Comparative Plant Ecology at Sheffield
1960-1980

Professor Roy Clapham, founder of the Sheffield-based research group later to become known as the Unit of Comparative Plant Ecology (UCPE).

Survey of old grasslands of the Sheffield region; Phillip Lloyd and Phil Grime at the Winnat's Pass.

Ian Rorison (centre left), first director of UCPE, leading a group sampling the mineral nutrient contents of the leaves of limestone grassland herbs.

John Hodgson, quadrat in hand, closes in on a rare species.

Comparative Plant Ecology at Sheffield
1981-2006

Rosemary Booth reviews progress in the Integrated Screening Programme.

Phil Grime, Joanna Mackey and Bruce Campbell exhibit at a Royal Society Soiree.

Some of the team that built the Buxton Climate Change Impacts Laboratory. Standing (left to right): Stuart Band, Rosemary Booth, George Hendry, Johnathan Kielty, David Corker, Greg Spring, Sarah Buckland, Lauchlan Fraser, Chris Thorpe, Sue Hillier. Kneeling: Phil Grime, Ric Colasanti, Ken Thompson. Inset: John Hodgson and Andrew Askew.

At Tapton Experimental Gardens, Jason Fridley and Chris Bennett examine the functioning of model plant communities of controlled genetic diversity.

Comparative Plant Ecology
a functional approach to common British species

SECOND EDITION

J. P. Grime   J. G. Hodgson   R. Hunt

assisted by

A. P. Askew   S. R. Band   R. E. Booth   S. M. Buckland   B. D. Campbell
H. C. Cornelissen   J. M. Fletcher   L. H. Fraser   G. A. F. Hendry   S. H. Hillier
S. Hubbard   J. M. L. Mackey   A. M. Neal   A. M. N. Ruttle   K. Thompson
P. C. Thorpe   M. van Ewijk   P. J. Wilson

Unit of Comparative Plant Ecology
Department of Animal and Plant Sciences   University of Sheffield UK

CASTLEPOINT PRESS

© J.P. Grime 2007

First published in Great Britain in 1988

Second edition published in 2007 by
Castlepoint Press
Colvend, Dalbeattie, Kirkcudbrightshire, DG5 4QB

Typeset in Minion by
MRM Graphics Ltd, Winslow, Bucks

British Library Cataloguing-in-Publication Data

A cataloguing record for this book is available from
the British Library

ISBN 978 1 8976 0430 4

All rights reserved.
No part of this publication may be reproduced, stored
in a retrieval system or transmitted in any form or by
any means, electronic, mechanical, photocopying,
recording or otherwise, without the prior written
consent of the copyright owner.

Castlepoint Press wishes to acknowledge the financial support
of the Botanical Society of the British Isles,
and also of Natural England.

Printed in Great Britain by CPI Antony Rowe, Chippenham.

Contents

List of panels	vii
List of figures	viii
List of tables	ix
Preface	xi
Acknowledgements	xii
Foreword to 1988 Edition	xii
Foreword	xiii

1 Introduction

Aims and methods of this book	1
The comparative approach to the British flora	1
Origins (1939-1970)	1
Two decades of 'creative conflict' (1970-1990)	2
Scaling up (1990-present)	2
Composition of the Species Accounts	3

2 Plant Functional Types

The need for a functional classification of plants	5
Tradeoffs: the basis for the evolution of widely-recurrent plant functional types	5
Data shortages: a world-wide problem	5
Different kinds of functional types for different ecological purposes	8
Primary plant functional types	9
The CSR model	9
The equilibrium between stress, disturbance and competition	11
Selection processes and evolutionary responses	12
Competition for resources	12
Resource shortage	12
Disturbance	12
Objections to CSR theory	12
Competition	13
Stress	13
Disturbance	13
Uses and limitations of CSR theory	13

3 Data sources for the Species Accounts

Introduction	17
Vegetation surveys	17
Survey field work	17
Features of the survey area	17
Methods used in Survey II	18
Outline results from Survey II	21
Habitats	21
Altitude	21
Slope	21
Aspect	21
Soil pH	22
Species diversity	22
Hydrology	23
Bare soil	23
Phenology, demography and regeneration	23
Rationale	23
Shoot phenology	23
Seed banks	25
Screening experiments	26
Rationale	26
Nuclear DNA amount	26
Germination characteristics	27
Relative growth rate	28
Variation in screened attributes	28
The scientific literature	29

4 The contents of the Species Accounts

Introduction	31
Nomenclature	31
Life-form	31
Phenology	31
Height	32
Seedling RGR	32
Specific leaf area (SLA) and leaf dry matter content (DMC)	32
Nuclear DNA content and chromosome number	33
Nuclear DNA amount (genome size)	33
Chromosome number	33
Established strategy	33
Regenerative strategies	33
Flowers	36
Germinule	36
Germination	36
Biological flora	37
Geographical distribution	37
Habitats	37
Gregariousness	37
Altitude	37
Slope	37
Aspect	37
Hydrology	38
Soil pH	38

Bare soil	38
△ ordination	38
Associated floristic diversity	39
Similarity in habitats	39
Commonest associated species	39
Synopsis	39
Current trends	39

5 Interpretation and use of the Species Accounts

Introduction	41
Autecological implications	41
Nomenclature	41
Life-form	41
Phenology	42
Seedling RGR	42
Nuclear DNA amount	44
Chromosome number	46
Established strategy	46
Competitors	46
Stress-tolerators	47
Ruderals	47
Intermediate strategies	47
Regenerative strategies	47
Vegetative expansion	48
Seasonal regeneration in vegetation gaps	48
Regeneration involving a persistent seed bank	48
Regeneration involving numerous widely-dispersed seeds or spores	49
Regeneration involving a bank of persistent juveniles	49
Flexibility in regeneration	49
Regenerative failure	49
Failure in plants with a single regenerative strategy	49
Failure in plants with more than one regenerative strategy	49
Flowers	50
Germinule	50
Germination	51
Initial germinability	51
Response to dry storage	51
Response to chilling	51
Response to scarification	51
Response to temperature	51
Response to light	52
Germination rate	52
Epigeal and hypogeal germination	53
Biological flora	53
Geographical distribution	53
Habitats	54
Gregariousness	54
Altitude	55
Slope	56
Aspect	57
Hydrology	57
Soil pH	57
Bare soil	57
Triangular ordination	58
Associated floristic diversity	58
Similarity in habitats	59
Commonest associates	60
Implications for communities and ecosystems	61
Introduction	61
Communities	61
Dominants	61
Subordinates	61
Transients	65
Co-existence mechanisms: theory	65
Co-existence mechanisms: real communities	65
Community dynamics	66
Ecosystems	73
The mass ratio hypothesis	73
Plant traits that drive ecosystem processes	75

6 The Species Accounts 81

7 Tables of attributes

Introduction	645
Explanation of tables	
Table 7.1 (Ecological attributes)	645
Table 7.2 (Attributes of the established phase)	647
Table 7.3 (Attributes of the regenerative phase)	648
Table 7.1	650
Table 7.2	664
Table 7.3	677

References 691

Subject Index 719

Species Index 735

List of panels

1	The research contribution of L. G. Ramenskii	5
2.	Tradeoffs in British plants; early speculations	6
3.	Protocols for screening plant traits and recognising plant functional types	7
4.	A hierarchy of plant functional types	8
5.	The Integrated Screening Programme : the main results	10
6.	Primary functional types: the debate	11
7.	Primary functional types: is the debate over?	12
8.	What is happening to the British Flora?	14
9.	Plant traits, communities and ecosystems	63
10.	The humped-back model: some refinements	67
11	The humped-back model: the cost of protecting grassland diversity	68
12.	Co-existence mechanisms: spatial niches	69
13.	Co-existence mechanisms: phenological niches	70
14.	Co-existence mechanisms: foraging niches	71
15.	Co-existence mechanisms: regeneration niches	72
16.	Co-existence mechanisms: genetic variation	72
17.	Community responses to extreme events	73
18.	Community responses to climate change	74
19.	Plant functional types control the tempo of ecosystems : evidence from Chernobyl	76
20.	Plant functional types control the tempo of ecosystems : evidence from butterflies	77

List of figures

2.1 A model describing the various equilibria between competition, stress and disturbance in vegetation and the location of primary and secondary strategies. 9

3.1 The UCPE survey area. 17

3.2 Numbers of quadrats in the primary, intermediate and terminal habitats of UCPE Survey II. 21

3.3 The frequency distribution of species by number of quadrats in UCPE Survey II. 23

3.4 A complete specimen quadrat record from UCPE 'Survey II'. 24

3.5 A scheme relating the pattern of seasonal change in shoot biomass to strategy type. 25

3.6 The relationship between maximum standing crop (plus litter) and species density at 14 sites. 25

3.7 Representation of the four types of seed bank described by Thompson and Grime (1979). 25

3.8 A chronology of the UCPE screening experiments. 26

3.9 Nuclear DNA amounts in 162 flowering plants occurring in the British Isles. 27

3.10 The distribution of germination rates. 28

3.11 Variation in mean relative growth rate within four levels of organization. 29

4.1 The chief types of life form, based upon Raunkiaer's classification. 31

5.1 The relationship between relative growth rate, specific leaf area and leaf dry matter content in the flora of the Sheffield region. 43

5.2 The relationship between nuclear DNA amount and the mean length times breadth of epidermal cells in mature leaves of a range of herbaceous species. 45

5.3 The relationship between nuclear DNA amount and the time of shoot expansion in 24 plant species commonly found in the Sheffield region. 45

5.4 A prediction of the range of combinations of regenerative strategies in relation to the established phase of the life-cycle. 48

5.5 The relationship between chlorophyll concentration of the investing structures and dark-germination of mature seeds in various species of flowering plants. 53

5.6 The contribution of rapidly-germinating and slow-germinating species to groups classified with respect to estimates of the maximum potential rate of seedling dry matter production. 53

5.7 The distribution of three floristic elements within the triangular model. 62

5.8 The relationship between DNA amount and the mean rate of leaf extension over the period 5 April- 25 May in 14 grassland species co-existing in the same turf. 66

5.9 Interacting characteristics of plant strategies and some ecosystem processes under conditions of high and low nutrient availability (From Chapin 1980). 76

List of tables

2.1 The basis for the evolution of three strategies in plants (From Grime 1977). 9

3.1 Facts and figures concerning the three UCPE vegetation surveys. 18

3.2 Climatic data for the Sheffield region for the period 1972-82. 19

3.3 Variations in land-use and in the potential quality of land for agriculture within the Sheffield region. 19

3.4 A full list of the coded habitat characteristics which were recorded for quadrats in UCPE Survey II. 20

3.5 Numbers of quadrats in the seven primary habitats of UCPE Survey II, classified with respect to **altitude**. 22

3.6 Numbers of quadrats in the seven primary habitats of UCPE Survey II, classified with respect to **slope**. 22

3.7 Numbers of quadrats in the seven primary habitats of UCPE Survey II, classified with respect to **aspect**. 22

3.8 Numbers of quadrats in the seven primary habitats of UCPE Survey II, classified with respect to **soil surface pH**. 23

3.9 Numbers of quadrats in the seven primary habitats of UCPE Survey II, classified with respect to **species diversity**. 24

3.10 Numbers of quadrats in the seven primary habitats of UCPE Survey II, classified with respect to **bare soil cover**. 24

4.1 An explanation of life-history classes. 31

4.2 A summary of the experimental design and environmental conditions used in the screening for mean relative growth rate under a standardised, productive regime. (From Grime and Hunt 1975). 32

4.3 Some characteristics of competitive, stress-tolerant and ruderal plants. 34

4.4 Species typical of seven ecological strategies. 35

4.5 Five regenerative strategies of widespread occurrence in terrestrial vegetation. 3

4.6 Classes used in the **'Hydrology'** histogram of the Species Accounts, with numbers of quadrats involved. 38

4.7 Classes used in the **'Bare soil'** histogram of the Species Accounts. 38

4.8 An explanation of the numerical data provided by the **'Triangular ordination'** diagrams of the Species Accounts. 39

5.1 The 20 most common vascular plant species of spoil habitats in the Sheffield region (From Grime 1986b). 54

5.2 The distribution of **altitude** classes, by percentage of records in each of the 32 'terminal' habitats of UCPE Survey II. 55

5.3 The distribution of **slope** classes, by percentage of records in each of the 32 'terminal' habitats of UCPE Survey II. 56

5.4 The distribution of classes of **bare soil** cover, by percentage of records in each of the 'terminal' habitats of UCPE Survey II. 59

5.5 Means and standard deviations (across all species) for the percentage occurrences recorded in each of the 32 'terminal' habitats of UCPE Survey II. 60

5.6 The species most frequently included in the 'Commonest Associates' tables and their % occurrence in all of the tables. 61

5.7 Attributes of the most common species in two contrasted grassland communities. 66

Publication supported by the
Botanical Society of the British Isles

The Society takes great pleasure in facilitating the publication of the second edition of *Comparative Plant Ecology*. The recording of plant distributions and the detection of changes in them is central to the activities of the Society. The accurate interpretation of our results relies on the availability of codified ecological information such as that presented in this volume, and this is increasingly important now that species are beginning to show the impacts of climate change. The first edition proved immensely helpful to us, especially during the writing of the *New Atlas of the British & Irish Flora* (Preston *et al.* 2002). In this second edition, not only have the accounts of the 281 species treated in full been updated and revised, but also an additional 320 species have been added to the ecological summary tables, bringing the total number of species with autecological information to 821. In recent times the emphasis in academic circles on molecular biology has led to the comparative neglect of whole-organism studies, and it is volumes like the present one that remind us of how important the latter are. It also shows the immense value of the results that can be obtained from carefully planned, long-term research programmes. Such data are essential both to help understand changing distribution patterns and to inform management plans for conservation.

About the BSBI
Tracing its origins back to 1836, the *Botanical Society of the British Isles* (BSBI) is a vibrant organisation that combines a body of professional scientific expertise with an active and committed corpus of skilled amateur members. It welcomes anyone who is interested in the flora of Britain and Ireland. The BSBI aims to:
- promote the study of British and Irish vascular plants and charophytes;
- support, conduct and participate in research into the taxonomy, biogeography, ecology and conservation of the British and Irish flora;
- co-operate with European and other botanists in matters of mutual interest and concern.

There are three main strands of activity within the Society:
1. *Survey work* in BSBI is managed by a network of vice-county recorders, and involves thousands of individuals who collect the millions of botanical records that underpin almost all botanical and nature conservation studies in the British Isles.
2. *Research activities* include conducting and funding specific research projects, publishing scientific works and hosting scientific conferences; much of this is undertaken in partnership with academic institutions, in many of which the Society has members.
3. *Educational activities* include field meetings and training sessions, often in partnership with other organisations, as well as the production of a variety of publications and web-based materials.

For more information on what we do and how to get involved, go to the Society's website (www.bsbi.org.uk).

Richard Gornall, President BSBI
University of Leicester, January 2007

Preface

We begin this preface with a quotation from an experienced commentator on scientific research:

If one believes in science, one must accept the possibility- even the probability- that the great era of scientific discovery is over. By science I mean not applied science, but science at its purest and grandest, the primordial human quest to understand the universe and our place in it. Further research may yield no more great revolutions, but only incremental, diminishing returns.

From John Horgan's 1996 book *The End of Science: facing the limits of Knowledge in the Twilight of the Scientific Age*

For all scientists this is a chastening diagnosis but for ecologists Horgan's book is particularly devastating; in the chapter devoted to Biology there is no mention of Ecology at all, either in terms of past achievements or future prospects! This raises the question as to whether ecological research (including that of the kind we offer in this book) can have any part in Horgan's 'primordial human quest.' Our response to this question is an emphatic 'Yes' but, at the same time, we can understand why a detached observer might presume that ecologists should not aspire to contribute to science 'at its purest and grandest.'

A clue to Horgan's dismissal of Ecology lies in what he quite properly selects in his book to represent the crowning achievement of Biology, namely the theory of evolution by natural selection (Darwin 1859). We suspect that Horgan, in the company of many biologists, regards ecological research as an applied science with its foundations mainly rooted in evolutionary biology. We believe that evolutionary concepts will remain indispensable to ecological research but we are of the opinion that a greater degree of independence from our sister discipline is now appropriate. In a new chapter (Chapter 2) added to this second edition we conclude that, to fulfil its true destiny among the sciences, it will be necessary for Ecology to establish research goals, theories and methods that are different from those of Evolutionary Biology. In Chapter 2, we argue that the defining objectives in a modern approach to ecology are to recognise and elucidate, at a range of spatial scales from the individual organism to the ecosystem, recurring patterns of functional specialisation. Crucial to this activity is a developing ecological vocabulary that recognises similarity and difference between organisms in terms of function rather than evolutionary affiliation. This, as we explain in Chapters 2 and 5, brings the opportunity for unifying concepts relating to the ecology, management and conservation of species, communities and ecosystems.

In order to understand why such an approach is necessary and effective we must look beyond the micro-evolutionary processes that determine that each species is different in it's biology and detailed ecology. A deeper analysis reveals that, in core functioning, the adaptive pathways available to plants and animals are severely constrained by the laws of physics and chemistry. For vascular plants in particular, there is now experimental confirmation that these constraints are ubiquitous and powerful enough to allow a world-wide functional classification. We believe that this system is capable of supporting generalising principles with regard to variation in the structure and dynamics of vegetation and the functioning of communities and ecosystems. One of the main purposes of this second edition is to illustrate more fully the opportunities that now exist to use functional classification of the British flora as an aid to ecological research and management.

The parts of this book that approach what John Horgan might recognize as 'science at its purest and grandest' are those relating to the CSR model of plant functional types. For the three authors of this book, 2005 marks the 31st year of our association with this controversial model. We often wonder what will be the final outcome of our odyssey. Perhaps we should draw some reassurance from another quotation from John Horgan's book.

Science's success stems in large part from its conservatism, its insistence on high standards of effectiveness. Quantum mechanics and general relativity were as new, as surprising, as anyone could ask for. But they were believed ultimately not because they imparted an intellectual thrill but because they were effective: they accurately predicted the outcome of experiments. Old theories are old for good reason. They are robust, flexible. They have an uncanny correspondence to reality. They may even be True.

Sheffield 2005
Phil Grime
John Hodgson
Rod Hunt

Acknowledgements

The first edition of this book, published in 1988, was based on researches conducted over a period of 25 years with the full support of, first, the Nature Conservancy then, later, the Natural Environment Research Council. The scale of the research effort during this period may be judged from the fact that no less than fifteen names appeared on the title page. Eighteen years later our ambition remains undiminished but our financial resources are much reduced. This places a mighty burden on the few who have elected to carry forward the distinctive type of studies conducted at the Unit of Comparative Plant Ecology since 1961. My purpose here therefore is to offer heartfelt thanks to those who in recent times, often at some cost to themselves, have become involved in the large-scale, long-term investigations of UCPE. It is a particular pleasure to acknowledge the vital and substantial contributions of co-author John Hodgson, my colleagues Ken Thompson, Rosemary Booth and Andrew Askew at Tapton Experimental Gardens and our multi-talented collaborator at Syracuse University, Jason Fridley.

This new edition of Comparative Plant Ecology has been modified to contain a large number of estimates of specific leaf area and leaf dry matter content, traits now widely used as ecological predictors. This has been possible as a result of the doctoral studies of Peter Wilson and recent collaborations with researchers Amy Bogaard, Carol Palmer, Mike Charles and Glynis Jones in the Department of Archaeology at the University of Sheffield. Despite expansions to accommodate more species and to document progress in the development of functional ecology, several features of the first edition of the book remain prominent in the second and here special mention must be made of the excellent foundations laid by Rod Hunt in analyses of the field survey data presented in the Species Accounts and by Judith Fletcher who created subject and species indexes of extraordinary coverage and precision.

Production of this new edition has depended to a large extent on the tenacity, high standards and unfailing good humour of our publisher, to whom we extend our warm appreciation. Finally, we thank the Journal of Vegetation Science for permission to reproduce in Panel 9, arguments first expressed in volume 17, pages 255–260 of that journal and we also acknowledge permission from the Journal of Ecology to include in Panel 12 a figure appearing as Figure 8 in the paper by Grubb, Green and Merrifield in volume 57, pages 175-212.

Philip Grime

Foreword to the 1988 Edition

In the third week of September 1986 I attended the 25th Anniversary Symposium of Sheffield University's **Unit of Comparative Plant Ecology** by kind invitation of the Director of the Unit, Professor Ian H. Rorison, and his staff. During the one full day and two half-days of the meeting we heard a series of papers on **Frontiers of Comparative Plant Ecology 1961–86**, and numbers of poster abstracts and video displays were on view. On the evening of Tuesday, 23 September there was a well-attended and much appreciated Symposium Dinner, and the whole constituted a highly successful and memorable sequence of events greatly enjoyed by all who took part.

I make special reference to it in this Foreword because it made me so much more acutely aware of the very large amount of first-class scientific research that had been completed and published during the 25 years since the establishment of the Unit, and also how outstandingly that research continues to revolutionize our ways of viewing, describing and, in due course, classifying natural vegetation as a basis for 'causal analysis, informed management and effective conservation'. It is already clearly insufficient merely to list component species and estimate their relative abundance. There is still a long way to go, but the provision in this volume of standardized Autecological Accounts of so many of our common wild vascular plants was undoubtedly a worthy objective.

The authors, and indeed the whole staff of the Unit, are much to be congratulated on their achievement.

A. R. Clapham

Foreword

by T.F.H. Allen, Ph. D.
Professor of Botany and Environmental Studies,
University of Wisconsin, Madison, USA

Academic life has changed over the last few decades. There used to be a large suite of different ethics: including medical, ecclesiastical and scholarly ethics. Now all have been replaced with the mercantile ethic, such that success in academe more or less requires vulgar careerism; we have to run it like a business. But then there are a few left who still deserve the appellation "gentleman and scholar," and the authors here are in that number. First the scholarship in this book is breathtaking, but more of that later. Equally impressive is the poise and relentless argument herein, as it deals with the points of contention arising as different schools of thought take this intellectual posture and that. Even when the disagreements are substantial and have involved heated debate elsewhere, the authors here have consistently handled each situation with a grace and fairness that only helps us all advance the science. Gentlefolk they are indeed, and relentless scholars too. Academic life has changed, but the best of what it used to be survives, and is found between these covers.

My first college lecture was given by John Harper, and I confess to model some of my lecturing style on his. Over a decade before Harper's great tome, I had already seen many of those ideas in lecture and seminar at Bangor. The papers that lie behind the push for plant demography were written by some of my colleagues in graduate school. So I do understand what the austerity of plant demography was about. Also in Bangor in those heady days of the 1960s was Paul Richards, with his tropical rain forest work, and Peter Greig-Smith, with his community analysis, so I am well aware of postures alternative to Harper's. Those of us in the multivariate faction of community ecology in Bangor were, much like the present authors, a bit bemused by the reductionism of demographic experiment, but we respected a bunch of clever folk doing rigorous work. I have tried to unify my whole experience in Bangor in my own work, but stand humbled here by this book, as the authors work a middle ground that is not a compromise, but is rather a massive statement in its own right.

A central tension here is competition. My dissertation was on the aerial algae of North Wales, and that phycological background leads me to understand the alternative position taken by others central in the intellectual tension raised by this book. David Tilman, like me, was a phycologist, and that leads both of us to give a heavy weight to mineral nutrients as the first cause to consider. And Tilman, the present authors and I have all focused on strategies arising from interaction between organisms in the face of environmental challenges. The solution offered by the present volume not only presses the competition issue to resolution, but it does so in a way that captures the basis of the other strategies wherein competition plays a small role. The pivot point in all of this is how many different constraints are there. The phycologist in me leads me to intuit that there are many limiting factors, and most of them are mineral in nature. Hutchinson's paradox of the plankton has some of the same gestalt. But the biological systems analyst I have become tells me that the patterns and regularity that we see must come from only a few constraints. The huge amount of experimentation performed in Sheffield, combined with a wide knowledge of comparisons across Britain, has led the authors of this book to the conclusion that it is a only a few factors acting on all plants. And they are right.

Recently I have been working on strategies of biological and social systems, and so have some thoughts on strategies as a general issue. It is very invigorating for me to see an alternative approach to strategies worked through so throughly and convincingly in the triangular ordinations herein. In my approach to strategies I have come to the conclusion that it is worthwhile to separate two crucial aspects of bio-social systems. All biological and social systems have these basic characteristics. First there is always a thermodynamic part of the system, as it consumes resource in a process of degradation. High quality material and energy enters, and low quality is extruded. That difference in quality is what drives the ordering processes in all of life. Second are semiotic-linguistic components that come as codes such as DNA or regulations in a society. The codes are inside the system, but external to the thermodynamics. All living systems have both thermodynamics and codes, and the effects of both sets of causes can be seen in all systems. But there are some interesting limitations. Our examples come from a wide set of systems from contemporary and historical societies, colonial insects, birds, sphagnum versus fen vegetation, and diseases. We have found that prediction of system behavior is either driven by the thermodynamics or the linguistic elements, but not both in a given analysis. When the thermodynamics dominate there is self-organized emergence, which requires no plan. Alternatively, when the planned coded elements dominate, the fluxes are so constrained that energy and resource quantity are not predictors. Strategies appear in the switch between the two sides of control. The reader should be impressed with how different is my approach to strategies compared to the ordinations in this book, but I can nevertheless show below a certain unity between the two sets of conclusions.

The critical distinction in the present context notes that there are severe restrictions as to the number of types of thermodynamically driven emergent behaviors that appear possible. By contrast, there is a huge number of different ways for a coded plan to affect efficiency. High temperature and humidity usually creates thunder heads, whereas water over a hole only makes a whirlpool to constrain system behavior. Thermodynamic ordering is limited. Meanwhile, there is an infinity of ways to plan to economize or become more efficient. Coding efficient strategies has a huge space in which to work. In the present book there is an emphasis on the physical limitations that appear responsible for much of the pattern. Physical limitations pertain to the thermodynamic causes of pattern, not the effects of planning and coded control. Thus from the completely different line of reasoning, and an almost entirely different set of examples, the conclusion is that one would expect plants limited by the physicality of their environment to be constrained by only a small number of factors. The Tilman-style models for competition expect multiple limiting factors, each directing a different outcome for specific situations. By contrast, the model for competition and the other drivers of strategies in the present book suggest that it is always the same small set of drivers, and outcomes are all easily related in a small triangular parameter space. And as the data come in, particularly as presented in the present volume, the short odds are on a small number of causal factors.

There is always a place for the flash of insight coming from some special contrived situation, but at least as important, although less glamorous, is the systematic exploration of the full space, to make sure no part of the large pattern is missing. Being systematic is very hard work, and is unfairly judged as drudgery. But the rewards of a systematic treatment of any exploration are huge, and have no substitute. The body of work behind the present volume is indeed immense, but the rewards are clearly there. In his tutorial books on chess, the British Grandmaster, Harry Golombek, often warned against changing the plan of attack so as to take advantage of some happenstance in the game. Do not dismantle a King-side attack and move to the Queen-side just because your opponent has left a minor piece unprotected, susceptible to an easy hit. Stick with the plan. In the Foreword to the first edition of Comparative Plant Ecology, Professor A. R. Clapham took delight in the solid achievements theretofore of the Natural Environment Research Council's Comparative Plant Ecology unit. And delighted he should have been. So here we are, some twenty years later, and the pace of achievement has never slackened. And in no small way, this is because they never changed the plan. Gaps needed to be filled. Experiments and calibration had to be done. And the scope of the whole needed to be expanded in an orderly systematic way. Phil Grime was instrumental in my writing this forward. I am always open to new ideas, grabbing from anywhere, and he wanted the book to be put in a large setting. He contrasted himself with me, saying that he could rarely afford to take on some new position, because, every time he does it, he has twenty years of experimentation to do. The day to day the work on this project must have been crushing, a full twenty years and more of experimentations. But here we are with the crowning glory of the autecology of all the common British species. There was a time when the vegetation of Wisconsin was the best described in the world. Not any more, and one reason is the wealth of information in this volume. And how timely is this second edition. The crisis of species and habitat loss presses ever harder. Theorists hold forth about the situation and what to do about it, and that is all well and good. But there is no substitute for hard data and grand synthesis about the prevailing situation. While loss of diversity is a good enough preliminary signal, a proper treatment of the losses needs to understand each player as an individual species. When I took plant systematics, Greig-Smith made a clever move. He said we could not do justice to all the Angiosperm families in just one course, but we could study the Monocotyledons systematically, covering all the families. And so it is with species under siege. We cannot do all the species in the world, and have time left to take action, but apparently it is possible to do it for just Britain, as a microcosm. I remain filled with admiration for the size of the effort that appears in this volume, but even more important is that there is a certain closure by focusing particularly on the British flora.. This book shows that it really is possible to get properly on top of a flora, without collapsing everything down to crude diversity measures that suppress species differences. A proper rich treatment of Britain shows the bonuses of accumulating a proper base species by species. Clearly there are many more ideas and summaries that can be extracted from the information in the present work. But the patterns and potential for synthesis already shows through, and promises to show much more. Perhaps the most important message to take from this work is the worth of courage to take on systematic treatments of large coherent subjects, all in the cause of conservation and restoration.

I referred earlier to the delight in the first edition of Professor Clapham. And I am delighted too. But my delight is a bit different. This is a book that gives comfort to a lover of the British flora. I can never resist looking up familiar places on a map. On the Google world map I can focus in on my childhood home in London, and can even see the conservatory still there, tucked between the kitchen and the breakfast room. The small track of Epping Forest is still there, although I know the holly has grown up to cut off the paths I used to take to school. Then I might google off to Bangor to find my research sites. And then over to Malham Tarn, near where my mother still lives, and where Peter Greig-Smith led me on an undergraduate field trip over forty years ago. Then a skip over to Sheffield to focus in on the Department of Plant and Animal Sciences, where I gave a colloquium on a recent sabbatical. Somehow time spent just probing organized information is never wasted, and it is always pleasing. And so it is with this book. What a joy to sit and hop around the body of this book to see old friends from the British flora. While one knows them well, it is so much fun to guess what the page for Viola riviniana has to say about slope. Oh, so it is almost as common of very steep terrain as it is on flat ground, while occurring mostly on middling slopes. This is a book for the laboratory, so we can calibrate ideas about the complexity of plant communities. But it is also a fireside book, for Botanists to "google" around the British flora.

Tim Allen
Madison, Wisconsin
August 2006

1 INTRODUCTION

Aims and methods of this book

The main purpose of this book is to provide standardized accounts of the biology and ecology of common vascular plants of the British flora. The structure and content of each account, and in particular, the balance between data presentation and interpretation, has been strongly influenced by our concern to meet the specific requirements of academic research, vegetation management and nature conservation. Consequently, the book contains a rather lengthy section (Chapter 5) which describes in some detail the various ways in which the accounts may be used in efforts to understand or manipulate the ecology of individual species or to control the composition of plant communities or the properties of the ecosystems in which they occur.

The form of the book is a reflection of two additional intentions. The first is our wish to reaffirm the value of a broadly-based comparative approach to the ecological study of the British flora. The second is the commitment of all three authors to the view that the future of ecology as a rigorously predictive science will benefit considerably from the development of a universal functional classification of organisms, which by complementing the Linnaean system can be central to an understanding of community and ecosystem processes. The remainder of this chapter attempts to clarify these aims and explains how they are addressed by the **Species Accounts** that are a dominant feature of the book.

As explained in the Preface, this book appears at a critical time in the history of ecological research when, after protracted debate, some major philosophical and procedural barriers to progress are breaking down. In this new edition we have therefore added a new feature to mark this development. This takes the form of a set of **Panels** added to Chapters 2 and 5. Each panel occupies one or two pages and highlights a particular topic that we believe to be relevant to the understanding and effective use of the information provided in this book. Most of the panels refer to subjects where a definitive statement is possible with regard to either the science or its implications for vegetation management and here an effort has been made to address the topic in non-technical language. Inevitably some key problems in plant ecology remain unresolved: here our approach has been to use the panels to identify remaining difficulties, to review the prospects for their resolution and to cite relevant publications. Panel 9 is unusually long reflecting the key importance of distinguishing between plant traits that drive ecosystem functioning and those that are more exclusively concerned with the struggle between populations and species to gain membership of the plant community.

The comparative approach to the British flora

Origins (1939 –1970)

Although the British Isles present the aspiring plant ecologist with a considerable diversity of habitats, their geographical area is small and their flora is mercifully depauperate. At a comparatively early stage in the development of ecology as an organized science it was thus possible for A. G. Tansley to produce a general synopsis in the form of his highly influential book *The British Islands and their Vegetation* (1939). These circumstances quickly established among amateur and professional botanists alike a strong educational tradition for 'getting to know the British flora'. For many students, and more particularly those with a critical interest in the functioning of vegetation, the practical outcome of this tradition was a challenging apprenticeship of travel, taxonomy and, sometimes also, premature exposure to insolubly complex ecological problems. The authors know personally several plant biologists who were diverted away from plant ecology by such early traumas.

Those who fell at the first hurdle were not alone in succumbing to the dangers inherent in this broad approach to plant ecology. Among those who persisted to become professional ecologists, the effort of grappling with the whole flora, or a major part of it, sometimes resulted in an overextension of both data and powers of interpretation. From its inception to the 1960's, *The Journal of Ecology* displayed a strong reliance upon description and spatial correlation. Perusal reveals many instances where the attempt to maintain progress on a wide front spawned unsatisfactory assumptions about the nature, integrity and dynamic relationships of the vegetation in question. The attempt also generated a great deal of 'facile guesswork' (Harper 1982) concerning the identity of the processes controlling the distribution and abundance of individual species.

In the current era of increasingly refined research methods it has been easy to misunderstand the motives and to undervalue the achievements of many of the earlier plant ecologists who adopted a broad approach to the British flora. A corrective to this unfortunate tendency is available from the more philosophical contributions of E. J. Salisbury, A. S. Watt and A. R. Clapham. Each fostered and drew upon the broad comparative tradition, but at the same time was committed to a critical reductionist analysis of the mechanisms underlying the ecological behaviour of species and the structure and dynamics of vegetation.

The influence of the comparative approach upon Salisbury is nowhere more obvious than in his monograph on plant reproductive biology (Salisbury 1942) and in his much-underestimated 'infection pressure' theory of plant colonising behaviour (Salisbury 1953). The benefits of Watt's secure grasp of the British flora are evident throughout his many pioneering investigations of vegetation dynamics, not least in his classic paper 'Pattern and process in the plant community' (Watt 1947). It is with Clapham, however, that we find an active promulgation of the broad comparative tradition in terms that seem specifically designed to shape the future course of plant ecology in Britain. This occurs first in his contributions as taxonomist and ecologist to the *Flora of the British Isles* and in his initiation of the *Biological Flora of the British Isles* published in *The Journal of Ecology*. Most revealing, however, is the set of ideas contained in his Presidential Address to the British Ecological Society (Clapham 1956). Here he suggests that 'Our main concern as plant ecologists is to know why a plant of this species, and not of that, is growing in a given spot; that whatever views we might entertain about the community as organism, or quasi-organism, or as a mere assemblage of individuals, we should show ourselves to be interested primarily in autecological problems, that is in enlarging our knowledge and understanding of the biology of individual species of plants.'

The address went on to advocate the use of controlled-environment facilities to elucidate key factors in the ecology of British species and to suggest how the information might be brought together and utilized for the conservation and management of vegetation.

The collection and comparison of standardized information on the species of the British flora, recommended by Clapham as a means of understanding vegetation, followed closely the philosophies prevailing in the physical sciences. Perhaps the most obvious of these is the role played by the Periodic Table of the Elements in classifying, analysing, and even predicting, the structure and properties of chemical elements and compounds.

Two decades of 'creative conflict' (1970–1990)

Over the next two decades research on the British flora did not advance on the broad front envisaged by Clapham. The developments that might have been expected to follow the appearance of the *Flora of the British Isles* (Clapham et al. 1952) and the *Atlas of the British Flora* (Perring & Walters 1962) were slow to materialize. Research efforts became more detailed and specialized. Under the charismatic leadership of John Harper, the demographic approach to plant population ecology, presaged by the work of E. J. Salisbury, A. S. Watt and C. O. Tamm, came into its own (Harper 1967, 1977), latterly fortified by modern techniques of data analysis and computer modelling. This was a welcome development that began to add temporal and genetic components to our thinking about vegetation. However, there can be little doubt that the introduction of more critical and reductionist methods in plant population biology occurred at the cost of a drastic narrowing of research objectives and subject material. Both in population studies and in physiological experimentation on the British Flora, precision tended to assume priority over ecological perspective. One consequence of this shifting emphasis was a widening gap between academic research and the activities and objectives of the agencies and voluntary organisations engaged in monitoring, conserving and managing natural resources. For those who were keenly aware of the continuing damage to our botanical heritage arising from habitat losses and changing land use (eg Ratcliffe 1984) the academic focus on demography was inexplicable. However, this situation did not arise from a callous disregard for the future of the British Flora. As the following quotations illustrate, the unity of purpose that characterised this remarkable, if rather myopic, episode in the history of academic plant ecological research in Britain arose from a conviction that demography through its close acquaintance with the births and deaths of plants brought the plant scientist closer to an understanding of the underlying mechanisms of vegetation and would eventually have practical benefits.

'…the search for generalities in ecology has been disappointing – more so in plant than in animal ecology. The few generalities that have emerged come from studies of stands of single species.'
'…it is from the work of many individuals working scattered over a variety of parts of the world, but concentrating their attention over long periods on the behaviour of individual plants, that the development of ecology as a generalising and predictive science may be possible.'

Harper (1982)

Among those plant ecologists familiar with the practical and pressing problems of vegetation conservation and management this diagnosis of what was required to drive plant ecology forward drew a mixed reception. It was widely acknowledged that:

'…in communities with a rapid turnover of populations and individuals, demography can provide strong clues to the processes which influence species composition, control relative abundance and determine cyclical and successional change. Even in these most favourable circumstances, however, doubts may persist concerning the underlying mechanisms of control. In those plants where the life-span of individuals is long and both death and successful recruitment are rare and protracted events, the status of the demographer is likely to be reduced to that of a puzzled observer of events occurring several years downstream from the complex and unknown phenomena that caused them.'

Grime (1993).

In view of Clapham's long association with Sheffield it was fitting, and in retrospect perhaps inevitable, that this was where the national trend towards specialized and/or detailed studies in plant ecology encountered the strongest resistance. Initially under the guidance of Clapham and funded by the Nature Conservancy (and latterly by the Natural Environment Research Council), a small team of scientists including the authors of this book embarked upon a long-term research programme designed to characterize and compare the field ecology and laboratory characteristics of many of the more common plants of inland Britain and to use the resulting data to develop guidelines for vegetation management. The methods and scale of the operations undertaken by the Sheffield group were intended to complement the more specialised and detailed approach of the Harperian school but there is no doubt that they were regarded with suspicion by those preferring finer levels of analysis with more explicit connections to the demography of particular plant populations. Some insight into the potential for misunderstandings that surrounded our approach during this period is apparent in the following comment by a neutral observer:

'We are currently in a reductionist, population-dominated era…Once we spent a lot of time looking at the behaviour of species, then ecosystems became important. Now, nearly any ecologist who feels he must be respectable will work on populations. Although an historian may be able to see a proximal cause for what happens – the arguments of a persuasive scientist or a novel and interesting discovery – the ultimate causes may be little other than those which give the world its flourishing fashion industry.'

A.D. Bradshaw (1987)

Against this uncertain background there were many temptations to abandon our broad comparative approach to the British flora. However, as we entered the last decade of the 20th century new factors began to make such thoughts of retreat unnecessary and encouraged an even wider perspective.

Scaling up (1990 – present)

From satellite imaging and other forms of environmental surveillance it has become apparent that floristic changes are occurring across the world and at an unprecedented rate. These are driven by changes in land-use and climate and involve species extinctions and functional shifts as sets of species with particular characteristics are replaced by other sets with attributes more attuned to the new conditions. Because plants capture and process the resources later utilised by the other organisms it is inevitable that such changes in the functional characteristics of the vegetation will have profound repercussions on ecosystem structure and functioning. Many of these changes are taking place in circumstances where there has been little research conducted on the flora and where there is an urgent need to predict the consequences for both vegetation and ecosystems.

The British Isles have not escaped these processes and from two main sources, The Countryside Survey (Barr et al. 1993, Bunce et al 1999, Firbank et al. 2000, Haines-Young et al. 2000) and The New Atlas of the British and Irish Flora (Preston et al. 2002a) there is confirmation that widespread and substantial changes are now occurring in the distribution and abundance of plant species. These changes, first reported by Hodgson (1986 a,b,c, 1987a ; see Panel 8, page 14) are most pronounced in lowland areas of the British Isles. Many vascular plants have declined in both the number and size of their populations and some appear to face extinction. On a wide geographical scale familiar vegetation types of the British countryside have changed radically in species composition and there has been a general decline in species-richness (number of species per unit area) in many plant habitats.

This is clear evidence that a transformation of the British flora is now occurring at an unprecedented rate with immediate and long-term consequences for the floristic composition of the countryside and the conservation of natural resources. We suggest that it is in relation to this scenario of widespread and profound vegetation change that we should assess the current state of plant ecological research in Britain. Has our science achieved the state of maturity and rigour that allows recognition of the causes of these changes? Can we forecast the rate and direction of future changes? Are we now capable of recommending changes in management to arrest or reverse losses of particular species? Do we know enough of what will be required to re-establish lost plant populations?

Faced with such large-scale problems, it is clearly impracticable to study each species in detail following the protocol advocated by Harper (1982). This situation has prompted the search for criteria that will support a functional classification of plants that can be applied to all terrestrial vegetation and can provide a basis for understanding how it is maintained at present and will respond to future changes in management and climate.

Viewed in this context the various activities of the Sheffield group can be regarded as a contribution to the development of such a classification. However, in order to pursue this objective it has been necessary, in alignment with laboratories around the world, to address several fundamental issues that, until recently, blocked progress towards the consensus required to provide the basis for a universal functional classification of plants. These have included arguments about the circumstances in which competition for resources controls vegetation composition, the role of herbivores and their predators as vegetation determinants and the relationships between plant functional types and ecosystem functioning. As we explain in Chapter 3, the diversity of environmental conditions and land-uses present in the Sheffield region has provided the basis for our contribution to the task of trying to resolve these issues. More specifically, in a programme of field surveys and laboratory screening experiments, now extending over a period of 45 years, the vegetation of the South Pennines has been used as a test-bed for the development and validation of generalising principles in plant ecology. The 1999-2003 Report of the Unit of Comparative Plant Ecology contains an historical account of this research programme and is available at http://www.sheffield.ac.uk/~nuocpe/.

Composition of the Species Accounts

As explained above, much of our research in recent years has been concerned with the effort to devise a useful and universal functional classification of terrestrial plants and it was inevitable that the results of this research would be prominent in this book. However it is important to recognise that in this monograph we have used a variety of additional sources providing information of specific relevance to the inland vascular plants of the British Isles. The Species Accounts in Chapter 6 and the associated tables of synopses in Chapter 7 are a compilation of original field and laboratory data together with additional information from a wide range of publications. Their form owes a great deal to earlier 'species biographies' written by taxonomists or ecologists. Although the primary purpose of the *Flora of the British Isles* (Clapham *et al.* 1952), the *New Flora of the British Isles* (Stace 1997) and the *Flora Europaea* (Tutin *et al.* 1964-80) is taxonomic, such publications also contain ecological information in the form of notes appended to the descriptions of species. These function as aids to taxonomy, but often they also provide a synoptic account of the habitats exploited by the species. There have been several attempts to augment these sources, and accounts of the field ecology of species which have appeared using standardized procedures of data collection and presentation. Examples here include the studies of Hundt (1966) in Central Europe, of Zarzycki (1976) in Poland, of Grime and Lloyd (1973) in N England and of Hansen and Jensen (1974) on road-verge vegetation in Denmark. All of these investigations have been consulted in preparing our own accounts, and they have been each brought with them the considerable advantages (Bradshaw 1987) which accrue from a large common database. Foremost among these is the opportunity for direct ecological comparisons between species.

The accounts produced as contributions to the *Biological Flora of the British Isles* published in *The Journal of Ecology* represent a much more ambitious approach to formalized description of the ecology of plant species. These contain not only field data, but the results of laboratory investigations and reviews of published literature relating to the species. The *Biological Flora* accounts are exceedingly valuable sources of information. We have used them extensively in this book but, as Grime and Lloyd (1973) noted, they 'have been designed to provide an account of the ecology of a species over a large geographical area and it has not been found practicable or perhaps desirable to standardize in detail the collection and presentation of field data. In particular, little or no account has been taken of the field situations in which a given species *fails* to occur'. Further limitations arise because *Biological Flora* accounts are not available for many of our subject species, even some of the common ones, and many of the earlier accounts no longer provide an adequate summary of the current state of knowledge.

The accounts in this book attempt to carry the 'species biography' into a further stage of its evolution by introducing an additional degree of standardization in the collection, analysis and presentation of laboratory and field data. Earlier in this introduction we have explained that the development of a functional classification of plants has played a major part of our effort to understand the ecology of individual species and the composition of communities. In this book we have found this classification to be a useful framework for comparisons between species and a helpful source in those parts of the species accounts that are concerned with prediction and management. In view of the extent of our reliance on functional classification, in this new edition we have added a new chapter (Chapter 2) providing a general introduction to this subject. Chapter 3 reviews the principal sources of data for the accounts. Chapter 4 then explains in detail the contents of the accounts. This is followed in Chapter 5 by illustrations of the use of this autecological data in ecological research and in the monitoring and management of vegetation.

2 PLANT FUNCTIONAL TYPES

The need for a functional classification of plants

For more than a century it has been recognised that there is a need for a dual classification of organisms. In the first place, a system is required that can describe evolutionary relationships; following the massive contribution of Carl Linnaeas and, much more recently, through the widespread application of molecular techniques, this classificatory activity is now in an advanced state of development. For the rather different purposes of ecology there is a requirement for a second classification that reflects the way in which the current ecology of each species or population is determined by its functional characteristics.

In certain cases, such as the preservation of rare populations, the purposes of conservation and management are served by research with a very detailed focus on the distinguishing functional characteristics of individual species and populations. Quite clearly, we cannot hope to conduct this type of research on more than a tiny minority of the worlds' flora. However, it is much more common for the task of the ecologist to be that of explaining or predicting the impacts of a changed environment or management regime on the fate of groups of species in communities or even larger units of vegetation and landscape. This raises the question 'Can we recognise general rules concerning the ways in which the present ecology and abundance of types of species are controlled and will respond to change?' For some ecologists, particularly those with detailed knowledge of the field ecology of individual species, the notion of such general guidelines to floristic patterns and functional shifts has been controversial (e.g. Grubb 1992) but others have embraced the concept with enthusiasm. In particular, Raunkiaer (1934), Holdridge (1947), Box (1981), and Woodward (1987) have championed the search for generalisations based upon relationships between climate, plant form, plant physiology and plant function. In a world experiencing rapid climate change there is an obvious requirement to sustain this line of research.

The need for a comprehensive framework for interpretation and prediction of vegetation characteristics and dynamics extends beyond relationships with climate. We also require a basis upon which to understand how vegetation responds to soil conditions and is affected by competition between neighbours and by animal populations, pests and pathogens. Perhaps most important of all, there is an urgent need to analyse the mechanisms by which, at a regional and global scale, floras are responding to changing exploitation and management by humankind. It is not immediately obvious why we should expect such ambitious objectives to be attainable through the development of a classification of plant functional types. From taxonomic (evolutionary) classifications of plants it is abundantly clear that the Plant Kingdom is the result of a long and complex history resulting in an immense variety of forms and ecologies. Challenged by such variety, how can we recognise criteria that are sufficiently clear and comprehensive to provide a key to plant functional types and general guidelines for management and conservation? In order to address this question it is necessary to recognise that, despite their apparent diversity, the evolutionary and ecological pathways available to plants (and animals) have been severely constrained, in some quite fundamental respects, by the laws of physics and chemistry.

Tradeoffs; the basis for the evolution of widely-recurrent plant functional types

On first inspection, the notion of widely-recurrent plant functional types does not seem to be compatible with what we know of the process of evolution by natural selection. We are all familiar with the occurrence of variation both between and within many plant populations. This ever-present reminder of the lability of plant form and function and of the genetic potential for local 'fine-tuning' of the ecological potential of species by natural selection presents a picture of plant evolution as a continuous and infinitely diversifying process hardly consistent with the idea of a functional typology. **But this is to allow the micro-evolutionary processes**

PANEL 1 THE RESEARCH CONTRIBUTION OF L.G. RAMENSKII

The legacy of Darwin's theory did not contain any general perspective about the ways in which the process of natural selection was constrained by the laws of physics and chemistry. It was apparent that each animal and plant species occupied a restricted range of habitats, but there was no general theory to explain the basis of such restriction. It is only during the last twenty years that ecologists in the West have been made aware (Rabotnov 1985) of the pioneering contribution of L.G. Ramenskii (1938) on this subject.

Ramenskii's achievement was to begin the process of recognising fundamental and inescapable constraints in plant design. Certain avenues of adaptive specialisation, in particular circumstances, would inevitably lead to the sacrifice of fitness in certain others. We now know that there are numerous such constraints that limit (to varying extents) facets of the ecology of individual species and populations. It is remarkable that Ramenskii was not content to merely comment on the possibility of tradeoffs in plant design. Without the insights now available from comparative physiology and working mainly by field observation, he proposed that some constraints were so integral to the core functioning of plants that they surfaced throughout the world and predetermined the paths of ecological specialisation available to The Plant Kingdom.

References and additional reading: Rabotnov (1985), Ramenskii (1938)

PANEL 2 TRADEOFFS IN BRITISH PLANTS: EARLY SPECULATIONS

Some of the first clues to the importance of trade-offs in restricting the ecological ranges of native plants coincided with the development of glasshouses and, at a later stage, plant growth chambers, that permitted direct comparison of the performance of species of contrasted ecology under standardised conditions. Particularly where experiments included enough species to allow a broad perspective it was frequently observed that some plants consistently failed to exploit conditions that allowed others to prosper. The word 'trade-off' was not widely used in ecology until later in the last century but the origins of the concept are apparent in publications that began to establish connections between success in one set of environmental conditions and failure in certain others. The examples in Table A are drawn from papers in which various such connections were made by speculating about the advantages and disadvantages associated with particular traits in some common plants.

Table A Early speculations concerning tradeoffs in herbaceous plants (adapted from Grime 1965).

Species	Traits	Advantage	Disadvantage
1 *Bromopsis erectus*	Slow growth, low transpiration	Drought tolerance	Weak competitor on moist soils
2 *Deschampsia flexuosa*	Resistance to metal toxicity	Tolerant of acid soil conditions	Susceptible to lime chlorosis
3 *Pilosella officinarum*	Flattened rosette	Can exploit dry rocky terrain	Cannot survive in tall vegetation
4 *Arenaria serpyllifolia*	Numerous small seeds	Rapidly colonises bare soil	Fails to establish in closed turf

References and additional reading: Grime (1965), Grime & Hodgson (1969), Grime & Jeffrey (1965), Olsen C (1958).

---

that undoubtedly exist and should not be ignored to dominate our perspective. For too long, such emphasis has obscured the constrained and repetitive nature of plant macro-evolution and has deprived Plant Ecology of the conceptual framework essential to its development as a predictive science.

Darwin's theory of natural selection (Darwin 1859) represented a giant stride for ecology as it did for all areas of biology but this did not immediately provide a general perspective about the ways in which the processes of natural selection would be constrained and channelled by the limited potentiality of the organisms themselves. However, at a comparatively early stage, some ecologists (e.g. Macleod 1894; Ramenskii 1938) had described widely-recurrent patterns of functional specialisation in the life-histories of plants and there can be little doubt that these authors recognised that particular plant types coincided with particular ecological conditions and forms of natural selection. It is also probable that they had also begun to connect success in one set of circumstances with failure in others. This is very likely in the case of Ramenskii (1938, see Panel 1) who had recognised fundamentally different mechanisms of resource capture representing tradeoffs between priority of access ('plants like lions'), opportunism ('plants like jackals') and tolerance of shortage ('plants like camels'). By these rather graphic analogies Ramenskii had clearly begun the search for the internal constraints and inevitable tradeoffs that would later (Grime 1965, see Panel 2; Stearns 1976; Shugart 1997) provide the basis for a more formalised search for plant functional types.

Here we define a tradeoff as follows:

'A tradeoff is an evolutionary dilemma whereby genetic change conferring increased fitness in one circumstance inescapably involves sacrifice of fitness in another.'

(Grime 2001)

In recent years, building on the pioneering ideas of Macleod and Ramenskii, evidence from diverse schools of research has begun to confirm the existence in animals and plants of recurring functional types. It is also becoming clear that their recognition can often provide a key to understanding the structure and dynamics of communities and ecosystems. However in order to avoid any confusion it is necessary to point out that it is only recently that the terminology in this field of research has tended to be dominated by use of 'functional types'. Over the period 1960-1990, ecologists tended to use the word 'strategy' to describe a recurrent type of specialisation associated with particular habitat conditions or niches.

Scientists differ sharply in their attitudes to the use of the term 'strategy' in ecology. Some theorists (Hutchinson 1959, MacArthur & Wilson 1967, Pianka 1970, Southwood 1977, Maynard-Smith 1982, May & Seger 1986) have used the term freely, while others (Harper 1982, Godwin 1985) have taken strong exception to it. With its teleological and anthropomorphic connotations the term is not ideal, and it is understandable that some biologists have preferred to use neutral expressions such as 'functional type', 'set of traits' or 'syndrome' (Stebbins 1974).

We feel no special commitment to the term 'strategy' but in this book we will use it occasionally and interchangeably with 'functional type' as a mark of respect for those who first used it in this context. Their achievement was to recognize that organisms may exhibit sets of traits which are predictably related to their ecology. Here a functional type (strategy) is defined as 'a grouping of similar or analogous genetic characteristics which recurs widely among species or populations and causes them to exhibit similarities in ecology' (Grime 1979)

Data shortages: a world-wide problem

Functional types can be detected or validated by conducting an objective search in which sets of values measured on variable traits such as leaf nitrogen concentration, seedling growth rate and leaf decomposition rate have been found to vary in concert with each other and to coincide with similar plant ecologies. Ideally a large number of species of contrasted ecology should be screened using standardised methods (Grime *et al.* 1997, see Panel 3), the traits investigated should relate to as many variable aspects of plant functional specialisation as possible and multivariate analysis techniques should be used both to recognise positive and negative associations between values for particular traits and to detect circumstances where sets of trait values recur so consistently that functional types can be recognised. Unfortunately, this

PANEL 3 PROTOCOLS FOR SCREENING PLANT TRAITS AND RECOGNISING PLANT FUNCTIONAL TYPES

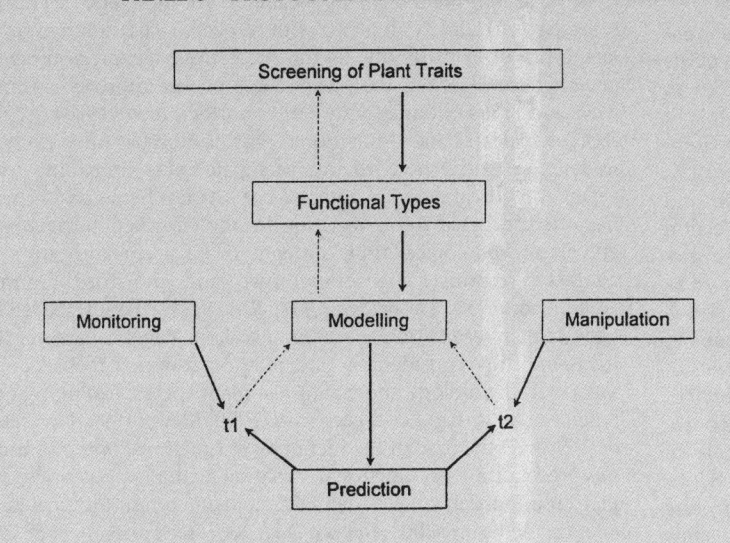

Figure A Protocol for testing the predictive value of plant functional types. Discrepancies revealed in the tests at t1 and t2 initiate further modelling cycles, each of which may necessitate refinement of the functional types or even additional screening. (Reproduced from Grime 1993 by permission of *Oikos*.)

With remarkable foresight, the renowned Victorian artist and philosopher John Ruskin explained the purposes of screening plant traits as follows:

Now what we especially need at present for educational purposes is to know, not the anatomy of plants, but their biography – how and where they live and die, their tempers, benevolences, malignities, distresses, and virtues. We want them drawn from their youth to their age, from bud to fruit. We ought to see the various forms of their diminished but hardy growth in cold climates, or poor soils; and their rank or wild luxuriance, when full-red, and warmly nursed. And all this we ought to have drawn so accurately, that we might at once compare any given part of a plant with the same part of any other, drawn on the like conditions.

John Ruskin *Aratra Pentelici*, 1872

A more strictly scientific case for screening is made by Grime and Hunt (1975) and, more recently, specific protocols have been suggested (Diaz & Cabido 1997, Knevel *et al.* 2003) in attempts to provide an objective basis for the recognition of plant functional types. These protocols (e.g. Figure A) also may contain suggestions for the use of screening results in attempts to interpret vegetation responses to changes in land use and climate.

References and additional reading: Diaz & Cabido (1997), Grime (1993), Grime & Hunt (1975), Knevel *et al.* (2003)

---

experimental protocol is time-consuming, costly and initially unrewarding in terms of scientific publications. Hence, at a time when many ecologists and managers of natural resources are eager to use plant functional types as a basis for interpretation and prediction of the changing state of vegetation and ecosystems, implementation is often difficult because we lack the critical data required for functional classification.

Plant ecologists have responded in a variety of ways to the problem of developing a functional approach in circumstances where national, regional or local floras have not been characterised on the scale required:

Research has been conducted to find quick and easy measurements that are sufficiently informative and reliable to replace more costly or time-consuming procedures. Used with care, these so-called 'soft tests' (Hodgson *et al.* 1999) hold considerable promise. Particular attention has focussed (Wilson *et al.* 1999) on two very easily measured leaf traits, specific leaf area (SLA) and leaf dry matter content (DMC) which are often correlated with other plant attributes such as seedling growth rate, defence against herbivores and rate of leaf decomposition. Determinations of SLA and DMC have been added to the Species Accounts and in Chapter 5 there is a brief review of the advantages and limitations of these two 'soft' predictors.

In a much simplified extension of the 'soft' approach Westoby (1998) has advocated a scheme in which each species is characterised by reference to only three soft traits (SLA, leaf canopy height and seed size): whilst recognising that each of these traits provides some clues to particular aspects of plant ecology we feel impelled to point out that the Westoby scheme leaves untouched many aspects of vegetation and ecosystem functioning in which ecologists and resource managers have a vital interest.

In those floras where very little screening of functional traits has been carried out quite rapid progress can be achieved by restricting attention to a relatively small number of carefully-selected traits of proven informational value in other floras. This approach has been applied on a massive scale, embracing approximately 3000 plant species of North-West Europe (Knevel *et al.* 2003). A further example of this research tactic is provided by Diaz *et al.* (2004) where traits previously used to characterise a local flora in England were found to be equally useful as predictors of variation in ecosystem properties under very different conditions in Argentina, Iran and Spain.

Not all plant ecologists have been persuaded of the need to develop comprehensive databases of plant traits as a basis for interpreting or predicting vegetation and ecosystem properties. In the United States and in Australia, there has been comparatively little activity beyond reviews that have searched for patterns of variation in plant traits based upon existing data sources in the published literature (Reich *et al.* 1992, Wright *et al.* 2004). Whilst this is a legitimate line of enquiry it has inherent weaknesses in terms of the lack of standardisation in methods when such diverse sources are consulted and the inevitable focus upon those traits and species for which most data are available.

In view of the large volume of ecological research conducted in the United States it is perhaps surprising that the screening of plant traits and the recognition of plant functional types have not been pursued more vigorously in this part of the world. Differences in history and philosophy appear to be taking many European and North American plant ecologists along rather different paths at the present time. As explained in Panels 6 and 7, one consequence of this divergence has been a protracted international debate (Grime 1974, Tilman 1982) that involves very different characterisations of the primary plant strategies and only recently appears to be approaching a resolution.

Different kinds of functional types for different ecological purposes

Plant functional types have been used for a range of different purposes using a variety of criteria. In each of the Species Accounts in this book particular attention has been given to the need to distinguish between the strategy of the established plant and that of the juvenile phase. The need for this separation became apparent through the work of several ecologists, including Stebbins (1951, 1971,1974), Wilbur *et al.* (1974) Grubb (1977), Gill (1978) and Grime (1979), who recognized the peculiar nature of the selection forces and design constraints which determine the characteristics of the offspring as distinct from those of the adult. This has led to the conclusion that there are distinct regenerative strategies which differ in such respects as degree of resource investment, mobility and dormancy, and which confer different but predictable sets of ecological capacities and limitations upon the organism. In each of the Species Accounts we have characterised the regenerative strategies and Chapter 5 contains a discussion of the role of different types of offspring in the ecology of plants. In many plants, the same genotypes may be capable of regenerating by several quite different mechanisms such as by developing stolons that allow local expansion and by releasing wind-dispersed seeds or spores that may simultaneously colonise distant habitats. This leads to the hypothesis (Grime 1979) that ecological amplitude is determined not only by genetic variation and by phenotypic plasticity, but also by regenerative flexibility, a function of the number and identity of the regenerative strategies exhibited by the species.

A second major source of variation in plant functional types is related to the geographical scale at which they are applied. Panel 4 classifies functional types arbitrarily into those with *global, regional, local* or *within-population* applicability. At the *global* scale the objective is to classify all plants with respect to variation in a set of key traits of consistently useful predictive value regardless of geographical location. Early exponents of this approach include MacLeod (1894), Ramenskii (1938) and Rabotnov (1983). At the *regional* scale in Panel 4 functional classification has often evolved in relation to the need to interpret patchiness coinciding with vegetation changes across climatic boundaries: here classification has often focussed upon variation in photosynthetic mechanism and in morphological traits thought to be useful indicators of adaptive responses to temperature and moisture regimes (Raunkiaer 1934, Holdridge 1947, Box 1981, Woodward 1987, Cunningham *et al.* 1999). The *local* scale in Panel 4 narrows the focus of study considerably and may be confined to particular aspects of the biology and ecology of plant species exploiting one type of habitat (e.g. Landsberg *et al.* 1999, Kleyer 1999, Lavorel *et al.* 1999). At the base of the hierarchy of functional types we must now recognise that functional variation within populations can play an important role in maintaining plant community structure (Booth and Grime 2003, Reusch *et al.* 2005, see Panel 16, page 72).

Finally, it is necessary to acknowledge that the ecologists and managers who are using plant functional types may have very specific objectives that necessitate careful choice of functional types. In other cases however the same system has been shown to be applicable to a diversity of tasks including the detection of the impacts of landuse changes and the measurement of resistance and resilience following extreme climatic events (MacGillivray *et al.* 1995, Hunt *et al.* 2004).

PANEL 4   A HIERARCHY OF PLANT FUNCTIONAL TYPES

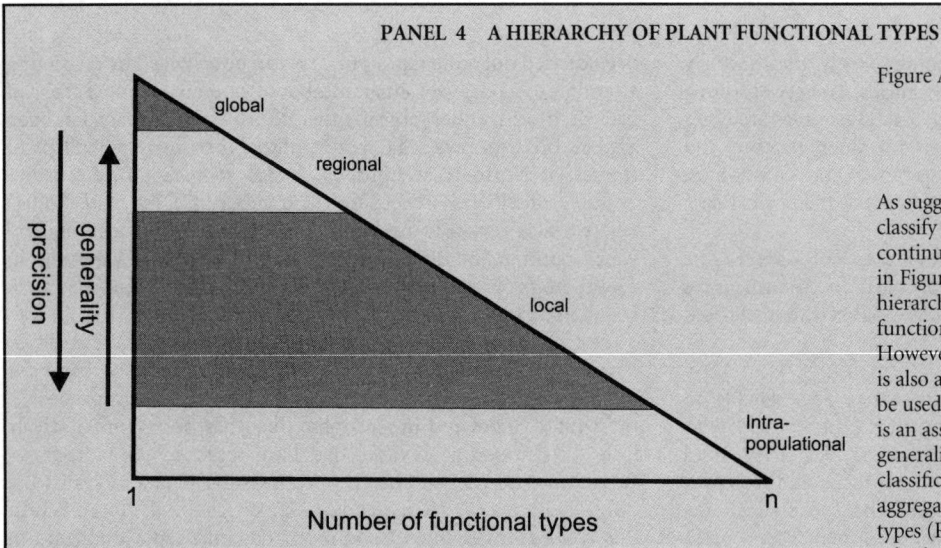

Figure A. A hierarchy of plant functional types.

As suggested by Day *et al.* (1988), it is useful to classify plant functional types within a continuous hierarchy. In the scheme illustrated in Figure A the most obvious feature of the hierarchy is the geographical scale at which the functional types are most likely to be useful. However, as we move from top to bottom there is also an inevitable shift in the criteria that can be used to define the functional types and there is an associated change in emphasis from generality to precision. At the **global** scale are classifications that include all plants and aggregate them into a small number of primary types (Raunkiaer 1934, Ramenskii 1938) using criteria that are readily available for most plant species. In early studies this led to heavy reliance upon plant morphologies and life histories but, with the advent of screening tests (see Panel 3), greater weight has been placed upon variation in the capture and utilisation of resources and the characteristics of juvenile plants (seeds and seedlings).

At the **regional** scale, plant functional types are often defined in ways that help us to recognise adaptive solutions to the challenges associated with each of the world's major climatic zones. Here, there is usually a strong focus upon traits that reflect plant specialisations related to particular combinations of temperature and moisture supply (Box 1981, Woodward 1987). In addition, plant functional types at this intermediate geographical scale may be chosen to enable ecologists to recognise circumstances where variation in geology and soils are suspected to be acting as a dominant factor.

In work at a **local** scale (e.g. Lavorel *et al.* 1999) it has been found possible to define plant strategies more narrowly, even to the extent of examining variation in only a small number of traits in one type of habitat.

Finally at the base of the hierarchy of plant functional types exploratory studies are in progress to measure the extent to which we can recognise functional types within local populations of individual species (see Panel 16).

References and additional reading: Booth & Grime (2003), Box (1981), Day *et al.* (1988), Grime (2001), Lavorel *et al.* (1999), Woodward (1987)

Primary plant functional types

In the preceding section of this chapter and later (in Chapter 5), we have emphasised the great variety of possibilities that are now available to use variable plant traits, either individually or by aggregating them into plant functional types, to analyse patterns and processes in vegetation and ecosystems at many different geographical scales. However, it is implicit in Panel 4 that some plant functional types are more important than others. At the apex of the hierarchy depicted in this figure there is a clear assertion that some axes of variation in plant traits occur universally and affect all members of the Plant Kingdom. For this to be true it is necessary that there should be certain constraints and tradeoffs that operate inescapably upon the evolution and current ecology of all plants regardless of their detailed ecology and taxonomic identity. In a search for primary plant functional types, this usefully narrows our search to traits and tradeoffs in components of plant functioning that are common to all plants such as the capture of resources and their utilisation in growth, defence and reproduction.

On the basis of researches by many ecologists over the last forty years we conclude that such universal constraints and tradeoffs do, in fact, occur and have determined that, with respect to core physiology, biochemistry and life-history, all plant evolution is predictably channelled into one of a small number of alternative pathways each predictably related to key habitat characteristics. As explained in greater detail elsewhere (Grime 2001), this conclusion has two implications of potentially revolutionary importance for the future conduct of ecological science:

While Linnaean taxonomy will remain essential to ecology as an unequivocal identifier of field and experimental material, we should not expect this form of classification to map reliably in relation to the ecology of species. It is useful, of course, for ecologists to be aware that the current ecology of certain plant taxa continues to be constrained by ancestral specialisations (Stebbins 1951, Hodgson 1986c) but we suggest that this kind of general insight is no substitute for quantitative determinations of species traits. Some plant families, for example, show extremely wide variation with respect to traits of fundamental importance in vegetation and ecosystem functioning (Diaz et al. 2004).

Until quite recently efforts to recognise functional types of plants and animals (e.g. MacArthur and Wilson 1967, Pianka 1970, Harper 1977, Caswell 1989) have focussed on their population biology and demography with the objective of understanding the evolution and ecology of individual species and populations. More recently the perspective has expanded to include differences in the capture, utilisation and release of resources. To a crucial extent this has transformed the study of plant functional types from an activity concerned exclusively with the ecology of plant populations to an engagement with global changes in vegetation and study of the role of plants as controllers of ecosystem structure and dynamics.

In succeeding sections of this chapter, attention is focussed on the CSR model of primary plant functional types. We believe that this emphasis is now justified in the light of the large quantity of supporting evidence (Panels 5-8) and in view of the extent to which the model has been implemented. However, it is important to recognise that, over the last 20 years there have been other models and much debate. In particular, as explained in Panels 6 and 7, David Tilman has developed an approach to this subject that is distinct from the CSR system in proposing competition for resources as the dominant mechanism structuring all plant communities (Tilman 1982) and in the extent to which trade-offs between root and shoot competitive abilities have been granted a defining role in recognising primary strategies (Tilman 1988).

The CSR model

The CSR model, to which we refer in each of the Accounts, originates from the suggestion (Grime 1974) that it is useful to classify the external factors which affect vegetation into two broad categories. The first, which we may describe as *stress*, consists of the phenomena which restrict photosynthetic production, such as shortages of light, water and mineral nutrients, or sub-optimal temperatures. The second, here referred to as *disturbance*, is associated with the partial or total destruction of the plant biomass

Table 2.1 The basis for the evolution of three strategies in plants. (From Grime 1977.)

Intensity of disturbance	Intensity of stress Low	High
Low	Competitors	Stress-tolerators
High	Ruderals	(No viable strategy)

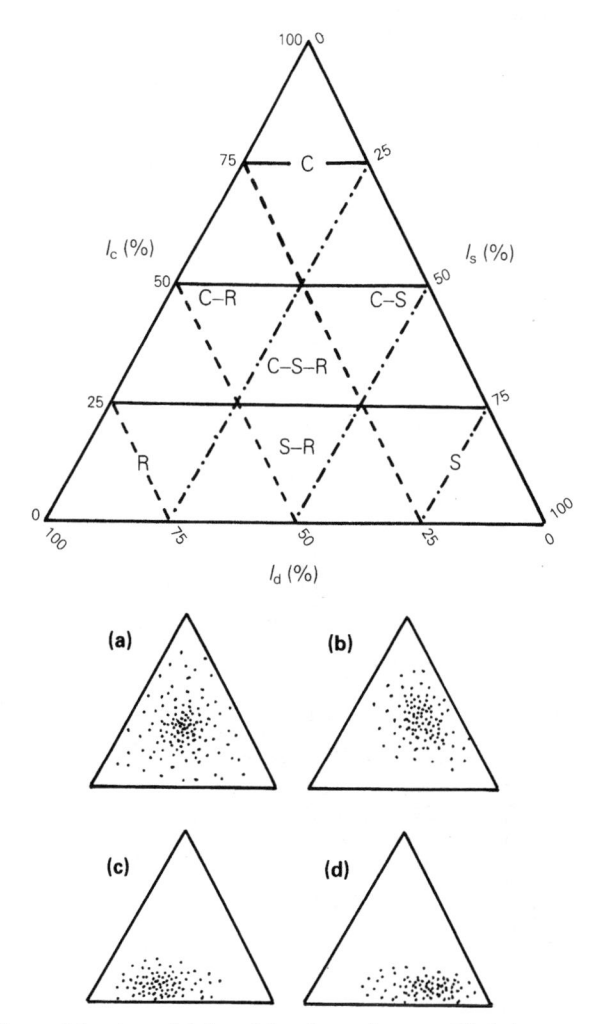

Figure 2.1 A model describing the various equilibria between competition, stress and disturbance in vegetation and the location of primary and secondary strategies. C, competitor; S, stress-tolerator; R, ruderal; C-R, competitive-ruderal; S-R, stress-tolerant ruderal; C-S, stress-tolerant competitor; C-S-R, 'C-S-R strategist'. I_c, relative importance of competition (—); I_s, relative importance of stress (– . –); I_d, relative importance of disturbance (---). The strategic range of four life forms is also shown: (a) herbs, (b) trees and shrubs, (c) bryophytes and (d) lichens.

PANEL 5 THE INTEGRATED SCREENING PROGRAMME (ISP): THE MAIN RESULTS

Already in Panel 3 it has been explained that, if we are to have confidence in plant functional types as a basis for explaining or predicting variation in vegetation and ecosystem structure and functioning, it is necessary to devise an experimental protocol that validates their existence. In the special case where our objective is to test for the existence of primary plant strategies(see Panels 6 and 7), there are additional requirements. It is essential that the screening is applied to a wide range of species of contrasted ecology and it is also necessary that the selected traits refer to all aspects of the core functioning of the plant and include both juvenile and mature stages of the life-history.

The final list of 43 species and 63 traits involved in the Integrated Screening Programme (ISP) conducted in Sheffield in the 1990's emerged from many discussions between participating scientists and resulted in an extensive programme of laboratory procedures extending over a period of five years. The main results of the ISP can be summarised as follows:

(1) Multivariate analysis revealed that many of the screened traits varied widely across the 43 species and were positively or negatively associated with variation in other traits. The strongest axis (Axis 1) accounted for 22% of the total variation in the ISP dataset and revealed strong linkage between leaf nutrient concentrations, seedling growth rate, root and shoot foraging responses, the longevity, tensile strength, and palatability of leaves and the rate of decomposition of leaf litter. This axis of variation confirmed the occurrence across the species of a trade-off, predicted from CSR theory between a set of trait values conferring an ability for high rates of resource acquisition (and loss) in productive habitats and another set dictating retention of resources in unproductive conditions.
(2) The second axis of variation in the ISP data was essentially taxonomic in character and separated monocotyledons from dicotyledons. More interesting for the purposes of ecology, the third axis reflected variation in length of life-history and when plotted against Axis 1 (see Figure A) produced a distribution of species within a triangular space that was highly consistent with CSR theory.
(3) Many of the traits responsible for the distribution of species in the figure below (e.g. palatability to generalist herbivores and rates of litter decomposition) not only involve aspects of the core functioning of all plants but are also deeply implicated in the functioning of ecosystems. This points unmistakably to a key role for primary plant strategies as controllers of ecosystem productivity and the circulation and storage of resources (Panel 9) and pollutants (Panel 19).
(4) Variation in many other ecologically-important traits, particularly those relating to seeds and seedlings, was found to be unrelated to the axes of the CSR model. This illustrates the way in which a trait-based analysis can allow all plants to be classified with respect to primary functional type without compromising the idea that each may possess a distinctive ecology. Arising from this duality is the suggestion (Panel 9), that whilst the traits that define the primary strategy have a determining role in relation to ecosystem functioning there are others that merely control admission and persistence in the plant community.

References and additional reading
Grime *et al.* (1997), Grubb (1977), Diaz *et al.* (2004), Thompson *et al.* (1996), Wright *et al.* (2004)

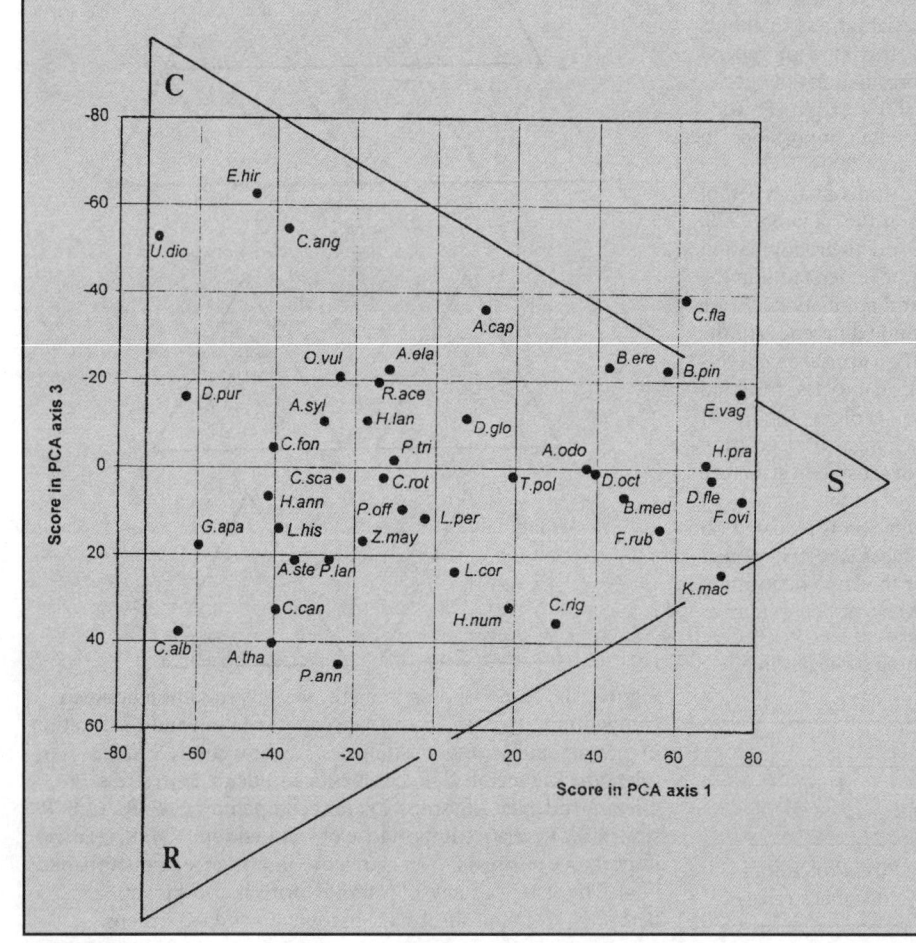

Figure A A scatter-diagram of species positions on Axes 1 and 3 of the PCA conducted on the established phase in the Integrated Screening Programme. The superimposed equilateral triangle has been inserted to indicate the consistency of the distribution of species with the three poles of the CSR system (Grime 1974, 1977, 1979). The species are: *Agrostis capillaris, Anisantha sterilis, Anthoxanthum odoratum, Anthriscus sylvestris, Arabidopsis thaliana, Arrhenatherum elatius, Brachypodium pinnatum, Briza media, Bromopsis erecta, Campanula rotundifolia, Carex flacca, Catapodium rigidum, Centaurea scabiosa, Cerastium fontanum, Chamerion angustifolium, Chenopodium album, Conyza canadensis, Dactylis glomerata, Dryas octopetala, Epilobium hirsutum, Eriophorum vaginatum, Festuca ovina, Festuca rubra, Galium aparine, Helianthemum nummularium, Helianthus annuus, Helictotrichon pratense, Holcus lanatus, Koleria macrantha, Leontodon hispidus, Lolium perenne, Lotus corniculatus, Origanum vulgare, Pilosella officinarum, Plantago lanceolata, Poa annua, Poa trivialis, Rumex acetosella, Thymus polytrichus, Urtica dioica, Zea mays.* (Reproduced from Grime *et al.* 1997 by permission of *Oikos*.)

and arises from the activities of herbivores, pathogens and humans (trampling, mowing and ploughing), and from phenomena such as wind-damage, frosting, droughting, soil erosion and fire. When the four permutations of high and low stress with high and low disturbance are examined (Table 2.1) it is apparent that only three of these are viable as plant habitats. This is because the effect of continuous and severe stress in highly-disturbed habitats is to prevent a sufficiently rapid recovery or re-establishment of the vegetation. It is suggested that the three remaining contingencies in Table 2.1 have been associated with the evolution of primary strategies of the established phase which conform to three distinct types (Ramenskii 1938, Grime 1974). These are the *competitors*, exploiting conditions of low stress and low disturbance, the *stress-tolerators*, associated with high stress and low disturbance, and the *ruderals*, characteristic of low stress and high disturbance. The three strategies are, of course, extremes of evolutionary specialization. There are others which exploit the various intermediate conditions which correspond to particular equilibria between stress, disturbance and competition. These can be displayed in a triangular diagram, which can be used also to indicate the strategic range of various life-forms. At the present time the strongest evidence supporting the CSR model relates to the results from the Integrated Screening Programme (Panel 5) but, as explained in Panel 7, investigations of various kinds have provided information consistent with the model. At this point it is necessary to clarify two particular aspects of the model; first, the nature of the equilibrium between stress, disturbance and competition and, second, the selection processes associated with these three phenomena.

The equilibrium between stress, disturbance and competition

The CSR model proposes that the vegetation which develops in a particular place and at a particular time is the result of an equilibrium which is established between the intensities of stress (constraints on production), disturbance (physical damage to the vegetation) and competition (the attempt by neighbours to capture the same unit of resource). In this model stress and disturbance control the intensity of competition by restricting the density and vigour of the vegetation. In the short term this control is exerted through immediate impacts on the established plants by either restricting the rate of plant growth (stress) or by destroying plant biomass (disturbance). Over a longer time-span control occurs by modification of the species and genotypic composition of the vegetation through selective effects on extinctions and immigrations. Where stress and disturbance remain low, large, fast-growing plants of exceedingly high competitive ability will invade, eventually occupying the site and, as an inevitable consequence of competitive exclusion, cause a drift towards monoculture. Where severe stress and frequent disturbance coincide, no vegetation is possible. The viable equilibria between stress, disturbance and competition occupy a triangular area and, as explained earlier on this page, the characteristics of the plants expected to occupy a particular position within the triangle are predictable from their location at specific intersections between the three co-ordinates of the triangle.

In the real world the CSR equilibrium varies from place to place, even within a plant community, and on diurnal, seasonal and

---

PANEL 6   PRIMARY FUNCTIONAL TYPES: THE DEBATE

Attempts to recognise, on a world-scale, a small number of recurring types of plants that inform us about variation in the structure and dynamics of vegetation and the functioning of ecosystems began more than a century ago (MacLeod 1894) and this search continues to the present day. The purpose of this panel is to classify these attempts according to the different methods and sources of information that have inspired them; in this way it is possible to identify some of the reasons why this subject has stimulated debate over such a long period of time.

Approach 1: Functional types based on field observations and study of plant life-histories
Even to a casual observer it is obvious that plant species may differ consistently with respect to life-span, size at maturity and the timing and extent of reproductive activity. This prompted MacLeod (1894) to attach primary significance to the distinction between 'capitalists' that were long-lived and monopolised large areas of habitat and 'proletarians' that were short-lived and exhibited a fugitive existence. Much later, this idea was formalised as the theory of r- and K-selection (MacArthur and Wilson 1967) and more recently it has surfaced again as a proposed trade-off between competitive and colonising abilities (Hastings 1980, Tilman 1994).

Approach 2: Functional types based on theories of plant competition for resources
Tilman (1982) proposed that theories based upon the competitive abilities of phytoplankton populations in surface waters (Tilman 1976) could be extended to vascular plants rooted in soil. On this basis he suggests that the outcome of competition for particular mineral nutrients is determined by differences in the external concentrations below which competing species fail to extract a sufficient supply for growth. Later this approach was applied to the task of defining primary strategies (Tilman 1988) by proposing a trade-off between the above- and below-ground competitive abilities of plants. Through the work of researchers on plant mineral nutrition and soil physics over a long period (Panel 7) it has become apparent that the Tilman models are in conflict with our current understanding of the behaviour of nutrients and plant roots within the soil and with our knowledge of the interdependence of root and shoot competitive abilities (Donald 1958, Colasanti & Hunt 1997, see also Panel 7).
A more restricted role for resource competition is proposed in the theory advanced by Ramenskii (1938) and Grime (1974). Here it is suggested that:
(a) competition for resources reaches maximum intensity and importance as a determinant of vegetation in circumstances where resource abundance imposes few constraints on plant production and large, fast-growing plants monopolise the habitat and competitive exclusion drives the vegetation toward monoculture
(b) resource competition declines in importance in habitats that have fewer resources or are subject to a high intensity of vegetation destruction.

Approach 3: Functional types based upon screening plant traits
As explained in Panels 2, 3 and 5, the search for primary plant strategies has not always proceeded as a test of specific theories. With growing momentum over the past 50 years, empirical studies have generated a fund of knowledge about the way in which plant traits vary in relation to each other and in relation to environmental factors and vegetation management. In retrospect we can now see that this pragmatic approach has played an influential role, sometimes prompting theory (Panels 2 and 3) and at other times (Panel 7) allowing falsification of existing ideas.

References and additional reading: Colasanti and Hunt (1997), Donald (1958), Grace (1990), Grime (1974), Hastings (1980), MacArthur & Wilson (1967), MacLeod (1894), Ramenskii (1938), Thompson (1987a), Thompson & Grime (1988), Tilman (1976, 1982, 1987, 1988, 1994).

successional timescales. For this reason communities often contain species of widely differing strategy. In modern, floristically-depauperate landscapes many species may be under-dispersed, and there may be delay in adjustment of the species composition of the vegetation to changes in the intensities of stress or disturbance. Here we suspect that the role of intraspecific variation in the occupying species will be enlarged.

Selection processes and evolutionary responses

In the preceding section and in Grime (1979, 2001) it has been argued that the vegetation that develops in any particular place is the result of three interacting forces each of which achieves greatest potency in different but well-defined and easily-recognisable circumstances. These can be summarised as follows:

Competition for resources

The struggle between neighbours for preemptive capture of the resources needed for plant growth exerts maximum impact in productive, relatively undisturbed habitats where large, fast-growing plants enjoy a competitive advantage. Often a consequence of the struggle for resources is competitive exclusion and suppression of diversity to a point where local areas may be occupied by only a single species. (note, however, as explained in Panels 10 and 11, pages 67 and 68, that there are mechanisms apart from competition that are capable of reducing species diversity).

Resource shortage

This phenomenon is most evident on droughted or nutrient-impoverished soils and is associated with selection of long-lived, inherently slow-growing plants. Plant vigour may be so reduced that individuals grow in isolation from neighbours or exhibit only weak competitive interactions. In Panel 7, on this page, recent evidence is cited that documents the decline in competition in unproductive habitats and demonstrates experimentally that under very unproductive (stressed) conditions plants frequently benefit from the presence of neighbours.

Disturbance

This process is most pronounced in circumstances where potential productivity is high but development is frequently interrupted by cultivation techniques or climatic events that partially or completely destroy the vegetation. Under such conditions where plants are repeatedly killed soon after their establishment there is strong selection for short-lived individuals that reproduce early.

The idea that all plants can be classified with regard to the same fundamental tradeoffs resulting from their exposure during their evolutionary history to particular equilibria between these three selective mechanisms has generated controversy over a long period of time. However, as explained in Panel 7, this debate is now approaching a resolution as new data becomes available. Before proceeding further it may be helpful to identify the main objections to CSR theory and the reasons why we have remained persuaded of its validity and usefulness from 1974 to the present day.

Objections to CSR theory

Perhaps the greatest, but often unstated, objection to CSR theory arises from disbelief that the selection processes and evolutionary responses associated with only three phenomena (competition, stress and disturbance) are sufficiently predictable to form the basis for a universal functional classification of plants. Following the theories of Stebbins (1974), it may be supposed that, according to

---

PANEL 7 PRIMARY FUNCTIONAL TYPES: IS THE DEBATE OVER?

In Panel 6 it is suggested that there have been three main mechanisms for the development of ideas concerning primary plant strategies. In the first and second, general theories emerged at an early stage either from direct observation of plants in the field or by development of mathematical or verbal models based upon principles of life-history evolution or resource competition. The third mechanism has been more empirical; here ideas have developed more slowly as interpretations of patterns detected in data obtained by screening traits in a large number of species of similar and contrasted ecology.

So what is the current status of theories resulting from the interaction of these three approaches? In order to address this question it is useful first to be reminded of an observation made by Steven Stearns (Stearns 1976). He noted that even at an early stage of its history there was an imbalance in this research field between model development and model testing. He complained that ideas often achieved wide currency without ever being subjected to rigorous comparison against the realities of the field or laboratory. Until recently this situation has continued with the theoretical 'hare' far outstripping the empirical 'tortoise'. Perhaps we should not be surprised by this state of affairs: models are easier to construct and appreciate than multi-trait and multi-species databases and their appeal to teachers and textbook writers is more immediate and compact.

The problem of imbalance between model development and model testing has particular relevance to the ideas of primary plant functional types suggested by Tilman (1982, 1988). These have achieved wide exposure despite rather obvious conflicts with both theory and empirical data, some of which have been apparent for two decades or longer. Specifically, the failure of the Tilman models to accommodate the severe nutrient gradients surrounding competing roots, recently re-emphasised by Craine *et al.* (2005), has been widely understood by experts in soil physics and root functioning (Bhat & Nye 1973, Nye & Tinker 1977, Thompson 1987a, Huston & DeAngelis 1994) but, until recently, it appears to have escaped the attention of many ecologists.

We can conclude therefore on the relatively narrow but theoretically crucial issue of variation in root competitive ability for mineral nutrients that the mechanism proposed by Tilman (1982,1988) is not supported.

On the broader question of which models have been validated or falsified by trait screening there is rapidly accumulating support for CSR theory from the large comparative databases (Field & Mooney 1986, Reich *et al.* 1992, Grime *et al.* 1997, Diaz *et al.* 2004, Wright *et al.* 2004) that are becoming available as ecologists in many different parts of the world collaborate to identify the most consistent patterns of trait variation in plants ; all of these place strong emphasis on the transition from 'acquisitive' to 'retentive' functional types along gradients in ecosystem productivity and first validated in the Integrated Screening Programme (Panel 5). Additional support for the CSR model has appeared in a range of more specialised investigations (e.g. Aerts & van der Peijil 1993, Bolker & Pacala 1999, Burt-Smith *et al.* 1998, Henry *et al.* 2004, Mustard *et al.* 2003, Pearce *et al.* 2006, Ladd & Facelli 2005).

A related issue that has divided ecologists for a long time arises from the contention of Tilman that competition for resources remains equally important as a vegetation determinant in productive and unproductive habitats. Recently this interpretation has been falsified by experimental studies comparing the impacts of neighbours on target transplants in a variety of contrasted alpine and low altitude situations around the world (Callaway *et al.* 2002). The results of this investigation not only confirmed that competition declined in low productivity habitats but also revealed positive benefits from close proximity to established vegetation in alpine situations.

References and additional reading: Aerts & van der Peijil (1993), Bhat & Nye (1973), Bolker & Pacala (1999), Brooker & Callaghan (1998), Burt-Smith *et al.* (1998), Callaway *et al.* (2004), Colasanti & Hunt (1997), Craine *et al.* (2005), Diaz *et al.* (2004), Field and Mooney (1986), Fraser & Grime (1997), Grime et al. (1997), Henry *et al.* (2004), Huston & DeAngelis (1994), Ladd & Facelli (2005), Mustard *et al.* (2003), Nye and Tinker (1977), Pearce *et al.* (2005), Reich *et al.* (1992), Stearns (1976), Thompson (1987), Wright *et al.* (2004).

their evolutionary history, taxa will respond in different ways to the same selection pressure. It could be argued that against the different climatic, edaphic and biotic backgrounds afforded by contrasted biomes and habitats, evolutionary responses to competition, stress and disturbance will differ radically. With these complications in mind, Grubb (1995) has suggested that only complex arrays of sets of plant traits can adequately describe the range of functional types actually occurring in nature. Ultimately, these discussions can be resolved only by further research of the kinds reviewed by Grime (1987, 2001). Already, however, two important issues can be clarified:

The existence of great variety in the characteristics of species and populations is not in dispute, nor is the relevance of this detailed variation to our understanding of the 'fine-tuning' of plant ecologies. The critical question refers to a higher level of organization, and concerns the existence (or not) of recurrent predictable patterns of specialization, recognition of which assists functional analysis of communities and ecosystems.

There can be little doubt that evolutionary histories can strongly influence paths of contemporary ecological specialization. The question to be addressed here is whether different evolutionary histories can lead to radically different solutions to the same ecological problem, or whether they merely determine the extent to which particular taxa have been capable of approaching the same basic solution. From diverse schools of research there is already available a considerable amount of evidence to support the view that with respect to basic characteristics of life-history and physiology the paths of ecological and evolutionary specialization in response to high intensities of competition, stress and disturbance are remarkably circumscribed. The nature of these overarching constraints will now be examined.

Competition

Where productive, undisturbed habitats are colonized by robust, perennial plants of high potential growth rate, the zones immediately above and below the ground surface are occupied by a dense, rapidly-expanding biomass. In the aerial environment, a shaded stratum extends upward beneath a rapidly-ascending layer of foliage; mortality is high among individuals which are outstripped by their neighbours and become trapped in the shaded zone. Here the high respiratory burden and etiolated tissues render the suppressed individuals particularly susceptible to pathogens (Vaartaja 1952). However, mortalities in such dense vegetation are not simply the result of events above ground. Competition for light imposes a severe drain on the carbon and energy reserves of the plant, which may lead to starvation of the root system and its confinement to zones of nutrient depletion within the rhizosphere (Nye 1966,1969, Bhat & Nye 1973, Fitter & Hay 1981). Equally important, however, is the demand for mineral nutrients and water imposed by the rapid growth and turnover of foliage characteristic of plants competing within an ascending canopy. The interdependence of root and shoot is therefore a crucial part of the analysis of the selection forces that operate upon competing plants. Elsewhere (Donald 1958, Mahmoud & Grime 1976, Grime 1987, Kadmon & Shmida 1990, Van der Werf *et al.* 1993, Ryser & Lambers 1995) the implications of this scenario have been considered in relation to the phenomenon of covariance in the resource 'foraging' abilities of leaves and roots, (Panel 14, page 71). At this point, however, the essential argument is that competition generates severe spatial gradients in resources, and that selection is likely to favour those genotypes in which rapid morphological adjustments facilitate escape from the depletion zones, sustain resource capture and maintain reproductive fitness.

Stress

A marked contrast is immediately evident when attention is turned to the selection processes operating in habitats where severe stress restricts plant production to a continuously low level.

Here only low rates of capture of the limiting resources are possible, and both survival and reproduction depend crucially upon the capacity of the plant to remain viable through long periods in which little growth occurs. This may confer a selective advantage upon species which uncouple growth from resource intake (Grime 1977). As suggested later (page 47), even where other stresses are conspicuous (e.g. the low temperatures of the tundra and the low precipitation of arid deserts) the ultimate determinants of the selection processes and adaptive responses associated with severe stress are usually chronic shortages of mineral nutrients such as nitrogen or phosphorus, or both. The protection against herbivory of plant tissues (and hence the safeguarding of captured mineral nutrients) is also conspicuous in plants of chronically unproductive habitats (Bryant & Kuropat 1980, Coley *et al.* 1985, Grime *et al.* 1996, Pearce *et al.* 2006).

Disturbance

Where frequent and severe disturbance becomes the dominant influence upon vegetation, natural selection is likely to favour those species and genotypes in which rapid growth and early reproduction increase the probability that sufficient offspring will be produced to allow the survival and continuous re-establishment of the population. There is little difficulty in recognizing the relevance of this pattern of natural selection to circumstances where fertile habitats are subject to frequent mechanical disturbance (e.g. arable fields, gardens and heavily-trampled paths). Here, through human activities, brief opportunities are created which can be exploited by plants with a condensed life-history. However, more-careful analysis is required where the agents of disturbance are climatic. Where, for example, the effect of summer desiccation is to lower the water-level of a pond, the resulting exposure of bare mud may create a temporary but productive habitat which can be colonized by ephemeral plants (Salisbury 1967, Furness & Hall 1981).

An essentially similar opportunity for ephemerals occurs during the vernal phase in some deciduous woodlands on fertile soils and is especially conspicuous in parts of North America (Baskin & Baskin 1985). A further example is provided in certain desert areas where soil fertility is sufficient to allow extremely rapid development of ephemerals following rain showers (Went 1949). It is of vital significance in assessing the predictive value of the CSR theory to recognize that, in all of these instances, ephemerals are exploiting a temporal niche of short duration but relatively high productivity. At those pond margins, woodlands or dry habitats where mineral nutrients are strongly limiting we may predict that the low quality of the 'growth window' will preclude exploitation by ephemerals. Here conditions are more likely to favour an uncoupling of resource-capture from growth, a characteristic of stress-tolerators.

Uses and limitations of CSR theory

In various publications (Grime 1979, 1980, 1986b, Leps *et al.* 1982), and in Chapter 5 of this book, there are accounts of the numerous opportunities for ecological prediction and manipulation of vegetation, which follow from a recognition not only of the strategies of the established phase of plant life-histories, but of the regenerative phase also (p. 47). Of particular importance here are the procedures (Grime 1984) which have begun to take

PANEL 8 WHAT IS HAPPENING TO THE BRITISH FLORA?

In 1986, John Hodgson published four papers in the journal *Biological Conservation (Hodgson (1986a-d)* based upon two decades of botanical fieldwork in an extensive area of North Central England. The main purpose of these publications was to recognise the causes of commonness and rarity in a local flora that included a wide range of habitats representative of most of those occurring in upland and lowland areas of Inland Britain. The conclusions drawn from the investigation followed analysis of a database described in Chapter 3 of this book and comprising approximately 10,000 metre square samples of herbaceous vegetation in each of which floristic and environmental data was collected using standardised methods. To an unusual extent, Hodgson employed sampling procedures (see page 18) that ensured that comparable ecological information was obtained for both common and rare species.

From this unique source two main conclusions were drawn:
1. Although patterns related to geology, climate and early land-use history were detected, the most influential factors determining commonness and rarity in vascular plants are recent and current changes in land-use. Expanding and declining species can be distinguished in functional terms. Greatest vulnerability to extinction is associated with long-lived, slow-growing plants of unproductive habitats (stress-tolerators). Expansion is most evident in fast-growing perennial or ephemeral species (competitors and ruderals) that are associated with fertile soils and frequently disturbed plant communities of agricultural, industrial or urbanised habitats.
2. Commonness and rarity are related to taxonomy and evolutionary history. Current benefits from intensifying and disruptive land exploitation are evident in many members of plant families e.g. Brassicaceae that emerged late in the Angiosperm record whereas more ancient taxa e.g. Rosaceae were more likely to be experiencing decline.

Twenty years after these conclusions were published, it is interesting to consider how far they have been vindicated by subsequent research. Strong support for Hodgson's interpretation of the current trajectories of floristic change in Britain is contained in Figure A, which presents the results of a search for functional shifts in the national floras of several countries in Western Europe. In this study, Thompson (1994) compared the functional characteristics of expanding and declining components and found that in heavily populated countries (England, The Netherlands and the former West Germany) the increasing species included a high proportion of 'weedy' species (ruderals and competitors) whereas the declining species were mainly stress-tolerators. Comparable shifts were not detected in the national vegetation statistics of predominantly rural and less densely populated countries (Scotland, Wales, The Irish Republic and Northern Ireland). This interesting study not only substantiated the conclusions of Hodgson but also showed how human population densities could be used as a predictor of the functional shifts that occur as floras suffer the consequences of intensifying changes in land-use. Later the same approach was used successfully to predict patterns at a more local scale in the loss of rare species in Britain (Thompson & Jones 1999).

There are now two additional sources of information concerning vegetation changes taking place across the British Isles and there is an obvious need to consult these to determine whether the functional shifts reported to be taking place in the earlier studies of Hodgson and Thompson can be confirmed. In the *Countryside Survey* (Barr *et al.* 1993, Bunce *al.* 1999, Firbank *al.* 2000) data have been collected at 10 year intervals from a network of permanent plots to record changes in habitat and vegetation composition within the most abundant habitats of rural landscapes. A very different approach has been applied in the *New Atlas of the British and Irish Flora* (Preston *al.* 2002a) where data is provided for all species, both native and introduced. Here, however, data are on a coarse scale, with presence or absence recorded on a grid of 10 x 10 km squares. Although neither of these surveys is an ideal basis upon which to analyse floristic change, it is notable that both have highlighted dramatic recent changes in the floristic composition of the countryside and have identified the major role of changing land-use in this phenomenon. Using data on plant traits in this book and soon to be available from the LEDA screening programme (Knevel *et al.* 2003) it will be possible to perform definitive tests to establish the extent to which the functional shifts reported in earlier studies are now of general occurrence in heavily exploited areas of the British Isles.

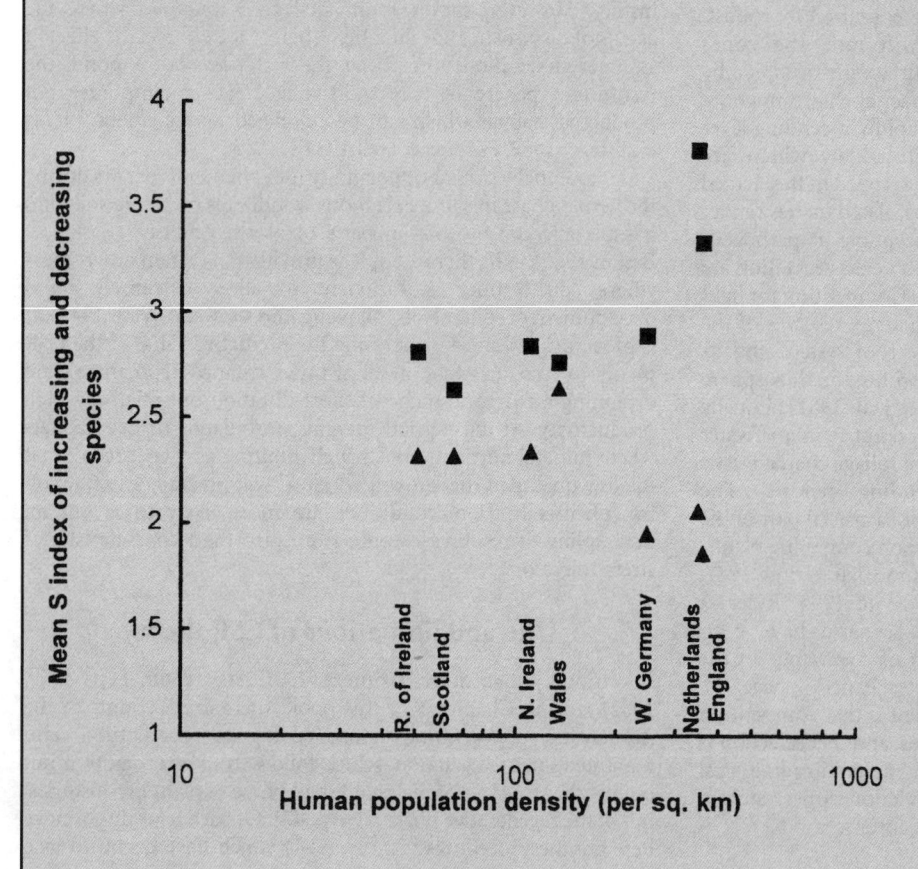

References and additional reading: Barr *et al.* (1993), Bunce *et al.* (1999), Firbank *et al.* (2000), Hodgson (1986a-d), Knevel *et al.* (2003), Preston *et al.* (2002a), Thompson (1994), Thompson & Jones (1999).

Figure A. Relationship between mean S index of increasing (▲) and decreasing (■) species and human population density in seven European countries where the S index reflects the proximity of the strategy of the species to the S corner of the CSR model calculated following the procedures of Hodgson *et al.* (1998). S class of the two groups was not significantly different in Scotland, N. Ireland, or Wales, but the two groups do differ significantly in the R. of Ireland ($p = 0.049$), England, W. Germany, and The Netherlands (all <0.001). Reproduced from Thompson 1994 by permission of Springer-Verlag.)

account of intraspecific variation in strategy in variable species such *Poa annua* and *Agrostis capillaris*.

As explained in Chapter 5, the classification of plant *communities* with respect to the strategies of their component species is also valuable, because it provides a means of predicting the rates and direction of any cyclical or successional change. Knowledge of strategies also allows prediction of the resilience of vegetation to climatic fluctuations, herbivory or human disruptive influences.

Strategy concepts have proved useful in analysing major effects of management within large areas of landscape. In one investigation (Hodgson 1986a,b; see Panel 8, page 14), the CSR model was used to explain the mechanisms controlling the rarity and abundance of species in the vascular flora of an area of intensively-developed landscape in N England. The general conclusion here is that land-use is the dominant influence on current levels of species abundance. Agricultural inputs of fertilizer, coupled with severe habitat disturbances, have resulted in the majority of rare plants being relatively stress-tolerant species confined to islands of aboriginal, less-productive countryside. Such areas are surrounded by landscapes that are either productive or disturbed, or both, and are particularly exploited by species exhibiting competitive or ruderal characteristics.

Recent studies have revealed scope for using the CSR model in relation to the ecological analysis, management and conservation of larger taxa, including some families of vascular plants. In this research, particular emphasis has been placed upon the concept of 'channelling' (Stebbins 1951,1974), whereby recent or current evolution appears to be strongly influenced by ancestral patterns of ecological specialization. Grime (1984) and Hodgson (1986c) have linked this principle to strategy concepts in analysing the mechanisms controlling the ecology of pteridophytes, and in an attempt. to explain why families of angiosperms have shown very different response to recent changes in land-use in Britain. It was concluded from these studies that the decline of most pteridophytes, and of many species drawn from relatively primitive families, e.g. Rosaceae and Fabaceae is related to possession of inflexibly stress-tolerant traits (e.g. low potential relative growth rate and delayed reproduction) and propagule characteristics which are inappropriate (either too small, e.g. Pteridophyta, or too large, e.g. many Rosaceae and Fabaceae) for effective colonization of the productive or disturbed vegetation which is created by modern forms of land use.

Further research is required to identify the attributes which, in particular taxa, have obstructed evolution towards ruderal or competitive strategies. Constraints upon the rapid evolutionary modification of offspring size and shape are more obvious (Stebbins 1971, 1974, Hodgson & Mackey 1986), and may be related to ancestral commitments to characteristic types of flower and fruit architecture on the part of individual families.

Although many inferences follow from the classification of a species within the CSR model and the identification of regenerative strategies, it would be unreasonable to expect all aspects of the ecology of a species to be predictable from this approach. For this reason each Species Account also contains additional information and draws attention, as required, to gaps in our knowledge and understanding. The uncertain role of many historical factors, the occurrence of genetic variation over the geographical or habitat range of a species, and the complexity of the processes responsible for the 'fine-tuning' of distributions and abundances within individual sites all set obvious lower limits to the scale at which strategy concepts are able to explain phenomena and generate predictions.

The failure of strategy theory to address all of the minutiae of proximal causes and effects has prompted some ecologists to reject the approach in favour of research programmes in which attempts at synthesis are *non grata* until the results of very many detailed studies of the field behaviour of plant populations become available (e.g. Harper 1982). With respect to this latter approach we remain unconvinced of its potential for providing a direct route to a general, predictive theory which is capable of development into guidelines for the conservation and management of vegetation. The fluctuating and stochastic nature of plant population processes, viewed locally and over short periods, is fascinating in its own right. However and this is especially true in the case of long-lived plants, they do not appear to provide a practicable and systematic basis for the recognition of the critical events and underlying causes which determine the essential ecology of large numbers of species.

3 DATA SOURCES FOR THE SPECIES ACCOUNTS

Introduction

To compile the Species Accounts presented in this book, four principal sources of information have been employed: vegetation surveys, other field observations, laboratory screening experiments and published sources. The first three originate from Sheffield, but in quantity of data the Accounts are dominated by the findings of just one of these: information from a vegetation survey performed in the Sheffield region by UCPE and hitherto unpublished. Most of the material derived from this source remains unchanged from that used in the first edition of this book. In this chapter we provide background information on each of these principal sources. Chapter 7 consists of tables providing synopses of these data, not only for the species dealt with in the Accounts themselves but also for others which are important in the British flora.

Vegetation surveys

Survey fieldwork

In many ways Sheffield is ideally situated for the broad botanical fieldwork which is necessary for the support of generalizing hypotheses. Many local species are at the northern limit of their British and European distributions, and some are at their southern limits. The surrounding area also offers a wide range in geological strata, altitude, aspect and land use, with a corresponding diversity of plant life. The regional geology, climate and land use have been described by Linton (1956), Edwards (1966) and Anderson and Shimwell (1981).

The UCPE vegetation surveys were conducted within an area of 3000 km² surrounding Sheffield (Fig. 3.1). In all, three surveys have been performed. Table 3.1 lists some of their principal features. Survey I (semi-natural grassland only) has been described by Lloyd *et al.* (1971, 1972), Lloyd (1972a) and Grime and Lloyd (1973). Survey II (all common herbaceous communities) is the chief source of field data for the present Accounts. Survey III (rarer plants and communities) provides an ancillary source of information, and has also been the foundation of a separate investigation into the causes of commonness and rarity in the Sheffield flora (Hodgson 1986a,b,c,d , see also Panel 8).

Features of the survey area

Some features of the climate, soils and land use of the area, which is shown in Figure 2.1, are given in Tables 3.2 and 3.3. Higher rainfall and lower temperatures are associated with the western portion of the region, with the climate becoming progressively drier and warmer to the east (Table 3.2). The major geological substrata outcrop as bands which run N-S through the survey area. The older rocks and higher land are both associated with the western part of the region, and the younger rocks and lower altitudes with the east. Since discontinuities in climate, geology and topography follow a similar pattern, the Sheffield region may be subdivided into two approximately equal portions (Table 3.3). An 'upland' area (Carboniferous Limestone, Millstone Grit and Lower Coal Measures) has the majority of its land lying above 200 m and rainfall generally exceeds 850 mm year^{-1}. A drier, 'lowland' area consists of the Middle and Upper Coal Measures, Magnesian Limestone, Bunter Sandstone and Keuper Marl. The lowland is also generally warmer. Both regions have geological similarities, each including a calcareous substratum, together with both separate and shared strata which are non-calcareous. The one minor anomaly is the Keuper Marl, associated with the lowland only, and having no counterpart within the upland.

A wide range of soil conditions is associated with each

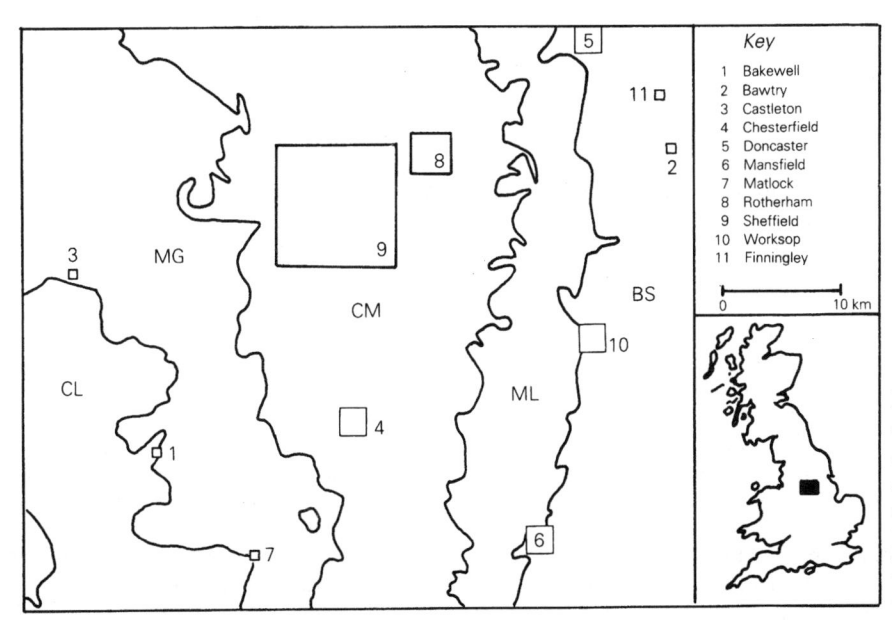

Figure 3.1 The UCPE survey area. The approximate extent of urban development is shown by numbered squares. Geological substrata are: CL, Carboniferous Limestone; MG, Millstone Grit; CM, Coal Measures; ML, Magnesian Limestone; BS, Bunter Sandstone. (From Grime & Lloyd 1973.)

Table 3.1 Facts and figures concerning the three UCPE vegetation surveys.

	Survey I	Survey II	Survey III
Objects	(a) An objective description of the ecology of grassland species (b) Description and classification of the vegetation of semi-natural grassland	(a) A description of the ecology of the most common species of the region (b) Recognition of the main types of vegetation in each of the major habitats	(a) A description of the ecology of species recorded in <20 quadrats in Survey II (b) Records of the less common vegetation
Year commenced	1965	1967	1972
Sampling policy	Random within uniformly selected localities	Subjective within subjectively selected vegetation	Subjective within subjectively selected vegetation
Sampled area*		National Grid references	
NW corner	SE 100020	SE 100020	SE 100050
NE corner	SE 670020	SE 670020	SE 700050
SW corner	SK 100550	SK 100550	SK 100550
SE corner	SK 670550	SK 670550	SK 700550
Number of 1 m² quadrats	657	2748	7324
Main habitat types		Quadrats per thousand	
Wetland	–	156.1	237.6
Skeletal	–	130.6	86.3
Grassland	1000	80.8	43.8
Arable	–	44.4	41.2
Spoil	–	172.1	111.7
Wasteland	–	149.6	326.1
Woodland	–	266.4	116.2
Minor habitats	–	–	37.1
Solid geology*		Quadrats per thousand	
Carboniferous limestone	377.5	193.2	162.6
Millstone grit	249.6	245.3	165.2
Coal measures	187.2	283.4	207.7
Magnesian limestone	129.4	129.2	197.2
Bunter sandstone	56.3	141.9	254.4
Keuper marl	–	7.0	12.9

*See also Figure 3.1.

substratum. Many of the soils associated with the non-calcareous substrata tend to be acidic, while those of calcareous substrata often contain free calcium carbonate and are of high pH. This is discussed further in relation to the soils associated with semi-natural grasslands of the region by Lloyd et al. (1971).

A majority of the region is subjected to agricultural management (Table 3.3). Urban development and forestry account for most of the land not utilized for farming. The lowland has a potential quality of land for agriculture of Grade 2 or 3, while in the upland the less-productive Grades 4 and 5 prevail. This difference in land quality is reflected in the type of farming practised, the lowland being predominantly arable and the upland consisting mainly of grassland.

The proportion of land utilized for farming is lower in the more fertile areas, reflecting the competition which exists within the lowland between contending types of land-use. In Britain urban development tends to occur most markedly within areas of high agricultural potential (Engledow & Amey 1980).

Notwithstanding the large area of urban land associated with the Lower Coal Measures, this national trend is also repeated within the Sheffield region. The difference between the two regions is further accentuated by the presence within the upland of the Peak District National Park, established in 1950 (Edwards 1962).

Methods used in Survey II

Important differences exist between the sampling procedures of the three vegetation surveys, as is clear from Table 3.1. In Survey II, our principal source of distributional data, the objective was to survey a full range of examples of each of the main herbaceous vegetation types present within the survey area. Localities were therefore selected subjectively, according to prior knowledge of the region.

Table 3.2 Climatic data for the Sheffield region for the period 1972-82: Buxton (near the western margin of the survey area, altitude 307 m), Sheffield (135 m) and Finningley (near the eastern margin, 10 m). Comparable data for the period 1943-65 are reported by Grime and Lloyd (1973). (From Hodgson 1986a.)

	J	F	M	A	M	J	J	A	S	O	N	D	Annual
(a) Mean daily maximum temperature													
Buxton	4.7	4.2	6.9	9.3	13.1	16.4	17.7	17.9	14.8	10.9	7.8	5.4	
Sheffield	6.1	6.3	8.8	11.6	15.4	18.5	20.1	20.4	17.3	13.3	9.3	6.8	
Finningley	6.4	7.0	9.6	12.0	15.9	19.3	21.0	21.3	18.3	13.6	9.7	7.1	
(b) Mean daily minimum temperature (T)													
Buxton	0.1	0.0	1.0	2.7	5.3	8.7	10.5	10.4	8.1	5.2	2.7	1.0	
Sheffield	1.6	1.5	2.7	4.5	6.8	10.0	12.1	12.0	10.0	7.0	4.3	2.3	
Finningley	0.7	0.8	2.0	3.9	5.9	9.3	11.3	11.2	9.1	6.1	3.3	1.4	
(c) Mean monthly and annual rainfall (mm)													
Buxton	143	108	131	69	74	83	82	95	107	113	146	139	1290
Sheffield	77	74	80	48	57	75	54	62	61	61	75	89	813
Finningley	46	45	51	33	42	60	43	52	52	46	40	55	565

Table 3.3 Variations in land-use and in the potential quality of land for agriculture within the Sheffield region: data abstracted from Sheets S102, S103, S111 and S112 of Ministry of Agriculture, Fisheries and Food (1966). (From Hodgson 1986a.)

Solid geology	Area (% Sheffield region)	Land use (% geological substratum)			Potential quality of land for agriculture* (% agricultural land)				
		Agriculture	Urban	Other	1	2	3	4	5
Lowland									
Magnesian limestone	12	73	16	11	0	76	23	<1	<1
Keuper marl	2	86	3	11	0	41	59	0	0
Bunter sandstone	20	65	12	24	<1	6	90	3	<1
Coal measures (upper and middle)	18	52	29	10	0	1	69	30	<1
All lowland strata	52	67	18	15	<1	22	66	12	<1
Upland									
Coal measures (lower)	17	61	29	9	0	0	24	62	14
Carboniferous limestone	11	92	1	6	0	0	19	67	14
Millstone grit	21	86	2	11	0	0	6	37	57
All upland strata	48	79	11	10	0	0	15	53	32

*Key to land quality.
Grade 1 – Very minor limitations to agricultural use; most crops may be grown.
Grade 2 – Some limitations; usually edaphic, but a wide range of agricultural and horticultural crops may be grown.
Grade 3 – Moderate limitations (edaphic, climatic, topographic or a combination); most suitable for grass and cereals, but some horticultural and root crops are grown.
Grade 4 _ Severe limitations; suitable for grass with some oats, barley and forage crops.
Grade 5 - Very severe limitations; generally under grass or rough grazing; some pioneer forage crops.

Within each locality quadrats were positioned deliberately to provide examples of each of the major plant assemblages within each major habitat. A guide to the habitats recognized in Survey II is given in Figure 3.2. Fieldwork extended over 6 years, in each of which sampling was restricted to the period April to September inclusive. Because winter annuals and vernal geophytes tend to disappear completely during the summer months, vegetation likely to contain these floristic elements was not sampled after the end of May.

Had the object of the sampling been vegetation description, then the sample size would have needed to vary according to vegetation type so as to include the majority of species present within each community. At one extreme, for example, a sampling area of <0.25 m² might be appropriate (Clark 1974) for a community of winter annuals exploiting a small pocket of soil on a rock outcrop; at the other, in wetland or woodland habitats, a sampling area of 4 m² might be more satisfactory (Wheeler 1975; Sydes 1981). However, our main aim has been to focus upon the distribution and ecology of species. For this purpose a rigorously standardized sample size is more appropriate. Accordingly, the area of each of our quadrats was fixed at 1 m^2. This size was considered to be the maximum advisable in view of the small-scale variation in soil type and vegetation cover known to occur in grassland and some other habitats (Balme 1953, Grime 1963a), and was also the size used by Grime and Lloyd (1973). This size of sample has also proved to be informative in studies of the role of species interactions in the control of species diversity in herbaceous vegetation (Grime 1973a,b, see also Panels 10 and 11, pages 67 and 68).

For each species present in the quadrat a figure representing percentage frequency was obtained, being the number of 100 x 100 mm subdivisions of the quadrat in which rooted shoots of the subject species was found to occur. Though a laborious technique, this minimises 'observer error', and has more constancy through the summer than simpler estimates of cover. In practice the metre-square quadrat was divided into four quarters of 25 subdivisions, each quarter being examined by a single observer. Comparison of the four subtotals provided a rough guide to the homogeneity of the area. Two observers, always including one or both of the senior authors of this book, normally worked on each quadrat.

The mean slope and aspect were recorded at each quadrat site, and soil samples were taken at depths of 0-30 mm ('surface') and 90-120 mm, for subsequent pH determinations, which were performed on the fresh soil saturated with distilled water, using a glass electrode. Environmental measurements formed part of a set of 39 habitat characteristics which were recorded on each occasion. A full list of these appears in Table 3.4. In addition, informal notes were often made, recording features such as site ownership and management. A complete specimen quadrat record is illustrated later, in Figure 3.4. The computerized storage of the data, and the

Table 3.4 A full list of the coded habitat characteristics which were recorded for quadrats in UCPE Survey 11. The characters later referred to in most detail in the Species Accounts are shown in bold type.

Number	Title	Explanation
1	**Habitat type**	A 3-digit code number indicating the principal intermediate and terminal habitat type (for an explanation see Fig. .2.2)
2	Quadrat number	A 3-digit number indicating the accession sequence within the habitat
3	NGR easting	The 100 m National Grid value
4	NGR northing	The 100 m National Grid value
5	Locality	A code identifying the 5-km grid square in which the quadrat is located
6	**Altitude**	Altitude to the nearest 25 ft (7.6 m)
7	**Aspect**	Degrees of mean aspect, clockwise direction, magnetic north 0
8	**Slope**	Degrees of mean slope from the horizontal
9	pH (i)	Surface soil pH, sampled at 0-30 mm depth
10	pH (ii)	Soil pH sampled at 90-120 mm depth
11	**Species richness**	Number of vascular species present in the quadrat
12	Gramineae	Number of species of Gramineae present in the quadrat
13	Total RF (i)	Sum of all rooted frequencies (thousands)
14	Total RF (ii)	Sum of all rooted frequencies (hundreds, tens and units)
15	Sphagnum	Scored 0 if absent, 1 if present
16	Bare rock	Six categories of percentage cover from 0 to 100
17	**Bare -soil**	Six categories of percentage cover from 0 to 100
18	Plant litter	Six categories of percentage cover from 0 to 100
19	Bryophyte mat	Six categories of percentage cover from 0 to 100
20	Refuse (waste paper, etc.) or algal scum	Six categories of percentage cover from 0 to 100
21	Water surface	Six categories of percentage cover from 0 to 100 (includes area occupied by floating vegetation)
22	Exposed water surface	Six categories of percentage cover from 0 to 100 (excludes area occupied by floating vegetation)
23	Water depth	Mean water depth, cm (aquatic habitats only)
24	Geology	Code for geological substratum
25	Pedology	Code for pedological substratum
26	Spoil	Code for type of spoil material, if present
27	Vegetation type	Coded subsidiary information on vegetation type
28	Habitat (i)	
29	Habitat (ii)	Coded information on habitat type
30	Habitat (iii)	
31	Management (i)	Coded information on management
32	Management (ii)	
33	**Hydrology**	Coded information on water movement and depth
34	Neighbourhood	Coded information on adjacent habitats -
35	Canopy (i)	Coded information identifying tree species forming canopy, if any, above quadrat
36	Canopy(ii)	
37	Miscellaneous	Coded minor information, very diverse
38	Strategy	A (superseded) habitat classification
39	Status	Existence of NNR, SSSI, etc.

features of its general-purpose retrieval program, closely followed the principles described by Lloyd *et al.* (1972). More-specialized calculations and tabulations were performed either by means of purpose-written programs or by standard statistical packages.

Outline results from Survey II

Habitats

After a preliminary programme of field sampling a simple, branched key to major habitat types was devised (Fig. 3.2). This was based almost exclusively upon self-evident physical and physionomical attributes of the environment and its exploitation, and was constructed in terms designed to be accessible to non-specialist users of the data. The key contained seven 'primary', eight 'intermediate' and 32 'terminal' categories. Knowing the full range of habitats which were likely to be present within the whole survey, it was then possible to carry out the fieldwork in such a way that all of the terminal categories were adequately represented. As far as possible, each type of habitat was sampled throughout the Sheffield region. In order to avoid bias, samples were taken from the first example of each habitat type encountered within each sector of the survey area.

Figure 3.2 gives the number of quadrats recorded in each of the primary, intermediate and terminal habitats. Most of the terminal categories are common British habitats. Lead-mine spoil, the British distribution of which has been mapped by Halliday (1960), is an exception. The hay-meadows of the region are also unusual, in that most are grazed after the hay crop has been harvested. Generally, the number of records required reflected the area and geographical spread of the habitats within the survey area and the degree of vegetation heterogeneity encountered. A minimum of about 40 records per terminal habitat was considered desirable, though 11 habitats required 100 or more records for satisfactory representation.

The distributions of the major environmental attributes, including those presented diagrammatically in the Species Accounts, are now discussed with reference to the seven primary habitats mentioned in Table 3.1 (wetland, skeletal, grassland, arable, spoil, wasteland and woodland). Salient points involving the 32 terminal habitats are also mentioned. In Chapter 4 the ecological significance of certain of these attributes is discussed more fully at the level of the terminal habitats.

Altitude

The lowland-upland dichotomy within the survey area has been noted already (p. 17). Information on the distribution of altitude with respect to the seven primary habitats, (Table 3.5), confirms the pronounced lowland bias in arable habitats mentioned earlier, with 85% of these records lying below 200 m. Wetlands and woodlands are also somewhat lowland-biased, whereas skeletal habitats, spoil and wasteland are well represented throughout the whole altitudinal range. Only grassland has an upland bias: low-grade agricultural land (Table 3.3) lying at or above 200 m supports 78% of all grassland recorded in the survey. Within the three wide-ranging primary habitats, coal-mine heaps, cinders, bricks and mortar, and sewage heaps are particularly lowland-biased within the spoil habitats whereas, in contrast, quarry heaps of all kinds and lead-mine heaps exhibit an upland bias. River and stream banks, paths and walls all exhibit a lowland bias.

Slope

Table 3.6 reveals that records in all of the seven primary habitats, except the skeletal group, are strongly biased toward low angles of slope. In each case records with values <20° outnumber all other slope classes combined. Within skeletal habitats rock outcrops are biased towards low slope, but 92% of all scree records lie within the range 20-39°. Naturally enough, very steep slopes predominate on cliffs and walls, with 21 and 14%, respectively, of records in these two terminal categories consisting of vertical or overhanging sites (of slope not less than 90°).

Aspect

About half of the survey samples are of <10° mean slope, and thus have no effective topographical aspect at all (Table 3.7). Wetlands and arable habitats are each almost entirely non-sloping (93% of records in each case), while in skeletal and spoil habitats the majority of sites have a definite aspect, and non-sloping records account for only 11 and 41% respectively. Within grasslands, wastelands and woodlands the mean proportion of sites possessing no definite aspect is 47%. Throughout the survey, where aspect is exhibited its distribution between the four quadrants of the compass (delimited by aspects of 45, 135, 225 and 315°) is relatively even.

HABITATS

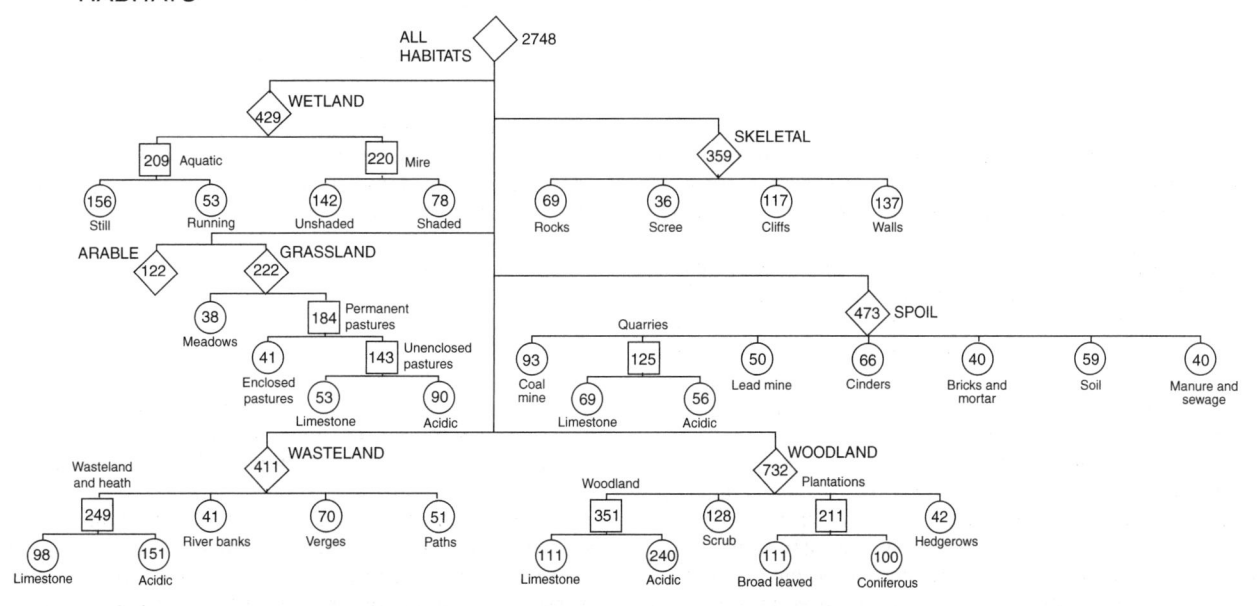

Figure 3.2 Numbers of quadrats in the primary, intermediate and terminal habitats of UCPE 'Survey II'.

Table 3.5 Numbers of quadrats in the seven primary habitats of UCPE Survey II, classified with respect to **altitude**.

Habitat	<100	100–199	200–299	300–399	400+	Habitat totals
Wetland	166	147	40	60	16	429
Skeletal	82	110	130	32	5	359
Grassland	16	32	74	87	13	222
Arable	51	53	13	5	0	122
Spoil	154	140	104	73	2	473
Wasteland	152	110	92	49	8	411
Woodland	221	269	199	42	1	732
Class totals	842	861	652	348	45	2748

Altitude (m)

Table 3.6 Numbers of quadrats in the seven primary habitats of UCPE Survey 11, classified with respect to **slope**.

Habitat	<20	20–39	40–59	60–79	80+	Habitat totals
Wetland	420	7	1	1	0	429
Skeletal	63	59	22	90	125	359
Grassland	139	69	14	0	0	222
Arable	119	2	1	0	0	122
Spoil	278	161	33	1	0	473
Wasteland	279	107	24	0	1	411
Woodland	521	190	21	0	0	732
Class totals	1819	595	116	92	126	2748

Slope (degrees from the horizontal)

Table 3.7 Numbers of quadrats in the seven primary habitats of UCPE Survey II, classified with respect to **aspect**.

Habitat	Below 10° slope	N-facing	E-facing	S-facing	W-facing	Habitat totals
Wetland	400	13	2	6	8	429
Skeletal	40	101	56	83	79	359
Grassland	105	33	27	20	37	222
Arable	114	1	3	2	2	122
Spoil	195	71	77	72	58	473
Wasteland	200	70	46	61	34	411
Woodland	337	107	95	101	92	732
Class totals	1391	396	306	345	310	2748

Soil pH

In keeping with the varied geological composition of the survey area (Fig. 3.1 and Table 3.1), all of the principal habitats are widely represented with respect to range of soil surface pH (Table 3.8). The one exception is the arable group, from which pH values below 5.0 are absent. On few occasions, very shallow soil profiles in terminal habitats such as scree, cliffs, limestone quarry heaps and limestone woodlands made it impossible for soil surface pH to be measured. The most notably extreme biases within the terminal habitats were those created by definition: 85% of pastures on base-poor strata lay below pH 4.0. Conversely, 38% of bricks-and-mortar heaps had a pH of 8.0 or above.

Because soil pH at 90-100 mm depth was also measured, though not presented here, differences in pH between the two depths could be calculated. When the surface pH was less than the deeper pH, some form of leaching was indicated; when greater, some form of surface accumulation of calcareous material or acidifying microbial activity at depth. Examples of the former occurred in most of the grassland habitats (with the exception of meadows) where, on average, 18% of records showed a pH gradient of at least 0.5 units in favour of the deeper measurement. On coal-mine heaps, cinders, and limestone wasteland such gradients occurred, on average, in 38% of each habitat. Examples of reversed pH gradients of similar magnitude were seen in mires (22% of all records), manure and sewage waste (23%) and in coniferous plantations (30%).

Species diversity

In the 2748 quadrats which comprised Survey II (Table 3.1), a total of 629 vascular plant species were recorded within 188 genera. As many as 86 of the species were recorded only once, but the most common, *Poa trivialis*, was present in 605 quadrats. A frequency distribution showing numbers of species by numbers of quadrats is given in Figure 3.3.

The predominantly leftwards skew of this diagram, even with its geometrically scaled class intervals, indicates that the greater part of the vegetation of the region consists of a few, relatively common species. A total of 281 of these are dealt with in detail in the Species Accounts (Chapter 6).

Table 3.8 Numbers of quadrats in the seven primary habitats of UCPE Survey II, classified with respect to **soil surface pH**.

Habitat	>4.0	4.0–4.9	5.0–5.9	6.0–6.9	7.0+	Totals	No info.	Habitat totals
Wetland	22	49	57	144	156	428	1	429
Skeletal	48	39	24	69	168	348	11	359
Grassland	86	32	39	44	20	211	1	222
Arable	0	0	35	44	42	121	1	122
Spoil	40	79	54	87	208	467	5	473
Wasteland	93	78	57	55	128	411	0	411
Woodland	208	281	56	76	100	721	11	732
Class totals	497	558	322	519	788	2718	30	2748

The *criteria for inclusion* in our selection were as follows. Fifteen records or more in the whole survey guaranteed the inclusion of a species in the Accounts. For less than 15 records a value-judgement was made as to the ecological importance of the species; the minimum number of records for any species included was normally 10, but three exceptions were made to this rule because of exceptional importance or interest: *Phragmites australis*, (9 quadrats); *Carlina vulgaris*, (7); *Fallopia japonica*, (5). The accounts for these three species rely heavily upon the support of information gained from the literature and from Survey III (p. 18).

Because most of the species included are very widespread, it has been possible to analyse a very large proportion of the region's floristics (as defined by the exhaustively complete Survey III) in terms of the distribution and attributes of no more than 36% of its local species, 51% of its local genera and 57% of its local families. Nationally, these represent 10, 26 and 46%, respectively, of the native British flora. The coverage of Britain's established alien flora is less complete, with only 3% of species, 5% of genera and 29% of families being included in the Accounts.

Within the 1 m² sample, 135 quadrats contained no more than a single species. At the other end of the scale, one quadrat (illustrated as a specimen record in Fig. 3.4) contained 40. Table 3.9 shows how floristic diversity was distributed among the seven primary habitats. Grassland, spoil and wasteland were the most diverse of these, averaging 15% of records with 21 species or more per square metre. Woodlands were least diverse, with 99% of records lying below this figure. Of the terminal habitats, by far the most diverse was limestone pasture, with 51% of records at 21 species or above, followed by the other (always limestone) habitats -quarry heaps (30%) and wasteland and heath (28%). Least diverse were coniferous plantations with 85% of records at 5 species m^{-2} or less, followed by aquatic habitats (average of both 81%) and broad-leaved plantations (74%).

Hydrology
The 429 wetland records have been classified according to a composite hydrological index which not only takes into account the mean depth of the free water, if any, but also incorporates information on slope when water is absent. This classification is explained in detail on p. 38. However, because it applies only to wetland records, a discussion of its distribution between the seven primary habitats is inappropriate.

Bare soil
In Table 3.10 the samples drawn from various habitats are classified according to amount of bare soil surface. As might be expected, arable habitats and spoil heaps contain large quantities compared with grassland, wasteland and woodland. Bare soil exposure in mire was considerably higher in shaded situations.

Phenology, demography and regeneration

Rationale
The vegetation surveys belong to a class of ecological research which is high in realism (being wholly field-based), high in generality (because of the very large number of samples and species included) but low in precision (because of the simple nature of the floristic observations and their lack of continuity, or even repetition, in time). [See Harper (1982) and Hunt & Doyle (1984) for further discussions of these concepts.] A second line of fieldwork, involving a moderate sacrifice of generality to precision, has concerned itself with more detailed investigations of the phenology, demography and regenerative biology of a more limited selection of species. The two examples summarized below play particularly prominent parts in providing data for the Species Accounts.

Shoot phenology
Variation in standing crop and litter was measured by seasonal sampling and sorting of the herbaceous vegetation at 13 sites in the survey area (Al-Mufti *et al.* 1977). The vegetation types examined comprised tall herb, woodland floor and grassland communities.

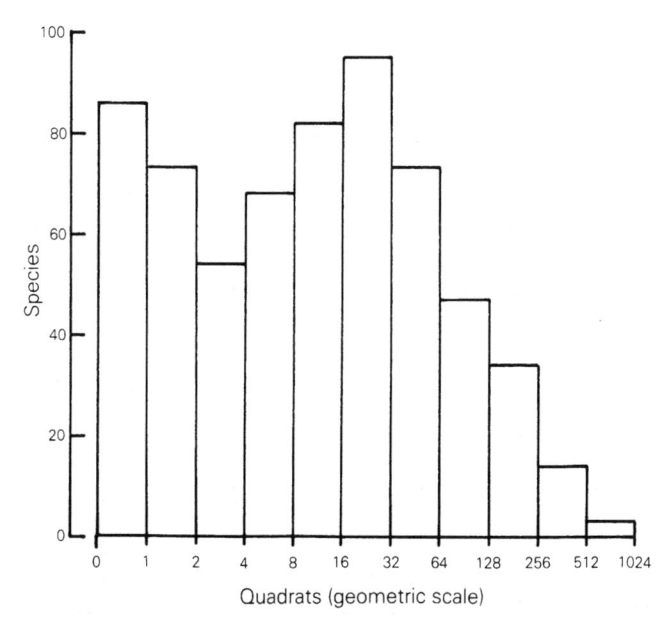

Figure 3.3 The frequency distribution of species by number of quadrats in UCPE 'Survey II'; n = 629.

```
HAB CODE    500   QUADRAT      17   NGR EAST    501   NGR NORTH  667   LOCALITY    96   ALT/10      50   ASPECT     160   DEG SLOPE   9
PH[0]*10     59   PH[10]*10    69   NO SPP       40   NO GRAMIN   14   TRF THOU     0   TRF UNITS  685   SPHAGNUM     0   BARE ROCK   0
BARE SOIL     0   PL LITTER     3   BRYOPHYTE     0   REFUSE       0   WATER        0   EXP WATER    0   WAT DEPTH    0   GEOLOGY     5
PEDOLOGY      0   SPOIL         0   VEGETN        3   HABITAT A    0   HABITAT B    0   HABITAT C    0   USAGE A      9   USAGE B     3
HYDROLOGY     0   MISCELL A     0   MISCELL B     0   MISCELL C    0   MISCELL D    0   STRATEGY     2   STATUS       0

ACHIL MIL    16   AGROS TEN    43   ANTHO ODO    14   ARRHE ELA    3   BRACH PIN   92   BRIZA MED    8   BROMU ERE   64   CAMPA ROT  10
CAREX CAR    10   CAREX FLA    44   CENTA NIG    62   CENTA SCA    7   CERAS FON    1   DACTY GLO   18   DAUCU CAR    2   FESTU PRA   5
FESTU RUB    82   GALIU VER    15   HELIA CHA     8   HELIC PRA    2   HELIC PUB   31   HIERA SPP    1   KNAUT ARV    2   KOELE CRI   3
LATHY PRA    15   LEONT HIS     3   LOLIU PER     1   LOTUS COR   52   PIMPI SAX    7   PLANT LAN   15   PLANT MED   12   POA   PRA   9
POTER SAN     6   PRIMU VER     1   SCABI COL     1   TARAX OFF    1   TRIFO MED    2   TRIFO PRA    1   TRISE FLA    7   VIOLA HIR   9
```

Figure 3.4 A complete specimen quadrat record from UCPE 'Survey II'. The first part of this facsimile record consists of coded habitat characteristics (see Table 3.4). The second part lists the percentage rooted frequency of each of the 40 plant species present (names abbreviated from Clapham *et al.* 1981). This particular quadrat was from Magnesian Limestone wasteland near Pleasley, North Nottinghamshire. Floristically, it was the richest encountered in the survey.

At each site the living fraction in replicated 0.25 m² quadrats was separated into its components, and graphs were plotted to illustrate the shoot phenology of the more common species.

For the sites at which herbaceous vegetation was growing under fertile conditions, there was a large peak in standing crop during the summer (>1000 g m^{-2} dry mass). The vegetation here contained fewer species. The low species densities at such sites appeared to be related to the ability of certain members of the plant community (e.g. *Chamaerion angustifolium*, *Petasites hybridus*, *Urtica dioica*) to exercise competitive dominance. This involved the rapid expansion of a dense leaf canopy during the period June-August, allied to the production of a high density of litter.

At the woodland and grassland sites examined, the sums of weights of the maximum standing crop and the litter never exceeded 800 g m^{-2} and a variety of phenological behaviours was encountered. At two of the woodland sites perennial vernal species (including geophytes) were prominent (e.g. *Anemone nemorosa*, *Hyacinthoides non-scripta*, *Ranunculus ficaria*). These exhibited phenologies in which leaves expanded during a brief period immediately following the marked decline in density of tree litter in early spring, and senesced soon after the full expansion of the tree canopy.

At the sites where the standing crop was severely restricted by low soil fertility, the most common phenological pattern was that of the evergreen. In certain of these species (e.g. *Carex caryophyllea*, *Carex flacca* and *Lamiastrum galeobdolon*) a seasonal peak of shoot expansion was difficult to detect. In two of the limestone grasslands investigated forbs with midsummer peaks of shoot expansion were prominent. The majority of these (e.g. *Leontodon hispidus*, *Lotus corniculatus*, *Sanguisorba minor*) had relatively deep root systems, and appeared to exploit reserves of moisture during periods when many grasses were subjected to desiccation.

A consistent feature of the results was the marked amplitude of bimodal seasonal variation in the abundance of bryophytes, expansion of which coincided with the moist cool conditions of spring and autumn.

Two general conclusions may be drawn from this study. The

Table 3.9 Numbers of quadrats in the seven primary habitats of UCPE Survey 11, classified with respect to **species diversity**.

Habitat	<6	6–10	11–15	16–20	21–25	26–30	31+	Habitat totals
			No. of species in the 1 ml quadrat					
Wetland	250	111	43	11	6	5	3	429
Skeletal	188	87	41	21	15	6	1	359
Grassland	61	48	34	32	20	13	14	222
Arable	16	28	39	19	10	9	1	122
Spoil	119	128	109	51	43	17	6	473
Wasteland	121	103	89	54	22	15	7	411
Woodland	411	214	80	22	5	0	0	732
Class totals	1166	719	435	210	121	65	32	2748

Table 3.10 Numbers of quadrats in the seven primary habitats of UCPE Survey II, classified with respect to **bare soil cover**.

Habitat	0	1–10	11–25	26–50	51–75	76+	Totals	No info.	Habitat totals
			Bare soil cover						
Wetland	271	36	27	22	21	52	429	0	429
Skeletal	213	71	26	15	9	11	345	14	359
Grassland	105	31	8	9	1	3	157	65	222
Arable	1	1	1	3	1	115	122	0	122
Spoil	256	58	39	34	11	71	469	4	473
Wasteland	166	54	31	24	24	39	388	73	411
Woodland	490	89	47	36	18	47	727	5	732
Class totals	1502	340	179	143	85	338	2587	161	2748

first is that there is a consistent relationship between shoot phenology and the established strategies of plants; this is summarized in Figure 2.5. The second is that there is a control of species density in herbaceous vegetation, such that the potential for high species density corresponds approximately to the range 350–750 g m^{-2} in the sum of maximum standing crop and litter (Fig. 3.6).

Seed banks

Seasonal variation in the density and composition of the reservoir of germinable seeds present in surface soil samples was measured at 6-weekly intervals in 10 ecologically-contrasted sites in the survey area (Thompson & Grime 1979). The techniques adopted sought particularly to detect persistent seed banks (i.e. those in which some of the component seeds were at least 1 year old). They also allowed the recognition of species in which there is a transient accumulation of detached germinable seeds during the summer.

The results obtained for populations of the same species in different types of habitat suggested that seasonal variation in seed number is a function of the species, rather than of the environment. It seems likely that, for the Sheffield region at least, the major ecological significance of seed bank type is the selective advantage derived from mechanisms of seed dormancy and germination which allow seedlings to evade either seasons unfavourable to seedling establishment or the potentially-dominating effects of established plants.

From the data collected in this study, four types of seed bank (I–IV) were recognized (Fig. 3.7). The transient seed banks (Types I and II) appear to facilitate seedling colonization of the gaps created by seasonally predictable damage and mortality in the vegetation,

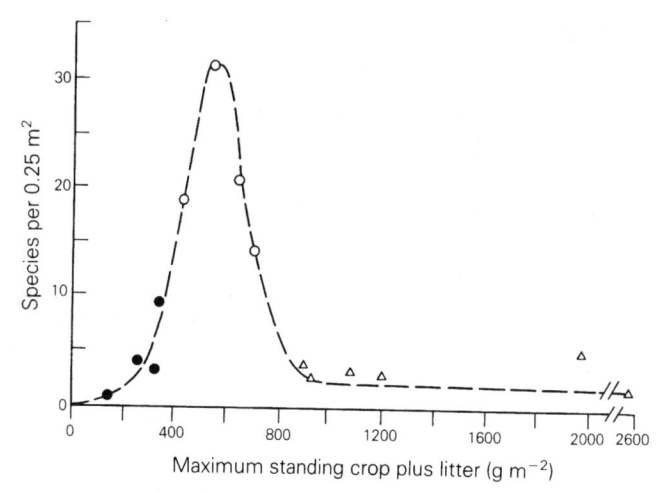

Figure 3.6 The relationship between maximum standing crop (plus litter) and species richness at 14 sites: ○, grasslands, ●, woodlands; △, tall herb communities. The curve is fitted by eye. (From Al-Mufti *et al.* 1977.)

whereas the persistent seed bank (Type IV) confers the potential for regeneration in circumstances where disturbance of the established vegetation is temporally unpredictable. A second category of persistent seed bank (Type III) has characteristics intermediate between those of Types I and IV. It contains some seeds which germinate soon after release, and others which are more persistent in the soil. This type is often associated with the capacity for both rapid population expansion and persistence, and is characteristic of some of the most widely successful species of the British flora (e.g. *Agrostis* spp., *Holcus lanatus* and *Poa trivialis*). As explained (p. 50), it is often possible to predict the type of seed bank developed by a species from the morphology of the seed and from its germination responses in the laboratory. An important source of evidence, strongly corroborating the seed bank classification reviewed here, has become available in the form of an

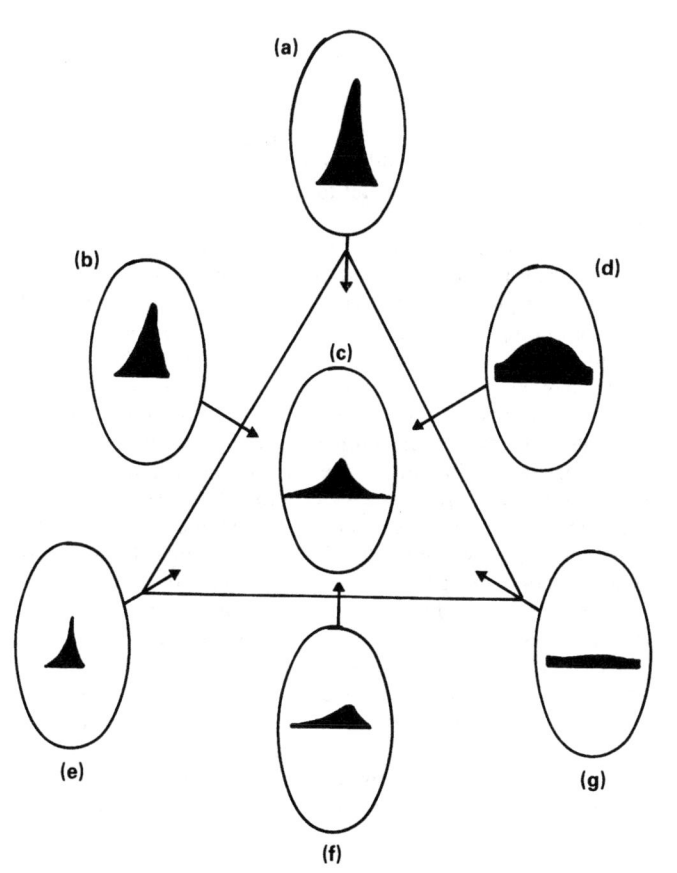

Figure 3.5 A scheme relating the pattern of seasonal change in shoot biomass to strategy type: (a) competitor, (b) competitive ruderal, (c) C–S–R strategist, (d) stress-tolerant competitor, (e) ruderal, (f) stress-tolerant ruderal and (g) stress-tolerator.

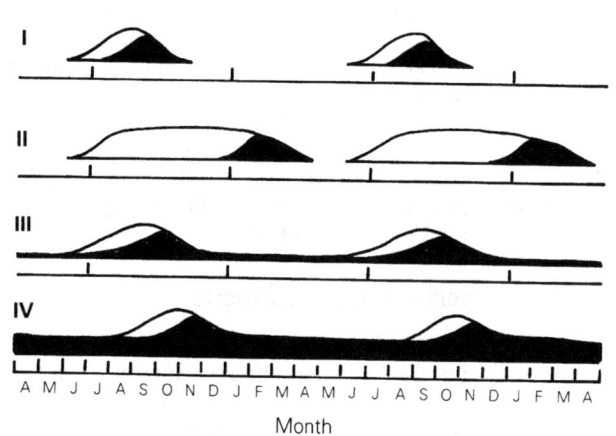

Figure 3.7 Representation of the four types of seed bank described by Thompson and Grime (1979). Shaded areas: seeds capable of germinating immediately after removal to suitable laboratory conditions. Unshaded areas: seeds viable but not capable of immediate germination. Type I, annual and perennial grasses of dry or disturbed habitats. Type II, annual and perennial herbs colonizing vegetation gaps in early spring. Type III, species mainly germinating in the autumn but maintaining a small presistent seed bank. Type IV, annual and perennial herbs and shrubs with large persistent seed banks.

DATA SOURCES FOR THE SPECIES ACCOUNTS

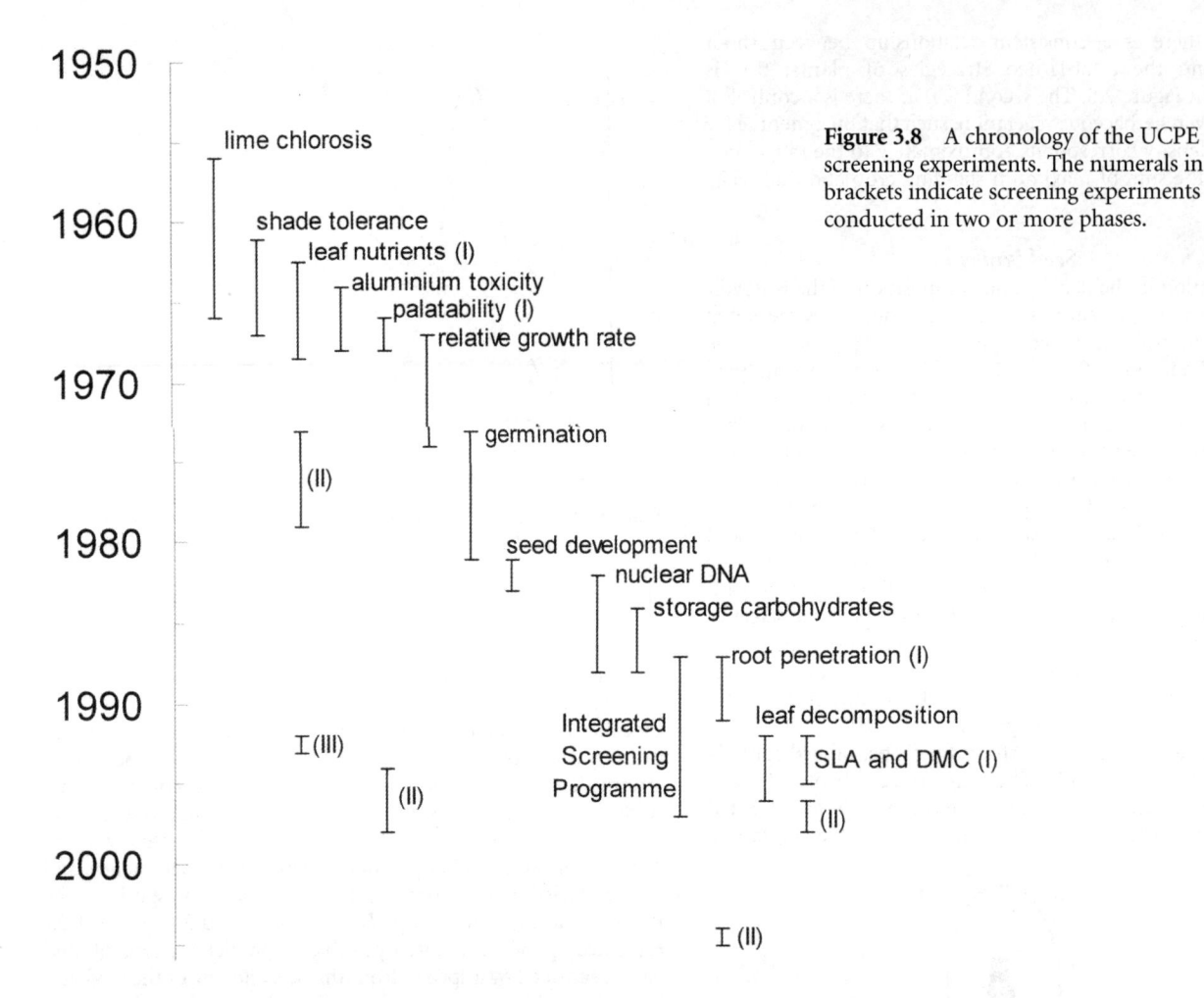

Figure 3.8 A chronology of the UCPE screening experiments. The numerals in brackets indicate screening experiments conducted in two or more phases.

extensive experiment monitoring the fate of experimentally-buried seeds (Roberts 1986a,b).

A feature of the Sheffield seed bank investigations was the inconsistent correspondence between the species composition of the seed flora and that of the associated vegetation. At certain sites substantial, persistent seed banks were detected for species which were either extremely scarce or did not occur at all in the vegetation actually established there.

Both transient and persistent types of seed banks were represented at each of the ten sites. As explained later (p. 65), this is consistent with the hypothesis (Panel 15) that complementary mechanisms of regeneration are involved in the maintenance of floristic diversity within plant communities.

Screening experiments

Rationale

The reasoning behind the UCPE screening programme has been dealt with at length in Chapter 2 (Panels 3 and 5) because it is a fundamental component of our comparative and predictive approach to plant ecology.

In such experiments high generality and precision may be maintained, but realism is sacrificed. An additional and important advantage of the approach is that many growth-room investigations can be reproduced or extended wherever there are adequate facilities. Data collected on different species or genotypes and in various laboratories can therefore be compared directly. When comparable data are available for a large number of species and populations drawn from a wide range of habitats, it is possible to estimate the limits of variation of the plant attribute itself. Hence, it becomes possible to place subsequent individual measurements into context and to begin to assess their ecological significance.

A continuous series of screening experiments (Fig. 3.8) has been conducted over a 35-year period, commencing with a comparative study of susceptibility to lime chlorosis (Grime & Hodgson 1969), and continuing most recently with screening of rooting depth (Shanta, Grime and Thompson, unpublished). All of these sources of data have been drawn upon in writing the Species Accounts.

Although most data have been obtained from only one field population per species, the number of species is large and includes representatives from all of the major habitats of the area. Uncertainty concerning the extent to which each sample is representative of the species does not, therefore, invalidate the screening, either as an estimate of the overall range of variation or as an attempt to recognize differences between groups of species of contrasted ecology. The problem of variation within the species is considered further in the final part of this section (p. 28). The next sections provide background information on the ranges and patterns of variation evident within three particularly important plant attributes: nuclear DNA amount, germination characteristics and potential growth rate. In each case Chapter 4 includes more details of how the measurements were made and Chapter 5 discusses their relevance to the ecology of the species considered in this book.

Nuclear DNA amount (Genome size)

Using the technique of Feulgen microdensitometry, Grime and Mowforth (1982) and Band (unpubl.) examined the 2C DNA amounts of cell nuclei in root-tip preparations drawn from a large

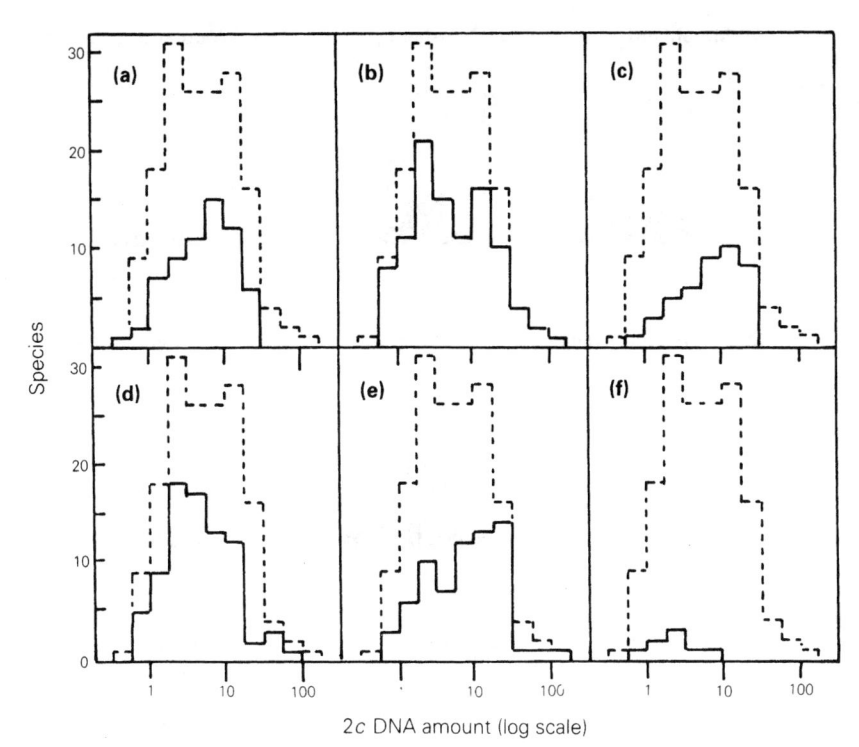

Figure 3.9 Nuclear DNA amounts in 162 flowering plants occurring in the British Isles: (a) annuals, (b) perennials (including biennials), (c) grasses, and trees and shrubs, (d) species of widespread distribution, (e) of southern distribution and (f) of northern distribution. The background distribution for all 162 species is also reproduced. (From Grime & Mowforth 1982.)

sample of the Sheffield flora. Data were augmented by determinations made elsewhere on other British species (Fig. 3.9, see also Bennett and Leitch 2004). Large DNA amounts were found to be particularly associated with Mediterranean geophytes and certain grasses in which growth is confined to the cool conditions of winter and early spring. As explained on p. 45, this observation prompted further studies relating nuclear DNA amount to the rate and timing of spring growth. DNA amounts have also been found to provide a useful index of temporal niche differentiation in plant communities (Grime et al. 1985).

Germination characteristics

Using a standardized procedure (p.36), Grime et al. (1981) made a laboratory study of the germination characteristics of seeds collected from a wide range of habitats in the survey area. Measurements were conduct in freshly-collected seeds and on samples subjected to dry storage, chilling and scarification. Responses to temperature and light flux were also examined.

The data were used to compare the germination biology of groups of species classified with respect to various criteria including life-form, family (Fig. 3.10), geographical distribution, ecology, and seed shape, weight and colour.

Marked differences were observed in the capacity of freshly-collected seeds for immediate germination. Of the 403 species examined, 158 failed to exceed 10% germination, but 128 attained values greater than 80%. Germination was high in the majority of grasses and low in many annual forbs and woody species. With respect to initial germinability, major families could be arranged in the series:

Poaceae > Asteraceae > Fabaceae = Cyperaceae > Apiaceae.

Seeds capable of germinating immediately after collection were often small in size, elongated with antrorse hairs (Poaceae), or conical with teeth on the dispersule (Asteraceae). High initial germinability was conspicuous among species of great abundance in the Sheffield flora.

In the majority of species germination percentage increased during dry storage; this effect was most marked in small-seeded species. Among the 75 species which responded to chilling, some germinated at low temperature in darkness (many of these are known to form Type II seed banks, see p. 25) whereas others were dependent upon subsequent exposure to light or higher temperature, or both (many of these have Type IV seed banks). Responses to chilling were characteristic of the Apiaceae. In all of the legumes examined, rapid germination to a high percentage was brought about by scarification.

Under the experimental conditions all of the annual grasses showed the potential for rapid germination. High rates were also observed in many of the annual forbs and perennial grasses. Low rates of germination occurred in the majority of sedges, shrubs and trees, and were particularly common in species of northern distribution in Britain. Rapid germination was characteristic of the species of greatest abundance in the Sheffield flora. Rate of germination showed a progressive decline with increasing seed weight, and, with some exceptions, there was a positive correlation between rate of germination and the relative growth rate of the seedling.

In 16 species germination in the light was found to be dependent upon exposure to diurnal fluctuations in temperature; germination responses to temperature fluctuations are analysed in detail in Thompson and Grime (1983). Under constant temperature conditions the majority of grasses, legumes and composites germinated over a wide range of temperature, and the same feature was evident in species of ubiquitous or southern distribution in the British Isles. A requirement for relatively high temperature was apparent in sedges, plants of northern distribution and the majority of marsh plants. The range of constant temperatures conducive to germination tended to be wider in grassland plants than in woodland species. Rapid germination over a wide range of temperature occurred in many of the species which attain great abundance in the Sheffield flora.

Although germination in most species was promoted by light, in a few there was inhibition under relatively high light flux. In 104 species a marked reduction in germination occurred if seeds were kept in the dark, and in many species this inhibitory effect could be intensified by excluding temperature fluctuations. A capacity for germination in darkness was observed in all of the legumes and

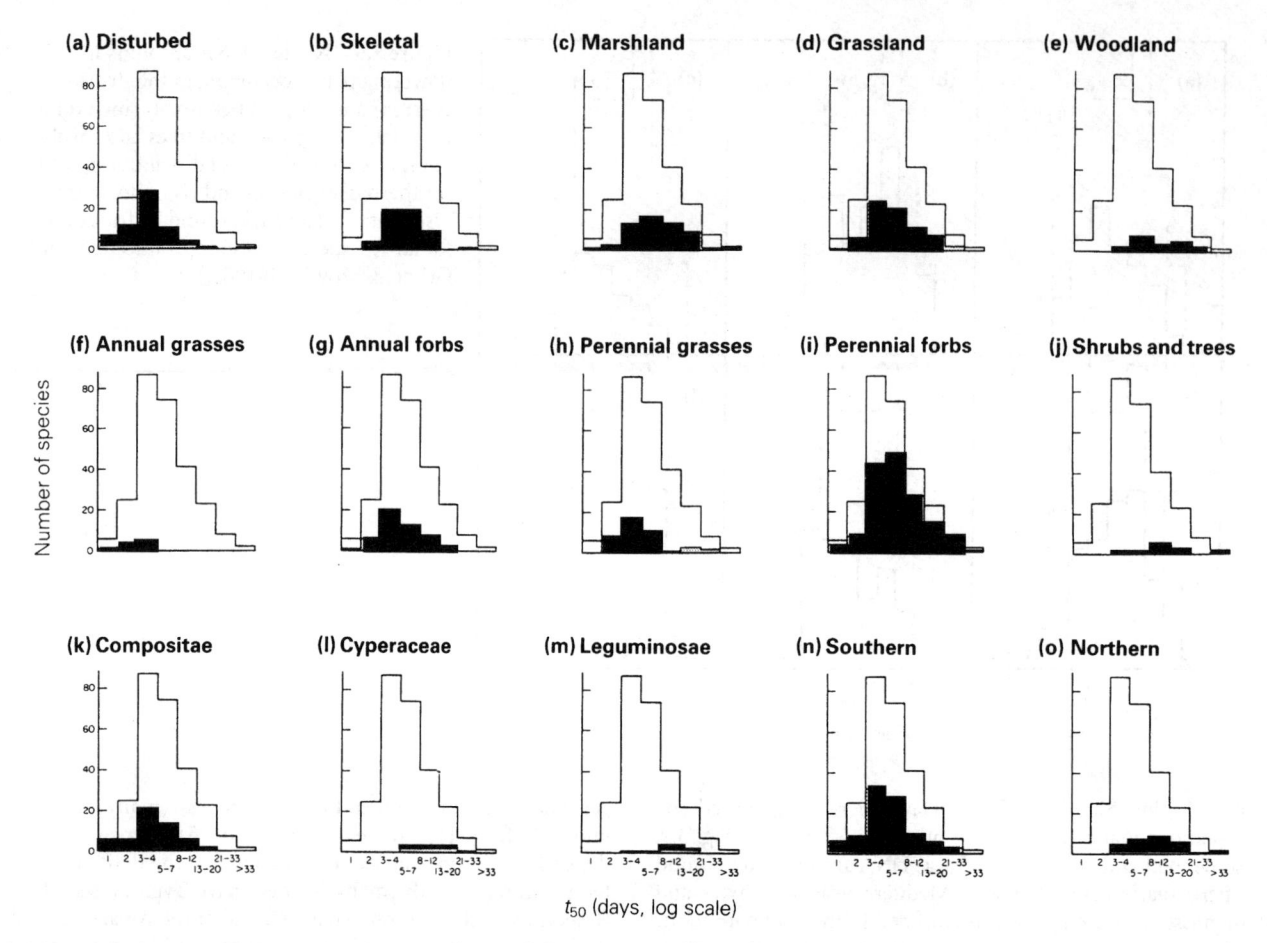

Figure 3.10 The distribution of germination rates. In each histogram the contribution of the specified group of species to a percentage class is indicated by the 'filled' part of the column. The complete column is the total for all species. (From Grime et al. 1981.)

many of the grasses. Dark germination did not occur in the Cyperaceae, and was uncommon in the Asteraceae. The inhibitory effect of darkness was characteristic of many of the species known to form reserves of buried seeds, but it also occurred in certain species with more transient seed banks.

Relative growth rate

Grime and Hunt (1975) estimated the maximum potential relative growth rate (RGR) attained in the exponential phase by 132 species of flowering plant from the UCPE survey area. The period of growth between 2 and 5 weeks after germination was studied in a standardized, productive environment (p. 32), and fitted curves were used to derive various growth analytical parameters (Hunt 1979, 1981, 1982). Additional (unpublished) measurements which have subsequently become available have also been drawn upon in the Accounts (Fig. 3.11).

Woody species exhibited a bias towards low values of RGR, and a similar trend was evident among fine-leaved grasses. Annuals were most frequent in the high RGR category. Grasses and forbs included a wide range of growth rates and, in both, high values of RGR were associated with a variety of growth forms. With the exception of the woody plants and some biennial herbs, all of the low RGRs recorded were from species which as seedlings and mature plants tended to be small in stature.

The possibility that RGR is of adaptive significance in the field was tested by examining the frequency of species of low or high RGR in vegetation samples from a range of habitat types. In several habitats that are disturbed or productive, or both, fast-growing species were predominant and species of low RGR were virtually or completely absent. The converse was true of several stable, unproductive habitats. Species of moderate RGR were ubiquitous.

Measurements of RGR are time-consuming and considerable effort has been expended in a search (e.g. Garnier & Laurent 1994, Westoby 1998, Wilson et al. 1999) for alternative traits that are easier to measure but have similar predictive power. In Chapters 4 and 5 two alternative candidates to fulfil this role, specific leaf area (SLA), and leaf dry matter content (DMC) are evaluated (see pp. 32 and 43).

Variation in screened attributes

As explained on pp. 7 and 26, the purpose of the screening experiments was to recognize broad patterns of ecological specialization associated with habitat, rather than to establish with precision the mean and range of variation with respect to particular attributes displayed by individual species. Despite ecotypic variation, the majority of species do show consistent and distinct ecologies, and it is not unreasonable to expect that, for many of the more important attributes investigated by the screening experiments, variation within species is unlikely to exceed that observed between species. Naturally, no complete validation of this prediction can be made without obtaining full information on all major attributes of each species, in which case the 'broad-picture' approach would itself be redundant. However, variation in selected variables may be explored in moderate detail and at different levels of organization, to gain at least some insight into the relative magnitude of inter- and intraspecific variation.

Such an exploration was undertaken by Hunt (1984), using a

	(a) Between-species	n	Ref.
		132	4
	(b) Between-population		
	Agrostis capillaris	3	3
	Dactylis glomerata	4	3
	Festuca ovina	5	1
	Festuca rubra	5	3
	Holcus lanatus	7	1
	Lolium perenne	6	3
	Trifolium repens	6	3
	Bromopsis erecta	6	5
	(c) Between-genet		
	Trifolium repens	48	2
	(d) Between-ramet		
	Carex flacca	18	6

Mean relative growth rate (day^{-1})

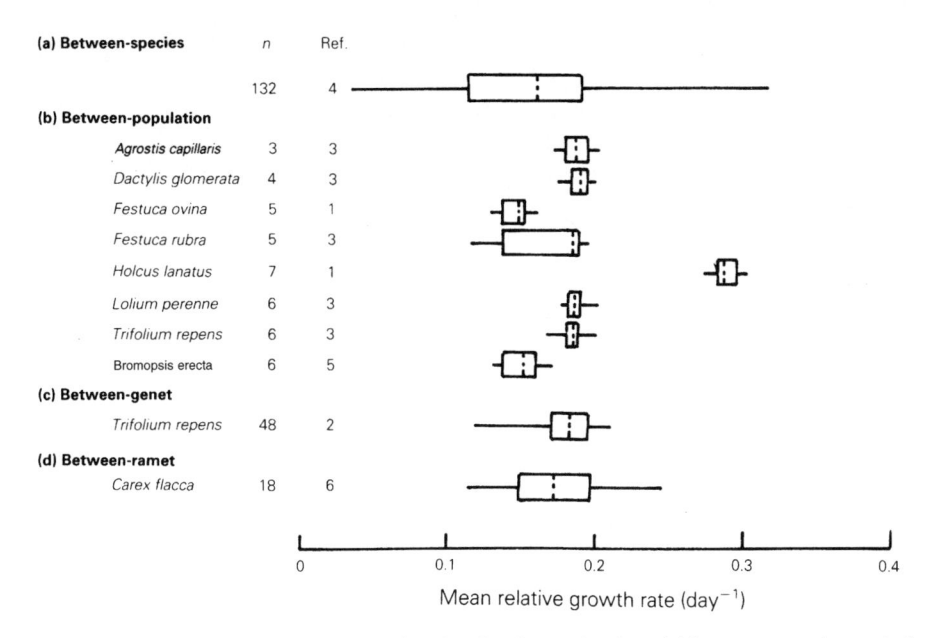

Figure 3.11 Variation in mean realtive growth rate within four levels of organization: (a) between-species variation within a large sample of species representative of each of the dry terrestrial habitats of the Sheffield region; (b) between-population variation between natural or 'commercial' populations of eight different species; (c) between-genet variation between individuals of one population of one species; (d) between-ramet varitation between individuals cloned from one genet of one species. Figures (b), (c) and (d) are scaled to correspond with (a) to allow for small differences in the conditions of measurement and stages of growth. The results are presented in the form of 'box' and 'whisker' diagrams, which identify the ranges and quartiles (Tukey 1977, p. 39). References: (a) A.J.M. Baker (unpubl.); (2) Burdon and Harper (1980); (3) Elias and Chadwick (1979); (4) Grime and Hunt (1975); (5) Law (1974); (6) Hunt 1984).

combination of information already published and new experimental data. It showed that, at least in respect of relative growth rate attained under productive conditions, variation between species is about ten-fold, whereas variation between populations, genets and ramets is perhaps no more than two-fold within any single level (Fig. 3.11). Information gained on the statistical frequency distributions of values within these levels of organization was also encouraging, with normal or log-normal distributions appearing to be the rule. This gives increased confidence to programmes of research which draw their material randomly from natural situations and attempt a preliminary characterization of the attributes of species.

Longer-term experiments reported by Hunt & Lloyd (1987) have shown that the innate variations in growth potential which may exist between species at the juvenile stage are also preserved into at least a second season of growth, though naturally at a reduced amplitude. The authors also present firm evidence for innate variations between species emerging within just a few days of germination.

An early concern in plant ecology exemplified by the classic studies of Clausen, Keck and Hiesey (1940) was to survey the extent of trait variation *between* populations of the same species occurring across wide geographical and environmental gradients. This approach proved conclusively that substantial inter-specific variation can occur at this scale. It is only relatively recently, however that genetic and functional variation *within* local populations has become a focus of research activity as ecologists seek to understand the contribution of such variation to the maintenance of plant community structure and species composition. On the basis of studies such as that of Booth and Grime (2003, Panel 16, page 72) it seems likely that trait variation below the level of the species is poised to emerge as a significant and undervalued factor in the management and conservation of plant communities.

The scientific literature

The fourth source of information for the Species Accounts is the literature already published concerning work done outside Sheffield. Because of the large number of species involved, an exhaustive literature search of the type which is mandatory for *Biological Flora* accounts was not attempted. Also, some reference books in foreign languages, e.g. Hegi (1907-1981), were consulted only infrequently because of the difficulty in obtaining a translation. Nevertheless, in preparing the first edition of this book, a formal review of the modern ecological and taxonomic literature was undertaken for each species with the full-time assistance of a qualified library researcher. Believing that as many of the relevant published sources should be consulted as possible, we included in this all manner of formal and informal reports, papers, theses and books, as the 'References' indicate. These sources covered an extremely wide range of plant science, with geography, ecology, agronomy, environmental physiology, cytogenetics, plant anatomy, taxonomy, systematics and evolutionary biology all being represented. It is a matter of considerable regret that in producing this second edition without financial support or assistance we have been unable to repeat the exhaustive search that was achieved for the first edition.

In addition to the large number of publications which have yielded points of individual detail, a small number of special sources have been used repeatedly to provide items of standard information. These are all identified in Chapter 4, and some are listed on the endpapers.

The next chapter explains in detail the contents of the Accounts themselves, and is followed by a chapter concerned with the ecological interpretation of the data which appear in them.

4 THE CONTENTS OF THE SPECIES ACCOUNTS

Introduction

Each Account follows a standardized format occupying two pages. Original field data are represented in the form of diagrams which are interpreted in notes located on the left-hand page. Also included on the left-hand page is information relating to the morphology, life-history, phenology and cytology of the species, derived either from published sources or from our own field observations and laboratory studies.

A third major element in each Account is the 'Synopsis', which reviews the distinctive features of the species and its ecology. This section incorporates references to autecological studies and to other specialized datasets, if relevant. Wherever possible the Synopsis also includes comments on the responses of the species to biotic factors and to forms of vegetation management. Immediately following the Synopsis, 'Current trends' contains a brief account of changes which may be predicted or are known to be taking place in the distribution or ecology of the species within the British Isles.

What follows here is an explanation of each part of the Species Account. Sources of information are identified, and the methods of presentation are described. For quicker reference a brief guide to this information is provided on pages 79 and 80.

Nomenclature

Each species is identified by the scientific and common names cited in the *New Flora of the British Isles*, (Stace 1997). The scientific name of species of introduced origin is preceded by an asterisk, which is included in parenthesis if the species is either doubtfully native or has an increased geographical or ecological range as a result of escaping from cultivation. Any synonym used in *Flora Europaea* (Tutin *et al.* 1964-80) is added in italics. Where a different name has been used for a *Biological Flora* account (*q. v.*), or in the New Atlas of the British and Irish Flora (Preston *et al.* 2002a), this is explained, within the body of the text. Other species are sometimes mentioned even though they are not the subject of the Account; these are distinguished by the addition of an authority to the scientific name. Some closely related taxa could not be separated readily during fieldwork. In the few instances where the taxonomy is complex or unresolved (e.g. *Rubus fruticosus* agg. or *Taraxacum* agg.), the Account refers to the aggregate species. Where two species are of similar ecology and morphology and hybridize frequently (e.g. *Betula pendula* and *B. pubescens*, or *Quercus petraea* and *Q. robur*), only one Account is presented for each pair. However, wherever possible, and particularly within the Synopsis, any characteristics which distinguish the separate species are emphasized.

[Some frequently used general references are usually cited in an abbreviated form. *CTW: Flora of the British Isles*, 2nd edn (Clapham *et al.* 1961); *ExcurFl: Excursion Flora of the British Isles*, 3rd edn (Clapham *et al.* 1981); *FE 1(-5): Flora Europaea* (5 Vols) (Tutin *et al.* 1964-80).]

Life-form

The form and life-history of the species is described using standardized terminology. This section is an elaboration of the life-form classification of Raunkiaer (1934) in which plants are placed in categories according to the position of the growing points of the shoots over the 'inactive season' (Fig. 4.1). In Britain this varies according to habitat and vegetation type, and may occur either during the winter or, less commonly, in summer.

Introducing each life-form description are terms relating to life-span (see Table 4.1) and the arrangement of the leaf canopy (see Fig. 4.1).

In the later part of the life-form description, further phrases are added to characterize first the shoot, then the root system. Standard botanical terms, defined by *CTW*, are used to characterize leaf shape; leaf size is also assessed. Wherever possible, information on depth of root penetration is added, either from our own observations or from published sources (e.g. Anderson 1927).

We next comment on the nature and extent of mycorrhizal infection of the root system making particular reference to the literature survey of Harley & Harley (1986).

For a guide to the further interpretation and use of information under this heading, see page 41.

Phenology

Here the timing of major events or phases *in the life of the established plant* is described in relation to season. Particular attention is given to the state of the plant in winter and during the period in which the main onset of shoot expansion takes place (winter, early spring, late spring or summer). For many species quantitative data are available describing the seasonal development of the shoot biomass (Williamson 1976, Al-Mufti *et al.* 1977, Al-Mashhadani 1980, Furness 1980), at least within particular plant

Table 4.1 An explanation of life-history classes.

Description	Life-span (years)	Frequency of reproduction
Monocarpic annual	<1	Once, then dies
Monocarpic biennial	2	Once, then dies
Monocarpic perennial	>1	Once, then dies
Polycarpic perennial	>1	More than once (perhaps many times in long-lived species)

Table 4.2 A summary of the experimental design and environmental conditions used in the screening for mean relative growth rate under a standardised, productive regime. (From Grime and Hunt 1975.)

Medium: Sand + Hewitt nutrient solution	Visible radiation at plant height: 38.0 W m^{-2} (0.0545 cal cm^{-2} min^{-1})
Day temperature: 20 ± 0.5°C	Night temperature: 15± 0.5°C
Day length: 18 h (06.00 to 00.00 h)	Relative humidity: always above 60%

Dates of experiments: 24 November 1967 to 24 May 1973

Design of each experiment: 12-18 species × 4 harvests × 5 replicates

---

communities. These data have provided valuable reference points in cases where the classification of shoot phenology has, of necessity, relied upon more informal field observations. The months over which flowers and fruits appear are noted (and for pteridophytes the season of spore release is identified). Where a high proportion of propagules is known to be retained on the parent plant (e.g. in *Rumex obtusifolius* and *R. crispus*) this is reported.

Phenological events within a single species may vary considerably according to latitude, altitude, aspect, habitat, management and year of observation. In recognition of this, the ranges quoted in the Accounts are relatively wide, and take heed both of published information and of our own observations over an extended period of years.

For a guide to the further interpretation and use of information under this heading, see page 42.

Height

Here the upper limit of the leaf canopy of the established plant is recorded, together with that attained by the flowering shoot.

Seedling RGR

Estimates of the mean relative growth rate in total dry weight per plant, measured during the seedling phase, are based upon the screening procedure devised by Grime and Hunt (1975). The period of growth between 2 and 5 weeks after germination was investigated by means of weekly harvesting of sets of individuals selected randomly from a population growing in a standardized, productive, growth-room environment. A summary of the experimental design and environmental conditions used in this screening procedure is given in Table 4.2. A full description of the methods was given by Grime and Hunt (1975). See also the notes provided in Chapter 3 (p. 28).

It is unlikely that the conditions selected were optimal for all of the species examined; in some at least the values of RGR may be underestimates. Moreover, relative growth rates vary ontogenetically during the seedling phase. Within species, genets and even ramets, there may be variations in potential RGR (Bradshaw *et al*. 1964, Elias & Chadwick 1979, Burdon & Harper 1980, Hunt 1984, Hunt & Lloyd 1987). In view of this we have not quoted exact estimates of RGR. Instead, each value has been classified using a class-interval of 0.5 week^{-1}.

For a guide to the further interpretation and use of information under this heading, see page 42.

Specific leaf area (SLA) and leaf dry matter content (DMC)

Values of SLA, expressed as mm^2, and DMC, as percentage of the fresh weight of the leaf, were obtained using methods described in Wilson *et al.* (1999) and relate to material collected from unshaded field conditions within the Sheffield region by Peter Wilson and other members of UCPE and the Department of Archaeology, Sheffield University. Much of this information is unpublished but is incorporated into the LEDA database (Knevel *et al*. 2003). There is a close correspondence between values from UK samples and those measured on leaves from continental populations of the same species. In a few species for which no UK samples were collected we have used data from continental material; such cases are identified by an asterisk.

Leaves were wrapped in moist tissue paper and sealed in a polythene bag then left overnight in a refrigerator at 4°C in the

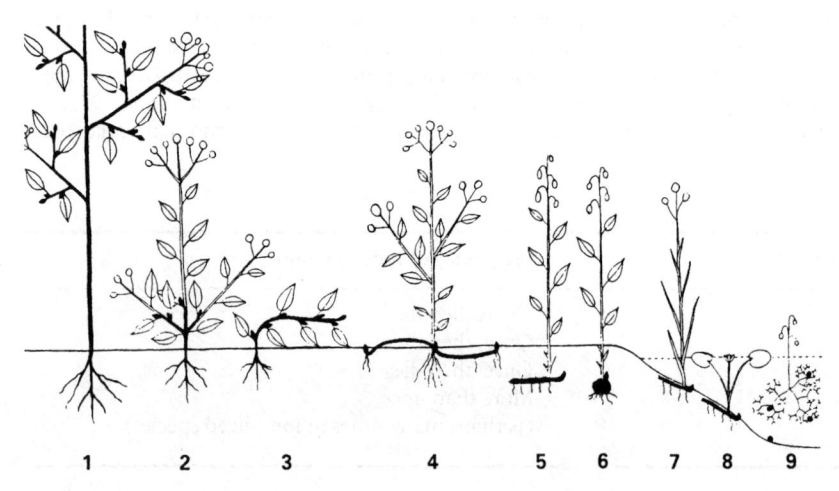

Figure 4.1 The chief types of life form, based upon Raunkiaer's classification. Light lines denote the parts of the plants which die in the unfavourable season, heavy lines the parts which persist. Terms in use are: 1, phanerophytes; 2 and 3, chamaephytes; 4, hemicryptophytes; 5 and 6, geophytes; 7, helophytes; 8 and 9, hydrophytes. (From Raunkiaer 1934.)

dark to ensure full hydration. Subsequently, they were blotted dry and fresh mass was measured. After scanning of leaf area, the samples were oven-dried at 80°C for 2-3 days prior to reweighing. SLA was calculated as area divided by dry mass and DMC as dry mass divided by fresh mass and expressed as a percentage.

Nuclear DNA amount and chromosome number

(For brevity this title is shortened in the Accounts, but the section deals with the following two topics.)

Nuclear DNA amount (genome size)

The values for nuclear DNA cited here are weights per nucleus, based upon Feulgen staining and microdensitometry of root-tip preparations. Values were available from published sources (e.g. Bennett & Smith 1976, Grime & Mowforth 1982), or contained in the web-site maintained *at Bennett and Leich (2004)* but some are the result of new measurements made on freshly-germinated seedlings or on established plants removed from the field and grown in nutrient solution (S. R. Band unpubl.).

The procedure for DNA measurement at UCPE followed that described by Bennett and Smith (1976), with refinements by Goldstein (1981). Root tips of the species under study were processed simultaneously with root tips of two standard species, *Senecio vulgaris* and *Allium cepa* (cv. Ailsa Craig), for which the accurate nuclear DNA amounts are known from chemical determinations. The root tips were fixed, hydrolysed, stained with Schiffs reagent, squashed and mounted on glass in DePeX. The density of stain in the early prophase nuclei of all three species was then estimated using a Vickers M86 scanning microdensitometer. For each species 30 nuclei (ten on each of three slides) were measured.

The DNA amounts are expressed as the 2C value, that is the total amount of DNA present in the nucleus in the early prophase of cell division, measured in picograms. Where polyploidy occurs (e.g. in *Ranunculus ficaria*) separate determinations for the different karyotypes are quoted, if available. In the rare instance of *Poa annua*, where interpopulation variation in DNA amount has been observed at constant chromosome number, the overall range of values is recorded under 'Synopsis'.

Chromosome number

Chromosome numbers for each species are based upon those quoted by published sources. Most fortunately, we have been able to draw upon the extensive unpublished compilation for the British flora of Professor D. H. Valentine. These data have been augmented by European records taken from D. M. Moore (1982), R. J. Moore (1973, 1974, 1977) and Federov (1974), but a comprehensive scan of the modem cytological literature was beyond our scope.

Estimates of ploidy are also included. Normally, for convenience, the base number for the genus is treated as *2x*. However, in the case of some polyploid genera, ploidy is also assessed relative to the base number of the family. No estimate of ploidy is given for the Cyperaceae, the members of which have a diffuse centromere and, in many instances, form an aneuploid series (Stebbins 1971).

[General references cited in abbreviated form in this section are *BC: Compilation of British Chromosome Numbers* (D. H. Valentine unpubl.) and *FE1* (-5): *Flora Europaea* (cited above).]

For a guide to the further interpretation and use of information under this heading, see page 44.

Established strategy

Each species is ascribed a position (or in genetically and ecologically variable species, a range of positions) within the C-S-R model of primary ecological strategies (Ramenskii 1938, Grime 1974, 1977, 1979). Already in Chapter 2 (p. 9), and again in Chapter 5 (p. 46), the nature of this theoretical model has been examined and its implications for ecology reviewed. Current efforts to test its validity are described elsewhere (Grime 2001). Here we simply state the main assumptions upon which the theory is based, and describe the criteria used to classify species with respect to strategy. An account of the method used to classify species with respect to primary strategy is contained in Hodgson *et al.* (1999) and illustrations of the use of CSR strategies to interpret the results of monitoring studies and ecological experiments are presented in Hunt *et al.* (2004).

A plant strategy may be, defined as a grouping of similar or analogous genetic characteristics which recurs widely among species or populations, such that they show similarities in ecology. The CSR theory (p. 9) argues that there are three primary strategies in plants, because there are three distinct threats to existence: competitive exclusion, chronic stress and repeated severe disturbance. Each threat occurs under particular types of environmental conditions, and each confers a selective advantage upon a different type of ecological specialization: competitors (C) prevail under the threat of competitive exclusion, stress-tolerators (S) under severe stress, and ruderals (R) in conditions of frequent and severe disturbance.

The extreme conditions favouring C, S or R strategists form only part of a range of environments exploited by plants. The full spectrum of habitat conditions and associated plant strategies can be represented by an equilateral triangle (see Fig. 2.1, p. 9) within which variation in the relative importance of competition, stress and disturbance controls not only the three types of plant specialization described above, but also a range of intermediate strategies (C-R, S-C, S-R and C-S-R), each associated with a less extreme equilibrium between competition, stress and disturbance.

Table 4.3 lists attributes of morphology, life-history and physiology which appear to be associated with the three primary strategies, and identifies those which have proved most useful in classifying the strategies of the species for which Species Accounts are provided in Chapter 6. It might be considered unlikely that so many variable attributes of plants would conform to these patterns. However, it is the cardinal assertion of the C-S-R model of primary plant strategies that under the distinctive selection pressures associated with the extremes of competition, stress and disturbance (see p. 13) the range of adaptive possibilities is extremely constrained and few evolutionary solutions are viable. Further conformity arises from the fact that each of these solutions depends upon an integrated response involving most of the fundamental activities of the plant.

In order to illustrate the system, typical examples of each strategy have been assembled in Table 4.4, together with references to published investigations relevant to their classification as such. For a guide to the further interpretation and use of information under this heading, see page 46.

Regenerative strategies

Many previous treatments of regeneration in plants (e.g. Harper 1977) have been particularly concerned with the genetic structure of populations, placing strong emphasis upon the relative importance of sexual and vegetative processes in the regenerative biology of plants. We share this concern, and this is particularly evident in the 'Synopsis' of many Accounts. However, under 'Regenerative

Table 4.3 Some characteristics of competitive, stress-tolerant and ruderal plants. Characteristics shown in bold type have proved particularly useful in classifying plant strategies.

	Competitive	Stress-tolerant	Ruderal
(i) Morphology			
1. Life-forms	Herbs, shrubs, trees	Lichens, bryophytes, herbs, shrubs and trees	Herbs, bryophytes
2. Morphology of shoot	**High dense canopy of leaves. Extensive lateral spread above and below ground**	Extremely wide range of growth forms	**Small stature, limited lateral spread**
3. Leaf form	Robust, often mesophytic	**Often small or leathery, or needle-like**	Various, often mesophytic
4. Canopy	**Rapidly-ascending monolayer**	Often multilayered; if monolayer, not rapidly ascending	Various
(ii) Life-history			
5. Longevity of established phase	Long or relatively short	**Long to very long**	**Very short**
6. Longevity of roots	Relatively short	Long	**Short**
7. Leaf phenology	**Well-defined peaks of leaf production coinciding with periods of maximum potential productivity**	Evergreens, with various patterns of leaf production	Short phase of production in period of high potential productivity
8. Phenology of flowering (or sporulation in ferns)	Flowers produced after (or more-rarely before) periods of maximum potential productivity	No general relationship between time of flowering and season	**Flowers produced early in the life-history**
9. Frequency of flowering	Established plants usually flower each year	**Intermittent flowering over a long life-history**	High frequency of flowering
10. Proportion of annual production devoted to seeds	Small	Small	**Large**
11. Perennation	Dormant buds and seeds	**Stress-tolerant leaves and roots**	Dormant seeds
12. Most commonly associated regenerative* strategies	V, S, W, B,	V, W, B$_j$	S, W, B,
(iii) Physiology			
13. Mean potential relative growth-rate	High	Low	**High**
14. Response to resource depletion	Rapid morphogenetic responses in the form and distribution of leaves and roots	Morphogenetic responses slow and small in magnitude	Rapid curtailment of vegetative growth, diversion of resources into flowering
15. Photosynthesis and uptake of mineral nutrients	Strongly seasonal coinciding with long continuous period of vegetative growth	Opportunistic, often uncoupled from vegetative growth	Opportunistic, coinciding with vegetative growth

*Key to regenerative strategies (see Table 3.5): V, vegetative expansion; S, seasonal regeneration in vegetation gaps; W, numerous small, wind-dispersed seeds or spores; B$_s$, persistent bank of seeds or spores; B$_j$, persistent bank of juveniles.

	Competitive	Stress-tolerant	Ruderal
16. Acclimation of photosynthesis, mineral nutrition and tissue hardiness to seasonal change in temperature, light and moisture supply	Weakly developed	Strongly developed	Weakly developed
17. Storage of photosynthate mineral nutrients	**Most photosynthate and mineral nutrients are rapidly incorporated into vegetative structure, but a proportion is stored and forms the capital for expansion of growth in the following growing season**	Storage systems in leaves, and in stems or roots, or both	**Confined to seeds**
(iv) **Miscellaneous**			
18. Litter	Copious, not usually persistent	Sparse, but often persistent	Sparse, not usually persistent
19. Palatability to unspecialized herbivores	Various	Low	Various, often high
20. Nuclear DNA amount	Usually small	Various	Small to very small

*Key to regenerative strategies (see Table 4.5): V, vegetative expansion; S, seasonal regeneration in vegetation gaps; W, numerous small, wind-dispersed seeds or spores; B_s, persistent bank of seeds or spores; B_j, persistent bank of juveniles.

Table 4.4 Species typical of seven ecological strategies (see Fig. 2.1).

Strategy	Typical species
Competitor	*Chamerion angustifolium*[1–4] *Urtica dioica*[1, 5–10] *Phalaris arundinacea*[1, 11]
Ruderal	*Capsella bursa-pastoris*[12, 13] *Senecio vulgaris*[1, 12–14] *Urtica urens*[7, 9, 15]
Stress-tolerator	*Danthonia decumbens*[1, 16, 17] *Primula veris*[18] *Sanicula europaea*[1, 19]
Competitive-ruderal	*Impatiens glandulifera*[20] *Ranunculus repens*[1, 21–23] *Tussilago farfara*[1, 2, 24, 25]
Stress-tolerant ruderal	*Carlina vulgaris*[26–28] *Catapodium rigidum*[1, 29] *Linum catharticum*[26, 30]
Stress-tolerant competitor	*Dryopteris filix-mas*[31–33] *Mercurialis perennis*[5, 34] *Vaccinium myrtillus*[1, 35, 36]
'C-S-R strategist'	*Holcus lanatus*[1, 37, 38] *Hypochaeris radicata*[4, 39, 40] *Rumex acetosa*[1, 41]

Publications relevant to this classification are:
1, Grime and Hunt 1975; 2, Myerscough and Whitehead 1966, 1967; 3, Myerscough 1980; 4, Van Andel and Vera 1977; 5, Al-Mufti et al. 1977; 6, Pigott and Taylor 1964; 7, Boot et al. 1986; 8, Wheeler 1981; 9, Greig-Smith 1948a; 10, Bassett et al. 1977; 11, Buttery and Lambert 1965; 12, Fryer and Makepeace 1977; 13, Salisbury 1942; 14, Harper and Ogden 1970; 15, Salisbury 1964; 16, Higgs and James 1969; 17, Furness 1980; 18, Tamm 1972; 19, Inghe and Tamm 1985, 20, Al-Mashhadani 1979; 21, Sarukhan 1974; 22, Doust 1981a,b; 23, Sarukhan and Harper 1973; 24, Bakker 1960; 25, Ogden 1974; 26, Verkaar et al. 1983; 27, Watt 1981; 28, Verkaar and Schenkeveld 1984b; 29, Clark 1974; 30, During et al. 1985; 31, Pogorelova. and Rabotnov 1978; 32, Willmott 1985; 33, Page 1982; 34, Hutchings 1978; 35, Ritchie 1956; 36, Pigott 1983; 37, Watt 1978; 38, Beddows 1961; 39, Aarssen 1981, 40, Turkington and Aarssen 1983; 41, Putwain and Harper 1970.

Table 4.5 Five regenerative strategies of widespread occurrence in terrestrial vegetation.

Strategy	Functional characteristics	Conditions under which the strategy appears to enjoy a selective advantage
Vegetative expansion (V)	New shoots vegetative in origin and remaining attached to parent plant until well established	Productive or unproductive habitats subject to low intensities of disturbance
Seasonal regeneration (S)	Independent offspring (seeds or vegetative propagules) produced in a single cohort	Habitats subjected to seasonally predictable disturbance by climate or biotic factors
Persistent seed or spore bank (B_s)	Viable but dormant seeds or spores present throughout the year; some persisting more than 12 months	Habitats subjected to temporally unpredictable disturbance
Numerous widely dispersed seeds or spores (W)	Offspring numerous and exceedingly buoyant in air; widely dispersed and often of limited persistence	Habitats subjected to spatially unpredictable disturbance or relatively inaccessible (cliffs, walls, tree trunks, etc.)
Persistent juveniles (B_j)	Offspring derived from an independent propagule but seedling or sporeling capable of long-term persistence in a juvenile state	Unproductive habitats subjected to low intensities of disturbance

strategies' our object is to recognize, and where possible classify, mechanisms of regeneration in *functional* terms by considering the size, number, dispersal, dormancy and degree of independence of the offspring, and the conditions affecting its establishment.

Five major types of regenerative strategies of common occurrence in the British Isles are listed and described in Table 4.5. A shortage of reliable demographic information has made it impossible to recognize the strategy involving banks of persistent juveniles with certainty. Few references to this strategy have therefore been made in the Accounts. Also, it has been difficult to distinguish between vegetative expansion and seasonal regeneration by vegetative fragmentation in the case of certain species; in such cases of uncertainty we have used the symbol (V).

For a guide to the further interpretation and use of information under this heading, see page 47, Table 4.3 (page 34) and Fig. 5.4 (page 48).

Flowers

Here we note flower colour, structure of the inflorescence, mode of pollination and breeding system. Where available, information also appears on the distribution of male and female parts between flowers, on inflorescences and on any marked differences in the time of emergence of stigmas and stamens. The number of flowers in an inflorescence is very variable even within a single species. However, a series of classes is used to allow a crude separation of species with few-flowered inflorescences from those with many-flowered ones. These classes represent averages of ⩽10, 11-25, 26-50, 51-100, 101-1000 and >1000 flowers per inflorescence.

In the case of pteridophytes the distribution of the sori is described.

For a guide to the further interpretation and use of information under this heading, see pages 41 and 42.

Germinule

This word is used to identify the structure which germinates to give rise to the seedling (or prothallus in the case of ferns). Each Account notes the mean weight and linear dimensions (length x breadth) of air-dried, viable germinules sampled from natural populations in the British Isles. Most of these values are taken from Grime *et al.* (1981). Information is also included as to whether the germinule is a seed or a fruit (such as an achene). Notes follow on the presence and dimensions of any appendage to the seed such as a pappus, lemma, bristle or an elaiosome. The instances where, for example, the seed is enclosed in a berry or a capsule are also recorded. Such features have relevance to mechanisms of dispersal.

Finally, an estimate of the number of seeds per flower is given. This, like the value for number of flowers per inflorescence, is necessarily imprecise, and can be used only to differentiate between extremes.

For a guide to the further interpretation and use of information under this heading, see page 50.

Germination

The germination data are mainly drawn from the standardized screening procedures reported by Grime *et al.* (1981), augmented by subsequent unpublished measurements conducted in Sheffield using the same techniques.

First, the pattern of germination is described in terms of whether the cotyledons are emergent (epigeal) or remain within the seed (hypogeal). This is followed by a note of the most rapid rate of germination recorded for the species, expressed as the number of days following imbibition of air-dried seeds required for half of the final percentage germination to be attained, t_{50}. For the majority of species this rate was attained in experiments where seeds, ripened by dry storage in the laboratory, were incubated on moist Whatman No.1 filter paper in Petri dishes in the light (under a regime illuminated by a mixture of fluorescent tubes and tungsten lamps providing 40 W m^{-2} of short-wave radiation, at 20°C over an 18-h day, with a night temperature of 15°C).

Next, the response of the seed to temperature is described in terms of the upper and lower limits of the range over which seeds achieved half of the maximum percentage germination attained under the most favourable temperature regime. These data refer exclusively to thermogradient bar experiments, in which seeds were exposed to constant temperatures over the range 5-40°C under fluorescent light and with ample water supply and high atmospheric humidity.

The final piece of standardized information is derived from a comparison of germination under fluorescent light or in darkness against that observed under constant background conditions of day length (18 h), temperature (20/15°C) and moisture supply. Here the species are broadly classified into those in which germination always occurred to a high percentage and was relatively unaffected by irradiance (L = D), and those in which germination was to some extent inhibited by darkness (L > D) or inhibited by light (L < D).

Where appropriate, additional information relating to germination has been added, drawing either upon our own studies or upon other published investigations. In particular, we highlight any evidence implicating hard seed-coats or chilling requirements in mechanisms of seed dormancy.

For a guide to the further interpretation and use of information under this heading, see page 51.

Biological flora

The *Biological Flora* accounts published in *The Journal of Ecology* and *The Canadian Journal of Botany* are a rich source of information on many of the species described in Chapter 6. Relevant accounts are identified by date and author and we also refer to any other publications which may provide substantial synopses of the ecology of the subject species. The abbreviation *Biol Fl* is used when information abstracted from a *Biological Flora* is cited.

Geographical distribution

Notes on the British, European and world distributions of the species are derived mainly from *CTW* (Clapham *et al.* 1962), *ABIF* (Preston, *et al.* 2002a, *New Atlas of the British and Irish Flora*) and *FE1* (-5) (Tutin *et al.*, 1964-80, *Flora Europaea*, 5 Vols). More limited references are made to *CSABF* (Perring & Sell 1968, *Critical Supplement to the Atlas of the British Flora*) and *AF* (Jermy *et al.* 1978, *Atlas of Ferns of the British Isles*).

Habitats

The first notes which appear under this heading indicate the pattern of occurrence of the species in the UCPE 'Survey II'. As explained already (p. 18), this was a survey of all of the major herbaceous vegetation types of the Sheffield region. The data to which these notes refer are presented in the centre of the left-hand page in the form of a simple branched key in which habitats are classified according to physical features of the environment and to forms of land-use. The key contains seven 'primary', eight 'intermediate' and 32 'terminal' categories (indicated as lozenges, squares and circles, respectively). The value which is quoted in each box indicates the percentage of survey samples of that habitat which contains the subject species. The survey base for this habitat key consists exclusively of inland habitats sampled in north-central England; the diagram thus indicates where each species is most commonly found within the Sheffield region.

The latter part of the notes given under 'Habitats' draws attention to aspects of the habitat distribution which are not reflected in the habitat key. In particular, information has been taken from the unpublished 'Survey III' (p. 18), involving rarer plant species and communities and performed by JGH with the assistance of S. R. Band and the co-operation of many local naturalists. Data from this source are identifiable from our use of the word 'observed' in place of 'recorded' or 'found'. Finally, other aspects of habitat distribution within Britain, particularly those relating to maritime and montane habitats, are noted within square brackets.

For a guide to the further interpretation and use of information under this heading, see page 54.

Gregariousness

In the table which appears at the foot of the right-hand page, 'Abundance in quadrats', and in the brief interpretative notes under 'Gregariousness' on the left-hand page, we use survey data to review the range in abundance of the subject species in vegetation samples associated with each of the seven primary habitats. For each habitat the table lists the number of occasions on which the species attains each of ten classes of abundance. The ten classes are distributed over the range 1-100, where each unit corresponds to one 100 x 100 mm subsection of the 1 m^2 quadrat. The subsection is counted if it is occupied by rooted shoots of the subject species. The bottom line in the table provides totals for each abundance class, expressed as a percentage of the total number of occurrences of the species in the whole survey. In some cases where there is a 'carry-over' of text from the left-hand page, this is the only line which appears. In other cases the table is omitted entirely.

For a guide to the further interpretation and use of information under this heading, see page 55.

Altitude

The information interpreted here occurs mainly in the diagram of this title situated in the top, left-hand corner of the right-hand page. This figure, based upon survey data from the Sheffield region, describes for each of five altitude classes the percentage of quadrats containing at least one subsection in which the species is rooted. This is shown as an unstippled area. The figure also indicates the percentage of samples in which 20% or more of the subsections are occupied (the stippled area). In the notes under 'Altitude' on the left-hand page we draw attention to any consistent features of these data and, where possible, note the upper altitudinal limit of the species within the Sheffield region. In some instances the maximum value is derived from the survey of rarer plant species (see 'Habitats'), with data from this source again being quoted as 'observed'. In a majority of the Accounts a further note is added (in square brackets) describing the upper altitudinal limit of the subject species in the British Isles, from Wilson (1956).

For a guide to the further interpretation and use of information under this heading, see page 55.

Slope

Here the main source of information is the slope histogram centred at the top of the right-hand page. The unstippled columns in this figure record the percentage of simple occurrences of the subject species over each of five classes of slope. The stippled columns correspond to the percentage of samples in which 20% or more of the quadrat subsections contained rooted shoots, as under 'Altitude'.

For a guide to the further interpretation and use of information under this heading, see page 56.

Aspect

Two circular diagrams, situated at the top right of the right-hand page, use UCPE survey data to provide comparisons of the frequency of occurrence of the species on north- and south-facing slopes in unshaded and shaded habitats. These analyses exclude

Table 4.6 Classes used in the 'Hydrology' histogram of the Species Accounts, with numbers of quadrats involved (this classification applies to wetlands only, see Fig. 3.2).

Class	Slope (degrees)	Water depth above soil surface (mm)	Quadrants
A	>5	Nil	33
B	≤5	Nil*	82
C	≤5	Nil†	107
D	Nil	<100	83
E	Nil	101–250	58
F	Nil	>250	66

* Not marginal to open water; † marginal to open water.

samples from wetlands and from slopes of less than 10°. In each diagram the results of chi-squared statistical tests are indicated, showing the probability of any observed difference between north- and south-facing slopes being due solely to chance, either (above) with respect to the proportion of samples simply containing the species or (below) with respect to the proportion of the occurrences in which rooted shoots of the species occupy 20% or more of the quadrat.

On the left-hand page the notes under 'Aspect' describe salient features of the two aspect diagrams. Where appropriate we refer to the results of other investigations on the aspect distributions of plant species (e.g. Grime & Lloyd (1973) *An Ecological Atlas of Grassland Plants*, abbreviated *Ecol Atl*, and the UCPE Survey III).

For a guide to the further interpretation and use of information under this heading, see page 57.

Hydrology

The distribution of the species in relation to site hydrology is examined in the 'Hydrology' histogram near the bottom left of the right-hand page. In this figure 'Survey II' samples from the Sheffield region have been arranged into six classes, forming a series of decreasing probability of surface drainage (explained in detail in Table 4.6). The unstippled columns record the percentage occurrence of the subject species in each class, and the stippled columns correspond to the percentage of samples in which 20% or more of the 1 m² quadrat subsections contain rooted shoots. Notes summarizing this distribution appear under 'Hydrology' on the left-hand page. Notes on the hydromorphology of wetland sites are also included, based on notes made during fieldwork. For simplicity, only three categories of mire are separated: *ombrogenous mire*, with a peaty soil which is irrigated mainly by precipitation (e.g. raised bogs, blanket peat); *topogenous mire*, with a high water table which is maintained primarily by topography (e.g. pond and river margins), and *soligenous mire*, where the water table is fed mainly by lateral movement of ground water (e.g. spring fens). This classification follows that of McVean and Ratcliffe (1962).

For a guide to the further interpretation and use of information under this heading, see page 57.

Soil pH

In the 'Soil pH' histogram in the centre left of the right-hand page, survey samples from the Sheffield region have been divided into five classes, of width 1 pH unit, according to the reaction of a soil paste prepared from an homogenized sample taken from top *c.* 30 mm of the soil profile. The unstippled columns refer to the percentage occurrence of the subject species in each class, and the stippled columns describe the percentage of samples in which 20% or more of the 1 m² quadrat subsections contained rooted shoots. Notes summarizing this distribution appear under 'Soil pH' on the left-hand page.

For a guide to the further interpretation and use of information under this heading, see page 57.

Bare soil

In the 'Bare soil' histogram near the bottom right of the right-hand page, survey samples from the Sheffield region have been divided into six classes (explained in detail in Table 4.7), arranged according to the percentage of subsections of the 1 m² quadrat which individually exhibited more than 50% of their area as exposed soil. The unstippled columns record the percentage occurrence of the species in each class, and the stippled columns indicate the percentage of samples in which 20% or more of the quadrat subsections contained rooted shoots. Notes summarizing this distribution appear under 'Bare soil' on the left-hand page.

The Sheffield surveys also included estimates of exposure of bare rock, spoil (e.g. demolition rubble and cinders) and cover by herbaceous litter, tree litter, bryophytes and lichens. Although these data are not presented here, we comment on them where the subject species shows marked changes in abundance in relation to such factors.

For a guide to the further interpretation and use of information under this heading, see page 57.

△ ordination

In the triangular diagram located in the centre of the right-hand page, an ordination of vegetation samples drawn from all the major habitats of the Sheffield region has been used to estimate the ecological amplitude of the subject species and the 'mean strategy' of the floristic components of the communities with which the species is associated.

The initial step in this ordination was to classify certain of the more common herbaceous species of the Sheffield region with

Table 4.7 Classes used in the 'Bare soil' histogram of the Species Accounts (this classification applies only to non-aquatic habitats, see Fig. 3.2 and Table 3.10).

Class	Percentage cover of bare soil
A	Nil
B	1–10
C	11–25
D	26–50
E	51–75
F	76–100

respect to established strategy. The next step was to ordinate vegetation samples drawn from the range of habitats represented in the Sheffield region. Each 1 m² sample was located within the triangle according to the position and frequency of any marker species that it contained. This was achieved by calculating mean stress and disturbance co-ordinates, with the contribution of each marker species being weighted according to its frequency in the vegetation sample. Classification of all Survey II samples in this way was not possible: first, because criteria for dealing with many aquatic plants were not available and, second, because it was not considered useful to calculate means for vegetation samples containing less than four species. Nevertheless, 2008 vegetation samples were successfully ordinated by these means. They were distributed fairly evenly within the central circular area of the triangle.

A calculation of the percentage occurrence of each subject species within 91 different zones of the circle was then performed. These zones appear as small hexagons within the outer framework of the diagrams presented in the Accounts. The distribution of subject-species percentages permitted contours to be constructed, describing range and abundance in relation to strategy type. These percentages are represented in the published figure on a scale of nil (-), 1, 2, …, 10, explained fully in Table 4.8.

The small scale of this figure does not permit a detailed reproduction of all of the underlying contour diagram. Instead, a simplified version is presented in which the centre of the contoured distribution of the subject species is indicated by a cross, with a single broken contour to delimit that area of the diagram in which the percentage occurrence of the species is at least 25% of the value at its centre, i.e. in the region of the cross. Occasionally, rather diffuse patterns emerge which do not allow contours, or sometimes even centres, to be defined. In a small number of other cases species may show a bimodal pattern. Some are not eligible for contouring at all, as in the case of most tree species, where the individuals recorded in the survey quadrats were seedlings and thus not legitimate members of the established plant community. The significance of these and of the more orthodox patterns is examined later (p. 58).

In each Account a brief interpretation of the analysis is presented in notes under the heading '△ ordination' on the left-hand page.

For a guide to the further interpretation and use of information under this heading, see page 58.

Associated floristic diversity

In the diagram situated at the centre right of each right-hand page all 1 m² quadrats from the vegetation survey containing the species are examined with respect to their species richness. Each sample is allocated to one of five classes, according to the abundance of the subject species in the quadrat at 20% intervals, and in the histogram the mean number m⁻² of species of vascular plants in each class is represented. Thus, the diagram allows an assessment of the general level of species richness associated with the species, and exposes instances in which variation in the abundance of the subject species coincides with changes in floristic diversity. A brief summary and interpretation of these results is presented under the heading 'Associated floristic diversity' on the left-hand page. Relevant observations from Survey III also appear in some cases.

For a guide to the further interpretation and use of information under this heading, see page 58.

Similarity in habitats

Under this heading on the left-hand page, five species are listed which, in terms of their distribution in the terminal categories of the habitat key, are the ones most similar to the subject-species. The species were identified by means of a similarity analysis conducted with the aid of the CLUSTAN statistical package (Wishart 1978). The values quoted for the 'similar' species are similarity coefficients, expressed as percentages. The primary and intermediate habitat categories were not included in this analysis, since this would involve duplication of data. All of the variables were of the continuous, numerical type and the multivariate dataset thus consisted of 281 cases × 32 variables = 8992 elements. All 32 variables were standardized to zero mean and unit variance.

In some less-common species, particularly those from wetlands (which are poorly represented in the survey area), the similarity coefficients were low and the species list was ecologically heterogeneous. In such cases the results of the similarity analysis would have been potentially misleading, and have therefore been omitted from the Account, with a note to that effect.

Commonest associates

Under the triangular ordination on the right-hand page, the five species which co-occur most frequently with the subject–species are listed. The values given are percentages of vegetation samples for the subject species that also contain the associate.

Table 4.8 An explanation of the numerical data provided by the 'Triangular ordination' diagrams of the Species Accounts. Values given indicate the relative constancy of the subject species in groups of vegetation samples ordinated with respect to the established strategies and abundance of component 'marker' species (see p. 48).

Value given	Percentage occurrence of subject species
–	Nil
1	1–10
2	11–20
3	21–30
4	31–40
5	41–50
6	51–60
7	61–70
8	71–80
9	81 and above

Synopsis

The aim of each Synopsis is to review the distinctive features of the subject species and its ecology. This involves an interpretation of the standardized field and laboratory data presented in the Account, with an incorporation of the sometimes more fragmentary information which is available from the various published sources. Unpublished field observations from 'Survey III' (p. 18) have also been drawn upon.

Inevitably the amount of useful data available from the scientific literature varies widely according to species. Agricultural and ecological research workers have dealt exhaustively with commercial forage plants and arable weeds, such as *Lolium perenne*, *Trifolium repens*, *Poa annua* and *Stellaria media*, but at the other extreme there has been great neglect of many common plants of waysides, marginal land and wooded habitats, such as *Leontodon autumnalis*, *Melica uniflora* and *Stellaria holostea*. Hence, for some species our task has been to distil relevant data from an extensive literature, whereas in others we have simply exhausted a woefully incomplete database. In cases where knowledge is exceptionally fragmentary (e.g. *Vicia sepium*) the Synopsis, by its brevity and shallow penetration into ecological analysis, serves mainly as a reminder of our great ignorance of the subject species and its ecology.

However, though the length and content of the Synopses may vary considerably, we have tried to review what knowledge there is of the species in a standardized and logical sequence. Hence, the initial sections are mainly concerned with the geography and habitat range of the subject, and with its niche within plant communities. This is followed, where possible, by an assessment of the species' tolerance of ecological factors, including those under direct or indirect human control. Current or past patterns of human utilization of the species in, for example, agriculture are noted, and any well-defined responses to vegetation management are described. The regenerative mechanisms of the species are considered, both in relation to the maintenance of its natural populations and in relation to its potential for colonizing new habitats. Finally, where information is available, we comment on the genetic integrity and taxonomic affinities of the species and cite any evidence of ecotypic differentiation.

In keeping with the main title of this book, many of the Synopses include passages in which comparisons have been drawn between the subject species and plants of similar ecology within the British flora. Where such plants occur together, as in the case of the guilds of small winter annuals of droughted, calcareous soils, we have tried to recognize any differences which may permit their coexistence within plant communities (see, for example, the discussion of niche differentiation between *Myosotis arvensis* and smaller winter annuals, p. ??).

[Certain common general references in this section are again cited in abbreviated form. *PGE*: *Park Grass Experiment* (Brenchley & Warington 1958, Thurston 1969); *SIEF*: *Survey of Inorganic Elements in Foliage* (J. P. Grime & S. R. Band unpubl.); *WCH*: *Weed Control Handbook* (Fryer & Makepeace 1977).]

Current trends

Under this heading any conspicuous changes which are currently taking place in the abundance of the species are reported and, based upon more-extensive studies of commonness and rarity (Hodgson 1986a,b, Bunce *et al.* 1999, Preston *et al.* 2002a), we briefly assess the current and future status of the species in the British flora.

5 INTERPRETATION AND USE OF THE SPECIES ACCOUNTS

Introduction

Much of the information contained in the Species Accounts is presented in a form already familiar to plant ecologists. It thus requires little interpretation, especially as the written sections of each Account (notably the Synopsis) assess the ecological significance of the more important attributes of the subject species. The main purpose of the notes given in this explanatory chapter is therefore to examine critically some of the less familiar analyses and uses of the data which may be possible using the large body of standardized autecological information which the Accounts provide.

Most of this chapter is concerned with the implications of the Accounts for our understanding of the autecology of the subject species. Later, however, we also explore the use of the same data to investigate the structure and functioning of plant communities and ecosystems.

Autecological Implications

In general terms the most obvious contribution of the Accounts to the autecological study of plants is to provide a review of the essential features of the species and an introduction to the major literary sources regarding its biology and ecology. This may furnish clues to the plant attributes and external factors contributing to the geographical and habitat range of the subject species, to its level of abundance in Britain and to the nature and range of its contributions to British vegetation. At an applied level, the Accounts are likely to be useful to those attempting to control the distribution and abundance of particular species. They may be helpful also in guiding the selection of plants or management techniques for use in the reclamation of landscape and when re-introducing species to nature reserves or to areas maintained for amenity and recreation.

Nomenclature

Characters utilized in plant taxonomy have not been used extensively in ecological analysis. This is because ecological specialization is not as immediately evident in, or characteristic of, families of plants as in the case of many of the higher taxa of animals. Lack of specialization is suggested by the wide range of life-forms adopted and diversity of habitats exploited by some families. However, recent research has brought about an increasing recognition that in characters such as association with nitrogen-fixing organisms (in the Fabaceae), seed size and structure (Hodgson & Mackey 1986), seed germination (Guppy 1912, Grime *et al.* 1981), plant strategy (Hodgson 1986c) and wood structure and its effects on early leafing (Lechowicz 1984) taxonomy coincides with ecological specialization. It seems most likely that as research at the interface of ecology and taxonomy expands, the predictions of Stebbins (1974) and Grime and Hodgson (1987) will be confirmed and family history will emerge as an important influence on the present characteristics and contemporary ecology of individual species.

Life-form

The classification of plants according to life-form was originally conceived by Raunkiaer (1934), as a means whereby floras in different parts of the world could be equitably compared in terms of their 'morphological spectra'. Following elaboration, and computerization, the life-form approach is now the basis for a comprehensive classification of major world vegetation types (Box 1981). The strength of this system lies in its reliance upon morphological attributes which are readily available for most species, and upon the strong global correlations which have been established between dominant life-forms and climate. Despite its increasing sophistication, the life-form approach remains essentially a 'broad-brush' exercise, of greatest value in the hands of phytogeographers and climatologists.

In the notes included under 'Life-form' in the Accounts we have made additions to the Raunkiaer classification in an attempt to increase the usefulness of such descriptions for ecological purposes. One amendment applied to all perennials has been to distinguish between (a) species which flower once and then die ('monocarpic'), (b) those which survive to flower at intervals ('polycarpic') and (c) those which, according to genotype or habitat conditions, vary in the extent to which survival and flowering continue after the' first reproductive season ('facultatively polycarpic'). Used in conjunction with other information in the Account (pp. 46 to 53), this description of the life-history also provides an insight into the population dynamics of the species. Also following Raunkiaer (1934) we have included information on leaf size. However, the ecological significance of this plant attribute is imperfectly understood. In a few studies (Loveless 1961, Beadle 1966), leaf size has been correlated with site productivity, and even with the phosphorus status of the habitat, but most theories (e.g. Gates 1968, Parkhurst & Loucks 1972, Givnish & Vermeij 1976) suggest that leaf size is determined by trade-offs between transpiration cost and potential photosynthetic gain. Givnish and Vermeij (1976) provide predictions of optimal leaf size dependent upon the degree of illumination and water status of the habitat. Nevertheless, attempts to correlate leaf size directly with climate are frequently unsuccessful (e.g. Dolph & Dilcher 1980), and our knowledge is too incomplete to attempt a synthesis for the British flora.

Wherever available, additional information is provided with respect to the root system. The depth of root penetration into the soil is known for many species and often shows interrelationships with habitat and phenology. A high proportion of the species with shallow root systems are either plants of continuously moist habitats (e.g. *Cardamine flexuosa*, *Lamiastrum galeobdolon* and *Oxalis acetosella*), or have growth compressed into the cool damp seasons of the year (e.g. *Aira praecox*, *Myosotis ramosissima* and *Ranunculus ficaria*). In a restricted number of species (e.g. *Asplenium ruta-muraria*, *Festuca ovina* and *Sedum acre*), a shallow root system coincides with an ability on the part of the plant to persist as established mature individuals or seedlings in droughted situations, thus providing circumstantial evidence of physiological tolerance of moisture stress. In a much larger number of perennial plants of desiccated habitats (e.g. *Centaurea scabiosa*, *Hypochaeris radicata* and *Sanguisorba minor*) the root system penetrates deeply and is associated with a capacity for sustained shoot development during the summer. As might be expected, deep root systems are vulnerable to waterlogging and anaerobiosis; several Accounts (e.g. those for *Achillea millefolium*, *Hyacinthoides non-scripta* and *Pteridium aquilinum*) implicate this phenomenon in the exclusion of the subject species from wetland habitats.

Under 'Life-form' the root system is further characterized by reference to the nature and extent of its mycorrhizal infection. It is well known that mycorrhizas are of major significance in the ecology of many native species (Harley 1969), and it is clear that their importance is greatest on undisturbed, infertile, aerobic soils.

In assessing the possible significance of vesicular-arbuscular (VA) mycorrhizal infections, it is important to recognize that there is growing evidence within some herbaceous communities of infertile soils that the roots of individual plants of the same, or even different, species may become connected through a common network of investing hyphae through which mineral nutrients and, perhaps also, assimilate appears to move from 'source' to 'sink' plants, (Whittingham & Read 1982, Chiariello *et al*. 1982, Brownlee *et al*. 1983). The autecological significance of this phenomenon may lie in relation to the process of seedling establishment on infertile soils.

As explained (p. 49), a mechanism of regeneration in plants exploiting unproductive habitats characteristically involves persistent juveniles. Here, often in close proximity to the parent plant, and beneath other established individuals of the same or different species, seedlings and small juveniles may exist for extended periods before eventually either failing or graduating into the established, reproductive population (Tamm 1956, Inghe & Tamm 1985). Such persistence under adverse conditions owes much to various tolerance mechanisms, including the conservative deployment of resources and resistance to pathogens (Vaartaja 1952, Loach 1970, Mahmoud & Grime 1974).

However, it now seems necessary to recognize the strong possibility that, in some species at least, persistent juveniles are subsidized by resources imported from neighbouring established plants through interconnecting VA mycorrhizal hyphae. Evidence in support of this hypothesis is contained in the Accounts for a number of species including *Centaurea nigra*, *Centaurium erythraea*, *Galium verum*, *Pilosella officinarum* and *Scabiosa columbaria*. Usually these are all heavily mycorrhizal, and in the 'Synopsis' for each it is reported that seedling survival or yield in experimental turf microcosms showed a considerable benefit from the introduction of VA mycorrhizas. In a striking corollary, moreover, it is recorded in the Account for *Rumex acetosa* that seedlings of this species, which normally fail to develop significant infections, showed no benefit from the presence of VA mycorrhizas in the same microcosm experiment (Grime *et al*. 1987).

Phenology

Already, under 'Life-form' (p. 41), it has been convenient to examine phenological characteristics in relation to other plant attributes. This approach is carried much further under 'Nuclear DNA amount' (p. 44), 'Established strategy' (see particularly Fig. 5.3, p. 45) and 'Regenerative strategies' (p. 47). Here we simply confine attention to those interpretations and applications which follow directly from field observation of seasonal patterns in the development of the established plant. Detailed treatments of the relationships between phenology and other features of the plant, such as shoot architecture, leaf form, leaf longevity and rooting pattern are available in the studies of Williamson (1976), Al-Mufti *et al*. (1977), Givnish (1982, 1986), Sydes (1984) and Sydes and Grime (1984).

The most obvious implications of phenological data relate to the timing of management procedures. Where control measures need to be applied to restrict the vigour or spread of an aggressive, noxious or otherwise undesirable species (e.g. *Fallopia japonica*, *Rumex obtusifolius* and *Senecio jacobaea*), information relating to the time of shoot expansion, flowering and seed release is clearly essential to achieving maximal effect with minimal effort and cost. This argument applies to the eradication of weeds from arable land, where in timing herbicide applications it is helpful to distinguish between (a) weeds which develop over winter and in early spring (e.g. *Elytrigia repens*, *Galium aparine* and *Poa trivialis*), (b) weeds which are usually delayed in appearance (e.g. *Chenopodium album*, *Matricaria discoidea* and *Urtica urens*) and (c) weeds which are present throughout the year (e.g. *Poa annua*, *Senecio vulgaris* and *Stellaria media*).

Similar considerations are important where the object of management is to manipulate the relative abundance of species in perennial grasslands and tall-herb communities through seasonal grazing, mowing or burning. Here it is imperative not merely to foresee the specific effects of the chosen management procedure, but also to anticipate the interaction of phenology with time of application. Hence, grazing designed to reduce the potentially dominant impact of *Bromopsis erectus* in chalk grassland would be expected to exert a strong selective effect against this species if applied during the winter and early spring, since *B. erectus* is unusual in the degree to which leaf expansion continues at low temperatures (Wells 1971, Law 1974). In contrast, winter grazing and trampling would be unlikely to suppress vegetation dominants such as *Pteridium aquilinum* and *Fallopia japonica*, in which the frost-sensitive canopy is reconstructed annually during the summer months.

The times of flowering, seed maturation and dehiscence are particularly relevant to vegetation management in species which are unusually dependent upon regeneration by seed. Examples here include meadow species such as *Bromus hordeaceus* and *Rhinanthus minor*, where the established plants are short-lived and there is no persistent seed bank. In these circumstances population size is strongly determined by the output of viable seeds in the preceding year, and it seems very likely that some of the major fluctuations in abundance observed in these species are due to year-to-year variations in the rates of seed maturation and dates of mowing. In the majority of meadow species, population numbers are buffered by the longer life-spans of established individuals or their propagules, but even here the timing of the haycut may be expected, over a number of years, to affect the relative abundance of the species through its selective effects on seed production and biomass removal.

Seedling RGR

In exploring the predictive value of laboratory measurements of seedling RGR in plant species it may be helpful to recall the arguments used by Grime and Hunt (1975), whose publication forms the source of most of the RGR values quoted in the Accounts.

The extrinsic and intrinsic constraints on RGR in plants growing under natural conditions are such that it is pointless to extrapolate to the field in absolute terms from measurements made in the laboratory (see also Hunt & Lloyd 1987). Even so, the possibility exists that values of RGR still remain a clue to the relative productivity of species growing under natural conditions. Such a link is suggested because species associated with fertile habitats, such as arable fields or manure heaps (e.g. *Chenopodium album*, *Persicaria* spp., *Poa annua*, *Stellaria media* and *Urtica dioica*), have higher values of RGR than species characteristic of unproductive habitats such as unfertilized pastures (e.g. *Helictotrichon pratense*, *Danthonia decumbens*, *Festuca ovina* and *Nardus stricta*).

Correlations between potential relative growth rate and site fertility have been commented upon, or are evident in, a number of publications (e.g. Bradshaw *et al.* 1964, Grime 1966). The extensive data collected by Grime and Hunt (1975) provided an opportunity for testing the strength of this correlation, and confirmed that productive and unproductive vegetation types indeed differed with respect to the potential maximum growth rates of their major constituents. The habitat types used in these comparisons were broad and rather heterogeneous, both in vegetation and in productivity. More-precise comparisons on the basis of individual vegetation samples were performed by Grime (1974), using prepublication data from the RGR screening programme. Grime used the frequency of each species within the sample to calculate weighted means for RGR, with an outcome which suggested that in natural vegetation there may be a sensitive adjustment of general levels of potential maximum RGR prevailing in a community in response to variation in site fertility.

Developments over the 12 years since these conclusions were drawn (Chapin 1980, Hunt 1984, Hunt and Lloyd 1987) have confirmed the strong predictive value of seedling RGR in relation to habitat productivity, and have narrowed causal analysis of the correlation to studies of the very different mechanisms of resource capture and utilization in habitats differing in productivity (Grime *et al.* 1986).

Under 'Established strategy' the ecological significance of seedling RGR is re-examined in a broader context and further reference to the underlying mechanisms is deferred until later (p. 47). However, it is necessary at this point to draw attention to groups of species which provide conspicuous exceptions to the general relationship between seedling RGR and site productivity. The first group consists of woody plants of relatively fertile soils (e.g. *Betula* spp. and *Salix cinerea*) in which seedling growth rates are low, due to the expenditure of photosynthate on woody tissue at the expense of leaf area (Jarvis & Jarvis 1964, Grime & Hunt 1975). A second group is exemplified by the umbellifers *Anthriscus sylvestris* and *Heracleum sphondylium* in which potential RGRs below those usually associated with fertile soils are associated with the translocation of photosynthate into the rootstock, a process which begins at a very early stage of seedling development, and which provides the capital for production of the shoot in the subsequent growing season.

Specific leaf area (SLA) and leaf dry matter content (DMC).
In unproductive ecosystems plant species have inherently low relative growth rates, low leaf nutrient concentrations, slow litter decomposition and low palatability to generalist herbivores while in productive conditions we encounter rapid growth, high nutrient concentrations, high palatability and rapid decomposition (Grime *et al.* 1997, Diaz *et al.* 2004). As explained in Panels 5, page 10 and 9, page 63) these relationships indicate a tradeoff in leaf structure and functioning that reflects a shift in plant adaptive biology from rapid resource acquisition to effective conservation of captured resources in well-defended tissues. Specific leaf area and leaf dry matter content are predicted (Reich *et al.* 1992, Garnier and Laurent 1994, Diaz *et al.* 2004, Wright *et al.* 2004) to vary in tandem with this transition in leaf traits as soft, mesophytic leaves of fertile habitats give way to tough, fibrous structures in unproductive conditions.

It is relatively easy, without need of specialised equipment and facilities, to amass data on SLA and DMC for large numbers of species. The resulting databases now appear to provide unprecedented opportunities to study patterns of variation in vegetation characteristics and habitat productivity at various scales and to analyse current processes of change (e.g. Hodgson *et al.* 2005a,c). In anticipation of an explosion in the use of SLA and DMC in ecological research (see for example Hodgson 2005a,c) **it is essential to assess the accuracy with which these two leaf traits can be used as surrogates for other traits of established reliability. The need for caution in the use of SLA and DMC is only too apparent in Figure 5.1 where their relationship to seedling relative growth rate is examined. These data confirm an overall statistical correlation but reveal great variability at the level of individual species and groups of species.**

Some of the main sources of this variability are as follows:

(1) Sun and shade leaves
Although DMC is generally consistent between different collections of the same species, values for SLA are particularly variable in material collected from natural populations of species that commonly occur in both shaded and unshaded conditions (Wilson *et al.* 1999). This phenomenon is probably caused by plasticity in leaf structure particularly in relation to light intensity. Sun leaves are thicker than shade leaves (Lewis 1972) and SLA is a function of both DMC and leaf thickness (Vile *et al.* 2005). In consequence the thinner leaves taken from shaded conditions will exhibit higher values of SLA. This problem can be recognised where samples collected from different localities and vegetation types are available.

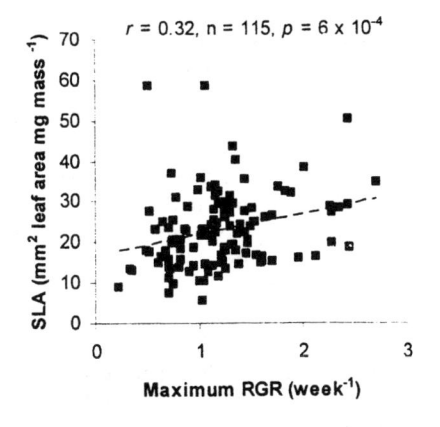

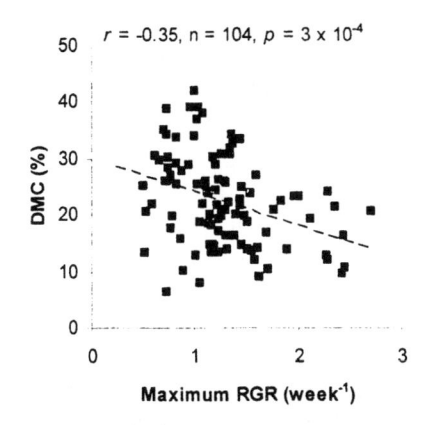

Figure 5.1. The relationship between relative growth rate (RGR), specific leaf area (SLA) and dry matter content (DMC) within the Sheffield flora.
Only species measured in the growth rate analyses of Grime & Hunt (1975) have been included. [Pearson correlation coefficient *r* between SLA and DMC = 0.41, no of species = 104, p = 2×10^{-4}.]

(2) Shade plants

A rather different problem arises in the species that habitually occupy shaded habitats either on the forest floor or as submerged aquatics. Here, light may be more influential than mineral nutrients in constraining plant growth. With low supplies of photosynthate for leaf construction, SLA is consistently high and DMC is always low, irrespective of growth rate. *Oxalis acetosella*, for example, is a slow-growing species (RGR c.0.5 day^{-1}) but the high values of SLA (58 mm^2 g^{-1}) and low estimate of DMC (13%) quoted in this book (page 450) are similar to those of fast-growing species such as *Stellaria media*. This is a very different circumstance from that considered under (1) above in that it is not associated with a plastic response. Values for SLA and DMC tend to remain unaltered even when the plant is growing in unshaded conditions. For certain genera that exploit wet habitats with extremely low light intensities (e.g. *Hymenophyllum* and *Elodea*) the leaf lamina is only one cell thick. We conclude that it is inappropriate to use SLA and DMC as predictors of nutrient status and growth rate in shaded and aquatic habitats.

(3) Succulence

In many species of arid habitats, the presence of water-storage tissue within the leaf increases thickness and mass and affects the relationships between RGR, SLA and DMC. In contrast to the relationship in Figure 5.1, succulents may combine low RGR, low SLA and low DMC. For example, *Sedum acre* combines an RGR of 0.7 day^{-1} with an SLA of 14 mm^2 g^{-1} and a DMC of 6%. As recommended by Vendramini et al. (2002), we suggest that SLA is the preferred ecological predictor for succulent species.

[A somewhat similar situation exists for wetland plants where the leaf lamina may perform the additional function of providing a conduit for oxygen diffusion to the roots through the production of aerenchymatous tissue (Armstrong 1964). The consequences of this additional function on the indicator value on SLA and DMC has not been investigated. However, since aerenchyma, unlike water-storage tissue, usually contains chloroplasts, the modifying effects on SLA and DMC may be minor].

Support tissue

The viability of the leaf as a photosynthetic organ is crucially dependent upon supporting structures that ensure effective display of chlorophyll and transport of materials into and out of the leaf. In most dicotyledonous species these functions are performed by petioles and stems neither of which are included in measurements of SLA and DMC. However, particularly in grasses and sedges, the supporting structure is often located almost entirely within the leaf lamina and will have effects on both SLA and DMC. These may be more pronounced in species e.g. *Deschampsia cespitosa* where support is also related to high silica content. It is also apparent that the greater contribution of supporting veins in highly dissected leaves will dictate lower values of SLA and higher DMC.

We conclude that, despite their usefulness in making broad comparisons between ecologically-contrasted groups of species, interpretation of SLA and DMC values will be affected, in some cases, by the extent to which ecologists are able and prepared to take into consideration variation between species that arises from differences in the balance between photosynthetic and support functions in the sampled leaf tissue. This is likely to be of critical importance if the objective of research is to search for minor differences in resource capture and utilisation between species co-existing in a plant community. The need, in such cases, to take the support function into account is well illustrated by comparing the determinations of SLA and DMC quoted in this book for the two frequent co-habitants, *Lolium perenne* (RGR 1.3 day^{-1}; SLA 28 mm^{2}g^{-1}; DMC 21%) and *Trifolium repens* (RGR 1.3 day^{-1}; SLA 35 mm^{2}g^{-1}; DMC 18%); here we may suspect that the slightly higher SLA and lower DMC of the latter species is due to the inevitable exclusion of a petiolar support function from an assay and calculation based exclusively upon the leaf lamina.

In view of the extent to which both SLA and DMC are subject to these various complicating factors we conclude that they should be used judiciously. Applied with caution, both have considerable potential where the objective is to rapidly map and interpret ecological processes at the landscape scale. Even for this purpose, however, we suspect that it may be unwise to rely on single leaf characters, each with its known limitations. For rapid survey we advocate the use of a set of leaf assays including other leaf traits such as linear dimensions, thickness and toughness (e.g. Diaz et al. 2004, Hodgson et al. 2005a,b,c) and, where they can be afforded, leaf chemistry.

Nuclear DNA amount (genome size)

In both plants and animals the quantity of DNA residing in the nucleus varies considerably according to species (Van't Hof & Sparrow 1963, Britten & Davidson 1971, Bennett & Smith 1976). The vascular plants of the British flora are no exception to this, and exhibit a range extending from 0.03 pg of DNA in the diploid genome of *Thlaspi caerulescens* J. S. & C. Presl to 141.4 pg in that of *Fritillaria meleagris* L.

The 'extra' DNA of organisms of high DNA amount is not related to organizational complexity nor to evolutionary advancement; much of it appears to consist of highly-repeated base-sequences which are not transcribed into protein (Davidson & Britten 1973). This has prompted various theories (e.g. Bennett 1971, 1972, Cavalier-Smith 1978, 1980, Doolittle & Sapienza 1980, Orgel & Crick 1980) which attempt to explain in cytological or genetic terms the origin and functional significance of variation in nuclear DNA amount. However, no consensus has emerged from recent reviews (e.g. Dover & Flavell 1982, Olmo 1983, Hardman 1986).

In parallel, (but, unfortunately, largely in isolation from, the work of molecular biologists and geneticists) is the ecological research on nuclear DNA amount. This has now advanced to the stage where there are data available for the majority of common British vascular plants [see the website database maintained by M.D.Bennett at *Bennett and Leich (2004)*], and relationships between DNA amount and ecology can be examined. However, before commenting on the significance of the data cited in our Accounts, we must refer to earlier studies which established the principles relevant to an ecological understanding of variation in nuclear DNA amount in plants.

Though within one individual plant or animal the sizes and shapes of cells vary considerably according to type of tissue and conditions experienced during differentiation, it is clear that species with large nuclei and high DNA amounts also exhibit generally larger cell sizes (Bradley 1954, Van't Hof & Sparrow 1963, Commoner 1964, Darlington 1965). Evidence of the generality of this phenomenon as applied to plants is provided in Figure 5.2, which relates nuclear DNA amount to the linear dimensions of the leaf epidermal cells in a range of common herbaceous species. It is also worth noting that plants and animals appear to employ a range of elaborate developmental processes whereby nuclear DNA amount and cell size are adjusted in concert, producing cells which are abnormally large or small for the species and which perform specialized functions (e.g. germ cells, xylem vessels, root hairs and endosperm). The occurrence of such variants is reviewed by Cavalier-Smith (1978, 1982), who suggests that there is a strong functional integration between the size of the nucleus and that of the cell.

Variation in cell size and nuclear DNA amount is directly correlated with differences in the minimum length of the cell-cycle

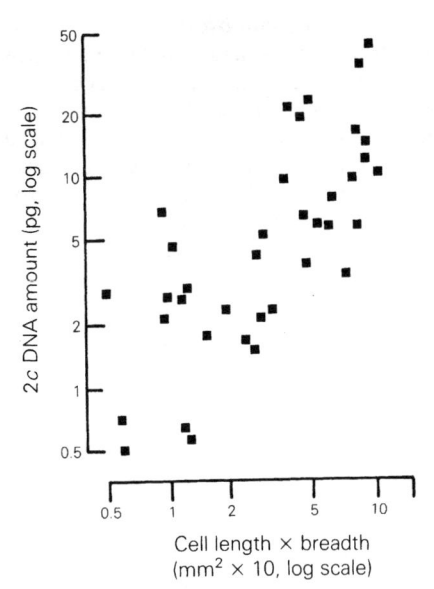

Figure 5.2 The relationship between nuclear DNA amount and the mean length times breadth of epidermal cells in mature leaves of a range of herbaceous species. (Modified by Bennett 1987 from Grime 1983.)

(Darlington 1965, Bennett 1971), an observation which led Bennett (1972) to suggest that by shortening the duration of the mitotic and meiotic cycles, the reductions in cell and nuclear volumes have played a key role in the evolution of rapidly-growing ephemeral species such as arable weeds and some annual crops.

An additional 'geographical' line of enquiry into the significance of variation in nuclear DNA amount in plants is available from comparative studies such as those of Avdulov (1931), Stebbins (1956), Bennett (1976), Bennett and Smith (1976), Jones and Brown (1976) and Levin and Funderburg (1979), who have shown that genome size in tropical species is consistently small when compared with the wide range observed in temperate floras. In crop plants Bennett (1976) has drawn attention to a latitudinal trend in DNA amounts, whereas for native geophytes and grasses of the British flora there is a tendency for species with enlarged nuclear DNA amounts to be concentrated in southern England and to have European distributions centred on the Mediterranean (Grime & Mowforth 1982).

The phenomenon of interspecific variation in genome size has been placed into an ecological context by evidence that, in a range of species common in northern England, nuclear DNA amounts are correlated with shoot phenology (Grime & Mowforth 1982, Grime 1983). Evidence of the predictive value of nuclear DNA amount with respect to phenology is provided in Figure 5.3, which is based upon measurements of seasonal change in the shoot biomass of naturally-established species in various plant communities in the Sheffield region (Al-Mufti et al. 1977, Al-Mashhadani 1980, Furness 1980). These data show that from early spring to midsummer changes in the identity of the most actively expanding species range from *Hyacinthoides non-scripta* (2C DNA = 42.4 pg) in March to *Chamerion angustifolium* (2C DNA = 0.7 pg) in June, with a continuous reduction in nuclear DNA amount between the two.

In an attempt to explain the correlation between nuclear DNA amount and the timing and rate of shoot growth in the spring, it has been suggested by Grime and Mowforth (1982) that climatic selection has operated upon nuclear DNA amount through the differential sensitivity of cell division and cell expansion to low temperature. According to this hypothesis, species from continuously warm climates, or species which exploit the summer season in cool temperate climates such as that of the British Isles, have been subject to natural selection for low nuclear DNA amount and small cell size. These are features which coincide with a potentially short cell-cycle, and which are conducive to rapid production of plant tissue in conditions that allow continuous division and expansion of cells. This strategy of growth is attuned to summer conditions, and is clearly vulnerable to the potentially limiting effect of cold spring temperatures upon mitosis. It seems reasonable to suggest that this could explain the delayed shoot growth of the species of low DNA amount shown in Figure 5.3.

A quite different strategy of growth is proposed to explain the capacity of many geophytes and grasses of high DNA amount for expansion of shoots in the spring. Here it is suggested that rapid early growth is achieved by expansion of large cells (with large nuclei) which are formed, but not expanded, during preceding warmer conditions. This growth strategy circumvents the limitations imposed upon mitosis by low spring temperatures, but it may be expected to restrict the performance of the plant towards the latter part of the growing season. This is because, in species with large cells and large nuclear DNA amounts, relatively slow rates of summer growth may be expected to arise from the extended cell-cycle (Bennett 1971). Moreover, in the geophytes with massive nuclei (e.g. *Allium ursinum* and *Hyacinthoides non-scripta*) and in some species of moderately high DNA amount (e.g. *Bromopsis erectus*, *Lolium perenne* and *Ranunculus bulbosus*), the summer is marked by a quiescent phase in which development appears to be mainly restricted to meristematic activity, preparatory to future episodes of cell expansion. This phenomenon has been particularly well documented in geophytes (Hartsema 1961), but there is also evidence of its importance in grasses and cereals, where it is commonly described by agriculturalists as 'stored growth' (Salter & Goode 1967).

Further research into the ecological value of measurements of nuclear DNA amount is required. It is clear, nevertheless, that data of this kind can be helpful in predicting and interpreting seasonal patterns in shoot development of the types already discussed under 'Phenology' (p. 42), and, as explained later (p. 66), variation in

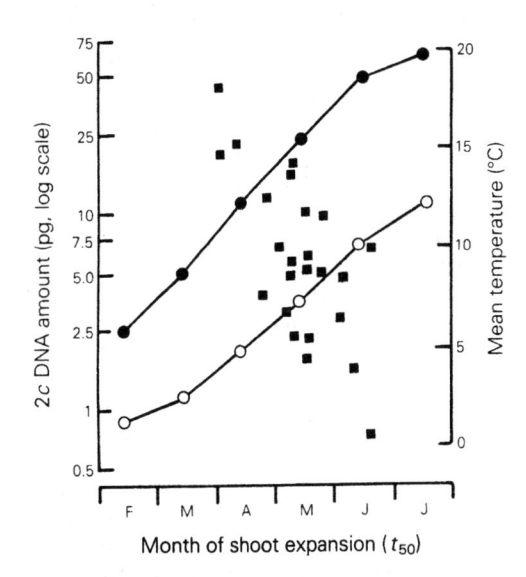

Figure 5.3 The relationship between nuclear DNA amount and the time of shoot expansion in 24 plant species commonly found in the Sheffield region. Temperature at Sheffield is expressed as the long-term averages for each month of daily minima (○) and maxima (●) in air temperature at 1.5 m above the ground. (Redrawn from Grime and Mowforth 1982.)

INTERPRETATION AND USE OF THE SPECIES ACCOUNTS

DNA amount within plant communities appears to provide an index of temporal niche differentiation.

Despite the general relationship between nuclear DNA amount and shoot phenology, some exceptions occur. 'Cool season' species, which grow in both the autumn and the spring, often have a low DNA amount. This group includes many small winter annuals (e.g. *Arabidopsis thaliana* and *Arenaria serpyllifolia*) and species such as *Alliaria petiolata*, *Galium aparine*, *Poa trivialis* and *Taraxacum* spp. In marked distinction to the geophytes and grasses of high nuclear DNA amounts, many of the 'cool season' species are extremely variable from year to year, both in phenology and in the final size attained by the shoot. This strongly suggests that spring growth in these plants is highly opportunistic, depending primarily upon exploitation of unseasonably warm days and sunny locations. This conclusion is supported by the extreme stunting of seedlings of species such as *Galium aparine* in consistently cold springs (such as that of 1986) and by the restriction of many of these species to south-facing slopes.

A further anomaly lies in the delayed appearance of the shoots of *Pteridium aquilinum*, despite their belonging to a species of relatively high nuclear DNA amount. The explanation for this apparent deviation remains obscure, although it may be significant that such a phenology involves a component of 'hidden' activity, with fronds emerging from rhizomes deep in the soil.

Chromosome number

The functions of chromosomes are (a) the 'storage, replication and transmission of hereditary information', (b) the regulation of gene action and (c) the regulation of gene recombination (Stebbins 1971). However, some characters related to ploidy are of more overt ecological significance. For example, during their early evolutionary history many polyploids show a high level of heterozygosity and have a wide ecological range (Stebbins 1971,1980). Comparison of polyploids within the Sheffield flora with their extant diploid relatives reveals that the polyploids tend to be more common either than the diploids or than any taxonomically- isolated polyploids (Hodgson 1987b). Polyploids are particularly characteristic of fertile sites (e.g. *Agrostis stolonifera*, *Chenopodium album*, *Poa annua* and *Urtica dioica*), but they are also well represented among the common species of less-productive habitats (e.g. *Agrostis capillaris* and *Festuca rubra*).

Established strategy

As explained in Chapter 2 (p. 9), the CSR model of primary strategies consists of an attempt to recognize the main avenues of ecological specialization in the established phase of plant life-histories. The sets of attributes associated with the primary strategies (Table 4.3, p.34) suggest many opportunities to analyse the distribution and population dynamics of species, and to predict the consequences of changes in their environment or management regime. In fact, analyses published by Hodgson in 1986 (see Panel 8, page 14) strongly suggest that profound changes are taking place in the flora of the British Isles as a consequence of changes in land use and can be interpreted as shifts in the relative abundances of C-, S- and R-strategists. Furthermore, as proposed by Whittaker and Goodman (1979), Grime (1979, 1986a) and Leps et al. (1982), identification with respect to strategy often provides the basis for explaining the role of particular species in succession (Caccianaga et al. 2006) and for predicting sensitivity to vegetation perturbation ('resistance') and capacity to recover from disturbance ('resilience').The scope for use of CSR theory also extends to predictions of how vegetation will be affected by climate change (Panel 18).

In order to explain more precisely why CSR theory is useful in devising testable ecological predictions it is now necessary to identify the defining characteristics of C-, S- and R-strategists.

Competitors

The list of plant attributes associated with the competitive strategy (see Table 4.3, page 34) contains several which are obviously related to the capacity of C-strategists to monopolize resource capture in productive, relatively undisturbed environments. These include a high potential relative growth rate, tall stature and a tendency to form a consolidated growth form by vigorous lateral spread above and below ground. These characteristics are all evident in what we may describe as the 'classical competitors' of the British landscape (e.g. *Chamerion angustifolium*, *Epilobium hirsutum*, *Fallopia japonica*, *Petasites hybridus*, *Phalaris arundinacea* and *Urtica dioica*).

The high rates of resource capture achieved by these plants are also due to other less-obvious but equally important characteristics. These include the formation of substantial underground storage organs which fuel the initial surge in root and shoot growth in the spring and allow a very large peak in shoot biomass to be developed in summer (Fig. 3.5, page 25). In any attempt to understand the functional characteristics of C-strategists it is vitally important to recognize that these plants have the capacity to withdraw resources from the environment at rapid rates. The effect of this phenomenon is to create a spatially patchy environment in which, despite the general abundance of resources within the habitat at large, zones of severe depletion develop rapidly above and below ground during each growing season in close proximity to functional leaves and roots (this subject has been discussed already on p. 13). It is for this reason that we may suspect that the most important characteristic of the C-strategist is the rapidity and scale of the morphological changes that occur in the root and shoot systems over each growing season. This feature, coupled with the normally short life-span of individual leaves and fine roots, brings about a constant readjustment in the spatial distribution of the leaf canopy and in the actively absorbing part of the root system. The consequence of this constant deployment of new leaves and roots into unexploited, resource-rich zones of the patchy environments created by competing plants is the phenomenon of 'active-foraging' (Grime 1977, 1979, Panel 14). This has been analysed experimentally and, in terms of resource capture from productive environments, has been shown to be superior to mechanisms involving less-dynamic root and shoot systems (Crick 1985, Grime et al. 1986, Campbell and Grime 1989).

Compared with ephemeral plants of frequently-disturbed productive habitats (ruderals), species exhibiting the competitive strategy often show a temporary delay in the onset of seed production. In terms of Darwinian fitness, the advantage of this developmental pattern is clear in the case of habitats normally exploited by these species; in *undisturbed* productive conditions rapid vegetative monopoly is an essential prelude to sustained heavy reproductive output over many years (Bolker and Pacala 1999). Equally, however, there can be little doubt that the delayed reproduction of the competitors is one of the major factors restricting their resilience and abundance in *severely disturbed* habitats.

The costs involved in the active foraging for light, mineral nutrients and water, which is characteristic of the competitive strategy, are considerable because of the high rates of reinvestment of captured resources necessary for the construction of new leaves and roots following their rapid senescence. To these costs we must add those associated with the high rates of herbivory experienced by the weakly defended tissues of many competitors (for evidence of this phenomenon, with evolutionary interpretations, see Grime (1979), Coley (1983), Coley et al. (1985) and Edwards & Wratten (1985). We suspect, therefore, that there are severe penalties attached to active foraging and that these will greatly restrict the success of the competitive strategy in chronically unproductive habitats.

Stress-tolerators

Under the previous heading we suggested that heavy expenditure of captured resources in new leaves and roots would be of selective advantage only in circumstances where active foraging gains access to large reserves of light energy, water and mineral nutrients. From this argument we predict that competitors will fail in habitats where productivity is low and where resource availability is brief and unpredictable (e.g. where light occurs as sunflecks and/or where mineral nutrients become available as short, rich pulses, as from intermittent decomposition processes, e.g. Davison (1964) and Gupta & Rorison (1975). The experimental study of Campbell and Grime (1989) provides confirmatory evidence of this effect: here the stress-tolerator, *Festuca ovina* was found to be superior in nitrogen capture and growth-rate to the competitor, *Arrhenatherum elatius*, when nitrogen supply was confined to pulses of short (<12h.) duration.

In unproductive habitats, conservation of captured resources is of primary significance and successful plants are likely to be those with the ability to capture *and retain* scarce resources in a continuously hostile physical environment. Conservation of resources is apparent in the low potential relative growth rates and delayed onset of reproduction in many species of unproductive habitats. Also in keeping with these predictions, we find that the leaves of species of unproductive environments tend to be comparatively long-lived, morphologically implastic structures which are strongly defended against herbivory (Grime 1979, Bryant & Kuropat 1980, Coley 1983, Cooper-Driver 1985, Grime et al. 1996). Additional significance has been attached to the strong anti-herbivore defences of many stress-tolerators by the suggestion (Grime & Anderson 1986, Grime et al. 1996) that physical defences which protect the living foliage often remain operational after senescence, retarding the breakdown of litter by decomposing-organisms. This is consistent with the high organic content of the superficial horizons of infertile soils, and perhaps explains also the dense accumulation of litter under trees of comparatively low growth rate (e.g. *Fagus sylvatica* and *Quercus petraea*). As explained in several Accounts, this phenomenon has pronounced effects on the distribution of herbaceous plants on the woodland floor (Sydes & Grime 1981a,b).

Stress-tolerators, despite sharing many common features of life-history and physiology (Table 4.3, p. 34), are associated with a wide range of life-forms and ecological behaviour. As the Accounts illustrate, there are stress-tolerators characteristic of calcareous soils (e.g. *Koeleria macrantha* and *Primula veris*), of acidic soils (e.g. *Juncus squarrosus* and *Nardus stricta*), of droughted habitats (e.g. *Sedum acre* and *Thymus polytrichus*), of wetland or damp grassland (e.g. *Carex panicea* and *Succisa pratensis*), of shaded situations (e.g. *Sanicula europaea* and *Viola riviniana*) and of heavy-metal-contaminated spoil (e.g. *Minuartia verna*). Quite clearly, there are important distinctions to be drawn within the general category 'stress-tolerator', and the Accounts contain an abundance of references to evidence for mechanisms of tolerance which are specific to particular types of stress-tolerators. This diversity prompts the question:

'*What is/are the nature of the selection force(s) responsible for the features common to stress-tolerators?*'

Two possible answers deserve consideration:
(1) stress-tolerant traits are an inevitable evolutionary response to chronically low productivity, and are relatively independent of the nature of the stresses constraining production.

(2) despite superficial differences all the habitats exploited by stress-tolerators share a common underlying stress.

Debate continues with regard to these two rival explanations. However, the Accounts suggest that not only is the balance of evidence currently in favour of the *latter* hypothesis, but that the common underlying stress is likely to be low availability of mineral nutrients, especially phosphorus and nitrogen. The implications of this tentative conclusion are considerable and have been discussed at length elsewhere (Grime 2001). It is already evident that the notion of limiting mineral nutrient elements as the ultimate selective pressure determining the stress-tolerant strategy is not incompatible with the concept of a recurrent pattern of evolutionary specialization involving all major aspects of the plant. As explained by Chapin (1980), most plant activities depend upon the level of supply of mineral nutrients and, as proposed in Figure 5.9, page 76, it seems likely that conservative mechanisms of mineral nutrient capture and utilization will be invariably associated with constraints upon both carbon assimilation and reproductive activity.

Ruderals

Two plant characteristics in particular are relevant to an analysis of the population dynamics and ecology of plants exhibiting the ruderal strategy. The first is a potentially high RGR during the seedling phase and the second is an early onset of reproduction, a feature which in Angiosperms often coincides with self-pollination and rapid maturation and release of seeds. In arable weeds, and in ephemeral plants of frequently and severely disturbed habitats such as paths, these characteristics undoubtedly confer resilience and explain the remarkably rapid population fluctuations observed in species such as *Matricaria discoidea*, *Polygonum aviculare*, *Senecio vulgaris* and *Stellaria media*. As shown clearly by the data of Salisbury (1942), cited in many of the Accounts, it is also a characteristic of the ruderal strategy that allocation to seed production is sustained (as a proportion of the total biomass) in plants stunted by drought, by mineral nutrient stress or as a consequence of growth at high population densities.

The reproductive imperative evident in the life-histories, physiology and breeding systems of plants classified as ruderals in the Accounts also plays an inevitable part in these strategists' failure to exploit relatively undisturbed habitats. The early diversion of captured resources into flowers and seeds is not compatible with the development of the extensive root and shoot systems necessary for dominance and extended occupation of productive, stable habitats, nor is it conducive to survival in undisturbed but highly-stressed environments where success is usually associated with very conservative patterns of resource utilization.

Intermediate strategies

In Chapter 4 (p. 33) we list the full range of established strategies recognized in the Accounts. In addition to the three primary strategies these include the four intermediates (C-R, S-R, S-C and C-S-R) corresponding to particular intermediate positions within the triangular model (see Fig. 2.1, p. 9). With some slight modification of the C-S-R category, these intermediate strategies conform to the detailed descriptions provided by Grime (1979). In Table 4.4, (page 35), we give examples of each intermediate strategy, together with publications relevant to their classification as such, which relies upon their 'intermediacy' with respect to the criteria used to recognize the primary strategies (Table 4.3, page 34). A particularly clear example of 'intermediacy' is provided in Figure 3.5, (page 25), which summarizes the proposed relationships between plant strategy and seasonal change in shoot biomass.

Regenerative strategies

Seedlings and vegetative offspring are comparatively small. They are therefore exposed to hazards which may be much more severe than, and often quite different from, those experienced by

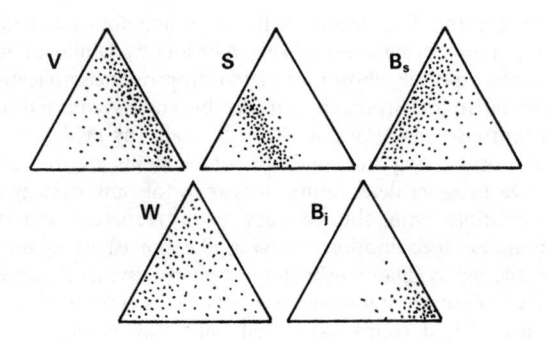

Figure 5.4 A prediction of the range of combinations of regenerative strategies in relation to the established phase of the life-cycle: V, combinations involving vegetative expansion; S, combinations involving seasonal regeneration; B$_s$, combinations involving a persistent seed bank; W, combination involving numerous widely-dispersed seeds or spores; B$_j$, combinations involving persistent juveniles. The location of the established strategies within the triangular model is shown in Figure 2.1.

established plants. Successful regeneration in many species depends upon exploitation of locally favourable sites (Williams & Harper 1965, Grubb 1977), and plants appear to have evolved a variety of regenerative mechanisms which exploit particular niches. Five major types of regenerative strategy are distinguished in Table 4.5, (page 36) which also contains a brief description of the conditions under which particular strategies appear to enjoy a selective advantage. The five strategies will now be considered in turn.

Vegetative expansion

Under this heading it is convenient to assemble many of the regenerative mechanisms which involve the expansion and subsequent fragmentation of the vegetative plant through the formation of persistent rhizomes, stolons or suckers. The most consistent feature of vegetative expansion is the relatively low risk of mortality to the offspring. This is achieved through a prolonged attachment to the parent plant and the mobilization of resources from parent to offspring.

The outstanding feature of the distribution of vegetative expansion in the triangular model (see Figure 5.4) is its coincidence with established strategies characteristic of relatively undisturbed habitats (competitors, stress-tolerant competitors and stress-tolerators). This pattern may exist because vegetative expansion involves a period of attachment between parent and offspring, and is therefore not a viable mechanism when vegetation is affected by frequent and severe disturbance. It seems likely that the role of vegetative expansion may be rather different in competitors and stress-tolerators. In productive, relatively undisturbed habitats, vegetative expansion is often an integral part of a mechanism whereby competitive herbs (e.g. *Fallopia japonica*, *Phragmites australis* and *Pteridium aquilinum*) rapidly monopolize the environmental resources, and thus suppress the growth and regeneration of neighbouring plants. Under the very different conditions of highly-stressed environments, however, the main advantage of vegetative expansion in species such as *Anemone nemorosa*, *Carex caryophyllea* and *Sanicula europaea* arises from its capacity to sustain the offspring under conditions in which establishment from an independent propagule is a lengthy and hazardous process.

Seasonal regeneration in vegetation gaps

In a wide range of habitats herbaceous vegetation is subjected to seasonally-predictable damage from phenomena such as temporary drought, flooding, trampling and grazing. Under these conditions the most common regenerative strategy is that in which the areas of bare ground or sparse vegetation cover created every year are recolonized annually during a particularly favourable season. The propagules involved in seasonal regeneration are relatively large, lack long-term dormancy, and may be sexual or asexual in origin. They include cohorts of synchronously germinating seeds (e.g. in *Bromus hordeaceus*, *Heracleum sphondylium* and *Impatiens glandulifera*) and populations of offsets or bulbils (e.g. in *Anthriscus sylvestris* and *Ranunculus ficaria*), or the tips of rapidly-fragmenting rhizomes or stolons (e.g. in *Poa trivialis*, *Ranunculus repens* and *Trifolium repens*).

Seasonal regeneration is particularly associated with competitive ruderals and stress-tolerant ruderals, both of which are restricted to sites at which the intensity of disturbance is sufficient to cause seasonal and temporary gaps in the vegetation. In habitats subjected to forms of disturbance that are more-severe or less-predictable, or both, dependence upon propagules which lack long-term dormancy may be expected to limit regenerative capacity. It is probably for this reason that the incidence of exclusively seasonal forms of regeneration is low among ruderals *sensu stricto*.

The strategy of seasonal regeneration therefore represents a rather unsophisticated mechanism of gap exploitation in which, after propagules have been dispersed locally within the habitat and have germinated simultaneously (or recommenced growth, in the case of fragments of rhizomes or stolons), survival is usually limited to those individuals which occur in gaps. It is evident that this method of regeneration depends upon the creation in the same habitat and in each year of a high density of suitable gaps. Where gaps occur more rarely or less predictably, seasonal regeneration tends to give way to other types of regenerative strategy.

Regeneration involving a persistent seed bank

As explained on page 25, when flowering plants are compared with respect to the fate of their seeds, two contrasting groups may be recognized. In one, most if not all of the seeds germinate soon after release. In the other, many become incorporated into a bank of dormant seeds which is detectable in the habitat at all times during the year and may represent an accumulation of many years' production. These two groups are, of course, extremes, and between them there are species and populations in which the seed bank, although present throughout the year, shows pronounced seasonal variations in size. Nevertheless, it is convenient to draw an arbitrary distinction between 'transient' and 'persistent' seed banks. **A *transient seed bank* may therefore be defined as one in which none of the seed crop remains in the habitat in a viable condition for more than 1 year, and a *persistent seed bank* as one in which at least some of the component seeds are 1 year old or more.**

Regeneration involving a persistent seed bank commonly occurs in association with each of the established strategies, with the exception of the stress-tolerators, in which the role of a seed bank is often played by a bank of persistent seedlings (see below).

Although some of the plants which develop persistent seed banks have mechanisms which facilitate seed dispersal by animals (e.g. *Danthonia decumbens*, *Rubus fruticosus* and *Vaccinium myrtillus*), most seed banks are located close to their parent plants and may even be situated directly beneath them. It seems therefore likely that the main functional significance of a persistent seed bank is the extent to which it allows regeneration *in situ*. This interpretation is consistent with the fact that seed banks are particularly common in proclimax vegetation, such as grassland, heathland and disturbed marsh, in which the process of vegetational change is cyclical rather than successional. Many forms of proclimax vegetation, especially those in farmland, are subject to alternating patterns of land use, with the result that at any one location conditions suitable for the establishment of each potential member of the community are available only intermittently. In this situation the presence in the soil of a persistent seed bank allows the survival of populations during periods in which the management regime is temporarily unfavourable to the established plants.

Persistent seed banks occur in association with a wide range of primary and secondary strategies, and it is clear that in certain respects the role of a seed bank changes according to the established strategy with which it is associated. In ruderals such as the annuals of arable fields and marshland (e.g. *Juncus bufonius, Papaver rhoeas, Rorippa palustris* and *Veronica persica*) the seed bank may permit rapid recovery from catastrophic mortalities inflicted by cultivation, herbicide treatments or natural phenomena such as flooding. It also allows survival of unfavourable seasons and longer periods of temporary stability during which the ruderals are excluded by perennial species. In competitive ruderals (e.g. *Juncus articulatus* and *Rumex obtusifolius*), in stress-tolerant ruderals (e.g. *Arenaria serpyllifolia* and *Geranium molle*), in C-S-R strategists (e.g. *Anthoxanthum odoratum* and *Holcus lanatus*) and in stress-tolerators (e.g. *Danthonia decumbens* and *Thymus polytrichus*), the role of the seed bank is similar to that in ruderals, except that it appears to be more concerned with regeneration within gaps in perennial vegetation.

In habitats dominated by competitors and stress-tolerant competitors, the intervals between major disturbances may be very long (perhaps more than 15 years). Competitors and stress-tolerant competitors with persistent seed banks include many of the most familiar herbs of vegetation subject to occasional disturbance, for example by burning or flooding (e.g. *Calluna vulgaris, Epilobium hirsutum, Juncus effusus* and *Urtica dioica*).

Regeneration involving numerous widely-dispersed seeds or spores
This mechanism of regeneration occurs in association with the complete range of established strategies, and is conspicuous in two types of situation. The first consists of localized and relatively inaccessible habitats such as crevices in cliffs, walls and tree trunks, where pteridophytes (e.g. *Asplenium trichomanes* and *Cystopteris fragilis*) and wind-dispersed composites (e.g. *Mycelis muralis* and *Hieracium* spp.) are strongly represented. The second arises in circumstances where landscape is subject to spatially-unpredictable disturbance. Where disturbance occurs as an exceptional event in an environment of moderate to high productivity, the site often presents a small target for colonization, and usually remains open only for a relatively brief period. Here the most successful invaders include those herbs, shrubs and trees (e.g. *Tussilago farfara, Chamerion angustifolium, Salix* spp. and *Betula* spp.) which produce locally-saturating densities of mobile seeds.

Because of their prolific seed production and effective dispersal, species in the 'widely dispersed' category are frequently recorded as seedlings or small plants in habitats which are quite unsuitable for the established plant. In only a few instances do species in this grouping form a persistent seed bank (e.g. *Epilobium* spp.).

Regeneration involving a bank of persistent juveniles
Alternatively described as 'advance reproduction' (Marks 1974), this mechanism of regeneration appears to be characteristic of species exploiting circumstances where seedling establishment is lengthy and uncertain, and consists essentially of a process of replacement of occasional mortalities in the population of established plants (e.g. *Quercus* spp. and *Sanicula europaea*). Banks of persistent juveniles are especially common among stress-tolerators, many of which tend to produce seeds rather intermittently over the long life-span of the established plant. It may be significant in the maintenance of populations of these plants that the persistent juveniles provide the possibility of recruitment to the breeding population between seed crops.

On page 31 we refer to evidence suggesting that some persistent seedlings may benefit from mineral nutrients and assimilate imported by means of mycorrhizal connections from established plants.

Flexibility in regeneration

In contrast to the strategies associated with the established phases of plants' life-histories, the five regenerative strategies are not primarily determined by inflexible 'design constraints'. They are not, therefore, mutually exclusive. For this reason it is not uncommon for the same genotype to exhibit two or more regenerative strategies simultaneously. This led Grime (1979) to suggest that the ecological amplitude of a species may be determined, to some degree, by the *number of regenerative strategies* which it exhibits. Although limited by incomplete data, the Accounts lend strong support to this hypothesis. At one extreme it is possible to recognize species in which a narrowly defined ecological range coincides with a restricted capacity for regeneration (e.g. *Euphrasia officinalis, Fagus sylvatica* and *Impatiens glandulifera*). The other extreme is exemplified by species such as *Epilobium hirsutum, Holcus lanatus, Juncus effusus* and *Poa trivialis*. In these, the ecological versatility evident from a capacity for persistence and rapid population expansion under a variety of forms of vegetation management is in turn supported by a flexibility in mechanisms of regeneration.

Polymorphism in seed characters which is particularly associated with plants of disturbed habitats, e.g. *Atriplex, Chenopodium, Rumex* and many composites, provides a further dimension to regenerative flexibility.

Regenerative failure

As explained on page 36, the success of particular regenerative strategies appears to be strongly dependent upon habitat conditions. We predict that failure of regeneration is a major factor limiting the ecological and geographical range of species.

Failure in plants with a single regenerative strategy
In the British flora there are many species in which regeneration depends exclusively upon the production of seeds which lack dormancy and which produce a single cohort of seedlings each year. Such reliance upon seasonal regeneration is particularly characteristic of autumn-germinating annual grasses (e.g. *Aira praecox, Anisantha sterilis* and *Hordeum murinum*) and of a number of spring-germinating trees and shrubs (e.g. *Acer pseudoplatanus, Fraxinus excelsior* and *Sambucus nigra*). For both of these groups there is evidence that dependence upon seasonal regeneration is often a potent limiting factor in the field. In *Hordeum murinum*, for example, it has been shown by Davison (1977) that the altitudinal limit of the species is related to its failure to produce sufficient seed to maintain viable populations. A quite different example of failure is represented by the inability of common trees such as *A. pseudoplatanus* and *F. excelsior* to regenerate in pastures and unfenced woodlands. Here seed production is prolific, but cannot usually compensate for the losses of seedlings inflicted by domestic grazing animals, rodents and deer.

Failure in plants with more than one regenerative strategy
Among the many perennial species which have an ability both to spread vegetatively and to regenerate by seed, it is not uncommon to find populations in which one of these two mechanisms

predominates. At the northern limits of their distribution in Britain many species (e.g. *Brachypodium pinnatum* and *Cirsium acaulon*) establish from seed only rarely, but are capable of vigorous clonal expansion (Pigott 1968). A similar shift away from regeneration by seed often results from management practices: pastures, meadows and frequently mown road verges, lawns and sports fields usually contain populations of grasses (e.g. *Agrostis capillaris* and *Festuca rubra*) in which persistence is relatively independent of seed production. In contrast, situations occur where vegetative expansion is constrained and seed production remains as the only effective means of regeneration. An example of this is provided by the stunted, but flowering, individuals of *Chamerion angustifolium* which are frequently observed in crevices on cliffs and walls.

Flowers

Sexual reproduction is of fundamental importance in creating genetic diversity, both within and between populations, and is exhibited by most species (Richards 1986). Flowers, which provide the machinery for this process, are thus of vital significance. There is, of course, a considerable diversity of floral structure in evidence, even within the British Isles. Although much of this diversity is a legacy of the evolutionary history of the major taxa, a great deal of structural variety can be related to pollination mechanisms, even though most British species rely upon generalist pollinators.

The floral character which is most generally regarded as being of highest ecological importance is the breeding system, since this regulates the amount of variation between offspring. In particular, ruderals and stress-tolerant ruderals in the British flora (and elsewhere (Richards 1986)) are often in-breeding. This may be of selective importance in short-lived plants of disturbed habitats, either (a) because it reduces the probability that pollination failure will limit the rate or yield of reproduction, or (b) because offspring which are homozygous have, like their parents, a relatively high probability of reproductive success. By contrast, in the longer-lived competitors and stress-tolerators there tends to be a higher incidence of out-breeding.

Kay (1987) considers that the effects of flowering upon the growth of the vegetative plant are considerable. With Thompson and Stewart (1981), he points out that the cost of reproduction by seed (in terms of resources sequestered from the growing vegetative plant) is much larger than the simple dry-weight yields of seed would imply. In addition, competition for pollinators, or other demands made by the flowering process, may necessitate an allocation of resources into floral development at a time when resources would otherwise be available for vegetative growth. This implies an important role for flowering in the determination of the ecological distributions of species. However, despite much detailed and innovative recent work on the subject (e.g. Charnov 1979, Willson & Burley 1983, Kay 1987), we have yet to establish useful generalisations regarding the ecological consequences of variation in the form, number and pollination of flowers and the production of seeds.

Germinule

The size and shape of the germinule often provides valuable information about function both before and after germination. As pointed out by Salisbury (1942), Grime and Jeffrey (1965), Harper et al. (1970) and Baker (1972), a conflict can be recognized between selection forces favouring the production of numerous offspring and those promoting germinules of large size. In the Accounts the advantages and limitations associated with dependence upon small germinules are evident within the pteridophytes, and in species of *Betula*, *Epilobium* and *Salix*, in which efficient long-range dispersal has been achieved at the expense of the extreme vulnerability of the small offspring to dominance by established plants. However, it would be a mistake to assume that the functional significance of small germinule size is invariably related to dispersal and to 'escape through space' from the influence of the dominant species of the established vegetation. As recognized by Thompson and Grime (1979), the production of numerous tiny germinules is often associated with species in which the offspring 'escape through time' in the relative sanctuary of a persistent seed bank. Through the agency of rainwater percolation and earthworm activity, the small compact germinules of a wide range of species are incorporated into banks of seeds or spores in the soil. Germination tends to be delayed until disturbances of the vegetation or the soil expose the seeds to unfiltered sunlight or large diurnal fluctuations in temperature, both of which often appear to function as cues for germination. Such 'gap-detection mechanisms' (Thompson et al. 1977) restrict the appearance of seedlings to microsites where conditions may be relatively favourable for establishment. The Accounts provide numerous examples of persistent seed banks occurring among species of marshland (e.g. *Juncus* spp., *Rorippa palustris* and *Stellaria uliginosa*), heathland (e.g. *Calluna vulgaris* and *Erica cinerea*), wasteland (e.g. *Hypericum perforatum* and *Origanum vulgare*), disturbed woodland (e.g. *Digitalis purpurea* and *Moehringia trinervia*), grassland (e.g. *Agrostis capillaris* and *Deschampsia cespitosa*) and arable land (e.g. *Capsella bursa-pastoris* and *Chenopodium album*). Tiny seeds are less vulnerable to predation than large ones; Thompson (1987) has suggested that this may have provided an additional selection force, explaining the strong correlation observed between small seed size and seed bank persistence.

As recognized by Salisbury (1942), the offspring of species producing large germinules appear to be capable of establishing successfully in closed herbaceous vegetation. This is evident from the Accounts for large-seeded tree species (e.g. *Acer pseudoplatanus*, *Fagus sylvatica* and *Quercus* spp.) and from those describing the large-seeded annual herbs (e.g. *Galeopsis tetrahit*, *Galium aparine* and *Impatiens glandulifera*), the obligately or facultatively monocarpic perennials (e.g. *Anthriscus sylvestris*, *Arctium minus* and *Heracleum sphondylium*) and the polycarpic perennials (e.g. *Allium ursinum*, *Myrrhis odorata* and *Vicia cracca*). In these species, however, it is important to recognize that in their natural habitats the benefit of large seed reserves may arise in relation to emergence through plant litter rather than in relation to competition with established plants (Sydes & Grime 1981a,b, Gross 1984, Thompson 1987).

Baker (1972) noted a tendency for large seed size to be associated with plants of droughted habitats in the flora of California. The Accounts show evidence of a similar pattern in the British Isles. Comparatively large seeds are particularly a feature of those spring-germinating plants which, even in dry habitats, continue to expand their shoot biomass during the late spring and summer. Examples of these include obligate or facultative annual species such as *Medicago lupulina* and *Torilis japonica*, and perennials such as *Centaurea scabiosa*, *Lotus corniculatus*, *Pimpinella saxifraga* and *Sanguisorba minor*, all of which have taproots, the initial extension of which is presumably dependent upon mobilization of a large fraction of the considerable seed reserves into the radicle.

The shape of the germinule also provides clues to function. Specializations increasing buoyancy in air or in water, or facilitating dispersal by animals, are well documented (Ridley 1930, van der Pijl 1972). These are commented upon in the Accounts. A more contentious area of speculation arises in relation to the morphology of germinules in those species which form persistent buried seed banks. The majority in this category have a germinule which is compact and usually without hairs or larger projections. This suggests that there may have been selection for rapid incorporation into the surface layers of the soil, either by rainwater percolation or by the activity of invertebrates. Evidence for this hypothesis is particularly strong in two families - Asteraceae and Poaceae – where the species forming persistent seed banks (e.g. *Agrostis* spp., *Danthonia decumbens*, *Lapsana communis*, *Matricaria discoidea*, *Milium effusum* and *Tripleurospermum inodorum*) are, atypically, bereft of projections which might aid dispersal by wind or by animals. It is also interesting that in two species which form persistent seed banks, *Anthoxanthum odoratum* and *Cirsium arvense*, the paraphernalia assisting initial dispersal often become detached, releasing a compact germinule.

Germination

The germination data presented in the Accounts, together with information on species of more-restricted abundance in the Sheffield region. have been analysed in full by Grime *et al.* (1981). The following general conclusions may be drawn.

Initial germinability

Many of the species which show the capacity for immediate germination have other characteristics, such as specialized temperature requirements, which suggest that the germination of freshly-dispersed seeds may be prevented by the intervention of limiting factors which operate in the field but are excluded from simple laboratory germination tests. This hypothesis is supported by the fact that many of the species which display high initial germinability (e.g. *Deschampsia cespitosa*, *Epilobium hirsutum* and *Holcus lanatus*) are known to accumulate large and persistent banks of buried seeds in the field.

Response to dry storage

In the majority of species the germination percentage increases during dry storage. This effect is most marked at warm temperatures and in small-seeded species. The requirement for warm dry storage is most pronounced in winter annuals (e.g. *Aira praecox*, *Myosotis ramosissima* and *Saxifraga tridactylites*), where the phenomenon has been recognized widely (e.g. Ratcliffe 1961, Newman 1963) as a mechanism reducing the risk of premature germination in habitats subject to summer desiccation. The same explanation may be applied to the characteristic but less-pronounced responses to dry storage which are evident in certain of the autumn-germinating perennial grasses such as *Festuca ovina* and *Koeleria macrantha*.

Of the perennial species which display a marked improvement in germinability during dry storage, the majority are small-seeded, with many forming persistent seed banks in the soil (e.g. *Danthonia decumbens*, *Plantago lanceolata* and *Rumex crispus*). The possibility must therefore be considered that in certain species delayed ripening and delayed germination are important in facilitating seed burial.

Response to chilling

Among the 75 species which responded to chilling in the tests of Grime *et al.* (1981), some, including all of the Umbelliferae examined, germinated at low temperatures whether in light or in darkness. The majority of species in this group are known to produce large, even-aged populations of seedlings in the early spring (e.g. *Heracleum sphondylium*, *Hyacinthoides non-scripta* and *Rhinanthus minor*). There is no evidence to suggest that any of these maintains a persistent seed bank in the soil, and the most probable explanation for this germination behaviour is that it facilitates the spring colonization of vegetation gaps in perennial vegetation. We suspect that the large seeds and the capacity for early germination allow seedlings of these species to compete effectively with shoots from neighbouring established plants.

A different significance may be attached to the chilling responses in some arable weeds (e.g. *Atriplex patula*, *Chenopodium album*, *Papaver rhoeas* and *Polygonum aviculare*). Here the requirement for chilling is allied to a need for subsequent exposure to light, or to higher temperatures. All of the species in this category form persistent seed banks, suggesting that the chilling requirement may be implicated, initially in delayed germination and seed burial, and subsequently in mechanisms of secondary dormancy, whereby germination of the buried seeds is favoured only by conditions obtaining in the early spring (Courtney 1968).

Response to scarification

All of the legumes for which Accounts are presented, and certain other species such as *Convolvulus arvensis*, *Geranium molle* and *Helianthemum nummularium*, are capable of rapid germination to a high percentage only after scarification. Although the role of impermeable seed-coats under field conditions is still uncertain (Ballard 1973), there can be little doubt that this characteristic is often conducive to delayed germination and incorporation into persistent seed banks. A further insight into the possible ecological significance of the hard seed-coat of legumes is provided by studies such as those of Hamly (1932), Hagon and Ballard (1969) and Hagon (1971), who showed that permeability depends upon the condition of the strophiole, the only point of entry of water into the unscarified seeds of many legumes.

More suggestive still are the results of Quinlivan (1961,1968) and Quinlivan and Millington (1962), who established that in some Australian legumes the permeability of the strophiole can be increased under field conditions by diurnal temperature fluctuations of a magnitude commonly experienced by buried seeds lying on, or close to, a soil surface devoid of vegetation. More recently it has been shown (Moreno-Casasola 1985) that the breaking of hard-coat dormancy in legume seed by diurnal temperature fluctuations and changes in soil-water content varies considerably between species. This, therefore, is circumstantial evidence suggesting that seed-coat impermeability in the legumes, and perhaps also in other species, facilitates the persistence of seed reserves in the habitat and may also provide a mechanism of 'gap-detection' (Thompson *et al.* 1977) by surface-lying or superficially buried seeds.

Response to temperature

In the Accounts the germination response to temperature has been described by reference to the upper and lower limits of the range over which at least 50% of the maximum percentage germination occurs. This ignores the fact that some germination can occur outside this range. It is also important to recognize that these results refer to constant temperatures, and that in many species the germination range, particularly at its lower extremity, may be extended by imposing a fluctuating temperature regime (Thompson & Grime 1983).

Among annual and perennial grasses a high proportion of the species are capable of germinating over a wide range of temperatures, perhaps of more than 20°C. This is especially true of species of dry habitats and southern distribution (e.g. *Bromopsis erectus*, *Catapodium rigidum* and *Trisetum flavescens*), and suggests that a relative insensitivity to temperature is characteristic of seeds in which water supply acts as the primary determinant of the timing of germination in the field.

Several sources of variation in temperature-response appear to be associated with taxonomic differences. Scarified seeds of legumes tend to germinate over a very wide range of temperatures. In both the Fabaceae and the Poaceae, wide range is related to the capacity for germination at both high and low temperatures. In the Asteraceae, however, wide germination range is due mainly to the high germinability maintained at elevated temperatures. A quite distinct pattern of response is characteristic of the Cyperaceae (e.g.

Carex flacca, *C. nigra* and *Eriophorum angustifolium*), in which germination tends to be restricted to a narrow range of relatively high temperatures.

When groups of species belonging to various geographical elements within the British flora are compared, several features become apparent. The most striking of these is the failure of a majority of plants of northern distribution (e.g. *Carex pilulifera*, *Molinea caerulea* and *Vaccinium vitis-idaea*) to germinate at temperatures below 7°C. This is in agreement with earlier studies (Billings & Mooney 1968, Thompson 1968, 1970, Wein & Maclean 1973), and is consistent with the hypothesis that the high temperature requirement of these plants is the result of natural selection at high latitudes for a germination-response restricting the appearance of seedlings to the short but relatively favourable summer.

Species concentrated in southern Britain include a high proportion capable of germinating at low temperature, and tend to have a temperature range exceeding that of the northern species. Species germinating over a wide range of temperatures reach their highest frequency among plants which are ubiquitous within the British Isles.

Relationships between temperature response and habitat are also evident from the Accounts. Marshland and disturbed habitats contain a large proportion of the species which do not germinate below 7°C. Some of these species (e.g. *Chenopodium rubrum*, *Juncus effusus* and *Rumex crispus*) are also unusually responsive to diurnal fluctuations in temperature (Thompson & Grime 1983). In Britain seedling establishment in marshes and at the margins of lakes, rivers, ponds and ditches is usually restricted to the late spring, when the water table recedes and muddy surfaces become exposed and aerobic. It seems likely, therefore, that the requirement for high or fluctuating temperature in these marsh plants provides a mechanism delaying spring germination until such time as the declining water table no longer insulates the soil from the increasing radiation load (Thompson *et al.* 1977).

Germination over a wide range of temperatures is particularly common in grassland plants (e.g. *Bellis perennis*, *Festuca rubra*, *Holcus lanatus* and *Trifolium repens*) and species from skeletal habitats (e.g. *Arabidopsis thaliana*, *Helianthemum nummularium* and *Medicago lupulina*). In woodland plants, however, (e.g. *Brachypodium sylvaticum*, *Geum urbanum* and *Luzula pilosa*) germination is often restricted to a narrow range of intermediate temperatures.

Response to light

The information on the germination responses to light contained in the Accounts agrees with the comprehensive screening data of Kinzel (1920), inasmuch as germination in most species is promoted by light. Both sources, however, reveal the existence of some seed populations in which germination is inhibited by light. In the data drawn upon in our Accounts this effect is most pronounced in *Urtica urens*, but appears also in *Cardamine hirsuta*. The tendency of certain species to be inhibited from germinating 'in light' has been observed by numerous earlier investigators (Toole 1973). The inhibitory effect is most severe under prolonged high light flux {Black & Wareing 1960, Schultz & Klein 1963, Chen & Thiman 1964, Rollin & Maignan 1967, Kendrick & Frankland 1969) or at high temperature (Black & Wareing 1960, Chen & Thiman 1964). This suggests that under field conditions the inhibitory effects of high light flux will usually coincide with strongly desiccating conditions, which may pose a severe threat to seedling establishment. It seems, therefore, possible that sensitivity to high irradiance provides a safeguard reducing the risk of seedling death caused by drought.

The proportion of species exceeding 50% germination in darkness is considerably higher in grasses than in either annual or perennial forbs. After scarification all of the Leguminosae described in the Accounts germinated rapidly to a high percentage in darkness. Species with the capacity to germinate to a high percentage in the dark are especially common among plants of ubiquitous or widespread occurrence in the British Isles. These also form a high proportion of the species of common occurrence in the Sheffield-based sample.

Members of the Compositae and Cyperaceae exhibit a strong light requirement. When the species appearing in the Accounts are broadly classified with respect to habitat, they can be arranged into a series, through which the proportion of species attaining 50% germination in darkness declines as follows:

grassland > skeletal habitats > woodland > disturbed habitats > marshland

In seeking a causal explanation for variation between plant species in their capacity for germination in darkness, it is necessary to recognize that this characteristic is phenotypically variable and depends upon the experience of the seed during the later stages of maturation. From laboratory studies and experiments with natural leaf canopies (Pigott 1971, Gorski 1975, King 1975, Grime & Jarvis 1976, Panetta 1979, Fenner 1980a,b, Silvertown (1980), it has been demonstrated that sunlight which has been filtered through a leaf canopy may be depleted in red light to an extent which causes germination to be inhibited. In view of this evidence for post-dispersal effects of leaf canopies upon germinability, it seems likely that similar effects could occur at an earlier stage during the maturation of the seeds on the parent plant.

Experimental studies (Cresswell & Grime 1981) suggest that the induction of a light requirement in developing seeds occurs not only in inflorescences shaded by neighbouring plants, but also in the very large number of plants in which seed maturation is completed within green parental structures. From data such as those in Figure 5.5, it seems, therefore, that the most likely explanation for the light requirement noted in many of the Accounts is to be found in the light-filtering properties of the tissues which surround the developing seeds of these species. Failure to germinate in darkness is most commonly observed among small-seeded species of disturbed habitats and marshland (e.g. *Cardamine flexuosa*, *Juncus articulatus* and *Sagina procumbens*). Many of these are known to develop large and persistent seed reserves in the soil. This suggests that the light requirement of the freshly-dispersed seed (perhaps reinforced in some cases by the influence of leaf-canopies prior to burial) prevents the germination of the buried seed and initiates germination only if the seed is unearthed by some form of disturbance.

Germination rate

Wherever possible, each Account contains an estimate of the maximum rate of germination achieved by the species. For some, at least, the data may be an underestimate of the true maximum, because the experimental conditions were not optimal for the species. The rates recorded in the Accounts are based upon experiments in which germination was initiated by hydration of air-dried seeds. This procedure reproduces the common circumstance in which the germination of many grasses, members of the Asteraceae and winter annuals coincides with the onset of wet conditions at the end of the summer, but it is scarcely relevant to the variety of instances in which fully imbibed seeds are induced to germinate by environmental stimuli such as changes in temperature or illumination.

Despite these uncertainties, some interesting patterns of variation in germination rate (analysed in more detail by Grime *et al.* (1981)) are apparent from the Accounts. A wide range of

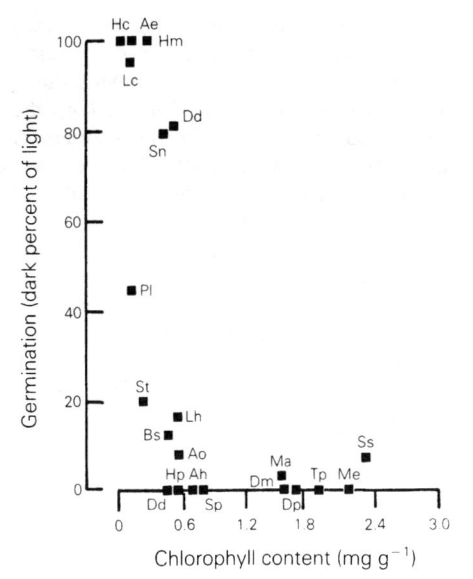

Figure 5.5 The relationship between chlorophyll concentration of the investing structures and dark-germination of mature seeds in various species of flowering plants. Values for chlorophyll refer to the concentration at the midpoint in the range of seed moisture contents. Ae, *Arrhenatherum elatius*; Ah, *Arabis hirsuta*; Ao, *Anthoxanthum odoratum*; Bs, *Brachypodium sylvaticum*; Dd, *Danthonia decumbens*; Dm, *Draba muralis* L.; Dp, *Digitalis purpurea*; He, *Helianthemum nummularium*; Hm, *Hordeum murinum* L.; Hp, *Hypericum perforatum*; Lc, *Lotus corniculatus*; Lh, *Leontodon hispidus*; Ma, *Myosotis arvensis*; Me, *Milium effusum*; Pl, *Plantago lanceolata*; Sd, *Silene dioica*; Sn, *Silene nutans* L.; Sp, *Succisa pratensis*; Ss, *Senecio squalidus*; St, *Serratula tinctoria* L.; Tp, *Tragopogon pratensis* L. (From Cresswell & Grime 1981.)

germination rates occurs in perennial forbs, but there is a clear bias towards low rates in the shrubs and trees. Germination rates are high in many Asteraceae, but low in all of the Cyperaceae and many of the Fabaceae. There is a well-defined tendency for species of northern distribution in Europe (e.g. *Danthonia decumbens*, *Luzula pilosa* and *Vaccinium vitis-idaea*) to germinate relatively slowly. Rapid germination is characteristic of the majority of the species of fertile disturbed habitats. Marshland and woodland plants include a high proportion of slowly germinating species.

With increasing seed weight there is a progressive decline in the proportion of rapidly germinating species. Rapid germination is correlated with the presence of a pappus or conical seed shape (Asteraceae), hygroscopic awns (Poaceae) and antrorse (backwardly-projecting) hairs or teeth (Asteraceae and Poaceae).

A very clear relationship exists between rate of germination and rate of seedling dry-matter production (Fig. 5.6). It is not unreasonable to propose that in many species the two attributes will show parallel responses to natural selection, which in some cases may be linked physiologically. Caution is needed, however, in extending this generalization; exceptions are apparent in such as *Helianthemum nummularium*, *Lotus corniculatus* and several other tap-rooted forbs of infertile soils, in which the capacity of, scarified seeds to germinate rapidly coincides with an inherently low rate of seedling growth.

Epigeal and hypogeal germination

In most species the cotyledons have a photosynthetic function in the young seedling. However, in a few larger-seeded species, mainly in the Papilionaceae, the cotyledons remain below ground, fuelling the germination process (and radicle penetration in particular) by means of seed storage products. In some species the ecological role of hypogeal germination is uncertain, particularly since often only a small proportion of the reserves is actually utilized by the seedling (Fenner 1985). However, in at least one local example, *Quercus* spp., an excess of food reserve allows recovery if the juvenile is grazed (see p. 498).

Biological flora

Biological flora accounts lack the rigorous standardization of the Accounts presented in this book. However, they provide a large volume of additional data of considerable value to the specialist. This includes notes on important predators and pollinators, together with a wide range of detailed information which is generally beyond the scope of the Accounts.

Geographical distribution

The moist tropics are considered to be the aboriginal home of the angiosperms (Cronquist 1968, Takhtajan 1969). Perhaps related to this, the number of angiosperm species per unit area decreases progressively from the tropics to polar regions (Rejmanek 1976). Superimposed on this general relationship are the facts that the British flora is of recent origin (c.1 000 000 BC onwards (West 1969), and that for the past 7000 years Britain has been geographically isolated from mainland Europe.

Although historical factors have radically influenced the distribution of species and the composition of the British flora it is the influence upon plant distribution of the present climate that has tended to receive greatest attention in efforts to explain the geographical limits of plant species. As already noted on page 30, there is a strong global correlation between climate and life-form (Box 1981). The British flora can be usefully classified into elements such as' Arctic-Alpine', ' Atlantic' and 'Continental', which relate to both geographical and climatic distributions (Matthews 1937). However, despite current concerns related to global warming, in only a few instances have experimental data been produced to support the hypothesis that the current climate regulates plant boundaries (Grace 1987). Perhaps the best demonstration relates to *Cirsium acaulon* and to *Tilia cordata*, where regenerative failure at the edge of their range appears to be due to low summer temperature (Pigott 1974, Pigott & Huntley 1978,1980,1981). An inability to tolerate low winter temperatures can also limit distribution, as in the case of *Ilex aquifolium* and *Hedera helix* (Godwin 1975), and the same seems true in spring for

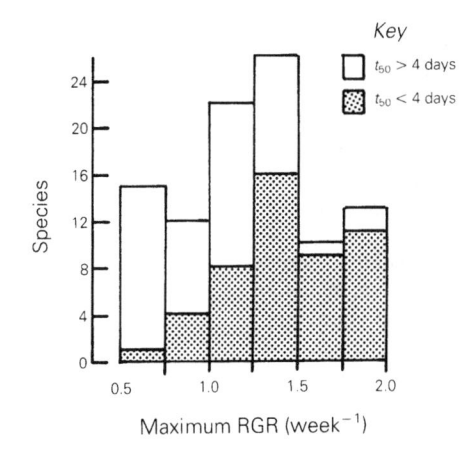

Figure 5.6 The contribution of rapidly-germinating (t_{50} <1.4 days) and slow-germinating (t_{50} >4 days) species to groups classified with respect to estimates of the maximum potential rate of seedling dry-matter production (R_{max}). (From Grime *et al.* 1981.)

the maritime subspecies of *Pyrola rotundifolia* (Hunt *et al.* 1985).

Several factors account for the failure of climatic studies to explain plant distribution satisfactorily. Measurement of microclimatic variables is beset with difficulties. There are, for example, considerable variations in humidity and temperature associated with microtopography (Rorison *et al.* 1986), and the climate experienced at the leaf surface of the plant may be very different from that of the surrounding air (Geiger 1957, Gates 1962, Grace 1977, 1987). Limits may be imposed by extreme climatic events such as severe droughts or very late frosts which only occur once or twice a century. Moreover, because temperature affects a whole range of metabolic processes, even actual failure due to climate may have many possible mechanisms (Grace 1987). In particular it must be remembered that many of the populations near to the geographical limits of a species are small, isolated and genetically impoverished and these factors alone can restrict reproductive vigour. These complications make analysis by direct field study extremely uncertain and, often, unrewarding. We conclude that simple comparative experiments defining the climatic tolerances of large numbers of species of contrasted geographical distribution (e.g. Larcher 1980) may provide the most reliable way forward.

Habitats

Even on the basis of a casual inspection, the habitat diagram in each Account provides useful ecological information. The value for 'All habitats' at the top of the diagram is a measure of the subject species' overall abundance in the whole area investigated. A crude index of the ecological amplitude of the species may be obtained by comparing the frequencies attained in the seven intermediate categories, each of which represents a distinctive type of habitat. Using the habitat diagram in this way, we are able to conclude, for example, that *Agrostis stolonifera* and *Holcus lanatus* are species of relatively wide ecological amplitude and, in this respect, contrast strongly with certain other common species of narrowly-defined ecology such as *Rubus fruticosus* agg. and *Holcus mollis*. 'Ecological amplitude' must be used with caution, however, since it is being inferred from the presence or absence of the species in habitats classified very broadly in terms of environment and land-use. As explained in the Account for *Poa trivialis* (p. 478), the possibility must be considered that some subordinate members of plant communities achieve 'wide amplitude' merely by exploiting essentially similar microhabitats or temporal niches within otherwise distinct types of habitats and vegetation.

An additional problem in estimating ecological amplitude from the habitat diagrams arises when considering species regenerating by means of numerous, widely dispersed offspring. As explained (p. 65), the dispersal efficiency of these species is often such as to produce large populations of juveniles in habitats in which there is little prospect of successful establishment. *Betula* spp., *Chamerion angustifolium*, *Dryopteris filix-mas* and *Epilobium hirsutum* provide examples of species for which the inclusion of such transients in the survey records has created a spuriously high index of ecological amplitude.

Although a general impression of the habitat distribution of a species is available directly from the 'Habitats' diagram, further calculations may be necessary to examine the affinity of a species for particular intermediate or terminal categories. This is especially desirable where finer comparisons between species are required, since the absolute values for percentage occurrence presented in the habitat diagram are clearly influenced by the abundance of the species in the surveyed area as a whole. In order to take this factor into account in comparisons between species it is convenient to calculate, for any intermediate or terminal category under scrutiny (H), an index of affinity (A), where:

A = (% occurrence in *H*)/(% occurrence in all habitats)

As an illustration of this approach, Table 5.1 provides a list of certain species occurring in the intermediate category 'Spoil'. Comparisons of percentage occurrences identify *Agrostis stolonifera*, *Chamerion angustifolium* and *Holcus lanatus* as the most common species of spoiled land. However, when the index of affinity is used as the basis of comparison, *Chaenorhinum minus*,

Table 5.1 The 20 most common vascular plant species of spoil habitats in the Sheffield region. (From Grime 1986b.)

Rank	Species	F*	A	P	G	C	W
1	*Holcus lanatus*	36.4	+	+			
2	*Chamerion angustifolium*	32.8		+			+
3	*Agrostis stolonifera*	32.0		+	+		
4	*Dactylis glomerata*	29.4		+	+		
5	*Festuca rubra*	28.5		+	+		
6	*Agrostis capillaris*	26.0		+	+		
7	*Poa annua*	24.9	(+)	(+)	+		
8	*Poa pratensis*	24.8		+	+		
9	*Arrhenatherum elatius*	24.1		+	+		
10	*Poa trivialis*	23.5		+	+		
11	*Tussilago farfara*	23.5		+		+	+
12	*Taraxacum* agg.	22.0		+		+	+
13	*Deschampsia flexuosa*	18.3		+	+		
14	*Cerastium fontanum*	18.2		+			
15	*Plantago lanceolata*	17.3		+			
16	*Cirsium arvense*	16.1		+		+	+
17	*Elymus repens*	14.4		+	+		
18	*Trifolium repens*	14.2		+			
19	*Senecio vulgaris*	14.2	+			+	+
20	*Senecio squalidus*	13.7	+			+	+

*F, percentage of spoil samples containing the species; A, annual; P, perennial; G, grass; C, composite; W, wind-dispersed. The parentheses for *Poa annua* denote the fact that annual and perennial phenotypes of this species occur in the Sheffield region.

Minuartia verna and *Senecio viscosus* emerge as the species which are most *particularly associated* with Spoil.

The categories used in the 'Habitats' diagram are all extremely broad and some (e.g. river and stream banks) are very heterogeneous in vegetation. It is possible, nevertheless, to use the diagram as a broad basis upon which to assess the present and future status of a species. In particular, we can differentiate between those (e.g. *Galium sterneri*, *Mercurialis perennis* and *Primula veris*) which have remained confined to habitats that are remnants of the more ancient countryside (unenclosed pastures, woodland, rock outcrops, screes, cliffs) and those (e.g. *Digitalis purpurea*, *Elytrigia repens* and *Senecio squalidus*) which have successfully colonized recently created or radically transformed habitats (various types of spoil, plantations and arable land). With caution, therefore, the diagram may be used to detect both threatened and expanding species.

Gregariousness

Inspection of the table 'Abundance in quadrats', classifying the occurrences of the subject species in terms of the proportion of quadrat subsections occupied by rooted shoots, permits a broad assessment of population structure. Species in which records are concentrated towards the left-hand side of the table ('Type 1') include not only the tree seedlings (e.g. *Acer pseudoplatanus*, *Alnus glutinosa* and *Quercus* spp.), but also various herbaceous species originating from widely scattered propagules and possessing little capacity for vegetative spread (e.g. *Cirsium vulgare*, *Cystopteris fragilis* and *Hypochaeris radicata*).

Species with a large proportion of occurrences in the higher frequency classes ('Type 3') are those capable of developing dense populations. These populations may consist of a large number of discrete individuals originating as seedlings in open habitats (e.g. *Aira praecox*, *Linum catharticum* and *Matricaria discoidea*) or they may be the result of vegetative expansion (e.g. *Agrostis stolonifera*, *Brachypodium pinnatum* and *Mercurialis perennis*).

However, in some large clonal species (e.g. *Pteridium aquilinum*, *Fallopia japonica* and *Petasites hybridus*) the capacity to dominate herbaceous vegetation is not reflected in a high frequency of rooted shoots. This is because these species produce a relatively low numerical density of individually massive fronds, shoots or leaves.

In addition to the two extremes in gregariousness, an intermediate category can be recognized, 'Type 2'. This includes a high proportion of species (e.g. *Briza media*, *Campanula rotundifolia* and *Carex caryophyllea*) which are of restricted height and capacity for lateral spread and which normally play a subordinate role in perennial herbaceous vegetation.

Altitude

The altitude histograms in the Accounts form a continuous array

Table 5.2 The distribution of **altitude** classes, by percentage of records in each of the 32 'terminal' habitats of UCPE Survey II (see Fig. 2.2).

Habitat	<100	100–199	200–299	300–399	400+
Still aquatic	55.1	21.2	0.6	21.8	1.3
Running aquatic	26.4	45.3	20.8	7.5	0.0
Unshaded mire	29.6	35.2	10.6	14.8	9.9
Shaded mire	30.8	51.3	16.7	1.3	0.0
Rock outcrops	23.2	30.4	27.5	13.0	5.8
Scree	0.0	2.8	97.2	0.0	0.0
Cliffs	21.4	20.5	41.9	15.4	0.9
Walls	29.9	46.7	19.7	3.6	0.0
Arable	41.8	43.4	10.7	4.1	0.0
Meadows	10.5	39.5	34.2	15.8	0.0
Enclosed pastures	29.3	34.1	22.0	12.2	2.4
Limestone pastures	0.0	5.7	50.9	39.6	3.8
Acidic pastures	0.0	0.0	27.8	61.1	11.1
Coal-mine heaps	48.4	51.6	0.0	0.0	0.0
Limestone quarry heaps	29.0	14.5	30.4	26.1	0.0
Acidic quarry heaps	0.0	16.1	37.5	42.9	3.6
Lead-mine heaps	0.0	0.0	60.0	40.0	0.0
Cinders	59.1	30.3	6.1	4.5	0.0
Bricks and mortar	30.0	47.5	20.0	2.5	0.0
Soil heaps	35.6	28.8	28.8	6.8	0.0
Manure and sewage waste	42.5	42.5	7.5	7.5	0.0
Limestone woodlands	43.2	23.4	33.3	0.0	0.0
Acidic woodlands	28.3	41.3	27.1	3.3	0.0
Scrub	22.7	43.7	24.2	9.4	0.0
Broad-leaved plantations	32.4	38.7	18.9	9.9	0.0
Coniferous plantations	32.0	19.0	38.0	10.0	1.0
Hedgerows	19.0	61.9	16.7	2.4	0.0
Limestone wasteland	27.6	31.6	37.8	3.1	0.0
Acidic wasteland	46.4	14.6	15.9	19.9	3.3
River banks	43.9	43.9	9.8	2.4	0.0
Verges	25.7	28.6	27.1	17.1	1.4
Paths	36.3	37.3	15.7	5.9	3.9

of distributions with the two extremes being represented by exclusively lowland plants (e.g. *Artemisia absinthium*, *Senecio squalidus* and *Tamus communis*) and by species which are most frequent at high altitudes (e.g. *Dryopteris dilatata*, *Empetrum nigrum* and *Vaccinium vitis-idaea*). Differences in altitude coincide with major changes in climate which may be expected to impose limitations on the distribution of individual species, either directly by affecting vegetative growth or seed production (Davison 1977, Pigott & Huntley 1981), or indirectly by modifying the vigour of competing species (Carter & Prince 1981).

There is a danger of overestimating the importance of climatic factors in relation to altitudinal ranges. In the Sheffield region, as in the rest of Britain, changes in altitude are confounded with differences in geology and land-use, and the information presented in Table 5.2 confirms that many of the terminal habitats contribute unequally to the altitude classes used to construct the histogram. Evidently, therefore, distinctive plant distributions with respect to altitude may arise simply because of their restriction to particular habitats. An excellent example of this is provided by the large number of calcicolous pasture plants (e.g. *Helictotrichon pratense*, *Polygala vulgaris*, *Sanguisorba minor* and *Scabiosa columbaria*) which attain maximum frequency of occurrence in the intermediate altitude range 200-300 m. We may be fairly certain that this pattern is merely a reflection of the distribution of old calcareous pastures in the Sheffield region, since elsewhere in Britain all of these species are common at both higher and lower altitude in ancient chalk and limestone pasture.

Another striking example of the impact of habitat availability on the altitudinal distribution of species is available from a number of heathland plants (e.g. *Galium saxatile*, *Calluna vulgaris* and *Nardus stricta*). Reference to Lees (1888), Howitt and Howitt (1963) and Preston *et al.* (2002a) reveals that the present upland bias of these species in Britain is in part a result of the destruction of formerly suitable sites in lowland heathlands.

Slope

As in the case of altitude, species occurrences in relation to angle of slope are strongly determined by the distribution of habitats in relation to slope. Table 5.3 indicates that records in all of the primary habitats except the skeletal group are strongly biased towards shallow gradients. In each case records of slope <20° outnumber all other classes combined. Within skeletal habitats, rock outcrops are biased towards low angles of slope, while 92% of all scree records lie within the range 20-39°. Naturally enough, very steep slopes predominate on cliffs and walls, with 21 and 14%, respectively, of records in these two terminal categories consisting of vertical or overhanging sites (of slope not less than 90°). We conclude that investigations of relationships between species distributions and slope *per se* offer little to the ecologist unless they are reviewed in the context of the particular habitats involved.

Table 5.3 The distribution of **slope** classes, by percentage of records in each of the 3 2 'terminal' habitats of UCPE Survey II (see Fig. 2.2).

Habitat	<20	20–39	40–59	60–79	80+
Still aquatic	100.0	0.0	0.0	0.0	0.0
Running aquatic	100.0	0.0	0.0	0.0	0.0
Unshaded mire	96.5	3.5	0.0	0.0	0.0
Shaded mire	94.9	2.6	1.3	1.3	0.0
Rock outcrops	44.9	29.0	15.9	10.1	0.0
Scree	5.6	91.7	2.8	0.0	0.0
Cliffs	0.0	2.6	5.2	47.8	44.4
Walls	21.9	2.2	3.0	19.7	53.3
Arable	97.5	1.6	0.8	0.0	0.0
Meadows	100.0	0.0	0.0	0.0	0.0
Enclosed pastures	92.7	7.3	0.0	0.0	0.0
Limestone pastures	26.4	60.4	13.2	0.0	0.0
Acidic pastures	54.4	37.8	7.8	0.0	0.0
Coal-mine heaps	55.9	38.7	5.4	0.0	0.0
Limestone quarry heaps	29.0	53.6	17.3	0.0	0.0
Acidic quarry heaps	35.7	53.5	10.7	0.0	0.0
Lead-mine heaps	46.0	40.0	12.0	2.0	0.0
Cinders	89.4	9.1	1.5	0.0	0.0
Bricks and mortar	85.0	12.5	2.5	0.0	0.0
Soil heaps	66.1	30.5	3.4	0.0	0.0
Manure and sewage waste	77.5	22.5	0.0	0.0	0.0
Limestone woodlands	56.7	38.7	4.5	0.0	0.0
Acidic woodlands	74.1	23.0	2.9	0.0	0.0
Scrub	70.3	25.8	3.9	0.0	0.0
Broad-leaved plantations	75.7	24.3	0.0	0.0	0.0
Coniferous plantations	71.0	25.0	4.0	0.0	0.0
Hedgerows	83.4	16.7	0.0	0.0	0.0
Limestone wasteland	37.7	52.1	10.2	0.0	0.0
Acidic wasteland	65.5	25.9	8.6	0.0	0.0
River banks	80.5	17.1	0.0	0.0	2.4
Verges	84.3	14.3	1.4	0.0	0.0
Paths	100.0	0.0	0.0	0.0	0.0

Aspect

The data presented in the two aspect diagrams in each Species Account allow comparisons of frequency of occurrence and abundance between north- and south-facing slopes both in open and in shaded situations.

From an earlier detailed study of distribution in relation to aspect in grasslands plants of the Sheffield region (Grime & Lloyd 1973), it is clear that the simple frequency of occurrence on north- and south-facing slopes often varies independently from the vigour attained at such sites. For example, in a number of species (e.g. *Agrostis capillaris*, *Briza media* and *Carex flacca*) the frequency of occurrence does not vary significantly in relation to aspect, but the level of abundance is considerably greater at sites situated on north-facing slopes. Similar results are evident elsewere in the Accounts, suggesting that in some plants aspect-related phenomena affecting establishment and persistence may be rather different from those which determine vigour and population size.

A useful generalisation in interpreting plant distributions in relation to aspect in unshaded habitats is that species of predominantly northern distribution in Britain (e.g. *Carex pilulifera*, *Galium saxatile* and *Molinea caerulea*) tend to be more common on north-facing slopes. Also, a bias towards sites of northerly aspect is often apparent in situations where species of marshland and damp grassland (e.g. *Carex panicea*, *Cirsium palustre* and *Succisa pratensis*) or woodland (e.g. *Anemone nemorosa*, *Mercurialis perennis* and *Oxalis acetosella*) occur in the open.

A high proportion of the species which exhibit an association with unshaded south-facing slopes are plants of southern distribution in the British Isles. Many are small annuals or monocarpic perennials associated with open habitats subject to summer desiccation (e.g. *Arabidopsis thaliana*, *Carlina vulgaris* and *Trifolium dubium*) or are xeromorphic or tap-rooted perennials (e.g. *Centaurea scabiosa*, *Lotus corniculatus* and *Thymus polytrichus*).

When aspect distributions in shaded and unshaded habitats are compared, a very commonly recurring pattern is that in which species which occur more frequently on north-facing slopes in the open are associated with south-facing slopes when under shade. Examples of these include *Angelica sylvestris*, *Galium saxatile* and *Epilobium montanum*, all of which, typically, are restricted to continuously damp habitats. A further nuance in plant-aspect relations is evident in the 'Synopsis' for species such as *Anemone nemorosa*, *Hyacinthoides non-scripta*, *Mercurialis perennis* and *Oxalis acetosella*. These are largely restricted to shaded habitats of all aspects in lowland areas, but frequently occur on unshaded north-facing sites within upland regions.

A large proportion of the shaded samples used in the analysis of aspect relationships refer to plantations, small fragments of disturbed woodland, scrub and hedgerows. In many of these habitats the shading is of recent origin and is low in intensity. The result of this is that much of the vegetation is not composed of 'true' woodland species. It is interesting to note, however, that some of the species which are exclusive to woodland (e.g. *Lamiastrum galeobdolon*, *Sanicula europaea* and *Veronica montana*) show a strong association with north-facing shaded slopes.

Hydrology

The hydrology diagrams allow the species to be characterized in terms of its affinity for wetland habitats. At one extreme are the 'dryland specialists' (e.g. *Arenaria serpyllifolia*, *Asplenium trichomanes* and *Sedum acre*) and deep-rooted species (e.g. *Medicago lupulina*, *Pteridium aquilinum* and *Tamus communis*), which are all wholly excluded from wetland habitats. At the other extreme, as we might expect, are the anatomically and physiologically specialised strict hydrophytes (e.g. *Potamogeton* spp. and *Ranunculus penicillatus*). Such species, discussed by Sculthorpe (1967), occupy the 'submerged' end of the hydrology diagram.

Among the species with peaks in occurrence at the left-hand side of the hydrology diagram a distinction can be drawn between plants which, although not exclusive to wetland habitats, are of frequent occurrence in mire (e.g. *Cardamine flexuosa*, *Poa trivialis* and *Ranunculus repens*), those which are restricted to wetlands with a retreating water table during the summer (e.g. *Ranunculus sceleratus* and *Rorippa palustris*) and those which are capable *of* exploiting either exposed soil or shallow water during the growing season (e.g. *Apium nodiflorum*, *Callitriche stagnalis* and *Persicaria amphibia*). A diversity of physiological specializations has been implicated in tolerance of mire conditions. Many species exploiting topogenous mire have well-developed aerenchyma in their shoots and roots. This facilitates oxygen diffusion into the rhizosphere, a process which reduces the levels of ferrous iron and other toxins (Crawford 1982). However, the importance of this detoxification in other types of mire is uncertain. For example, spring fens have high redox potentials (B. D. Wheeler pers. comm.), but their assemblages of species frequently resemble grassland communities as much as those of wetland. Clearly, much additional work is needed before a satisfactory explanation of the factors controlling species distribution can be achieved for all wetland habitats.

Soil pH

Distinctive distributions with respect to surface soil pH are evident in extreme calcifuges (e.g. *Calluna vulgaris*, *Deschampsia flexuosa* and *Erica cinerea*) and in species restricted to calcareous soils (e.g. *Galium sterneri*, *Helictotrichon pratense* and *Scabiosa columbaria*). It is interesting to observe that plants with occurrences extending up to very high pH values include species with some populations rooted in the mortar and cement of old walls (e.g. *Asplenium ruta-muraria*, *Cystopteris fragilis* and *Sedum acre*) or exploiting brick rubble on urban demolition sites (e.g. *Artemisia absinthium*, *Senecio squalidus* and *Tussilago farfara*). Some of the widest distributions with respect to soil pH occur in grasses such as *Anthoxanthum odoratum*, *Deschampsia cespitosa* and *Festuca ovina*. The Synopses of the Accounts for these three species cite published evidence for the existence of edaphic ecotypes.

In view of the complex changes in habitat factors and management regime across the soil pH spectrum (Table 3.8), few generalizations can be offered with regard to the causal mechanisms responsible for the correlations observed between soil pH and species distribution. It is particularly hazardous to speculate about the phenomena operating within soils of high pH, which are exceedingly heterogeneous in terms of productivity, vegetation and management. However, as noted by Grime and Hodgson (1969), a recurring feature of many of the relatively wide pH distributions (e.g. those of *Cerastium fontanum*, *Festuca rubra*, *Holcus lanatus* and *Rumex acetosa*) is a sharp 'cut-off below pH 4.0. This truncation coincides with the change from the mull to mor type of humus, and occurs at a point where there is often a marked increase in exposure to the potentially toxic aluminium and ammonium ions (Magistad 1925, Rorison 1960, Gigon & Rorison 1972).

Bare soil

Under the edaphic and climatic conditions prevailing in most of Britain the extent of bare soil exposure provides an index of vegetation disturbance both by 'natural' factors such as drought, flood damage, soil creep and animal activity, and by human interventions in agriculture, industry, domestic activities or recreation. The 'Bare soil' histogram in each Account records the frequency and abundance of the species in relation to quantity of exposed soil, and thus provides a useful index of vulnerability to, or dependence upon, habitat disturbance.

However, in order to avoid too facile an approach to relationships between plant distribution and disturbance, it is necessary to warn that measurements of soil exposure at one instant (as here) are not a direct measure of overall intensity of disturbance. The quantity of bare soil amidst vegetation in the terminal habitats (Table 5.4) is determined also by the resilience of the vegetation itself, which is in turn dependent upon various other factors of which the most important is productivity. Hence, in the Accounts some associations with large amounts of bare soil (e.g. in *Poa annua*, *Senecio vulgaris* and *Stellaria media*) arise from the frequent and severe disturbance of productive vegetation, whereas in others (e.g. *Arenaria serpyllifolia*, *Minuartia verna* and *Myosotis ramosissima*) the associations are due to occasional and less-extensive damage occurring within much less-productive (and hence more-slowly 'repaired') communities.

High, exposures of bare soil are also common at productive sites dominated by trees (e.g. *Fraxinus excelsior*), shrubs (e.g. *Sambucus nigra*) or herbs (e.g. *Petasites hybridus*) which produce an extensive and dense leaf canopy and deposit large quantities of rapidly decomposing litter.

Just as the significance of positive correlations with bare soil can vary according to habitat, so also can that of negative correlations. Species may be concentrated in sites with little exposed soil, either because they are themselves capable of creating a dense cover of foliage or litter (e.g. *Brachypodium pinnatum*, *Carex acutiformis* and *Pteridium aquilinum*) or because, they are species (e.g. *Anthriscus sylvestris*, *Cruciata laevipes* and *Potentilla erecta*) which are able to coexist with dominants. Alternatively, they may be species (e.g. *Asplenium trichomanes*, *Geranium robertianum* and *Mycelis muralis*) capable of exploiting skeletal habitats in which there is very little bare soil because the ground surface is occupied by rock, bryophytes or lichens or the spoil of cinder tips and demolition sites (e.g. *Chaenorhinum minus*, *Senecio squalidus* and *S. viscosus*).

Triangular ordination

As explained on page 42, the 'Habitats' diagram provides an estimate of the ecological range of each species, which is limited in its precision by the arbitrary nature of the criteria used to define the habitat categories and by the resulting heterogeneity of the vegetation within each grouping. The 'Triangular ordination' represents an attempt to avoid some of these difficulties by measuring the ecological amplitude of each species across a matrix of functionally-defined vegetation types.

The procedure used to construct the matrix (p. 38) was to ordinate a large number of individual vegetation samples, drawn from all of the major terrestrial habitats of the Sheffield region, with reference to the importance of particular traits in the component species. These traits were selected on the basis of their consistent relationship to the axes recognized in the CSR model of primary plant strategies (p. 9). Quite clearly, the usefulness of the ordination therefore rests upon the validity of the model, which in turn depends upon the arguments discussed on pages 46–47 and and ultimately upon experimental tests of the kind described in Grime (1987).

In most of the Accounts the triangular ordination has allowed the position of greatest concentration of the species to be identified (by a cross on the diagram) and, as described on page 39, 'ecological amplitude' has been calculated and identified by a single contour. In a small number of species it has not been possible to follow these procedures because of the diffuse nature of the underlying data. The lack of any coherent pattern in certain species is related to the small number of data available. In trees regenerating by widely dispersed seeds (e.g. *Acer pseudoplatanus*, *Betula* spp. and *Salix* spp.), the pattern is obscured by an abundance of records arising from the seedlings which occur as ultimately-abortive colonists of a wide range of habitats. Also, on cliffs, walls and similar rocky habitats an ability to colonize, often by small wind-borne propagules, is critical and species of dissimilar strategy may as a result coexist in the same habitat. The lack of pattern on the triangular ordination diagrams of species of rocky habitats, e.g. *Mycelis muralis*, may indicate a dominant effect of this distributional factor over other strategies.

More interesting, however, is the small group of plants containing *Geranium robertianum*, *Geum urbanum* and *Silene dioica*, in which no clear pattern has emerged despite the availability of a considerable number of samples, the majority of which contain established individuals of the subject species. All of the species in this group are associated with habitats subject to intermittent disturbance. We suspect that the absence of well-defined patterns in the triangular ordinations for these species is due to their restriction to parts of a micro-successional mosaic which is too fine-grained to be detected by 1 m² samples.

A comparison of the position of the centre of the ordination with the strategy classification given under 'Established strategy' reveals that in the majority of subjects there is a close correspondence between the two. As might be expected, this confirms that plants tend to occupy vegetation containing other species of similar strategy. There are, however, some important exceptions to this general relation. Perhaps the most striking examples are provided by species such as *Cardamine flexuosa*, *Galium aparine* and *Ranunculus ficaria*, all of which are to varying degrees ruderal in strategy, but nevertheless occur most commonly in vegetation dominated by C-strategists. The evidence assembled in the Synopses of the Accounts for these species leaves little doubt that their elevated contour centres derive from morphological and phenological specializations which permit the exploitation of microhabitats and coexistence with robust perennials.

The contours in the triangular ordinations provide an index of ecological amplitude which accords well with published evidence. The compact distributions of species such as *Arabidopsis thaliana*, *Campanula rotundifolia* and *Veronica persica* are predictable on the basis of the apparent lack of major variations in their life-history and essential ecology. In contrast, the literature reviewed in the Accounts for *Agrostis capillaris*, *Arrhenatherum elatius*, *Poa annua* and *Poa pratensis* provides evidence of considerable genetic variation in life-history, morphology or physiology. We should not be surprised, therefore, to find that these species have relatively widely-spread contours.

Associated floristic diversity

The most obvious value of the data analysed under this heading is that they provide a broad indication of the level of species richness (i.e. number of species of vascular plants m²) in the vegetation exploited by the species. The histograms in the Accounts suggest a number of broad correlations between the habitat conditions exploited by a plant and the floristic diversity of the communities in which it occurs. It is quite clear, for example, that the extreme calcifuges (e.g. *Deschampsia flexuosa*, *Calluna vulgaris* and *Galium saxatile*) are consistently associated with species-poor vegetation (< 10 species m²), whereas plants of calcareous pastures (e.g. *Koeleria macrantha*, *Polygala vulgaris* and *Primula veris*) are usually found in communities where the number of associated species exceeds 20 m². It seems possible that this difference may be a consequence of the nationally depauperate calcifuge flora of the British Isles (Grime 1973b).

The histogram in each Account also allows us to recognize

Table 5.4 The distribution of classes of **bare soil** cover, by percentage of records in each of the 'terminal' habitats of UCPE Survey II (see Fig. 2.2).

	Bare soil class (percentage cover)						
Habitat	0	1–10	11–25	26–50	51–75	76+	No Info.
Still aquatic	91.0	4.5	1.9	2.6	0.0	0.0	0.0
Running aquatic	83.0	11.3	1.9	1.9	1.9	0.0	0.0
Unshaded mire	41.5	9.9	13.4	9.2	9.2	16.9	0.0
Shaded mire	33.3	11.5	5.1	5.1	9.0	35.9	0.0
Rock outcrops	18.8	18.8	15.9	17.4	13.0	15.9	0.0
Scree	58.3	2.8	0.0	0.0	0.0	0.0	38.9
Cliffs	52.1	35.0	11.1	1.7	0.0	0.0	0.0
Walls	86.1	11.7	1.5	0.7	0.0	0.0	0.0
Arable	0.8	0.8	0.8	2.5	0.8	94.3	0.0
Meadows	81.6	5.3	2.6	2.6	2.6	5.3	0.0
Enclosed pastures	63.4	12.2	4.9	9.8	0.0	2.4	7.3
Limestone pastures	43.4	15.1	3.8	0.0	0.0	0.0	37.7
Acidic pastures	27.8	17.8	3.3	4.4	0.0	0.0	46.7
Coal-mine heaps	41.9	10.8	11.8	3.2	0.0	32.3	0.0
Limestone quarry heaps	37.7	24.6	11.6	13.0	4.3	8.7	0.0
Acidic quarry heaps	58.9	19.6	10.7	8.9	0.0	1.8	0.0
Lead-mine heaps	32.0	22.0	10.0	16.0	8.0	12.0	0.0
Cinders	89.4	3.0	3.0	1.5	0.0	3.0	0.0
Bricks and mortar	72.5	5.0	10.0	7.5	5.0	0.0	0.0
Soilheaps	23.7	8.5	5.1	8.5	3.4	44.1	6.8
Manure and sewage waste	100.0	0.0	0.0	0.0	0.0	0.0	0.0
Limestone woodlands	36.9	19.8	12.6	9.9	4.5	15.3	0.9
Acidic woodlands	77.9	9.6	3.8	2.5	2.5	2.9	0.8
Scrub	68.0	10.2	6.3	5.5	2.3	7.0	0.8
Broad-leaved plantations	63.1	17.1	3.6	8.1	2.7	5.4	0.0
Coniferous plantations	89.0	4.0	4.0	0.0	0.0	3.0	0.0
Hedgerows	38.1	19.0	19.0	7.1	2.4	11.9	2.4
Limestone wasteland	38.8	10.2	4.1	5.1	4.1	2.0	35.7
Acidic wasteland	50.3	13.9	4.0	3.3	2.6	0.7	25.2
River banks	31.7	12.2	14.6	7.3	14.6	19.5	0.0
Verges	47.1	22.9	17.1	4.3	1.4	7.1	0.0
Paths	11.8	3.9	5.9	15.7	17.6	45.1	0.0

circumstances in which increasing abundance of the subject species is correlated positively or negatively with floristic diversity. Various alternative explanations are possible for such effects, and these have been reviewed extensively in relation to the so-called 'humped-back' model (Grime 1973a, see also Figure 3.6, page 25 and Panels 10 and 11, pages 67 and 68). Here it is necessary only to discriminate between four major types of relationship.

In the first, a rise in abundance of the subject species brings about a decline in species diversity through a clear capacity to exercise dominance (e.g. on the part of *Pteridium aquilinum*, *Phalaris arundinacea* and *Brachypodium pinnatum*). In the second, a similar decline also coincides with increasing abundance, but here the subject species (e.g. *Anthriscus sylvestris* and *Galium aparine*) is not itself a potential dominant. The relationship instead depends upon its capacity to coexist with dominants. In the third relationship a progressive decrease in floristic diversity again coincides with a rising abundance of the subject species, but this is a reflection of the capacity of the subject to develop large populations in habitats which are inhospitable to the majority of other species (as with *Minuartia verna*, *Pilosella officinarum* and *Senecio viscosus*).

The remaining category involves a rise in floristic diversity in concert with an increasing abundance of the subject species.

Here, at least in some instances, the rise in diversity appears to be due to escape from the dominance of large clonal perennials along a gradient of decreasing habitat productivity or increasing disturbance, or both (e.g. in *Catapodium rigidum*, *Festuca ovina* and *Sedum acre*). In theory, this fourth pattern could also arise where diversity increases with the abundance of the subject species, as part of a transition from an extremely hostile environment with few species to one of slightly higher carrying capacity. No clear examples of this phenomenon occur in the data.

Similarity in habitats

The procedure used to identify the five species most similar to the subject species in distribution between habitats depends upon a calculation of 'percentage similarity' (p. 39). This returns a high value when there is a good 'match' between the percentage occurrences of species in a large number of the 32 terminal habitat categories (Fig. 3.2, page 21) and a low value when there is not. Thus calculated, the percentage similarity simply indicates a similarity *in habitats*. It implies neither that species thus identified are necessarily similar in ecology to the subject species, nor that they are necessarily to be found in association with the subject species. Table 5.5 lists means and standard deviations, across all species, for the percentage occurrences of species recorded in each

Table 5.5 Means and standard deviations (across all species) for the percentage occurrences recorded in each of the 32 'terminal' habitats of UCPE Survey II (see also Fig. 2.2).

Habitat	Mean (%)	S.D.
Still aquatic	1.199	4.341
Running aquatic	1.100	4.213
Unshaded mire	3.199	5.549
Shaded mire	2.815	5.860
Rocks	4.548	7.848
Scree	3.787	11.004
Cliffs	1.943	3.879
Walls	2.064	4.642
Arable	4.441	11.521
Meadows	6.352	16.607
Enclosed pastures	1.676	3.942
Limestone pastures	7.406	13.401
Acidic pastures	2.064	9.109
Soilheaps	5.146	10.570
Coal mine	2.804	6.899
Lead mine	4.754	10.038
Cinders	4.534	9.274
Bricks and mortar	4.975	11.229
Manure and sewage	2.851	7.873
Limestone quarries	5.662	11.259
Acidic quarries	1.705	7.701
Scree	2.509	5.627
Hedgerows	3.114	8.417
Limestone woodlands	3.061	8.125
Acidic woodlands	2.068	5.194
Broad-leaved plantations	1.737	4.988
Coniferous plantations	1.399	4.618
River and stream banks	4.338	8.215
Verges	5.004	11.117
Paths	3.449	8.949
Limestone wasteland	6.025	11.089
Acidic wasteland	2.132	6.224

of the 32 terminal habitat categories. High means, as in meadows and limestone habitats, can indicate the presence of relatively more species (i.e. non-zero records) in the habitat; high standard deviations can indicate a relatively 'non-specialist' flora (as in the case of scree slopes, in comparison with unshaded mire).

Three general types of result may emerge from the similarity analysis; these should be borne in mind when interpreting the percentage similarity in habitats reported in the Accounts.

First, where the subject species has a relatively specialised and restricted habitat distribution there will be a large number of 'zero matches' in the calculation of percentage similarity. Provided that other species exist with broadly the same limited pattern of positive occurrences, these will emerge as very 'similar' indeed to the subject species. In such cases even the fifth most similar species can exhibit a percentage similarity of almost 90%. For example, the sequence of five similarities to the wetland species *Apium nodiflorum* is *Nasturtium officinale* agg. *Veronica beccabunga*, *Callitriche stagnalis*, *Equisetum palustre* and *Glyceria fluitans*, with high percentage similarities ranging from 98.7 down to 89.3. On the same basis, in arable habitats the sequence of five species most similar to *Veronica persica* is *Spergula arvensis*, *Fallopia convolvulus*, *Myosotis arvensis*, *Persicaria maculosa* and *Papaver rhoeas*, with percentage similarities from 99.2 down to 88.7.

Second, in this analysis it is difficult for species of genuinely widespread distribution to attain high percentage similarities to their statistical neighbours. Values in the range 60-80% are more common here. For example, the sequence for *Dactylis glomerata* is *Taraxacum* agg., *Poa pratensis*, *Festuca rubra*, *Leucanthemum vulgare* and *Plantago lanceolata*, with percentage similarities ranging from 79.1 down to 67.1. Another example is provided by *Taraxacum* itself, for which the sequence is *Dactylis glomerata*, *Poa pratensis*, *Festuca rubra*, *Plantago lanceolata* and *Leucanthemum vulgare*, with percentage similarities from 79.1 down to 57.2. These two sequences also illustrate the high incidence of networks *of* mutual similarity among species of widespread distribution.

Third, some species of restricted occurrence have no counterparts capable of returning high percentage similarities in this kind of analysis. Even their closest statistical neighbours may have percentage similarities which lie only in the range 50-60. For example, the sequence for *Cruciata laevipes* is *Lathyrus linifolius*, *Bromopsis erectus*, *Trifolium medium*, *Stachys officinalis* and *Brachypodium pinnatum*, with percentage similarities ranging from 56.8 to 47.5. *Vicia sepium* has the sequence *Arctium minus*, *Fallopia japonica*, *Torilis japonica*, *Vicia cracca* and *Trifolium medium*, with percentage similarities from 52.5 to 41.4. In extreme cases of this kind we have omitted the listing of 'similar species' altogether. This has also been the practice where very few records exist for the subject species.

Commonest associates

The procedure used to identify the five species most commonly associated with the subject species (page 39) would be expected to

Table 5.6 The species most frequently included in 'Commonest Associates' tables and their percentage contribution to the listings for all species. [In total 147 species are recorded as 'Commonest Associates'.]

Species	Rank 1	Rank 2	Rank 3	Rank 4	Rank 5	Total	% of dataset	Cumulative % of dataset
1 Poa trivialis	66	31	14	13	8	132	8.4	8.4
2 Festuca rubra	59	22	15	23	8	127	8.1	16.6
3 Holcus lanatus	8	24	18	22	15	87	5.6	22.1
4 Dactylis glomerata	10	17	20	18	13	78	5.0	27.1
5 Festuca ovina	28	18	15	10	3	74	4.7	31.9
6 Arrhenatherum elatius	13	15	19	8	10	65	4.2	36.0
7 Agrostis stolonifera	19	18	9	4	10	60	3.8	39.9
8 Plantago lanceolata	6	11	14	14	8	53	3.4	43.3
9 Ranunculus repens	1	8	17	12	12	50	3.2	46.4
10 Urtica dioica	1	17	8	10	7	43	2.8	49.2
11 Agrostis capillaris	6	11	8	6	4	35	2.2	51.4
12 Trifolium repens	2	7	9	3	14	35	2.2	53.7
13 Poa pratensis		6	9	6	12	33	2.1	55.8
14 Poa annua	17	6	1	4	2	30	1.9	57.7
15 Deschampsia flexuosa	16	3	7	1	2	29	1.9	59.6
16 Anthoxanthum odoratum	4	8	4	7	6	29	1.9	61.4
17 Holcus mollis	5	6	6	2	2	21	1.3	62.8
18 Mercurialis perennis	4	4	4	7	2	21	1.3	64.1
19 Campanula rotundifolia		2	10	5	3	20	1.3	65.4
20 Galium aparine		4	7	2	7	20	1.3	66.7
21 Taraxacum agg.	1	3	3	7	6	20	1.3	67.9
22 Polygonum aviculare	6	4	2	4	2	18	1.2	69.1
23 Stellaria media	4	6	1	4	1	16	1.0	70.1
24 Galium palustre	3	3	4	2	3	15	1.0	71.1
25 Rubus fruticosus agg.		5	2	2	6	15	1.0	72.0
26 Lotus corniculatus	1	1		9	4	15	1.0	73.0
27 Hyacinthoides non-scripta	4	2	3	3	1	13	0.8	73.8
28 Carex flacca	1	3	4	2	3	13	0.8	74.7
29 Juncus effusus	3	4	1	3	1	12	0.8	75.4
30 Geranium robertianum	4			5	3	12	0.8	76.2
31 Koeleria macrantha		3	2	3	4	12	0.8	77.0
32 Cardamine pratensis	3	1	2	1	4	11	0.7	77.7
33 Vaccinium myrtillus		4	2	2	2	10	0.6	78.3
34 Plantago major	1	1	2	5	1	10	0.6	79.0
35 Myosotis scorpioides	1		4	1	4	10	0.6	79.6
36 Lemna minor	8			1		9	0.6	80.2
37 Potentilla erecta		2	4	1	2	9	0.6	80.7
38 Elytrigia repens	I	1	1	4	2	9	0.6	81.3
39 Brachypodium sylvaticum			3	2	4	9	0.6	81.9
40 Heracleum sphondylium			1	3	5	9	0.6	82.5
41 Lolium perenne	1	1	2	4		8	0.5	83.0
42 Leontodon hispidus	2	1		5	8		0.5	83.5

result in a bias towards common and ecologically wide-ranging species: this is confirmed in Table 5.6. Although in total 147 species appear in the lists, eleven species contribute more than 50% of the records. The commonest associates are plants of pasture and wasteland.

Some practical uses may arise from the listing of commonest associated species in each of the Species Accounts. These are particularly obvious in relation to the choice of species for reclamation projects. Where, for example, dereliction of formerly species-rich pastures has resulted in losses of subordinates it may be helpful to consult the accounts for the dominants that remain in order to identify their missing associates. Similarly, in attempts to create sustainable and attractive plant assemblages for the purposes of amenity, reference to the lists may assist the choice of species.

The lists of commonest associates also have some relevance to academic studies of the mechanisms maintaining plant community structure. As explained in Panel 9, one of the main challenges in this field of research is to recognise and characterise the forces that promote trait convergence and those that are responsible for trait divergence. Used in tandem with trait databases, the lists of commonest associates are likely, therefore, to provide useful insights into the role of similarity and difference in the combining ability of species pairs. Such analysis will also suggest which of the co-existence mechanisms outlined in Panels 12–16 are implicated in particular common associations.

Implications for communities and ecosystems

Introduction

The standardized form of the information contained in the Species Accounts creates the possibility of analyses designed to recognize the functional characteristics and dynamics of plant communities. Here we are seeking to understand the properties of the plant community as an expression of measureable traits of its component species and populations. It is important to recognize that autecological data can be used not merely to forecast the different responses of component species to the conditions of climate, soil and management prevailing in the habitat, but also to predict the role of species interactions and regenerative processes in determining the structure and dynamics of vegetation and the relative abundance of species.

But green plants and their communities do not exist in isolation from the ecosystems they occupy. They provide the point of entry of resources into the ecosystem and through variation in the way that they utilise these resources different functional types exercise major effects on all the other ecosystem components and processes. In this rapidly developing area of basic and applied ecological research it has become a matter of vital importance to identify how variation in plant traits and plant functional types influences the structure and functioning of ecosystems. More specifically, we need to know which plant traits play an integral role in the ecosystem and which are mainly concerned with the composition and functioning of the plant community. It is a mistake to assume that all traits that affect the fitness of a plant species must also exert effects on the ecosystem; it is likely that much of the Darwinian struggle between contending plant populations to enter and persist in the plant community occurs between species with closely similar effects on ecosystem processes such as primary production, herbivory, decomposition, nutrient cycling, and carbon storage.

This is such an important issue and is so integral to the objectives of this book that it will now be addressed separately as a heavily-referenced topic in Panel 9.

Communities

Against the theoretical background set out in Panel 9, it is now possible to begin to consider what mechanisms are involved in the assembly of plant communities. Few plant assemblages consist of a single species although this can occur locally where all others are excluded by large plants such as *Fallopia japonica*, *Pteridium aquilinum* or *Urtica dioica*. It is much more common, even where the vegetation is dominated by potentially monopolistic species, for several species to co-exist and in some communities of the British Isles (Figure 3.4, page 24) as many as 40 herbaceous species may be found in each square metre of turf. Later in this section we will consider in turn the various factors conducive to species co-existence in plant communities. First, however, it will be useful to recognise three different roles that species can play in the plant community-dominants, subordinates and transients.

Dominants

Many of the Accounts draw attention to the fact that the subject species has a capacity for functioning as a dominant within plant communities. This usage of 'dominance' implies that the species not only makes a major contribution to the total biomass of the plant community, but also tends (as an individual or a population) to ramify throughout the edaphic and aerial environments and to influence the identity, quantity and local distributions of the associated flora and fauna. This argument leads to the proposition (Grime 1977) that dominance has two major components:

(a) The mechanism whereby the dominant plant achieves a biomass greater than that of its associates —this will vary according to the species and habitat conditions.

(2) The mechanism by which the dominant plant influences the fitness of its neighbours —this includes deleterious effects (e.g. resource depletion and release of phytotoxins) and promotional effects (e.g. the provision of surfaces for epiphytes, the release of substrates to parasites, symbionts and decomposing organisms and the production of food and shelter exploited by herbivores and their predators).

These definitions provide for considerable variety in types of vegetation dominant. The CSR model of primary strategies (p. 9) permits potential dominants to be distinguished from other plants (Fig. 5.7) with a subdivision into three functional classes which, for simplicity, may be described respectively as competitive, ruderal and stress-tolerant dominants. The differences between these three types are described by Grime (1979) and will not be reviewed in detail here. However, certain of the distinguishing characteristics of the three classes of dominant are essential to an analysis of their role in community dynamics. It is necessary to point out that although species exhibiting the competitive strategy (p. 46) are conspicuous among vegetation dominants, they achieve this status only in highly productive, undisturbed conditions. The full spectrum of vegetation dominants includes two other major groups of plants. The first consists of competitive-ruderals (e.g. *Anisantha sterilis* and *Impatiens glandulifera*) which, despite being annual, are capable of dominating productive habitats if they are subjected each year to a single, predictable major disturbance, such as winter flooding or autumn ploughing. The second group consists of stress-tolerant competitors; these may be herbaceous plants (e.g. *Carex acutiformis* and *Mercurialis perennis*), shrubs (e.g. *Ulex europaeus* and *Vaccinium myrtillus*) or trees (e.g. *Fagus sylvatica* and *Quercus* spp.) which, although relatively slow-growing, are capable of assuming dominance in moderately productive, undisturbed environments.

Subordinates

In Figure 5.7 it is suggested that species which are not capable of dominance fall into two broad classes. First, there are those (e.g. *Asplenium ruta-muraria*, *Saxifraga tridactylites*, and *Sedum acre*) which occupy extreme habitats where the intensities of stress or disturbance, or of various combinations of the two, are sufficient to exclude dominants. Second, there are those which by various means coexist with dominants. Some of these (e.g. *Anemone nemorosa*, *Galium aparine* and *Ranunculus ficaria*) have phenologies and morphologies which permit many of their individuals to escape the main impacts of the stresses generated by the dominants. However, others (e.g. *Campanula rotundifolia*, *Sanicula europea* and *Viola riviniana*) are suspected of being relatively tolerant of the shading and mineral nutrient depletions imposed by their larger neighbours. Complementing the definition of dominance (see above), therefore, we may describe subordinates

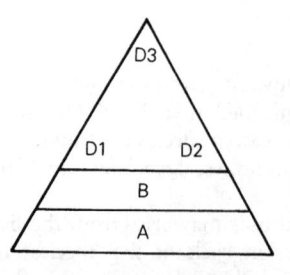

Figure 5.7 The distribution of three floristic elements within the triangular model (Fig. 2.2). Potential dominants are: D1, ruderal dominants; D2, competitive dominants; D3, stress-tolerant dominants. Area A is occupied by plants highly adapted to conditions that are extremely disturbed or unproductive, or both, and area B by subordinates. (From Grime 1985.)

PANEL 9 PLANT TRAITS, COMMUNITIES AND ECOSYSTEMS

Researchers in this field have inherited insights from two schools with very different philosophies. One, traceable to Darwin (1859) and implemented by Diamond (1975), has emphasised coexistence between organisms with differing traits or trait values as the key to the coherence and predictable composition of both animal and plant communities. The second, with roots extending back to the pioneers of plant geography, sociology and physiology, has drawn attention to the extent to which members of the same plant community often tend to exhibit similarity in plant traits. The objective in this panel is to suggest that research has now reached a stage where these two conflicting sources of evidence can be reconciled.

Trait convergence

Plant species coexisting in communities usually exhibit some conspicuous similarities in life history, morphology and physiology (Tansley 1939, Ellenberg 1963, Box 1981, Chabot & Mooney 1985, Rodwell 1992) but, this evidence of trait similarity within the community has been slow to impact upon theories of community assembly. An exception to this general pattern has been the early and widespread acceptance of a controlling effect of soil fertility on trait similarity within communities. Pearsall (1950), for example, observed that in productive and unproductive grasslands in Northern Britain coexisting species were similar with respect to the concentrations of mineral nutrients in their leaves and Parsons (1968), working in Australia, noted that the potential growth rates of sets of species coexisting in plant communities appeared to be similar and positively correlated with the fertility of the soils of their natural habitats. These relationships were later confirmed by investigations that involved many more species and by comparative experiments in which species from contrasted communities were examined under controlled laboratory conditions (Higgs & James 1969, Grime & Hunt 1975, Field & Mooney 1986, Lambers & Poorter 1992, Thompson et al. 1996b). At the same time the perspective has widened beyond leaf nutrient concentrations and plant growth rates to include other traits that vary in relation to vegetation productivity: these include leaf longevity, defence against generalist herbivores and rate of leaf litter decomposition (Reich et al. 1992, Sydes 1984, Rogers & Clifford 1993, Grime et al. 1997, Cornelissen et al. 1996, 1999). From these surveys it is apparent that, as foreshadowed in the reviews of Grime (1977) and Chapin (1980), the transition from productive to unproductive vegetation is associated with parallel shifts in a set of traits that are deeply embedded in the core physiology of plants, the structure and dynamics of the community and the functioning of the ecosystem. The consistency with which these predictable and parallel changes in the trait values of component species occur across the productivity gradient is apparent in three recent studies (Grime et al. 1997, Diaz et al. 2004, Wright et al. 2004) in each of which it is concluded that they represent a shift that simultaneously imposes upon plant individuals, plant communities and ecosystems a transition in resource dynamics from the acquisitive ('fast and leaky') to the retentive ('slow and tight'). Further support for the hypothesis that productivity acts as a filter selecting predictable sets of traits is available from experiments in which plant communities have been allowed to assemble from a large and functionally diverse pool of candidate species under contrasted conditions of soil fertility (Fraser & Grime 1999, Buckland & Grime 2000).

Although trait convergence within the community is most fully documented with respect to the impact of productivity examples can also be found with respect to other variables. For example, in circumstances where vegetation on fertile soils is completely destroyed at least once every year (e.g. arable fields) there is a strong tendency for communities to consist exclusively of ephemerals with early allocation to seed production.

Trait divergence

The idea that coexisting species within a plant community may be expected to have different biologies whereby resources are captured and exploited in different ways had an auspicious beginning:

'The truth of the principle, that the greatest amount of life can be supported by great diversification of structure, is seen under many natural circumstances. In an extremely small area especially if freely open to immigration, and where the contest between individual and individual must be severe, we always find great diversity in its inhabitants. For instance, I found that a piece of turf, three feet by four in size, which had been exposed for many years to exactly the same conditions, supported twenty species of plants , and these belonged to eighteen genera and to eight orders, which shows how much these plants differed from each other....the advantages of diversification of structure, with the accompanying differences of habit and constitution, determine that the inhabitants, which thus jostle each other most closely shall, as a general rule, belong to what we call different genera and orders.'

(Darwin 1859)

This idea that competition for resources has resulted in trait divergence and encouraged stable coexistence between organisms has been formalised as 'limiting similarity' (Diamond 1975, Pacala and Tilman 1994). According to this line of reasoning, past competition for the same resources between organisms has resulted in trait divergences that, now, as a legacy or 'ghost of competition past' (Connell 1980) reduce competition and permit coexistence often leading to more exhaustive use of resources and enhanced productivity.

It is, to say the least, unfashionable to question the conclusions of Charles Darwin but there can be little doubt that the Darwin-Diamond model of competition as the mainspring of trait variation within communities is not supported by empirical study of plant communities. Further, one may suspect that it is this hypothesis and the unresolved debates that it has provoked (Weiher & Keddy 1999) that provide an explanation for what Lewontin (1974) has graphically described as 'the agony of community ecology'. Trait variation is a common feature of species-rich plant communities but there is no convincing evidence that this phenomenon is a consequence of competitive interactions. To the contrary, field observations and experiments conducted over a long period of time (Tansley & Adamson 1925, Donald 1958, Mahmoud & Grime 1976, Kadmon & Shmida 1990, Janssens et al. 1998) confirm that trait variation and species-richness are suppressed by competition and promoted by natural disturbance events, animal activities and forms of vegetation management that reduce the vigour and competitive ability of potentially dominant species. As explained later, there is a large body of evidence implicating vegetation disturbance rather than competition as the most potent mechanism creating and sustaining trait variation in plant communities

Mechanisms responsible for trait convergence and divergence

There remains an acute shortage of standardised data on the responses of plants to specific climatic and edaphic factors and this contrasts strongly with the wealth of information with respect to seeds, germination and leaf characteristics. Despite this patchiness in the supply of reliable comparative data some clues are now available in the quest to distinguish between the circumstances conducive to trait convergence and those responsible for trait divergence within plant communities. In particular, insights can be obtained by examining the causes of the relative uniformity within the community of productivity-related traits and comparing this with the diversity often observed in regenerative traits. The value of such a comparison is underlined by the investigation of Thompson et al. (1996a) where screening of many traits in the same community at one particular site revealed that trait convergence in the established phase of the life-cycle coincided with trait divergence in the regenerative phase.

In seeking an explanation for convergent and divergent patterns of trait variation within the community it is informative to recognise a difference in the spatial and temporal scales at which productivity and disturbance impact on the plant community. With some well-defined exceptions such as the dissected communities of limestone pavement and rock outcrops, most semi-natural plant communities are relatively homogeneous and constant with respect to productivity. In marked contrast, a great variety of phenomena are capable of inflicting disturbance and there are wide ranges in the temporal and spatial scales at which particular forms of disturbance may operate. Thus it is possible, even within the same small area and over the passage of only a few years for vegetation to be exposed to many contrasted types of disturbance. As explained by Grubb (1977) vegetation disturbance has acted as a powerful evolutionary force, restricting the competitive effects of established plants and promoting the development of a rich variety of regenerative mechanisms capable of exploiting different opportunities for recruitment of juveniles. The traits involved include the size, shape, dispersal mechanism and dormancy of seeds (Grime & Jeffrey 1965, Thompson et al. 1993, Moles et al. 2000) and the physiological characteristics that influence the time and place of germination (Grime et al. 1981, Baskin & Baskin 1998). Examples of the extent to which variation in the agency, scale and timing of disturbance can bring together a diversity of regenerative traits within the same community are apparent in investigations such as those of Thompson and Grime (1979), Peart 1984, Masuda and Washitani (1990), Gigon and Leutert (1996) and Thompson et al. (1996b), in which plant species with contrasted regenerative mechanisms were found to be coexisting in communities.

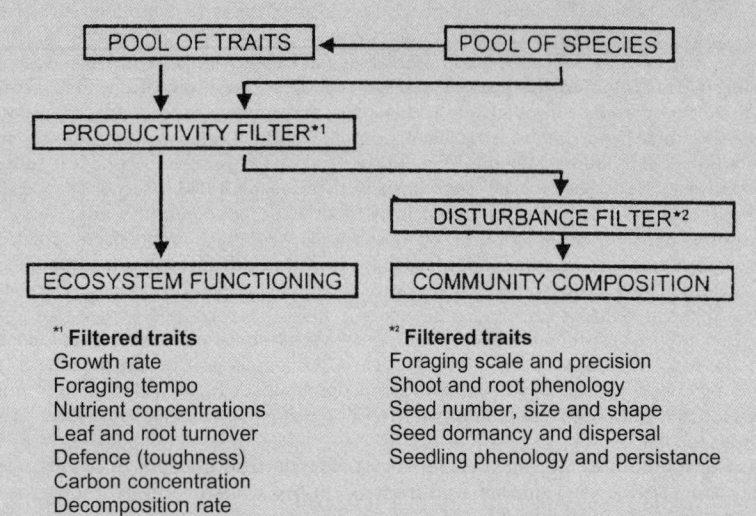

Figure A Scheme representing two of the filters and some of the traits involved in the assembly of plant communities and the functioning of ecosystems.

[*1] **Filtered traits**
Growth rate
Foraging tempo
Nutrient concentrations
Leaf and root turnover
Defence (toughness)
Carbon concentration
Decomposition rate

[*2] **Filtered traits**
Foraging scale and precision
Shoot and root phenology
Seed number, size and shape
Seed dormancy and dispersal
Seedling phenology and persistance

The scope for trait diversification as a consequence of disturbance extends beyond the regenerative phase. Variation in the intensity and spatial distribution of disturbance may permit coexistence between ephemeral and perennial life histories (Crawley & May 1987). There is experimental evidence (Campbell et al. 1991, Grime & Mackey 2002) that occasional grazing and mowing events are sufficient to allow coexistence between potential dominants and smaller subordinates in pastures and meadows It is also suspected that variation in the seasonal distribution of grazing, mowing and burning events over a number of years promotes trait and species diversity in grassland communities by sustaining differences in the phenology of leaf growth, photosynthesis, flowering and seed production (Al-Mufti et al. 1977, Grime et al. 1985, Bakker1989, Kahlert et al. 2005).

In this comparison of trait convergence and divergence strong emphasis has been placed on the relative constancy of productivity in space and time over the area occupied by a community and this has been contrasted with the capacity of disturbance to generate spatial and temporal heterogeneity within the community and to allow regeneration and persistence of a diversity of traits, trait values and species. It may be a mistake, however, to seek to explain the mechanisms responsible for trait convergence and divergence simply in terms of the diversity and stochasticity of disturbance events. It seems necessary also to recognise differences in the severity of the physiological and genetic constraints that operate in different stages of the life cycle and in different components of the plant's biology. The scope for variation in evolutionary responses to disturbance, through minor changes in germination biology or shoot architecture is likely to be considerably greater than that associated with responses to productivity which appear to involve relatively intractable, multigenic linkages and tradeoffs between different aspects of the core physiology of the plant.

Consequences for communities and ecosystems

Following the initiative of Schulze and Mooney (1993), many plant ecologists have been prompted to investigate the impacts of declining plant diversity upon the functioning of ecosystems. In a recent paper (Grime 2002) it has been argued that the random deletions of species often applied in diversity/ecosystem functioning experiments do not resemble the functional shifts in species composition and ecosystem properties that are the main processes by which the biosphere is changing under the impact of human activities. This criticism about the unsatisfactory nature of 'diversity' as an experimental variable connects with a more general unease about the way in which its use has sometimes diverted attention away from the primary task of elucidating the mechanism by which specific plant traits, individually or as sets, influence the development and functioning of communities and ecosystems. With these problems in mind, Figure A, applies our current understanding of the productivity and disturbance filters to examine their combined effects on community and ecosystem properties. It is expected that the productivity filter will exert a profound effect on both the community and the ecosystem by admitting or excluding traits, trait values and species that are directly implicated in dry matter production, carbon storage, nutrient cycling, trophic interactions and litter decomposition. In contrast, the influence of the disturbance filter will be a divergence in traits that are mainly confined to the Darwinian struggle for entry, persistence and relative abundance in the plant community. This lower impact of the disturbance filter on ecosystem functioning is predicted on the basis that it mainly selects between regenerative, architectural, life-history and phenological traits that have a lesser influence on the physical and chemical processes that drive ecosystems.

References and additional reading: Al-Mufti et al. (1977), Bakker (1989), Baskin & Baskin (1998), Box (1981), Buckland & Grime (2000), Campbell et al. (1991), Chabot & Mooney (1985), Chapin (1980), Connell (1980), Cornelissen (1996, 1999), Crawley & May (1987), Cunningham et al. (1999), Darwin (1859), Diamond (1975), Diaz et al. (2004), Donald (1958), Ellenberg (1963), Field & Mooney (1986), Fraser & Grime (1999), Fukami et al. (2005), Gigon & Leutert (1996), Grime (1977, 2002, 2006), Grime & Hunt R (1975), Grime & Jeffrey (1965), Grime & Mackey (2002), Grime et al. (1981, 1985, 1997), Grubb (1977), Higgs & James (1969), Janssens et al. (1998), Kadmon & Shmida (1990), Kahlert et al. (2005), Keddy (1992), Knevel et al. (2003), Lambers & Poorter (1992), Lewontin (1974), Mahmoud & Grime (1976), Masuda & Washitani (1990), Moles et al. (2000), Parsons (1968), Pacala & Tilman D (1994), Pearsall (1950), Peart (1984), Reich et al. (1992), Rodwell (1992), Rogers & Clifford (1993), Schulze & Mooney (1993), Stubbs & Wilson (2004), Sydes (1984), Tansley (1939), Tansley & Adamson (1925), Thompson & Grime (1979), Thompson et al. (1993, 1996a,b), Weiher & Keddy (1999), Wright et al. (2004)

as plants with specialisations of many different kinds which limit their ability to monopolize the edaphic and aerial environments. These specialisations may be in morphology, phenology or life-history and, at their most conspicuous, restrict the activity of their subject to parts only of the environmental mosaic or of the growing season, or both.

It is interesting to note that many of the most common plants of the British flora are species which are restricted to a subordinate role in plant communities. Grime (1985b) identifies the 24 most commonly occurring herbaceous plants of the Sheffield region, and classifies them into potential dominants and subordinates. From these data it is clear that some subordinates (notably *Cerastium fontanum, Plantago lanceolata* and *Poa trivialis*) are as widely successful as the most familiar dominants. Although many factors are likely to contribute to the fitness of these subordinates (e.g. genetic variation within and between populations, flexibility in mechanisms of regeneration, capacity to exploit the most extensive habitats and communities), an examination of the Accounts suggests that two additional phenomena should be taken into consideration in an attempt to explain their success:

Many subordinate species are capable of exploiting similar niches in each of a variety of habitats, and often appear in association with a variety of dominants. *Poa trivialis*, for example, occurs as a late vernal constituent in eutrophic woodlands, grasslands and marshes, in association with a very wide range of taller grasses, broad-leaved herbs and woody species.

The reduced morphology and restricted phenology of many of the most widely successful subordinate species reduces both the risk and severity of damage caused by factors such as drought, flooding, grazing, mowing and burning. Consequently, populations of subordinates may be expected to show a degree of homoeostasis which is greater than that of the dominants when vegetation is exposed to fluctuating conditions of climate and management.

Transients

In the vegetation surveys drawn upon in this book (pages 17–23), a strenuous effort was made to make a complete list of all the species present in each vegetation sample. When the resulting data were examined it became apparent that in addition to dominants and subordinates, most samples contained species that were represented as scattered seedlings and immature individuals. Many of these species were dominants or subordinates of different types of neighbouring plant communities. It seems necessary therefore to recognise a third vegetation component made up of transients originating from the seed rain from the surrounding landscape or from seed banks occurring as a legacy of previous occupants of the site.

In comparison with the dominants and subordinates, the role of the transients is ill-defined and it may be justified to accept Janzen's description of them as 'lost plants' (Janzen 1986) destined for an early death and failure to gain admission to the community. However, it is possible to view the presence of these 'misfits' in a more positive light. The modern landscape of Europe consists of a dynamic patchwork of habitats in which plant communities are constantly destroyed and reassembled. Where a rich diversity of transients are detectable this signifies the potential for rapid vegetation adaptation to new conditions and changes in management. Decline in the wealth of transients, on the other hand, may be taken as an indication of falling plant diversity at the landscape scale and a warning that processes of plant community reassembly may increasingly result in the recruitment of dominants and subordinates with poor fit to the prevailing habitat conditions.

Co-existence mechanisms: theory

It is now appropriate to move beyond broad categorisations and to ask how it is often possible for some plant communities to contain several dominants and a rich diversity of subordinates. A general response to this question is implicit in models such as the humped-back model (Grime 1973a, see Panels 10 and 11) and the centrifugal model (Keddy 1990, Wisheu and Keddy 1992) which seek to explain the creation of species-rich communities as a consequence of the operation of stress or disturbance factors that reduce the vigour and competitive impacts of dominants and simultaneously introduce a diversity of opportunities for other species through various processes of niche differentiation. Argument continues about the mechanisms whereby niches are created and community diversity is sustained in species-rich vegetation.

To provide an overview of this very active and diverse area of ecological research and to identify emerging lessons for conservation and management Panels 12-16 illustrate five distinct mechanisms that are implicated in niche differentiation and species co-existence. We can summarise the essential features of these mechanisms by recognising that plants may co-exist within the community by exploiting different places above or below ground (Panel 12), by growing or flowering at different times in the year (Panel 13), by developing different root and shoot architectures that result in differences in the scale and precision with which resources are captured (Panel 14), by exploiting different opportunities for regeneration (Panel 15) and by maintaining a high level of genetic diversity within component populations (Panel 16).

Evidence supporting these various theoretical mechanisms is included in the panels, which also contain some specific pointers for management, but we believe that it is helpful here to reiterate a general argument of crucial importance in Panel 9 and in the first paragraph of this section. **None of the mechanisms of co-existence described above (Panels 12-16), are by themselves capable of sustaining diversity. All, in various ways, depend upon the operation of stress or disturbance factors (many under human control) that constrain the competitive vigour of potential dominants and, in consequence, expose niches and permit them to be exploited. Where the objective in managing semi-natural vegetation is to maintain diversity it is therefore essential (see Panels 10 and 11) to restrict dominance.**

Co-existence mechanisms: real communities

In real communities it is likely that several of the mechanisms illustrated in Panels 12-16 operate simultaneously to sustain species diversity. For both academic and applied purposes it would be very helpful if the identity of the mechanisms promoting species co-existence in particular communities could be recognised by examining the traits of constituent species. An example of this type of analysis is provided in Table 5.7 in which two samples of vegetation from grassland communities of contrasted productivity are compared with respect to the identity, potential relative growth rates, nuclear DNA amounts and regenerative strategies of their principal constituent species. As might be expected, the species of the two communities differ consistently with respect to potential relative growth rate but within each growth rates are fairly constant. In marked contrast, however, it is apparent that within both communities there is considerable variety in nuclear DNA amount and in type of seed bank. On this basis the tentative hypothesis may be advanced that differences in phenology (Panel 13) and in mechanisms of regeneration (Panel 15) but not in growth rate contribute to niche differentiation and species coexistence in both communities. Thompson *et al.* (1996) describe a similar situation in which the basis of species-richness in a

Table 5.7 Attributes of the most common species in two contrasted grassland communities.

Species	Frequency	RGR (week^{-1}) <1.0	RGR (week^{-1}) >1.0	DNA (pg) <7.5	DNA (pg) >7.5	Seed bank Transient	Seed bank Persistent
Productive pasture							
Festuca rubra	98		+		+	+	
Poa trivialis	92		+	+			+
Lolium perenne	90		+		+	+	
Agrostis capillaris	49		+	+			+
Dactylis glomerata	49		+		+	+	
Holcus lanatus	49		+	+			+
Cynosurus cristatus	32		+	+		+	
Trifolium repens	15		+	+			+
Unproductive pasture							
Vaccinium myrtillus	97	+		+			+
Deschampsia flexuosa	67	+			+	+	
Festuca ovina	63	+			+	+	
Nardus stricta	63	+		+		+	
Luzula campestris	15		+	+			+
Galium saxatile	8	+		+			+
Anthoxanthum odoratum	6	+			+		+

limestone grassland appeared to reside in a complex array of regenerative mechanisms.

In seeking to explain co-existence mechanisms in particular communities it can be profitable to compare the life-histories and established strategies of co-occurring species. This will often indicate communities in which floristic diversity is maintained by the presence of a sub-structure of microhabitats. In situations where ephemeral species are recorded within perennial herbaceous communities this is particularly obvious. In unproductive calcareous grasslands, for example, the ephemerals may be exploiting small areas of droughted shallow soil (Grubb 1976) which provide semi-permanent gaps of fixed position within a matrix of perennial grass cover. On deeper and more-fertile soils the co-occurrence of ephemerals and perennials is more likely to be associated with the successional dynamics of patches which arise, for example, from the spatially unpredictable impacts such as those caused by moles and the larger grazing animals.

As explained in Panel 13, additional insights into mechanisms of species coexistence in plant communities may be gained from the phenological data contained within the Accounts. Many communities can be identified in which the component species exhibit major differences in the timing of shoot growth, flowering, seed dispersal or germination. Where evidence of this type is obtained, it often points to the existence of temporal partitioning of resources or of opportunities for regeneration.

Nuclear DNA amount provides a useful indicator of the timing of shoot growth, and differences in DNA amount between species in a plant community may provide an index of the extent to which differences in the phenology of shoot growth contribute to niche differentiation and species coexistence. Figure 5.8 shows a clear association between DNA amount and leaf extension rate in the early spring in a species-rich grassland community.

Within individual Accounts further examples of inferences based upon nuclear DNA amounts are considered. Here it suffices to draw attention to the strong possibility that differences in DNA amount provide a clue to the role of phenological differences in the 'combining ability' of some familiar grassland co-dominants. The complementary seasonal peaks of growth in ryegrass-white clover swards coincide with a considerable difference between these species (*L. perenne* DNA = 9.9 pg, *T. repens* = 3.0 pg). A similar pattern is evident (Al-Mashhadani 1980) within the red fescue-brown bent grasslands of upland Britain (*Festuca rubra* DNA = 13.9 pg, *Agrostis capillaris* = 5.6 pg). In lowland limestone or chalk grassland phenological differences consistent with those predictable from nuclear DNA amounts have been observed between the co-dominants *Bromopsis erectus* (DNA = 22.5 pg) and *Brachypodium pinnatum* (2.3 pg) (Law 1974).

Community dynamics

As explained in the preceding section, the Accounts can provide

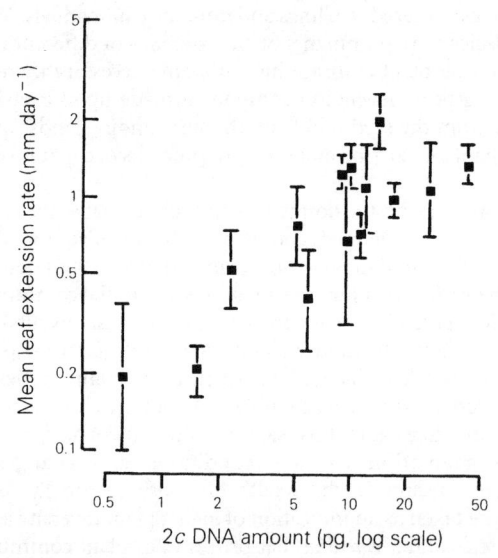

Figure 5.8 The relationship between DNA amount and the mean rate of leaf extension over the period 5 April–25 May in 14 grassland species coexisting in the same turf. The 95% confidence limits are indicated by vertical lines. (Redrawn from Grime 1983.)

PANEL 10 THE HUMPED-BACK MODEL: SOME REFINEMENTS

The humped-back model (Grime 1973 a,b, Figure A) originated as an attempt to summarise in one diagram the interaction of five factors (abiotic stress, vegetation destruction, dominance, niche differentiation and the species pool) in controlling the diversity of vascular plant species in herbaceous vegetation. One of the main implications of the model was the prediction that species richness would exhibit a unimodal pattern of variation where gradients occurred in the above-ground mass of living and dead plant material. The first attempt to quantify this relationship was reported in the large-scale investigation of Al-Mufti et al. (1977). Grace (1999) provides an account of subsequent investigations testing the validity and generality of the model. His review notes that, while many authors found a unimodal relationship between species-richness and the mass of living and dead plant material above the ground, other studies failed to observe this pattern. There are several possible explanations for these results. First, and this applies particularly to cases where measurements have been restricted to one type of habitat, the range of conditions may have been too narrow, corresponding to only the up- or down-slope of the model. Alternatively, the sampling may have been applied to immature or erratically-managed vegetation where conditions did not allow an equilibrium to be established between the local species pool and the conditions of resource supply and management. Thompson et al. (2005) suggest that a similar situation is arising in some diversity experiments when plant communities synthesised from seed are terminated and their results published prematurely.

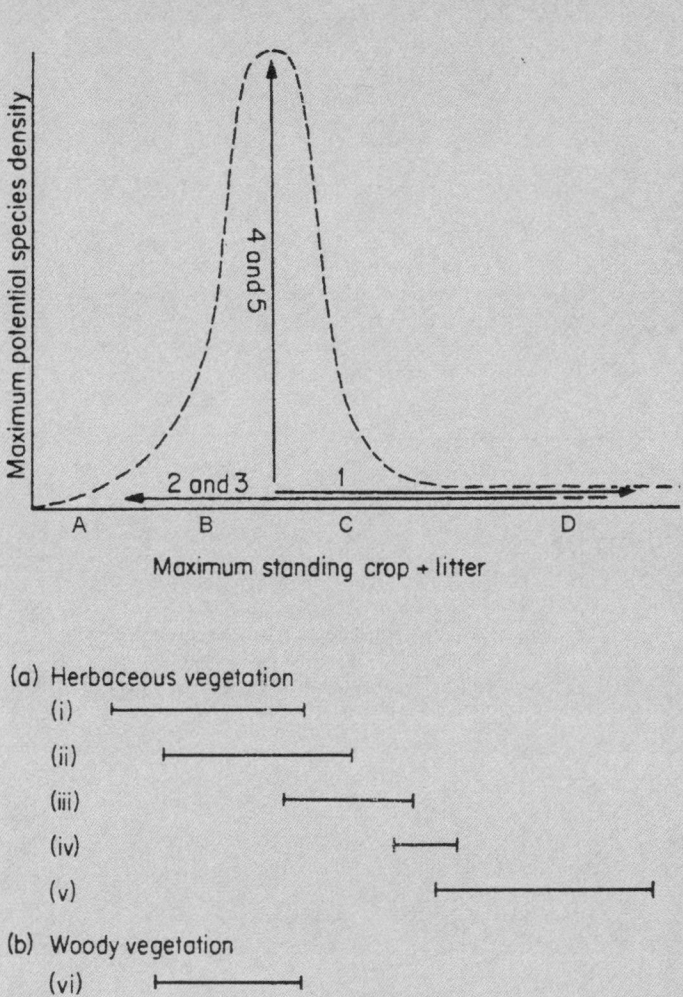

In a recent investigation (Shanta et al. 2007), an additional factor has been found to affect the precise form of the unimodal relationship. This phenomenon was observed in a survey of species-richness conducted in a local area of ancient landscape in North Derbyshire containing a wide diversity of habitats and plant functional types. The results are consistent with the model but show differences between habitats with respect to the number of vascular plant species persisting in conditions that allow the accumulation of high biomass and litter. On deep, fertile soils, dominance by tall, fast-growing grasses is associated with low species-richness. In neighbouring habitats on shallow, infertile soils the number of vascular plant species remains higher despite the fact that the mass of living and dead material, mainly composed of bryophytes, is closely similar in amount and more persistent than that of the fertile habitats. It would appear from these data that the decline in species-richness seen on the right-hand slope of the model is only partially explained by the physical effects of the accumulated mass and the capture of resources by the dominant species. An additional factor, specifically associated with high soil fertility (Henry et al. 2004), is the dynamic and hazardous nature, for smaller individuals, of the competitive neighbourhoods dominated by large, fast growing perennials.

Figure A Model summarising the impact of five processes upon species richness in vegetation. Key to processes: 1, dominance; 2, stress; 3, disturbance; 4, niche differentation; 5, ingress of suitable species or genotypes. The horizontal lines describe the range of contingencies encompassed by a number of familiar herbaceous and woody vegetation types. Key to vegetation types: (i) paths, (ii) grazed outcrops with discontinuous soil cover, (iii) infertile pastures, (iv) fertilised pastures, (v) derelict fertile pastures, (vi) arctic scrub, (vii) temperate scrub, (viii) temperate hedgerows, (ix) species-rich tropical rainforest, (x) mature temperate or tropical forest on fertile soil. (Reproduced from Grime 1979.)

References and additional reading: Al-Mufti et al. (1977), Grace (1999), Grime (1973), Henry et al. (2004), Shanta et al. (2007), Thompson et al. (2005)

---

some understanding of the complementary seasonal patterns of growth and regeneration which permit species coexistence in certain plant communities. The same information is helpful also in attempts to analyse longer-term (successional) change in vegetation and to predict community responses to perturbations arising from climatic events, herbivory, pathogens and impacts of management.

At an extremely simple level, the identification of the established strategies of the species represented within a community can provide a basis for predictions of community stability (Leps et al. 1982, MacGillivray et al. 1995, see Panel 17) and has particular relevance to forecasts of the sensitivity of plant communities to climate change (Panel 18). In particular, an essential distinction can be made between *community resistance* to perturbation (which is low where most of the component species are ruderals or competitors, but higher where stress-tolerators predominate) and *community resilience* (which is greatest in ruderals and lowest in stress-tolerators).

Further insights into responses to perturbation are available by reference to the regenerative strategies of plants. In general,

PANEL 11 THE HUMPED-BACK MODEL: THE COST OF PROTECTING GRASSLAND DIVERSITY

Efforts to protect declining populations of animals and plants often involve management agreements that include financial compensation to farmers but there is accumulating evidence (Vitousek *et al.* 1997) that these measures are failing to arrest the losses of wildlife that arise from continuing intensification of land use.

In a recent collaboration between ecologists and economists in England, Spain and Argentina (Hodgson *et al.* 2005c), this problem has been investigated by calculating the economic benefits of grassland intensification across the widest possible range in productivities and farming systems and matching these profits against the coincident losses in plant diversity.

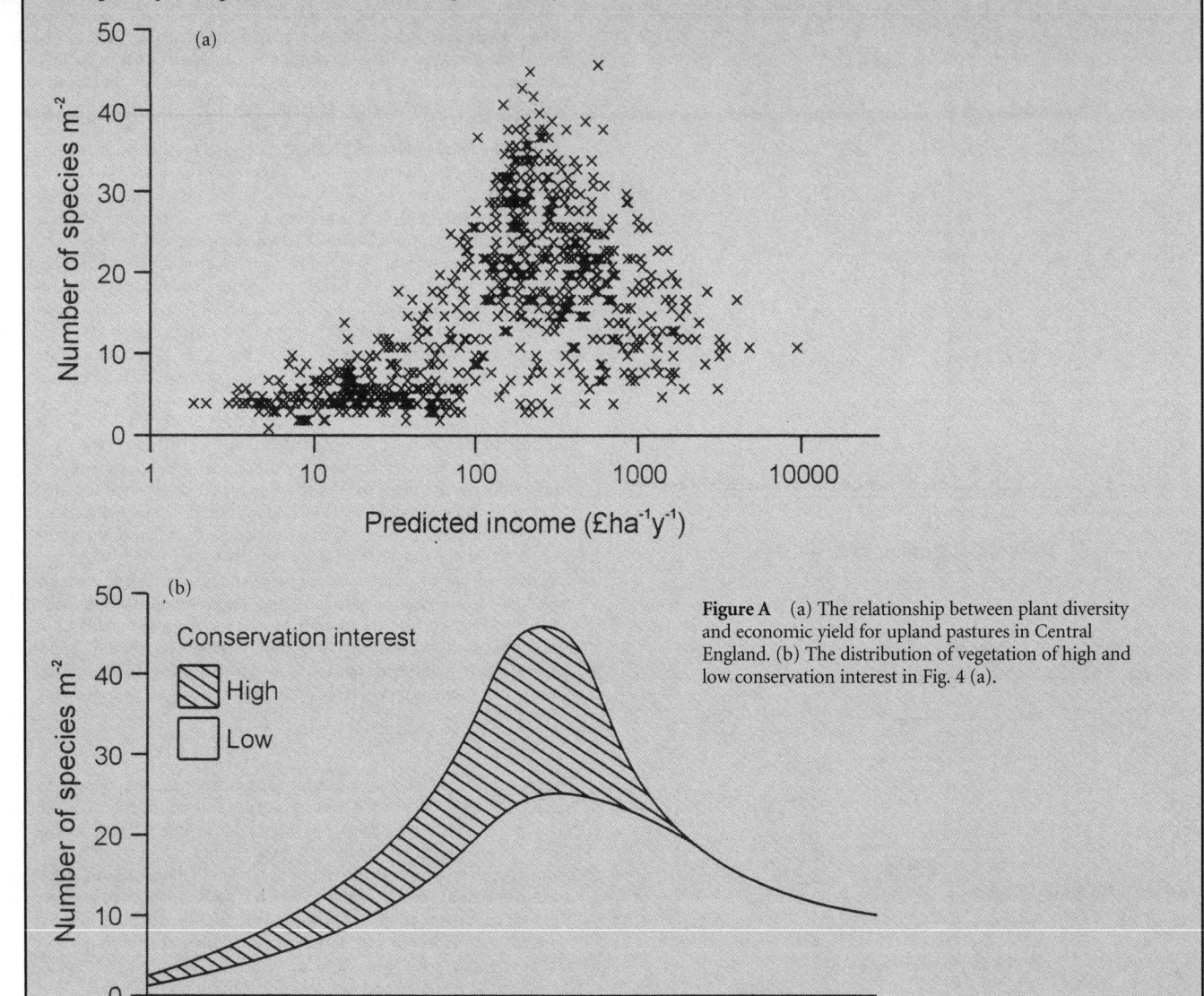

Figure A (a) The relationship between plant diversity and economic yield for upland pastures in Central England. (b) The distribution of vegetation of high and low conservation interest in Fig. 4 (a).

The results, here represented by data collected from pastures in North Central England (Figure A), show remarkable consistency between the three areas of study. It is shown that increasing fertility coincides with an exponential increase in livestock carrying capacity and in marginal returns. Across the same gradient, however, high levels of species diversity are confined to less fertile soils, with low stocking rates and low income to the farmer.

This demonstration of the negative relationship between agricultural production and diversity is consistent with the extensive European grassland survey of Janssens *et al.* (1998) but departs from the conclusions drawn by Hector *et al.* (1999) who claimed to have detected benefits to productivity in diverse plots of grassland recently synthesised from seed (Thompson *et al.* (2005) explain this anomaly as a consequence of the extreme immaturity of the experimental plots).

The current losses of diversity through intensification are a relatively new phenomenon. In earlier centuries pastoral agriculture operated primarily on the left-hand (ascending) slope of the humped-back model where rising productivity and biodiversity coincide. However, in Britain and in many other parts of the world, technology has taken farming on to the right-hand (descending) slope with damaging consequences for diversity. This process is difficult to arrest by financial compensation because the returns to the farmer from intensification appear to continue to rise exponentially across the range of soil fertilities corresponding to the right-hand slope.

References and additional reading: Hector *et al.* (1999), Hodgson *et al.* (2005c), Janssens *et al.* (1998), Thompson *et al.* (2005), Vitousek *et al.* (1997)

PANEL 12 CO-EXISTENCE MECHANISMS: SPATIAL NICHES

The most species-rich communities of vascular plants in the British Isles occur in unfertilised chalk or limestone pastures with shallow soils. Early investigations (Grime 1963a,b, Grubb *et al.* 1969) provide circumstantial evidence that a significant contribution to the diversity of these communities arises from the presence of both shallow- and deeply-rooting species. In Northern Britain the shallow-rooted component often includes marshland species such as *Succisa pratensis, Carex panicea* and *Carex pulicaris*. As illustrated below (Figure A), in the relatively high rainfall conditions prevailing in Britain, it is not unusual for the surface layers of calcareous soils to become acidic and for shallow-rooted calcifuge species such as *Calluna vulgaris, Hypericum pulchrum* and *Potentilla erecta* to be locally abundant.

In an investigation conducted at Buxton in North Derbyshire in the drought year of 1995 (Buckland *et al.* 1997) it was established that both marsh plants and calcifuge species were differentially susceptible to desiccation. More recently in experiments using rain shelters to impose even more severe droughts at the same site it has been confirmed (Figure B) that shallow-rooted marsh plants such as *Carex panicea* and calcifuges such as *Potentilla erecta* suffer high rates of mortality in conditions in which deeper-rooted species such as *Festuca ovina* remain relatively unaffected and may even increase in abundance.

References and additional reading: Buckland *et al.* (1997), Grime (1963a,b), Grubb *et al.* (1969)

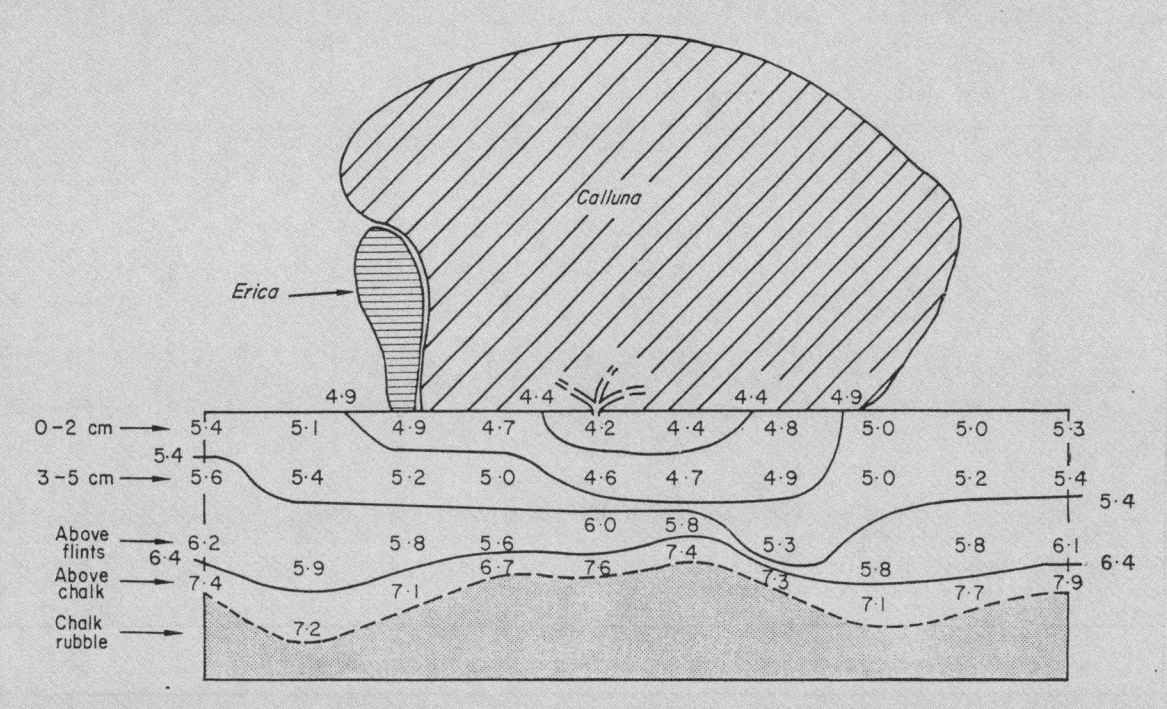

Figure A The pH of the mineral soil in a transect through an isolated bush of *Calluna* with a suppressed plant of *Erica cinerea*. Horizontal scale same throughout and same as vertical scale above soil. Vertical scale within soil exaggerated up to 2.5 times the horizontal. Height of bush was 55 cm and pH determination was 15 cm apart (Grubb *et al.* 1969).

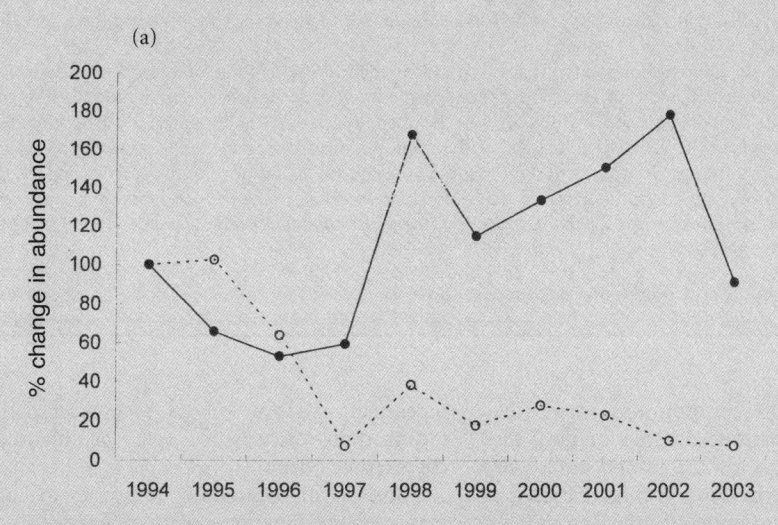

Figure B Changes in the abundance of (a) *Carex panicea*. (Figure and caption continued on page 70.)

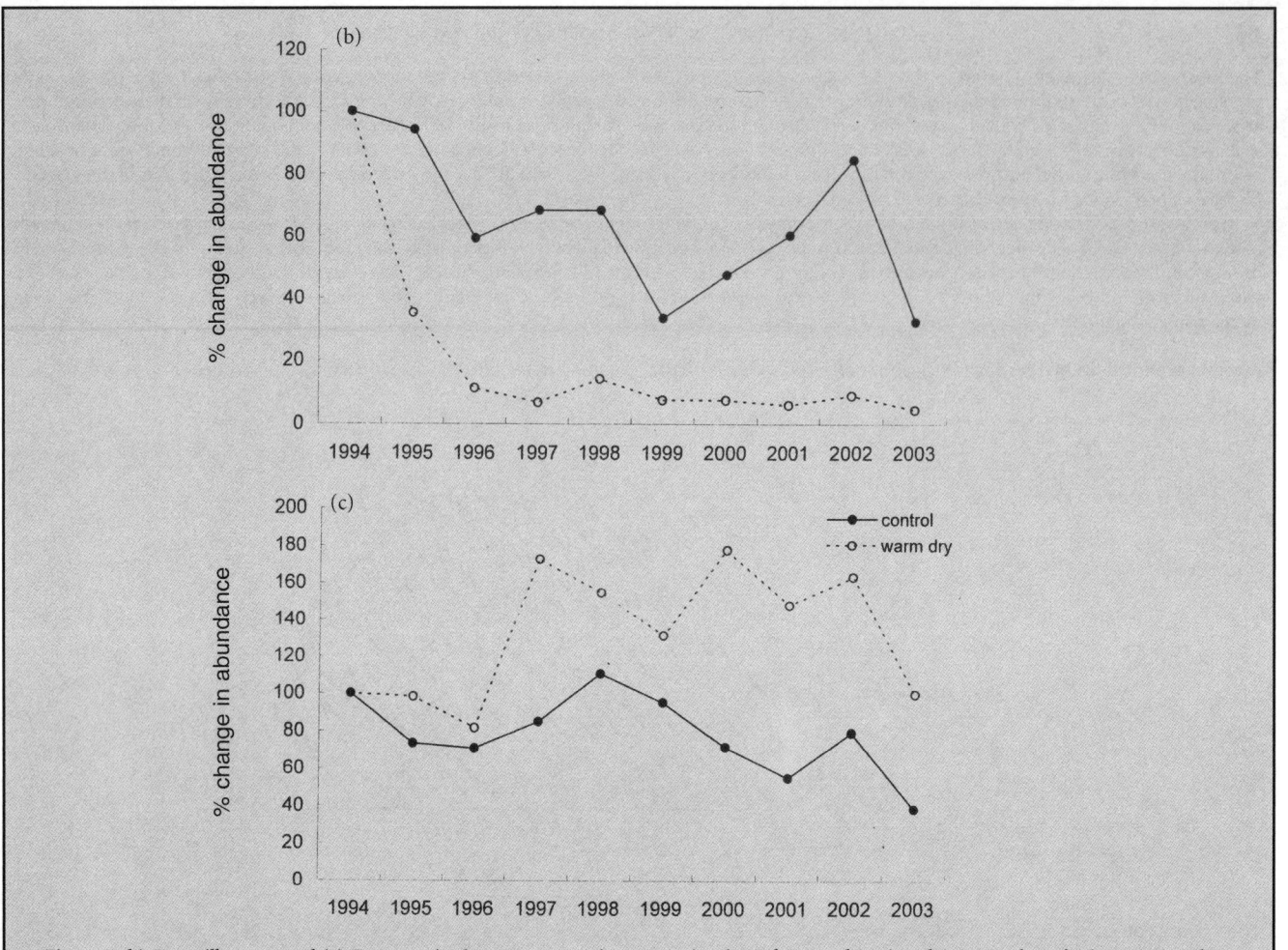

Figure B b) *Potentilla erecta* and (c) *Festuca ovina* in response to winter warming (+3° above ambient) and summer drought (no rain in July and August) in limestone grassland at Buxton in North Derbyshire.

PANEL 13   CO-EXISTENCE MECHANISMS : PHENOLOGICAL NICHES

In cool temperate climates such as those of Western Europe shoot expansion and flowering vary widely even between plant species co-existing within the same community. However, it is important to recognise (Al-Mufti *et al.* 1977) that such phenological differences cannot by themselves guarantee that a community will become species-rich. As discussed in Panels 10 and 11, in circumstances which support the growth of robust 'monopolists' and allow the accumulation of a large mass of plant material, diversity is likely to fall despite the fact that the climatic conditions are suitable for a wide range of both spring and summer phenologies.

Co-existence between plants that grow at different times of the year is most likely therefore where the total biomass of the community remains low either because the soils are too poor or because the vegetation is subject to grazing, cutting or some other form of biomass removal.

With the onset of global warming, there is an obvious need to consider how the phenologies of co-existing species will be altered. How will the balance between spring- and summer-growing species be affected? Research on this topic requires that we develop a physiological understanding of the differences in growth processes that allow some species to expand their tissues under the cold conditions of the early spring whilst in others growth is delayed until temperatures rise. Earlier in this chapter it is suggested that differences in cell and genome size provide a valuable clue to the different mechanisms of growth of spring and summer species (see Figure 5.8, page 66) and we forecast that eventually databases of plant genome size will play a useful role in predicting sensitivity to climate change.

References and additional reading: Al-Mufti *et al.* (1977), Grime and Mowforth 1982.

---

resilience will be most marked in communities where the component species maintain persistent seed banks or have the capacity to re-sprout from buds at or below the ground surface. As might be expected, both of these mechanisms of regeneration are strongly represented in the weed communities of annual crops (e.g. *Agrostis gigantea*, *Elytrigia repens*, *Papaver rhoeas* and *Persicaria* spp.), and persistent seed banks are particularly important in perennial vegetation subject to cyclical changes resulting from intermittent disturbance (e.g. heathland communities and coppiced woodland).

Forecasts of community responses to changes in vegetation management may be possible through judicious use of the Accounts. For example, reference to the potential relative growth rates, phenologies and established strategies of constituent species

PANEL 14   CO-EXISTENCE MECHANISMS: FORAGING NICHES

Since the pioneering studies of mature, perennial plant community structures by RH Whittaker (Whittaker 1965) it has been recognised that plant species tend to occupy a characteristic position within the biomass hierarchy of the community, allowing classification into dominant and subordinate roles. Progress has also been made in identifying some of the plant traits that confer the potential to dominate a plant community. From measurements in the field (Grime 1973a, Grubb et al. 1982) and in laboratory experiments (Gaudet and Keddy 1988) it is evident that dominant plants are usually the tallest but this leaves unresolved the question as to whether this is a cause or consequence of competitive success. Even more contentious is the need to explain why many species consistently occupy subordinate positions in communities. Are these plants 'also-rans' in the struggle for existence or are there fitness benefits associated with subordinate status?

Two possible mechanisms, neither exclusive of the other, have been suggested (Grime 1985b) whereby populations of subordinates may achieve high fitness. First, they may compensate for their lower abundance in the community by an ability to co-exist with a wide range of dominants in a variety of plant communities. Second, many subordinates appear to be able to flourish intermittently when the vigour of the potential dominants is temporarily reduced by impacts of weather or management. As research has continued we have come to rely on a further hypothesis-that the dominant/subordinate distinction is genetically predetermined and relates to a trade-off between the scale (high in dominants) and precision (high in subordinates) of foraging for resources by the shoots and roots. Experimental support for this hypothesis is available (Figure A) from experiments (Campbell et al. 1991, Grime and Mackey 2002) in which the foraging behaviour in a standardised patchy environment has been assayed in the shoot and root systems of 43 species as part of the Integrated Screening Programme (Panel 5).

The results from these studies reveal that, with some well-defined exceptions, there is a close correspondence between the shoots and the roots of each species in foraging characteristics. In vegetation dominants e.g. *Arrhenatherum elatius* the scale of foraging is high, indicating the potential to exploit, and eventually monopolise, a large volume of habitat above-and below-ground. In contrast, the smaller shoot- and root-systems of subordinates e.g. *Campanula rotundifolia* exhibit fine-grained foraging in which the surfaces of the plant are concentrated in local areas of higher resource quality both above and below ground. In Figure A, the scale of shoot and root foraging is compared in 8 species which were subsequently allowed to compete in equi-proportional mixture for 16 weeks. The resulting data confirm the correlation between shoot and root foraging behaviour and the ability of foraging information to predict the status of species in a community. The same conclusion has been drawn from the comprehensive screening of root development in patchy nutrient conditions conducted on more than 50 species from the Great Plains Grasslands (Johnson & Biondini 2001, Levang-Brilz & Biondini 2002).

We conclude, on the basis of the surveys of foraging now available for a substantial number of species, that a tradeoff between scale and precision does exist and can influence co-existence within communities. For the purposes of management, however, it is essential to recognise that, by itself, the tradeoff cannot sustain diversity for an indefinite period. Particularly under conditions of high productivity, suppression of the subordinates will occur unless the vigour of the dominants is constrained by biomass removal.

References and additional reading: Campbell et al. (1991), Gaudet & Keddy (1988), Grime & Mackey (2002), Grime (1973a, 1987), Grubb et al. (1982), Hutchings & deKroon (1994), Johnson & Biondini (2001), Levang-Brilz & Biondini (2002), Whittaker (1965)

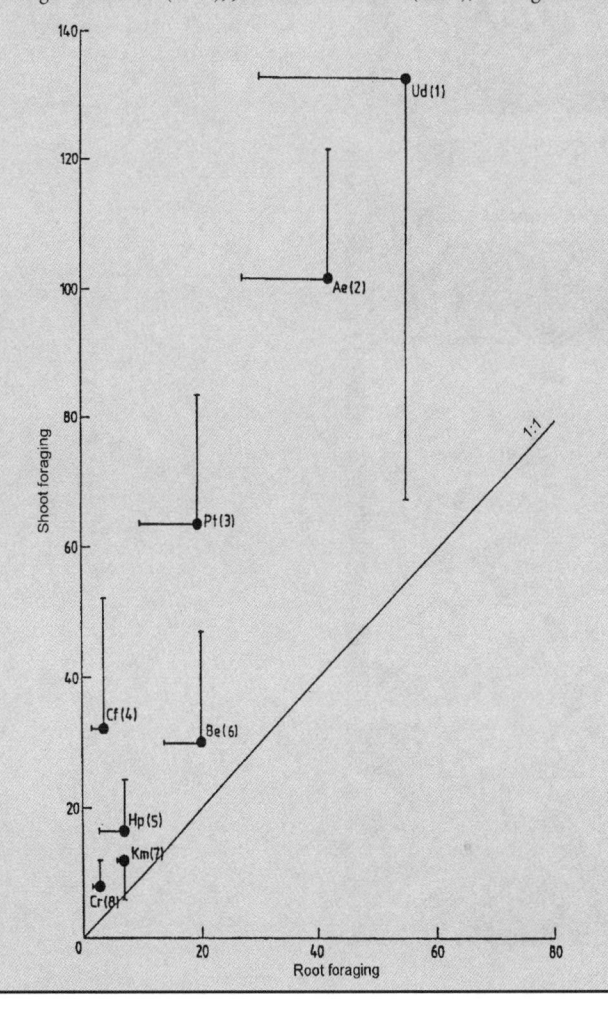

Figure A An examination of the relationship between root and shoot responses to resource heterogeneity in eight herbaceous species of contrasted ecology. A description of the methods used to expose the plants to resource patchiness is provided in Grime (2001). Scales of foraging by the roots and shoots in the foraging assays are expressed as the respective increments of biomass (mg) to two undepleted quadrants which in both assays constitute 50% of the available volume. The numbers in brackets refer to the species ranking in a conventional competition experiment in which all eight species were grown together in an equiproportional mixture on fertile soil for 16 weeks. The vertical and horizontal lines refer to 95% confidence limits. The species are: Ae, *Arrhenatherum elatius*; Be, *Bromopsis erecta*; Cf, *Cerastium fontanum*; Cr, *Campanula rotundifolia*; Hp, *Hypericum perforatum*; Km, *Koeleria macrantha*; Pt, *Poa trivialis*; Ud, *Urtica dioica*. (Reproduced from Campbell et al. 1991a by permission of *Oecologia*.)

INTERPRETATION AND USE OF THE SPECIES ACCOUNTS

PANEL 15 CO-EXISTENCE MECHANISMS : RENERATION NICHES

More than a quarter of a century has passed since the publication by Peter Grubb (Grubb 1977) of his admirable and much-cited paper explaining the role of juvenile plant strategies in the maintenance of diversity within plant communities.

A remarkable feature of the ecological literature is the way in which his interpretation has withstood the test of time; it is hard to find any authors who have rejected his hypothesis and in Panel 9 considerable space is devoted to placing the regeneration niche within the wider context of the mechanisms of plant community assembly.

With the continuing accumulation of standardised information on the seeds, seed banks and germination biology of plant species there are increasing opportunities for vegetation managers to control the composition and diversity of plant communities by influencing regeneration. A particularly good illustration is provided by Thompson *et al.* (1996) in which laboratory and field information was available on the regeneration characteristics of most of the component species in a species-rich grassland community. This analysis revealed dimensions of variation in seed bank type and germination requirements within the community that, in theory, could provide a basis for manipulation of both floristic composition and species-richness.

References and further reading: Grubb (1977), Lavorel *et al.* (1994), Thompson & Grime (1979), Thompson *et al.* (1996)

PANEL 16 CO-EXISTENCE MECHANISMS: GENETIC VARIATION

Antonovics (1976) commented that 'Forces maintaining species diversity and genetic diversity are similar' and he forecast that 'An understanding of community structure will come from considering how these kinds of diversity interact'. In order to test this idea it is necessary to conduct experiments that involve controlled manipulation of genetic-diversity in plant communities and study of it's consequences for vegetation composition and species diversity. This kind of experiment necessitates the use of populations of long-lived individuals with low rates of replacement by sexually derived progeny. To be in a position to synthesise communities with controlled levels of genetic diversity it is necessary first to create large stocks (a 'living library') of genetically identical individuals of each of the species and genotypes to be used. Current experiments in Sheffield utilise a library resulting from clonal propagation of 176 randomly selected cuttings (16 of each of 11 species) all from within the same 10m x 10m area in an ancient species-rich calcareous pasture in North Derbyshire. The results of the first experiment using these materials (Booth & Grime 2003) have revealed a surprising amount of functional diversity within coexisting plant populations and have confirmed (Figure A) that genetic diversity can have strong effects in maintaining the composition and species diversity of a grassland community.

References: Antonovics (1976), Booth & Grime (2003)

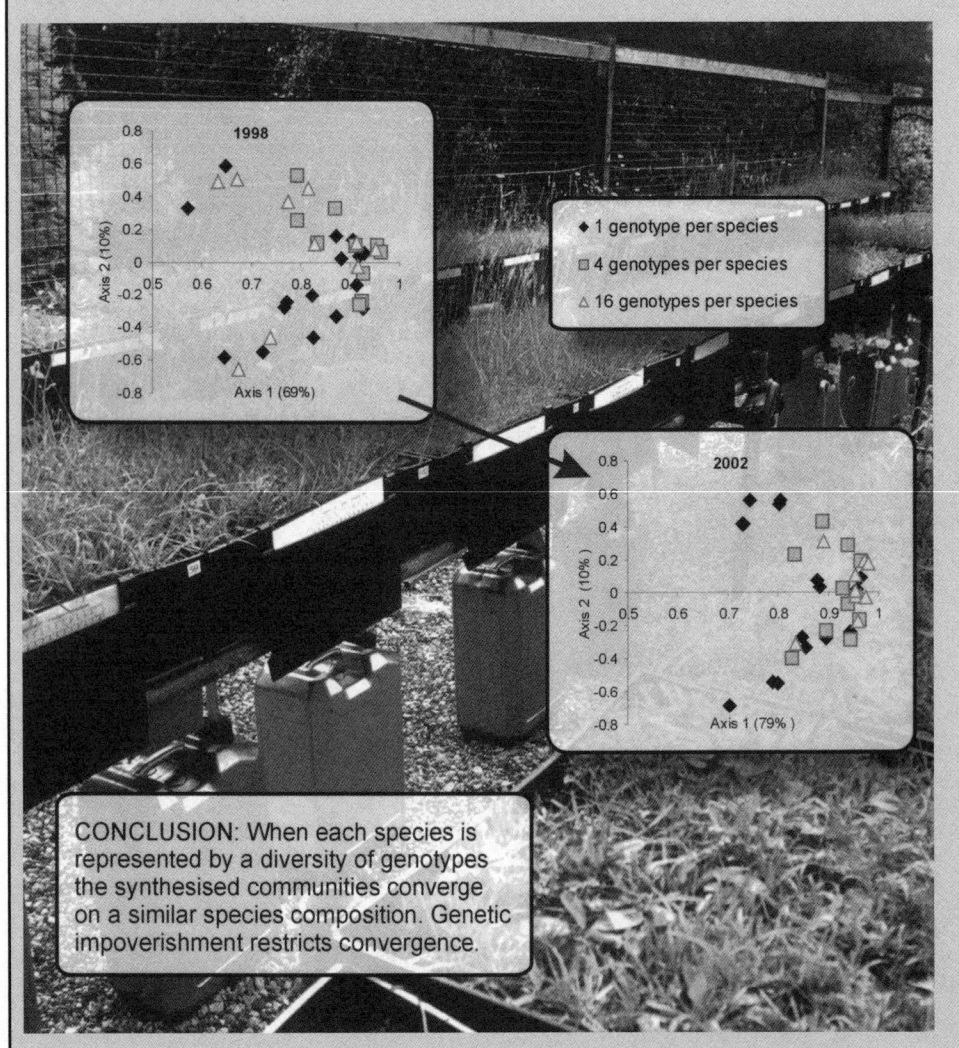

Figure A Changes in the species composition of synthesised plant communities of differing genetic diversity in all component species over a period of five years.

CONCLUSION: When each species is represented by a diversity of genotypes the synthesised communities converge on a similar species composition. Genetic impoverishment restricts convergence.

> **PANEL 17   COMMUNITY RESPONSES TO EXTREME EVENTS**
>
> For ecological purposes it is useful to define extreme events as episodes in which biological systems are briefly exposed to mechanical or climatic conditions that exceed their recent experience. At the level of both the plant community and the whole ecosystem it is essential to develop a broad understanding of the impacts of extreme events. This has become an urgent issue for two main reasons:
>
> 1. As explained in Panel 8, modern landscapes with high densities of humans are subjected to a high frequency of disruptive impacts of changing patterns of land use.
> 2. Theoretical models concerning the future course of climate change (Mearns *et al.* 1984, Wigley 1985) predict that there will be an increasing frequency of extreme climatic events in the 21st century.
>
> If extreme events can arise from such a diversity of causes it might be expected that attempts to develop general guidelines to the sensitivity of plant communities will be fraught with difficulties. This rather pessimistic assumption relies on the hypothesis that plant sensitivity will vary widely and unpredictably according to the nature of each extreme event. However, field observations (Leps *et al.* 1982) and experiments (MacGillivray *et al.* 1995) suggest that a basis for prediction is available if the CSR model of plant functional traits is employed to forecast both the resistance of vegetation to an extreme event and the speed and completeness of recovery (resilience). The predictions arising from CSR theory are specific and propose
>
> 1. a trade-off between the two sets of traits associated, respectively, with resistance and resilience.
> 2. a predictable relationship between resistance and the set of traits identified with stress tolerance and between resilience and the set of traits associated with ruderals and, to a lesser extent, competitors.
>
> Why should we expect stress-tolerators to exhibit greater resistance to extreme events? Here, a key factor relates to the ecological and evolutionary penalties that would follow if stress-tolerators had low resistance to damage. These would arise from the inevitably low capacity for replacement of lost tissues and lost individuals in these slow-growing plants. On this basis we can expect that many stress-tolerators will be resistant to several kinds of extreme events, particularly those to which they have been exposed during their evolutionary history. In marked contrast to the stress-tolerators, greater resilience is likely to occur in ruderal and competitive plants as a simple consequence of their faster growth rates.
>
> It would be a mistake to rely exclusively on CSR theory to predict the resistance or resilience of vegetation. In particular circumstances other criteria will be available and should be utilised. Seed bank composition, for example, can provide useful predictions (van der Valk & Davis 1976) of the speed and direction of vegetation recovery following an extreme event.
>
> **References and further reading:** Buckland *et al.* (1997), De Angelis (1980), Leps *et al.* (1982), MacGillivray *et al.* (1995), Mearns *et al.* (1984), Tilman & Downing (1994) van der Valk and Davis (1976), Wigley (1985)

may allow a prediction of the rate and direction of the changes in species composition which would occur as a consequence of the removal of grazing animals from pastures. By the same approach it may be possible also to recognize those species which are most likely to increase in abundance when the productivity of a community is increased by the addition of fertilizers. Where management changes lead to loss of diversity and the disappearance of particular species, reference to the Accounts should help to identify which of the missing species are likely to be persistent in the habitat as populations of buried seeds. Where the aim of management is to restore such species, the Accounts will in some cases provide pointers to the measures necessary to encourage their re-establishment.

The Accounts may also be useful where the objective is to achieve a more subtle manipulation of the relative abundance of constituent species. Reference to phenological information can ensure that the seasonal distribution of grazing, mowing, burning or growth-regulator applications is optimal when efforts are being made to suppress particular dominants. In a similar way, the timing of management procedures can be arranged to minimize harmful effects on seed production and seedling establishment in species for which population expansions are desirable. Where vegetation contains a high proportion of relatively short-lived species there may be additional scope for 'informed' manipulation of species composition by creating patches of bare soil at times propitious for the germination and establishment of particular species.

Ecosystems
The mass ratio hypothesis

Each plant population, however localised or small in total biomass, will eventually contribute to the functioning of the ecosystems in which it occurs or into which it deposits mobile plant products in the form of litter, leachates, fruits or seeds. This has been evident since the earliest days of ecological enquiry. Much more problematic and lacking in definition are questions relating to the ways in which variation in the functional characteristics of a plant community influences specific ecosystem processes.

Before addressing these issues, however, it is necessary to consider the more general question:

'Is there any way to recognise which members of a plant community are exerting the greatest impact on ecosystem processes?'

In order to answer this question it is useful first to be reminded that ecosystem processes such as primary production, transfers of resources to herbivores and their predators and parasitoids, nutrient cycling, energy flows, carbon storage and various decomposition processes are all strongly dependent on the quantity of material and energy captured and released from the plant biomass. This leads to the proposition (Grime 1998) that the extent to which a plant species affects ecosystem functions is likely to be closely predictable from its contribution to the total plant biomass. This 'mass ratio' hypothesis is implicit in many commentaries and models relating to ecosystem function (Shugart 1984, Pastor and Post 1986, Huston and Smith 1987, Grime 1987, Sala *et al.* 1992, Huston 1997) and is dictated by laws that require that large effects of autotrophs within ecosystems involve major participation in syntheses, and in inputs to resource fluxes and degradative processes. It follows that ecosystem properties should be determined to a large extent by the characteristics of the dominants and will be relatively insensitive to variation in species richness in circumstances where this is attributable to changes in the number of subordinates and transients. It is important to specify that the mass ratio hypothesis is restricted in application to the role of autotrophs in ecological processes. It is conceivable that in other trophic elements, such as parasites, herbivores, predators and symbionts, impacts on ecosystems may be less predictably related to abundance.

PANEL 18 PLANT COMMUNITY RESPONSES TO CLIMATE CHANGE

In addition to impacts arising from the increased frequency of extreme weather events such as droughts, floods and windstorms (Panel 17), the long-term consequences of climate change for vegetation are expected to include those due to slow but progressive changes in atmospheric carbon dioxide concentration, temperature and rainfall (IPCC, 2001). Plant communities will differ in their responses to these changes. In the search for guidelines for management strategies, difficulties are arising from the need to understand long-term responses across a wide range of environments and land-uses. It is inevitable that, faced with such complexity, we will rely upon theoretical models. This highlights the desirability that such models are tested and supported by reference to carefully-planned and strategically-located long-term experiments on real ecosystems.

An example of the potential of field experiments to assist research on the impacts of climate change is available from identical temperature and rainfall manipulations (Grime *et al.* 2000) conducted simultaneously on two contrasted calcareous grasslands situated at Buxton in North Derbyshire (Fig. A) and at Wytham near Oxford. Results from the first 5 years of this experiment (Figure B) revealed a profound difference between the two sites in vegetation responsiveness to the temperature and rainfall treatments. Under the conditions of high soil fertility at Wytham, transformations of the flora occurred through rapid changes in species abundance but only minor shifts occurred in the ancient, unproductive grassland at Buxton. This evidence is consistent with theory in the sense that two vegetation types occupying opposite extremities of Axis 1 of the CSR model (Panel 5) would be expected to differ profoundly (Panel 17) in their rates of response to the experimental manipulations. These results are also in agreement with the broader interpretation of current vegetation changes in the British Isles presented in Panel 8.

Figure A Manipulating of rainfall and temperature at the Buxton Climate Change Impacts Laboratory in North Derbyshire, England. Top: the imposition of summer drought by automatic rainshelters; these are activated by a rain-sensor to slide over the plots during rain showers and conduct water downslope through pipework. Bottom: elevation of winter temperatures by 3°C using heating cables fastened to the soil surface.

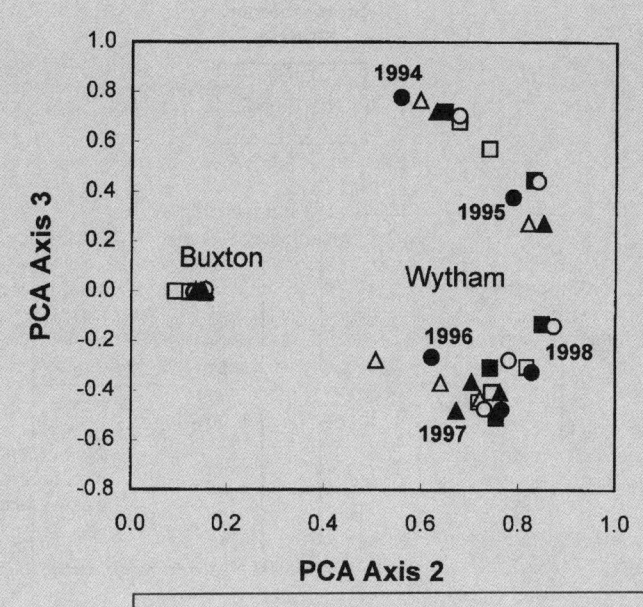

Figure B Principal components analysis of the combined floristic data collected from two contrasted grasslands subjected to identical climate manipulations over a five-year period (1994–98). Proportion of variance accounted for by the PCA axes: Axis 1 (not shown), 46%; Axis 2, 27%; Axis 3, 12%. Treatments: ■, control; ▲, winter warming; ●, summer drought; □, supplementary summer rainfall; △, winter warming + summer drought; ○, winter warming + supplementary summer rainfall. (Reproduced from Grime et al. 2000 by permission of Science.)

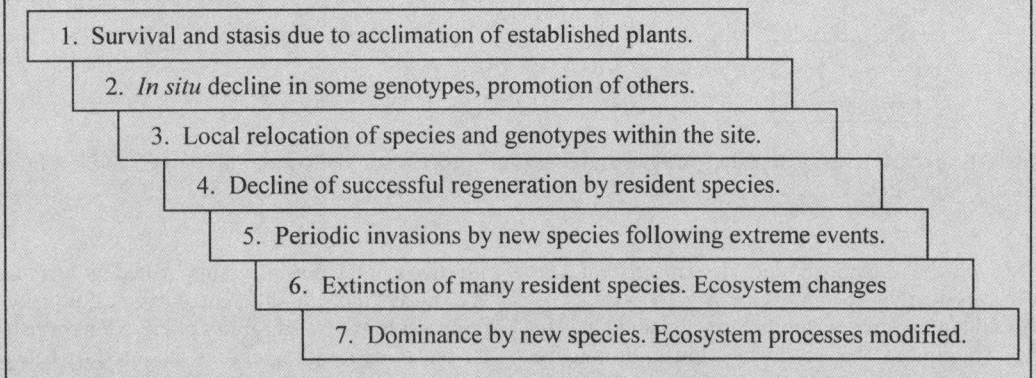

1. Survival and stasis due to acclimation of established plants.
2. *In situ* decline in some genotypes, promotion of others.
3. Local relocation of species and genotypes within the site.
4. Decline of successful regeneration by resident species.
5. Periodic invasions by new species following extreme events.
6. Extinction of many resident species. Ecosystem changes
7. Dominance by new species. Ecosystem processes modified.

Figure C Conceptual framework of the processes predicted to characterise the response of a perennial plant community to climate change.

The resistance to climate variation of the vegetation at Buxton raises important questions concerning the eventual fate in a changing climate of the large areas of Upland Britain that are currently occupied by ancient unproductive vegetation and provide refuges for a high proportion of rare and endangered plants and animals. In attempts to forecast the future of this major resource, key considerations are the long life-span and climatic tolerance of the stress-tolerators that dominate the vegetation of infertile soils. In the scheme presented in Figure C an attempt is made to identify some of the processes that are likely to characterise the long-term responses of unproductive perennial vegetation to a changing climate. It is suggested that the initial impacts will be inconspicuous as long-lived individuals survive change by the seasonally-reversible molecular adaptations that are a characteristic feature of their tissues. This phase may be expected to give way eventually to differential promotion of favoured genotypes and local redistribution of species to less extreme microhabitats. Later still it is forecast that extreme events may cause extinctions, facilitate invasions and cause changes in both vegetation structure and ecosystem processes.

But vital unanswered questions persist. What will be the rates of these various processes and what will be the synergies between them? Who will pay for the experiments and who will have the dedication to see them through?

References and further reading: Grime *et al.* (2000, 2003), IPCC, 2001.

A growing body of experimental evidence supports the mass ratio hypothesis. In a study of resistance and resilience to drought, late frost and fire (MacGillivray *et al.* 1995) predictions using traits measured in the laboratory were found to be accurate when calculations were based upon means weighted according to the relative abundance of each plant species in the experimental plots. Similar conclusions were drawn from recent investigations (Wardle *et al.* 1997, Hooper and Vitousek 1997, Garnier *et al.* 2004) in which various ecosystem properties were found to be strongly correlated with the functional characteristics of the dominant contributors to the biomass.

Plant traits that drive ecosystem processes
One of the main conclusions in the argument advanced in Panel 9, page 63, is that the plant traits that have the most powerful controlling effects on ecosystem processes are those that vary in relation to vegetation productivity and are subject to selection forces exerted by soil fertility. More specifically it is concluded from comparative studies (e.g. Grime *et al.* 1997, Diaz *et al.* 2004, Wright *et al.* 2004) that as we move from productive to unproductive vegetation (ie. from side C-R to corner S in the CSR triangular model, see Panel 5) there are parallel shifts within a set of traits including potential growth rate, leaf longevity, leaf chemistry, defence against generalist herbivores and litter decomposition rate. These traits all have a deep involvement in the

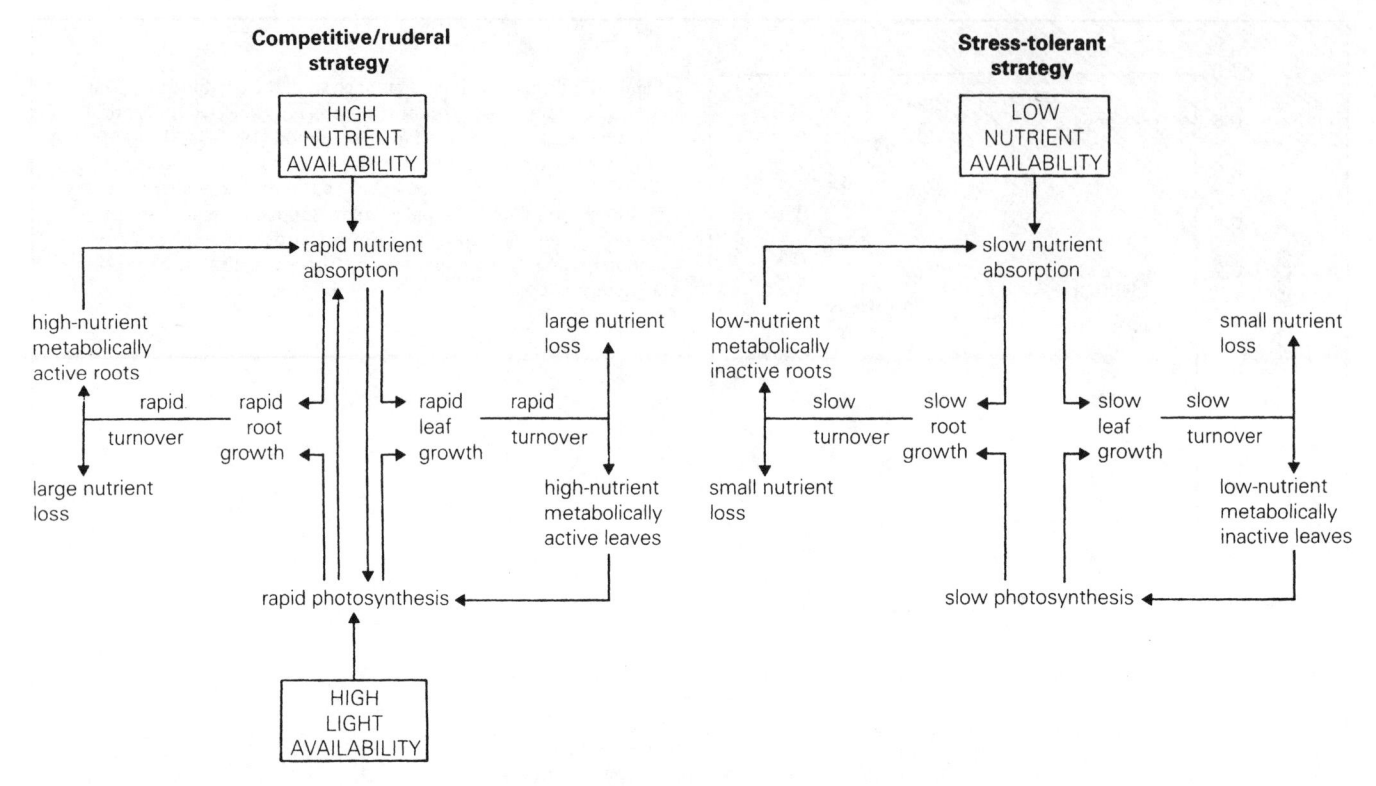

Figure 5.9 Interacting characteristics of plant strategies and ecosystem processes under conditions of high and low nutrient availability. (From Chapin 1980.)

central processes of the ecosystem and determine that the transition from productive to unproductive ecosystems is associated with a shift in resource dynamics from 'fast and leaky' to 'slow and tight'. This means that productive and unproductive habitats act as filters selecting contrasted sets of plant trait values that have very different effects that, respectively, sustain high and low productivity. The operation of the contrasted feedbacks of widely different plant trait values on productivity at opposite ends of the first axis of the CSR model (see Panel 5) was captured effectively in the flow diagrams created by Terry Chapin (Fig. 5.9). We suggest that the differences in rates of processes and flux of resources in the two diagrams of Fig. 5.9 have general implications for our understanding of how the interplay between productivity and plant traits sets the tempo for aspects of ecosystem functioning as diverse as the retention of radionucleides (Panel 19) and the characteristics of butterflies (Panel 20).

One of the most important implications of Figure 9 concerns the quantity and quality of carbon stored in the soil. The rapid

---

PANEL 19 PLANT FUNCTIONAL TYPES CONTROL THE TEMPO OF ECOSYSTEMS: EVIDENCE FROM CHERNOBYL

In 1986, a large cloud of radioactive material released from an explosion at the Chernobyl nuclear facility in the Soviet Union, deposited radiocaesium (^{137}Cs) at various locations including some hill farms in areas of Northern Britain. Unusually high levels of ^{137}Cs were detected in vegetation and sheep carcasses and the marketing of animals was suspended. Predictions of the time that it would take for the contamination to fall to safe levels proved to be highly optimistic and following a British Government Inquiry there has been recognition of the long residence time of pollutants such as ^{137}Cs in living and dead organic components of unproductive ecosystems.

Table A identifies several features of upland pastures that may account for the persistence and slow dynamics of ^{137}Cs following this incident. It seems clear that the characteristics of stress-tolerant plants and their residues have played a key role in the slow tempo of ^{137}Cs circulation and release from the contaminated areas.

Table A Some characteristics of the stress-tolerant vegetation of upland pastures and their potential effects on ^{137}Cs sequestration.

Plant characteristics	Implications for capture and persistence of ^{137}Cs
1 Long-lived, slow-growing, lichens, bryophytes, herbs and low shrubs	All capable of incorporating ^{137}Cs into long-lived tissues
2 Mycorrhizal infection present in the roots of most of the vascular plants	Mycorrhizas have a high affinity for metals and form an effective network throughout the surface soil
3 Low palatability of most of the species	Low sheep densities and slow rate of depletion of ^{137}Cs in plants
4 Slow decomposition of plant litter	Slow rate of depletion of ^{137}Cs especially in mycorrhizal residues

Reference: Grime (1988)

PANEL 20   PLANT FUNCTIONAL TYPES CONTROL THE TEMPO OF ECOSYSTEMS: EVIDENCE FROM BUTTERFLIES

There is a large volume of circumstantial evidence implicating plant functional types in the control of ecosystem processes. The consistent patterns of variation in leaf palatability and decomposition rates strongly suggest how plant traits might exercise control over the identity and activity of organisms in other trophic levels of the ecosystem. Two papers (Hodgson 1993, Dennis et al. 2004) provide rare, unusually detailed and comprehensive support for the hypothesis that plant traits can influence the functional characteristics of the animals that exploit them. In the second of these two investigations the authors summarised their findings as follows:

Butterfly biology is linked to host-plant strategies. An increasing tendency of a butterfly's host plants to a particular strategy biases that butterfly species to functionally linked life-history attributes and resource breadth and type. In turn, population attributes and geography are significantly and substantially affected by host choice and the strategies of these host plants…..high host-plant C and R strategy scores bias butterflies to rapid development, short early stages, multivoltinism, long flight periods, early seasonal emergence, higher mobility, polyphagy, wide resource availability and biotope occupancy….. Increasing host-plant S strategy scores have reversed tendencies, biasing those butterfly species to extended development times, fewer broods, short flight periods, smaller wing expanse and lower mobility, monophagy, restricted resource exploitation and biotope occupancy, closed, areally limited populations with typical metapopulation structures, sparce distributions, and limited geographical ranges, range retractions, and increased rarity.

References: Hodgson (1993), Dennis *et al.* (2004)

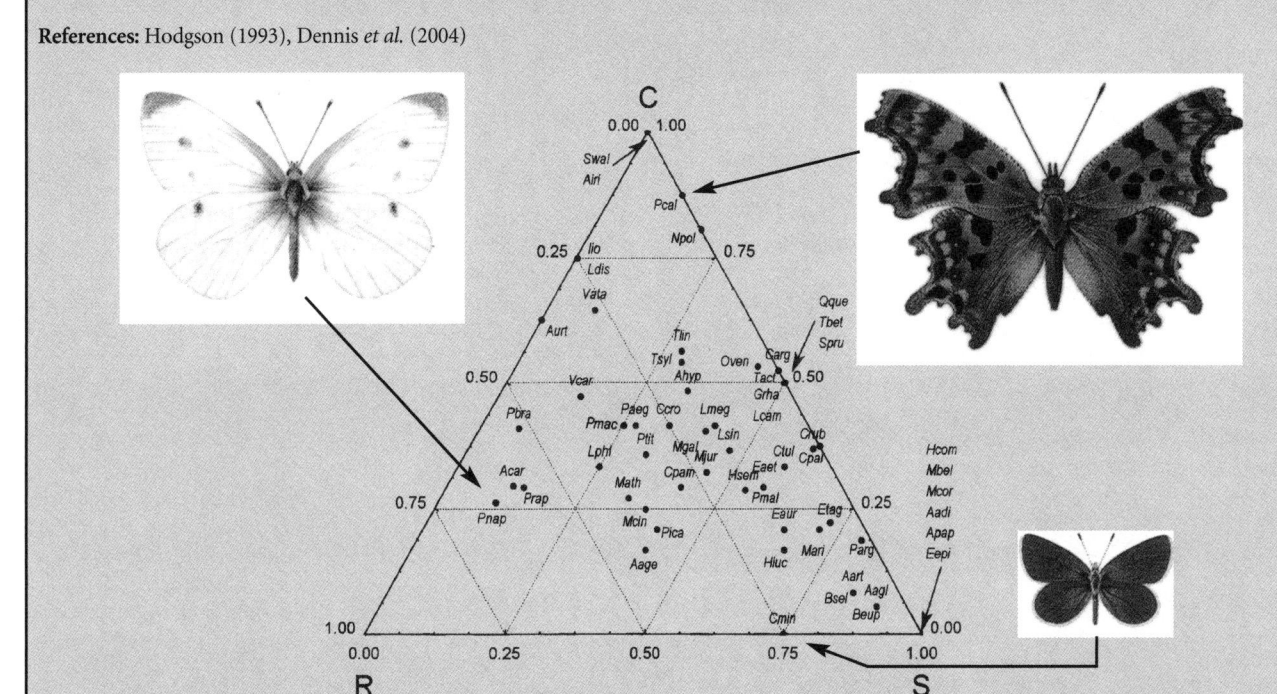

Figure A   British butterflies in positions corresponding to the functional type of the food plant exploited by their caterpillars (Dennis et al. 2004).

replacement of mature vegetation dominated by stress-tolerators by early-successional communities of R- and C-strategists is destined to significantly reduce the capacity of soils to maintain a large stock of relatively inactive carbon. This development may be expected to release carbon dioxide to the atmosphere and to accelerate the rate of global warming.

The most immediate consequences of the feedbacks between plant traits and ecosystem functioning in Fig. 5.9 will be their tendency to conserve particular combinations of ecosystem properties and trait values in the dominant components of the vegetation. We know, however, that both ecosystems and their plant communities are subject to functional shifts during primary and secondary successions and following various changes in climate and management regime. For the purposes of ecosystem management, therefore, it is necessary to identify under what circumstances the natural tendency for stasis will be overcome and both vegetation and ecosystem functioning will move together toward a new equilibrium point. At the present time there is a developing consensus among ecologists about the mechanisms most responsible for functional shifts in vegetation across the world. Clearly some of the changes are autogenic in the sense that they arise from sequestration of resources in the plant biomass or enrichment of the total stock of resources though processes such as nitrogen fixation. To an increasing extent, however, change is driven by external impacts of climate and man either through effects on resource supply or vegetation disturbance. This is a large and complex subject beyond the scope of this book but we cannot resist one observation with particular reference to the British Isles. In the search for a general perspective with regard the current agents of change in ecosystem structure and functioning we urge upon fellow ecologists a theoretical framework in which, following the scheme in Figure 5.9 and embracing the neglected philosophy of TCR White (1993) we grant to mineral nutrients (rather than energy) the seminal role in our attempts to understand the changes in resource currencies that are currently driving change in terrestrial ecosystems in Britain and further afield.

In this book we have argued that it is now feasible to unify our understanding of vegetation processes by implementing functional classifications of plants. If this target is achieved we believe that the development of functional typologies for heterotrophs is inevitable and will provide the next step toward the greater goal of a universal functional classification of ecosystems for the purposes of conservation and sustainable exploitation.

A brief guide to the Species Accounts

Nomenclature Principal name of the subject species; authority; any alternative names(s); family and common name(s). (Explanation p. 31, interpretation p. 34.)
Life-form Plant form (e.g. tufted/ stoloniferous); longevity (e.g. annual/perennial); Raunkiaer life-form; shoot, leaf and root characters (including size and mycorrhizal status) (Explanation p. 32, interpretation p. 41.)
Phenology Whether winter green; period of flowering and fruiting; leaf longevity (if known). (Explanation p. 31, interpretation p. 42.)
Height Maximum height of leaf canopy and inflorescences. (Explanation p. 32.)
Seedling RGR Mean relative growth rate of subject's seedlings under standardised, productive conditions (six classes of 0.5 week^{-1}). (Explanation p. 32, interpretation p. 42.) Estimations of SLA and DMC. (Explanation p. 32, interpretation p. 43.)
Nuclear DNA content Values of 2C nuclear DNA amount assessed by Feulgen microdensitometry; chromosome numbers and estimated ploidy. (Explanation p. 33, interpretation p. 44.)
Established strategy Classification with respect to the seven primary or secondary CSR strategies of the established phase: C, competitive; S, stress-tolerant; R, ruderal; C-R, competitive ruderal; S-R, stress-tolerant ruderal; C-S, stress-tolerant competitor; C-S-R, C-S-R 'strategist'. (Explanation p. 32, interpretation p. 46.)

Regenerative strategies Listing of regenerative strategies exhibited by the subject, any of: V, vegetative expansion; S, seasonal regeneration; B$_s$, regeneration from a persistent seed bank; W, regeneration involving numerous widely dispersed propagules; additional descriptive text; classification, into seed bank types I-IV. (Explanation p. 33, interpretation p. 47.)
Flowers Floral structure; pollination; number and arrangement of flowers. (Explanation p. 36; interpretation p. 50.)
Germinule Air-dry weight, physical dimensions and form of the germinule and (where different) the dispersule; fruit characters (including number of seeds per fruit). (Explanation p. 36, interpretation p. 50.)
Germination Epigeal/hypogeal form; time taken for half of the sample to germinate in the most successful test; temperature range over which at least 50% of the maximum observable germination occurs in the light; where appropriate, method of breaking dormancy; comparison of germination in light and darkness. (Explanation p. 36, interpretation p. 51.)
Biological flora References to published Biological Floras and to other important autecological works. (Explanation p. 37, interpretation p. 53.)
Geographical distribution Notes on the British, European and world distributions of the subject. (Explanation p. 37, interpretation p. 53.)

Habitats Interpretation of data presented in the 'Habitats diagram' (q.v.), with additional information drawn from local [2] and national [3] sources. (Explanation p. 37, interpretation p. 54.)
Gregariousness Interpretation of data presented in the 'Abundance in quadrats' diagram (q.v.), discriminating between: (1) sparse, (2) intermediate, and (3) patch-forming species; inferences regarding competitive exclusion. (Explanation p. 37, interpretation p. 55.)
Altitude Interpretation of information presented in the 'Altitude diagram' on the right-hand page (q.v.) [2] with maximum altitude in Britain [3]. (Explanation p. 37, interpretation p. 55.)
Slope Interpretation of information presented in the 'Slope diagram' on the right-hand page (q.v.). (Explanation p. 37, interpretation p. 56.)
Aspect Interpretation of the information presented in the 'Aspect diagram' on the right-hand page (q.v.). (Explanation p. 37, interpretation p. 57.)
Hydrology Interpretation of information presented in the 'Hydrology diagram' on the right-hand page (q.v.). (Explanation p. 38, interpretation p. 57.)
Soil pH Interpretation of information presented in the 'Soil pH diagram' on the right-hand page (q.v.). (Explanation p. 38, interpretation p. 57.)
Bare soil Interpretation of information presented in the 'Bare soil diagram' on the right-hand page (q.v.). (Explanation p. 38, interpretation p. 57.)

Associated floristic diversity Interpretation of information presented in the 'Associated floristic diversity diagram' on the right-hand page (q.v.). (Explanation p. 39, interpretation p. 58.)
△ ordination Interpretation of information presented in the 'Triangular ordination diagram' on the right-hand page (q.v.). (Explanation p. 38, interpretation p. 58.)
Habitats diagram Percentage occurrence of the subject in each of seven primary categories (lozenge-shaped), eight intermediate categories (square) and 32 terminal categories (circular) within the habitat classification explained in Figure 3.2. Except in 'All habitats', values are rounded up. (Explanation p. 37).
Similarity in habitats Listing of the five species most similar to the subject in distribution and abundance in the 32 terminal categories of the habitat classification (Figure 3.2), percentage similarity scale. (Explanation p. 39, interpretation p. 59.)
Commonest associates Listings of the five species which co-occur most frequently with the subject species. (Explanation p. 39, interpretation p. 60.)
Synopsis Notes providing an ecological overview of the more important field, screening and published information. (Explanation p. 40.)
Current trends Estimates of the response of the subject, and of its future status, in relation to current changes in land use in Britain. (Explanation p. 40.)

Altitude diagram Percentage occurrence (i.e. simple presence, unstippled) and abundance (i.e. presence at 20% frequency or more, stippled) of the subject within five classes of altitude. (Explanation p. 37.)
Slope diagram Percentage occurrence (i.e. simple presence, unstippled) and abundance (i.e. presence at 20% frequency or above, stippled) of the subject within five classes of slope. (Explanation p. 37.)
Aspect diagram Percentage occurrence (i.e. simple presence, unbracketed) and percentage of that occurrence which constitutes 'abundance' (i.e. at 20% frequency or above, bracketed) of the subject in four contingencies of aspect and shading, excluding wetlands and slopes of <10. Where sufficiently clear data exist the significance of any N-S difference is stated, either as the appropriate P-value, e.g. $P > 0.05$, or as a dash. (Explanation p. 37).
Associated floristic diversity diagram Mean species m^2 associated with five groups of samples classified according to rooted frequency of the subject. Indicates the general level of species-richness associated with the subject and (in some cases) the apparent influence of the subject on species diversity. (Explanation p. 39.)
Hydrology diagram Percentage occurrence (i.e. simple presence, unstippled) and abundance (i.e. presence at 20% frequency or more, stippled) of the subject in a sequence of six hydrological classes:

Class	Slope	Water depth (mm)
A	5+	nil
B	<5	nil
C	marginal to open water	
D	nil	<100
E	nil	101-250
F	nil	>250

(Explanation p. 27.)

△ **ordination diagram** Relative constancy of the subject in a matrix of vegetation types classified according to the strategies of the component species. Numbers quoted indicate the percentage of samples of each vegetation type in which the subject occurs: –, absent; 1, 1-10%; 2, 11-20%; 3, 21-30%, etc., to 9, 80% or above. The cross indicates the approximate position of the maximum percentage occurrence of the subject and a dashed contour delimits the region within which it achieves at least one-quarter of its maximum percentage occurrence. (Explanation p. 38.)

Soil pH diagram Percentage occurrence (i.e. simple presence, unstippled) and abundance (i.e. presence at 20% frequency or more, stippled) of the subject in five classes of surface soil pH. (Explanation p. 38.)

Bare soil diagram Percentage occurrence (i.e. simple presence, unstippled) and abundance (i.e. presence at 20% frequency or more, stippled) of the subject within six classes of percentage bare soil cover:

Class	Bare soil
A	nil
B	1- 10%
C	11-25
D	26- 50
E	51-75
F	76-100

(Explanation p. 38.)

Abundance in quadrats table Listing of the number of occasions within each primary habitat on which the subject attains each of ten classes of frequency in the sample. Occurrences in each habitat are classified into one of three 'distribution types' if they number at least ten in all: (1), occurrences at 1-10% frequency five or more times the sum of all other occurrences; (3), occurrences at 1-50% frequency less than three times the sum of all other occurrences; (2), intermediate between (1) and (3). The bottom line provides the same on a whole-survey basis. *When text carries over from the left-hand page of the Account only the bottom line of the table appears.*

6 THE SPECIES ACCOUNTS

Acer pseudoplatanus L.

Aceraceae
Sycamore, Great Maple

Field data analysed in this account refer to seedlings and small saplings

Life-form Deciduous tree with broad and spreading crown and smooth bark. Leaves broad, palmately lobed, <20 000 mm²; roots with VA or rarely ecto-mycorrhizas.
Phenology Flowers April to June, when leaves begin to expand. Seed set September to November. Leaves relatively long-lived.
Height Foliage and flowers to 30 m.
Seedling RGR <0.5 week^{-1}. SLA 13.3 DMC. No data.
Nuclear DNA amount 2.4 pg; $2n = 52$ (*FE2*); $4x$.
Established strategy Intermediate between competitor and stress-tolerant competitor.
Regenerative strategies W, S. Seeds germinate synchronously in spring and are not incorporated into a persistent seed bank; SB II.

Flowers Yellowish-green, monoecious, insect-pollinated; male and female flowers mixed in pendulous 60-100-flowered panicle.
Germinule 30.60 mg, 6.4 × 4.7 mm, seed in a samara, 34.47 mg, with a long wing of *c*. 35 mm (2 per ovary).
Germination Epigeal; dormancy broken by chilling.
Biological flora Jones (1945).
Geographical distribution Native of the mountains of S and C Europe and Asia, but widely planted and well naturalized throughout the British Isles (100% of British vice-counties), *ABIF* page 439; also widely naturalized in the lowlands of C and N Europe (56% of European territories and 10% naturalized).

Habitats Seedlings and saplings recorded or observed in most habitats, both shaded and unshaded, but particularly frequent in woodland. A familiar seedling weed in many lawns.
Gregariousness Seedlings may occur at high densities.
Altitude Most frequent in lowland regions, but seedlings observed to 310 m. [Up to 480 m (*Biol Fl*).]
Slope Juveniles widely distributed.

Aspect Frequent on both N- and S-facing slopes.
Hydrology A non-persistent colonist of soligenous and topogenous mire.
Soil pH Juveniles widely distributed, but most-frequently recorded in the pH range 3.5-4.5, unlike the mature tree which is equally frequent between pH 4.0 and 8.0.
Bare soil Juveniles widely distributed.
△ **ordination** Juveniles occur in all vegetation types except for severely disturbed situations, but they are uncommon in very unproductive sites.
Associated floristic diversity Often low.

Habitats

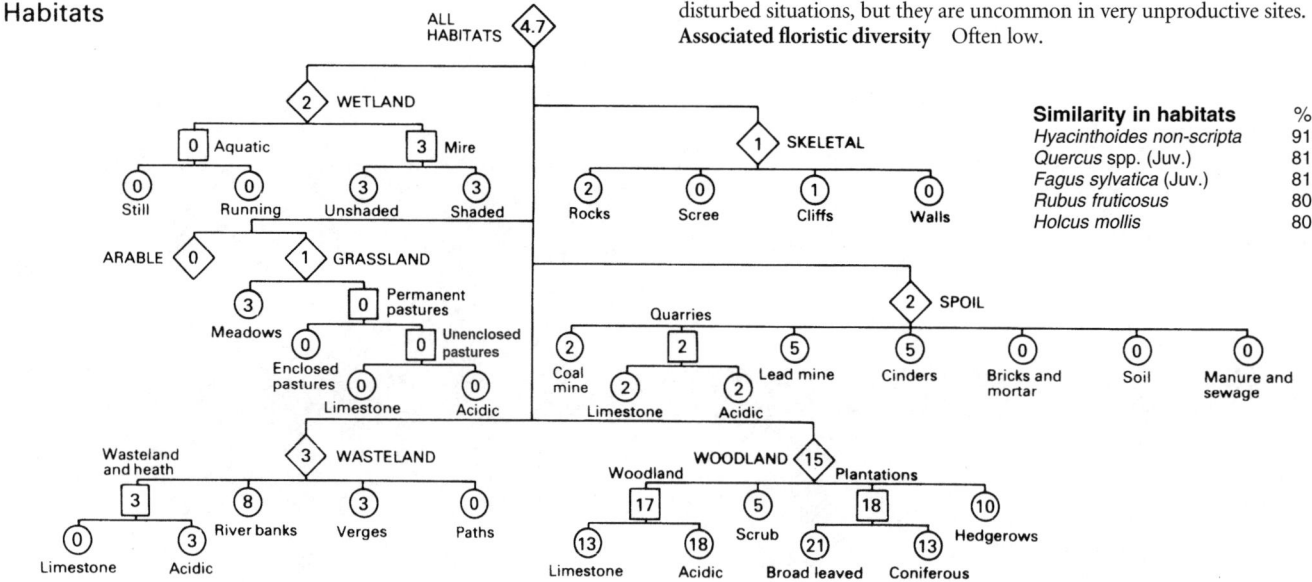

Similarity in habitats	%
Hyacinthoides non-scripta	91
Quercus spp. (Juv.)	81
Fagus sylvatica (Juv.)	81
Rubus fruticosus	80
Holcus mollis	80

Synopsis *A.p.* is a relatively frost-tolerant, deciduous timber tree with diffuse porous wood, up to 30 m in height, and exceptionally living for 400-600 years (*Biol Fl*). The species is a native of C and S Europe, particularly in mountainous regions (*FE2*), and might have been expected to colonize Britain but for the barrier to migration provided by the English Channel (Green 1981). In fact, *A.p.* was introduced into Britain during the 15th or 16th century (*Biol Fl*) and has been planted widely from the later 18th century onwards (Rackham 1980). In 1947 *A.p.* covered *c*. 31 000 ha of woodland (>4% of the total), making it the fifth most common tree of broad-leaved woodland (Steele & Peterken 1982). *A.p.* produces numerous winged, wind-dispersed fruits and, apart from *Fraxinus excelsior*, is the tree species most commonly recorded as seedlings or saplings within the Sheffield region. The species is also the most frequent contributor to the shrub or tree canopy, and is infiltrating many older woodlands. *A.p.* is tolerant of coppicing (*Biol Fl*). The largest trees are particularly associated with moist, base-rich, fertile soils, but *A.p.* is widespread on all but the most acidic sites. Damage to trees by grey squirrels may be extensive and causes problems in the management of *A.p.* for timber (Stern 1982). Seeds germinate in spring and no persistent seed bank is formed. The seedling is intolerant of competition from herbaceous species (*Biol Fl*) and sensitive to grazing; consequently, saplings tend to occur either beneath a tree canopy or in inaccessible rocky sites. A feature which *A.p.* shares with many other trees is the capacity of the saplings to expand leaves before bud-break of the adult trees. As a seedling, *A.p.* is found in sites similar to those exploited by *Fraxinus excelsior*. However, perhaps through mechanisms involving the ring porous wood, bud-break is considerably earlier. Saplings are also moderately shade-tolerant, but exhibit a higher frequency of occurrence on S-facing slopes and tend not to persist under a canopy of *A.p.* (*Biol Fl*). Saplings are capable of more rapid growth than those of most other British trees (*Biol Fl*) and in the Derbyshire Dales establishment of *A.p.* tends to be restricted to highly disturbed sites. The species establishes particularly freely in lightly-shaded sites in disturbed

Altitude

Slope

Aspect

n.s.	2.4	0.0
—	1.8	0.0 Unshaded

n.s.	11.4	0.0
—	12.2	0.0 Shaded

Soil pH

Associated floristic diversity

Hydrology

Triangular ordination

Bare soil

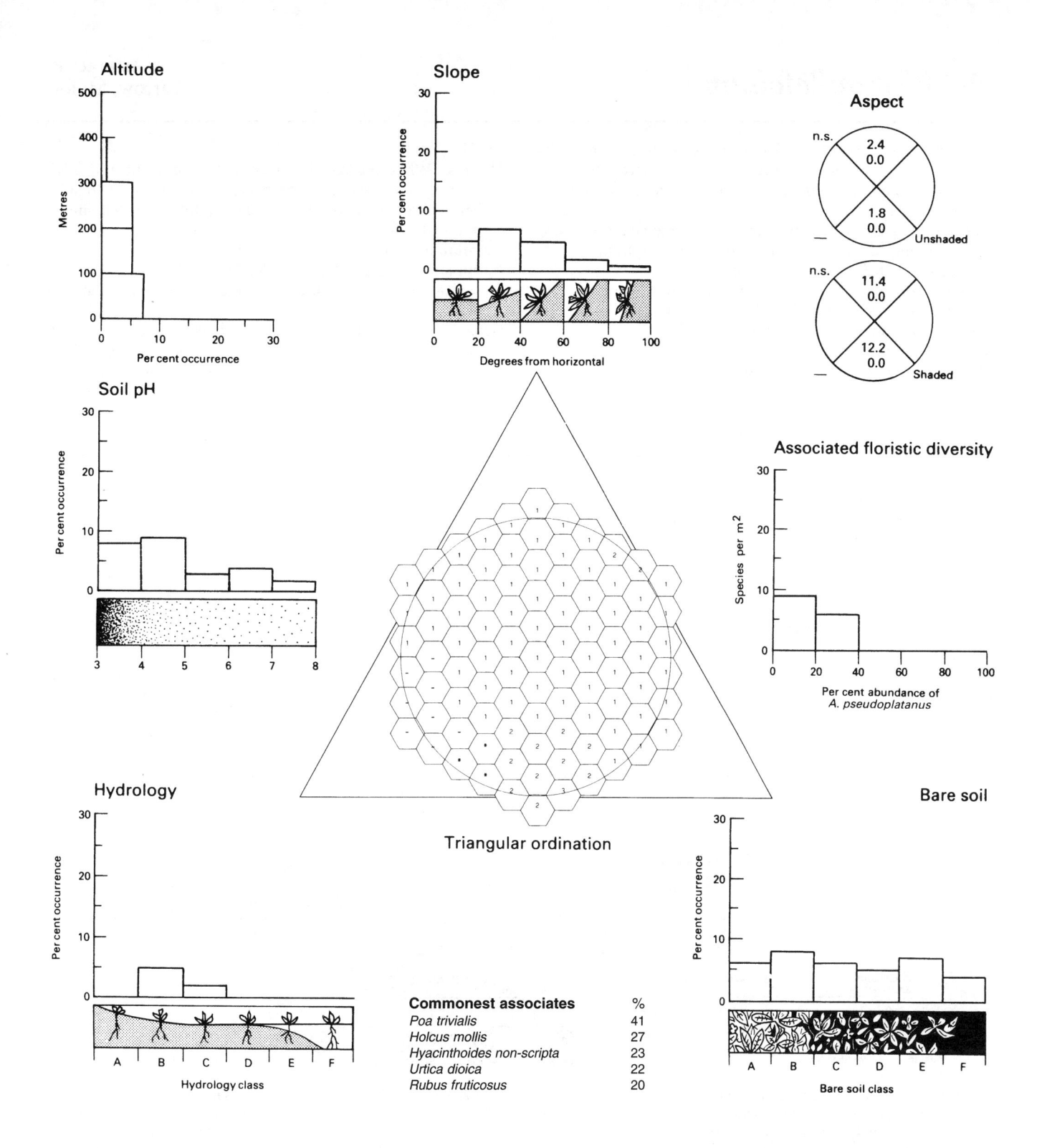

Commonest associates %
Poa trivialis 41
Holcus mollis 27
Hyacinthoides non-scripta 23
Urtica dioica 22
Rubus fruticosus 20

woodland. Leaves expand early in summer to form a rather dense canopy and leaf litter is relatively persistent, though less so than that of *Fagus sylvatica* (Sydes & Grime 1981b). Dense accumulations of *A.p.* litter may reduce the floristic diversity of the ground flora of woodlands (Anderson 1979, Sydes & Grime 1981b). The aphid *Drepanosiphum platanoides* Schr. may cause considerable reduction in the growth of *A.p.* (Dixon 1971) but very few insects are associated with *A.p.* (Southwood 1961). However, Stern (1982) observes that *A.p.* has been little-studied and suggests that its insect fauna has been underestimated. A list of diseases and predators is included in *Biol Fl*. Populations from different European climatic zones show different phenological patterns (*Biol Fl*).

Current trends A more economic crop than oak or beech, arguably under-used in modern British forestry (Stern 1982). Formerly much planted for forestry and, because of its ease of transplanting, much in demand for amenity use. However, *A.p.* is a weedy species and its rapid and effective regeneration is in contrast to that of *Tilia* spp. in ancient woodland (see Pigott 1969). The relationship between *A.p.* and the species of mixed ancient woodland may be similar to that between *Betula* and *Quercus* (see *Betula* account), and concerted efforts have been made to remove *A.p.* from some ancient woodlands. In this context the indiscriminate planting of *A.p.* for amenity may be viewed as highly irresponsible.

Achillea millefolium L.

Asteraceae
Yarrow, Milfoil

Life-form Polycarpic perennial, semi-rosette chamaephyte with far-creeping rhizomes. Stems erect; leaves lanceolate, finely dissected, often <1000 mm²; roots often deep with VA mycorrhizas.
Phenology Leaf rosettes overwinter and some give rise during late spring to leafy flowering shoots. Flowers from June to August and fruits from July onwards.
Height Foliage to 300 mm; flowering shoot to 450 mm.
Seedling RGR 1.5-1.9 week^{-1}. SLA 26.7. DMC 19.6
Nuclear DNA amount 15.3 pg; $2n = 54^*$ (*BC*); $6x^*$.
Established strategy C-S-R.
Regenerative strategies V, ?S. Regeneration *in situ* mainly by clonal expansion. Seeds, which are only briefly persistent in the soil (Roberts 1986a), germinate in autumn or spring, and may be important in the colonization of new sites (see 'Synopsis'); ?SB I.
Flowers White, pollinated by insects, self-incompatible (*Biol Fl*), each capitulum with *c.* 5 outer female ray-florets and *c.* 10 inner hermaphrodite disk-florets; >100 capitula crowded into a terminal corymb.
Germinule 0.16 mg, 2.1 × 1.1 mm, achene.
Germination Epigeal, t_{50} 1 day; 6-33°C; L = D.
Biological flora Warwick and Black (1982), see also Zarzycki (1976).
Geographical distribution The British Isles (100% of British vice-counties), *ABIF* page 647; most of Europe, although rare in the Mediterranean (85% of European territories and 3% naturalized); Asia; introduced in N America, Australia and New Zealand.

Habitats Most characteristically a species of grazed or ungrazed grassland. Also prominent on waysides, spoil heaps and rock outcrops. Absent from wetland and woodland. [Found in maritime habitats such as sand dunes, cliff-tops and stabilized shingle. Also extends into montane grassland.]
Gregariousness Intermediate, often represented by isolated clones.
Altitude Suitable habitats for this grassland species are more abundant in the uplands.
Slope Mainly confined to slopes of <50°.
Aspect Found over the full range of aspects.
Hydrology Absent from wetland, like many species with a deep root system.
Soil pH Found on soils varying in pH between 4.0 and 8.0. Most abundant in the middle of this range.
Bare soil A wide range of occurrences, but only achieving abundance where there is a relatively small proportion of bared ground.

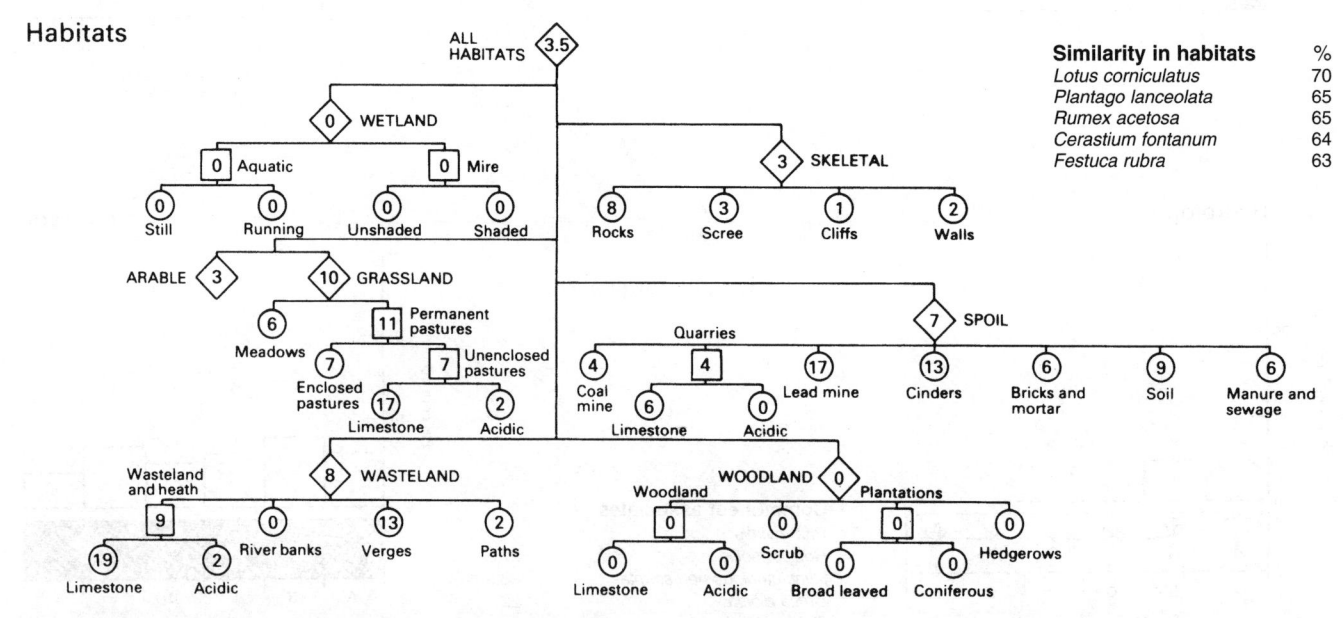

Habitats

Similarity in habitats	%
Lotus corniculatus	70
Plantago lanceolata	65
Rumex acetosa	65
Cerastium fontanum	64
Festuca rubra	63

△ **ordination** Distribution centred on vegetation associated with moderately disturbed and relatively fertile conditions. Absent from very unproductive conditions and from heavily disturbed vegetation. Flourishes in grazed pastures where the disturbance is sufficient to debilitate larger and potentially dominant perennial herbs.
Associated floristic diversity Intermediate to species-rich.
Synopsis *A.m.* is a winter-green herb of wide edaphic tolerance found in a range of grassy and open habitats. *A.m.* is equally common and ecologically wide-ranging in N Europe (Ellenburg 1978) and is included among the six most common species in a survey of Danish roadside habitats (Hansen & Jensen 1974). In grazed turf the leaves are appressed to the ground in basal rosettes which may be aggregated into large clonal patches. In tall grassland foliage is longer and held erect and, although not a woodland plant, *A.m.* may persist in shade, where it tends not to flower (Bourdot *et al.* 1984). However, *A.m.* is relatively intolerant of competition from taller and more robust herbs. The root system may descend more than 200 mm (Anderson 1927, *Biol Fl*). This rooting habit coincides both with a degree of drought tolerance and with vulnerability to waterlogging. The relatively high nuclear DNA amount, and the presence of high concentrations of fructan in the rhizomes during winter (Hendry & Brocklebank unpubl.), suggests that the species may be capable of growth at low temperatures; this hypothesis needs to be tested. The plants have a pungent odour and bitter taste, and may taint milk. *A.m.* is usually considered to be a grassland weed (*Biol Fl*). In the *PGE* the species is most abundant in plots of low productivity and is suppressed by applications of ammonium salts or by liming. In poorer pasture *A.m.* may be an important food source for sheep. *A.m.* has also been used as a source of food, flavouring and medicine (*Biol Fl*). *A.m.* reproduces vegetatively by means of long rhizomes to form clonal patches which expand at a rate of up to 200 mm year^{-1} (*Biol Fl*). The eventual breakdown of the older rhizomes results in the production of a number of daughter plants. *A.m.* may also become established from pieces of rhizome detached as a result of

Altitude
Slope
Aspect

	n.s.	6.7	
		41.2	
		6.3	
	n.s.	35.7	Unshaded

—	0.0	—
—	0.0	—

Shaded

Soil pH

Associated floristic diversity

Hydrology

Absent from wetland

Triangular ordination

Commonest associates %
Festuca rubra 85
Plantago lanceolata 63
Dactylis glomerata 62
Holcus lanatus 52
Poa pratensis 52

Bare soil

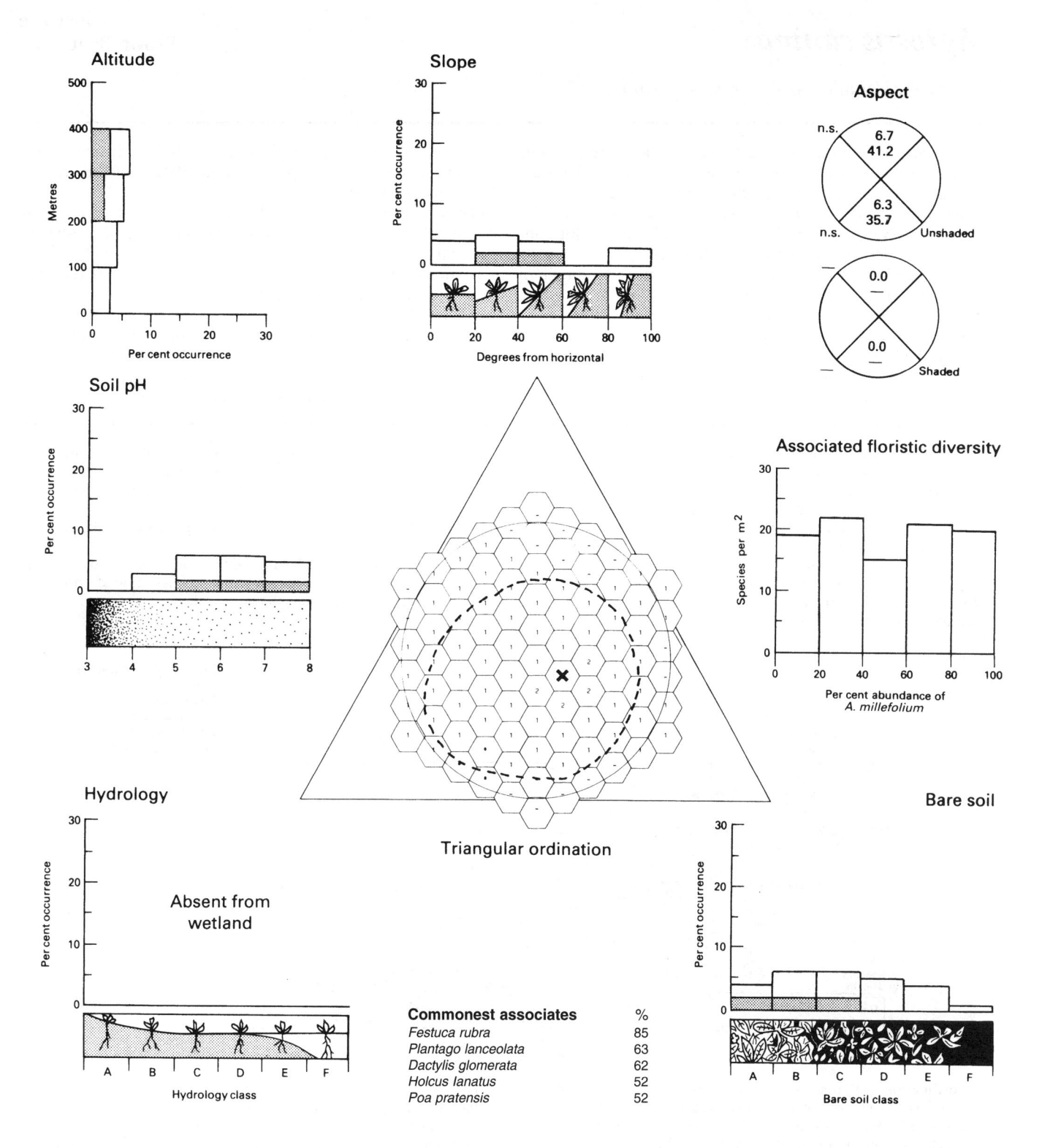

soil disturbance. In ungrazed communities seed is often produced in large quantities (1500 per flowering stem (*Biol Fl*)), and *A.m.* has potentially a high colonizing ability in ungrazed or lightly grazed habitats. The species is considered an important weed of arable land in New Zealand (Kannangara & Field 1985), but it does not appear to exploit farmland to the same extent in Britain. Seedlings, which are small, tend to establish only in rather open sites and in Britain regeneration by seed may be less important in established populations but may be critical during initial colonization. Seeds germinate during autumn or the following spring (see Bostock 1978, Roberts 1986a). A transient seed bank has been reported in New Zealand (Kannagara & Field 1985, Roberts 1986a). The flattened achenes are dispersed locally by wind and may survive ingestion by some animals (Ridley 1930). However, human activities probably provide the most important means of dispersal. The species is polymorphic and in Europe *A.m.* is included in a taxonomically complex grouping of eight species. Heavy-metal-tolerant races have been identified from lead-mine spoil (Shaw 1984).

Current trends *A.m.* is seldom an important grassland weed in Britain and current agricultural land-use does not favour its increase. Nevertheless, *A.m.* remains a common plant of pastures, waysides and lawns.

Agrostis canina L.

Poaceae
Brown Bent-grass

Formerly classified as *A. canina* ssp. *canina* (*CTW*)

Life-form Polycarpic perennial, protohemicryptophyte with creeping stolons. Shoots decumbent; leaves linear, often <250 mm^2; roots nodal, shallow, with VA mycorrhizas.

Phenology Some young leaves overwinter, and plant is capable of some winter growth (Philipson 1937). Flowers in June and July and sets seed in August and September.

Height Foliage <100 mm (except when stolons held upright by other vegetation); flowering shoots to 300 mm.

Seedling RGR No data. SLA 37.8. DMC 27.3.

Nuclear DNA amount 7.0 pg; $2n = 14*$ (*BC*); $2x*$.

Established strategy Intermediate between SR and C-S-R.

Regenerative strategies V, B$_s$. Regenerates effectively both by vegetative means and by seed (see 'Synopsis'); SB III or IV.

Flowers Green, hermaphrodite, wind-pollinated, self-pollination initially impeded (Knuth 1906); >100 in an open panicle (1 per spikelet)

Germinule 0.05 mg, 1.3 × 0.4 mm, caryopsis dispersed with an attached lemma sometimes bearing an hygroscopic bristle (1.7 × 0.4 mm).

Germination Epigeal; t_{50} 2 days; 9-30°C germination rate increased after dry storage at 20°C; L = D.

Biological flora None.

Geographical distribution Widely distributed in the British Isles (*A.c.* and *A. vinealis* are together found in 98% of British vice-counties), *ABIF* page 776; most of Europe (79% of European territories); temperate Asia; introduced in N America.

Habitats Virtually confined to wetland habitats, and in particular to unshaded and shaded mire.

Gregariousness Intermediate.

Altitude Occurs at all altitudes, but because its habitats are more common in upland areas *A.c.* becomes much more frequent and abundant on higher ground, particularly that over 400 m. [Up to 1040 m for aggregate species including *A.v.*; the lists of McVean & Ratcliffe (1962) suggest that each species extends to over 700 m.].

Slope No significant trends.

Aspect No significant trends.

Hydrology Frequent in soligenous mire, but also associated with topogenous mire including sites adjacent to water. In aquatic systems generally rooted on the bank but capable of extending floating shoots over the water surface.

Soil pH Associated particularly with acidic peaty soils in the pH range 3.5-6.0.

Bare soil Found mainly in habitats with a relatively low exposure of bare soil. Many sites have a large proportion of the ground covered by *Sphagnum* spp. or *Polytrichum commune* Hedw.

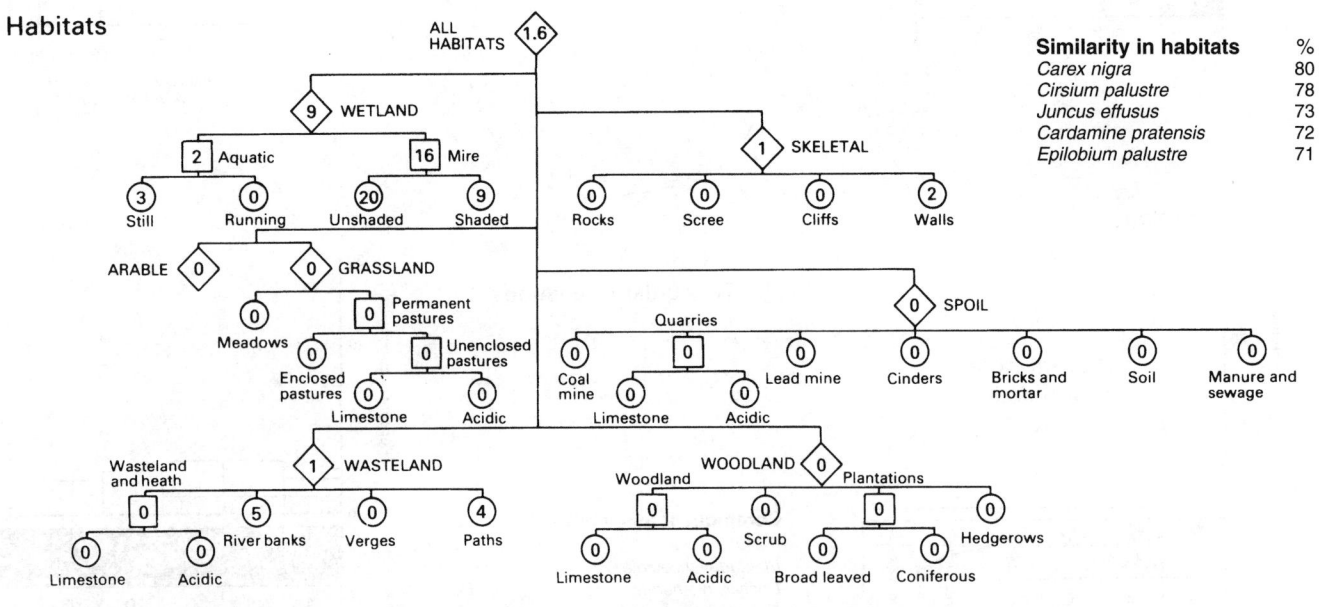

Habitats

Similarity in habitats	%
Carex nigra	80
Cirsium palustre	78
Juncus effusus	73
Cardamine pratensis	72
Epilobium palustre	71

△ **ordination** Occurs mainly in association with somewhat unproductive and relatively undisturbed vegetation types. Not present in heavily disturbed situations; unable to compete in highly productive vegetation and unable to survive in highly disturbed sites.

Associated floristic diversity Low to high, depending upon the degree of dominance exerted by its associated species.

Synopsis The most ecologically distinct of the common *Agrostis* spp., *A.c.* is a low-growing, stoloniferous wetland grass virtually restricted to infertile, acidic, peaty soils. The plant is essentially a gap-colonizing, subordinate species rather than a potential dominant and shoot densities may fluctuate widely from year to year. Vegetative spread occurs by means of long creeping stolons. In the growing season *A.c.* extends aerial stems into gaps in the canopy of stands dominated by more-robust species, such as *Juncus effusus*. *A.c.* may form floating rafts in peaty pools from plants rooted on the bank, and can form diffuse mats over *Sphagnum* patches.

The species may increase in response to disturbance such as ditch-cutting and light trampling. Locally, *A.c.* forms virtual monocultures in shaded habitats, but it seldom flowers or sets seed in heavily-shaded sites. Fragments of stolons detached as a result of disturbance readily root to form new plants. *A.c.* also regenerates in vegetation gaps by means of freshly shed seed in autumn or spring and a persistent seed bank is formed. The species is frequently an effective colonist of sites linked by water and colonization of, for example, upland reservoirs from water-borne plant fragments is suspected. *A.c.* may also colonize more-isolated sites such as gravel pits. No hybrids have been reliably reported for Britain (Stace 1975).

Current trends Apparently fairly mobile and potentially favoured by human activities such as peat-cutting and sowing as a turf grass (Shildrick 1980). Nevertheless, *A.c.* is probably decreasing as the result of habitat destruction by drainage and eutrophication.

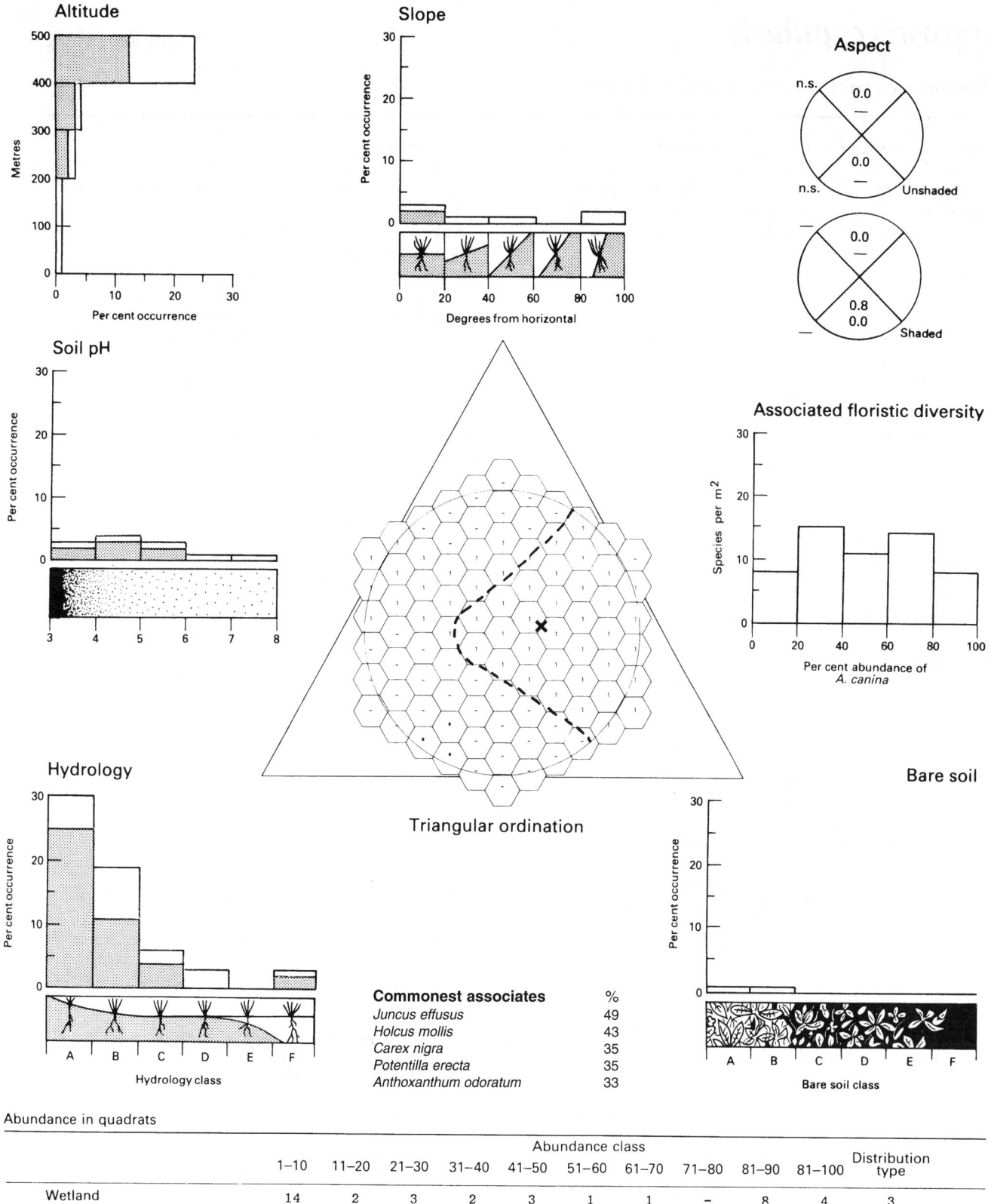

Altitude

Slope

Aspect

Soil pH

Associated floristic diversity

Species per m^2 vs Per cent abundance of *A. canina*

Hydrology

Triangular ordination

Bare soil

Commonest associates	%
Juncus effusus	49
Holcus mollis	43
Carex nigra	35
Potentilla erecta	35
Anthoxanthum odoratum	33

Abundance in quadrats

	\multicolumn{10}{c}{Abundance class}	Distribution type
	1–10	11–20	21–30	31–40	41–50	51–60	61–70	71–80	81–90	81–100	
Wetland	14	2	3	2	3	1	1	–	8	4	3
Skeletal	1	–	–	–	–	–	1	–	–	–	–
Arable	–	–	–	–	–	–	–	–	–	–	–
Grassland	4	4	2	2	–	1	1	–	1	1	2
Spoil	2	2	–	–	–	–	–	–	–	–	–
Wasteland	14	7	3	2	–	–	2	2	1	7	3
Woodland	1	–	–	–	–	–	–	–	–	–	–

Agrostis capillaris L.

Poaceae
Common Bent-grass

Taxonomic problems are outlined in 'Synopsis'

Life-form Polycarpic perennial protohemicryptophyte with spreading rhizomes or more rarely stolons. Shoots erect; leaves linear, typically <500 mm²; roots with VA mycorrhizas.

Phenology Winter green. Leaf expansion in late spring followed by flowering from June to August. Seed set August to October. Some winter growth.

Height Foliage usually <150 mm; flowering shoots to 500 mm.

Seedling RGR 1.0–1.4 week⁻¹. SLA 30.8. DMC 25.1.

Nuclear DNA amount 7.1 pg; $2n = 28^*$ (*BC*); $4x^*$. Values of 9.3 pg nuclear DNA and $2n = 42$ have also been recorded from a non-European source (Bennett *et al.* 1982).

Established strategy C-S-R.

Regenerative strategies V, B_s. Regenerative biology is complex (see 'Synopsis'); SB III or IV.

Flowers Brownish, hermaphrodite, wind-pollinated, outbreeding (Bradshaw 1959(a)); >100 aggregated into a loose panicle.

Germinule 0.06 mg, 1.0 × 0.5 mm, caryopsis dispersed with an attached lemma (1.9 × 0.5 mm).

Germination Epigeal; t_{50} 2 days; 7–31°C; L > D.

Biological flora None, but see Zarzycki (1976).

Geographical distribution The British Isles (100% of British vice-counties), *ABIF* page 775 (as *A. tenuis* Sibth.); Europe (82% of European territories); temperate Asia; introduced and established in N and S America, Australia and New Zealand.

Habitats Present in all major types of habitat. Particularly abundant in permanent pasture, on heaths and in waste places, and also on spoil heaps of coal mines, lead mines and sandstone quarries. Other common habitats are grassy paths, road verges, building rubble, railway ballast, rock outcrops and open or grazed areas in plantations, scrub and woodland. Represented more locally on river banks and walls, and in marshes, arable fields and hedgerows. [An agronomically important species of montane grassland.]

Gregariousness A patch-forming species.

Altitude Widespread, but suitable habitats are most common in upland areas. [Up to 1040 m.]

Slope Associated with a wide range of slopes.

Aspect Significantly more occurrences on unshaded S-facing slopes but achieves abundance more consistently when sites are N-facing.

Hydrology Essentially a dryland species, but found on moist soils and occasionally in soligenous mire or growing on tussocks of other species. Only of casual occurrence in submerged habitats.

Soil pH Greatest frequency and abundance on moderately acidic soils in the pH range 4.0–6.0 but occurs more locally on strongly acidic and on base-rich soils.

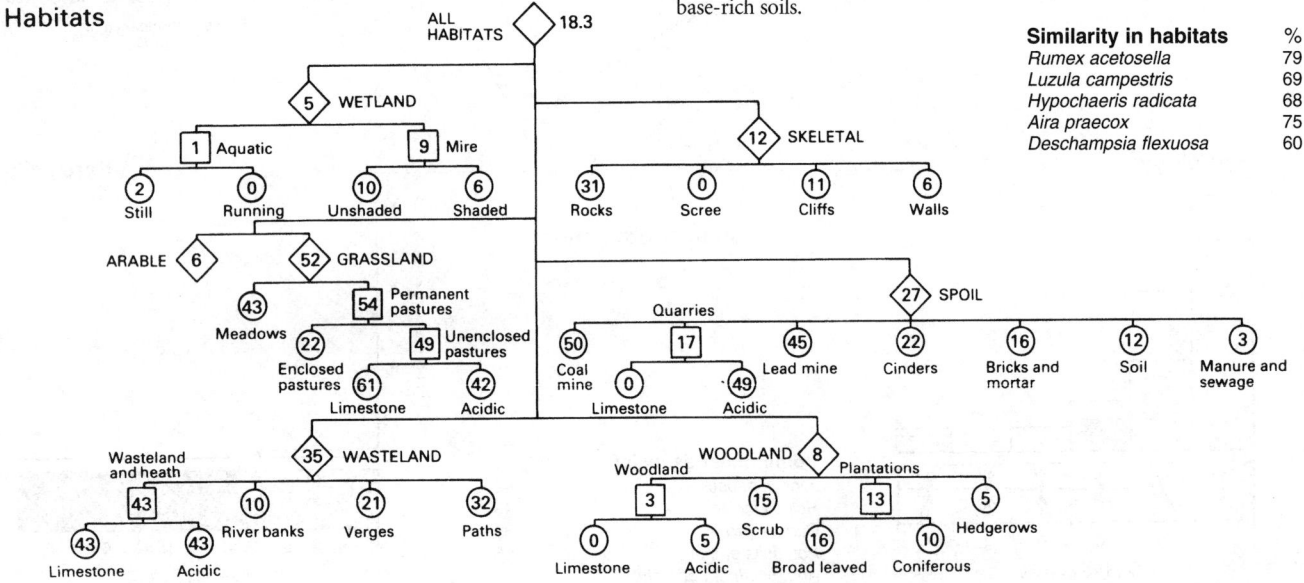

Habitats

Similarity in habitats	%
Rumex acetosella	79
Luzula campestris	69
Hypochaeris radicata	68
Aira praecox	75
Deschampsia flexuosa	60

Bare soil Associated with all contingencies but more frequent and abundant in habitats having low to moderate amounts of exposed soil.

△ **ordination** Distribution extremely wide-ranging but centred on vegetation of intermediate to low productivity subject to moderate or low intensities of disturbance.

Associated floristic diversity Typically low to intermediate and not varying in relation to the abundance of *A.c.*

Synopsis *A.c.* is a tufted, often patch-forming grass with an exceptionally wide ecological range. Grazing land and amenity grassland dominated by *A.c.* cover extensive areas of upland and lowland Britain, but *A.c.* is most characteristic of short turf in upland pastures on moderately acidic, brown earth soils. In this habitat the species is represented by genotypes of moderately high relative growth rate, and is capable of establishing a dense low leaf canopy which under heavy grazing is rapidly renewed during late spring and summer. Defoliation experiments (Mahmoud 1973) reveal that under fertile conditions the resilience of *A.c.* under frequent cropping is related to the ability to produce very high densities of small tillers, many of which are situated close to the ground surface and escape damage. Although largely restricted to mildly acidic soils, *A.c.* extends on to calcareous N-facing slopes, perhaps for reasons similar to those suggested for *Potentilla erecta*. In the *PGE*, yields of *A.c.* were promoted by ammonium fertilizer and the species was 'very much discouraged by lime', by treatments leading to tall productive turf and by supply of superphosphate alone. Leaves have the low concentrations of Ca typical of many grasses (*SIEF*). *A.c.* shows some tolerance of burning. *A.c.* regenerates by means of rhizomes or stolons to form extensive patches, and both in upland pasture and elsewhere *A.c.* may assume the role of dominant due in large measure to an ability for rapid lateral spread. However, in keeping with the relatively low nuclear DNA amount leaf growth is delayed compared with many other turf grasses, especially

Altitude

Slope

Aspect

P<0.05

| 20.1 | 62.7 |
| 29.1 | 49.2 |

n.s. / Unshaded

n.s.

| 3.8 | 40.0 |
| 5.7 | 14.3 |

n.s. / Shaded

Soil pH

Associated floristic diversity

Triangular ordination

Commonest associates	%
Festuca rubra	53
Anthoxanthum odoratum	47
Festuca ovina	46
Holcus lanatus	45
Dactylis glomerata	36

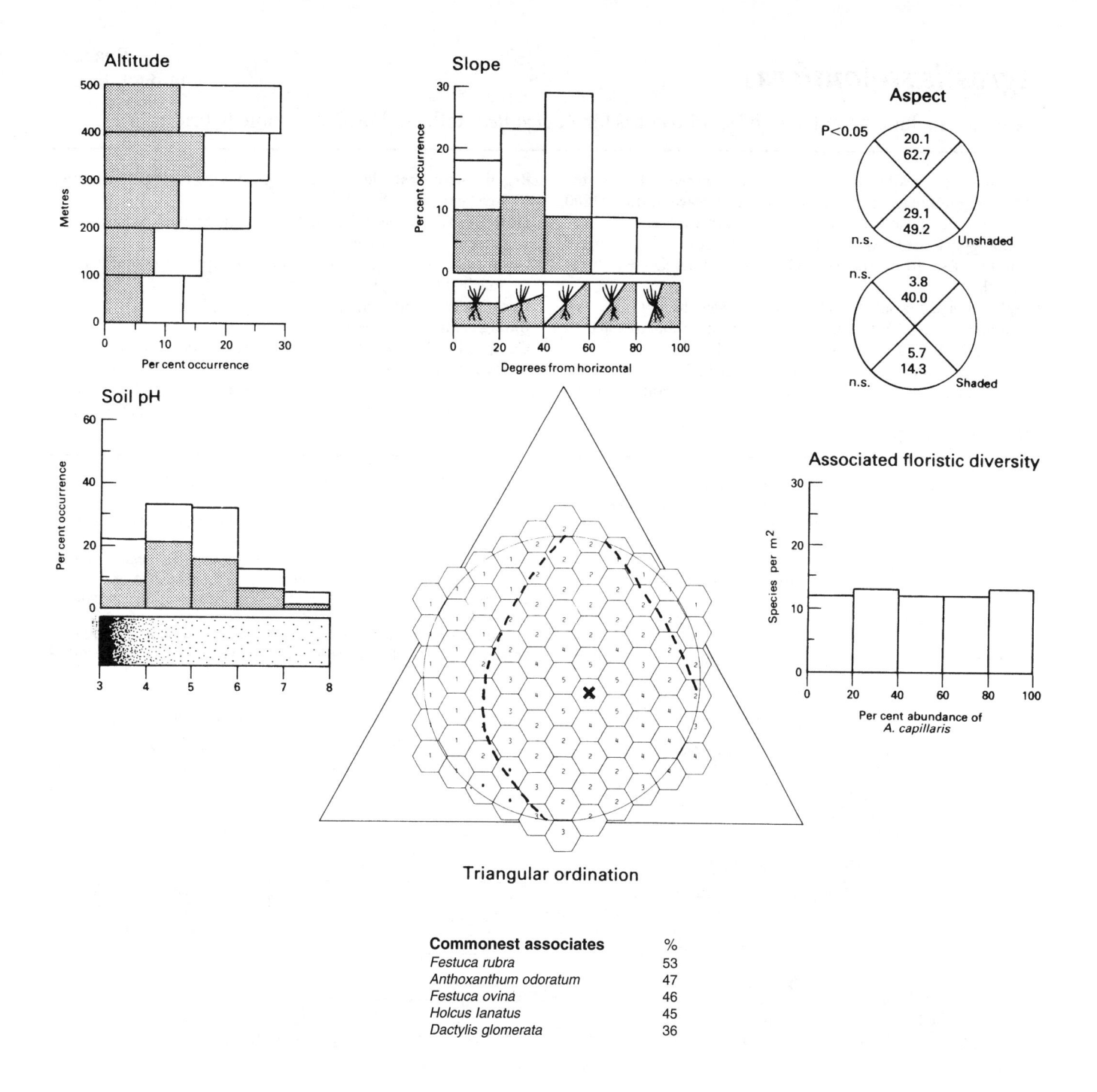

Festuca rubra with which the species frequently coexists. Shoots detached as a result of disturbance may give rise to new plants. *A.c.* also regenerates by seeds which germinate in autumn, in spring, or after incorporation into a persistent seed bank. Seed-set, though normally prolific, is irregular at higher altitudes (Bradshaw 1959a). The persistence and colonizing ability of *A.c.* is in part a product of the diversity of regenerative strategies and of the mobility of seeds which are presumably mainly transported in soil. However, the wide success of the outbreeding *A.c.* is also related to its remarkable ability to develop genetically specialized populations. Populations differ in capacity for lateral spread and seed production (Bradshaw 1959a). Upland populations and those from sandy areas are mainly rhizomatous, while those from lowland grassland are stoloniferous (Bradshaw 1959a). Individuals with an inherently slower growth rate have been identified in unproductive habitats such as lead-mine spoil (Bradshaw 1959b). *A.c.* is both genetically variable and phenotypically plastic in growth form; in consequence it is often difficult to identify taxa by vegetative characteristics. The ecological range of *A.c.* overlaps with all other native British species of *Agrostis* and *A.c.* hybridizes with other species. In particular, the hybrid with *A. stolonifera*, which has been said to be better attuned to certain grazed environments than its parents (Bradshaw 1958), may be common and some field records included under *A.c.* refer to hybrid taxa or to the recently introduced *A. castellana* Boiss. and Reuter.

Current trends *A.c.* has declined, particularly in lowland areas, with the destruction of permanent pasture (Hopkins 1979). The species remains economically important in upland Britain. However, *A.c.* remains a common species and is an effective colonist of artificial habitats, particularly on mildly acidic soils. The development of ecotypes specialized towards new artificial habitats is resulting in an increase in the ecological range and genetic diversity of *A.c.* The disturbance associated with modern land-use is likely to reduce the ecological isolation further and to increase the degree of hybridization between species of *Agrostis*.

Agrostis stolonifera L.

Poaceae
Creeping Bent, Fiorin

Field data include a limited number of records for *A. gigantea* Roth. and probably some hybrid taxa

Life-form Stoloniferous, polycarpic perennial proto-hemicryptophyte. Stems prostrate to erect; leaves linear, <1000 mm²; roots shallow, often with VA mycorrhizas.
Phenology Usually winter green. Flowers from July to August and seed set from August to October. Leaves relatively short-lived.
Height Foliage usually <200 mm and flowers to 400 mm. Var. *palustris* (Huds.) Farw. may attain 600 mm (Hubbard 1984).
Seedling RGR 1.0-1.4 week^{-1}. SLA 35.6. DMC 26.0.
Nuclear DNA amount 7.0 pg; $2n$ = 28*, 30, 32, 35, 42, 44, 46 (*FE5*, *BC*); 4x*-7x, 4x plants appear most common in Europe (*FE5*).
Established strategy Competitive-ruderal.

Regenerative strategies V, B$_s$. Regenerative biology is complex (see 'Synopsis'); SB III.
Flowers Green, hermaphrodite, wind-pollinated; >100 in each panicle (1 per spikelet).
Germinule 0.02 mg, 1.0 × 0.5 mm, caryopsis dispersed with an attached lemma (1.5 × 0.5 mm).
Germination Epigeal; t_{50} 3 days; 11-36°C; L > D.
Biological flora None.
Geographical distribution The British Isles, especially in lowland districts (100% of British vice-counties), *ABIF* page 776; Europe (97% of European territories); C Asia, Japan and N America; introduced in Australia, New Zealand and S Africa.

Habitats Recorded on fertile soils in almost every habitat type listed, and particularly abundant on spoil heaps, road verges, paths, meadows, pastures, mire at the margin of streams and on arable land. [Widespread on sea cliffs, in salt marshes, dune slacks and wet montane habitats.]
Gregariousness Potentially patch-forming.
Altitude Suitable habitats are more frequent in lowland areas. [Up to 750 m.]
Slope Wide-ranging.

Aspect Slightly more frequent on unshaded S-facing slopes.
Hydrology Common in both soligenous and topogenous mire. Also frequent at the edge of streams and ditches, floating in shallow water but usually rooted to the bank.
Soil pH Becoming progressively more frequent and abundant with increasing soil pH, with highest values in the pH range 5.5 to >8.0.
Bare soil Distribution biased towards habitats having a large proportion of exposed soil.
Δ ordination Widely distributed, but with distribution centred on vegetation associated with relatively productive and moderately disturbed conditions.

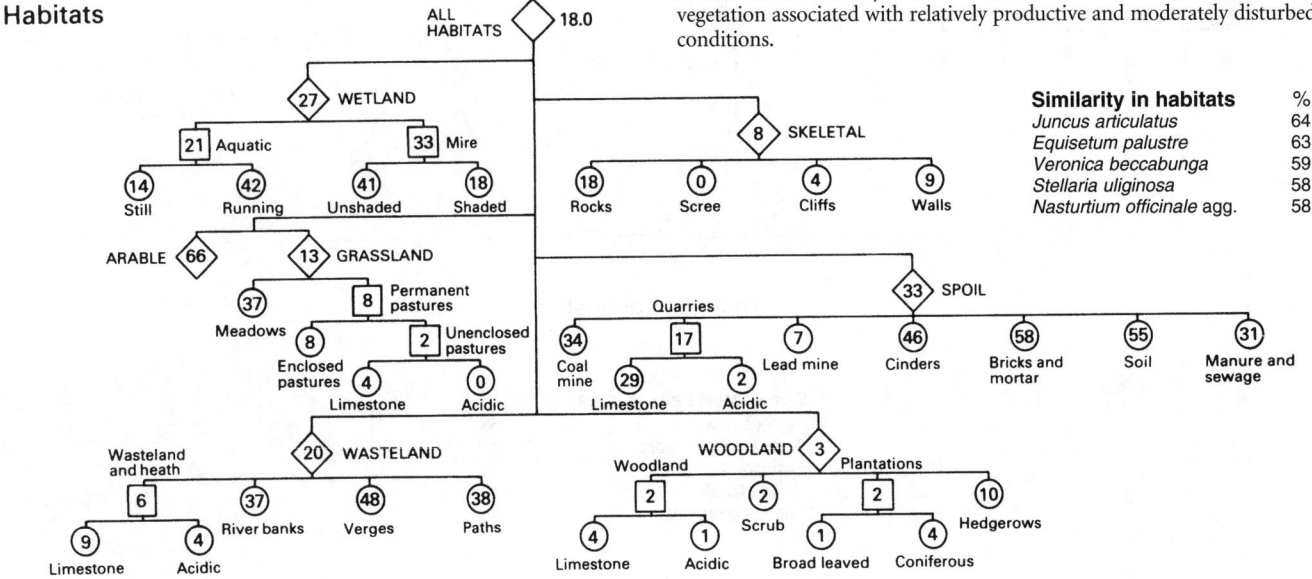

Habitats

Similarity in habitats	%
Juncus articulatus	64
Equisetum palustre	63
Veronica beccabunga	59
Stellaria uliginosa	58
Nasturtium officinale agg.	58

Associated floristic diversity Intermediate; lower at high frequencies of *A.s.*, indicating the capacity of the species to exert local dominance in some circumstances.
Synopsis *A.s.* is a short, fast-growing, stoloniferous, patch-forming grass which is common in an exceptionally wide range of fertile habitats. It is most abundant where the growth of tall dominants has been restricted by disturbance. *A.s.* exploits aquatic habitats and mire, woodland margins, most types of spoil heaps, moist grassland and arable land. *A.s.* is equally frequent in a range of maritime habitats. The roots and shoots show 'active foraging' in space (Grime *et al.* 1986); this explains the marked capacity of *A.s.* to exploit pockets of nutrient enrichment and canopy gaps (Crick 1985). *A.s.* is moderately tolerant of ferrous iron toxicity (R. E. D. Cooke unpubl.). *A.s.* forms extensive clonal patches by means of long stolons. Detached shoots, which may be transported in soil, readily root to form a new plant, and this mode of regeneration appears to be important in disturbed habitats such as arable fields. The species is more palatable to stock than *A. capillaris* (Spedding & Diekmahns 1972). *A.s.* also regenerates by seed in autumn, in spring or after incorporation into a persistent seed bank. This diversity of regenerative strategies doubtless contributes considerably to the success of *A.s.* as both a colonist of new sites and a survivor in disturbed habitats. The capacity to exploit a wide range of habitats probably also arises from the capacity of *A.s.* to form ecologically specialized populations in a manner similar to that of *A.c.* Ecotypes tolerant of heavy metals have been detected in N Derbyshire on contaminated soils (Shaw 1984), and less-winter-green salt-tolerant populations occur on roadsides. Philipson (1937) has suggested that *A.s.* may be subdivided into var. *palustris* (most freshwater mire populations) and var. *stolonifera* (the remainder). No relationships between morphology and chromosome number have been detected (*FE5*). *A.g.* ($2n = 42*$ (*BC*), 6x*) differs in morphology from typical *A.s.* in that (a) shoots are taller and more robust (foliage to 400 mm; flowers to 800 mm) and (b) it produces rhizomes and, consequently, tends to have a somewhat deeper root system. The first of these attributes enhances the capacity of *A.g.* to persist in tall vegetation, while the second may tend to restrict *A.g.* to well-drained soils.

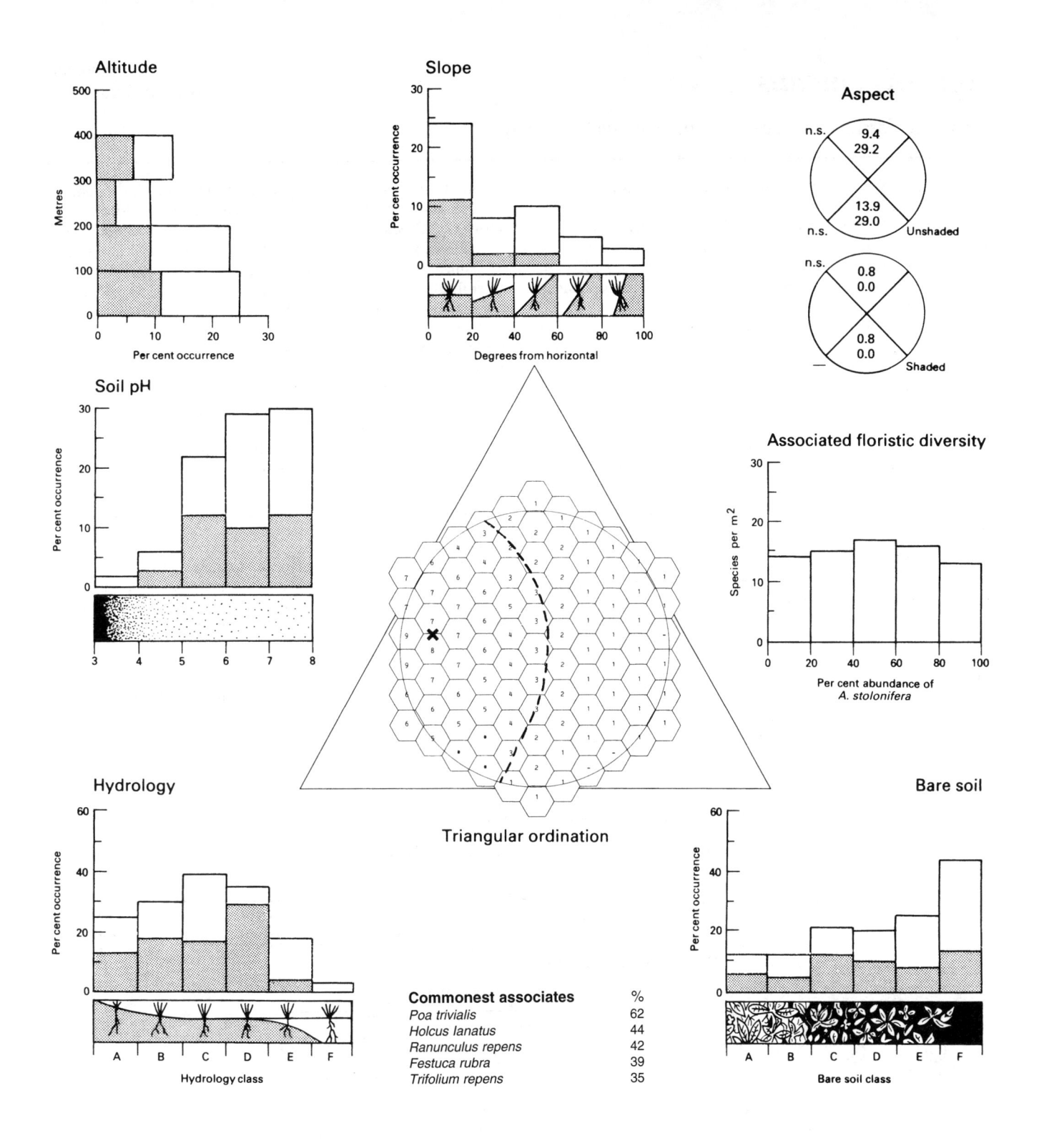

A.g. reaches its greatest frequency and abundance in cereal crops on sandy soils and exploits lighter poorer soils than those on which *Elytrigia repens* is abundant (Thurston & Williams 1968). *A.g.*, unlike *A.s.*, is predominantly a lowland plant, occurring in 31% of British vice-counties and 74% of European territories, and is most abundant in arable fields and on wasteland on dry soils. *A.g.*, produces *c.* 1000 seeds per panicle (Williams 1973) each weighing *c.* 0.09 mg, and has a range of regenerative strategies similar to that of *A.s.* The temperature range over which seeds germinate (7-34°C) suggests that *A.g.* has a greater potential to germinate in autumn than *A.s.* Populations from wetlands are less frequent and have been afforded varietal status by Philipson (1937), and a further ssp. is recorded from Russia (*FE5*). *A.g.*, *A.s.* and *A.c.* are closely related (Jones 1956) and form hybrids with one another, perhaps quite frequently (Stace 1975). *A.s.* also hybridizes with *Polypogon monspeliensis* (L.) Desf. (Stace 1975).

Current trends Despite better prospects for their control by herbicides (*WCH*), both *A.s.* and *A.g.* are apparently increasing as a result of the intensification of cereal growing and the adoption of minimal cultivation. *A.s.* is in addition a common weed of productive pasture systems, a useful turf-grass species (Shildrick 1980) and has greatly expanded its habitat range.

Agrostis vinealis Schreber

Poaceae
Brown Bent-grass

Formerly classified as *A. canina* ssp. *montana* Hartman

Life-form Polycarpic perennial protohemicryptophyte with shortly creeping rhizomes. Shoots erect with linear leaves, typically <500 mm²; root system with VA mycorrhizas.
Phenology Winter green. Capable of some winter growth but little leaf extension until late spring (Philipson 1937). Flowers June to August and sets seed from September. Leaves relatively long-lived.
Height Foliage typically <150 mm; flowering shoots to 500 mm.
Seedling RGR 1.0-1.4 week^{-1} (for *A.c.* ssp. *m.*). SLA 25.7. DMC 34.1.
Nuclear DNA amount 6.9 pg; $2n = 28*$ (*BC*); $4x*$.
Established strategy C-S-R.
Regenerative strategies B_s, V. Regeneration by seed occurs either directly in autumn or after incorporation into a persistent bank of seeds. There is some capacity for lateral vegetative spread by means of rhizomes, but expansion is slight compared with other locally-occurring species of *Agrostis*; the extent of separation into daughter plants remains uncertain; SB III or IV.
Flowers Green, hermaphrodite, wind-pollinated, self-pollination initially impeded (Knuth 1906); >100 in an open panicle (1.1 per spikelet).
Germinule 0.06 mg, 1.7 × 0.5 mm, caryopsis dispersed with an attached lemma (2.5 × 0.5 mm), sometimes bearing an hygroscopic bristle.
Germination Epigeal; t_{50} 3 days; L > D.
Biological flora None.
Geographical distribution Generally distributed in the British Isles (*A. canina* and *A.v.* are together found in 98% of British vice-counties), *ABIF* page 777; much of Europe (41% of European territories) but distribution imperfectly known; temperate Asia (Hubbard 1984).

Habitats Most common on dry acidic grasslands, on both limestone and more-acidic strata. Occurs in both grazed and ungrazed grassland, and is observed regularly as a colonist of disturbed artificial habitats such as sand pits and forestry rides in parts of the Sheffield region. Virtually absent from wetland and skeletal habitats. [A frequent constituent of montane grassland.]

Gregariousness Intermediate.
Altitude Occurs at all altitudes, but habitats suitable for *A.v.* are restricted in the lowlands and *A.v.* is most frequent above 200 m. [Up to 1040 m for aggregate species including *A. canina*; the lists of McVean & Ratcliffe (1962) suggest that both species extend to over 700 m.]

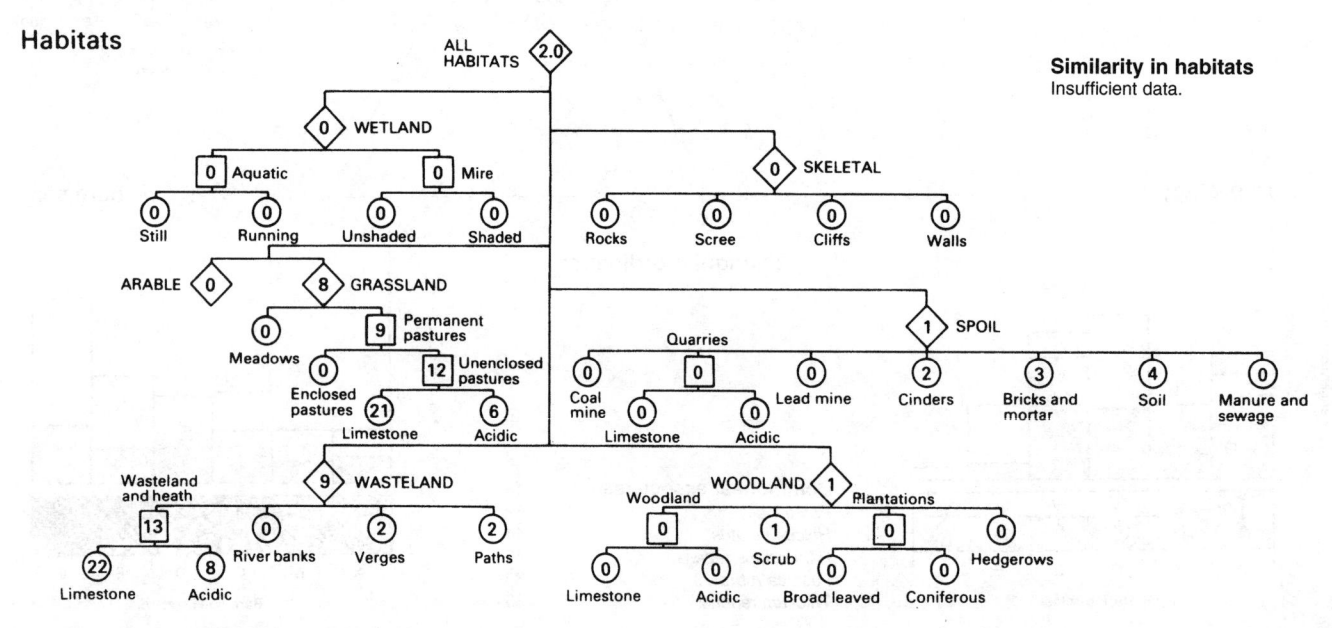

Habitats

Similarity in habitats
Insufficient data.

Slope Largely absent from steep slopes.
Aspect Found over a wide range of aspects.
Hydrology Not recorded in wetland habitats, but has been observed adjacent to soligenous mire.
Soil pH Wide-ranging with respect to pH, but most frequent and abundant on acidic soils within the range 3.5-5.0.
Bare soil Most typically recorded from well-vegetated ground with little exposed soil.
△ ordination Most frequent in undisturbed and relatively unproductive vegetation, but also present in more-productive and in some disturbed vegetation types.
Associated floristic diversity Associates usually few or intermediate in number.
Synopsis *A.v.* is a shortly rhizomatous winter-green grass of acidic grassland and open habitats, generally found growing in close association with *A. capillaris*. Apart from the much narrower ecological range of *A.v.*, the biology of the two species is so similar and the overlap in their ecological and geographical distributions so complete that ecological reasons for the separation of the two species are hard to find. *A.v.*, which has lesser capacity for lateral spread than *A.c.*, and an earlier, and possibly less-protracted, flowering period (Philipson 1937), is considered drought-tolerant (Hubbard 1984). Though obviously responsive to variations in nutrient levels (Miles 1974), *A.v.* may have a greater ability to persist on acidic, relatively unproductive soils. Hybrids with *A.c.* and *A. stolonifera* have been reported (Stace 1975). Listed as a turf-grass species (Shildrick 1980) but seldom seen in amenity grassland. The brevity of these comments is a reflection of our ignorance of the ecology of this elusive species.
Current trends Decreasing as a result of pasture improvement and upland dereliction.

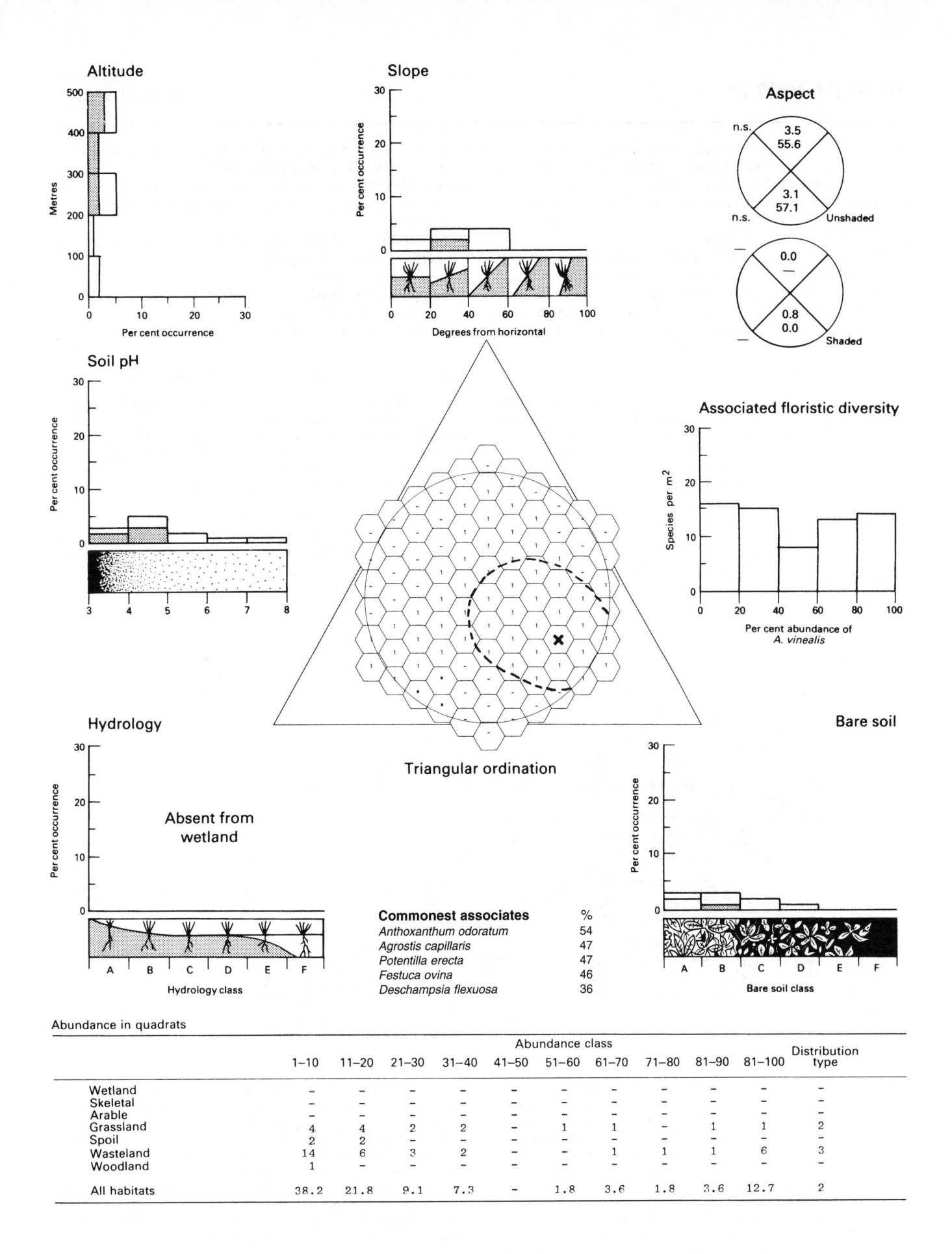

Altitude

Slope

Aspect

	n.s.	3.5	
n.s.		55.6	
	3.1		Unshaded
	57.1		

	—	0.0	
—		—	
	0.8		Shaded
	0.0		

Soil pH

Associated floristic diversity

Hydrology

Absent from wetland

Triangular ordination

Bare soil

Commonest associates	%
Anthoxanthum odoratum	54
Agrostis capillaris	47
Potentilla erecta	47
Festuca ovina	46
Deschampsia flexuosa	36

Abundance in quadrats

	Abundance class										Distribution type
	1–10	11–20	21–30	31–40	41–50	51–60	61–70	71–80	81–90	81–100	
Wetland	–	–	–	–	–	–	–	–	–	–	–
Skeletal	–	–	–	–	–	–	–	–	–	–	–
Arable	–	–	–	–	–	–	–	–	–	–	–
Grassland	4	4	2	2	–	1	1	–	1	1	2
Spoil	2	2	–	–	–	–	–	–	–	–	–
Wasteland	14	6	3	2	–	–	1	1	1	6	3
Woodland	1	–	–	–	–	–	–	–	–	–	–
All habitats	38.2	21.8	9.1	7.3	–	1.8	3.6	1.8	3.6	12.7	2

Aira praecox L.

Poaceae
Early Hair-grass

Life-form Tufted, semi-rosette, winter-annual therophyte. Stems erect, spreading or prostrate; leaves hair-like, <100 mm²; roots sometimes with VA mycorrhizas.
Phenology Seeds germinate in autumn. Flowers April to May and sets seed June to July.
Height Foliage <50 mm; flowers to 120 mm.
Seedling RGR 0.5–0.9 week^{-1}. SLA 20.1. DMC 27.7.
Nuclear DNA amount 5.9 pg, $2n = 14$ (Bennett *et al*. 1982); $2x$.
Established strategy Stress-tolerant ruderal.
Regenerative strategies S, ?B$_s$. Normally germinates in vegetation gaps during autumn but some seeds may persist briefly in the soil (Roberts 1986b); SB I or ?III.

Flowers Silvery-green, hermaphrodite, wind-pollinated; often *c*. 30 in a spike-like panicle (2 per spikelet).
Germinule 0.18 mg, 1.6 × 0.6 mm, caryopsis with an attached lemma, an hygroscopic bristle and a basal tuft of hairs, 2.6 × 0.6 mm.
Germination Epigeal; t_{50} 4 days; 10–31°C; germination rate increased after dry storage at 20°; L > D.
Biological flora None.
Geographical distribution The British Isles (100% of British vice-counties), *ABIF* page 772; S, W and C Europe (49% of European territories); introduced in N America.

Habitats Occurs, often with other winter annuals, in dry, rocky or freely-drained soils, particularly those of low pH. Frequent in sandstone quarries, and observed in sand and gravel workings. Also widespread in habitats such as rock outcrops and spoil tips, and found at the margin of paths, on roadsides, wasteland and in pasture on dry sandy soils. [Found also on lowland heaths, sand dunes and on soil bared by erosion in mountain areas.]
Gregariousness Intermediate.
Altitude Found over a wide range of altitudes to >400 m. [Up to 690 m.]
Slope Generally associated with flat ground or moderate slopes.
Aspect Significantly more frequent on unshaded S-facing slopes.
Hydrology Absent from wetlands.

Soil pH Most widely distributed and abundant on soils in the pH range 3.5–6.0. Less frequent at higher values.
Bare soil *A.p.* often exploits sites with large quantities of bare soil. Where the exposure of bare soil is low this usually coincides with the presence of an extensive carpet of bryophytes such as *Polytrichum juniperinum* Hedw.
△ **ordination** Distribution centred on vegetation subject to some disturbance.
Associated floristic diversity The low diversity associated with *A.p.* appears to be mainly due to the paucity of species exploiting acidic soils. High values coincide with the occasional excursions of *A.p.* to soils of higher pH.

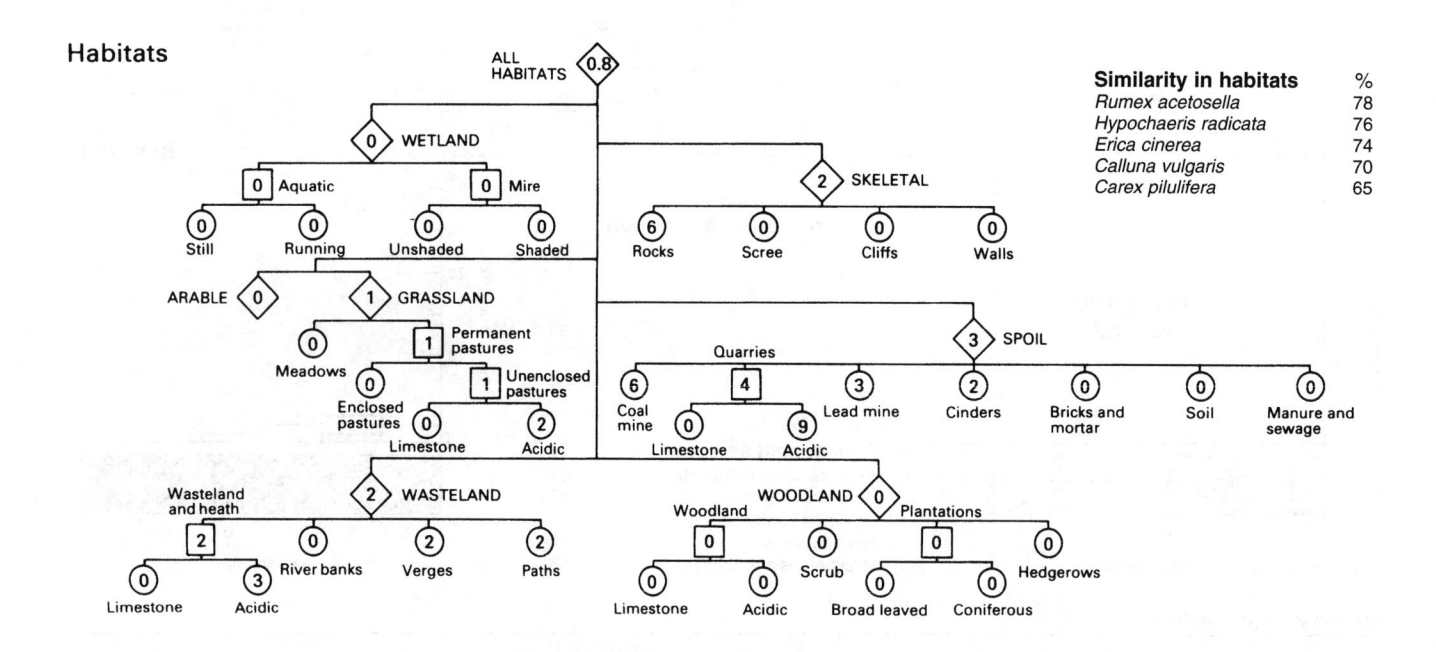

Habitats

Similarity in habitats	%
Rumex acetosella	78
Hypochaeris radicata	76
Erica cinerea	74
Calluna vulgaris	70
Carex pilulifera	65

Synopsis *A.p.* is an early-flowering, winter-annual grass of short, open vegetation on thin rocky or dry sandy soils where the vigour of perennial species is restricted by the combined effects of summer drought and low soil fertility. Thus, the species is characteristic of habitats such as gritstone outcrops, cinder tips and sand pits, and other habitats with semi-permanent winter-annual communities. *A.p.* is the most common winter annual of calcifuge distribution, and in many acidic upland sites the species is the only winter annual present. In lowland habitats two winter annuals which occur most commonly with *A.p.* on acidic soils are *Aphanes australis* and *Ornithopus perpusillus* L. Both flower and fruit later than *A.p.* and are tap-rooted, extending, for example, into short mown turf. Of the three, *A.p.* appears to adopt the most drought-avoiding strategy; germination occurs in mid-autumn, and the seedlings, which have a short root system, appear to be drought-sensitive (Newman 1963).

(The ecologically similar *Teesdalia nudicaulis* (L.) R.Br. germinates a little earlier but produces a tap-root (Newman 1963)). The persistence of seeds in the soil is uncertain. *A.p.* is widely distributed, e.g. beside railways and in sand pits, and we suspect that the awned caryopses of *A.p.* are readily dispersed over long distances, probably through human agency.
Current trends Uncertain. *A.p.* is an effective colonist of artificial sites but has remained restricted to relatively infertile areas.

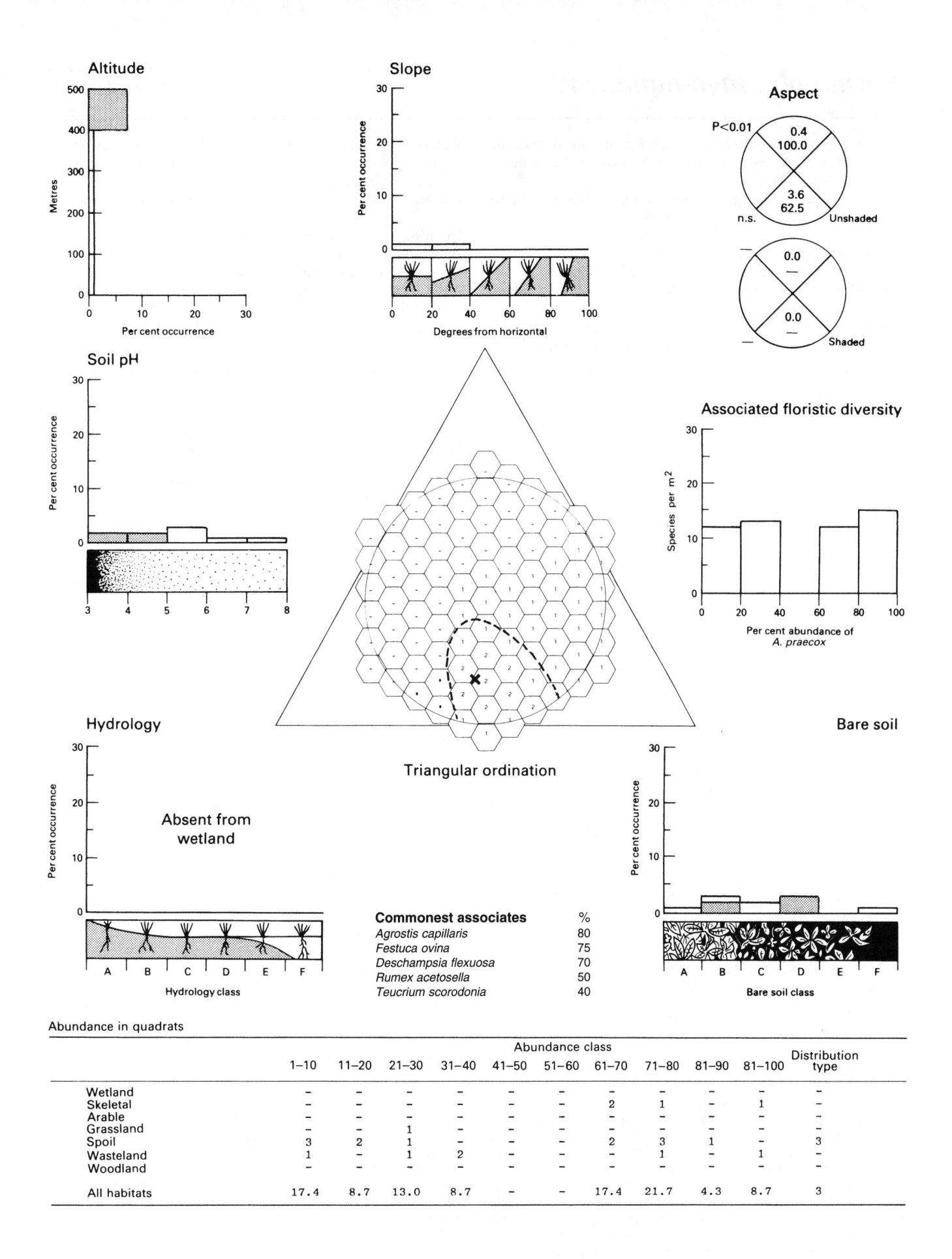

Altitude

Slope

Aspect

P<0.01   0.4
         100.0
n.s.   3.6
         62.5   Unshaded

         0.0

         0.0   Shaded

Soil pH

Associated floristic diversity

Hydrology

Absent from wetland

Triangular ordination

Commonest associates	%
Agrostis capillaris	80
Festuca ovina	75
Deschampsia flexuosa	70
Rumex acetosella	50
Teucrium scorodonia	40

Bare soil

Abundance in quadrats

	Abundance class										Distribution type
	1–10	11–20	21–30	31–40	41–50	51–60	61–70	71–80	81–90	81–100	
Wetland	–	–	–	–	–	–	–	–	–	–	–
Skeletal	–	–	–	–	–	–	2	1	–	1	–
Arable	–	–	–	–	–	–	–	–	–	–	–
Grassland	–	–	1	–	–	–	–	–	–	–	–
Spoil	3	2	1	–	–	–	2	3	1	–	3
Wasteland	1	–	1	2	–	–	–	1	–	1	–
Woodland	–	–	–	–	–	–	–	–	–	–	–
All habitats	17.4	8.7	13.0	8.7	–	–	17.4	21.7	4.3	8.7	3

Alisma plantago-aquatica L.

Alismataceae
Water Plantain

Life-form Polycarpic perennial, rosette-forming hydrophyte or helophyte. Scape erect; leaves typically ovate, <20 000 mm^2; roots non-mycorrhizal.
Phenology Leaves emerge in spring and plant flowers June to August. Seed set from August to October.
Height Foliage to 600 mm; flowers to 1 m.
Seedling RGR No data. SLA 31.0. DMC 14.3.
Nuclear DNA amount 20.6 pg; $2n = 14^*$ (*BC*); $2x^*$.
Established strategy Intermediate between competitive-ruderal and ruderal.
Regenerative strategies B$_s$, V. Regeneration occurs by seed and by lateral spread, and is poorly understood (see 'Synopsis'); ?SB IV.

Flowers Pale lilac, hermaphrodite, self-compatible, insect- or wind-pollinated (Sculthorpe 1967); >100 in a whorled, branched inflorescence.
Germinule 0.27 mg, 2.4 × 1.6 mm, achene, up to *c.* 20 produced per flower.
Germination Epigeal; hard-coat dormancy.
Biological flora None.
Geographical distribution The British Isles, although rarer in the N and absent from the N of Scotland (94% of British vice-counties), *ABIF* page 668; widespread in most of Europe (87% of European territories); N and E Africa, W Asia, N and S America, Australia.

Habitats Exclusively a wetland species, occurring at the edge of ponds, ditches, canals and lakes, either in mire at the margin of open water or less frequently in shallow water. Occasionally in shaded wetland habitats.
Gregariousness Sparse.
Altitude Not recorded above 200 m. [Up to 380 m.]
Slope Like most wetland species, mainly on flat ground and gentle slopes.
Aspect No data.
Hydrology Typical of the transitional area between mire and open water.

A.p-a. as an aquatic is restricted to stagnant or slow-flowing water, and has been observed at depths of up to 450 mm. However, it is most typical of topogenous mire which is temporarily submerged in winter.
Soil pH Most frequent on neutral soils and absent from acidic soils of pH <4.0.
Bare soil Most frequent in sites with high exposures of bare soil.
△ **ordination** Distribution confined to vegetation associated with fertile habitats subject to moderate disturbance.
Associated floristic diversity Typically low.

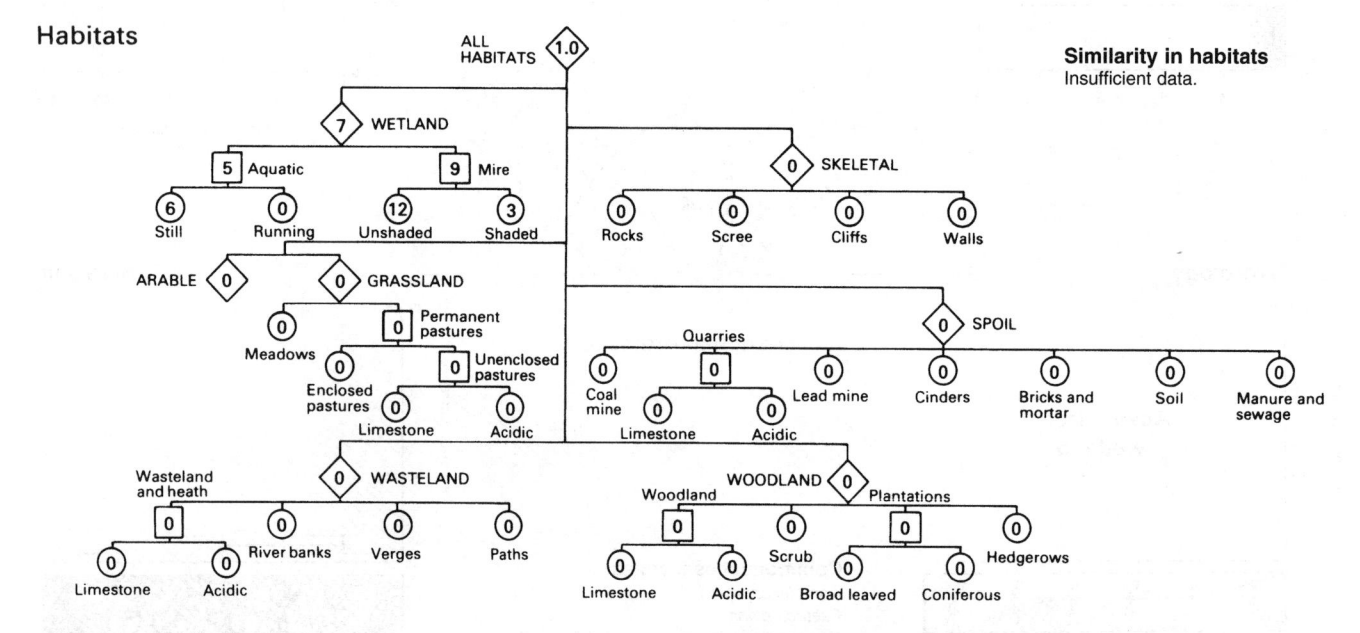

Habitats

Similarity in habitats
Insufficient data.

Synopsis *A.p-a.* exploits the boundary between aquatic and mire systems, e.g. ditch and pond margins. The leaves, though moderately large, are all basal, an arrangement which renders the species susceptible to competitive exclusion at sites where taller, productive, clonal, wetland dominants are established. Consequently, *A.p-a.* is promoted by ditch clearance and other disturbance factors. *A.p-a.* is particularly characteristic of sites with a fluctuating water table. Here, young plants survive inundation as phenotypes in which all the leaves are unusually narrow and submerged. In shallow water young plants with floating leaves are also frequent. Except in the case of some young submerged plants, leaves die back in autumn (Bjorkqvist 1967). The nuclear DNA amount in *A.p-a.* is relatively high and suggests that leaf growth in the spring may benefit from the expansion of tissues developed during the previous summer (cf. *Ranunculus bulbosus*). *A.p-a.* produces an acrid sap which may confer some defence against herbivory; this was formerly administered medically with dire consequences for the patient (Sculthorpe 1967). In common with other members of the Alismataceae, a family of wetland species, *A.p-a.* possesses a poorly developed water conduction system, with vessels confined to the roots (Dahlgren *et al.* 1985). *A.p-a.* regenerates freely by seed, and an average plant produces *c.* 3600 achenes (Salisbury 1942). Freshly-shed seed exhibit hard-coat dormancy. Germination, in autumn or spring (Bjorkqvist 1967), occurs mainly on bare mud, and massive populations have emerged subsequent to summer droughting (Salisbury 1942). The comments of Salisbury (1942) imply the existence of a persistent seed bank. *A.p-a.* is also capable of vegetative propagation through branching of the vegetative axis (Bjorkqvist 1967, Charlton 1973), but observations suggest that regeneration by seed is more important except perhaps in closed communities. Seeds may be effectively transported by water, can survive ingestion by birds and fishes, and may adhere to birds or to clothing (Ridley 1930). *A.p-a.* is frequently an early colonist of newly-created wetland habitats.

Current trends Because of the extensive destruction of wetlands, *A.p-a.* cannot be considered an increasing species, but the species is not under threat and appears to be favoured by the combined effect of eutrophication and disturbance.

Altitude

Slope

Aspect

Unshaded

Shaded

Soil pH

Associated floristic diversity

Species per m²

Per cent abundance of *A. plantago-aquatica*

Triangular ordination

Hydrology

Commonest associates	%
Lemna minor	73
Callitriche stagnalis	36
Holcus lanatus	36
Juncus effusus	36
Poa trivialis	36

Bare soil

Abundance in quadrats

	Abundance class										Distribution type
	1–10	11–20	21–30	31–40	41–50	51–60	61–70	71–80	81–90	81–100	
Wetland	23	4	–	–	–	–	–	–	–	–	1
Skeletal	–	–	–	–	–	–	–	–	–	–	–
Arable	–	–	–	–	–	–	–	–	–	–	–
Grassland	–	–	–	–	–	–	–	–	–	–	–
Spoil	–	–	–	–	–	–	–	–	–	–	–
Wasteland	–	–	–	–	–	–	–	–	–	–	–
Woodland	–	–	–	–	–	–	–	–	–	–	–
All habitats	85.2	14.8	–	–	–	–	–	–	–	–	1

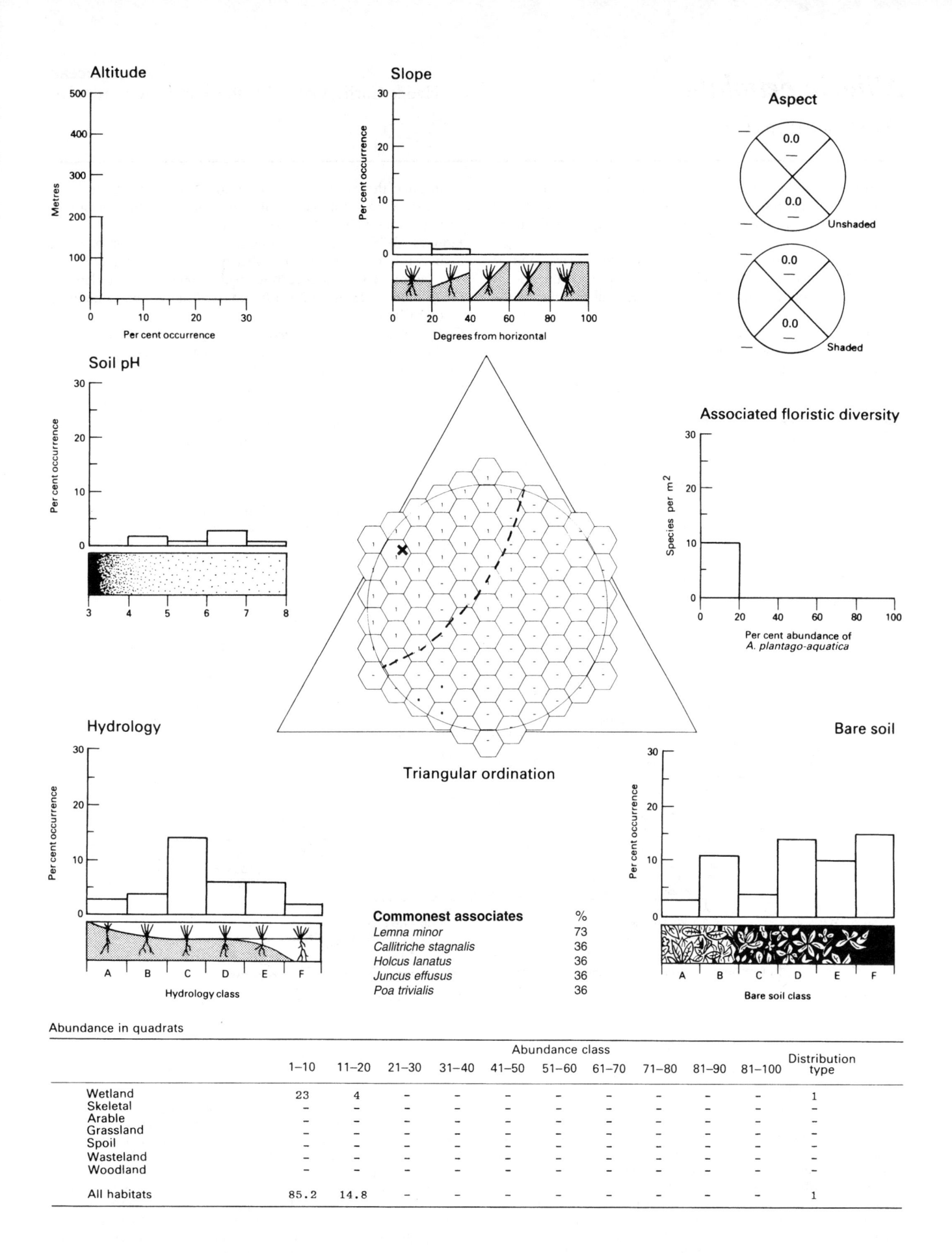

Alliaria petiolata
(Bieb.) Cavara and Grande

Brassicaceae
Hedge Garlic, Garlic Mustard, Jack-by-the-hedge

Life-form Winter-annual, monocarpic (or perhaps rarely polycarpic) perennial, semi-rosette hemicryptophyte. Shoot erect; leaves rounded, <10 000 mm^2; tap-rooted, non-mycorrhizal.

Phenology Overwintering as a rosette of small leaves and as dormant seeds. Flowers produced from April to June. Seeds shed from July onwards and some often still present on the dead plant in August.

Height Foliage and flowers to 1200 mm.

Seedling RGR No data. SLA 35.2. DMC 14.2.

Nuclear DNA amount 3.9 pg; $2n = 36^*$ (Bennett & Smith, 1976); $2n = 36^*, 42^*$ (*BC*); $2x^*$, genus polyploid in origin.

Established strategy Competitive-ruderal.

Regenerative strategies B_s. Regenerates exclusively by seed, which germinates in late winter or spring (*Biol Fl*). Seeds have a limited persistence in the soil (*Biol Fl*). *A.p.* may perennate through the formation of adventitious buds, but there is no evidence that this constitutes an important mechanism of regeneration (*Biol Fl*); SB IV.

Flowers White, hermaphrodite, homogamous, visited by small insects but apparently automatically self-pollinated (*Biol Fl*); on a racemose inflorescence, often with >100 flowers per plant.

Germinule 2.25 mg, 3.2 × 1.0 mm, seed in a siliqua containing up to 20 seeds.

Germination Epigeal; seed-coat tough and impermeable, dormancy broken by prolonged chilling (*Biol Fl*).

Biological flora Cavers *et al.* (1979).

Geographical distribution The British Isles, but less common in the N and W (91% of British vice-counties), *ABIF* page 251; Europe except for the extremes of N and S (85% of European territories); N Africa and Asia Minor; naturalized in N America.

Habitats Most frequent in lightly shaded habitats beside rivers and on road verges. Also at the margin of woodland, in hedgerows and beside paths. Infrequent in skeletal habitats and on spoil, and absent from agricultural land and wetlands.

Gregariousness Because of their robust habit, established plants usually occur at rather low densities.

Altitude Suitable habitats are most frequent in lowland areas, but populations occur up to 350 m, the British altitudinal limit for the species.

Slope *A.p.* is restricted to comparatively deep, moist soils which are concentrated in areas of gentle slope.

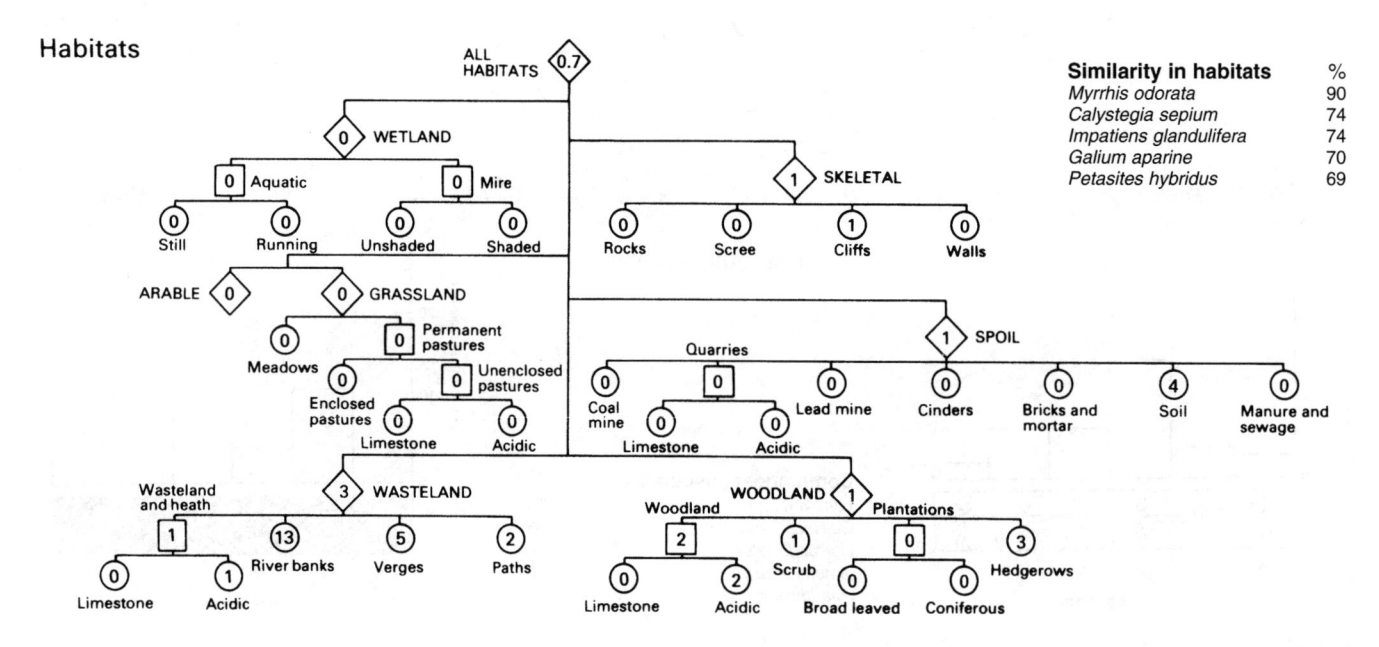

Habitats

Similarity in habitats	%
Myrrhis odorata	90
Calystegia sepium	74
Impatiens glandulifera	74
Galium aparine	70
Petasites hybridus	69

Aspect Exhibits a slight bias towards N-facing slopes in both shaded and unshaded habitats.

Hydrology Found on moist rather than waterlogged soils.

Soil pH Absent from highly acidic soils.

Bare soil Most frequently associated with moderate to high exposure of bare soil.

△ ordination Distribution widely scattered with high values occurring in disturbed, moderately productive vegetation; also represented in perennial communities of relatively productive sites.

Associated floristic diversity Intermediate.

Synopsis *A.p.* is a tall, short-lived herb characteristic of river banks, woodland and hedgerows on relatively moist fertile soils where the vigour of more-robust species is restricted by a combination of light to moderate shade and disturbance by factors such as occasional flooding on river banks. Although absent from grazed or trampled sites, *A.p.* responds to cutting by producing lateral flowering shoots and may persist on infrequently mown roadside verges. The leaves are characterized by unusually high concentrations of N, P, Ca, Fe and Na (*SIEF*) and smell of garlic when crushed. Their use in salads has been suggested (*Biol Fl*). *A.p.* is usually monocarpic, an unusual feature among species of shaded habitats, and occurs in habitats in which tree litter is rapidly broken down. The seed contains a large embryo but a long period of chilling appears necessary to break dormancy (*Biol Fl*). Seed often remains dormant for 18 months, and sometimes longer. Thus, *A.p.* can be considered to have a moderately persistent seed bank (*Biol Fl*). Seeds germinate in late winter or early spring and plants flower in the following year (*Biol Fl*). Regeneration through the formation of adventitious buds on the roots has also been recorded, but appears to be of rare occurrence (*Biol Fl*). Leaves are formed

Altitude
Slope
Aspect
Soil pH
Associated floristic diversity
Triangular ordination
Hydrology

Absent from wetland

Commonest associates	%
Poa trivialis	86
Galium aparine	66
Urtica dioica	51
Arrhenatherum elatius	31
Ranunculus ficaria	31

Bare soil

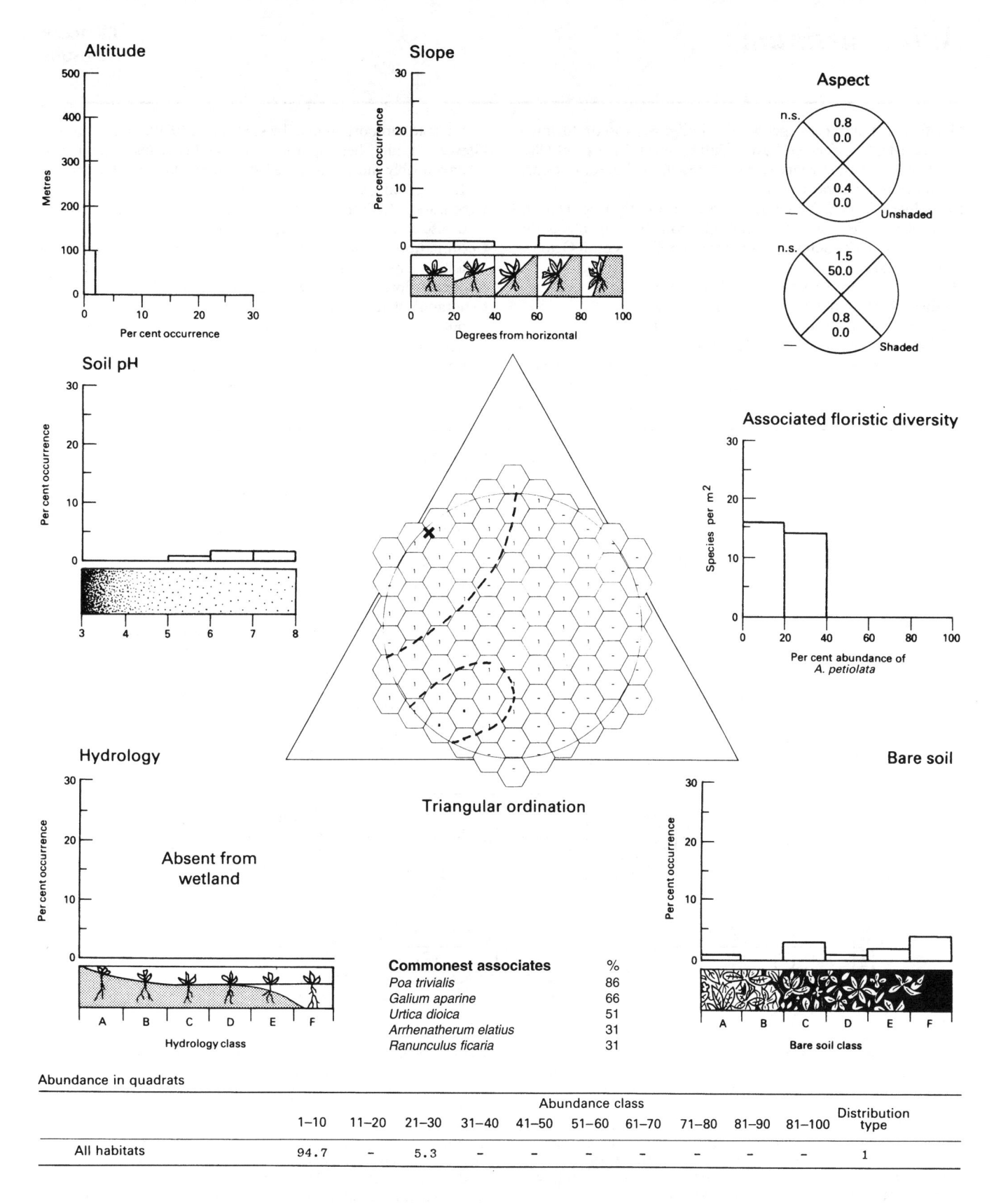

Abundance in quadrats

	\multicolumn{10}{c	}{Abundance class}	Distribution type
	1–10	11–20	21–30	31–40	41–50	51–60	61–70	71–80	81–90	81–100	
All habitats	94.7	–	5.3	–	–	–	–	–	–	–	1

in both autumn and spring and leaf area is at a minimum in midsummer, when the tree canopy is fully expanded. This phenology is the same as that of *Anthriscus sylvestris*, which exploits a similar range of habitats. However, A.p. has a greater dependence upon seed for regeneration, and perhaps for this reason is restricted to more-disturbed sites. The species is also more frequent in sites subject to winter flooding, and extends less frequently into drier and unshaded habitats. Two chromosome numbers are widely recorded, $2n = 36$ and $2n = 42$ (*Biol Fl*); these have not been linked to variation in morphology or ecology.

Current trends A.p. is found in disturbed fertile habitats and is more characteristic of secondary than of primary woodland (Peterken 1981). Furthermore, despite its comparatively large seeds and lack of well-defined mechanism of dispersal, A.p. frequently colonizes artificial habitats and is probably increasing.

Allium ursinum L.

Liliaceae
Ramsons

Life-form Polycarpic perennial, bulbous, rosette-forming geophyte. Scapes erect; leaves elliptic, usually 2 per plant (*Biol Fl*), <10 000 mm²; roots with VA mycorrhizas, but some deeper and weakly contractile (*Biol Fl*).
Phenology Vernal. New leaves appear in early spring. Flowers April to June. Seeds shed gradually from June to July as the plant dies back. Summer dormancy of bulb lasts until October (Ernst 1979).
Height Foliage to 250 mm; scape to 450 mm.
Seedling RGR No data. SLA 36.7. DMC 11.4.
Nuclear DNA amount 71.4 pg; $2n = 14^*$ (*BC*); $2x^*$.
Established strategy Stress-tolerant ruderal.
Regenerative strategies S. (V.). Produces numerous seeds which germinate during winter and spring. Larger plants, with three leaves, may produce two bulbs, but vegetative regeneration is much less important than by seed (Ernst 1979); SB II.
Flowers White, hermaphrodite, protandrous, insect-pollinated or probably more usually selfed (Knuth 1906, *Biol Fl*); often *c.* 15 in an umbel.
Germinule 4.27 mg, 2.5 × 2.4 mm, seed dispersed from a dehiscent three- to six-seeded capsule.
Germination Epigeal; seeds require warm-moist incubation followed by chilling.
Biological flora Tutin (1957), see also Ernst (1979).
Geographical distribution Widely distributed in the British Isles but rather local in the S Midlands and East Anglia (97% of British vice-counties), *ABIF* page 821; most of Europe northward to Norway, but absent from much of the NE and rare in the Mediterranean (67% of European territories).

Habitats Predominantly a plant of woodland, but also found in other moist shaded habitats such as shaded river banks and mires, with isolated records from scree and rock crevices. Occasionally in the open, especially on river banks, but seldom far from woodland. [Occurs in grikes of limestone pavement (Silvertown 1983).]
Gregariousness Patch-forming.
Altitude Distributed mainly in lowland areas but also observed to 270 m. [Up to 430 m.]
Slope As a result of its virtual absence from skeletal habitats *A.u.* is also absent from the steepest slopes.
Aspect Slightly more frequent on N-facing slopes, but more abundant in S-facing shaded habitats.
Hydrology Occurs on moist rather than waterlogged or submerged soils, and not usually a wetland species.
Soil pH Largely absent from soils of pH <4.5, and most abundant within the pH range 6.0–7.5.

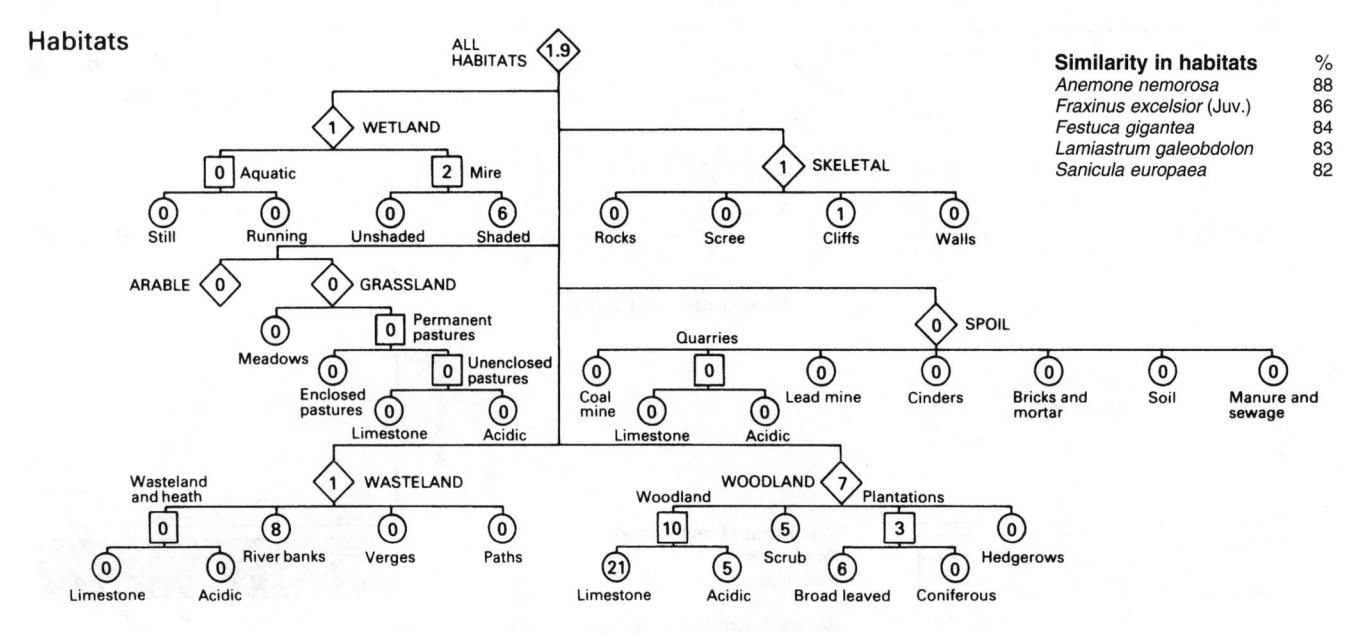

Habitats

Similarity in habitats	%
Anemone nemorosa	88
Fraxinus excelsior (Juv.)	86
Festuca gigantea	84
Lamiastrum galeobdolon	83
Sanicula europaea	82

Bare soil Associated with a wide range of values, but generally in areas with discontinuous cover of herbaceous vegetation. Sites with low exposure of bare soil frequently have large quantities of bryophyte mat or tree litter.
△ **ordination** Occurs in most vegetation types except those which are heavily disturbed. Distribution centred on vegetation of moderate productivity subject to some disturbance. Most typically associated with alluvial terraces subject to winter flood damage and usually also experiencing disruptive impacts of seasonal shading by trees.
Associated floristic diversity Typically low to intermediate.
Synopsis *A.u.* is a bulb-forming woodland plant smelling pungently of garlic when crushed. *A.u.* has a specialized phenology and a narrow ecological range, and no ecotypes have been described (*Biol Fl*). *A.u.* is characterized by a very high nuclear DNA amount (71.4 pg), consistent with the vernal phenology and determinate growth habit. Bulbs and roots analysed in midwinter have exceptionally high concentrations of the reserve carbohydrate fructan (Hendry & Brocklebank unpubl.). Like other vernal species, *A.u.* is essentially a shade-avoiding species: under continuously dense canopies growth is poor and flowering is often inhibited (*Biol Fl*). *A.u.* is typical of moist, well-drained, base-rich soils and, particularly on alluvial terraces, the foliage contains relatively high concentrations of P (*SIEF*). *A.u.* is sensitive both to waterlogging and to drought (*Biol Fl*). Susceptibility to drought and restriction to relatively mild oceanic climates (*Biol Fl*) may be related in part to the poor system of water conduction in which vessels are confined to the roots (Dahlgren *et al.* 1985). Furthermore, leaves of *A.u.* persist longer in moist habitats (*Biol Fl*), and during the cool moist summer of 1985 some leaves were still green in a Sheffield woodland during August, nearly 2 months after normal senescence. In Britain *A.u.* extends into unshaded habitats in the moist climate of the W (*Biol Fl*). The young shoot of *A.u.* can spear its way through dense litter and *A.u.* is one of the most successful exploiters of beech plantations, where litter accumulation is considerable and shade is

Altitude

Slope

Aspect

Soil pH

Associated floristic diversity

Hydrology

Commonest associates	%
Poa trivialis	59
Ranunculus ficaria	40
Mercurialis perennis	38
Festuca gigantea	36
Galium aparine	32

Triangular ordination

Bare soil

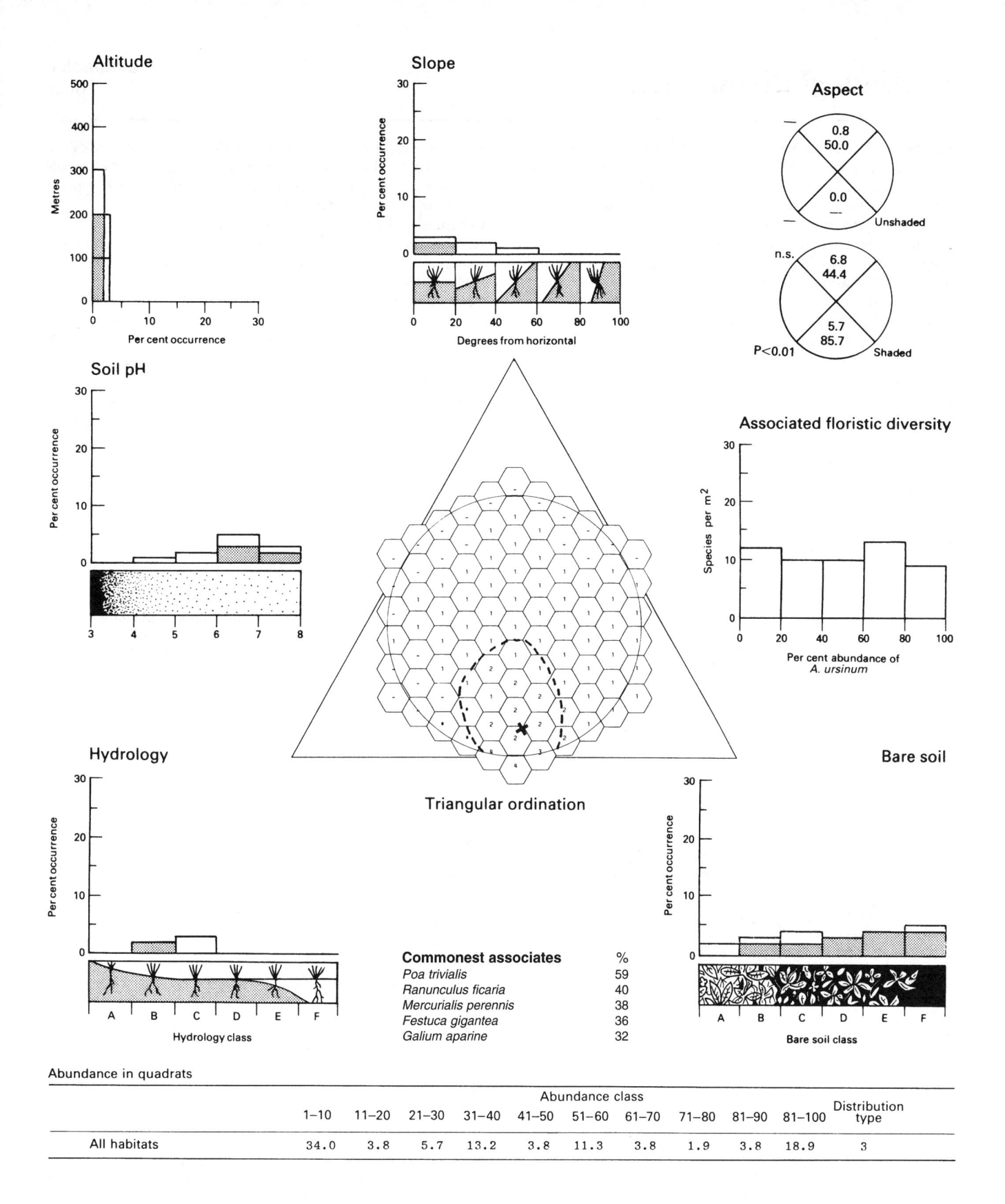

Abundance in quadrats

	\multicolumn{10}{c}{Abundance class}	
	1–10	11–20	21–30	31–40	41–50	51–60	61–70	71–80	81–90	81–100	Distribution type
All habitats	34.0	3.8	5.7	13.2	3.8	11.3	3.8	1.9	3.8	18.9	3

dense in summer. Foliage is poisonous to stock if consumed in large quantities (Cooper & Johnson 1984). *A.u.* regenerates mainly by seed and annual production may exceed 9000 seeds m^{-2} in dense stands (*Biol Fl*). However, the seeds normally fall within 2.5 m of the parent (Ernst 1979). Plants usually take 3 years to flower and persist for *c.* 8 years (*Biol Fl*, Ernst 1979). Roots of seedlings are close to the surface, where supplies of macronutrients are high, while those of plants of reproductive age are concentrated in the less fertile A$_2$ horizon (Ernst 1979). *A.u.* is exceptional among species forming monospecific stands in that dominance does not result in differential mortality of the smallest individuals (Ernst 1979). A further ssp. is recorded from Europe (*FE5*).

Current trends Probably decreasing, since *A.u.* has poor mobility and is not a common colonist of sites distant from existing populations.

(*) *Alnus glutinosa* (L.) Gaertner

Betulaceae
Alder

Life-form Deciduous tree with a fissured bark and an unusually narrow crown. Leaves ovate, sometimes <5000 mm²; roots nodulated and normally with ecto-mycorrhizas.
Phenology Flowers February to April before the leaves expand; seed set October onwards.
Height Foliage and flowers to 20 m.
Seedling RGR No data. SLA 14.6. DMC. No data.
Nuclear DNA amount 1.1 pg; $2n = 28$ (*FE1*); $2x$, genus of polyploid origin.
Established strategy Stress-tolerant competitor.
Regenerative strategies W. Reproduces by means of wind-dispersed or water-borne seeds which germinate during spring. No evidence of a persistent seed bank; SB II.

Flowers Monoecious, wind-pollinated, self-sterile (McVean 1955a); >100 male flowers in pendulous catkins, <50 female flowers in small upright to spreading catkins; each type in groups of 3-6 at end of branches.
Germinule 1.3 mg, 2.7 × 2.7 mm, winged nutlet.
Germination Epigeal, t_{50} 3 days; germination improved by dry storage (McVean 1955b); L = D (McVean 1955b).
Biological flora McVean (1953).
Geographical distribution The British Isles, but increasingly frequent in the high-rainfall areas of the NW (99% of British vice-counties), *ABIF* page 134; Europe except the extreme N and S (85% of European territories and 3% naturalized); W Asia and N Africa.

Habitats Seedlings and saplings of *A.g.* are largely restricted to moist or waterlogged habitats. They are recorded from river banks, mire, walls and cliffs beside streams and from damp woodland.
Gregariousness Sparse.
Altitude Suitable wetland habitats are more frequent in lowland areas and seedlings recorded only to 240 m. [Up to 510 m (Wilson 1956), but good seed not produced above 305 m (McVean 1955a).]
Slope Juveniles observed over most angles of slope.

Aspect Slightly more frequent on N-facing slopes in unshaded habitats and on S-facing slopes in shaded situations.
Hydrology Consistent with their requirement for high oxygen tensions for effective establishment (McVean 1955b), juveniles are most frequent on soligenous mire or on moist, freely-drained alluvial soils on river floodplains.
Soil pH Occurs over a wide range of values, but less frequent on strongly acidic soils.

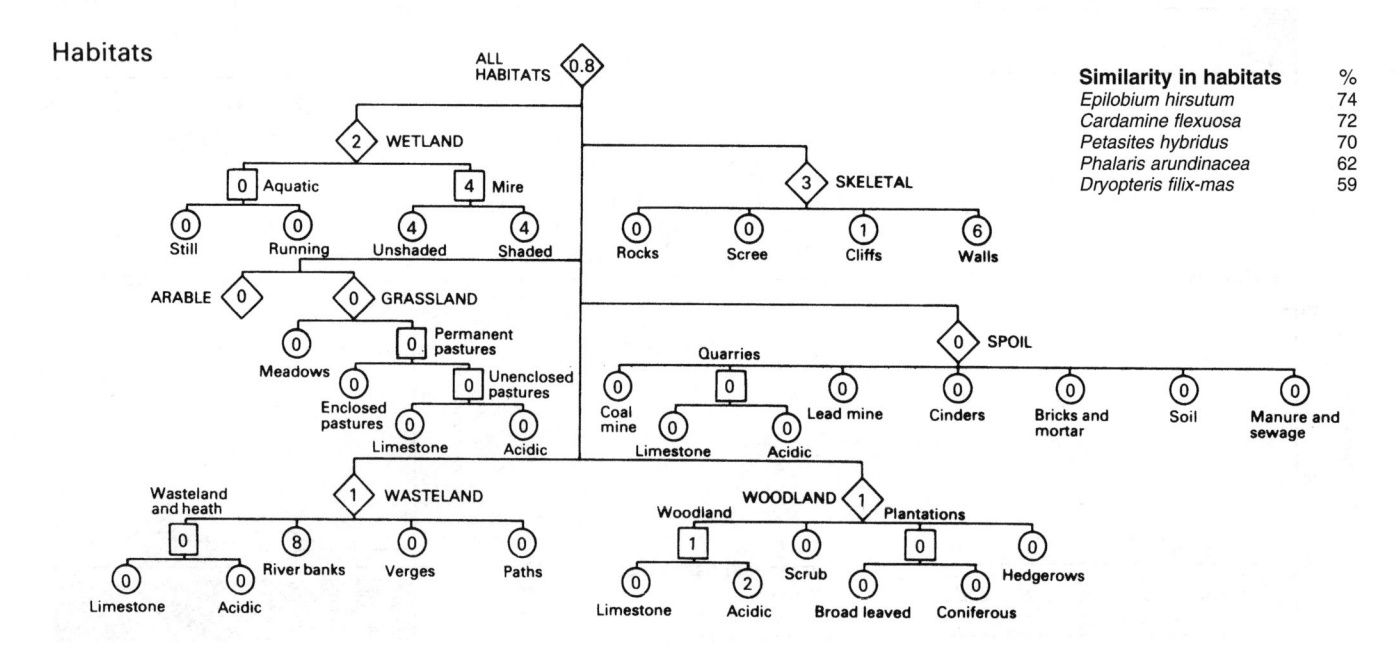

Habitats

Similarity in habitats	%
Epilobium hirsutum	74
Cardamine flexuosa	72
Petasites hybridus	70
Phalaris arundinacea	62
Dryopteris filix-mas	59

Bare soil Mainly associated with habitats having intermediate values for bare soil. Sites are often characterized by moderate amounts of litter or bryophyte mat.
△ **ordination** Juveniles associated with relatively fertile and infrequently disturbed vegetation types. Absent from vegetation subject to severe stress or disturbance.
Associated floristic diversity Typically intermediate.
Synopsis *A.g.* is a deciduous tree with diffuse porous wood, growing up to *c.* 20 m in height and living for a maximum of *c.* 300 years (Bernatzky 1978) (*c.* 120 years (*Biol Fl*)). Alders cover *c.* 11 000 ha of British woodland (>1% of the total) (Steele & Peterken 1982). Seedlings are found in the same types of habitat as the adult tree, i.e. on river and stream banks, in soligenous mire or forming more-extensive woodland on river floodplains. Elsewhere the species occurs with other trees in moist plateau woodland (Rackham 1980). This close parallel between the distribution of seedlings and that of the canopy tree is related to the conditions required for seedling establishment. Radical elongation is slow in seedlings (McVean 1955b). As a result, seedlings are very susceptible to droughting as well as to cold periods in early spring (McVean 1955b). Further, high humidity and high oxygen tension are required for germination (McVean 1955b), and seedling establishment is dependent upon high light intensity (McVean 1956a). These critical factors limit the ecological range of the adult tree. Saplings have the potential to survive on substrates, e.g. colliery spoil, where surface layers may become severely desiccated. In general, mature alder woods do not provide conditions suitable for regeneration; even-aged populations of seedlings often appear, but these usually die off, to be replaced by other species (McVean 1956a). Thus, *A.g.* may be regarded as a seral species. Succession of wetlands involving *A.g.* and *Salix cinerea* tend to occur under more nutrient-rich conditions than those associated with *B.pub.* (Tucker & Fitter 1981, Wiegers 1985). Patterns of succession are at present poorly understood (McVean 1956c, Wheeler 1980, Wiegers 1985). Seeds are effectively dispersed either by wind-drift over standing water or in water currents (McVean 1955b), and consequently linear populations are common along the strand-line.

Altitude
Slope
Aspect
Soil pH
Associated floristic diversity
Triangular ordination

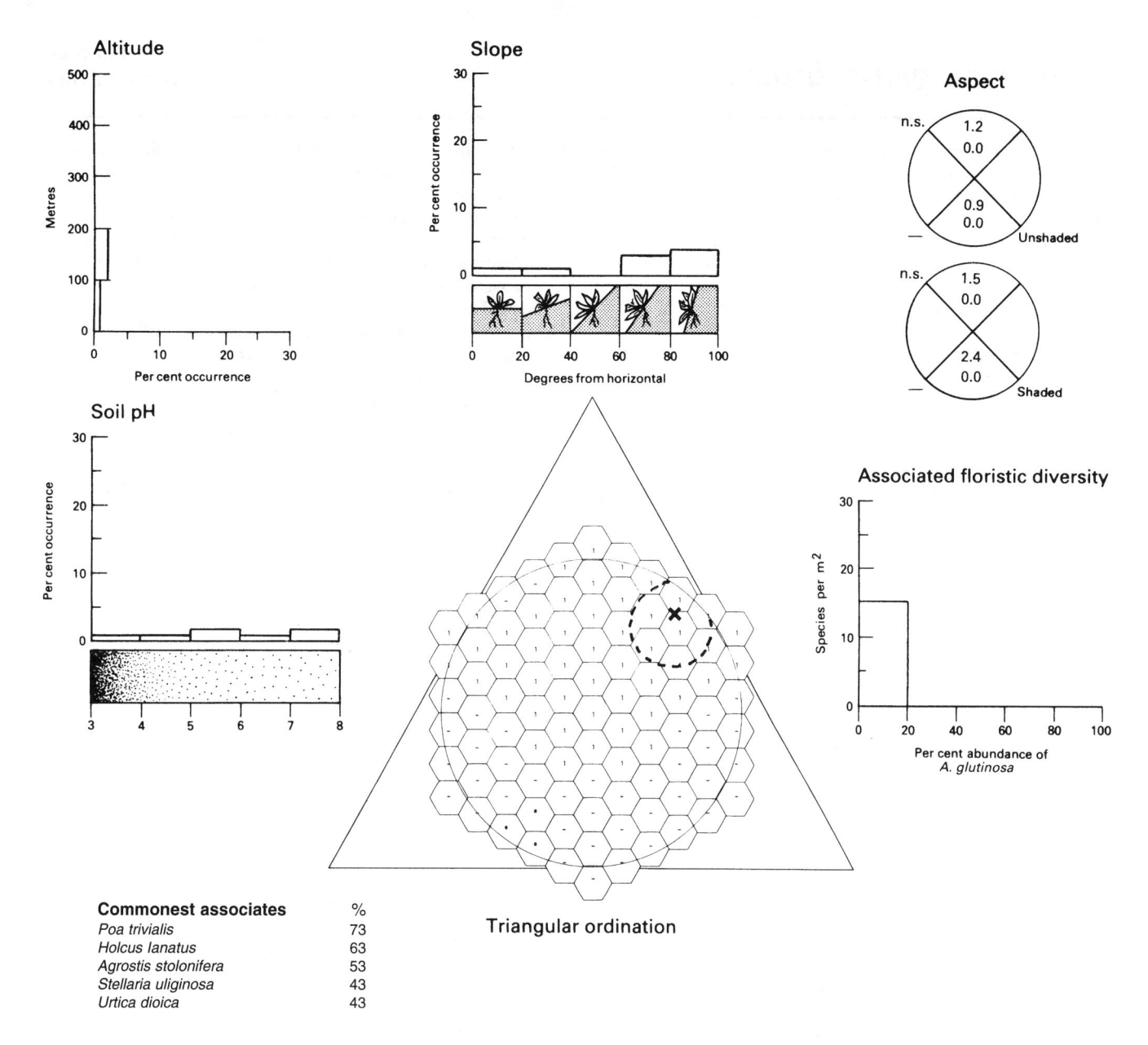

Commonest associates %
Poa trivialis 73
Holcus lanatus 63
Agrostis stolonifera 53
Stellaria uliginosa 43
Urtica dioica 43

However, McVean (1955b) suggests that the number of seeds dispersed by wind >30 m from the parent tree is normally insufficient to ensure establishment. Seedlings are more tolerant of waterlogging than those of *Betula pubescens*, but they are more restricted to continuously moist habitats (McVean 1956a). Cold treatment reduces the minimum temperature for germination to 7°C and seeds germinate in early spring (McVean 1955b). No persistent seed bank is formed (McVean 1955b). The tree roots contain the nitrogen-fixing actinomycete bacterium *Frankia* spp. within root nodules, and the nitrogen content of the tree may double between February and May (Kirchner *et al.* 1908). The fixation of nitrogen is an oxygen-demanding process, and nodules are restricted to the upper regions of the soil (McVean 1956b). The roots have the capacity to diffuse oxygen (McVean 1956b) and are at a wide range of depths, including some which are deeply penetrating and tolerant of reducing conditions (McVean 1956b). As a result *A.g.* can survive in conditions of changing water table better than most other trees, and produces adventitious roots from the bole in response to flooding (McVean 1956b). As in many trees, the developmental processes of seed formation are slow. Catkins are formed in July; pollination occurs the following February to March; ovules appear in June; fertilization occurs in July to August; seed ripens at end of September (McVean 1955a). Unlike *Betula* spp., the female catkins are not enclosed in leaf buds during winter, and this, coupled with the early pollination and extended period before fertilization, may restrict the capacity of *A.g.* to set seed at high altitudes and under severe winter or early spring conditions (McVean 1955a). The percentage of good seeds may be low, but mast years do occur (McVean 1955a,b). Although found over a wide range of pH (3.4-7.2 (Peterken 1981)), the species is more frequently observed on non-calcareous strata. This distribution appears to hold for continental Europe, where *A. incana* (L.) Moench. prevails on calcareous soils (McVean 1956a). Compared with large willows such as *Salix fragilis* L., the species appears to be more tolerant of acidic, less-fertile soils and shaded conditions, and is less capable of exploiting fertile, base-rich wetland sites. However, further studies are required to elucidate the factors regulating the distribution of alders and willows. *A.g.* forms a relatively dense canopy and its litter is not usually persistent. The number of insects associated with *A.g.* is rather low for a native deciduous tree (Southwood 1961), suggesting that the leaves are relatively palatable (see Wratten *et al.* 1981). No inducible defence system against insect predators has been identified (Edwards & Wratten 1985). *A.g.*, which is one of the faster-growing trees (Jones 1945), can be used for reclamation of coal-mine spoil, but certain introduced species which grow even more rapidly are generally preferred. The hybrid with the introduced *A.i.* is widely planted and may have arisen spontaneously (Stace 1975).

Current trends Plantations of *A.g.* occur in certain low-lying areas but the species is no longer widely planted and is probably decreasing through the destruction of habitats in many lowland districts. Remains relatively common in river- and stream-side habitats.

Alopecurus geniculatus L.

Poaceae
Marsh Foxtail

Life-form Polycarpic perennial protohemicryptophyte. Stems decumbent, rooting at the nodes; leaves linear, often <500 mm^2; roots shallow, with VA mycorrhizas.
Phenology Shoot extension in spring, followed by flowering in June and July. Seed set mainly in July and August.
Height Foliage usually <100 mm; flowering shoots and some leaves to 250 mm.
Seedling RGR 1.0–1.4 week^{-1}. SLA 33.4. DMC 19.3.
Nuclear DNA amount 14.2 pg; $2n = 28$ (*FE5*); $4x$.
Established strategy Competitive-ruderal.
Regenerative strategies V, ?S. Capable of regeneration by vegetative means and by seed (see 'Synopsis'); ?SB II or IV.

Flowers Grey-purple, hermaphrodite, wind-pollinated, self-compatible (Wiemarck 1968); >100 in a spike-like panicle (1 per spikelet).
Germinule 0.38 mg, 1.5 × 0.7 mm, caryopsis dispersed with an attached lemma and an hygroscopic bristle (3.0 × 1.4 mm).
Germination Epigeal; t_{50} 5 days; dormancy broken by dry storage at 20°C.
Biological flora None.
Geographical distribution The British Isles (99% of British vice-counties), *ABIF* page 782; most of Europe, although rare in the extreme S (74% of European territories); temperate Asia; introduced in N America.

Habitats A wetland species occurring in damp meadows, on marshy ground and at the edges of ponds, streams and ditches. Less frequent in other disturbed unshaded habitats, particularly those which are both wet and fertile. Occasional records from manure heaps, mining waste, arable land and wasteland.
Gregariousness Intermediate.
Altitude Wide-ranging. [Up to 600 m.]
Slope As with most wetland species, largely restricted to flat and gently sloping ground.

Aspect Insufficient data.
Hydrology Largely restricted to habitats which are flooded in winter. These include topogenous mire, hollows in pasture and the margins of ponds, ditches and reservoirs.
Soil pH Most records from soils in the pH range 5.5–7.0.
Bare soil Mainly in habitats with a large exposure.
△ **ordination** Present only in relatively productive and moderately disturbed vegetation types.
Associated floristic diversity Typically intermediate.

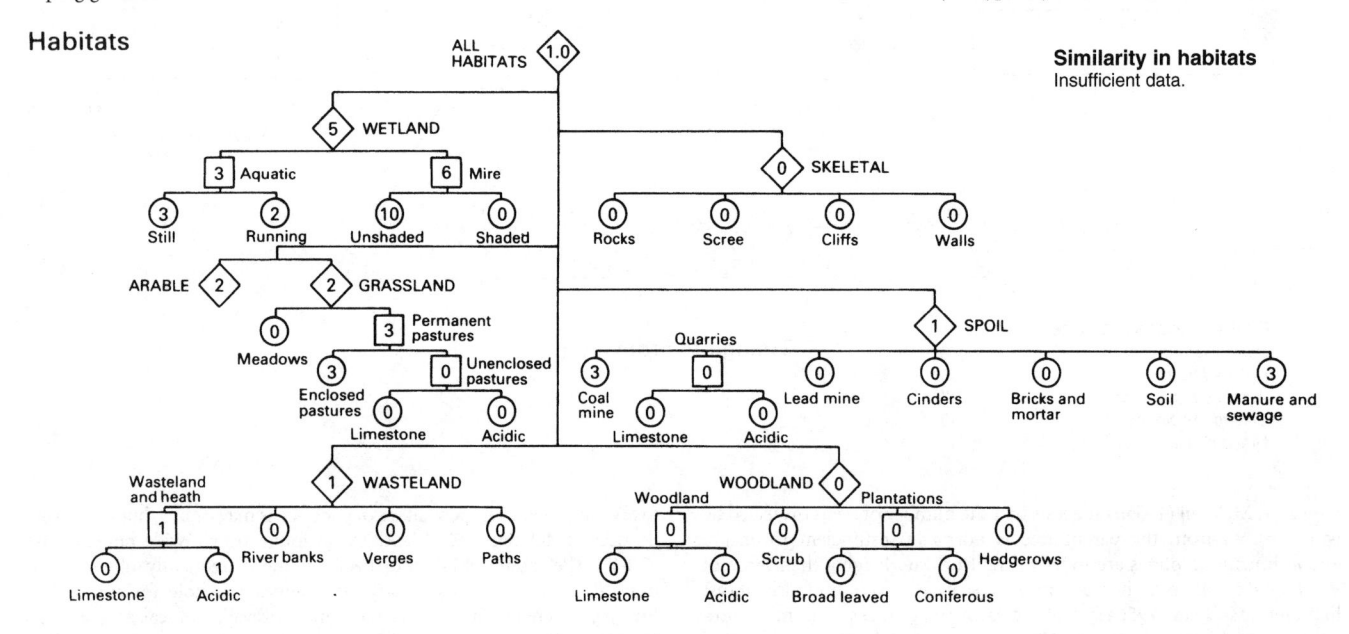

Habitats

Similarity in habitats
Insufficient data.

Synopsis A relatively common species which has not received sufficient attention from plant ecologists. *A.g.* is a low-growing grass, usually associated with fertile sites in which the growth of potential dominants is restricted by inundation by flood water in winter and sometimes also by exposure to summer grazing. *A.g.* closely resembles *Agrostis stolonifera* in growth form, and the two frequently occur together. There is considerable difficulty in distinguishing between seedlings of *A.g.* and *A.s.* under field conditions. However, apart from the fact that *A.s.* is ecologically wide-ranging, the differences between the two species are slight. Within the Sheffield region the only common habitat which *A.g.* regularly exploits to a greater extent than *A.s.* is the margin of reservoirs. *A.g.* is often a dominant species on bare mud exposed as a result of a fall in water level during summer. We suspect that the higher nuclear DNA amount of *A.g.* (14.2 pg) relative to that of *A.s.* (7.0 pg) is associated with an ability to exploit wetland sites while they are still cool and wet, during spring and early summer. Consistent with this hypothesis is the fact that *A.g.* flowers and sets seed earlier than *A.s.* Whether the two species differ also in rooting depth and tolerance of anaerobic soil conditions is not known. *A.g.* has also been observed as a dominant of wet mowing meadows, where its abundance may be related to a capacity to set seed early, before the grass is cut. Like *A.s.*, *A.g.* often forms extensive carpets by means of long stolons (up to 450 mm). Shoots detached as a result of disturbance also readily re-root to form new plants. Seeds germinate from April to September (Roberts 1986b). A persistent seed bank might be predicted in view of the disturbed nature of the habitats exploited by *A.g.* Although one has been recorded by Stieperaere and Timmerman (1983) and Roberts (1986a), several studies, including those of Thompson (1977), have failed to confirm these observations. It is also not clear to what extent the shoots and foliage of *A.g.* are capable of overwintering underwater along reservoir margins. Hybrids have been recorded with *A. aequalis* Sobol. and *A. pratensis* (Stace 1975).
Current trends Uncertain. In view of its wetland distribution the species is likely to decrease as a consequence of habitat destruction. It is not clear to what extent this trend is offset by the capacity of *A.g.* to function as an effective colonist of disturbed artificial habitats.

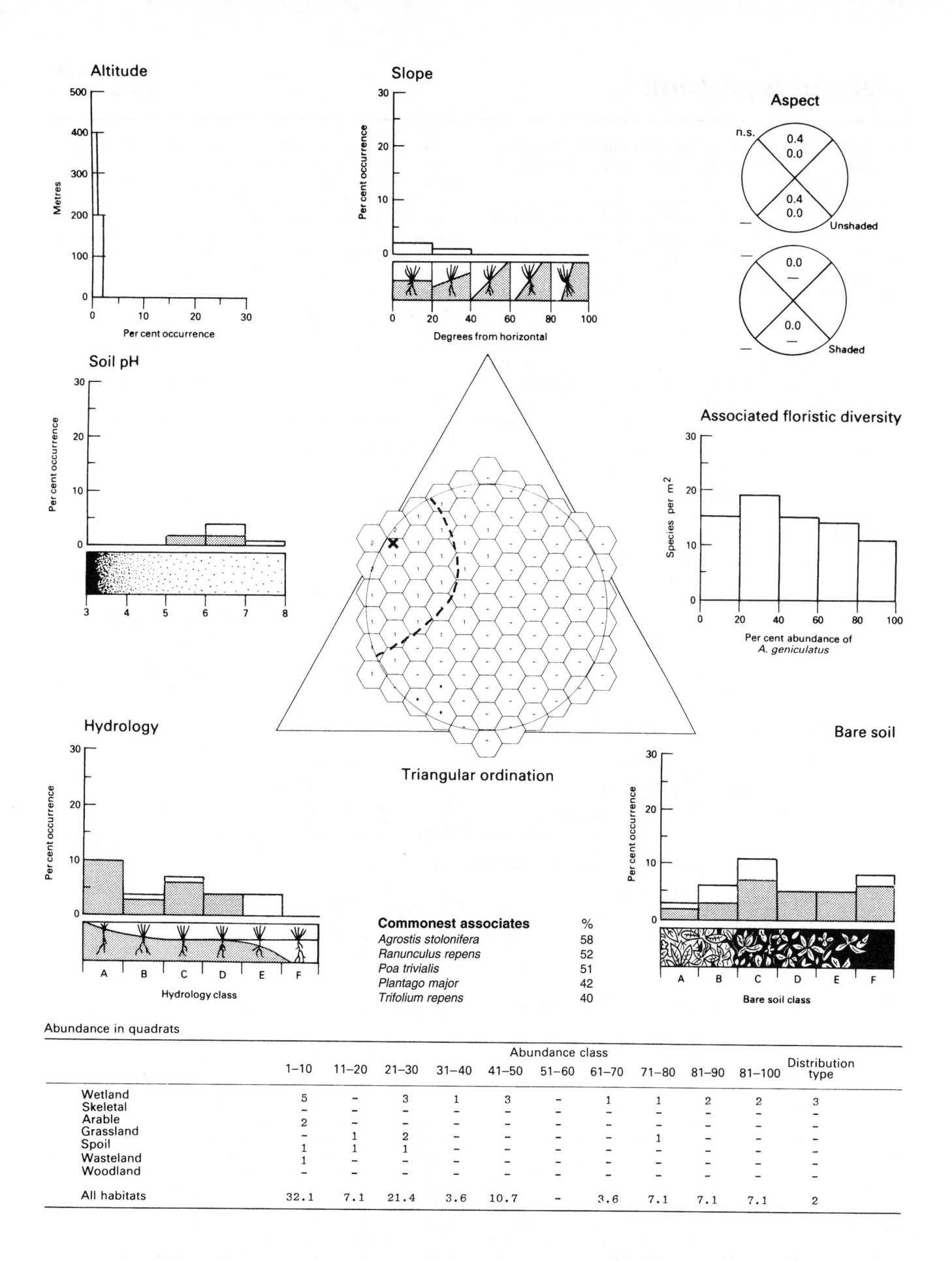

Altitude
Slope
Aspect
Soil pH
Associated floristic diversity
Species per m²
Per cent abundance of *A. geniculatus*

Hydrology
Triangular ordination
Bare soil

Commonest associates	%
Agrostis stolonifera	58
Ranunculus repens	52
Poa trivialis	51
Plantago major	42
Trifolium repens	40

Abundance in quadrats

	1–10	11–20	21–30	31–40	41–50	51–60	61–70	71–80	81–90	81–100	Distribution type
Wetland	5	–	3	1	3	–	1	1	2	2	3
Skeletal	–	–	–	–	–	–	–	–	–	–	–
Arable	2	–	–	–	–	–	–	–	–	–	–
Grassland	–	1	2	–	–	–	–	1	–	–	–
Spoil	1	1	1	–	–	–	–	–	–	–	–
Wasteland	1	–	–	–	–	–	–	–	–	–	–
Woodland	–	–	–	–	–	–	–	–	–	–	–
All habitats	32.1	7.1	21.4	3.6	10.7	–	3.6	7.1	7.1	7.1	2

Alopecurus pratensis L.

Poaceae
Meadow Foxtail

Life-form Tufted, polycarpic perennial protohemicryptophyte. Shoots erect; leaves linear, <4000 mm²; roots with VA mycorrhizas.
Phenology Winter green and showing early spring growth (Spedding & Diekmahns 1972). Flowers April to June and seed set in June and July. Summer leaves short-lived.
Height Foliage mainly <300 mm; flowering stem to 1 m.
Seedling RGR 1.0-1.4 week^{-1}. SLA 27.3. DMC 25.9.
Nuclear DNA amount 13.6 pg (Olszewska & Olsiecka 1982); $2n = 28$ (*FE5*); $4x$.
Established strategy Intermediate between C-S-R and competitor.
Regenerative strategies S, ?V. By freshly shed seed in vegetation gaps during autumn; available evidence (e.g. Roberts 1986b) suggests that *A.p.* does not have a buried seed bank. *A.p.* has a restricted capacity to increase the size of tuft by producing new shoots, but rarely forms extensive stands; SB I.
Flowers Grey-purple, hermaphrodite, protogynous, wind-pollinated, usually self-incompatible (Weimarck 1968); >100 in spike-like panicles (1 per spikelet).
Germinule 0.71 mg, 2.7 × 1.4 mm, caryopsis dispersed with an attached lemma (5.5 × 2.0 mm) bearing an hygroscopic bristle.
Germination Epigeal; t_{50} 1 day; <5-35°C; L = D.
Biological flora None.
Geographical distribution The British Isles except in mountainous regions, local in Ireland (100% of British vice-counties), *ABIF* page 781; most of Europe except for the extreme NE and the Mediterranean region (87% of European territories); N Asia; introduced in Australia, New Zealand and the USA.

Habitats Essentially a species of productive meadows and pastures. *A.p.* also occurs less frequently on road verges, in hedgerows, and in a range of other grassy habitats. *A.p.* is not found in wetlands or in arable land, and is of rare occurrence in woodland and skeletal habitats.
Gregariousness Sparse to intermediate.
Altitude Occurs at altitudes of up to 400 m. Because permanent pasture and hay meadows are more abundant in the upland region *A.p.* is more abundant at higher altitudes. [Up to 450 m.]

Slope The most suitable habitats are situated on gentle slopes.
Aspect Insufficient data.
Hydrology Present in moist, rather than waterlogged sites.
Soil pH Most frequent and abundant on mildly acidic soils in the pH range 5.0-7.0.

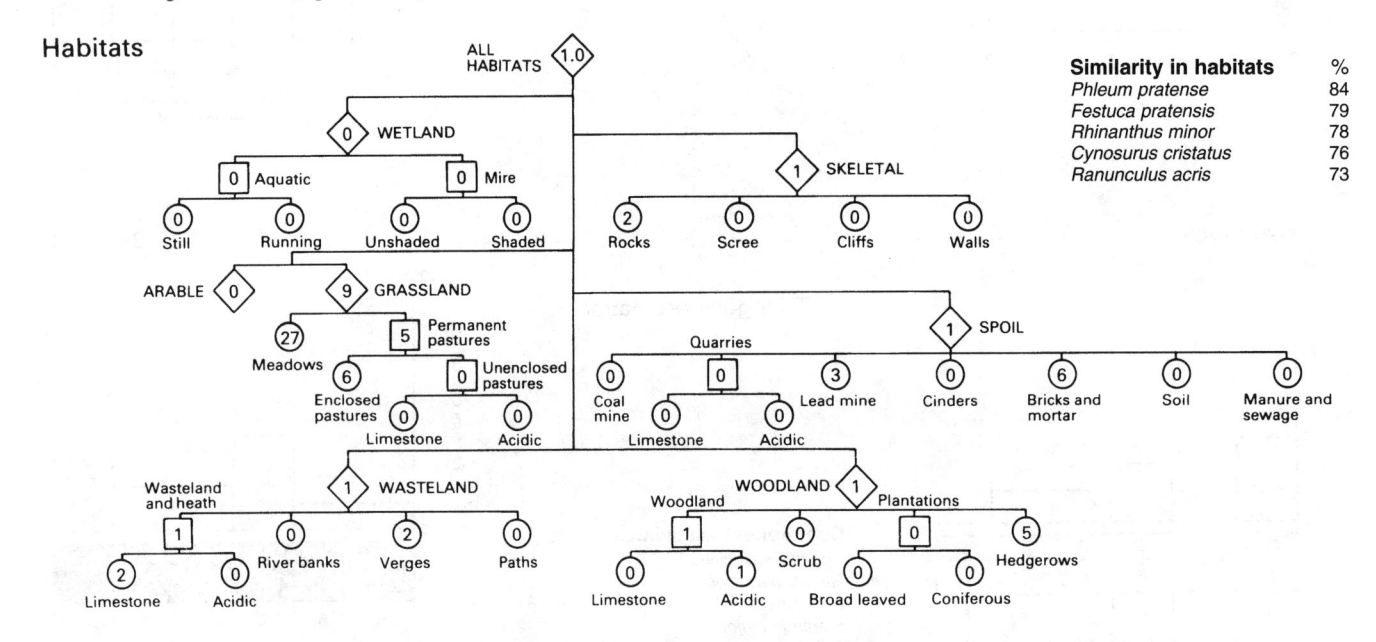

Habitats

Similarity in habitats	%
Phleum pratense	84
Festuca pratensis	79
Rhinanthus minor	78
Cynosurus cristatus	76
Ranunculus acris	73

Bare soil Associated with a wide range of values, but most strongly so in habitats with intermediate exposures of bare soil.
△ **ordination** Distribution centred on relatively productive vegetation subject to some disturbance. Absent from unproductive vegetation.
Associated floristic diversity Associated with a moderately diverse flora, declining where *A.p.* is more abundant, suggesting that *A.p.* has a limited capacity to exercise dominance.
Synopsis *A.p.* is a loosely-tufted winter-green grass most characteristic of fertile hay meadows on moist soils. *A.p.* is moderately productive and may on occasion outyield *Lolium perenne* (Spedding & Diekmahns 1972). In the *PGE* the species 'responds readily to a plentiful supply of nutrients provided sufficient lime is applied. It requires abundant nitrogen and thrives on the unlimed areas of plots receiving sodium nitrate but requires lime before it can take advantage of nitrogen supplies as ammonium sulphate'. *A.p.* grows rapidly in early spring and, except for *Anthoxanthum odoratum*, is the earliest flowering of the common perennial grasses. Early spring growth and the ability for growth under moderate shade (Spedding & Diekmahns 1972) are probably important dimensions of the niche of *A.p.* in grasslands, and may also enable *A.p.* to persist, for example, at woodland margins. *A.p.* has a rather narrow ecological distribution and is very much restricted to moist, well-drained soils, seldom occurring in sites which are waterlogged during summer, and excluded from very dry soils. In a classical competition experiment conducted along an experimental gradient of soil water supply (Ellenberg 1953), it was shown that in monoculture *A.p.* attained maximum vigour in moist but not waterlogged soils, but tended to be displaced from such optimal conditions by competition from other species. *A.p.* is moderately tolerant of grazing and cutting, but not of heavy trampling. *A.p.* regenerates by seeds which are shed in summer, often before hay is cut. No buried seed bank has been reported. *A.p.* was formerly sown in seed mixtures to a considerable extent and cultivars were developed, but now *A.p.* has largely been replaced by *Phleum pratense* (Spedding & Diekmahns 1972). In tall vegetation *A.p.* is,

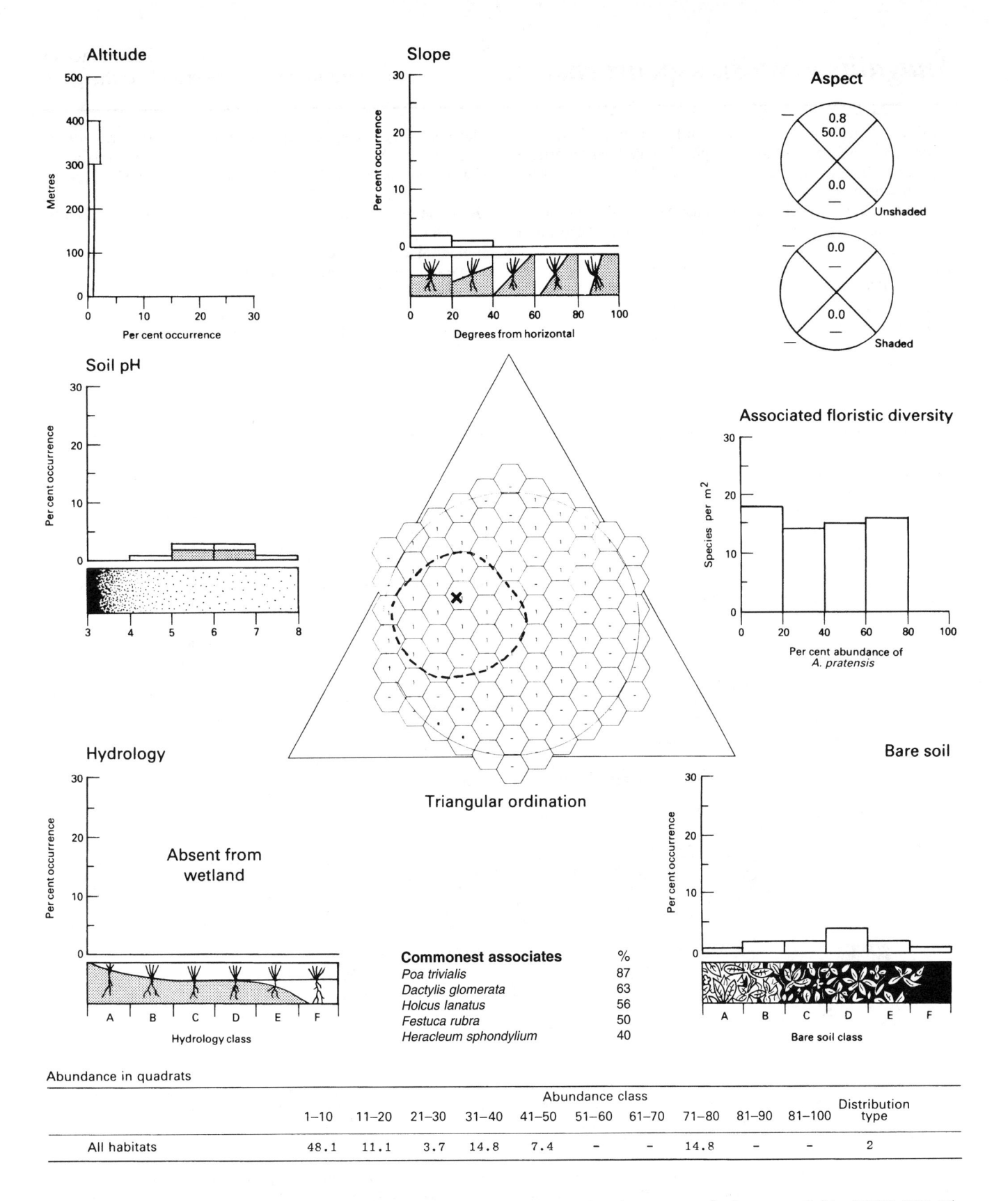

Altitude
Slope
Aspect
Soil pH
Associated floristic diversity
Triangular ordination
Hydrology
Absent from wetland
Bare soil

Commonest associates	%
Poa trivialis	87
Dactylis glomerata	63
Holcus lanatus	56
Festuca rubra	50
Heracleum sphondylium	40

Abundance in quadrats

	1–10	11–20	21–30	31–40	41–50	51–60	61–70	71–80	81–90	81–100	Distribution type
All habitats	48.1	11.1	3.7	14.8	7.4	–	–	14.8	–	–	2

however, more persistent than *Phleum pratense*, a capacity which appears to involve some vegetative regeneration. In common with several other aspects of the biology of *A.p.*, this phenomenon has not been adequately investigated. *A.p.* hybridizes with *A. geniculatus* (Stace 1975) and a further ssp. is recognized in Europe (*FE5*).

Current trends *A.p.* has decreased through a decline both in its agricultural importance and in the acreage of suitable habitats. The amount of seed sown annually is now negligible (MAFF 1980-84). Although found to some extent in new artificial environments, the capacity of *A.p.* to exploit recently created environments is likely to be restricted by the absence of both a persistent seed bank and by the lack of an effective method of vegetative spread.

Anagallis arvensis ssp. *arvensis* L.

Primulaceae
Scarlet Pimpernel, Shepherd's Weatherglass

Life-form Usually a summer annual in Britain, but behaves as a winter annual in parts of its geographical range. Stems prostrate to erect; leaves ovate, with glands, often <100 mm^2; roots with VA mycorrhizas.
Phenology Variable. Probably germinating in spring, but plants capable of overwintering. Flowers mainly June to August and sets seed July to October.
Height Foliage and flowers usually <200 mm.
Seedling RGR No data. SLA 37.6. DMC 12.5.
Nuclear DNA amount 3.5 pg; $2n = 40$ (*FE5*); $4x$.
Established strategy Intermediate between ruderal and stress-tolerant ruderal.
Regenerative strategies B$_s$. Regenerates by seed, which germinates mainly in spring. Forms a persistent seed bank; SB IV.

Flowers Usually red, hermaphrodite, self- or more rarely insect-pollinated (Marsden-Jones & Weiss 1959); solitary axillary.
Germinule 0.40 mg, 1.5 × 1.1 mm, seed, in a 12-45-seeded dehiscent capsule (*FE3*).
Germination Epigeal; t_{50} 1 day; <5-35°C; dormancy broken by chilling (but see 'Synopsis').
Biological flora None, but see Holm *et al.* (1977).
Geographical distribution The British Isles, but rare in the extreme N (100% of British vice-counties); *ABIF* page 303; the aggregate species is found in most of Europe (92% of European territories) and is relatively cosmopolitan with the exception of the tropics.

Habitats Most commonly occurring on arable land, particularly in cereal crops, but also recorded or observed from open or rocky habitats (soil and cinder heaps, quarry spoil and rock outcrops), and from bared soil on road verges and disturbed wasteland. [Found on sand dunes.]
Gregariousness Typically occurring as scattered individuals.
Altitude Confined to lowlands and not recorded above 200 m. [Up to 320 m.]
Slope Absent from the steepest slopes.
Aspect Found on both N- and S-facing aspects.

Hydrology Absent from wetlands.
Soil pH Observed at pH 4.5, but mainly restricted to the pH range 5.5-8.0.
Bare soil Only recorded from open sites. Sites with low exposures of bare soil are usually rocky.
△ **ordination** Distribution strongly centred on ruderal vegetation types associated with disturbed, highly fertile conditions.
Associated floristic diversity Typically intermediate to species-rich.

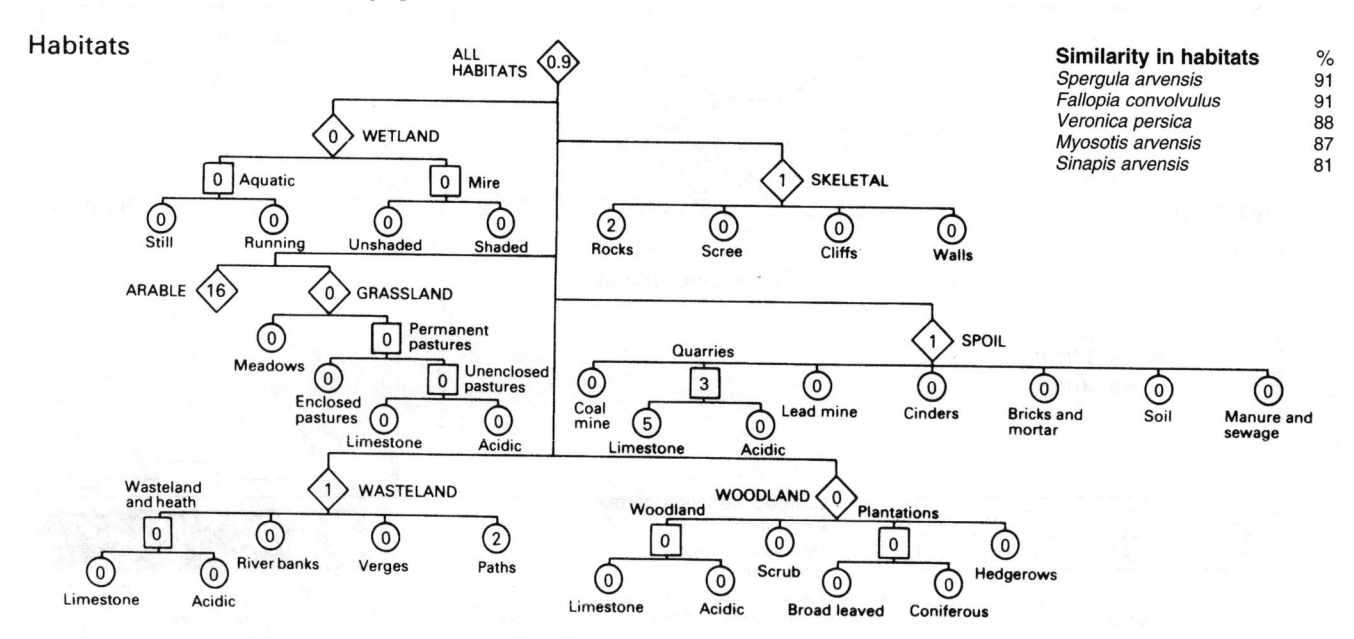

Habitats

Similarity in habitats	%
Spergula arvensis	91
Fallopia convolvulus	91
Veronica persica	88
Myosotis arvensis	87
Sinapis arvensis	81

Synopsis Although restricted to lowland sites in Britain (Wilson 1956), *A.a.* is relatively cosmopolitan and is classified as one of the world's worst weeds (Holm *et al.* 1977). In warm climates *A.a.* usually behaves as a winter annual, but in cooler regions the species is a summer annual (Holm *et al.* 1977). In Britain *A.a.* is probably a summer annual (see Roberts & Boddrell 1983), and in arable fields, a favoured habitat, it is more frequent in spring-sown crops (Chancellor & Froud-Williams 1984). The population dynamics of *A.a.* may be complex, since polycarpic perennial individuals have been observed in fallow fields. The possibility that the species may behave locally as a winter annual deserves investigation, particularly since germination may occur in the autumn (*WCH*). *A.a.* is highly toxic to stock (Cooper & Johnson 1984). Because *A.a.* can germinate and grow at low temperatures, and its seedlings persist under moderate shade, this low-growing species can exploit the phase before the cereal canopy is fully established (Holm *et al.* 1977). *A.a.* is excluded from most broad-leaved crops, perhaps because of shading by the crop and by other weed species. Like *Veronica persica*, *A.a.* typically produces procumbent indeterminant shoot growth, with flowers and fruits formed sequentially in the axils of the leaves. The flowers of *A.a.* tend to open in full sunlight and remain closed on dull or rainy days, earning the species its common name 'Shepherd's Weatherglass'. On average, one plant produces less than 1000 seeds (Salisbury 1942); this appears to be rather few for a common arable weed. Seeds are released close to the soil surface; this probably facilitates their incorporation into a persistent seed bank (Chippindale & Milton 1934, Jalloq 1974). Seed is widely dispersed as a result of agricultural management, particularly as a contaminant of crop seeds (Holm *et al.* 1977), and can survive ingestion by birds (Ridley 1930). Consequently, *A.a.* often becomes established in new open habitats such as quarry spoil, road

Altitude

Slope

Aspect

Soil pH

Associated floristic diversity

Triangular ordination

Hydrology

Absent from wetland

Commonest associates	%
Poa annua	81
Polygonum aviculare	68
Agrostis stolonifera	67
Elytrigia repens	59
Stellaria media	45

Bare soil

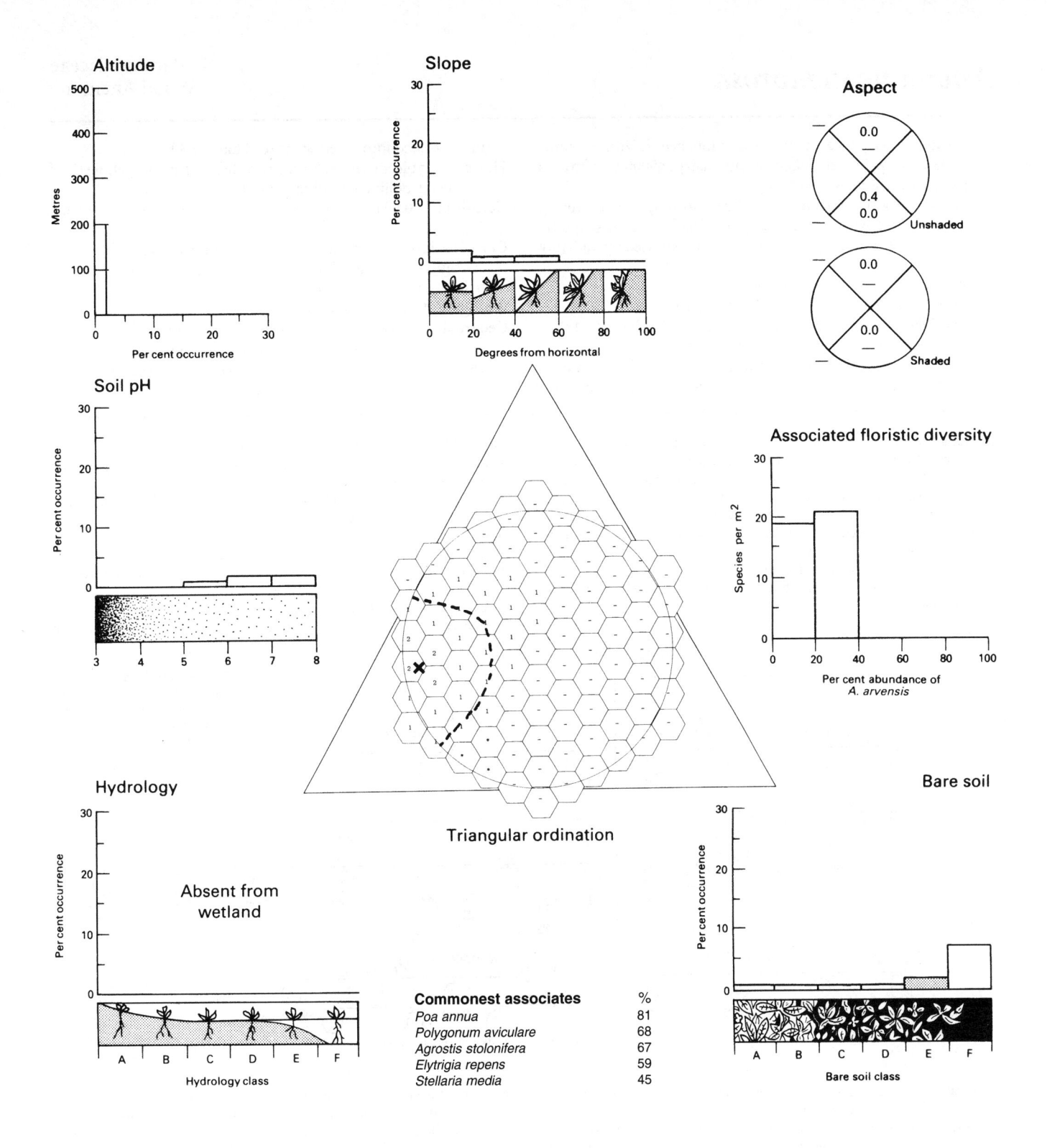

and rail cuttings and, more rarely, disturbed road verges and other transient habitats. Germination of freshly-shed seeds is said to be prevented by the presence of a water-soluble inhibitor (Holm *et al.* 1977). A range of other environmental controls of germination appear to exist, including a requirement for light (Holm *et al.* 1977), but their exact ecological significance remains unclear. Except when growing at low temperatures, *A.a.* is a long-day plant (Holm *et al.* 1977) and the main flowering season, June to August, is more restricted than that of most arable weeds. Exploitation of some sites may be restricted by the strongly seasonal pattern of flowering, which may explain the rarity of *A.a.* in gardens. Factors controlling the distribution of *A.a.* in rocky habitats and sand dunes (the only semi-natural habitat with which the species is commonly associated) are poorly understood. The species is genetically variable, and plants with blue flowers predominate near the Mediterranean (*FE3*). Populations from sand dunes are deep-rooted (Salisbury 1952), fleshy-leaved and usually form compact tufts. The rare ssp. *foemina* (Miller) Schinz and Thell. (*A. foemina* Miller), with blue flowers, produces fewer, heavier seeds and is summer annual (Salisbury 1964). Hybrids between the two taxa are reported (Stace 1975).

Current trends Often associated with rarer arable weeds, e.g. *Euphorbia exigua* L., and probably decreasing as weed-control measures on arable land become more effective.

Anemone nemorosa L.

Ranunculaceae
Wood Anemone

Life-form Perennial geophyte with branched rhizome. Stems erect; leaves palmately lobed, slightly hairy, <5000 mm^2; roots shallow with VA mycorrhizas.
Phenology Shoots emerge from March to May and flowering occurs only 2 weeks after emergence (*Biol Fl*). Fruits ripe by May or June and above-ground parts have senesced by mid-July (*Biol Fl*).
Height Foliage and flowers to 350 mm.
Seedling RGR No data. SLA 26.3. DMC 20.5.
Nuclear DNA amount 38.1 pg; usually $2n = 30^*$ (*Biol Fl*); $4x^*$.
Established strategy Stress-tolerant ruderal.
Regenerative strategies V, S. Produces clonal patches by means of rhizomes. Seeds, which germinate in spring, are not incorporated into a persistent seed bank; SB II.
Flowers Typically white, hermaphrodite, self-incompatible (*Biol Fl*), insect-pollinated; solitary and terminal.
Germinule 0.99 mg, 4.9 × 1.9 mm, achene, 16-31 per flower (*Biol Fl*).
Germination Hypogeal; period of moist preferably cool conditions during which the embryo enlarges are required before germination (*Biol Fl*).
Biological flora Shirreffs (1985).
Geographical distribution Most of the British Isles, but less common in Ireland (99% of British vice-counties), *ABIF* page 98; N temperate zone of Europe (74% of European territories); W Asia.

Habitats Most frequent and abundant in woodlands, particularly on limestone strata. Also frequent in damp calcareous pastures in the limestone dales of the Derbyshire Peak District, and recorded for scree, wasteland, shaded mire and cliffs. Observed more-locally in acidic pastures, hedgerows and on roadsides. Absent from aquatic, arable and spoil habitats. [Also occurs under *Calluna vulgaris*, in limestone pavement and in rocky montane habitats (*Biol Fl*).]
Gregariousness Often patch-forming.
Altitude Extending to 400 m, but frequency and abundance decrease with altitude. [Up to 1190 m.]

Slope Most frequent and only abundant on slopes of <30°.
Aspect Slightly more frequent and abundant on N-facing slopes; significantly so in grasslands (*Ecol Atl*).
Hydrology Not a wetland species, but occurs in moist ground beside soligenous and topogenous mire.
Soil pH Tolerant of all but the most acidic or alkaline soils; highest frequency associated with soils in the pH range 4.5-7.5.
Bare soil Some woodland sites often have a high exposure of bare soil, but the species also occurs in association within a continuous cover of persistent closed grassland vegetation.
△ **ordination** Distribution centred on vegetation associated with infertile and undisturbed habitat types.

Similarity in habitats	%
Mercurialis perennis	94
Brachypodium sylvaticum	92
Sanicula europaea	90
Allium ursinum	88
Melica uniflora	86

Habitats

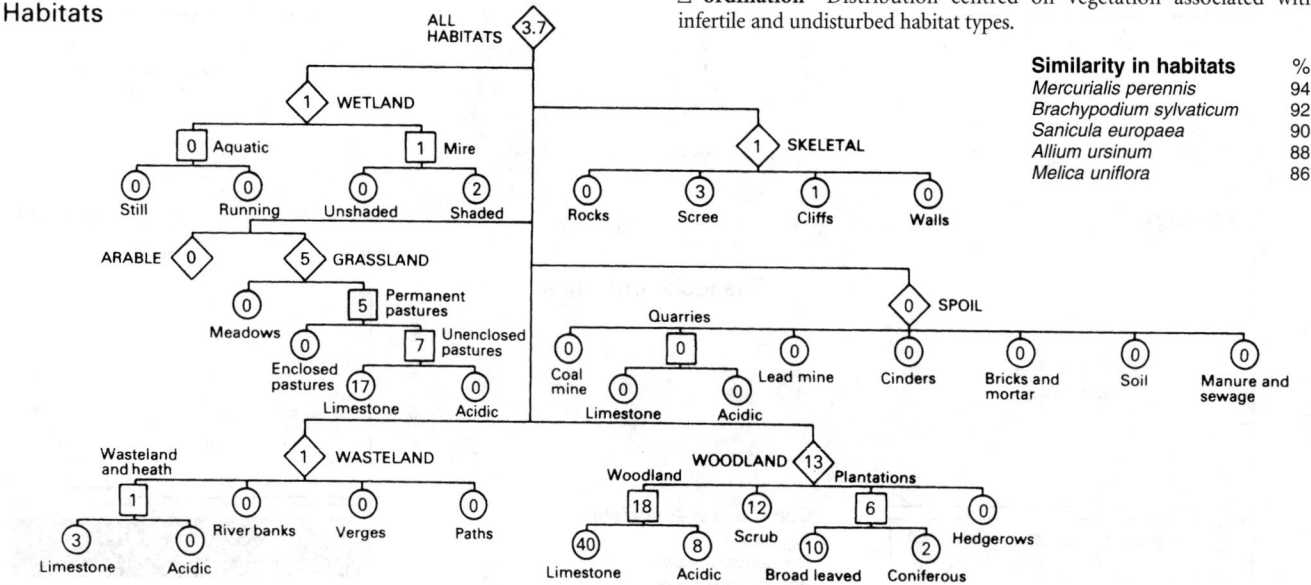

Associated floristic diversity Typically low to intermediate in woodland, but often high in calcareous grassland.
Synopsis *A.n.* is typically a rhizomatous, patch-forming vernal herb of less-fertile woodlands. As might be predicted from its high nuclear DNA amount, the phenology of *A.n.* involves shoot expansion during the late winter and early spring and exploitation of the light phase before canopy development by neighbouring trees and shrubs; by mid-July the above-ground portions of the plant have died back. The emergent shoot is crozier-shaped. This affords protection to the flower bud, which was formed in the previous autumn (*Biol Fl*), and often enables the shoot to push its way through tree litter. The same phenomenon occurs in some derelict grasslands, and high population densities have been observed in coarse turf with dense accumulations of *Brachypodium pinnatum* litter. Differences in ecology between *A.n.* and the other common, shallow-rooted vernal species, *Ranunculus ficaria*, are in part attributable to the more erect morphology of *A.n.*: this permits the species to penetrate litter and to rise through closed canopies much more effectively. The species has a shallow root system and is most vigorous under moist conditions where leaves may be produced up to 10 days earlier than on adjacent dry soils (*Biol Fl*). Further, in the moist climates of the N and W the ecological range of *A.n.* is greater, and the species extends into grassland and other open habitats (*Biol Fl*). *A.n.* may occur in lightly grazed habitats. The plant contains protoanemonin and is unpalatable to stock (Cooper & Johnson 1984), but it can survive occasional defoliation. *A.n.* is also capable of exploiting some grasslands managed by burning (*Biol Fl*) and may escape the impact of summer fires. The distribution of *A.n.* shows a considerable overlap with that of *Hyacinthoides non-scripta* and *Mercurialis perennis*. However, *A.n.* is more tolerant of woodland disturbance and responds to coppicing by a marked increase in flowering (*Biol Fl*). In marked contrast to *H.n-s.*, the roots are superficial and *A.n.* frequently exploits soils where the water table is close to the surface. In common with *H.n-s.*, but in contrast with *M.p.*, the species extends onto both acidic and calcareous

Altitude

Slope

Aspect

Soil pH

Associated floristic diversity

Hydrology

Triangular ordination

Commonest associates	%
Mercurialis perennis	48
Festuca rubra	45
Viola riviniana	42
Deschampsia cespitosa	40
Dactylis glomerata	37

Bare soil

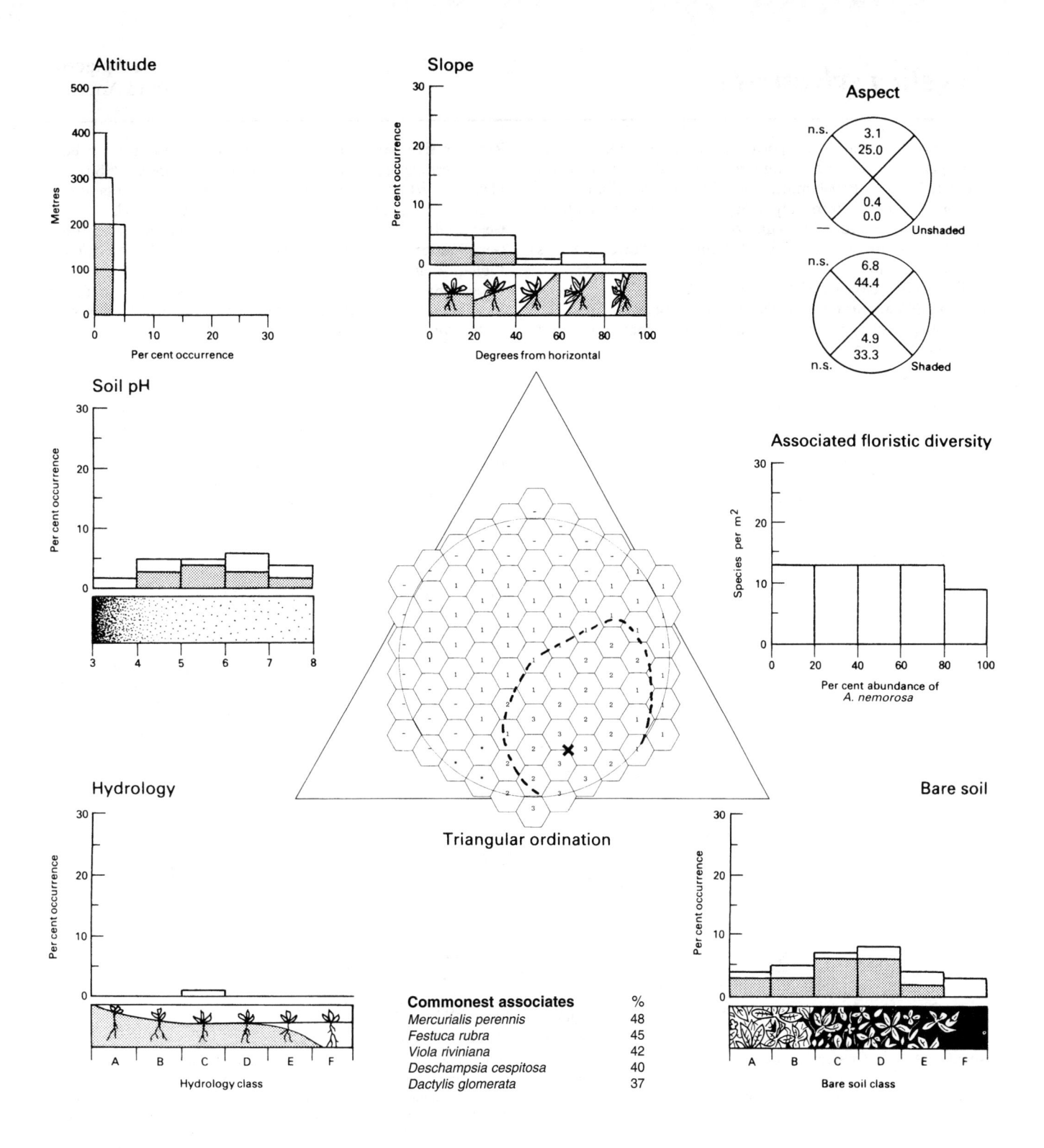

soils. With Rackham (1980), we postulate that A.n. is confined to situations where relatively infertile soils coincide with factors (e.g. management regime, hydrology) which restrict the vigour of both H.n-s. and M.p. The leaves of A.n. are relatively low in P, but often contain rather high concentrations of Ca, Mg, Fe and Al (SIEF). A.n. regenerates mainly by means of rhizome growth to form large but slow-growing clonal patches. It is suspected that in fragmented form genets may survive for hundreds of years (Biol Fl) and may thus persist for as long as some of the trees and shrubs in their overstorey. Although A.n. is self-incompatible, seed set occurs regularly, but it is higher in woodland than in grassland sites. The achene contains a minute embryo, and has a prolonged requirement for moist conditions before germination will take place. Germination is particularly high after severe winters. No persistent seed bank has been reported (Biol Fl). Development of the seedling and young plant is very slow, and individuals may not flower until 10 years old (Biol Fl). The seed is not dispersed far from the parent plant. A maximum distance of 130 mm has been reported in one population (Biol Fl). Thus, A.n. has a low colonizing ability and is most typically associated with older areas of woodland or grassland (Ecol Atl, Rackham 1980). Several varieties and complements of chromosome number have been reported (Biol Fl).

Current trends A poor colonist. Remains relatively common, but is decreasing, certainly in grassland habitats and probably in some woodlands.

Angelica sylvestris L.

Apiaceae
Wild Angelica

Life-form Monocarpic or polycarpic perennial, semi-rosette hemicryptophyte with a stout rootstock. Stem erect; leaves divided, some >10 000 mm^2; roots with VA mycorrhizas.
Phenology Flowers from July to September and sets seed from September onwards. Shoots die back in autumn.
Height Majority of foliage below 600 mm, but flowering stems may reach 2 m.
Seedling RGR No data. SLA 23.7. DMC 22.0.
Nuclear DNA amount No data; $2n = 22$ (*FE2*); $2x$.
Established strategy Intermediate between competitor and competitive-ruderal.

Regenerative strategies ?S. More information is needed concerning regeneration from seed (see 'Synopsis'); ?SB II.
Flowers White or pink-tinged, hermaphrodite, protandrous, insect-pollinated (Tutin 1980); >1000 in each umbel.
Germinule 1.15 mg, 4.6 × 3.4 mm, winged mericarp (2 per fruit).
Germination Epigeal; t_{50} 15 days; dormancy broken by chilling.
Biological flora None.
Geographical distribution The British Isles (100% of British vice-counties), *ABIF* page 473; Europe (92% of European territories); temperate Asia.

Habitats Occurs in a wide range of habitats, but is most frequent in moist rough grassland (particularly in the limestone dales), in shaded mire, and on the banks of rivers and streams. Also found at low frequency and often only as seedlings in a variety of other habitats, including woodland, spoil heaps, waysides and pasture. Absent from aquatic, arable and most skeletal habitats. [Also found on mountain ledges and sheltered sea cliffs.]
Gregariousness Sparse. Mature individuals may cover more than 0.25 m^2. Records with high frequencies of *A.s.* usually signify the presence of seedlings and young plants.
Altitude Occurring only at altitudes of <300 m. [Up to 850 m.]
Slope Found both on relatively level habitats in wetland and on steeper-sloping river banks and dalesides.

Aspect More frequent on N-facing slopes in unshaded situations and S-facing slopes in shade; this pattern is consistent with a requirement for moist, moderately shaded conditions.
Hydrology Typical of moist rather than submerged soils. Particularly common at sites adjacent to open water, and on river banks, but found in both topogenous and soligenous mire.
Soil pH Occurs over a wide range of soil pH values >4.0.
Bare soil Occurs over a wide range of values, but not abundant where exposure is high.
△ **ordination** Distribution centred on vegetation associated with fertile, relatively undisturbed conditions, but locally recorded as seedlings in some unproductive sites. Excluded from heavily disturbed vegetation.
Associated floristic diversity Intermediate

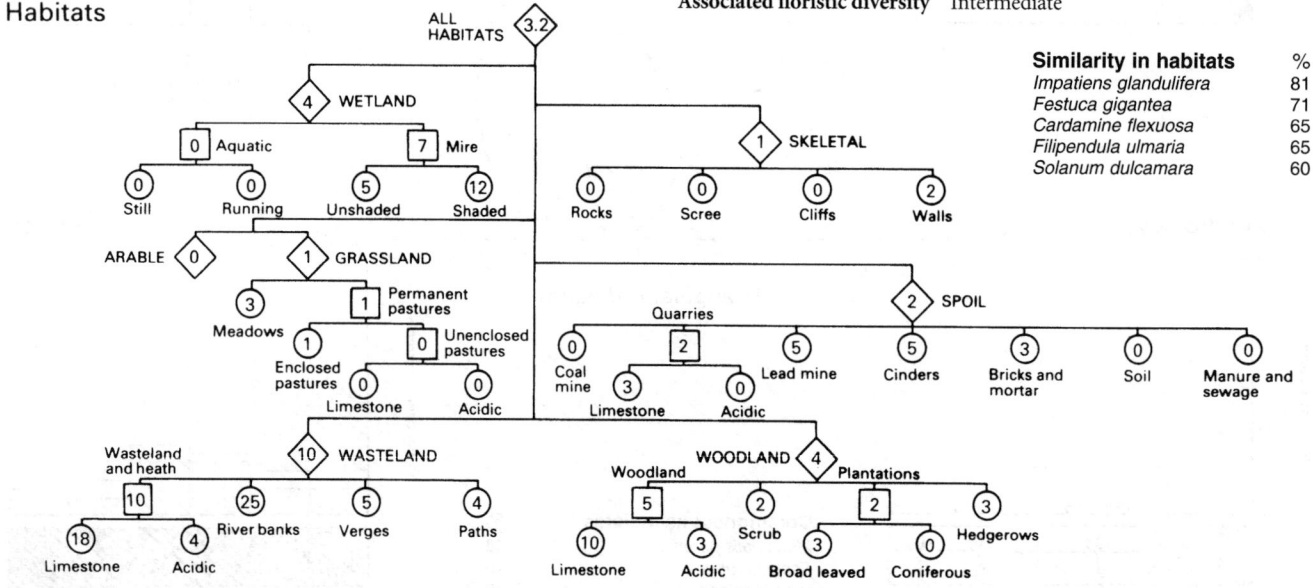

Similarity in habitats	%
Impatiens glandulifera	81
Festuca gigantea	71
Cardamine flexuosa	65
Filipendula ulmaria	65
Solanum dulcamara	60

Habitats

Synopsis *A.s.* is a robust species of moist, often relatively fertile, sites and has a wide habitat range which includes damp grassland, mire and woodland margins. Although generally considered to be a polycarpic perennial, we suspect that in many habitats part of the population is monocarpic. Observations of the species in winter indicate that a large proportion (sometimes >50%) of the plants die after flowering. At maturity, individuals produce a small number of massive basal or near-basal leaves up to 600 mm in length and a large flowering stem which may reach 2 m. Despite the large size, the capacity for dominance is low compared with wetland species such as *Epilobium hirsutum* which have the added advantages of rapid vegetative spread and a taller, more densely packed leaf canopy. *A.s.* achieves rapid early growth through the mobilization of reserves from the stout rootstock, and benefits where management allows the development of a larger summer peak in shoot biomass. Leaf tissue is rich in Ca (*SIEF*). In sites subject to frequent mowing or grazing *A.s.* is usually represented by seedlings or stunted individuals. Sites conducive to large populations of *A.s.* are either periodically disturbed, e.g. river and ditch banks, woodland margins and mire subject to variable winter flooding, or situated in habitats such as coarse grassland where grazing has recently been relaxed. It is not known to what extent the perpetuation of *A.s.* in grassland is due to cutting regimes, but we suspect a dependence similar to that described by Harvey & Meredith (1981) for another, much rarer, wetland umbellifer *Peucedanum palustre* (L.) Moench. Regeneration is entirely by seed. Populations often consist of a few mature flowering individuals associated with many young plants, each with 1 or 2 small leaves and existing in an apparently suppressed condition beneath the vegetation canopy. Studies are needed to ascertain whether these young plants constitute a bank of persistent seedlings which are dependent upon the subsequent disturbance of the vegetation in order to reach maturity. Seeds, which contain a rather small embryo, have a chilling requirement and germinate in the spring. No persistent seed bank has been reported. Seeds float (Ridley 1930) and long-distance dispersal by this means along waterways may be important in the survival of the species.
Current trends Uncertain, but the future of the species in habitats such as river banks appears secure.

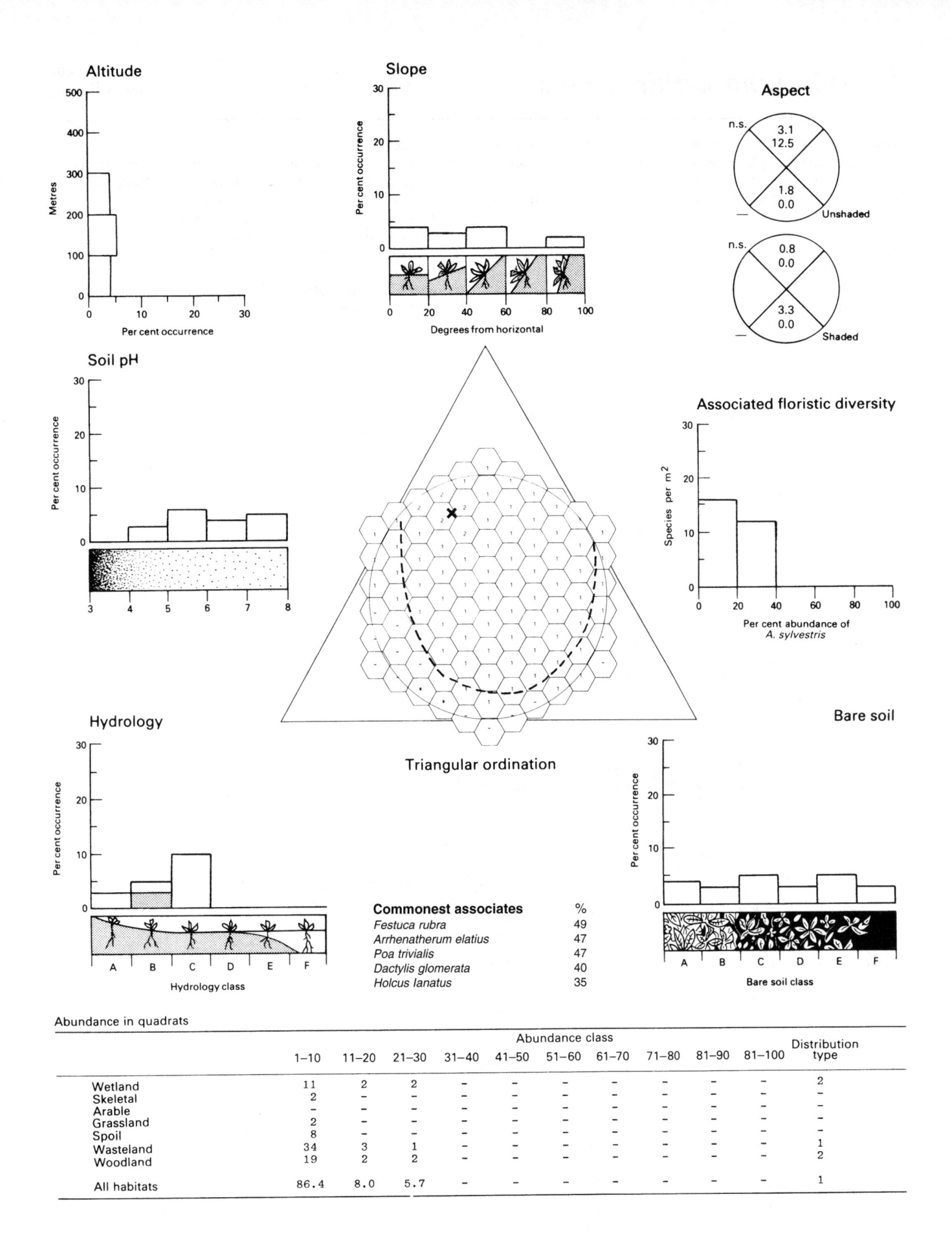

Altitude

Slope

Aspect

	n.s.	3.1	
		12.5	
		1.8	Unshaded
		0.0	

	n.s.	0.8	
		0.0	
		3.3	Shaded
		0.0	

Soil pH

Associated floristic diversity

Hydrology

Triangular ordination

Commonest associates

	%
Festuca rubra	49
Arrhenatherum elatius	47
Poa trivialis	47
Dactylis glomerata	40
Holcus lanatus	35

Bare soil

Abundance in quadrats

	\multicolumn{10}{c}{Abundance class}	Distribution type
	1–10	11–20	21–30	31–40	41–50	51–60	61–70	71–80	81–90	81–100	
Wetland	11	2	2	–	–	–	–	–	–	–	2
Skeletal	2	–	–	–	–	–	–	–	–	–	–
Arable	–	–	–	–	–	–	–	–	–	–	–
Grassland	2	–	–	–	–	–	–	–	–	–	–
Spoil	8	–	–	–	–	–	–	–	–	–	–
Wasteland	34	3	1	–	–	–	–	–	–	–	1
Woodland	19	2	2	–	–	–	–	–	–	–	2
All habitats	86.4	8.0	5.7	–	–	–	–	–	–	–	1

(+) *Anisantha sterilis* (L.) Nevski

Poaceae
Barren Brome

Life-form A winter- or more rarely summer-annual, semi-rosette therophyte. Stems erect or spreading; leaves linear, hairy, <2000 mm^2; roots non-mycorrhizal.
Phenology Typically winter green. Flowers from May to July and sets seed from June to October.
Height Foliage mainly <400 mm; flowers to 1 m.
Seedling RGR 1.0-1.4 week^{-1}. SLA 31.4. DMC 24.1.
Nuclear DNA amount 6.7 pg, $2n = 14^*$ (Bennett & Smith 1976); $2x^*$.
Established strategy Intermediate between ruderal and competitive-ruderal.
Regenerative strategies S. By seed in autumn, or more rarely early spring. No persistent seed bank is formed (ADAS 1983b, Roberts 1986b); SB I.
Flowers Green, hermaphrodite, mainly selfed but cross-fertilization by wind may occasionally occur (Beddows 1931); often >100 in a drooping panicle (4-10 per spikelet).
Germinule 8.4 mg, 11.2 × 1.3 mm, caryopsis dispersed with a long-awned lemma (37.6 × 1.5 mm).
Germination Epigeal; t_{50} 1 day; <5-37°C; L = D.
Biological flora None.
Geographical distribution Widely distributed in the British Isles but less common in Scotland and Ireland (92% of British vice-counties), *ABIF* page 791; SW and C Europe extending to S Sweden (79% of European territories and 5% naturalized); SW Asia; introduced in N America.

Habitats Most commonly found in hedgerows, but occurs also in a range of disturbed and moderately fertile artificial habitats, e.g. demolition sites and the edges of arable fields. Also present in semi-natural skeletal habitats such as outcrops, where it behaves strictly as a winter annual. Particularly abundant on sandy and other dry, freely-drained soils.
Gregariousness Typically sparse, but can form dense patches at arable field margins.
Altitude Observed to 275 m, but suitable habitats are more frequent in lowland areas. [Up to 370 m.]
Slope Wide-ranging.

Aspect Widely distributed but additional, more extensive, sampling (Hodgson unpubl.) suggests a bias towards S-facing slopes.
Hydrology Absent from wetlands.
Soil pH Most frequent and abundant at sites of pH >6.5.
Bare soil Found over a wide range of values, but most abundant at intermediate exposures.
△ **ordination** Distribution centred on vegetation associated with disturbed fertile habitats.
Associated floristic diversity Generally relatively low; highest values associated with sites that have recently been subjected to an extremely high level of disturbance.

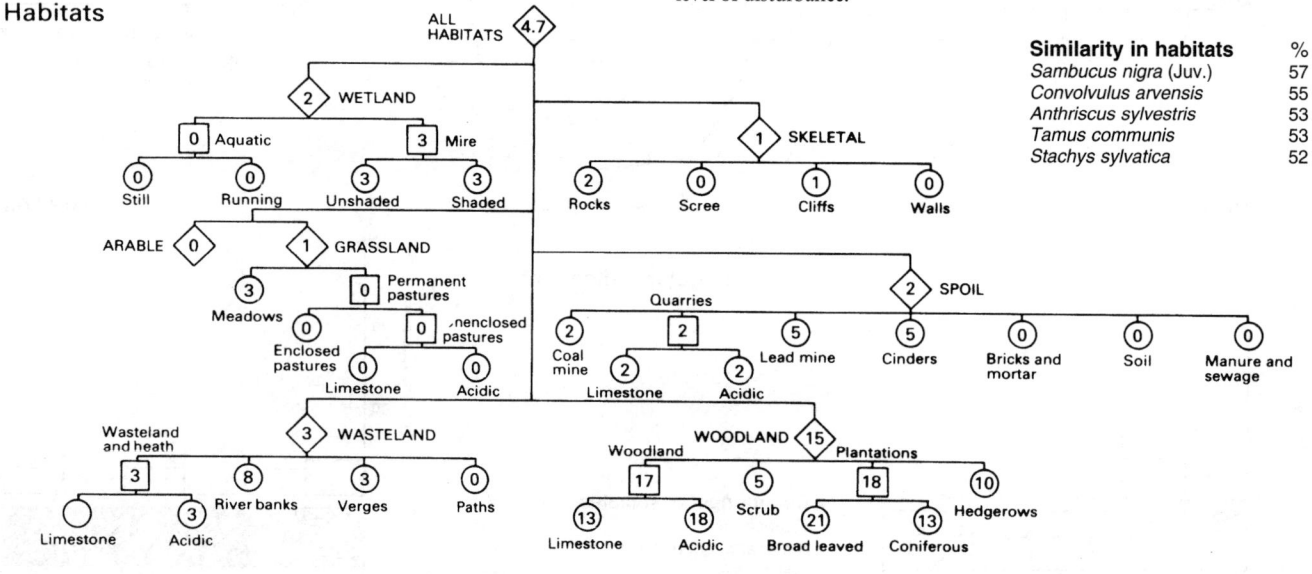

Habitats

Similarity in habitats	%
Sambucus nigra (Juv.)	57
Convolvulus arvensis	55
Anthriscus sylvestris	53
Tamus communis	53
Stachys sylvatica	52

Synopsis *A.s.* is a winter- or more rarely summer-annual grass which is similar in many ecological respects to *Galium aparine*, though it is unable to persist in tall herb communities and tends to germinate earlier, before the onset of winter. *A.s.* is found in fertile habitats where, as a result of summer shade in the case of hedgerows or summer drought in the case of rock outcrops, the vigorous growth of polycarpic perennial herbs is prevented. The species is also becoming increasingly important as a weed of winter-sown cereals (Chancellor & Froud-Williams 1984). Growth of *A.s.* is restricted to the period between autumn and early summer, and *A.s.* flowers before the onset in summer of the period, with greatest risk of disturbance in these and other related habitats. Flowering is controlled by a weak vernalization requirement, and even seedlings which are produced in early spring may flower (Froud-Williams 1983). Panicle extension takes place only in long days (Froud-Williams 1983). The seeds are exceptionally large and the number produced (often *c.* 200 per plant (ADAS 1983b)), is lower than in most ruderal species. Germination of seeds, which are shed from June to August, is frequently delayed until autumn, apparently through the absence of adequate moisture, or sufficiently close contact between the seed and the soil (ADAS 1983a). Seedlings can establish successfully through up to 130 mm of soil (ADAS 1983a). However, at 150 mm. depth, although seeds still germinate, the seedlings fail to survive. Germination is known to be inhibited by white and by red light (Hilton 1984), but the ecological significance of this behaviour remains obscure. A persistent seed bank is not formed (Froud-Williams 1983). In arable fields seed is dispersed within the crop (Froud-Williams 1983), but in other habitats the caryopses, which have a long barbed awn, adhere to the coats of animals and to clothing (Ridley 1930). *A.s.* is a contaminant of pasture-grass seed (*WCH*), and seeds may survive ingestion by stock (Ridley 1930). The species has an Atlantic distribution and is largely restricted to lowland sites. It is considered by Webb (1985) to have been introduced.
Current trends Encouraged by disturbance and by high fertility; apparently increasing. *A.s.* is also becoming a serious weed in fields where winter cereals are grown repeatedly using minimum cultivation techniques (ADAS 1983b).

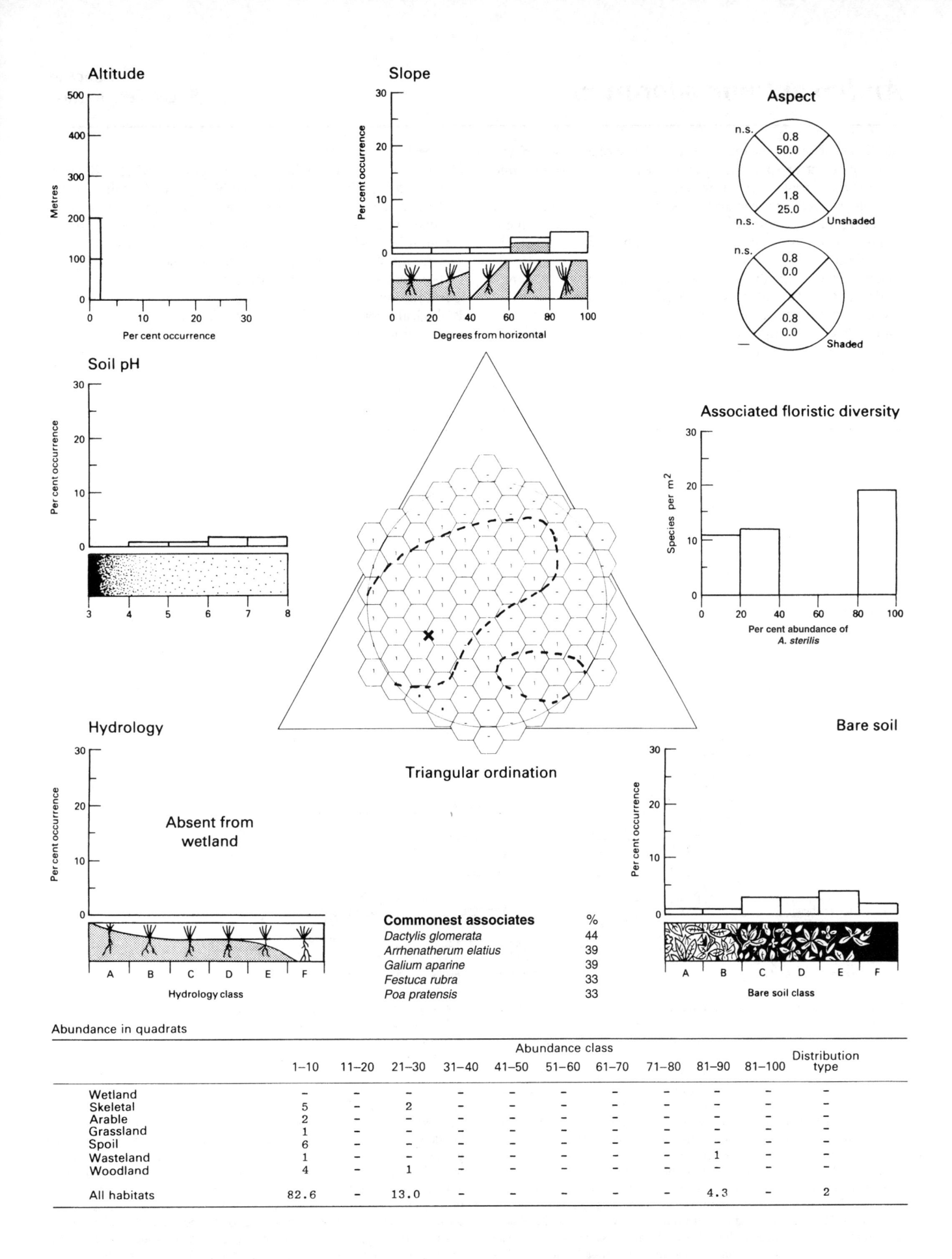

Altitude

Slope

Aspect

	n.s.	0.8	
		50.0	
n.s.		1.8	Unshaded
		25.0	

	n.s.	0.8	
		0.0	
		0.8	Shaded
		0.0	

Soil pH

Associated floristic diversity

Species per m² vs Per cent abundance of *A. sterilis*

Triangular ordination

Hydrology

Absent from wetland

Commonest associates %
Dactylis glomerata 44
Arrhenatherum elatius 39
Galium aparine 39
Festuca rubra 33
Poa pratensis 33

Bare soil

Abundance in quadrats

	1–10	11–20	21–30	31–40	41–50	51–60	61–70	71–80	81–90	81–100	Distribution type
Wetland	–	–	–	–	–	–	–	–	–	–	–
Skeletal	5	–	2	–	–	–	–	–	–	–	–
Arable	2	–	–	–	–	–	–	–	–	–	–
Grassland	1	–	–	–	–	–	–	–	–	–	–
Spoil	6	–	–	–	–	–	–	–	–	–	–
Wasteland	1	–	–	–	–	–	–	–	1	–	–
Woodland	4	–	1	–	–	–	–	–	–	–	–
All habitats	82.6	–	13.0	–	–	–	–	–	4.3	–	2

Anthoxanthum odoratum L.

Poaceae
Sweet Vernal-grass

Life-form Tufted, polycarpic perennial protohemicryptophyte. Shoots erect or spreading; leaves linear, sparsely hairy, often <500 mm²; roots with VA mycorrhizas.
Phenology Winter green and probably growing to some extent during winter. Plants flower from April to June and seed ripens June to August. There is a second flush of shoot growth in autumn.
Height Foliage usually <150 mm; flowering shoots to 500 mm.
Seedling RGR 0.5-0.9 week^{-1}. SLA 27.3. DMC 22.0.
Nuclear DNA amount 11.8 pg; $2n = 10, 20^*$ (*BC, FE5*); $2x, 4x^*$.
Established strategy Intermediate between stress-tolerant ruderal and C-S-R strategist.
Regenerative strategies S, B$_s$. Primarily by seed directly during summer and autumn or after incorporation into a persistent seed bank. Also shows a limited capacity for vegetative spread (see 'Synopsis'); SB III.
Flowers Green, hermaphrodite, protogynous, wind-pollinated, typically highly self-incompatible (Wu & Jain 1980). Inflorescence a spike-like panicle usually with <50 or more spikelets, each composed of one fertile and two sterile florets.
Germinule 0.45 mg, 2.1 × 1.0 mm, caryopsis dispersed either without attached structures or in a spikelet (4.3 × 1.3 mm) bearing long hairs and hygroscopic bristles.
Germination Epigeal; t_{50} days; 7-31°C; L = D.
Biological flora None, but see Zarzycki (1976).
Geographical distribution The British Isles (100% of British vice-counties), *ABIF* page 772; Europe (97% of European territories); temperate Asia and northern Africa, although only on mountains in the S; introduced in N America, Australia and New Zealand.

Habitats Abundant in a wide variety of grassy habitats especially hill pastures, meadows, moors and wasteland. Common on stabilized scree slopes in the limestone dales, near old lead mines, and on marshy ground, rock outcrops, river banks, road verges and other waste places. Occurs less frequently in the herb layer of scrub and open woodlands, on spoil heaps (especially quarry waste), and is of casual occurrence in arable fields. Absent from aquatic habitats.
Gregariousness Intermediate.

Altitude Recorded at all altitudes, but most frequent and abundant in the upland region, where suitable habitats are most common. [Up to 1030 m.].
Slope Associated with a full range of slopes.
Aspect More frequent and abundant on N-facing slopes, particularly in unshaded situations.
Hydrology Frequent on moist soils and in soligenous mire on non-calcareous strata.

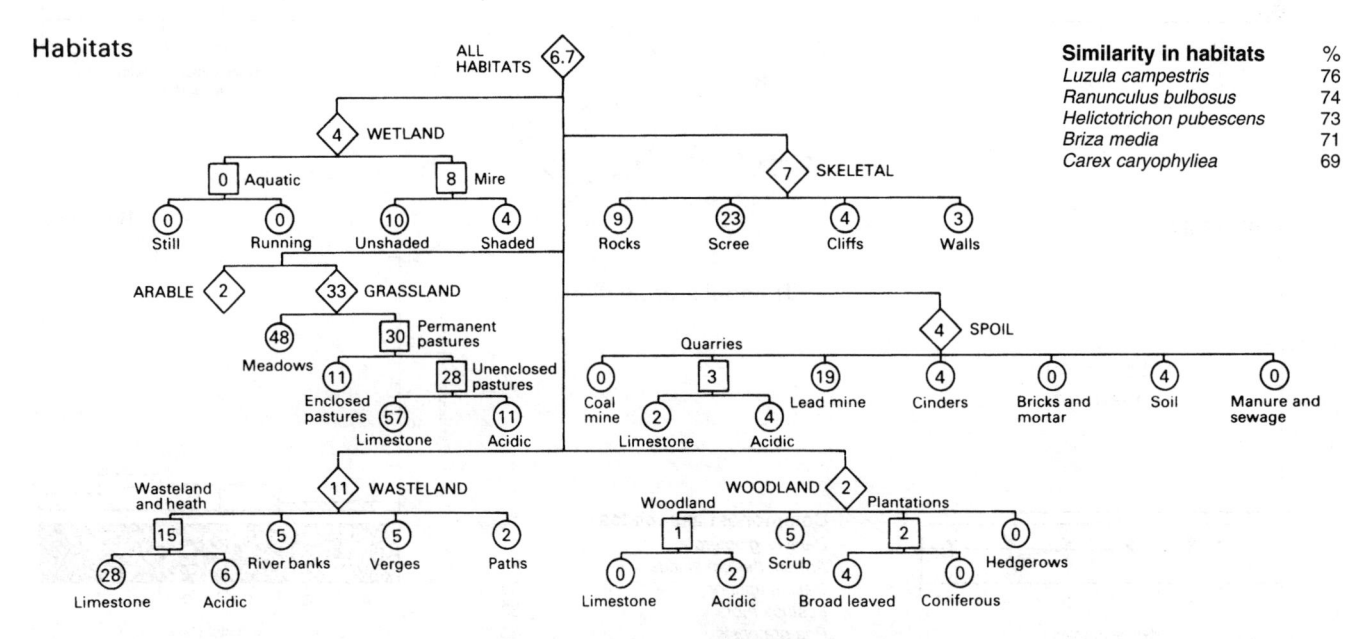

Habitats

Similarity in habitats	%
Luzula campestris	76
Ranunculus bulbosus	74
Helictotrichon pubescens	73
Briza media	71
Carex caryophyliea	69

ALL HABITATS 6.7
WETLAND 4
Aquatic 0; Mire 8
Still 0; Running 0; Unshaded 10; Shaded 4
SKELETAL 7
Rocks 9; Scree 23; Cliffs 4; Walls 3
ARABLE 2
GRASSLAND 33
Meadows 48; Permanent pastures 30
Enclosed pastures 11; Unenclosed pastures 28
Limestone 57; Acidic 11
Quarries 4; SPOIL 4
Coal mine 0; 3; Lead mine 19; Cinders 4; Bricks and mortar 0; Soil 4; Manure and sewage 0
Limestone 2; Acidic 4
Wasteland and heath; WASTELAND 11; WOODLAND 2
River banks 15; Verges 5; Paths 5; 2
Limestone 28; Acidic 6
Woodland 1; Scrub 5; Plantations 2; Hedgerows 0
Limestone 0; Acidic 2; Broad leaved 4; Coniferous 0

Soil pH Distributed over the full range in soil pH but most common in the range 4.5-6.0.
Bare soil Distribution wide ranging, but biased towards habitats having a relatively small exposure of bare soil.
△ **ordination** Distribution centred on vegetation associated with low to intermediate productivity and subject to slight or moderate disturbance; scattered occurrences in a wide range of other vegetation types.
Associated floristic diversity Typically intermediate, but often associated with species-poor vegetation on acidic soils and also not uncommon in species-rich communities.
Synopsis *A.o.* is a relatively slow-growing, short, loosely-tufted, winter-green grass typically occurring as scattered individuals in a wide range of grasslands and, to a lesser extent, in open habitats particularly on slightly acidic soils. The species is said to flourish best on 'well-manured and acid soil' and is 'usually reduced by lime' (*PGE*). From the same source it is also evident that *A.o.* is not encouraged by ammonium salts alone and is suppressed by heavy applications of nitrogenous fertilizer. *A.o.* is thought to be relatively unpalatable but was formerly included in commercial seed mixtures and is preferred by stock to species such as *Festuca ovina*, despite the presence of coumarin (Spedding & Diekmahns 1972). *A.o.* is the earliest-flowering common pasture grass and has distinct spring and autumn peaks of vegetative growth. Perhaps as a result of this phenology and its relatively short life-span (Grant & Antonovics 1978), *A.o.* is sensitive to defoliation (Spedding & Diekmahns 1972) and to shading in tall derelict grassland. Thus, *A.o.* reaches maximum abundance in damp pastures and meadows of low to moderate fertility. Regeneration is mainly by seed, which is shed during early summer and may be incorporated into a persistent seed bank. Under favourable conditions effective seed production occurs over a wide range of temperatures, and we suspect that seedlings originating from surface-lying or buried seeds play a key role in colonization of gaps where these are caused by factors such as trampling, grazing and urine scorching. Less importantly, *A.o.* has a limited capacity

Altitude
Slope
Aspect
$P < 0.01$

15.4	48.7
6.3	35.7

n.s. — Unshaded

2.3	66.7
1.6	0.0

n.s. — Shaded

Soil pH

Associated floristic diversity

Hydrology

Triangular ordination

Commonest associates %
Festuca rubra 67
Holcus lanatus 58
Agrostis capillaris 56
Festuca ovina 49
Plantago lanceolata 43

Bare soil

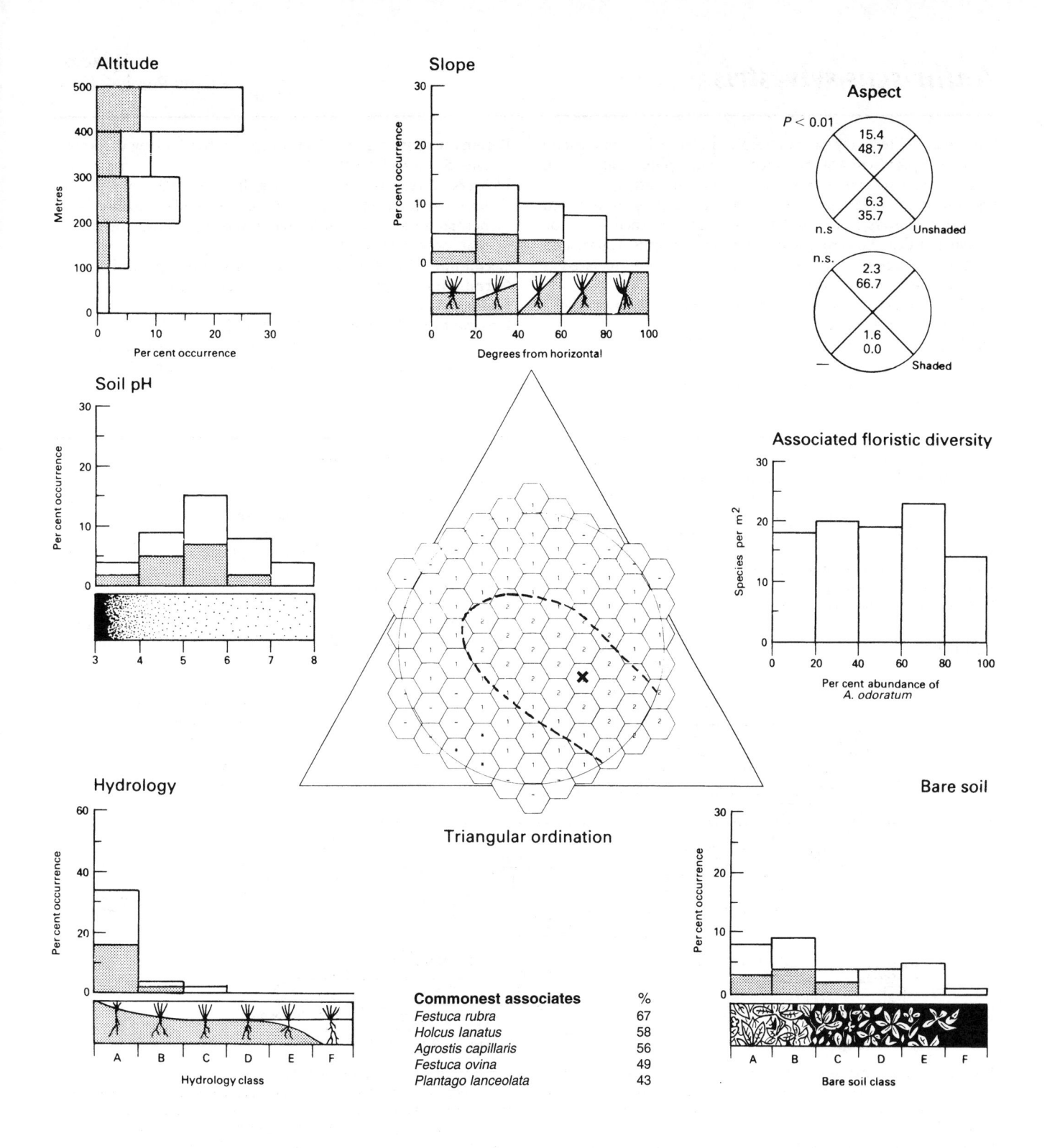

for tillering but individuals retain a distinctly tufted habit. In laboratory conditions detached vegetative shoots of *A.o.* readily produce roots and can form new plants, but there is no evidence that this occurs in the field. Ecotypes differing in morphology and mineral nutrition have been reported (Snaydon & Davies 1972, Davies & Snaydon 1974) and evolutionary changes in populations may occur extremely rapidly (Snaydon & Davies 1972). Consistent with classical theory, the genetic variability of *A.o.* and its capacity to exploit a fairly wide range of habitats coincide with a short life-span, rapid population turnover and the tendency to be strongly outbreeding (Wu & Jain 1980). British plants appear to be tetraploid (*BC*). Plants with $2n = 10$, ($2x$), occur mainly in the N part of the European range of *A.o.* (*FE5*).

Current trends Continued abundance is favoured by the capacity of *A.o.* to evolve rapidly in new types of habitat. Opposing this, however, is the association of the species with less fertile habitats. Thus, a maintained genetic diversity, coupled with a decrease in the overall abundance of *A.o.*, is predicted.

Anthriscus sylvestris (L.)

Apiaceae
Cow Parsley, Keck

Life-form Effectively a polycarpic perennial, semi-rosette hemicryptophyte. Stems erect; leaves pubescent, much dissected, <10 000 mm^2; tap-rooted, non-mycorrhizal.

Phenology Winter green. Foliage produced during autumn and spring. Flowers from April to June and seed shed from the dying or dead flowering stems from July onwards. Foliage dies at the same time as the flowering stem but is soon replaced as new offsets begin to grow.

Height Flowering stems to 1500 mm, but a majority of the foliage below 400 mm.

Seedling RGR 0.5-0.9 week^{-1}. SLA 24.1. DMC 20.7.

Nuclear DNA amount 4.2 pg; $2n = 16^*$ (*BC*); $2x^*$.

Established strategy Intermediate between competitor and competitive-ruderal.

Regenerative strategies S. (V.) Regenerative biology is complex (see 'Synopsis'); SB II.

Flowers Creamy-white, insect-pollinated, >50 per umbel; terminal umbel has male flowers in the middle, surrounded by hermaphrodite flowers; lateral flowers with mostly or entirely male flowers (Tutin 1980).

Germinule 5.18 mg, 3.4 × 1.4 mm, mericarp (2 per fruit).

Germination Epigeal; dormancy broken by chilling.

Biological flora None.

Geographical distribution The British Isles (100% of British vice-counties), *ABIF* page 456; Europe except in the Mediterranean region (74% of European territories); temperate Asia and N Africa.

Habitats Most abundant on road verges, in hedgerows, in meadows and on river banks. *A.s.* also occurs at lower frequencies in a wide range of other types of tall herbaceous vegetation in derelict pasture and on waste ground and spoil heaps, as well as in woodlands and shaded skeletal habitats. Only occurs as seedlings in arable land and absent from wetland. [Also rarely recorded from damp cliffs on mountains (Clapham 1953).]

Gregariousness Intermediate. High values often indicate the incidence of large numbers of seedlings.

Altitude Suitable habitats are more frequent at lower altitudes, but extends to 380 m. [Up to 460 m.]

Slope Wide-ranging.

Aspect Marginally higher occurrence on N-facing slopes but abundance greater on S-facing slopes in both shaded and unshaded situations.

Hydrology Absent from wetland habitats.

Soil pH Most frequent and abundant at pH >7.0. Infrequent below pH 4.5.

Bare soil Greatest occurrence and maximum abundance at intermediate values.

△ **ordination** Distribution centred on vegetation associated with fertile, moderately disturbed conditions. Excluded from very unproductive vegetation.

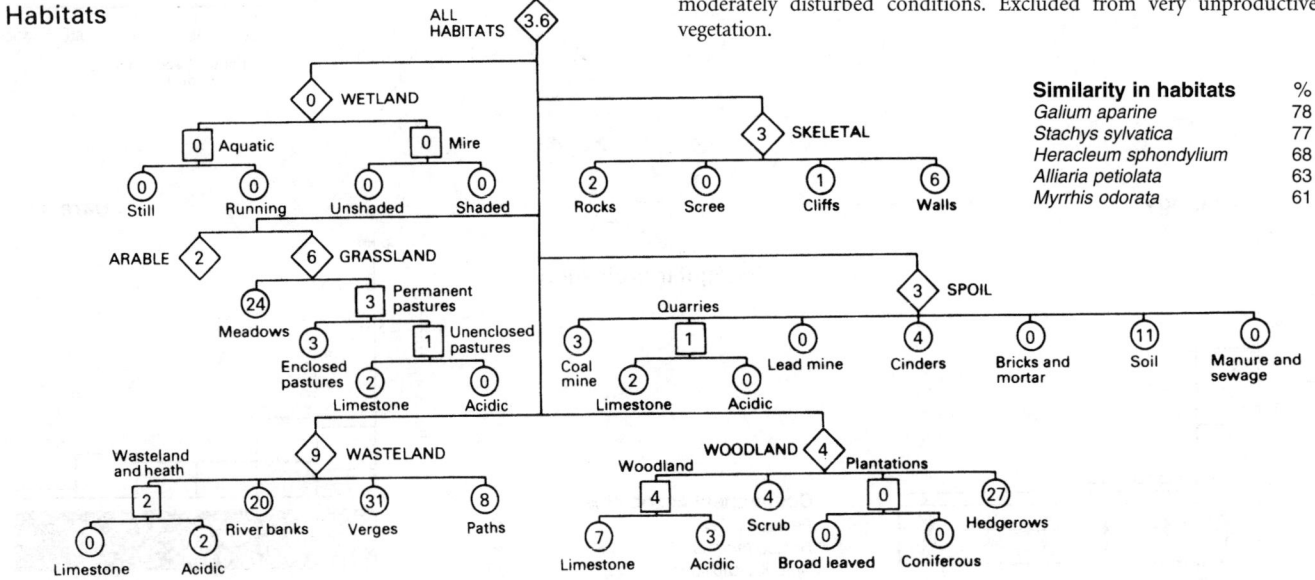

Habitats

Similarity in habitats	%
Galium aparine	78
Stachys sylvatica	77
Heracleum sphondylium	68
Alliaria petiolata	63
Myrrhis odorata	61

Associated floristic diversity Intermediate or relatively low and tending to decline with increasing abundance of *A.s.* This may reflect the impact of dominant grasses (e.g. *Arrhenatherum elatius*) which tend to flourish at sites conducive to *A.s.*

Synopsis *A.s.* is a winter-green, tap-rooted competitive-ruderal which exploits a range of moist or shaded fertile habitats. *A.s.* has a cool-season phenology with leaf canopy reaching a minimum during summer. This phenology, coupled with early seed-set, appears to contribute to the ability of *A.s.* to persist in open woodland, a habitat in which *A.s.* is an important constituent in parts of N Europe (Clapham 1953, Ellenberg 1978). *A.s.* occurs in hay meadows and is able to coexist on roadsides and in hedgerows with tall grasses such as *Arrhenatherum elatius* and *Elytrigia repens*, which have later (summer) peaks in biomass. However, *A.s.* is sensitive to regular cutting, and in roadsides and hedgerows – the major habitats of the species – is often restricted to the less-intensively managed areas close to the hedge, where it may form a conspicuous white zone of flowering shoots in early spring. *A.s.*, which is eaten by cattle and rabbits (Clapham 1953), is also intolerant of heavy grazing and trampling, and does not occur on heavily droughted or waterlogged soils, although it is characteristic of productive flooded meadows in reclaimed land in Holland (Joenje 1978). Regeneration is rather specialized. *A.s.* is a 'biennial' which perennates by buds in the axils of the basal leaves (Tutin 1980). These produce new plants when the flowering stem dies, and thus represent a means both of perennation and of vegetative reproduction. The tall flowering stems dry out during the summer and seed is lost over a prolonged period. Seeds contain a poorly differentiated embryo when shed, have a chilling requirement and germinate synchronously in spring. Under productive conditions the relative growth rate of the seedling is slow, but this has been shown to result from the early allocation of photosynthate to the swollen root-stock which forms the capital for shoot expansion during the second growing season (Hunt 1970). The leaves are relatively rich in P and Ca (*SIEF*). *A.s.* is a 'follower of humans' and river banks are perhaps the only semi-natural habitat in which the species occurs. Clapham (1953) speculates that the common northern form,

Altitude

Slope

Aspect

	n.s.	2.8	
n.s.		14.3	
		2.7	Unshaded
n.s.		16.7	

	n.s.	3.8	
n.s.		20.0	
		2.4	Shaded
n.s.		33.3	

Soil pH

Associated floristic diversity

Triangular ordination

Hydrology

Absent from wetland

Bare soil

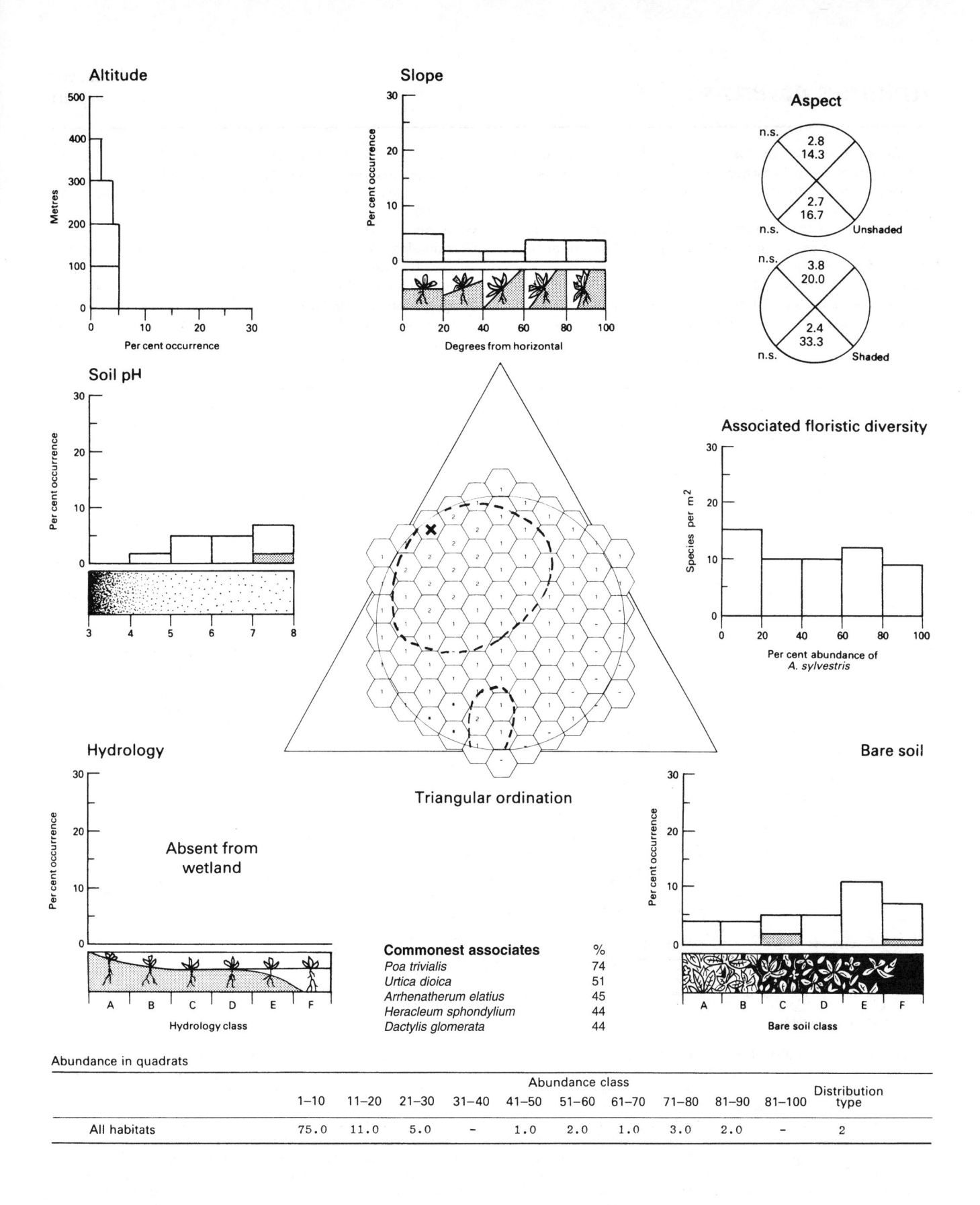

Commonest associates %
Poa trivialis 74
Urtica dioica 51
Arrhenatherum elatius 45
Heracleum sphondylium 44
Dactylis glomerata 44

Abundance in quadrats

	Abundance class									Distribution type
	1–10	11–20	21–30	31–40	41–50	51–60	61–70	71–80	81–90	81–100	
All habitats	75.0	11.0	5.0	–	1.0	2.0	1.0	3.0	2.0	–	2

which has been called var. *angustisecta* Druce, may be indigenous, while the southern plant (var. *latisecta* Druce) may have immigrated later as a result of human activities.

Current trends The specialized phenology and regenerative biology, including the absence of a persistent seed bank, renders *A.s.* rather vulnerable to seasonally unpredictable disturbance. Nevertheless, *A.s.* is common and likely to increase due to its capacity to exploit artificial linear sites such as roadsides, which provide corridors for population expansion.

Aphanes arvensis L.

Rosaceae
Parsley-piert

Life-form Winter- or less frequently summer-annual therophyte. Stems branched, decumbent or ascending; leaves ovate, trisect, hairy, often <50 mm²; roots with VA mycorrhizas.
Phenology Variable. Flowers April to October (early if a winter- and later if a summer-annual). Seed set from May onwards.
Height Foliage and flowers <80 mm.
Seedling RGR No data. SLA 26.0. DMC 26.4.
Nuclear DNA amount 1.1 pg; $2n = 48$ (*FE2*); $6x$.
Established strategy Intermediate between ruderal and stress-tolerant ruderal.
Regenerative strategies S, B_s. Seeds, which germinate in autumn and to a lesser extent in spring (*WCH*), may also be incorporated into a persistent seed bank; SB III.
Flowers Green, hermaphrodite, apomictic; <10 in leaf-opposed clusters.
Germinule 0.18 mg, 2.3 × 1.1 mm, dry fruit.
Germination Epigeal; t_{50} 4 days; <5–25°C; germination rate increased after dry storage at 20°C; L > D.
Biological flora None.
Geographical distribution The British Isles (100% of British vice-counties), *ABIF* page 343; S, W and C Europe (72% of European territories); N America.

Habitats Mainly a weed of arable land (particularly in cereal crops) or a component of winter-annual communities on rock outcrops and other droughted soils. Recorded also from spoil tips, dry grassland and beside paths. Observed as a weed of short turf in lawns, cricket fields and on putting greens. Absent from woodland and wetlands. [Occurs in winter-annual communities on fixed dunes.]
Gregariousness Sparse.
Altitude Suitable habitats are most frequent in the lowlands, but extends to 300 m. [Up to 520 m.].
Slope Absent from the steepest terrain.
Aspect Distribution biased towards S-facing slopes.
Hydrology Absent from wetlands.
Soil pH Rarely observed below pH 5.0.

Bare soil Virtually confined to open sites, mainly with high exposures of bare soil. Low values are generally associated with rocky habitats.
△ **ordination** Distribution centred on vegetation associated with disturbed environments on soils of moderate fertility. Absent from highly productive undisturbed sites.
Associated floristic diversity Typically species-rich.
Synopsis *A.a.* is an annual herb associated with a wide range of disturbed habitats. These include relatively infertile, shallow calcareous and dry, sandy soils on which the cover of perennials is restricted by summer drought. Immediate germination is prevented by an after-ripening requirement, and germination is delayed until autumn. The overwintering plant flowers apomictically in spring and early summer. Like *Trifolium dubium*, *A.a.* appears to occupy microsites with deeper soil, and is one of

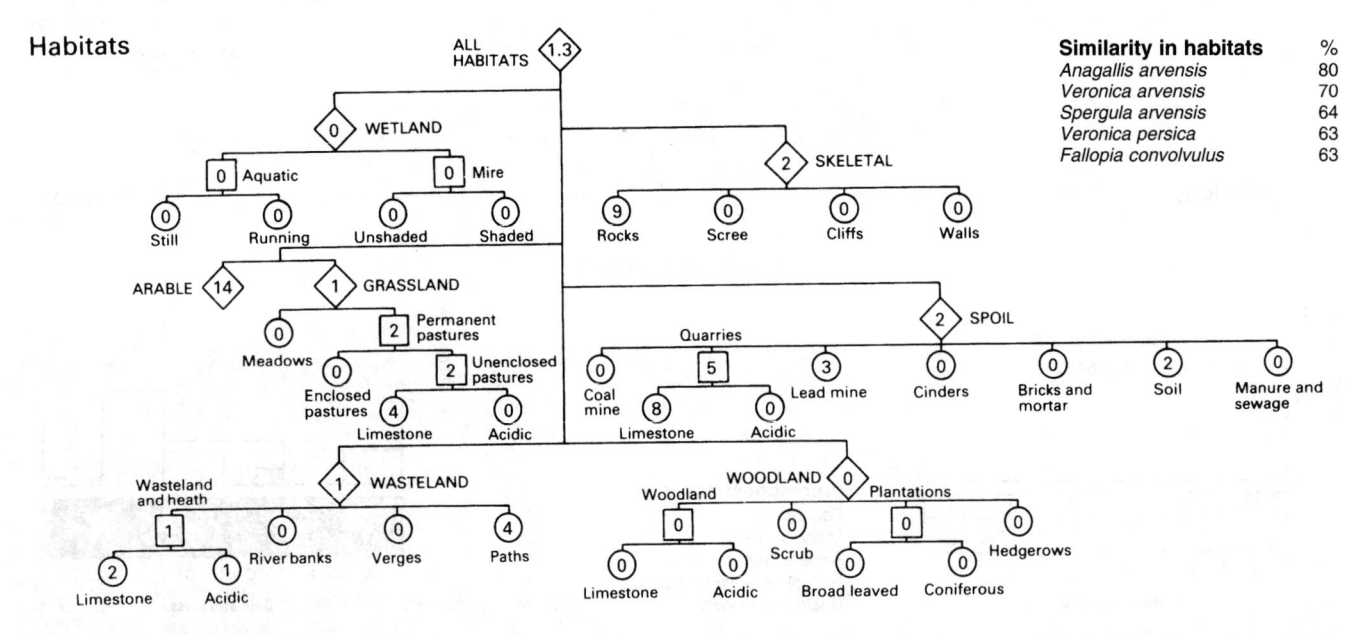

Habitats

Similarity in habitats	%
Anagallis arvensis	80
Veronica arvensis	70
Spergula arvensis	64
Veronica persica	63
Fallopia convolvulus	63

the last species to succumb to drought. Some plants may even persist until autumn during wet summers. With *T.d.*, *A.a.* extends into more-fertile, managed habitats, e.g. lawns, where the vigour of perennials is restricted by frequent close mowing. *A.a.* also exploits sites where the cover of perennials is restricted by annual cultivation, particularly for winter cereals. In this habitat *A.a.* appears to occupy a low, shaded stratum during summer, and it seems probable that growth occurs mainly in the autumn and spring before the crop is fully established, and again in late summer after the crop has been harvested. In further contrast with many other small winter annuals, *A.a.* often colonizes heavy, water-retaining clays. The capacity to exploit lawns and arable fields, and habitats subject to summer drought, may stem from the fact that unlike (for example) *Myosotis ramosissima*, *A.a.* is not obligately winter-annual. A secondary peak in germination often occurs during spring on arable land (*WCH*). Even so, *A.a.* is probably discouraged by the cultivation of spring-sown crops (Roberts & Neilson 1981). *A.a.* is entirely dependent upon seed for regeneration, and produces seeds which are large compared with those of most winter-annuals but smaller than average among arable weeds. Buried seeds of *A.a.* show seasonal changes in germination requirements and the species forms a large and persistent seed bank (Roberts & Neilson 1981). It is not known whether the ability of *A.a.* to exploit three very different types of habitat in Britain is under genotypic or phenotypic control. The related sexual species, *A. australis* Rydb. ($2n = 16$, $2x$), a recently described segregate of *A. microcarpa* auct. (non (Boiss. and Reuter) Rothm.) occurs on dry sandy soils. Except in arable fields, where the distributions of the two overlap, *A.au.* largely replaces *A.a.* on the (non-calcareous) Bunter sandstone of NE England. *A.au.* occurs within the pH range 4.0–7.5 and is more frequent towards the middle to upper end of this range.
Current trends Uncertain, but apparently favoured by reduced cultivation and direct drilling (Pollard & Cussans 1981).

Altitude
Slope
Aspect

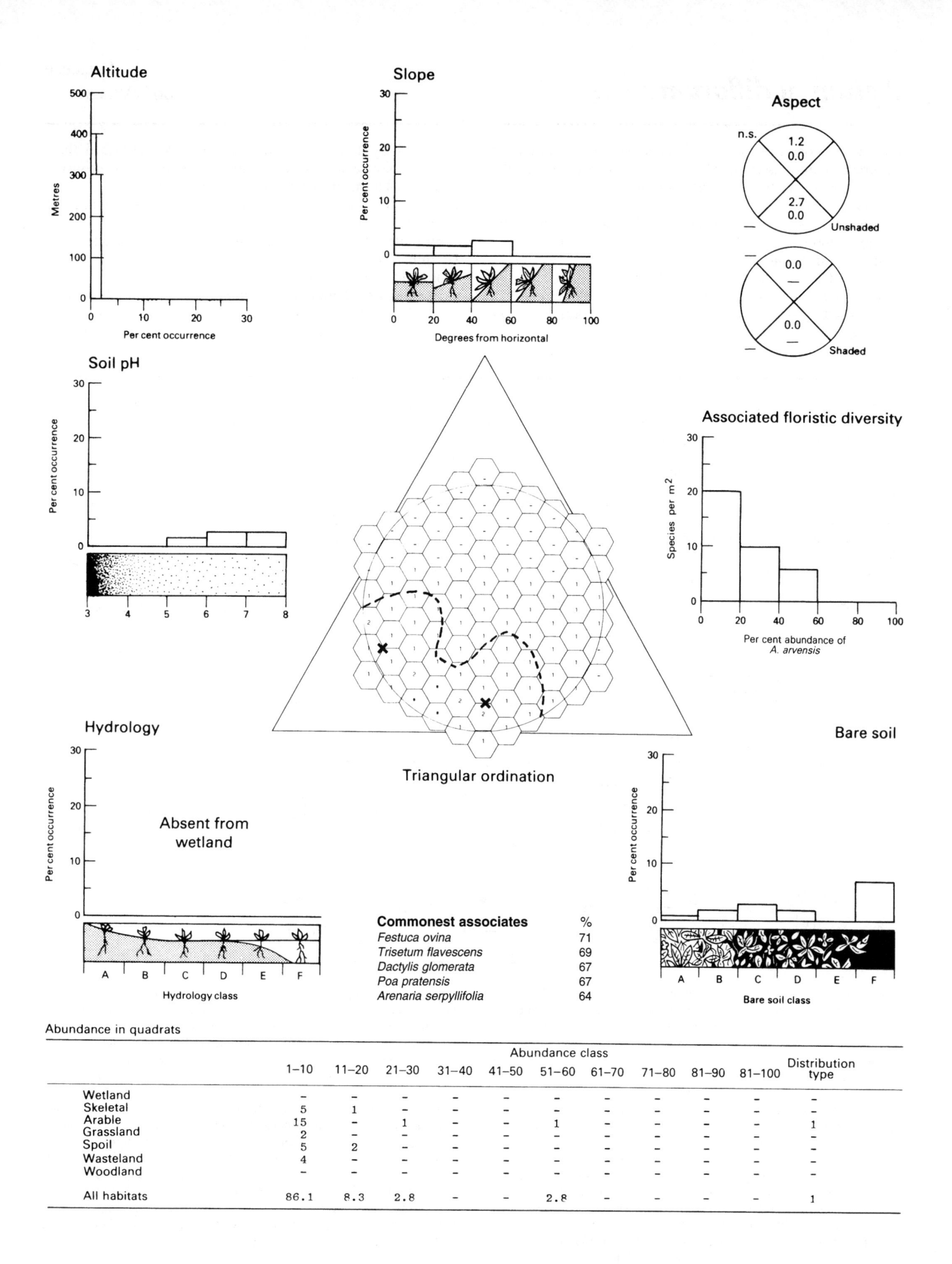

	n.s.	1.2	0.0	
	—	2.7	0.0	Unshaded

		0.0	
	—	0.0	Shaded

Soil pH

Associated floristic diversity

Hydrology

Absent from wetland

Triangular ordination

Commonest associates	%
Festuca ovina	71
Trisetum flavescens	69
Dactylis glomerata	67
Poa pratensis	67
Arenaria serpyllifolia	64

Bare soil

Abundance in quadrats

	1–10	11–20	21–30	31–40	41–50	51–60	61–70	71–80	81–90	81–100	Distribution type
Wetland	–	–	–	–	–	–	–	–	–	–	–
Skeletal	5	1	–	–	–	–	–	–	–	–	–
Arable	15	–	1	–	–	1	–	–	–	–	1
Grassland	2	–	–	–	–	–	–	–	–	–	–
Spoil	5	2	–	–	–	–	–	–	–	–	–
Wasteland	4	–	–	–	–	–	–	–	–	–	–
Woodland	–	–	–	–	–	–	–	–	–	–	–
All habitats	86.1	8.3	2.8	–	–	2.8	–	–	–	–	1

Apium nodiflorum (L.) Lag.

Apiaceae
Fool's Watercress

Life-form Polycarpic perennial helophyte. Stems branched, procumbent or ascending; leaves pinnate, sometimes >10 000 mm^2; roots shallow, adventitious on lower part of stem, non-mycorrhizal.

Phenology Shoot growth in late spring is followed by flowering in July and August and seed set September to October.

Height Usually <400 mm.

Seedling RGR No data. SLA 61.8. DMC 11.0.

Nuclear DNA amount 2.2 pg; $2n = 22*$ (*BC*); $2x*$.

Established strategy Intermediate between competitor and competitive-ruderal.

Regenerative strategies V, ?S. Forms extensive patches by means of trailing shoots which root at the nodes. Detached shoot pieces also capable of regeneration. Seedlings occur in spring, but establishment is probably infrequent; ?SB II.

Flowers Greenish-white, hermaphrodite, insect-pollinated; often >100 in each umbel.

Germinule 0.53 mg; 1.9 × 1.0 mm, mericarp (2 per fruit).

Germination Epigeal; t_{50} <7 days.

Biological flora None.

Geographical distribution Much of the British Isles, but rare in Scotland (91% of British vice-counties), *ABIF* page 469; much of Europe, particularly in W (56% of European territories); W and C Asia, N Africa.

Habitats A wetland species of rivers and streams (particularly in their upper reaches on calcareous strata), ditches and ponds, and shaded and unshaded mire. Found to a lesser extent on river banks and in other waterside habitats.

Gregariousness Recorded as intermediate but sometimes stand-forming.

Altitude Extends up to 340 m, above the British maximum recorded by Wilson (1956), but most abundant at <100 m.

Slope As with many wetland species, occurrences restricted to gentle slopes.

Aspect Insufficient data.

Hydrology A semi-aquatic species. Found in both slow-flowing and stagnant shallow water, but extending to >70 mm. Also frequent in topogenous and, to a lesser extent, soligenous mire, particularly in sites adjacent to open water.

Soil pH Widely distributed, but most abundant in the pH range 6.5-8.0.

Bare soil Insufficient data.

△ ordination Distribution centred on vegetation associated with relatively fertile, moderately disturbed conditions. Absent from highly disturbed vegetation and from very unproductive sites.

Associated floristic diversity Typically low.

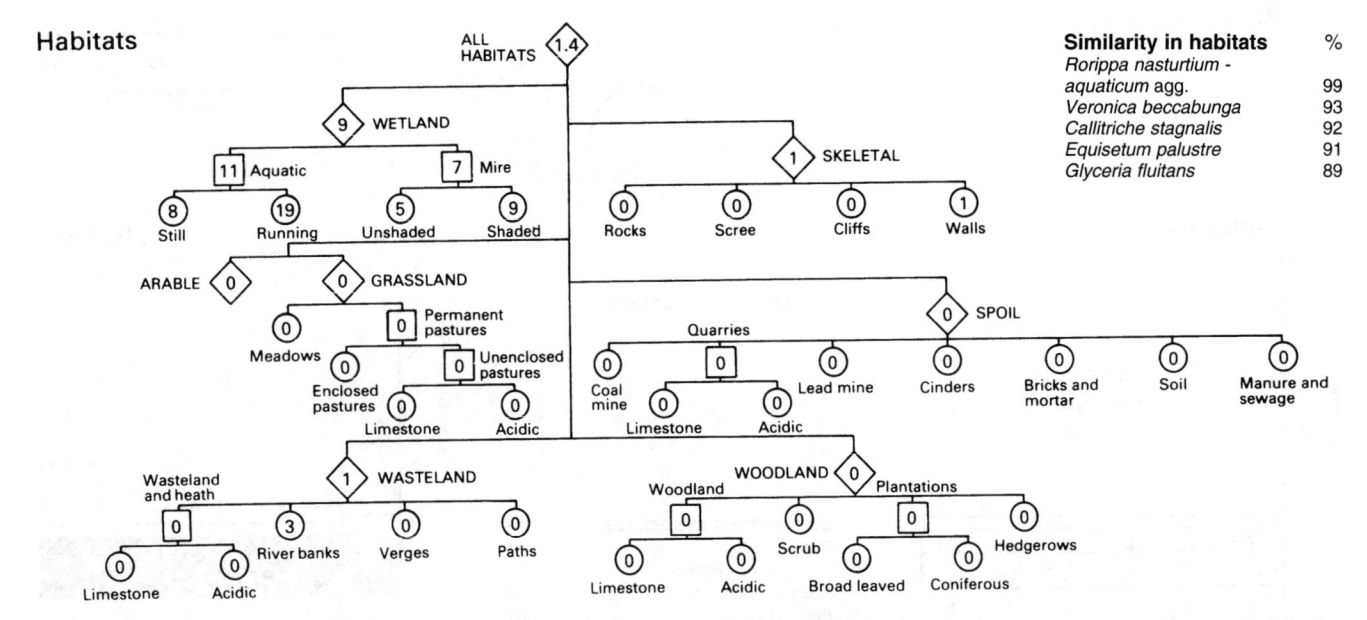

Habitats

Similarity in habitats	%
Rorippa nasturtium-aquaticum agg.	99
Veronica beccabunga	93
Callitriche stagnalis	92
Equisetum palustre	91
Glyceria fluitans	89

Synopsis *A.n.* is a moderately robust, prostrate to ascending stand-forming species which straddles the boundary between aquatic and mire habitats, and may behave as an emergent or, more rarely, submerged aquatic. The species occurs in fertile sites where the growth of potential dominants is restricted by disturbances such as ditch clearance and erosion by water currents. *A.n.* is frequent in ditches, at the margin of ponds, lakes, canals and slow-moving rivers and streams, on both calcareous and non-calcareous soils. Roots, mainly located at the basal internodes, provide good anchorage (Thommen & Westlake 1981), and the species is also characteristic of shallow, potentially fast-flowing streams on chalk and limestone. Many of these streams often dry up, partly or completely, during summer, providing conditions which *A.n.* appears better able to exploit than *Rorippa nasturtium-aquaticum* agg. (Thommen & Westlake 1981). The species is infrequent in grazed sites. Typically *A.n.* develops an erect growth form, but shoots in contact with the ground readily produce adventitious roots. Seedlings may become established either in mud or under water from early spring onwards, but seldom survive except in very open areas and appear to be vulnerable to competition from established plants (Thommen & Westlake 1981). Consequently, effective regeneration is mainly by vegetative means and detached vegetative shoots, which in water develop new roots within 2 days under laboratory conditions, are probably widely dispersed to new sites in water currents. The commonest semi-aquatics of similar stature and ecological distribution to *A.n.* are *Rorippa nasturtium-aquaticum* agg. and *Veronica beccabunga*. *A.n.* appears more commonly associated with fluctuating water levels and still-water habitats than *V.b.* Perhaps associated with this latter difference, *A.n.* shows much greater die-back of shoots in autumn than *V.b.* Differences from *N.o.* are described in the account for this species. *A.n.* is phenotypically plastic, and phenotypes resembling *A. repens* (Jacq.) Lag. may occur in short turf. *A.n.* hybridizes with *A. inundatum* (L.) Reichenb. fil.

Current trends Uncertain. Tolerant of eutrophic conditions, but many habitats formerly occupied by the species have been destroyed by drainage and other factors related to land-use.

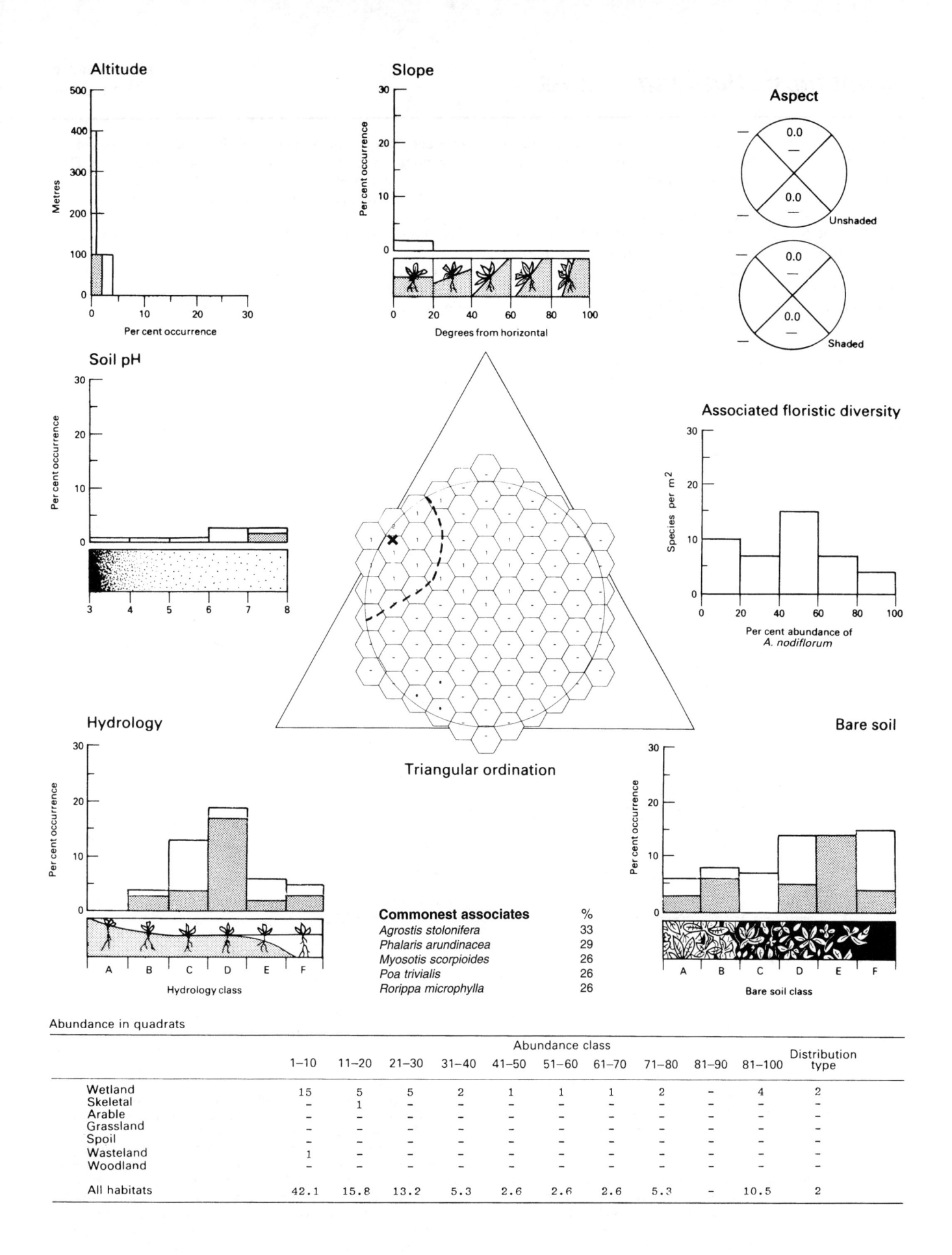

Altitude

Slope

Aspect

Unshaded

Shaded

Soil pH

Associated floristic diversity

Species per m^2 vs Per cent abundance of *A. nodiflorum*

Hydrology

Hydrology class: A B C D E F

Triangular ordination

Commonest associates %
Agrostis stolonifera 33
Phalaris arundinacea 29
Myosotis scorpioides 26
Poa trivialis 26
Rorippa microphylla 26

Bare soil

Bare soil class: A B C D E F

Abundance in quadrats

	1–10	11–20	21–30	31–40	41–50	51–60	61–70	71–80	81–90	81–100	Distribution type
Wetland	15	5	5	2	1	1	1	2	–	4	2
Skeletal	–	1	–	–	–	–	–	–	–	–	–
Arable	–	–	–	–	–	–	–	–	–	–	–
Grassland	–	–	–	–	–	–	–	–	–	–	–
Spoil	–	–	–	–	–	–	–	–	–	–	–
Wasteland	1	–	–	–	–	–	–	–	–	–	–
Woodland	–	–	–	–	–	–	–	–	–	–	–
All habitats	42.1	15.8	13.2	5.3	2.6	2.6	2.6	5.3	–	10.5	2

Arabidopsis thaliana (L.) Heynh.

Brassicaceae
Thale Cress

Life-form Winter- or perhaps occasionally summer-annual, semi-rosette therophyte. Shoots erect; leaves lanceolate to ovate, hairy, <500 mm^2; with a slender tap-root and sometimes VA mycorrhizas.
Phenology Usually winter-annual. Flowers from April often until the death of the plant in summer. Most seed set from May onwards. Plants flowering during September or October are perhaps summer annuals.
Height Foliage mainly <20 mm; flowering stem to 500 mm.
Seedling RGR No data. SLA 30.4. DMC 10.4.
Nuclear DNA amount 0.5 pg; $2n = 10$ (*FE1*), $2x$.
Established strategy Stress-tolerant ruderal.
Regenerative strategies S, B$_s$. Regenerates entirely by seedlings which establish in vegetation gaps, mainly during autumn. Forms a persistent bank of buried seeds; SB III.

Flowers White, hermaphrodite, homogamous, sometimes gynomonoecious, visited by small insects (*CTW*), predominantly but not exclusively self-fertilizing (Lawrence 1976) in a racemose inflorescence often with >200 flowers per plant.
Germinule 0.02 mg, 0.5 × 0.5 mm, seed, in a dehiscent 25-40-seeded siliqua.
Germination Epigeal; t_{50} 2 days; 7-31°C; L > D; germination rate increased after dry storage at 20°C.
Biological flora None, but see Lawrence (1976).
Geographical distribution Most of the British Isles, but less common in N and W and in Ireland (93% of British vice-counties), *ABIF* page 252; Europe except the extreme N (89% of European territories); E Africa and Asia; introduced in N America, S Africa and Australia.

Habitats Recorded most frequently from rock outcrops, quarry heaps and walls, with scattered occurrences in areas of shallow soil in pastures and derelict grassland on limestone. Observed with equal frequency on freely-drained sandy soils, on waysides, railway banks, soil heaps and disturbed wasteland. Also a weed of tree nurseries, park flower-beds and gardens.
Gregariousness Sparse.

Altitude Widespread in lowland and upland regions extending to 320 m. [Up to 690 m.]
Slope Over a wide range of slopes. The high frequency of *A.p.* on steep slopes is a reflection of its occurrence in winter-annual communities on rock outcrops.
Aspect Significantly more frequent on unshaded S-facing slopes.
Hydrology Absent from wetlands.
Soil pH Most frequent on relatively neutral soils in the pH range 6.0-8.0. Virtually absent from acidic soils of pH <4.5.

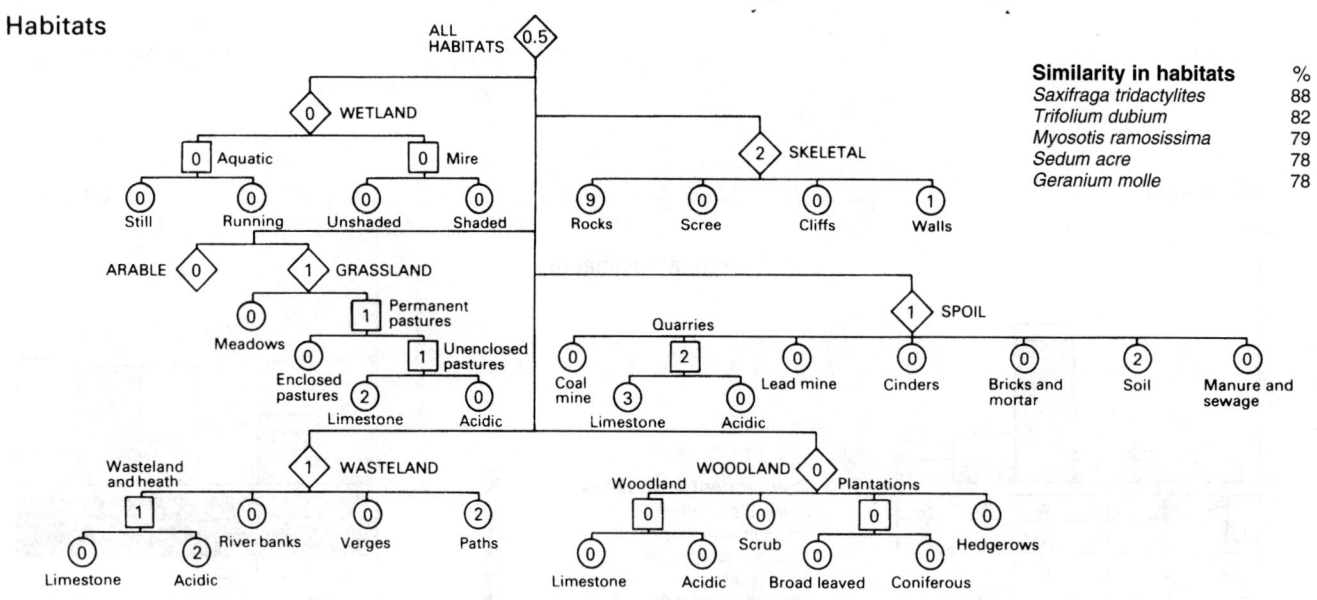

Habitats

Similarity in habitats	%
Saxifraga tridactylites	88
Trifolium dubium	82
Myosotis ramosissima	79
Sedum acre	78
Geranium molle	78

Bare soil Frequently associated with low to moderate exposures of bare earth and with the presence of bare rock. Occurrences on the Bunter sandstone and in horticultural habitats usually coincide with large amounts of bare soil.
△ **ordination** Distribution centred on vegetation associated with disturbed, relatively infertile conditions.
Associated floristic diversity Intermediate to species-rich.
Synopsis *A.t.*, which is normally a winter annual, is most characteristic of shallow rocky or sandy soils subject to summer drought. Both the association with S-facing slopes and the extremely low nuclear DNA amount suggest that vegetative development, which occurs in the cool parts of the year, depends upon the capacity for opportunistic growth in brief warmer periods. Flowering and seed set occurs later than in some other winter annuals, such as *Cardamine hirsuta, Erophila verna (L.)* DC., *Myosotis ramosissima, Saxifraga tridactylites* and *Veronica arvensis*.

Accordingly, *A.t.* persists only in less-droughted areas, often on deeper soil, and in open microsites within perennial grassland. Populations in grasslands fluctuate from year to year in inverse relation to the cover of perennials. In years following a severe drought large populations, often containing individuals of unusually large size, may be observed. Elsewhere *A.t.* occurs sporadically as a weed of gardens and park flower-beds. The species also occurs in tree nurseries and on railway ballast, two habitats in which colonization by perennials is often restricted by herbicide treatment. Seed is released in early summer and is prevented from germinating in summer by dormancy, which is removed by exposure to warm temperatures for several weeks (Ratcliffe 1961, Lawrence 1976). A persistent seed bank is also formed, and this appears to act as a buffer against local extinction. According to season and site, individuals vary considerably in size and seed production; much of this variation appears to be phenotypic (Ratcliffe 1961, Myerscough & Marshall 1973) but equally

Altitude
Slope
Aspect
P < 0.01

Soil pH
Associated floristic diversity
Hydrology
Absent from wetland

Triangular ordination

Bare soil

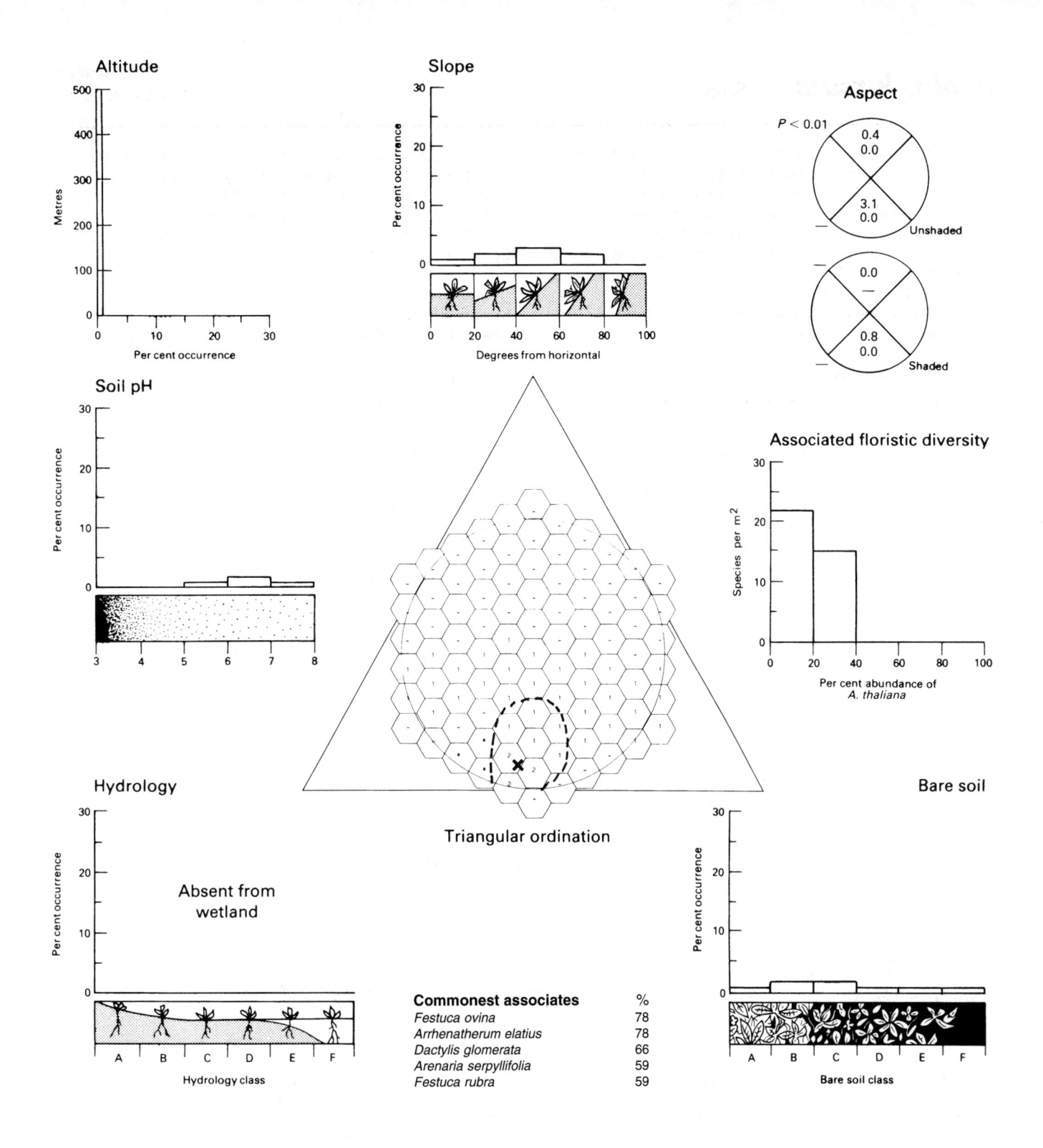

Commonest associates	%
Festuca ovina	78
Arrhenatherum elatius	78
Dactylis glomerata	66
Arenaria serpyllifolia	59
Festuca rubra	59

there is much genetic variation, particularly between populations (Lawrence 1976). Thus, plants from a limestone dale-side, a habitat where the period suitable for growth varies little from year to year, have been shown to require vernalization before flowering whereas a park population showed no such requirement. Plants of *A.t.* occasionally seen in flower during late summer on cultivated ground in Britain may belong to this latter type of population, arising from seed germinating from a buried seed bank following spring or summer disturbances of the soil. Summer-annual races of *A.t.* have been recorded (Laibach 1951), and it is suspected that in some tree nurseries *A.t.* may have been introduced from continental Europe along with container plants. However, in some populations (perhaps even a majority) seed dormancy is induced during winter, and as a result germination only occurs in autumn (Baskin & Baskin 1983). The small size and short life-cycle of some genotypes, coupled with the low chromosome number, have made *A.t.* a useful subject for genetic studies, and a range of genotypes and mutants circulates widely between laboratories.

Current trends *A.t.* is likely to increase further, both in abundance and in the range of habitats exploited.

Arabis hirsuta (L.) Scop.

Brassicaceae
Hairy Rock Cress

Life-form Polycarpic perennial, or perhaps more rarely monocarpic, semi-rosette hemicryptophyte. Stems erect; leaves ovate, hairy, <500 mm^2; tap-rooted, roots non-mycorrhizal.
Phenology Winter green. Flowers in June to August; seed set August to October. Leaves relatively long-lived.
Height Much leaf area located in an appressed basal rosette; flowering shoot to 600 mm in fruit.
Seedling RGR 1.0-1.4 week^{-1}. SLA 23.7. DMC 16.6.
Nuclear DNA amount No data; $2n = 32^*$ (*BC*); $4x^*$.
Established strategy Intermediate between stress-tolerator and C-S-R.
Regenerative strategies S, ?B$_s$. Reproduces exclusively by seed, which germinates mainly in autumn. On the basis of the responses of the seed in the laboratory the presence of a persistent seed bank is suspected; ?SB III.
Flowers White, hermaphrodite, homogamous, normally self-pollinated (Stace 1975); often >50 in a racemose inflorescence.
Germinule 0.09 mg, 1.7 × 0.9 mm, seed in a siliqua containing *c.* 40 seeds.
Germination Epigeal; t_{50} 3 days; 10-31°C; L > D.
Biological flora None.
Geographical distribution Scattered throughout the British Isles (86% of British vice-counties), *ABIF* page 263; Europe (85% of European territories); N Africa, N Asia, Japan and N America.

Habitats Widespread on skeletal soils overlying calcareous strata. Most frequent on rock outcrops and scree, but also on cliffs, quarry spoil, rocky pasture and wasteland. Rare in shaded habitats and almost absent in recent artificial habitats (the record from a demolition site is exceptional).
Observed on old limestone walls and rarely also on well-drained road verges off the limestone, where the road foundations provide a calcareous soil. Absent from wetlands. [Also found on calcareous sand dunes and on mountain ledges.]
Gregariousness Sparse.

Altitude Suitable habitats are most frequent on the carboniferous limestone and this is reflected in an altitude range which extends to 400 m. [Up to 920 m.]
Slope The greater frequency of *A.h.* on relatively steep slopes results from its common association with rocky habitats.
Aspect Equally widespread on N- and S-facing slopes but in grassland shows a S-facing bias (*Ecol Atl*). Insufficient data for shaded sites.
Hydrology Absent from wetlands.
Soil pH All records are from soils of pH >6.0.

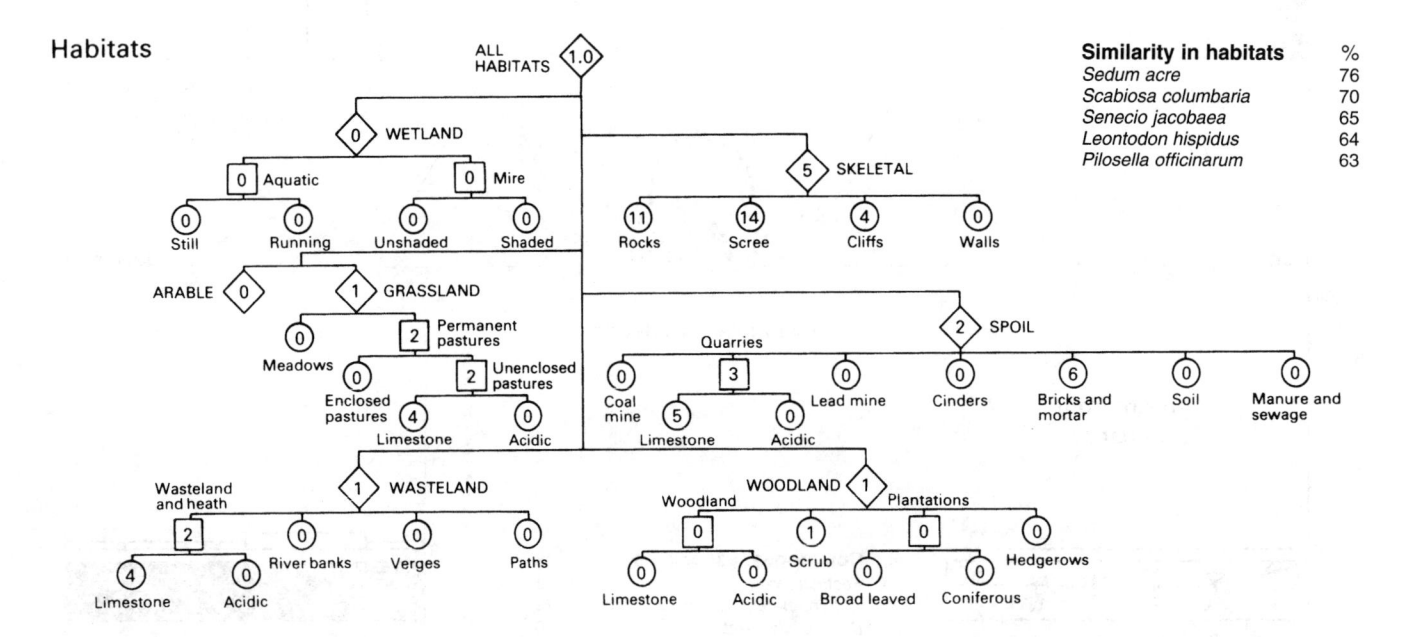

Habitats

Similarity in habitats	%
Sedum acre	76
Scabiosa columbaria	70
Senecio jacobaea	65
Leontodon hispidus	64
Pilosella officinarum	63

Bare soil Occurs mainly in habitats with small amounts of bare soil, and large quantities of bare rock.
Δ ordination Distribution centred on unproductive and relatively undisturbed vegetation. Absent from highly disturbed conditions and from undisturbed fertile situations.
Associated floristic diversity Typically intermediate.
Synopsis *A.h.* is a winter-green herb of infertile, calcareous soils. *A.h.* forms a low-growing rosette and is confined to relatively open microsites. The species lacks the capacity for lateral vegetative spread and regeneration of this short-lived perennial is entirely by seed, which germinates mainly in autumn. As in many other plants of disturbed habitats, large quantities of seed are produced and a robust plant may release in excess of 2000 seeds. *A. h.* may increase in abundance following burning and the development of a persistent seed bank is suspected. However, flowers and seeds are borne on a tall stem, up to *c.* 600 mm. Thus, flowering individuals of *A.h.* are vulnerable to the effects of grazing and in pasture the species is particularly associated with those outcrops and scree margins which are less accessible to stock. The low stature of the leaf canopy renders *A.h.* liable to dominance by taller species, and probably accounts in part for the restriction to rocky sites and grassland on shallow soils. *A.h.* attains peak above-ground biomass in summer and we suspect that in drier sites growth is sustained during this period through the capacity of the tap-root to exploit subsoil water in rock crevices. Unusually among species of infertile calcareous habitats, *A.h.* lacks mycorrhizas, (Harley & Harley 1986). Locally *A.h.* extends into lightly shaded habitats, where flowering and seed set are reduced. *A.h.* is mainly self-pollinated and populations are often topographically isolated (Stace 1975). As a result, *A.h.* is genetically polymorphic. Populations from Irish sand dunes were formerly separated as *A. brownii* Jordan (*FE1*). On a broader European scale *A. h.* is part of a closely related group in which five species have been recognized (*FE1*).
Current trends Likely to decrease. The capacity of *A.h.* to exploit artificial habitats is probably restricted more by narrow ecological niche than by powers of seed production and dispersal.

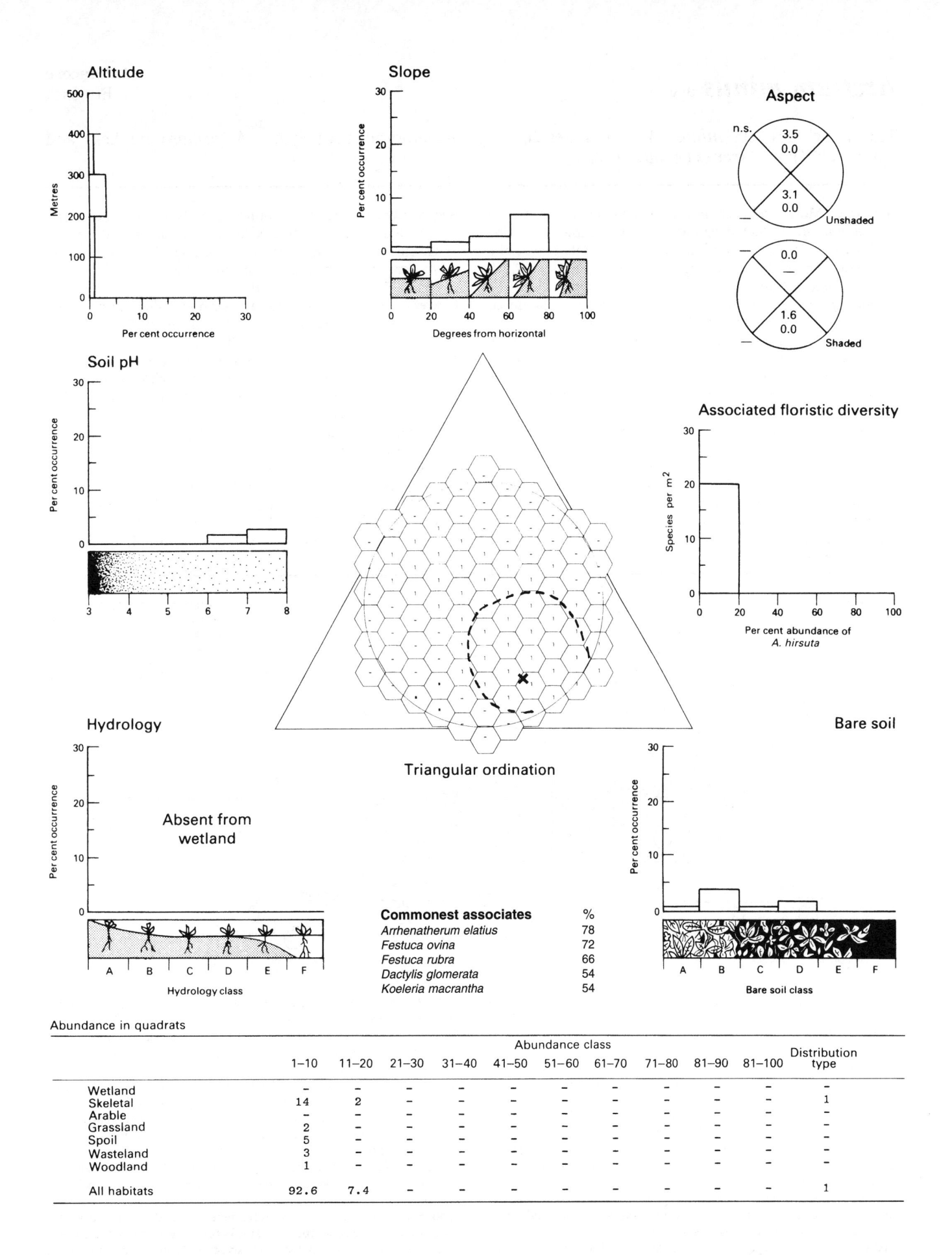

Altitude

Slope

Aspect

n.s.
3.5 / 0.0
3.1 / 0.0
Unshaded

0.0 / —
1.6 / 0.0
Shaded

Soil pH

Associated floristic diversity

Species per m² vs Per cent abundance of *A. hirsuta*

Hydrology

Absent from wetland

Triangular ordination

Bare soil

Commonest associates	%
Arrhenatherum elatius	78
Festuca ovina	72
Festuca rubra	66
Dactylis glomerata	54
Koeleria macrantha	54

Abundance in quadrats

	Abundance class									Distribution type
	1–10	11–20	21–30	31–40	41–50	51–60	61–70	71–80	81–90	81–100	
Wetland	–	–	–	–	–	–	–	–	–	–	–
Skeletal	14	2	–	–	–	–	–	–	–	–	1
Arable	–	–	–	–	–	–	–	–	–	–	–
Grassland	2	–	–	–	–	–	–	–	–	–	–
Spoil	5	–	–	–	–	–	–	–	–	–	–
Wasteland	3	–	–	–	–	–	–	–	–	–	–
Woodland	1	–	–	–	–	–	–	–	–	–	–
All habitats	92.6	7.4	–	–	–	–	–	–	–	–	1

Arctium minus agg.

Asteraceae
Burdock

Taxon includes ssp. *minus* (*A. minus* Bernh.), ssp. *nemorosum* (Lej.) Syme (*A. nemorosum* Lej.) and ssp. *pubens* (Bab.) Arenes (*A. pubens* Bab.)

Life-form Monocarpic perennial hemicryptophyte arising from a stout tap-root. Root system with VA mycorrhizas. Flowering shoot erect, branched and leafy. Vegetative plants consist of a rosette composed of a few large (<0.1 m^2) ovate leaves which are tomentose below.

Phenology Seeds germinate in spring, and the leaves produced persist until autumn then die back. More-mature plants, which may be more than 2 years old, produce leaves in late spring, followed in early summer by an upright shoot. Flowers produced from July to September and fruits from September onwards. A few achenes often still present on the dead plant in early winter.

Height Stems bearing leaves and flower heads to 1500 mm (to 2500 mm in ssp. *nemorosum*). Basal leaves of juvenile plants to 700 mm.

Seedling RGR No data. SLA 24.1. DMC 17.7.

Nuclear DNA amount No data; $2n = 36$ * (*BC*); $2x$*. (Tribe also with a polyploid base number; $n = 17$.)

Established strategy Intermediate between competitor and competitive-ruderal.

Regenerative strategies S. Entirely by seeds which are either shed close to the plant or dispersed as a burr, by animals or humans. Germination occurs mainly in spring. Indirect evidence suggests that seed does not remain viable in the soil for more than 1-3 years (*Biol Fl*); ?SB II.

Flowers Florets purple tubular, hermaphrodite, protandrous, with the capacity for self- or insect- pollination (*Biol Fl*), <50 in each capitulum in a racemose inflorescence.

Germinule 11.47 mg, 8.0 × 3.7 mm, achene with a pappus of short bristles up to 3.5 mm in length, shed either singly or in a burr consisting of fruits enclosed within a detached capitulum covered in hooked bracts.

Germination Epigeal; t_{50} 11 days; some populations show hard-coat dormancy; L > D.

Biological flora Gross *et al.* (1980).

Geographical distribution The British Isles, most common in England (95% of British vice-counties), *ABIF* page 609; most of Europe (87% of European territories); N Africa and Asia. Delimitation is difficult due to hybridization with *A. lappa* L.

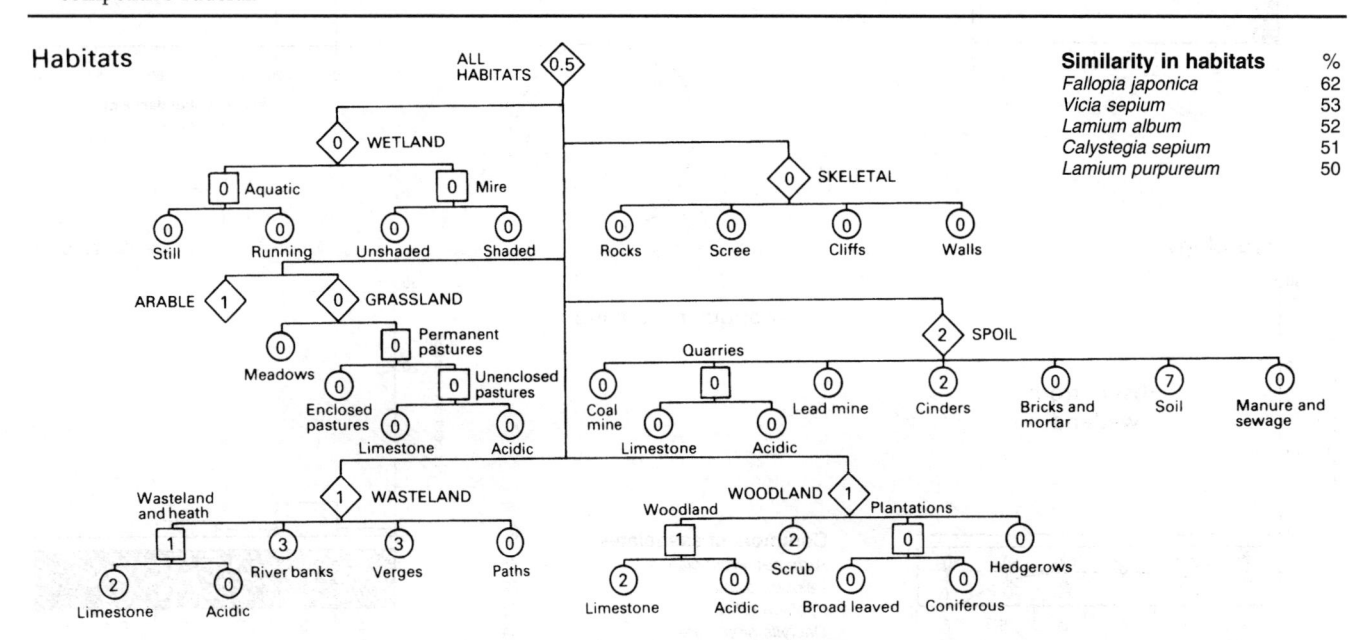

Habitats

Similarity in habitats	%
Fallopia japonica	62
Vicia sepium	53
Lamium album	52
Calystegia sepium	51
Lamium purpureum	50

Habitats Frequent on soil heaps, river banks and roadsides. Also characteristic of woodland clearings and ride edges, and found on wasteland and as a casual in arable fields. Absent from wetlands, pasture and skeletal habitats.

Gregariousness Sparse. An adult plant may shade up to 1 m^2 of ground surface and values in all frequency classes other than 1-10 usually refer to stunted plants or seedlings.

Altitude Most common in lowland areas. Relatively infrequent in the upland region, but extending to c. 350 m. Exact altitudinal limit difficult to assess because *A.m.* is impermanent in many of its sites. [Up to 380 m.]

Slope Not recorded from the steepest slopes as a consequence of absence from skeletal habitats.

Aspect Widely distributed. On S-facing slopes most records refer to juveniles. More-recent and extensive data (Hodgson unpubl.) indicate a N-facing bias in plants surviving to maturity.

Hydrology Like most species with a deep tap-root, absent from wetlands.

Soil pH Most records are from neutral soils in the pH range 6.0-8.0, but the species is known to occur locally at values down to pH 4.5.

Bare soil Associated with a wide range of values, but largely absent from completely closed vegetation; this is a reflection of the capacity of *A.m.* to colonize either large bared areas or vegetation gaps.

△ ordination Attains highest frequencies in productive vegetation subject to moderate levels of disturbance, but represented also in relatively stressed and undisturbed vegetation. These latter occurrences correspond to shaded habitats, where *A.m.* may persist for several years as small immature plants.

Associated floristic diversity Typically intermediate. When forming extensive stands *A.m.* presents a dense canopy and few other species are present.

Synopsis *A.m.* is a tall, monocarpic, perennial herb associated with a wide range of fertile, disturbed habitats. When close to maturity after two, or usually more, years, *A.m.* forms an exceptionally large and robust plant.

Altitude
Slope
Aspect
P<0.05, n.s. / Unshaded: 0.0, —, 0.4, 0.0
—, 0.0, —, 0.8, 0.0 / Shaded

Soil pH

Associated floristic diversity

Triangular ordination

Hydrology
Absent from wetland

Bare soil

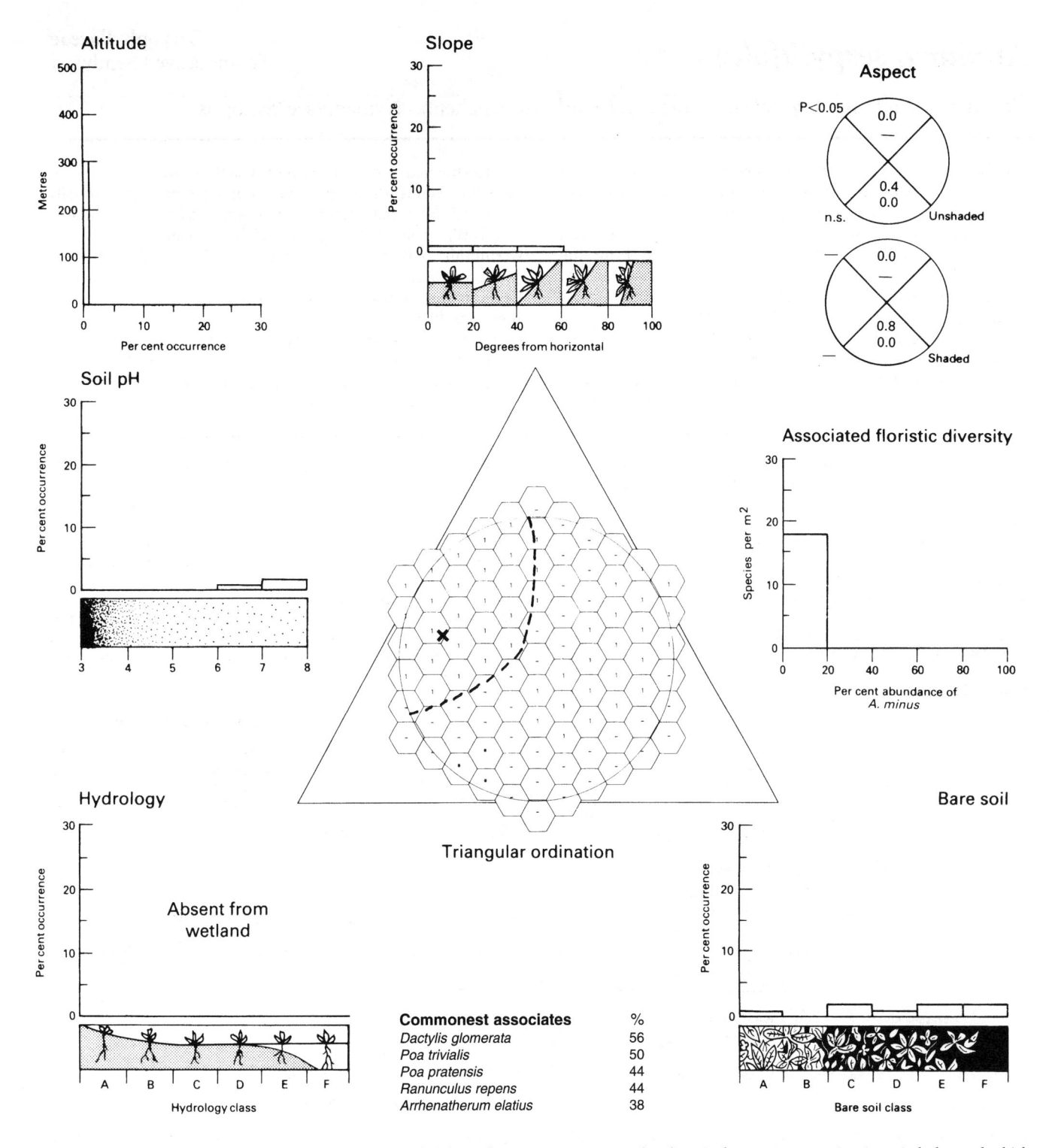

Commonest associates | %
Dactylis glomerata | 56
Poa trivialis | 50
Poa pratensis | 44
Ranunculus repens | 44
Arrhenatherum elatius | 38

Plants observed in Michigan, USA (Gross & Werner 1983) did not flower until the five-leaved stage, not attained for at least 4 years. The leaves of the mature plant are reminiscent of and only slightly smaller than those of rhubarb (*Rheum* × *hybridum* Murray) and cast a dense shade on neighbouring plants. The openings in the vegetation created by adult plants may be important as sites for colonization by their own seedlings, as described for *Dipsacus fullonum* L. by Werner (1977) and *Senecio jacobaea* by McEvoy (1984). However, *A.m.* is perhaps most characteristically an opportunist colonizer of fertile disturbed sites within or adjacent to woodland. *A.m.* is not, however, a true woodland species and under moderate shade plants remain in a suppressed state consisting of only a few, relatively small, basal leaves. Flowering occurs only in light shade or in the open. *A.m.* extends almost to its British altitudinal limit of 380 m within the Sheffield region. Thus, its distribution is probably limited to some extent by climatic factors. Regeneration is entirely by seed which appears to be shortly persistent in the soil (*Biol Fl*). Like many woodland species and unlike most common Asteraceae, which are wind-dispersed, the hooked fruits are dispersed by animals. Three ssp. are recognized (*Excur Fl*) but these are inter-fertile, both with each other and with *A. lappa* L. (Stace 1975) and are often difficult to distinguish. Consequently, differences in ecology between the ssp. remain obscure. In Britain *A.l.* largely restricted to disturbed, unshaded alluvial sites in S England, where it is represented by large, persistent populations with less-ephemeral seed banks (*Biol Fl*). The burrs of *A.l.* open at maturity and seeds fall near the parents. The seeds of *A.l.* are also less heavily predated (Rollo *et al.* 1985).

Current trends Likely to become more common in woodland areas as a result of clear felling and other disturbances related to modern forestry; perhaps also increasing on roadsides and in wasteland generally.

Arenaria serpyllifolia L.

Caryophyllaceae
Thyme-leaved Sandwort

Field data include ssp. *serpyllifolia* L. and ssp. *leptoclados* (Reichenb.) Nyman. (see 'Synopsis')

Life-form Winter- or perhaps more rarely summer-annual therophyte. Stems decumbent or ascending; leaves ovate, hairy, often >10 mm²; roots non-mycorrhizal.
Phenology Usually overwinters as a seedling or small plant. Flowers from May to August and sheds seed from July onwards.
Height Foliage and flowers to 250 mm.
Seedling RGR 1.0-1.4 week^{-1}. SLA 21.6. DMC 22.0.
Nuclear DNA amount 1.7 pg; $2n = 40^*$ (*BC*); $4x^*$.
Established strategy Stress-tolerant ruderal.
Regenerative strategies S, B$_s$. Regenerates by means of seed, which germinates mainly in autumn and which may be incorporated into a persistent seed bank; SB III.
Flowers White, hermaphrodite, homogamous, occasionally visited by small insects but automatically self-pollinated (*CTW*); often >100 per plant in dichasial cymes.
Germinule 0.06 mg, 0.5 × 0.5 mm, seed in a dehiscent *c.* 15 seeded capsule.
Germination Epigeal; t_{50} 2 days; 7-28°C; L > D.
Biological flora None.
Geographical distribution The British Isles (100% of British vice-counties), *ABIF* page 153; Europe northwards to Scandinavia (89% of European territories); temperate Asia, New Zealand and N America.

Habitats Recorded most frequently from limestone rock outcrops, walls, cliffs and scree slopes. Also found on spoil heaps including mining and quarry waste, and observed at similar frequencies on the dry sandy soils of the Bunter sandstone. Occasionally present on waysides, arable land and wasteland. Absent from shaded habitats and wetlands. [Also found on sand dunes.]
Gregariousness Intermediate.
Altitude Found over a wide range in altitude up to 400 m. [Up to 600 m.]
Slope Wide-ranging.
Aspect A significant positive association with unshaded S-facing slopes.
Hydrology Absent from wetlands.

Soil pH Mainly recorded on soils of pH >7.0. Not observed below pH 5.0.
Bare soil Widely distributed, but most frequently associated with moderate exposures.
△ **ordination** Distribution centred on vegetation associated with lightly disturbed, unproductive habitats. Virtually absent from undisturbed situations.
Associated floristic diversity Often species-rich.
Synopsis *A.s.* is typically a small winter-annual herb of droughted or disturbed habitats. Most characteristically the species exploits rock outcrops and dry sandy soils but, compared with co-occurring winter

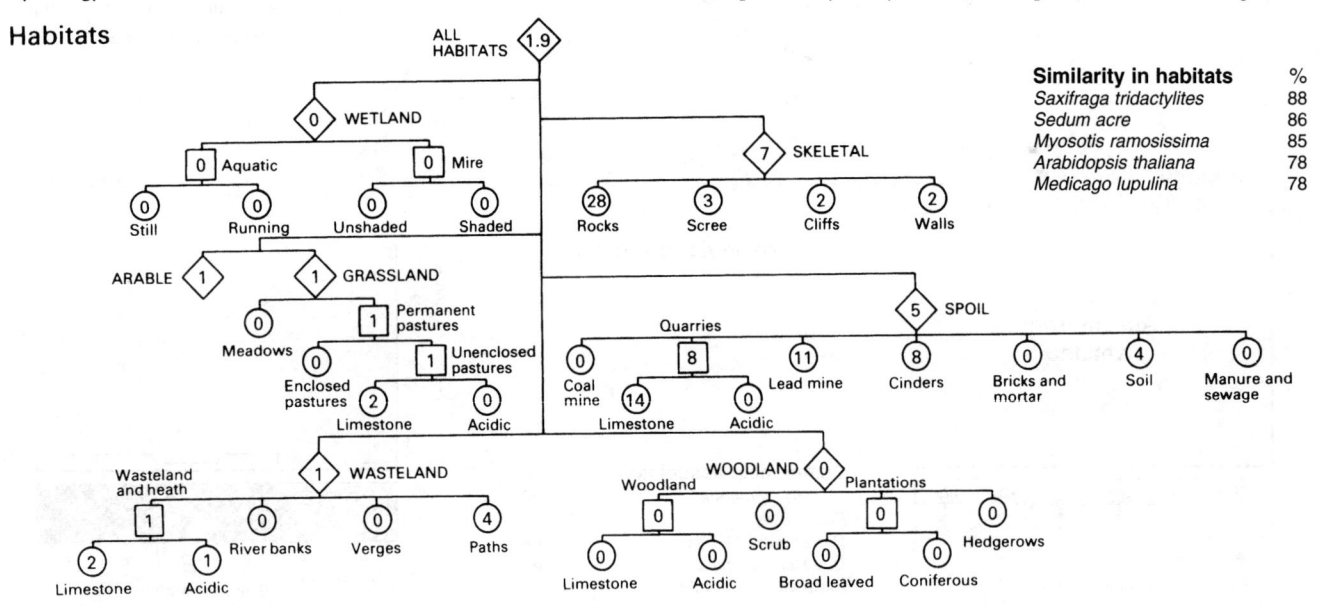

Habitats

Similarity in habitats	%
Saxifraga tridactylites	88
Sedum acre	86
Myosotis ramosissima	85
Arabidopsis thaliana	78
Medicago lupulina	78

annuals such as *Saxifraga tridactylites*, *A.s.* is relatively late-flowering (May or June to August) and tends to exploit less-droughted microsites. The species also requires relatively high levels of soil moisture for germination (Ratcliffe 1961). *A.s.* does not persist in shaded sites, is absent from mown grassland and in pasture is restricted to rocky sites inaccessible to grazing stock. The seedlings are grazed by sheep and snails (King 1977b) but are apparently little predated by rabbits (*CTW*). The species extends into some relatively fertile but highly disturbed habitats, e.g. arable fields. Flowering is induced by vernalization and long days (Ratcliffe 1961, Krekule & Hajkova 1972) and it is not certain whether the late-flowering plants of *A.s.* frequently observed in the field originate by spring germination. Freshly-shed seed has an after-ripening requirement and germination is inhibited by canopy shade (King 1975), providing a possible mechanism of 'gap detection' and imposed dormancy. These characteristics facilitate the incorporation of seeds into a persistent seed bank (Roberts 1986a).

Populations fluctuate widely from year to year and the presence of the persistent seed bank probably enables populations in more 'stable' sites, such as rock outcrops, to be maintained following years of poor seed set. Although as few as 100 seeds (King 1977b) may be borne on small plants on ant hills, seeds are frequently produced in much larger numbers. Salisbury (1942) suggests a mean number per plant of >3000 and, although lacking obvious dispersal mechanisms, *A.s.* appears highly mobile. Thus, in addition to exploiting sites which remain open for many years, the species frequently occupies transient areas of exposed soil in grassland or wasteland. On ant hills, a favoured habitat on chalk in S England, seedling mortality is very high and has a variety of causes (King 1977b). In particular, seedlings formed late in the autumn are very vulnerable to winter fatalities and drought (King 1977b). A large-seeded form, var. *macrocarpa* Lloyd, is found on the Atlantic coast (*CSAFB* p. 13, *CTW*, *FE1*). Ssp. *leptoclados* with $2n = 20$, $2x$ and a more-lowland distribution

Altitude

Slope

Aspect

$P < 0.05$

	2.4	
33.3		5.8
	15.4	

n.s. — Unshaded

	0.0	
—		—
	0.8	
	0.0	

— Shaded

Soil pH

Associated floristic diversity

Triangular ordination

Hydrology

Absent from wetland

Bare soil

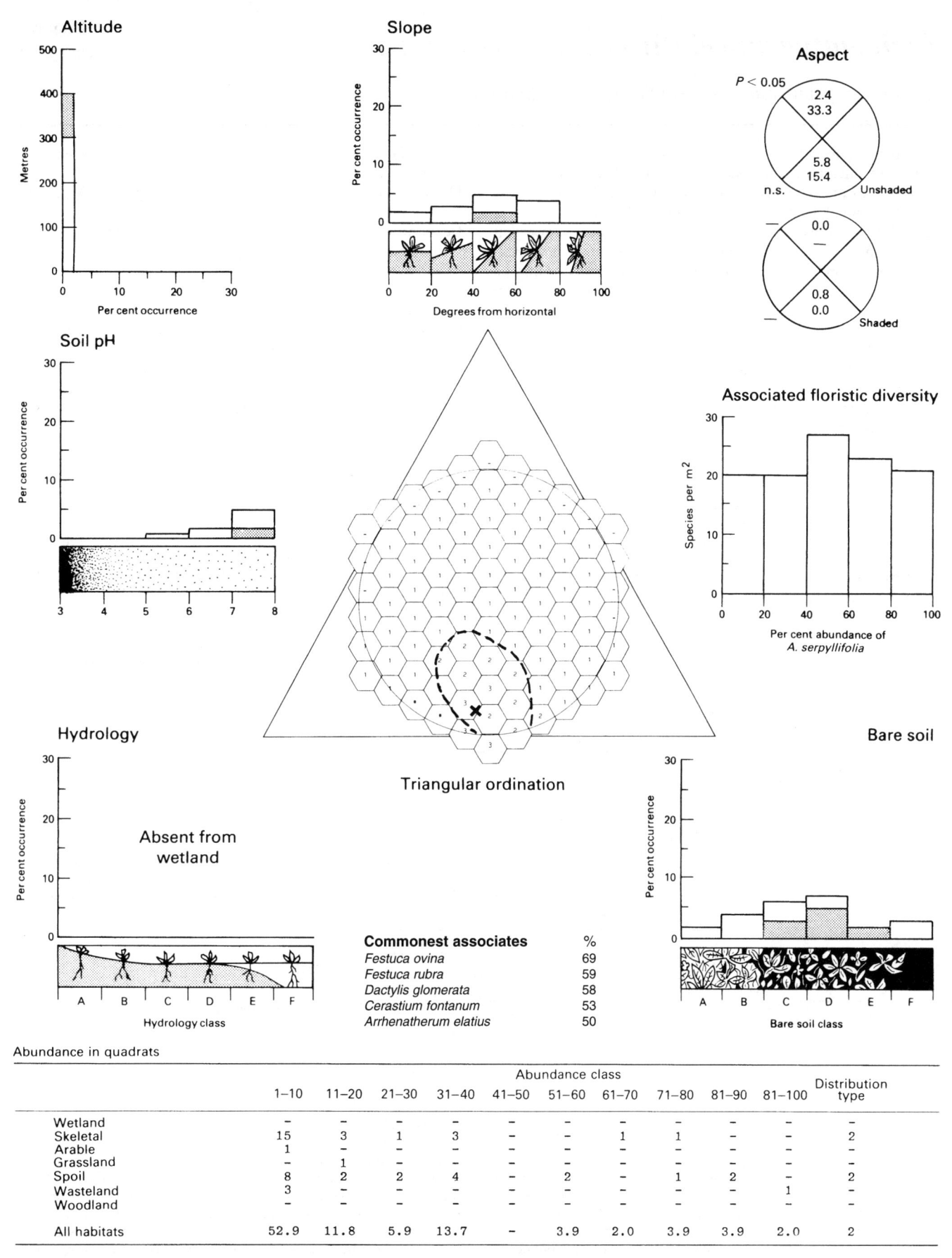

Commonest associates %
Festuca ovina 69
Festuca rubra 59
Dactylis glomerata 58
Cerastium fontanum 53
Arrhenatherum elatius 50

Abundance in quadrats

	1–10	11–20	21–30	31–40	41–50	51–60	61–70	71–80	81–90	81–100	Distribution type
Wetland	–	–	–	–	–	–	–	–	–	–	–
Skeletal	15	3	1	3	–	–	1	1	–	–	2
Arable	1	–	–	–	–	–	–	–	–	–	–
Grassland	–	1	–	–	–	–	–	–	–	–	–
Spoil	8	2	2	4	–	2	–	1	2	–	2
Wasteland	3	–	–	–	–	–	–	–	–	1	–
Woodland	–	–	–	–	–	–	–	–	–	–	–
All habitats	52.9	11.8	5.9	13.7	–	3.9	2.0	3.9	3.9	2.0	2

(*CSABF* p. 13, *FE1*), also occurs in Britain. Ssp. *l.* is much less common than ssp. *s.*, and in the Sheffield region is only recorded with certainty from lowland sites on the Bunter sandstone.

Current trends Has decreased on arable land, but is a common plant, particularly beside railways and in quarries. On balance it is probably increasing.

Arrhenatherum elatius (L.) Beauv. ex J. and C. Presl

Poaceae
Oat-grass

Including ssp. *bulbosum* (Willd.) Schubler and Martens

Life-form Loosely tufted, polycarpic perennial protohemicryptophyte. Shoots erect or slightly spreading; leaves linear, glabrous or slightly hairy, <4000 mm^2; often deeply rooted, with or without VA mycorrhizas.

Phenology Shoots produced from March onwards. Flowers in June and July. Seed shed while still green in July and August. Leaves generally short-lived but ssp. *elatius* in particular may be winter green (*Biol Fl*), especially in shorter turf. In ssp. *b*. stem bases become swollen in autumn.

Height Foliage and flowers to 1200 mm.

Seedling RGR 1.0-1.4 week^{-1}. SLA 24.3. DMC 30.7.

Nuclear DNA amount 16.0 pg; $2n = 28*$ (*BC*); $4x*$.

Established strategy Ssp. *b*. competitor; ssp. *e*. less robust and extends into less-productive habitats.

Regenerative strategies S, (V). By freshly shed seed in autumn.

Populations vary in their capacity to regenerate by vegetative means (see 'Synopsis'); SB I.

Flowers Green, wind-pollinated; self-incompatible (Weimarck 1968); usually <50 in a loose panicle (2 per spikelet).

Germinule 2.39 mg, 2.6 × 1.2 mm, caryopsis dispersed with an attached lemma, an hygroscopic bristle and a basal tuft of hairs (10.8 × 1.8 mm).

Germination Epigeal; t_{50} 3 days; <5-30°C; L = D.

Biological flora Pfitzenmeyer (1962).

Geographical distribution The British Isles (100% of British vice-counties), *ABIF* page 766; most of Europe, but absent from much of the NE and only on mountains in the S (89% of European territories); N Africa and W Asia; introduced in N America, New Zealand and Australia.

Habitats Most commonly recorded from limestone scree, where the species is an important colonizer and stabilizer. Other major habitats include limestone quarry heaps, hedgerows, unmown grass verges, soil heaps, cinder tips and other spoil, wasteland, meadows, pastures, river banks, rock outcrops, hedge banks and walls. Recorded at lower frequencies in clearings and margins of woodlands and plantations. The two ssp. may differ in their habitat range (see 'Synopsis').

Gregariousness Intermediate.

Altitude Widespread but especially common on the carboniferous limestone between 200-300 m. [Up to 550 m.]

Slope Wide ranging.

Aspect No marked effect of aspect in unshaded habitats. In shade, the species is significantly more frequent on S-facing slopes.

Hydrology Generally associated with moist rather than waterlogged soils.

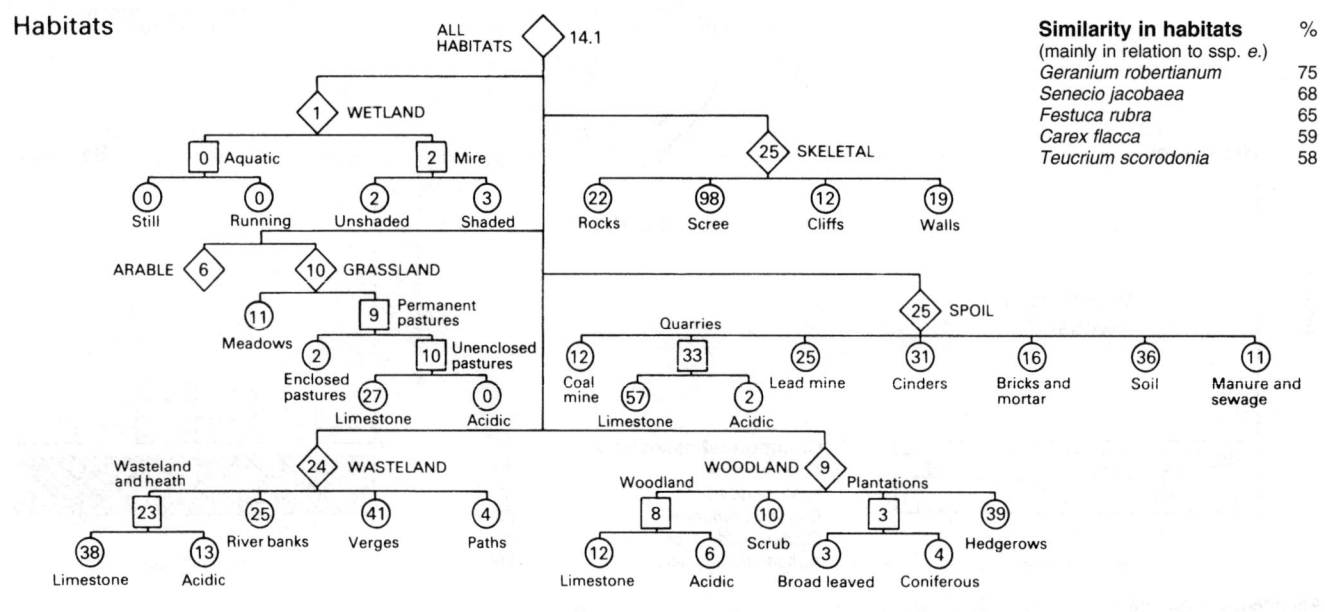

Habitats

ALL HABITATS 14.1

- WETLAND 1
  - Aquatic 0
    - Still 0
    - Running 0
  - Mire 2
    - Unshaded 2
    - Shaded 3
- ARABLE 6
- GRASSLAND 10
  - Meadows 11
  - Permanent pastures 9
    - Enclosed pastures 2
      - Limestone 27
      - Acidic 0
    - Unenclosed pastures 10
- SKELETAL 25
  - Rocks 22
  - Scree 98
  - Cliffs 12
  - Walls 19
- Quarries
  - Coal mine 12
  - Limestone 33 (57)
  - Lead mine 25
  - Acidic 2
- SPOIL 25
  - Cinders 31
  - Bricks and mortar 16
  - Soil 36
  - Manure and sewage 11
- WASTELAND 24
  - Wasteland and heath 23
    - Limestone 38
    - Acidic 13
  - River banks 25
  - Verges 41
  - Paths 4
- WOODLAND 9
  - Woodland 8
    - Scrub 10
      - Limestone 12
      - Acidic 6
    - Broad leaved 3
    - Coniferous 4
  - Plantations 3
  - Hedgerows 39

Similarity in habitats (mainly in relation to ssp. *e*.)	%
Geranium robertianum	75
Senecio jacobaea	68
Festuca rubra	65
Carex flacca	59
Teucrium scorodonia	58

Soil pH In lightly-managed grassland present on all but the most strongly acidic soils, and becoming more frequent and abundant with increasing pH up to values of 8.0.

Bare soil Widely distributed but not abundant at high exposures.

△ **ordination** Distribution extremely wide ranging. Highest frequencies occur in association with highly productive and undisturbed vegetation, indicating that *A.e.* is capable of competitive dominance. However, the descending distribution indicates a well-recognized ability to persist in less-productive conditions (Lloyd 1972b, Ellenberg 1978) and to establish as seedlings in disturbed habitats. We predict that ssp. *b*. may be more centred on fertile and disturbed vegetation while ssp. *e*. may be confined to parts of the diagram representing less-fertile conditions (see 'Synopsis')

Associated floristic diversity Low and declining slightly at the highest frequencies of *A.e.*

Synopsis *A.e.* is a tall, tussock-forming grass found in a wide range of unshaded or lightly shaded habitats. It has extensive roots and exploits habitats with a deep water table better than grasses such as *Alopecurus pratensis* (Ellenberg 1953). However, both ssp. appear to have a high moisture demand and the constancy in limestone screes may be in part a result of continuously moist soil maintained by the 'mulch effect' of talus. In meadows the species thrives best when 'receiving heavy complete manures whether as organic or artificial fertilisers' (*PGE*). The leaves of *A.e.* have unusually high P levels and their Ca status is greater than that of the foliage of most common grasses (*SIEF*). Ecologically, the two subspecies are poorly distinguished. Ssp. *b*., with bulbous stem bases during winter, grows predominantly on fertile soils and was formerly a common arable weed (*Biol Fl*). Ssp. *e*. occurs on limestone scree and on other less fertile sites and shows a greater tendency to winter greenness (*Biol Fl*). Plants

Altitude
Slope
Aspect

	n.s.
n.s.	28.3 / 37.5
n.s.	24.2 / 40.7 Unshaded

$P < 0.05$	7.6 / 20.0
n.s.	15.4 / 26.3 Shaded

Soil pH
Associated floristic diversity
Hydrology
Triangular ordination
Bare soil

Commonest associates %
Festuca rubra 61
Dactylis glomerata 47
Poa pratensis 41
Poa trivialis 32
Festuca ovina 30

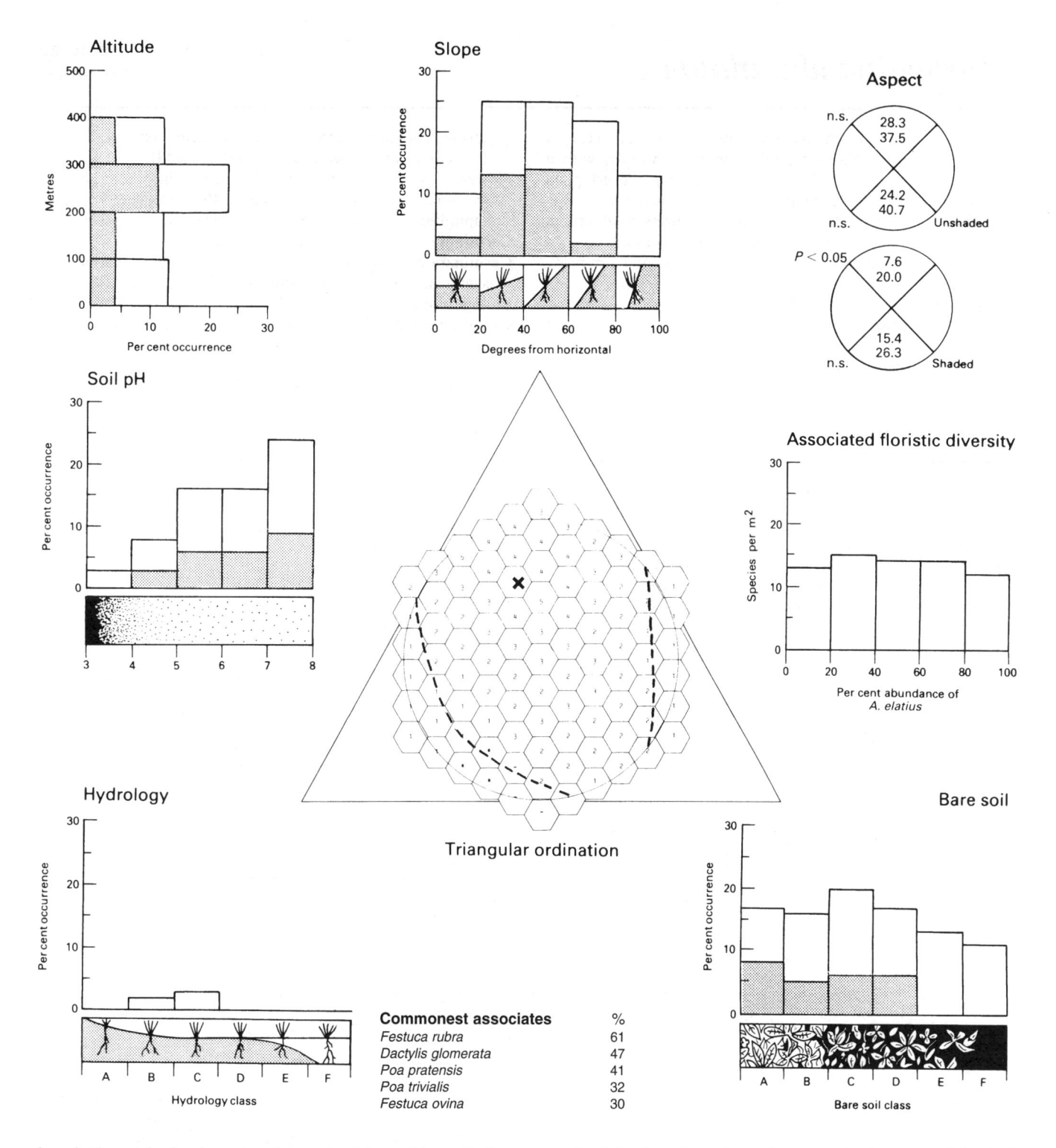

from fertile wasteland and uncut road verges (mainly ssp. *b.*) are tall; they may achieve co-dominance with broad-leaved herbs. On less-fertile soils (e.g. scree) *A.e.* (mainly ssp. *e.*) also may have a robust morphology and may replace shorter, slow-growing grasses, such as *Festuca ovina,* when grazing ceases. Both ssp. are particularly susceptible to grazing since (a) stems are upright and nodal, bearing several leaves, (b) only a small number of basal axillary buds are available for regeneration (*Biol Fl*) and (c) regeneration by seed is almost entirely from summer and autumn germination (Thompson & Grime 1979, Roberts 1986b), no seed bank is formed and the seed is large and perhaps poorly dispersed. The seed of *A.e.* is moderately large and under shade the first leaf is capable of considerable extension. This may facilitate establishment on deep block scree (Grime & Jeffrey 1965). *A.e.* is shortly rhizomatous, with a limited capacity for lateral vegetative spread, and rhizome connections may break, resulting in closely associated daughter plants. In ssp. *b.* regeneration by means of detached stem bases can occur during autumn or winter disturbance, e.g. in ploughed fields. This ssp., 'Onion Couch', used to be a frequent agricultural weed (*Biol Fl*). In some lightly grazed pastures semi-prostrate phenotypes or genotypes that are more resilient under defoliation have been recorded (Mahmoud *et al.* 1975). Plants perhaps referable to *A.e.* with $2n = 14$, $2x$ are recorded from SW Europe (*FE5*).

Current trends *A.e.* exploits a wide range of derelict, fertile environments. It has, for example, increased in response to a relaxation of management of road verges and railway banks, and to withdrawal of sheep grazing on some limestone dales. Thus, given current trends, *A.e.* is likely to become even more common. With the greater use of minimum tillage and direct drilling, an increase in ssp. *b.* on arable land has also been predicted (Ayres 1977).

(+)*Artemisia absinthium* L.

Asteraceae
Wormwood

Life-form Perennial chamaephyte or, in its first year, a hemicryptophyte. Stems typically somewhat woody, with a crown of hairy, dissected leaves (often c. 4000 mm^2) overtopped by erect leafy flowering stems; roots non-mycorrhizal.

Phenology Crown of leaves overwinters. Flowers produced in July and August and seed set from September onwards.

Height Foliage and flowers to 900 mm but above 450 mm most leaves small.

Seedling RGR No data. SLA 24.8. DMC 22.1.

Nuclear DNA amount 7.3 pg, $2n = 18*$ (*BC*, Bennett & Smith 1976); $2x*$.

Established strategy Intermediate between competitor and C-S-R.

Regenerative strategies S, B$_s$. Seed germinates when shed in autumn or, if conditions are unfavourable, during the following spring. Persistent buried seeds have been reported (Odum 1978), but further investigation is required; ?III.

Flowers Florets yellow, central hermaphrodite, outer female, wind-pollinated (*Biol Fl*); often c. 40 aggregated into each capitulum; >1000 capitula may occur in each racemose panicle.

Germinule 0.10 mg, 1.2 × 0.7 mm, achene.

Germination Epigeal; t_{50} 3 days; 12-31°C; germination rate slightly increased after dry storage at 20°C; L > D.

Biological flora Maw *et al.* (1985).

Geographical distribution Largely restricted to England; most Scottish and Irish records are near the coast; perhaps only naturalized (73% of British vice-counties), *ABIF* page 645; most of Europe (74% of European territories and 3% naturalized); temperate Asia; introduced into N and S America and New Zealand.

Habitats Most characteristic of areas of urban dereliction. Particularly frequent on demolition sites. Also found on a wide range of other types of spoil and on waysides, sometimes growing in cracks at the edges of pavements. Sometimes present in skeletal habitats on cliffs and outcrops, but even here always close to roads or urban settlements. [Even in coastal areas is mainly found on derelict land and is often impermanent.]

Gregariousness Usually represented by scattered individuals.

Altitude A lowland species observed to 350 m. The greater frequency between 100 and 200 m reflects the concentration of sites in the industrial conurbation comprising Sheffield and Rotherham. [Up to 370 m.]

Slope Wide-ranging.

Aspect No trend detected.

Hydrology Absent from wetlands.

Soil pH Mainly found at high pH values. The peak above pH 8.0 reflects a strong association with demolition sites.

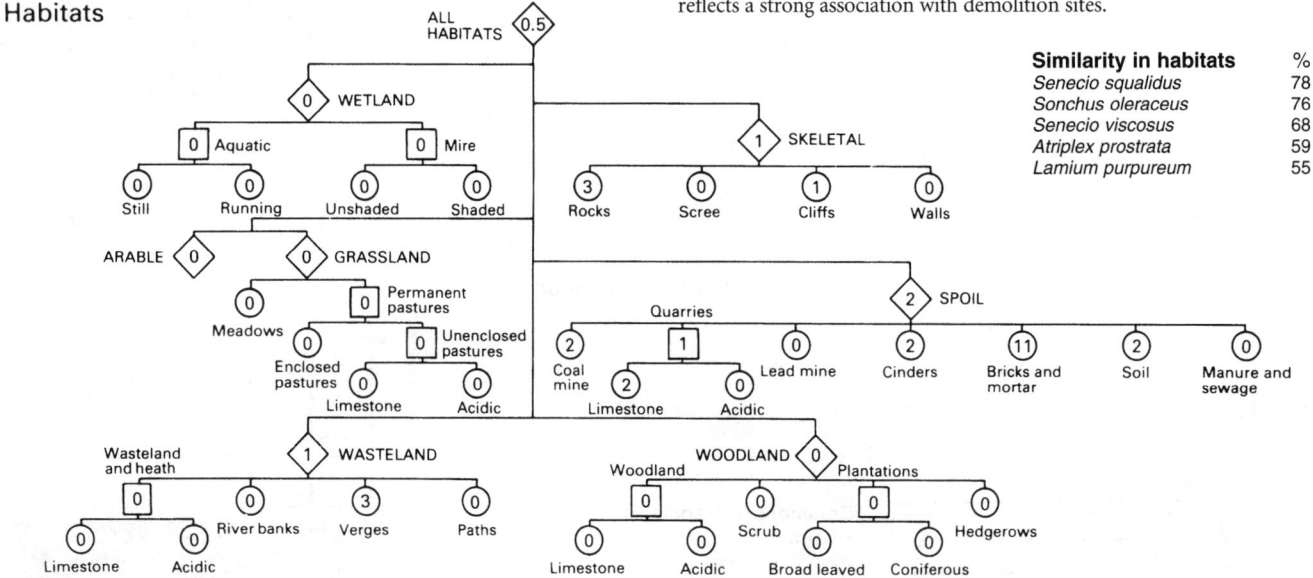

Similarity in habitats	%
Senecio squalidus	78
Sonchus oleraceus	76
Senecio viscosus	68
Atriplex prostrata	59
Lamium purpureum	55

Habitats

Bare soil Usually associated with habitats where the exposure of soil or spoil is high.

△ **ordination** Present only in vegetation which is relatively fertile; distribution centred on vegetation which is subject to some disturbance.

Associated floristic diversity Typically intermediate.

Synopsis *A.a.* is a tall, bushy, aromatic herb formerly much cultivated for cuisine and medicine. *A.a.* is probably as abundant within the Sheffield region as anywhere in Britain; it is largely restricted to areas of urban dereliction. *A.a.* does not exploit semi-natural habitats and is probably not native to Britain. Colonizes relatively fertile soils and reproduces entirely by seed in vegetation gaps. Populations tend to be common in recently disturbed areas, and in situations where opportunities for vegetation consolidation are restricted by the terrain. Demolition sites, where pieces of brick or concrete rubble reduce access to the soil and pavement edges, are frequently exploited. Elsewhere individuals of *A.a.* may occur in isolation in, for example, tall grassland at sites of previous disturbance. Little else is known of the ecological foundation of the distribution of *A.a.*, but its occurrence in demolition sites and in areas close to the sea (*ABF*) suggests some nutritional specialization. *A.a.* is known to extract large quantities of N from the soil, possibly to the detriment of potential competitors (*Biol Fl*). The association of *A.a.* with large urban areas and maritime regions may also indicate a sensitivity to harsh winter climates. *A.a.* has a predominantly S distribution in Britain and plants often show signs of winter damage. The significance of the winter growth form, typically an aggregation of leaves at the top of a woody stem, is obscure, but it is a form shared with the maritime *Brassica oleracea* L. *A.a.* is a weed of overgrown pasture in Canada (*Biol Fl*). This is not the case in Britain, although *A.a.* has been observed to persist on derelict rubbish tips reclaimed for horse-grazing. *A.a.* may taint milk and is rejected by cattle in Canada (*Biol Fl*) and can withstand infrequent cutting.

Current trends *A.a.* appears to have increased as a result of urban dereliction and industrial spoilage. Whether this is likely to continue is uncertain, but it seems improbable that *A.a.* will become an important grassland weed in Britain.

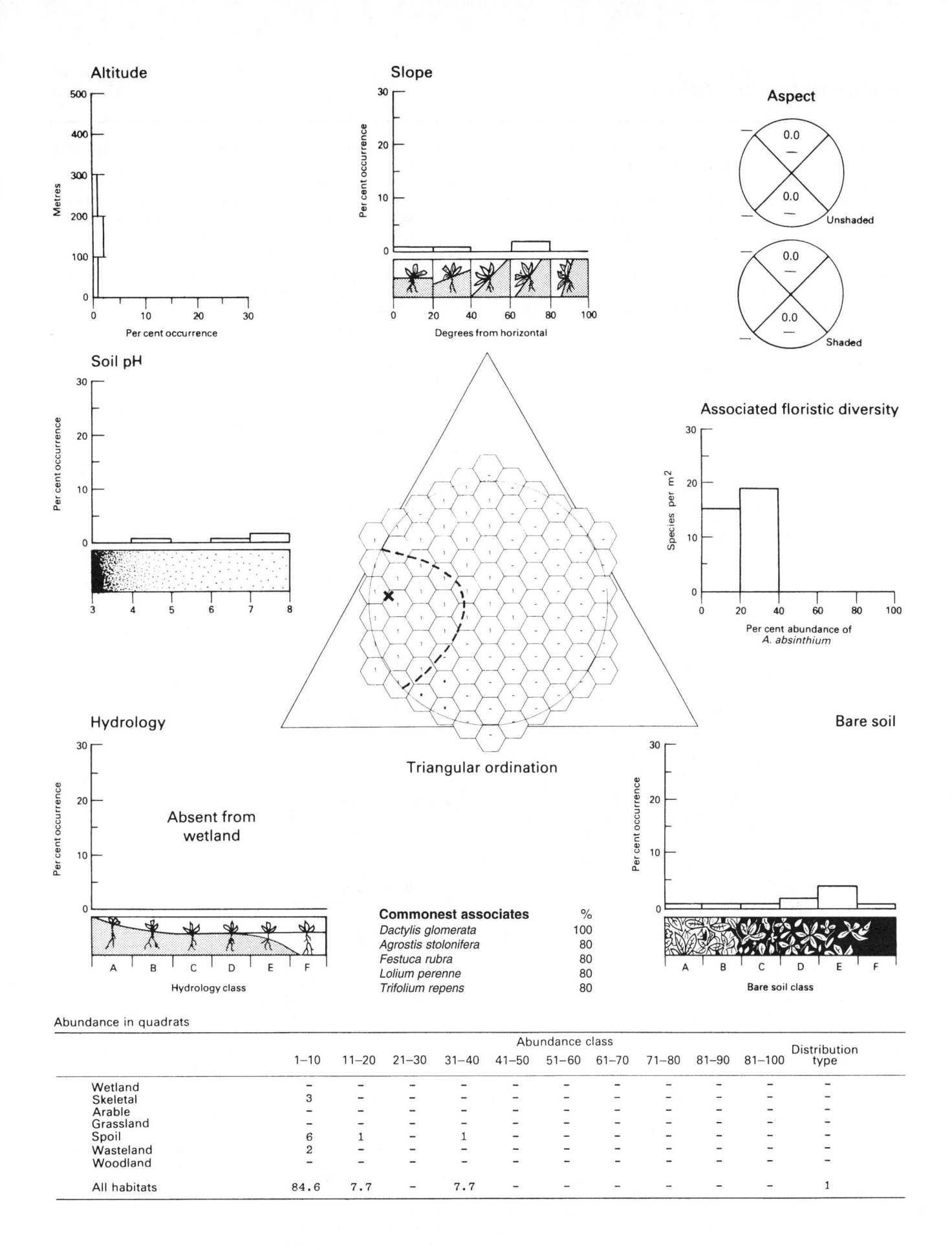

Altitude

Slope

Aspect

Unshaded

Shaded

Soil pH

Associated floristic diversity

Species per m²

Per cent abundance of *A. absinthium*

Hydrology

Absent from wetland

Hydrology class

Triangular ordination

Commonest associates	%
Dactylis glomerata	100
Agrostis stolonifera	80
Festuca rubra	80
Lolium perenne	80
Trifolium repens	80

Bare soil

Bare soil class

Abundance in quadrats

	Abundance class									Distribution type
	1–10	11–20	21–30	31–40	41–50	51–60	61–70	71–80	81–90	81–100	
Wetland	–	–	–	–	–	–	–	–	–	–	
Skeletal	3	–	–	–	–	–	–	–	–	–	
Arable	–	–	–	–	–	–	–	–	–	–	
Grassland	–	–	–	–	–	–	–	–	–	–	
Spoil	6	1	–	1	–	–	–	–	–	–	
Wasteland	2	–	–	–	–	–	–	–	–	–	
Woodland	–	–	–	–	–	–	–	–	–	–	
All habitats	84.6	7.7	–	7.7	–	–	–	–	–	–	1

Artemisia vulgaris L.

Asteraceae
Mugwort

Life-form Tufted perennial hemicryptophyte with branching root stock. Stems erect; leaves ovate, pinnatifid, usually <2000 mm^2; roots with VA mycorrhizas.

Phenology New shoots overwinter as low rosette. Flowers July to September and sets seed September to October. Stems then die.

Height Foliage and flowers to 1200 mm but above 600 mm most leaves are small.

Seedling RGR No data. SLA 21.7. DMC 26.1.

Nuclear DNA amount 6.1 pg; $2n = 16^*$ (BC); $2x^*$.

Established strategy Intermediate between competitor and competitive-ruderal.

Regenerative strategies S, B$_s$, V. By seed which normally germinates in vegetation gaps during spring or after incorporation into a bank of buried seed (Bostock 1978, Roberts 1986a). Capacity for vegetative spread is limited; SB II or ?IV.

Flowers Yellow or purplish, tubular, wind-pollinated, c. 15 in each capitulum, marginal florets female, central florets hermaphrodite (see also Garnock-Jones 1986); often >1000 capitula in a racemose panicle.

Germinule 0.12 mg, 1.3 × 0.5 mm, achene.

Germination Epigeal; t_{50} 1 day; 7-34°C; germination rate increased after dry storage at 20°C; L > D.

Biological flora None.

Geographical distribution The British Isles, but local in Scotland and Ireland, *ABIF* page 645; most of Europe (82% of European territories); other temperate regions of the N hemisphere except for the extremes of N and S.

Habitats Predominant in disturbed urban habitats such as spoil heaps of mineral soil, cinders, slag and building rubble. Also present on banks of rivers and streams, and path edges, and frequently observed on road verges and beside railways in some lowland areas. Only of casual occurrence on arable land. Absent from woodlands, pastures and wetland.

Gregariousness Sparse to intermediate.

Altitude Most-suitable habitats are found in lowland areas, but extends to 350 m. [Up to 420 m.]

Slope Widely distributed.

Aspect Associated with a wide range of aspects. Only recorded as abundant on unshaded S-facing slopes.

Hydrology Absent from wetland.

Soil pH Seldom recorded below pH 5.5.

Bare soil Most frequent and abundant in habitats with high exposures.

△ **ordination** Distribution centred on vegetation associated with relatively fertile but heavily disturbed situations. Present but less common in some relatively undisturbed productive vegetation types. Absent from unproductive habitats.

Associated floristic diversity Varying widely, but typically low to intermediate.

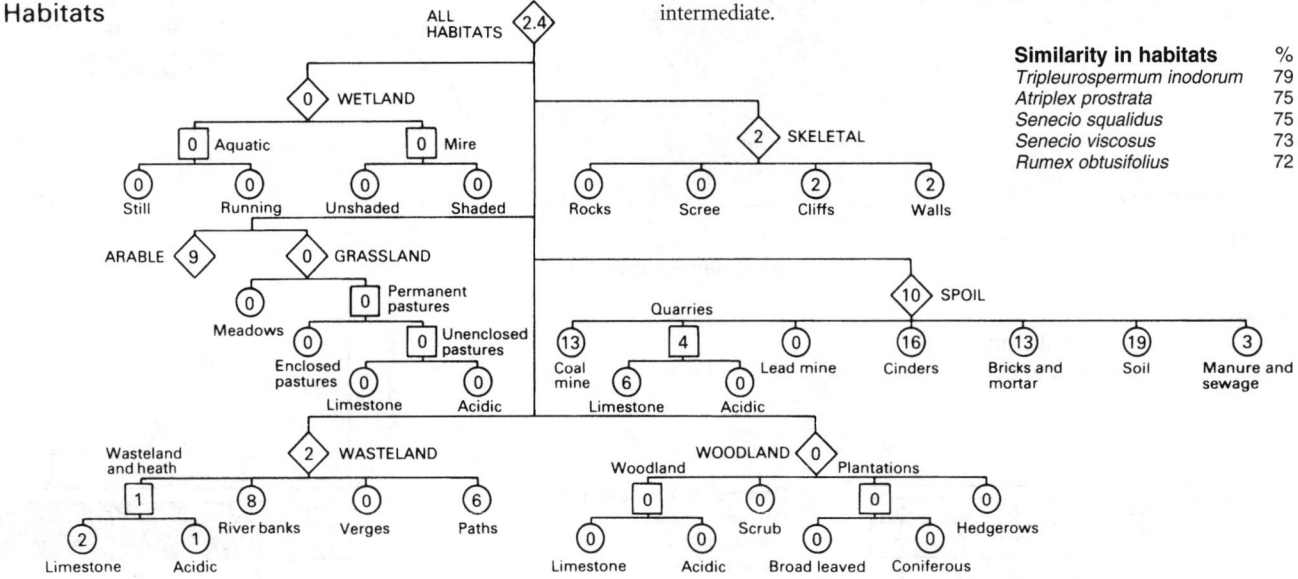

Similarity in habitats	%
Tripleurospermum inodorum	79
Atriplex prostrata	75
Senecio squalidus	75
Senecio viscosus	73
Rumex obtusifolius	72

Habitats

[Habitat diagram: ALL HABITATS 2.4; WETLAND 0 (Aquatic 0: Still 0, Running 0; Mire 0: Unshaded 0, Shaded 0); SKELETAL 2 (Rocks 0, Scree 0, Cliffs 2, Walls 2); ARABLE 9; GRASSLAND 0 (Permanent pastures 0: Meadows 0, Enclosed pastures 0 Limestone/Acidic 0/0; Unenclosed pastures 0); Quarries (Coal mine 13, Lead mine 0, Limestone 4, Acidic 6/0); SPOIL 10 (Cinders 16, Bricks and mortar 13, Soil 19, Manure and sewage 3); WASTELAND 2 (Wasteland and heath: Limestone 2, Acidic 1; River banks 8, Verges 0, Paths 6); WOODLAND 0 (Woodland 0 Limestone/Acidic 0/0; Plantations 0: Scrub 0, Broad leaved 0, Hedgerows 0, Coniferous 0)]

Synopsis *A.v.* is an erect, tufted, aromatic herb which occurs most frequently in disturbed, relatively fertile, urban sites. The species produces a lower canopy than that of species such as *Cirsium arvense* and *Urtica dioica*, and does not form extensive stands. Thus, *A.v.* is particularly associated with sites where consolidation by more-robust species is prevented by the rocky substrata, e.g. cinders, brick rubble and paving stones. In such sites the deep root system of *A.v.* may allow access to mineral soil beneath the spoil. Colonies may also persist for many years on roadsides, particularly on steep slopes or sandy soils, which are liable to some soil movement. However, in many habitats, e.g. soil heaps, *A.v.* may be an early colonist which is later replaced by taller species with greater potential for dominance. *A.v.* is absent from shaded, grazed and waterlogged sites, and is potentially toxic to livestock (Cooper & Johnson 1984). The species has a lowland distribution in Britain (Wilson 1956); this is determined in part by patterns of land-use, but a restriction by climatic factors is also likely. The colonizing potential of *A.v.* is related to its high level of seed production, which may exceed 9000 per flowering stem (Bostock & Benton 1979). Achenes are not plumed but may be wind-dispersed on a local scale (Ridley 1930). Seeds also appear to be widely distributed through human activities and many field records from bared ground refer to isolated seedlings which may not reach maturity. In particular, *A.v.* appears to exploit road and railway systems as corridors of dispersal. Survival of disturbance in sites where *A.v.* is established is facilitated by the presence of a persistent seed bank (Bostock 1978, Roberts 1986a) and rhizomes and patches of *A.v.* may expand at a rate of 300 mm year^{-1} by vegetative means (Bostock & Benton 1979). Seed germination, which occurs mainly in spring, is partially inhibited in the presence of a leaf canopy (Gorski *et al.* 1977), and a light requirement can be induced in some seeds by storage in the soil (Bostock 1978). There is much apparently genotypic variation within the species, e.g. in leaf dissection and flower colour. The extent to which this variation is ecologically important is uncertain.

Current trends Probably still increasing in urban habitats and in other disturbed sites in lowland Britain.

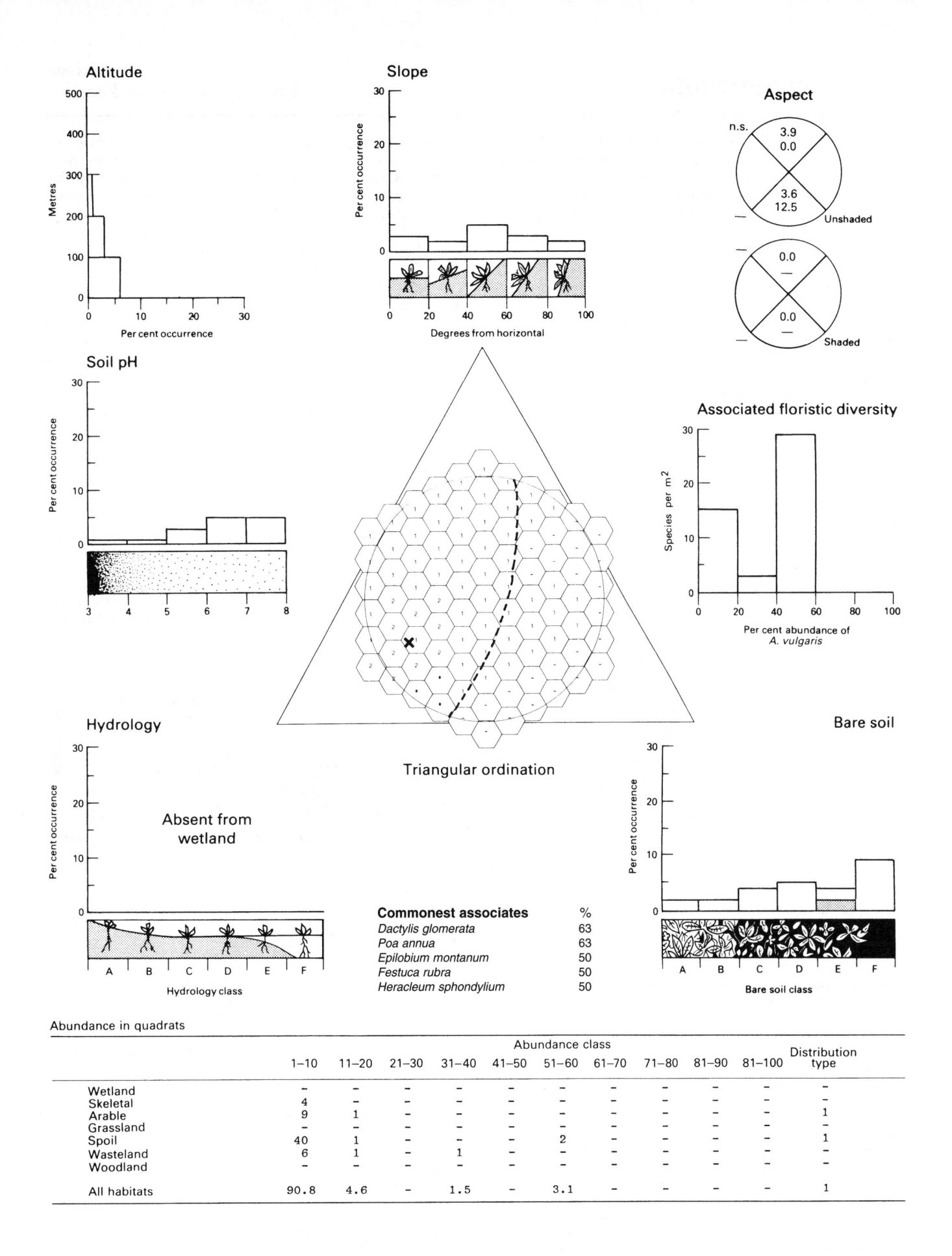

Altitude

Slope

Aspect

	n.s.	3.9
		0.0
	3.6	12.5 Unshaded

		0.0
		—
	0.0	— Shaded

Soil pH

Associated floristic diversity

Hydrology

Absent from wetland

Triangular ordination

Bare soil

Commonest associates	%
Dactylis glomerata	63
Poa annua	63
Epilobium montanum	50
Festuca rubra	50
Heracleum sphondylium	50

Abundance in quadrats

	1–10	11–20	21–30	31–40	41–50	51–60	61–70	71–80	81–90	81–100	Distribution type
Wetland	–	–	–	–	–	–	–	–	–	–	–
Skeletal	4	–	–	–	–	–	–	–	–	–	1
Arable	9	1	–	–	–	–	–	–	–	–	1
Grassland	–	–	–	–	–	–	–	–	–	–	–
Spoil	40	1	–	–	–	2	–	–	–	–	1
Wasteland	6	1	–	1	–	–	–	–	–	–	–
Woodland	–	–	–	–	–	–	–	–	–	–	–
All habitats	90.8	4.6	–	1.5	–	3.1	–	–	–	–	1

(*) *Arum maculatum* L.

Araceae
Lords-and-Ladies, Cuckoo-pint

Life-form Polycarpic perennial, rosette-forming geophyte with a deep underground corm. Scape erect; leaves triangular-hastate, <10 000 mm^2; roots with VA mycorrhizas.
Phenology Vernal. New shoots emerge in late winter or early spring. Flowers in April to May but fruit not ripe until autumn. Leaves usually senescent by July.
Height Foliage and inflorescence to 400 mm.
Seedling RGR No data. SLA 32.3. DMC 11.8.
Nuclear DNA amount 21.8 pg (Bennett & Smith 1976); $2n = 56^*, 84^*$ (*BC*); $4x^*, 6x^*$.
Established strategy Stress-tolerant ruderal.
Regenerative strategies V, S. Regenerates by seed which germinates in early spring. Seed is short-lived and there is no evidence of a persistent seed bank. Vegetative regeneration results in local colonies; SB II.

Flowers Small unisexual flowers at basal portion of floral column (spadix) with zone of 30-50 female flowers below c. 100 male flowers, the whole enclosed within a yellow-green spathe; cross-pollinated by small insects by an elaborate mechanism described by Church (1908).
Germinule 31.93 mg, 8.1 × 5.9 mm, seed dispersed by birds within a 1- (to 4-) seeded red berry (*Biol Fl*).
Germination Hypogeal; dormancy broken by chilling and promoted by darkness.
Biological flora Sowter (1949), see also Prime (1960).
Geographical distribution Throughout England, Wales, S Scotland (introduced further N) and Ireland (86% of British vice-counties), *ABIF* page 684; W, C and S Europe (67% of European territories and 3% naturalized); N Africa.

Habitats Only recorded in or adjacent to shaded habitats. Most frequent in deciduous woodlands, scrub and hedgerows, on base-rich soils, but also found in broad-leaved plantations and more rarely beneath conifers. Virtually absent from wetlands.
Gregariousness Sparse.
Altitude Up to 300 m, but more frequent and abundant in lowland areas despite the fact that suitable upland habitats are plentiful. [Up to 410 m (*Biol Fl*).]
Slope Absent from skeletal habitats, and hence absent from slopes of more than 50°.

Habitats

Aspect Rather more occurrences on shaded N-facing slopes, but significantly more abundant on shaded S-facing slopes.
Hydrology Like other species with a deep underground corm and root system, absent from strongly waterlogged soils.
Soil pH Most frequent and abundant on soils in the pH range 5.5-7.5, and absent from acidic soils of pH <4.5.
Bare soil Often associated with habitats with an incomplete ground cover of herbaceous vegetation. Sites often contain a mixture of bare soil, tree litter and bryophyte mat.
Δ ordination Distribution relatively wide-ranging, although less frequent in heavily disturbed vegetation types. Distribution centred both on vegetation associated with undisturbed, moderately fertile habitats and on vegetation of infertile, slightly disturbed areas.

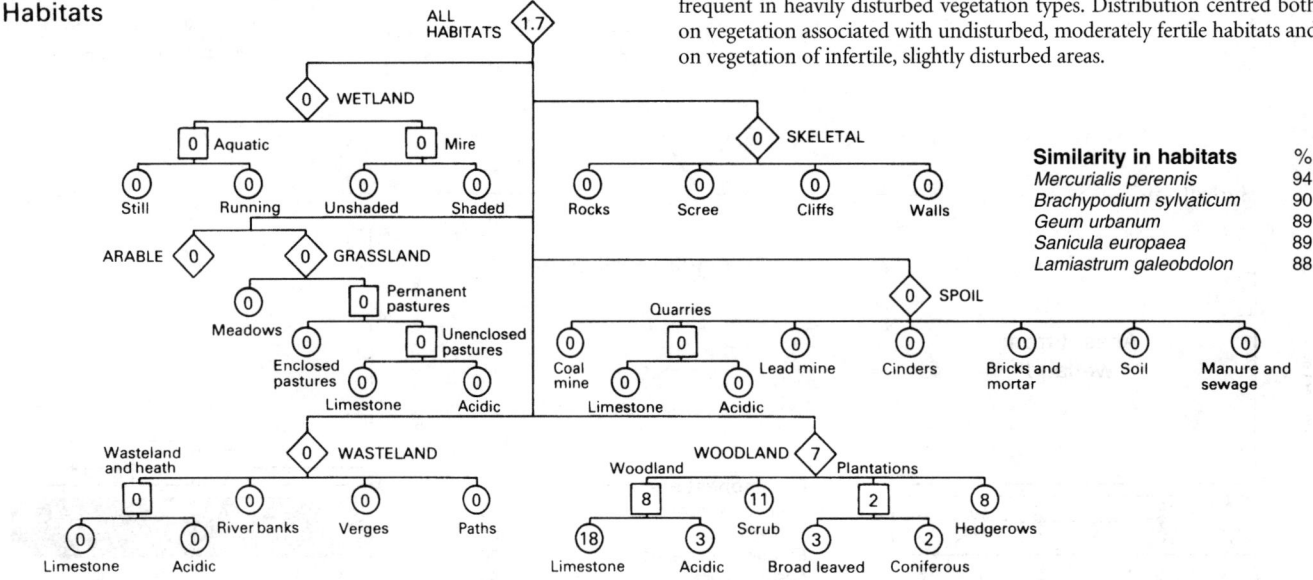

Similarity in habitats	%
Mercurialis perennis	94
Brachypodium sylvaticum	90
Geum urbanum	89
Sanicula europaea	89
Lamiastrum galeobdolon	88

Associated floristic diversity Usually low to intermediate.
Synopsis *A.m.* is a vernal, monocotyledonous herb of relatively fertile soils in woodland and hedgerows, and shows poor growth in full sunlight (*Biol Fl*). *A.m.* reaches the N edge of its distribution in Britain and is primarily a lowland species in N England. The young shoot is said to be tolerant of low temperatures (*Biol Fl*), but inflorescences are sometimes damaged by frost. The corm becomes buried to depths of 200-300 mm by means of contractile roots (*Biol Fl*). *A.m.* is largely confined to moist but well-drained soils. Foliage of *A.m.* is toxic and is generally avoided by stock (Cooper & Johnson 1984). The buried corm and deep root system are disadvantaged by anaerobic soils, and the poorly differentiated conductive tissue (Dahlgren et al. 1985) may preclude establishment in droughted sites. *A.m.* is slow to establish from seed and is heavily dependent upon its mycorrhizal associate. Foliage is not produced until the second or third year's growth; seed is rarely produced before the seventh year (Prime 1960). In view of this extended and vulnerable juvenile phase, and the fact that plants often produce <30 seeds (Salisbury 1942), it is surprising that *A.m.* is frequent in habitats subject to intermittent disturbance, such as plantations and hedgerows. It seems likely that the reduced competition from other species following disturbance promotes *A.m.*, as it does many orchids which are similarly dependent upon mycorrhizas and which exploit bare areas of derelict quarries and gravel pits. Regeneration by vegetative means occurs through proliferation of daughter tubers (*Biol Fl*). However, regeneration by seed dispersed by birds is equally important. The floral structure and mechanisms promoting cross-pollination by insects, described by Lamarck (1778), Church (1908) and Meeuse (1975), involve thermogenic respiration and release of an attractive odour. Flowers are as elaborate and spectacular as any within the British Flora. Considerable morphological variation exists (Prime 1960), but the extent of ecotypic differentiation in uncertain. *A.m.* hybridizes with *A. italicum* Miller (FE5).
Current trends Uncertain. Vulnerable to sustained and frequent disturbance, but able to exploit broad-leaved plantations if subject to normal forestry management.

Altitude

Slope

Aspect

Unshaded

0.0
0.0

n.s. 9.1 n.s.
     8.3
     4.9
     16.7

Shaded

Soil pH

Associated floristic diversity

Hydrology

Absent from wetland

Triangular ordination

Bare soil

Commonest associates %
Mercurialis perennis 73
Poa trivialis 35
Urtica dioica 34
Galium aparine 31
Ranunculus ficaria 30

Abundance in quadrats

	\multicolumn{10}{c	}{Abundance class}	Distribution type
	1–10	11–20	21–30	31–40	41–50	51–60	61–70	71–80	81–90	81–100	
Wetland	–	–	–	–	–	–	–	–	–	–	–
Skeletal	–	–	–	–	–	–	–	–	–	–	–
Arable	–	–	–	–	–	–	–	–	–	–	–
Grassland	–	–	–	–	–	–	–	–	–	–	–
Spoil	–	–	–	–	–	–	–	–	–	–	–
Wasteland	–	–	–	–	–	–	–	–	–	–	–
Woodland	42	2	1	1	–	–	–	–	–	–	1
All habitats	91.3	4.3	2.2	2.2	–	–	–	–	–	–	1

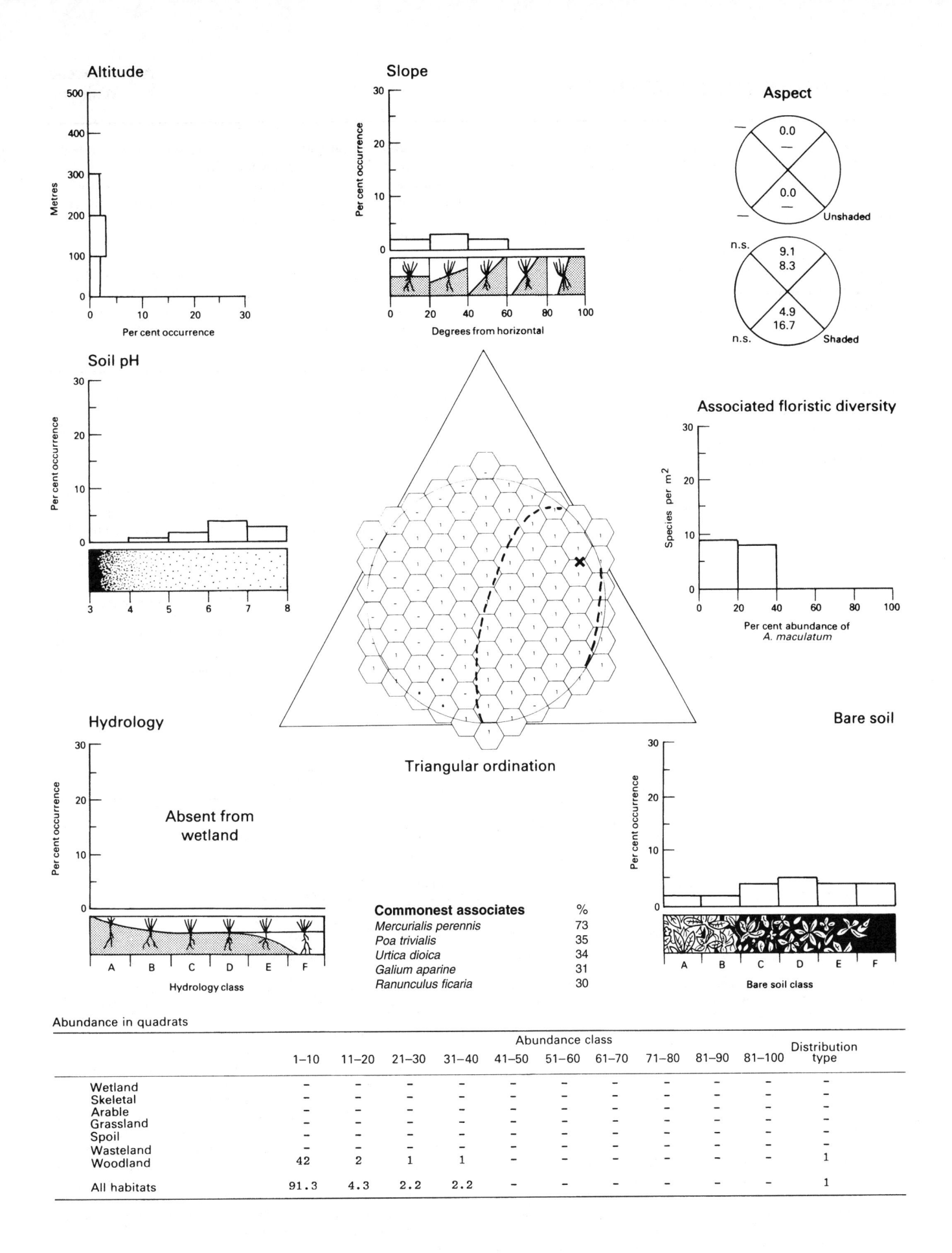

Asplenium ruta-muraria L.

Aspleniaceae
Wall-rue

Life-form Polycarpic perennial, rosette hemicryptophyte with very shortly-creeping rhizome. Fronds ascending, divided, with a glandular stipe, usually <500 mm^2; VA mycorrhizas present only if rooted in soil.
Phenology Winter green, with fronds replaced in spring and early summer (Page 1982). Spores produced June to August.
Height Fertile and sterile fronds to 80 mm.
Sporeling RGR No data. SLA 12.4. DMC 29.2.
Nuclear DNA amount 10.5 pg; $2n = 144^*$ (*BC*); $4x^*$ (Page 1982).
Established strategy Stress-tolerator.

Regenerative strategies W, ?B$_s$. Regenerative strategies insufficiently studied (see 'Synopsis').
Sporangia Aggregated into sori on the underside of the fronds.
Germinule 46 × 33 µm, spore, brown, explosively dispersed from a many-spored sporangium.
Germination t_{50} 8 days; L > D.
Biological flora None.
Geographical distribution Widely distributed in the British Isles, but local in many eastern areas (100% of British vice-counties), *ABIF* page 71, *AF* page 62; most of W, C and S Europe (87% of European territories); N and S Asia, E N America.

Habitats Confined to base-rich crevices in walls, cliffs and rock outcrops with one record for limestone quarry rubble.
Gregariousness Sparse to intermediate.
Altitude More common in upland areas, where suitable habitats are more abundant. Observed to 340 m. [Up to 610 m.]
Slope As a result of its association with walls and cliffs *A.r-m.* is most frequent and abundant on slopes of more than 80°.
Aspect Rather more frequent and only abundant on unshaded S-facing slopes.
Hydrology Absent from wetlands.
Soil pH Not recorded from soils of pH <6.0.

Bare soil Restricted to rocky habitats with little or no exposed soil.
△ **ordination** Insufficient data, but in an ordination based upon a more recent survey in the Sheffield region the species was found mainly in lightly disturbed, unproductive habitats.
Associated floristic diversity Typically low, but occasionally higher due to occurrences on species-rich rock outcrops.
Synopsis *A.r-m.* is a small, winter-green fern with thick leathery leaves (Page 1982). The species is characteristic of crevices, usually in unshaded calcareous cliffs and mortared walls. The species is the only common fern in which the distribution shows a S-facing bias. *A.r-m.* is also the fern which most consistently survives to maturity on walls in the lowland parts

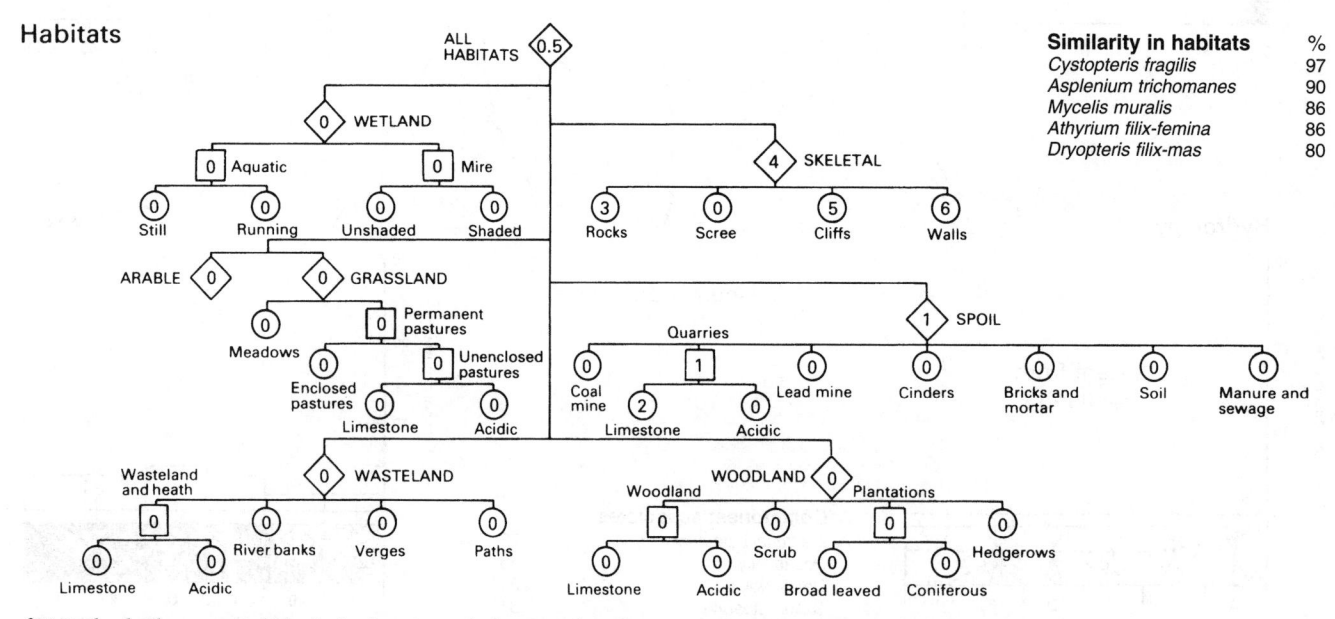

Habitats

Similarity in habitats	%
Cystopteris fragilis	97
Asplenium trichomanes	90
Mycelis muralis	86
Athyrium filix-femina	86
Dryopteris filix-mas	80

of N England. There can be little doubt that *A.r-m.* is the most drought-tolerant of the common ferns. However, in the context of the vascular flora of Britain, this represents only a modest degree of tolerance. In common with other widespread ferns of skeletal habitats, the species is more frequent in the moist W part of the British Isles (*ABF*). Further, the frond canopy, or even the whole plant, may be periodically destroyed during summer drought. This vulnerability may be partly compensated for by the facts that the species produces large quantities of spores from an early age and that most of its fronds are fertile (Page 1982). It is not known whether a persistent spore bank is formed. As in *Asplenium trichomanes* and *Cystopteris fragilis*, spores are slow to germinate (see *C.f.*). It is not known whether the prothallus itself is exceptionally drought-resistant, but the spores of *A.r-m.* have a higher temperature requirement for germination than those of *A.t.* (Young 1985), and *A.r-m.* tends to be found in shallower crevices than *A.t.* (Young 1985). *A.r-m.* has a strictly limited capacity for lateral vegetative spread. The species is phenotypically plastic, with leaves produced early in the season being considerably less divided than those produced later (Page 1982). The species hybridizes, very rarely, with *A.t.* and with one other species (Stace 1975). Diploid populations of this predominantly tetraploid species have been reported from Italy (*FE1*).
Current trends Although a very effective colonist by virtue of its wind-dispersed spores, *A.r-m.* appears to have declined considerably during the last 50 years in industrial areas, possibly due to atmospheric pollution (Page 1982). However, there are signs, in Sheffield at least, of rehabilitation, and *A.r-m.* is occurring with increasing frequency on stone and even on relatively modern brick walls in moister residential parts of the city. With respect to *A.r-m.* and other ferns, two possible causal factors may be suggested; first, the level of the atmospheric toxin sulphur dioxide has declined, and, second, the concentration of oxides of nitrogen has increased (M. H. Unsworth pers. comm.). The latter, by increasing the nitrogen status of the substratum, may at least in theory accelerate the process of colonization. However, the species is probably still declining in dry lowland areas, due to the loss of old mortared stone walls, a favoured habitat.

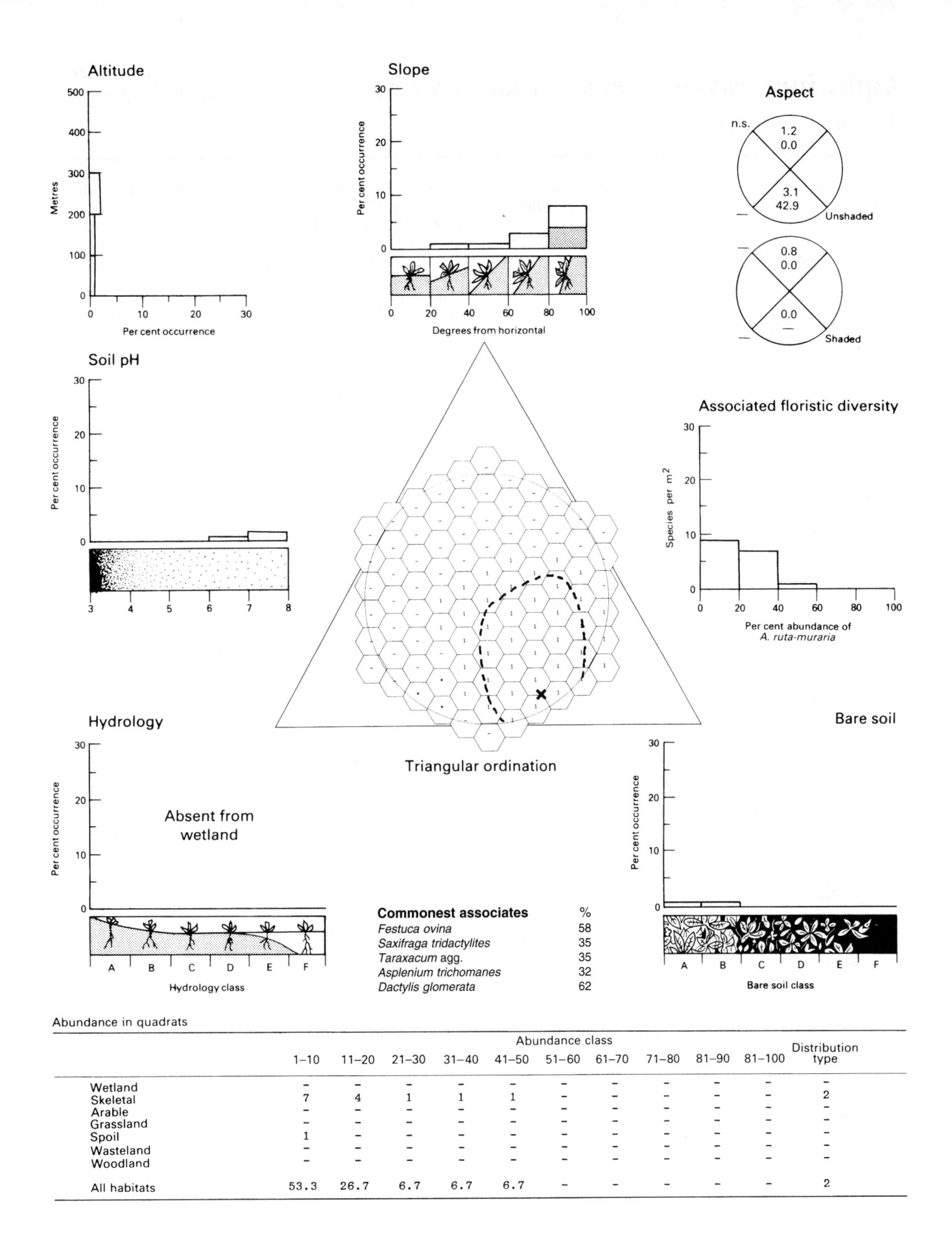

Altitude

Slope

Aspect

	n.s.	1.2	
		0.0	
		3.1	
		42.9	Unshaded

		0.8	
		0.0	
		0.0	
		—	Shaded

Soil pH

Associated floristic diversity

Species per m^2

Per cent abundance of *A. ruta-muraria*

Hydrology

Absent from wetland

Hydrology class

Triangular ordination

Commonest associates %
Festuca ovina — 58
Saxifraga tridactylites — 35
Taraxacum agg. — 35
Asplenium trichomanes — 32
Dactylis glomerata — 62

Bare soil

Bare soil class

Abundance in quadrats

	Abundance class									Distribution type
	1–10	11–20	21–30	31–40	41–50	51–60	61–70	71–80	81–90	81–100	
Wetland	–	–	–	–	–	–	–	–	–	–	–
Skeletal	7	4	1	1	1	–	–	–	–	–	2
Arable	–	–	–	–	–	–	–	–	–	–	–
Grassland	–	–	–	–	–	–	–	–	–	–	–
Spoil	1	–	–	–	–	–	–	–	–	–	–
Wasteland	–	–	–	–	–	–	–	–	–	–	–
Woodland	–	–	–	–	–	–	–	–	–	–	–
All habitats	53.3	26.7	6.7	6.7	6.7	–	–	–	–	–	2

Asplenium trichomanes ssp. *quadrivalens*
D. E. Meyer emend. Lovis

Aspleniaceae
Maidenhair Spleenwort

Life-form Shortly rhizomatous, polycarpic perennial, rosette hemicryptophyte. Fronds erect or spreading, linear, pinnately divided, <1000 mm^2; roots with VA mycorrhizas.
Phenology Winter green with fronds replaced in early summer (Page 1982). Spores produced August to October.
Height Fertile and sterile fronds to 200 mm in height in vertical crevices but usually less.
Sporeling RGR No data. SLA 10.9. DMC 33.7.
Nuclear DNA amount 13.7 pg; $2n = 144^*$, $4x^*$ (Page 1982).
Established strategy Stress-tolerator.

Regenerative strategies W, B$_s$. Regenerative strategies insufficiently studied (see 'Synopsis').
Sporangia Aggregated into sori on the underside of the fronds.
Germinule 37 × 28 µm, spore, brown, explosively dispersed from a many-spored sporangium.
Germination t_{50} 8 days; L > D.
Biological flora None.
Geographical distribution The British Isles, especially in the N and W (99% of British vice-counties), *ABIF* page 69, *AF* pages 57-9; much of Europe (97% of European territories); N and S temperate zones and on mountains in the tropics.

Habitats Virtually confined to shaded or unshaded rocky habitats on the limestone, particularly cliffs and outcrops, and in the crumbling mortar of old walls. Occasionally observed on wet rocks. [Found on base-rich mountain rocks (Page 1982).]
Gregariousness Sparse to intermediate.
Altitude The upland bias is related to the abundance of *A.t.* on the carboniferous limestone of N Derbyshire. Observed up to 370 m. [Extends beyond 900 m (Page 1982).]
Slope Mainly found in steeply sloping skeletal habitats.

Aspect More frequent and only abundant on N-facing slopes in both shaded and unshaded habitats.
Hydrology Absent from wetlands.
Soil pH Rarely observed in the region below pH 6.5, reflecting an association of *A.t.* with limestone crevices and mortar in walls.
Bare soil All records from rocky habitats with little bare soil.
△ **ordination Distribution** somewhat diffuse, reflecting the heterogeneous nature of the vegetation in crevices on cliffs and walls. Centred mainly on undisturbed and unproductive vegetation, and absent from highly disturbed or highly productive communities.
Associated floristic diversity Usually low, but higher values are recorded from outcrops.

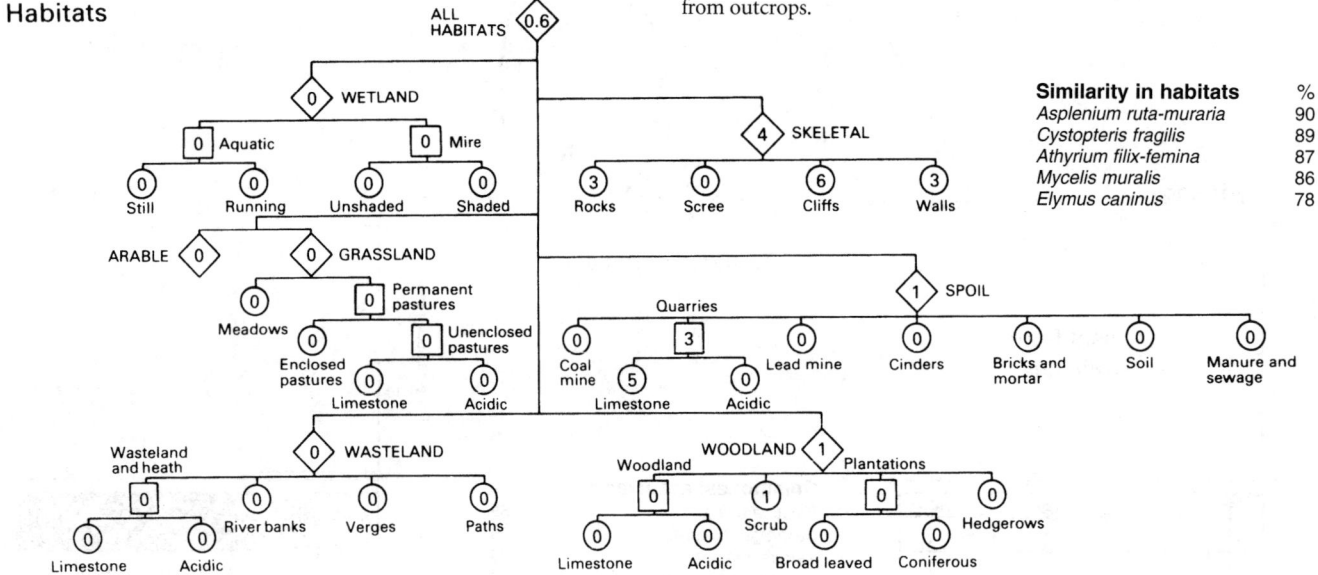

Habitats

Similarity in habitats	%
Asplenium ruta-muraria	90
Cystopteris fragilis	89
Athyrium filix-femina	87
Mycelis muralis	86
Elymus caninus	78

Synopsis *A.t.* is a small winter-green fern which is most characteristically associated with crevices in calcareous cliffs and mortared walls. In form and distribution the species resembles *A. ruta-muraria*, with which it frequently occurs. However, *A.t.* is more frequent on N-facing slopes, regularly occurs in shaded sites and is infrequent within the lowland parts of the region. The species appears to require moister habitats and exhibits greater shade-tolerance than *A.r-m*. In N Derbyshire *A.t.* is frequent on walls of carboniferous limestone but, in contrast with *A.r-m.*, the species is unusually scarce on adjacent walls of mortared gritstone. It remains uncertain whether this distribution pattern is a reflection of a more calcicolous habit on the part of *A.t.* or is related to other factors such as the potential of limestone to counteract the acidity associated with sulphur dioxide pollution. *A.t.* has only a restricted capacity for lateral spread (by means of short rhizomes) and regeneration is mainly from spores. A single frond of *A.t.* may produce *c.* 750 000 spores (Page 1982), and the species is an effective long-distance colonist. As with *A.r-m.*, spores are slow to germinate. Studies are required to measure the persistence of spores in natural environments. It is unclear to what extent the distribution of the sporophyte generation of *A.t.* (and other British ferns) is restricted by the ecological requirements of the prothallus. The smaller ssp. *pachyrachis* (H. Christ) Lovis and Reichst ($2n = 144$) occurs mainly in the West and on less-calcareous rocks in montane regions the diploid ssp. *trichomanes* ($2n = 72$) largely replaces the more widespread ssp. Both can hybridize with ssp. of *A.r-m.* and with *Phyllitis scolopendrium* (L.) Newman.
Current trends In lowland areas now largely restricted to old walls and natural rock faces. Remains an effective colonist in upland areas, and here the species is not under threat. Like *A.r-m.*, the species may have decreased in industrial areas as a result of the effects of atmospheric pollution. Currently, as with *A.r-m.*, *A.t.* appears to be increasing in some parts of residential Sheffield.

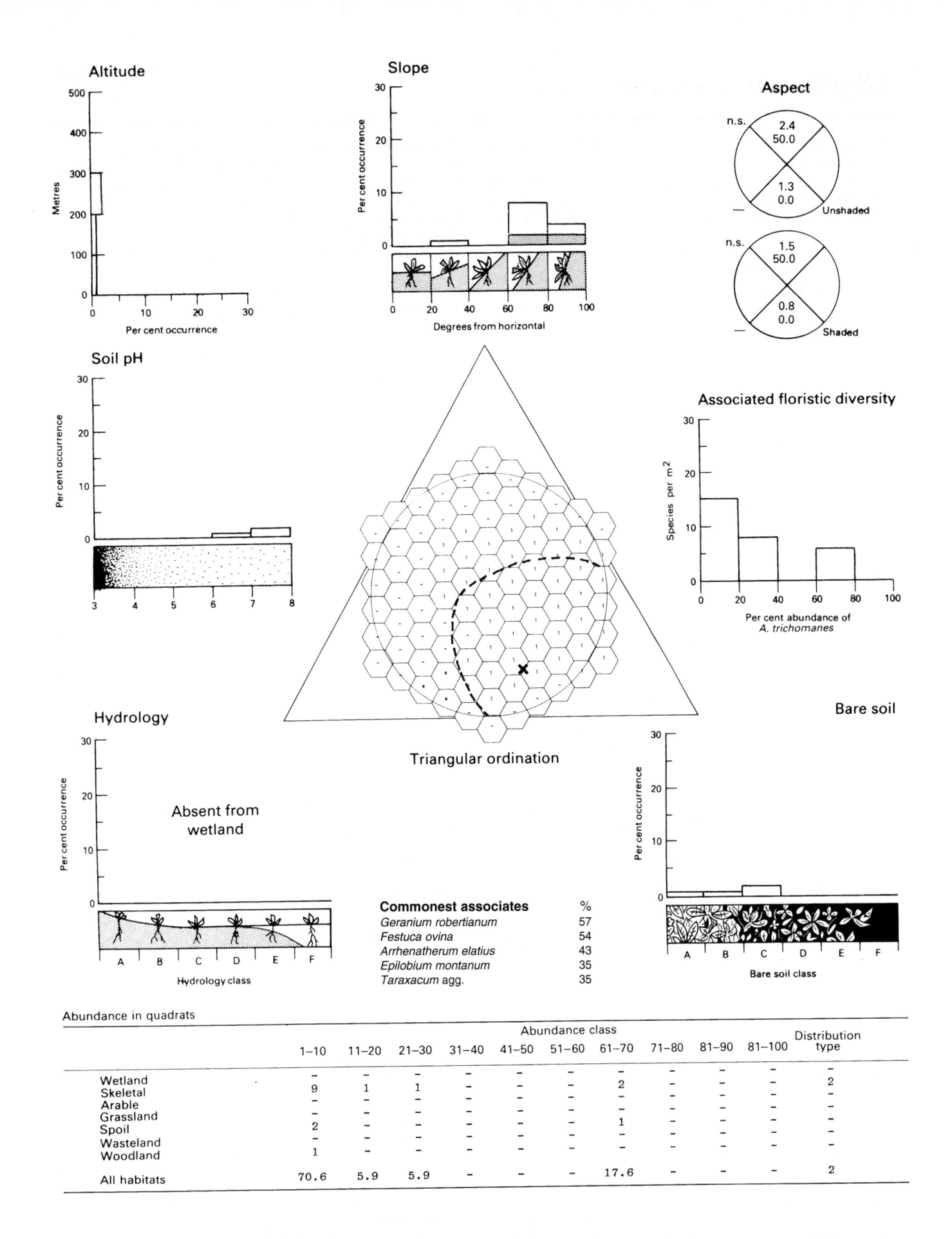

Altitude

Slope

Aspect

	n.s.	2.4	50.0	
		1.3	0.0	Unshaded

	n.s.	1.5	50.0	
		0.8	0.0	Shaded

Soil pH

Associated floristic diversity

Species per m² vs Per cent abundance of *A. trichomanes*

Triangular ordination

Hydrology

Absent from wetland

Bare soil

Commonest associates	%
Geranium robertianum	57
Festuca ovina	54
Arrhenatherum elatius	43
Epilobium montanum	35
Taraxacum agg.	35

Abundance in quadrats

	1–10	11–20	21–30	31–40	41–50	51–60	61–70	71–80	81–90	81–100	Distribution type
Wetland	–	–	–	–	–	–	–	–	–	–	–
Skeletal	9	1	1	–	–	–	2	–	–	–	2
Arable	–	–	–	–	–	–	–	–	–	–	–
Grassland	–	–	–	–	–	–	–	–	–	–	–
Spoil	2	–	–	–	–	–	1	–	–	–	–
Wasteland	–	–	–	–	–	–	–	–	–	–	–
Woodland	1	–	–	–	–	–	–	–	–	–	–
All habitats	70.6	5.9	5.9	–	–	–	17.6	–	–	–	2

Athyrium filix-femina (L.) Roth

Woodsiaceae
Lady-fern

Life-form Polycarpic perennial, rosette hemicryptophyte with a branched rhizome. Fronds erect, lanceolate, much dissected, with a scaly stipe, <0.1 m²; roots usually with VA mycorrhizas.
Phenology Fronds all emerge together during April and May and spores are produced from July until autumn (Page 1982), after which the fronds die.
Height Fertile and sterile fronds to 1200 mm.
Sporeling RGR No data. SLA 12.1. DMC 29.2.
Nuclear DNA amount 5.9 pg; $2n = 80*$; $2x*$ (Page 1982).
Established strategy Intermediate between stress-tolerant competitor and competitor.

Regenerative strategies W, V, ?B_s. See 'Synopsis'.
Sporangia Aggregated into sori on the underside of the frond.
Germinule 41 × 25 μm, spore, brown, explosively dispersed from a many-spored sporangium.
Germination t_{50} 4 days; L > D.
Biological flora None.
Geographical distribution Most of the British Isles (99% of British vice-counties), *ABIF* page 72, *AF* page 67; Europe, although rare in the Mediterranean (95% of European territories); N temperate zones and in tropical forests and mountains.

Habitats A species of moist or shaded habitats, most frequently recorded from cliffs, walls, outcrops, woodland, river banks and shaded or unshaded mire. Often not attaining reproductive maturity in skeletal habitats. [Occurs on mountain rocks and in lighter shade within coniferous plantations (Page 1982).]
Gregariousness May achieve high canopy density and dominance at low frequencies. High-frequency values refer exclusively to colonies of young plants.
Altitude Most frequent in the moist sheltered valleys of upland regions; extends to 430 m. [Up to c. 915 m (Page 1982).]
Slope Found both on the level ground of woodland floors and on the vertical slopes of cliffs.

Aspect Mainly confined to shaded habitats, where it is more frequent on N-facing slopes.
Hydrology On moist soils, in soligenous mire and wet ground beside streams; otherwise absent from flooded ground.
Soil pH Associated with a wide range of values. Most records lie between pH 3.5 and 6.5. Rare on calcareous strata, and only present when rock is overlain by non-calcareous soils.
Bare soil Widely distributed.
△ **ordination** Absent from heavily disturbed habitats. Distribution centred on vegetation types associated with undisturbed, relatively fertile conditions.
Associated floristic diversity Species-poor.

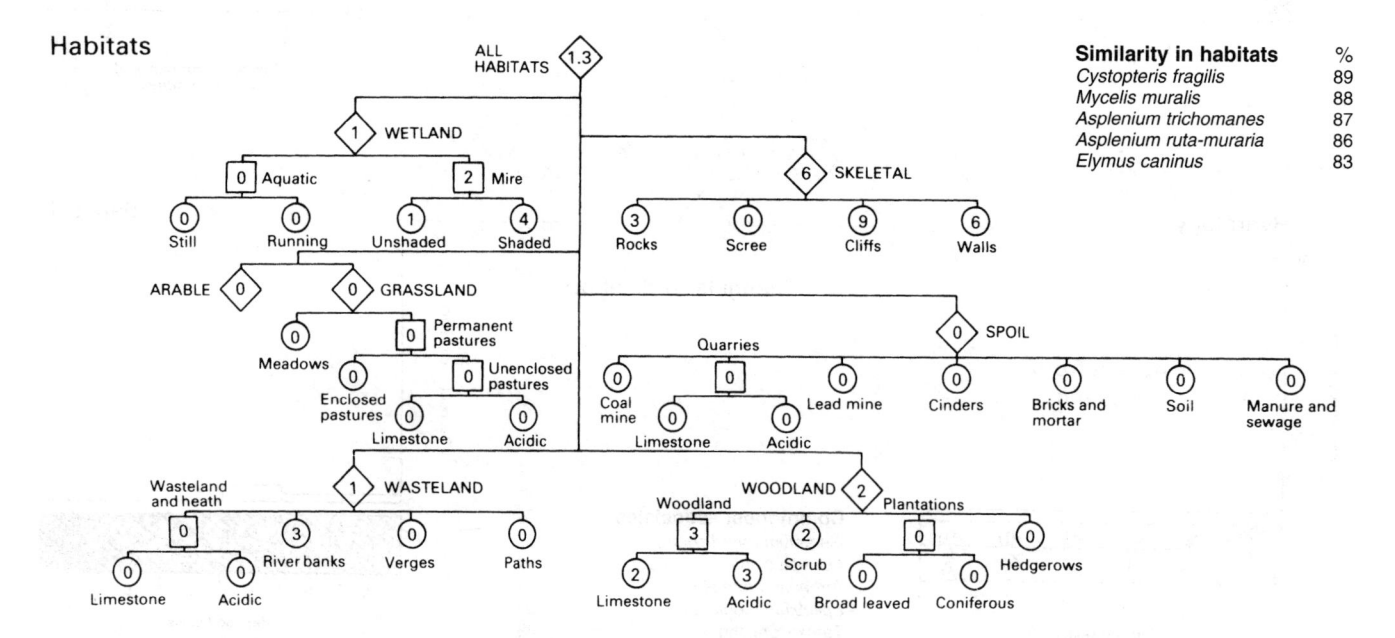

Habitats

Similarity in habitats	%
Cystopteris fragilis	89
Mycelis muralis	88
Asplenium trichomanes	87
Asplenium ruta-muraria	86
Elymus caninus	83

Synopsis *A.f-f.* is a potential dominant of moist woodland, and shares many of the attributes of *Dryopteris dilatata*. However, the species commonly occurs on river and stream banks, and on woodland floodplains, and is thus more frequently associated with fertile, regularly disturbed habitats than *D.d.* Indeed, *A.f-f.* has a shoot phenology typical of the 'competitive strategy' (*sensu* Al-Mufti *et al.* 1977). The fronds expand mainly during a single flush in early summer, attain maximum size in July, and die back with the onset of autumn frosts (Page 1982). Again, unlike *D.d.*, old plants of *A.f-f.* may consist of a branched rhizome and a clump of numerous crowns (Page 1982); this morphology, coupled with the tendency to accumulate a dense layer of litter, confers upon *A.f-f.* a strong capacity for vegetation dominance. Vigorous stands of *A.f-f.* contain few associated species. The possibility that rhizomes detached by disturbance may be transported in water to form new colonies elsewhere deserves investigation. The species is not recorded from highly calcareous soils, but occurs at its most luxuriant between pH 5.0 and 7.0, the upper end of its pH range. Unlike *D.d.* and *D. filix-mas*, the fronds are not cited as toxic by Cooper and Johnson (1984). In rocky sites many plants die before reaching reproductive maturity, often apparently because crevices colonized by *A.f-f.* provide insufficient soil and moisture to sustain an adult plant. A single frond of *A.f-f.* may produce 2 million spores (Page 1982). It is not known whether *A.f-f.* forms a persistent spore bank, but the spores are widely dispersed and the species is a common colonist of ditch banks and wet walls. Much of the considerable morphological variation observed between populations is maintained in cultivation and the presence of heavy-metal-tolerant populations is suspected (Page 1982).

Current trends Appears to be decreasing in many lowland areas as a consequence of habitat destruction.

Altitude

Slope

Aspect

Unshaded
- n.s. 0.4
- 0.0
- 1.3
- 0.0

Shaded
- n.s. 5.3
- 0.0
- 2.4
- 0.0

Soil pH

Associated floristic diversity

Hydrology

Triangular ordination

Bare soil

Commonest associates %
Holcus mollis 42
Poa trivialis 30
Dryopteris dilatata 25
Oxalis acetosella 25
Deschampsia flexuosa 22

Abundance in quadrats

	\multicolumn{10}{c}{Abundance class}	Distribution type
	1–10	11–20	21–30	31–40	41–50	51–60	61–70	71–80	81–90	81–100	
Wetland	4	–	–	–	–	–	–	–	–	–	–
Skeletal	19	1	–	–	–	–	–	–	–	–	1
Arable	–	–	–	–	–	–	–	–	–	–	–
Grassland	–	–	–	–	–	–	–	–	–	–	–
Spoil	–	–	–	–	–	–	–	–	–	–	–
Wasteland	1	–	–	–	–	–	–	–	–	–	–
Woodland	9	2	–	–	–	–	–	–	–	–	2
All habitats	91.7	8.3	–	–	–	–	–	–	–	–	1

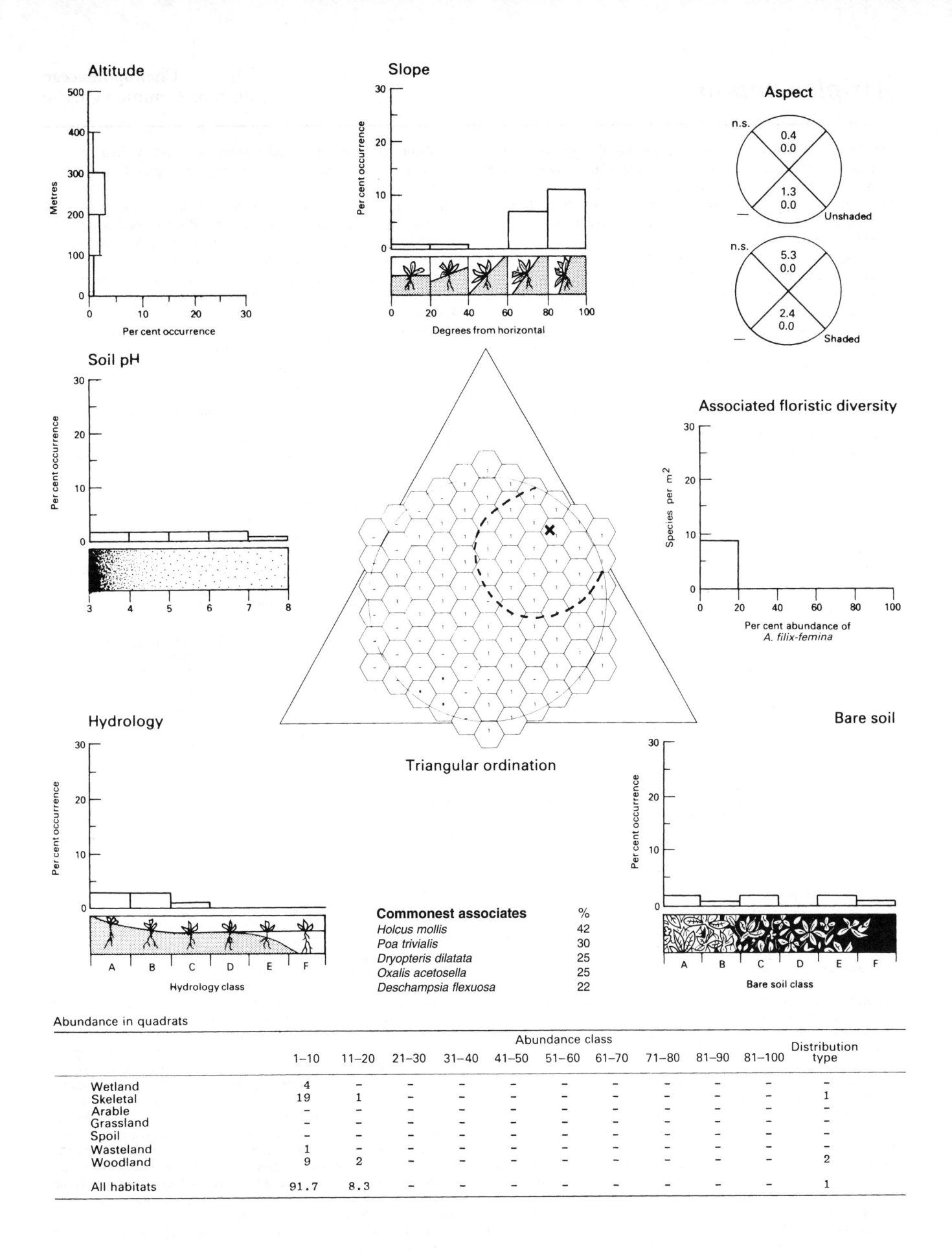

Atriplex patula L.

Chenopodiaceae
Iron-root, Common Orache

Life-form Semi-rosette, summer-annual therophyte. Stems prostrate to erect, branched; leaves lanceolate, usually <2000 mm²; roots with VA mycorrhizas.
Phenology Seeds germinate in spring. Flowers from June to October and seed set from September to November.
Height Foliage and flowers to 1 m.
Seedling RGR No data. SLA 17.5. DMC 15.8.
Nuclear DNA amount 4.3 pg; $2n = 36^*$ (Taschereau 1985); $4x^*$.
Established strategy Intermediate between ruderal and competitive-ruderal.
Regenerative strategies B_s. Entirely by seed which normally germinates in spring. Also forms a long-persistent bank of buried seeds (Brenchley & Warington 1930); SB II or ?IV.
Flowers Green, monoecious, facultatively autogamous but slightly protogynous, wind-pollinated but also visited by insects (Taschereau 1985); often >100 in a spike-like inflorescence.
Germinule 1.33 mg, 1.8 × 1.6 mm, dry fruit (remains of ovary wall thin and papery) often dispersed enveloped within two fleshy bracteoles. Polymorphic with small black seeds (1.5-2.5 mm) and larger brown seeds (2.0-3.5 mm).
Germination Epigeal; t_{50} 3 days; 10-23°C; dormancy broken by chilling (but see 'Synopsis'); L > D.
Biological flora None.
Geographical distribution Most of the British Isles but less common in N Scotland (97% of British vice-counties), *ABIF* page 142; Europe except the extreme N (89% of European territories); N Africa, W Asia; naturalized in N America.

Habitats A plant of disturbed, artificial habitats. Common as an arable weed, and on manure heaps and demolition sites. Also frequent on soil heaps, with sporadic occurrences on other types of spoil, river banks, waysides, muddy pond margins, walls and in gardens. Absent from grassland, woodland and aquatic habitats. [Frequent in submaritime habitats.]
Gregariousness Sparse to intermediate.
Altitude Up to 380 m but suitable sites most frequent in lowland areas. [Up to 440 m.]
Slope Mainly on gentle and moderate slopes.
Aspect Slightly more frequent on N-facing slopes, and only abundant where the aspect is to the north.
Hydrology Rare in wetlands, and restricted to exposed mud beside ponds and to wet areas on mine heaps.
Soil pH Mainly restricted to soils of pH >5.0.
Bare soil Particularly frequent and abundant in habitats having a relatively large amount of exposed soil.
△ **ordination** Distribution strongly centred on productive and highly disturbed vegetation types.
Associated floristic diversity Low to intermediate.

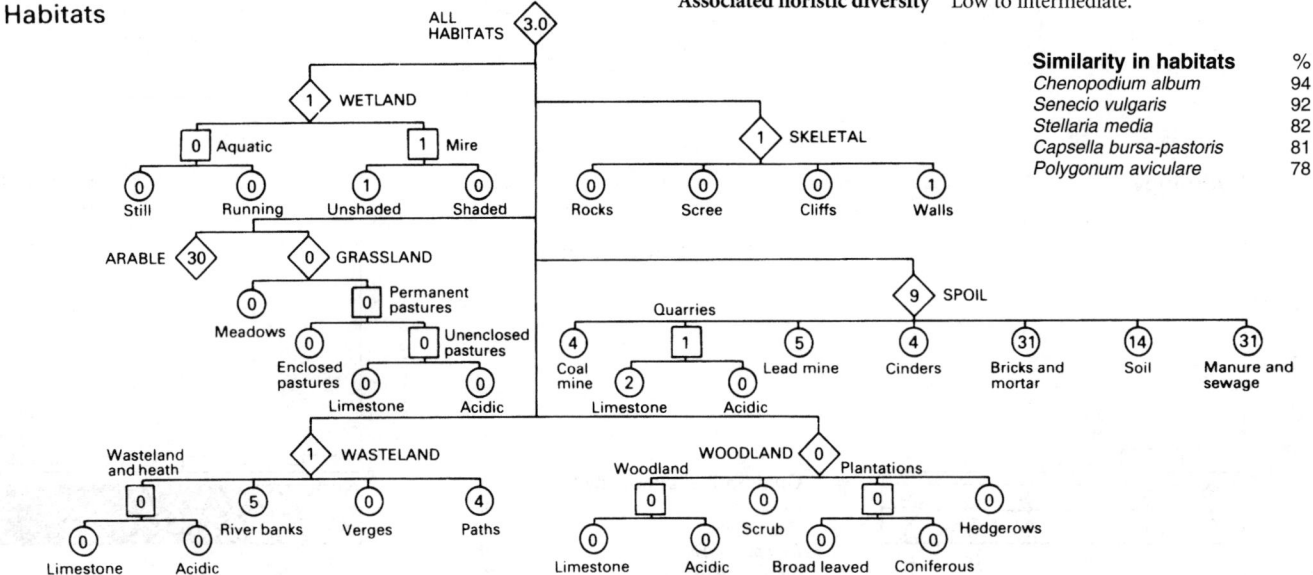

Habitats

Similarity in habitats	%
Chenopodium album	94
Senecio vulgaris	92
Stellaria media	82
Capsella bursa-pastoris	81
Polygonum aviculare	78

Synopsis *A.p.* is an erect or prostrate summer-annual with a growth form intermediate between those of *Polygonum aviculare* and *Chenopodium album*. The species colonizes a range of artificial, moist, unshaded, relatively fertile sites following periods of severe disturbance. The three main habitats with which the species is associated are demolition sites, manure heaps and arable land, particularly in broad-leaved crops or fallow areas. *A.p.* is usually much branched from near the base, and control measures involving cutting may encourage the species at the expense of taller, more-dominant weeds (*WCH*). The species is largely absent from trampled and from grazed sites. *A.p.* shows a greater restriction to fertile mineral soils and is less gregarious than *A. prostrata*. Like other ephemerals, plant size and reproductive output are very variable. Plants with less than 10 and nearly 1000 flowers have been observed at a single site. Regeneration is entirely by seed, which germinates between late winter and early summer (*WCH*), and a very persistent seed bank may be formed (Brenchley & Warington 1930). Most common weeds have two peaks of germination, in autumn and in spring, and the restricted germination period of *A.p.* must limit its regenerative flexibility compared with these species, even though seed is polymorphic, both with respect to size and to germination requirements. Brown seeds are produced, which are typically present in small quantities, and will germinate in the laboratory within 2 weeks, whereas black seeds require scarification and alternating temperatures (Taschereau 1985). The disadvantages of this strongly seasonal pattern of germination are perhaps offset by the widespread occurrence of *A.p.* as an impurity in crop seed (*WCH*); in this respect *Fallopia convolvulus*, *Persicaria* spp. and *Polygonum aviculare* show similar characters. Seeds are probably dispersed through human activities and they can survive ingestion by a range of mammals and birds (Ridley 1930) and may be dispersed through these agencies. The hybrid with *A. littoralis* L. is recorded from Britain (Taschereau 1985).
Current trends Remains an effective colonist with persistent populations in artificial habitats.

Altitude

Slope

Aspect

Soil pH

Associated floristic diversity

Triangular ordination

Hydrology

Commonest associates %
Polygonum aviculare 69
Poa annua 68
Poa trivialis 63
Stellaria media 61
Persicaria maculosa 54

Bare soil

Abundance in quadrats

	\multicolumn{9}{c	}{Abundance class}	Distribution type
	1–10	11–20	21–30	31–40	41–50	51–60	61–70	71–80	81–90	81–100	
Wetland	1	–	–	–	–	–	–	–	–	–	–
Skeletal	1	–	–	–	–	–	–	–	–	–	1
Arable	32	4	–	–	–	–	–	–	–	–	–
Grassland	–	–	–	–	–	–	–	–	–	–	–
Spoil	33	5	1	–	–	–	–	1	–	–	2
Wasteland	4	–	–	–	–	–	–	–	–	–	–
Woodland	–	–	–	–	–	–	–	–	–	–	–
All habitats	86.6	11.0	1.2	–	–	–	–	1.2	–	–	1

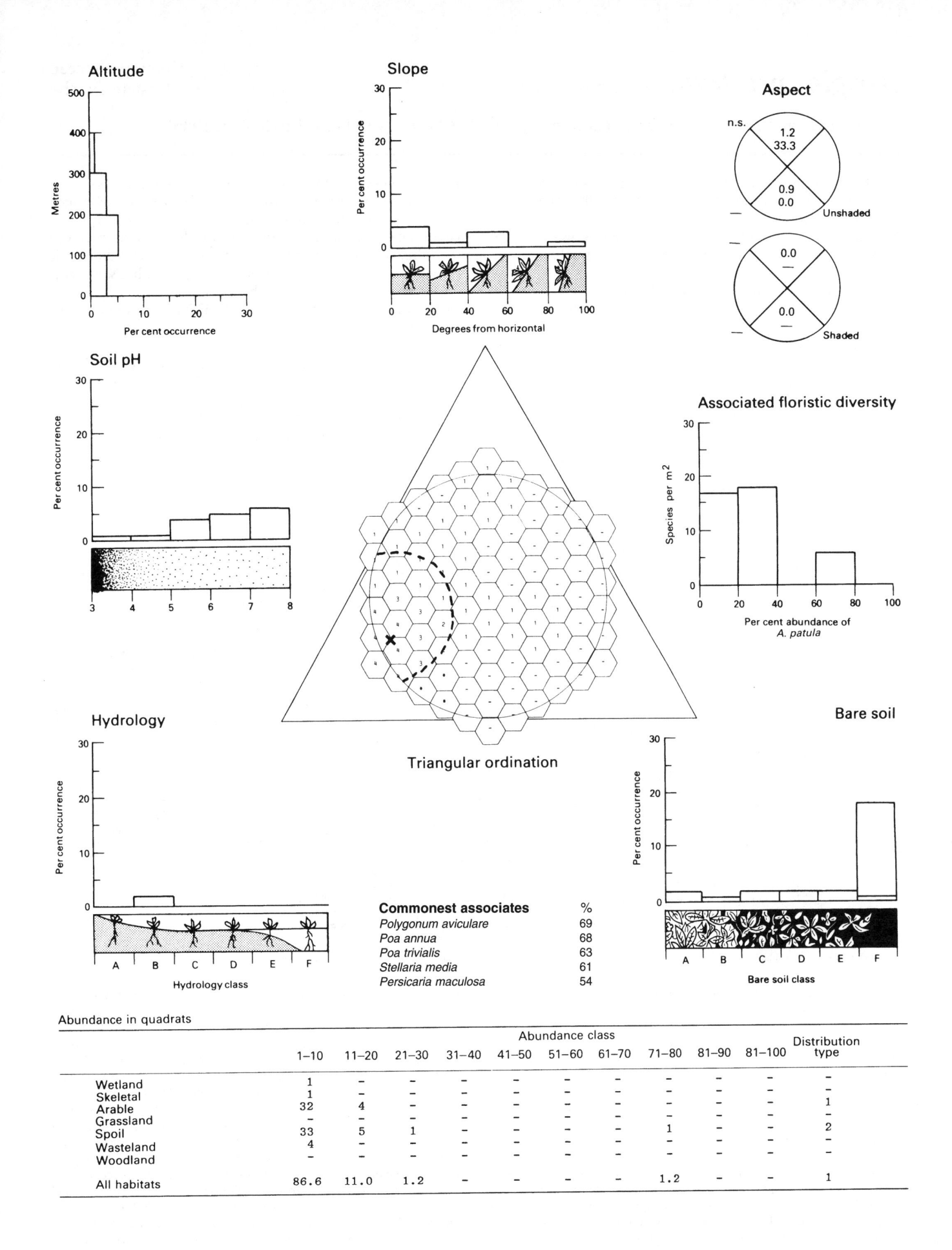

147

Atriplex prostrata Boucher ex DC.

Chenopodiaceae
Hastate Orache

A. hastata L. auct. angl. non L. This taxonomic group has been revised (see Taschereau 1985)

Life-form Semi-rosette, summer-annual therophyte. Stems erect or procumbent; leaves triangular, usually <4000 mm^2; roots non-mycorrhizal.

Phenology Seeds germinate in spring. Flowers from July to September and seed set from September to November.

Height Foliage and flowers to 1 m.

Seedling RGR No data. SLA 19.0. DMC 14.4.

Nuclear DNA amount No data; $2n = 18^*$ (Taschereau 1985); $2x^*$.

Established strategy Internediate between ruderal and competitive-ruderal.

Regenerative strategies B$_s$. Exclusively by seed, which germinates in spring. Seeds may remain dormant in the soil for up to 3 years (Chepil 1946), and a buried seed bank is suspected; SB IV.

Flowers Green, monoecious, facultatively autogamous, wind-pollinated but also visited by insects (Taschereau 1985); often >100 in a spike-like inflorescence.

Germinule 0.86 mg, *c.* 1.5 mm, seed usually dispersed enveloped within two bracteoles up to 5 mm in length. Seed dimorphic with most seeds small and black (1.0-2.5 mm) but some large and brown (1.5-3 mm) (Taschereau 1985).

Germination Epigeal; t_{50} 3 days; 19-33°C; dormancy broken by chilling; L > D.

Biological flora None.

Geographical distribution Most coastal regions of the British Isles except N Scotland, but local inland except in S and E England (Taschereau 1985); *ABIF* page 141, Europe northwards to Scandinavia (89% of European territories); N Africa, Asia; probably introduced in N America.

Habitats Particularly associated with disturbed spoil heaps such as building rubble, coal-mine spoil and soil heaps. Also found on exposed mud beside ponds and ditches, and on river banks, road verges and paths. Of casual occurrence in pasture and absent from woodland, skeletal habitats and arable land. [Occurs on sand, silt and shingle beaches, and in maritime and inland salt marshes (Taschereau 1985).]

Gregariousness Intermediate.

Altitude Confined to altitudes of <300 m. [Up to 460 m.]

Slope *A.p.* is absent from most steeply sloping sites.

Aspect More frequently recorded on N-facing slopes, but abundant in a greater proportion of S-facing sites.

Hydrology Recorded from waterside habitats which are flooded in winter but which present areas of bare mud in summer.

Soil pH Restricted to soils of pH >4.5.

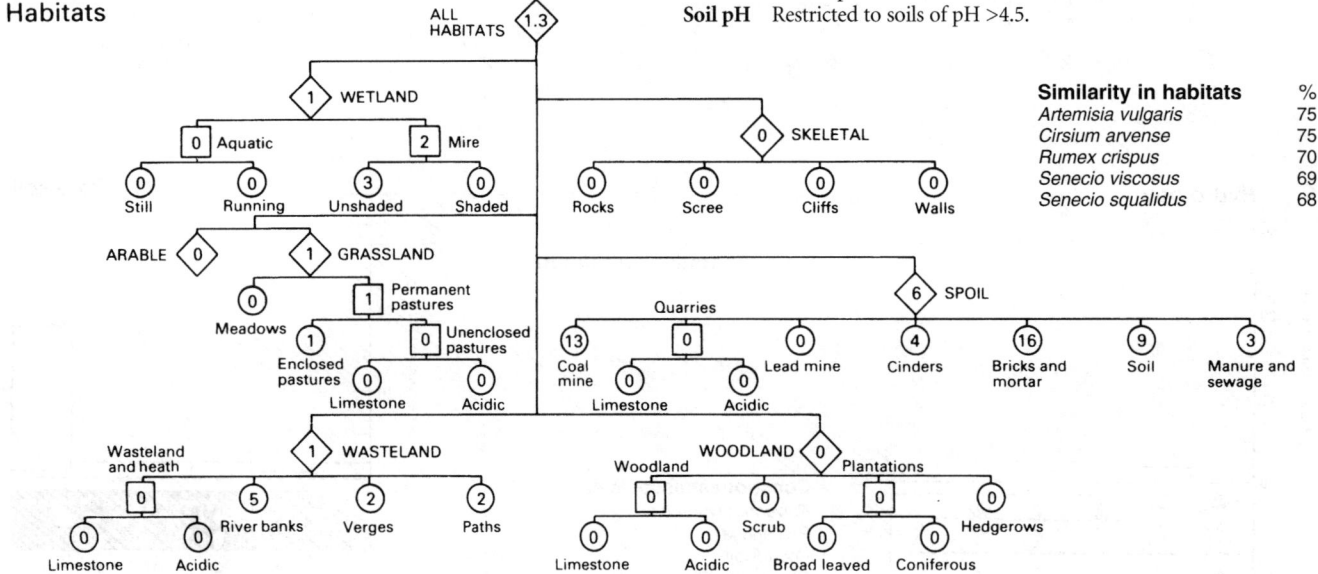

Habitats

Similarity in habitats	%
Artemisia vulgaris	75
Cirsium arvense	75
Rumex crispus	70
Senecio viscosus	69
Senecio squalidus	68

Bare soil Widely distributed, but most frequent and abundant in habitats having a large exposure of bare soil.

Δ ordination Distribution centred on disturbed and relatively productive vegetation, and absent from unproductive vegetation.

Associated floristic diversity Typically intermediate. Because the species frequently behaves as a pioneer colonist, high values of *A.p.* are often associated with low species diversity.

Synopsis *A.p.* is an erect or prostrate summer annual which exploits open, moist, unshaded sites, either (a) as an early colonist following disturbance and before the establishment of potentially-dominant perennial species, e.g. on demolition sites and coal-mine spoil, or (b) as a persistent component of habitats where the growth of perennials is restricted by regular disturbance, e.g. on the sea shore. Although perhaps originally a maritime species exploiting the upper reaches of salt marshes and sand, silt and shingle beaches (Taschereau 1985), *A.p.* is now widespread inland. The origin of *A.p.* as a plant of maritime soils high in available sodium presumably explains the exceptional ability of the species to exploit certain nutritionally-extreme habitats; *A.p.* is particularly common on demolition sites and coal-mine spoil. Demolition sites frequently exhibit very high pH (>8.0) and coal-mine spoil is often characterized by high sodium and chloride concentrations (Caswell *et al.* 1984a,b). *A.p.* also exploits pulverized fly-ash, another substrate with a very high pH (Rees & Sidrak 1955, Weston 1964). Not all of the sites colonized by *A.p.* are fertile, and the species can effectively utilize the low levels of nitrogen available in fly-ash (Weston 1964). Although growth is better where nitrogen is supplied in the form of nitrate, *A.p.* is also tolerant of the ammonium ion (Weston 1964). This may be important in enabling the species to exploit manure heaps and shorelines where detritus accumulates at the high-tide mark. When grown in solution culture *A.p.* tends to absorb a relatively low concentration of chloride ion (Black 1956). However, the presence of increased salt does lead to increased succulence (Black 1958). *A.p.* is not exclusively confined to extreme habitats of the type described

Commonest associates

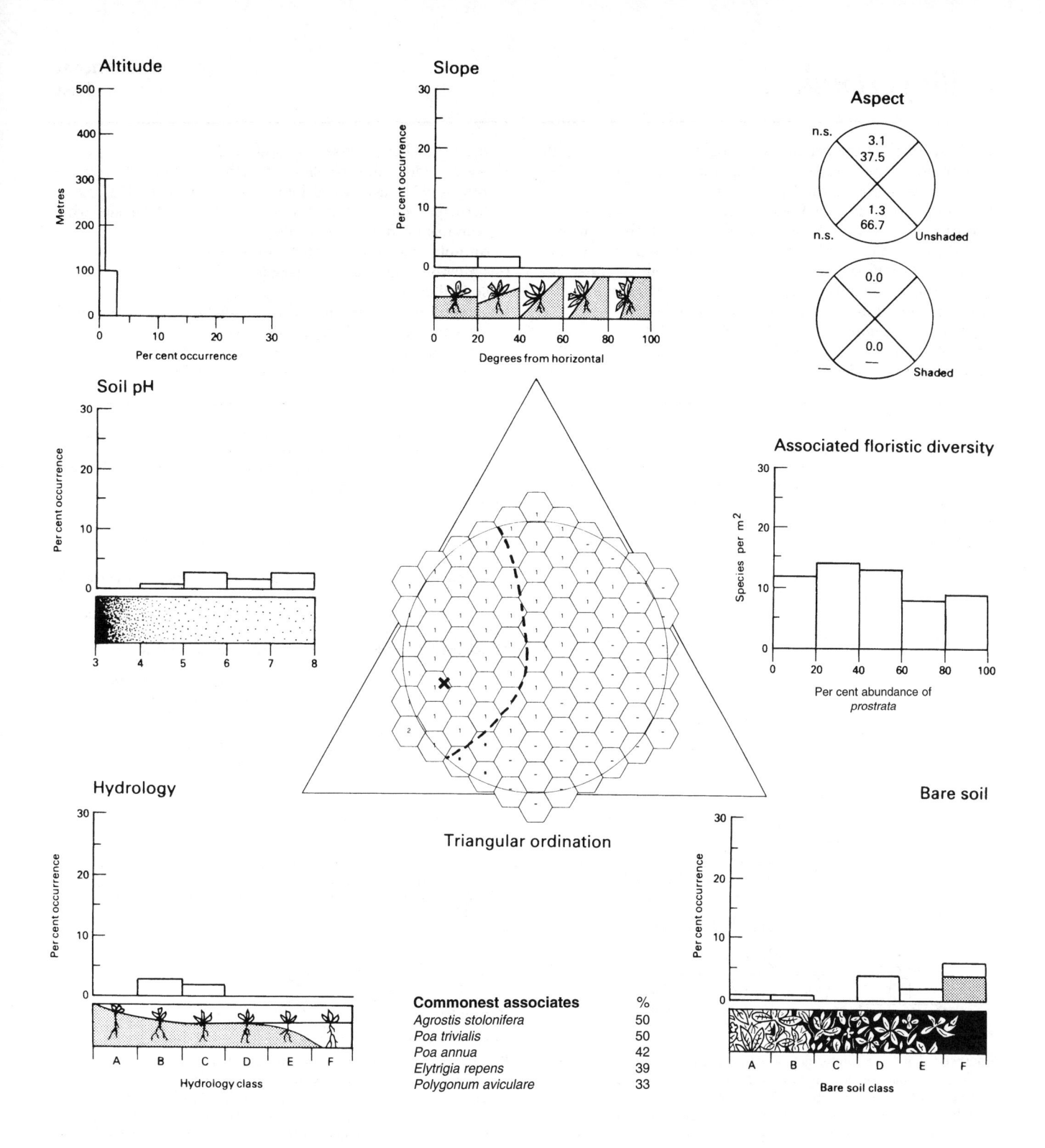

	%
Agrostis stolonifera	50
Poa trivialis	50
Poa annua	42
Elytrigia repens	39
Polygonum aviculare	33

above. The factors regulating the distribution of *A.p.* in more 'normal' ruderal habitats require investigation. *A.p.* is a C3 species (Slatyer 1970) and, of British species of *Atriplex*, only *A. laciniata* L. exhibits the Kranztypus leaf venation of C4 plants (Taschereau 1985). *A.p.* is largely absent from trampled and from grazed sites. The species regenerates entirely by seed, which germinates in spring. Mature seeds of *A.p.* are shed from within green tissues and, consequently, all require subsequent exposure of the imbibed seed to unfiltered sunlight before germination can occur (Cresswell & Grime 1981). The seeds are polymorphic but, unlike those of *A. patula*, the large brown and the (more frequent) small black seeds germinate at the same rate (Taschereau 1985). A buried seed bank is suspected. Seeds can survive ingestion by cattle (Ridley 1930) and are presumably spread by mammals and birds as well as by human activities. *A.p.* is polymorphic, often with several genotypes occurring within one population, and highly plastic (Gustafsson 1976). A genetically-fixed, prostrate form is restricted to certain exposed maritime sites (Gustafsson 1976). Three other British species are recognized within the *prostrata* group. *A.p.* forms hybrids with two of these, and with one other species also (Taschereau 1985).

Current trends An apparently increasing colonist of artificial habitats such as spoil heaps and derelict land. Perhaps increasing also in maritime areas and on verges beside roads salted in winter. Disturbance of maritime areas may be expected to increase the extent of hybridization with other species.

Bellis perennis L.

Asteraceae
Daisy

Life-form Polycarpic perennial, rosette hemicryptophyte with stout erect stock. Scapes ascending; leaves hairy, ovate, <500 mm^2; roots shallow, with VA mycorrhizas.

Phenology Winter green. We suspect that *B.p.* grows appreciably during the winter. Flowers over much of the year but particularly from April to June. Seed shed from May onwards.

Height Foliage usually appressed to the ground, but may extend to 30 mm in height in taller vegetation; scapes to 120 mm in fruit.

Seedling RGR 1.0-1.4 week^{-1}. SLA 30.9. DMC 11.3.

Nuclear DNA amount 3.9 pg; $2n = 18^*$ (*BC*); $2x^*$.

Established strategy Intermediate between ruderal and C-S-R.

Regenerative strategies V. Spreads both by short stolons to form patches and by seed (see 'Synopsis'); ?SB I.

Flowers Florets insect-pollinated, self-compatible (Warwick & Briggs 1979); aggregated into a capitulum of 14-152 central, hermaphrodite, yellow disc-florets and 18-59 female, white marginal ray florets (Hull 1961).

Germinule 0.09 mg, 1.7×0.8 mm, achene.

Germination Epigeal; t_{50} 3 days; 6-35°C; L = D.

Biological flora None.

Geographical distribution The British Isles (100% of British vice-counties), *ABIF* page 643; S, W and C Europe (82% of European territories and 8% naturalized); W Asia; New Zealand and N America, introduced from Europe.

Habitats Occurring primarily in those grasslands which for at least part of the year are low in height, e.g. meadows, pastures, mown grass verges and paths, also on some calcareous rock outcrops and quarry heaps. More occasional in arable fields, spoil heaps, marshy ground and wasteland overlying limestone. Very common in lawns and sports fields. Virtually absent from shaded habitats. [Often frequent in dune grassland.]

Gregariousness Intermediate.

Altitude More frequent in the upland region, where managed grassland habitats are more abundant. Observed up to 410 m. [Up to 920 m.]

Slope Found on slopes of up to 60°, but most habitats suitable for *B.p.* are flat or have a slope of <20°.

Aspect Associated with a wide range of values, but only recorded as abundant on S-facing slopes and level ground.

Hydrology Appearances in wetland habitats restricted to moist but well-aerated soils or rarely to soligenous mire.

Soil pH Clearly associated with base-rich soils of pH >5.5. Most frequent and abundant in the range 7.0-8.0.

Bare soil Widely distributed, but most abundant at low or moderate levels of bare ground.

△ ordination Distribution centred on vegetation associated with relatively fertile, moderately disturbed conditions. Absent from very heavily disturbed vegetation, and also from undisturbed vegetation in which *B.p.* cannot coexist with more-robust perennial species.

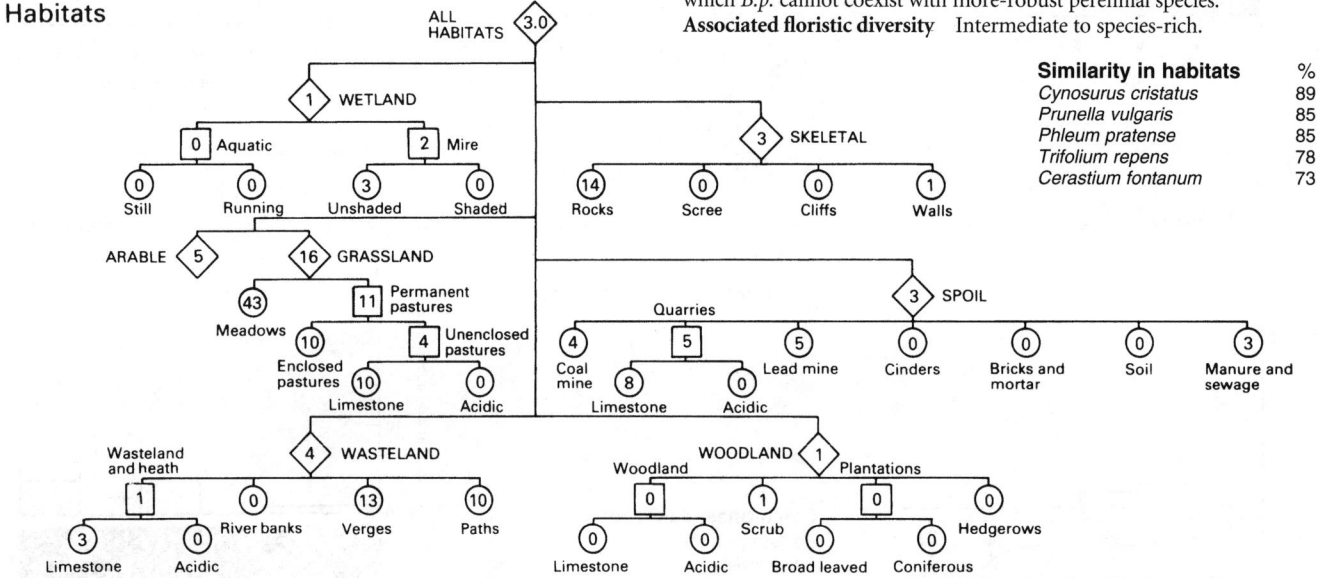

Associated floristic diversity Intermediate to species-rich.

Similarity in habitats	%
Cynosurus cristatus	89
Prunella vulgaris	85
Phleum pratense	85
Trifolium repens	78
Cerastium fontanum	73

Habitats

[Habitat hierarchy diagram: ALL HABITATS 3.0 → WETLAND 1 (Aquatic 0: Still 0, Running 0; Mire 2: Unshaded 3, Shaded 0); ARABLE 5; GRASSLAND 16 (Meadows 43; Permanent pastures 11: Enclosed pastures 10 (Limestone 10, Acidic 0), Unenclosed pastures 4); SKELETAL 3 (Rocks 14, Scree 0, Cliffs 0, Walls 1); Quarries (Coal mine 4 (Limestone 8, Acidic 0), Lead mine 5 (Limestone 0, Acidic 0), Cinders 0); SPOIL 3 (Bricks and mortar 0, Soil 0, Manure and sewage 3); WASTELAND 4 (Wasteland and heath: Limestone 3, Acidic 0; River banks 1; Verges 13; Paths 10); WOODLAND 1 (Woodland 0: Limestone 0, Acidic 0; Scrub 1: Broad leaved 0, Coniferous 0; Plantations 0; Hedgerows 0)]

Synopsis *B.p.* is a winter-green, low-growing rosette-forming herb producing clonal patches in short turf on fertile soils, and is strongly associated with trampled habitats. *B.p.* is essentially a 'cool season' plant capable of flowering throughout most of the year, particularly in lawns. During winter the roots of *B.p.* contain large amounts of the storage carbohydrate fructan (Hendry & Brocklebank unpubl.). Summer growth is often checked by drought. Despite the S-facing bias in field distribution, *B.p.* is most characteristic of moist soils, particularly where the effects of trampling and other forms of disturbance create local patches of bare soil. *B.p.* is most familiar as a weed of lawns and sports fields, but is also capable of persisting at reduced population densities in meadows, provided that there is a post-harvest grazing period. In meadows, before the hay is cut, *B.p.* produces upright leaves. However, when the field is grazed the leaves of *B.p.*, which are palatable to stock (Milton 1933), form an appressed rosette. *B.p.* is phenotypically plastic in many morphological features, including the number of florets within each capitulum (Hull 1961), but the genetic differences between British populations appear to be insufficient to allow the recognition of distinct ecotypes (Warwick & Briggs 1979, 1980).

B.p. regenerates vegetatively by means of stolons to form patches. Breakdown of stolons results in the isolation of daughter plants from the parent. Regeneration by seed is generally less important, but is more common in hay meadows than in lawns (Foster 1964). Seed germinates over an extended period during summer and autumn, reflecting the long flowering season and the capacity for germination over a wide range of temperatures. Indeed, flowers are present in most months of the year; during the spring and summer they develop extremely rapidly and consequently seeds can escape predation even in heavily grazed and frequently mown grasslands. Buried viable seeds have been detected (Chippindale & Milton 1934), but this report should be treated with caution as many other workers have failed to find them and, according to Chancellor (1985) and Roberts (1986a), the seed of *B.p.* is only briefly persistent. Seeds appear to be dispersed in mud on human feet and on animals and vehicles.

Current trends The abundance of *B.p.* is maintained at a high level through human activities and, as a result of modern land use, *B.p.* may be expected to increase still further.

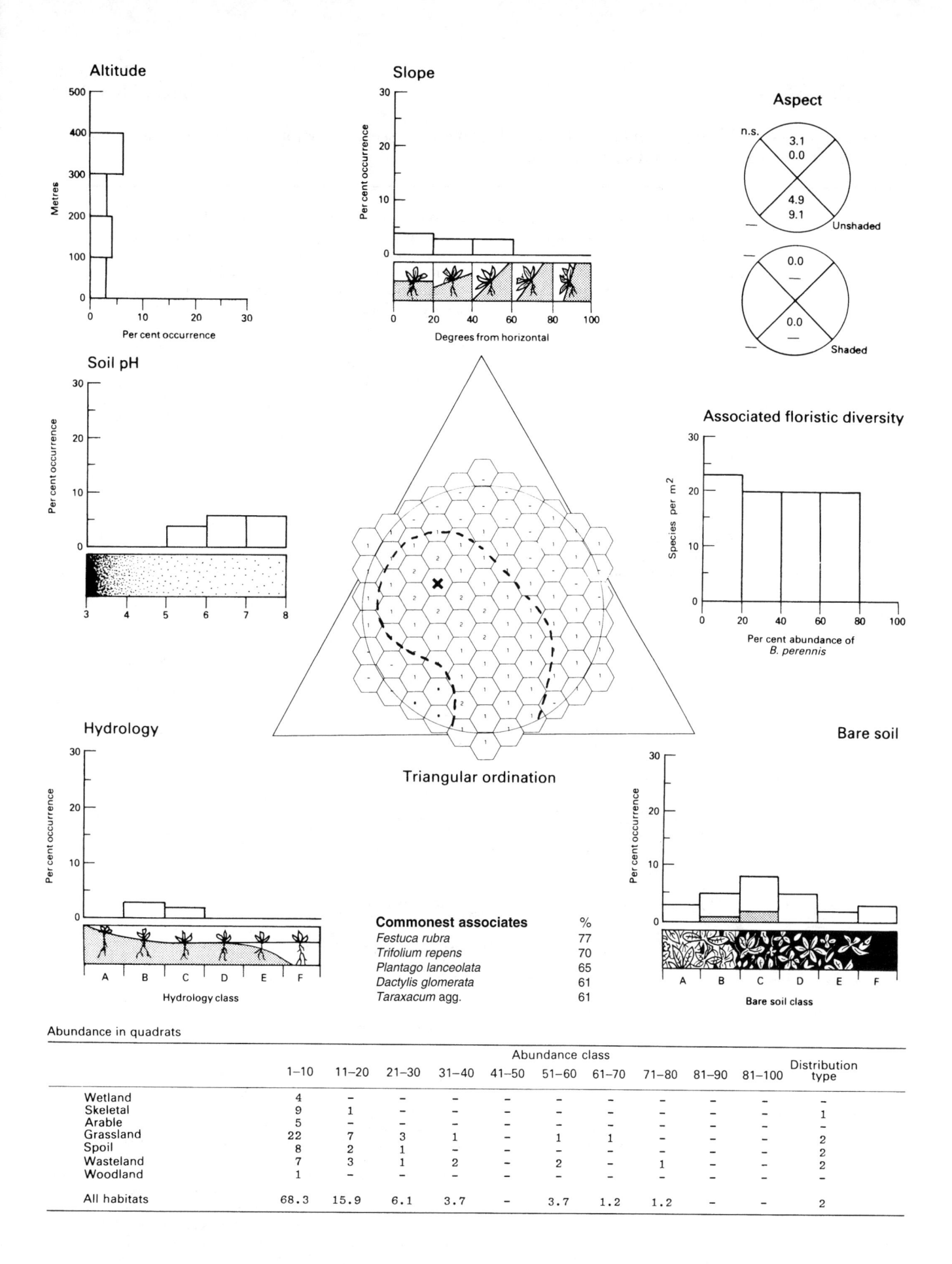

Altitude

Slope

Aspect

Soil pH

Associated floristic diversity

Triangular ordination

Hydrology

Bare soil

Commonest associates	%
Festuca rubra	77
Trifolium repens	70
Plantago lanceolata	65
Dactylis glomerata	61
Taraxacum agg.	61

Abundance in quadrats

	\multicolumn{10}{c	}{Abundance class}	Distribution type
	1–10	11–20	21–30	31–40	41–50	51–60	61–70	71–80	81–90	81–100	
Wetland	4	–	–	–	–	–	–	–	–	–	–
Skeletal	9	1	–	–	–	–	–	–	–	–	1
Arable	5	–	–	–	–	–	–	–	–	–	–
Grassland	22	7	3	1	–	1	1	–	–	–	2
Spoil	8	2	1	–	–	–	–	–	–	–	2
Wasteland	7	3	1	2	–	2	–	1	–	–	2
Woodland	1	–	–	–	–	–	–	–	–	–	–
All habitats	68.3	15.9	6.1	3.7	–	3.7	1.2	1.2	–	–	2

(*) *Betula* agg.

Betulaceae
Birch, Silver Birch

Includes *B. pubescens* Ehrh., *B. pendula* Roth. and their hybrids (see also 'Synopsis'). Field data analysed in this account refer to seedlings and small saplings

Life-form Deciduous tree or multi-stemmed shrub. Leaves ovate, sometimes >1000 mm^2; roots usually with ectomycorrhizas.
Phenology Flowers April and May before the leaves have expanded. Seeds shed from September onwards.
Height Foliage and flowers to 25 m.
Seedling RGR 0.5-0.9 week^{-1}. SLA 13.9. DMC. No data.
Nuclear DNA amount *B.pub.*: 1.5 pg; $2n = 56*$ (*BC*); $4x*$. *B.pen.*: $2n = 28*$; $2x*$.
Established strategy Intermediate between competitor and stress-tolerant competitor.
Regenerative strategies W, S. Regenerates by wind-dispersed seed, which germinates in spring. Seed shows some limited capacity for persistence on or near the soil surface.

Flowers Green, monoecious, wind-pollinated; >50 male flowers grouped in drooping catkin, and >50 female flowers in shorter, erect or spreading, more-compact catkins.
Germinule 0.12 mg, 3.4 × 2.3 mm, flattened winged nutlet.
Germination Epigeal; t_{50} 3 days; L > D (Harding 1981).
Biological flora Atkinson (1992) and see Harding (1981) and Henderson and Mann (1984).
Geographical distribution *B.pub.*: throughout the British Isles, but more common in the N and W (95% of British vice-counties), *ABIF* page 133; N and C Europe and locally on mountains southwards to Spain (64% of European territories); temperate Asia. *B.pen.*: similar, but 90% of British vice-counties and 74% of European territories and 3% naturalised.

Habitats Seedlings and young saplings common in shaded habitats, particularly scrub and woodland on acidic strata. They are frequent also in rocky open habitats such as limestone quarry spoil, cinders and rock outcrops. Also recorded at lower frequencies from all but the following habitats: enclosed pasture, hedgerows, paths, manure heaps, scree and aquatic habitats.

Gregariousness Seedlings may occur at high densities.
Altitude Juveniles observed above 400 m. [Up to 760 m.]
Slope Juveniles widely distributed.
Aspect Juveniles equally frequent on N- and S-facing slopes in unshaded sites, but in shaded habitats significantly more frequent on N-facing slopes and only achieving abundance in this situation.
Hydrology Seedlings occasionally recorded in topogenous, soligenous and ombrogenous mires.

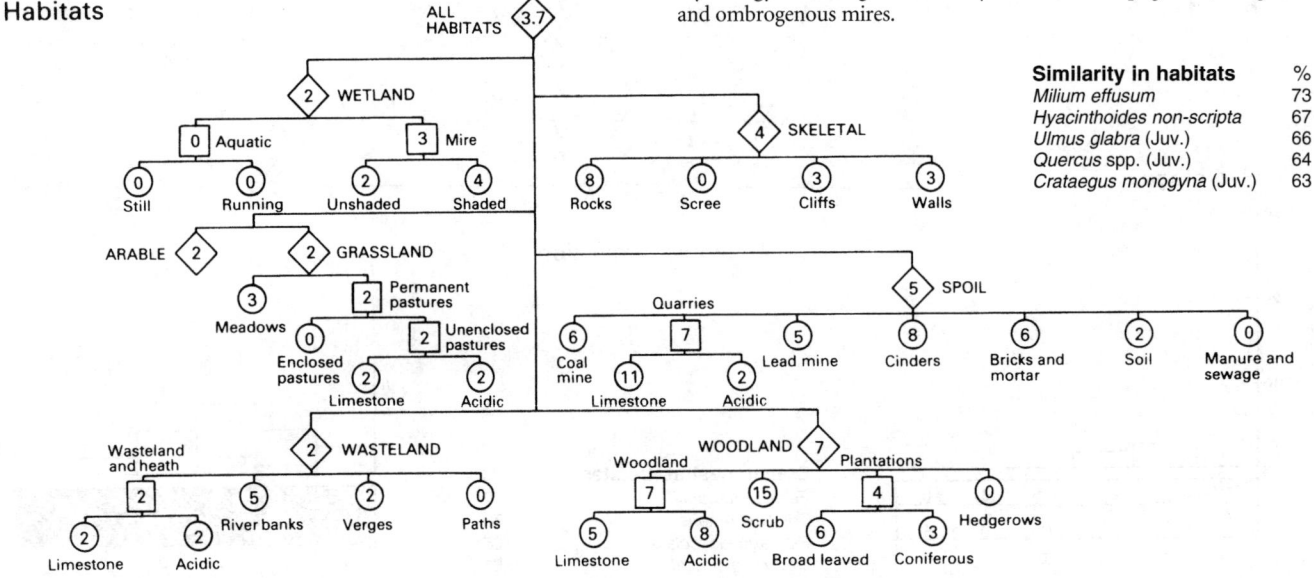

Habitats

Similarity in habitats	%
Milium effusum	73
Hyacinthoides non-scripta	67
Ulmus glabra (Juv.)	66
Quercus spp. (Juv.)	64
Crataegus monogyna (Juv.)	63

Soil pH Widely distributed, but most frequent and abundant below pH 5.0.
Bare soil Widely distributed. Sites with an abundance of juveniles tend to have small amounts of exposed soil, but often include a continuous bryophyte mat or a thin layer of tree litter. [Establishment is much better on bare soil.]
△ ordination Highest values clustered at the right-hand side of the distribution, a reflection of an association with relatively undisturbed vegetation. Seedlings recorded in association with various levels of productivity.
Associated floristic diversity Relatively low, particularly in sites containing high densities of seedlings.
Synopsis *B.* agg. is a fast-growing deciduous tree which may reach 25 m in height and has a life-span of c. 60-70 years (Harding 1981). Wood is diffuse-porous and bud-break comparatively early. In a forestry survey summarized by Steele and Peterken (1982), *B.* spp. was found to cover 157 000 ha of woodland (>21% of the total), making it the second most common broad-leaved tree of British woodland and the most common in Scotland. *B.* agg. was exceedingly common in Britain in the pre-Boreal period (Godwin 1956). The extremely high abundance of *B.* agg. particularly *B.pub.*, in Scotland today probably relates in part to a tolerance of cold climate (McVean & Ratcliffe 1962). *B.* agg. may commence flowering within its first 5-10 years (Harding 1981). Small, winged, wind-dispersed fruits are produced in large quantities (up to 43 000 m^{-2} of ground (Kinnaird 1974)), although a good seed year is generally followed by a number of years with lower production (Kinnaird 1968). Seed viability is particularly high in good seed years (Harding 1981). Seeds have a light requirement but may germinate in darkness after chilling (Henderson and Mann 1984). However, germination may be inhibited by placement under a leaf canopy (Black 1969, Sidhu & Cavers 1977, Wareing 1966). These findings are consistent with the frequent observation that young birch seedlings tend to be restricted to areas of bare soil, especially

Altitude

Slope

Aspect

n.s.
2.4 / 0.0
2.2 / 0.0
Unshaded

$P<0.05$
10.6 / 35.7
2.4 / 0.0
Shaded

Soil pH

Triangular ordination

Associated floristic diversity

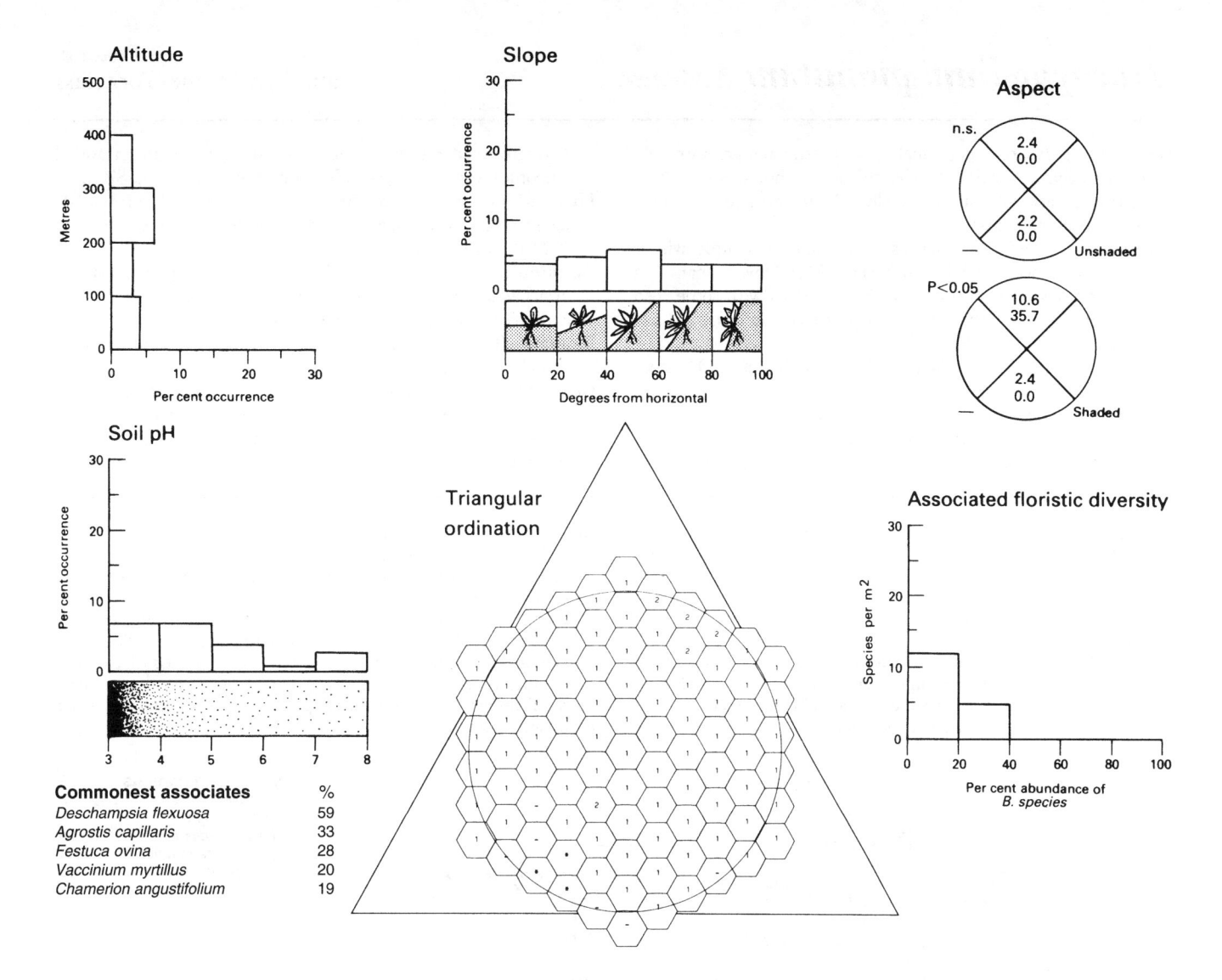

Commonest associates	%
Deschampsia flexuosa	59
Agrostis capillaris	33
Festuca ovina	28
Vaccinium myrtillus	20
Chamerion angustifolium	19

where there are canopy gaps in grassland or woodland. Seedlings are small and very susceptible to drought (Harding 1981), although less so than those of *Alnus glutinosa* (McVean 1956a). In sharp contrast with oak, the seedling root:shoot ratio never exceeds 1 (Shaw 1974). It has been suggested that root competition from heather plays a major role in the prevention of *B.* agg. from establishing in Callunetum (Dimbleby 1953). Establishment is often also unsuccessful in the vicinity of mature birch trees (Harding 1981). In soils of low nutrient status seedling establishment is much enhanced by mycorrhizal infection of the roots (Harding 1981). Many fatalities occur during the first winter, and subsequently due to the effects of shade (Harding 1981). A list of common pathogens is given by Harding (1981). Establishment from seed is poor in heavily-grazed sites. A few viable seeds are often found in soils but, with the exception of Brown and Oosterhuis (1981), there are no reports of a persistent seed bank. Seedlings and young saplings have been observed in most of the major habitats of the Sheffield region, often in sites quite remote from established trees and in areas unsuitable for subsequent establishment of trees. *B.pen. is* often found at high pH, and may be an early colonist of limestone quarries. Nevertheless, birch is most conspicuous as a colonist of open ground, burned areas and forest clearance on acidic soils. The soil under *B.* agg. may be converted from mor to mull humus following the penetration of the iron pan by the root system and the uptake of calcium and other nutrients from lower horizons of the soil profile (Harding 1981). Breakdown of litter is relatively rapid. *B.pen.* survives swift-moving fires (Beckett & Beckett 1979) and is tolerant of coppicing. *B.pub.* is relatively palatable to the snail *Helix aspersa* Muller, and supports a diverse insect fauna (Wratten *et al.* 1981). An inducible defence against foliar predation has been described (Haukioja & Niemela 1976, 1977, Edwards & Wratten 1982) in which decreased palatability is correlated with increased levels of phenolics (Haukioja & Niemela 1979, Niemela *et al.* 1979). Often, birch scrub does not give rise to oak woodland, as might have been predicted, but appears in many places to be able to perpetuate itself (Rackham 1980). Indeed, Rackham. suggests that the high reproductive capacity of *B.* agg. offsets its short life-span, and that oak-to-birch succession frequently occurs. The two segregates *B.pen.* and *B.pub.* frequently occur together. *B.pub. is* more characteristic of moist soils (Walters 1968), but the distribution of the two in woodlands may be determined by chance factors of colonization (Rackham 1980). This is probably due to the high level of disturbance associated with British woodlands; fertile hybrids between the two species are frequent (Stace 1975). Rackham. (1980) notes a greater association of *B.pub.* with ancient woodland. *B.pub.* shows some evidence of less efficient seed mobility, is shorter-lived, and is more tolerant of cold, of acidic and of nutrient-deficient conditions (Harding 1981). The relative growth rate of *B.pen.* exceeds that of *B.pub.* over a wide range of light intensities (Kinnaird & Kemp 1971). In Scotland *B.pub.* hybridizes also with *B. nana* L. (Stace 1975).

Current trends The capacity for effective long-distance dispersal and regeneration by seed enabled birch to spread rapidly at the end of the last glaciation. This aggressive species is showing a similar capacity for expansion in the present disturbed landscapes, both in areas of former heath or grassland and (as a weed) in forestry. Marrs *et al.* (1986) now consider birch a major threat to loss of open heathland in the Brecklands of East Anglia. Rackham (1980) makes two suggestions: (a) that birch should not normally be planted for conservation (or amenity) purposes, nor should tree planting be undertaken in ancient woodland, since, whatever is planted, birches normally appear, and (b) allowing for the fact that birch is likely to increase further, attempts should be made to find an economic use for it. (The species was probably formerly of considerable economic importance in N England and Scotland.) Both suggestions seem eminently reasonable.

Brachypodium pinnatum (L.) Beauv.

Poaceae
Heath False-brome, Tor Grass

Life-form Polycarpic perennial, semi-rosette hemicryptophyte or chamaephyte with extensive rhizomes. Shoots erect; leaves coarse, occasionally hairy, usually <2000 mm^2; roots usually with VA mycorrhizas.

Phenology Growth commences in late spring (a little winter growth may occur in S England; Hall 1971). Flowers mainly in July and seed shed from September onwards with some seed overwintering on the plant. Most leaves die back in late autumn.

Height Foliage usually <300 mm; flowering shoots to 600 mm.

Seedling RGR 1.0-1.4 week^{-1}. SLA 23.2. DMC 36.9.

Nuclear DNA amount 2.5 pg; $2n = 28*$ (*BC*); $4x*$.

Established strategy Stress-tolerant competitor.

Regenerative strategies V, ?S. Relies almost exclusively on vegetative generation forming extensive patches by means of rhizomes; seed set is generally poor. (See 'Synopsis'); ?SB II.

Flowers Green, hermaphrodite, wind-pollinated; >100 aggregated into a spike-like inflorescence composed of sessile, 8-22-flowered spikelets.

Germinule 2.85 mg, 7.6 × 1.6 mm, caryopsis dispersed with an attached short-awned lemma (12.3 × 1.9 mm).

Germination Epigeal; t_{50} 2 days; 6-34°C; L > D.

Biological flora None published, but one is in preparation (T. C. Hutchinson, pers. comm.).

Geographical distribution Mainly in S and E of Britain with scattered distribution in Ireland (41% of British vice-counties), *ABIF* page 793; Europe except for the extreme N (87% of European territories); N Africa.

Habitats Predominantly a species of dry grassland, particularly in little-grazed areas overlying calcareous strata. Other habitats include quarry heaps and rock outcrops, hedgebanks and road verges, scrub, woodland and plantations. Absent from wetlands and arable land. [Isolated records from sea cliffs and sand dunes (T. C. Hutchinson, pers. comm.)]

Gregariousness A stand-forming species.

Altitude Largely restricted to the magnesian limestone and to altitudes of <200 m. Often introduced in its upland sites and observed up to 380 m, which is above the altitudinal maximum given by Wilson (1956).

Slope Most records from slopes in the range 10-60°.

Aspect Significantly more frequent on S-facing slopes.

Hydrology Absent from wetlands.

Soil pH Mainly on soils of pH >5.0, with a peak at 7.5-8.0 consistent with the calcicolous distribution of *B.p.*

Bare soil Associated with a wide range of exposures. In ungrazed and unburnt habitats *B.p.* may form a dense and persistent layer of litter.

△ **ordination** Distribution centred on vegetation associated with undisturbed conditions of intermediate fertility. Also extends into rather more disturbed vegetation, although absent from sites at which disturbance is severe.

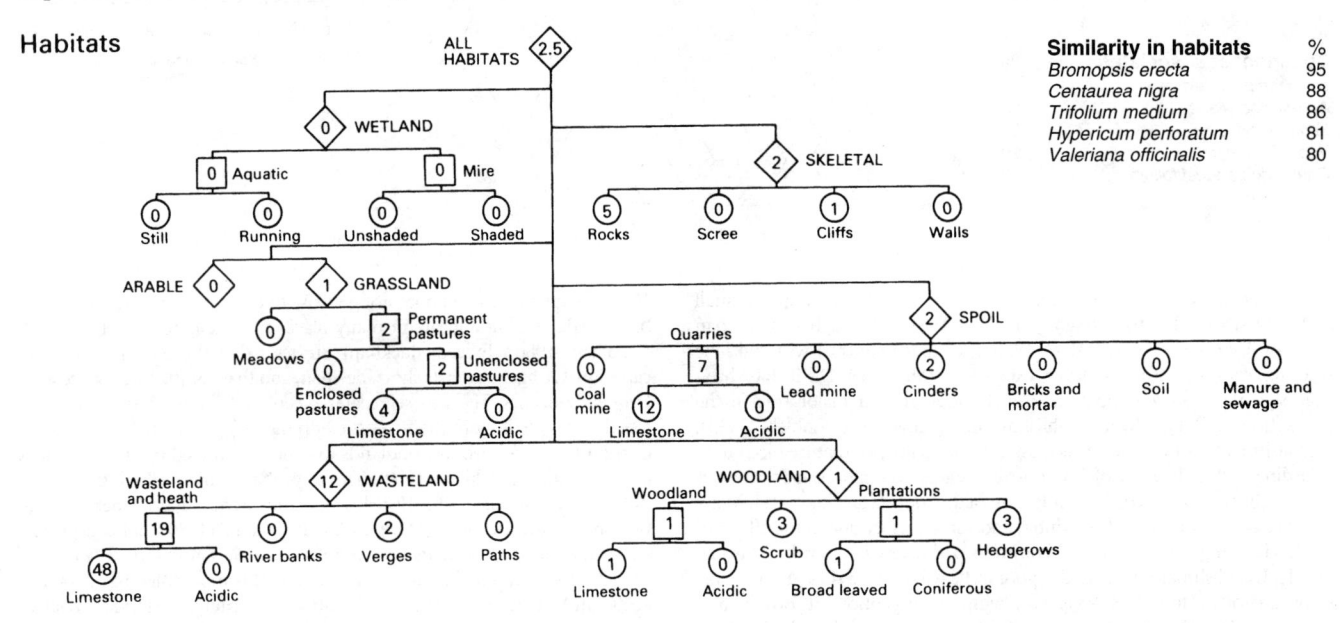

Habitats

Similarity in habitats	%
Bromopsis erecta	95
Centaurea nigra	88
Trifolium medium	86
Hypericum perforatum	81
Valeriana officinalis	80

Associated floristic diversity Usually intermediate, but higher in calcareous pastures. In ungrazed and unburnt habitats *B.p.* can form dense stands with persistent litter. Here associated floristic diversity is invariably low.

Synopsis *B.p.* is an aggressively rhizomatous, potentially dominant, stand-forming grass of infertile lowland pasture and wasteland, usually on calcareous soil. As might be predicted from its low nuclear DNA amount, *B.p.* has a summer peak in growth. This is in marked contrast to *Bromopsis erecta* with which the species often grows. *B.p.* has a deep root system and tends to attain maximum vigour on terraces with an accumulation of mineral subsoil. In N and upland Britain *B.p.* is rare and is often confined to S-facing slopes. This geographical restriction appears to arise from problems in initial establishment. In common with certain other grasses which form large clonal patches (e.g. *Elytrigia repens* and *Glyceria maxima*), *B.p.* often produces little viable seed. A seed set of 20% is considered a good average in lowland sites (T. C. Hutchinson, pers. comm.) although, somewhat paradoxically, generally high amounts of good seed were observed in the cool summer of 1985. In upland sites seed is not normally produced at all; sometimes even the inflorescences are not formed. Accordingly, at higher altitudes new colonies of *B.p.* are often the result of unwitting human introduction; however, the species is frost-resistant (T. C. Hutchinson, pers. comm.) and, once established, may be locally invasive even in N-facing grasslands. Seeds germinate in the spring and do not form a persistent seed bank (Roberts 1986b). Regeneration by means of transported rhizome fragments requires investigation, but may have been important during the colonization of quarries. *B.p.* is resistant to trampling, and only the young leaves are eaten by sheep and cattle. In many unmanaged areas *B.p.* has become more prominent following myxamatosis and the decline in rabbit populations (Thomas 1960).

Increases in area and biomass of *B.p.* have brought about a decrease in

Altitude
Slope
Aspect
P<0.01
4.7
91.7
n.s.
11.2
88.0
Unshaded

0.0
1.6
100.0
Shaded

Soil pH

Associated floristic diversity

Hydrology
Absent from wetland

Triangular ordination

Bare soil

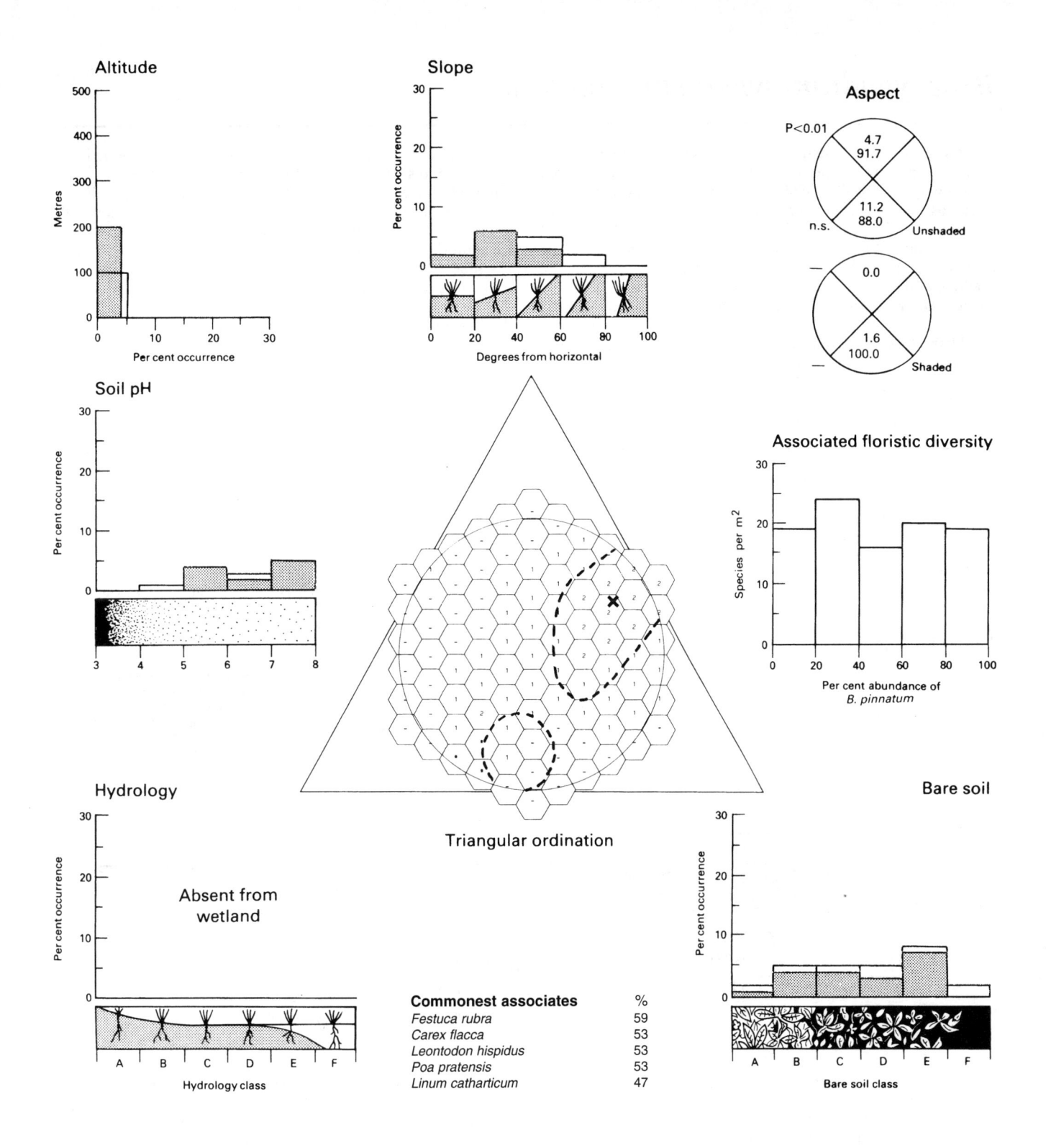

Commonest associates	%
Festuca rubra	59
Carex flacca	53
Leontodon hispidus	53
Poa pratensis	53
Linum catharticum	47

the floristic diversity of the associated vegetation, since the relatively dense canopy of *B.p.* and the often thick layer of persistent leaf litter which accumulates beneath it inhibit the survival of many of the smaller herbaceous species of calcareous grassland. *B.p.*, with rhizomes down to 20 mm below the soil surface, is tolerant of burning; this frequently allows *B.p.* to displace the more fire-sensitive *Bromopsis erecta* (Grime 1955). Although a relatively strict calcicole in native British sites, *B.p.* also occurs on more-acidic serpentine soils in France (T. C. Hutchinson, pers. comm.) and, as an adventive, is not restricted to calcareous soils. Another subspecies, ssp. *rupestre* (Host) Schubler & Martens with $2n = 14$ $(2x)$ occurs in Europe (*FE5*).

Current trends *B.p.* is likely to decrease in frequency of occurrence in lowland areas as a result of the further destruction of semi-natural calcareous grassland. Nevertheless, the species has increased in abundance within many of its extant sites due to a relaxation of grazing pressure (Hope-Simpson 1941) and, possibly, an increased nitrogen input from air pollution (R. Bobbink, pers. comm.). In uplands *B.p.* has spread along railway banks and roadsides, sometimes on non-calcareous strata, and has also expanded following quarrying activity. Thus, the range of *B.p.* appears to be slowly increasing. The species is apparently also expanding in Europe (Willems 1983).

Brachypodium sylvaticum (Hudson) Beauv.

Poaceae
Slender False-brome

Life-form Densely tufted, polycarpic perennial, semi-rosette hemicryptophyte. Shoots erect; leaves linear, hairy, often yellow-green, usually <4000 mm^2; roots with VA mycorrhizas.
Phenology Some winter growth of young, autumn-produced shoots (Szujko-Lacza & Feketa 1974). Flowers in July and seed set in August to October. Shoots die back in autumn.
Height Foliage to 350 mm; flowers to 900 mm.
Seedling RGR 1.0-1.4 week^{-1}. SLA 41.2. DMC 30.6.
Nuclear DNA amount 1.0 pg; $2n = 18*, 28,42,56$ (*FE5*); $2x*$, $4x$, $6x$ and $8x$.
Established strategy Intermediate between C-S-R and stress-tolerant competitor.
Regenerative strategies S, V. Seeds shed in autumn are dormant and most germinate in spring (Cresswell 1982), although autumn germination is also possible (Hutchinson pers. comm.). Plants can increase in size through the production of new shoots, but *B.s.* rarely forms large patches; ?SB II.
Flowers Green, hermaphrodite, wind-pollinated, probably self-compatible; sometimes >100 aggregated into a spike-like inflorescence composed of sessile, 8-16-flowered spikelets.
Germinule 0.62 mg, 7.5 × 1.4 mm, caryopsis dispersed with an attached long-awned lemma (19.7 × 1.6 mm).
Germination Epigeal; t_{50} 5 days; 9-24°C; dormancy broken by dry storage at 20°C; L > D.
Biological flora In preparation (T. C. Hutchinson).
Geographical distribution The British Isles, particularly in S and C England (99% of British vice-counties), *ABIF* page 793; Europe northwards to Scandinavia (89% of European territories); N Africa and W Asia.

Habitats Mainly in shaded habitats, particularly base-rich woodland, but also frequent in scrub and hedgerows. Widespread, too, in unshaded habitats, especially in upland areas; recorded from base-rich grassland, scree slopes, lead-mine spoil, limestone quarry heaps, and to a lesser extent other rocky and grassy habitats. [Also found on sea cliffs (Burnett 1964) and fixed sand dunes (Salisbury 1952).]
Gregariousness Intermediate.

Altitude Within the lowland part of its range (below 200 m) *B.s.* is predominantly a plant of shaded habitats. In more-upland regions it occurs in both open and shaded habitats. The large number of occurrences above 200 m reflects this greater habitat range at higher altitudes. [Up to 380 m.]
Slope Found on both flat and sloping ground.
Aspect In shaded habitats significantly more frequent on S-facing slopes; otherwise no significant trends.
Hydrology Absent from wetlands.

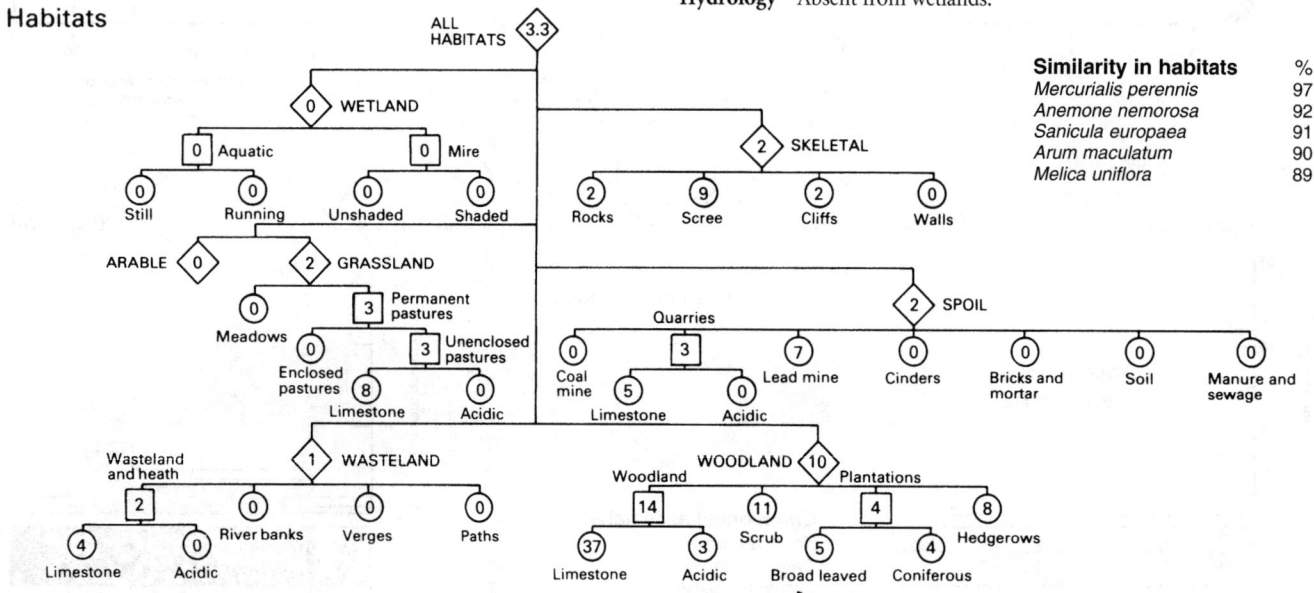

Habitats

Similarity in habitats	%
Mercurialis perennis	97
Anemone nemorosa	92
Sanicula europaea	91
Arum maculatum	90
Melica uniflora	89

Soil pH Most frequent and abundant on neutral soils in the pH range 6.0-8.0. Absent from very acidic soils.
Bare soil Widely distributed, but highest frequency and abundance in woodland sites with intermediate amounts of exposed soil.
△ **ordination** Distribution centred on vegetation associated with undisturbed and relatively unproductive vegetation. Absent from heavily disturbed communities.
Associated floristic diversity Typically intermediate.
Synopsis *B.s.* is typically a tufted grass of woodland habitats. However, in the Sheffield region the species shows a significant S-facing bias in woodlands, and is generally absent from deep shade (Hutchinson pers. comm.). Further, tussocks of *B.s.* in woodland may produce as few as 30 seeds or may even fail to flower (Hutchinson pers. comm.). In keeping with the DNA amount, which is exceptionally low among native grasses, leaf expansion in *B.s.* is somewhat delayed, and shoot biomass reaches its maximum in summer (Szujko-Lacza & Feketa 1974). *B.s.* is a common constituent of coppiced woodland. Particularly in upland areas, *B.s.* is also frequent in grassland and scree habitats, and here one vigorous plant may produce as many as 2700 seeds (Hutchinson pers. comm.). Plants in open sites show signs of photo-bleaching and, as a result, their leaves are distinctly more yellow-green than those of adjacent grasses. It is not clear whether this yellowing is associated with any diminution of growth rate. Hubbard (1984) suggests that in grassland sites *B.s.* is a relict of former woodland. The species is found under a wide range of edaphic conditions, and is particularly characteristic of less fertile, calcareous soils. *B.s.* loses water more rapidly than *B. pinnatum* during both the stomatal and the cuticular phases of transpiration (Hutchinson pers. comm.), and its frequency on scree slopes in the Derbyshire dales may be, in part, a result of the continuously moist soil maintained by the 'mulch effect' of talus. *B.s.* may be eaten by sheep and rabbits (Hutchinson pers. comm.), and is absent from heavily grazed grassland. However, the species is frequently found in sites subject to burning (*Ecol Atl*). *B.s.* probably regenerates mainly by seed, but its limited capacity for lateral vegetative spread may be of significance in shaded sites, where seed set is frequently poor. No

Altitude

Slope

Aspect

	2.0	
	0.0	
n.s.	1.8	Unshaded
	50.0	

	4.5	
	33.3	
P<0.01	14.6	Shaded
	33.3	

Soil pH

Associated floristic diversity

Triangular ordination

Hydrology

Absent from wetland

Commonest associates

	%
Arrhenatherum elatius	56
Mercurialis perennis	50
Festuca rubra	42
Dactylis glomerata	41
Viola riviniana	34

Bare soil

Abundance in quadrats

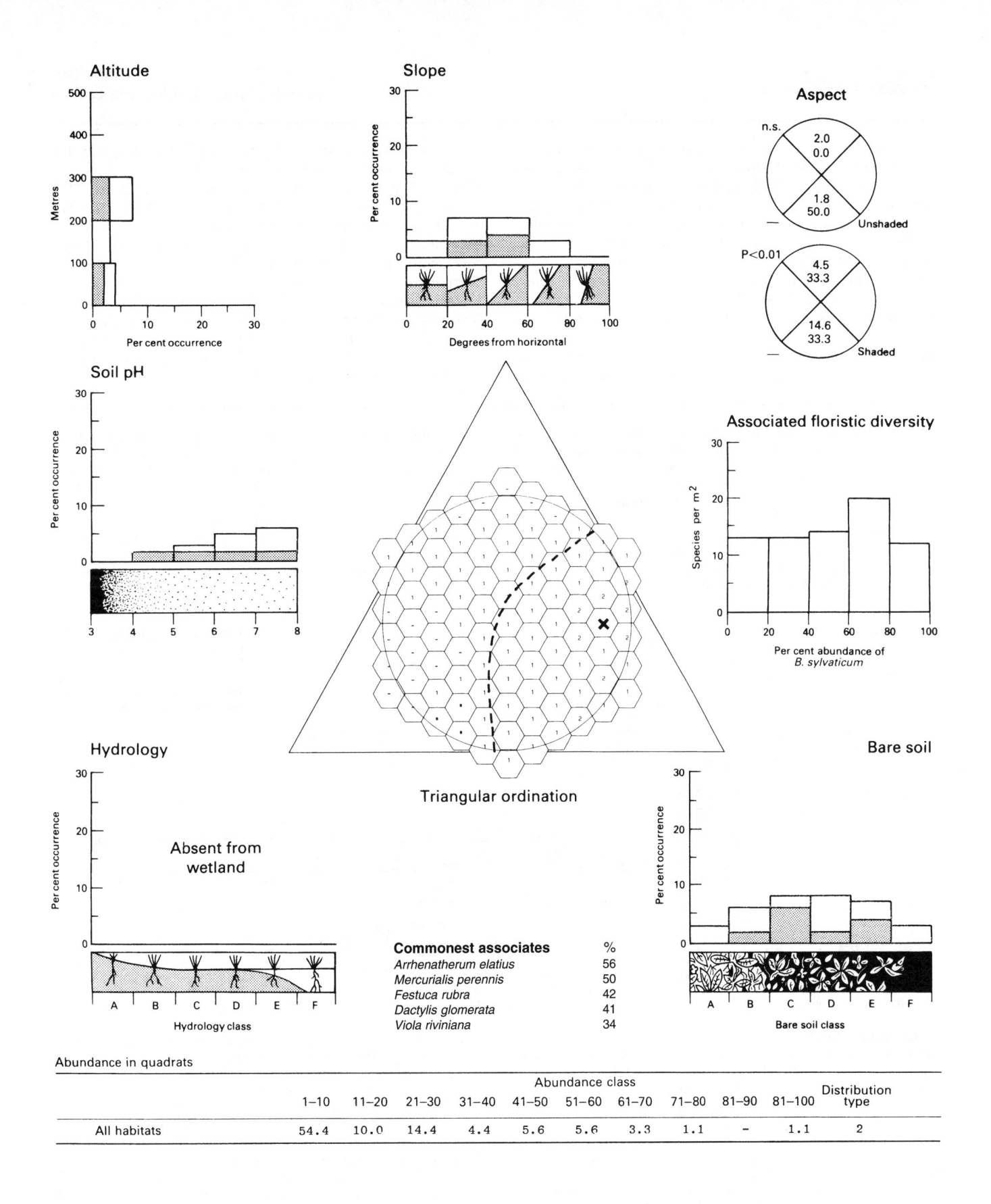

	\multicolumn{9}{c	}{Abundance class}	Distribution type
	1–10	11–20	21–30	31–40	41–50	51–60	61–70	71–80	81–90	81–100	
All habitats	54.4	10.0	14.4	4.4	5.6	5.6	3.3	1.1	–	1.1	2

persistent seed bank has been reported. Despite low seed mobility, *B.s.* is also frequently found along the edges of woodland rides in modern forests. The species is morphologically uniform in Britain (Hutchinson pers. comm.). Elsewhere the species is more variable, and a further ssp. is recorded from Europe (*FE5*).

Current trends Persistence in woodland does not appear to be severely threatened by modern forestry practice. However, *B.s.* is not a regular colonist of artificial habitats, and a continuing decrease in its abundance is predicted.

Briza media L.

Poaceae
Quaking Grass, Doddering Dillies

Life-form Polycarpic perennial, semi-rosette hemicryptophyte with short rhizomes. Shoots erect; leaves linear, typically <500 mm^2; roots often relatively shallow, with VA mycorrhizas.
Phenology Winter green. Leaf expansion occurs in early spring. Flowers in June and July and seed shed from July to September.
Height Foliage usually <100 mm; flowering shoots to 500 mm.
Seedling RGR 1.0-1.4 week^{-1}. SLA 19.8. DMC 30.4.
Nuclear DNA amount 16.9 pg, $2n = 14*$ (Bennett & Smith 1976); $2n = 14*, 28*$ (*BC*); $2x*, 4x*$.
Established strategy Intermediate between stress-tolerator and C-S-R.
Regenerative strategies S, V. Mainly by freshly shed seed in vegetation gaps during autumn. *B.m.* has a limited capacity for lateral vegetative spread by means of rhizomes, but the importance of this mode is uncertain; SB I.
Flowers Purplish, hermaphrodite, homogamous (Knuth 1906), wind-pollinated, strongly self-incompatible (Murray 1974); >100 in a loose panicle (4-10 per spikelet).
Germinule 0.23 mg, 1.7 × 1.0 mm, caryopsis dispersed with an attached lemma (3.0 × 1.8 mm).
Germination Epigeal; t_{50} 4 days; L = D.
Biological flora Dixon (2002) and see Zarzycki (1976).
Geographical distribution Widely distributed throughout the British Isles except N Scotland (95 % of British vice-counties), *ABIF* page 754; Europe except the Arctic and parts of the S (85% of European territories); temperate Asia.

Habitats Particularly associated with semi-natural calcareous pastures, but also widespread in a range of calcareous ungrazed grassy habitats such as wasteland, quarry spoil, scree and road verges. Also observed in older meadows and enclosed pastures, sometimes on non-calcareous soils. Occasional records from mire. Absent from aquatic and woodland habitats.
Gregariousness Intermediate.
Altitude Found over a wide range of altitudes up to 400 m. Most commonly found at higher altitudes where unimproved calcareous grassland is most abundant. [Up to 660 m.]

Slope Grows on both flat and sloping ground but particularly frequent on slopes in the range 20-40°, most of which are Derbyshire dale sides.
Aspect Frequent and locally abundant on both N- and S-facing slopes. Previously recorded (*Ecol Atl*) as more abundant on N-facing grassland.
Hydrology Largely absent from wetlands, but occasionally observed in soligenous mire and on tussocks of other wetland species.
Soil pH Shows a calcicolous distribution with greatest frequency and abundance on soils of pH >7.0.
Bare soil A clear bias towards habitats with low to moderate amounts of exposed earth.

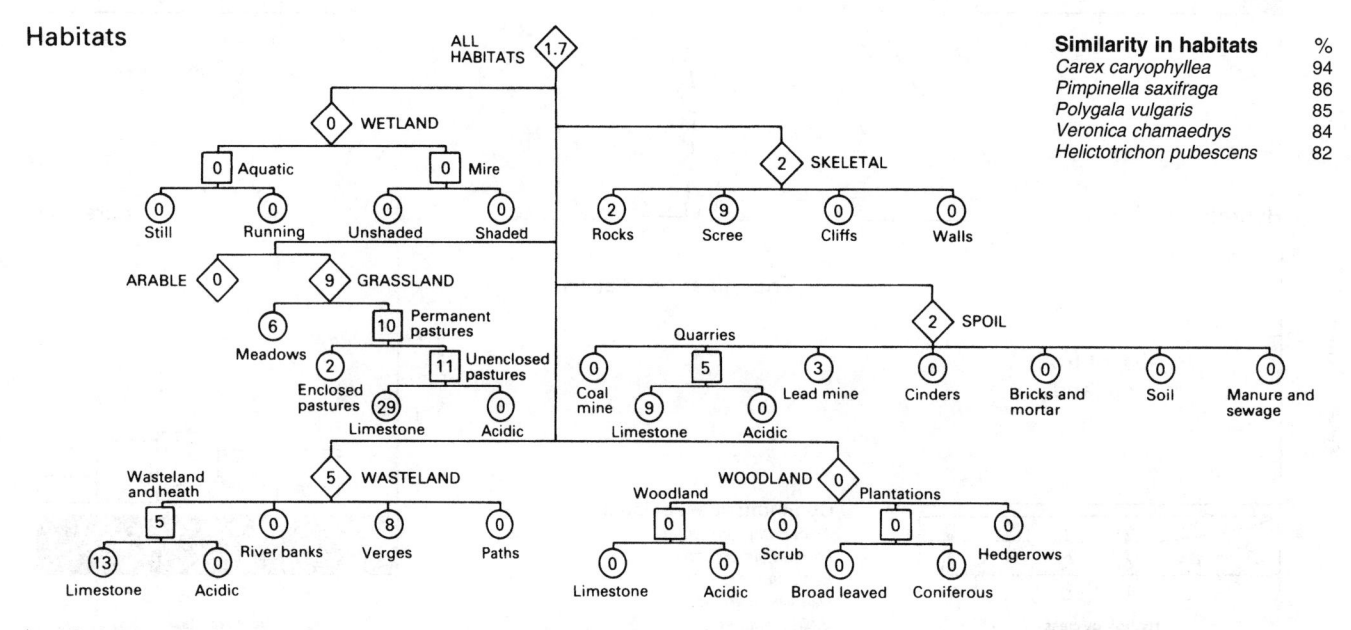

Habitats

Similarity in habitats	%
Carex caryophyllea	94
Pimpinella saxifraga	86
Polygala vulgaris	85
Veronica chamaedrys	84
Helictotrichon pubescens	82

△ **ordination** Distribution centred on vegetation of infertile, relatively undisturbed habitats. Also modest representation within vegetation of moderate productivity and subject to intermediate levels of disturbance. Absent entirely from heavily disturbed vegetation and from communities dominated by highly productive perennial species.
Associated floristic diversity Typically species-rich.
Synopsis *B.m.* is a slow-growing, winter-green grass largely restricted to unfertilized, often species-rich grassland, particularly on calcareous soils. Agricultural experience (*PGE*) suggests that *B.m.* is 'generally an indicator of poverty or exhaustion of soil and disappears when conditions are improved. It is usually increased by lime'. In turf microcosms, roots of this species became heavily colonized by VA mycorrhizas with increases of up to 4-fold in seedling yield (Grime, Mackey, Hillier & Read 1987). *B.m.* produces little foliage (Hubbard 1984), and typically in short turf the leaves are held close to the ground surface. In this respect *B.m.* resembles *Cynosurus cristatus*. The species is susceptible to submergence by litter and to shading from taller species, and is a poor competitor, failing to persist in derelict grassland. The low growth habit may afford some protection from mammalian grazing. Experiments with snails (Grime *et al.* 1968) suggest that the leaves possess no major chemical deterrent to herbivory. In keeping with the fairly high nuclear DNA amount, *B.m.* has a vernal phenology and rapid leaf extension occurs during early spring (Grime *et al.* 1985). This early growth contrasts with the delayed phenologies of many of the species, e.g. *Carex flacca*, with which *B.m.* frequently occurs. *B.m.* is significantly more abundant in N-facing grassland (*Ecol Atl*), is characteristic of moist but unproductive habitats and often extends into soligenous mire. Seeds germinate soon after release (Thompson & Grime 1979) in late summer and autumn, and the species does not develop a persistent seed bank. In marked contrast with many other grasses of calcareous habitats, the seeds and young seedlings are relatively small and the latter suffer heavy mortalities during the initial phase of establishment (Davison 1964). It is not clear whether regeneration by seed or by

Altitude

Slope

Aspect

n.s.	5.5 42.9
n.s.	4.5 40.0 Unshaded

—	0.0
—	0.0 Shaded

Soil pH

Associated floristic diversity

Hydrology

Absent from wetland

Triangular ordination

Commonest associates %
Carex flacca 76
Festuca ovina 74
Plantago lanceolata 73
Festuca rubra 72
Lotus corniculatus 66

Bare soil

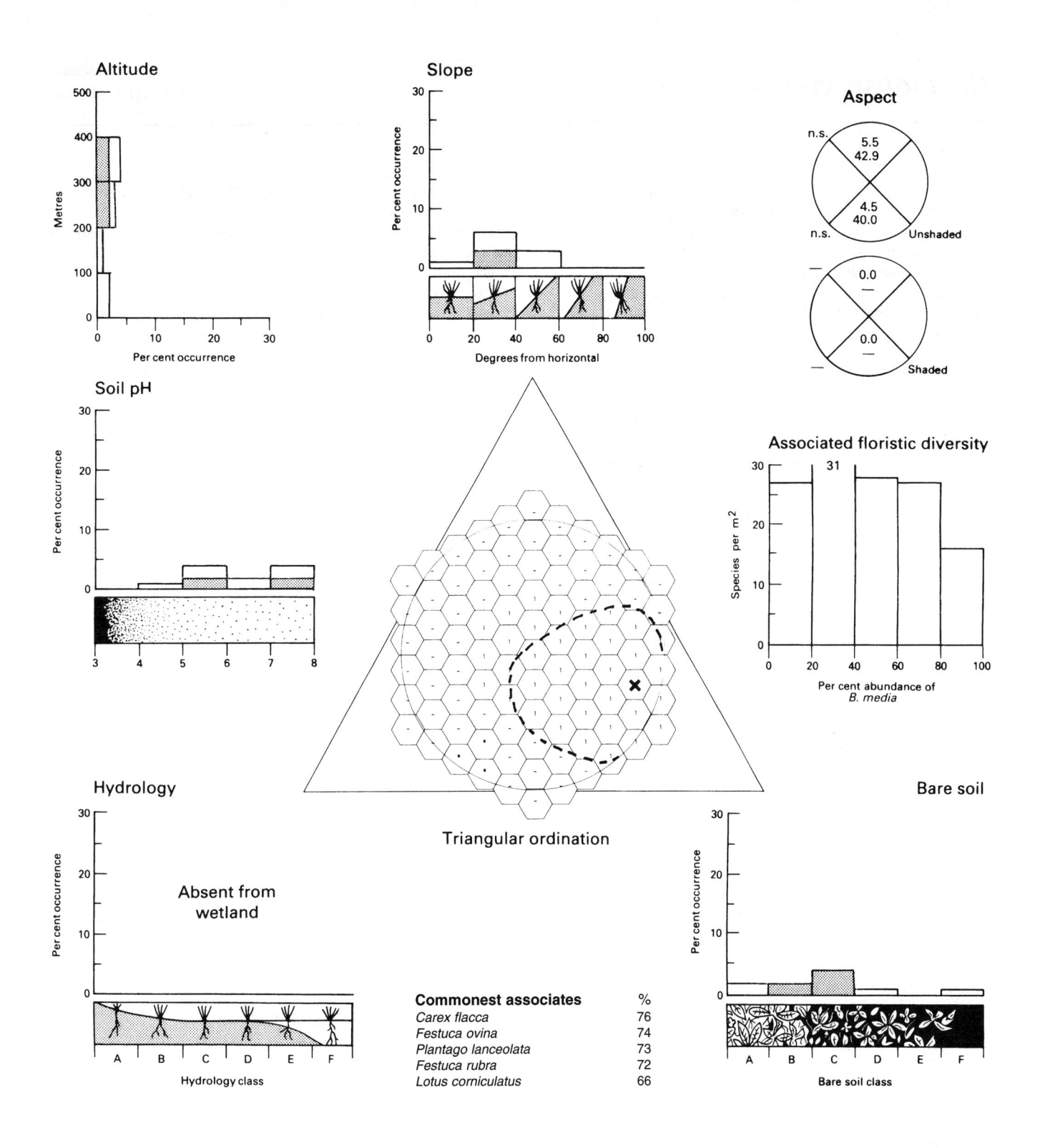

vegetative means is the more important. Cytological variation occurs with $4x$ cytotypes having a predominantly E and $2x$ plants W European distributions, a pattern which recurs in Britain (Murray 1976). Although mainly restricted to calcareous soils and to short turf, *B.m.* is wide-ranging with respect to soil type and may occur on heavy non-calcareous clays and on sandy soils. Ecotypes differing in their mineral nutrition have been identified (Cooper 1976), and it is suspected that much of the edaphic tolerance of this outbreeding species results from the formation of ecotypes. Another subspecies is reported from SE Europe (*FE5*).

Current trends *B.m. is* largely restricted to older, unproductive, managed grasslands and a further contraction in abundance of the species seems likely, particularly in lowlands. In the past *B.m.* has colonized artificial habitats such as roadsides and old quarries, but the species does not appear to be effectively colonizing recently-created environments. With the continuing fragmentation of suitable habitats the evolution of inbreeding populations of *B.m. is* predicted.

Bromopsis erecta (Huds.) Fourr.

Poaceae
Upright Brome

Life-form Densely tufted, polycarpic perennial, semi-rosette hemicryptophyte. Shoots erect; leaves coarse, hairy along margins, usually <1000 mm^2; roots with VA mycorrhizas.
Phenology Winter green with some winter growth. Flowers in June or July. Seed set in July or August and soon shed. Leaves relatively long-lived, but many die back in autumn.
Height Foliage 0-300 mm; flowering shoots to 1 m.
Seedling RGR 1.0-1.4 week^{-1}. SLA 14.0. DMC 35.3.
Nuclear DNA amount 22.6 pg; $2n$ = 42, 56 (*FE5*); $6x$, $8x$.
Established strategy Intermediate between C-S-R and stress-tolerant competitor.
Regenerative strategies S, V. Principally by seed directly in autumn; some seed may also overwinter at the soil surface. Size of clump is increased by the production of additional vegetative shoots, and old plants may in some sites form a spreading 'fairy ring' (Austin 1968); SB I.
Flowers Green, hermaphrodite, wind-pollinated; self-incompatible (Frankel & Galun 1977); >100 in a loose panicle (4-14 per spikelet).
Germinule 4.23 mg, 6.1 × 1.4 mm, caryopsis dispersed with an awned lemma attached (17.5 × 1.8 mm).
Germination Epigeal; t_{50} 2 days; 7-35°C; L = D.
Biological flora None.
Geographical distribution Common in SE Britain, rather local in the N and doubtfully native in Scotland, in C Ireland only native to central regions (50% of British vice-counties), *ABIF* page 790; S, W and C Europe northwards to Scandinavia (56% of European territories and 18% naturalized); W Africa and W Asia; introduced in N America.

Habitats Commonest on dry ungrazed semi-natural grassland on calcareous soils. Also occurs on limestone quarry heaps and pastures, and more rarely in open habitats such as paths and rock outcrops, and in hedgerows, again on calcareous soils. Absent from wetlands and arable fields. Observed more rarely as an introduction at sites on road verges over non-calcareous strata and on cindery ground beside railways; the soil at these sites may contain limestone.
Gregariousness A tussock or stand-forming species.
Altitude On the magnesian limestone, where it is frequent, *B.e.* occurs at up to 150 m, but the species is most frequent and abundant below 100 m. (Also observed rarely on the carboniferous limestone up to 300 m.) [Not found above 3 10 m.]
Slope In the Sheffield region semi-natural grassland is mainly found on valley sides. Consequently, most records are from sloping ground in the range 20-50°.
Aspect Occurrences significantly more frequent on unshaded S-facing slopes. In shaded habitats all records are from S-facing slopes.
Hydrology Not present in wetlands.

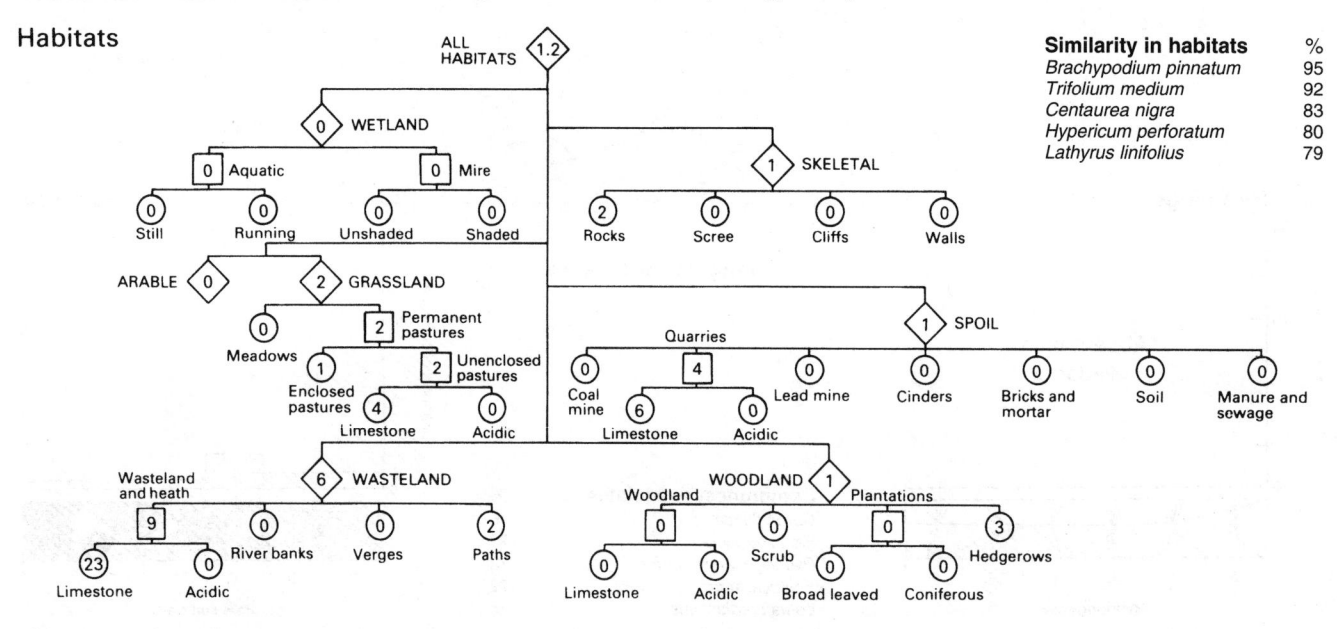

Habitats

Similarity in habitats	%
Brachypodium pinnatum	95
Trifolium medium	92
Centaurea nigra	83
Hypericum perforatum	80
Lathyrus linifolius	79

Soil pH Absent from acidic soils below pH 4.5. Most frequent and abundant in the range 7.0-8.0.
Bare soil Most strongly associated with habitats having low to intermediate exposure of bare soil, although where *B.e.* is abundant there is usually little bare soil.
△ ordination Distribution centred on vegetation subject to little or no disturbance and of intermediate productivity.
Associated floristic diversity Most frequently associated with species-rich vegetation.
Synopsis *B.e.* is a tufted winter-green grass which, like *Brachypodium pinnatum*, is largely restricted to semi-natural infertile lowland grassland on dry calcareous soils. *B.e.* is not rhizomatous and consequently is a less effective dominant than *B.p*. Despite the relatively shallow root system (Anderson 1927), *B.e.* tends to be restricted to drier, shallow soils which *B.p.* is less able to exploit. As might be predicted from its high nuclear DNA amount, *B.e.* shows rapid extension growth in early spring. The capacity to exploit dry habitats during a cool, moist period of the year (*B.e.* also grows to some extent during winter) constitutes a critical phenological difference between *B.e.* and *B.p.*. In studies of European populations (Bornkamm 1961) *B.e.* has been found to be favoured by a moist spring, but the balance in mixed communities is shifted in favour of *B.p.* if the weather is dry. More rarely, *B.e. is* a local dominant on moist N-facing slopes; its status in these sites requires investigation. Because of its morphology, *B.e.* is less protected from, and less tolerant of, burning than *B.p.* and, although unpalatable, tends to be replaced by *Festuca ovina* under heavy grazing (Hope-Simpson 1940, Willems 1985). *B.e.* regenerates effectively by seed, and appears to form large colonies only in habitats in which open areas are intermittently available for seedling establishment. Thus, although locally *B.e.* and *B.p.* appear equally well-dispersed in derelict lowland calcareous grasslands, *B.e. is* less common as a vegetation dominant, probably because

Altitude
Slope
Aspect

P<0.05
2.4 / 83.3
6.7 / 73.3
n.s. — Unshaded

— / 0.0
0.8 / 100.0
— Shaded

Soil pH

Associated floristic diversity

Hydrology

Absent from wetland

Triangular ordination

Commonest associates	%
Lotus corniculatus	100
Carex flacca	83
Centaurea nigra	83
Festuca rubra	83
Leontodon hispidus	83

Bare soil

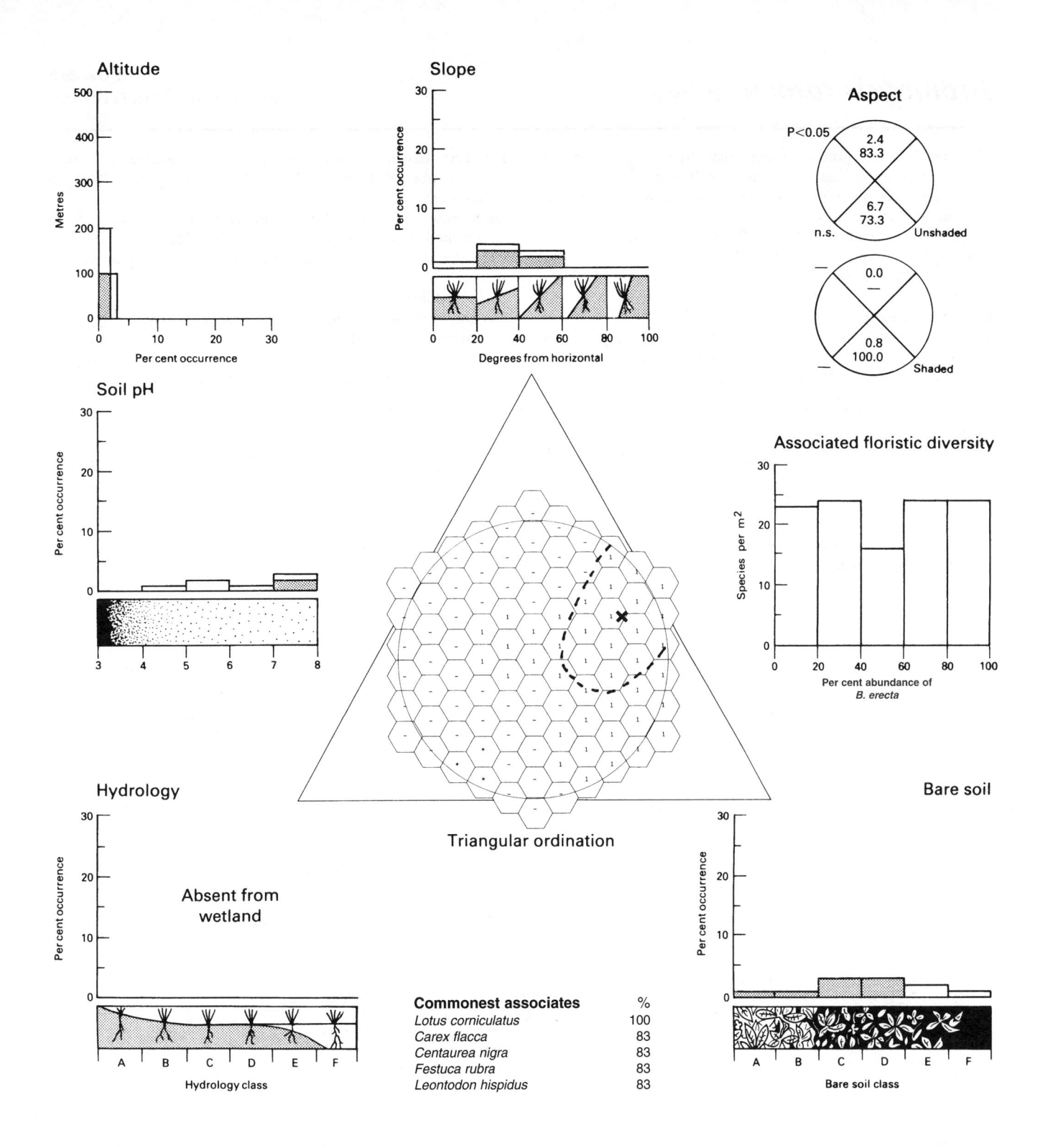

of its higher level of dependence upon regeneration by seed and its greater sensitivity to burning. A small number of native upland populations of *B.e.* occur on the carboniferous limestone where they set good seed, suggesting that restriction of *B.e.* at higher altitudes may arise from problems of seedling establishment rather than seed production. In Europe *B.e.* is more ecologically wide ranging (Austin 1968) and four ssp. are recognized (*FE5*).
Current trends Like most other species of lowland calcareous semi-natural grassland, *B.e. is* decreasing in abundance as a result of habitat destruction. However, relaxation of management has resulted in the formation of dense, species-poor swards of *B.e.* Where *Brachypodium pinnatum* is present, dereliction often results in the suppression of *B.e.* The edaphic range of *B.e.* appears to be increasing, at least temporarily, as a result of its establishment on roadsides and on railway banks on non-calcareous strata.

Bromopsis ramosa Hudson

Poaceae
Hairy Brome, Wood Brome

Life-form Tufted, polycarpic perennial protohemicryptophyte. Stem erect; leaves linear, hairy, up to *c.* 6000 mm².

Phenology Winter green. Flowers July and August and sets seed from August to October.

Height Foliage mainly <600 mm; flowers to 1400 mm.

Seedling RGR No data. SLA 24.3. DMC 25.1.

Nuclear DNA amount 24.0 pg; $2n = 14, 28, 42*$ (*BC, FE5*); $2x, 4x, 6x*$.

Established strategy C-S-R.

Regenerative strategies S. By seed in spring. Although additional tillers may be produced, plants remain tufted and compact, and there is no effective method of vegetative regeneration; SB II.

Flowers Green, hermaphrodite, wind-pollinated, self-compatible (Beddows 1931); >100 in a loose panicle (4-11 per spikelet).

Germinule 7.37 mg, 9.2×1.5 mm, caryopsis dispersed with an attached long-awned lemma (14.5×2.2 mm).

Germination Epigeal; t_{50} 2 days; 7-35°C; dormancy broken by chilling.

Biological flora None.

Geographical distribution The British Isles except parts of N Scotland (97% of British vice-counties), *ABIF* page 789; W, C and S Europe northwards to Scandinavia (67% of European territories and 3% naturalized); N Africa and temperate Asia.

Habitats Largely confined to base-rich shaded habitats, especially woodland and hedgerows. Absent from grazed grassland and arable land.

Gregariousness Sparse to intermediate.

Altitude Only recorded from <300 m. [Up to 370 m.]

Slope Wide-ranging.

Aspect Widely distributed with respect to aspect.

Hydrology Recorded from moist but not from waterlogged soils.

Soil pH Widely distributed, but mainly found at pH values >5.0.

Bare soil Over a wide range of values, but typically occurring on open ground in woodland with moderate amounts of bare soil and often some tree leaf litter and/or bryophyte mat.

△ **ordination** Excluded from highly disturbed situations. Distribution centred on vegetation of intermediate productivity and subject to low levels of disturbance.

Associated floristic diversity Typically low to intermediate.

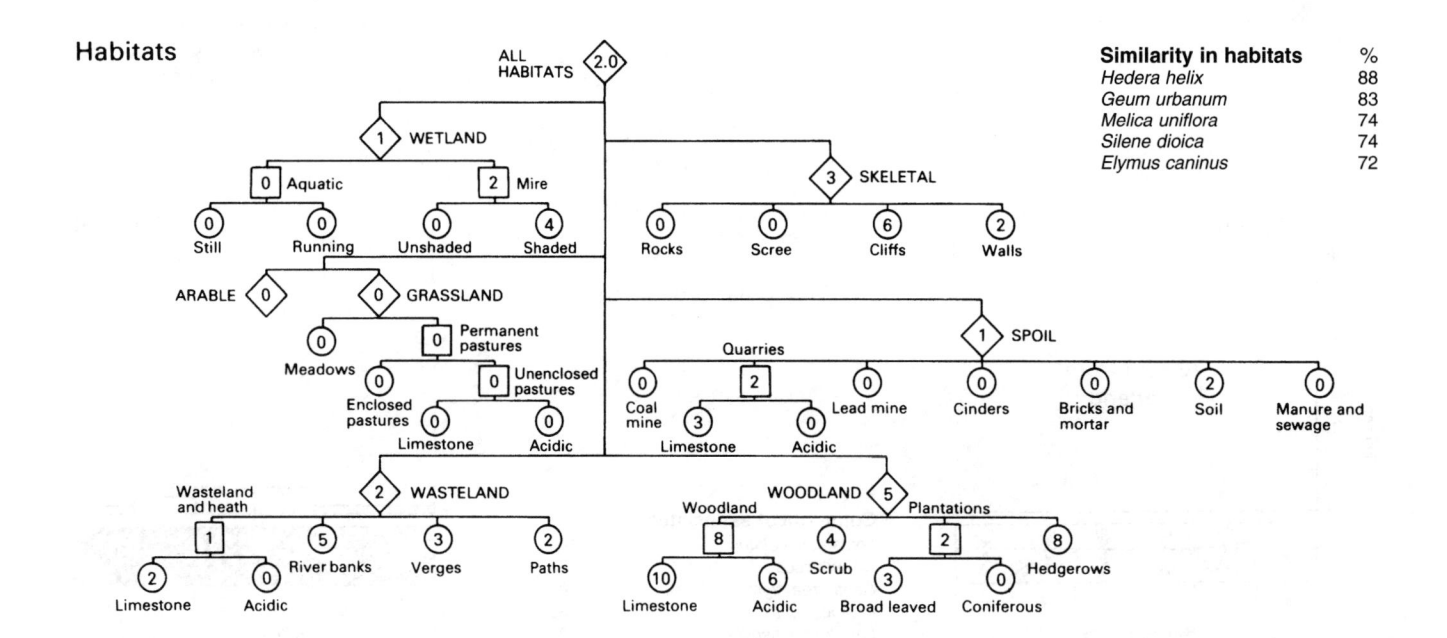

Habitats

Similarity in habitats	%
Hedera helix	88
Geum urbanum	83
Melica uniflora	74
Silene dioica	74
Elymus caninus	72

Synopsis *B.r.* is a tufted, winter-green grass of base rich lowland woodlands and has a Sub-Atlantic distribution within Europe. *B.r.* shares many attributes with *Festuca gigantea*. The species are remarkably similar in their field distributions, and frequently occur together. The main differences between the two are that *B.r.* tends to occur in drier habitats than *F.g.* and is less abundant in deep shade. Consistent with these observations *B.r.* is recorded more frequently from cliff ledges and less commonly from mire than *F.g.*, and its distribution shows a greater bias towards S-facing woodland. These differences in distribution coincide with a higher DNA amount in *B.r.*, allowing differences in shoot phenology between the two species to be predicted. The regenerative strategies of *B. r.* are also similar to those of *F.g. B.r.* shows a similar dependence on seedlings, which appear in spring and rarely establish in sites with a large accumulation of tree litter. The seed of *B.r.* is larger than that of *F.g.*, but the awn is shorter both proportionally and in absolute terms than that of *F.g.*. Nevertheless, the seed is effectively dispersed by attachment to fur or clothing, and *B.r.*, like *F.g.*, is a frequent colonist of secondary woodland in Lincolnshire (Peterken 1981). In data collected from the Sheffield region *F.g.* is recorded in 70% more quadrats than *B.r.*, and the possibility that the seed of *B.r.* is less mobile requires investigation. Both *B.r.* and *F.g.* are in urgent need of thorough ecological study.

Current trends Status uncertain but appears favoured by modern methods of management of broad-leaved woodland.

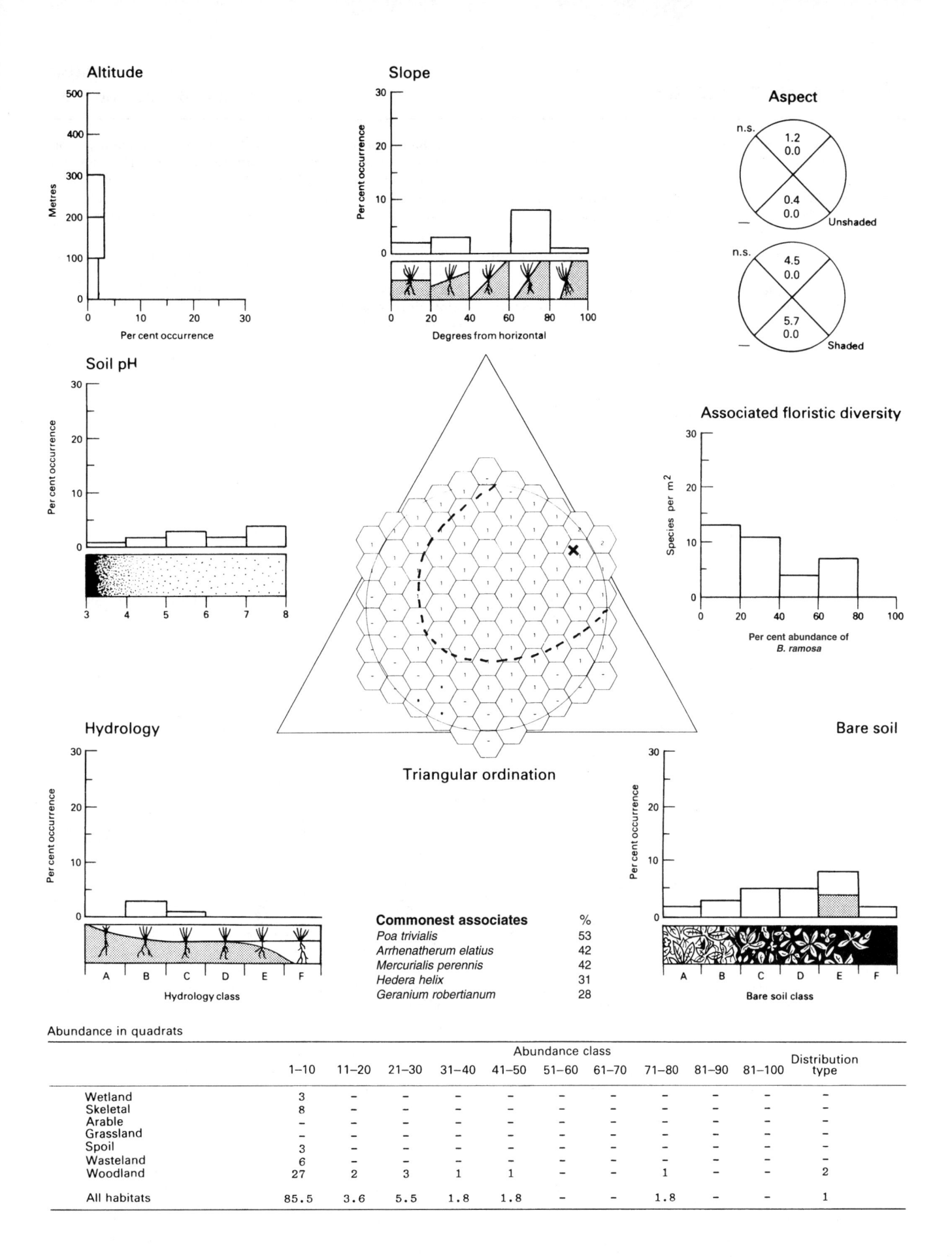

Altitude

Slope

Aspect

	n.s.	
	1.2 / 0.0	
	0.4 / 0.0	Unshaded

	n.s.	
	4.5 / 0.0	
	5.7 / 0.0	Shaded

Soil pH

Associated floristic diversity

Hydrology

Triangular ordination

Bare soil

Commonest associates	%
Poa trivialis	53
Arrhenatherum elatius	42
Mercurialis perennis	42
Hedera helix	31
Geranium robertianum	28

Abundance in quadrats

	Abundance class									Distribution type
	1–10	11–20	21–30	31–40	41–50	51–60	61–70	71–80	81–90	81–100	
Wetland	3	–	–	–	–	–	–	–	–	–	–
Skeletal	8	–	–	–	–	–	–	–	–	–	–
Arable	–	–	–	–	–	–	–	–	–	–	–
Grassland	–	–	–	–	–	–	–	–	–	–	–
Spoil	3	–	–	–	–	–	–	–	–	–	–
Wasteland	6	–	–	–	–	–	–	–	–	–	–
Woodland	27	2	3	1	1	–	–	1	–	–	2
All habitats	85.5	3.6	5.5	1.8	1.8	–	–	1.8	–	–	1

Bromus hordeaceus L.

Poaceae
Soft Brome, Lop Grass

Life-form Tufted, winter-annual therophyte. Shoots erect or spreading; leaves linear, hairy, <2000 mm^2; roots usually non-mycorrhizal.
Phenology Winter green. Vegetative growth occurs primarily in autumn and spring. Flowers from May to July and seed set from May to early August.
Height Foliage usually <400 mm overtopped by inflorescences to 800 mm.
Seedling RGR No data. SLA 22.7. DMC 26.9.
Nuclear DNA amount 18.4 pg; $2n = 28$* (Bennett & Smith 1976); $4x$*.
Established strategy Intermediate between ruderal and competitive-ruderal.

Regenerative strategies S. By freshly shed seed in autumn; perhaps a small amount of seed also germinates in spring. Does not form a persistent seed bank (Lewis 1973, Roberts 1986b); SB I.
Flowers Green, hermaphrodite, wind-pollinated, facultatively autogamous (Smith 1981); >100 in a loose panicle (6-12 per spikelet).
Germinule 2.9 mg, 6.0 × 1.6 mm, caryopsis dispersed with an attached long-awned lemma (9.0 × 5.0 mm).
Germination Epigeal; t_{50} 3 days; <5->30°C; L = D.
Biological flora None.
Geographical distribution The British Isles, but less common in the N (100% of British vice-counties), *ABIF* page 787; most of Europe (92% of European territories and 5% naturalised); Asia; introduced in N and S America.

Habitats Associated with a range of grassy and open habitats, most of which are managed or disturbed by humans. Common in meadows and more sporadic on rock outcrops (where it behaves as a winter annual and is associated with species of similar phenology and life-history), spoil tips, paths and in pastures. Occurs as a contaminant of grass seed, and is commonly observed on road verges, particularly when these have been recently sown. Absent from aquatic, mire and woodland habitats. [Occurs in sandy places near the sea and in cliff-top turf.]
Gregariousness Intermediate.
Altitude No suitable habitats for *B.h.* exist at high altitudes within the region, and the species is not recorded above 400 m. [Up to 550 m; unclear whether this represents an established or casual occurrence.]
Slope Found in habitats of level ground (e.g. meadows) and of moderate slope (outcrops).
Aspect Has a significant bias towards S-facing slopes, consistent with the tendency to behave as a winter annual.
Hydrology Absent from wetland habitats.
Soil pH Absent from acidic soils of pH <5.0.
Bare soil Associated with a wide range of values. In its meadow habitats the proportion of bare ground is frequently low; on outcrops and in other disturbed habitats it is often high.
△ ordination Distribution confined to vegetation subject to some disturbance but of varying productivity.
Associated floristic diversity Usually intermediate to species-rich.

Similarity in habitats	%
Rhinanthus minor	93
Phleum pratense	78
Festuca pratensis	74
Alopecurus pratensis	73
Trifolium pratense	72

Habitats

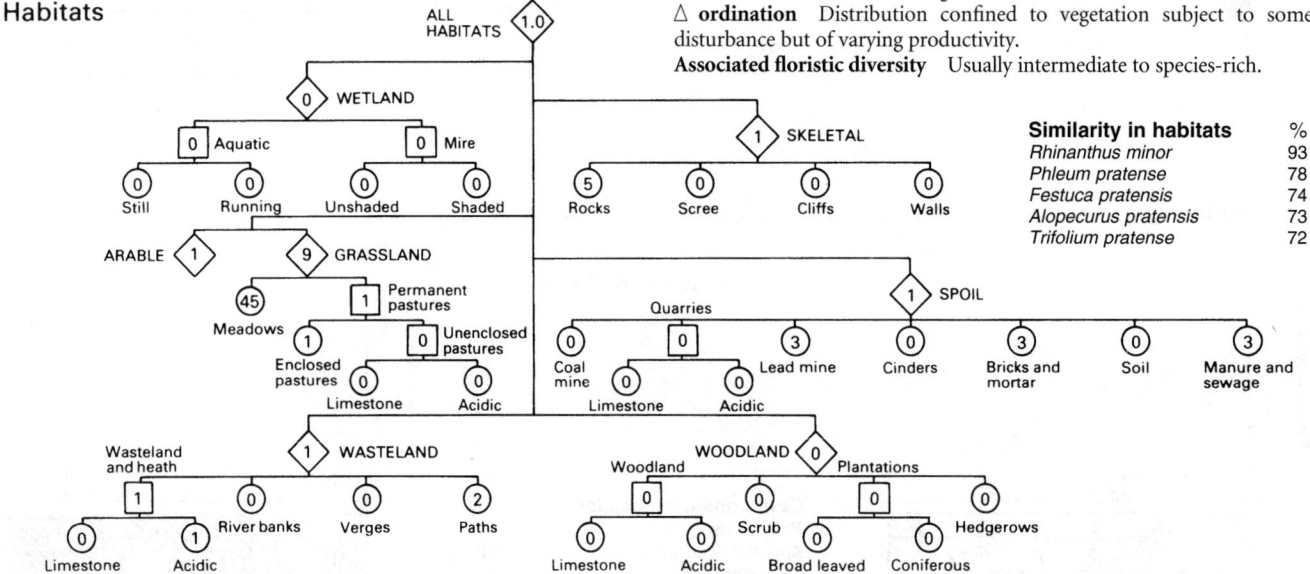

Synopsis *B.h.* is a tufted winter-annual grass of hay meadows, roadsides, waste places and rock outcrops, particularly on moderately fertile soils. *B.h.* is very much a 'follower of humans' and is often dispersed as an impurity in grass seed (Smith 1968). In the *PGE*, *B.h.* was increased by applications of sodium nitrate and minerals but declined in plots where the soils were untreated or received ammonium salts. Brenchley & Warington (1958) also record that the species 'shows a specially close connection with certain Leguminosae, notably *Lathyrus pratensis*'. *B.h.* is dependent upon seed for regeneration, and no persistent seed bank has been reported (Lewis 1973, Roberts 1986b). *B.h.* is particularly susceptible to heavy grazing, frequent mowing and trampling, and seldom persists in pastures or lawns. Seeds ripen in midsummer at or slightly before the time at which hay meadows are cropped. The seeds, which have no dormancy, germinate equally rapidly in darkness or in light, and under constant or fluctuating temperatures. The caryopses are large, and germination occurs underneath the grass canopy during summer and autumn. *B.h. is* phenotypically highly plastic; robust plants produce many flowering culms, but in unfavourable sites an individual may bear only a solitary spikelet with just six florets. *B.h.* is also genotypically variable, but much of its variation appears to stem from hybridization and introgression with *B. lepidus* Holmberg (Stace 1975), producing the hybrid, *B.* × *pseudothominei* P. M. Sm. In upland areas of the Sheffield region the smaller-seeded *B.* × *p.* shows little persistence in older hay meadows, and is found at higher altitudes only as a rarity on rock outcrops. In lowland areas, however, *B.* × *p.* is well established, particularly in dry, open sites, and *B.h.* var. *leiostachys* Hartm. with hairless spikelets is also relatively frequent. In Britain hybridization may be less frequent than in much of Europe, where self-fertilization is less common (Stace 1975). Various ecotypes, which do not occur in the Sheffield region, have been given subspecific rank (Smith 1981). Ssp. *thominei* (Hard.) Braun-Blanq. includes dwarf sand-dune populations, ssp. *ferronii* (Mabille) P. M. Sm. occurs on cliff tops in S England, and ssp. *divaricartus* (Bonnier & Layens) Kerquelen. is a casual from S Europe.
Current trends Uncertain. In lowland areas *B.* × *p.* may be increasing relative to *B.h.*, and the ecological differences between the two taxa are becoming increasingly obscure

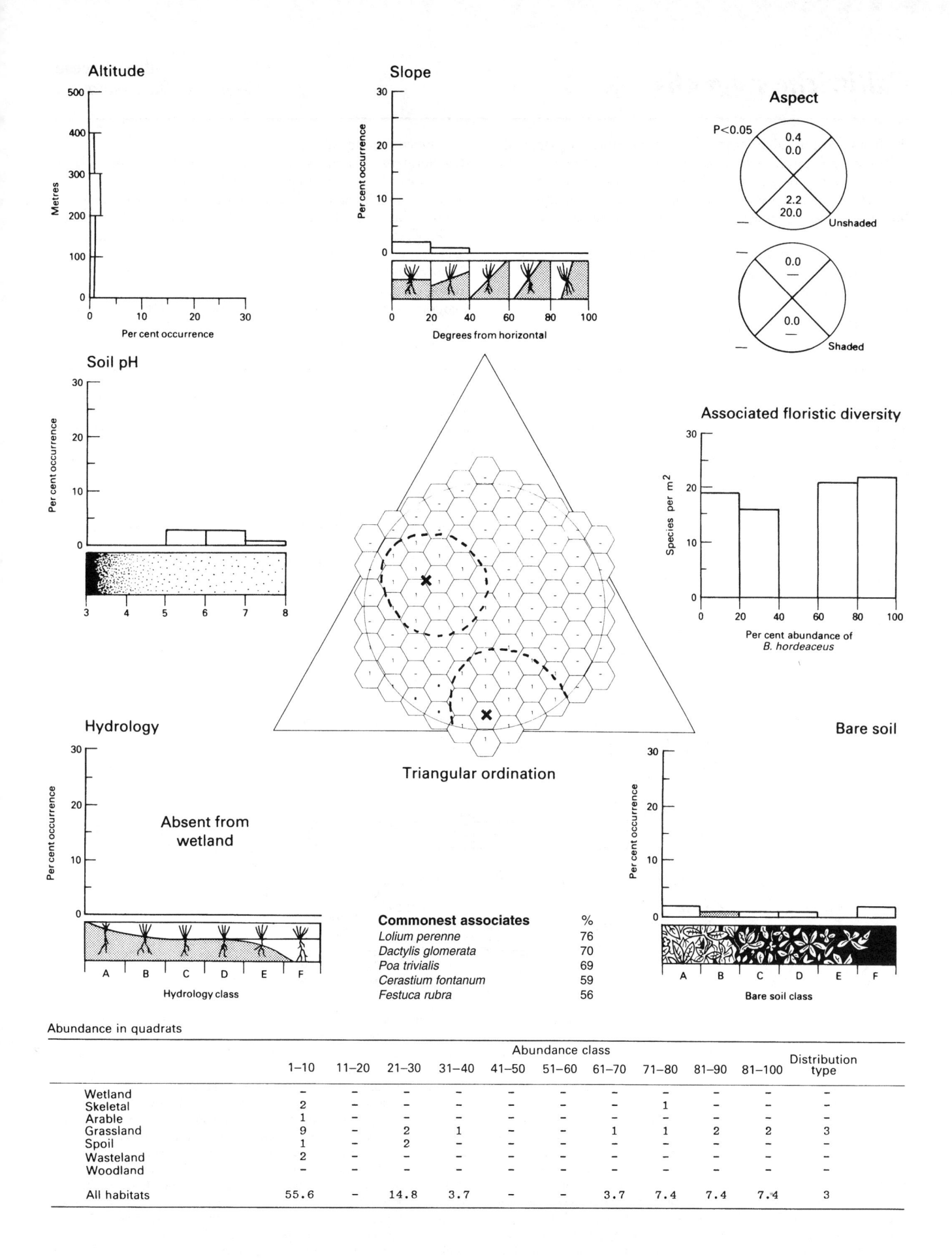

Altitude

Slope

Aspect

$P<0.05$

Unshaded: 0.4 / 0.0 / 2.2 / 20.0

Shaded: — / 0.0 / 0.0 / —

Soil pH

Associated floristic diversity

Hydrology

Absent from wetland

Triangular ordination

Commonest associates	%
Lolium perenne	76
Dactylis glomerata	70
Poa trivialis	69
Cerastium fontanum	59
Festuca rubra	56

Bare soil

Abundance in quadrats

	1–10	11–20	21–30	31–40	41–50	51–60	61–70	71–80	81–90	81–100	Distribution type
Wetland	–	–	–	–	–	–	–	–	–	–	–
Skeletal	2	–	–	–	–	–	–	1	–	–	–
Arable	1	–	–	–	–	–	–	–	–	–	–
Grassland	9	–	2	1	–	–	1	1	2	2	3
Spoil	1	–	2	–	–	–	–	–	–	–	–
Wasteland	2	–	–	–	–	–	–	–	–	–	–
Woodland	–	–	–	–	–	–	–	–	–	–	–
All habitats	55.6	–	14.8	3.7	–	–	3.7	7.4	7.4	7.4	3

Callitriche stagnalis Scop.

Callitrichaceae
Common Water-starwort

Life-form Annual therophyte, or perhaps more frequently in Britain, polycarpic perennial helophyte or hydrophyte with erect floating or submerged stems in aquatic habitats and prostrate creeping stems on mud; leaves ovate, <250 mm².

Phenology Winter green. Flowers May to September; seed set July onwards.

Height Foliage and flowers 0-10 mm, borne on prostrate stems in terrestrial forms; in aquatic habitats at or below surface of water (stems to 600 mm).

Seedling RGR No data. SLA 42.6. DMC 11.3.

Nuclear DNA amount 2.5 pg; $2n = 10^*$ (BC); $2x^*$.

Established strategy Intermediate between ruderal and competitive-ruderal.

Regenerative strategies (V), ?B_s. Mainly vegetative (see 'Synopsis'). Studies of regeneration by seed are required.

Flowers Green, monoecious, protandrous, perhaps pollinated by wind, but possibly also by water, or perhaps self-pollinated (Philbrick 1984); solitary or one of each sex in axils of leaves.

Germinule 0.09 mg; 1.5×0.5 mm, dry fruit (4 per ovary).

Germination Epigeal; t_{50} 4 days.

Biological flora None.

Geographical distribution The British Isles (100% of British vice-counties), *ABIF* page 532; Europe (82% of European territories); N Africa.

Habitats Confined to waterlogged or submerged habitats. Most common in stagnant or flowing water, but also frequent both in unshaded and in shaded mire. Impermanent in disturbed terrestrial habitats such as soil heaps and arable fields close to water.

Gregariousness Potentially patch-forming.

Altitude Found over a wide range of altitudes to 470 m. [Up to 890 m.]

Slope As with most wetland species, *C.s.* is generally associated with flat or gently sloping ground.

Aspect No data.

Hydrology Occurs in a range of wetland habitats. *C.s.* is widespread as a floating or submerged aquatic, particularly in ditches, rooted in shallow water of <250 mm or more rarely associated with deeper water to 1 m. Not recorded from fast-flowing water. Found also in both soligenous and topogenous mire.

Soil pH Widely distributed on soils of pH >4.0, but most commonly recorded from neutral soils in the pH range 6.5-7.5.

Bare soil Terrestrial populations are generally associated with moderate to high exposures of bare soil.

△ ordination Most frequently associated with relatively productive vegetation subject to some disturbance.

Associated floristic diversity Relatively low and usually declining further at high frequencies of *C.s.*, particularly where these are drawn from dense submerged stands in aquatic systems.

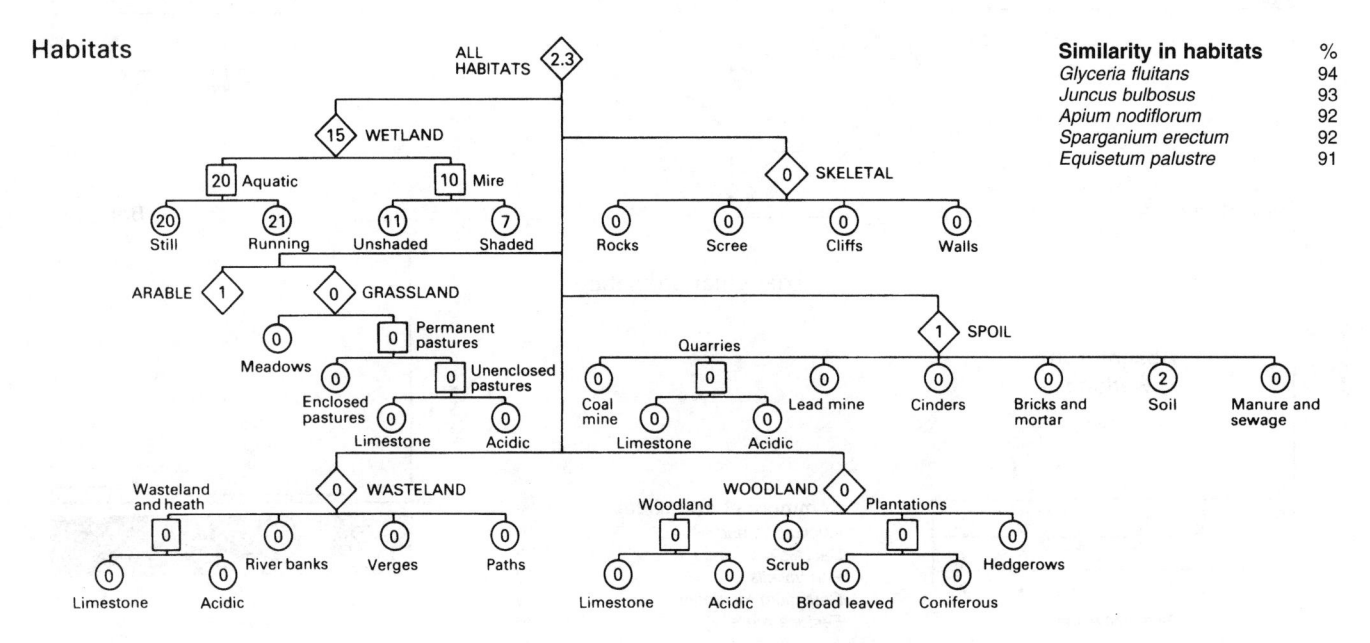

Habitats

Similarity in habitats	%
Glyceria fluitans	94
Juncus bulbosus	93
Apium nodiflorum	92
Sparganium erectum	92
Equisetum palustre	91

ALL HABITATS 2.3
WETLAND 15
Aquatic 20, Mire 10
Still 20, Running 21, Unshaded 11, Shaded 7
SKELETAL 0
Rocks 0, Scree 0, Cliffs 0, Walls 0
ARABLE 1
GRASSLAND 0
Permanent pastures 0
Meadows 0, Unenclosed pastures 0
Enclosed pastures 0
Limestone 0, Acidic 0
SPOIL 1
Quarries 0
Coal mine 0, Limestone 0, Acidic 0, Lead mine 0, Cinders 0, Bricks and mortar 0, Soil 2, Manure and sewage 0
WASTELAND 0
Wasteland and heath 0
Limestone 0, Acidic 0, River banks 0, Verges 0, Paths 0
WOODLAND 0
Woodland 0, Plantations 0
Limestone 0, Acidic 0, Scrub 0, Broad leaved 0, Coniferous 0, Hedgerows 0

Synopsis *C.s.* is a patch-forming herb associated with relatively fertile wetland habitats. In shallow ditches and muddy cart-tracks *C.s.* occurs with species such as *Juncus bufonius* and in soligenous mire with *Montia fontana*. The species also occurs in aquatic habitats (deeper ditches and ponds) typically in the shallows but sometimes in water up to 1 m deep. In mire and in sites normally flooded only in winter *C.s.* is a small, prostrate plant with short leaves and internodes, while in aquatic situations the species produces long leafy stems, often terminating in a floating leaf rosette. In both types of site *C.s.* often displays considerable resilience, populations recovering from major impacts of disturbance whether mechanical or resulting from large fluctuations in watertable. Thus, in aquatic systems *C.s.* remains abundant in large drainage ditches which are regularly cleared of foliage. Like many other aquatic macrophytes, *C.s.* is heterophyllous. The first leaves formed are narrowly elliptical and are often short-lived. Subsequent leaves are ovate and the terminal ones, in the leaf rosette, may be almost circular. The form of these mature leaves is not greatly altered by immersion or emergence. Even by the standard of aquatic plants (Sculthorpe 1967), the vascular tissue of *C.s.* is very much reduced and terrestrial phenotypes are persistent only in very wet habitats. The apparent absence of epidermal chloroplasts in *Callitriche is* unusual amongst aquatic macrophytes (Sculthorpe 1967). *C.s.* is wide-ranging with respect to soil type, and a degree of salt tolerance is evident (Schotsman 1958). The species is most frequent on non-calcareous soils and may achieve abundance with *Juncus bulbosus* and *Potamogeton polygonifolius* Pourr. in peaty drainage ditches. Regeneration is mainly by vegetative means. Stems readily root on contact with mud, and clonal patches are formed in both terrestrial and aquatic systems. Detached shoots readily regenerate to form new plants. Seed production does not occur in submerged plants (*FE3*), and appears to be greater in plants growing on mud than in aquatic specimens (with floating leaves). We suspect that

Altitude
Slope
Aspect
Soil pH
Associated floristic diversity
Hydrology
Triangular ordination
Bare soil

Commonest associates	%
Agrostis stolonifera	41
Poa trivialis	38
Holcus lanatus	31
Juncus articulatus	31
Juncus effusus	31

Abundance in quadrats

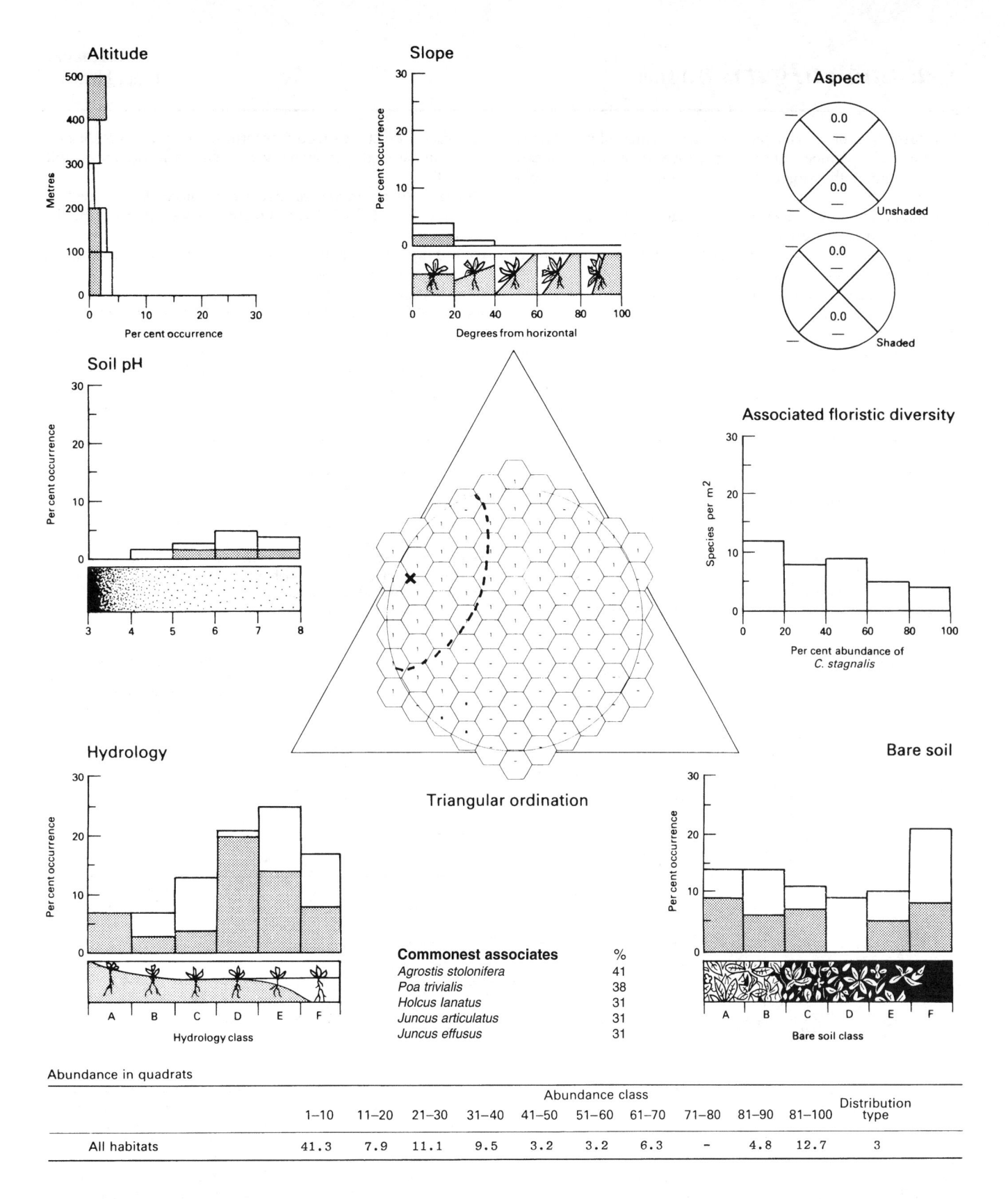

	1–10	11–20	21–30	31–40	41–50	51–60	61–70	71–80	81–90	81–100	Distribution type
All habitats	41.3	7.9	11.1	9.5	3.2	3.2	6.3	–	4.8	12.7	3

further research will confirm the existence of a persistent seed bank, and on the basis of laboratory responses it seems likely that germination will usually occur in spring. Flowers are perhaps typically wind-pollinated, but a remarkable form of self-pollination, where the pollen tube grows from the anther of the male flower through the vegetative plant to a female flower, has been described for several species of *Callitriche* (Philbrick 1984). *C.s.* is replaced in relatively base-poor (often peaty) aquatic habitats particularly in upland areas by *C. hamulata* Kutz. ex Koch ($2n = 38$; $8x$) and in lowland water enriched by agricultural run-off by *C. platycarpa* Kutz. ($2n = 20$; $4x$). Both species are European endemics and are more consistently aquatic than *C. stagnalis*.

Current trends Apparently a highly mobile species of disturbed fertile wetlands. Not under threat and perhaps even increasing.

Calluna vulgaris (L.) Hull

Ericaceae
Ling, Heather

Life-form Shrubby chamaephyte or phanerophyte. Stems branched, decumbent or ascending; leaves xeromorphic, linear, <10 mm², with margins strongly revolute; roots with ericoid mycorrhizas.

Phenology Evergreen. Flowers from August to September. Seed shed from September onwards and some may overwinter on the plant. Leaves on the main shoot survive for 1 year, those on short lateral shoots up to 3 years (*Biol Fl*).

Height Foliage and flowers to 800 mm.

Seedling RGR <0.5 week^{-1}. SLA 13.2. DMC No data.

Nuclear DNA amount No data; $2n = 16$ (*FE3*); $2x$.

Established strategy Intermediate between stress-tolerator and stress-tolerant competitor.

Regenerative strategies B$_s$. A large and persistent bank of buried seed is formed. Large populations of seedlings appear after fires (*Biol Fl*). Prostrate shoots may form adventitious roots (*Biol Fl*), but in the Sheffield region *C.v.* typically has an ascending habit; SB IV.

Flowers Purple, hermaphrodite, weakly protandrous, insect- or wind-pollinated (*Biol Fl*); numerous flowers in a raceme-like inflorescence.

Germinule 0.03 mg, 0.7 × 0.5 mm, seed in a dehiscent capsule containing up to 32 seeds.

Germination Epigeal; t_{50} 14 days; 10-28°C; L > D.

Biological flora Gimingham (1960), see also Gimingham (1972).

Geographical distribution Widespread in the British Isles, but scarce in E and C England (100% of British vice-counties), *ABIF* page 291; most common in N and W Europe, rare in the Mediterranean region and the SE (77% of European territories); eastern N America, rare and probably introduced.

Habitats Abundant in acidic pastures, on heathland and moorland, in sandstone quarries, and on outcrops and cliffs. Occasional in lightly shaded scrub and unshaded mire and beside paths, rivers and roads. Also recorded from walls and from lead-mine and other toxic metal workings on acidic soils. There are isolated records from species-rich calcareous grasslands. [Prominent in a range of montane habitats, particularly on heaths and in bogs, but also in sites of high exposure on mountains and on scree slopes (*Biol Fl*). Occurs on fixed sand dunes and sea heaths.]

Gregariousness Produces large spreading plants, which may exceed 0.5 m² in diameter, and is a stand-forming species. High scores for frequency usually refer to seedlings and young plants.

Altitude Recorded at all altitudes, but because of the destruction of lowland heaths the records are strongly biased towards upland habitats. [Up to 1100 m (*Biol Fl*).]

Slope Recorded at all angles, but abundant only on shallower slopes.

Aspect In unshaded habitats slightly more frequent and abundant on S-facing slopes; in shaded situations there is a N-facing bias. The significance of this distribution pattern is obscure.

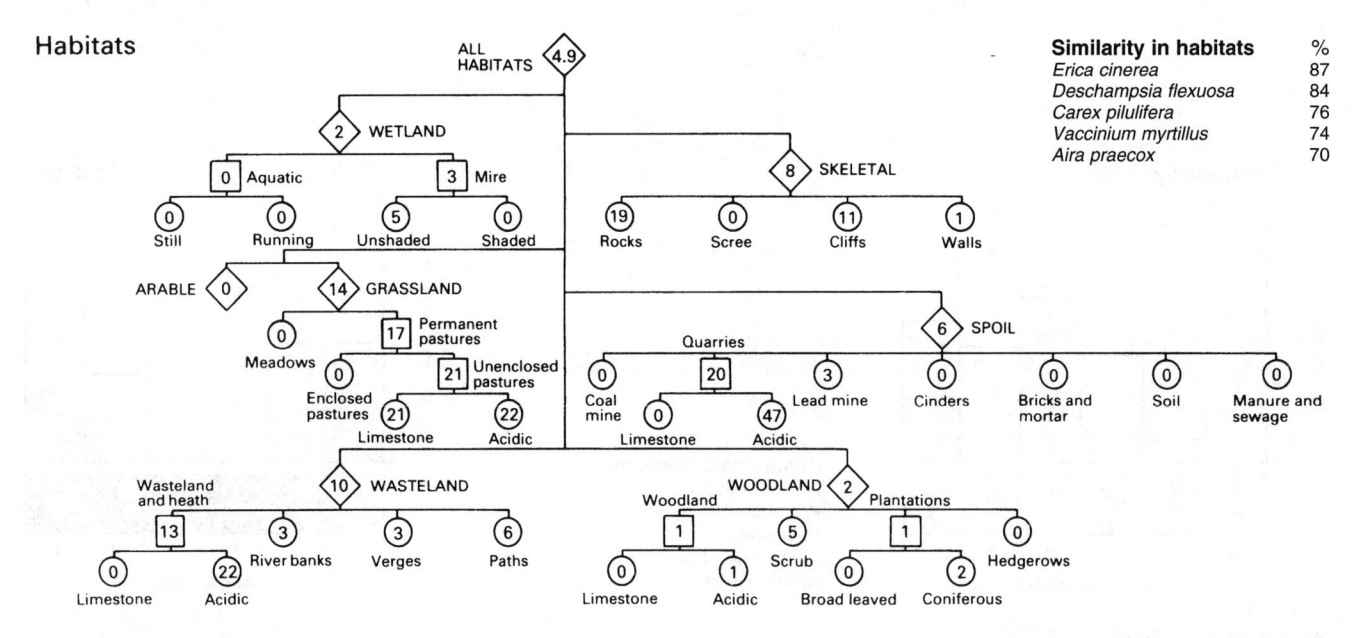

Habitats

Similarity in habitats	%
Erica cinerea	87
Deschampsia flexuosa	84
Carex pilulifera	76
Vaccinium myrtillus	74
Aira praecox	70

Hydrology Not typically a wetland species, but often found growing on tussocks of other species in ombrogenous mire and at the edges of soligenous mire.

Soil pH Most frequent and abundant in strongly acidic habitats and seldom recorded at pH >5.0. There are isolated occurrences from highly calcareous, N-facing slopes on the carboniferous limestone.

Bare soil Present over a wide range of values.

Δ ordination Distribution centred on relatively unproductive and undisturbed vegetation types. Absent from heavily disturbed sites.

Associated floristic diversity Usually low.

Synopsis *C.v.* is a relatively short-lived (normally under 30 years (*Biol Fl*)), evergreen shrub found in a wide range of habitats on acidic, usually well-drained soils. It is by far the most important species of heathland in Britain. *C.v.* is a valuable food plant for hill sheep and grouse, particularly during the first 15-20 years of its life, when its canopy is increasing (*Biol Fl*). Leaves are low in P and Ca, but often contain relatively high amounts of Mn and Al (*SIEF*). Later, the major shoot branches tend to spread apart, leaving a gap in the canopy above the centre of the plant, and 'leggy' older plants are little-grazed (*Biol Fl*). Controlled burning, preferably in spring, is used to rejuvenate older plants which subsequently resprout, particularly if <15 years old (*Biol Fl*). However, *C.v.* is killed by fierce fire and is sensitive to overgrazing. *C.v.* is sometimes regarded as a weed of forestry, since its presence restricts the growth and establishment of some coniferous species. *C.v.* shows physiological adaptations to an oceanic type of climate (Gimingham 1972). *C.v.* is potentially taller than most heathland species, has a dense canopy during much of its life and forms persistent litter. In consequence, sites with vigorous, mature *C.v.* are often virtually devoid of other herbaceous species. *C.v.* also has some more

Altitude

Slope

Aspect

	n.s.	7.9	
		50.0	
	n.s.	11.7	Unshaded
		53.8	

	n.s.	2.3	
		0.0	
	—	0.8	Shaded
		0.0	

Soil pH

Associated floristic diversity

Triangular ordination

Hydrology

Commonest associates %
Deschampsia flexuosa 87
Vaccinium myrtillus 38
Festuca ovina 27
Agrostis capillaris 26
Nardus stricta 23

Bare soil

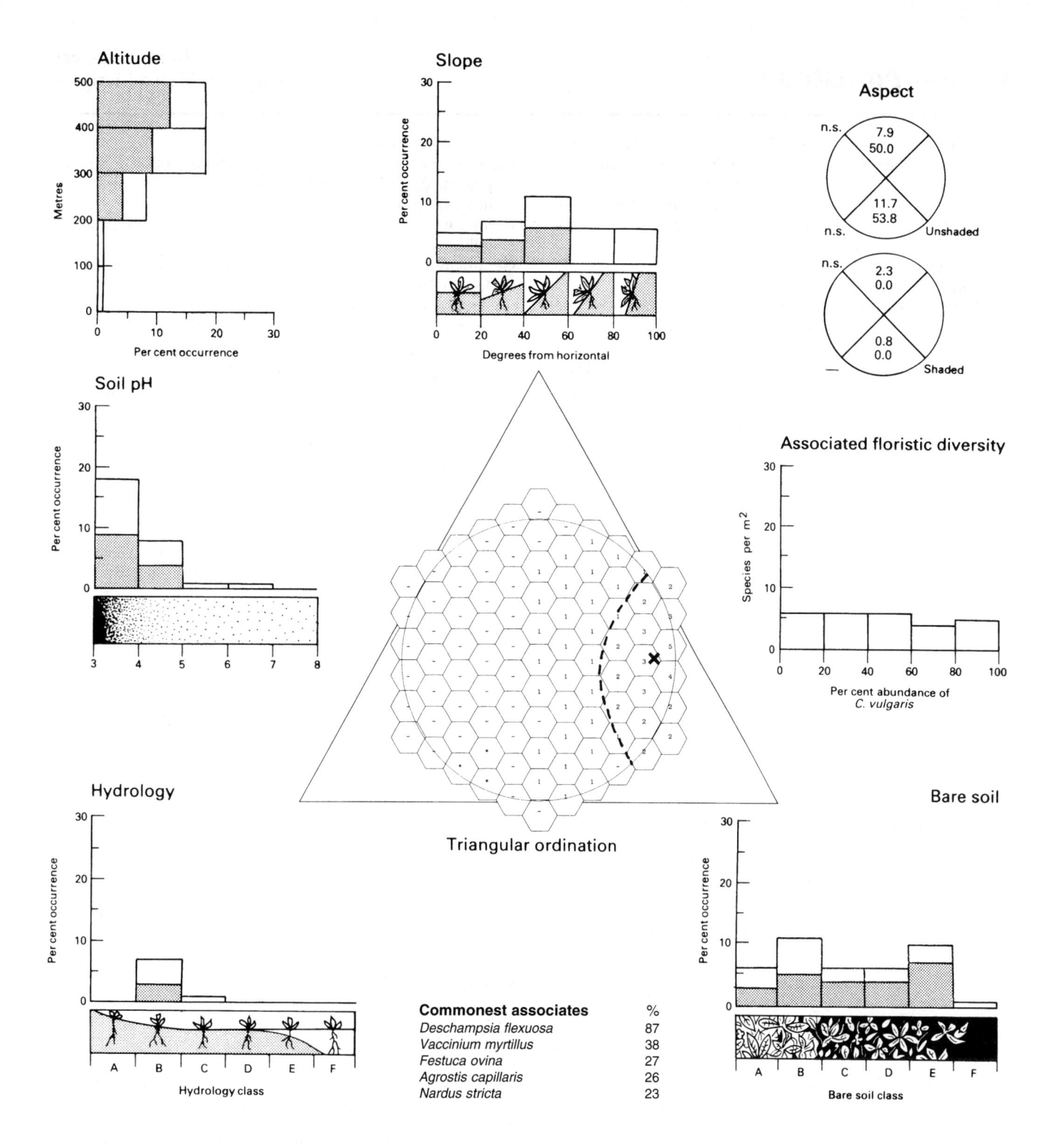

ruderal characteristics: the life-span is relatively short for a woody species, there is prodigious seed output and under favourable conditions flowering may occur in the second year. *C.v.* produces numerous minute seeds (up to 1 million in^{-2} (*Biol Fl*)). These vary in morphology and germination characteristics (Helpsper & Klerken 1984), and may be incorporated into a buried seed bank in which they are capable of long-term survival (Odum 1978), eventually germinating at any time from spring until autumn. Seedling establishment appears to depend upon the creation of bare ground. Seeds may be wind-dispersed (*Biol Fl*). The growth form shows a degree of plasticity (e.g. dwarf prostrate forms occur on exposed sites). Some populations exhibit pronounced genotypic variation in morphology (*Biol Fl*). On acidic soils contaminated by lead and other heavy metals *C.v.* may be virtually the only colonist. Survival here depends heavily upon the capacity of the mycorrhizal associate to bind heavy metals rather than the tolerance of *C.v.* itself (Bradley *et al.* 1982). The soil around the roots is frequently more acidic than the surrounding soil (Grubb & Suter 1971). The mechanism of this acidification is examined by Read and Bajwa (1985).

Current trends Many lowland heaths in Britain have now been destroyed. In some upland areas *C.v.* has decreased through the conversion of moorland to commercial forests, and locally through overgrazing and other practices. Thus, although still an abundant upland species, *C.v.* is decreasing in its original habitats. However, the species appears to be extending its range into acidic artificial habitats such as roadsides and a variety of types of spoil.

Caltha palustris L.

Ranunculaceae
Kingcup, Marsh Marigold

Life-form Polycarpic perennial, semi-rosette helophyte with creeping rhizome. Shoots ascending; leaves rounded, sometimes >6000 mm²; roots with VA mycorrhizas.

Phenology New shoots elongate in early spring. Flowers from April to June and seed shed gradually in July and August. Leaves relatively long-lived.

Height Foliage and flowers to 300 mm.

Seedling RGR No data. SLA 19.6. DMC 16.6.

Nuclear DNA amount 33.0 pg; $2n = 16, 24, 32, 48, 52\text{-}64^*$ (*BC*, Kootin-Sanwu & Woodell 1970); $2x, 3x, 4x, 6x, 7x^*, 8x^*$.

Established strategy C-S-R.

Regenerative strategies V, ?S. Regenerates both vegetatively and by seed (see 'Synopsis'); ?SB II.

Flowers Yellow, hermaphrodite, homogamous, insect-pollinated, self-incompatible (Kootin-Sanwu & Woodell 1971); in a few-flowered cymose panicle.

Germinule 0.99 mg, 2.5×1.3 mm, seed in a dehiscent *c.* 8 seeded follicle; *c.* 7 follicles per fruit.

Germination Epigeal; t_{50} 12 days; dormancy broken by chilling.

Biological flora None.

Geographical distribution The British Isles (99% of British vice-counties), *ABIF* page 95; Europe, although rare in the Mediterranean (79% of European territories); temperate and Arctic Asia and N America.

Habitats Exclusively a wetland species. Most frequent in mire, particularly in shaded habitats. Also recorded from river banks, ditches and walls adjoining open water, and observed in wet pastures.

Gregariousness Usually occurring as isolated individuals.

Altitude Mainly recorded from lowland areas, where suitable sites are most frequent, but observed to 400 m. [Up to 1100 m.]

Slope As with most wetland species, most records are from relatively flat ground.

Aspect Insufficient data.

Hydrology Recorded commonly in topogenous mire, often in sites marginal to open water and with winter flooding. Also frequent in soligenous mire. Restricted to shallow water (60 mm) and not found in permanently flooded sites.

Soil pH Not observed from acidic soils below pH 4.5 and most frequent on soils of pH 6.0-7.0.

Bare soil Recorded from a wide range of conditions, but most frequent in sites with high exposures.

Δ ordination Absent from vegetation which is relatively unproductive and also from heavily disturbed communities, otherwise present in a rather wide range of vegetation types.

Associated floristic diversity Often in species-poor communities but may occur in soligenous mire with as many as 38 species m⁻².

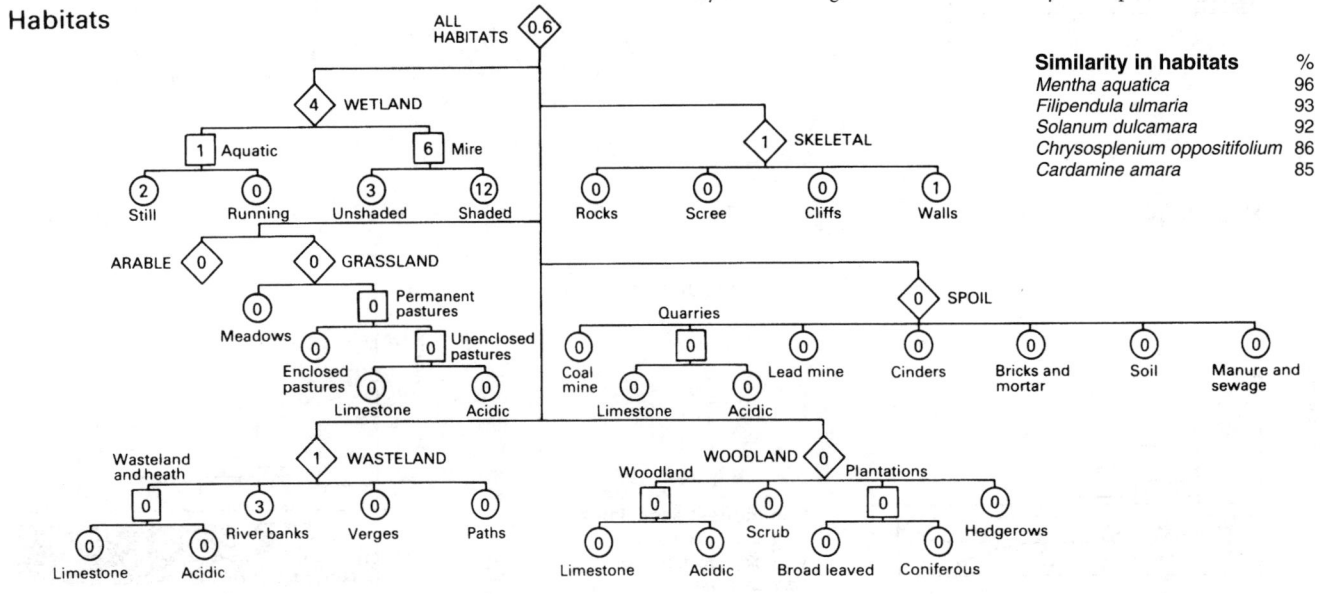

Habitats

Similarity in habitats	%
Mentha aquatica	96
Filipendula ulmaria	93
Solanum dulcamara	92
Chrysosplenium oppositifolium	86
Cardamine amara	85

Synopsis *C.p.* is a wide-ranging, wetland herb with a variable but restricted capacity for vegetative spread. Individual plants appear to be capable of surviving for at least 50 years (Panigrahi 1955). The species is particularly characteristic of topogenous mire on moderately fertile soils in more-open, tall-herb communities in sites where the vigour of potential dominants is restricted by summer shade and winter flooding. *C.p.* also occurs on river banks close to the water level, a habitat which is frequently too disturbed to allow consolidated growth of stand-forming dominants. With increasing altitude *C.p.* becomes particularly frequent in unshaded situations; the species extends into montane habitats and, in Europe, into Arctic regions (see Panigrahi 1955). Consistent with the high DNA value, shoot extension commences in late winter and plants produce leaves and flowers in early spring before the expansion of deciduous tree canopies. As in other early flowering species of shaded wetlands, such as (*Cardamine amara* and *Chrysosplenium oppositifolium*), the leaves of *C.p.* persist throughout the summer; they are characterized by high concentrations of N and Na, and low levels of P and Fe (*SIEF*). *C.p.* occurs in grazed sites and contains protoanemonin, which is toxic to stock (Cooper & Johnson 1984). Shoots detached during disturbance are capable of regeneration. Although typically plants occur in discrete clumps, forms with prostrate stems rooting at the nodes are also observed, particularly in upland areas (var. *radicans* (Forst.) Beck) (Woodell & Kootin-Sanwu 1971). Viable seed is produced regularly and has a well-defined chilling requirement; germination occurs in spring. It is uncertain whether seeds persist in the soil, but it is known that they may float for up to 4 weeks (Ridley 1930), and widespread dispersal by water is possible. Although *C.p.* is variable in the field in Europe and N America (Woodell & Kootin-Sanwu 1971), there are no major morphological discontinuities between British populations, and many differences between populations disappear on cultivation. *C.p.* is represented by a polyploid complex with in addition aneuploidy, but chromosome races are not morphologically distinct (Kootin-Sanwu & Woodell 1971). Plants with $2n = 56$ are by far the commonest in Britain (Kootin-Sanwu & Woodell 1971).

Current trends Decreasing with the destruction of wetlands.

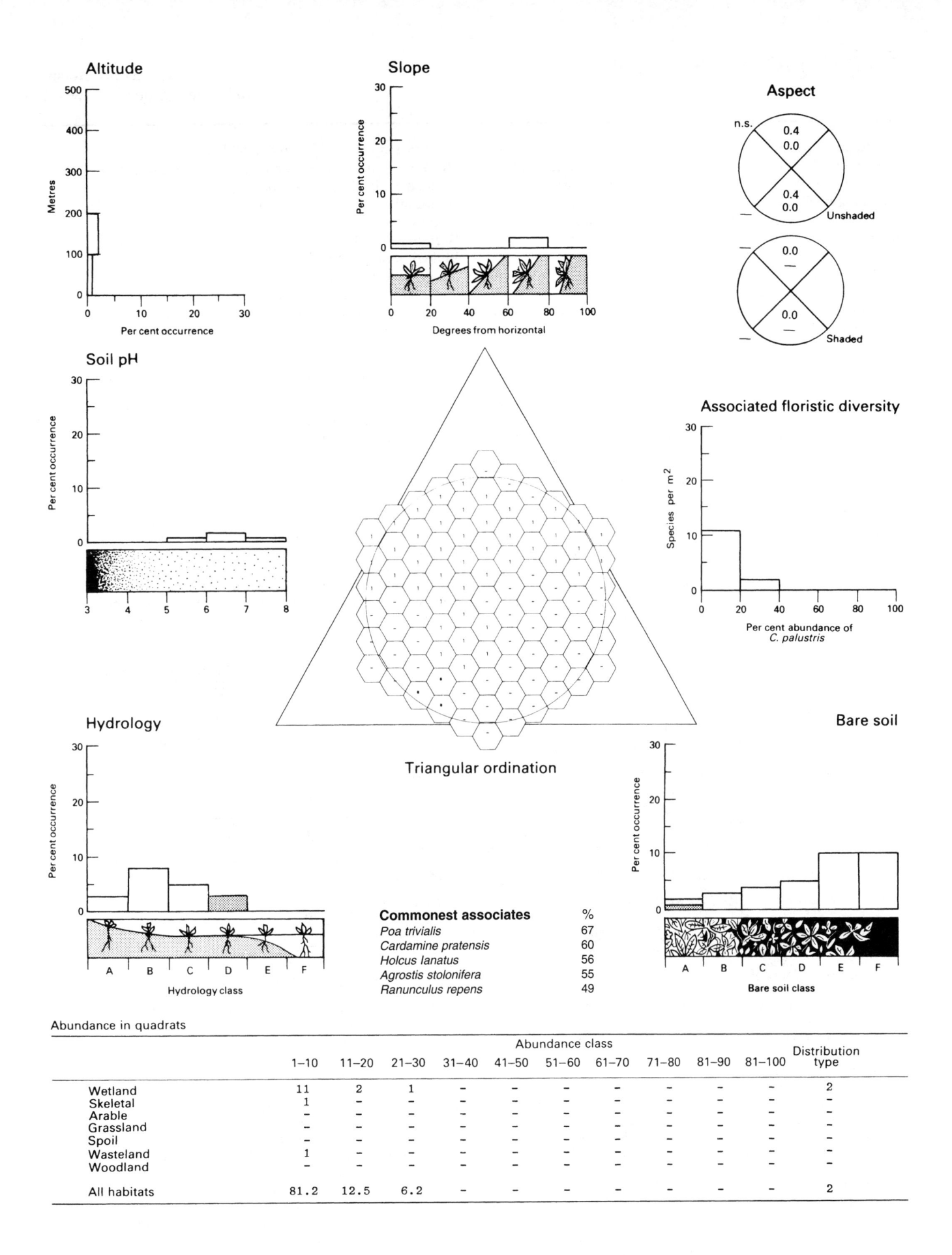

Altitude

Slope

Aspect

Soil pH

Associated floristic diversity

Species per m² vs Per cent abundance of *C. palustris*

Hydrology

Hydrology class: A B C D E F

Triangular ordination

Commonest associates	%
Poa trivialis	67
Cardamine pratensis	60
Holcus lanatus	56
Agrostis stolonifera	55
Ranunculus repens	49

Bare soil

Bare soil class: A B C D E F

Abundance in quadrats

	1–10	11–20	21–30	31–40	41–50	51–60	61–70	71–80	81–90	81–100	Distribution type
Wetland	11	2	1	–	–	–	–	–	–	–	2
Skeletal	1	–	–	–	–	–	–	–	–	–	–
Arable	–	–	–	–	–	–	–	–	–	–	–
Grassland	–	–	–	–	–	–	–	–	–	–	–
Spoil	–	–	–	–	–	–	–	–	–	–	–
Wasteland	1	–	–	–	–	–	–	–	–	–	–
Woodland	–	–	–	–	–	–	–	–	–	–	–
All habitats	81.2	12.5	6.2	–	–	–	–	–	–	–	2

Calystegia sepium (L.) R. Br.

Convolvulaceae
Bellbine, Larger Bindweed

Life-form Polycarpic perennial geophyte often with deep rhizomes. Stems twining, procumbent or upright; leaves ovate, usually <5000 mm²; roots deep, with VA mycorrhizas.
Phenology Flowers June to October and seed set from September to October. Shoots die back in autumn.
Height Foliage and flowers at *c.* 150 mm if prostrate or up to >2 m if supported.
Seedling RGR No data. SLA 35.3. DMC 18.4.
Nuclear DNA amount 1.7 pg; $2n = 22*$ (*BC*); $2x*$.
Established strategy Intermediate between competitor and competitive-ruderal.
Regenerative strategies (V), B_s. Regeneration by seed requires further study (see 'Synopsis'); SB IV.
Flowers Usually white, hermaphrodite, homogamous, insect-pollinated; self-incompatible (Stace 1965).
Germinule 33.9 mg, 4-7 mm, seed in a dehiscent, ≤4-seeded capsule.
Germination Epigeal; t_{50} 3 days; hard-coat dormancy; L = D.
Biological flora None.
Geographical distribution Common in England, Wales and Ireland, becoming rarer northwards and usually introduced in Scotland *ABIF* page 535; Europe except for the extreme N (87% of European territories); W Asia; N Africa and N America.

Habitats Recorded on the banks of rivers and streams, road verges and other waste places. Also present on spoil such as manure heaps, mineral soil, building rubble and mining waste. Other habitats include hedgerows and mire. Not recorded from woodland, skeletal habitats, pasture, arable and permanently flooded habitats. A troublesome garden weed. [One variant, ssp. *roseata* Brummitt, is found in brackish marshes, maritime sands and waste places (*FE3*).]
Gregariousness Even a few individuals have the potential to form a dense canopy.
Altitude Suitable habitats are most frequent at low altitudes; 380 m is the British altitudinal limit for the species.
Slope Typically associated with flat or gently sloping ground.

Aspect In the data presented here only recorded on N-facing slopes, but more-extensive studies (Hodgson *et al.* unpubl.) have revealed a more wide-ranging distribution.
Hydrology Frequent in moist habitats and occasionally found in topogenous mire.
Soil pH Virtually restricted to soils with pH >5.5.
Bare soil Occurs over a wide range of values.
△ **ordination** Associated with productive vegetation subjected in some cases to occasional disturbance, but usually dominated by robust competitive herbs on which *C.s.* is capable of gaining physical support for its climbing shoots.
Associated floristic diversity Low to intermediate.
Synopsis *C.s.* is a robust, rhizornatous, twining herb of moist, fertile, disturbed habitats. In the less disturbed and more semi-natural of its

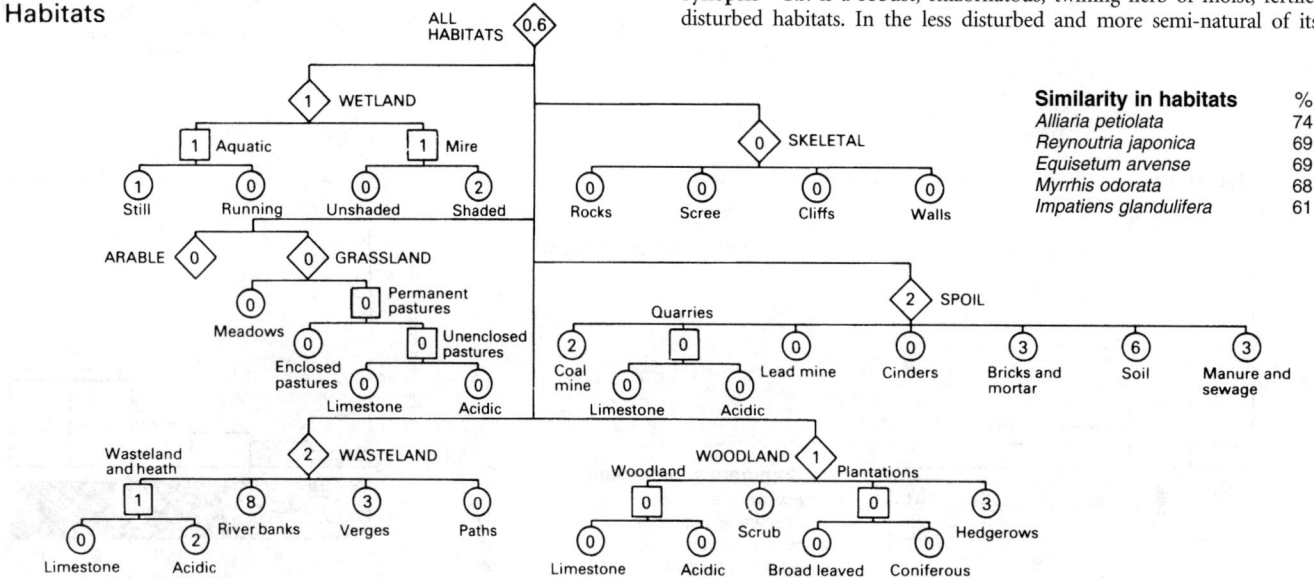

Habitats

Similarity in habitats	%
Alliaria petiolata	74
Reynoutria japonica	69
Equisetum arvense	69
Myrrhis odorata	68
Impatiens glandulifera	61

habitats, e.g. shaded mire and reed beds, *C.s.* is often an inconspicuous, non-flowering component of the vegetation. However, on river banks, roadsides and other, more-frequently disturbed, productive sites the shoots of very large plants may occur as extensive curtains supported by tall herbs, fences or hedges, sometimes ascending to >2 m. In recently abandoned or neglected gardens the species has been observed to develop near-monocultures in a short time. Field observations indicate that moist conditions and a degree of disturbance are necessary to sustain populations of the species; where populations of *C.s.* remain undisturbed there is usually a decline in vigour and an expansion of dominance by other clonal herbs or shrubs. *C.s.* is a lowland species and its range is presumably restricted by climatic factors. *C.s.* is also vulnerable to regular grazing or cutting. The species shows aggressive vegetative spread by means of a network of rhizomes which extend to a depth of *c.* 0. 3 m in the *soil* (*Excur Fl*). In disturbed sites vigorous regeneration occurs through both rhizome and root fragments, both of which readily form new plants (Salisbury 1964). The species frequently colonizes roadsides and spoil tips by means of detached plant fragments in dumped soil. Regeneration by seed appears to be of lesser importance. *C.s.* is self-incompatible (Stace 1961), and in small colonies, which are often single clones, seed set is frequently poor. In larger colonies dependence on vegetative regeneration may be less extreme. The large seeds germinate in autumn or in spring (Muller 1978) and may retain a high level of viability even when buried for as long as 39 years (Salisbury 1964). Seed of riverside populations may be transported by water (Ridley 1930). In the Sheffield region pink-flowererd genotypes are restricted to mire. *C. pulchra* Brummitt and Heywood and *C. silvatica* Kit. Griseb. are aliens occurring in waste places, mainly in urban areas. They both hybridize locally with the native taxon and with each other, and form extensive and at least partially fertile clones (Stace 1975). *C. silvatica* is the most robust taxon and is particularly characteristic of the most fertile sites.
Current trends Uncertain. A common colonist of artificial disturbed habitats; may be increasing locally.

Altitude
Slope
Aspect

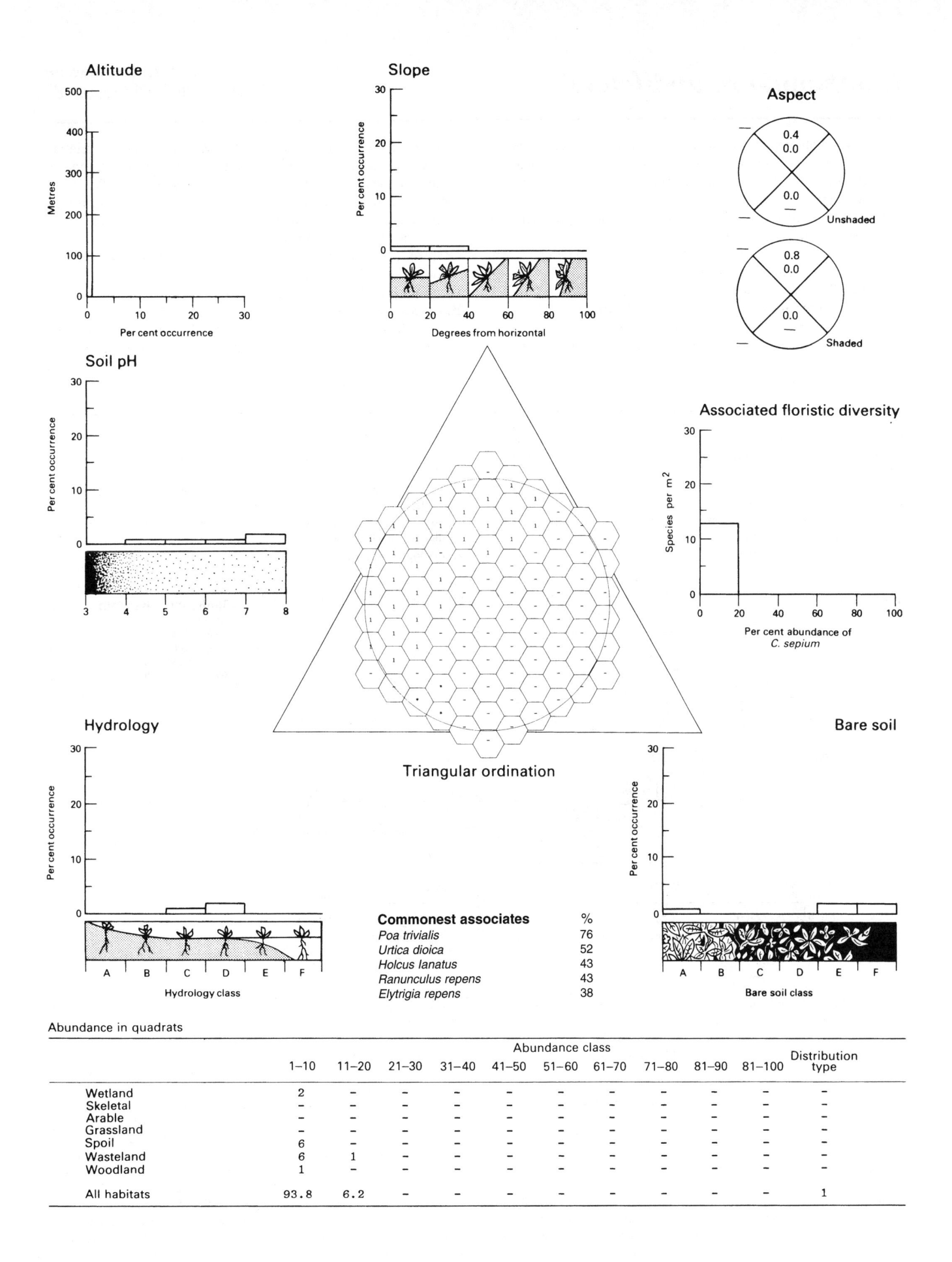

Unshaded

Shaded

Soil pH

Associated floristic diversity

Species per m²

Per cent abundance of *C. sepium*

Hydrology

Hydrology class

Triangular ordination

Commonest associates	%
Poa trivialis	76
Urtica dioica	52
Holcus lanatus	43
Ranunculus repens	43
Elytrigia repens	38

Bare soil

Bare soil class

Abundance in quadrats

	\multicolumn{10}{c}{Abundance class}	
	1–10	11–20	21–30	31–40	41–50	51–60	61–70	71–80	81–90	81–100	Distribution type
Wetland	2	–	–	–	–	–	–	–	–	–	–
Skeletal	–	–	–	–	–	–	–	–	–	–	–
Arable	–	–	–	–	–	–	–	–	–	–	–
Grassland	–	–	–	–	–	–	–	–	–	–	–
Spoil	6	–	–	–	–	–	–	–	–	–	–
Wasteland	6	1	–	–	–	–	–	–	–	–	–
Woodland	1	–	–	–	–	–	–	–	–	–	–
All habitats	93.8	6.2	–	–	–	–	–	–	–	–	1

Campanula rotundifolia L.

Campanulaceae
Harebell, Bluebell (Scotland)

Life-form Polycarpic perennial, semi-rosette hemicryptophyte with slender rhizomes. Stems erect; leaves with chalk glands, basal ones ovate, those on the flowering stem linear, <250 mm^2; roots bearing VA mycorrhizas.

Phenology Winter green. Flowers from July to September and sets seed from August onwards.

Height Non-flowering plants often <30 mm; flowering stems 150-400 mm, leafy to about half-way.

Seedling RGR 0.5-0.9 week^{-1}. SLA 22.2. DMC 25.5.

Nuclear DNA amount 5.3 pg; $2n = 34^*, 68^*, 102^*$ (*BC*); $2x^*, 4x^*, 6x^*$ (base number for genus strictly $n = 8$ or 10).

Established strategy Intermediate between stress-tolerator and C-S-R.

Regenerative strategies V, B$_s$. Vegetative spread effected by means of slender underground rhizomes. Also regenerates in vegetation gaps by means of seed either in autumn or in spring, dependent upon habitat, and forms a buried seed bank; SB III or IV.

Flowers Blue, rarely white, hermaphrodite, protandrous, insect-pollinated, self- incompatible (Laane 1983); either solitary or <10 in a branched panicle.

Germinule 0.07 mg, 1.2 × 0.5 mm, seed dispersed from a dehiscent >50-seeded capsule.

Germination Epigeal; t_{50} days; 10-28°C; L > D.

Biological flora None.

Geographical distribution Most of the British Isles but local in Ireland (91% of British vice-counties), *ABIF* page 587; most of Europe but rare in the S (69% of European territories); N temperate regions.

Habitats Most abundant in dry grassland such as calcareous sheep pasture, wasteland, lead-mine spoil and on scree slopes. Also on rock outcrops, quarry spoil and cliffs. Occurs at waysides, on walls and cinder tips, and in habitats such as road verges, paths and railway ballast. Almost totally absent from wooded habitats, wetlands and arable fields. [Widespread in montane areas on rock ledges and in short turf, and in dune grassland and on cliff tops.]

Gregariousness Intermediate.

Altitude Recorded at altitudes of up to 400 m. Particularly frequent and abundant on the carboniferous limestone (between 200 and 400 m). [Up to 1170 m.]

Slope Particularly common on slopes of between 20 and 40°, as a result of its abundance on limestone dalesides.

Aspect Slightly more frequent and abundant on N-facing slopes. [A similar pattern has been recorded in Dorset (Perring 1959).]

Hydrology Absent from wetlands.

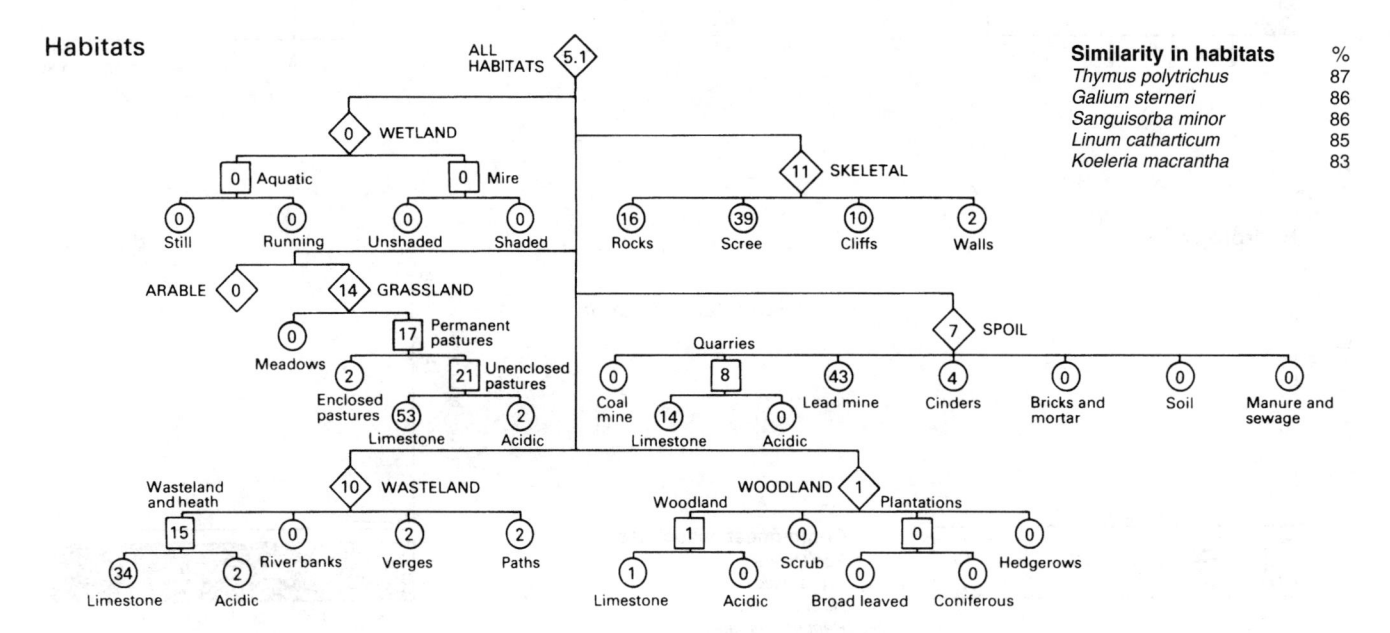

Habitats

Similarity in habitats	%
Thymus polytrichus	87
Galium sterneri	86
Sanguisorba minor	86
Linum catharticum	85
Koeleria macrantha	83

Soil pH Frequency and abundance show a progressive increase as soil pH rises. Most common on base-rich soils in the pH range 7.0-8.0. Also frequently observed on mildly acidic soils and quite characteristic of such soils in some parts of the country, but generally absent from strongly acidic soils.

Bare soil Wide ranging, but most strongly associated with habitats having a relatively small amount of bare soil. Some sites containing *C.r.* include exposed rock.

△ **ordination** Distribution centred on vegetation associated with relatively unproductive habitats subject to little disturbance. Absent from heavily disturbed or fertile environments.

Associated floristic diversity Often high.

Synopsis *C.r.* is a diminutive, slow-growing, rhizomatous herb found usually as a minor component of vegetation in a wide range of unproductive grassy and rocky habitats. *C.r.* usually occupies the lower levels of the canopy, and we suspect that it is shade-tolerant and occupies a niche in infertile habitats not dissimilar to that of *Poa trivialis* in moist, fertile ones. *C.r.* is heterophyllous with leaf form dependent largely on illumination (Goebel 1896, 1908). Ovate, shade leaves are found on non-flowering shoots, while more-linear sun leaves are produced on the disproportionately robust flowering stems. During the winter roots of *C.r.* contain high concentrations of the reserve carbohydrate fructan (Hendry & Brocklebank in prep.). In turf microcosms, roots of this species became colonized by VA mycorrhizas with increases of up to 6-fold in seedling yield (Grime, Mackey, Hillier & Read 1987). Regeneration amongst closed communities is facilitated by the production of creeping stems which ramify locally at the soil surface and within the bryophyte layer. Seed germination is stimulated by the creation of vegetation gaps (Cresswell 1982). A persistent seed bank may be formed. The season of germination varies with latitude and aspect. At one site in N Derbyshire germination on a N-facing slope was delayed until spring whereas seedlings appeared in autumn and spring on an adjacent slope of S-facing aspect (Hillier 1984).

Altitude
Slope
Aspect
Soil pH
Associated floristic diversity
Hydrology
Absent from wetland
Bare soil

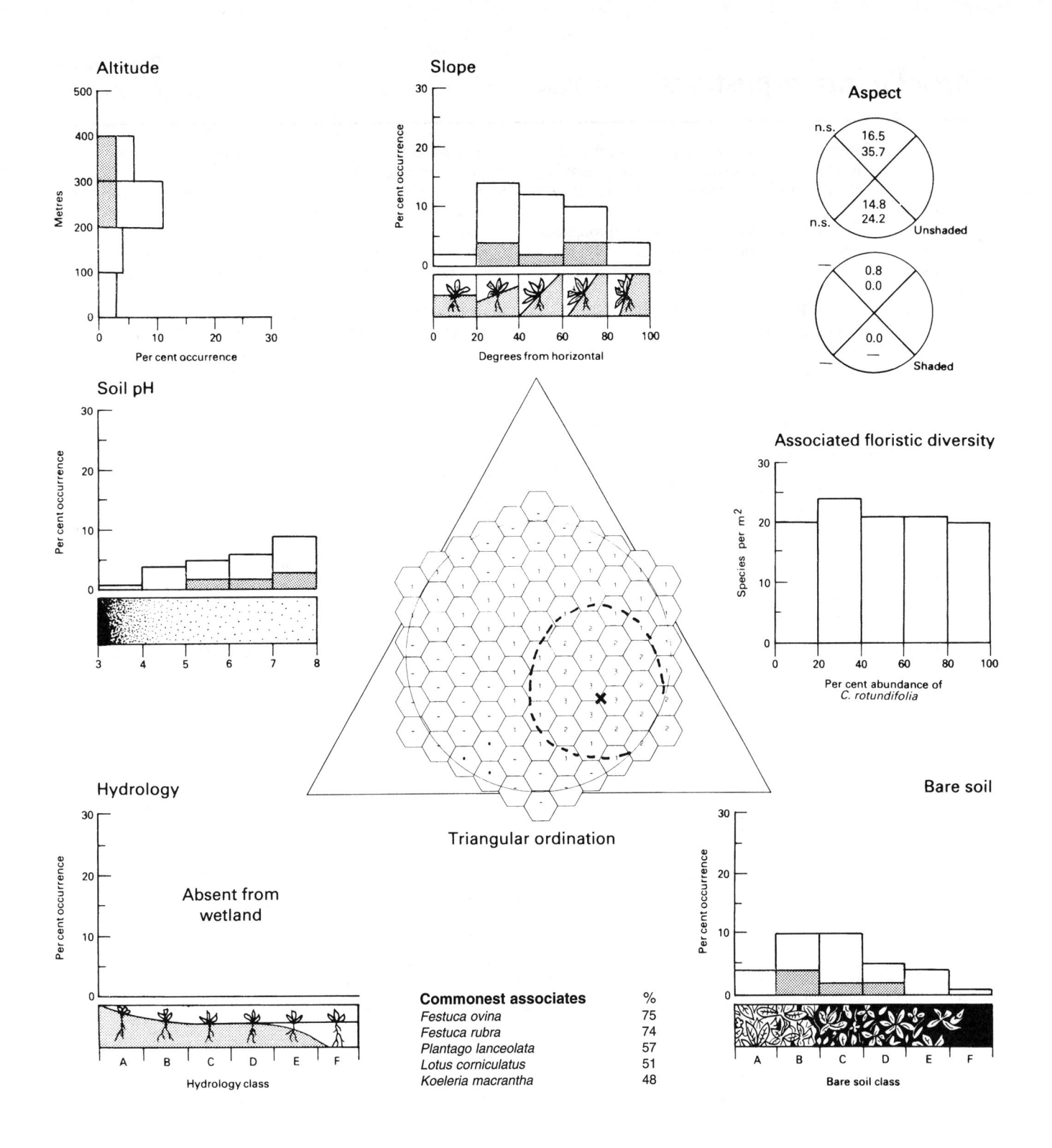

Triangular ordination

Commonest associates	%
Festuca ovina	75
Festuca rubra	74
Plantago lanceolata	57
Lotus corniculatus	51
Koeleria macrantha	48

In some habitats seeds of C.r. are said to be dispersed by ants (Segal 1969). Most British populations are $4x$ except in the W and N, where they are apparently replaced by $6x$ plants (McCallister 1970). Plants with higher ploidy levels have larger seeds (FE4) but ecological differences associated with the two ploidy levels have not been detected. C.r. is an outbreeding species which is extremely variable within Europe and is associated with a fairly wide range of edaphic and climatic conditions in Britain. Heavy-metal-tolerant races have been recognized within the Sheffield region (Shaw 1984).

Current trends C.r. is a plant of infertile habitats. Therefore, despite its widespread distribution and apparently high mobility, we suspect that C.r. is likely to continue to decrease unless current trends in land use are reversed.

Capsella bursa-pastoris (L.) Medicus

Brassicaceae
Shepherd's Purse

Life-form A summer- or winter-annual, semi-rosette therophyte. Stem erect; leaves lanceolate, entire to pinnatifid, hairy or not, <4000 mm^2; tap-rooted, with or without mycorrhizas.
Phenology Variable, since germination may occur throughout most of the year. Plants capable of overwintering. Flowers most abundant from May to October and seed set mainly from June to October.
Height Foliage mostly <100 mm; flowers to 400 mm.
Seedling RGR No data. SLA 24.5. DMC 13.2.
Nuclear DNA amount 1.4 pg; $2n = 32$ (*BC*); $4x^*$.
Established strategy Ruderal.
Regenerative strategies B$_s$. Exclusively by seed in vegetation gaps, either from spring onwards or after incorporation into a persistent seed bank; SB IV.
Flowers White, hermaphrodite, homogamous, mainly self-pollinating (Hurka *et al.* 1976); >100 in a racemose inflorescence.
Germinule 0.11 mg, 1.0 × 0.5 mm, seed in *c.* 30-seeded silicula.
Germination Epigeal; dormancy broken by chilling or by scarifying.
Biological flora Aksoy, Dixon and Hale (1998).
Geographical distribution The British Isles, although less common in the N (100% of British vice-counties), *ABIF* page 271; Europe (100% of European territories); worldwide distribution.

Habitats A species of disturbed, fertile ground. Particularly abundant as an arable weed, mainly in broad-leaved crops and fallow margins. Also found in a range of disturbed artificial habitats such as demolition sites, paths, soil spoil and manure heaps. A familiar garden weed. Infrequent and restricted to bared areas in grassland, and only of casual occurrence in woodland. Observed as a winter-annual on rock outcrops. Absent from wetlands.
Gregariousness Typically occurring as scattered individuals.
Altitude Recorded at altitudes of up to 400 m, but suitable habitats are more frequent and abundant on low-lying ground. [Up to 530 m.]
Slope Mostly restricted to flat or gently sloping ground.
Aspect Significantly more frequent on S-facing slopes.
Hydrology Absent from wetlands.
Soil pH Absent from acidic soils of pH <5.0.
Bare soil Most frequent and abundant in habitats with a high exposure of bare soil.
△ **ordination** Distribution virtually confined to vegetation associated with disturbed, fertile conditions.
Associated floristic diversity Typically intermediate.

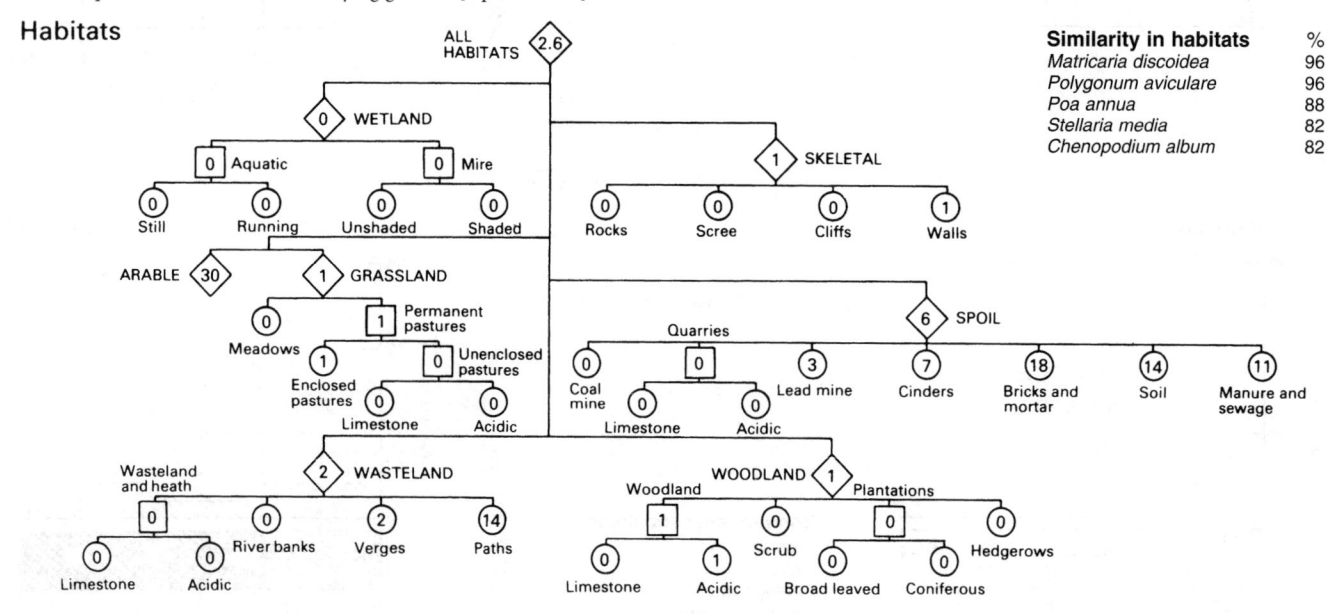

Habitats

Similarity in habitats	%
Matricaria discoidea	96
Polygonum aviculare	96
Poa annua	88
Stellaria media	82
Chenopodium album	82

Synopsis *C.b-p.* is an annual herb which exploits disturbed, fertile, artificial habitats. In arable fields it is most frequent in broad-leaved crops. The status of *C.b-p.* as one of the 'world's worst weeds' (Holm *et al.* 1977) and as an ardent 'follower of man' may be related to the short life-span, coupled with the capacity to germinate over much of the year. Seed of this facultatively long-day plant (Hurka *et al.* 1976) may be produced within 6 weeks, and up to three generations may be formed each year (*WCH*). *C.b-p.* thrives on soils of high mineral status, and an average plant may produce *c.* 3000 seeds (*WCH*). However, seed number is variable, with robust plants producing *c.* 17 000 (Salisbury 1942) or more (Hurka & Haase 1982) and stunted plants from droughted or nutrient-stressed habitats forming less than 50 seeds. Seeds are shed slowly from each siliqua and have a variety of mechanisms for long-distance dispersal (see Hurka & Haase 1982). They are small and flattened, and may be dispersed locally by wind. The seed testa is mucilaginous, and transport in mud on footwear and agricultural machinery appears to be the major form of seed dispersal. Seeds may survive ingestion by stock, birds (Ridley 1930) and earthworms (Hurka & Haase 1982). An exceptionally persistent seed bank is formed in the soil. Seeds require chilling before germination is possible (see Popay & Roberts 1970a,b for a detailed discussion of the complex germination biology). Because freshly shed seed is dormant, *C.b-p.* is limited in its capacity rapidly to exploit impermanent areas of bare soil. *C.b-p.* is more characteristic of continuously disturbed sites, e.g. gardens and arable fields. In these habitats, moreover, the semi-rosette growth form enables the species to survive regimes of trampling, grazing and mowing to a much greater extent than erect ruderals such as *Senecio vulgaris*. *C.b-p.* is very variable and largely in-breeding; many morphologically distinctive local populations were formerly separated (*CTW*). Although in less-disturbed sites particular genotypes tend to predominate, in highly disturbed sites where selection pressures may be weaker populations are genetically variable (Bosbach *et al.* 1982). *C.b-p.* hybridizes with the alien *C. rubella* Reut. (Stace 1975).
Current trends Very common in a range of artificial habitats, and a further increase in its abundance and ecological range appears likely.

Altitude

Slope

Aspect

$p < 0.05$

	0.8	
—	0.0	
	3.1	
	0.0	Unshaded

	0.0	
—	—	
	0.8	
	0.0	Shaded

Soil pH

Associated floristic diversity

Hydrology

Absent from wetland

Triangular ordination

Commonest associates	%
Poa annua	84
Stellaria media	70
Poa trivialis	66
Matricaria discoidea	60
Plantago major	60

Bare soil

Abundance in quadrats

	\multicolumn{10}{c	}{Abundance class}	Distribution type
	1–10	11–20	21–30	31–40	41–50	51–60	61–70	71–80	81–90	81–100	
Wetland	–	–	–	–	–	–	–	–	–	–	–
Skeletal	1	–	–	–	–	–	–	–	–	–	–
Arable	30	4	2	–	–	–	–	–	–	–	1
Grassland	1	–	–	–	–	–	–	–	–	–	–
Spoil	23	1	–	–	–	–	–	–	–	–	1
Wasteland	4	3	–	1	–	–	–	–	–	–	–
Woodland	1	–	–	–	–	–	–	–	–	–	–
All habitats	84.5	11.3	2.8	1.4	–	–	–	–	–	–	1

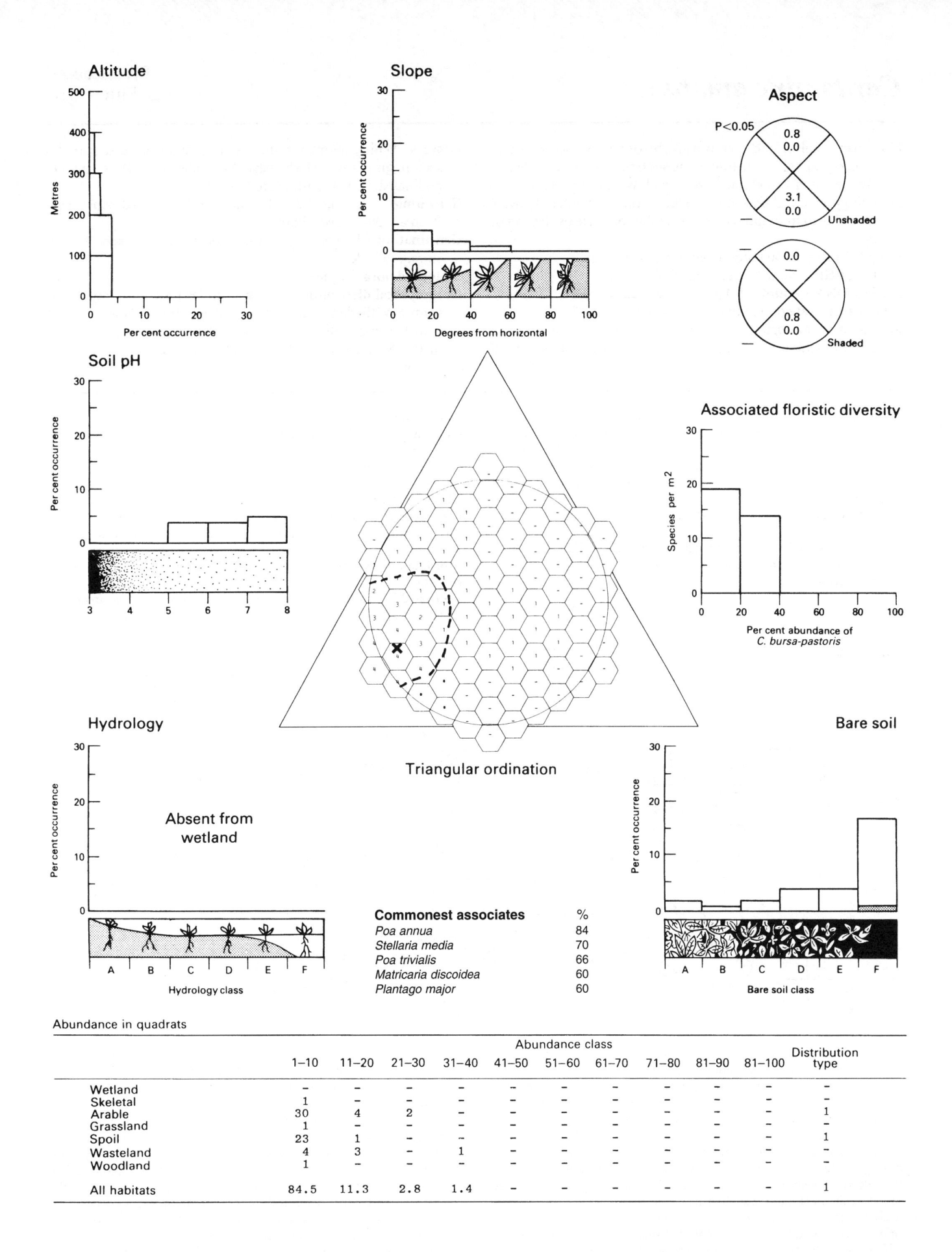

Cardamine amara L.

Brassicaceae
Large Bitter-cress

Life-form Polycarpic perennial, rhizomatous, semi-rosette hemicryptophyte with numerous stolons. Shoots erect; leaves pinnate, often <1000 mm^2; roots shallow, non-mycorrhizal.

Phenology New shoots are produced during winter. Flowers from May to June and seed set generally complete by the end of July.

Height Foliage to 400 mm, overtopped by flowers.

Seedling RGR No data. SLA 49.6. DMC 12.5.

Nuclear DNA amount 1.0 pg; $2n = 16^*$, 32 (BC, FE1); $2x^*$, $4x$.

Established strategy Competitive-ruderal.

Regenerative strategies (V.), ?Bs. Mainly by vegetative means; establishment by seed probably a rare event (see 'Synopsis'); ?SB III.

Flowers White, hermaphrodite, homogamous, at least partially self-incompatible (Urbanska-Worytkiewicz 1980), insect-pollinated; in a 10-20-flowered raceme.

Germinule 0.26 mg, 1.3 × 1.0 mm, seed dispersed explosively from a *c.* 30-seeded siliqua.

Germination Epigeal; t_{50} 7 days; dormancy broken by dry storage at 20°C.

Biological flora None.

Geographical distribution Most of the British Isles, but absent from SW England and much of Wales and Ireland (56% of British vice-counties), *ABIF* page 260; most of Europe, but rare in the N and extreme S (72% of European territories); Asia Minor.

Habitats Exclusive to wetland and semi-aquatic habitats. Frequent in shaded and unshaded mire and on river banks. Also found in ditches and beside open water, and on moist soils adjacent to wetlands. More-recent observations indicate that in the lowland region *C.a.* is mainly found in shaded habitats, whereas in the uplands it is equally frequent in shaded and unshaded situations.

Gregariousness Recorded as intermediate, but potentially patch-forming.

Altitude Observed up to 310 m but most sites <200 m. [Up to 460 m.]

Slope Like most wetland species, largely restricted to flat or gently sloping ground.

Aspect Insufficient data.

Hydrology Found in soligenous or, more rarely, topogenous mire. Also occurs in shallow, usually flowing water, but in this situation colonies straddle the boundary between aquatic habitats and mire.

Soil pH Mainly confined to soils of pH >5.0, but more frequent on non-calcareous strata.

Bare soil Observed in association with a wide range of values, but particularly frequent and abundant where there are moderate exposures of bare soil.

△ ordination Mainly associated with fertile and moderately disturbed vegetation.

Associated floristic diversity Relatively low, especially at high-frequency values of *C.a.* where the species exhibits the capacity for dominance.

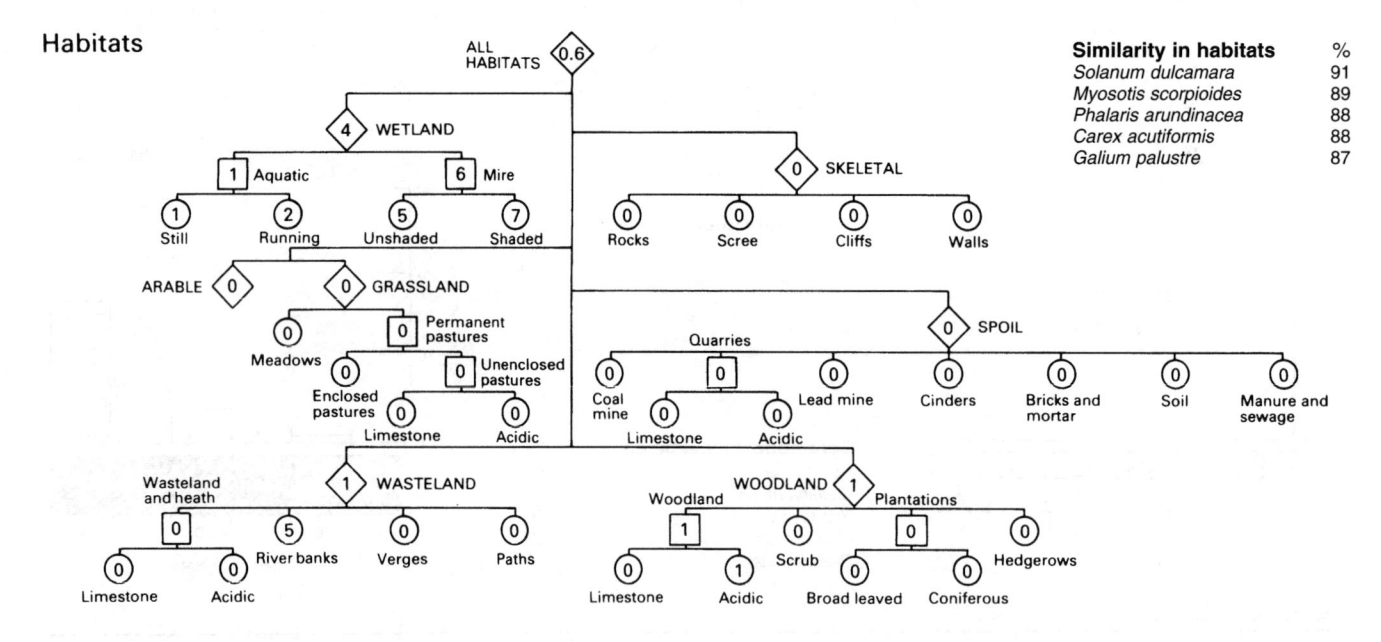

Habitats

Similarity in habitats	%
Solanum dulcamara	91
Myosotis scorpioides	89
Phalaris arundinacea	88
Carex acutiformis	88
Galium palustre	87

Synopsis *C.a.* occurs typically in soligenous mire or 'semi-aquatic' habitats, and is characteristic of moderately fertile soils where the vigour of the more robust wetland species is restricted by summer shade and often also by winter flood damage. The species exhibits rapid shoot growth in spring, and flowering is completed before the tree canopy has fully formed. In upland areas *C.a.* frequently occurs in the open. Unlike other native members of its genus, *C.a.* is not truly winter green, but leaves are formed during winter. *C.a.* forms extensive patches by means of stolons. Breakdown of older stolons by mechanical damage during winter flooding results in the isolation of daughter plants. In the laboratory shoot pieces readily re-root to form new plants, and this process may enable *C.a.* to colonize whole stream or river systems. The species is variously reported as self- incompatible (Urbanska-Worytkiewicz 1980) and partly autogamous (Lovkvist 1975), and it has been reported that stream-side populations, which may consist of a single clone, often produce virtually no seed (Urbanska-Worytkiewicz 1980). Little is known about regeneration by seed, although Urbanska-Worytkiewicz records it as a rare event, probably involving a bank of persistent seeds. The laboratory characteristics of the seed suggest that germination occurs in spring and, like a number of other wetland species, fluctuating temperatures are required for germination (UCPE unpubl.). It remains uncertain to what extent *C.a.* is mobile between sites which are unconnected by water, nor is it clear whether the rarity of *C.a.* in W Britain and Ireland is related to ineffective dispersal of the species or to other ecological factors. Two other members of the *C. amara* group are recognized in Europe (*FE1*).

Current trends Probably decreasing as a result of habitat destruction.

Altitude
Slope
Aspect

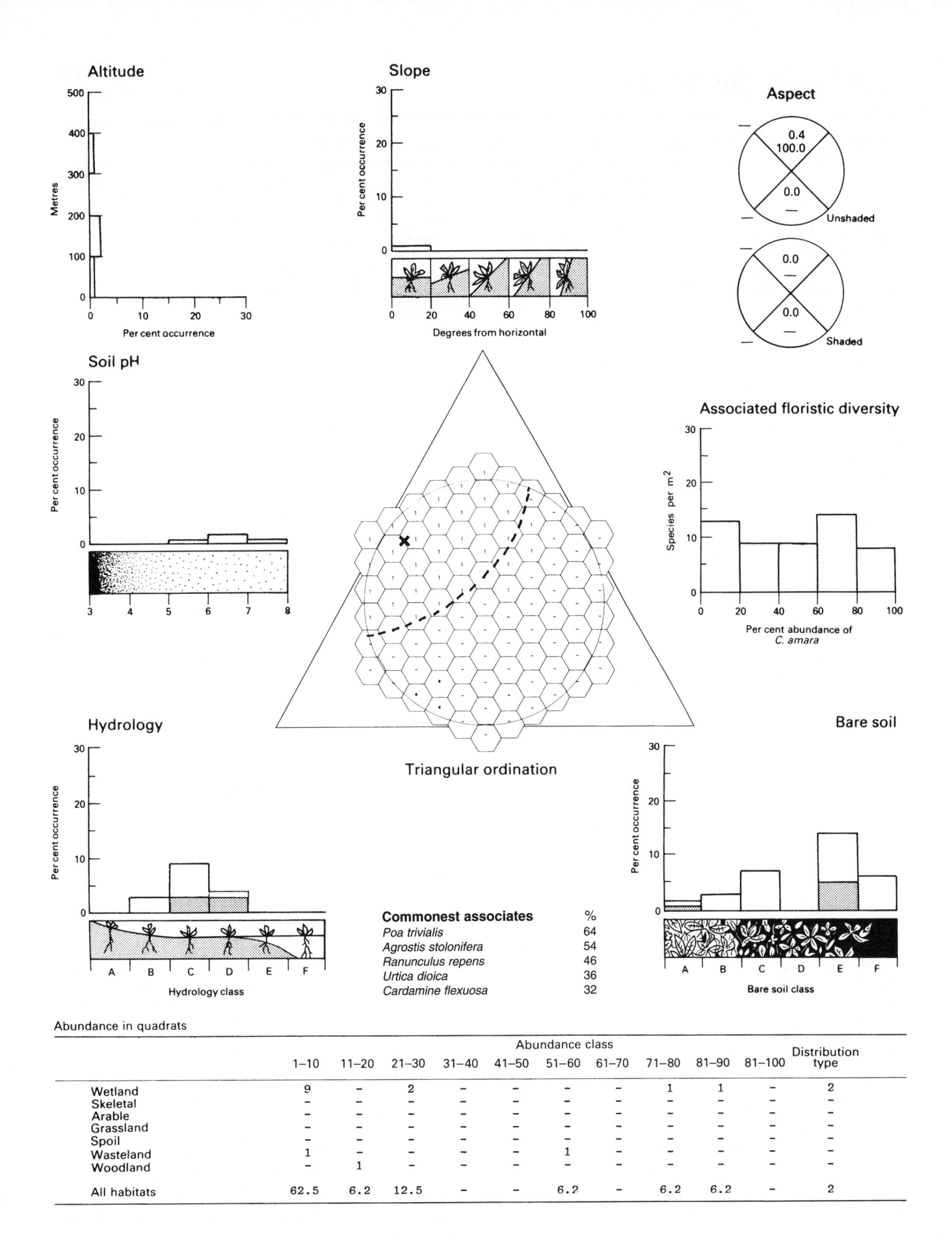

0.4
100.0
0.0
Unshaded

0.0
0.0
Shaded

Soil pH

Associated floristic diversity

Per cent abundance of *C. amara*

Hydrology

Hydrology class

Triangular ordination

Commonest associates	%
Poa trivialis	64
Agrostis stolonifera	54
Ranunculus repens	46
Urtica dioica	36
Cardamine flexuosa	32

Bare soil

Bare soil class

Abundance in quadrats

	Abundance class										Distribution type
	1–10	11–20	21–30	31–40	41–50	51–60	61–70	71–80	81–90	81–100	
Wetland	9	–	2	–	–	–	–	1	1	–	2
Skeletal	–	–	–	–	–	–	–	–	–	–	–
Arable	–	–	–	–	–	–	–	–	–	–	–
Grassland	–	–	–	–	–	–	–	–	–	–	–
Spoil	–	–	–	–	–	–	–	–	–	–	–
Wasteland	1	–	–	–	–	1	–	–	–	–	–
Woodland	–	1	–	–	–	–	–	–	–	–	–
All habitats	62.5	6.2	12.5	–	–	6.2	–	6.2	6.2	–	2

Cardamine flexuosa With.

Brassicaceae
Wood Bitter-cress

Life-form Variously a winter- or summer-annual or monocarpic to polycarpic perennial, semi-rosette hemicryptophyte with a short, branched ascending stock. Stems erect; leaves hairy, pinnate, <1000 mm²; roots shallow, sometimes with VA mycorrhizas.
Phenology Rosette winter green. Flowers mainly in April and May and sets seed in June to July, but a second flowering may occur during summer.
Height Foliage usually <150 mm; flowering stems exceptionally to 500 mm.
Seedling RGR 1.0-1.4 week^{-1}. SLA 53.6. DMC 12.9.
Nuclear DNA amount 1.7 pg; 2n = 32*, c. 50 (*BC, FE1*); 4x*, 6x.
Established strategy Intermediate between stress-tolerant ruderal and ruderal.
Regenerative strategies S, B$_s$. Regenerating almost entirely by seed although vegetative spread is theoretically possible (see 'Synopsis'); SB III.
Flowers White, hermaphrodite, inbreeding (Ellis & Jones 1969); usually >20 in a raceme.
Germinule 0.13 mg, 1.3 × 0.6 mm, seed explosively discharged from a *c.* 30-seeded siliqua.
Germination Epigeal; t_{50} 2 days; <5-30°C; L > D.
Biological flora None.
Geographical distribution The British Isles (99% of British vice-counties), *ABIF* page 261; W Europe (71% of European-territories); Asia and Japan; probably introduced in N America.

Habitats Most common in shaded marshland and disturbed banks of rivers and streams. Also present on the exposed mud of river beds, disturbed woods and embankments, and other moist or shady habitats. Recorded, too, in a variety of damp, shaded situations, on walls, cliffs, soil heaps and cinder tips. Observed as a frequent weed of moist, shaded gardens. Absent from grassland and from dry, unshaded habitats.
Gregariousness Sparse to intermediate.
Altitude Observed to 320 m and most records from altitudes of between 100 and 200 m. [Up to 1190 m.].
Slope Associated with a wide range of values.
Aspect More frequent and abundant on N-facing slopes.
Hydrology Typical of soligenous mire and moist ground beside rivers or streams. Less characteristic of topogenous mire and absent from permanently submerged habitats.
Soil pH Most frequent and abundant on base-rich soils in the pH range 6.0-7.0. Absent from strongly acidic soils.
Bare soil Distribution includes open areas of bare soil and mud, and sites with a virtually complete cover of vegetation or plant litter.
Δ **ordination** An exceedingly diffuse pattern with the species excluded from heavily disturbed habitats, declining in unproductive vegetation and reaching highest frequencies in moderately disturbed fertile sites. This distribution appears to be related to the ability of *C.f.* to exploit damp, shaded microsites in a wide range of habitats and vegetation types (see 'Synopsis').

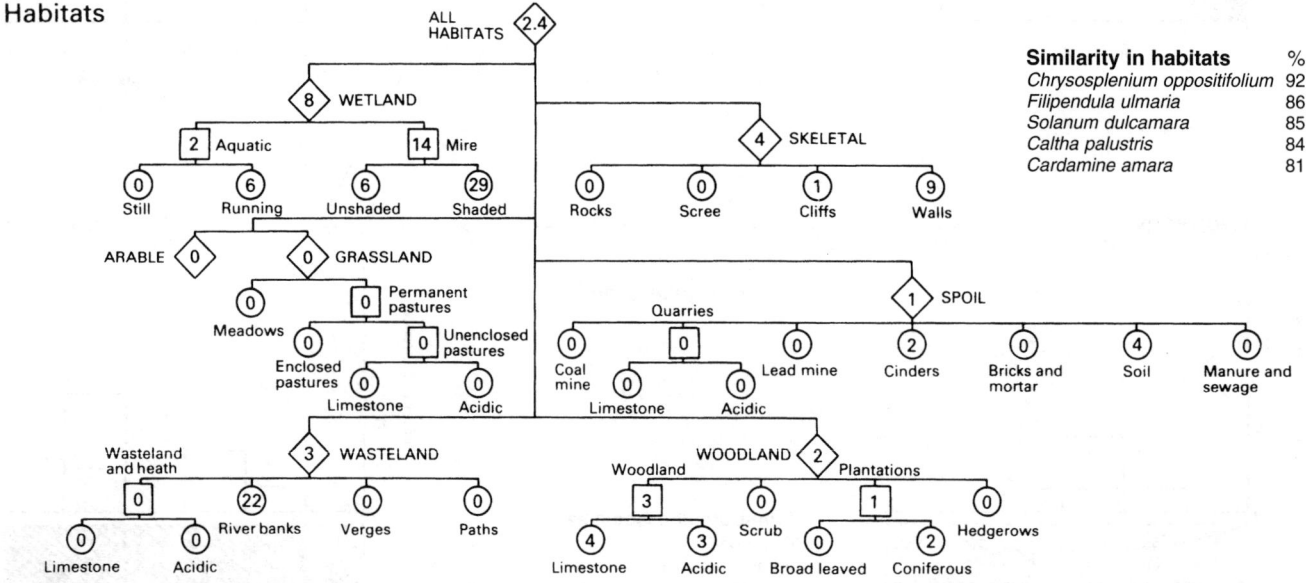

Habitats

Similarity in habitats	%
Chrysosplenium oppositifolium	92
Filipendula ulmaria	86
Solanum dulcamara	85
Caltha palustris	84
Cardamine amara	81

Associated floristic diversity Relatively low.
Synopsis *C.f.* is a small, rosette-forming herb of moist, usually shaded habitats. The species is particularly characteristic of those fertile sites in which the vigour of the more robust colonizing perennials is restricted by seasonal shade or some form of physical disturbance. Thus, *C.f.* occurs in mire, on stream banks and at river margins which may be subject to scouring and flooding in winter. The species is found in gardens, and also colonizes crevices in brickwork and cliffs. The explosive discharge of the narrowly-winged seeds may be important as a means of locating these relatively inaccessible microsites. As illustrated by the triangular ordination and the list of species 'most similar in habitats', despite the relatively small stature and rather ephemeral life-history, the species occurs most consistently in vegetation dominated by competitive ruderals. Several features of *C.f.* appear to determine this ecology. They include (a) the ability of seedlings and overwintering shoots to commence growth and produce seeds early in the season before consolidation of the leaf canopy, (b) flexibility in life-history which allows swift fruiting in temporary situations and continued reproduction in more secure locations, (c) tolerance of moderate intensities of shade and (d) flexibility in seed germination biology. Seeds germinate over a wide range of temperatures, and are often prominent in the buried seed bank, in locations where *C.f.* is only a minor component of the established vegetation. Shoot fragments detached as a result of disturbance are capable of rooting to form new plants; the significance of this mode of regeneration remains uncertain. In Japan, ssp. are recognized which differ in their longevity, capacity of the stem segments to root and distance to which seed is explosively dispersed (Kimata 1983). *C.f.* hybridizes with *C. hirsuta* and with *C. pratensis* (Stace 1975), and may itself be an allotetraploid resulting from hybridization between *C.h.* and *C. impatiens* L. (Ellis & Jones 1969).
Current trends An effective colonist of disturbed fertile habitats, and probably increasing.

Altitude
Slope
Aspect

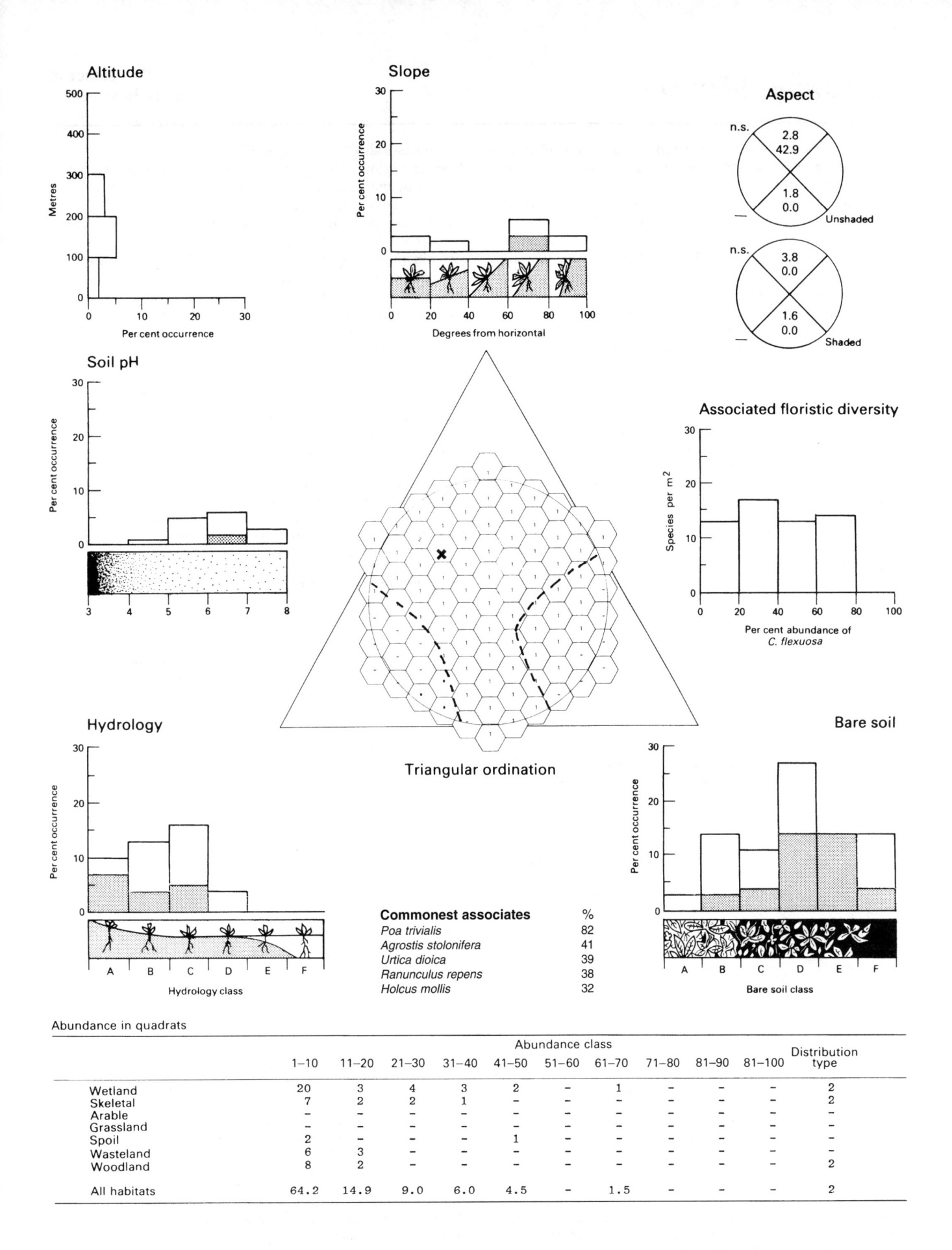

	2.8
n.s.	42.9
	1.8
	0.0	Unshaded

	3.8
n.s.	0.0
	1.6
	0.0	Shaded

Soil pH

Associated floristic diversity

Hydrology

Triangular ordination

Bare soil

Commonest associates	%
Poa trivialis	82
Agrostis stolonifera	41
Urtica dioica	39
Ranunculus repens	38
Holcus mollis	32

Abundance in quadrats

	Abundance class										Distribution type
	1–10	11–20	21–30	31–40	41–50	51–60	61–70	71–80	81–90	81–100	
Wetland	20	3	4	3	2	–	1	–	–	–	2
Skeletal	7	2	2	1	–	–	–	–	–	–	2
Arable	–	–	–	–	–	–	–	–	–	–	–
Grassland	–	–	–	–	–	–	–	–	–	–	–
Spoil	2	–	–	–	1	–	–	–	–	–	–
Wasteland	6	3	–	–	–	–	–	–	–	–	–
Woodland	8	2	–	–	–	–	–	–	–	–	2
All habitats	64.2	14.9	9.0	6.0	4.5	–	1.5	–	–	–	2

Cardamine hirsuta L.

Brassicaceae
Hairy Bitter-cress

Life-form Winter- or more rarely summer-annual, semi-rosette therophyte. Stem erect; leaves pinnate, <500 mm²; roots typically non-mycorrhizal.
Phenology Usually winter annual. Flowers mainly in April and seed shed in May and June. In moist habitats a further generation may occur which flowers and sets seed in autumn.
Height Foliage largely confined to a flattened rosette; flowering stems to 70-300 mm.
Seedling RGR No data. SLA 30.1. DMC 12.6.
Nuclear DNA amount 0.5 pg; $2n = 16^*$ (*BC*); $2x^*$.
Established strategy Stress-tolerant ruderal.
Regenerative strategies S, B_s. Regenerates entirely by seed, which typically germinates in autumn and forms a persistent seed bank (see also 'Synopsis'); SB III.
Flowers White, hermaphrodite, homogamous, rarely visited by insects, and automatically self-pollinated (*CTW*); typically <30 in a racemose inflorescence.
Germinule 0.09 mg, 1.0 × 1.0 mm, seed explosively discharged from a *c*. 20-seeded siliqua.
Germination Epigeal; t_{50} 5 days; 11-28°C; germination rate increased after dry storage at 20°C; L < D.
Biological flora None.
Geographical distribution The British Isles (99% of British vice-counties), *ABIF* page 262; Europe except the extreme N (95% of European territories); most of the N hemisphere.

Habitats Occurs commonly as a winter annual on limestone quarry spoil, rock outcrops, cinder tips and, to a lesser extent, in rocky pastures, and on cliffs and walls. In many moist, often shaded, habitats, gardens, waysides, soil heaps and wasteland *C.h.* may produce more than one generation per year. Commonly observed as a weed of tree nurseries, where the species is often winter-annual. [Also recorded from sand dunes.]
Gregariousness Sparse.
Altitude Wide-ranging, but suitable habitats are particularly frequent on the carboniferous limestone between 200 and 400 m. [Up to 1160 m.]
Slope Extending on to steep slopes.
Aspect Widely distributed.

Hydrology Absent from wetlands.
Soil pH Mainly recorded from soils in the pH range 7.0-8.0.
Bare soil Widely distributed, although most frequently associated in this survey with rocky habitats having intermediate amounts of bare earth.
△ **ordination** Most common in vegetation associated with relatively infertile disturbed habitats. Absent from highly disturbed conditions and from circumstances conducive to intense competition for resources. Occurrences outside the main centre of distribution in the centre of the triangular model appear to correspond to excursions of *C.h.* into artificial disturbed habitats.
Associated floristic diversity Varying widely. Often high in winter-annual communities; frequently low in gardens.

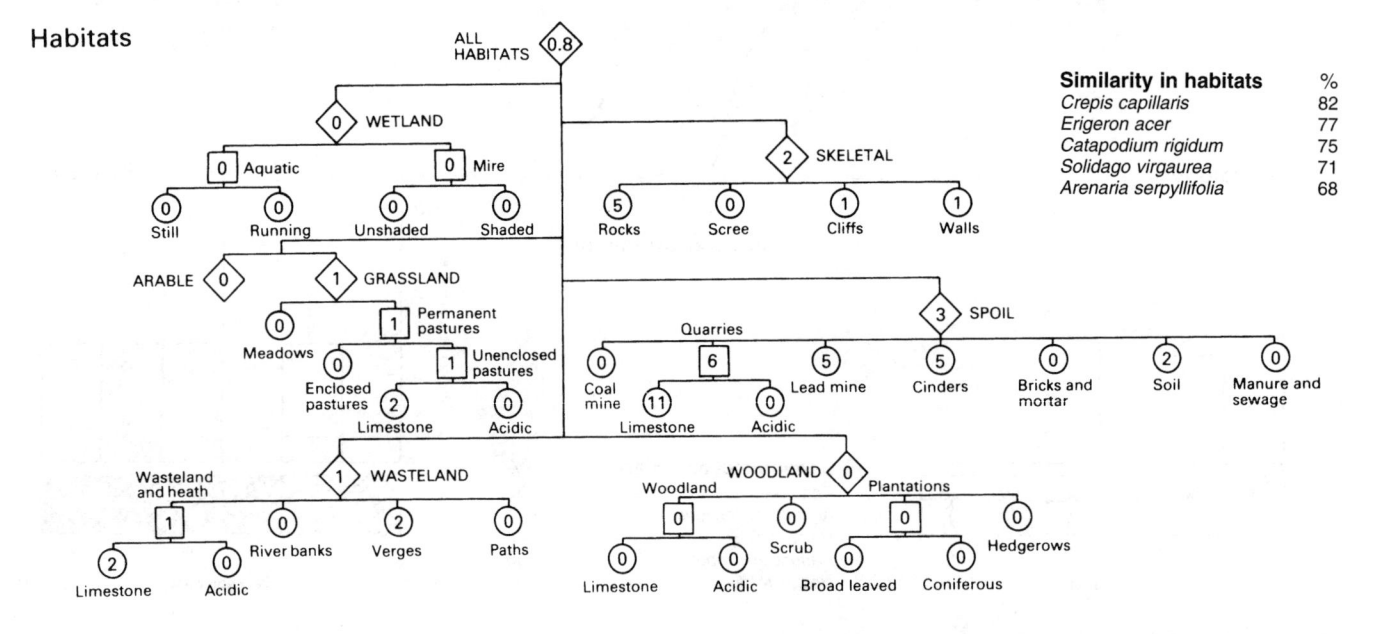

Habitats

Similarity in habitats	%
Crepis capillaris	82
Erigeron acer	77
Catapodium rigidum	75
Solidago virgaurea	71
Arenaria serpyllifolia	68

Synopsis *C.h.* is most characteristically a winter annual on well-drained sandy or calcareous soils subject to summer drought. After *Erophila verna* (L.) DC., *C.h.* is the earliest flowering of the common winter annuals, and occupies the shallowest soils and other particularly drought-prone sites. Regeneration is entirely by seed, which is explosively discharged up to *c*. 1 m from the plant in still conditions (Salisbury 1964). Plant size and seed production vary considerably with soil fertility and prevailing climate; Salisbury (1942) quotes an average of *c*. 100 seeds per plant for a sand-dune population, but >50 000 seeds were found on a nearby garden plant. Seeds are often shed as early as May. However, germination is delayed until autumn as a result of an after-ripening requirement and, typically of many winter annuals, *C.h.* forms a persistent seed bank (Chippindale & Milton 1934, Jalloq 1974). Fragments of leaves with pieces of stem attached are capable of rooting (Salisbury 1965); however this is seldom important as a means of regeneration. Like *Arabidopsis thaliana*, the species has expanded its range to become an abundant weed of tree nurseries. Here the factor restricting the incursion of perennials and favouring this ephemeral is not drought but the herbicide applied during early summer. *C.h.* also occurs as a winter annual on shaded rock outcrops and is also found in moist, often shaded, gardens and in other waste places; here two generations may occur within a single year, one in spring and another in autumn. Thus, *C.h.* is associated with a wide range of edaphic conditions and produces populations which differ in phenology. Whether these differences have a genetic basis, as in the case of *A.t.*, requires investigation. Hybrids have been recorded with *C. flexuosa* (Stace 1975).
Current trends An effective colonist of artificial habitats, this species appears to be increasing.

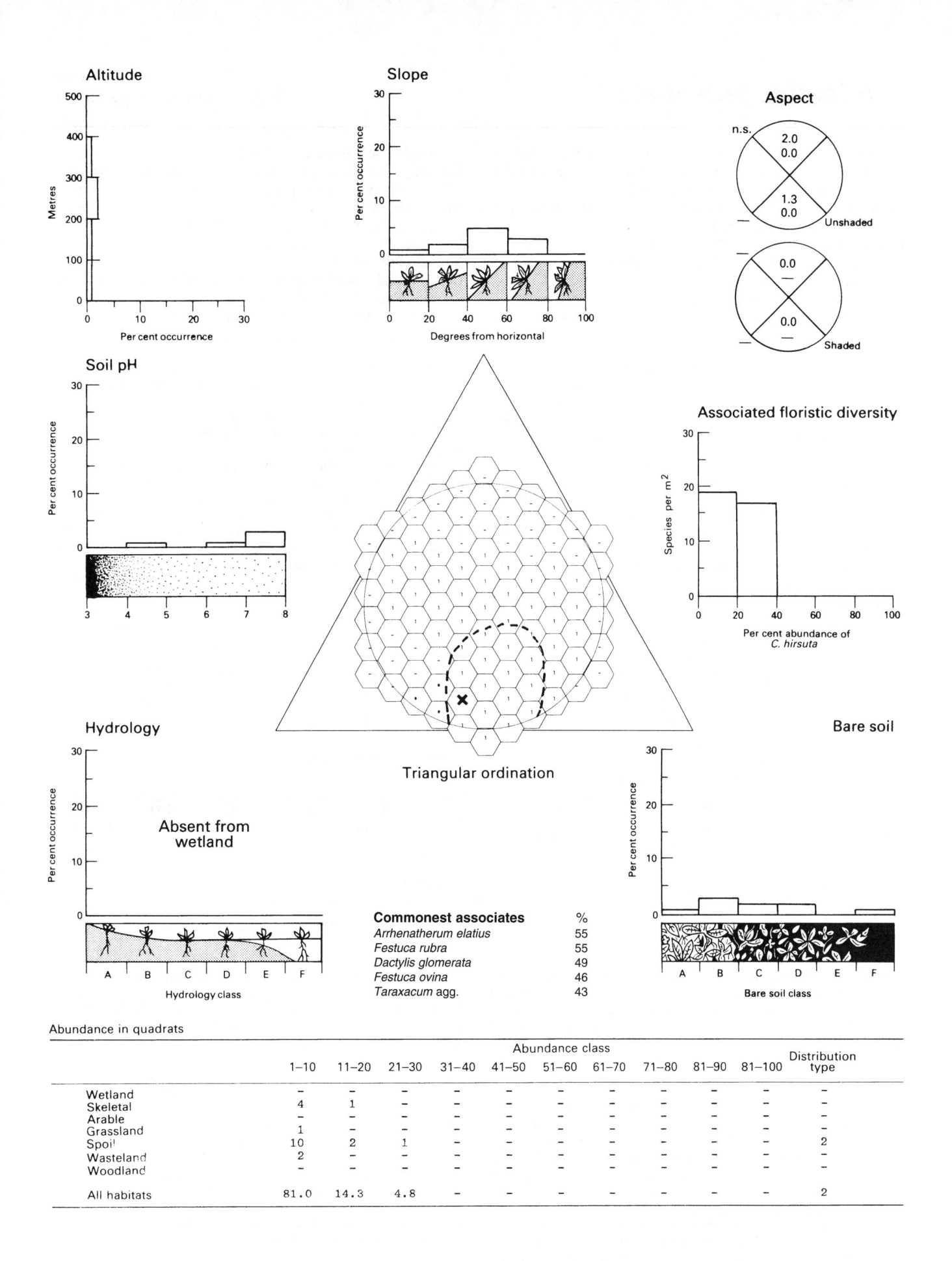

Altitude

Slope

Aspect

Unshaded

Shaded

Soil pH

Associated floristic diversity

Per cent abundance of *C. hirsuta*

Hydrology

Absent from wetland

Hydrology class

Triangular ordination

Commonest associates	%
Arrhenatherum elatius	55
Festuca rubra	55
Dactylis glomerata	49
Festuca ovina	46
Taraxacum agg.	43

Bare soil

Bare soil class

Abundance in quadrats

	\multicolumn{10}{c	}{Abundance class}	Distribution type
	1–10	11–20	21–30	31–40	41–50	51–60	61–70	71–80	81–90	81–100	
Wetland	–	–	–	–	–	–	–	–	–	–	–
Skeletal	4	1	–	–	–	–	–	–	–	–	–
Arable	–	–	–	–	–	–	–	–	–	–	–
Grassland	1	–	–	–	–	–	–	–	–	–	–
Spoil	10	2	1	–	–	–	–	–	–	–	2
Wasteland	2	–	–	–	–	–	–	–	–	–	–
Woodland	–	–	–	–	–	–	–	–	–	–	–
All habitats	81.0	14.3	4.8	–	–	–	–	–	–	–	2

Cardamine pratensis L.

Brassicaceae
Cuckoo Flower, Lady's Smock

Life-form Polycarpic perennial, semi-rosette hemicryptophyte with short rhizomes. Stems erect; leaves pinnate, typically <1000 mm^2; roots shallow, non-mycorrhizal.
Phenology Overwintering as a small rosette. Flowers from April to June and sets seed from June to July.
Height Foliage mainly <150 mm; flowering shoot to 600 mm.
Seedling RGR 0.5-0.9 week^{-1}. SLA 16.6. DMC 20.0.
Nuclear DNA amount 3.3 pg; $2n = 30^*, 56^*$ (*BC*); $4x^*, 8x^*$.
Established strategy Intermediate between C-S-R and ruderal.
Regenerative strategies (V), ?B$_s$. Regenerative strategies are complex (see 'Synopsis'); ?SB III.
Flowers Lilac, hermaphrodite, protogynous, self-incompatible (Lovkvist 1956), insect-pollinated; typically <50 in a raceme.
Germinule 0.60 mg, 1.6 × 1.0 mm, seed explosively dispersed from a siliqua often containing *c.* 10 seeds.
Germination Epigeal; t_{50} 5 days; 12-18°C; L > D.
Biological flora None.
Geographical distribution The British Isles (100% of British vice-counties), *ABIF* page 261; much of Europe except the S (79% of European territories); N-temperate regions.

Habitats Primarily a wetland species associated with both shaded and unshaded mire, but also frequent in moist pastures. There are also a few records from river banks and from aquatic habitats. May persist on moist soils as a garden weed, and occurs in lawns on heavy soils. Not recorded from spoil, ungrazed grassland and arable land, and virtually absent from skeletal habitats and woodland. [Occurs in soligenous mire in mountain areas (McVean & Ratcliffe 1962) and in dune slacks.]
Gregariousness Sparse to intermediate.
Altitude Found over a wide range of values to 480 m. [Up to 1000 m.]
Slope Like most wetland species, *C.p.* is most frequent and abundant on flat ground and gentle slopes, and is largely absent from steeply sloping ground.
Aspect *C.p.* shows a bias towards N-facing slopes, presumably due to its restriction to moist or waterlogged soils.
Hydrology Found on moist soils, in sites subject to winter flooding, soligenous and, to a lesser extent, also topogenous mire, but absent from permanently submerged sites.
Soil pH Mainly restricted to soils of pH >5.0.
Bare soil Found over a wide range of values, but more typical of sites with continuous vegetation cover.
△ **ordination** Distribution centred on productive and moderately disturbed vegetation but also present in a wide range of other vegetation types. Absent from heavily disturbed situations.
Associated floristic diversity Typically intermediate, although a frequent constituent of highly species-rich vegetation.
Synopsis *C.p.* is an early-flowering, winter-green, rosette-forming herb characteristic of wet grassland and mire on moderately fertile soils. The species, like several others characteristic of soligenous mire (e.g. *Carex*

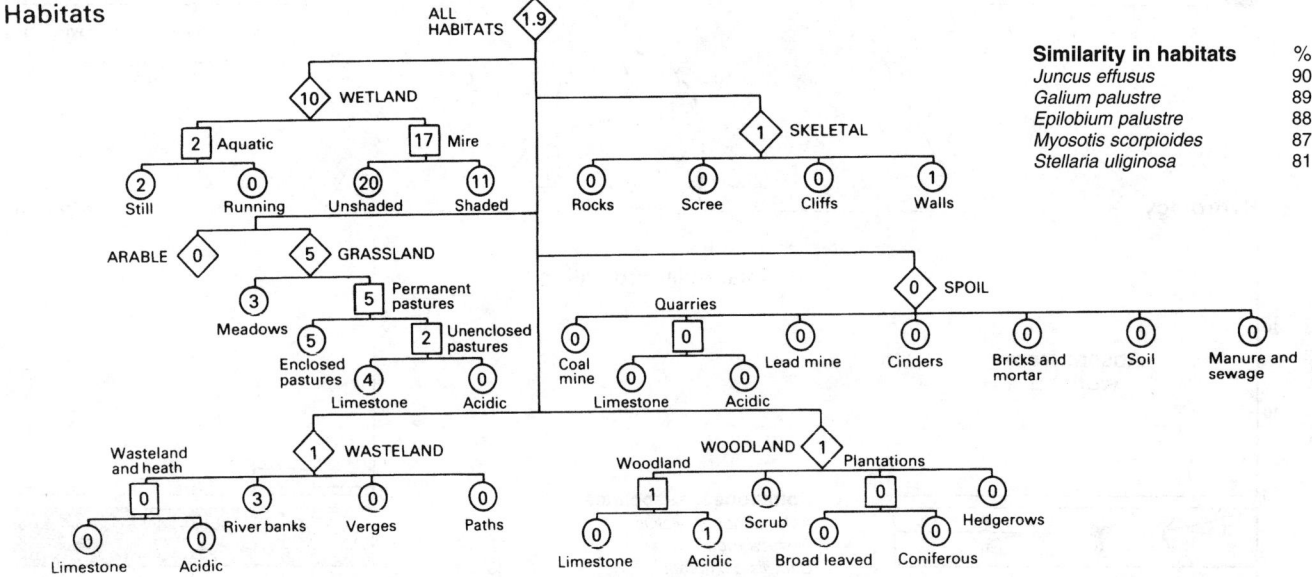

Habitats

Similarity in habitats	%
Juncus effusus	90
Galium palustre	89
Epilobium palustre	88
Myosotis scorpioides	87
Stellaria uliginosa	81

panicea), is also frequent on N-facing limestone grasslands in upland areas. The basal leaves generally occupy the lower levels of the leaf canopy and *C.p.* may persist in open woodland. Thus, we suspect that *C.p.* is to some extent shade-tolerant. However, the species is less common in tall vegetation and is most frequent where the vigour of potentially dominant species is restricted by grazing. *C.p.* exploits vegetation gaps in a manner similar to *Ranunculus repens* except that new plants arise from leaflets in contact with the soil rather than from stolons. The resulting plantlets may be as much as 200 mm from the centre of the parent plant, and reproduction by this means may constitute the main method of regeneration in moist sites (Salisbury 1965). Leaves and shoot pieces detached as a result of disturbance are capable of producing new plants and in the absence of competition *C.p.* may also extend through the development of branched rhizomes at the rate of 40-50 mm year^{-1} (Salisbury 1965). Seed set is frequently low in grazed sites, and is probably most important in drier sites where vegetative regeneration is less easily effected (Salisbury 1965). Seeds are very variable in size and are explosively discharged up to 2.5 m from the parent plant (Salisbury 1965).

Germination occurs mainly in spring, and a persistent seed bank occurs. Despite being an outbreeding species (Lovkvist 1956) much of the locally-occurring morphological variation within the species is attributable to phenotypic plasticity (Dale & Elkington 1974). Six species are included within the *C. pratensis* group in Europe (*FE1*): *C. pratensis* $2n$ = 16, 28-34, 38-44, 48 and *C. palustris* (Wimmer and Grab.) Peterm $2n$ = 56, 64, 72, 76, 80, 84, *c.* 96 are recognized for Britain (*FE1*). They cannot be recognized by morphological characters in Britain (Dale & Elkington 1974) and have not been distinguished, either in British floras or here. While populations in Europe often have a range of chromosome numbers, those from the British Isles are more constant (Dale & Elkington 1974). The $8x$ karyotype tends to be the commonest within the Sheffield region (Dale & Elkington 1974), with the $4x$ plant appearing to be largely restricted to S England (Hussein 1955). Hybridizes with *C. flexuosa* (Stace 1975).
Current trends Probably not as common as formerly in lowland pastures, but likely to remain as a frequent component of marshland and damp grassland in Upland Britain.

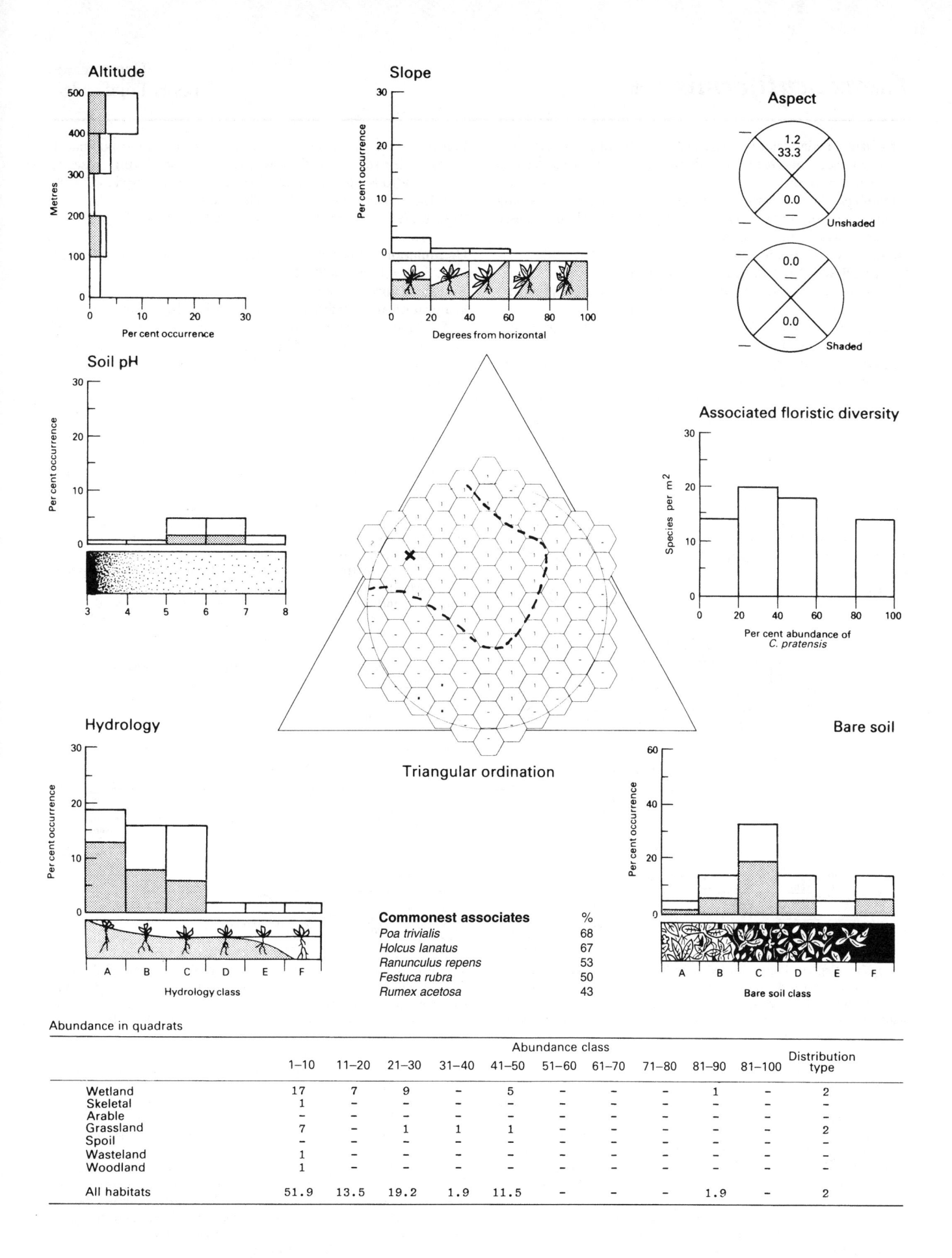

Altitude

Slope

Aspect

Unshaded

Shaded

Soil pH

Associated floristic diversity

Hydrology

Hydrology class

Commonest associates	%
Poa trivialis	68
Holcus lanatus	67
Ranunculus repens	53
Festuca rubra	50
Rumex acetosa	43

Triangular ordination

Bare soil

Bare soil class

Abundance in quadrats

	\multicolumn{10}{c}{Abundance class}	Distribution type
	1–10	11–20	21–30	31–40	41–50	51–60	61–70	71–80	81–90	81–100	
Wetland	17	7	9	–	5	–	–	–	1	–	2
Skeletal	1	–	–	–	–	–	–	–	–	–	–
Arable	–	–	–	–	–	–	–	–	–	–	–
Grassland	7	–	1	1	1	–	–	–	–	–	2
Spoil	–	–	–	–	–	–	–	–	–	–	–
Wasteland	1	–	–	–	–	–	–	–	–	–	–
Woodland	1	–	–	–	–	–	–	–	–	–	–
All habitats	51.9	13.5	19.2	1.9	11.5	–	–	–	1.9	–	2

Carex acutiformis Ehrh.

Cyperaceae
Lesser Pond-sedge

Life-form Perennial, semi-rosette helophyte with stout creeping rhizomes. Stems erect; leaves linear, <2000 mm^2; roots lacking VA mycorrhizas.

Phenology Shoot growth in late spring is followed by flowering in June to July and seed is shed August onwards. Shoots senesce during winter. Leaves long-lived.

Height Foliage and flowers to 1.5 m.

Seedling RGR No data. SLA 16.5. DMC 34.3.

Nuclear DNA amount No data; $2n = c.$ 38, 78 (*FE5*); Cyperaceae with diffuse centromeres, base number obscure.

Established strategy Intermediate between competitor and stress-tolerant competitor.

Regenerative strategies V, B$_s$. Regeneration by seed requires further study (see 'Synopsis').

Flowers Red or purple-brown, monoecious, wind-pollinated and also occasionally visited by pollen-eating insects, protandrous (Jermy *et al.* 1982); 3-4 female spikes (>100 flowers) overtopped by 2-3 male spikes.

Germinule 1.11 mg, 1.7 × 1.0 mm, nut in an utricle (3.5 × 1.7 mm).

Germination Epigeal; t_{50} 10 days; germination rate increased after dry storage at 20°C.

Biological flora None.

Geographical distribution The British Isles except for N Scotland, but rather more common in the S and E (83% of British vice-counties), *ABIF* page 722; most of Europe except for the extremes of N and S (79% of European territories); N and S Africa, temperate Asia and N America.

Habitats Exclusively a wetland or semi-aquatic species, occurring most frequently on river banks and in ungrazed mire adjacent to open water, in both unshaded and shaded situations. Also extends into aquatic habitats at the margins of ponds, ditches, rivers and streams.

Gregariousness Data suggest an intermediate degree of clumping, but field observations confirm that *C.a.* is often a stand-forming species.

Altitude Suitable habitats are more common in lowland areas but observed up to 320 m. [Up to 370 m.]

Slope As with many wetland species, largely restricted to flat ground and gentle slopes.

Aspect Insufficient data.

Hydrology Most frequent in topogenous mire adjacent to open water, but also found in soligenous mire. Occurs in shallow water, but in these circumstances part of the colony usually extends on to ground above the water level.

Soil pH Largely absent from acidic soils.

Bare soil Stands of *C.a.* often accumulate some persistent litter, and observations indicate that *C.a.* is typical of closed communities rather than those with large exposures of bare soil.

△ **ordination** Distribution centred on undisturbed vegetation types of moderately high productivity.

Associated floristic diversity Typically low. Recent unpublished studies indicate a decrease in species number at high-frequency values, indicating that *C.a.* exhibits competitive exclusion.

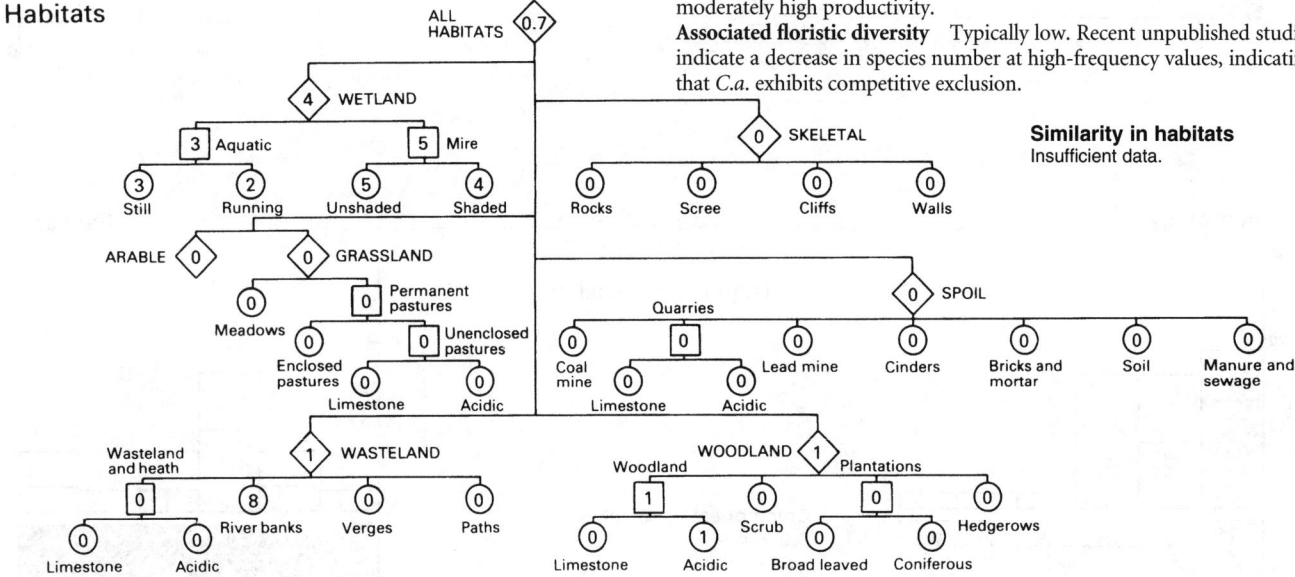

Similarity in habitats Insufficient data.

Synopsis A species in urgent need of further study. *C.a.* is a tall (to *c.* 1.5 m), wetland sedge forming extensive stands where the vigour of other potential dominants is reduced by intermittent grazing (Wheeler 1983), by light shade from a tree canopy or by low soil fertility. *C.a.* also frequently grows at the water's edge, particularly beside large ponds, lakes and rivers, and the possibility that it may persist in these sites because the growth of other dominants is suppressed by the disturbance associated with wave action or water currents requires investigation. *C.a.* is not normally found in water systems subject to large fluctuations in water level and is absent from reservoir margins. Although extending from mire into aquatic systems, in no site is *C.a.* exclusively restricted to submerged soil (see notes on *C. riparia* Curtis below). *C.a.* is primarily a lowland species, with a maximum altitude in the British Isles of only 170 m (Wilson 1956). *C.a.*, which is moderately tolerant of ferrous iron toxicity (R. E. D. Cooke unpubl.), is found in a wide range of edaphic conditions including calcareous mire but is absent from acidic peat. In lowland Britain *C.a.* is primarily found in water-margin communities and fen-meadows, whereas in the uplands it is almost exclusively a species of soligenous mire. It is assumed that this difference relates to the availability of suitable habitats at different altitudes, rather than to any major ecological discontinuity in the behaviour of *C.a.* Species of *Carex* often contain cyanogenic glycosides and other toxic principles (Cooper & Johnson 1984). The role of these constituents in defence against herbivory remains uncertain for *C.a.* and for other British species of *Carex*. *C.a.* shows effective vegetative regeneration, forming extensive patches by means of rhizomatous growth. Rhizome fragments detached during periods of disturbance may be transported to other sites and form new colonies. This may be the most efficient method of long-distance dispersal, since seed set is frequently poor. We suspect that germination takes place in spring. The size of the seed bank is highly variable. The dispersule floats on water and may also be dispersed by waterfowl (Ridley 1930), but there is no evidence that *C.a.* is an early colonist of artificial habitats. A frequent associate of *C.a.* in water-side habitats is *C. riparia* Curtis ($2n = 72$ (*FE5*)). The distribution of the two, overlap but *C.r.*, which commences growth and flowers slightly earlier, consistently extends further into the water than *C.a.*

Current trends Probably decreasing.

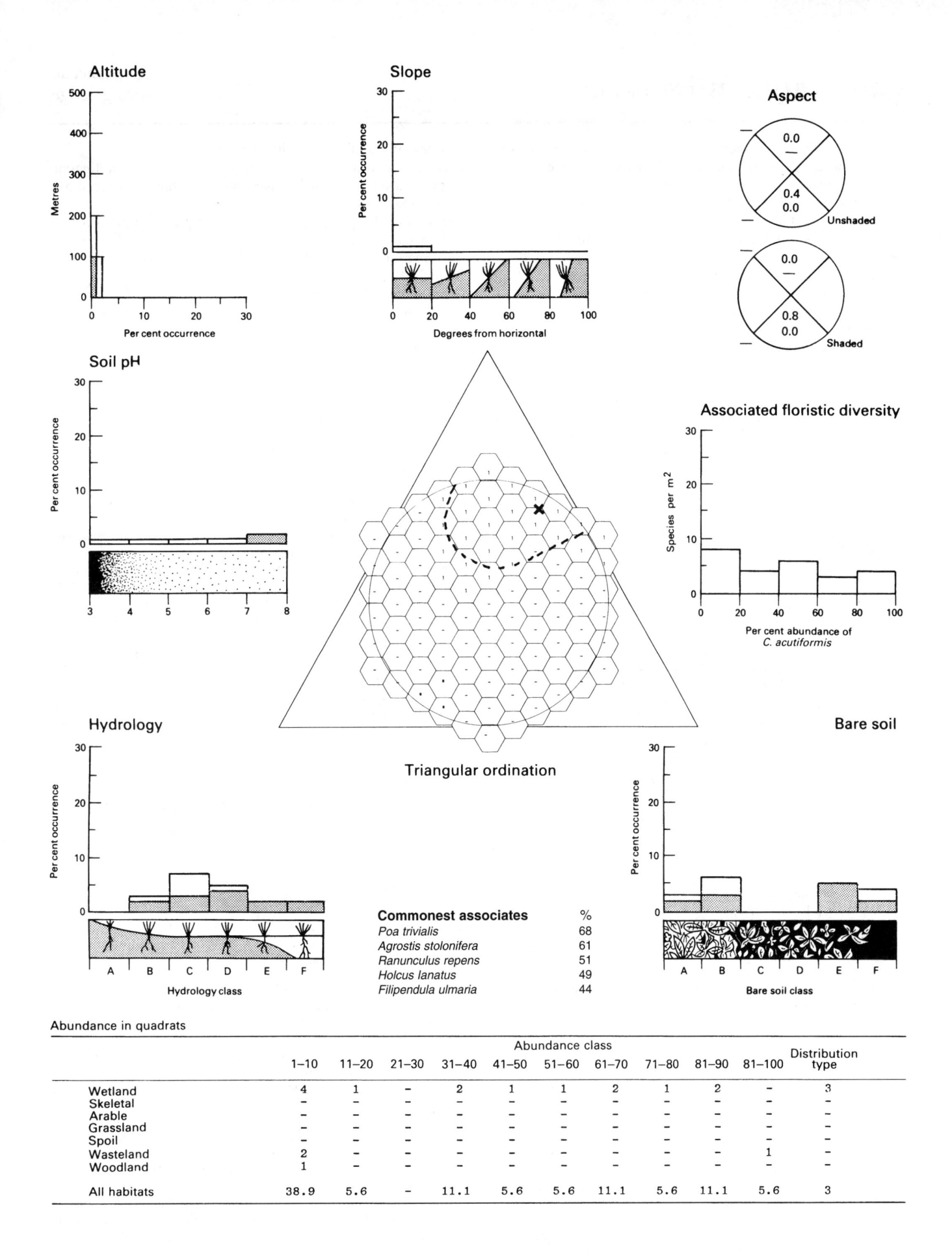

Altitude

Slope

Aspect

Unshaded

Shaded

Soil pH

Associated floristic diversity

Species per m² vs Per cent abundance of *C. acutiformis*

Hydrology

Triangular ordination

Commonest associates	%
Poa trivialis	68
Agrostis stolonifera	61
Ranunculus repens	51
Holcus lanatus	49
Filipendula ulmaria	44

Bare soil

Abundance in quadrats

	1–10	11–20	21–30	31–40	41–50	51–60	61–70	71–80	81–90	81–100	Distribution type
Wetland	4	1	–	2	1	1	2	1	2	–	3
Skeletal	–	–	–	–	–	–	–	–	–	–	–
Arable	–	–	–	–	–	–	–	–	–	–	–
Grassland	–	–	–	–	–	–	–	–	–	–	–
Spoil	–	–	–	–	–	–	–	–	–	–	–
Wasteland	2	–	–	–	–	–	–	–	–	1	–
Woodland	1	–	–	–	–	–	–	–	–	–	–
All habitats	38.9	5.6	–	11.1	5.6	5.6	11.1	5.6	11.1	5.6	3

Carex caryophyllea Latour.

Cyperaceae
Spring Sedge

Life-form Polycarpic perennial, semi-rosette hemicryptophyte with shortly creeping rhizomes. Shoots erect; leaves linear, tough, typically <500 mm^2; roots non-mycorrhizal.
Phenology Winter green. Flowers April to May. Seed set August to September. Leaves relatively long-lived.
Height Foliage to 100 mm; flowering shoots to 150 mm.
Seedling RGR No data. SLA 23.0. DMC 31.0.
Nuclear DNA amount 1.6 pg; 2n = 62, 64, 66, 68* (*FE5*, *BC*); base number obscure.
Established strategy Stress-tolerator.
Regenerative strategies V. Regeneration by seed requires further study. Regeneration primarily by means of rhizomes to form extensive but diffuse patches (see 'Synopsis').
Flowers Red-brown, monoecious, wind-pollinated, protandrous; 1-3 female spikes (often <50 flowers) overtopped by 1 male spike.
Germinule 1.19 mg, 1.7 × 1.1 mm, nut dispersed in an utricle (2.5 × 1.1 mm).
Germination Epigeal; dormancy broken by chilling.
Biological flora None.
Geographical distribution Most of the British Isles (96% of British vice-counties), *ABIF* page 732; Europe northward to Scandinavia (82% of European territories); temperate Asia and N America.

Habitats Primarily a species of dry grasslands such as calcareous pastures, waste ground and road verges, but also present in skeletal habitats such as screes, rock outcrops and spoil heaps from disused quarries and lead mines located on limestone. The use of limestone as a foundation in roadbuilding has facilitated the spread of *C.c* from the limestone areas on to adjacent road-verge sites overlying non-calcareous rock types. In a few localized upland areas *C.c.* is a natural, local constituent of grassland on acidic, podsolized soils.
Gregariousness Intermediate.
Altitude Most frequent and abundant on the carboniferous limestone (200-300 m) extending to over 350 m. [Up to 690 m.]

Slope Particularly associated with moderate slopes, a reflection of its frequency on Derbyshire dale-sides.
Aspect Slightly more frequent on N-facing slopes.
Hydrology Absent from wetlands.
Soil pH Recorded on soils in the pH range 4.0-8.0.
Bare soil Associated with habitats having little exposed soil.
△ **ordination** Distribution confined to vegetation associated with infertile, relatively undisturbed habitats.
Associated floristic diversity Typically species-rich; even sites with soil of pH 4.0 are of intermediate species-richness.
Synopsis *C.c.* is an inconspicuous, but locally abundant, winter-green creeping sedge of unproductive low turf, and usually only reaches high frequencies in situations where dominance by taller and more-robust species is prevented by mineral nutrient stress and grazing. In pastures the

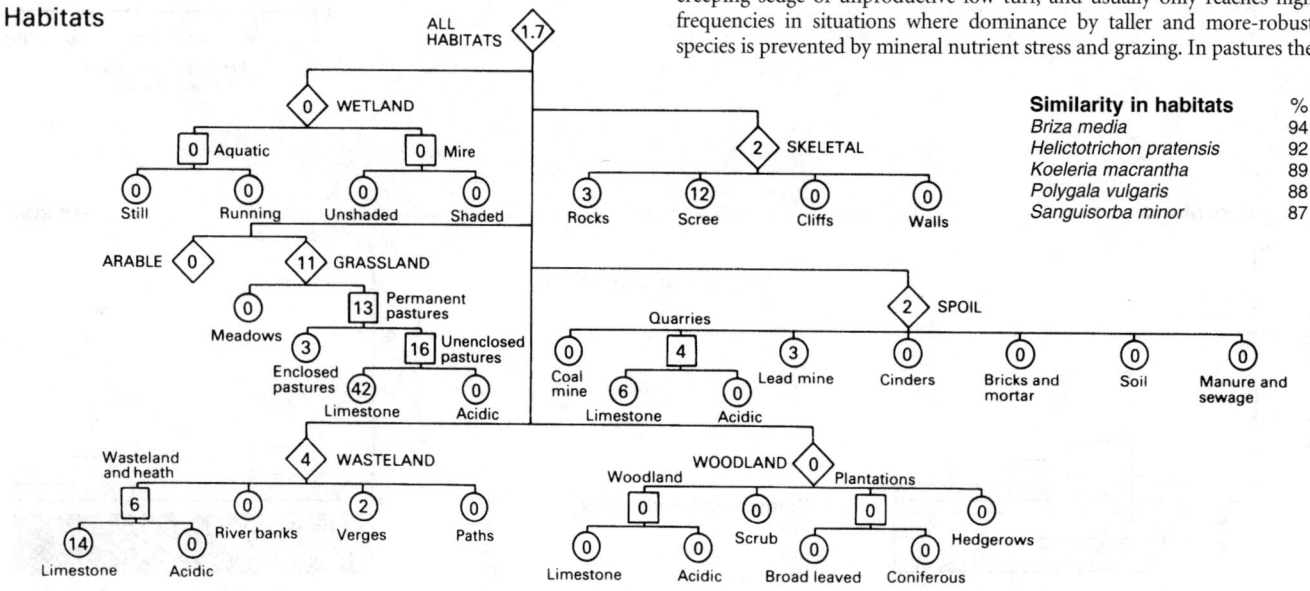

Habitats

Similarity in habitats	%
Briza media	94
Helictotrichon pratensis	92
Koeleria macrantha	89
Polygala vulgaris	88
Sanguisorba minor	87

canopy of *C.c.* may become submerged between grazing episodes, and the species shows considerable shade-tolerance. However, in derelict calcareous pastures where the leaves of *C.c.*, which do not exceed 200 mm (Jermy *et al.* 1982), become heavily shaded, the species tends slowly to disappear. In common with most sedges, *C.c.* is non-mycorrhizal (Harley & Harley 1986); this dependence upon the root system alone for the uptake of mineral nutrients is unusual among species from infertile calcareous grasslands. Though largely restricted to calcareous soils, *C.c.* shows wide edaphic tolerance. In the Sheffield region the species is restricted to calcareous sites within the lowland area; additional fieldwork has revealed that in the uplands *C.c.* extends onto moderately acidic podsolized soils, of pH as low as 4.0. This accords with the British distribution, where *C.c.* is frequent on calcareous soils in the S and E but extends onto more acidic soils in mountain areas (Jermy *et al.* 1982). This pattern is reminiscent of that of *Helianthemum nummularium* and *Thymus polytrichus*. With *Viola hirta*, *C.c.* is the earliest-flowering common species of calcareous grassland. Nevertheless, like other sedges *C.c.* has small cells and a low DNA amount and leaf growth is delayed until late spring (Grime *et al.* 1985). Seed production is often extremely poor and fruits, which are ripe by May to July, are frequently retained on the plant until August to September, leading to losses during grazing. Seedlings are rarely seen in the field, and time of seed germination is uncertain, although the known behaviour of similar grassland sedges predicts spring germination. The information needed to characterize the seed bank is lacking. However, it is clear that regeneration is mainly by vegetative means from clonal patches which may fragment to form daughter colonies. Although *C.c.* has colonized lime-enriched roadsides across acidic moorland, the species appears to have a low potential for colonizing new sites and is characteristic of old grasslands (Wells *et al.* 1976). There is no information on the range of ecotypes formed by *C.c.*

Current trends *C.c.* shows low seed production and poor mobility and is vulnerable both to increased fertility and to the relaxation of grazing pressure. Thus, *C.c.* is likely to decrease still further, particularly in lowland areas.

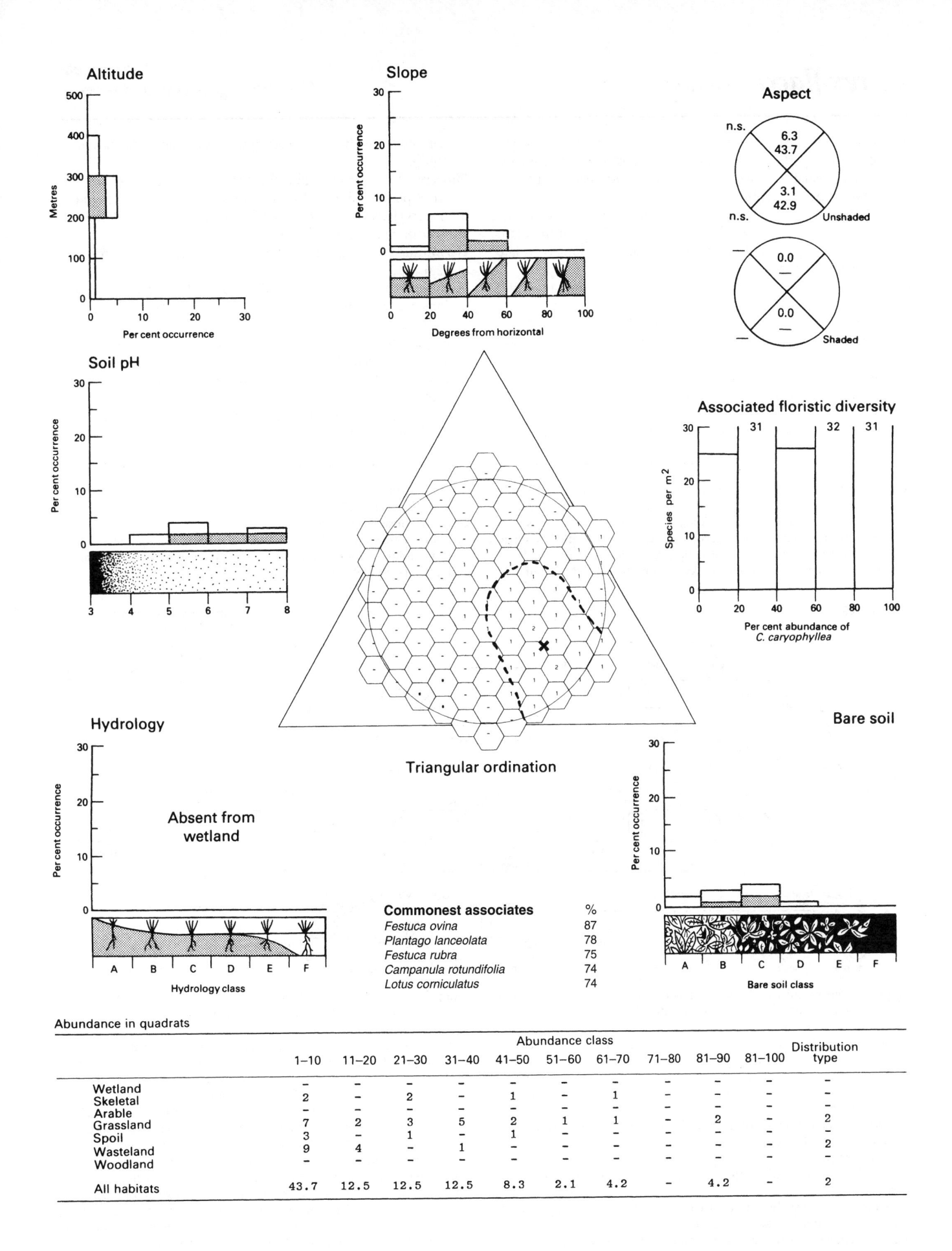

Altitude

Slope

Aspect

	n.s.	6.3	
		43.7	
	n.s.	3.1	Unshaded
		42.9	

	—	0.0	
	—	0.0	Shaded

Soil pH

Associated floristic diversity

Species per m²

Per cent abundance of *C. caryophyllea*

Hydrology

Absent from wetland

Triangular ordination

Bare soil

Commonest associates %
Festuca ovina 87
Plantago lanceolata 78
Festuca rubra 75
Campanula rotundifolia 74
Lotus corniculatus 74

Abundance in quadrats

	\multicolumn{10}{c}{Abundance class}	Distribution type
	1–10	11–20	21–30	31–40	41–50	51–60	61–70	71–80	81–90	81–100	
Wetland	–	–	–	–	–	–	–	–	–	–	–
Skeletal	2	–	2	–	1	–	1	–	–	–	–
Arable	–	–	–	–	–	–	–	–	–	–	–
Grassland	7	2	3	5	2	1	1	–	2	–	2
Spoil	3	–	1	–	1	–	–	–	–	–	2
Wasteland	9	4	–	1	–	–	–	–	–	–	–
Woodland	–	–	–	–	–	–	–	–	–	–	–
All habitats	43.7	12.5	12.5	12.5	8.3	2.1	4.2	–	4.2	–	2

Carex flacca Schreber
Cyperaceae
Carnation-grass, Glaucous Sedge

Life-form Polycarpic perennial, semi-rosette hermcryptophyte with creeping rhizomes. Shoots erect; leaves linear, glaucous beneath, usually <1000 mm^2; roots rather shallow, usually without VA mycorrhizas.

Phenology Winter green. Flowers from May to June. Fruits formed from June onwards, but usually retained on the plant until autumn. Leaves long-lived and replaced annually in late spring (Sydes 1981).

Height Foliage usually <300 mm, but flowering shoots to 600 mm.

Seedling RGR 1.0-1.4 week^{-1}. SLA 16.8. DMC 31.8.

Nuclear DNA amount 0.6 pg; $2n = 76*$, 90 (*BC, FE5*); base number obscure.

Established strategy Stress-tolerator.

Regenerative strategies V, B$_s$. Mainly vegetative, by means of rhizomes forming extensive patches. Seedlings appear in the spring but are only rarely observed in the field (*Biol Fl*). A persistent seed bank has been reported (Donelan & Thompson 1980).

Flowers Dark brown, purplish-brown or black, monoecious, wind-pollinated and also occasionally visited by pollen-eating insects (Jermy *et al.* 1982); 1-5 female spikes (often >50 flowers) overtopped by 2-3 male spikes.

Germinule 0.37 mg, 1.6 × 1.3 mm, nut dispersed in an utricle (2.6 × 1.5 mm).

Germination Epigeal; t_{50} 7 days; 16-28°C; germination rate increased by chilling and to a lesser extent by dry storage at 20°C; L > D.

Biological flora Taylor (1956), see also Zarzycki (1976).

Geographical distribution Britain, but less frequent in the N and W (100% of British vice-counties), *ABIF* page 725; most of Europe, but absent from parts of the NE, also absent from the extreme S (92% of European territories); N Africa; introduced in N America and New Zealand.

Habitats Recorded most frequently in pasture and on stable scree slopes in the limestone dales. Also widespread on calcareous strata in pastures, wasteland and old lead-mine workings. Less frequent in open scrub, woodland clearings and rides. Observed in calcareous mire and in a range of older artificial habitats, e.g. cinder tips, coal-mine spoil and enclosed pasture. [Frequent in sand-dune slacks and on cliff tops. Also found in montane habitats and estuarine marshes, where it can withstand some salinity (Jermy *et al.* 1982).]

Gregariousness Intermediate.

Altitude Distribution shows an upland bias reflecting the fact that habitat conditions suitable for *C.f.* are most frequent on the carboniferous limestone. [Up to 790 m.]

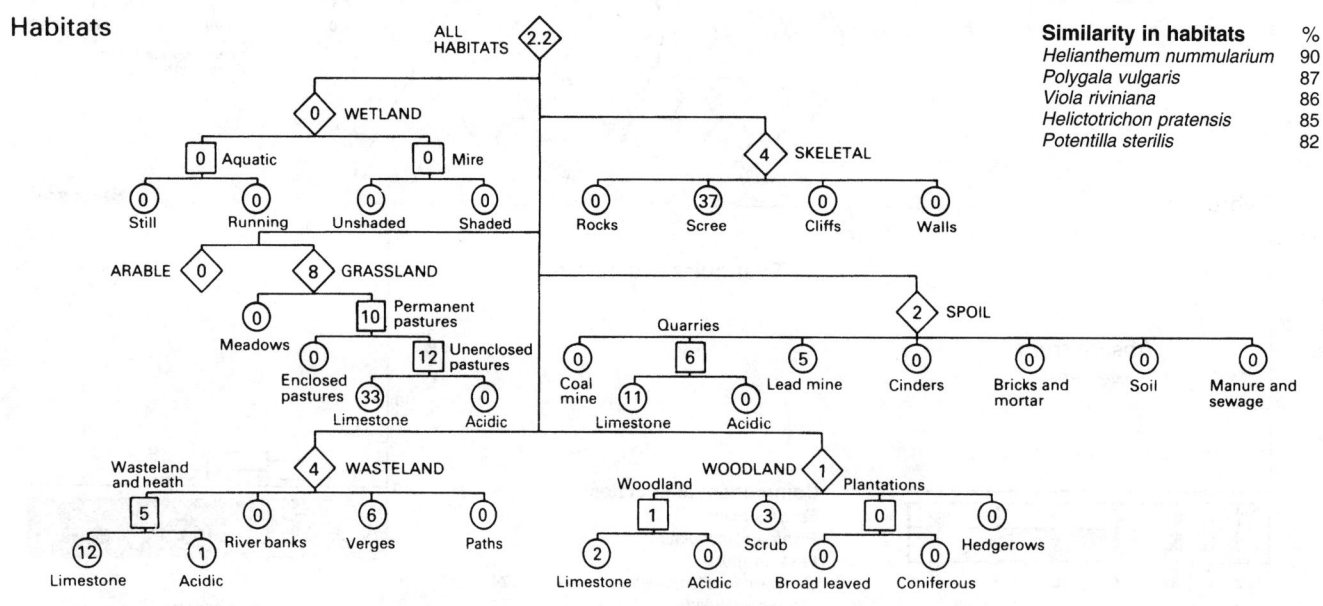

Habitats

Similarity in habitats	%
Helianthemum nummularium	90
Polygala vulgaris	87
Viola riviniana	86
Helictotrichon pratensis	85
Potentilla sterilis	82

Slope Found on all but the steepest slopes. Because *C.f.* is especially frequent on the Derbyshire dale sides, the distribution is biased towards moderate angles of slope.

Aspect More frequent and abundant on unshaded N-facing slopes but this trend is not statistically significant.

Hydrology In wetland habitats *C.f.* is largely restricted to soligenous mire. [Also occurs in dune slacks which may be flooded for 4-6 months of the year (Willis *et al.* 1959).]

Soil pH Shows a broadly calcicolous distribution with most records in the range 7.0-8.0, but also common on slightly more acidic soils. Rare in habitats with pH <5.0.

Bare soil Associated with habitats having low to moderate amounts of exposed soil.

△ **ordination** Distribution centred on vegetation associated with unproductive habitats. Absent from disturbed habitats.

Associated floristic diversity Most typical of species-rich vegetation.

Synopsis *C.f.* is the commonest British sedge of nutrient-deficient calcareous soils and is particularly frequent in short turf. The species is widespread but generally less abundant on non-calcareous strata. Studies of this presumably long-lived, stress-tolerant species in derelict limestone grassland revealed little seasonal change in shoot biomass (Al-Mufti *et al.* 1977). This is due to leaves surviving for 1 year and being replaced annually in late spring (Sydes 1981). Thus the leaves appear to have a longer life-span than many of the species with which *C.f.* is commonly associated (Sydes 1981), and we predict that *C.f.* is among the less palatable species of calcareous grassland. *C.f.* exhibits a N-facing bias in grassland (*Ecol Atl*). Infrequent occurrence in shaded habitats may be related more to susceptibility to fungal attack than to intolerance of low light *per se*, since experimental plants grow successfully in sand-culture at 1.6% daylight while failing at this light intensity under the more humid conditions associated with solution culture (Hunt 1970). Once established on a site *C.f.* forms large clonal patches. A persistent seed bank has been reported

Altitude
Slope
Aspect
Soil pH
Associated floristic diversity
Hydrology
Triangular ordination
Bare soil

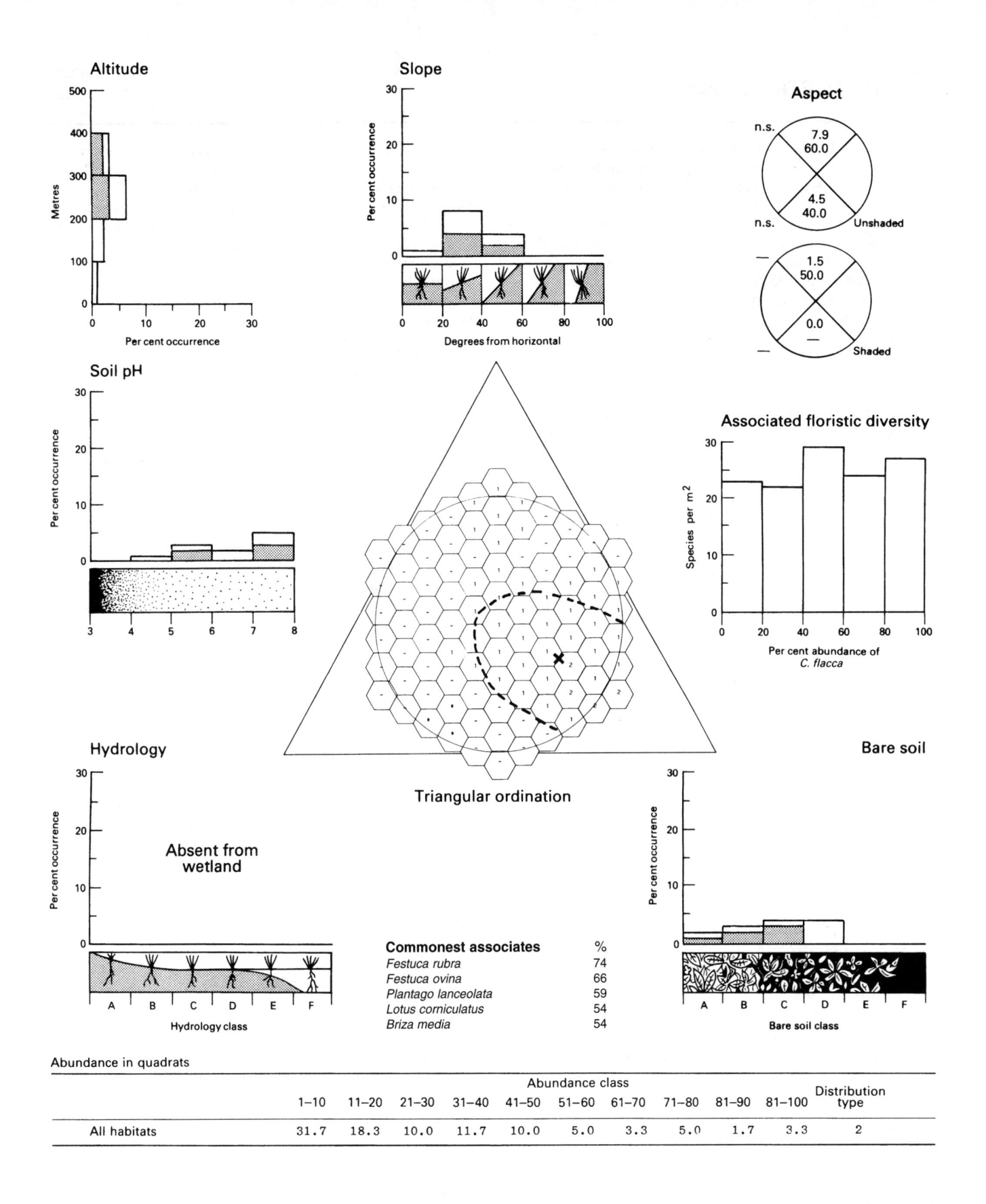

Commonest associates	%
Festuca rubra	74
Festuca ovina	66
Plantago lanceolata	59
Lotus corniculatus	54
Briza media	54

Abundance in quadrats

	Abundance class										Distribution type
	1–10	11–20	21–30	31–40	41–50	51–60	61–70	71–80	81–90	81–100	
All habitats	31.7	18.3	10.0	11.7	10.0	5.0	3.3	5.0	1.7	3.3	2

(Donelan & Thompson 1980) but regeneration by seed appears to be a rare event. *C.f.* extends into a wide range of calcareous habitats including soligenous mire (often with *C. panicea*). Survival in wetland habitats may be related to the capacity of the roots to diffuse oxygen from shoot to root to soil (Jones 1975). The extent to which ecotypes have been formed in this morphologically variable and ecologically wide-ranging species appears not to have been investigated. However, relative growth rates of comparable ramets cloned from a single genet of *C.f.* have been shown to vary by more than twofold (Hunt 1984).

Current trends   *C.f.* is a species of infertile, often calcareous soils and is declining, particularly in lowland areas, as a result of the replacement of infertile pasture by a range of disturbed and fertile habitats.

Carex nigra (L.) Reichard

Cyperaceae
Common Sedge

Life-form Polycarpic perennial, semi-rosette hemicryptophyte or helophyte, typically with a creeping rhizome but sometimes forming dense tufts. Stems erect; leaves linear, usually <1000 mm^2; roots lacking VA mycorrhizas.

Phenology New shoots elongate in spring. Flowers from May to July and seed shed August onwards. Shoots senesce during winter. Leaves long-lived.

Height Foliage and flowers usually <400 mm; in tufted forms, shoots may extend a further 100 mm above the soil surface.

Seedling RGR No data. SLA 11.9. DMC 34.0.

Nuclear DNA amount 1.1 pg; $2n$ = 82, 83*-85 * (*BC*, *FE5*); base number obscure.

Established strategy Intermediate between stress-tolerant competitor and stress-tolerator.

Regenerative strategies V, B$_s$. Typically forms large patches as a result of rhizome extension. These patches often split with the disintegration of old rhizomes. Regeneration by seeds, which germinate in spring, is less important. A persistent seed bank has been reported (K. Thompson unpubl.); ?SB IV.

Flowers Black, monoecious, wind-pollinated and also visited by pollen-eating insects, almost completely self-incompatible (Faulkner 1973); 1-4 female spikes (often >100 flowers) overtopped by 1-2 male spikes.

Germinule 0.81 mg, 1.8 × 1.1 mm, nut in an utricle (2.7 × 1.1 mm).

Germination Epigeal; t_{50} 4 days; 19-37°C; dormancy broken by chilling and less effectively by dry storage at 20°C; L > D.

Biological flora None.

Geographical distribution The British Isles (100% of British vice-counties), *ABIF* page 737; most of Europe, although rare in the extreme S (87% of European territories); W Asia, N America and Australia.

Habitats A wetland species. Though most frequent in unshaded and shaded mire, *C.n.* is also found in moist pasture and wasteland. Generally absent from permanently aquatic habitats. Recorded from, but not persisting on, arable land. [Found in dune slacks and very common in marshy ground in montane areas (Jermy *et al.* 1982).]

Gregariousness Intermediate to stand-forming.

Altitude Widespread, but suitable habitats more frequent in the uplands. [Up to 990 m.]

Slope Like most wetland species, mainly restricted to relatively flat ground.

Aspect Insufficient data.

Hydrology Most frequent and abundant in soligenous mire, but also widespread in topogenous mire and on moist soils.

Soil pH Most commonly and abundantly recorded on acidic soils of pH <5.0, but distribution extends locally to highly calcareous soils.

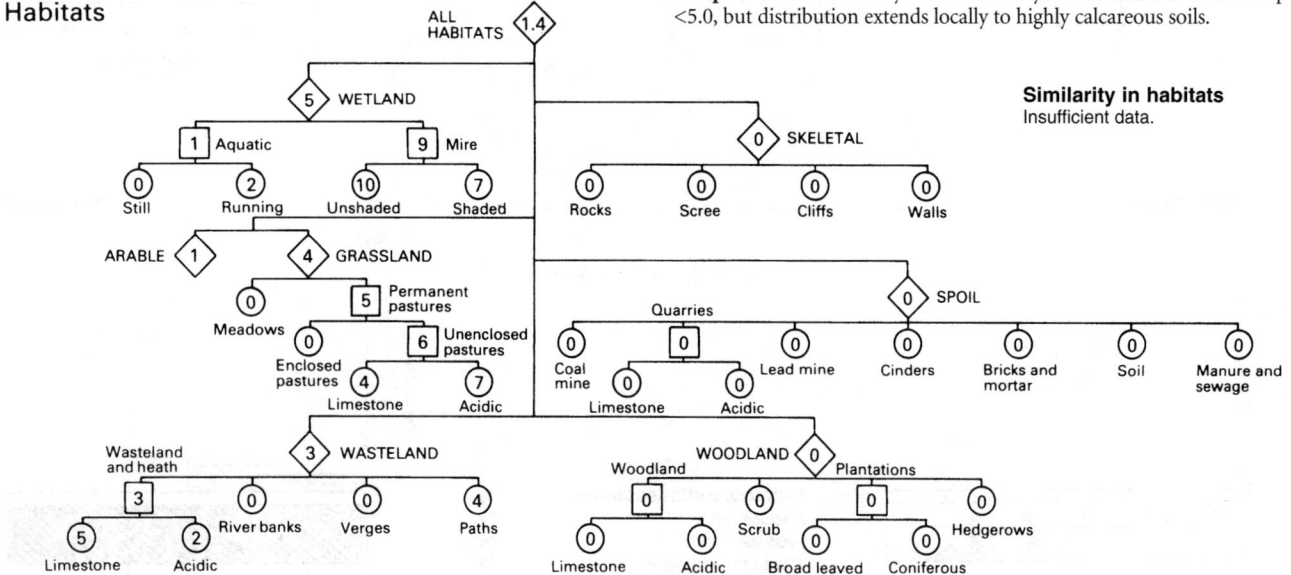

Habitats

Similarity in habitats
Insufficient data.

Bare soil Mainly occurs in habitats with little exposed soil. Often associated with a carpet of bryophytes, particularly *Sphagnum* spp.

△ **ordination** Distribution centred on relatively undisturbed and unproductive vegetation. Absent from heavily disturbed and from undisturbed, productive vegetation.

Associated floristic diversity On acidic soils generally associated with low species diversity. In sites of high soil pH *C.n.* frequently occurs in species-rich vegetation.

Synopsis An ecologically wide-ranging species in need of further investigation. *C.n.* is a rhizomatous, stress-tolerant sedge most typical of soligenous mire in sites where the growth of potential dominants is suppressed by low fertility (and often also by grazing pressure). *C.n.* also occurs beside upland streams, in moist grassland and flood hollows, but is virtually absent from mire which is adjacent to open water. Although typically a minor component of vegetation, *C.n.*, which is taller than *C. panicea* and shows winter die-back, may exert local dominance particularly on acidic peaty soils. *C.n.* also extends into calcareous mire. Regeneration is mainly by vegetative means, and large clonal patches may be formed as a result of rhizome growth. A persistent seed bank has been reported (K. Thompson pers. comm.) but regeneration by seed is probably a rare event. Further, *C.n.* rarely colonizes recent artificial habitats. *C.n.* hybridizes with four other members of the *Carex nigra* group (Stace 1975). Some of the variation within *C.n.* in Britain may be due to subsequent introgression (Jermy *et al.* 1982); this, in turn, may be responsible in part for the wide habitat range and the lack of cohesive ecological identity in the species. In topogenous mire subject to large fluctuations in water table, *C.n.* occasionally forms large tussocks, perhaps for reasons similar to those advanced for *Molinea caerulea*. In dune slacks and around sheep shelters the leaves are short and broad, while those in nutrient-poor mire are long and narrow (Jermy *et al.* 1982).

Current trends Still common. but decreasing as a result of habitat destruction.

Altitude
Slope
Aspect

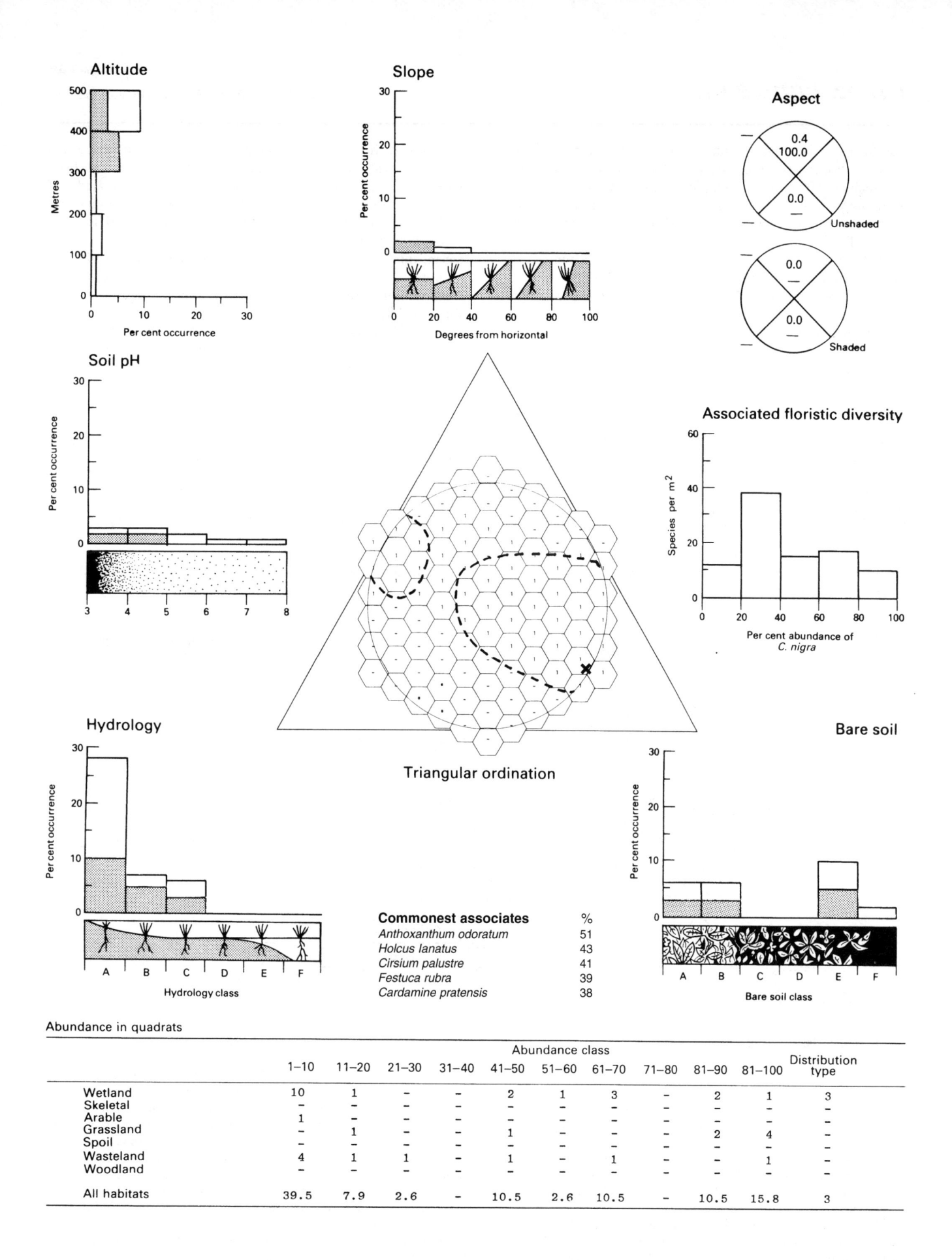

Unshaded

Shaded

Soil pH

Associated floristic diversity

Hydrology

Triangular ordination

Commonest associates	%
Anthoxanthum odoratum	51
Holcus lanatus	43
Cirsium palustre	41
Festuca rubra	39
Cardamine pratensis	38

Bare soil

Abundance in quadrats

	Abundance class										Distribution type
	1–10	11–20	21–30	31–40	41–50	51–60	61–70	71–80	81–90	81–100	
Wetland	10	1	–	–	2	1	3	–	2	1	3
Skeletal	–	–	–	–	–	–	–	–	–	–	–
Arable	1	–	–	–	–	–	–	–	–	–	–
Grassland	–	1	–	–	1	–	–	–	2	4	–
Spoil	–	–	–	–	–	–	–	–	–	–	–
Wasteland	4	1	1	–	1	–	1	–	–	1	–
Woodland	–	–	–	–	–	–	–	–	–	–	–
All habitats	39.5	7.9	2.6	–	10.5	2.6	10.5	–	10.5	15.8	3

Carex panicea L.

Cyperaceae
Carnation-grass, Carnation Sedge

Life-form Polycarpic perennial, semi-rosette hemicryptophyte with shortly creeping rhizomes. Stems erect; leaves linear, usually <1000 mm^2; roots typically lacking VA mycorrhizas.
Phenology Winter green. Leaf growth in late spring followed by flowering in May and June. Seed shed July to October. Leaves long-lived.
Height Foliage <300 mm; flowering shoots to 600 mm.
Seedling RGR 0.5-0.9 week^{-1}. SLA 15.5. DMC 35.1.
Nuclear DNA amount 2.0 pg; $2n = 32^*$ (*BC*); base number obscure.
Established strategy Intermediate between stress-tolerator and C-S-R.
Regenerative strategies V, B$_s$. Mainly vegetative, by means of creeping rhizomes. B$_s$ seedlings, which appear in spring and summer (Muller 1978), are seen rather infrequently in the field.
Flowers Purplish or brown, monoecious, wind-pollinated and also occasionally visited by pollen-eating insects, protandrous (Jermy *et al.* 1982); 1-3 female spikes (often <50 flowers) overtopped by 1 male spike.
Germinule 1.88 mg, 2.1 × 1.5 mm, nut in an utricle (3.7 × 2.1 mm).
Germination Epigeal; dormancy broken by chilling.
Biological flora None.
Geographical distribution Widespread in the British Isles, particularly in N and W (100% of British vice-counties), *ABIF* page 726; most of Europe, although rare in the Mediterranean region (82% of European territories); much of the USSR, but sometimes observed on the steep slopes of wet cliffs and outcrops.
Aspect All records are from unshaded N-facing slopes.
Hydrology Most characteristic of soligenous mire. Also observed on moist soils and tussocks of other wetland species but infrequent in topogenous mire.
Soil pH Occurs on soils within the pH range 4.0-7.5. Slightly more frequent and abundant in the range 5.0-6.0.
Bare soil Exposures usually low.
△ **ordination** Distribution confined to vegetation associated with unproductive and relatively undisturbed habitats.

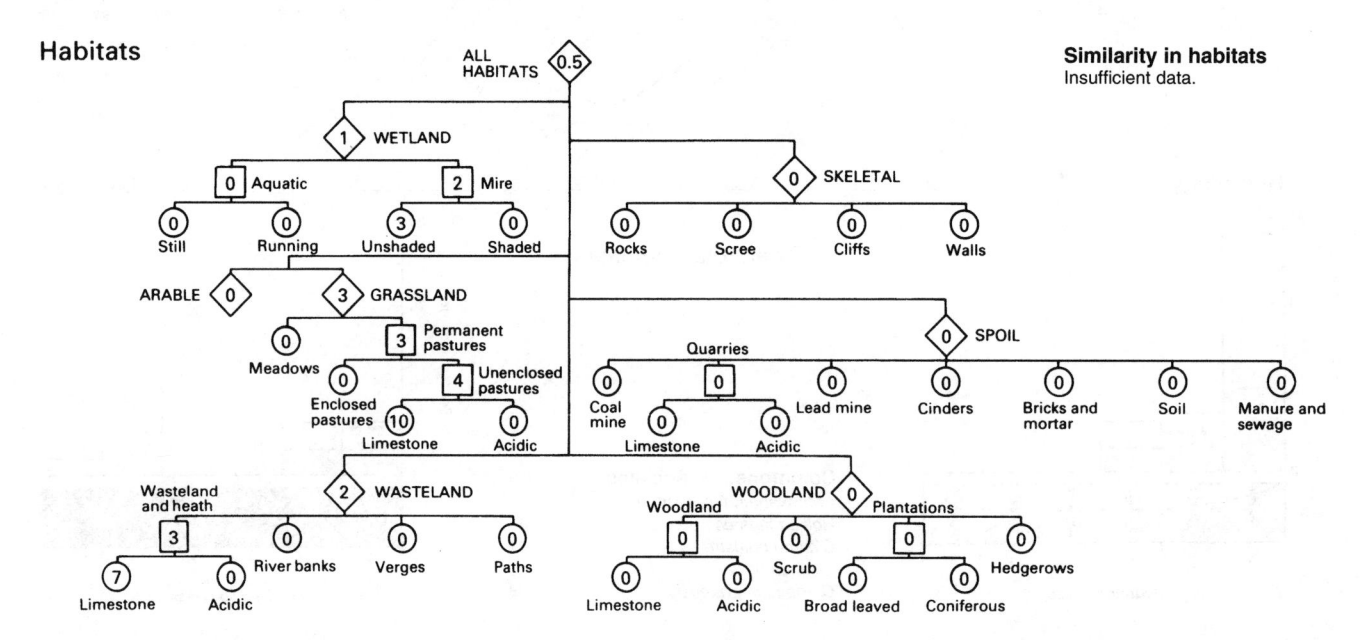

Habitats Found in damp pastures and grassy wasteland over limestone. Although only encountered infrequently in unshaded wetland sites in this survey, *C.p.* is locally an important constituent of certain types of mire, both on calcareous and non-calcareous strata. [In Britain as a whole *C.p.* is most characteristically a wetland species.]
Gregariousness Typically intermediate, but locally patch-forming in mire.
Altitude Observed over a wide range of altitudes to 450 m but suitable habitats more frequent in the upland region. [Up to 1130 m.]
Slope Most records are from flat ground or gentle to moderate slopes,

Habitats

Similarity in habitats
Insufficient data.

Associated floristic diversity Typically species-rich.
Synopsis *C.p.* is a winter-green, low-growing, rhizomatous sedge most characteristic of moist or waterlogged sites where the growth of potential dominants is suppressed by low soil fertility and by grazing or cutting. The species, which is virtually absent from shaded habitats, is most typical of soligenous mire, both on moderately base-poor soils with *C. viridula* Michx. and *C. echinata* Murray and in highly calcareous sites, often with the smaller-seeded *C. flacca*. *C.p.* also occurs in wet pasture and dune slacks, and within the Sheffield region is particularly characteristic of calcareous grassland on moist N-facing slopes. A similar capacity to exploit both soligenous mire and limestone grassland in Northern Britain is exhibited by a number of grassland species, e.g. *Anthoxanthum odoratum*, *Briza media*, *C.f.* and *Festuca rubra*, and by some predominantly wetland species (*Anagallis tenella* (L.) L., *C. hostiana* DC, *C. pulicaris* L., *Parnassia palustris* L., *Pinguicula vulgaris* L. and *Valeriana dioica* L.). At small springs within soligenous mire, *C.p.* is often replaced close to the water's edge by species more exclusively restricted to wetland, such as *C.v.* and *C.e.* in base-poor sites, or by *Eleocharis quinqueflora* (F. X. Hartmann) O. Schwartz in base-rich areas. *C.p.* frequently forms diffuse but extensive clonal patches, usually in species-rich vegetation. Regeneration by means of the rather large seeds, which germinate in the spring, is probably infrequent. The utricle surrounding the nut gives the dispersule prolonged buoyancy in water (Ridley 1930), and where the species occurs close to water this may be an important method of colonizing new sites. However, in lowland areas at least, *C.p.* appears a poor colonist and is indicative of older vegetation types.
Current trends Decreasing as a result of habitat destruction. Already becoming rare in some lowland areas.

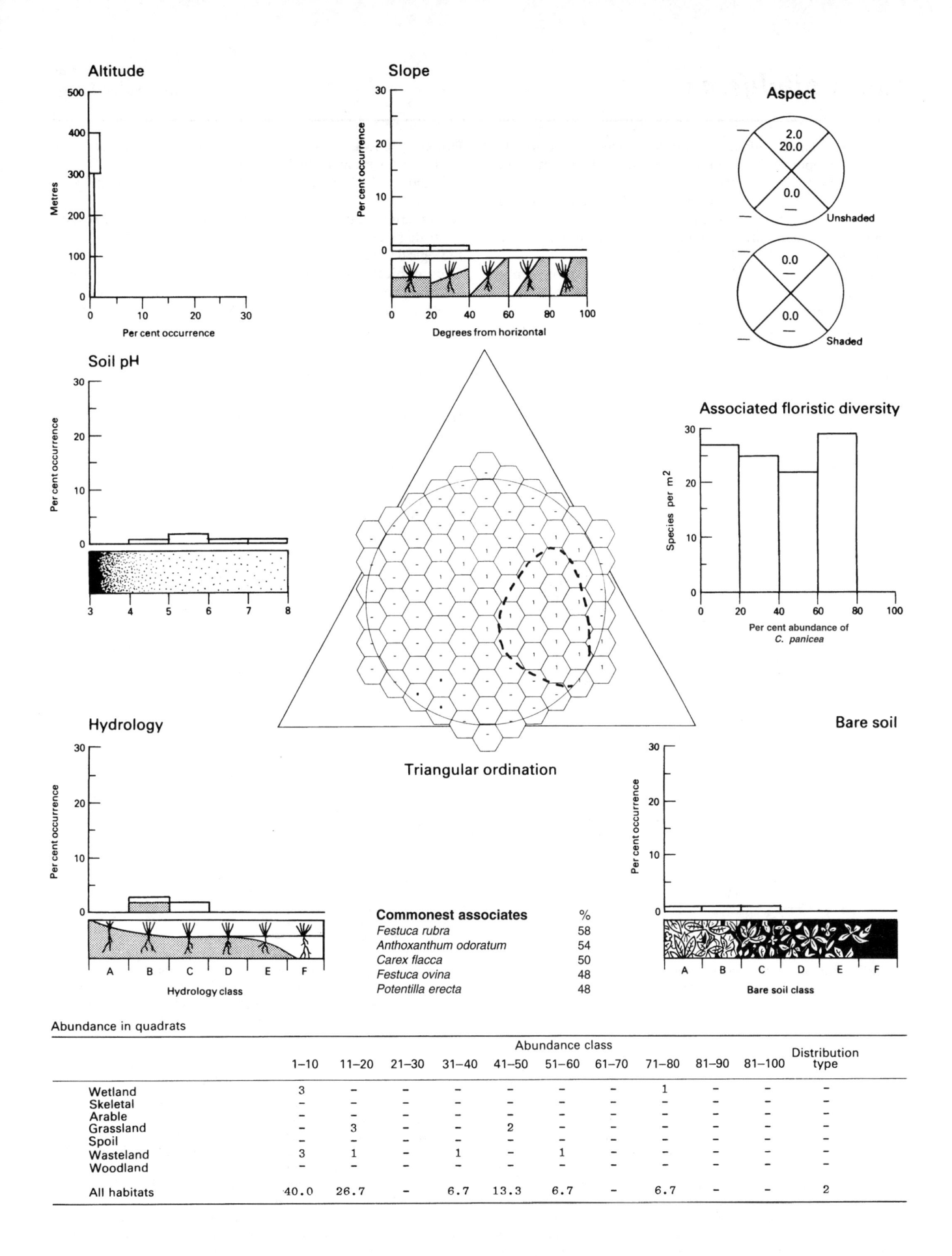

Altitude

Slope

Aspect

Soil pH

Associated floristic diversity

Hydrology

Triangular ordination

Bare soil

Commonest associates	%
Festuca rubra	58
Anthoxanthum odoratum	54
Carex flacca	50
Festuca ovina	48
Potentilla erecta	48

Abundance in quadrats

	1–10	11–20	21–30	31–40	41–50	51–60	61–70	71–80	81–90	81–100	Distribution type
Wetland	3	–	–	–	–	–	–	1	–	–	–
Skeletal	–	–	–	–	–	–	–	–	–	–	–
Arable	–	–	–	–	–	–	–	–	–	–	–
Grassland	–	3	–	–	2	–	–	–	–	–	–
Spoil	–	–	–	–	–	–	–	–	–	–	–
Wasteland	3	1	–	1	–	1	–	–	–	–	–
Woodland	–	–	–	–	–	–	–	–	–	–	–
All habitats	40.0	26.7	–	6.7	13.3	6.7	–	6.7	–	–	2

Carex pilulifera L.

Cyperaceae
Pill-headed Sedge

Life-form Densely tufted, polycarpic perennial, semi-rosette hemicryptophyte. Shoots erect; leaves linear, <1000 mm^2; roots apparently non-mycorrhizal.
Phenology Winter green. Flowers May and June and fruits June to July. Leaves relatively long-lived.
Height Foliage typically <100 mm; flowering stems to 300 mm.
Seedling RGR No data. SLA 20.4. DMC 32.8.
Nuclear DNA amount No data; 2n = 18* (*BC*); base number obscure.
Established strategy Stress-tolerator.
Regenerative strategies B$_s$, V. Timing of seed germination uncertain (see 'Synopsis'); SB IV.

Flowers Brown, mainly wind-pollinated; 2-4 female spikes (<50 flowers) overtopped by 1 male spike.
Germinule 1.17 mg, 1.5 × 1.2 mm, nut in an utricle (2.5 × 1.3 mm).
Germination Epigeal; dormancy breaking mechanism requires further study.
Biological flora None.
Geographical distribution A European endemic found throughout the British Isles (98% of British vice-counties), *ABIF* page 733; mainly N and W Europe (41% of European territories).

Habitats Typically found in short turf on acidic soils. Most frequent in upland pasture, but also occurs on quarry spoil, waste land, paths and outcrops, and extends into lightly shaded woodland habitats. [Found in acidic montane grassland.]
Gregariousness Sparse to intermediate.
Altitude Most-suitable habitats occur on the millstone grit uplands. [Up to 1010 m.]
Slope Generally absent from skeletal habitats and from the steepest slopes.

Aspect A slight bias towards N-facing slopes.
Hydrology Absent from wetland habitats.
Soil pH Distribution strongly biased towards acidic soils even when growing on calcareous strata.
Bare soil Not recorded from sites with high exposures of bare soil.
Δ ordination Distribution strongly centred on unproductive, undisturbed types of vegetation.
Associated floristic diversity Low.

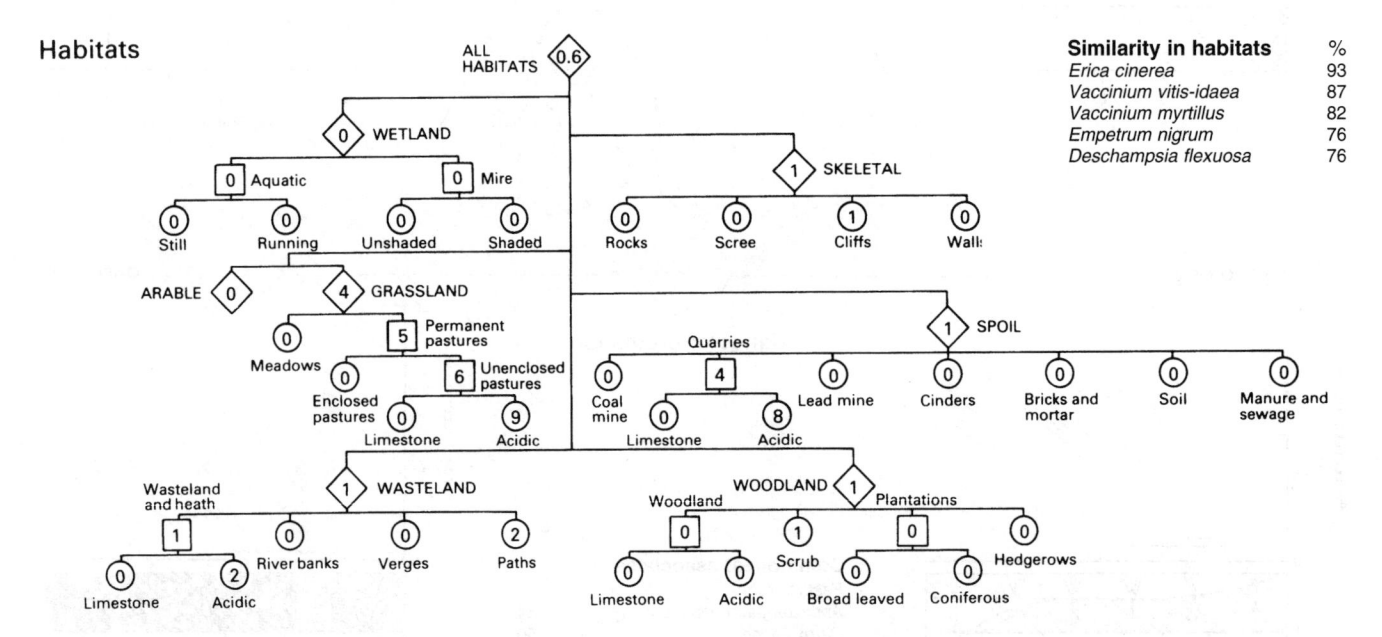

Habitats

Similarity in habitats	%
Erica cinerea	93
Vaccinium vitis-idaea	87
Vaccinium myrtillus	82
Empetrum nigrum	76
Deschampsia flexuosa	76

Synopsis *C.p.* is a tufted winter-green sedge most typically found in infertile grassland and heathland on leached, moderately acidic soils. The species has also been observed, albeit rarely, on calcareous clay soils on limestone, and its distribution extends into dry habitats to a greater extent than that of any other common sedge. The tussock growth form of *C.p.* affords the young shoots of the mature plant some protection against burning and the shoot apices are protected by leaf bases (Mallik & Gimingham 1985). Although moderately resistant to trampling and able to persist and set seed in light shade, *C.p.* is more susceptible to close grazing than more-productive species such as *Agrostis capillaris*, and is often replaced in little-managed sites by more-robust species such as *Nardus stricta*. However, *C.p.* produces a persistent seed bank (Hill & Stevens 1981, Mallik *et al.* 1984, Kjellsson 1985b) and would be expected to be most abundant following extensive disturbance. This suggestion is consistent with the behaviour of *C.p.* described by Mallik *et al.* (1984) in *Calluna* heath burned about every 15 years. Here, abundant seed is produced only during the early phase of the *Calluna* regeneration cycle, at which time the seed bank of *C.p.* rapidly increases in size with stand age. The common occurrence of *C.p.* in old gritstone quarries is also strongly indicative of the importance of past disturbance events, in allowing the species to establish as an opportunist colonist of bare areas. Some lateral vegetative expansion of established plants occurs and patches may occasionally exceed 0.5 m in diameter. However, *C.p.* retains its densely tufted habit and there is little evidence of extensive clonal development. Thus, effective regeneration is by seed. Both the mechanisms breaking seed dormancy and the seasonality of natural germination require further study. Seeds temporarily possess an elaiosome and are dispersed by ants up to *c.* 1.4 m away from the immediate canopy of the parent plant, to sites which may be more conducive to seedling establishment (Kjellsson 1985a). Heavy seed predation by mice and beetles has been noted (Kjellsson 1985b). Geographically isolated plants are sometimes found by paths and occasional long-distance dispersal by humans or vehicles is suspected. Another ssp. of *C.p.* is endemic to the Azores (*FE5*).

Current trends Although *C.p.* can colonize some artificial habitats, few suitable habitats for *C.p.* are being created, and the species is probably decreasing as a result of the destruction of heaths and acidic grassland.

Altitude
Slope
Aspect

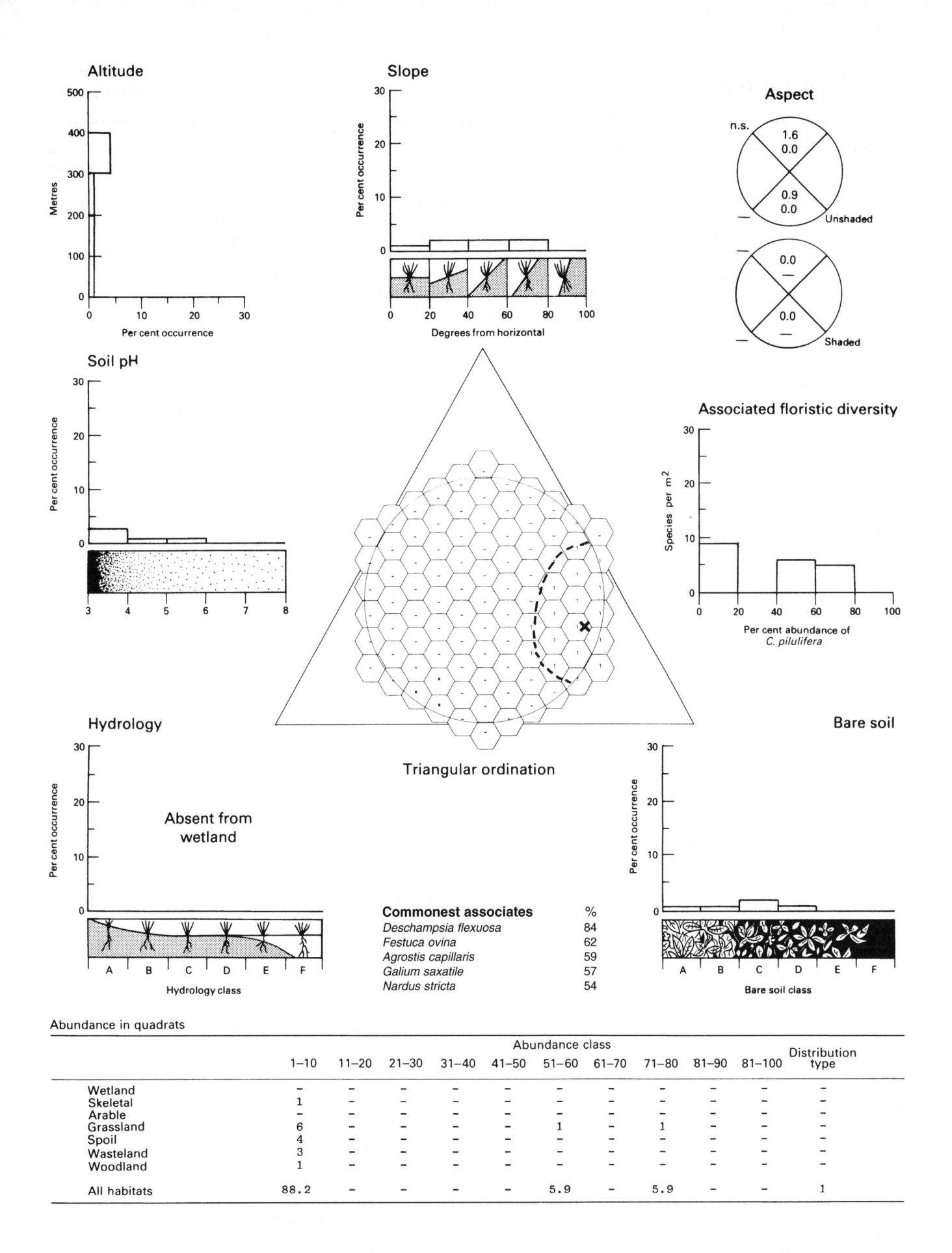

Aspect circles:
- Unshaded: n.s. 1.6 / 0.0 / 0.9 / 0.0
- Shaded: — / 0.0 / 0.0 / —

Soil pH

Associated floristic diversity

Species per m² vs Per cent abundance of *C. pilulifera*

Hydrology

Absent from wetland

Triangular ordination

Bare soil

Commonest associates	%
Deschampsia flexuosa	84
Festuca ovina	62
Agrostis capillaris	59
Galium saxatile	57
Nardus stricta	54

Abundance in quadrats

	1–10	11–20	21–30	31–40	41–50	51–60	61–70	71–80	81–90	81–100	Distribution type
Wetland	–	–	–	–	–	–	–	–	–	–	–
Skeletal	1	–	–	–	–	–	–	–	–	–	–
Arable	–	–	–	–	–	–	–	–	–	–	–
Grassland	6	–	–	–	–	1	–	1	–	–	–
Spoil	4	–	–	–	–	–	–	–	–	–	–
Wasteland	3	–	–	–	–	–	–	–	–	–	–
Woodland	1	–	–	–	–	–	–	–	–	–	–
All habitats	88.2	–	–	–	–	5.9	–	5.9	–	–	1

Carlina vulgaris L.

Asteraceae
Carline Thistle

Life-form Spiny, monocarpic perennial hemicryptophyte, taprooted and with VA mycorrhizas. Immature plant with a rosette of lanceolate, cottony, spiny leaves, usually <500 mm². Flowering stems erect and leafy.

Phenology Flowers are produced from July to October. Seed shed during winter (Verkaar & Schenkeveld 1984b). Dead flowering shoots long-persistent, and seeds may be retained on the plant for considerable periods. Leaves relatively long-lived but not overwintering (Verkaar & Schenkeveld 1984a).

Height Foliage in immature plants, often <20 mm. In flowering plant foliage and flowers to 600 mm.

Seedling RGR No data. SLA 13.3. DMC 21.9.

Nuclear DNA amount No data; $2n = 20*$ (*BC*); $2x*$.

Established strategy Stress-tolerant ruderal.

Regenerative strategies Limited dispersal by wind. Some evidence of short-term persistence near the soil surface. See 'Synopsis'.

Flowers Florets yellow-brown, tubular, homogamous, insect-pollinated or selfed if insect-pollination fails (Knuth 1906); >100 in terminal capitula (Salisbury 1942).

Germinule 1.53 mg, 3.2 × 1.2 mm, achene dispersed with an easily detached pappus *c.* 8 mm in length.

Germination Epigeal; t_{50} 3 days; 8-30°C.

Biological flora None.

Geographical distribution Lowland Britain, extending northwards to Ross (77% of British vice-counties), *ABIF* page 608; most of Europe northwards to S Norway (77% of European territories); Asia Minor.

Habitats In areas of discontinuous turf associated with calcareous pastures, rock outcrops, lead-mine heaps and derelict wasteland. Present locally on calcareous quarry spoil and roadsides, and observed in one instance on non-calcareous sandy wasteland. [Frequent on sand dunes.]

Gregariousness Occurs as a few scattered individuals.

Altitude Recorded at altitudes of up to 340 m. [Up to 460 m.]

Slope Most frequent on dale-sides on moderate slopes. [Elsewhere, however, the species is equally characteristic of flat ground, e.g. on sand dunes.].

Aspect More frequent on S-facing slopes.

Hydrology Absent from wetland.

Soil pH Strongly calcicole, save for one site where *C.v.* has been observed on acidic sand of pH 4.8; most common in pH range 7.5-8.0.

Bare soil Widely distributed, but highest frequencies associated with moderate amounts of bare soil. Habitats frequently contain significant exposures of bare rock.

△ **ordination** Restricted to vegetation of infertile habitats.

Associated floristic diversity Typically intermediate to species-rich.

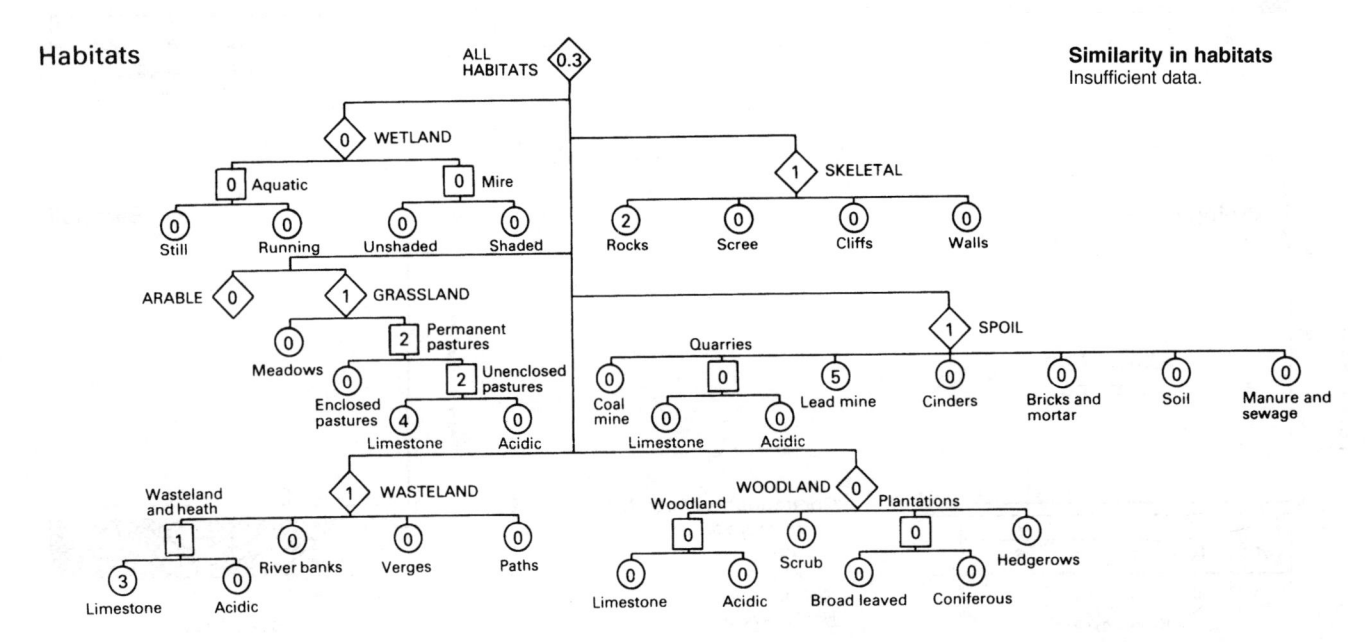

Habitats

Similarity in habitats
Insufficient data.

Synopsis *C.v.* is a monocarpic, perennial thistle of unproductive calcareous grassland, open rocky habitats and stabilized sand dunes. Generally only a small proportion of individuals flower in any one season, and rosettes usually persist in the vegetative state for several years, some needing 6 years to flower (Watt 1981). The low-growing rosettes appear particularly vulnerable to overshading, and tend to be restricted to open microsites (Verkaar *et al.* 1983). All parts of the shoot, including the bracts protecting the inflorescence, bear stiff spines which are effective deterrents to mammalian herbivores. *C.v.* occurs in dry habitats, and we suspect that its relatively deep tap-root (Anderson 1927) may allow *C.v.* to utilize subsoil moisture during periods of drought. Plants growing in the Sheffield region are typically small and with few capitula. Those from, for example, S England are often taller, and under very favourable conditions *C.v.* may produce 1-30 capitula which may each contain up to 300 achenes (Greig-Smith & Sagar 1981). Regeneration is entirely by seeds, which are shed late in the year and germinate mainly during spring. Studies in Holland (Verkaar & Schenkeveld 1984b) suggest that low numbers of seeds persist in the seed bank. Population densities in some sites may be regulated primarily by severe seed predation (Greig-Smith & Sagar 1981). The size of the achene is large relative to that of the easily detachable pappus, and *C.v.* has a relatively short dispersal distance (Greig-Smith & Sagar 1981). Two other subspecies are recognized from Europe, including one which exploits damp grassland (*FE4*).

Current trends The creation of bare infertile areas through quarrying and lead mining has probably led to an increase in *C.v.* However, *C.v.* rarely shows evidence of colonization over long distances, and appears to have a narrow ecological range. *C.v.* is likely to decrease as many of the unproductive calcareous pasture sites in which it occurs continue to become overgrown or are destroyed.

Altitude
Slope
Aspect

Soil pH

Associated floristic diversity

Hydrology

Absent from wetland

Triangular ordination

Commonest associates	%
Festuca ovina	95
Linum catharticum	95
Carex flacca	81
Lotus corniculatus	76
Euphrasia officinalis agg.	71

Bare soil

Abundance in quadrats

	1–10	11–20	21–30	31–40	41–50	51–60	61–70	71–80	81–90	81–100	Distribution type
Wetland	–	–	–	–	–	–	–	–	–	–	–
Skeletal	–	–	1	–	–	–	–	–	–	–	–
Arable	–	–	–	–	–	–	–	–	–	–	–
Grassland	2	–	–	–	–	–	–	–	–	–	–
Spoil	2	–	–	–	–	–	–	–	–	–	–
Wasteland	2	–	–	–	–	–	–	–	–	–	–
Woodland	–	–	–	–	–	–	–	–	–	–	–
All habitats	85.7	–	14.3	–	–	–	–	–	–	–	–

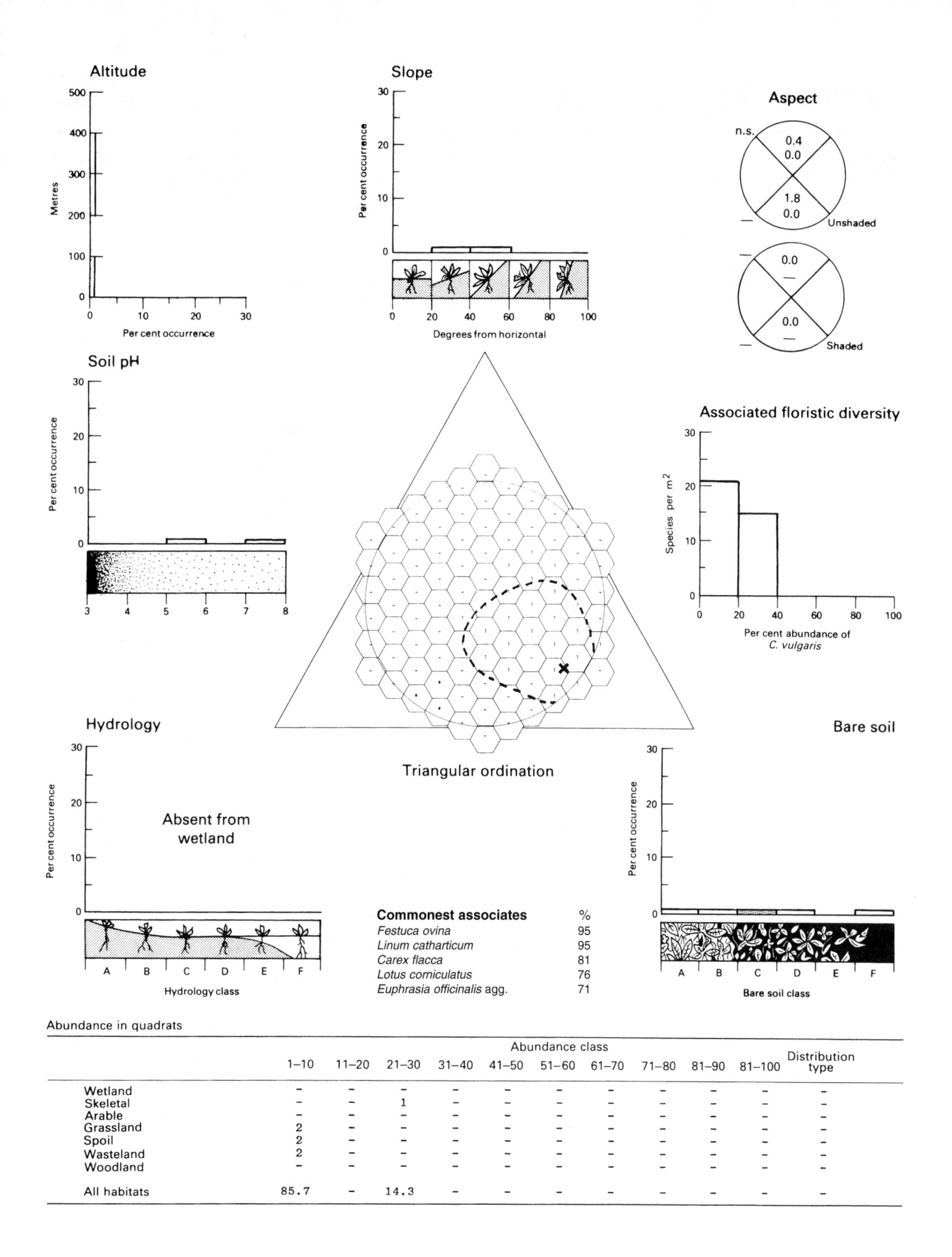

Catapodium rigidum (L.) C.E. Hubb.

Poaceae
Hard Poa, Fern Grass

Life-form Winter- or more rarely summer-annual therophyte. Stems spreading; leaves linear, usually <100 mm²; roots with VA mycorrhizas.
Phenology Germinates mainly in autumn, and capable of some winter growth (*Biol Fl*). Flowers May to June and seeds shed from July onwards.
Height Foliage to 100 mm; flowers to 200 mm.
Seedling RGR 1.0-1.4 week^{-1}. SLA 26.2. DMC 27.1.
Nuclear DNA amount 5.9 pg; $2n = 14*$ (*BC*); $2x*$.
Established strategy Stress-tolerant ruderal.
Regenerative strategies S. Principally by freshly-shed seed in autumn. Some ungerminated seeds persist over winter (*Biol Fl*); SB I.

Flowers Green, hermaphrodite, wind-pollinated; populations consisting of 'sexual but generally self-fertilizing lines' (*Biol Fl*) often <50 in a one-sided panicle (spikelets 3-10-flowered).
Germinule 0.19 mg, 1.8 × 0.5 mm, caryopsis dispersed with an attached lemma.
Germination Epigeal; t_{50} 2 days; 6-33°; germination rate increased after dry storage at 20°C; L = D.
Biological flora Clark (1974).
Geographical distribution The British Isles, but mainly in S and E England with coastal extensions of its range to the N and W (75% of British vice-counties), *ABIF* page 761; W and S Europe (59% of European territories); N Africa and W Asia; introduced to N and S America, Australia and S Africa.

Habitats Mainly in skeletal or disturbed calcareous habitats, particularly on quarry spoil and rock outcrops. More rarely observed on wall tops, cinders, ballast and foundry spoil tips. [Common on semi-stabilized sand dunes (Clark 1974) and in a range of artificial habitats near the sea.]
Gregariousness Intermediate.
Altitude Mainly a lowland species, but observed to 340 m. [Up to 350 m.]
Slope Recorded from all but the steepest slopes.
Aspect More frequent on unshaded S-facing slopes. This trend has been confirmed in a subsequent survey.

Hydrology Absent from wetlands.
Soil pH Mainly confined to calcareous soils within the pH range 7.0-8.0.
Bare soil Always in very open vegetation, and usually in habitats with a moderate to high exposure of bare soil. Occurrences in sites with very little bare soil correspond to rock outcrops.
△ **ordination** Distribution centred on disturbed vegetation types of low to moderate productivity.
Associated floristic diversity Highly variable. Often intermediate to species-rich, but in sites with much bare rock and little soil, low values for number of spp. per m² are observed.

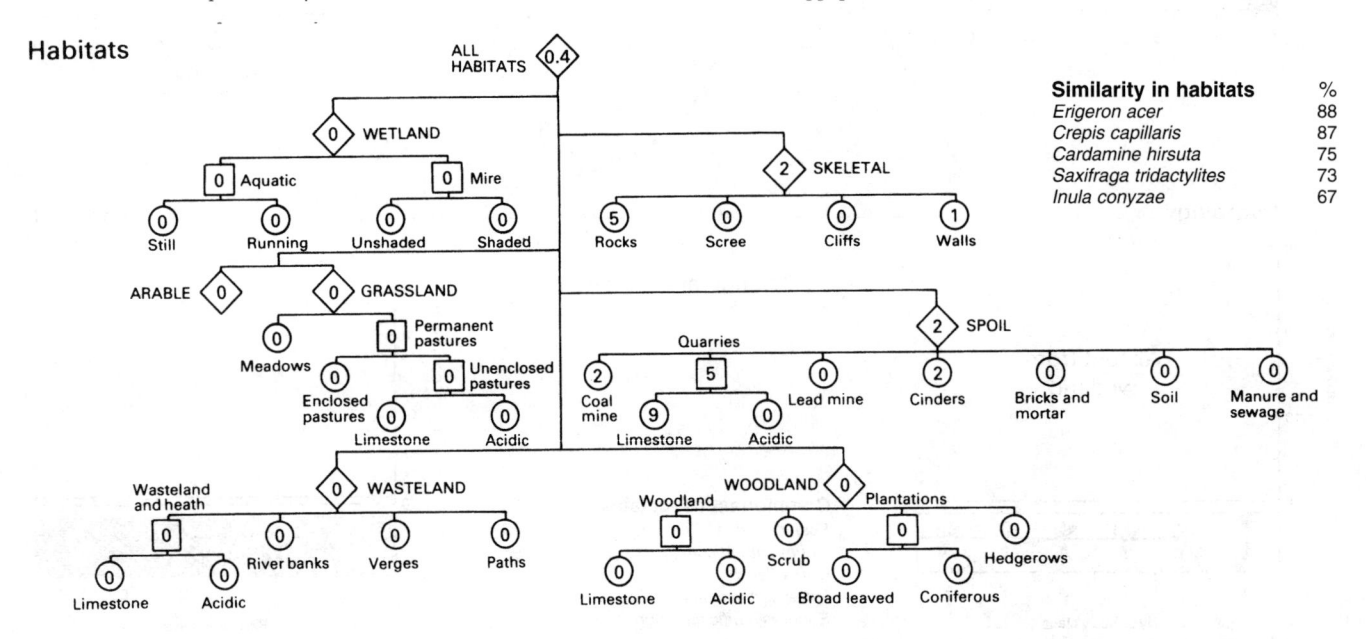

Habitats

Similarity in habitats	%
Erigeron acer	88
Crepis capillaris	87
Cardamine hirsuta	75
Saxifraga tridactylites	73
Inula conyzae	67

Synopsis *C.r.* is a small, mainly winter-annual grass of relatively infertile droughted soils, either on calcareous strata or near the sea. Often the species produces a single inflorescence containing less than 10 caryopses per plant (*Biol Fl*). In view of this exceptionally low level of seed production it is not surprising that *C.r.* tends to be restricted to continuously open sites, e.g. rock outcrops and sand dunes. As might be expected for a relatively late-flowering winter annual, floral development shows some requirement for vernalization and long days (*Biol Fl*). However, *C.r.* shows regenerative flexibility, in that (a) the species is not an obligate winter annual, although the spring-germinating plants which occur in certain populations produce even fewer seeds (*Biol Fl*), (b) the after-ripening requirement is short and seedlings, many of which do not survive, may be found after any wet period between spring and autumn (*Biol Fl*), and (c) additional later-flowering, secondary inflorescences may be formed (*Biol Fl*). These characteristics suggest that the habitats exploitable by *C.r.* are not necessarily heavily droughted every year and, in the most common habitat for the species, limestone quarries, *C.r.* often occurs in the absence of other winter annuals. The species, which often grows to some extent during winter, is susceptible to low winter and spring temperatures, and has a predominantly lowland distribution in Britain which lies at the N limit of its European range (*Biol Fl*). The majority of plants overwinter as small individuals and are particularly vulnerable to frost-heave (*Biol Fl*). Because of the small and localized nature of many colonies, the species is also potentially vulnerable to the effects of heavy grazing (*Biol Fl*). Like other stress-tolerant ruderals, *C.r.* responds to added nutrients. Thus, on ground manured by rabbits, seed set per plant may be twice as great as on unfertilized sites and cultivation on fertile soil has been shown to cause nearly a 30-fold increase in the number of inflorescences (*Biol Fl*). More-robust populations, formerly treated as ssp. *majus* (C. Presl) F. H. Perring and P. D. Sell, occur in the S and W of the British Isles (*CSABF* p. 149) but appear to deserve only varietal status (*FE5*). Other studies have failed to find evidence of local specialization between

Altitude
Slope
Aspect
Soil pH
Associated floristic diversity
Hydrology
Absent from wetland

Triangular ordination

Commonest associates %
Festuca rubra 88
Dactylis glomerata 75
Poa pratensis 75
Arenaria serpyllifolia 63
Arrhenatherum elatius 63

Bare soil

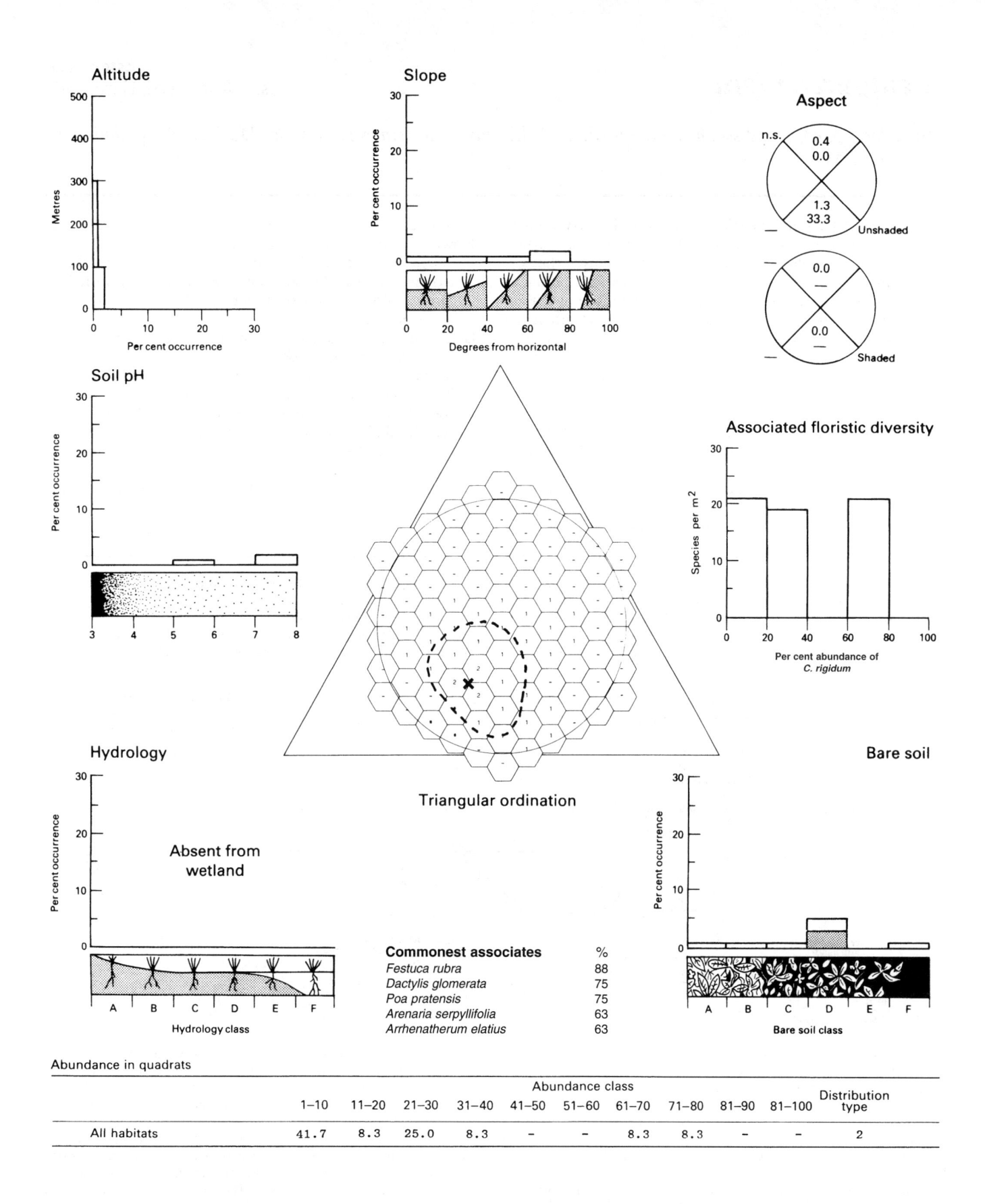

Abundance in quadrats

	1–10	11–20	21–30	31–40	41–50	51–60	61–70	71–80	81–90	81–100	Distribution type
All habitats	41.7	8.3	25.0	8.3	–	–	8.3	8.3	–	–	2

populations (Clark 1980a,b). Another ssp. has been identified from maritime sand in the Mediterranean region *(FE5)*. *C. marinum* (L.) C. E. Hubb replaces *C.r.* in areas closest to the sea and a sterile hybrid between the two has once been recorded (Stace 1975).

Current trends *C.r.* has effectively colonized many calcareous quarries in lowland England; despite the low seed production, examples occur where the species appears to have been dispersed to sites at least 10 km from established populations. On this basis, a further expansion of the species is anticipated.

Centaurea nigra L.

Asteraceae
Lesser Knapweed, Hardheads

Includes ssp. *nigra* and ssp. *nemoralis* Jordan (*C. debeauxii* ssp. *nemoralis* (Jordan) Dostal and ssp. *thuillieri* Dostal) CTW.

Life-form Polycarpic perennial, semi-rosette hemicryptophyte with stout oblique stock. Shoots often branched; leaves hairy, lanceolate, typically <4000 mm^2; roots with VA mycorrhizas.
Phenology New shoots emerge in spring. Flowers produced during June to September dependent upon the population. Fruits shed from July onwards, and some may be retained in the inflorescence after the death of the shoot in autumn. In S England basal leaves may appear in autumn (Marsden-Jones & Turrill 1954) but local populations do not appear winter green.
Height Large leaves below 300mm; flowers to 600 mm.
Seedling RGR 1.0-1.4 week^{-1}. SLA 26.3. DMC 15.5.
Nuclear DNA amount 3.7 pg; 2n = 22*, 44* (*BC*); 2x, 4x*. (Base number for tribe is polyploid; n = 17 (Moore & Frankton 1962).)
Established strategy Variable, with a majority of populations C-S-R.
Regenerative strategies S, V. Mainly by means of seed but with some slight capacity for vegetative spread (see 'Synopsis').
Flowers Florets purple, tubular, hermaphrodite, insect-pollinated, typically self-incompatible; *c.* 100 aggregated into terminal capitula (Marsden-Jones & Turrill 1954). In radiate forms, which are restricted to S England, the outer florets are longer and neuter.
Germinule 2.55 mg, 3.1 × 1.7 mm, achene dispersed with a rudimentary pappus *c.* 0.8 mm in length.
Germination Epigeal; t_{50} 3 days; 11-34°C; germination rate increased after dry storage at 20°C; L > D.
Biological flora None, but see Marsden-Jones & Turrill (1954).
Geographical distribution Endemic to Europe. Throughout the British Isles (99% of British vice-counties), *ABIF* page 617; Europe eastwards to Sweden and Italy (33% of European territories and 8% naturalized); introduced in New Zealand and N America.

Habitats Recorded most frequently from ungrazed limestone grassland. Also present on rock outcrops, limestone quarry heaps, pastures and meadows. Occurs at lower frequencies on cliffs, spoil, banks of rivers and streams, road verges and paths. Virtually absent from wetland, and not recorded from woodland.
Gregariousness Intermediate.
Altitude Up to 400 m, but rather more abundant at lower altitudes despite grassland habitats being more abundant in upland regions. [Up to 580 m.]

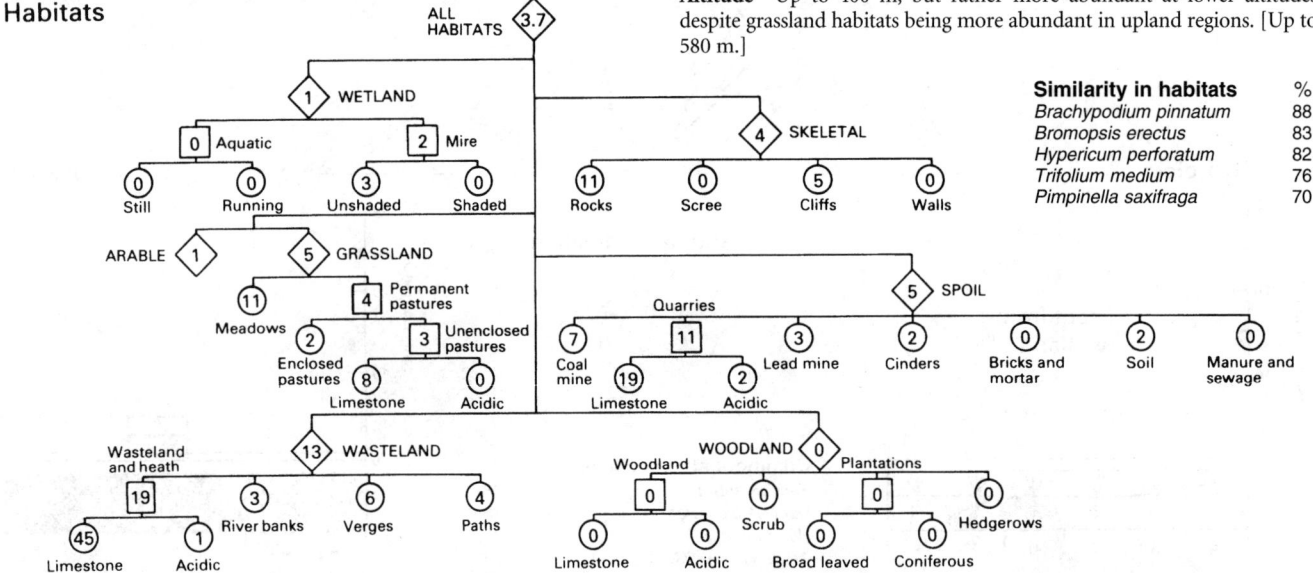

Similarity in habitats	%
Brachypodium pinnatum	88
Bromopsis erectus	83
Hypericum perforatum	82
Trifolium medium	76
Pimpinella saxifraga	70

Habitats

Slope Widely distributed, but not recorded from near-vertical slopes.
Aspect Rather more common on S-facing slopes.
Hydrology Recorded from moist soils peripheral to mire but, like many deeper-rooted species, *C.n.* is generally absent from wetlands.
Soil pH Widely distributed except below pH 4.0. Histogram has two peaks (5.0-6.0 and 7.5-8.0).
Bare soil Widely distributed though most frequently at low to moderate exposures.
△ ordination A relatively wide distribution centred on vegetation associated with moderately fertile and relatively undisturbed habitats. Strongly suppressed in heavily disturbed vegetation and, in circumstances conducive to dominance, by robust, clonally-expansive species.
Associated floristic diversity Intermediate to high.
Synopsis *C.n.* is a genetically-variable, short- or longer-lived, tufted herb found in a wide range of grasslands on moderately fertile or infertile soils.

In the *PGE*, the species occurs 'in the very mixed associations of plots receiving no manure or incomplete fertilisers' and 'is seldom found on well-manured soils'. In turf microcosms, roots of this species became heavily colonized by VA mycorrhizas with increases of up to 8-fold in seedling yield (Grime, Mackey, Hillier & Read 1987). Frequency, vigour and flowering are much greater in grassland experiencing occasional burning or infrequent mowing than in pastures. However, *C.n.* can persist in grazed turf, where its leaves may be eaten by sheep, or more rarely by cattle (Marsden-Jones & Turrill 1954). The tough wiry stems are normally avoided by stock (Marsden-Jones & Turrill 1954). *C.n.* shows a summer peak in shoot production and in more-droughted sites such as shallow calcareous soils on S-facing slopes; we suspect that this phenology is sustained by access to subsoil moisture. The species may be partially winter green at low altitudes but not in upland areas. Regeneration is primarily by seeds, and a robust plant may produce >1000 each year. Much seed is

Altitude
Slope
Aspect

n.s.	7.9	
	25.0	
n.s.	11.7	Unshaded
	30.8	

—	0.0	
—	0.0	Shaded

Soil pH
Associated floristic diversity
Hydrology
Triangular ordination
Bare soil

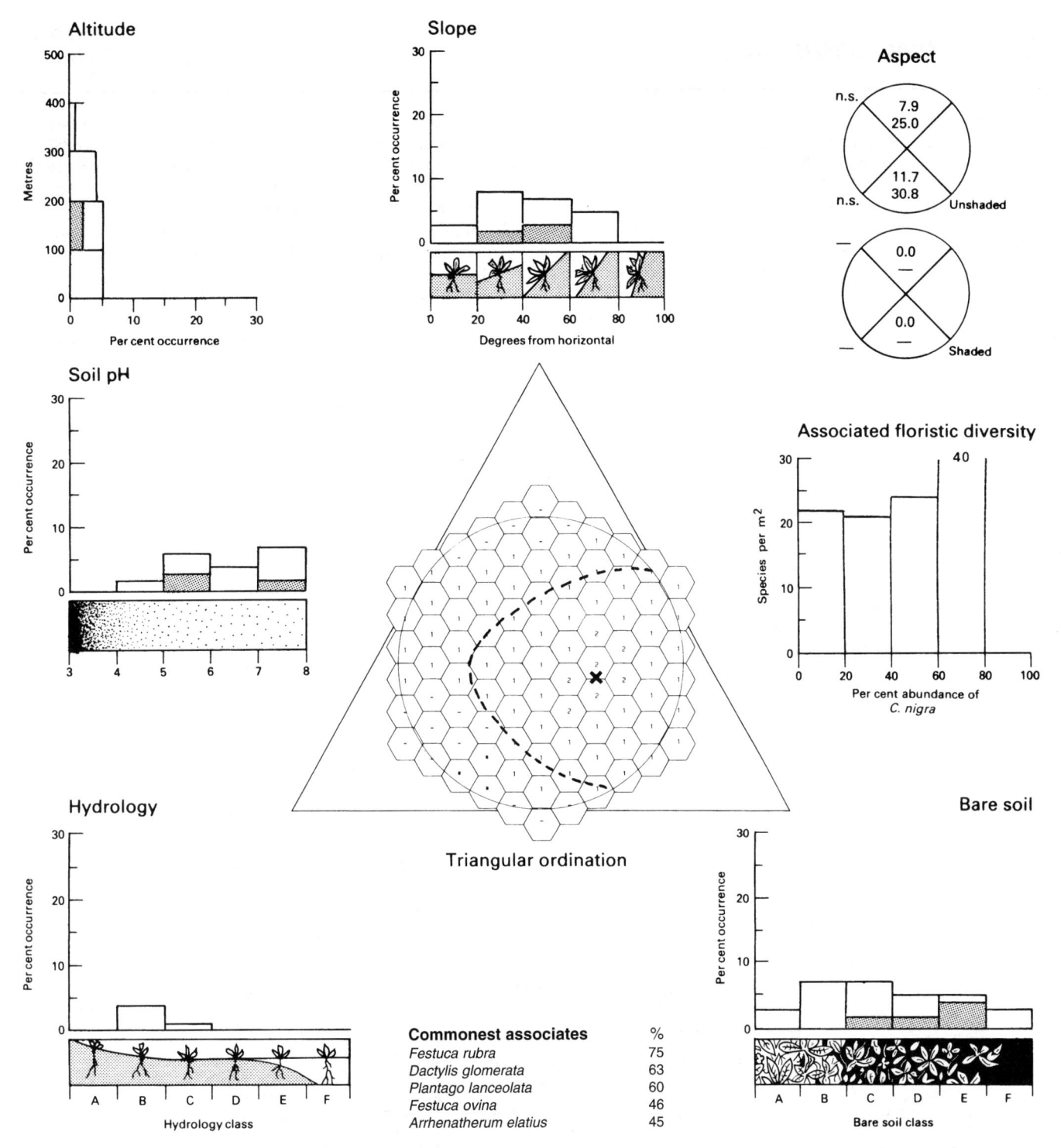

Commonest associates	%
Festuca rubra	75
Dactylis glomerata	63
Plantago lanceolata	60
Festuca ovina	46
Arrhenatherum elatius	45

retained in the fruit head after ripening and is shed at irregular times during autumn and winter. Seeds are heavily predated by insects in the capitula and by small rodents on the ground. Germination probably takes place mainly in the spring (Roberts 1986a) but some autumn germination may occur (Marsden-Jones & Turrill 1954). No substantial persistent seed bank is formed but a few ungerminated seeds may survive for several years (Roberts 1986a).

Vegetative spread seldom appears important. However, new side-shoots may be produced up to 25 mm from the main stem and these may become detached from the parent plant within 12 years, occasionally giving rise to a 'fairy ring' type of spread (Marsden-Jones & Turrill 1954). *C.n.* comprises a complex group in need of taxonomic revision and ecological study. Early work suggested that segregates with different ecological and geographical ranges could be identified (Marsden-Jones & Turrill 1954). Ssp. *nigra* is considered the prevalent northern form, found particularly on heavy moist soils, while ssp. *nemoralis* has a southern distribution and is associated with lighter, often calcareous soils (*CTW*). These segregates intergrade within the Sheffield area and their validity has been questioned (Elkington & Middlefell 1972). *C.n.* has not been subdivided in this account or in *Excur Fl*, but in *FE4* a large number of spp. and ssp. have been separated.

Doubtless the considerable genetic variation within the aggregate taxa contributes to the ecological range of *C.n.* and to the conspicuously wide variation in flowering times observed between populations. In S England populations with radiate heads (*C. debeauxii* ssp. *thuillieri* (*FE4*)) have distributions similar to that of *C. scabiosa* (Lack 1976).

Current trends More common in older pastures, and probably decreasing in farmland. However, this outbreeding and genetically variable species has perhaps increased its ecological amplitude by colonizing a range of artificial grassy or forb-rich communities.

Centaurea scabiosa L.

Asteraceae
Greater Knapweed

Life-form Tufted, polycarpic perennial, semi-rosette hemicryptophyte with stout oblique woody stock. Shoots erect, branched; leaves pinnatifid, hairy, <4000 mm^2; roots deep, with VA mycorrhizas.
Phenology Leaves normally dying back in winter. Flowers mainly in July and August. Seed shed in Aug to Nov. Leaves relatively long-lived.
Height Foliage mainly <250 mm; flowering shoots to 900 mm.
Seedling RGR No data. SLA 20.7. DMC 19.7.
Nuclear DNA amount 3.5 pg; $2n = 20^*$, 40 (*FE4, BC*); $2x^*$, $4x$ (base number for tribe is polyploid, $n = 17$).
Established strategy Intermediate between stress-tolerant competitor and C-S-R..
Regenerative strategies S. Regeneration by seed requires further study. Unlike *C. nigra*, seed is not retained for long in the fruit head, and appears to germinate in spring (Roberts 1986a). No evidence of a buried seed bank. Forms tufts from a branching stock, but has no marked capacity for vegetative spread; ?SB II.
Flowers Florets purple, insect-pollinated; self-incompatible (Lack 1982c); capitula terminal with <25 outer large neuter and *c*. 50 central, tubular, hermaphrodite florets.
Germinule 7.46 mg, 5.0 × 2.4 mm, achene dispersed with a pappus *c*. 6 mm in length. Pappus not effective in long-distance dispersal by wind.
Germination Epigeal; t_{50} 3 days; 13-34°C; L > D.
Biological flora None.
Geographical distribution Lowland Britain but much more common in the SE, rare in Scotland and Ireland (70% of British vice-counties), *ABIF* page 615; Europe from Spain and Italy northwards to Scandinavia (64% of European territories and 3% naturalized); W Asia.

Habitats Occurs predominantly in ungrazed calcareous grassland and on limestone cliffs. Local populations also exist on rock outcrops, quarry heaps and railway ballast. Also observed on roadsides and railway banks, sometimes away from limestone. [Often found near the sea, particularly on cliff tops, but also sand dunes (Valentine 1980).]
Gregariousness Occurs as isolated individuals.
Altitude Up to 300 m. More abundant at low altitudes, even though apparently suitable grassland habitats are more frequent on the upland, carboniferous limestone. [Up to 320 m.]
Slope A wide distribution.

Habitats

Aspect More common on S-facing slopes.
Hydrology Absent from wetland.
Soil pH Occurs mainly on base-rich soils of pH >6.0.
Bare soil A wide distribution, but most frequently found in association with relatively small exposures of bare soil.
△ **ordination** A diffuse pattern, probably arising from the occurrence of the species in rather different upland and lowland habitats. At higher altitudes on the Derbyshire limestone *C.s.* is usually associated with sparse vegetation on skeletal S-facing slopes. On the magnesian limestone the species is often represented by robust phenotypes, and has a habitat range which includes sites in more productive derelict grasslands.
Associated floristic diversity Usually intermediate to high.
Synopsis *C.s.* is an apparently long-lived, tufted, summer-green herb, largely restricted to dry and, to a lesser extent, open grassland habitats and rock outcrops, all nearly always on calcareous soils. *C.s.* is more common on S-facing slopes and, despite occupying habitats subject to drought, *C.s.* develops its maximum leaf canopy during midsummer. It seems likely that the deep tap root (Anderson 1927) allows evasion of drought by exploitation of moisture in deep crevices. *C.s.* is uncommon on the upland carboniferous limestone of Derbyshire, despite an abundance of apparently suitable habitats. We suspect that the rarity of *C.s.*, which has a southerly distribution, may be determined at least in part by climatic factors. *C.s.* is infrequent in heavily grazed or trampled areas, but appears to be tolerant of burning. Regeneration is primarily by seed, which is produced in abundance. In Derbyshire populations seed production may be affected by heavy infestation of the flowering stems by aphids, and the seeds are heavily predated by small rodents. The seed is large and the temperatures required for germination are high compared with those of the autumn-germinating grasses with which *C.s.* is often associated. This suggests that the strategy of establishment in *C.s.* involves spring germination (Roberts 1986a) and rapid penetration of the radicle into the subsoil before summer desiccation of the surface soil. Observations by Lack (1982a-c) suggest that *C.s.* is a better competitor for insect pollinators than *C. nigra*. *C.s.* may be subdivided into a large number of varieties (Britton 1923), and recently variation with respect to growth form and leaf characters has been shown to have both a genetic and an ecological basis, although *C.s.* also shows some phenotypic plasticity (Valentine 1980). Plants from the upland carboniferous limestone are smaller and less robust than those from the magnesian limestone of N England and tend to be restricted to more skeletal soils.
Current trends *C.s.* has some capacity to exploit new disturbed habitats as an impurity in grass seed (see *WCH*). Despite this, a decrease in the abundance of *C.s.* is predicted as a consequence of habitat destruction. As populations decrease in size the incidence of self-compatibility may be expected to increase.

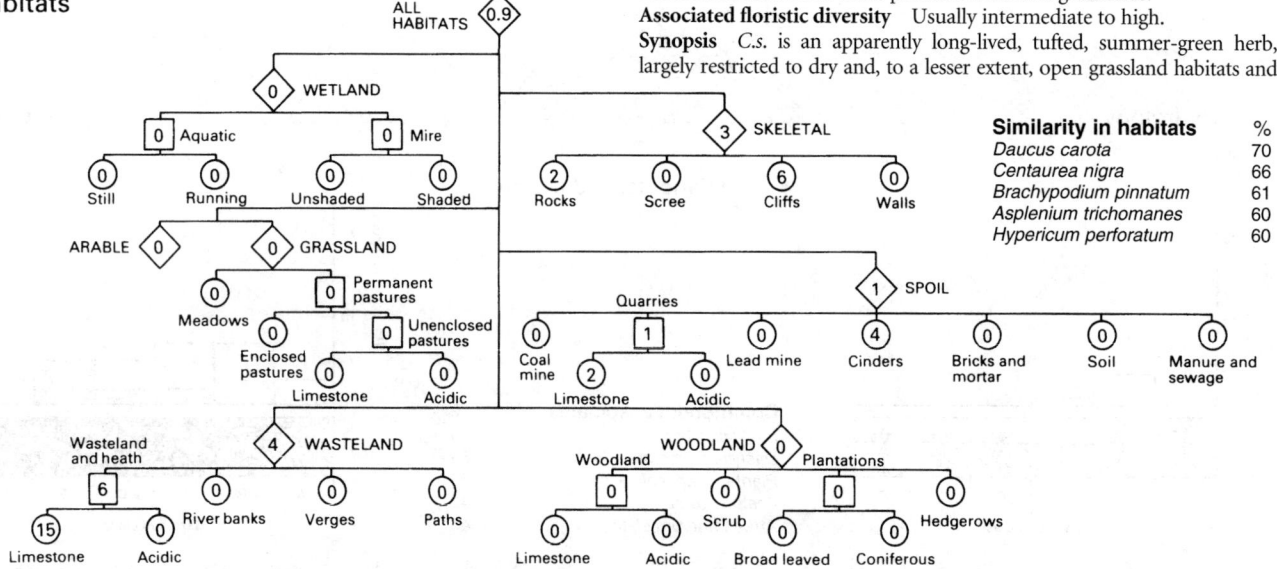

Similarity in habitats	%
Daucus carota	70
Centaurea nigra	66
Brachypodium pinnatum	61
Asplenium trichomanes	60
Hypericum perforatum	60

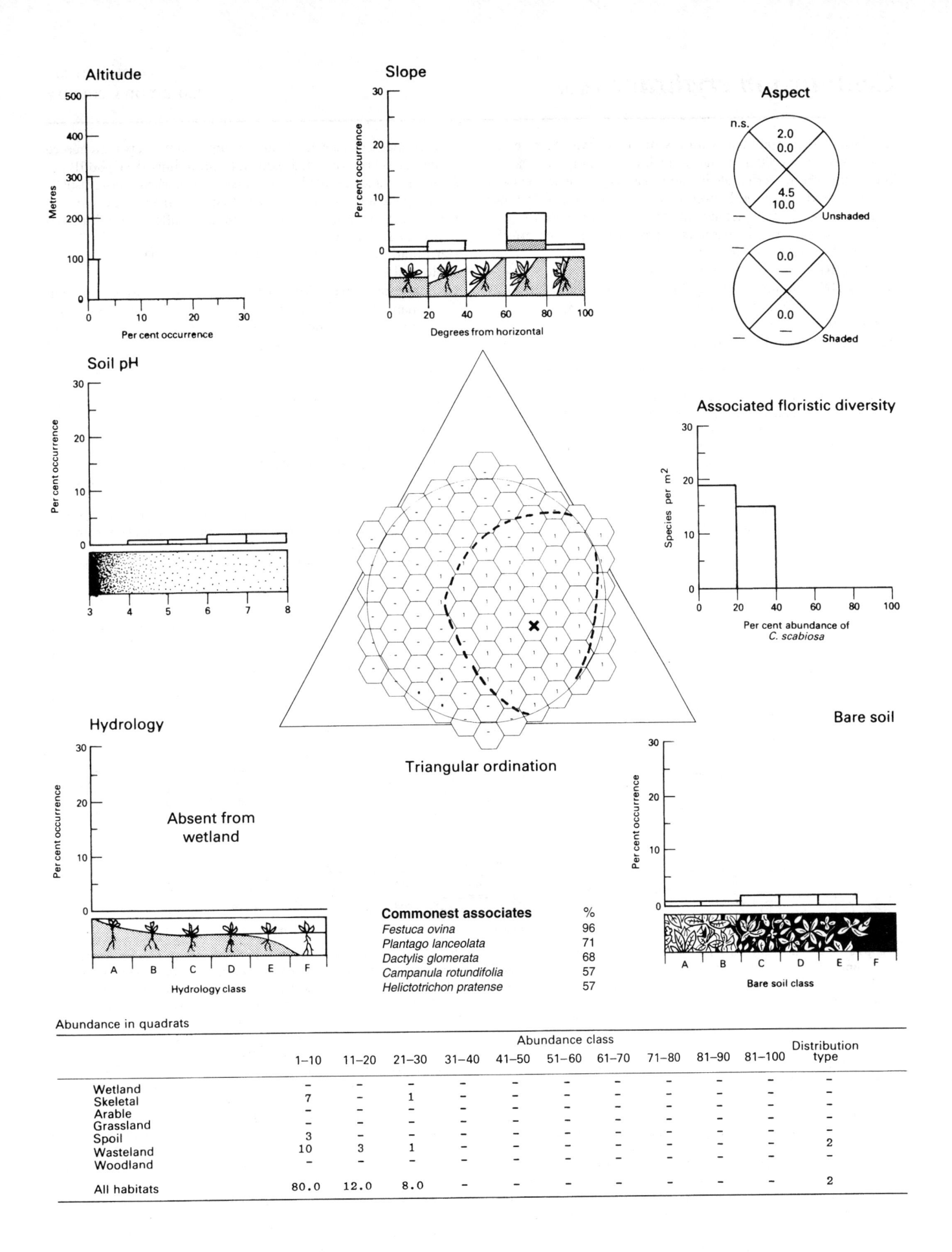

Altitude

Slope

Aspect

	n.s.	2.0
		0.0
—	4.5	10.0 Unshaded

		0.0
—		—
	0.0	— Shaded

Soil pH

Associated floristic diversity

Hydrology

Absent from wetland

Triangular ordination

Bare soil

Commonest associates	%
Festuca ovina	96
Plantago lanceolata	71
Dactylis glomerata	68
Campanula rotundifolia	57
Helictotrichon pratense	57

Abundance in quadrats

	Abundance class										Distribution type
	1–10	11–20	21–30	31–40	41–50	51–60	61–70	71–80	81–90	81–100	
Wetland	–	–	–	–	–	–	–	–	–	–	–
Skeletal	7	–	1	–	–	–	–	–	–	–	–
Arable	–	–	–	–	–	–	–	–	–	–	–
Grassland	–	–	–	–	–	–	–	–	–	–	–
Spoil	3	–	–	–	–	–	–	–	–	–	2
Wasteland	10	3	1	–	–	–	–	–	–	–	
Woodland	–	–	–	–	–	–	–	–	–	–	–
All habitats	80.0	12.0	8.0	–	–	–	–	–	–	–	2

Centaurium erythraea Rafn.

Gentianaceae
Common Centaury

Life-form Winter-annual, semi-rosette therophyte. Stem erect; leaves ovate, <1000 mm²; tap-rooted, with VA mycorrhizas.

Phenology Seeds germinate in autumn and *C.e.* overwinters as a small rosette of leaves. Flowering occurs from June to October and seed set from September, after which the plant dies.

Height Foliage mainly basal and typically appressed to the ground; flowering stems up to 500 mm.

Seedling RGR No data. SLA 20.7. DMC 24.7.

Nuclear DNA amount 2.5 pg; $2n = 40*$ (*BC*); $4x*$. Subspecies with $2n = 20$ ($2x$) occur elsewhere in Europe (*FE3*).

Established strategy Stress-tolerant ruderal.

Regenerative strategies S, B_s. Reproduces entirely by seed, which germinates in vegetation gaps during autumn. A persistent seed bank has been reported (Brown & Oosterhuis 1981); SB III.

Flowers Pink, hermaphrodite, insect- or perhaps more usually self-pollinated (Ubsdell 1979); often *c.* 25 in a dense cyme.

Germinule 0.01 mg, 0.4 × 0.3 mm, seed; often <200 dispersed from each capsule.

Germination Epigeal; t_{50} 6 days; 13–29°C; L > D.

Biological flora None.

Geographical distribution The British Isles, but less common in Scotland (93% of British vice-counties), *ABIF* page 479; Europe northwards to Scandinavia (87% of European territories); SW Asia; naturalized in N America.

Habitats Recorded from open ground on wasteland, quarry spoil, lead mine heaps, rock outcrops, cliffs, paths and road verges. Mainly on soils overlying limestone. Also observed in sand and gravel pits, and more rarely on cinders and coal-mine spoil. Absent from woodlands, wetlands, arable land and pasture. [Also found in maritime habitats such as sand dunes and cliff tops, though often replaced by related taxa in these habitats.]

Gregariousness Sparse.

Altitude Most frequent and abundant on the magnesian limestone at altitudes of <100 m. Extends to 250 m, but infrequent in upland areas despite an abundance of apparently suitable habitats. [Up to 280 m.]

Slope Found both in habitats such as wasteland with gentle slopes, and on steeply sloping cliffs and outcrops.

Aspect Recorded from S- but not N-facing slopes.

Hydrology Absent from wetlands.

Soil pH Most records from pH >5.5; minimum observed value pH 4.8.

Bare soil Widely distributed, but most frequently recorded in sites with a high exposure of bare soil.

△ **ordination** Associated mainly with relatively unproductive vegetation types which are not subject to intense disturbance. Distribution centred on moderately disturbed vegetation of intermediate productivity.

Associated floristic diversity Frequently associated with species-rich vegetation.

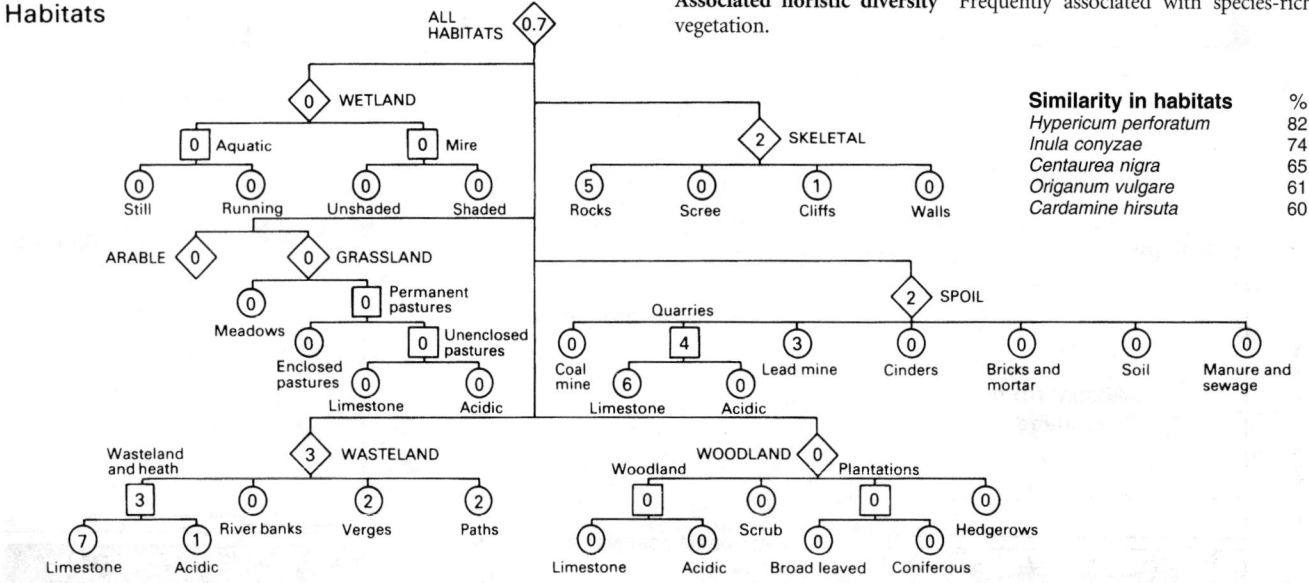

Habitats

Similarity in habitats	%
Hypericum perforatum	82
Inula conyzae	74
Centaurea nigra	65
Origanum vulgare	61
Cardamine hirsuta	60

Synopsis *C.e.* is a rosette-forming winter-annual, colonizing a range of open unproductive habitats, particularly on calcareous soils. Overwinters as a flattened rosette which contains exceptionally high concentrations of glucose and fructose. As might be predicted from its late-summer flowering period, *C.e.* is absent from frequently droughted soils, and the low-growing rosettes are susceptible to the effects of shade from surrounding species. A virtual absence of the species from pasture presumably reflects the vulnerability of its erect flowering stems to grazing by stock. *C.e.* is mainly found in lowland sites and on S-facing slopes, and the distribution in upland parts of Britain may be limited by climatic factors. *C.e.* regenerates exclusively by means of minute seeds and a single plant of average size may produce *c.* 40 000. A persistent seed bank has been recorded (Brown & Oosterhuis 1981). Establishment requires the existence of bare areas in which the seedling experiences minimal competition. Disturbed unproductive sites associated with calcareous quarry spoil or sand and gravel pits, or even the clay overburden of coal-mine heaps, may support large colonies of *C.e.* during the earlier stages of colonization. Similarly, less-transient habitats such as outcrops, sand-dune grassland and other sites subject to a degree of soil erosion or disturbance, may have sizeable populations of *C.e.* In laboratory experiments with turf microcosms, seedling establishment on infertile calcareous soil fails completely in the absence of VA mycorrhizas, whereas infection leads to the development of robust plants (Grime, Mackey, Hillier & Read 1987). *C.e.* is phenotypically plastic with respect to size and flower number. It also shows much genotypic variation in Europe, where six ssp. and numerous varieties have been recognized (*FE3*). British material has been subdivided into five varieties (Ubsdell 1976a,b). Two refer to sea-cliff populations, and only one, var. *erythraea*, is typical of inland situations (Ubsdell 1976a,b). Some dwarf populations from sea cliffs have been separated as var. *capitatum* (Willd.) Melderis. The ecology of *C.e.* is further obscured near the sea by the presence of other species of *Centaurium*, and hybrids have been reported (Stace 1975). The ecology and life-history of *C.e.* is also very similar to that of *Blackstonia perfoliata* (L.) Hudson, another annual member of the Gentianaceae; mixed populations of *C.e.* and *B.p.* are frequently observed in lowland calcareous habitats.

Current trends A relatively mobile species which is ecologically more wide-ranging than formerly and which may become more common.

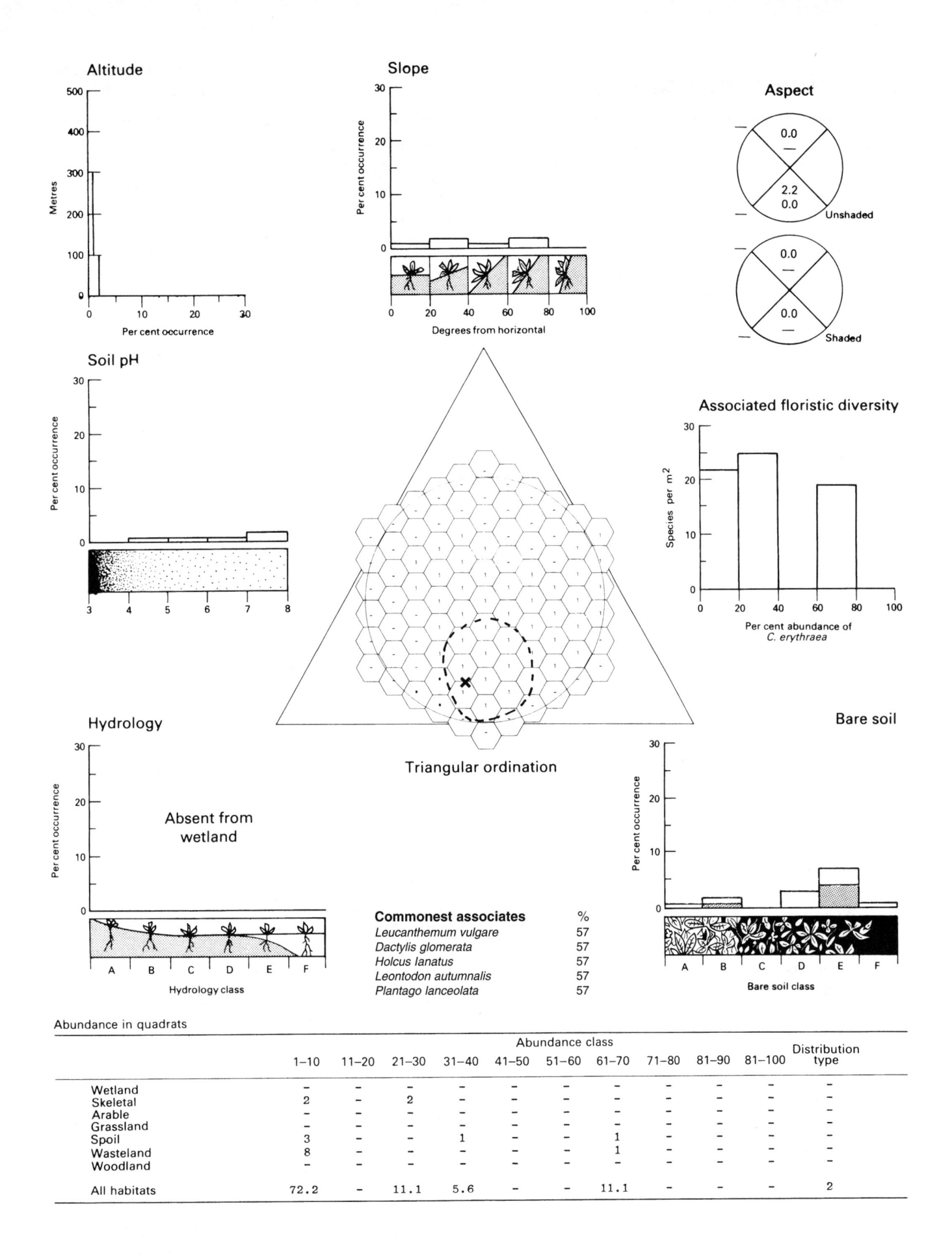

Altitude

Slope

Aspect

Unshaded

Shaded

Soil pH

Associated floristic diversity

Species per m²

Per cent abundance of *C. erythraea*

Hydrology

Absent from wetland

Hydrology class

Triangular ordination

Commonest associates	%
Leucanthemum vulgare	57
Dactylis glomerata	57
Holcus lanatus	57
Leontodon autumnalis	57
Plantago lanceolata	57

Bare soil

Bare soil class

Abundance in quadrats

	Abundance class										Distribution type
	1–10	11–20	21–30	31–40	41–50	51–60	61–70	71–80	81–90	81–100	
Wetland	–	–	–	–	–	–	–	–	–	–	–
Skeletal	2	–	2	–	–	–	–	–	–	–	–
Arable	–	–	–	–	–	–	–	–	–	–	–
Grassland	–	–	–	–	–	–	–	–	–	–	–
Spoil	3	–	–	1	–	–	1	–	–	–	–
Wasteland	8	–	–	–	–	–	1	–	–	–	–
Woodland	–	–	–	–	–	–	–	–	–	–	–
All habitats	72.2	–	11.1	5.6	–	–	11.1	–	–	–	2

Cerastium fontanum Baugm.

Caryophyllaceae
Common Mouse-ear Chickweed

Life-form Polycarpic perennial chamaephyte, but often a therophyte on arable land. Shoots decumbent or if flowering ascending; leaves lanceolate, hairy, usually <250 mm²; roots shallow, usually lacking VA mycorrhizas.
Phenology Winter green. Flowers April to September. Seed shed from June onwards. Flowering shoots relatively short-lived.
Height Foliage mainly <100 mm, overtopped by flowers.
Seedling RGR 1.0–1.4 week^{-1}. SLA 29.2. DMC 12.9.
Nuclear DNA amount 5.9 pg; $2n = 144^*$ (108^*–152) (*BC*, *FE1*); $16x^*$.
Established strategy Intermediate between ruderal and C-S-R.

Regenerative strategies B_s, (V). Regenerative biology is complex (see 'Synopsis'); SB III.
Flowers White, hermaphrodite, protandrous, selfed or insect-pollinated; <20 in terminal cymes.
Germinule 0.16 mg, 0.9 × 0.8 mm, seed in a dehiscent *c.* 40-seeded capsule (Salisbury 1964).
Germination Epigeal; t_{50} 2 days; <5–30°C; L = D.
Biological flora None, but see Peterson (1969).
Geographical distribution The British Isles (100% of British vice-counties), *ABIF* page 163; Europe (97% of European territories and 3% naturalized); temperate regions worldwide.

Habitats In all types of unshaded dryland habitat, though often at relatively low frequencies. Most commonly recorded in meadows and pastures, on outcrops and on lead-mine, cinder and quarry spoil. Also frequent on waysides, demolition sites, soil heaps and wasteland. Infrequent in mire and in shaded habitats, and often only a casual on arable land. Absent from aquatic habitats.
Gregariousness Sparse to intermediate, but occasionally forming patches.

Altitude Widely distributed, extending to 470 m, but more abundant in the upland region where the area of grassland is greater. [Up to 1220 m.]
Slope Wide-ranging.
Aspect Slightly more frequent in occurrence and locally abundant on N-facing slopes.
Hydrology Associated with moist rather than waterlogged soils, but not infrequent in soligenous mire.
Soil pH Found over a wide range of values, but infrequent below pH 4.5.
Bare soil Occurring in association with a wide range of values.

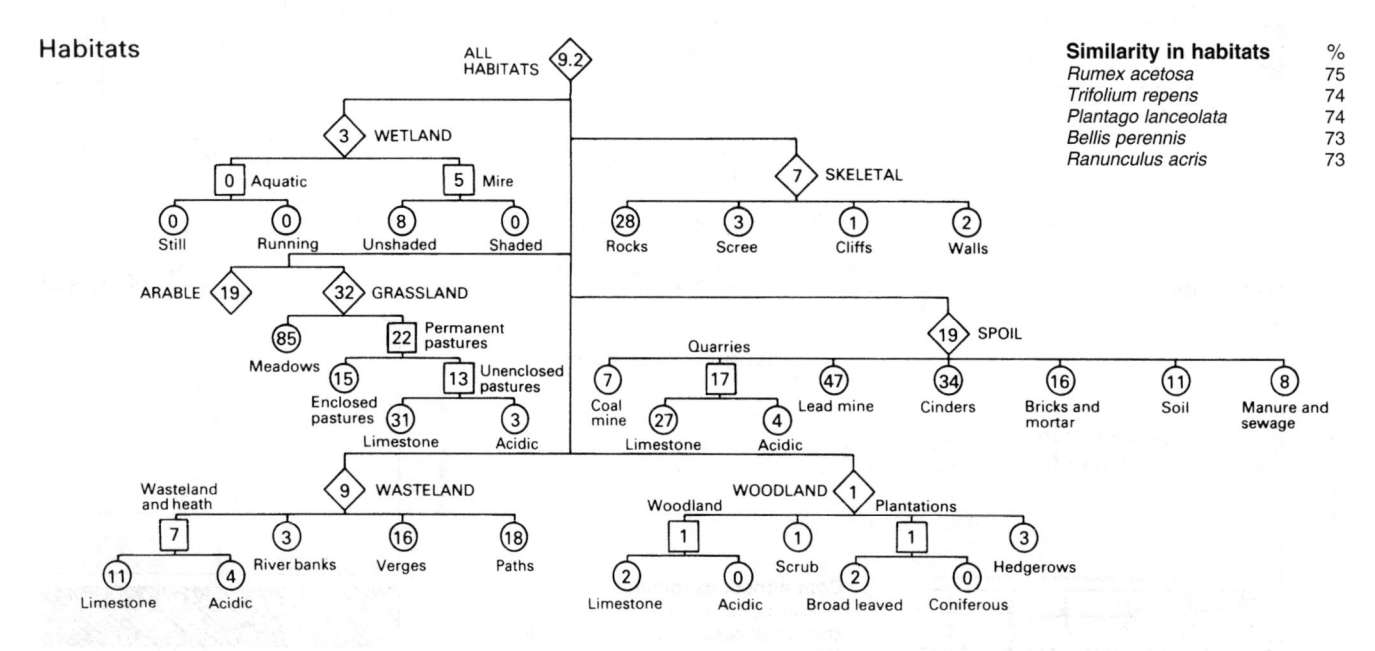

Habitats

Similarity in habitats	%
Rumex acetosa	75
Trifolium repens	74
Plantago lanceolata	74
Bellis perennis	73
Ranunculus acris	73

△ **ordination** Distribution centred on vegetation experiencing moderately severe levels of disturbance, but extending also into relatively unproductive habitats. In view of the creeping growth form it is not surprising to find the species strongly excluded from vegetation dominated by C and S-C strategists.
Associated floristic diversity Typically intermediate to species-rich.
Synopsis *C.f.* is a low-growing, winter-green herb typically of moderately fertile but disturbed habitats. In many respects its ecological characteristics are similar to those of *Sagina procumbens*, although the latter is a much smaller plant. *C.f.* is a perennial which may form patches in excess of 0.5 m² in lawns. These may persist for periods in excess of 6 years. However, in some circumstances *C.f.* is capable of flowering within 9 weeks of germination (Peterson 1969) and can function as an annual in highly disturbed habitats, e.g. arable fields. Field observations suggest that, particularly in dry years, the species behaves as a winter annual in habitats such as rock outcrops. Invariably *C.f.* is a subordinate component of plant communities and is dependent upon the absence or debilitation of potential dominants for survival; accordingly *C.f.* is most frequent in grazed or mown grassland, and in the *PGE* the species is prominent only in unproductive plots. *C.f.* is widespread on forms of wasteland liable to occasional mechanical disturbance and shoots can grow through the heaped soil of ant hills (King 1977a–c). *C.f.* is consumed by cattle (Peterson 1969) and appears relatively sensitive to trampling. In the absence of competition *C.f.* persists in shaded conditions (Peterson 1969) and the capacity to exploit hay-meadows, which are grazed after cutting, suggests also a modest degree of shade tolerance in the early part of the growing season. Much of the success of *C.f.* relates to its regenerative biology. An average plant may produce 6500 seeds in a year (Salisbury 1964), although plants on infertile soils produce many fewer (King 1977c). *C.f.* forms a persistent seed bank, in which some buried seeds may survive for 40 years (Salisbury 1964). As in the case of *Agrostis capillaris*, freshly-dispersed seeds often show no evidence of a light-requirement for germination (Grime *et al.* 1981) and it seems likely that this characteristic is acquired during the phase in which the seed lies on the soil surface beneath a leaf canopy before burial (Fenner 1978). Germination occurs from early spring to late autumn (Roberts 1986a). In addition, *C.f.* also has the capacity for lateral

Altitude

Slope

Aspect

	n.s.	
	14.2	
	8.3	
—	11.2	Unshaded
	0.0	

	n.s.	
	0.8	
	0.0	
—	0.8	Shaded
	0.0	

Soil pH

Associated floristic diversity

Hydrology

Triangular ordination

Commonest associates %
Festuca rubra 73
Holcus lanatus 57
Trifolium repens 52
Plantago lanceolata 48
Dactylis glomerata 47

Bare soil

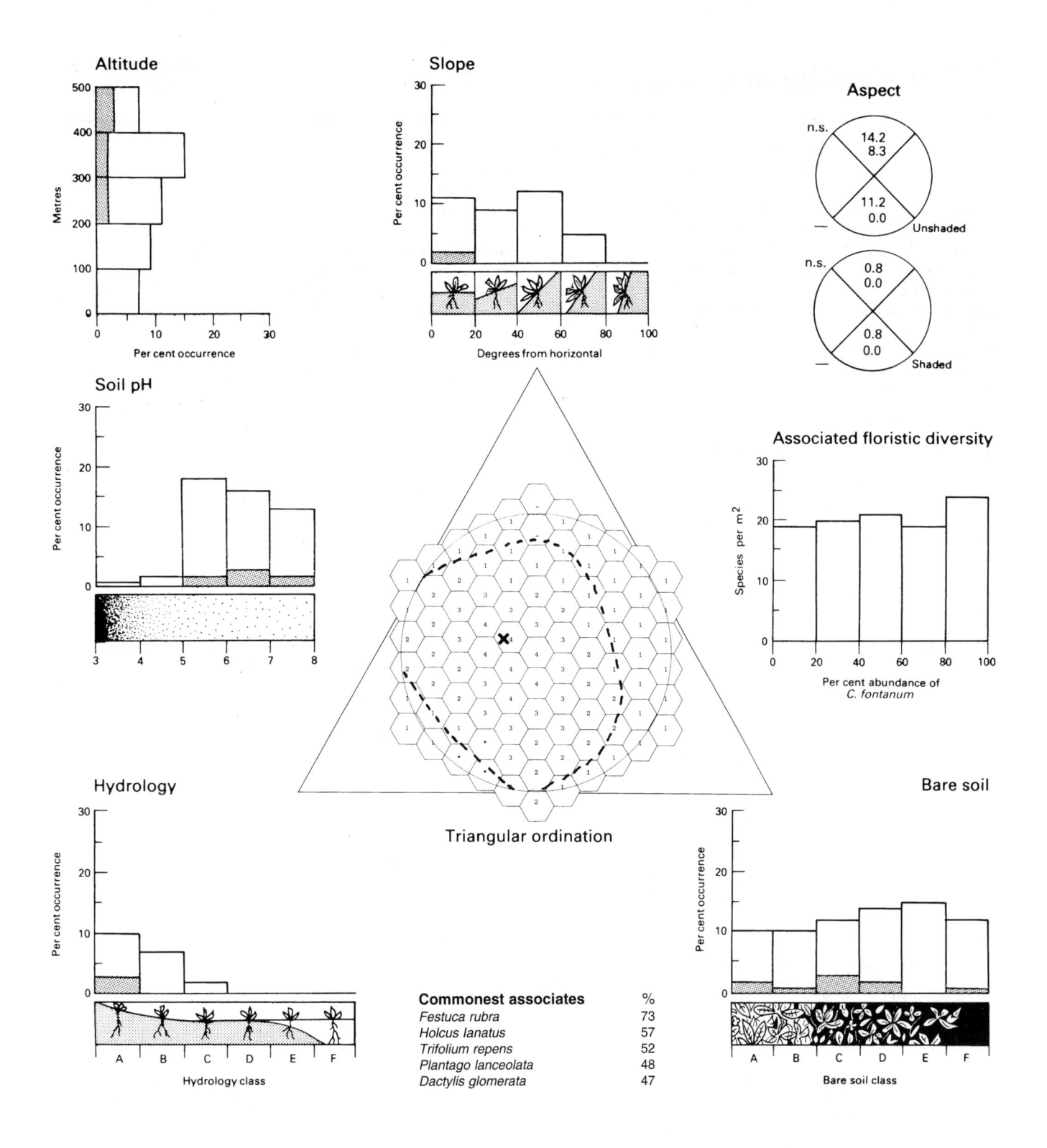

spread since its decumbent vegetative shoots root freely. Detached shoot apices are also capable of regeneration to form a new plant. The small seeds appear to be very mobile, and *C.f.* is a listed impurity of agricultural seed of *Phleum pratense* and *Trifolium repens* (WCH). Although the distribution is centred on moist, fertile disturbed sites, *C.f.* is ecologically wide-ranging. The distribution overlaps those of winter-annual *Cerastium* spp. in areas subject to summer drought, that of the more robust species *C. arvense* L. in dry, less fertile, grassland and that of the arctic *Cerastium* spp. in montane areas. Hybrids with species in the latter two groups have been recorded (Stace 1975). The species is very variable in Europe, where it is subdivided into five ssp. (*FE1*). Three subspecies have been recognised in the UK (Stace 1991).

Current trends An effective and probably increasing colonist of artificial habitats. Likely to be encouraged on arable land by reduced cultivation and direct drilling (Pollard & Cussans 1981).

(+)*Chaenorhinum minus* (L.) Lange

Scrophulariaceae
Small Toadflax

Life-form Summer-annual, often glandular-hairy, therophyte. Stems erect, branching; leaves linear, usually <100 mm^2; roots without mycorrhizas.

Phenology Seeds germinate in spring. Plants flower from May to October and set seed from July to October.

Height Foliage and flowers to 250 mm.

Seedling RGR No data. SLA 20.3. DMC 12.9.

Nuclear DNA amount No data; $2n = 14$ (Federov 1974); $2x$.

Established strategy Intermediate between ruderal and stress-tolerant ruderal.

Regenerative strategies S. Regenerates exclusively by seed which germinates in spring; SB II.

Flowers Purple, hermaphrodite, usually selfed; often >20 in a racemose inflorescence.

Germinule 0.07 mg, 1.1 × 0.7 mm, seed in a dehiscent capsule which may contain >50 seeds.

Germination Epigeal; t_{50} 4 days; 12-35°C; dormancy broken by chilling; L > D.

Biological flora None.

Geographical distribution Doubtfully native. Common in S England, becoming rarer northwards, a casual in Scotland, widespread in Ireland but less common in the N (58% of British vice-counties), *ABIF* page 547; most of Europe, but possibly not native in much of the N (native or naturalized in 79% of European territories); West Asia; introduced in N America.

Habitats Only recorded from railway ballast and cinder tips, but observed in arable fields on clayey soils and on wasteland.

Gregariousness Sparse to intermediate.

Altitude Most suitable habitats occur in the lowland but *C.m.* has been observed up to 340 m, its British altitudinal limit.

Slope Mainly restricted to flat ground and gentle slopes.

Aspect Observed on both N- and S-facing slopes.

Hydrology Absent from wetlands.

Soil pH Mostly on soils in the pH range 6.0-8.0, and absent from soils of pH <4.5.

Bare soil Restricted to very open sites, many with high exposures of bare soil.

△ ordination Associated with wind-dispersed species growing in heavily disturbed vegetation.

Associated floristic diversity Varying widely from low on railway ballast sprayed with herbicides, to relatively high on some cinder tips and arable fields.

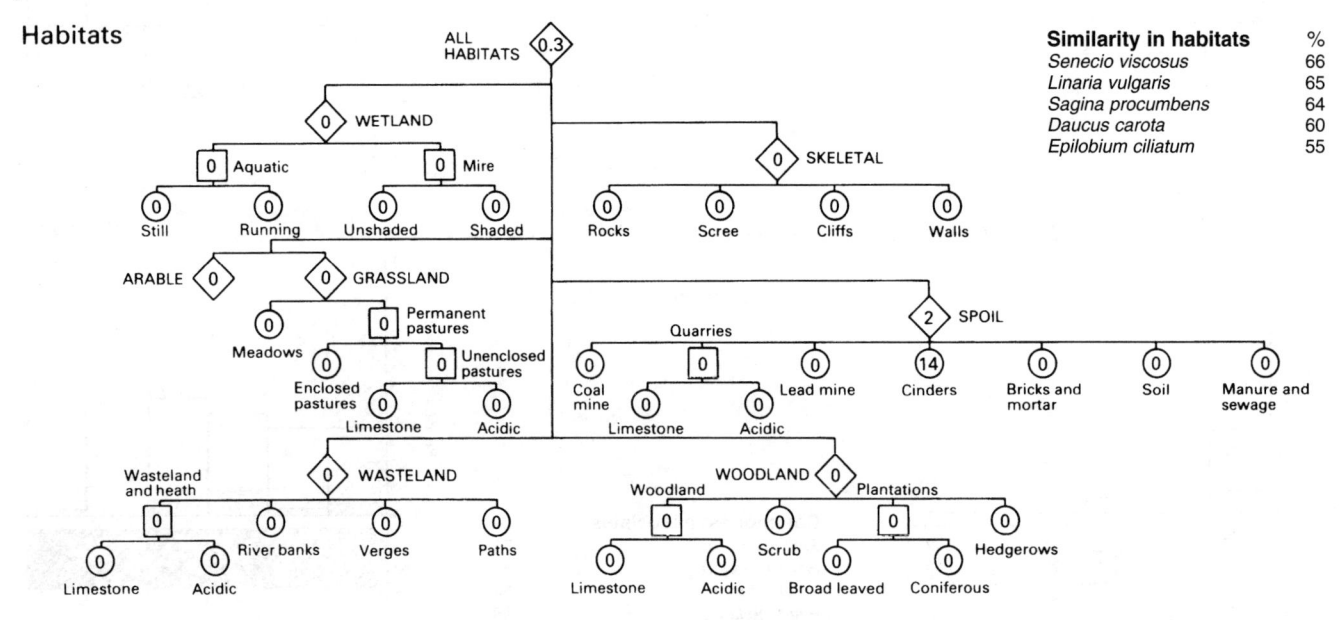

Habitats

Similarity in habitats	%
Senecio viscosus	66
Linaria vulgaris	65
Sagina procumbens	64
Daucus carota	60
Epilobium ciliatum	55

Synopsis *C.m.* is a small, summer-annual herb, which most typically grows on ballast, particularly cinders, beside railway lines in sites where the lateral vegetative spread of perennials is restricted by the rocky nature of the substrate. In habitats on actively-used track the growth of perennials may be further constrained by the application of herbicides. *C.m.* can persist in such sprayed sites because it is capable of setting seed relatively early in summer (see Arnold 1981), but in untreated areas, where plants may persist until autumn, seed-set and population density are generally greater. *C.m.* produces seeds abundantly, Salisbury (1942) suggesting that an average plant produces >2000. The seeds are minute and ribbed, and are shed from pores in a many-seeded capsule. Along railway tracks the seeds are dispersed most effectively as a result of air currents caused by passing trains (Arnold 1981). However, the mobility of the seeds is generally low (Arnold 1981), and the dispersal corridors provided along this linear habitat appear critical for the spread of the species. The plants remain erect for a while after their death, and thus the period of dispersal may be prolonged. Seeds may also be dispersed by the plant behaving as a 'tumble-weed' (Widlechner 1983). *C.m.* is also found more occasionally in a range of intermittently disturbed situations, but usually these populations are impermant, a characteristic which almost certainly arises from the transient nature of the seed bank (Brenchley & Warington 1930, Roberts & Boddrell 1983). The lack of persistent seeds in the soil, coupled with restriction of the germination period, may have contributed to the present rarity of *C.m.* on arable land, a habitat in which it was formerly widespread. In farmland *C.m.* is now restricted to scattered occurrences in cereal fields on calcareous soils. The species is restricted to lowland habitats in the British Isles (Wilson 1956). It is not known whether arable populations differ genetically from those on railways, but further investigation seems justified in a species capable of exploiting two such contrasting habitats. Two additional subspecies are recorded from Europe (*FE3*).

Current trends Formerly a frequent arable weed but becoming increasingly rare in this habitat. Now an expanding species on cinders beside railways, both in Britain and in North America (Widlechner 1983).

Altitude
Slope
Aspect

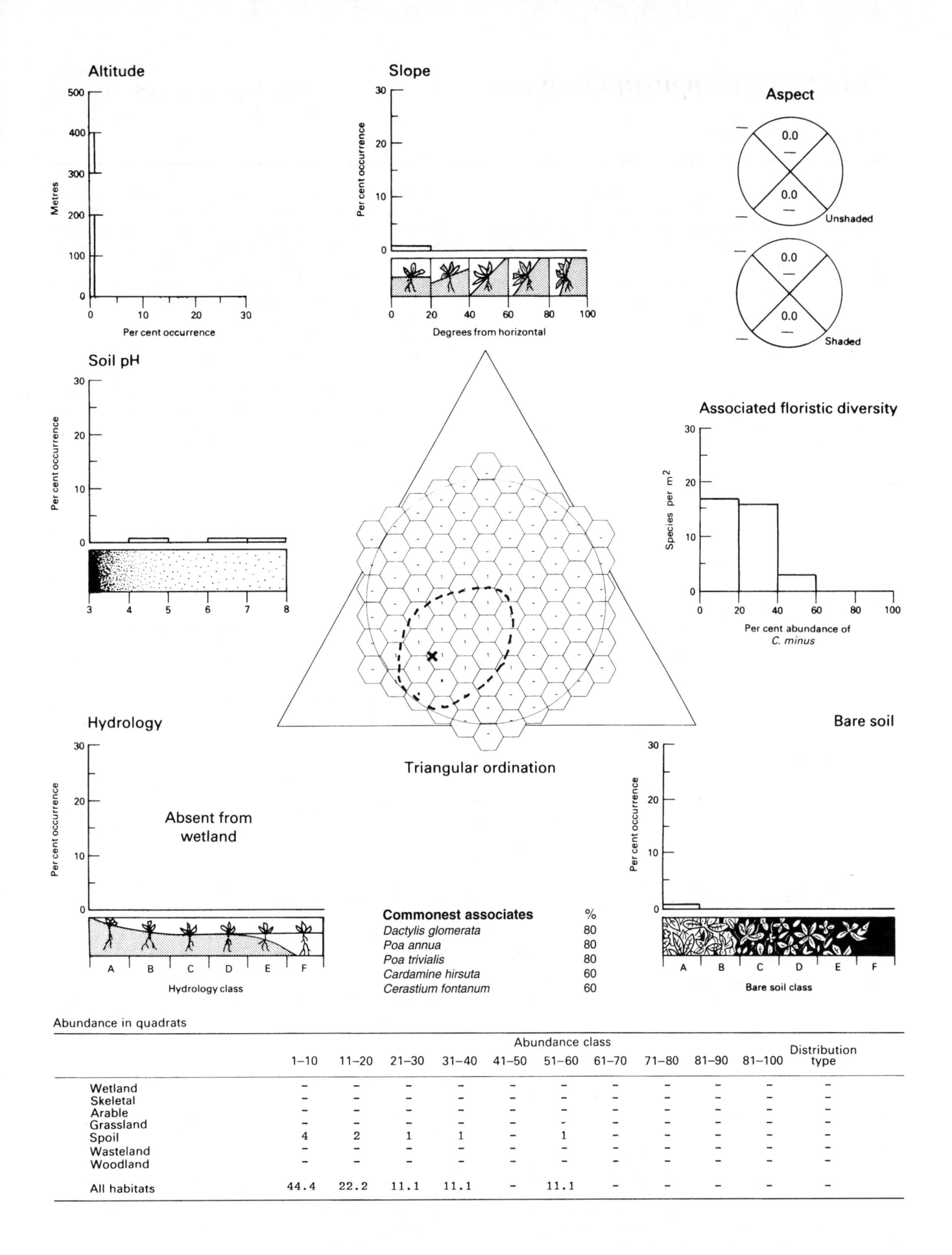

Unshaded

Shaded

Soil pH
Associated floristic diversity

Triangular ordination

Hydrology

Absent from wetland

Bare soil

Commonest associates	%
Dactylis glomerata	80
Poa annua	80
Poa trivialis	80
Cardamine hirsuta	60
Cerastium fontanum	60

Abundance in quadrats

	\multicolumn{10}{c	}{Abundance class}	Distribution type
	1–10	11–20	21–30	31–40	41–50	51–60	61–70	71–80	81–90	81–100	
Wetland	–	–	–	–	–	–	–	–	–	–	–
Skeletal	–	–	–	–	–	–	–	–	–	–	–
Arable	–	–	–	–	–	–	–	–	–	–	–
Grassland	–	–	–	–	–	–	–	–	–	–	–
Spoil	4	2	1	1	–	1	–	–	–	–	–
Wasteland	–	–	–	–	–	–	–	–	–	–	–
Woodland	–	–	–	–	–	–	–	–	–	–	–
All habitats	44.4	22.2	11.1	11.1	–	11.1	–	–	–	–	–

(*)*Chamerion angustifolium* (L.) J. Holub

Onagraceae
Rose-bay Willow-herb, Fireweed

Epilobium angustifolium

Life-form Polycarpic perennial geophyte (rarely protohemicryptophyte in mild winters) with horizontal roots and old stem bases giving rise to erect simple or branched stems with many lanceolate leaves, <4000 mm^2; roots relatively deep with VA mycorrhizas.
Phenology Shoots emerge in spring (or rarely autumn). Flowers from July to September and seeds dispersed from August onwards. Shoots die back completely in autumn. Leaves short-lived.
Height Foliage to 900 mm overtopped by inflorescence.
Seedling RGR 1.0-1.4 week^{-1}. SLA 18.9. DMC 22.8.
Nuclear DNA amount 0.8 pg; $2n = 36^*$ (*BC*), $2x^*$.
Established strategy Competitor.
Regenerative strategies W, V. Forms large clonal patches. In unshaded conditions produces numerous wind-dispersed seeds which germinate mainly in autumn. Some ungerminated seeds survive until the following spring (Thompson & Grime 1979), but they are not incorporated into a persistent seed bank; SB I.
Flowers Purple, hermaphrodite, protandrous, self-compatible, insect-pollinated (*Biol Fl*); often *c.* 200 in a raceme.
Germinule 0.05 mg, 1.1 × 0.3 mm, seed dispersed with plume *c.* 14 mm from a dehiscent often *c.* 500-seeded capsule.
Germination Epigeal; t_{50} 1 day; <5-35°C; germination rate increased after dry storage at 20°C; L > D.
Biological flora Myerscough (1980), see also Van Andel & Vera (1977) (as *E.a.*).
Geographical distribution The British Isles, scarce only in NW Scotland and Ireland (78% of British vice-counties), *ABIF* page 418; most of Europe, but rare in the S (82% of European territories); N Africa, N America and Asia.

Habitats Occurs in a wide variety of habitats, but is especially common in fertile, derelict environments such as urban clearance sites, cinders, building rubble, mining and quarry waste, and other spoil heaps. Also abundant in woodland clearings, wood margins, scrub and young plantations. Widespread in open habitats such as walls, cliffs and outcrops, waysides and wasteground, and river banks, with scattered occurrences of seedlings in wetlands, arable fields and pastures. Also frequent in neglected gardens. [On montane cliffs and mobile sand dunes (*Biol Fl*).]
Gregariousness Recorded as intermediate, but capable of forming dense stands at frequency values as low as 50%; also occurs widely as scattered seedlings and small rosettes.

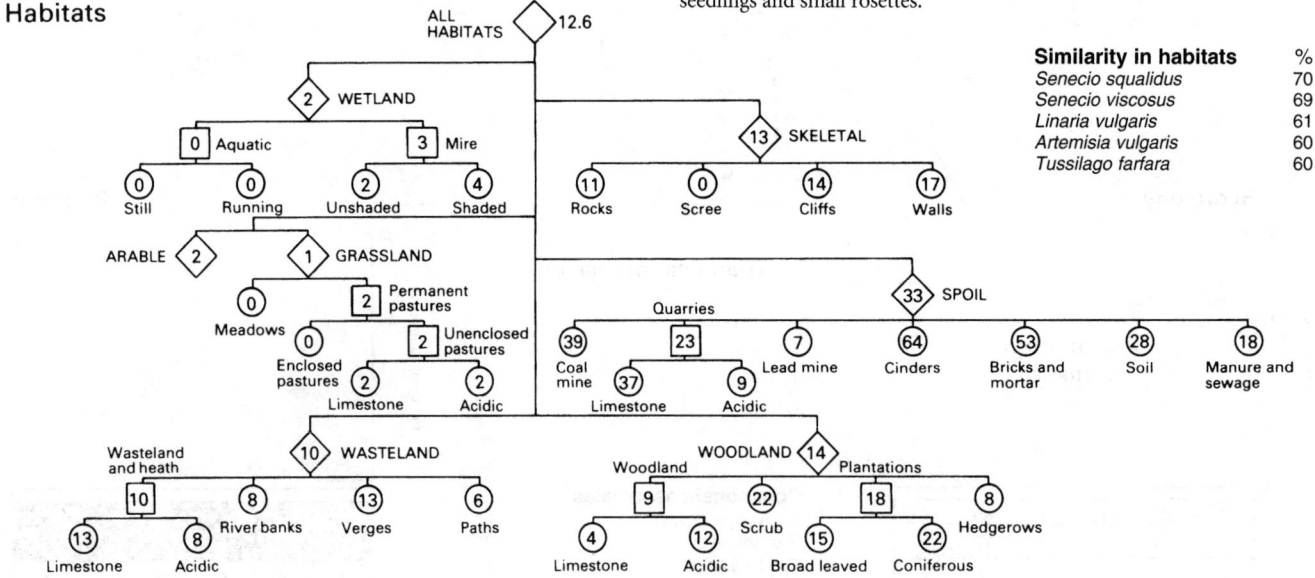

Habitats

Similarity in habitats	%
Senecio squalidus	70
Senecio viscosus	69
Linaria vulgaris	61
Artemisia vulgaris	60
Tussilago farfara	60

Altitude Most frequent and abundant below 100 m, and becoming progressively less common at higher elevations as suitable habitats become less common. [Up to 980 m.]
Slope Occurs over a wide range of slope; abundant only on slopes <50°.
Aspect No clear trends.
Hydrology Occasionally recorded in wetlands on moist but not submerged soils.
Soil pH Widely distributed with a peak in the pH range 4.5-5.0. A further peak at pH >7.5 reflects the abundance of *C.a.* on brick and mortar rubble in urban demolition sites.
Bare soil Associated with a wide range of values.
△ ordination A diffuse pattern, reflecting the presence of recently germinated seedlings throughout the landscape. Distribution centred in two distinct areas, the first associated with fertile undisturbed conditions in which the species is a widespread dominant, and the second in less-productive disturbed habitats (crevices in cliffs, walls, building rubble), where *C.a.* is often represented by isolated individuals or small populations of stunted plants.
Associated floristic diversity Relatively low, and declining with increased abundance of *C.a.*, confirming the potentially dominant impact of the species.
Synopsis *C.a.* is a tall, potentially dominant colonizing herb rarely found in long-established plant communities (*Biol Fl*) and most characteristically associated with derelict land. A peak of shoot biomass occurs in midsummer and involves a rapidly-ascending monolayer of short-lived leaves and results in a dense canopy. *C.a.* can form local monocultures with considerable accumulation of stem litter. The litter is not colonized by bryophytes, and new robust shoots of *C.a.* may penetrate it annually for many years. The habitat range of *C.a.* is restricted by the palatability of the shoots to stock and *C.a.* is also susceptible to shoot loss resulting from cutting or even frequent burning (*Biol Fl*). *C.a.* is found under a wide range of edaphic conditions, from sandy soils to clay and from low to high pH. However, it is usually absent from waterlogged sites and, other than on demolition sites (of pH >7.5), is most characteristic of slightly acidic soils.

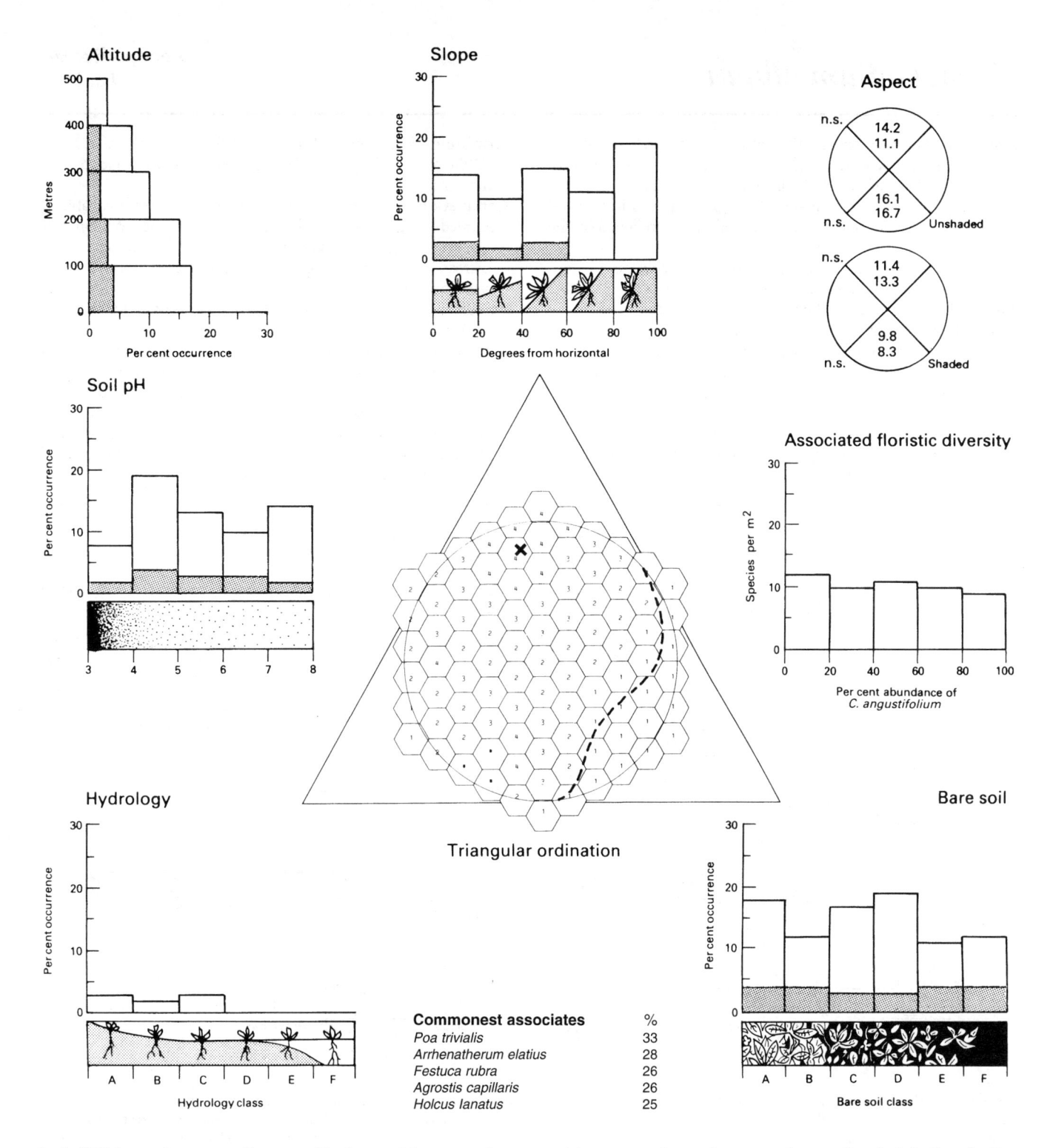

Commonest associates	%
Poa trivialis	33
Arrhenatherum elatius	28
Festuca rubra	26
Agrostis capillaris	26
Holcus lanatus	25

In the *PGE* the species temporarily appeared in plots receiving ammonium salts and the 'inhibiting effect of lime on establishment' was noted. On soils of higher pH we suspect that *C.a.* is often replaced as an early colonist by *Tussilago farfara*. The foliage contains high concentrations of N and P (*SIEF*). The current widespread abundance may be attributed to the numerous widely-dispersed seeds which frequently approach 'saturating densities' in lowland habitats and provide access not only to relatively inaccessible regeneration niches on cliffs and walls, but also to open sites recently created by local fires or earth-moving operations. The seeds are relatively short-lived and do not persist in the soil. The small seedling is highly susceptible to dominance by established vegetation. Seedling establishment is usually restricted to circumstances where, after the creation of bare soil, the habitat remains relatively undisturbed. During the first year the seedling develops into a flattened rosette, and in the second and later years of secondary succession on disturbed fertile soils *C.a.* rosettes rapidly consolidate colonies by the development of horizontal roots generating tall leafy stems. At woodland margins and in clearings *C.a.* produces phenotypes with an efficient display of thin leaves and fewer flowers. Another non-European ssp. has $2n = 72$ ($4x$) and $2n = 108$ ($6x$) (*Biol Fl*). *C.a.* was formerly a rare native species and was grown in gardens for ornament (where it may still be seen in Derbyshire). Several early local records are described as garden escapes (Linton 1903). A North American origin for the later invasive spread in this century is suggested by Heslop-Harrison (1953), but this has not been verified.

Current trends *C.a.* has increased as a result of such factors as mechanized disturbance and industrial and urban dereliction. Further increases in abundance appear likely.

Chenopodium album L.

Chenopodiaceae
Fat Hen

Life-form Summer annual therophyte. Stem erect, simple, or more usually well-branched; leaves ovate, mealy or glabrous, often <1000 mm^2; roots usually non-mycorrhizal.

Phenology A summer annual which begins to flower in July and fruits from August onwards. Some seed may be retained on the plant after its death in autumn.

Height Foliage to 1 m or more, overtopped by flowers.

Seedling RGR 1.0–1.4 week^{-1}. SLA 16.6. DMC 16.4.

Nuclear DNA amount 3.8 pg; $2n = 54*$ (*BC*); $6x*$.

Established strategy Competitive-ruderal.

Regenerative strategies B_s. Reproduces entirely by seed, which germinates mainly in spring. Seedlings originating in autumn do not overwinter (*Biol Fl(a)*). A persistent seed bank is formed (*Biol Fl(a)*); SB IV.

Flowers Green, hermaphrodite or female, wind- or self-pollinated (*Biol Fl*); aggregated into terminal heads; often many thousand in a branched inflorescence.

Germinule Dimorphic; brown *c.* 1.5 mg, black *c.* 1.2 mg (Williams & Harper 1965); seed enclosed within a papery pericarp.

Germination Epigeal; t_{50} 2 days; 8–36°C, brown seeds capable of immediate germination, black seeds dormant (Williams & Harper 1965); germination rate of freshly-collected seeds increased after dry storage at 5 or 20°C; dormancy of some seeds broken by chilling; L > D.

Biological flora (a) Williams (1963), (b) Bassett and Crompton (1978).

Geographical distribution The British Isles, but less frequent in the N and W (100% of British vice-counties), *ABIF* page 140; originally a native of Europe (92% of European territories and 5% naturalized) and a N temperate species, but has spread with cultivation to all inhabited parts of the world except where there are extreme desert conditions; widely distributed as a weed in N and C America, Africa and Australia.

Habitats Most common as an arable weed of broad-leaved crops or fallow field margins, and also a colonist of disturbed habitats, particularly manure heaps, sewage sludge, soil heaps and building rubble. Also recorded, too, from open habitats such as cinder tips, paths and walls. Not recorded from grassland, woodland and wetlands.

Gregariousness Tends to occur as robust scattered individuals.

Altitude Recorded up to 310 m, but most suitable habitats occur within the lowland region. [Up to 460 m.]

Slope Widely distributed but most records are from flat or gently sloping ground.

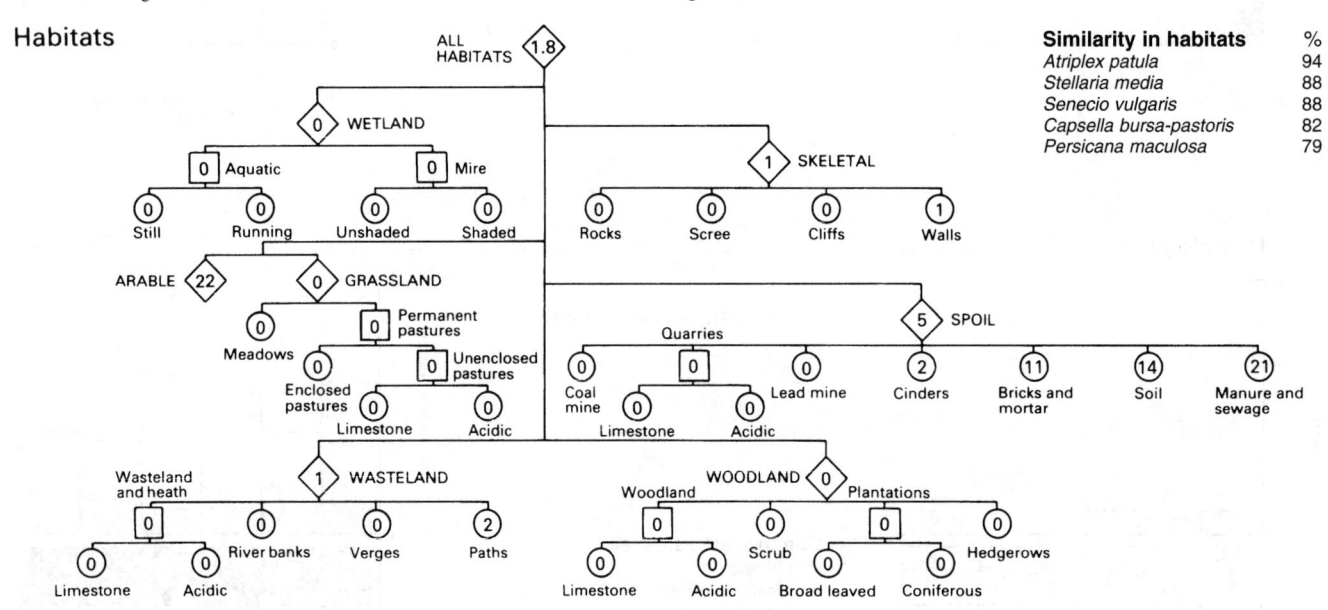

Habitats

Similarity in habitats	%
Atriplex patula	94
Stellaria media	88
Senecio vulgaris	88
Capsella bursa-pastoris	82
Persicana maculosa	79

Aspect Slightly more frequent on unshaded N-facing slopes.

Hydrology Absent from wetlands.

Soil pH Absent from acidic soils below pH 5.0.

Bare soil Strongly represented in habitats with a large exposure of bare soil.

Δ ordination Distribution mostly confined to vegetation associated with heavily disturbed and fertile habitats.

Associated floristic diversity Typically intermediate.

Synopsis *C.a.* is an often robust summer annual found in a wide range of disturbed relatively fertile habitats. It is ranked by Holm *el al.* (1977) as the world's worst weed of potatoes and sugar beet. *C.a.* shares with many other troublesome weeds, e.g. *Sinapis arvensis*, (a) the ability to form a buried and highly persistent seed bank (Roberts & Feast 1972, Odum 1978), (b) a tendency for some seed to be retained on the plant at harvest time, leading to contamination of crop seed, and (c) the tendency for dispersal by birds and mammals (*Biol Fl(a)*). *C.a.* is characteristic of nitrogen-rich habitats (*Biol Fl(b)*), a feature even more marked in *C. rubrum*. *C.a.* is palatable to livestock, though toxic in large quantities (*Biol Fl(b)*). Both the foliage and the seeds have been eaten by humans and the shoot has exceptionally high protein concentrations (Hendry & Brocklebank unpubl.). *C.a.* is not, however, found generally in grazed habitats, as it does not readily recover from damage to the shoot. Regeneration is entirely by seed, and *C.a.* takes several months to set seed, which prevents its successful establishment in the most frequently disturbed sites. Although some seed germination occurs in autumn, only seedlings originating in spring survive to flower (*Biol Fl(a)*). A number of genetic variants of *C.a.* have been identified, including populations resistant to atrazine (*Biol Fl(b)*). In common with other ruderals *C.a.* also shows considerable phenotypic plasticity; this attribute is particularly marked with respect to seed characters. Under long days many small, black dormant seeds are produced whereas under short days most seeds are brown and non-dormant (*Biol Fl(a)*). The ecological significance of this particular seed polymorphism remains unclear, but the proportion of each type of seed is known to vary from year to year within a single population (*Biol Fl(b)*), and the differences in germination

Altitude

Slope

Aspect

Soil pH

Associated floristic diversity

Triangular ordination

Hydrology

Absent from wetland

Bare soil

Commonest associates	%
Atriplex patula	74
Polygonum aviculare	74
Stellaria media	74
Matricaria discoidea	58
Poa annua	58

Abundance in quadrats

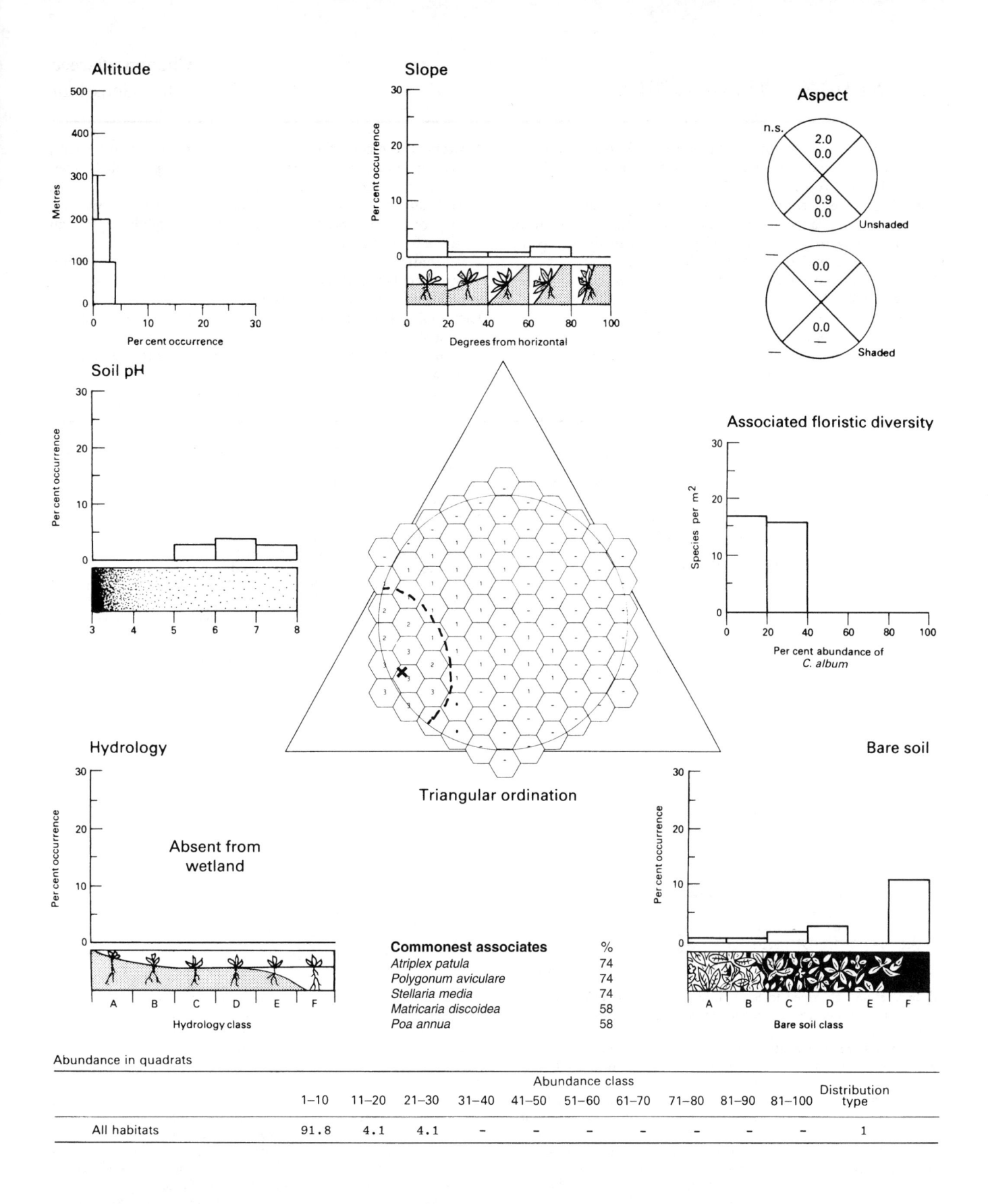

	\multicolumn{10}{c}{Abundance class}	Distribution type
	1–10	11–20	21–30	31–40	41–50	51–60	61–70	71–80	81–90	81–100	
All habitats	91.8	4.1	4.1	–	–	–	–	–	–	–	1

requirements between the different types are as great as those occurring between some species (Williams & Harper 1965).

Current trends Now primarily a species of broad-leaved crops, although formerly widespread in barley and other cereals (*Biol Fl(b)*). The range of *C.a.* in agricultural habitats thus appears to be declining, but the species remains common on urban wasteland. On balance *C.a.*, though still a very common and troublesome weed, may be decreasing in abundance.

Chenopodium rubrum L.

Chenopodiaceae
Red Goosefoot

Life-form Summer annual therophyte. Shoots simple or branched, ascending or erect; leaves toothed, ovate, often fleshy, < 1000 mm²; roots non-mycorrhizal.

Phenology Seeds germinate mainly during spring and early summer and plants flower from July onwards. Seed shed from August onwards, and some still present when the plants die back in autumn. Leaves relatively short-lived.

Height Foliage to 700 mm overtopped by flowers.

Seedling RGR 2.0–2.4 week⁻¹. SLA 21.0. DMC 12.7.

Nuclear DNA amount No data; $2n = 36*$ (*BC*); $4x*$.

Established strategy Competitive-ruderal.

Regenerative strategies B_s. Regenerates exclusively by seed which remains viable in the soil for up to 50 years (*Biol Fl*); SB IV.

Flowers Hermaphrodite or more rarely predominantly male, usually protogynous, mainly wind- or self-pollinated; often many thousand in a branched inflorescence.

Germinule 0.09 mg, 0.6 × 0.6 mm, seed, enclosed within a papery pericarp. Seeds of terminal flowers slightly larger than those from lateral flowers (*Biol Fl*).

Germination Epigeal; t_{50} <1 day; 25–35°C; germination promoted by fluctuating temperatures; L > D.

Biological flora Williams (1969).

Geographical distribution Mainly in S and E Britain, but often a casual and probably not native in N England, Wales, Scotland and Ireland (63% of British vice-counties), *ABIF* page 137; W, C and S Europe but rare in the Mediterranean (74% of European territories); C Asia; introduced to N and S America.

Habitats Most common on manure heaps and sewage spoil but found in a variety of other disturbed artificial habitats such as demolition sites, coal mine spoil and soil heaps. Occasionally on exposed mud beside ponds and reservoirs, or on their retaining walls. Also found in arable fields associated with broad-leaved crops. Absent from grassland, woodland and aquatic habitats. [Grows in sand-dune slacks and on the upper reaches of salt marshes (*Biol Fl*).]

Gregariousness Intermediate. High frequencies are associated with dense populations of stunted plants. A robust individual may shade an area in excess of 0.5 m².

Altitude Most frequent and abundant below 200 m, but recorded up to 300 m, the British altitudinal limit for the species.

Slope Virtually absent from skeletal habitats and consequently restricted to flat or moderately sloping ground.

Aspect Found over a wide range of aspects.

Hydrology Recorded from exposed mud marginal to open water.

Soil pH Very few records from acidic soils, and largely restricted to soils of pH >6.0.

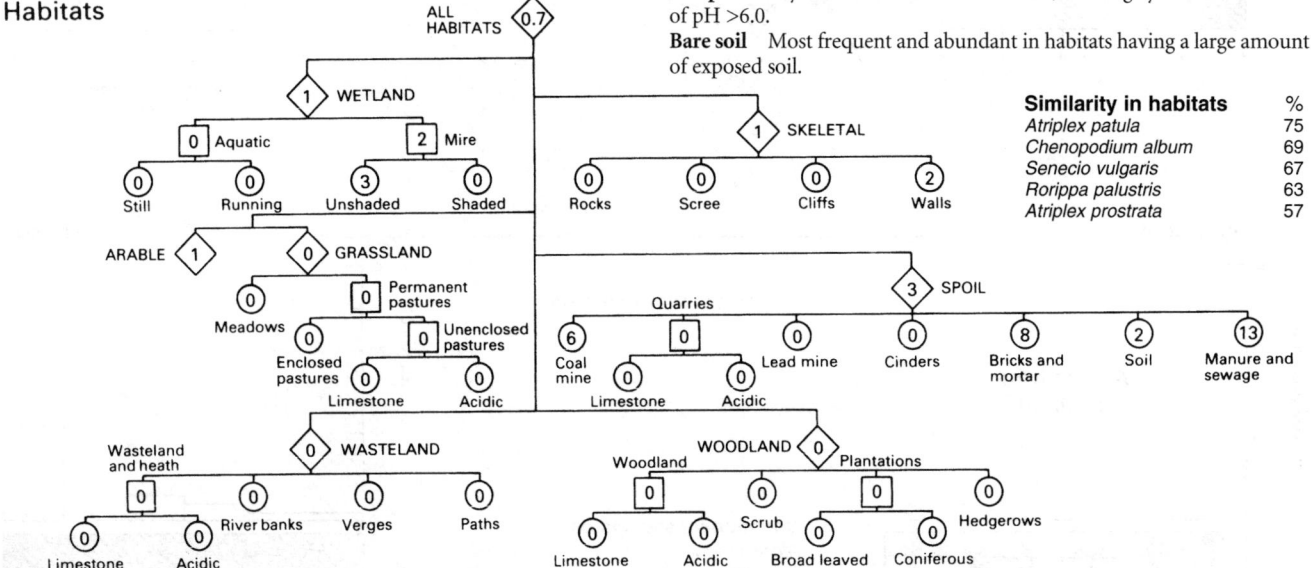

Bare soil Most frequent and abundant in habitats having a large amount of exposed soil.

Habitats

Similarity in habitats	%
Atriplex patula	75
Chenopodium album	69
Senecio vulgaris	67
Rorippa palustris	63
Atriplex prostrata	57

Δ ordination Confined to fertile habitats subject to varying levels of disturbance, and centred on vegetation associated with disturbed, fertile habitats.

Associated floristic diversity Typically low to intermediate.

Synopsis Although widely distributed in fertile disturbed habitats, *C.r.* has a more restricted ecology than *C. album*. *C.r.* is most characteristic of nitrogen-rich environments, and in the Sheffield region reaches its maximum abundance on farmyard manure and sewage residues. Occurrences with *Lycopersicon esculentum* Miller (tomato) on river shingle may provide a useful indicator of pollution of waterways; both species are often observed where raw sewage is discharged into the water. *C.r.* is also common beside eutrophic lakes and ponds, and may develop high even-aged populations on mud exposed during the summer months. Occurrences at the upper reaches of salt marshes (*Biol Fl*) imply that at least certain populations of *C.r.* exhibit salt tolerance. Like *C. album*, *C.r.* is absent from continuously and severely disturbed sites. *C.r.* rarely occurs in grazed habitats. *C.r.* shows extreme phenotypic plasticity in reproductive effort (Cumming 1969). Stunted plants may reach only 30 mm in height and produce *c.* 20 seeds, while robust plants may be nearly 30 times as tall and produce 10 000 times as many seeds (*Biol Fl*). The seeds have a hard coat and may pass unchanged through the digestive tract of grazing animals (see *Biol Fl*), and presumably also resist digestion by seed-eating birds. Seeds are persistent in the soil (Odum 1978).

Germination is strongly promoted by fluctuating temperatures (Thompson & Grime 1983); this effect is most pronounced in darkness and probably functions as a 'depth-sensing' and 'gap-detecting' mechanism in buried seeds. *C.r.*, a short-day plant, is often used as an experimental subject in physiological investigations of flowering. Floral induction can be effected at the cotyledon stage, and races differ in the range of day-length under which flowering is initiated. Most other aspects of the biology have not been studied intensively, although *C.r.* clearly shares many of the distinctive features of *C. album*, a species with which it shows marked ecological similarities.

Current trends Likely to increase in lowland regions, particularly in urban and probably in some agricultural areas. *C.r.* is also a regular colonist of artificial habitats in upland, and an increase in its altitudinal maximum is predicted.

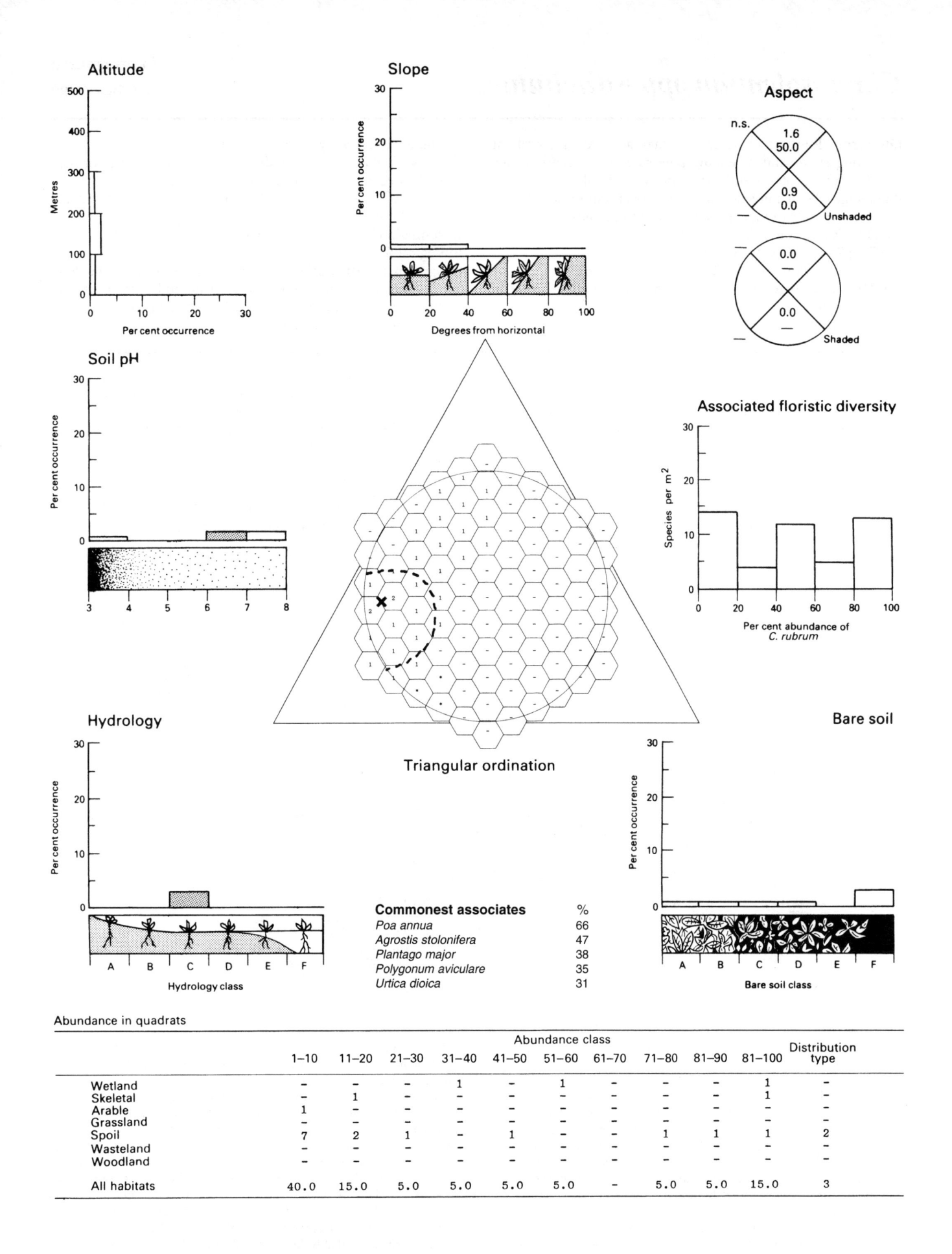

Altitude

Slope

Aspect

Soil pH

Associated floristic diversity

Species per m^2 vs Per cent abundance of *C. rubrum*

Triangular ordination

Hydrology

Bare soil

Commonest associates %
Poa annua 66
Agrostis stolonifera 47
Plantago major 38
Polygonum aviculare 35
Urtica dioica 31

Abundance in quadrats

	1–10	11–20	21–30	31–40	41–50	51–60	61–70	71–80	81–90	81–100	Distribution type
Wetland	–	–	–	1	–	1	–	–	–	1	–
Skeletal	–	1	–	–	–	–	–	–	–	1	–
Arable	1	–	–	–	–	–	–	–	–	–	–
Grassland	–	–	–	–	–	–	–	–	–	–	–
Spoil	7	2	1	–	1	–	–	1	1	1	2
Wasteland	–	–	–	–	–	–	–	–	–	–	–
Woodland	–	–	–	–	–	–	–	–	–	–	–
All habitats	40.0	15.0	5.0	5.0	5.0	5.0	–	5.0	5.0	15.0	3

Chrysosplenium oppositifolium L.

Saxifragaceae
Golden Saxifrage

Life-form Polycarpic perennial chamaephyte. Decumbent rooting stems and ascending flowering stems; leaves ovate; <500 mm^2; roots adventitious, non-mycorrhizal.
Phenology Winter green. Plants flower mainly in April and May. Seed shed May to June. Leaves relatively long-lived.
Height Foliage and flowers <40 mm.
Seedling RGR No data. SLA 23.3. DMC 12.5.
Nuclear DNA amount No data; $2n = 42*$ (*BC*); $6x*$.
Established strategy Intermediate between competitive-ruderal and C-S-R.
Regenerative strategies (V), B$_s$. Regenerates both vegetatively and by seed (see 'Synopsis'); ?SB III.
Flowers Yellow, hermaphrodite, protogynous, insect- or self-pollinated; often <30 in a cymose inflorescence.
Germinule 0.04 mg, 0.7 × 0.5 mm, seed in a dehiscent *c.* 30-seeded capsule.
Germination Epigeal; t_{50} 11 days.
Biological flora None.
Geographical distribution A European endemic. Throughout the British Isles, but local in E England and Ireland (98% of British vice-counties), *ABIF* page 320; W and part of C Europe (38% of European territories).

Habitats Most frequent in unshaded mire, but also found on river banks, where it may be temporarily immersed, and in flood-plain woodland. Also found on cliffs and walls beside open water, or on dripping rocks.
Gregariousness Recorded as intermediate, but potentially carpet-forming.
Altitude Wide-ranging, but in lowland sites largely restricted to shaded habitats. [Up to 1100 m.]
Slope As a result of its capacity to grow on wet rocks, extends on to steep slopes.
Aspect No trends evident in the data presented opposite, but additional studies suggest that in both shaded and unshaded habitats *C.o.* shows a N-facing bias.
Hydrology Largely confined to waterlogged river margins, soligenous mire and moist ground on river-floodplains. Some sites liable to flooding but *C.o.* absent from permanently submerged sites.
Soil pH Most records on soils with a pH of between 6.0 and 7.5 and absent from acidic soils of pH <4.5.
Bare soil Wide-ranging.
△ **ordination** Forms an understorey in a wide range of communities, and distribution diffuse. Absent from heavily disturbed vegetation, and less frequent in sites of moderate fertility and with little or no disturbance.
Associated floristic diversity Low to intermediate.

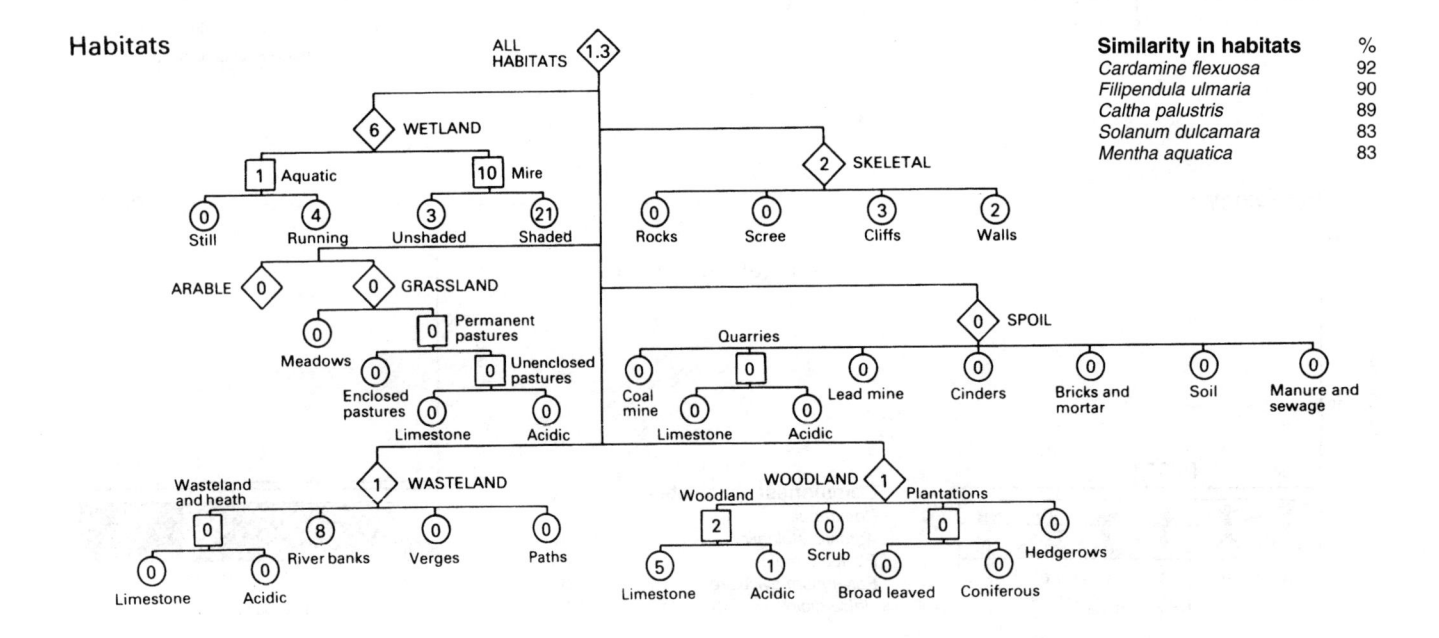

Habitats

Similarity in habitats	%
Cardamine flexuosa	92
Filipendula ulmaria	90
Caltha palustris	89
Solanum dulcamara	83
Mentha aquatica	83

Synopsis *C.o.* is a low-growing, mat-forming, winter-green herb most typically associated with soligenous mire and river banks. The species is a European endemic and part of the sub-Atlantic element of the British flora (Birks 1973). Typically *C.o.* forms an understorey beneath taller vegetation, and in lowland habitats *C.o.* is largely restricted to sites shaded by trees. On river floodplains in upland areas *C.o.* is most frequent beneath tall-herb communities, whereas at higher altitudes the species forms continuous carpets in the open on wet rocks and along mountain stream sides associated with species such as *Epilobium obscurum* and *Montia fontana* L. Foliage of *C.o.* is characterized by low concentrations of N and P and elevated levels of Na and Fe (*SIEF*). *C.o.* maintains an extensive leaf canopy throughout the year and is the earliest flowering species of shaded mire. These two attributes are consistent with the ability of the low-growing *C.o.* to coexist with tall herbs such as *Epilobium hirsutum* and *Petasites hybridus*, which do not expand their foliage fully until early summer. The species has a wide edaphic range, occurring both on limestone cliffs and in base-poor flushes on gritstone. *C.o.* forms large clonal patches by means of creeping stoloniferous growth. In addition, shoots detached as a result of disturbance readily re-establish, and this probably constitutes an important means of colonizing river systems. The species also produces numerous small seeds, which may be incorporated into a buried seed bank (UCPE unpubl. data). However, regeneration by seed appears to be infrequent, and the species shows a poor colonizing ability in new artificial habitats. *C.o.* often coexists with *C. alternifolium* L. ($2n = 48$ (*FE1*)) which is rhizomatous and tends to occur on slightly better drained soils.
Current trends Probably decreasing, at least in lowland areas.

Altitude
Slope
Aspect

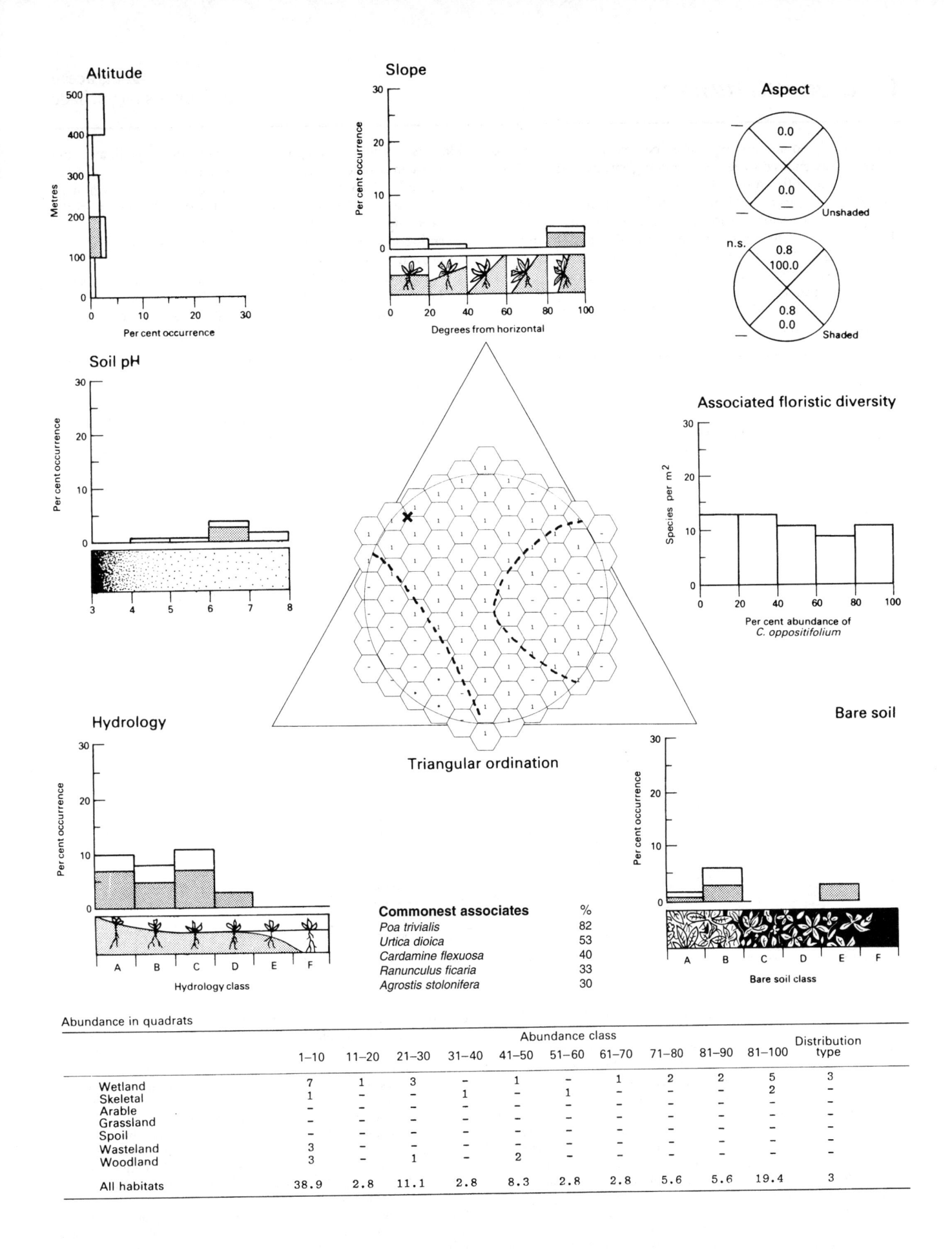

	Unshaded
—	0.0
—	0.0

	Shaded
n.s.	0.8
	100.0
	0.8
—	0.0

Soil pH

Associated floristic diversity

Species per m²

Per cent abundance of *C. oppositifolium*

Triangular ordination

Hydrology

Bare soil

Commonest associates	%
Poa trivialis	82
Urtica dioica	53
Cardamine flexuosa	40
Ranunculus ficaria	33
Agrostis stolonifera	30

Abundance in quadrats

	1–10	11–20	21–30	31–40	41–50	51–60	61–70	71–80	81–90	81–100	Distribution type
Wetland	7	1	3	–	1	–	1	2	2	5	3
Skeletal	1	–	–	1	–	1	–	–	–	2	–
Arable	–	–	–	–	–	–	–	–	–	–	–
Grassland	–	–	–	–	–	–	–	–	–	–	–
Spoil	–	–	–	–	–	–	–	–	–	–	–
Wasteland	3	–	–	–	–	–	–	–	–	–	–
Woodland	3	–	1	–	2	–	–	–	–	–	–
All habitats	38.9	2.8	11.1	2.8	8.3	2.8	2.8	5.6	5.6	19.4	3

Circaea lutetiana L.

Onagraceae
Enchanter's Nightshade

Life-form Stoloniferous, polycarpic perennial geophyte. Stems erect or ascending; leaves ovate, often glabrous, usually <2000 mm^2; roots with VA mycorrhizas.
Phenology Plants flower from June to August and set seed from September to October. Shoots die back in autumn.
Height Foliage <300 mm, overtopped by flowers to 600 mm.
Seedling RGR No data. SLA 36.7. DMC 19.3.
Nuclear DNA amount No data; $2n = 22*$ (*BC*); $2x*$.
Established strategy Competitive-ruderal.
Regenerative strategies (V), S. Mainly vegetative (see 'Synopsis'); SB II.

Flowers White, hermaphrodite, usually insect-pollinated but self-compatible (Raven 1963); >50 in a raceme.
Germinule 0.88 mg, 2.6 × 0.9 mm, seed dispersed within an indehiscent 1-2-seeded fruit covered with stiff, hooked bristles (1.97 mg, 3.0 × 2.3 mm).
Germination Epigeal; a chilling requirement is suspected.
Biological flora None.
Geographical distribution Throughout much of the British Isles, but infrequent in N Scotland (92% of British vice-counties), *ABIF* page 421; most of Europe except the NE (79% of European territories); N Africa.

Habitats Almost totally confined to moist shaded habitats, particularly woodland, scrub and shaded mire. Occasionally observed in hedgerows, on shaded road verges and as a weed of shrubberies and shaded gardens.
Gregariousness Potentially patch-forming.
Altitude Widely distributed to 300 m. [Up to 370 m.]
Slope Absent both from skeletal habitats and steep slopes.
Aspect Although recorded only from N-facing slopes in this survey, additional studies have revealed a wide distribution with respect to aspect.
Hydrology Occasionally in shaded soligenous mire, but more typically in moist woodland, sometimes at the boundary of topogenous mire.
Soil pH Absent from acidic soils of pH <4.5.
Bare soil Largely confined to shaded habitats having a relatively high proportion of exposed soil or tree litter.
△ **ordination** Present in all types of vegetation except those which are very disturbed. Distribution centred on relatively productive vegetation subject to intermediate levels of disturbance.
Associated floristic diversity Low to intermediate.

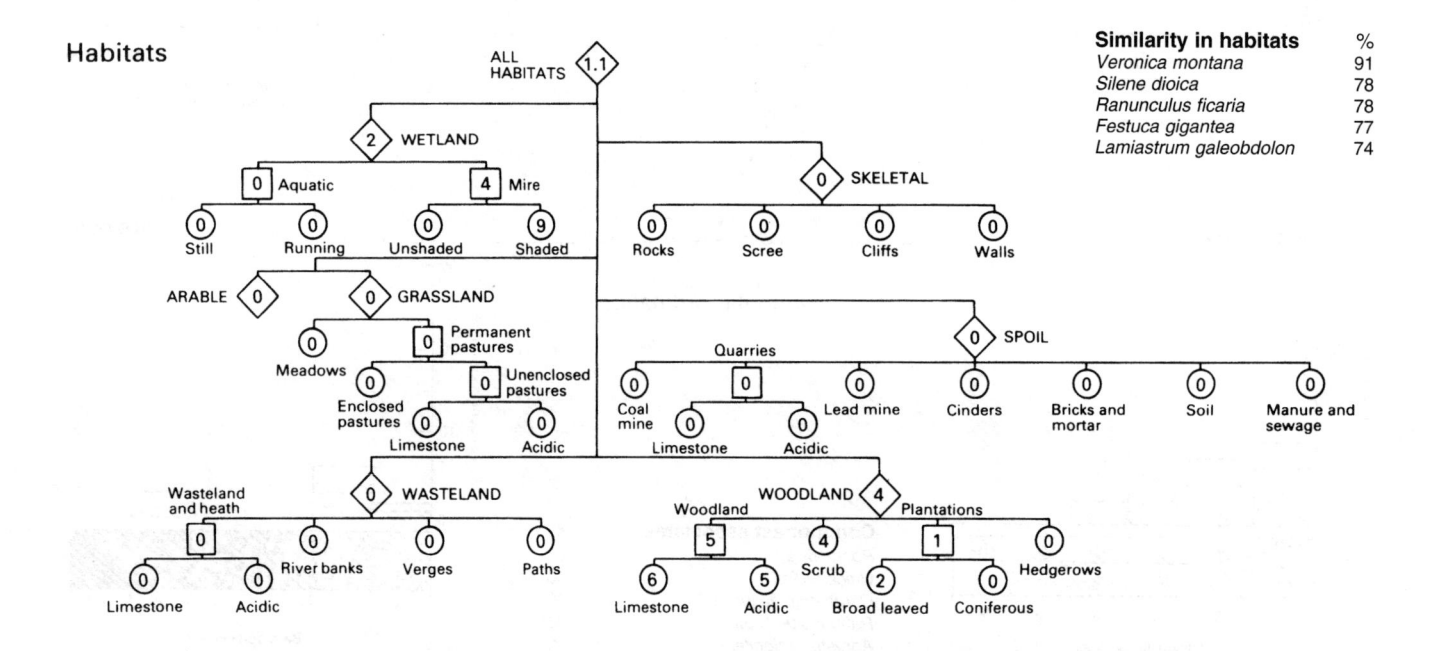

Habitats

Similarity in habitats	%
Veronica montana	91
Silene dioica	78
Ranunculus ficaria	78
Festuca gigantea	77
Lamiastrum galeobdolon	74

Synopsis *C.l.* is a medium-sized stand-forming herb of fertile woodland in sites where the vigour of potential dominants is limited by shade. The species produces a network of short-lived shoots and stolons, and is classified by Salisbury (1942) as a 'pseudo-annual'. The rapid growth and long stolons (to 300 mm) enable the species to persist as a weed of shrubberies and shaded gardens. More importantly, these attributes may also enable *C.l.* to achieve rapid local dominance during the formation of secondary woodland, and may facilitate the active foraging for optimal sites within a changing or disturbed environment. Unlike most woodland species, *C.l.* is not evergreen and shows no pronounced vernal growth; it achieves its peak in above-ground biomass during summer. Thus, *C.l.* is similar to two other species which exploit disturbed woodland: *Chamerion angustifolium* and *Stachys sylvatica*. All three exhibit a similar phenology, coupled with a well-developed capacity for vegetative spread. Their occurrence within secondary woodland is likely to be related to a high colonizing ability in disturbed sites. We suspect that their persistence during more-stable late phases of woodland development depends in part upon the poor colonizing ability of many of the woodland herbs with vernal or winter-green phenologies which are complementary to those of trees. *C.l.* does, however, differ from *C.a.* and *S.s.* in two important respects: (a) it is virtually restricted to shaded habitats and is thus more truly a woodland species and (b) the fruits are clothed in hooked bristles and are, like a number of other woodland plants, dispersed by attachment to animals (and to clothing). *C.l.* does not form a persistent seed bank (Brown & Oosterhuis 1981). A hybrid, *C.* × *intermedia* Ehrh., is formed with *C. alpina* L. Vegetatively, this plant is even more vigorous than *C.l.* and is widespread in N Britain.

Current trends A woodland species capable of exploiting disturbed habitats and of surviving in deep shade. An expansion in plantations and urban woodland is forecast.

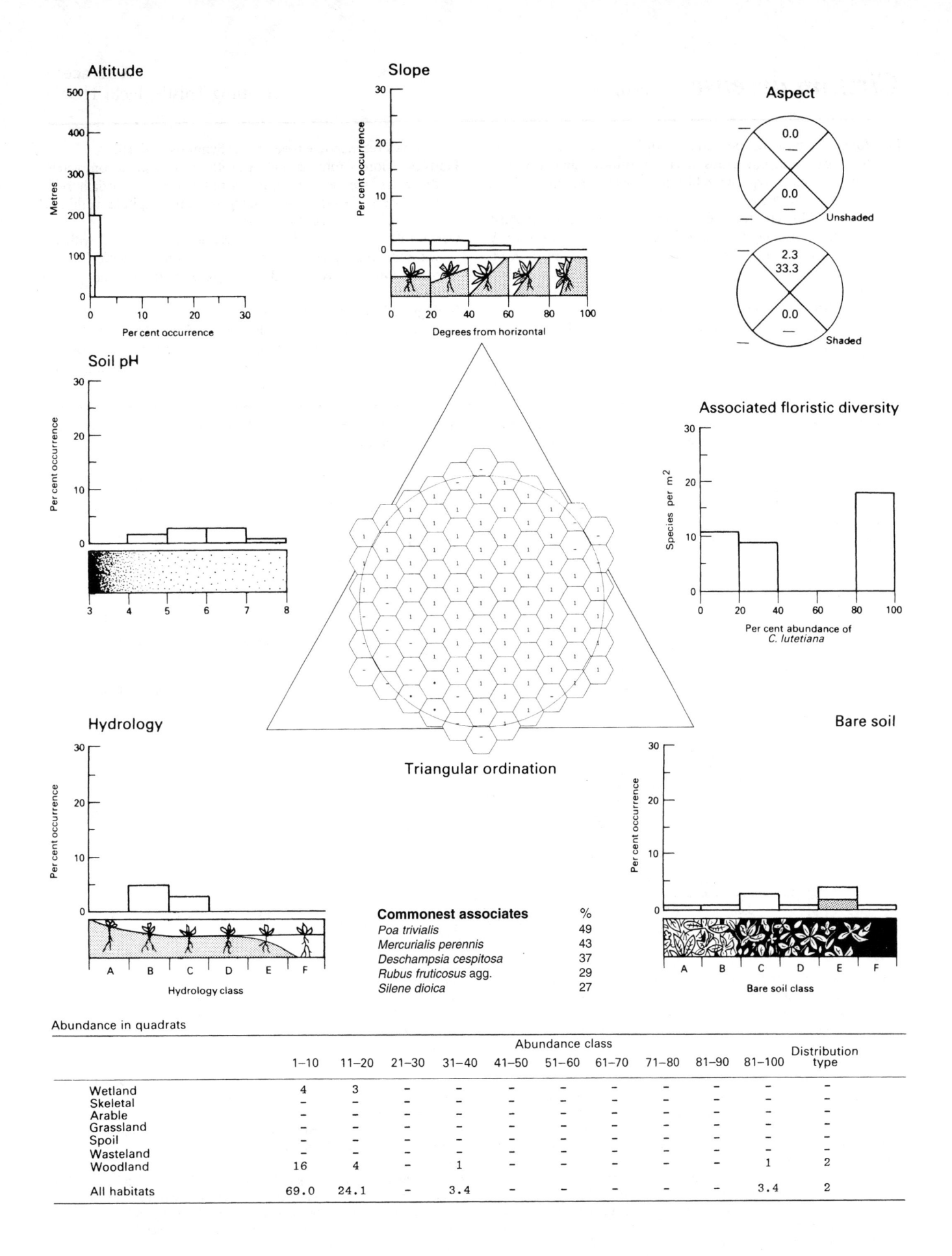

Altitude

Slope

Aspect

	Unshaded
0.0	0.0
—	—

	Shaded
2.3	33.3
0.0	—

Soil pH

Associated floristic diversity

Hydrology

Triangular ordination

Bare soil

Commonest associates	%
Poa trivialis	49
Mercurialis perennis	43
Deschampsia cespitosa	37
Rubus fruticosus agg.	29
Silene dioica	27

Abundance in quadrats

	Abundance class										Distribution type
	1–10	11–20	21–30	31–40	41–50	51–60	61–70	71–80	81–90	81–100	
Wetland	4	3	–	–	–	–	–	–	–	–	–
Skeletal	–	–	–	–	–	–	–	–	–	–	–
Arable	–	–	–	–	–	–	–	–	–	–	–
Grassland	–	–	–	–	–	–	–	–	–	–	–
Spoil	–	–	–	–	–	–	–	–	–	–	–
Wasteland	–	–	–	–	–	–	–	–	–	–	–
Woodland	16	4	–	1	–	–	–	–	–	1	2
All habitats	69.0	24.1	–	3.4	–	–	–	–	–	3.4	2

Cirsium arvense (L.) Scop.

Asteraceae
Creeping Thistle, Field Thistle

Life-form Polycarpic perennial geophyte with far-creeping lateral roots and numerous erect adventitious shoots. Leaves spiny, lanceolate, typically <4000 mm², with VA mycorrhizas.
Phenology Flowers July to September. Seed shed August onwards; some may be retained in the seed head until early winter, by which time the above-ground shoot is dying or dead.
Height Leafy shoots with terminal inflorescences attaining 900 mm.
Seedling RGR No data. SLA 16.1. DMC 13.2.
Nuclear DNA amount 3.1 pg; $2n = 34^*$ (*BC*); $2x^*$ Base number for tribe is also $n = 17$.
Established strategy Competitor.
Regenerative strategies V, W, B_s. For an explanation of the complex regenerative biology see 'Synopsis'; ?SB III.
Flowers Florets tubular, pale purple or whitish, incompletely dioecious (see 'Synopsis'), insect-pollinated and partially self-fertile (Kay 1985); >100 aggregated into capitula within a branched terminal inflorescence.
Germinule 1.17 mg, 3.5 × 1.6 mm, achene dispersed with or without a readily-detached pappus *c.* 25 mm in length.
Germination Epigeal; t_{50} 3 days (seed often dormant - see 'Synopsis').
Biological flora Moore (1975).
Geographical distribution The British Isles (100% of British vice-counties), *ABIF* page 614; Europe northwards to Scandinavia (87% of European territories and 5% naturalized); N Africa, W Asia; introduced in N America.

Habitats Widely distributed, but most abundant on heaps of coal-mine spoil, mineral soil, cinders and building rubble, in meadows and pastures, on river and stream banks and road verges, and in unmanaged grasslands. Also frequent in arable fields, on quarry heaps and organic spoil, and in hedgerows. Less common in skeletal and wetland habitats, and largely absent from woodland.
Gregariousness Potentially stand-forming.
Altitude Up to 400 m, but more abundant at lower altitudes. [Up to 700 m.]

Slope Largely absent from both skeletal habitats and the steepest slopes. Abundant only on flat ground.
Aspect No pattern detected.
Hydrology Occasionally present on moist soils marginal to open water but, as with many deeply rooted species, infrequent in wetlands.
Soil pH Widely distributed, but most common on base-rich soils of pH >5.0.
Bare soil Widely distributed with respect to amount of bare soil, a reflection of the occurrence of *C.a.* in tall herb communities as well as in open sites.
△ ordination Distribution centred on vegetation associated with fertile, moderately disturbed conditions. Absent only from very unproductive habitats.
Associated floristic diversity Typically low or intermediate.

Habitats

Similarity in habitats	%
Holcus lanatus	81
Atriplex prostrata	75
Artemisia vulgaris	71
Tussilago farfara	63
Cirsium vulgare	61

Synopsis *C.a.* is a tall, locally dominant broad-leaved herb. It is designated a noxious weed (Weeds Act 1959) and is one of the world's worst weeds (Holm *et al.* 1977). In grassland the spiny leaves are avoided by grazing animals, and the species regenerates aggressively by means of an extensive system of branched, lateral horizontal roots. Each of these may give rise to shoots, resulting in the formation of large clonal patches which may expand at up to 6 m year^{-1} (*Biol Fl*). Roots degenerate after about 2 years to produce isolated daughter colonies (*Biol Fl*). The success of *C.a.* on arable land, and in other disturbed environments, arises from its capacity to regenerate from root fragments. *C.a.* is also difficult to eradicate since its root system may extend several metres in depth (*Biol Fl*). *C.a.* is generally believed to be dioecious, but Kay (1985) indicates that some 'male' clones in Britain are hermaphrodite and some are subhermaphrodite. *C.a.* is also self-fertile, although seed set is relatively low (Kay 1985). Where the species is dioecious, seed set usually requires male and female plants to be within a few hundred metres of each other (*Biol Fl*). Since the pappus is readily detached, seed dispersal is often poor. Nevertheless, colonization of new sites by seedlings is frequently seen in the field, and this largely out-breeding species exhibits a range of morphological and ecotypic variations (*Biol Fl*). The main germination period is probably spring. Recently-collected seeds often germinate well but some authors, including Grime *et al.* (1981), record poor germination in fresh seed (*Biol Fl*). Hodgson (unpubl.) has found scarification improves the germination of some thistles, suggesting that the diversity in reported germination behaviour may relate to variation in the condition of the seed coat when the test was carried out (cf. many legumes which germinate well if tested before hardening of the seed coat). *C.a.* may form a bank of persistent seeds (*Biol Fl*). Both the root and young shoot were formerly eaten by humans (*Biol Fl*). The species hybridizes with *C. palustre* and one further species in Britain (Stace 1975).
Current trends *C.a.* successfully exploits agricultural land and urban wasteground. Despite its status as a noxious weed and the fact that it may be controlled by combinations of herbicides, ploughing and crop rotation (Holm *et al.* 1977), *C.a.* remains common and may even be increasing in grassland and wasteland.

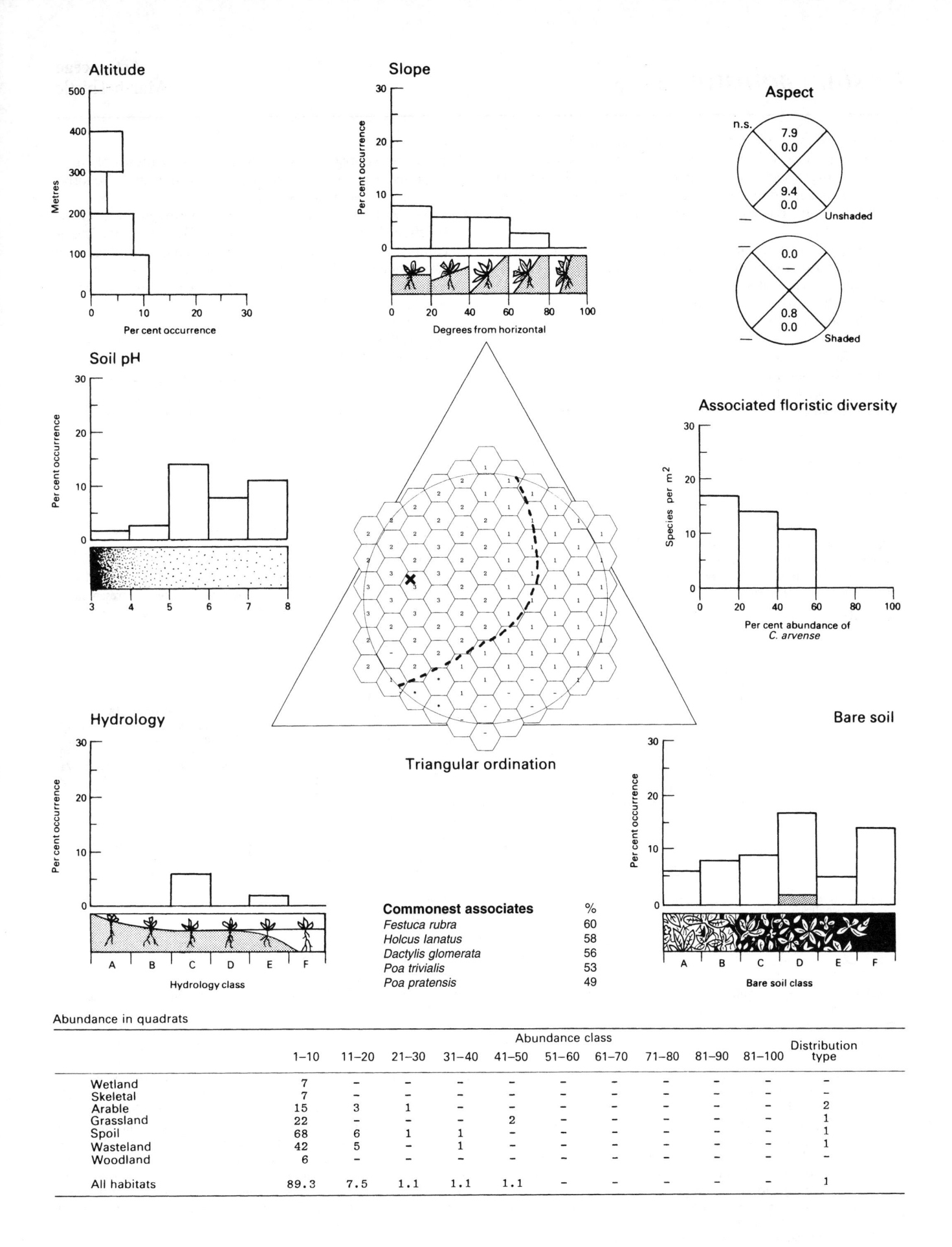

Altitude

Slope

Aspect

Soil pH

Associated floristic diversity

Hydrology

Triangular ordination

Bare soil

Commonest associates %
Festuca rubra 60
Holcus lanatus 58
Dactylis glomerata 56
Poa trivialis 53
Poa pratensis 49

Abundance in quadrats

	1–10	11–20	21–30	31–40	41–50	51–60	61–70	71–80	81–90	81–100	Distribution type
Wetland	7	–	–	–	–	–	–	–	–	–	–
Skeletal	7	–	–	–	–	–	–	–	–	–	–
Arable	15	3	1	–	–	–	–	–	–	–	2
Grassland	22	–	–	–	2	–	–	–	–	–	1
Spoil	68	6	1	1	–	–	–	–	–	–	1
Wasteland	42	5	–	1	–	–	–	–	–	–	1
Woodland	6	–	–	–	–	–	–	–	–	–	–
All habitats	89.3	7.5	1.1	1.1	1.1	–	–	–	–	–	1

Cirsium palustre (L.) Scop.

Asteraceae
Marsh Thistle

Life-form Monocarpic perennial, semi-rosette hemicryptophyte with an erect stock. Immature plant consisting of a leaf rosette of hairy, ovate, spiny leaves, each of which is sometimes >4000 mm². Flowering stems erect with spiny, pinnatifid, decurrent leaves; roots relatively shallow, with or without VA mycorrhizas.
Phenology Winter green. Flowers from July to September. Seed shed from August to October. After flowering, plant dies in late autumn. Leaves relatively long-lived.
Height Foliage usually <100 mm in non-flowering plants; in adult plants stem bearing leaves overtopped by flowers to 1500 mm.
Seedling RGR No data. SLA 16.8. DMC 9.6.
Nuclear DNA amount 2.7 pg; $2n = 34^*$ (*BC*); $2x^*$, $n = 17$ also base number for the tribe.
Established strategy Intermediate between competitive-ruderal and C-S-R.
Regenerative strategies W, B_s. Regenerates by means of wind-dispersed seeds, many of which germinate in spring. Also forms a persistent seed bank; ?SB III.
Flowers Florets purple, rarely white, hermaphrodite (some functionally male, at least in Europe (*CTW*)), moderately self-compatible (Mogford 1974), insect-pollinated; >50 in each capitulum; capitula solitary or 2-4 in terminal clusters.
Germinule 2.0 mg, 3.0 × 1.6 mm, achene dispersed with a pappus *c.* 9 mm in length.
Germination Epigeal; t_{50} 4 days; 12-33°C; L > D.
Biological flora None.
Geographical distribution The British Isles (100% of British vice-counties), *ABIF* page 613; most of Europe northwards to Scandinavia, although rare in the Mediterranean (69% of European territories); New Zealand, N Africa and W Asia.

Habitats Common in mire and damp calcareous pastures, and on wasteland and the banks of rivers and streams. Also found in enclosed pastures, on road verges and in open scrub or woodland margins. Of more-localized occurrence in moist skeletal habitats. Absent from arable and from all habitats on highly acidic soils.
Gregariousness Usually represented by scattered individuals.
Altitude Widely distributed. [Up to 730 m.]
Slope Occurs in mire on relatively flat ground and in flushes and damp grassland on sloping dale-sides.
Aspect Records are concentrated on unshaded N-facing sites.
Hydrology Found over a wide range of hydrological conditions, but most frequent and abundant in soligenous mire or on soils which remain moist in summer.
Soil pH Most frequent within the pH range 4.5-6.5, and absent from strongly acidic soils.
Bare soil Associated with a wide range of values.
△ **ordination** Distribution centred on vegetation associated with moderate levels of disturbance and fertility. Scarcely represented in vegetation dominated by strongly competitive species, and excluded completely from highly disturbed habitats.
Associated floristic diversity Usually relatively high.

Habitats

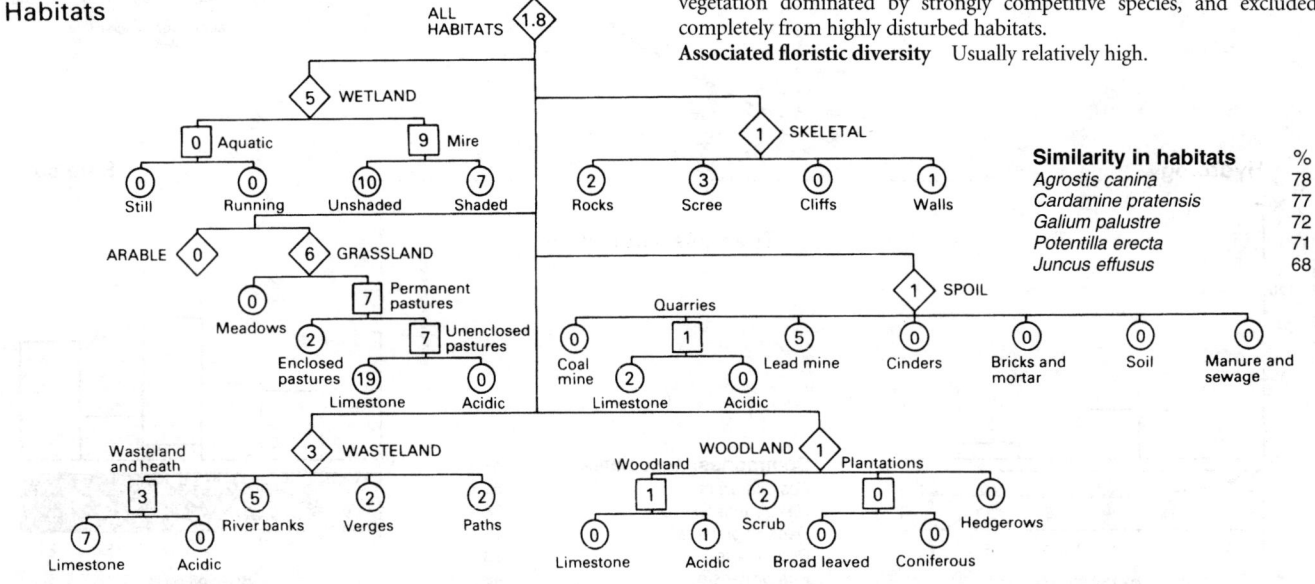

Similarity in habitats	%
Agrostis canina	78
Cardamine pratensis	77
Galium palustre	72
Potentilla erecta	71
Juncus effusus	68

Synopsis *C.p.* is a spiny, winter-green, monocarpic, perennial thistle of mire and moist grassland, particularly on mildly acidic soils of moderate fertility. *C.p.* is infrequent in permanently waterlogged topogenous mire, and is more characteristic of soligenous mire or sites where the surface soil is moist in summer and flooded in winter. In upland regions *C.p.* extends into damp dale-side grassland, and we suspect that *C.p.* does not persist in a strongly anaerobic rooting medium. The low-growing rosettes of *C.p.*, which contrast with the tall flowering stem, are very dependent upon short turf or open vegetation for establishment, and may develop for several years before flowering. Seed production is often high (*c.* 700 per plant (Salisbury 1942)). *C.p.* has some shade-tolerant attributes (Pons 1977) and may persist without flowering in lightly shaded woodland. *C.p.* is little grazed by stock, and is particularly frequent in grazed habitats and on the banks of rivers, ditches, etc. The capacity of *C.p.* to exploit these regularly disturbed sites is enhanced by the formation of a persistent seed bank. Buried seeds are particularly important during the 'closed' phases of coppice cycles (Pons 1977). In coppice, seed burial and persistence appears to involve canopy-induced inhibition of germination in surface-lying seeds and thermal reversion of phytochrome in buried seeds (Pons 1983, 1984). Van Leeuwen (1981) has observed that self-pollinated plants produce fewer, larger achenes than cross-pollinated individuals. The seedlings from larger achenes have a lower risk of mortality, and it is suggested that this may facilitate the establishment of colonies from single isolated individuals (van Leeuwen 1981). Mogford (1974) has observed white-flowered plants amongst normal purple-flowered individuals in sea-cliff and mountain populations. This white-flowered form is preferentially pollinated, and, in theory, may enjoy a selective advantage under conditions of low pollination. *C.p.* forms hybrids with other species of *Cirsium* (Stace 1975).
Current trends Uncertain. *C.p.* is favoured by the increased disturbance associated with modern land-use, but not by the effects of increased drainage on wetland communities.

Altitude
Slope
Aspect

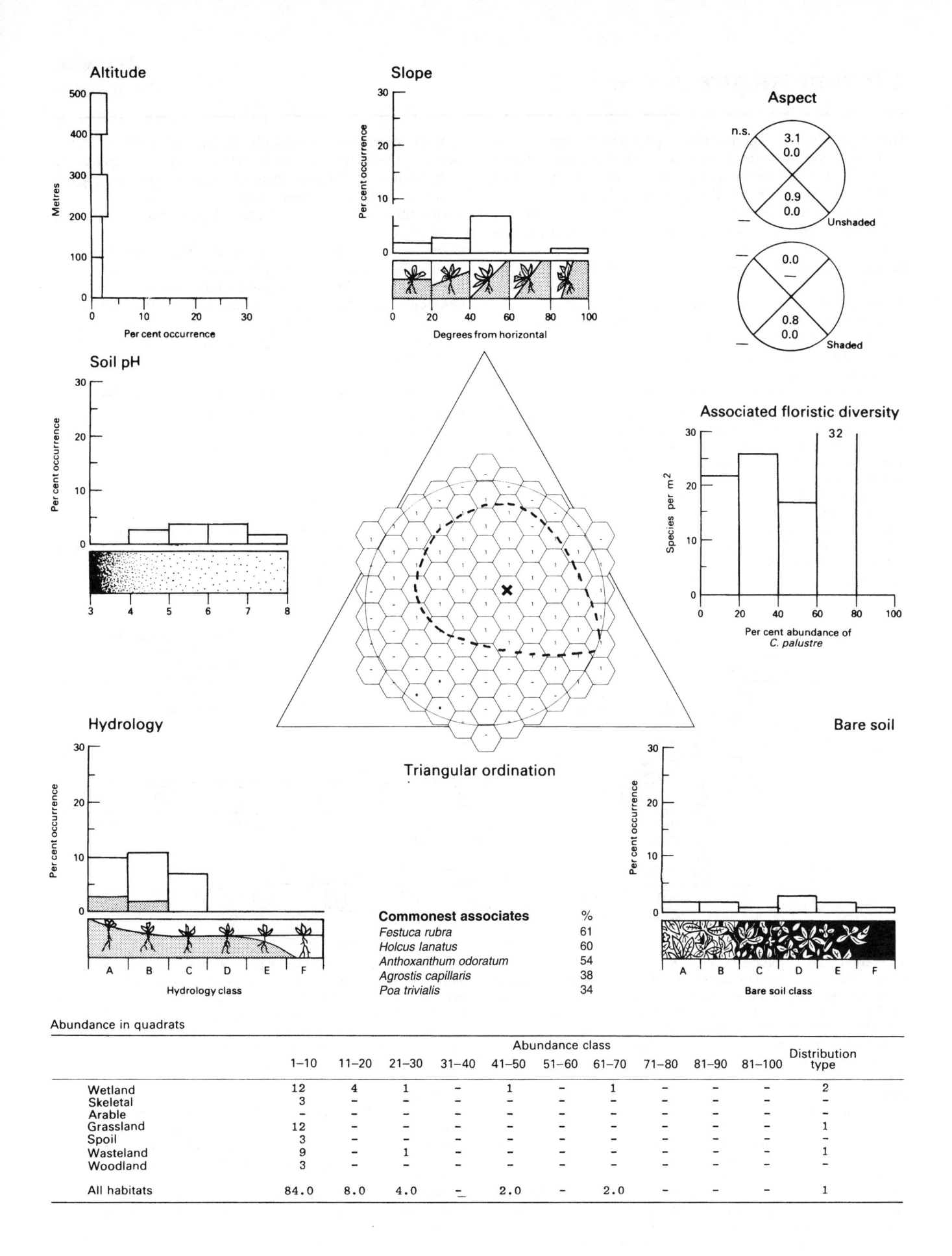

n.s.
| 3.1 | 0.0 |
| 0.9 | 0.0 | Unshaded

| 0.0 | — |
| 0.8 | 0.0 | Shaded

Soil pH

Associated floristic diversity

32

Per cent abundance of *C. palustre*

Hydrology

Triangular ordination

Bare soil

Commonest associates	%
Festuca rubra	61
Holcus lanatus	60
Anthoxanthum odoratum	54
Agrostis capillaris	38
Poa trivialis	34

Abundance in quadrats

	\multicolumn{10}{c}{Abundance class}	Distribution type
	1–10	11–20	21–30	31–40	41–50	51–60	61–70	71–80	81–90	81–100	
Wetland	12	4	1	–	1	–	1	–	–	–	2
Skeletal	3	–	–	–	–	–	–	–	–	–	–
Arable	–	–	–	–	–	–	–	–	–	–	–
Grassland	12	–	–	–	–	–	–	–	–	–	1
Spoil	3	–	–	–	–	–	–	–	–	–	–
Wasteland	9	–	1	–	–	–	–	–	–	–	1
Woodland	3	–	–	–	–	–	–	–	–	–	–
All habitats	84.0	8.0	4.0	–	2.0	–	2.0	–	–	–	1

Cirsium vulgare (Savi) Ten.

Asteraceae
Spear Thistle

Life-form Spiny, monocarpic perennial, semi-rosette hemicryptophyte with a long tap root. Stem erect; leaves lanceolate, prickly, hairy above, cottony below, sometimes >10 000 mm^2; roots with or without VA mycorrhizas.

Phenology Immature plant with an overwintering rosette. Flowers from July to October and seed set from August to October, after which the plant dies.

Height Foliage and flowers to 1500 mm; non-flowering rosette <100 mm.

Seedling RGR 0.5-0.9 week^{-1}. SLA 13.5. DMC 12.1.

Nuclear DNA amount 5.1 pg; $2n = 68*$ (*BC*); $4x*$ (Tribe also with a polyploid base number; $n = 17$.)

Established strategy Competitive-ruderal.

Regenerative strategies W, ?B$_s$. Regenerates exclusively by means of wind-dispersed seeds (see 'Synopsis'); SB I.

Flowers Florets tubular, purple, hermaphrodite, aggregated into heads, insect-pollinated, often >200 in each capitulum; capitula solitary or 2-3 in terminal clusters.

Germinule 2.64 mg, 3.5 × 1.5 mm, achene with a pappus 23 mm in length.

Germination Epigeal; t_{50} 2 days; 11-30°C; germination rate increased after dry storage at 20°C; L > D.

Biological flora Klinkhamer and De Jong (1989).

Geographical distribution The British Isles (100% of British vice-counties), *ABIF* page 611; Europe northwards to Scandinavia (89% of European territories); W Asia and N Africa; introduced in N America.

Habitats Most common in pastures and on road verges, productive spoil heaps, building rubble and cinder tips. Also present (mainly as seedlings) on rock outcrops, arable fields, banks of rivers and streams, hedgerows and paths. Largely absent from shaded and from waterlogged habitats.

Gregariousness Sparse.

Altitude Up to 400 m, but suitable habitats are more common at lower altitudes. [Up to 690 m.]

Slope Widely distributed.

Aspect No significant trends in unshaded habitats. Occurrence in shade restricted to S-facing slopes.

Hydrology Not a wetland species, but sometimes a casual on exposed mud at the margin of open water.

Soil pH Widely distributed, but only common on soils of pH >5.0.

Bare soil Histogram has two peaks coinciding with low and high values of bare ground. The former corresponds mainly to occurrences in perennial communities such as grasslands; the latter to those in more-disturbed habitats.

A ordination Distribution centred on vegetation associated with productive, heavily disturbed habitats, but also present (perhaps mainly as seedlings) in many other situations.

Associated floristic diversity Typically intermediate.

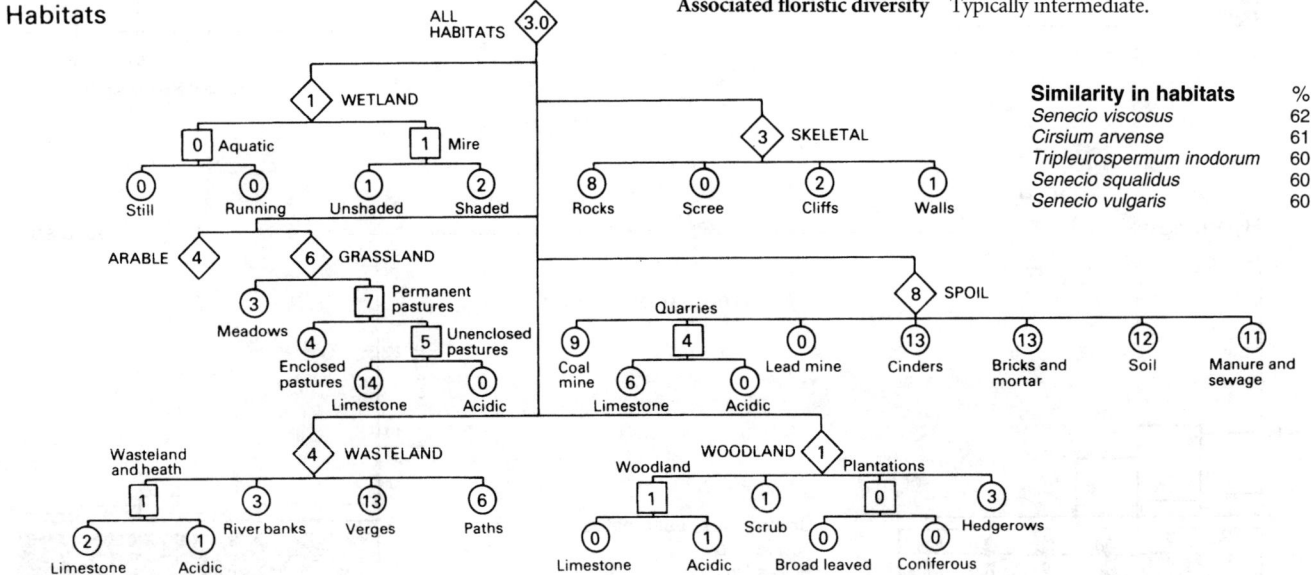

Habitats

Similarity in habitats	%
Senecio viscosus	62
Cirsium arvense	61
Tripleurospermum inodorum	60
Senecio squalidus	60
Senecio vulgaris	60

Synopsis *C.v.* is a typically robust monocarpic perennial herb which, because of its spiny stems and leaves, is unpalatable to stock. It can colonize the bare areas created by overgrazing in pasture and, for this reason, *C.v.* is listed under the 1959 Weeds Act as a noxious weed which landowners are obliged to control (*WCH*). The species is among the earlier colonists of a range of other fertile disturbed habitats, e.g. spoil heaps. The species has a deep tap-root and is intolerant of waterlogging and does not prosper in shaded habitats. *C.v.* is less persistent in droughted sites, e.g. rock outcrops, than *Carduus nutans* L. However, young plants can withstand a degree of damage and can persist in lawns. Immature plants consist of a rosette, up to 1.5 m or more in diameter, of spiny leaves and with a stout tap-root; flowering is delayed until the plant attains a size threshold (De Jong & Klinkhamer 1986). The dry matter allocated to seeds is c. 10% of the total plant dry weight, irrespective of plant size (De Jong & Klinkhamer 1986). Where productivity is lower or defoliation is experienced, flowering may be delayed for several years. A single plant may produce up to 8000 large seeds on a tall, erect flowering shoot (De Jong & Klinkhamer 1986). Because of the large pappus, seeds are very buoyant in air. However, De Jong and Klinkhamer (1986) found that only 10% were dispersed outside the parental population, and seed predation was between 72 and 93%. *C.v.* seeds show little innate dormancy, may germinate in autumn or spring (De Jong & Klinkhamer 1986) and are highly effective colonists of transient vegetation gaps in a disturbed landscape. One source (Jalloq 1974) suggests the formation of a persistent seed bank, but this was not confirmed by other studies (van Breeman & van Leeuwen 1983, De Jong & Klinkhamer 1986). In fertile sites the species may even behave as a winter annual, completing its lifecycle in 10 months (De Jong & Klinkhamer 1986). As with *C. palustre*, it has been suggested that self-pollinated plants produce a smaller number of larger achenes than cross-pollinated individuals (van Leeuwen 1981). The species is very variable, but the taxon *C.v.* has not yet been satisfactorily subdivided (*FE4*). *C.v.* forms hybrids in Britain with two other species of *Cirsium* (Stace 1975).

Current trends A common colonist of disturbed artificial habitats. Perhaps still increasing overall, despite the control measures applied to populations on agricultural land.

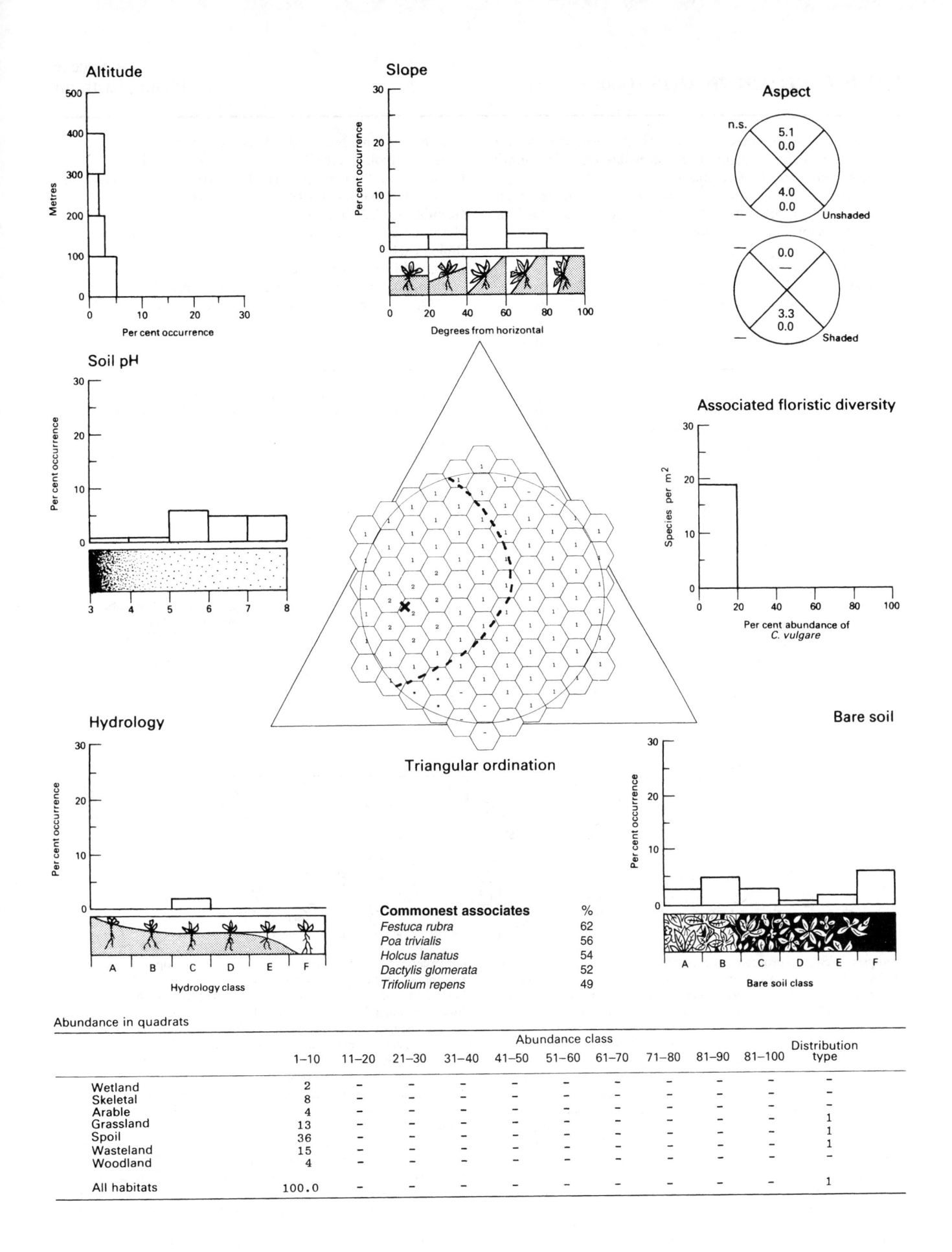

Altitude

Slope

Aspect

	n.s.	5.1	
		0.0	
		4.0	Unshaded
	—	0.0	

		0.0	
	—	—	
		3.3	Shaded
	—	0.0	

Soil pH

Associated floristic diversity

Triangular ordination

Hydrology

Commonest associates

	%
Festuca rubra	62
Poa trivialis	56
Holcus lanatus	54
Dactylis glomerata	52
Trifolium repens	49

Bare soil

Abundance in quadrats

	Abundance class										Distribution type
	1–10	11–20	21–30	31–40	41–50	51–60	61–70	71–80	81–90	81–100	
Wetland	2	–	–	–	–	–	–	–	–	–	–
Skeletal	8	–	–	–	–	–	–	–	–	–	–
Arable	4	–	–	–	–	–	–	–	–	–	–
Grassland	13	–	–	–	–	–	–	–	–	–	1
Spoil	36	–	–	–	–	–	–	–	–	–	1
Wasteland	15	–	–	–	–	–	–	–	–	–	–
Woodland	4	–	–	–	–	–	–	–	–	–	1
All habitats	100.0	–	–	–	–	–	–	–	–	–	1

Conopodium majus (Gouan) Loret.

Apiaceae
Pignut, Earthnut

Life-form Polycarpic perennial, semi-rosette geophyte, with a deeply-buried tuber. Stem erect; leaves dissected, <500 mm^2; root system with VA mycorrhizas.

Phenology Vernal. Leaves elongate below ground during winter and emerge in early spring. Flowers formed in May and June and seed set in June to July, by which time the shoot is dead.

Height Initial foliage height usually <100 mm, but flowering stem may attain 600 mm.

Seedling RGR 0.5–0.9 week^{-1}. SLA 24.0. DMC 18.9.

Nuclear DNA amount 1.6 pg, $2n = 22$ (Silvestre 1972); $2x$.

Established strategy Stress-tolerant ruderal.

Regenerative strategies S. Regenerates by seed which germinates in early spring. No seed bank has been reported and vegetative regeneration is rare (Doust & Doust 1982); SB II.

Flowers White, insect-pollinated, flowers mostly hermaphrodite but a few male (Tutin 1980); >50 in each umbel.

Germinule 2.26 mg, 4.6×1.4 mm, mericarp (2 per ovary).

Germination Epigeal; dormancy broken by chilling.

Biological flora None.

Geographical distribution A European endemic found throughout the British Isles (100% of British vice-counties), *ABIF* page 459; W Europe extending E to Italy (23% of European territories and 3% naturalised).

Habitats Found in two distinct habitat types: (a) in pasture, meadows and to a lesser extent grassy wasteland, road verges and lead-mine spoil, and (b) in woodland and scrub, including shaded cliffs. Absent from most spoil and skeletal habitats, arable land and wetlands.

Gregariousness Intermediate.

Altitude Widely distributed to 400 m. [Up to 700 m.]

Slope Wide-ranging, but infrequent on the steepest slopes.

Aspect In unshaded sites, frequency is greater on N-facing slopes; in shaded situations the distribution shows a slight S-facing bias.

Hydrology Absent from wetland habitats.

Soil pH Most frequent in the pH range 4.5–7.0.

Bare soil Associated with a wide range of values due to occurrence in two types of situation: (a) grassland with little exposure of bare soil and (b) woodland including sites with high exposures of soil.

Δ ordination Distribution centred on vegetation associated with relatively undisturbed and relatively unproductive conditions. Absent from severely disturbed situations.

Associated floristic diversity Low to species-rich, dependent upon pH and habitat.

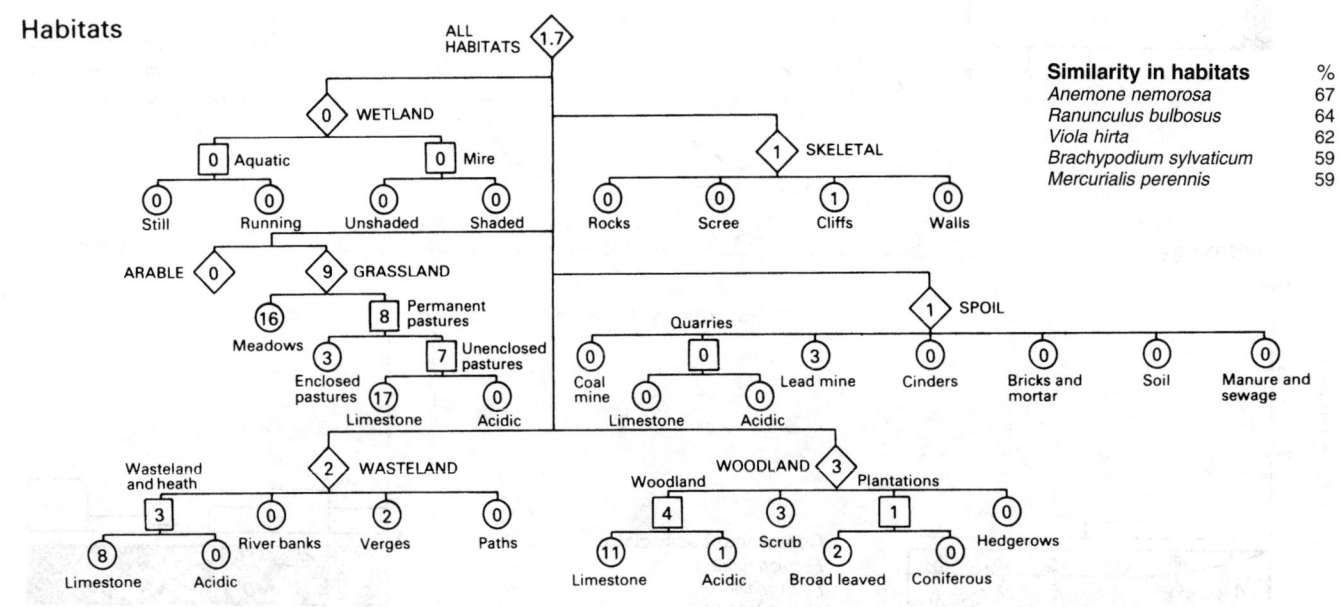

Habitats

Similarity in habitats	%
Anemone nemorosa	67
Ranunculus bulbosus	64
Viola hirta	62
Brachypodium sylvaticum	59
Mercurialis perennis	59

Synopsis *C.m.*, a European endemic with a sub-Atlantic distribution (Birks 1973), is a vernal herb with a long-lived (Doust 1980), edible, underground tuber. The species occurs in woodland and grassland habitats. In both *C.m.* is associated with relatively infertile, deep, mildly acidic, mineral soils. Leaves are high in N and Ca but low in P *(SIEF)*. Shoots emerge from the soil in early spring, flowering and fruiting are essentially over by midsummer; by July only the tuber remains alive. This phenology closely parallels that of *Hyacinthoides non-scripta*, and in woodlands shoots die soon after full tree-canopy expansion. The nuclear DNA is lower than would be expected for a vernal species, and may reflect a tendency for the shoot to grow underground throughout most of the winter and be dependent upon warm intervals and microsites for spring growth. However, the species is not exceptionally thermophilous and seedling growth rate under 20°C day and 10°C night was little different from that under a 20/15°C regime (UCPE unpubl.). While the vernal phenology represents a shade-avoidance mechanism in woodland, its role in grassland systems is less clear. As suggested for *Oxalis acetosella*, *C.m.* may be a woodland relict in these sites. The tendency for grassland occurrences to be restricted to N-facing upland slopes may indicate a requirement for continuously moist conditions. *C.m.* is seldom found in burned sites and the basal leaves, which usually senesce by May or early June, are vulnerable to spring fires. In common with *Lathyrus linifolius*, *C.m.* is susceptible to heavy grazing during early spring. The species relies upon regeneration by seed and does not form a persistent seed bank. Only c. 7% of biomass was allocated to ripe fruits in one study (Doust 1980). Clonal growth is rare (Doust & Doust 1982). Like many members of the Apiaceae, the embryo is relatively small and chilling is required to break dormancy. Seeds germinate in spring and the seedlings observed in the Sheffield region possess a single cotyledon (cf. Tutin 1980). This unusual feature for a dicot herb is also shared by *Ranunculus ficaria*. *C.m.* is seldom found as a colonist of new habitats, and is usually indicative of sites with habitat continuity. Whether this restriction is exercised through failure of seedling establishment or lack of seed mobility is unknown. Studies carried out in the Iberian Peninsula suggest the existence of two ssp. (Silvestre 1973).

Current trends Apparently decreasing, and likely to decline still further.

Altitude
Slope
Aspect

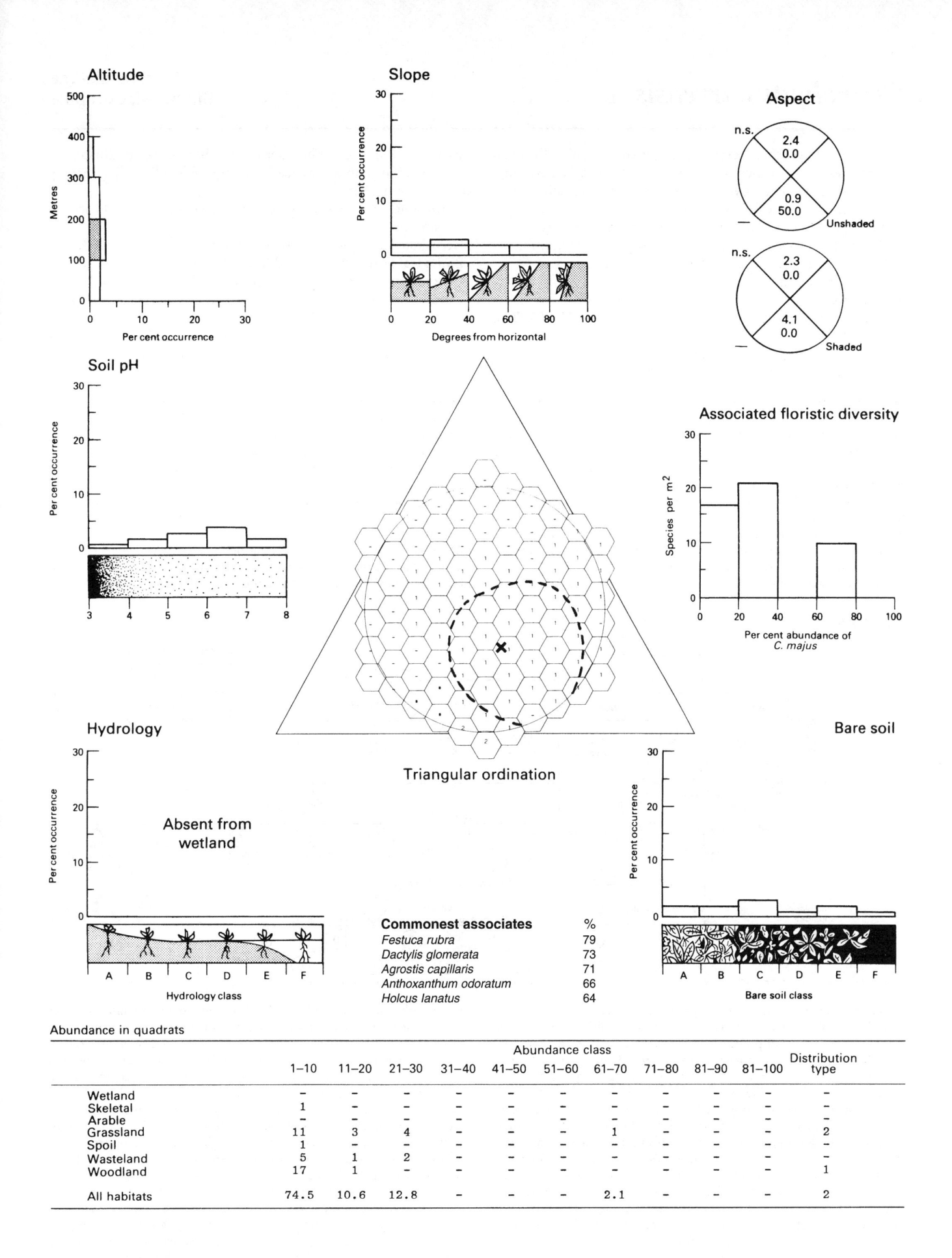

	n.s.	2.4
		0.0
		0.9
	—	50.0 Unshaded

	n.s.	2.3
		0.0
		4.1
	—	0.0 Shaded

Soil pH

Associated floristic diversity

Hydrology

Absent from wetland

Triangular ordination

Bare soil

Commonest associates %
Festuca rubra 79
Dactylis glomerata 73
Agrostis capillaris 71
Anthoxanthum odoratum 66
Holcus lanatus 64

Abundance in quadrats

	Abundance class										Distribution type
	1–10	11–20	21–30	31–40	41–50	51–60	61–70	71–80	81–90	81–100	
Wetland	–	–	–	–	–	–	–	–	–	–	–
Skeletal	1	–	–	–	–	–	–	–	–	–	–
Arable	–	–	–	–	–	–	–	–	–	–	–
Grassland	11	3	4	–	–	–	1	–	–	–	2
Spoil	1	–	–	–	–	–	–	–	–	–	–
Wasteland	5	1	2	–	–	–	–	–	–	–	–
Woodland	17	1	–	–	–	–	–	–	–	–	1
All habitats	74.5	10.6	12.8	–	–	–	2.1	–	–	–	2

Convolvulus arvensis L.

Convolvulaceae
Bindweed, Cornbine

Life-form Polycarpic perennial geophyte, often with deep rhizomes. Stems procumbent or erect, shoots often twining; leaves lanceolate to ovate, sometimes hairy, usually <1000 mm^2; roots deep with VA mycorrhizas.
Phenology Flowers from June to September and sets seed from August to October. Shoots die back completely in autumn.
Height Foliage and flowers either <100 mm if prostrate or to 750 mm if supported.
Seedling RGR 1.0-1.4 week^{-1}. SLA 33.4. DMC 13.9.
Nuclear DNA amount 3.6 pg; $2n = 50$, (48*) (*Biol Fl, FE5, BC*); ?$16x$*.
Established strategy Competitive-ruderal.
Regenerative strategies (V), B$_s$. See 'Synopsis'; SB IV.

Flowers Pink or white, hermaphrodite, insect-pollinated; a Canadian population self-incompatible (*Biol Fl*); usually solitary, axillary.
Germinule 10.1 mg, 3.1 × 1.6 mm, seed in dehiscent <14 seeded capsule.
Germination Epigeal; hard-coat dormancy; L = D (*Biol Fl*).
Biological flora Weaver and Riley (1982).
Geographical distribution Throughout England and Wales, but rarer in Scotland, generally distributed although probably introduced in Ireland (98% of British vice-counties), *ABIF* page 490; Europe except the extreme N (92% of European territories); temperate regions throughout the world.

Habitats A weed of arable land. Also frequent in a variety of other less disturbed habitats such as hedgerows, wasteland, soil heaps, cinders, railway banks, quarry and mining waste, banks of rivers and streams, road verges, rock outcrops and cliffs. Not recorded in wetlands, meadows, pastures or woodland. [Occurs on disturbed sand dunes.]
Gregariousness Capable of forming an extensive leaf canopy at low frequency values.
Altitude Suitable habitats are more common in lowland areas. Sufficiently uncommon above 200 m, however, to suggest climatic restriction at higher altitudes. [Up to 230 m.]
Slope Widely distributed.

Aspect Significantly more frequent on unshaded S-facing slopes.
Hydrology Absent from wetlands.
Soil pH Not recorded from acidic soils of pH <4.5.
Bare soil Particularly associated with high exposures.
△ ordination Most frequent in fertile habitats subject to disturbance, but also associated with a wide variety of conditions including tall productive vegetation subject to occasional disturbance. Absent from very infertile sites.
Associated floristic diversity Intermediate, but declining at higher frequencies of *C.a.*; this may be due in part to the dominant effects of *C.a.* itself or to those arising from more-robust species associated with *C.a.*

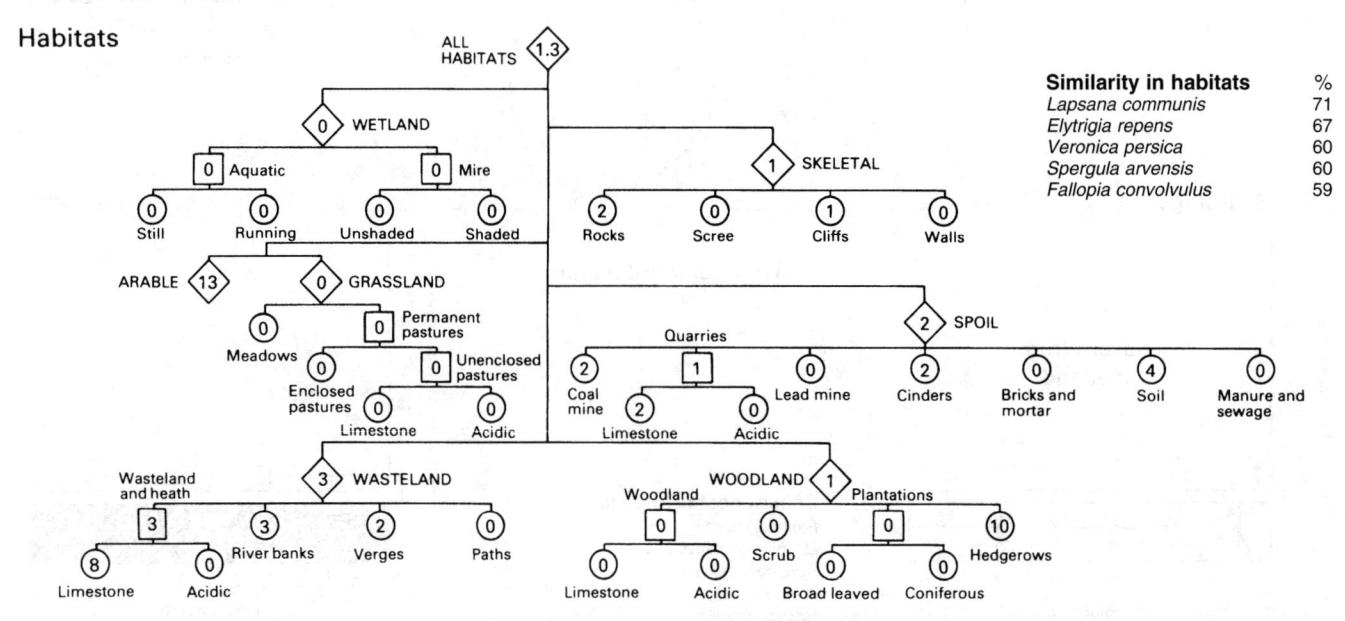

Habitats

Similarity in habitats	%
Lapsana communis	71
Elytrigia repens	67
Veronica persica	60
Spergula arvensis	60
Fallopia convolvulus	59

Synopsis *C.a.* is a rhizomatous, prostrate or erect twining herb of disturbed fertile habitats. Like *Calystegia sepium*, which it resembles in habit, *C.a.* exploits both productive open sites, e.g. arable land, and less-disturbed terrain such as unmanaged roadsides and railway or field banks, where it may twine around the stems of the taller grasses such as *Arrhenatherum elatius*. The species is most conspicuous, and probably most abundant, in recently disturbed sites, and is particularly troublesome as an arable weed. Crop yield may be severely reduced (*Biol Fl*) and, where the shoots envelop the stems of the crop plant, difficulties in harvesting may arise (*WCH*). The species, formerly widespread in all crops (Brenchley 1920), is now mainly associated with cereals and is most common in fields which are winter-sown (Chancellor & Froud-Williams 1984). However, it has been suggested (*Biol Fl*) that, because above-ground shoot elongation is delayed until late spring, *C.a.* is an ineffective competitor with those crops which grow in early spring. Many of the roots are in the top 600 mm of soil, but some may penetrate to 9 m (*Biol Fl*) and the species is notoriously difficult to eradicate. The leaves and stems are difficult to wet with herbicide sprays (*Biol Fl*), and control by herbicides is often ineffective in sites where shoots are afforded protection by rocky ground (*WCH*). As a result *C.a.* is also especially frequent on railway ballast. *C.a.* is a lowland species with a bias towards S-facing slopes, and is seldom recorded above 200 m. Overwintering roots may suffer damage if the ground freezes (*Biol Fl*) and both distribution and seed set appear to be limited by climatic factors. As we might expect from the deep root system, *C.a.* is absent from wetlands and is strongly persistent on droughted cliffs and outcrops. *C.a.* is absent from regularly grazed or mown situations, and is also excluded from heavily shaded habitats. In common with *C.s.*, the species exhibits aggressive vegetative spread in gardens; this is due in large measure to the ability to regenerate from a rapidly extending network of horizontal roots, which also regenerate effectively if fragmented by, for example, digging (*Biol Fl*). *C.a.* produces seeds which are long-persistent in the soil and have the potential to germinate during the warmer part of the year, particularly

Altitude
Slope
Aspect
P < 0.05

0.8	0.0
2.7	0.0

Unshaded

0.0	—
1.6	0.0

Shaded

Soil pH

Associated floristic diversity

Hydrology
Absent from wetland

Triangular ordination

Bare soil

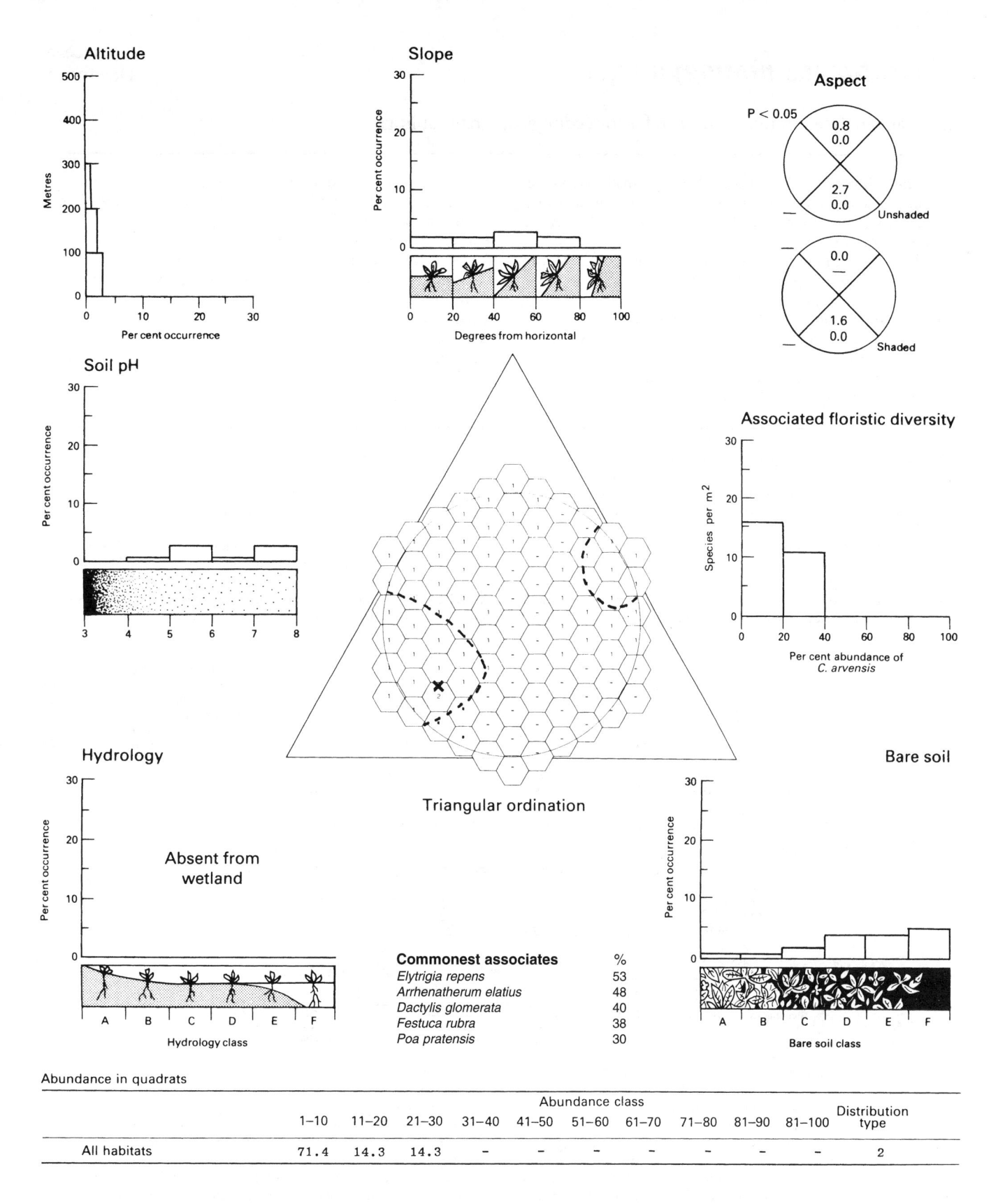

Commonest associates	%
Elytrigia repens	53
Arrhenatherum elatius	48
Dactylis glomerata	40
Festuca rubra	38
Poa pratensis	30

Abundance in quadrats

	\multicolumn{10}{c}{Abundance class}	Distribution type
	1–10	11–20	21–30	31–40	41–50	51–60	61–70	71–80	81–90	81–100	
All habitats	71.4	14.3	14.3	–	–	–	–	–	–	–	2

in late spring (*Biol Fl*). Flowers persist for only 1 day (*Biol Fl*), and in Britain regeneration by seed is probably infrequent since seed is set only in good summers (Thurston 1960). However, seedling establishment is very rapid and the importance of seed in colonizing new sites is often underestimated (*Biol Fl*). Rhizome fragments and seeds are probably transported to new sites in soil, and seeds are also known to be dispersed by birds (Ridley 1930). Populations vary in leaf shape and flower colour, and within Europe ecotypes have been recognized (*Biol Fl*). Some plants growing beside railway tracks may be referable to var. *linearifolius* Choisy., which is native to S Europe. Populations of *C.a.* resistant to 2,4-D have been reported (Whitworth & Muzik 1967).

Current trends Probably decreasing on arable land as a result of weed control but remaining common in other types of disturbed habitats.

(*) *Crataegus monogyna* Jacq.

Rosaceae
Hawthorn

Field data analysed in this account refer to seedlings or small saplings

Life-form Thorny, much-branched shrub or small tree. Leaves mainly on side-branches (short shoots), leathery, ovate, lobed, with few hairs, often <500 mm^2; roots with VA or ectomycorrhizas.
Phenology Deciduous. Leaves expand in spring. Flowers May to June and sets seed September to November.
Height Foliage and flowers to 10 m.
Seedling RGR No data. SLA 10.4. DMC. No data.
Nuclear DNA amount No data; $2n = 34$ (*FE2*); $2x$. Subfamily also of polyploid origin.
Established strategy Stress-tolerant competitor.
Regenerative strategies S. By seeds which germinate in spring. Does not appear to form a persistent seed bank, although because of their germination requirements seeds are shortly persistent in the soil.
Flowers Usually white, hermaphrodite, self-incompatible (Stace 1975), insect-pollinated; up to *c*. 16 in a corymb.
Germinule 8.1 × 5.1 mm, pyrene; in a usually 1-seeded fleshy fruit.
Germination Epigeal.
Biological flora None.
Geographical distribution The British Isles (100% of British vice-counties), *ABIF* page 371; most of Europe except for the N and E margins (82% of European territories); Asia.

Habitats The most widely-planted hedgerow shrub. Seedlings and saplings are widely distributed and recorded frequently both in unshaded habitats such as meadows, pastures, wasteland, outcrops and limestone quarry heaps, and in woodlands. Not recorded from aquatic habitats.
Gregariousness Sparse.
Altitude Distribution of juveniles predominantly lowland, but extending to 350 m. [Up to 550 m.]
Slope Widely distributed.
Aspect Juveniles widely distributed, but showing some bias towards S-facing (often rocky) slopes in the open and N-facing slopes in woodland, suggesting some restriction to sites where the vigour of herbs is reduced.
Hydrology Seedlings are recorded, but do not persist at the margin of soligenous and topogenous mire.
Soil pH Throughout the full range of pH, but less common below pH 4.0.
Bare soil Widely distributed, but most frequent at intermediate values.
△ **ordination** Juveniles wide-ranging but infrequent in highly disturbed and very fertile sites.
Associated floristic diversity Intermediate to high in grasslands, low to intermediate in skeletal habitats and low in woodlands.

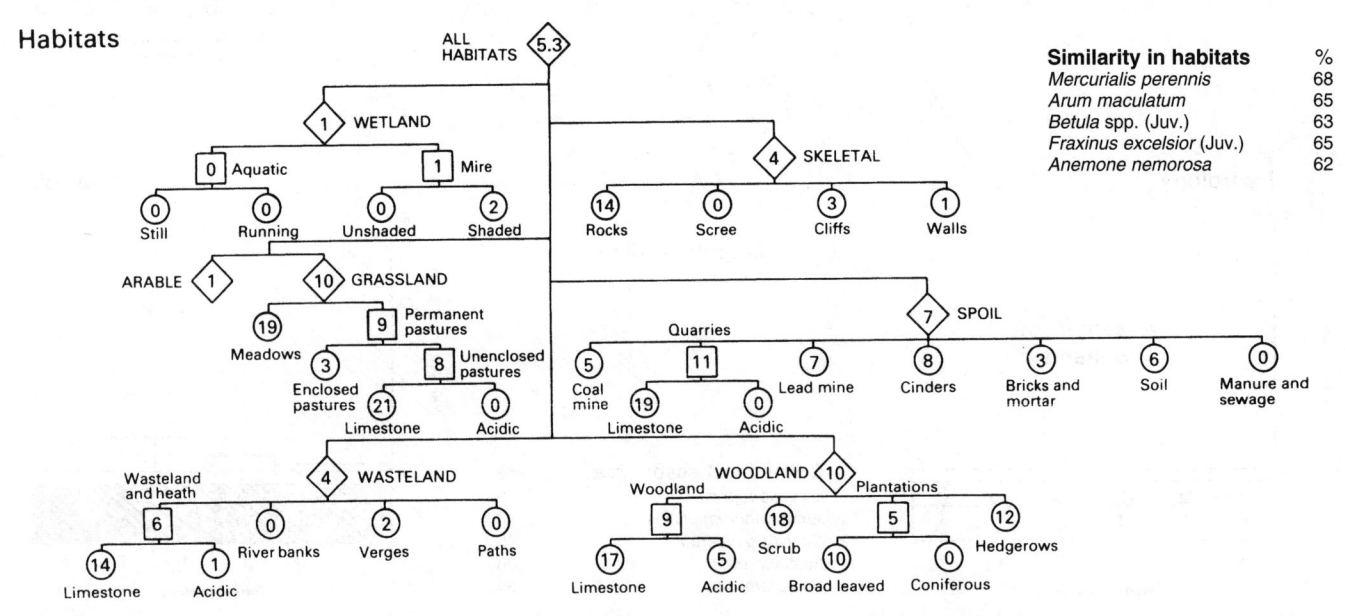

Habitats

ALL HABITATS ⟨5.3⟩

1 WETLAND
  0 Aquatic
    0 Still
    0 Running
  1 Mire
    0 Unshaded
    2 Shaded

4 SKELETAL
  14 Rocks
  0 Scree
  3 Cliffs
  1 Walls

ARABLE ⟨1⟩

10 GRASSLAND
  19 Meadows
  9 Permanent pastures
  3 Enclosed pastures
  8 Unenclosed pastures
  21 Limestone
  0 Acidic

7 SPOIL
  Quarries
    5 Coal mine
    11
    7 Lead mine
    8 Cinders
    3 Bricks and mortar
    6 Soil
    0 Manure and sewage
  19 Limestone
  0 Acidic

4 WASTELAND
  Wasteland and heath
    6
    14 Limestone
    1 Acidic
  0 River banks
  2 Verges
  0 Paths

WOODLAND ⟨10⟩
  Woodland
    9
    17 Limestone
    5 Acidic
  18 Scrub
    10 Broad leaved
    0 Coniferous
  5 Plantations
  12 Hedgerows

Similarity in habitats	%
Mercurialis perennis	68
Arum maculatum	65
Betula spp. (Juv.)	63
Fraxinus excelsior (Juv.)	65
Anemone nemorosa	62

Synopsis *C.m.* is a deciduous shrub or small tree with diffuse, porous wood and, predictably from this feature (Lechowicz 1984), bud-break is relatively early in spring. One specimen in the Derbyshire Dales was found to be 165 years old (Merton 1970). The species occurs in a wide range of habitats as a seedling or young sapling. As a taller plant contributing to the tree or under-shrub canopy, *C.m.* is less widespread and is mainly found on soils of pH >4.5, showing a peak within the pH range 6.5-7.5 (UPCE data). In Britain *C.m.* is the most common colonizing shrub of grazed and ungrazed grassland on base-rich soils. Pollard *et al.* (1974) even suggest that hawthorn is the commonest woody species in Britain. *C.m.* is by far the most abundant hedgerow species. Apart from *Betula* spp., which exploit more acidic soils, *C.m.* is the most frequently recorded canopy species in scrub. However, *C.m.* is less frequent as a constituent of woodland, and the species is only the fourth most common canopy species of limestone woodlands of the Sheffield region. It is scarcely recorded in the remaining woodland categories. Classical studies, summarized by Tansley (1939), indicate that invasion by *C.m.* following a relaxation in grazing pressure represents a first stage in the development of secondary woodland. The distribution of *C.m.* within the Derbyshire Dales consistently suggests that *C.m.* is a seral species of developing woodland (see Pigott 1969, Merton 1970). Typically, in wooded habitats *C.m.* is restricted to the understorey of secondary woodland (Pigott 1969, Merton 1970) and does not appear to regenerate effectively under shaded conditions. However, this is rather an oversimplification, since some hawthorn thickets are long-persistent (Rackham 1980) and the species shows some capacity to regenerate in gaps in long-established woodland. Sites dominated by *C.m.* seldom accumulate persistent litter. The species is thorny and may even be toxic to stock (Cooper & Johnson 1984). The foliage is also relatively unpalatable to the snail *Helix aspersa* and, like other unpalatable species, *C.m.* supports a large insect flora (Wratten *et al.* 1981).

Altitude
Slope
Aspect

	5.9	
n.s.	0.0	
	7.2	
	0.0	Unshaded

	7.6	
n.s.	0.0	
	4.1	
	0.0	Shaded

Soil pH

use as is

Associated floristic diversity

Hydrology

Triangular ordination

Commonest associates	%
Dactylis glomerata	53
Festuca rubra	51
Festuca ovina	48
Plantago lanceolata	43
Carex flacca	35

Bare soil

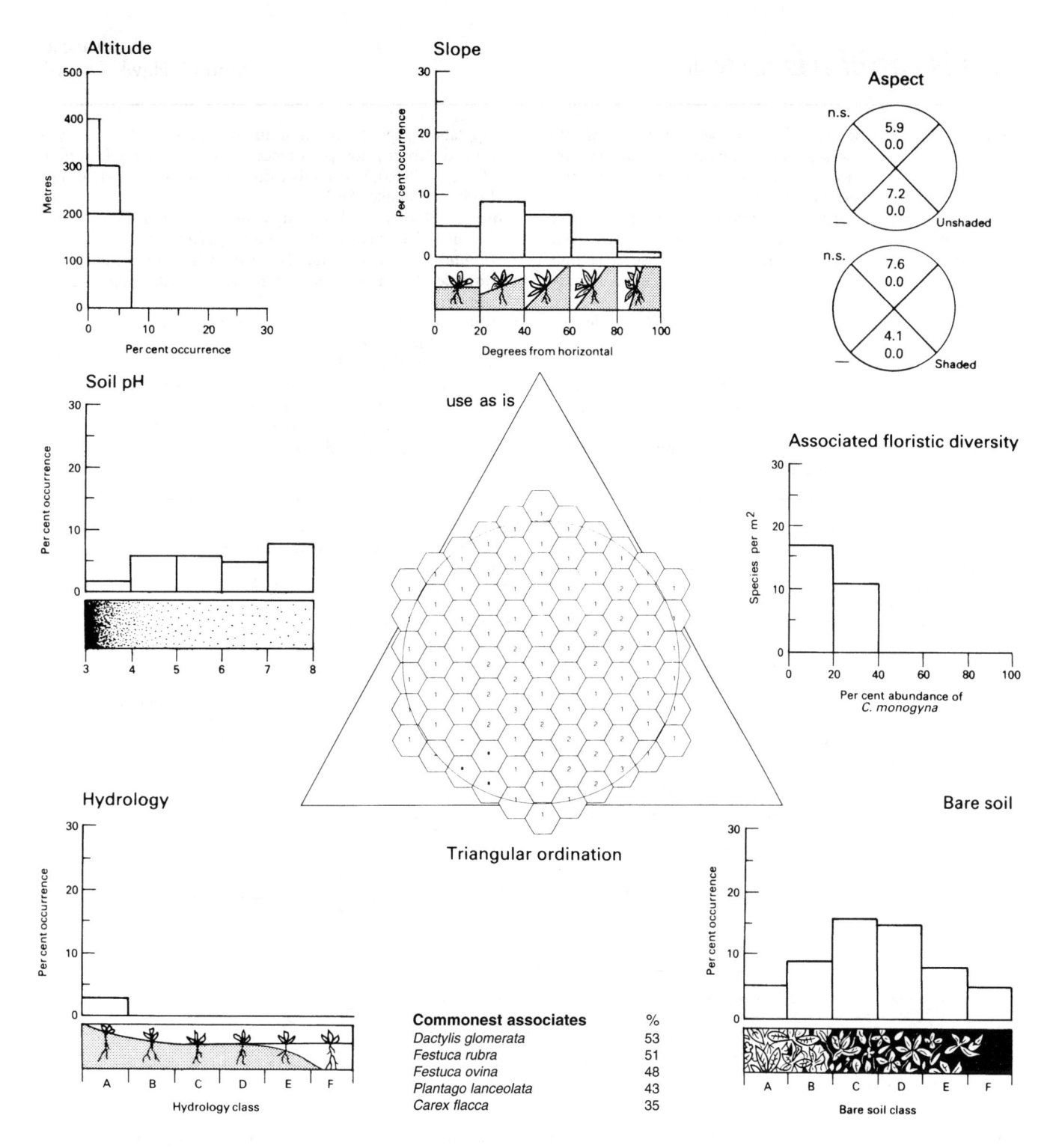

Experimental evidence (Edwards & Wratten 1985) suggests that *C.m.* possesses a mechanism of inducible defence against foliar predation by insects. Consequently, *C.m.* is better able to establish from seed in grazed areas than many other woody species (Linhart & Whelan 1980). However, heavy predation of foliage has been observed (Port & Thompson 1980) and may contribute to the transitory nature of the dominance exerted by the species. *C.m.* is self-incompatible (Stace 1975). Immature fruits may be predated (Manzur & Courtney 1984) and shaded bushes tend not to flower. Although these factors reduce seed set, the species displays an impressive capacity for dispersal and regeneration by seed. The fleshy fruits are consumed both by resident and migrating birds, and by mammals, and seedlings are common in most types of herbaceous vegetation. Seeds are enclosed in a lignified endocarp and show embryo dormancy (Dickinson 1985). Germination requires exposure to warm moist conditions to break hard-coat dormancy, followed by chilling (Bean 1970). As a result germination in the field does not occur until 18 months after seeds are shed (Beckett & Beckett 1979). No persistent seed bank has been reported. The variability of the species in Europe is such that six ssp. are recognized (*FE2*). In parts of S England *C.m.* shows extensive introgression with *C. laevigata* (Poiret) DC (Byatt 1975). The latter is typical of woodland conditions and heavy soils (Byatt 1975, Stace 1975). *C.a.* also hybridizes with *Mespilus germanica* L. (*FE2*).

Current trends Aerial photographs indicate substantial increases in this species during recent years in the semi-natural grasslands of the Derbyshire Dales. *C.m.* has shown a similar increase in many other grassland areas when the severity of grazing has been relaxed. The species is also increasingly prominent in the landscape as a result of activities such as planting of *C.m.* on motorway verges and in new hedgerows.

Crepis capillaris (L.) Wallr.

Asteraceae
Smooth Hawk's-beard

Life-form Semi-rosette therophyte, probably usually a winter annual but perhaps also sometimes a summer annual or longer-lived. Stems erect; leaves lanceolate, pinnatifid, sometimes hairy, <2000 mm^2; tap-rooted and with VA mycorrhizas.

Phenology Variable, but typically germinating in autumn and overwintering with a rosette of leaves. Flowers mainly from June to July and seed set from July to August. A second flowering in August to September is characteristic of some populations.

Height Foliage mainly <300 mm; large leaves <100 mm; flowers to 500 mm (to 900 mm in some genotypes).

Seedling RGR No data. SLA 30.5. DMC 12.6.

Nuclear DNA amount 4.1 pg, $2n = 6*$ (Bennett *et al.* 1982); $2x*$.

Established strategy Intermediate between ruderal and stress-tolerant ruderal.

Regenerative strategies W, B$_s$. Reproduces entirely by wind dispersed seed (see 'Synopsis'), typically germinating in vegetation gaps during autumn or spring. Seed has only a limited capacity for persistence at or near the soil surface (Roberts 1986a), but whether this constitutes a persistent seed bank is debatable; ?SB III.

Flowers Florets yellow, ligulate, hermaphrodite, self-incompatible (Fryxell 1957), insect-pollinated; *c.* 40 in each capitulum; often >10 capitula in a branched inflorescence.

Germinule 0.21 mg, 2.0 × 0.5 mm, achene with a pappus *c.* 3.0 mm long.

Germination Epigeal; t_{50} 1 day; 13-30°C; L > D.

Biological flora None.

Geographical distribution The British Isles (100% of British vice-counties), *ABIF* page 627; W, C and S Europe extending northwards to Scandinavia (56% of European territories and 15% naturalized); Canary Islands; introduced in N America, Australia and New Zealand.

Habitats A species of dry open ground, most frequently recorded from limestone quarry heaps, rock outcrops, coal-mine waste, cinder tips and building rubble. A less frequent constituent of grasslands and road verges. Widely observed beside railways, both on ballast and on the banks themselves. Absent from woodland and wetland habitats.

Gregariousness Sparse to intermediate.

Altitude Occurs up to 400 m, but most abundant on the dry sandy soils of the low-lying Bunter sandstone. [Up to 450 m.]

Slope Widely distributed.

Aspect Frequency of occurrence significantly greater on unshaded S-facing slopes.

Hydrology Absent from wetland.

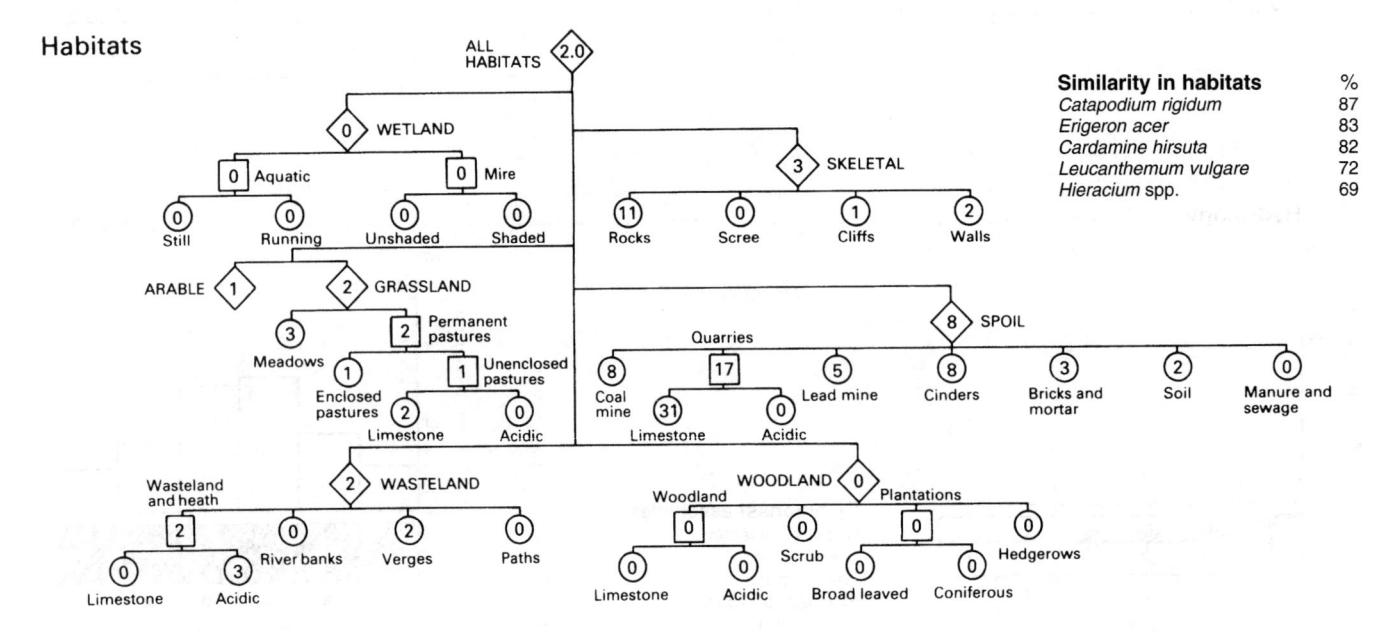

Similarity in habitats	%
Catapodium rigidum	87
Erigeron acer	83
Cardamine hirsuta	82
Leucanthemum vulgare	72
Hieracium spp.	69

Soil pH Restricted to soils with pH values >4.5 and most frequent on base-rich soils of pH >6.0.

Bare soil Widely distributed but tending to associate with a high proportion of bare ground.

Δ ordination A rather unusual distribution centred on vegetation associated with disturbed and only moderately productive conditions.

Associated floristic diversity Typically intermediate.

Synopsis In view of its widespread occurrence, *C.c.* deserves more study than it has so far received. The species is most typically a relatively robust winter-annual herb, often exploiting open, moderately productive habitats where the cover of perennials is reduced by drought or mechanical disturbance. However, some populations appear to contain, and may even consist entirely of, polycarpic perennial individuals. Damaged stems may also re-flower in autumn. The species is associated with two main types of habitat: (a) semi-permanent habitats such as droughted rock outcrops, sandy wasteland, and steep road and railway cuttings or embankments subject to soil creep, and (b) disturbed, often transient habitats such as soil heaps and localized areas of bare ground in pasture and on road verges. *C.c.* appears to avoid heavily droughted sites and does not normally occur in association with the early-flowering winter annuals, such as *Cardamine hirsuta* and *Saxifraga tridactylites*. Seed is set later than that of most other winter annuals, and it is capable of rapid and immediate germination. The species is frequently the only ruderal present in relatively closed vegetation dominated by perennials, and on sandy soils it may even persist in lawns. Plants without a main central stem (var. *capillaris*) are distinguishable from those with one (var. *agrestis* (Waldst. and Kit) Dalla Torre and Sarnth) (*FE4*), though the ecologies of these varieties have not been compared. In theory, var. *capillaris* would be at a selective advantage in the short open vegetation of habitats similar to those exploited by *Geranium molle*, *Medicago lupulina* and *Trifolium dubium*, while the taller and more erect

Altitude

Slope

Aspect

$P < 0.01$

	1.2	
33.3		4.9
n.s.		9.1 Unshaded

	0.0	
—		—
	0.8	
—	0.0	Shaded

Soil pH

Associated floristic diversity

Hydrology

Absent from wetland

Triangular ordination

Bare soil

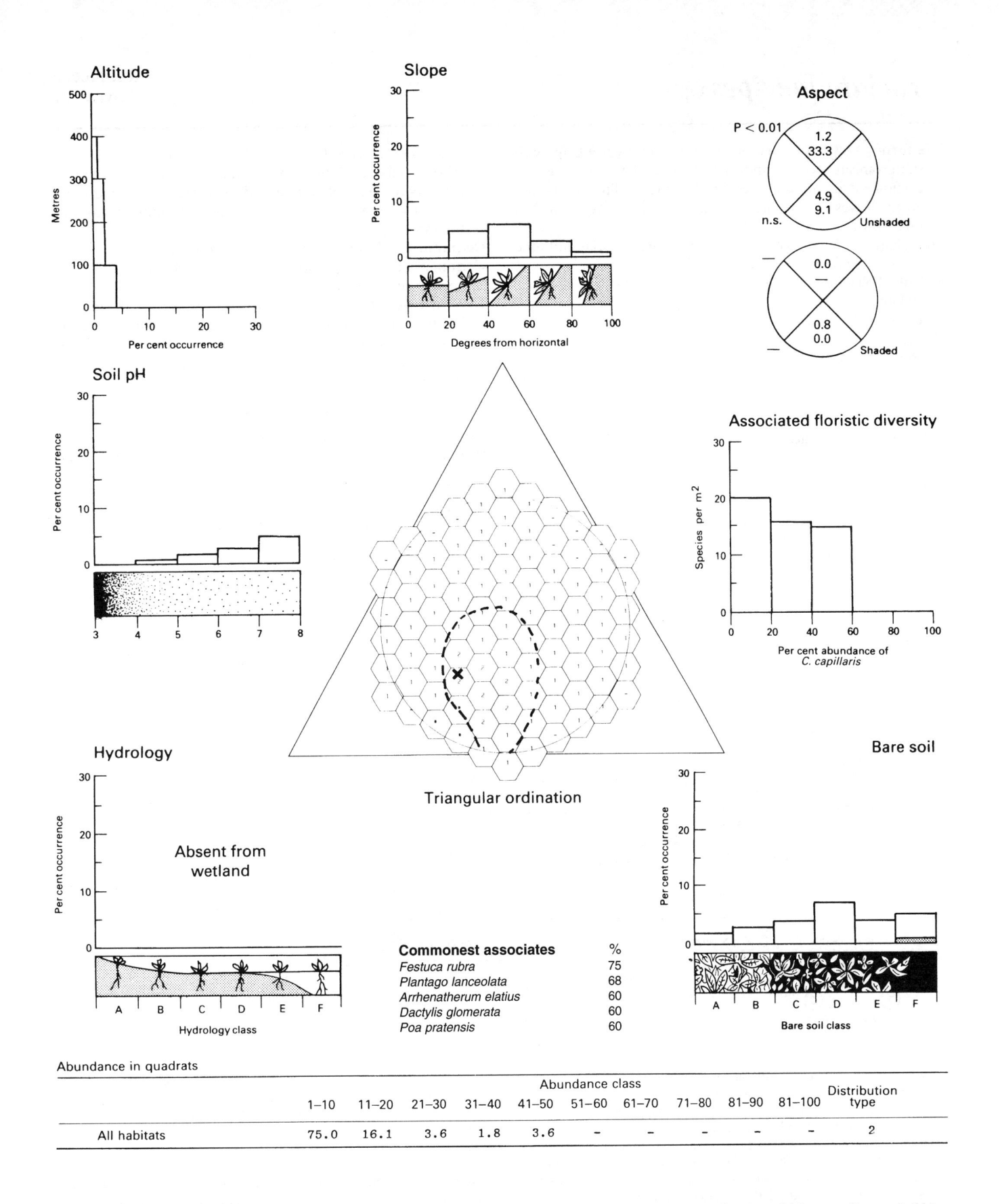

Commonest associates	%
Festuca rubra	75
Plantago lanceolata	68
Arrhenatherum elatius	60
Dactylis glomerata	60
Poa pratensis	60

Abundance in quadrats

	Abundance class									Distribution type
	1–10	11–20	21–30	31–40	41–50	51–60	61–70	71–80	81–90	81–100	
All habitats	75.0	16.1	3.6	1.8	3.6	–	–	–	–	–	2

var. *agrestis*, which also has larger involucres (*FE4*), would show a stronger affinity with annuals of taller vegetation such as *Torilis japonica* and *Vicia sativa* ssp. *nigra* (L.) Ehre. The species displays a strong colonizing ability, which is related to the production of numerous wind-dispersed seeds with a limited capacity for persistence at the soil surface (Grime 1978, Roberts 1986a). *C.c.* has fewer chromosomes than any other common vascular plant in the British flora, a specialization which, according to Stebbins (1974), facilitates rapid evolution. Perhaps therefore, we should expect considerable ecotypic differentiation within this species.

Current trends Often a rapid colonist of dry, relatively unproductive, disturbed habitats, and probably still increasing.

Cruciata laevipes Opiz

Rubiaceae
Crosswort

Life-form Polycarpic perennial protohemicryptophyte with subterranean stolons and slender rhizome. Shoots either decumbent or, if supported by the surrounding vegetation, erect; leaves ovate, hairy, <250 mm^2; roots with VA mycorrhizas.
Phenology Shoots apices winter green. Flowers in May and June and seed set from August to November.
Height Foliage and flowers either procumbent or ascending to *c.* 500 mm.
Seedling RGR No data. SLA 30.3. DMC 23.3.
Nuclear DNA amount No data; $2n = 22$ (*FE4*); $2x$.
Established strategy C-S-R.
Regenerative strategies (V), S. Regenerative biology is complex (see 'Synopsis'); SB II.
Flowers Yellow, insect-pollinated, terminal flowers hermaphrodite, protandrous and the lateral flowers male, 5-9 in each axillary cyme; cymes in whorls forming a terminal inflorescence.
Germinule 1.5 mm; dry mericarp (2 per ovary).
Germination Epigeal; dormancy broken by chilling.
Biological flora None.
Geographical distribution Most of Britain except N Scotland, less frequent in Wales, local and probably only introduced in Ireland (63% of British vice-counties), *ABIF* page 596; W, C and S Europe (67% of European territories and 3% naturalized); Asia Minor.

Habitats Most frequent in rough grazed or ungrazed grassland, mainly on deep calcareous soils. Widespread in occurrences on waysides, river banks and in hedgerows. Recorded also from lead-mine and quarry spoil, and the edges of unshaded mire and scrub. Absent from aquatic and arable habitats.
Gregariousness Occasionally patch-forming.
Altitude Most frequent on the carboniferous limestone above 200 m. Extends to 380 m. [Up to 470 m.]
Slope Absent from the steepest slopes.
Aspect Recorded slightly more frequently on N-facing slopes.
Hydrology Not a wetland species, though found on moist ground adjacent to mire.
Soil pH Occurrences scattered over the pH range 4.0-8.0; frequency and abundance greatest on base-rich soils.
Bare soil Distribution wide ranging, but generally biased towards habitats with little or no bare soil.
△ **ordination** Insufficient data.
Associated floristic diversity Ranging from intermediate to relatively high values.

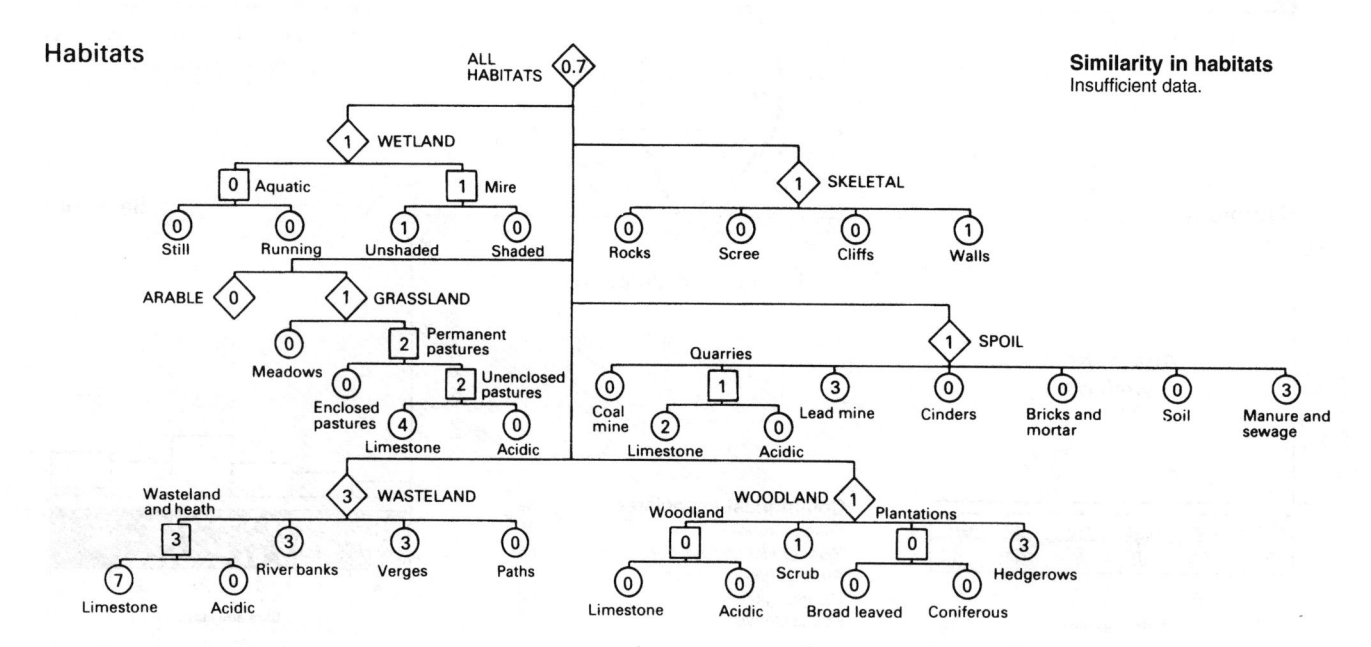

Habitats

Similarity in habitats
Insufficient data.

Synopsis *C.l.* is a weakly scrambling winter-green herb on moist, often calcareous soils in vegetation of intermediate productivity. In lowland parts of the Sheffield region *C.l.* is mainly found at woodland margins and in hedgerows, while in upland areas it is characteristic of grasslands, especially on steep N-facing slopes subject to topographic shade. Thus, we suspect that *C.l.* is, like *Stellaria holostea*, a 'semi- shade' mesophyte. In common with *S. holostea*, the leaves persist through the winter, becoming increasingly senescent, and are then replaced by a new flush of leaves in spring, followed by flowers in early summer. The canopy of *C.l.* is very low and, despite a capacity for lateral spread, only persists in sites where the growth of taller species is debilitated through the operation of disturbance factors such as occasional grazing. *C.l.* is susceptible to heavy grazing, and probably also to burning, and may become prominent for a number of years in grassland sites released from grazing. *C.l.* has a slender rhizome and subterranean stolons, and can form large clonal patches, particularly in shaded habitats. Detached shoots regenerate readily, and *C.l.* may spread along roadsides and river flood plains by this means. *C.l.* also regenerates by means of rather large seeds which germinate in spring and appear to be poorly dispersed. Lack of seed mobility may explain the failure of *C.l.* to colonize Ireland in Postglacial times. Seed production appears to be rather erratic; heavy fruiting in Derbyshire was observed in the unusually cool summer of 1985. No seed bank has yet been detected. *C.l.* is phenotypically plastic; tall scrambling colonies when grazed become prostrate with smaller leaves and shorter internodes.
Current trends Despite a limited capacity to colonize river banks and roadsides, *C.l.* is mainly restricted to older habitats and appears to be decreasing.

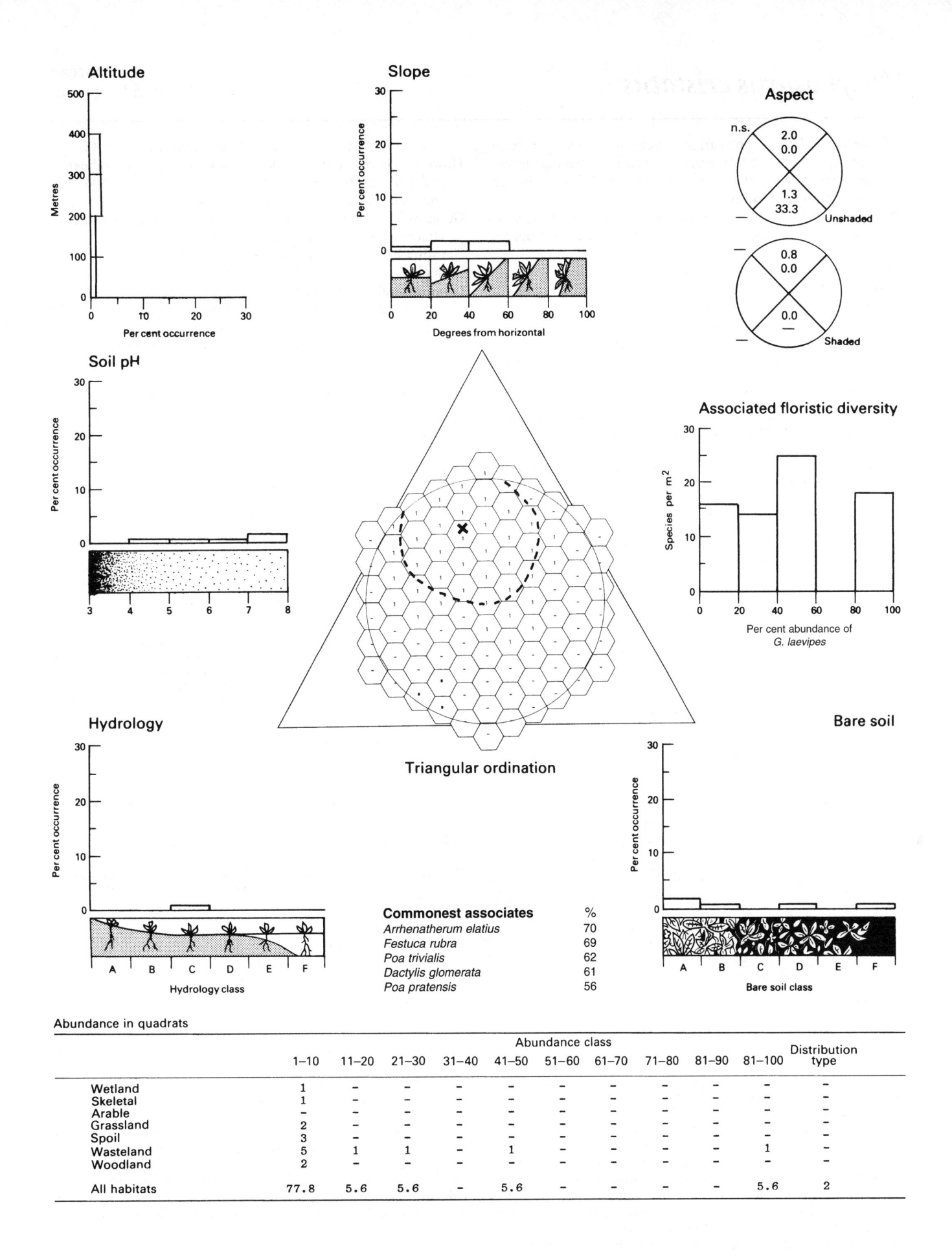

Altitude

Slope

Aspect

Soil pH

Associated floristic diversity

(Species per m², Per cent abundance of *G. laevipes*)

Hydrology

Triangular ordination

Bare soil

Commonest associates	%
Arrhenatherum elatius	70
Festuca rubra	69
Poa trivialis	62
Dactylis glomerata	61
Poa pratensis	56

Abundance in quadrats

	\multicolumn{10}{c}{Abundance class}	Distribution type
	1–10	11–20	21–30	31–40	41–50	51–60	61–70	71–80	81–90	81–100	
Wetland	1	–	–	–	–	–	–	–	–	–	–
Skeletal	1	–	–	–	–	–	–	–	–	–	–
Arable	–	–	–	–	–	–	–	–	–	–	–
Grassland	2	–	–	–	–	–	–	–	–	–	–
Spoil	3	–	–	–	–	–	–	–	–	–	–
Wasteland	5	1	1	–	1	–	–	–	–	1	–
Woodland	2	–	–	–	–	–	–	–	–	–	–
All habitats	77.8	5.6	5.6	–	5.6	–	–	–	–	5.6	2

(*)*Cynosurus cristatus* L.

Poaceae
Crested Dog's-tail

Life-form Tufted, polycarpic perennial, semi-rosette hemicryptophyte. Shoots erect or slightly spreading; leaves linear, typically <500 mm²; roots usually shallow, with or without VA mycorrhizas.
Phenology Winter green. Flowers mainly in June. Seeds ripen in July-August and shed in August (*Biol Fl*). Summer leaves short-lived.
Height Foliage typically <100 mm; flowering shoots to 750 mm.
Seedling RGR 1.5-1.9 week^{-1}. SLA 26.4. DMC 23.9.
Nuclear DNA amount 5.6 pg; $2n = 14$* (*BC*); $2x$*.
Established strategy Intermediate between ruderal and C-S-R.
Regenerative strategies S. Principally by freshly shed seed in vegetation gaps in autumn; some germination may be delayed until spring. Rarely shows clonal expansion (*Biol Fl*); SB I.
Flowers Green, hermaphrodite, homogamous, wind-pollinated, self-incompatible (Ennos 1985, *Biol Fl*); often c. 100 in a spike-like panicle (2-5 per spikelet).
Germinule 0.70 mg, 2.6 × 1.0 mm, caryopsis dispersed with an attached short-awned lemma (3.5 × 0.8 mm).
Germination Epigeal; t_{50} 3 days; <5-33°C; L = D.
Biological flora Lodge (1959).
Geographical distribution The British Isles (100% of British vice-counties), *ABIF* page 752; most of Europe except the extreme N (87% of European territories); N Asia and N Africa; introduced to N America, Australia and New Zealand.

Habitats Commonest in productive, sown, agricultural grassland (meadows and enclosed pastures). Also frequent on verges and grassy paths. A less important constituent of vegetation of rock outcrops and a number of artificial habitats (building rubble, mining spoil, walls). Also occurs occasionally in grazed marshes and as a non-persistent colonist of arable land. Absent from woodland. [Found in base-rich grassy places near the sea and locally in montane areas.]
Gregariousness Intermediate.
Altitude Present at all altitudes, although less frequent and abundant below 100 m, reflecting the replacement of grassland by arable land at low altitudes. [Up to 580 m.]
Slope Particularly common on level ground, due to its association with sown grasslands.
Aspect No significant trends.
Hydrology Uncommon in wetlands, but recorded from soligenous mire.
Soil pH Not recorded on soils of pH <4.0. Most frequent and abundant within the range 5.0-7.5.
Bare soil Occurs in association with widely different amounts of bare soil.
△ ordination Distribution centred on vegetation associated with relatively fertile and moderately disturbed habitats; less common in undisturbed vegetation.
Associated floristic diversity Intermediate to species-rich.

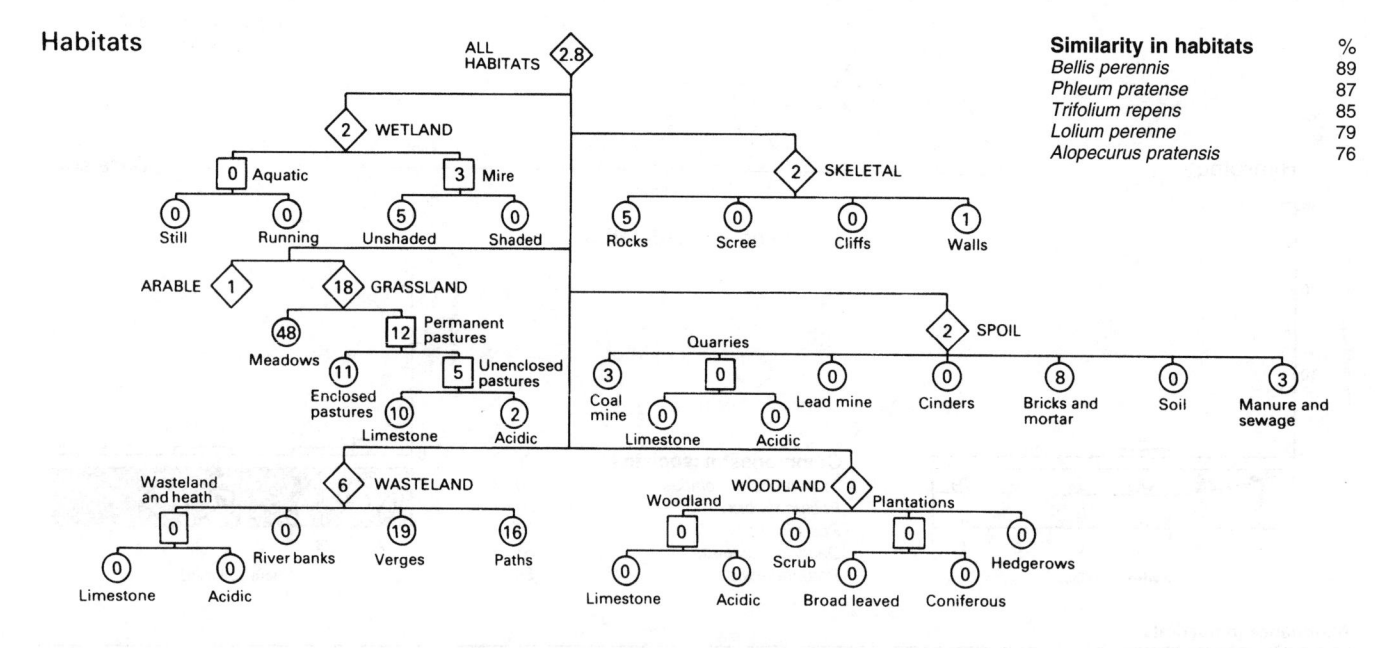

Habitats

Similarity in habitats	%
Bellis perennis	89
Phleum pratense	87
Trifolium repens	85
Lolium perenne	79
Alopecurus pratensis	76

Synopsis *C.c.* is a small, short-lived, tufted grass of relatively fertile grasslands, but is now little used in agriculture. Morphologically, *C.c.* is rather uniform and of narrow ecological range, although ecotypes may occur which are distributed in relation to land management (Lodge 1962a). The species is susceptible both to drought and to waterlogging, and exhibits low tolerance of shade (*Biol Fl*). *C.c.* shows a marked restriction to heavily grazed short turf. Although the foliage is very palatable, the wiry flowering stem is rejected by stock and the conspicuous dead flowering stems often overwinter (*Biol Fl*). Flowering may occur in the first growing season and seed production may reach 1100 year^{-1} in a mature plant. Provided that the plant is not subject to heavy spring grazing, seed production may be maintained at a high level each year (*Biol Fl*). However, as in a number of other sown pasture grasses (including *Dactylis glomerata*, *Lolium perenne* and *Phleum pratense*), there is no persistent seed bank. With its rather low canopy *C.c.* performs better in pastures than in permanent hay fields (*Biol Fl*). The abundance of *C.c.* in productive pastures appears to be limited by the dominant effects of more robust species.
Current trends Formerly sown for pasture on soils of relatively low fertility, but now probably decreasing as a result of reductions in the acreage of permanent pasture. Still grown in small quantities, chiefly in lawn-grass mixtures (*Biol Fl*), but since flower stems are only partially resistant to mowing the capacity for *C.c.* to persist in lawns is probably limited. Sowings for amenity are unlikely to offset the declining abundance of *C.c.* in pastures.

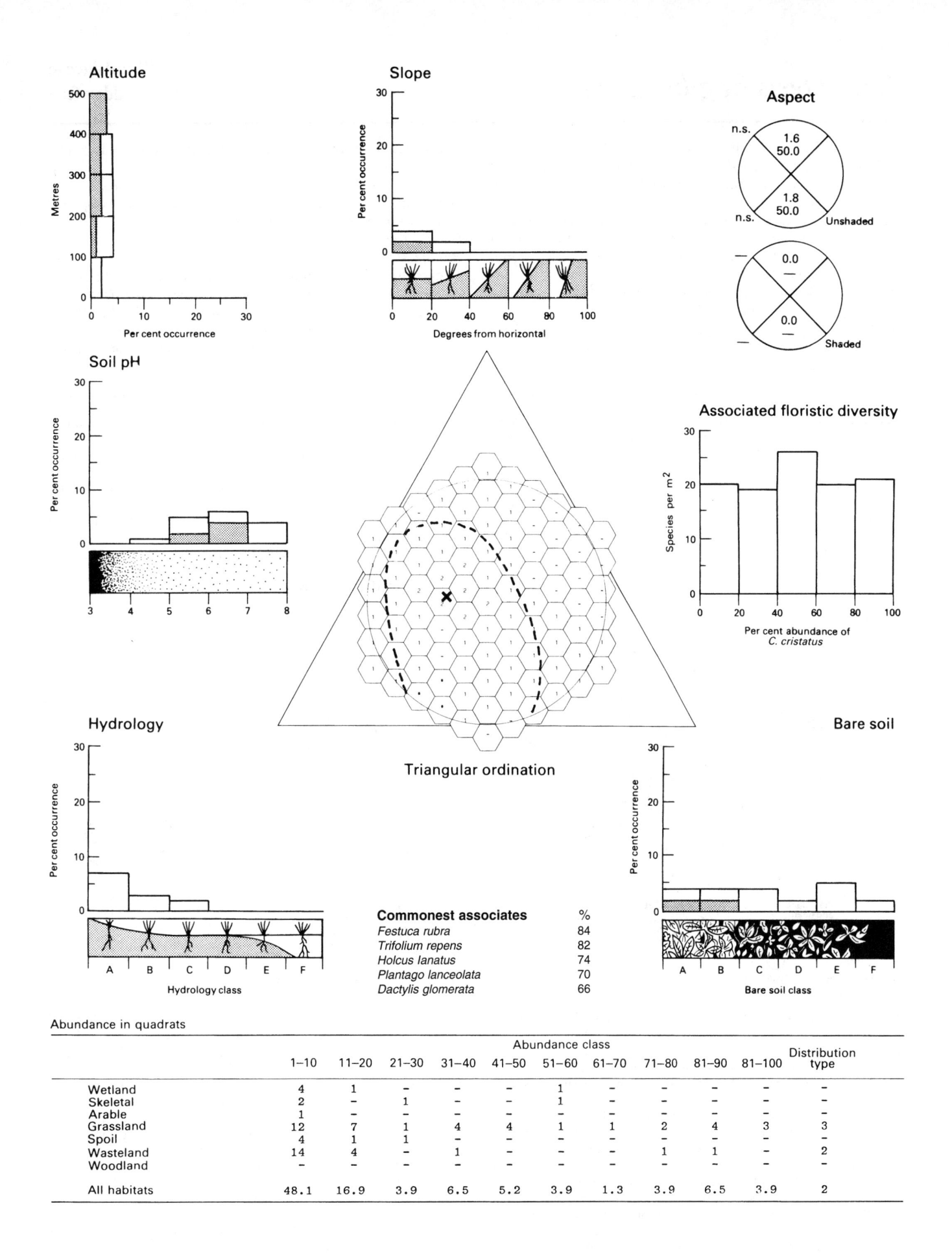

Altitude

Slope

Aspect

n.s.	1.6 / 50.0	
n.s.	1.8 / 50.0	Unshaded

—	0.0 / —	
—	0.0 / —	Shaded

Soil pH

Associated floristic diversity

Hydrology

Triangular ordination

Bare soil

Commonest associates %
Festuca rubra — 84
Trifolium repens — 82
Holcus lanatus — 74
Plantago lanceolata — 70
Dactylis glomerata — 66

Abundance in quadrats

	\multicolumn{10}{c	}{Abundance class}	Distribution type
	1–10	11–20	21–30	31–40	41–50	51–60	61–70	71–80	81–90	81–100	
Wetland	4	1	–	–	–	1	–	–	–	–	–
Skeletal	2	–	1	–	–	1	–	–	–	–	–
Arable	1	–	–	–	–	–	–	–	–	–	–
Grassland	12	7	1	4	4	1	1	2	4	3	3
Spoil	4	1	1	–	–	–	–	–	–	–	–
Wasteland	14	4	–	1	–	–	–	1	1	–	2
Woodland	–	–	–	–	–	–	–	–	–	–	–
All habitats	48.1	16.9	3.9	6.5	5.2	3.9	1.3	3.9	6.5	3.9	2

Cystopteris fragilis (L.) Bernh.

Woodsiaceae
Brittle Bladder-fern

Life-form Tufted, shortly rhizomatous, polycarpic perennial hemicryptophyte. Fronds erect or spreading, deeply dissected with a slightly scaley stipe, often <1000 mm^2; roots with VA mycorrhizas.
Phenology Fronds elongate in early spring. Spores are produced from June to September and fronds die with the onset of autumn frosts (Page 1982). Fronds produced in early autumn are short-lived (Page 1982).
Height Fertile and sterile fronds to 200 mm.
Sporeling RGR No data. SLA 21.4. DMC 24.8.
Nuclear DNA amount 16.6 pg; $2n = 168^*, 252^*, 336$ (BC, Page 1982); $4x^*, 6x^*, 8x$.
Established strategy Intermediate between SR and C-S-R.

Regenerative strategies W, ?B$_s$. Insufficiently studied (see 'Synopsis').
Sporangia Aggregated into sori on underside of the frond.
Germinule 44×29 μm, spore, black, spiny, explosively dispersed from a many-spored sporangium.
Germination t_{50} 7 days; L > D.
Biological flora None.
Geographical distribution Frequent in NW Britain, becoming local in the SE (82% of British vice-counties), *ABIF* page 74, *AF* page 72; Europe but only on mountains in the Mediterranean region (97% of European territories); the most widespread of all British ferns in distribution, although only on mountains in the tropics.

Habitats Restricted to shaded or unshaded skeletal habitats. Most frequent on limestone cliffs or in mortar-filled crevices in walls. Also observed on outcrops, stabilized scree, rocky lead-mine spoil and wet gritstone cliffs. [Occurs on mountain cliffs (Page 1982).]
Gregariousness Usually sparse.
Altitude Observed to 415 m, but mainly found on the carboniferous limestone between 200 and 300 m. [To 915 m (Page 1982).]
Slope A species of cliffs and walls, and consequently mainly restricted to very steep slopes.

Aspect More frequent and only abundant on N-facing slopes, both in shaded and unshaded sites.
Hydrology Absent from wetlands, but sometimes observed on rocks with dripping water.
Soil pH Most records and observations are within the pH range 6.5-8.0 and not recorded below pH 5.5.
Bare soil All records from habitats with little exposed soil but high exposures of bare rock.
△ **ordination** Distribution diffuse, as in most species from cliffs and walls.
Associated floristic diversity Usually low.

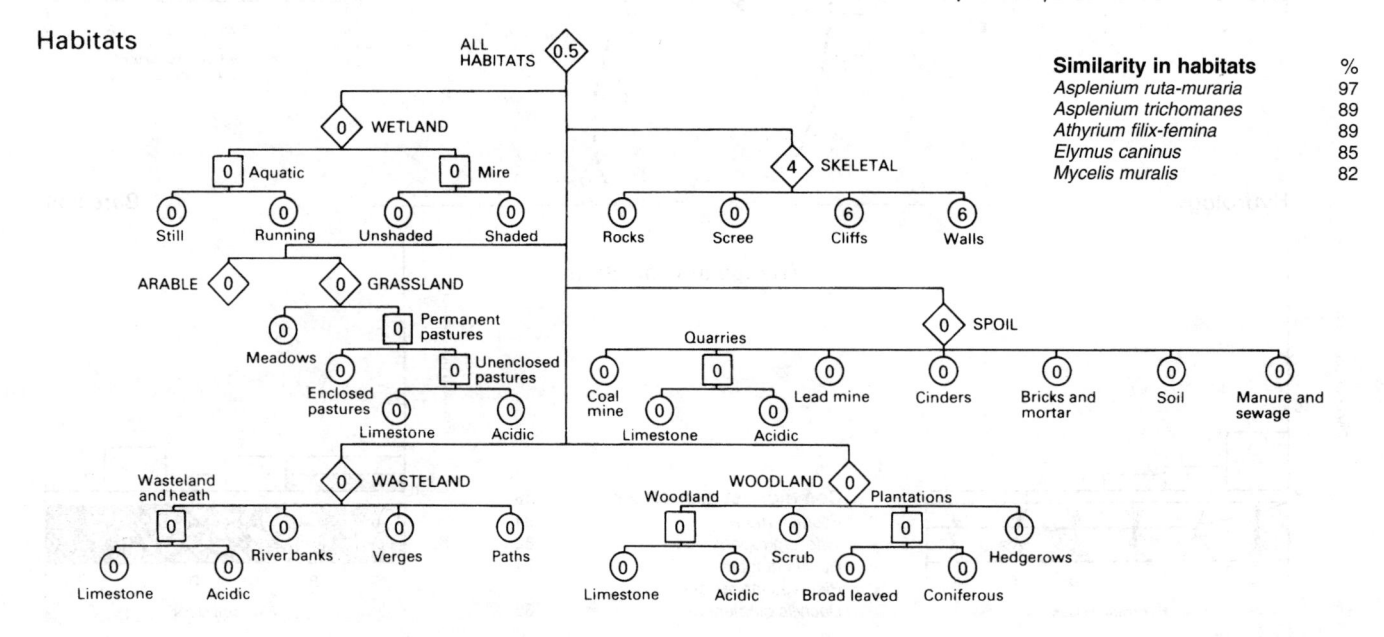

Habitats

Similarity in habitats	%
Asplenium ruta-muraria	97
Asplenium trichomanes	89
Athyrium filix-femina	89
Elymus caninus	85
Mycelis muralis	82

Synopsis A species which has not been studied intensively. *C.f* is a small fern of moist skeletal habitats. Growth is more rapid than that of most ferns (Page 1982). The fronds are unusually thin and the species can survive at exceptionally low light intensities in, for example, cave entrances. *C.f.* is also frequent on unshaded cliffs and walls. In N England the species is almost totally restricted to upland areas, and nationally *C.f.* is distributed mainly in the N and W (*ABF*). It is not known whether this restriction is related to the susceptibility of established plants to summer drought (Page 1982) or to problems of establishment from spores in drier climates. Each fertile frond produces an estimated 1 500 000 wind-dispersed spores (Page 1982), which are the most important means of regeneration. Even under ideal conditions, germination rate of the spores is lower than that recorded for *Athyrium filix-femina, Dryopteris dilatata* or *D. filix-mas*. It is possible that this feature, which is shared by *Asplenium ruta-muraria* and *A. trichomanes*, may be a specialization reducing the risk of premature spore germination following summer showers in otherwise droughted skeletal habitats. It is also possible that the spiny spore is hydrophobic, and thus less easily hydrated under intermittently dry conditions. It is not known whether a persistent spore bank is formed in the soil. *C.f.* produces a short branching rhizome but has little capacity for lateral, vegetative spread. Cf. has brittle stipes, hence its English name 'Brittle Bladder-fern'. The species is capable of regeneration from detached fronds. There is much genetically determined morphological variation within the species, and tetraploid and hexaploid chromosome races have also been identified (Page 1982). These form a sterile hybrid (Stace 1975). It unclear to what extent the distribution of *C.f.* is determined by the (unknown) ecology of the prothallial stage and it is not known whether plants from different habitats, and of different cytotypes, are ecotypically distinct.
Current trends Uncertain, but apparently decreasing on walls through destruction or alteration of this habitat.

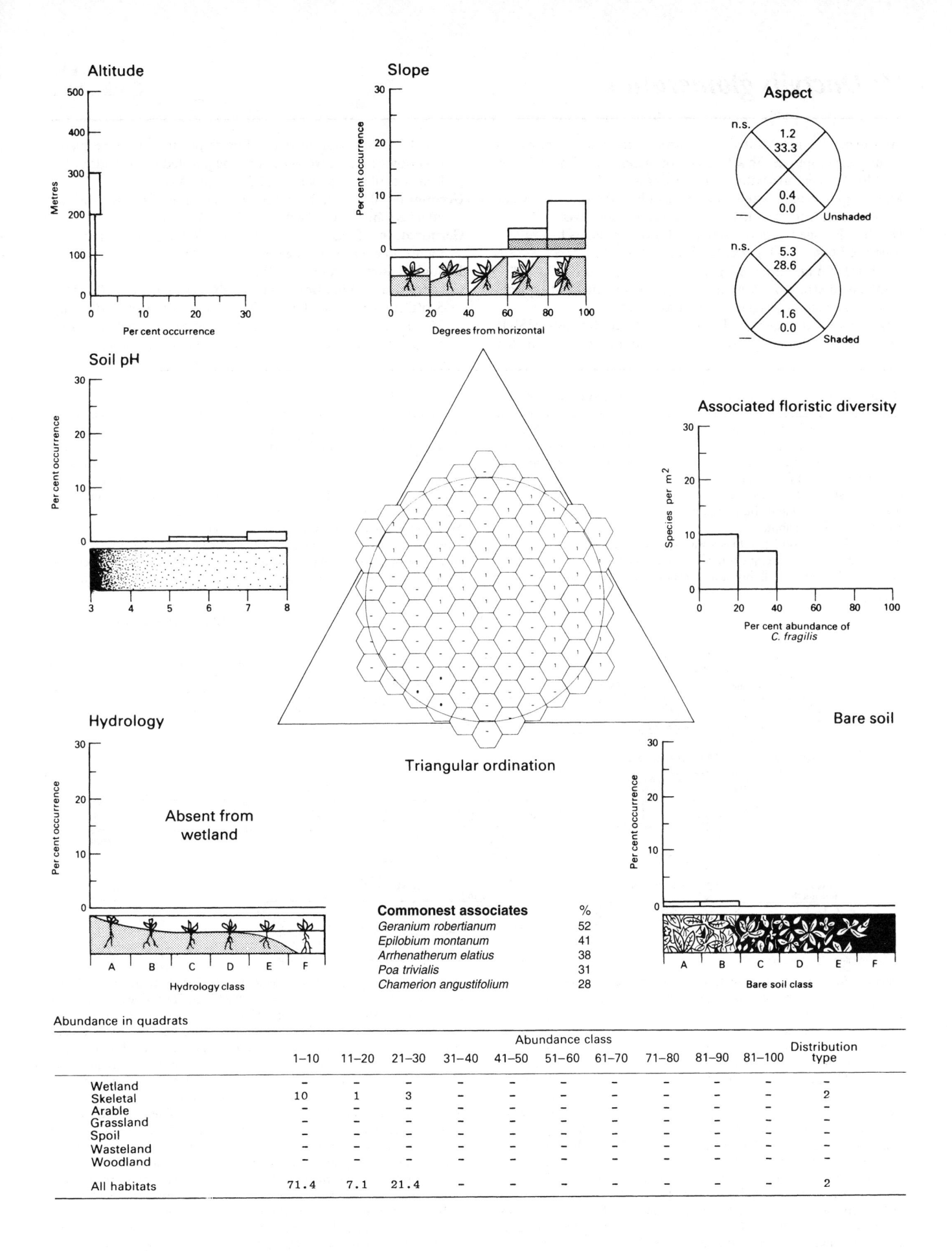

Altitude

Slope

Aspect

	n.s.	
	1.2	
	33.3	
—	0.4	Unshaded
	0.0	

	n.s.	
	5.3	
	28.6	
—	1.6	Shaded
	0.0	

Soil pH

Associated floristic diversity

Hydrology

Absent from wetland

Triangular ordination

Bare soil

Commonest associates	%
Geranium robertianum	52
Epilobium montanum	41
Arrhenatherum elatius	38
Poa trivialis	31
Chamerion angustifolium	28

Abundance in quadrats

	1–10	11–20	21–30	31–40	41–50	51–60	61–70	71–80	81–90	81–100	Distribution type
Wetland	–	–	–	–	–	–	–	–	–	–	–
Skeletal	10	1	3	–	–	–	–	–	–	–	2
Arable	–	–	–	–	–	–	–	–	–	–	–
Grassland	–	–	–	–	–	–	–	–	–	–	–
Spoil	–	–	–	–	–	–	–	–	–	–	–
Wasteland	–	–	–	–	–	–	–	–	–	–	–
Woodland	–	–	–	–	–	–	–	–	–	–	–
All habitats	71.4	7.1	21.4	–	–	–	–	–	–	–	2

(*) *Dactylis glomerata* L.

Poaceae
Cock's-foot

Life-form Densely tufted, polycarpic perennial, semi-rosette hemicryptophyte. Shoots erect or ascending; leaves linear, <4000 mm²; roots with VA mycorrhizas.
Phenology Winter green. Flowers mainly from May to July and seed set July to October. Leaves relatively short-lived.
Height Foliage typically <400 mm; flowering shoots to 1 m.
Seedling RGR 1.0-1.4 week^{-1}. SLA 27.7. DMC 24.3.
Nuclear DNA amount 8.7 pg; $2n = 14, 28*$ (*BC, FE5*); $2x, 4x*$.
Established strategy Varying with genotype but most typically intermediate between C-S-R and competitor.
Regenerative strategies S. Largely by seed (see 'Synopsis'); ?SB II.
Flowers Green, hermaphrodite, homogamous, wind-pollinated, usually cross-fertilized but selfing is possible in some plants (Beddows 1959); several hundred aggregated at the ends of the branches of a loose panicle (2-5 per spikelet).
Germinule 0.51 mg, 2.9 × 1.3 mm, caryopsis dispersed with an attached short-awned lemma (6.7 × 1.4 mm).
Germination Epigeal; t_{50} 1 day; <5-36°C; germination rate increased after dry storage at 20°C; L > D.
Biological flora Beddows (1959).
Geographical distribution The British Isles except in parts of the Scottish Highlands (100% of British vice-counties), *ABIF* page 760; Europe except the Arctic (97% of European territories); temperate Asia and N Africa; naturalized in N America.

Habitats Occurs in most dryland habitats, but most frequent as a constituent of meadows, pastures, rough grassland, waste places and road verges. Also abundant on spoil such as mining and quarry waste, building rubble, cinders, manure heaps, on rock outcrops, cliffs, scree slopes and walls, along hedgerows, river banks, paths and occasionally in open woodlands. Absent from aquatic habitats.
Gregariousness Intermediate.
Altitude Frequent and abundant up to 400 m. [Up to 690 m.]
Slope Widely distributed.
Aspect Shows a consistant bias towards S-facing slopes.
Hydrology Rather infrequent in wetland habitats.
Soil pH Found on all but the most acidic soils; most commonly associated with soils in pH range 5.0-8.0.
Bare soil Occurs in association with widely differing amounts of bare soil but declines in abundance at the two extremes.
△ **ordination** Distribution very wide-ranging, possibly due to the considerable genetic variation occurring in the species (*Biol Fl*). Centred on relatively productive, moderately disturbed vegetation, but the species often occurs as a minor constituent (sometimes only as seedlings and small tussocks) in unproductive or heavily disturbed vegetation.
Associated floristic diversity Typically intermediate.
Synopsis *D.g.* is a potentially robust, winter-green grass forming clumps on a wide range of soil types and under management regimes extending from intensive to lax. *D.g.* is palatable to stock and tolerant of summer

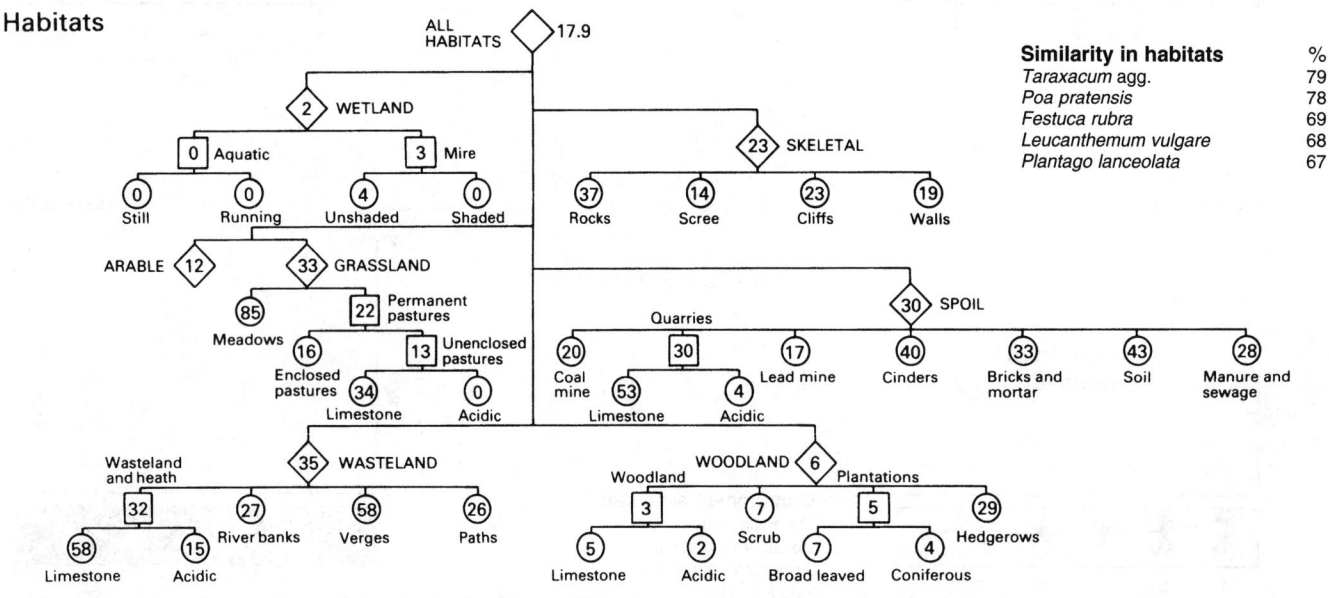

Habitats

Similarity in habitats	%
Taraxacum agg.	79
Poa pratensis	78
Festuca rubra	69
Leucanthemum vulgare	68
Plantago lanceolata	67

grazing. However, *D.g.* is most successful in grasslands where defoliation takes place infrequently and a tall canopy is established. In derelict grasslands, e.g. neglected meadows and roadsides, *D.g.* may develop relatively massive tussocks which exert local dominance upon smaller herbs. In infertile, base-rich pastures and particularly on S-facing slopes *D.g.* is represented by small-tussocked phenotypes. In turf microcosms, roots of this species became heavily colonized by VA mycorrhizas with increases of up to 80 per cent in seedling yield (Grime *et al.* unpubl.). *D.g.* exhibits some tolerance of drought and also survives light burning. Despite the widespread occurrence on paths (see 'Habitats') the species is usually considered to be susceptible to trampling (*Biol Fl*). In the *PGE*, *D.g. is* described as widespread and plentiful across the range of treatments 'except those inducing very acid or starved conditions'. Foliage is characterized by high Na and low Ca and Fe (*SIEF*). The tissues of *D.g.* are rich in fructan (Pollock 1979). In common with *Lolium perenne* and several other sown species, *D.g.* has only a restricted capacity for lateral tussocking and regenerates almost exclusively by seed in the spring (Roberts 1986b), primarily in vegetation gaps. Seed may persist for 2 to 3 years near the soil surface (Roberts 1986b) but *D.g.* does not form a substantial persistent seed bank. This feature appears to reduce the resilience of *D.g.* in heavily grazed and trampled pastures where bare ground is quickly colonized by species such as *Poa annua*. However, seed set is often high and the stiff cilia on the dispersule may enhance mobility. The capacity to exploit a wide range of habitats appears to stem both from phenotypic plasticity and ecotypic differentiation. Particularly in fertile sites, the distribution of *D.g.* has been profoundly influenced by the use of its seed in agriculture since the early 19th century (*Biol Fl*). Over 250 t are now sown annually (MAFF 1980-84). Plants with $2n = 14$ $(2x)$ occur in Europe and are sometimes given subspecific rank (*FE5*).
Current trends Uncertain, owing to the difficulty of distinguishing between native and introduced populations and in predicting the fate of a grass which regenerates solely by seed but which has little capacity for seed persistence. The use of *D.g.* in agriculture appears to be declining (Hopkins 1979), and the abundance of *D.g.* in farmland is likely to decrease.

Altitude
Slope
Aspect

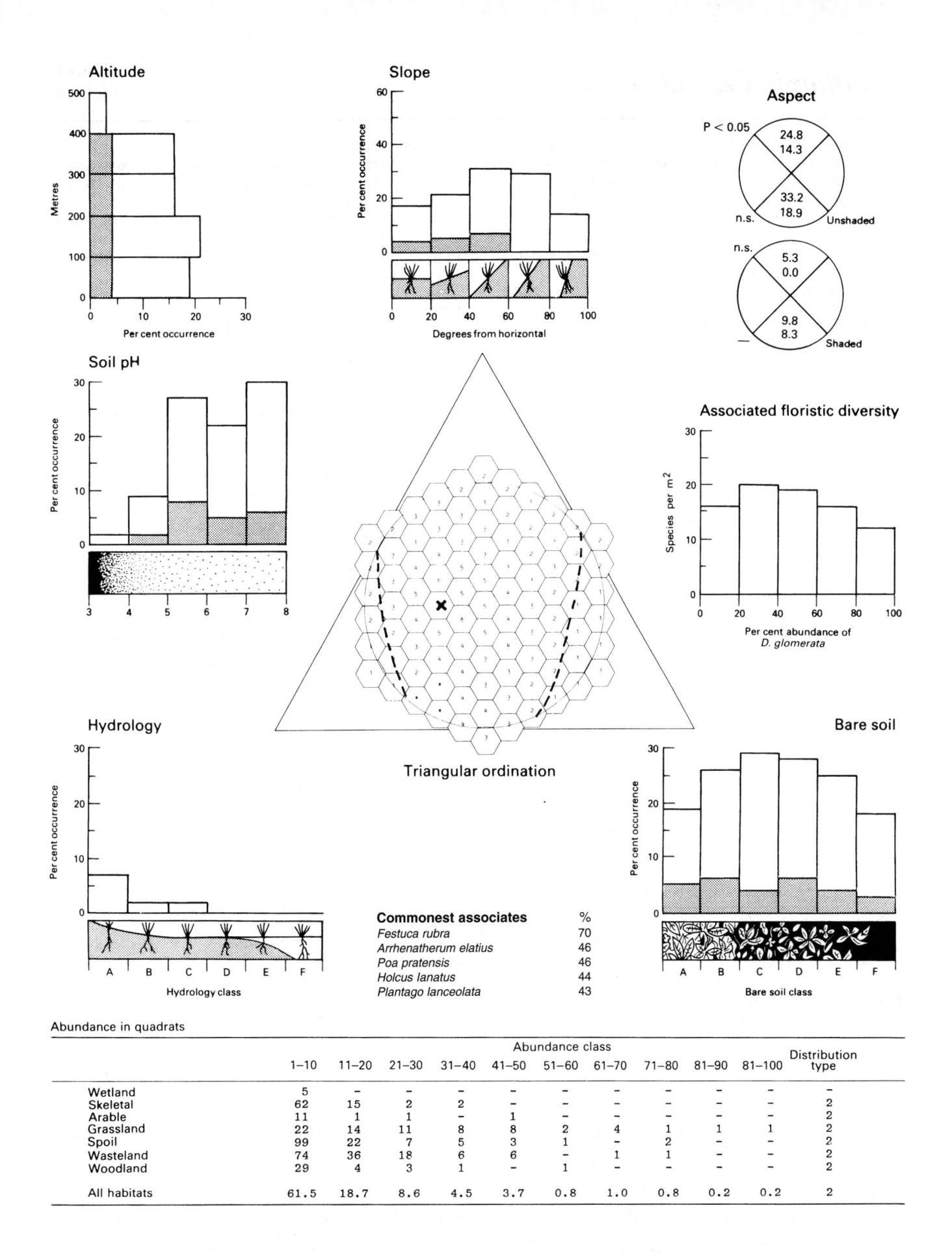

P < 0.05

24.8	14.3
33.2	18.9

n.s. — Unshaded

n.s.

5.3	0.0
9.8	8.3

— Shaded

Soil pH

Associated floristic diversity

Per cent abundance of *D. glomerata*

Hydrology

Hydrology class: A B C D E F

Triangular ordination

Commonest associates   %
Festuca rubra   70
Arrhenatherum elatius   46
Poa pratensis   46
Holcus lanatus   44
Plantago lanceolata   43

Bare soil

Bare soil class: A B C D E F

Abundance in quadrats

	1–10	11–20	21–30	31–40	41–50	51–60	61–70	71–80	81–90	81–100	Distribution type
Wetland	5	–	–	–	–	–	–	–	–	–	–
Skeletal	62	15	2	2	–	–	–	–	–	–	2
Arable	11	1	1	–	1	–	–	–	–	–	2
Grassland	22	14	11	8	8	2	4	1	1	1	2
Spoil	99	22	7	5	3	1	–	2	–	–	2
Wasteland	74	36	18	6	6	–	1	1	–	–	2
Woodland	29	4	3	1	–	1	–	–	–	–	2
All habitats	61.5	18.7	8.6	4.5	3.7	0.8	1.0	0.8	0.2	0.2	2

243

Danthonia decumbens (L.) DC.

Poaceae
Heath Grass

Life-form Tufted, polycarpic perennial, semi-rosette hemicryptophyte. Shoots decumbent to erect; leaves sparsely hairy, linear, typically <1000 mm^2; roots with VA mycorrhizas.

Phenology Winter green. Leaf expansion occurs in late spring. Flowers produced in July and seed set from August to October. Leaves relatively long-lived.

Height Foliage usually <100 mm; flowering shoots to 400 mm.

Seedling RGR 0.5-0.9 week^{-1}. SLA 20.6. DMC 30.6.

Nuclear DNA amount 5.9 pg; $2n$ = 24, 36*, 124* (*BC, FE5*); 4x, 6x*, 20x*.

Established strategy Intermediate between S and C-S-R.

Regenerative strategies B$_s$. By seed directly in autumn or after incorporation into a persistent seed bank. Plants increase in size as the result of the formation of new shoots but retain their tufted character; SB IV.

Flowers Green to purple, hermaphrodite, usually cleistogamous (wind-pollinated flowers infrequent); often >20 in a compact panicle; (4-6 per spikelet).

Germinule 0.87 mg, 2.5 × 1.4 mm, caryopsis dispersed with an attached lemma and a basal tuft of hairs (3.2 × 2.1 mm).

Germination Epigeal; t_{50} 5 days; 13-29°C; dormancy broken by dry storage at 20°C; L > D.

Biological flora None.

Geographical distribution Widespread in the British Isles especially in the W (100% of British vice-counties), *ABIF* page 799; most of Europe except the extreme N (87% of European territories); SW Asia and NW Africa.

Habitats In pastures, heaths and moorland and other waste places with occasional appearances on road verges, grassy paths and scree slopes. Occasionally in wetlands and on rock outcrops. [Frequent in montane grassland (Burnett 1964).]

Gregariousness Sparse to intermediate.

Altitude Found over a wide range of altitudes, but most suitable habitats for *D.d.* are situated within the upland region. [Up to 1040 m.]

Slope Because of its greater abundance in upland areas *D.d.* is most generally associated with sites of moderate slope (20-30°).

Aspect Frequent on slopes of both N- and S-facing aspect, but only recorded in abundance from N-facing slopes.

Hydrology Occasionally observed in soligenous mire.

Soil pH Shows a bias towards mildly acidic soils in the pH range 4.0-6.0, with a secondary peak above pH 6.5 associated with occurrences in calcareous grassland (this peak has been confirmed in a more recent vegetation survey). [The occurrence of *D.d.* in calcareous habitats appears most frequent in the N and W of Britain.]

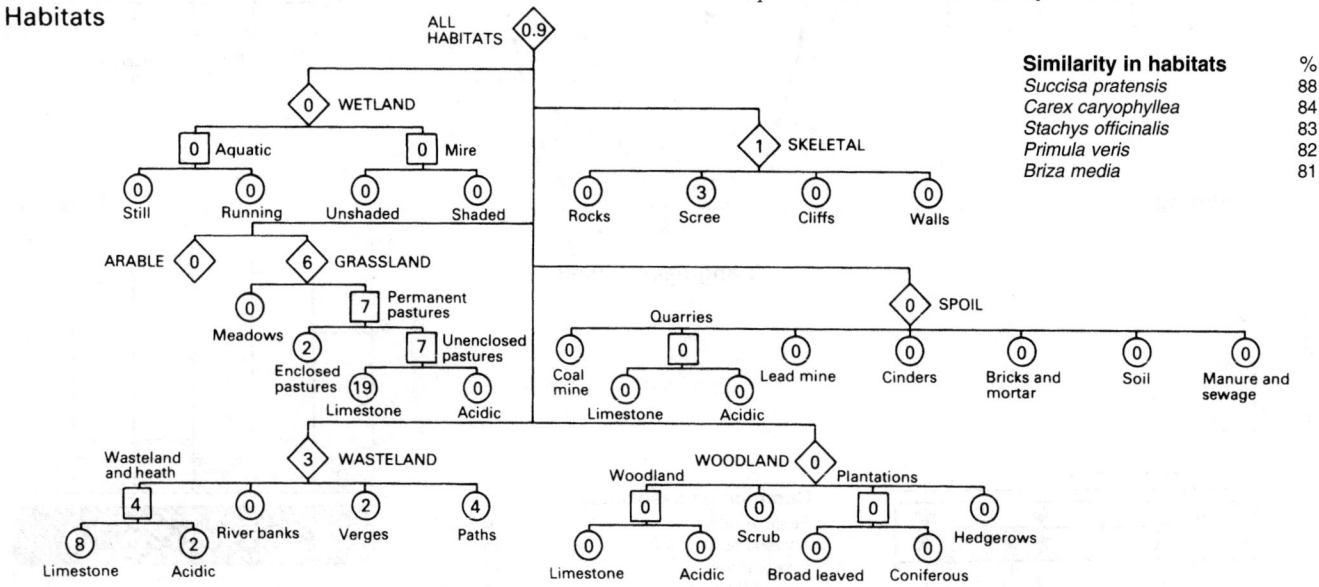

Habitats

Similarity in habitats	%
Succisa pratensis	88
Carex caryophyllea	84
Stachys officinalis	83
Primula veris	82
Briza media	81

Bare soil Usually associated with low levels of soil exposure.

△ **ordination** Distribution clearly centred on vegetation associated with infertile, relatively undisturbed habitats.

Associated floristic diversity Widely variable. Vegetation with >30 species m^{-2} is associated with a soil pH of >6.0, while species-poor vegetation tends to be of pH <4.5.

Synopsis *D.d.* is a slow-growing, loosely-tufted, winter-green grass restricted to infertile grassland and heathland, and present on both mildly acidic and calcareous soils. *D.d.* is vulnerable to dominance by more-productive species (Higgs & James 1969) and is totally excluded from fertile grassland. Milton (1933) suggests that the palatability of *D.d.* is greater than that of either *Agrostis capillaris* or *Festuca ovina*. In pastures damage by grazing may have severe effects as a result of the slow rate of replacement of the relatively long-lived leaves. Where pasture productivity is high *D.d.* is rapidly displaced by species such as *A.c.* which have a more dynamic leaf canopy. *D.d.* is more abundant in N-facing grassland, a pattern consistent with its frequent association with moist soils and local excursions into soligenous mire. *D.d.* shows only modest resistance to trampling and its distribution appears little affected by the presence or absence of burning. Compared with those of other common grasses forming persistent seed banks, the caryopses of *D.d.* are unusually large. Buried seed populations of *D.d.* are significantly clumped (Thompson 1986). This may be related to the role of ants in dispersal of the seed, which has an elaiosome (Ridley 1930). It seems possible that burial and persistence of seeds in the soil may be related to their late maturation, delayed release and high temperature requirement for germination, all of which militate against immediate germination. *D.d.* exhibits a limited capacity for clonal growth. Little is known of the genetic and physiological factors underlying the bimodal distribution with respect to soil pH. It appears that populations from calcareous soils may refer to ssp. *decipiens* Sch. and Bass., which has a more slender habit and $2n$ = 24 (Hubbard 1984), while those from mildly acidic soils belong to ssp. *decumbens*, but

Altitude
Slope
Aspect

	n.s.	2.4
	16.7	
	2.2	Unshaded
	0.0	

	0.0
	0.0	Shaded

Soil pH

Associated floristic diversity

Hydrology

Absent from wetland

Triangular ordination

Bare soil

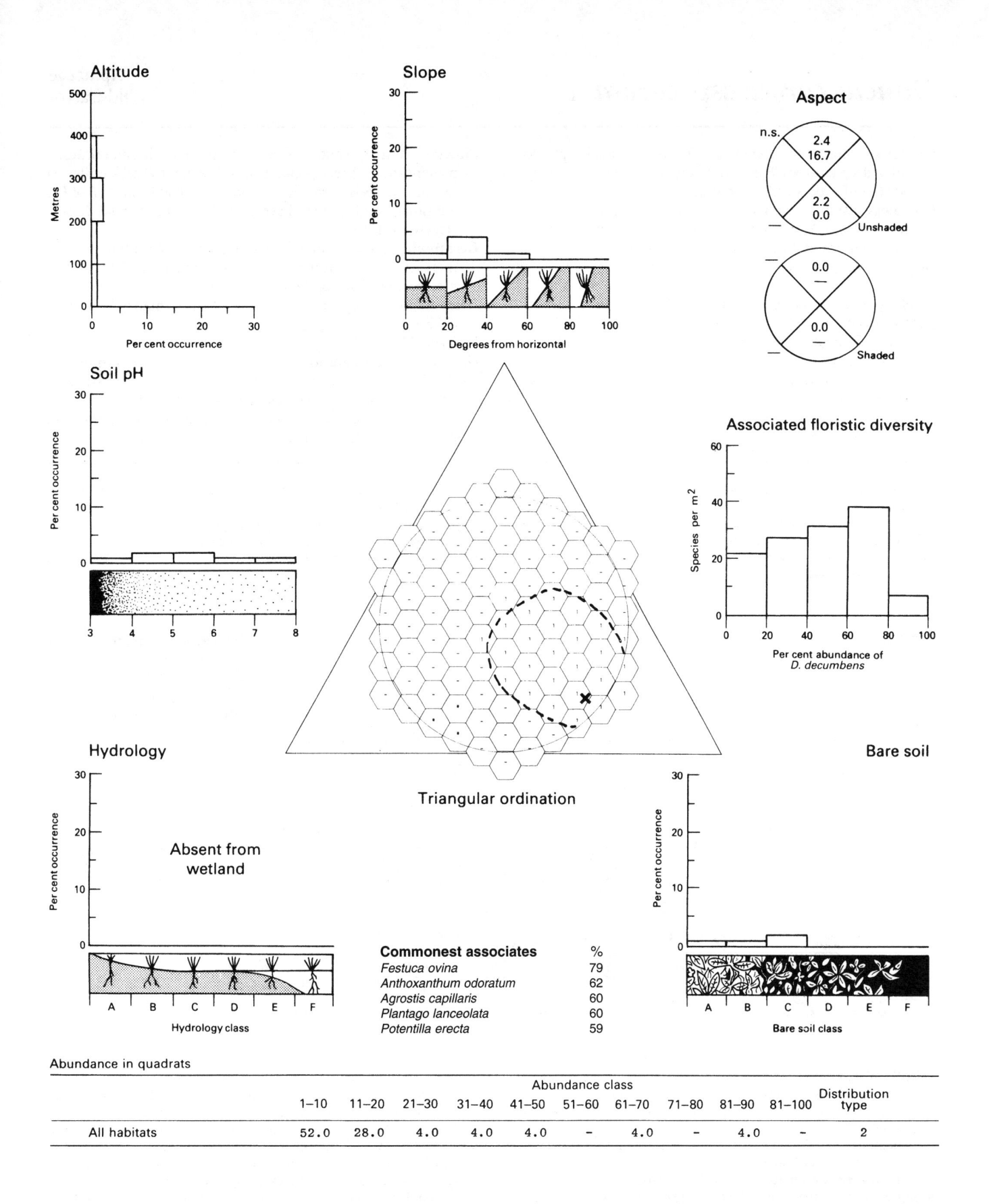

Commonest associates	%
Festuca ovina	79
Anthoxanthum odoratum	62
Agrostis capillaris	60
Plantago lanceolata	60
Potentilla erecta	59

Abundance in quadrats

	Abundance class									Distribution type
	1–10	11–20	21–30	31–40	41–50	51–60	61–70	71–80	81–90	81–100	
All habitats	52.0	28.0	4.0	4.0	4.0	–	4.0	–	4.0	–	2

the ecological distribution of these taxa requires further study. Flowers are often cleistogamous and it is tempting to conclude that the rather uniform morphology and restricted ecological distribution of the species may be related to in-breeding. The measurements of seedling RGR and nuclear DNA amount cited in this account are based on populations from calcareous soils; those for seed size and germination requirements refer to populations associated with acidic soils.

Current trends *D.d.* is confined to infertile grassland. The species is indicative of old or semi-natural grassland and has decreased, particularly in lowland areas as a result of intensive methods of farming.

Daucus carota ssp. carota L.

Apiaceae
Wild Carrot

Life-form Monocarpic perennial, semi-rosette hemicryptophyte with a deep tap-root. Stem erect; leaves dissected, often hispid, <4000 mm^2; roots with VA mycorrhizas.

Phenology Winter green. Flowers produced in June and July and seed set from August onwards. Some seed retained on plant after the death of the flowering stem in autumn.

Height Foliage mainly <200 mm. Flowering stems exceptionally to 1 m.

Seedling RGR No data. SLA 29.1. DMC 25.1.

Nuclear DNA amount 2.0 pg; $2n = 18*$ *(BC)*; $2x*$.

Established strategy Intermediate between stress-tolerant ruderal and C-S-R.

Regenerative strategies ?S, B$_s$. Regenerates entirely by seed which in Britain tends to germinate mainly in spring (Roberts 1986a). There is evidence of a persistent seed bank (*Biol Fl*, During *et al.* 1985, Roberts 1986a); SB III.

Flowers White, central one may be red, hermaphrodite, protandrous, insect-pollinated, subsequently self-fertile; in terminal umbels normally of more than 1000 flowers (*Biol Fl*). Subsidiary umbels may have large numbers of male flowers (Heywood 1983).

Germinule From terminal umbel, 0.88 mg, 2.7 × 1.9 mm, spiny mericarp (2 per fruit); from lateral umbels *c.* 75% of this maximum weight (*Biol Fl*).

Germination Epigeal; t_{50} 4 days; 6-29°C; dormancy broken by chilling; L > D.

Biological flora Dale (1974).

Geographical distribution The aggregate species widespread in Britain, but the distribution mainly southern and coastal (99% of British vice-counties), *ABIF* page 478; most of Europe (87% of European territories and 5% naturalized); naturalized in most other temperate and many tropical countries.

Habitats Recorded most frequently from ungrazed limestone grassland. Also found on cinders, rock outcrops and cliffs, and more locally in pastures, coal-mine spoil and waysides. Frequently observed beside railways. Absent from wetland, woods and arable land. [Also found near the sea but here often replaced by ssp. *gummifer* Hooker fil.]

Gregariousness Sparse.

Altitude Mainly restricted to lowland habitats, but observed up to 270 m. [Up to 400 m.]

Slope Associated with a wide range of slope.

Aspect Slightly more frequent on N-facing slopes.

Hydrology Absent from wetlands.

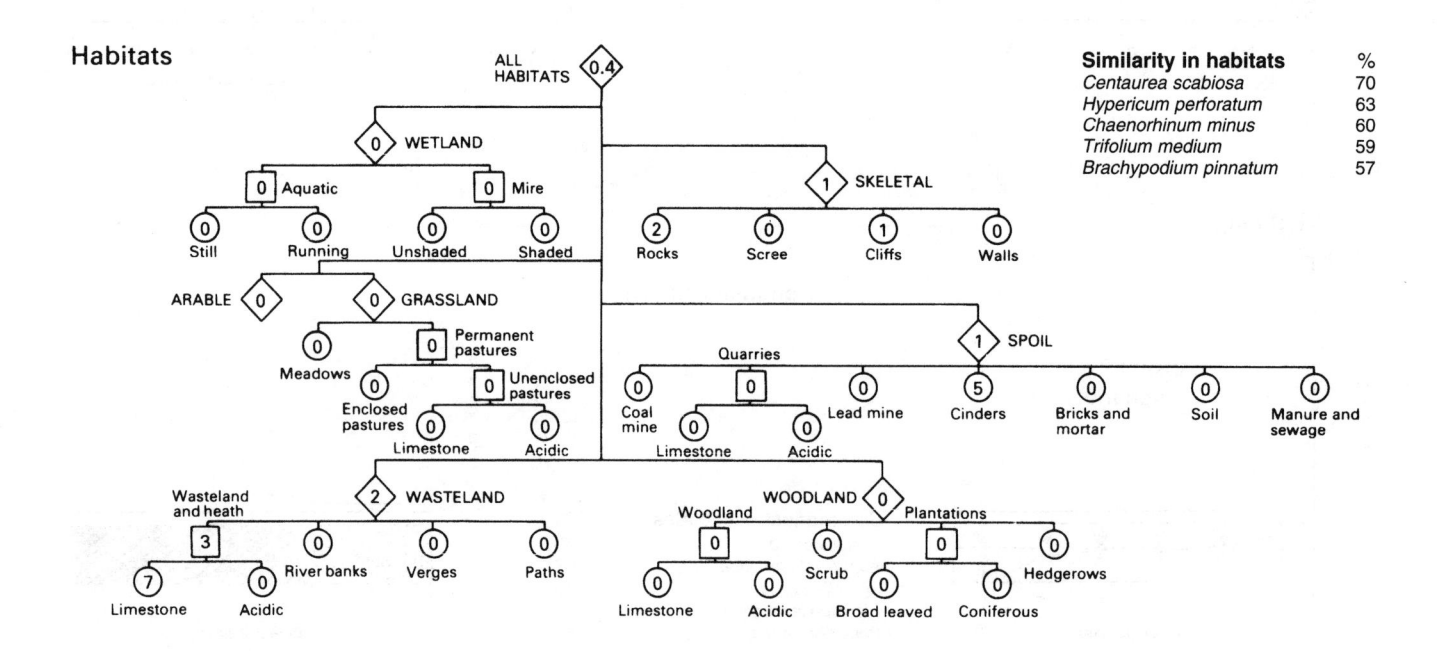

Habitats

Similarity in habitats	%
Centaurea scabiosa	70
Hypericum perforatum	63
Chaenorhinum minus	60
Trifolium medium	59
Brachypodium pinnatum	57

Soil pH Not recorded from soils of pH <5.0 and most frequent above pH 7.0, a reflection of its frequent association with calcareous strata.

Bare soil Often associated with bare soil and sites with discontinuous vegetation cover.

△ **ordination** No clear-cut relationship is apparent, but distribution shows some bias towards infertile habitats.

Associated floristic diversity Typically species-rich.

Synopsis *D.c.* is a monocarpic herb of dry, relatively infertile, grassland and open habitats, often taking 3-4 years to reach maturity (During *et al.* 1985). *D.c.* has sometimes been considered a 'moisture-loving' plant (*Biol Fl*) but, presumably because of its stout tap-root, can exploit drier environments (Anderson 1927). In Britain *D.c.* is characteristic of calcareous soils (*Biol Fl*) but it is also widely distributed as an adventive on other well-drained soil types. In Europe the calcicolous tendency is less pronounced (*Biol Fl*). *D.c.* is vulnerable to ploughing and cutting, particularly in late summer, and grows poorly with delayed flowering in shaded habitats (*Biol Fl*). *D.c.* taints milk and is not grazed preferentially by cattle (*Biol Fl*). Nevertheless, it is eaten to some extent by cows, horses and sheep, and the flowering shoot is particularly vulnerable to grazing (*Biol Fl*). *D.c.* regenerates entirely by seed and until maturity consists of a compact rosette of basal leaves easily overgrown by more-robust vegetation dominants. Thus, *D.c.* tends to be persistent only in short turf or open sites in ungrazed or under-utilized pasture. There is a close correlation between rosette size and probability of flowering (Lacey 1986). Seed production in large plants has been estimated at between 1000 and 40 000 (*Biol Fl*), but is often considerably less. Numbers as low as *c.* 80 are reported by During *et al.* (1985), who also record low rates of seedling mortality. Seeds vary in size and germination characteristics even when

Altitude
Slope
Aspect

	n.s.	1.6
	0.0	
	0.4	0.0 Unshaded

| | 0.0 | |
| | 0.0 | Shaded |

Soil pH

Associated floristic diversity

Triangular ordination

Hydrology
Absent from wetland

Commonest associates %
Plantago lanceolata 88
Poa pratensis 75
Arrhenatherum elatius 63
Cynosurus cristatus 63
Festuca ovina 63

Bare soil

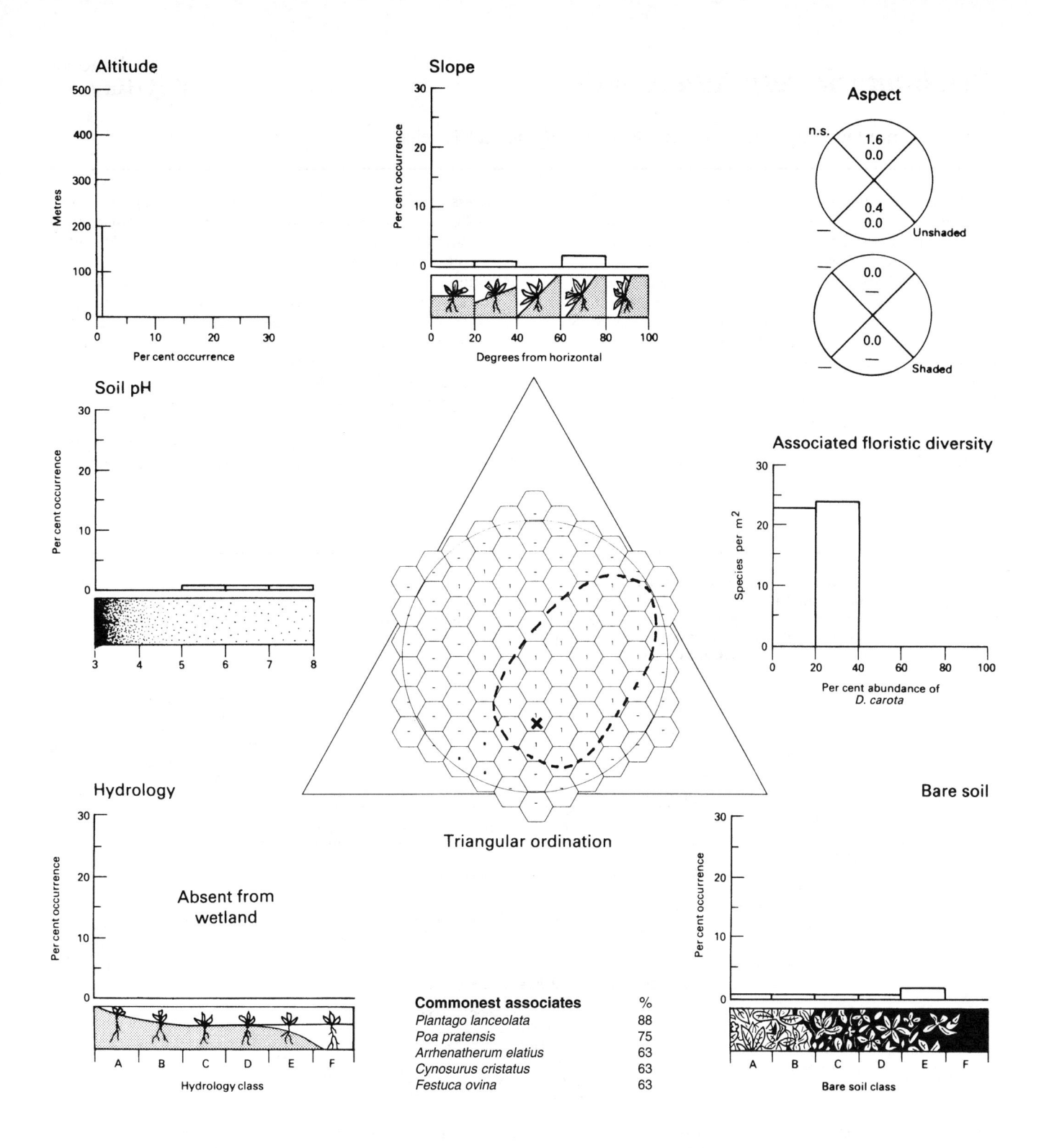

derived from the same plant *(Biol Fl)*. In chalk grassland, detached germinable seeds are detectable on or near the soil surface throughout the year, although their numbers fall in summer (Schenkeveld & Verkaar 1984). Seedlings tend to survive better on more-productive microsites (Verkaar *et al.* 1983). Seed is dispersed long distances, largely by humans although the spiky mericarps are carried locally by wind and in animal fur (Lacey 1982) and may survive ingestion *(Biol Fl)*. The longevity of plants of *D.c.* is under both environmental and genetic control, and annual, biennial and longer-lived populations have been identified (Lacey 1986). *D.c.* is polymorphic and 11 ssp. have been described for Europe *(FE2)*. Another taxon, ssp. *gummifer* (Syme) Hook. f. is found on sea cliffs and sand dunes on Atlantic coasts (Tutin 1980, *CSABF* p. 43). Aspects of the ecology of the ssp. *g.* are described in El-Sheikh (1973), Malloch and Okusanya (1979) and Okusanya (1979a-c). Ssp. *carota* is capable of interbreeding with ssp. *g.* and the cultivated carrot (ssp. *sativus* (Hoffm.) Arcangeli) (Heywood 1983). Whether there is any appreciable gene flow from cultivated carrot crops is doubtful, however, since crops and wild populations are seldom in close contact.

Current trends Probably decreasing as a consequence of the declining acreage of infertile habitats. *D.c.* may have increased in genetic diversity as a result of the introduction of alien genotypes along railways and in grass seed.

Deschampsia cespitosa (L.) Beauv.

Poaceae
Tufted Hair-grass

Excluding montane populations referrable to ssp. *alpina* (L.) Tzveler

Life-form Densely tufted, polycarpic perennial, semi-rosette hemicryptophyte. Shoots erect; leaves linear, tough, with barbed ridges on upper surface, sometimes >2000 mm^2; roots with VA mycorrhizas.
Phenology Winter green with maximum above-ground biomass in August and minimum in February. Some growth occurs during winter (*Biol Fl*). Flowers from June to August. Seeds shed from August onwards. Leaves relatively long-lived.
Height Foliage usually <500 mm; flowers to 2 m.
Seedling RGR 1.0-1.4 week^{-1}. SLA 18.5. DMC 30.8.
Nuclear DNA amount 18.0 pg, $2n = 52$ (non-European material; Bennett *et al.* 1982); $2n = 26^*, 52^*$ (*Biol Fl*); $2x^*, 4x^*$. (Base number of genus strictly $n = 7$.)
Established strategy Intermediate between C-S-R and stress-tolerant competitor.
Regenerative strategies S, B$_s$. Regeneration biology is complex (see 'Synopsis'); SB III.

Flowers Silvery-green, hermaphrodite, protandrous, wind-pollinated, self-incompatible (*Biol Fl*); often >1000 in an open panicle (2 per spikelet); vivipary is recorded (*Biol Fl*) but is mostly unimportant.
Germinule 0.31 mg, 2.1 × 0.8 mm, caryopsis dispersed with an attached lemma, a short hygroscopic bristle and a basal tuft of hairs (2.9 × 0.9 mm). Germinule weight of $2x$ and $4x$ plants 0.25 and 0.31 mg, respectively (*Biol Fl*).
Germination Epigeal; t_{50} 3 days; 7-33°C; L > D.
Biological flora Davy (1980).
Geographical distribution The British Isles (99% of British vice-counties), *ABIF* page 769; Europe except for the extreme S (87% of European territories); widely distributed in temperate and Arctic regions and on mountains in tropical Africa and Asia.

Habitats Occurs particularly frequently on soils with impeded drainage. Most-regularly recorded from base-rich woodland and unimproved pasture, but also common in a range of grazed and ungrazed grassland systems, river banks, both shaded and unshaded mire, and spoil heaps. Infrequent in plantations, in aquatic habitats, and on cliffs and outcrops. Scarce and of casual occurrence on arable land. [Also found in montane grassland (*Biol Fl*).]
Gregariousness Recorded as intermediate, but may occasionally form near monocultures.

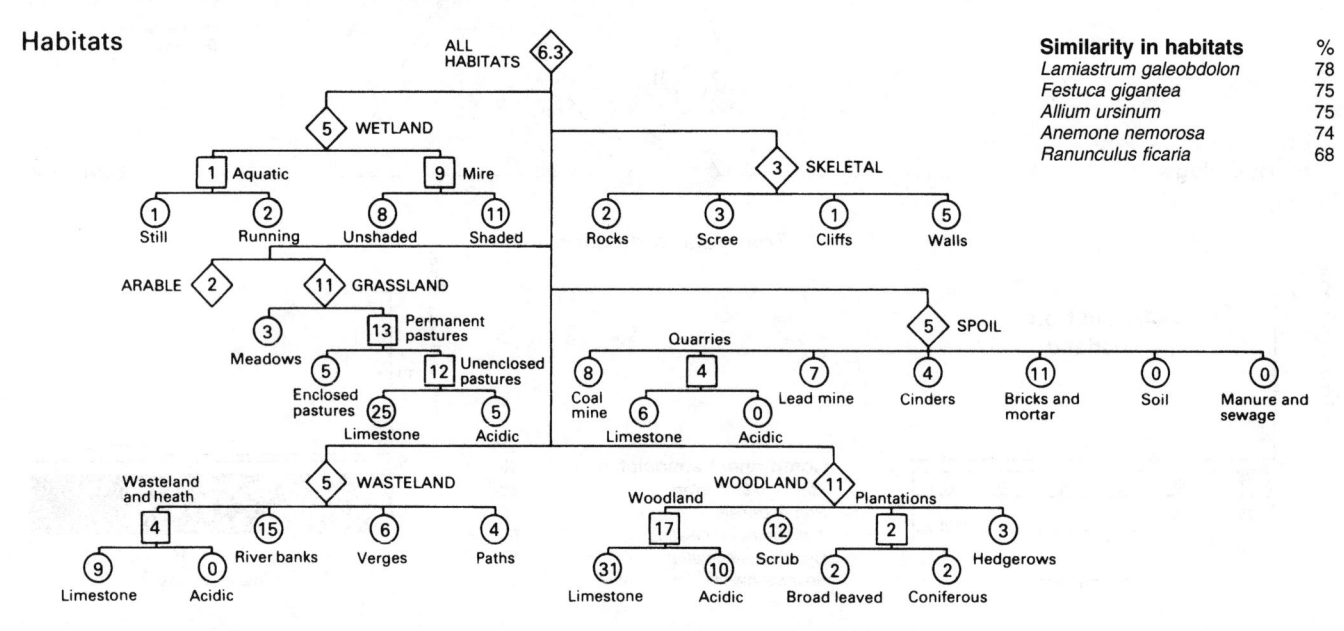

Habitats

Similarity in habitats	%
Lamiastrum galeobdolon	78
Festuca gigantea	75
Allium ursinum	75
Anemone nemorosa	74
Ranunculus ficaria	68

Altitude Common at all altitudes. [Up to 1130 m.]
Slope Found over a wide range of slopes, but only abundant on slopes of less than 40°.
Aspect More frequent on N-facing slopes, in both open and shaded sites.
Hydrology Common in both topogenous and soligenous mire but not usually found in permanently flooded habitats.
Soil pH Present over a wide range in soil pH; most frequently associated with soils of pH 4.5-6.5.
Bare soil Associated with a wide range of values, but distribution biased towards habitats having a relatively small exposure of bare soil.
△ **ordination** Widely distributed, but centred on two types of relatively undisturbed vegetation differing in productivity; these reflect the two major habitat types in which *D.c.* occurs, (woodland and pasture land). Absent from heavily disturbed vegetation.

Associated floristic diversity Typically low or intermediate.
Synopsis *D.c.* is a relatively persistent tufted grass often forming a dense tussock with persistent leaf litter. These characteristics often allow local dominance of communities by the species. The principal habitats are grassland and woodland. In both *D.c.* is most frequent on poorly drained, slightly acidic soils. The coarse leaves have a high silica content, and mature leaves are usually avoided by herbivores (*Biol Fl*). As a result *D.c.* may thrive as a weed of lowland cattle pasture. However, young foliage may be eaten by horses and rabbits, and in upland areas, where leaves tend to have less silica, the species is grazed freely by cattle, sheep and deer, and is sometimes considered a useful pasture species (*Biol Fl*). Foliar levels of N, P and Ca are low (*SIEF*). Survival in woodland, where flowering and seed production is often poor, may be related to (a) the considerable longevity of established plants (>30 years), (b) the capacity to produce phenotypes

Altitude
Slope
Aspect

	Unshaded
n.s. 4.7	16.7
n.s. 2.7	16.7

	Shaded
n.s. 12.1	31.3
n.s. 6.5	25.0

Soil pH

Associated floristic diversity

Hydrology

Triangular ordination

Bare soil

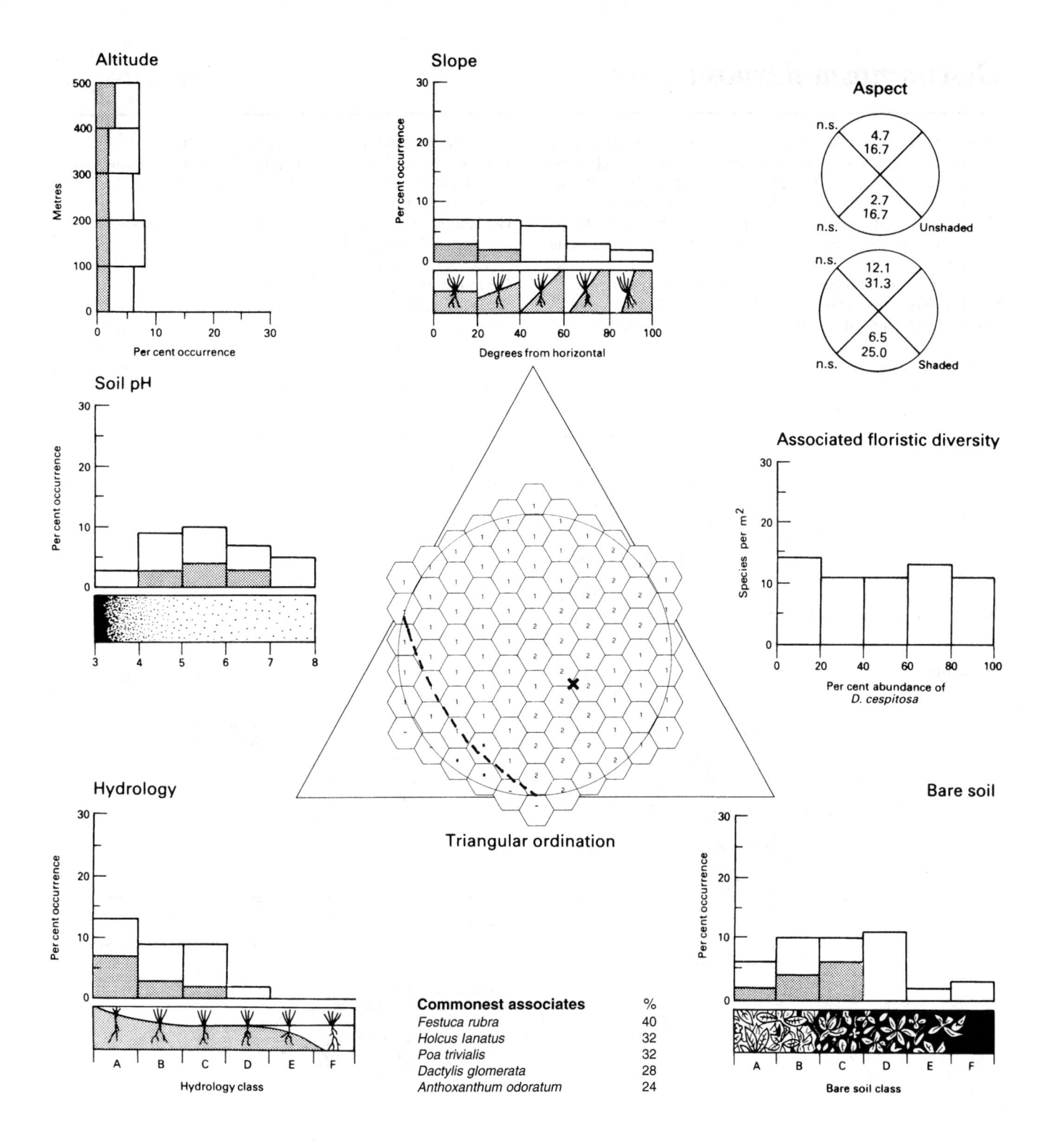

Commonest associates %
Festuca rubra 40
Holcus lanatus 32
Poa trivialis 32
Dactylis glomerata 28
Anthoxanthum odoratum 24

with long narrow leaves, reducing the degree of self-shading usually characteristic of tussock grasses (*Biol Fl*) and (c) persistence as seeds through the densely-shaded phase of coppice cycles. Under favourable conditions *D.c.* may flower in its second year, producing up to 500 000 small seeds per plant (*Biol Fl*) and forming a persistent buried seed bank (Roberts 1986b). Regeneration by seed appears to occur only in areas of open ground mainly during the period February to June (Roberts 1986b). *D.c.* has only a restricted capacity for lateral vegetative spread, although as a result of breakdown of the rhizomes at the centre of the tussock, *D.c.* may occasionally fragment into a number of daughter plants (*Biol Fl*). Edaphic and climatic ecotypes have been reported (*Biol Fl*) and heavy-metal-tolerant populations have been identified (*Biol Fl*). There are variations in ploidy and plants with $2n = 52$ ($4x$) are more common and diploids are mainly restricted to semi-natural woodland (Rothera & Davy 1986). This taxonomically complex species has been subdivided into seven ssp., many of which hybridize freely (*FE5*).

Current trends *D.c.* may have decreased in grasslands as a result of improved drainage and management, and the conversion of pasture to arable in lowland areas. However, in woodlands and many other artificial habitats a high level of mechanized disturbance appears beneficial to *D.c.*, and tetraploids of the species may be increasing. The future status of diploids, which have a restricted habitat range, remains uncertain. *D.c.* is frequently a colonist of many artificial habitats and is common on derelict land. Changes in the status of the species are therefore more likely to be expressed as fluctuations in relative abundance in particular habitats rather than more generally.

Deschampsia flexuosa (L.) Trin.

Poaceae
Wavy Hair-grass

Life-form Tufted or rhizomatous and mat-forming, polycarpic perennial, semi-rosette hemicryptophyte. Shoots erect; leaves hair-like, sometimes >100 mm^2; roots mainly shallow but some extending to >500 mm (*Biol Fl*), with VA mycorrhizas.
Phenology Winter green. Flowers June and July and sheds seed shortly after ripening in August or September. Leaves long-lived.
Height Foliage up to 200 mm; flowering shoots to 400 mm.
Seedling RGR 0.5-0.9 week^{-1}. SLA 15.0. DMC 33.7.
Nuclear DNA amount 11.0 pg; $2n = 26, 28*, 56$ (*BC, FE5*); $4x*, 8x$.
Established strategy Intermediate between stress-tolerator and stress-tolerant competitor.
Regenerative strategies V, S. Regenerates effectively both vegetatively and by seed (see 'Synopsis'); SB I.
Flowers Silvery green, hermaphrodite, homogamous, wind-pollinated, self-incompatible (Weimarck 1968); >100 in an open panicle (2 per spikelet).
Germinule 0.43 mg, 1.9 ×1.0 mm, caryopsis dispersed with an attached lemma, an hygroscopic bristle and a basal tuft of hairs (3.5 × 1.3 mm).
Germination Epigeal; t_{50} 4 days; 6-28°C; L = D.
Biological flora Scurfield (1954).
Geographical distribution The British Isles, especially in the N and W (98% of British vice-counties), *ABIF* page 770; most of Europe, but rarer in the S and absent from much of the SE (85% of European territories); N Asia, N America and on mountains in warmer climates, e.g. America and E Africa.

Habitats Of widespread occurrence in a range of unproductive, base-poor habitats. Abundant and widely distributed in upland pastures, moors, acidic heaths, scrub, woodlands and plantations. Occurs on the acidic soils of rock ledges in sandstone quarries, rock outcrops, cliffs and spoil heaps. Successful also in neglected hill pastures, the sides of railway cuttings and other waste places. Relatively frequent in bogs, in which it often grows on the top of tussocks of other species, e.g. *Eriophorum vaginatum*. Uncommon in unleached calcareous habitats, and absent from arable land. [An important component of many montane plant communities on acidic soils.]

Gregariousness A patch-forming species.
Altitude Widely distributed, but suitable habitats are more frequent in upland areas. [Up to 1150 m.]
Slope Found at all angles of slope.
Aspect More frequent on N-facing slopes, with a significant concentration in N-facing shaded habitats.
Hydrology Occurs in wetland habitats but infrequent on waterlogged soils.
Soil pH Mostly restricted to sites with pH <5.0.

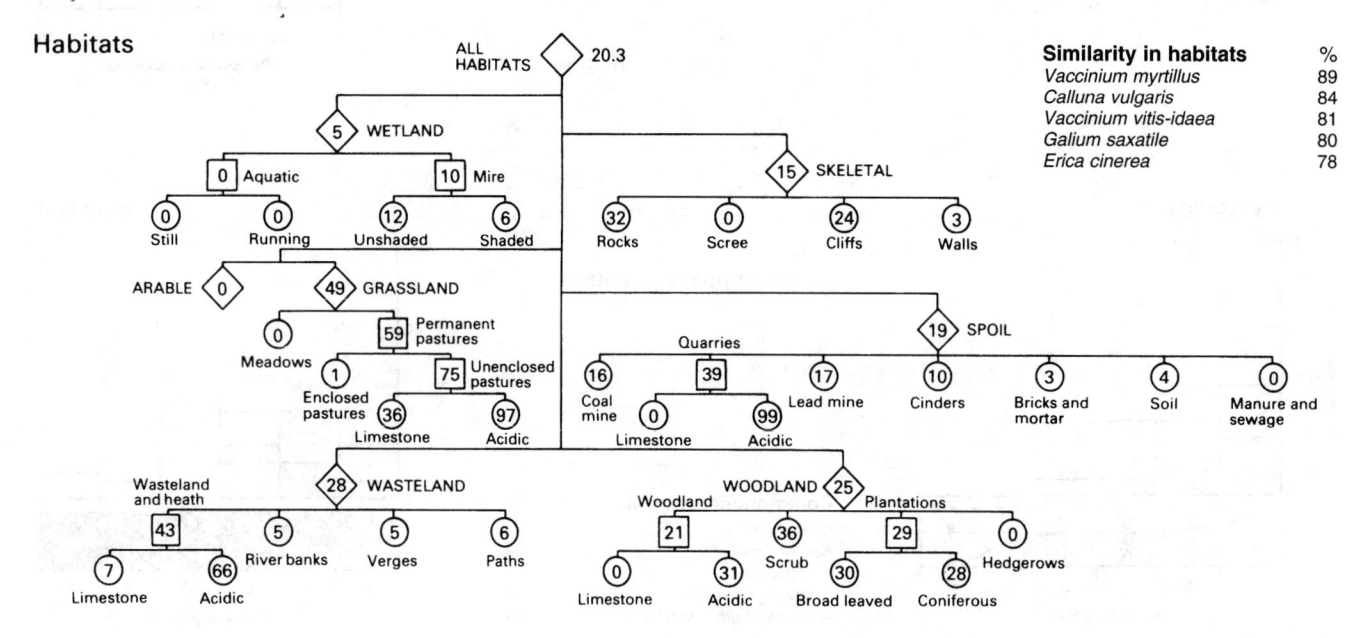

Habitats

ALL HABITATS 20.3

WETLAND 5
- Aquatic 0
  - Still 0
  - Running 0
- Mire 10
  - Unshaded 12
  - Shaded 6

SKELETAL 15
- Rocks 32
- Scree 0
- Cliffs 24
- Walls 3

ARABLE 0

GRASSLAND 49
- Meadows 0
- Permanent pastures 59
  - Enclosed pastures 1
  - Unenclosed pastures 75
    - Limestone 36
    - Acidic 97

Quarries
- Coal mine 16
- Limestone 39 / 0
- Lead mine 17 / Acidic 99
- Cinders 10

SPOIL 19
- Bricks and mortar 3
- Soil 4
- Manure and sewage 0

WASTELAND 28
- Wasteland and heath 43
  - Limestone 7
  - Acidic 66
- River banks 5
- Verges 5
- Paths 6

WOODLAND 25
- Woodland 21
  - Limestone 0
  - Acidic 31
- Scrub 36
  - Broad leaved 30
- Plantations 29
  - Coniferous 28
- Hedgerows 0

Similarity in habitats	%
Vaccinium myrtillus	89
Calluna vulgaris	84
Vaccinium vitis-idaea	81
Galium saxatile	80
Erica cinerea	78

Bare soil Frequency and abundance decline progressively with increasing exposure of bare soil.
△ **ordination** Distribution strongly concentrated on the right-hand side of the diagram, indicating restriction to unproductive, relatively undisturbed vegetation. Excursions into other types of vegetation are extremely rare and probably correspond to seedlings.
Associated floristic diversity Low and declining slightly with increased abundance of *D.f.*
Synopsis *D.f.* is a slow-growing, evergreen, clump- or carpet-forming grass tolerant of low pH, low mineral nutrient supply and high external concentrations of aluminium and manganese (Hackett 1965, Grime & Hodgson 1969, Mahmoud & Grime 1974, Rorison 1985). The leaves contain low amounts of N, P, Mg and Ca (*SIEF*) and *D.f.* responds to low levels of nitrogen by more-efficient utilization and redistribution of dry matter in the form of finer roots and root hairs (Robinson & Rorison 1983). The species is typical of cold climates and podzolic soils and produces humus which is persistent and inhibitory to plant growth (Grime 1963b, Jarvis 1964). *D.f.* is the most successful calcifuge grass in Britain, and combines a very wide geographical distribution with the capacity to exploit a diversity of acidic dryland habitats. The species is particularly shade-tolerant, and seedlings are capable of persistence in total darkness for 3 months (Hutchinson 1967). However, vegetative vigour and flowering are both inhibited by dense shade, and *D.f.* is also rather vulnerable to submergence beneath deciduous tree litter. In unshaded habitats in Britain differences in performance may be associated with aspect. On N-facing slopes vegetative expansion may result in continuous monocultures, whilst on adjacent S-facing slopes tussocks are often stunted but produce abundant inflorescences. *D.f.* is eaten by sheep and rabbits, but in moorland habitats new shoots of *Calluna vulgaris* are preferred and grazing here can lead to an increase in *D.f.* (*Biol Fl*).

Altitude
Slope
Aspect

n.s.	31.1 / 87.3	
n.s.	24.7 / 80.0	Unshaded
$P < 0.05$	30.3 / 70.0	
$P < 0.001$	18.7 / 60.9	Shaded

Soil pH

Associated floristic diversity

Hydrology

Triangular ordination

Bare soil

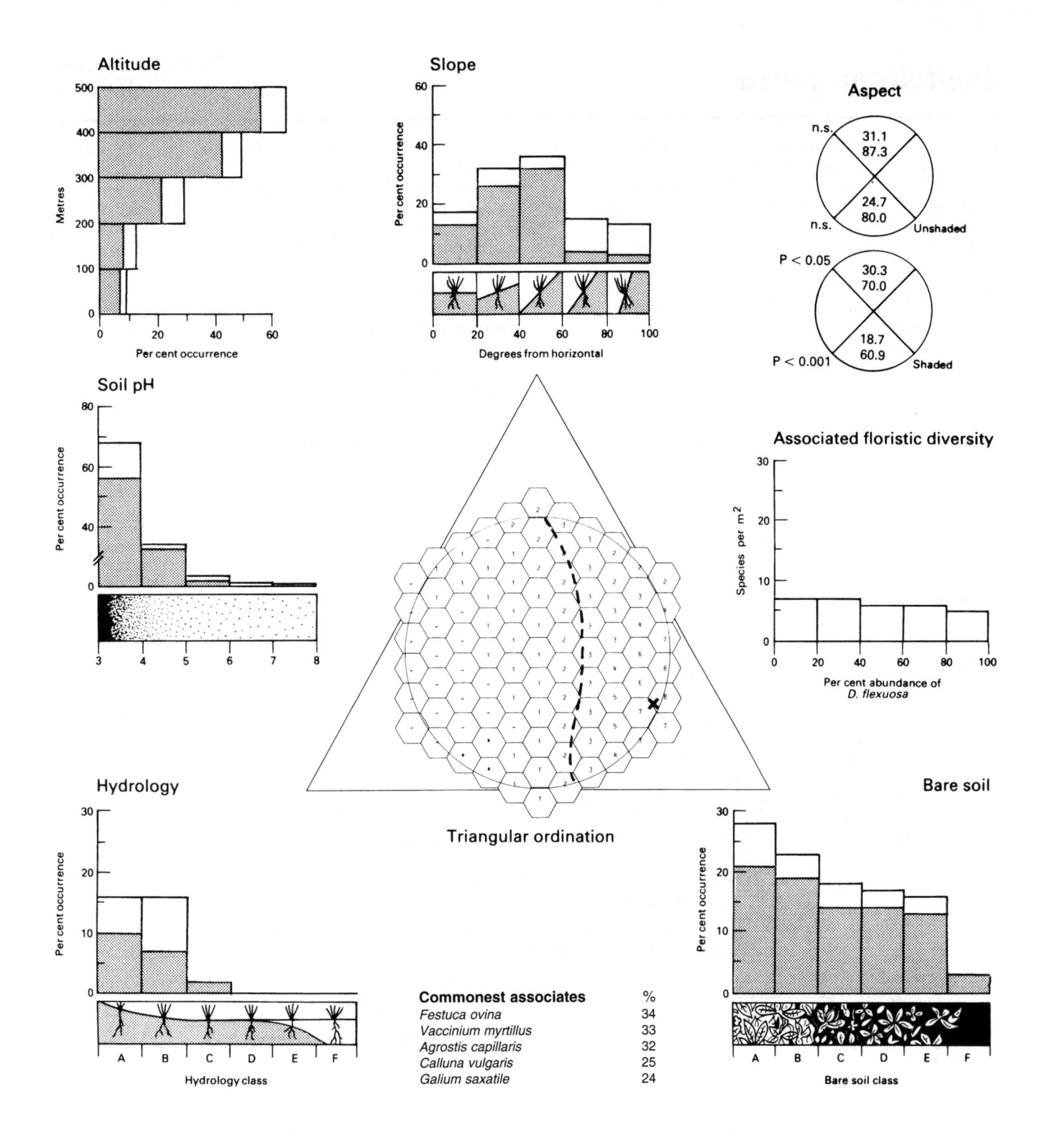

Commonest associates %
Festuca ovina 34
Vaccinium myrtillus 33
Agrostis capillaris 32
Calluna vulgaris 25
Galium saxatile 24

However, under conditions of intense grazing, mowing or trampling, or if the soil is dry, *Festuca ovina* tends to be more prevalent. The species is capable of forming large clonal patches as a result of rhizome growth. Freshly shed seed germinates in autumn. The seedlings so formed may be long-lived, often persisting in a stunted form until conditions become more favourable for growth, but no persistent seed bank is accumulated. In consequence, *D.f.* is not an effective colonist after severe fires. However, the species often persists vegetatively in habitats subject to regular fires of low intensity. The species is occasionally planted as a shade-tolerant amenity grass (Shildrick 1980).

Current trends Although *D.f.* colonizes a number of artificial habitats, e.g. railway banks and coal-mine spoil, the species is decreasing in lowland areas, due to the destruction of heathland and acidic grassland. *D.f.* is likely to remain abundant in upland areas.

Digitalis purpurea L.

Scrophulariaceae
Foxglove

Life-form Polycarpic or more rarely monocarpic, semi-rosette perennial. Stems erect; leaves ovate, usually glandular hairy, <30 000 mm²; tap-rooted, often with VA mycorrhizas.

Phenology Winter green. Plants flower from June to September and seed set from July to October.

Height Foliage to *c.* 1 m (or a basal rosette in the case of juveniles); flowers to 1.5 m.

Seedling RGR 1.0-1.4 week⁻¹. SLA 13.8. DMC 24.3.

Nuclear DNA amount 2.4 pg; $2n = 56$ (*FE3*); $2x$. (Base number of tribe probably 7.)

Established strategy Intermediate between stress-tolerant ruderal and C-S-R.

Regenerative strategies S, B_s. By seed which may germinate directly in autumn or in spring. A persistent seed bank is formed. *D.p.* has a limited capacity to form daughter rosettes; SB III.

Flowers Usually purple, hermaphrodite, protandrous, insect-pollinated and automatically selfed if this fails (Knuth 1906); in a 20- to 80-flowered raceme (*Excur Fl*).

Germinule 0.07 mg, 1.0 × 0.5 mm, seed in a dehiscent capsule containing <1000 seeds.

Germination Epigeal; t_{50} 6 days; 10-29°C; L > D.

Biological flora None.

Geographical distribution The British Isles (99% of British vice-counties), *ABIF* page 551; W, SW and WC Europe, and also cultivated and widely naturalized further east (33% of European territories and 13% naturalized); USA, Australia and New Zealand, probably introduced.

Habitats Mainly restricted to disturbed shaded habitats, particularly woodland, river banks and hedgerows. Recorded, sometimes only as seedlings, from a variety of spoil habitats, particularly soil heaps, quarry spoil and cinder tips. Also present in skeletal habitats including cliffs and walls, and in shaded mire. Frequently observed on roadsides and disturbed wasteland. Only a casual on arable land, and absent from managed grassland and from aquatic habitats.

Gregariousness Sparse.

Altitude Observed to 350 m, and recorded more widely from upland areas. [Up to 890 m.]

Slope Wide-ranging.

Aspect In shaded habitats species shows a S-facing bias.

Hydrology Found occasionally on moist soils at the edge of soligenous mire.

Soil pH Recorded on soils varying in pH from 3.5 to 7.5, but most frequent on acidic soils of pH <5.0.

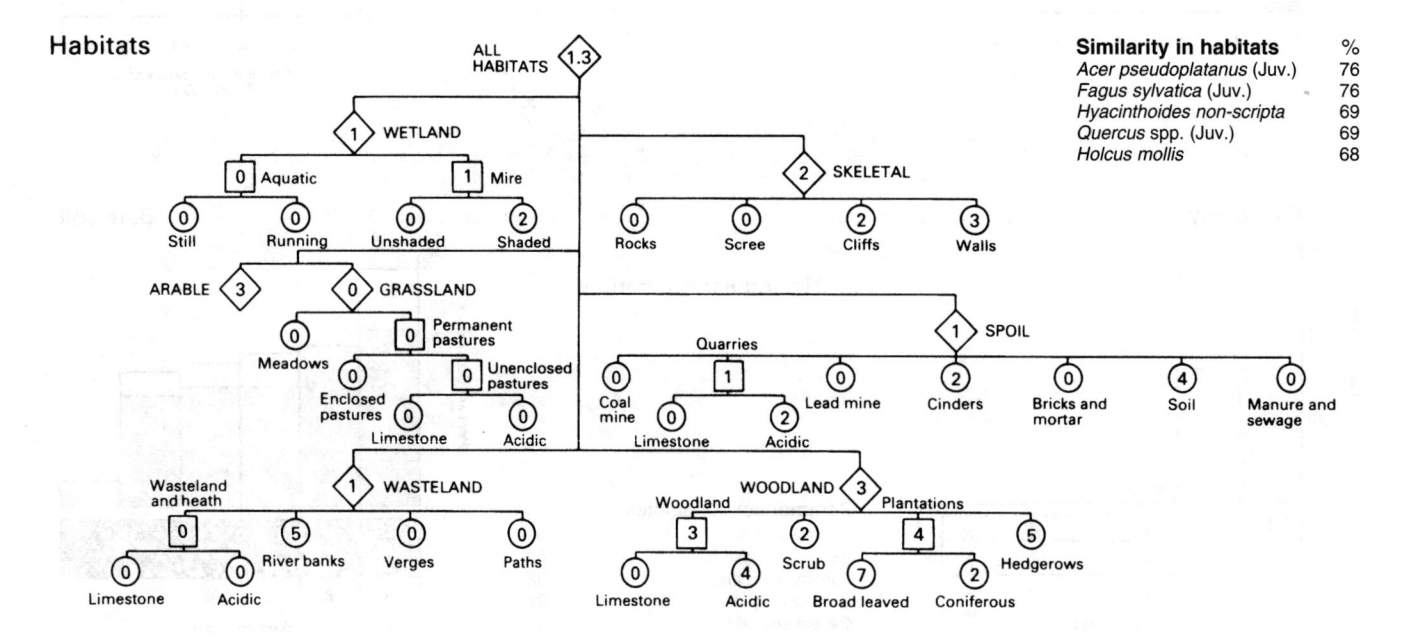

Habitats

Similarity in habitats	%
Acer pseudoplatanus (Juv.)	76
Fagus sylvatica (Juv.)	76
Hyacinthoides non-scripta	69
Quercus spp. (Juv.)	69
Holcus mollis	68

Bare soil Associated with various degrees of exposure, but most frequent in habitats with low to moderate amounts of bare soil.

△ **ordination** A rather diffuse pattern resulting from a capacity to exploit local patches of disturbed soil in a wide range of shaded vegetation types.

Associated floristic diversity Typically low to intermediate.

Synopsis *D.p.* is a winter-green, rosette-forming, polycarpic (or more rarely monocarpic) perennial with an Atlantic distribution. It exploits lightly-shaded (often S-facing) sites usually on disturbed acidic soils of moderate fertility. Particularly in upland Britain, *D.p.* is characteristic of perennial plant communities on river banks, rocky habitats and steep hedgebanks. Here flooding and soil creep at intervals give rise to areas of bared soil suitable for regeneration by seed, and dictate that the growth of potential dominants is to some extent restricted. *D.p.* also occurs in recent plantations, burned areas in woods and around the exposed roots of windthrown trees. Rackham (1980) has also observed expansions in *D.p.* after coppicing. Here populations are short-lived, becoming replaced in a few years by taller, more-dominant herbaceous species. In comparison with most calcifuge herbs, *D.p.* is relatively fast-growing, although under unfavourable conditions development may be slow. On fertile soils the species may form a large rosette within 1 year and flower in its second season. Dense populations on acidic soils may be indicative of locally enhanced soil fertility (Rackham 1980). Foliage of *D.p.* is characterized by high concentrations of P, Na, Fe and Mn (*SIEF*). Both fresh and after drying and storage, foliage of the species is toxic to stock and humans (Cooper & Johnson 1984). The poisonous principle (digitalin) is used in the treatment of heart complaints, and *D.p.* is grown as a crop (MAFF 1980). Despite its toxicity *D.p.* is infrequent in grazed habitats. The species may occasionally produce daughter rosettes, but regenerates mainly by seed. An average plant may produce >70 000 seeds (Salisbury 1942). Freshly-shed seed has a light requirement, and germination is inhibited by

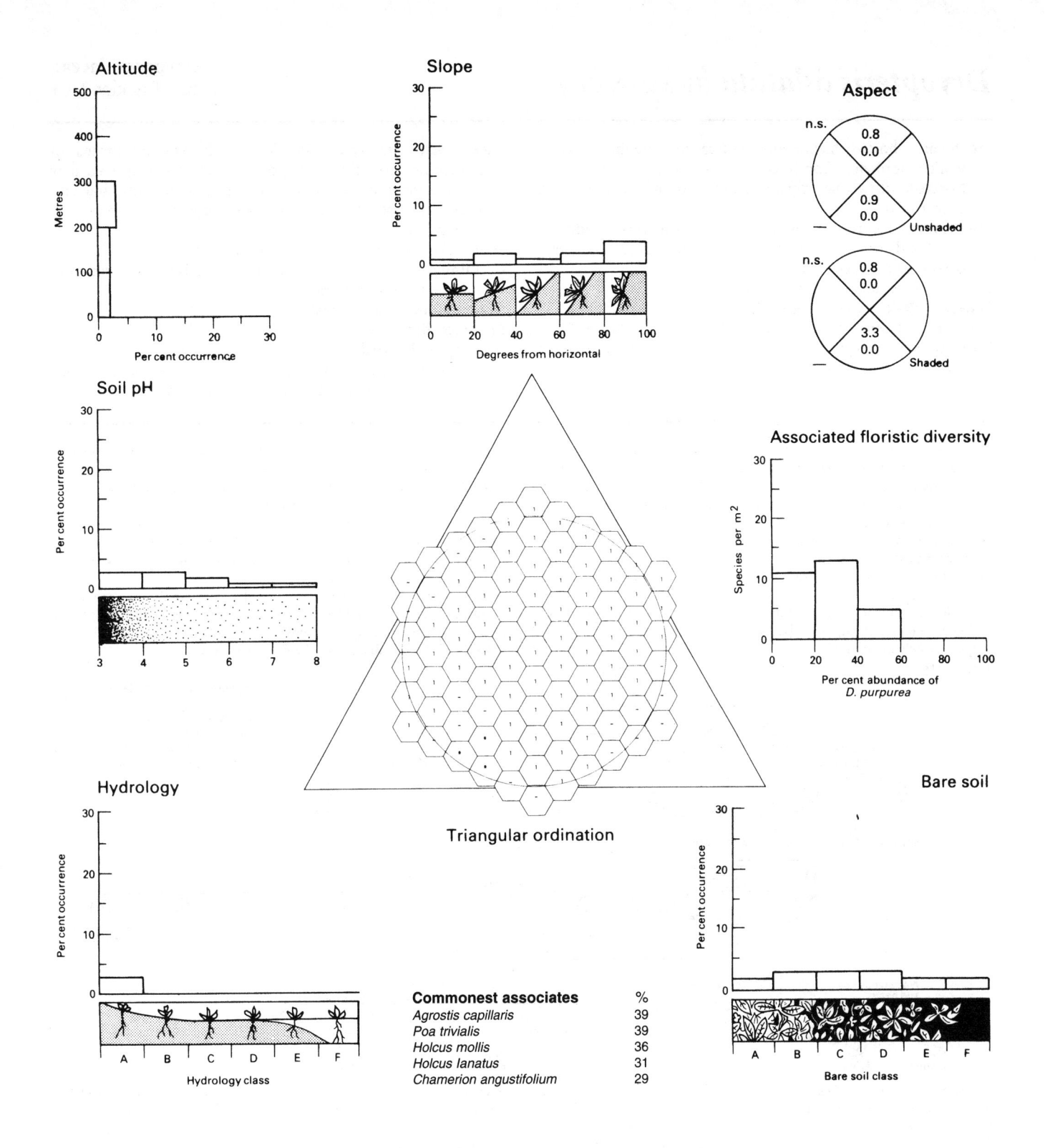

Altitude
Slope
Aspect
Soil pH
Associated floristic diversity
Hydrology
Triangular ordination
Bare soil

Commonest associates	%
Agrostis capillaris	39
Poa trivialis	39
Holcus mollis	36
Holcus lanatus	31
Chamerion angustifolium	29

a leaf canopy (Van Baalen 1982). As a result large quantities of seed are incorporated into a persistent seed bank, with the result that *D.p.* may remain in sites subject to intermittent disturbance, and often develops high population densities following disturbance. In an outdoor pot experiment, germination was observed in all months except December (Roberts 1986a). Seed may be dispersed by wind (Ridley 1930) and other agencies, and the species is a frequent colonist of a range of artificial habitats. *D.p.* is extremely variable in several of its morphological characters, and dwarf alpine ecotypes occur (*FE3*). A further two ssp. are recognized in Europe (*FE3*).

Current trends Remains common in uplands, but perhaps decreasing in the lowland region since the increased level of habitat disturbance (a feature favouring *D.p.*) is probably outweighed by the loss of habitat to agriculture and coniferous woodland.

Dryopteris dilatata (Hoffm.) A. Gray

Dryopteridaceae
Broad Buckler-fern

Life-form Shortly rhizomatous, polycarpic perennial, rosette hemicryptophyte. Fronds erect or spreading, ovate, much dissected, with a scaley stipe, sometimes >0.1 m^2; roots with VA mycorrhizas.
Phenology Fronds often overwinter in a senescent condition. New fronds produced in April and May and spores from June to September (Page 1982).
Height Fertile and sterile fronds to 1200 mm.
Sporeling RGR No data. SLA 23.9. DMC 27.7.
Nuclear DNA amount 16.1 pg; $2n = 164^*$ (*BC*); $4x^*$ (Page 1982).
Established strategy Intermediate between stress-tolerant competitor and C-S-R.

Regenerative strategies W, V, ?B$_s$. Mainly by means of numerous wind-dispersed spores, which it is suspected may become incorporated into a persistent spore bank, but also shows a restricted capacity for vegetative spread (see 'Synopsis').
Sporangia Sori on the underside of the fronds.
Germinule 50 × 33 µm, spore, black, explosively dispersed from a many-spored sporangium.
Germination No information.
Biological flora None.
Geographical distribution The British Isles (100% of British vice-counties), *ABIF* page 80, *AF* page 93; N and C Europe (74% of European territories); temperate Asia.

Habitats Mainly in two habitats, (a) woodland, particularly coniferous plantations and woodland on acidic strata, and (b) skeletal habitats, particularly cliffs. Also found in shaded and (mainly at higher altitudes) unshaded mire. Frequently observed on gritstone-quarry heaps, but rare on other types of spoil. [Occurs in rock crevices in montane areas, and behaves as an epiphyte on rough-barked trees (Page 1982).]
Gregariousness May achieve high canopy densities and dominance at low population density; high frequency-values refer exclusively to colonies of young plants.
Altitude Extends to >400 m, and most commonly recorded from moist sheltered valleys in the uplands. [Up to 1130 m (Page 1982).]
Slope Widely distributed.

Aspect Slightly more frequent in N-facing situations and significantly more abundant in populations on N-facing slopes.
Hydrology Frequent on moist soils and soligenous mire; less so on topogenous mire, and absent from aquatic habitats.
Soil pH Widely distributed, but most frequent and abundant on acidic soils below pH 5.0 and infrequent above pH 6.5. Scarce on limestone.
Bare soil Most sites have little bare soil, due to high exposures of rock or dense cover of tree or fern litter.
△ **ordination** Distribution mainly confined to undisturbed relatively productive vegetation, but there is an outlier at the base of the triangle (see opposite) corresponding to scattered occurrences on cliffs and walls.
Associated floristic diversity Typically very low.

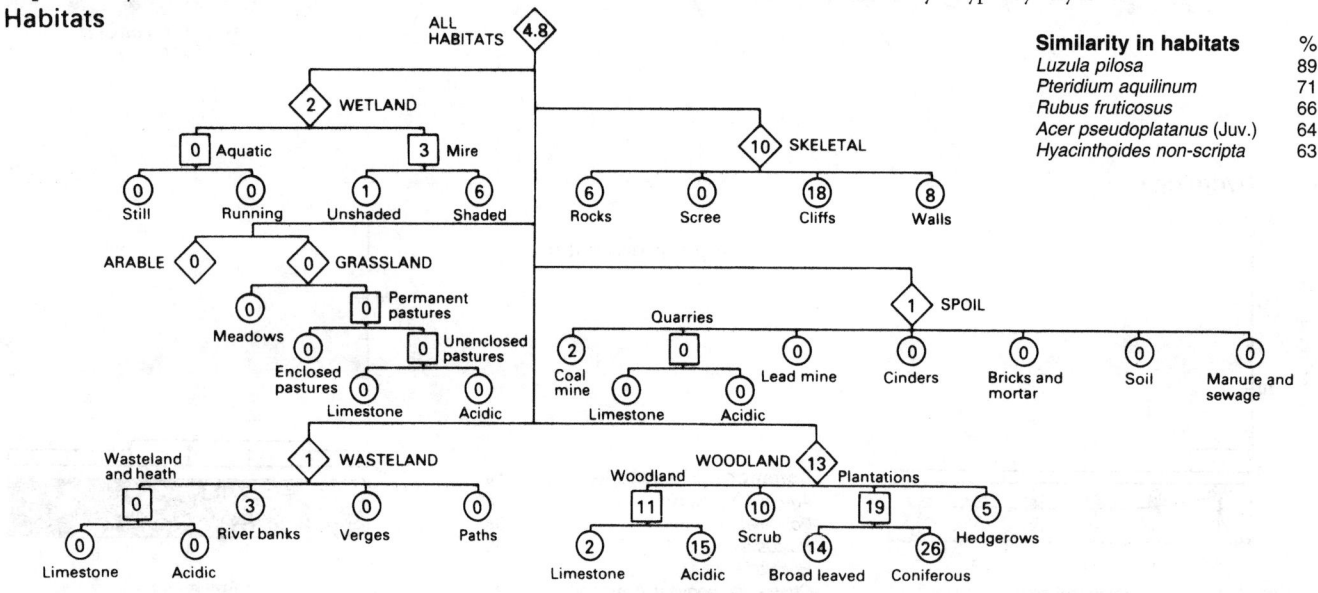

Habitats

Similarity in habitats	%
Luzula pilosa	89
Pteridium aquilinum	71
Rubus fruticosus	66
Acer pseudoplatanus (Juv.)	64
Hyacinthoides non-scripta	63

Synopsis Long-lived (Page 1982) and highly shade-tolerant, this fern forms a tall crown of fronds from an erect or ascending rhizome, and often functions as a woodland herb layer dominant on moist, acidic, moderately fertile, peaty soils. *D.d.* extends more frequently into unshaded habitats at higher altitudes. The fronds, particularly when young, contain thiamase, which induces thiamine deficiency (Ottosson & Anderson 1983). This provides an effective defence against a majority of insects (Ottosson & Anderson 1983), and the fronds are potentially toxic to stock (Cooper & Johnson 1984). Occasionally plants produce a number of small crowns from a slender, long-creeping rhizome (Page 1982), but it is more typical for vegetative spread to be strongly confined. An estimated 13.5 million minute wind-borne spores are produced by each frond (Page 1982), and these are released at 'saturating densities' into the landscape. Large numbers of viable spores, which have a light requirement for germination, occur throughout the top 120 mm of the soil profile in many woodland sites, and the presence of a shortly-persistent spore bank is suspected.

Interpretation of the ecological range of *D.d* is complicated by the fact that the distribution of the long-lived sporophyte is determined to a considerable extent by the habitat requirements of the ephemeral gametophytic prothallus. Since the prothallus has little effective control of water-loss and requires moisture for the fertilization of the gametes, establishment of sporelings is largely confined to damp microsites. In the event of soil movement or litter decomposition, prothalli and sporelings are often killed, and they are also susceptible to competition from bryophytes and higher plants. Young sporelings tend to be observed mainly on moist, bare soil on the woodland floor, in rock crevices and on fallen trees or rotting stumps, particularly in heavy shade. The specialized nature of the microsites required for the establishment of the sporeling may account for the tendency of *D.d.* to occur as widely-spaced individuals rather than as a stand-forming species. Moreover, as a result of the ecological dissimilarity between the gametophyte and sporophyte generations, *D.d.* often persists as a young sporeling in sites which are

Altitude
Slope
Aspect
Soil pH
Associated floristic diversity
Hydrology
Triangular ordination
Bare soil

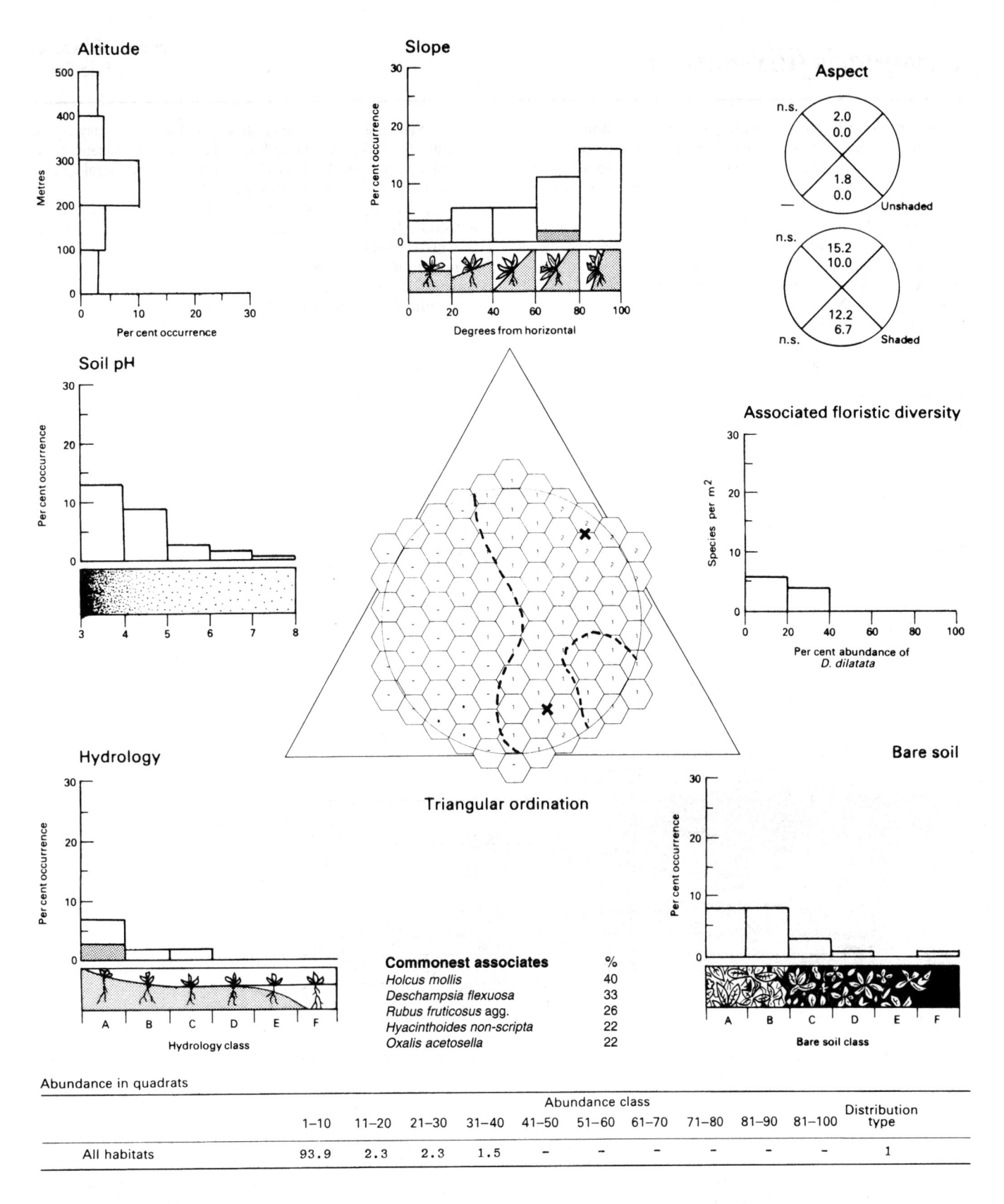

Commonest associates	%
Holcus mollis	40
Deschampsia flexuosa	33
Rubus fruticosus agg.	26
Hyacinthoides non-scripta	22
Oxalis acetosella	22

Abundance in quadrats

	1–10	11–20	21–30	31–40	41–50	51–60	61–70	71–80	81–90	81–100	Distribution type
All habitats	93.9	2.3	2.3	1.5	–	–	–	–	–	–	1

unsuitable for the adult sporophyte. In a study in Derbyshire woodlands (Willmot 1985) it was found that populations of *D.d* exhibited a much greater level of recruitment and a higher rate of mortality than *D. filix-mas*. This 'weedy' behaviour is most marked in the cooler and moister conditions of N Britain (Page 1982). Consistent with its wide habitat range, including sites as distinct as montane rock ledges and lowland woodland, the species is morphologically very variable and is believed to be a tetraploid derived as a result of hybridization between *D. expansa* (C. Presl) Fraser-Jenkins and Jermy (*D. assimilis* S. Walker), and a montane species *D. intermedia* (Muhl.) A. Gray, or related species (Gibby & Walker 1977). *D.d.* hybridizes with four other related species in Britain (Stace 1975).

Current trends A rapid colonist of new plantations and rocky habitats; apparently increasing.

Dryopteris filix-mas (L.) Schott

Dryopteridaceae
Male Fern

Life-form Shortly rhizomatous, polycarpic perennial, rosette hemicryptophyte. Fronds erect or spreading, lanceolate, much dissected with a scaley stipe, sometimes >0.1 m²; roots with VA or rarely ectomycorrhizas. Prothallus may also have VA mycorrhizas.
Phenology Fronds often overwintering in a semi-senescent form. New fronds appear from May to June and spores produced from August to November.
Height Fertile and sterile fronds to 1400 mm.
Sporeling RGR No data. SLA 24.2. DMC 29.5.
Nuclear DNA amount 17.4 pg; $2n = 164^*$ (*BC*); $4x^*$ (Page 1982).
Established strategy Stress-tolerant competitor.
Regenerative strategies W, V. Regenerates by means of wind-dispersed spores. Large quantities of viable spores may occur within the soil profile (UCPE unpubl.), and a persistent spore bank is suspected. There is a limited capacity for lateral spread through branching of the rhizome; ?SB III.
Sporangia Aggregated into sori on the underside of the fronds.
Germinule 44×30 µm, spore, black, explosively dispersed from a many-spored sporangium.
Germination t_{50} 3 days; L > D.
Biological flora None.
Geographical distribution The British Isles (100% of British vice-counties), *ABIF* page 77, *AF* pages 82–83; most of Europe except the extreme N and S (85% of European territories); temperate Asia, N America.

Habitats *D.f-m.* is restricted to two very different habitats, woodland and skeletal habitats, the latter mainly cliffs and walls, but also quarry spoil, outcrops and screes, river banks and some cinder tips. In some skeletal habitats, especially walls, *D.f-m.* often fails to reach reproductive maturity.
Gregariousness Potentially a stand-forming dominant.
Altitude Frequency increasing up to 300 m, thereafter decreasing. [Extends to over 610 m (Page 1982).]
Slope Recorded from all slopes.
Aspect More frequent on N-facing slopes both in shaded and unshaded sites.
Hydrology Recorded from moist ground at the edge of mire, but not present in strongly waterlogged conditions.
Soil pH Found over the full range of values from pH 3.0 to 8.0. In woodlands mainly in the range pH 3.5–5.0; in skeletal habitats largely above pH 6.5.
Bare soil Mainly found in habitats with little exposed soil. In woodland most of the soil is covered by tree leaf litter whilst in skeletal habitats most of the substratum consists of bare rock.

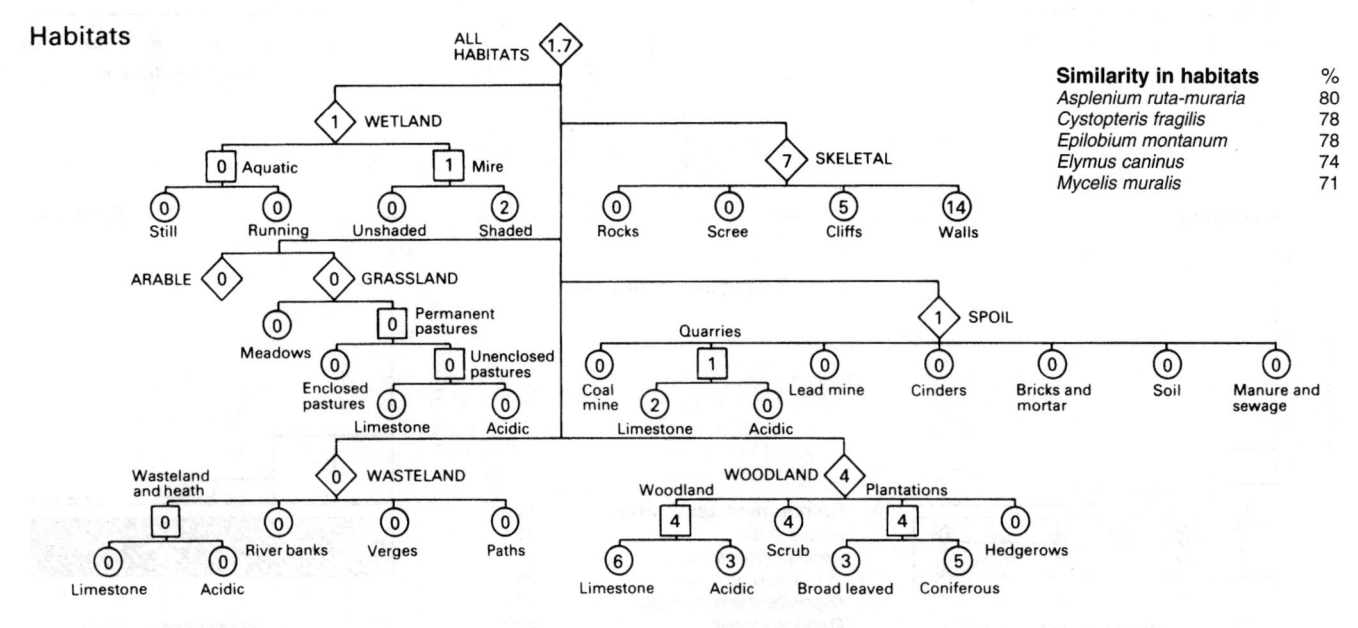

Habitats

Similarity in habitats	%
Asplenium ruta-muraria	80
Cystopteris fragilis	78
Epilobium montanum	78
Elymus caninus	74
Mycelis muralis	71

△ **ordination** Most frequently associated with vegetation of undisturbed and relatively fertile habitats. Absent from heavily disturbed sites.
Associated floristic diversity Typically low.
Synopsis In common with *D. dilatata*, with which it frequently occurs, *D.f-m.* has the potential for functioning as a woodland-floor dominant. However, the species is less restricted to soils of low pH than is *D.d.*, and is frequent on calcareous rocks and in limestone woodland. The fronds of *D.f-m.* are less persistent in winter, and expand slightly later in spring, than those of *D.d.* (Page 1982). A further ecological difference from *D.d.* may be recognized from the fact that *D.f-m.* occurs primarily on drier, well-drained soils and is completely absent from wetlands. In addition, *D.f-m.* is more common than *D.d.* in lightly or even unshaded habitats. *D.f-m.* is better able than *D.d.* to colonize mortared walls which are the major habitat for the species in some urban areas. *D.f-m.* is grown in rockeries, and some populations in urban areas are likely to be naturalized genotypes of cultivated stock. On free-standing walls plants often do not grow large enough to sporulate, either because the crevice in which *D.f-m.* is rooted is too small to allow a large plant to establish or because development is interrupted by pointing. However, the species produces spores moderately frequently in other rocky habitats such as screes, cliffs and outcrops. Studies in Russian woodland (Pogorelova & Rabotnov 1978) suggest that plants of *D.f-m.* may live for at least 30–40 years and do not sporulate until 6–7 years old. Willmot (1985) suggests that establishment in Derbyshire woodland is comparatively uncommon but that, once established, *D.f-m.* is long-lived. This contrasts with the more ruderal population dynamics of *D.d.*. The species declines in abundance at high altitudes. In upland woodlands on acidic strata *D.f-m.* is largely restricted to nutrient-enriched areas of higher pH close to the streams in the valley bottoms. In these woodlands *D.f-m.* is replaced on more acidic soils by *D. affinis* (Lowe) Fraser-Jenkins (*D. borreri* Newman), a species with a more oceanic distribution. In unshaded acidic sites at high altitudes replacement of *D.f-m.* by *D.d.* is almost complete. It is not known whether the ecological

Altitude

Slope

Aspect

Soil pH

Associated floristic diversity

Hydrology

Triangular ordination

Bare soil

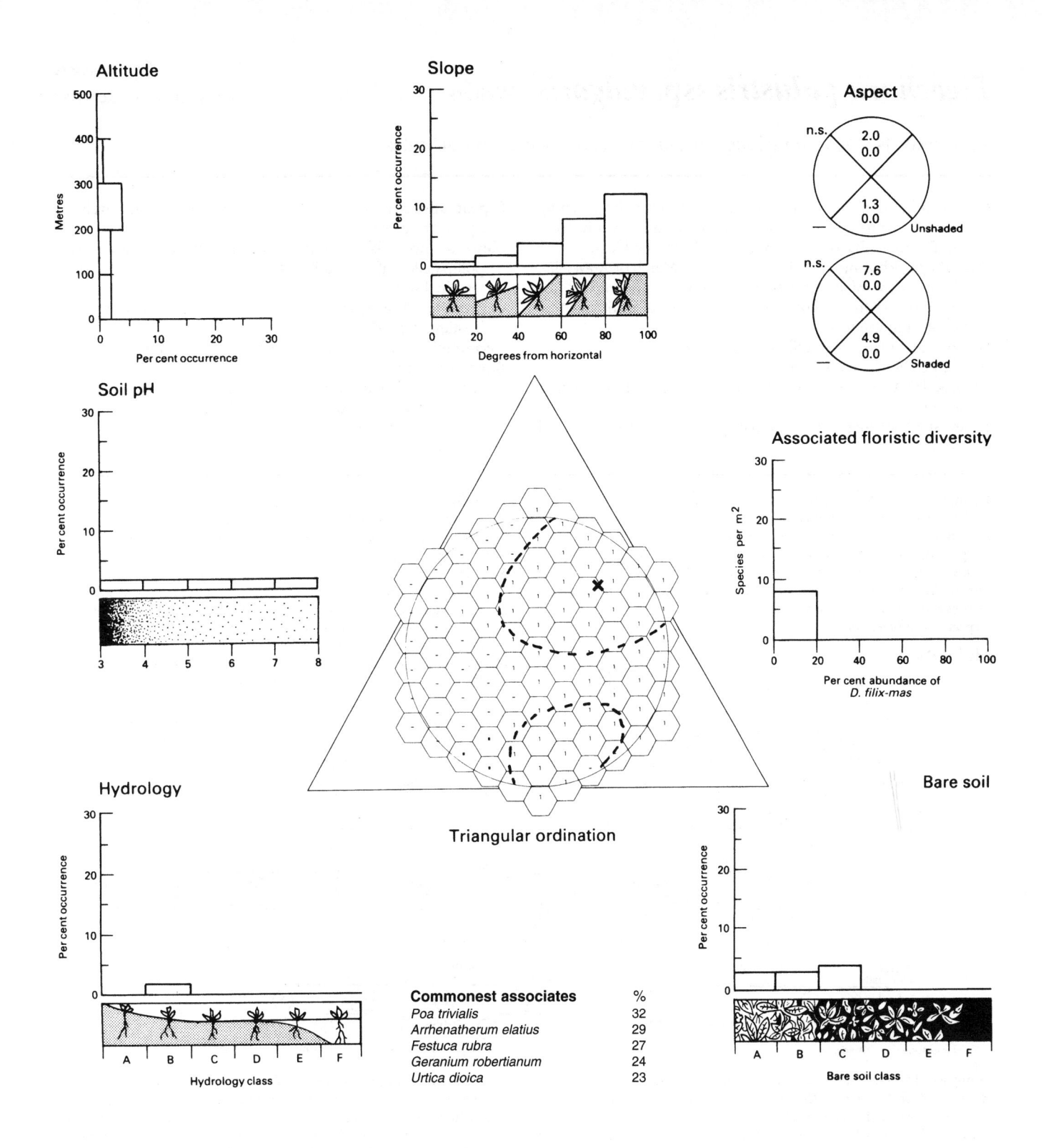

Commonest associates	%
Poa trivialis	32
Arrhenatherum elatius	29
Festuca rubra	27
Geranium robertianum	24
Urtica dioica	23

differences between *D.f-m.* and *D.d.* arise from characteristics of the sporophyte or relate to differences in requirements for prothallial establishment. It may be significant that whereas *D.a.* is apomictic *D.f-m.* is sexual (Schneller 1975); more research is required to examine the effect of this difference during the establishment of new colonies. There is often considerable variation, presumably genetic in origin, between populations (Page 1982). *D.f-m.* hybridizes frequently with *D.a.* and rarely with *D.d.* and two other species.

Current trends Considered by Page (1982) to have been decreasing since Roman times through the destruction of woodland habitats and to be becoming still rarer for the same reason. Page suggests that much of the genetic variation still present in the species is attributable to the former existence of more extensive populations; this characteristic is likely to be lost as *D.f-m.* becomes increasingly restricted to plantations and rocky habitats.

Eleocharis palustris ssp. *vulgaris* Walters

Cyperaceae
Common Spike-rush

Field records may also include ssp. *palustris* (L.) Roemer and Schultes

---

Life-form Polycarpic perennial helophyte with far-creeping rhizome. Stems green with leaves reduced to brownish, basal sheaths; roots shallow, usually lacking VA mycorrhizas.
Phenology Rhizome growth commences in March or April (*Biol Fl*) and some of the shoots formed towards the end of this period flower from May to July. Shoots persist until killed by the late autumn frosts.
Height Flowering and sterile shoots to 600 mm.
Seedling RGR No data. SLA 10.1. DMC 32.6.
Nuclear DNA amount 11.1 pg; $2n = 38*$ (*BC*); base number obscure.
Established strategy Intermediate between competitor and C-S-R.

Regenerative strategies V, ?B_s. Regenerates both vegetatively and by seed (see 'Synopsis').
Flowers Brownish, hermaphrodite, protogynous, wind-pollinated, self-compatible (Stace 1975); 15-30 in a short, terminal spike.
Germinule 0.96 mg, 2.4 × 1.5 mm, nut dispersed with 4 short bristles attached.
Germination Epigeal; dormancy broken by chilling.
Biological flora Walters (1949).
Geographical distribution For the aggregate species, the British Isles (100% of British vice-counties), *ABIF* page 703; Europe (97% of European territories); Asia, N America and N Africa and India.

Habitats Exclusively a wetland species. Most common in unshaded mire or shallow water at the margin of lakes, ponds and ditches. Less frequent in soligenous mire and beside streams. [Occurs in dune slacks (*Biol Fl*).]
Gregariousness Often patch-forming.
Altitude Recorded up to 400 m. [Up to 460 m.]
Slope As with most marsh and semi-aquatic species, *E.p.* is largely restricted to flat or gently sloping ground.
Aspect Insufficient data.

Hydrology Most frequent in topogenous mire marginal to open water with a narrowly fluctuating water table. Often also extending into shallow water, but rarely observed from water depths >300 mm. Found occasionally in soligenous mire and sites that flood only during winter.
Soil pH Mainly confined to soils in the pH range 5.0-7.5.
Bare soil Associated with a wide range in amount of exposed soil.
Δ ordination Distribution centred on productive vegetation subject to moderate intensity of disturbance, but sometimes persisting in situations which are only infrequently disturbed.
Associated floristic diversity Usually low, but in soligenous mire *E.p.* has been observed in comparatively species-rich habitats.

Habitats

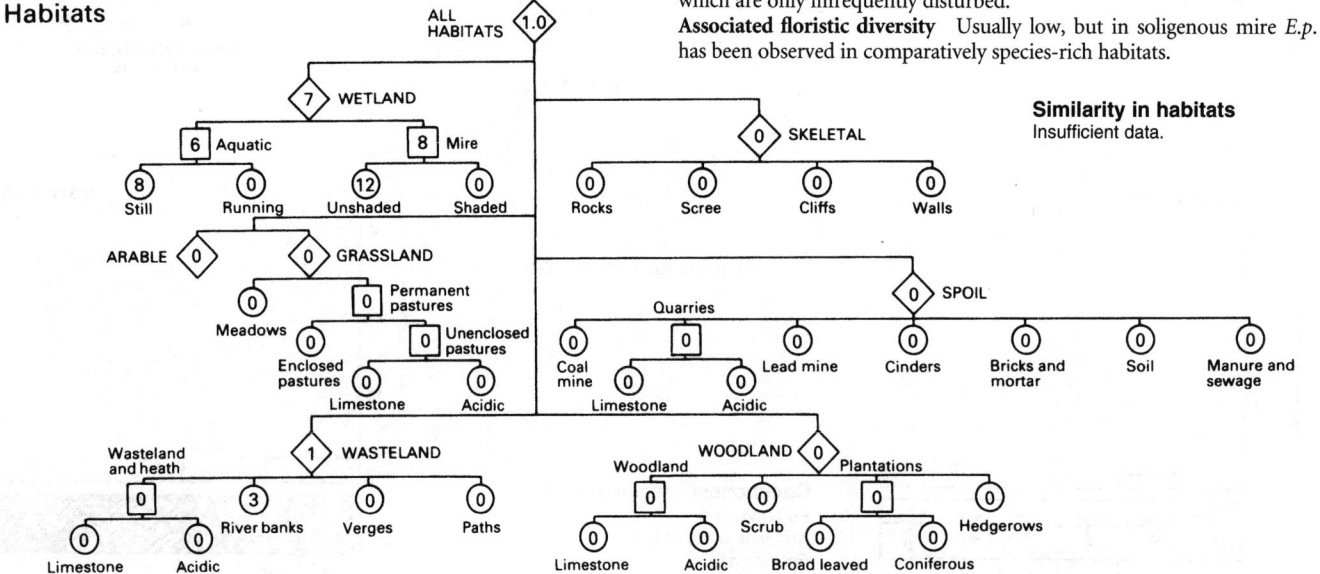

Similarity in habitats
Insufficient data.

Synopsis In common with all other common angiosperms of the British flora, in which the stem is the only major photosynthetic organ, *E.p.* is confined to wetland. *E.p.* is much shorter than many other water-margin plants, and tends to be restricted to relatively open situations where the growth of more-robust species is restricted by factors such as disturbance by water currents and a degree of soil infertility. Its tolerance of these factors is particularly marked at the margin of Scottish lochs, where *E.p.* is characteristic of shallow water on exposed sandy shores (Spence 1964). According to habitat and phenotype, the stems vary between 1 and 8 mm in diameter (*Biol Fl*), and their height is also very variable. *E.p.* does not extend far into drier habitats where the growth form tends to become tufted and less vigorous (*Biol Fl*). Thus, the species is generally absent from small ponds with a large fluctuation between summer and winter water table. *E.p.* may persist in a non-flowering condition in lightly grazed habitats. Vegetative growth by means of rhizomes gives rise to extensive stands. These may break into daughter plants either through the breakdown of old rhizomes during the second year (*Biol Fl*) or through natural disturbance caused by, for example, flood damage. Detached plantlets can re-root to form new colonies. Seedlings are rarely seen, and regeneration by seed appears to play little part in the maintenance of populations, but is presumably involved in the initial colonization of sites. The nut of *E.p.* has barbed bristles which appear to facilitate dispersal by animals. *E.p.* is eaten by water-fowl (Ridley 1930), and seed discharged in faeces may be responsible for long-distance dispersal. No seed bank has been detected during the limited studies so far undertaken. Field records may include ssp. *palustris* with 20-40 smaller seeds in a terminal spike, $2n = 16$ (*BC*) and a restricted British distribution (*CSABF* p. 146). This ssp. is self-incompatible. Ssp. *vulgaris* may have arisen through hybridization between ancestors similar to *E. palustris* ssp. *palustris* and *E. uniglumis* (Link) Schultes (Strandhede 1965). The two ssp. of *E.p.* form hybrids, and ssp. *v.* also hybridizes with *E. uniglumis* (Link) Schultes.
Current trends Not as regular a colonist of new artificial wetland sites as *Juncus* spp. and *Typha latifolia*. Because of its small stature, *E.p.* is vulnerable to dominance by larger, fast-growing clonal species in habitats subject to eutrophication. Probably decreasing.

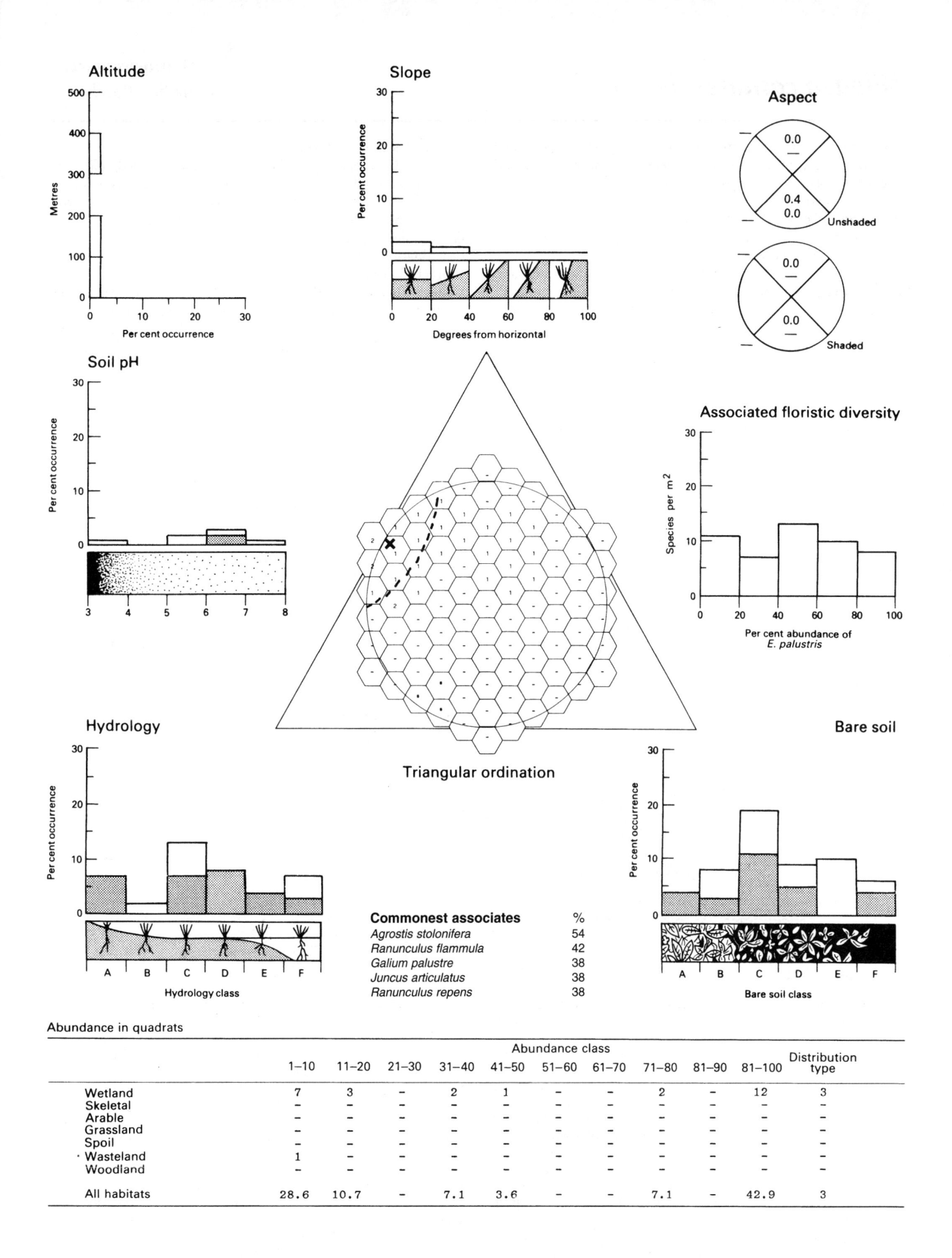

Altitude

Slope

Aspect

Unshaded

Shaded

Soil pH

Associated floristic diversity

Per cent abundance of *E. palustris*

Hydrology

Triangular ordination

Commonest associates	%
Agrostis stolonifera	54
Ranunculus flammula	42
Galium palustre	38
Juncus articulatus	38
Ranunculus repens	38

Bare soil

Abundance in quadrats

	Abundance class										Distribution type
	1–10	11–20	21–30	31–40	41–50	51–60	61–70	71–80	81–90	81–100	
Wetland	7	3	–	2	1	–	–	2	–	12	3
Skeletal	–	–	–	–	–	–	–	–	–	–	–
Arable	–	–	–	–	–	–	–	–	–	–	–
Grassland	–	–	–	–	–	–	–	–	–	–	–
Spoil	–	–	–	–	–	–	–	–	–	–	–
Wasteland	1	–	–	–	–	–	–	–	–	–	–
Woodland	–	–	–	–	–	–	–	–	–	–	–
All habitats	28.6	10.7	–	7.1	3.6	–	–	7.1	–	42.9	3

Elodea canadensis Michx.

Hydrocharitaceae
Canadian Pondweed

Life-form Submerged, mat-forming, polycarpic perennial hydrophyte with an extensive system of prostrate stems and erect shoots; leaves lanceolate, <50 mm^2; roots unbranched, non-mycorrhizal.

Phenology Flowers June to September but no fruits formed. Shoots die back in autumn in association with the production of overwintering dormant buds.

Height Leafy stems submerged; flowers floating on water surface.

Seedling RGR No data. SLA 25.8. DMC 23.5.

Nuclear DNA amount 10.0 pg; $2n = c.$ 24*, 48 (Federov 1974, Simpson 1986); $2x^*$, $4x$.

Established strategy Competitive-ruderal.

Regenerative strategies V. Forms extensive clonal patches by means of prostrate stems. Detached shoots, either as a result of disturbance or during autumn die-back, readily form new colonies. No seed formed.

Flowers Greenish-purple, dioecious, water-pollinated; solitary in a tubular spathe in the leaf axil. As a result of extension of the axis, female flowers float at surface of water. Male flowers recorded only once from Britain (Sculthorpe 1967).

Germinule None.

Germination Not relevant.

Biological flora None.

Geographical distribution Native of N America and naturalized in much of Britain, although absent from the extreme N (82% of British vice-counties), ABIF page 670; much of Europe except for the Arctic and some islands (67% of European territories). Also established as a weed in parts of S America, Asia and Africa (Holm et al. 1979).

Habitats An aquatic species found in lakes, ponds, and canals, and also in slow-flowing rivers and streams. The plant may become exposed if the water level drops, but E.c. is never a permanent component of mire vegetation.

Gregariousness Patch-forming.

Altitude More common in lowland areas, where suitable habitats are more frequent, extending to 220 m.

Slope No data.

Aspect No data.

Hydrology Most frequent and abundant in deeper waters. Extending to depths of 750 mm.

Soil pH Always in a substrate of pH 4.5, and most frequent around pH 7.0.

Bare soil Occasionally stranded on exposed mud after a drop in the water level, otherwise always submerged.

△ **ordination** No data.

Associated floristic diversity Usually low, as with most aquatic species.

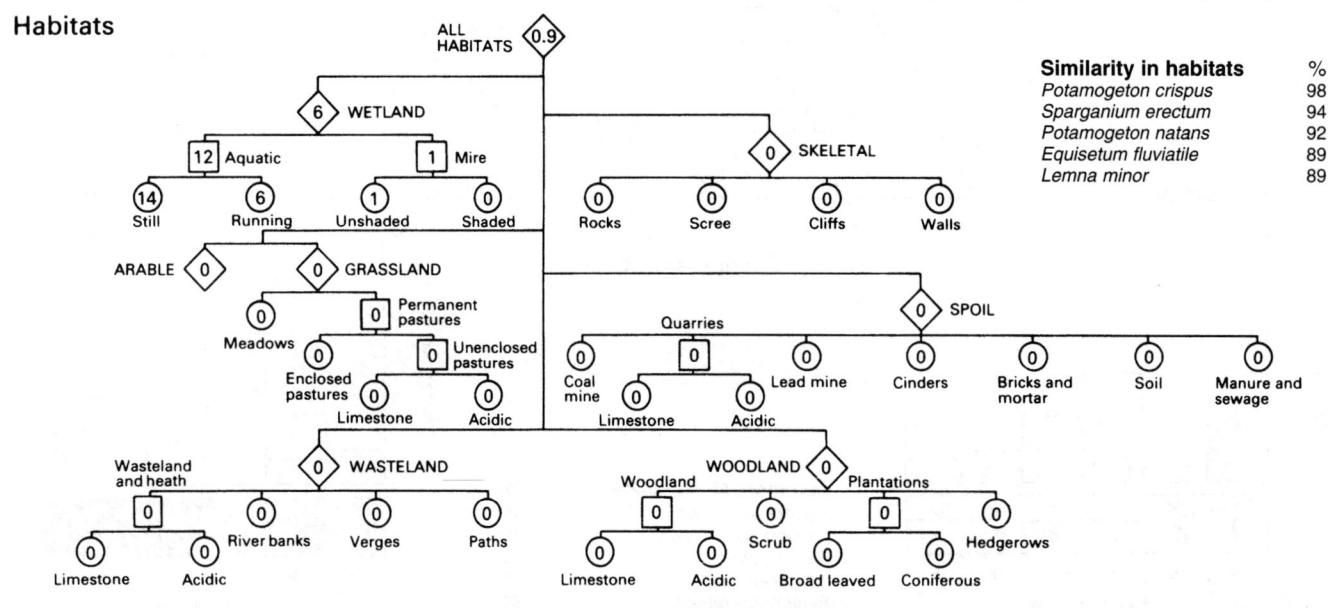

Habitats

Similarity in habitats	%
Potamogeton crispus	98
Sparganium erectum	94
Potamogeton natans	92
Equisetum fluviatile	89
Lemna minor	89

Synopsis A native of N America, E.c. is now naturalized and widespread in the British Isles, where it was first recorded in the early 19th century (Simpson 1984). E.c. is a submerged aquatic of still or slow-moving water, and the plant is often rooted in silt. In some habitats E.c. is the dominant, or even the only, hydrophyte present; in others it forms a lower layer beneath the canopy of larger aquatic macrophytes such as *Potamogeton natans* and *Sparganium erectum*. However, the species is usually suppressed if the water has a floating carpet of *Lemna minor* or if there is a high level of water turbulence (Simpson 1986). The pattern of colonization is described thus by Simpson (1984): (a) build-up of the initial colony, often to the exclusion of other aquatics (perhaps over a period of 3-4 years), (b) attainment of an equilibrium state (next 3-10 years), (c) decline (next 7-15 years), (d) the presence of a small residual population, or complete disappearance, of E.c., with possibly a return later. E.c. became a serious weed following its naturalization, choking waterways before declining to its present less-abundant (but still common) level. It has been suggested that the decline may be associated with some form of nutrient depletion (Sculthorpe 1967), but the exact cause is unknown and vulnerability of the genetically uniform populations to some form of pathogen attack seems worthy of consideration. The plant consists of an extensive system of prostrate stems and erect leafy shoots. The leaves are only two cells in thickness and do not bear stomata (Simpson 1986). E.c. can utilize both dissolved carbon dioxide and the bicarbonate ion for photosynthesis (Sculthorpe 1967). The plant lacks vessels. The roots are adventitious and unbranched, and usually contribute less than 3% of the dry weight of the plant (Sculthorpe 1967). Nevertheless, inorganic nutrients are absorbed through the roots and are translocated to the shoot (Denny 1980). The roots are characterized by large polyploid root hairs (Dosier & Riopel 1978). The species does not exhibit aquatic acid metabolism (Keeley & Morton 1982) to any appreciable extent (see account for *Ranunculus flammula*). In a Canadian study (Haag 1979) shoots were found to lack dormancy in winter and although the light-compensation point was low, a

Altitude

Slope

Aspect

Unshaded: 0.0 / 0.0
Shaded: 0.0 / 0.0

Soil pH

Commonest associates %
Lemna minor 54
Lemna trisulca 35
Potamogeton natans 35
Ceratophyllum demersum 27
Potamogeton crispus 20

Associated floristic diversity

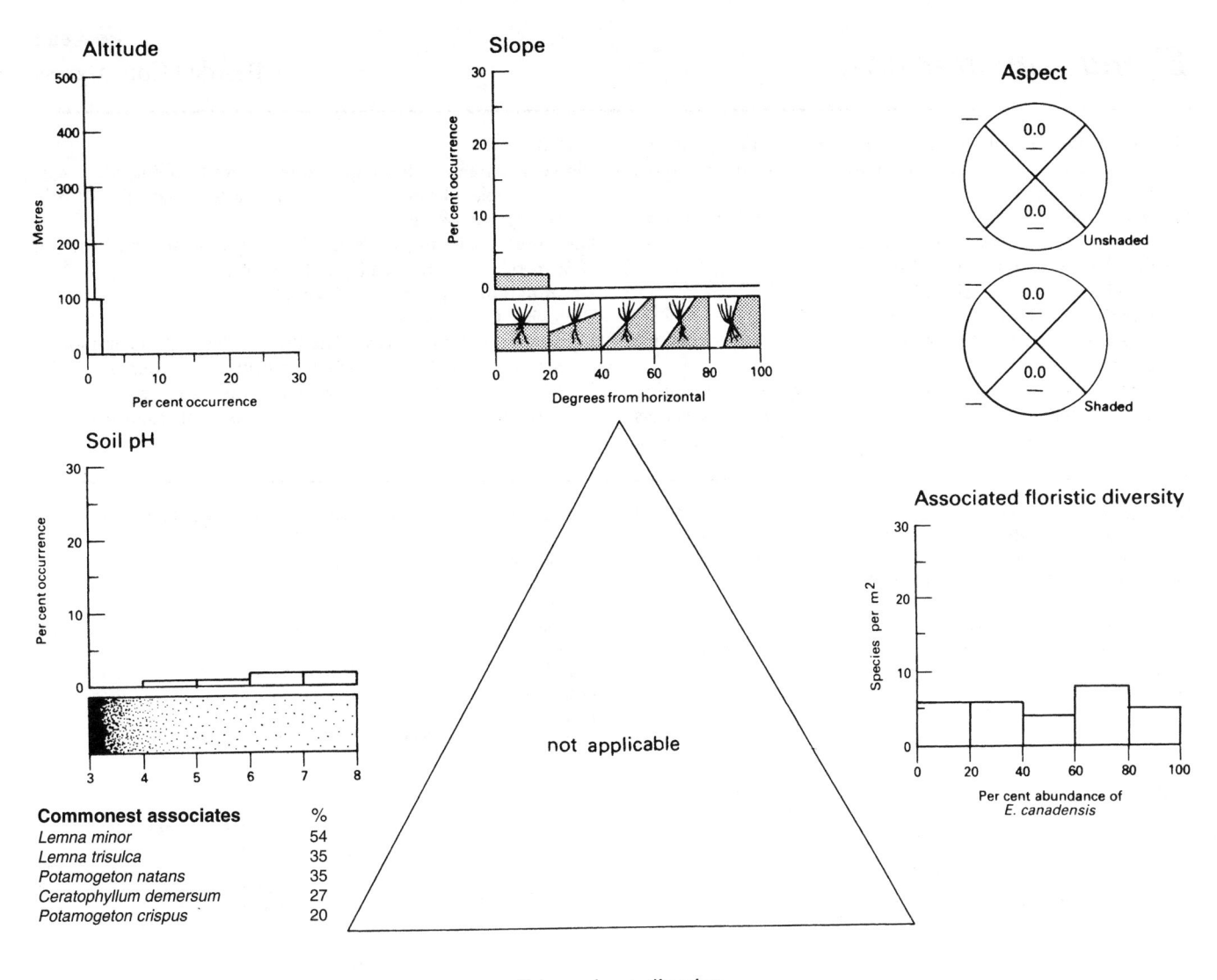

Triangular ordination — not applicable

Abundance in quadrats

	1–10	11–20	21–30	31–40	41–50	51–60	61–70	71–80	81–90	81–100	Distribution type
All habitats	24.0	–	20.0	8.0	8.0	8.0	–	4.0	12.0	16.0	3

thick covering of ice in winter resulted in some shoot death (Haag 1979). Re-establishment in iced-up areas occurred by means of rapid growth in summer, but the probability of survival, and the longevity, of such populations was low. The mechanism of the apparent restriction of E.c. to lowland sites in Britain may have a similar climatic component. Regeneration is entirely vegetative. Patches increase in size as a result of shoot extension *in situ* and shoot pieces, carried by water or attached to wild-fowl (Ridley 1930), may form new colonies. Shoots may continue growing even when not rooted (Salisbury 1961). Plants overwinter as dormant buds which separate as the old stems decay. These dormant buds, which normally sink, may be transported to new sites. *E.c.* is shy-flowering, and no populations in very deep or in flowing water have been observed to flower. The production of flowers has little ecological significance in any event, since the species is dioecious and the male plant has been reported only once in Britain (Sculthorpe 1967). *E.c.* has a wide latitudinal range both as a native species and as an established alien (Sculthorpe 1967). The species extends from peaty waters to calcareous sites and, judging by its associated species, the plant is found in both mesotrophic and eutrophic waters, an observation which is consistent with the data of Spence (1964) and Simpson (1986). *E. nuttallii* (Planchon) St John, another N American species first recorded in 1966 (Simpson 1984) is at an earlier stage in colonizing Britain, and is still spreading rapidly within the Sheffield region. As in *E.c.*, populations of *E.n.* appear to consist exclusively of female plants. *E.n.* ($2n = c.\ 48^*$, $4x^*$ (Simpson 1986)) differs from *E.c.* in a number of ways. First, it is usually more robust, and shoots often extend to the water surface. Second, *E.n.* is more plastic phenotypically; growth forms in flowing water differ from those in still water, and the species also adopts an unusual growth form in deep lakes (Simpson 1984). Third, *E.n.* flourishes in nutrient-rich situations and is not recorded from peaty waters. The two species often occur together, with *E.n.* usually assuming the dominant role. Both locally and nationally, *E.n.* is replacing *E.c.* in many sites (Simpson 1984). The continued spread of *E.n.* is likely to be encouraged by eutrophication.

Current trends Probably decreasing as a result of the destruction of aquatic habitats and the rapid spread of *E.n.* (see above).

Elymus caninus (L.) L.

Poaceae
Bearded Couch-grass

Life-form Tufted, polycarpic perennial protohemicryptophyte. Shoots erect; leaves linear, sometimes hairy, <4000 mm^2; roots with VA mycorrhizas.
Phenology Flowers in July and sets seed from September to October. Shoots are replaced in autumn.
Height Foliage and flowers to 1 m.
Seedling RGR No data. SLA 27.8. DMC 30.9.
Nuclear DNA amount 17.1 pg; $2n = 28$ (*FE5*); $4x$.
Established strategy Intermediate between stress-tolerant competitor and C-S-R.
Regenerative strategies S. By seed, mainly in early spring (Roberts 1986b). Forms loose tufts but growth of lateral shoots insufficient to constitute a strategy of vegetation regeneration; ?SB II.
Flowers Green, hermaphrodite, wind-pollinated, self-compatible (Beddows 1931); >50 in a terminal spike (2-6 in each sessile spikelet).
Germinule 4.04 mg, 6.0 × 1.9 mm, caryopsis dispersed with an awned lemma attached (10.5 × 2.0 mm).
Germination Epigeal; t_{50} 6 days; L > D.
Biological flora None.
Geographical distribution The British Isles, but uncommon in N Scotland and Ireland (81% of British vice-counties), *ABIF* page 793; most of Europe except the extreme S (85% of European territories); temperate Asia; introduced in N America.

Habitats More typically associated with wood and scrub margins. Found also in a range of shaded habitats which includes road verges, cliffs and walls. Frequently observed on tree-lined river banks and occasionally in unshaded habitats either adjacent to woodland or hedgerows, or as a relict of former woodland.
Gregariousness Intermediate.
Altitude Not recorded above 300 m. [Normally up to 410 m, but a montane variant extends to 610 m.].
Slope Wide-ranging.
Aspect Most frequently recorded on S-facing slopes both here and in a subsequent survey.
Hydrology Absent from wetlands.
Soil pH Most commonly found on soils of pH <6.0 and only rarely observed at pH <5.0.
Bare soil Wide-ranging, but most abundant in sites with relatively high exposures.
△ **ordination** Displays a relatively wide strategic range, but with distribution centred on relatively productive and little disturbed vegetation types.
Associated floristic diversity Typically low to intermediate.

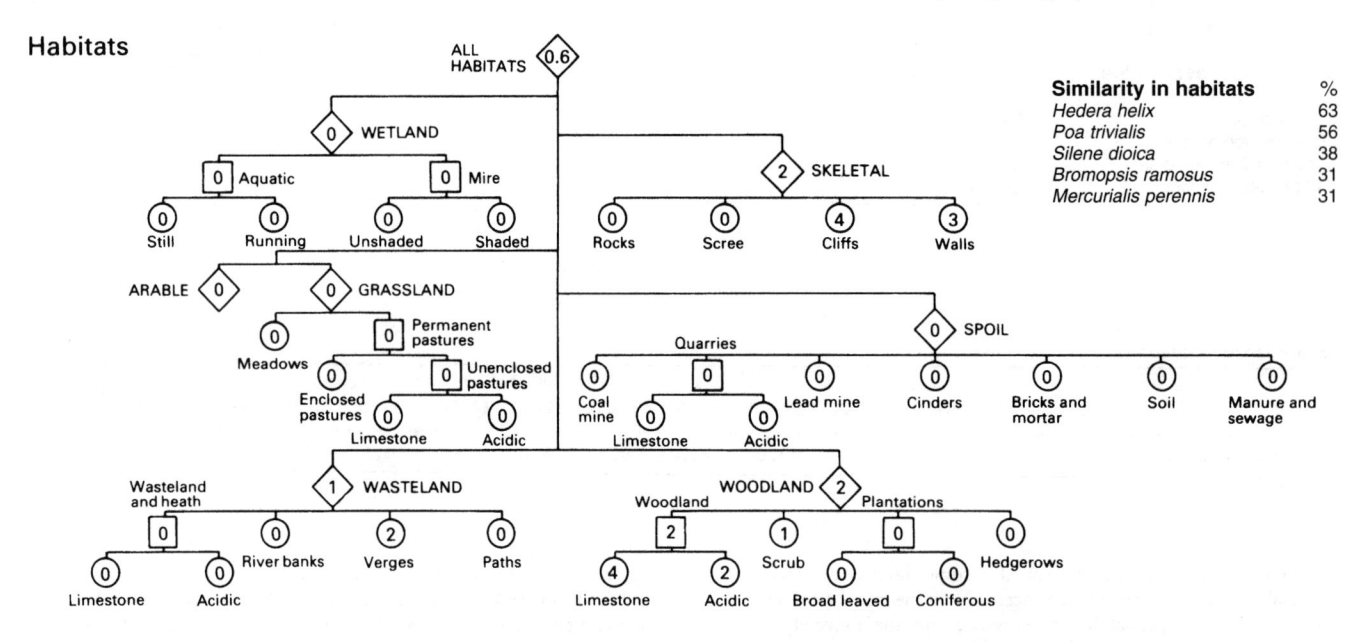

Habitats

Similarity in habitats	%
Hedera helix	63
Poa trivialis	56
Silene dioica	38
Bromopsis ramosus	31
Mercurialis perennis	31

Synopsis *E.c.* is a tufted grass most typical of base-rich woodland margins in sites where the vigour of potential dominants is reduced not only by light-to-moderate shade but also by infertility (as in some woodland sites) or by disturbance (as on river banks and roadsides), or more usually by a combination of the two. The tendency to occur on S-facing slopes is a reflection of an association with more-open woodland sites. The annual increase in shoot biomass continues into the shaded phase, resulting from expansion of the tree canopy. The species produces new shoots in autumn and is winter green. *E.c.* is characteristic of freely-drained soils as the frequent occurrences on rock ledges and the absence from mires testify. *E.c.* is non-rhizomatous and largely dependent upon seed for regeneration (see Tripathi & Harper 1973). The species is virtually absent from sites with large quantities of tree litter, and the possibility that regeneration by seed is poor in the presence of tree litter requires investigation. Some seed germinates directly in autumn and the remainder in the following spring (Dorph-Petersen 1924). No persistent seed bank has been reported, although a small number of seeds germinated after 2 years of burial in a pot experiment (Roberts 1986b). It is not known whether the caryopsis floats, but the species extends along many kilometres of tree-lined river bank within the upland portion of the Sheffield region. This is the habitat which *E.c.* occupies most consistently here, and in which the species reaches its maximum abundance in N England. A number of variants have been described (*FE5*) and distinctive populations recorded from montane habitats in Scotland were formerly separated as *Agropyron donianum* F. B. White (*CTW*).
Current trends *E.c.* does not appear to be a very effective colonist of new artificial habitats, including secondary woodland, and is probably decreasing.

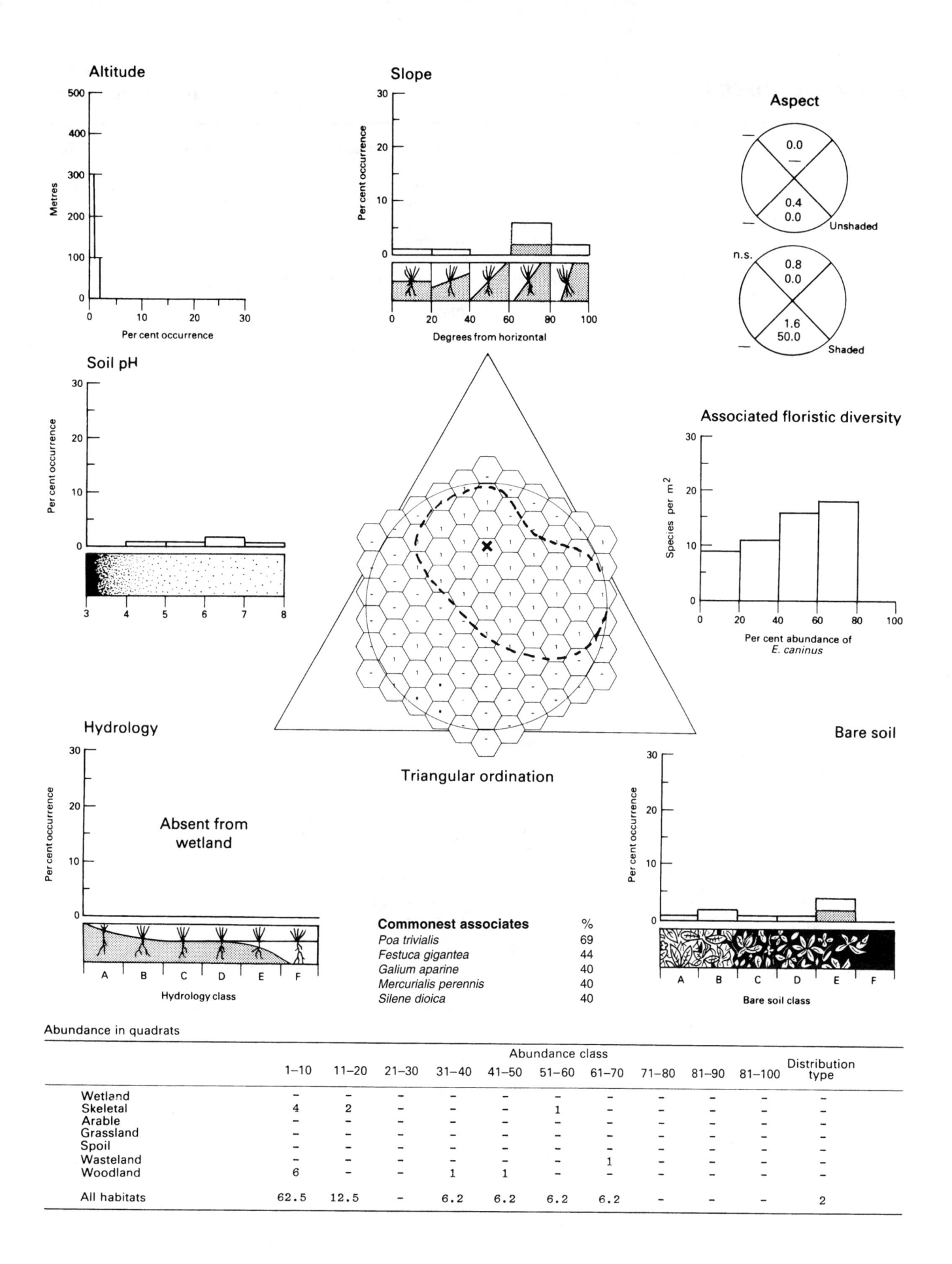

Altitude

Slope

Aspect

Unshaded: — 0.0 / — — / 0.4 0.0
Shaded: n.s. 0.8 / — 0.0 / — 1.6 50.0

Soil pH

Associated floristic diversity

Hydrology

Absent from wetland

Triangular ordination

Bare soil

Commonest associates %
Poa trivialis 69
Festuca gigantea 44
Galium aparine 40
Mercurialis perennis 40
Silene dioica 40

Abundance in quadrats

	1–10	11–20	21–30	31–40	41–50	51–60	61–70	71–80	81–90	81–100	Distribution type
Wetland	–	–	–	–	–	–	–	–	–	–	–
Skeletal	4	2	–	–	–	1	–	–	–	–	–
Arable	–	–	–	–	–	–	–	–	–	–	–
Grassland	–	–	–	–	–	–	–	–	–	–	–
Spoil	–	–	–	–	–	–	–	–	–	–	–
Wasteland	–	–	–	–	–	1	–	–	–	–	–
Woodland	6	–	–	1	1	–	–	–	–	–	–
All habitats	62.5	12.5	–	6.2	6.2	6.2	6.2	–	–	–	2

Elytrigia repens (L.) Desv. ex Nevski

Poaceae
Couch-grass, Scutch, Twitch

Life-form Rhizomatous, polycarpic perennial protohemicryptophyte. Shoots erect; leaves linear, hairy on upper surface, sometimes >2000 mm^2; roots often deep, with VA mycorrhizas.

Phenology In relatively undisturbed sites young shoots produced in autumn overwinter. Plants flower from June to September and seed set from August to October. Mature shoots senesce in autumn. In other habitats phenology varies according to the timing of disturbance.

Height Foliage to 600 mm; flowering shoots to 1 m.

Seedling RGR 1.0–1.4 week^{-1}. SLA 24.6. DMC 26.7.

Nuclear DNA amount 24.2 pg; $2n = 42^*$ (Bennett *et al.* 1982); $6x^*$.

Established strategy Intermediate between competitor and competitive-ruderal.

Regenerative strategies (V), ?B$_s$. Mainly by vegetative means (see 'Synopsis'); SB III.

Flowers Green, hermaphrodite, largely self-sterile, wind-pollinated (*Biol Fl* (a)); often *c.* 100 in a terminal spike (3–8 per spikelet).

Germinule 2.02 mg, 5.0 × 1.4 mm, caryopsis dispersed with an attached, sometimes awned lemma (12.9 × 1.8 mm).

Germination Epigeal; shows a requirement for alternating temperatures (*Biol Fl* (a)); L < D.

Biological flora (a) Palmer and Sagar (1963), (b) Werner and Rioux (1977) (both as *Agropyron repens* (L.) Beauv.).

Geographical distribution The British Isles, particularly in lowland areas (100% in British vice-counties), *ABIF* page 794; Europe (92% of European territories); most temperate regions.

Habitats Abundant in a wide range of fertile, disturbed habitats particularly arable land where it is often a troublesome weed, on road verges, in hedgerows and on various types of spoil heaps. Also frequent in meadows, pastures and various categories of wasteland. Infrequent in marshy, skeletal and shaded sites. Absent from aquatic habitats. [Occurs at the coast, but in sites closest to the sea is generally replaced by other species of *Elymus*.]

Gregariousness Intermediate to stand-forming.

Altitude Over a wide range of altitudes up to 400 m, but most frequent and abundant in the lowland arable region. [Up to 450 m.]

Slope Most suitable habitats associated with level or gently sloping ground.

Aspect No obvious trends.

Hydrology Recorded infrequently beside lakes and ponds, but not in situations which are permanently waterlogged.

Soil pH Widely distributed, but most frequent at pH >7.0.

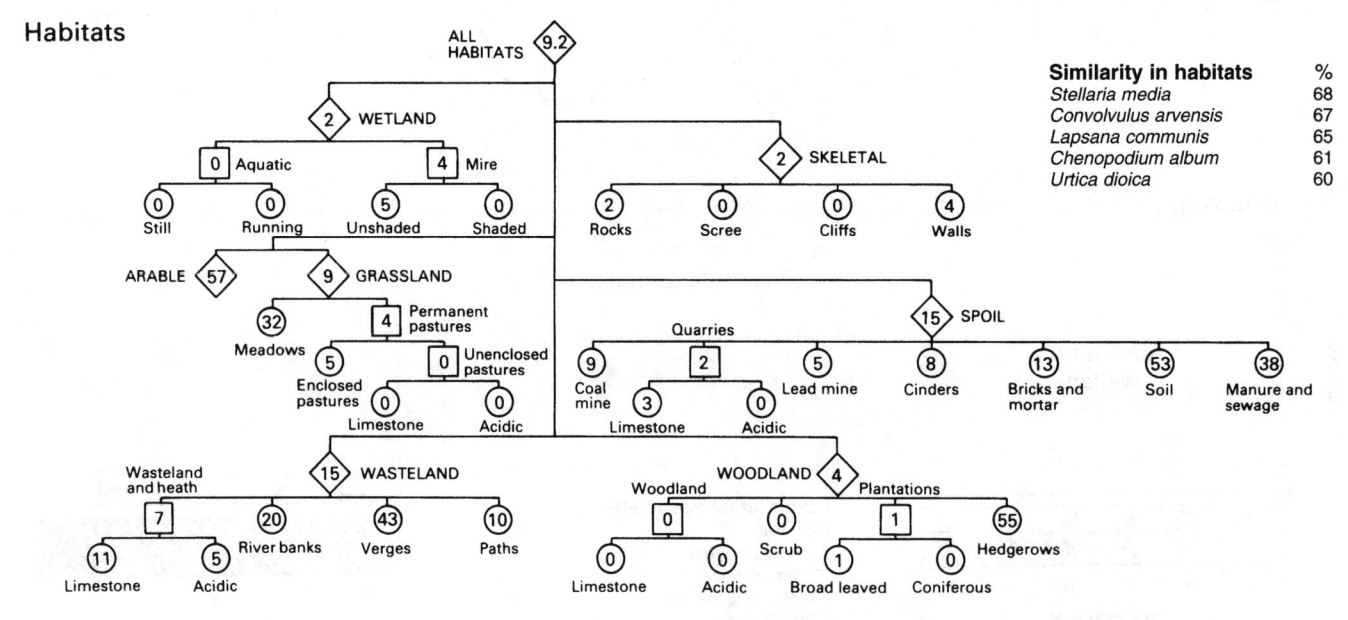

Habitats

Similarity in habitats	%
Stellaria media	68
Convolvulus arvensis	67
Lapsana communis	65
Chenopodium album	61
Urtica dioica	60

Bare soil Most frequent and abundant in habitats with high exposure of bare soil, but can also persist in intermittently disturbed areas with little bare ground.

△ **ordination** Distribution centred on vegetation of disturbed fertile habitats but *E.r.* is wide-ranging and absent only from very unproductive situations.

Associated floristic diversity Intermediate, declining with increasing abundance of *E.r.*

Synopsis *E.r.* is a rhizomatous grass associated with a wide range of fertile, disturbed habitats. *E.r.*, classified as one of the world's worst weeds, is perhaps the most serious perennial weed of the cooler regions of the N temperate zone (Holm *et al.* 1977). *E.r.* is recorded from 32 crops in more than 40 countries, and in Britain is most important as a weed of cereals. The success of *E.r.* on arable land arises particularly from the following characteristics. (a) An extensive system of rhizomes which in undisturbed sites give rise to large stands with a dense tall leaf canopy and which, after disturbance by ploughing, cause extensive regeneration from rhizome fragments. In some places rhizomes may reach 400 mm in depth (*Biol Fl* (a)) and are thus difficult to eradicate. Rhizome fragments may also be transplanted to new sites as a result of human activities. (b) *E.r.*, which has a high nuclear DNA amount and contains abundant fructans in winter (Smith 1967), is a 'cool season' grass; tillering and photosynthesis are most active during spring and autumn (*Biol Fl*(b)). This phenology enables *E.r.* to compete with the young cereal crop and to exploit the site after the crop has been harvested (*Biol Fl* (b)). Furthermore *E.r.* is a 'luxury consumer' of major nutrients (*Biol Fl* (b)) although, in common with most grasses, the species has a low Ca content in the leaves (*SIEF*). (c) *E.r.* is unaffected by many types of herbicides used to control dicotyledonous weeds (*Biol Fl* (b)). Its control is hindered by growing cereals continually on one site rather than in rotation. (d) Its capacity to exploit many field banks and road verges enables *E.r.* to re-invade arable land following effective weed control. Seed production is often poor, and regeneration by seed generally

Altitude
Slope
Aspect

n.s.	8.7 / 40.9	n.s.
n.s.	5.8 / 38.5	Unshaded

n.s.	3.8 / 60.0	n.s.
$P < 0.05$	3.3 / 75.0	Shaded

Soil pH
Associated floristic diversity
Hydrology
Triangular ordination
Bare soil

Commonest associates %
Poa trivialis 56
Poa pratensis 43
Dactylis glomerata 42
Urtica dioica 37
Festuca rubra 36

Abundance in quadrats

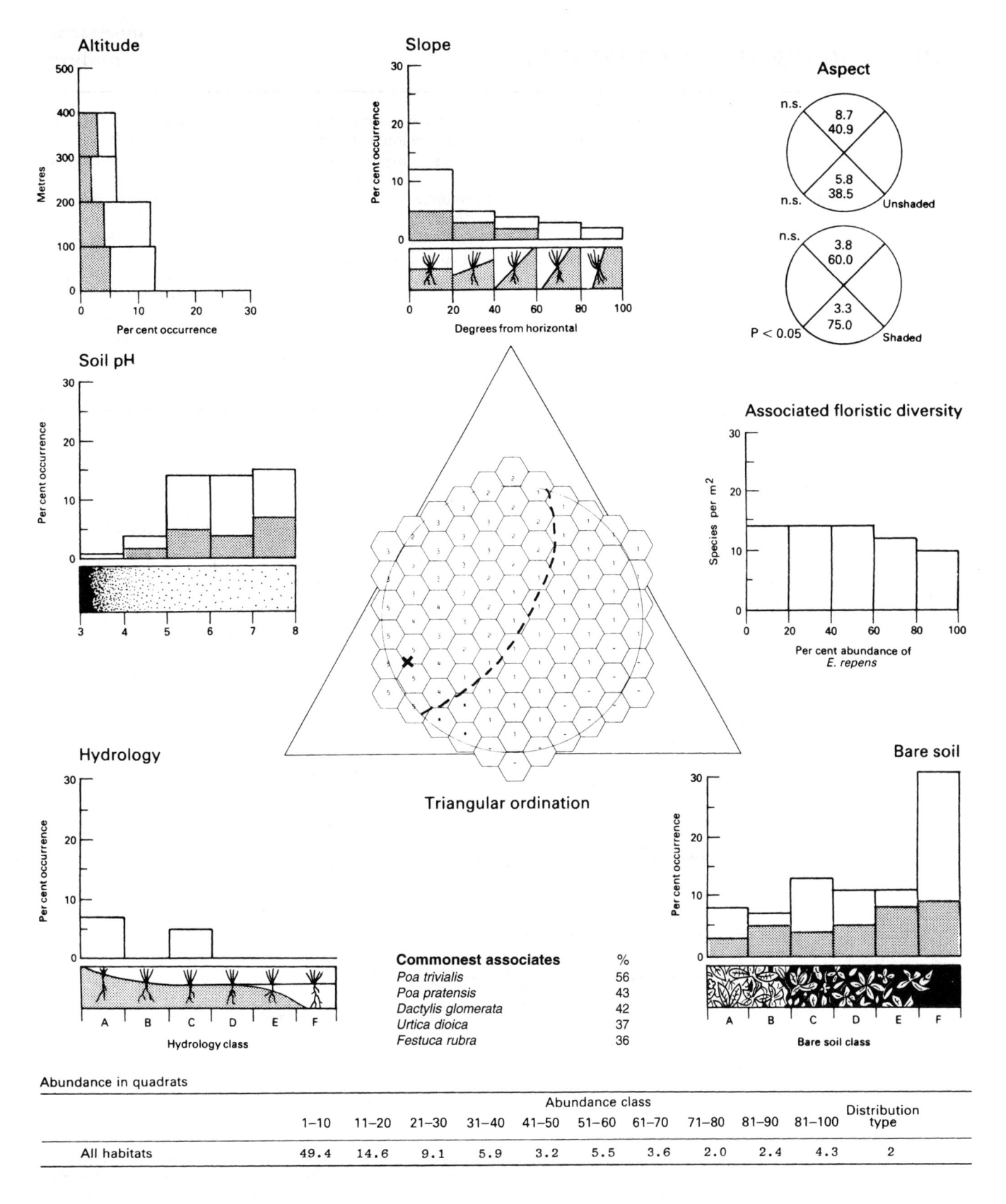

	1–10	11–20	21–30	31–40	41–50	51–60	61–70	71–80	81–90	81–100	Distribution type
All habitats	49.4	14.6	9.1	5.9	3.2	5.5	3.6	2.0	2.4	4.3	2

has a subsidiary role to rhizome production. E.r. may form a short-lived seed bank (Biol Fl (b)). Like other relatively palatable grasses with tall nodal stems, E.r. is usually absent from heavily grazed habitats and although relatively cold-tolerant, the young autumn shoots are sometimes killed under severe winter conditions (Biol Fl (b)). Claims that E.r. is allelopathic require further investigation. E.r. is out-breeding and genetic variation in several important ecological attributes has been described (Biol Fl (a)). Five ssp. are recognized from Europe (FE5). Known to hybridize with two other species of Elytrigia and one of Hordeum (Stace 1975).

Current trends Despite its poor ability to regenerate from seed, E.r. is, thanks to human agency, abundant and probably still increasing. Genotypes exploiting arable land and those tolerant of salt-spray on roadsides appear particularly favoured at present.

Empetrum nigrum ssp. nigrum L.

Empetraceae
Crowberry

Life-form Phanerophyte or woody chamaephyte. Older shoots procumbent, some of younger shoots ascending; leaves xeromorphic, strongly revolute, < 100 mm²; roots shallow, bearing ericoid mycorrhizas.

Phenology Evergreen. New shoot growth commences in April and flowering occurs at about the same time. Fruits ripe from July onwards, but may overwinter on the plant. Shoot extension ceases in September or October. Leaves persist for several years (*Biol Fl*).

Height Foliage and flowers usually <300 mm.

Seedling RGR No data. SLA 8.1. DMC. No data.

Nuclear DNA amount No data; $2n = 26^*$ (*BC*); $2x^*$.

Established strategy Intermediate between stress-tolerator and stress-tolerant competitor.

Regenerative strategies V, S, ?B$_s$. Mainly vegetative, but regeneration by seed possible (see 'Synopsis'); SB III.

Flowers Purple, usually dioecious, wind-pollinated, but occasionally visited by insects (Knuth 1906); axillary.

Germinule 0.75 mg, 5.9 × 5.3 mm, in a black berry with 6-9 seeds (*Biol Fl*).

Germination Epigeal; dormancy broken by chilling (*Biol Fl*).

Biological flora Bell and Tallis (1973).

Geographical distribution W and N Britain, but absent from SE England (73 % of British vice-counties), *ABIF* page 286; N and C Europe except for the extreme N and on mountains further S (69% of European territories); N Asia and N America and the cooler regions of the N hemisphere.

Habitats Restricted to grazed or ungrazed moorland areas. Occasionally found rooted in pockets of soil on gritstone outcrops and quarry spoil, and also on blanket peat and in mire. Found more rarely in lightly shaded habitats. [Although typically a plant of acidic upland or montane areas, *E.n.* is also recorded from limestone pavement, from calcium carbonate-rich sand and from dune heaths (*Biol Fl*).]

Gregariousness Sometimes patch-forming.

Altitude Formerly found locally at *c.* 100 m (Howitt & Howitt 1963) but now restricted to the uplands (to 550 m). [Up to 1270 m.]

Slope Not recorded from slopes of >40°.

Aspect Distribution has a N-facing bias.

Hydrology Essentially a species of moist rather than waterlogged habitats. Recorded from ombrogenous mire, sometimes rooted on tussocks of *Eriophorum vaginatum* and also more rarely from soligenous mire at the edge of blanket peat.

Soil pH Particularly frequent and abundant on the most acidic soils. Seldom recorded above pH 4.5. [Elsewhere there are also a few records from calcareous soils of high pH (*Biol Fl*).]

Bare soil Mainly in habitats having relatively little exposed soil.

△ ordination Distribution confined to vegetation associated with undisturbed and relatively infertile habitats.

Associated floristic diversity Low or, more rarely, intermediate. At high frequencies of *E.n.* other species tend to be excluded.

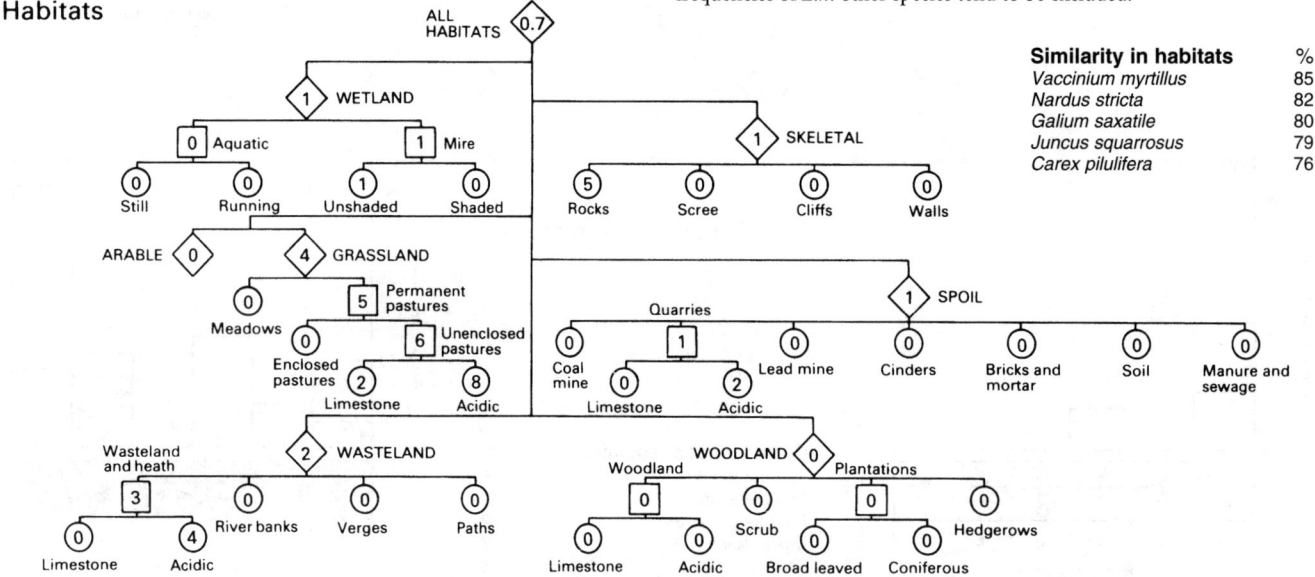

Habitats

Similarity in habitats	%
Vaccinium myrtillus	85
Nardus stricta	82
Galium saxatile	80
Juncus squarrosus	79
Carex pilulifera	76

Synopsis *E.n.* is a low-growing, long-lived (to 140 years *Biol Fl*), evergreen shrub restricted to areas with a cool, moist climate. In N England, which lies at the S edge of its distribution, the species is confined to upland sites. The lower altitudinal limit decreases in N Britain (*Biol Fl*) and occurrences in the Sheffield region show a bias towards N-facing slopes. *E.n.* is relatively tolerant of snow cover and severe winter conditions. Shoot growth commences in early spring and involves the emergence of flowers which were fully formed in the preceding autumn (*Biol Fl*). In the nutrient-deficient habitats of which *E.n.* is characteristic, the evergreen habit of *E.n.* may be important in the retention and efficient utilization of limiting elements such as N and P (Small 1972). The species is perhaps most typical of acidic peaty soils, but is also recorded on highly calcareous soils in N Britain (*Biol Fl*) and W Ireland (Grime 1963b). Though frequent in moist habitats, *E.n.* is intolerant of severe waterlogging, other than in soligenous mire. *E.n.* is tolerant of controlled burning but is destroyed by severe fires (*Biol Fl*). The foliage is eaten by grouse but not by sheep and seed is dispersed by a variety of birds and animals and is eaten by humans (*Biol Fl*). Regeneration is mainly by means of vegetative growth. Prostrate stems may root on contact with the ground, and as a result *E.n.* may form large clonal patches which may increase at a rate of up to 100 mm year−1 (*Biol Fl*). Male and female plants are often found in association and fruiting is of frequent occurrence. Flowering is usually confined to open habitats. Most seeds germinate in the first spring following their release. The remainder germinate in small numbers for several years (*Biol Fl*). However, establishment from seed appears to be infrequent (*Biol Fl*). Work in Alaska (McGraw 1980) suggests that *E.n. sensu lato* may form a small bank of persistent buried seeds. This combination of regenerative strategies, involving extensive clonal growth and the production of only moderate numbers of relatively large seeds, is not unlike that of *Vaccinium myrtillus* and *V. vitis-idaea*, either of which *E.n.* may replace in moister,

Altitude

Slope

Aspect

Soil pH

Associated floristic diversity

Triangular ordination

Hydrology

Commonest associates	%
Deschampsia flexuosa	74
Vaccinium myrtillus	62
Eriophorum angustifolium	43
Calluna vulgaris	34
Rubus chamaemorus	28

Bare soil

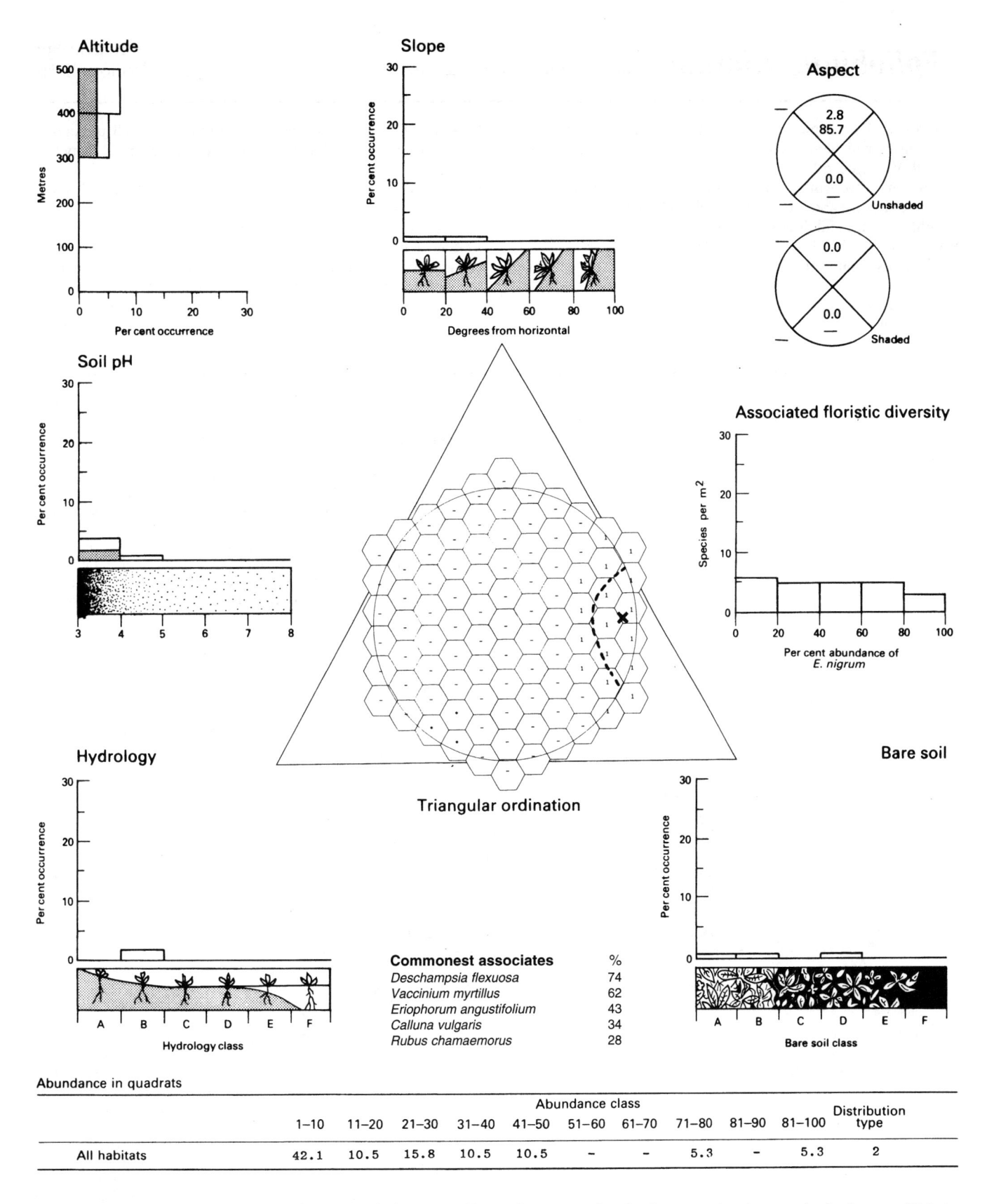

Abundance in quadrats

	1–10	11–20	21–30	31–40	Abundance class 41–50	51–60	61–70	71–80	81–90	81–100	Distribution type
All habitats	42.1	10.5	15.8	10.5	10.5	–	–	5.3	–	5.3	2

more northerly or montane sites. However, there is a marked contrast with the biology of *Calluna vulgaris* and *Erica cinerea*, which tend to form relatively short-lived plants and are dependent upon the production of numerous minute seeds and large persistent seed banks. Ssp. *hermaphroditum* (Hagerup) Bocher replaces ssp. *nigrum* in some mountainous regions of N Britain, parts of the Alps and the extreme N of Europe. It is monoecious and has $2n = 52$.

Current trends *E.n.* is now extinct in many lowland parts of Britain through the destruction of habitats, but perhaps also through an increase in winter temperatures since the 19th century (*Biol Fl*). In upland areas *E.n.* appears more common than formerly. This may be due to overgrazing by sheep, and to other forms of land management which have resulted in increased bare ground for seedling establishment and have reduced the vigour of some potential competitors.

Epilobium ciliatum Rafin. *E. adenocaulon* Hausskn.

Onagraceae
American Willow-herb

Life-form Polycarpic perennial protohemicryptophyte or chamaephyte. Stem erect; leaves lanceolate, <2000 mm^2; roots with VA mycorrhizas.
Phenology Overwinters by means of leaf rosettes. Plants flower July to August and set seed from July to September.
Height Foliage and flowers to 900 mm.
Seedling RGR No data. SLA 30.9. DMC 16.3.
Nuclear DNA amount 1.1 pg; $2n = 36*$ (*BC*); $2x*$ (or perhaps $n = 9$).
Established strategy Intermediate between ruderal and C-S-R.
Regenerative strategies W, ?B$_s$. Regenerates both vegetatively and by means of seed (see 'Synopsis'); SB III.
Flowers Pink, hermaphrodite, automatically self-pollinated (Myerscough & Whitehead 1967); often >50 in a branched racemose inflorescence.
Germinule 0.06 mg, 1.3 × 0.5 mm, seed with a plume of hairs of *c.* 5 mm in a dehiscent, >100-seeded capsule.
Germination Epigeal; t_{50} 2 days; 8-36°C; L > D.
Biological flora None.
Geographical distribution Native of N America. Common in SE England and in scattered localities northwards to Scotland (>20% of British vice-counties), *ABIF* page 415; much of Europe, mainly in the NW (naturalized in 38% of European territories).

Habitats Occurs particularly frequently on cinder tips, soil heaps, building rubble and other spoil. Also recorded or observed from all habitats except aquatic and enclosed pastures.
Gregariousness Sparse.
Altitude Suitable sites are more abundant in lowland areas, but *E.c.* extends to 400 m.
Slope Wide-ranging.
Aspect Widely distributed but only recorded in abundance on S-facing slopes during this survey.
Hydrology Typical of moist rather than waterlogged sites, e.g. exposed mud beside reservoirs. Absent from submerged habitats.
Soil pH Seldom recorded from sites of pH <5.0.
Bare soil Wide-ranging. Most typically abundant in sites with high exposures of soil or spoil.
A ordination A diffuse pattern indicating that *E.c.* is present in a wide range of vegetation types.
Associated floristic diversity Typically intermediate.
Synopsis A native of N America, *E.c.* is a tall, fast-growing polycarpic perennial herb (Myerscough & Whitehead 1967), which colonizes a range of disturbed, often relatively fertile sites. In some of the species' habitats, such as gardens, stony river margins and railway ballast beside the permanent way, the spread of potential dominants is limited by regular disturbance in the form of weeding, flooding and the application of herbicides respectively. In others, e.g. demolition sites, coal-mine spoil and

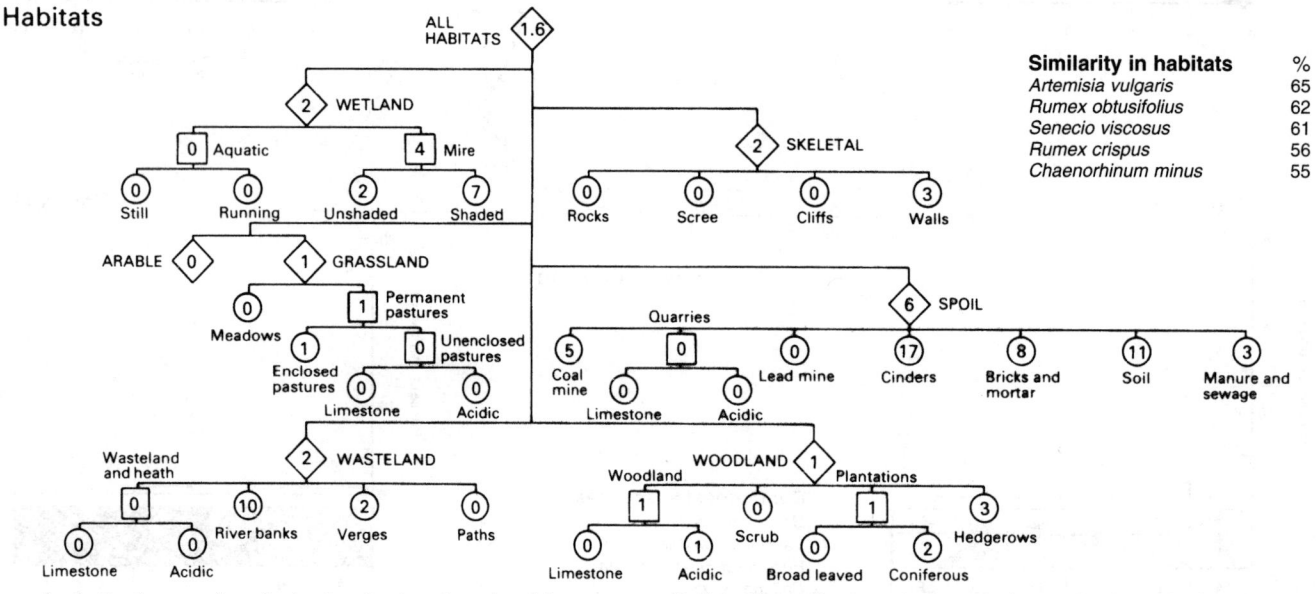

Habitats

Similarity in habitats	%
Artemisia vulgaris	65
Rumex obtusifolius	62
Senecio viscosus	61
Rumex crispus	56
Chaenorhinum minus	55

wasteland, *E.c.* is an early colonist, but in time is replaced by more-dominant species. *E.c.* was first recorded in Britain in 1891, and since 1932 has spread rapidly (Salisbury 1964). First observed in the Sheffield region in 1959 (Sledge 1959), *E.c.* has since spread to such an extent that it is now the most abundant willow-herb in urban areas. The success of *E.c.* appears to be related to the considerable habitat range. Despite being an inbreeding species, *E.c.* is able to colonize a wide variety of environments and appears better able than most native species to exploit artificial habitats such as cinders, demolition sites, gravel pits and disturbed ground beside forestry rides. Effective colonization appears to relate to the exceptional regenerative versatility of *E.c.* Although normally a polycarpic perennial, seedlings develop rapidly and under long day conditions may set seed within 10 weeks (Myerscough & Whitehead 1966); thus the species has the potential to function as an annual. Seed production is prolific; a large plant may produce over 10 000 plumed and wind-dispersed seeds. Germination occurs over a wide range of temperatures, and establishment from seed may be initiated over much of the year (Myerscough & Whitehead 1967). Seeds accumulate in large numbers on or near the soil surface, and a seed bank is often still present in the soil after *E.c.* has been displaced by later successional species (UCPE unpubl.). Although lacking the capacity for extensive lateral vegetative spread, the species is able to regenerate vegetatively to a limited extent by the leafy stolons, which are produced at the base of the stem in autumn; these may become detached as a result of disturbance and give rise to new plants. However, in undisturbed sites the leafy stolons appear to function primarily as organs of perennation. The stolons may grow to some extent during winter (Myerscough & Whitehead 1967), suggesting that in upland regions the distribution of *E.c.* may be restricted by climate as well as by the availability of suitable habitats. Like *Senecio squalidus*, which also produces plumed seeds, *E.c.* has spread widely along the network of railways. *E.c.* forms hybrids with a wide range of other *Epilobium* species (Stace 1975) and *E.c.* × *E. montanum* appears to be the commonest hybrid in N England.
Current trends An alien which appears to be still increasing, particularly in urban and industrial areas. Now the most abundant and ecologically wide-ranging species of *Epilobium* in lowland Britain.

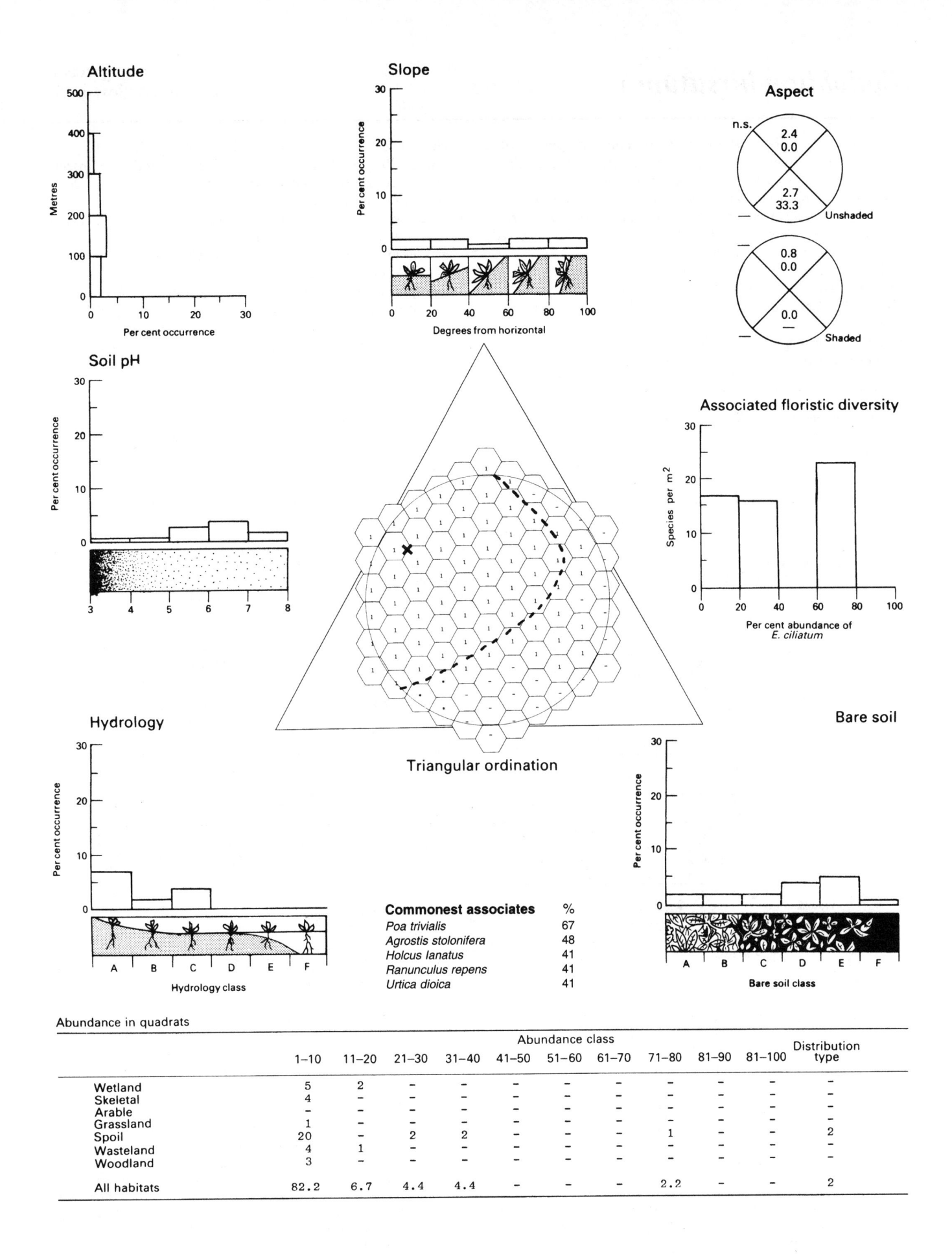

Altitude

Slope

Aspect

	n.s.	2.4	0.0	
		2.7	33.3	Unshaded

		0.8	0.0	
		0.0	—	Shaded

Soil pH

Associated floristic diversity

Species per m² vs Per cent abundance of *E. ciliatum*

Hydrology

Triangular ordination

Bare soil

Commonest associates	%
Poa trivialis	67
Agrostis stolonifera	48
Holcus lanatus	41
Ranunculus repens	41
Urtica dioica	41

Abundance in quadrats

	1–10	11–20	21–30	31–40	41–50	51–60	61–70	71–80	81–90	81–100	Distribution type
Wetland	5	2	–	–	–	–	–	–	–	–	–
Skeletal	4	–	–	–	–	–	–	–	–	–	–
Arable	–	–	–	–	–	–	–	–	–	–	–
Grassland	1	–	–	–	–	–	–	–	–	–	–
Spoil	20	–	2	2	–	–	–	1	–	–	2
Wasteland	4	1	–	–	–	–	–	–	–	–	–
Woodland	3	–	–	–	–	–	–	–	–	–	–
All habitats	82.2	6.7	4.4	4.4	–	–	–	2.2	–	–	2

Epilobium hirsutum L.

Onagraceae
Great Hairy Willow-herb

Life-form Polycarpic perennial protohemicryptophyte with stolons and extensive rhizomes. Stems erect; leaves lanceolate, hairy, <4000 mm^2; roots with VA mycorrhizas.
Phenology Produces leafy stolons during the latter part of winter. After an extended period of vegetative growth in spring and summer, plants flower from late July to August and set seed from September to October.
Height Foliage and flowers to 2 m.
Seedling RGR 1.5-1.9 week^{-1}. SLA 28.1. DMC 20.8.
Nuclear DNA amount 0.6 pg; $2n = 36^*$ (*BC*); ?$2x^*$ (or perhaps $n = 9$).
Established strategy Competitor.
Regenerative strategies V, W, B$_s$. Regenerates effectively both vegetatively and by means of seed (see 'Synopsis'); SB III.
Flowers Purple, hermaphrodite, often protandrous, insect-pollinated, generally out-breeding (Shamsi 1970); often >20 in a racemose inflorescence.
Germinule 0.05 mg, 1.0 × 0.5 mm, seed with plume of hairs *c.* 4 mm; often >100 in a dehiscent capsule.
Germination Epigeal; t_{50} 2 days; 12-33°C; L > D.
Biological flora None.
Geographical distribution The British Isles, but absent from much of Scotland (92% of British vice-counties), *ABIF* page 412; Europe except the extreme N (89% of European territories); temperate Asia, N, E and S Africa; introduced in N America.

Habitats Essentially a wetland species. Most frequently recorded from river banks and unshaded and shaded mire. Less frequent in aquatic habitats and on various types of spoil tip. Also found on damp walls and as seedlings on cliffs and outcrops. Absent from pasture, arable land and woodland.
Gregariousness A stand-forming dominant even at intermediate frequencies.
Altitude Only recorded up to 300 m, and frequency and abundance much greater in the lowlands. [Up to 370 m.]
Slope Largely restricted to and only abundant on gentle inclines, but a few occurrences (mainly seedlings) on steep slopes.
Aspect No significant effects.
Hydrology Frequent on the moist soils of river banks, topogenous mire beside ponds and ditches and canals, and in soligenous mire.
Soil pH Most frequent and abundant in the pH range 6.0-8.0. Less common on acidic soils down to pH 4.0.
Bare soil Widely distributed, but most frequently associated with wetland habitats having a high exposure of bare mud.
△ **ordination** Only recorded in productive vegetation. Distribution centred on vegetation associated with highly productive and only occasionally or superficially disturbed conditions.
Associated floristic diversity Relatively low, and declining with increased abundance of *E.h.*, confirming the capacity of the species for dominance.

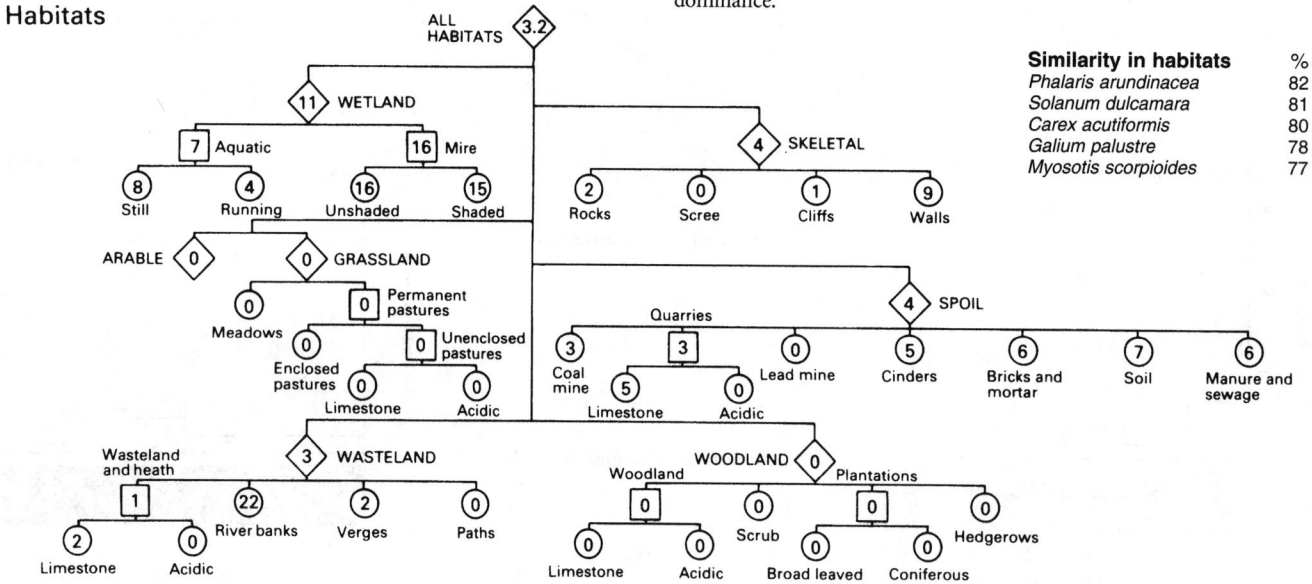

Habitats

Similarity in habitats	%
Phalaris arundinacea	82
Solanum dulcamara	81
Carex acutiformis	80
Galium palustre	78
Myosotis scorpioides	77

Synopsis *E.h.* is a tall, stand-forming herb particularly associated with river banks and productive mire. Although characteristic of sites with high net production (Wheeler & Giller 1982), *E.h.* is sometimes capable of dominance in sites of slightly lower fertility, but it is sensitive to grazing (Wheeler 1983) and mowing, both of which lead to rapid displacement by shorter species. The germination rate and total germination percentage of seeds is increased by ethene, and in response to flooding, shallow adventitious roots are formed at the expense of the deep primary root (Etherington 1983). Both responses may be interpreted as specializations associated with waterlogged habitats, especially since neither is shown by the dryland species *Chamerion angustifolium* (Etherington 1983). Soils colonized by *E.h.* are often characterized by low levels of extractable iron and manganese (Al-Farraj 1983), and *E.h.* is, within the context of wetland species, very susceptible to ferrous ion toxicity (Hendry & Brocklebank 1985, R. E. D. Cooke unpubl.). This sensitivity may explain the greater frequency of *E.h.* on river banks (flooded mainly during winter) than in mire (which is permanently waterlogged). The plant produces leafy stolons during the later part of winter, and leaves damaged by winter conditions may be replaced in early spring (Shamsi & Whitehead 1977). Whether this behaviour plays a part in the tendency for *E.h.* to be restricted to lowland areas (maximum British altitude 370 m (Wilson 1956)) requires investigation. In this context it may be significant that seed set was very low in the cool wet summer of 1985. The species has a formidable array of regenerative strategies, including the production of numerous wind-dispersed seeds which have the capacity to germinate over a wide range of temperatures. Some seeds germinate in autumn soon after dispersal, but in many others germination may be delayed until the following spring. *E.h.* also accumulates considerable reserves of buried seeds, and forms large clones by means of rhizomes, which may extend at rates of 0.5 m year^{-1} and die in the year following their initiation (Shamsi & Whitehead 1974).

Altitude
Slope
Aspect

	n.s.	1.6	
		0.0	
		2.2	Unshaded
		0.0	

		0.0	
		0.0	Shaded

Soil pH

Associated floristic diversity

Hydrology

Triangular ordination

Commonest associates	%
Poa trivialis	72
Urtica dioica	52
Agrostis stolonifera	41
Ranunculus repens	39
Holcus lanatus	35

Bare soil

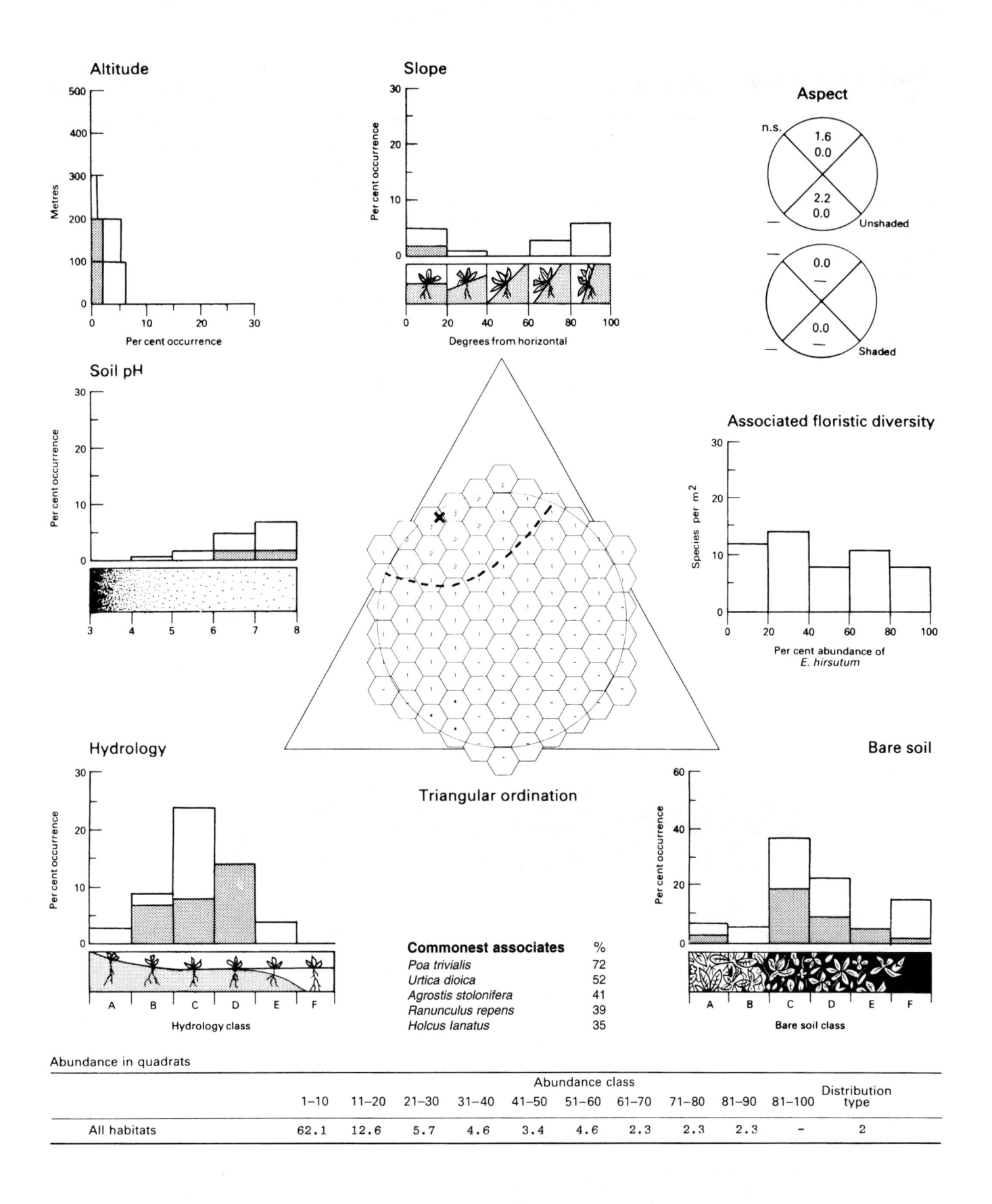

Abundance in quadrats

	1–10	11–20	21–30	31–40	41–50	51–60	61–70	71–80	81–90	81–100	Distribution type
All habitats	62.1	12.6	5.7	4.6	3.4	4.6	2.3	2.3	2.3	–	2

Regeneration may also occur from detached rhizome fragments. These attributes, coupled with the high potential growth rate and capacity to form a dense, rapidly ascending, leaf canopy during summer, confer upon *E.h.* the ability for effective colonization of fertile habitats, high rates of resource capture and dominance of perennial communities in productive habitats. Hybrids with other species of *Epilobium* are infrequent, and only the crosses with *E. ciliatum* and *E. montanum* have been observed within the Sheffield region.

Current trends Increasing. An effective colonist of artificial habitats, highly responsive to eutrophication.

Epilobium montanum L.

Onagraceae
Broad-leaved Willow-herb

Life-form Polycarpic perennial protohemicryptophyte or chamaephyte. Stems erect; leaves ovate, often slightly hairy, usually <2000 mm²; roots with VA mycorrhizas.
Phenology Overwintering as short above- or below-ground stolons. Flowers June to August and sets seed from July to September.
Height Foliage and flowers to 600 mm.
Seedling RGR No data. SLA 28.4. DMC 22.9.
Nuclear DNA amount No data; $2n = 36*$ (*BC*); $2x*$ (or perhaps $n = 9$).
Established strategy Intermediate between ruderal and C-S-R.

Regenerative strategies W, $?B_s$. Regenerates vegetatively and by means of seed (see 'Synopsis'); ?SB III.
Flowers Pink, hermaphrodite, homogamous, commonly self-pollinated; often <20 in a terminal raceme.
Germinule 0.13 mg, 1.1×0.5 mm, seed with a plume of hairs of *c.* 6 mm; often >100 in a dehiscent capsule.
Germination Epigeal; t_{50} 4 days; 8-22°C; L > D.
Biological flora None.
Geographical distribution The British Isles (100% of British vice-counties), *ABIF* page 412; Europe except for the extreme N (85% of European territories); W Asia.

Habitats Most frequent on coal-mine spoil, cinders and walls. A casual of arable land, and largely absent from pasture and wetland habitats, but otherwise widely distributed on rocky, disturbed, shaded ground.
Gregariousness Sparse.
Altitude Wide-ranging. [Up to 790 m.]
Slope Found over the full range.
Aspect Wide-ranging, but tending to be more abundant on N-facing slopes in unshaded situations, and S-facing slopes in shaded habitats.

Hydrology Typical of moist rather than waterlogged soils.
Soil pH Seldom recorded below pH 5.5.
Bare soil Widely distributed.
△ **ordination** Insufficient data.
Associated floristic diversity Typically intermediate.

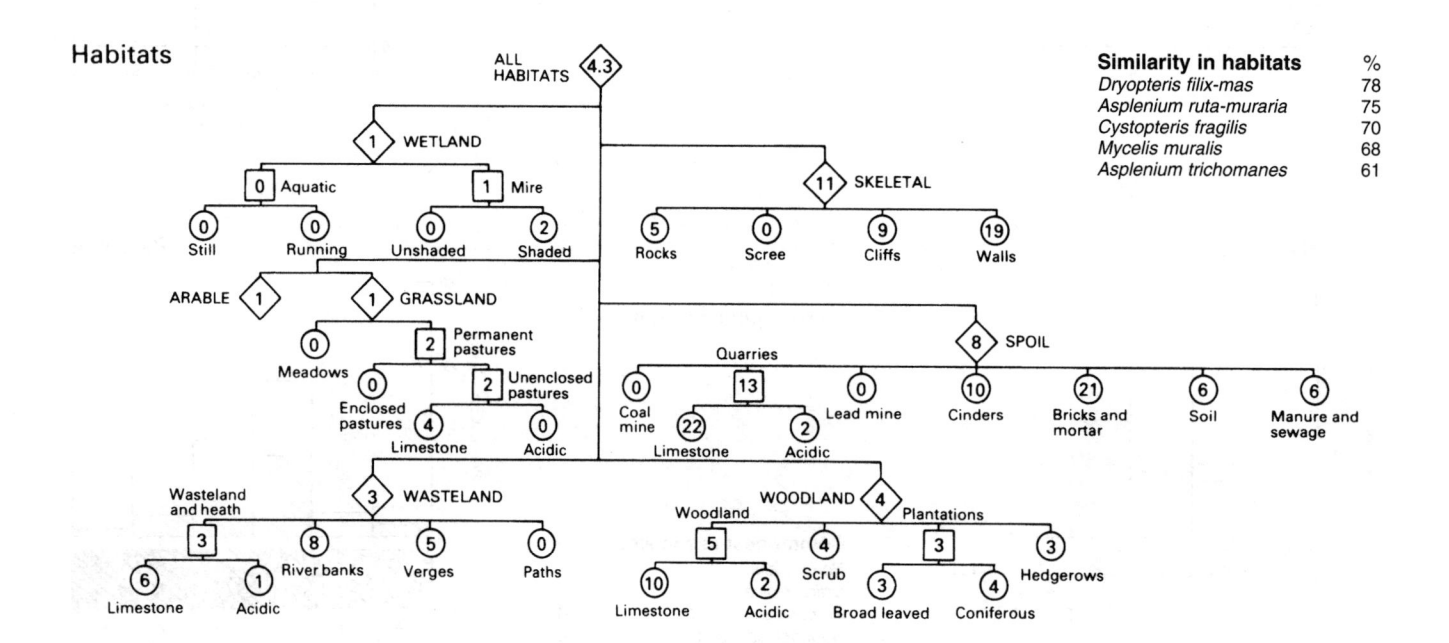

Habitats

Similarity in habitats	%
Dryopteris filix-mas	78
Asplenium ruta-muraria	75
Cystopteris fragilis	70
Mycelis muralis	68
Asplenium trichomanes	61

Synopsis *E.m.* is an erect, relatively fast-growing (Myerscough & Whitehead 1967) polycarpic perennial herb which colonizes a range of disturbed and skeletal sites. In habitat range and life-history *E.m.* is very similar to *E. ciliatum*. However, for a number of reasons *E.m.* is much less abundant. First, its biology and ecology is less ruderal in character. Thus, the species takes *c.* 10 days longer to set seed (Myerscough & Whitehead 1966); plants are smaller and few-flowered, and therefore produce smaller numbers of seeds; these germinate over a narrower range of temperatures. Second, *E.m.* is most characteristic of lightly shaded conditions and only attains abundance as a colonist of moist habitats. Thus, *E.m.* has a narrower habitat range than *E.c.* However, *E.m.*, whose stolons bear less leaf area and grow less during winter than those of *E.c.* (Myerscough & Whitehead 1967), extends extensively into upland regions. Also, judging by its habitat range, *E.m.* exhibits greater shade-tolerance. The production of numerous wind-dispersed propagules and the association with moist, sometimes shaded conditions are shared with many ferns and with *Mycelis muralis*, all of which exploit walls and other steep, rocky habitats (see 'Similarity in habitats'). A degree of ecotypic differentiation has been reported in this predominantly in-breeding species. Garden populations tend to allocate more reserves to seeds, whereas woodland populations produce more stolons (Atkinson & Davison 1971). As in the case of *E.c.*, *E.m.* forms many hybrids (Stace 1975).
Current trends Uncertain, although as an effective colonist of artificial habitats, *E.m.* is likely to remain common.

Altitude
Slope
Aspect

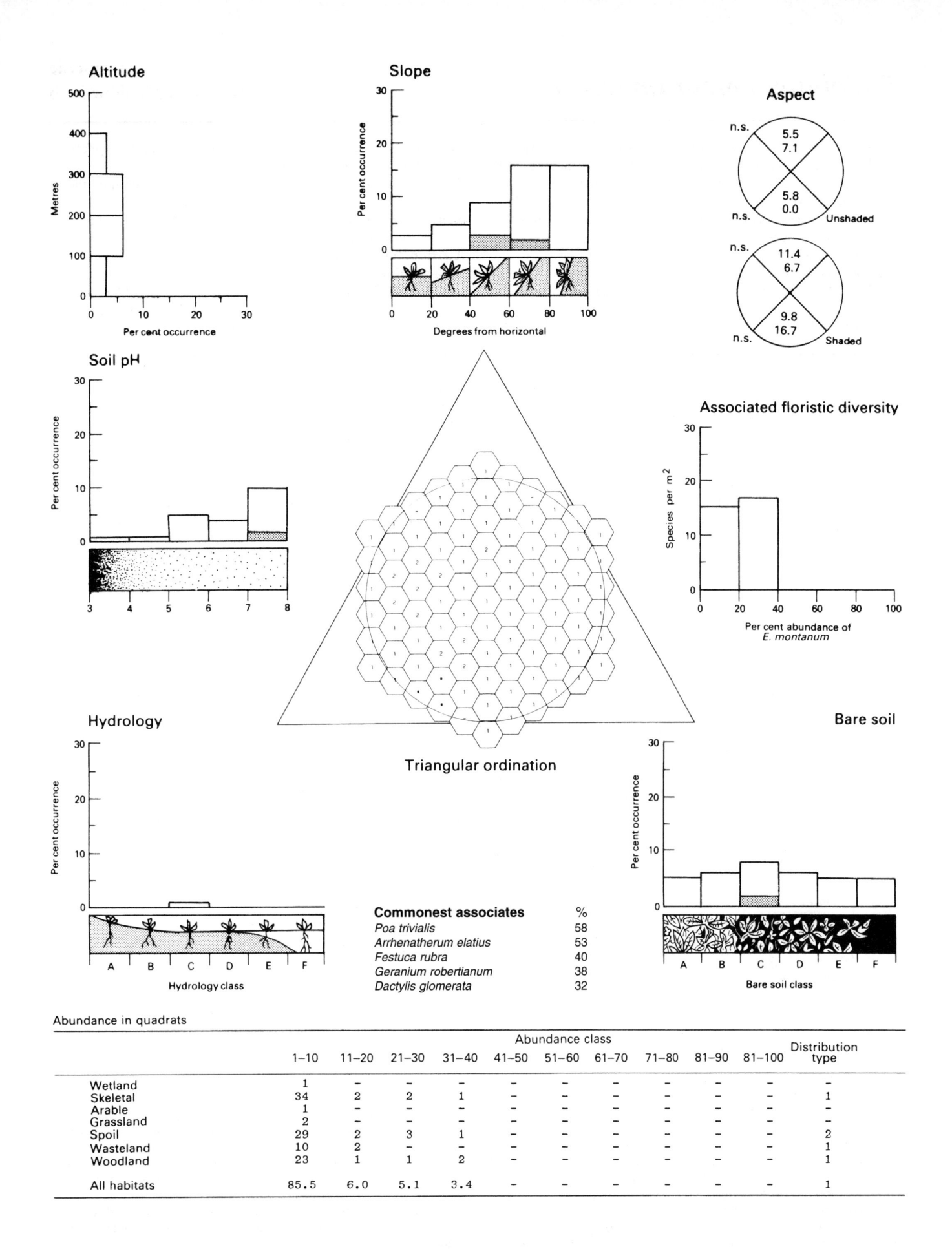

	n.s.	5.5	
		7.1	
		5.8	
	n.s.	0.0	Unshaded

	n.s.	11.4	
		6.7	
		9.8	
	n.s.	16.7	Shaded

Soil pH

Associated floristic diversity

Hydrology

Triangular ordination

Bare soil

Commonest associates %
Poa trivialis 58
Arrhenatherum elatius 53
Festuca rubra 40
Geranium robertianum 38
Dactylis glomerata 32

Abundance in quadrats

	Abundance class										Distribution type
	1–10	11–20	21–30	31–40	41–50	51–60	61–70	71–80	81–90	81–100	
Wetland	1	–	–	–	–	–	–	–	–	–	–
Skeletal	34	2	2	1	–	–	–	–	–	–	1
Arable	1	–	–	–	–	–	–	–	–	–	–
Grassland	2	–	–	–	–	–	–	–	–	–	–
Spoil	29	2	3	1	–	–	–	–	–	–	2
Wasteland	10	2	–	–	–	–	–	–	–	–	1
Woodland	23	1	1	2	–	–	–	–	–	–	1
All habitats	85.5	6.0	5.1	3.4	–	–	–	–	–	–	1

Epilobium obscurum Schreber

Onagraceae
Dull-leaved Willow-herb

Life-form Polycarpic perennial protohemicryptophyte or chamaephyte. Stems erect; leaves lanceolate, sometimes with hairs, often < 1000 mm^2; roots with VA mycorrhizas.

Phenology Overwintering by means of leafy stolons. After shoot extension in late spring, flowers in July and August and sets seed July to September.

Height Foliage and flowers to 600 mm.

Seedling RGR No data. SLA 31.8. DMC 13.5.

Nuclear DNA amount 0.5 pg; $2n = 36*(BC)$; $2x*$ (or perhaps $n = 9$).

Established strategy Intermediate between ruderal and C-S-R.

Regenerative strategies W, ?B$_s$. Regenerates both vegetatively and by seed (see 'Synopsis'); ?SB III.

Flowers Pink, hermaphrodite, homogamous, self-pollinated; often >20 in a branched racemose inflorescence.

Germinule 0.05 mg, 1.0 × 0.5 mm seed with a plume of hairs of c. 5 mm, in a dehiscent >100-seeded capsule.

Germination Epigeal t_{50} 3 days; L > D.

Biological flora None.

Geographical distribution The British Isles (99% of British vice-counties), *ABIF* page 146; Europe except the extreme N (97% of European territories); much of the USSR and N Africa.

Habitats Recorded only as a species of unshaded and, to a lesser extent, shaded mire. Also present on wet rocks with running water, and may experience temporary submergence at the margins of streams. Observed on disturbed wasteland, on soil heaps and as a garden weed. Absent from woodland, grassland habitats and most spoil categories.

Gregariousness Sparse.

Altitude Observed from 15 to 400 m. [Up to 780 m.]

Slope Typical of gentle slopes, but also recorded from wet walls and cliffs.

Aspect Only recorded from S-facing slopes, but additional studies incorporating data from a larger number of samples suggest that the species is more widely distributed.

Hydrology Typical of soligenous mire, but some records from topogenous mire. Also found temporarily submerged in flowing water.

Soil pH Recorded on soils with pH values varying between pH 5.0 and 7.0 but more occurrences and greatest abundance within the pH range 6.0-7.0.

Bare soil Wide-ranging. In mire often occurs in closed vegetation; in more-disturbed habitats high exposures are frequently observed.

△ **ordination** Associated with productive vegetation subject to slight or moderate levels of disturbance. Not recorded from heavily disturbed vegetation.

Associated floristic diversity Typically intermediate.

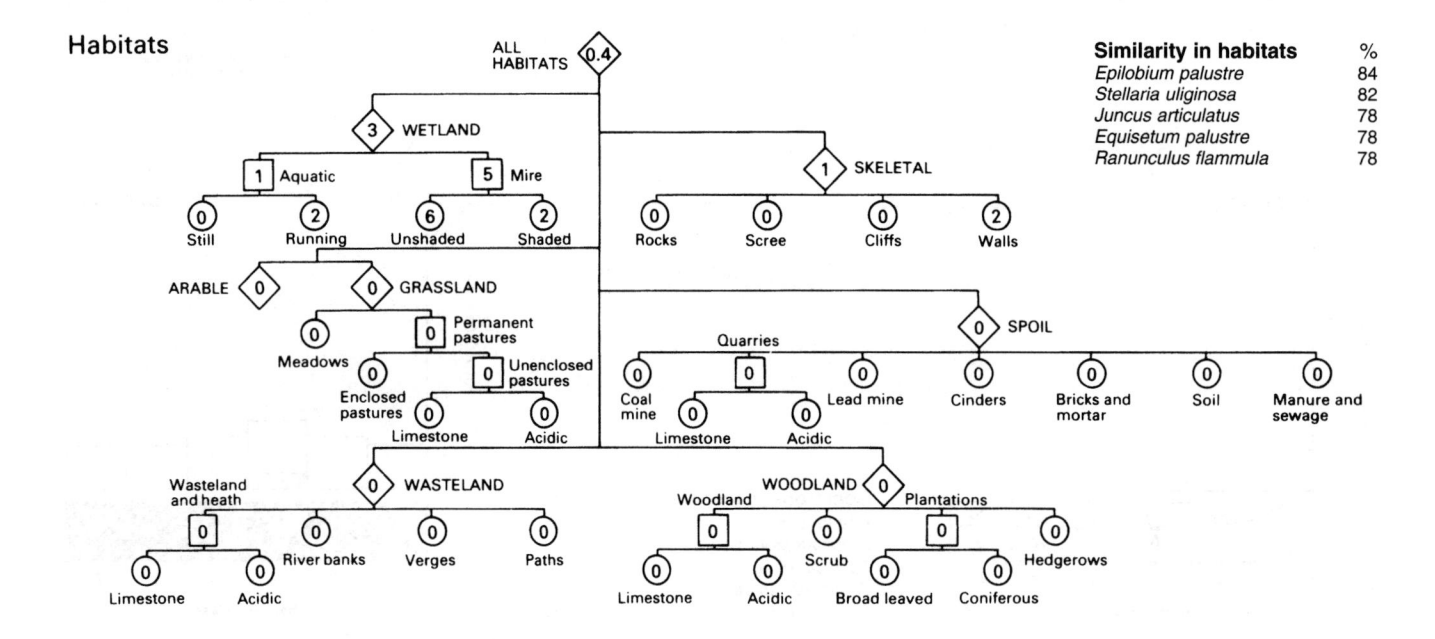

Habitats

Similarity in habitats	%
Epilobium palustre	84
Stellaria uliginosa	82
Juncus articulatus	78
Equisetum palustre	78
Ranunculus flammula	78

Synopsis *E.o.* is an erect herb which produces numerous wind-dispersed seeds, and is an effective colonist of a wide range of moist habitats. The species is most common in two relatively fertile types of environment: (a) mire, river banks and wet rocks, often in situations where the vigour of potential dominants is restricted by winter-flooding and sometimes by shade; and (b) disturbed ground, including garden plots. In wetlands the species is absent from highly calcareous soils and is most frequent in soligenous mire. *E.o.* frequently exploits sites which are intermediate in soil fertility between productive habitats capable of supporting *E. hirsutum* and the impoverished sites of *E. palustre*. In highly disturbed wetland habitats, including those with wide fluctuations in water table, e.g. river shingle, *E.o.* is invariably replaced by *E. ciliatum*. In gardens and waste places the species is restricted to moist, often shaded sites which are infrequently disturbed. In such sites *E.o.* is less frequent than either *E.c.* or *E. montanum*. As in the case of other species of *Epilobium*, the tall stems of *E.o.* are vulnerable to grazing. A robust plant of *E.o.* produces many thousands of small, wind-dispersed seeds. These do not float (Ridley 1930) and may form a bank of persistent seeds in the soil. The species is stoloniferous, and additional plants may be formed by this means in closed vegetation. However, despite this capacity for vegetative spread, *E.o.* is sparsely distributed in vegetation. Detached shoot segments are also capable of regeneration, and plant fragments washed away during flooding may prove to be of importance in colonizing new sites downstream. The distribution of *E.o.* overlaps those of several other species of *Epilobium* and hybrids occur (Stace 1975).

Current trends Uncertain. Perhaps increasing as a result of disturbance in artificial habitats, but decreasing as a wetland plant due both to drainage and to eutrophication.

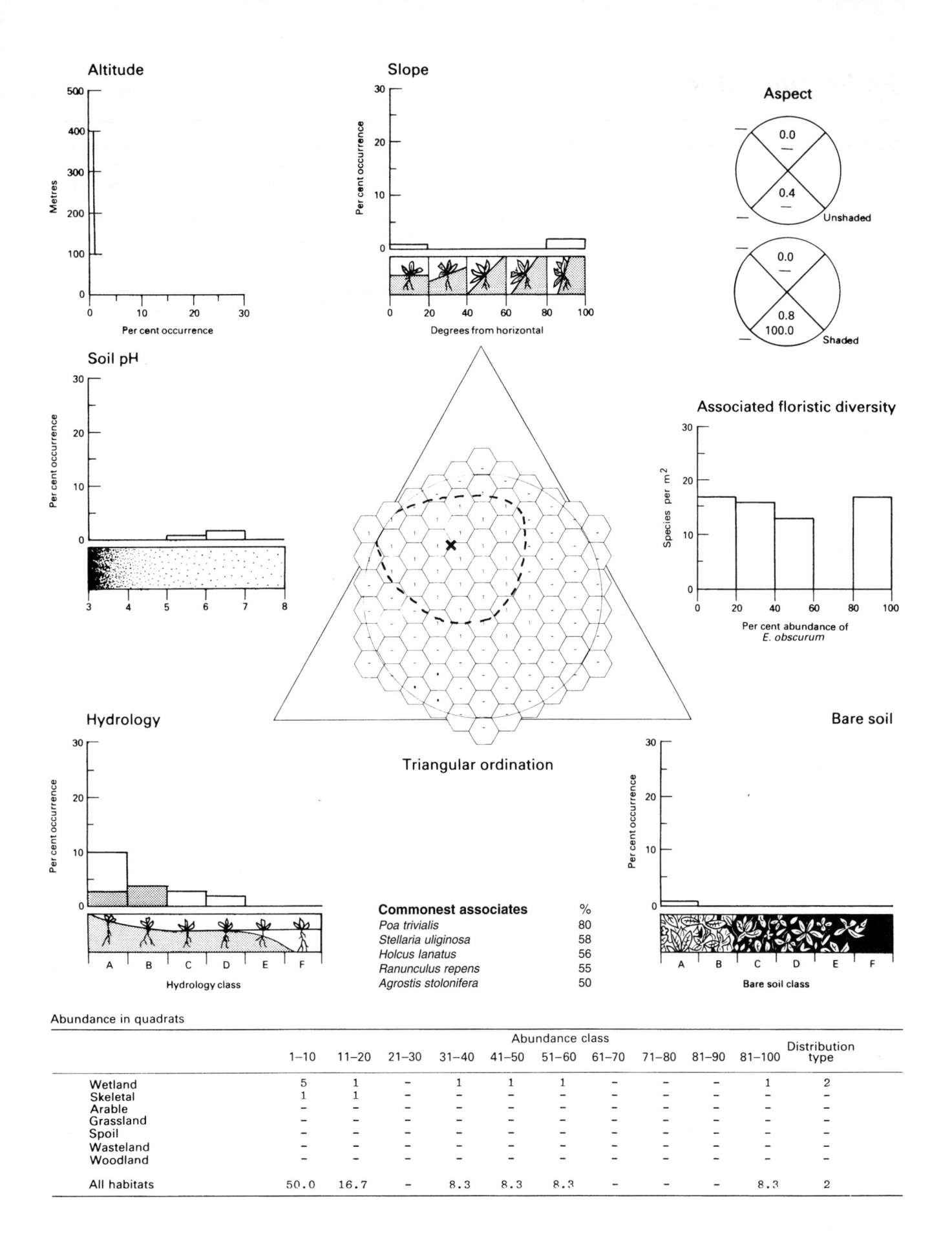

Altitude

Slope

Aspect

Unshaded
Shaded

Soil pH

Associated floristic diversity

Species per m²

Per cent abundance of *E. obscurum*

Hydrology

Hydrology class

Triangular ordination

Commonest associates	%
Poa trivialis	80
Stellaria uliginosa	58
Holcus lanatus	56
Ranunculus repens	55
Agrostis stolonifera	50

Bare soil

Bare soil class

Abundance in quadrats

	Abundance class										Distribution type
	1–10	11–20	21–30	31–40	41–50	51–60	61–70	71–80	81–90	81–100	
Wetland	5	1	–	1	1	1	–	–	–	1	2
Skeletal	1	1	–	–	–	–	–	–	–	–	–
Arable	–	–	–	–	–	–	–	–	–	–	–
Grassland	–	–	–	–	–	–	–	–	–	–	–
Spoil	–	–	–	–	–	–	–	–	–	–	–
Wasteland	–	–	–	–	–	–	–	–	–	–	–
Woodland	–	–	–	–	–	–	–	–	–	–	–
All habitats	50.0	16.7	–	8.3	8.3	8.3	–	–	–	8.3	2

Epilobium palustre L.

Onagraceae
Marsh Willow-herb

Life-form Polycarpic perennial protohemicryptophyte. Stems erect; leaves lanceolate, hairy, <1000 mm^2; roots with VA mycorrhizas.
Phenology Overwinters by means of swollen, terminal buds at the end of stolons. After shoot extension in late spring, plants flower in July and August and set seed from July to September.
Height Foliage and flowers to 600 mm.
Seedling RGR No data. SLA 36.1. DMC 16.1.
Nuclear DNA amount 0.6 pg; $2n = 36^*$ (*BC*); $2x^*$ (or perhaps $n = 9$).
Established strategy Intermediate between C-S-R and stress tolerator.

Regenerative strategies W, (V), ?B$_s$. Regenerates both vegetatively and by means of seed (see 'Synopsis'); ?SB III.
Flowers Pink, hermaphrodite, usually selfed; up to 10 in a raceme.
Germinule 0.04 mg, 1.8 × 0.5 mm, seed with a plume of hairs of *c*. 6 mm, in a dehiscent >50-seeded capsule.
Germination Epigeal; t_{50} 3 days; 6-34°C; L > D.
Biological flora None.
Geographical distribution The British Isles (100% of British vice-counties), *ABIF* page 416; Europe except for parts of the Mediterranean region (79% of European territories); Asia and N America.

Habitats Confined to mire, usually on base-poor soils, and infrequent in shaded sites. Also observed on wet cliffs and outcrops.
Gregariousness Sparse.
Altitude Wide-ranging, *c*. 5-470 m, but most-suitable habitats are found in upland areas. [Up to 790 m.]
Slope Like all wetland species, largely restricted to flat or gently sloping ground.
Aspect Insufficient data.
Hydrology Most frequent in soligenous mire, but found also in topogenous mire. Uncommon in sites marginal to open water, and absent from aquatic sites.

Soil pH All records in the pH range 4.5-7.5; most frequent and abundant on mildly acidic soils of pH 5.0-6.0, and not recorded from highly calcareous soils.
Bare soil Mainly restricted to sites with little exposed soil. Vegetation often comparatively open and with large quantities of bryophyte mat.
△ ordination Distribution centred on vegetation associated with relatively undisturbed habitats of intermediate fertility. Absent from disturbed vegetation types and also from both extremes of the productivity range.
Associated floristic diversity Intermediate to species-rich.

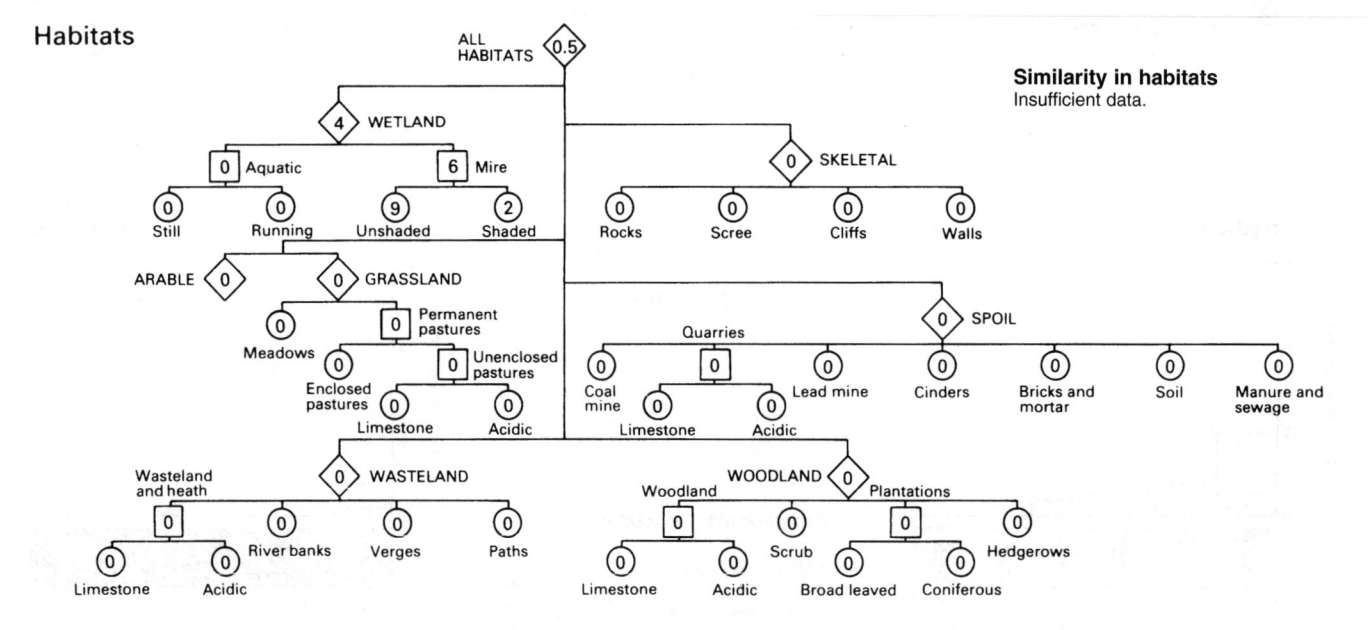

Habitats

Similarity in habitats
Insufficient data.

Synopsis *E.p.* is an erect herb growing at low densities in a range of wetland habitats, where the growth of potential dominants is restricted by an intermediate level of soil fertility and sometimes also by grazing. Sites with *E.p.* are further characterized by low levels of extractable macronutrients and low values of net above-ground production (Al-Farraj 1983). Within the Sheffield region the species is most characteristic of soligenous mire, but is also found in topogenous mire, flood hollows and ditch banks. Although recorded from calcareous sites in some parts of Britain (Wheeler 1980), and indifferent in its response to calcium level when grown experimentally (Al-Farraj 1983), *E.p.* is virtually restricted to mildly acidic soils. On calcareous strata the species is largely confined to acidic drift or to leached soils, a pattern similar to that shown by *Ranunculus flammula*. On highly calcareous soils *E.p.* appears to be replaced by *E. parviflorum*. At lowland sites the plant has a tall leafy stem (to 600 mm) and, when growing in pasture, is often found in areas of *Juncus effusus* where it may receive some protection from grazing. In upland areas plants are smaller (often <150 mm) and few-flowered, and tend to occur in lightly grazed habitats. In the more base-rich and productive non-calcareous mire sites (characterized, for example, by *Cirsium palustre* and *Rumex acetosa*) the species is apparently replaced by *E. obscurum*. Though less susceptible than *E. hirsutum*, *E.p.* is relatively intolerant of ferrous iron toxicity (R. E. D. Cooke unpubl.) and is characteristic of soils with moderate levels of extractable iron (Al-Farraj 1983). Effective regeneration is probably mainly by means of stolons. Swollen terminal buds are formed at the end of thread-like stolons, and these give rise to new plants, often over 100 mm away from the parent plant. Detached segments of stem are also capable of regeneration (Hodgson unpubl.). *E.p.* produces small, plumed seeds and annually an average plant may release *c*. 14 000 (Kytovuori 1969), although the output from the smaller plants frequently observed in upland populations is likely to fall short of 1000. No seed bank studies appear to have been carried out, but *E.p.* would be exceptional for the genus if it did not maintain some

Altitude
Slope
Aspect
Unshaded

Shaded

Soil pH

Associated floristic diversity

Hydrology

Triangular ordination

Commonest associates	%
Cardamine pratensis	71
Holcus lanatus	60
Galium palustre	51
Cirsium palustre	50
Poa trivialis	49

Bare soil

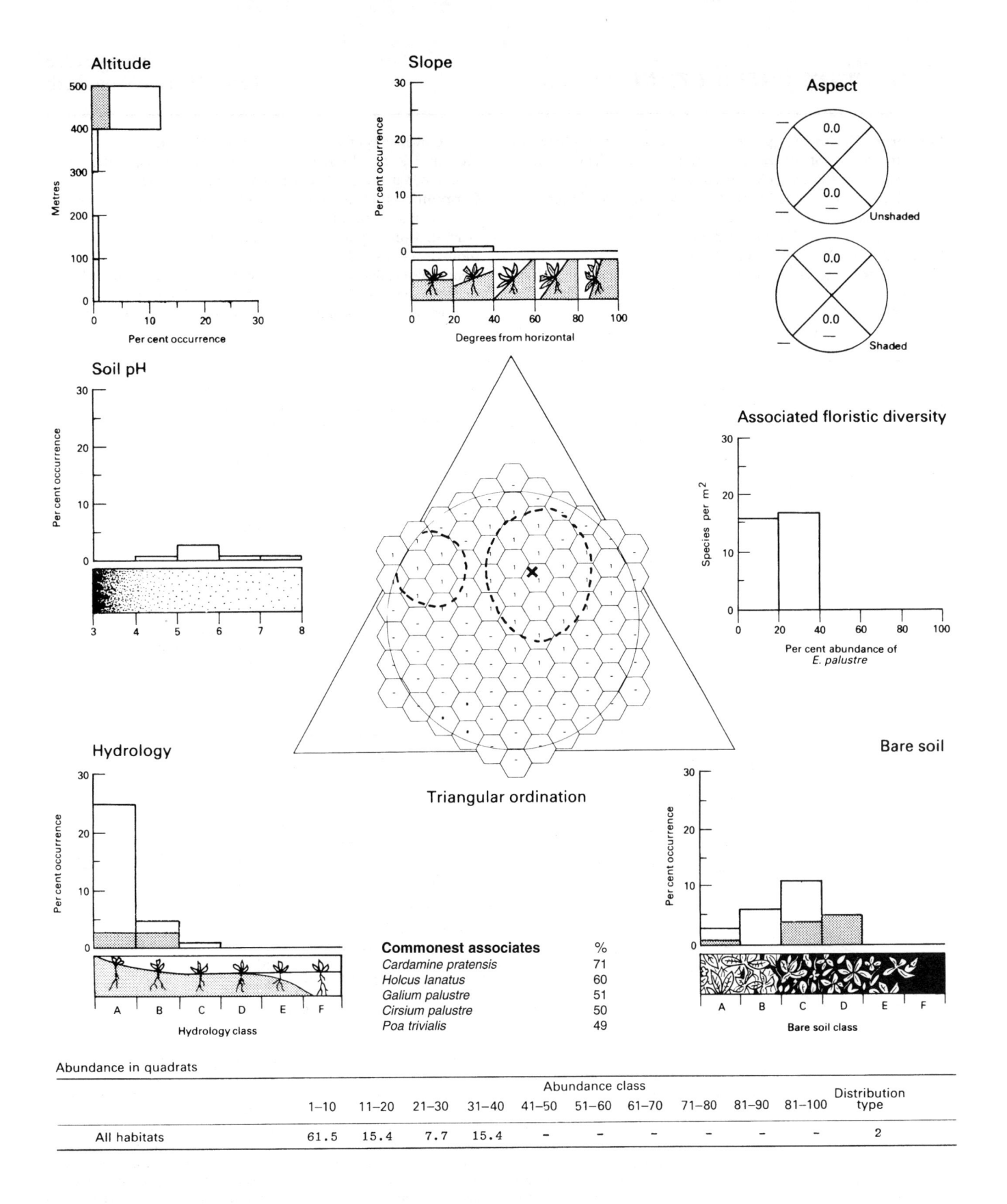

Abundance in quadrats

	1–10	11–20	21–30	31–40	41–50	51–60	61–70	71–80	81–90	81–100	Distribution type
All habitats	61.5	15.4	7.7	15.4	–	–	–	–	–	–	2

dormant seeds in the surface soil horizons. The seeds do not float (Ridley 1930) but wind-dispersal in *E.p.* appears to be effective, and the species has colonised gravel pits and colliery subsidence areas remote from some established populations. Although *E.p.* is known to form hybrids with most other British members of the genus (Stace 1975), the species is often ecologically isolated from other species, and no hybrids have been detected within the Sheffield region.

Current trends Although an effective colonist of new sites, *E.p.* is largely restricted to less-fertile soils and, at least in lowland areas, is likely to decline as a consequence of drainage, eutrophication and habitat destruction.

Epilobium parviflorum Schreber

Onagraceae
Lesser Hairy Willow-herb

Life-form Polycarpic perennial protohemicryptophyte or chamaephyte. Stems erect; leaves lanceolate, hairy, usually <2000 mm^2; roots with VA mycorrhizas.
Phenology Overwinters by means of leafy stolons. Flowers July and August and seed set August to October.
Height Foliage and flowers to 600 mm.
Seedling RGR No data. SLA 31.9. DMC 16.1.
Nuclear DNA amount No data; $2n = 36*$ (*BC*); $2x*$ (or perhaps $n = 9$).
Established strategy C-S-R.
Regenerative strategies W, ?B$_s$. Regenerates both vegetatively and by means of seed (see 'Synopsis'); ?SB III.
Flowers Rose, hermaphrodite, homogamous, often self-pollinated; often >20 in a racemose inflorescence.
Germinule 0.11 mg, 1.0 × 0.5 mm, with a plume of hairs of *c.* 5 mm, in a dehiscent <100-seeded capsule.
Germination Epigeal; t_{50} 3 days; L > D.
Biological flora None.
Geographical distribution The British Isles, but less abundant in the N (99% of British vice-counties), *ABIF* page 412; Europe northwards to Scandinavia (89% of European territories); N Africa, W Asia.

Habitats Most records are from shaded or unshaded calcareous mire, or from limestone stream beds which dry out in summer. Also recorded on limestone quarry spoil on the magnesian limestone and from damp walls. Observed on cinders and beside railways, and as an occasional weed of gardens. Absent from grassland, woodland and arable land.
Gregariousness Sparse to intermediate.
Altitude Extending to 350 m, but recorded most frequently below 100 m. [Up to 370 m.]
Slope Found over a wide range of values.
Aspect Insufficient data.
Hydrology Most frequent in soligenous mire, but also recorded from topogenous mire and rarely from temporarily submerged sites.
Soil pH Mainly restricted to soils of pH >5.5.
Bare soil Associated with widely different values, a reflection of its ecology both as a colonist of disturbed ground and as a component of species-rich mire.
△ **ordination** Present in a relatively wide range of vegetation types but absent from very unproductive vegetation and from the most heavily disturbed habitats.
Associated floristic diversity Often species-rich in wetland and usually intermediate in disturbed artificial habitats.

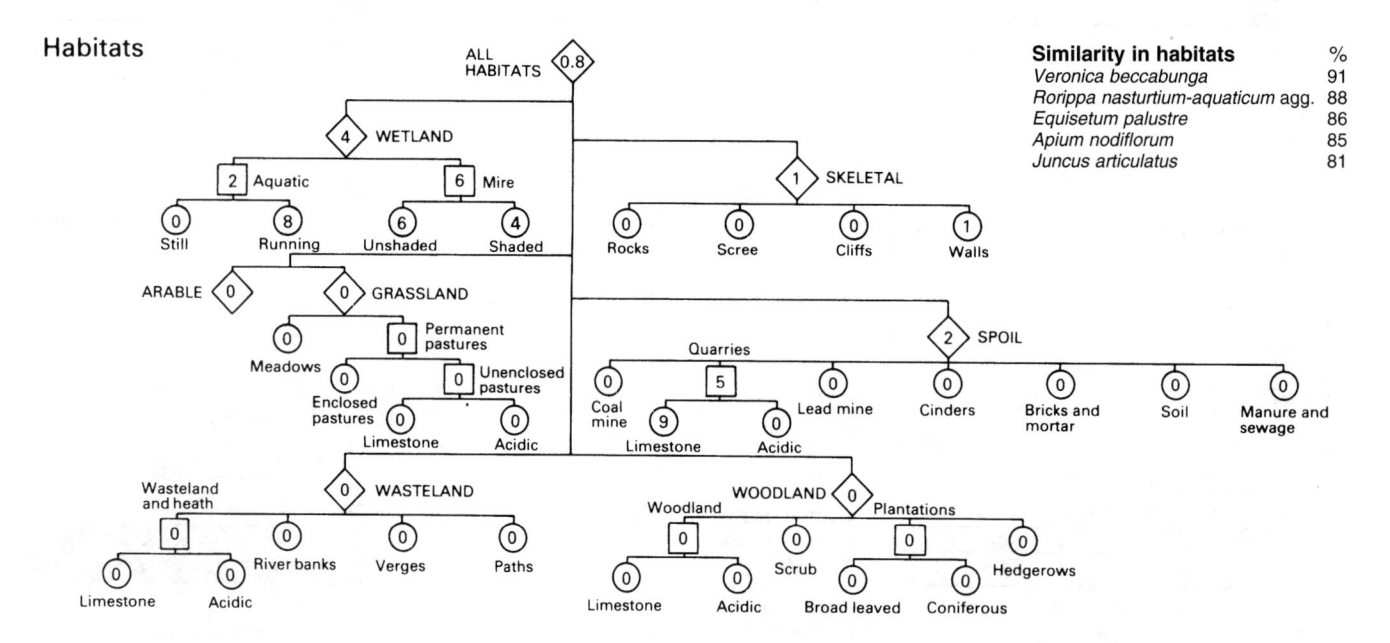

Habitats

Similarity in habitats	%
Veronica beccabunga	91
Rorippa nasturtium-aquaticum agg.	88
Equisetum palustre	86
Apium nodiflorum	85
Juncus articulatus	81

Synopsis A species which requires further study. *E.p.* colonizes a wide range of relatively moist habitats, and is persistent in circumstances where the vigour of potential dominants is restricted by intermediate levels of soil fertility and occasional disturbance. The species occurs primarily in two types of habitat: (a) derelict land and (b) soligenous mire and seasonally exposed river beds on calcareous strata. On derelict land the species is associated with a wide range of soil conditions and is unusual in its capacity to grow on finely-powdered magnesian limestone quarry spoil of pH 8.0. In wetlands, the species' commonest habitat, *E.p.* is rare on non-calcareous soils and, consequently there is little overlap with either *E. obscurum* or *E. palustre*. Moreover, in regions where *E. palustre* occurs frequently on calcareous soils, *E.p.* exploits the more productive areas (B. D. Wheeler pers. comm.). In topogenous mire *E.p.* tends to occupy drier upper reaches of the wetland habitat. *E.p.* is relatively tall, and the shoots are vulnerable to trampling and heavy grazing. However, *E.p.* is not infrequent in grazed mire, perhaps because trampling results in the formation of open microsites suitable for colonization by seedlings. In common with *E. hirsutum*, *E.p.* is a lowland species which is close to its altitudinal limit over much of N Britain. *E.p.* produces numerous wind-dispersed seeds which germinate directly in vegetation gaps or after incorporation into a persistent seed bank. The species also forms short overwintering stolons, but these are produced close to the base of the flowering stem and it is uncertain whether these operate primarily as a means of regeneration or merely allow perennation. *E.p.* hybridizes with many other species (Stace 1975), but none of the crosses achieves widespread abundance.
Current trends Overall and as a colonist of artificial habitats, *E.p.* is probably increasing. However, the species may be decreasing in wetlands as a result of habitat destruction.

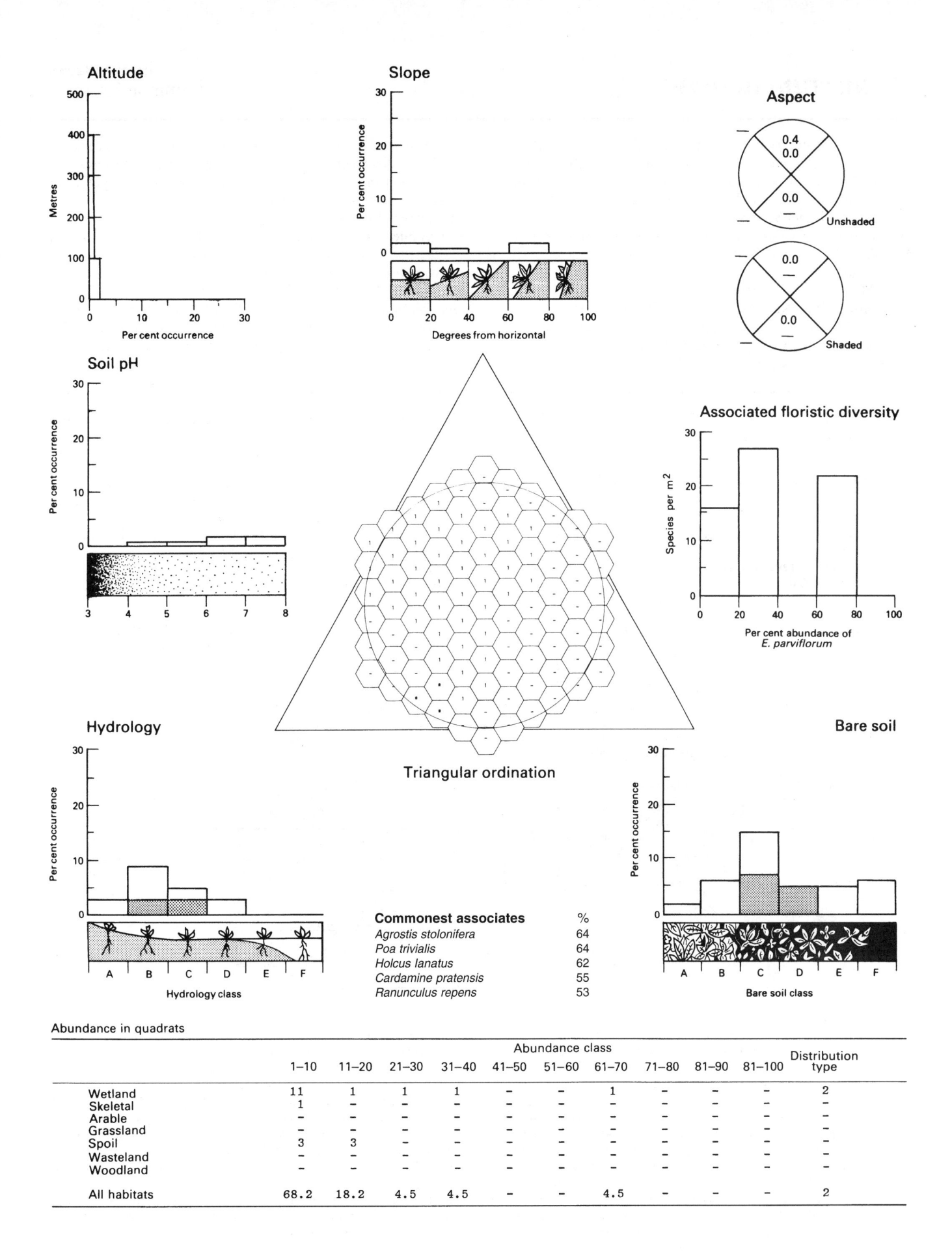

Altitude

Slope

Aspect

Unshaded

0.4	—
0.0	0.0
—	

Shaded

—	—
0.0	0.0
—	

Soil pH

Associated floristic diversity

Hydrology

Triangular ordination

Bare soil

Commonest associates %
Agrostis stolonifera 64
Poa trivialis 64
Holcus lanatus 62
Cardamine pratensis 55
Ranunculus repens 53

Abundance in quadrats

	\multicolumn{10}{c	}{Abundance class}	Distribution type
	1–10	11–20	21–30	31–40	41–50	51–60	61–70	71–80	81–90	81–100	
Wetland	11	1	1	1	–	–	1	–	–	–	2
Skeletal	1	–	–	–	–	–	–	–	–	–	–
Arable	–	–	–	–	–	–	–	–	–	–	–
Grassland	–	–	–	–	–	–	–	–	–	–	–
Spoil	3	3	–	–	–	–	–	–	–	–	–
Wasteland	–	–	–	–	–	–	–	–	–	–	–
Woodland	–	–	–	–	–	–	–	–	–	–	–
All habitats	68.2	18.2	4.5	4.5	–	–	4.5	–	–	–	2

Equisetum arvense L.

Equisetaceae
Common Horsetail

Life-form Polycarpic perennial geophyte with extensive rhizomes and small tubers. Sterile stems branched, erect or prostrate, green with scale-like leaves, <10 mm², fused into a sheath; fertile stems, unbranched, light brown; roots often deep, with or without VA mycorrhizas.
Phenology Fertile shoots produced in April and persist *c*. 10 days (Page 1982). Sterile shoots develop slightly later and die back in late autumn.
Height Green sterile stems to 800 mm; sporangia to 250 mm.
Sporeling RGR No data. SLA. DMC. Leaves absent.
Nuclear DNA amount 28.4 pg; $2n = 216^*$ (*BC*); $2x^*$.
Established strategy Competitive-ruderal.
Regenerative strategies V, W, S. Forms extensive rhizomatous, clonal patches. Rhizome fragments and detached sterile stems also readily regenerate. Establishment by means of spores may be infrequent since moist bare soil appears to be required for colonization (Page 1967); SB I.
Sporangia In a terminal cone.
Germinule 34 × 33 μm, spore, short-lived, green, explosively dispersed from a many-spored sporangium.
Germination t_{50} 1 day; L > D.
Biological flora Cody and Wagner (1980).
Geographical distribution The British Isles (100% of British vice-counties), *ABIF* page 58, *AF* page 20; most of Europe (97% of European territories); Arctic and N temperate zones.

Habitats Frequent on river banks, spoil heaps, waysides, both shaded and unshaded mire, and wasteland. Occurs at the edge of pastures and arable land, often as a result of vegetative spread from the adjoining hedge or field bank. Only of casual occurrence in aquatic systems, and absent from skeletal habitats. Particularly abundant beside railways. [Occurs on fixed dunes and cliff-tops, in mountain areas beside streams and in flushes (Page 1982).]
Gregariousness Recorded as sparse to intermediate, but can form extensive patches.
Altitude Widely distributed and observed up to 400 m in the Sheffield region. [Up to 950 m.]
Slope Absent from skeletal habitats, and hence also from the steepest slopes.
Aspect Recorded slightly more frequently on unshaded S-facing slopes, but more often abundant in N-facing situations.
Hydrology Mainly found on moist soils or in soligenous mire, where it may replace *E. palustre*. Not present in permanently submerged aquatic habitats.
Soil pH Widely distributed, but largely absent from acidic soils. Most frequent and abundant in the pH range 5.5-7.5.
Bare soil Widely distributed with greatest frequency in habitats having intermediate amounts.

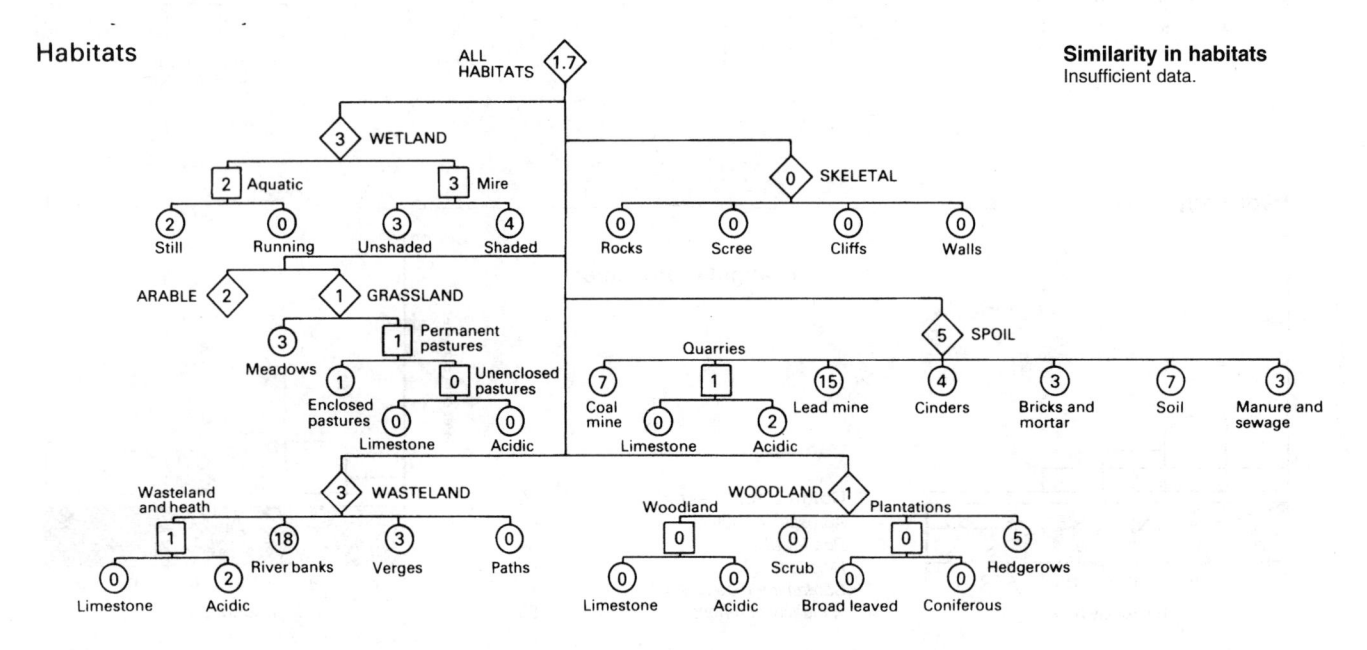

Habitats

Similarity in habitats
Insufficient data.

△ **ordination** Distribution relatively widespread, with contours centred on productive types of vegetation subject to some disturbance.
Associated floristic diversity Varying widely, from species-poor in sites where *E.a.* is an early colonist to species-rich in, for example, soligenous mire.
Synopsis *E.a.* resembles *Tussilago farfara* in both ecology and life-history. Each is a colonist of moist disturbed areas, and both are capable of persisting long after perennial vegetation has been re-established. Morphological similarities include an extensive, potentially fast-growing and often deep rhizome system and the production of large photosynthetic organs (in the case of *E.a.*, branched stems) which persist throughout the summer. As with *T.f.*, the rhizome system exploits a wide range of soil depths, enabling *E.a.* to persist in sites prone to soil slippage. However, the most remarkable parallel lies in regeneration. *E.a.* produces cones on short-lived achlorophyllous stems in early spring before the green shoots have developed (similarly *T.f.* flowers before the leaves appear). Like the seeds of *T.f.*, the numerous spores soon lose viability. Cones are usually only produced in abundance in open habitats, and are seldom observed in well-vegetated sites. *E.a.*, which is toxic to stock (Cooper & Johnson 1984), is a weed of over 25 broad-leaved, cereal and pasture crops, and is regarded as one of the world's worst weeds (Holm *et al.* 1977). However, this designation seems unjustified within the Sheffield region (see 'Habitats'). Distinctive features which relate generally to the genus *Equisetum* are described under *E. palustre*.
Current trends Perhaps increasing in disturbed artificial habitats, particularly beside railways, roads and ditches, and to a lesser extent in gardens and on arable land.

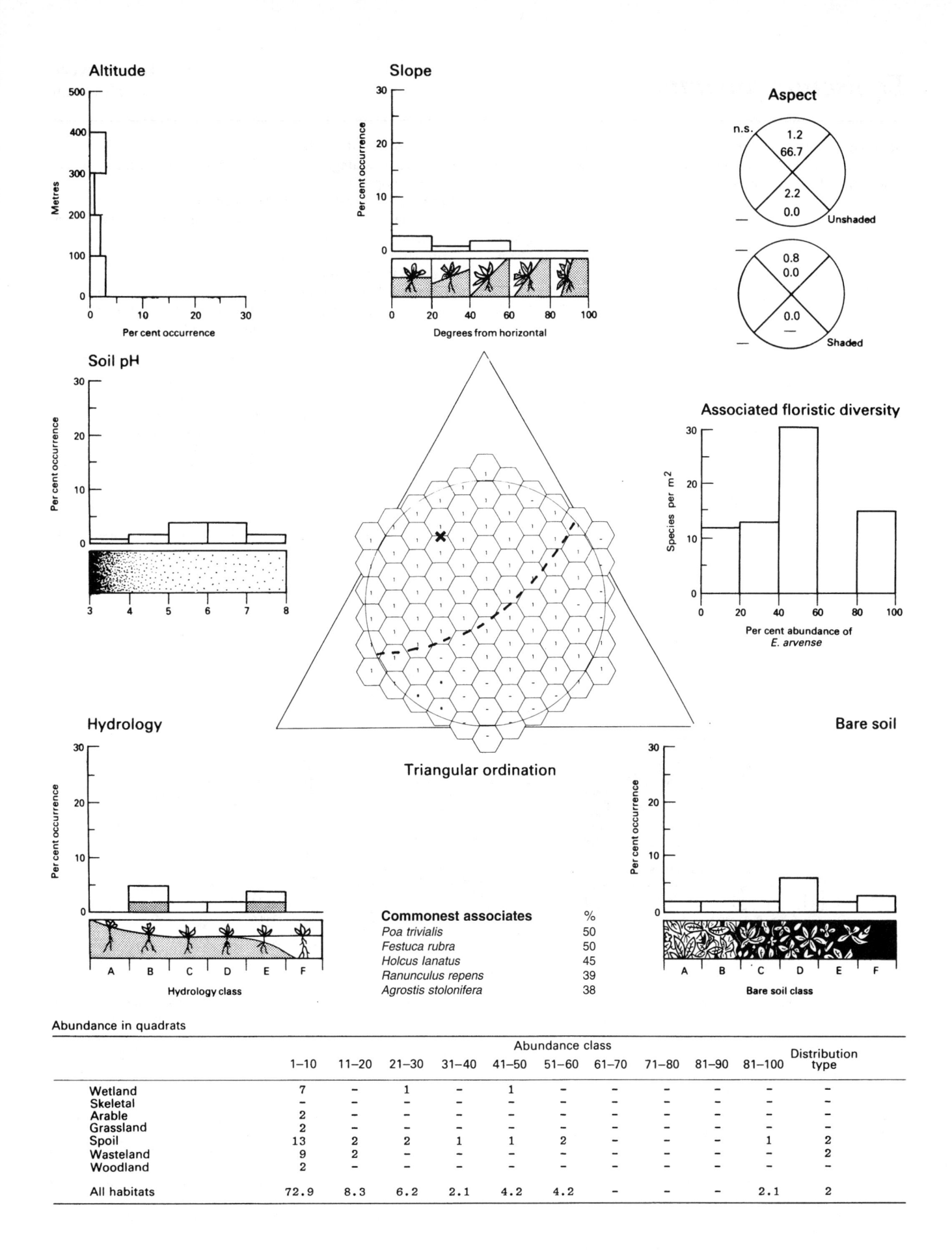

Altitude

Slope

Aspect

n.s. 1.2 / 66.7 / 2.2 / 0.0 Unshaded

0.8 / 0.0 / 0.0 / — Shaded

Soil pH

Associated floristic diversity

Hydrology

Triangular ordination

Bare soil

Commonest associates	%
Poa trivialis	50
Festuca rubra	50
Holcus lanatus	45
Ranunculus repens	39
Agrostis stolonifera	38

Abundance in quadrats

	1–10	11–20	21–30	31–40	41–50	51–60	61–70	71–80	81–90	81–100	Distribution type
Wetland	7	–	1	–	1	–	–	–	–	–	–
Skeletal	–	–	–	–	–	–	–	–	–	–	–
Arable	2	–	–	–	–	–	–	–	–	–	–
Grassland	2	–	–	–	–	–	–	–	–	–	–
Spoil	13	2	2	1	1	2	–	–	–	1	2
Wasteland	9	2	–	–	–	–	–	–	–	–	2
Woodland	2	–	–	–	–	–	–	–	–	–	–
All habitats	72.9	8.3	6.2	2.1	4.2	4.2	–	–	–	2.1	2

Equisetum fluviatile L.

Equisetaceae
Water Horsetail

Life-form Polycarpic perennial helophyte or hydrophyte with extensive rhizomes bearing erect green stems with scale-like leaves, <10 mm² and fused into a sheath; roots non-mycorrhizal.

Phenology Shoots expand in spring and cones produced in June to July. In mire shoots die back in autumn, but in aquatic habitats they may remain green over winter, at least at the base.

Height Sterile and fertile stems to 1200 mm.

Sporeling RGR No data. SLA. DMC. Leaves absent.

Nuclear DNA amount 27.0 pg; $2n = 216*$ (*BC*); $2x*$ (Page 1982).

Established strategy Intermediate between stress-tolerant competitor and C-S-R.

Regenerative strategies V, W, S. Primarily by vegetative means (see 'Synopsis').

Sporangia Aggregated into a terminal cone.

Germinule Spore, short-lived, green, explosively dispersed from a many-spored sporangium.

Germination t_{50} c. 2 days; L > D.

Biological flora None.

Geographical distribution The British Isles (99% of British vice-counties), *ABIF* page 58, *AF* page 18; most of Europe (74% of European territories); temperate Asia and N Africa.

Habitats A wetland species occurring most commonly in shallow water at the edges of lakes and ponds; also frequently recorded or observed in shaded or unshaded mire.

Gregariousness Stand-forming.

Altitude Found over a wide range of altitudes to 400 m, but suitable habitats are rare on some strata (carboniferous and magnesian limestones), leading to discontinuities in altitudinal range. [Up to 1050 m.]

Slope Like most wetland species, mainly confined to flat or gently sloping ground.

Aspect No data.

Hydrology An emergent aquatic growing in shallow water at the edge of ponds and ditches. Often extending from mire at the water's edge to water up to 650 mm in depth. [Down to 1.5 m (Spence 1964).] Also found in topogenous mire adjacent to open water and in soligenous mire.

Soil pH Recorded within the range pH 4.0-7.5 (rarely observed at lower or higher values). Most frequent and abundant on mildly acidic soils in the middle of this range.

Bare soil Mainly associated with low exposures.

△ **ordination** Insufficient data.

Associated floristic diversity Typically low, but not infrequent in species-rich soligenous mire.

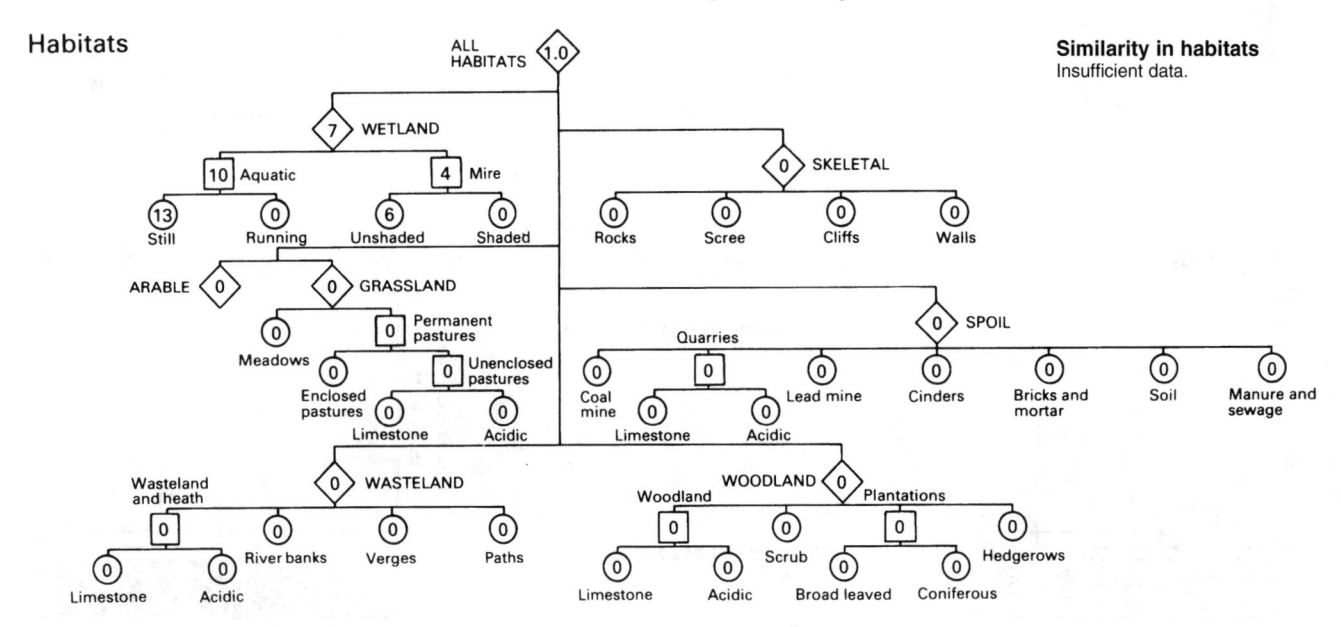

Habitats

Similarity in habitats
Insufficient data.

Synopsis (See. also that of *E. palustre*.) *E.f.* is a rhizomatous, often stand-forming species which most characteristically behaves as an emergent aquatic from water at depths of up to c. 1.5 m (Spence 1964) at the margins of lakes or ponds and in ditches. In such sites the growth of other potential dominants is usually restricted by soil infertility and sometimes by shade. In N England the species is typically associated with mildly acidic soils, though it may occur locally on highly calcareous soils. The species is most frequent where there are comparatively minor fluctuations in water table during summer. *E.f.* shows some vulnerability to the effects of wave action, but may occur on relatively exposed shores where it replaces *Eleocharis palustris* as the water becomes deeper (Spence 1964). Although most abundant as an aquatic, *E.f.* may persist in mire during pond siltation, and may increase again when a tree canopy is formed. *E.f.* also occurs as a usually minor constituent in soligenous mire and wet grazed meadows. The species lacks tubers (Page 1982), but the rhizome system is extensive and up to 80% (by weight) of plant biomass may be below ground (Sculthorpe 1967). The length of shoot projecting above water is proportional to the depth of the water (Spence 1964). The aerial stems have a central hollow which may reach 90% of the diameter of the shoot. Unusually, this central cavity extends into the rhizome (Page 1982) and may facilitate oxygen diffusion to the roots, apparently allowing *E.f.* to persist in highly anaerobic soils. Cones are frequently produced, but establishment of new plants from spores is probably very rare. *E.f.* can form large clonal patches by means of rhizome growth, and detached rhizomes or pieces of shoot may regenerate either *in situ* or after being transported by water to a new site. The species is unusually plastic. Typically the stems are unbranched, but in sheltered or shaded sites plants may have whorls of lateral branches (Page 1982). Hybrids are formed with *E. arvense* and probably also with *E.p.* (Stace 1975).

Current trends Apparently decreasing in lowland areas (Page 1982) as a result of habitat destruction and eutrophication.

Altitude

Slope

Aspect

Unshaded 0.0

Shaded 0.0

Soil pH

Associated floristic diversity

Not applicable

Hydrology

Triangular ordination

Bare soil

Commonest associates	%
Cardamine pratensis	39
Epilobium palustre	38
Carex nigra	36
Ranunculus repens	36
Galium palustre	35

Abundance in quadrats

	1–10	11–20	21–30	31–40	Abundance class 41–50	51–60	61–70	71–80	81–90	81–100	Distribution type
Wetland	8	7	4	2	–	1	–	3	–	3	2
Skeletal	–	–	–	–	–	–	–	–	–	–	–
Arable	–	–	–	–	–	–	–	–	–	–	–
Grassland	–	–	–	–	–	–	–	–	–	–	–
Spoil	–	–	–	–	–	–	–	–	–	–	–
Wasteland	–	–	–	–	–	–	–	–	–	–	–
Woodland	–	–	–	–	–	–	–	–	–	–	–
All habitats	28.6	25.0	14.3	7.1	–	3.6	–	10.7	–	10.7	2

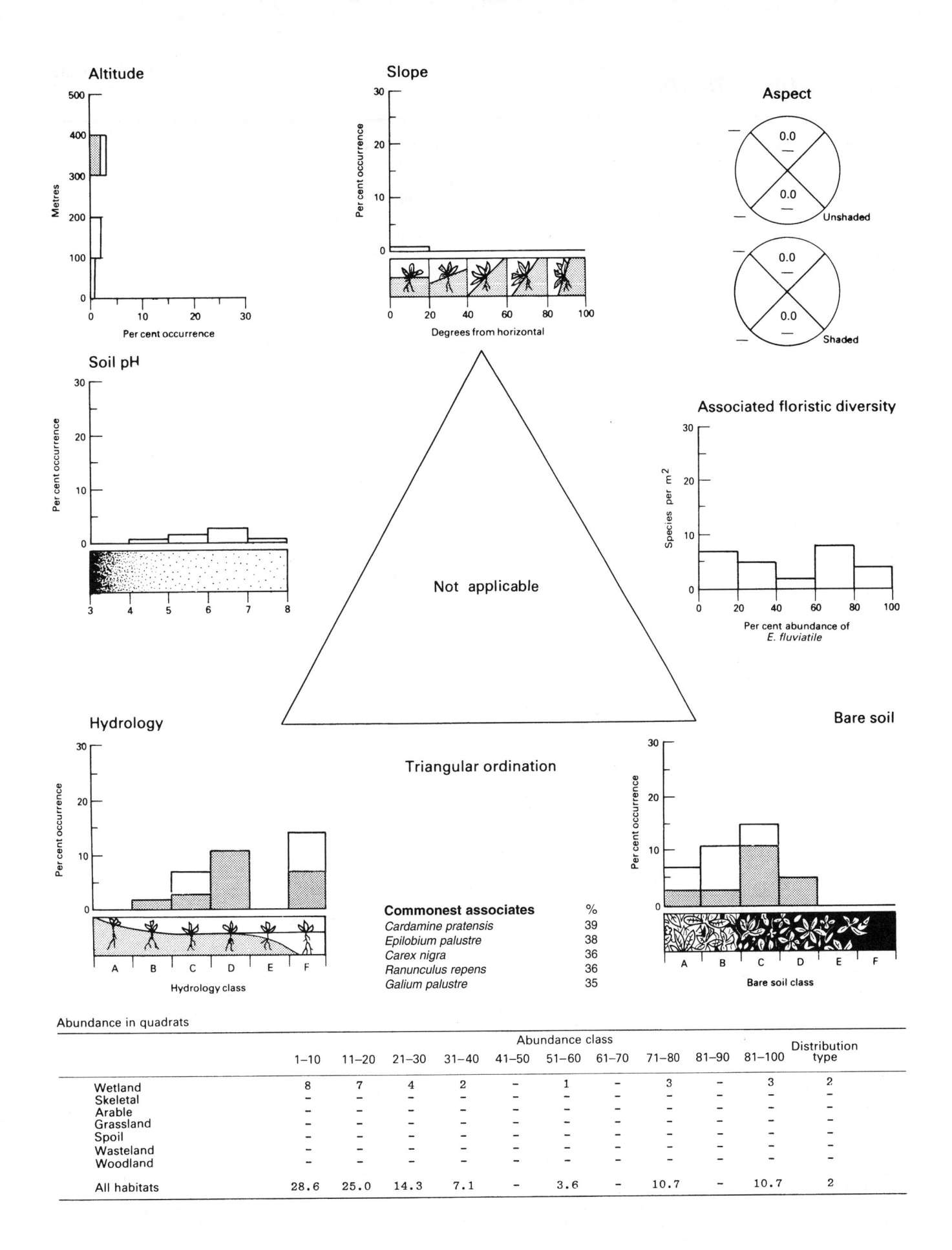

Equisetum palustre L

Equisetaceae
Marsh Horsetail

Life-form Polycarpic perennial geophyte with extensive system of rhizomes bearing tubers. Both fertile and sterile stems green, branched, typically erect with scale-like leaves <10 mm^2 and fused into a sheath; roots often with VA mycorrhizas.
Phenology Cones produced from June to July and plants die back in autumn.
Height Sterile and fertile stems to 500 mm.
Sporeling RGR No data. SLA. DMC. Leaves absent.
Nuclear DNA amount 25.0 pg; $2n = 216^*$ (*BC*); $2x^*$.
Established strategy Intermediate between competitive-ruderal and C-S-R.

Regenerative strategies V, W, S. Primarily by vegetative means (see 'Synopsis'); SB I.
Sporangia Brown, aggregated into a terminal cone.
Germinule Spore, short-lived, green, explosively dispersed from a many-spored sporangium.
Germination t_{50} 2 days; L > D.
Biological flora None, but see Borg (1971).
Geographical distribution The British Isles (99% of British vice-counties), *ABIF* page 59, AF page 25; most of Europe, less common in the S (89% of European territories); temperate Asia and N America.

Habitats Restricted to wetland habitats. Most common in unshaded mire, including the beds of streams which dry up in summer. Also recorded from aquatic habitats and from bare soil or rocky habitats adjacent to open water. Observed as an abundant colonist of gravel pits and in moist pastures. [Found in dune slacks, though *E. variegatum* Schleich. ex Weber and Mohr is more common in this habitat. Extends onto mountains beside streams and in flushes (Page 1982).]
Gregariousness Typically sparse to intermediate, but sometimes stand-forming.
Altitude Recorded over a wide range of altitudes and without the pronounced lowland bias evident in the majority of common wetland species in Britain. [Up to 920 m.]

Slope As with most wetland species, largely confined to level or gently sloping ground.
Aspect Insufficient data. Subsequent fieldwork has failed to detect any aspect bias.
Hydrology Frequent in both soligenous and topogenous mire. Often found in sites marginal to open water and in temporarily flooded situations, but absent from permanently submerged sites.
Soil pH Largely restricted to soils of pH >5.0.
Bare soil Associated with a wide range of values, but most frequent at intermediate values.
△ ordination Distribution centred on relatively fertile disturbed habitats. Absent from highly disturbed or very unproductive situations.
Associated floristic diversity Low to high. Occurs both as an early colonist of bared ground with few other associated species and in species-rich wet meadows.

Similarity in habitats	%
Juncus articulatus	95
Glyceria fluitans	92
Callitriche stagnalis	91
Apium nodiflorum	91

Habitats

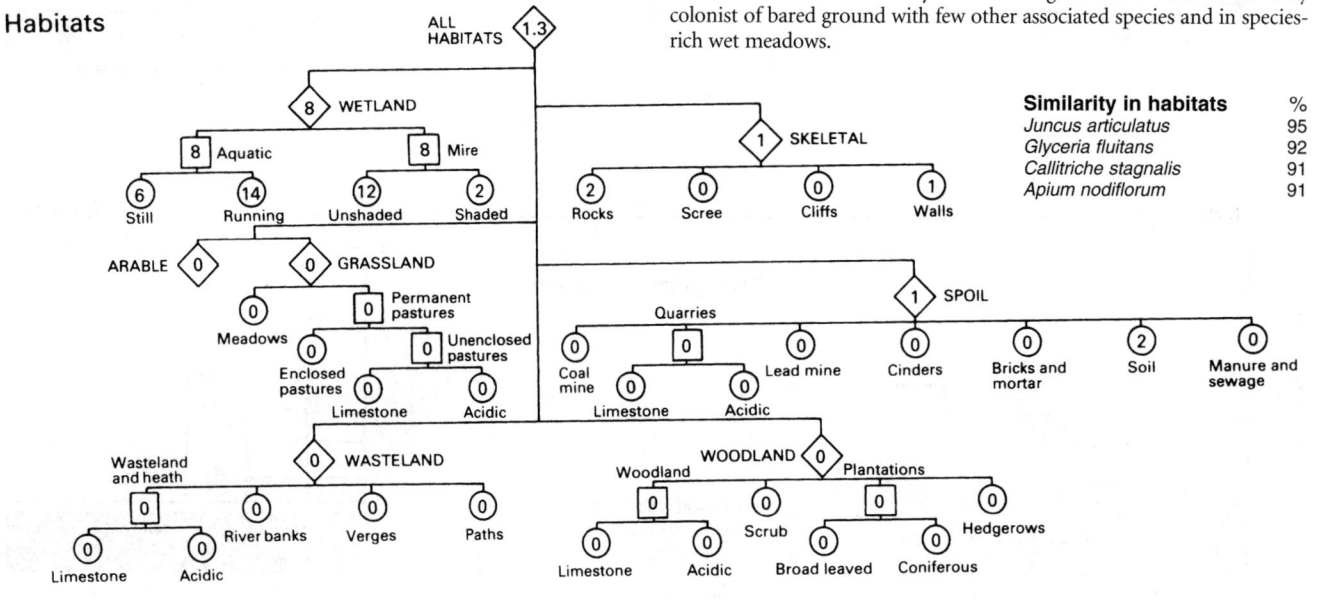

Synopsis The genus *Equisetum* belongs to an ancient section of the Pteridophyta, and has no closely related extant relatives. Over 30% of the world taxa are found in the British Isles (Page & Barker 1985). Plants lack secondary thickening (Bell 1985) and shoots are xeromorphic in character, with the leaves reduced to scales and the stem, which is often highly siliceous and unpalatable to stock (Cooper & Johnson 1984), acting as the main photosynthetic organ. The green spores are short-lived, and development of the green, delicate gametophyte is contingent upon dispersal to moist, bared ground. The gametophytes of the three commonest species, *E. arvense*, *E. fluviatile* and *E. palustre*, are seldom recorded, but may form large populations on bare mud beside lakes and reservoirs (Duckett & Duckett 1980). They are very sensitive to competition, and apparently also to allelopathic substances produced by the sporophyte, but mature rapidly and are typically dioecious (Duckett & Duckett 1980). In the three common species, *E.a.*, *E.f.* and *E.p.*, the vulnerability of the gametophyte is offset by vigorous vegetative regeneration. This is unusual in the Pteridophytes; here, as in *Pteridium*, it is associated with the presence of vessels in the vascular tissue (White 1963). *E.p.* exploits a variety of moderately fertile wetland habitats. The rhizomes are often deep, enabling *E.p.* to persist on drained land and to locate roots in soil horizons deeper than those exploited by most of its associates (Borg 1971). The species has a rather sparse canopy and is not normally a dominant of plant communities. Instead, the species tends to occur either as a colonist of disturbed habitats, such as ditch sides and recently excavated gravel pits, or as a subordinate component of communities in which the growth of potential dominants is restricted by grazing or cutting. Apart from being unpalatable, *E.p.* is also toxic to stock and 2 g day^{-1} in hay may reduce the milk yield of cattle (Holm *et al.* 1977). The toxic principle involved appears to be an alkaloid rather than the enzyme thiaminase, which is very important in other species (Holm *et al.* 1977, Cooper & Johnson 1984). Both *E.a.* and *E.p.* are relatively frequent on metalliferous spoil. However, reports that these species accumulate gold

Altitude
Slope
Aspect
Unshaded
Shaded

Soil pH

Associated floristic diversity
Species per m² vs Per cent abundance of *E. palustre*

Triangular ordination

Hydrology

Commonest associates	%
Holcus lanatus	59
Poa trivialis	56
Ranunculus repens	53
Agrostis stolonifera	50
Cardamine pratensis	45

Bare soil

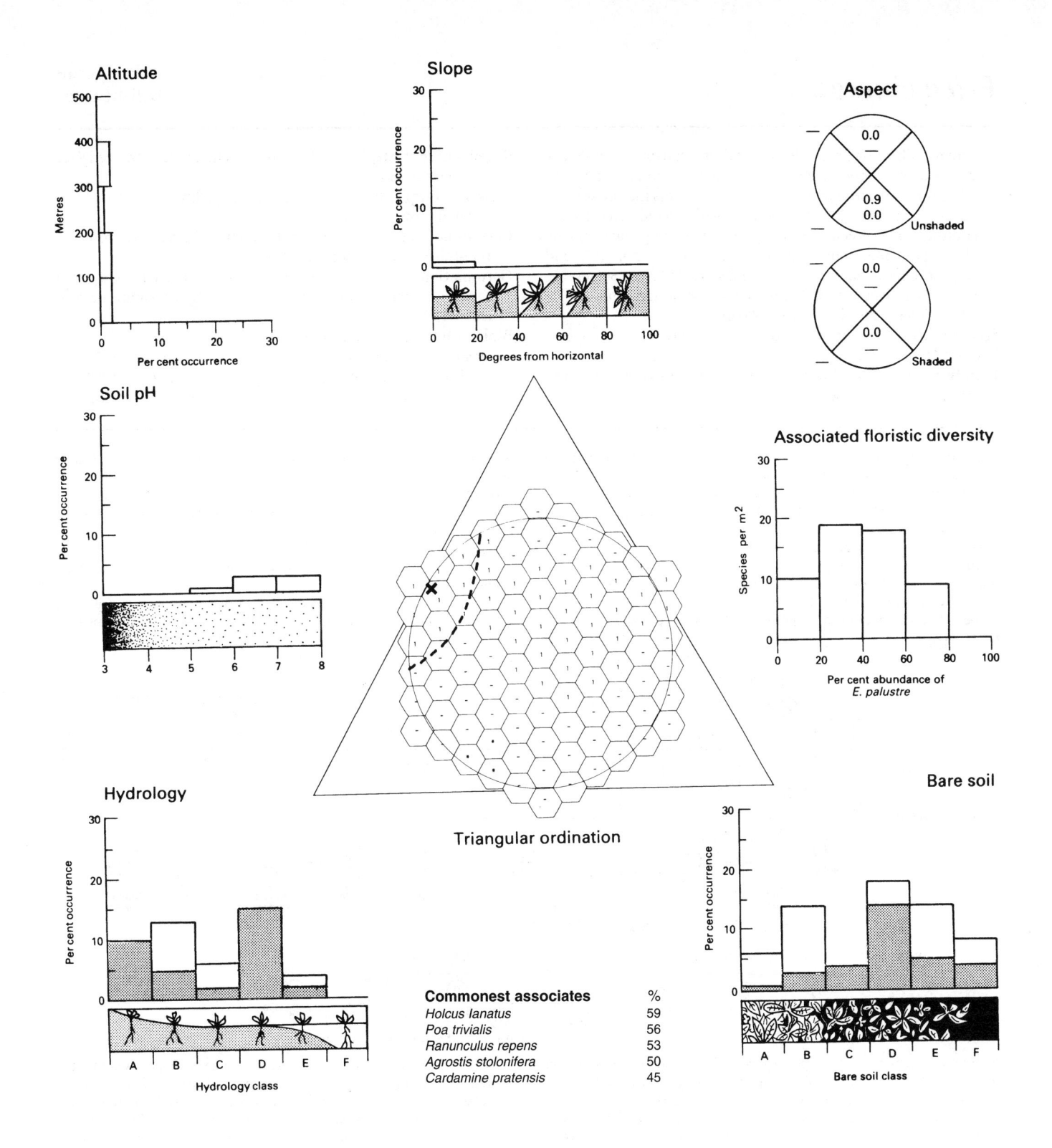

in their tissues are grossly exaggerated, resulting from a failure to take account of the presence of arsenic, which affects some analytical procedures and may lead to overestimation of gold content. *E.p.* spreads by means of stout rhizomes to form extensive clonal patches, and may also regenerate from plant fragments. It is suspected that plant parts are occasionally transported by water to form new colonies. In low-growing vegetation in soligenous mire, particularly on less-fertile calcareous soils, *E.p.* may be replaced by *E.a.*. The species, which is included by Holm *et al.* (1977) as one of the world's worst weeds, varies markedly in habit and branch form according to habitat, and hybridizes with three other species of *Equisetum* (Stace 1975).

Current trends Uncertain. Capable of long-distance dispersal, and both invasive and persistent once established, but not usually amongst the first colonists of new wetland habitats, presumably because of problems in establishment from spores.

Erica cinerea L.

Ericaceae
Bell-heather

Life-form Shrubby phanerophyte or chamaephyte. Stems ascending, branched, rooting at the base; leaves xeromorphic, linear, <10 mm² with strongly revolute margins; roots mainly shallow but may extend to 240 mm (*Biol Fl*), with ericoid mycorrhizas.

Phenology Evergreen. Shoots expand during late spring although limited winter growth occurs. Flowers in July and August (before *Calluna vulgaris*) and seed shed from September onwards. Leaves last for 2 years, sometimes longer (*Biol Fl*).

Height Foliage and flowers to *c.* 600 mm.

Seedling RGR No data. SLA 10.9. DMC. No Data.

Nuclear DNA amount No data; $2n = 24$ (*FE3*); $2x$.

Established strategy Intermediate between stress-tolerator and stress-tolerant competitor.

Regenerative strategies B_s. Effective regeneration is by seed (see 'Synopsis'); SB IV.

Flowers Purple, hermaphrodite, insect-pollinated or selfed; up to 50 in terminal racemes (*Biol Fl*).

Germinule 0.04 mg, seed in a dehiscent, <34-seeded capsule (*c.* one-third of seeds abortive) *Biol Fl*.

Germination Epigeal; low germination percentage for untreated seeds; germination promoted by chilling, scarification or heat shock (*Biol Fl*).

Biological flora Bannister (1965).

Geographical distribution Most of the British Isles (97% of British vice-counties), *ABIF* page 292; W Europe (28% of European territories); introduced into parts of N America.

Habitats Largely confined to well-drained, acidic soils. Recorded from heathland, grazed moorland, old quarries, rock outcrops and cliffs. In lowland areas largely restricted to roadsides, golf-courses and railway banks. [Also recorded from oak, birch and pine woodland, mire, sand dunes and maritime and low-alpine heaths (*Biol Fl*).]

Gregariousness Intermediate.

Altitude Formerly widespread, but suitable habitats are now more frequent in the uplands. Extends to 350 m. [Up to 1240 m in Britain, but not above 620 m in N England (*Biol Fl*).]

Slope Recorded both from flat ground and from steep slopes.

Aspect Significantly more frequent on S-facing slopes, but locally abundant in both N- and S-facing situations.

Hydrology Absent from wetlands within survey area. [Recorded infrequently from hummocks in ombrogenous mire and other wetland areas (*Biol Fl*).]

Soil pH Confined to soils of pH <5.0.

Bare soil Distribution biased towards habitats with low to moderate exposures of bare soil.

△ ordination Absent from frequently disturbed vegetation. Distribution centred on relatively undisturbed communities of low or intermediate productivity.

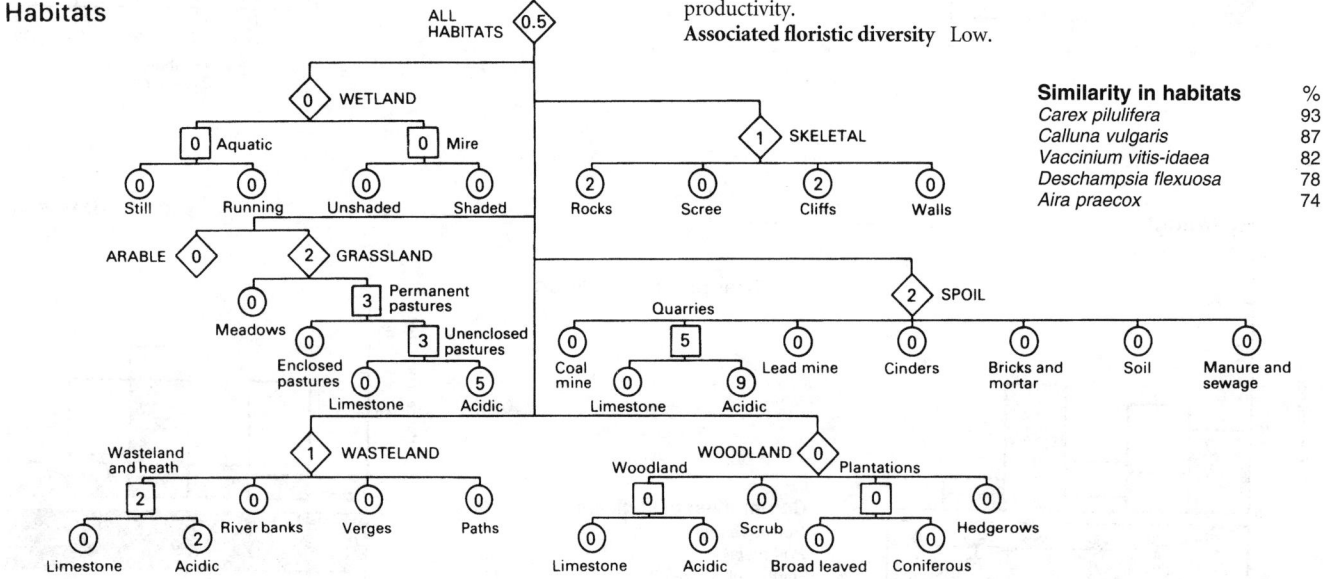

Associated floristic diversity Low.

Similarity in habitats	%
Carex pilulifera	93
Calluna vulgaris	87
Vaccinium vitis-idaea	82
Deschampsia flexuosa	78
Aira praecox	74

Habitats

Synopsis *E.c.* is a small evergreen shrub which is found primarily on well-drained acidic mineral soils, with *Calluna vulgaris*. Unlike *C.v.*, *E.c.* has a narrow ecological range and tends not to vary greatly in morphology (*Biol Fl*). The species is usually shorter in stature than *C.v*, and has a lower altitudinal limit which decreases with increasing latitude (*Biol Fl*). *E.c.* is associated with oceanic climates, has a capacity greater than that of *C.v.* to exploit dry habitats (*Biol Fl*), and shows a significant bias towards S-facing slopes in the Sheffield region. *E.c.* is also more shade-tolerant (*Biol Fl*). Elsewhere in heathland *E.c.* is less successful, tending to survive under the canopy of *C.v.* or as scattered individuals (*Biol Fl*). *E.c.* often expands following burning and in these circumstances may achieve temporary dominance (*Biol Fl*). The stem bases appear to re-sprout better than those of *C.v.* Germination is stimulated by heat treatment (a short period at 100°C may overcome the requirement for vernalization (Mallik & Gimingham 1985)) and after burning is frequently superior to that of *C.v.* (*Biol Fl*). If sites are grazed *E.c.* is less successful despite its low palatability to sheep (*Biol Fl*) and a majority of our survey records are from ungrazed sites. *E.c.* shows some winter growth and is more frost-sensitive than *C.v.* (*Biol Fl*). Like *C.v.*, the species is unusual among woody plants of infertile sites in being relatively short-lived (perhaps *c.* 20 years *Biol Fl*) and in its potential to flower in the second or third year (*Biol Fl*). More research is needed to examine the extent to which the life histories and reproductive effort of *E.c.* and *C.v.* are dependent upon the pulses of mineral nutrient release associated with heathland burning. The species produces numerous minute seeds which germinate in vegetation gaps during autumn and intermittently during the following growing season (*Biol Fl*). Seeds are also incorporated into a persistent seed bank (*Biol Fl*). The plant forms compact tufts, but vegetative spread is not sufficiently extensive to be regarded as an important means of regeneration. Populations may exhibit some attunement to the climate of their site of origin (Bannister 1978) and some are physiologically adapted to alkaline heath soils (Davies 1984).

Current trends *E.c.* has been drastically reduced in the lowlands by destruction of heathland and *E.c.* is now uncommon and largely restricted to roadsides, railway banks and golf courses. Trends within the uplands are uncertain.

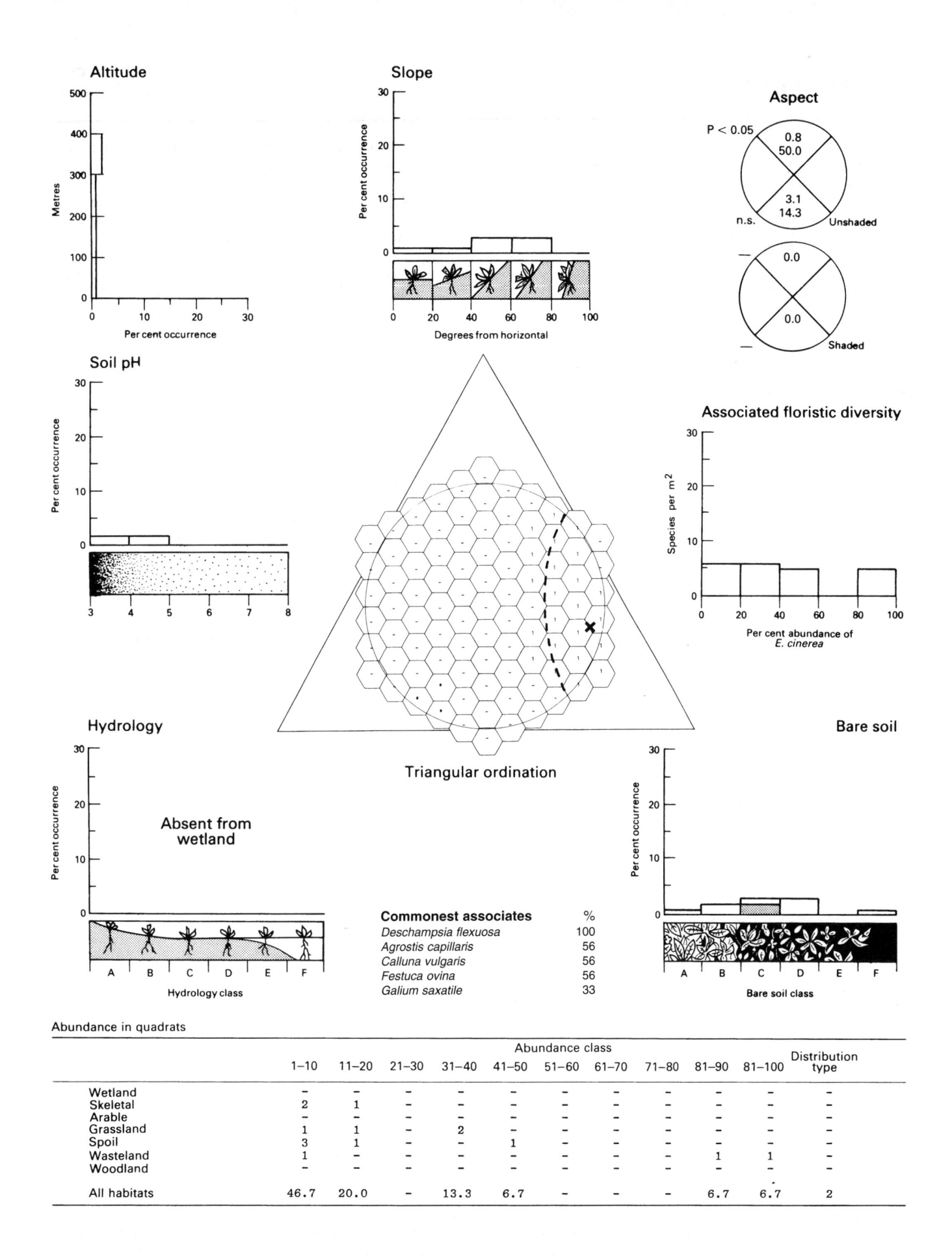

Altitude

Slope

Aspect

P < 0.05
0.8 | 50.0
3.1 | 14.3
n.s. — Unshaded

— | 0.0
— | 0.0
— Shaded

Soil pH

Associated floristic diversity

Species per m²

Per cent abundance of *E. cinerea*

Hydrology

Absent from wetland

Hydrology class

Triangular ordination

Commonest associates	%
Deschampsia flexuosa	100
Agrostis capillaris	56
Calluna vulgaris	56
Festuca ovina	56
Galium saxatile	33

Bare soil

Bare soil class

Abundance in quadrats

	1–10	11–20	21–30	31–40	41–50	51–60	61–70	71–80	81–90	81–100	Distribution type
Wetland	–	–	–	–	–	–	–	–	–	–	–
Skeletal	2	1	–	–	–	–	–	–	–	–	–
Arable	–	–	–	–	–	–	–	–	–	–	–
Grassland	1	1	–	2	–	–	–	–	–	–	–
Spoil	3	1	–	–	1	–	–	–	–	–	–
Wasteland	1	–	–	–	–	–	–	–	1	1	–
Woodland	–	–	–	–	–	–	–	–	–	–	–
All habitats	46.7	20.0	–	13.3	6.7	–	–	–	6.7	6.7	2

Erigeron acer L.

Asteraceae
Blue Fleabane

Life-form Semi-rosette monocarpic to polycarpic perennial. Stem erect; leaves ovate, hairy, usually <500 mm^2; roots reported to lack mycorrhizas.
Phenology Occasionally winter green. Flowers July to August and sets seed from July to September.
Height Foliage mainly <100 mm; flowers to 400 mm.
Seedling RGR No data. SLA 19.3. DMC 19.5.
Nuclear DNA amount No data; 2n = 18* (*BC*); 2x*.
Established strategy Intermediate between stress-tolerant ruderal and C-S-R.
Regenerative strategies W. Regeneration biology requires further study (see 'Synopsis'); ?SB I.
Flowers Outer florets purple, ligulate, female; disk florets yellow, outer female, inner hermaphrodite; insect-pollinated, self-incompatible (Gray 1982); >100 florets in each capitulum; capitula in a few- to 36-flowered panicle (Salisbury 1942).
Germinule 0.11 mg, 2.2 × 0.5 mm, achene with a pappus of 4.2 mm.
Germination Epigeal; t_{50} 3 days; 13-31°C; L = D.
Biological flora None.
Geographical distribution Mainly in S and E England and in the W mainly near the coast, local in Ireland and virtually absent from Scotland (59% of British vice-counties), *ABIF* page 642; much of Europe (77% of European territories); temperate regions of the N hemisphere.

Habitats Recorded in unshaded, calcareous, rocky habitats such as quarry spoil and rock outcrops. Widely observed in calcareous wasteland and over limestone and siliceous strata and colonizes soils of high pH on cinders, refractory sand spoil, sand pits and road cuttings. [Frequent on fixed dunes.]
Gregariousness Sparse to intermediate.
Altitude A predominantly lowland species but observed to 400 m. [Up to 430 m.]
Slope Occurs on all but the steepest slopes.
Aspect Records indicate a S-facing bias.
Hydrology Absent from wetlands.
Soil pH Only recorded from calcareous soils above pH 6.5; not observed below pH 5.5.
Bare soil Distribution wide-ranging with highest frequencies associated with rocky sites with intermediate exposures of bare soil.
△ **ordination** Particularly associated with sites experiencing intermediate intensities of both stress and disturbance.
Associated floristic diversity Typically intermediate to species-rich.

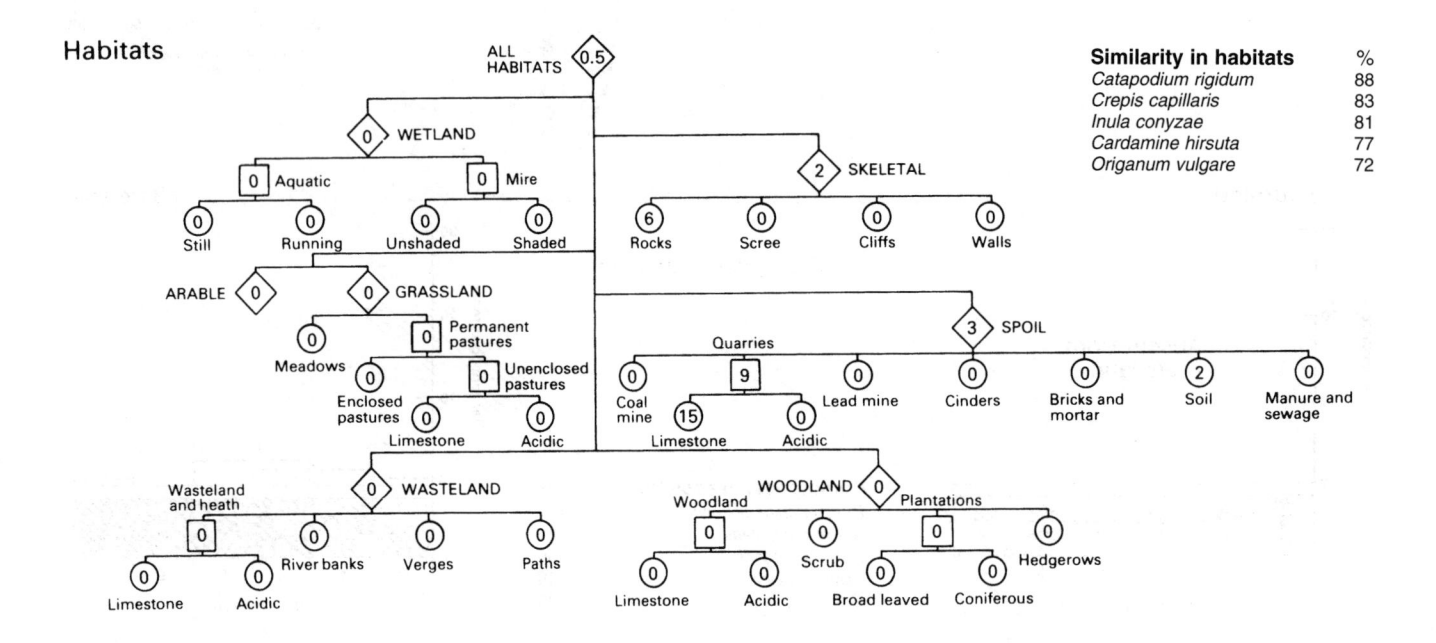

Habitats

Similarity in habitats	%
Catapodium rigidum	88
Crepis capillaris	83
Inula conyzae	81
Cardamine hirsuta	77
Origanum vulgare	72

Synopsis A species which requires further study in field and laboratory. Populations appear to include monocarpic and polycarpic individuals. *E.a.* is a short-lived herb which is found on sand dunes and on shallow calcareous soils where the cover of polycarpic perennials is restricted by soil infertility and summer drought. Some sites are also liable to soil creep or to sand blow, but only seldom are other annual species well represented in the vegetation. The species is restricted, presumably by climatic factors, to lowland sites and is very scarce, for example, on the upland carboniferous limestone of Derbyshire, despite being frequent at the same latitude on the lowland magnesian limestone of S Yorkshire. In favourable conditions individuals may become large and survive to flower in a second season, but the capacity for vegetative spread is slight. Perhaps because of the relatively erect form of the flowering shoot and the dependence upon seed for regeneration, *E.a.* is generally absent from grazed sites. An average plant with 6-8 capitula may produce over 1000 small achenes, and robust specimens may produce five times this number (Salisbury 1942). Achenes have a large plume, and individuals of *E.a.* have been observed both on roadsides and beside railways 15 km from the nearest known populations. The consistent occurrence of *E.a.* in lowland limestone quarries further illustrates the high mobility of the species. However, in a majority of sites with large populations *E.a.* is long-established. No seed bank studies appear to have been undertaken. Four ssp. are described from other parts of Europe (*FE4*), and *E.a.* hybridizes with *Conyza canadensis* (L.) Cronq. (Stace 1975).
Current trends Uncertain. An effective colonist, but narrowly restricted to dry, unproductive, usually calcareous habitats.

Altitude
Slope
Aspect
Unshaded
Shaded

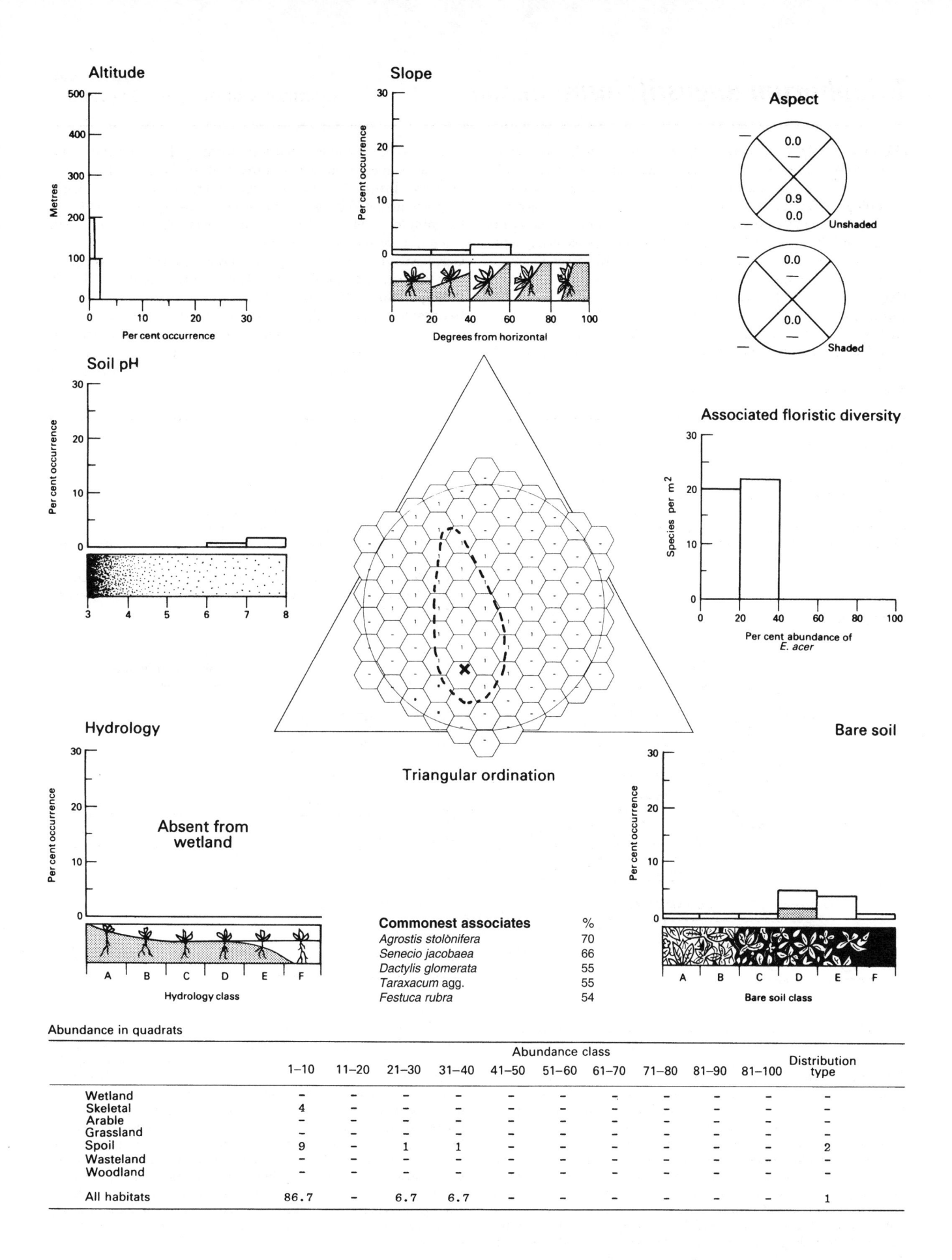

Soil pH

Associated floristic diversity
Species per m^2
Per cent abundance of *E. acer*

Hydrology
Absent from wetland

Triangular ordination

Commonest associates	%
Agrostis stolonifera	70
Senecio jacobaea	66
Dactylis glomerata	55
Taraxacum agg.	55
Festuca rubra	54

Bare soil

Abundance in quadrats

	\multicolumn{10}{c}{Abundance class}	Distribution type
	1–10	11–20	21–30	31–40	41–50	51–60	61–70	71–80	81–90	81–100	
Wetland	–	–	–	–	–	–	–	–	–	–	–
Skeletal	4	–	–	–	–	–	–	–	–	–	–
Arable	–	–	–	–	–	–	–	–	–	–	–
Grassland	–	–	–	–	–	–	–	–	–	–	–
Spoil	9	–	1	1	–	–	–	–	–	–	2
Wasteland	–	–	–	–	–	–	–	–	–	–	–
Woodland	–	–	–	–	–	–	–	–	–	–	–
All habitats	86.7	–	6.7	6.7	–	–	–	–	–	–	1

Eriophorum angustifolium Honckeny

Cyperaceae
Common Cotton-grass, Bog Cotton

Life-form Polycarpic perennial, semi-rosette helophyte with extensive rhizomes. Shoots erect; leaves linear, coarse, <2000 mm^2; root systems shallow, usually without VA mycorrhizas.

Phenology New shoots are produced from spring onwards. Flowering typically occurs in May and June and seed is set in June or July. Leaves, which survive the growing season, begin to die back in late autumn, but may remain partly green through much of the winter.

Height Foliage and flowers to 600 mm.

Seedling RGR 0.5-0.9 week^{-1}. SLA 8.9. DMC 34.2.

Nuclear DNA amount 1.3 pg; $2n = 58^*$ (*BC*); ?$2x^*$.

Established strategy Intermediate between stress-tolerator and stress-tolerant competitor.

Regenerative strategies V, W. Regeneration mainly vegetative by means of rhizome extension, forming large patches, and establishment from seed is infrequent (see 'Synopsis').

Flowers Brownish-green, hermaphrodite, protogynous, wind-pollinated; <50 in each inflorescence of 3-7 spikes (*Biol Fl*).

Germinule 0.44 mg, 2.8 × 1.1 mm, wind-dispersed nut, with cottony bristles up to 35 mm.

Germination Epigeal; t_{50} 4 days; 16-31°C; percentage germination drastically reduced after dry storage at 20°C; L > D.

Biological flora Phillips (1954).

Geographical distribution Locally abundant throughout the British Isles, particularly in the N and W (100% of British vice-counties), *ABIF* page 701; Europe except for the S Mediterranean (79% of European territories); N America and Arctic regions.

Habitats Exclusively a wetland species. Most common on ombrogenous mire, but grows on moorland associated with moist rather than waterlogged conditions, and in the shallow water of moorland pools. Also observed in highly calcareous soligenous mire, with isolated records from wet areas in limestone quarries and coal-mine spoil and from shaded sites.

Gregariousness Intermediate to stand-forming.

Altitude Although wide-ranging, *E.a.* is most common on high moorland areas of the millstone grit and lower coal-measure plateaux. [Up to 2000 m.]

Slope Like most wetland species, largely absent from steep slopes.

Aspect No suitable data.

Hydrology Found most frequently in ombrogenous mire, but also recorded from shallow pools and soligenous mire.

Soil pH Mostly confined to acidic soils of pH <5.0, but with isolated occurrences on neutral and calcareous soils of high pH.

Bare soil Occurs in vegetation with a relatively closed canopy, in eroded peatland areas with much bare soil and in areas carpeted with bryophyte, particularly *Sphagnum* spp.

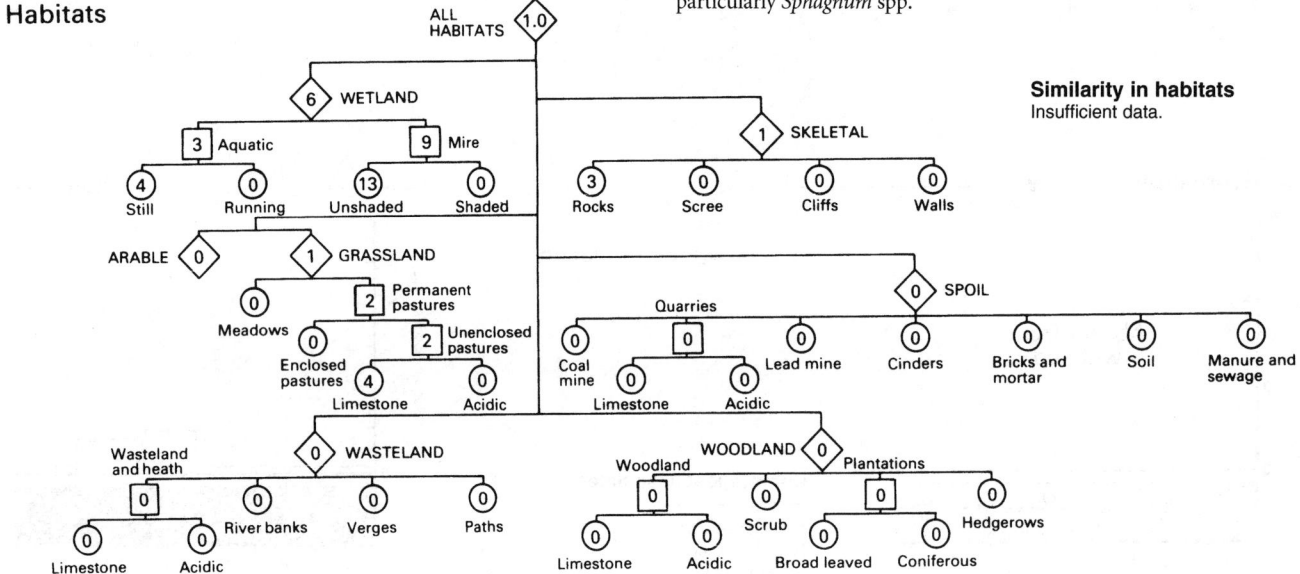

Habitats

Similarity in habitats
Insufficient data.

△ **ordination** Distribution centred on undisturbed and relatively unproductive vegetation types. Absent from highly disturbed vegetation and from undisturbed productive communities.

Associated floristic diversity On acidic soils, low, especially at high-frequently values of *E.a.* On calcareous soils >30 species m^{-2} recorded.

Synopsis *E.a.* is most typical of infertile ombrogenous mire, although its habitat range includes soligenous mire and shallow water. *E.a.* forms large, often diffuse, clonal patches by means of rhizomes which in exceptional cases may extend by up to 1 m year^{-1} (*Biol Fl*). As a result of this rhizome growth *E.a.* is able to colonize eroding peat in overgrazed habitats, and in this situation may have a niche not unlike that of *Carex arenaria* L. in sand dunes. Breakdown of older rhizomes after 3-4 years (*Biol Fl*) leads to the isolation of daughter plants. Detached shoots readily re-root and may form new plants in eroding peat or areas subject to peat cutting. Although *E.a.* produces large amounts of wind-dispersed fruit, establishment by seed in spring is very rare (*Biol Fl*), and there is no evidence of a persistent seed bank. *E.a.* is tolerant of grazing by sheep and of burning, but rhizome growth is poor in dry habitats (*Biol Fl*). *E.a.* is one of the very few species which exploit both highly acidic and calcareous sites, and there is evidence that clones from plants on acidic soils show normal growth on calcareous soils (Hodgson 1972). Leaves have very low Ca contents (*SIEF*).

Current trends In Britain as a whole the destruction of wetlands has led to a marked decline of *E.a.* in lowland areas, and in some regions the species is close to extinction. However, *E.a.* remains abundant in many upland areas, and is particularly abundant on the millstone grit within the Sheffield region. This abundance, which is greater than is typical for British ombrogenous mires (B. D. Wheeler pers. comm.), may be due to an ability to colonize the extensive areas of eroded peat arising from overgrazing and burning.

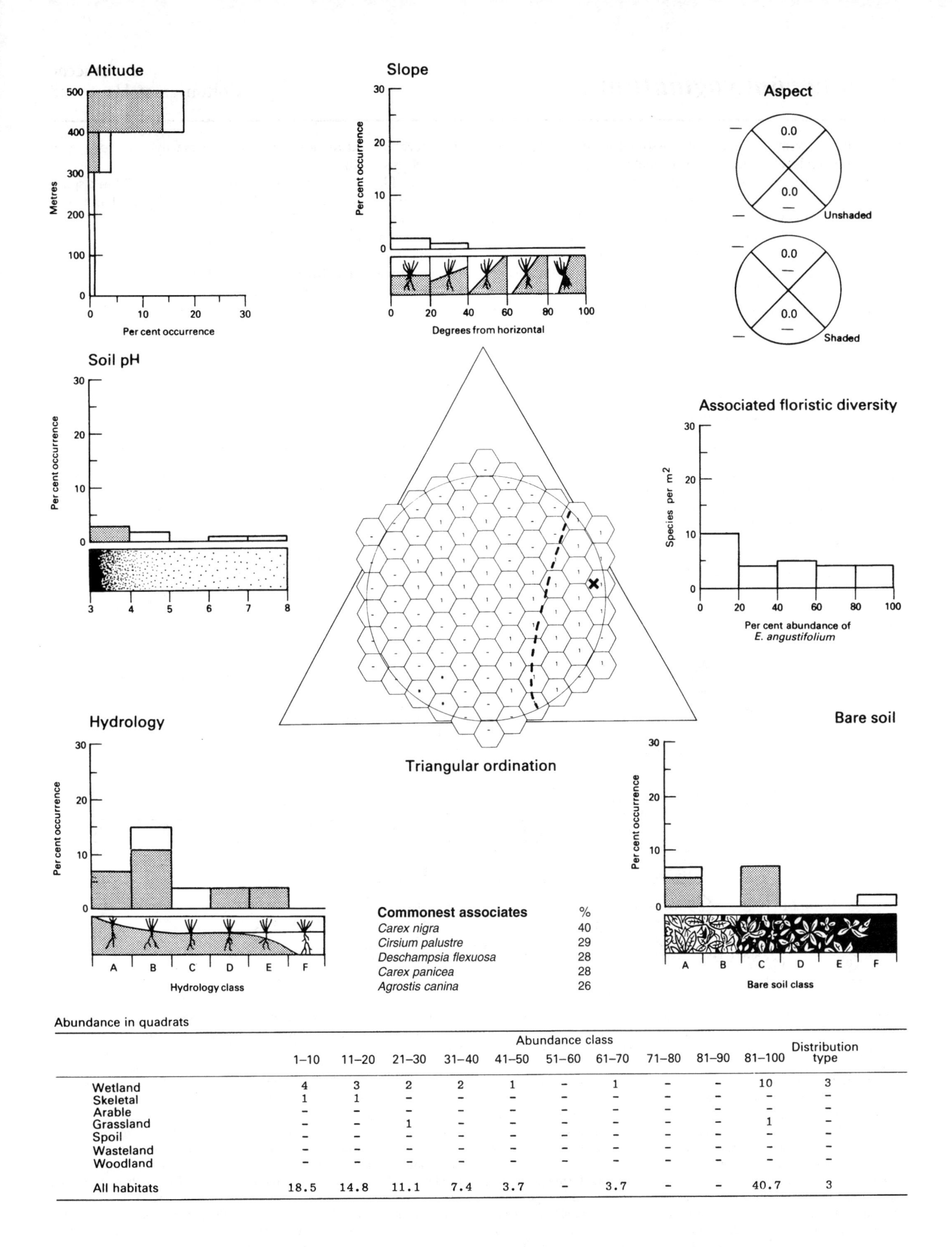

Altitude

Slope

Aspect

Unshaded

Shaded

Soil pH

Associated floristic diversity

Species per m²

Per cent abundance of *E. angustifolium*

Hydrology

Hydrology class

Triangular ordination

Commonest associates	%
Carex nigra	40
Cirsium palustre	29
Deschampsia flexuosa	28
Carex panicea	28
Agrostis canina	26

Bare soil

Bare soil class

Abundance in quadrats

	Abundance class										Distribution type
	1–10	11–20	21–30	31–40	41–50	51–60	61–70	71–80	81–90	81–100	
Wetland	4	3	2	2	1	–	1	–	–	10	3
Skeletal	1	1	–	–	–	–	–	–	–	–	–
Arable	–	–	–	–	–	–	–	–	–	–	–
Grassland	–	–	1	–	–	–	–	–	–	1	–
Spoil	–	–	–	–	–	–	–	–	–	–	–
Wasteland	–	–	–	–	–	–	–	–	–	–	–
Woodland	–	–	–	–	–	–	–	–	–	–	–
All habitats	18.5	14.8	11.1	7.4	3.7	–	3.7	–	–	40.7	3

Eriophorum vaginatum L.

Cyperaceae
Cotton-grass, Hare's tail

Life-form Polycarpic perennial, tussock-forming, helophyte. Stems erect; leaves long, narrow, triquetrous, <500 mm²; roots often deep, with or without VA mycorrhizas.
Phenology New growth commences in spring and flowering occurs shortly afterwards, in April or May (earlier than in *E. angustifolium*). Fruit formed *c.* 1 month later (*Biol Fl*). Leaves, which survive the growing season, die back in late autumn, but many are still green at the base in the following spring. Roots replaced annually (*Biol Fl*).
Height Foliage and flowering shoots to 500 mm from tussocks up to 300 mm in height.
Seedling RGR No data. SLA 5.6. DMC 38.7.
Nuclear DNA amount 1.1 pg; 2*n* = 58 (*FE5*); base number obscure.
Established strategy Intermediate between stress-tolerant competitor and stress-tolerator.

Regenerative strategies W, ?B$_s$. Primarily by seed (see 'Synopsis'); ?SB III.
Flowers Black, hermaphrodite, wind-pollinated; >50 in a spike.
Germinule 1.02 mg, 2.5 × 1.2 mm, wind-dispersed nut with cottony bristles up to 35 mm.
Germination Epigeal; t_{50} 3 days; 10-22°C; L > D.
Biological flora Wein (1973).
Geographical distribution Most of the British Isles, but decreasing towards the S and E and extinct in a number of counties (87% of British vice-counties), *ABIF* page 702; N, NE and C Europe (69% of European territories) and other N temperate zones, extending southwards locally and mainly in mountainous regions.

Habitats A species of unshaded wetlands on highly acidic peat. Also common in grazed and ungrazed areas of moist, peaty moorland, and observed at one atypical site consisting of a small calcareous marsh with a clay soil in a limestone quarry. [*E.v.* extends into montane bog communities (Burnett 1964).]
Gregariousness Stand-forming.
Altitude Recorded almost exclusively from the millstone grit uplands, particularly above 400 m. Extinct in most of its lowland sites but still surviving locally almost down to sea level. [Up to 940 m.]
Slope Like most wetland species, largely restricted to shallow slopes.
Aspect Insufficient data.
Hydrology Most frequent in ombrogenous mire, but also occurs in soligenous mire and beside or even within shallow moorland pools. Sometimes found on drained moorland sites as a relict of former wetland communities.
Soil pH Largely restricted to acidic peaty soils. Rarely observed at pH values >6.0.

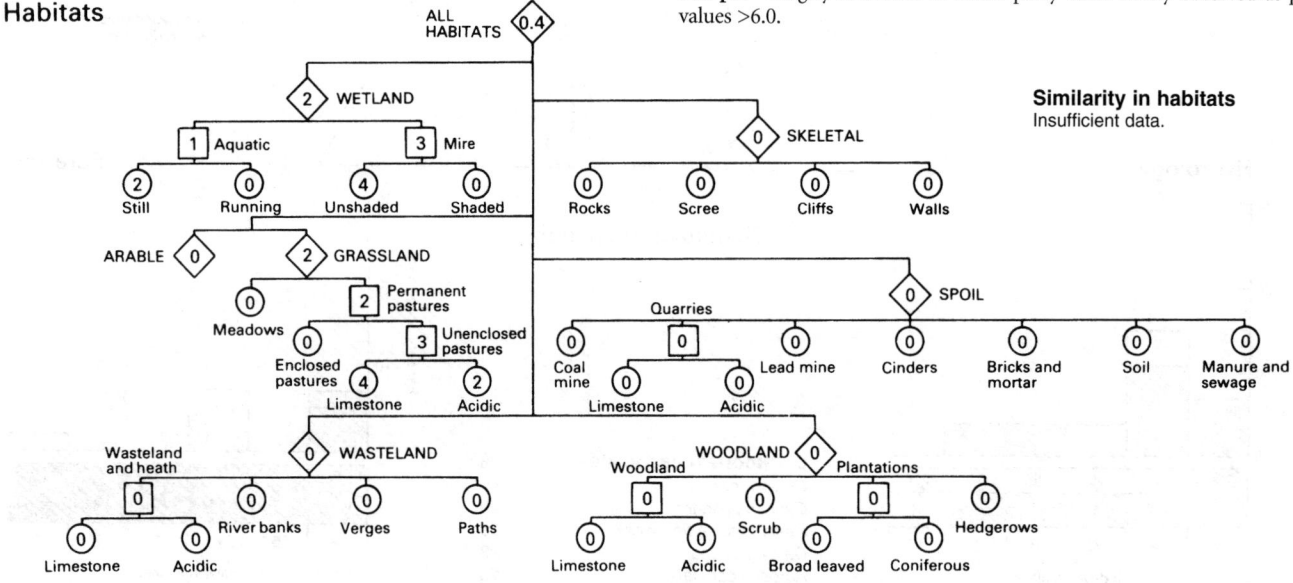

Habitats

Similarity in habitats
Insufficient data.

Bare soil Most frequent in habitats with little exposed bare soil but observed as an early colonist of cut peat and other disturbed sites.
△ **ordination** Largely confined to vegetation types which are subject to little or no disturbance in situations of moderate to low fertility.
Associated floristic diversity Usually low, but a value of 21 species m⁻² has been observed. *E.v.* forms large tussocks, and at higher-frequency values might be expected to exhibit competitive exclusion. However, there is little evidence of a decrease in species number with increasing frequency of *E.v.*, probably because the tops of tussocks are frequently colonized by dryland species such as *Deschampsia flexuosa*.
Synopsis *E.v.* is a long-lived (>100 years) tussock-forming species developing extensive stands on wet, acidic, peaty soils of low potential productivity. The species is particularly characteristic of Arctic tundra (*Biol Fl*). In Britain *E.v.* has a northerly and an upland bias in its distribution, and is infrequent S of the Midlands. *E.v.* is particularly prominent in sites which are waterlogged in spring but drier in summer (*Biol Fl*). The species is tolerant of light grazing by sheep or cattle, and survives superficial burning. In Alaska growth is severely limited by nutrient availability, particularly nitrogen (Shaver *et al.* 1986), and marked seasonal changes in the levels of nutrients and carbohydrates are described in Chapin *et al.* (1986). In Alaskan populations (ssp. *spissum* (Fern.) Hult.) leaf production has been found to be coupled with the remobilization of mineral nutrients from senescing leaves (Jonasson & Chapin 1985). This minimizes the dependence of *E.v.* upon the current supply of soil nutrients, and may be important in facilitating dominance in the chronically unproductive habitats in which the species grows (Jonasson & Chapin 1985). The leaves are particularly low in Ca (*SIEF*). Early flowering is facilitated by development of the inflorescence during the previous year's growth, and seeds germinate in spring (Muller 1978). A bank of persistent seed has been recorded in Alaska (McGraw 1980, Gartner *et al.* 1983). Under favourable

Altitude

Slope

Aspect

Unshaded

Shaded

Soil pH

Associated floristic diversity

Hydrology

Triangular ordination

Bare soil

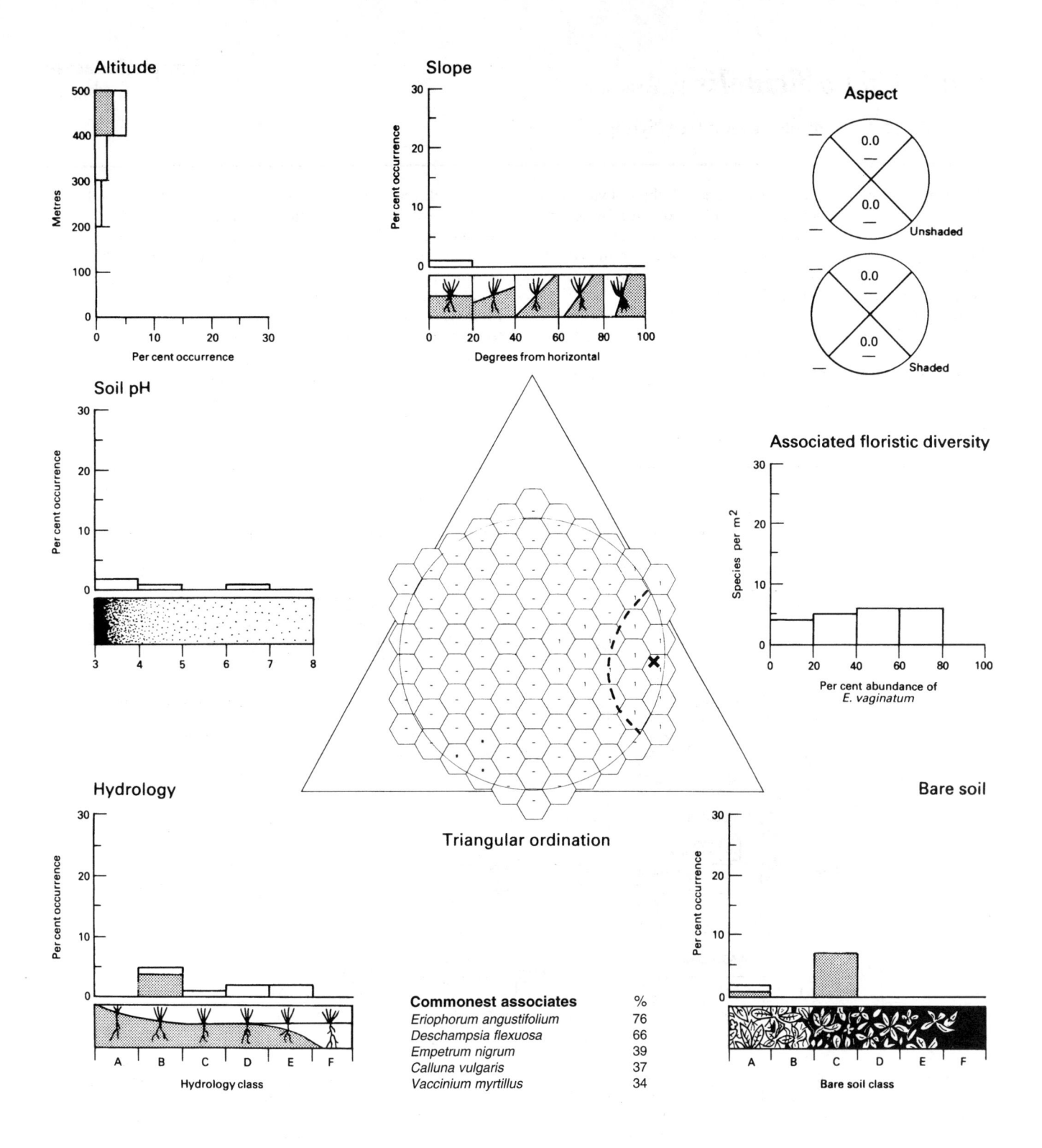

Commonest associates	%
Eriophorum angustifolium	76
Deschampsia flexuosa	66
Empetrum nigrum	39
Calluna vulgaris	37
Vaccinium myrtillus	34

conditions *E.v.* may produce over 1000 wind-dispersed seeds m^{-2} and is an effective colonist of areas of peat-cutting and peat-erosion both in Alaska and in the S Pennines. Disturbances such as fire, grazing, human activity and frost action result in the formation of microsites suitable for establishment by seed (Gartner *et al.* 1986). Seedling establishment in tundra is particularly associated with microsites covered with liverworts, since bared areas are subject to a high level of soil instability and in other sites there are dominant impacts by vascular plants and mosses (Gartner *et al.* 1986). There is no evidence of budding of tussocks (*Biol Fl*). Detached plantlets can be grown in the field (*Biol Fl*), and may afford a means of regeneration in the eroding peat that is common in many higher moorland areas. However, at least one authority, *Biol Fl*, contends that the only effective mechanism of regeneration is by seed.

Current trends In the lowland areas of Britain *E.v.* is decreasing as a result of habitat destruction. However, the macro-fossil record in peat suggests that *E.v.* has flourished during past episodes of artificial disturbance (*Biol Fl*). *E.v.* may expand at the expense of *Calluna vulgaris* in response to increased grazing (Rawes & Hobbs 1979), and may be colonizing in areas of eroding peat in upland Britain.

Euphrasia officinalis L. sensu lato

A taxonomically complex group (see 'Synopsis')

Scrophulariaceae
Eyebright

Life-form Hemiparasitic, summer-annual therophyte. Stem erect or ascending, branched or unbranched; leaves ovate, with or without hairs, <250 mm^2; roots non-mycorrhizal.
Phenology Seeds germinate in spring. Flowers from June to September and sets seed from September onwards.
Height Foliage and flowers to 300 mm.
Seedling RGR No data. SLA 12.9. DMC 24.7.
Nuclear DNA amount No data; $2n = 22^*, 44^*$ (*BC*); $2x^*, 4x^*$.
Established strategy Stress-tolerant ruderal.
Regenerative strategies S. Regenerates entirely by seed which germinates in late winter or spring. No buried seed bank is formed (Kelly 1986); SB II.
Flowers White, hermaphrodite, moderately self-fertile, insect-, or in the case of smaller-flowered taxa, usually self-pollinated (Yeo 1966, Stace 1975); often >10 in a terminal spike.
Germinule 0.13 mg, 1.0 × 0.6 mm, seed; up to *c.* 20 shed from a dehiscent capsule (Yeo 1966); L = D (During *et al.* 1985).
Germination Epigeal; chilling requirement (but not present in all populations (Yeo 1961)).
Biological flora None.
Geographical distribution Widespread, particularly in N and W Britain, *ABIF* page 563, *CSABF* pages 55-62; related taxa are found in 65% of European territories; and *Euphrasia* spp. occur throughout the temperate N hemisphere, in temperate S America, Australia and New Zealand.

Habitats Frequently recorded from calcareous pasture and lead-mine spoil heaps. Also found in a range of other rocky habitats including scree, quarry spoil and rock outcrops, and from waysides and unproductive meadows. Scattered occurrences in mire and in grasslands on non-calcareous strata. [Occurs on heathland, by mountain streams and other montane habitats, sea cliffs, sand dunes and salt marshes (*CSABF*).]
Gregariousness Sparse to intermediate.

Altitude Most frequent on the carboniferous limestone in the range 200-400 m. [Up to 1210 m.]
Slope Most frequent on intermediate slopes, a reflection of an association with the limestone dales.
Aspect Widely distributed on both N- and S-facing slopes.
Hydrology Observed in soligenous mire.
Soil pH Most frequent and abundant on base-rich soils of pH >6.5, but observed at values down to pH 4.4 [and recorded from heathland].
Bare soil Widely distributed, but with greatest frequency and abundance in relatively open, but vegetated, sites.

Habitats

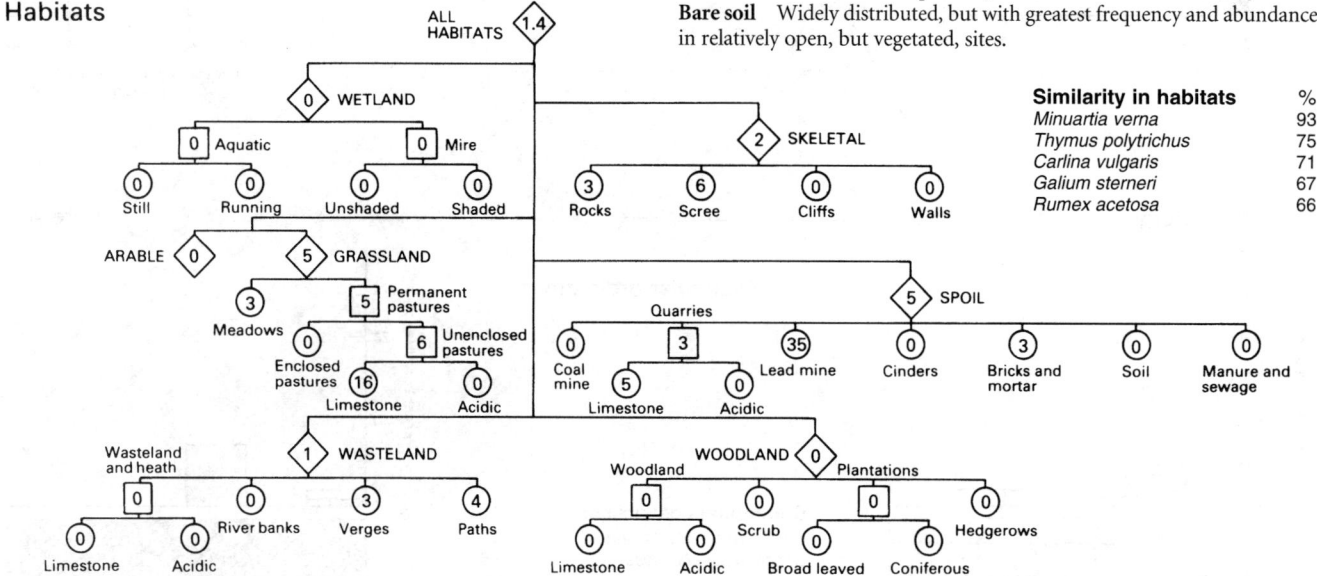

Similarity in habitats	%
Minuartia verna	93
Thymus polytrichus	75
Carlina vulgaris	71
Galium sterneri	67
Rumex acetosa	66

△ **ordination** Distribution centred on unproductive and undisturbed vegetation types. Absent from very disturbed or highly productive sites.
Associated floristic diversity Typically species-rich.
Synopsis *E.o.* is a root hemiparasite found in a wide range of less-fertile habitats. *E.o.* is represented in Britain by 19 microspecies, 7 of which are endemic to Britain and a further 9 endemic to Europe. Highest concentrations of species occur in the N and W (*CSABF*). All have a narrow habitat range coupled with a well-defined geographical distribution (see *CSABF*, Yeo 1966). Thus, each is potentially distinct ecologically, except where hybridization occurs when taxa of the same ploidy level are found in close proximity (Stace 1975). The commonest taxa in N England (*E. confusa* Pugsl. and *E. nemorosa* (Pres.) Wallr.) are particularly associated with dry habitats on calcareous strata. However, other taxa are widespread in heathland and on other moderately acidic sites, particularly in the N and W. Specialization also occurs with respect to land management (tall microspecies in meadows, short in sheep pastures), with respect to pollination mechanism (large-flowered forms insect-pollinated, small often selfed) and with respect to habitat (maritime, montane and wetland taxa are recognized (Yeo 1966)). Although capable of autotrophic existence, *E.o.* is essentially hemiparasitic and may be attached experimentally to the roots of a wide range of hosts (Yeo 1964). *E.o.* probably also has a wide host range in the field, perhaps even parasitizing more than one species simultaneously. No single potential host is present with all occurrences of *E.o.*, but the commonest hosts may be members of the Poaceae and Fabaceae (Hodgson 1973). The early germination of the seed, reported by Yeo (1961), and the complementary early spring growth of many grasses, also point to the importance of the Poaceae as host plants. *E.o.*, which has a poorly-developed root system, receives carbohydrates (and presumably also water and mineral nutrients) from its host; it grows considerably better attached than unattached (Hodgson 1973). The possibility that the relatively high levels of mannitol and galactitol in *E.o.* (Hodgson 1973) play an important osmotic role in the maintenance of hemiparasitism deserves investigation (D. H. Lewis pers. comm.). However, the origin and ecological advantages of this species'

Altitude
Slope
Aspect

3.1 62.5 n.s. Unshaded
5.4 50.0 n.s.
0.0 —
0.0 — Shaded

Soil pH

Associated floristic diversity

Triangular ordination

Hydrology

Absent from wetland

Commonest associates %
Festuca ovina 69
Festuca rubra 69
Plantago lanceolata 69
Campanula rotundifolia 64
Linum catharticum 63

Bare soil

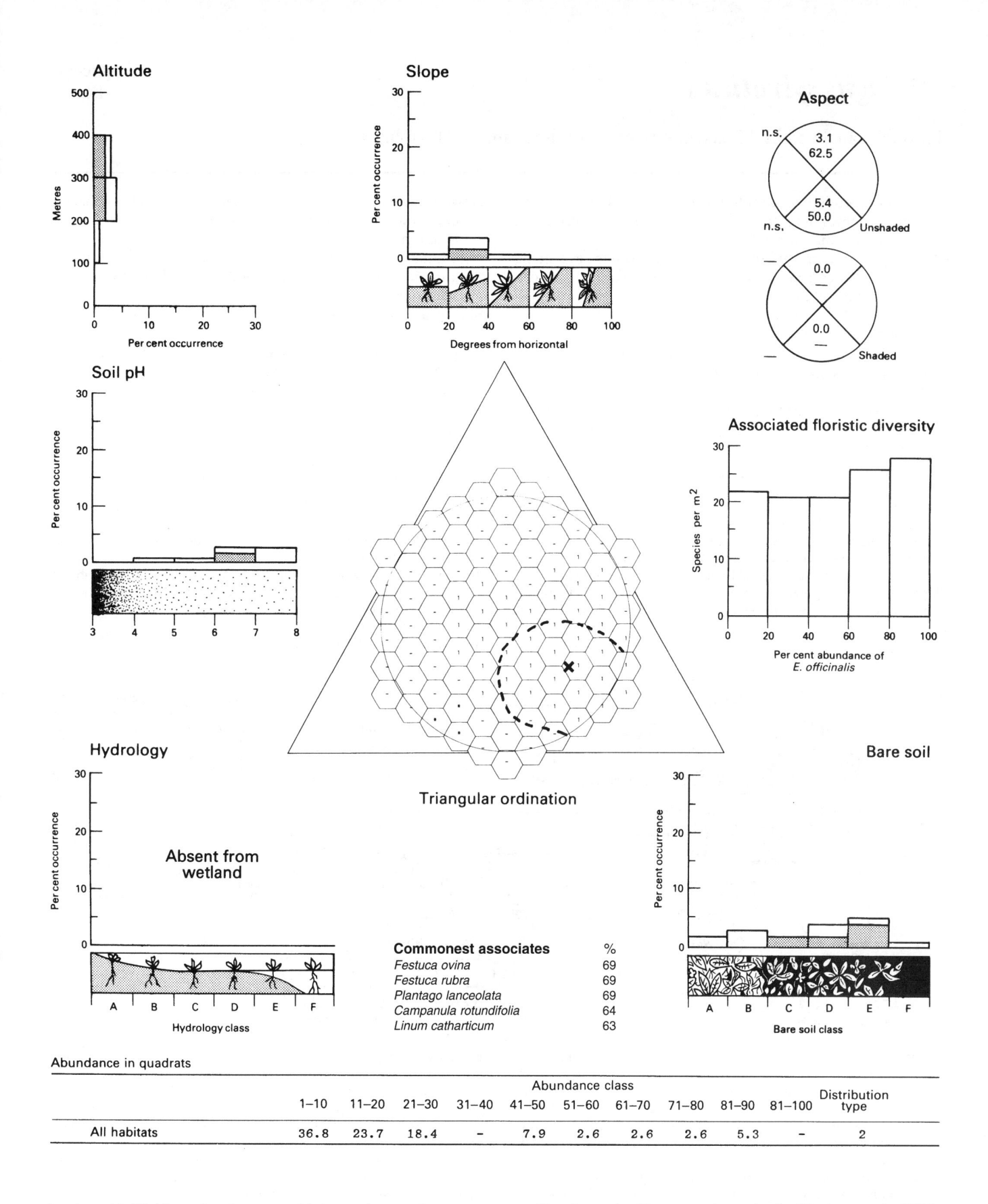

Abundance in quadrats

	1–10	11–20	21–30	31–40	41–50	51–60	61–70	71–80	81–90	81–100	Distribution type
All habitats	36.8	23.7	18.4	–	7.9	2.6	2.6	2.6	5.3	–	2

hemiparasitic life-history is unknown, and it is not clear to what extent host plants are debilitated by its presence. Like most ruderals, *E.o.* is phenotypically plastic and its size and reproductive effort vary considerably according to habitat and host (Wilkins 1963, Yeo 1964). Large vigorous plants form on a good host; under less-favourable conditions a small stunted plant is produced with, on occasion, just a single capsule. *E.o.* is considered unpalatable to grazing animals (During *et al.* 1985). Regeneration is entirely by seed, and seedling mortalities are high (During *et al.* 1985). There is apparently no persistent seed bank (Kelly 1986).

Current trends Like other species of less-fertile habitats, *E.o.* is probably decreasing.

(*) *Fagus sylvatica* L.

Fagaceae
Beech

Field data analysed in this account refer to seedlings and small saplings

Life-form Deciduous tree with broad crown, smooth bark and dense canopy. Leaves ovate, slightly hairy, usually <5000 mm²; roots with ectomycorrhizas.
Phenology Flowers April to May, coinciding with the expanding leaves. Seed set from September to October (earlier than oak (Newbold & Goldsmith 1981)).
Height Foliage and flowers to 30 m.
Seedling RGR No data. SLA 12.3. DMC. No data.
Nuclear DNA amount 0.8 pg; $2n = 24$ (*FE1*); $2x$.
Established strategy Stress-tolerant competitor.
Regenerative strategies S. Reproduces by seed, which germinates during spring. There is no evidence of a persistent seed bank, but seedlings and saplings are relatively persistent in shade; SB II.

Flowers Green, monoecious, protogynous, wind-pollinated; male flowers numerous on tassel-like heads, female flowers usually in pairs, wind-pollinated.
Germinule 22.5 mg, 15.0×10.0 mm, nut; 1-2 in a scaly capsule.
Germination Epigeal; dormancy broken by chilling (Newbold & Goldsmith 1981).
Biological flora None, but see Brown (1953) and Newbold and Goldsmith (1981).
Geographical distribution Native of SE England, but planted for forestry or ornament and naturalized to some extent in the remainder of the British Isles, *ABIF* page 129; W and C Europe extending northwards to SE Norway (62% of European territories and 3% naturalized).

Habitats Seedlings and saplings only recorded from shaded habitats, and much more frequently in beech plantations than elsewhere.
Gregariousness Seedlings sometimes in small clusters.
Altitude Juveniles recorded to 320 m. [Up to 370 m.]
Slope Not recorded from skeletal habitats or from steep slopes.
Aspect Slightly more frequent on N-facing slopes, possibly a reflection of the vulnerability of the seed to desiccation (Newbold & Goldsmith 1981).

Hydrology Not a wetland species, but seedlings recorded from topogenous mire.
Soil pH Most frequently recorded on acidic soils of pH <5.0, but also present within the pH range 6.0-8.0.
Bare soil Juveniles mainly under beech trees, and therefore in habitats with relatively little exposed soil but containing large quantities of persistent tree litter.
△ **ordination** Insufficient data.
Associated floristic diversity Low.

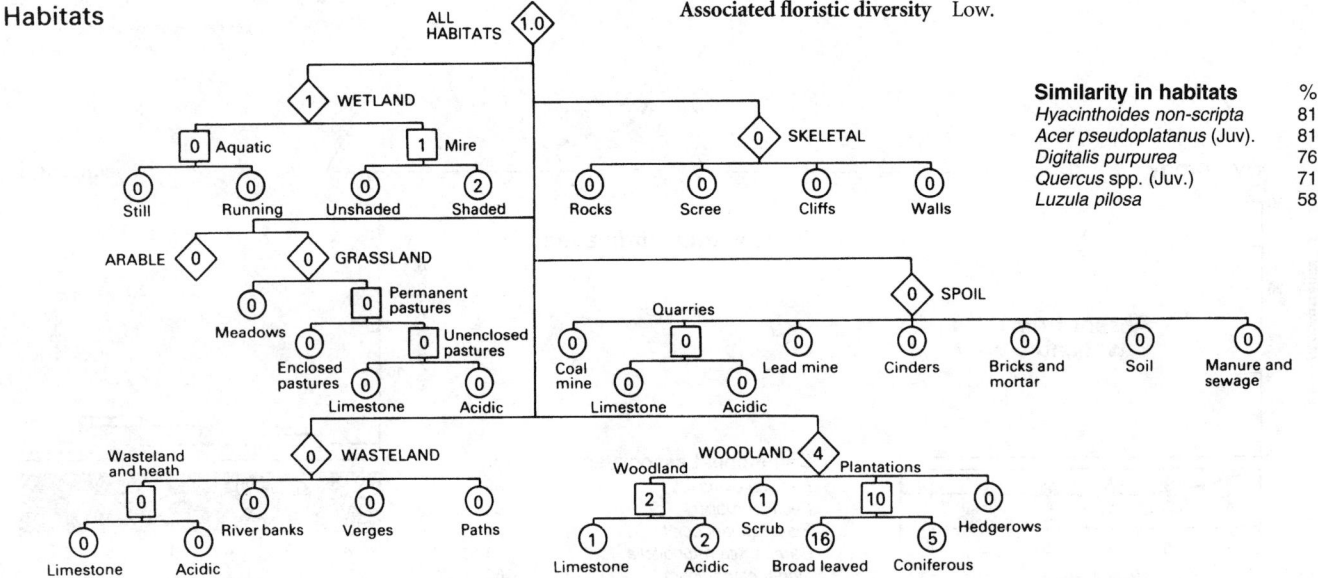

Habitats

Similarity in habitats	%
Hyacinthoides non-scripta	81
Acer pseudoplatanus (Juv).	81
Digitalis purpurea	76
Quercus spp. (Juv.)	71
Luzula pilosa	58

Synopsis *F.s.* is a tall, deciduous, dominant forest timber tree which extends to 30 m in height and exploits fertile conditions more effectively than oak (Newbold & Goldsmith 1981). *F.s.* has a life-span which rarely exceeds 300 years (Rackham 1980). As a native tree the species is restricted, apparently by climatic factors, to SE England (Rackham 1980). Indigenous populations occur on a wide range of soils particularly over acidic gravels, sands and sandstones and only rarely over calcareous strata (Rackham 1980). *F.s.* occurs over an exceptionally wide range of soil pH (<3.5 to >7.5), but maximum height growth tends to occur at *c.* pH 5.0 (Rackham 1980). *F.s.* is relatively shallow-rooted and vulnerable to strong winds (Brown 1953). The species is also unable to exploit dry sites with deep subsoil water (Rackham 1980). The species is widely planted for commercial forestry and for ornament, and occupies *c.* 74 000 ha of woodland (>10% of total), making it the third most common tree of British broad-leaved woodlands (Steele & Peterken 1982). The species is not resilient to coppice management (Beckett & Beckett 1979). Planted *F.s.* can regenerate outside the present geographical range of the species. Despite the role of birds and squirrels as vectors (Ridley 1930), beech nuts are poorly dispersed in native populations (Watt 1934). In the Sheffield region the distribution of seedlings is closely correlated with the presence of *F.s.* in the woodland canopy. Annual seed production is often low. Mast years, in which heavy crops of seed are produced, tend to follow warm summers and are dependent also upon the absence of spring frost, to which the young shoots are sensitive (Brown 1953). Heavy seed crops tend to occur once every 5-10 years, are more frequent in S England than the N, and may be associated with a narrow annual growth ring (Newbold & Goldsmith 1981). Seeds may be killed by excessive water loss, but in this respect *F.s.* is less vulnerable than oak (Newbold & Goldsmith 1981). Moist conditions are required for establishment, but nuts may loose viability under waterlogged conditions (Newbold & Goldsmith 1981). Subsequent

Altitude

Slope

Aspect

Soil pH

Commonest associates — %
- Hyacinthoides non-scripta — 62
- Holcus mollis — 48
- Deschampsia flexuosa — 38
- Dryopteris dilatata — 24
- Hedera helix — 19

Associated floristic diversity

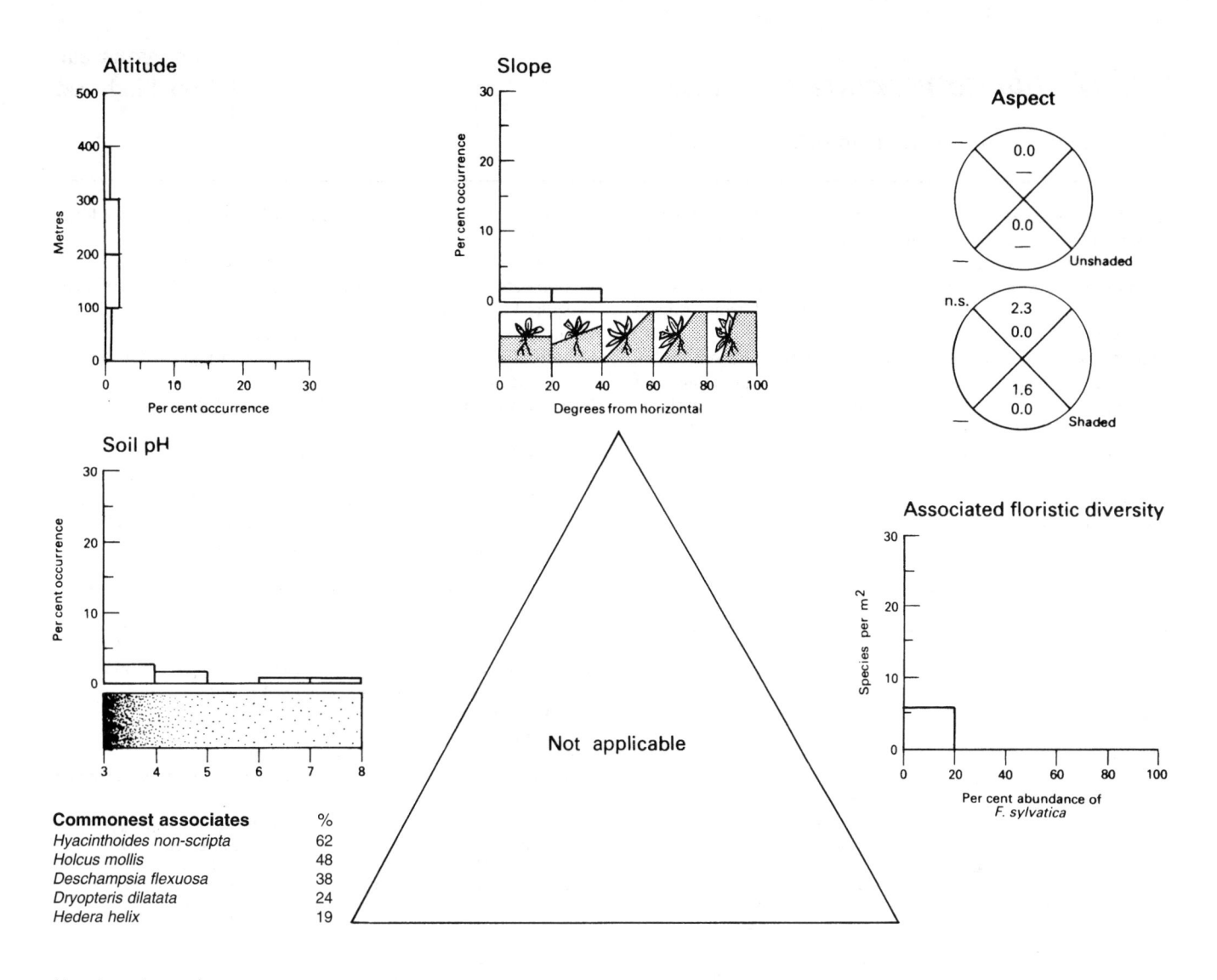

Not applicable

Abundance in quadrats

	1–10	11–20	21–30	31–40	41–50	51–60	61–70	71–80	81–90	81–100	Distribution type
Wetland	1	–	–	–	–	–	–	–	–	–	–
Skeletal	–	–	–	–	–	–	–	–	–	–	–
Arable	–	–	–	–	–	–	–	–	–	–	–
Grassland	–	–	–	–	–	–	–	–	–	–	–
Spoil	–	–	–	–	–	–	–	–	–	–	–
Wasteland	–	–	–	–	–	–	–	–	–	–	–
Woodland	26	–	–	–	–	–	–	–	–	–	1
All habitats	100.0	–	–	–	–	–	–	–	–	–	1

establishment of seedlings, which germinate in spring, usually takes place in the shelter of bushes or trees (Rackham 1980). Seedlings are highly shade-tolerant (Watt 1923). However, under a full canopy of *F.s.* saplings only persist for a few years. Under more-favourable conditions they can produce seed when only 28 years old, although 60 years may be a more typical age for first fruiting (Brown 1953). Seed production may continue for more than 200 years (Brown 1953). No persistent seed bank is formed, and nuts are toxic to stock (Cooper & Johnson 1984). The insect fauna of *F.s.* is of intermediate diversity compared with that associated with other broad-leaved species (Southwood 1961). Seed predation is high but the synchronization of leaf and fruit fall may afford the seed some protection from predators (Newbold & Goldsmith 1981). Variation in annual amount of seed set may also minimize any build-up of potential predators. A sufficient number of seedlings are normally produced in mast years to ensure regeneration of the woodland, but seedlings and saplings are subsequently exposed to a variety of environmental and biotic hazards (Newbold & Goldsmith 1981). Monocultures of *F.s.* are particularly characteristic of modern forestry, and are characterized by heavy shade and accumulation of the exceptionally persistent leaf litter (Sydes 1981), with the result that the ground flora is often sparse and species-poor. *F.s.* is more productive than oak on fertile soils, and the concept of oak woodland as the climax vegetation of much of the British Isles appears ill-founded; indeed the notion of a single predictable stable forest climax is becoming increasingly suspect.

Current trends Many beech woods were originally planted, but there is some evidence, e.g. Pigott (1983), that this potentially dominant species is increasing. Rackham (1980) considers native beech woods to be less under threat than many other woodland types, although their composition may be altering through the cessation of traditional methods of woodland management. Much of the seed used in cultivation in this country originated in Rumania (Gordon & Fraser 1982). Thus, the genetic diversity of *F.s.* in Britain has presumably been increased, although the fitness of these 'foreign' genotypes in Britain remains uncertain.

(+)*Fallopia convolvulus* (L.) A. Love

Polygonaceae
Black Bindweed

Bilderdykia convolvulus (L.) Dumort.

Life-form Summer-annual therophyte. Stems branched, scrambling or twining; leaves ovate, usually <2000 mm^2; taprooted and usually non-mycorrhizal.

Phenology Seeds germinate in spring (*WCH*), flowers are formed from July to October and seed shed from August onwards.

Height Determined by the height of the supporting vegetation. Foliage and flowers usually <400 mm, even though the stems of robust plants may exceed 1 m in length; more rarely scrambling at ground level.

Seedling RGR 1.0-1.4 week^{-1}. SLA 25.7. DMC 20.0.

Nuclear DNA amount 1.5 pg; $2n = 40$ (*FE1*), $4x$.

Established strategy Intermediate between ruderal and competitive-ruderal.

Regenerative strategies B_s. Regenerates exclusively by seed, which germinates in spring and may become incorporated into a persistent seed bank (*Biol Fl*); SB IV.

Flowers Greenish-white, hermaphrodite, insect- or self-pollinated, sometimes cleistogamous (*Biol Fl*); often <30 in lax, spike-like inflorescences.

Germinule 1.28 mg, 4.0 × 2.4 mm, nut.

Germination Epigeal; chilling requirement; germination also enhanced by scarification (*Biol Fl*).

Biological flora Hume *et al.* (1983) (as *Polygonum convolvulus* L).

Geographical distribution The British Isles, but less common in upland regions and Ireland (99% of British vice-counties), *ABIF* page 190; Europe, but absent or introduced in the extreme N (97% of European territories); N Africa and temperate Asia; introduced in N America, S Africa and New Zealand.

Habitats Frequent on arable land, particularly in cereal crops. Occasionally recorded in other disturbed habitats such as building rubble, soil and manure heaps, cinders and on waysides. Also occurs as a garden weed.

Gregariousness Usually occurs as scattered individuals.

Altitude Mainly a lowland species but extending locally to 340 m. Often occurs as a casual, and upper altitudinal limit uncertain. [Up to 450 m.]

Slope Most frequent and only abundant on flat ground and gentle slopes, a reflection of an association with arable land.

Aspect Insufficient data.

Hydrology Absent from wetlands.

Soil pH Recorded only from soils of pH >5.0.

Bare soil Clearly associated with habitats having large exposures of bare soil.

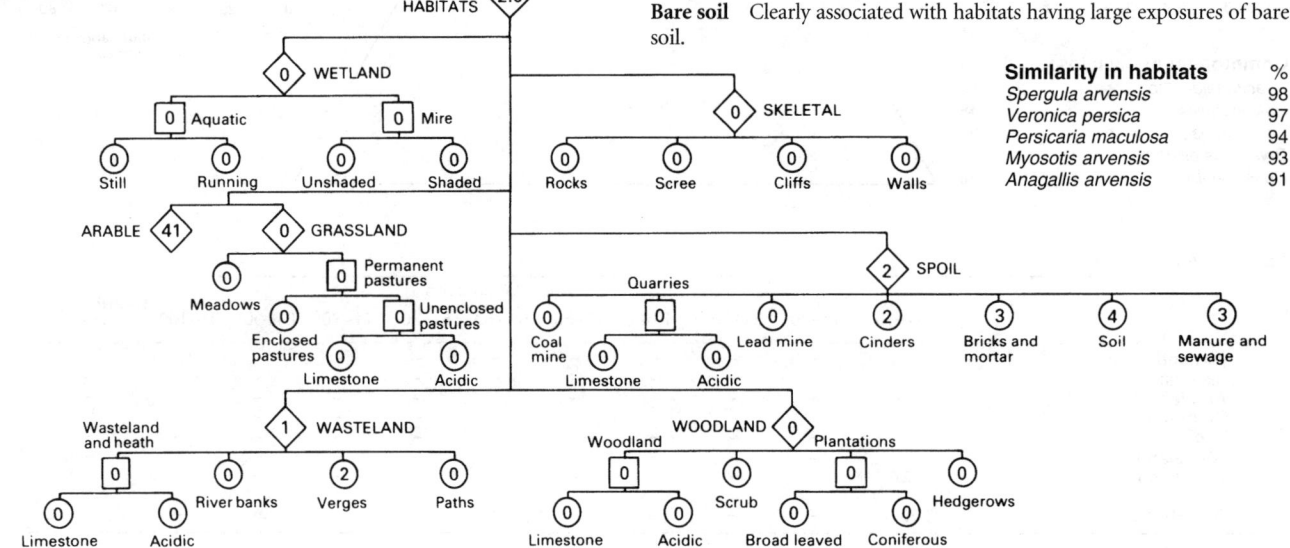

Similarity in habitats	%
Spergula arvensis	98
Veronica persica	97
Persicaria maculosa	94
Myosotis arvensis	93
Anagallis arvensis	91

Habitats

[Habitat diagram: ALL HABITATS 2.0; WETLAND 0 (Aquatic 0: Still 0, Running 0; Mire 0: Unshaded 0, Shaded 0); ARABLE 41; GRASSLAND 0 (Permanent pastures 0: Meadows 0, Unenclosed pastures 0; Enclosed pastures 0: Limestone 0, Acidic 0); SKELETAL 0 (Rocks 0, Scree 0, Cliffs 0, Walls 0); Quarries 0 (Coal mine 0, Lead mine 0: Limestone 0, Acidic 0); SPOIL 2 (Cinders 2, Bricks and mortar 3, Soil 4, Manure and sewage 3); WASTELAND 1 (Wasteland and heath 0: Limestone 0, Acidic 0; River banks 0, Verges 2, Paths 0); WOODLAND 0 (Woodland 0: Limestone 0, Acidic 0; Scrub 0: Broad leaved 0, Coniferous 0; Plantations 0, Hedgerows 0)]

△ **ordination** Distribution concentrated on vegetation of highly fertile and heavily disturbed sites.

Associated floristic diversity Typically intermediate.

Synopsis *F.c.* is a summer-annual species which is considered only doubtfully native by Webb (1985) and is in need of further study. Unlike other annual arable weeds, the species produces weak stems which scramble over or twine around adjacent vegetation. This growth form is likely to minimize the usage of carbohydrates for structural tissue, thus allowing *F.c.* to grow exceedingly rapidly and to allocate resources first to flowers and then to the unusually large seeds at an early stage of development. *F.c.* occurs frequently, but at low population densities on arable land, mainly in cereal crops, and is equally frequent in spring- and autumn-sown cereals (Chancellor & Froud-Williams 1984). *F.c.* is also found to a lesser extent in other unshaded, disturbed habitats on moist, relatively fertile soils. Seedlings are large and have the potential for a high relative growth rate. In favourable situations early-germinating seedlings may form large, much-branched individuals producing several thousand seeds. Reductions of up to 25% in crop yield have been reported where *F.c.* is the dominant weed (*Biol Fl*). Under water- or mineral nutrient-stress, vegetative growth is curtailed, and small plants bearing flowers and seed are frequently observed. Seeds, which are incorporated into a buried seed bank, survive particularly well at great depths in the soil (Roberts & Feast 1972). The seeds have a chilling requirement, with germination occurring mainly in spring. Seed banks of *F.c.* tend to increase if cereals are grown for several years in the same field (*Biol Fl*). We suspect that, in common with several other British arable weeds in the Polygonaceae (Courtney 1968), dormancy is re-imposed annually in response to rising summer temperatures. Although surviving ingestion by birds (Ridley 1930), seeds are probably dispersed mainly by human activity. The species is a widespread contaminant of wheat, other cereal grains and a variety of broad-leaved crops (*Biol Fl*, *WCH*), and its frequency in arable fields is perhaps due more to the difficulty in cleaning crop seeds than to any other aspects of the biology of *F.c.*.

Current trends A common impurity of crop seeds, but producing fewer, larger seeds than many arable weeds, and consequently potentially vulnerable to improved methods of cultivation and seed cleaning.

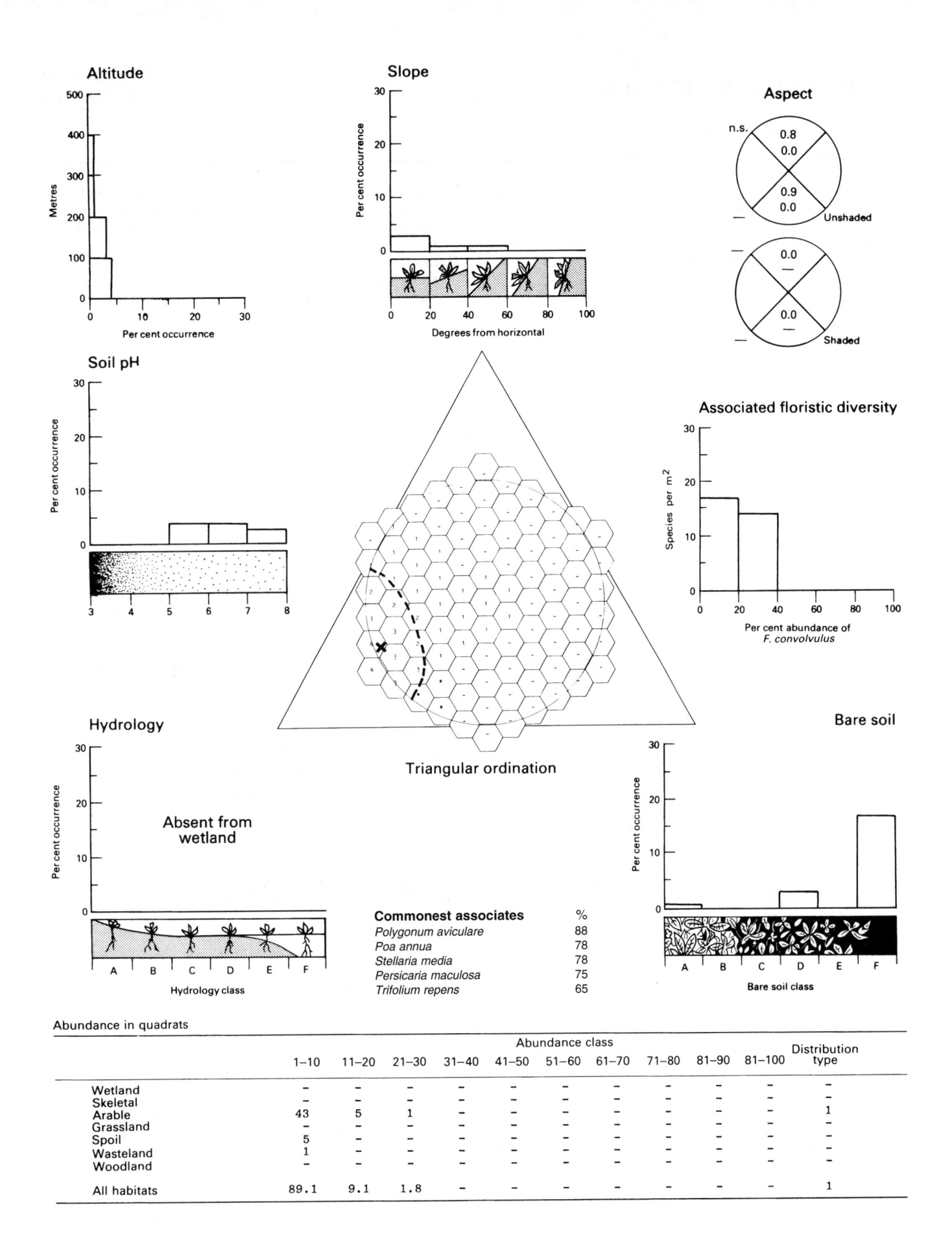

Altitude

Slope

Aspect

	n.s.	0.8
—	0.0	0.9
	0.0	Unshaded

	—	0.0
—	0.0	—
		Shaded

Soil pH

Associated floristic diversity

Hydrology

Absent from wetland

Triangular ordination

Commonest associates	%
Polygonum aviculare	88
Poa annua	78
Stellaria media	78
Persicaria maculosa	75
Trifolium repens	65

Bare soil

Abundance in quadrats

	Abundance class										Distribution type
	1–10	11–20	21–30	31–40	41–50	51–60	61–70	71–80	81–90	81–100	
Wetland	–	–	–	–	–	–	–	–	–	–	–
Skeletal	–	–	–	–	–	–	–	–	–	–	–
Arable	43	5	1	–	–	–	–	–	–	–	1
Grassland	–	–	–	–	–	–	–	–	–	–	–
Spoil	5	–	–	–	–	–	–	–	–	–	–
Wasteland	1	–	–	–	–	–	–	–	–	–	–
Woodland	–	–	–	–	–	–	–	–	–	–	–
All habitats	89.1	9.1	1.8	–	–	–	–	–	–	–	1

Fallopia japonica Adans.

Polygonaceae
Japanese Knotweed

Data refer to var. *japonica*

Life-form Rhizomatous, polycarpic perennial geophyte. Stems stout, branched; leaves ovate, glandular along margins, <10 000 mm^2.
Phenology Flowers August to October. Seed may be set in October, after which shoots die.
Height Foliage and flowers to 2 m.
Seedling RGR No data. SLA 16.7. DMC 26.2.
Nuclear DNA amount 3.4 pg; $2n = 88^*$, $8x^*$ (Bailey & Conolly 1985).
Established strategy Competitor.
Regenerative strategies V. Probably exclusively by vegetative means (see 'Synopsis').
Flowers Greenish-white, functionally dioecious (male flowers with reduced ovary and female flowers with reduced stamens), insect-pollinated; often >50 in each axillary panicle.
Germinule 0.61 mg (in Japan, 0.82 mg (Hayashi 1977)), 4.0×2.0 mm, nut.
Germination Epigeal; dormancy broken by dry storage at room temperature (Justice 1941). A low percentage germination under these conditions effected for a population from the Sheffield region.
Biological flora Beerling, Bailey and Conolly (1994).
Geographical distribution Native of Japan, Taiwan and N China but now extensively naturalized; as a garden escapee in the British Isles (99% of British vice-counties (Conolly 1977)), *ABIF* page 189 and Europe (41% of European territories).

Habitats Recorded or observed from a variety of relatively productive artificial habitats, namely soil heaps, river banks, road verges and railway banks, wasteland, coal-mine spoil and cinder tips. Also found occasionally in woodland. Absent from wetland, skeletal habitats and managed grassland.

Habitats

Similarity in habitats Insufficient data.

Gregariousness A stand-forming species which, even at low-frequency values, can form a dense canopy.
Altitude Largely associated with sites of urban dereliction, and therefore predominantly a lowland species, but observed to 320 m.
Slope Most records are from flat or gently sloping ground.
Aspect Widely distributed.
Hydrology Absent from wetland habitats.
Soil pH Most records and observations fall within the pH range 6.5–8.0, but the species does occur to some extent on acidic soils, both in the Sheffield region and in its native Japan (Conolly 1977).

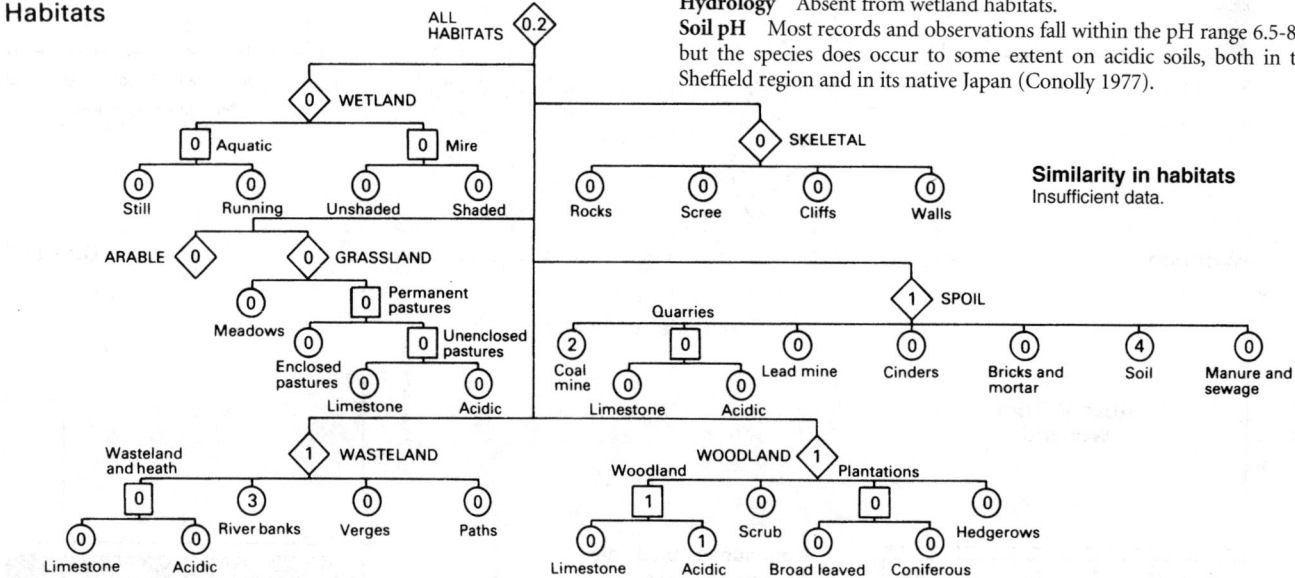

Bare soil In established stands of *F.j.* a majority of the soil is covered by plant litter. During the colonization phase the amount of bare soil is variable and may be high.
△ **ordination** Insufficient data.
Associated floristic diversity Generally low, and can form virtual monocultures.
Synopsis Probably the most aggressive of the relatively common herbaceous dominants of the British flora. It is now an offence to allow *F.j.* to escape into the wild (Bailey & Conolly 1985). *F.j.* is a native of Japan, Taiwan and N China, and was first introduced into Britain in 1825 as an ornamental plant. It was first recorded as a naturalized alien in 1886 (Conolly 1977). In Japan *F.j.* is often an early colonist of volcanic soils, and may occur on soils of pH <4.0 (Maruta 1976, Conolly 1977). In Britain the species is a widespread colonist of habitats such as river banks, road verges and railway banks, with coal-mine spoil and cinder tips being perhaps the nearest equivalent of its native habitat. Many of its habitats are undoubtedly productive but coal-mine sites, some of which have a low pH, are likely to be less so. Moreover, *F.j.* has a well-defined capacity to grow on soils of low N status in Japan (Hirose & Tateno 1984). The concentration of *F.j.* in urban areas in Britain is a reflection of its horticultural origin. *F.j.* is an exceedingly tall (to 2 m), polycarpic perennial, and a dense canopy develops each summer beneath which few species are capable of surviving. Coexistence with *F.j.* is restricted further by the persistent stem litter which accumulates within established stands. The shoots die back each autumn, and early growth of new shoots in the spring is sustained by reserves in the thick fleshy rhizome. *F.j.* is toxic to at least some species of livestock (Cooper & Johnson 1984). The species is functionally dioecious, and only female flowers (with reduced anthers) have been observed within the Sheffield region. A small amount of seed may be set, and some has been induced to germinate in the laboratory (UCPE unpubl.). However, no seedlings have been recorded in the field in Britain (Conolly 1977). Thus, effective regeneration appears to be entirely by means of the extensive and rapidly-growing rhizome system. Clones many metres in diameter are common. Where *F.j.* occurs adjacent to rivers

Altitude
Slope
Aspect

	Unshaded
0.8	50.0
0.4	0.0

	Shaded
0.0	—
0.0	—

Soil pH

Associated floristic diversity

Not applicable

Hydrology

Absent from wetland

Triangular ordination

Bare soil

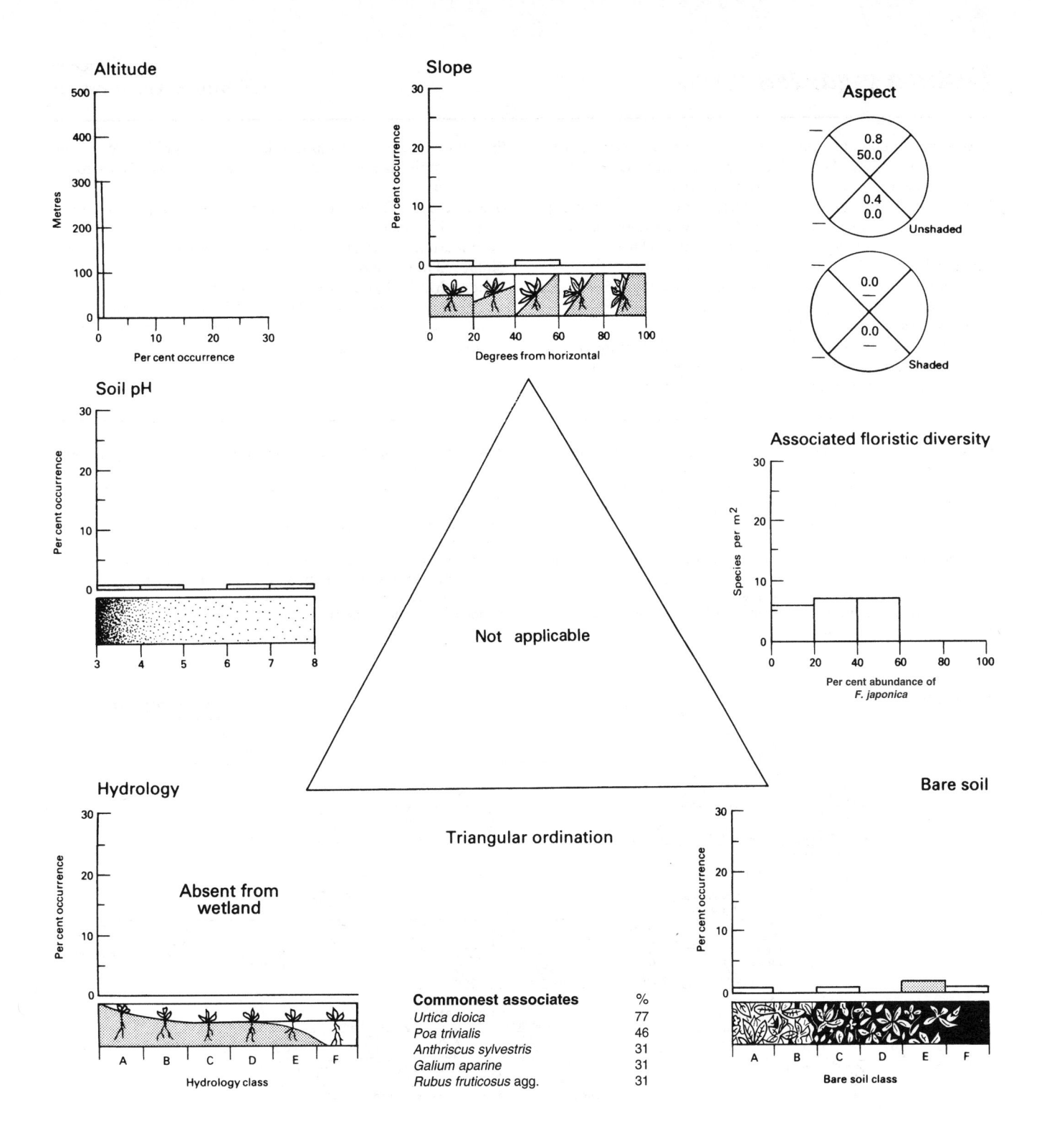

Commonest associates	%
Urtica dioica	77
Poa trivialis	46
Anthriscus sylvestris	31
Galium aparine	31
Rubus fruticosus agg.	31

it may be widely dispersed by natural means, but on most forms of wasteland long-distance dispersal appears to be sporadic and dependent upon human disturbance, such as the transport and tipping of soil contaminated with rhizomes. Apart from a geographical restriction resulting from low mobility, the spread of *F.j.* may be constrained further by climatic factors (Conolly 1977). Shoots are vulnerable both to late frosts and to summer drought, and *F.j.* has spread most rapidly in regions where these phenomena are less severe, particularly in the W of Britain and in relatively frost-free urban areas (Conolly 1977). If it were not for these limiting factors *F.j.*, which is the tallest common polycarpic herbaceous species (excluding twining plants) in the British flora, would be even more common, particularly since the species is resistant to a variety of herbicides (Fuchs 1957, Harper & Scott 1966). The robust shoot apices can penetrate asphalt (Conolly 1977). Var. *compacta* (J. D. Hook.) Buchheim, a native of India (Lousley & Kent 1981), with stems <700 mm tall and $2n = 44$ ($4x$) is less frequent in Britain (Conolly 1977, Bailey & Conolly 1985). Plants with $2n = 66$ ($6x$) may be referable to the hybrid between *F.j.* and the more robust (to 3 m) and less-frequently naturalized *F. sachalinensis* (F. Schmidt ex Maxim) Ronse Decr. (with $2n = 44$ (Bailey & Conolly 1985)).

Current trends An invasive, highly competitive, polycarpic perennial weed, which is likely to become even more troublesome.

Festuca gigantea (L.) Vill.

Poaceae
Tall Brome, Giant Fescue

Life-form Tufted, polycarpic perennial, semi-rosette hemicryptophyte. Stems erect or spreading; leaves linear, usually <6000 mm^2; roots non-mycorrhizal.
Phenology Winter green. After extension growth in early spring, flowers in June and July. Seed set from August to October.
Height Foliage mainly <600 mm; flowers to 1500 mm.
Seedling RGR 1.0-1.4 week^{-1}. SLA 27.1. DMC 21.6.
Nuclear DNA amount 14.5 pg, $2n = 56$ (Bennett *et al.* 1982), not the number normally recorded; $2n = 42$ (*FE5*); $6x$.
Established strategy C-S-R.
Regenerative strategies S. Regenerates by seed in spring. Forms loose tufts but growth of lateral shoots insufficient to constitute an effective mechanism of vegetative regeneration; SB II.

Flowers Green, hermaphrodite, wind-pollinated, self-compatible (Beddows 1931); >100 in a loose panicle (3-10 per spikelet).
Germinule 3.12 mg, 5.0 × 1.5 mm, caryopsis dispersed with an awned lemma attached (19.5 × 1.5 mm).
Germination Epigeal; t_{50} 10 days; dormancy broken by chilling.
Biological flora None.
Geographical distribution The British Isles except for the extreme N of Scotland (94% of British vice-counties), *ABIF* page 742; Europe except for the extreme N, and rare S of the Alps (74% of European territories); Asia.

Habitats Primarily a plant of woodland margins and scrub, but also found in a range of other shaded habitats such as river banks and mire. Occasionally in unshaded habitats, but usually confined to sites adjacent to wooded areas. Observed on shaded road verges and hedgebanks. Absent from skeletal habitats and pasture.
Gregariousness Usually represented by scattered individuals.
Altitude More frequently found in lowland habitats and not recorded above 300 m. [Up to 370 m.]
Slope Occurs on a wide range of slopes.
Aspect Recorded more frequently in N-facing sites, but only seen in abundance on S-facing slopes.

Hydrology Frequently recorded from soligenous mire, but largely absent from topogenous mire.
Soil pH Widely distributed, but with greatest frequency and abundance on soils of pH >5.0.
Bare soil Wide-ranging, but most frequent and abundant at high exposures.
△ **ordination** Most frequently associated with relatively fertile sites in which disturbance is slight or moderate, but widely distributed and only absent from vegetation subject to heavy disturbance.
Associated floristic diversity Typically intermediate.

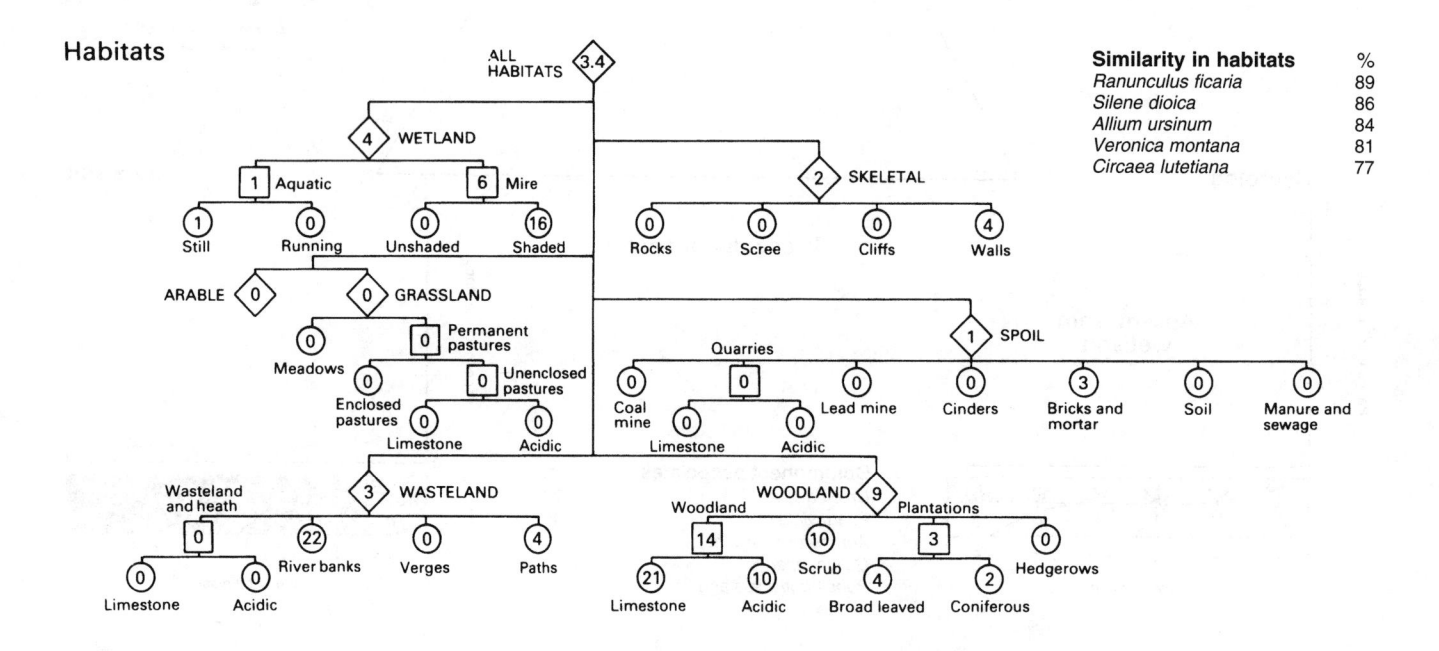

Habitats

Similarity in habitats	%
Ranunculus ficaria	89
Silene dioica	86
Allium ursinum	84
Veronica montana	81
Circaea lutetiana	77

Synopsis *F.g.* is a tufted grass of moist, base-rich sites in lowland woodland, often growing in association with *Bromopsis ramosus*. The species occurs in moderate shade and is particularly frequent beside streams and other disturbed fertile sites in woodland margins and clearings. *F.g.* is not a potential dominant and is usually a minor component of vegetation. *F.g.* is winter green, with an annual increase in shoot biomass beginning in early spring and continuing into the shaded phase. Germination occurs in spring and no persistent seed bank has been reported. *F.g.* is largely restricted to sites with little accumulation of tree litter, perhaps because the species is non-rhizomatous and is largely dependent upon seed for regeneration. The species is a widespread colonist of secondary woodland in Lincolnshire (Peterken 1981) and appears to be relatively mobile. The dispersule has much longer awns than those of other British species of *Festuca* (Hubbard 1984) and fruits adhere tenaciously to fur and clothing. *F.g.* often occurs close to streams and rivers, and the possibility that seeds may be water-dispersed requires investigation. The species is rather isolated ecologically from other species of *Festuca*, but hybrids with *F. arundinacea* Schreb., *F. pratensis* and *Lolium perenne* have been recorded within the British Isles (Stace 1975).
Current trends Status uncertain. Certainly not under threat, and apparently favoured by modern methods of managing broad-leaved woodland.

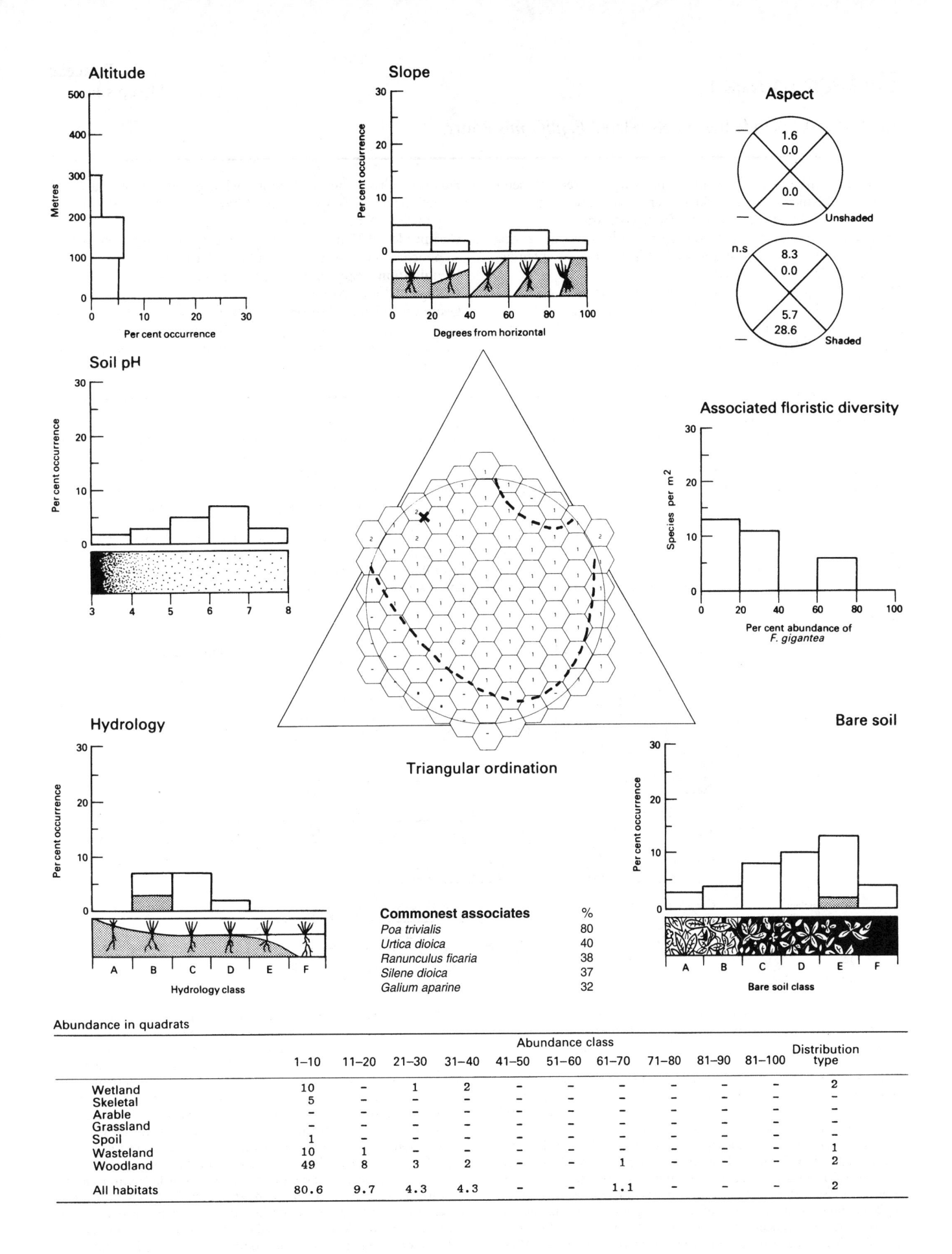

Altitude

Slope

Aspect

	1.6	
	0.0	
	0.0	
	—	Unshaded

n.s
	8.3	
	0.0	
	5.7	
	28.6	Shaded

Soil pH

Associated floristic diversity

Hydrology

Triangular ordination

Commonest associates %
Poa trivialis 80
Urtica dioica 40
Ranunculus ficaria 38
Silene dioica 37
Galium aparine 32

Bare soil

Abundance in quadrats

	1–10	11–20	21–30	31–40	41–50	51–60	61–70	71–80	81–90	81–100	Distribution type
Wetland	10	–	1	2	–	–	–	–	–	–	2
Skeletal	5	–	–	–	–	–	–	–	–	–	–
Arable	–	–	–	–	–	–	–	–	–	–	–
Grassland	–	–	–	–	–	–	–	–	–	–	–
Spoil	1	–	–	–	–	–	–	–	–	–	1
Wasteland	10	1	–	–	–	–	–	–	–	–	1
Woodland	49	8	3	2	–	–	1	–	–	–	2
All habitats	80.6	9.7	4.3	4.3	–	–	1.1	–	–	–	2

Festuca ovina L.

Poaceae
Sheep's Fescue

Field records include the closely related *F. filiformis* Pourr.

Life-form Tufted or mat-forming, polycarpic perennial, semi-rosette hemicryptophyte. Shoots erect or spreading; leaves hair-like, <250 mm²; roots with VA mycorrhizas.

Phenology Winter green, producing new tillers in early spring. Flowers from May to July and seed set from June to September. Leaves relatively long-lived.

Height Foliage usually <150 mm; flowering shoots to 450 mm.

Seedling RGR 0.5-0.9 week^{-1}. SLA 10.9. DMC 41.0.

Nuclear DNA amount 9.5 pg; $2n = 14*, 28*$ (*FE5, BC*); $2x*, 4x*$.

Established strategy Stress-tolerator.

Regenerative strategies S, V. See 'Synopsis'; SB I.

Flowers Green, hermaphrodite, wind-pollinated, strongly self-incompatible (Snaydon & Bradshaw 1961); often >50 in a loose panicle (3-9 per spikelet).

Germinule 0.35 mg, 2.7 × 1.0 mm, caryopsis dispersed with an attached short-awned lemma (4.3 × 1.1 mm).

Germination Epigeal; t_{50} 4 days; 7-33°C; L = D.

Biological flora None.

Geographical distribution The British Isles (100% of British vice-counties), *ABIF* page 746; N and C Europe (59% of European territories and 3% naturalized); other N temperate zones.

Habitats Occurs in a wide variety of habitats, and particularly abundant in upland pastures and on heaths and moors. Also common on scree slopes, rock outcrops, cliffs, lead-mine heaps and quarry spoil. Recorded at low frequency in a variety of other habitats, including grass verges and scrub. Absent from arable land. [An agronomically important species in montane grassland.]

Gregariousness A patch-forming species.

Altitude Suitable habitats are most frequent in the uplands, and distribution shows an upland bias. [Up to 1310 m.]

Slope Widely distributed.

Aspect Slightly more frequent on S-facing aspects and slightly more abundant on N-facing slopes.

Hydrology Recorded from moist ground, and in wetland occasionally observed on tussocks of larger species, e.g. *Eriophorum vaginatum*. Occasionally in soligenous mire, sometimes growing with *Sphagnum*.

Soil pH Frequent and abundant over the full range of soil pH, but declining slightly over the range pH 5.0-7.5, which includes the majority of the soils of high fertility.

Bare soil Wide-ranging, but most frequent and abundant at low exposures.

△ **ordination** Distribution concentrated in the lower right-hand side of the diagram, indicating an association with unproductive, relatively undisturbed vegetation. Excluded from heavily disturbed habitats.

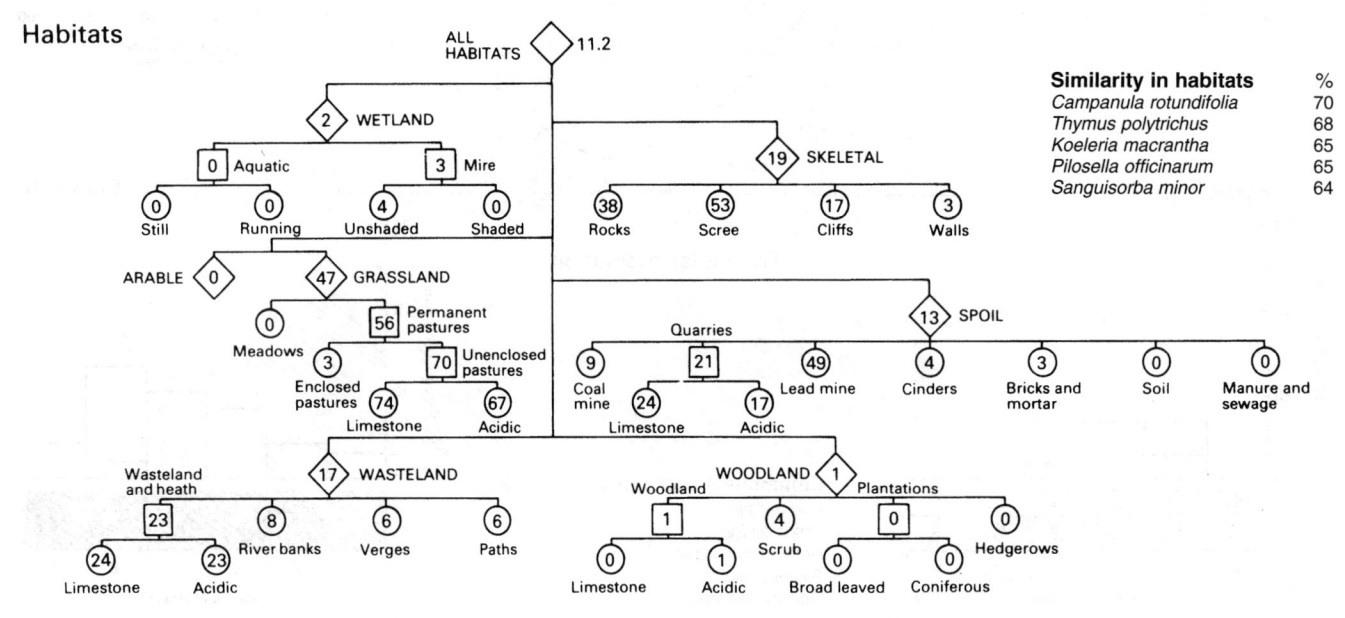

Habitats

Similarity in habitats	%
Campanula rotundifolia	70
Thymus polytrichus	68
Koeleria macrantha	65
Pilosella officinarum	65
Sanguisorba minor	64

Associated floristic diversity Low to intermediate, but rising with increasing frequency of *F.o.*. More rarely *F.o.* is a component of very species-rich vegetation.

Synopsis *F.o.* is a slow-growing, tufted to mat-forming, winter-green grass found in a wide range of unproductive grasslands and rocky habitats. In particular, *F.o.* is the most consistent component species of infertile pasture in Britain, and is unique amongst our grassland species in its capacity to achieve high frequency and abundance in both highly acidic and strongly calcareous sites. However, the species most similar in habitat range are calcicolous. In sites of greater fertility *F.o.* declines in importance relative to faster-growing and broader-leaved turf grasses, e.g. *Agrostis capillaris* and *F. rubra*. The failure of *F.o.* in competition with *A.c.* has been demonstrated experimentally (Mahmoud & Grime 1976). *F.o.* is less palatable than some of the pasture grasses with which it occurs (Milton 1933). Nevertheless, the species is an important food plant in hill pastures extensively grazed by sheep, comparing very favourably with, for example, *Nardus stricta*. *F.o.* can withstand moderate trampling and is often found along the edges of paths and sheep walks, but it is sensitive to burning. *F.o.* is cold-tolerant but tends to be replaced in moist montane habitats by the proliferous *F. vivipara* (L.) Sm. Calcium status of the leaves is low (*SIEF*). Dereliction of pastures leads to the suppression of *F.o.* by taller species, but on shallow, nutrient-deficient soils *F.o.* may persist indefinitely. Despite the relatively shallow root system (Grime & Curtis 1976) of *F.o.*, many of the habitats it exploits are subject to severe drought. Features of *F.o.* which appear to contribute to its success in dry habitats include xerophylly and an early shoot phenology and seed-set. On shallow soils tussocks are frequently killed by drought, and in very dry years populations may be severely reduced. However, re-establishment occurs from the autumn cohorts of seedlings. Seeds lack dormancy and tend to germinate

Altitude

Slope

Aspect

Soil pH

Associated floristic diversity

Species per m^2

Per cent abundance of *F. ovina*

Hydrology

Triangular ordination

Bare soil

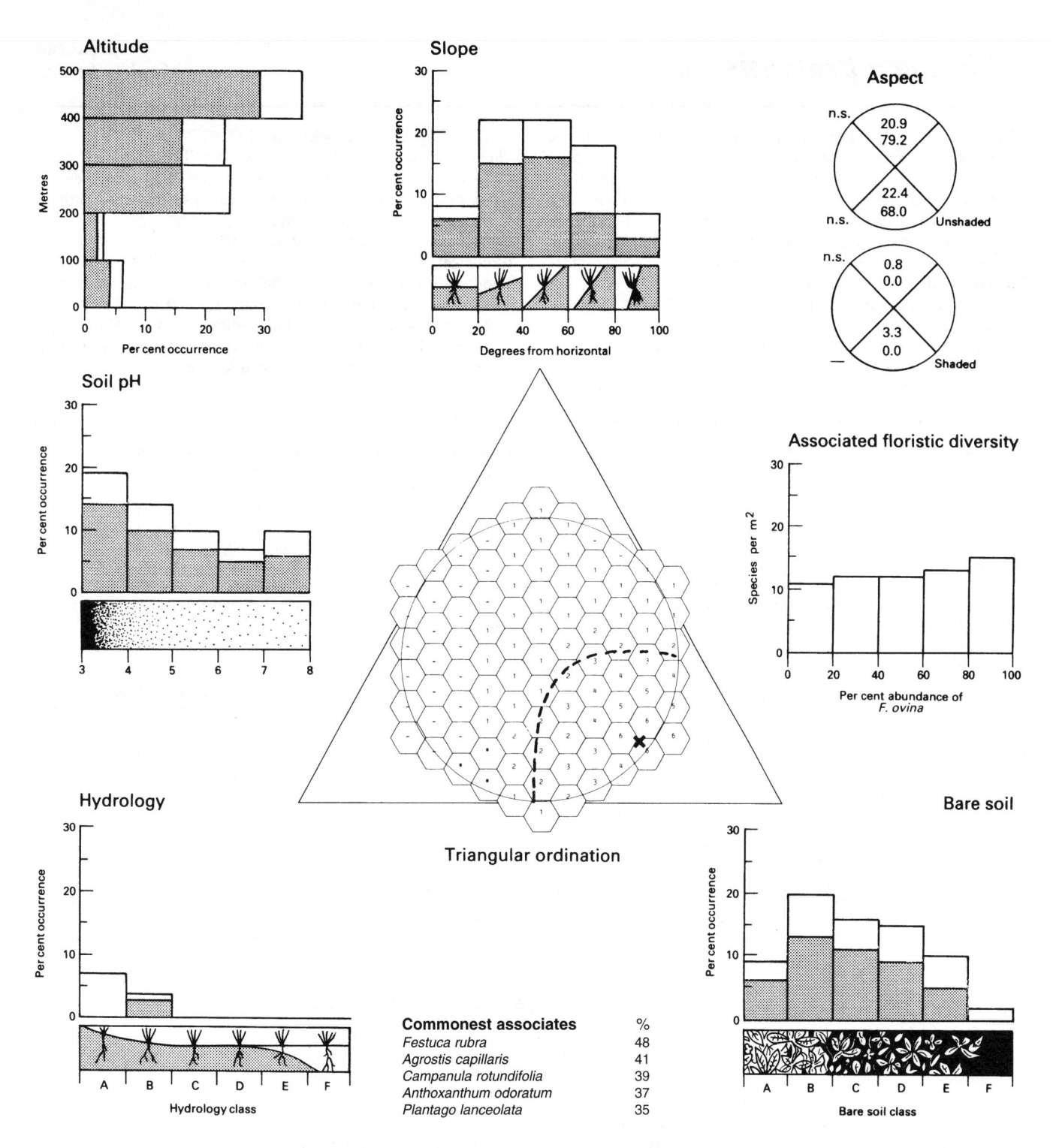

Commonest associates	%
Festuca rubra	48
Agrostis capillaris	41
Campanula rotundifolia	39
Anthoxanthum odoratum	37
Plantago lanceolata	35

synchronously in late summer, although at northern stations a small proportion may remain ungerminated until spring (Harmer & Lee 1978). In moister habitats, where longevity of plants is likely to be greater, the capacity for slow vegetative expansion may also be important and, exceptionally, clones *c.* 10 m in extent and apparently hundreds of years old may develop (Harberd 1962). However, in *F.o.* there is no evidence of the well-defined capacity for rapid vegetative proliferation exhibited by *F.r.* ssp. *rubra*. *F.o.* has colonized a number of artificial habitats, but does not appear to be very mobile. The capacity of *F.o.* to exploit a wide range of habitats appears to be related to its out-breeding habit, which promotes the emergence of distinct ecotypes. Thus ssp. *ovina* and ssp. *hirtula* (Hack. ex Travis) M. J. Wilk. are mainly restricted to acidic soils while ssp. *ophioliticola* (Kerguelen) M. J. Wilk. is found on calcareous and serpentine soils. The genetic and ecological implications of the existence of long-lived individuals in some habitats and relatively short-lived ones in others requires investigation. *F.f.*, $2n = 14, 28$ ($2x, 4x$), is largely, or possibly exclusively, restricted to acidic soils, and has a smaller germinule (*FE5, CSABF* p. 148). Tetraploids are moderately self-fertile (Watson 1958). *F.o.* hybridizes, though rarely, with *F.f.*, with *F. rubra* and with one other species (Stace 1975).

Current trends *F.o.* (and *F. filiformis*) are sown in small quantities (less than 50 t year^{-1}, MAFF 1980-84), primarily for turf or amenity grassland. However, this does not compensate for the decrease in abundance of *F.o.* due to the destruction of older grassland systems (Hopkins 1979), and *F.o.* is becoming uncommon in some lowland areas. As its lowland sites become smaller and more isolated *F.o.* might be expected to evolve a greater capacity for in-breeding.

(*) *Festuca pratensis* Hudson

Poaceae
Meadow Fescue

Life-form Loosely tufted, polycarpic perennial, semi-rosette hemicryptophyte or chamaephyte. Shoots erect or spreading; leaves linear, <4000 mm²; roots often with VA mycorrhizas.
Phenology Winter green. Some growth in winter and appreciable growth in early spring (Spedding & Diekmahns 1972). Flowers in June and seed is shed from July to September.
Height Foliage predominantly <300 mm; flowers to 800 mm.
Seedling RGR No data. SLA 21.1. DMC 26.9.
Nuclear DNA amount 4.4 pg; $2n = 14^*$ (*BC*); $2x^*$.
Established strategy C-S-R.
Regenerative strategies S, V. By seed in autumn. No persistent bank of seeds is formed (Lewis 1973). Usually present as small compact tufts; older plants may fragment into short-lived clones (Spedding & Diekmahns 1972); SB I.
Flowers Green, hermaphrodite, wind-pollinated, rarely selfed (Beddows 1931); >100 in an open panicle (5-14 per spikelet).
Germinule 1.53 mg, 5.0 × 1.3 mm, caryopsis dispersed with an attached lemma (7.5 × 1.5 mm).
Germination Epigeal; t_{50} 2 days; 8-30°C.
Biological flora None.
Geographical distribution The British Isles, but rare in N Scotland (93% of British vice-counties), *ABIF* page 741; most of Europe but rare in parts of the Mediterranean and SW Europe, doubtfully native in parts of the N (87% of European territories); temperate Asia; introduced in N America.

Habitats Found in moist grassland, both managed (meadows, pastures and road verges) and unmanaged (wasteland), and also locally frequent in some marshy habitats, reaching maximum abundance on clayey and alluvial soils with high water retention. Observed on paths, but only of casual occurrence in highly disturbed habitats.
Gregariousness Intermediate or occurring as scattered individuals.
Altitude Recorded over a wide range of altitudes but rare on freely drained soils of the low-lying Bunter sandstone. [Up to 500 m.]
Slope Largely confined to habitats with gentle slopes.
Aspect Insufficient data.
Hydrology Frequent on moist soils and widely observed in base-rich soligenous mire. Only of casual occurrence in aquatic habitats.
Soil pH Restricted to soils pH >5.0.
Bare soil Most typical of vegetation stands with little exposed soil.
△ **ordination** Distribution confined to vegetation subject to some disturbance and of intermediate productivity.
Associated floristic diversity Found over a wide range of floristic diversity. Grassland sites are usually intermediate or species-rich, whereas some wetland sites are very species-rich and *F.p.* has been observed in vegetation with >40 species m⁻².

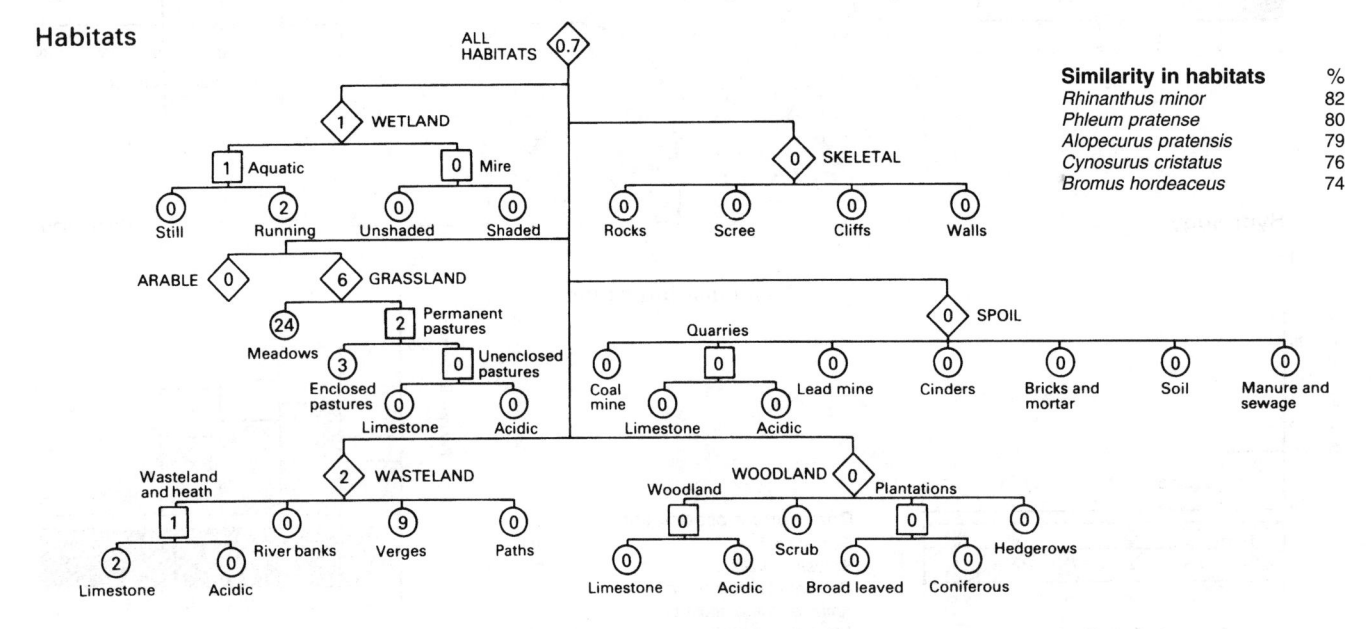

Habitats

Similarity in habitats	%
Rhinanthus minor	82
Phleum pratense	80
Alopecurus pratensis	79
Cynosurus cristatus	76
Bromus hordeaceus	74

Synopsis *F.p.* is a tufted, often relatively short-lived grassland species which is palatable to livestock. The species is of agricultural importance, particularly in meadows (Spedding & Diekmahns 1972), and over 100 t are sown annually (MAFF 1980-84). *F.p.* is, in many respects, similar to *Lolium perenne* and frequently forms the sterile hybrid × *Festulolium loliaceum* (Huds.) P. Fourn. Unlike *L.p.*, however, *F.p.* is most frequent on moist to waterlogged soils and, in keeping with the lower nuclear DNA amount, the shoot phenology is delayed in comparison with that of *L.p.*. Also, *F.p.* is less productive than *L.p.* (Spedding & Diekmahns 1972) and as an apparently native species frequently occurs on soils of intermediate fertility often in relatively species-rich vegetation. Characteristically *F.p.* occurs as a subordinate component of turf, and is usually represented by scattered individuals. *F.p.*, which does not persist either in tall grassland or in intensively grazed pasture, has been most frequently recorded from hay meadows. Like *L.p.*, *F.p.* appears largely dependent upon seed for regeneration and does not form a persistent seed bank. As in the case of the annual grass, *Bromus hordeaceus*, the capacity to persist in hay meadows probably results from early shedding of the seed, a feature facilitating colonization of the open ground available immediately after the hay has been cropped. The species hybridizes with *F. gigantea* and *F. arundinacea* Schreb., as well as with *L.p.* The ecology of the ssp. *apennina* (De Not.) Hegi, with $2n = 28$, $4x$, from continental Europe makes an interesting contrast with that of the more widespread and British ssp. *pratensis*. Ssp. *apennina* occurs at higher elevations in the mountains than ssp. *pratensis*, where it survives the severe winters by dying back in autumn to become winter-dormant (Tyler et al. 1978). Seeds of ssp. *apennina* require cold-treatment for germination.
Current trends Decreasing as a native species through the destruction of alluvial grasslands and other suitable habitats. Also becoming agronomically less important, and expected to decrease further in popularity (Parry & Butterworth 1981).

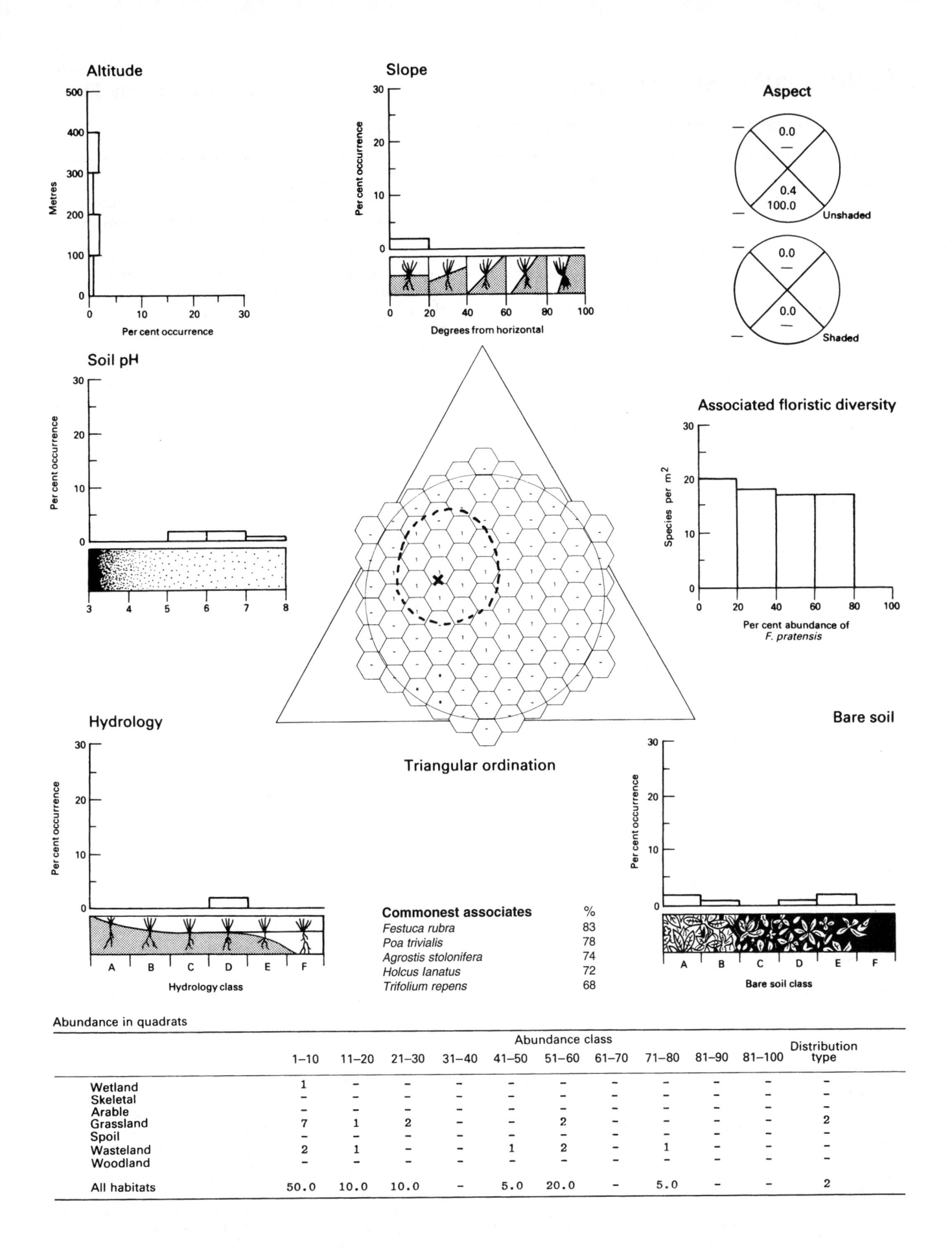

Altitude

Slope

Aspect

Unshaded

Shaded

Soil pH

Associated floristic diversity

Species per m²

Per cent abundance of *F. pratensis*

Hydrology

Hydrology class

Triangular ordination

Commonest associates	%
Festuca rubra	83
Poa trivialis	78
Agrostis stolonifera	74
Holcus lanatus	72
Trifolium repens	68

Bare soil

Bare soil class

Abundance in quadrats

	\multicolumn{10}{c	}{Abundance class}	Distribution type
	1–10	11–20	21–30	31–40	41–50	51–60	61–70	71–80	81–90	81–100	
Wetland	1	–	–	–	–	–	–	–	–	–	–
Skeletal	–	–	–	–	–	–	–	–	–	–	–
Arable	–	–	–	–	–	–	–	–	–	–	–
Grassland	7	1	2	–	–	2	–	–	–	–	2
Spoil	–	–	–	–	–	–	–	–	–	–	–
Wasteland	2	1	–	–	1	2	–	1	–	–	–
Woodland	–	–	–	–	–	–	–	–	–	–	–
All habitats	50.0	10.0	10.0	–	5.0	20.0	–	5.0	–	–	2

Festuca rubra ssp. *rubra* L.

Poaceae
Red Fescue, Creeping Fescue

Field records may include data for ssp. *commutata* Gaud in dry grassland

Life-form Polycarpic perennial, semi-rosette hemicryptophyte with long creeping rhizomes. Shoots erect or spreading; leaves linear, hair-like, <500 mm^2; roots normally with VA mycorrhizas.

Phenology Winter green, with considerable early spring and late autumn growth. Flowers from May to July and seed set from July to August.

Height Foliage usually <300 mm; flowering shoots to 700 mm.

Seedling RGR 1.0–1.4 week^{-1}. SLA 17.7. DMC 26.7.

Nuclear DNA amount 13.9 pg, $2n$ = 28, 42*, 56, 70, including non-British ssp. (*FE5, BC*); $4x$, $6x$*, $8x$, $10x$.

Established strategy C-S-R, but some populations tending towards C, S-C or S.

Regenerative strategies V, S. See 'Synopsis'. SB I.

Flowers Green, hermaphrodite, wind-pollinated, mainly self-incompatible (Weimarck 1968); often <100 in a loose panicle (3–9 per spikelet); SB I.

Germinule 0.79 mg, 3.5 × 1.1 mm, caryopsis dispersed with an attached usually short-awned lemma (7.3 × 1.1 mm).

Germination Epigeal; t_{50} 2 days; 6–35°C; L = D.

Biological flora None, but see Zarzycki (1976).

Geographical distribution The *F.r.* aggregate occurs throughout the British Isles (100% of British vice-counties), *ABIF* page 743; Europe (85% of European territories and 5% naturalized); N Africa, temperate Asia and N America, but only on mountains in the southern part of its range.

Habitats Particularly characteristic of base-rich grassland and very common on road verges and wasteland and in meadows and pastures. Equally frequent in rocky habitats such as scree slopes, quarry spoil and rock outcrops, and common and abundant in a wide variety of other habitats including mining and industrial spoil, waste places such as paths and river banks, and open habitats such as cliffs and walls. Occurs more rarely in woods, hedgerows and scrub, and in unshaded wetlands. [*F.r. sensu lato* is also found on sand dunes, shingle ridges, cliffs, salt marshes and mountains.]

Gregariousness Intermediate to stand-forming.

Altitude Common at all altitudes, but particularly abundant in the limestone dale-sides of the upland region. [Up to 1080 m.]

Slope Widely distributed.

Aspect In unshaded sites significantly more frequent on N-facing slopes.

Hydrology Found on moist soils, in soligenous mire and on tussocks of other wetland species. Not associated with permanently submerged habitats.

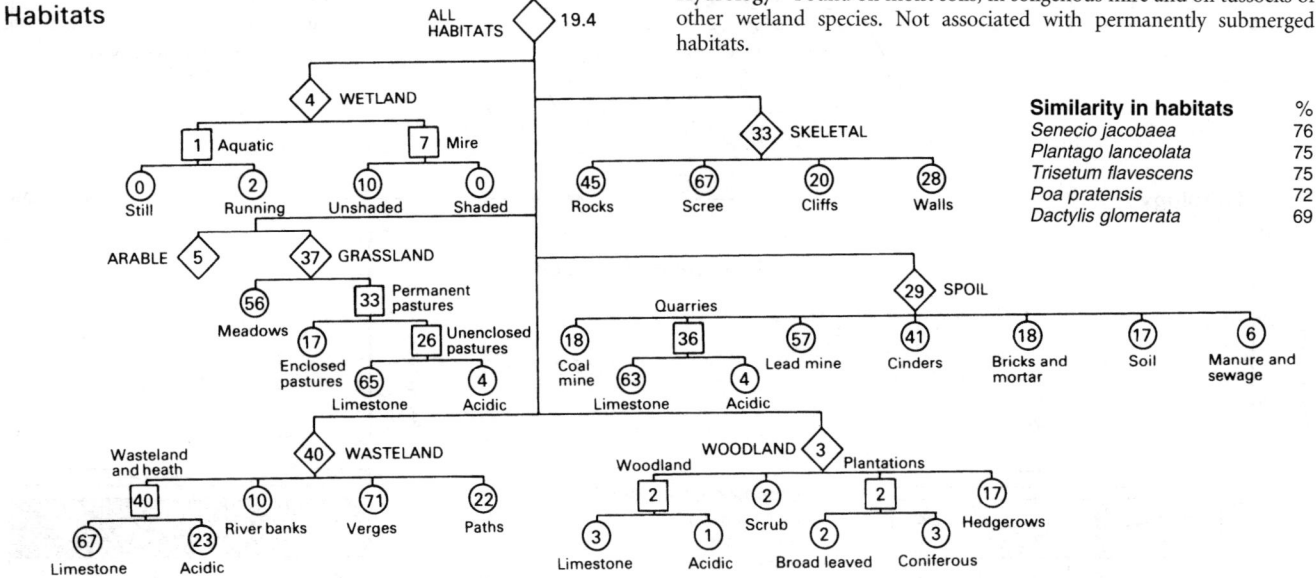

Similarity in habitats	%
Senecio jacobaea	76
Plantago lanceolata	75
Trisetum flavescens	75
Poa pratensis	72
Dactylis glomerata	69

Soil pH Absent from only the most acidic soils. More frequent and abundant on soils of pH >5.0.

Bare soil Wide-ranging, but less common in habitats with large amounts of exposed soil.

Δ ordination Occurs in a very wide range of vegetation types, with distribution centred on vegetation in which competition is reduced to moderate intensities by disturbance or low productivity.

Associated floristic diversity Typically intermediate. In sites where *F.r.* is the most abundant grass the vegetation is frequently species-rich.

Synopsis Ssp. *r.* is a rather slow-growing, winter-green grass forming large clones by means of extensive rhizome growth in a wide range of grasslands and other habitats, particularly on base-rich soils. *F.r.* occurs in 'every plot, limed and unlimed' in the *PGE*, where the species' abundance is increased at low levels of fertilizer input and decreased by heavy nitrogenous dressings'. Comparatively rapid rates of leaf growth are achieved in early spring (Grime *et al.* 1985), suggesting that temporal niche differentiation plays a critical role in coexistence with *Agrostis capillaris*, which has a much later pattern of shoot development in the agriculturally important 'Red Fescue-Brown Bent' pastures of intermediate productivity in Britain. Ssp. *r.* can also occupy a somewhat different niche in more-productive tall-grass communities, forming the non-flowering understorey described by Grubb (1982). Levels of N and Ca in the leaves are usually low (*SIEF*). In turf microcosms, roots of *F.r.* became heavily colonized by VA mycorrhizas but infection brought about no significant increase in seedling yield (Grime, Mackey, Hillier & Read 1987). Although readily grazed by sheep, ssp. *r.* is considered less palatable than pasture grasses such as *Lolium perenne* (Spedding & Diekmahns 1972). This may contribute to the ability of ssp. *r.* to persist as scattered individuals in many productive pastures. The distribution of ssp. *r.* in limestone grassland strongly overlaps with that of *F. ovina*, essentially a species of infertile soils and readily grazed by sheep. Unlike some other ssp., *r.* is tolerant of cutting but does not persist under close (5 mm) mowing (Shildrick 1984). Ssp. *r.*

Altitude

Slope

Aspect

P < 0.05

	41.3	
n.s.	62.9	Unshaded
	30.9	
	60.9	

	3.8	
n.s.	40.0	
	4.1	Shaded
—	0.0	

Soil pH

Associated floristic diversity

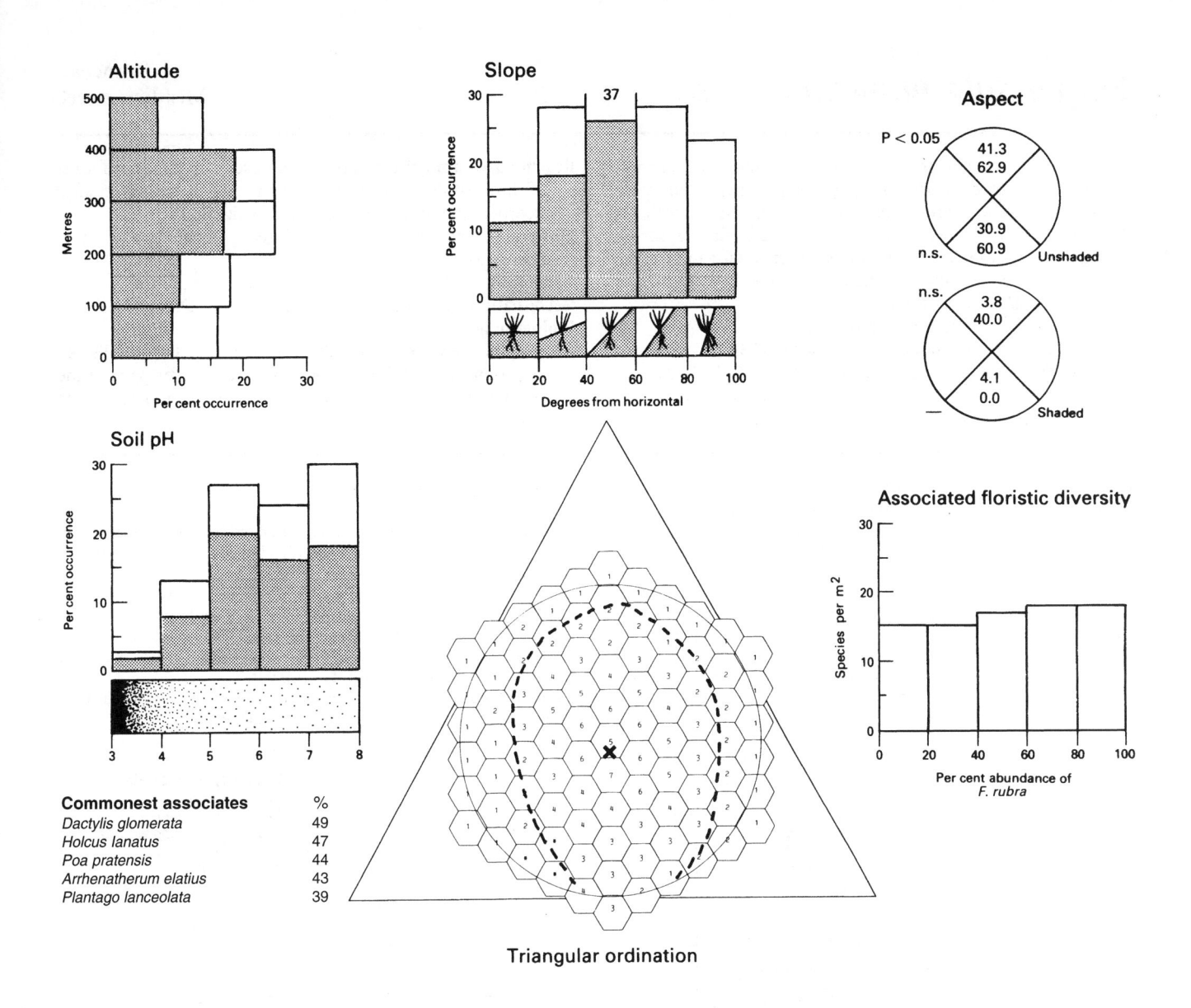

Commonest associates %
Dactylis glomerata 49
Holcus lanatus 47
Poa pratensis 44
Arrhenatherum elatius 43
Plantago lanceolata 39

Triangular ordination

Abundance in quadrats

	1–10	11–20	21–30	31–40	41–50	51–60	61–70	71–80	81–90	81–100	Distribution type
All habitats	29.3	9.4	7.5	5.8	6.0	4.9	4.7	5.6	6.0	20.8	3

is an important stabilizer of limestone screes onto which it encroaches vegetatively. Rhizome growth also plays an important role in regeneration, and clones perhaps 1000 years old have been recorded (Harberd 1961). Rhizomes break down rapidly, leaving nutritionally-independent daughter plants (Hall 1971). Seeds are released in summer and germinate mainly in autumn. Ssp. *r.* only forms a transient seed bank (Thompson & Grime 1979, Roberts 1986b). The *F.r.* group is extremely complex, and a number of morphologically distinct taxa of *F.r. sensu stricto*, with specialized ecological distributions, have been recognized (Hubbard 1984). These include ssp. *arctica* (Hack.) Govor (mountains above 600 m), ssp. *commatata* (dry grassland), ssp. *litoralis* (G. F. W. Meyer) Aquier (salt marsh) and ssp. *pruinosa* (Hack.) Piper (sea cliffs, salt marshes and stony sea shores). This taxonomic treatment may prove unsatisfactory. In most ecological studies it is not clear which segregates have been studied. The maritime ssp. are presumably salt-tolerant. Others, found in sown grassland, are more robust than native populations, and may contribute to the extensive monospecific stands on motorway verges in which there is a dense build-up of persistent litter. However, evolution within the species is potentially rapid; lead-tolerant ecotypes have been detected on motorway verges (Atkins et al. 1982). It seems probable that the exceptionally wide range of habitats exploited by *F.r.* may result from ecotypic differentiation. The problems concerning ecological distribution and taxonomy are compounded by the fact that cultivars of two maritime ssp., as well as *F. nigrescens*, are now recommended for use in turfgrass (Shildrick 1980). *F.r.* hybridizes frequently with *Lolium perenne* and crosses with three species of *Vulpia* (Stace 1975).

Current trends Though still common, *F.r.* has decreased in abundance due to the destruction of older grassland systems (Hopkins 1979). However, this effective colonist of artificial habitats is probably found in a more diverse range of environments than was formerly the case. Seeds of *F.r.* agg. are sown at over 1500 t year^{-1} (MAFF 1980-84). It is predicted that even larger quantities will be sown in the future, since *F.r.* is still underused in amenity grassland. *F.r.* may also have more agronomical potential than is at present appreciated.

Filipendula ulmaria (L.) Maxim.

Rosaceae
Meadow-sweet

Life-form Rhizomatous, polycarpic perennial, semi-rosette hemicryptophyte. Stems erect; leaves pinnate, often hairy, usually <6000 mm²; roots shallow, often with VA mycorrhizas.
Phenology Leaves develop from winter buds during spring. Plants flower June to August and seed set August to October. Leaves produced until September (Yapp 1912), but by November shoots are dead.
Height Foliage and flowers <1200 mm.
Seedling RGR 0.5–0.9 week^{-1}. SLA 18.6. DMC 39.1.
Nuclear DNA amount No data; $2n = 14* \ 16$ (*BC, FE3*); $2x*$.
Established strategy Intermediate between competitor and stress-tolerant competitor.

Regenerative strategies V, ?B$_s$. Forms extensive stands by means of rhizomatous growth. Seeds germinate in spring or after incorporation into a persistent seed bank; ?SB IV.
Flowers Cream, hermaphrodite, insect-pollinated; >100 in a cymose panicle.
Germinule 0.99 mg, 4.2 × 1.9 mm, achene; *c.* 6 per fruit.
Germination Epigeal; t_{50} 3 days; 11–30°C; L = D.
Biological flora None.
Geographical distribution The British Isles (100% of British vice-counties), *ABIF* page 324; Europe except for parts of the Mediterranean region (77% of European territories); temperate Asia.

Habitats Occurs mainly on damp and marshy ground. Particularly common in shaded mire and on river and ditch banks. Also widespread in unshaded mire and in open woodland and hedgerows. Also recorded from temporarily flooded sites and on soil dredged from ponds. More rarely as a constituent of species-rich herb communities of N-facing calcareous terraces (Pigott 1958). Rarely observed from wet skeletal habitats, meadows and pastures, and absent from most spoil categories.
Gregariousness Forms dense stands in which dominance may be exerted at relatively low densities of shoots.
Altitude A majority of suitable wetland sites occur in the lowland but observed to 400 m. [Up to 890 m.]

Slope Mainly restricted to gentle slopes.
Aspect Insufficient data.
Hydrology Shows a clear association with moist soils near or marginal to open water. Also found in soligenous mire, but absent from permanently flooded sites.
Soil pH Absent from soils of pH <4.5, but frequent above this value.
Bare soil Widely distributed, but most frequently associated with moderate levels of bare soil and often associated with persistent litter.
△ ordination Distribution clearly centred on vegetation associated with productive and relatively undisturbed habitats.
Associated floristic diversity Low and declining slightly with increasing abundance of *F.u.*, indicating a capacity for dominance.

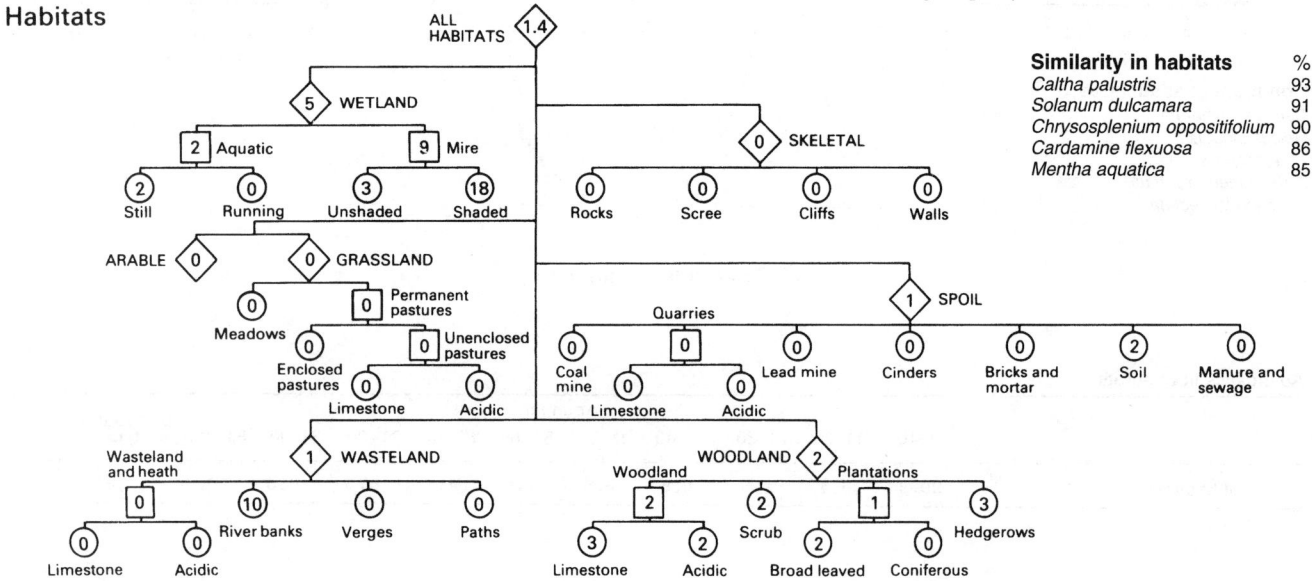

Habitats

Similarity in habitats	%
Caltha palustris	93
Solanum dulcamara	91
Chrysosplenium oppositifolium	90
Cardamine flexuosa	86
Mentha aquatica	85

Synopsis *F.u.* is a tall herb, particularly characteristic of moderately fertile sites at the transition between wetland and dryland. Habitats include river banks and floodplain terraces. The absence from sites which remain waterlogged throughout the year may be related to the sensitivity of *F.u.* to ferrous ion toxicity (R. E. D. Cooke unpubl.). The aerenchyma of the roots is relatively poorly developed compared with that of many wetland species (Smirnoff & Crawford 1983), and for a wetland species *F.u.* shows an unusually high level of drought resistance in the field (Rackham 1980). Grazing and frequent mowing eliminate the species. The plant shows rapid extension growth in the late spring and reaches a peak in above-ground biomass in July (Al-Mufti *et al.* 1977). Unlike many stand-forming species (e.g. *Epilobium hirsutum* and *Urtica dioica*) which have a high relative growth rate and the capacity to form a rapidly ascending canopy of short-lived foliage, the RGR of *F.u.* is low, and most of the leaves survive until the end of the growing season. Leaf persistence in shade is particularly marked in the case of the foliage of non-flowering shoots, which occur as a basal rosette, often submerged beneath a dense canopy. The above-ground biomass in stands of *F.u.* tends to be less than that in other commonly occurring tall-herb communities (Al-Mufti *et al.* 1977), and *F.u.* tends to occur in sites where the vigour of these more productive species is reduced by mineral-nutrient stress or shade. Thus, *F.u.* is frequent in shaded mire and widespread in open woodland on wet clayey soils. Although this is not apparent from the triangular ordination, we also suspect that some of the sites exploited by *F.u.* are less rich in nutrients than those occupied by, for example, *E.h.* Foliage of *F.u.* contains high concentrations of N, P and Mg and a relatively low level of Fe (*SIEF*). *F.u.* regenerates mainly by means of creeping rhizomes to form extensive patches. In open habitats an abundance of seed may be produced, but in shaded sites flowering and seed set are frequently reduced. Germination occurs in spring, and we suspect that seeds may be incorporated into a large persistent seed bank (UCPE unpubl.). Seeds are capable of floating for several weeks (Ridley 1930); this may enable *F.u.* to colonize water

Altitude

Slope

Aspect

Soil pH

Associated floristic diversity

Hydrology

Triangular ordination

Commonest associates %
Poa trivialis 65
Festuca rubra 46
Holcus lanatus 46
Ranunculus repens 37
Arrhenatherum elatius 34

Bare soil

Abundance in quadrats

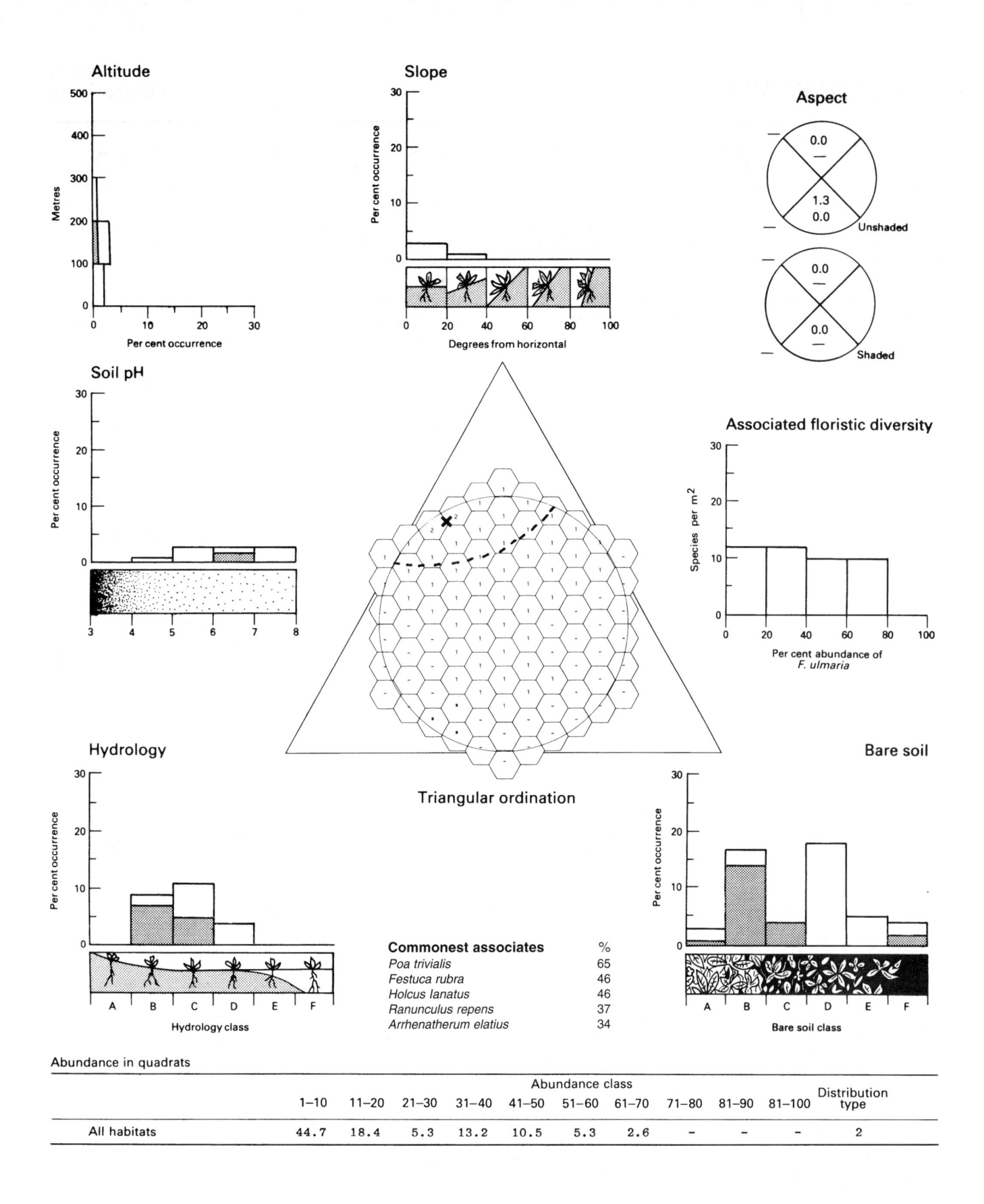

	\multicolumn{9}{c	}{Abundance class}	Distribution type
	1–10	11–20	21–30	31–40	41–50	51–60	61–70	71–80	81–90	81–100	
All habitats	44.7	18.4	5.3	13.2	10.5	5.3	2.6	–	–	–	2

systems. It is suspected that vegetative portions detached by disturbance are also capable of regeneration. Leaves are plastic with respect to hairiness, hairs being developed under dry atmospheric conditions (Yapp 1912). In Europe the existence of numerous ecotypes has been demonstrated (Turesson 1920) and *F.u.* has been formally subdivided into three ssp. (*FE2*).

Current trends Not generally a rapid colonist of artificial habitats and, like most species associated with wetlands, *F.u.* is probably decreasing.

Fragaria vesca L.

Rosaceae
Wild Strawberry

Life-form Polycarpic perennial, rosette hemicryptophyte with woody stock and long arching runners. Flowering scapes erect; leaves ternate, hairy, usually <2000 mm²; roots with VA mycorrhizas.

Phenology Winter green. Flowers from April to June and bears fruits from June to July. New stolons produced after flowering.

Height Foliage and flowers usually <200 mm.

Seedling RGR No data. SLA 17.5. DMC 38.8.

Nuclear DNA amount No data; $2n = 14*$ (BC); $2x*$.

Established strategy Intermediate between stress-tolerator and C-S-R.

Regenerative strategies V, B_s. Regenerates effectively by long arching runners to form large clones. Also produces seeds which germinate in spring and summer (Muller 1978) and which may be incorporated into a persistent seed bank; ?SB IV.

Flowers White, hermaphrodite, insect-pollinated and selfed automatically if insect visits fail (Knutt 1906); <10 in a cyme.

Germinule 0.31 mg, 1.3 × 1.0 mm, achene, often >40 embedded on the surface of the red fleshy receptacle.

Germination Epigeal; t_{50} 6 days; 11-30°C; L > D.

Biological flora None.

Geographical distribution The British Isles (100% of British vice-counties), *ABIF* page 333; most of Europe (89% of European territories); Asia, N America.

Habitats Recorded from well-drained, skeletal habitats, particularly scree, quarry and lead-mine spoil. Also found on wasteland and pasture. Frequently associated with shallow soils, and widely observed on railway ballast and cinder tips. Also frequent in open woodland and scrub and present in some hedgerows. Absent from wetlands and arable land.

Gregariousness Intermediate.

Altitude Observed to 310 m. [Up to 640 m.]

Slope Absent from cliffs and from the steepest slopes.

Aspect Wide-ranging.

Hydrology Absent from wetlands.

Soil pH Not recorded below pH 5.0, and most records fall in the pH range 6.5-8.0.

Bare soil Widely distributed.

△ ordination Distribution centred on vegetation associated with relatively infertile conditions. Absent from highly disturbed or productive communities.

Associated floristic diversity Typically intermediate to species-rich. Lower in woodland,

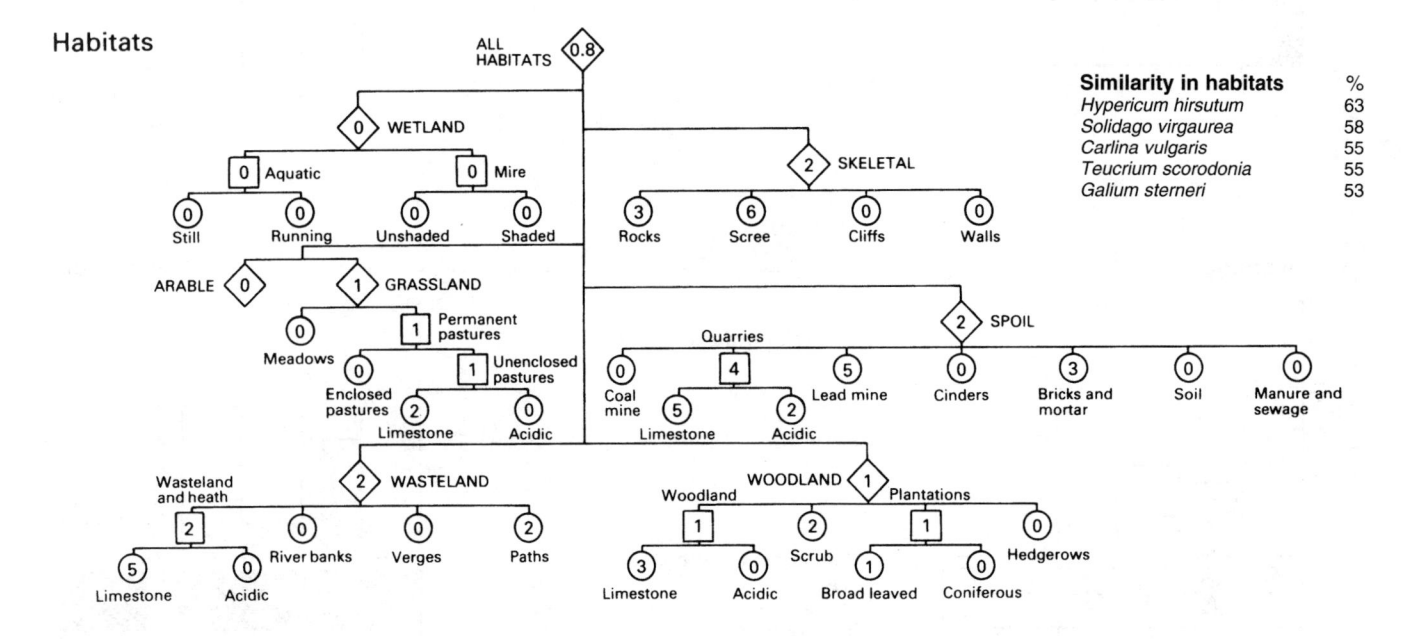

Habitats

Similarity in habitats	%
Hypericum hirsutum	63
Solidago virgaurea	58
Carlina vulgaris	55
Teucrium scorodonia	55
Galium sterneri	53

Synopsis *F.v.* is a low-growing, winter-green herb which exploits sites where the vigour of potential dominants is suppressed by moderately low fertility and where there is often a rocky substrate or moderate shade from a tree canopy, or both. The species is particularly characteristic of open turf in situations where, despite the shallow soil, the incidence of droughting is slight. In this respect *F.v.* resembles several other broadleaved, rosette plants such as *Potentilla sterilis*, *Prunella vulgaris*, *Primula veris* and *Ajuga reptans* L. Thus, the habitat range of *F.v.* includes sites such as stabilized scree, quarry spoil and ballast beside the permanent way, and woodland margins, scrub and hedgerows. In upland areas the species is more frequent in unshaded sites, and throughout the habitat range flowers and fruits are more prolific in unshaded habitats. In N Britain this edaphically wide-ranging species is particularly frequent on limestone strata. Regeneration is mainly vegetative and *F.v.* produces long stolons. The runners also allow ramification over rocky substrata and experiments are required to examine the extent to which they enable *F.v.* to exploit sites of uneven soil depth or with a patchy light environment. The turnover of rosettes is relatively slow (Angevine 1983). Plantlets detached as a result of disturbance may also play a role in the spread of the species. Vigorous clonal spread is frequently observed, and regeneration by seed is probably mainly of importance during the colonization of new sites. Strawberries are dispersed by birds (Ridley 1930) and seed is incorporated into a persistent seed bank (UCPE unpubl.). There is a high level of ecotypic differentiation in N American populations (Hancock & Bringhurst 1978), and a large number of varieties and forms, including cultivars, have been described (Staudt 1962). However, the most commonly cultivated strawberry is *F.* × *ananassa* Duchesne (*F. chiloensis* (L.) Duchesne × *F. virginiana* Duchesne; $2n = 56$, $6x$ (*FE2*)), which is more robust in all its parts and, when naturalized, occurs on deeper moister soils than *F.v.*, usually in unshaded sites.

Current trends Has spread in the past beside railways and hedgerows but is not locally a rapid colonist of new artificial habitats and is perhaps decreasing.

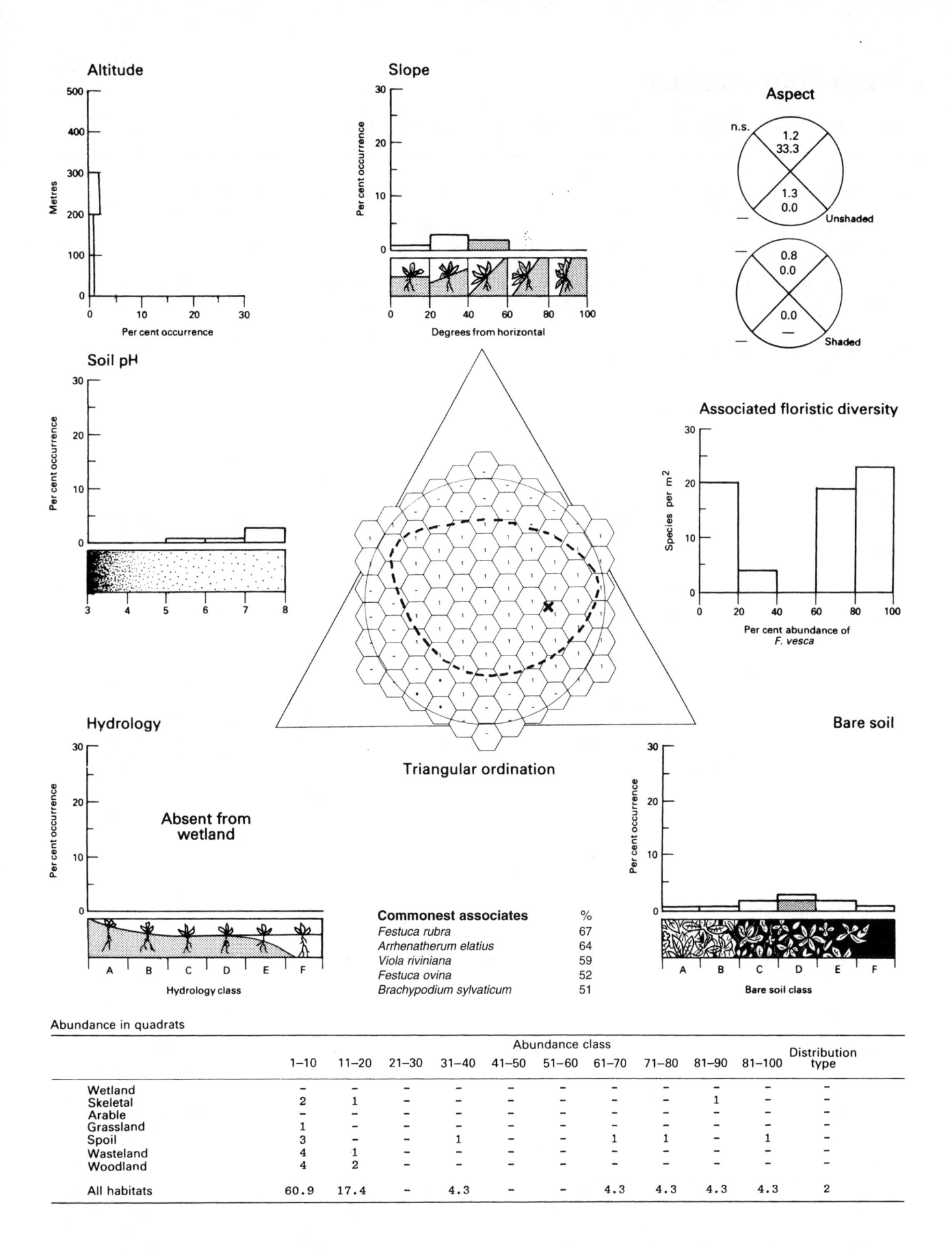

Altitude

Slope

Aspect

n.s.
| 1.2 |
| 33.3 |
| 1.3 |
| 0.0 |
Unshaded

| — |
| 0.8 |
| 0.0 |
| 0.0 |
| — | — |
Shaded

Soil pH

Associated floristic diversity

Hydrology

Absent from wetland

Triangular ordination

Commonest associates	%
Festuca rubra	67
Arrhenatherum elatius	64
Viola riviniana	59
Festuca ovina	52
Brachypodium sylvaticum	51

Bare soil

Abundance in quadrats

	\multicolumn{10}{c}{Abundance class}	Distribution type
	1–10	11–20	21–30	31–40	41–50	51–60	61–70	71–80	81–90	81–100	
Wetland	–	–	–	–	–	–	–	–	–	–	–
Skeletal	2	1	–	–	–	–	–	–	1	–	–
Arable	–	–	–	–	–	–	–	–	–	–	–
Grassland	1	–	–	–	–	–	–	–	–	–	–
Spoil	3	–	–	1	–	–	1	1	–	1	–
Wasteland	4	1	–	–	–	–	–	–	–	–	–
Woodland	4	2	–	–	–	–	–	–	–	–	–
All habitats	60.9	17.4	–	4.3	–	–	4.3	4.3	4.3	4.3	2

(*)*Fraxinus excelsior* L.

Oleaceae
Ash

Field data analysed in this account refer to seedlings and small saplings

Life-form Deciduous tree with a narrow crown and a smooth bark. Leaves pinnate, typically <5000 mm^2; roots usually with VA mycorrhizas.
Phenology Flowers April to May before leaves and sets seed from September onwards. Leaves shed October to November.
Height Foliage and flowers typically <18 m (to 35 m).
Seedling RGR <0.5 week^{-1}. SLA 12.8. DMC. No data.
Nuclear DNA amount 2.0 pg; $2n$ = 46 (*FE3*); $2x$, genus of polyploid origin.
Established strategy Intermediate between competitor and stress-tolerant competitor.
Regenerative strategies W, S. Regenerates by seeds, which germinate in spring usually in the second year after shedding (Gardner 1977). Seeds have a limited persistence at the soil surface. Suckering has been observed (*Biol Fl*).
Flowers Purplish, partially dioecious, wind-pollinated; >100 in axillary panicles.
Germinule 29.3 mg, 14.4 × 3.7 mm; seed dispersed in a samara (46.19 mg, 28.2 × 7.9 mm).
Germination Epigeal; seeds have both an 'immature' embryo and a chilling requirement (*Biol Fl*).
Biological flora Wardle (1961).
Geographical distribution The British Isles except for parts of the Scottish Highlands (99% of British vice-counties), *ABIF* page 538; most of Europe except the N, S and E margins (64% of European territories); N Africa and W Asia.

Habitats A frequent constituent of hedgerows. Seedlings and saplings recorded or observed in all but aquatic habitats. Particularly frequent in woodlands and meadows, and on shaded river banks, lead-mine spoil and scree.
Gregariousness Seedlings and saplings, often in clumps, in shaded habitats.
Altitude Observed to 400 m. [Up to 585 m.]

Slope Juveniles wide-ranging.
Aspect Seedlings and saplings frequently recorded from shaded slopes of both northern and southern aspect, although only abundant on S-facing slopes. In unshaded habitats, the species was recorded slightly more frequently on N-facing slopes.
Hydrology Not persistent in soligenous and topogenous mire.
Soil pH Juveniles widely distributed with slightly greater abundance in the pH range 6.0-8.0.
Bare soil Associated rather more frequently with habitats having low to moderate exposures of bare earth, often in the additional presence of litter or bare rock.

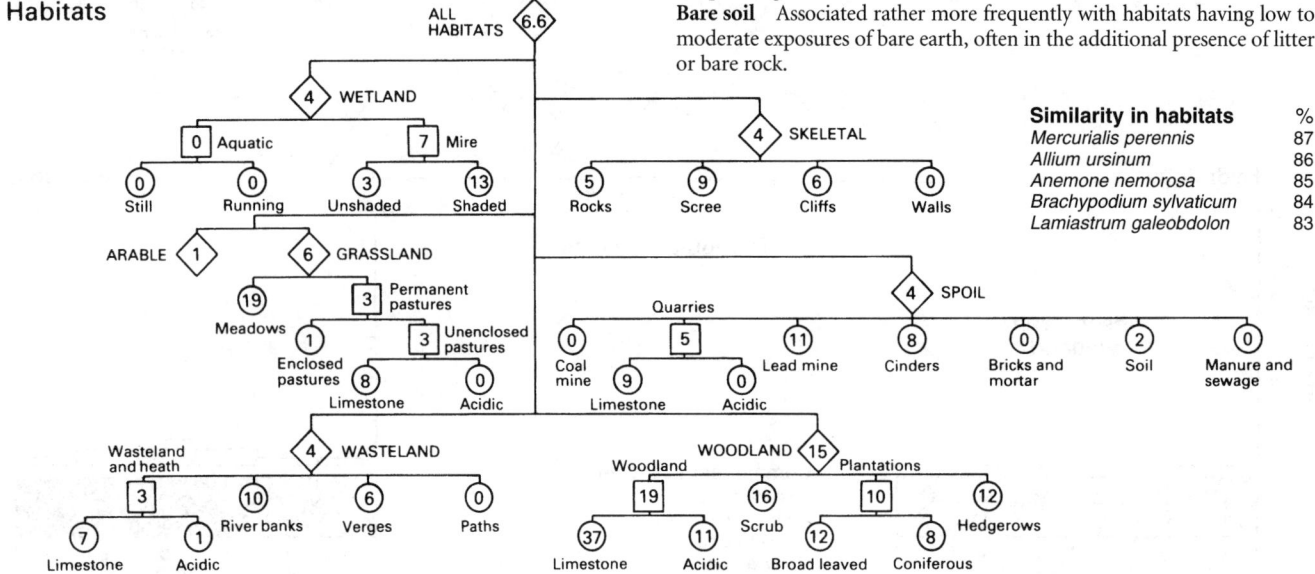

Habitats

Similarity in habitats	%
Mercurialis perennis	87
Allium ursinum	86
Anemone nemorosa	85
Brachypodium sylvaticum	84
Lamiastrum galeobdolon	83

△ **ordination** Distribution of juveniles extremely wide-ranging, but concentrated in vegetation of relatively unproductive and moderately disturbed habitats.
Associated floristic diversity Typically low in woodland habitats; more frequently intermediate to species-rich in unshaded sites.
Synopsis *F.e.* is a timber tree or shrub with ring-porous wood. New shoots are vulnerable to frost (*Biol Fl*), and bud-break is delayed until late spring. The species occupies 56 000 ha and is the fourth most common broad-leaved tree in British woodlands (Steele & Peterken 1982). In optimal sites, which are sheltered, moist, well-drained and fertile, the species at maturity may extend to 35 m (Beckett & Beckett 1979) but lives for only *c*. 180 years (Rackham 1980). However, coppiced specimens may survive for over 300 years (Rackham 1980), and in less-favourable habitats *F.e.* may persist as a shrub (*Biol Fl*). The species regenerates freely through the production, after 30-40 years, of numerous wind-dispersed fruits (up to 100 000 every second year (*Biol Fl*)). Although fruits are less mobile than those of *Betula* spp. (McVean 1955b), seedlings are widely distributed. However, many of the sites in which seedlings germinate are unsuitable for establishment. Seedlings are sensitive to shade and summer-green woodland herbs, e.g. *Mercurialis perennis*, may provide an effective barrier against establishment of saplings (Wardle 1959, Gardner 1976). In turf *F.e.* becomes established only in vegetation gaps (*Biol Fl*), and effective increase in height is dependent upon high light intensities (*Biol Fl*). However, once the sapling has penetrated the herb layer the species may be long-persistent (to at least 28 years (Gardner 1976)) in shaded conditions. Seedlings are vulnerable to grazing, but establishment may take place within the protection of bushes of *Crataegus monogyna* or on screes if inaccessible to grazing animals. Many of the ash-woods of the Derbyshire Dales can be attributed to a relaxation of grazing pressure in the 19th century (Merton 1970). Seedlings are susceptible to damping off in moist, shaded sites and

Altitude

Slope

Aspect

Soil pH

Associated floristic diversity

Commonest associates	%
Festuca rubra	39
Arrhenatherum elatius	37
Mercurialis perennis	32
Dactylis glomerata	31
Poa trivialis	30

Triangular ordination

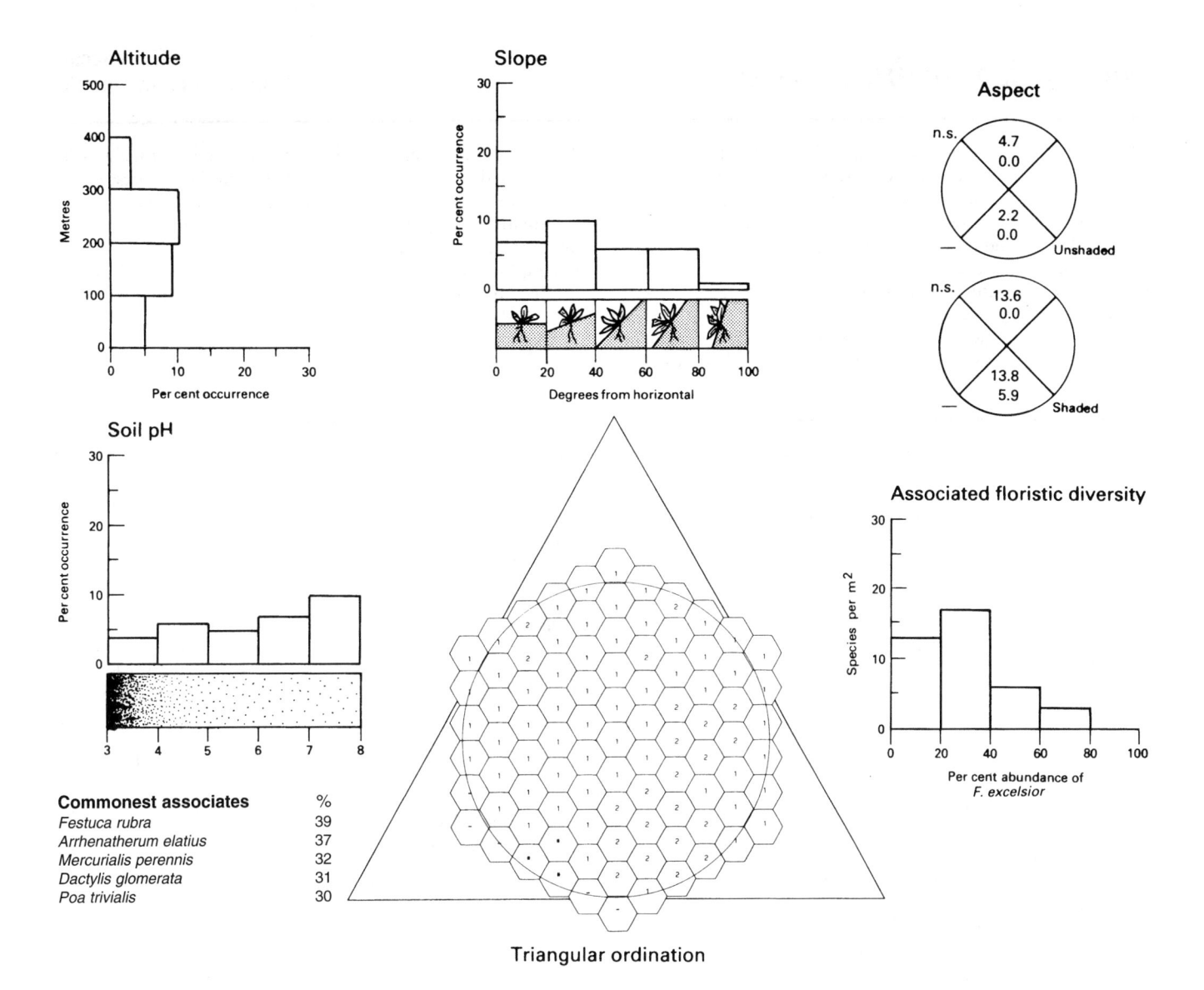

to desiccation on shallow soils (*Biol Fl*). These constraints on seedling establishment dictate that *F.e. is* most typically a seral species colonizing relatively open, little-grazed, base-rich sites. In historical times an increase in the abundance of *F.e.* was associated with forest clearance (*Biol Fl*), with the result that in the Sheffield region *F.e. is* the most frequently recorded canopy species of limestone woodland, is the second most common hedgerow tree and is the third most commonly occurring solitary tree or bush. The species shows a bias towards N-facing slopes, and is virtually absent from acidic soils. In addition to its role in the formation of secondary woodland (see Scurfield 1959, Merton 1970), *F.e.* is also a frequent constituent of the canopy of ancient woodland (Pigott 1969). However, although it attains its greatest size on fertile soils, *F.e.* reaches maximum abundance, and exerts maximum ecological impact, under conditions much less favourable for its growth, namely on relatively infertile calcareous soils where most individuals form stunted trees or shrubs. *F.e.* regenerates almost entirely by seed. Flowers are variously male, female and hermaphrodite, and may occur in various combinations on each tree (*Biol Fl*). The proportion of male and female flowers can vary from year to year (*Biol Fl*), but Gardner (1977) concluded that some trees, which did not fruit, are consistently male-flowered. Female trees have thinner trunks than males, apparently because of their greater allocation of resources to seed production (Harper 1977). Fruits reach their full length by July and their maximum seed size by August (*Biol Fl*). However, embryos continue to develop until August or September (*Biol Fl*) and when the seed is shed (from September onwards) the embryo is still only half the length of the seed (*Biol Fl*). Germination inhibitors are present both in endosperm and in embryo, and most fruits lie on the ground for two winters, during which time the embryo develops (*Biol Fl*). In one Derbyshire ash-wood a majority of seeds were consumed by small mammals whereas others were predated by caterpillars (Gardner 1977). Germination of the few survivors occurs in April and May. Viable fruits may lie on the ground for up to 6 years (*Biol Fl*), forming a shortly-persistent seed bank. This may facilitate rapid colonization following the formation of gaps in the woodland canopy. Seedlings and saplings expand a greater part of the leaf area before canopy expansion of the mature tree and, unlike the adults, which produce only sun-leaves, both sun- and shade-leaves are produced (*Biol Fl*). Rapid growth appears to be dependent upon high rainfall in May and June (*Biol Fl*). Roots do not penetrate below the level of the permanent watertable, but the species may persist in wetland habitats provided there is a shallow zone of well-drained soil for establishment. The canopy of *F.e.* produces a relatively light shade and the leaves, which are still green when shed (*Biol Fl*), do not form persistent litter (Sydes & Grime 1981a,b). *F.s.* combines relatively low palatability to the snail *Helix aspersa* Muller with a low diversity of associated insects (Wratten *et al.* 1981). As a forest crop *F.e.* is only economic under optimal soil conditions. In Germany populations exploiting dry limestone soils are physiologically different from those exploiting moist fertile soils (*Biol Fl*). A further ssp. has been recorded for SE Europe (*FE3*).

Current trends Regeneration of *F.e.* is favoured by disturbance, and the species may be increasing, at least on less-fertile soils.

Galeopsis tetrahit L. sensu lato

Lamiaceae
Common Hemp-nettle

Life-form Summer-annual therophyte. Stem erect; leaves ovate, <4000 mm², with glandular hairs; roots non-mycorrhizal.
Phenology Germination occurs in spring. Flowers from July to September and sets seed from August to October.
Height Foliage and flowers to 1 m.
Seedling RGR No data. SLA 29.0. DMC 17.6.
Nuclear DNA amount No data; $2n = 32^*$ (*BC*); $4x^*$.
Established strategy Competitive-ruderal.
Regenerative strategies B_s. Entirely by seeds, which germinate in spring and may form a persistent seed bank (Roberts 1986a); ?SB IV.

Flowers Pink, purple or white, hermaphrodite, usually self-pollinated (Stace 1975); often >50 in a whorled inflorescence.
Germinule 4.83 mg, 3.0×2.3 mm, nutlet (4 per ovary).
Germination Epigeal; dormancy broken by chilling.
Biological flora None.
Geographical distribution The British Isles (100% of British vice-counties), ABIF page 515; most of Europe, but rare in the SE (79% of European territories); N Asia; naturalized in N America.

Habitats Occurs in two main habitats: (a) disturbed ground, particularly arable land amongst broad-leaved crops and (b) closed vegetation of moist shaded sites of river banks, in damp woodland and in hedgerows. More rarely observed in shaded mire and in stands of *Phragmites*. Absent from managed grassland and most skeletal and wetland habitats.
Gregariousness Sparse.
Altitude Extending to 400 m. [Up to 450 m.]
Slope Not recorded from most skeletal habitats or from steep slopes.
Aspect Slightly more frequent on S-facing slopes in shaded habitats, and on N-facing slopes in unshaded habitats.
Hydrology Not a wetland species, but occasionally observed from drier parts of topogenous mire, where the water table is well below the soil surface during summer.
Soil pH Widely-distributed, but mainly recorded within the pH range 4.5–7.0.
Bare soil Most frequent at high exposures.
△ **ordination** Widely distributed, but centred on heavily disturbed, productive vegetation.
Associated floristic diversity Typically low to intermediate.
Synopsis The aggregate species is a tall annual herb which exploits two main types of disturbed ground; (a) arable land and other open disturbed sites and (b) moist, moderately shaded habitats. In both types of habitat the species usually functions as a summer-annual, although there may be a second period of germination in autumn (*WCH*). The aggregate species consists of two taxa, *G. bifida* Boenn. and *G. tetrahit, sensu stricto*. Extremely limited observations suggest that *G.b. is* restricted to arable and other similar habitats, and is thus more overtly than *G.t.* a follower of humans. The species flourishes in open fertile ground, and plants are considerably more robust in broad-leaved crops and on fallow ground than in cereal crops. In woodlands *G.t.* occurs in clearings, and in moist habitats exploits a niche similar to that of *Impatiens glandulifera* but, unlike this species, is not stand-forming. In view of the tall stature, it is not surprising to find that the species does not occur in sites which are regularly grazed, cut or trampled. Usually only one generation per year is produced, and consequently the species is vulnerable to the effects of weeding and is seldom found in gardens. *G.t.* can be toxic to stock (Cooper & Johnson 1984). As with other common ruderals, *G.t.* is prolific (on average *c.* 2000 seeds per plant (Salisbury 1942)), but the seeds are much larger and are produced in smaller numbers than those of other common weeds of similar size and morphology (e.g. *Chenopodium album*); this, coupled with the strong seasonality of germination (*WCH*, Roberts 1986a), appears to reduce the capacity of the species to colonize new disturbed sites of transient duration. Nutlets may float on water for 2 days, and can survive ingestion by stock and by birds (Ridley 1930). It is not known whether a persistent seed bank is formed in woodland, but one is recorded from arable land (Chippindale & Milton 1934), the commonest habitat of the aggregate species. In arable habitats seed is also effectively dispersed by humans since some seed is still retained on the plant when the crop is harvested and the species may be an impurity in commercial crop seed (*WCH*). However, it is probably significant that whereas seed is now mainly found as an impurity in harvested cereals, the species performs best in association with broad-leaved crops. *G.b.* and *G.t.* form a sterile hybrid (Stace 1975).
Current trends Possibly decreasing overall (*WCH*), especially on arable land. The high levels of disturbance associated with modern management are probably causing an increase of the species in its woodland habitats.

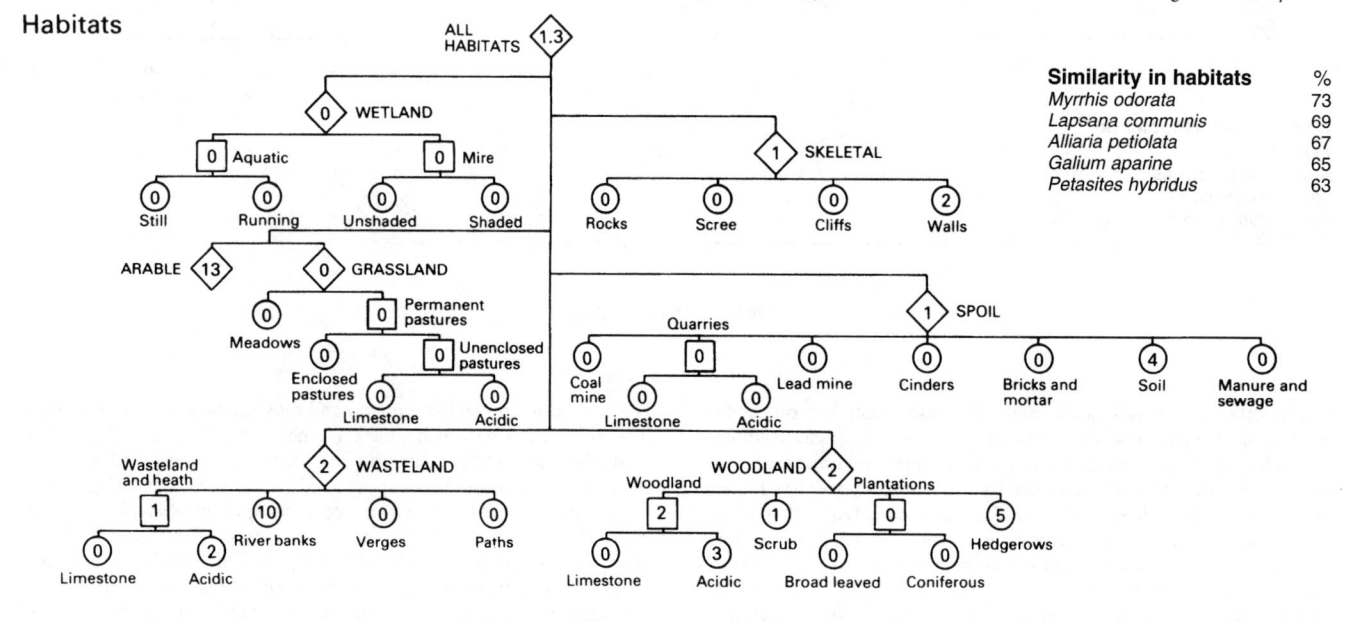

Similarity in habitats	%
Myrrhis odorata	73
Lapsana communis	69
Alliaria petiolata	67
Galium aparine	65
Petasites hybridus	63

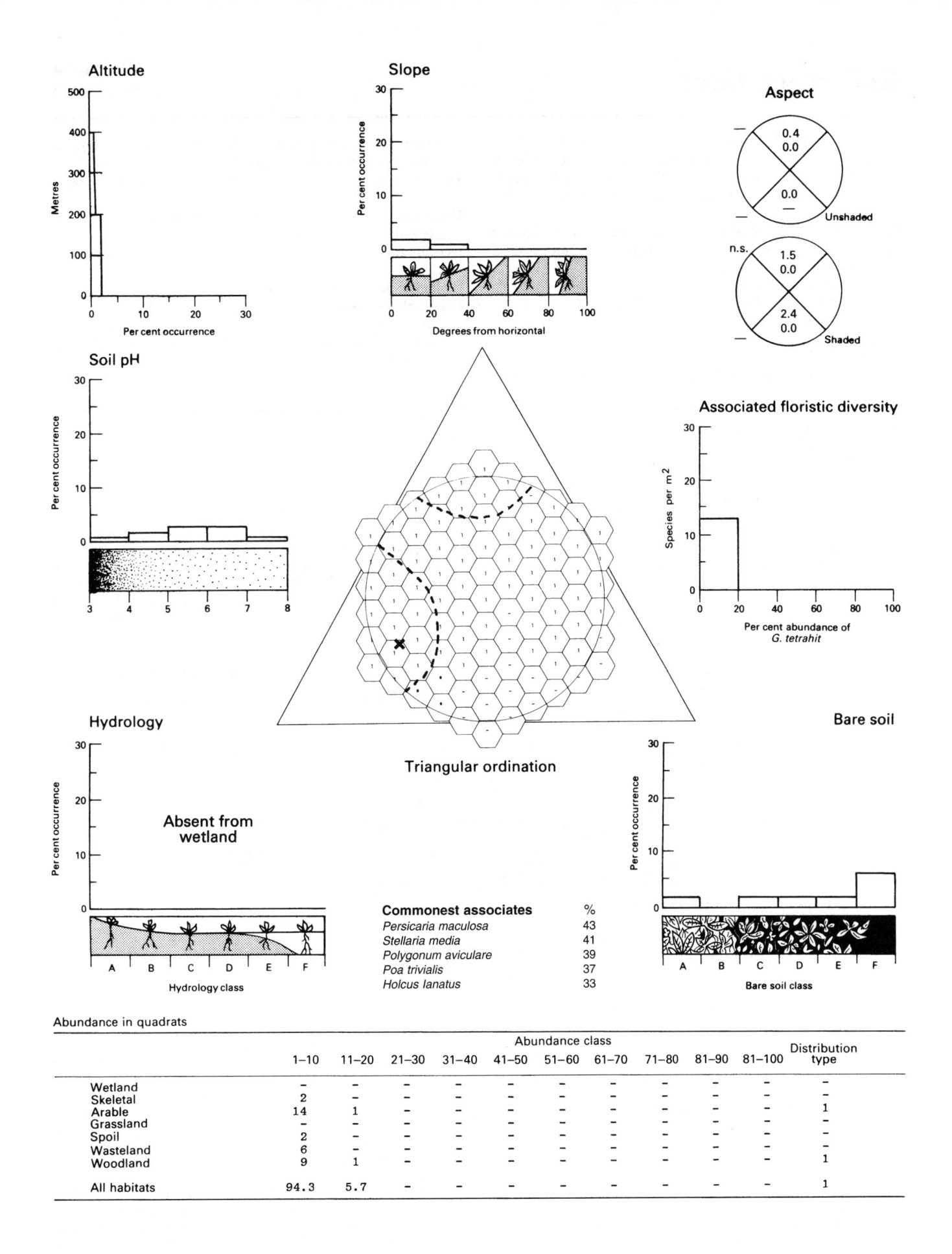

Altitude

Slope

Aspect

Unshaded: 0.4 / 0.0 / 0.0 / —
Shaded: n.s. 1.5 / 0.0 / 2.4 / —

Soil pH

Associated floristic diversity

Per cent abundance of *G. tetrahit*

Hydrology

Absent from wetland

Triangular ordination

Commonest associates	%
Persicaria maculosa	43
Stellaria media	41
Polygonum aviculare	39
Poa trivialis	37
Holcus lanatus	33

Bare soil

Abundance in quadrats

	1–10	11–20	21–30	31–40	41–50	51–60	61–70	71–80	81–90	81–100	Distribution type
Wetland	–	–	–	–	–	–	–	–	–	–	–
Skeletal	2	–	–	–	–	–	–	–	–	–	–
Arable	14	1	–	–	–	–	–	–	–	–	1
Grassland	–	–	–	–	–	–	–	–	–	–	–
Spoil	2	–	–	–	–	–	–	–	–	–	–
Wasteland	6	–	–	–	–	–	–	–	–	–	–
Woodland	9	1	–	–	–	–	–	–	–	–	1
All habitats	94.3	5.7	–	–	–	–	–	–	–	–	1

Galium aparine L.

Rubiaceae
Goosegrass, Cleavers

Life-form Winter- to summer-annual therophyte. Stems prostrate or scrambling and ascending, often branched; leaves lanceolate, with hairs and, like the stem, with backward-pointing prickles, <250 mm²; roots with or without VA mycorrhizas.

Phenology Capable of winter growth. Flowers April to June and seed is set from July to October, often after the plant has died. Leaves relatively short-lived.

Height Foliage and flowers to 1.2 m.

Seedling RGR 1.0-1.4 week^{-1}. SLA 27.9. DMC 14.0.

Nuclear DNA amount 2.0 pg; $2n = 42, 44, 48, 62, 66, 68$ (*FE4*); $4x, 6x$.

Established strategy Competitive-ruderal.

Regenerative strategies S, but germination often occurring in autumn or winter. Regenerates entirely by seed, which is only briefly persistent in the soil; SB II.

Flowers Greenish-white, hermaphrodite, protandrous, usually self-pollinated; in few-flowered axillary cymes.

Germinule 7.25 mg, 2.3 × 2.0 mm, dry mericarp, covered in hooked bristles (2 per ovary).

Germination Epigeal; t_{50} 5 days; 6-26°C; dormancy broken by chilling; L = D.

Biological flora Taylor (1999).

Geographical distribution The British Isles (100% of British vice-counties), *ABIF* page 594; Europe except the NE (92% of European territories); N and W Asia, widely introduced and now virtually cosmopolitan.

Habitats Widely distributed and, with the exception of aquatic habitats, recorded or observed in every terminal category of the habitat key. Particularly frequent in shaded or unshaded tall-herb communities, especially those associated with hedgerows, river and stream banks, and road verges, but also a major arable weed, particularly of cereals. Generally scarce on spoil and in wetland, grazed and skeletal habitats. [Also occurs on maritime shingle.].

Gregariousness Typically intermediate.

Altitude More abundant in lowland habitats, decreasing progressively at higher elevations.

Slope Wide-ranging, but most frequent on level ground.

Aspect Slightly more frequent on S-facing slopes, especially in shaded habitats.

Hydrology Recorded in association with moist but not submerged soils.

Soil pH Most frequent and abundant between pH 5.5 and 8.0, but seedlings are recorded down to pH 3.5.

Bare soil Widely distributed, but more frequent at higher exposures.

△ **ordination** Distribution very wide-ranging, but centred on vegetation associated with fertile undisturbed habitats, where the species survives by scrambling over the shoots of the dominant herbs. As an annual *G.a.* is also able to flourish in some disturbed fertile habitats (e.g. arable). The few occurrences in unproductive habitats consist largely of seedlings.

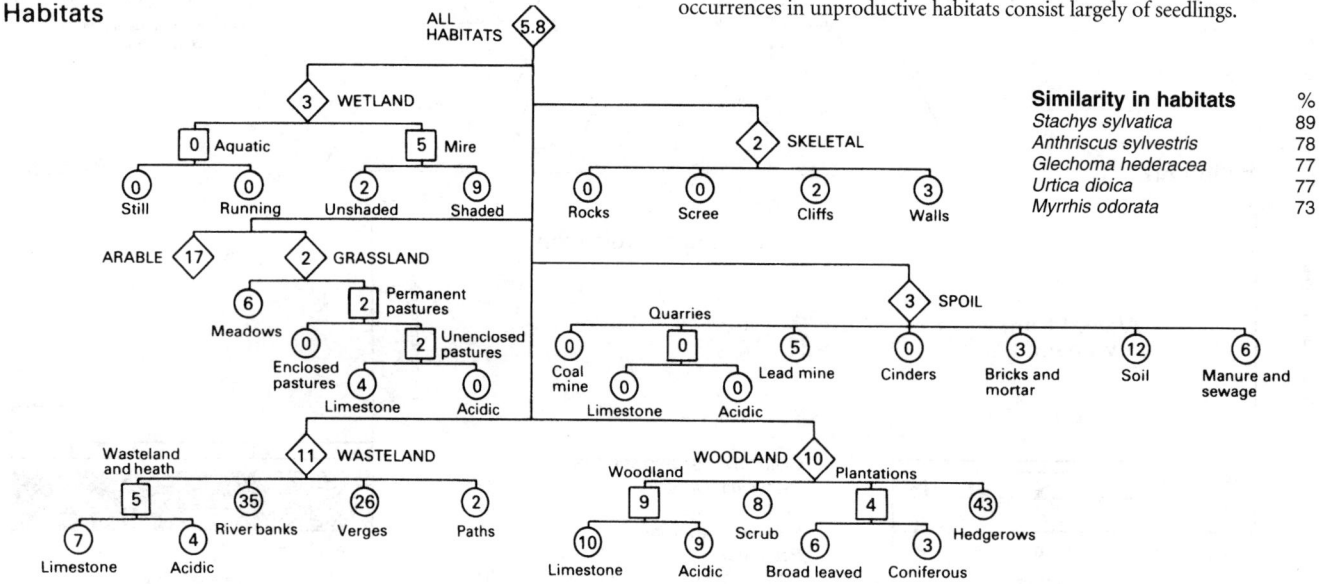

Habitats

Similarity in habitats	%
Stachys sylvatica	89
Anthriscus sylvestris	78
Glechoma hederacea	77
Urtica dioica	77
Myrrhis odorata	73

Associated floristic diversity Relatively low and declining at higher frequencies (see 'Synopsis').

Synopsis *G.a.* is an often tall, scrambling, facultatively winter- or summer-annual herb. The commonness of *G.a.* is related to an ability to occur in two different types of vegetation: tall-herb communities and ephemeral assemblages of disturbed land. In stands of tall herbs the large seed capital and ability to make appreciable growth during milder winters allow *G.a.* to keep pace in height growth with established perennials. The phenology of *G.a.* is semi-vernal, in that flowering commences in April, seed is set from June onwards and the plant dies in late summer or autumn, before the onset of frosty weather. *G.a.* flourishes under conditions of high soil fertility, and may form extensive monocultures in exceptionally enriched sites such as derelict sewage beds. The foliage of *G.a.* contains high concentrations of P and Ca (*SIEF*), and the species is sensitive to ferrous iron toxicity (R. E. D. Cooke unpubl.), which may explain the restriction of *G.a.* to the drier parts of wetland systems. The stems and leaves have backward-pointing prickles which provide adherence to and gain support from the shoots of canopy dominants. The stems also stick to clothing and fur, as do the mericarps, which are covered with hooked bristles. In this way *G.a.* is dispersed very effectively by humans and animals. Dispersal by water or after ingestion by birds or stock is also possible (Ridley 1930). Of the ruderal habitats with which *G.a.* is characteristically associated, the most important is arable land, where the species is classified by Holm *et al.* (1977) as one of the world's worst weeds; it is the most frequent weed of winter wheat in Central S England (Chancellor & Froud-Williams 1984). In arable land *G.a.* has been encouraged by the use of phenoxy-herbicides which are ineffective against the species (Holm *et al.* 1977). Control of *G.a.* is particularly difficult since the species is a contaminant of many species of crop seeds (*WCH*), and its adhesive seeds are readily dispersed in sacking, clothing and in mud

Altitude

Slope

Aspect

Unshaded
n.s.	2.8
14.3	
3.6	n.s.
12.5	

Shaded
n.s.	6.1
25.0	
11.4	n.s.
28.6	

Soil pH

Associated floristic diversity

Hydrology

Triangular ordination

Bare soil

Commonest associates — %
Poa trivialis — 67
Urtica dioica — 41
Arrhenatherum elatius — 48
Heracleum sphondylium — 40
Anthriscus sylvestris — 25

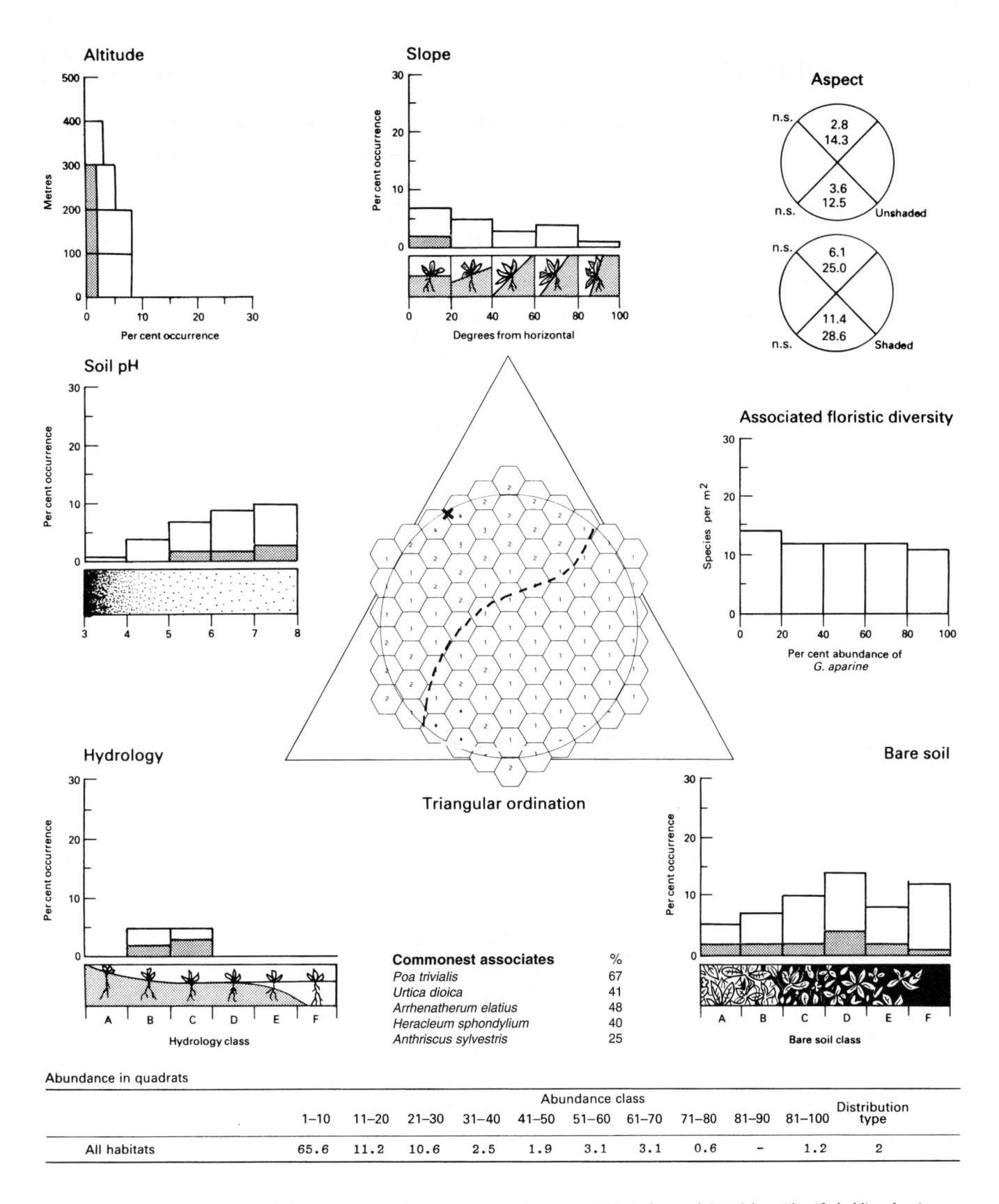

Abundance in quadrats

	1–10	11–20	21–30	31–40	41–50	51–60	61–70	71–80	81–90	81–100	Distribution type
All habitats	65.6	11.2	10.6	2.5	1.9	3.1	3.1	0.6	–	1.2	2

(WCH, Ridley 1930). Seeds require a brief exposure to chilling, but are then capable of germination at low temperatures. Germination frequently occurs over a protracted period, and populations in spring may include both large, well-developed seedlings and others with the radicle just emergent. The stage of development is also dependent upon climate, with seedlings being more advanced in spring after mild winters. G.a. forms a persistent seed bank, but seeds seldom persist beyond 2 years (Brenchley & Warington 1930). Bocher et al. (1955) have identified obligately winter-, obligately summer- and facultatively polycarpic races within Europe, all with the same chromosome number ($2n = 64$). The distribution of other cytological races is not known.

Current trends Probably still increasing in all habitats except arable, where herbicides designed for specific use on G.a. have recently been introduced.

Galium palustre L.

Rubiaceae
Marsh Bedstraw

Includes ssp. *elongatum* C. Presl Arcang. and ssp. *palustre* L. Most of our field records refer to ssp. *palustre*.

Life-form Polycarpic perennial, protohemicryptophyte or helophyte with slender creeping stock. Stems decumbent or ascending; leaves lanceolate, often <100 mm² (to 250 mm² in ssp. *e.*); roots typically non-mycorrhizal.

Phenology Overwinters as small shoots. Flowers June to July (ssp. *p.* 2-3 weeks earlier than ssp. *e.* (*CTW*)). Seed shed from August to October.

Height ssp. *p.* foliage and flowers to 500 mm; ssp. *e.* to 1200 mm.

Seedling RGR (ssp. *p.*) 1.0-1.4 week^{-1}. SLA 33.9. DMC 18.1.

Nuclear DNA amount ssp. *p.* 2.6 pg; $2n = 24^*, 48^*$ (*BC*); $2x^*, 4x^*$. ssp. *e.* $2n = 96^*, 144$ (*BC, FE4*); $8x^*, 12x$.

Established strategy Intermediate between C-S-R and competitive-ruderal.

Regenerative strategies V, B_s. Regenerates both vegetatively and by means of seed (see 'Synopsis'); SB IV.

Flowers White, hermaphrodite, protandrous, insect-pollinated; usually <100 in a loose panicle.

Germinule 0.91 mg, 1.7 × 1.5 mm, dry mericarp (2 per fruit).

Germination Epigeal; t_{50} 2 days; 6-34°C.

Biological flora None.

Geographical distribution The aggregate species occurs throughout the British Isles (100% of British vice-counties), *ABIF* page 592; most of Europe northwards to Scandinavia (82% of European territories); Asia Minor.

Habitats Exclusively a wetland species. Equally common in unshaded and shaded mire. Also frequent on river banks. Elsewhere recorded from the retaining walls of a reservoir, and found on soil heaps derived from pond dredgings.

Gregariousness Intermediate.

Altitude Wide-ranging. [Up to 600 m.] ssp. *p.*, the most common segregate throughout the region, more frequent in upland areas, and ssp. *e.* more abundant in the lowlands.

Slope Mainly restricted to flat or gently sloping sites.

Aspect Insufficient data.

Hydrology ssp. *p.* occurs in topogenous mire with water standing only in winter and in soligenous mire. ssp. *e.* is particularly common in sites with standing water (*CTW*).

Soil pH Virtually absent from soils of pH <4.5, and most common in the pH range 5.0-7.0.

Bare soil Insufficient data.

△ **ordination** Absent from unproductive or heavily disturbed vegetation types; most frequent in fertile sites subject to some disturbance.

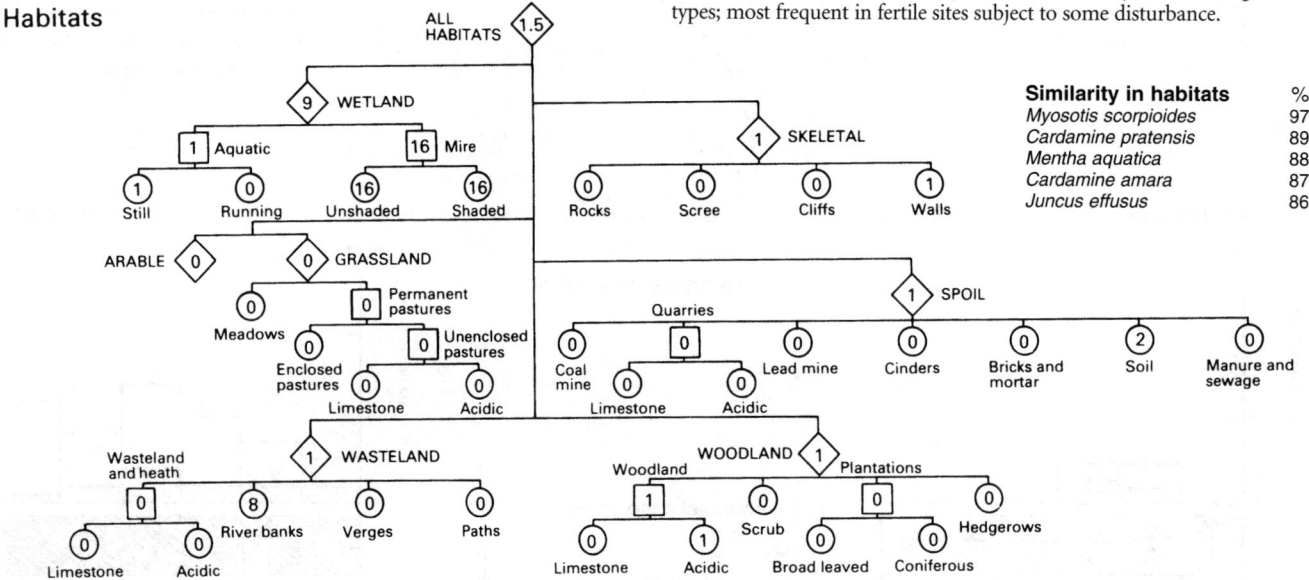

Habitats

Similarity in habitats	%
Myosotis scorpioides	97
Cardamine pratensis	89
Mentha aquatica	88
Cardamine amara	87
Juncus effusus	86

Associated floristic diversity Typically intermediate.

Synopsis The aggregate species is a tall, scrambling wetland herb which is mainly restricted to sites where the vigour of potential dominants is to some extent suppressed (a) by shade from trees, (b) by disturbance factors such as light grazing or flooding, or (c) by a degree of nutrient stress. The aggregate species is found on a wide range of soil types, but is largely absent from calcareous strata. ssp. *p.*, in particular, is very frequent on mildly acidic peat and exhibits a moderate tolerance of ferrous ion (R. E. D. Cooke unpubl.) The aggregate species normally provides only a minor proportion of the total biomass, and frequently much of its foliage is in the lower layers of the leaf canopy. Survival appears to depend upon an opportunistic expansion into small gaps in the canopy of tall wetland vegetation. The capacity for lateral vegetative spread by means of a slender creeping stock and decumbent shoots may be important in this respect. As is the case for many other wetland species, detached cuttings root freely. This is in marked contrast with the response of cuttings of the grassland Bedstraws, *G. saxatile*, *G. sterneri* and *G. verum*, in which adventitious roots are poorly developed, even under controlled laboratory conditions (J. G. Hodgson unpubl.). Detached fragments may be important in enabling the species to colonize more-distant sites beside water. The flowering shoots are tall and usually attain the height of the surrounding vegetation. Seed may survive ingestion by cattle, and may float for several weeks (Ridley 1930). Germination occurs mainly in spring and seeds may be incorporated into a persistent seed bank (Thompson & Grime 1979). The ecological differences between ssp. *e.* and ssp. *p.* have been little studied. ssp. *e.* is the taller and is found in wetter sites (*CTW*, see 'Hydrology'), but the two are connected by intermediates (*FE4*). Both morphologically and ecologically, *G. uliginosum* L. shows close affinities with ssp. *p.* Both occur in soligenous mire, but *G.u.* consistently occupies the drier sites. *G.u.* also differs from ssp. *p.* in its greater capacity to exploit highly calcareous mire, and its lower potential to form rooted cuttings.

Current trends Like most wetland species, probably decreasing as a result of habitat destruction.

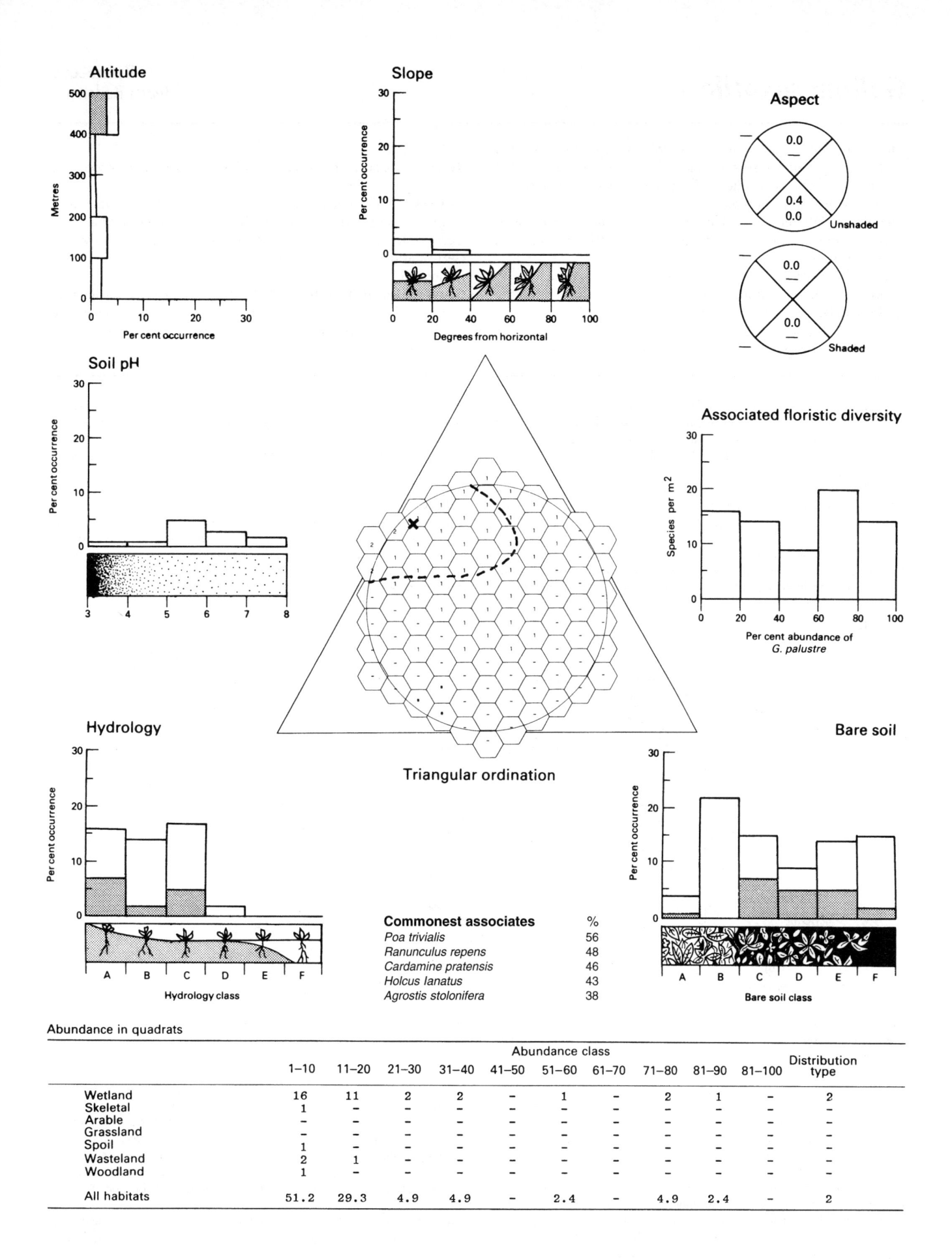

Altitude

Slope

Aspect

Unshaded

Shaded

Soil pH

Associated floristic diversity

Species per m²

Per cent abundance of *G. palustre*

Hydrology

Hydrology class

Triangular ordination

Commonest associates	%
Poa trivialis	56
Ranunculus repens	48
Cardamine pratensis	46
Holcus lanatus	43
Agrostis stolonifera	38

Bare soil

Bare soil class

Abundance in quadrats

	\multicolumn{10}{c}{Abundance class}	
	1–10	11–20	21–30	31–40	41–50	51–60	61–70	71–80	81–90	81–100	Distribution type
Wetland	16	11	2	2	–	1	–	2	1	–	2
Skeletal	1	–	–	–	–	–	–	–	–	–	–
Arable	–	–	–	–	–	–	–	–	–	–	–
Grassland	–	–	–	–	–	–	–	–	–	–	–
Spoil	1	–	–	–	–	–	–	–	–	–	–
Wasteland	2	1	–	–	–	–	–	–	–	–	–
Woodland	1	–	–	–	–	–	–	–	–	–	–
All habitats	51.2	29.3	4.9	4.9	–	2.4	–	4.9	2.4	–	2

Galium saxatile L.

Rubiaceae
Heath Bedstraw

Life-form Polycarpic perennial protohemicryptophyte with a slender tap root and rooted stolons. Leaves ovate, less than 250 mm² in whorls on slender trailing or weakly ascending stems; roots with VA mycorrhizas.

Phenology Winter green. Flowers June to August. Seed shed August onwards, but some retained on the inflorescence until late autumn.

Height Non-flowering shoots normally prostrate. Leafy flowering shoots seldom exceeding 80 mm.

Seedling RGR 0.5-0.9 week^{-1}. SLA 25.5. DMC 18.8.

Nuclear DNA amount 2.9 pg; $2n = 22, 44^*$ (*BC, FE4*); $2x, 4x^*$.

Established strategy Intermediate between stress-tolerator and C-S-R.

Regenerative strategies V, B$_s$. Both vegetatively and by seed (see 'Synopsis'); ?SB IV.

Flowers White, hermaphrodite, protandrous, self-incompatible, insect-pollinated (*CTW*), usually <50 in a compound panicle.

Germinule 0.56 mg, 1.5 × 2.2 mm, dry mericarp (2 per ovary).

Germination Epigeal; t_{50} 13 days; 7-22°C.

Biological flora None.

Geographical distribution A European endemic; throughout most of Britain (99% of British vice-counties), *ABIF* page 594; confined to NW and CE Europe (51% of European territories and 3% naturalized).

Habitats *G.s.* is most abundant in pastures, heathland and moorland on acidic strata and also has a high incidence on disused quarry heaps, in derelict grassland and in scrub. Occurs locally in mire, open woods and plantations, and on rock outcrops and upland stream banks. Absent from arable land and aquatic habitats.

Gregariousness Intermediate to patch-forming.

Altitude Distribution shows an upland bias, a reflection of the greater abundance of acidic habitats at higher elevations. Extends to 470 m. [Up to 1010 m.]

Slope Virtually absent from very steeply sloping terrain.

Aspect In unshaded habitats *G.s.* is slightly more abundant on N-facing slopes (significantly so in grasslands (*Ecol Atl*)) while in shaded habitats the species tends to occur at higher levels of abundance on S-facing slopes.

Hydrology Typically a dryland species, but recorded in soligenous mire and more rarely in ombrogenous mire.

Soil pH Most frequent and abundant at pH values of <5.0, with isolated occurrences up to pH 7.5.

Bare soil Distribution clearly skewed towards habitats having little exposed soil.

△ **ordination** Absent entirely from disturbed vegetation and centred on unproductive vegetation. Increasing occurrence on the right provides evidence of some capacity to coexist with more robust species of higher competitive ability (e.g. *Vaccinium myrtillus, Pteridium aquilinum*).

Associated floristic diversity Typically low.

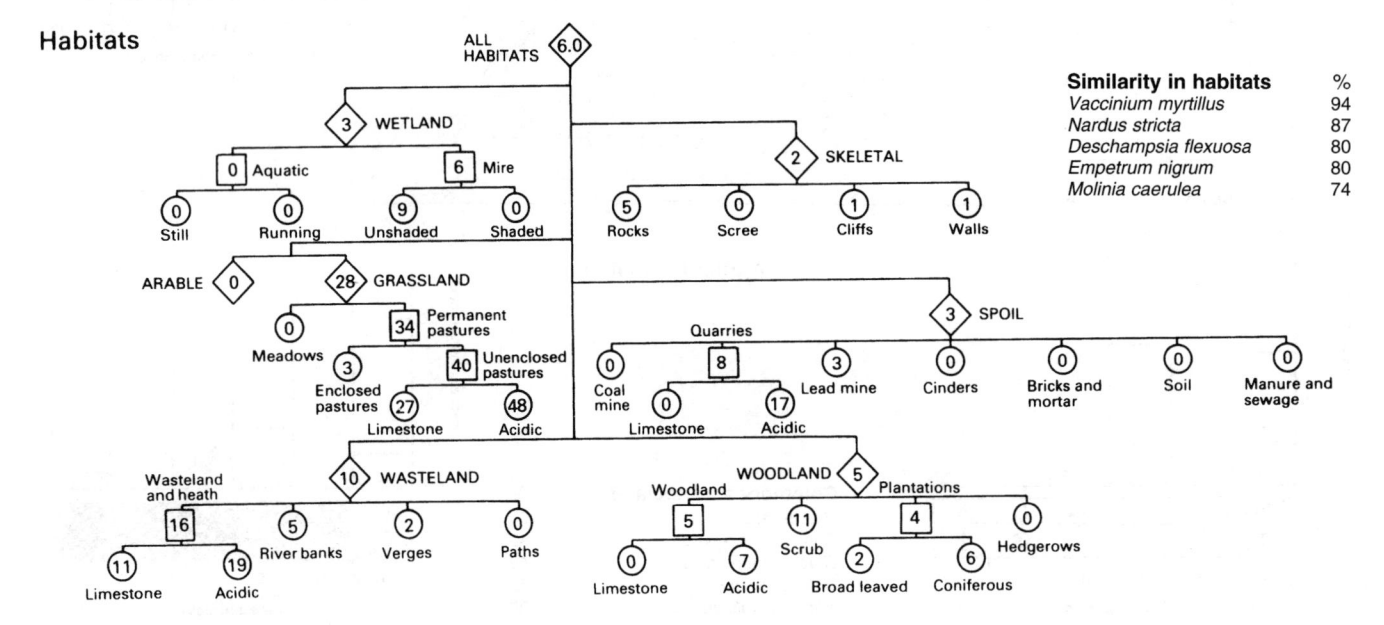

Habitats

Similarity in habitats	%
Vaccinium myrtillus	94
Nardus stricta	87
Deschampsia flexuosa	80
Empetrum nigrum	80
Molinia caerulea	74

Synopsis *G.s.* is a slow-growing, prostrate, mat-forming herb found in a wide range of rocky habitats on acidic soils. A classic calcifuge, when planted on calcareous soils it grows poorly and exhibits root stunting and lime-chlorosis (Tansley 1917). The foliage of plants in the Sheffield region is high in Mn and Al (*SIEF*). We suspect that the few field records at high soil pH represent calcareous soils containing an undetected superficial leached horizon. *G.s.* appears to be moderately shade-tolerant and occurs in open woodland and as a diffuse understorey beneath grass or heather canopies. However, the low mats of *G.s.* tend to be suppressed in derelict grassland and heathland where they are often submerged beneath grass litter and low shrub canopies. As a result *G.s.* tends to attain its highest frequencies in sheep pasture. The affinity for N-facing slopes suggests that in grassland the distribution of *G.s.* may be affected by drought, an hypothesis which is consistent with known fluctuations in abundance corresponding to rainfall variation from year to year (Watt 1960, Van den Bergh 1979). The prostrate stems of *G.s.* readily form roots and are thus conducive to the development of large clonal patches. Unlike other species of similar growth form, *G.s.* has little capacity for regeneration from detatched, rooted shoot tips (Hodgson unpubl.). Seed is set regularly in unshaded sites and persistent seed banks may develop. However, regeneration by seed is probably infrequent. The significance of the seed bank in relation to the survival of *G.s.* in burned habitats requires investigation. Apart from the observed differences between individuals growing in shaded and unshaded habitats, little is known concerning the degree of phenotypic plasticity and ecotypic differentiation within *G.s.*. Slender, earlier-flowering diploid populations occur in Europe (Kliphuis 1967).

Current trends Destruction of habitats has led to a considerable decline in *G.s.* in many lowland regions, but *G.s.* is still common in the uplands.

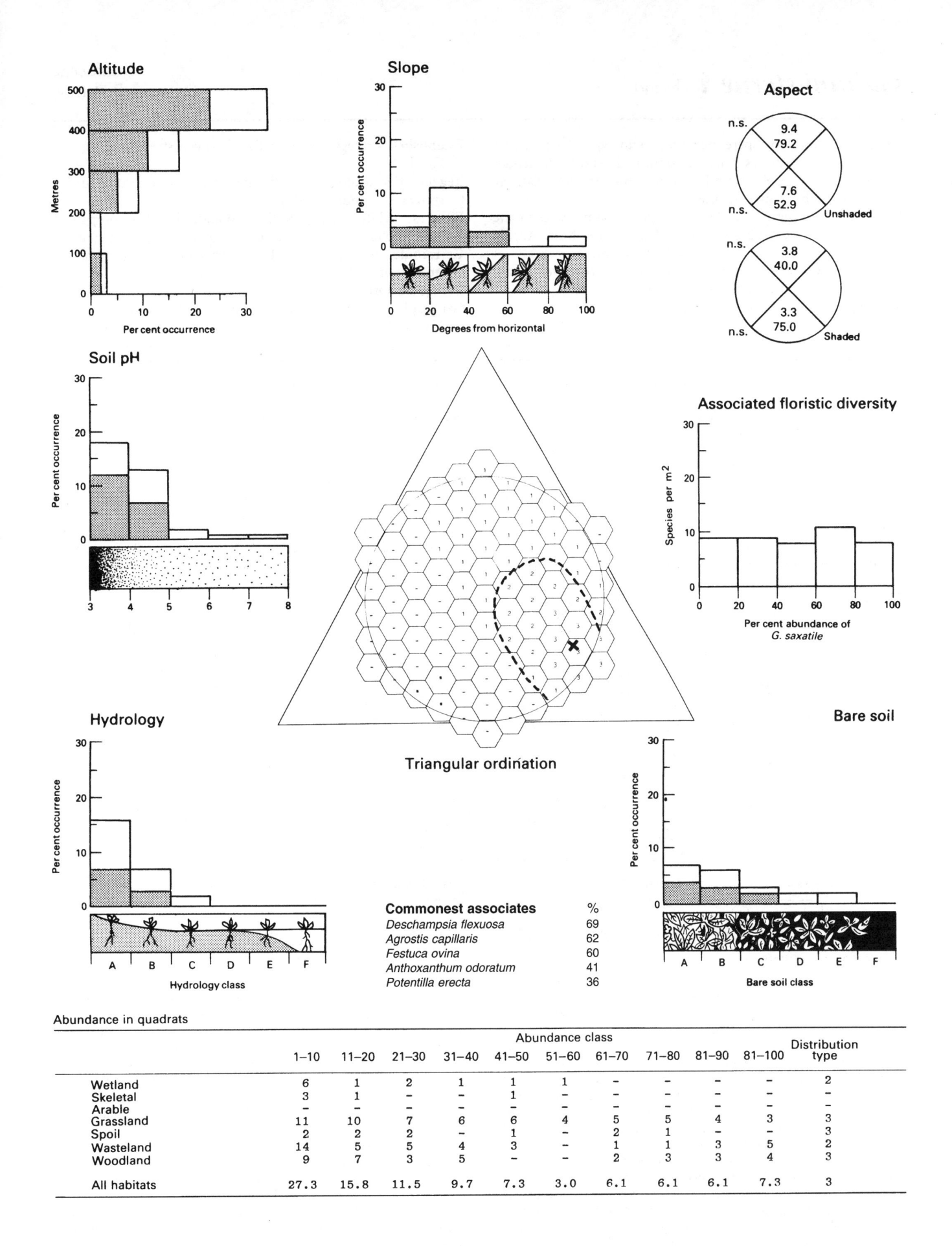

Altitude

Slope

Aspect

	n.s.	
n.s.	9.4 / 79.2	
	7.6 / 52.9	Unshaded

	n.s.	
n.s.	3.8 / 40.0	
	3.3 / 75.0	Shaded

Soil pH

Associated floristic diversity

Hydrology

Triangular ordination

Bare soil

Commonest associates	%
Deschampsia flexuosa	69
Agrostis capillaris	62
Festuca ovina	60
Anthoxanthum odoratum	41
Potentilla erecta	36

Abundance in quadrats

	1–10	11–20	21–30	31–40	41–50	51–60	61–70	71–80	81–90	81–100	Distribution type
Wetland	6	1	2	1	1	1	–	–	–	–	2
Skeletal	3	1	–	–	1	–	–	–	–	–	–
Arable	–	–	–	–	–	–	–	–	–	–	–
Grassland	11	10	7	6	6	4	5	5	4	3	3
Spoil	2	2	2	–	1	–	2	1	–	–	3
Wasteland	14	5	5	4	3	–	1	1	3	5	2
Woodland	9	7	3	5	–	–	2	3	3	4	3
All habitats	27.3	15.8	11.5	9.7	7.3	3.0	6.1	6.1	6.1	7.3	3

Galium sterneri Ehrend.

Rubiaceae
Sterner's Bedstraw

Life-form Polycarpic perennial protohemicryptophyte. Leaves lanceolate, sometimes hairy, <250 mm² in whorls on slender prostrate or weakly ascending stems; root system relatively shallow, with VA mycorrhizas.

Phenology Winter green. Vegetative growth commences in late spring. Flowers June and July and seed shed from August onwards. Flowering shoots die back in autumn.

Height Non-flowering shoots normally prostrate. Leafy flowering shoots seldom exceeding 80 mm.

Seedling RGR No data. SLA 24.4. DMC 22.8.

Nuclear DNA amount 2.0 pg; $2n = 22^*, 44^*$ (*BC*, *FE4*); $2x^*, 4x^*$.

Established strategy Intermediate between stress-tolerator and C-S-R.

Regenerative strategies V, ?B_s. Regeneration in need of further study (see 'Synopsis'); ?SB IV.

Flowers White, hermaphrodite, protandrous, self-incompatible, insect-pollinated (*CTW*); usually <50 in a compound panicle.

Germinule 0.39 mg, 1.1 × 1.0 mm, dry mericarp (2 per ovary).

Germination Epigeal; t_{50} 4 days; 12-29°C.

Biological flora None.

Geographical distribution A European endemic confined to the N and W of the British Isles (25% of British vice-counties, *ABIF* page 594) and NW Europe (15% of European territories).

Habitats Confined to limestone. Most abundant on calcareous screes, rocky pastures and lead-mine spoil. Also recorded from cliffs and outcrops, quarry heaps and wasteland.

Gregariousness Intermediate.

Altitude Recorded only on the carboniferous limestone, and hence restricted to altitudes of between 200 and 400 m.

Slope Largely restricted to rocky limestone dales, and therefore, although wide-ranging, most frequent and abundant on slopes of between 20 and 60°.

Aspect Slightly more frequent and abundant on N-facing slopes.

Hydrology Absent from wetlands.

Soil pH Most frequent and abundant on calcareous soils of pH >7.0. Shows a progressive decrease as soils become more acidic, and not recorded at pH values <4.0.

Bare soil Associated with widely different amounts of bare soil and bare rock.

△ ordination Distribution centred on vegetation associated with relatively undisturbed, unproductive conditions.

Associated floristic diversity Typically intermediate to high.

Synopsis Despite its rarity in Britain and Europe as a whole, *G.s.* is a common plant on the carboniferous limestone of N Derbyshire. It is a low-growing herb restricted to open grassland and rocky habitats on calcareous soils. It is a classic calcicole (Tansley 1917 as *G. sylvestre* Poll.), but in many other facets of its ecology *G.s.* is similar to the calcifuge *G. saxatile*, with

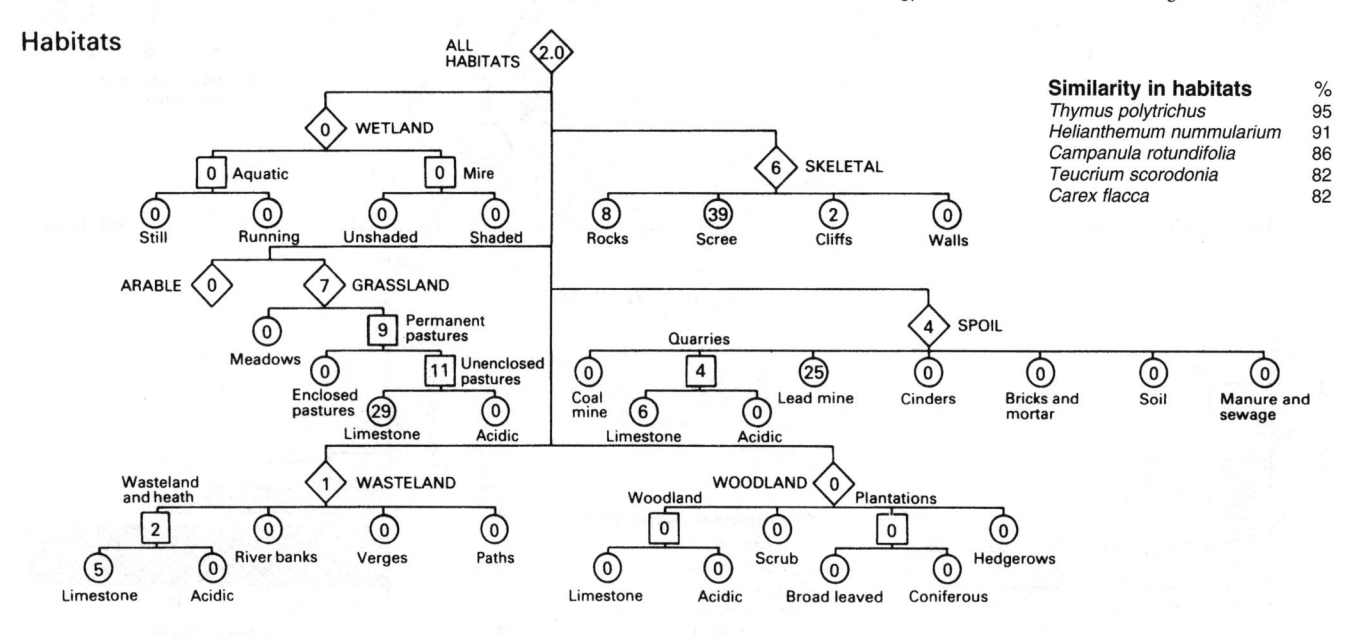

Habitats

Similarity in habitats	%
Thymus polytrichus	95
Helianthemum nummularium	91
Campanula rotundifolia	86
Teucrium scorodonia	82
Carex flacca	82

which it forms a sterile hybrid in Wales and Scotland (Stace 1975). In common with *G. saxatile*, *G.s.* tends to be absent from droughted sites, a feature perhaps related to the shallow root system and the delay of shoot growth until late spring. *G.s.* is also a poor competitor, and generally occurs as an inconspicuous vegetation component. *G.s.* is generally found in relatively open habitats, and appears to be less shade-tolerant than *G. saxatile*. *G.s.* is most conspicuous at the margin of screes and outcrops, where its shoots often form mats which extend over rocks and bared areas. However, these mats are less extensive than those often associated with *G. saxatile*, and their component shoots are less consistently rooted. Thus, regeneration by the seeds, which are commonly produced except where the plant is shaded, may assume a greater importance than in *G. saxatile*. A buried seed bank is suspected. Field studies are required to determine the phenology of seed germination and the circumstances conducive to seedling establishment. Genotypes differing in level of hairiness are often found close to each other, and some populations exploit lead-mine spoil. Otherwise nothing is known of the genetic variability of populations and of the capacity of *G.s.* to form ecotypes. Derbyshire populations are tetraploid ($2n = 44$). Slender diploid populations ($2n = 22$) from the W coast of Britain and from Ireland also occur (*FE4*). These populations, although geographically and cytologically isolated, are morphologically indistinguishable from $4x$ plants (Goodway 1957). One closely related species is the octoploid, *G. pumilum* Murray ($2n = 88$), found on calcareous soils in S Britain (*Excur Fl*, *FE4*). Its ecological separation from the tetraploid *G.s.* is uncertain. Both this species and *G.s.* are part of a closely interrelated polyploid complex of 27 European species centred around *G. pusillum* L. (*FE4*).

Current trends Generally restricted to semi-natural. or at least relatively ancient habitats. Changes in land management, such as the cessation of grazing in many of the limestone dales, are likely to have resulted in a decrease in the abundance of *G.s.*, but its survival on screes and rock outcrops seems assured.

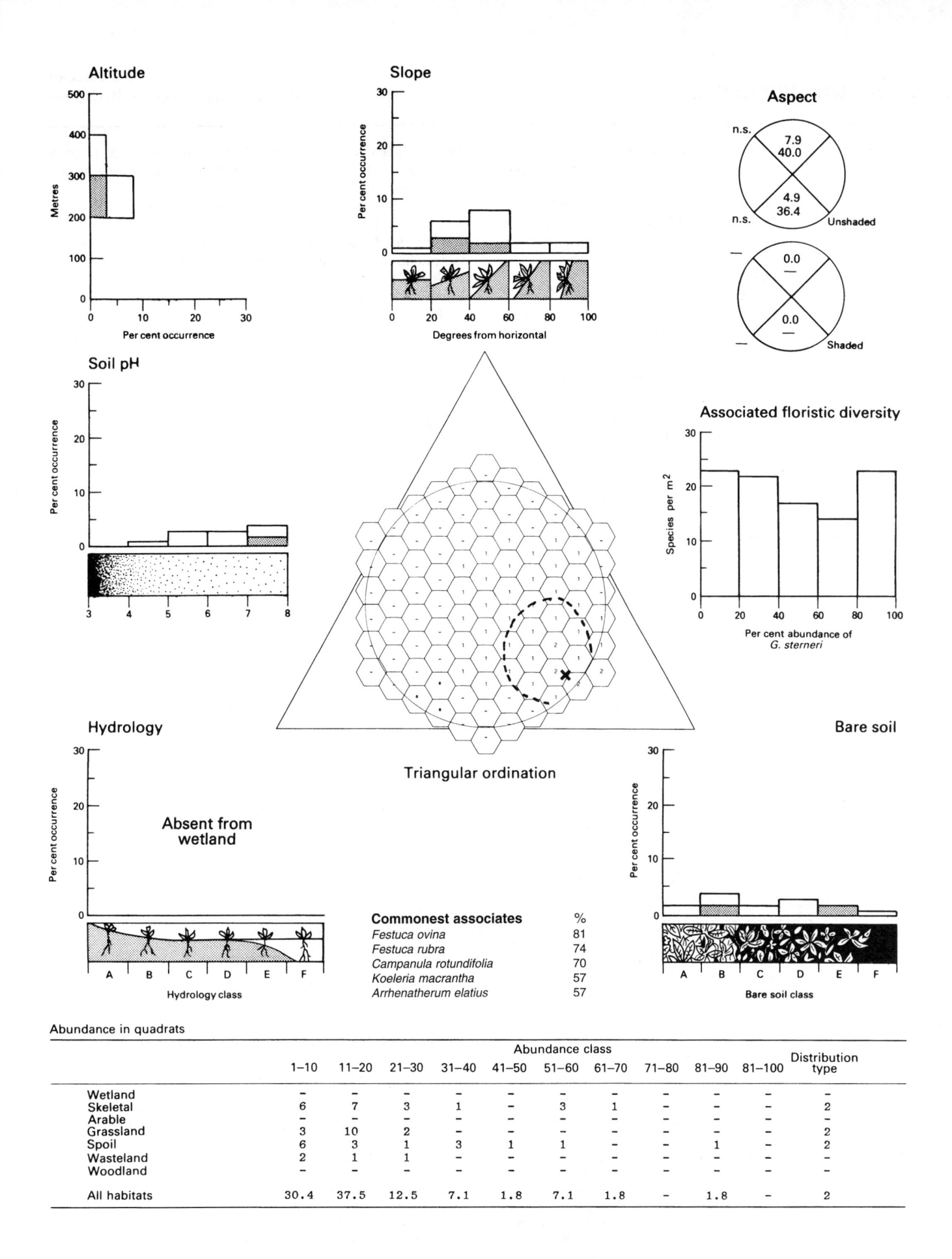

Altitude

Slope

Aspect

n.s.	7.9	
	40.0	
	4.9	
n.s.	36.4	Unshaded

—	0.0	
	—	
	0.0	
—	—	Shaded

Soil pH

Associated floristic diversity

Hydrology

Absent from wetland

Triangular ordination

Bare soil

Commonest associates	%
Festuca ovina	81
Festuca rubra	74
Campanula rotundifolia	70
Koeleria macrantha	57
Arrhenatherum elatius	57

Abundance in quadrats

	1–10	11–20	21–30	31–40	41–50	51–60	61–70	71–80	81–90	81–100	Distribution type
Wetland	–	–	–	–	–	–	–	–	–	–	–
Skeletal	6	7	3	1	–	3	1	–	–	–	2
Arable	–	–	–	–	–	–	–	–	–	–	–
Grassland	3	10	2	–	–	–	–	–	–	–	2
Spoil	6	3	1	3	1	1	–	–	1	–	2
Wasteland	2	1	1	–	–	–	–	–	–	–	–
Woodland	–	–	–	–	–	–	–	–	–	–	–
All habitats	30.4	37.5	12.5	7.1	1.8	7.1	1.8	–	1.8	–	2

Galium verum L.

Rubiaceae
Lady's Bedstraw, Yellow Bedstraw

Life-form Polycarpic perennial hemicryptophyte with slender creeping stock. Shoots usually decumbent with whorls of hairy linear leaves <25 mm^2; root system deep, with VA mycorrhizas.

Phenology Winter green. Flowers July and August and sets seed from September to November. Leaves relatively long-lived. Flowering shoots die back in early winter, some still bearing fruit.

Height Foliage and flowers typically <300 mm.

Seedling RGR 1.0–1.4 week^{-1}. SLA 24.1. DMC 23.4.

Nuclear DNA amount 3.9 pg; $2n = 22, 44^*$ (*BC, FE4*); $2x, 4x^*$.

Established strategy Intermediate between C-S-R and stress-tolerant competitor.

Regenerative strategies V, ?B$_s$. Regeneration from seed requires further study. Primarily vegetatively by means of stolons (see 'Synopsis'); ?SB IV.

Flowers Yellow, hermaphrodite, protandrous, insect-pollinated, usually >50 in a compound panicle.

Germinule 0.4 mg, 1.5 × 1.4 mm, dry mericarp (2 per ovary).

Germination Epigeal; t_{50} 7 days; 11–29°C; L = D.

Biological flora None.

Geographical distribution Most of the British Isles (100% of British vice-counties), *ABIF* page 592; Europe (85% of European territories); New Zealand and N America, probably introduced.

Habitats Recorded mainly in grazed or ungrazed calcareous grasslands. Also occurs on rock outcrops, waysides and on limestone quarry tips. Absent from arable, wetland and woodland habitats. [Occurs on stabilized sand dunes and on sea cliffs.]

Gregariousness Intermediate, although can form patches over 1 m^2 in extent.

Altitude Recorded at altitudes up to 400 m. Considerably more frequent and abundant in the range 200–300 m, corresponding to the distribution of the carboniferous limestone. [Up to 670 m.]

Slope Most records are from limestone dale-sides in the range 20–50°. Absent from cliffs and from steeper slopes.

Aspect Widely distributed, but rather more abundant on N-facing sites.

Hydrology Absent from wetlands.

Soil pH Recorded within the pH range 4.0–8.0, but recorded at maximum frequency and abundance between pH 5.5 and 6.5.

Bare soil Not present in habitats having large amounts of exposed soil.

△ **ordination** Confined to relatively undisturbed habitats, and concentrated in vegetation associated with unproductive conditions.

Associated floristic diversity Typically species-rich.

Habitats

Similarity in habitats	%
Pimpinella saxifraga	90
Stachys officinalis	85
Helictotrichon pubescens	82
Carex caryophyllea	80
Briza media	77

Synopsis *G.v.* is an often low-growing, winter-green herb of unproductive grassland and, to a lesser extent, open rocky habitats. Within the Sheffield region *G.v.* is particularly characteristic of soils of intermediate base-status over limestone, conditions which usually coincide with surface-leached calcareous soils. In Britain as a whole *G.v.* is also common on sandy soils, and here the unusually deep root system (Anderson 1927, Salisbury 1952) may be of critical significance. In turf microcosms, roots of this species became colonized by VA mycorrhizas with increases of up to 9-fold in seedling yield (Grime, Mackey, Hillier & Read 1987). *G.v.* persists in dune grasslands in which it exhibits a limited ability to withstand burial, which may also explain the capacity to colonize ant-hills (King 1977a-c). Field observations suggest that *G.v.* has a midsummer peak in biomass. *G.v.* is relatively unpalatable to stock (Milton 1933). The species is much more robust when released from grazing, and extensive mats of ascending shoots can develop in derelict grassland as a result of stolon growth. However, where grassland remains derelict and unburnt, *G.v.* is eventually suppressed by taller species. The seed ecology of *G.v.* has been studied little, but we suspect that regeneration by seed is rather rare and that a persistent seed bank is formed. Heavy-metal-tolerant populations have been identified within the flora of Derbyshire (Shaw 1984). In N Europe populations of *G.v.* ssp. *verum* are $4x$ and in S Europe they are $2x$ (*FE4*). A large number of ecological and geographical races occur in S Europe, but the extent of genotypic variation within British populations appears not to have been investigated. The hybrid with *G. mollugo* L. is frequent in S England.

Current trends An infrequent colonist of new habitats and, in common with most species of less-fertile habitats, *G.v.* is probably decreasing.

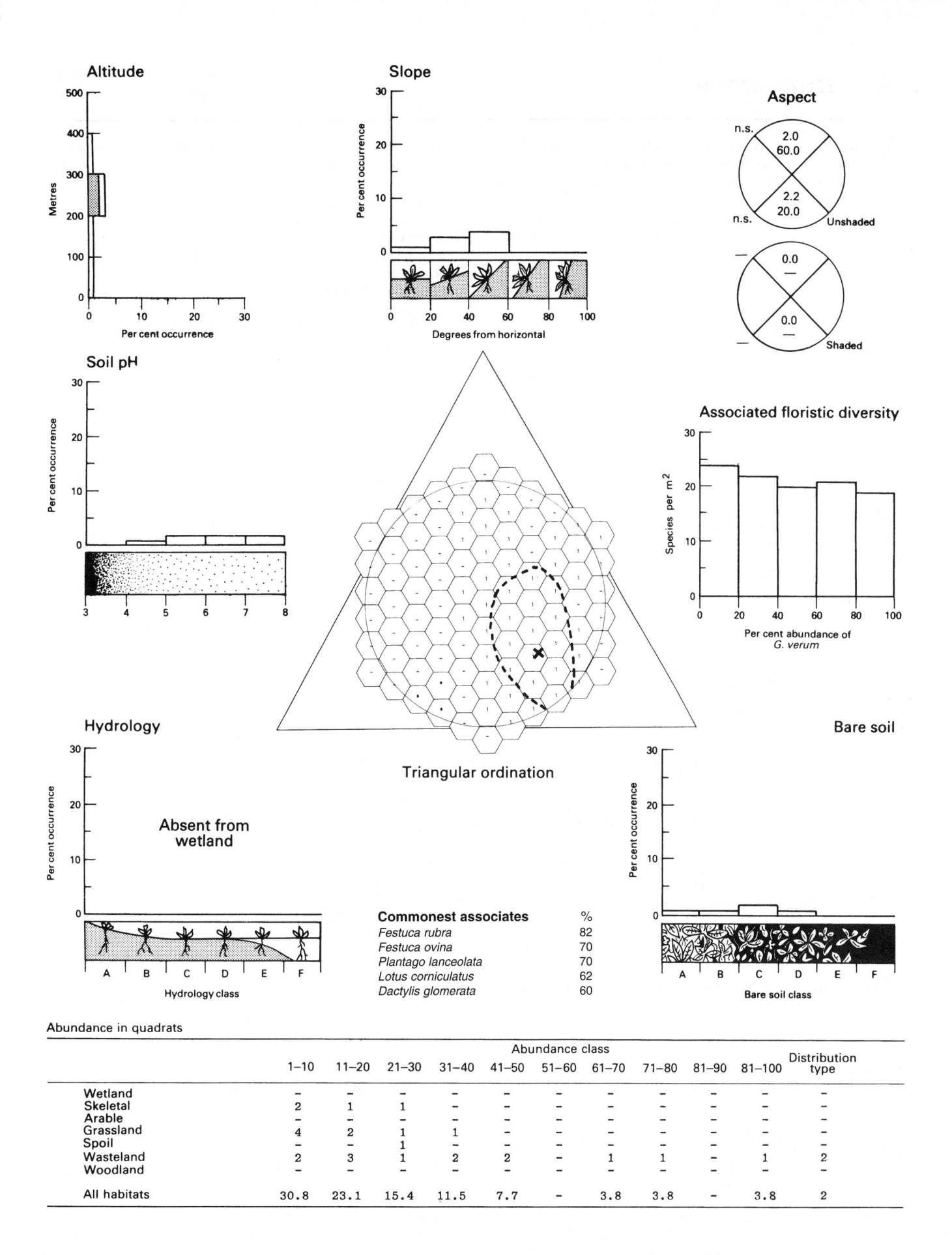

Altitude
(histogram: Metres vs Per cent occurrence)

Slope
(histogram: Per cent occurrence vs Degrees from horizontal)

Aspect

n.s.	2.0	60.0	n.s.
n.s.	2.2	20.0	Unshaded

—	0.0	—
—	0.0	—	Shaded

Soil pH
(histogram: Per cent occurrence vs pH 3–8)

Associated floristic diversity
(histogram: Species per m² vs Per cent abundance of *G. verum*)

Triangular ordination

Hydrology
Absent from wetland
(Per cent occurrence vs Hydrology class A–F)

Bare soil
(Per cent occurrence vs Bare soil class A–F)

Commonest associates   %
Festuca rubra   82
Festuca ovina   70
Plantago lanceolata   70
Lotus corniculatus   62
Dactylis glomerata   60

Abundance in quadrats

	1–10	11–20	21–30	31–40	41–50	51–60	61–70	71–80	81–90	81–100	Distribution type
Wetland	–	–	–	–	–	–	–	–	–	–	–
Skeletal	2	1	1	–	–	–	–	–	–	–	–
Arable	–	–	–	–	–	–	–	–	–	–	–
Grassland	4	2	1	1	–	–	–	–	–	–	–
Spoil	–	–	1	–	–	–	–	–	–	–	–
Wasteland	2	3	1	2	2	–	1	1	–	1	2
Woodland	–	–	–	–	–	–	–	–	–	–	–
All habitats	30.8	23.1	15.4	11.5	7.7	–	3.8	3.8	–	3.8	2

Geranium molle L.

Geraniaceae
Dove's-foot Cranesbill

Life-form Winter- or more rarely, summer-annual therophyte. Stems decumbent or ascending; leaves rounded, divided, hairy, sometimes >1000 mm²; tap-rooted, often with VA mycorrhizas.

Phenology Variable, dependent upon season of germination. Autumn-germinating plants overwinter as rosettes. Flowers within the period April to September and sets seed from June to October.

Height Foliage mainly <200 mm; flowers to 400 mm.

Seedling RGR No data. SLA 26.6. DMC 18.6.

Nuclear DNA amount No data; $2n = 26*$ (*BC*); $2x*$.

Established strategy Intermediate between ruderal and C-S-R.

Regenerative strategies S, B_s. By seed, either in autumn soon after seed set or after incorporation into a persistent seed bank (Dorph-Petersen 1924); SB III.

Flowers Purple, hermaphrodite, insect-pollinated but probably often selfed (Knuth 1906); in pairs within a cymose inflorescence.

Germinule 1.24 mg, 1.9 × 1.5 mm, seed dispersed from an explosively dehiscent up to 5-seeded fruit.

Germination Epigeal; t_{50} 3 days; hard-coat dormancy; germination enhanced by dry storage at 20°C.

Biological flora None.

Geographical distribution The British Isles, but less frequent in the N and W (100% of British vice-counties), *ABIF* page 448; Europe except for the extreme N (89% of European territories and 3% naturalized); SW Asia and N Africa; naturalized in N and S America and New Zealand.

Habitats Mainly recorded from limestone outcrops, but widely distributed in other habitats with bare ground. Recorded from wasteland, pasture, scree, soil heaps, coal-mine spoil and manure heaps. Observed on sandy soils beside roads and railways and on cinder tips. Rare in lightly-shaded sites; otherwise absent from woodlands and from wetlands and only a casual of arable land.

Gregariousness Sparse.

Altitude Widely distributed, but most frequently recorded on the carboniferous limestone. Observed equally commonly on the sandy soils of the Bunter sandstone below 100 m. [Up to 530 m.]

Slope Not recorded from cliffs and walls or from the steepest slopes.

Aspect Data presented opposite suggest a N-facing bias. However, *G.m.* is often found in droughted habitats with other winter annuals, and in a further more extensive survey (Hodgson *et al.* in prep.) *G.m.* is much more common on S-facing slopes.

Hydrology Absent from wetlands.

Soil pH Only recorded from soils of pH >5.0.

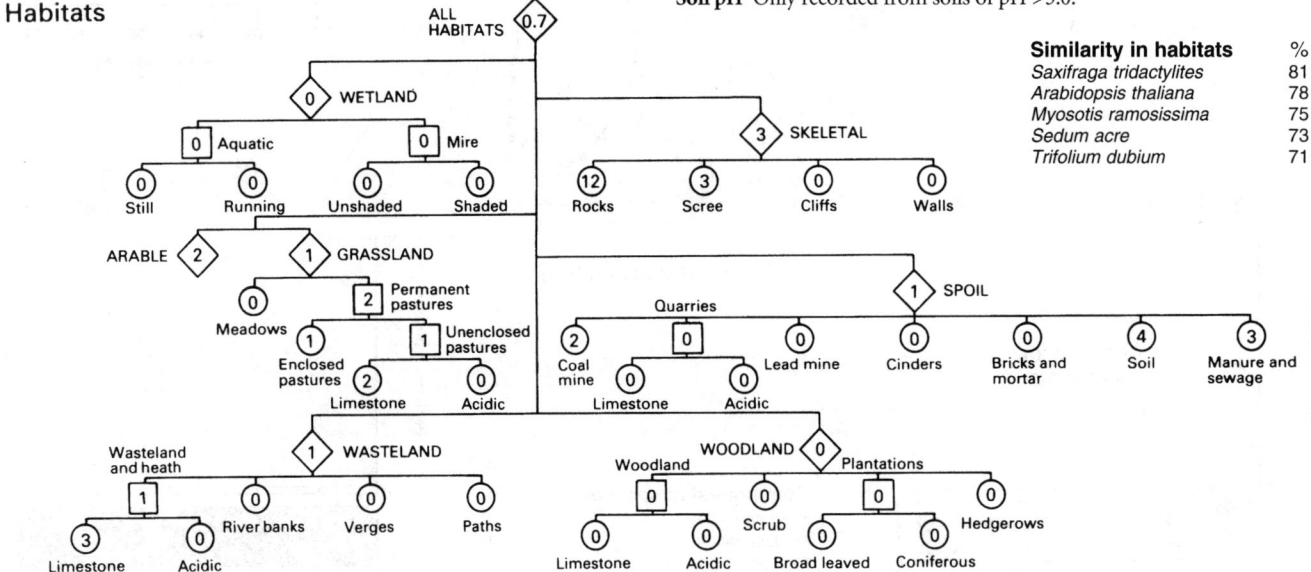

Habitats

Similarity in habitats	%
Saxifraga tridactylites	81
Arabidopsis thaliana	78
Myosotis ramosissima	75
Sedum acre	73
Trifolium dubium	71

Bare soil Recorded in habitats having widely different amounts of bare soil. Many of the occurrences in sites with low exposures of bare soil are situated on rock outcrops.

△ **ordination** Associated with vegetation subject to moderate amounts of disturbance with productivity ranging from relatively productive to unproductive. Most frequent in disturbed unproductive vegetation.

Associated floristic diversity Typically intermediate or species-rich.

Synopsis *G.m.* is most typically a relatively robust, winter-annual herb exploiting open habitats where encroachment by perennials is checked annually by drought and their vigour reduced by nutrient stress. The species is associated with permanently open habitats, such as droughted rock outcrops and sandy wasteland, but occurs also in transient habitats, e.g. new soil heaps and localized areas of recently-bared ground in pasture and on road verges. Thus, its habitat range is remarkably similar to that of *Medicago lupulina*, with which *G.m.* is often associated in disturbed lawns and pastures. *G.m.* frequently extends into heavily-droughted habitats and, like the most ephemeral winter-annuals of these sites, flowers early and produces dormant seeds. However, in less-droughted sites *G.m.* often has an extended flowering period, perhaps because the long tap-root enables the species to exploit subsoil moisture. The possibility that the prolonged flowering may also be due to the production of more than one cohort of seedlings within a season (as in the case of *G. robertianum*) also merits investigation. *G.m.* is absent from woodland and wetland. Regeneration relies upon seeds, which exhibit hard-coat dormancy. A persistent seed bank may be formed (Dorph-Petersen 1924). Large plants may produce 1500 seeds (Salisbury 1961), but small plants may form less than 100. Short-distance dispersal of seed (up to c. 6 m) is effected by means of explosive dehiscence of the fruit (Salisbury 1961). Seeds survive ingestion by cattle (Ridley 1930) and are an impurity in clover seed (*WCH*). The species of *Geranium* which are most closely related in ecology to *G.m.* are *G. dissectum* L. and *G. pusillum* L. *G.d* exploits moister, often impermanent, sites, while *G.p.* is more strictly a winter annual of dry lowland habitats.

Current trends Uncertain. Combines an attribute favourable to future expansion, namely the capacity for wide dispersal, with an unfavourable one, the dependence upon less-fertile habitats.

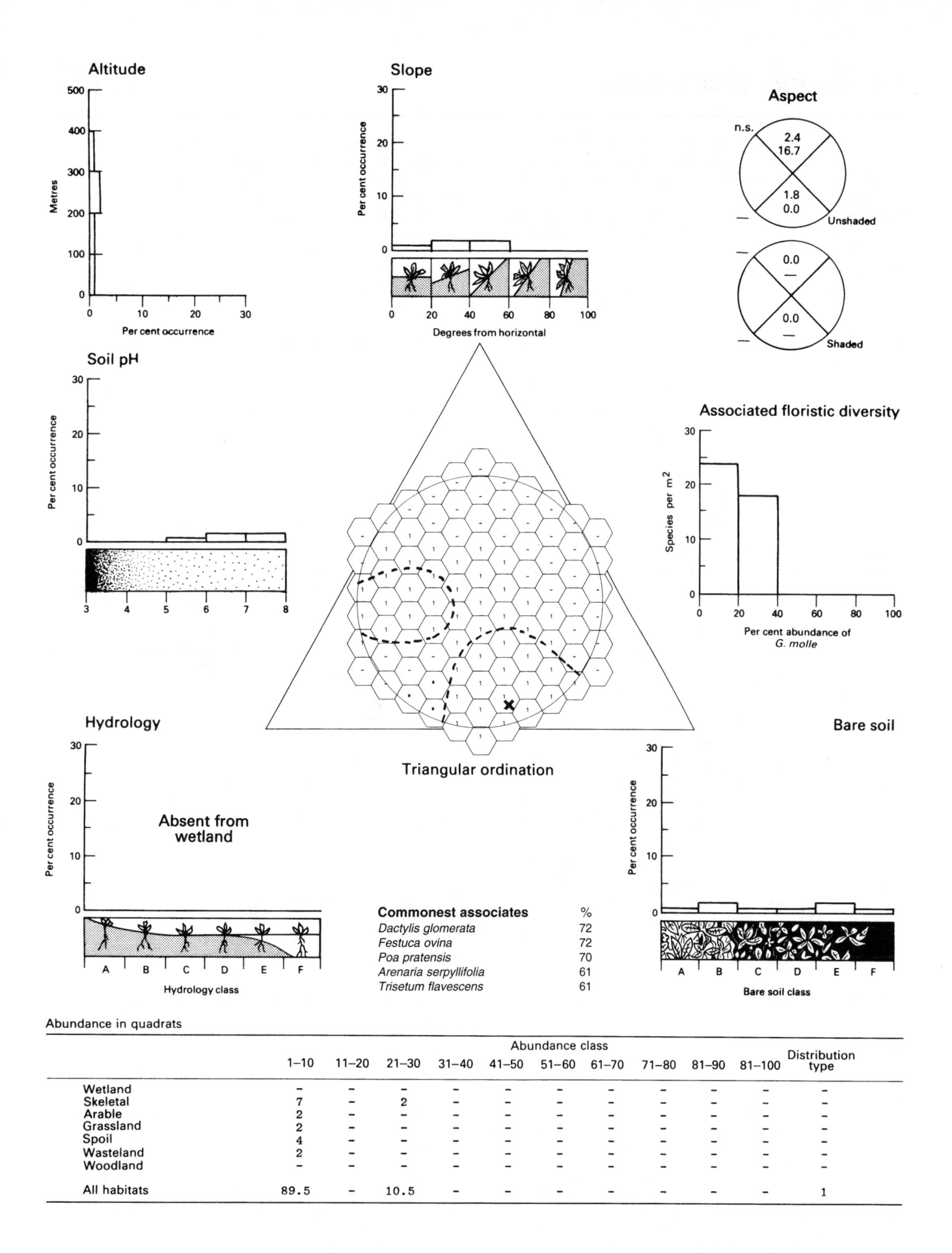

Altitude

Slope

Aspect

	n.s.	
—	2.4	16.7
—	1.8	0.0

Unshaded

—	0.0	—
—	0.0	—

Shaded

Soil pH

Associated floristic diversity

Hydrology

Absent from wetland

Triangular ordination

Commonest associates	%
Dactylis glomerata	72
Festuca ovina	72
Poa pratensis	70
Arenaria serpyllifolia	61
Trisetum flavescens	61

Bare soil

Abundance in quadrats

	1–10	11–20	21–30	31–40	41–50	51–60	61–70	71–80	81–90	81–100	Distribution type
Wetland	–	–	–	–	–	–	–	–	–	–	–
Skeletal	7	–	2	–	–	–	–	–	–	–	–
Arable	2	–	–	–	–	–	–	–	–	–	–
Grassland	2	–	–	–	–	–	–	–	–	–	–
Spoil	4	–	–	–	–	–	–	–	–	–	–
Wasteland	2	–	–	–	–	–	–	–	–	–	–
Woodland	–	–	–	–	–	–	–	–	–	–	–
All habitats	89.5	–	10.5	–	–	–	–	–	–	–	1

Geranium robertianum L.

Geraniaceae
Herb Robert

Life-form A semi-rosette species most typically flowering in its second season (see 'Synopsis'). Stems decumbent or ascending; leaves dissected, hairy, <2000 mm²; roots usually with VA mycorrhizas.

Phenology Variable (see 'Synopsis'). Winter green. Populations in flower from May to September and in seed from June to October.

Height Foliage and flowers to 500 mm.

Seedling RGR 1.0-1.4 week^{-1}. SLA 30.0. DMC 18.6.

Nuclear DNA amount No data; $2n = 32^*, 64^*$ (*BC*); $2x^*, 4x^*$. (Genus with a polyploid base number; $n = 14$.)

Established strategy Intermediate between ruderal and C-S-R.

Regenerative strategies ?B_s Regenerates exclusively by seed. Germinates over an extended period (see 'Synopsis'). Seed bank studies have not yet been undertaken; ?SB III.

Flowers Pink, hermaphrodite, protandrous, insect- or self-pollinated (Knuth 1906); leaf-opposed, in pairs.

Germinule 1.14 mg, 1.3 × 1.1 mm, seed forcibly dispersed enclosed within a mericarp; up to 5 seeds per ovary.

Germination Epigeal; t_{50} 10 days; hard-coat dormancy; germination rate also increased after dry storage at 20°C; L = D.

Biological flora Tofts (2004).

Geographical distribution The British Isles except in N of Scotland (100% of British vice-counties), *ABIF* page 449; Europe northwards to Scandinavia (89% of European territories); temperate Asia and N Africa; naturalized in N and S America.

Habitats A very common species of limestone scree slopes. Also frequent in woodland on limestone soils, in hedgerows, on river banks, limestone quarry heaps, lead-mine spoil and in skeletal habitats; more occasionally in woodland on non-calcareous strata, scrub, plantations, waysides, wasteland, cinders, demolition sites and mine debris, and a casual of arable land. Absent from aquatic habitats and pasture. [Characteristic of maritime shingle.]

Gregariousness Sparse to intermediate.

Altitude Observed at altitudes of up to 460 m, with frequency and abundance greatest in moist rocky habitats on the carboniferous limestone, within the range 200-300 m. [Up to 700 m.]

Slope Wide-ranging.

Aspect More frequent and abundant on S-facing slopes in shaded habitats and on N-facing slopes in unshaded situations.

Hydrology Occurs predominantly on moist soils. The few records for wetlands are mainly from sloping ground, reflecting an association with soligenous mire.

Soil pH Absent from soils of pH <4.5, and only abundant above pH 5.5.

Bare soil Associated with a wide range of values, but most frequent and abundant in sites with low exposures, a reflection of the abundance of *G.r.* on scree slopes.

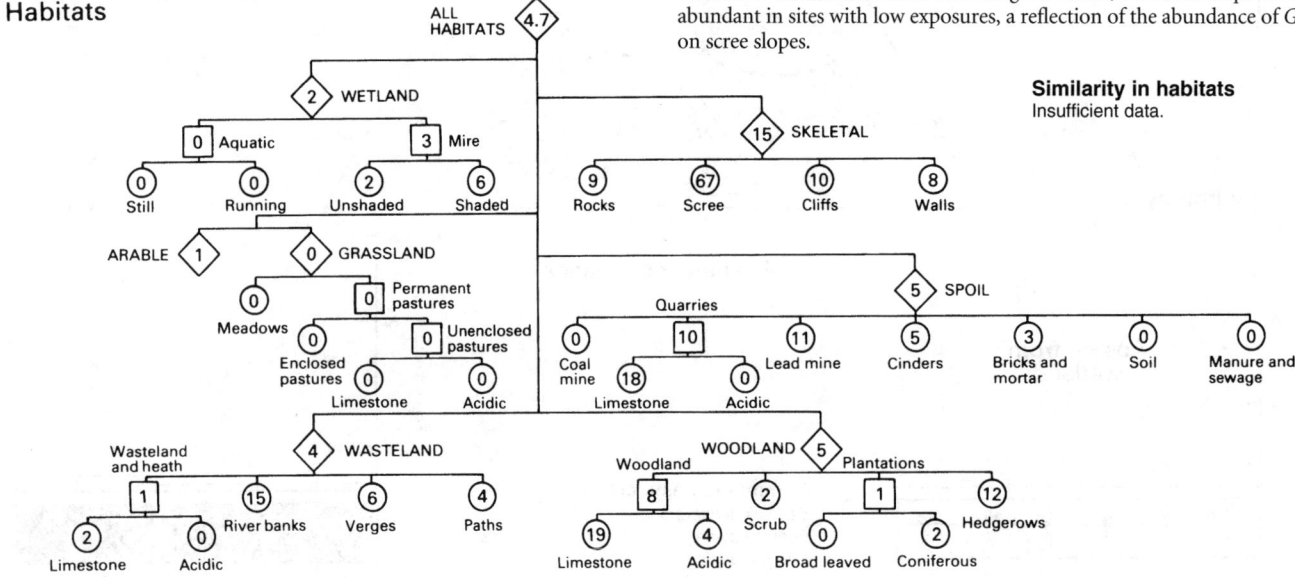

Similarity in habitats
Insufficient data.

Habitats

△ **ordination** Widely distributed, but centred on unproductive and relatively undisturbed vegetation.

Associated floristic diversity Low to intermediate.

Synopsis *G.r.* is a much-branched, usually monocarpic herb, growing in base-rich sites where vegetation density is reduced by fairly frequent disturbance, coupled with either shade, low soil fertility, or both. The species occurs in such diverse habitats as shaded woodland beside rivers and streams, and slightly unstable limestone scree subject to intermittent disturbance, and is frequently the only ruderal in the habitats in which it occurs. Within the Sheffield region suitable habitats are particularly well represented on limestone. In dry lowland areas *G.r.* is largely confined to moist shaded habitats. However, in damp upland climates *G.r.* occurs in a wide range of open and skeletal habitats and in maritime areas (where atmospheric moisture is, also high) *G.r.* is also characteristic of shingle beaches. The species is relatively shade-tolerant (Hegi 1964), and flowers and fruits during the expanded phase of the woodland canopy. Foliage of *G.r.* sampled from the Sheffield region contains high concentrations of Ca, Na and Fe (*SIEF*). When crushed, the foliage has an unpleasant smell; this characteristic may act as a deterrent to grazing. Throughout the summer months most populations contain a mixture of flowering individuals, juveniles and seedlings. Work in Poland by Falinska and Piroznikow (1983) suggests that this is because often three cohorts of seedlings are produced over the course of the year. Seedlings arising in spring flower and fruit the following summer, and have the highest survival rates and the highest reproductive potential. Thus, the spring cohort is the most important in the population dynamics of the species. The summer cohort of seedlings (July to August) overwinters as small plants which frequently do not flower until their third growing season. A few, flowering in the autumn of their second year, may also flower in their third, and are thus polycarpic. The autumn cohort is not always produced and few of the seedlings overwinter successfully. Nevertheless, such staggering of germination time appears likely to facilitate survival of a short-lived

Altitude

Slope

Aspect

Soil pH

Associated floristic diversity

Hydrology

Absent from wetland

Triangular ordination

Bare soil

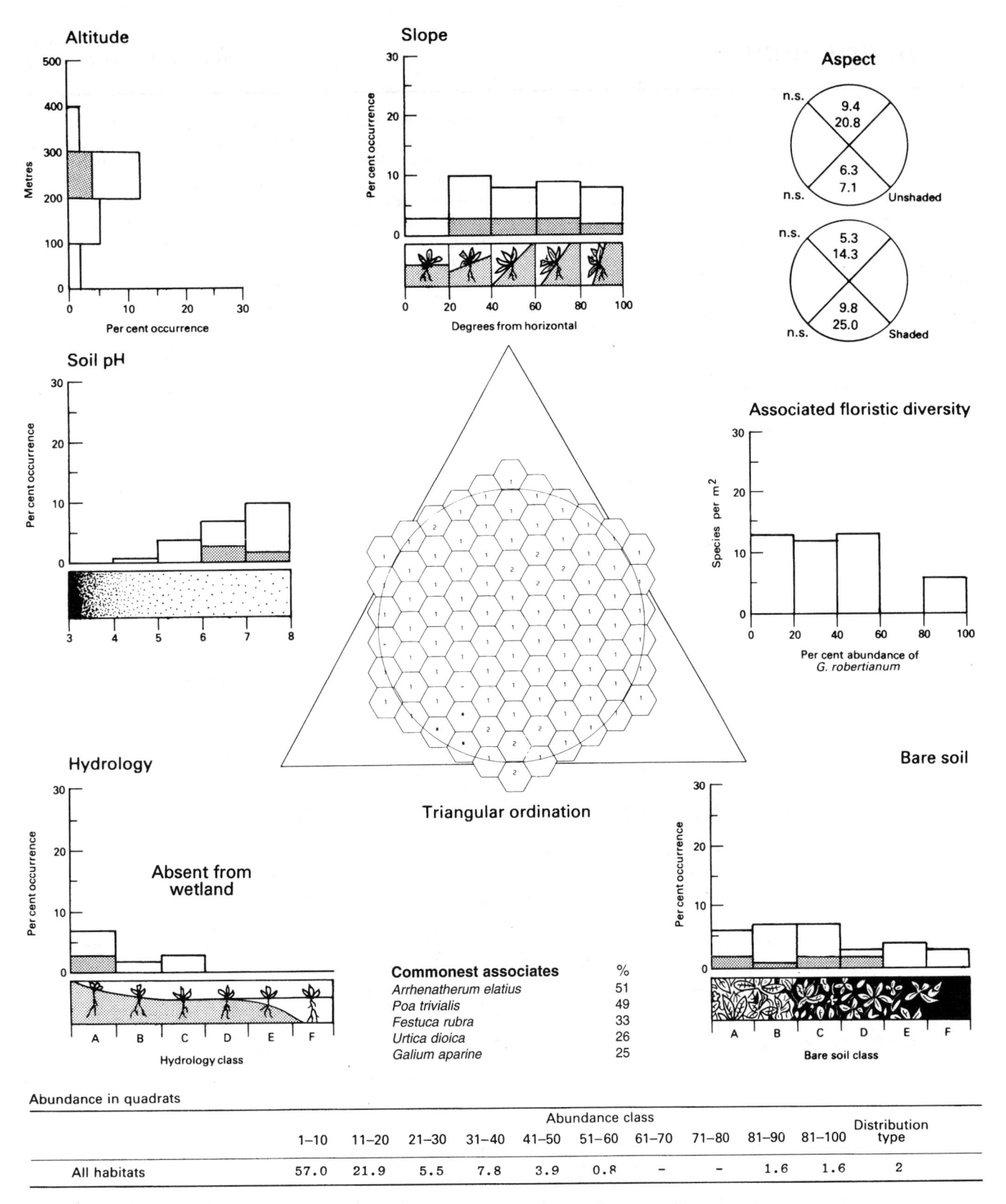

Commonest associates	%
Arrhenatherum elatius	51
Poa trivialis	49
Festuca rubra	33
Urtica dioica	26
Galium aparine	25

Abundance in quadrats

	1–10	11–20	21–30	31–40	41–50	51–60	61–70	71–80	81–90	81–100	Distribution type
All habitats	57.0	21.9	5.5	7.8	3.9	0.8	–	–	1.6	1.6	2

monocarpic species in environments subject to occasional, seasonally unpredictable disturbance. The seeds of the spring cohorts show the highest germination percentage (Falinska & Piroznikow 1983). The seed, which exhibits hard-coat dormancy, is comparatively large and supports considerable height growth in the hypocotyl and cotyledon stalks. As in the case of the seedling of *Arrhenatherum elatius*, the resulting capacity for height growth in shade is important in the colonization of limestone screes (Grime & Jeffrey 1965). The presence of a persistent seed bank is predicted.

The mericarps of *G.r.* float in water and are capable of attachment to animals and clothing (Ridley 1930). Both methods of dispersal may contribute to the relatively high colonizing ability of *G.r.* both locally and within the secondary woodland of Lincolnshire (Peterken 1981). The species is variable and three ecological races were formerly separated as ssp. (*CTW*). *G.r.* may hybridize with *G. purpureum* Vill.

Current trends Not under threat, and perhaps even increasing in disturbed artificial habitats.

Geum urbanum L.

Rosaceae
Herb Bennet, Wood Avens

Life-form Polycarpic perennial, semi-rosette hemicryptophyte with a short thick rhizome. Stem erect; lower leaves pinnate, hairy, sometimes >4000 mm^2, upper leaves less divided; roots with VA mycorrhizas.
Phenology Winter green. Flowers from June to August. Seed set July to September.
Height Foliage mainly <400 mm; flowers to 600 mm.
Seedling RGR 0.5–0.9 week^{-1}. SLA 22.6. DMC 30.2.
Nuclear DNA amount No data; 2n = 42 (*FE2*); 6x (2x for section, (Gajewski 1959).
Established strategy Intermediate between competitive-ruderal and C-S-R.

Regenerative strategies S. Regenerates by means of seeds which germinate in spring (Roberts 1986a). No persistent seed bank has been detected. *G.a.*, which is only shortly rhizomatous, appears to have little capacity for vegetative spread; SB II.
Flowers Yellow, hermaphrodite, self- or more rarely insect-pollinated (*CTW*); in few-flowered cymose inflorescences.
Germinule 0.73 mg, 8.3 × 1.3 mm, achene with a hooked awn, *c.* 100 flower^{-1} in a sessile head.
Germination Epigeal; t_{50} 5 days; 10–22°C; L > D.
Biological flora Taylor (1997).
Geographical distribution The British Isles (100% of British vice-counties), *ABIF* page 335; most of Europe except the extreme N (85% of European territories); W Asia, N Africa.

Habitats A species of woodland and hedgerows, which also extends on to river banks, waysides, cliffs and walls, particularly under shaded conditions. Also occurs more rarely in shaded mire and on shaded spoil tips.
Gregariousness Sparse to intermediate.
Altitude Observed to 310 m. [Up to 510 m.]
Slope Wide-ranging.
Aspect More frequent and abundant on shaded S-facing slopes.
Hydrology Not a wetland species but can persist on moist soil at the edge of soligenous mire.

Soil pH No records below pH 4.0, and only abundant above pH 5.0.
Bare soil Associated with a wide range of exposures, but sites generally with little herbaceous cover and various amounts of bare soil, litter, bryophyte mat and rock.
△ **ordination** A wide and diffuse distribution. Least frequent in highly stressed and highly disturbed habitats.
Associated floristic diversity Typically intermediate.

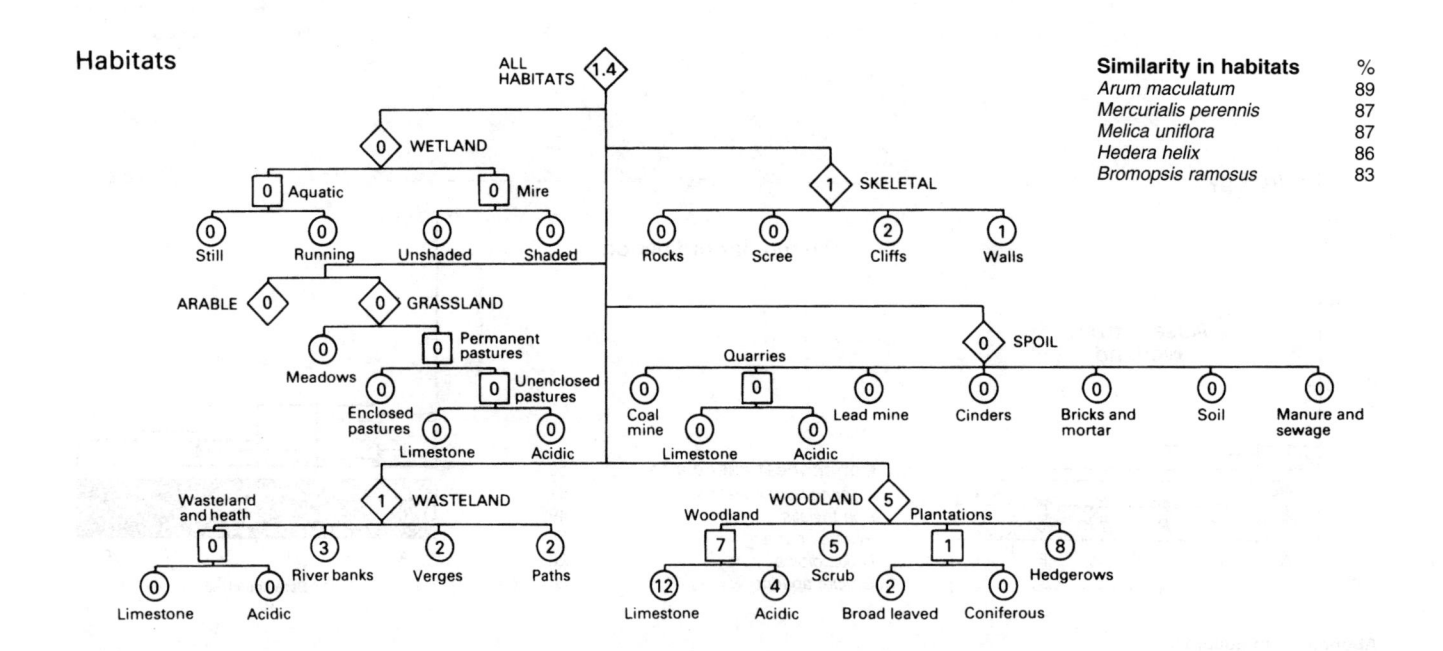

Habitats

Similarity in habitats	%
Arum maculatum	89
Mercurialis perennis	87
Melica uniflora	87
Hedera helix	86
Bromopsis ramosus	83

Synopsis *G.u.* is a winter-green herb producing a rosette of ascending compound leaves. The species is frequent in situations of moderate shade in woodland, particularly on moist S-facing slopes, and is also widespread in the partial shade of hedgerows. Flowering and seed set are greatest in unshaded conditions. In coppiced woodland *G.u.* shows an intermediate degree of shade-tolerance comparable with that of *Cirsium palustre* (Pons 1977). *G.u.* is perhaps best regarded as a 'semi-shade' species occurring in sites where the vigour of potential dominants of the herb layer is restricted by intermittent disturbance. *G.u.* is particularly frequent in hedgerows, in secondary woodland (Peterken 1981) and along woodland rides. The species is virtually absent from waterlogged ground and from grazed habitats. *G.u.* has little capacity for vegetative spread. Like *Anthriscus sylvestris*, another hedgerow species of similar ecology, *G.u.* produces numerous seeds, sometimes >1000 per plant, and apparently lacks a persistent seed bank (Roberts 1986a). However, the achene has a strong hooked awn and is dispersed on clothing and fur. This mechanism may contribute to the ability of *G.u.* to occur widely in woodlands and hedgerows. *G. rivale* L., a species of N-facing grassland, wet meadows and soligenous mire, forms hybrid swarms with *G.u.*, typically in open woodland (see Stace 1975). Usually introgression is towards *G.r.*, but in shaded, drier sites some hybrids may resemble *G.u.* more closely.
Current trends Favoured by the increasing level of disturbance of woodland and scrub, which may be compensating for losses of hedgerow habitats.

Altitude
Slope
Aspect

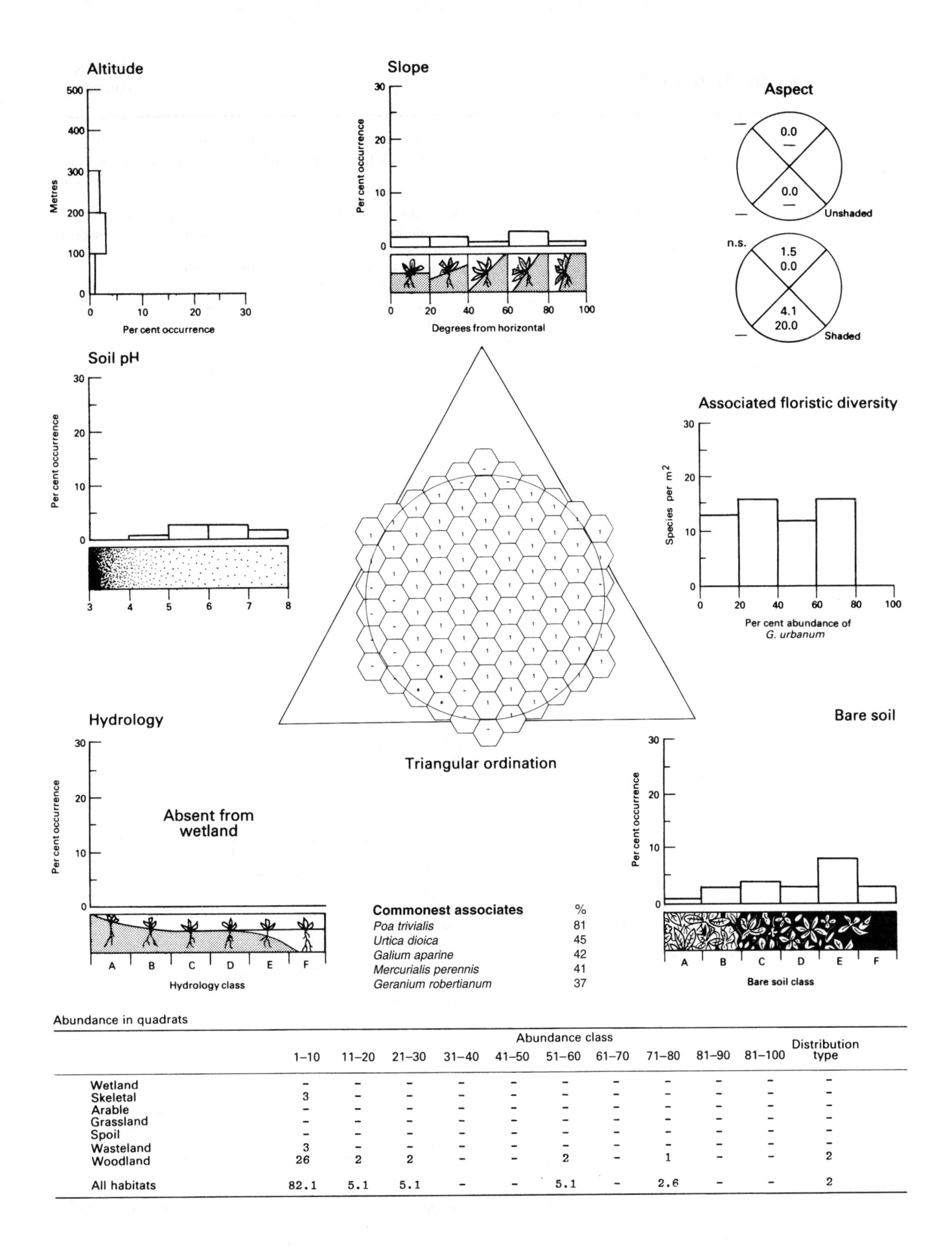

Unshaded
0.0 / 0.0

n.s.
1.5 / 0.0
4.1 / 20.0
Shaded

Soil pH

Associated floristic diversity

Per cent abundance of *G. urbanum*

Hydrology

Absent from wetland

Triangular ordination

Commonest associates	%
Poa trivialis	81
Urtica dioica	45
Galium aparine	42
Mercurialis perennis	41
Geranium robertianum	37

Bare soil

Abundance in quadrats

	1–10	11–20	21–30	31–40	41–50	51–60	61–70	71–80	81–90	81–100	Distribution type
Wetland	–	–	–	–	–	–	–	–	–	–	–
Skeletal	3	–	–	–	–	–	–	–	–	–	–
Arable	–	–	–	–	–	–	–	–	–	–	–
Grassland	–	–	–	–	–	–	–	–	–	–	–
Spoil	–	–	–	–	–	–	–	–	–	–	–
Wasteland	3	–	–	–	–	–	–	–	–	–	–
Woodland	26	2	2	–	–	2	–	1	–	–	2
All habitats	82.1	5.1	5.1	–	–	5.1	–	2.6	–	–	2

Glechoma hederacea L.

Lamiaceae
Ground Ivy

Life-form Polycarpic perennial protohemicryptophyte with long creeping, rooted stems. Flowering branches ascending; leaves reniform, hairy, often <500 mm^2; roots often with VA mycorrhizas.
Phenology Winter green. Following early spring growth the plant flowers from March to May. Main period of seed set June.
Height Foliage and flowers usually <200 mm.
Seedling RGR No data. SLA 39.5. DMC 16.0.
Nuclear DNA amount No data; $2n = 18, 36*$ (*BC, FE3*); $2x, 4x*$.
Established strategy Intermediate between competitive-ruderal and C-S-R.
Regenerative strategies V. ?B$_s$. Regeneration by seed probably uncommon and requiring further study (see 'Synopsis').
Flowers Violet, hermaphrodite (small female flowers also commonly occur), insect-pollinated; often <20 in a whorled inflorescence.
Germinule 0.69 mg, 1.9 × 1.1 mm, nutlet (4 per ovary).
Germination Epigeal; t_{50} 5 days; germination rate increased after dry storage at 20°C; L > D.
Biological flora Hutchings and Price (1999).
Geographical distribution The British Isles, but rare in W Ireland and N Scotland (97% of British vice-counties), *ABIF* page 520; W and N Europe (85% of European territories); W and N Asia, New Zealand and USA.

Habitats Typically a plant of shaded habitats, particularly hedgerows. Frequent also in mire and on soil heaps, river banks, in woodlands and on waysides. Recorded also from walls, wasteland and from spoil tips. Very rarely observed at the edges of pasture, or encroaching into arable fields from their margins. Absent from aquatic habitats.
Gregariousness Patch-forming.
Altitude Extends to 380 m. [Up to 410 m.]
Slope Wide-ranging.
Aspect Slightly more frequent on S-facing slopes in both this and a subsequent survey, but achieves greater abundance on N-facing slopes in unshaded situations.
Hydrology Recorded occasionally from soligenous mire and from moist soils at the edge of topogenous mire.
Soil pH Most frequent and abundant in the pH range 5.5-7.5, but extends down to pH 4.0.
Bare soil Wide-ranging, but most abundant in sites with moderate to high exposures.
△ **ordination** Distribution centred on vegetation associated with slightly disturbed, fertile conditions.
Associated floristic diversity Most commonly intermediate.

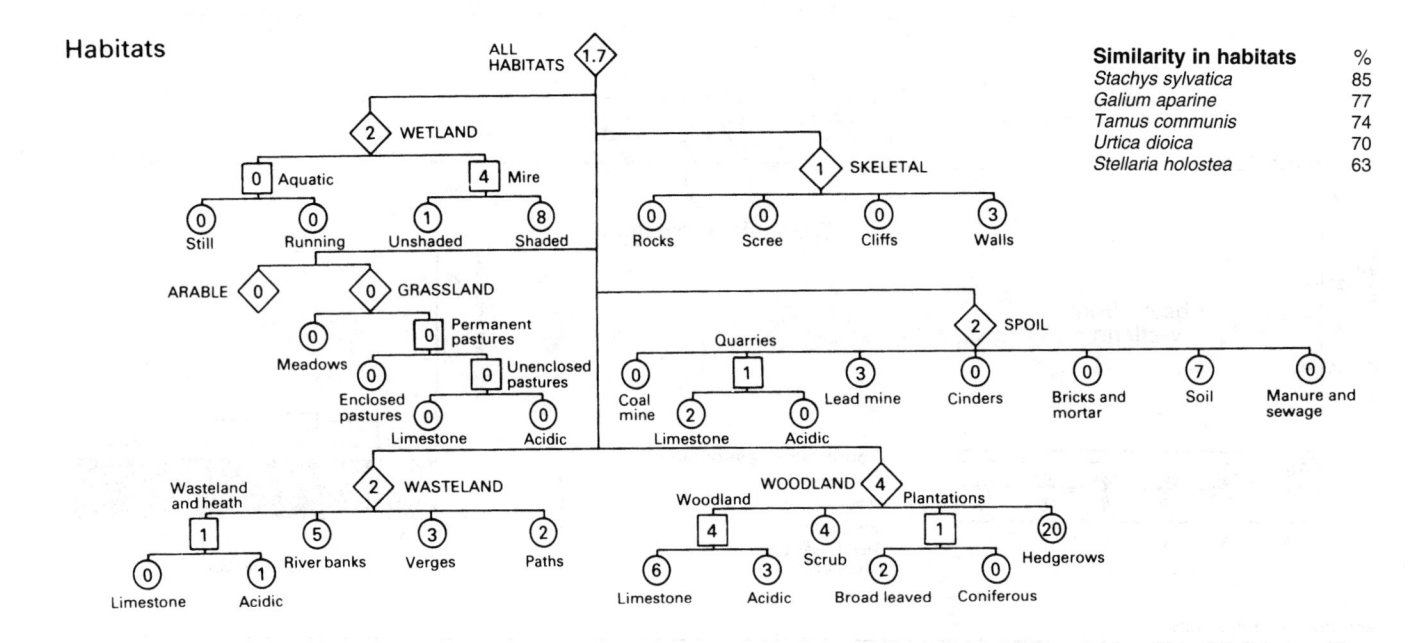

Habitats

Similarity in habitats	%
Stachys sylvatica	85
Galium aparine	77
Tamus communis	74
Urtica dioica	70
Stellaria holostea	63

Synopsis *G.h.* is a carpet-forming herb most typical of fertile habitats where the vigour of potential dominants is restricted by a combination of shade and disturbance. *G.h.* is frequent in hedgerows, on shaded roadsides and in woodland, and tends to occur beside rides, on floodplains, beside rabbit scrapes or in clear-felled areas. Peterken (1981) considers the species to be characteristic of secondary woodland. Although generally absent from pasture, *G.h.* is known to be toxic to stock (Cooper & Johnson 1984). When growing in the open *G.h.* produces more flowers and has a higher level of seed set, forming leaves which are small relative to those seen in shaded sites. In common with many other species from shaded habitats, *G.h.* flowers early in the year; however, the leaf canopy reaches a maximum in summer. The species has a limited capacity to persist under tall herbs or beneath a continuous tree canopy, and is most characteristic of lightly or patchily shaded habitats. When released from canopy shade, *G.h.* often spreads rapidly and extensively by means of long creeping stems. Thus, at an early stage in colonization, the floor of a quarry adjacent to woodland has been seen covered with *G.h.*. Effective regeneration appears to be primarily vegetative, and it is likely that, as in the case of *Stachys sylvatica*, detached shoot fragments are important for long-distance colonization. Seed set is often extremely poor, and studies by Chancellor (1985) suggest that seed is only shortly persistent in the soil.
Current trends A regular colonist of secondary woodland (Peterken 1981), probably by vegetative means, and likely to remain relatively common in shaded habitats subject to occasional disturbance.

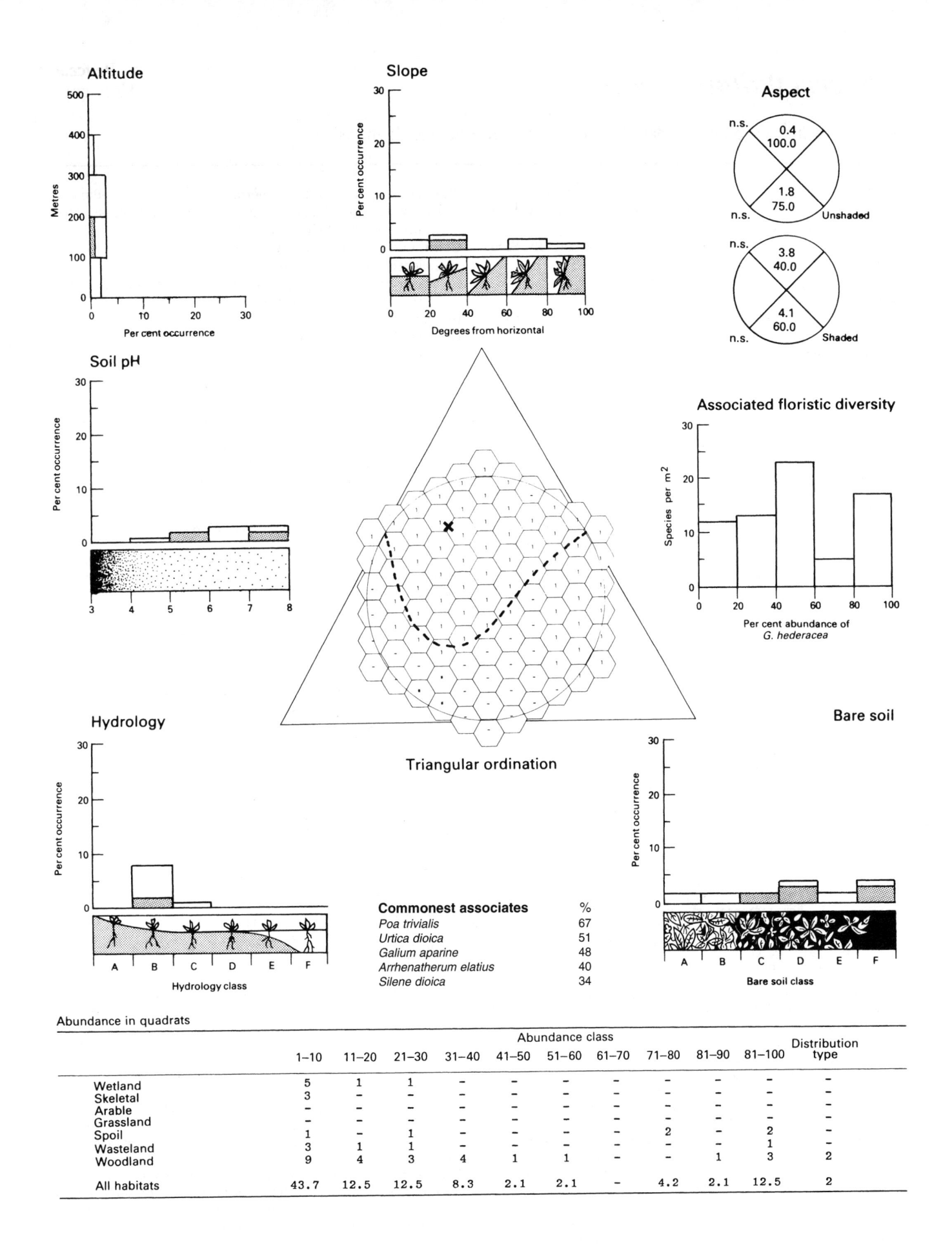

Altitude

Slope

Aspect

	n.s.	0.4 / 100.0	n.s.
	n.s.	1.8 / 75.0	Unshaded

	n.s.	3.8 / 40.0	n.s.
	n.s.	4.1 / 60.0	Shaded

Soil pH

Associated floristic diversity

Species per m² vs Per cent abundance of *G. hederacea*

Hydrology

Hydrology class: A B C D E F

Triangular ordination

Commonest associates	%
Poa trivialis	67
Urtica dioica	51
Galium aparine	48
Arrhenatherum elatius	40
Silene dioica	34

Bare soil

Bare soil class: A B C D E F

Abundance in quadrats

	\multicolumn{10}{c}{Abundance class}	Distribution type
	1–10	11–20	21–30	31–40	41–50	51–60	61–70	71–80	81–90	81–100	
Wetland	5	1	1	–	–	–	–	–	–	–	–
Skeletal	3	–	–	–	–	–	–	–	–	–	–
Arable	–	–	–	–	–	–	–	–	–	–	–
Grassland	–	–	–	–	–	–	–	–	–	–	–
Spoil	1	–	1	–	–	–	–	2	–	2	–
Wasteland	3	1	1	–	–	–	–	–	–	1	–
Woodland	9	4	3	4	1	1	–	–	1	3	2
All habitats	43.7	12.5	12.5	8.3	2.1	2.1	–	4.2	2.1	12.5	2

Glyceria fluitans (L.) R. Br.

Poaceae
Flote-grass

Perhaps including some field data for *Glyceria plicata* Fries (see 'Synopsis')

Life-form Polycarpic perennial helophyte or hydrophyte. Shoots creeping or floating; leaves linear, <2000 mm^2; roots with or without VA mycorrhizas.
Phenology Winter green. Flowers mainly in June and July. Seed shed in August and September.
Height Foliage and flowers to 600 mm.
Seedling RGR 1.0-1.4 week^{-1}. SLA 43.5. DMC 22.2.
Nuclear DNA amount 3.3 pg; $2n = 40*$ (*BC*); $4x*$ (For the tribe the base number is 5.).
Established strategy Competitive-ruderal.
Regenerative strategies V, ?B$_s$. Regeneration most commonly by vegetative means (see 'Synopsis'); ?SB IV.

Flowers Green or purple, hermaphrodite, wind-pollinated; largely self-incompatible (Borrill 1958); >100 in an open panicle (spikelets 8-16-flowered).
Germinule 1.20 mg, 1.8×0.9 mm, caryopsis dispersed with an attached lemma (6.9×1.6 mm).
Germination Epigeal; t_{50} 3 days; 8-34°C; germination rate increased after dry storage at 20°C; L > D.
Biological flora None.
Geographical distribution The British Isles (100% of British vice-counties), *ABIF* page 763; most of Europe except the Arctic (87% of European territories); temperate Asia and N America.

Habitats Exclusively a wetland species, frequently forming floating rafts at the margin of ponds and ditches. Equally common in unshaded or shaded marsh. Less frequent in fast-flowing water, and seldom recorded from moist terrestrial habitats.
Gregariousness Sometimes patch-forming.
Altitude Frequent and abundant at altitudes of up to 400 m, but suitable habitats most common at the upper end of this range. [Up to 670 m.]
Slope Mainly on gently sloping or level ground.
Aspect Insufficient data.

Hydrology Typically occurring in mud and shallow water at the edge of ponds, ditches and, to a lesser extent, streams. Also widespread in topogenous and, to a lesser extent, soligenous mire.
Soil pH Absent from very acidic or highly calcareous soils, and most abundant within the pH range 5.0-7.0.
Bare soil Associated with a wide range of exposures of bare soil and of exposed water.
△ **ordination** Distribution centred on productive vegetation subject to moderate levels of disturbance; also persists, at least temporarily, in little-disturbed, productive vegetation.

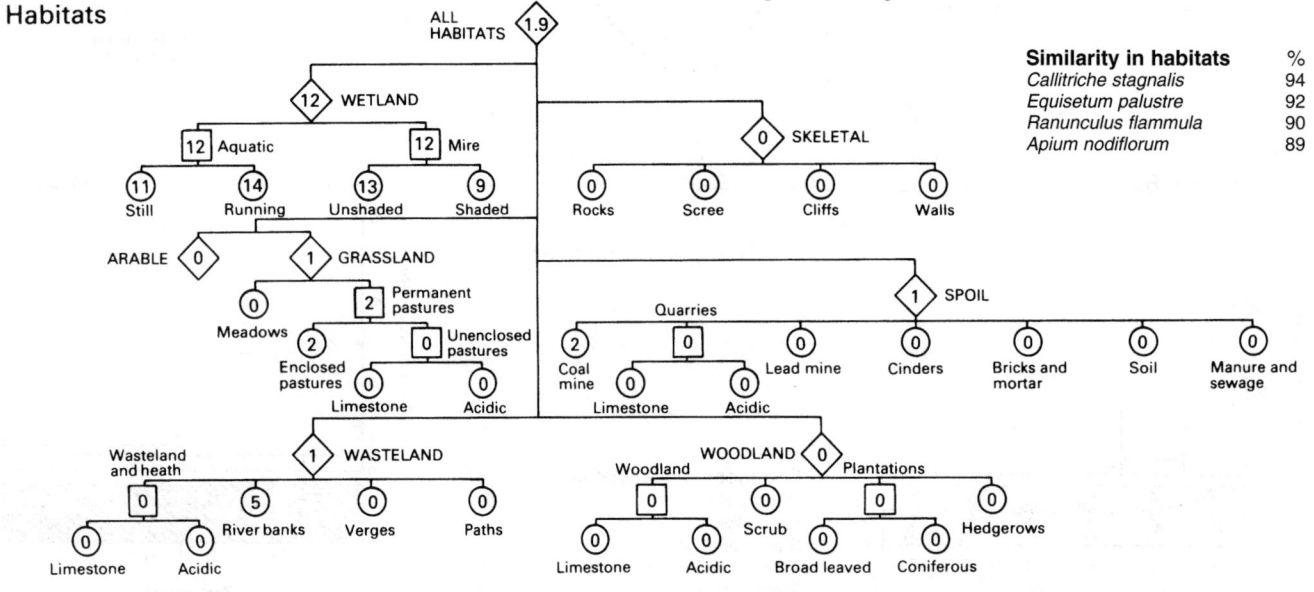

Habitats

Similarity in habitats	%
Callitriche stagnalis	94
Equisetum palustre	92
Ranunculus flammula	90
Apium nodiflorum	89

Synopsis *G.f* is a winter-green stand-forming grass of relatively fertile mire and semi-aquatic habitats. The species is usually rooted in mud, and frequently occurs with the shoots floating on shallow water at the edge of small ponds and drainage ditches. *G.f.* is one of the species which is most capable of exploiting sites where the water table fluctuates extremely widely during the year. Although associated with a wide range of soil conditions, the species is absent from calcareous soils. Both in mud and in water *G.f.* reproduces vegetatively to form extensive patches by means of creeping shoots. Detached shoots readily re-root and may be transported by water and colonize new sites. Although populations of this clonal outbreeding species regularly set seed, regeneration by this means is probably not a regular occurrence. A persistent seed bank is recorded by Chippindale and Milton (1934). The ecological amplitude of *G.f.* is wide, and overlaps with that of a number of other *Glyceria* spp. However, *G.f.* tends to be replaced by the more calcicolous *G. notata* Chevall ($2n = 40$) in base-rich soligenous mire. Hybrids are found in flushes where the species co-occur. The hybrid *G.* × *pedicellata* Townsend ($2n = 40$), which is sterile but vegetatively vigorous, is locally common, in the absence of its parents, beside rivers on the upland carboniferous limestone; this may constitute a habitat which is intermediate in terms of hydrology and base status between those occupied by the parents. *G. declinata* Breb ($2n = 20$) is smaller, does not form extensive stands, and tends to replace *G.f.* in heavily trampled sites and base-poor soligenous mire. *G.d.* rarely hybridizes with *G.f.* (Stace 1975). Unlike *G.f.*, both *G.d.* and *G.n.* are inbreeding (Borrill 1958), and each has smaller seeds than *G.f.* In contrast with those of *G.d.* and *G.f.* seeds of *G.n.* do not exhibit any dormancy. This may facilitate establishment in soligenous mire during autumn.
Current trends Activities such as ditch-making favour *G.f.*, and the species may be increasing in ponds, perhaps as a result of eutrophication. However, destruction of mire has been considerable and, in common with many other wetland plants, the future status of *G.f.* remains uncertain.

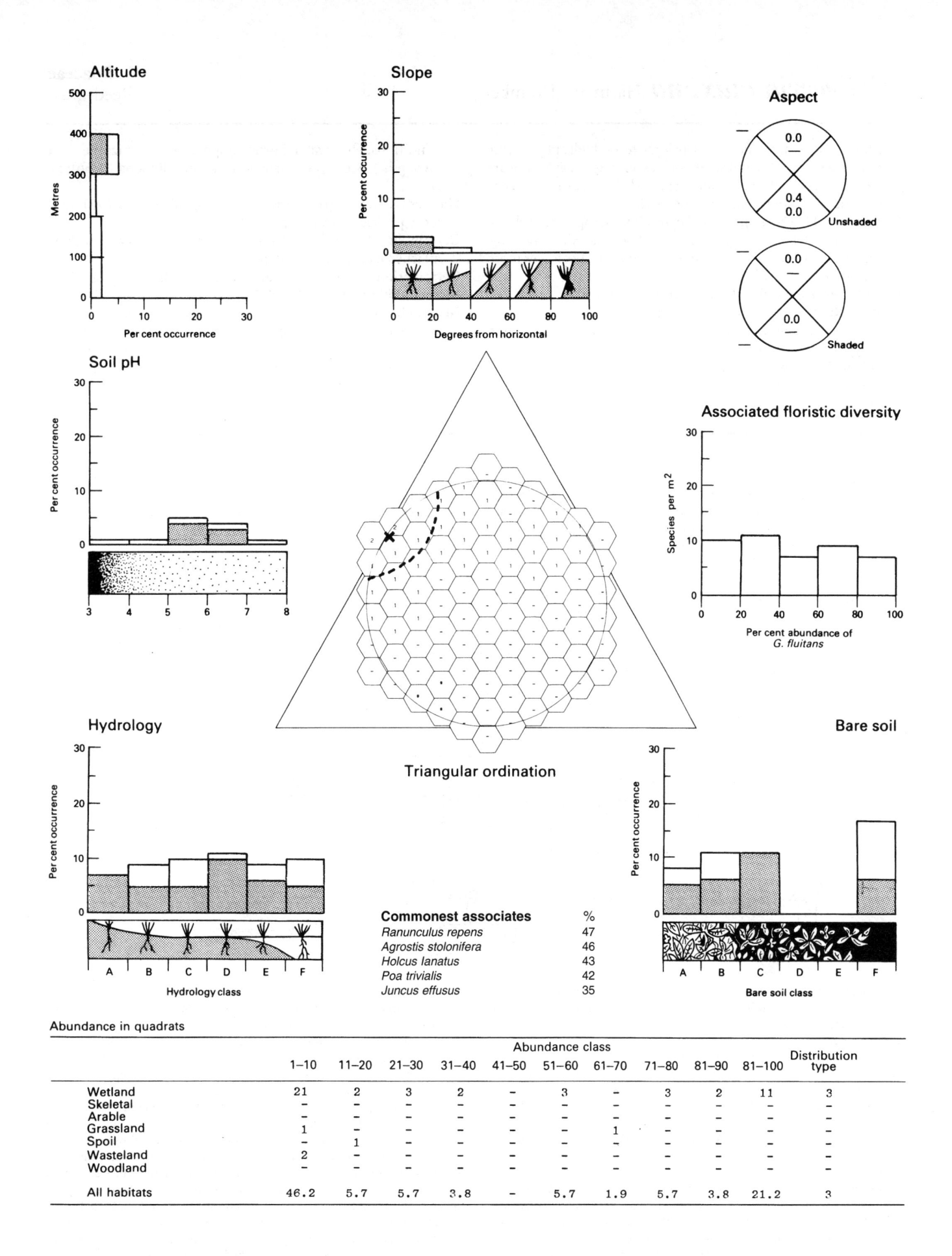

Altitude

Slope

Aspect

Unshaded

Shaded

Soil pH

Associated floristic diversity

Species per m² vs Per cent abundance of *G. fluitans*

Hydrology

Hydrology class: A B C D E F

Triangular ordination

Commonest associates	%
Ranunculus repens	47
Agrostis stolonifera	46
Holcus lanatus	43
Poa trivialis	42
Juncus effusus	35

Bare soil

Bare soil class: A B C D E F

Abundance in quadrats

	\multicolumn{10}{c	}{Abundance class}	Distribution type
	1–10	11–20	21–30	31–40	41–50	51–60	61–70	71–80	81–90	81–100	
Wetland	21	2	3	2	–	3	–	3	2	11	3
Skeletal	–	–	–	–	–	–	–	–	–	–	–
Arable	–	–	–	–	–	–	–	–	–	–	–
Grassland	1	–	–	–	–	–	1	–	–	–	–
Spoil	–	1	–	–	–	–	–	–	–	–	–
Wasteland	2	–	–	–	–	–	–	–	–	–	–
Woodland	–	–	–	–	–	–	–	–	–	–	–
All habitats	46.2	5.7	5.7	3.8	–	5.7	1.9	5.7	3.8	21.2	3

(*)*Glyceria maxima* Hartman (Holmberg)

Poaceae
Reed-grass

Life-form Polycarpic perennial helophyte or hydrophyte with long creeping rhizomes. Shoots ascending; leaves linear, sometimes >4000 mm^2; roots mainly shallow but some deep, providing anchorage; non-mycorrhizal.

Phenology Flowers in July and August, but caryopses ripe about 1 month after flowering (*Biol Fl*) and soon shed. In floating reed-swamp, leaves remain winter green; in erect colonies most leaves die back (*Biol Fl*).

Height Foliage mostly <1 m; flowering shoots to 2 m.

Seedling RGR 1.0-1.4 week^{-1}. SLA 22.1. DMC 21.9.

Nuclear DNA amount 12.2 pg; $2n = 60^*$ (*BC*); $6x^*$. (Genus polyploid; base number of tribe is 5 (*FE5*).)

Established strategy Competitor.

Regenerative strategies V, ?B$_s$. Primarily by means of rhizome extension to form extensive stands and, in disturbed conditions, by water-borne fragments of rhizome (see 'Synopsis'). Seeds germinate in the spring (Roberts 1986b); SB III.

Flowers Green, hermaphrodite, wind-pollinated, >1000 in a loose panicle (8-16 per spikelet).

Germinule 0.74 mg, 1.6 × 1.0 mm, caryopsis dispersed with an attached lemma (3.0 × 1.0 mm).

Germination Epigeal, dormancy broken by dry storage at 20°C.

Biological flora Lambert (1947).

Geographical distribution Most of the British Isles, but most abundant in lowland areas, rare in Scotland (73% of British vice-counties), *ABIF* page 763; N temperate zone of Europe (72% of European territories); Asia; sown commercially in S Africa and N America.

Habitats Exclusively a wetland species. Frequent in unshaded mire beside canals, ditches, lakes and ponds, and at the edge of open water, either rooted on the bank or in the water itself. Following drainage, may persist as a relict of former wetland in dryland habitats. Observed in lightly shaded sites. [Extends into slightly brackish water and even into sites which may rarely undergo inundation by sea water (*Biol Fl*).]

Gregariousness Often stand-forming.

Altitude A lowland species which is most frequent and abundant at altitudes of less than 150 m, a value given by Wilson (1956) as its British maximum.

Slope Distribution reflects the gentle slopes generally associated with wetland habitats.

Aspect Insufficient data.

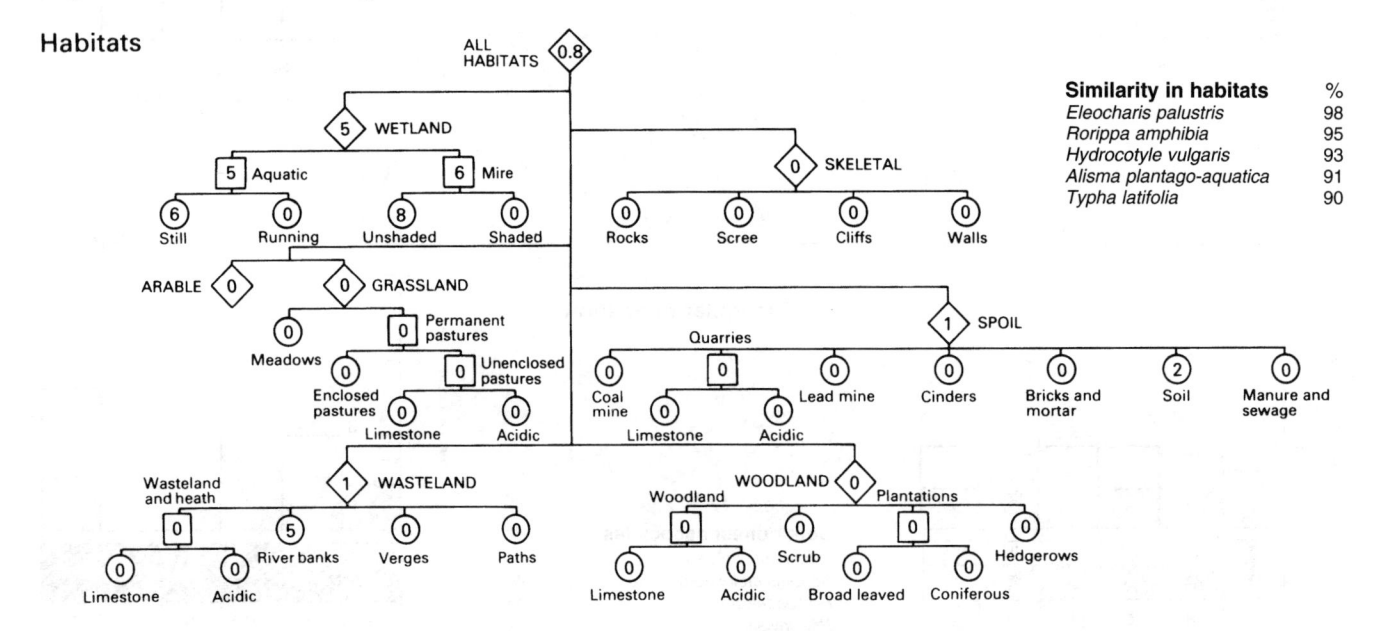

Habitats

Similarity in habitats	%
Eleocharis palustris	98
Rorippa amphibia	95
Hydrocotyle vulgaris	93
Alisma plantago-aquatica	91
Typha latifolia	90

Hydrology Tolerant of a range of hydrological conditions, but most frequent in topogenous mire beside open water. Also commonly rooted in shallow water and can spread from the margin to form floating rafts over deeper water.

Soil pH Only recorded within the pH range 5.0-7.5.

Bare soil Associated with widely differing amounts of bare soil, persistent litter and exposed water surface.

Δ ordination Confined to vegetation associated with fertile and infrequently-disturbed habitats.

Associated floristic diversity Variable, depending on site disturbance, but usually low. At very high values of *G.m.* competitive exclusion of other species is indicated.

Synopsis *G.m.* is a tall robust grass producing a dense canopy and often forming a virtual monoculture as a result of expansion of an extensive and close-packed rhizome system. *G.m.* is highly aerenchymatous (*Biol Fl*) and is characteristic of waterlogged alluvial soils of high fertility, particularly in sites close to slow-flowing water. The species may also form floating rafts of prostrate stems which can choke canals. *G.m.* may compete successfully with *Phragmites australis* under productive conditions, since it produces a dense cover in spring before the development of *P.a.* (Buttery & Lambert 1965), a phenomenon consistent with the higher nuclear DNA amount of *G.m.*. However, factors such as increased anaerobiosis may tilt the balance in favour of *P.a.* (Buttery *et al.* 1965). *G.m.* is a lowland species mainly found below 150 m. The species is probably restricted in part by the scarcity and fragmented nature of suitable habitats in upland areas, but has been observed as a persistent introduction at 290 m. *G.m.* shows a low level of morphological variation and a modest ecological amplitude (*Biol Fl*). *G.m.* is palatable to stock when young and was formerly cultivated as a fodder crop (*Biol Fl*). It is tolerant of cutting, more so than *P.a.* and its relative abundance may be increased by mowing (*Biol Fl*). Seed set is poor and establishment from the seeds which persist in the soil (Roberts 1986b) appears to be a rare event. Thus, *G.m.* is largely dependent upon vegetative means for regeneration. The species shows aggressive rhizome growth to form large clonal stands and detached rhizome pieces, which float,

Altitude

Slope

Aspect

Unshaded

Shaded

Soil pH

Associated floristic diversity

Hydrology

Triangular ordination

Bare soil

Commonest associates	%
Lemna minor	52
Galium palustre	29
Agrostis stolonifera	24
Holcus lanatus	24
Myosotis scorpioides	24

Abundance in quadrats

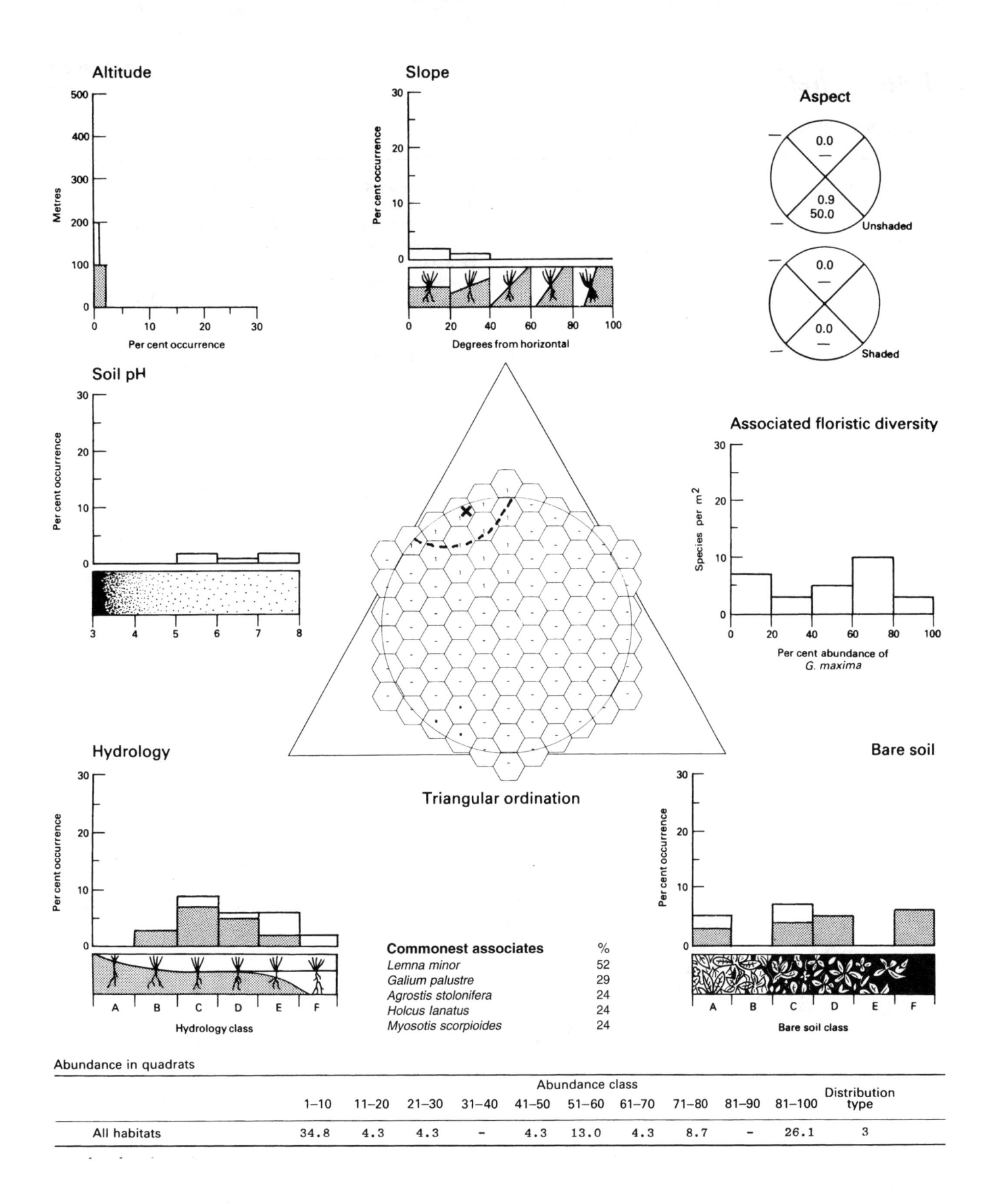

	1–10	11–20	21–30	31–40	41–50	51–60	61–70	71–80	81–90	81–100	Distribution type
						Abundance class					
All habitats	34.8	4.3	4.3	–	4.3	13.0	4.3	8.7	–	26.1	3

regenerate readily to form new colonies. By these means whole water systems may be colonized. However, such dependence upon large, water-borne vegetative propagules for regeneration severely limits the capacity of G.m. to colonize land-locked sites. Thus, G.m. may be abundant in a canal system but absent from nearby ponds, and we suspect that in many of its more isolated sites G.m. is an ornamental introduction.

Current trends Appears to have increased, apparently at the expense of P.a., in the ditches beside arable land in lowland areas, and has choked a number of disused canals. However, since much wetland has been drained, on balance G.m. is probably decreasing.

Hedera helix L.

Araliaceae
Ivy

Taxonomy under revision (see 'Synopsis')

Life-form Woody, perennial phanerophyte or chamaephyte. Stems long, either creeping or climbing, attached to the substrate by numerous small roots, which bear VA mycorrhizas. Leaves evergreen, heterophyllous, mostly palmately lobed and <5000 mm^2; those on flowering branches ovate and larger.
Phenology Evergreen. Flowers September to November; berries ripen April to June the following year. Leaves long-lived.
Height Foliage <150 mm in creeping (sterile) stems. Climbing stems on tree trunks and walls bear foliage and flowers to 30 m.
Seedling RGR 0.5–0.9 week^{-1}. SLA 14.39. DMC 32.1.
Nuclear DNA amount 3.6 pg; $2n = 48^*$; $?2x^*$ (generic base number possibly $n = 12$).

Established strategy Stress-tolerant competitor.
Regenerative strategies V, S. Has the capacity for extensive vegetative spread. Establishment from seed appears to be infrequent (see 'Synopsis').
Flowers Yellow-green, hermaphrodite, sometimes protandrous, insect-pollinated; often *c.* 20 in terminal umbels.
Germinule 20.43 mg, 5.7 × 3.7 mm, dispersed in a black 2-3-seeded berry.
Germination Epigeal; t_{50} 6 days; L < D.
Biological flora Metcalfe (2005).
Geographical distribution For aggregate species: the British Isles (100% of British vice-counties), *ABIF* page 453; W, C and S Europe extending as far N as Norway (85% of European territories); Asia Minor.

Habitats Most characteristic of shaded habitats, and commonly occurring in woodlands and hedgerows, either carpeting the ground or growing vertically up the trunks of trees. Also common on cliffs, walls and rock outcrops. Less frequent on river banks, road verges, path sides and spoil heaps. Not found on agricultural land, and virtually absent from wetlands. [On rocks and walls near the sea, but absent from the most exposed sites; on sheltered rocks in submontane areas.]

Gregariousness Carpet-forming.
Altitude Observed to 340 m. [Up to 610 m.]
Slope Found over the full range of slopes.
Aspect Slightly more frequent and abundant on N-facing slopes.
Hydrology A species of moist rather than waterlogged soils.
Soil pH Most frequent and abundant above pH 6.0, but extending onto mildly acidic soils though seldom found below pH 4.0.
Bare soil Wide-ranging.
△ ordination Distribution centred on vegetation associated with moderately fertile, undisturbed habitats. Absent from highly disturbed vegetation.

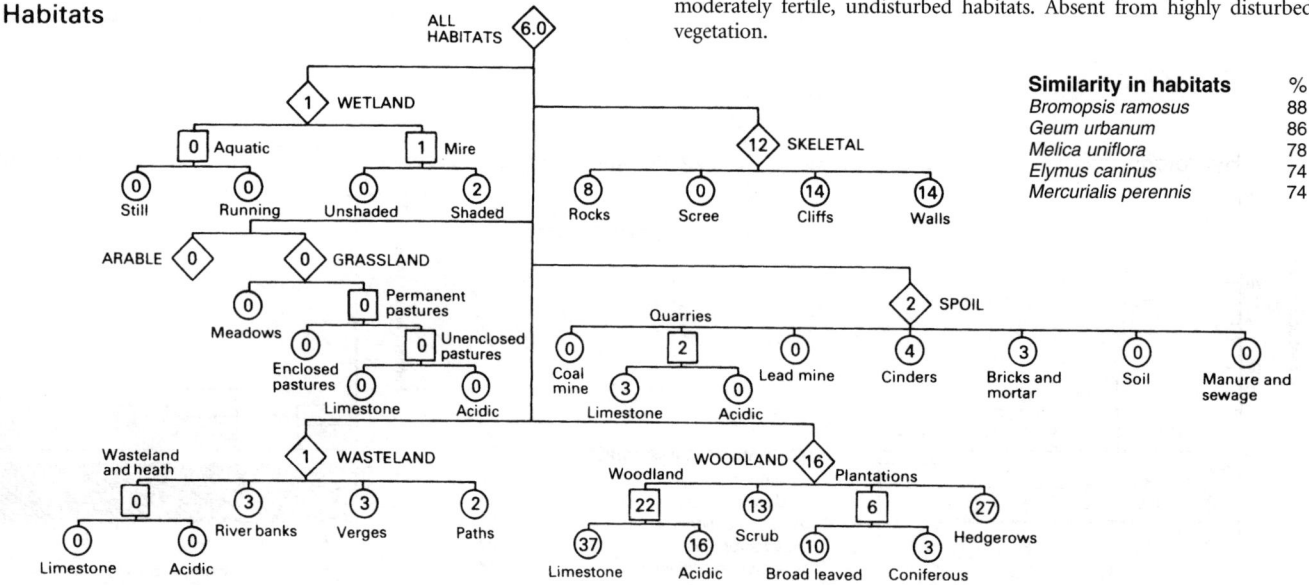

Similarity in habitats	%
Bromopsis ramosus	88
Geum urbanum	86
Melica uniflora	78
Elymus caninus	74
Mercurialis perennis	74

Habitats

Associated floristic diversity Typically low.
Synopsis *H.h.* is a long-lived (to 400 years (Rose 1980)), evergreen species. Its long woody stems grow either vertically up tree trunks, cliffs or walls, attached by means of numerous short roots, or horizontally to form continuous carpets on the woodland floor. Its capacity to extend over soil-less habitats from a base rooted locally in soil is unique within the British flora. The vertical stems are vital to the reproductive capacity of the species, since they alone bear flowering shoots and do so only in relatively unshaded sites. Locally, plants in woodland or hedgerows seldom escape the shade of their accompanying trees and shrubs, and so rarely flower. Thus, like *Lonicera periclymenum* and *Tamus communis*, *H.h.* mainly sets seed in habitats which are less shaded than those in which it occurs most abundantly. The large, berried seeds are dispersed by birds, and germination probably occurs mainly in spring, though very young seedlings can be observed during winter. The seed is short-lived and does not form a persistent seed bank. Flowers which are produced in late autumn are susceptible to the effects of frost. Problems of initial colonization apppear to limit the woodland distribution of *H.h.*, and the species is characteristic of secondary rather than ancient woodland (Rackham 1980, Peterken 1981). Thus, colonies frequently radiate inwards from the woodland edge or are found on areas of former quarrying. Rackham (1980) suggests that invasion is associated with hawthorn succession from open ground. The possibility that *H.h.* requires unvegetated, relatively unshaded sites for establishment from seed requires investigation. However, once established in woodland *H.h.* can form extensive clonal patches by means of rooted horizontal stems. *H.h.* is also often abundant in woodland close to active limestone quarries, apparently due to its resistance to the effects of limestone dust. *H.h.* is heterophyllous

Altitude

Slope

Aspect

n.s. 1.2 / 66.7 n.s. / Unshaded

n.s 22.7 / 70.0 n.s / 15.4 / 68.4 Shaded

P < 0.001

Soil pH

Associated floristic diversity

Hydrology

Bare soil

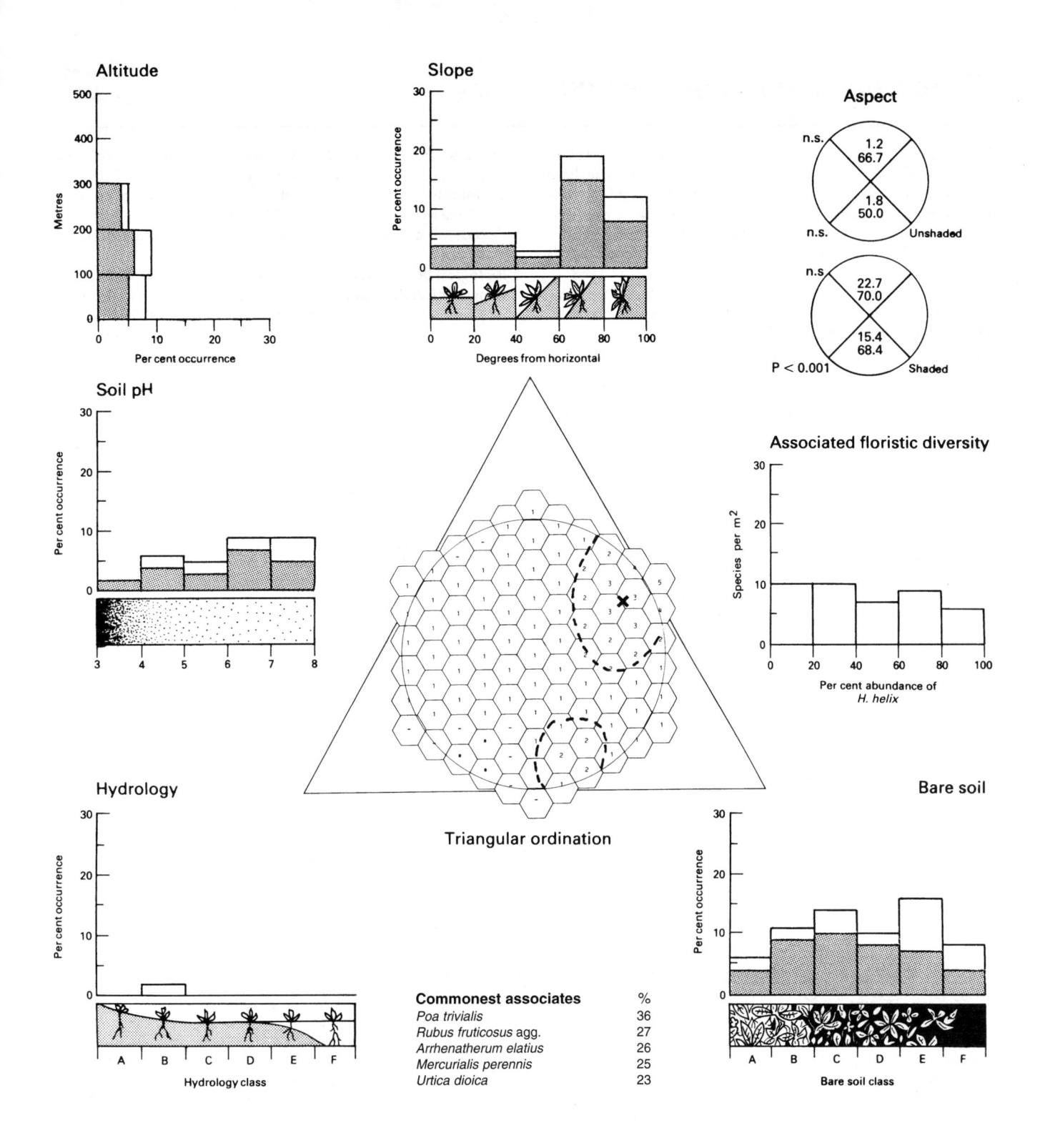

Triangular ordination

Commonest associates	%
Poa trivialis	36
Rubus fruticosus agg.	27
Arrhenatherum elatius	26
Mercurialis perennis	25
Urtica dioica	23

with palmately lobed 'shade leaves' except on flowering branches where larger, ovate sun-leaves occur. These two types of shoot have very different characteristics. 'Juvenile' shoots, with shade leaves, root readily but do not flower; in contrast, 'mature' shoots, with sun-leaves can be rooted only with great difficulty (Zimmermann et al. 1985). Ivy is the only British member of the predominantly tropical family Araliaceae, and is susceptible to low winter temperatures (Godwin 1975). For this reason the species has been used by palaeobotanists as an indicator of mild winters (Godwin 1975). *H.h.* was gathered as fodder from Neolithic times until the 16th century (Rackham 1980). However, the plant is toxic if consumed in quantity (Cooper & Johnson 1984) and is rejected by gastropods (Grime, Blythe & Thornton 1970). Foliage of *H.h.* appears to be subject to extremely low rates of herbivory, and provides excellent cover for some early-nesting birds. McAllister (1979) recognizes two taxa, *H. helix* and *H. hibernica*. The latter, with $2n = 96$ ($4x$), replaces *H.h.* in SW coastal regions of the British Isles, Ireland and the Iberian peninsula. Two further ssp. are recognized from S Europe (*FE2*).

Current trends Uncertain. Favoured by artificial disturbances, but some habitats, e.g. dead standing trees and old walls, are less frequent than formerly.

Helianthemum nummularium (L.) Miller

Cistaceae
Common Rockrose

Life-form Shrubby, mat-forming, polycarpic perennial chamaephyte. Stems branched, woody; leaves lanceolate, hairy, leathery, often <10 mm^2; tap-rooted and with VA mycorrhizas.

Phenology Evergreen. Shoots never completely dormant (*Biol Fl*). Flowers June and July and seed shed in July and August.

Height Foliage usually <200 mm high, overtopped by the inflorescence.

Seedling RGR 0.5–0.9 week^{-1}. SLA 17.3. DMC 25.3.

Nuclear DNA amount 4.5 pg; $2n = 20^*$ (*BC*); $2x^*$.

Established strategy Stress-tolerator.

Regenerative strategies ?B$_s$. See 'Synopsis'; ?SB IV.

Flowers Yellow, usually hermaphrodite, insect- or self-pollinated; up to 12 in a terminal cyme (*Biol Fl*).

Germinule 1.38 mg, 1.8 × 1.0 mm, seed dispersed from a 10–20-seeded dehiscent capsule (*Biol Fl*).

Germination Epigeal; t_{50} 2 days; <5–36°C, hard-coat dormancy; L = D.

Biological flora Proctor (1956) (as *H. chamaecistus* Mill.).

Geographical distribution Widespread, particularly in the S and SE of England, perhaps introduced in its one Irish locality 61% of British vice-counties), *ABIF* page 223; most of Europe except the extreme N (85% of European territories).

Habitats Most abundant in dry calcareous pastures and on stabilized scree slopes. Also recorded on old lead-mine heaps, rock outcrops, calcareous waste ground and in disused limestone quarries. [Also occurs on calcareous brown sand, base-rich, well-drained river shingle and even on montane heath (*Biol Fl*).]

Gregariousness Intermediate.

Altitude Extends to 360 m, but most abundant between 200 and 300 m, a reflection of its abundance on the carboniferous limestone of Derbyshire. [Up to 660 m. (*Biol Fl*).]

Slope Mainly restricted to limestone dale-sides, and therefore particularly frequent and abundant on moderate slopes.

Aspect Recorded more frequently on unshaded N-facing slopes, but abundance significantly greater on S-facing slopes. Found to be more frequent on S-facing slopes in a previous grassland survey (*Ecol Atl*).

Hydrology Absent from wetlands.

Soil pH Shows a progressive increase in frequency and abundance with increasing pH from 4.5 to 8.0. [Can occur at pH 3.8 or less in Scotland (*Biol Fl*).]

Bare soil Occurs in association with low to moderate amounts of exposed soil or bare rock.

△ ordination Distribution strongly centred on vegetation associated with relatively undisturbed, infertile habitats.

Associated floristic diversity Typically species-rich.

Synopsis *H.n.* is a slow-growing, prostrate, evergreen undershrub most typically associated with species-rich, short turf on well-drained, infertile soils. *H.n.* is strongly calcicolous in SE England and in N Derbyshire.

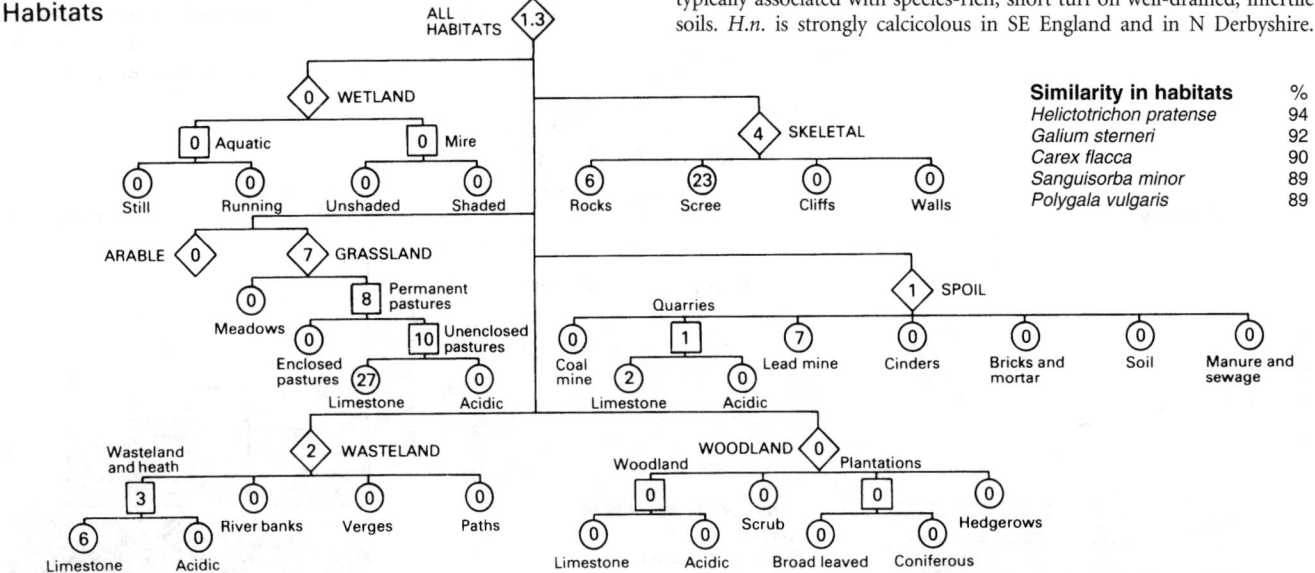

Habitats

Similarity in habitats	%
Helictotrichon pratense	94
Galium sterneri	92
Carex flacca	90
Sanguisorba minor	89
Polygala vulgaris	89

However, in E Scotland *H.n.* is characteristic of dry *Agrostis-Festuca* grasslands and may even occur with *Calluna vulgaris* and other calcifuges on soils of pH 3.8 (*Biol Fl*). This pattern of distribution with respect to soil reaction is similar to, though less extreme than, that shown by *Thymus polytrichus*. Leaves of plants established on calcareous soils in the Sheffield region have high Ca and Al concentrations (*SIEF*). On the southern chalk *H.n.* may be particularly frequent on ant hills (King 1977a-c). The root systems of plants examined in chalk downland are described as intermediate in depth (Anderson 1927), but our own observations on fissured limestone suggest that the tap-root penetrates deeply and facilitates drought-avoidance. The species also exhibits some desiccation tolerance involving sclerophylly, and under extreme drought *H.n.* has the capacity to shed leaves (P. J. Grubb pers. comm.). The adult plant is also frost-resistant. *H.n.* can persist under moderate trampling and grazing, but cannot compete successfully in taller grasslands of the type which develops in response to dereliction or fertilizer application. Benefits from disturbance also arise from the creation of local areas of bare soil necessary for seedling establishment, which is more important than the vegetative expansion which occurs by the layering of old branches (*Biol Fl*). Seeds have hard-coat dormancy, but further research is required to recognize the mechanisms which break dormancy under natural conditions and to examine the persistence of the seed bank. Germination tends to occur in spring, at least in N England, and seedlings are susceptible to droughting. Seeds, which are mucilaginous when moistened, appear to be poorly dispersed. As a result *H.n.* is restricted to old or semi-natural habitats. This low dispersability may have contributed to the failure of *H.n.* to colonize Ireland and to its replacement in the Burren by *H. oelandicum* (L.) Dum. Cours. which, elsewhere in Britain, is a species of xeric grassland (Proctor 1958). Some morphological variation exists and appears to be genetic in origin; a number of ecotypes have been identified (*Biol Fl*). The existence of more-extreme morphological discontinuities allow the recognition of 7 additional ssp. within the mountainous regions of C and S Europe (*FE2*). *H.n.* also hybridizes with *H.o.* (Stace 1975).

Current trends Decreasing and close to extinction in some intensively-managed lowland areas, in which *H.n.* is becoming increasingly restricted to local refugia such as rock outcrops.

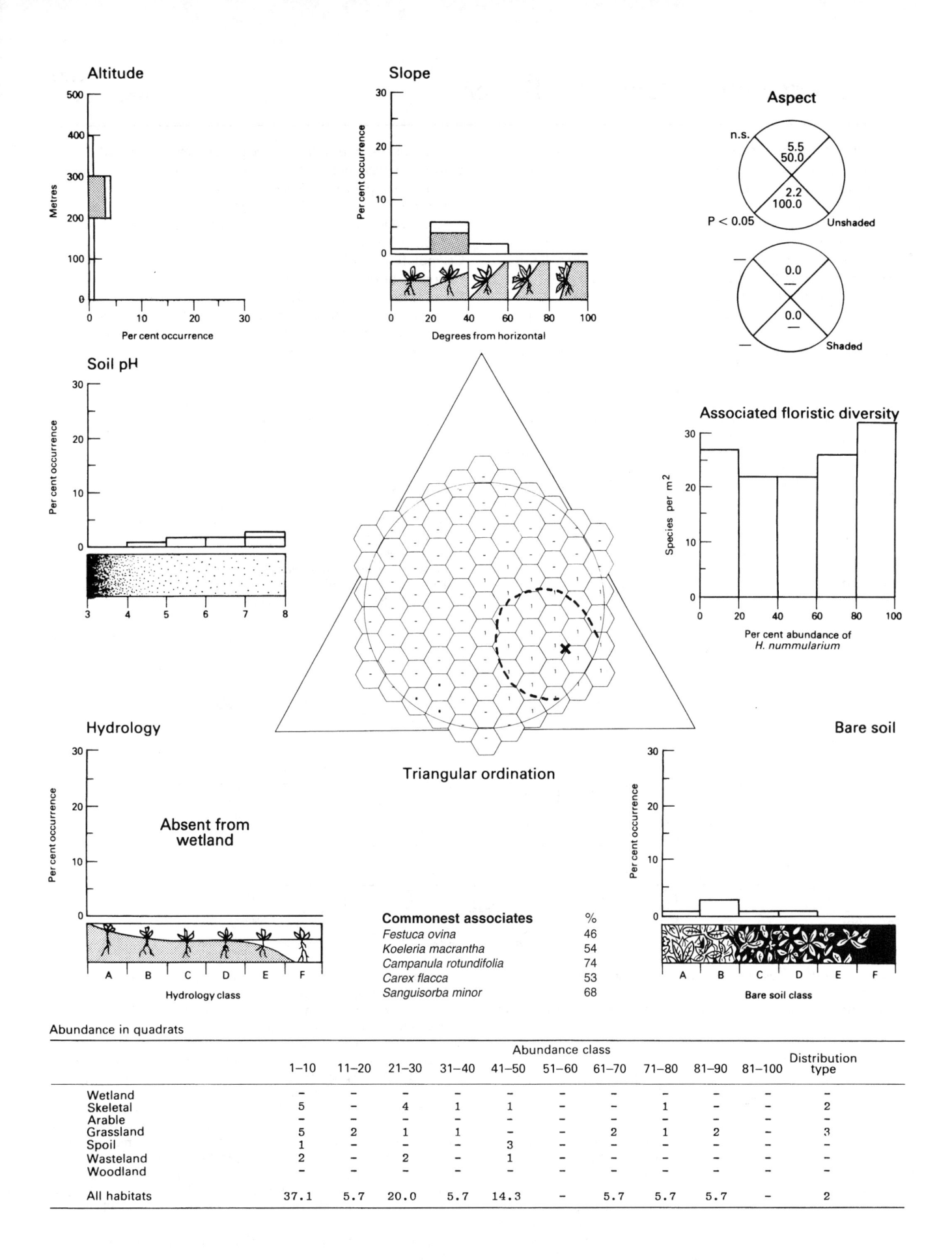

Altitude

Slope

Aspect

n.s.

5.5	
50.0	
2.2	
100.0	

$P < 0.05$ Unshaded

0.0	
—	
0.0	
—	

— Shaded

Soil pH

Associated floristic diversity

Hydrology

Absent from wetland

Triangular ordination

Bare soil

Commonest associates %
Festuca ovina 46
Koeleria macrantha 54
Campanula rotundifolia 74
Carex flacca 53
Sanguisorba minor 68

Abundance in quadrats

	\multicolumn{10}{c	}{Abundance class}	Distribution type
	1–10	11–20	21–30	31–40	41–50	51–60	61–70	71–80	81–90	81–100	
Wetland	–	–	–	–	–	–	–	–	–	–	–
Skeletal	5	–	4	1	1	–	–	1	–	–	2
Arable	–	–	–	–	–	–	–	–	–	–	–
Grassland	5	2	1	1	–	–	2	1	2	–	3
Spoil	1	–	–	–	3	–	–	–	–	–	–
Wasteland	2	–	2	–	1	–	–	–	–	–	–
Woodland	–	–	–	–	–	–	–	–	–	–	–
All habitats	37.1	5.7	20.0	5.7	14.3	–	5.7	5.7	5.7	–	2

Helictotrichon pratense (L.) Besser.

Poaceae
Meadow Oat

Life-form Densely tufted, polycarpic perennial, semi-rosette hemicryptophyte. Shoots erect; leaves tough, linear, usually <2000 mm^2; roots with VA mycorrhizas.
Phenology Winter green. Rapid rates of leaf expansion occur in early spring. Flowers in June and seed shed in July to August. Leaves relatively long-lived.
Height Foliage usually <300 mm; flowering stems to 600 mm.
Seedling RGR 0.5-0.9 week^{-1}. SLA 10.4. DMC 32.3.
Nuclear DNA amount 35.9 pg; $2n = 126*$ (*BC*); $18x*$.
Established strategy Intermediate between stress-tolerant competitor and C-S-R.
Regenerative strategies S. Primarily by seed freshly shed in autumn. Plants increase in size through the production of additional shoots to form large but discrete clumps; SB I.
Flowers Green, hermaphrodite, wind-pollinated; often <50 in a contracted panicle (3-6 per spikelet).
Germinule 2.08 mg, 5.2 × 1.3 mm, caryopsis dispersed with an attached lemma, an hygroscopic bristle and a basal tuft of hairs (18.5 × 1.6 mm).
Germination Epigeal; t_{50} 7 days; 8-26°C; L = D.
Biological flora Dixon (1991).
Geographical distribution A European endemic found throughout much of the British Isles but absent from Ireland (54% of British vice-counties), *ABIF* page 765; N and C Europe (51% of European territories).

Habitats Restricted to dry calcareous habitats, particularly semi-natural, calcareous pastures. Present also in derelict grassland and on scree slopes, rock outcrops and cliff ledges. Recorded locally on spoil heaps of old lead mines, but absent from areas of highly toxic material. Only infrequently observed in shaded habitats. [Found in a range of base-rich habitats in montane areas (McVean & Ratcliffe 1962). However, further taxonomic studies may lead to the classification of these and other montane populations of *H.p.* as a separate species (*FE5*).]
Gregariousness Intermediate.

Altitude Occurs in both upland and lowland areas, but most records within the range 200-300 m corresponding to the distribution of the carboniferous limestone in the survey area. [Up to 840 m for the montane variant.]
Slope Since *H.p.* is most frequent in the grasslands of the Derbyshire dale sides, most records are from slopes in the range 20-40°.
Aspect Slightly more frequent and abundant on N-facing slopes, but in a previous survey of older grasslands the abundance was found to be greater on S-facing slopes (*Ecol Atl*).
Hydrology Absent from wetlands.
Soil pH Common only on soils of pH > 4.5, and most frequently associated with soils of high base-status.

Habitats

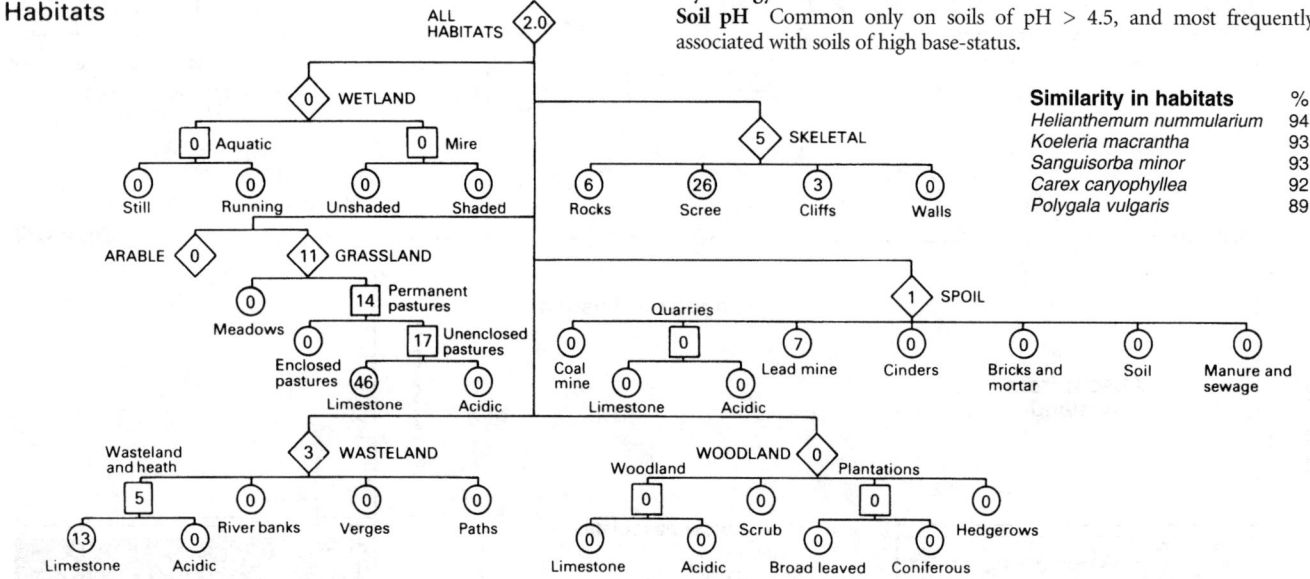

Similarity in habitats	%
Helianthemum nummularium	94
Koeleria macrantha	93
Sanguisorba minor	93
Carex caryophyllea	92
Polygala vulgaris	89

Bare soil Most records are from habitats having low or intermediate amounts of exposed soil.
△ **ordination** Distribution confined to vegetation types associated with infertile undisturbed sites.
Associated floristic diversity *H.p.* is typically associated with species-rich vegetation.
Synopsis *H.p.* is a potentially robust but slow-growing, tufted, wintergreen grass restricted to infertile, calcareous grassland and rocky habitats. *H.p.* shows some ecological affinities with calcareous races of *Festuca ovina* with which it is often associated. *H.p.* shares several attributes with *Festuca ovina* and *Koeleria macrantha*, including (a) high leaf extension rate in early spring, which in the case of *H.p.* is predictable from the exceptionally high value for nuclear DNA amount, (b) a short, and early, flowering period, (c) a dependence upon regeneration by seed, which germinate in autumn, soon after dispersal, (d) a wide temperature range for seed germination and (e) the absence of a persistent seed bank. However, *H.p.* is potentially taller and more robust than either of the other two species, and has a greater capacity to dominate derelict limestone grassland where, as in many other situations inaccessible to grazing animals, *H.p.* can develop large winter-green tussocks with long spreading leaves. Compared with many other grasses, including *F.o.*, *H.p.* suffers less damage during fires, probably because meristems near the centre of tussocks are insulated by green leaf tissue (Lloyd 1972a). In grazed sites *H.p.* is reduced in stature and remains a relatively inconspicuous component. We suspect that its leaves are more palatable to sheep than those of *F.o.* Foliar concentrations of N and Ca are relatively low (*SIEF*). *H.p.* appears to have low mobility and is restricted to grasslands of some antiquity (Wells *et al.* 1976) and absent from Ireland. It is one of only four common calcicolous species which are strictly confined to limestone strata within the Sheffield area, the other three being *Helianthemum nummularium*, *Scabiosa columbaria* and *Viola hirta*. The full extent of ecotypic differentiation within *H.p.* remains to be investigated although montane populations with larger spikelets may be taxonomically distinct (*FE5*).
Current trends Already close to extinction in many lowland areas and likely to decrease further as a result of habitat destruction.

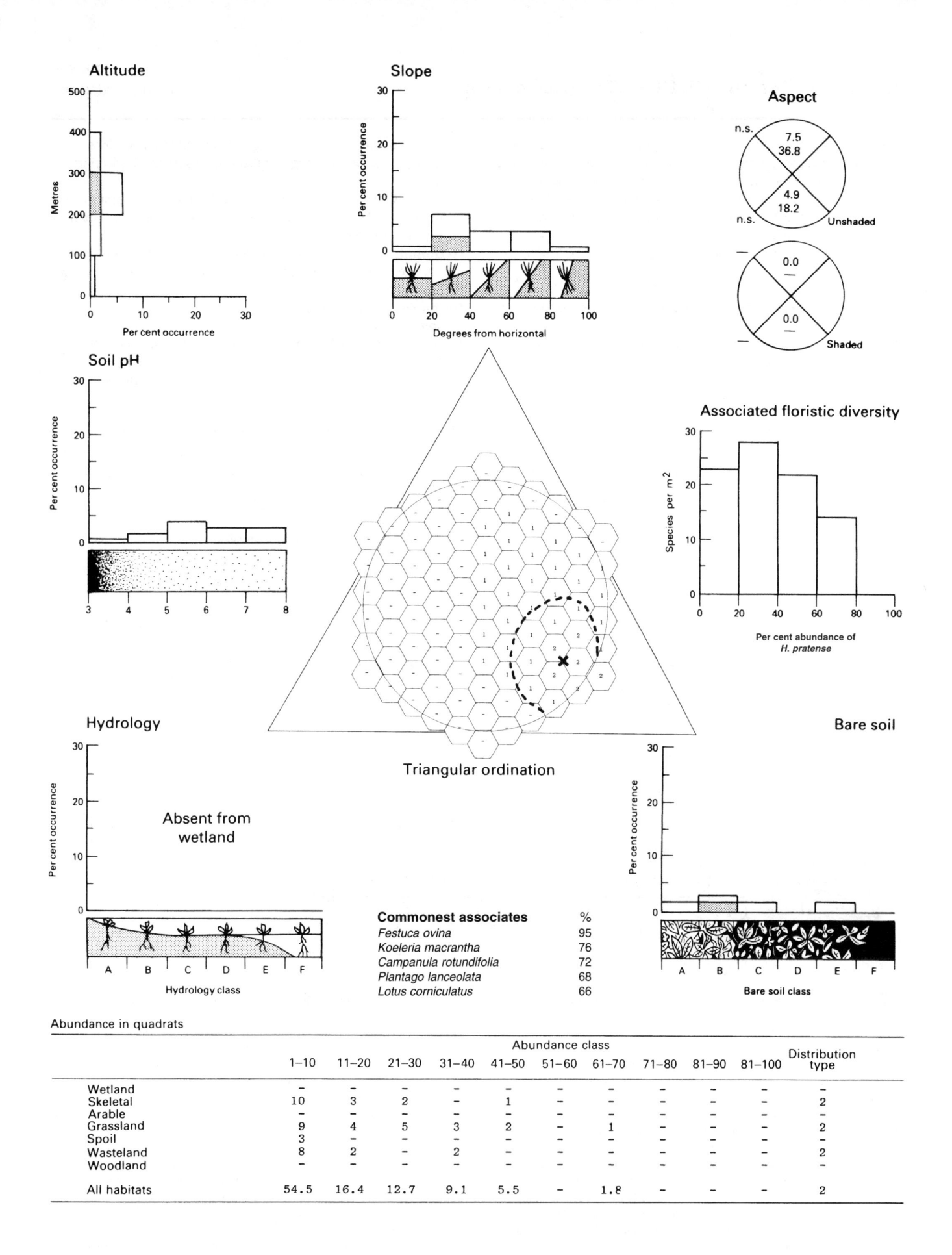

Altitude
(histogram: Metres vs Per cent occurrence)

Slope
(histogram: Per cent occurrence vs Degrees from horizontal)

Aspect
n.s. 7.5 / 36.8 / 4.9 / 18.2 n.s. Unshaded
0.0 / — / 0.0 / — Shaded

Soil pH
(histogram: Per cent occurrence vs pH 3–8)

Associated floristic diversity
(histogram: Species per m² vs Per cent abundance of *H. pratense*)

Triangular ordination

Hydrology
Absent from wetland
(Hydrology class A–F)

Bare soil
(Bare soil class A–F)

Commonest associates	%
Festuca ovina	95
Koeleria macrantha	76
Campanula rotundifolia	72
Plantago lanceolata	68
Lotus corniculatus	66

Abundance in quadrats

	1–10	11–20	21–30	31–40	41–50	51–60	61–70	71–80	81–90	81–100	Distribution type
Wetland	–	–	–	–	–	–	–	–	–	–	–
Skeletal	10	3	2	–	1	–	–	–	–	–	2
Arable	–	–	–	–	–	–	–	–	–	–	–
Grassland	9	4	5	3	2	–	1	–	–	–	2
Spoil	3	–	–	–	–	–	–	–	–	–	–
Wasteland	8	2	–	2	–	–	–	–	–	–	2
Woodland	–	–	–	–	–	–	–	–	–	–	–
All habitats	54.5	16.4	12.7	9.1	5.5	–	1.8	–	–	–	2

Helictotrichon pubescens (Hudson) Pilg.

Poaceae
Hairy Oat-grass

Life-form Loosely tufted, polycarpic perennial protohemicryptophyte. Shoots erect or spreading; leaves linear, softly hairy <2000 mm², root system with VA mycorrhizas.

Phenology Winter green. Leaf expansion occurs in early spring. Flowers in June and July and seed shed mainly in July. Leaves relatively long-lived.

Height Foliage usually <300 mm; flowering stems to 700 mm.

Seedling RGR No data. SLA 23.4. DMC 26.4.

Nuclear DNA amount 10.1 pg; $2n = 14^*$ (*BC*); $2x^*$.

Established strategy Intermediate between stress-tolerator and C-S-R.

Regenerative strategies S, (V). Principally by freshly-shed seed in autumn, but some seed may persist on the soil surface and germinate in spring. *H.p.* has short rhizomes which give rise to diffuse patches. Rhizome connections between shoots may break, making daughter plants nutritionally independent, but it is uncertain whether regeneration by this method is important; SB I.

Flowers Green, hermaphrodite, wind-pollinated, possibly self-incompatible (Weimarck 1968); >100 in a contracted panicle (2-3 per spikelet).

Germinule 1.92 mg, 5.5 × 0.9 mm, caryopsis dispersed with an attached lemma, an hygroscopic bristle and a basal tuft of hairs (14.9 × 1.1 mm).

Germination Epigeal; t_{50} 5 days; 11-30°C.

Biological flora Dixon (1991).

Geographical distribution Most of the British Isles (95% of British vice-counties), *ABIF* page 765; N, C and E Europe (72% of European territories); temperate Asia; introduced into N America.

Habitats Common and locally abundant on calcareous strata in a range of grassy habitats, particularly pasture and wasteland. Also found locally in meadows, on grass verges, and on spoil heaps from disused limestone quarries and lead mines. Less frequent elsewhere, and mainly confined to road verges, on non-calcareous strata.

Gregariousness Intermediate.

Altitude Suitable habitats are most abundant on the carboniferous limestone, and in consequence *H.p.* is mainly found at higher altitudes, particularly within the range 200-300 m. Occasionally extending to 420 m. [Up to 520 m.]

Slope Found on both flat and moderately sloping ground, but absent from the steepest inclines.

Aspect Rather more records and greater abundance on unshaded N-facing slopes.

Hydrology Absent from wetlands.

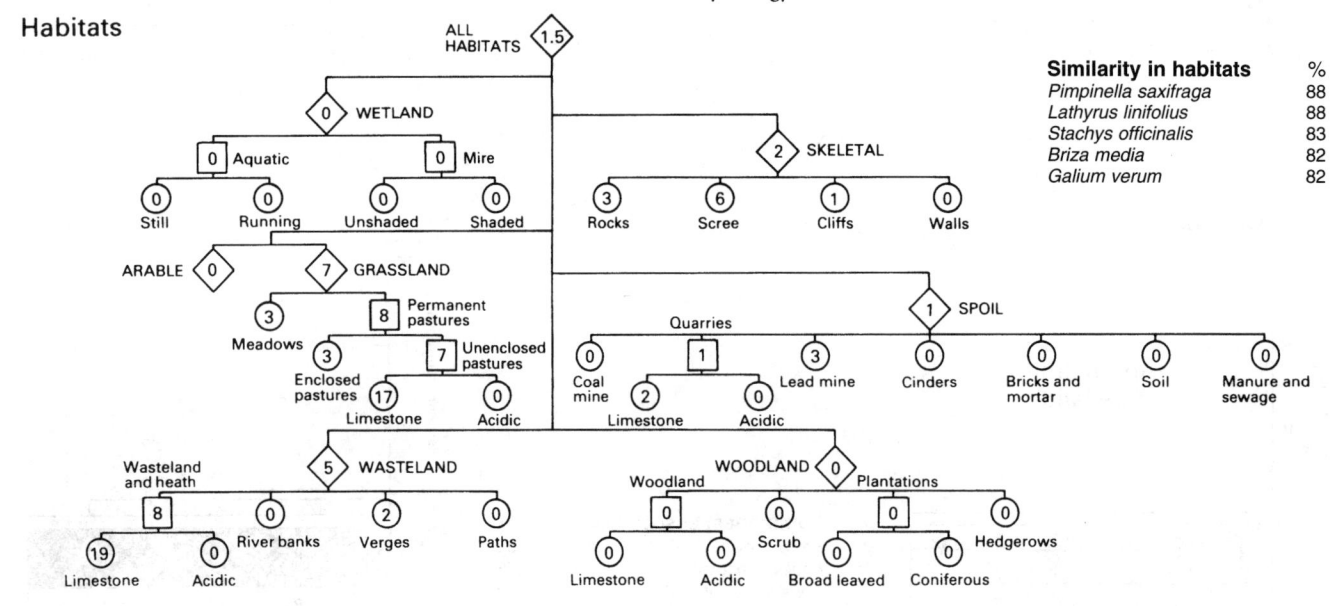

Habitats

Similarity in habitats	%
Pimpinella saxifraga	88
Lathyrus linifolius	88
Stachys officinalis	83
Briza media	82
Galium verum	82

Soil pH Most abundant on calcareous soils of pH >7.0, but also frequent in the range 5.5-7.0. Absent from strongly acidic soils of pH <4.0.

Bare soil Least frequent at high exposures.

△ ordination Distribution mainly centring on relatively undisturbed and unproductive vegetation with some evidence of persistence in more-productive sites. Absent from highly disturbed vegetation and suppressed by more-competitive species in undisturbed fertile habitats.

Associated floristic diversity Typically species-rich.

Synopsis *H.p.* is a loosely-tufted winter-green grass mainly restricted to moist but well-drained, relatively unproductive, calcareous grassland. Considered worthless for agriculture (Hubbard 1984). The calcicolous habit of the species is evident in the *PGE*, where it is 'very intolerant of ammonium salts' and 'considerably increased by lime even to the extent of becoming one of the three chief grass species'. In chalk downland *H.p.* was found to be relatively shallow-rooted (Anderson 1927). *H.p.* produces tillers with long ascending leaves not unlike those found in some forms of *Poa pratensis*. This appears to allow the persistence of *H.p.* in relatively tall grassland. However, *H.p.*, which has little capacity for lateral vegetative spread, is usually a minor component of such vegetation. With similar frequency *H.p.* also occurs as a minor component in grazed and burned sites. Thus, *H.p.* is a morphologically 'intermediate' species with no obvious specialization for any one regime of grassland management and in the *PGE* is described as 'a rather insignificant member of various mixed associations'. Thus, *H.p.* may be regarded as occupying a subordinate niche within moist, less-productive grassland analogous to that of *Trisetum flavescens* in drier sites. Like those of other 'intermediate' species, populations may be expected, under fluctuating conditions of climate, soil fertility and management, to show a greater degree of homeostasis than those of potential dominants. In this context it is interesting to note that *H.p.* reaches maximum local abundance in upland parts of the Sheffield region, growing on roadsides which are infrequently cut and which in many instances are subject to light sheep-grazing during summer. *H.p.*

Altitude
Slope
Aspect

	Unshaded
n.s.	4.7
	41.7
	3.6
n.s.	25.0

	Shaded
—	0.0
	0.8
—	0.0

Soil pH
Associated floristic diversity
Triangular ordination
Hydrology

Absent from wetland

Bare soil

Commonest associates %
Festuca rubra | 90
Dactylis glomerata | 75
Plantago lanceolata | 68
Festuca ovina | 63
Anthoxanthum odoratum | 61

Abundance in quadrats

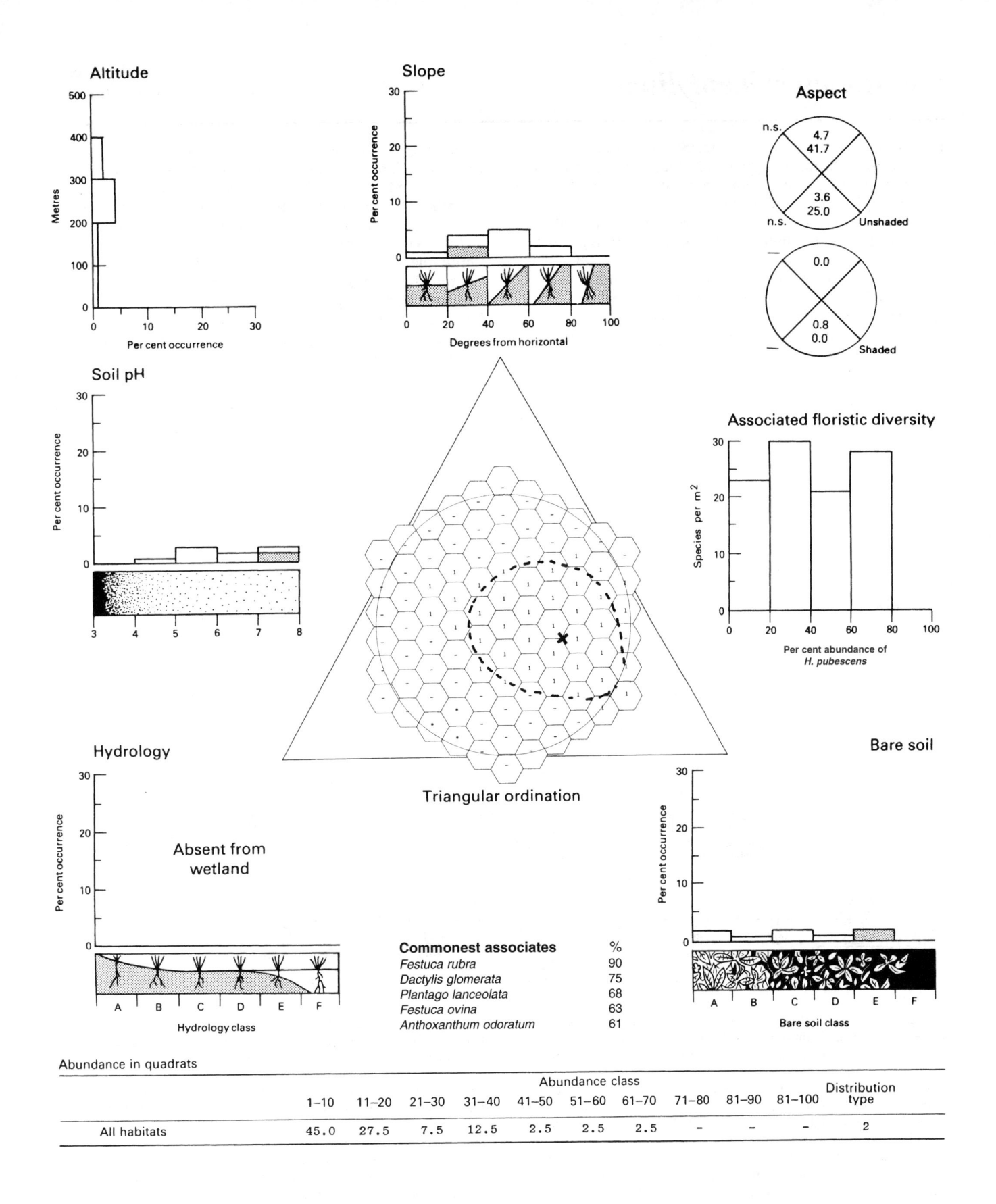

	1–10	11–20	21–30	31–40	41–50	51–60	61–70	71–80	81–90	81–100	Distribution type
All habitats	45.0	27.5	7.5	12.5	2.5	2.5	2.5	–	–	–	2

forms a transient seed bank with germination in late summer and early autumn. Within the Sheffield region *H.p.* has spread extensively from calcareous geological strata, but only along roadsides where, because of the use of limestone in road-making, edaphic conditions are closely similar to those in calcareous grassland. An additional subspecies is recorded from Europe (*FE5*).

Current trends *H.p.* is largely restricted to older grassland although it has colonized road verges extensively in the past. *H.p.* shows no sign of spreading into new artificial habitats, and seems destined to decline further.

Heracleum sphondylium L.

Apiaceae
Cow Parsnip, Hogweed, Keck

Life-form Polycarpic or more rarely monocarpic perennial, semi-rosette hemicryptophyte with stout tap-root. Stem erect; leaves pinnate, hairy, sometimes >0.25 m²; root system generally lacking VA mycorrhizas.

Phenology Flowers from June to September and seed shed slowly from August until winter. Foliage and stems die back in autumn.

Height Foliage mainly <500 mm; flowers to *c.* 2 m.

Seedling RGR 0.5–0.9 week^{-1}. SLA 21.0. DMC 22.0.

Nuclear DNA amount 3.8 pg; $2n = 22^*$ (*BC*); $2x^*$.

Established strategy Intermediate between competitor and C-S-R.

Regenerative strategies S. Regenerates by seed which germinates in spring. No persistent seed bank is formed; SB II.

Flowers White, protandrous but self-compatible (Stewart & Grace 1984), insect-pollinated; usually several hundred in an umbel. The terminal umbels with hermaphrodite flowers, lateral umbels with male and hermaphrodite or only male flowers (Tutin 1980).

Germinule 5.52 mg, 8.6 × 6.3 mm, winged mericarp (2 ovary^{-1}).

Germination Epigeal; dormancy broken by chilling.

Biological flora Sheppard (1991).

Geographical distribution The British Isles (100% of British vice-counties), *ABIF* page 476; Europe except for the extreme N and the Mediterranean region (79% of European territories); N-temperate regions.

Habitats Recorded or observed in all but aquatic habitats but particularly frequent on wasteland, road verges and river banks and in hedgerows and meadows. Contribution reduced to seedlings and stunted, isolated individuals in most other habitats.

Gregariousness Reproductive individuals occur as scattered large plants; all records of *H.s.* in abundance refer to seedlings and young plants.

Altitude Common at all altitudes to 400 m. [Up to 1010 m.]

Slope Wide-ranging.

Aspect No consistent trends.

Hydrology Found in moist ground at the edge of soligenous and topogenous mire, but not typically a wetland species.

Soil pH Most frequent and only abundant above pH 5.0.

Bare soil Wide-ranging.

△ **ordination** Distribution centred on vegetation associated with relatively fertile and moderately disturbed conditions, but scattered individuals found in all but the most unproductive or heavily disturbed vegetation.

Associated floristic diversity Intermediate.

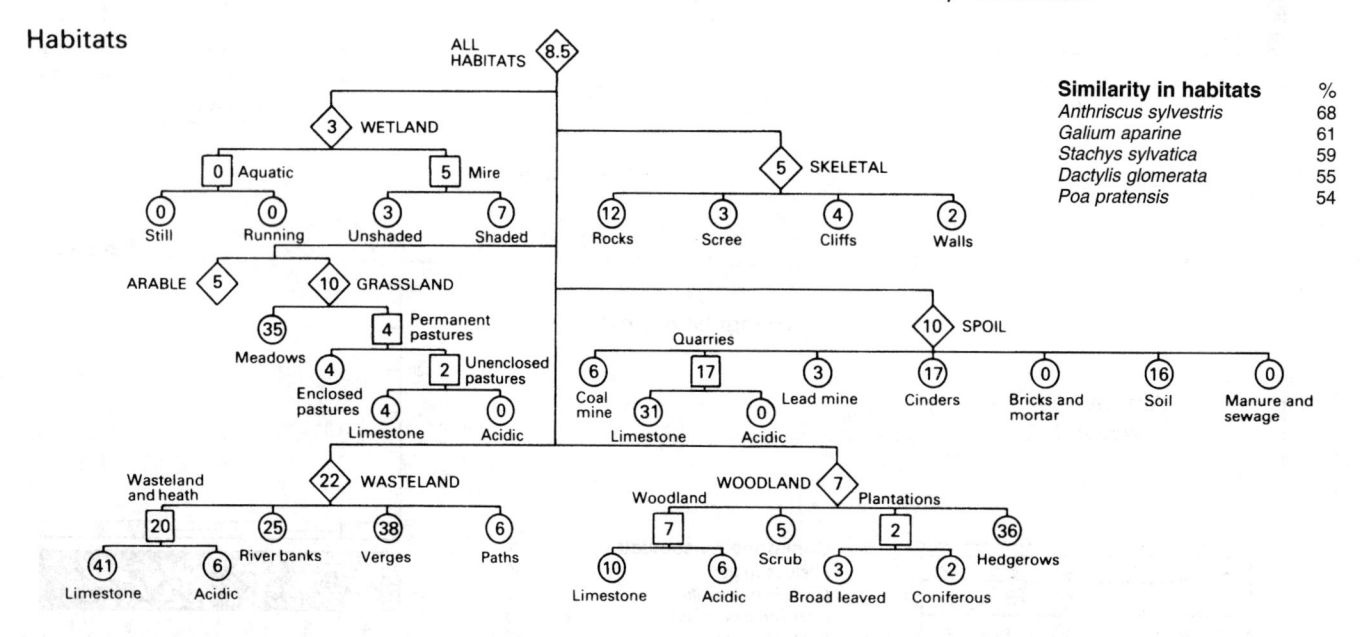

Habitats

Similarity in habitats	%
Anthriscus sylvestris	68
Galium aparine	61
Stachys sylvatica	59
Dactylis glomerata	55
Poa pratensis	54

Synopsis *H.s.* is a robust, tap-rooted herb of rough grassland and wasteland, particularly on moist, fertile soils. The leaves are high in P, Ca and Fe (*SIEF*), and when sampled in winter the plant contains unusually high concentrations of protein and fructose (Hendry & Brocklebank unpubl.). *H.s.* is perhaps typically a short-lived perennial, but often persists for over 5 years (Stewart & Grace 1984). The species has no capacity for vegetative spread, and its ecology is limited by the need for regular establishment from seed. Following from this, *H.s.* is often found in sites within roadside and hedgerow vegetation where robust competitive perennials with a capacity for clonal expansion are debilitated by occasional cutting. *H.s.* also occurs in hay-meadows. The leaf canopy consists of a few large leaves which are vulnerable to heavy grazing. Consequently, *H.s.* is infrequent in pastures. *H.s.* has a low seedling RGR primarily because at an early stage of seedling establishment photosynthate is allocated to the tap-root at the expense of new leaf development. However, the expansion of leaves and the growth of the stout stem in spring are relatively rapid in established plants due to a mobilization of these below-ground reserves. The tap-root may provide access to moisture in deep crevices, enabling *H.s.* to colonize rocky habitats. Like many ruderal species, *H.s.* has a protracted flowering period. Mericarps, which are both winged and flattened, are initially scattered short distances by wind. Like many members of the Umbelliferae, seeds contain a poorly differentiated embryo whose development requires a period of chilling (Stokes 1952) before germination can take place. Despite their large size and conspicuous presence on the soil surface over the winter, large numbers of the aromatic seeds escape predation and germinate in cohorts in spring. Populations often consist of a few flowering individuals and young plants, which may persist for several years at the one-leafed stage, often lie partially hidden beneath litter and the established plants' canopy. Many of our field records refer to plants in this state, and the apparently wide habitat range of *H.s.* is due in large measure to the presence of seedlings and small plants in situations which will not be conducive to their further development. Like several other large-seeded species, e.g. *Anthriscus sylvestris* and *Galeopsis tetrahit,* there is circumstantial evidence

Altitude

Slope

Aspect

Soil pH

Associated floristic diversity

Triangular ordination

Commonest associates	%
Festuca rubra	57
Dactylis glomerata	52
Arrhenatherum elatius	50
Poa trivialis	48
Poa pratensis	42

Hydrology

Bare soil

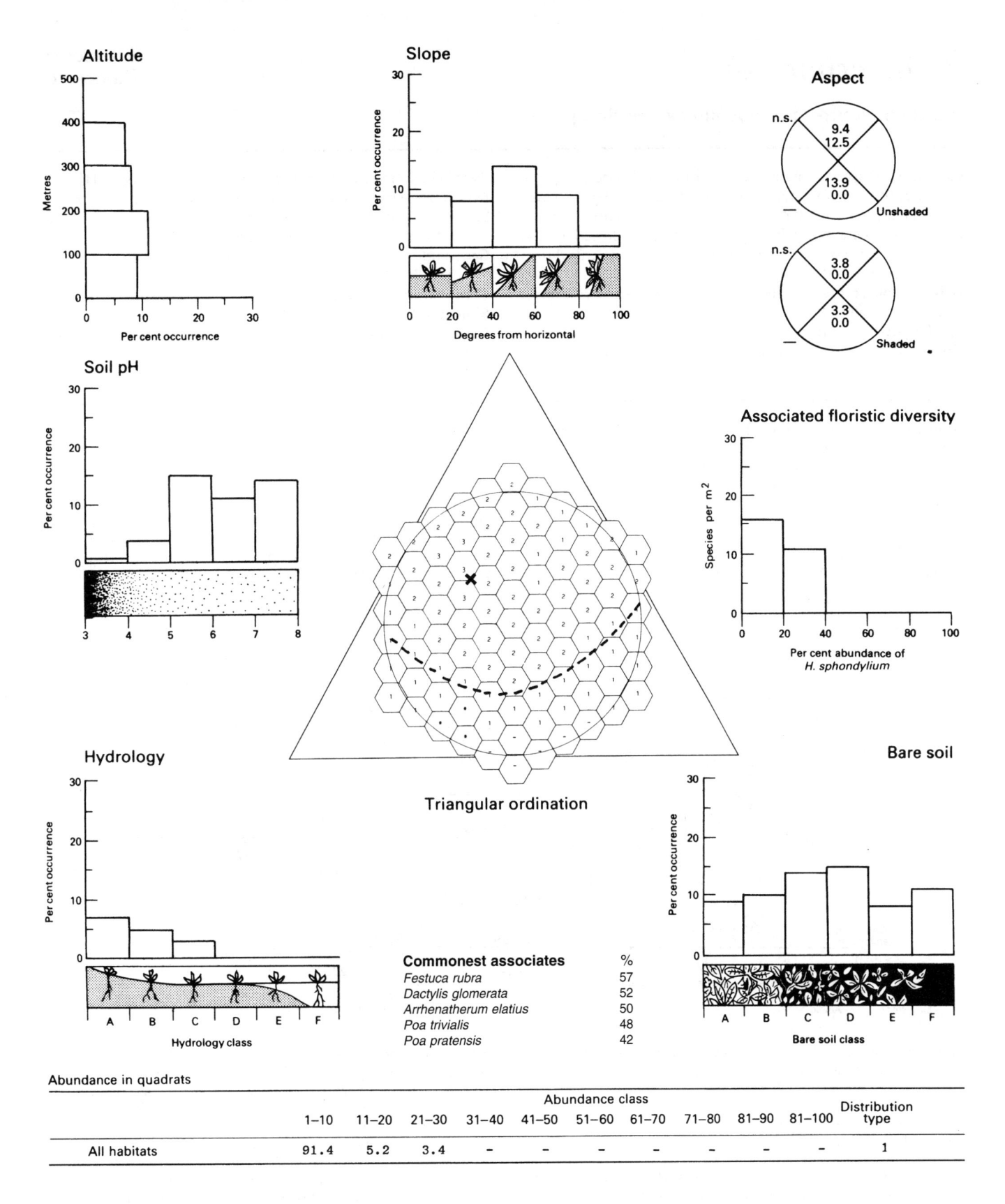

Abundance in quadrats

	\multicolumn{10}{c	}{Abundance class}	Distribution type
	1–10	11–20	21–30	31–40	41–50	51–60	61–70	71–80	81–90	81–100	
All habitats	91.4	5.2	3.4	–	–	–	–	–	–	–	1

that seedlings of *H.s.* can establish, at least on a local scale, within closed perennial communities. Further research is required on this phenomenon. *H.s.* is phenotypically plastic in response to management treatment, and flowering stems developed after cutting are much shorter than those produced otherwise. *H.s.* is a very variable species, and nine geographical variants have been recognized within Europe (*FE2*). British material refers to ssp. *sphondylium* except in East Anglia, where ssp. *sibiricum* (L.) Simonkai of NE and EC Europe also occurs (Tutin 1980). The distribution of *H.s.* overlaps with that of *H. mantegazzianum* Sommier & Levier and the two hybridize (Stace 1975).

Current trends A common and probably increasing species. *H.s.* may become more important as an arable weed as a consequence of the increased popularity of minimum tillage (*WCH*).

(*)*Hieracium* agg. L.

Asteraceae
Hawkweed

A taxonomically-difficult grouping (see 'Synopsis')

Life-form Polycarpic perennial, semi-rosette hemicryptophyte or a geophyte with a stout stock. Leaves linear to ovate, usually with long hairs, <10 000 mm^2; roots with VA mycorrhizas.
Phenology Some taxa winter green, others not. Flowering time variable, dependent upon taxon, mainly June onwards. Fruits from July onwards.
Height Foliage basal (0-200 mm) or cauline (to *c.* 700 mm); flowers usually <1 m.
Seedling RGR No data. SLA. DMC. No data.
Nuclear DNA amount No data; 2*n* = 18*, 27*, 36* (*BC*); 2*x**, 3*x**, 4*x**.

Established strategy Varying between stress-tolerator and C-S-R.
Regenerative strategies W, V (a few taxa only). See 'Synopsis'.
Flowers Yellow, ligulate, mostly apomictic; in a many-flowered capitulum; several capitula in a cymose panicle; ?SB I.
Germinule Section *Vulgata* (*H. exotericum* agg.) 0.40 mg, achene with a detached pappus *c.* 6 mm in length.
Germination Epigeal; no dormancy.
Biological flora None.
Geographical distribution The British Isles, *ABIF* page 631; temperate, Sub-Arctic and Arctic regions of Europe and Asia; SW Greenland, N Africa and N America.

Habitats *H.* spp. is most frequently recorded from spoil, wasteland and skeletal habitats. Infrequent in grazed grassland and woodland. Absent from wetlands and only a casual on arable land. [Also found in montane habitats, sand dunes and cliffs near the sea (*CSABF*).]
Gregariousness Typically intermediate.
Altitude Recorded to 380 m. [Up to 1220 m.]
Slope Wide-ranging.
Aspect Slightly more frequent and abundant in S-facing sites.
Hydrology Absent from wetland habitats.

Soil pH Present on all but the most acidic soils.
Bare soil Wide-ranging.
△ **ordination** Infrequent in the most disturbed and the most productive sites.
Associated floristic diversity Typically intermediate.
Synopsis *H.* agg. is a perennial herb associated with a wide range of less fertile, often rocky habitats. Habitat diversity coincides with the existence of many ecotypes within the grouping. *H.* agg. is a taxonomically difficult grouping and has been subdivided into a number of Sections, with over

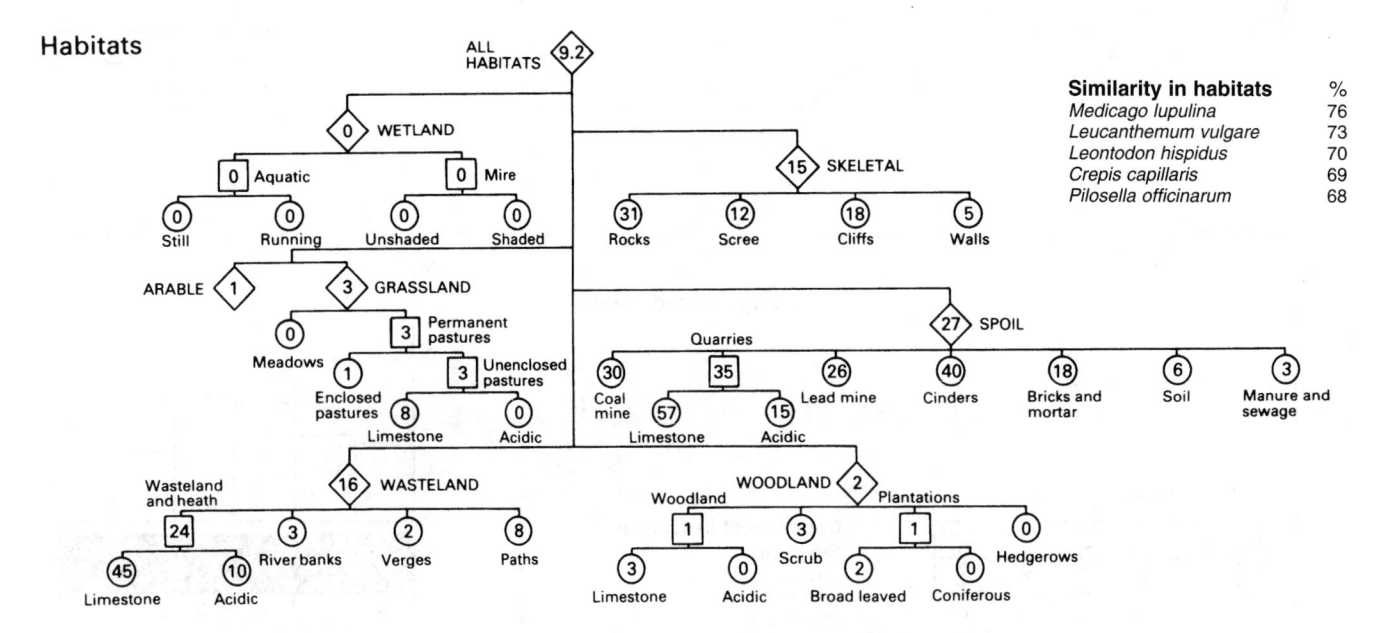

Habitats

Similarity in habitats	%
Medicago lupulina	76
Leucanthemum vulgare	73
Leontodon hispidus	70
Crepis capillaris	69
Pilosella officinarum	68

240 microspecies (*CSABF*). The Sections are often characterized by taxa of distinctive morphology and habitat range (*CSABF, Excur Fl*), but unfortunately it was not feasible to separate taxa during the fieldwork for this study. Over 66% of British taxa are either semi-rosette or rosette species and, consistent with this basal arrangement of the leaves, a majority are found in rocky (presumably little-vegetated and often desiccated) habitats (*CSABF*). Taxa frequently have a deep tap-root, and appear able to exploit subsoil moisture in droughted periods. They are perhaps mainly stress-tolerators. In Britain, 'semi-rosette' species of *Hieracium* are most abundant on calcareous rocks, but some records are from siliceous strata (*CSABF*). The stock may occasionally branch to give two rosettes, but effective regeneration is by seed which is produced in quantity and which is, in these apomictic species, a means of asexual reproduction. The plumed seed is widely dispersed and 'semi-rosette' species regularly colonize artificial habitats such as railway margins, road verges and old walls. Further, the presence of a number of alien species in these habitats (*CSABF*) suggests that the taxa are sometimes capable of dispersal over long distances, perhaps through the human agency. Flowering stems are tall and the leaves relatively long-lived. Consequently, the taxa are very susceptible to grazing (*CSABF*) and are seldom recorded from pasture. In a minority of taxa persistent leaves only occur on the stem. Some, but not all, of such plants are associated with taller closed vegetation including the herb layers of open woodland (*CSABF*). One of the commonest 'leafy' taxa both in the survey area and in much of Britain is *H. vagum* Jordan of Section Sabauda (F. N. Williams), which colonizes railway banks and coal mines and extends onto acidic soils. *H.v.* also has some capacity for lateral vegetative spread.
Current trends Uncertain. Reported to have decreased in the 20th century, perhaps as a result of sheep-grazing (*CSABF*). However, the reduced levels of grazing in areas such as the Derbyshire dales, and the capacity to colonize artificial habitats, e.g. quarry heaps, suggests the potential for an increase in the abundance of certain taxa.

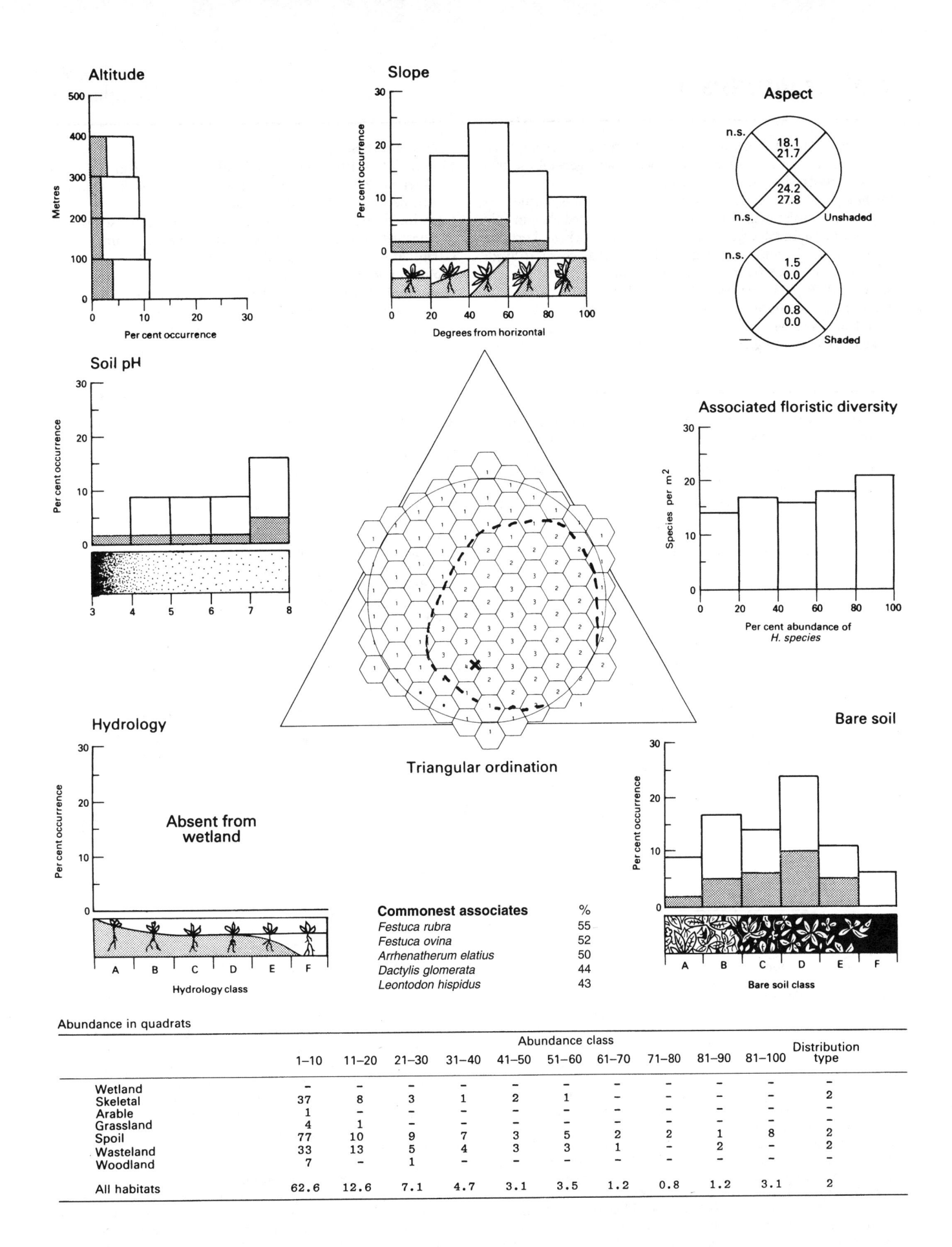

Altitude

Slope

Aspect

	n.s.	18.1	
		21.7	
		24.2	
		27.8	
	n.s.		Unshaded

	n.s.	1.5	
		0.0	
		0.8	
		0.0	
	—		Shaded

Soil pH

Associated floristic diversity

Triangular ordination

Hydrology

Absent from wetland

Bare soil

Commonest associates

	%
Festuca rubra	55
Festuca ovina	52
Arrhenatherum elatius	50
Dactylis glomerata	44
Leontodon hispidus	43

Abundance in quadrats

	\multicolumn{10}{c}{Abundance class}	Distribution type
	1–10	11–20	21–30	31–40	41–50	51–60	61–70	71–80	81–90	81–100	
Wetland	–	–	–	–	–	–	–	–	–	–	–
Skeletal	37	8	3	1	2	1	–	–	–	–	2
Arable	1	–	–	–	–	–	–	–	–	–	–
Grassland	4	1	–	–	–	–	–	–	–	–	–
Spoil	77	10	9	7	3	5	2	2	1	8	2
Wasteland	33	13	5	4	3	3	1	–	2	–	2
Woodland	7	–	1	–	–	–	–	–	–	–	–
All habitats	62.6	12.6	7.1	4.7	3.1	3.5	1.2	0.8	1.2	3.1	2

Holcus lanatus L.

Poaceae
Yorkshire Fog

Life-form Tufted, polycarpic perennial protohemicryptophyte. Shoots erect or ascending; leaves linear, hairy, generally <2000 mm^2; roots with VA mycorrhizas.
Phenology Little winter growth, and new shoots formed in spring. Flowers mainly in June and July and seed shed in July and August. In tall vegetation *H.l.* largely dies back; in short turf it remains obviously winter-green. Leaves relatively short-lived.
Height Foliage <350 mm; flowers to 600 mm.
Seedling RGR 1.5-1.9 week^{-1}. SLA 33.5. DMC 23.2.
Nuclear DNA amount 3.4 pg; $2n = 14*$ (*BC*); $2x*$.
Established strategy C-S-R.
Regenerative strategies S, B$_s$, V. Principally by seed (see 'Synopsis'); SB III.
Flowers Greyish-green, wind-pollinated, self-incompatible (Weimarck 1968); >100 in a panicle. Each spikelet usually of one hermaphrodite and one male flower.
Germinule 0.32 mg, 1.8 × 0.8 mm, caryopsis dispersed with an attached lemma bearing a short bristle (4.8 × 2.0 mm).
Germination Epigeal; t_{50} 3 days; <5-35°C; L > D.
Biological flora Beddows (1961).
Geographical distribution The British Isles (100% of British vice-counties), *ABIF* page 770; Europe except the Arctic (89% of European territories and 5% naturalized); temperate Asia and NW Africa; introduced to N and S America, Australia, New Zealand and S Africa.

Habitats Recorded, at least as seedlings, in every type of habitat. Achieves greatest abundance in meadows and pastures, but is also prominent on spoil heaps, waste ground, grass verges, paths and in hedgerows. Less common on stream banks, arable land, marshy ground and skeletal habitats such as walls and rock outcrops. Occurs at low frequencies in scrub and in woodland clearings. [Found also in grassy habitats near the sea and on mountains.]
Gregariousness Intermediate.
Altitude Frequent and abundant at all altitudes. [Up to 610 m.]
Slope Wide-ranging.
Aspect Widely distributed, but rather more records from S-facing slopes, especially in shaded habitats.
Hydrology Although most typical of moist soils, *H.l.* is common in wetland, particularly in soligenous mire. Recorded from temporarily submerged sites (water depth up to 1 m).
Soil pH Most frequent and abundant on soils of intermediate base-status (pH 5.0-6.0), but relatively common over the pH range 4.5-8.0. Few records from highly acidic soils.
Bare soil Widely distributed, consistent with the broad spectrum of habitat types occupied by the species.
△ **ordination** An extremely wide-ranging distribution, but with distribution concentrated on vegetation associated with moderately disturbed and relatively fertile conditions.

Habitats

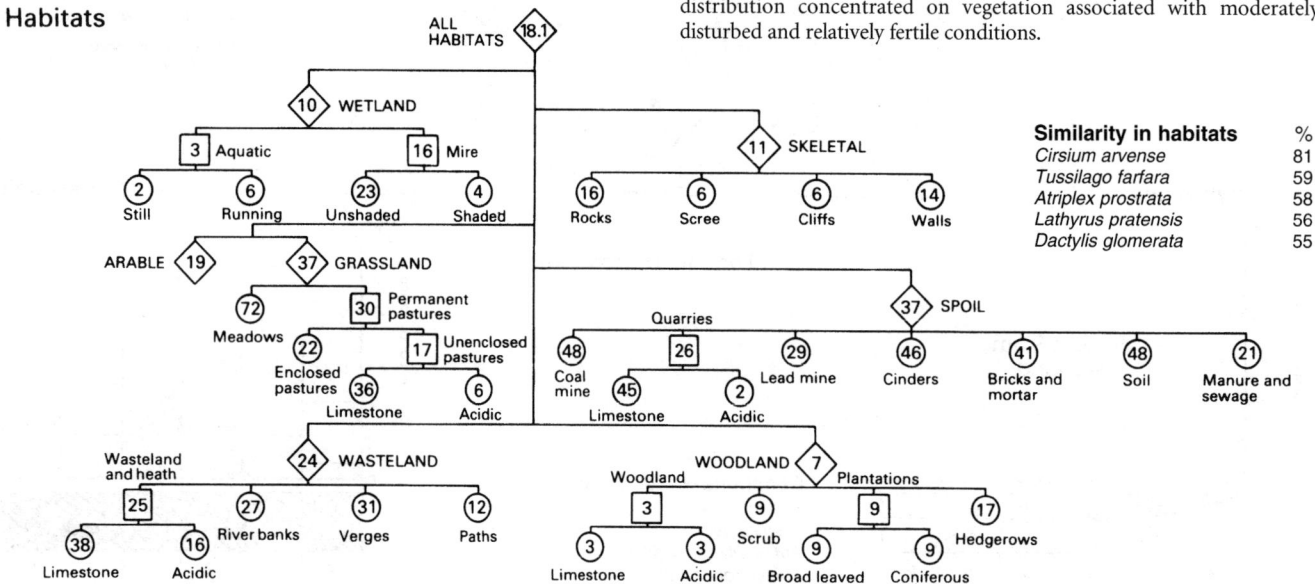

Similarity in habitats	%
Cirsium arvense	81
Tussilago farfara	59
Atriplex prostrata	58
Lathyrus pratensis	56
Dactylis glomerata	55

Associated floristic diversity Typically intermediate.
Synopsis *H.l.* is a tufted, partially winter-green grass. The ecological range is unusually wide, but is limited by five major attributes. First, *H.l.* shows lax tillering and has relatively few shoot buds (*Biol Fl*); consequently, the species is not tolerant of close grazing or heavy trampling and tends to be rather sparsely distributed in pasture. Second, although present on soils of widely different soil pH and fertility, *H.l.* is most frequent on relatively fertile soils of pH 5.0-6.0. In the PGE it is reported that 'high nutrition associated with soil acidity gives it great encouragement'. Third, the species achieves maximum vigour in moist rather than waterlogged habitats, although survival on anaerobic soils can occur through the capacity of the species to produce a network of fine roots at the soil surface (Watt 1978). Fourth, the species has restricted tolerance of shaded habitats. Fifth, *H.l.* is sensitive to winter damage, and above-ground biomass is reduced after severe winters (Al-Mashhdani 1980). Despite these constraints *H.l.* is an exceptionally versatile grass, rivalled only by *Festuca rubra* and *Agrostis capillaris* in its capacity to exploit widely different soil conditions, and forms of vegetation management. Foliar levels of N, Ca and Fe are relatively low in the Sheffield region, but elevated concentrations of Na have been found (*SIEF*). Seed production, from the plant's second year onward, is often prolific (*Biol Fl*) and, since many seeds are shed before harvesting, *H.l.* may be abundant in hay-meadows. Caryopses are released early and are capable of rapid germination over a wide range of temperatures. Seedling establishment is the main source of population expansion. Many seedlings appear in autumn, but large numbers of ungerminated seeds remain on the soil surface and a persistent seed bank is formed (Thompson 1977, Roberts 1986b). Germination is stimulated by the presence of light and by fluctuating temperatures and is largely restricted to gaps in the vegetation (Thompson & Whatley 1983). The copious seed production and wide seed dispersal, often by human agency, coupled with flexibility of germination and high growth rate, makes *H.l.* a highly efficient colonist of open habitats and confers some persistence

Altitude
Slope
Aspect
Soil pH
Associated floristic diversity
Hydrology
Triangular ordination
Bare soil

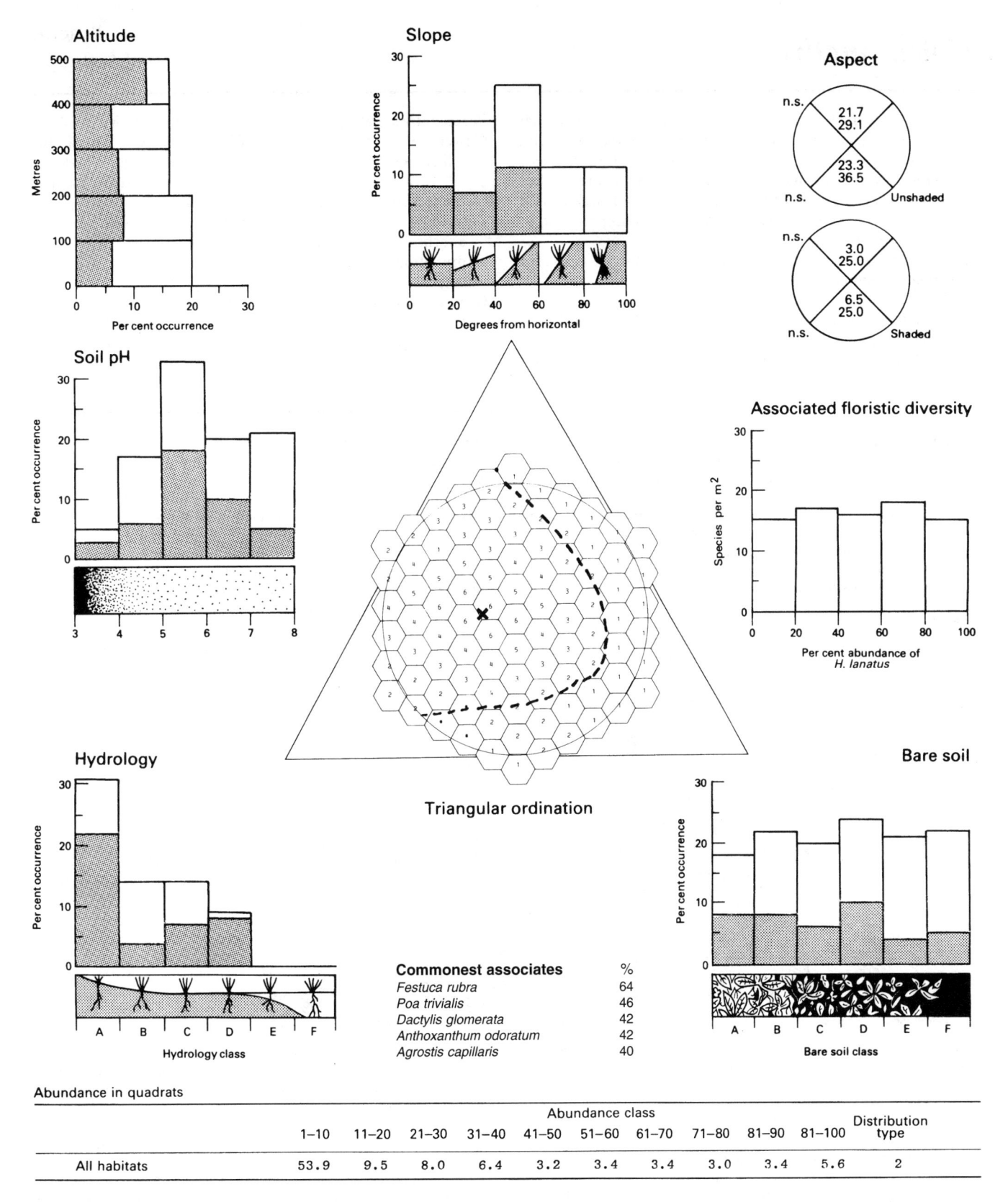

Commonest associates	%
Festuca rubra	64
Poa trivialis	46
Dactylis glomerata	42
Anthoxanthum odoratum	42
Agrostis capillaris	40

Abundance in quadrats

	\multicolumn{10}{c	}{Abundance class}	Distribution type
	1–10	11–20	21–30	31–40	41–50	51–60	61–70	71–80	81–90	81–100	
All habitats	53.9	9.5	8.0	6.4	3.2	3.4	3.4	3.0	3.4	5.6	2

under fluctuating types of land management. There is a restricted capacity for lateral spread through the production of prostrate shoots which root at the nodes (Watt 1978). This phenomenon probably only attains importance in sites where seed set is prevented. *H.l.* is possibly allelopathic (Al-Mashhadani 1980). *H.l.* shows a high degree of ecotypic differentiation, and edaphic ecotypes are recorded (McGrath 1979). *H.l.* also hybridizes with *H. mollis* (Stace 1975).

Current trends *H.l.* has many attributes which confer success in common artificial habitats. The species, which is a major problem in herbage seed crops (*WCH*), and one of the later colonists during sward deterioration (Morrison 1979), is probably still increasing. Further, although little grown, its yield of herbage in field experiments compares favourably with those of sown cultivars of agricultural grasses (Watt & Haggar 1980); we predict *H.l.* will be cultivated more frequently.

Holcus mollis L.

Poaceae
Creeping Soft-grass

Life-form Strongly rhizomatous, polycarpic perennial protohemicryptophyte. Shoots spreading or erect; leaves linear, often hairy, <2000 mm^2; roots shallow, with VA mycorrhizas.
Phenology Young shoots are produced mainly in autumn, but do not elongate until the following spring. Flowers in June and July and sets seed from July to September. Leaves produced in summer relatively short-lived.
Height Foliage mainly <150 mm (to 300 mm in unshaded habitats); flowers to 450 mm.
Seedling RGR 1.0-1.4 week^{-1}. SLA 40.7. DMC 21.6.
Nuclear DNA amount 7.8 pg; $2n$ = 14, 28*, 35*, 42*, 49* (*BC*, *FE5*); $2x$, $4x^*$, $5x^*$, $6x^*$, $7x^*$.
Established strategy Intermediate between competitor and C-S-R.

Regenerative strategies V. Mainly vegetative (see 'Synopsis').
Flowers Greyish-green, wind-pollinated, highly self-sterile but self-fertility increased with increasing polyploidy (Beddows 1971); >150 in a loose panicle. Each spikelet usually includes one hermaphrodite and one male flower.
Germinule 0.32 mg, 2.0 × 0.5 mm, caryopsis, some with an attached lemma and an hygroscopic bristle.
Germination Epigeal.
Biological flora Ovington and Scurfield (1956).
Geographical distribution The British Isles, but less common in the S Midlands and in parts of Ireland (99% of British vice-counties), *ABIF* page 771; most of Europe except for the extreme SE, only as a casual in parts of the N (75% European territories and 3% naturalized); introduced into N America.

Habitats Predominantly a species of acidic woodland, particularly in clearings, forest rides and other relatively open situations, along hedge banks and in scrub. Also abundant on heaths, waste ground, grass verges, along the banks of drainage ditches and hillside flushes, on sandy alluvial terraces of streams and in shaded mire. Other habitats include spoil heaps, such as slag heaps and sandstone quarry waste, rock outcrops and other open habitats and hill pastures. Also found occasionally as a weed at the edge of arable fields or in tree nurseries. [*H.m.* can colonize talus (*Biol Fl*).]
Gregariousness A stand-forming species.

Altitude Widely distributed. [Up to 580 m.]
Slope Wide ranging.
Aspect In the open, *H.m.* is more abundant in sites on N-facing slopes; in shaded habitats slightly more abundant on S-facing slopes.
Hydrology Frequent in soligenous base-poor mire and also forms floating rafts in upland rills.
Soil pH Distribution calcifuge; particularly frequent on soils in the pH range 3.5-5.0, and becoming scarcer as pH increases above this range.
Bare soil Widely distributed, but frequency and abundance declining towards habitats having little exposed soil.

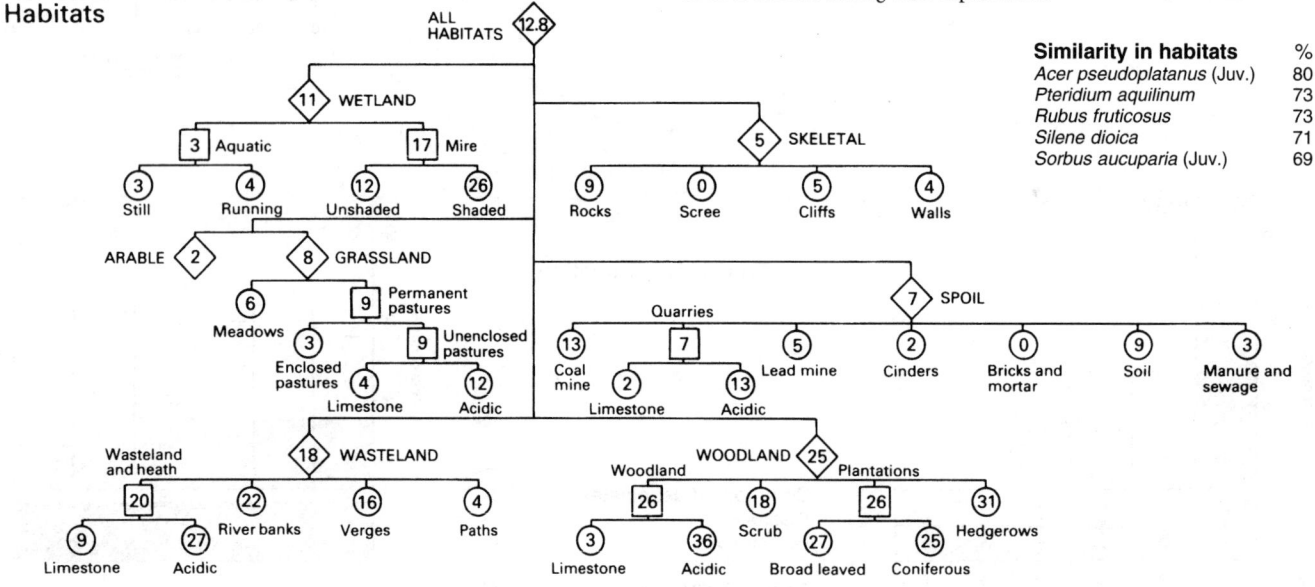

Habitats

Similarity in habitats	%
Acer pseudoplatanus (Juv.)	80
Pteridium aquilinum	73
Rubus fruticosus	73
Silene dioica	71
Sorbus aucuparia (Juv.)	69

ALL HABITATS 12.8
- WETLAND 11
  - Aquatic 3: Still 3, Running 4
  - Mire 17: Unshaded 12, Shaded 26
- SKELETAL 5: Rocks 9, Scree 0, Cliffs 5, Walls 4
- ARABLE 2
- GRASSLAND 8
  - Meadows 6
  - Permanent pastures 9: Enclosed pastures 3, Unenclosed pastures 9 (Limestone 4, Acidic 12)
- SPOIL 7: Quarries (Coal mine 13, Limestone 7, Acidic 13), Lead mine 5, Cinders 2, Bricks and mortar 0, Soil 9, Manure and sewage 3
- WASTELAND 18: Wasteland and heath (Limestone 9, Acidic 27) 20, River banks 22, Verges 16, Paths 4
- WOODLAND 25: Woodland 26 (Limestone 3, Acidic 36), Scrub 18 (Broad leaved 27, Coniferous 25), Plantations 26, Hedgerows 31

△ **ordination** Distribution centred on vegetation of undisturbed, moderately fertile habitats, but there is evidence of an ability to extend into unproductive sites.
Associated floristic diversity Low, reflecting the smaller number of species exploiting acidic soils. Values decline somewhat at high frequencies of *H.m.*
Synopsis *H.m.* produces extensive and sometimes quite ancient clonal stands, or even monocultures, on acidic soils. In exceptional circumstances old clones may form diffuse patches *c.* 1 km in diameter (Harberd 1967). Compared with most calcifuges, *H.m.* is capable of rapid dry-matter production and appears to be relatively nutrient-demanding. The distribution is generally restricted to brown earths and does not normally extend to strongly podsolized soils. Foliar concentrations of Ca are low (*SIEF*). *H.m.* exploits a diversity of habitats from woodland to wetlands, grasslands and even arable land, but in many *H.m.* is subject to disturbance of one form or another. This is true even in woodland, its major habitat. *H.m.* is exceedingly vigorous in lightly shaded and open habitats in woodland, and may spread explosively after coppicing. However, where there is dense shade the expansion of the canopy in summer usually brings about etiolation and a premature decline in vigour and the shoots frequently succumb to fungal attack (Grime 1981). It is thus difficult to understand why *H.m.* is not usurped from its woodland habitats by more-shade-adapted species. One possible reason may be the ability of *H.m.* to emerge through tree leaf litter (Sydes & Grime 1981 a,b) and to complement the deeper rooting depth of *Hyacinthoides non-scripta*. *H.m.*, though frequent in grassland, does not persist in heavily grazed pasture, as its few robust shoot stems are eaten more quickly than they are replaced. But if grazing ceases the erect growth form which the species adopts in unshaded habitats is competitively very advantageous. Most commonly, *H.m.* is $5x$ in W Europe (*FE5*) and produces little seed (Jones 1958). Plants in shade flower sparingly. Thus, the reproductive capacity of many populations of *H.m.* depends upon clonal expansion by means of long, spreading rhizomes and a facility for rapid regeneration from rhizome, and even shoot, fragments. The importance of regeneration from fragments in

Altitude

Slope

Aspect

	n.s.	7.1 / 55.6	
n.s.		9.4 / 33.3	Unshaded

	n.s.	28.0 / 54.1	
P< 0.001		20.3 / 64.0	Shaded

Soil pH

Associated floristic diversity

Hydrology

Triangular ordination

Bare soil

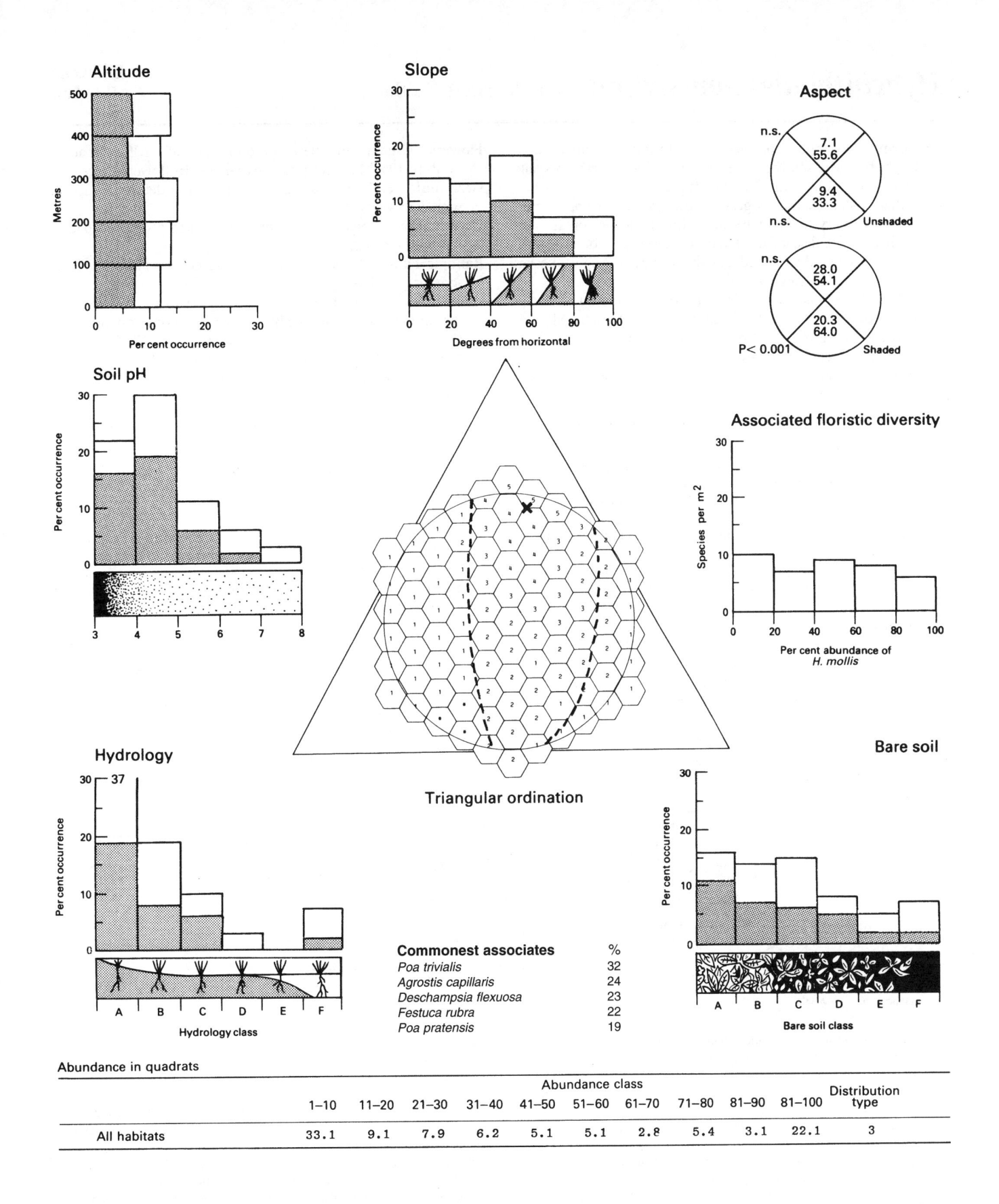

Commonest associates	%
Poa trivialis	32
Agrostis capillaris	24
Deschampsia flexuosa	23
Festuca rubra	22
Poa pratensis	19

Abundance in quadrats

	Abundance class										Distribution type
	1–10	11–20	21–30	31–40	41–50	51–60	61–70	71–80	81–90	81–100	
All habitats	33.1	9.1	7.9	6.2	5.1	5.1	2.8	5.4	3.1	22.1	3

mechanically-disturbed sites is illustrated by this species' persistence in tree nurseries and, to a lesser extent, on arable land. There is no evidence of a persistent seed bank, but sites with 4x plants may not have been sampled. Greater vegetative vigour has been noted in 5x clones; 4x plants tend to occur on drier soils, but otherwise the ecological differences between the cytological races remain obscure (Jones 1958, Carroll & Jones 1962). An additional ssp. is recorded from Spain (FE5). H.m. forms hybrids with H. lanatus. These resemble H.m., and tetraploids of H.m. may have arisen as a result of past hybridization between H.l. and H.m. (Stace 1975).

Current trends Uncertain. H.m. is able to exploit some disturbed conditions but is confined to moderately base-rich acidic soils.

Hyacinthoides non-scripta (L.) Chouard ex Rothm.

Liliaceae
Bluebell

Life-form Polycarpic perennial, rosette-forming geophyte with a deeply buried bulb. Stems erect; leaves linear, fleshy, usually <4000 mm^2; roots with VA mycorrhizas.

Phenology Vernal. Leaves grow below ground during winter, and by January to February their tips reach the soil surface. Further growth delayed until March. Flowers produced in April to June. Seed gradually shed, mainly in July and August. Leaves normally dead by July (*Biol Fl*).

Height Foliage typically <250 mm; flowers to 450 mm.

Seedling RGR <0.5 week^{-1} (Al-Mufti 1978). SLA 22.7. DMC 8.8.

Nuclear DNA amount 42.4 pg; $2n = 16^*$ (*BC*); $2x^*$.

Established strategy Stress-tolerant ruderal.

Regenerative strategies S, V. Regenerative biology is complex (see 'Synopsis'); SB II.

Flowers Usually blue, hermaphrodite, partially self-compatible (Wilson 1959b), insect-pollinated; in a 4-16-flowered raceme.

Germinule 6.17 mg, 2.8 × 2.1 mm, seed in a dehiscent capsule with 2-35 seeds (*Biol Fl*).

Germination Epigeal; requirement for warm moist incubation followed by chilling.

Biological flora Blackman and Rutter (1954) (as *Endymion non-scriptus* (L.) Gareke).

Geographical distribution A European endemic widely distributed throughout the British Isles, less common in Ireland (99% of British vice-counties), *ABIF* page 817; N and W Europe, locally naturalized from gardens in C Europe (21% of European territories and 5% naturalized).

Habitats A common woodland species. Recorded most frequently in broad-leaved plantations. Also common in scrub and woodland overlying either acidic or limestone strata, but less frequent in coniferous plantations. Occurs also in upland areas on waste ground and heaths, and occasionally in unproductive pastures, on spoil heaps and on cliffs. [Elsewhere recorded from sea cliffs, montane areas and rarely, from sand dunes (*Biol Fl*).]

Gregariousness Typically intermediate but sometimes stand-forming.

Altitude Extends to 340 m. [Up to 660 m. (*Biol Fl*.)]

Slope Largely absent from skeletal habitats and steep slopes.

Aspect In woodland habitats more frequent and significantly more abundant on S-facing slopes. A N-facing bias has been observed in unshaded habitats.

Hydrology Absent from wetlands.

Soil pH Wide-ranging, but more frequent and abundant in the pH range 3.5-4.5. Less common on soils of high base status, and seldom recorded at pH >7.5.

Bare soil Widely distributed, but most frequent and abundant in habitats where there is much tree litter and little exposed soil.

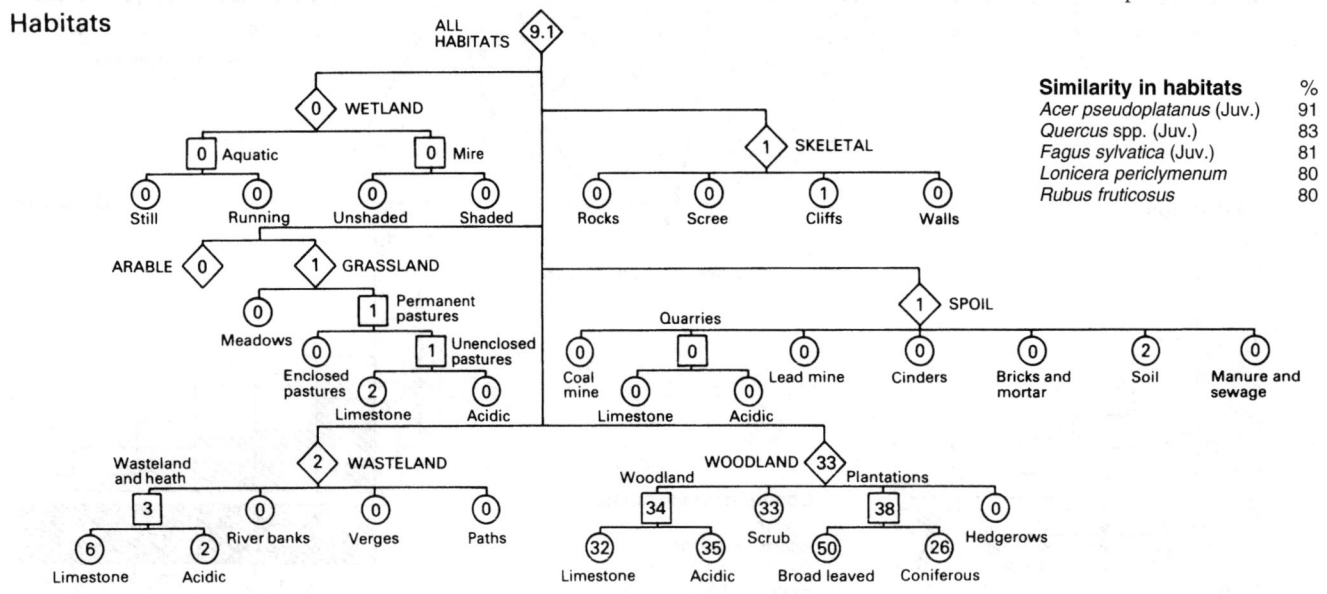

Habitats

Similarity in habitats	%
Acer pseudoplatanus (Juv.)	91
Quercus spp. (Juv.)	83
Fagus sylvatica (Juv.)	81
Lonicera periclymenum	80
Rubus fruticosus	80

△ **ordination** Distribution wide-ranging, although absent from heavily disturbed vegetation. Most frequently associated with relatively unproductive woodland herb layers subject to the disruptive influences (*sensu* Grime 1979) of tree canopy expansion in the late spring and tree leaf litter deposition in the autumn.

Associated floristic diversity Low and declining slightly with increased abundance of *Hn-s*.

Synopsis *Hn-s*. is a woodland, bulb-forming, monocotyledonous herb with a vernal phenology, exploiting the light phase before the development of a full summer canopy, and restricted to sites where the light intensity does not fall below 10% of daylight between April and mid-June (*Biol Fl*). This involves shoot expansion during late winter and early spring, a capacity consistent with its high nuclear DNA amount and the presence of exceptionally high concentrations of fructan and fructose in the bulb in winter (Hendry & Brocklebank unpubl.). *H.n-s*. then undergoes a period of bulb aestivation prior to the expansion of a new set of roots in autumn (*Biol Fl*). The number of leaves produced is predetermined and, since damaged leaves cannot be replaced, *Hn-s*. is very vulnerable to grazing, cutting or trampling (*Biol Fl*). However, foliage contains toxic glycosides and, though eaten by cattle and sheep, is not grazed by rabbits (Knight 1964, Cooper & Johnson 1984). *H.n-s*. has a poorly-developed system for water conduction (vessels are confined to the roots (Dahlgren *et al.* 1985)), and seeds released in summer require exposure to warm moist conditions followed by chilling in order to become germinable (Thompson & Cox 1978). These attributes appear to be important in restricting *H.n-s*. to mesic sites, and perhaps also in confining this 'Atlantic' species to shaded habitats in areas of low rainfall. As a result of the action of contractile roots, the mature bulb of *H.n-s*. may be buried to a depth of 250 mm or more, rendering the species vulnerable to waterlogging and subject to replacement by the more superficial bulbs of *Allium ursinum* on alluvial terraces. Thus, *H.n-s*. is characteristic of moist, freely-drained environments. The location of the bulb in the subsoil may limit the ability

Altitude

Slope

Aspect

Soil pH

Associated floristic diversity

Triangular ordination

Hydrology

Commonest associates %
Holcus mollis 39
Rubus fruticosus agg. 29
Oxalis acetosella 25
Mercurialis perennis 24
Deschampsia cespitosa 23

Bare soil

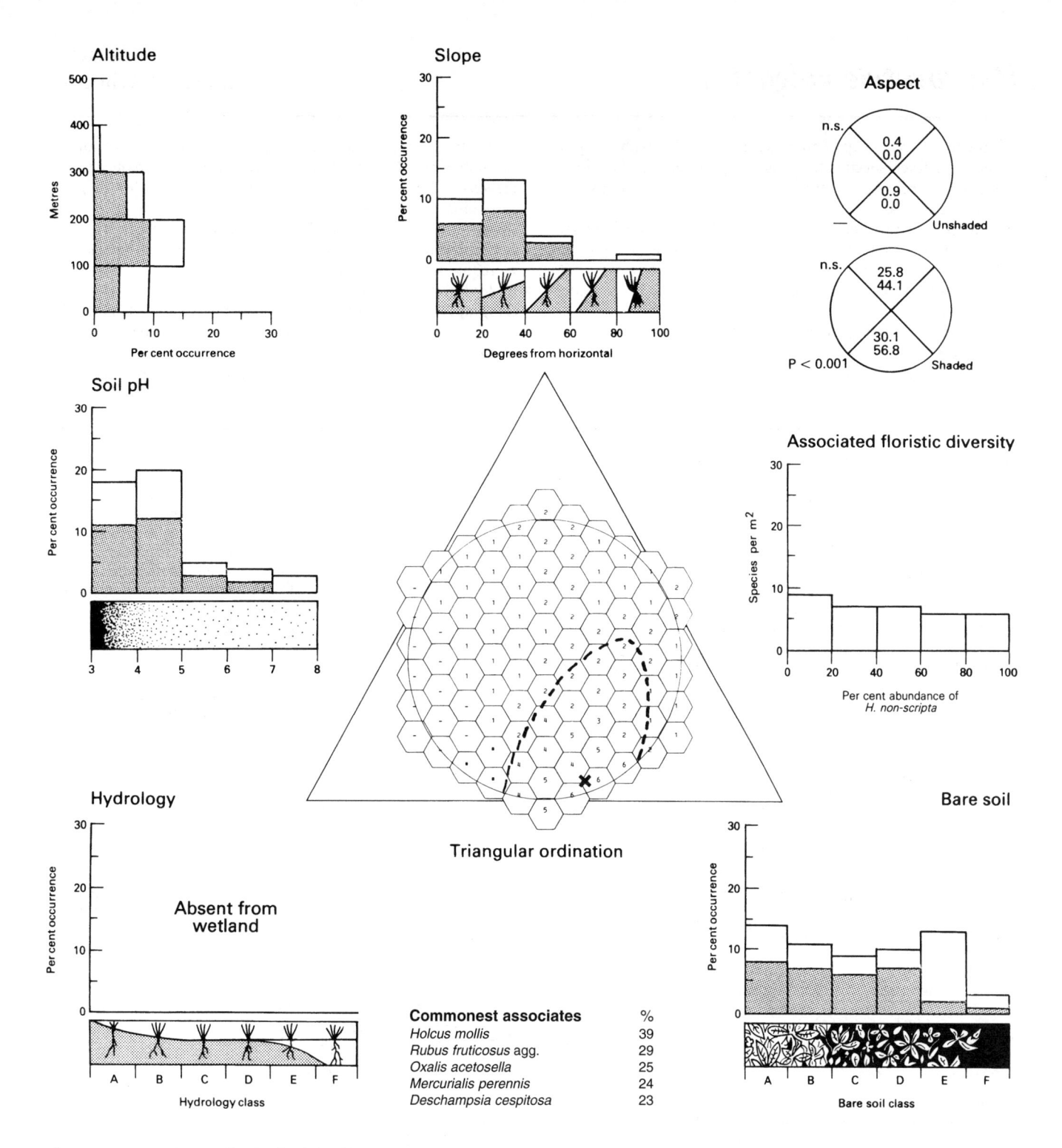

of *H.n-s.* to occupy the woodlands of stabilized limestone scree. However, the deep root system often allows co-existence with strongly competitive woodland species. *Hn-s.* often exploits the subsoil below the rhizomatous growth of *Pteridium aquilinum* (Knight 1964, Grabham & Packham 1983). Leaves are high in Na and Mn, but low in P and Mg (*SIEF*). During the early phase of shoot emergence the leaves form a compact spear-shoot (Salisbury 1916), capable of penetrating through considerable depths of deciduous tree litter (Sydes & Grime 1981a,b). In favourable situations dense stands of robust individuals may exert local dominance of the ground flora during the vernal phase, and probably restrict the vigour of other species. Various studies (Wilson 1959a, Knight 1964, Grabham & Packham 1983) consistently indicate that *H.n-s.* regenerates primarily by seeds, which germinate from October onwards, usually in microsites protected by a covering of leaf litter and within 0.4 m of the parent plant (Knight 1964). A variable number of seeds remain ungerminated until spring (Grabham & Packham 1983). Seedlings suffer high mortality in close proximity to parent plants and in sites with deep tree litter. Plants do not flower for *c.* 5 years (Wilson 1959a). At maturity they may produce daughter bulbs but, with the exception of (*Biol Fl*), most authorities have concluded that vegetative regeneration is of minor importance. Genotypic variation between and within populations has been reported (Wilson 1959a), but few taxonomic variants have been described (*Biol Fl*). *H.n-s.* forms hybrids with the related *H. hispanica* (Miller) Rothm. (Stace 1975).
Current trends Uncertain. In moist upland areas *H.n-s.* is a frequent colonist of plantations, and may be among those woodland species which are less vulnerable to changes in forestry management. In some lowland areas, however, the species is largely restricted to ancient woodland (Rackham 1980), and may therefore be declining.

Hydrocotyle vulgaris L.

Apiaceae
Pennywort, White-rot

Life-form Polycarpic perennial hemicryptophyte with creeping shoots. Leaves succulent, peltate, sparsely hairy, often long-stalked, usually <1000 mm^2; roots shallow, adventitious with VA mycorrhizas.

Phenology Overwintering with minute leaves. Growth in late spring is followed by flowering in June to August and seed set from August to October.

Height Foliage to 250 mm, but generally <120 mm; flowers often <50 mm.

Seedling RGR No data. SLA 32.6. DMC 13.6.

Nuclear DNA amount 1.9 pg; 2n = 96 (*FE2*); 8x.

Established strategy Intermediate between ruderal and C-S-R.

Regenerative strategies V. ?S. Forms extensive patches by means of long creeping stems. Detached shoots capable of re-establishment. Regeneration by seed probably infrequent.

Flowers Greenish-white tinged with pink, hermaphrodite, self-pollinated, 3-6, clustered.

Germinule 0.31 mg, 1.7 × 1.0 mm, mericarp (2 per fruit).

Germination Epigeal; seed dormancy breaking mechanism requires investigation.

Biological flora None.

Geographical distribution The British Isles (100% of British vice-counties), *ABIF* page 454; W, C and S Europe northwards to Scandinavia and southwards to Portugal and Greece (64% of European territories); N Africa; introduced in New Zealand.

Habitats Largely restricted to unshaded mire, but extending into aquatic sites and into various habitats adjacent to open water.

Gregariousness Patch-forming.

Altitude Wide-ranging but occurrences more abundant within the upland region. [Up to 530 m.]

Slope As with most wetland species largely restricted to gentle slopes.

Aspect Insufficient data.

Hydrology Most characteristic of soligenous mire, but also found on reservoir margins and ditch banks. May occur in ditches as an aquatic with submerged leaves. Only rarely observed in shaded habitats.

Soil pH Mainly restricted to soils of pH 4.5-6.0, and absent from calcareous soils.

Bare soil Only recorded from vegetation providing an almost continuous vegetation cover and often associated with *Sphagnum* carpets. However, the stoloniferous growth does allow *H.v.* to spread onto bare disturbed ground adjacent to established colonies.

△ **ordination** Present in vegetation types which are subject to slight or moderate levels of disturbance, with distribution centred on moderately productive vegetation. Absent from situations in which disturbance is very intense, and also excluded from tall, dense communities.

Associated floristic diversity Typically low.

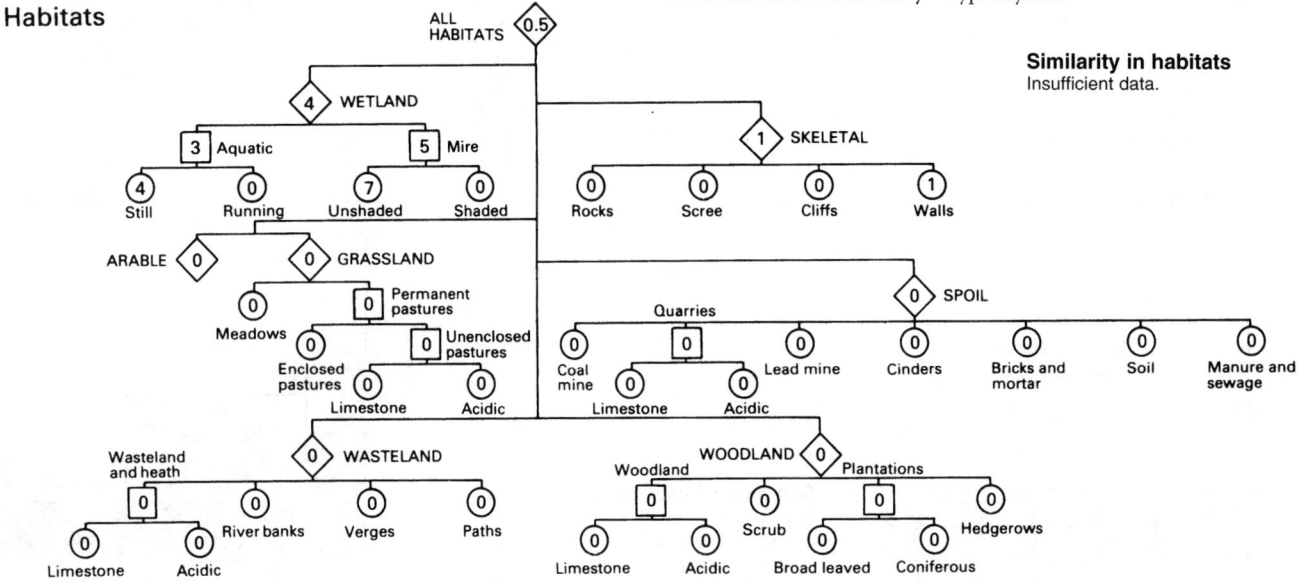

Habitats

Similarity in habitats
Insufficient data.

Synopsis *H.v.* is a wetland species with creeping rooted stems bearing leaves held aloft on long petioles. The species is perhaps most characteristic of soligenous mire on peaty, mildly acidic soils where the growth of potential dominants is suppressed by an intermediate level of fertility and by occasional grazing. To a lesser extent, *H.v.* occurs in topogenous mire and plants rooted on, for example, ditch banks may extend into aquatic habitats. However, persistence and growth on reservoir margins down only to positions corresponding to the winter water level suggest an inability to exploit sites which are waterlogged throughout the year. The species may occur in local monocultures or in a turf with other species. *H.v.* may form an understorey to species such as *Juncus effusus*, but is usually absent from shaded mire. Canopy height is determined by the length of the petiole, and plants growing in dry or grazed habitats have much shorter petioles and smaller leaves than those in taller vegetation. Marked changes in morphology are often associated with plants growing beside water; individuals rooted under water often have a disproportionately long petiole and lack the hairs present on the leaves of terrestrial forms (Sculthorpe 1967). *H.v.* regenerates vigorously by the creeping shoots which often form extensive patches. Thus, the species may be a particularly successful colonist of areas bared by disturbance. Detached fragments, which root readily, are probably important in the colonization of stream- and ditch-side sites following disturbance. Fruits, which float and may adhere to the feet of birds (Ridley 1930), may also play a role in the colonization of more-distant sites. However, regeneration by seed is probably of infrequent occurrence, particularly since *H.v.* appears rather shy-flowering when growing amongst tall vegetation. No persistent seed bank has been detected. The subfamily to which *H.v.* belongs, the Hydrocotyloideae (formerly treated as the separate family, Hydrocotylaceae (*CTW*) is mainly distributed in the S hemisphere, and *H.v.* is the only British member (Tutin 1980).

Current trends Decreasing as a result of habitat destruction. Not under any great threat at present in upland regions, but becoming rare in many lowland districts.

Altitude
Slope
Aspect
Unshaded
Shaded

Soil pH
Associated floristic diversity
Species per m²
Per cent abundance of *H. vulgaris*

Hydrology
Hydrology class

Triangular ordination

Commonest associates	%
Carex nigra	59
Cirsium palustre	50
Juncus bulbosus	45
Anthoxanthum odoratum	43
Juncus effusus	43

Bare soil
Bare soil class

Abundance in quadrats

	1–10	11–20	21–30	31–40	41–50	51–60	61–70	71–80	81–90	81–100	Distribution type
Wetland	2	1	1	1	–	1	2	2	–	4	3
Skeletal	–	–	1	–	–	–	–	–	–	–	–
Arable	–	–	–	–	–	–	–	–	–	–	–
Grassland	–	–	–	–	–	–	–	–	–	–	–
Spoil	–	–	–	–	–	–	–	–	–	–	–
Wasteland	–	–	–	–	–	–	–	–	–	–	–
Woodland	–	–	–	–	–	–	–	–	–	–	–
All habitats	13.3	6.7	13.3	6.7	–	6.7	13.3	13.3	–	26.7	3

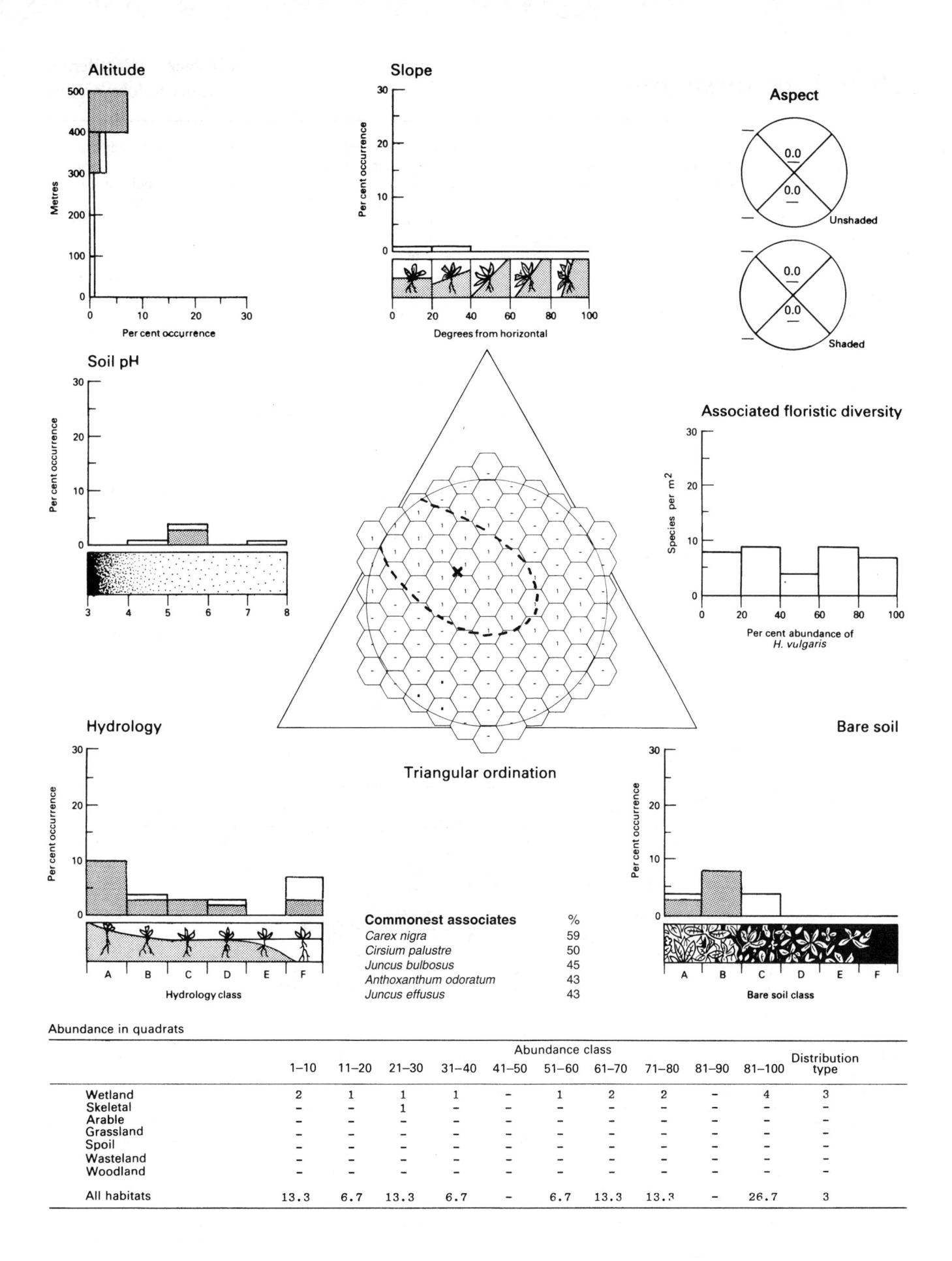

Hypericum hirsutum L.

Clusiaceae (Guttiferae)
Hairy St John's Wort

Life-form Loosely-tufted, polycarpic perennial protohemicryptophyte. Stems erect; leaves glandular, ovate, hairy, <1000 mm^2; roots often bearing VA mycorrhizas.

Phenology May overwinter as small shoots produced in autumn. Flowers July to August and sheds seed from September onwards, although some may be retained on the plant over winter. Shoots normally die back in autumn, but may persist, albeit in a senescent form, in sheltered sites.

Height Foliage up to *c*. 700 mm, overtopped by flowers.

Seedling RGR No data. SLA 27.2. DMC 27.6.

Nuclear DNA amount No data; $2n = 18$ (*FE2*); $2x$.

Established strategy Intermediate between stress-tolerant competitor and C-S-R.

Regenerative strategies B$_s$, V. Regenerative biology is complex (see 'Synopsis'); SB IV.

Flowers Yellow, hermaphrodite, insect- or self-pollinated in bad weather (Knuth 1906); >50 in a cymose inflorescence.

Germinule 0.07 mg, 1.3 × 0.5 mm, seed; >50 in a dehiscent capsule.

Germination Epigeal; t_{50} 5 days; 12–29°C; germination rate increased after dry storage at 20°C or after washing out inhibitors from the seed coat (UCPE unpubl.); L > D.

Biological flora None.

Geographical distribution Widely scattered throughout the British Isles, very local in Ireland (63% or British vice-counties), *ABIF* page 215; most of Europe except the NE and the extreme S (69% of European territories); W Asia.

Habitats Almost exclusively restricted to calcareous soils. Found on scree and in lightly-grazed or ungrazed grassland. Tolerant of light shade in scrub or at the margin of woods and recorded from cliffs and lead-mine spoil. Observed on ditch banks, waysides, railway banks, outcrops, old limestone quarries, coal-mine debris and cinder tips. Absent from arable or wetland habitats.

Gregariousness Occurs typically as isolated clumps.

Altitude Widely distributed, but mainly restricted to the carboniferous limestone and most frequently between 200 and 300 m. [Up to 410 m.]

Slope Most records are from dale-side slopes on the carboniferous limestone, but frequently observed on more-level ground on the magnesian limestone.

Aspect Predominantly on N-facing slopes in both shaded and unshaded habitats.

Hydrology Absent from wetlands.

Soil pH Only recorded on soils of pH >6.0.

Bare soil Associated with widely different amounts of exposed soil.

△ **ordination** Only recorded in relatively infertile undisturbed sites.

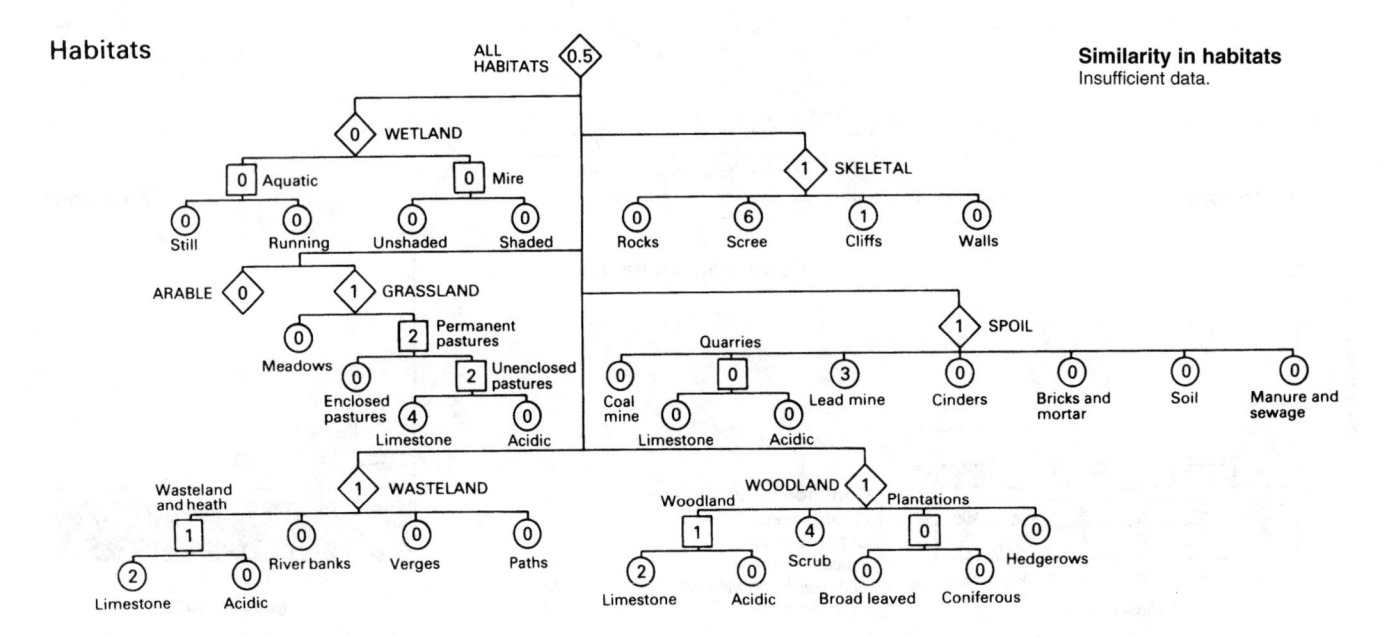

Habitats

Similarity in habitats
Insufficient data.

Associated floristic diversity Typically intermediate to species-rich.

Synopsis *H.h.* is a tall, tuft-forming herb characteristic of relatively unproductive wasteland on moist, often clayey, calcareous soils. Although tall like *H. perforatum*, its capacity for dominance is restricted by the limited extent of lateral vegetative spread. Persistence of *H.h.* appears to be dependent upon infertility and occasional disturbance, factors which restrict encroachment by taller clonal species. *H.h.* exhibits a limited tolerance of burning. The species contains hypericin (Knox & Dodge 1985) which, as in *H.p.*, confers some protection against herbivory. It is unusual to find extensive populations of *H.h.*, and there are grounds to suspect that the species is a plant of ecotones. In common with *Teucrium scorodonia*, *Fragaria vesca* and *Potentilla sterilis*, the species is most common at the boundary between shaded and unshaded habitats. *H.h.* is less overtly a colonizing species than *H.p.* and, although found on quarry spoil and shaded rock outcrops, *H.h* is more typical of tall grassland, particularly at wood margins. Further contrasts with *H.p.* are evident in its greater restriction to calcareous strata and failure to exploit droughted habitats. *H.h.* produces numerous very small seeds which appear to require areas of bared ground for establishment. Seeds probably germinate mainly in spring after the inhibitor in the seed coat has been washed out. *H.h.* also forms a persistent seed bank (Roberts 1987a), and this almost certainly plays an important role in allowing persistence during the 'closed' phases of coppice cycles. *H.h.* has only a limited capacity for vegetative reproduction, and detached shoots root much less readily than those of *H.p.*

Current trends *H.h.* is restricted to a relatively narrow range of infertile habitats and, despite some capacity for persistence and dispersal, is probably decreasing.

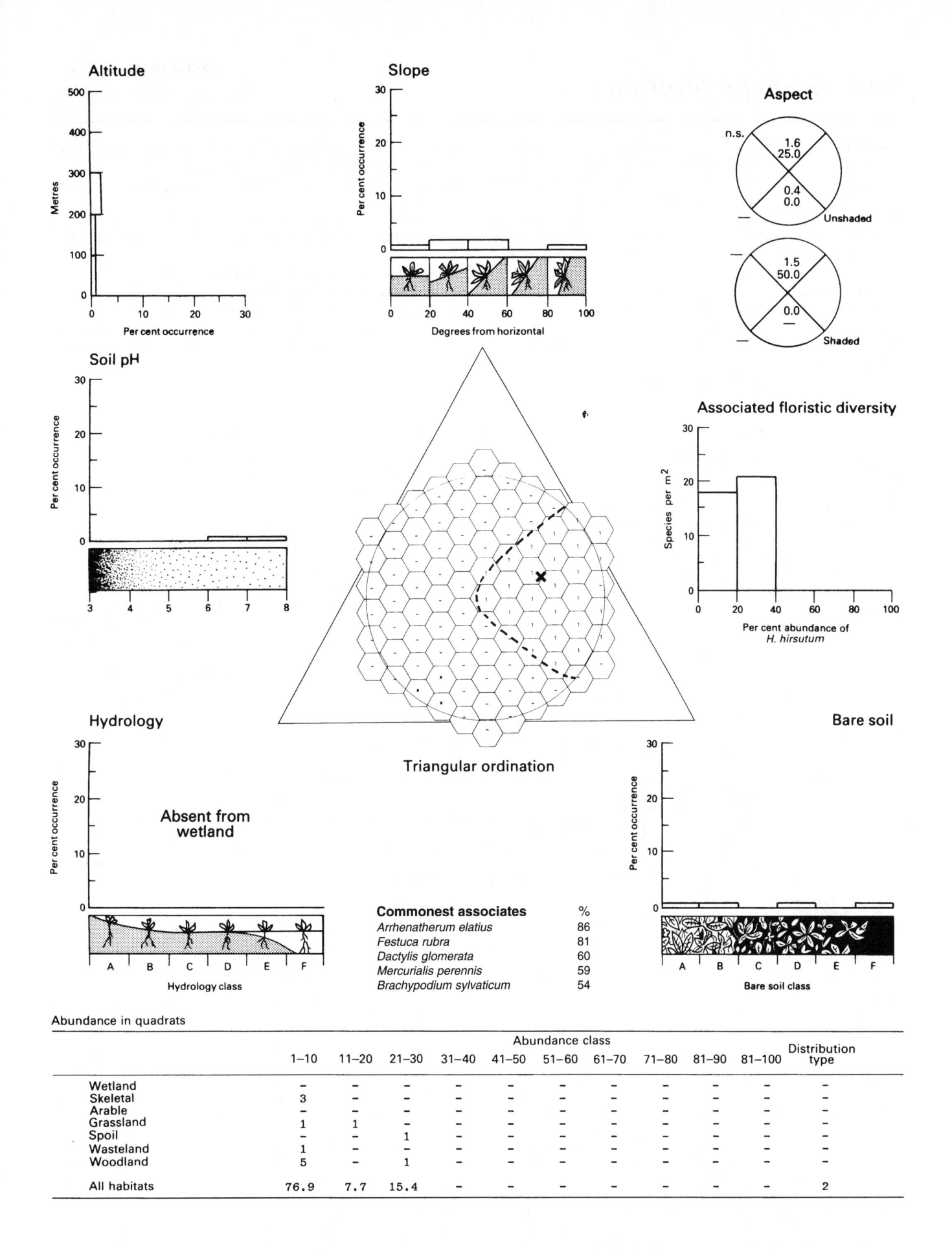

Altitude

Slope

Aspect

	n.s.	1.6
	25.0	
—	0.4	0.0
		Unshaded

	1.5	
	50.0	
—	0.0	—
		Shaded

Soil pH

Associated floristic diversity

(Species per m², Per cent abundance of *H. hirsutum*)

Triangular ordination

Hydrology

Absent from wetland

Bare soil

Commonest associates	%
Arrhenatherum elatius	86
Festuca rubra	81
Dactylis glomerata	60
Mercurialis perennis	59
Brachypodium sylvaticum	54

Abundance in quadrats

	1–10	11–20	21–30	31–40	41–50	51–60	61–70	71–80	81–90	81–100	Distribution type
Wetland	–	–	–	–	–	–	–	–	–	–	–
Skeletal	3	–	–	–	–	–	–	–	–	–	–
Arable	–	–	–	–	–	–	–	–	–	–	–
Grassland	1	1	–	–	–	–	–	–	–	–	–
Spoil	–	–	1	–	–	–	–	–	–	–	–
Wasteland	1	–	–	–	–	–	–	–	–	–	–
Woodland	5	–	1	–	–	–	–	–	–	–	–
All habitats	76.9	7.7	15.4	–	–	–	–	–	–	–	2

Hypericum perforatum L.

Clusiaceae (Guttiferae)
Common St John's Wort

Life-form Rhizomatous, polycarpic perennial protohemicryptophyte. Shoots erect, branched; leaves lanceolate, glandular, >250 mm²; some roots penetrating deeply, other superficial, bearing VA mycorrhizas.

Phenology Flowers from June to September and sheds seed from late August onwards with some possibly overwintering on the plant. Older shoots die back in autumn, but young unexpanded shoots may overwinter.

Height Foliage to *c.* 600 mm; overtopped to 900 mm by flowers.

Seedling RGR No data. SLA 23.1. DMC 24.8.

Nuclear DNA amount 1.2 pg; $2n = 32$ (*FE2*); $4x$.

Established strategy Intermediate between competitive-ruderal and C-S-R strategist.

Regenerative strategies B_s, V. Regenerative biology is complex (see 'Synopsis'); SB IV.

Flowers Yellow, hermaphrodite, usually apomictic (*FE2*); often >100 aggregated in a cymose inflorescence.

Germinule 0.1 mg, 1.3 × 0.6 mm, seed; >50 in a dehiscent capsule.

Germination Epigeal; t_{50} 3 days; 13–35°C; dormancy influenced by the presence of inhibitors in the seed coat (Campbell 1985); L > D.

Biological flora Campbell and Delfosse (1984).

Geographical distribution Most of the British Isles, but less common in the N and W (94% of British vice-counties), *ABIF* page 212; Europe except for the extreme N (92% of European territories); W Asia and N Africa, Australia and New Zealand.

Habitats Frequent on calcareous wasteland and limestone quarries, on both spoil and ledges in the quarry face itself. Found also in a variety of spoil habitats and frequently observed on cinders and ballast beside railways. Occasionally recorded in lightly grazed pasture, waysides and as an impermanent colonist of arable land and woodland margins. Observed in sand and gravel workings and in woodland rides. Absent from wetland. [Found on fixed dunes (Salisbury 1952).]

Gregariousness May form diffuse patches.

Altitude Mainly associated with disturbed artificial habitats which, like *H.p.* itself, decrease with increasing altitude. Extends to 310 m. [Up to 460 m.]

Slope Found over a wide range of slopes.

Aspect Significantly more frequent and only abundant on S-facing slopes.

Hydrology Absent from wetlands.

Soil pH Restricted to soils of pH >5.0, and particularly frequent and abundant above pH 7.0.

Bare soil Widely distributed, with highest values of frequency and abundance occurring at intermediate levels of exposed soil.

△ **ordination** A constituent of many types of vegetation except where high productivity is combined with low or moderate levels of disturbance. Distribution centred on disturbed vegetation in rather unproductive sites.

Similarity in habitats	%
Centaurea nigra	82
Centaurium erythraea	82
Brachypodium pinnatum	81
Bromopsis erectus	80
Inula conyzae	73

Habitats

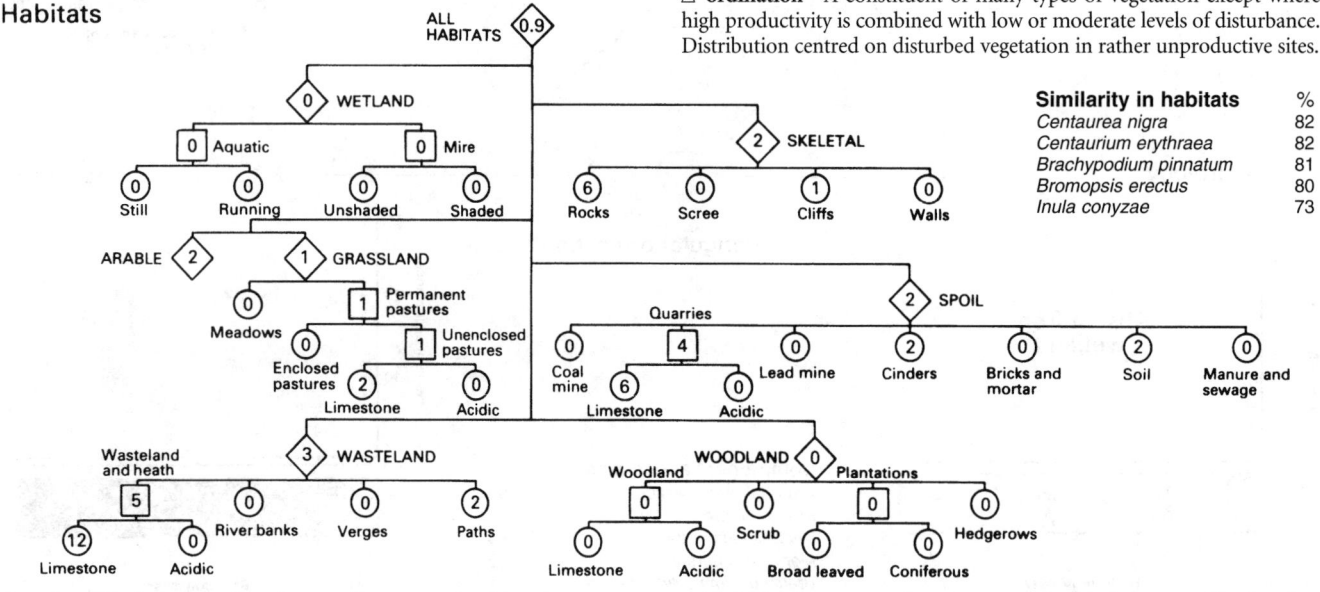

Associated floristic diversity Intermediate to species-rich.

Synopsis *H.p.* is a tall herb characteristic of relatively unproductive wasteland on freely drained soils. Although a tall species, the potential to exercise dominance is restricted by the limited capacity to form dense stands. *H.p.* is most characteristic of dry rocky or sandy ground where, as a result of factors such as disturbance and low fertility, competition from more-robust species is restricted. In particular, *H.p.* is associated with chalk and limestone quarries and with railway ballast. *H.p.* is tolerant of, or even favoured by, fire (*Biol Fl*) and its grassland distribution is strongly biased towards burned sites. However, *Hp.* is ecologically wide-ranging and occurs on a diversity of soil types, and in association with widely contrasted levels of disturbance. It seems likely that the deep tap-root (*Biol Fl*) is of critical importance in allowing *H.p.* to sustain a summer peak in shoot biomass. In Australia and New Zealand, where the species is much more common than in Britain, the glandular shoots are a major hazard to farm animals, and heavy economic losses are experienced through poisoning of stock. This phenomenon is not common in Britain because *H.p.* is only a minor weed of grassland, and the major effects of hypericin toxicity are induced only in the presence of bright sunlight (*Biol Fl*, Cooper & Johnson 1984), an elusive factor during many British summers. Vegetative regeneration of *H.p.* takes two forms. First, many axillary shoots are produced from the decumbent rooting base of the clump; this enables a limited degree of lateral spread and tends to be particularly well developed in plants growing on deep soils (*Biol Fl*). Second, and mainly in shallow soils, new clumps may develop from lateral roots at some distance from the parent shoots (*Biol Fl*). This mode of regeneration is very similar to that of *Linaria vulgaris* with which *H.p.* is frequently associated on railway ballast. Detached shoot branches root fairly readily but in many of the habitats exploited by *H.p.* soil conditions are probably normally too dry to allow their establishment. Regeneration by seed which is produced in large quantities (often >30 000 per plant, according to Salisbury (1942)), is probably dependent upon the existence of relatively large areas of bare

Altitude

Slope

Aspect

$p < 0.01$

0.8	0.0
4.0	22.2

Unshaded

—	0.0
0.0	—

Shaded

Soil pH

Associated floristic diversity

Hydrology

Absent from wetland

Triangular ordination

Bare soil

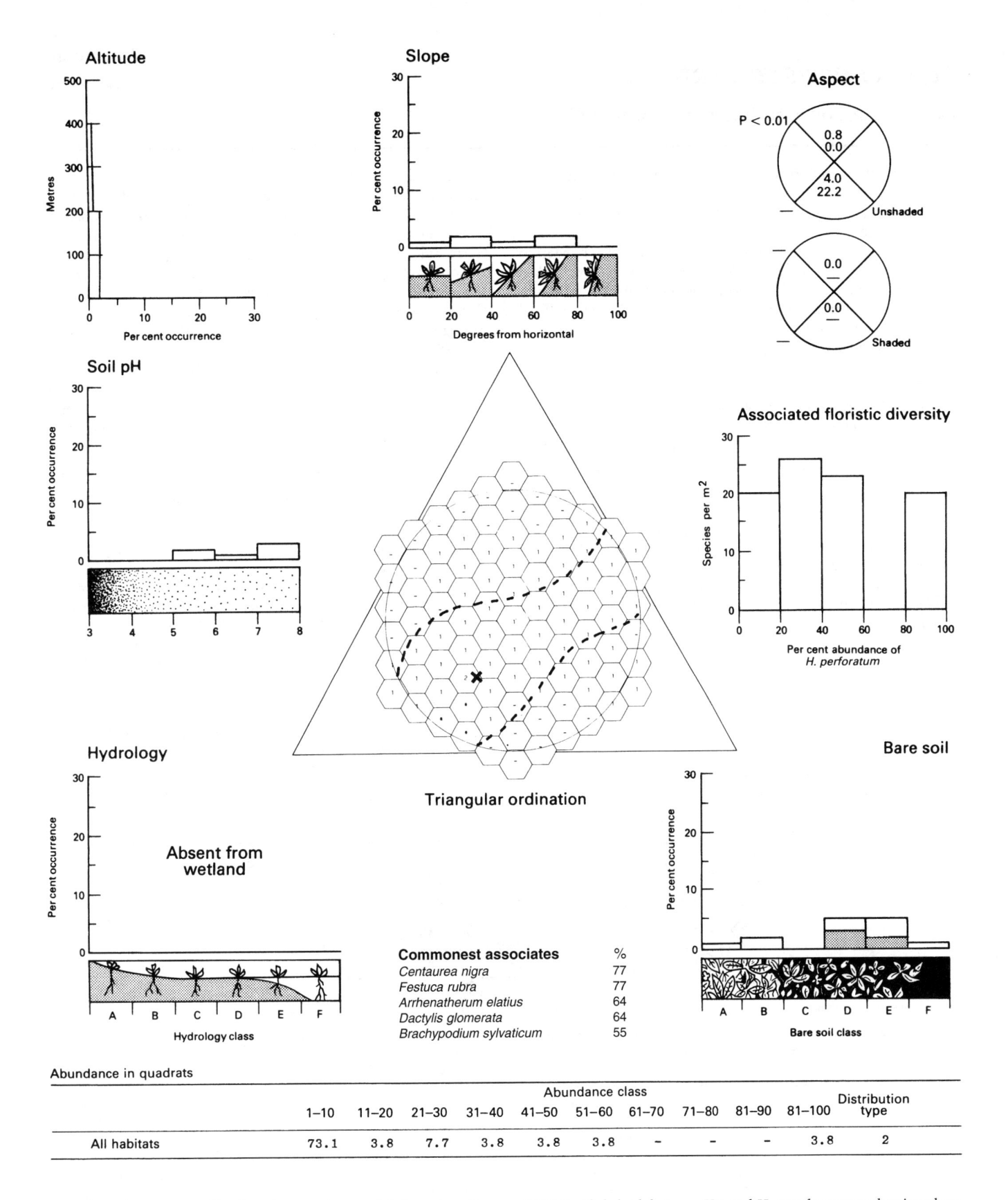

Commonest associates	%
Centaurea nigra	77
Festuca rubra	77
Arrhenatherum elatius	64
Dactylis glomerata	64
Brachypodium sylvaticum	55

Abundance in quadrats

	Abundance class										Distribution type
	1–10	11–20	21–30	31–40	41–50	51–60	61–70	71–80	81–90	81–100	
All habitats	73.1	3.8	7.7	3.8	3.8	3.8	–	–	–	3.8	2

ground. Seeds are very small and tend to germinate mainly in spring, after an inhibitor in the seed coat (Campbell 1985) has been washed out. *H.p.* regularly forms a persistent buried seed bank (Grime 1978, Roberts 1986a). The extent of ecotypic differentiation remains uncertain, but geographical races differing in leaf morphology have been identified (*FE2*) and Pritchard (1960) suggests the existence of weed, woodland and sand-dune ecotypes. *H.p.* shows signs of hybrid origin and is sometimes male-sterile (Robson 1981). Fertile hybrids between *H.p.* and *H. maculatum* ssp. *obtusiusculum* (Tourlet) Hayek are recorded from Britain (Stace 1975). *H.m.* tends to occur on moist, less calcareous soils, but its ecological distribution frequently overlaps with that of *H.p.*

Current trends Possibly still increasing due to its capacity to exploit a number of artificial habitats, particularly those associated with railways.

Hypochaeris radicata L.

Generic name often spelt *Hypochoeris*, as in *FE4*

Asteraceae
Cat's Ear

Life-form Polycarpic perennial, rosette hemicryptophyte. Stock stout, erect and branching; scape branched, erect; leaves lanceolate, hairy, <4000 mm²; roots deep, with VA mycorrhizas.

Phenology Overwinters as a small rosette (or in cold climates no leaves overwinter (*Biol Fl(b)*). Main period of vegetative growth probably until June (*Biol Fl(b)*). Flowers mainly from June to September particularly during the early part of this period. Seed set from July onwards.

Height Foliage typically appressed to the ground but rarely up to 200 mm. Scapes to 600 mm.

Seedling RGR No data. SLA 23.3. DMC 11.6.

Nuclear DNA amount No data; $2n = 8*$ (*BC*); $2x*$.

Established strategy C-S-R.

Regenerative strategies W, V. Regenerates effectively both by seed and by vegetative means (see 'Synopsis'); SB I.

Flowers Florets yellow, ligulate, insect-pollinated; generally considered to be self-incompatible, but some self-pollination may occur (*Biol Fl(b)*); 50 aggregated in each capitulum.

Germinule 0.96 mg, 6.0 × 0.8 mm, achene dispersed with a beak about as long as the fruit itself and a pappus *c*. 11 mm in length. Outer fruits smaller, often unbeaked (*FE4*).

Germination Epigeal; t_{50} 3 days; 11-30°C; L = D.

Biological flora (a) Aarssen (1981), (b) Turkington and Aarssen (1983).

Geographical distribution The British Isles (100% of British vice-counties), *ABIF* page 619; most of Europe except the NE (82% of European territories); widely distributed and frequently introduced in cooler, temperate parts of the world.

Habitats Primarily a plant of short turf in pastures, meadows, heaths, derelict pastures and waysides. Also found amongst rocks, and on cliffs and on raw soils developed from spoil. Only of casual occurrence on arable land, and absent from woodland and wetland habitats. [Found in cliff-top grassland and dry areas on sand dunes.]

Gregariousness Typically sparse to intermediate.

Altitude Found up to 400 m, but because of its association with dry, well-drained soils and open vegetation, *H.r.* is particularly common on the low lying Bunter sandstone. [Up to 610 m.]

Slope Frequent over a wide range in slope.

Aspect Occurrences and abundance greater on S-facing slopes.

Hydrology Absent from sites subject to prolonged waterlogging.

Soil pH Widely distributed, but most common and abundant in the pH range 4.5-5.5. Absent from most strongly acidic soils. [In areas with a less oceanic climate *H.r.* is absent from calcareous soils (*Biol Fl(b)*).]

Bare soil Most frequently associated with intermediate exposures.

△ **ordination** Distribution indicates conditions of moderate disturbance and intermediate fertility, such as those associated with rock outcrops, spoil

Habitats

Similarity in habitats %
Aira praecox 76
Agrostis capillaris 68
Rumex acetosella 67
Hieracium spp. 59
Calluna vulgaris 56

and heavily exploited pastures. Absent from severely disturbed situations. On the most fertile soils the species is confined to sites which have been recently disturbed or are subject to frequent cutting or grazing.

Associated floristic diversity Typically intermediate to species-rich.

Synopsis *H.r.* is found in a wide range of short grasslands and open habitats, but is particularly characteristic of dry, sandy, slightly acidic soils. Though highly palatable (*Biol Fl(b)*), *H.r.* often persists in grazed turf, probably because the leaves are restricted to a basal rosette which is closely appressed to the ground. In mown turf the tough, springy flowering stems often evade damage. *H.r.* is very tolerant of drought (*Biol Fl(b)*). We suspect that this is because the deep root system (Salisbury 1952) is able to tap subsoil water during dry periods. In much of Britain *H.r.* overwinters with a small rosette of leaves, but in very cold regions the species is not winter green (*Biol Fl(b)*). In heavily grazed or mown turf *H.r.* regenerates vegetatively to form diffuse clonal patches, whereas in lightly grazed sites numerous, wind-dispersed seeds are produced. Germination may occur over much of the year, with peaks in spring and autumn, but no persistent seed bank is formed (Roberts 1986a). Flexibility of regeneration and transport in commercial grass seed samples have contributed to the ability of *H.r.* to become an important weed of grassland in many parts of the world (*Biol Fl(a)*). This colonizing ability may have been assisted by the capacity of plants under favourable conditions to flower within 2 months of establishment (*Biol Fl(a)*). Considerable variation between populations has also been reported (*Biol Fl(a,b)*), Van der Dijk 1981). Heukels and Oostroom (1968), working in Holland, have suggested the existence of two ssp., one from fertile habitats and with long fruits, the other from less-fertile, often acidic soils and with smaller fruits and a more prostrate growth habit. *H.r.* has a very low chromosome number ($2n = 8$); Stebbins (1974) discusses the possible advantages of this characteristic in facilitating rapid evolution. The species hybridizes with *H. glabra* L.

Future trends The data compiled in *Biol Fl(a)* suggest that *H.r.* is probably still increasing worldwide. However, *H.r.* is not an important constituent of heavily fertilized pasture, and the area of habitat suitable for this species has declined in response to modern methods of grassland management. However, *H.r.* remains a highly effective colonist of new habitats, and its future as a common plant appears assured.

Altitude
Slope
Aspect

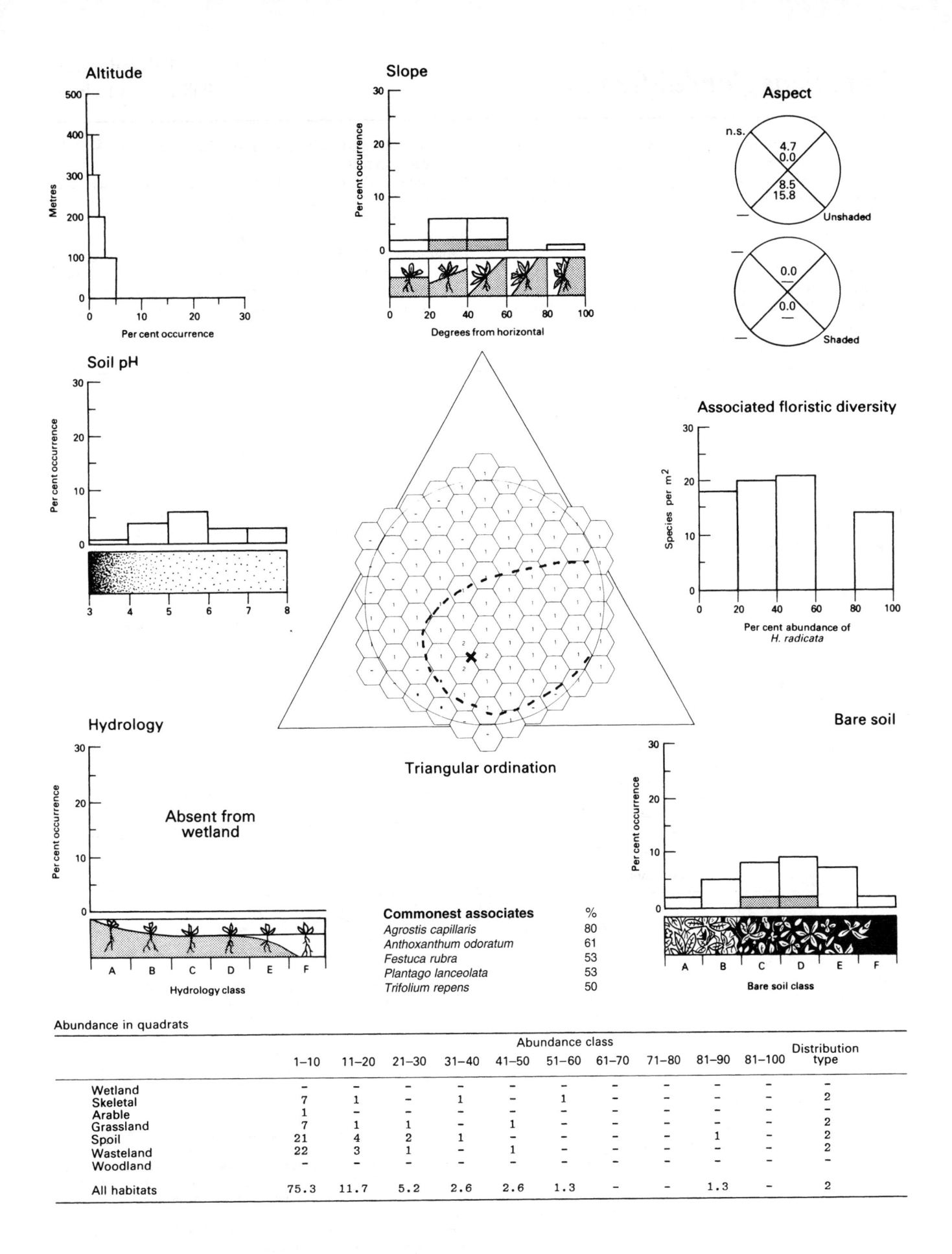

	n.s.	4.7
	—	0.0
		8.5
		15.8

Unshaded

		0.0
	—	0.0

Shaded

Soil pH

Associated floristic diversity

Hydrology

Absent from wetland

Triangular ordination

Commonest associates	%
Agrostis capillaris	80
Anthoxanthum odoratum	61
Festuca rubra	53
Plantago lanceolata	53
Trifolium repens	50

Bare soil

Abundance in quadrats

	1–10	11–20	21–30	31–40	41–50	51–60	61–70	71–80	81–90	81–100	Distribution type
Wetland	–	–	–	–	–	–	–	–	–	–	–
Skeletal	7	1	–	1	–	1	–	–	–	–	2
Arable	1	–	–	–	–	–	–	–	–	–	–
Grassland	7	1	1	–	1	–	–	–	–	–	2
Spoil	21	4	2	1	–	–	–	–	1	–	2
Wasteland	22	3	1	–	1	–	–	–	–	–	2
Woodland	–	–	–	–	–	–	–	–	–	–	–
All habitats	75.3	11.7	5.2	2.6	2.6	1.3	–	–	1.3	–	2

Impatiens glandulifera Royle

Balsaminaceae
Policeman's Helmet

Life-form Summer-annual therophyte. Shoot erect; leaves lanceolate to *c.* 4000 mm^2; roots non-mycorrhizal.

Phenology Seeds germinate in spring. Flowers from July until October and seed set from August to October, at which time plants are killed by frost.

Height Foliage and flowers to 2 m.

Seedling RGR No data. SLA 38.3. DMC 13.7.

Nuclear DNA amount 2.3 pg; $2n = 18^*, 20^*$ (*BC*); $2x^*$.

Established strategy Competitive-ruderal.

Regenerative strategies S. Exclusively by seed which germinates in spring. No persistent seed bank occurs; SB II.

Flowers Usually purplish-pink, hermaphrodite, protandrous, insect-pollinated, self-compatible (Valentine 1978); 5-12 in axillary racemes.

Germinule 7.32 mg, 4.8 × 3.5 mm, seed explosively discharged from a 1-6-seeded capsule (Salisbury 1964).

Germination Epigeal; t_{50} 4 days; chilling requirement.

Biological flora Beerling and Perrins (1993).

Geographical distribution Native of Himalaya. Introduced and locally common in N and W England and Wales, less common in SE England, Scotland and Ireland, but still increasing (39% of British vice-counties), *ABIF* page 453; most of Europe (naturalized in 54% of European territories).

Habitats Abundant on river banks, shaded marshland and disturbed lightly-shaded areas in woodland and scrub.

Gregariousness A robust stand-forming species capable of dominance at quite low densities of established plants.

Altitude Confined to lowland habitats and only observed to 210 m.

Slope Recorded from gently sloping terrain.

Aspect Insufficient data.

Hydrology Typical of moist soils on river banks and occurring less commonly in topogenous mire.

Soil pH All records in the pH range 5.0-8.0.

Bare soil Particularly associated with habitats having a moderate to high exposure of bare earth.

△ **ordination** Distribution fairly wide-ranging, but centred on vegetation of high productivity and experiencing a moderate degree of disturbance. Absent from highly disturbed habitats and from very unproductive vegetation.

Associated floristic diversity Relatively low.

Synopsis *I.g.* is a tall, summer-annual herb, typically forming linear stands along river banks on fertile alluvial soils. In these sites populations are re-established annually from large seeds which are dispersed

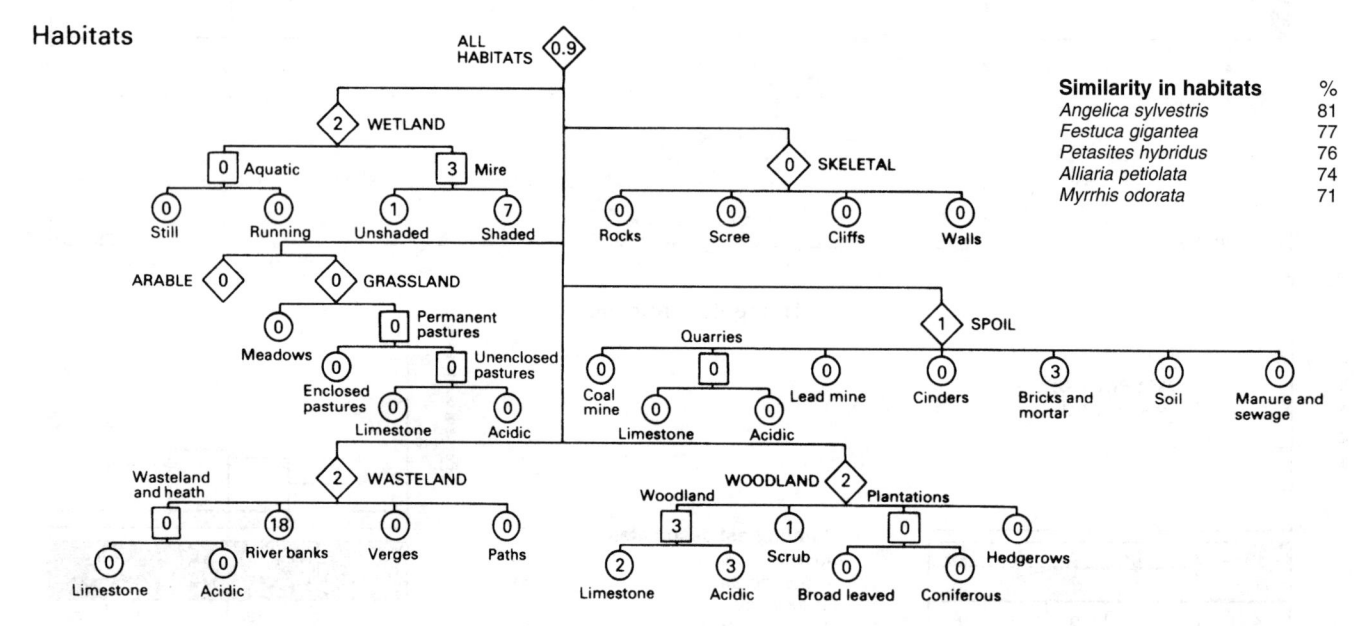

Habitats

Similarity in habitats	%
Angelica sylvestris	81
Festuca gigantea	77
Petasites hybridus	76
Alliaria petiolata	74
Myrrhis odorata	71

explosively in autumn. About 800 seeds are produced by each medium-sized plant (Salisbury 1964). Seeds must experience chilling temperatures in order to become capable of germination. Germination of the entire seed population occurs synchronously in spring, and leads to the establishment of dense even-aged stands in which other species, including perennials, may be submerged. River banks colonized by *I.g.* frequently have reduced floristic diversity. In the British flora ruderal dominance such as that exercised by *I.g.* is relatively uncommon except in weed-infested arable land. *I.g.*, a native of Himalaya, is at 2 m the tallest annual in the British flora. It is interesting to note, too, that the tallest monocarpic and polycarpic perennial herbs in the flora (excluding twining plants), respectively *Heracleum mantegazzianum* Sommier and Levier (SW Asia, 3.5 m) and *Fallopia sachalinensis* (F. Schmidt ex Maxim.) Nakai (Sakhalin Island, 3 m), are also aliens. The capacity of *I.g.* to exploit riverside sites appears to depend strongly upon overwintering by seed. Thus, unlike the perennials of alluvial terraces *I.g.* is relatively unaffected by winter flooding, silt deposition and soil erosion. *I.g.* is tolerant of moderate shade, and is able to flower and set seed in open woodland. We assume that seeds are frequently transported by water, since colonies on river banks extend much more rapidly than others. The species appears to be restricted to very moist habitats; shoots wilt rapidly after picking, and plants of drier sites have short internodes and small leaves. Like other British *Impatiens* spp., *I.g.* has very brittle, hollow stems with a high water content and a low proportion of vascular tissue, and relies upon buttressing by adventitious roots (cf. *Zea mays* L.) to maintain its erect stature. Shoots are very sensitive to autumn frosts. Flowers are insect-pollinated (Valentine 1978). Insect-pollinated species often experience problems with respect to pollination when introduced into foreign countries, and most British-introduced annual species are self-pollinated (Valentine 1978). *I.g.* is an exception in this respect. The degree of genetic heterogeneity in populations is uncertain, but flower colour frequently varies within populations.

Current trends *I.g.*, which was first recorded in the wild in Britain in 1848 (Valentine 1978), is still increasing.

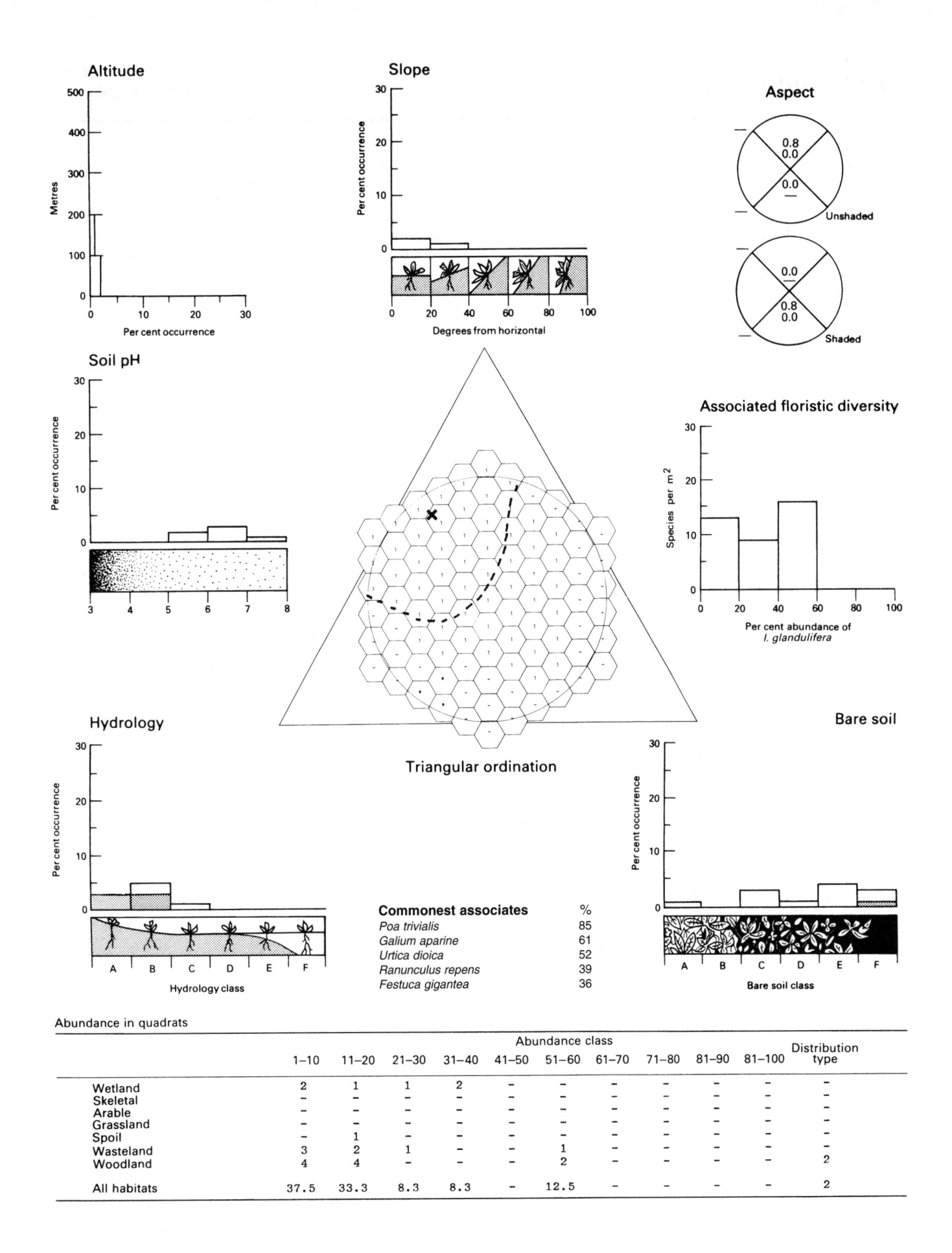

Altitude

Slope

Aspect

Unshaded

Shaded

Soil pH

Associated floristic diversity

Species per m^2 vs Per cent abundance of *I. glandulifera*

Hydrology

Hydrology class: A B C D E F

Triangular ordination

Commonest associates	%
Poa trivialis	85
Galium aparine	61
Urtica dioica	52
Ranunculus repens	39
Festuca gigantea	36

Bare soil

Bare soil class: A B C D E F

Abundance in quadrats

	1–10	11–20	21–30	31–40	41–50	51–60	61–70	71–80	81–90	81–100	Distribution type
Wetland	2	1	1	2	–	–	–	–	–	–	–
Skeletal	–	–	–	–	–	–	–	–	–	–	–
Arable	–	–	–	–	–	–	–	–	–	–	–
Grassland	–	–	–	–	–	–	–	–	–	–	–
Spoil	–	1	–	–	–	–	–	–	–	–	–
Wasteland	3	2	1	–	–	1	–	–	–	–	–
Woodland	4	4	–	–	–	2	–	–	–	–	2
All habitats	37.5	33.3	8.3	8.3	–	12.5	–	–	–	–	2

Inula conyzae DC.

Asteraceae
Ploughman's Spikenard

Life-form Monocarpic or polycarpic perennial, semi-rosette hemicryptophyte with oblique stock. Stems erect or ascending; leaves ovate, hairy, sometimes >4000 mm^2; root system relatively deep, with VA mycorrhizas.
Phenology Overwintering as a basal rosette. Flowers July to September and seed shed from September to November. Leaves relatively long-lived.
Height Flowering stems to 1300 mm, leafy to about half-way.
Seedling RGR No data. SLA 15.0. DMC 24.5.
Nuclear DNA amount 6.7 pg; $2n = 32^*$ (*BC*); $4x^*$.
Established strategy Intermediate between stress-tolerant ruderal and C-S-R.
Regenerative strategies W, ?B$_s$. Entirely by means of wind-dispersed seeds which germinate in vegetative gaps, probably mainly during spring. A persistent seed bank is suspected but has not been reported.
Flowers Yellow, *c.* 90 in each capitulum (Salisbury 1942). Central florets tubular, hermaphrodite, marginal florets shortly ligulate, female; visited by insects (Knuth 1906); usually >50 capitula in a terminal corymb.
Germinule 0.26 mg, 2.5 × 0.5 mm, achene dispersed with a pappus *c.* 10 mm in length.
Germination Epigeal; t_{50} 3 days; 13-30°C; L > D.
Biological flora None.
Geographical distribution Much of England and Wales, but absent from Scotland and Ireland (43% of British vice-counties), *ABIF* page 636; W, C and S Europe (67% of European territories); E Asia and N Africa.

Habitats Occurs mainly on the skeletal, calcareous soils of rock outcrops, quarry heaps, pastures and wasteland. Also observed on calcareous, sandy soils and known to be a frequent colonist of foundry spoil tips. A few rare occurrences in lightly shaded scrub or woodland. Absent from wetlands. [Found on coastal sand dunes.]
Gregariousness Sparse.
Altitude Most frequent at low (<100 m) altitudes and exhibiting an exceptionally low British altitudinal limit of 260 m.
Slope Found over a wide range of slopes.
Aspect Significantly more frequent on S-facing slopes.
Hydrology Absent from wetlands.
Soil pH Associated typically with base-rich soils, particularly in the pH range 7.0-8.0, but observed down to pH 5.0.
Bare soil Widely distributed, though most frequently associated with a moderate proportion of bare soil.
△ **ordination** A diffuse and uninformative pattern reflecting the limited amount of data and the inability of m^2 samples to characterize the microhabitats exploited by *I.c.*
Associated floristic diversity Often associated with relatively species-rich vegetation.

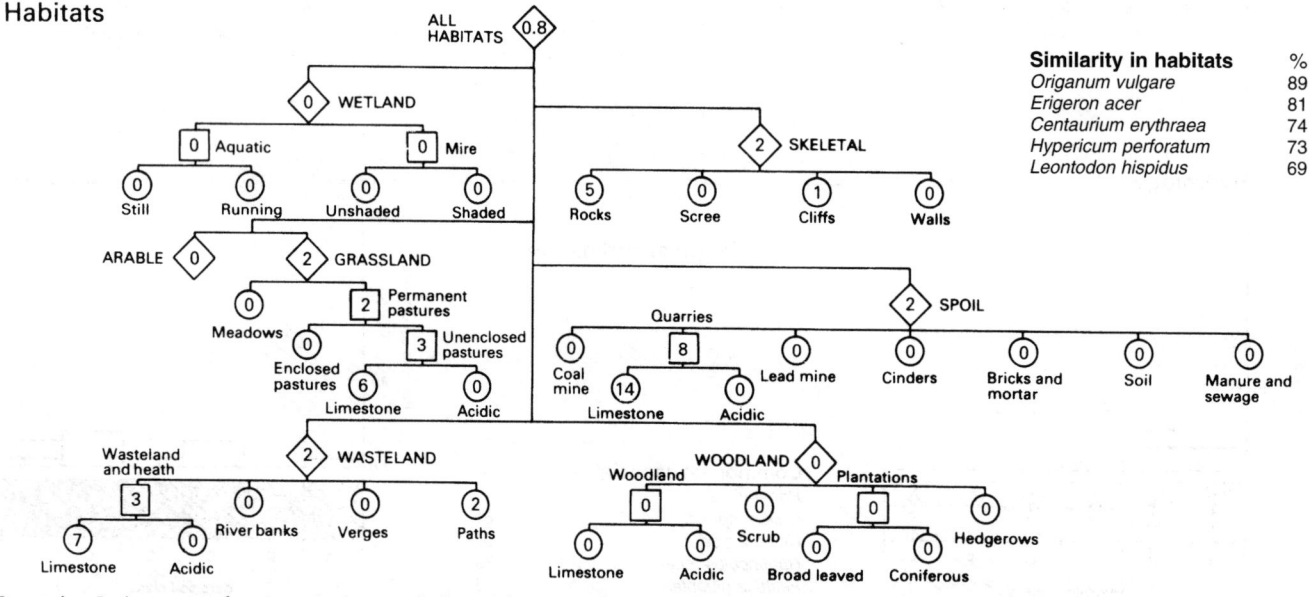

Habitats

Similarity in habitats	%
Origanum vulgare	89
Erigeron acer	81
Centaurium erythraea	74
Hypericum perforatum	73
Leontodon hispidus	69

Synopsis *I.c.* is a rosette-forming, winter-green herb, mainly restricted to dry infertile, calcareous soils. The ecology of the species is poorly understood. *I.c.* appears to function as a polycarpic perennial at the centre of its distribution in S Europe, but behaves as a monocarpic perennial at sites in Holland (van Gils & Huits 1978). A consistent feature of sites with *I.c.* in Britain is that they are habitats with discontinuous vegetation cover. This is doubtless because regeneration is entirely dependent upon small seeds. Seedlings exploit vegetation gaps and appear mainly during autumn. During establishment *I.c.* is a rosette species with little capacity for lateral vegetative spread. *I.c.* may be fatally toxic to sheep and cattle and is little grazed (Cooper & Johnson 1984). However, as in Holland (van Gils & Huits 1978), *I.c.* is most abundant in unmanaged sites and we suspect that the species is particularly characteristic of dry wasteland at a successional stage intermediate between open ground and tall grassland. *I.c.* is mainly restricted to S Britain, and the distribution is biased towards low altitudes and S-facing slopes. Thus, despite an apparent abundance of suitable habitats, *I.c.* shows a restricted distribution on the upland carboniferous limestone and the species reaches its British altitudinal limit on this substratum. As in the case of *Centaurea scabiosa*, therefore, there are indications that some stages in the life-history may be unusually thermophilous, and that the geographical and ecological distribution of *I.c.* is restricted by climatic factors. Despite an association with S-facing slopes, *I.c.* is largely absent from the most heavily droughted sites and in dry situations, we suspect that the deep root system may sustain water supply during the peak of growth in summer. A persistent seed bank is suspected but has not been demonstrated. The relative size of the achene and pappus appear well suited for long-distance dispersal (Salisbury 1964) and *I.c.* has colonized a number of foundry spoil tips situated at a considerable distance from the calcareous strata to which the species is normally restricted.

Current trends *I.c.* is probably increasing as a colonist of dry, disturbed, usually calcareous habitats.

Altitude
Slope
Aspect
$p < 0.05$

	0.8	
	50.0	
3.1		0.0

Unshaded

	0.0	
0.8		0.0

Shaded

Soil pH

Associated floristic diversity

Hydrology
Absent from wetland

Triangular ordination

Bare soil

Commonest associates %
Arrhenatherum elatius 82
Teucrium scorodonia 82
Brachypodium sylvaticum 64
Festuca ovina 64
Viola riviniana 64

Abundance in quadrats

	1–10	11–20	21–30	31–40	41–50	51–60	61–70	71–80	81–90	81–100	Distribution type
Wetland	–	–	–	–	–	–	–	–	–	–	–
Skeletal	3	–	–	–	1	–	–	–	–	–	–
Arable	–	–	–	–	–	–	–	–	–	–	–
Grassland	2	1	–	–	–	–	–	–	–	–	–
Spoil	8	1	–	–	–	–	–	–	–	–	–
Wasteland	7	–	–	–	–	–	–	–	–	–	–
Woodland	–	–	–	–	–	–	–	–	–	–	–
All habitats	87.0	8.7	–	–	4.3	–	–	–	–	–	1

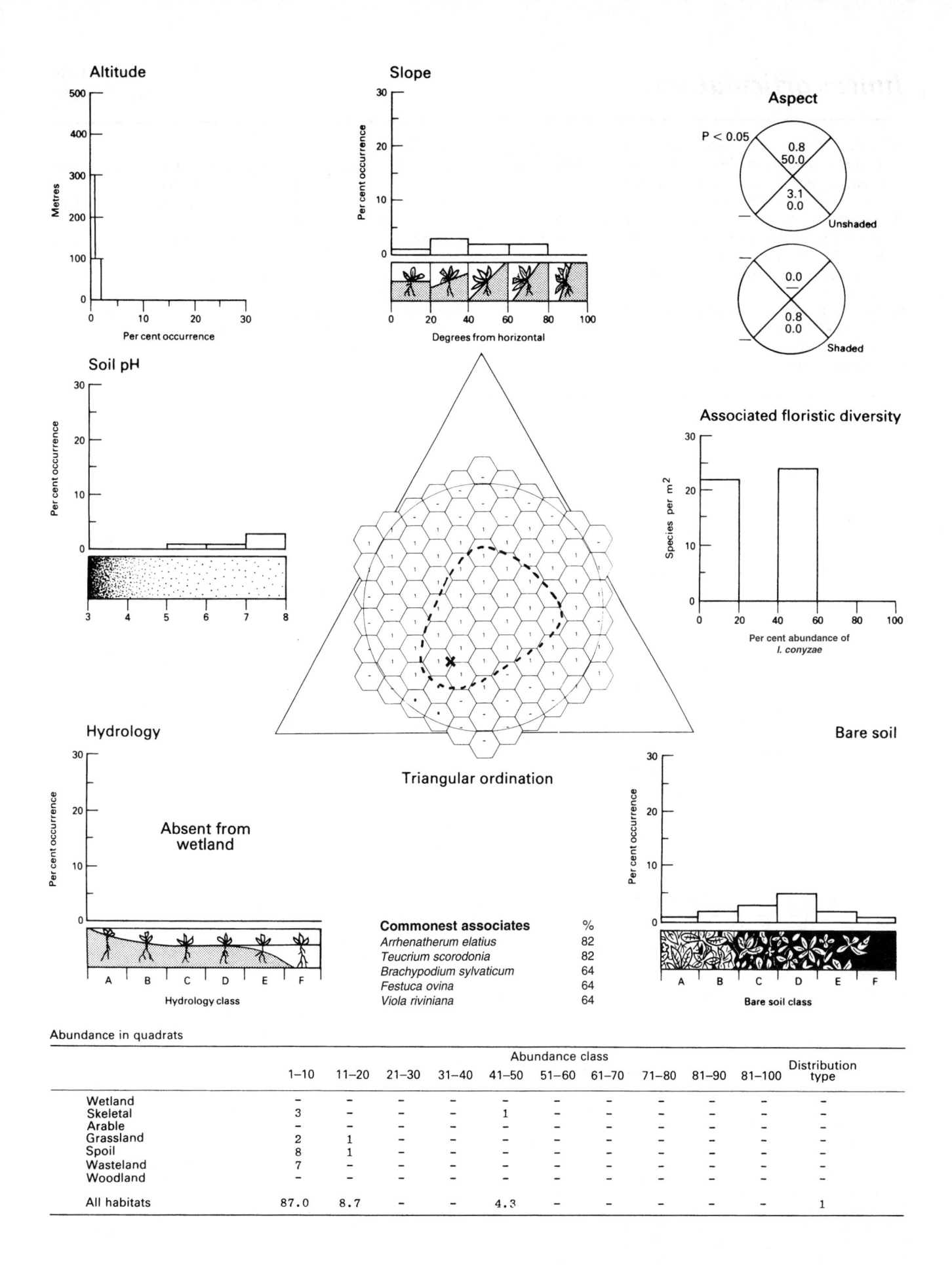

Juncus articulatus L.

Juncaceae
Jointed Rush

Life-form Tufted or rhizomatous, creeping, polycarpic perennial, semi-rosette hemicryptophyte or helophyte. Stems ascending, decumbent or prostrate; leaves linear, <500 mm²; roots sometimes with VA mycorrhizas.

Phenology Winter green. Flowers June to September and seed set from September to October. Leaves relatively long-lived.

Height Variable, from 20 mm for both foliage and inflorescence in prostrate forms to *c.* 600 mm in the case of robust erect plants.

Seedling RGR No data. SLA 32.7. DMC 18.7.

Nuclear DNA amount 3.6 pg; $2n = 80^*$ (*BC*, *FE5*); $4x^*$ (see Zandee 1981).

Established strategy Intermediate between competitive ruderal and C-S-R.

Regenerative strategies V, B_s. Regenerates effectively both vegetatively and by seed (see 'Synopsis'); SB IV.

Flowers Brown, hermaphrodite, protogynous, wind-pollinated; up to >100 in a cymose inflorescence.

Germinule 0.02 mg, 0.7 × 0.3 mm, seed dispersed from a dehiscent capsule containing *c.* 40 seeds (CTW).

Germination Epigeal; t_{50} 4 days; 17-31°C; L > D.

Biological flora None.

Geographical distribution The British Isles (100% of British vice-counties), *ABIF* page 692; Europe except the extreme N (97% of European territories); Asia, eastern N America and N Africa; introduced in S Africa, Australia and New Zealand.

Habitats A species of permanently or temporarily waterlogged habitats. Most commonly recorded from unshaded mire. Also found on stream- or river-sides and at the margin of ponds, lakes and ditches. Recorded too from wet pastures, wasteland, walls by water and from paths. Absent from dry and from shaded habitats. [Found in dune slacks and montane habitats.]

Gregariousness Intermediate.

Altitude Found over a wide range of values to 400 m, but suitable habitats more frequent in upland areas. [Up to 790 m.]

Slope As with most wetland species, confined to flat or gently sloping ground.

Aspect Insufficient data.

Hydrology Typically a species of soligenous mire, but also found as an emergent or floating aquatic in flowing water, in topogenous mire and in sites which are flooded in winter and moist in summer.

Soil pH Not recorded below pH 4.5.

Bare soil Found over a wide range of values; in some sites with low exposures there are extensive areas of bryophyte mat.

△ ordination Occurs mainly in vegetation types of moderate to high productivity which are subject to some disturbance.

Associated floristic diversity Typically intermediate to species-rich.

Synopsis *J.a.* is a morphologically variable (low-growing to relatively tall), winter-green rush which is most typically found in base-rich mire in situations where disturbance, and often also a degree of soil infertility, limit the growth of potential dominants. *J.a.* is moderately resistant both to grazing and to trampling, and is often found in grazed mire, where it adopts a more prostrate growth form and is usually smaller than when growing in ungrazed sites. *J.a.* is also found beside streams in close proximity to fast-flowing water, and may even behave as a semi-aquatic, forming floating rafts in upland rills or drainage ditches but usually attached to the bank. In its pronounced morphological plasticity and wide ecological amplitude in wetland habitats, *J.a.* shows ecological affinities with the more shade-tolerant *Ranunculus flammula*. *J.a.* is edaphically wide-ranging, occurring on mildly acidic to highly calcareous soils. The leaves and stems have, like most wetland species, well-developed air spaces but *J.a.* exhibits only moderate tolerance of ferrous ion (R. E. D. Cooke unpubl.). *J.a.*, particularly in its prostrate forms, may root at the nodes to form extensive clonal patches. This is especially evident close to water, and appears to be important in the colonization of stream margins. Even in grazed sites plants produce large numbers of minute seeds. These germinate mainly during spring in vegetation gaps, or are incorporated into a persistent seed bank. The seeds, which are mucilaginous and adhesive when moist, are widely dispersed even between landlocked sites, and are probably transported in mud. The two other frequent jointed rushes, *J. acutiflorus* Ehrh. ex Hoffm. and *J. subnodulosus* Schrank, are both taller, stand-forming dominants of relatively undisturbed mire of moderate productivity. Both are often found on peat; the former on mildly acidic, the latter on base-rich, frequently calcareous substrata. *J.a.* hybridizes in Britain with three other rushes, the commonest being the hybrid with *J. acutiflorus* which forms clonal patches (Stace 1975). A large anomalous form of *J.a.* described by Timms and Clapham (1940) may also prove to be of hybrid origin.

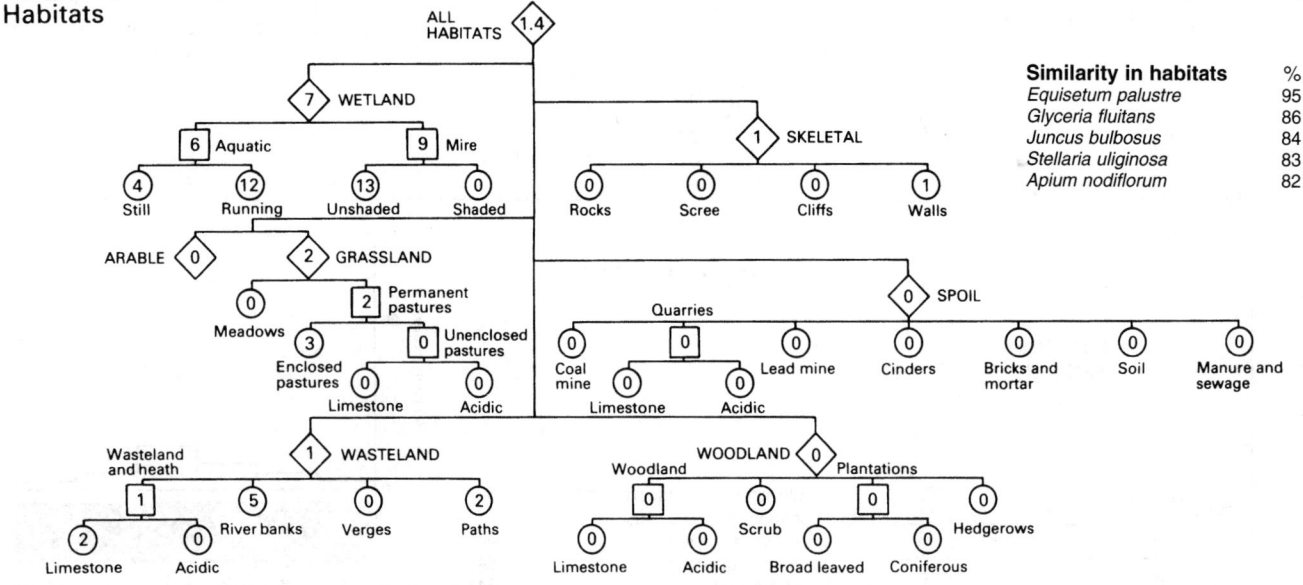

Current trends Although an effective colonist, probably decreasing as a result of wetland habitat destruction, and perhaps also declining under the impacts of eutrophication.

Similarity in habitats	%
Equisetum palustre	95
Glyceria fluitans	86
Juncus bulbosus	84
Stellaria uliginosa	83
Apium nodiflorum	82

Altitude
Slope
Aspect

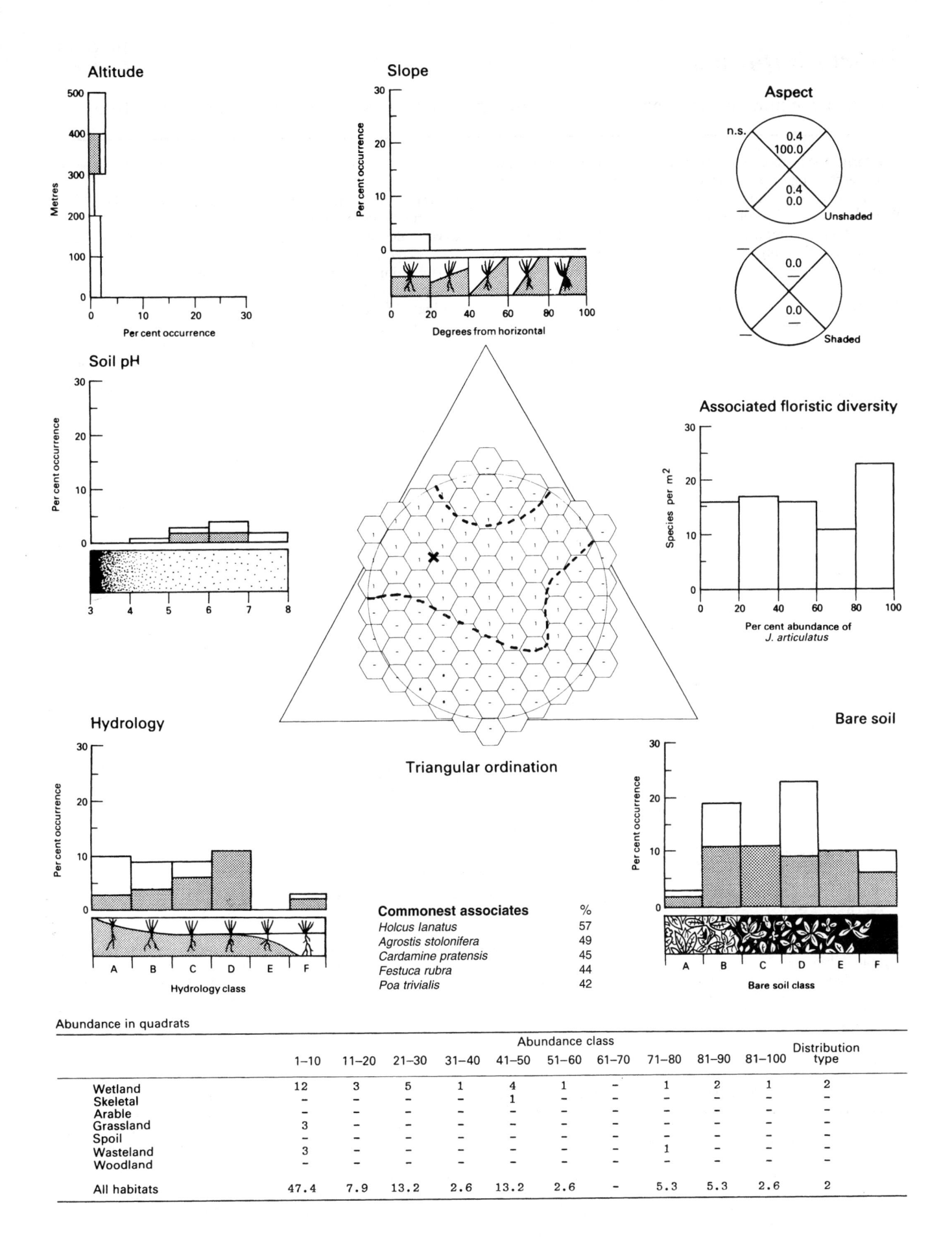

Soil pH
Associated floristic diversity

Species per m^2 / Per cent abundance of *J. articulatus*

Hydrology
Triangular ordination
Bare soil

Commonest associates	%
Holcus lanatus	57
Agrostis stolonifera	49
Cardamine pratensis	45
Festuca rubra	44
Poa trivialis	42

Abundance in quadrats

	1–10	11–20	21–30	31–40	41–50	51–60	61–70	71–80	81–90	81–100	Distribution type
Wetland	12	3	5	1	4	1	–	1	2	1	2
Skeletal	–	–	–	–	1	–	–	–	–	–	–
Arable	–	–	–	–	–	–	–	–	–	–	–
Grassland	3	–	–	–	–	–	–	–	–	–	–
Spoil	–	–	–	–	–	–	–	–	–	–	–
Wasteland	3	–	–	–	–	–	–	1	–	–	–
Woodland	–	–	–	–	–	–	–	–	–	–	–
All habitats	47.4	7.9	13.2	2.6	13.2	2.6	–	5.3	5.3	2.6	2

Juncus bufonius L.

Juncaceae
Toad Rush

Including *J. minutulus* Albert and Jahordiez, now considered conspecific with *J.b.* (Cope & Stace 1985)

Life-form Semi-rosette, summer-annual therophyte. Stems erect or ascending; leaves linear, <100 mm^2; roots shallow, sometimes with VA mycorrhizas.

Phenology Germination in spring or summer, varying according to habitat conditions; flowering period May to September and fruiting period July to October. By November plants normally dead.

Height Foliage <100 mm; flowering stems to 250 mm.

Seedling RGR No data. SLA 17.1. DMC 13.6.

Nuclear DNA amount 2.6 pg; $2n$ = 70, 100–110, 108* (*FE5*, Cope & Stace 1985); probably *c.* $4x$, $6x$* (Cope & Stace 1985).

Established strategy Intermediate between ruderal and stress-tolerant ruderal.

Regenerative strategies B_s. By seed; successful establishment mainly in spring and early summer. A large buried seed bank is formed; ?SB IV.

Flowers Green, hermaphrodite, cleistogamous; usually <50 in a cymose inflorescence.

Germinule 0.02 mg, 0.6 × 0.4 mm, seed in a dehiscent <100-seeded capsule.

Germination Epigeal; t_{50} 7 days; dormancy broken by dry storage at 20°C; L > D.

Biological flora None.

Geographical distribution The British Isles (100% of British vice-counties), *ABIF* page 689; Europe (97% of European territories); most temperate zones worldwide, but probably only native in Eurasia, N Africa and N America (Cope & Stace 1978).

Habitats Restricted to wet, disturbed habitats. Most frequent in unshaded mire, on rutted tracks or paths and on river banks. Also found in areas of arable land with impeded drainage, and in trampled areas in pastures. More rarely recorded from wet cliffs and walls and observed occasionally on cinders, coal-mine spoil and demolition sites. Absent from woodland and aquatic habitats. [Extends into dune and salt-marsh communities (Cope & Stace 1978).]

Gregariousness Sparse to intermediate.

Altitude Widely distributed and observed to 400 m. [Up to 600 m.]

Slope Recorded over a wide range of values although, like other wetland species, more frequent and abundant on relatively flat ground.

Aspect No significant trends.

Hydrology Most frequent and only abundant in mire adjacent to water. More rarely observed from trampled soligenous mire. Not recorded from submerged habitats.

Soil pH Absent from acidic soils below pH 4.0 but also infrequent on calcareous strata.

Similarity in habitats Insufficient data.

Habitats

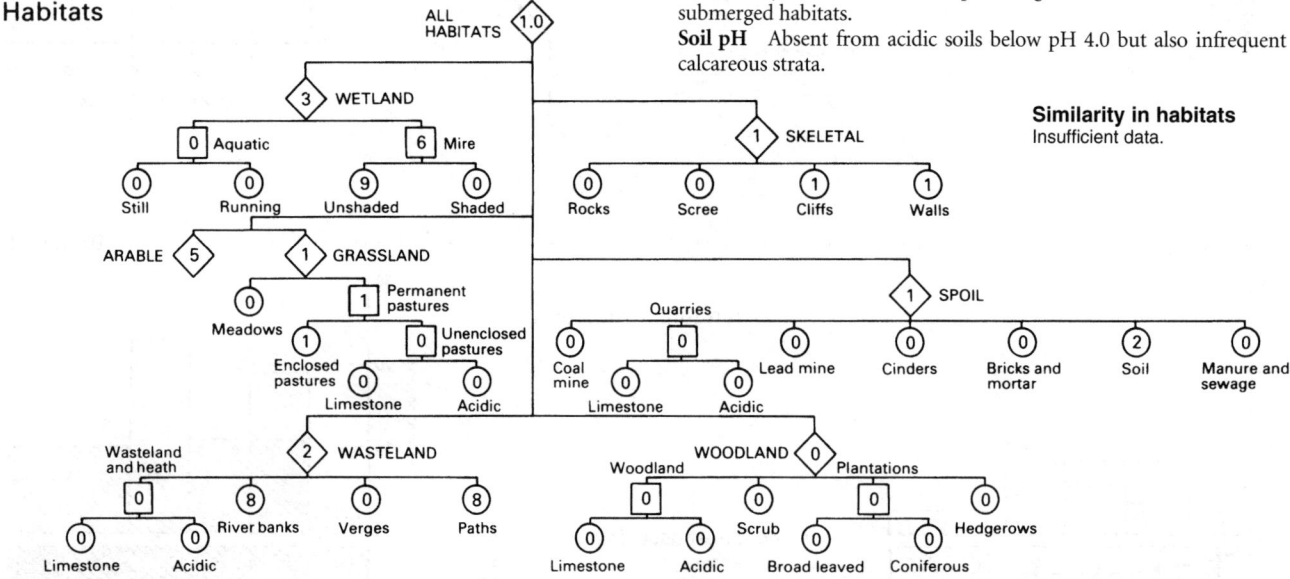

ALL HABITATS 1.0
- WETLAND 3
  - Aquatic 0
    - Still 0
    - Running 0
  - Mire 6
    - Unshaded 9
    - Shaded 0
- ARABLE 5
- GRASSLAND 1
  - Meadows 0
  - Permanent pastures 1
    - Enclosed pastures 1
      - Limestone 0
      - Acidic 0
    - Unenclosed pastures 0
- SKELETAL 1
  - Rocks 0
  - Scree 0
  - Cliffs 1
  - Walls 1
- Quarries 0
  - Coal mine 0
    - Limestone 0
    - Acidic 0
  - Lead mine 0
- SPOIL 1
  - Cinders 0
  - Bricks and mortar 0
  - Soil 2
  - Manure and sewage 0
- WASTELAND 2
  - Wasteland and heath 0
    - Limestone 0
    - Acidic 0
  - River banks 8
  - Verges 0
  - Paths 8
- WOODLAND 0
  - Woodland 0
    - Limestone 0
    - Acidic 0
  - Scrub 0
    - Broad leaved 0
    - Coniferous 0
  - Plantations 0
  - Hedgerows 0

Bare soil Most frequently associated with habitats having a high exposure of bare soil.

△ ordination Absent from unproductive vegetation and excluded from communities dominated by perennial plants. Most frequently associated with disturbed fertile habitats.

Associated floristic diversity Varies widely, but typically intermediate to species-rich.

Synopsis The only common obligately annual, non-graminaceous monocotyledon in the British flora, *J.b.* is most typical of fertile sites which are submerged during winter and remain moist during summer. The species is low-growing, short-lived and without the capacity for vegetative expansion. Thus, *J.b.* is restricted to (a) continuously open sites such as exposed mud beside ponds and reservoirs, and (b) vegetated sites subject to moderate intermittent disturbance such as rutted cart-tracks and path margins. Plants may flower within *c.* 4 weeks and in dry habitats may be small and short-lived. However, in favourable sites where the level of competition is low, numerous flowering stems may be produced over a protracted period. *J.b.* regenerates by means of numerous minute seeds, and forms a persistent seed bank. Seeds can survive ingestion by cattle and horses, are mucilaginous and adhesive but do not float. However, seedlings float (Ridley 1930) and may be transported in water to new sites. Detached shoots, when young, are capable of rooting to form new plants; this capacity is probably significant mainly in enabling the plant to recover *in situ* following disturbance. The *J. bufonius* aggregate is represented in Europe by four near-diploids, differing in ecological and geographical distribution, and *J.b. sensu stricto*, which is *c.* $4x$ and $6x$ (Cope & Stace 1985). It is suggested that *J.b.* has arisen as a result of hybridization amongst these or related species, a process which may be continuing, and *J.b.* is the most morphologically variable and ecologically diverse of the group (Cope & Stace 1985). One of the near-diploids, *J. ambiguus* Guss. (*J. ranarius* Song. & Perr.) occurs in saline habitats and is allogamous (Loenhoud & Sterk 1976). *J.a.* might be expected to occur on salted roadsides, but has not yet been recorded in this habitat.

Current trends A highly mobile species which exploits disturbed, fertile, artificial habitats. *J.b.* is one of the few wetland species which is clearly increasing.

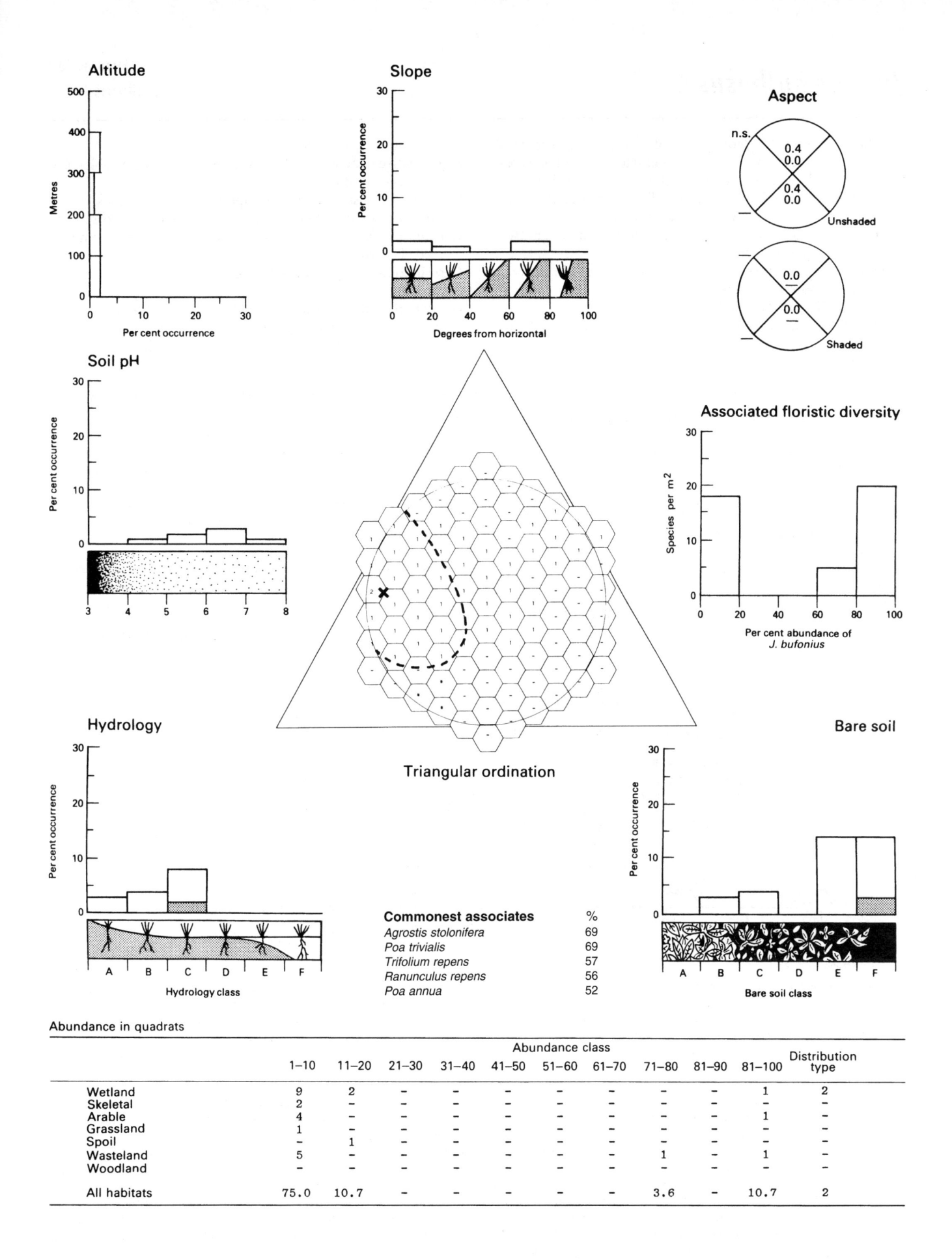

Altitude

Slope

Aspect

n.s.
| 0.4 | 0.0 |
| 0.0 | 0.4 |
Unshaded

| 0.0 | — |
| — | 0.0 |
Shaded

Soil pH

Associated floristic diversity

Hydrology

Triangular ordination

Bare soil

Commonest associates	%
Agrostis stolonifera	69
Poa trivialis	69
Trifolium repens	57
Ranunculus repens	56
Poa annua	52

Abundance in quadrats

	Abundance class										Distribution type
	1–10	11–20	21–30	31–40	41–50	51–60	61–70	71–80	81–90	81–100	
Wetland	9	2	–	–	–	–	–	–	–	1	2
Skeletal	2	–	–	–	–	–	–	–	–	–	–
Arable	4	–	–	–	–	–	–	–	–	1	–
Grassland	1	–	–	–	–	–	–	–	–	–	–
Spoil	–	1	–	–	–	–	–	–	–	–	–
Wasteland	5	–	–	–	–	–	–	1	–	1	–
Woodland	–	–	–	–	–	–	–	–	–	–	–
All habitats	75.0	10.7	–	–	–	–	–	3.6	–	10.7	2

Juncus bulbosus L.

Juncaceae
Bulbous Rush

Life-form Either a densely tufted, polycarpic perennial, semirosette, hemicryptophyte, a helophyte with a bulbous base and spreading shoots, rooted at the nodes, or a floating or submerged hydrophyte with long trailing stems. Leaves hairlike, <100 mm²; roots shallow, non-mycorrhizal.

Phenology Winter green. Flowers June to September and seed set from August to October.

Height Foliage to 100 mm and flowering stems to 150 mm, or a floating or submerged aquatic.

Seedling RGR No data. SLA 19.2. DMC 21.3.

Nuclear DNA amount No data; $2n = 40$ (*FE5*); $2x$ (see Zandee 1981).

Established strategy Intermediate between stress-tolerator and stress-tolerant ruderal.

Regenerative strategies ?B_s, V. Variable according to habitat (see 'Synopsis'); ?SB IV.

Flowers Reddish-brown, hermaphrodite, protogynous, wind-pollinated (Muller 1883); often < 50 in a cymose inflorescence.

Germinule 0.03 mg, 0.5 × 0.1 mm, seed dispersed from a many-seeded capsule.

Germination Epigeal; t_{50} 5 days; L > D.

Biological flora None.

Geographical distribution The British Isles although rare in parts of C, S and E England (100% of British vice-counties), *ABIF* page 693; much of Europe except the SE (74% of European territories); N Africa.

Habitats Occurs as a submerged or floating aquatic in peaty pools, ditches and at the edge of streams on non-calcareous strata. Equally frequent in unshaded mire, and also found on stream banks in moorland areas. Extends locally into habitats with well-drained but continuously moist soils.

Gregariousness Sparse to intermediate.

Altitude Most frequent in wet peaty habitats on gritstone over 300 m. Below this altitude apparently restricted by a shortage of habitats. [Up to 1070 m.]

Slope As with most wetland species, mainly confined to flat ground or gentle slopes.

Aspect Insufficient data.

Hydrology May occur as partially floating mats in ditches, lakes and streams up to 750 mm in depth. Also frequent in soligenous mire and recorded in peat cuttings (ombrogenous mire). May become quite abundant on mire at the margin of reservoirs and pools.

Soil pH Not recorded below pH 4.0 or above pH 7.0. Most records from within the pH range 5.0-5.5. Absent from calcareous strata.

Bare soil Insufficient data.

△ **ordination** Insufficient data.

Associated floristic diversity Relatively low, particularly in aquatic habitats where *J.b.* occurs at greatest frequency.

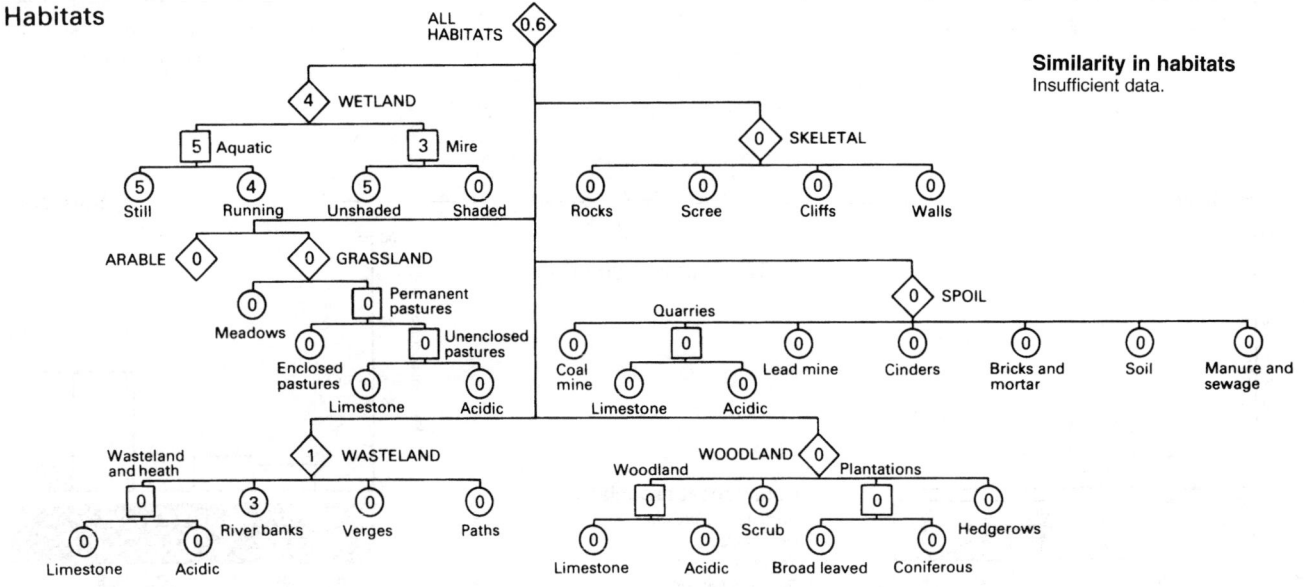

Habitats

Similarity in habitats
Insufficient data.

Synopsis *J.b.* is a low-growing rush of base-poor, often peaty wetland habitats where the growth of potential dominants is suppressed by the combined effects of infertility, relatively low pH and the various forms of wetland-habitat disturbances. Each plant consists of a tuft of basal leaves with axillary stems arising near the base. In mire, these stems are often short, whereas in aquatic systems they may reach 1 m (*FE5*). Leaves are also longer on submerged shoots, and the aquatic and mire forms are morphologically so dissimilar that they are scarcely recognizable as the same species. This phenotypic plasticity appears to render a single genotype capable of exploiting two different habitats (mire and aquatic) and may be a reflection of fluctuating watertable in many of the habitats with which *J.b.* is associated. Thus, *J.b.* often occurs in sites subject to seasonal variation in water depth (shallow pools, ditches, moorland reservoirs), and even in soligenous mire *J.b.* is often found in close proximity to running water. Populations may vary markedly in size from year to year (e.g. Macan 1977), and it seems likely that, as in other species of *Juncus*, populations are buffered by the presence of a persistent seed bank. However, *J.b.* occurs in several apparently more-stable habitats such as mountain lakes, and many aspects of its ecology are poorly understood. The leaves and stems, like those of most wetland species, have well-developed air spaces. In aquatic systems *J.b.* forms large floating patches, and detached portions of shoot readily regenerate and may even be carried some distance in the water. Unlike individuals growing in mire, aquatic forms do not flower and semi-aquatic plants are often viviparous. No information is available concerning the existence of ecotypes and *J. kochii* F. W. Schultz, which was formerly treated as a separate taxon, is no longer recognized.

Current trends Appears relatively mobile, with a capacity to colonize even land-locked sites; it may be encouraged by habitat disturbance. However, since *J.b.* is associated with infertile wetlands, it is decreasing rapidly in many lowland areas. In upland areas its status is more uncertain. *J.b.* is increasing in Dutch water systems, apparently as a consequence of 'acid rain' (Roelofs 1983). A similar trend should be looked for in upland Britain.

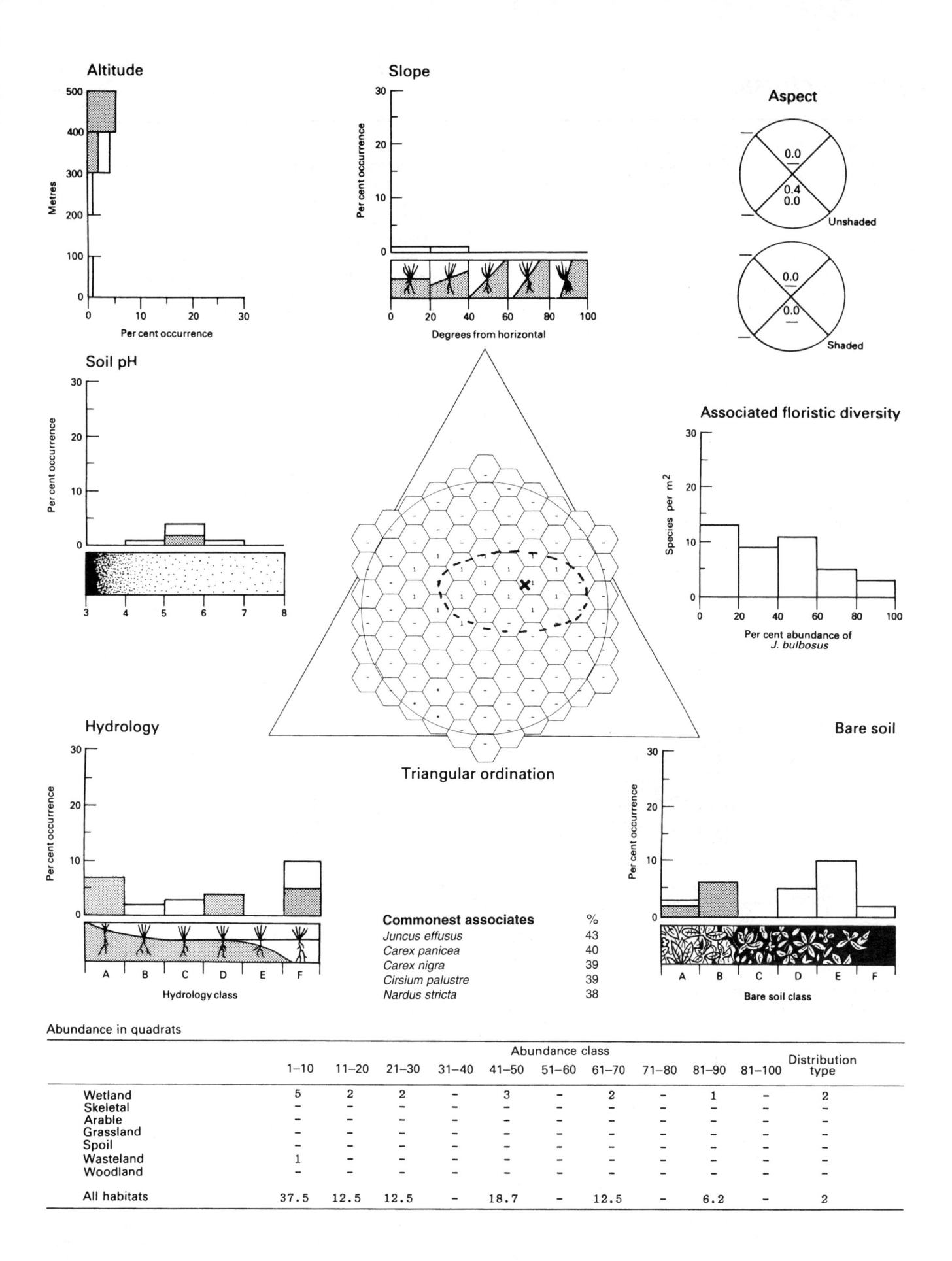

Altitude

Slope

Aspect

Unshaded

0.0	
0.4	0.0

Shaded

0.0	
0.0	

Soil pH

Associated floristic diversity

Species per m^2 vs Per cent abundance of *J. bulbosus*

Hydrology

Triangular ordination

Commonest associates   %
Juncus effusus   43
Carex panicea   40
Carex nigra   39
Cirsium palustre   39
Nardus stricta   38

Bare soil

Abundance in quadrats

	1–10	11–20	21–30	31–40	41–50	51–60	61–70	71–80	81–90	81–100	Distribution type
Wetland	5	2	2	–	3	–	2	–	1	–	2
Skeletal	–	–	–	–	–	–	–	–	–	–	–
Arable	–	–	–	–	–	–	–	–	–	–	–
Grassland	–	–	–	–	–	–	–	–	–	–	–
Spoil	–	–	–	–	–	–	–	–	–	–	–
Wasteland	1	–	–	–	–	–	–	–	–	–	–
Woodland	–	–	–	–	–	–	–	–	–	–	–
All habitats	37.5	12.5	12.5	–	18.7	–	12.5	–	6.2	–	2

Juncus effusus L.

Juncaceae
Soft Rush

Life-form Rhizomatous, polycarpic perennial hemicryptophyte. Stems erect with brownish sheathing leaves; roots lacking mycorrhizas. Green leaves, similar to those found in *J. bufonius*, restricted to very young plants.
Phenology Flowers from June to August and seed shed gradually from July or August onwards. Stems overwinter, but become brown towards their tips.
Height Fertile and sterile stems 300-1500 mm.
Seedling RGR 1.0-1.4 week^{-1}. SLA 5.5. DMC 25.4.
Nuclear DNA amount 0.6 pg; $2n$ = 40, 42 (*FE5*); probably $2x$.
Established strategy Intermediate between competitor and stress-tolerant competitor.
Regenerative strategies V, B_s. Effective regeneration both by seed and vegetative means (see 'Synopsis'); SB IV.

Flowers Brown, hermaphrodite, protogynous, wind-pollinated or occasionally cleistogamous (*Biol Fl*); >100 in a terminal inflorescence subtended by a long stem-like bract.
Germinule 0. 01 mg, 0.5 × 0.3 mm, seed; in a dehiscent, <100-seeded capsule.
Germination Epigeal; t_{50} 5 days; L > D.
Biological flora Richards and Clapham (1941).
Geographical distribution The British Isles (100% of British vice-counties), *ABIF* page 696; much of Europe except the Arctic (92% of European territories); widespread in N temperate regions, Australia and New Zealand.

Habitats Characteristic of a wide range of damp or waterlogged habitats, particularly unshaded mire. Also frequent on river banks, in pasture and in shaded mire. Occasional in aquatic habitats, wasteland and open woodland, and on waysides. Only recorded here from one type of spoil habitat (soil heaps), but observed in damp areas on cinder tips and coalmine spoil, and in gravel and sand pits. Absent from arable land and skeletal habitats except where adjacent to water.
Gregariousness Patch-forming.
Altitude Common at all altitudes, but suitable habitats are most abundant in the millstone grit and coal measure uplands. Extends to 470 m. [Up to 850 m.]
Slope Most frequent and abundant on flat ground and gentle slopes.
Aspect Few data are available, but apparently equally common on N- and S-facing slopes.
Hydrology Found in both soligenous and topogenous mire. Although frequent in sites marginal to open water, *J.e.* rarely occurs in permanently submerged habitats, except in recently flooded sites such as subsidence hollows where it may occur in >500 mm of water (B. D. Wheeler pers. comm.).
Soil pH Widely distributed, even occurring at pH values of 6.5, but less common above pH 7.0.
Bare soil Wide-ranging. Mature stands are often associated with persistent litter; young plants tend to be found on bare mud.

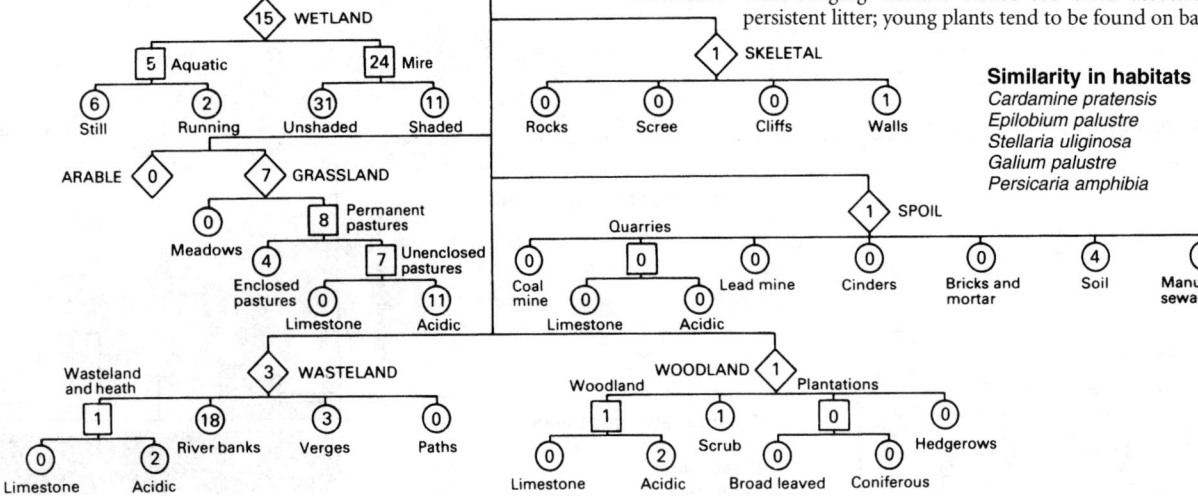

Habitats

Similarity in habitats	%
Cardamine pratensis	90
Epilobium palustre	87
Stellaria uliginosa	86
Galium palustre	85
Persicaria amphibia	83

△ ordination Absent from vegetation types which are both unproductive and subject to high levels of disturbance. Distribution centred on vegetation of moderate productivity and subjected to low intensities of disturbance.
Associated floristic diversity Typically low to intermediate.
Synopsis *J.e.* is a tall perennial with photosynthetic stems and scale-like basal leaves. *J.e.* is one of the most abundant and wide-ranging of wetland species, and is by far the most common British rush. *J.e.* is found on a wide range of moist or waterlogged soils, and is particularly widespread on the more-acidic mineral soils. Levels of N, P and Ca in the leaves are low, but Na and Fe are relatively high (*SIEF*). *J.e.* is frequent in grazed habitats, where it tends to be avoided by sheep or cattle, and is moderately tolerant of annual cutting and trampling (*Biol Fl*). *J.e.* also occurs in some shaded habitats, often without flowering. The success of *J.e.* in wetland appears to be related to (a) the dense basal structure of the tussocks and persistent litter, both of which tend to prevent the establishment of other species, (b) the extreme tolerance of ferrous ion toxicity (R. E. D. Cooke unpubl.), and (c) a regeneration mechanism which allows seed persistence in the soil and rapid colonization of disturbed habitats. Over 13 000 seeds may be produced by a single inflorescence (Salisbury 1964). Many are incorporated *in situ* into a buried seed bank, and small seed banks have been detected at sites where there is no historical record of *J.e.*, suggesting that there is effective dispersal of the seeds through animals or some other agency. Thus, *J.e.* is, in many mire habitats, among the first species to establish on soil bared by disturbance. Germination occurs particularly in spring. Seeds, which are mucilaginous when moist, sink in water but colonization of mud at the edge of drainage ditches and larger water bodies can arise through 'beaching' of seedlings which may be free-floating (Ridley 1930). However, seedlings do not appear to establish under water. Thus, *J.e.* normally colonizes only those sites which are above the water table for part of the year. *J.e.* will survive in shallow water if established plants are subject to a change in water table, e.g. as a result of land subsidence. Once established, regeneration by seed is much less important. *J.e.* is strongly rhizomatous and may form extensive clonal patches;

Altitude

Slope

Aspect

Soil pH

Associated floristic diversity

Hydrology

Triangular ordination

Bare soil

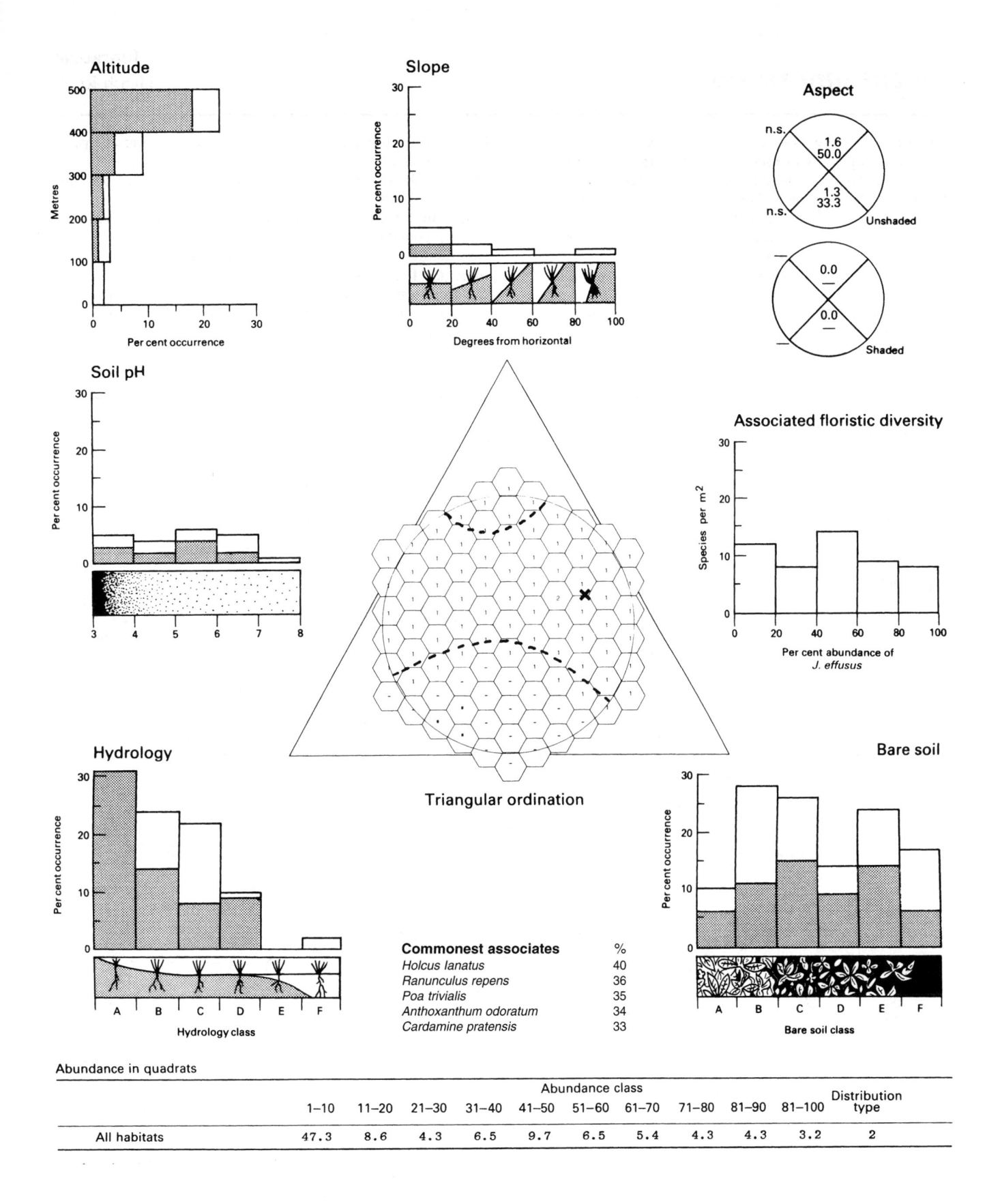

Commonest associates	%
Holcus lanatus	40
Ranunculus repens	36
Poa trivialis	35
Anthoxanthum odoratum	34
Cardamine pratensis	33

Abundance in quadrats

	1–10	11–20	21–30	31–40	41–50	51–60	61–70	71–80	81–90	81–100	Distribution type
All habitats	47.3	8.6	4.3	6.5	9.7	6.5	5.4	4.3	4.3	3.2	2

breakdown of the rhizome results in the isolation of daughter plants. Several genotypes have been recognized (*Biol Fl*), including var. *compactus* Hoppe, which is a variant of reduced stature normally found on upland acidic soils. *J.e.* is morphologically similar to *J. conglomeratus* L. and *J. inflexus* L., both of which are more restricted than *J.e.* to drier habitats or to soligenous mire. *J.c.* is less robust than *J.e.*, flowers earlier, and is particularly characteristic of sites of intermediate soil pH. *J.i.* is most abundant in base-rich, particularly calcareous conditions, and is much less tolerant of ferrous ion toxicity than *J.e.* (R. E. D. Cooke, unpubl.). *J.e.* hybridizes both with *J.c.* and with *J.i.* (Stace 1975).

Current trends One of the most rapid colonists of artificial wetland habitats and moist disturbed ground, and probably increasing.

Juncus squarrosus L.

Juncaceae
Heath Rush

Life-form Tufted or mat-forming, polycarpic perennial, rosette hemicryptophyte with a short branched rhizome. Leaves linear, often <250 mm^2; roots sometimes with VA mycorrhizas.
Phenology Winter green. Flowers mainly in June and July and seed shed from August onwards, but some may remain in the capsule until the following spring (*Biol Fl*). All leaves from previous year dead by middle of summer (*Biol Fl*).
Height Foliage generally <100 mm, flowering stems to 500 mm.
Seedling RGR 1.0-1.4 week^{-1}. SLA 4.7. DMC 36.8.
Nuclear DNA amount 1.1 pg; $2n = 42$ (*FE5*); genus with diffuse centromere.
Established strategy Intermediate between stress-tolerant and C-S-R.
Regenerative strategies V, B$_s$. Regenerates both vegetatively and by seed (see 'Synopsis'); ?SB IV.
Flowers Brown, hermaphrodite, protogynous, wind-pollinated (*Biol Fl*); <50 in a terminal cyme.
Germinule 0.05 mg, 0.8 × 0.5 mm, seed dispersed from a capsule often containing *c.* 50 seeds (*Biol Fl*).
Germination Epigeal; t_{50} 3 days; 10-30°C; L > D.
Biological flora Welch (1966).
Geographical distribution The British Isles, although only common in the N and W due to a shortage of suitable habitats elsewhere (97% of British vice-counties); *ABIF* page 688; C, W and NW Europe (51% of European territories); Morocco.

Habitats Mainly restricted to non-calcareous soils. Recorded from grazed or ungrazed moorland, moist heathland, unshaded mire, paths and outcrops. Found on quarry heaps and in sand pits, but otherwise absent from spoil. Absent from shaded, aquatic, arable and most skeletal habitats. [Widespread in montane areas, and recorded from dune-slacks (*Biol Fl*).]
Gregariousness Typically sparse to intermediate in the Sheffield region, but can form patches up to 5 m in diameter (*Biol Fl*).
Altitude Observed from 2 to >400, m but suitable habitats are most frequent in the gritstone uplands above 300 m. [Up to 1040 m.]
Slope Only recorded from slopes of <20°.
Aspect Insufficient data.
Hydrology A species of moist soils which also occurs in soligenous mire and at the edge of blanket peat.
Soil pH Only recorded from soils of pH <5.0 [although not exclusive to acidic soils (*Biol Fl*)].
Bare soil Associated with a wide range of values, but typically in sites with an incomplete cover of vegetation.
△ **ordination** Restricted mainly to undisturbed and unproductive vegetation types.
Associated floristic diversity Typically low.
Synopsis *J.s.* is a winter-green clump-forming rush developing flattened rosettes and exhibiting a rather restricted ecological and geographical range. The species is virtually confined to infertile acidic soils, and seedlings from sites of low pH grow very poorly and may be highly

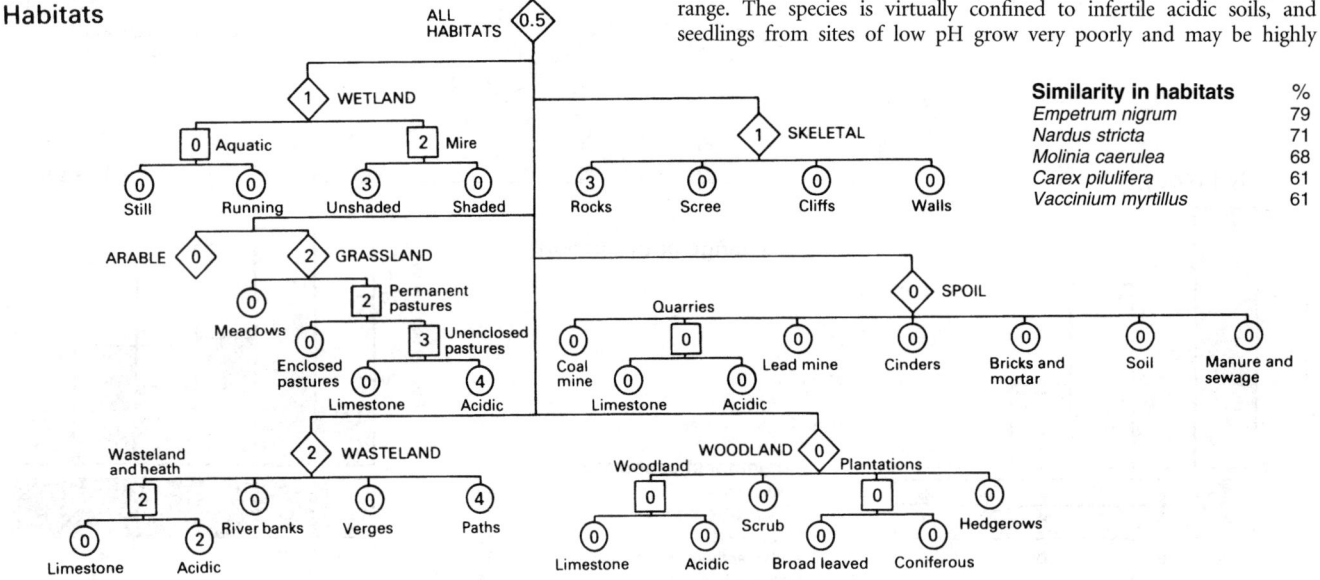

Habitats

Similarity in habitats	%
Empetrum nigrum	79
Nardus stricta	71
Molinia caerulea	68
Carex pilulifera	61
Vaccinium myrtillus	61

chlorotic when planted on calcareous soils (*Biol Fl*). Growth of *J.s.* is strongly inhibited by high external concentrations of calcium (Jefferies & Willis 1964). Leaves are rich in Na but have low Ca content (*SIEF*). *J.s.* is occasionally eaten by beef cattle and horses and, in winter and early spring when food is scarce, by sheep (*Biol Fl*). However, *J.s.* is generally avoided by stock, and is increasing in those upland areas which are now only grazed by sheep in summer. Thus, in the upland areas of N and W Britain, *J.s.* exploits a niche similar to, but wetter than, that of *Nardus stricta*. *J.s.* grows slowly and tends to be a very subordinate component of plant communities, particularly since its foliage is held close to the ground surface. *J.s.* is also intolerant of shade and is dependent upon the creation of short turf or bare ground by grazing or trampling (*Biol Fl*). *J.s.* is tolerant of occasional burning. Like *N.s.*, *J.s.* regenerates in established communities mainly by rhizomatous growth, to form larger clumps. This is particularly effective on moist soils, and die-back of older rhizomes in the centre of a patch results in the isolation of daughter plants. *J.s.* may release up to 10 000 seeds m^{-2} (*Biol Fl*), which germinate in vegetation gaps during spring or are incorporated into a persistent seed bank. *J.s.* can expand locally within sites bared by fire or other disturbances. Seeds are mucilaginous and adhesive, and the species shows some potential for long-distance colonization of new sites. Regeneration by seeds is probably infrequent in established vegetation (*Biol Fl*). The species exhibits little morphological variation (*Biol Fl*).
Current trends *J.s.* is relatively mobile and appears to have colonized a number of acidic artificial habitats from a long distance. However, the rate of appearance of these new habitats is much lower than the rate of destruction of heathland. Consequently *J.s.* is decreasing in many lowland areas. However, current management regimes in upland areas (i.e. reduction in the extent of winter and early spring sheep grazing) probably favour *J.s.* (see 'Synopsis').

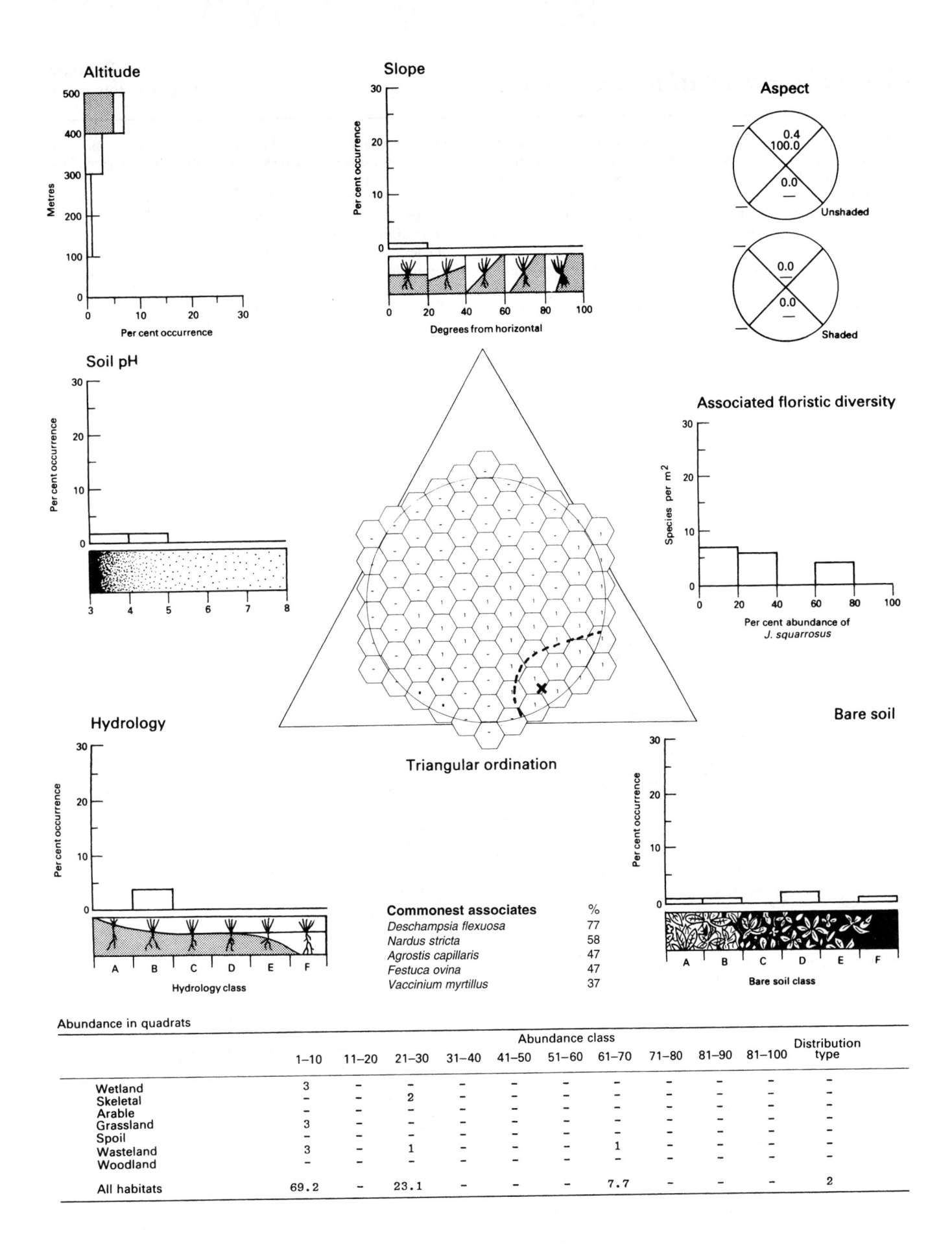

Altitude

Slope

Aspect

Unshaded

Shaded

Soil pH

Associated floristic diversity

Hydrology

Triangular ordination

Commonest associates	%
Deschampsia flexuosa	77
Nardus stricta	58
Agrostis capillaris	47
Festuca ovina	47
Vaccinium myrtillus	37

Bare soil

Abundance in quadrats

	\multicolumn{10}{c	}{Abundance class}	Distribution type
	1–10	11–20	21–30	31–40	41–50	51–60	61–70	71–80	81–90	81–100	
Wetland	3	–	–	–	–	–	–	–	–	–	–
Skeletal	–	–	2	–	–	–	–	–	–	–	–
Arable	–	–	–	–	–	–	–	–	–	–	–
Grassland	3	–	–	–	–	–	–	–	–	–	–
Spoil	–	–	–	–	–	–	–	–	–	–	–
Wasteland	3	–	1	–	–	–	1	–	–	–	–
Woodland	–	–	–	–	–	–	–	–	–	–	–
All habitats	69.2	–	23.1	–	–	–	7.7	–	–	–	2

Koeleria macrantha (Ledeb.) Schultes

Poaceae
Crested Hair-grass

Life-form Loosely tufted, polycarpic perennial protohemicryptophyte. Shoots erect or creeping; leaves linear, grooved on the upper surface, typically <500 mm^2; roots with VA mycorrhizas.
Phenology Winter green. Exhibits early spring growth. Flowers produced June and July and seed shed July and August. Leaves relatively long-lived.
Height Foliage usually <100 mm; flowering stems to 400 mm.
Seedling RGR 0.5-0.9 week^{-1}. SLA 9.7. DMC 32.7.
Nuclear DNA amount 9.3 pg; $2n$ = 14, 28, 42, 70 (FE5); $2x$, $4x$, $6x$, $10x$.
Established strategy Stress-tolerator.
Regenerative strategies S. Mainly by seed in vegetation gaps directly after shedding during autumn. Tufted and with only restricted capacity for lateral vegetative spread; SB I.

Flowers Green, hermaphrodite, homogamous, wind-pollinated, often automatically selfed (Knuth 1906); >100 aggregated into a spike-like panicle (2-3 per spikelet).
Germinule 0.3 mg, 2.7 × 0.5 mm, caryopsis dispersed with an attached lemma (4.9 × 1.0 mm).
Germination Epigeal; t_{50} 2 days; 6-31°C; L = D.
Biological flora Dixon (2000).
Geographical distribution Generally distributed throughout the British Isles (86% of British vice-counties), ABIF page 768; Europe except for the extremes of N and S (64% of European territories); temperate Asia and N America.

Habitats Frequently recorded in dry, semi-natural pasture, scree slopes in the limestone dales, rock outcrops, old lead-mine workings, derelict limestone pastures and limestone quarry heaps. Absent from woodland, wetland and arable land. [Occurs on coastal sand dunes and cliff-top turf.]

Gregariousness Intermediate.
Altitude Not recorded above 400 m, with most records predictably occurring in the range 200-400 m, corresponding to the distribution of the carboniferous limestone stratum in the survey area. [Up to 550 m.]
Slope Most common on daleside slopes in the range 20-60°.
Aspect Slightly more frequent and more abundant on N-facing slopes.
Hydrology Absent from wetlands.

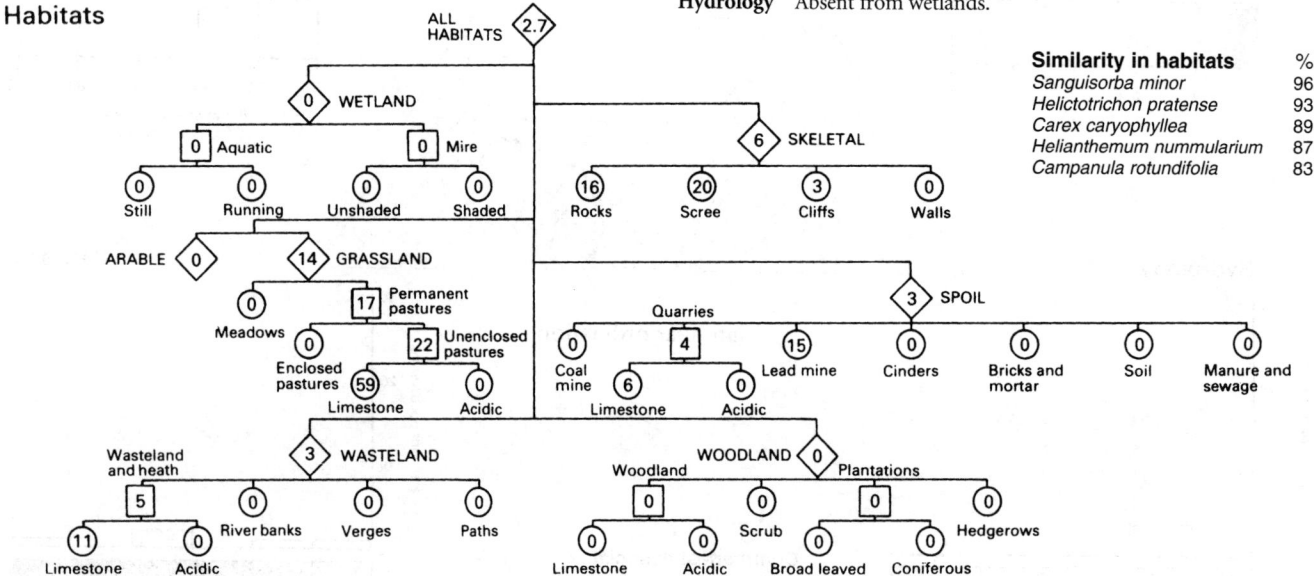

Habitats

Similarity in habitats	%
Sanguisorba minor	96
Helictotrichon pratense	93
Carex caryophyllea	89
Helianthemum nummularium	87
Campanula rotundifolia	83

Soil pH Frequent on calcareous and relatively base-rich soils, declining in frequency of occurrence below pH 4.5. Although not a strict calcicole in all parts of Britain, all records in the Sheffield region are from calcareous substrata.
Bare soil Mostly recorded with relatively low exposures, but in many habitats bare rock is also present, and the species is less frequent where there are no gaps in the vegetation.
△ **ordination** Distribution fairly clearly centred on vegetation associated with unproductive, relatively undisturbed conditions. Notably absent from heavily disturbed situations and circumstances conducive to intense competition.
Associated floristic diversity Generally high.
Synopsis *K.m.* is a slow-growing, loosely tufted, usually short, wintergreen grass most characteristic of infertile calcareous grassland and rock outcrops. *K.m.* shares several attributes with calcareous races of *Festuca ovina* including (a) high leaf-extension rates in early spring, (b) a short, early flowering period occurring before the onset of summer drought, (c) a dependence for regeneration on seed, which germinates soon after release, resulting in the appearance of cohorts of seedlings during autumn, (d) intermediate seed size, (e) a wide temperature range for seed germination and (f) the absence of a persistent seed bank. However, unlike *F.o.*, *K.m.* is never a dominant contributor to the shoot biomass. It seems likely that this may be related to the many aspects of its biology in which *K.m.* exhibits an intermediate degree of specialization compared with the various dominants with which it is able to coexist. Thus, *K.m.* is less xeromorphic than *F.o.* and appears more palatable to sheep, but is less productive and resilient under close-grazing than *Agrostis capillaris*. *K.m.* is also less robust than *Avenula pratensis*. Hence, because *K.m.* is an 'intermediate', its populations may be expected, under fluctuating conditions of climate, soil fertility and management, to show a greater degree of homeostasis than those of potential dominants. In this respect the ecology of *K.m.* in calcareous pasture is somewhat analogous to that of *Trisetum flavescens* in meadows. In some areas *K.m.* is less strongly calcicolous and may occur, for example, on sea cliffs in essentially base-poor soil which has been nutritionally enriched by sea spray. Plants with various ploidy levels are recorded from Europe (FE5), but cytological

Altitude
Slope
Aspect

	Unshaded
n.s.	8.7 / 45.5
n.s.	6.7 / 20.0

	Shaded
—	0.0
—	0.0

Soil pH

Associated floristic diversity

Hydrology

Absent from wetland

Commonest associates

	%
Festuca ovina	89
Festuca rubra	68
Campanula rotundifolia	66
Plantago lanceolata	65
Lotus corniculatus	57

Bare soil

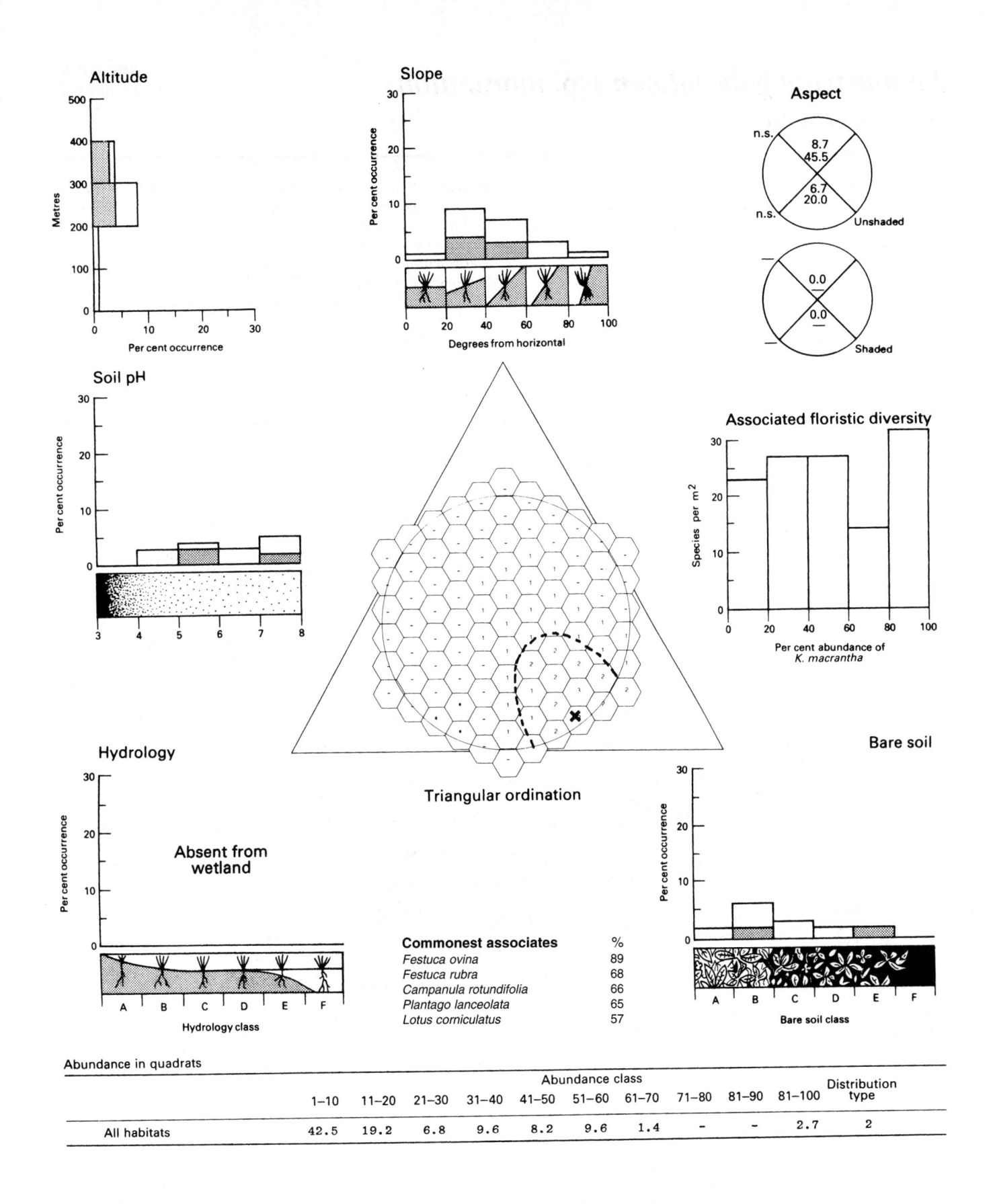

Triangular ordination

Abundance in quadrats

	1–10	11–20	21–30	31–40	41–50	51–60	61–70	71–80	81–90	81–100	Distribution type
All habitats	42.5	19.2	6.8	9.6	8.2	9.6	1.4	–	–	2.7	2

studies have not been carried out in Britain. Although *K.m.* has a relatively narrow ecological range in the Sheffield region, the existence of ecotypic differences between local populations has been demonstrated (Shaw 1984, Pearce unpubl.). Heavy-metal-tolerant populations have been recorded (Shaw 1984) and *K.m.* appears to be an 'excluder' *sensu* Baker (1981).

Hybridizes with *K. vallesiana* (Honckeny) Gaudia (Stace 1975).

Current trends Mainly restricted to semi-natural vegetation, and hence decreasing as a result of habitat destruction. Now close to extinction in some lowland areas.

Lamiastrum galeobdolon ssp. *montanum*
(Pers.) Ehrend. & Polatschek.

Lamiaceae
Yellow Archangel

Life-form Stoloniferous polycarpic perennial chamaephyte. Flowering stems ascending; leaves ovate, hairy, sometimes >2000 mm^2; roots shallow, with or without VA mycorrhizas.
Phenology Winter green. Flowers from May to June and nutlets shed from August onwards. Leaves long-lived.
Height Foliage mostly <300 mm, flowers to 600 m.
Seedling RGR 0.5-0.9 week^{-1}. SLA 22.9. DMC 27.0.
Nuclear DNA amount 6.5 pg; $2n = 36^*$ (*BC*), $4x^*$.
Established strategy Intermediate between stress-tolerator and stress-tolerant competitor.
Regenerative strategies V, S. Mainly vegetative (see 'Synopsis'); SB II.

Flowers Yellow, hermaphrodite, homogamous, insect-pollinated, rarely self-fertilized, often *c.* 40 in a whorled inflorescence (*Biol Fl*).
Germinule 1.98 mg, 3.5 × 1.9 mm, nutlet (4 per ovary).
Germination Epigeal; dormancy broken by chilling.
Biological flora Packham (1983).
Geographical distribution Common in much of England and Wales, rare and often introduced in Scotland, and only in SE Ireland (43% of British vice-counties), *ABIF* page 511; most of Europe, but rare in the Mediterranean region and the N (65% of European territories and 5% naturalized); Iran.

Habitats Common in mixed deciduous woodland and scrub, with scattered occurrences on banks of rivers and streams, and other shady or damp places, usually on heavier soils. [Rarely occurs in limestone pavement (*Biol Fl*).]
Gregariousness Intermediate.
Altitude Up to 300 m. [Up to 430 m (*Biol Fl*).]
Slope Virtually absent from skeletal habitats and from steep slopes.
Aspect In shaded habitats more frequent and significantly more abundant on N-facing slopes.

Hydrology Not typically a wetland species, but occurs on moist or even waterlogged soils, usually beside woodland streams.
Soil pH Occurs over a wide range of soil pH.
Bare soil A wide range of occurrences, but highest frequencies associated with intermediate levels of bare ground, often in association with the presence of tree litter.
Δ ordination Distribution centred on vegetation associated with unproductive and relatively undisturbed conditions; *L.g.* is occasionally found in more-productive situations. Absent from highly disturbed habitats.
Associated floristic diversity Relatively low.

Similarity in habitats	%
Arum maculatum	88
Mercurialis perennis	84
Anemone nemorosa	84
Sanicula europaea	83
Fraxinus excelsior (Juv.)	83

Habitats

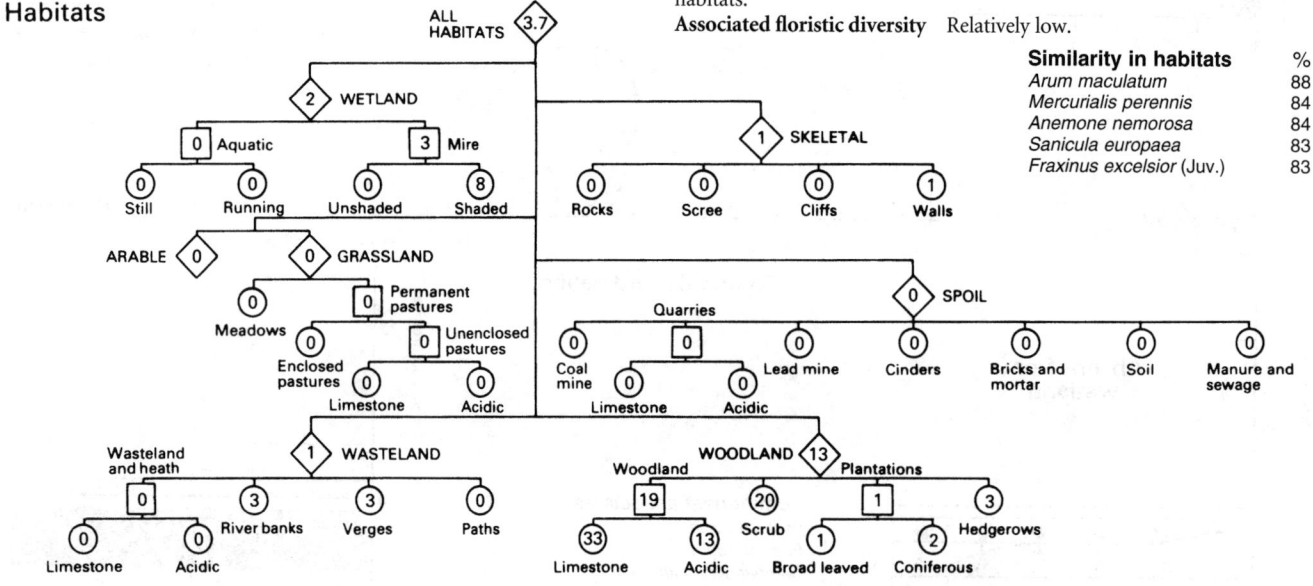

Synopsis *L.g.* is a slow-growing herb restricted to moist woodland and other shaded habitats. *L.g.*, which is winter green and produces both sun and shade leaves, is tolerant of deep shade but under these conditions flowering is suppressed (*Biol Fl*). *L.g.* is morphologically well-suited to exploit deep accumulations of tree litter in which it may form large, often diffuse stands as a result of stolon growth (Sydes & Grime 1981a,b). *L.g.* is cold-tolerant but susceptible to drought, and although found on a wide range of soils is scarce on sandy substrata. The greater abundance in N-facing woodland (see 'Aspect') probably relates both to the sensitivity to desiccation and to the shade tolerance exhibited by *L.g.* *L.g.* is susceptible to trampling (*Biol Fl*). The mineral nutrition of *L.g.* deserves attention since, despite a low growth rate, *L.g.* is regarded by some ecologists as a plant of relatively nutrient-rich soils (*Biol Fl*). Leaves contain high concentrations of N and Fe, but appear to be low in P and Mg (*SIEF*). *L.g.* produces long spreading stolons which root at the nodes and may form large clonal patches. Breakdown of the internodes results in the formation of daughter plants. We suspect that shoot pieces may be transported along woodland streams to form new colonies. Also regenerates by means of seeds, which germinate in spring. The balance between seed production and stolon development appears to be regulated by climate. In drier, open sites, and in unusually warm summers, more seed is set and stolons are shorter (*Biol Fl*). Thus, *L.g.* may be more dependent upon vegetative regeneration towards the N edge of its range, particularly in upland areas, than in S England. Seeds of *L.g.* have elaiosomes and are dispersed by ants (*Biol Fl*). Long-distance dispersal is generally poor and *L.g.* has been identified as an indicator species of older woodlands (*Biol Fl*). Ssp. *galeobdolon* with $2n = 18^*$ ($2x^*$), which is less vigorous and flowers earlier, occurs in N Lincolnshire (*Biol Fl*). Ssp. *m.* ($4x$), the main British taxon, is believed to have originated through hybridization between ssp. *g.* ($2x$) and ssp. *flavidum* (F. Hermann) Ehrend. & Polatschek ($2x$), and occupies an intermediate geographical range.
Current trends There appears to be no significant recruitment of *L.g.* into the wild from garden populations, and *L.g.* is presumably decreasing as a result of habitat destruction.

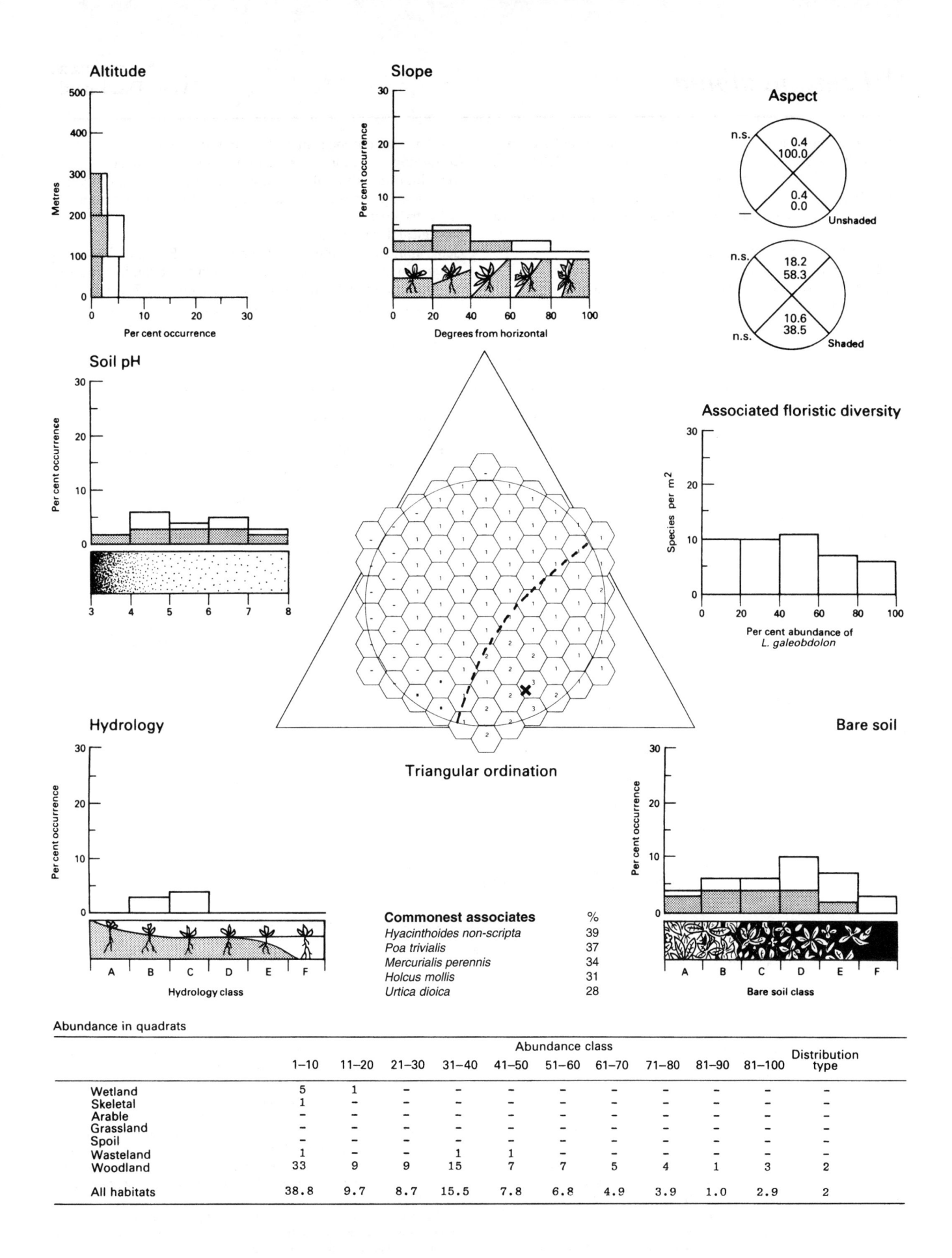

Altitude

Slope

Aspect

	n.s.	0.4 100.0	
		0.4 0.0	Unshaded

	n.s.	18.2 58.3	
	n.s.	10.6 38.5	Shaded

Soil pH

Associated floristic diversity

Species per m² vs Per cent abundance of *L. galeobdolon*

Hydrology

Triangular ordination

Bare soil

Commonest associates %
Hyacinthoides non-scripta 39
Poa trivialis 37
Mercurialis perennis 34
Holcus mollis 31
Urtica dioica 28

Abundance in quadrats

	1–10	11–20	21–30	31–40	41–50	51–60	61–70	71–80	81–90	81–100	Distribution type
Wetland	5	1	–	–	–	–	–	–	–	–	–
Skeletal	1	–	–	–	–	–	–	–	–	–	–
Arable	–	–	–	–	–	–	–	–	–	–	–
Grassland	–	–	–	–	–	–	–	–	–	–	–
Spoil	–	–	–	1	1	–	–	–	–	–	–
Wasteland	1	–	1	1	1	–	–	–	–	–	–
Woodland	33	9	9	15	7	7	5	4	1	3	2
All habitats	38.8	9.7	8.7	15.5	7.8	6.8	4.9	3.9	1.0	2.9	2

(+)*Lamium album* L.

Lamiaceae
White Dead-nettle

Life-form Polycarpic perennial protohemicryptophyte with creeping rhizome. Stems erect; leaves ovate, hairy, <2000 mm²; roots with VA mycorrhizas.
Phenology Winter green. Following early spring growth the plant flowers mainly in May to June. Seed set from June to August.
Height Foliage and flowers to 600 mm.
Seedling RGR No data. SLA 31.0. DMC 18.3.
Nuclear DNA amount 2.2 pg; $2n = 18*$ (*BC*); $2x*$.
Established strategy Competitive-ruderal.
Regenerative strategies V, ?B_s. Regeneration by seed requires further study (see 'Synopsis').
Flowers White, hermaphrodite, homogamous, insect-pollinated; often >50 terminally in a whorled inflorescence.
Germinule 1.43 mg, 2.4 × 1.2 mm, nutlet (4 per ovary).
Germination Epigeal; in one test only 20% germination achieved by freshly collected seed, and this was reduced to 0% by dry storage.
Biological flora None.
Geographical distribution Common in England, becoming rarer in the W of the British Isles and in N Scotland, introduced into Ireland (82% of British vice-counties), *ABIF* page 512; most of Europe, but rare in the S and absent from many islands (67% of European territories and 5% naturalized); the Himalayas and Japan.

Habitats Most commonly recorded from soil heaps, but most characteristic of shaded roadsides and hedgerows. Also found in a variety of spoil categories, on wasteland and walls, and as a casual of arable land. Rarely observed in mire, meadows and pastures, woodland and skeletal habitats. Absent from aquatic habitats.
Gregariousness Intermediate.
Altitude Up to 340 m, the British altitudinal limit for *L.a.*
Slope Wide-ranging.
Aspect Distribution shows a slight S-facing bias.
Hydrology Not a wetland species.
Soil pH Observed down to pH 4.5, but most frequent at values >7.0.
Bare soil Associated with a wide range of values, but abundant only where there are low exposures.
△ **ordination** Wide-ranging, but distribution centred on vegetation associated with disturbed fertile habitats. Absent from stressed environments.
Associated floristic diversity Low to intermediate.

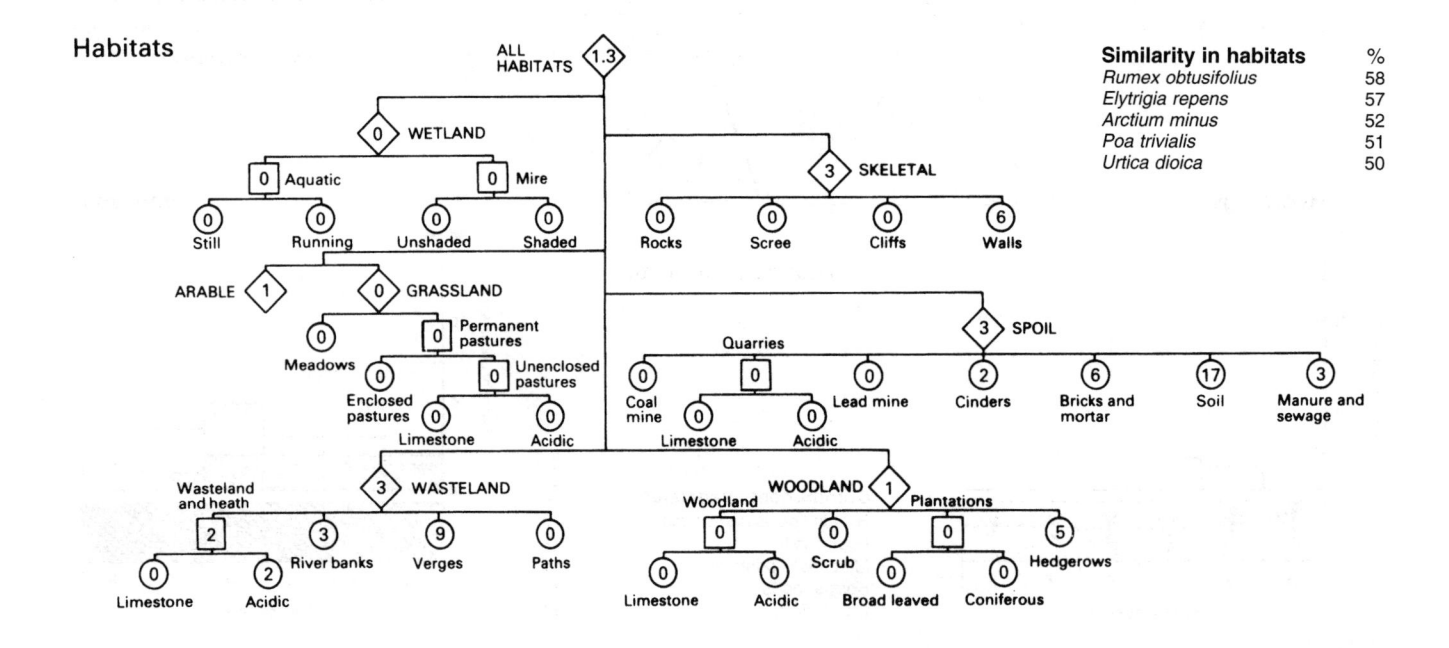

Habitats

Similarity in habitats	%
Rumex obtusifolius	58
Elytrigia repens	57
Arctium minus	52
Poa trivialis	51
Urtica dioica	50

Synopsis *L.a.* is a sprawling, relatively low-growing rhizomatous, wintergreen herb of fertile, well-drained and often lightly-shaded sites, where the vigour of taller potential dominants is suppressed by occasional disturbance. Thus, *L.a.* is particularly characteristic of hedgerows and road verges which are infrequently disturbed by verge cutting and hedge trimming and where, as a result, species such as *Urtica dioica* can only achieve incomplete dominance. The species flowers in early summer, at which time it is a conspicuous component of the flora. Subsequently, the growth of taller species usually renders *L.a.* a subordinate member of the community. Although attaining maximum shoot biomass in summer, the phenology of *L.a.* involves much shoot growth in early spring. This may in part explain the restriction of *L.a.* to lowland sites in the British Isles. Effective regeneration appears to be vegetative; *L.a.* forms loose clonal patches by means of rhizome growth. Seed production is occasionally prolific (up to 2400 per plant (Salisbury 1964)), but regeneration by seed has been little studied. A persistent seed bank is suspected to occur. The seed has an elaiosome and may be dependent upon ants for dispersal. Long-distance seed dispersal during Postglacial colonizing episodes appears to have been slow, and may account for the absence of *L.a.* from the native Irish flora. However, Peterken (1981) describes *L.a.* as a colonist of secondary woodland, and the possibility must be considered that some long-distance dispersal of seeds or detached portions of the vegetative plant is taking place.

Current trends Colonizes roadsides and hedgerows, and present in some secondary woodlands (Peterken 1981). Favoured both by fertile conditions and by disturbance; probably increasing.

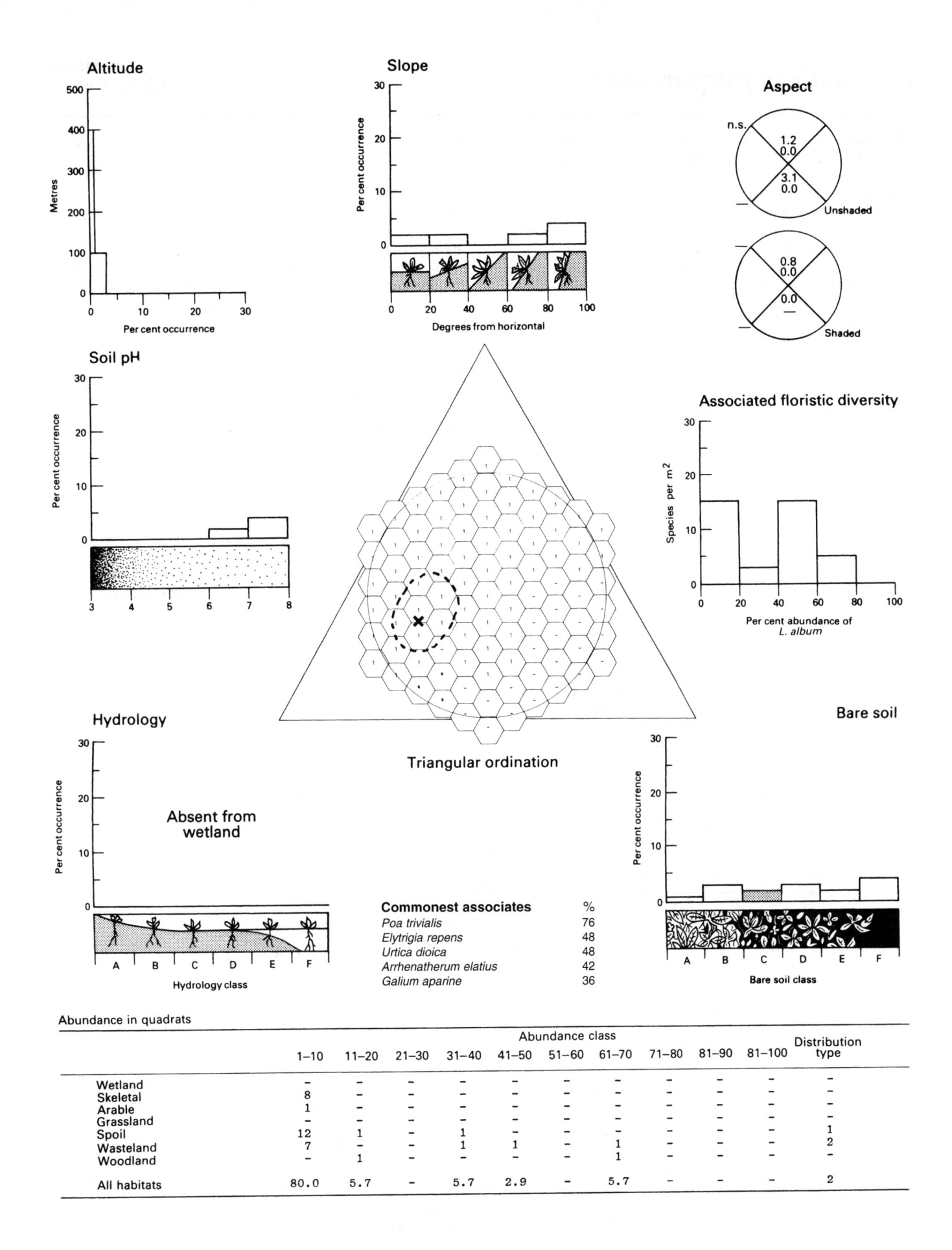

Altitude

Slope

Aspect

	n.s.	
	1.2	
	0.0	
	3.1	
	0.0	Unshaded

	0.8	
	0.0	
	0.0	
	—	Shaded

Soil pH

Associated floristic diversity

Hydrology

Absent from wetland

Triangular ordination

Commonest associates	%
Poa trivialis	76
Elytrigia repens	48
Urtica dioica	48
Arrhenatherum elatius	42
Galium aparine	36

Bare soil

Abundance in quadrats

	Abundance class										Distribution type
	1–10	11–20	21–30	31–40	41–50	51–60	61–70	71–80	81–90	81–100	
Wetland	–	–	–	–	–	–	–	–	–	–	–
Skeletal	8	–	–	–	–	–	–	–	–	–	–
Arable	1	–	–	–	–	–	–	–	–	–	–
Grassland	–	–	–	–	–	–	–	–	–	–	–
Spoil	12	1	–	1	–	–	–	–	–	–	1
Wasteland	7	–	–	1	1	–	1	–	–	–	2
Woodland	–	1	–	–	–	1	–	–	–	–	
All habitats	80.0	5.7	–	5.7	2.9	–	5.7	–	–	–	2

(+)*Lamium purpureum* L.

Lamiaceae
Red Dead-nettle

Life-form Winter- to summer-annual, semi-rosette therophyte. Stems erect, branching from the base; leaves ovate, hairy, <1000 mm^2; roots with VA mycorrhizas.

Phenology Variable, since germination may occur from early summer to autumn. Plants capable of overwintering and flower from March to October but particularly from April to June. Seed set mainly May to July (to November).

Height Foliage and flowers to 450 mm.

Seedling RGR No data. SLA 31.3. DMC 14.6.

Nuclear DNA amount 2.2 pg; $2n = 18*$ (*BC*); $2x*$.

Established strategy Ruderal.

Regenerative strategies B$_s$. See 'Synopsis'; SB IV.

Flowers Purple, hermaphrodite, homogamous, pollinated by bees or selfed (*CTW*); often >50 in a dense whorled inflorescence.

Germinule 0.70 mg, 2.3 × 1.3 mm, nutlet (4 per ovary).

Germination Epigeal; t_{50} 3 days; germination rate increased after dry storage at 20°C; L = D.

Biological flora None.

Geographical distribution The British Isles (100% of British vice-counties), *ABIF* page 513; most of Europe, but only on mountains in the S (87% of European territories and 3% naturalized); SW Asia.

Habitats Recorded most frequently from arable land, soil heaps and demolition sites. Also found on cinder tips. Observed on waysides and as a winter-annual of rock outcrops, disturbed roadsides and hedgerows. A common garden weed. Only of casual occurrence in grassland, and absent from wetlands and woodlands.

Gregariousness Sparse.

Altitude Mainly a lowland species, but observed up to 340 m. [Up to 610 m.]

Slope Not recorded from skeletal habitats, and rarely observed on steep slopes.

Aspect Distribution biased towards S-facing slopes.

Hydrology Absent from wetlands.

Soil pH Virtually restricted to sites of pH >5.0, and most frequent in the pH range 6.0-8.0.

Bare soil Typically associated with high exposures.

Δ ordination Largely confined to fertile disturbed conditions.

Associated floristic diversity Typically intermediate.

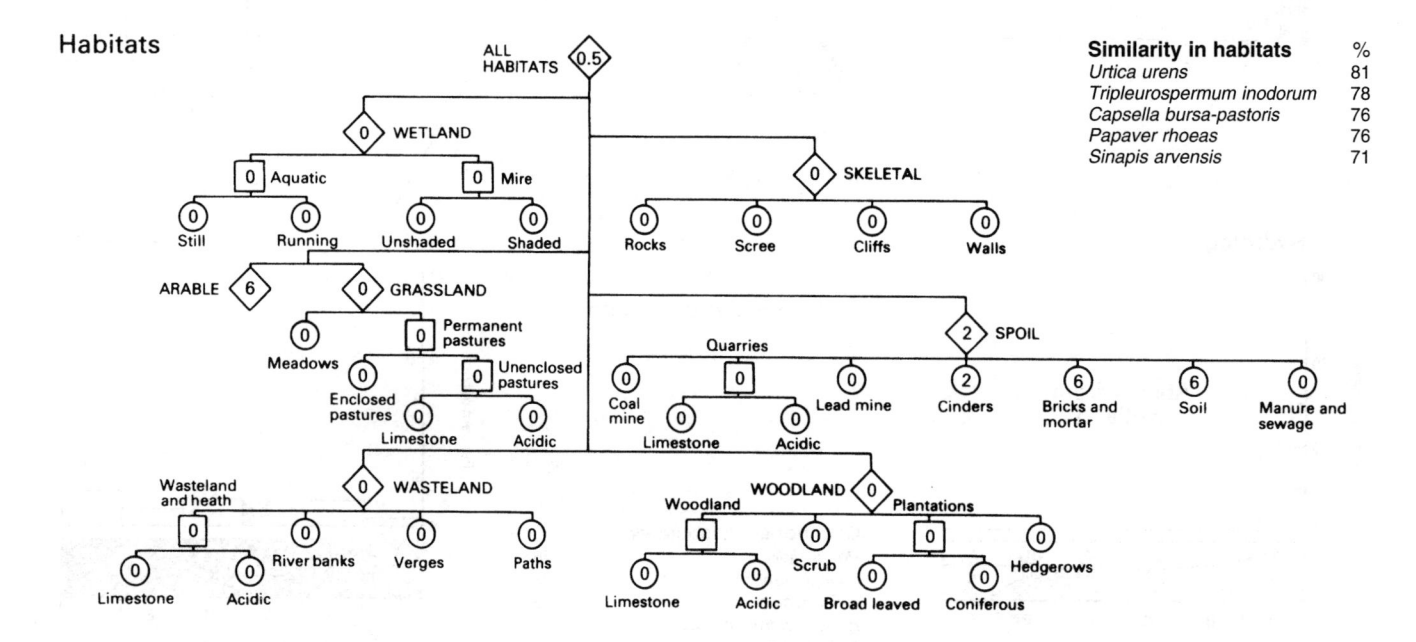

Habitats

Similarity in habitats	%
Urtica urens	81
Tripleurospermum inodorum	78
Capsella bursa-pastoris	76
Papaver rhoeas	76
Sinapis arvensis	71

Synopsis *L.p.*, regarded by Webb (1985) as doubtfully native, is a short-lived annual herb which exploits gardens and a wide range of other disturbed, relatively fertile habitats. In some sites *L.p.* is facultatively winter-annual. Thus, the species occurs on rock outcrops liable to summer drought and on steep or sandy hedgebanks, where seed set precedes the major growth period of associated woody and perennial herbaceous species. In winter barley, too, *L.p.* sets seed early, before the crop has developed a dense canopy. Important controls upon the phenology of *L.p.* are exercised by its germination biology. Summer- and winter-annual forms with different requirements for germination may occur (Lauer 1953), and most local populations appear facultatively winter- or summer-annual. In Britain the germination period in the field is mainly from May to October, both on arable land (Roberts & Boddrell 1983) and in a vegetable garden population examined in N Derbyshire, where two generations regularly occur in a season. In some N American populations seed is dormant when shed in early summer, and remains so until autumn (Baskin & Baskin 1984). Any residual seeds are returned to a state of secondary dormancy by low temperatures in winter (Baskin & Baskin 1984). *L.p.* often produces >1000 seeds per plant and shows some persistence in the soil (Roberts & Boddrell 1983). The species also has a restricted capacity for vegetative spread. During the colder months *L.p.* may produce short, rooted, prostrate shoots, forming a clump (Salisbury 1964). Non-flowering shoot tips readily re-root, and plants can establish and set seed following spring rotavation (Hodgson unpubl.). Little is known about the dispersal of *L.p.*, but the species appears to be mainly dependent for its long-range transport upon movement of seeds contained in soil, although some local dispersal by ants (Ridley 1930) has been observed.

Current trends Not under threat, and perhaps increasing locally in lowland areas.

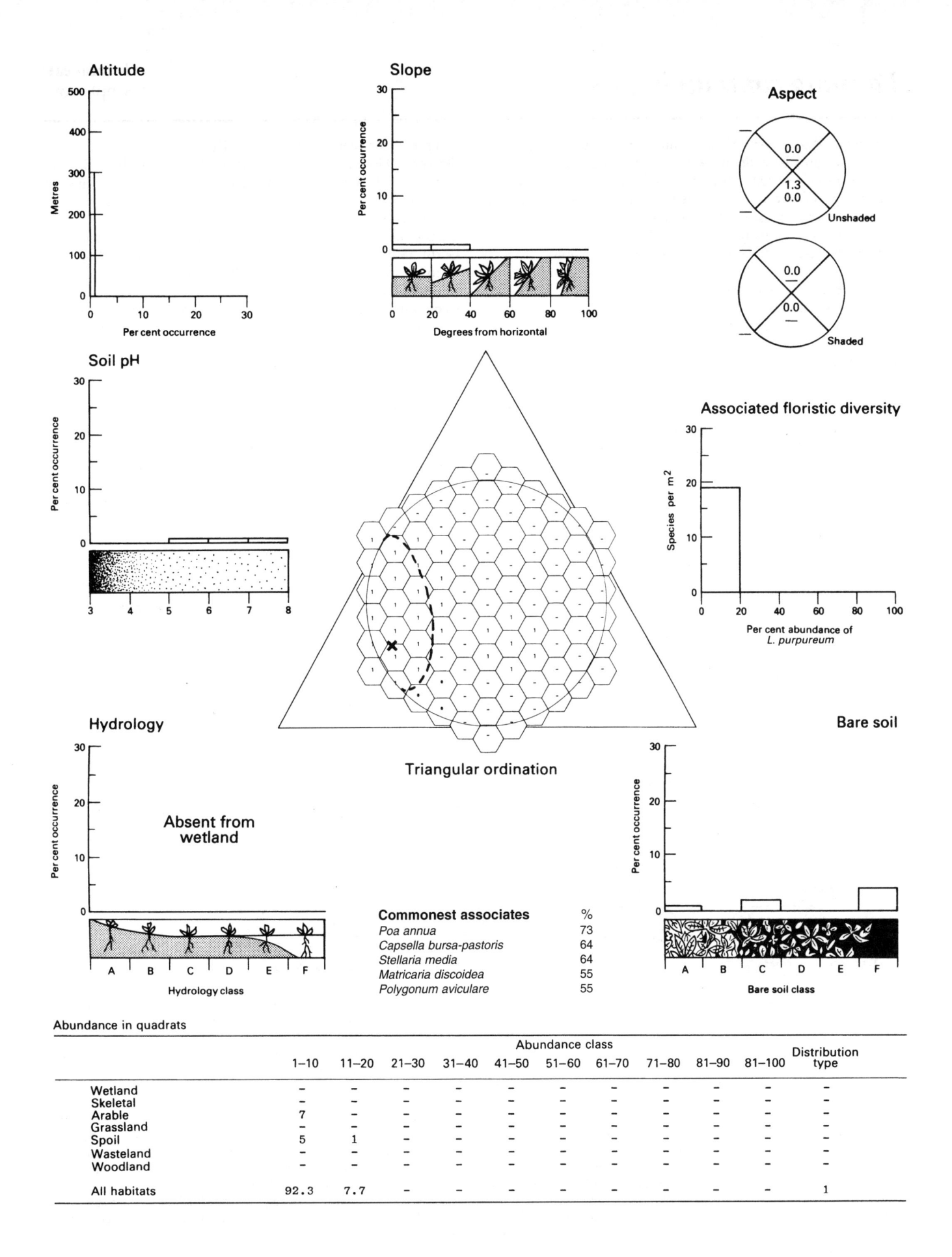

Altitude

Slope

Aspect

Unshaded

Shaded

Soil pH

Associated floristic diversity

Species per m^2 vs Per cent abundance of *L. purpureum*

Hydrology

Absent from wetland

Triangular ordination

Commonest associates	%
Poa annua	73
Capsella bursa-pastoris	64
Stellaria media	64
Matricaria discoidea	55
Polygonum aviculare	55

Bare soil

Abundance in quadrats

	\multicolumn{10}{c	}{Abundance class}	Distribution type
	1–10	11–20	21–30	31–40	41–50	51–60	61–70	71–80	81–90	81–100	
Wetland	–	–	–	–	–	–	–	–	–	–	–
Skeletal	–	–	–	–	–	–	–	–	–	–	–
Arable	7	–	–	–	–	–	–	–	–	–	–
Grassland	–	–	–	–	–	–	–	–	–	–	–
Spoil	5	1	–	–	–	–	–	–	–	–	–
Wasteland	–	–	–	–	–	–	–	–	–	–	–
Woodland	–	–	–	–	–	–	–	–	–	–	–
All habitats	92.3	7.7	–	–	–	–	–	–	–	–	1

Lapsana communis L.

Asteraceae
Nipplewort

Life-form Winter- or summer-annual, semi-rosette therophyte, hairy towards base. Stem erect; lower leaves pinnatifid, upper ovate, usually <6000 mm^2; tap-rooted, with VA mycorrhizas.

Phenology Overwintering as rosettes from autumn-germinating seeds or germinating in spring. Flowers from July to September and sets seed gradually from July to October.

Height Foliage mainly <450 mm, overtopped by flowers to 900 mm.

Seedling RGR No data. SLA 52.6. DMC 12.7.

Nuclear DNA amount 2.4 pg; $2n = 14^*$ (*BC*); $2x^*$.

Established strategy Intermediate between ruderal and competitive-ruderal.

Regenerative strategies B_s. Regeneration entirely by seed, mainly germinating in autumn. Seeds appear to have the capacity for limited persistence in the soil; SB III.

Flowers Florets yellow, ligulate, hermaphrodite, insect-pollinated, but regularly self-fertilized in absence of insects (Muller 1883); *c.* 30 in each capitulum; often 15-20 capitula in each corymbose panicle (*CTW*).

Germinule 1.27 mg, 4.5 × 1.4 mm, achene, without a pappus.

Germination Epigeal: t_{50} 4 days; 13-35°C; germination rate increased after dry storage at 20°C; L > D.

Biological flora None.

Geographical distribution The British Isles except for the N of Scotland (100% of British vice-counties), *ABIF* page 618; Europe northwards to Scandinavia (87% of European territories and 3% naturalized); Australia, N Africa, W and C Asia; introduced in N America.

Habitats Predominantly on arable land, especially in cereal crops, but also in other disturbed habitats such as soil heaps, sewage waste and quarry heaps. In addition, recorded from hedgerows, woodlands, plantations, banks of rivers and streams, cliffs and walls. Absent from pasture and wetland habitats.

Gregariousness Sparse.

Altitude Up to 300 m. [Up to 440 m.]

Slope Wide-ranging.

Aspect Slightly more frequent on unshaded, S-facing slopes.

Hydrology Absent from wetland.

Soil pH Recorded only from soil of pH >5.0.

Bare soil Widely distributed, but most strongly associated with habitats having a large proportion of bare ground.

△ **ordination** Distribution centred on vegetation associated with fertile heavily disturbed conditions, but evidence of some capacity to persist in less-disturbed sites.

Associated floristic diversity Typically intermediate.

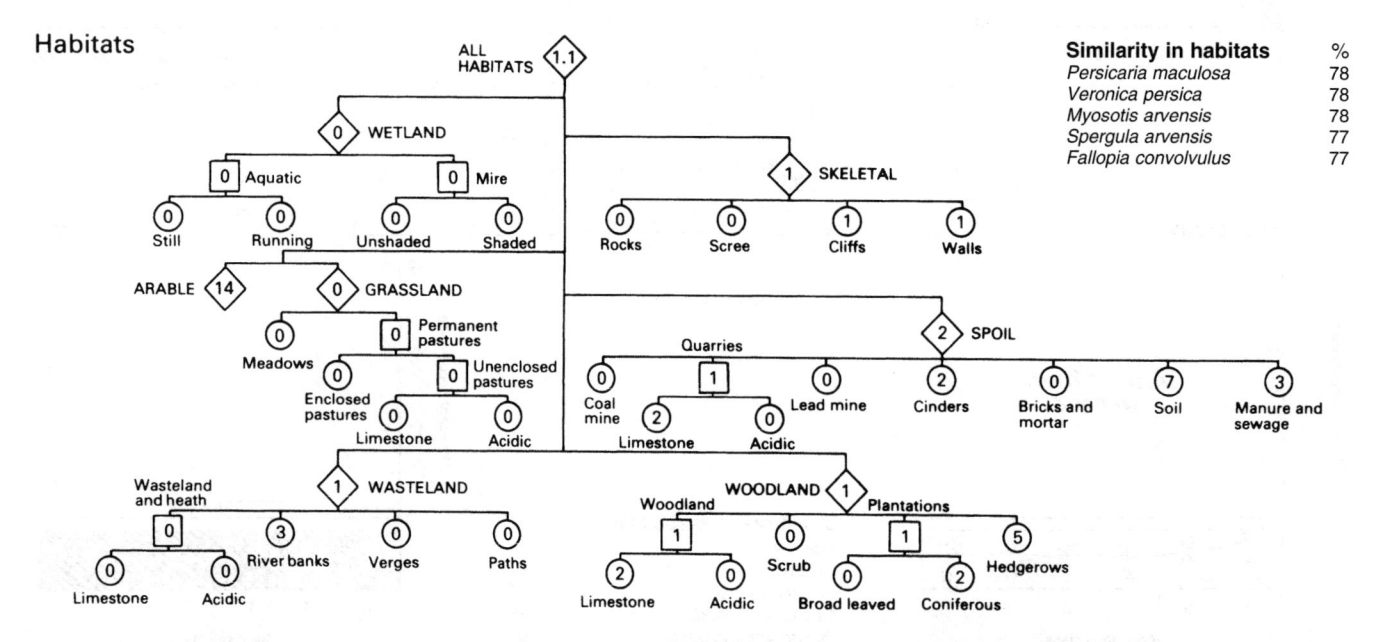

Habitats

Similarity in habitats	%
Persicaria maculosa	78
Veronica persica	78
Myosotis arvensis	78
Spergula arvensis	77
Fallopia convolvulus	77

Synopsis *L.c.* is a tall winter- or summer-annual herb which exploits fertile habitats, e.g. (a) arable land and soil heaps where the growth of perennial herbs has been checked or prevented by major disturbance before the growing season or (b) where the growth of perennials is restricted by light-to-moderate shade, and by disturbances such as soil creep (e.g. in hedgebanks) or the creation of woodland clearings. *L.c.* occurs in a range of habitats similar to that exploited by *Galeopsis tetrahit*, although *L.c.* is less frequent in wetter sites. The two species are of similar height and under favourable conditions, both become very tall plants capable of dominating neighbours, including smaller perennials. In autumn each produces relatively large seeds, which may contaminate crop seeds (*WCH*) and may become incorporated into a shortly persistent seed bank. On average plants produce *c.* 1000 seeds (Salisbury 1942). Seeds appear to germinate with equal facility in autumn and spring and this, coupled with the lesser vulnerability to weeding of the non-flowering juvenile rosette (compared with that of the erect leafy *G.t.*), may contribute to the persistence of *L.c.* in shaded gardens and explain the absence of *G.t.* from this habitat. The semi-rosette growth form may also prevent *L.c.*, which is found mainly in cereals, from exploiting sites where taller dicotyledonous crops (e.g. *Brassica* spp. and potatoes) are grown. The outer achenes in each capitulum are much longer than the inner (*FE4*). Whether they also differ in germination characteristics is not known. The extent to which ecotypes have developed in this wide-ranging species, which exploits both fugacious and semi-stable habitats, is uncertain, but three further ssp. have been recorded from Europe (*FE4*). One, ssp. *intermedia* (Bieb.) Hayek., from SE Europe, is introduced into Britain. Ssp. *i.* is an annual to polycarpic perennial and has a larger involucre (and presumably also larger achenes) (*Excur Fl*).

Current trends Perhaps increasing in shaded habitats, as a result of the greater level of mechanized disturbance in land management, but probably decreasing on arable land.

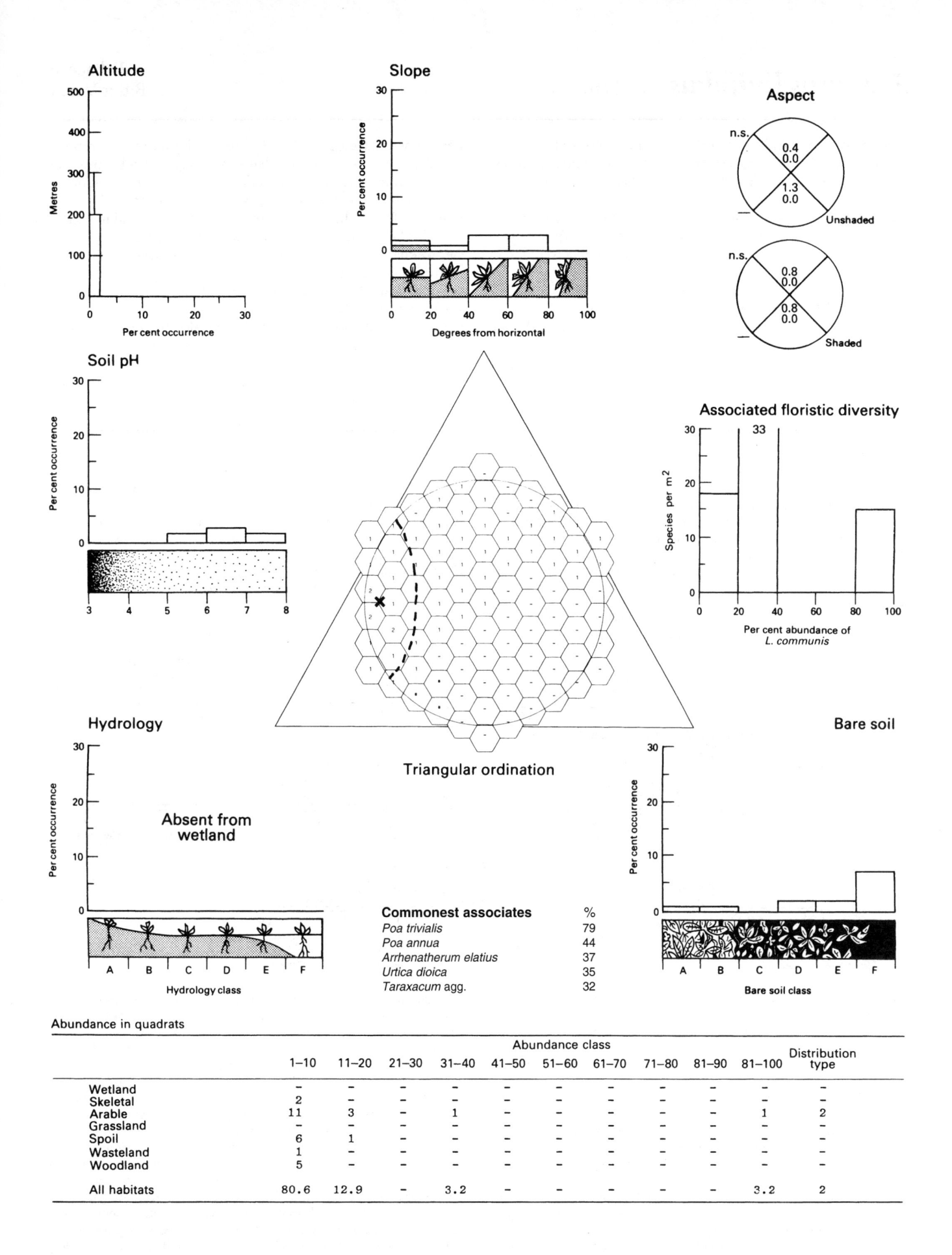

Altitude

Slope

Aspect

Soil pH

Associated floristic diversity

Triangular ordination

Hydrology

Absent from wetland

Commonest associates	%
Poa trivialis	79
Poa annua	44
Arrhenatherum elatius	37
Urtica dioica	35
Taraxacum agg.	32

Bare soil

Abundance in quadrats

	1–10	11–20	21–30	31–40	41–50	51–60	61–70	71–80	81–90	81–100	Distribution type
Wetland	–	–	–	–	–	–	–	–	–	–	–
Skeletal	2	–	–	–	–	–	–	–	–	–	–
Arable	11	3	–	1	–	–	–	–	–	1	2
Grassland	–	–	–	–	–	–	–	–	–	–	–
Spoil	6	1	–	–	–	–	–	–	–	–	–
Wasteland	1	–	–	–	–	–	–	–	–	–	–
Woodland	5	–	–	–	–	–	–	–	–	–	–
All habitats	80.6	12.9	–	3.2	–	–	–	–	–	3.2	2

Lathyrus linifolius (Reichard) Bässler

Fabaceae
Bitter Vetch

Life-form Polycarpic perennial protohemicryptophyte with creeping tuberous rhizome. Stems erect; leaves pinnate, <2000 mm², with a terminal tendril; roots shallow, bearing VA mycorrhizas.

Phenology New shoots expand in late winter or early spring. Flowers produced from April to July and seed shed from July to October. Shoots die back in autumn. Leaves relatively long-lived.

Height Foliage and flowers to 400 mm.

Seedling RGR <0.5 week^{-1}. SLA 22.7. DMC 26.3.

Nuclear DNA amount 16.5 pg; $2n = 14$* (Bennett & Smith 1976); $2x$*.

Established strategy Intermediate between C-S-R and stress-tolerator.

Regenerative strategies V, ?B$_s$. Spreads by means of creeping rhizomes to form diffuse patches. Also regenerates by means of seeds which germinate in autumn or spring. No information on the presence or absence of a persistent seed bank; ?SB III.

Flowers Purple, hermaphrodite, insect-pollinated and not selfed if this fails (Knuth 1906); 2-6 in axillary racemes.

Germinule 12.49 mg, 2.6 × 2.2 mm, seed dispersed explosively from a 4-6-seeded pod.

Germination Hypogeal; t_{50} 6 days; hard-coat dormancy; L = D.

Biological flora None.

Geographical distribution The British Isles, but more common in the W and N, virtually absent from East Anglia, the SE Midlands and C Ireland (95% of British vice-counties), *ABIF* page 385; Europe except the extreme N and rare in the SE (59% of European territories).

Habitats Recorded from grazed and ungrazed semi-natural grassland, including that associated with lead-mine spoil. Occasionally observed at margins of woodland and scrub in light shade, particularly in lowland areas, and more rarely at the edge of mire, in roadside vegetation and on damp cliffs. Otherwise absent from wetlands, woodland, spoil and skeletal habitats, and not recorded from arable land.

Gregariousness Intermediate.

Altitude In the present survey only recorded from the carboniferous limestone uplands, but more-detailed observations (Hodgson *et al.* unpubl.) indicate an altitudinal range of 20-350 m. [Up to 760 m.]

Slope Virtually absent from skeletal habitats and therefore absent from the steepest slopes.

Aspect Only recorded from unshaded N-facing slopes. More-recent studies have confirmed the existence of a strong N-facing bias in the field distribution in the Sheffield region.

Hydrology Not a wetland species, but typical of moist soils and also observed rarely at the edge of soligenous mire.

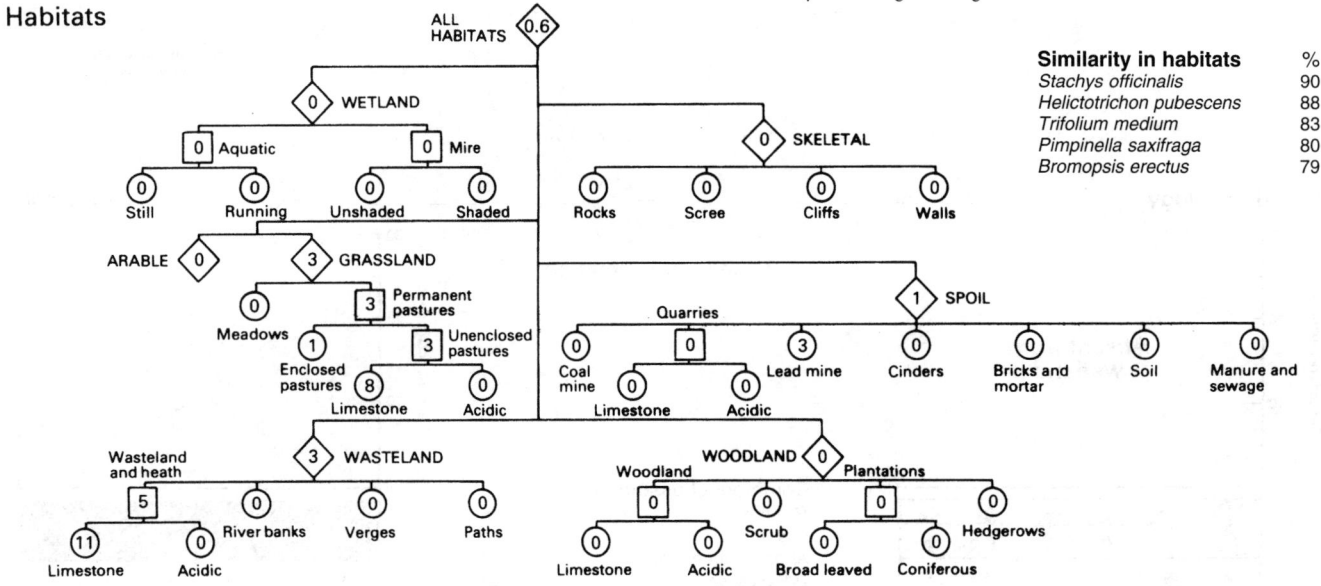

Habitats

Similarity in habitats	%
Stachys officinalis	90
Helictotrichon pubescens	88
Trifolium medium	83
Pimpinella saxifraga	80
Bromopsis erectus	79

Soil pH Largely restricted to soils of pH >4.0, but most frequently in soils of intermediate base status and uncommon at pH >7.0.

Bare soil Associated with habitats with little or no exposed soil.

Δ ordination Distribution centred on relatively unproductive and undisturbed vegetation types, with evidence of some capacity to persist in more-productive situations.

Associated floristic diversity Usually associated with either intermediate or species-rich vegetation.

Synopsis *L.l.* is a slow-growing, rhizomatous legume of unproductive grassland, usually on soils of intermediate pH. Nitrogen-fixing root nodules are formed in conjunction with *Rhizobium leguminosarum*. The species appears to be restricted to continuously moist soils and is characteristic of neutral grasslands dominated by *Agrostis capillaris*. In these circumstances *L.l.* is often associated with a well-defined group of species of similar ecology (*Stachys officinalis*, *Hypericum pulchrum* L., *Potentilla erecta* and *Viola lutea* Hudson), all of which resemble *L.l.* in sensitivity to moisture stress and in susceptibility to lime-chlorosis (Grime & Hodgson 1969). *L.l.* is observed at woodland margins in lowland areas. In common with species such as *Ophrys insectifera* L. and *Primula vulgaris* Hudson, the species is predominantly in shaded sites in lowland Britain and in more-open localities in upland areas. Occurrence in woodlands appears to be correlated with the fact that the phenology of *L.l.* is 'semi-vernal'. Consistent with its high nuclear DNA value, *L.l.* produces a rapid flush of growth in early spring by mobilizing the reserves in its tuberous rhizome. Flowering commences in April, and little further shoot growth takes place during summer. The distribution of *L.l.* within pastures is strongly affected by its morphology and phenology. *L.l.* has an erect stature (150-400 mm) and is confined to lightly-grazed sites. The species appears to be absent from sites which are regularly grazed during the early spring. *L.l.* persists through summer with short vegetative shoots at or a little below the canopy height of the pasture. Effective regeneration is mainly by means of rhizomatous growth to form diffuse patches. Usually

Altitude
Slope
Aspect

Unshaded: 3.5 / 33.3 / 0.0 / —
Shaded: — / 0.0 / 0.0 / —

Soil pH

Associated floristic diversity

Triangular ordination

Hydrology

Absent from wetland

Bare soil

Commonest associates	%
Agrostis capillaris	83
Anthoxanthum odoratum	72
Potentilla erecta	69
Festuca rubra	66
Festuca ovina	64

Abundance in quadrats

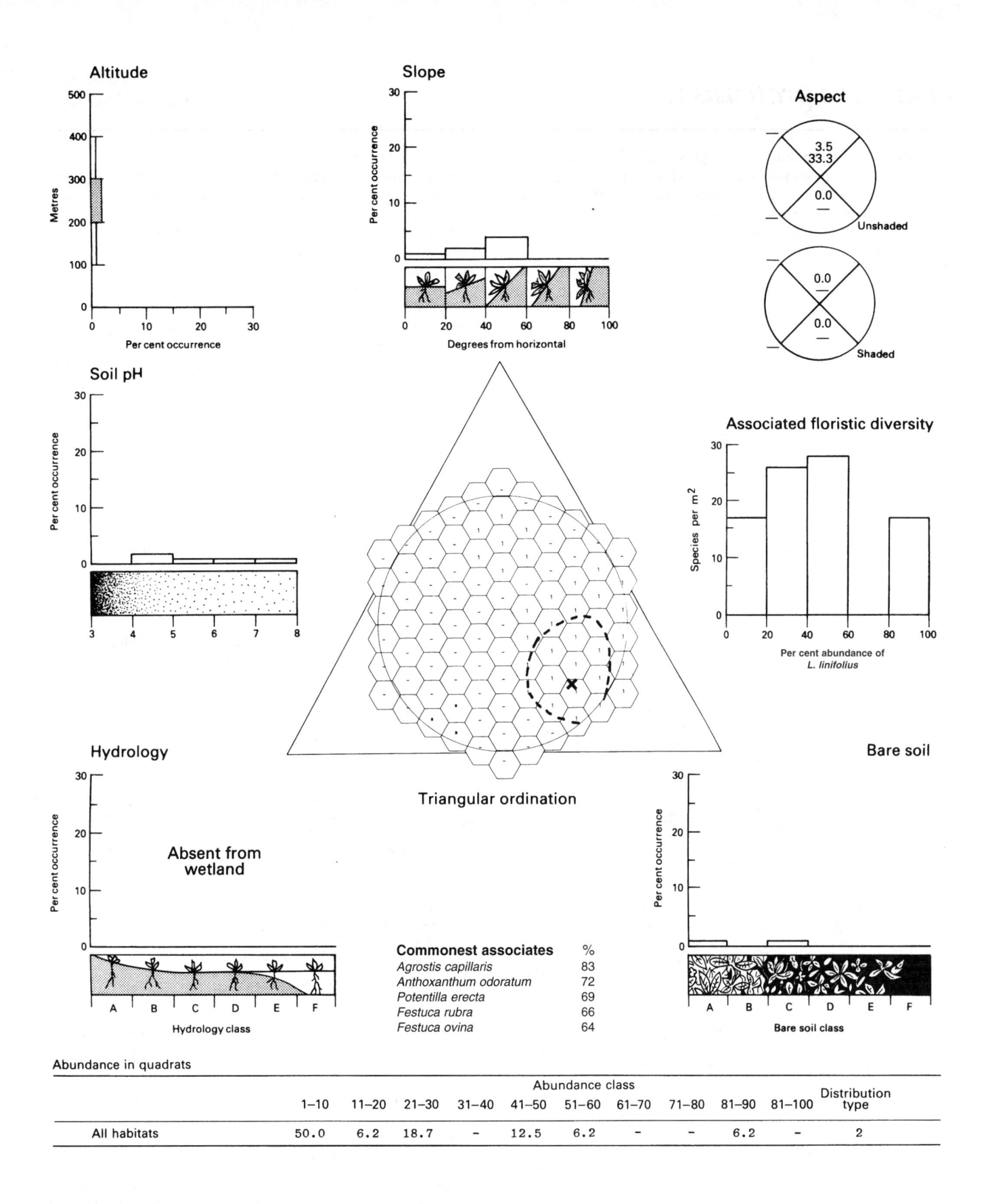

	1–10	11–20	21–30	31–40	41–50	51–60	61–70	71–80	81–90	81–100	Distribution type
All habitats	50.0	6.2	18.7	–	12.5	6.2	–	–	6.2	–	2

the large seed of *L.l.* is released with a hard seed coat which requires scarification before germination is possible. However, under wet summer conditions seeds have been observed to germinate in pods which are still attached to the parent plant. Germination in autumn or spring is hypogeal and the seedling is presumably dependent upon its seed reserves for a considerable period during establishment. A species of low colonizing ability considered by Peterken (1981) to be characteristic of ancient woodland in Lincolnshire.

Current trends *L.l.* is rather narrowly restricted to an increasingly scarce set of conditions of soil type and management. A further reduction in abundance seems inevitable.

Lathyrus pratensis L.

Fabaceae
Meadow Vetchling

Life-form Rhizomatous polycarpic perennial protohemicryptophyte. Shoots branched and climbing; leaves tendrillate with one pair of leaflets, usually <1000 mm²; root system nodulated, sometimes with VA mycorrhizas.
Phenology Flowers from May to August. Seed set from August to October and shoots die in autumn.
Height Stems bearing leaves and flowers to 900 mm.
Seedling RGR 0.5–1.0 week^{-1}. SLA 23.7. DMC 27.3.
Nuclear DNA amount 14.7 pg; $2n = 14*, 28*$ (*BC*, Brunsberg 1977); $2x*, 4x*$.
Established strategy C-S-R.
Regenerative strategies V, ?B$_s$. Regenerates vegetatively by means of rhizome growth and by seed which apparently germinates in spring. The extent to which seed persists in the soil is not known.
Flowers Yellow, hermaphrodite, insect-pollinated; in a 5–10-flowered axillary raceme. Self-incompatibility mechanism incomplete (Brunsberg 1977).
Germinule 12.85 mg, 3.3 × 2.8 mm, seed in an explosively dehiscent usually 3–6-seeded pod (Brunsberg 1977).
Germination Hypogeal; t_{50} 4 days; <5–28°C; hard-coat dormancy; L = D.
Biological flora None, but see Brunsberg (1977).
Geographical distribution The British Isles except for N Scotland (100% of British vice-counties), *ABIF* page 386; Europe except for the extremes of N and S (89% of European territories); N Africa, temperate Asia; introduced in N America.

Habitats Occurs most frequently in grassland habitats, particularly in meadows, wasteland and waysides. Also found in pasture, in a variety of different types of spoil habitats, scree and river banks. Infrequent in mire, and only a casual of arable land. Present in lightly shaded habitats, but absent from woodland, aquatic and most skeletal habitats.
Gregariousness Intermediate.
Altitude Recorded from a wide range of altitudes to 380 m. [Up to 460 m.]
Slope Only rarely occurring in skeletal habitats, and not recorded from the steepest slopes.
Aspect Slightly more frequent and more abundant on N-facing slopes.
Hydrology Recorded on moist soils at the edge of topogenous or soligenous mire.
Soil pH Virtually absent from soils of pH <4.5, and most frequent between pH 5.0 and 7.0.
Bare soil Widely distributed, but most frequent and abundant in closed communities with little exposed soil.

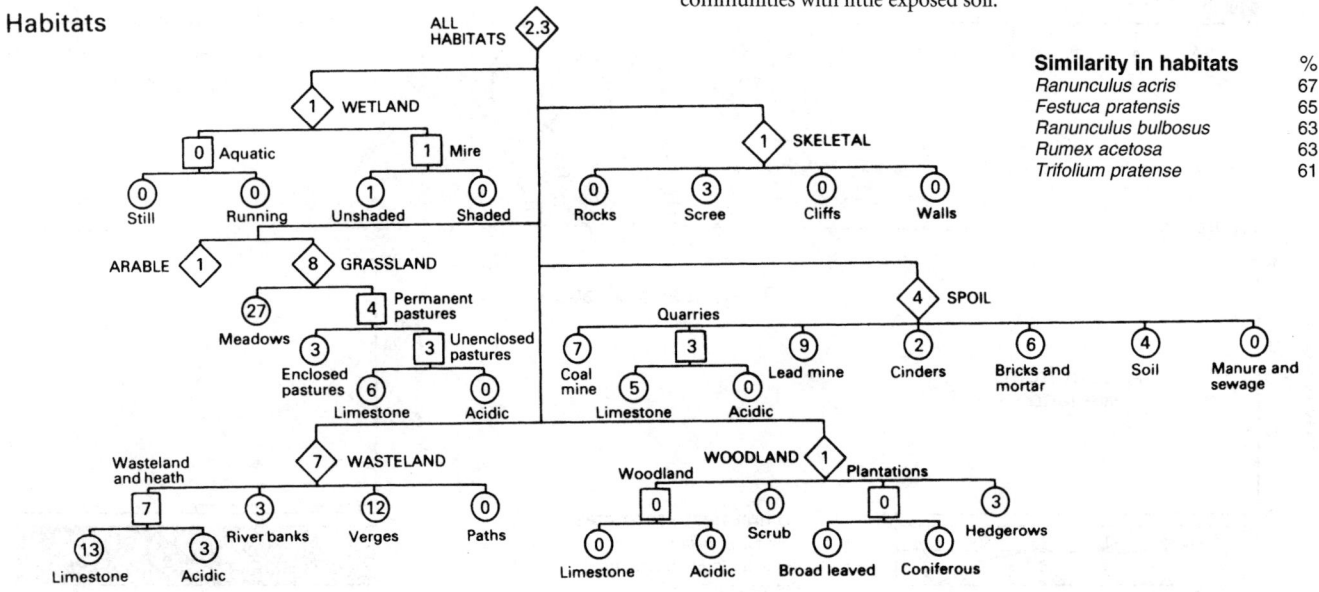

Habitats

Similarity in habitats	%
Ranunculus acris	67
Festuca pratensis	65
Ranunculus bulbosus	63
Rumex acetosa	63
Trifolium pratense	61

△ **ordination** Absent from sites subject to high levels of disturbance and excluded from vegetation of low productivity, but nevertheless present in a wide range of communities. Distribution is centred on vegetation of moderate productivity and experiencing an intermediate level of disturbance.
Associated floristic diversity Typically associated with intermediate or species-rich communities.
Synopsis *L.p.* is a rhizomatous legume supported on its leaf tendrils by the surrounding vegetation. It is most typical of meadows, roadsides and other tall vegetation on soils of moderate fertility. In the *PGE L.p.* benefited from treatments involving applications of minerals alone or of sodium nitrate and occasional farmyard manure. The species was suppressed by 'starvation, ammonium salts or sodium nitrate alone'. *L.p.* is common on calcareous soils, and much more resistant to lime-chlorosis than *Lathyrus linifolius* (Grime 1965). N-fixing root nodules are formed in association with *Rhizobium leguminosarum*. As a consequence of its ascending growth form, the occurrences of *L.p.* in pastures are restricted to lightly grazed situations. *L.p.* is often tall and has the capacity for lateral vegetative spread, but tends to exercise only localized and temporary dominance in undisturbed vegetation. The species tends to be restricted to sites where the vigour of potential dominants, e.g. *Arrhenatherum elatius*, is debilitated by occasional mowing or by some other form of infrequent disturbance. The capacity of *L.p.* to behave as a dominant, appears to be restricted by (a) its scrambling habit and dependence upon support from other vegetation and (b) its capacity to fix nitrogen which, we suspect, may ameliorate soil infertility to an extent that faster-growing species may dominate. Regeneration is predominantly vegetative; *L.p.* can produce rhizomes up to 7 m in length (Brunsberg 1977). The importance of vegetative regeneration is particularly evident in habitats such as pasture, where grazing often prevents flowering and seed set, and in hay meadows which are often cut before seeds are ripe. This vigorous vegetative reproduction is coupled with rather modest seed set (usually 10–12 ovules, 3–6 seeds) (Brunsberg 1977). The large seeds are generally shed close to the plant and so the species has only limited colonizing ability. However, *L.p.* is widespread and in the past

Altitude
Slope
Aspect

Unshaded: n.s. 5.1 / 23.1 / 3.6 / 0.0
Shaded: — / 0.0 / 0.0 / —

Soil pH

Associated floristic diversity

Hydrology

Triangular ordination

Bare soil

Commonest associates %
Festuca rubra 87
Holcus lanatus 62
Dactylis glomerata 60
Plantago lanceolata 50
Poa pratensis 47

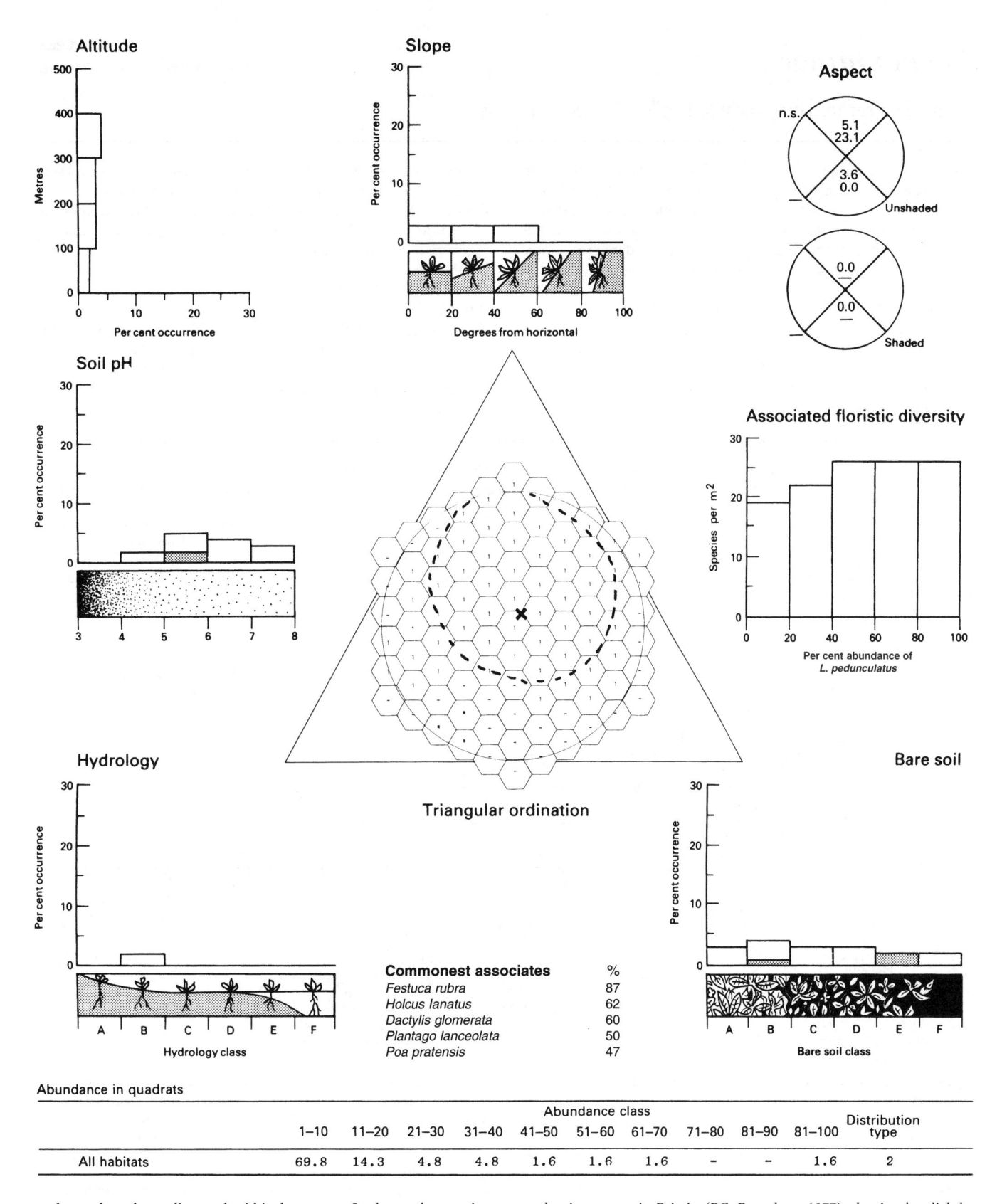

Abundance in quadrats

	1–10	11–20	21–30	31–40	41–50	51–60	61–70	71–80	81–90	81–100	Distribution type
All habitats	69.8	14.3	4.8	4.8	1.6	1.6	1.6	–	–	1.6	2

seed may have been dispersed within hay-crops. Seed can also survive ingestion by animals (Brunsberg 1977). Many seeds germinate in spring. Seed-bank studies have not been undertaken. More recently, mechanized disturbance resulting in the movement of soil containing rhizomes may have allowed the spread of *L.p.* in disturbed habitats. Two main cytotypes occur in Europe; $2n = 14$ (2x), which is widespread in Europe but with a predominantly E distribution, and $2n = 28$ (4x) in W and C Europe (Brunsberg 1977). This apparent autotetraploid appears to be the predominant type in Britain (BC, Brunsberg 1977); despite the slightly heavier seeds and lower pollen fertility, it is morphologically indistinguishable from diploid plants, and has a similar habitat range (Brunsberg 1977). However, the two are genetically isolated. Other closely related but ecologically separated species occur in Europe (Brunsberg 1977).

Current trends Remains common, but is decreasing in farmland.

Lemna minor L.

Lemnaceae
Duckweed, Duck's-meat

Some field records may include *L. gibba* L. (see 'Synopsis')

Life-form Floating hydrophyte. Thallus flat, ovate, up to 4 mm in diameter with a solitary, unbranched non-mycorrhizal root.

Phenology In early summer thalli start budding off, and may continue to do so throughout the summer. Flowering, which is rare, occurs in June and July (*CTW*). Frost-damaged shoots sink during winter, some with a small living bud (Landolt 1982).

Height At the water surface; on land height *c*. 1 mm.

Thallus RGR 2.0-2.4 week^{-1}. SLA 30.7. DMC 10.7.

Nuclear DNA amount 1.2 pg; $2n = 40, 42, 50$ (*FE5*); 4-5x (see also 'Synopsis').

Established strategy Competitive-ruderal.

Regenerative strategies (V). Forms extensive floating mats by means of vegetative buds. Flowers are rarely produced, and seed set has not been observed within the Sheffield region.

Flowers Minute, with no well-defined pollination mechanism (Sculthorpe 1967); inflorescence containing 2 stamens and 1 ovary enclosed in a sheath.

Germinule *c*. 0.6 mm, seed.

Germination No data.

Biological flora None.

Geographical distribution The British Isles, except for the Scottish Highlands (98% of British vice-counties), *ABIF* page 686; Europe (92% of European territories); cosmopolitan in temperate regions of the world.

Habitats An aquatic species of still or slow-moving water, especially in ponds, canals and ditches; sometimes in moderately shaded situations. Also occurs beached above the water level in mire and in other habitats adjacent to open water, but not persistent in these sites.

Gregariousness Stand-forming.

Altitude Suitable habitats are most frequent in the lowlands, but *L.m.* has been observed to 440 m. [Up to 500 m.]

Slope Not relevant.

Aspect Not relevant.

Hydrology Found floating on water of various depths. Most frequent in moderately shallow water (101-250 mm deep). Also occurs beached on wet mud.

Soil pH Over a wide range in soil pH, but more common above pH 4.5.

Bare soil An aquatic species. Often beached on bare mud.

△ **ordination** Absent from unproductive wetland vegetation.

Associated floristic diversity Typically low, particularly at high values of *L.m.*

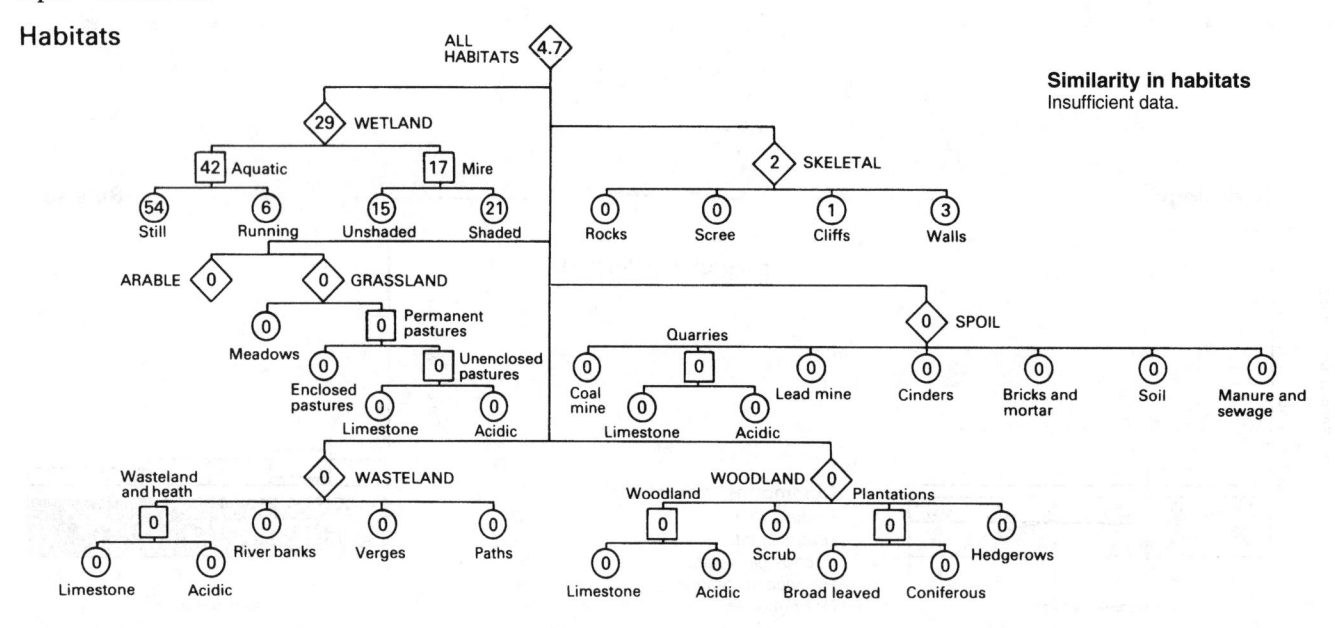

Habitats

Similarity in habitats
Insufficient data.

Synopsis *L.m.* is the most widespread free-floating vascular plant in the world (Sculthorpe 1967) and is by far the most common aquatic in Britain. *L.m.* is absent from base-poor, and perhaps also highly calcareous, waters but elsewhere is frequent on still waters where these are of sufficiently small extent to escape the major effects of wave action and wind blow. Despite the poorly protected epidermis and a water content of *c*. 96%, beached thalli are capable of active growth (Sculthorpe 1967). *L.m.* persists, albeit with reduced vigour, on water or on mud shaded by taller species. Each plant is minute, consisting of a single green, flat thallus up to 4 mm in diameter and a single unbranched root. The vascular tissue lacks vessels (Sculthorpe 1967). Plants contain oxalate raphids (Sculthorpe 1967), but it is not clear to what extent these deter predation. Vegetative buds are produced from two pouches on each frond, and new fronds may separate or remain joined to the parent. Each thallus may survive for 5-6 weeks during summer. By this means the water surface may be quickly covered, and in exceptional circumstances watercourses may become choked. Seed is rarely set, but high rates of flowering occurred during Postglacial Boreal and Atlantic periods, when the climate was warmer and milder (Sculthorpe 1967). Whole plants may adhere to the feathers or feet of birds (Sculthorpe 1967) and *L.m.* is an effective colonist of still water. In Europe *L.m.* is predominantly tetraploid and more rarely pentaploid, but diploid, triploid and octaploid counts have been recorded within the extra-European range of the species (Urbanska-Worytkiewicz 1975). The distribution of two other related free-floating species overlaps with that of *L.m.*: *L. gibba* L. and *Spirodela polyrhiza* (L.) Scheid. *L.g.*, which does not normally sink in winter and is thus vulnerable to the effects of low air temperatures (Landolt 1982), is largely confined to canals and ditches in lowland areas. *L.g.* exhibits a high phosphorus requirement (Landolt 1982) and a higher relative growth rate, and in mixed populations at high density may exclude *L.m.* through the ability of the buoyant thalli to occupy a superior position

Altitude

Slope

Aspect

Soil pH

Associated floristic diversity

Not applicable

Hydrology

Triangular ordination

Commonest associates	%
Potamogeton natans	26
Agrostis stolonifera	22
Lemna trisulca	21
Elodea canadensis	17
Sparganium erectum	14

Bare soil

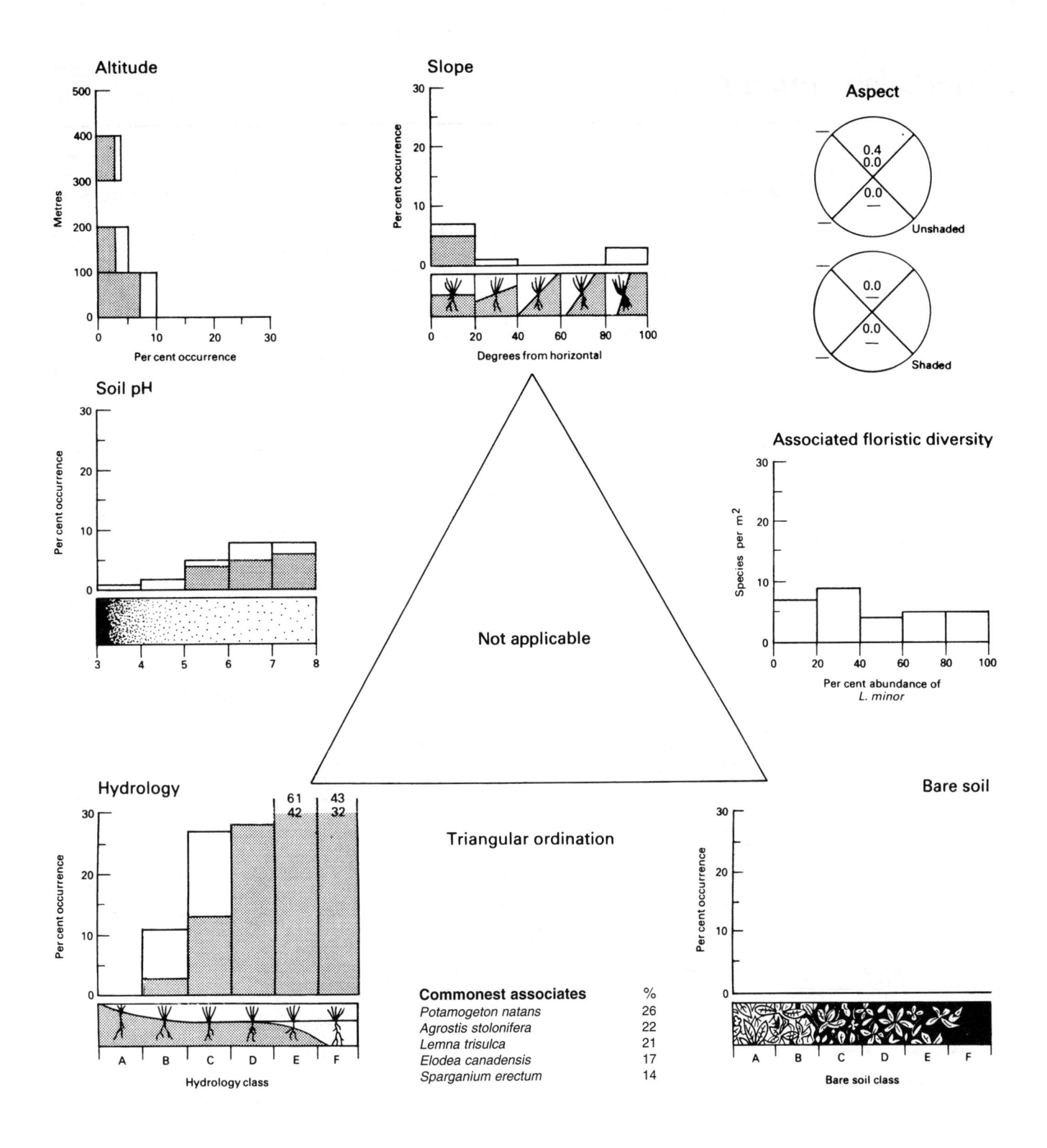

in the canopy (Clatworthy & Harper 1962, Rejmankova 1975). Gibbosity, is however, perhaps primarily a specialization for water-retention, of value in more-arid climates in the event of the water body drying up (see Landolt 1982). This character is often poorly developed in local populations, leading to confusion with *L.m*. Indeed, some field records of *L.m*. may include *L.g*. or even consist entirely of this species. (Records may also include some data for *L. minuscula* Herter., which is also readily mistaken for *L. minor* (Leslie & Walters 1983), but this species has not yet been recorded for the Sheffield region.) The larger *S.p*. has a high nitrogen and phosphorus requirement and forms turions (Landolt 1982). *S.p*. survives winter well but, in contrast with *L.m*. and *L.g*., has a high minimum temperature for growth (Landolt 1982). Perhaps for this reason, within the Sheffield region *S.p*. is mainly found in shallow water, which may be expected to warm up more rapidly.

Current trends A colonist of eutrophic water, *L.m*. may still be increasing despite the continued destruction of aquatic habitats. However, Rejmankova (1975) suggests that *L.g*. is the potential dominant in warmer eutrophic waters. In lowland regions eutrophication may be creating habitats more suitable for the latter species.

Leontodon autumnalis L.

Asteraceae
Autumn Hawkbit

Life-form Polycarpic perennial, rosette hemicryptophyte with branched stock. Scapes erect, or ascending, branched; leaves lanceolate, pinnatifid, glabrous or hairy, <2000 mm^2; roots with VA mycorrhizas.
Phenology Small rosettes overwinter. Flowers from June to October and sets seed from July to October.
Height Leaves usually appressed to the ground, but attaining up to 150 mm in height in tall vegetation; scapes to 300 mm.
Seedling RGR No data. SLA 26.1. DMC 15.2.
Nuclear DNA amount 2.7 pg, $2n = 12$ (Bennett & Smith 1976); $2n = 12^*$, 24 (*BC, FE4*); $2x^*$, $4x$.
Established strategy Intermediate between C-S-R and ruderal.
Regenerative strategies W, ?B$_s$. Regenerative biology is complex (see 'Synopsis'); SB ?III.

Flowers Florets yellow, ligulate, hermaphrodite, self-incompatible (Finch 1967), pollinated by insects; >50 aggregated into each capitulum; up to 7 capitula on a branched scape.
Germinule 0.70 mg, 4.5 × 0.8 mm, achene dispersed with a pappus of hairs *c.* 6 mm in length.
Germination Epigeal; t_{50} 2 days; 18-32°C; germination rate increased after dry storage at 20°C; L > D.
Biological flora None.
Geographical distribution The British Isles (100% of British vice-counties), *ABIF* page 620; most of Europe, although rather local in the S (82% of European territories); N and W Asia and NW Africa; introduced in N America.

Habitats Common on various types of spoil heaps, in enclosed pastures and on road verges and paths. Also frequently recorded from rock outcrops and meadows. Uncommon in ungrazed grassland and in most skeletal habitats. Observed as an infrequent component of soligenous mire in uplands. Only casual in arable fields. Absent from woodland. [Extends to high altitudes in wetland habitats (McVean & Ratcliffe 1962).]
Gregariousness Intermediate.
Altitude Over a wide range of altitudes up to 400 m. [Up to 980 m.]
Slope Not recorded at values >50°.

Aspect Associated with a wide range of aspect.
Hydrology Extending into moist rather than waterlogged soils. Occasionally observed in soligenous mire, particularly in upland areas.
Soil pH Largely confined to soils of pH >5.0.
Bare soil Found over a wide range of values, reflecting its occurrence in both relatively unvegetated disturbed habitats and in short turf.
△ **ordination** Present in disturbed vegetation types of varying levels of productivity, with distribution centred on productive vegetation subject to high or moderate levels of disturbance.

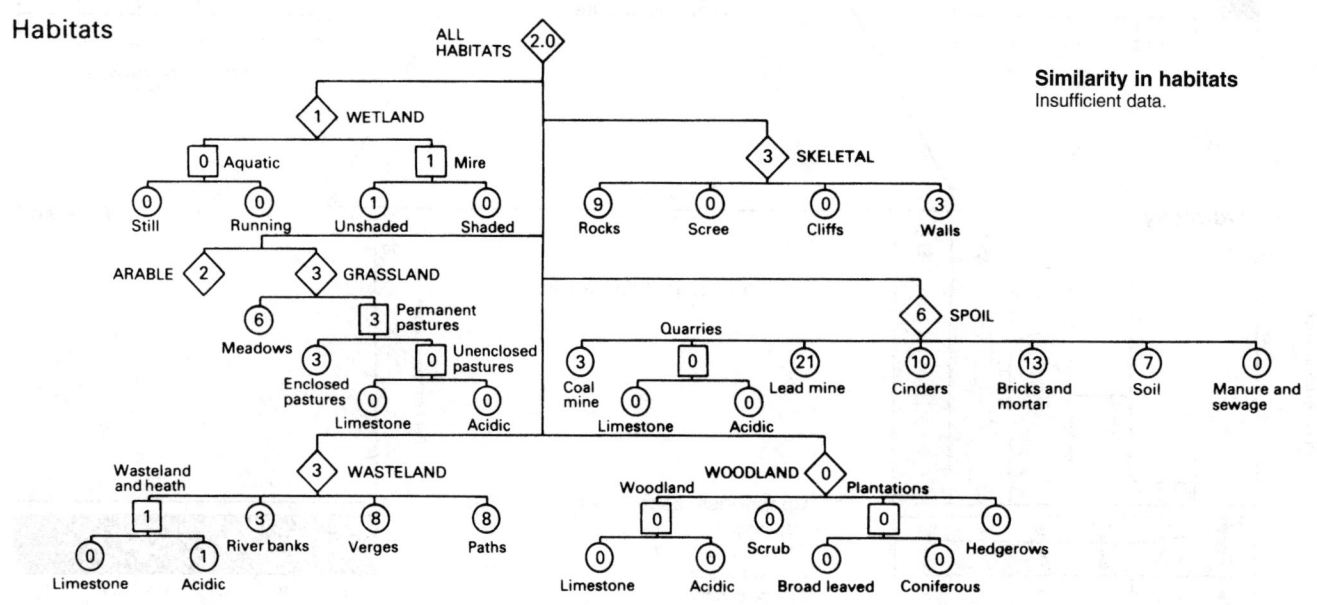

Habitats

Similarity in habitats
Insufficient data.

Associated floristic diversity Typically associated with intermediate values.
Synopsis *L.a.* is a rosette-forming herb of short turf and open ground. Found under a range of edaphic conditions, but particularly characteristic of road verges and pasture on moist fertile soil. *L.a.* is highly plastic in morphology, and is capable of exploiting sites subject to periodic disturbance by grazing, cutting or trampling. In short vegetation the leaves, which are palatable to stock, tend to be held close to the soil and may be pinnatifid. In taller grassland the leaves are generally long, ascending and relatively undissected. *L.a.* is absent from woodland and tall herbaceous vegetation, and appears to be intolerant of shading. The shoot phenology contrasts markedly with that of many of the grasses, such as *Lolium perenne*, with which it grows, since there is a pronounced peak in leaf area during summer. *L.a.* has a limited capacity for vegetative spread, and may produce a branched stock, with each branch terminating in a rosette. There is, however, no evidence for the development of extensive clones. *L.a.* regenerates by means of wind-dispersed seeds. Germination is dependent upon warm temperatures (Grime *et al.* 1981), and is partially inhibited by leaf canopies (Gorski *et al.* 1977). Some seedlings appear in vegetation gaps during autumn, but there is evidence, for the Sheffield region and elsewhere (Wesson & Wareing 1969, Roberts 1986a), that *L.a.* forms a transient reserve of surface-lying or superficially-buried seeds. The pappus remains stiff when wet, an unusual feature among wind-dispersed composites; this appears to provide anchorage during seedling establishment (Sheldon 1974). Fruit production is often extremely limited in summer-grazed sites. As well as being phenotypically plastic *L.a.* is also out-breeding and genetically variable (Finch 1967). Populations of shorter, few-flowered plants have been separated recently as ssp. *pratensis* (Koch) Arcangelis (*FE4*) and occur mainly in montane areas.
Current trends *L.a.* exploits some common artificial habitats, is relatively mobile and appears to be increasing, perhaps most conspicuously in continuous bands close to the metalled surface of many major roads and motorways.

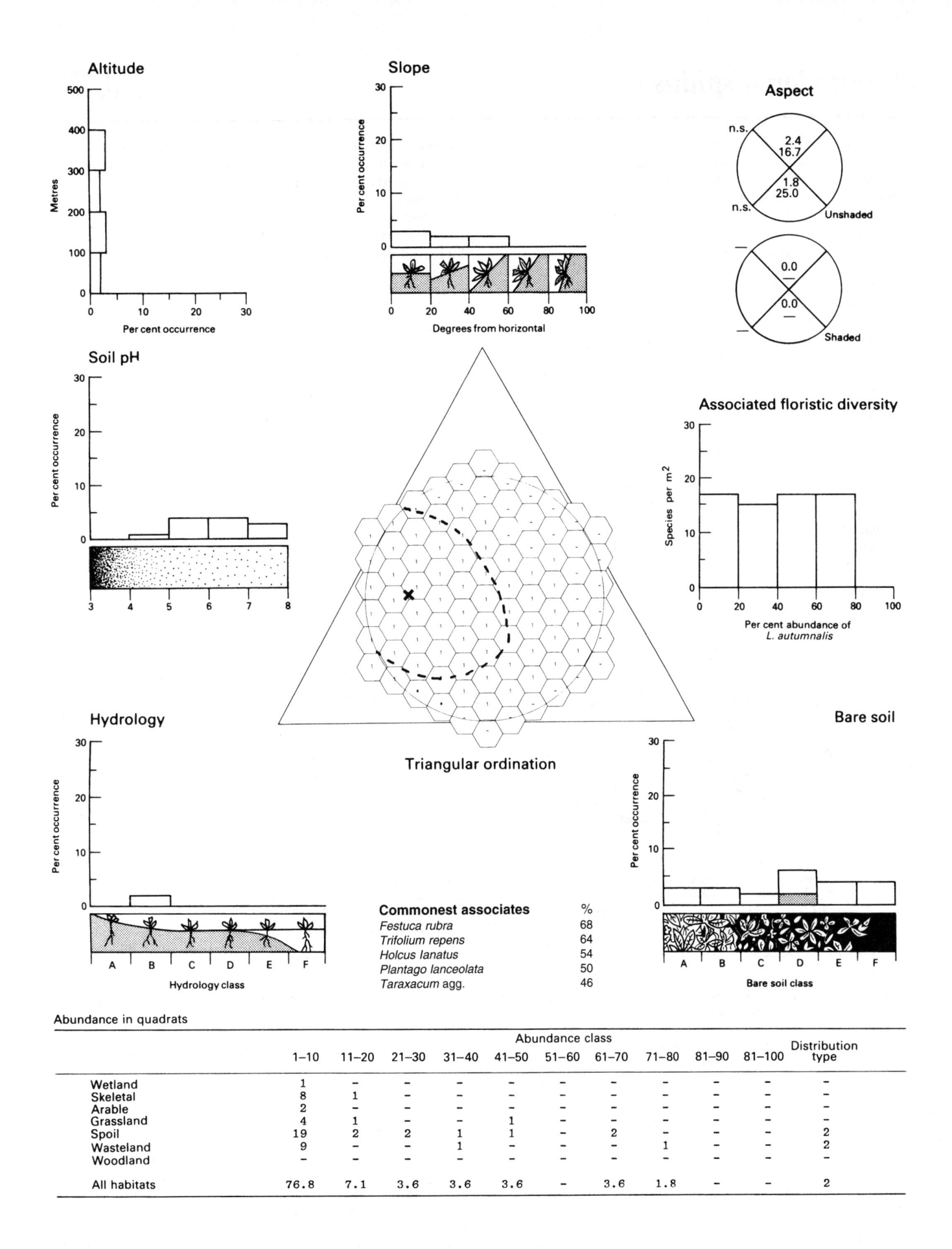

Altitude

Slope

Aspect

	n.s.	2.4
		16.7
	n.s.	1.8
		25.0

Unshaded

		0.0
		0.0

Shaded

Soil pH

Associated floristic diversity

Hydrology

Triangular ordination

Bare soil

Commonest associates %
Festuca rubra 68
Trifolium repens 64
Holcus lanatus 54
Plantago lanceolata 50
Taraxacum agg. 46

Abundance in quadrats

	Abundance class										Distribution type
	1–10	11–20	21–30	31–40	41–50	51–60	61–70	71–80	81–90	81–100	
Wetland	1	–	–	–	–	–	–	–	–	–	–
Skeletal	8	1	–	–	–	–	–	–	–	–	–
Arable	2	–	–	–	–	–	–	–	–	–	–
Grassland	4	1	–	–	1	–	–	–	–	–	–
Spoil	19	2	2	1	1	–	2	–	–	–	2
Wasteland	9	–	–	1	–	–	–	1	–	–	2
Woodland	–	–	–	–	–	–	–	–	–	–	–
All habitats	76.8	7.1	3.6	3.6	3.6	–	3.6	1.8	–	–	2

Leontodon hispidus L.

Asteraceae
Rough Hawkbit

Life-form Polycarpic perennial, rosette hemicryptophyte with branched stock. Scapes erect or ascending; leaves lanceolate, hairy, up to 4000 mm² or more; roots with VA mycorrhizas.
Phenology New leaves expand in late spring. Flowers from June to September and fruits from July to October. Shoots die back in autumn.
Height Foliage often appressed to the ground, not exceeding 300 mm; scape to 600 mm.
Seedling RGR 0.5-0.9 week^{-1}. SLA 28.8. DMC 10.5.
Nuclear DNA amount 5.6 pg; $2n = 14*$ (*BC*); $2x*$.
Established strategy C-S-R.
Regenerative strategies W, V, ?B_s. Regenerative biology is complex (see 'Synopsis'); ?SB III.

Flowers Florets yellow, ligulate, hermaphrodite, self-incompatible (Finch 1967), insect-pollinated; often >100 aggregated into a terminal capitulum.
Germinule 0.85 mg, 6.5 × 1.0 mm, achene dispersed with a pappus *c.* 7 mm in length.
Germination Epigeal; t_{50} 4 days; 12-32°C; L > D.
Biological flora None.
Geographical distribution Widespread in England, Wales and C Ireland, infrequent elsewhere (77% of British vice-counties), *ABIF* page 620; W, C and S Europe northwards to Scandinavia (72% of European territories); Asia Minor.

Habitats Predominantly in calcareous pastures and quarries and on screes, rock outcrops and less-toxic lead-mine spoil. Also present on limestone cliffs, wasteland, road verges, railway banks and various types of spoil heap. Absent from woodland and only observed in wetland on tussocks of other species. Most common on calcareous soils, but by no means restricted to them. [Frequent in calcareous sand dunes.]
Gregariousness Intermediate.
Altitude Up to 400 m. Suitable habitats are most frequent on the carboniferous limestone and in consequence the distribution shows an upland bias. [Up to 580 m.]

Slope Widely distributed with respect to slope, but most abundant on slopes of 20-70° in the limestone dales.
Aspect No influence of aspect detected.
Hydrology Absent from wetland other than its rare occurrence on tussocks of other species.
Soil pH Clearly associated with base-rich soils and considerably more frequent on soils of pH >7.0. Absent from acidic soils of pH <4.5.
Bare soil Widely distributed, but most frequently associated with low or moderate amounts of bare ground.
△ **ordination** Distribution centred on vegetation associated with relatively undisturbed and infertile habitats. Absent from heavily disturbed situations and excluded from productive perennial communities.
Associated floristic diversity Relatively high.

Similarity in habitats	%
Linum catharticum	89
Scabiosa columbaria	86
Pilosella officinarum	85
Senecio jacobaea	85
Solidago vigaurea	77

Habitats

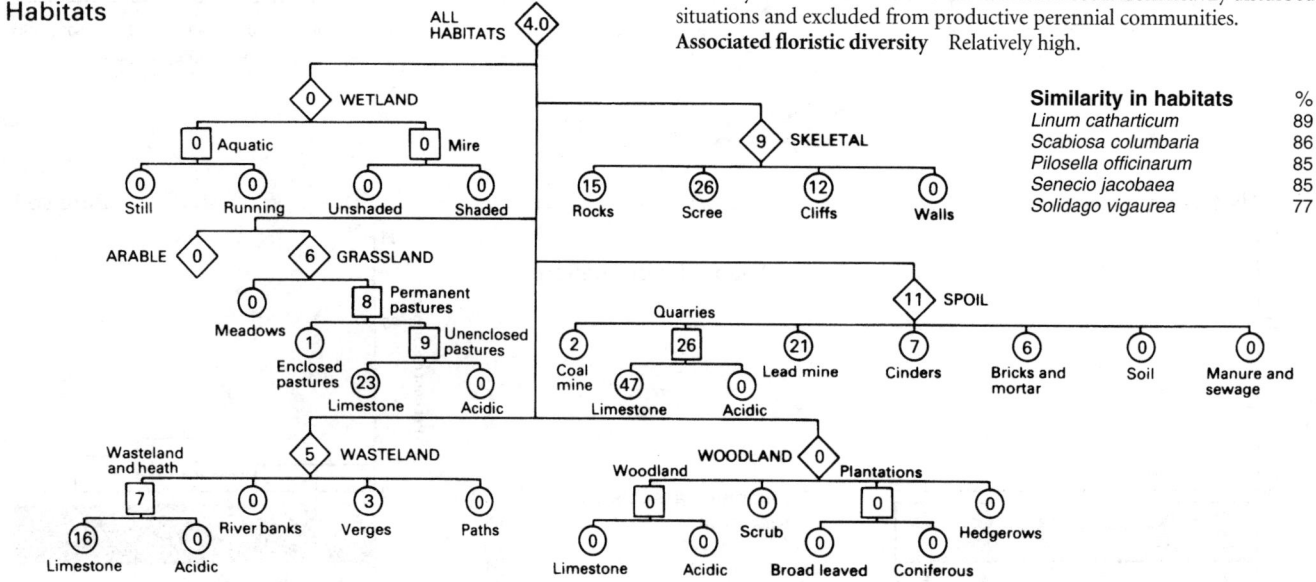

Synopsis *L.h.* is a slow-growing rosette-forming species which grows in a range of unproductive grasslands and open habitats, particularly on calcareous soils. It is moderately palatable to sheep (Milton 1933) and tends to subtend its leaves close to the ground surface in grazed turf. In the *PGE* the species is mainly confined to unproductive plots. In turf microcosms, roots became heavily colonized by VA mycorrhizas, with increases of up to 14-fold in seedling yield (Grime, Mackey, Hillier & Read 1987). The marked capacity to regenerate after close grazing and cutting (Hillier 1984) is related to the ability to re-sprout from buds on the stock. In tall grassland, leaves are more upright but *L.h.* is suppressed by larger species in rank vegetation. Leaf production in spring is considerably delayed compared with that of many of the grasses with which the species grows. The tap root penetrates to a considerable depth (Salisbury 1916) and in dry habitats appears to sustain water supply during the peak of shoot growth, which occurs in summer (Al-Mufti *et al.* 1977). We suspect that in dry habitats the dependence upon water provided via the tap root severely limits the capacity of *L.h.* for vegetative spread, since in moist habitats *L.h.* can form clonal patches through the production of a branching stock with adventitious roots. These clones may reach 200 mm in diameter. Otherwise, as in quarries (Finegan & Harvey 1982), *L.h.* regenerates mainly by means of wind-dispersed seeds which germinate in vegetation gaps. Ungerminated seeds of *L.h.* have a limited capacity for persistence on the soil surface (Grime 1978, Roberts 1986a). However, seed-set is often low in summer in grazed turf and establishment from seed is usually slow (King 1977b,c, Elzebroek 1981) although seedlings invading large vegetation gaps in Derbyshire have been observed to reach the flowering stage 2 years after germination (Hillier 1984). The field distribution of *L.h.* reveals that the species often behaves as a colonizer. *L.h.* is particularly common in open communities on unstable calcareous substrata, e.g. recently abandoned quarry heaps; this capacity for invasion is clearly related to the production of wind-dispersed seeds. *L.h.* is subdivided into six subspecies in Europe (*FE4*) and is essentially

Altitude

Slope

Aspect

n.s. 9.4 / 33.3
n.s. 8.5 / 31.6 Unshaded

0.0 / 0.0 Shaded

Soil pH

Associated floristic diversity

Species per m²
Per cent abundance of *L. hispidus*

Hydrology

Absent from wetland

Triangular ordination

Bare soil

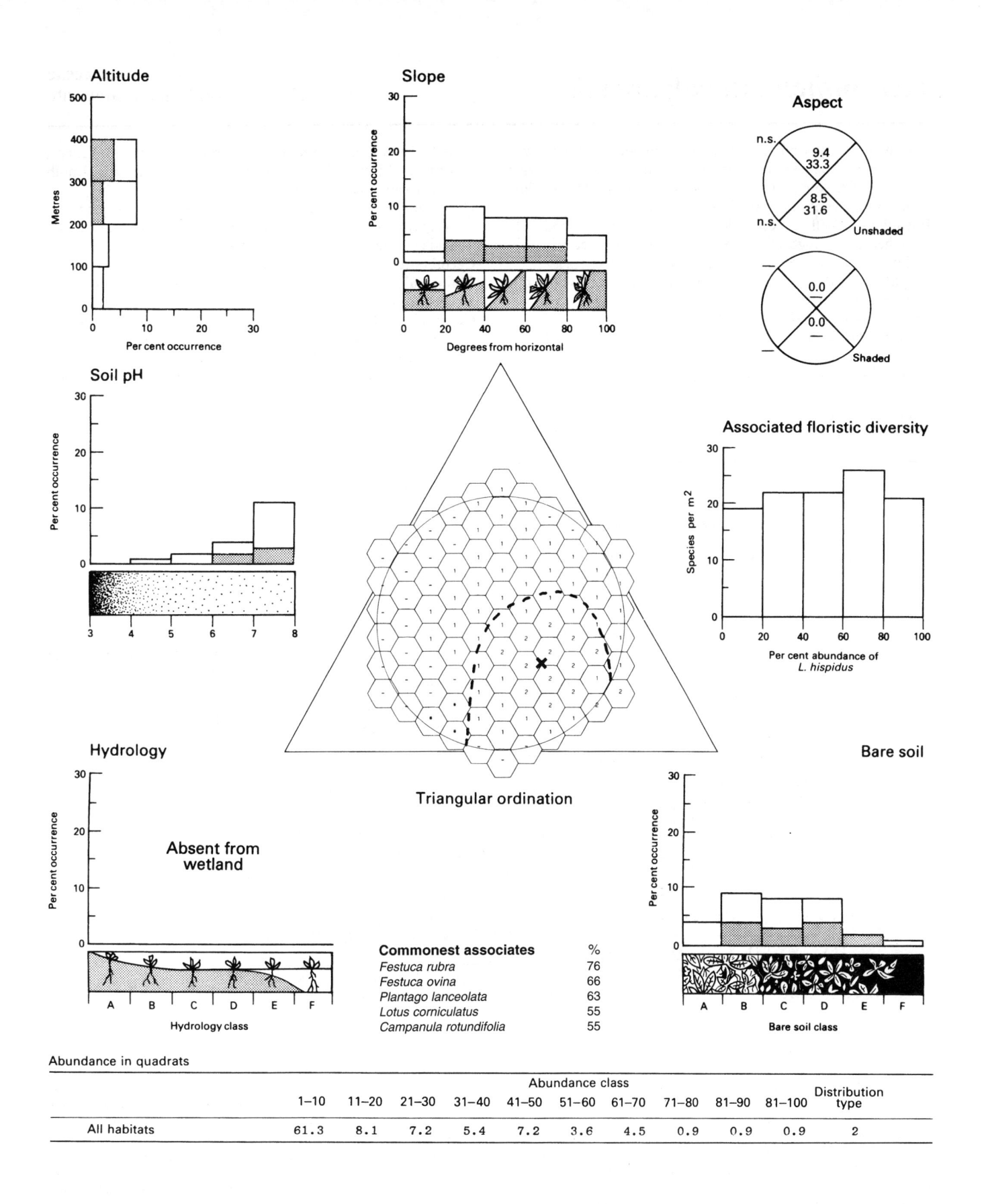

Commonest associates	%
Festuca rubra	76
Festuca ovina	66
Plantago lanceolata	63
Lotus corniculatus	55
Campanula rotundifolia	55

Abundance in quadrats

	1–10	11–20	21–30	31–40	41–50	51–60	61–70	71–80	81–90	81–100	Distribution type
All habitats	61.3	8.1	7.2	5.4	7.2	3.6	4.5	0.9	0.9	0.9	2

outbreeding (Finch 1967). *L.h.* hybridizes with *L. saxatilis* (Lam.) Merat.
Current trends Uncertain, since although it is mobile and exploits artificial habitats such as quarries and railway banks, *L.h.* is largely restricted to less-fertile habitats. As a result of the ability to colonize derelict land, it is nevertheless more secure than the majority of species characteristic of species-rich pastures.

Leucanthemum vulgare Lam.

Asteraceae
Ox-eye Daisy, Moon Daisy, Marguerite

Life-form Polycarpic perennial protohemicryptophyte with branched stock. Shoots erect; leaves oblong to ovate; sparsely hairy when young, usually <1000 mm^2; roots rather shallow, mainly adventitious, with VA mycorrhizas.

Phenology Winter green. Flowers mainly in June and July. Achenes, although often ripe before then, are shed while the inflorescences are dying in August and September (*Biol Fl*).

Height Non-flowering rosettes <100 mm. Foliage to 400 mm; flowers to 700 mm.

Seedling RGR No data. SLA 22.1. DMC 13.0.

Nuclear DNA amount 19.0 pg; $2n = 18*, 36*, 54$ (*BC, FE4*); $2x*, 4x*, 6x$.

Established strategy Intermediate between C-S-R and competitive-ruderal.

Regenerative strategies S, B$_s$, V. See 'Synopsis'; SB III.

Flowers Central florets, often >150 per capitulum, tubular, yellow and hermaphrodite, outer florets usually <25, ligulate, white and female; insect-pollinated, selfed if this fails (Knuth 1906); capitulum solitary, terminal.

Germinule 0.33 mg, 2.4 × 0.5 mm, achene, ribbed, pappus a scaly rim or absent. Seed of disk and ray florets of similar size and germination characteristics (*Biol Fl*).

Germination Epigeal; t_{50} 2 days; 9-31°C; L > D.

Biological flora Howarth and Williams (1968) (as *Chrysanthemum leucanthemum* L.), see also Zarzycki (1976).

Geographical distribution The British Isles, but less frequent in the N (100% of British vice-counties), *ABIF* page 649; Europe to N Scandinavia but rare in the extreme N (85% of European territories and 5% naturalized); also distributed with cultivation to various parts of the world, e.g. USA and New Zealand.

Habitats Common on spoil heaps, particularly limestone quarry waste. Also frequent in meadows and abandoned pasture, and particularly characteristic of railway banks and areas of cinders close to the track. More scarce in skeletal habitats and on road verges, river banks and paths; only a casual on arable land, and absent from wetland and woodland habitats. [Distinct maritime and high mountain ecotypes are recognized (*Biol Fl*).]

Gregariousness Intermediate.
Altitude Up to 400 m. [Up to 800 m.]
Slope Widely distributed.
Aspect Recorded more frequently on S-facing slopes, but in semi-natural grassland *L.v.* shows a N-facing bias (*Ecol Atl*), although more abundant on S-facing slopes.

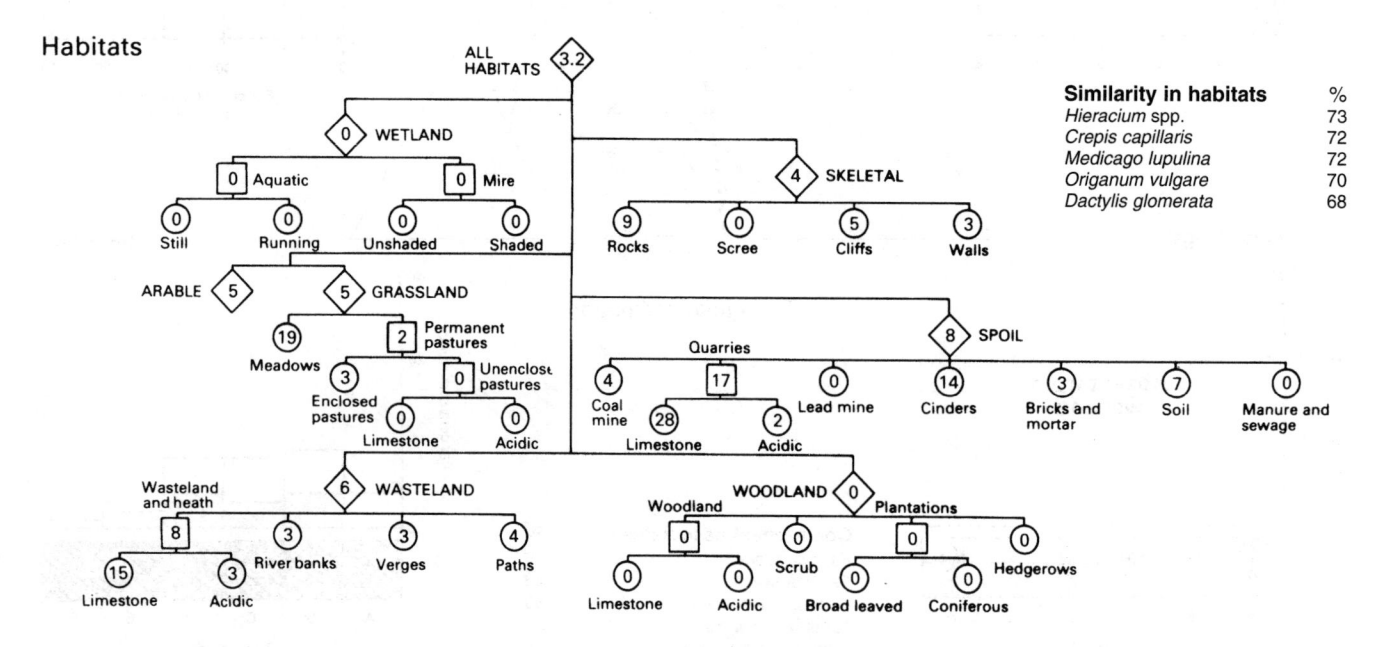

Habitats

Similarity in habitats	%
Hieracium spp.	73
Crepis capillaris	72
Medicago lupulina	72
Origanum vulgare	70
Dactylis glomerata	68

Hydrology Generally absent from wetlands, although recorded from fen in *Biol Fl*.

Soil pH Widely distributed on base-rich soils of pH >5.0, showing a progressive increase in frequency and abundance with rising soil pH.

Bare soil Widely distributed, but most common at moderate exposures.

△ **ordination** Distribution wide-ranging, but centred on vegetation associated with intermediate levels of disturbance and productivity. Evidence of competitive exclusion from productive undisturbed communities.

Associated floristic diversity Typically associated with species-rich vegetation.

Synopsis *L.v.* is a loosely-tufted winter-green herb of grassy and open habitats where the growth of potential dominants is restricted by a degree of soil infertility and disturbance factors such as cutting for hay and light grazing. The leaves of *L.v.* are basal and situated close to the ground over the winter period but, associated with a high value for nuclear DNA, there is rapid shoot expansion in spring. In ungrazed sites during the summer, functional leaves are largely confined to the flowering stems. These features, together with the production of seeds in early summer, enables *L.v.* to occur as a weed of hay meadows. In common with many arable weeds, *L.v.* retains many seeds on the plant at harvest time, is capable of germinating in autumn or spring, and forms a persistent seed bank (Chippindale & Milton 1934, Roberts 1986a). *L.v.* also exploits pasture grazed by horses, sheep and goats, and was formerly eaten as a salad plant, but is said to be generally avoided by cows and pigs (*Biol Fl*). Although tolerant of grazing and trampling, *L.v.* is generally less abundant in pastures; this may be a consequence of the limited capacity for clonal spread (patches usually <50 mm in diameter) and low seed-set under grazed conditions. Ungrazed plants may produce up to 4000 achenes (*Biol Fl*), which may explain why *L.v.* is often an effective colonist of areas of

Altitude
Slope
Aspect
Soil pH
Associated floristic diversity
Hydrology
Absent from wetland
Triangular ordination
Bare soil

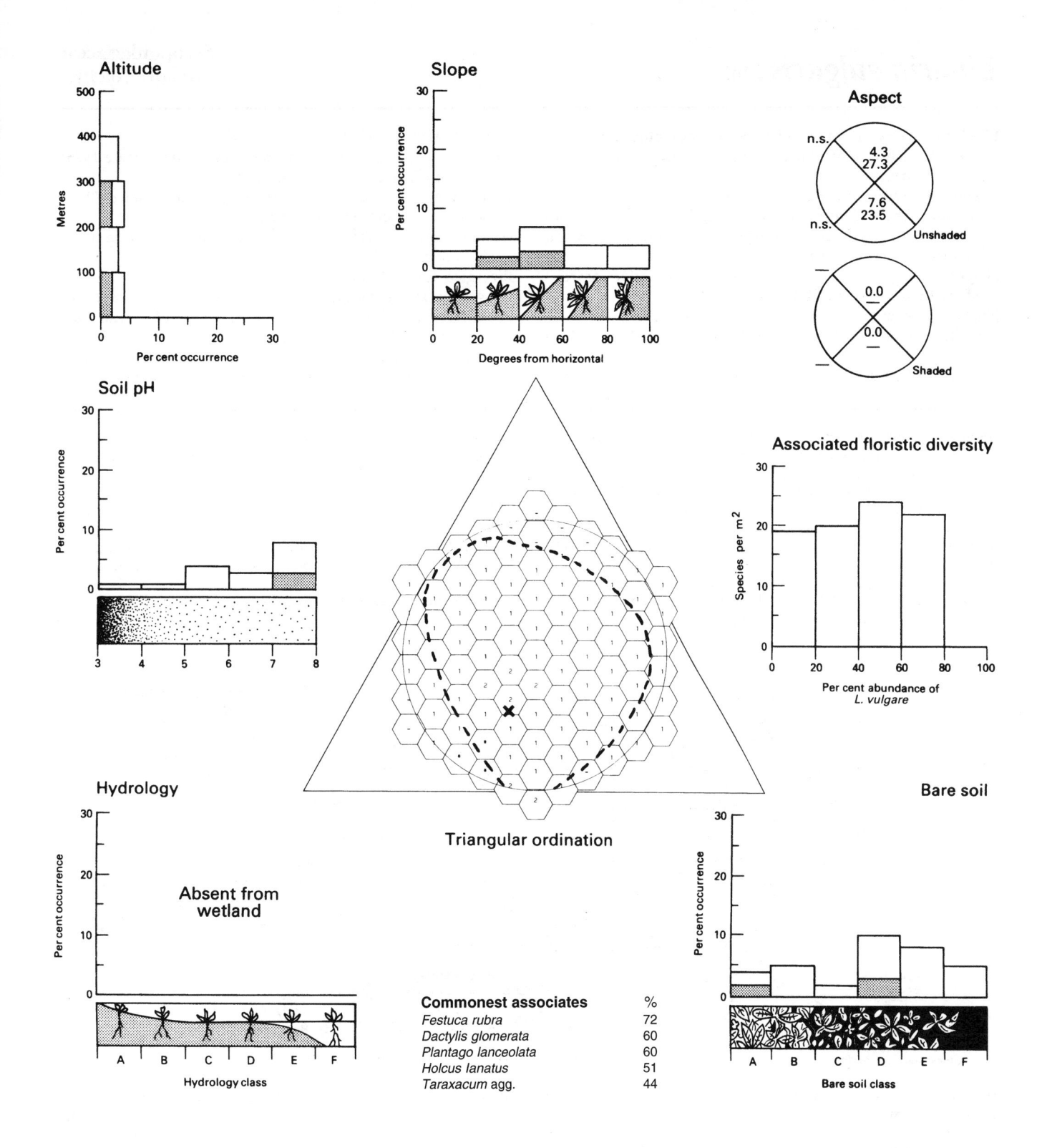

Commonest associates	%
Festuca rubra	72
Dactylis glomerata	60
Plantago lanceolata	60
Holcus lanatus	51
Taraxacum agg.	44

discontinuous vegetation cover on spoil heaps, roadsides and railway banks. *L.v.* is considered drought tolerant (*Biol Fl*); in outdoor pot experiments substantial seedling emergence occurred between March and September (Roberts 1986a) and the characteristics of the achene allow rapid germination at the surface of relatively dry soils (Oomes & Elberse 1976). *L.v.* forms a polyploid series and numerous ecotypes have been recognized (*Biol Fl*). Tetraploids replace diploids in N and E Britain (Pearson 1967), while in S and W they tend to occur only in sites subject to heavy-metal pollution (Grant & Whitebrook 1983). 2x plants are also commonest in N and W Europe, with 4x plants most frequent in the N and E, and in the S on mountains. Both occur in C Europe (Bocher & Larsen 1957). Drought or low N- supply can induce leaf succulence, a character under genetic control in maritime populations (*Biol Fl*).

Current trends The wide distribution of *L.v.* owes much to human activities (*Biol Fl*). However, *L.v.* is uncommon in highly fertilized sites, and appears to be decreasing in meadows and pastures. As a result *L.v.* is becoming more characteristic of areas such as railway and motorway banks, and less common in farmland.

Linaria vulgaris Miller

Scrophulariaceae
Yellow Toadflax

Life-form Polycarpic perennial protohemicryptophyte. Stems erect, branched; leaves linear, <250 mm^2; deeply rooted, with VA mycorrhizas.

Phenology Overwintering as short leafy shoots produced in autumn. These elongate in late spring. Flowers from July to October and seed set from September onwards. Flowering shoots die back during autumn.

Height Foliage to *c.* 500 mm, overtopped by the inflorescence.

Seedling RGR No data. SLA 23.4. DMC 18.3.

Nuclear DNA amount No data; $2n = 12$ (*FE3*); $2x$.

Established strategy Intermediate between competitive-ruderal and C-S-R.

Regenerative strategies V, B$_s$. Regenerative biology is complex (see 'Synopsis'), SB III.

Flowers Yellow, hermaphrodite, self-incompatible (Stace 1975), insect-pollinated; often *c.* 20 in a raceme.

Germinule 0.14 mg, 2.1 × 1.8 mm, winged seeds, often *c.* 70 in a dehiscent capsule (Salisbury 1964).

Germination Epigeal; t_{50} 5 days; germination rate increased after dry storage at 20°C; L > D.

Biological flora None.

Geographical distribution England, Wales and S Scotland, local in Ireland and possibly introduced (83% of British vice-counties), *ABIF* page 549; most of Europe except for the extreme N and much of the Mediterranean region (77% of European territories); W Asia; naturalized in N America.

Habitats Mainly restricted to disturbed artificial habitats, and particularly abundant on cinder tips and beside railways. Frequent on coal-mine spoil and recorded from demolition sites, hedgerows and walls. Also observed on mining waste and building rubble. Absent from meadows, pasture, arable land, wetlands and most skeletal habitats.

Gregariousness Sparse to intermediate.

Altitude Most suitable habitats occur in the lowlands but *L.v.* has been observed up to 260 m. [Up to 320 m.]

Slope Wide-ranging, but only abundant on flat ground or gentle slopes.

Aspect Occurrences slightly more frequent on N-facing slopes, but significantly more abundant on S-facing aspects.

Hydrology Absent from wetlands.

Soil pH Most records within the pH range 5.5-8.0, with occasional occurrences on more-acidic soils down to pH 3.5.

Bare soil Sometimes in well-vegetated sites with little bare soil, but mainly found in association with open areas of cinders, ballast or other types of spoil.

△ **ordination** Usually associated with moderately productive vegetation and most frequent in sites experiencing occasional disturbance. (see 'Synopsis').

Associated floristic diversity Low to intermediate.

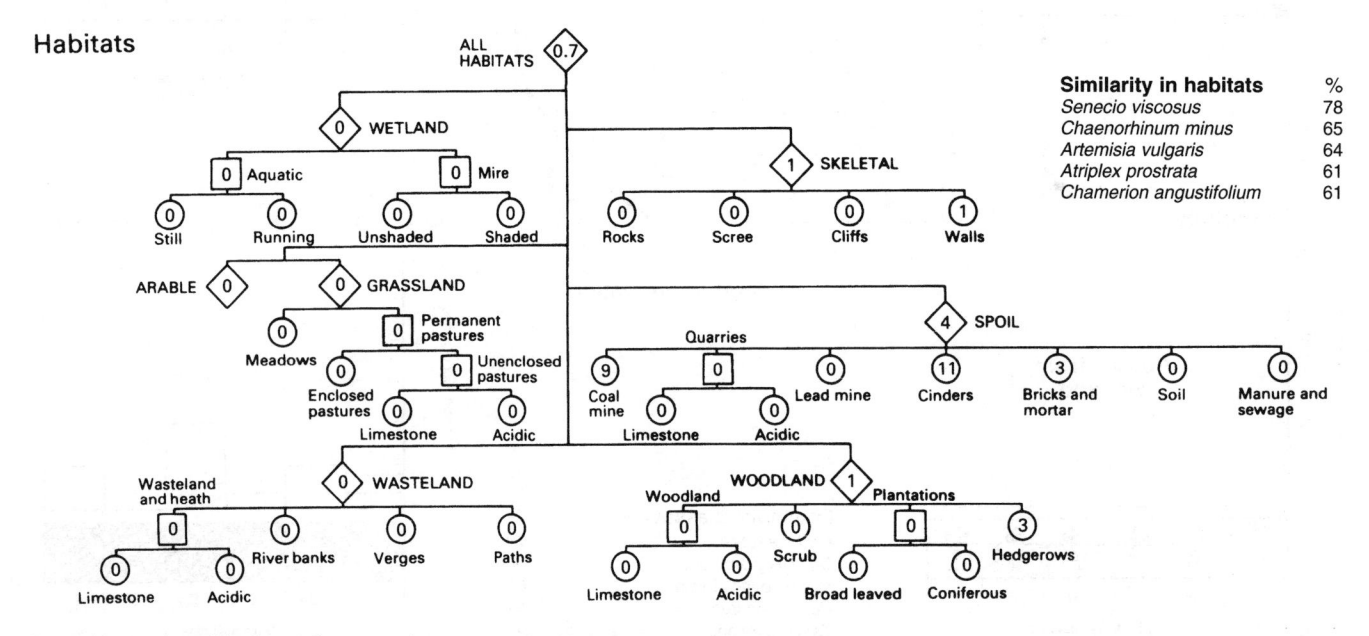

Habitats

Similarity in habitats	%
Senecio viscosus	78
Chaenorhinum minus	65
Artemisia vulgaris	64
Atriplex prostrata	61
Chamerion angustifolium	61

Synopsis *L.v.* is a tall herb mainly restricted to open, artificial habitats on dry soils. *L.v.* is particularly common in situations where deep mineral soils are overlain by coarse debris, and in this respect the species resembles *Tussilago farfara* and *Equisetum arvense*. Although tall and with an extensive root system, *L.v.* has only a limited capacity to spread laterally and to develop a dense leaf canopy; consequently the species is generally restricted to sites where the growth of more-robust perennials is debilitated. Thus, *L.v.* is most frequent in sites of relatively low annual productivity and some disturbance, and is particularly characteristic of open cinders and ballast beside railways, where the vigour of other potentially fast-growing, colonizing species is limited by the low nutritional status of the substratum. Persistence beside the permanent way may also arise from a resistance to selective herbicides (Salisbury 1964), and we suspect that *L.v.* is also tolerant of burning. *L.v.* has a relatively deep root system which may sustain water supply during the peak of growth, which occurs during summer. The foliage is said to be toxic to stock (Cooper & Johnson 1984) but *L.v.* is rarely present in pasture. Like *Hypericum perforatum*, with which it often grows, *L.v.* regenerates by means of adventitious buds produced on the roots (Bakshi & Coupland 1960), and forms large but diffuse clonal patches by this means. Disturbance appears to have a deleterious effect on vegetative spread (Bakshi & Coupland 1960). Regeneration by seed is also important and each inflorescence has the potential to produce over 1000 small, winged seeds, but *L.v.* is self-incompatible and seed set is often poor (Salisbury 1964). Seeds may persist for several years in the soil and germinate in spring and early summer (Roberts 1986a). *L.v.* is part of a taxonomically complex grouping centred on S Russia (*FE3*) and hybridizes in Britain with *L. repens* (L.) Miller.

Current trends Probably increasing due to a capacity to exploit railways, industrial spoil and other artificial habitats.

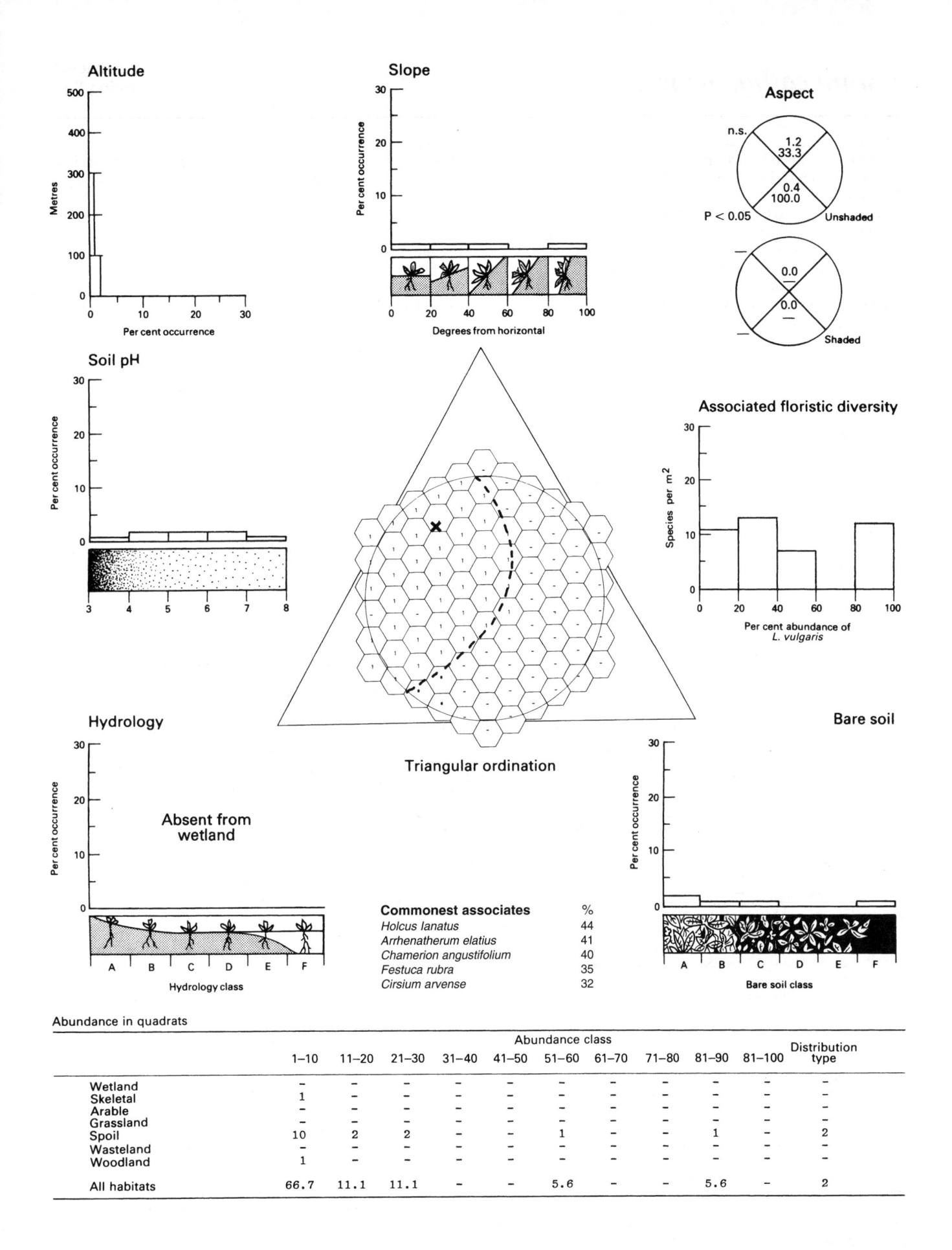

Altitude

Slope

Aspect

n.s.
1.2 / 33.3
0.4 / 100.0
$p < 0.05$ — Unshaded

0.0 / 0.0 — Shaded

Soil pH

Associated floristic diversity

Species per m²

Per cent abundance of *L. vulgaris*

Hydrology

Absent from wetland

Triangular ordination

Bare soil

Commonest associates	%
Holcus lanatus	44
Arrhenatherum elatius	41
Chamerion angustifolium	40
Festuca rubra	35
Cirsium arvense	32

Abundance in quadrats

	\multicolumn{10}{c}{Abundance class}	Distribution type
	1–10	11–20	21–30	31–40	41–50	51–60	61–70	71–80	81–90	81–100	
Wetland	–	–	–	–	–	–	–	–	–	–	–
Skeletal	1	–	–	–	–	–	–	–	–	–	–
Arable	–	–	–	–	–	–	–	–	–	–	–
Grassland	–	–	–	–	–	–	–	–	–	–	–
Spoil	10	2	2	–	–	1	–	–	1	–	2
Wasteland	–	–	–	–	–	–	–	–	–	–	–
Woodland	1	–	–	–	–	–	–	–	–	–	–
All habitats	66.7	11.1	11.1	–	–	5.6	–	–	5.6	–	2

Linum catharticum L.

Linaceae
Purging Flax

Life-form Biennial or more rarely annual (During *et al.* 1985). Stems erect, often branched; leaves ovate to lanceolate, usually <10 mm^2; roots shallow, normally bearing VA mycorrhizas.
Phenology Winter green. Flowers from June to September. Seed shed from July to October. Leaves relatively long-lived.
Height Foliage typically <30 mm in non-flowering plants, but flowering stems up to 150 mm.
Seedling RGR No data. SLA 37.5. DMC 21.0.
Nuclear DNA amount 1.2 pg; $2n = 16$ (*FE2*); $2x$.
Established strategy Stress-tolerant ruderal.
Regenerative strategies S, B$_s$. Regenerates entirely by seed, which germinates predominantly during spring. Evidence of a persistent seed bank (Roberts 1986a, J. Willems pers. comm.); SB II-IV.
Flowers White, hermaphrodite, homogamous, pollinated by various insects or selfed (Knuth 1906); often >20 in an open cyme.
Germinule 0.15 mg, 1.4 × 0.9 mm, seed within an up to 10-seeded dehiscent capsule.
Germination Epigeal; t_{50} 3 days; 7-36°C; dormancy broken by chilling.
Biological flora None, but see Zarzycki (1976).
Geographical distribution The British Isles (100% of British vice-counties), *ABIF* page 435; Europe except the extreme N (85% of European territories); Asia Minor, Iran.

Habitats Recorded mainly from calcareous strata. Particularly frequent on limestone quarry heaps, lead-mine spoil and in dale-side pastures, but also present on rock outcrops, screes, wasteground and road verges. Other habitats include cliffs, cinder tips, soil heaps and building rubble. Also observed in mire. Not recorded in woodland or arable fields.
Gregariousness Intermediate, but capable of developing high population densities.
Altitude Widely distributed, but most frequent on the carboniferous limestone between 200 and 400 m. [Up to 840 m.]

Habitats

Slope Wide-ranging, but suitable habitats particularly frequent in limestone dales of moderate slope.
Aspect Frequent on N- and S-facing slopes, but slightly more abundant on the former.
Hydrology Observed in calcareous soligenous mire and more rarely on tussocks of other species; otherwise restricted to dryland habitats.
Soil pH Most frequent and abundant above pH 7.0. Not recorded from acidic soils below pH 5.0.
Bare soil Associated mainly with habitats with intermediate exposures. Usually declining in abundance at the two extremes, but capable of developing large populations in open habitats which remain moist during the summer.
△ **ordination** A rather wide distribution, with contours centred on

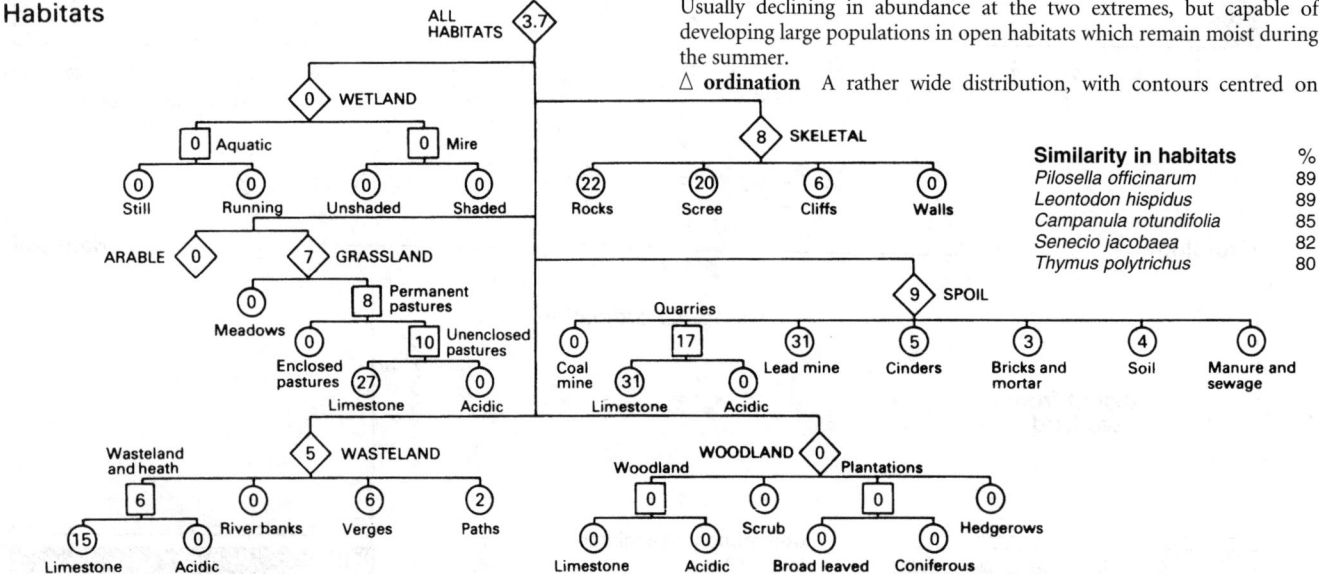

Similarity in habitats	%
Pilosella officinarum	89
Leontodon hispidus	89
Campanula rotundifolia	85
Senecio jacobaea	82
Thymus polytrichus	80

relatively undisturbed unproductive vegetation; also some capacity to persist in heavily disturbed sites. Suppressed in vegetation dominated by robust species.
Associated floristic diversity Typically species-rich.
Synopsis *L.c.* is a diminutive, slow-growing, biennial or annual herb most characteristic of open, infertile, calcareous grassland and damp skeletal habitats. The root system of *L.c.* is both delicate and shallow (Anderson 1927). The occurrences of *L.c.* within the *PGE* are restricted to unproductive plots. In grassland *L.c.* is associated with open turf, but germination and seedling survival is greatest in microsites providing some cover by perennial plants (Verkaar *et al.* 1983, Hillier 1984). *L.c.* reaches greater abundance on N-facing slopes and can develop large populations in wet skeletal habitats, such as quarry floors, and in open calcareous mire. *L.c.* usually overwinters as small plants, flowering in the following summer. In winter the tissues contain high concentrations of fructan (Hendry & Brocklebank unpubl.). The tendency to exploit the margins of transient vegetation gaps and to persist beneath the turf canopy suggest both a vulnerability to moisture stress and a degree of shade tolerance. *L.c.* is potentially toxic to stock (Cooper & Johnson 1984) but is predated by slugs. Emergence through the sward and effective display of flowers is facilitated by the erect growth form of the shoot. *L.c.* regenerates entirely by means of seeds, which are released in summer or autumn and require chilling at *c.* 5°C for 6-8 weeks before germination is possible (King 1977a). Recent investigations in England (Roberts 1986a) and Holland (Willems pers. comm.) reveal that some seeds of *L.c.* persist in the habitat and revert to a dormant state during the warmer summer months (cf. *Polygonum aviculare*). Seedlings appear in large populations mainly in spring (Roberts 1986a) but occasionally in autumn. There are high mortalities (During *et al.* 1985). The timing of spring germination has been found to be relatively constant from year to year, and it has been suggested (Kelly 1986) that a daylength cue may be implicated. In Dutch grassland, biennial plants produce *c.* 75 seeds per plant and annuals only 20-30 (During *et al.* 1985). The annual forms may represent a different ecotype (During *et al.* 1985).
Current trends *L.c.* exploits infertile artificial habitats such as quarry spoil and railway banks. It is not clear to what extent this compensates for the losses occurring through destruction of more-ancient grassland habitats.

Altitude

Slope

Aspect

	n.s.	10.6	
		51.9	
		10.3	
	n.s.	43.5	Unshaded

	—	0.0	
		0.0	
	—	—	Shaded

Soil pH

Associated floristic diversity

Hydrology

Absent from wetland

Triangular ordination

Commonest associates	%
Festuca ovina	78
Festuca rubra	70
Campanula rotundifolia	65
Plantago lanceolata	65
Lotus corniculatus	63

Bare soil

Abundance in quadrats

	Abundance class										Distribution type
	1–10	11–20	21–30	31–40	41–50	51–60	61–70	71–80	81–90	81–100	
Wetland	–	–	–	–	–	–	–	–	–	–	–
Skeletal	14	4	1	2	–	–	4	–	–	3	2
Arable	–	–	–	–	–	–	–	–	–	–	–
Grassland	5	–	1	1	–	1	3	3	–	–	3
Spoil	25	4	3	1	3	2	2	2	–	–	2
Wasteland	8	3	1	1	1	–	1	1	1	2	3
Woodland	–	–	–	–	–	–	–	–	–	–	–
All habitats	50.5	10.7	5.8	4.9	3.9	2.9	9.7	5.8	1.0	4.9	2

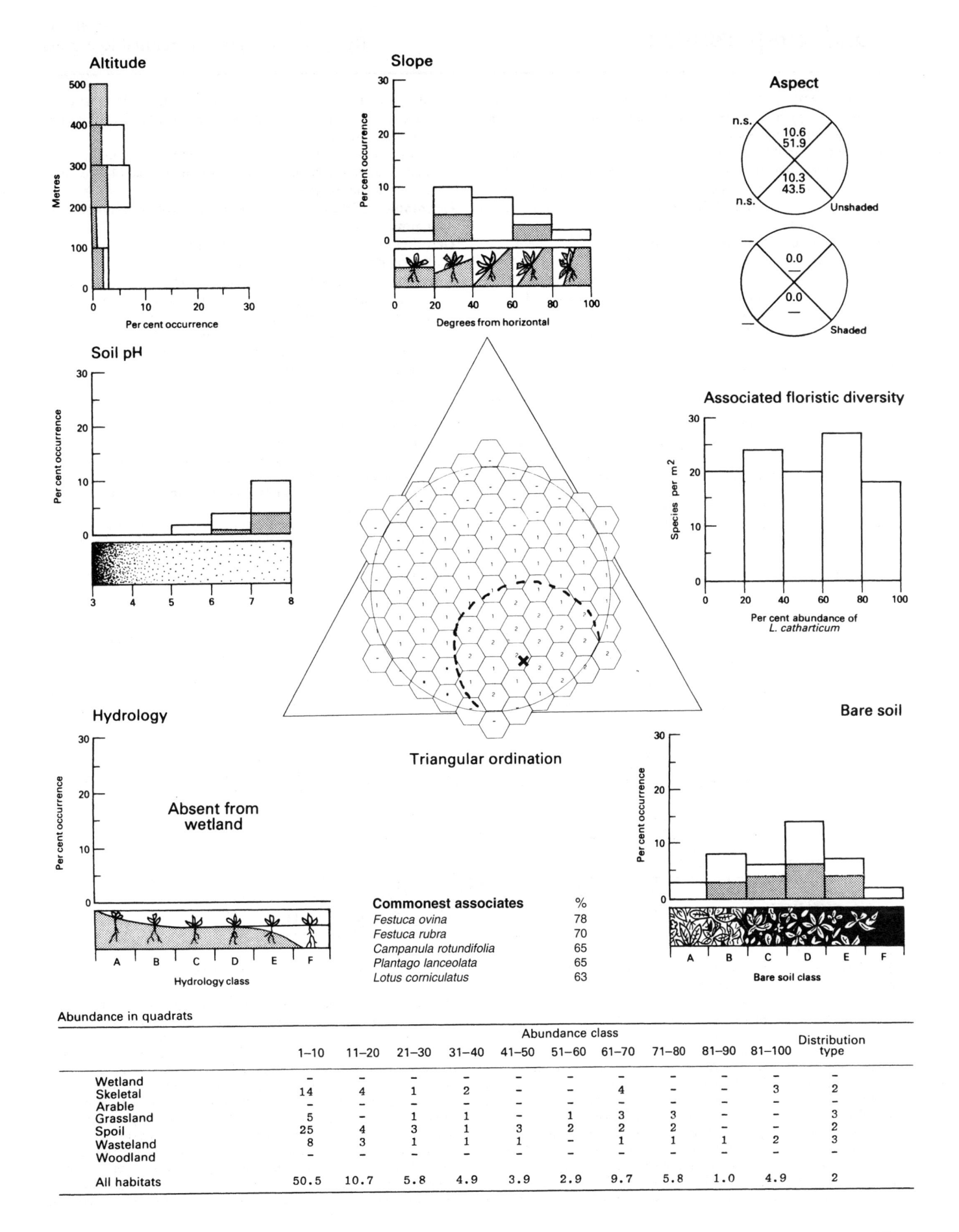

(*)*Lolium perenne* L.

Poaceae
Rye-grass, Ray-grass, Perennial Rye-grass

Life-form Tufted, usually polycarpic perennial, semi-rosette hemicryptophyte. Shoots erect or spreading; leaves linear, often <1000 mm^2; roots with VA mycorrhizas.

Phenology Variable according to genotype. Some genotypes grow more than others during the winter, but all are winter green. Certain varieties flower mainly in early summer (May), others are late heading (July-August). Seed sets after a further month (*Biol Fl*).

Height Foliage to 250 mm; flowers to 500 mm.

Seedling RGR 1.0-1.4 week^{-1}. SLA 26.4. DMC 21.2.

Nuclear DNA amount 9.9 pg, $2n = 14$* (Bennett & Smith 1976); $2x$*. Cultivars with $2n = 28$ ($4x$) are included in lists of recommended varieties (ADAS 1984).

Established strategy Intermediate between competitive-ruderal and C-S-R.

Regenerative strategies S. Principally by seed directly in autumn. Lateral tillers may increase the size of the plant, but these do not appear to constitute an effective means of reproduction; SB I.

Flowers Green, hermaphrodite, wind-pollinated, self-incompatible (Weimarck 1968); >100 in a spike (4-14 in each distant sessile spikelet).

Germinule 1.79 mg, 3.8 × 1.0 mm, caryopsis with an attached lemma, 6.0 × 1.8 mm.

Germination Epigeal; t_{50} 2 days; <5-34°C; L = D.

Biological flora Beddows (1967).

Geographical distribution The British Isles (100% of British vice-counties), *ABIF* page 749; Europe except the Arctic (87% of European territories and 10% introduced); temperate Asia and N Africa; introduced into N America, Australia and New Zealand and most parts of the temperate world, including higher altitudes in the tropics.

Habitats Common in a range of fertile, artificial, grassy or disturbed habitats, particularly where there is grazing or trampling. Abundant in meadows, enclosed pastures and road verges, where it is often sown. Also very common on paths, demolition sites, and manure and sewage spoil. Less frequent in droughted habitats, e.g. cliffs, and in waterlogged situations, and not persistent in infertile grassland. Absent from woodland. [Occurs in a range of grassy or open habitats near the sea. Will grow in suitably fertilized uplands (*Biol Fl*).]

Gregariousness Intermediate.

Altitude Less common at higher altitudes where productive pasture becomes increasingly scarce. Absent above 400 m. [Up to 490 m under certain conditions (*Biol Fl*).]

Slope Most frequent and abundant on flat ground and gentle slopes.

Aspect Widely distributed.

Hydrology A species of moist rather than waterlogged soils.

Soil pH Absent from highly acidic soils; frequent and abundant in the pH range 5.0-8.0.

Bare soil Widely distributed, but most frequently associated with relatively high exposures.

Habitats

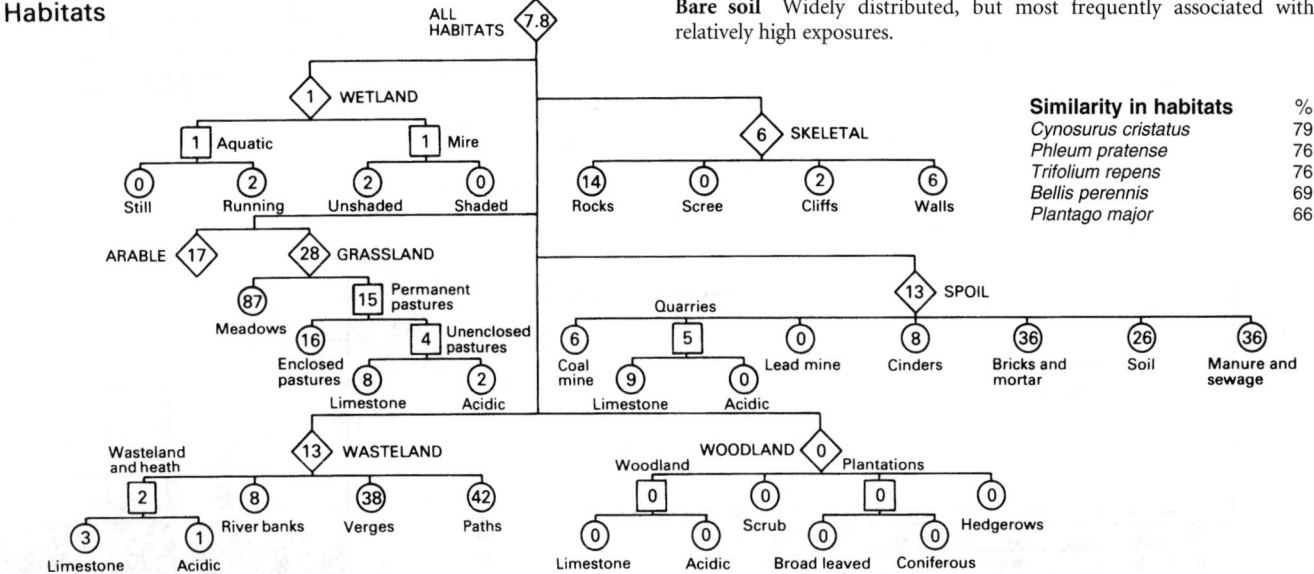

Similarity in habitats	%
Cynosurus cristatus	79
Phleum pratense	76
Trifolium repens	76
Bellis perennis	69
Plantago major	66

△ **ordination** Distribution wide-ranging but clearly centred on vegetation associated with habitats which are fertile and subject to relatively frequent disturbance.

Associated floristic diversity Low to intermediate.

Synopsis *L.p.* has been cultivated since at least the 17th century (*Biol Fl*) and is the most economically important forage grass in Britain. The management of productive grasslands, for both agriculture and amenity, at present depends upon the introduction and maintenance of a sward of *L.p.* There is a current annual sowing of more than 20 000 t of *L.p.* and *L. multiflorum* Lam. – the ecology of which is described by Beddows (1973) – and *L. multiflorum* × *L. perenne*. This is almost 10 times the amount sown of any other grass (MAFF 1980-84). Many genotypes have been bred or identified, but *L.p.* may be conveniently subdivided into two groups: early-heading types, which tend to be short-lived and to have few tillers and an erect growth form, and late-heading types, which are often longer-lived and with many tillers (*BiolFl*). Both are extensively sown. The former are considered in historical terms to have been 'followers of man' (*Biol Fl*), while the latter are characteristic of old fattening pasture and may have entered Britain before humans (*Biol Fl*). Leaf extension commences under cool conditions in the early spring and, consistent with this phenology, the tissues of *L.p.* are rich in fructan (Pollock & Jones 1979). *L.p.* has a high nutrient requirement, and is highly productive on fertile soils, very tolerant of trampling, mowing and grazing, and palatable to stock. Foliage has a high content of P but, in common with most grasses, tends to be low in Ca (*SIEF*). Despite the high densities of seeds sown, *L.p.* is rapidly replaced by other grasses unless carefully managed. The low persistence in many pastures relates to the transient nature of the seed bank (Thompson & Grime 1983, Roberts 1986b) and to ineffective vegetative spread. Both of these seriously reduce the cost-effectiveness of *L.p.* for agricultural and amenity use, and often necessitate frequent resowing.

Altitude

Slope

Aspect

	n.s.	7.5	
		36.8	
		10.8	Unshaded
	n.s.	20.8	

		0.0	
		—	
		0.0	Shaded
	—	—	

Soil pH

Associated floristic diversity

Hydrology

Triangular ordination

Bare soil

Commonest associates	%
Trifolium repens	68
Poa trivialis	65
Holcus lanatus	57
Festuca rubra	56
Dactylis glomerata	53

Abundance in quadrats

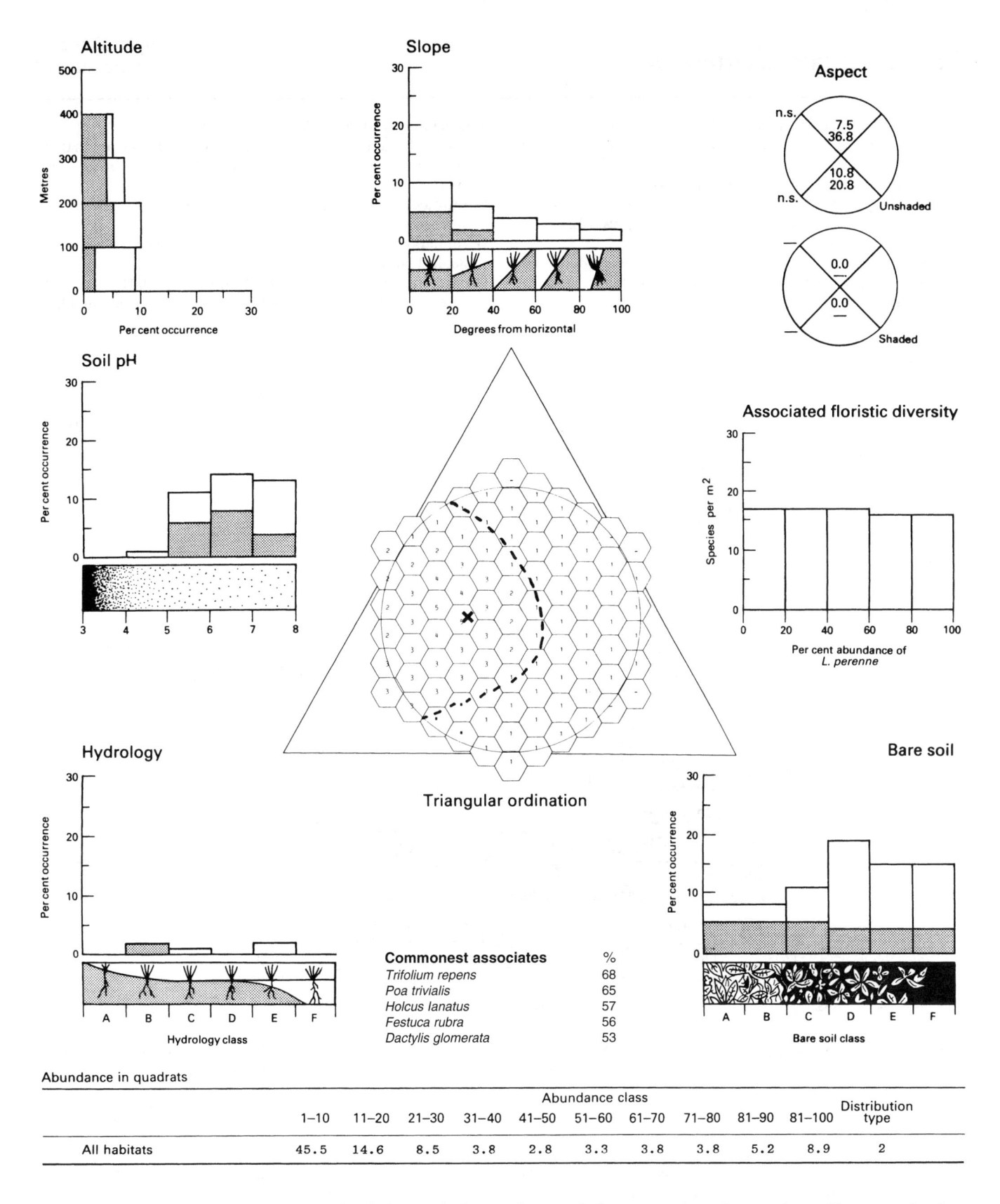

	\multicolumn{10}{c}{Abundance class}	Distribution type
	1–10	11–20	21–30	31–40	41–50	51–60	61–70	71–80	81–90	81–100	
All habitats	45.5	14.6	8.5	3.8	2.8	3.3	3.8	3.8	5.2	8.9	2

Heavy-metal-tolerant populations have been identified in Derbyshire (Shaw 1984) and, apart from hybrids with *L.m.*, *L.p.* also crosses with various species of *Festuca* (Stace 1975).

Current trends *L.p.* is abundant, and perhaps increasing since vast amounts of seed are sown annually, both for agriculture and for amenity. Seed is also shed on tracks and roadsides during harvesting. Commercially, however, there is an over-reliance upon *L.p.*, and when the ecological features of other grass species are better understood by agronomists, *L.p.* will be less-consistently recommended as a grassland crop. Also, the advantages of using slower-growing, less nutrient-demanding and more-persistent species on roadsides and in some amenity grasslands are now being more generally appreciated. In the near future the abundance of *L.p.* is thus likely to decrease.

Lonicera periclymenum L.

Caprifoliaceae
Honeysuckle

Life-form Twining or prostrate phanerophyte. Stems woody, much branched; leaves ovate, hairy, <2000 mm²; roots with VA mycorrhizas.
Phenology Some leaves and young shoots overwintering, particularly in sheltered areas. Flowers from June to July (to September) and berries are ripe from August onwards. Leaves relatively long-lived.
Height Foliage and flowers may exceed 3 m in hedgerows; non-flowering prostrate forms occur on woodland floors.
Seedling RGR No data. SLA 14.2. DMC 25.6.
Nuclear DNA amount 5.5 pg; $2n = 18^*, 36^*, 54$, (*CTW*, *FE4*); $2x^*, 4x^*, 6x$.
Established strategy Stress-tolerant competitor.
Regenerative strategies V, S. See 'Synopsis'; ?SB II.
Flowers Creamy-yellow, hermaphrodite, insect-pollinated; sometimes selfed if this fails (Knuth 1906); <10 in terminal heads.
Germinule 5.21 mg, 3.7×2.7 mm, dispersed within a few seeded, red berry.
Germination Epigeal; dormancy broken by chilling.
Biological flora None.
Geographical distribution The British Isles (100% of British vice-counties), *ABIF* page 601; W, C and S Europe extending NE to Sweden (49% of European territories and 3% naturalized); N and C Morocco.

Habitats Virtually restricted to woodland and hedgerows, and absent from established coniferous plantations. In upland areas observed occasionally in open ungrazed habitats. [Extends into unshaded, rocky montane habitats and, more rarely, sheltered sea cliffs.]
Gregariousness Forms stands or patches even at low frequencies.
Altitude Recorded more frequently from lowland areas. [Up to 610 m.]
Slope Recorded only from slopes of <40°, but observed on near-vertical, wooded cliffs.
Aspect In shaded habitats more frequent on S-facing slopes. Observations in unshaded habitats suggest a N-facing bias.
Hydrology Largely restricted to moist rather than waterlogged soils.
Soil pH Over a wide range of values, but most frequent and abundant on acidic soils of pH <5.0.
Bare soil Only found in habitats with relatively little bare soil, but often associated with persistent tree litter.
△ **ordination** Distribution confined to undisturbed habitats.
Associated floristic diversity Typically low.

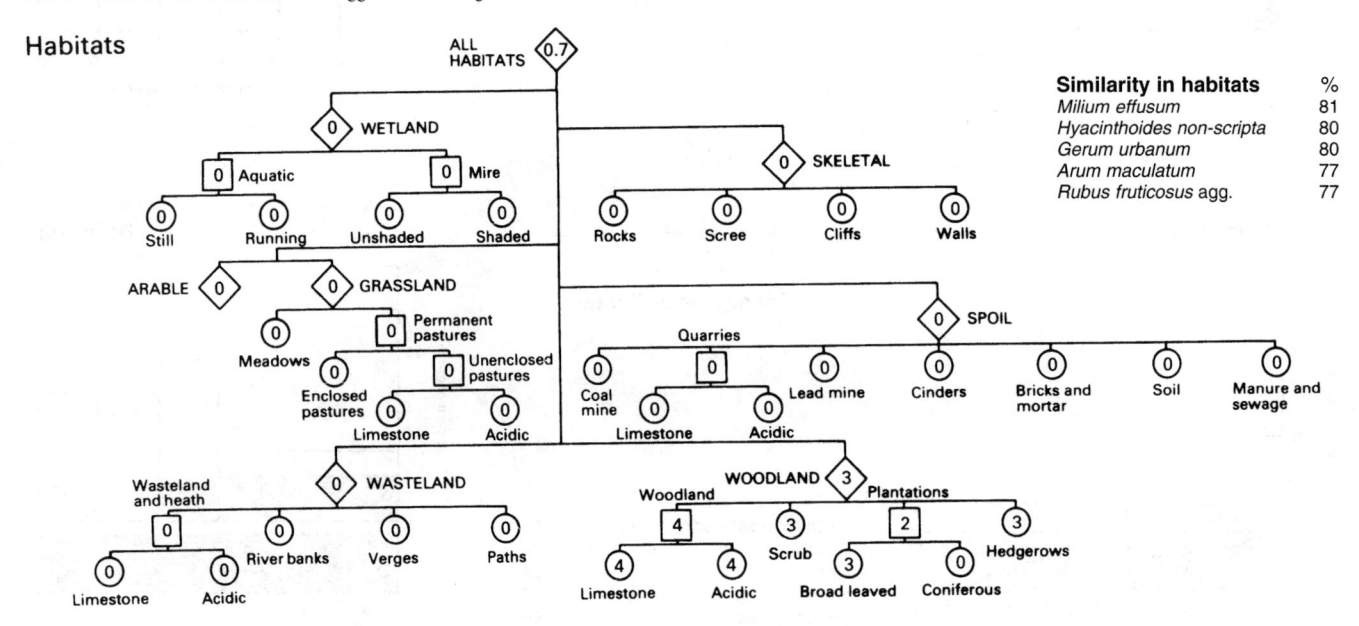

Habitats

Similarity in habitats	%
Milium effusum	81
Hyacinthoides non-scripta	80
Geum urbanum	80
Arum maculatum	77
Rubus fruticosus agg.	77

Synopsis *L.p.*, which has an sub-Atlantic distribution (Birks 1973), is represented by two main phenotypes: (a) a tall (often >3 m), woody, freely-flowering, deciduous climber of hedgerows and other more-open habitats and (b) a prostrate, non-flowering undershrub of woodland, dependent upon the creation of gaps in the canopy by tree-fall or woodland clearance before flowering can take place. In these respects the species resembles several other climbing plants of woodland habitats, e.g. *Hedera helix*, *Tamus communis* and *Clematis vitalba* L. In N England the species exhibits a bias towards S-facing slopes in shaded habitats and a tendency to occur on N-facing slopes in the open. In shoot phenology, shade-tolerance and reproductive biology *L.p.* resembles *Rubus fruticosus*, and the two species have similar ecologies in both woodlands and hedgerows. However, the distribution of *L.p.* is strongly biased towards less-fertile acidic soils, and *R.f.* has a greater potential to dominate herbaceous vegetation. Also, of course, *L.p.* is without spines and, judging by its rarity in grazed sites, is more vulnerable to grazing. In woodland sites the prostrate stems, which root at irregular intervals to form large clonal patches, are the only means of regeneration. In hedgerow sites, where plants are erect, vegetative reproduction is absent, or at best restricted, and regeneration appears to be mainly by seeds, which germinate in spring. There is no information on the presence or otherwise of a bank of persistent seed, or any direct evidence concerning the frequency of establishment by seed. However, *L.p.* is an effective colonist of secondary woodland, both in the Sheffield region and within Lincolnshire (Peterken 1981), presumably because its berries are widely dispersed by birds (Ridley 1930). It is not known whether stems of *L.p.* cut during, for example, hedge-trimming, regenerate to form new plants. A further European ssp. is recorded (*FE4*).
Current trends Uncertain. A frequent colonist of broad-leaved plantations on freely-drained acidic soils, but not an early recruit in hedgerows, where most seed is set.

Altitude

Slope

Aspect

Unshaded 0.0 / 0.0 / — / —

Shaded 0.8 / 0.0 / 2.4 / 0.0 — n.s.

Soil pH

Associated floristic diversity

Species per m² vs Per cent abundance of *L. periclymenum*

Hydrology

Absent from wetland

Triangular ordination

Bare soil

Commonest associates	%
Holcus mollis	73
Rubus fruticosus agg.	38
Deschampsia flexuosa	35
Hyacinthoides non-scripta	35
Deschampsia cespitosa	31

Abundance in quadrats

	1–10	11–20	21–30	31–40	41–50	51–60	61–70	71–80	81–90	81–100	Distribution type
Wetland	–	–	–	–	–	–	–	–	–	–	–
Skeletal	–	–	–	–	–	–	–	–	–	–	–
Arable	–	–	–	–	–	–	–	–	–	–	–
Grassland	–	–	–	–	–	–	–	–	–	–	–
Spoil	–	–	–	–	–	–	–	–	–	–	–
Wasteland	–	–	–	–	–	–	–	–	–	–	–
Woodland	7	6	1	2	1	1	1	–	–	–	2
All habitats	36.8	31.6	5.3	10.5	5.3	5.3	5.3	–	–	–	2

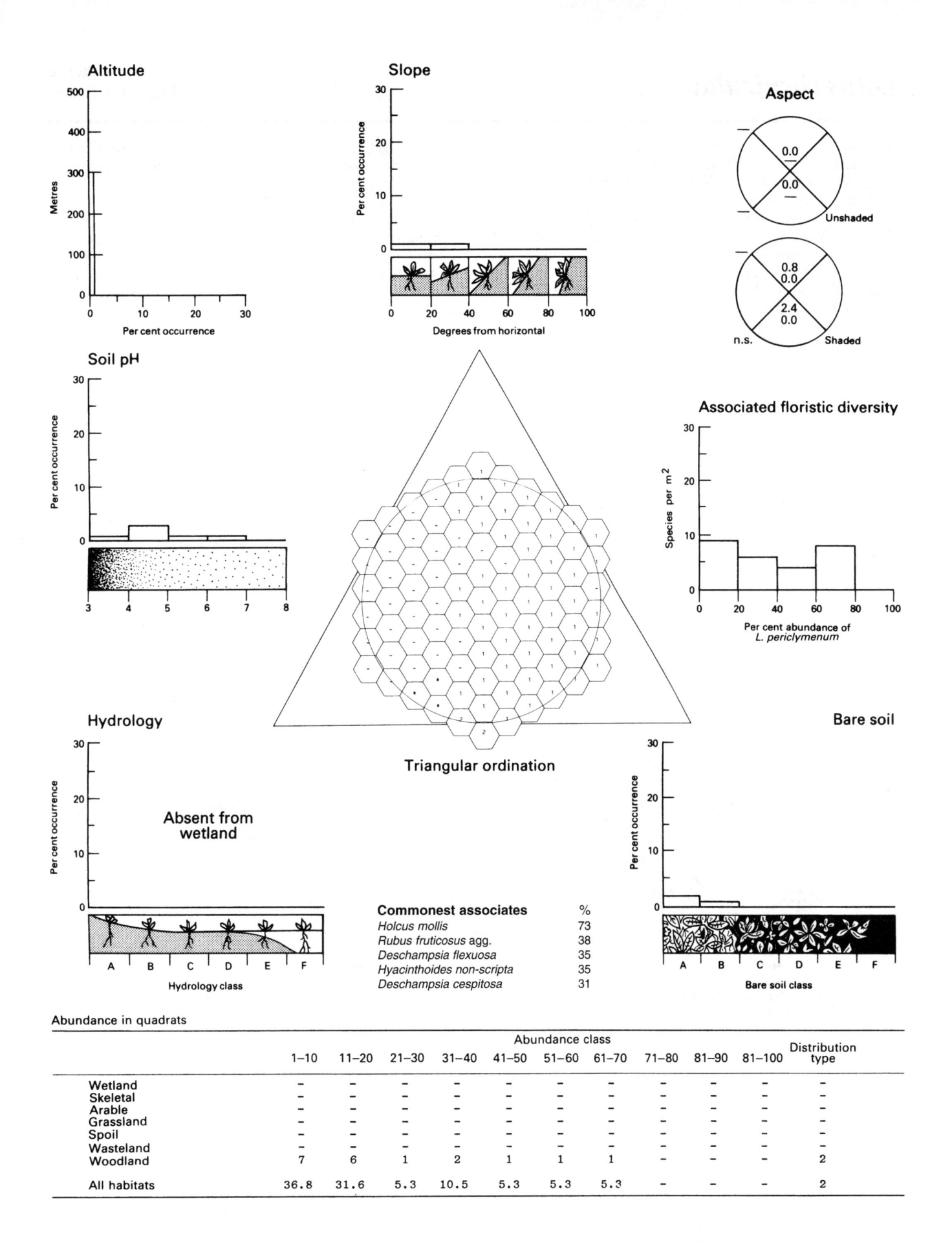

Lotus corniculatus L.

Fabaceae
Birdsfoot-trefoil, Eggs-and-Bacon

Life-form Polycarpic perennial protohemicryptophyte. Shoots decumbent; leaves 5-foliate, usually <500 mm^2; with long taproot, nodules and usually with VA mycorrhizas.

Phenology Flowers mainly June and July, longer in grazed situations. Seeds produced c. 1 month after pollination (*Biol Fl(a)*). Most shoots die back in late autumn, although occasionally in sheltered sites some may persist until the new flush of growth in spring.

Height Foliage usually <200 mm overtopped by flowers.

Seedling RGR 1.0-1.4 week^{-1}. SLA 29.3. DMC 16.7.

Nuclear DNA amount 2.2 pg; $2n = 24^*$ (*BC*), $4x^*$.

Established strategy Intermediate between stress-tolerator and C-S-R.

Regenerative strategies B_s. Regenerates mainly by seeds, which usually germinate in spring or become incorporated into a persistent seed bank. There is also a limited capacity for vegetative regeneration (see 'Synopsis'); ?SB III.

Flowers Yellow, hermaphrodite, insect-pollinated; partially self-compatible (*Biol Fl(a)*); 2-6 in each axillary umbel.

Germinule 1.67 mg, 1.9 × 1.5 mm, seed; up to >20 per pod, explosively dispersed.

Germination Epigeal; t_{50} 1 day; <5-34°C; hard-coat dormancy; L = D.

Biological flora (a) Turkington and Franko (1980), (b) Jones and Turkington (1986). See also Zarzycki (1976).

Geographical distribution The British Isles (100% of British vice-counties), *ABIF* page 377; Europe except for the extreme N; Asia, N and E Africa, but only on mountains in the tropics.

Habitats Widely dispersed in grassy and waste places. Particularly common in limestone pastures and wasteland, and on quarry waste and lead-mine heaps. Also found on various other types of spoil heap, on rock outcrops or screes, and in meadows, enclosed pastures and on road verges. Only a casual of arable land, and absent from woods and from wetland. [Found on shingle, sand dunes, cliff-top turf, sea heaths and a range of montane habitats.]

Gregariousness Intermediate.

Altitude Most suitable habitats are found on the carboniferous limestone, and the distribution of *L.c.* shows a corresponding upland bias. [Up to 915 m.]

Slope Wide-ranging, but absent from near-vertical surfaces.

Aspect Rather more frequent and abundant on S-facing slopes.

Hydrology Absent from wetlands.

Soil pH Excluded from very acidic soils. Histogram shows two peaks in the pH ranges 5.5-6.0 and 7.5-8.0.

Bare soil Not frequent in sites with closed vegetation or those with very little plant cover.

△ ordination Distribution fairly wide-ranging, but absent from habitats in which competition is intense. Centred on vegetation associated with habitats of moderate to low productivity and occasional disturbance.

Associated floristic diversity Typically high.

Similarity in habitats	%
Pimpinella saxifraga	82
Pilosella officinarum	81
Linum catharticum	74
Campanula rotundifolia	73
Briza media	72

Habitats

(Habitat diagram: ALL HABITATS 5.6; WETLAND 0 [Aquatic 0 (Still 0, Running 0), Mire 0 (Unshaded 0, Shaded 0)]; SKELETAL 4 [Rocks 11, Scree 9, Cliffs 3, Walls 0]; ARABLE 1; GRASSLAND 14 [Meadows 6, Permanent pastures 15 (Enclosed pastures 3: Limestone 42, Acidic 3; Unenclosed pastures 17)]; SPOIL 13 [Quarries (Coal mine 10, Lead mine 19: Limestone 32, Acidic 6; 26), Cinders 14, Bricks and mortar 3, Soil 0, Manure and sewage 3]; WASTELAND 13 [Wasteland and heath 18 (Limestone 40, Acidic 4), River banks 0, Verges 12, Paths 0]; WOODLAND 0 [Woodland (Limestone 0, Acidic 0), Scrub 0 (Broad leaved 0, Coniferous 0), Plantations 0, Hedgerows 0])

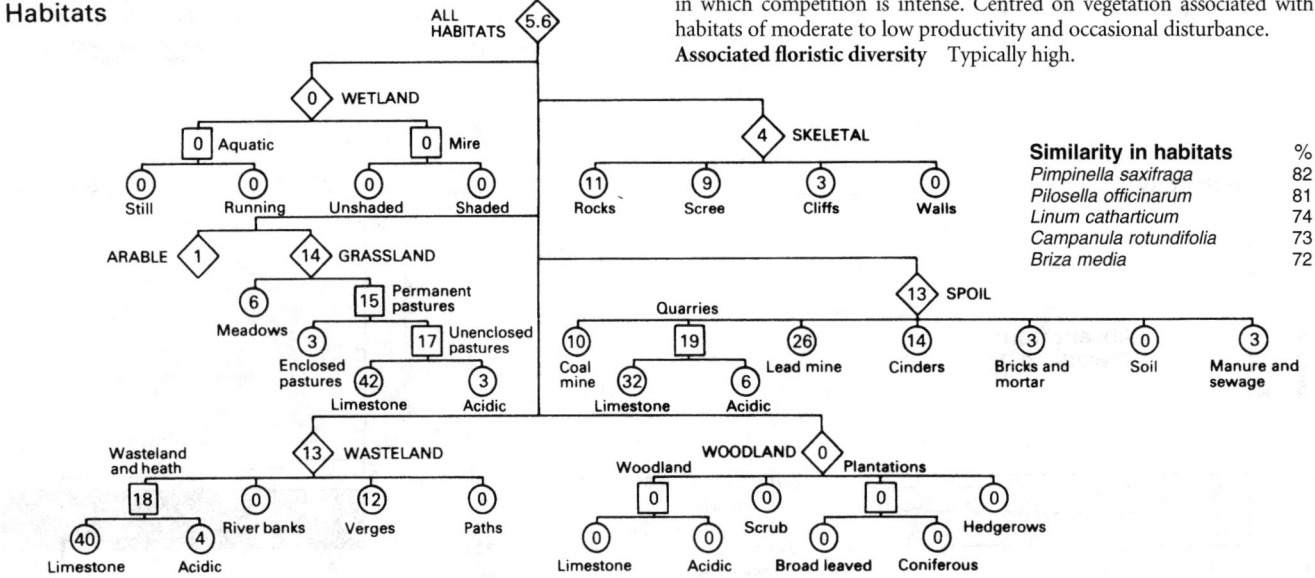

Synopsis *L.c.*, which forms nitrogen-fixing nodules (in conjunction with *Rhizobium lupini*), is the commonest legume of unproductive semi-natural grassland, and is perhaps the most ecologically wide-ranging legume in Britain. The species extends from maritime to montane environments, from moderately low to high soil pH, from infertile to moderately fertile soils, and from spoil and open habitats to grassland. *L.c.* is restricted to unproductive plots of the *PGE* and is strongly suppressed by nitrogenous manures. *L.c.* is generally regarded as relatively tolerant of grazing, is palatable to stock, and attains agricultural importance in less-productive grasslands. Grant and Marten (1985) even suggested that *L.c.* is the second most important forage legume in N America. Some populations of *L.c.* produce cyanogenic glucosides, which are an effective deterrent against small herbivores (*Biol Fl(a)*) and may also be toxic to stock (Cooper & Johnson 1984). The presence of *L.c.* produces local enrichment of soil nitrogen. This can result in invasion and dominance of the patch by more-N-demanding species, e.g. *Festuca rubra* (A. J. Willis pers. comm.).

L.c. is not a drought-tolerant species (*Biol Fl(a)*) and the establishment and persistence of *L.c.* on desiccated, often S-facing limestone soils, is probably related to its relatively large seed and the capacity of its roots to penetrate fissures deeply (Anderson 1927, Salisbury 1952). In ungrazed sites up to 100 seeds may be produced in each inflorescence (*Biol Fl(a)*), and this moderate seed output is apparently sufficient to allow *L.c.* to colonize successfully many infertile habitats, e.g. quarry-spoil heaps, where an ability to fix nitrogen may be important. However, in many grazed habitats seeds are produced infrequently, and population expansion may be dependent upon the persistent seed bank (Donelan & Thompson 1980) and the limited capacity for clonal expansion. Populations from S England frequently have a greater potential for vegetative regeneration than those from N upland sites (A. J. Willis pers. comm.). *L.c.* can also regenerate from root or stem cuttings (*Biol Fl(a)*), but this is probably only rarely important under natural conditions. *L.c.* is genotypically and probably to a lesser extent phenotypically variable. Chrtkova-Zertova (1973b) identifies

Altitude
Slope
Aspect

n.s.	10.6 / 40.7
n.s.	16.1 / 52.8 Unshaded

—	0.0
—	0.0 Shaded

Soil pH

Associated floristic diversity

Hydrology

Absent from wetland

Triangular ordination

Commonest associates %
Festuca rubra 78
Plantago lanceolata 73
Festuca ovina 67
Campanula rotundifolia 57
Dactylis glomerata 55

Bare soil

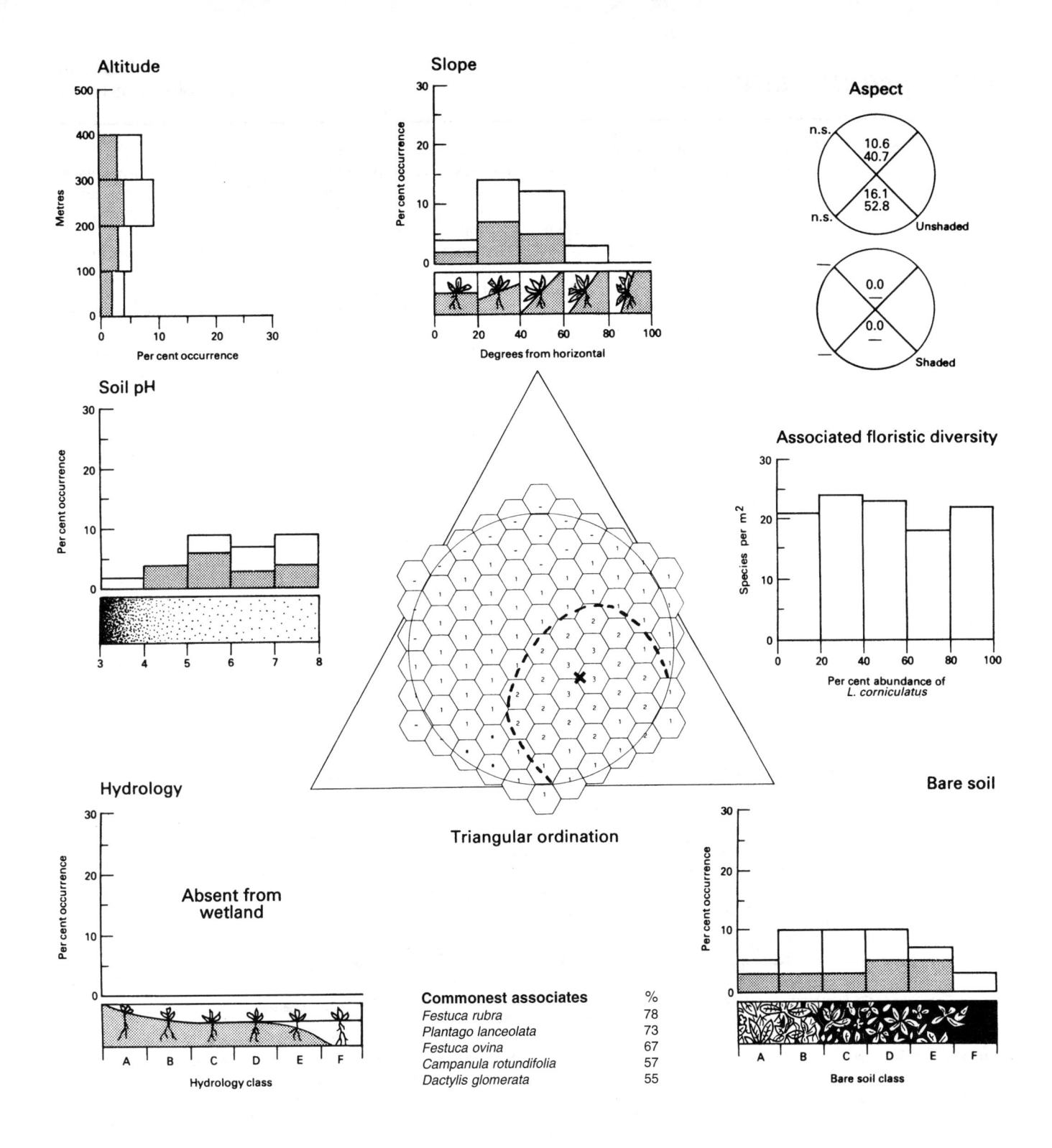

14 varieties from Europe and suggests that such variability has been crucial in determining the ability of *L.c.* to exploit a range of different habitats. A heavy-metal-tolerant race of *L.c.* has been described within the Sheffield area (Shaw 1984). As a result of its N-fixing ability *L.c.* is potentially a valuable species for the reclamation of derelict land (Dancer *et al.* 1977). In Britain seed of *L.c.* was formerly sown for agricultural and amenity use at the rate of <5 t year^{-1} (MAFF 1980-84). More-erect, alien genotypes, with larger leaflets, have been introduced in seed mixtures sown onto new road verges; these are capable of persistence in relatively productive swards (Bonnemaison & Jones 1986). *L.c.* and *L. pedunculatus* are part of a taxonomically complex group which is represented by at least 12 spp. in Europe.

Current trends In improved pasture *L.c.* is rapidly suppressed, and the species is a poor competitor in the tall grasslands which result from the cessation of grazing in marginal land. Thus, although *L.c.* exploits a number of new artificial habitats, e.g. limestone and cinder spoil, a further decline in its abundance seems inevitable.

Lotus pedunculatus Cav.

Fabaceae
Large Birdsfoot-trefoil

Life-form Polycarpic perennial protohemicryptophyte with spreading stoloniferous shoots; leaves 5-foliate, often hairy, usually <1000 mm^2; roots shallow nodulated and with VA mycorrhizas.
Phenology New shoots are produced in spring. Flowers from June to August and sets seed from August to October.
Height Foliage and flowers to 600 mm.
Seedling RGR No data. SLA 24.3. DMC 22.1.
Nuclear DNA amount 1.1 pg; 2*n* = 12* (*BC*); 2*x**.
Established strategy Intermediate between competitor and C-S-R.
Regenerative strategies V, B$_s$. Regenerates both by vegetative means and by seed (see 'Synopsis'); ?SB IV.
Flowers Yellow, hermaphrodite, protandrous, insect-pollinated, self-incompatible (Duke 1981); 5-12 in axillary cymose heads.
Germinule 0.4 mg, 1.1 × 1.0 mm, seed explosively dispersed from a *c.* 14-seeded pod.
Germination Epigeal; t_{50} 1 day: 7-37°C; hard-coat dormancy; L > D.
Biological flora None.
Geographical distribution The British Isles except the extreme N (98% of British vice-counties), *ABIF* page 377; W, C and S Europe, and often as a casual elsewhere (67% and naturalized in a further 10% of European territories); Asia and N Africa.

Habitats Primarily a plant of mire or wet grassland. Most frequent in unshaded mire and moist wasteland. Less common in pasture and on waysides. Also recorded from aquatic habitats. Absent from skeletal habitats, spoil, woodland and arable land.
Gregariousness Recorded as sparse to intermediate, but occasionally stand-forming.
Altitude Widely distributed and observed to 395 m. [Up to 490 m.]
Slope Like most wetland species, occurs mostly on flat ground and gentle slopes.
Aspect Insufficient data.

Hydrology Found over a wide range of slopes, reflecting the capacity of *L.p.* to grow in soligenous and the drier parts of topogenous mire. Not found in permanently submerged habitats.
Soil pH Largely absent from soils of pH <4.5, infrequent on calcareous strata and most common between pH 4.5 and 6.5.
Bare soil Most records are from sites with relatively closed vegetation and little exposed soil.
△ **ordination** Found under both moderately productive and unproductive conditions, but absent from heavily disturbed sites.
Associated floristic diversity Typically intermediate to species-rich.

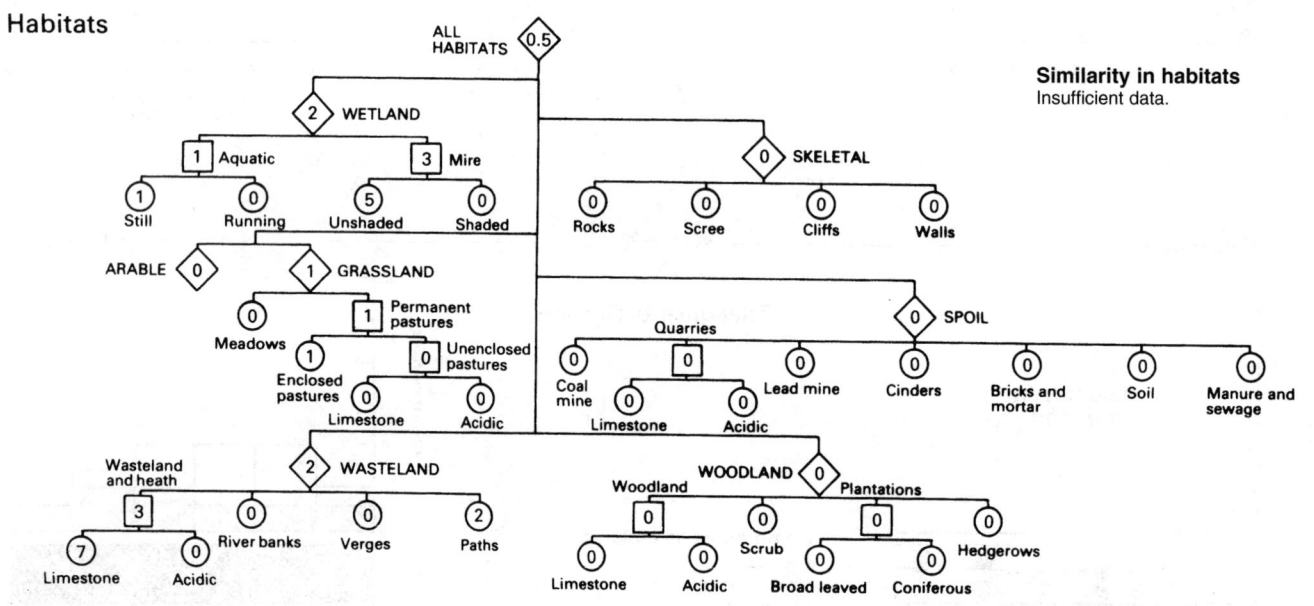

Habitats

Similarity in habitats
Insufficient data.

Synopsis *L.p.* is a relatively tall, patch-forming legume of the drier parts of topogenous mire and of soligenous mire. The species is the only common wetland legume in the British flora, and forms nitrogen-fixing nodules in association with *Rhizobium lupini*. The general absence of legumes from wetlands may be a reflection of the conflict between a requirement for oxygen during nitrogen-fixation (Sprent 1984) and a possible requirement for oxygen diffusion from the root as part of the detoxification of the anaerobic soil environment. Consistent with these constraints, *L.p.* is, within topogenous mire, largely restricted to drier habitats, and is fairly sensitive to ferrous ion toxicity (R. E. D. Cooke unpubl.). *L.p.* occurs both as a colonist of infertile habitats (e.g. on stony ground beside reservoirs and in gravel pits) and in tall wet grassland and on soligenous mire, where the species appears to be a component of relatively stable and moderately productive vegetation. We suspect that *L.p.* is restricted to sites where the growth of potential dominants is limited to some extent by soil infertility. However, the extent to which the capacity to fix nitrogen affects the status and persistence of the species in communities is not known. *L.p.* is tolerant of light grazing and may be cut for hay (Duke 1981). The species is, like other wetland species, sensitive to drought but is moderately salt-tolerant (Duke 1981). *L.p.* can form clonal patches by means of spreading stolons, and breakdown of stolons results in the isolation of daughter colonies. Shoot pieces root relatively slowly compared with those of many wetland species (Hodgson unpubl.), and regeneration by means of shoot fragments is probably uncommon. Reproduces also by seeds which appear to germinate in spring (Muller 1978) and a persistent seed bank is reported (Turner 1933). Despite the fact that seeds are relatively large and lacking in any obvious dispersal mechanism, *L.p.* is a not-infrequent colonist of moist, disturbed sites. The degree of ecotypic differentiation within this out-breeding species is not known. *L.p.* is part of the *Lotus corniculatus* complex, which consists of 12 species.

Current trends *L.p.* has only a restricted colonizing ability, and is decreasing through the destruction of wetland habitats, particularly in lowland areas. Considered agriculturally superior to *L.p.* due to its higher protein content (Duke 1981), but not at present used in any great quantity for agricultural purposes or in landscape reclamation.

Altitude

Slope

Aspect

Unshaded

Shaded

Soil pH

Associated floristic diversity

Species per m^2 vs Per cent abundance of *L. pedunculatus*

Hydrology

Absent from wetland

Triangular ordination

Bare soil

Commonest associates	%
Holcus lanatus	72
Festuca rubra	60
Cirsium palustre	59
Anthoxanthum odoratum	58
Ranunculus repens	56

Abundance in quadrats

	Abundance class										Distribution type
	1–10	11–20	21–30	31–40	41–50	51–60	61–70	71–80	81–90	81–100	
Wetland	7	–	–	–	–	–	–	–	–	–	–
Skeletal	–	–	–	–	–	–	–	–	–	–	–
Arable	–	–	–	–	–	–	–	–	–	–	–
Grassland	1	–	–	–	–	–	–	–	–	–	–
Spoil	–	–	–	–	–	–	–	–	–	–	–
Wasteland	3	1	1	–	1	–	–	1	–	–	–
Woodland	–	–	–	–	–	–	–	–	–	–	–
All habitats	73.3	6.7	6.7	–	6.7	–	–	6.7	–	–	2

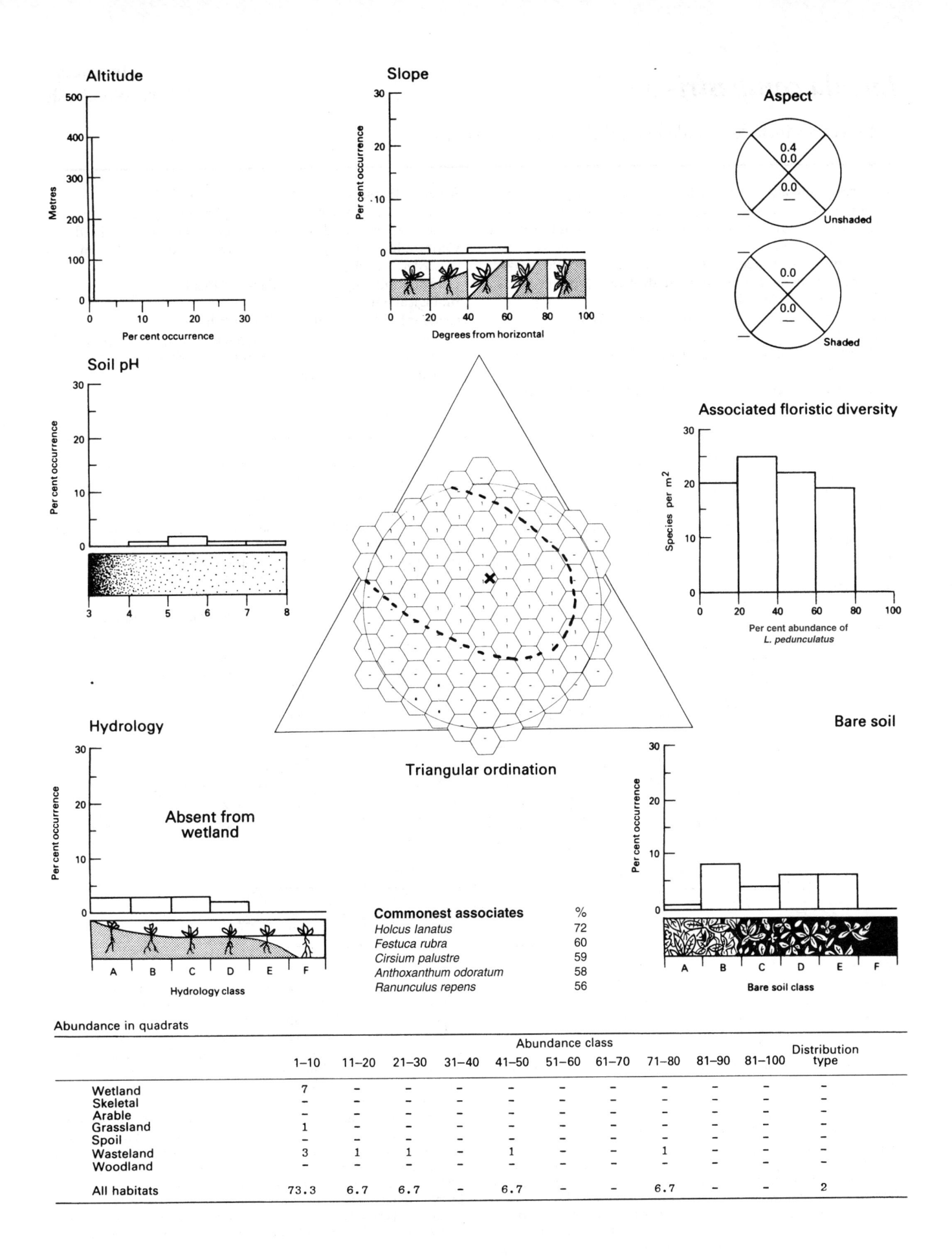

Luzula campestris (L.) DC.

Juncaceae
Field Woodrush

Field records include *L. multiflora* (Ehrh.) Lej. (see 'Synopsis')

Life-form Polycarpic perennial, semi-rosette hemicryptophyte with short stock and shortly creeping stolons. Shoots erect; leaves linear, hairy, typically <500 mm^2; roots with or without VA mycorrhizas.

Phenology Winter green, but leaf expansion in spring delayed. Flowers from March to June and seeds shed in July and August. Leaves relatively long-lived.

Height Foliage usually <100 mm; flowers to 150 mm.

Seedling RGR 1.0-1.4 week^{-1}. SLA 21.6. DMC 21.4.

Nuclear DNA amount 1.2 pg; $2n = 12^*$ (*BC*); $2x^*$ (with a diffuse centromere).

Established strategy Intermediate between stress-tolerator and C-S-R.

Regenerative strategies V, B$_s$. See 'Synopsis'; ?SB IV.

Flowers Chestnut-brown, hermaphrodite, protogynous but autogamous, wind-pollinated; often <50 in a cymose head.

Germinule 0.64 mg; 1.7 × 1.1 mm, seed dispersed from a 3-seeded capsule.

Germination Epigeal; t_{50} 8 days; 7-24°C; L = D.

Biological flora None.

Geographical distribution The British Isles but less common in Ireland (100% of British vice-counties), *ABIF* page 699; Western Europe (85% of European territories); N Africa, N America and New Zealand.

Habitats A pasture species which is also common in, for example, meadows and on wasteland, grass verges and lead-mine heaps. Also present on mining and quarry waste, with scattered occurrences in open rocky habitats and on scree slopes. Recorded in scrub but not in woodland. Absent from arable land, and records for wetlands probably refer to *L.m*.

Gregariousness Intermediate.

Altitude Increasing in frequency and abundance at higher elevations up to 460 m, a reflection in part of the greater amount of permanent pasture in upland regions. [Up to 1010 m.]

Slope Wide-ranging, but particularly frequent in dale-side grasslands which are concentrated on intermediate slopes.

Aspect Widely-distributed, but somewhat more abundant on N-facing slopes. In grasslands *L.c.* has a pronounced N-facing bias (*Ecol Atl*).

Hydrology Not typically a wetland species. Occasionally recorded from soligenous mire, but here records probably refer to *L.m*.

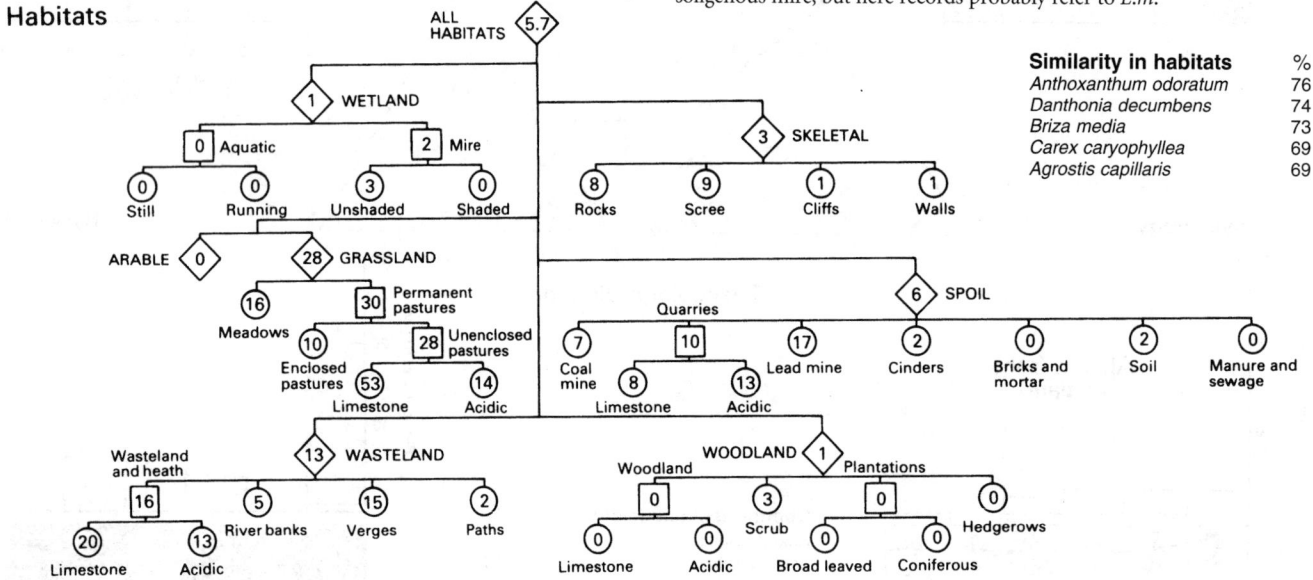

Habitats

Similarity in habitats	%
Anthoxanthum odoratum	76
Danthonia decumbens	74
Briza media	73
Carex caryophyllea	69
Agrostis capillaris	69

Soil pH Very wide-ranging, but more often associated with acidic than with alkaline soils. Most frequent and abundant in the pH range 4.5-6.0. (Records from the most acidic soils are predominantly of *L.m*.)

Bare soil Wide-ranging, but not abundant where there is a high exposure.

△ ordination Distribution relatively wide-ranging, but tending to centre on vegetation associated with undisturbed and rather infertile habitats. Absent from heavily disturbed vegetation and suppressed in communities dominated by robust fast-growing species.

Associated floristic diversity Typically intermediate.

Synopsis *L.c.* is a winter-green, patch-forming species which is in need of further study. Although occurring in association with a relatively wide range of habitats and soil types, *L.c.* typically occupies the lower stratum in short turf on moist, less fertile, often mildly acidic soils. The species is primarily a pasture plant and, despite its delayed leaf growth, is one of the earliest-flowering of the common pasture species. In short turf *L.c.* often forms conspicuous dense patches as a result of stolonifery. Although absent from most of the taller types of herbaceous vegetation, *L.c.* displays a marked ability to persist under a discontinuous canopy of pasture grasses, and occasionally occurs as non-flowering colonies in scrub. This strongly suggests that the species is shade-tolerant. Palatability to stock is similar to that of *Festuca ovina* and *Agrostis* spp. (Milton 1933). Seed is produced regularly. However, the seeds have an unusually low temperature range for germination, and the timing of germination is uncertain. A persistent seed bank is also formed, but regeneration by seed appears to be infrequent. The mucilaginous seeds have an elaiosome and may be dispersed by ants (Handel 1978). *L. multiflora* is a densely-tufted taller plant, virtually restricted to acidic peaty and sandy soils. In Britain, *L.m.* includes ssp. *multiflora* ($2n = 36$, ssp. *congesta* (Thuill.) Arcang. ($2n = 48; 8x$) and ssp. *hibernica* Kirschner and T. C. G. Rich ($2n = 24; 4x$).

Current trends Uncertain. Most common in less-fertile habitats, but capable of persisting as a rather inconspicuous component of some relatively productive, artificial habitats such as lawns and road verges.

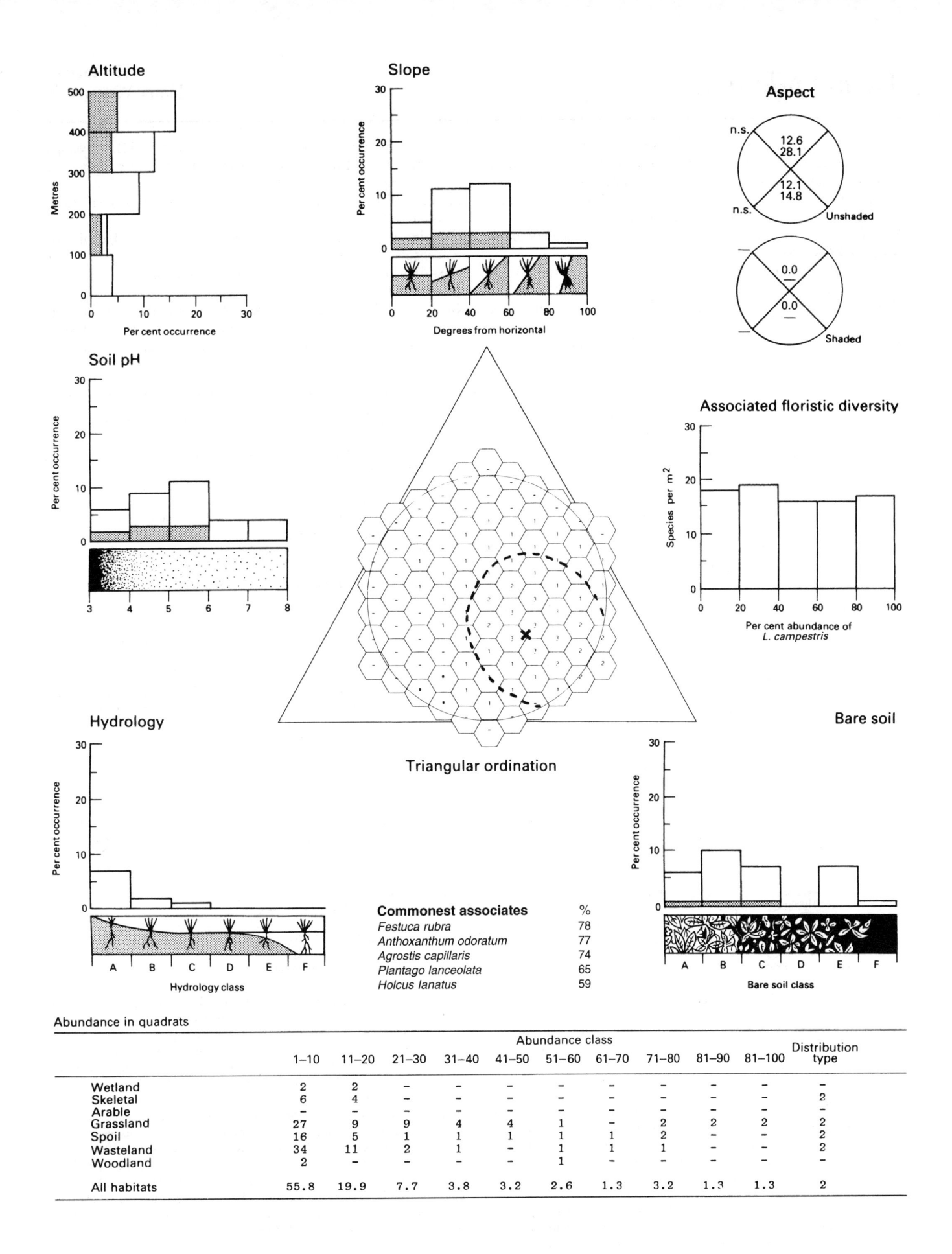

Altitude

Slope

Aspect

	n.s.	12.6	
		28.1	
		12.1	
	n.s.	14.8	Unshaded

	—	0.0	
		0.0	
	—		Shaded

Soil pH

Associated floristic diversity

Hydrology

Triangular ordination

Bare soil

Commonest associates %
Festuca rubra 78
Anthoxanthum odoratum 77
Agrostis capillaris 74
Plantago lanceolata 65
Holcus lanatus 59

Abundance in quadrats

	\multicolumn{10}{c}{Abundance class}	Distribution type
	1–10	11–20	21–30	31–40	41–50	51–60	61–70	71–80	81–90	81–100	
Wetland	2	2	–	–	–	–	–	–	–	–	–
Skeletal	6	4	–	–	–	–	–	–	–	–	2
Arable	–	–	–	–	–	–	–	–	–	–	–
Grassland	27	9	9	4	4	1	–	2	2	2	2
Spoil	16	5	1	1	1	1	1	2	–	–	2
Wasteland	34	11	2	1	–	1	1	1	–	–	2
Woodland	2	–	–	–	–	1	–	–	–	–	–
All habitats	55.8	19.9	7.7	3.8	3.2	2.6	1.3	3.2	1.3	1.3	2

Luzula pilosa (L.) Willd.

Juncaceae
Hairy Woodrush

Life-form Tufted, shortly-stoloniferous, polycarpic perennial, semi-rosette, hemicryptophyte. Stems erect; leaves linear, with hairs, usually <1000 mm^2; roots often without VA mycorrhizas.
Phenology Winter green. Flowers from April to June and sets seed from June to July. Leaves long-lived.
Height Foliage mainly <100 mm; flowering stems to 300 mm.
Seedling RGR No data. SLA 25.1. DMC 22.8.
Nuclear DNA amount 0.6 pg; $2n = 66*$ (*BC*, Bennett & Smith 1976), 72 (*FE5*); $6x^*$.
Established strategy Intermediate between stress-tolerator and C-S-R.
Regenerative strategies S, B$_s$, V. Seed normally germinate directly in summer (Muller 1978), but may form a very persistent, buried seed bank (Grandstrom 1982). Develops small tufts, but there is no evidence that *L.p.* forms substantial clonal patches; SB III or IV.
Flowers Brown, hermaphrodite, wind-pollinated (Muller 1883); often >10 in a lax cyme.
Germinule 0.81 mg, 1.4 × 1.0 mm, seed in a 3-seeded, dehiscent capsule.
Germination Epigeal; to 7 days; 9-23°C; L > D.
Biological flora None.
Geographical distribution The British Isles, although local in Ireland (91% of British vice-counties), *ABIF* page 698; most of Europe except the extreme S (74% of European territories); introduced into N America.

Habitats Mainly restricted to moist, shaded habitats, particularly in lowland areas. Most frequent in woodland, but also recorded from shaded outcrops, quarry spoil and shaded and unshaded cliffs. Observed from pasture, wasteland, waysides and railway banks, with isolated records from lead-mine spoil and mire. Absent from aquatic and arable habitats.
Gregariousness Sparse to intermediate.
Altitude More widespread in upland areas, but recorded from 10 to 320 m. [Up to 670 m.] In lowland areas almost exclusively a plant of shaded habitats, but in upland areas extends into open areas.
Slope Found over a wide range of values, but infrequent on the steepest slopes.
Aspect More frequent on slopes of northern aspect, particularly in unshaded habitats.
Hydrology Not a wetland species, but there are isolated records for *L.p.* at the edge of soligenous mire.
Soil pH Infrequent on calcareous strata, and largely restricted to acidic soils within the pH range 3.5-5.0.
Bare soil Habitats usually have little or no bare soil; often this is due to the presence of an accumulation of tree litter.
△ **ordination** Absent from disturbed vegetation. Distribution centred on stable and unproductive vegetation types.
Associated floristic diversity Typically low, but has been observed in pasture with >40 species m^{-2}.

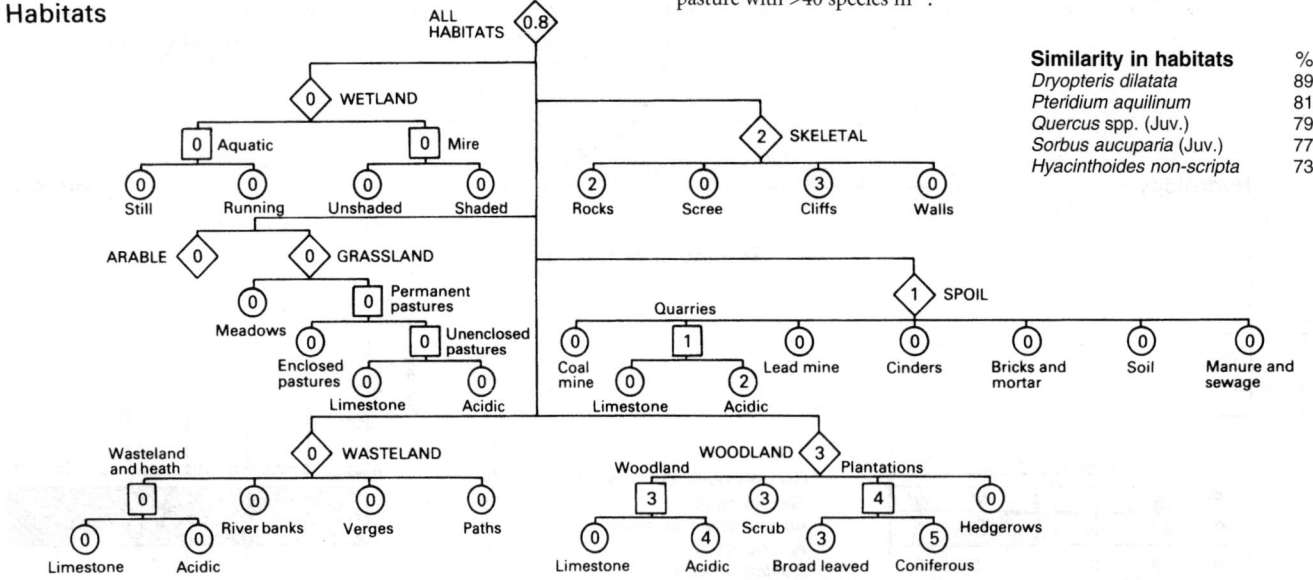

Similarity in habitats	%
Dryopteris dilatata	89
Pteridium aquilinum	81
Quercus spp. (Juv.)	79
Sorbus aucuparia (Juv.)	77
Hyacinthoides non-scripta	73

Habitats

Synopsis *L.p.* is a tufted winter-green species most typically associated with situations in woodlands where the vigour of potential dominants is restricted by shade and by soil infertility. *L.p.* is particularly characteristic of moist acidic soils. The species is widespread on clay but virtually absent from the freely-drained acidic soils, such as those of the Bunter sandstone of South Yorkshire and North Nottinghamshire. In lowland parts of Britain *L.p.* is largely confined to broad-leaved woodland, and Peterken (1981) considers that in Lincolnshire the plant is characteristic of ancient woodland. In upland areas the species is ecologically more wide-ranging; here *L.p.* may occur in old pastures, often with *Oxalis acetosella* and is a frequent colonist of gritstone quarries, particularly where these are adjacent to woodland. Presumably because of its shade tolerance, *L.p.* is also a common constituent of upland conifer plantations in the Sheffield region, and is recorded more frequently from this habitat than from any other. The short stolons confer only limited capacity for vegetative spread, and regeneration by seed is important in population expansion and persistence. On average a plant produces *c.* 200 seeds year^{-1} (Salisbury 1942); these have a well-developed elaiosome and are frequently dispersed by ants (Ridley 1930). Even under heavy shade *L.p.* produces seed, but it is in open areas after forest fire or clear felling that seed set is greatest (Grandstrom 1982). These are also the circumstances most favourable for seedling establishment, either from freshly-shed seed or from the long-persistent seed bank. On this basis Grandstrom (1982) concludes that infrequent periods of catastrophic disturbance play an essential role in the maintenance of populations of *L.p.* in woodland habitats. This regenerative strategy is remarkably similar to that associated with *Carex pilulifera* on burned heathland. *L.p.* forms hybrids with *L. forsteri* (Sm) DC, and the distributions of the two species overlap in S England (Stace 1975).
Current trends Probably decreasing in lowland areas, where it shows a low capacity to colonize new sites, but with perhaps a stable or even an increasing distribution in forested upland areas.

Altitude
Slope
Aspect

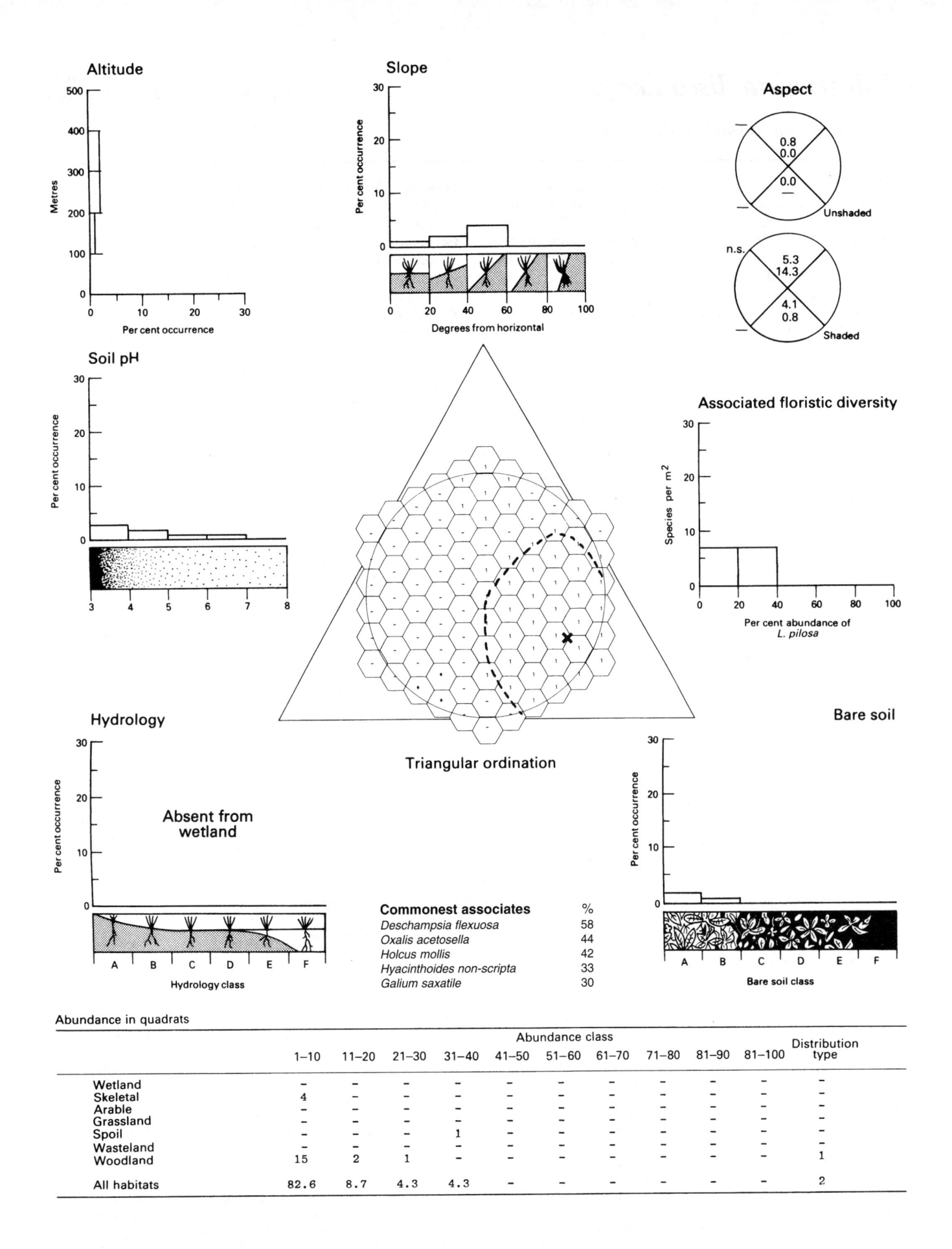

	0.8	
	0.0	
	0.0	
	—	Unshaded

	n.s.	
	5.3	
	14.3	
	4.1	
	0.8	Shaded

Soil pH

Associated floristic diversity

Species per m² vs Per cent abundance of *L. pilosa*

Hydrology

Absent from wetland

Triangular ordination

Bare soil

Commonest associates	%
Deschampsia flexuosa	58
Oxalis acetosella	44
Holcus mollis	42
Hyacinthoides non-scripta	33
Galium saxatile	30

Abundance in quadrats

	1–10	11–20	21–30	31–40	41–50	51–60	61–70	71–80	81–90	81–100	Distribution type
Wetland	–	–	–	–	–	–	–	–	–	–	–
Skeletal	4	–	–	–	–	–	–	–	–	–	–
Arable	–	–	–	–	–	–	–	–	–	–	–
Grassland	–	–	–	–	–	–	–	–	–	–	–
Spoil	–	–	–	1	–	–	–	–	–	–	–
Wasteland	–	–	–	–	–	–	–	–	–	–	–
Woodland	15	2	1	–	–	–	–	–	–	–	1
All habitats	82.6	8.7	4.3	4.3	–	–	–	–	–	–	2

Matricaria discoidea DC.

Chamomilla suaveolens (Pursh) Rydb.

Asteraceae
Pineapple Weed, Rayless Mayweed

Life-form Winter- or summer-annual therophyte. Stems erect, often branched; leaves finely divided, often <100 mm^2; roots with VA mycorrhizas.
Phenology Overwintering from autumn-germinating seed or seed germinating in spring. Flowers June to July and sets seed from July onwards.
Height Foliage and flowers to 300 mm.
Seedling RGR 1.0-1.4 week^{-1}. SLA 28.7. DMC 13.5.
Nuclear DNA amount 4.6 pg, $2n = 18$* (Bennett & Smith 1976); $2x$*.
Established strategy Ruderal.
Regenerative strategies B_s. Regenerates entirely by seed either in vegetation gaps during autumn or the following spring or after incorporation into a persistent seed bank; SB III.

Flowers Florets greenish-yellow, hermaphrodite, tubular, seldom visited by insects; often *c*. 200 in each capitulum; capitula solitary at the end of each branch.
Germinule 0.08 mg, 1.2 × 0.5 mm, achene; pappus a short membranous rim.
Germination Epigeal; t_{50} 3 days; 11-23°C; germination rate increased after dry storage at 20°C; L > D.
Biological flora None.
Geographical distribution A native of NE Asia and perhaps also W N America; widely naturalized in the British Isles (86% of British vice-counties), *ABIF* page 650; Europe, although absent from much of the S (69% of European territories); established in Chile and New Zealand.

Habitats A common weed of cereals and broad-leaved arable crops in field margins on compacted soils. Also found at high frequencies on paths and tracks, particularly around farms, and on urban spoil, especially demolition sites and other areas subjected to heavy trampling. Scarce in skeletal habitats and mire, and absent from aquatic, grassland and wooded habitats.
Gregariousness Intermediate.

Altitude Extends to 400 m. but suitable habitats are more frequent in lowland areas. [Up to 530 m.]
Slope Not recorded from the most skeletal habitats or from the steepest slopes.
Aspect Wide-ranging.
Hydrology Not a wetland species, but occasionally on bare mud exposed during the summer months at the edges of ponds and reservoirs.

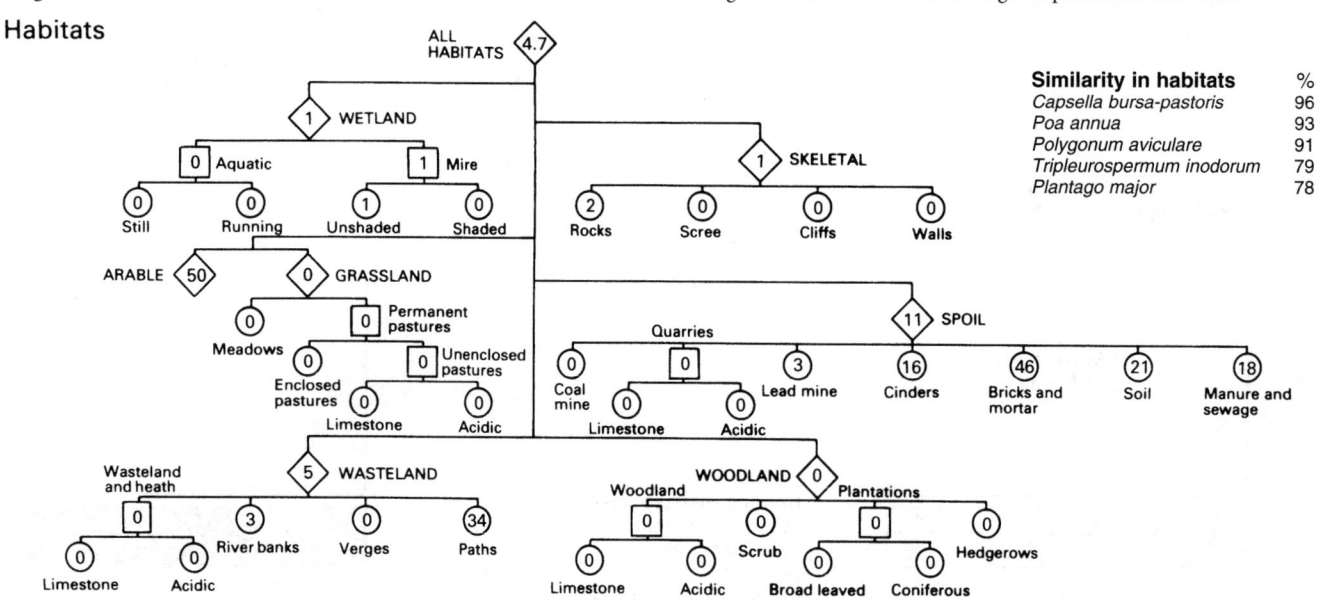

Habitats

Similarity in habitats	%
Capsella bursa-pastoris	96
Poa annua	93
Polygonum aviculare	91
Tripleurospermum inodorum	79
Plantago major	78

Soil pH Restricted to soils of pH >5.0.
Bare soil Particularly common in habitats having a high proportion of bare ground.
△ **ordination** Distribution centred on vegetation associated with highly disturbed, fertile habitats. Absent almost entirely from unproductive sites and from productive but undisturbed vegetation.
Associated floristic diversity Typically intermediate.
Synopsis A native of NE Asia and perhaps also of NW America, first recorded from Britain in 1871 (*FE4*, Salisbury 1964). Strongly aromatic when crushed. *M.d.* is facultatively a summer or winter annual, and exploits compacted soils where dominance by perennial species is prevented by trampling and by vehicular pressure. When flowering the species is upright with leafy stems, and at this stage is potentially more vulnerable to damage from trampling than more-decumbent species such as *Poa annua*; thus, *M.d.* tends to be abundant only on road margins and in the central grassy areas of cart-tracks, and scarce on the most heavily trampled areas of paths. The success of the species is related in part to the production of numerous seeds. Although a long-day plant, seed may be set within 40-50 days in spring-germinating plants, a much shorter period than that required by *Tripleurospermum inodorum* (Roberts & Feast 1974). A plant from little-disturbed sites produces on average over 6000 small seeds; an exceptionally large capitulum may release over 400 achenes (Salisbury 1942). Even small plants with a single flower head, such as may be found in heavily trampled sites, may produce over 50 seeds. Seeds may be carried in mud on tyres, and the rapid spread of *M.d.* at the beginning of the 20th century relates to the increased use of motor cars (Salisbury 1964). Seeds may also survive ingestion by horses (Ridley 1930). The spread of the species to distant sites in England, Ireland and Scotland in the 19th century (Salisbury 1964) illustrates the capacity of *M.d.* for long-distance dispersal. *M.d.* forms a persistent seed bank and has the capacity to germinate over most of the year, although spring and autumn peaks are evident which undoubtedly enable populations to survive

Altitude

Slope

Aspect

Soil pH

Associated floristic diversity

Triangular ordination

Hydrology

Commonest associates	%
Poa annua	83
Polygonum aviculare	66
Poa trivialis	58
Lolium perenne	51
Agrostis stolonifera	48

Bare soil

Abundance in quadrats

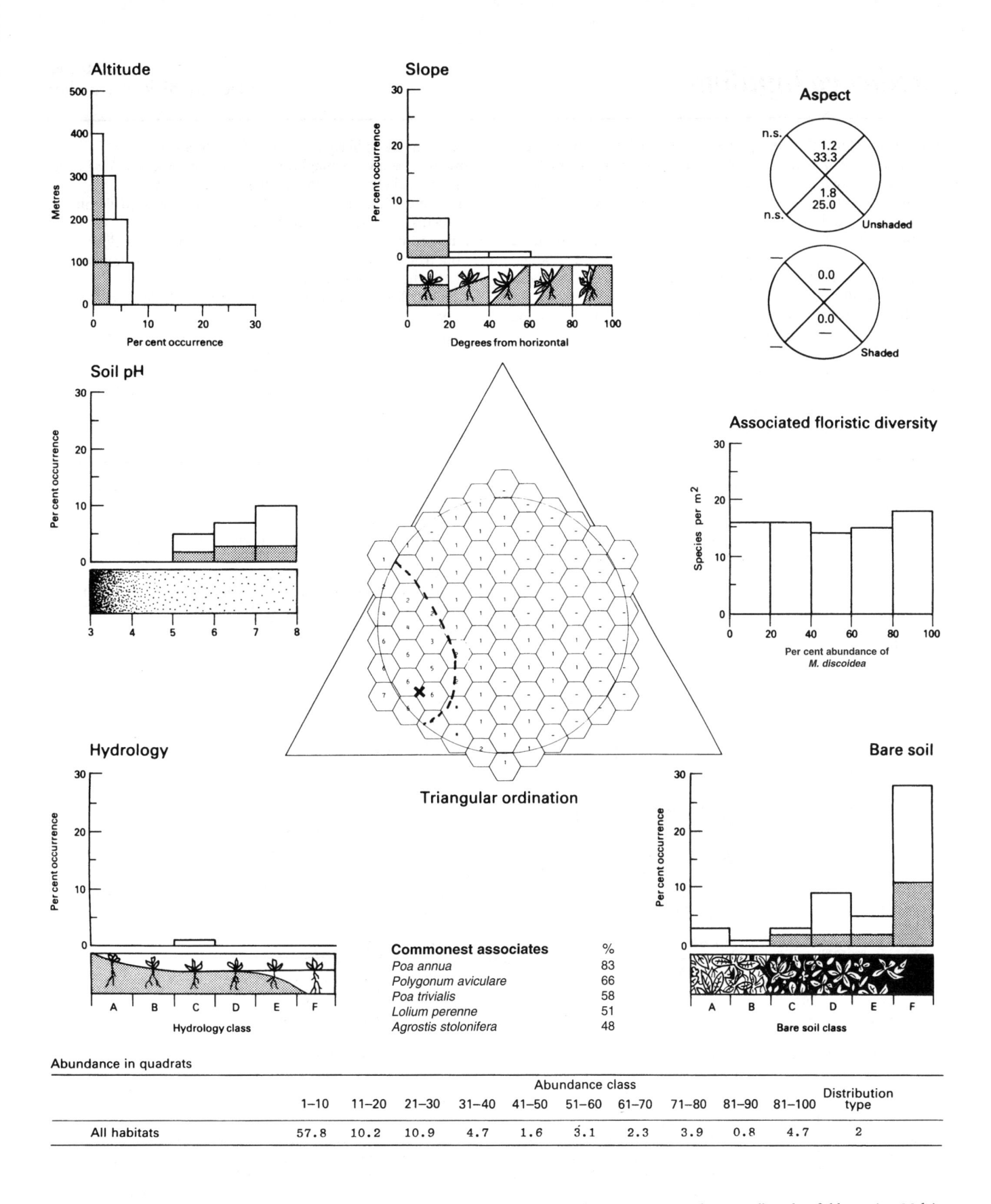

	Abundance class										Distribution type
	1–10	11–20	21–30	31–40	41–50	51–60	61–70	71–80	81–90	81–100	
All habitats	57.8	10.2	10.9	4.7	1.6	3.1	2.3	3.9	0.8	4.7	2

repeated disturbance. The occurrence of $M.d.$ on bare areas nearest to the metalled road surface suggest that the species possesses some salt tolerance. On the basis of the association with sites subject to soil compaction, a modest tolerance of waterlogging is predicted, but this hypothesis has not been tested. The high frequency with which $M.d.$ is recorded from arable fields in the Sheffield region probably results from sampling bias since, to avoid crop damage, most records were collected at field margins. $M.d.$ is particularly abundant in gateways and in other access areas for tractors, but is largely absent from gardens.

Current trends Remains abundant on fertile paths, tracks and road margins at low to moderate altitudes. Perhaps still increasing in range and abundance.

Medicago lupulina L.

Fabaceae
Black Medick, Black Hay

Life-form Annual therophyte to short-lived, polycarpic perennial, protohemicryptophyte. Shoots often much branched from the base, procumbent or ascending; leaves ternate, hairy, usually <500 mm^2; deeply tap-rooted and with VA mycorrhizas and root nodules.

Phenology Winter green. New shoots and seedlings produced mainly in spring. Plants flower from May to August, although this period may be curtailed by droughting, and seed set about 1 month after flowering.

Height Typically procumbent and <200 mm, but ascending plants may exceed 400 mm.

Seedling RGR 1.0-1.4 week^{-1}. SLA 29.3. DMC 19.6.

Nuclear DNA amount 1.8 pg; $2n = 16^*$ (BC), $2x^*$.

Established strategy Intermediate between ruderal and C-S-R.

Regenerative strategies B$_s$. Regenerates only by seed. A persistent seed bank is formed. Seedlings appearing in late autumn are unlikely to survive (*Biol Fl*); SB IV.

Flowers Yellow, hermaphrodite, homogamous, self- or insect-pollinated (*Biol Fl*), often *c*. 30 in an axillary raceme.

Germinule 2.01 mg, 1.6 × 1.3 mm, seed dispersed within an indehiscent 1-seeded pod, 2.4 × 2.0 mm.

Germination Epigeal; t_{50} 2 days; <5-36°C, hard-coat dormancy, but fruits detached before maturity may germinate immediately in the field (Sidhu & Cavers 1977); L = D.

Biological flora Turkington and Cavers (1979), see also Zarzycki (1976).

Geographical distribution Widespread, but infrequent in much of Scotland and parts of Ireland (99% of British vice-counties), ABIF page 391; Europe except the extreme N (92% of European territories and 5% naturalized); Australia, New Zealand, USA, N Africa and W Asia.

Habitats Mainly in open habitats such as rock outcrops, limestone quarry waste and cinder tips and tracks. Less frequent on wasteland over limestone, paths, demolition sites and coal-mine spoil. Also associated with calcareous pastures, cliffs, soil heaps, meadows and walls. Only a casual on arable land, and absent from woodland and wetland habitats. [A frequent plant of calcareous sand dunes.]

Gregariousness Intermediate.

Altitude Most abundant below 300 m but extending to 370 m. [Up to 400 m.]

Slope Widely distributed.

Aspect Significantly more frequent and only abundant on S-facing slopes.

Hydrology Absent from wetlands.

Soil pH Recorded mainly on soils overlying limestone and absent from strongly acidic soils. Most frequent and abundant where soil pH is >7.0.

Bare soil Associated with widely different amounts of bare ground.

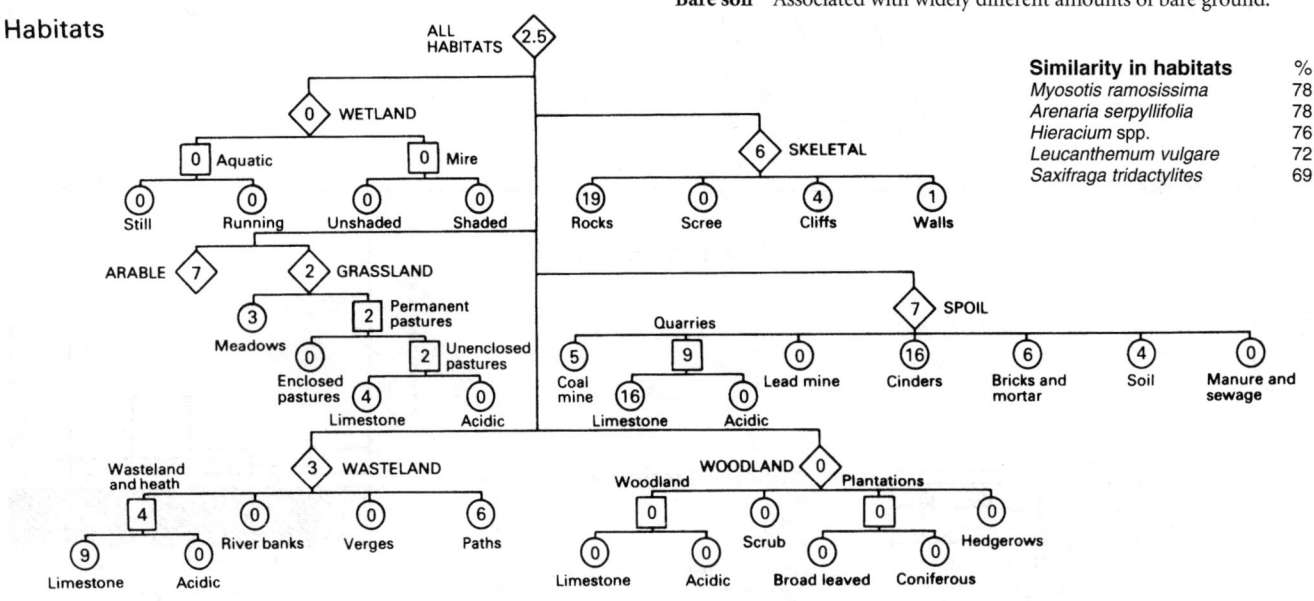

Habitats

Similarity in habitats	%
Myosotis ramosissima	78
Arenaria serpyllifolia	78
Hieracium spp.	76
Leucanthemum vulgare	72
Saxifraga tridactylites	69

△ **ordination** Shows a rather wide distribution, but absent from vegetation dominated by robust perennial species. Distribution centred on vegetation associated with moderately disturbed and relatively infertile habitats.

Associated floristic diversity Typically species-rich.

Synopsis *M.l.* is a short-lived legume forming nitrogen-fixing nodules in conjunction with *Rhizobium meliloti*, and is characteristic of relatively disturbed, infertile sites, particularly on calcareous soils. *M.l.* produces a long tap-root (Anderson 1927) and frequently exploits grasslands on S-facing slopes and other habitats where summer droughting of perennial vegetation creates areas of bare ground. The species has a predominantly lowland distribution in Britain, extending only to 400 m (Wilson 1956). *M.l.* is grazed by sheep but is not very palatable to cattle (*Biol Fl*). *M.l.* can withstand repeated cutting and may persist in prostrate form as a lawn weed. In short turf, fruits are borne close to the ground and are thus protected to some extent from predation. Regeneration is entirely by seed, which is often set over a prolonged period. Under moisture stress *M.l.* tends to sustain seed production at the expense of vegetative development (Foulds 1978), and in droughted habitats *M.l.* usually functions as a winter annual. In moist fertile sites a life-span of at least 4 years is probably common and robust plants may produce over 2000 seeds annually (*Biol Fl*). Some genetic differences exist with respect to life-history; diploid and tetraploid races have been recorded (*Biol Fl*). *M.l.* may set seed within 9 weeks of germination (*Biol Fl*), forms a persistent buried seed bank (Roberts & Feast 1972, Odum 1978) and is essentially a colonist of bare areas. Because of its capacity to fix nitrogen, *M.l.* is sometimes recommended for use in land reclamation (see Dancer *et al.* 1977).

Current trends Uncertain. An effective colonist of disturbed habitats and spoil, but potential range is limited by the association with infertile soils.

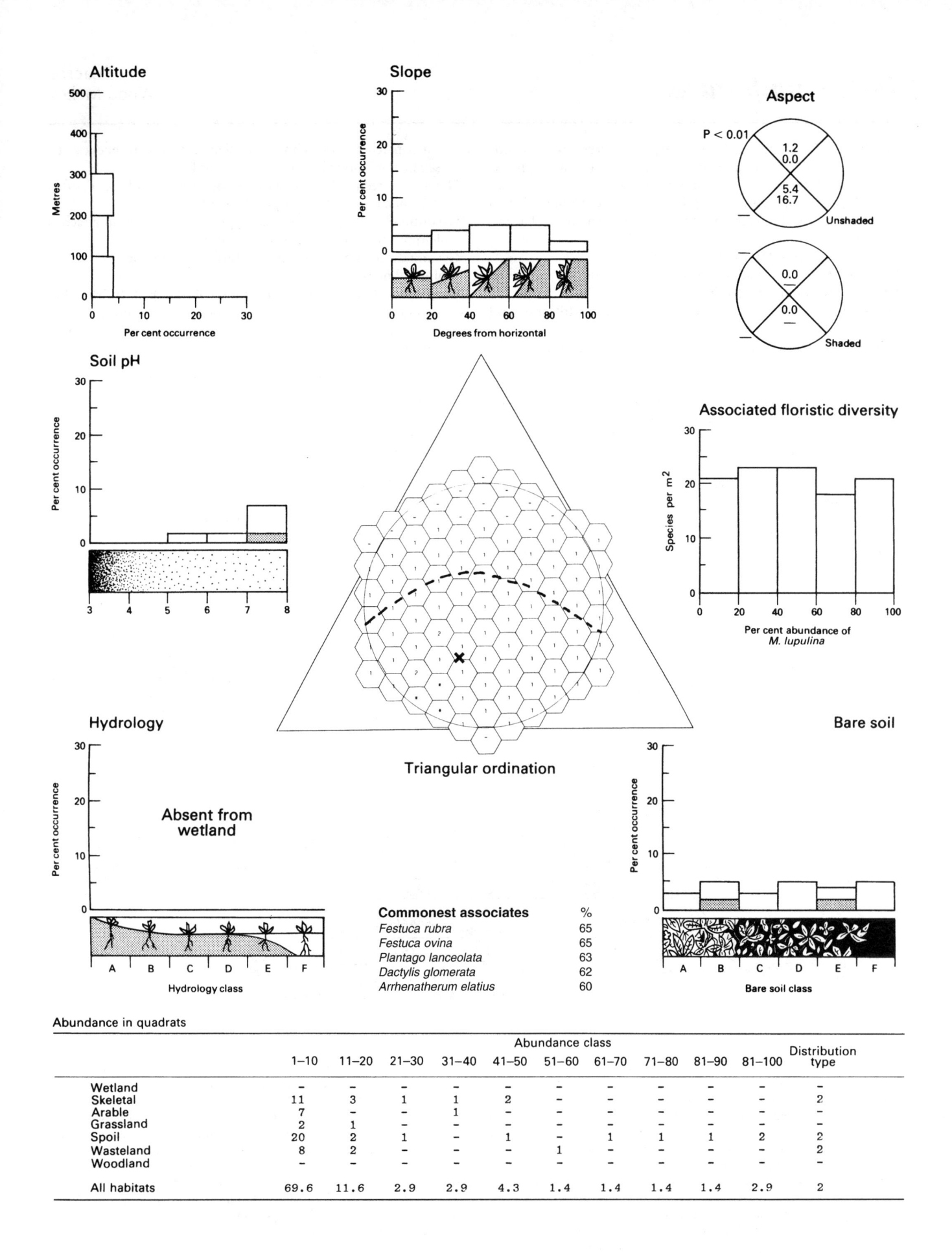

Altitude

Slope

Aspect

$P < 0.01$

	1.2	
	0.0	
—	5.4	Unshaded
	16.7	

	0.0	
—	0.0	Shaded

Soil pH

Associated floristic diversity

Triangular ordination

Hydrology

Absent from wetland

Bare soil

Commonest associates	%
Festuca rubra	65
Festuca ovina	65
Plantago lanceolata	63
Dactylis glomerata	62
Arrhenatherum elatius	60

Abundance in quadrats

	\multicolumn{10}{c	}{Abundance class}	Distribution type
	1–10	11–20	21–30	31–40	41–50	51–60	61–70	71–80	81–90	81–100	
Wetland	–	–	–	–	–	–	–	–	–	–	–
Skeletal	11	3	1	1	2	–	–	–	–	–	2
Arable	7	–	–	1	–	–	–	–	–	–	–
Grassland	2	1	–	–	–	–	–	–	–	–	–
Spoil	20	2	1	–	1	–	1	1	1	2	2
Wasteland	8	2	–	–	–	1	–	–	–	–	2
Woodland	–	–	–	–	–	–	–	–	–	–	–
All habitats	69.6	11.6	2.9	2.9	4.3	1.4	1.4	1.4	1.4	2.9	2

Melica uniflora Retz.

Poaceae
Wood Melick

Life-form Rhizomatous, polycarpic perennial proto-hemicryptophyte. Stems erect or spreading; leaves linear, hairy, usually <1000 mm²; roots non-mycorrhizal.
Phenology Shoot growth in spring is followed by flowering in May and June. Seed set from July to August. Shoots die back in autumn.
Height Foliage and flowers to 700 mm.
Seedling RGR No data. SLA 39.6. DMC. No data.
Nuclear DNA amount No data; $2n = 18$ (*FE5*); $2x$.
Established strategy Intermediate between stress-tolerator and stress-tolerant competitor.
Regenerative strategies V, S. Spreads by means of rhizomes to form patches 1 m or more in diameter. Also regenerates by seed, germinating in spring. SB I, II or IV.
Flowers Green or brown, hermaphrodite, wind-pollinated; often *c.* 20 in a loose panicle (1 per spikelet).
Germinule 2.78 mg, caryopsis dispersed with both an attached lemma and a club-shaped sterile floret.
Germination Epigeal; dormancy broken by chilling.
Biological flora None.
Geographical distribution In most of the British Isles, but absent from N Scotland (91% of British vice-counties), *ABIF* page 765; Europe northwards to Scandinavia (77% of European territories); Asia Minor.

Habitats Exclusively a species of woodland and other shaded habitats. Most common in woodland and scrub on base-rich soils, but also found in hedgerows (possibly as a woodland relict), on shaded cliffs and observed on shaded walls and outcrops. Occurs more rarely in unshaded habitats close to woodland.
Gregariousness Stand-forming.
Altitude Recorded at altitudes of up to 300 m. [Up to 490 m.]
Slope Wide-ranging.
Aspect No significant trends.
Hydrology Absent from waterlogged soils, but occasionally observed on moist ground adjacent to mire.

Soil pH Widely distributed, but most common on soils in the pH range 5.5-7.5. Absent from highly acidic soils.
Bare soil Widespread, but most abundant at moderately low exposures.
△ **ordination** Distribution centred on undisturbed and relatively unproductive vegetation.
Associated floristic diversity Relatively low and declining further at higher frequencies of *M.u.*, suggesting that the species may be exerting a significant degree of dominance in some communities.

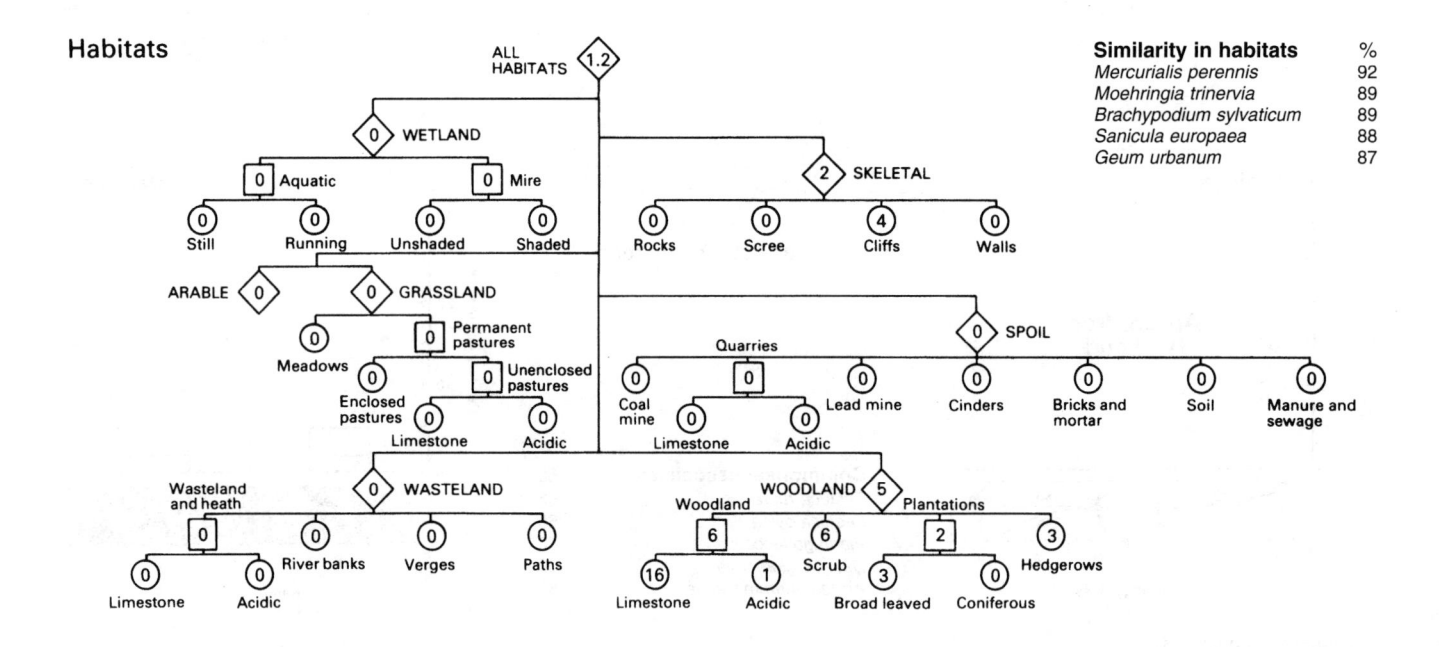

Habitats

Similarity in habitats	%
Mercurialis perennis	92
Moehringia trinervia	89
Brachypodium sylvaticum	89
Sanicula europaea	88
Geum urbanum	87

Synopsis *M.u.* is a stand-forming woodland grass more rarely found in hedgerows and restricted to moderately base-rich but relatively unproductive sites where the vigour of taller potential dominants is reduced both by shade and by low soil fertility. The annual increase in shoot biomass continues into the shaded phase but, unusually amongst woodland grasses, *M.u.* is not winter green. *M.u.* is characteristic of freely-drained sites; frequent occurrence on shaded rock ledges and the absence of the species from mire suggest intolerance of waterlogging. Although associated with a wide range of soil types, the species is recorded most frequently from calcareous strata. *M.u.* regenerates in situ by means of rhizomatous growth to form large clonal patches. Stands are often extremely localized, suggesting that seedling colonization is infrequent either because the comparatively large seeds are underdispersed or because they are heavily predated. Seeds germinate in spring and a persistent seed bank has been detected. Spikelets of *M.u.* consist of one fertile floret and 2-3 reduced to form a club-shaped mass (Hubbard 1984). This appendage encourages dispersal by ants (Ridley 1930). *M.u.* is only rarely among the recruits to recent artificial habitats, and within the context of Lincolnshire is regarded as a species of ancient woodland (Peterken 1981). It is surprising that *M.u.* produces a persistent seed bank. Presumably, like *Milium effusum*, it exploits sites of localized disturbance. This neglected species is in urgent need of ecological investigation.
Current trends Probably decreasing.

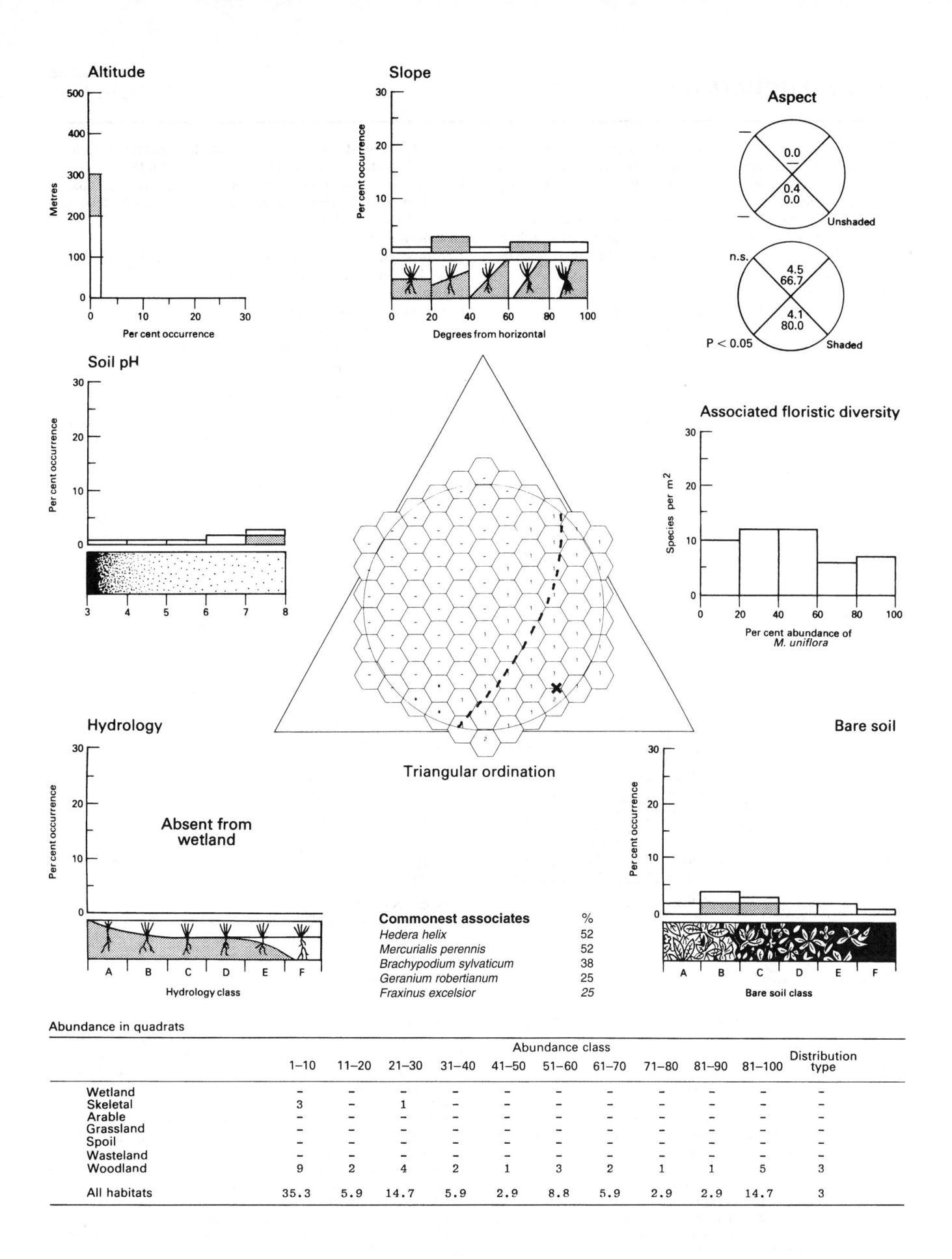

Altitude

Slope

Aspect

	0.0	
—	0.4	Unshaded
	0.0	

	4.5	
n.s.	66.7	
	4.1	Shaded
$P < 0.05$	80.0	

Soil pH

Associated floristic diversity

Species per m^2 vs Per cent abundance of *M. uniflora*

Hydrology

Absent from wetland

Triangular ordination

Bare soil

Commonest associates	%
Hedera helix	52
Mercurialis perennis	52
Brachypodium sylvaticum	38
Geranium robertianum	25
Fraxinus excelsior	25

Abundance in quadrats

	1–10	11–20	21–30	31–40	41–50	51–60	61–70	71–80	81–90	81–100	Distribution type
Wetland	–	–	–	–	–	–	–	–	–	–	–
Skeletal	3	–	1	–	–	–	–	–	–	–	–
Arable	–	–	–	–	–	–	–	–	–	–	–
Grassland	–	–	–	–	–	–	–	–	–	–	–
Spoil	–	–	–	–	–	–	–	–	–	–	–
Wasteland	–	–	–	–	–	–	–	–	–	–	–
Woodland	9	2	4	2	1	3	2	1	1	5	3
All habitats	35.3	5.9	14.7	5.9	2.9	8.8	5.9	2.9	2.9	14.7	3

Mentha aquatica L.

Lamiaceae
Water Mint

Life-form Polycarpic perennial protohemicryptophyte or helophyte with creeping rhizome. Stems erect; leaves ovate, usually hairy, <2000 mm²; roots with VA or, very rarely, ectomycorrhizas.

Phenology Overwinters as short shoots with green, scale-like leaves. Shoots elongate in late spring. Flowers from July to September and seed set from August to October, after which the shoots die.

Height Foliage to *c.* 600 mm, overtopped by flowers to 900 mm.

Seedling RGR 1.0–1.4 week^{-1}. SLA 32.0. DMC 16.4.

Nuclear DNA amount 3.0 pg; $2n = 96^*$ (*BC*); $8x^*$.

Established strategy Intermediate between competitor and competitive-ruderal.

Regenerative strategies V, B_s. Regenerates vegetatively to form clonal patches and by seed (see 'Synopsis'); ?SB IV.

Flowers Lilac, gynomonoecious, protandrous, insect-pollinated, self-compatible (Harley & Brighton 1977); usually <100 in a whorled terminal head.

Germinule 0.14 mg, 1.1 × 0.6 mm, nutlet (4 per ovary).

Germination Epigeal; t_{50} 10 days; dormancy broken by chilling and by dry storage at 20°C; L > D.

Biological flora None.

Geographical distribution The British Isles, but rarer in N Scotland (100% of British vice-counties), *ABIF* page 526; Europe except for the extreme N (92% of European territories and 3% naturalised); SW Asia and Africa.

Habitats Exclusively a wetland species. Found in both shaded and unshaded mire, and to a lesser extent in standing water in ditches or pond margins.

Gregariousness Recorded as intermediate, but sometimes stand-forming.

Altitude Suitable sites are most frequent in lowland areas, but *M.a.* has been observed to 270 m. [Up to 460 m.]

Slope As with most wetland species, only occurring on flat ground and gentle slopes.

Aspect No data.

Hydrology Occurring most frequently and in greatest abundance in topogenous mire marginal to open water; less characteristic of soligenous mire. Also found in temporarily aquatic habitats in up to 30 mm of water.

Soil pH Largely absent from soils of pH < 5.0, and most frequent in neutral soils.

Bare soil Associated with a wide range of values.

A ordination Distribution centred on vegetation associated with fertile, infrequently disturbed conditions. Absent from heavily disturbed or unproductive situations.

Associated floristic diversity Typically low or intermediate.

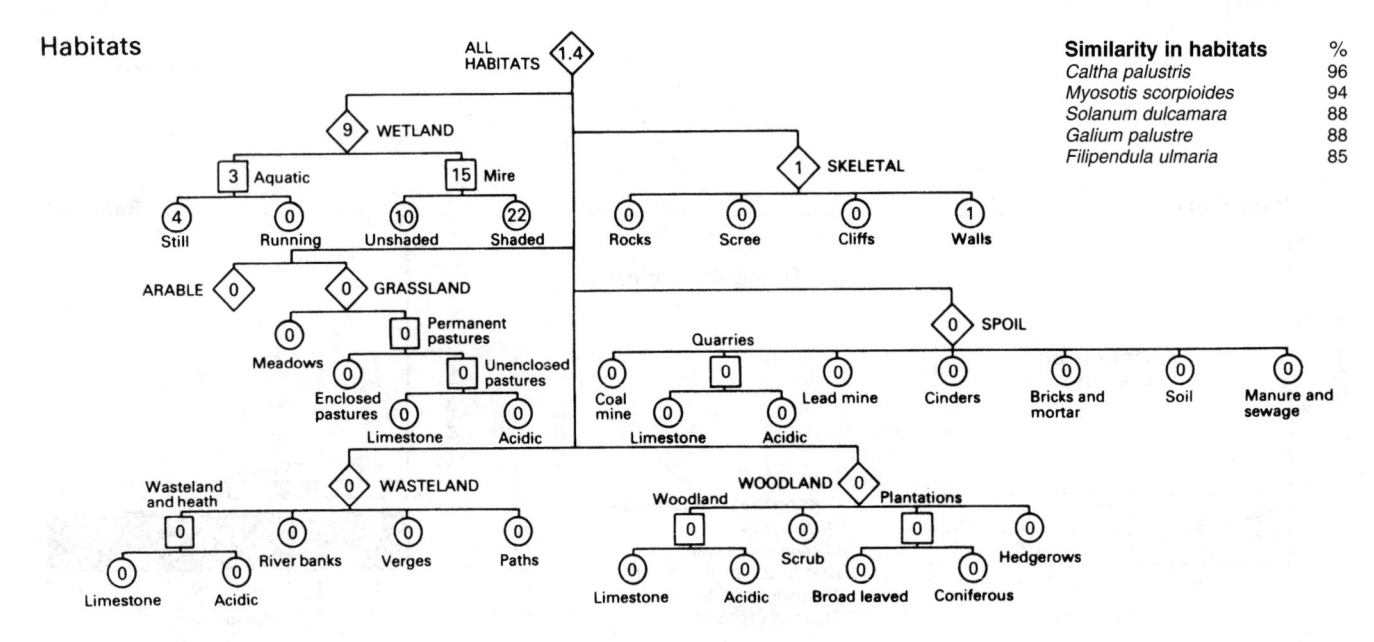

Habitats

Similarity in habitats	%
Caltha palustris	96
Myosotis scorpioides	94
Solanum dulcamara	88
Galium palustre	88
Filipendula ulmaria	85

Synopsis *M.a.* is a rhizomatous, aromatic, stand-forming herb typically associated with mire adjacent to open water but more common in mid-marsh positions than those close to the water-line. Less frequently *M.a.* may occur wholly or partially submerged in ditches. The species appears to be restricted to situations where the vigour of taller waterside dominants is reduced by a degree of soil infertility. *M.a.* is also found in moderately shaded habitats. Because of its erect stature, *M.a.* is sensitive to trampling and grazing, and despite the apparent toxicity of *Mentha* species to stock (Cooper & Johnson 1984) the species is suppressed in grazed marshes. *M.a.* is found in a wide range of edaphic conditions, and is particularly frequent in soligenous mire on calcareous soils. The species forms extensive patches by means of creeping rhizomes, breakdown of which often results in the formation of daughter clones. Rhizome or shoot fragments detached as a result of disturbance during, for example, flooding readily regenerate and *M.a.* may be transported in water to colonize distant sites. Various sterile hybrid mints have colonized water systems extensively in this way. Regeneration by seeds, which may float on water for long periods (Ridley 1930), is probably infrequent. Seeds normally germinate in vegetation gaps during spring or after incorporation into a persistent seed bank. *M.a.* varies greatly in habit, leaf shape and degree of hairiness, both as a result of genetic differences and phenotypic plasticity. In drier sites where the water table fluctuates widely, e.g. woodland rides and reservoir margins, *M.a.* is frequently replaced by *M. arvensis* L. ($2n = 72$, $6x$). The usually sterile but vegetatively vigorous hybrid between the two (*M.* × *verticillata* L.; $2n = 84$. $7x$ (Harley & Brighton 1977)) is also frequent, often in the absence of one or both parents, and usually in slightly drier habitats than those exploited by *M.a.* Two further hybrids involving *M.a.* occur in Britain (Harley & Brighton 1977); these are often of horticultural origin.

Current trends Still common, but perhaps decreasing as a result of eutrophication and habitat destruction.

Altitude
Slope
Aspect
Unshaded
Shaded

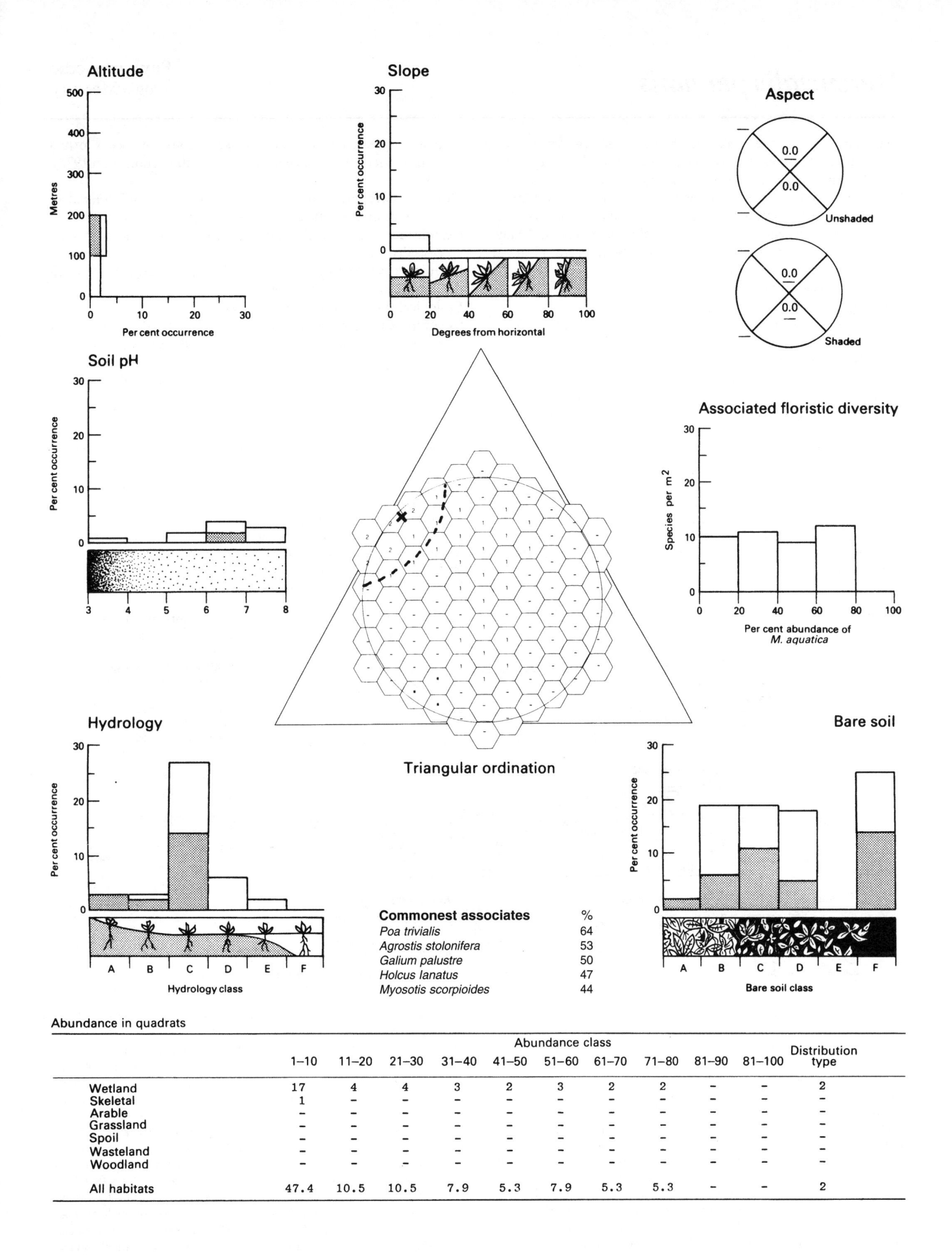

Soil pH
Associated floristic diversity
Species per m^2
Per cent abundance of *M. aquatica*

Triangular ordination

Hydrology
Hydrology class

Commonest associates	%
Poa trivialis	64
Agrostis stolonifera	53
Galium palustre	50
Holcus lanatus	47
Myosotis scorpioides	44

Bare soil
Bare soil class

Abundance in quadrats

	1–10	11–20	21–30	31–40	41–50	51–60	61–70	71–80	81–90	81–100	Distribution type
Wetland	17	4	4	3	2	3	2	2	–	–	2
Skeletal	1	–	–	–	–	–	–	–	–	–	–
Arable	–	–	–	–	–	–	–	–	–	–	–
Grassland	–	–	–	–	–	–	–	–	–	–	–
Spoil	–	–	–	–	–	–	–	–	–	–	–
Wasteland	–	–	–	–	–	–	–	–	–	–	–
Woodland	–	–	–	–	–	–	–	–	–	–	–
All habitats	47.4	10.5	10.5	7.9	5.3	7.9	5.3	5.3	–	–	2

Mercurialis perennis L.

Euphorbiaceae
Dog's Mercury

Life-form Polycarpic perennial protohemicryptophyte with extensive rhizomes. Stems erect; leaves ovate, hairy, typically <1000 mm^2; roots usually with VA mycorrhizas.

Phenology Some leafy shoots often persist until late winter or early spring, the time at which new shoots emerge. Plants flower from February, before the leaves are fully expanded, until May, by which time there is a full leaf canopy. Seed set from June to August. Leaves long-lived.

Height Foliage and flowers to 400 mm.

Seedling RGR No data. SLA 25.3. DMC 22.4.

Nuclear DNA amount 4.7 pg; $2n$ = 42, 64 (female), 66 (male) c. 80, 84 (*FE3*); 6x, 8x, 10x.

Established strategy Stress-tolerant competitor.

Regenerative strategies V, S. Mainly vegetatively by means of rhizomes to form large clonal patches. Although rarely reproducing by seed (Hawkes 1966), most populations contain male and female plants, and seed set is common. Seeds form a transient seed bank and germinate in spring (Thompson 1977); SB II.

Flowers Green, dioecious, rarely monoecious, wind-pollinated; <20 male or 1-3 female in axillary spikes.

Germinule 2.15 mg, 2.7 × 2.6 mm, seed shed explosively from a 2-seeded capsule.

Germination Mainly but not exclusively hypogeal (Mukerji 1936); dormancy broken by chilling.

Biological flora None, but see Mukerji (1936).

Geographical distribution Most of the British Isles except for N Scotland and rare in Ireland (80% of British vice-counties), *ABIF* page 426; most of Europe northwards to Norway, but only on mountains in southern parts (77% of European territories); SW Asia.

Habitats Predominantly a species of shaded habitats, but extending into open habitats, particularly in upland areas. Particularly abundant in limestone woodlands, but frequent in all types except coniferous plantations and in hedgerows. Widespread in upland limestone grassland and scree. Less frequent on river banks, waysides, lead-mine spoil, outcrops, cliffs and walls. Absent from most types of spoil heaps, wetlands and arable land.

Gregariousness Stand-forming.

Altitude Widely distributed but less frequent above 300 m. Extends to 390 m. [Up to 1010 m.]

Slope Found over a wide range of values.

Aspect Slightly more frequent on N-facing slopes in both open and shaded sites but achieving greatest abundance on shaded S-facing slopes.

Hydrology Absent from wetlands.

Soil pH Largely absent from acidic soils below pH 4.0 and most frequent and abundant on neutral soils.

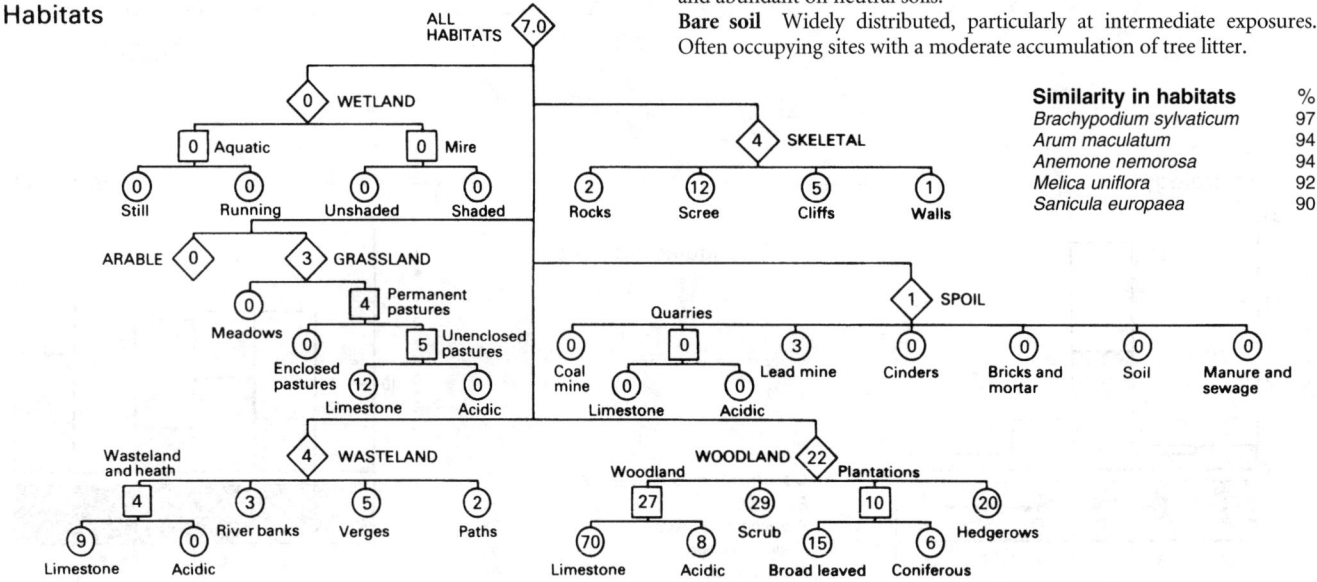

Bare soil Widely distributed, particularly at intermediate exposures. Often occupying sites with a moderate accumulation of tree litter.

Similarity in habitats	%
Brachypodium sylvaticum	97
Arum maculatum	94
Anemone nemorosa	94
Melica uniflora	92
Sanicula europaea	90

Habitats

△ **ordination** Absent from highly disturbed vegetation. Distribution centred on undisturbed vegetation types of intermediate productivity.

Associated floristic diversity Relatively low, and declining at high-frequency values of *M.p.* where we conclude that the species is exerting competitive exclusion.

Synopsis *M.p.* is a rhizomatous, stand-forming woodland herb exploiting relatively unproductive base-rich sites where the growth of more-robust dominants is suppressed. In lowland habitats the species is largely restricted to shaded sites. However, in the uplands *M.p.* is found in unshaded herbaceous vegetation, particularly on moist N-facing slopes; in some populations established in the open *M.p.* forms the upper layer of the canopy and may suffer considerable loss of chlorophyll. It is quite common, however, for *M.p.* to occupy the shaded zone beneath the canopy of a taller herb, e.g. *Filipendula ulmaria* (Al-Mufti et al. 1977). *M.p.* is relatively deeply rooted (normally 100-150 mm (Mukerji 1936)) and is very susceptible to waterlogging and to the effects of Fe^{2+} toxicity (Martin 1968). On wet soils roots are shallower and *M.p.* is susceptible to summer drought (Martin 1968). The foliage of *M.p.* contains high concentrations of N and Ca, but levels of P are low (*SIEF*). The leaves have an unpleasant odour when bruised and rarely show signs of extensive predation. They contain, amongst other substances, toxic volatile oils, and are highly poisonous both to stock and to humans (Cooper & Johnson 1984). The species combines winter green and vernal aspects in its rather unusual phenology. As with many vernal species, new shoots are formed in the autumn and expand in late winter (February-March). The emergent shoot is bent over at the tip, facilitating emergence through leaf litter (Salisbury 1916). The plant flowers before the leaves are fully expanded. *M.p.* reaches a peak in biomass at the end of May at the termination of the light phase (Al-Mufti et al. 1977, Hutchings 1978). However, unlike vernal species, most shoots and leaves persist through the summer and gradually senesce

Altitude

Slope

Aspect

Soil pH

Associated floristic diversity

Triangular ordination

Hydrology

Bare soil

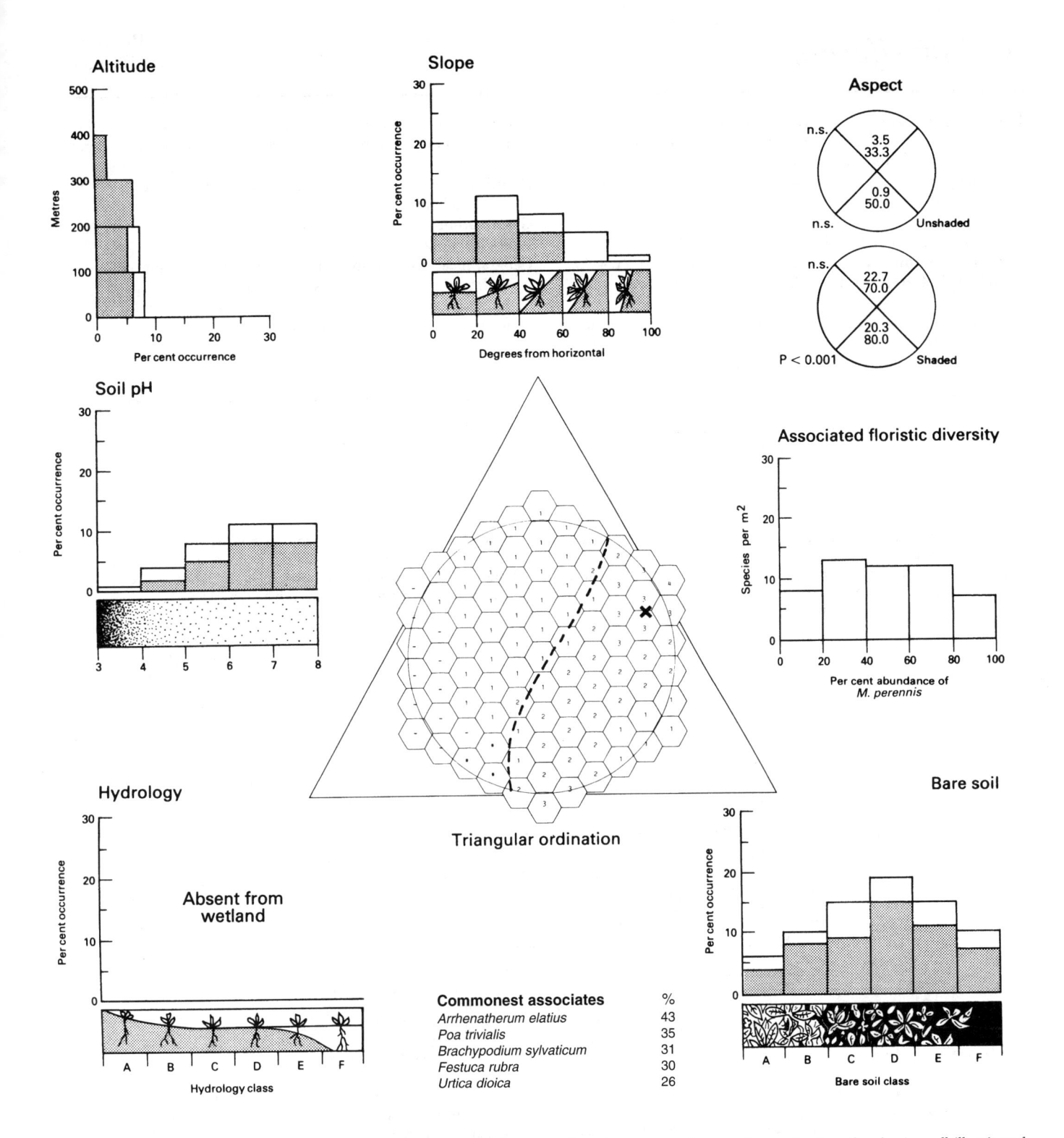

Commonest associates	%
Arrhenatherum elatius	43
Poa trivialis	35
Brachypodium sylvaticum	31
Festuca rubra	30
Urtica dioica	26

during winter. Regeneration is mainly by vegetative means, and *M.p.* forms extensive long-lived clones which may extend by 100-150 mm. year^{-1} (Mukerji 1936). The species, which can survive at very low light intensities (Mukerji 1936), responds little to the additional light present during clear felling or coppicing, but survives beneath other species and returns to prominence after the regrowth of the tree canopy. The interaction between individual shoots and the control of shoot density are described by Hutchings and Barkham (1976). Although normally dioecious, female populations usually set seed (up to 300 per plant (Mukerji 1936)), but many may be non-viable (Mukerji 1936) and seedling establishment is uncommon (Hawkes 1966). No persistent seed bank is formed. Male and female plants appear to differ in their ecological distribution. Male plants have been recorded more frequently than females in well-illuminated conditions (Mukerji 1936, Wade 1981a,b, Wade *et al*. 1981) and at high pH (Cox 1981). *M.p.* is a common species of ancient woodland in Lincolnshire (Peterken 1981), and its large seeds, which have an elaiosome and are said to be specialized for dispersal by ants (Ridley 1930), appear to be poorly dispersed. *M.p.* shows some morphological variations which have been given varietal rank. The species is also phenotypically plastic, and can form sun or shade leaves which differ in morphology and anatomy (Mukerji 1936).

Current trends Shows poor mobility and is probably decreasing with the destruction of older woodland.

Milium effusum L.

Poaceae
Wood Millet

Life-form Loosely tufted, polycarpic perennial protohemicryptophyte. Stems erect or ascending; leaves linear, usually <1000 mm^2; roots often with VA mycorrhizas.
Phenology Winter green. Flowers in June and seed set from July to September.
Height Foliage usually <350 mm; flowers to 1200 mm.
Seedling RGR 1.0-1.4 week^{-1}. SLA 33.3. DMC 25.1.
Nuclear DNA amount 9.4 pg; $2n = 14, 28^*$ (*BC, FE5*); $2x, 4x^*$.
Established strategy Intermediate between C-S-R and stress-tolerator.
Regenerative strategies B$_s$, V. By seed in spring or after incorporation into a bank of buried seeds. Also shows a restricted capacity for lateral vegetative spread; SB III or ?IV.

Flowers Green, hermaphrodite, wind-pollinated, feebly protogynous (Knuth 1906); >100 in a loose panicle (1 per spikelet).
Germinule 1.2 mg, 2.1 × 1.2 mm, caryopsis dispersed with an attached lemma (2.7 × 2.1 mm).
Germination Epigeal; t_{50} 22 days; 11-22°C; L > D.
Biological flora None.
Geographical distribution The British Isles but more frequent in England (79% of British vice-counties), *ABIF* page 741; Europe except for the Mediterranean region (79% of European territories); temperate Asia and N America.

Habitats Mainly a plant of woodland and scrub, particularly on base-poor strata. Less frequent in hedgerows and plantations, and not recorded from wetland, grassland and skeletal habitats.
Gregariousness Intermediate.
Altitude Restricted to <300 m and particularly abundant on the coal measures. [Up to 340 m.]
Slope Most frequent on flat ground and gentle slopes, and not recorded from steep slopes.
Aspect More frequently found on S-facing slopes, but abundance showing no marked response to aspect.
Hydrology Absent from wetlands.

Soil pH Recorded mainly on acidic soils and most common in the pH range 4.0-4.5. Occasionally observed at pH >7.0. [In some areas more characteristic of limestone soils (M. B. Usher pers. comm.).]
Bare soil Occurs in association with widely different amounts of bare soil and variable amounts of tree litter, but absent from habitats which have a very high exposure of bare soil.
△ **ordination** Distribution wide ranging, though noticeably absent from heavily disturbed vegetation. Highest frequencies concentrated in communities of low productivity experiencing slight disturbance.
Associated floristic diversity Low, with higher values where *M.e.* is least abundant.

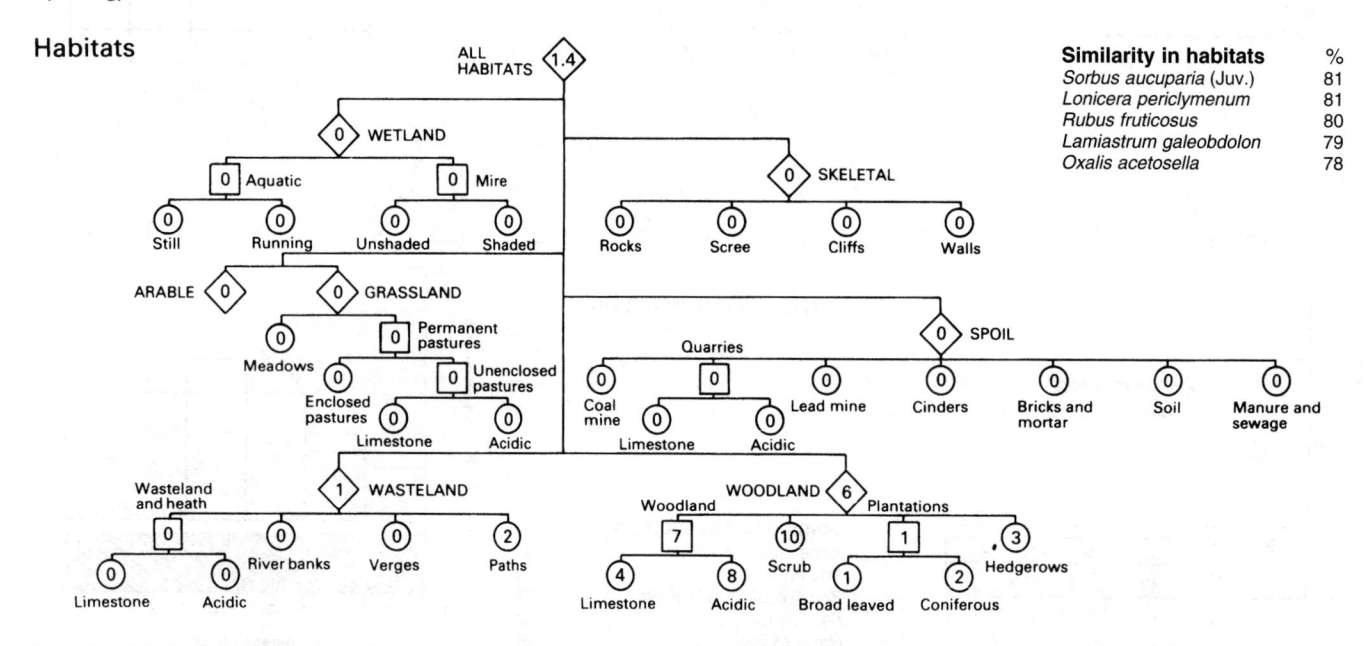

Habitats

Similarity in habitats	%
Sorbus aucuparia (Juv.)	81
Lonicera periclymenum	81
Rubus fruticosus	80
Lamiastrum galeobdolon	79
Oxalis acetosella	78

Synopsis *M.e.* is a loosely-tufted woodland grass occurring in sites where the vigour of potential dominants is reduced both by shade and by low site fertility. The species is winter green, with an annual increase in shoot biomass beginning in spring and continuing into the shaded phase (Al-Mufti *et al.* 1977). The robust shoots are relatively efficient at penetrating deciduous layers of tree-litter (Sydes & Grime 1981a,b). *M.e.* is restricted to moist soils, particularly clays, and locally is virtually absent from the Bunter sandstone. Within the Sheffield region *M.e.* is often recorded from mildly acidic soils. However, elsewhere *M.e.* may be characteristic of calcareous sites (Hubbard 1984, M. B. Usher pers. comm.). The species has only a limited capacity for vegetative spread, but numerous seeds are produced. These show low rates of germination when first transferred to the laboratory (Grime *et al.* 1981). *M.e.* exhibits variation in germination biology over its geographical range (Thompson 1980). At southern stations *M.e.* flowers early and seeds germinate predominantly in autumn (Thompson 1980). Nearer to its northern limit seed set is later and at least some germination may be delayed until spring (Thompson 1980). A similar pattern has been reported for *Festuca ovina* (Harmer & Lee 1978). *M.e.* forms a persistent seed bank, which suggests that the species may have the capacity to exploit sites where periodic disturbance occurs in the form of either tree-felling or fire. Caryopses, which can survive ingestion by some animals (see Ridley 1930), are shed without an attached lemma. They show no obvious specialization for dispersal, and in Lincolnshire *M.e.* is regarded as a species of ancient woodland (Peterken 1981). However, *M.e.* has colonized more-recent woodland in some upland sites. Hubbard (1984) also states that *M.e.* was formerly planted for ornament and that its seeds provided food for game birds, but there is no evidence that *M.e.* has been introduced in the Sheffield region (see also Rackham 1980).
Current trends Uncertain, but probably decreasing, particularly in lowland areas.

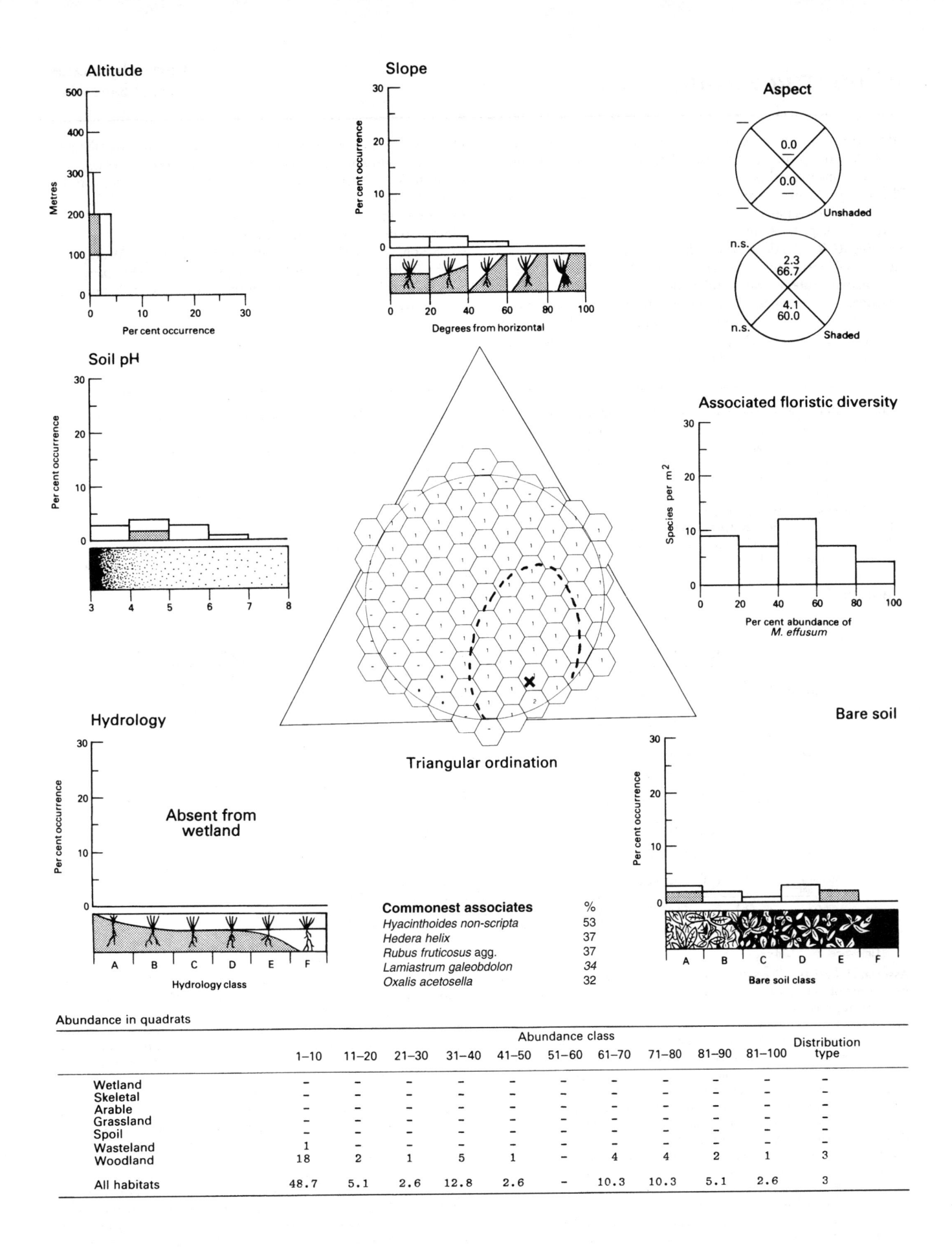

Altitude

Slope

Aspect

Unshaded
0.0
0.0
n.s. — —

Shaded
2.3 66.7
4.1 60.0
n.s.

Soil pH

Associated floristic diversity

Species per m² vs Per cent abundance of *M. effusum*

Hydrology

Absent from wetland

Triangular ordination

Bare soil

Commonest associates	%
Hyacinthoides non-scripta	53
Hedera helix	37
Rubus fruticosus agg.	37
Lamiastrum galeobdolon	34
Oxalis acetosella	32

Abundance in quadrats

	1–10	11–20	21–30	31–40	41–50	51–60	61–70	71–80	81–90	81–100	Distribution type
Wetland	–	–	–	–	–	–	–	–	–	–	–
Skeletal	–	–	–	–	–	–	–	–	–	–	–
Arable	–	–	–	–	–	–	–	–	–	–	–
Grassland	–	–	–	–	–	–	–	–	–	–	–
Spoil	–	–	–	–	–	–	–	–	–	–	–
Wasteland	1	–	–	–	–	–	–	–	–	–	–
Woodland	18	2	1	5	1	–	4	4	2	1	3
All habitats	48.7	5.1	2.6	12.8	2.6	–	10.3	10.3	5.1	2.6	3

Minuartia verna (L.) Hiern

Caryophyllaceae
Vernal Sandwort

Life-form Polycarpic perennial chamaephyte with a branching stock and tap-root. Shoots ascending; leaves linear, <10 mm^2; roots non-mycorrhizal.
Phenology Winter green. Flowers May to September and sets seed July to October.
Height Foliage to 60 mm; flowers to 120 mm.
Seedling RGR No data. SLA 19.2. DMC 18.6.
Nuclear DNA amount 2.8 pg; $2n = 14^*$ (*BC*); $2x^*$.
Established strategy Stress-tolerator.
Regenerative strategies V, B$_s$. Forms cushions or mats through the slow growth and rooting of prostrate shoots. Also produces numerous seeds which germinate particularly in autumn and form a persistent seed bank; SB III.

Flowers White, hermaphrodite, protandrous, insect-pollinated or self-fertilized (Halliday 1960); in a 2-6-flowered cyme.
Germinule 0.08 mg, 1.0 × 0.7 mm, seed in a 5-25-seeded capsule (Halliday 1960).
Germination Epigeal; t_{50} 7 days; 10-27°C; L > D.
Biological flora None, but see Halliday (1960).
Geographical distribution Scattered throughout the W and N of the British Isles (19% of British vice-counties), *ABIF* page 156; S, W and C Europe (59% of European territories); N temperate regions.

Habitats Abundant on the rocky debris of old lead workings. Occurs locally and at lower frequencies in calcareous pastures, usually in the vicinity of lead-mine heaps. The marshland records all refer to one exceptional site in which a spring occurs within lead workings. [Found on sea-heath on serpentine soils and on basic montane rocks, sometimes in flushes (Halliday 1960).]
Gregariousness Forms small patches.
Altitude Suitable habitats restricted to the carboniferous limestone and millstone grit, and consequently *M.u.* is an upland species. [Occurs from almost sea level to 875 m.]

Slope Not recorded from steep slopes [but occurs on cliffs in montane areas].
Aspect Only in unshaded habitats and more frequently encountered on S-facing slopes.
Hydrology Virtually absent from wetland.
Soil pH Restricted to calcareous soils in the pH range 5.0-8.0 and most abundant in the centre of this range [but extends onto more acidic soils (Halliday 1960)].
Bare soil No clear-cut trends, but rather more abundant in habitats having a moderately high exposure of bare soil, almost certainly resulting from the interaction of soil toxicity and summer drought.
△ **ordination** Distribution confined to vegetation associated with unproductive undisturbed conditions.

Habitats

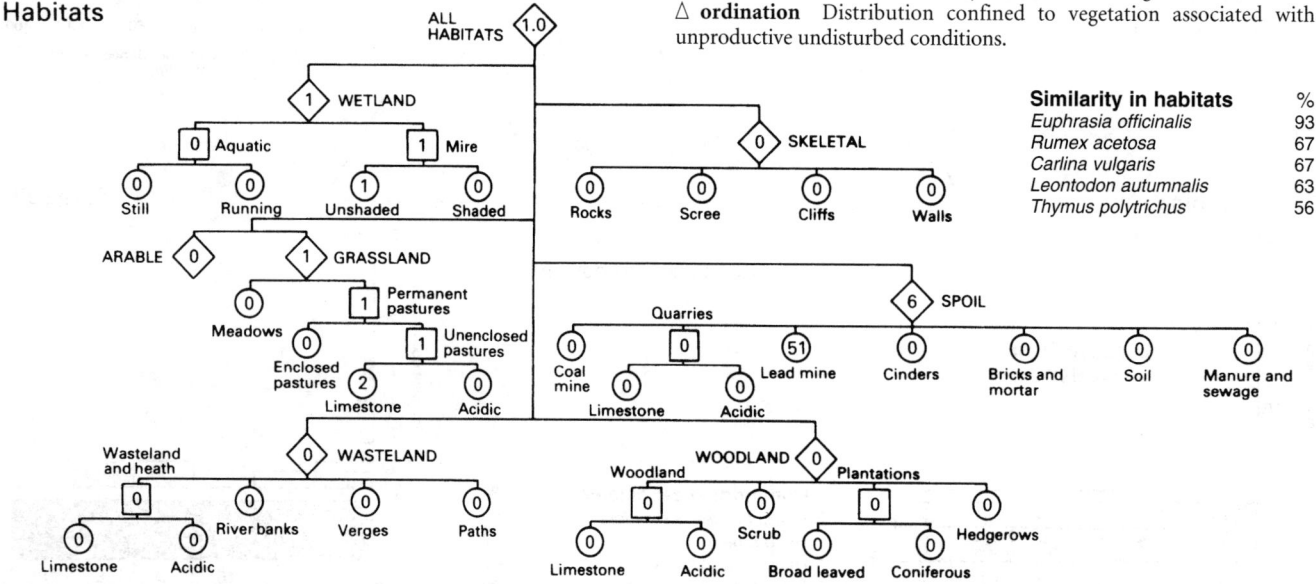

Similarity in habitats	%
Euphrasia officinalis	93
Rumex acetosa	67
Carlina vulgaris	67
Leontodon autumnalis	63
Thymus polytrichus	56

Associated floristic diversity Species diversity declining at highest frequencies of *M.v.*, presumably as a result of heavy-metal toxicity effects on associated species.
Synopsis *M.v.* is a low-growing, cushion-forming herb, which is relatively uncommon and has a very disjunct distribution in Britain (ABF). The species may perhaps best be regarded as a late glacial relict (see Antonovics *et al.* 1971). Nearly all populations in N Derbyshire occur on lead-mine spoil; the remainder, in natural rock outcrops and limestone quarries, also occur on contaminated soils (A. J. M. Baker pers. comm.). *M.v.* also exploits serpentine soils (Halliday 1960). In rocky montane sites, the other main habitat type for the species, heavy-metal concentrations are assumed to be low. The wide edaphic tolerance of *M.v.* is further illustrated by its capacity to grow on mine spoil on acidic soils in addition to its more typical calcareous habitats. *M.v.* is, after *Thlaspi caerulescens* J. and C. Presl., the most heavy-metal-tolerant species in the Sheffield flora (Hajar 1986) and is an accumulator of heavy metals (*sensu* Baker 1981). This may be a reflection of the absence of mycorrhizas in the root system; since all the other heavy-metal- tolerant plants mentioned in this volume are mycorrhizal and behave as excluders, *sensu* Baker (1981) and Shaw (1984). The foliage of plants sampled from lead mine heaps in Derbyshire contains unusually low concentrations of N but high levels of Ca and Fe (*SIEF*). *M.v.* is found over a wide range of climatic conditions in Britain, extending from sites near sea-level which are virtually frost-free to montane areas where the number of days with frost may exceed 150 (Halliday 1960). The distribution is mainly restricted to sites with at least 1000 mm of rainfall each year (Halliday 1960). Halliday also suggests that summer humidity may be important in the survival of this oceanic montane species. This is consistent with local observations of heavy fatalities during summers of exceptional drought (A. R. S. M. Hajar pers. comm.). Many Derbyshire sites for *M.v.* are highly toxic and, as a result, have an incomplete vegetation cover. The species is particularly associated with sites liable to soil erosion (Halliday 1960); plants detached as a result of soil disturbance are frequently observed in the field. *M.v.* has an unusually long flowering period and produces numerous very small seeds. The presence of a

Altitude

Slope

Aspect

	n.s.	1.6 / 100.0	
	3.1 / 71.4	Unshaded
	n.s.		

	—	0.0 / —	
	0.0 / —	Shaded
	—		

Soil pH

Associated floristic diversity

Hydrology

Triangular ordination

Bare soil

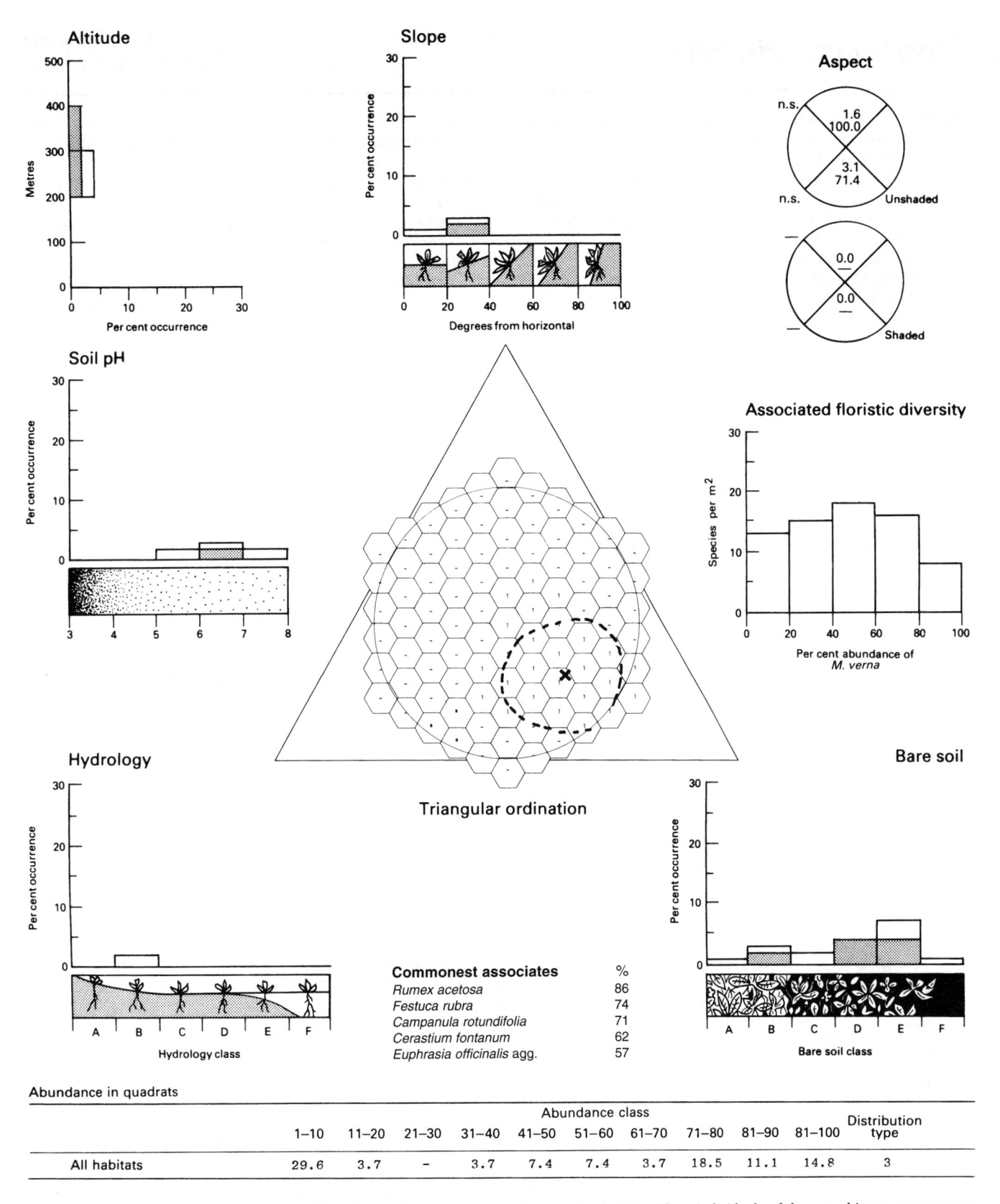

Commonest associates	%
Rumex acetosa	86
Festuca rubra	74
Campanula rotundifolia	71
Cerastium fontanum	62
Euphrasia officinalis agg.	57

Abundance in quadrats

	1–10	11–20	21–30	31–40	41–50	51–60	61–70	71–80	81–90	81–100	Distribution type
All habitats	29.6	3.7	–	3.7	7.4	7.4	3.7	18.5	11.1	14.8	3

persistent seed bank is probably critical in enabling the species to recover after drought. Under more-favourable conditions *M.v.* also spreads by clonal expansion but field observations and recent experiments (Hajar & Grime unpubl.) have shown that here the species is exceedingly vulnerable to competition from metal-tolerant races of *Festuca rubra*. *M.v.* is polymorphic and five further ssp. have been identified in Europe. Some populations are tetraploid ($2n = 48$) *FE1*; their recognition is to some extent confounded by the interaction between genetic diversity and phenotypic plasticity. Thus, individuals of the same biotype may appear very different in contrasted environments whereas genetically distinct races may be indistinguishable through the effects of climate (Halliday 1960). Crosses between diploid populations from widely differing geographical areas are successful, but diploids are genetically isolated from tetraploids (Halliday 1960).

Current trends Reported as decreasing in S Scotland (Halliday 1960) and decreasing in Derbyshire also, as a result of habitat destruction.

Moehringia trinervia (L.) Clairv.

Caryophyllaceae
Three-nerved Sandwort

Life-form Usually a summer-annual therophyte, but sometimes a short-lived polycarpic perennial chamaephyte. Stems much branched, prostrate or weakly ascending; leaves ovate, ciliate, usually <250 mm^2; roots without mycorrhizas.
Phenology Winter green when perennial. Germinates mainly in spring. Flowers May and June and sheds seed June and July.
Height Foliage and flowers often < 150 mm.
Seedling RGR No data. SLA 36.1. DMC 18.0.
Nuclear DNA amount No data; 2n = 24* (*BC*); 2x*.
Established strategy Intermediate between ruderal and C-S-R.
Regenerative strategies B$_s$ Regenerates entirely by seedlings which appear in spring. A persistent seed bank is reported (Donelan & Thompson 1980, Brown & Oosterhuis 1981); ?SB III.

Flowers White, hermaphrodite, visited by insects and automatically self-pollinated (*CTW*); solitary and axillary or in terminal cymes.
Germinule 0.22 mg, 1.0 × 0.8 mm, seed in a *c.* 12-seeded dehiscent capsule.
Germination Epigeal; t_{50} 60 days; 9-21°C; germination rate increased after dry storage at 20°C; L > D.
Biological flora None.
Geographical distribution The British Isles, but less common in N Scotland and Ireland (95% of British vice-counties), *ABIF* page 155; most of Europe (87% of European territories); W Asia.

Habitats A species of shaded habitats, particularly open disturbed woodland. Also recorded from hedgerows, shaded cliffs and outcrops, and disturbed mineral soil. Rarely observed in mire and grassland, and absent from aquatic and arable habitats.
Gregariousness Usually occurs as scattered individuals.
Altitude Observed to 290 m. [Up to 430 m.]
Slope Found over a wide range of slopes.
Aspect Slightly more common on S-facing slopes.
Hydrology Not strictly a wetland species, but there are isolated observations from moist soils and at the edge of soligenous and topogenous mire.
Soil pH Widely distributed above pH 3.5, but most frequent on acidic soils of pH 4.0-5.0.
Bare soil Most commonly associated with habitats having a moderate amount of bare ground.
△ **ordination** Concentrated in vegetation experiencing moderate intensities of stress and disturbance. Excluded from sites subject to heavy disturbance or intensive competition.

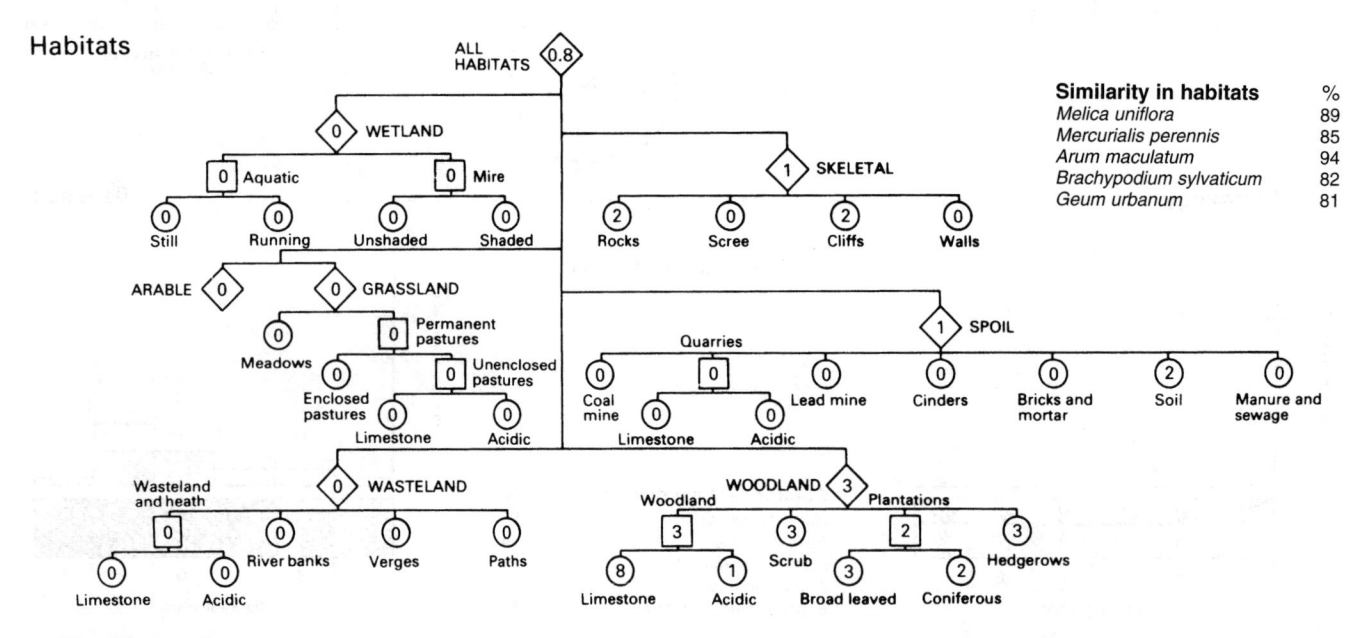

Habitats

Similarity in habitats	%
Melica uniflora	89
Mercurialis perennis	85
Arum maculatum	94
Brachypodium sylvaticum	82
Geum urbanum	81

Associated floristic diversity Typically low.
Synopsis *M.t.* is one of only two relatively common annual species which are exclusive to British woodlands, the other being *Ceratocapnos claviculata* (L.) Liden. As with *C.c.*, *M.t.* is a summer-annual. In a majority of sites with *M.t.* the cover of tree litter is less than 25%, and it seems likely that the presence of large amounts is prejudicial to establishment; indeed, the general absence from woodland of winter annuals and other autumn-germinating species may be due to problems of seedling establishment arising from the impacts of leaf-fall and litter persistence. More rarely, and perhaps only in light shade, *M.t.* may behave as a short-lived, winter-green polycarpic perennial. The species exploits disturbed ground, e.g. rabbit scrapes or sites of tree falls, and in appearance and in ecology *M.t.* is 'the *Stellaria media* of woodland'. The distribution of *M.t.* shows a slight bias towards acidic soils, and the slight tendency for the species to occur on S-facing slopes in woodland suggests that, in common with many other ephemerals, *M.t.* requires well-insolated, relatively warm sites for completion of the life-cycle. Salisbury (1942) suggests that the average output per plant is *c.* 2500 seeds, and although this estimate appears rather high *M.t.* is undoubtedly a prolific seed producer. Germination is inhibited by high temperatures and is delayed until spring, and, unusually for a woodland species (and typically for species exploiting unpredictably disturbed habitats), a persistent seed bank is formed (Donelan & Thompson 1980, Brown & Oosterhuis 1981). In this latter specialization an obvious parallel exists with another woodland opportunist, *Digitalis purpurea*. The seed has an oily elaiosome (*CTW*) and dispersal is often influenced by ants although, judging by the distribution of the species, dispersal by humans is much more important.
Current trends Peterken (1981) regards *M.t.* as a characteristic species of secondary woodland. *M.t.* is favoured by the high level of mechanized disturbance associated with modern forestry, and is probably increasing.

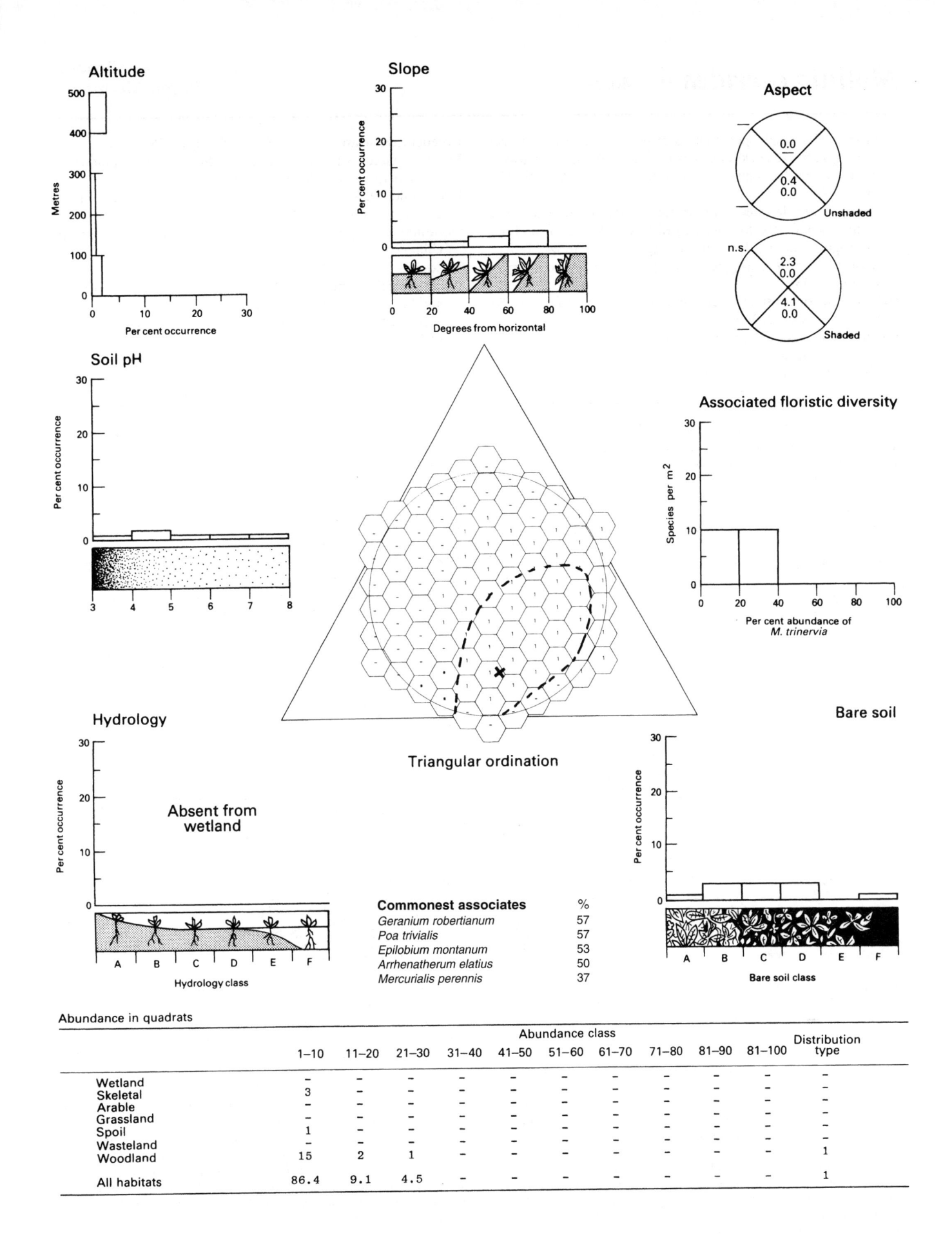

Altitude

Slope

Aspect

Unshaded
0.0
0.4 —
0.0
—

Shaded
n.s.
2.3
0.0 —
4.1
0.0
—

Soil pH

Associated floristic diversity

Species per m²

Per cent abundance of *M. trinervia*

Hydrology

Absent from wetland

Hydrology class: A B C D E F

Triangular ordination

Bare soil

Bare soil class: A B C D E F

Commonest associates	%
Geranium robertianum	57
Poa trivialis	57
Epilobium montanum	53
Arrhenatherum elatius	50
Mercurialis perennis	37

Abundance in quadrats

	1–10	11–20	21–30	31–40	41–50	51–60	61–70	71–80	81–90	81–100	Distribution type
Wetland	–	–	–	–	–	–	–	–	–	–	–
Skeletal	3	–	–	–	–	–	–	–	–	–	–
Arable	–	–	–	–	–	–	–	–	–	–	–
Grassland	–	–	–	–	–	–	–	–	–	–	–
Spoil	1	–	–	–	–	–	–	–	–	–	–
Wasteland	–	–	–	–	–	–	–	–	–	–	–
Woodland	15	2	1	–	–	–	–	–	–	–	1
All habitats	86.4	9.1	4.5	–	–	–	–	–	–	–	1

Molinia caerulea (L.) Moench

Poaceae
Purple Moor-grass

Life-form Polycarpic perennial, semi-rosette hemicryptophyte forming either tussocks or extensive swards. Shoots erect; leaves linear, somewhat hairy, <4000 mm^2; roots deep with VA mycorrhizas.

Phenology Swollen stem bases remaining green throughout the winter. New shoots elongate in April or May. Flowers from June to August and seed set from August to October. Leaves deciduous in autumn, but roots persist for up to 3 years (Proffitt 1985).

Height Tussocks usually 0-200 mm; foliage up to a further 450 mm; flowers to 1200 mm.

Seedling RGR No data. SLA 21.6. DMC 31.4.

Nuclear DNA amount 4.9 pg; $2n = 36^*$ (*BC*); $4x^*$.

Established strategy Stress-tolerant competitor.

Regenerative strategies V, ?B$_s$. See 'Synopsis'; ?SB II-IV.

Flowers Green or purplish, hermaphrodite, wind-pollinated; >100 in an open or dense panicle (1-4 per spikelet).

Germinule 0.53 mg, 1.9 × 0.8 mm, caryopsis dispersed with an attached lemma (4.1 × 1.0 mm).

Germination Epigeal; t_{50} 2 days; 19-39°C; dormancy broken by chilling; L = D.

Biological flora Taylor, Rowland and Jones (2001).

Geographical distribution The British Isles, especially in the N and W (100% of British vice-counties), *ABIF* page 800; Europe except for some islands, but mainly on mountains in the S (77% of European territories); SW and N Asia; introduced into N America.

Habitats Recorded from a range of wet, acidic habitats, usually on peaty soils. These include moist grazed or ungrazed grassland, moorland, mire and river banks. There are also isolated records for moist woodland and walls beside water. Also observed in calcareous mire and calcareous grassland. [Often abundant on wet acidic soils in montane areas. Common on limestone pavement in W Ireland (Grime 1963b).]

Gregariousness Intermediate to patch-forming.

Altitude Found over a wide range of values, but suitable sites are more frequent at higher altitudes. [Up to 920 m.]

Slope Occurs on flat ground and on gentle to moderate slopes.

Aspect In drier habitats *M.c.* tends to be associated with N-facing slopes.

Hydrology Found in soligenous mire and on moist ground at the edge of ombrogenous and, to a lesser extent, topogenous mire.

Soil pH Commonly recorded on acidic soils of pH <4.0, with a decrease in frequency as pH rises. However, additional sampling reveals that the distribution is bimodal, with another peak at pH >7.0 [a general feature of the British distribution of *M.c.* (B. D. Wheeler pers. comm.)].

Bare soil Produces an abundance of persistent litter, and consequently sites tend to contain little bare soil.

△ **ordination** Distribution centred on undisturbed vegetation types with low productivity, but *M.c.* also shows some capacity to persist in disturbed sites.

Similarity in habitats
Insufficient data.

Habitats

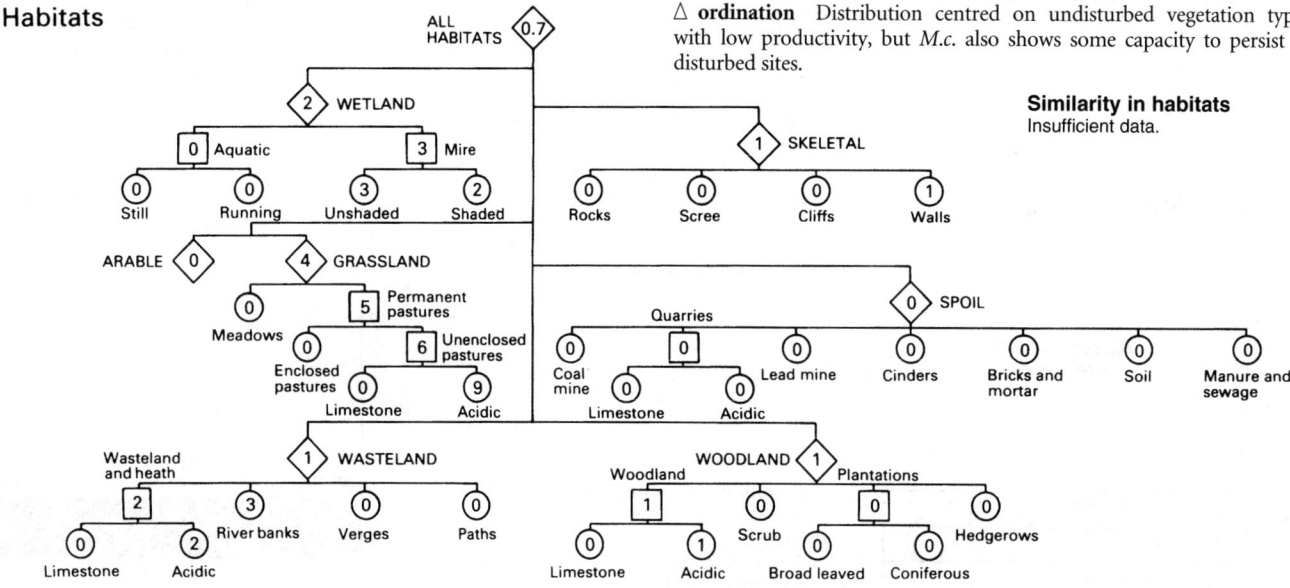

Associated floristic diversity In this survey *M.c.* is associated with vegetation of low species diversity, but in calcareous mire values of up to 35 species m^{-2} have been recorded.

Synopsis *M.c.* is a tufted, or turf-forming, deciduous grass which, like *Festuca ovina*, has a bimodal pH distribution with peaks of abundance on both highly acidic and calcareous soils. At both high and low pH *M.c.* tends to be associated either with moist grassland or with soligenous mire possessing a well-oxygenated soil profile (Armstrong & Boatman 1967), but it is tolerant of ferrous ion (R. E. D. Cooke unpubl.). The lack of tolerance of waterlogging may in part stem from the deep root system (down to at least 800 mm (Webster 1962)) and in areas with a relatively sharp transition between wetland and dryland *M.c.* often occupies the transition zone. In drier moorland habitats *M.c.* gives way to *Nardus stricta*, whereas in waterlogged habitats it is replaced by species such as *Juncus effusus* and *Eriophorum vaginatum* (Jefferies 1916). *M.c.* is tolerant of grazing and burning, and is particularly important in acidic upland areas where the species has a grazing value intermediate between that of *Agrostis-Festuca* grassland and impoverished *Nardus* grassland (McVean 1952). In *Molinia*-dominated communities there are high rates of nutrient turnover (Loach 1968). This contrasts with the behaviour of *Eriophorum vaginatum*, which exploits impoverished soils and appears to show more-efficient mineral retention and recycling (see *E.v.*). *M.c.* litter was formerly harvested for animal bedding (Godwin 1929). The species persists in moderate shade, but tends to produce few flowers under these conditions. Leaves are deciduous with an abscission zone, but new growth, fuelled by below-ground carbohydrate reserves, begins in April or May. *M.c.* is vulnerable to damage in cold springs, particularly in upland areas (Jefferies 1915). In Derbyshire *M.c.* has a tussock growth form in moorland areas, but forms more-uniform carpets in moist calcareous grassland and in soligenous mire with a pronounced lateral water flow. This behaviour is consistent with the findings of Rutter (1955), who suggested that the height of tussocks is greatest when both the growth of *M.c.* and the rate of decomposition of its dead remains is slow. He equates this combination with the conditions associated with a fluctuating water table. *M.c.* regenerates *in situ* through lateral vegetative spread, particularly in habitats favouring the non-tufted growth form. However, tussocks may include more than one genotype (Proffitt 1985). Break-up of patches of *M.c.* is generally associated with patch degeneration rather than with regeneration, and is often accompanied by invasions by other species, e.g. bryophytes and *Deschampsia flexuosa*, on top of the tussocks (Jefferies 1915). *M.c.* produces much seed which germinates in spring and colonizes bare ground. A persistent seed bank has been recorded (Chippindale & Milton 1934, Proffitt 1985). In common with various grasses and sedges of wet northerly habitats (e.g. *E.v.*) there is a high-temperature requirement for seed germination (Grime *et al.* 1981). This characteristic is likely to reduce the risk of excessive damage to seedlings by spring frosts. Taller plants with larger caryopses and $2n = ?36, 90 (4x, 10x)$ have been separated as ssp. *arundinacea* (Schrank) H. Paul, but many intermediates occur (*FE5*). Ssp. *a.* is recorded from rather base-rich mineral soils with a fluctuating watertable (*FE5*) and has (as *M. litoralis* Host) been recorded from S England and Wales (Hubbard 1984).

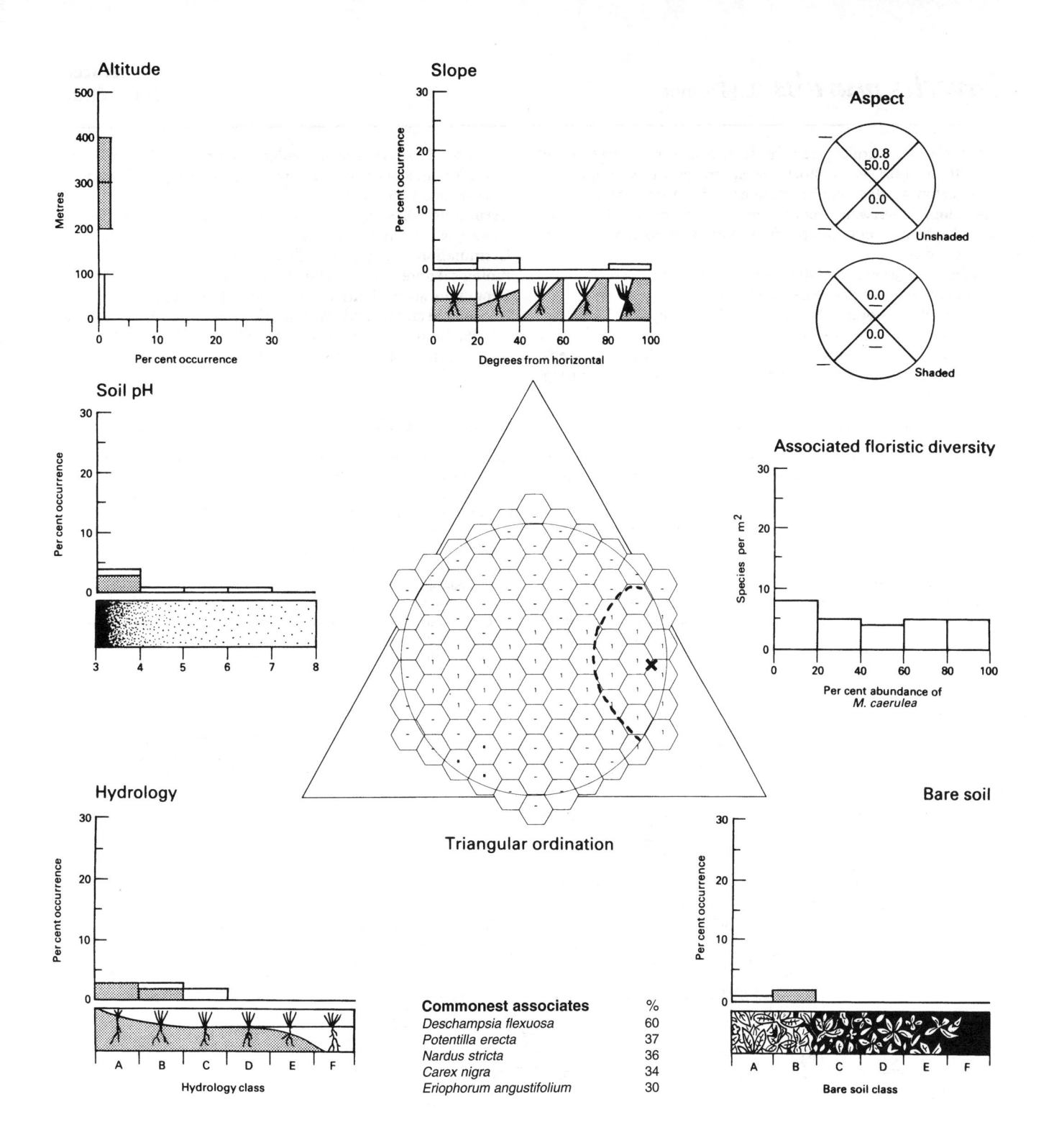

Current trends Although no longer of agricultural importance *M.c.* continues to be widespread and abundant in upland regions. The species is declining in many lowland areas as a result of habitat destruction.

Mycelis muralis (L.) Dumort.

Asteraceae
Wall Lettuce

Life-form Polycarpic perennial, semi-rosette hemicryptophyte with a short erect stock. Stem erect; leaves pinnatifid, sometimes >4000 mm^2; roots often with VA mycorrhizas.
Phenology Overwintering as a small rosette. Flowers from July to September and sets seed from July to October, after which shoot dies.
Height Foliage mainly <400 mm; flowers to 1 m.
Seedling RGR No data. SLA. DMC. No Data.
Nuclear DNA amount No data; $2n = 18^*$ (*BC*); $2x^*$.
Established strategy C-S-R.
Regenerative strategies W. Produces wind-dispersed seeds which germinate in vegetation gaps probably mainly during autumn. Seed bank studies are needed; ?SB I.

Flowers Florets yellow, ligulate, hermaphrodite; *c.* 5 per capitulum and often >50 capitula in an open panicle. Mode of pollination uncertain.
Germinule 0.34 mg, 3.5 × 0.9 mm, achene dispersed with a pappus *c.* 5.5 mm in length.
Germination Epigeal; t_{50} 4 days; 10-24°C; L > D.
Biological flora Clabby and Osborne (1999).
Geographical distribution The British Isles, but less common in Scotland and Ireland (62% of British vice-counties), *ABIF* page 626; Europe northwards to Scandinavia (77% of European territories and 3% naturalized); NW Africa.

Habitats Typically a plant of moist, shaded, skeletal habitats on calcareous soils (cliffs, rock outcrops and walls). Also frequent at the edge of woodland, particularly where associated with stony ground, and occurs in open rocky habitats (quarry spoil and cinders). Occasionally observed on road verges and wasteland, and recorded only as a transient in limestone pasture and shaded marsh. Excluded from aquatic and arable habitats.
Gregariousness Sparse.
Altitude Occurs at altitudes of up to 400 m. [Up to 460 m.]
Slope Associated with a wide range of values. The bias towards steep slopes reflects the frequent occurrence of *M.m.* in skeletal habitats.

Aspect Slightly more frequent and abundant on shaded N-facing slopes.
Hydrology Not a wetland species and virtually excluded from waterlogged soils.
Soil pH Associated with base-rich soils, particularly those above pH 6.0.
Bare soil Frequently low. The majority of sites have an incomplete vegetation cover and large exposures of bare rock, with the species rooted in crevices.
△ **ordination** The diffuse pattern reflects the strong association of *M.m.* with assemblages of other wind-dispersed species which are often heterogeneous in established strategy and frequently include casual non-persistent individuals.
Associated floristic diversity Typically intermediate.

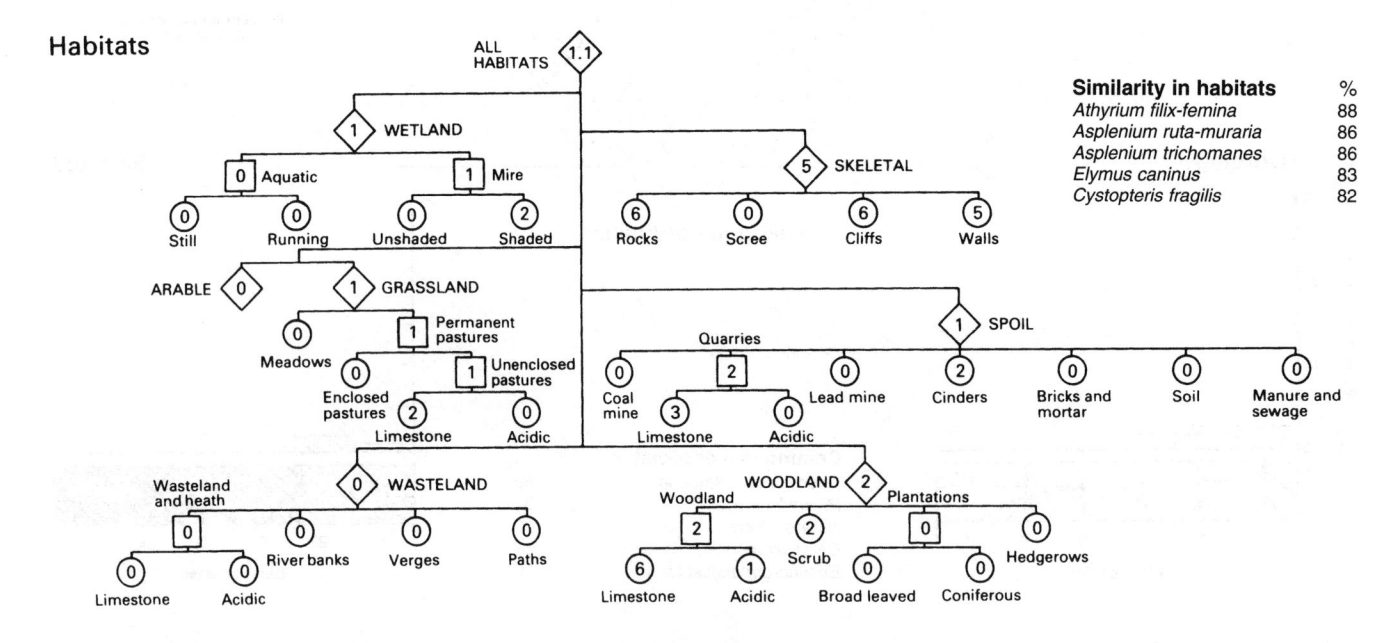

Habitats

Similarity in habitats	%
Athyrium filix-femina	88
Asplenium ruta-muraria	86
Asplenium trichomanes	86
Elymus caninus	83
Cystopteris fragilis	82

Synopsis *M.m.* is a tall winter-green herb. In the Sheffield flora four of the five species most similar in habitat range are ferns. This highlights the distinctive ecology of the species, which arises from the unusual combination (among British herbs) of shade tolerance with effective seed dispersal by wind. These two attributes undoubtedly confer a characteristic ability to exploit shaded but relatively inaccessible situations on cliffs, walls and rock outcrops. Despite its tolerance of shade, *M.m.* is not a common constituent of woodland herb layers; this may be related to sensitivity to dominance by more-robust species and submergence of the basal rosette by tree litter. The species is highly calcicolous within the Sheffield flora, occurring mainly on limestone rocks or mortared walls. *M.m.* is a drought-avoiding species whose roots penetrate fissures in the rock or stonework. Seed is produced throughout late summer and autumn, with only five seeds in each capitulum, a remarkably low number for a member of the Asteraceae. The species lacks effective vegetative spread and is restricted to small rocky or disturbed microsites which are large enough to sustain the established plant but too small to allow the vegetative expansion of more-dominant species. It is not known whether the species forms a persistent seed bank.
Current trends Uncertain, but not under threat. A frequent associate of artificial habitats, but some populations, particularly on old walls, are decreasing in abundance.

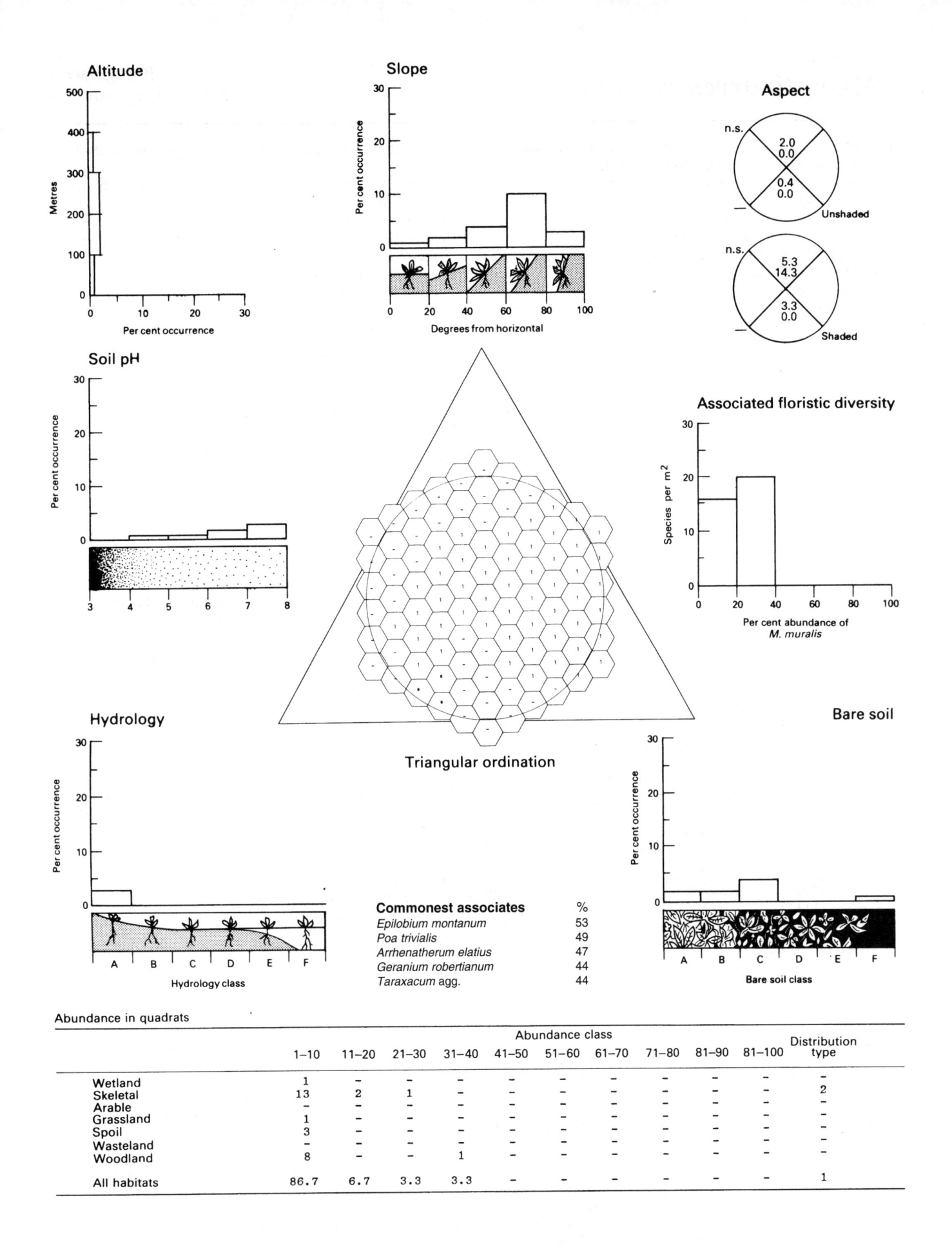

Altitude

Slope

Aspect

	n.s.	
—	2.0 / 0.0	Unshaded
	0.4 / 0.0	

	n.s.	
—	5.3 / 14.3	Shaded
	3.3 / 0.0	

Soil pH

Associated floristic diversity

Triangular ordination

Hydrology

Bare soil

Commonest associates	%
Epilobium montanum	53
Poa trivialis	49
Arrhenatherum elatius	47
Geranium robertianum	44
Taraxacum agg.	44

Abundance in quadrats

	1–10	11–20	21–30	31–40	41–50	51–60	61–70	71–80	81–90	81–100	Distribution type
Wetland	1	–	–	–	–	–	–	–	–	–	2
Skeletal	13	2	1	–	–	–	–	–	–	–	
Arable	–	–	–	–	–	–	–	–	–	–	–
Grassland	1	–	–	–	–	–	–	–	–	–	–
Spoil	3	–	–	–	–	–	–	–	–	–	–
Wasteland	–	–	–	–	–	–	–	–	–	–	–
Woodland	8	–	–	1	–	–	–	–	–	–	–
All habitats	86.7	6.7	3.3	3.3	–	–	–	–	–	–	1

+*Myosotis arvensis* (L.) Hill

Boraginaceae
Common Forget-me-not

Life-form A semi-rosette, winter- or perhaps occasionally summer-annual therophyte. Stem erect; leaves ovate, hairy, usually <1000 mm²; roots non-mycorrhizal.
Phenology Overwintering as a small rosette. Flowers mainly from April to July and sets seed from May onwards.
Height Foliage to *c.* 200 mm; flowers to 300 mm.
Seedling RGR No data. SLA 26.7. DMC 12.3.
Nuclear DNA amount No data; $2n = 52$ (*FE3*); $4x$.
Established strategy Ruderal.
Regenerative strategies S, B_s. Exclusively by seed which germinates in autumn. *M.a.* forms a persistent seed bank (Brenchley & Warington 1930; Wesson & Wareing 1969); SB III.
Flowers Blue, hermaphrodite, usually self-pollinated; >10 in each cyme.
Germinule 0.29 mg, 1.8×1.0 mm, nutlet (4 per ovary).
Germination Epigeal; t_{50} 3 days; <5–27°C; germination rate increased after dry storage; L > D.
Biological flora None.
Geographical distribution The British Isles (100% of British vice-counties), *ABIF* page 505; Europe (92% of European territories); W Asia.

Habitats Recorded mainly as weed of arable land, particularly in cereal crops. Also found in disturbed situations such as soil heaps, rocky areas associated with limestone grassland and waste ground. Observed on rock outcrops, scree, walls and also rarely at woodland margins, in sand and gravel pits, and on disturbed roadsides. Absent from wetlands.
Gregariousness Sparse.
Altitude Most frequent in lowland arable areas but extends to 350 m. [Up to 560 m.]
Slope Mainly restricted to flat ground and gentle slopes.
Aspect More frequent on unshaded S-facing slopes.
Hydrology Absent from wetlands.
Soil pH Absent from acidic soils of pH <5.0.
Bare soil Mainly associated with high exposures of soil.
△ ordination Distribution centred on productive, disturbed vegetation types. Also occurs in disturbed microhabitats within otherwise relatively undisturbed vegetation.
Associated floristic diversity Typically intermediate.

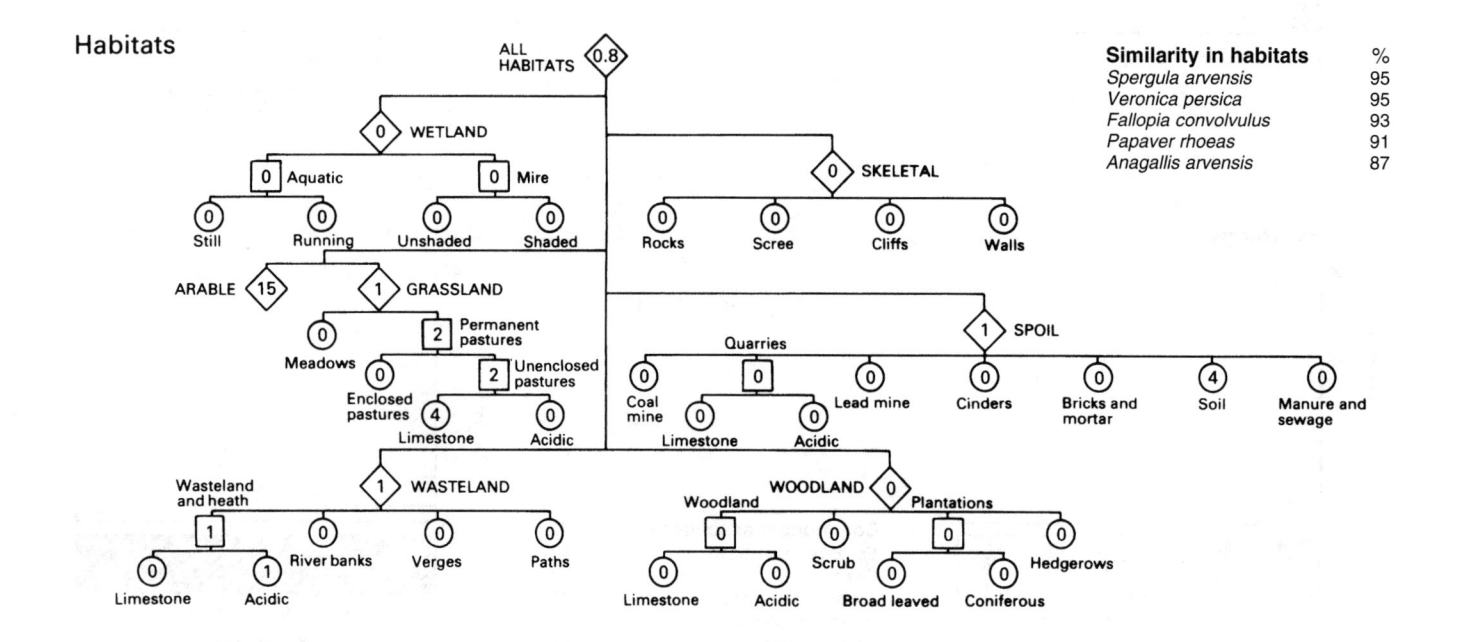

Habitats

Similarity in habitats	%
Spergula arvensis	95
Veronica persica	95
Fallopia convolvulus	93
Papaver rhoeas	91
Anagallis arvensis	87

Synopsis *M.a.* is a relatively tall, annual herb which occurs in disturbed artificial habitats and occupies a wide range of soil types. The species tends to occur on relatively shallow calcareous soils or dry sandy soils in sites where, as a result of summer drought and often also some physical disturbance of the soil, the dominance of polycarpic perennials is incomplete. *M.a.* usually occurs as scattered individuals, and tends to occupy sites which have both deeper soils and are more transient than those of the majority of small winter annuals. However, the species does normally behave as a winter-annual, and germination of the seeds, which are shed in summer, is usually delayed until autumn. Other transient habitats exploited by *M.a.* include waste ground and woodland clearings. The species appears to be highly mobile. The seeds can survive ingestion by cattle and horses, and may be dispersed also by humans or in the fur of animals, enclosed in the hispid spiky calyx (Ridley 1930). Seeds are also incorporated into a persistent seed bank (Brenchley & Warington 1930, Wesson & Wareing 1969), which is particularly important in the main habitat, arable land. Here the species may behave to a limited extent as a summer annual, but *M.a.* is more frequent in winter wheat and winter barley than in spring barley (Chancellor & Froud-Williams 1984). Responses to agricultural management have been little studied, but *M.a.* has been observed to re-sprout and flower following defoliation. *M.a.* varies greatly in habit according to environment (*FE3*), but the role of ecotypic differentiation in determining the wide ecological amplitude of *M.a.* is uncertain. It seems likely that local populations belong to var. *arvensis* ($2n = 52$) rather than the larger fruited var. *sylvestris* Schitdl. with $2n = 66$.

Current trends Uncertain and strongly dependent upon developments in agricultural practice.

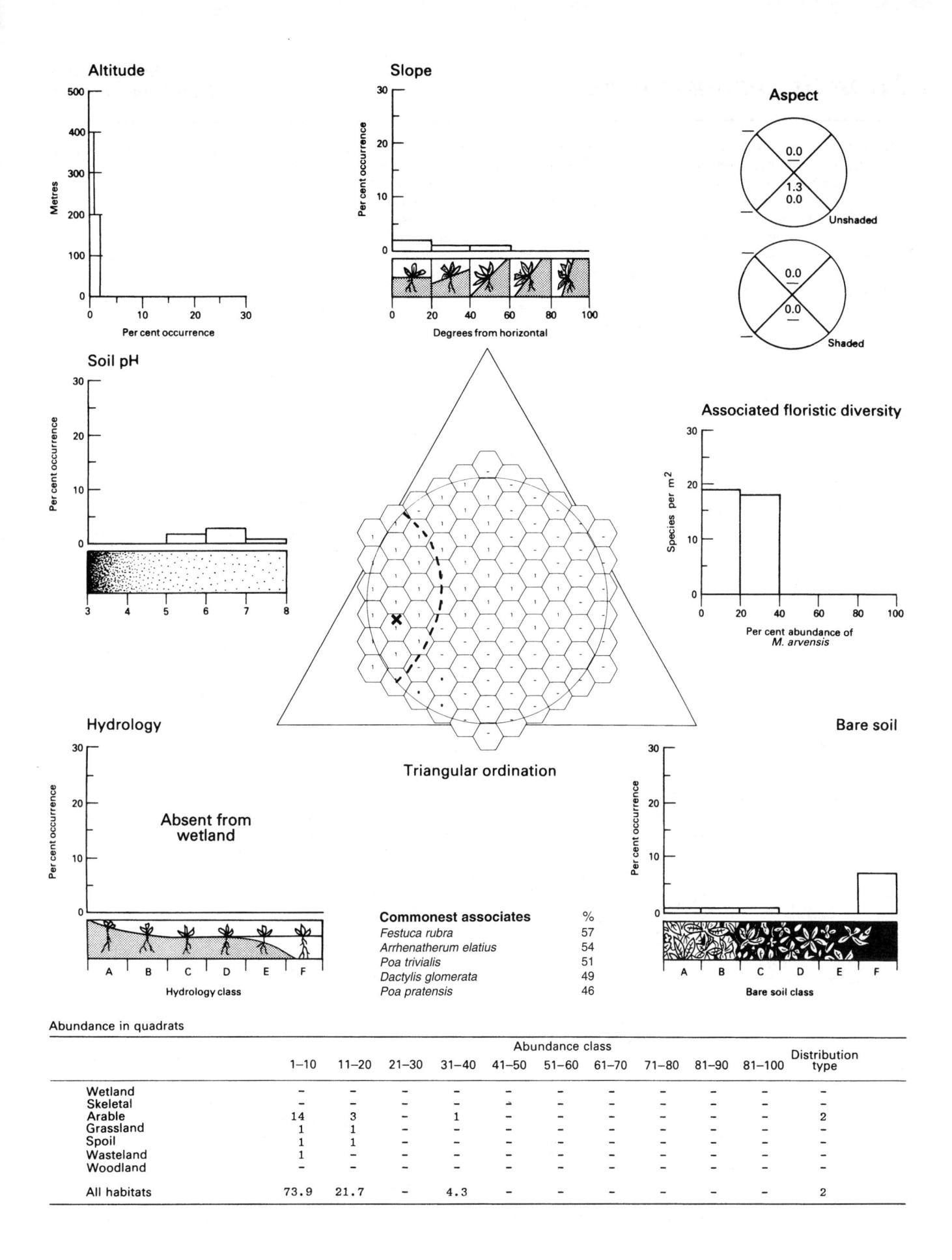

Altitude

Slope

Aspect

Unshaded
0.0 | 0.0
1.3 | 0.0

Shaded
0.0 | 0.0
— | —

Soil pH

Associated floristic diversity

Hydrology

Absent from wetland

Triangular ordination

Bare soil

Commonest associates	%
Festuca rubra	57
Arrhenatherum elatius	54
Poa trivialis	51
Dactylis glomerata	49
Poa pratensis	46

Abundance in quadrats

	1–10	11–20	21–30	31–40	41–50	51–60	61–70	71–80	81–90	81–100	Distribution type
Wetland	–	–	–	–	–	–	–	–	–	–	–
Skeletal	–	–	–	–	–	–	–	–	–	–	–
Arable	14	3	–	1	–	–	–	–	–	–	2
Grassland	1	1	–	–	–	–	–	–	–	–	–
Spoil	1	1	–	–	–	–	–	–	–	–	–
Wasteland	1	–	–	–	–	–	–	–	–	–	–
Woodland	–	–	–	–	–	–	–	–	–	–	–
All habitats	73.9	21.7	–	4.3	–	–	–	–	–	–	2

Myosotis ramosissima Rochel

Boraginaceae
Early Forget-me-not

Life-form Semi-rosette, winter-annual therophyte. Stems branched, ascending; leaves lanceolate, hairy, <500 mm^2; roots sometimes with VA mycorrhizas.
Phenology Winter annual. Flowers from April to June and sets seed from May to July.
Height Foliage mainly <30 mm; flowers to 250 mm.
Seedling RGR No data. SLA 26.9. DMC 14.7.
Nuclear DNA amount No data; $2n = 48$ (*FE3*); $4x$.
Established strategy Stress-tolerant ruderal.
Regenerative strategies S, ?B$_s$. Exclusively by seed, which germinates in vegetation gaps during autumn. The existence of a persistent seed bank is predicted; ?SB III.

Flowers Blue, hermaphrodite, perhaps usually self-pollinated; sometimes >20 in each cyme.
Germinule 0.11 mg, 1.0×0.8 mm, nutlet (4 per ovary).
Germination Epigeal; t_{50} 5 days; germination rate greatly increased by dry storage; L = D.
Biological flora None.
Geographical distribution The British Isles, but absent from most of Wales, NW England, N Scotland and W Ireland (74% of British vice-counties), *ABIF* page 505; most of Europe except the extreme N (87% of European territories); SW Asia and N Africa.

Habitats Exclusively on dry rocky or sandy soils. Frequent on rock outcrops and, to a lesser extent, on stony pasture or wasteland. Also found on quarry spoil and cinders, with one record atypically from a demolition site. Also observed widely on bare sandy soils, particularly in sand pits and beside roads and railways. [Grows with other winter annuals on sand dunes.]
Gregariousness Often loosely clumped in small patches.
Altitude Recorded over a wide range of altitude. The peak in distribution between 200 and 300 m corresponds to occurrences on the carboniferous limestone. [Up to 430 m.]
Slope Wide-ranging.
Aspect Slightly more frequent on S-facing slopes, a trend which is even more pronounced in a more detailed recent study of the species (Hodgson unpubl.).
Hydrology Absent from wetlands.
Soil pH Rarely observed below pH 5.5.
Bare soil Recorded only from sites which have high amounts of bare rock and little exposed soil. However, on sandy soils the species is frequently associated with large exposures.
△ **ordination** Distribution centred on vegetation associated with infertile habitats subject to some disturbance. Absent from the most highly disturbed areas and from undisturbed fertile sites.
Associated floristic diversity Typically associated with species-rich vegetation.

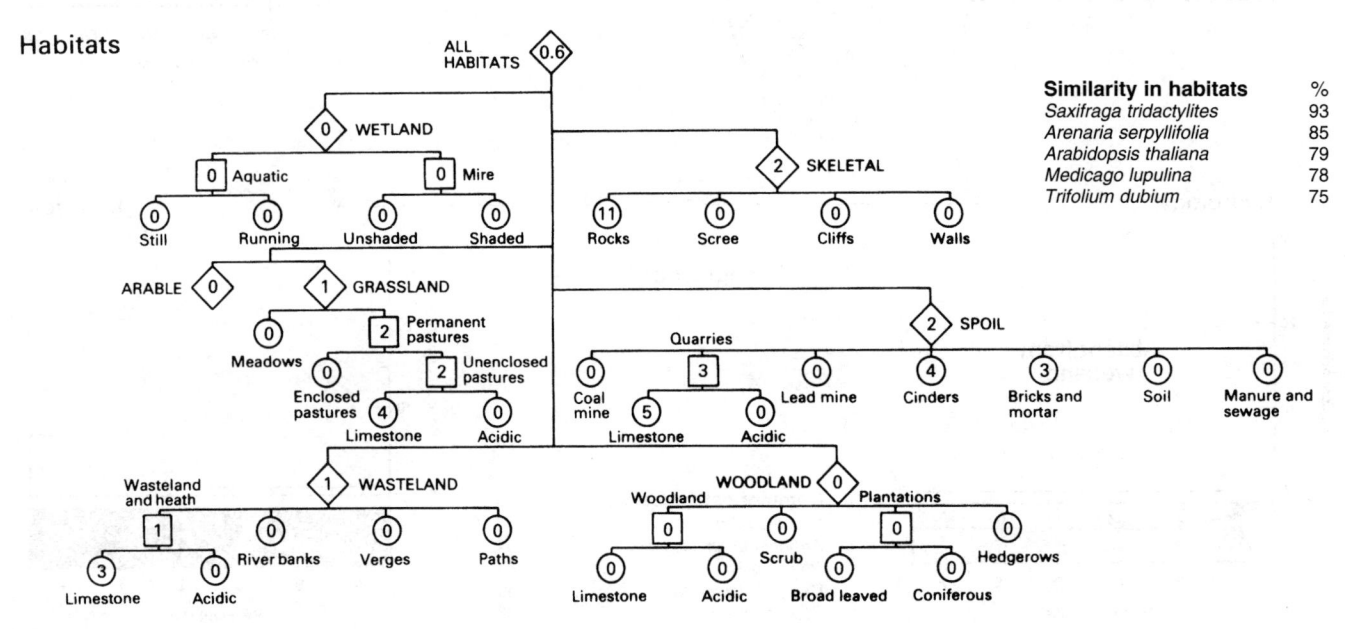

Habitats

Similarity in habitats	%
Saxifraga tridactylites	93
Arenaria serpyllifolia	85
Arabidopsis thaliana	79
Medicago lupulina	78
Trifolium dubium	75

Synopsis *M.r.* is a small, winter-annual herb restricted to shallow organic soils on limestone, or to dry sandy soils, where, in both, the cover of perennials is restricted by summer drought. *M.r.* may also occur on ant hills. Seeds are released during summer, and immediate germination is prevented by an after-ripening requirement. Seeds germinate in autumn and plants, which overwinter as rosettes, require vernalization before flowering (Krekule & Hajkova 1972). The flowering period commences in early spring and continues until the onset of drought. *M.r.* is much more common in long-established, winter-annual communities than in fugacious sites. The species shows a similar restriction to more-predictably open sites in Holland, where its complex germination biology is regarded as more conservative and less opportunistic than that of *Veronica arvensis* (Janssen 1973). The existence of a persistent seed bank is suspected. Common species of similar ecology in Britain are *Catapodium rigidum*, *Erophila verna* (L.) DC. and *Saxifraga tridactylites*. The former is later-flowering and is presumably associated with less drought-prone sites. *E.v.* is earlier-flowering. In addition, *E.v.* and (particularly) *S.t.*, produce numerous small seeds, while *M.r.* forms fewer (often <100), larger seeds. The reasons for the restriction of *M.r.* to predictably open sites are unknown. However, there is no evidence implicating low mobility and one local colony on railway cinders in N Derbyshire lies at least 10 km from the next known site. Populations from the limestone strata of N England appear to refer to ssp. *ramosissima*. In sandy places, especially near the sea in S England, the species is represented by ssp. *globularis* (Samp.) Grau (*FE3*).
Current trends As a colonist of quarries, sand pits and railway banks, *M.r.* has probably increased in the recent past. However, current trends in land management are destroying many of its sites, and the species appears to be decreasing.

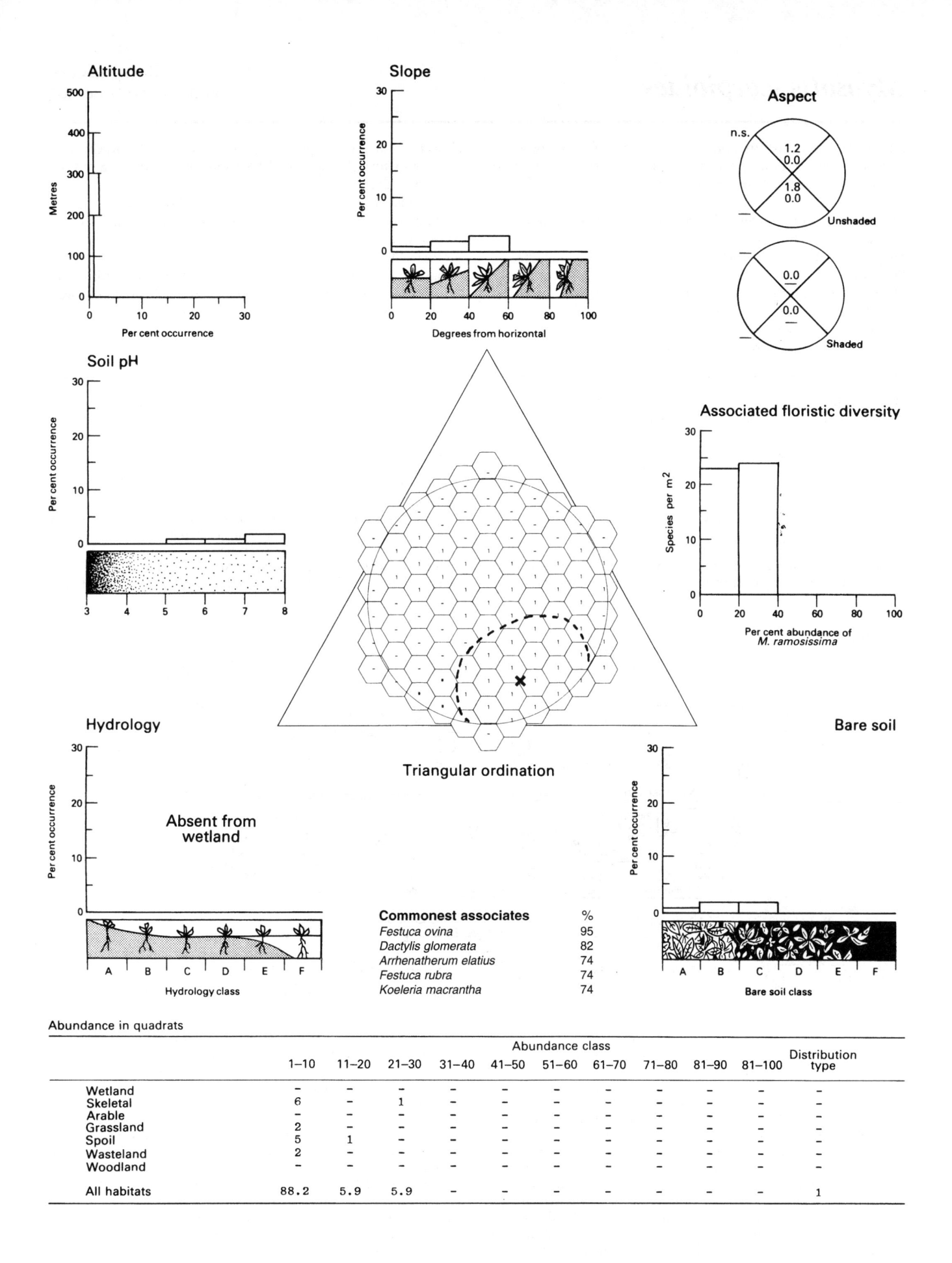

Altitude

Slope

Aspect

n.s.
1.2 0.0
1.8 0.0
Unshaded

0.0
0.0
Shaded

Soil pH

Associated floristic diversity

Per cent abundance of *M. ramosissima*

Triangular ordination

Commonest associates	%
Festuca ovina	95
Dactylis glomerata	82
Arrhenatherum elatius	74
Festuca rubra	74
Koeleria macrantha	74

Hydrology

Absent from wetland

Bare soil

Abundance in quadrats

	\multicolumn{10}{c	}{Abundance class}	Distribution type
	1–10	11–20	21–30	31–40	41–50	51–60	61–70	71–80	81–90	81–100	
Wetland	–	–	–	–	–	–	–	–	–	–	–
Skeletal	6	–	1	–	–	–	–	–	–	–	–
Arable	–	–	–	–	–	–	–	–	–	–	–
Grassland	2	–	–	–	–	–	–	–	–	–	–
Spoil	5	1	–	–	–	–	–	–	–	–	–
Wasteland	2	–	–	–	–	–	–	–	–	–	–
Woodland	–	–	–	–	–	–	–	–	–	–	–
All habitats	88.2	5.9	5.9	–	–	–	–	–	–	–	1

Myosotis scorpioides L.

Boraginaceae
Water Forget-me-not

Life-form Rhizomatous, polycarpic perennial helophyte or semi-rosette hemicryptophyte with creeping runners. Stems erect; leaves lanceolate, hairy, <1000 mm^2; roots often with VA mycorrhizas.
Phenology Winter green. Flowers May to September and sets seed from August to October.
Height Foliage mainly <200 mm; flowering stems to 450 mm.
Seedling RGR No data. SLA 34.0. DMC 13.8.
Nuclear DNA amount 2.7 pg; $2n = 66$ (*FE3*); $6x$.
Established strategy Competitive-ruderal.
Regenerative strategies V, ?B$_s$. Regenerates vegetatively to form clonal patches and by seed (see 'Synopsis'); ?SB III.

Flowers Blue, hermaphrodite, insect-pollinated, largely self-incompatible (Varapoulos 1979); sometimes >100 in a cymose inflorescence.
Germinule 0.28 mg, 2.0 ×1.2 mm, nutlet (4 per ovary).
Germination Epigeal; t_{50} 10 days; dormancy broken by dry storage at 20°C; L > D.
Biological flora None.
Geographical distribution The British Isles (97% of British vice-counties), *ABIF* page 502; C and N Europe (69% of European territories); N Africa, Asia and N America.

Habitats Exclusively a wetland species, equally abundant in shaded and unshaded mire. Less frequent on river banks or as an aquatic at the edge of still or flowing water. There are also isolated records from a wall and a soil heap, both beside water.
Gregariousness Intermediate to stand-forming.
Altitude Observed up to 260 m, but suitable habitats are more frequent in the lowlands. [Up to 460 m.]
Slope As with most wetland species, largely confined to flat or gently sloping sites.
Aspect Insufficient data.
Hydrology Most frequent and abundant in mire on level ground or close to open water, indicating an association with topogenous mire adjacent to open water. Also found in soligenous mire, and forms floating mats at the edge of still or flowing water. Not recorded from water depths >450 mm.
Soil pH Mainly restricted to soils of pH >5.0.
Bare soil Found over a wide range of values, but most frequent at high exposures.
△ **ordination** Occurs in vegetation types which are relatively productive and subject to some disturbance; distribution centred on vegetation subject to moderate disturbance, but also occurs in some relatively undisturbed vegetation types. Absent from both heavily disturbed situations and from unproductive vegetation.
Associated floristic diversity Typically low to intermediate.
Synopsis *M.s.* is a low-growing, winter-green herb of fertile mire close to

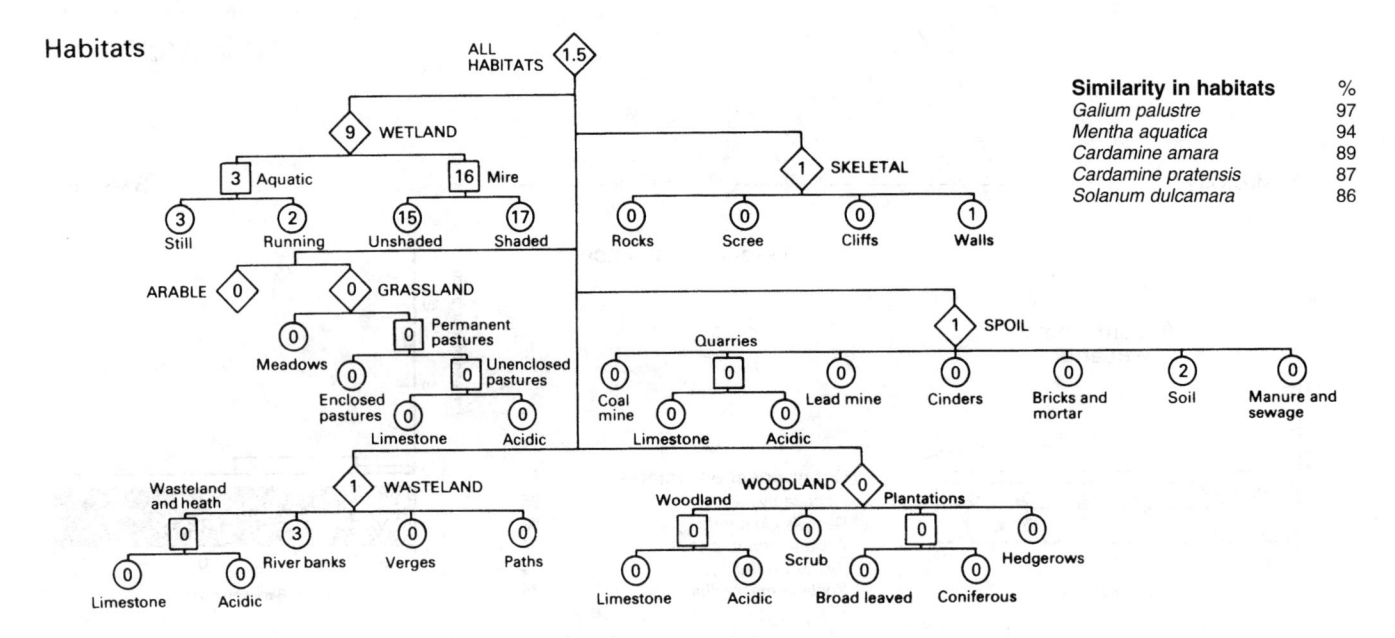

Habitats

Similarity in habitats	%
Galium palustre	97
Mentha aquatica	94
Cardamine amara	89
Cardamine pratensis	87
Solanum dulcamara	86

open water, usually in situations where the growth of potential dominants is restricted by disturbance associated with winter flooding. Thus, *M.s.* is characteristic of ditch, pond and river margins, and may sometimes form floating rafts. The species is perhaps more tolerant of a fluctuating water table and of submergence than the ecologically similar but smaller-seeded *Veronica beccabunga*. *M.s.* occurs in shaded sites, and may also persist at the edges of stands of taller dominant species. *M.s.* forms clonal patches by means of rhizomes and runners. Older stems break down to leave isolated daughter plants. Shoot fragments washed away by flooding readily regenerate to form new colonies, and may be important in colonizing water systems. Reproduction by seed which germinate in vegetation gaps during spring is probably of lesser importance in established colonies. However, the seeds do not float (Ridley 1930), and we suspect they are incorporated into a persistent seed bank. The germination requirements of *M.s.* have been studied little and there is a need to ascertain whether, as in many other wetland species of similar ecology, germination is promoted by fluctuating temperatures. The extent of ecotypic differentiation in this out-breeding species is uncertain, but the progeny from selfed plants may show in-breeding depression (Varapoulos 1979). *M.s.* has a wide edaphic range, extending from calcareous to mildly acidic soils. However, beside nutrient- and base-poor water, generally in upland areas, *M.s.* is often replaced by the morphologically similar polycarpic perennial, *M. secunda* A. Murray ($2n = 24, 48$; $2x, 4x$ (*FE3*)). The annual, *M. laxa* ssp. *cespitosa* (C. F. Schultz) Hyl ex Hordh ($2n = 22$, $2x$ (Stace 1975)) replaces *M.s.* in annual communities on bare mud, and the hybrid between these two species is perennial and vegetatively vigorous (Stace 1975).
Current trends Tolerant of eutrophic waters, and relatively mobile. Likely to remain common.

Altitude
Slope
Aspect

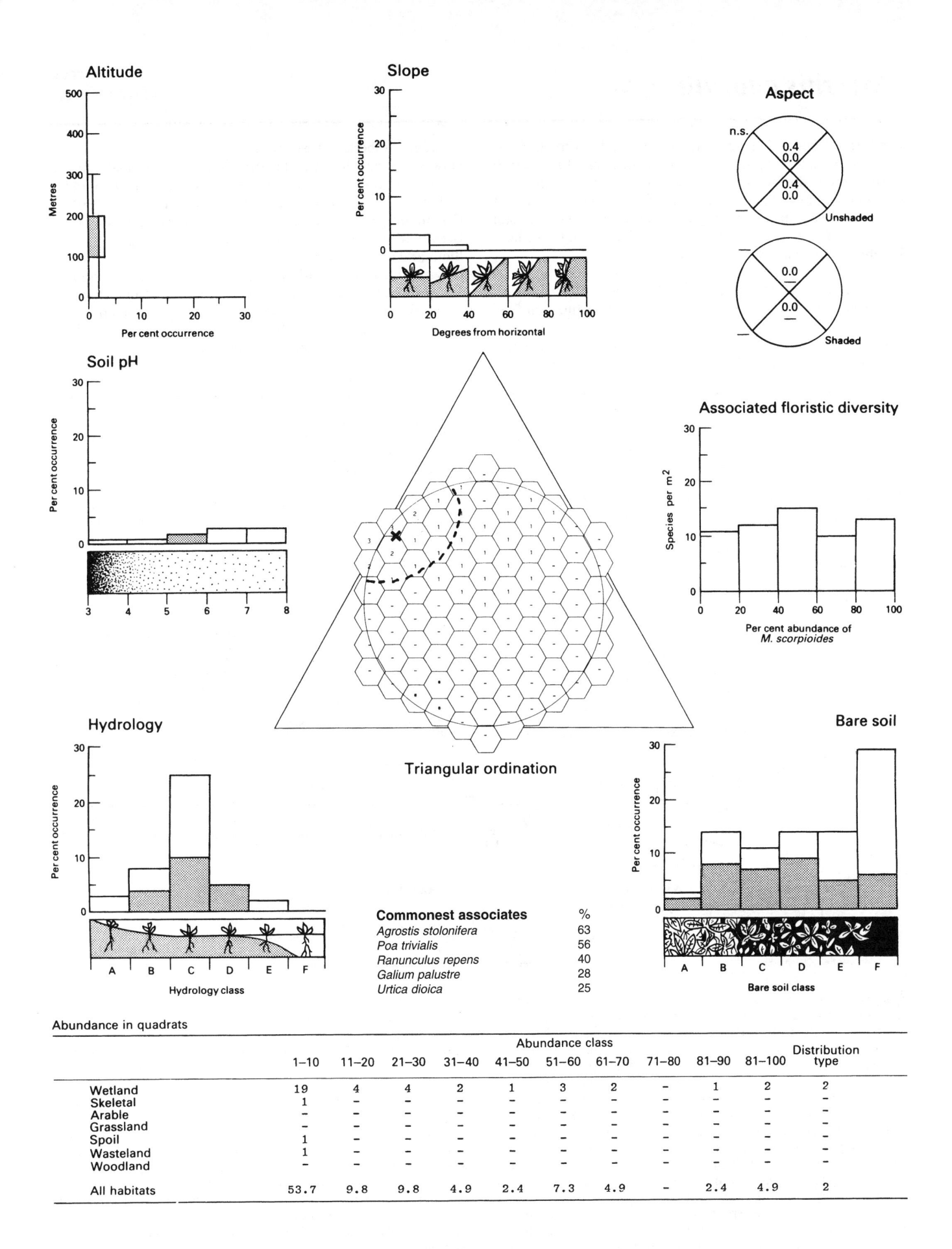

Soil pH

Associated floristic diversity

Species per m^2

Per cent abundance of *M. scorpioides*

Hydrology

Triangular ordination

Bare soil

Commonest associates	%
Agrostis stolonifera	63
Poa trivialis	56
Ranunculus repens	40
Galium palustre	28
Urtica dioica	25

Abundance in quadrats

	\multicolumn{10}{c}{Abundance class}	Distribution type
	1–10	11–20	21–30	31–40	41–50	51–60	61–70	71–80	81–90	81–100	
Wetland	19	4	4	2	1	3	2	–	1	2	2
Skeletal	1	–	–	–	–	–	–	–	–	–	–
Arable	–	–	–	–	–	–	–	–	–	–	–
Grassland	–	–	–	–	–	–	–	–	–	–	–
Spoil	1	–	–	–	–	–	–	–	–	–	–
Wasteland	1	–	–	–	–	–	–	–	–	–	–
Woodland	–	–	–	–	–	–	–	–	–	–	–
All habitats	53.7	9.8	9.8	4.9	2.4	7.3	4.9	–	2.4	4.9	2

443

Myrrhis odorata (L.) Scop.

Apiaceae
Sweet Cicely

Life-form Polycarpic perennial, semi-rosette hemicryptophyte. Stems erect; leaves pinnatifid, hairy, sometimes >10 000 mm^2; deeply rooted with a stout tap-root and VA mycorrhizas.
Phenology Flowers in May to June before the full expansion of leaves. Seeds ripen from August onwards, and many are still present on the dead flowering stems during part of the winter.
Height Foliage mainly <1.5 m; flowers to 2 m.
Seedling RGR No data. SLA 33.8. DMC 23.8.
Nuclear DNA amount 1.7 pg; $2n = 22$ (Federov 1974); $2x$.
Established strategy Intermediate between competitor and C-S-R.
Regenerative strategies S. Regenerates by means of seed, which germinates in spring (Roberts 1986a). No persistent seed bank occurs (Roberts 1986a); SB II.
Flowers White, insect-pollinated (Tutin 1980); several hundred in an umbel. Terminal umbel bears male and hermaphrodite flowers, lateral umbel bears male flowers only (Tutin 1980).
Germinule 35.01 mg, 20.8 × 3.0 mm, mericarp (2 per fruit).
Germination Epigeal; dormancy broken by chilling.
Biological flora None.
Geographical distribution Native of the Alps, Pyrenees, Appenines and mountains of the W part of Balkan Penninsula, widely naturalized elsewhere (21% of European territories and 36% naturalized). In the British Isles largely restricted to the N (58% of British vice-counties), *ABIF* page 458.

Habitats Particularly characteristic of river banks and flood plains, both unshaded and shaded. Widespread on road verges, in hedgerows and at woodland margins. Rarely observed on spoil, wasteland and in pastures. Non-persistent seedlings have been recorded from arable land. Absent from wetlands and skeletal habitats.
Gregariousness Capable of forming stands even at low frequencies.
Altitude Observed within the range 10-350 m but shows a strong upland bias. [Up to 500 m.]
Slope All records from slopes of <30°.
Aspect Distribution concentrated in the N-facing sector.
Hydrology Absent from wetlands.
Soil pH Occurs at highest frequency and abundance within the pH range 6.0-7.0. Absent from soils of pH <4.0.
Bare soil Associated with a wide range of values.
△ ordination Occurs in a wide range of vegetation types, although absent from heavily disturbed and from unproductive vegetation. Distribution centred on vegetation associated with productive and lightly disturbed conditions.
Associated floristic diversity Low to intermediate.

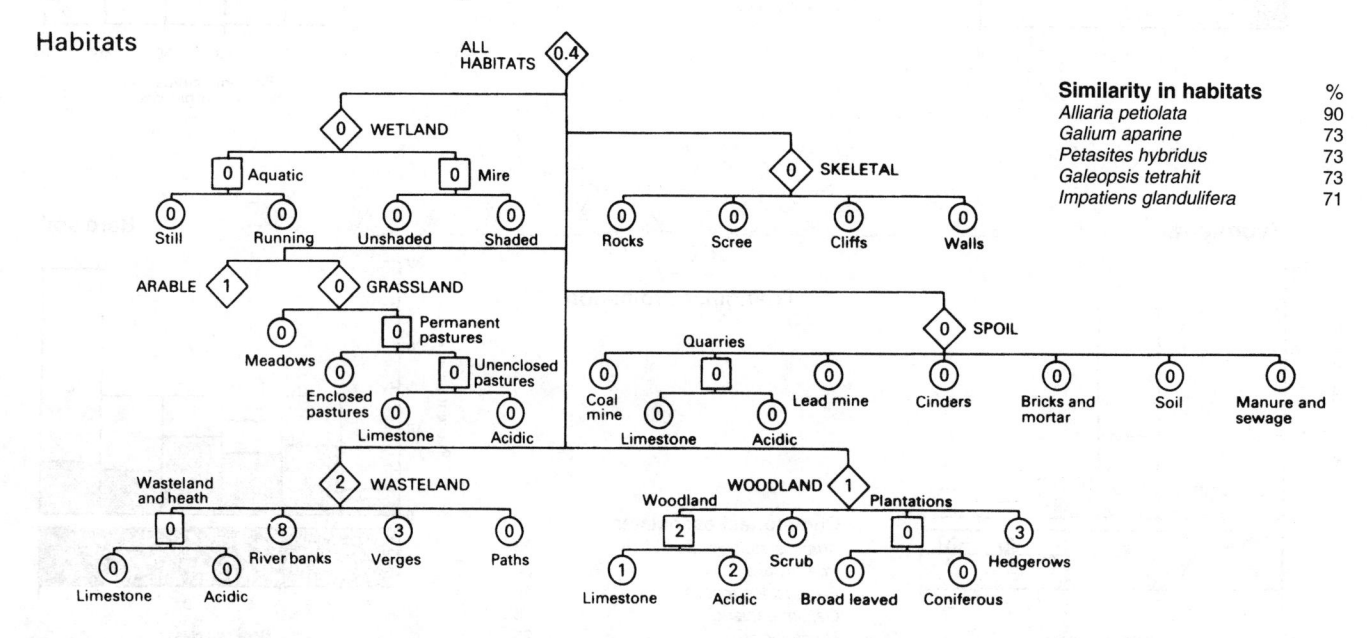

Habitats

Similarity in habitats	%
Alliaria petiolata	90
Galium aparine	73
Petasites hybridus	73
Galeopsis tetrahit	73
Impatiens glandulifera	71

Synopsis *M.o.* is a robust garden escape, smelling of aniseed and formerly cultivated as a pot-herb. The species is now thoroughly naturalized (*FE2*), particularly on shaded river banks, and has been established in semi-natural vegetation for so long that it has been regarded as possibly native (*CTW*). The species lacks the capacity for lateral vegetative spread and forms stands in moist, fertile sites only, where the growth of dominants is restricted by factors such as occasional flooding and light shade. *M.o.* is usually associated with deep mineral soils, and the foliage is characterized by high concentrations of nitrogen and phosphorus (*SIEF*). Despite its association with riversides, *M.o.* is not a wetland species, tending to occur on freely-draining soils at the tops of river banks. Where *M.o.* and *Impatiens glandulifera* occur together, *M.o.* occupies the areas further from the water's edge. On riverbanks the species benefits from the stout tap-root which affords very effective anchorage. Following floods it is not unusual to observe several centimetres of exposed root in stream-side populations. *M.o.* is frequently found on roadsides, where it survives occasional cutting (but not grazing). The ecology of the species in these sites may be similar to that of *Rumex pseudoalpinus* Hofft., also a garden escapee, which may persist near its site of introduction but shows little capacity for dispersal. In Britain *M.o.* is rare in lowland areas, a pattern matching the montane distribution in C and S Europe, where it is native. *M.o.* appears to be entirely dependent upon seed for regeneration. When shed, the seed, which is the largest in the herbaceous flora of the British Isles, has a poorly-differentiated embryo which occupies only 10% of the seed length (Lhotska 1977, J. G. Hodgson & J. M. L. Mackey unpubl.). The embryo grows while the seed is in a cold imbibed condition during the winter and, as a result, seed germination is delayed until spring. The large fruits float when dry, but when imbibed generally sink within 12 h (Lhotska 1977). Apart from an evident capacity to spread along river systems, *M.o.* is a poor colonist.
Current trends Uncertain. Does not appear immediately threatened, at least within upland Britain.

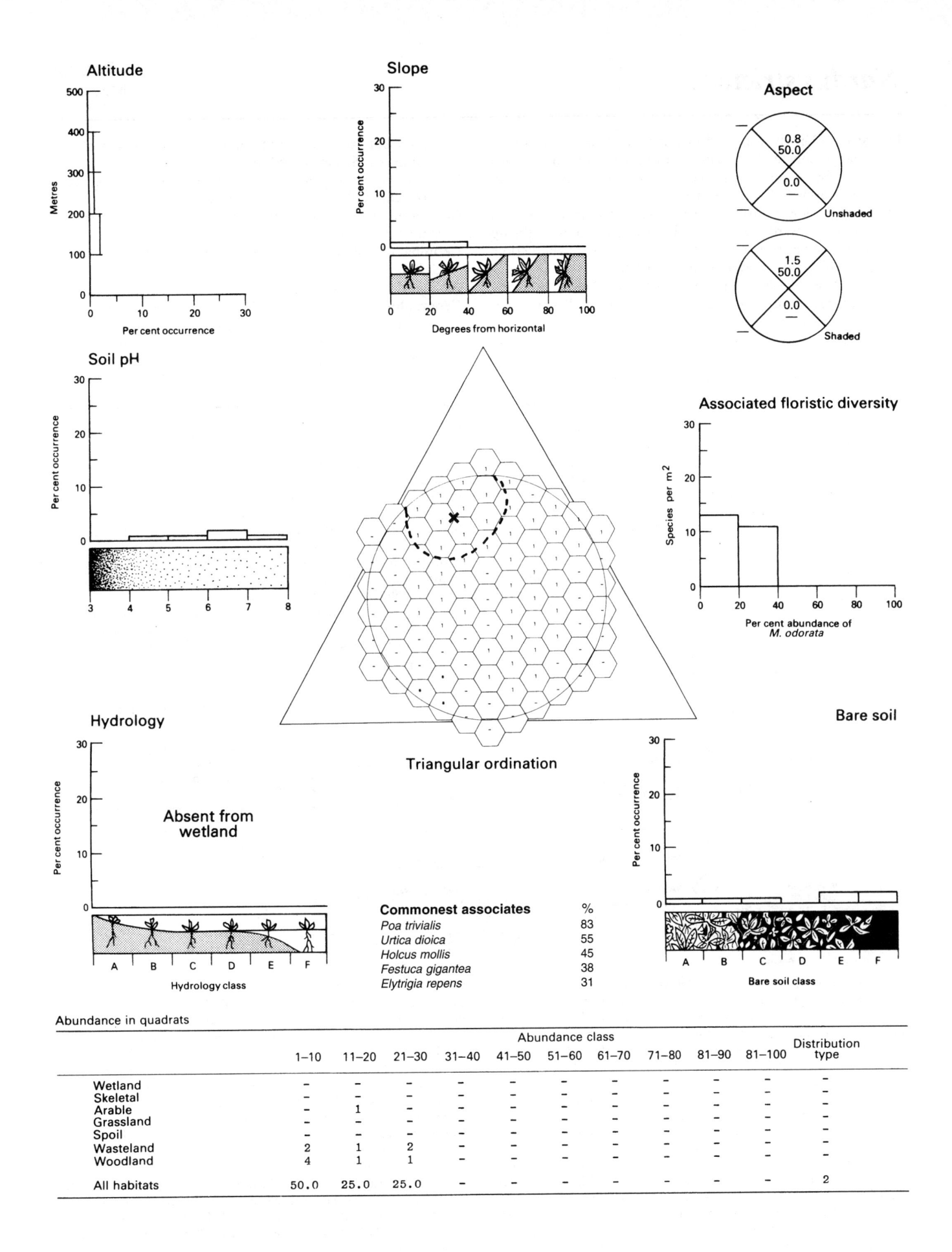

Altitude

Slope

Aspect

Unshaded

Shaded

Soil pH

Associated floristic diversity

Per cent abundance of *M. odorata*

Hydrology

Absent from wetland

Hydrology class

Triangular ordination

Commonest associates	%
Poa trivialis	83
Urtica dioica	55
Holcus mollis	45
Festuca gigantea	38
Elytrigia repens	31

Bare soil

Bare soil class

Abundance in quadrats

	\multicolumn{10}{c	}{Abundance class}	Distribution type
	1–10	11–20	21–30	31–40	41–50	51–60	61–70	71–80	81–90	81–100	
Wetland	–	–	–	–	–	–	–	–	–	–	–
Skeletal	–	–	–	–	–	–	–	–	–	–	–
Arable	–	1	–	–	–	–	–	–	–	–	–
Grassland	–	–	–	–	–	–	–	–	–	–	–
Spoil	–	–	–	–	–	–	–	–	–	–	–
Wasteland	2	1	2	–	–	–	–	–	–	–	–
Woodland	4	1	1	–	–	–	–	–	–	–	–
All habitats	50.0	25.0	25.0	–	–	–	–	–	–	–	2

Nardus stricta L.

Poaceae
Mat-grass

Life-form Densely tufted, polycarpic perennial, semi-rosette hemicryptophyte. Very short rhizomes are produced, each giving rise to numerous erect shoots; leaves hair-like, <250 mm^2; roots with VA mycorrhizas.

Phenology Flowers from June to August and caryopses shed from July onwards. Leaves long-lived, dying back from the tip but remaining green towards the base for a considerable period.

Height Foliage and flowering shoots to 400 mm.

Seedling RGR 0.5-0.9 week^{-1}. SLA 10.2. DMC 38.8.

Nuclear DNA amount 4.2 pg; $2n = 26^*$ (*BC*); $2x^*$. Base number for tribe also $n = 13$.

Established strategy Intermediate between stress-tolerator and stress-tolerant competitor.

Regenerative strategies V, S. Regenerates mainly by vegetative means, but seed may be important for colonisation of bare ground (see 'Synopsis'); ?SB II.

Flowers Green or purplish, hermaphrodite, protogynous, apomictic; <20 in a spike (1 per spikelet).

Germinule 0.38 mg, 4.2 × 0.6 mm, caryopsis with an attached lemma, 10.4 × 0.9 mm.

Germination Epigeal; t_{50} 12 days; 17-36°C; dormancy broken by chilling; L > D.

Biological flora Chadwick (1960).

Geographical distribution Most of the British Isles, but particularly abundant in the N and W (99% of British vice-counties), *ABIF* page 740; most of Europe (82% of European territories); Asia. Only on mountains in the southern part of its range.

Habitats Particularly frequent in grazed acidic grassland and on moorland. Also widespread on unmanaged heath, quarry spoil and river banks, and recorded from mire, cliffs, outcrops and lightly shaded woodland. Absent from most shaded, most skeletal and all flooded habitats. [Particularly abundant in acidic, montane grassland.]

Gregariousness Intermediate.

Altitude Widely distributed, but suitable habitats are more frequent in the upland regions. [Up to 1010 m.]

Slope Found over a wide range of slopes.

Aspect Frequent on both N- and S-facing slopes, but significantly more frequent and also more abundant on S-facing aspects in grasslands (*Ecol Atl*).

Hydrology Essentially a dryland species which occasionally extends into soligenous mire.

Soil pH Distribution strongly biased towards acidic habitats; particularly frequent and abundant on soils of pH <4.0.

Bare soil Most frequent in habitats with a relatively small proportion of exposed soil.

△ **ordination** Distribution centred on vegetation associated with infertile undisturbed conditions.

Associated floristic diversity Usually species-poor.

Synopsis *N.s.* forms long-lived, spreading tussocks, and is frequently a major constituent of unproductive grassland and heath vegetation on relatively free-draining acidic soils. *N.s.* is particularly abundant in upland

Habitats

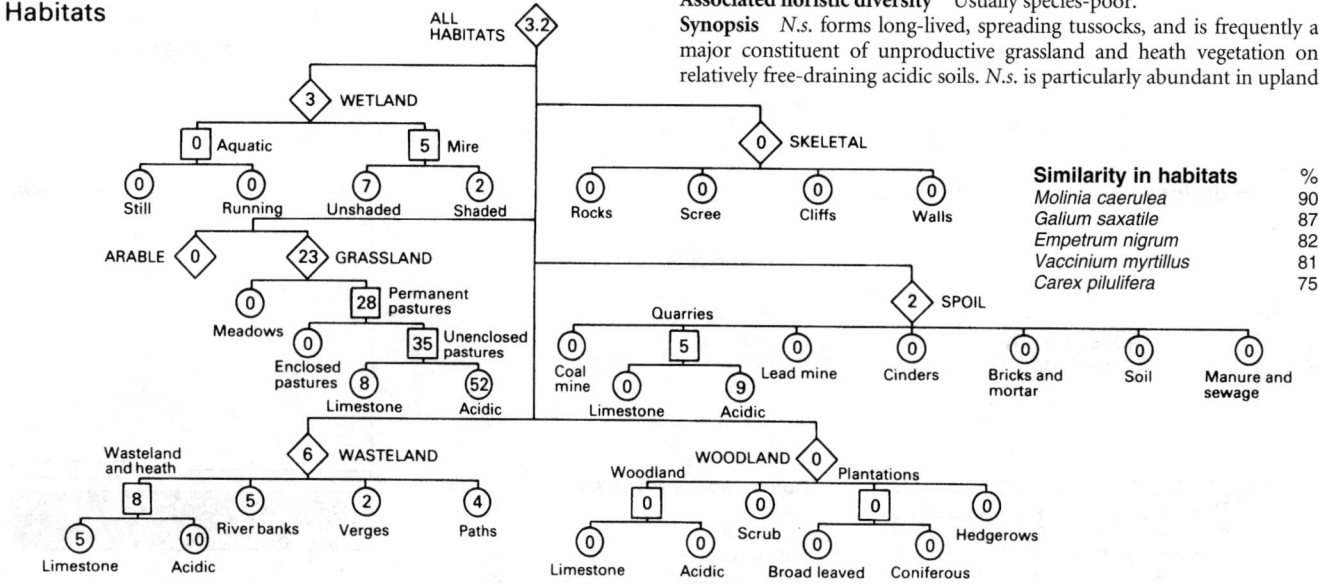

Similarity in habitats	%
Molinia caerulea	90
Galium saxatile	87
Empetrum nigrum	82
Vaccinium myrtillus	81
Carex pilulifera	75

areas of high rainfall (*Biol Fl*). The foliage is low in Ca (*SIEF*) and is unpalatable to sheep, though it may be eaten in winter when other food is scarce (*Biol Fl*). *N.s.* may re-colonize fired areas by seed, but adult plants are sensitive to burning and to cutting (*Biol Fl*). *N.s.* is, however, tolerant of trampling, and may adopt a flattened growth form beside paths. Regeneration is effected principally through a limited growth and branching of the rhizome system followed by decay of older rhizomes, producing daughter plants (*Biol Fl*). Tussocks, may expand at a rate of *c.* 20 mm year^{-1} (Jeffreys 1917). *N.s.* also regenerates from detached pieces of rhizome (*Biol Fl*). Regeneration by seed appears to be less important, particularly in closed vegetation, but may play a role in colonization of bare ground where incursions by other species have been delayed (King 1960, *Biol Fl*). Seed set may be low in poor summers, particularly at high altitudes, and the majority of caryopses are dormant when freshly shed (*Biol Fl*). However, a clump 200 mm in diameter may produce 2000 florets under favourable conditions (*Biol Fl*). Thus, under some circumstances *N.s.* may be an effective local colonist. There are isolated records suggesting that *N.s.* has a persistent seed bank (Chippindale & Milton 1934, K. Thompson pers. comm.), but negative results have been obtained in other investigations and there is no evidence of seed persistence in the Sheffield region. Since most seed is shed close to the parent plant (King 1960), *N.s.* is unlikely to be an effective long-distance colonist of new habitats. In parts of continental Europe *N.s.* is less-exclusively calcifuge (*Biol Fl*) and populations from acidic soil have been shown to grow successfully on highly calcareous soils following the addition of CaHPO$_4$ (James 1962). *N.s.* is morphologically relatively constant and apomictic.

Current trends In many upland areas of Britain the growth of *N.s.* was checked by winter grazing (*Biol Fl*). Now sheep are usually only released onto the hills during the summer months and, being selective feeders, they choose more-palatable species, which are often present in abundance. Thus, *N.s.* has increased dramatically in parts of N Britain and is replacing *Agrostis-Festuca* grassland on some sites in N Derbyshire (Evans 1977). In lowland areas habitat destruction for agriculture, forestry and urban development has diminished *N.s.*

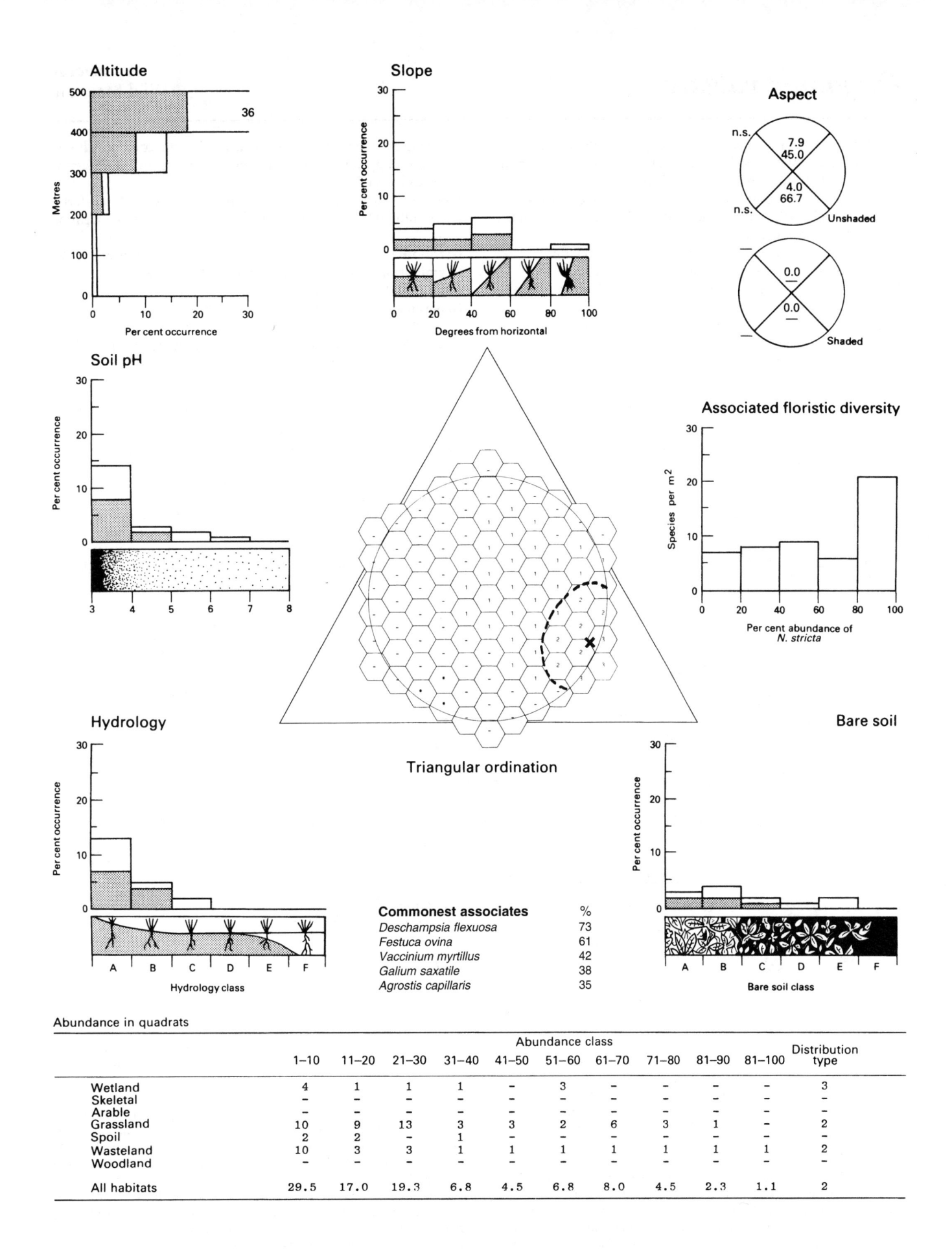

Altitude

Slope

Aspect

n.s. 7.9 / 45.0 / 4.0 / 66.7 n.s. Unshaded

— / 0.0 / 0.0 / — Shaded

Soil pH

Associated floristic diversity

Species per m² vs Per cent abundance of *N. stricta*

Hydrology

Hydrology class: A B C D E F

Triangular ordination

Commonest associates	%
Deschampsia flexuosa	73
Festuca ovina	61
Vaccinium myrtillus	42
Galium saxatile	38
Agrostis capillaris	35

Bare soil

Bare soil class: A B C D E F

Abundance in quadrats

	1–10	11–20	21–30	31–40	41–50	51–60	61–70	71–80	81–90	81–100	Distribution type
Wetland	4	1	1	1	–	3	–	–	–	–	3
Skeletal	–	–	–	–	–	–	–	–	–	–	–
Arable	–	–	–	–	–	–	–	–	–	–	–
Grassland	10	9	13	3	3	2	6	3	1	–	2
Spoil	2	2	–	1	–	–	–	–	–	–	–
Wasteland	10	3	3	1	1	1	1	1	1	1	2
Woodland	–	–	–	–	–	–	–	–	–	–	–
All habitats	29.5	17.0	19.3	6.8	4.5	6.8	8.0	4.5	2.3	1.1	2

Origanum vulgare L.

Lamiaceae
Wild Majoram

Life-form Rhizomatous, polycarpic perennial chamaephyte. Shoots erect, branched; leaves ovate, hairy, usually <1000 mm^2; deep-rooted and with VA mycorrhizas.
Phenology Some populations winter green (see 'Synopsis'). Flowers July to September and sheds seed August to November. Shoots die back in autumn.
Height Foliage overtopped by inflorescence to 800 mm.
Seedling RGR 1.0-1.4 week^{-1}. SLA 21.9. DMC 23.3.
Nuclear DNA amount 1.4 pg; $2n = 30*$ (*BC*); $2x*$, genus of polyploid origin.
Established strategy Intermediate between C-S-R and stress-tolerant competitor.
Regenerative strategies B$_s$, V. Regenerates by means of seeds which germinate in spring or become incorporated into a persistent seed bank. The capacity to form compact clumps seldom appears to constitute an effective means of vegetative spread; SB IV.
Flowers Purple, gynodioecious, protandrous, not self-fertile (Knuth 1906), insect-pollinated; several hundred in a terminal panicle.
Germinule 0.1 mg, 1.0 × 0.7 mm, nutlet (4 in each fruit).
Germination Epigeal; t_{50} 3 days; 13-33°C; L = D.
Biological flora None.
Geographical distribution The British Isles, but more local in Scotland and N Ireland (88% of British vice-counties), *ABIF* page 523; most of Europe (87% of European territories); N and W Asia and N America.

Habitats Occurs in rocky limestone habitats, particularly on quarry spoil heaps and derelict land, but also found in semi-derelict unenclosed pasture and skeletal habitats. Observed on roadsides, some situated on non-calcareous strata. Absent from arable land, woodland, most types of spoil heaps, enclosed pastures, meadows and wetlands.
Gregariousness Intermediate.
Altitude Extends up to 400 m. [Up to 410 m.]
Slope Found over a full range of slopes.
Aspect Occurrences significantly biased towards S-facing slopes, but slightly more abundant on slopes of N aspect.

Hydrology Absent from wetlands.
Soil pH Virtually restricted to calcareous soils and not recorded below pH 5.5.
Bare soil Found mainly in sites with high exposures of bare soil. Low values represent occurrences in skeletal habitats.
△ **ordination** Distribution centred on vegetation in which competition is restricted by mineral nutrient stress and occasional disturbance. Absent from heavily disturbed conditions and from very unproductive sites.
Associated floristic diversity Typically intermediate to species-rich.

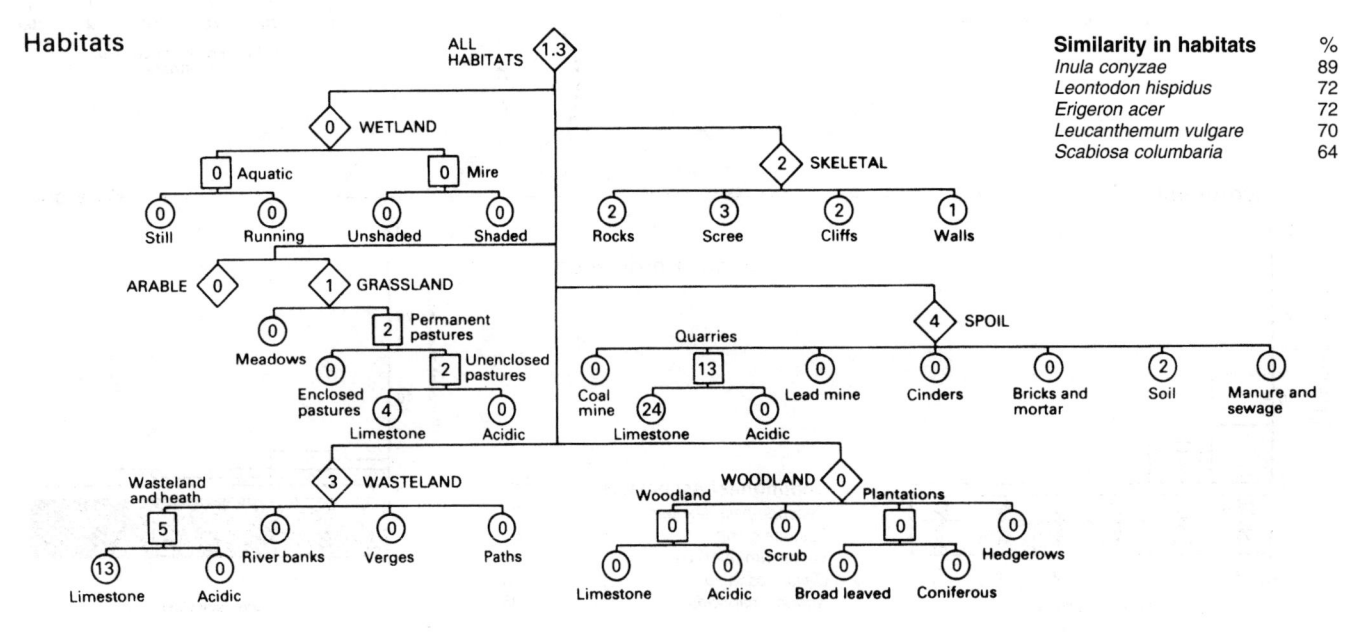

Habitats

Similarity in habitats	%
Inula conyzae	89
Leontodon hispidus	72
Erigeron acer	72
Leucanthemum vulgare	70
Scabiosa columbaria	64

Synopsis *O.v.* is a tall, tufted, aromatic herb of relatively dry and infertile, usually calcareous, soils. The species grows most vigorously at high external concentrations of Ca, and at low levels of supply exhibits Ca deficiency (Jefferies & Willis 1964). *O.v.* exhibits only a limited capacity for lateral spread and is unable to coexist with taller, fast-growing species. As a result of its stature *O.v.* is vulnerable to grazing and is infrequent in pasture. The species tends to occur either in grassland vegetation subject to periodic burning or on rock outcrops. *O.v.* has a relatively deep root system which often bears very long and numerous root hairs (Anderson 1927) and almost certainly allows the species to exploit subsoil water during periods of drought. The species is essentially of lowland distribution (Wilson 1956) and differences in phenology are observed at high and low altitude. Observations in the Sheffield region suggest that in lowland sites young green shoots overwinter, while during severe winters shoot growth in upland sites may be delayed until spring. *O.v.* produces numerous small seeds which germinate in vegetation gaps during spring (Cresswell 1982). *O.v.* also forms very large buried seed banks (Grime 1978, Thompson & Grime 1979). In samples from W Europe 1-62% of individuals are male-sterile plants (Kheyr-Pour 1981). Populations from disturbed vegetation show a higher incidence of male-sterility than those from more-stable sites (Iestewaart *et al.* 1984). It would seem that some populations have experienced high selection pressures for out-breeding, and that this has led to the formation of the many variants and ecotypes of *O.v.* which have been recorded (*FE3*, Bocher 1975). However, it should be pointed out that, at least in the British Isles, *O.v.* has a very narrow ecological range.
Current trends The species is capable of exploiting disused quarries and other artificial calcareous habitats, and has extended its range locally along roadsides. However, *O.v.* is restricted to rather infertile sites and may be expected to decline as these habitats diminish.

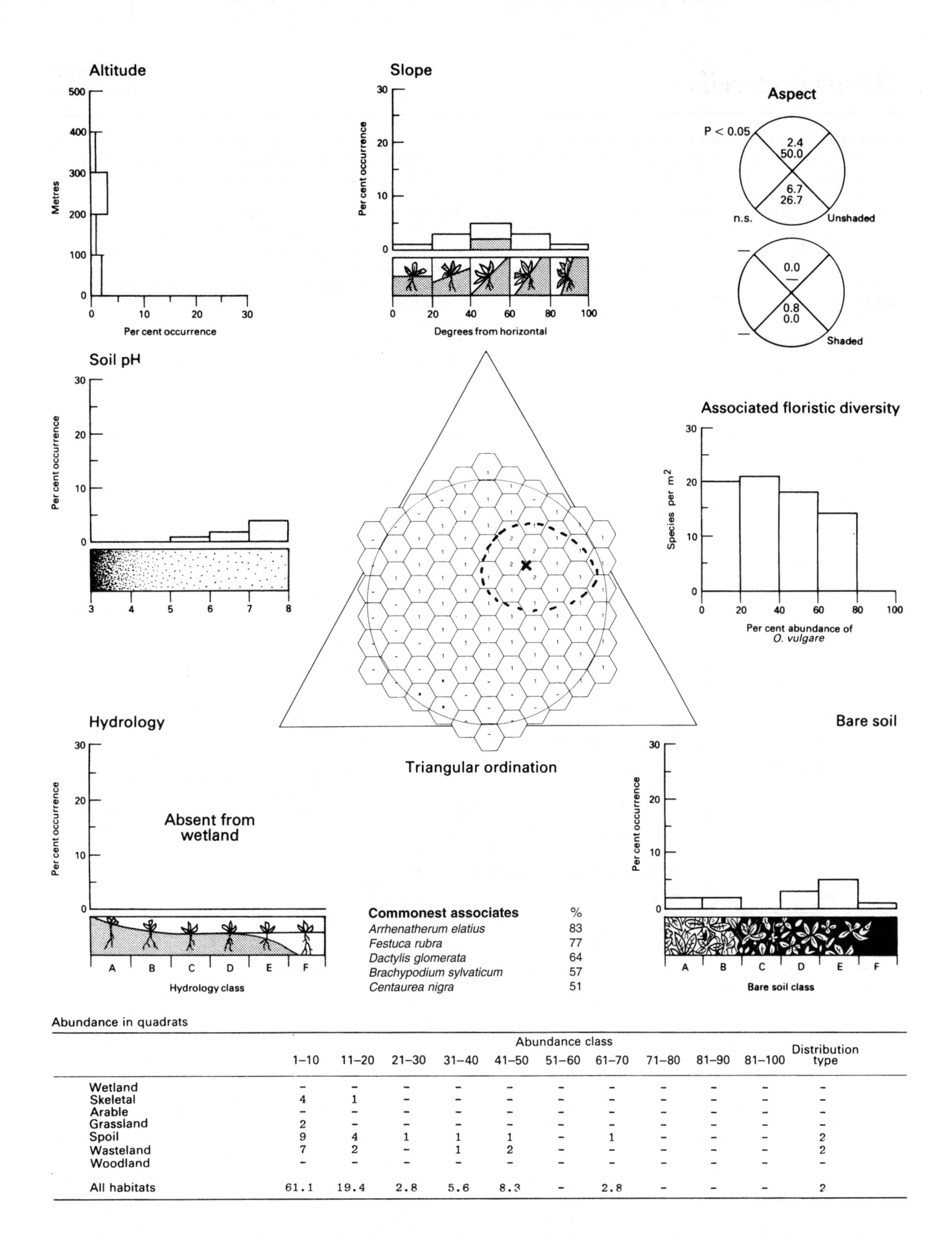

Altitude

Slope

Aspect

$P < 0.05$

	2.4	50.0	
n.s.	6.7	26.7	Unshaded

	—	0.0	
—	0.8	0.0	Shaded

Soil pH

Associated floristic diversity

Species per m² vs Per cent abundance of *O. vulgare*

Hydrology

Absent from wetland

Triangular ordination

Bare soil

Commonest associates	%
Arrhenatherum elatius	83
Festuca rubra	77
Dactylis glomerata	64
Brachypodium sylvaticum	57
Centaurea nigra	51

Abundance in quadrats

	1–10	11–20	21–30	31–40	41–50	51–60	61–70	71–80	81–90	81–100	Distribution type
Wetland	–	–	–	–	–	–	–	–	–	–	–
Skeletal	4	1	–	–	–	–	–	–	–	–	–
Arable	–	–	–	–	–	–	–	–	–	–	–
Grassland	2	–	–	–	–	–	–	–	–	–	–
Spoil	9	4	1	1	1	–	1	–	–	–	2
Wasteland	7	2	–	1	2	–	–	–	–	–	2
Woodland	–	–	–	–	–	–	–	–	–	–	–
All habitats	61.1	19.4	2.8	5.6	8.3	–	2.8	–	–	–	2

Oxalis acetosella L.

Oxalidaceae
Wood Sorrel

Life-form Polycarpic perennial, rosette hemicryptophyte or chamaephyte. Scapes erect; leaves ternate, hairy, typically <500 mm^2; base of petiole a swollen storage organ; roots shallow, usually with VA mycorrhizas.

Phenology Winter green, capable of growth throughout the winter in shaded sheltered sites (*Biol Fl*). Normal flowers produced April to May followed by cleistogamous flowers. Seed set from May onwards (*Biol Fl*). Leaves relatively long-lived, particularly in shaded sites.

Height Foliage and scapes to 100 mm.

Seedling RGR 0.5-0.9 week^{-1}. SLA 62.1. DMC 13.4.

Nuclear DNA amount 6.4 pg (Bennett *et al.* 1982); $2n = 22^*$ (*BC*); $2x^*$.

Established strategy Intermediate between stress-tolerator and stress-tolerant ruderal.

Regenerative strategies V, S, ?B$_s$. Forms clonal patches by means of rhizome extension. Die-back of older rhizome isolates daughter plants. Regenerates also by seeds, germinating in spring. May sometimes also form a persistent seed bank (Staaf *et al.* 1987).

Flowers Solitary, long-stalked, white, rarely pink, hermaphrodite, sparingly pollinated by insects and usually not producing seeds. Late flowers without petals, cleistogamous and fertile (*Biol Fl*).

Germinule 1.01 mg, 2.6 × 1.6 mm, seed explosively dispersed from a capsule containing up to 10 seeds.

Germination Epigeal; dormancy broken by chilling.

Biological flora Packham (1978).

Geographical distribution The British Isles, but less common in E England and also rare in C Ireland (100% of British vice-counties), *ABIF* page 442; most of Europe, although rarer in the S (79% of European territories); N and C Asia.

Habitats Predominantly a woodland species, but also found on unshaded scree slopes and more rarely in other skeletal habitats, on river banks and in pastures. Also present in both shaded and unshaded mire. Absent from aquatic, spoil and arable habitats. [Found in rock crevices and grassland in montane areas, and also in limestone pavement.]

Gregariousness Intermediate to patch-forming.

Altitude Recorded over a wide range of altitudes to 470 m, but less frequent on the drier soils of the Bunter sandstone and restricted to shaded habitats in lowland areas. [Up to 1150 m.]

Slope Widely distributed.

Aspect More frequent and abundant on N-facing slopes.

Hydrology Not typically a wetland species, but not infrequent in soligenous mire.

Soil pH Extending across the range of pH 3.5 to >8.0 but most frequent between pH 3.5 and 5.0.

Bare soil Most records from woodland habitats with little exposed soil but where depositions of tree litter may be high.

△ ordination Associated with a wide range of conditions, but distribution centred on undisturbed infertile sites. Absent from highly disturbed vegetation.

Associated floristic diversity Typically low, but in unshaded sites may occur in relatively species-rich vegetation.

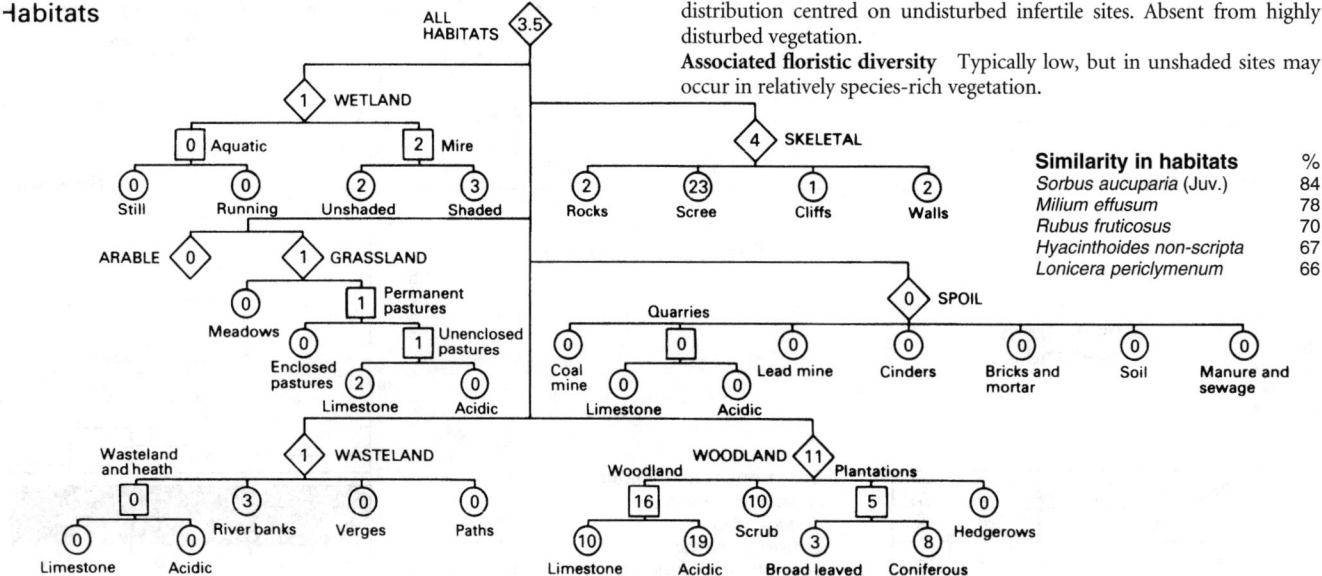

Similarity in habitats	%
Sorbus aucuparia (Juv.)	84
Milium effusum	78
Rubus fruticosus	70
Hyacinthoides non-scripta	67
Lonicera periclymenum	66

Synopsis *O.a.* is a rhizomatous, patch-forming woodland herb with a superficial resemblance to *Trifolium repens* both in habit and in the nastic 'sleep' movements of its trifoliate leaves. *O.a.* has a slow growth rate and the leaf canopy is lower than that of most other members of the woodland ground flora. Thus, *O.a.* tends to play a subordinate role in herbaceous communities. The species is suppressed by heavy depositions of deciduous tree litter, and for this reason is often restricted to raised areas of the woodland floor such as tree bases and small hummocks. *O.a.* is, however, very shade-tolerant, surviving in dimly-lit rock crevices and maintaining dry matter production and setting seed under moderate shade. *O.a.* is also physiologically attuned to sites of low illumination and relatively low temperatures (*Biol Fl*). The rhizome of *O.a.* lies on the soil surface, and the root system is typically shallow. Thus, *O.a.* exploits a relatively thin layer of soil and is largely restricted to continuously moist habitats. *O.a.* is found on a wide range of soil types from highly calcareous to acidic, but is more frequent on the latter. The leaves are unusually low in P, Ca, Mg and Fe (*SIEF*). It is suspected that in some calcareous habitats the niche exploited by *O.a.* is occupied by species such as *Hedera helix*; similarly, *O.a.* may be replaced by *Deschampsia flexuosa* in woodlands on the poorest acidic soils. *O.a.* also occurs in semi-natural grassland, usually on N-facing slopes, but tends to show less winter growth in such habitats than in more-sheltered woodland sites (*Biol Fl*). *O.a.* appears to be little grazed, perhaps because its leaves contain oxalic acid and are poisonous to livestock (*Biol Fl*). *O.a.* regenerates mainly as a result of clonal growth, rhizome extension in beech woods averaging 100 mm year^{-1} (*Biol Fl*), and through the production of cleistogamous flowers (*Biol Fl*). A persistent seed bank is reported (Staaf *et al.* 1987). *O.a.* is phenotypically plastic with respect to size and the proportion of resources allocated to rhizome and root. The leaves of plants from open sites differ in anatomy, colour and longevity from those of shaded sites (*Biol Fl*).

Future trends *O.a.* is one of the few woodland species which exploits coniferous plantations. However, as noted by Peterken (1981), *O.a. is* most characteristic of long-established woodland sites, and appears to be decreasing in abundance.

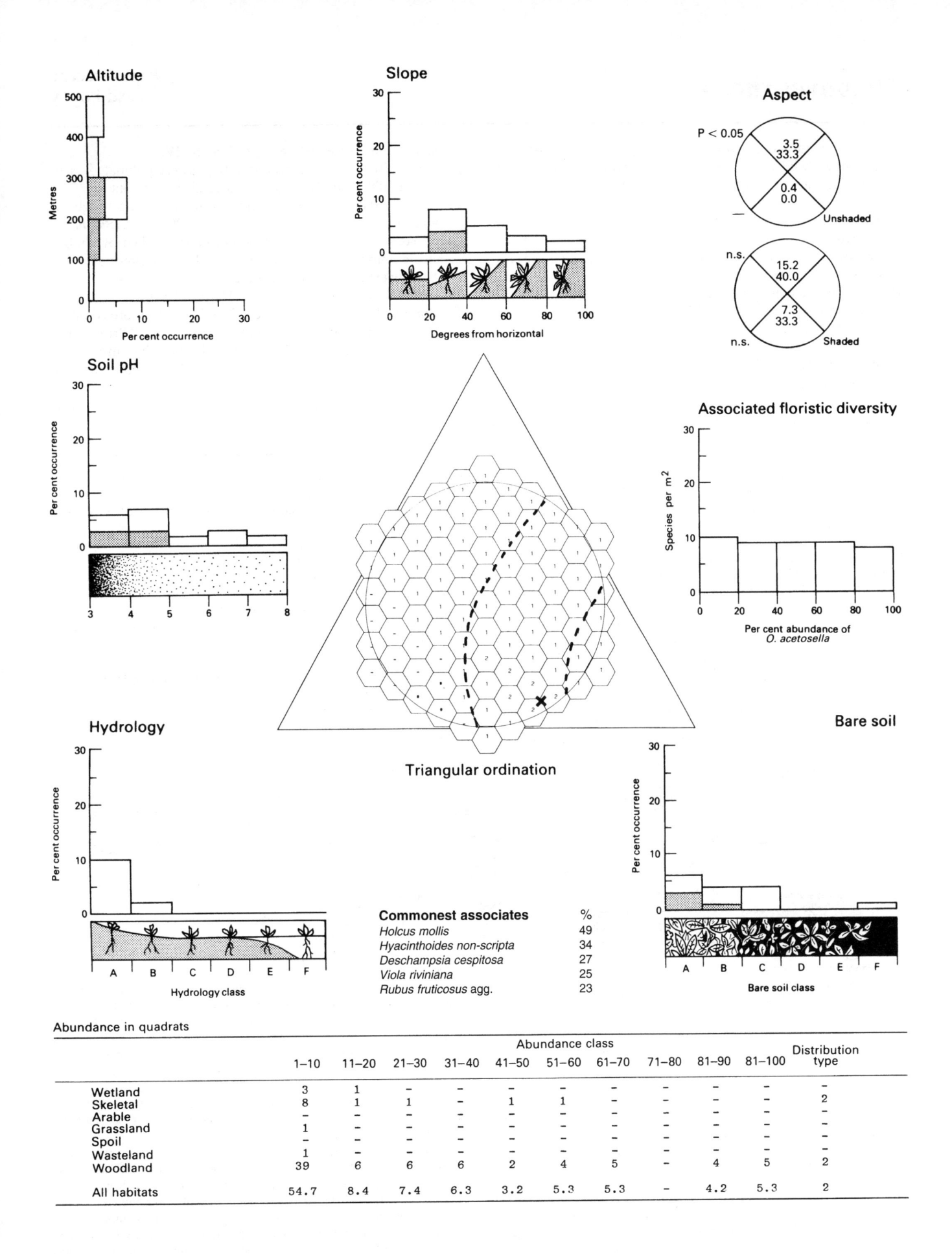

Altitude

Slope

Aspect

$P < 0.05$

	3.5	
	33.3	
	0.4	
	0.0	Unshaded

n.s.

	15.2	
	40.0	
	7.3	
	33.3	Shaded

n.s.

Soil pH

Associated floristic diversity

Per cent abundance of *O. acetosella*

Hydrology

Triangular ordination

Bare soil

Commonest associates

	%
Holcus mollis	49
Hyacinthoides non-scripta	34
Deschampsia cespitosa	27
Viola riviniana	25
Rubus fruticosus agg.	23

Abundance in quadrats

	Abundance class										Distribution type
	1–10	11–20	21–30	31–40	41–50	51–60	61–70	71–80	81–90	81–100	
Wetland	3	1	–	–	–	–	–	–	–	–	2
Skeletal	8	1	1	–	1	1	–	–	–	–	2
Arable	–	–	–	–	–	–	–	–	–	–	–
Grassland	1	–	–	–	–	–	–	–	–	–	–
Spoil	–	–	–	–	–	–	–	–	–	–	–
Wasteland	1	–	–	–	–	–	–	–	–	–	–
Woodland	39	6	6	6	2	4	5	–	4	5	2
All habitats	54.7	8.4	7.4	6.3	3.2	5.3	5.3	–	4.2	5.3	2

+*Papaver rhoeas* L.

Papaveraceae
Field Poppy

Life-form Summer- or more rarely winter-annual, semi-rosette therophyte with a slender tap-root. Stems branched erect or ascending; leaves dissected, hairy, <4000 mm^2; roots usually non-mycorrhizal.

Phenology Seedlings appear in spring or, more rarely, overwinter. Flowers from June to August. Seed-set from July onwards. A second flush of flowering may occur after the crop has been harvested.

Height Foliage mainly <300 mm. Flowers to 600 mm.

Seedling RGR No data. SLA 30.2. DMC 14.3.

Nuclear DNA amount 5.3 pg (Bennett & Smith 1976); $2n = 14^*$ (*BC*); $2x^*$.

Established strategy Ruderal.

Regenerative strategies B_s. Regenerates exclusively from seed in vegetation gaps during spring. In mild winters seeds germinating in autumn may also succeed (*Biol Fl*). Forms a bank of buried seed (Brenchley 1918); SB IV.

Flowers Usually red, hermaphrodite, insect-pollinated, self-incompatible (*Biol Fl*, Rogers 1969), axillary and solitary.

Germinule 0.09 mg, 0.9 × 0.7 mm, seed dispersed through the pores of a capsule which may contain > 1000 seeds (*Biol Fl*).

Germination Epigeal; t_{50} 4 days; dormancy broken by chilling and in common with *P. dubium* ssp. *lecoqii* (Lamotte) Syme, *P.r.* may exhibit hard-coat dormancy.

Biological flora McNaughton and Harper (1964).

Geographical distribution The British Isles, but more common in the S, rare and local in N Scotland (94% of British vice-counties), *ABIF* page 115; most of Europe, becoming rarer northwards (82% of European territories); temperate Asia and N Africa; introduced into N America, Australia and New Zealand. Perhaps not native in Britain and other parts of N Europe (*FE1*).

Habitats Typically associated with arable land, particularly in fallow field margins and to a lesser extent in cereal crops. Also found in a range of other disturbed artificial habitats such as soil heaps, waysides, cinder tips and wasteland. Infrequent in gardens. Absent from wetland, pasture, woodland and skeletal habitats.

Gregariousness Sparse to intermediate.

Altitude Principally an arable species, and therefore largely restricted to sites of less than 200 m but extends locally to 230 m. [Up to 300 m (*Biol Fl*).] Altitudinal range difficult to assess accurately and in many upland sites *P.r.* may only occur casually.

Slope As a result of its habitat range, *P.r.* is largely restricted to flat or gently sloping ground.

Aspect Only recorded in unshaded S-facing habitats. Additional sampling confirms this S-facing bias.

Hydrology Absent from wetlands.

Soil pH Absent from acidic soils and relating to its association with arable land records are all within the pH range 6.0–8.0.

Bare soil Associated almost exclusively with habitats having large amounts of bare soil.

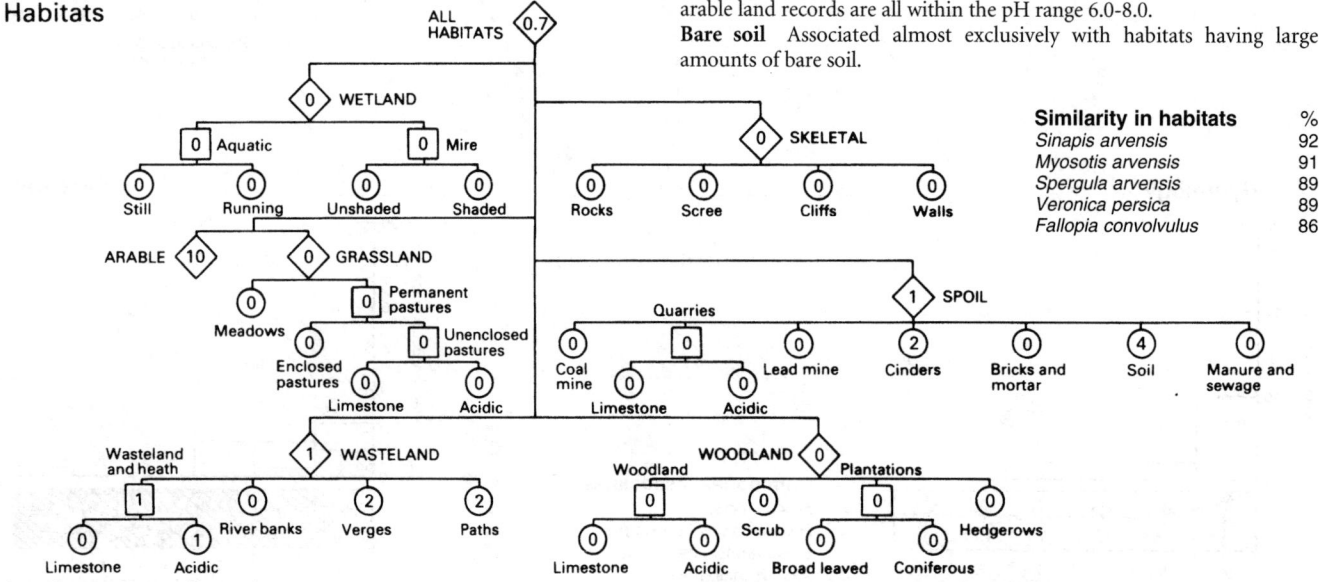

Similarity in habitats	%
Sinapis arvensis	92
Myosotis arvensis	91
Spergula arvensis	89
Veronica persica	89
Fallopia convolvulus	86

△ **ordination** Insufficient data.

Associated floristic diversity Typically associated with species-rich plant communities.

Synopsis *P.r.* is a summer-annual restricted to disturbed artificial habitats (particularly arable land), and may have been introduced from S Europe along with agricultural crops. Seed remains have been identified which date from the Late Bronze Age (*Biol Fl*). *P.r.* is more frequent in cereal than in root crops. With the advent of selective weed-killers *P.r.* is now mainly restricted to fallow ground and to disturbed sites not managed for agriculture. *P.r.* is infrequent in grazed habitats. Nevertheless, the species contains toxic alkaloids and is unpalatable to livestock. Seeds also contain toxins (*Biol Fl*). *P.r.* is sensitive to trampling, but may occasionally survive harvesting of the crop and produce new flowering shoots. Regeneration is entirely by seed which is rapidly shed from a many-seeded capsule by a censer mechanism (*Biol Fl*). Seedlings originating in autumn may survive mild winters, but most-effective regeneration is in spring (*Biol Fl*). *P.r.* is capable of survival in an unpredictably disturbed landscape, and dispersal in time is effected by the formation of a buried seed bank (Roberts & Feast 1972). Thus, the former boundaries of an arable field which had been grassed for over 20 years were recognized by the presence of poppies derived from a persistent seed bank (Chancellor 1985). *P.r.* may be dispersed considerable distances and occasionally occurs as an alien introduction from abroad, e.g. Webster (1978). *P.r.* shows considerable plasticity, both in form and in number of seeds set. Under unfavourable conditions *P.r.* may produce from a single small flower as few as 4 seeds. Robust plants are much branched, perhaps with over 400 flowers; the mean number of seeds per capsule has been given as 1360 (*Biol Fl*). Hybridizes with *P. dubium* L.

Current trends Formerly a common plant in lowland areas, often forming large populations in arable fields. Still common, but decreasing both in number of sites and in size of populations. *P.r.* is beginning to become characteristic of waste places rather than of arable fields. However, unlike the related *P.d.*, which has recently increased by colonizing dry, sandy and cindery railway banks, *P.r.* has remained strictly confined to productive ruderal sites. As *P.r.* becomes more restricted to small populations, we predict selection of increased in-breeding.

Altitude

Slope

Aspect

Unshaded 0.0 / 1.8 / 0.0

Shaded 0.0 / 0.0

Soil pH

Associated floristic diversity

Triangular ordination

Hydrology

Absent from wetland

Commonest associates

	%
Poa annua	100
Polygonum aviculare	100
Stellaria media	100
Elytrigia repens	80
Agrostis gigantea	60

Bare soil

Abundance in quadrats

	1–10	11–20	21–30	31–40	41–50	51–60	61–70	71–80	81–90	81–100	Distribution type
Wetland	–	–	–	–	–	–	–	–	–	–	–
Skeletal	–	–	–	–	–	–	–	–	–	–	–
Arable	9	1	–	–	1	–	–	–	1	–	2
Grassland	–	–	–	–	–	–	–	–	–	–	–
Spoil	2	1	–	–	–	–	–	–	–	–	–
Wasteland	3	–	–	–	–	–	–	–	–	–	–
Woodland	–	–	–	–	–	–	–	–	–	–	–
All habitats	77.8	11.1	–	–	5.6	–	–	–	5.6	–	2

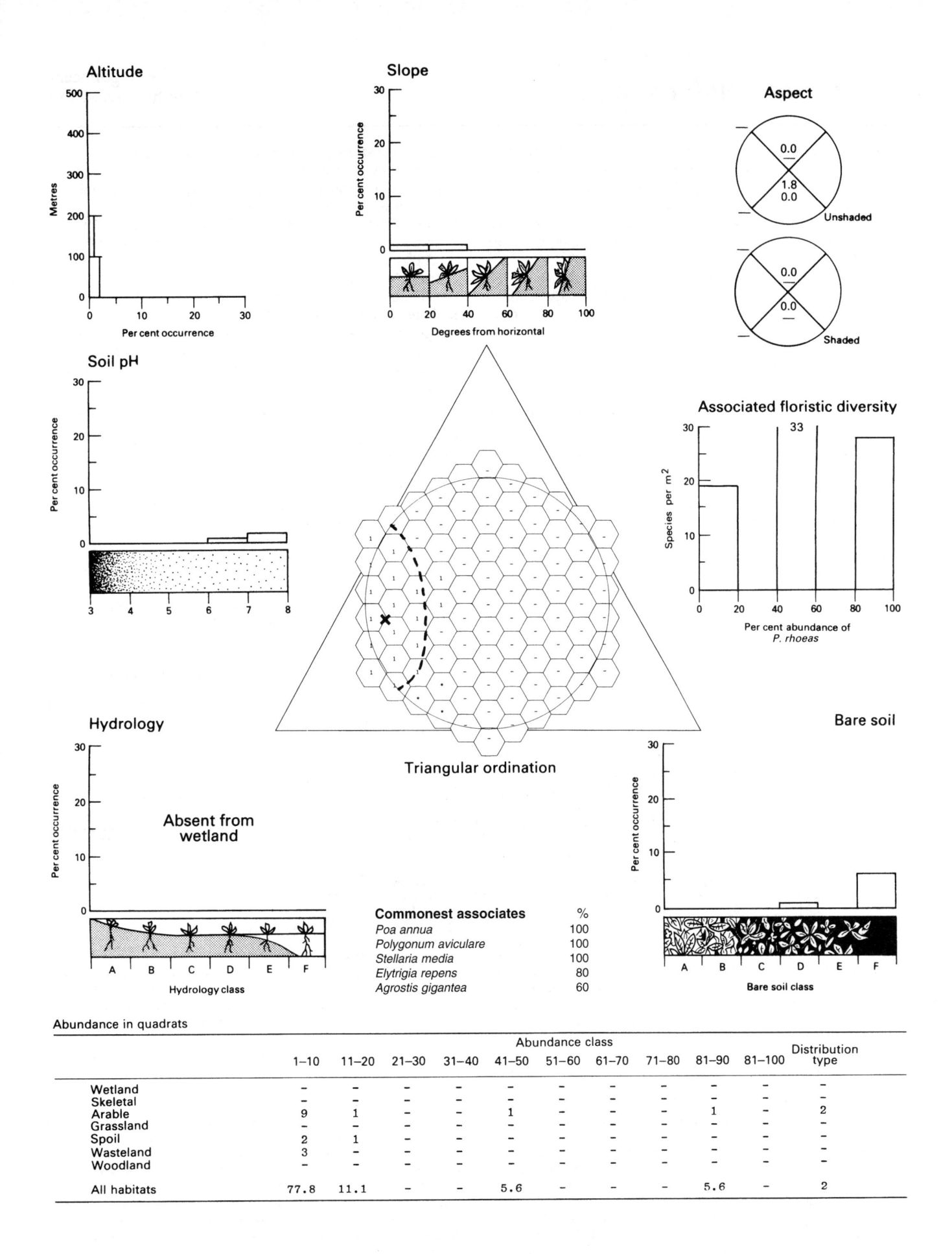

Persicaria amphibia (L.) Gray

Polygonaceae
Amphibious Bistort

Life-form Rhizomatous, polycarpic perennial floating hydrophyte or protohemicryptophyte with erect stems. Leaves ovate, often <2000 mm^2 (slightly larger in the aquatic form); roots non-mycorrhizal.

Phenology Flowers from July to September and seed is set from September to October. One of the earliest wetland species to die back in autumn.

Height Stems bearing leaves and overtopped by flowers to 750 mm in terrestrial forms. In aquatic forms the foliage floats on the surface of the water and the inflorescence is shortly emergent.

Seedling RGR No data. SLA 17.0. DMC 23.7.

Nuclear DNA amount 2.0 pg; $2n = 66, 88$ (*FE1*, Wcislo 1977); $6x$, $8x$.

Established strategy Intermediate between competitor and competitive-ruderal.

Regenerative strategies (V), ?B$_s$. Regenerates mainly by vegetative means (see 'Synopsis'); ?SB IV.

Flowers Pink, hermaphrodite, insect-pollinated; >50 in a terminal spike. Some inflorescences are functionally female (T. T. Elkington pers. comm.).

Germinule 4.48 mg, 2.7×1.9 mm, nut.

Germination Epigeal; dormancy broken by chilling.

Biological flora Partridge (2001).

Geographical distribution The British Isles, but more frequent in C and S areas (99% of British vice-counties), *ABIF* page 184: Europe except the extreme S (89% of European territories); Asia, N America and S Africa.

Habitats Restricted to wetland habitats or to sites which are either adjacent to, or were formerly, wetland. Most frequent in unshaded mire and as an aquatic in ponds. Also found in shaded mire and on river banks. In areas reclaimed from mire *P.a.* may occur in arable land, mowing meadows and waysides, and is frequently observed on ditch banks. [Also found in dune slacks (Lousley & Kent 1981).

Gregariousness Stand-forming.

Altitude Most suitable habitats are found in the lowlands; not observed above 200 m. [Up to 230 m.]

Slope Like a majority of wetland species, largely restricted to flat ground and to gentle slopes.

Aspect Insufficient data.

Hydrology Recorded mainly in topogenous mire marginal to open water. In aquatic systems *P.a.* is most common at depths greater than 100 mm, and occurrences down to 520 mm recorded.

Soil pH Virtually restricted to soils of pH >4.5.

Bare soil Associated with a wide range of values.

△ **ordination** Distribution centred on fertile sites subject to some disturbance.

Associated floristic diversity Typically low to intermediate, generally higher in terrestrial vegetation than in aquatic habitats.

Similarity in habitats	%
Alisma plantago-aquatica	98
Eleocharis palustris	97
Glyceria maxima	95
Hydrocotyle vulgaris	94
Typha latifolia	92

Habitats

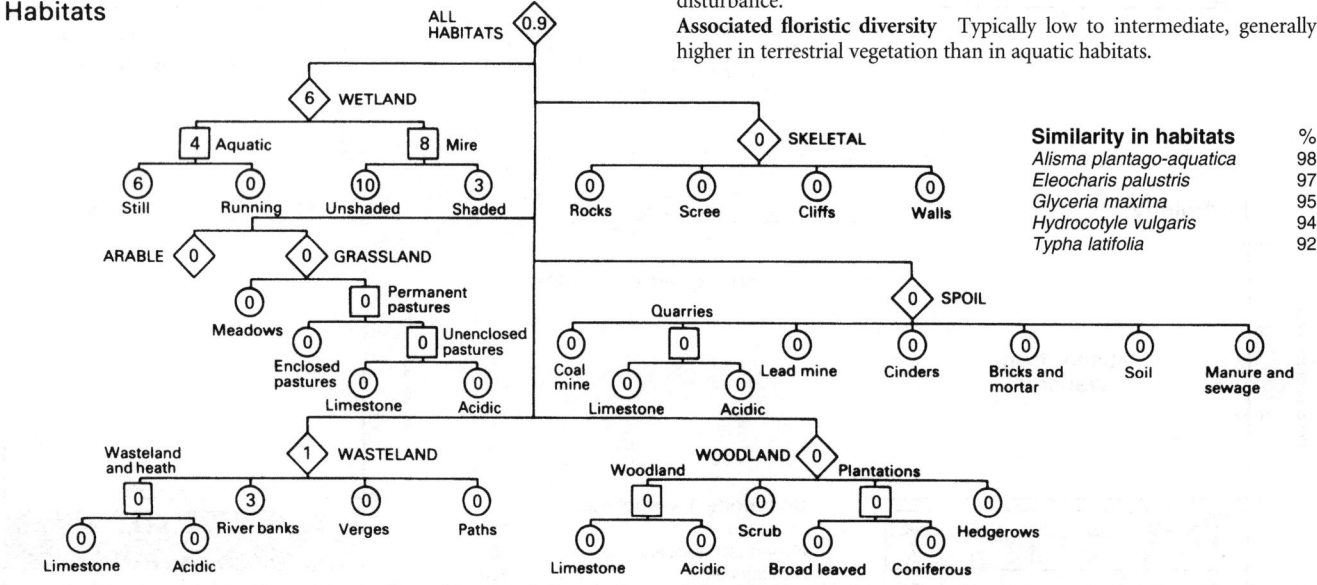

Synopsis *P.a.* is a robust, rhizomatous herb capable of exploiting both aquatic and terrestrial habitats apparently by virtue of its extreme phenotypic plasticity. The species behaves as an aquatic with glabrous, floating leaves with functional stomata restricted to the upper surface (Sculthorpe 1967). In mire or on ditch and river banks, however, the plant has hairy leaves with a much higher stomatal frequency on the lower surface. *P.a.* exploits fertile habitats, and the rhizomatous shoot system may extend to 13 m and spread laterally at a rate of >50 mm per day (Sculthorpe 1967). However, in terrestrial habitats *P.a.* reaches only 750 mm in height and only achieves abundance where the vigour of potential dominants is suppressed. Thus, *P.a.* is particularly abundant along reservoir margins and, to a lesser extent, river and ditch banks, where water levels are subject to wide fluctuations. The species is also frequent in areas of mining subsidence, which are now more waterlogged than formerly. Aquatic populations are less common than those of the terrestrial habitats, and there is circumstantial evidence that many, if not most, have been secondarily derived from populations of the land form following flooding. Although primarily a wetland species, *P.a.* may be long-persistent in drained habitats and has been observed as a weed at the edge of potato fields in the Sheffield region. The species has not been recorded from calcareous soils. *P.a.* is very frost-sensitive and is perhaps the first wetland species to die back in autumn; this may explain the restriction of *P.a.* to lowland sites <230 m (Wilson 1956). *P.a.* regenerates mainly by vegetative means, forming extensive patches by rhizome extension. Detached vegetative shoot fragments root within 2 days in the laboratory (Hodgson unpubl.), and under field conditions this form of regeneration is an important mechanism in the founding of new colonies; long-distance colonization along water courses may be effected by this means. *P.a.* also regenerates, perhaps only infrequently, by seeds which germinate during spring. A persistent seed bank is predicted but has not yet been demonstrated. Aquatic populations tend to flower moderately freely; terrestrial plants are shy-flowering.

Current trends *P.a.* exploits disturbed, fertile, moist conditions, but is not generally an early colonist of new sites. Despite a number of characteristics which make it potentially an aggressive weed, *P.a.* appears to be decreasing.

Altitude

Slope

Aspect

Unshaded

Shaded

Soil pH

Associated floristic diversity

Hydrology

Triangular ordination

Commonest associates %
Galium palustre 42
Agrostis stolonifera 38
Ranunculus repens 31
Poa trivialis 27
Urtica dioica 27

Bare soil

Abundance in quadrats

	Abundance class										Distribution type
	1–10	11–20	21–30	31–40	41–50	51–60	61–70	71–80	81–90	81–100	
Wetland	14	–	4	2	–	–	1	–	1	2	2
Skeletal	–	–	–	–	–	–	–	–	–	–	–
Arable	–	–	–	–	–	–	–	–	–	–	–
Grassland	–	–	–	–	–	–	–	–	–	–	–
Spoil	–	–	–	–	–	–	–	–	–	–	–
Wasteland	–	–	1	–	–	–	–	–	–	–	–
Woodland	–	–	–	–	–	–	–	–	–	–	–
All habitats	56.0	–	20.0	8.0	–	–	4.0	–	4.0	8.0	2

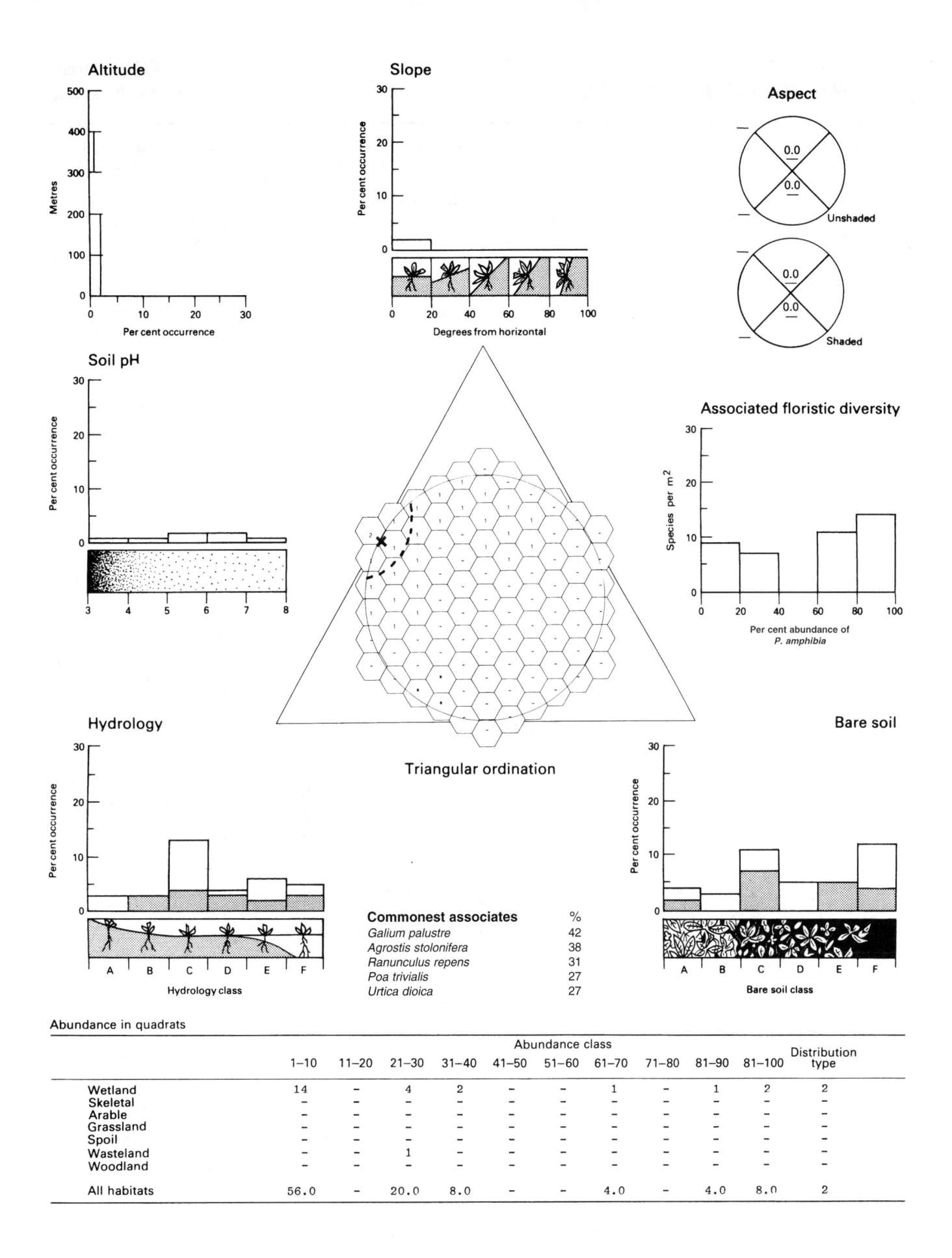

Persicaria maculosa Gray

Polygonaceae
Redshank, Persicaria

Life-form Summer-annual therophyte. Shoots erect or decumbent, usually branched; leaves lanceolate, <2000 mm^2, lower ones sometimes hairy; roots shallow, sometimes with VA mycorrhizas.

Phenology Seeds germinate in April and May. Flowers from June until the plant is killed by frost (October onwards). Seeds ripen July onwards.

Height Foliage and flowers to 750 mm.

Seedling RGR 1.0-1.4 week^{-1}. SLA 21.1. DMC 20.3.

Nuclear DNA amount 0.8 pg; $2n = 44^*$ (*BC*); $4x^*$.

Established strategy Intermediate between ruderal and competitive-ruderal.

Regenerative strategies B$_s$. Regenerates entirely by seed which germinates in spring. Forms a persistent seed bank; SB IV.

Flowers Pink or rarely white, hermaphrodite, self- or insect-pollinated, sometimes cleistogamous (*Biol Fl*); up to 50 or more in each terminal spike.

Germinule 2.12 mg, 3.2 × 2.3 mm, nut. Polymorphic (biconvex 2.71 mg; trigonous 3.21; tetragonous 4.04 (Hammerton 1967)).

Germination Epigeal; t_{50} 2 days; 30-40°C; dormancy broken by chilling and germination rate also increased after dry storage at 20°C; L > D.

Biological flora Simmonds (1945).

Geographical distribution The British Isles (100% of British vice-counties), *ABIF* page 184; Europe except for the extreme N and S (92% of European territories); widely distributed as a weed in Australia, New Zealand, N and S America, Asia and N Africa.

Habitats Most frequent and abundant on arable land, but also found in many other disturbed habitats, particularly soil heaps and manure and sewage spoil. Absent from aquatic and woodland habitats.

Gregariousness Typically as scattered individuals.

Altitude Most frequent at altitudes of <200 m, corresponding to the distribution of arable land in the survey area. Recorded up to 335 m. [Up to 450 m.]

Slope Absent from steeply sloping ground.

Aspect Widely distributed.

Hydrology Found on moist, exposed mud at the margin of open water, but not in permanently waterlogged sites.

Soil pH Most frequent in the pH range 5.0-7.0, and absent from acidic soil of pH <4.5.

Bare soil Mainly associated with high exposures of bare soil.

△ **ordination** Distribution centred on vegetation of disturbed, fertile conditions.

Associated floristic diversity Intermediate, occasionally species-rich.

Synopsis *P.m.* is an erect summer annual, exploiting disturbed fertile

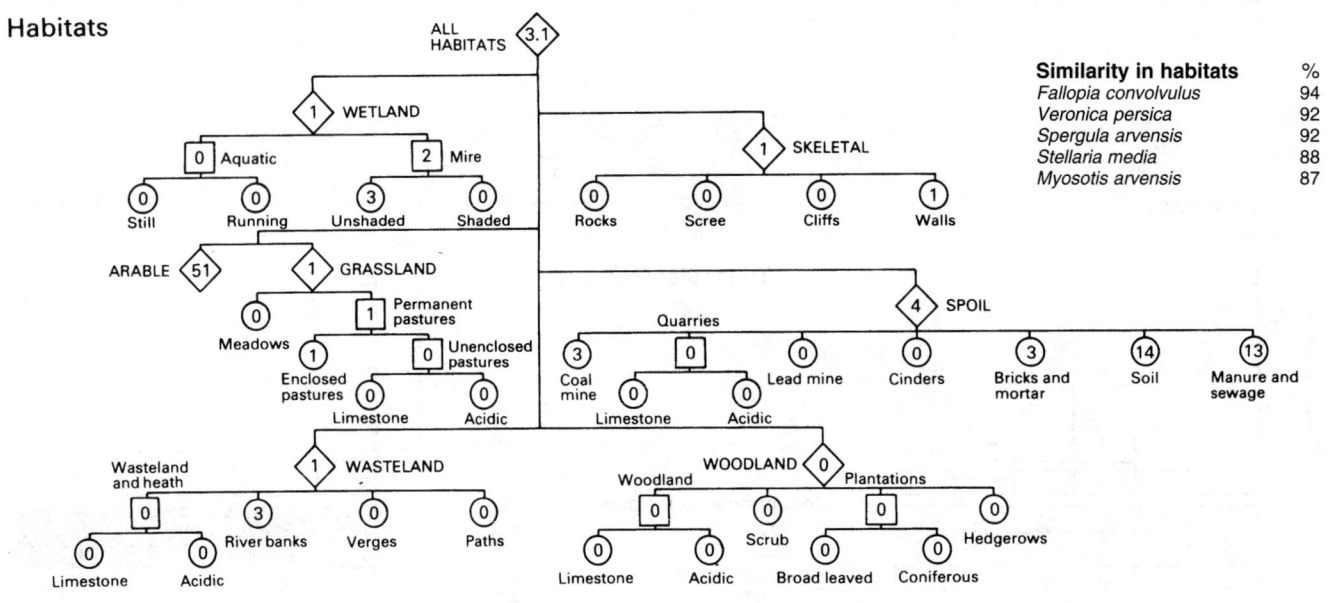

Habitats

Similarity in habitats	%
Fallopia convolvulus	94
Veronica persica	92
Spergula arvensis	92
Stellaria media	88
Myosotis arvensis	87

soils. *P.m.* is essentially a 'follower of humans' and is particularly characteristic of arable land in both cereal and dicotyledonous crops, although usually more vigorous and more frequent in the latter. *P.m.* is found on a wide range of moist soils, and occurs on exposed mud beside ponds, where it may be a native of long standing. Although there are no British records of poisoning, *P.m.* contains oxalate and is potentially toxic (Cooper & Johnson 1984). *P.m.* resembles many other arable weeds, e.g. *Sinapis arvensis*, in three respects: (a) seed is retained on the plant, leading to contamination of harvested crop seed (see *WCH*), (b) a persistent seed bank is formed and (c) seed may be ingested and dispersed by birds and animals (*Biol Fl*). Seeds are also polymorphic, and there is variation in seed weight and chilling requirement between populations and even individual plants (Hammerton 1967). Seeds in the soil have the potential to germinate each spring, but are returned to a state of secondary dormancy during the annual rise in late spring temperatures (Staniforth & Cavers 1979). This behaviour restricts the capacity of *P.m.* to exploit areas which are disturbed within the summer and to autumn. Cuttings of *P.m.* readily root at the node, and this facility may allow rapid re-establishment of plants during disturbance. *P.m.* is phenotypically very plastic (*Biol Fl*). Thus, *P.m.* may adopt a prostrate form on paths (*Biol Fl*) and, when grazed by water-fowl, can produce short, much-branched plants bearing seed. In unfavourable habitats as few as three flowers may be formed, whereas large specimens can produce 1200 (*Biol Fl*). The species shows a high level of in-breeding, and there are a large number of morphological differences between populations (Hammerton 1965). *P.m.* is very similar to *P. lapathifolia* (L.) Gray, both in form and in ecology. *P.l.*, which has $2n = 22$ and tends to be more robust, exhibits a more southerly distribution in Britain and is less common, but only a few ecological differences between the two species have been established (Hammerton 1967). *P.m.* hybridizes with *P.l.* and three other species in Britain (Stace 1975).

Current trends Apparently decreasing on arable land, particularly in cereal crops. Future status on wasteland uncertain.

Altitude
Slope
Aspect

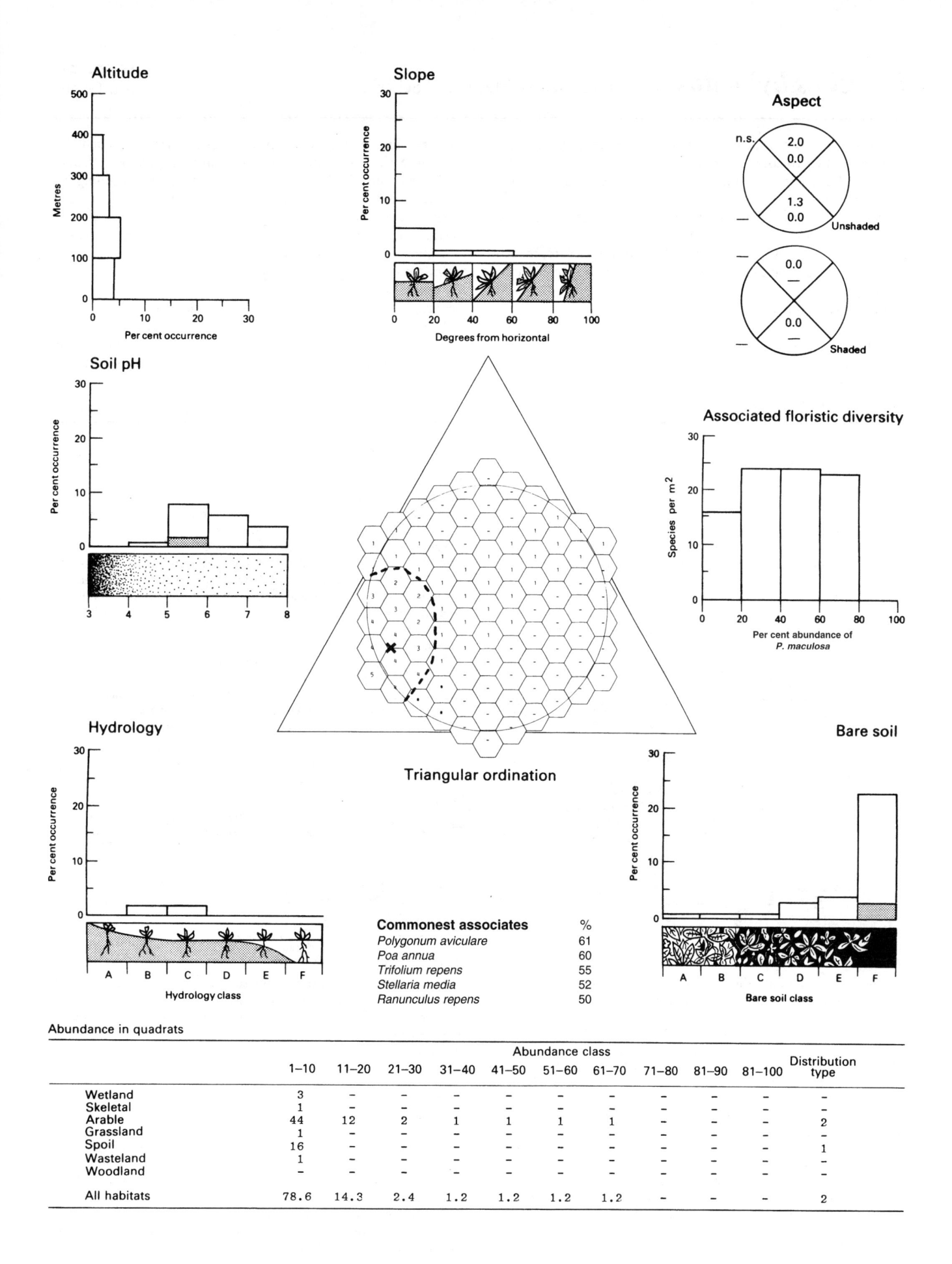

Unshaded
n.s.	2.0
—	0.0
—	1.3
—	0.0

Shaded
—	0.0
—	—
—	0.0
—	—

Soil pH

Associated floristic diversity

Hydrology

Triangular ordination

Bare soil

Commonest associates	%
Polygonum aviculare	61
Poa annua	60
Trifolium repens	55
Stellaria media	52
Ranunculus repens	50

Abundance in quadrats

	1–10	11–20	21–30	31–40	41–50	51–60	61–70	71–80	81–90	81–100	Distribution type
Wetland	3	–	–	–	–	–	–	–	–	–	–
Skeletal	1	–	–	–	–	–	–	–	–	–	–
Arable	44	12	2	1	1	1	1	–	–	–	2
Grassland	1	–	–	–	–	–	–	–	–	–	–
Spoil	16	–	–	–	–	–	–	–	–	–	1
Wasteland	1	–	–	–	–	–	–	–	–	–	–
Woodland	–	–	–	–	–	–	–	–	–	–	–
All habitats	78.6	14.3	2.4	1.2	1.2	1.2	1.2	–	–	–	2

Petasites hybridus (L.) P. Gaertner, B. Meyer and Scherb.

Asteraceae
Butterbur

Life-form Polycarpic perennial, rosette-forming geophyte with an extensive rhizome system. Scapes erect; leaves orbicular hairy, long-stalked, >0.5 m^2; roots sometimes with VA mycorrhizas.
Phenology Flowers March to May, before leaf expansion. Seed set May to July. Leaves die in late autumn.
Height Leaves up to 1.5 m; 'male' flowering spikes reach 400 mm, with the longer-lived 'female' spike extending to 800 mm in fruit.
Seedling RGR No data. SLA 18.4. DMC 20.1.
Nuclear DNA amount 1.8 pg; $2n = 60*$ (*BC*); $2x*$ (base number of tribe, $n = 10$).
Established strategy Competitor.
Regenerative strategies V, W. Regeneration biology is complex (see 'Synopsis'); SB I.

Flowers Florets pinkish-violet, dioecious, insect-pollinated; 16-55 in each 'male' capitulum (including 1-5 peripheral female flowers) and 32-130 in each 'female' capitulum (including 0-5 central sterile flowers) (*FE4*). Over 100 capitula in each spike-like panicle.
Germinule 0.26 mg, 2.5 × 0.6 mm, achene dispersed with a pappus of hairs up to 9 mm in length.
Germination Epigeal; t_{50} 1 day; L = D.
Biological flora None.
Geographical distribution The British Isles except for the N of Scotland (99% of British vice-counties), *ABIF* page 659; Europe northward to Scandinavia (62% of European territories and 13% naturalized); N and W Asia; introduced in N America. (See also 'Synopsis'.)

Habitats Most commonly associated with unshaded or lightly-shaded marshes on river floodplains. Present locally in open woods and at the margins of shallow ponds and ditches. Recorded as isolated occurrences on soil heaps and on arable land. Also observed on road verges and railway banks.
Gregariousness A stand-forming species whose large size precludes the possibility of high-frequency values.
Altitude Mainly in lowland habitats, but observed up to 340 m. [Up to 380 m.]

Habitats

Slope Mostly on flat or gently sloping sites, a reflection of an association with floodplains.
Aspect No significant trends.
Hydrology Associated with moist rather than waterlogged soils, but often occurs in sites which are flooded during the winter. Elsewhere observed most frequently in soligenous mire.
Soil pH Not recorded from soils of pH <6.0.
Bare soil Generally associated with high exposures, partly because the dense canopy of *P.h.* tends to exclude other species and partly because in many sites there is scouring and silt deposition during winter flooding.
Δ ordination Associated with relatively undisturbed and productive vegetation types.
Associated floristic diversity Relatively low as a result of dominance exercised by *P.h.*

Similarity in habitats	%
Impatiens glandulifera	76
Myrrhis odorata	73
Alnus glutinosa (Juv.)	70
Alliaria petiolata	69
Festuca gigantea	67

Synopsis *P.h.* is a robust rhizomatous herb of open or partially shaded river and stream terraces, usually growing on moist, fertile alluvial soils. The leaves, which may approach 1 m in diameter, are supported by petioles up to 2 m in length. They are by a considerable margin the most massive leaves found within the native British flora, and are in marked contrast with those of other competitive dominants (e.g. *Epilobium hirsutum*) which create dense shade by means of a large number of small leaves supported on an erect stem. Mobilization of reserves in the stout rhizome allows a dense leaf canopy to be produced rapidly during late spring and early summer. In many habitats *P.h.* is associated with species such as *Ranunculus ficaria* which have complementary vernal phenologies. Because of its morphology, *P.h.* is particularly vulnerable to grazing, trampling, cutting and other disturbances during summer and autumn. The roots of *P.h.* abscise at the end of the growing season and are replaced each spring. *P.h.* regenerates mainly by means of rhizome growth to form extensive patches. Older rhizomes break down to leave isolated daughter plants, and rhizome pieces may be transported to other sites during flooding of river banks and periods of erosion. *P.h.* is dispersed effectively along waterways in this manner and, since in some drainage systems all colonies are of the same sex, the possibility of a common vegetative parentage seems likely. The tendency for *P.h.* to be restricted to river-sides may arise from the poor mobility of its rhizome fragments. *P.h.*, which flowers in early spring, before leaf expansion, is dioecious. Where male and female plants coexist abundant seed is often produced. The seeds are short-lived and wind-dispersed, and germinate immediately after shedding in spring or in early summer. *P.h.* does not form a persistent seed bank, and establishment of new colonies by seed is probably rare. Indeed, since the female plant has a restricted geographical distribution (*CSABF* p. 72), regeneration by seed is unlikely over much of Britain. Female plants are also infrequent in parts of N Europe. Where only males are found, providing a good early source of nectar for bees, *P.h.* may be introduced (*FE4*). A further ssp. is recorded from Europe (*FE4*).
Current trends Although an occasional colonist of roadsides and railway banks, *P.h.* has rather specific habitat requirements, and seems likely to remain only locally abundant.

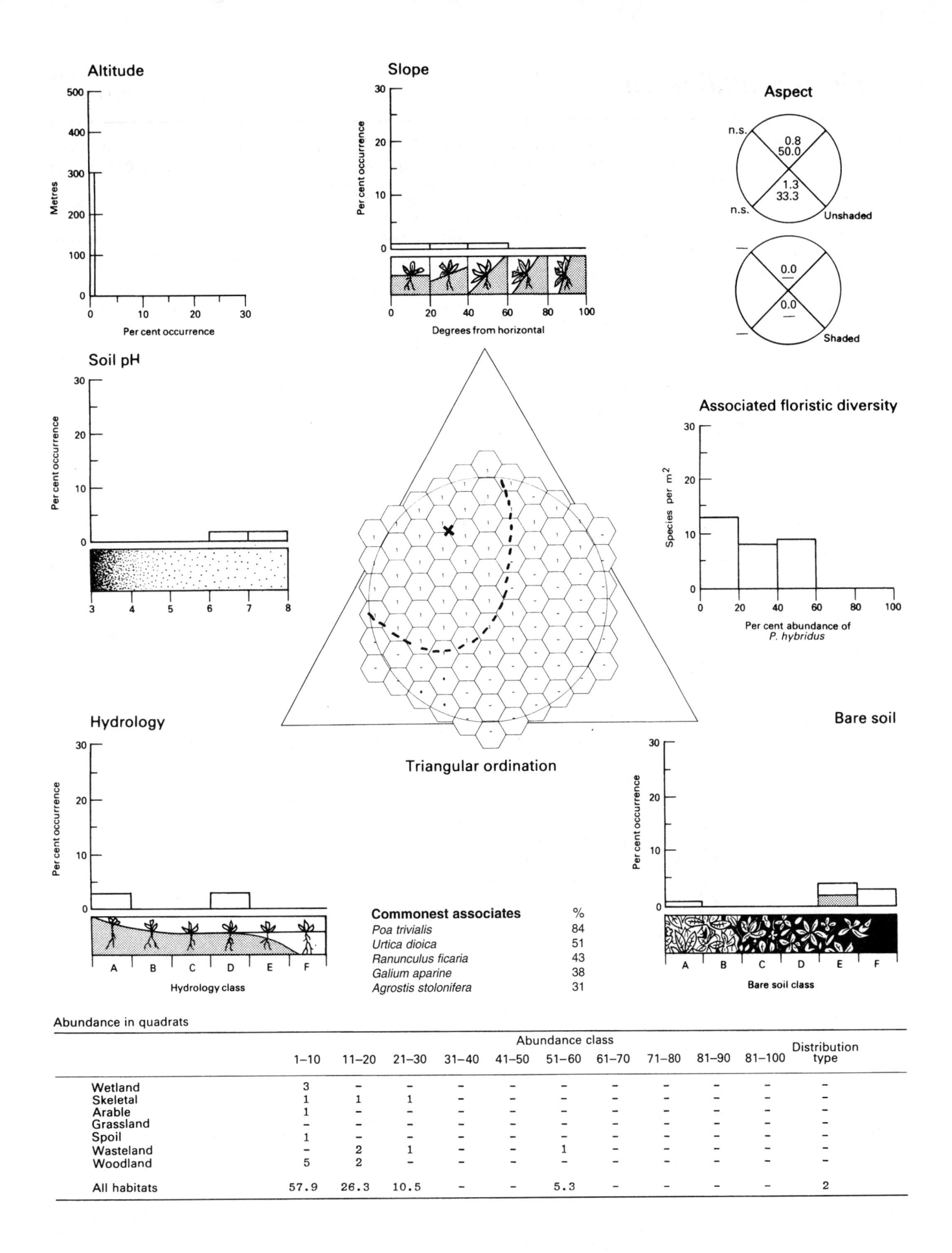

Altitude

Slope

Aspect

	0.8	
n.s.	50.0	
	1.3	
n.s.	33.3	Unshaded

	0.0	
—	—	
	0.0	
—	—	Shaded

Soil pH

Associated floristic diversity

Species per m^2 vs Per cent abundance of *P. hybridus*

Hydrology

Triangular ordination

Bare soil

Commonest associates	%
Poa trivialis	84
Urtica dioica	51
Ranunculus ficaria	43
Galium aparine	38
Agrostis stolonifera	31

Abundance in quadrats

	\multicolumn{10}{c}{Abundance class}	Distribution type
	1–10	11–20	21–30	31–40	41–50	51–60	61–70	71–80	81–90	81–100	
Wetland	3	–	–	–	–	–	–	–	–	–	–
Skeletal	1	1	1	–	–	–	–	–	–	–	–
Arable	1	–	–	–	–	–	–	–	–	–	–
Grassland	–	–	–	–	–	–	–	–	–	–	–
Spoil	1	–	–	–	–	–	–	–	–	–	–
Wasteland	–	2	1	–	–	1	–	–	–	–	–
Woodland	5	2	–	–	–	–	–	–	–	–	–
All habitats	57.9	26.3	10.5	–	–	5.3	–	–	–	–	2

Phalaris arundinacea L.

Poaceae
Reed-grass, Reed Canary-grass

Life-form Polycarpic perennial helophyte with extensive system of creeping rhizomes. Shoots erect; leaves linear, sometimes >4000 mm^2; roots deep (Hubbard 1984) with VA mycorrhizas.
Phenology New shoots elongate in spring. Flowers in June and July and sets seed from July to October. Shoots die in autumn.
Height Foliage and flowers to 1200 mm.
Seedling RGR 1.0-1.4 week^{-1}. SLA 26.2. DMC 26.3.
Nuclear DNA amount 11.4 pg, $2n = 42$* (Bennett & Smith 1976); $2n = 28, 42$ (*FE5*); $4x, 6x$*.
Established strategy Competitor.
Regenerative strategies V, B$_s$. Forms extensive stands by means of long rhizomes, and capable of regeneration by seed (see 'Synopsis').

Flowers Green, hermaphrodite, wind-pollinated, self-incompatible (Frankel & Galun 1977); often >1000 in a dense panicle (1 per spikelet).
Germinule 0.67 mg, 1.9 × 1.0 mm, caryopsis with an attached lemma (4.2 × 1.2 mm).
Germination Epigeal; t_{50} 3 days; 7-35°C; germination improved by dry storage at 20°C; or according to Vose (1962), by scarification; L = D.
Biological flora None.
Geographical distribution The British Isles (100% of British vice-counties), *ABIF* page 773; Europe except for the Mediterranean region (85% of European territories); temperate Asia, N America and S Africa.

Habitats A wetland species of frequent occurrence at the margins of rivers and streams, lakes, ponds and ditches, and in both shaded and unshaded mire. Also occasionally found on damp walls and embankments adjacent to wetland habitats.
Gregariousness Stand-forming.
Altitude Extending to 250 m, but suitable habitats are more frequent at lower altitudes. [Up to 410 m.]
Slope As with other wetland species, largely restricted to shallow slopes.
Aspect Insufficient data.
Hydrology Frequent in both soligenous and topogenous mire. Extends in some waterside habitats into shallow (<250 mm) water.

Soil pH Recorded only from soils of pH >5.0.
Bare soil Associated with varying exposures, but stands of *P.a.*, which are not subject to flooding often accumulate a dense surface layer of leaf litter.
△ **ordination** Confined to productive, relatively undisturbed vegetation.
Associated floristic diversity Relatively low, particularly at high frequencies of *P.a*.
Synopsis *P.a.* is a tall, strongly rhizomatous grass which forms extensive stands in mire and along water-courses. This potential dominant produces a dense canopy of broad leaves in summer, and is highly productive in unmanaged, undisturbed sites (Buttery & Lambert 1965). The growth of other flowering plants may be suppressed also by the considerable

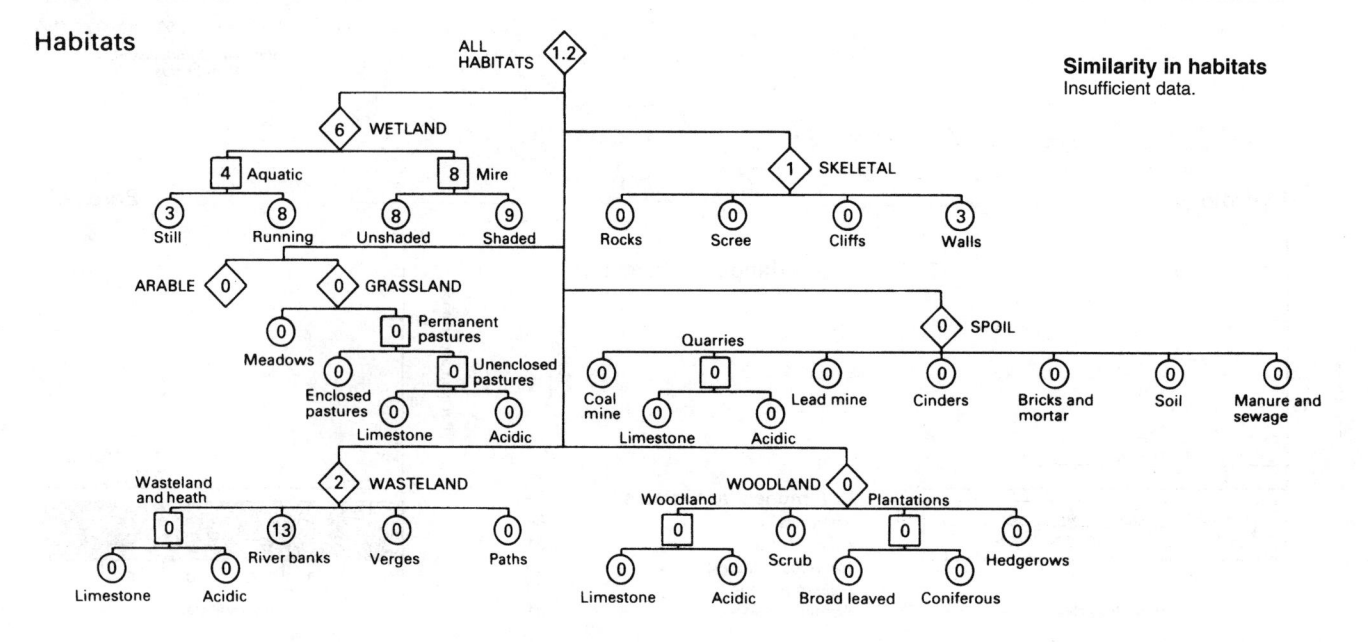

Habitats

Similarity in habitats
Insufficient data.

quantities of litter which accumulate when the shoots die back each autumn. *P.a.* is a flood-tolerant species, and an increased amount of aerenchyma is produced in the roots after inundation (Smirnoff & Crawford 1982). However, the species is deeply rooted (Hubbard 1984) and is most characteristic of habitats which are incompletely waterlogged in summer, e.g. river banks and the upper drier parts of topogenous mire. The species tends to be confined to drier sites than, for example, *Glyceria maxima* and *Phragmites australis*, and in aquatic habitats is mainly recorded from sites with flowing water. Although normally found in unmanaged sites, *P.a.* is tolerant of intermittent cutting. The young shoots, which are palatable to stock, may be exploited for hay and for grazing on land subject to flooding (Hubbard 1984). However, the plant contains alkaloids and may be highly toxic to sheep (Sculthorpe 1967). The species regenerates mainly by rhizomes to form large clonal patches. Nodal

portions of vegetative shoots are also capable of regeneration, and fragments which float offer a potential means of colonizing river systems. Despite being self-incompatible, seed production is often prolific, but there is frequent infection of the inflorescence by ergot (*Claviceps purpurea* (Fr.) Tul.). Distribution by means of floating seeds (Sculthorpe 1967) is probably only important in the colonization of new sites. Germination, which in the field occurs in spring, may be accelerated by scarification, dry storage, stratification or fluctuating temperatures (Colbury 1953, Vose 1962, Hoffman *et al.* 1980), and the formation of a persistent seed bank is suspected. Less-robust plants with $2n = 14$, $2x$, ssp. *rotgesii* (Fouc. & Mandon ex Husnot) Kerguelen also occur in Europe (*FE5*).
Current trends Uncertain. Not particularly imperilled by current patterns of land-use.

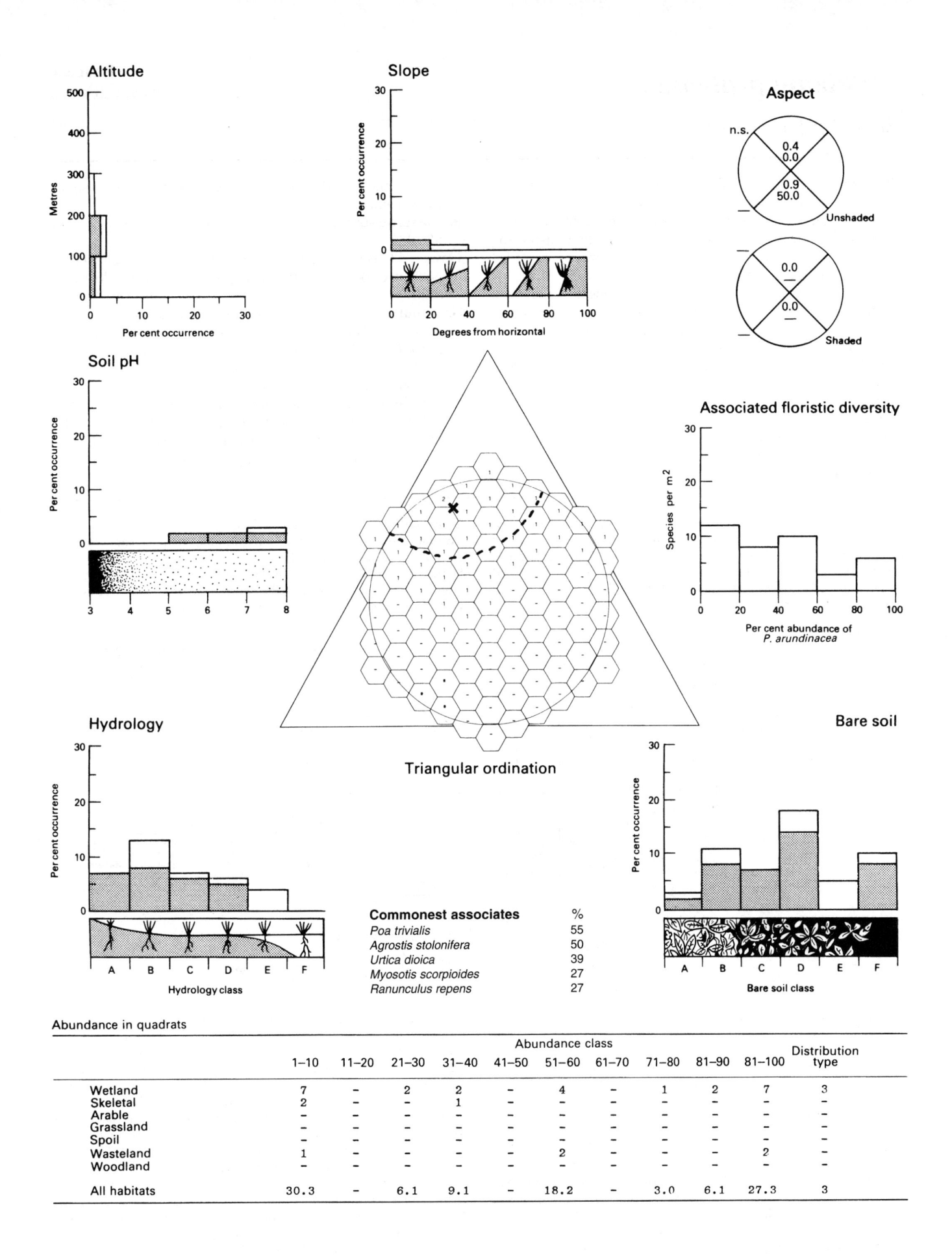

Altitude

Slope

Aspect

n.s.
0.4	—
0.0	0.9
—	50.0
Unshaded

—	—
0.0	—
0.0	—
Shaded

Soil pH

Associated floristic diversity

Species per m² vs Per cent abundance of *P. arundinacea*

Hydrology

Triangular ordination

Bare soil

Commonest associates	%
Poa trivialis	55
Agrostis stolonifera	50
Urtica dioica	39
Myosotis scorpioides	27
Ranunculus repens	27

Abundance in quadrats

	1–10	11–20	21–30	31–40	41–50	51–60	61–70	71–80	81–90	81–100	Distribution type
Wetland	7	–	2	2	–	4	–	1	2	7	3
Skeletal	2	–	–	1	–	–	–	–	–	–	–
Arable	–	–	–	–	–	–	–	–	–	–	–
Grassland	–	–	–	–	–	–	–	–	–	–	–
Spoil	–	–	–	–	–	–	–	–	–	–	–
Wasteland	1	–	–	–	–	2	–	–	–	2	–
Woodland	–	–	–	–	–	–	–	–	–	–	–
All habitats	30.3	–	6.1	9.1	–	18.2	–	3.0	6.1	27.3	3

(*)*Phleum pratense* L.

Poaceae
Timothy, Cat's-tail

Includes *P. bertolonii* DC.

Life-form Loosely tufted, polycarpic perennial proto-hemicryptophyte with a relatively shallow root system. *P.p.*: shoots typically erect with linear leaves up to *c.* 2000 mm². *P.b.*: often decumbent and sometimes with leafy non-rooting stolons; leaves typically <500 mm². Mycorrhizas are recorded for the aggregated species.

Phenology *P.p.* Tillers winter green and elongate from April onwards. Growth reaches maximum in spring, often with a second peak in July (Langer 1956). Flowers produced in June to July and sets seed from July to August. *P.b.* similar.

Height *P.p.*: foliage mainly <400 mm; flowers to 1200 mm. *P.b.*: foliage and flowers often <300 mm.

Seedling RGR No data. *P.p*: SLA 30.6. DMC 28.5. *P.b*: SLA 27.9. DMC 26.7.

Nuclear DNA amount *P.p.*: 8.3 pg; predominantly $2n = 42$ (FE5); $6x$. *P.b.*: 3.4 pg; $2n = 14$; $2x$.

Established strategy *P.p*: Intermediate between competitive-ruderal and C-S-R. *P.b*: Intermediate between stress-tolerant ruderal and C-S-R.

Regenerative strategies *P.p*: ?B_s, S. *P.b*: V, ?B_s (See 'Synopsis'); SB ?III.

Flowers Green, hermaphrodite, wind- and typically cross-pollinated (Beddows 1931); in a spike-like panicle (1 flower per spikelet).

Germinule *P.p.*: 0.45 mg, 1.6 × 0.8 mm, caryopsis dispersed within a spikelet bearing stiff hairs and short awns (3.5 × 1.2 mm). *P.b.*: 1.2 × 0.6 mm, dispersule (2.7 × 1.0 mm).

Germination *P.p.*: epigeal; t_{50} 2 days; 11-36°C; L > D.

Biological flora None.

Geographical distribution For aggregate spp.: the British Isles (100% of British vice-counties), *ABIF* page 784; N, W and C Europe (92% of European territories but *P.b.* rare in the N); introduced into most other temperate countries.

Habitats A majority of field records refer to *P.p.*. *P.p.* is a common sown constituent of meadows and pastures. Also found on waysides and various types of spoil heap. A frequent casual in arable fields. In less-productive conditions tends to be replaced by ssp. *b.*, which extends onto droughted rock outcrops, dry sandy places, paths and short turf. Both ssp. absent from heavily shaded or permanently waterlogged sites.

Gregariousness Sparse to intermediate.
Altitude Suitable habitats are more common in the uplands; both ssp. extend to c. 350 m. [Up to 500 m.]
Slope Widely distributed. *P.b.* is the more frequent on steeper slopes.
Aspect Widely distributed; *P.b.* has a bias towards S-facing aspects.
Hydrology Only of casual occurrence on permanently waterlogged soils.

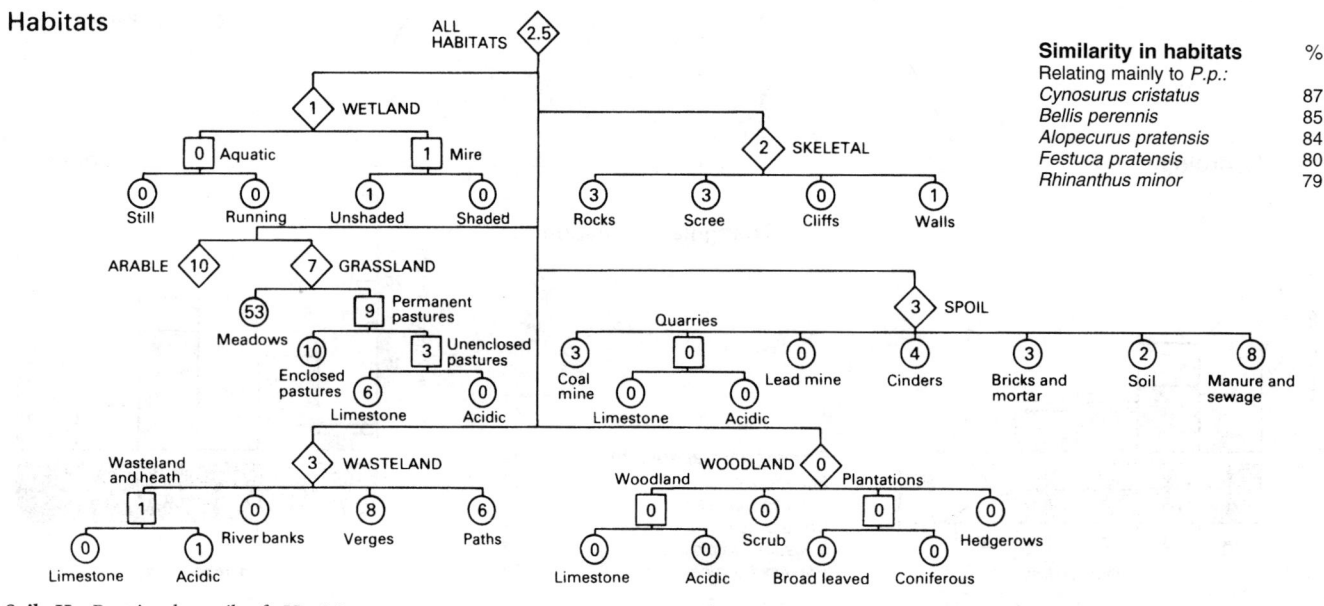

Habitats

Similarity in habitats	%
Relating mainly to *P.p.*:	
Cynosurus cristatus	87
Bellis perennis	85
Alopecurus pratensis	84
Festuca pratensis	80
Rhinanthus minor	79

ALL HABITATS 2.5
- WETLAND 1
  - Aquatic 0
    - Still 0
    - Running 0
  - Mire 1
    - Unshaded 1
    - Shaded 0
- ARABLE 10
- GRASSLAND 7
  - Permanent pastures 9
    - Unenclosed pastures 3
      - Limestone 6
      - Acidic 0
  - Meadows 10
    - Enclosed pastures 53
- SKELETAL 2
  - Rocks 3
  - Scree 3
  - Cliffs 0
  - Walls 1
- SPOIL 3
  - Quarries 3
    - Coal mine 0
    - Limestone 0
    - Lead mine 4
    - Acidic 0
  - Cinders 3
  - Bricks and mortar 3
  - Soil 2
  - Manure and sewage 8
- WASTELAND 3
  - Wasteland and heath 1
    - Limestone 0
    - Acidic 1
  - River banks 0
  - Verges 8
  - Paths 6
- WOODLAND 0
  - Woodland 0
    - Limestone 0
    - Acidic 0
    - Broad leaved 0
    - Coniferous 0
  - Plantations 0
    - Scrub 0
    - Hedgerows 0

Soil pH Restricted to soils of pH >5.0.
Bare soil Associated with habitats having widely different amounts of exposed soil.
△ **ordination** Present only in vegetation types of intermediate to high productivity, with distribution centred on moderately disturbed vegetation. Occurrences in less-productive vegetation correspond to *P.b.*
Associated floristic diversity Typically intermediate. Most of the occurrences in species-rich vegetation correspond to *P.b.*
Synopsis *P.p.* is a tall, tufted, winter-green grass now close to extinction in what are presumably its native habitats (water meadows and other low-lying moist grassland) (Hubbard 1984), but widely cultivated in fertile hay meadows and pastures, from which it escapes frequently. The species is very cold-tolerant (Spedding & Diekmahns 1972) and some growth may occur during winter, with tillers being produced in spring and, to a lesser extent, in autumn (Langer 1956). Because of this phenology *P.p.* can withstand some grazing during winter and early spring. Like other grasses producing tall nodal stems in summer, *P.p.* grows poorly if subjected to heavy summer grazing, and is perhaps best suited to regimes involving only occasional cutting or grazing (Spedding & Diekmahns 1972, Parry & Butterworth 1981). *P.p.* is typical of moist habitats, and can survive short periods of waterlogging in winter (Spedding & Diekmahns 1972). This taxon is intolerant of trampling. Although individual plants may survive for at least 20 years (Spedding & Diekmahns 1972), observations in the Sheffield region suggest that *P.p.* is only a transient constituent of meadows and pastures. This is probably related to the absence of a buried seed bank and to the virtual absence of vegetative spread in this laxly tufted plant, both features which limit the capacity to colonize gaps in vegetation. *P.p.* flowers during its first summer and, like *Arrhenatherum elatius*, is unusual among British pasture grasses in not having a chilling requirement for floral induction (Gardner & Loomis 1953).

Altitude
Slope
Aspect

	n.s.	
	1.6	
	0.0	
	1.8	
	25.0	Unshaded

	—	
	0.0	
	—	
	0.0	Shaded

Soil pH

Associated floristic diversity

Hydrology

Triangular ordination

Bare soil

Commonest associates	%
Poa trivialis	88
Trifolium repens	74
Lolium perenne	70
Holcus lanatus	67
Agrostis stolonifera	65

Abundance in quadrats

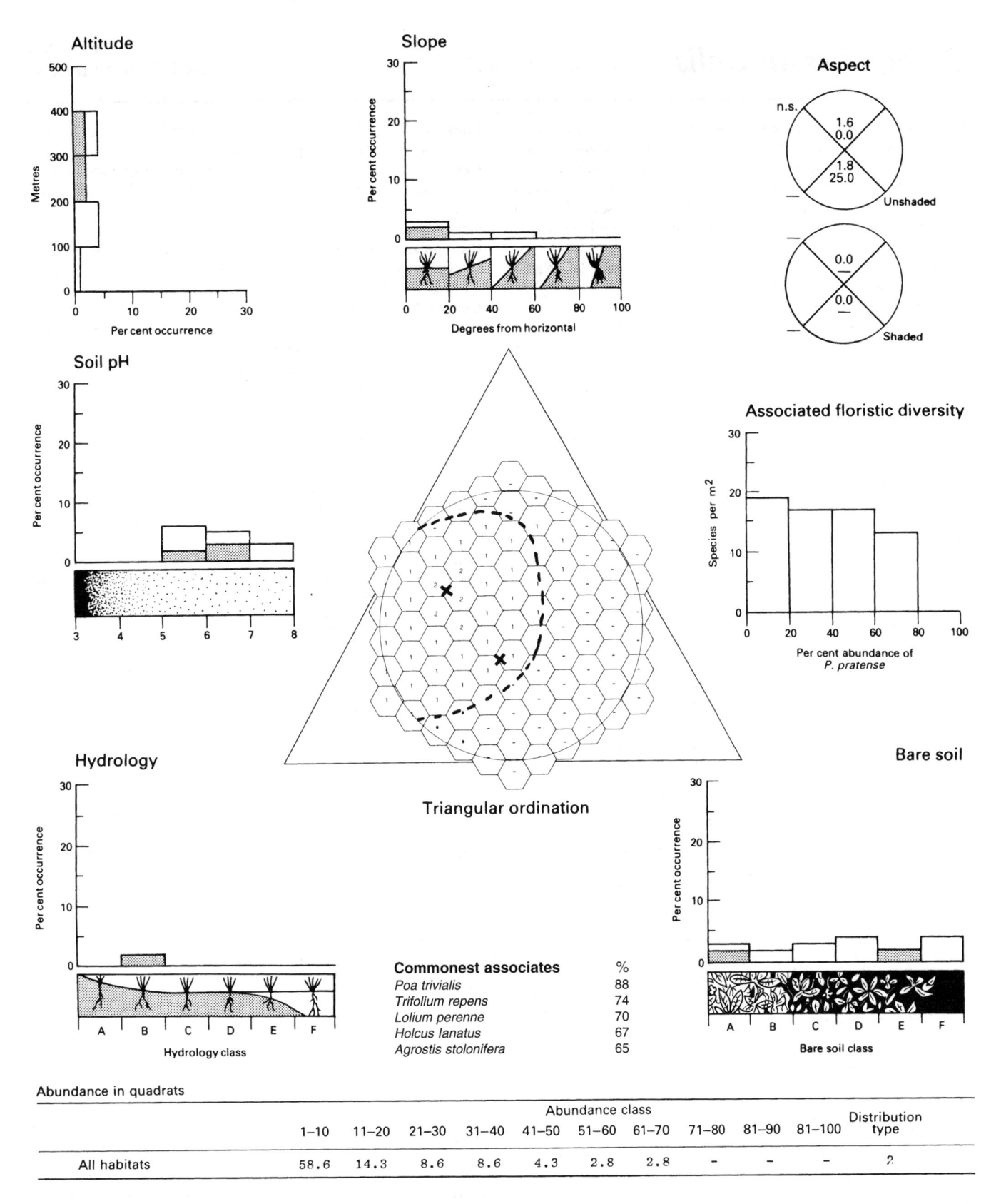

	1–10	11–20	21–30	31–40	41–50	51–60	61–70	71–80	81–90	81–100	Distribution type
All habitats	58.6	14.3	8.6	8.6	4.3	2.8	2.8	–	–	–	2

P.b. is a much smaller, relatively long-lived plant which, within the Sheffield region, is particularly characteristic of sandy soils and rocky ground in calcareous pasture. In many respects its ecology is similar to that of *P.p.*, but *P.b.* is generally found in less-fertile and drier habitats. Although sometimes cultivated, *P.b.* is native in many sites and shows some capacity for vegetative spread. Chancellor (1979) suggests that seed may persist in the soil for 20 years but this is at variance with data for *P.p.* and is in need of confirmation.

Current trends *P.p.*: nine cultivars are recommended (ADAS 1984) and over 1400 t of seed are purchased each year (MAFF 1980-84) for use in short-term grassland and, to a lesser extent, in turf. A decrease in the amount sown (predicted by Parry & Butterworth (1981)) would drastically reduce the abundance of *P.p.* Approximately 20 t of seed of *P.b.* are supplied annually (MAFF 1980-84) for use in permanent pasture or turf. It is not known to what extent the sowing of these cultivated strains compensates for the loss of native populations in semi-natural habitats.

Phragmites australis (Cav.) Trin. ex Steudel

Poaceae
Reed, Common Reed

Life-form Polycarpic perennial helophyte or hydrophyte with long woody rhizomes. Shoots erect, bearing linear leaves which may attain 10 000 mm^2; roots often with VA mycorrhizas.

Phenology New shoots elongate from April onwards. Flowers not produced until August or September. Ripe seed dispersed from November onwards, generally after the shoot has become senescent.

Height Inflorescence overtops foliage and may attain 2 or even 3 m in height.

Seedling RGR No data. SLA 15.5. 34.3.

Nuclear DNA amount 2.3 pg; $2n$ = 36, 44, 46, 48*, 49, 50, 51, 52, 54, 96 (*BC, FE5*); $3x$, $4x^*$, $5x$, $8x$.

Established strategy Competitor.

Regenerative strategies V, W. Primarily by vegetative means (see 'Synopsis').

Flowers Purplish, hermaphrodite, wind-pollinated; up to 1000 in a loose panicle (2-6 per spikelet).

Germinule 0.12 mg, 1.0 × 0.5 mm, caryopsis dispersed with an attached lemma and a basal tuft of long silky hairs (8.6 × 1.2 mm).

Germination Epigeal.

Biological flora Haslam (1972); see also Bjork (1967), Nikolajevskii (1971), van der Toorn (1972) and Rodewald-Rudescu (1974).

Geographical distribution The British Isles, particularly in the S and E (100% of British vice-counties), *ABIF* page 800; most of Europe (89% of European territories); virtually cosmopolitan in temperate regions of the world, but less abundant in America than in the Old World, uncommon in the tropics.

Habitats In a range of wetland habitats, including mire, drainage ditches, ditch sides and pond or lake margins. Also found in seepage zones of coalmine heaps. Observed beside slow-moving rivers and in wet areas resulting from mining subsidence. [Tolerant of brackish water and often occurs at the transition between fresh water and salt marsh (*Biol Fl*); also occasionally in wet places on sheltered sea cliffs.]

Gregariousness Characteristically forming extensive stands, a feature not immediately apparent from the data presented here, which include a high proportion of ditch-side and colonizing populations.

Altitude A lowland species observed up to 230 m. [Extends to 400-500 m (Haslam *et al*. 1975).]

Slope Associated with flat ground and slopes <20°.

Aspect Insufficient data.

Hydrology In topogenous mire beside water, and in sites liable to winter flooding; more sporadic in soligenous mire. Also present in shallow water (*c*. 250 mm) and occasionally persisting in >600 mm. [Occurs at depths of up to 2 m in still water (Haslam *et al*. 1975).]

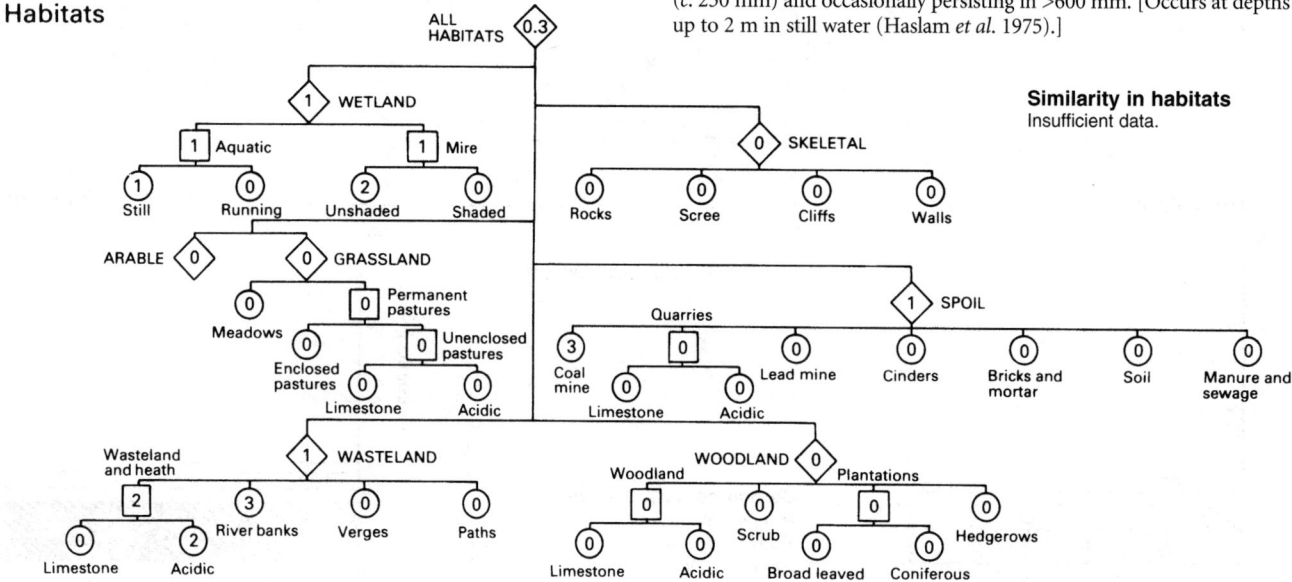

Habitats

Similarity in habitats
Insufficient data.

Soil pH Virtually confined to soils of pH >4.5.

Bare soil Associated with a wide range of values both for bare soil and persistent litter.

△ **ordination** Insufficient data.

Associated floristic diversity Intermediate at low frequencies of *P.a.* and dropping to extremely low values at higher frequencies, corresponding to stands where the species is functioning as a vegetation dominant.

Synopsis *P.a.* is a long-lived (perhaps to 1000 years (*Biol Fl*)), wetland grass with annual bamboo-like stems and rhizomes which may remain functional for *c*. 5 years (Fiala 1976). *P.a.* is the tallest non-woody species in the British flora (to over 3 m), forms a dense canopy with in places >100 shoots m^{-2} (*Biol Fl*) and shows extensive rhizome growth. A covering of persistent litter may also accumulate on the soil surface except in sites liable to cutting or flooding. As a result *P.a.* forms virtual monocultures by competitive exclusion in more fertile sites. *P.a.* is susceptible to intensive grazing, cutting or trampling. Often damaged by ground frost in spring (van der Toorn 1972) and showing its most vigorous growth in warmer climates (*Biol Fl*). *P.a.* has well-developed aerenchyma (*Biol Fl*) and shows slightly greater persistence in flowing water than *Glyceria maxima* (*Biol Fl*). *P.a.* can regenerate after occasional ploughing. *P.a.* spreads vigorously by means of rhizomes, and up to 25% of shoots may produce an inflorescence containing hundreds of spikelets. However, seed set is usually very low and the young seedlings are very vulnerable to flooding (*Biol Fl*). Shoots detached as a result of disturbance (e.g. maintenance work on drainage ditches and winter flooding) float readily and are capable of regenerating. Thus, dispersal and establishment often involves fragmentation rather than the wind-dispersal of seeds. However, effective long-distance dispersal, apparently by wind, does occur and *P.a.* was one of the first colonists of Krakatau (Ridley 1930). It is frequently an early colonist of spoilage in the S Yorks coal-field. Existence of a persistent seed bank remains uncertain. Apart from being one of the world's most widely distributed angiosperms, *P.a.* is also one of the world's worst weeds,

Altitude

Slope

Aspect

Unshaded

Shaded

Soil pH

Associated floristic diversity

Hydrology

Triangular ordination

Commonest associates	%
Myosotis scorpioides	43
Agrostis stolonifera	29
Carex acutiformis	29
Poa trivialis	29
Rorippa x sterilis	29

Bare soil

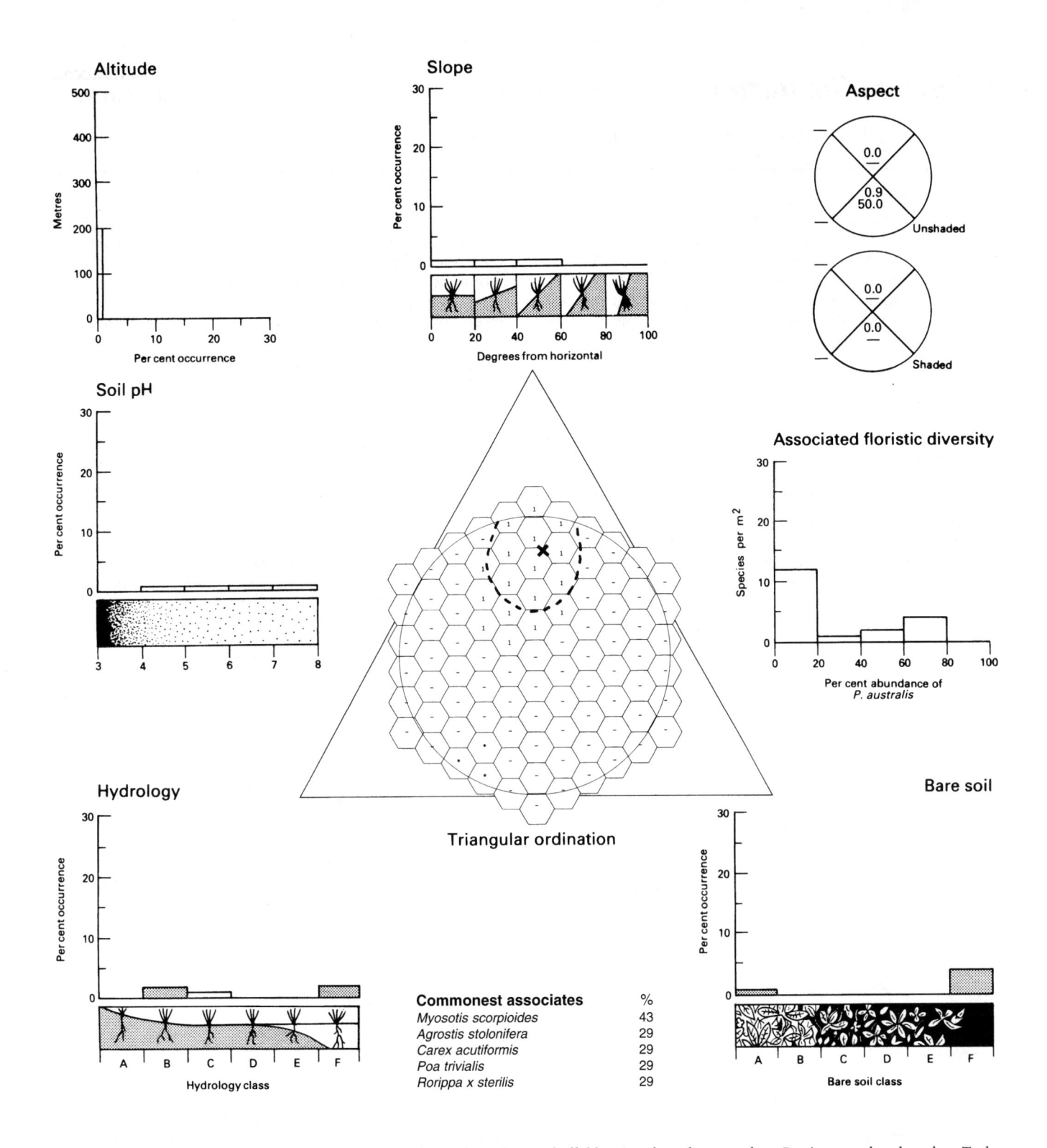

capable of blocking up waterways and drainage canals (Holm et al. 1977). The species is also of considerable value to humankind. In Britain the persistent stem litter was formerly extensively cropped during winter for thatching, and is still harvested locally. *P. a.* has a high protein content (Hendry & Brocklebank unpubl.) and the species is used for fodder, fuel, fertilizer and paper-making (Holm et al. 1977). Although ecologically wide-ranging and to some extent phenotypically plastic and existing as a number of cytotypes, *P.a.* shows comparatively little morphological variability in Britain (*Biol Fl*). In the Netherlands a peatland ecotype with short shoots and high shoot density has been distinguished from a river margin form which has fewer larger shoots and is more tolerant of salt (van der Toorn 1972). These genotypes of *P.a.* with long stems (van der Toorn 1969) replace other tall riverside species beside brackish water. Within the Sheffield region the only areas where *P.a.* is more abundant than *Typha latifolia* and *Glyceria maxima* are low-lying 'warp' lands formerly flooded with brackish water, colliery flashes, where Na and Cl concentrations may also be high (Caswell et al. 1984a,b), and on the magnesian limestone, where abnormally high levels of Mg in the water may be predicted.

Current trends *P.a.* is an effective colonist of some new habitats, particularly colliery waste, and may even have been planted in some of its extant sites. Severe losses of *P.a.* have occurred within recent historical times through the widespread drainage of wetlands. Also, in some areas *P.a.* appears to have been replaced by *Glyceria maxima*, which may perhaps exploit better than *P.a.* the lowered water table associated with increased drainage. A continuing decline in the abundance of *P.a.* is predicted.

Pilosella officinarum F. W. Schultz & Sch. Bip.

Asteraceae
Mouse-ear Hawkweed

A taxonomically difficult grouping (see 'Synopsis')

Life-form Polycarpic perennial, rosette hemicryptophyte with numerous prostrate leafy stolons. Scape erect; leaves lanceolate, hairy, often <1000 mm^2; roots with VA mycorrhizas.
Phenology Winter green. Flowers May and June (to August). Seed set within a few weeks of flowering.
Height Foliage normally appressed to the ground, but erect and up to 100 mm in taller vegetation. Scapes to 300 mm.
Seedling RGR 1.0-1.4 week^{-1}. SLA 20.4. DMC 18.8.
Nuclear DNA amount 7.9 pg; 2n = 18, 36*, 45*, 54*, 63 (*BC, FE4*); 2x, 4x*, 5x*, 6x*, 7x.
Established strategy Intermediate between stress-tolerator and C-S-R.
Regenerative strategies V, W. Regeneration biology is complex (see 'Synopsis'); SB I.

Flowers Florets yellow, ligulate, hermaphrodite; often c. 90-100 aggregated into a terminal capitulum (Bishop *et al.* 1978). Seeds produced by apomixis in the case of 5-7x plants or sexually (2x, 4x) and pollinated by insects.
Germinule 0.15 mg, 2.0 × 0.5 mm, achene with pappus c. 5.5 mm in length.
Germination Epigeal; t_{50} 3 days; 6-27°C; germination rate increased after dry storage at 20°C; L = D.
Biological flora Bishop and Davy (1994).
Geographical distribution The British Isles except Shetland (84% of British vice-counties), *ABF* page 298; temperate and sub-Arctic Europe (85% of European territories); W Asia; introduced in N America.

Habitats A plant of dry habitats, particularly on calcareous soils. Most frequent in calcareous pastures, and on scree, wasteland, rock outcrops and quarry, and lead-mine waste. Occurring more-locally on cliffs, road verges and paths, and in non-calcareous habitats such as coal-mine spoil and cinder tips. Observed only infrequently in non-calcareous pastures. Absent from wetlands and woodland. [Commonly found on sand dunes, and occurring in montane habitats.]
Gregariousness Sometimes patch-forming.
Altitude Widely distributed, but suitable habitats are particularly abundant on the carboniferous limestone, and distribution shows an upland bias. Extends to 415 m. [Up to 920 m.]
Slope Occurs most frequently on Derbyshire dale-sides, and hence distribution biased towards moderate slopes.

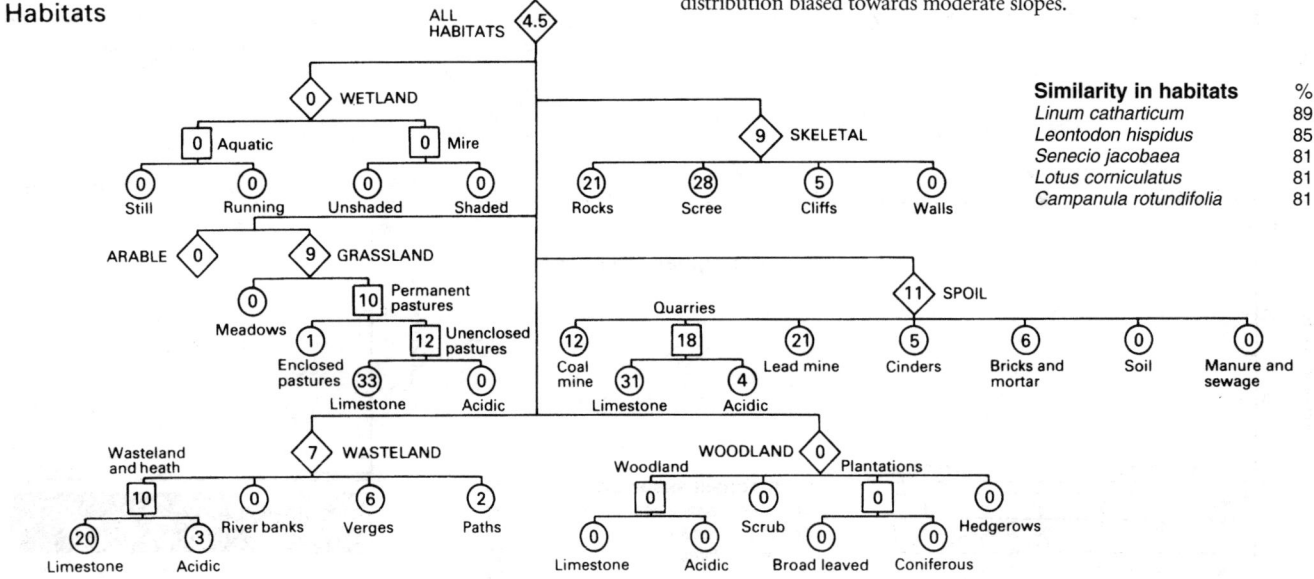

Habitats

Similarity in habitats	%
Linum catharticum	89
Leontodon hispidus	85
Senecio jacobaea	81
Lotus corniculatus	81
Campanula rotundifolia	81

Aspect Significantly more occurrences on S-facing slopes.
Hydrology Absent from wetland.
Soil pH Almost totally restricted to soils of pH >4.5.
Bare soil Widely distributed, but more frequently associated with moderate exposures of bare ground.
△ **ordination** Distribution fairly wide-ranging, but concentrated on vegetation associated with relatively unproductive habitats subject to some disturbance. Excluded completely from habitats which are highly disturbed, and scarcely represented in vegetation dominated by large, fast-growing species.
Associated floristic diversity Intermediate to species-rich, but declining in vegetation containing very high densities of *P.o.* (e.g. screes and rock outcrops).
Synopsis *P.o.* is a low-growing stoloniferous herb of dry, often calcareous soils. The leaves of *P.o.* are palatable to a range of herbivores, and *P.o.* typically forms appressed rosettes in grazed grassland. Although the foliage may be held upright in taller vegetation, the low stature of the shoot causes *P.o.* to be vulnerable to the shade of taller plants when pastures become derelict. In the *PGE P.o.* is restricted to unproductive grassland, and in experimental plots in chalk grassland, production and flowering were stimulated by addition of nitrogen (Lloyd & Pigott 1967). The leafy stolons allow ramification over rocky and unstable substrata, and on rock outcrops daughter rosettes are sustained at a considerable distance from the points at which roots penetrate into soil crevices. However, *P.o.* has shallow roots (Anderson 1927) and appears to show some susceptibility to drought (Bishop *et al.* 1978). In turf microcosms, roots of this species became heavily colonized by VA mycorrhizas with increases of up to 9-fold in seedling yield (Grime, Mackey, Hillier & Read 1987). *P.o.* regenerates largely by means of long stolons which give rise to rosettes and subsequently die back leaving unattached daughter plants. Stolons are produced only from rosettes which have produced a floral initial, even if this subsequently aborts (Bishop *et al.* 1978). Rosettes are frequently

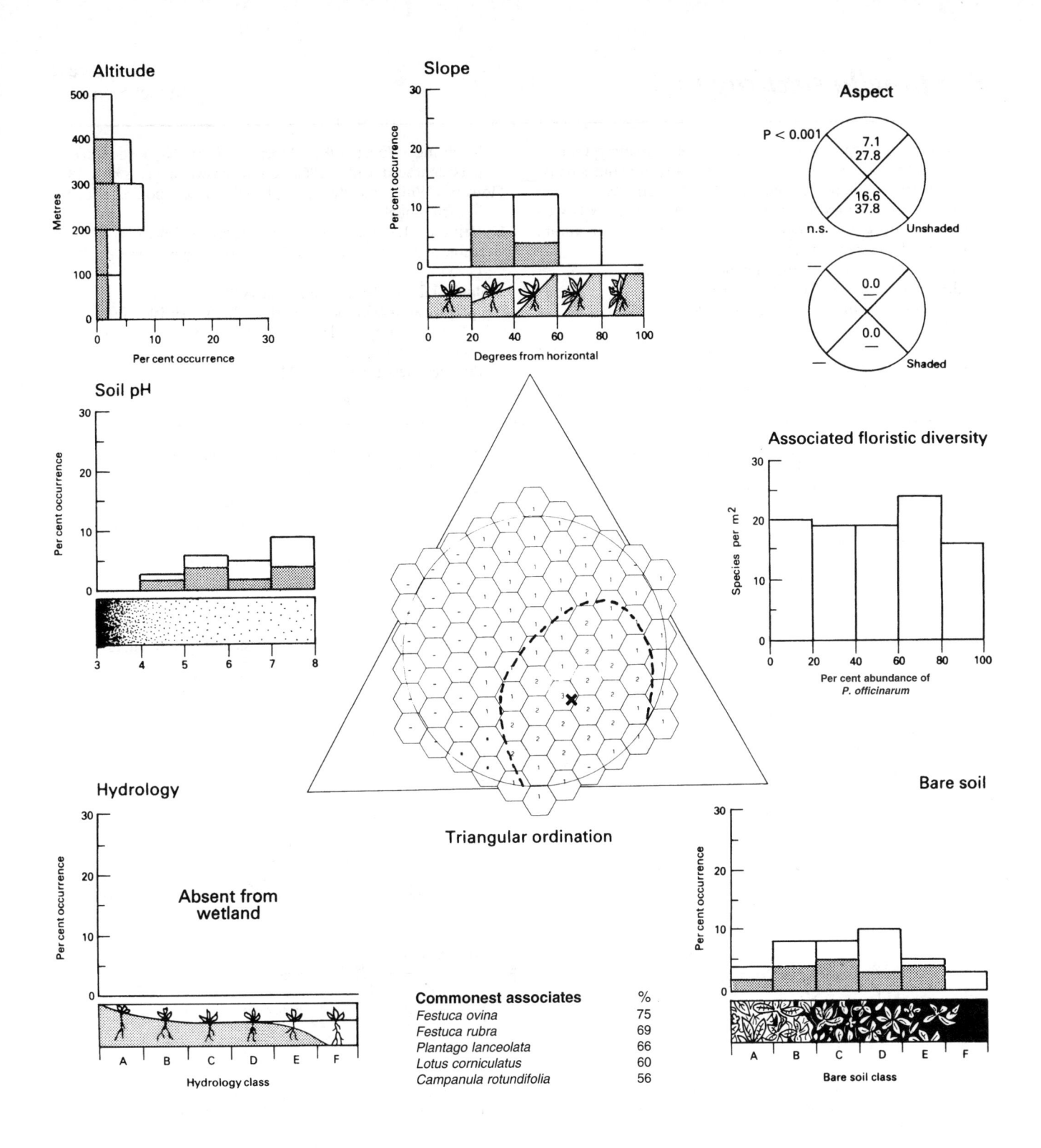

Altitude
Slope
Aspect
Soil pH
Associated floristic diversity
Hydrology
Triangular ordination
Bare soil

Commonest associates %
Festuca ovina 75
Festuca rubra 69
Plantago lanceolata 66
Lotus corniculatus 60
Campanula rotundifolia 56

short-lived, and considerable population fluctuations may occur as a result of the dynamics of stolon production and mortality (Bishop et al. 1978). *P.o.* also produces wind-dispersed seed, particularly in little-grazed sites. Seedlings have been observed in vegetation gaps mainly during autumn (Hillier 1984). A delay may occur between dispersal and germination but no persistent seed bank is developed (Roberts 1986a). In established grassland seedling establishment appears to be extremely rare, and Watt (1947, 1962) suggests that it takes place mainly where there is a low intensity of grazing and a wet spring. However, *P.o.* appears fairly widely on calcareous spoil, suggesting that regeneration from seed is effective in colonizing new habitats. $2x$ and $4x$ plants are sexual but higher ploidy levels are apomictic (Gadella 1972). Thus, in many cases reproduction by seed is an asexual process. *P.o.* has been subdivided into several ssp. differing in height and geographical distribution (*FE4*), and several occur in Britain. Data presented refer to the aggregate species, but both $4x$ (sexual) and $5x$ (apomictic) plants are recorded from the Sheffield region. The ecological differences between taxa and genotypes within the *P.o.* complex are poorly understood. A heavy-metal-tolerant race of *P.o.* has been described locally (Shaw 1984).

Current trends An effective colonist of artificial habitats such as quarries and railway banks, but perhaps decreasing due to the dependence upon infertile habitats.

Pimpinella saxifraga L.

Apiaceae
Burnet Saxifrage

Life-form Polycarpic perennial, semi-rosette hemcryptophyte with stout stock. Stems erect; leaves pinnate, sometimes hairy, <4000 mm^2; deep root system, with VA mycorrhizas.
Phenology Winter green. Flowers July to August and sets seed from August onwards. Flowering shoots die back in autumn. Leaves relatively long-lived.
Height Foliage typically <100 mm; flowers to 1 m.
Seedling RGR No data. SLA 15.1. DMC 28.5.
Nuclear DNA amount 10.2 pg; $2n$ = 20, 36, 40 (*FE2*, Moore 1973); $2x$, $4x$.
Established strategy Intermediate between stress-tolerant ruderal and C-S-R.
Regenerative strategies S. Regenerates by seed which germinates in spring (Hillier 1984, Roberts 1986a). No evidence of a persistent seed bank (1986a). Clonal growth appears rare; SB II.
Flowers White, mostly hermaphrodite, insect-pollinated, >100 in each umbel.
Germinule 1.2 mg, 2.7 × 1.7 mm, mericarp (2 per ovary).
Germination Epigeal; t_{50} 15 days; dormancy broken by chilling; L > D.
Biological flora None, but see Zarzycki (1976).
Geographical distribution Most of the British Isles, but local in N Scotland and N Ireland (94% of British vice-counties), *ABIF* page 460; most of Europe except the extreme S (77% of European territories); Asia Minor and E Asia.

Habitats Most common on calcareous strata. Frequent in grazed and ungrazed grassland, with local occurrences on limestone quarry heaps, outcrops, screes, lead-mine spoil, waysides and cliffs. Not recorded from arable land, woods or wetland.
Gregariousness Intermediate.
Altitude Most abundant at altitudes below 300 m, but extending to 350 m. [Up to 810 m.]
Slope Occurrences concentrated on limestone dale-sides of moderate slope.
Aspect Abundance slightly greater on N-facing slopes.
Hydrology Absent from wetlands.
Soil pH Frequent above pH 5.0, but rare below this value.
Bare soil Associated with low to moderately high exposures of bare soil.
△ **ordination** Distribution centred on vegetation associated with relatively undisturbed, unproductive habitats.
Associated floristic diversity Typically species-rich.

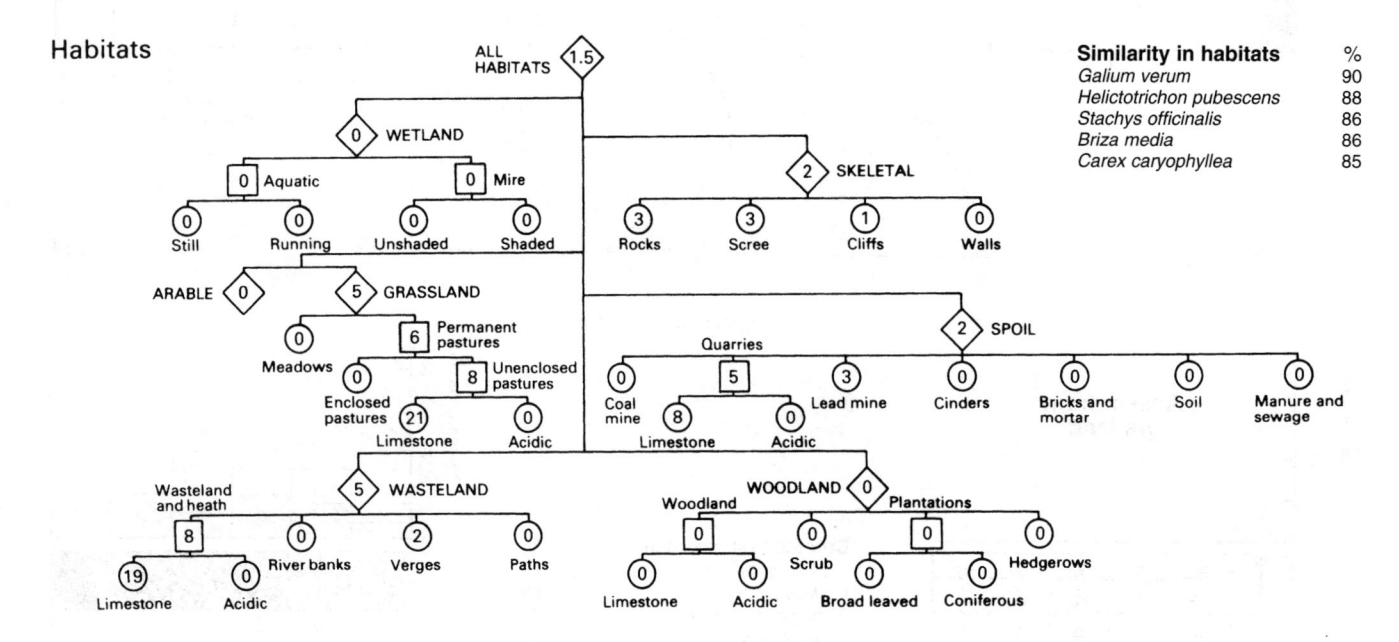

Habitats

Similarity in habitats	%
Galium verum	90
Helictotrichon pubescens	88
Stachys officinalis	86
Briza media	86
Carex caryophyllea	85

Synopsis *P.s.* is a rosette-forming, winter-green herb of short turf and rocky habitats on base-rich, infertile soils. *P.s.* is low-growing and tends to be suppressed in derelict or fertilized grassland. Restricted to unproductive plots in the PGE, *P.s.* exhibits a capacity for sustained growth during summer, and the distribution extends into dry, but not severely droughted, habitats, where the long tap-root provides access to moisture in soil crevices. In non-flowering plants there is a basal rosette consisting of a small number of leaves, but by the time flowers are produced these have normally withered and have been replaced by a small quantity of stem foliage. Thus, even where *P.s.* attains high population densities, the contribution to the total shoot biomass of the community is small. The capacity of *P.s.* to produce offsets is less than that of many other small tap-rooted species, e.g. *Sanguisorba minor* and *Scabiosa columbaria*. Consequently, *P.s.* is even more dependent than these species upon seedling establishment for the maintenance of populations and the colonisation of new sites. In common with those of *Succisa pratensis*, fruits develop late in the growing season and, in Britain, many are in a green state at the onset of winter. The seeds have the capacity to survive ingestion by cattle (Salisbury 1964), and this may be important in pasture sites. As with many other members of the Apiaceae, seeds contain a small embryo and have a chilling requirement for germination. In the field germination occurs in a synchronous burst in spring. Seedlings examined on a N-facing slope in N Derbyshire appeared at similar densities in closed turf and in gaps, and showed comparable mortalities (*c.* 50%) over the initial 2 years (Hillier 1984). *P.s.* has a low colonizing ability, and is regarded by Wells *et al.* (1976) as an indicator species of old grassland. Although mainly found over calcareous strata, *P.s.* has been observed occasionally in a number of grassland habitats on non-calcareous rocks. *P.s.* varies in size, pubescence, leaf shape and flower colour (*FE2*). The relationship, if any, between this polymorphism and the ecological distribution of *P.s.* is not known.
Current trends Decreasing, at least in many lowland areas. The fate of this species will be closely related to that of old calcareous grasslands.

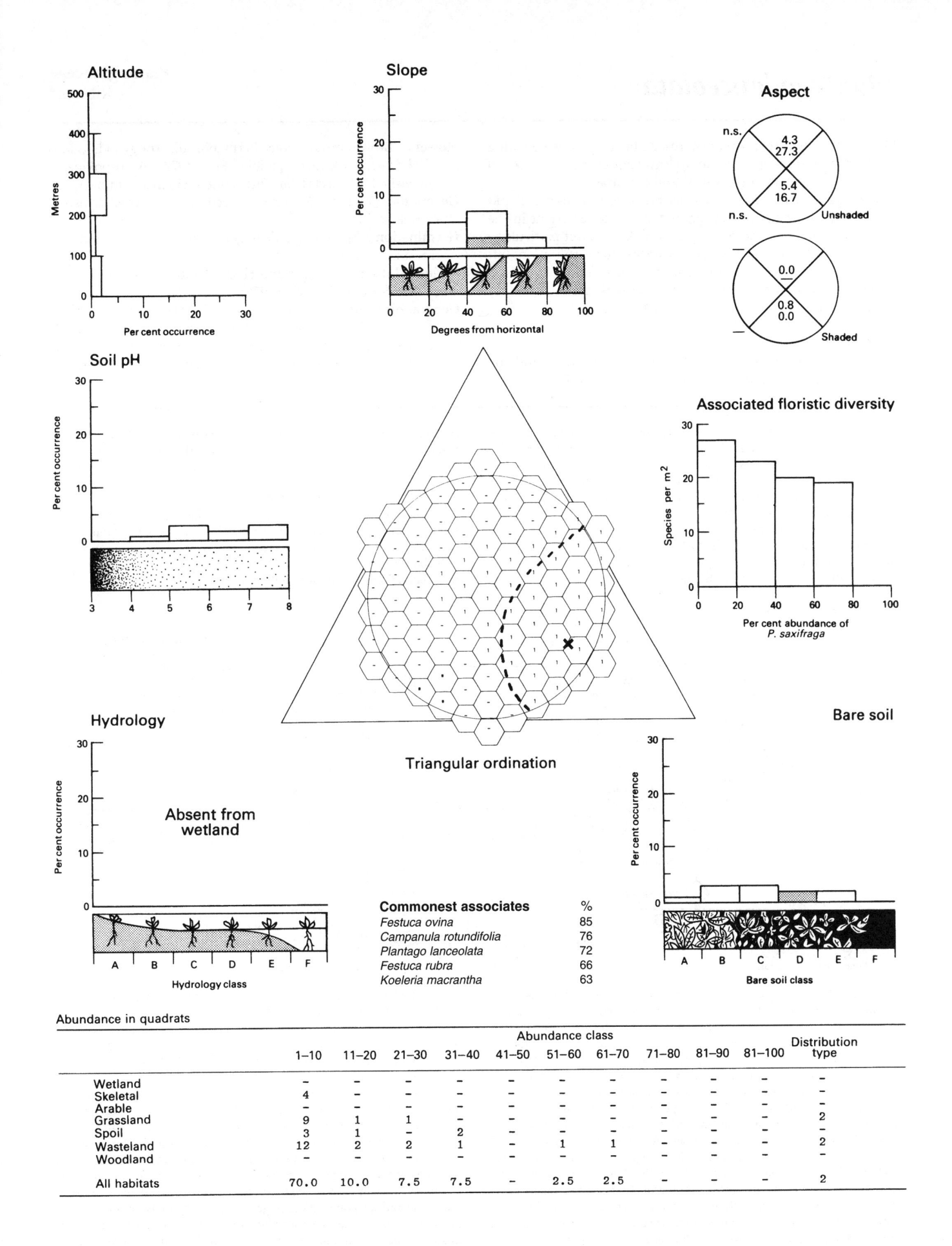

Altitude

Slope

Aspect

	n.s.	4.3	
		27.3	
	n.s.	5.4	Unshaded
		16.7	

	—	0.0	
		—	
	—	0.8	Shaded
		0.0	

Soil pH

Associated floristic diversity

Hydrology

Absent from wetland

Triangular ordination

Bare soil

Commonest associates %
Festuca ovina — 85
Campanula rotundifolia — 76
Plantago lanceolata — 72
Festuca rubra — 66
Koeleria macrantha — 63

Abundance in quadrats

	Abundance class										Distribution type
	1–10	11–20	21–30	31–40	41–50	51–60	61–70	71–80	81–90	81–100	
Wetland	–	–	–	–	–	–	–	–	–	–	–
Skeletal	4	–	–	–	–	–	–	–	–	–	–
Arable	–	–	–	–	–	–	–	–	–	–	–
Grassland	9	1	1	–	–	–	–	–	–	–	2
Spoil	3	1	–	2	–	–	–	–	–	–	
Wasteland	12	2	2	1	–	1	1	–	–	–	2
Woodland	–	–	–	–	–	–	–	–	–	–	–
All habitats	70.0	10.0	7.5	7.5	–	2.5	2.5	–	–	–	2

Plantago lanceolata L.

Plantaginaceae
Ribwort

Life-form Polycarpic perennial, rosette hemicryptophyte with a short rhizome. Scapes ascending; leaves lanceolate, hairy, <6000 mm^2; root system deep, with VA mycorrhizas.

Phenology Overwinters as a small broad-leaved rosette. Rapid leaf growth occurs in late spring and to a lesser extent in the autumn, and flowers from April until August, but particularly in June and July. Seed is ripe 2-3 weeks after fertilization, but some may overwinter in the seed head (*Biol Fl(a)*).

Height Leaves usually <150 mm; scapes to 450 mm.

Seedling RGR 1.0-1.4 week^{-1}. SLA 22.4. DMC 15.7.

Nuclear DNA amount 2.4 pg; $2n = 12^*$ (*BC*); $2x^*$.

Established strategy C-S-R.

Regenerative strategies B$_s$, V. Regenerates effectively both by vegetative means and by seed (see 'Synopsis'); SB III.

Flowers Brown, protogynous, hermaphrodite (or gynodioecious (*Biol Fl(a)*)), self-incompatible (Ross 1973), wind-pollinated and visited by insects (*Biol Fl(a)*); often <50 in a terminal spike.

Germinule 1.9 mg, 3.6 × 1.8 mm, seed in a 1-2-seeded dehiscent capsule.

Germination Epigeal; t_{50} 5 days; germination rate increased after dry storage at 20°C; L > D.

Biological flora (a) Sagar and Harper (1964a), (b) Cavers *et al.* (1980). See also Zarzycki (1976).

Geographical distribution The British Isles (100% of British vice-counties), *ABIF* page 536; Europe except for the extreme N (97% of European territories); N and C Asia; introduced into most other temperate countries, less frequent in the tropics.

Habitats Recorded abundantly from meadows, base-rich pastures, rock outcrops, quarry and lead-mine spoil, wasteland and waysides. Also frequent on other types of spoil and rocky habitats. A particular weed of lawns. Non-persistent in arable land, mire and scrub. Absent from closed woodland and aquatic habitats. [Present on sand dunes, cliff-top turf and in montane habitats (*Biol Fl(a)*).]

Gregariousness Intermediate.

Altitude Frequent at all altitudes to 450 m. [Up to 790 m.]

Slope Widely distributed.

Aspect Shows a bias towards S-facing slopes.

Hydrology Of only casual occurrence in permanently waterlogged sites.

Soil pH Particularly common on soils in the pH range 5.0-8.0, and rarely recorded from soils with pH <4.5. More common on soils overlying limestone.

Bare soil Associated with the full range of exposures.

△ **ordination** Extremely wide-ranging distribution, a reflection of the ability of the species to persist in many different types of vegetation. Most frequent in vegetation associated with reduced soil fertility and moderate disturbance.

Associated floristic diversity Often high.

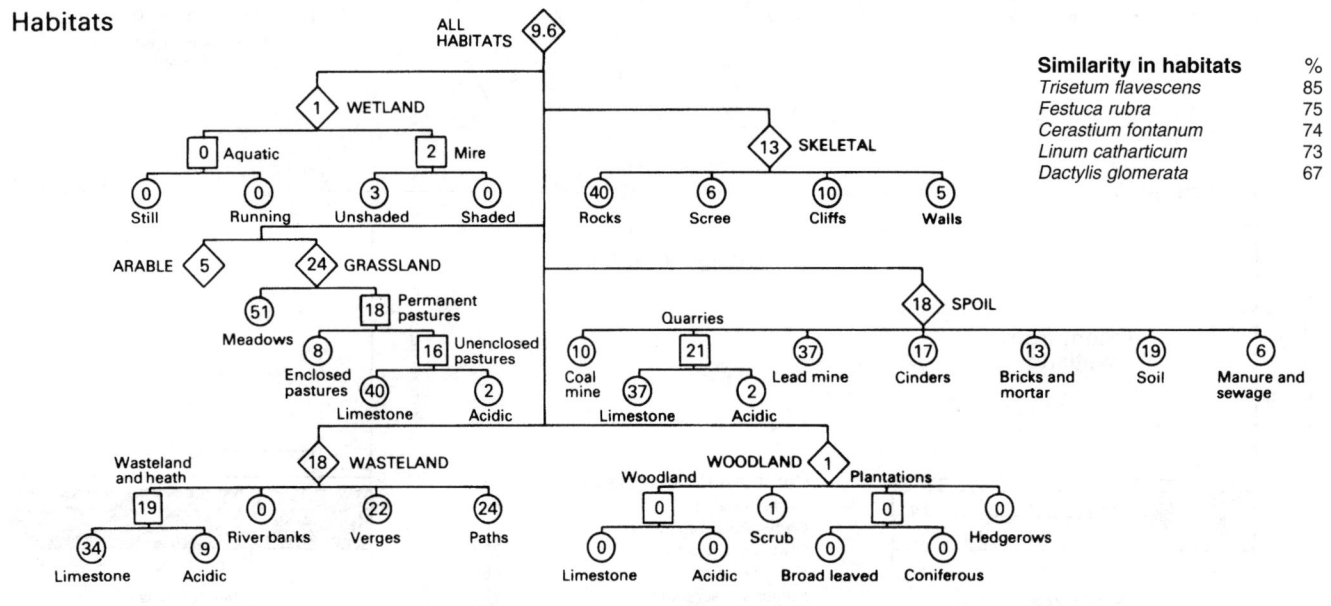

Habitats

Similarity in habitats	%
Trisetum flavescens	85
Festuca rubra	75
Cerastium fontanum	74
Linum catharticum	73
Dactylis glomerata	67

Synopsis *P.l.* is a frost-tolerant, winter-green, rosette-forming herb capable of flowering in its first year and, in some genotypes, exhibiting a potential life-span of >12 years (*Biol Fl(a)*). Although largely absent from woodland and wetlands, *P.l.* is found in a very wide range of habitats and the shoot is extremely plastic. There are ascending linear lanceolate leaves in dense vegetation, such as hay-meadows, and short ovate leaves forming prostrate rosettes in exposed or closely grazed sites. Although many roots are superficial (Anderson 1927), some penetrate deeply (*Biol Fl(a)*) and appear to confer drought avoidance in dry grassland and on S facing slopes. In the *PGE*, *P.l.* is 'chiefly associated with poor, exhausted soils' and is scarce in plots treated heavily with nitrogen. The foliage of plants collected from the Sheffield region is low in N, P and Fe, but high in Na and Ca (*SIEF*) and contains sorbitol (Hodgson 1973). In turf microcosms roots of *P.l.* were heavily colonized by VA mycorrhizas, and infection caused up to 12-fold increases in seedling yield (Grime, Mackey, Hillier & Read 1987). *P.l.* is palatable to sheep, but because of its appressed rosettes in short turf the foliage is not readily eaten in cattle pastures. The flowering heads are often heavily predated (*Biol Fl(a)*). *P.l.* has been described as one of the world's 12 most successful colonizing species, and as one of the world's most successful weeds. This status is achieved despite the fact that *P.l.* is seldom a principal weed of any one crop (Holm *et al.* 1977). The production of buds from the stem affords a method of vegetative regeneration in some established colonies (*Biol Fl(a)*). Connections between ramets subsequently decay. *P.l.* is also capable of regeneration from root fragments, but it is uncertain to what extent this is of importance in the field (*Biol Fl(a)*). Thus, *P.l.* has a limited capacity for vegetative spread and much flexibility in regeneration by seed. Some seed is shed in summer and some may overwinter on the parent plant. In one study (Blom 1979) a decline in mean seed weight was noted in response to trampling and soil compaction. Regeneration from seeds freshly dispersed into bare areas, such as those occurring in overgrazed grassland, may occur in spring and autumn. Seeds have no well-defined dispersal mechanism,

Altitude
Slope
Aspect

P < 0.05
14.6
18.9
22.0
32.7
n.s.   Unshaded

0.0
0.0
Shaded

Soil pH

Associated floristic diversity

Triangular ordination

Hydrology

Commonest associates %
Festuca rubra 77
Dactylis glomerata 59
Festuca ovina 48
Lotus corniculatus 47
Holcus lanatus 47

Bare soil

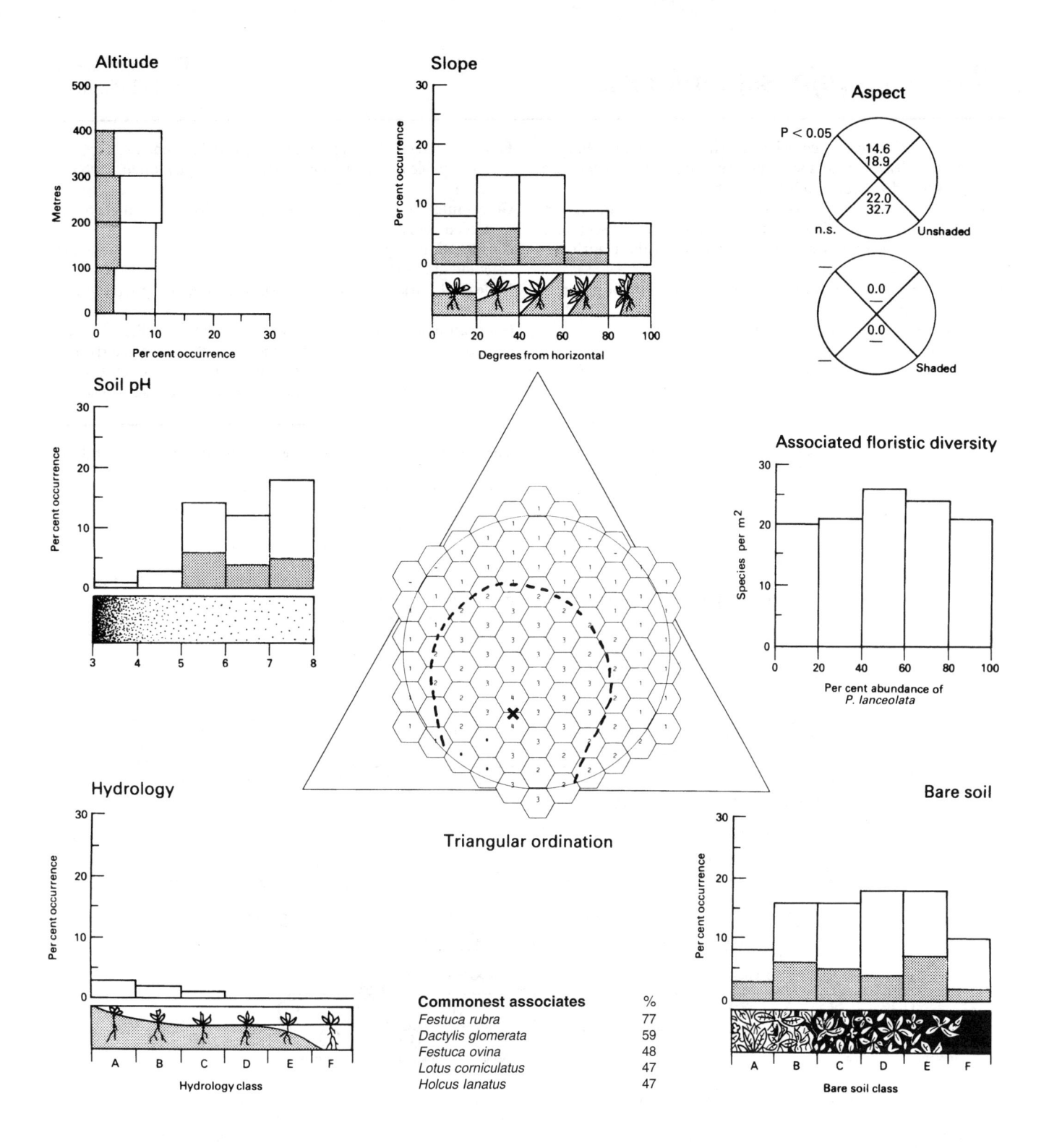

but may survive ingestion by animals; seeds also produce a sticky mucilage when wet (*Biol Fl(b)*). *P.l.* is mainly transported as a result of man's activities. The species was recorded among the 6 most common roadside plants in a large-scale survey in Denmark (Hansen & Jensen 1974), and seed may be a contaminant of hay crops. *P.l.* also forms a buried seed bank. In addition to the high level of morphogenetic plasticity, *P.l.* is also extremely variable genetically. Heavy-metal-tolerant populations occur in Derbyshire (Shaw 1984), and ecotypes differing in longevity, rosette number, leaf number and size, inflorescence shape and height, seed size and seed dormancy have been distinguished in comparisons between dune grassland and hay-meadow populations (van Groenendael 1985). Many other population studies have been carried out, and the evolution of ecotypes is potentially rapid (Pollard 1980). However, as the result of a large and detailed demographic study, Antonovics and Primack (1982) concluded that phenotypic variation was much more important than genotypic differences in determining the nature of grassland populations of *P.l.* The extremely wide range of morphological variation within this out-breeding species has resisted taxonomic classification (*FE4*).

Current trends *P.l.* has long been associated with human activities, and its abundance increased dramatically after early deforestation (*Biol Fl(a)*). *P.l.* is still very common, and appears capable of widespread persistence despite current changes in land-use.

Plantago major ssp. *major* L.

Plantaginaceae
Rat-tail Plantain

Life-form Polycarpic perennial, rosette hemicryptophyte. Scapes ascending; leaves ovate, sometimes hairy, typically <4000 mm^2; roots generally deep, with VA mycorrhizas.
Phenology Shoot apex overwinters either as a small rosette or below ground level. Flowers June to September and sets seed within *c.* 3 weeks, but seed often retained on the plant over winter. Shoots die in autumn.
Height Foliage usually <150 mm, often appressed to the ground; scape also to *c.* 150 mm.
Seedling RGR 1.5–1.9 week^{-1}. SLA 30.4. DMC 10.7.
Nuclear DNA amount 1.7 pg; $2n = 12^*$ (*BC*); $2x^*$.
Established strategy Intermediate between ruderal and C-S-R.
Regenerative strategies B$_s$. See 'Synopsis'; SB IV.

Flowers Green, hermaphrodite, wind-pollinated, protogynous but capable of full self-fertilization (*Biol Fl(a)*); often >100 in a spike.
Germinule 0.24 mg, 3.1×1.8 mm, seed dispersed from a 6–10 seeded dehiscent capsule (*FE4*).
Germination Epigeal; t_{50} 2 days; 15–34 °C dormancy broken by chilling; L > D.
Biological flora (a) Sagar and Harper (1964b), (b) Hawthorn (1974).
Geographical distribution Britain (100% of British vice-counties), *ABIF* page 535; Europe (95% of European territories and 3% naturalized); N and C Asia; naturalized in most other parts of the world, although rare in lowland tropical regions.

Habitats Most frequent on paths, tracks and at trampled field entrances. Also frequent in disturbed habitats such as demolition sites, soil heaps, cinder tips and on arable land. Occurs to a lesser extent in pasture and meadows, road verges, on river banks, in skeletal habitats, and on mire and other types of spoil. A common weed of lawns and sports fields. In woodland only in relatively unshaded areas on the rides. Absent from aquatic habitats and tall herb communities.
Gregariousness Intermediate.
Altitude Wide ranging up to 450 m, but particularly frequent in the lowlands. [Up to 630 m.]

Slope Over the full range, but most frequent and abundant in habitats associated with level ground.
Aspect Significantly more frequent on unshaded S-facing slopes, but achieves abundance in a high proportion of N-facing sites.
Hydrology Infrequent in wetlands, but occurs locally on wet poached ground beside ponds and on gravel and mud exposed during summer at the edge of ponds and reservoirs.
Soil pH Most frequent and abundant above pH 5.5, and not recorded from acidic soils of pH <4.5.
Bare soil Most frequent and abundant in disturbed habitats with large exposures, but shows some persistence once established in closed vegetation.

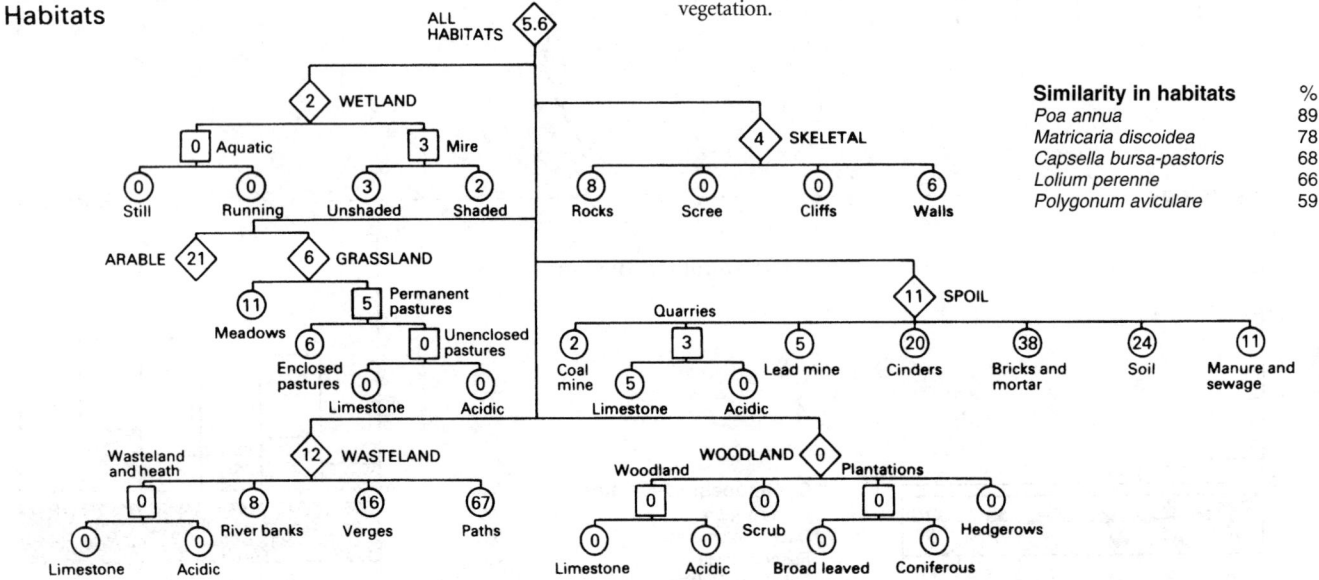

Habitats

Similarity in habitats	%
Poa annua	89
Matricaria discoidea	78
Capsella bursa-pastoris	68
Lolium perenne	66
Polygonum aviculare	59

△ **ordination** Present in a wide range of vegetation types, but distribution strongly centred on heavily disturbed, productive vegetation.
Associated floristic diversity Typically intermediate.
Synopsis *P.m.* is a rosette-forming herb found in a wide range of relatively fertile, disturbed habitats. It is exceptionally resistant to trampling and is characteristic of paths, tracks and gateways, where it usually forms appressed rosettes and is often associated with *Matricaria discoidea* and *Poa annua*. In closed vegetation the foliage is held erect. *P.m.* is frequent on compacted soils with impeded drainage, but is generally absent from sites subject to prolonged summer waterlogging. These two growth forms are to some extent under genotypic control. Erect genotypes are less tolerant of mowing and trampling than prostrate genotypes from lawns, and appear to be better attuned to tall vegetation (Warwick & Briggs 1979, 1980a,b). *P.m.* is less frequent on dry soils than *P. lanceolata*, and the foliage appears to be frost-sensitive. In grasslands *P.m.* is largely confined to short, heavily trampled turf, and in this habitat as elsewhere is strongly reliant upon seed for regeneration. *P.m.* germinates and shows effective root penetration in compacted soil (Blom 1979) but often fails to establish on loose or sandy soil. Establishment occurs in bare areas resulting from trampling or overgrazing in spring, but seedlings also appear sporadically during summer (*Biol Fl(a)*); *P.m.* often develops a large persistent seed bank. Although the leaves may be eaten by stock the fruiting heads are not consumed (*Biol Fl(a)*). Seed production may be prodigious, and an exceptionally large plant may release 14 000 seeds (*Biol Fl(b)*). Under optimal conditions seeds may be produced within 6 weeks (*Biol Fl(b)*). In a study reported by Blom (1979) an increase in average seed size was detected in plants subjected to trampling. Plants may produce daughter rosettes by means of lateral buds, and plants with several rosettes often have a higher seed output than those with a single rosette (*Biol Fl(b)*). On occasion the production of sideshoots follows mechanical damage, but vegetative regeneration is less important than in *P. lanceolata*. It is uncertain whether *P.m.* is capable of resprouting from shoot fragments

Altitude
Slope
Aspect
P < 0.01

2.4	33.3
7.2	12.5

n.s. — Unshaded

0.0	—
0.0	—

Shaded

Soil pH

Associated floristic diversity

Hydrology

Triangular ordination

Bare soil

Commonest associates	%
Poa annua	64
Trifolium repens	60
Agrostis stolonifera	58
Poa trivialis	58
Ranunculus repens	46

Abundance in quadrats

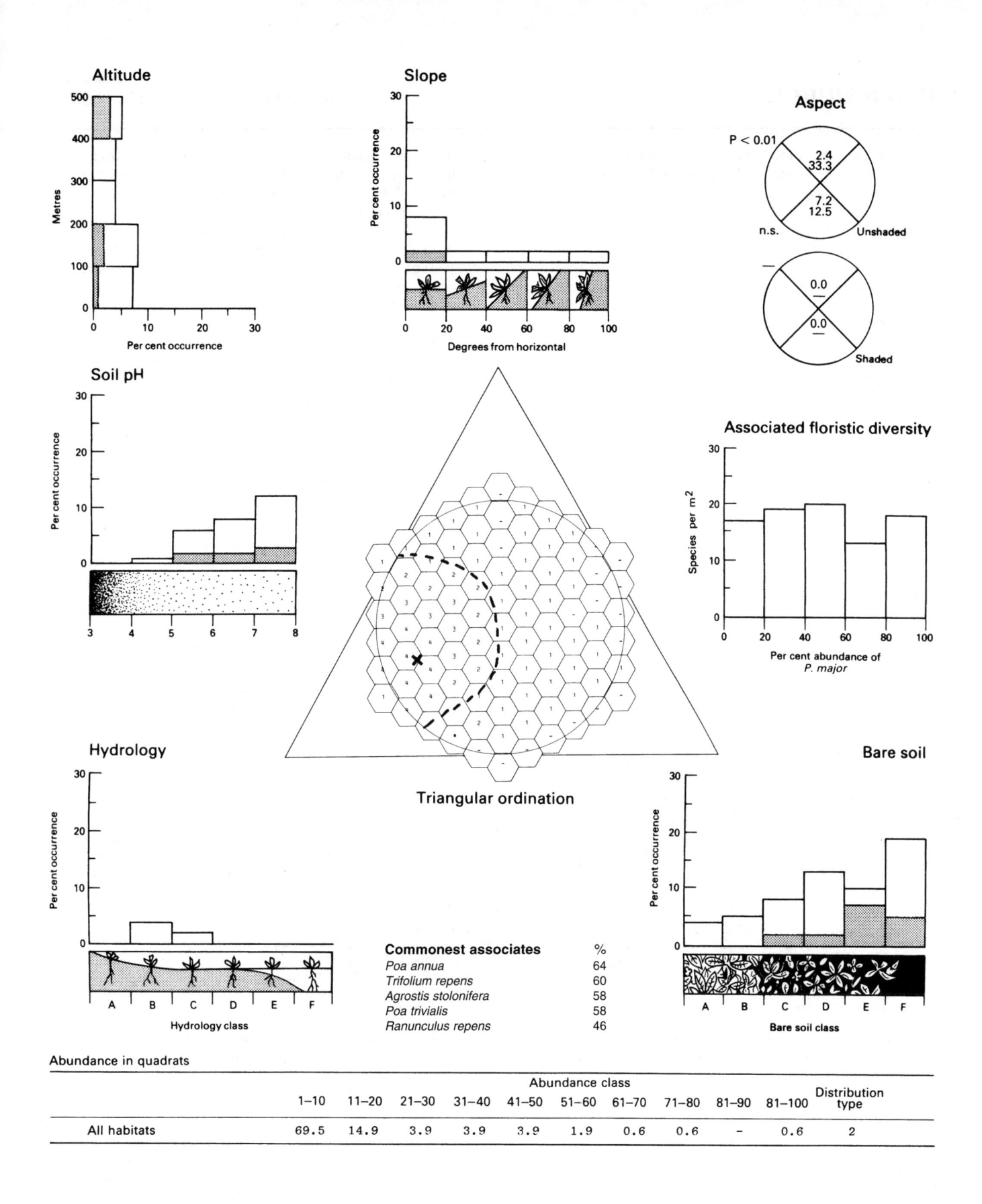

	\multicolumn{10}{c}{Abundance class}	Distribution type
	1–10	11–20	21–30	31–40	41–50	51–60	61–70	71–80	81–90	81–100	
All habitats	69.5	14.9	3.9	3.9	3.9	1.9	0.6	0.6	–	0.6	2

created by ploughing of arable land or by heavy trampling on paths. Seeds, which are mucilaginous when wet, are usually retained over winter on the parent plant (*Biol Fl(a)*) and are probably dispersed by humans, sometimes in crops (Holm *et al.* 1977). *P.m.* has colonized many parts of the world and, as a result, is identified by Holm *et al.* (1977) as one of the world's most successful weeds. Two other ssp. of *P.m.* are described in Europe, including ssp. *intermedia* (DC.) Arcangeli, which has a 14-34-seeded capsule and grows in damp, usually saline, sites (*Excur Fl*).

Current trends Further increases in abundance and ecological range may be expected.

Poa annua L.

Poaceae
Annual Poa, Annual Meadow-grass

Life-form Loosely tufted therophyte to short-lived polycarpic perennial, semi-rosette hemicryptophyte. Shoots erect, spreading or prostrate; leaves linear, typically <500 mm²; roots shallow (Salisbury 1952), with VA mycorrhizas.

Phenology Leaves, flowers and fruits may be found during all seasons in many disturbed habitats. Most typically a summer annual, but can also behave as a winter annual, particularly in droughted habitats. Perennial races tend to overwinter as leafy shoots and flower in summer. Leaves short-lived, at least in summer.

Height Foliage to 150 mm; flowers exceptionally to 300 mm.

Seedling RGR 1.5-1.9 week^{-1}. SLA 37.0. DMC 24.6.

Nuclear DNA amount 4.1 pg; $2n = 28*$ (*BC*); $4x*$.

Established strategy Typically ruderal, but some perennial forms exist.

Regenerative strategies S, B$_s$ and V in some biotypes. Annual races regenerate entirely by seed and may produce a persistent seed bank. Perennial forms also have a limited capacity for lateral vegetative spread by means of creeping rooted stems (Law *et al*. 1977), and new plants are potentially formed from shoots detached as a result of disturbance; SB III.

Flowers Green, hermaphrodite, homogamous, wind-pollinated; self-compatible and usually self-pollinated; often >100 in an open panicle (3-10 per spikelet). Cleistogamy frequent in winter (*Biol Fl*(*b*)).

Germinule 0.26 mg, 1.7 × 0.7 mm, caryopsis dispersed with an attached lemma (2.6 × 0.9 mm).

Germination Epigeal; t_{50} 2 days; 7-31°C; L > D.

Biological flora (a) Warwick (1979), (b) Hutchinson and Seymour (1982).

Geographical distribution The British Isles (100% of British vice-counties), *ABIF* page 756; Europe (97% of European territories); almost cosmopolitan in temperate zones of the world and on mountains in the tropics.

Habitats Occurs in a great variety of disturbed or trampled situations, but most common on arable land and disturbed fertile soil of spoiled land, paths and overgrazed fertile pastures. Also occurs locally in open situations on rock outcrops and walls, in marshland and on river banks. Absent from permanently flooded sites. [In maritime and montane regions mainly restricted to trampled sites.]

Gregariousness Intermediate.

Altitude Widely distributed, but suitable habitats are more frequent in lowland areas. [Up to 1200 m.]

Slope More common on gentle slopes of <20°, though recorded on all but the steepest slopes.

Aspect Wide-ranging but with slightly greater abundance on S-facing slopes.

Hydrology Found on exposed mud beside ponds and ditches, and in soligenous mire. Absent from aquatic habitats.

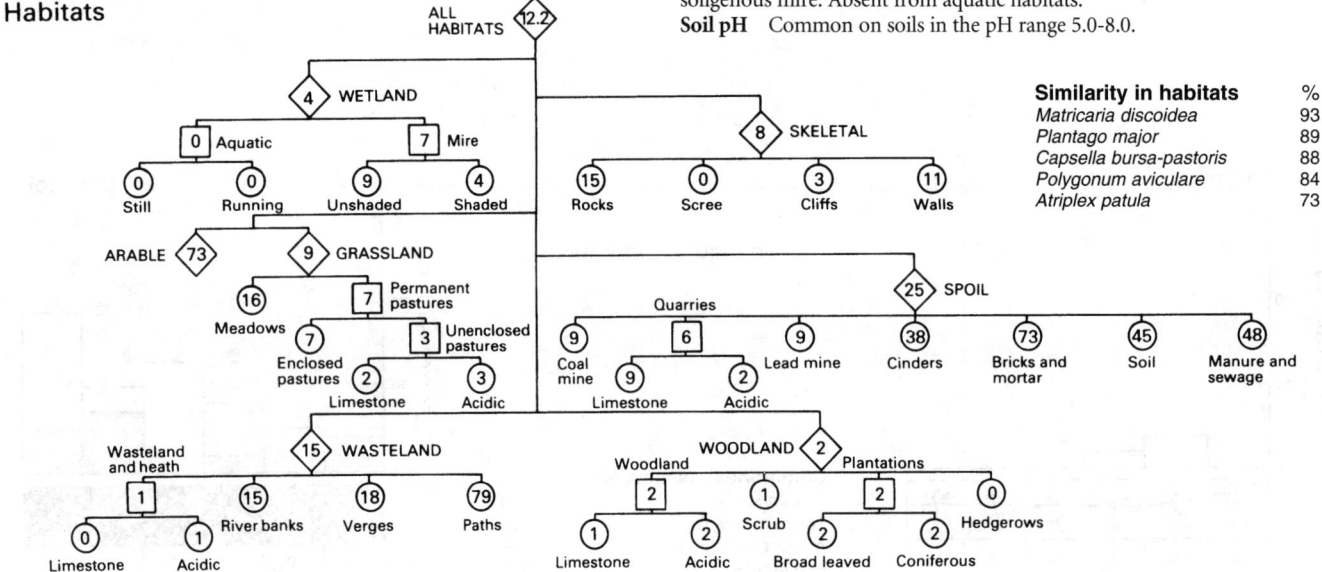

Soil pH Common on soils in the pH range 5.0-8.0.

Habitats

Similarity in habitats	%
Matricaria discoidea	93
Plantago major	89
Capsella bursa-pastoris	88
Polygonum aviculare	84
Atriplex patula	73

Bare soil Widely distributed, but showing a progressive increase in frequency and abundance towards habitats with large amounts of exposed earth.

△ ordination Distribution centred on vegetation associated with fertile, heavily disturbed conditions. However, also present in less-disturbed situations; this may be related to variation in genotype and to phenotypic plasticity displayed by the species (see 'Synopsis').

Associated floristic diversity Generally intermediate; a curvilinear relationship with maximum diversity corresponding to frequencies of *P.a.* in the range 10-70%. Frequencies for *P.a.* of <10% probably coincide with competitive dominance by other species, whereas those >70% are associated with severely disturbed species-poor habitats.

Synopsis *P.a.*, which is regarded by Webb (1985) as only doubtfully native to the British Isles, is arguably the most successful ruderal species in the British flora, with a habitat range which includes arable fields and agricultural and urban grassland. The species is among the first colonists during 'sward deterioration' of sown grassland (Morrison 1979) and is a common colonist of both winter- and summer-sown arable crops. *P.a.* also occurs in the second phase of colonization of new polders in Holland following desalination (Joenje 1978). Leaves of *P.a.* have high concentrations of N and P but are low in Ca (*SIEF*). Many factors contribute to the wide ecological amplitude, including: (a) Considerable genotypic variation and phenotypic plasticity. Most characteristically *P.a.* is a small annual plant of disturbed ground and in extreme circumstances may complete its life-cycle within 6 weeks. In contrast, perennial genotypes tend to occur in less disturbed, usually grassland, sites. Here they may not flower until their second year and can show stoloniferous growth (*Biol Fl*(*a*)). In some wetland sites *P.a.* may even approach *Catabrosa aquatica* (L.) Beauv. in size. Two-fold intraspecific variation in nuclear DNA amount (at constant ploidy) has been recorded within one population (Mowforth 1985). (b) Flexibility of regeneration. This has several components, including the capacity for immediate germination and the

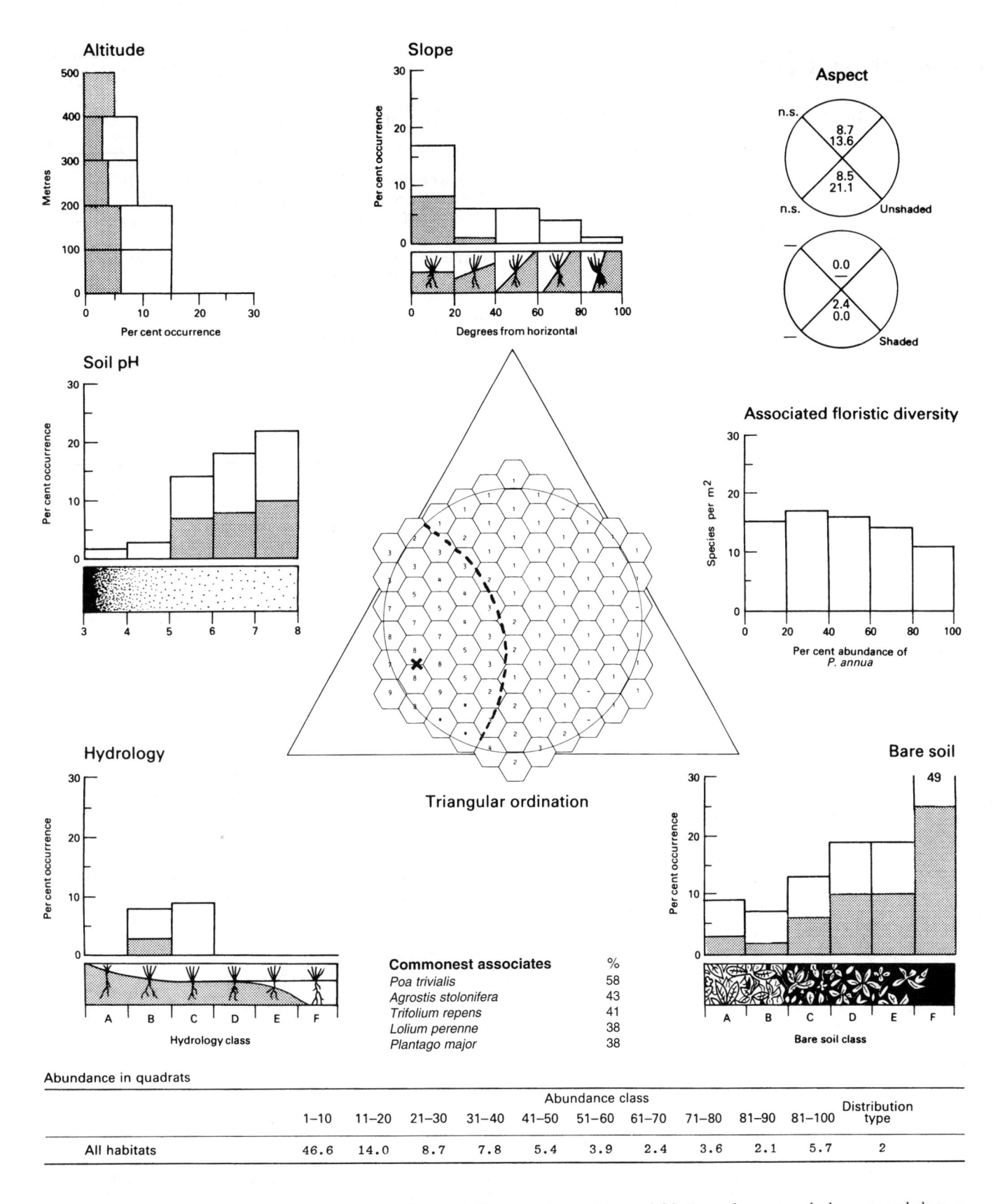

Altitude
Slope
Aspect
Soil pH
Associated floristic diversity
Hydrology
Bare soil
Triangular ordination

Commonest associates %
Poa trivialis 58
Agrostis stolonifera 43
Trifolium repens 41
Lolium perenne 38
Plantago major 38

Abundance in quadrats

	1–10	11–20	21–30	31–40	41–50	51–60	61–70	71–80	81–90	81–100	Distribution type
All habitats	46.6	14.0	8.7	7.8	5.4	3.9	2.4	3.6	2.1	5.7	2

formation of a buried seed bank (Roberts & Feast 1972, Roberts 1986b), germination throughout most of the year (Roberts 1986b), often with a peak in autumn (Thompson & Grime 1979), and survival of seedlings and young plants in winter conditions. Early-flowering individuals under unfavourable conditions may produce as few as 10 seeds, only 0.05% of the seed production possible under favourable conditions (*Biol Fl(b)*). (c) Tolerance of grazing, trampling and herbicides. Unlike most ruderals, *P.a.* is tolerant of severe defoliation and can set seed when cut regularly to as low as 6.5 mm (*Biol Fl(b)*). Also resistant to trampling, *P.a.* is the most common species on paths within the Sheffield region. *P.a.* is known to have developed resistance to some herbicides (*Biol Fl(a)*).

Current trends *P.a.* appears to be increasing. It combines success as a weed with palatability as a pasture grass. Now sown for amenity purposes.

Poa pratensis L.

Poaceae
Smooth-stalked Meadow-grass

Field data also include P. *angustifolia* L. and probably P. *humilis* Ehrh. ex Hoffm.

Life-form Polycarpic perennial, semi-rosette hemicryptophyte with creeping rhizomes. Shoots erect; leaves linear, typically <1000 mm^2; roots deeply penetrating, often with VA mycorrhizas.

Phenology Winter green. Some growth during winter but the main periods of vegetative growth are spring and perhaps autumn. Flowers May to July and seed normally shed by August.

Height Foliage usually <300 mm; flowers to 800 mm.

Seedling RGR No data. SLA 21.5. DMC 31.7.

Nuclear DNA amount 10.8 pg (Bennett *et al.* 1982), 2*n* = 84 (for non-European material); 2*n* = 42, 50-78, *c.* 86*, 91, 98 (*FE5, BC*); ?6*x*.-?14*x*.

Established strategy C-S-R.

Regenerative strategies V, B$_s$. See 'Synopsis'; ?SB III.

Flowers Green, hermaphrodite, wind-pollinated or apomictic; often >400 aggregated into an open branched panicle (2-5 per spikelet).

Germinule 0.25 mg, 1.6 × 0.6 mm, dispersed with an attached lemma (2.9 × 0.7 mm).

Germination Epigeal; t_{50} 4 days; 7-25°C; L > D.

Biological flora None.

Geographical distribution (Including *P.a.* and *P.h.*). The British Isles (100% of British vice-counties), *ABIF* page 756-7; Europe but only on mountains in the S (95% of European territories and 3% naturalised); temperate Asia; introduced into N America, New Zealand and Australia.

Habitats Widely distributed in grassy and open habitats, but particularly abundant on wasteland, road verges and limestone quarry spoil, and in enclosed pastures. Also common on limestone pastures, paths, spoil tips, outcrops, walls and in hedgerows. Scarce in marshes, on arable land and in woodland, and only of casual occurrence in aquatic systems.

Gregariousness Typically intermediate.

Altitude Found over the full range of altitudes. [Up to 950 m.]

Slope Wide-ranging.

Aspect Widely distributed but more abundant on S-facing slopes in semi-natural grassland (*Ecol Atl*).

Hydrology Extending onto moist rather than strongly waterlogged soils.

Soil pH Most frequent and abundant on soils of pH >4.5.

Bare soil Most frequent and abundant at lower exposures.

Δ ordination Found in a wide range of vegetation types, with highest frequencies associated with vegetation which is intermediate in terms of productivity and disturbance. Most of the records from the least disturbed vegetation correspond to *P.a.*

Associated floristic diversity Intermediate, declining slightly at the highest frequencies of *P.p.*

Habitats

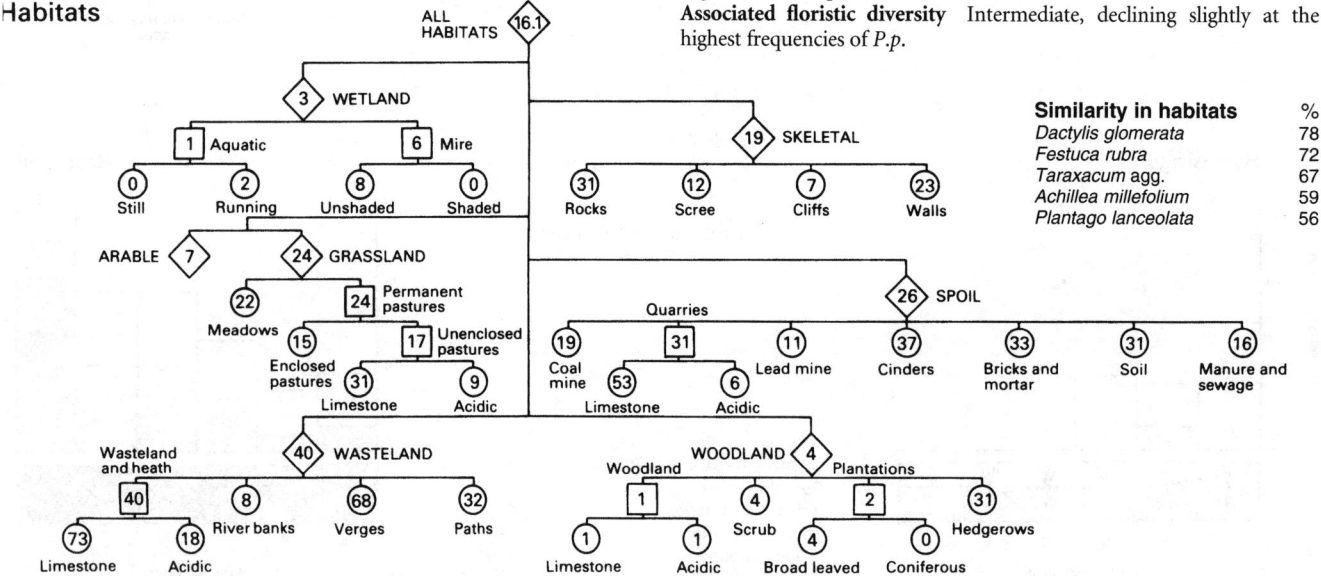

Similarity in habitats	%
Dactylis glomerata	78
Festuca rubra	72
Taraxacum agg.	67
Achillea millefolium	59
Plantago lanceolata	56

Synopsis *P.p.* is a wide-ranging grassland species exploiting both fertile and relatively infertile habitats, and occurring on moist or on dry soils. *P.p.* is frequent in pastures, is tolerant of grazing and trampling, and is potentially high-yielding (Spedding & Diekmahns 1972). The species rarely dominates vegetation, but is widespread as a minor component in grassland. In the PGE *P.p.* is 'present on most plots and is tenacious of its position in spite of the very small amounts that usually occur'. The attached tiller/rhizome system may be of benefit in these managed sites since, after cutting, the nutritional independence of the tillers may end following a reintegration of the shoot-rhizome system (Nyahza *et al.* 1973). Another important feature of the rhizomes is their ability to penetrate deeply into the soil. This may explain in part the capacity of *P.p.* to exploit dry soils and sites with uneven soil depth, e.g. rock outcrops, quarry heaps and cinder tips. *P.p.* can produce exceedingly long leaves in tall vegetation. This is apparent in productive tall herb communities of wasteland and unmanaged road verges, where *P.p.* frequently persists as a minor component dominated by, e.g. *Arrhenatherum elatius* and *Elytrigia repens*. *P.p.* is cold-tolerant and grows appreciably during winter (Spedding & Diekmahns 1972). The '*Poa pratensis* group' includes a number of ecologically and morphologically separable complexes which may be treated as species (*FE5*). In Britain these are *P.p.*, *P.a.* and *P.h.*. Each has arisen as a result of 'hybridization of various ancestral species followed by an increase in chromosome number and the disruption of the sexual breeding system' and inter-crossing between them is probably rare (*FE5*). Each 'species' has the capacity for extensive rhizomatous spread to form large clonal patches and is represented by an aneuploid series (*FE5*). No morphological separation of cytotypes of British species has been attempted and plants are in any event phenotypically plastic (Barling 1959, 1962). Different cytotypes may cohabit in the same site and *P.a.* and *P.h.* may sometimes coexist with *P.p.* (Barling 1959, 1962). Seed production is

Altitude
Slope
Aspect
Soil pH
Associated floristic diversity
Hydrology
Triangular ordination
Bare soil

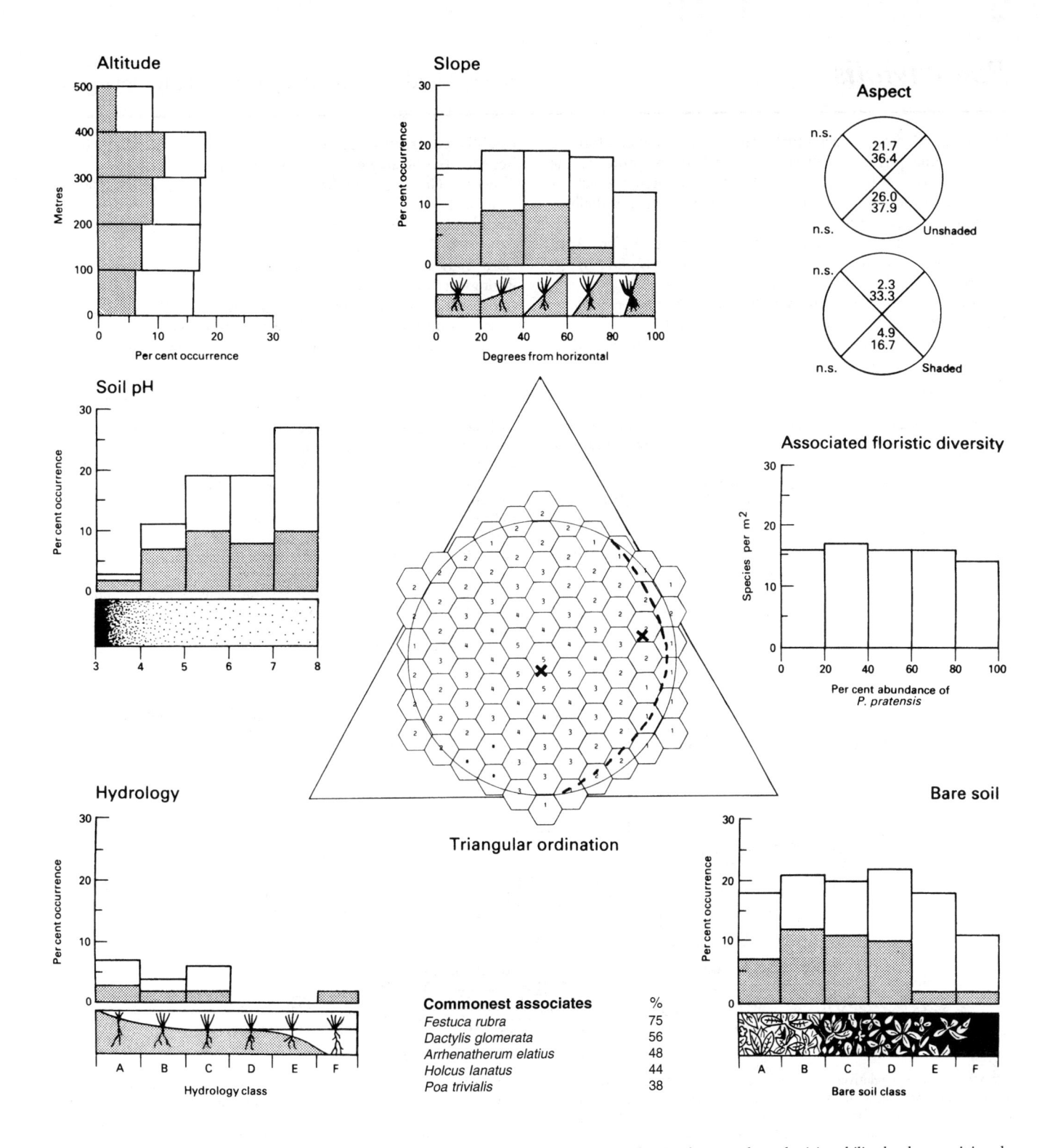

Commonest associates	%
Festuca rubra	75
Dactylis glomerata	56
Arrhenatherum elatius	48
Holcus lanatus	44
Poa trivialis	38

largely apomictic *(FE5)*. Persistent seeds occur in some situations. *P.p.* is by far the most common segregate, and the only one commercially sown (at a rate of *c.* 400 t year^{-1} (MAFF 1980-84) for permanent pasture or turf-grass). *P.a.* is found in dry, less-productive, mainly lowland sites, usually growing on calcareous soils with *Brachypodium pinnatum*, often on freely-drained railway banks. The period of spring growth is short (Barling 1959), and within the Sheffield region the species is absent from grazed sites, though tolerant of burning. *P.h.* is found primarily in two moist but unproductive types of habitat, namely montane grassland and dune slacks. *P.h.* is probably at its S geographical limit in Britain (Barling 1962). *P.h.* produces fewer larger seeds than *P.p.* (Barling 1959). Further ecological details of *P.a.* and *P.h.* are presented by Barling (1959, 1962) and Sargent *et al.* (1986).

Current trends *P.p.* has a modest colonizing ability, but because it is only sown in small quantities and not generally recommended for agricultural use (Spedding & Diekmahns 1972), the taxon is possibly decreasing. *P.h.* is mainly restricted to older habitats and may also be decreasing. The status of *P.a.* is uncertain. It appears to be increasing beside railways and motorways (Sargent, pers. comm.) but most of its semi-natural grassland habitats have been destroyed by modern land use. Changes in land-use which decrease the area of infertile grassland are likely to reduce the degree of geographical isolation between populations of *P.p.*, *P.a.* and *P.h.*. This brings the prospect of hybridization, with some breakdown of the ecological, genetic and morphological discontinuities between the three species. In some sites where *P.a.* and *P.p.* occur together the presence of intermediates is already suspected (Barling 1959).

Poa trivialis L.

Poaceae
Rough Meadow-grass, Rough-stalked Meadow-grass

Life-form Polycarpic perennial, semi-rosette hemicryptophyte or chamaephyte (effectively an annual in some arable situations) with creeping, leafy stolons. Stems erect or spreading; leaves linear, typically <500 mm^2; roots shallow, with VA mycorrhizas.
Phenology Winter green, but with little growth before April (Haggar 1971). Flowers in June and seed shed July and August. Leaves and stolons relatively short-lived.
Height Foliage usually <100 mm; flowering shoots to 600 mm.
Seedling RGR 1.0-1.4 week^{-1}. SLA 31.3. DMC 17.5.
Nuclear DNA amount 5.6 pg, $2n = 14$* (Bennett & Smith 1976); $2n = 14, 28$ (*FE5*); $2x$*, $4x$.

Established strategy Intermediate between ruderal and C-S-R.
Regenerative strategies V, B$_s$. See 'Synopsis'; SB III.
Flowers Green, hermaphrodite, wind-pollinated; often >1000 in a loose panicle.
Germinule 0.09 mg, 1.6×0.5 mm, caryopsis with an attached lemma (3.3×0.9 mm).
Germination Epigeal; t_{50} 2 days; 6-30°C; L > D.
Biological flora None.
Geographical distribution The British Isles (100% of British vice-counties), *ABIF* page 756; temperate Asia and N Africa; introduced to N and S America, Australia, New Zealand.

Habitats Recorded most frequently from damp, fertile and in some instances, shaded or highly disturbed habitats including meadows, productive pastures, river banks, road verges, mire, soil heaps, urban demolition sites and manure heaps. Frequent also on arable land and paths, and in hedgerows and disturbed woodlands and scrub. [In montane areas particularly associated with soligenous mire.]
Gregariousness Intermediate.
Altitude Widely distributed but suitable habitats are particularly abundant in lowland areas. [Up to 760 m.]
Slope Wide-ranging.
Aspect Associated with all aspects in both shaded and open habitats. In semi-natural grassland *P.t.* shows a N-facing bias (*Ecol Atl*).

Hydrology Abundant in wetlands, particularly in soligenous mire and on moist ground at the edge of topogenous mire. *P.t.* can also behave as a semi-aquatic on rafts of floating vegetation at the margins of streams and ponds. In such sites *P.t.* is usually rooted on the bank or on a boulder projecting above the water level.
Soil pH Most common on relatively base-rich soils of pH >5.0, but recorded on all but the most acidic soils.
Bare soil Evidence of an association with habitats having relatively high exposures.
△ **ordination** Distribution centred on productive vegetation subject to occasional disturbance, but also present at lower frequency in a wide variety of other vegetation types.
Associated floristic diversity Typically intermediate.

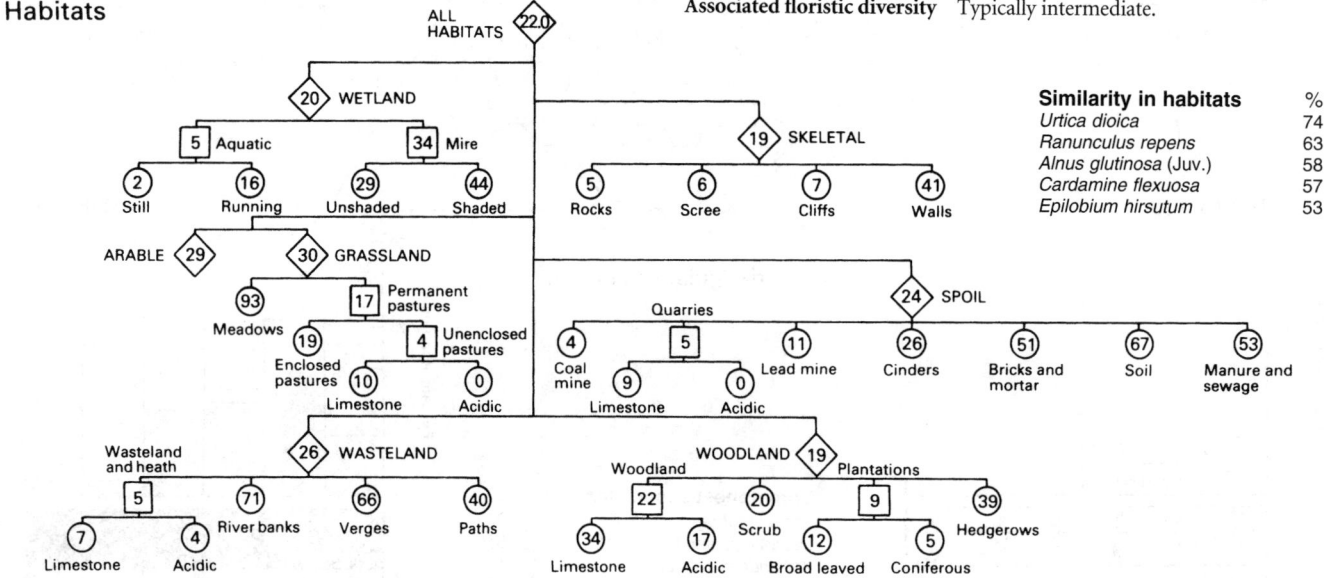

Habitats

Similarity in habitats	%
Urtica dioica	74
Ranunculus repens	63
Alnus glutinosa (Juv.)	58
Cardamine flexuosa	57
Epilobium hirsutum	53

Synopsis *P.t.* is a relatively fast-growing, winter-green grass with low-growing stolons, found in a wide variety of fertile habitats including wetland, woods, grassland and arable fields. *P.t.* is a palatable grass exploited for hay, although it is now little sown (Spedding & Diekmahns 1972). Despite low persistence under close mowing and susceptibility to trampling, the species is a useful turf-grass (Shildrick 1980) and about 40 t of seed are sown annually (MAFF 1980-84). *P.t.* appears to be moderately shade-tolerant (Al-Mufti 1978), typically forming an understorey below taller vegetation, and also occurring in woodland. In this latter habitat the low stature and weak physical structure are ill-suited for emergence through tree litter, which exerts a strong negative influence on *P.t.* (Sydes & Grime 1981a,b). *P.t.* also has a very shallow root system and is drought-sensitive. Foliage persists longer, and stolons are more robust, in unshaded but permanently moist habitats. The shoot phenology (75% of its annual yield is produced before June (Haggar 1971)) is complementary to that of tall herbs such as *Urtica dioica*, with which *P.t.* often coexists (Al-Mufti *et al.* 1977). In pasture *P.t.* usually occupies a stratum below the dominant grasses. Despite its cold-tolerance, *P.t.* exhibits little winter growth; in accordance with its relatively low nuclear DNA amount, little growth occurs before April (Haggar 1971). *P.t.* exhibits a degree of heterophylly, the culm leaves being much larger than those of the low-growing stolons. There is evidence of a light requirement for germination, and natural seed populations are least dormant in spring and autumn (Froud-Williams *et al.* 1986). However, seeds generally mature early and show little dormancy. Rapid germination is possible across a wide temperature range and huge cohorts of seedlings appear on bare soil each autumn. Individuals originating in autumn produce more-numerous and heavier seeds than those in spring (Budd 1970). Seeds which become buried in the soil are capable of persistence (Thompson & Grime 1979, Roberts 1986b). In lawns, and in other situations where seeding is prevented, *P.t.* frequently spreads vegetatively by means of stolons. Plants can withstand shallow burial and regeneration from shoot fragments occurs in arable land, though here (and elsewhere) regeneration by seed is important. *P.t.* is a weed of winter cereals, is a major problem in herbage

Altitude

Slope

Aspect

	n.s.	18.9 / 27.1	
n.s.	17.0 / 31.6		Unshaded

n.s.

	18.9 / 28.0	
	20.3 / 32.0	Shaded

n.s.

Soil pH

31 33 35

Associated floristic diversity

Triangular ordination

Hydrology

40

Commonest associates

	%
Holcus lanatus	37
Agrostis stolonifera	34
Ranunculus repens	34
Urtica dioica	33
Festuca rubra	33

Bare soil

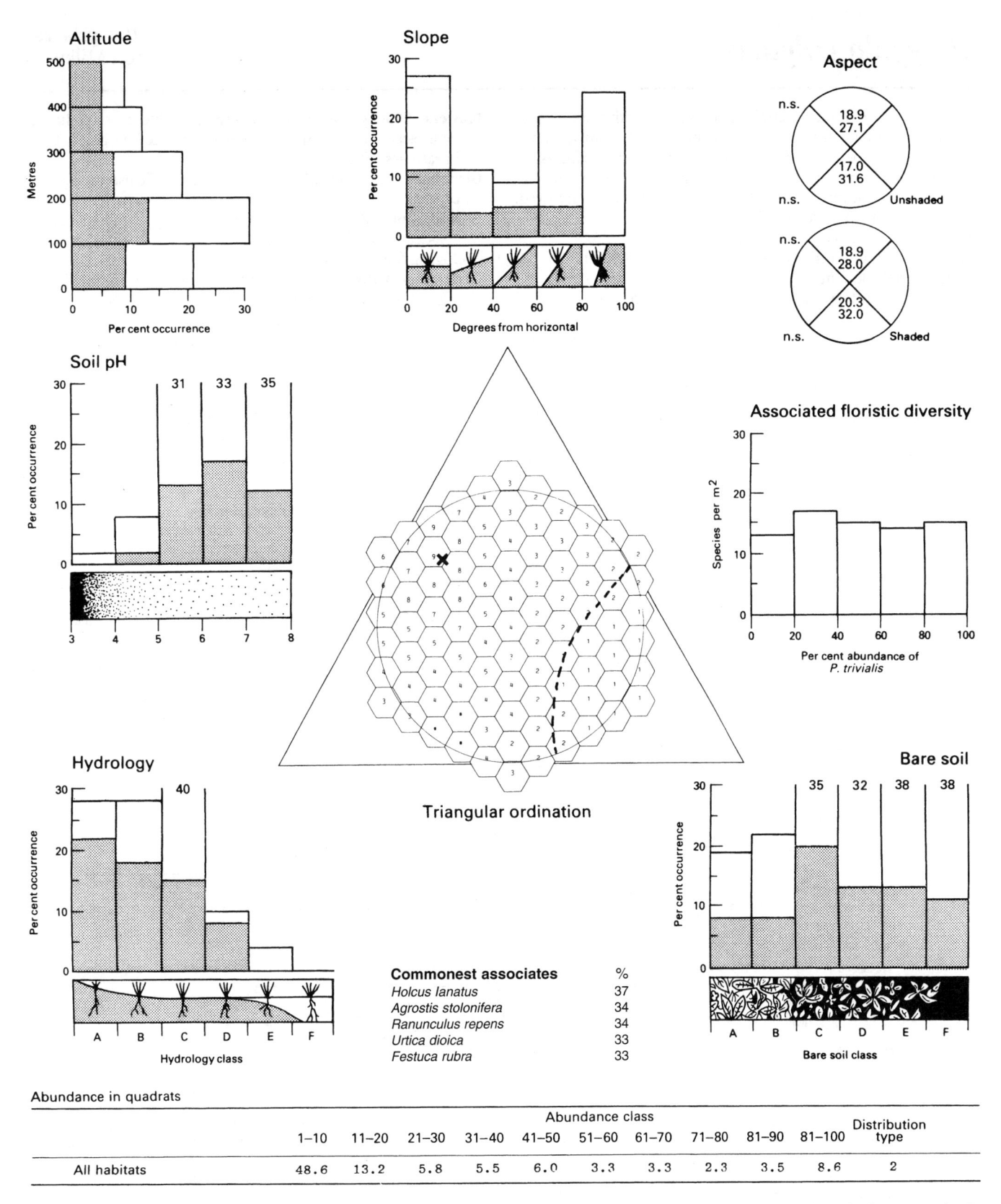

35 32 38 38

Abundance in quadrats

	\multicolumn{10}{c}{Abundance class}	Distribution type
	1–10	11–20	21–30	31–40	41–50	51–60	61–70	71–80	81–90	81–100	
All habitats	48.6	13.2	5.8	5.5	6.0	3.3	3.3	2.3	3.5	8.6	2

seed crops (WCH) and is often one of the first colonists during the 'sward deterioration' of cultivated grassland (Morrison 1979) which often has a deleterious effect on rye-grass yield (Haggar 1971). Although P.t. is rarely if ever a dominant, the species is arguably the most successful subordinate constituent of plant communities in the British Isles. At present it is unclear to what extent this is due to the evolution of ecotypes, to phenotypic plasticity or to the capacity of P.t. to occupy similar niches in a wide range of vegetation types. Seeds of arable populations appear to be markedly more dormant than seeds collected from grassland (Froud-Williams et al. 1986). Seeds from open arable habitats germinate poorly when exposed to a ratio of far-red/red irradiance similar to that found in natural shade, while seed from closed vegetation is less inhibited (Hilton et al. 1984). Another ssp. is recorded from Europe (FE5).

Current trends Apparently increasing in grassland and arable land and a frequent colonist of artificial habitats.

Polygala vulgaris L.

Polygalaceae
Common Milkwort

Life-form Polycarpic perennial chamaephyte with a number of erect or, more usually, trailing stems arising from a woody stock. Leaves ovate, <250 mm^2; roots with VA mycorrhizas.

Phenology Winter green. Flowers mainly from May to July and sets seed from July to August. Leaves comparatively long-lived.

Height Foliage and flowers usually <100 mm.

Seedling RGR No data. SLA 20.4. DMC 25.5.

Nuclear DNA amount No data; $2n = 68*$ (*BC*, Heubl 1984); $4x*$, genus polyploid in origin.

Established strategy Intermediate between stress-tolerator and C-S-R.

Regenerative strategies ?S. Regeneration requires further study (see 'Synopsis'); ?SB II.

Flowers Blue, purple or white, hermaphrodite, cross-pollinated by insects but predominantly selfed (Lack & Kay 1986); 10-40 in racemes (*FE2*).

Germinule 1.72 mg, 2.5 × 1.5 mm, seed in a dehiscent, 2 seeded capsule.

Germination Epigeal; dormancy broken by chilling (Heubl 1984); L > D (Heubl 1984).

Biological flora None.

Geographical distribution Widely distributed in the British Isles, but most common in S England (97% of British vice-counties), *ABIF* page 436; most of Europe (82% of European territories); W Asia and N Africa.

Habitats Most frequent in short, semi-natural grassland. Also found on the more vegetated areas of limestone scree, on wasteland, lead-mine and quarry spoil, and on waysides. Also recorded from cliffs and outcrops. Mainly restricted to calcareous strata. Absent from arable fields, wetland, woodland and most skeletal habitats.

Gregariousness Sparse.

Altitude Most habitats suitable for *P.v.* occur on the carboniferous limestone and the greater frequency of *P.v.* between 200 and 350 m reflects this. [Up to 730 m.]

Slope Most frequent on dale-sides on the carboniferous limestone, and therefore mainly restricted to moderate slopes. Absent from the most skeletal habitats and from the steepest slopes.

Aspect Significantly more frequent on unshaded S-facing slopes.

Hydrology Absent from wetlands although observed occasionally on cliffs with dripping water.

Soil pH Restricted in the data reported here to soils in the pH range 5.0-8.0.

Bare soil Found over a wide range of values, but most frequent in sites with intermediate exposures of bare soil.

△ **ordination** Distribution virtually confined to infertile, relatively undisturbed vegetation types.

Associated floristic diversity Associated with species-rich vegetation.

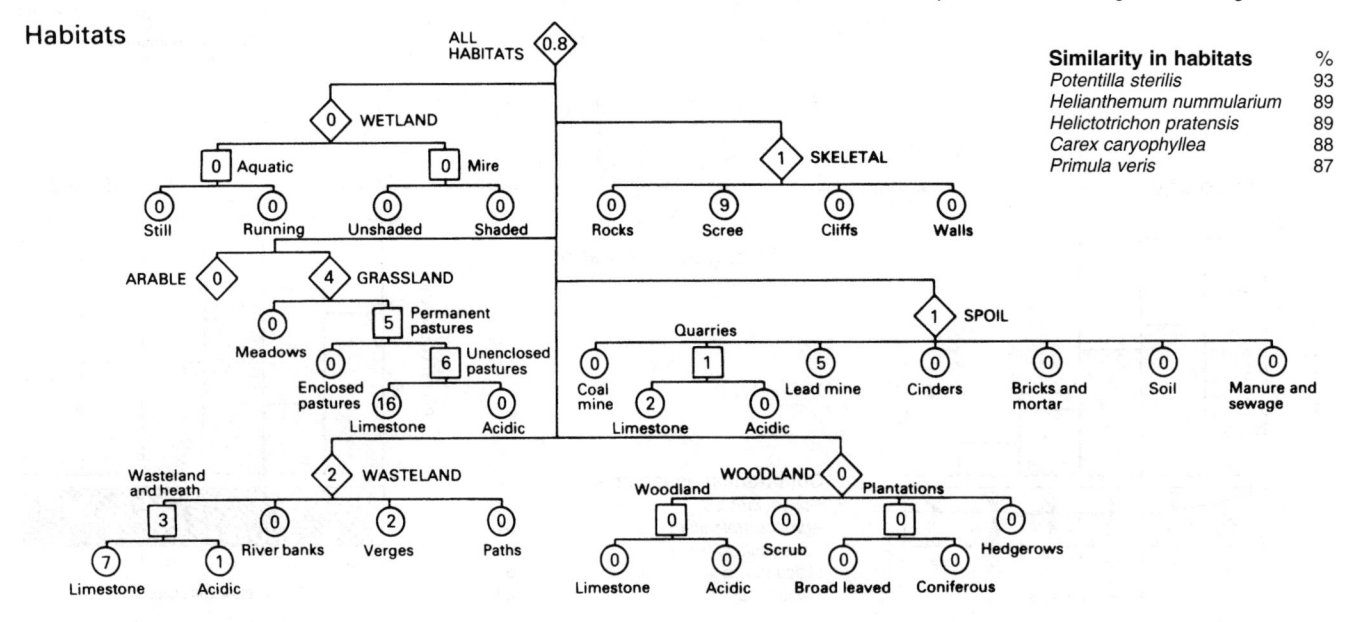

Habitats

Similarity in habitats	%
Potentilla sterilis	93
Helianthemum nummularium	89
Helictotrichon pratensis	89
Carex caryophyllea	88
Primula veris	87

Synopsis *P.v.* is a low-growing herb with a woody base, found in short, unproductive calcareous turf, where dominance by more-robust species is restricted by mineral-nutrient stress and grazing. The species is also found occasionally on base-rich soils associated with non-calcareous strata. In turf *P.v.* tends to occupy the lower levels of the leaf canopy, and we suspect that it is relatively shade-tolerant. However, *P.v.* reaches highest frequencies in more-open sites at the stabilized margins of limestone screes. The species forms small patches but the branches, although prostrate, do not root and there is no evidence of vegetative regeneration. Thus, effective regeneration occurs by seeds which germinate mainly in spring (Heubl 1984). No persistent seed bank has been detected. The seed is large, with an elaiosome and is dispersed by ants (Ridley 1930). However, as in the case of many other species with this dispersal mechanism, seed mobility appears to be localized. Although *P.v.* is found in many old quarries and on lime-enriched road verges across acidic moorland, the species seldom colonizes recent artificial habitats and is regarded by Wells *et al.* (1976) as an indicator species of ancient grassland. *P.v.* is the most common and ecologically wide-ranging species of the genus in Europe, and British material is genetically and biochemically diverse (Lack & Kay 1986). *P.v.* is also morphologically variable and has been subdivided recently into four ssp. (Heubl 1984); represented in the British Isles by ssp. *collina* (Reichenb.) Borbas. *P.v.* hybridizes with two other *Polygala* species in S England (Stace 1975). On acidic soils *P.v.* is replaced by *P. serpyllifolia* J. A. Hose, and in only one site in the Sheffield region, a species-rich, non-calcareous pasture of intermediate pH, has the ecological range of the two been observed to overlap.

Current trends Decreasing, particularly in lowland areas as a result of habitat destruction and the relaxation of grazing in old grassland.

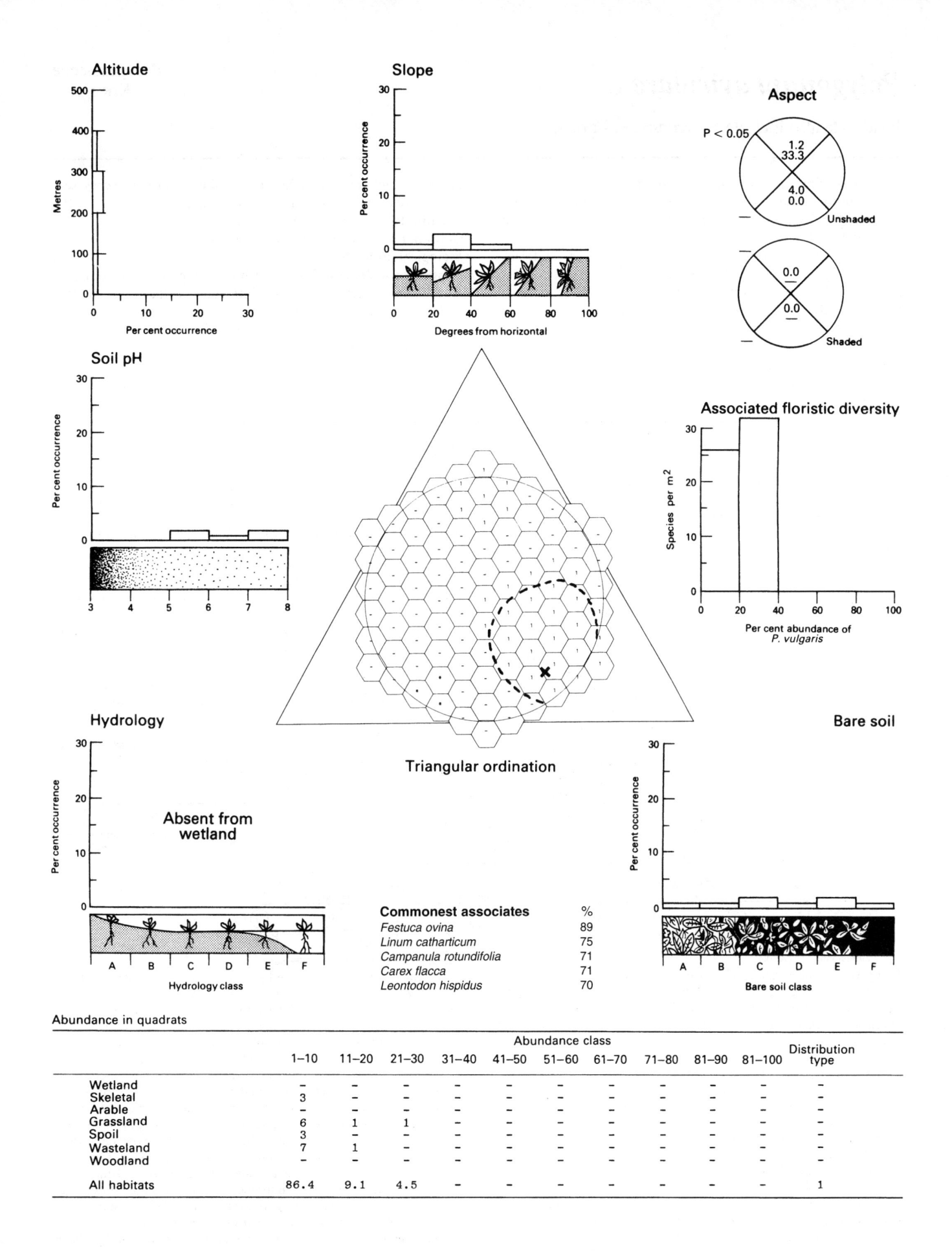

Altitude

(histogram: Metres vs Per cent occurrence)

Slope

(histogram: Per cent occurrence vs Degrees from horizontal)

Aspect

P < 0.05

Unshaded: 1.2 / 33.3 / 4.0 / 0.0

Shaded: 0.0 / 0.0

Soil pH

(histogram: Per cent occurrence vs pH 3–8)

Associated floristic diversity

(histogram: Species per m² vs Per cent abundance of *P. vulgaris*)

Triangular ordination

Hydrology

Absent from wetland

(Hydrology class A–F)

Bare soil

(Bare soil class A–F)

Commonest associates

	%
Festuca ovina	89
Linum catharticum	75
Campanula rotundifolia	71
Carex flacca	71
Leontodon hispidus	70

Abundance in quadrats

	1–10	11–20	21–30	31–40	41–50	51–60	61–70	71–80	81–90	81–100	Distribution type
Wetland	–	–	–	–	–	–	–	–	–	–	–
Skeletal	3	–	–	–	–	–	–	–	–	–	–
Arable	–	–	–	–	–	–	–	–	–	–	–
Grassland	6	1	1	–	–	–	–	–	–	–	–
Spoil	3	–	–	–	–	–	–	–	–	–	–
Wasteland	7	1	–	–	–	–	–	–	–	–	–
Woodland	–	–	–	–	–	–	–	–	–	–	–
All habitats	86.4	9.1	4.5	–	–	–	–	–	–	–	1

Polygonum aviculare L.

Polygonaceae
Knotgrass

Field data also include *P. arenastrum* Boreau

Life-form Summer-annual therophyte. Stems erect or procumbent; leaves heterophyllous, lanceolate, the larger typically <500 mm^2 (*c.* 3 × the smaller); deeply tap-rooted with or without VA mycorrhizas.
Phenology Germination occurs in spring. Flowers and fruits from July to November. Leaves short-lived.
Height Foliage and flowers exceptionally to 1 m.
Seedling RGR 1.0–1.4 week^{-1}. SLA 28.2. DMC 20.7.
Nuclear DNA amount 1.3 pg; $2n = 60*$ (*BC*); $6x*$.
Established strategy Ruderal.
Regenerative strategies B$_s$. Regenerates exclusively by seed, which germinates in spring. Also forms a persistent bank of buried seed; SB IV.

Flowers Pink or white, hermaphrodite, probably always self-fertilized (Styles 1962); 1–6 in axillary inflorescences.
Germinule 1.45 mg, 3.8 × 2.1 mm, nut.
Germination Epigeal; dormancy broken by chilling.
Biological flora None.
Geographical distribution Distribution imperfectly documented but almost certainly present in 100% British vice-counties (*ABIF* p. 87–8) and 100% European territories (but only an alien in the extreme N of Europe); Asia; introduced in N and S America and Australasia.

Habitats A plant of mechanically disturbed, or trampled, artificial habitats. Most commonly recorded from arable land, but also frequent on spoil heaps, particularly demolition sites, soil heaps, manure and sewage waste, and on paths and tracks. In pasture mainly restricted to gateways, and on road verges typically found close to the road margin. Infrequent on river banks, wasteland, rock outcrops and in mire. Absent from woodland and aquatic habitats. [Occurs in maritime habitats (Styles 1962).] (See 'Synopsis'.)
Gregariousness Intermediate.

Altitude Suitable habitats more frequent in lowland areas, but extends to 380 m. [Up to 550 m.]
Slope Not recorded from steeply sloping habitats.
Aspect Widely distributed with respect to aspect, but slightly more abundant on S-facing slopes.
Hydrology Recorded on exposed mud beside ponds, but not generally associated with waterlogged sites.
Soil pH Largely restricted to sites of pH >5.0.
Bare soil Most frequent and abundant at high exposures.
Δ **ordination** Distribution strongly centred in productive, disturbed vegetation.
Associated floristic diversity Varying widely but most typically intermediate.

Similarity in habitats	%
Capsella bursa-pastoris	96
Matricaria discoidea	91
Persicaria maculosa	86
Tripleurospermum inodorum	85
Stellaria media	84

Habitats

ALL HABITATS 6.1
- WETLAND 1
  - Aquatic 0
    - Still 0
    - Running 0
  - Mire 1
    - Unshaded 1
    - Shaded 2
- ARABLE 74
- GRASSLAND 2
  - Meadows 0
  - Permanent pastures 2
    - Enclosed pastures 3
      - Limestone 0
      - Acidic 0
    - Unenclosed pastures 0
- SKELETAL 1
  - Rocks 3
  - Scree 0
  - Cliffs 0
  - Walls 0
- SPOIL 12
  - Quarries
    - Coal mine 8
    - Lead mine 1 (Limestone 2, Acidic 0) 5
  - Cinders 16
  - Bricks and mortar 31
  - Soil 23
  - Manure and sewage 21
- WASTELAND 5
  - Wasteland and heath 1 (Limestone 0, Acidic 1)
  - River banks 3
  - Verges 8
  - Paths 24 (Limestone 0)
- WOODLAND 0
  - Woodland 0 (Acidic 0, Broad leaved 0)
  - Plantations 0 (Scrub 0, Coniferous 0, Hedgerows 0)

Synopsis The aggregate species, which exploits a wide range of open, disturbed, usually fertile, artificial habitats, is a summer-annual herb often with a deep tap-root. It is represented primarily by *P.are.* and *P.avi.*, both of which are very common. All other species in Section Polygonum are rare in Britain. *P.are.* is tetraploid ($2n = 40$), smaller in its parts, more usually prostrate, very tolerant of trampling and more frequent in drier sites (Styles 1962). *P.avi.* is particularly common in cereal crops, whereas *P.are.* predominates on tracks and roadsides, waste ground and spoil and, to some extent, in broad-leaved arable crops (Styles 1962). However, there is a considerable overlap in habitats between the two, and mixed populations are not uncommon (Styles 1962). The geographical distribution of the two is similar (Styles 1962). The aggregate species shares a number of attributes with other successful ruderals. (a) A relatively high potential for seed production; a robust plant may produce over 1000 relatively large seeds whereas, under conditions of stress, fewer than 10 may be produced. (b) Effective mechanisms of seed dispersal by humans or other agencies. Seeds occur both as an impurity in the harvested crop and as a contaminant of sown seed (*WCH*). They may be dispersed in mud on footwear or tyre treads, and can survive ingestion by stock or by birds (Ridley 1930). (c) Genetic variability. Both are polymorphic; morphologically separable distinct ecotypes of *P.are.* can be identified (Styles 1962). However, seeds germinate only in spring and are returned to a state of secondary dormancy during the rise in late spring temperatures (Courtney 1968). Consequently, the species produces only one generation each year, and is ill-equipped to exploit bared sites created by disturbance in summer. The major habitats of each (cereal fields in the case of *P.avi.*, trampled areas in the case of *P.are.*) are subject to a rather predictable pattern of disturbance. As with many other tap-rooted species, the aggregate is largely absent from wetlands. Both are also sensitive to shading, but can survive defoliation. Indeed, *P.are.* is frequent in mown or grazed grassland subject to trampling. Both, particularly *P.are.*, show a degree of seed polymorphism (Styles 1962), the ecological significance of which requires study. Each occurs by the sea and on road verges close to the metalled road surface, and a degree of salt tolerance is suspected in some populations.
Current trends An effective colonist of artificial habitats, and probably still increasing in range.

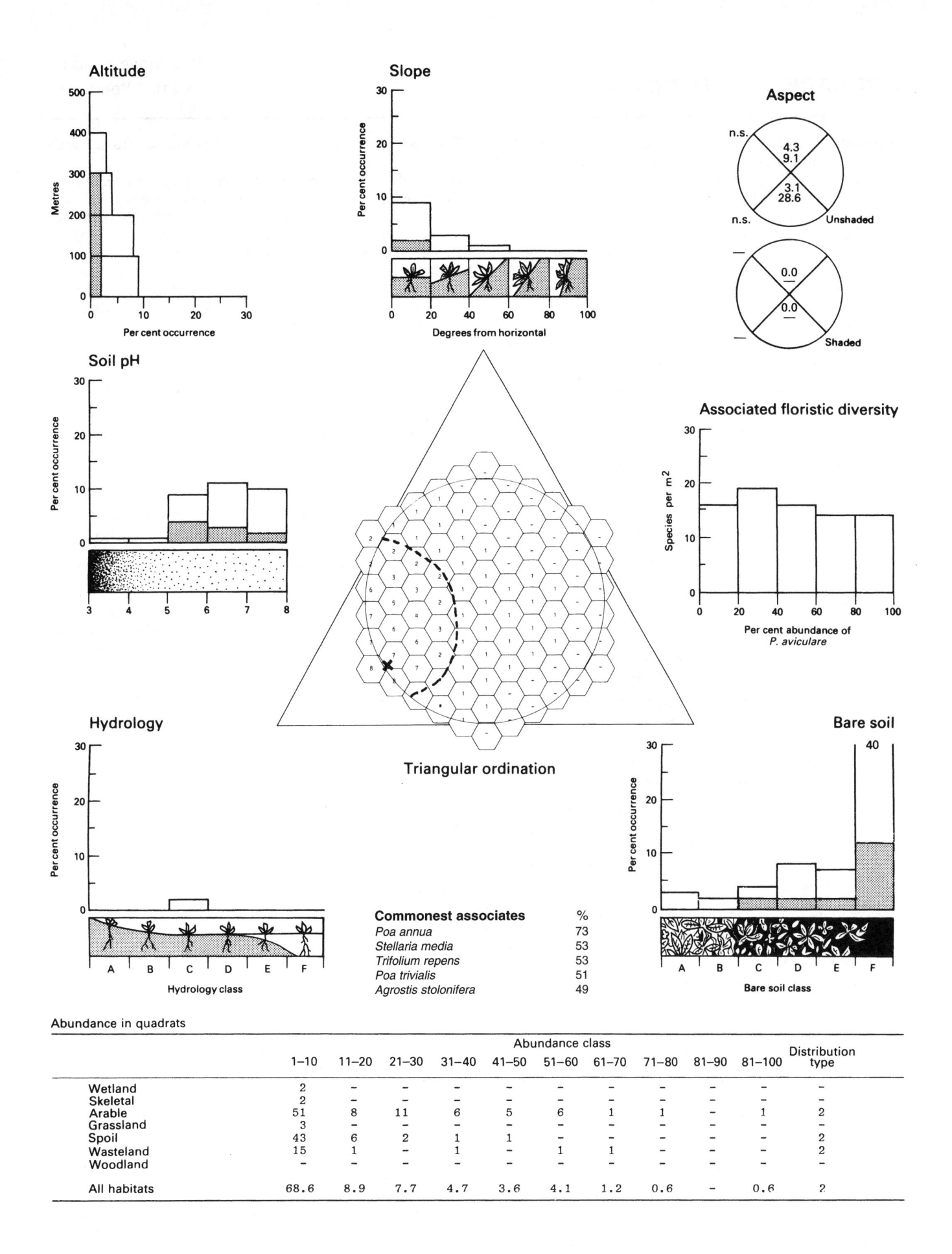

Altitude

Slope

Aspect

	n.s.	4.3
		9.1
n.s.	3.1	28.6
		Unshaded

	—	0.0
		—
—	0.0	—
		Shaded

Soil pH

Associated floristic diversity

Hydrology

Triangular ordination

Commonest associates	%
Poa annua	73
Stellaria media	53
Trifolium repens	53
Poa trivialis	51
Agrostis stolonifera	49

Bare soil

Abundance in quadrats

	1–10	11–20	21–30	31–40	41–50	51–60	61–70	71–80	81–90	81–100	Distribution type
Wetland	2	–	–	–	–	–	–	–	–	–	–
Skeletal	2	–	–	–	–	–	–	–	–	–	–
Arable	51	8	11	6	5	6	1	1	–	1	2
Grassland	3	–	–	–	–	–	–	–	–	–	–
Spoil	43	6	2	1	1	–	–	–	–	–	2
Wasteland	15	1	–	1	–	1	1	–	–	–	2
Woodland	–	–	–	–	–	–	–	–	–	–	–
All habitats	68.6	8.9	7.7	4.7	3.6	4.1	1.2	0.6	–	0.6	2

Potamogeton crispus L.

Potamogetonaceae
Curled Pondweed

Life-form Polycarpic perennial hydrophyte (but effectively a winter-annual with vegetative regeneration – see 'Synopsis'). Rhizome creeping; stems branched, ascending; leaves lanceolate to linear, translucent, typically <1000 mm^2; roots shallow, non-mycorrhizal.
Phenology Complex (see 'Synopsis').
Height All foliage below the water level; flowering spikes shortly emergent.
Seedling RGR No data. SLA 60.6. DMC 14.9.
Nuclear DNA amount 1.6 pg; $2n = 52$ (*FE5*); $4x$.
Established strategy Competitive-ruderal.
Regenerative strategies (V), S, ?B$_s$. Little studied and complex (see 'Synopsis'); ?SB IV.

Flowers Green, hermaphrodite, wind-pollinated; <10 in axillary spikes.
Germinule 4.0×2.5 mm, drupaceous fruit (up to 4 per flower).
Germination Cotyledon partially emergent from seed; weak hard-coat dormancy; disintegration of seed coat by micro-organisms may be important (Teltscherova & Hejny 1973).
Biological flora Catling and Dobson (1985).
Geographical distribution The British Isles, except in upland areas (94% of British vice-counties), *ABIF* page 679; most of Europe (85% of European territories) and other N-temperate regions; introduced in N America.

Habitats Restricted to aquatic habitats, particularly lakes, ponds, canals and ditches, but also in relatively still water in rivers and streams.
Gregariousness Occasionally patch-forming.
Altitude Observed to 240 m, but suitable habitats are more frequent in lowland areas. [Up to 400 m.]
Slope Not relevant, absent from terrestrial habitats.
Aspect Not relevant.
Hydrology Most frequent in deeper water and observed to 800 mm. In rivers and streams absent from water with a strong current.
Soil pH Rooted to substrates with a pH of between 5.5 and 7.5.
Bare soil Not relevant. Absent from terrestrial habitats.

Habitats

$\triangle$ **ordination** No data.
Associated floristic diversity Typically low, as with most aquatic species.
Synopsis *P.c.* is a submerged aquatic of relatively still waters. As a consequence of its winter-annual phenology (*Biol Fl*), *P.c.* experiences little competition from summer-green aquatics, and may even exert early season dominance (cf. *Poa trivialis* in some terrestrial communities). The production of summer foliage, characterized by strongly undulate, serrate leaves, is followed (about June or July) by the formation of fruiting spikes and dormant apices, the latter appearing in response to long days (*Biol Fl*). The plant dies back during July and the dormant apices, now black and thorny, sink to the bottom where they aestivate (*Biol Fl*). When the water body cools down in autumn, each germinates to give rise to a plant with small uncrisped leaves (*Biol Fl*). Typically this overwintering stage is deeply submerged. Sites with *P.c.* may develop a surface layer of ice in winter (*Biol Fl*). In the following spring the small winter leaves die and are replaced by

Similarity in habitats	%
Elodea canadensis	98
Potamogeton natans	93
Sparganium erectum	90
Equisetum fluviatile	89
Lemna minor	87

new summer foliage (*Biol Fl*). A single plant may produce 900 dormant apices in a season (*Biol Fl*). Vegetative regeneration of this type is extremely effective, as illustrated by the fact that sterile, hybrid taxa may nevertheless colonize large rivers or canal systems (see Stace 1975). Detached vegetative shoots do not develop roots readily, but dormant apices are formed irrespective of photoperiod (*Biol Fl*). Many fewer fruits are produced than dormant apices, and regeneration by seed, which appears to be stimulated by unusually low winter temperatures (Teltscherova & Hejny 1973) is probably a rare event. The exact conditions for optimal germination of the drupaceous fruit, which exhibits a slight degree of hard-coat dormancy (Teltscherova & Hejny 1973), remain to be elucidated. Fruits may survive ingestion by water-fowl and float in water for long periods (Ridley 1930). The shoots of *P.c.* are eaten by wildfowl, and they also support a large assemblage of aquatic invertebrates (*Biol Fl*). *P.c.* occurs in calcareous, brackish and relatively eutrophic environments, but the species is poorly represented in peaty water. In highly fertile ponds *P.c.* is often suppressed, apparently by the dominant effect of a floating raft of species such as *Lemna minor*. *P.c.* is not recorded from very stony substrata, and is usually rooted in silt. *P.c.* does not exhibit aquatic acid metabolism to any appreciable extent (Keeley & Morton 1982) (see *Ranunculus flammula*). The species normally utilizes carbon dioxide for photosynthesis, but is capable of using the bicarbonate ion. As with other submerged aquatics, the cuticle of the leaf is very thin (0.05 μm), allowing foliar absorption of inorganic nutrients (Denny 1980). However, vascular tissue in the stem is better developed than that of, for example, *P. pectinatus* L. (Denny 1980). As a result *P.c.* can take up more nutrients from the rooting substrate than *P.p.*, and can extend into less-eutrophic waters than *P.p.* (Denny 1980). *P.c.* is highly plastic in morphology, and the varieties described by early workers appear to have been phenotypic rather than genetic in origin. Hybridizes in Britain with six other species, some broad- and others narrow-leaved (Stace 1975).
Current trends A mobile species, but succumbs to dominance by *Lemna minor* and filamentous green algae under highly eutrophic conditions. Apparently decreasing.

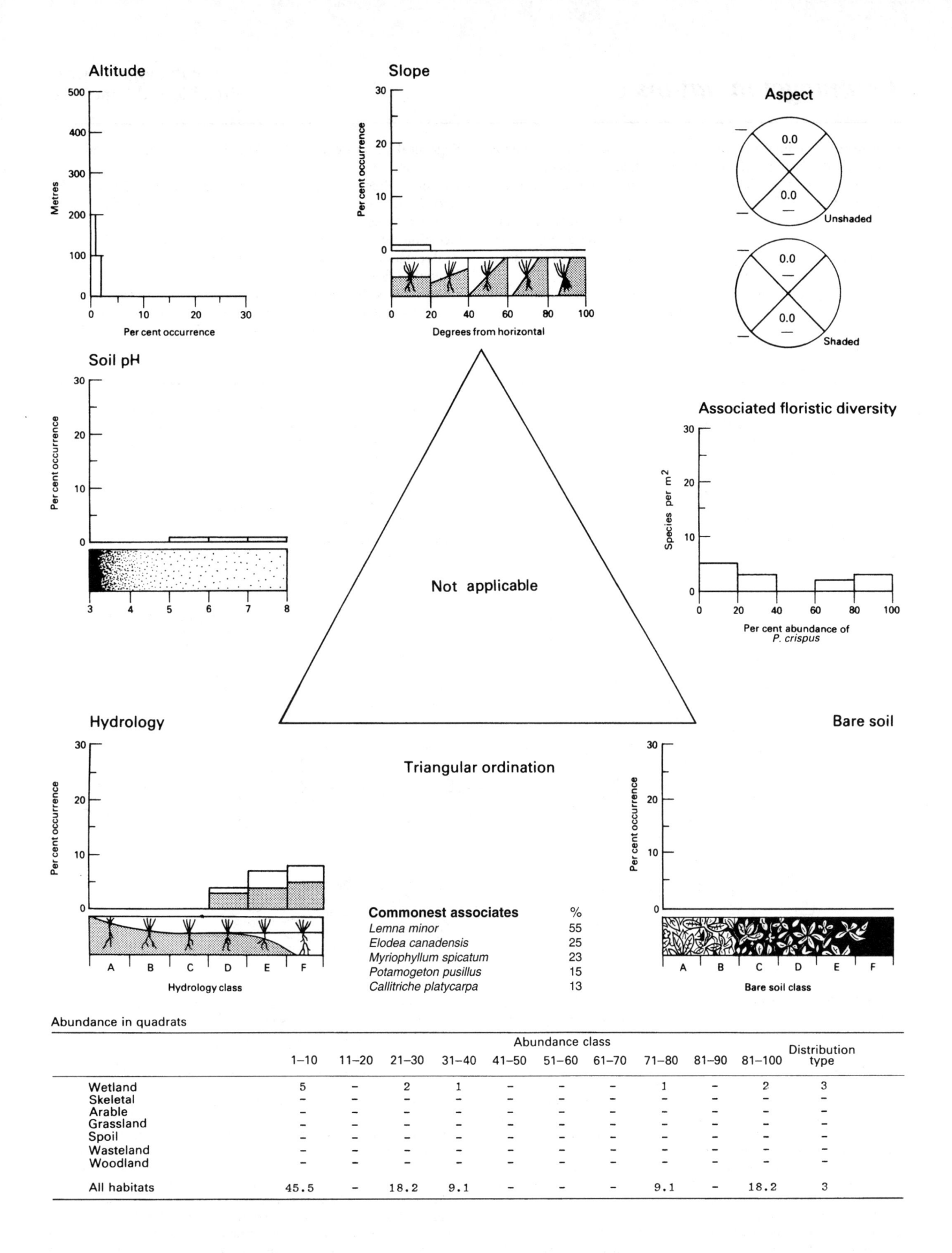

Altitude

Slope

Aspect

Unshaded 0.0

Shaded 0.0

Soil pH

Associated floristic diversity

Species per m²

Per cent abundance of *P. crispus*

Not applicable

Hydrology

Hydrology class: A B C D E F

Triangular ordination

Bare soil

Bare soil class: A B C D E F

Commonest associates	%
Lemna minor	55
Elodea canadensis	25
Myriophyllum spicatum	23
Potamogeton pusillus	15
Callitriche platycarpa	13

Abundance in quadrats

	1–10	11–20	21–30	31–40	41–50	51–60	61–70	71–80	81–90	81–100	Distribution type
Wetland	5	–	2	1	–	–	–	1	–	2	3
Skeletal	–	–	–	–	–	–	–	–	–	–	–
Arable	–	–	–	–	–	–	–	–	–	–	–
Grassland	–	–	–	–	–	–	–	–	–	–	–
Spoil	–	–	–	–	–	–	–	–	–	–	–
Wasteland	–	–	–	–	–	–	–	–	–	–	–
Woodland	–	–	–	–	–	–	–	–	–	–	–
All habitats	45.5	–	18.2	9.1	–	–	–	9.1	–	18.2	3

Potamogeton natans L.

Potamogetonaceae
Broad-leaved Pondweed

Life-form Perennial hydrophyte with creeping rhizome. Stems branched ascending; floating leaves with an ovate blade, <1000 mm^2, submerged leaves linear, often <500 mm^2; roots non-mycorrhizal.

Phenology Overwintering by means of dormant buds formed in autumn at the stem apices. Flowers from June to September and sets seed from August to October. Leaves and some stem apices die back in winter.

Height Foliage floating and to a lesser extent submerged; flowering spike shortly emergent.

Seedling RGR No data. SLA 25.5. DMC 20.5.

Nuclear DNA amount 1.7 pg; $2n = 52$ *(FE5)*; $4x$.

Established strategy Competitive-ruderal.

Regenerative strategies (V). Mainly by vegetative means (see 'Synopsis').

Flowers Green, hermaphrodite, wind-pollinated; >50 in a spike.

Germinule 4.25 mg, 4.5 × 3 mm; fleshy drupaceous fruit (up to 4 per flower).

Germination Cotyledon partially emergent from seed; hard-coat dormancy (Teltscherova & Hejny 1973).

Biological flora None.

Geographical distribution The British Isles, especially in C and E England (100% of British vice-counties), *ABIF* page 672; Europe (87% of European territories); much of the N hemisphere.

Habitats Only recorded from lakes, ponds, canals and ditches or impermanently as stranded shoots on bare mud adjacent to water. Observed also in river back-waters.

Gregariousness Patch-forming.

Altitude In lowland areas found in a wide range of aquatic habitats. In the upland regions, where there are fewer aquatic habitats, the distribution of *P.n.* is restricted and the species is particularly characteristic of dew ponds. Extends to 380 m. [Up to 700 m.]

Slope Not relevant.

Aspect Not relevant.

Hydrology Most frequent in deeper water extending to 750 mm [and particularly characteristic of the range <1-2 m (Spence & Chrystal 1970)]. Absent from fast-flowing water. Also found stranded on bare mud during droughts, but not usually persistent in mire habitats.

Soil pH Most frequent and abundant on soils within the pH range 5.5-7.0 and not recorded below pH 4.0.

Bare soil Confined to submerged habitats.

△ ordination Insufficient data.

Associated floristic diversity As with most aquatic species values are relatively low.

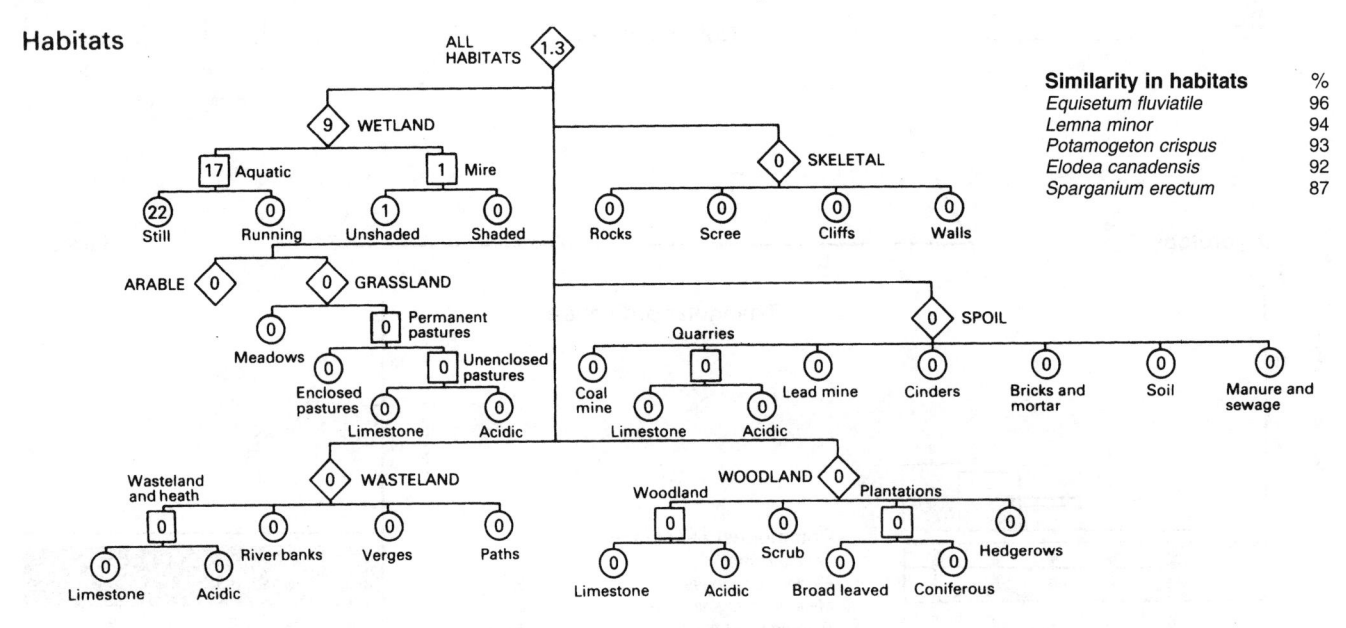

Habitats

Similarity in habitats	%
Equisetum fluviatile	96
Lemna minor	94
Potamogeton crispus	93
Elodea canadensis	92
Sparganium erectum	87

Synopsis *P.n.* is a large, heterophyllous aquatic with long-stalked, leathery, floating leaves and narrow translucent, inconspicuous, often short-lived, submerged leaves. The species is widespread in ditches, ponds, canals and sheltered lake-sides and is characteristic of relatively deep, stagnant water (exceptionally up to 2 m (Spence & Chrystal 1970)). However, *P.n.* is absent from very deep water, and Spence and Chrystal (1970) suggest that light may be an important factor controlling the zonation. In the Sheffield region *P.n.* is absent from reservoirs, suggesting that the species is vulnerable to wave action or has a limited tolerance of fluctuating water levels. In some areas of disused canal the growth of *P.n.* appears to be suppressed by a layer of *Lemna minor* and *L. gibba* L. at the water surface. Land forms are reported (Sculthorpe 1967) but, unlike *P. polygonifolius* Pourret, *P.n.* is of transient occurrence in mire. In running or deep water, phenotypes with leaves wholly submerged may occur (Sculthorpe 1967). The species overwinters by means of dormant buds formed at the stem apex. Some stems, unlike those of *P. crispus*, live for more than one season, and *P.n.*, in common with other *Potamogeton species*, is without vessels, the vascular tissue being less condensed than that of *P.c.* (Sculthorpe 1967). *P.n.* is unpalatable to the grass carp (*Ctenopharyngodon idella* Val.) (Fowler & Robson 1978), and may form extensive rhizomatous patches in fisheries. Breakdown of older rhizomes isolates daughter plants. Detached pieces of shoot (and presumably also rhizomes) are capable of rooting. These may either remain *in situ* or are carried some distance by water. Regeneration by the seeds, which show hard-coat dormancy (Teltscherova & Hejny 1973), is probably infrequent, even though seed is produced in abundance. It is suspected that germination occurs during spring. No information is available concerning the persistence of any seed bank. Seeds can survive ingestion by water-fowl and may be dispersed also in water, where they may float for up to 1 year (Ridley 1930). *P.n.* is an effective colonist, even of land-locked sites such as pools in quarries. However, it is uncertain whether such sites are colonized by seed or by transported vegetative fragments. The species extends from

Altitude
Slope
Aspect

Unshaded

Shaded

Soil pH

Associated floristic diversity

Not applicable

Hydrology

Triangular ordination

Commonest associates	%
Lemna minor	73
Lemna trisulca	35
Elodea canadensis	27
Ceratophyllum demersum	23
Equisetum fluviatile	23

Bare soil

Abundance in quadrats

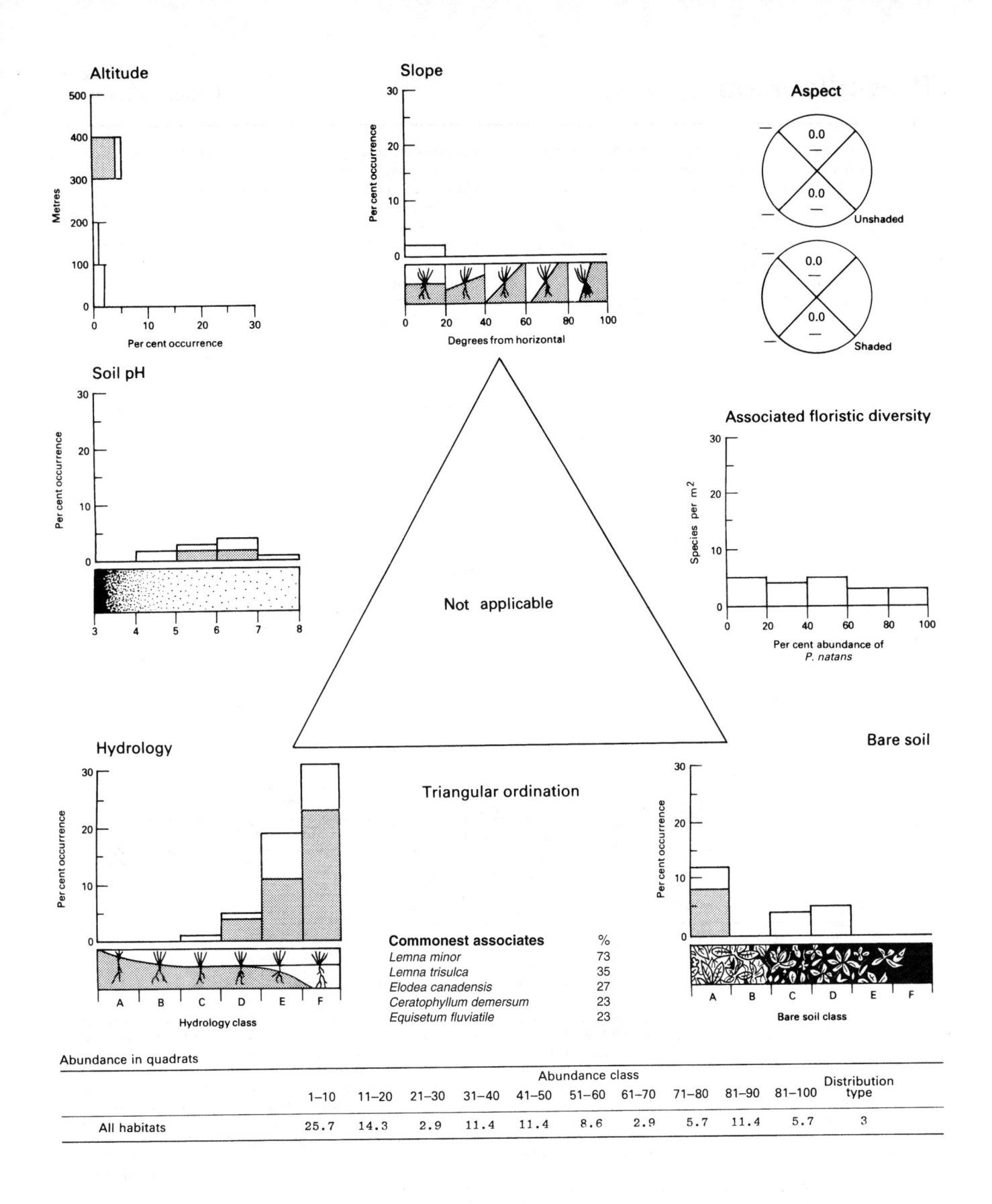

	\multicolumn{10}{c}{Abundance class}	Distribution type
	1–10	11–20	21–30	31–40	41–50	51–60	61–70	71–80	81–90	81–100	
All habitats	25.7	14.3	2.9	11.4	11.4	8.6	2.9	5.7	11.4	5.7	3

peaty, somewhat acidic sites to relatively base-rich conditions, and appears to exploit both nutrient-rich and more-nutrient-poor conditions (see Spence 1964). However, at high levels of acidity *P.n.* is replaced by *P.p.*, and in strongly calcareous waters *P. coloratus* Hornem becomes important. *P.n.* hybridizes, albeit rarely, with four other species in Britain (Stace 1975).

Current trends Despite being an effective colonist, *P.n.* appears to be decreasing through the destruction, and perhaps eutrophication, of aquatic habitats.

Potentilla erecta (L.) Rauschel

Rosaceae
Common Tormentil

Life-form Polycarpic perennial, semi-rosette hemicryptophyte with a woody stock. Stems decumbent or ascending; leaves ternate, <1000 mm^2; roots shallow, usually bearing VA mycorrhizas.

Phenology Flowers from June to September and sets seed from July onwards. Flowering stems die in autumn but some basal leaves may remain winter green. Leaves relatively long-lived.

Height Foliage and flowers usually <200 mm.

Seedling RGR 0.5-0.9 week^{-1}. SLA 21.7. DMC 31.8.

Nuclear DNA amount No data; $2n = 28^*$ (*BC*); $4x^*$.

Established strategy Intermediate between stress-tolerator and C-S-R.

Regenerative strategies V, B$_s$. Regenerative biology is complex (see 'Synopsis'); SB IV.

Flowers Yellow, hermaphrodite, insect-pollinated, self-incompatible (Watson 1969), usually <10 in loose terminal cymes.

Germinule 0.58 mg, 1.6 × 1.1 mm, achene (usually 4-8 in each fruit).

Germination Epigeal; t_{50} 21 days; seeds germinate more rapidly after dark incubation in warm moist conditions; L > D.

Biological flora None, but see Zarzycki (1976).

Geographical distribution The British Isles (100% of British vice-counties), *ABIF* page 331; Europe, although rare in the Mediterranean region (85% of European territories); W Asia.

Habitats Grows mainly in grassland and on grazed and ungrazed moorland and heaths, but recorded also from mire, waysides and quarry and lead-mine spoil. Rarely found in open woodland, and absent from aquatic, arable and skeletal habitats.

Gregariousness Intermediate.

Altitude More common in upland areas, where suitable habitats are more abundant, extending to 470 m. [Up to 1040 m.]

Slope Wide-ranging.

Aspect More frequent and only abundant on N-facing slopes.

Hydrology Not typically a wetland species, but frequent in soligenous mire and also recorded from turfs of *Sphagnum* spp.

Soil pH Occurs across a wide range of soil pH values from 3.5 to 7.5, but with fewer records above pH 6.0.

Bare soil Generally associated with low exposures.

△ **ordination** Distribution centred on vegetation associated with relatively infertile, undisturbed habitats. The range of occurrence extends into some vegetation types of higher productivity, but the species is entirely absent from heavily-disturbed habitats.

Associated floristic diversity Often low, particularly on acidic soils, but high in N-facing limestone grasslands.

Synopsis *P.e.* is a slow-growing, tufted herb typically found in a wide range of grassland and heathland habitats on acidic soils. *P.e.* reaches highest frequency and abundance in pastures of intermediate fertility and

Habitats

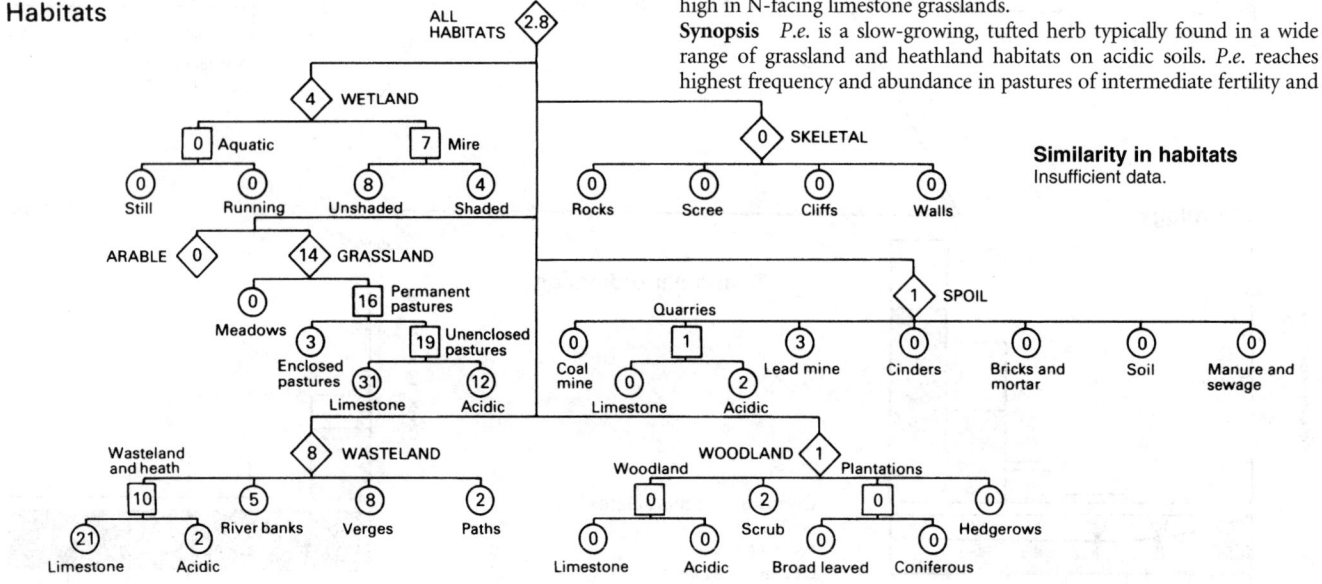

Similarity in habitats
Insufficient data.

is, for example, particularly common on brown earths and incipient podsols developed on loessic soils over limestone (Balme 1953). The distribution extends into calcareous grasslands, but is restricted here to soils of high moisture status such as those occurring on many N-facing dalesides of the Derbyshire limestone. Foliage of *P.e.* is low in P but unusually high in Fe, Mn and Al (*SIEF*). Field measurements and laboratory experiments show that vegetative growth is much more vigorous on non-calcareous soil (Grime 1963b). On calcareous soils it seems likely that *P.e.* may be excluded from dry S-facing sites where shallow root development renders seedlings vulnerable to summer drought; the same phenomenon could explain the apparently wider edaphic range of *P.e.* in the more humid climate of N and W Britain. The shallow root system of *P.e.* is also consistent with the occurrences of *P.e.* in soligenous mire. Flowering shoots of *P.e.* die back completely over winter, new ones being produced rapidly in the late spring by mobilization of reserves from the stout basal stock. This phenological pattern is very different from that of many of the species with which *P.e.* occurs on acidic

soils, and would perhaps not represent a viable strategy in an infertile grazed habitat but for the fact that *P.e.* is very unpalatable to stock (Milton 1933). *P.e.* is self-incompatible and has a long flowering period (Watson 1969). Achenes are often slow to ripen and many are released late in the summer, often in a green condition. Rapid germination is usually achieved only after warm moist incubation and there is no evidence of germination in autumn when many seeds become incorporated into a buried seed bank. Populations have been shown to differ genetically according to habitat and geographical location (Watson 1969). Hybrids occur with *P. anglica* Laicharding and *P. reptans* L., both of which regenerate vegetatively by producing long-rooted runners. The former is apparently an allo-octoploid hybrid between *P.e.* and *P.r.* (Matfield & Ellis 1972). *P.a.* which is self-compatible, tends to occur on soils of intermediate pH, whereas *P.r.*, which is self-incompatible, is confined to base-rich soils.

Current trends As a result of habitat destruction *P.e.* is decreasing in many lowland areas. The species remains abundant, however, in much of upland Britain.

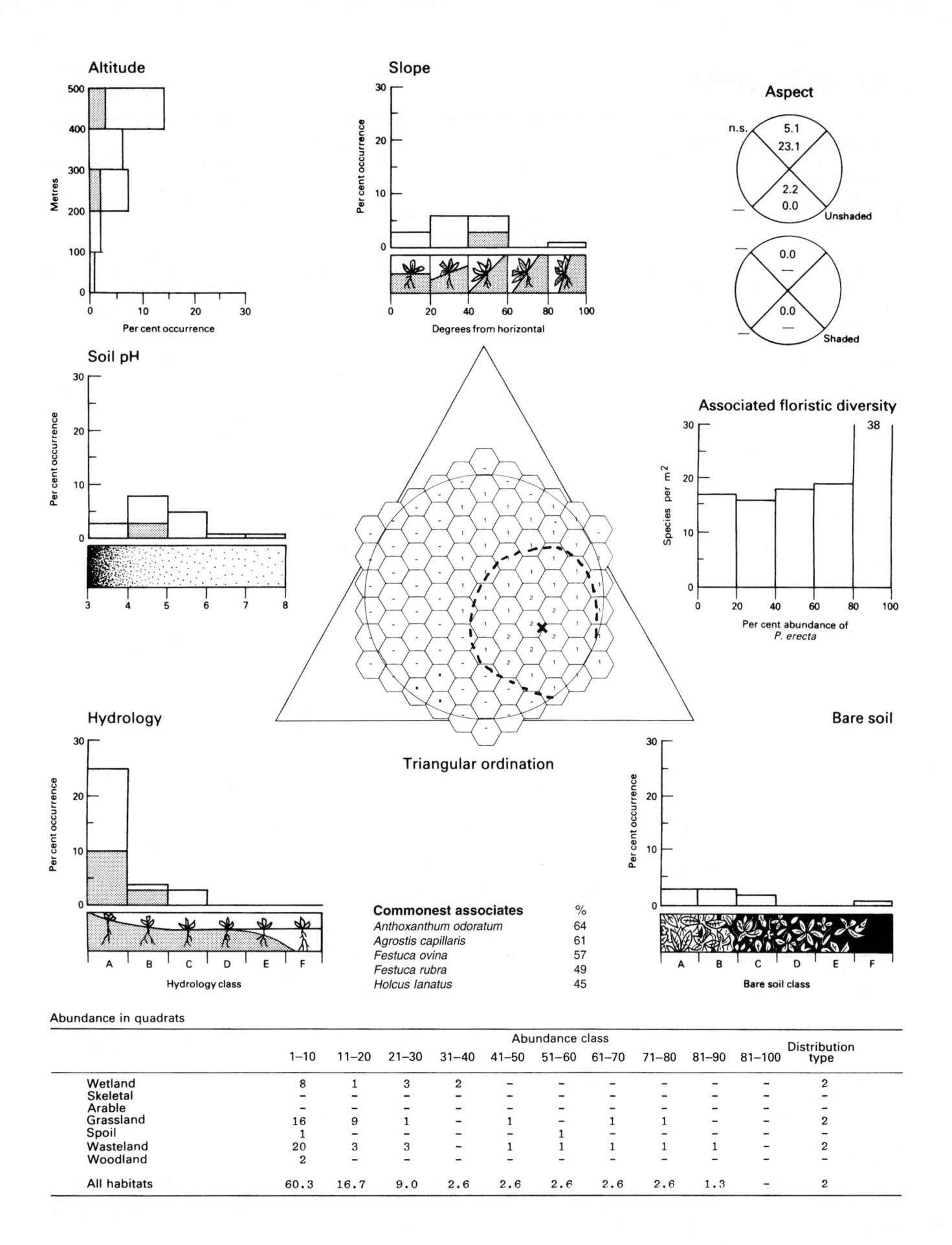

Altitude

Slope

Aspect

	n.s.	5.1
	23.1	
	2.2	
	0.0	Unshaded

	0.0	
	—	
	0.0	
	—	Shaded

Soil pH

Associated floristic diversity

Hydrology

Triangular ordination

Bare soil

Commonest associates	%
Anthoxanthum odoratum	64
Agrostis capillaris	61
Festuca ovina	57
Festuca rubra	49
Holcus lanatus	45

Abundance in quadrats

	Abundance class										Distribution type
	1–10	11–20	21–30	31–40	41–50	51–60	61–70	71–80	81–90	81–100	
Wetland	8	1	3	2	–	–	–	–	–	–	2
Skeletal	–	–	–	–	–	–	–	–	–	–	–
Arable	–	–	–	–	–	–	–	–	–	–	–
Grassland	16	9	1	–	1	–	1	1	–	–	2
Spoil	1	–	–	–	1	–	–	–	–	–	–
Wasteland	20	3	3	–	1	1	1	1	1	–	2
Woodland	2	–	–	–	–	–	–	–	–	–	–
All habitats	60.3	16.7	9.0	2.6	2.6	2.6	2.6	2.6	1.3	–	2

Potentilla sterilis (L.) Garcke

Rosaceae
Barren Strawberry

Life-form Polycarpic perennial, semi-rosette hemicryptophyte with a woody stock and prostrate stolons. Flowering stems decumbent; leaves ternate, hairy, <2000 mm^2; roots shallow, with VA mycorrhizas.
Phenology Winter green, showing appreciable growth in early spring. Flowers from February to May and seed shed May to June. Leaves relatively long-lived.
Height Foliage and flowers usually <60 mm.
Seedling RGR No data. SLA 24.6. DMC 25.0.
Nuclear DNA amount No data; $2n = 28^*$ (*BC*); $4x^*$.
Established strategy Intermediate between stress-tolerant ruderal and C-S-R.
Regenerative strategies V, ?B$_s$. Regeneration by seed requires further study (see 'Synopsis').
Flowers White, hermaphrodite, insect-pollinated or selfed if this fails; 1-3 on each axillary flowering stem.
Germinule 0.58 mg, 1.6 × 1.1 mm, achene; often *c.* 20 per fruit.
Germination Epigeal; no dormancy.
Biological flora None.
Geographical distribution A European endemic found throughout the British Isles except for N Scotland and around the Wash (98% of British vice-counties), *ABIF* page 333; W, C and S Europe extending N to Sweden (49% of European territories).

Habitats Most frequently recorded from the carboniferous limestone, where *P.s.* is found in pastures and scree, and to a lesser extent road verges, lead-mine spoil, quarry heaps and open woodland. Detailed observations elsewhere in the Sheffield region suggest that *P.s.* is most characteristic of lightly shaded roadsides, hedgebanks and wood margins. Absent from arable and wetland habitats.
Gregariousness Sparse.
Altitude Mainly between 200 and 400 m, a reflection of its frequency on the carboniferous limestone. [Up to 790 m.]
Slope Most common on carboniferous limestone dale-sides, and thus particularly associated with moderate slopes.
Aspect Slightly more frequent on N-facing slopes.
Hydrology Not recorded from wetlands.
Soil pH Absent from acidic soils below pH 4.5, and most frequent in the pH range 5.5-6.5.
Bare soil Widely distributed, but most frequent in habitats having low to moderate exposures.
△ ordination Distribution centred on vegetation which is relatively undisturbed and only moderately productive.
Associated floristic diversity In grassland, often associated with species-rich vegetation. In shaded habitats values are typically intermediate or even low.

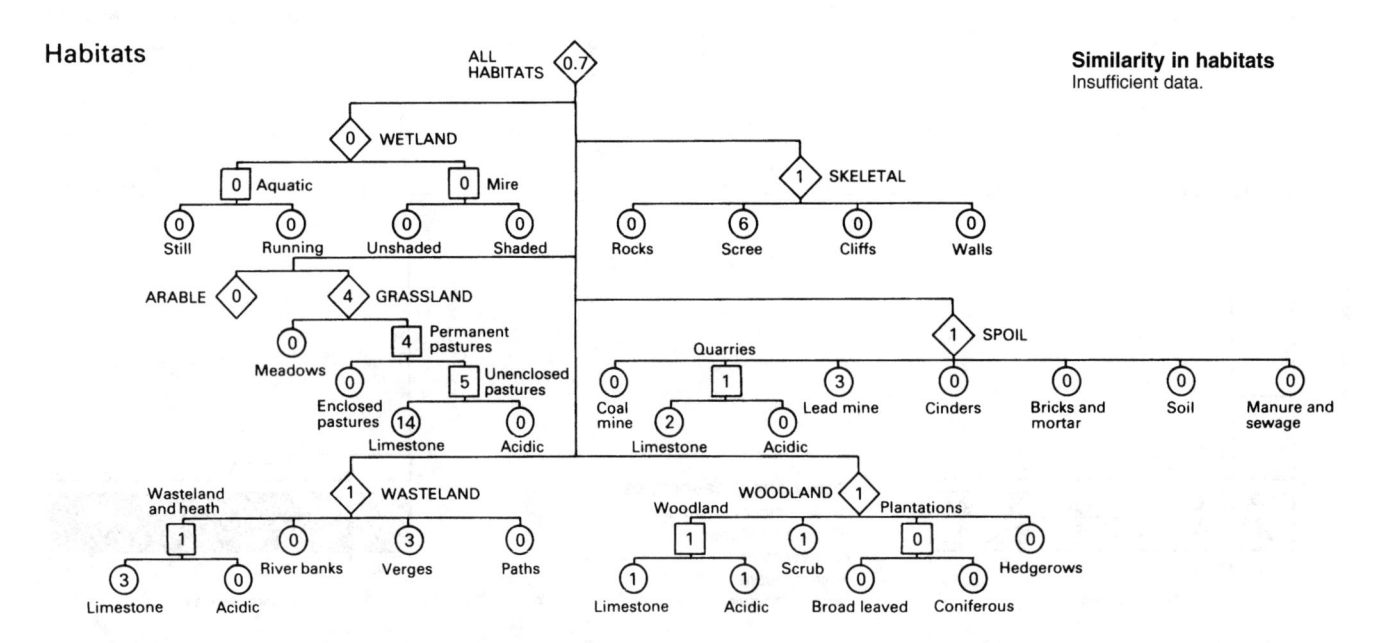

Habitats

Similarity in habitats
Insufficient data.

Synopsis *P.s.* is a tufted, low-growing winter-green herb of relatively infertile hedgebanks and pastures. *P.s.*, which has a relatively shallow root system, appears to be restricted to relatively moist base-rich soils or those of intermediate pH, and is absent from both waterlogged and droughted sites. In common with a number of other winter-green forbs, such as *Viola riviniana*, *Fragaria vesca* and *Ajuga reptans*, *P.s.* tends to be restricted to partially shaded habitats in lowland areas, but extends into more-open habitats in upland regions. Foliage persists throughout the summer and, though clearly shade-tolerant, *P.s.* is normally absent from deep shade. The occurrence of *P.s.* in pasture may in part reflect the low palatability of the leaves to stock and the ability of the foliage to exploit shaded layers low in the turf canopy. *P.s.* flowers and sets seed remarkably early, and appears capable of sustaining growth at low temperatures. It is not known whether a persistent seed bank is formed, and the season of seed germination remains uncertain. The species is often found in lightly disturbed sites such as steep banks subject to a degree of soil creep and the margin of woodland rides, and compared with other species of shaded habitats the achenes are relatively small. We predict that regeneration by seed is usually associated with disturbance events. *P.s.* forms extensive, usually diffuse, patches through the growth of rooted stolons. In shaded habitats plants tend to have larger leaves and longer petioles and a more erect habit than those from open sites. The extent of ecotypic differentiation in *P.s.* has not been investigated.
Current trends Regarded by Peterken (1981) as a plant of ancient woodland with a low colonizing ability. *P.s.* spreads relatively slowly on roadsides, and has a tendency to be restricted to ancient habitats. Probably decreasing.

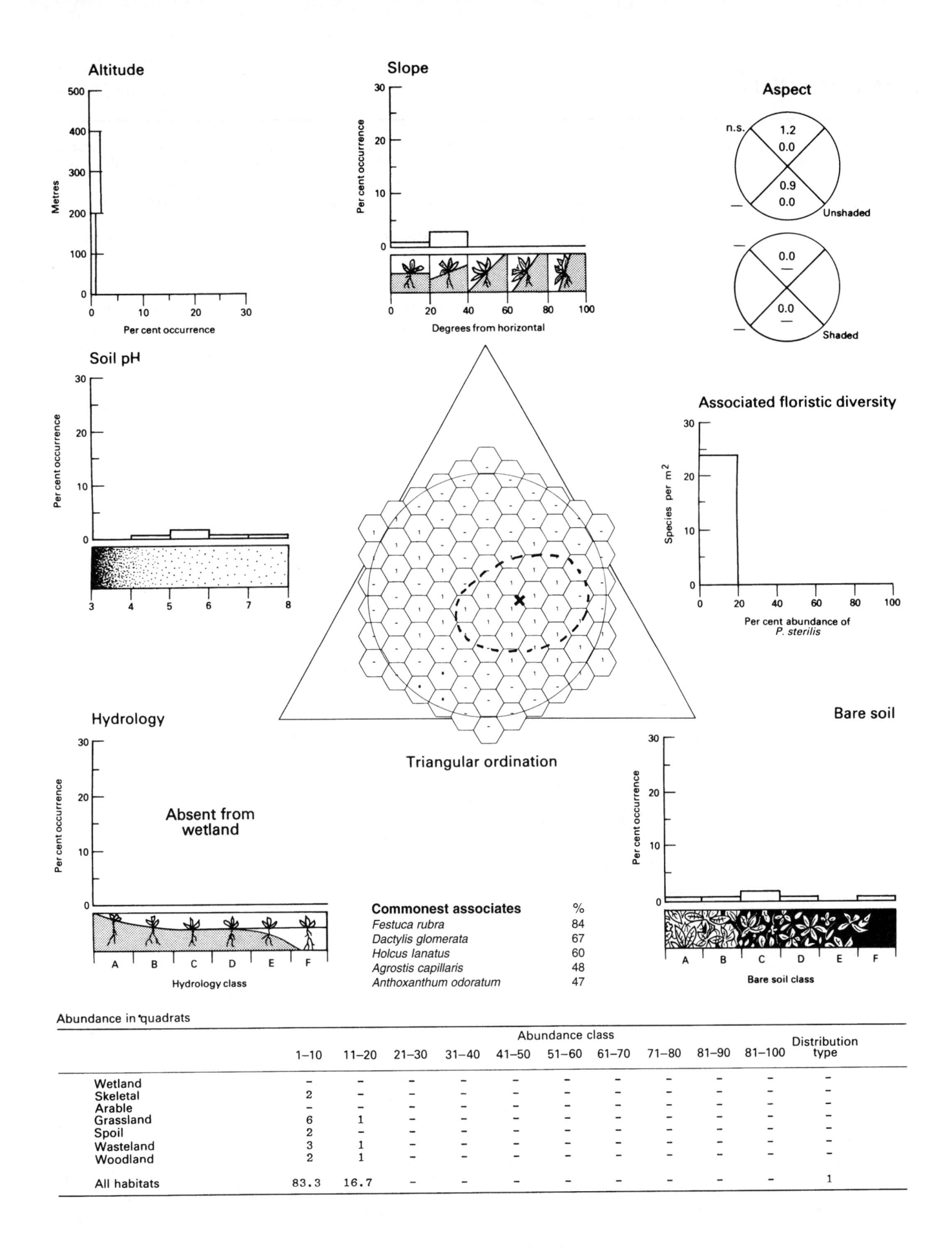

Altitude

Slope

Aspect

Soil pH

Associated floristic diversity

Hydrology

Absent from wetland

Triangular ordination

Commonest associates	%
Festuca rubra	84
Dactylis glomerata	67
Holcus lanatus	60
Agrostis capillaris	48
Anthoxanthum odoratum	47

Bare soil

Abundance in quadrats

	Abundance class									Distribution type
	1–10	11–20	21–30	31–40	41–50	51–60	61–70	71–80	81–90	81–100	
Wetland	–	–	–	–	–	–	–	–	–	–	–
Skeletal	2	–	–	–	–	–	–	–	–	–	–
Arable	–	–	–	–	–	–	–	–	–	–	–
Grassland	6	1	–	–	–	–	–	–	–	–	–
Spoil	2	–	–	–	–	–	–	–	–	–	–
Wasteland	3	1	–	–	–	–	–	–	–	–	–
Woodland	2	1	–	–	–	–	–	–	–	–	–
All habitats	83.3	16.7	–	–	–	–	–	–	–	–	1

Primula veris L.

Primulaceae
Cowslip, Paigle

Life-form Shortly rhizomatous, polycarpic perennial, rosette hemicryptophyte. Scapes erect; leaves in a basal rosette, hairy, ovate, usually <2000 mm², roots with VA mycorrhizas.

Phenology New leaves formed during late winter and spring. Flowers produced in April and May. Seed shed from July to September onwards. Leaves long-lived.

Height Foliage often appressed to the ground, but up to 150 mm in taller grassland; flowers to 300 mm.

Seedling RGR No data. SLA 17.7. DMC 21.5.

Nuclear DNA amount No data; $2n = 22$ (*FE3*); $2x$.

Established strategy Intermediate between stress-tolerator and C-S-R.

Regenerative strategies V, ?B_s Regeneration by seed requires further study. Regenerative biology is complex (see 'Synopsis').

Flowers Yellow, hermaphrodite, heterostylous, insect-pollinated; up to 30 aggregated into an umbel; self-incompatibility mechanism described by Lees (1971).

Germinule 0.69 mg, 1.7×1.4 mm, seed dispersed from a *c.* 50 seeded capsule.

Germination Epigeal; seeds contain water-soluble inhibitors (B. D. Wheeler pers. comm.), dormancy also broken by chilling; L > D.

Biological flora None.

Geographical distribution Most of the British Isles, but local outside England (92% of British vice-counties), *ABIF* page 298; Europe except the extreme N and much of the Mediterranean (74% of European territories); temperate Asia.

Habitats In grazed and ungrazed grassland, mainly on calcareous soils. Also recorded locally on stabilized scree slopes and limestone quarry heaps, lead-mine spoil, railway banks, road verges and, more rarely, in open scrub. Rarely observed in mire and on skeletal habitats. [Also occurs on stabilized sand dunes and sea cliffs.]

Gregariousness Sparse.

Altitude Most frequent on the carboniferous limestone, particularly between 200 and 300 m. [Up to 750 m.]

Slope Mainly recorded from intermediate slopes corresponding to daleside habitats.

Aspect Significantly more frequent on unshaded S-facing slopes. Additional, more-extensive studies within the Sheffield region (Hodgson unpubl.) suggest that *P.v.* is wide-ranging with respect to aspect and most characteristic of continuously moist soils.

Hydrology Absent from wetlands.

Soil pH All records in the pH range 5.0-8.0.

Bare soil Occurrences scattered, but abundant only in habitats with little bare soil.

△ **ordination** Restricted to vegetation of relatively unproductive habitats.

Associated floristic diversity Usually species-rich.

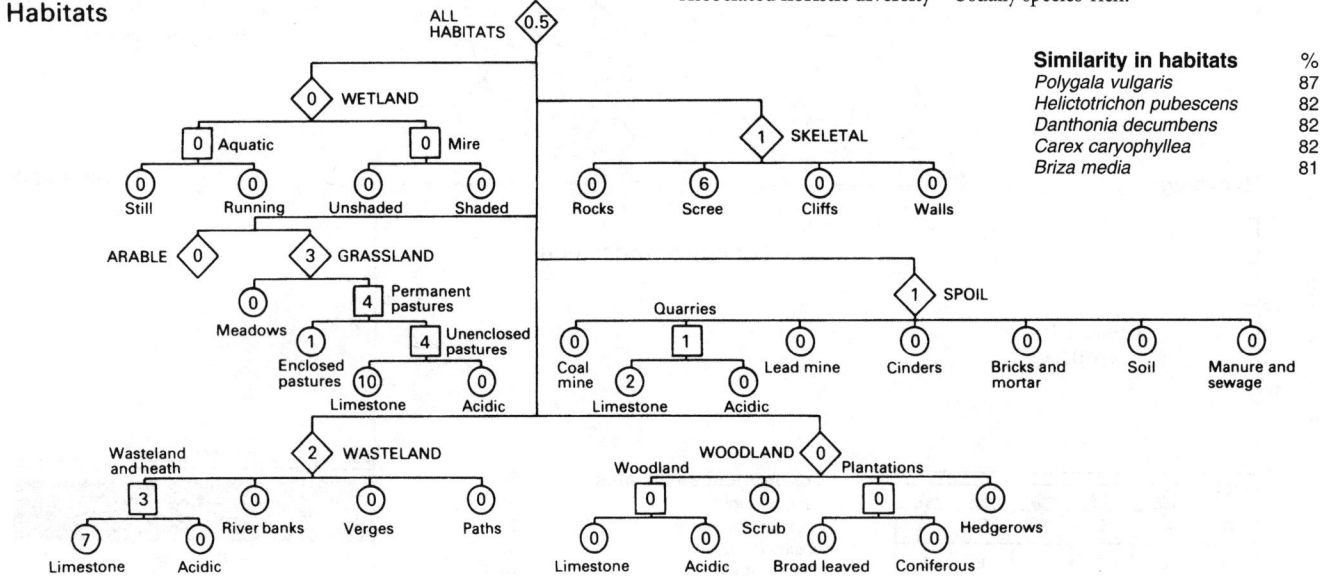

Habitats

Similarity in habitats	%
Polygala vulgaris	87
Helictotrichon pubescens	82
Danthonia decumbens	82
Carex caryophyllea	82
Briza media	81

Synopsis *P.v.* is a spring-flowering, long-lived (Tamm 1972), wintergreen, rosette-forming herb particularly characteristic of short, species-rich turf and only rarely occurring in shaded habitats. *P.v.* occurs predominantly on moist calcareous soils, but can be found occasionally on dry soils over non-calcareous strata. The species has an unusual phenology, in that it commences growth during winter and flowers in spring but achieves peak biomass in summer. This differs from, but overlaps, the phenology of summer-growing herbs such as *Leontodon hispidus*. The leaves, which are often held appressed to the ground, tend to be little predated by stock. *P.v.* has a very limited capacity for vegetative replication through branching of the rhizome, although this may be the most common mechanism of regeneration in stable communities (Tamm 1972). Elaborate mechanisms, including heterostyly (see Lees 1971), reduce the amount of self-fertilization. Nevertheless, *P.v.* produces considerable quantities of seed and in disturbed areas may establish relatively rapidly to form extensive populations. The timing of germination and nature of the seed bank is uncertain. *P.v.* forms four ssp. in Europe, of which only one, *ssp. veris*, occurs in Britain. The extent of genotypic variation and phenotypic plasticity between British populations is uncertain, though even within populations individual plants are known to vary considerably in leaf size and number of flowers per umbel.

Current trends *P.v.* has greatly decreased through the ploughing of old pastures and the relaxation of grazing pressure in semi-natural grassland. Although becoming ever rarer, *P.v.* is showing some resurgence along the verges of some motorways and trunk roads. Plants of the 'pin' form, with the long stigma, are able to self in the absence of 'thrum' plants, with the short stigma (Lees 1971). Normally the proportion of 'pin' types is about 60% (Lees 1971), but in small isolated populations a gradual increase in the proportion of 'pin' plants may be expected and, in unstable sites, and newly colonized areas, the greater heterozygosity associated with a high proportion of 'thrum' plants may be advantageous (Lees 1971).

Altitude

Slope

Aspect

$P < 0.05$

	0.8
	0.0
	2.7
	0.0

Unshaded

	0.0
	—
	0.0
	—

Shaded

Soil pH

Associated floristic diversity

35

Triangular ordination

Hydrology

Absent from wetland

Bare soil

Commonest associates

	%
Festuca ovina	80
Plantago lanceolata	80
Carex flacca	77
Festuca rubra	69
Campanula rotundifolia	66

Abundance in quadrats

	1–10	11–20	21–30	31–40	41–50	51–60	61–70	71–80	81–90	81–100	Distribution type
Wetland	–	–	–	–	–	–	–	–	–	–	–
Skeletal	2	–	–	–	–	–	–	–	–	–	–
Arable	–	–	–	–	–	–	–	–	–	–	–
Grassland	3	1	2	–	–	–	–	–	–	–	–
Spoil	1	–	–	–	–	–	–	–	–	–	–
Wasteland	6	–	–	–	–	–	–	–	–	–	–
Woodland	–	–	–	–	–	–	–	–	–	–	–
All habitats	80.0	6.7	13.3	–	–	–	–	–	–	–	2

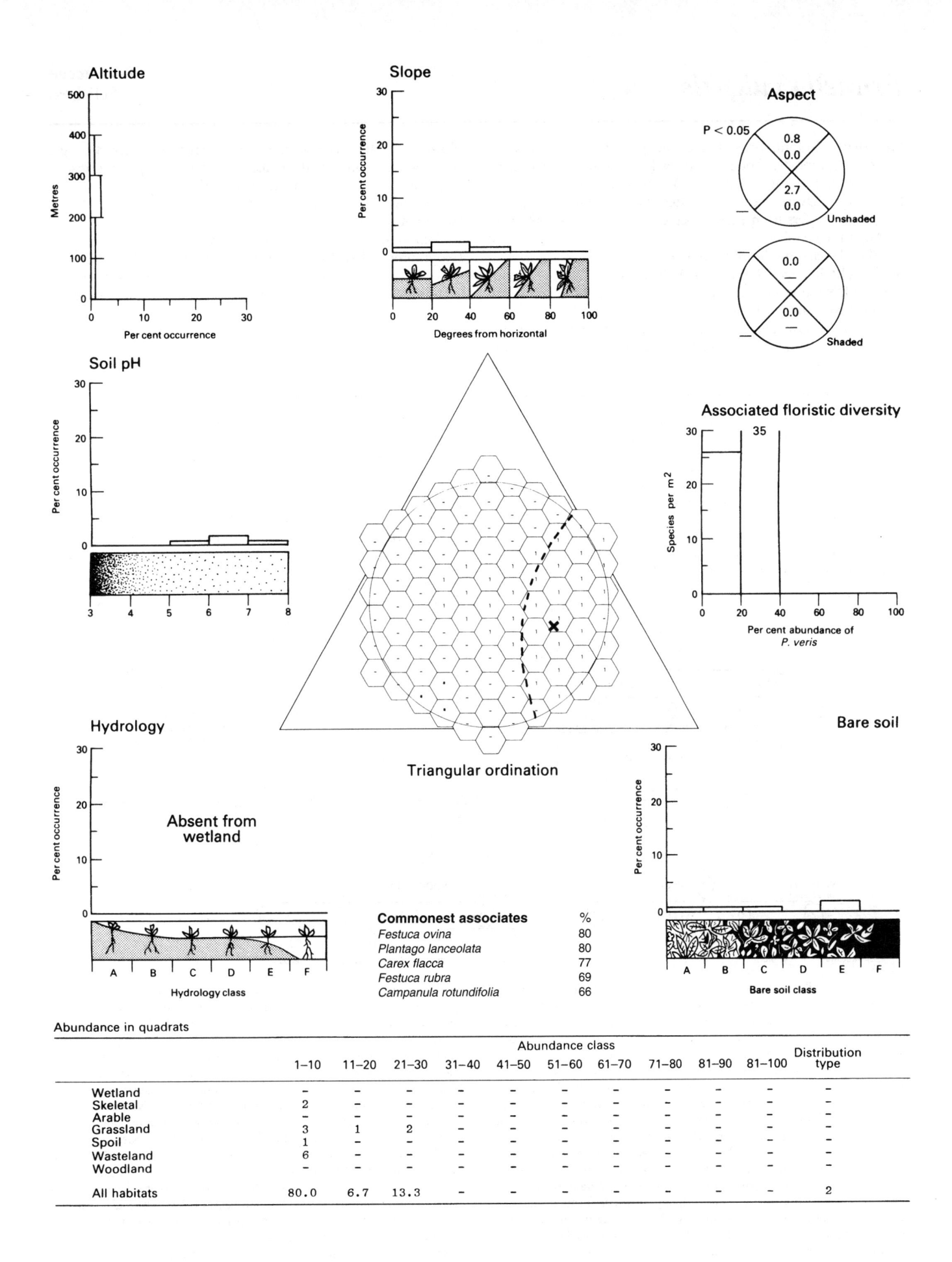

Prunella vulgaris

Lamiaceae
Self-heal

Life-form Stoloniferous, polycarpic perennial, semi-rosette hemicryptophyte. Shoots ascending; leaves ovate, hairy, usually <1000 mm^2; roots shallow, with VA mycorrhizas.
Phenology Overwinters as rosettes. Shoots elongate in late spring. Flowers from June to September and sheds seed from August to October.
Height Foliage and flowers <100 mm.
Seedling RGR 0.5-0.9 week^{-1}. SLA 30.3. DMC 15.1.
Nuclear DNA amount 1.3 pg; $2n = 28*$ (*BC*); $2x*$.
Established strategy C-S-R.
Regenerative strategies (V), ?B$_s$. Capable of regeneration by vegetative means and by seed. Further seed-bank studies are required (see 'Synopsis'); ?SB III.

Flowers Violet, hermaphrodite, self-sterile (Warwick & Briggs 1979), sometimes with smaller female flowers present, insect-pollinated; often <50 in a terminal inflorescence.
Germinule 0.73 mg, 2.0 × 1.0 mm, nutlet (4 per ovary).
Germination Epigeal; t_{50} 5 days; 13-34°C; L > D.
Biological flora None, but see Zarzycki (1976).
Geographical distribution The British Isles (100% of British vice-counties), *ABIF* page 521; Europe (95% of European territories); temperate Asia, N Africa, N America and Australia.

Habitats Occurs mainly in meadows and pastures. Also common on rock outcrops and waysides. Locally recorded from a number of other habitats, including limestone quarry spoil, arable fields (perhaps following the detachment of plants from the field margins during ploughing) and lead-mine spoil. Particularly frequent in the short or open turf of heavily grazed or trampled habitats, and a common lawn weed. Seldom persistent in woodlands and absent from aquatic habitats.
Gregariousness Sparse to intermediate.
Altitude Widely dispersed to 400 m. [Up to 760 m.]
Slope Scarce on both cliffs and slopes of >50°.

Aspect Overall slightly more abundant on S-facing slopes, but with a significant N-facing bias in grassland (*Ecol Atl*).
Hydrology Typical of moist soils, but occasionally occurring in soligenous mire.
Soil pH Largely restricted to soils of pH >5.0.
Bare soil Highest frequency of occurrence and abundance where there is a relatively large proportion of bare soil, but occurs in all categories.
△ ordination Scattered distribution centred on vegetation associated with moderate levels of stress and disturbance. Absent only from highly productive relatively undisturbed vegetation.
Associated floristic diversity Typically species-rich.
Synopsis *P.v.* is a short rather slow-growing, winter-green, patch-forming herb of grassland: typically associated with moist, moderately fertile soils. Like *Bellis perennis*, with which it often grows, *P.v.* is easily

Habitats

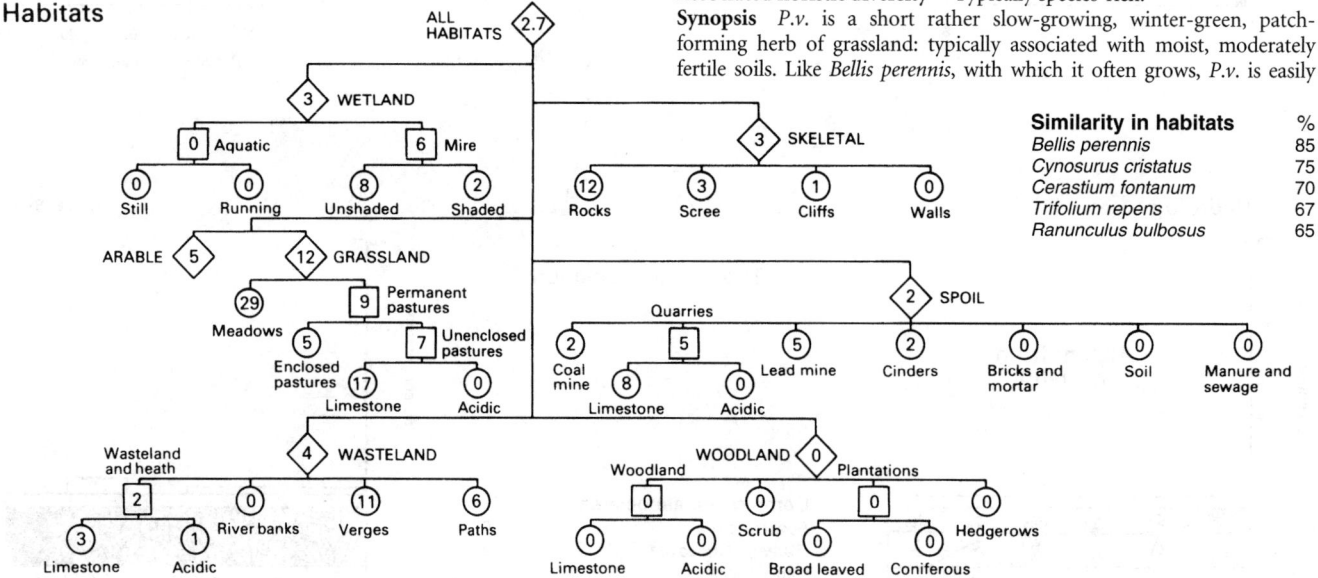

Similarity in habitats	%
Bellis perennis	85
Cynosurus cristatus	75
Cerastium fontanum	70
Trifolium repens	67
Ranunculus bulbosus	65

dominated by taller herbs and is abundant only in short turf, particularly in lawns and permanent pasture, where large clones may develop by vegetative expansion. *P.v.* is favoured by close grazing, presumably because of its creeping growth-form, moderate resistance to trampling and rather low palatability (Milton 1933). *P.v.* is infrequent in burned sites. The species exhibits some shade-tolerance and frequently occurs along woodland rides. It persists in meadows if grazing occurs after harvest. *P.v.* has a marked summer peak in above-ground biomass. Its parts are short-lived and, like *Trifolium repens*, *P.v.* is relatively mobile, expanding temporarily into gaps within turf before being replaced by more-robust species (see Schmid 1985a). The species forms loose clonal patches and connections with daughter ramets may decay in less than 1 year (Schmid & Harper 1985). We suspect that *P.v.* may regenerate from shoot fragments detached as a result of trampling, or other forms of damage. Regeneration also occurs by means of seeds, which germinate mainly in spring (Schmid 1985a, Winn 1985). Establishment from seed can occur in comparatively dry exposed microsites (Oomes & Elberse 1976). Plants developed from small lengths of shoot closely resemble seedlings in appearance, and the relative importance of seedings and vegetative fragmentation during the re-colonization of disturbed ground requires investigation. Winn (1985) contradicts Donelan and Thompson (1980) and Roberts (1986a) in suggesting that no seed bank is formed. The relative importance of regeneration by seed and by vegetative means varies greatly between populations (cf. Schmid 1985a, Winn 1985), and in Europe both annual and perennial races occur (Bocher 1963). *P.v.* is phenotypically plastic, and considerable change in morphology can be induced by temperature and by the intensity of grazing or mowing. Seed size is also variable, with plants from shaded sites producing heavier seeds (Winn 1985). Populations also vary in morphology and in growth rate according to habitat and geographical area. Bocher (1949), Warwick and Briggs (1979), Harper and Schmid (1985) and Schmid (1985a,b) make clear the ecological importance both of phenotypic plasticity and of genetic diversity in the widespread success of the species.
Current trends Common, and probably increasing.

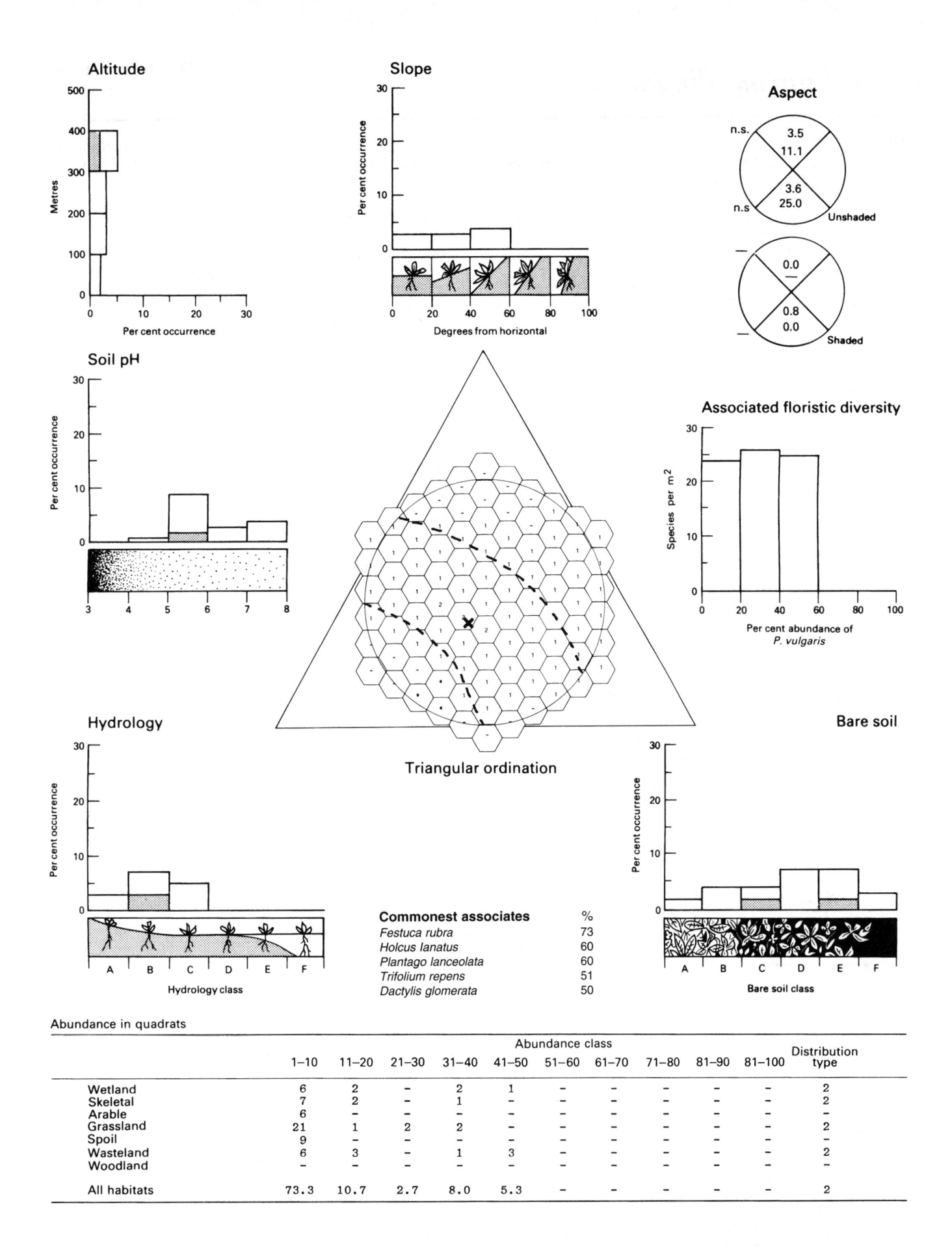

Altitude

Slope

Aspect

	n.s.	3.5	
		11.1	
	n.s	3.6	Unshaded
		25.0	

	—	0.0	
		—	
	—	0.8	Shaded
		0.0	

Soil pH

Associated floristic diversity

Hydrology

Triangular ordination

Commonest associates

	%
Festuca rubra	73
Holcus lanatus	60
Plantago lanceolata	60
Trifolium repens	51
Dactylis glomerata	50

Bare soil

Abundance in quadrats

	\multicolumn{10}{c	}{Abundance class}	Distribution type
	1–10	11–20	21–30	31–40	41–50	51–60	61–70	71–80	81–90	81–100	
Wetland	6	2	–	2	1	–	–	–	–	–	2
Skeletal	7	2	–	1	–	–	–	–	–	–	2
Arable	6	–	–	–	–	–	–	–	–	–	–
Grassland	21	1	2	2	–	–	–	–	–	–	2
Spoil	9	–	–	–	–	–	–	–	–	–	–
Wasteland	6	3	–	1	3	–	–	–	–	–	2
Woodland	–	–	–	–	–	–	–	–	–	–	–
All habitats	73.3	10.7	2.7	8.0	5.3	–	–	–	–	–	2

Pteridium aquilinum (L.) Kuhn

Dennstaedtiaceae
Bracken

Life-form Polycarpic perennial geophyte. Fronds arising directly from deep underground rhizome, much subdivided, sometimes >0.5 m^2; roots with VA mycorrhizas.
Phenology Fronds emerging from late spring onwards, persisting until autumn. Spores ripening August to September, shed August to October (Page 1982).
Height Both fertile and sterile fronds to 1800 mm.
Sporeling RGR No data. SLA 19.8. DMC 29.8.
Nuclear DNA amount 12.8 pg; $2n = 104*$ (BC); $2x*$, ($4x$ for family (Page 1982)).
Established strategy Competitor.
Regenerative strategies V, W, ?B$_s$. Mainly by vegetative means (see 'Synopsis'); ?SB III.

Sporangium Aggregated into sori on the underside of the frond.
Germinule 33 × 28 μm, spore, brown, explosively dispersed from a many-spored sporangium.
Germination t_{50} 4 days; L > D.
Biological flora Cody and Crompton (1975), see also Perring and Gardiner (1976) and Morton-Boyd *et al.* (1982).
Geographical distribution The British Isles (100% of British vice-counties), *ABIF* page 66, *AF* page 45; most of Europe although mainly on mountains in the S (92% of European territories); virtually worldwide except for temperate S America, the Arctic.

Habitats Typically found in woodland, hill pastures and on wasteland, particularly on acidic soils. Also frequent on waysides, river banks and spoil heaps, particularly in gritstone quarries, and recorded from mire and cliffs. Absent from arable and aquatic habitats. [Particularly abundant in montane areas. Also found on top of sheltered cliffs, and more rarely on stable sand dunes.]
Gregariousness Stand-forming even at low frond density.
Altitude Common at all altitudes, but suitable habitats are more frequent and abundant in upland areas. [Up to 590 m.]
Slope Largely absent from skeletal habitats, and thus less frequent on slopes >50°.

Aspect Slightly more frequent on S-facing slopes, both in shaded and unshaded sites. However, more abundant on N-facing slopes. The ecological significance, if any, of this pattern remains uncertain.
Hydrology Not a wetland species, but occasionally recorded from the margin of soligenous mire. Absent from topogenous mire and aquatic habitats.
Soil pH Recorded within the pH range 3.0-7.6, but most frequent and abundant below pH 4.5.
Bare soil Found over a wide range of values. Most sites have little exposed soil but a dense carpet of *P.a.* litter.
△ **ordination** Present in most undisturbed vegetation types, with distribution centred on highly productive vegetation.
Associated floristic diversity Low.

Habitats

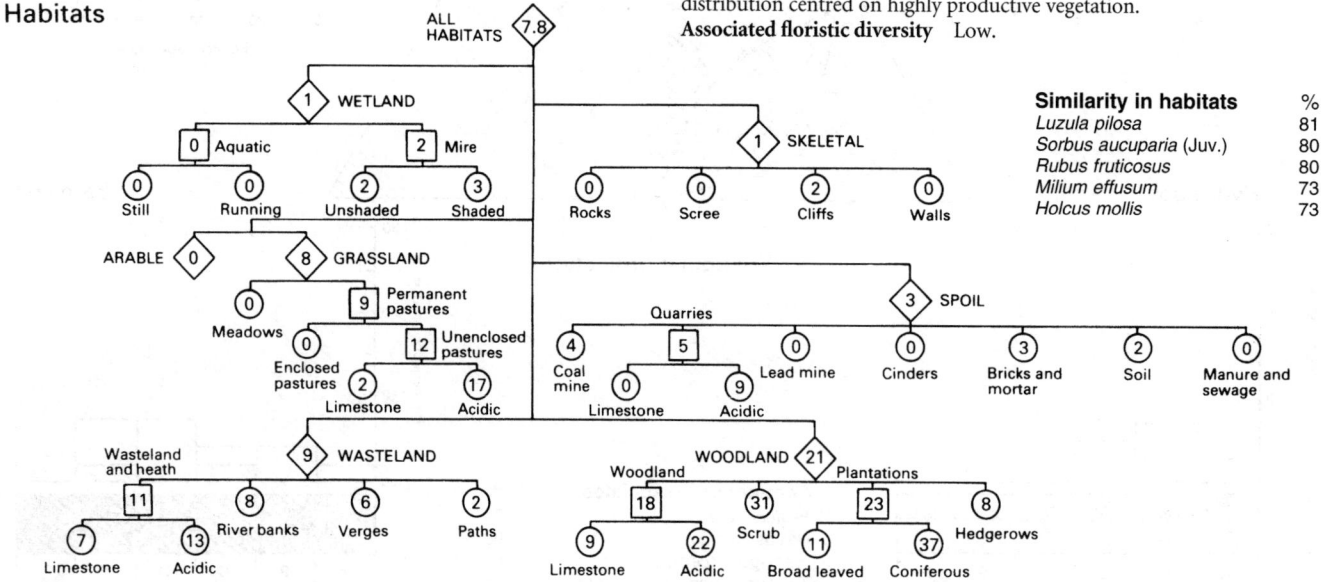

Similarity in habitats	%
Luzula pilosa	81
Sorbus aucuparia (Juv.)	80
Rubus fruticosus	80
Milium effusum	73
Holcus mollis	73

Synopsis *P.a.* is found in a wide range of shaded and unshaded habitats, particularly on deep acidic soils. In lowland areas *P.a.* is markedly more frequent in shaded habitats, while in upland areas it is equally, if not more, widespread in open sites. Whether this difference is brought about by climatic or land-use factors is not known. Unlike other common British ferns, *P.a.* possesses an extensive system of underground rhizomes and, in the context of the British flora, *P.a.* is a uniquely competitive and aggressively invasive fern. These attributes are associated with the presence of xylem vessels in the rhizome, an unusual feature in ferns (White 1963). Although *P.a.* is most characteristic of acidic strata, the species is usually restricted to deeper soils, and shows maximum vigour on productive brown earths. Over calcareous strata *P.a.* is often restricted to situations where there is overlying drift or loess. In vegetation subject to run-off from calcareous road foundations *P.a.* is often chlorotic and stunted. Leaves have low levels of Ca, but contain relatively high concentrations of P (*SIEF*) and the species is very efficient in recycling nutrients within the plant

(Petersen 1985). The whole plant is heavily defended chemically from mammalian and insect predators, and is toxic to livestock and to humans (Cooper-Driver 1976, Cooper & Johnson 1984). *P.a.* is also considered allelopathic by some biologists (Gliessman 1976). Young fronds produce extrafloral nectaries. These provide food for ants which may in turn rid *P.a.* of some of its potential insect predators (Lawton & Heads 1984, Heads & Lawton 1984). Young shoots of *P.a.*, the first of which emerge in April or May, are very sensitive to frost and trampling. *P.a.* was originally a woodland plant before the deforestations (Watt 1976). However, spore production and growth of the prothallus is greater in unshaded habitats (Conway 1949, Dring 1965). In open sites dominance by *P.a.* may be largely due to the presence of much surface-lying and standing litter. Litter accumulation may even suppress *P.a.* and this can lead eventually to re-invasion by other species. In woodland breakdown of litter is more rapid; here populations of *P.a.* are often more stable (Watt 1976). Once established, *P.a.* regenerates vegetatively to form large clonal patches,

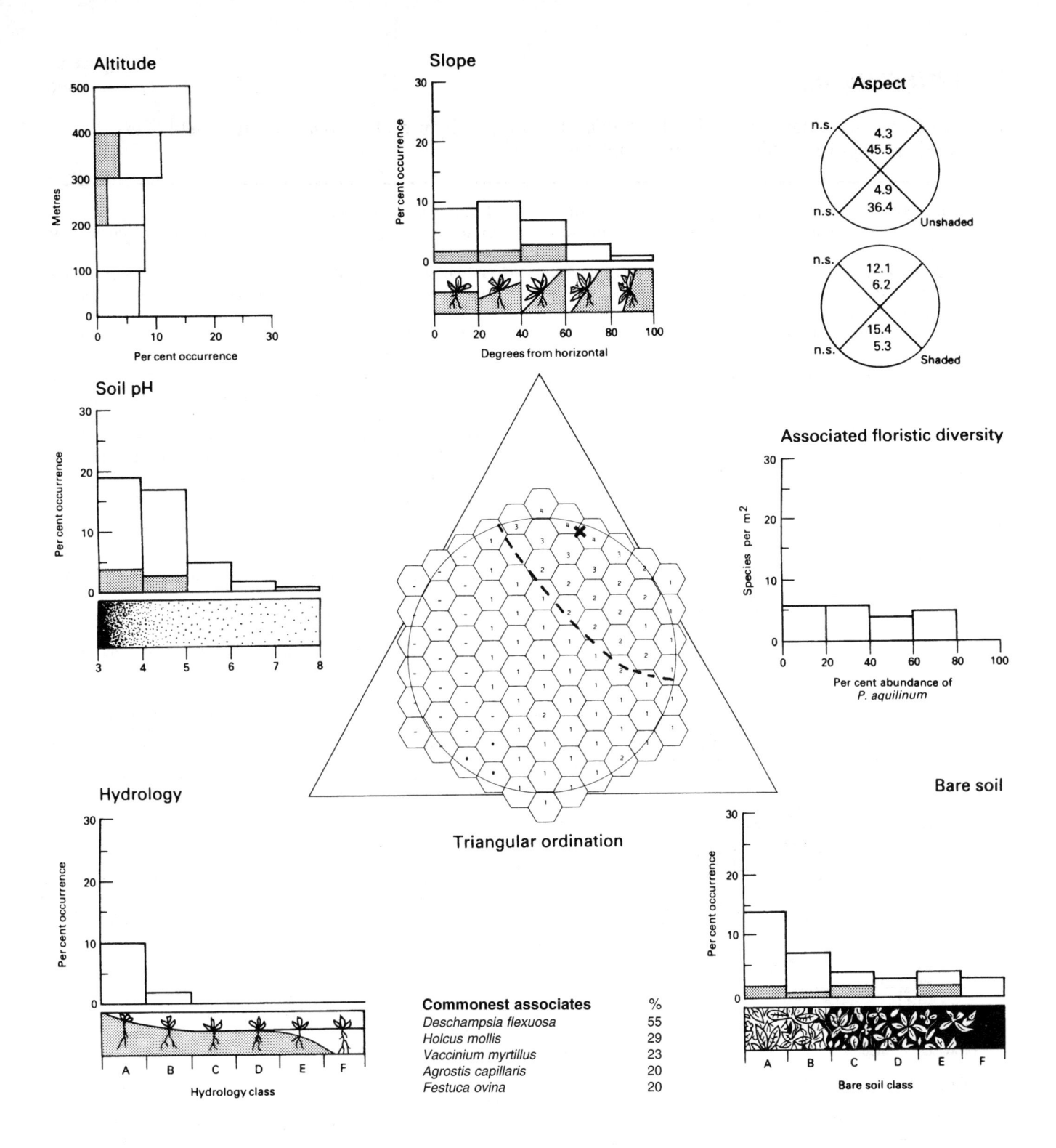

Altitude
Slope
Aspect
Soil pH
Associated floristic diversity
Hydrology
Triangular ordination
Bare soil

Commonest associates	%
Deschampsia flexuosa	55
Holcus mollis	29
Vaccinium myrtillus	23
Agrostis capillaris	20
Festuca ovina	20

bypassing the vulnerable prothallial stage that follows spore germination. Individual parts of the rhizome system may live for >50 years (Watt 1940) and their presence in the subsoil confers resistance to burning. Up to 30 million spores may be produced by a single fertile frond (Conway 1957) and, though initially wind-dispersed, viable spores are often found in abundance within the soil profile, and a buried spore bank is suspected. Spores retain their viability for up to 10 years (Conway 1949) and may even survive ingestion by insects (Scott *et al*. 1985). Although establishment of the sporeling is rapid (Conway 1953), regeneration from spores is mainly confined to areas of disturbed or burnt ground (Page 1976) and may occur in spring (*Biol Fl*). Thus, spores are probably only important in the colonization of new sites. The species shows genetic variation and, for example, exhibits polymorphism for cyanogenesis (Dyer & Hadfield 1985).

Formerly of economic importance as a source of fuel, thatch, litter, compost, food and potash, P.a. has some potential as a possible energy resource, either as a fuel or in the production of methanol or methane (Callaghan and Sheffield 1985).

Current trends Now a weed, primarily of grasslands (Rymer 1976) but also of forestry (Biggin 1982). Increasing in upland areas (Page 1982) and difficult to eradicate (McCreath 1982), despite recent efforts to reclaim grasslands by application of asulam and other mechanical or chemical means (ADAS 1983c). In lowland Britain many of its habitats have been destroyed as a result of modern land-usage, and P.a. may be decreasing in some areas. Nevertheless, P.a. remains common and is still an invasive plant in many lowland areas.

(*) *Quercus* agg.

Fagaceae
Oak

Includes *Q. petraea* (Mattuschka) Liebl., *Q. robur* L. and their hybrid *Q.* × *rosacea* Bechst. Field data analysed here refer to seedlings and small saplings

---

Life-form Deciduous tree with a broad crown and a fissured bark. Leaves ovate, pinnately lobed, >5000 mm^2; roots with ecto-mycorrhizas.
Phenology Flowers appear in April to May, with the expanding leaves and seed is shed from September to November. Leaves shed in late autumn.
Height Foliage and flowers up to 30 m.
Seedling RGR No data. SLA 13.2. DMC. No data.
Nuclear DNA amount Q.p. 1.5 pg; Q.r. 1.9, Q.p. and Q.r., 2n = 24 (*FE1*); 2x.
Established strategy Stress-tolerant competitor.
Regenerative strategies S. By seeds which germinate during autumn or winter. Seeds are not persistent (*Biol Fl*); SB I.
Flowers Green, monoecious, wind-pollinated; male flowers in drooping catkins; female spikes 1-5 -flowered (*CTW*). Q.r., at least, probably self-sterile (*Biol Fl*).
Germinule Q.p. c. 2500 mg, Q.r. c. 3500 mg (*Biol Fl*), 20.0 × 12.9 mm; nut.
Germination Hypogeal; radical capable of immediate extension but seedlings show epicotyl dormancy which is broken by chilling.
Biological flora Jones (1959a,b), see also Morris and Perring (1974) and Newbold and Goldsmith (1981).
Geographical distribution Q.p. more common in the N and W of British Isles (80% of British vice-counties), *ABIF* page 131; W, C and SE Europe extending northwards to Scandinavia (69% of European territories). Q.r. the most frequent oak in SE and C England, and less common and perhaps not native in N and W Britain (81% of British vice-counties) *ABIF* page 132; more widespread in Europe (77% of European territories).

Habitats Seedlings and saplings widely distributed in shaded habitats, particularly in woodland on acidic strata and in broad-leaved plantations. Also widespread in unshaded habitats, particularly quarry spoil, skeletal habitats and wasteland. Noticeably absent from grazed grassland.

Gregariousness Typically sparse.
Altitude Juveniles decreasing in frequency with increasing altitude. [Up to 460 m.]
Slope Occurs over a wide range of values.
Aspect Slightly more frequent on N-facing slopes in shaded and unshaded habitats, perhaps a reflection of the vulnerability of the seed to desiccation (Newbold & Goldsmith 1981).

Similarity in habitats	%
Hyacinthoides non-scripta	83
Acer pseudoplatanus (Juv.)	81
Luzula pilosa	79
Lonicera periclymenum	75
Sorbus aucuparia (Juv.)	72

Habitats

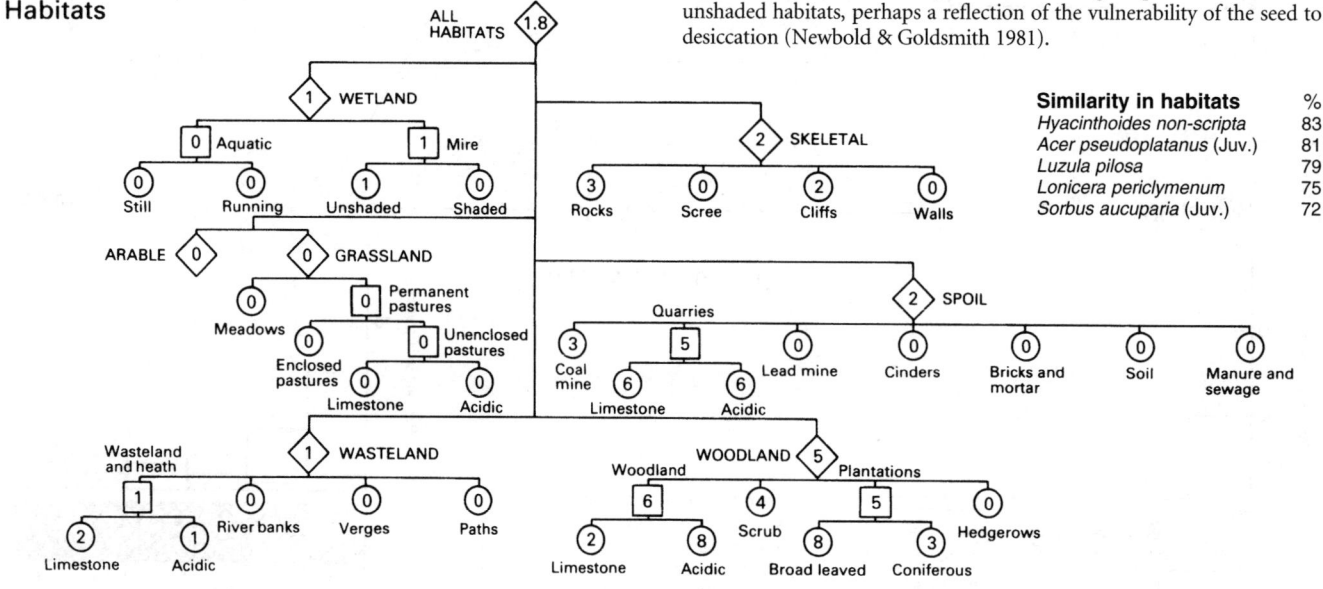

Hydrology Occurs on damp soils, but does not persist in strongly waterlogged soils.
Soil pH Over a wide range of values but most consistently recorded below pH 5.0.
Bare soil Widely distributed. Sites with little exposed soil include many with a covering of tree litter.
△ **ordination** No data.
Associated floristic diversity Juveniles restricted to species-poor vegetation.
Synopsis Tall but slow-growing, deciduous, timber trees, with ring porous wood and a late-emerging canopy. Q. spp. grow to 30 m, may live for >500 years (*Biol Fl*) and can persist in relatively infertile sites. Oak is the commonest tree of British broad-leaved woodlands (315 000 ha, 44% of total) except in Scotland, where birch is even more common (Steele & Peterken 1982). The concept of oak woodland as the climax vegetation of the British Isles appears unfounded (Shaw 1974, Rackham 1980). The production of new large xylem vessels in spring precedes bud-break by c. 1 week (Longman & Coutts 1974). Shoot growth and leaf expansion occur in April and May and subsequent shoot growth is intermittent with a resting period of 4-6 weeks (Longman & Coutts 1974). The genus Quercus has a predominantly warm-temperate distribution, and the two northern outlying species described here are susceptible to frost damage to the foliage in spring and show other signs of maladaptation to cool, temperate climates (Jones 1974). Q.p. is the more common species in ancient woodland and is found primarily on soils of low pH, particularly in the uplands of N and W Britain (Rackham 1980). In contrast, Q.r. is more generally associated with mixed woodland and hedgerows, and extends more regularly onto soils of higher pH (Rackham 1980) or of a moist clayey consistency (*Biol Fl*). Q.r. has larger acorns and the faster initial growth (Newbold & Goldsmith 1981). Ecological distinctions between the two species have been blurred, first as a result of hybridization, particularly in the N (Stace 1975, Rackham 1980), and second, because of the widespread planting of Q.r. for timber (Q.p. is now regarded as superior for this purpose (Newbold & Goldsmith 1981). The two species are highly

Altitude

Slope

Aspect

n.s. | 2.0
0.0
1.3 | 0.0
— | Unshaded

n.s. | 5.3
0.0
1.6 | 0.0
— | Shaded

Soil pH

Triangular ordination

Associated floristic diversity

Commonest associates %
Deschampsia flexuosa 71
Agrostis capillaris 43
Holcus mollis 31
Hyacinthoides non-scripta 23
Pteridium aquilinum 23

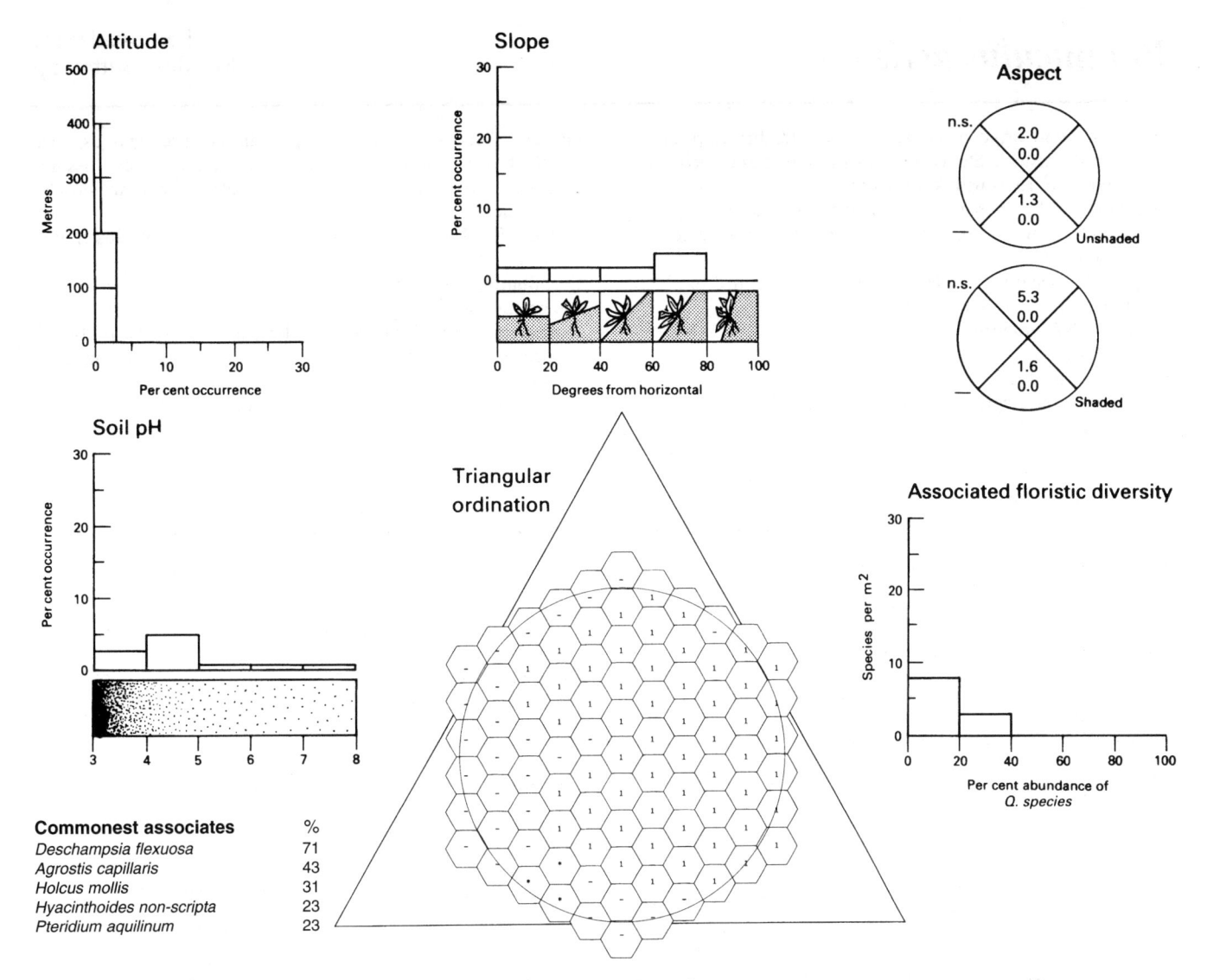

---

variable and show much overlap in their characteristics (*Biol Fl*) and were not separated during fieldwork. Regeneration is entirely by seed which is first produced when the tree is *c*. 40 years old, often slightly earlier in *Q.r.*, and seed production reaches a maximum at between 80 and 120 years (Newbold & Goldsmith 1981). Mast years, with up to 90 000 acorns per tree, only occur every 6-7 years, and even moderate yields may occur only every 3-4 years (*Biol Fl*). Good mast years are more frequent in S England and acorns are larger and tend to have higher viability under these circumstances (Newbold & Goldsmith 1981). Production of acorns may be erratic from one site to another (*Biol Fl*). A single tree may produce several million acorns in its lifetime (Newbold & Goldsmith 1981). Acorns are shed from September to November, but those falling before mid-October are usually non-viable (*Biol Fl*). Acorns are killed by excessive water-loss and are susceptible to frost damage (Newbold & Goldsmith 1981). Those of *Q.p.* are susceptible to waterlogging (*Biol Fl*). A covering of litter provides a suitably moist environment for seedling establishment (*Biol Fl*). The synchronization of fruit and leaf fall may aid this process and also reduce the accessibility of acorns to predators (Newbold & Goldsmith 1981). Seeds are heavily predated by insects, birds and mammals (*Biol Fl*). However, birds and mammals may aid dispersal by burying seed at some distance from the parent (*Biol Fl*), and acorns appear to be transported to greater distances than are beech nuts (Rackham 1980). Maximum growth is likely to occur at 20-40% of full daylight (Jarvis 1964). Acorns show epicotyl dormancy, which is broken by chilling. Germination is hypogeal and the seedling has a large tap-root which may constitute 75% of the weight of the plant (Rackham 1980). These reserves enable *Q.* spp. to colonize closed grassland (other common tree species are restricted to gaps) and confer resilience should the leading shoot be removed by grazing. The foliage is toxic and has an unpleasant taste (Cooper &

Johnson 1984), and trees are attacked by squirrels and rodents to a lesser extent than are *Fagus sylvatica* or *Acer pseudoplatanus* (*Biol Fl*). Although probably more insects predate *Q.* spp. than any other British tree, relatively few have a drastic effect (*Biol Fl*, Gradwell 1974, Morris 1974) and foliage is unpalatable to the snail *Helix aspersa* Muller (Wratten *et al.* 1981). However, two species of Lepidoptera may cause severe defoliation to mature trees during early summer (*Biol Fl*) and establishment of juveniles is often poor under oak, perhaps in part because defoliating caterpillars falling from the canopy predate the juveniles present (Shaw 1974). In long-lived species such as oak, with thousands of apical meristems, mutations may result in some branches being genetically distinct from the next (Cherfas 1985). The ecological significance of a 'genetic mosaic' produced in this way remains to be assessed but, in theory, interactions with pests and pathogens may well be modified (Cherfas 1985). The ability of *Q.* spp. to regenerate appears to have diminished greatly in the last 150 years (Rackham 1980). In the case of *Q.p.*, which appears the more shade-tolerant (*Biol Fl*), effective regeneration may be restored in many upland sites simply by excluding grazing animals from woodland (e.g. Pigott 1983). Rackham (1980) suggests that the combined effects of shade and predation by insects, mammals, birds and oak mildew may be too great to allow survival, and points to the fact that *Q.r.* frequently becomes established in unshaded habitats. In lowland areas the cessation of coppicing and pollarding, and thus the removal of a light phase, shows some correlation with the decline of oak regeneration (Rackham 1980).

Current trends Becoming less important as a timber tree and the bark is no longer utilized in the tanning industry (*Biol Fl*). However, as emphasized by Rackham (1980), *Q.r.* appears to be amongst the less vulnerable of our native trees. Indeed, *Q.r.*, and in many upland regions *Q.p.*, show some capacity for colonizing new, often unshaded, habitats.

Ranunculus acris L.

Ranunculaceae
Meadow Buttercup

Life-form Polycarpic perennial, semi-rosette hemicryptophyte with a short stock. Stems erect; leaves ovate, subdivided, hairy, <1000 mm^2; roots with VA mycorrhizas.

Phenology Winter green, but with little growth until early spring (*Biol Fl*). Flowers mainly from May to July and sets seed from July onwards.

Height Foliage usually <300 mm; flowering stem to 1 m.

Seedling RGR No data. SLA 23.8. DMC 19.1.

Nuclear DNA amount 10.7 pg, $2n = 14$* (Bennett & Smith 1976); $2x$*.

Established strategy C-S-R.

Regenerative strategies ?B$_s$, V. Regenerates by means of seed germinating in autumn or spring, and may form a persistent seed bank. Some vegetative regeneration occurs through the stock branching and by the subsequent daughter rosettes separating by decay (*Biol Fl*); SB III.

Flowers Yellow, typically hermaphrodite and protogynous, often self-sterile; typically <20 in a cymose inflorescence. In some populations many plants may be female and apomixis is also recorded (*Biol Fl*).

Germinule 1.9 mg, 3.0 × 2.0 mm, achene, up to 40 clustered in each fruit.

Germination Epigeal; t_{50} 41 days.

Biological flora Harper (1957a).

Geographical distribution The British Isles (100% of British vice-counties), *ABIF* page 101; widely distributed in temperate and Arctic Eurasia (79% of European territories and 3% naturalized); introduced into N America, S Africa and New Zealand.

Habitats Primarily a species of meadows and pastures. Also frequent in other grazed or mown habitats such as waysides and lead-mine spoil. Recorded at lower frequencies from various types of spoil heap, tall unmanaged grassland, mire, arable land, woodland and rock outcrops. Absent from most skeletal and all aquatic habitats. Records for arable land and some woodland and spoil habitats relate only to seedlings. [Found in dune grassland and in montane grassland and mire.]

Gregariousness Intermediate.

Altitude The area of pasture is appreciably greater within the upland region, and consequently *R.a.* is more frequent and abundant at higher altitudes. [Up to 1220 m (*Biol Fl*).]

Slope Absent from steepest slopes.

Aspect In unshaded sites slightly more frequent and abundant on N facing slopes. In shaded sites confined to S-facing slopes.

Hydrology Typical of moist rather than waterlogged soils, but characteristic of some grassy communities of soligenous mire.

Soil pH Most frequent and abundant within the pH range 5.5-7.0 and absent from acidic sites of pH <4.0.

Bare soil Found over the full range of values.

△ **ordination** Widely distributed, but absent from habitats which combine disturbance with low fertility.

Associated floristic diversity Intermediate to species-rich.

Similarity in habitats	%
Rumex acetosa	79
Cerastium fontanum	73
Alopecurus pratensis	73
Phleum pratense	72
Trifolium repens	71

Habitats

Synopsis *R.a.* is, like the other common buttercups (*R. bulbosus* and *R. repens*), a winter-green, frost-tolerant herb of grazed or mown grassland. *R.a.* is a plant of moist, but not waterlogged, habitats. In some ridge-and-furrow grasslands the three buttercups occupy different positions, with *R.b.* found on the more freely-drained ridge top, *R.a.* on the moist ridge side and *R.r.* in the wetter furrows (*Biol Fl*). *R.a.* is also taller than the other two and can compete effectively in moderately productive meadows. Detailed demographic studies have been carried out on all three species (Sarukhan & Harper 1973, Sarukhan 1974). Though often disadvantaged by being in full flower when hay is cut (*Biol Fl*), *R.a.* is more common in meadows than in any other habitat but, even here, its range is limited to relatively unproductive sites and its yield is suppressed by treatment with ammonium salts (PGE). Like *R.b.*, *R.a.* is unpalatable and it is avoided by stock (*Biol Fl*), in which it reduces yield. The claim that *R.a.*, which contains protoanemonin, is also highly toxic may be exaggerated (Cooper & Johnson 1984). Overgrazing produces areas of bare soil suitable for seedling establishment, but the seed has no obvious mechanism for dispersal. However, some vegetative regeneration occurs (see Sarukhan & Gadgil 1974). With respect both to habitat range and to dependence upon seed for regeneration, the three common buttercups may be ordered *R.b.* < *R.a.* << *R.r.*. Seed of *R.a.*, which is sometimes incorporated into a persistent seed bank (*Biol Fl*), is produced in much greater numbers than is the case in *R.r.* (Sarukhan 1974). *R.a.* is a variable and polymorphic species in need of taxonomic study in Britain and elsewhere (*CTW*). Two of the five ssp. tentatively recognized from Europe are recorded in Britain; ssp. *acris* and the montane ssp. *borealis* (Trautv.) Nyman (*FE1*). Studies on the origin and ecological significance of this variation are required.

Current trends *R.a.* exploits overgrazed, moderately fertile pastures. Since it is easily controlled, has poor dispersal and is slow to establish, *R.a.* is now mainly confined to permanent pastures subject to lax management. *R.a.* has almost certainly passed its peak of abundance and may be expected to decline further.

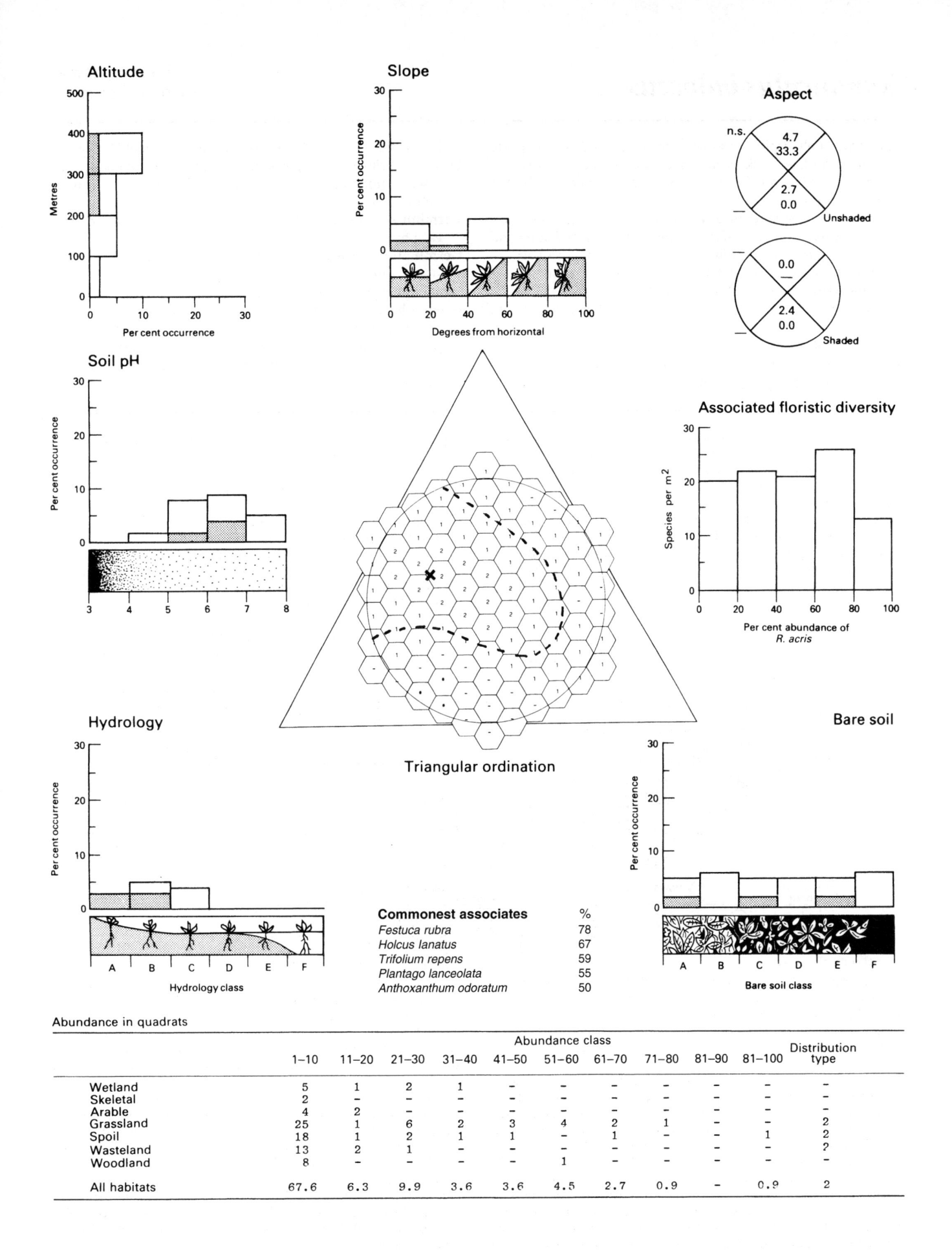

Altitude

Slope

Aspect

	n.s.	4.7	
		33.3	
—		2.7	Unshaded
		0.0	

		0.0	
—		—	
		2.4	Shaded
		0.0	

Soil pH

Associated floristic diversity

Hydrology

Triangular ordination

Bare soil

Commonest associates %
- Festuca rubra — 78
- Holcus lanatus — 67
- Trifolium repens — 59
- Plantago lanceolata — 55
- Anthoxanthum odoratum — 50

Abundance in quadrats

	1–10	11–20	21–30	31–40	41–50	51–60	61–70	71–80	81–90	81–100	Distribution type
Wetland	5	1	2	1	–	–	–	–	–	–	–
Skeletal	2	–	–	–	–	–	–	–	–	–	–
Arable	4	2	–	–	–	–	–	–	–	–	–
Grassland	25	1	6	2	3	4	2	1	–	–	2
Spoil	18	1	2	1	1	–	1	–	–	1	2
Wasteland	13	2	1	–	–	1	–	–	–	–	?
Woodland	8	–	–	–	–	1	–	–	–	–	–
All habitats	67.6	6.3	9.9	3.6	3.6	4.5	2.7	0.9	–	0.9	2

Ranunculus bulbosus L.

Ranunculaceae
Bulbous Buttercup

Life-form Polycarpic perennial, semi-rosette hemicryptophyte with corm-like stock. Stems erect; leaves ovate, subdivided, hairy, <1000 mm^2; some roots contractile, with VA mycorrhizas.
Phenology Overwintering as a rosette formed in autumn. New leaves expand in early spring. Flowers in May and June and sheds seed in June and July, after which the shoot dies back.
Height Leaves mainly <100 mm. Flowers to *c*. 400 mm.
Seedling RGR No data. SLA 18.2. DMC 18.8.
Nuclear DNA amount 11.2 pg; $2n = 16^*$ (Bennett & Smith 1976); $2x^*$.
Established strategy Stress-tolerant ruderal.
Regenerative strategies B$_s$. Regenerates in gaps by means of seeds which germinate in autumn. A persistent seed bank may be formed. Daughter 'corms' rarely produced (*Biol Fl*); SB III.
Flowers Yellow, hermaphrodite or more rarely gynodoecious, insect-pollinated (*Biol Fl*); in a few-flowered cymose inflorescence.
Germinule 3.1 mg; 3.0×2.3 mm, achene up to *c*. 30 clustered in each fruit.
Germination Epigeal; germination rate increased after dry storage (*Biol Fl*).
Biological flora Harper (1957c).
Geographical distribution The British Isles, but becoming less common in the N and W (99% of British vice-counties), *ABIF* page 101; Europe except for the extreme N, S and E (77% of European territories); introduced in N America and New Zealand.

Habitats A common grassland species, locally abundant in meadows and pastures. Also found occasionally in derelict pastures, on rock outcrops, limestone quarry heaps and waysides. Absent from woodland, arable land and wetlands. [Occurs on fixed sand dunes.]
Gregariousness Sparse.
Altitude Extends up to 400 m, with an upland bias reflecting the greater abundance of grasslands within the upland region. [Up to 580 m.]
Slope Found in pastures on relatively flat ground and on dale-sides, but not recorded from the steeper slopes associated with cliffs and other skeletal habitats.

Aspect In the Sheffield region more common on N-facing slopes, but *Biol Fl* suggests a S-facing bias.
Hydrology Absent from wetlands.
Soil pH Most frequent and abundant on soils above pH 5.0. Not present on strongly acidic soils of pH <4.0.
Bare soil Mainly associated with low exposures.
△ **ordination** Excluded from productive, undisturbed communities and those subject to severe disturbance. Distribution centred on vegetation associated with moderately high fertility and an intermediate level of disturbance, but extends into relatively undisturbed, infertile habitats.
Associated floristic diversity High.
Synopsis *R.b.* is a small rosette plant possessing a more-limited capacity for growth in height and in lateral spread than the other two common buttercups, *R. acris* and *R. repens*. Individual organs are replaced annually

Similarity in habitats	%
Anthoxanthum odoratum	74
Rhinanthus minor	71
Ranunculus acris	70
Phleum pratense agg.	68
Bellis perennis	67

Habitats

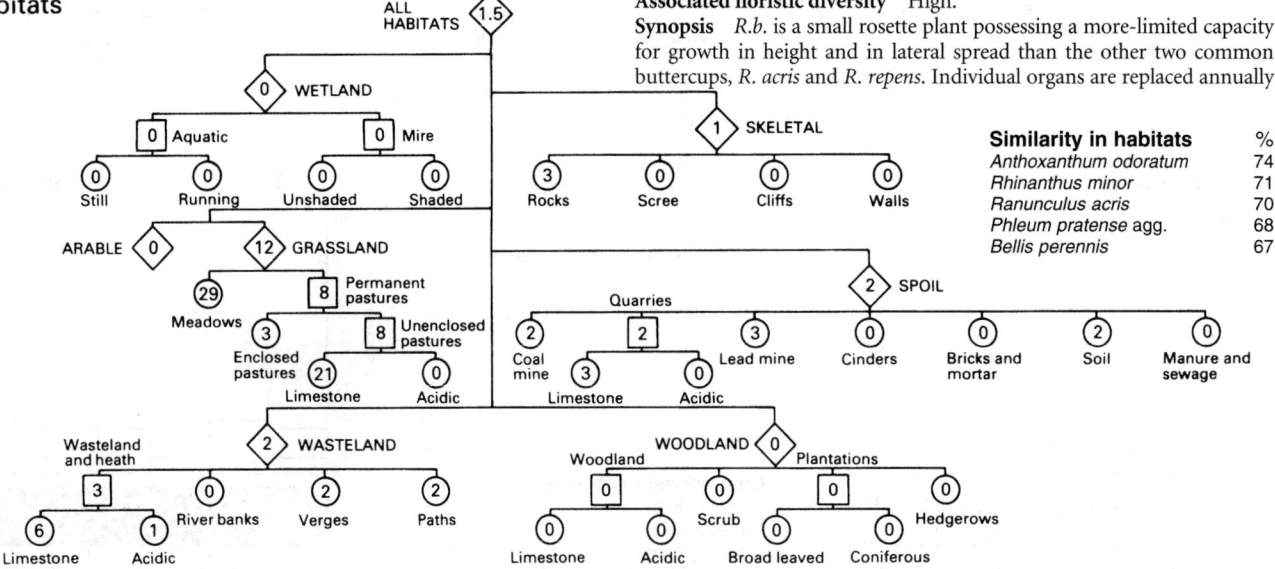

but the plant itself is relatively long-lived (*Biol Fl*). *R.b.* is particularly common in older, often species-rich, permanent pastures on freely-drained soils in sites where shading is prevented by heavy grazing or, in unproductive turf, by low soil fertility. The leaves of *R.b.* are mainly epinastic in a basal rosette, and the species tends to be suppressed in hay meadows (*Biol Fl*). Although data presented here and in *Ecol Atl* for the Sheffield region are rather contradictory, *R.b.* appears to be particularly characteristic of S and W slopes in sites where summer drought limits the growth of more-robust species (*Biol Fl*). *R.b.* exhibits an unusual phenology in which a small rosette is formed in autumn, with rapid leaf growth drawing upon the 'corm' reserves during the following spring. Flowering and seed set are earlier than in *R.a.* and *R.r.* and the plant usually aestivates as a below-ground 'corm' from mid-July (*Biol Fl*). This phenology affords a possible mechanism of drought-avoidance in drier sites and may also reduce sensitivity to the impacts of summer dominants in more-fertile sites. *R.b.* is encouraged by overgrazing, almost certainly because seedling establishment, which is critical for population increase (see Sarukhan 1974), occurs on the bare soil created by trampling (*Biol Fl*). No well-defined mechanism of dispersal has been described, and seeds are often shed close to the plant (*Biol Fl*). Seeds of this species, and of *R. acris*, are produced in greater numbers than is the case with *R. repens* but the seed bank of *R.b.* is less persistent (Sarukhan 1974). *R.b.* is unpalatable and, together with other herbage in its vicinity, is generally avoided by stock (*Biol Fl*). As a result, the agricultural productivity of pasture is reduced by infestations of *R.b.* and there are records of stock, and humans, being poisoned (Cooper & Johnson 1984). However, the level of seed predation is higher than that experienced by *R.a.*, though lower than in the case of *R.r.* (Sarukhan 1974). Ecotypes have been described (*Biol Fl*) and 6 ssp., including two found in Britain, have been tentatively recognized within Europe (*FE1*).
Current trends *R.b.* is mainly restricted to older areas of permanent pasture. The species reduces grassland production and is easily controlled by weed-killers (*Biol Fl*). A continuing decrease in its abundance is therefore inevitable.

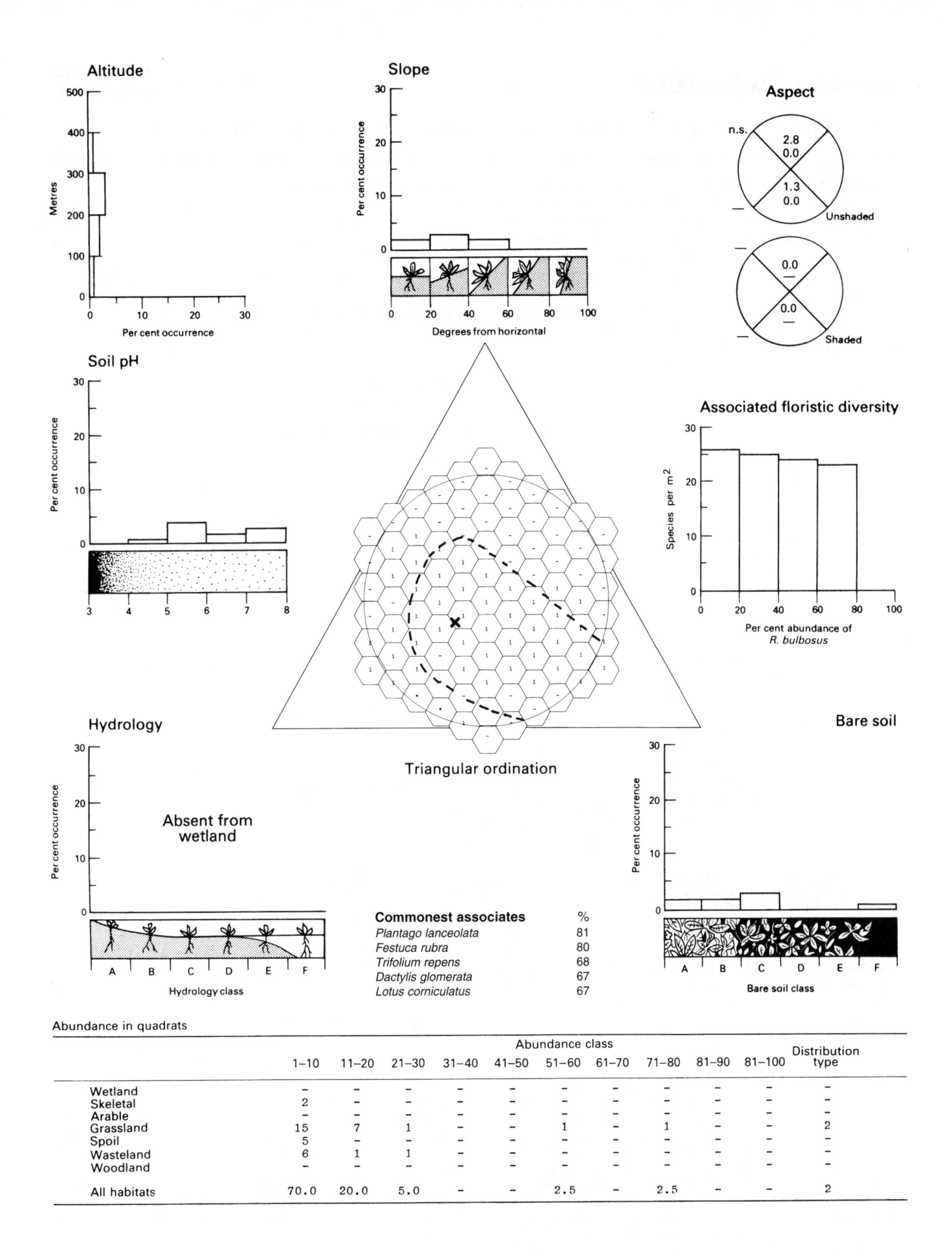

Altitude

Slope

Aspect

	n.s.	2.8	
		0.0	
	—	1.3	Unshaded
		0.0	

		0.0	
		—	
		0.0	Shaded
		—	

Soil pH

Associated floristic diversity

Species per m² vs Per cent abundance of *R. bulbosus*

Triangular ordination

Hydrology

Absent from wetland

Bare soil

Commonest associates %
Plantago lanceolata 81
Festuca rubra 80
Trifolium repens 68
Dactylis glomerata 67
Lotus corniculatus 67

Abundance in quadrats

	1–10	11–20	21–30	31–40	41–50	51–60	61–70	71–80	81–90	81–100	Distribution type
Wetland	–	–	–	–	–	–	–	–	–	–	–
Skeletal	2	–	–	–	–	–	–	–	–	–	–
Arable	–	–	–	–	–	–	–	–	–	–	–
Grassland	15	7	1	–	–	1	–	1	–	–	2
Spoil	5	–	–	–	–	–	–	–	–	–	–
Wasteland	6	1	1	–	–	–	–	–	–	–	–
Woodland	–	–	–	–	–	–	–	–	–	–	–
All habitats	70.0	20.0	5.0	–	–	2.5	–	2.5	–	–	2

Ranunculus ficaria L.

Ranunculaceae
Lesser Celandine

Includes ssp. *ficaria* and ssp. *bulbifer* (Marsden-Jones) Lawalree, not separated during fieldwork

---

Life-form Polycarpic perennial, semi-rosette geophyte with numerous root tubers. Stems branched, erect (ssp. *f.*) or spreading (ssp. *b.*); leaves ovate, almost succulent, typically <1000 mm^2; roots mainly shallow, with VA mycorrhizas.

Phenology Vernal. First leaves appear in January. Flowers from March to May and sheds seed mainly in June, by which time *R.f.* is often beginning to senesce. In ssp. *b.* leaf growth and senescence is slightly earlier than in ssp. *f.* (Nicholson 1983).

Height Foliage mainly <150 mm, overtopped by the flowers.

Seedling RGR No data. SLA 36.6. DMC 11.7.

Nuclear DNA amount Ssp. *f.* 18.6 pg, $2n = 16^*$, $2x^*$ (Bennett & Smith 1976); ssp. *b.* 34.3 pg, $2n = 32^*$, $4x^*$.

Established strategy Stress-tolerant ruderal.

Regenerative strategies S, (V.) (ssp. *b.* only). See 'Synopsis'.

Flowers Yellow usually hermaphrodite, protandrous, insect- or self-pollinated (*Biol Fl*), solitary at the end of each branch; SB II.

Germinule Ssp. *f.* achene; 1.04 mg; 2.5 × 1.4 mm, up to 70 in each fruit (*Biol Fl*).

Germination Epigeal; dormancy broken by chilling (*Biol Fl*).

Biological flora Taylor and Markham (1978).

Geographical distribution The British Isles, but less common in Scotland and Ireland (100% of British vice-counties), *ABIF* page 106; most of Europe (87% of European territories); W Asia; introduced in N America. Ssp. *b.* more widespread in Europe, and in Britain more abundant in the E; ssp. *f.* in W Europe and widespread in Britain.

Habitats Abundant on river and stream banks, and in woodland. Also on marshy ground and road verges, in hedgerows, meadows and pastures. Ssp. *b.* largely restricted to shaded sites (Nicholson 1983). [*R.f.* (probably ssp. *f.*) also a component of various maritime cliff-top communities.]

Gregariousness Intermediate.

Altitude Up to 400 m, but particularly frequent on the coal measures between 100 and 200 m. [Up to 730 m.]

Slope Largely confined to flat ground and gentle slopes.

Aspect Largely restricted to N-facing slopes in the open, and also slightly more frequent on N-facing slopes in shaded sites; abundance in shaded sites slightly greater on S-facing aspects.

Hydrology Recorded in damp situations and less frequently from soligenous mire.

Soil pH Found on soils with pH ranging between 4.0 and 8.0, and most commonly associated with pH values of 6.0-6.5.

Bare soil Widely distributed, but tending to associate most strongly with relatively unvegetated areas, often with intermediate exposures of bare soil and some bryophyte mat.

Habitats

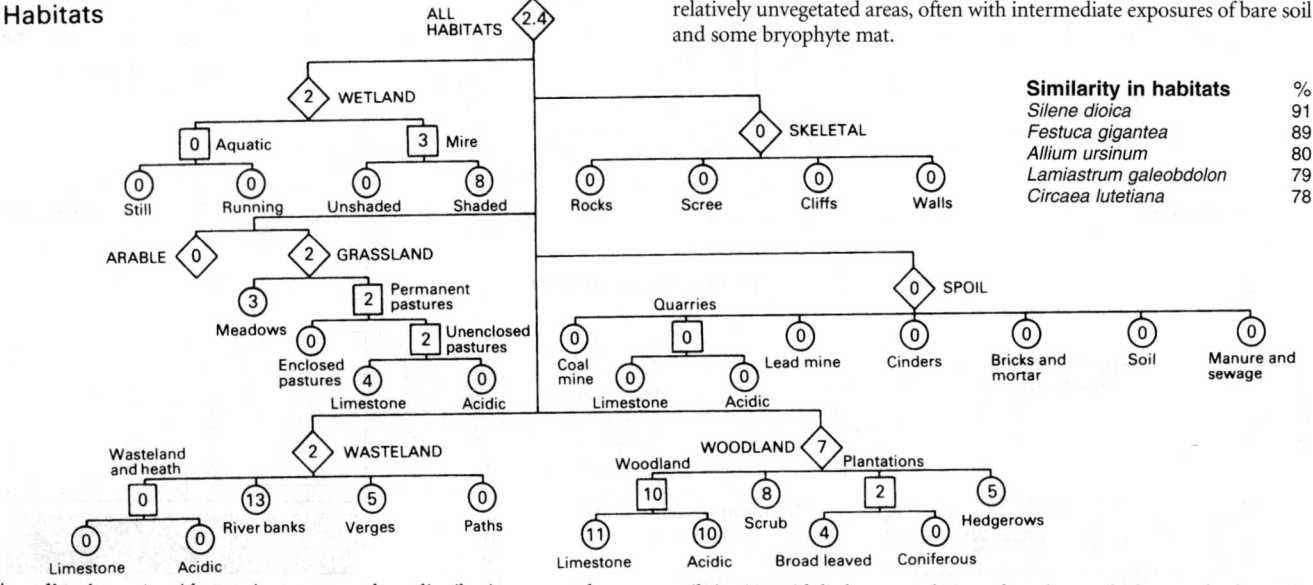

ALL HABITATS	2.4
WETLAND	2
Aquatic	0
Still	0
Running	0
Mire	3
Unshaded	0
Shaded	8
ARABLE	0
GRASSLAND	2
Meadows	3
Permanent pastures	2
Enclosed pastures	0
Unenclosed pastures	2
Limestone	4
Acidic	0
WASTELAND	2
Wasteland and heath	0
Limestone	0
Acidic	0
River banks	13
Verges	5
Paths	0
SKELETAL	0
Rocks	0
Scree	0
Cliffs	0
Walls	0
SPOIL	0
Quarries	0
Coal mine	0
Limestone	0
Acidic	0
Lead mine	0
Cinders	0
Bricks and mortar	0
Soil	0
Manure and sewage	0
WOODLAND	7
Woodland	10
Limestone	11
Acidic	10
Scrub	8
Broad leaved	4
Coniferous	0
Plantations	2
Hedgerows	5

Similarity in habitats	**%**
Silene dioica	91
Festuca gigantea	89
Allium ursinum	80
Lamiastrum galeobdolon	79
Circaea lutetiana	78

△ **ordination** A wide-ranging pattern, but distribution centred on vegetation subject to moderate disturbance and of moderate productivity. Also frequent in more-productive vegetation, but not strongly represented in undisturbed situations. Absent from conditions of severe disturbance. Ssp. *f.* tends to exploit less-disturbed sites than ssp. *b.* (*Biol Fl*).

Associated floristic diversity Relatively low, and declining slightly as *R.f.* achieves greater abundance.

Synopsis *R.f* is a tuber-forming herb which minimizes competition with summer-growing herbs, grasses and trees by virtue of its vernal phenology. The species grows in woods and, to a lesser extent, tall-herb and grassland communities. *R.f.* often also occurs on path sides and in grazed or mown habitats. Because of the vernal phenology, the growth of *R.f.* is often near completion in these sites before trampling, grazing or mowing reach any great intensity. Like other vernal geophytes, *R.f.* has a high nuclear DNA amount. *R.f.* is very cold-resistant (*Biol Fl*), and is the earliest common species to emerge, flower and senesce. As in *Hyacinthoides non-scripta*, unshaded sites are more favourable for growth, despite the fact that *R.f.* is frequent in woodland (*Biol Fl*). *R.f.* is common on moist, relatively fertile soils in sites with little accumulation of tree litter. The leaves of *R.f.* contain relatively high concentrations of Na (*SIEF*). They are toxic to stock (poisonous principle protoanemonin (Cooper & Johnson 1984)) and tend to escape predation. The seedling, like that of *Conopodium majus*, is monocotyledonous (*Biol Fl*). *R.f.* exists as two main ecotypes (*Biol Fl*). Ssp. *b.* is more vernal and more strictly a woodland plant, and regenerates both by means of bulbils (up to 24 per plant) borne in the axils of the lower leaves and by means of root tubers if these possess a bud. However, ssp. *b.* sets little seed. In contrast, ssp. *f.* produces seed but not bulbils. Thus, although the two overlap ecologically and can occur together, ssp. *b.* tends to be found in the more disturbed habitats, and may even be a garden weed (*Biol Fl*). The two occasionally hybridize (Stace 1975), and two further ssp. occur in Europe (*FE1*).

Current trends Ssp. *f.*, which tends to be found in less-disturbed sites, may be decreasing. Ssp. *b.* is probably more mobile both within and between woods, and the proportion of localities suitable for ssp. *b.* is likely to be increasing.

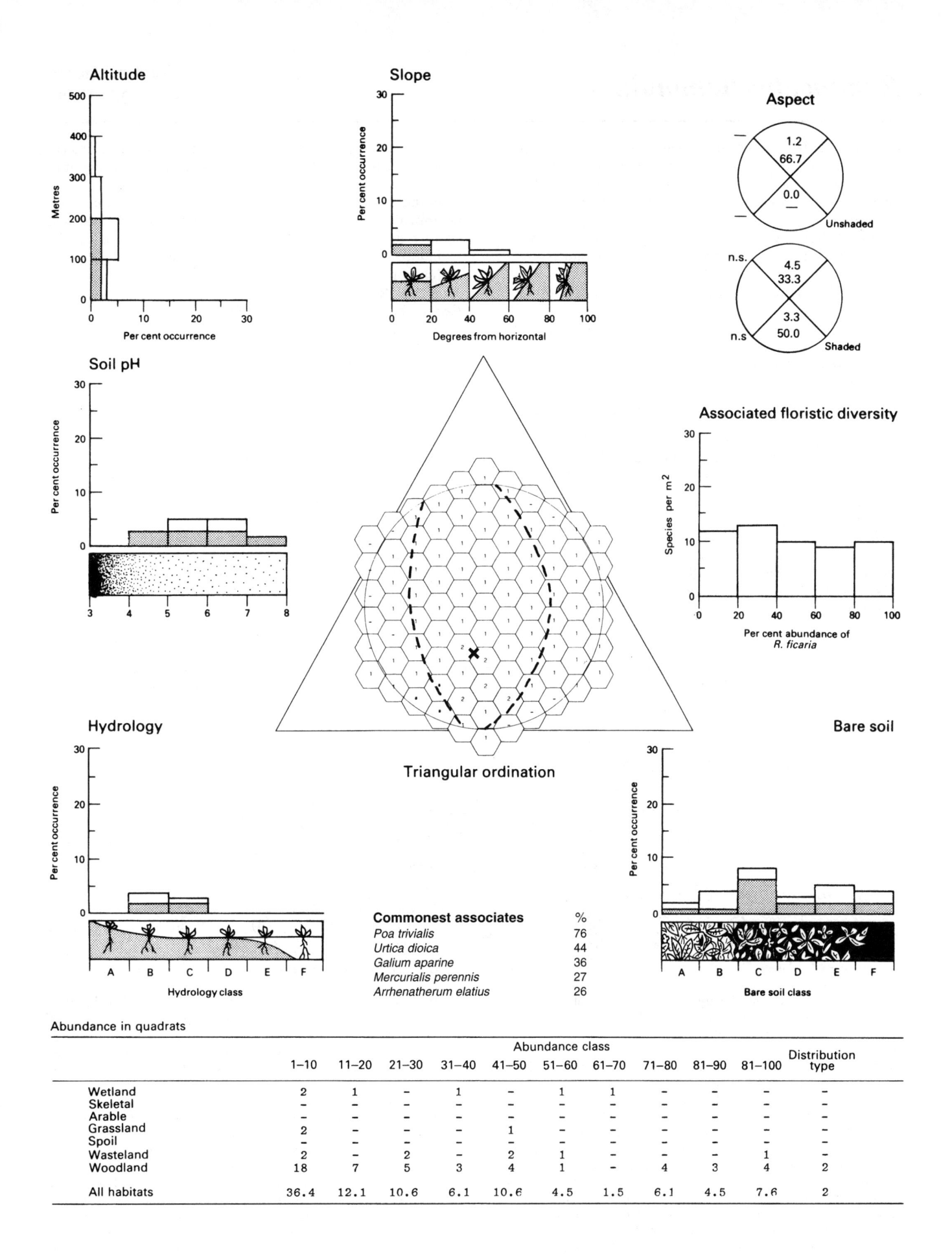

Altitude

Slope

Aspect

Soil pH

Associated floristic diversity

Hydrology

Triangular ordination

Commonest associates %
Poa trivialis 76
Urtica dioica 44
Galium aparine 36
Mercurialis perennis 27
Arrhenatherum elatius 26

Bare soil

Abundance in quadrats

	\multicolumn{10}{c}{Abundance class}	Distribution type
	1–10	11–20	21–30	31–40	41–50	51–60	61–70	71–80	81–90	81–100	
Wetland	2	1	–	1	–	1	1	–	–	–	–
Skeletal	–	–	–	–	–	–	–	–	–	–	–
Arable	–	–	–	–	–	–	–	–	–	–	–
Grassland	2	–	–	–	1	–	–	–	–	–	–
Spoil	–	–	–	–	–	–	–	–	–	–	–
Wasteland	2	–	2	–	2	1	–	–	1	–	–
Woodland	18	7	5	3	4	1	–	4	3	4	2
All habitats	36.4	12.1	10.6	6.1	10.6	4.5	1.5	6.1	4.5	7.6	2

Ranunculus flammula L.

Ranunculaceae
Lesser Spearwort

Life-form Polycarpic perennial, semi-rosette helophyte. Stems creeping and rooted or ascending; leaves mostly lanceolate, <1000 mm^2; roots shallow, with or without VA mycorrhizas.

Phenology Overwintering as very small rosettes. Flowers mainly from May to June and sets seed from July to August.

Height Foliage <20 mm in prostrate form to *c.* 300 mm when erect. Flowering stems to 500 mm, but usually shorter.

Seedling RGR No data. SLA 19.5. DMC 15.4.

Nuclear DNA amount 12.7 pg (Bennett & Smith 1976); 2*n* 32* (*BC*); 4*x**.

Established strategy Intermediate between competitive-ruderal and C-S-R.

Regenerative strategies V, B$_s$. Regenerates both vegetatively and by means of seeds (see 'Synopsis'); ?SB IV.

Flowers Yellow, hermaphrodite, protandrous, insect-pollinated, largely self-incompatible (Gibbs & Gornall 1976); solitary or in a <10-flowered cyme.

Germinule 0.37 mg, 1.8 × 1.2 mm, achene; up to 50 in each fruit.

Germination Epigeal; t_{50} 12 days; germination rate increased after dry storage at 20°C, more-rapid germination can be induced by alternate freeze-thaw pretreatments (Cook & Johnson 1968); L > D.

Biological flora None.

Geographical distribution The British Isles (100% of British vice-counties), *ABIF* page 104; most of Europe except the extreme N and the Mediterranean region (82% of European territories); temperate Asia.

Habitats Exclusively a wetland species found both in shaded or unshaded mire and at the margin of streams, ponds and ditches. Also found in grassland subject to winter flooding. [Particularly frequent in stony lakesides and reservoirs in upland areas. Found also on moist sea cliffs, in dune slacks and in wet montane habitats.]

Gregariousness Intermediate.

Altitude Wide-ranging with some discontinuities, a reflection of the virtual absence of *R.f.* from calcareous strata. Most suitable habitats found within the gritstone uplands. [Up to 950 m.]

Slope All records from slopes of <10°.

Habitats

Aspect No information.

Hydrology Frequent in soligenous mire, but also in topogenous mire beside lakes, ponds, ditches and reservoirs, and in flooded hollows. Not generally an aquatic species, but sometimes temporarily submerged in still or flowing water or floating with *Montia fontana* L., etc., at the edge of upland rills.

Soil pH Mainly on mildly acidic soils between pH 4.5 and 6.5 but extending locally to soils of pH up to 7.5. Rare on calcareous soils.

Bare soil Found over a wide range of values, but most frequent in areas with high exposure of bare soil, bare rock or bryophyte mat.

△ **ordination** Distribution centred on productive and slightly disturbed vegetation types. However, additional, more-extensive sampling (Hodgson *et al.* unpubl.) suggests an association with intermediate levels of fertility.

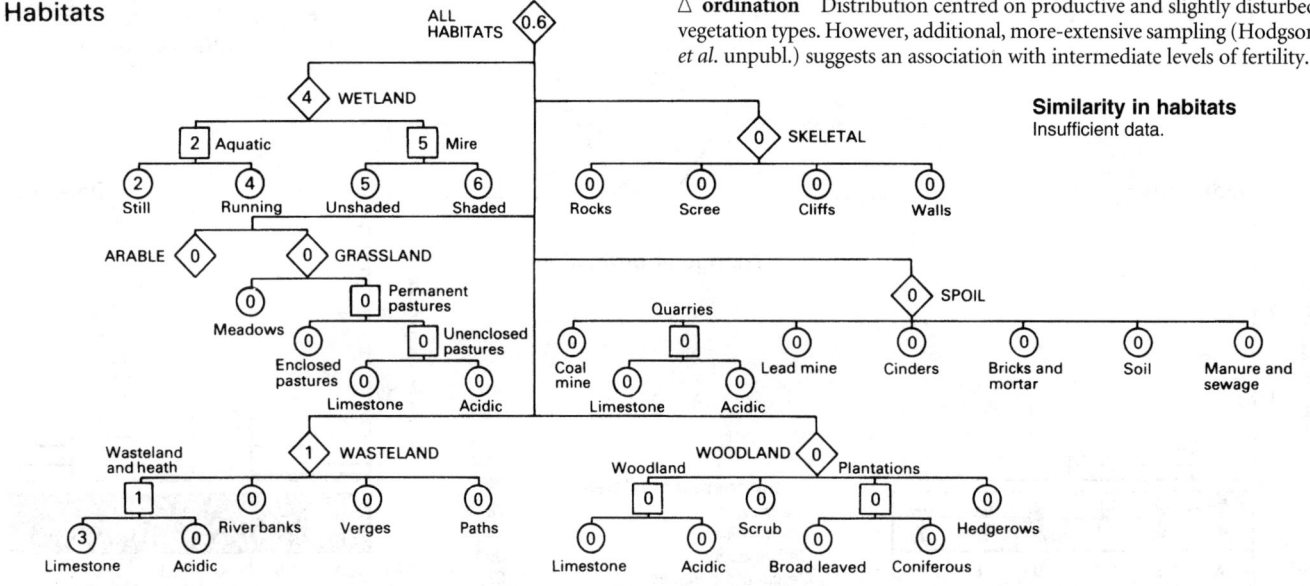

Similarity in habitats
Insufficient data.

Associated floristic diversity Varying widely, but typically intermediate.

Synopsis *R.f.* is found in a variety of wetland habitats where the growth of potential dominants is suppressed by disturbance and, in some sites, by infertility. The species is characteristic of mildly acidic, often peaty soils, but has also been recorded from highly calcareous sites within the Sheffield region. Much of the wide tolerance of wetland conditions coincides with phenotypic plasticity. Thus, under aquatic conditions *R.f.* exhibits nodal rooting; terrestrial and aquatic forms also differ in leaf shape (Gibbs & Gornall 1976) and prostrate rooted phenotypes are erect in cultivation (Stace 1975). In N America some genotypes are more heterophyllous than others (Cook & Johnson 1968). These are adaptable both to submergence and to desiccation, and are found in fluctuating habitats such as lake margins (Cook & Johnson 1968). Less-plastic genotypes are associated with more-stable conditions (Cook & Johnson 1968). However, despite the extent of phenotypic plasticity, single populations of this essentially out-breeding species may include morphologically distinct genotypes (Cook & Johnson 1968). Apart from a tendency to exploit disturbed habitats such as the stony edges of oligotrophic lakes and areas of soligenous mire, *R.f.* also occurs in vegetation kept low by grazing. Palatability has not been studied extensively, but the species contains protoanemonin and is toxic to stock (Cooper & Johnson 1984). *R.f.* is found in shaded mire and may persist in taller unshaded vegetation, where it adopts a more upright growth form. Aquatic acid metabolism (AAM) confers advantages where the concentration of CO_2 in the aquatic environment is low and diffusion rates are slow (Cockburn 1985). Several aquatic species of oligotrophic waters, e.g. *Littorella uniflora* (L.) Aschers. possess this adaptation, but *R.f.* is not one of them (Farmer & Spence 1985). The species shows some tolerance of ferrous iron (R. E. D. Cooke unpubl.). Prostrate stems root freely and diffuse clonal patches may be formed, particularly in open habitats. Shoots detached as a result of disturbance readily regenerate to form new plants; this may be an important mode of regeneration in lake- and stream-side populations.

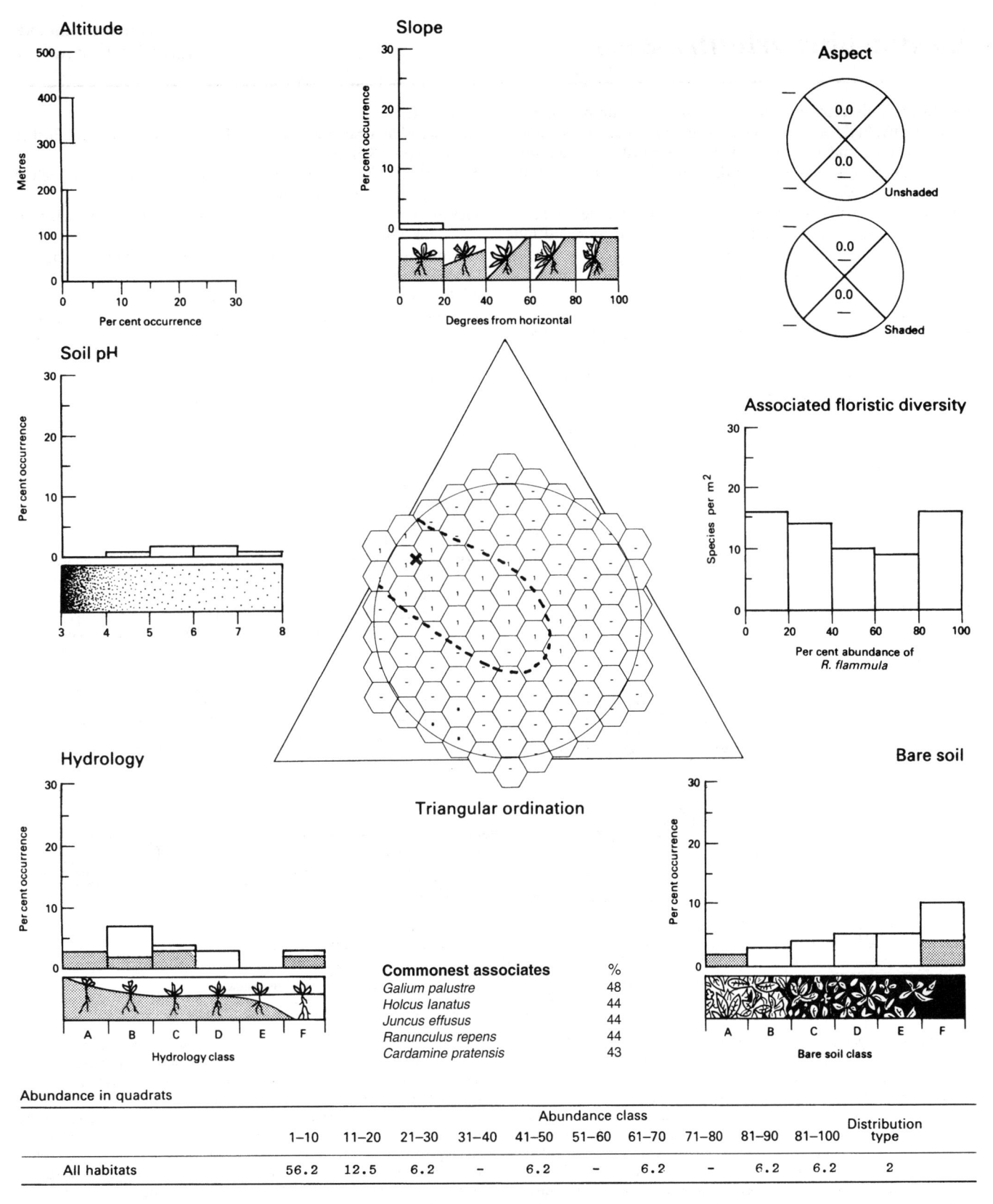

Altitude
Slope
Aspect
Soil pH
Associated floristic diversity
Triangular ordination
Hydrology
Bare soil

Commonest associates	%
Galium palustre	48
Holcus lanatus	44
Juncus effusus	44
Ranunculus repens	44
Cardamine pratensis	43

Abundance in quadrats

	\multicolumn{9}{c	}{Abundance class}	Distribution type
	1–10	11–20	21–30	31–40	41–50	51–60	61–70	71–80	81–90	81–100	
All habitats	56.2	12.5	6.2	–	6.2	–	6.2	–	6.2	6.2	2

Regeneration by seed is probably less important. Seeds may be incorporated into a persistent seed bank (UCPE unpubl.) and tend to germinate mainly in spring. Rapid germination can be induced if seeds are alternately frozen and thawed (Cook & Johnson 1968), but the significance of this and other possible germination cues remains uncertain. Seeds have been shown to survive ingestion by horses (Ridley 1930). Long-distance transport in water seems unlikely since the seeds lack buoyancy. Three ssp. are recorded from Britain, but only ssp. *flammula* occurs in the Sheffield region. The other two, ssp. *minimus* (Ar Benn.) Padmore and ssp. *scoticus* (E. S. Marshall) Clapham, are endemic to Scotland and Ireland (*FE1*). The former has a maritime and the latter a lake-side distribution. *R.f.* shows introgression with *R. reptans* L., which is now believed to be extinct in Britain (Stace 1975).

Current trends Occasionally colonizes new habitats, but probably decreasing as a result of wetland destruction and eutrophication.

Ranunculus peltatus Schrank
Ranunculaceae
Common Water-crowfoot

Life-form Polycarpic perennial or winter- or summer-annual hydrophyte (Cook 1966). Stems ascending in aquatic habitats or prostrate and rooted in mud; heterophyllous, often with both finely dissected and lobed, semi-orbicular leaves, <500 mm^2; roots shallow.

Phenology Capable of overwintering with dissected foliage. Flowers mainly in May and June and sets seed June and July onwards.

Height Typically a submerged or floating aquatic; on mud, foliage and flowers to *c.* 20 mm.

Seedling RGR No data. SLA 55.2. DMC 9.8.

Nuclear DNA amount 6.3 pg; $2n = 16, 32, 48*$ (*BC*, Cook 1966); $2x, 4x, 6x*$.

Established strategy Perhaps intermediate between ruderal and C-S-R.

Regenerative strategies V, ?B$_s$. Regeneration requires further study (see 'Synopsis').

Flowers White, hermaphrodite, insect-pollinated or selfed (Cook 1966); solitary, leaf-opposed.

Germinule 0.25 mg, 2.0 × 1.4 mm, achene; 30-40 in each fruit (Cook 1966).

Germination Epigeal; in one test freshly-collected seed yielded 30% germination.

Biological flora None, but see Cook (1966).

Geographical distribution Much of the British Isles, but absent from parts of Scotland (Cook 1966), *ABIF* page 109; most of Europe except N Scandinavia (87% of European territories); N Africa.

Habitats A still-water aquatic of ponds, reservoirs and ditches, which is also capable of surviving on wet mud near the water line in open situations or in partial shade.

Gregariousness Sparse to intermediate.

Altitude Observed up to 370 m.

Slope Not relevant.

Aspect Not relevant.

Hydrology Occurs in still-water habitats to a depth of 600 mm, and on exposed mud at the water margin.

Soil pH All records from sites with pH values in the range 5.0-7.5. Most frequent and abundant between pH 5.0 and 6.5.

Bare soil Found in both aquatic habitats and in sites with high exposures of bare mud.

Δ ordination Insufficient data.

Associated floristic diversity Low in aquatic sites; sometimes intermediate on exposed mud.

Synopsis *R.p.* is a still-water aquatic also capable of growing on wet mud and exploiting sites in which the fluctuation of water level prevents dominance by either obligate aquatics or by emergent 'semi-aquatic' and mire species. The species shows well-developed heterophylly; aquatic phenotypes usually have finely divided submerged leaves which collapse in air, and floating leaves with a well-defined blade. Terrestrial individuals

Similarity in habitats	**%**
Typha latifolia	96
Equisetum fluviatile	93
Lemna minor	92
Sparganium erectum	91
Potamogeton natans	87

Habitats

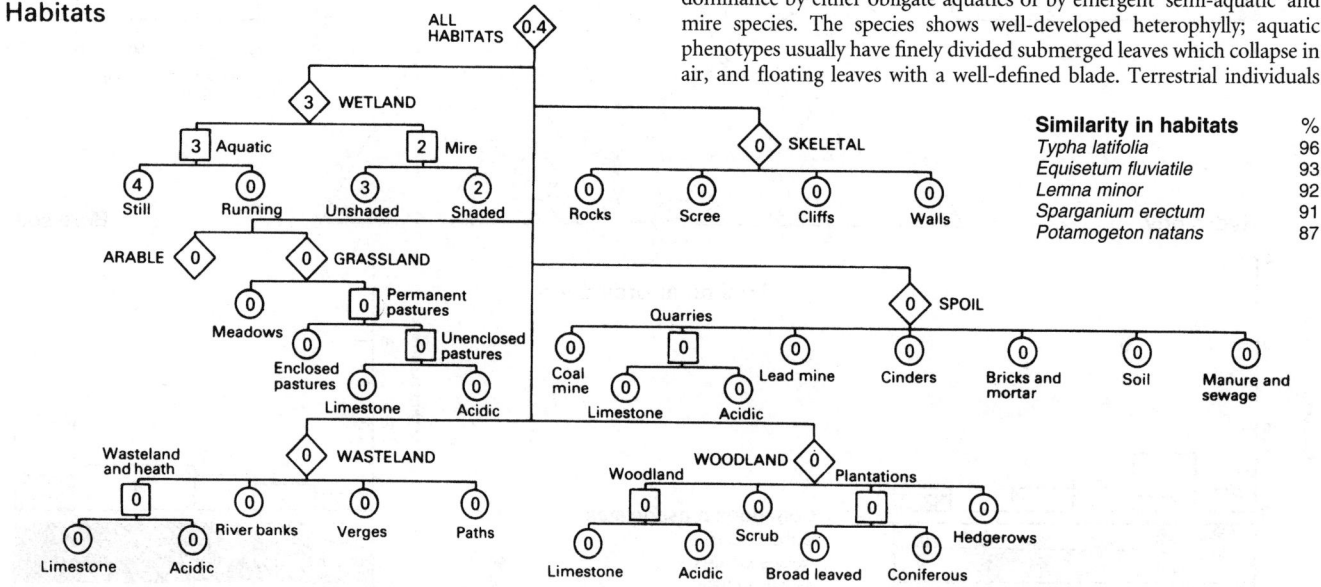

usually possess rigid leaves with short, rigid filiform. segments. Plants are susceptible to frost if not submerged, but can withstand being frozen in ice (Cook 1966). In *R. aquatilis* L., with which *R.p.* has close affinities, the type of leaf produced is dependent upon photoperiod and whether the shoot is submerged or terrestrial (Cook 1969), and the same appears to hold for *R.p.* (Cook 1966). This heterophylly appears to be of critical importance in allowing exploitation of water bodies subject to fluctuations in water level. In particular the capacity of laminar leaves to intercept light at the water surface may be important in competitive interactions with other aquatic species (Cook 1969). In its terrestrial form *R.p.* may occur beneath robust annuals, reflecting some capacity to tolerate reduced irradiance. Three other species of water-crowfoot are frequent in still water in N England; their distribution appears to be partially determined by hydrology. The non-heterophyllous *R. circinatus* Sibth. is virtually confined to aquatic systems, often occurring in relatively deep water and showing little capacity to exploit bare mud (Cook 1966). The two most common species in ditches and small pools, habitats liable to dry out completely during drought, are *R.a.* and the non-heterophyllous *R. trichophyllus* Chaix (although *R.t.*, and to a lesser extent *R.a.*, are also found in deep water). In contrast, *R.p.* is typical of aquatic systems with fluctuating levels of water, and is by far the most common water-crowfoot in reservoirs. The distribution of *R.p.* is biased towards mildly acidic (often peaty) habitats. The species is absent from calcareous waters, where it appears to be replaced by *R.t.*, which is also a species of 'brackish' waters associated with colliery workings. The regenerative strategies of *R.p.* are incompletely understood. Seedlings have been observed during autumn and spring, and mature plants may overwinter. Seeds float for short periods (Ridley 1930). No evidence of a persistent seed bank is available, but its existence is predicted from the ecology of the species. Vegetative shoot fragments root freely to form new plants, and this may be important in the dispersal of *R.p.* during the summer. The hybrid with *R. fluitans* Lam. has spread extensively along one river system, and *R.p.* also hybridizes with *R.t.* and with three further species (Stace 1975, Webster 1986).

Current trends Although apparently an effective colonist of recently created habitats, like most aquatics *R.p.* is probably now decreasing as a result of habitat destruction.

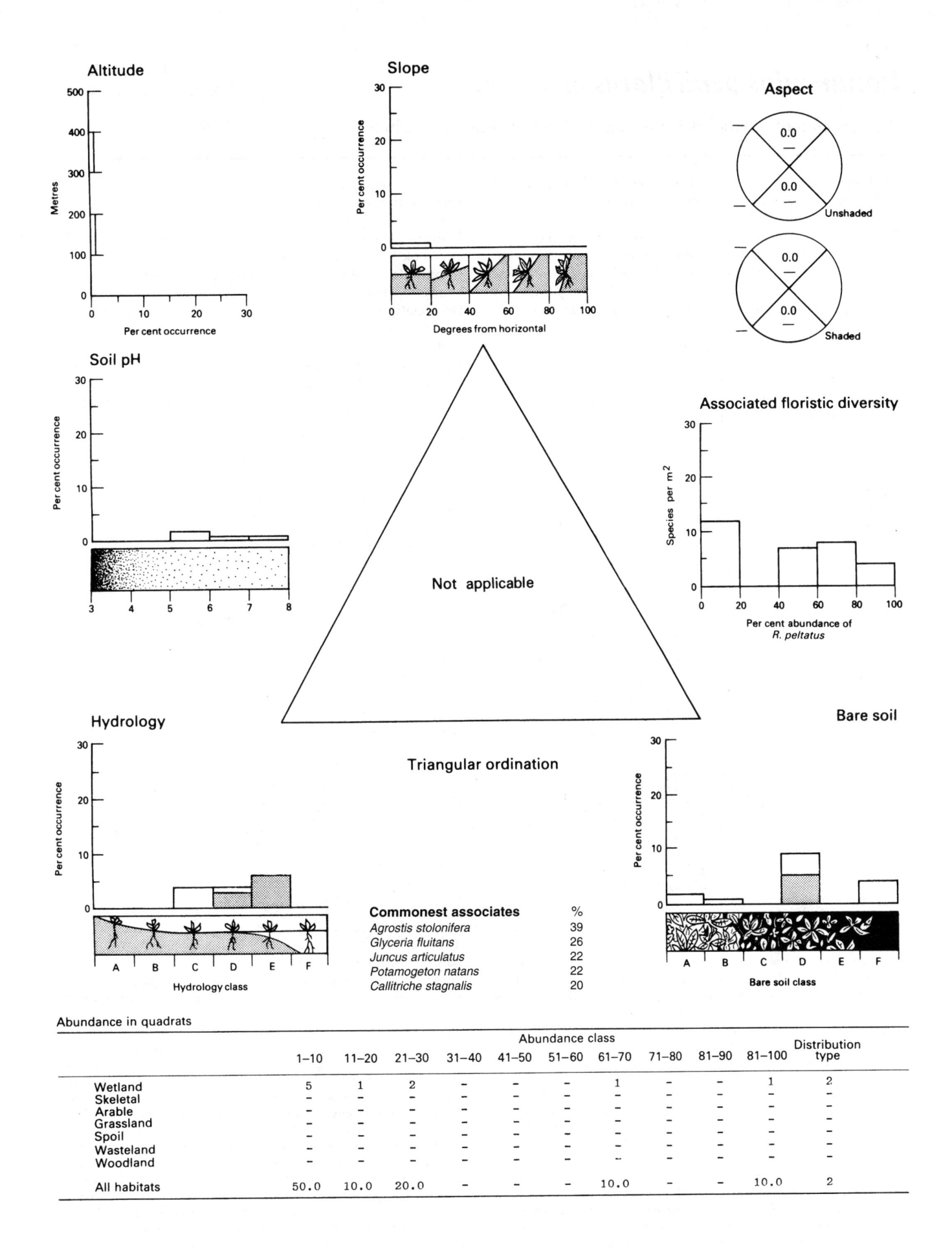

Altitude

Slope

Aspect

Unshaded

Shaded

Soil pH

Associated floristic diversity

Not applicable

Hydrology

Triangular ordination

Bare soil

Commonest associates	%
Agrostis stolonifera	39
Glyceria fluitans	26
Juncus articulatus	22
Potamogeton natans	22
Callitriche stagnalis	20

Abundance in quadrats

	Abundance class									Distribution type
	1–10	11–20	21–30	31–40	41–50	51–60	61–70	71–80	81–90	81–100	
Wetland	5	1	2	–	–	–	1	–	–	1	2
Skeletal	–	–	–	–	–	–	–	–	–	–	–
Arable	–	–	–	–	–	–	–	–	–	–	–
Grassland	–	–	–	–	–	–	–	–	–	–	–
Spoil	–	–	–	–	–	–	–	–	–	–	–
Wasteland	–	–	–	–	–	–	–	–	–	–	–
Woodland	–	–	–	–	–	–	–	–	–	–	–
All habitats	50.0	10.0	20.0	–	–	–	10.0	–	–	10.0	2

Ranunculus penicillatus (Dumort.) Bab.

Ranunculaceae
Stream Water-crowfoot

R. pseudofluitans (Syme) Baker & Foggitt. Field records may include *R. aquatilis* L. × *R. fluitans* Lam.

Life-form Polycarpic perennial hydrophyte with long, branched flexuous, submerged stems and finely-dissected leaves, <500 mm^2; roots adventitious.

Phenology Winter green. In spring or summer the stems increase in diameter and become hollow, and float at or near the water surface (Dawson 1976). Flowering occurs on these stems over a prolonged period but is locally particularly frequent in May and June and seed set from June onwards. Shoots tend to break up after fruiting.

Height In aquatic sites submerged or reaching the water surface; on land <20 mm.

Seedling RGR No data. SLA 37.8. DMC 11.0.

Nuclear DNA amount 9.8 pg; $2n = 32, 48*$ (*BC*); $4x, 6x*$.

Established strategy Intermediate between competitor and competitive-ruderal.

Regenerative strategies V, ?B$_s$. Regeneration requires further study (see 'Synopsis').

Flowers White, hermaphrodite, insect-pollinated or selfed (Cook 1966); solitary, leaf opposed.

Germinule 0.39 mg, 1.6 × 1.0 mm, achene, often 50-80 in each fruiting head (Cook 1966).

Germination Epigeal; in one test 20% germination of freshly-collected seed.

Biological flora None.

Geographical distribution Frequent in England, Wales, S Scotland and Ireland, *ABIF* page 109-10; much of Europe except the extreme N (49% of European territories); N Africa.

Habitats Virtually restricted to rivers and streams. Detached shoots beached on shingle or walls may form temporary colonies.

Gregariousness Patch-forming even at low values of shoot frequency.

Altitude Recorded up to 230 m, perhaps its British altitudinal limit.

Slope Not relevant.

Aspect Not relevant.

Hydrology Mainly restricted to flowing water, but also present in still water where streams have been dammed. Observed in up to 900 mm of water but extends to 2.5 m (Dawson 1976). Also shortly persistent on mud and gravel beside the water.

Soil pH All records from soils within the pH range 6.5-8.0.

Bare soil Typical of aquatic habitats, but rooted shoots may temporarily occupy exposed mud or gravel.

△ **ordination** Insufficient data.

Associated floristic diversity Normally very low.

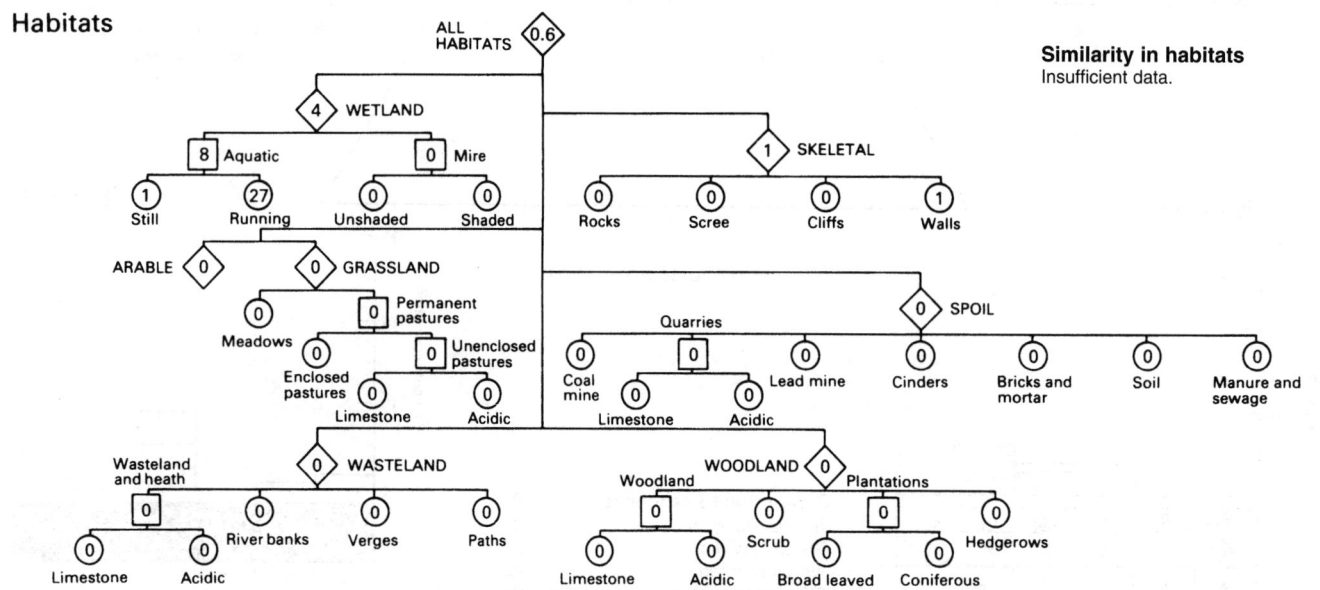

Habitats

Similarity in habitats
Insufficient data.

Synopsis *R.p.* is a dominant and productive submerged aquatic herb forming dense and extensive monocultures in rivers and streams but developing a standing crop which is usually much less than those of dominant emergent species such as *Typha latifolia* (Sculthorpe 1967). Peak biomass is attained in summer about 1 month after flowering (Dawson 1976), but there is also some photosynthetic activity during winter (Sculthorpe 1967). In water *R.p.* produces long, submerged, finely-dissected leaves which offer little mechanical resistance to the strong currents often experienced. Shoots produced in spring have thicker, hollow stems, produce flowers and float at or near the water surface (Dawson 1976). Photosynthesis is greater at faster flow rates and is possibly limiting in slow-moving water, as perhaps are nutrient uptake and the intake of oxygen for respiration (Sculthorpe 1967). Beached plants have the same dissected leaf form shown by *R. peltatus*. The control of *R.p.* by cutting in spring is ineffective and simply encourages further vegetative growth (Ham *et al.* 1982). In many river systems *R.p.* is the only aquatic macrophyte present, but in some regions, as in the Derbyshire Peak District, *R. fluitans* Lam. may also occur. Cook (1966) noted that stretches of water with a rocky or gravelly bottom are characterized by *R.p.*, which is replaced by *R.f.* where the substratum is more silty. This distribution appears unrelated to water chemistry, since both species occur on the base-poor millstone grit and on the carboniferous limestone. However, *R.p.* produces roots throughout the year and can thus remain attached to an unstable rocky substrate, whereas *R.f.* roots only during winter and is restricted to streams with a more stable silty bottom (Cook 1966). Regeneration is mainly vegetative. Stems may grow up to 4 m in the direction of the water flow, and single plants can form large clonal patches. Vegetative fragments root readily, and establishment in this manner has been observed. *R.p.* often produces abundant seed, which floats for only a short period (Ridley 1930). It is not known when seeds normally germinate. No persistent seed bank has so far been detected. It is likely that establishment by seed is rare. Populations in N England refer mostly to ssp. *pseudofluitans* (Syme) S. D. Webster, which occurs in slow-flowing water subject to frequent flooding Cook (1966). Ssp. *penicillatus* (Dumort.) Bab.,

Altitude
Slope
Aspect

	Unshaded
0.0 / — / 0.4 / 0.0	

Shaded: 0.0 / — / 0.0 / —

Soil pH
Associated floristic diversity

Not applicable

Hydrology

Triangular ordination

Commonest associates	%
Agrostis stolonifera	38
Poa trivialis	29
Apium nodiflorum	21
Myosotis scorpioides	21
Ranunculus repens	15

Bare soil

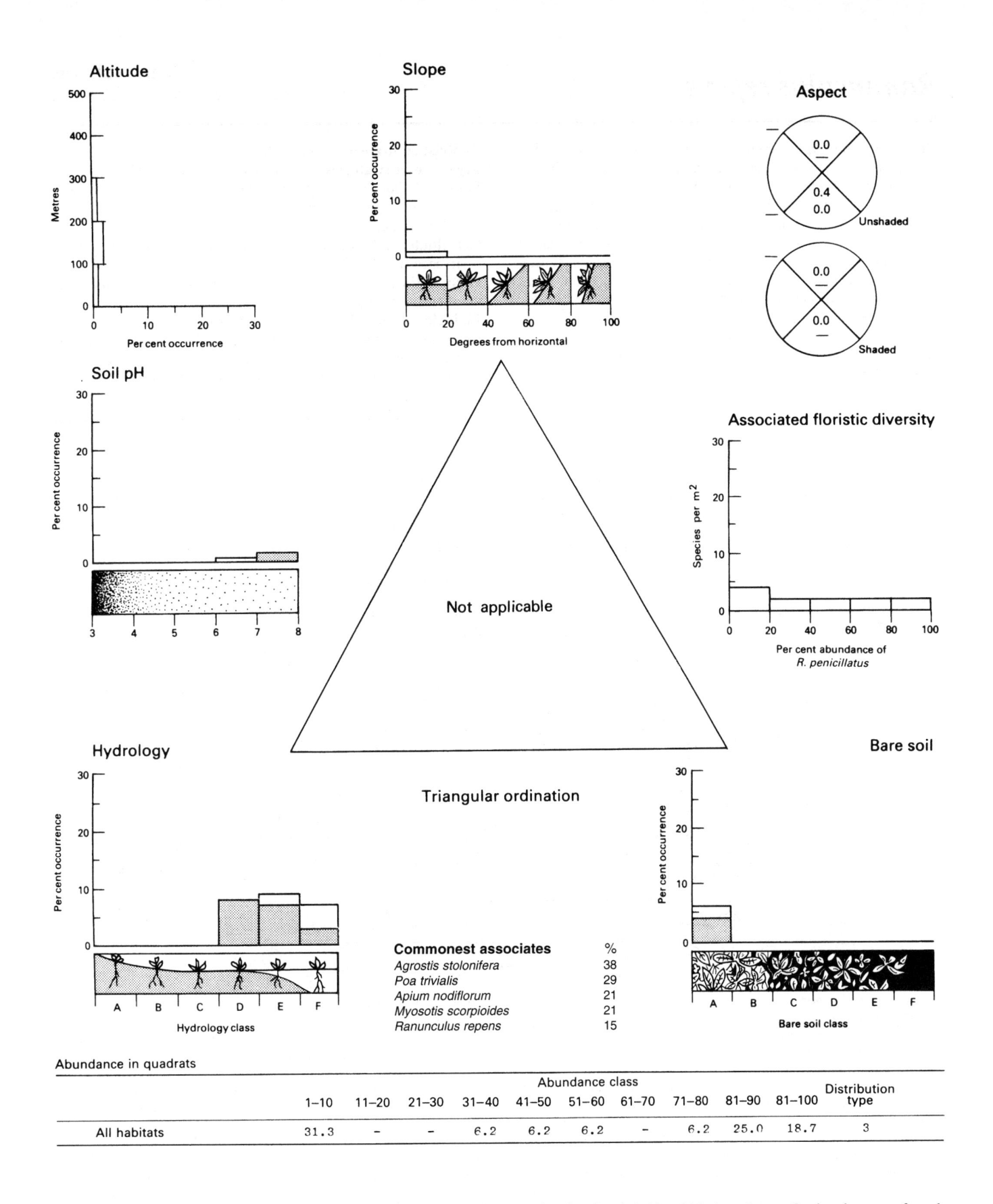

Abundance in quadrats

	1–10	11–20	21–30	31–40	41–50	51–60	61–70	71–80	81–90	81–100	Distribution type
All habitats	31.3	–	–	6.2	6.2	6.2	–	6.2	25.0	18.7	3

which may form floating laminar leaves, is found in fast-flowing water, and var. *vertumnus* Cook in clear, slow-flowing water, including canals not subject to regular flooding (Cook 1966). There is much additional variation within the species group, mostly ecotypic in origin (Cook 1966). The *R.p.* group is believed to have resulted from the hybridization of *R.f.* with *R. aquatilis*, *R. trichophyllus* Chaix and possibly *R. peltatus* (Stace 1975). The first hybrid, which is quite sterile, has been confirmed cytologically for Derbyshire (Stace 1975), and some field records for this hybrid may in error have been assigned to *R.p.*

Current trends Much less common than formerly, particularly in lowland areas, presumably as a result of increased eutrophication and/or industrial pollution.

Ranunculus repens L.

Ranunculaceae
Creeping Buttercup, Crowfoot

Life-form Polycarpic perennial, semi-rosette hemicryptophyte with epigeal stolons. Stems erect; leaves ovate, subdivided, hairy, often <2000 mm^2; roots with VA mycorrhizas.
Phenology Some growth in winter, but the main flush is in spring. Flowers mainly May to June and sets seed primarily in June and July. Stolons giving rise to newly formed ramets produced in early summer at the same time as the flowers. Stolons wither in late summer or autumn, and plants overwinter as rosettes of small leaves. Rosettes which produce flowers normally monocarpic (*Biol Fl*).
Height Foliage usually <250 mm; flowers to 600 mm.
Seedling RGR 0.5-0.9 week^{-1}. SLA 30.0. DMC 17.2.
Nuclear DNA amount 22.4 pg, $2n = 32$* (Bennett & Smith 1976); $4x$*.

Established strategy Competitive-ruderal.
Regenerative strategies (V), B$_s$. (See 'Synopsis'); ?SB IV.
Flowers Yellow, normally hermaphrodite, each day moving 'with the sun', insect-pollinated (*Biol Fl*); *c.* 5 in a cymose inflorescence.
Germinule 2.32 mg, 3.9 × 2.3 mm, achene, often *c.* 30 clustered in each fruit head (*Biol Fl*).
Germination Epigeal; t_{50} 19 days; germination rate increased after dry storage at 20°C; L > D.
Biological flora Harper (1957b), but see also Zarzycki (1976).
Geographical distribution The British Isles (100% of British vice-counties), *ABIF* page 101; Europe (87% of European territories and 5% naturalized); C Asia; introduced to N and S America and New Zealand.

Habitats Exceptionally wide-ranging and absent only from six terminal habitats. Most abundant in three major types: grassland (meadows and pastures), mire (and river banks) and disturbed habitats (soil heaps, the margins of arable land and gardens). Also frequent at waysides and on various types of spoil. In aquatic habitats usually at the water margin, often in recently flooded areas or rooted on the bank. Infrequent in skeletal habitats and woodland. [Found on sand dunes in moist microsites and on shingle (*Biol Fl*).]
Gregariousness Capable of forming patches, but usually interspersed with other species.

Altitude Perhaps because of its association with fertile and disturbed habitats, most frequent in lowland areas, but extending to 470 m. [Up to 1040 m (*Biol Fl*).]
Slope Most frequent and abundant on flat ground and gentle slopes, a reflection of an association with agricultural land and mire.
Aspect Widely distributed.
Hydrology Found most typically on moist soil beside water, but recorded from both topogenous and soligenous mire. Not recorded from permanently flooded sites.
Soil pH Most records from soils with pH in the range 5.0-8.0, and virtually absent below pH 4.5.
Bare soil Distribution wide-ranging, but biased towards habitats having a relatively large amount of exposed soil.

Habitats

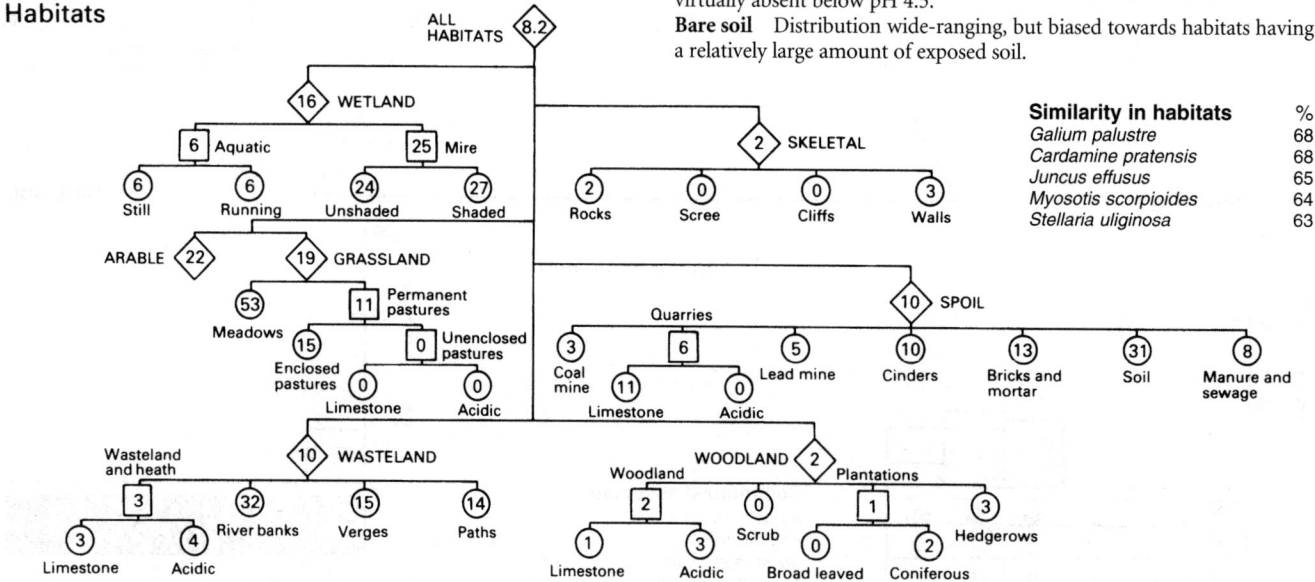

Similarity in habitats	%
Galium palustre	68
Cardamine pratensis	68
Juncus effusus	65
Myosotis scorpioides	64
Stellaria uliginosa	63

△ **ordination** Centred on fertile, moderately disturbed vegetation types. Absent from infertile habitats.
Associated floristic diversity Typically intermediate.
Synopsis *R.r.* is a stoloniferous, winter-green herb of fertile and disturbed habitats, particularly on poorly-drained soils, and in areas of high rainfall (*Biol Fl*). *R.r.* flourishes in mire, grassland and a range of disturbed habitats, and displays a wider habitat range than *R. acris* or *R. bulbosus*. *R.r.* is frequent in meadows, though less so than *R.a.*, and can survive infrequent digging or ploughing. The occurrence of *R.r.* is generally indicative of present or past human disturbance. *R.r.* appears to be less-effectively defended against predation than *R.a.* or *R.b.*, and it is frequently eaten by stock, apparently without ill-effect (*Biol Fl*, Cooper & Johnson 1984). Seeds, too, are more heavily predated than those of other common buttercups (Sarukhan 1974). *R.r.* is more sensitive to translocated herbicides, and its distribution relative to *R.a.* and *R.b.* may be restricted accordingly (*Biol Fl*). Demographic studies have been conducted on pasture populations (Sarukhan 1974, Doust 1981a,b). Like other buttercups, *R.r.* is dependent upon bare areas for regeneration by seed; open areas also encourage vigorous proliferation of stolons (*Biol Fl*). There is a rapid turnover of ramets (Sarukhan & Harper 1973) which may allow survival and spread of the species in sites subject to moderate disturbance. *R.r.* is the common buttercup most dependent upon vegetative regeneration. Some population biologists term the vegetative spread of *R.r.* (by means of widely-spaced offsets borne on stolons) 'the guerilla strategy', as opposed to 'the phalanx strategy' of tufted plants, where shoots are closer together (Doust 1981a). In closed turf, stolons may serve primarily to replace the parent by a daughter (*Biol Fl*), but in the absence of competition on fertile soils stolons up to 1.5 m in length are formed, producing several widely spaced ramets. Colonization of new sites appears to be effected both by seed and by detached plantlets, probably dispersed through the agency of humans or domestic animals. Although produced in smaller numbers, seeds of *R.r.* are more persistent in the soil than those of *R.a.* or *R.b.* (Odum 1978, Sarukhan 1974). The wide range of genetic variation within *R.r.* requires taxonomic treatment *(FE1)*.
Current trends *R.r.* colonizes and persists in a wide range of artificial habitats. A further increase in abundance is predicted.

Altitude
Slope
Aspect

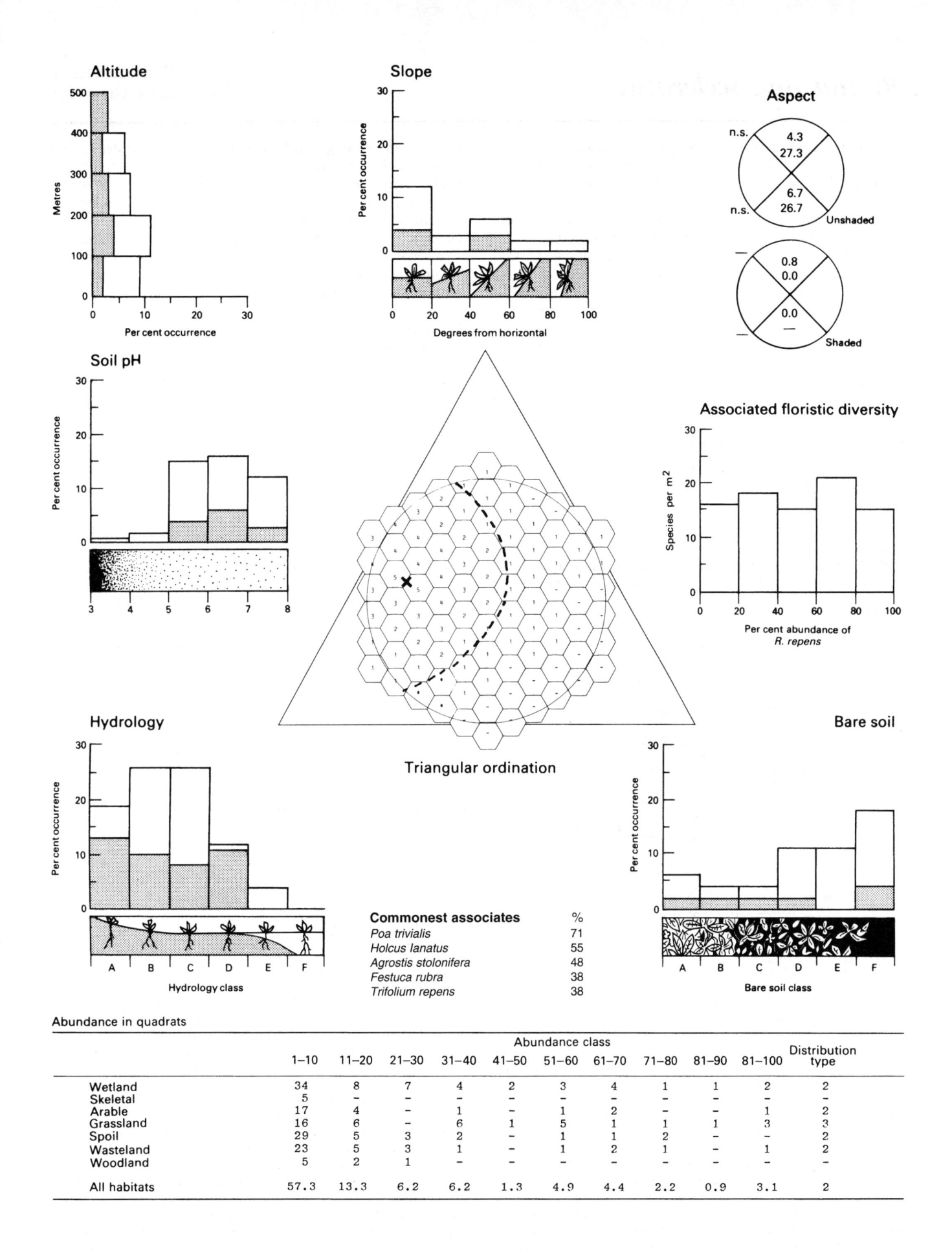

	n.s.	4.3	
		27.3	
	n.s.	6.7	Unshaded
		26.7	

	—	0.8	
		0.0	
	—	0.0	Shaded
		—	

Soil pH

Associated floristic diversity

Hydrology

Triangular ordination

Bare soil

Commonest associates

	%
Poa trivialis	71
Holcus lanatus	55
Agrostis stolonifera	48
Festuca rubra	38
Trifolium repens	38

Abundance in quadrats

	1–10	11–20	21–30	31–40	41–50	51–60	61–70	71–80	81–90	81–100	Distribution type
Wetland	34	8	7	4	2	3	4	1	1	2	2
Skeletal	5	–	–	–	–	–	–	–	–	–	–
Arable	17	4	–	1	–	1	2	–	–	1	2
Grassland	16	6	–	6	1	5	1	1	1	3	3
Spoil	29	5	3	2	–	1	1	2	–	–	2
Wasteland	23	5	3	1	–	1	2	1	–	1	2
Woodland	5	2	1	–	–	–	–	–	–	–	–
All habitats	57.3	13.3	6.2	6.2	1.3	4.9	4.4	2.2	0.9	3.1	2

Ranunculus sceleratus L.

Ranunculaceae
Celery-leaved Crowfoot

Life-form Summer- or occasionally winter-annual, semi-rosette therophyte or helophyte. Stem erect, often branched; leaves palmately divided, often <2000 mm²; roots shallow, with VA mycorrhizas.

Phenology Flowers mainly in June to July, with seed set from July to August, but life-span of plants may be as little as *c.* 2 months and a further generation is possible in summer (Bakker 1966), and *R.s.* is in flower and fruit through much of the summer. Plants which have not reached the flowering stage by autumn may overwinter in aquatic habitats as a rosette of floating leaves.

Height Foliage extending to *c.* 500 mm and overtopped by flowers in the mature plant.

Seedling RGR No data. SLA 30.4. DMC 18.5.

Nuclear DNA amount 8.0 pg, $2n = 32$ (Bennett & Smith 1976); $4x$.

Established strategy Intermediate between ruderal and competitive-ruderal.

Regenerative strategies B_s, Regenerates exclusively by seeds, which germinate mainly in spring and summer. Also forms a bank of persistent seeds (Thompson 1977); ?SB IV.

Flowers Yellow, hermaphrodite, protogynous, insect-pollinated, self-incompatible (van der Toorn 1980); in branching cymes, often >100 per plant.

Germinule 0.16 mg, 2.2 × 1.2 mm, achene; sometimes >100 in each fruit.

Germination Epigeal to 8 days; L > D.

Biological flora None.

Geographical distribution Throughout lowland Britain, rare in N Scotland (94% of British vice-counties), *ABIF* page 103; most of Europe, especially C and N parts (85% of European territories).

Habitats A wetland species particularly frequent at the edges of ponds and ditches. Found also on eroded river banks and as an impermanent colonist of moist soil heaps and sewage spoil. Rarely observed on poorly-drained arable land. Absent from grassland, woodland and skeletal habitats.

Gregariousness Sparse.

Altitude Most suitable habitats within the Sheffield region are associated with lowland areas, but the species occurs sparingly in the upland region and was observed at 320 m, a value above the maximum recorded by Wilson (1956) for the British Isles.

Slope As with most wetland species, largely confined to slopes of <10°.

Aspect Insufficient data.

Hydrology Most records are from topogenous mire adjacent to open water. *R.s.* typically occurs at the edge of ponds and ditches as a colonist of mud which becomes exposed as the water table becomes lower in spring or early summer. Seedlings produced in autumn may overwinter as submerged plants. Mature plants are not, however, found in permanently inundated sites.

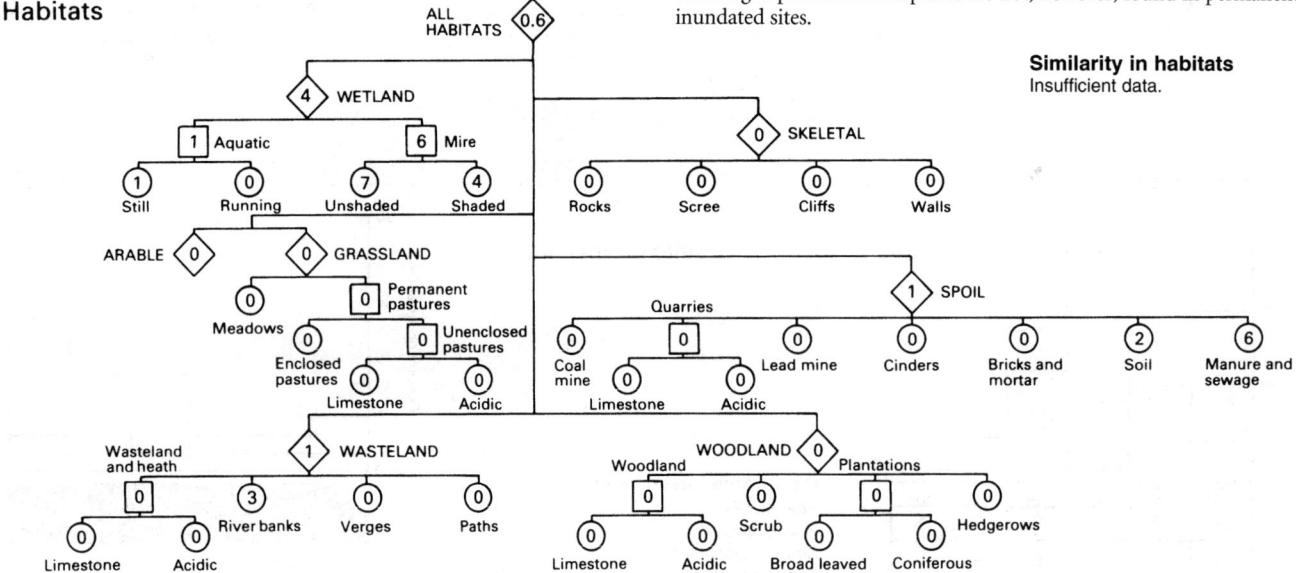

Habitats

Similarity in habitats
Insufficient data.

Soil pH Absent from acidic soils of pH <4.0.

Bare soil Found over a wide range of values, but more restricted to open sites than indicated here since many of the well-vegetated areas with *R.s.* have high densities of annual species.

△ **ordination** Distribution confined to vegetation associated with fertile habitats subject to some disturbance.

Associated floristic diversity Typically intermediate.

Synopsis *R.s.* is winter- or, more commonly, summer-annual and is characteristic of bare, fertile mud, particularly at the edges of ponds and ditches. *R.s.* and *Rorippa palustris* share a number of attributes which facilitate exploitation of transient water-side habitats. Under productive conditions each produces large quantities of small seeds (up to 56 000 per plant in the case of *R.s.* (van der Toorn 1980)). Like *R.p.*, *R.s.* forms a persistent seed bank (Thompson 1977) and submerged seeds remain viable for many years (van der Toorn 1980). In both species germination is enhanced by fluctuating temperatures and by exposure to sunlight (Salisbury 1942, Thompson 1977). Achenes are adhesive to animals and may float for several days (Ridley 1930). Both species are capable of rooting from detached stem pieces, although this does not appear to constitute a common mechanism of regeneration. The major differences between the two species are as follows. (a) In *R.s.* a proportion of the freshly dispersed seeds are readily germinable, and seedlings may become established in the field during autumn as well as in spring and summer. Seedlings are frost-tolerant (van der Toorn & ten Hove 1982). (b) *R.s.* has determinate growth and plants often set seed and die in *c.* 2 months. Thus, *R.s.* may produce two generations within a growing season (Bakker 1966). In contrast, the shorter *R.p.* tends to maintain flower and fruit production throughout the growing season if conditions remain favourable. (c) *R.s.* is more succulent and less woody than *R.p.*, and appears to be less persistent in drier sites. These differences accord with the tendency of *R.s.* to exploit fugacious sites close to the water's edge, whereas *R.p.* is particularly associated with areas of more-predictably exposed mud beside reservoirs.

Altitude
Slope
Aspect

Soil pH

Associated floristic diversity

Hydrology

Commonest associates	%
Juncus effusus	80
Callitriche stagnalis	60
Juncus bufonius	60
Plantago major	60
Poa trivialis	60

Triangular ordination

Bare soil

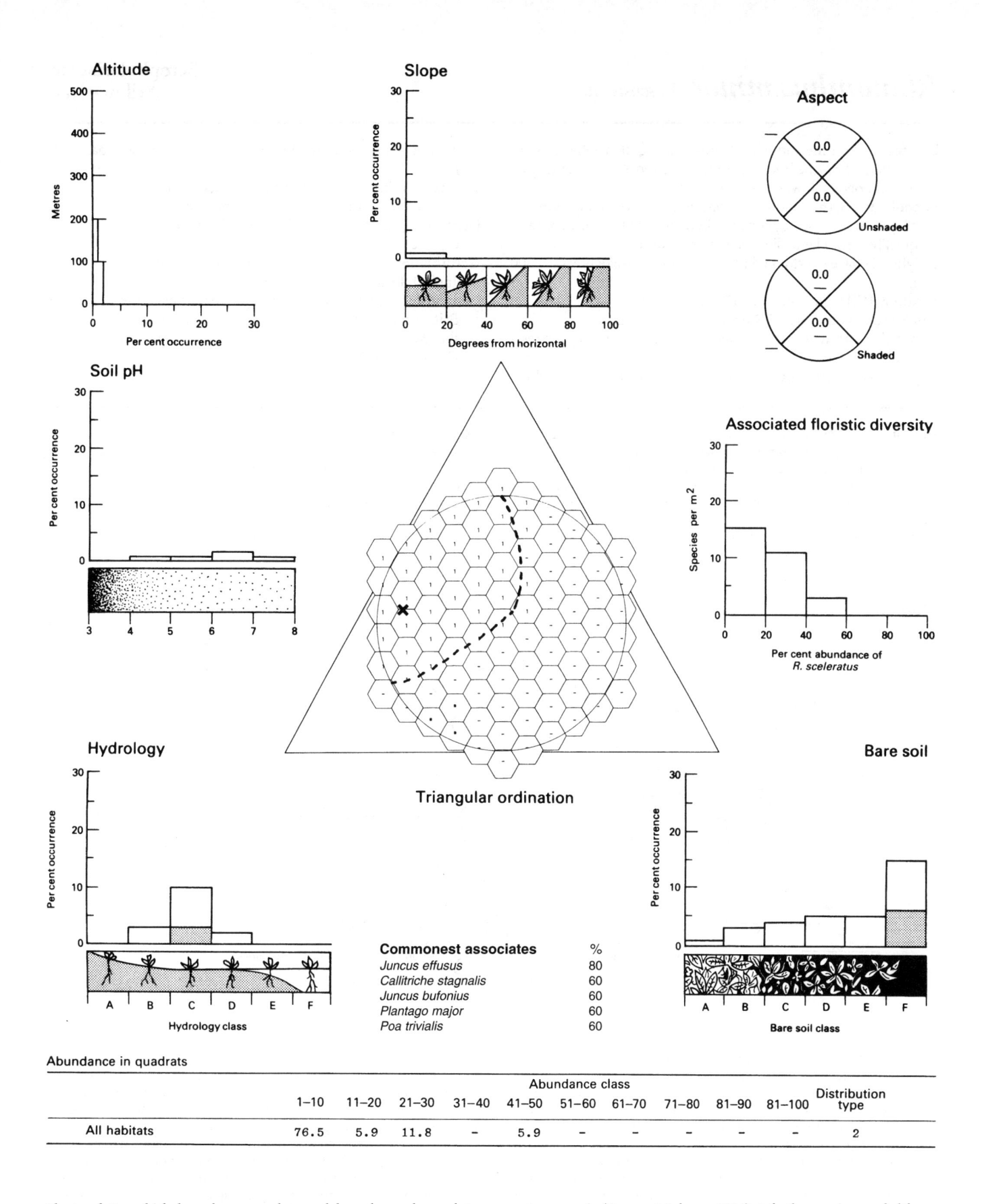

Abundance in quadrats

	1–10	11–20	21–30	31–40	41–50	51–60	61–70	71–80	81–90	81–100	Distribution type
All habitats	76.5	5.9	11.8	–	5.9	–	–	–	–	–	2

Plants of *R.s.* which have become submerged have been observed to develop floating leaves. Modified submerged leaves have also been recorded (Sculthorpe 1967). Like *R.p.*, *R.s.* has a southern and lowland distribution, and its sites seldom exceed 300 m. *R.s.* is potentially one of the more troublesome species of *Ranunculus* to farmers, since it may often be eaten by stock as a result of its succulent texture; the poisonous principle is protoanemonin (Cooper & Johnson 1984). A further ssp. is recorded from Europe (*FE1*).

Current trends An effective colonist of disturbed, fertile wetlands, which may be increasing. Perhaps also spreading to higher altitudes as a result of human activities.

Rhinanthus minor L. sensu lato

Scrophulariaceae
Yellow-rattle

Life-form Summer-annual, hemi-parasitic therophyte. Stem simple or branched; leaves ovate to lanceolate, typically <500 mm²; roots shallow, non-mycorrhizal.
Phenology Germination in spring is followed by flowering in May to August (see 'Synopsis'). Seed set from June onwards and quickly shed. Plants die in summer or autumn.
Height Foliage overtopped by flowers and leaf-like bracts to 500 mm.
Seedling RGR No data. SLA 17.8. DMC 17.9.
Nuclear DNA amount 7.9 pg; $2n = 22^*$ (*BC*); $2x^*$.
Established strategy Intermediate between ruderal and stress-tolerant ruderal.
Regenerative strategies S. Regenerates from seeds, which germinate in spring and do not form a persistent seed bank; SB II.
Flowers Yellow, hermaphrodite, insect-pollinated or selfed (Kwak 1979), usually <30 in each spike-like raceme.
Germinule 2.84 mg, 4.9 × 3.8 mm, winged seed in a *c.* 10 seeded capsule.
Germination Epigeal; dormancy broken by chilling.
Biological flora Westbury (2004).
Geographical distribution Most of the British Isles (100% of British vice-counties), *ABIF* pages 573-5; Europe, but rare in the Mediterranean region (79% of European territories); Greenland and Newfoundland.

Habitats Recorded mainly in meadows, with isolated records from pastures and atypically from rock outcrops. Also observed in ungrazed grassland, mire and on road verges. Absent from aquatic, arable, woods and several types of spoil habitats. [Found in sand-dune grassland.]
Gregariousness Frequently observed as dense populations.
Altitude Observed up to 350 m. [Up to 1020 m.]
Slope Absent from most skeletal habitats and steep slopes.
Aspect Insufficient data.
Hydrology Range includes meadows which become waterlogged in winter, and extends occasionally into the margins of soligenous mire.
Soil pH Absent from acidic sites below pH 5.0.
Bare soil No clear-cut trends.
△ **ordination** Only in vegetation of moderate productivity.
Associated floristic diversity Typically species-rich.

Habitats

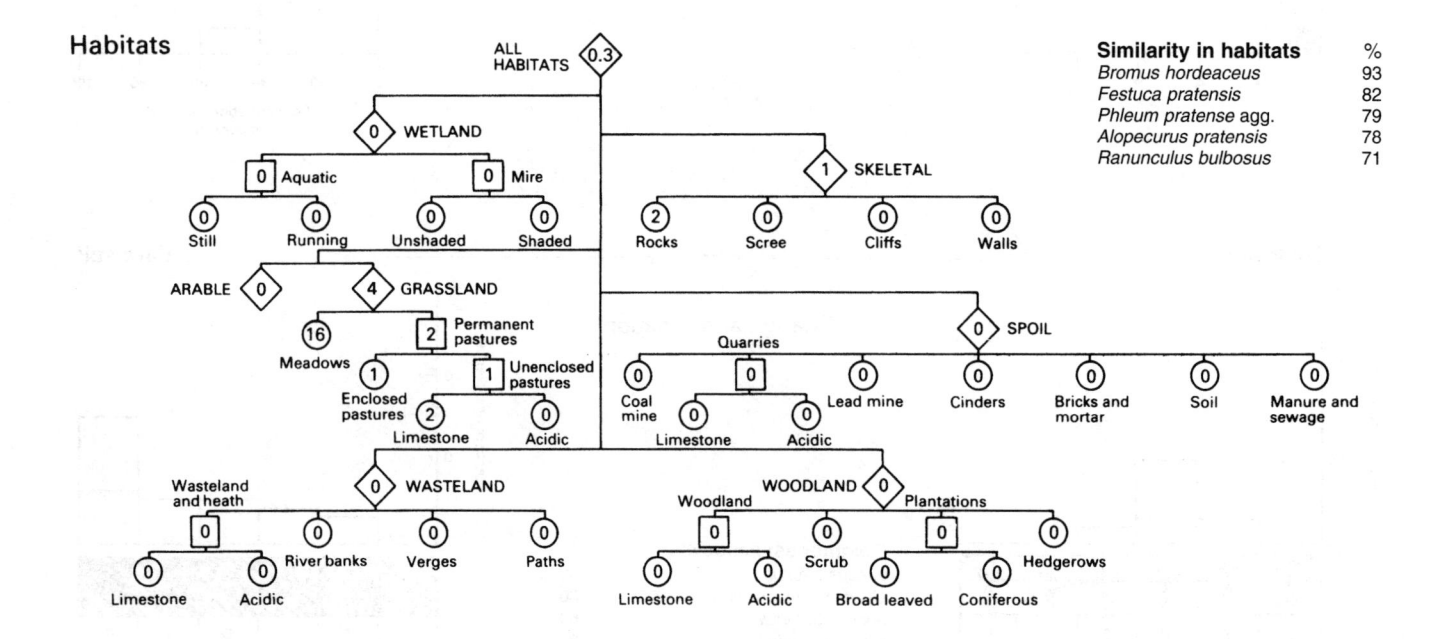

Similarity in habitats	%
Bromus hordeaceus	93
Festuca pratensis	82
Phleum pratense agg.	79
Alopecurus pratensis	78
Ranunculus bulbosus	71

Synopsis *R.m.* is a hemi-parasitic, summer-annual herb found in a wide range of grassland habitats on soils of moderate to low fertility. The species is most typical of hay meadows, where survival of *R.m.* depends critically upon the capacity to shed seed before harvest. *R.m.* is potentially toxic to stock (Cooper & Johnson 1984), and is extremely sensitive to the effects of heavy grazing, reflecting the vulnerability of the flowering shoots to predation and the total dependence upon seed production each year for regeneration and persistence of populations. *R.m.* is absent from droughted and from shaded habitats. Although capable of limited autotrophic growth, *R.m.* appears to be an obligate hemi-parasite in the field. *R.m.* has a wide host-range, particularly amongst the Poaceae, but grows considerably larger when attached to certain species, most notably *Trifolium repens* (Hodgson 1973). *R.m.* receives carbohydrates, and presumably also water and mineral nutrients, from its host (Hodgson 1973). The possibility that the relatively high levels of mannitol present in the plant (Hodgson 1973) plays an important osmotic role in the maintenance of hemi-parasitism deserves investigation (D. H. Lewis pers. comm.). *R.m.* produces moderately large winged seeds. These have a poorly differentiated embryo, and chilling is required to break dormancy. Germination occurs in spring and no persistent seed bank is produced (Roberts 1986a). *R.m.* is polymorphic, and the ecotypic variation present is not readily subdivided into taxa. The two commonest ecotypes are ssp. *stenophyllus* (Schur) Druce., which is found mainly in moist grassland, has a northern bias and flowers from July to August and ssp. *minor*, associated with drier sites, particularly in the S, and flowers from May to July (*ABIF, CTW, FE3*). Ecotypes from montane habitats and with an intermediate flowering period have also been separated. The origins of these variants, which are thought to have evolved recently (Hambler 1958), are uncertain.
Current trends Decreasing as a result of habitat destruction and the resowing and application of fertilizers to meadows.

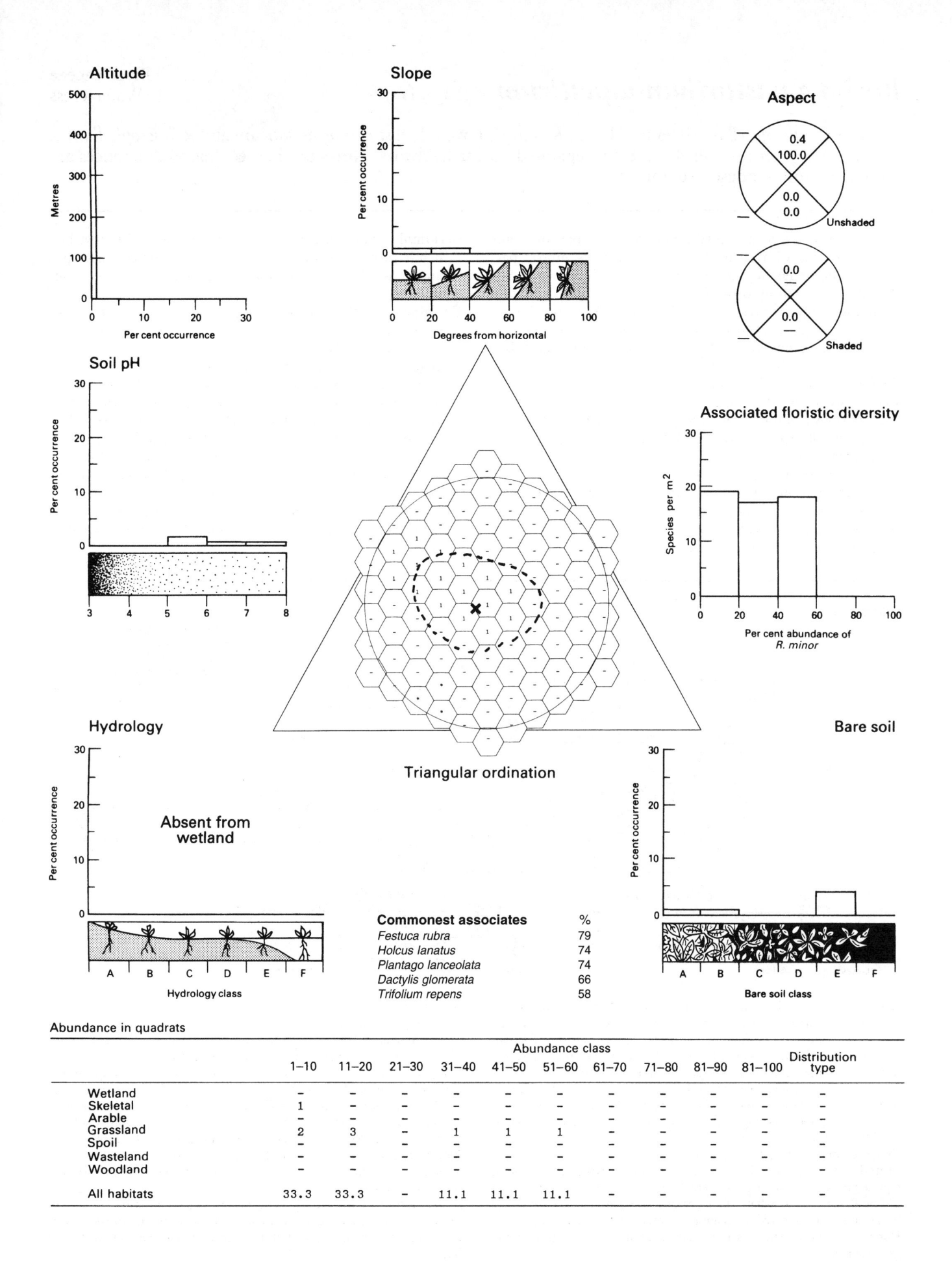

Altitude

Slope

Aspect

Unshaded

Shaded

Soil pH

Associated floristic diversity

Hydrology

Absent from wetland

Triangular ordination

Commonest associates	%
Festuca rubra	79
Holcus lanatus	74
Plantago lanceolata	74
Dactylis glomerata	66
Trifolium repens	58

Bare soil

Abundance in quadrats

	Abundance class										Distribution type
	1–10	11–20	21–30	31–40	41–50	51–60	61–70	71–80	81–90	81–100	
Wetland	–	–	–	–	–	–	–	–	–	–	–
Skeletal	1	–	–	–	–	–	–	–	–	–	–
Arable	–	–	–	–	–	–	–	–	–	–	–
Grassland	2	3	–	1	1	1	–	–	–	–	–
Spoil	–	–	–	–	–	–	–	–	–	–	–
Wasteland	–	–	–	–	–	–	–	–	–	–	–
Woodland	–	–	–	–	–	–	–	–	–	–	–
All habitats	33.3	33.3	–	11.1	11.1	11.1	–	–	–	–	–

Rorippa nasturtium-aquaticum (L.) agg. Hayek

Brassicaceae
Water-cress

Includes *R. microphyllum* (Boenn.) Hyl. ex A. and D. Love., *R. nasturtium-aquaticum* and *R. microphyllum* × *nasturtium-aquaticum*, which were not separated during fieldwork. Notes on their ecological differences are abstracted from Hodgson (unpubl.)

Life-form Polycarpic perennial helophyte. Older part of shoots procumbent, with shallow adventitious roots, upper part ascending or floating; leaves pinnately lobed, usually <2500 mm^2; roots non-mycorrhizal.

Phenology Overwinters as small plants. Flowering often commences during late May in *R.n-a.* or early June in *R.m.* and *R.m.* × *o.*, ceasing in July. Seed set about 2 months later (*Biol Fl*). Leaves relatively short-lived.

Height Foliage and flowering shoots to 450 mm.

Seedling RGR No data. SLA 44.7. DMC 9.8.

Nuclear DNA amount *R.m.*: 1.5 pg; $2n = 64*$ (*BC*); $4x*$. *R.m.* × *n-a.*: $2n = 48*$; $3x*$. *R.n-a.*: $2n = 32*$; $2x*$. Generic base number of polyploid origin.

Established strategy Competitive-ruderal.

Regenerative strategies V, B$_s$. Regeneration by seed requires further study (see 'Synopsis'); ?SB III.

Flowers White, hermaphrodite, homogamous, insect-pollinated, self-compatible; often >20 in each raceme (*Biol Fl*).

Germinule *R.m.*: 0.13 mg, 1.1 × 0.8 mm, seed dispersed in a dehiscent c. 30-seeded siliqua; *R.m.* × *n-a.*: mostly misshapen and non-viable, viable seed 0.14 mg, c. 0.8 × 0.7 mm; *R.n-a.*: 0.19 mg, 1.0 × 0.9 mm.

Germination *R.m.*: epigeal; t_{50} 6 days; *R.m.* × *o.*: poor germination; *R.n-a.*: t_{50} 8 days; germination rate increased after dry storage at 20°C.

Biological flora Howard and Lyon (1952).

Geographical distribution Distribution of segregates incompletely known. *R.m.*: common throughout the British Isles, *ABIF* page 257; W Europe (26% of European territories); W Asia. *R.m.* × *n-a.*: particularly in N England, Scotland and Ireland, *ABIF* page 257; France and C Europe. *R.n-a.*: common throughout lowland Britain, *ABIF* page 257; S and C Europe northwards to Scandinavia and extending eastward to the USSR (82% of European territories); N Africa, W Asia; introduced in many other parts of the world including New Zealand, where it is a serious river weed.

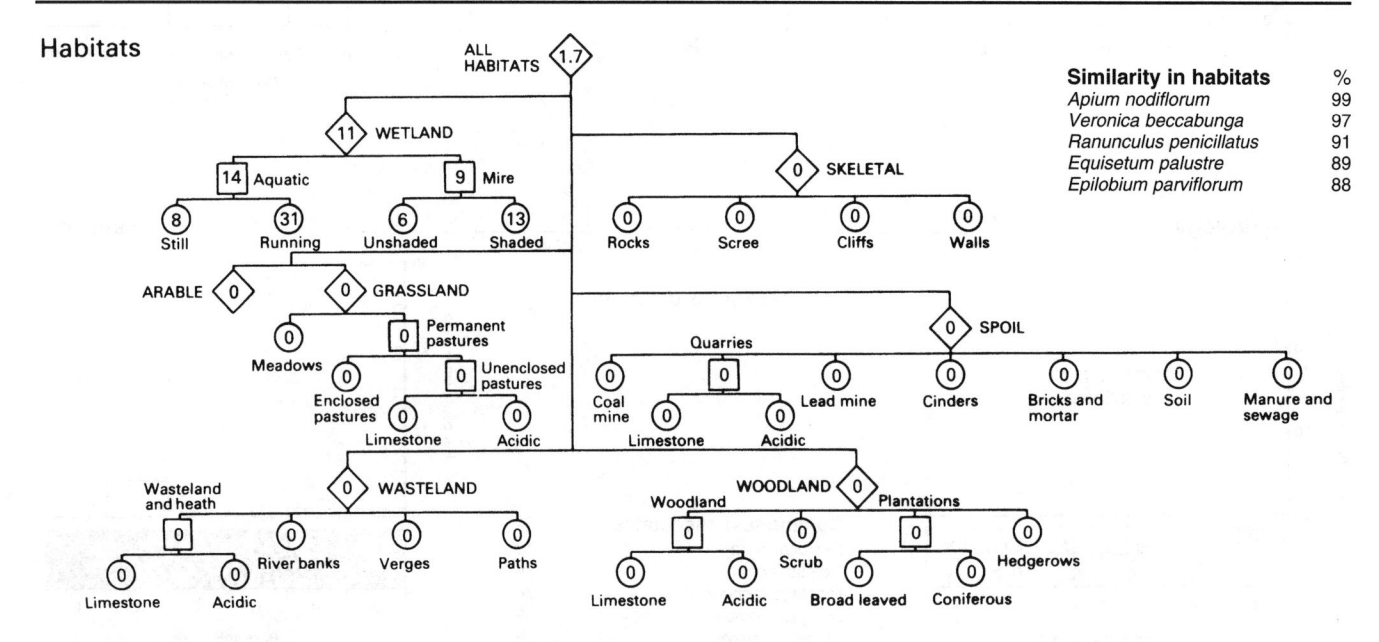

Habitats

Similarity in habitats	%
Apium nodiflorum	99
Veronica beccabunga	97
Ranunculus penicillatus	91
Equisetum palustre	89
Epilobium parviflorum	88

Habitats A wetland or semi-aquatic species. Most common at the margin of streams and to a lesser extent rivers, but also frequent in ditches and beside ponds, and in shaded or unshaded topogenous or soligenous mire, usually adjacent to water. In lowland parts of the Sheffield region *R.m.* is the most common segregate on non-calcareous soils, but is replaced to a considerable extent by *R.m.* × *n-a.* in upland areas, particularly on limestone and by *R.n-a.* or rarely *R.m.* × *n-a.* in or beside low-lying calcareous springs. [*R.n-a.* appears more tolerant of saline conditions (*Biol Fl*)]

Gregariousness Often forms patches.

Altitude Most frequent in lowland areas. In the upland region *R.m.* is replaced by *R.m.* × *n-a.* to a considerable extent. *R.n-a.* to 200 m; *R.m.* to 230 m; *R.m.* × *n-a.* to 340 m. [*R.m.* perhaps to 360 m; *R.n-a.* restricted to below 300 m and less frost-resistant than other two taxa (*Biol Fl*).]

Slope Like most wetland species, largely confined to flat ground and gentle slopes.

Aspect Not applicable.

Hydrology Most frequent and abundant in topogenous or soligenous mire marginal to water or rooted in shallow water. More commonly associated with flowing than with stagnant water (*Biol Fl*).

Soil pH Absent from strongly acidic situations and most frequent and abundant on neutral soils of the pH range 6.5-7.5. *R.m.*: 22% of records below pH 6.5; *R.m.* × *n-a.*: 14 %; *R.n-a.*: 11%. [In Britain as a whole *R.m.* × *n-a.* is also thought to occur at pH values, and calcium contents, intermediate between those associated with the parents (*Biol Fl*).]

Bare soil Associated with a wide range of exposures. Low values are associated with either a dense vegetation cover or the exposed water surface of aquatic habitats.

△ **ordination** Distribution centred on vegetation associated with productive, moderately disturbed conditions, but appears to show wide tolerance of varying levels of disturbance. Absent from unproductive habitats.

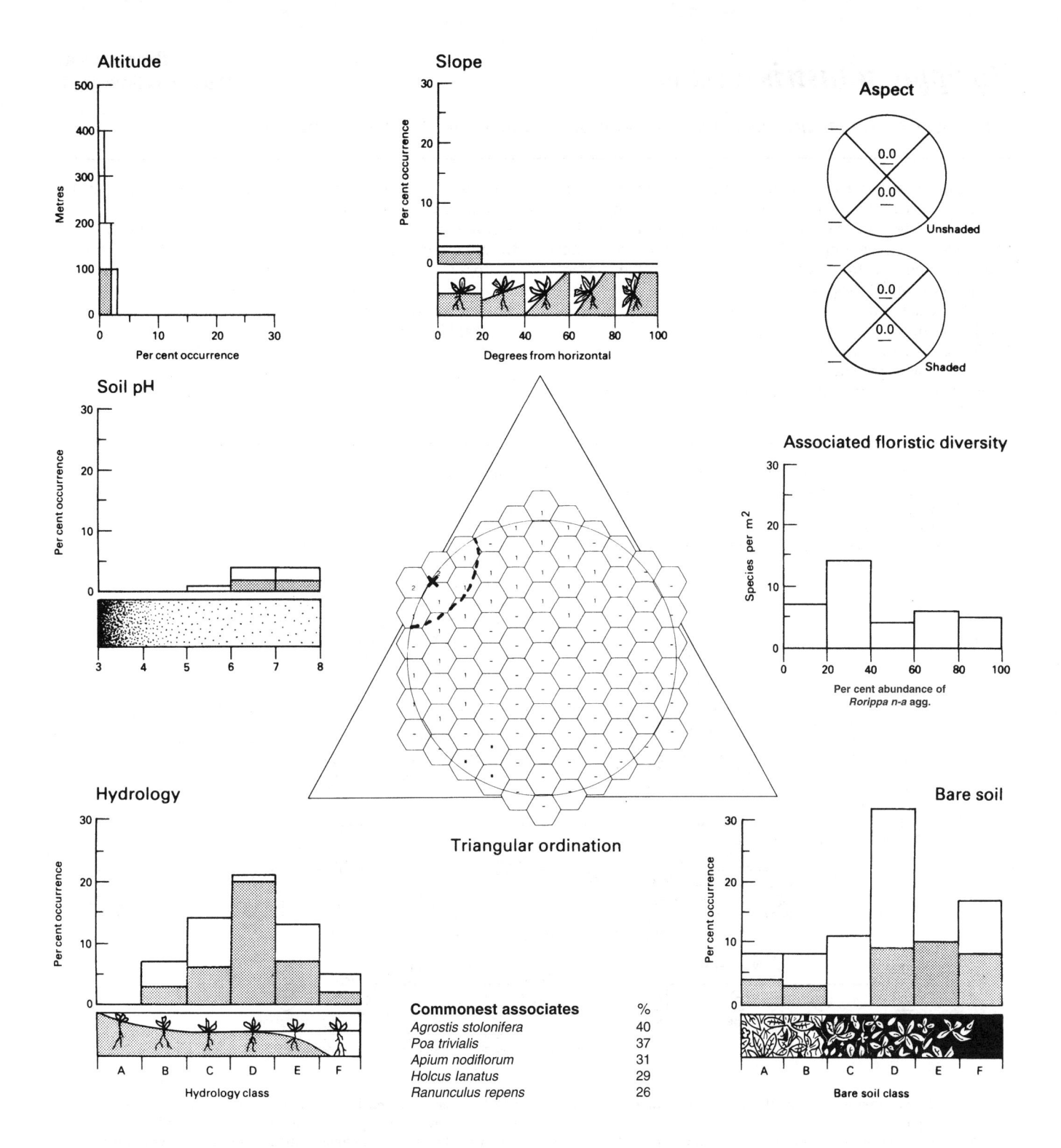

Altitude
Slope
Aspect
Soil pH
Associated floristic diversity
Hydrology
Triangular ordination
Bare soil

Commonest associates	%
Agrostis stolonifera	40
Poa trivialis	37
Apium nodiflorum	31
Holcus lanatus	29
Ranunculus repens	26

Associated floristic diversity Typically low. *R.m.* (and *R.m.* × *n-a.*) are occasionally found as monocultures and frequently form stands associated only with *Apium nodiflorum*.

Synopsis Despite belonging to a different family, *R.n-a.* agg. is ecologically and morphologically very similar to *Apium nodiflorum* but extends further N in Britain. Young shoots are eaten by cows (although this may result in tainted milk) and by water-fowl (*Biol Fl*), and the taxon survives heavy grazing. As in the case of *A.n.* procumbent stems root freely at the soil surface and give rise to large clones. Detached stem pieces have a remarkable ability to root and form new plants. *R.m.* × *n-a.* appears to have colonized whole water courses by this means. Seeds may be important for initial establishment, particularly in landlocked sites. Although seeds lack dormancy, germination probably takes place in the spring (*Biol Fl*).

R.n-a. agg. is apparently more ruderal than *A.n.* in the following respects: (a) roots are produced even more freely on stem fragments (often within 24 h); even the shoot of a seedling at the cotyledon stage (*Biol Fl*) and detached inflorescences are capable of rooting; (b) stems are exceedingly short-lived (Thommen & Westlake 1981); (c) the numerous small seeds are capable of either immediate or delayed germination, and a persistent seed bank is predicted on the basis of laboratory characteristics of the seed. *R.n-a.* and *R.m.* × *n-a.* are cultivated as salad plants in the S (*Biol Fl*) and both were formerly used as medicinal herbs; in some sites the two taxa may have been introduced for this purpose.

Future trends Still common, but decreasing in aquatic habitats as a result of water pollution. Probably lost from some mire sites through eutrophication and following invasion by taller wetland species.

Rorippa palustris (L.) Besser

Brassicaceae
Marsh Yellow-cress

A segregate of *R. islandica auct. mult. non* Borbas, separated since *FE1* (see 'Synopsis')

Life-form A summer-annual, semi-rosette therophyte. Stems erect, frequently branched; lower leaves deeply pinnatified, usually <2000 mm^2; roots relatively shallow, non-mycorrhizal.
Phenology Seeds germinate in spring or summer and plants flower from June to September. Seed is set from July onwards until the death of the plant.
Height Foliage and flowers to 600 mm.
Seedling RGR 2.0-2.4 week^{-1}. SLA 31.8. DMC 12.6.
Nuclear DNA amount 1.4 pg; $2n = 32^*$ (*BC*); $4x^*$.
Established strategy Ruderal.
Regenerative strategies B$_s$, S,. Effective regeneration entirely by seed (see 'Synopsis'); SB III.

Flowers Yellow, hermaphrodite, autogamous (Stace 1975); >100 in racemose inflorescence.
Germinule 0.07 mg, 0.75 mm, seed in a >50-seeded siliqua.
Germination Epigeal; t_{50} 4 days; dormancy broken by dry storage at 20°C and germination stimulated by large fluctuations of temperature, or scarification; L > D.
Biological flora None.
Geographical distribution The British Isles but rare in N Scotland (86% of British vice-counties), *ABIF* page 258; most of Europe, but less frequent in the Mediterranean region (74% of European territories); cosmopolitan.

Habitats A species of unshaded mire. Also found in moist disturbed situations such as spoil heaps and arable land, and on river banks and walls marginal to water.
Gregariousness Sparse.
Altitude Primarily associated with low-lying land, but observed to 275 m. [Up to 320 m.]

Slope Like most wetland species, largely confined to flat ground and gentle slopes.
Aspect Insufficient data.
Hydrology Mainly a colonist of mud exposed during summer but flooded in winter. Occurs beside both ponds and streams.
Soil pH Restricted to values of pH >5.0, and occasionally recorded from highly calcareous soils.
Bare soil Associated with a wide range of values, but distribution biased towards sites with high exposures of bare soil.
△ **ordination** Distribution confined to productive vegetation subject to moderate levels of disturbance.

Habitats

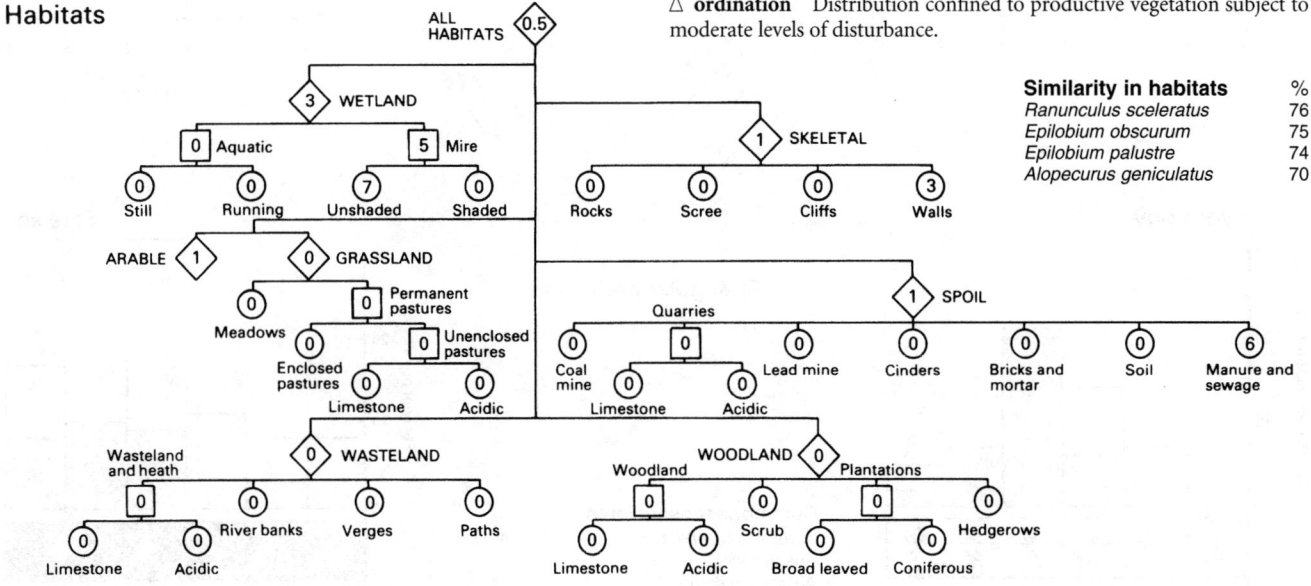

Similarity in habitats	%
Ranunculus sceleratus	76
Epilobium obscurum	75
Epilobium palustre	74
Alopecurus geniculatus	70

Associated floristic diversity Typically intermediate, varying little with increasing frequency of *R.p.*
Synopsis *R.p.* is a summer-annual which is characteristic of bare, fertile mud which becomes exposed in summer at the edge of reservoirs, and, to a lesser extent, ponds and ditches. Also rarely recorded from moist arable land and as a weed of horticulture. In drier sites, where the mud is exposed early, plants are frequently small and may produce <200 seeds whereas, according to Salisbury (1942), c. 130 000 may be formed on a plant of average size. *R.p.* has a southern and lowland distribution in Britain (*ABF*), and does not extend above 320 m (Wilson 1956). The species is associated with a wide range of edaphic conditions from mildly acidic silt flooded by acidic peaty water to exposed river beds in limestone dales. *R.p.* is absent from sites grazed by stock. Regeneration is almost entirely by seed, and involves the formation of a persistent seed bank (UCPE unpubl.). The germination of seeds is promoted by fluctuating temperatures and by light. Thus, seeds close to the surface of exposed mud may germinate, but those which are in submerged soil will not. Detached portions of vegetative shoots rapidly re-root. This capacity often causes the plant to re-anchor following flooding. Exceptionally, main stems dislodged onto the surface of the mud may produce roots and shoots along their entire length, to form a small clonal patches. Seeds float in water for several days and adhere to birds (Ridley 1930). They may also be transported by humans and *R.p.* is an early colonist of wetland sites. *R.p.* is ecologically very similar to *Ranunculus sceleratus*; differences between the two are described under *R.s.* Hybrids between *R.p.* and *R. amphibia* (L.) Bers. have been reported (Stace 1975). The closely related diploid ($2n = 16$) *R. islandica* (Oeder ex Murray) Borbas has been recognized since the publication of *FE1 R.i.* has a very restricted northern coastal distribution in Britain and Europe, with outlying stations in the Alps and Pyrenees (Randall 1974).
Current trends Although the distribution of *R.p.* is restricted at present, the association with fertile artificial habitats suggests the potential to expand into disturbed fertile wetlands; probably increasing.

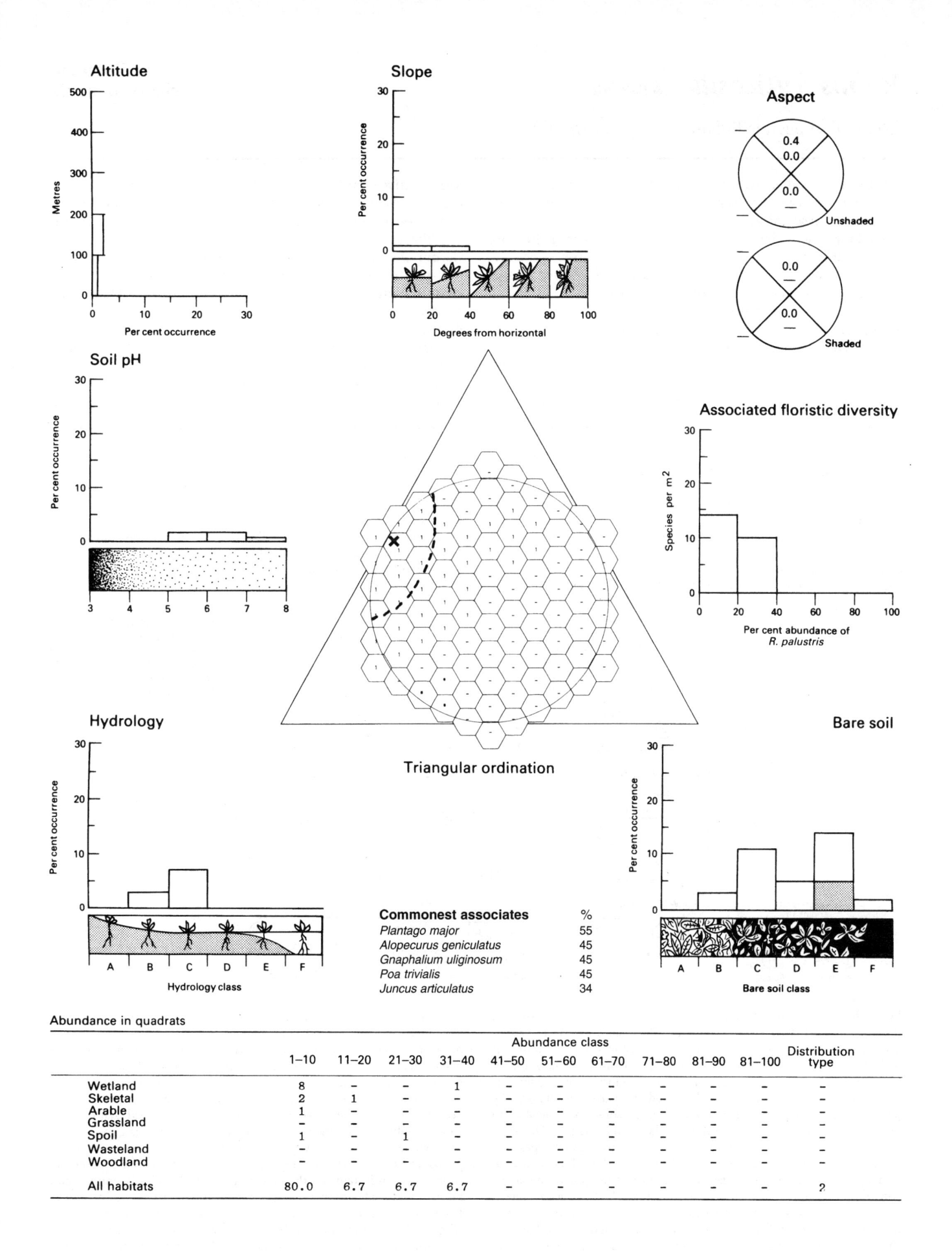

Altitude

Slope

Aspect

Unshaded

Shaded

Soil pH

Associated floristic diversity

Hydrology

Hydrology class

Triangular ordination

Commonest associates	%
Plantago major	55
Alopecurus geniculatus	45
Gnaphalium uliginosum	45
Poa trivialis	45
Juncus articulatus	34

Bare soil

Bare soil class

Abundance in quadrats

	1–10	11–20	21–30	31–40	41–50	51–60	61–70	71–80	81–90	81–100	Distribution type
Wetland	8	–	–	1	–	–	–	–	–	–	–
Skeletal	2	1	–	–	–	–	–	–	–	–	–
Arable	1	–	–	–	–	–	–	–	–	–	–
Grassland	–	–	–	–	–	–	–	–	–	–	–
Spoil	1	–	1	–	–	–	–	–	–	–	–
Wasteland	–	–	–	–	–	–	–	–	–	–	–
Woodland	–	–	–	–	–	–	–	–	–	–	–
All habitats	80.0	6.7	6.7	6.7	–	–	–	–	–	–	2

Rubus fruticosus L. *sensu lato*

Rosaceae
Blackberry, Bramble

(*Rubus* L. subgenus *Rubus*, except *R. caesius* L.)

Life-form Woody polycarpic perennial. Stems spiny, arching or erect; leaves ternate or palmate, sometimes >10 000 mm^2; roots with VA mycorrhizas.

Phenology Stems biennial to perennial. Flowers from June to September and sets seed from September to October. Leaves are relatively long-lived and in some taxa persist through the winter, when they are replaced by new foliage.

Height Foliage and flowers may exceed 3 m.

Seedling RGR No data. SLA 12.7. DMC. No data.

Nuclear DNA amount No data; $2n$ = 14*, 21*, 28*, 35*, 42*; $2x$* (<1% of British species), $3x$* (4%), $4x$* (91%), $5x$* (3%), $6x$* (<1%) (*Biol Fl*).

Established strategy Stress-tolerant competitor.

Regenerative strategies V, B$_s$. See 'Synopsis'; SB IV.

Flowers White or pink, hermaphrodite, cross-pollinated by insects or apomictic according to species; up to 20 present if inflorescence is compound.

Germinule 2.49 mg, 3.3 × 2.0 mm (minus fleshy outer layer), 1-seeded druplet; sometimes >20 in each 'blackberry'.

Germination Epigeal; up to 35% germination obtained by scarification followed by chilling (*Biol Fl*).

Biological flora Amor and Richardson 1980.

Geographical distribution The British Isles (100% of British vice-counties), *ABIF* page 327; Europe except the extreme N (87% of European territories); E Asia, Africa, S and N America; introduced into Australia and New Zealand.

Habitats Most characteristic of hedgerows, scrub and woodland, but recorded from all habitats except grazed and mown grassland and aquatic habitats.

Gregariousness A stand-forming species achieving dominance at very low densities of shoots.

Altitude Most frequent and abundant below 200 m, but extending to 400 m. [Up to 450 m.]

Slope Wide-ranging.

Aspect More frequent on S-facing slopes, both in the open and in shade.

Hydrology Not a wetland species, but occasionally found on moist ground beside soligenous or topogenous mire.

Soil pH Over the full range of values, but most frequent in the pH range 3.5–5.0.

Bare soil Widely distributed.

△ **ordination** Distribution including both fertile relatively undisturbed and unproductive undisturbed vegetation (see 'Synopsis').

Associated floristic diversity Typically low.

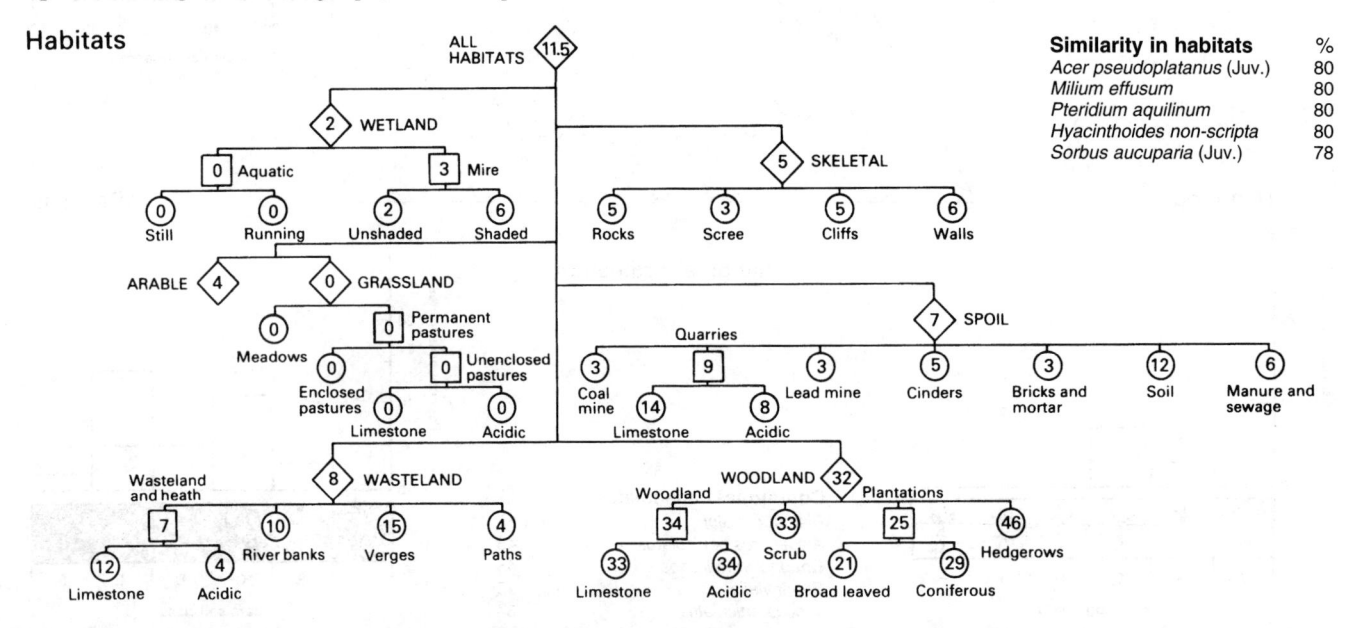

Habitats

Similarity in habitats	%
Acer pseudoplatanus (Juv.)	80
Milium effusum	80
Pteridium aquilinum	80
Hyacinthoides non-scripta	80
Sorbus aucuparia (Juv.)	78

ALL HABITATS 11.5
WETLAND 2; Aquatic 0 (Still 0, Running 0); Mire 3 (Unshaded 2, Shaded 6)
SKELETAL 5 (Rocks 5, Scree 3, Cliffs 5, Walls 6)
ARABLE 4
GRASSLAND 0; Meadows 0; Permanent pastures 0 (Enclosed pastures Limestone 0, Acidic 0); Unenclosed pastures 0
Quarries: Coal mine 3, Limestone 9 (14), Acidic 3 (8), Lead mine 3
SPOIL 7; Cinders 5, Bricks and mortar 3, Soil 12, Manure and sewage 6
WASTELAND 8; Wasteland and heath (Limestone 12, Acidic 4), River banks 7 (10), Verges 15, Paths 4
WOODLAND 32; Woodland 34 (Limestone 33, Acidic 34), Scrub 33, Broad leaved 21, Plantations 25 (Coniferous 29), Hedgerows 46

Synopsis *R.f.* is a robust, spiny, shrub or undershrub, capable of local dominance of herbaceous vegetation. Watt (1934) suggests that *R.f.* cannot achieve or maintain dominance in unshaded habitats. This is consistent with the predominantly woodland and hedgerow distribution, but takes insufficient account of the capacity of some taxa to form extensive thickets in unshaded sites or on derelict land and, in particular, on railway property. The greater frequency of *R.f.* on acidic soils, noted by Watson (1958), is confirmed by our own data for the Sheffield region. However, *R.f.* is remarkably wide-ranging and occurs in most habitat types. *R.f.* regenerates mainly by vegetative means. During the short days of autumn stem apices become positively geotropic, and when they establish contact with the soil produce roots and a resting bud (*Biol Fl*). More rarely, adventitious shoots (suckers) may be formed on the roots from a depth of up to 0.5 m (*Biol Fl*). *R.f.* may also reproduce from root fragments (*Biol Fl*). The woody stems usually persist for only 2–3 years, but stands do not degenerate in the centre as in the case of species such as *Pteridium aquilinum*. In their second year stems may produce flowers and seed production, often the result of apomixis, may reach 40 000 per 'plant' in open situations (*Biol Fl*). However, in very shaded sites flower and fruit production is low. Much of the seed is non-viable, and germination does not occur until the second spring after berries ripen (Nybom 1980). This is because there is double dormancy, imposed by a hard seed coat coupled with a chilling requirement (see Nybom 1980). Seedlings appear to be more vulnerable than new vegetative shoots to dominance by established plants (*Biol Fl*). Although regeneration by seed in dense perennial communities is infrequent, the berries are palatable to birds, animals and humans and seed is probably important in the colonization of new sites. Plants originating from seed do not flower until they are at least 3 years old (*Biol Fl*). *R.f.* is plastic, in growth form and morphology, to factors such as irradiance level, which makes some taxa difficult to identify (*Biol Fl*).

Altitude
Slope
Aspect
Soil pH
Associated floristic diversity
Hydrology
Triangular ordination
Bare soil

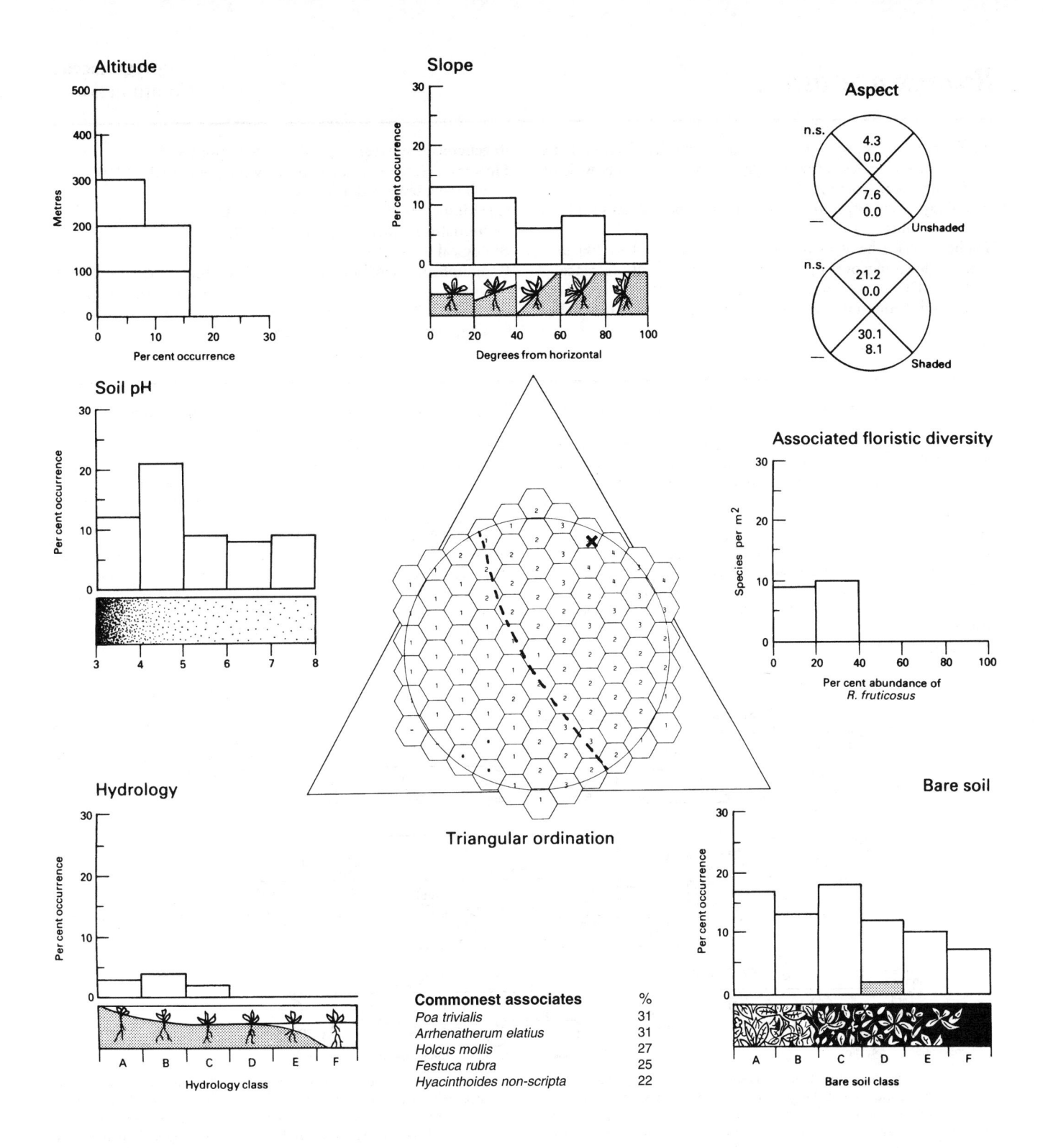

Commonest associates	%
Poa trivialis	31
Arrhenatherum elatius	31
Holcus mollis	27
Festuca rubra	25
Hyacinthoides non-scripta	22

However, the most important variations within *R.f.* are of genetic origin. Thus, *R.f.*, as defined here, is a complex group encompassing over 386 British microspecies, classified into 7 sections (Watson 1958). Most are apomictic, and many probably arose during the Pleistocene through the combined effects of hybridization and facultative apomixis. This speciation is still continuing (*FE2*). The wide habitat and strategic range of *R.f.* is in part a reflection of ecological differences between segregates. Thus, the commonest, and the only British diploid, *R. ulmifolius* Schott; is usually sexual, has a many-flowered compound, late-flowering inflorescence, has stems rooting at the apex, and is found on chalk and on heavy clay soils (*CTW*, *Excur Fl*). In contrast, members of Sect. *Rubus.* have a simple, early-flowering inflorescence, stems which do not root at their ends and are associated with acidic soils (*CTW*). The closely-related *R. caesius* (Sect Caesii Lej. and Courtois), in the Sheffield region typically is a plant of calcareous woodland. Hybrids between *R.c.* and *R.f.* are placed in Sect. *Corylifolii* Lindl. Species with late leaf-fall are restricted to areas where extreme winter conditions are short-lived; they have a smaller distributional area within Sweden (Oredsson 1975). Wide-ranging taxa in Sweden tend to have heavier seeds, form fewer seeds per berry, have early ripening of fruits and possess seeds with high germinability (Nybom 1980).
Current trends Uncertain. Reduction of populations has occurred through, for example, the removal of hedgerows but the species is encouraged by the increased disturbance associated with modern forestry.

Rumex acetosa L.

Polygonaceae
Common Sorrel

Life-form Polycarpic perennial, semi-rosette hemicryptophyte. Stems erect; leaves ovate, usually <2000 mm^2; roots with or without VA mycorrhizas.
Phenology Winter green. Flowers May to June and sets seed June to September.
Height Foliage basal in non-flowering plants and to 300 mm when flowering; overtopped to 500 mm by the inflorescence.
Seedling RGR 1.0–1.4 week^{-1}. SLA 28.1. DMC 10.3.
Nuclear DNA amount 3.3 pg; $2n = 14$ (female), $2n = 15^*$ (male) (*BC, FE1*); $2x^*$.
Established strategy C-S-R.

Regenerative strategies S, V. See 'Synopsis'; SB I.
Flowers Greenish, dioecious, wind-pollinated; >100 in a whorled, branched inflorescence.
Germinule 0.74 mg, 3.2 × 2.3 mm, nut.
Germination Epigeal; t_{50} 2 days; 6–24°C; L > D.
Biological flora None.
Geographical distribution The British Isles (100% of British vice-counties), *ABIF* page 192; most of Europe although rare in the S (77% of European territories and 3% naturalized); temperate Asia and N America.

Habitats Common in meadows and pastures, on lead-mine spoil, wasteland and waysides and in unshaded mire. Also recorded from cinders, quarry spoil, outcrops, and a range of other grassy and open habitats. Infrequent in woodland, only casual in arable land, and absent from aquatic habitats. [Found on maritime shingle and mountain ledges (Lousley & Kent 1981).]
Gregariousness Intermediate.
Altitude Recorded at all altitudes to 470 m. [Up to 1240 m.]
Slope Absent from cliffs and from the steepest slopes.
Aspect Slightly more abundant on unshaded N-facing slopes.
Hydrology Strongly associated with mire on sloping ground, a reflection of a high level of occurrence in soligenous mire in the millstone grit uplands.

Soil pH Shows relatively wide tolerance but most frequently associated with mildly acidic soils in the pH range 5.0–7.0.
Bare soil Found over the full range of values.
△ **ordination** Wide-ranging distribution, including types of vegetation from all except the most unproductive and disturbed conditions. However, it is clearly centred on vegetation in which competition is limited by moderate impacts of disturbance and reduced productivity.
Associated floristic diversity Typically intermediate.
Synopsis *R.a.* is found in a wide variety of habitats, but is particularly characteristic of meadows and pastures. The rosette of ascending leaves and the tall, flowering stem, coupled with the capacity for early seed set, facilitate the exploitation of hay-meadows. Under grazing *R.a.* persists as a small, low-growing rosette and sets seed less regularly. This plasticity of

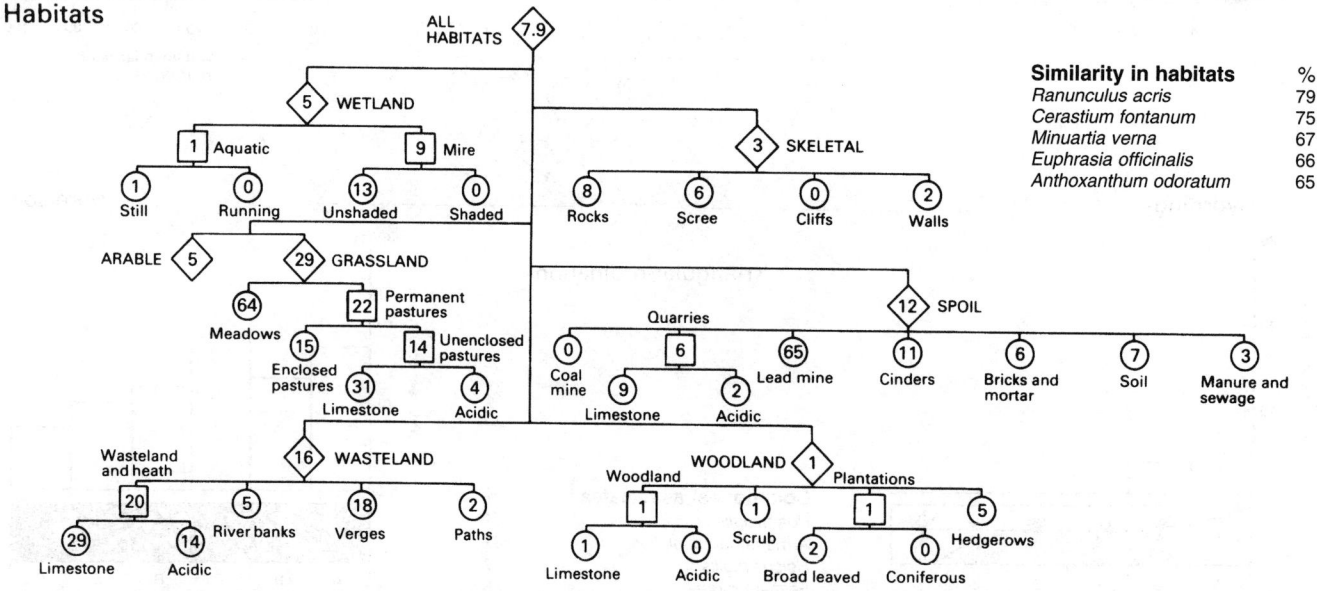

Habitats

Similarity in habitats	%
Ranunculus acris	79
Cerastium fontanum	75
Minuartia verna	67
Euphrasia officinalis	66
Anthoxanthum odoratum	65

response can operate within a single season, apparently enabling *R.a.* to exploit both the mown and the grazed phases of the 'hay and aftermath' management regime applied to many upland meadows. Much of the additional morphological variation within the species is also phenotypic in origin (Lousley & Kent 1981). *R.a.* extends occasionally onto calcareous soils, and is locally abundant on lead-mine spoil, where heavy-metal-tolerant races occur (Shaw 1984). In Europe races differing in shade tolerance (Bjorkman & Holmgren 1966) are recorded. The species most consistently achieves abundance on mildly acidic brown earths. The roots are not subject to heavy colonization by VA mycorrhizas, and in turf microcosms no consistent benefit from inoculation occurred in circumstances where other species, e.g. *Scabiosa columbaria*, were strongly promoted (Grime, Mackey, Hillier & Read 1987). The leaves contain oxalates and are potentially toxic to stock if eaten in large quantities (Cooper & Johnson 1984). Regeneration is mainly by means of seed, which is set abundantly despite the fact that *R.a.* is dioecious. No persistent seed bank is formed. Seeds can survive ingestion by cows (Ridley 1930).

However, *R.a.* is probably mainly dispersed by human activity as an impurity within hay and seed crops. *R.a.* exhibits a limited capacity to produce daughter ramets, and this may be of local significance in grazed and trampled habitats (Putwain & Harper 1970). Populations normally have about twice as many female as male ramets, despite the fact that the sex ratio of seed is c. 1:1 (Putwain & Harper 1972). Female inflorescences are formed later, but on taller stems, and leaves of female ramets are located at a greater height and are maintained later into the summer (Putwain & Harper 1972). On the basis of these results, Putwain and Harper suggest that male and female plants occupy different niches. A number of related species occur in Europe (*FE1*) but, in contrast with subgenus *Rumex* (including *R. crispus* and *R. obtusifolius*), hybridization is not of common occurrence (Stace 1975).
Current trends Since *R.a.* is most characteristic of mildly acidic soils, is dioecious, has little capacity for effective vegetative spread, and has no persistent seed bank, we predict that it will decline rapidly in areas converted to intensive agricultural management.

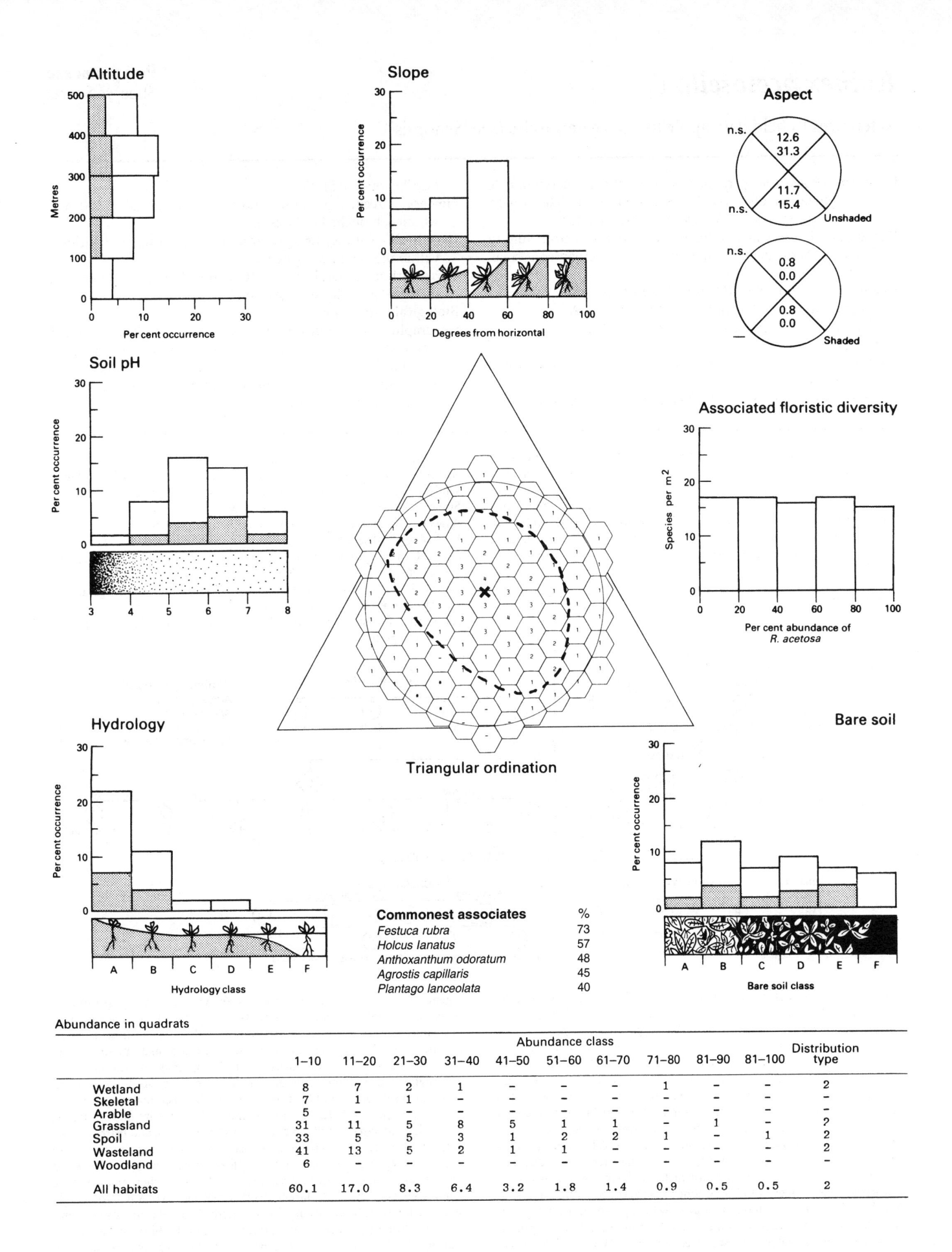

Altitude

Slope

Aspect

	Unshaded
n.s.	12.6 / 31.3
n.s.	11.7 / 15.4

	Shaded
n.s.	0.8 / 0.0
—	0.8 / 0.0

Soil pH

Associated floristic diversity

Per cent abundance of *R. acetosa*

Hydrology

Hydrology class

Triangular ordination

Commonest associates %
Festuca rubra 73
Holcus lanatus 57
Anthoxanthum odoratum 48
Agrostis capillaris 45
Plantago lanceolata 40

Bare soil

Bare soil class

Abundance in quadrats

	1–10	11–20	21–30	31–40	41–50	51–60	61–70	71–80	81–90	81–100	Distribution type
Wetland	8	7	2	1	–	–	–	1	–	–	2
Skeletal	7	1	1	–	–	–	–	–	–	–	–
Arable	5	–	–	–	–	–	–	–	–	–	–
Grassland	31	11	5	8	5	1	1	–	1	–	2
Spoil	33	5	5	3	1	2	2	1	–	1	2
Wasteland	41	13	5	2	1	1	–	–	–	–	2
Woodland	6	–	–	–	–	–	–	–	–	–	–
All habitats	60.1	17.0	8.3	6.4	3.2	1.8	1.4	0.9	0.5	0.5	2

Rumex acetosella L.

Polygonaceae
Sheep's Sorrel

References to field data apply to aggregate species (see 'Synopsis')

Life-form Polycarpic perennial, semi-rosette hemicryptophyte or geophyte. Shoots erect; leaves lanceolate, hastate, usually <500 mm^2; root system deep, without mycorrhizas.
Phenology Flowers in May to July and sets seed from July to October. Shoots die back in autumn. May overwinter as small rosettes.
Height Foliage mainly < 100 mm; flowers to 300 mm.
Seedling RGR 1.5-1.9 week^{-1}. SLA 21.5. DMC 11.9.
Nuclear DNA amount 3.4 pg (Bennett *et al.* 1982); 2n = 42* (*BC*); 6x*.
Established strategy Intermediate between C-S-R and stress-tolerant ruderal.
Regenerative strategies V, B$_s$. Regenerative biology is complex (see 'Synopsis'); SB IV.
Flowers Green, dioecious, wind-pollinated; usually >100 in a whorled, branched inflorescence.
Germinule 0.4 mg, 1.6 × 1.0 mm, nut often dispersed enclosed within the perianth segments.
Germination Epigeal; t_{50} 2 days; 16-36°C; germination rate increased after dry storage at 20°C; L = D.
Biological flora None.
Geographical distribution The British Isles (100% of British vice-counties), *ABIF* page 191; Europe southwards from Scandinavia (100% of European territories); temperate Asia, N and S Africa, temperate America and Australia, but possibly a recently introduced species in the S hemisphere.

Habitats A species of dry, well-drained habitats, largely restricted to non-calcareous strata. Frequent in both grazed and ungrazed grassland or heathland, on coal-mine and sandstone quarry waste, and on other types of spoil. Found also on outcrops and, more sparingly, in other open habitats, river banks, waysides and woodland. Only a casual of arable land, but observed as a weed of tree nurseries. Also observed frequently in sand and gravel pits, and on cinders beside railways. Absent from wetlands.

Habitats

Gregariousness Often patch-forming.
Altitude Suitable habitats are most frequent on the lowland Bunter sandstone, but *R.a.* has been observed to 410 m. [Up to 1050 m.]
Slope Occurs over a wide range in slopes.
Aspect Significantly more frequent on S-facing slopes.
Hydrology Absent from wetlands.
Soil pH Widely distributed, but frequency and abundance greatest on acidic soils in the pH range 3.5-5.5.
Bare soil Most frequent and abundant at low to intermediate values.
△ **ordination** Widely distributed.
Associated floristic diversity Relatively low, particularly where *R.a.* occurs at high frequency values.

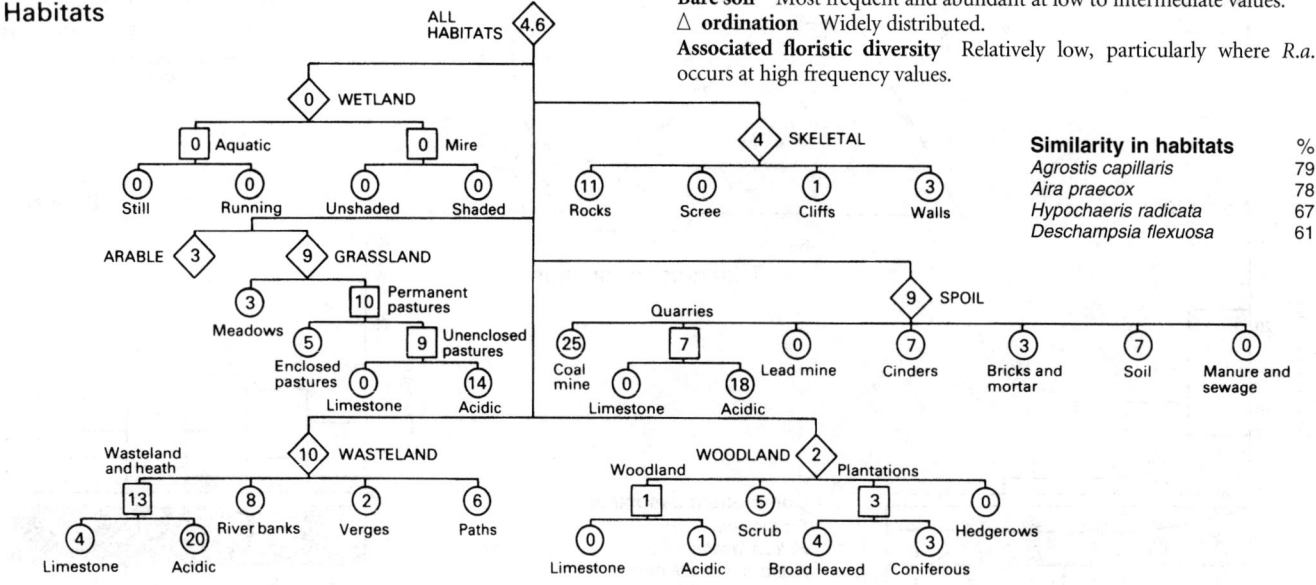

Similarity in habitats	%
Agrostis capillaris	79
Aira praecox	78
Hypochaeris radicata	67
Deschampsia flexuosa	61

Synopsis *R.a.* is a patch-forming herb characteristic of relatively infertile sandy or peaty soils. Because of its low growth habit, *R.a.* is readily dominated by taller herbs and grasses. Although capable of persistence in a non-flowering condition in open scrub and other shaded habitats, *R.a.* tends to be restricted to areas where, as a result of mechanical disturbance, fire or a rocky or sandy terrain, the vegetation is very open. *R.a.* also persists in poor pasture over acidic strata. The relatively deep root system affords access to subsoil water during summer and enables *R.a.* to coexist on dry sandy soils with winter annuals such as *Aira praecox*. The foliage is little grazed (Milton 1933), and the plant contains oxalates which are toxic to stock (Cooper & Johnson 1984). *R.a.* is characteristically calcifuge; seedlings grow very poorly and are highly chlorotic when grown on calcareous soils (Hodgson 1972). Among the species exploiting acidic soils, *R.a.*, like *Holcus mollis*, exhibits an unusually high potential relative growth rate. Perhaps for this reason both *R.a.* and *H.m.* are sometimes important weeds of ornamental or forestry tree nurseries (*WCH*). The species regenerates vegetatively, forming extensive patches by producing adventitious buds on horizontal roots. The subsequent breakdown or mechanical severing of these roots results in the formation of daughter colonies. *R.a.* also regenerates by seeds, which germinate in spring and form a persistent bank of buried seed (Chippindale & Milton 1934). Although dioecious, most populations appear to set seed and, though varying considerably, tend to show a 1:1 ratio between males and females (Putwain & Harper 1972). Zimmermann and Lechowitz (1982) suggest that males and females differ in response to water-stress, and Harris (1968) obtained evidence suggesting that males allocate more resources to vegetative regeneration and are therefore more persistent in closed vegetation. Male inflorescences are produced earlier than female inflorescences (Putwain & Harper 1972). The dispersal of *R.a.* is frequently aided by humans. Seeds occur as a commercially important contaminant of horticultural peat. Seeds can survive ingestion by birds, cattle, horses and pigs (Ridley 1930), and in some experiments germination has been

Altitude

Slope

Aspect

P < 0.01

3.9	50.0
10.3	56.5

n.s. Unshaded

0.0	—
—	0.0

Shaded

Soil pH

Associated floristic diversity

Hydrology

Triangular ordination

Bare soil

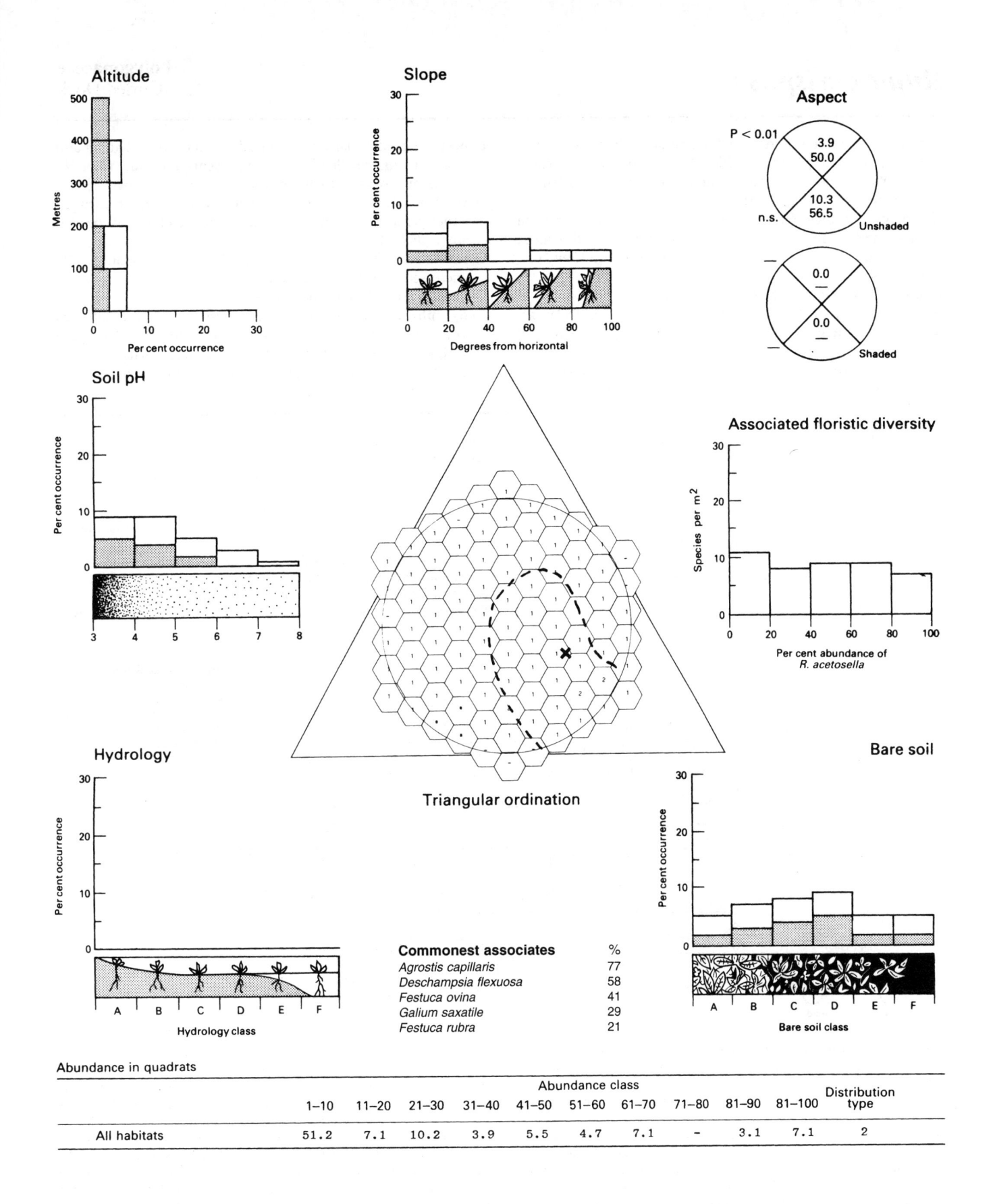

Commonest associates	%
Agrostis capillaris	77
Deschampsia flexuosa	58
Festuca ovina	41
Galium saxatile	29
Festuca rubra	21

Abundance in quadrats

	1–10	11–20	21–30	31–40	41–50	51–60	61–70	71–80	81–90	81–100	Distribution type
All habitats	51.2	7.1	10.2	3.9	5.5	4.7	7.1	–	3.1	7.1	2

found to be stimulated by nitrate (Steinbauer & Grigsby 1958). Most, and possibly all, of our field records refer to ssp. *acetosella*. *R.a.* consists of a polyploid series (ssp. *pyrenaicus* (Pourr.) Akeroid 2*x*, ssp. *a.* var. *tenuifolius* Wallr., 4*x*, and ssp *a.*, L., 6*x*) (Lousley & Kent 1981). In N England, forms resembling var. *t.* are found on some of the most acidic sandy soils of the Bunter sandstone, but many intermediate with typical ssp. *a.* occur and the two are not readily separable.

Current trends Population expansion can occur rapidly on disturbed acidic soils, and the species seems likely to remain locally common in upland and lowland Britain.

Rumex crispus L.

Polygonaceae
Curled Dock

Life-form Polycarpic perennial (or annual to biennial), semi-rosette hemicryptophyte (*Biol Fl*). Shoots erect; leaves lanceolate, <2000 mm^2; roots with a stout and often relatively deep tap-root, non-mycorrhizal.

Phenology Overwinters as a rosette of small leaves. Flowers from May onwards and sheds seed from late summer until winter. Large plants occasionally produce two crops of seed (*Biol Fl*).

Height Basal leaves to 300 mm; flowering stems to 1 m.

Seedling RGR No data. SLA 26.5. DMC 10.5.

Nuclear DNA amount 8.8 pg, $2n = 60$* (Bennett *et al.* 1982); $6x$*.

Established strategy Intermediate between competitive-ruderal and C-S-R.

Regenerative strategies B_s, (V). Regenerates mainly by seed (see 'Synopsis'); SB IV.

Flowers Green, usually hermaphrodite, wind-pollinated but also visited by insects, floral structure prevents autogamy (*Biol Fl*); numerous (often *c.* 1000), in whorls in a branched panicle.

Germinule 1.33 mg, 2.8 × 1.5 mm, nut enclosed within enlarged inner perianth segments; maritime forms tend to have heavier nuts (*Biol Fl*).

Germination Epigeal; t_{50} 3 days; 15-23°C; germination rate increased after dry storage; L > D.

Biological flora Cavers and Harper (1964a).

Geographical distribution The British Isles, although less common in the N (100% of British vice-counties), *ABIF* page 194; Europe (90% of European territories and 5% naturalized); most of Africa and now naturalized in most other parts of the world in association with human activities.

Habitats Particularly frequent on cinder tips, demolition sites, soil heaps and other types of spoil habitat. Also found on waysides, river banks, at the margin of meadows and pastures, and on wasteland. A casual of arable habitats and of seasonally flooded situations and here not usually reaching maturity. Not recorded from woodland or skeletal habitats. [A maritime variant is found on shingle ridges, often at the tidal drift-line, and on sand dunes; another is found on estuarine mud (*Biol Fl*).]

Gregariousness Mainly occurring as scattered individuals.

Altitude Distribution shows a lowland bias but extends locally to 400 m. [Up to 610 m (*Biol Fl*).]

Slope Absent from the steepest slopes.

Aspect Associated with the full range of aspects.

Hydrology Essentially a species of moist soils. Most records from wetlands refer to seedlings and juveniles, although mature plants are sometimes found in moist ground adjoining mire, or on exposed mud at the upper margins of ponds and reservoirs.

Soil pH Infrequent on acidic soils, and largely restricted to sites of pH >5.5.

Bare soil Most frequent in habitats with relatively high exposures.

△ ordination Largely confined to fertile sites subject to disturbance.

Associated floristic diversity Typically intermediate.

Synopsis *R.c.* is a short-lived perennial or, more rarely, annual herb, with

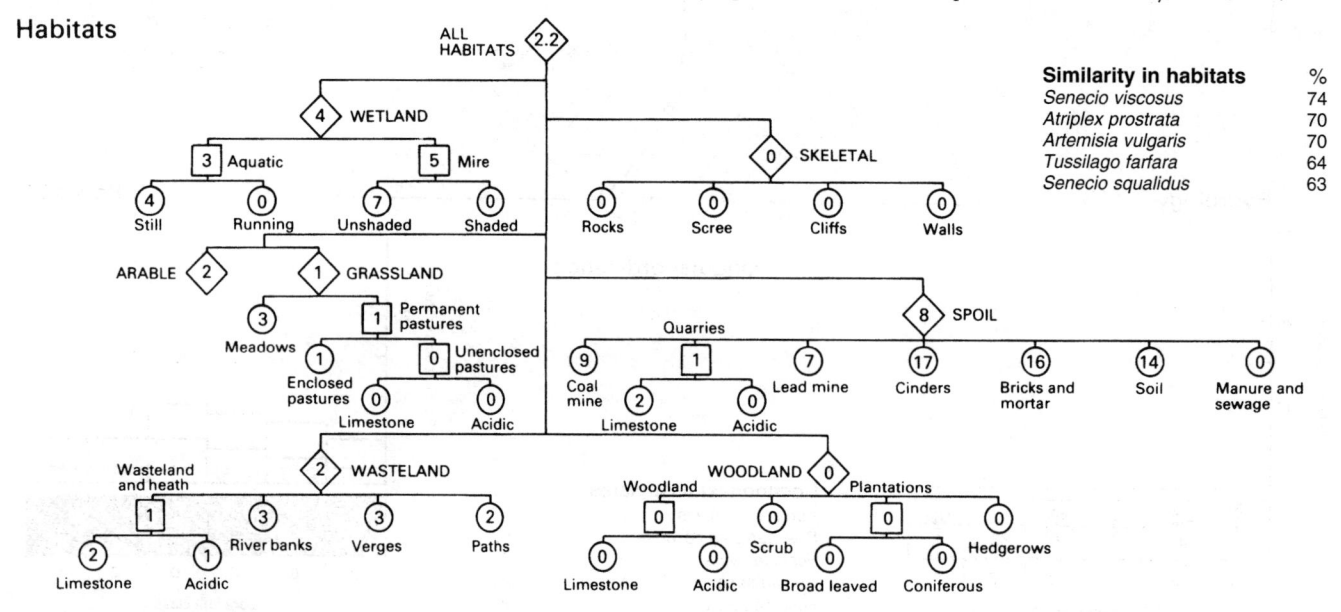

Habitats

Similarity in habitats	%
Senecio viscosus	74
Atriplex prostrata	70
Artemisia vulgaris	70
Tussilago farfara	64
Senecio squalidus	63

a tap-root and one to several clumped stems. *R.c.* is capable of flowering in its first year (*Biol Fl*), and is most frequently associated with sites that are both disturbed and fertile. As well as being one of the world's most widely distributed angiosperms (*Biol Fl*), *R.c.* is also considered to be one of the world's worst weeds and is recorded as important in 37 countries in 16 arable crops and in pastures (Holm *et al.* 1977). *R.c.* is toxic to stock, and in Britain its control is required under the 1957 Weeds Act (*Biol Fl*). In common with many species exploiting agricultural land and other artificial habitats, *R.c.* displays high phenotypic plasticity and forms well-defined ecotypes. Seeds, which are borne in numbers ranging from less than 100 to over 40 000 per plant, are polymorphic with respect both to size and to germination characteristics; seed characters may differ according to ecotype (*Biol Fl*). The complex germination of *R.c.* has been investigated by, among others, Cavers and Harper (1966), Le Deunff (1974), Baskin and Baskin (1978) and Vincent and Cavers, (1978). The probability of flowering is correlated with rosette size during the previous autumn (Weaver & Cavers 1980). *R.c.* is capable of regeneration from root fragments, but effective reproduction appears to be almost exclusively by seed, though establishment of seedlings is often rather slow (*Biol Fl*). A persistent seed bank is formed and seeds retained viability in soil for 80 years in one experiment (Darlington & Steinbauer 1961). Fruits survive ingestion by some birds (*Biol Fl*) and *R.c.* may be dispersed in this way. The most noteworthy ecotype is ssp. *littoreus* (J. Hardy) Akeroyd, which tolerates immersion in sea water and is often abundant on maritime shingle. This variety lacks the polymorphism with respect to seed dormancy of typical ruderal populations and may not form a persistent seed bank (Cavers & Harper 1966, 1967a,b). In addition this ecotype has three tubercles on its perianth segments (most inland races and typical British *R. obtusifolius* have one), and this characteristic, which is shared by many European docks that grow near water, may aid buoyancy and enable

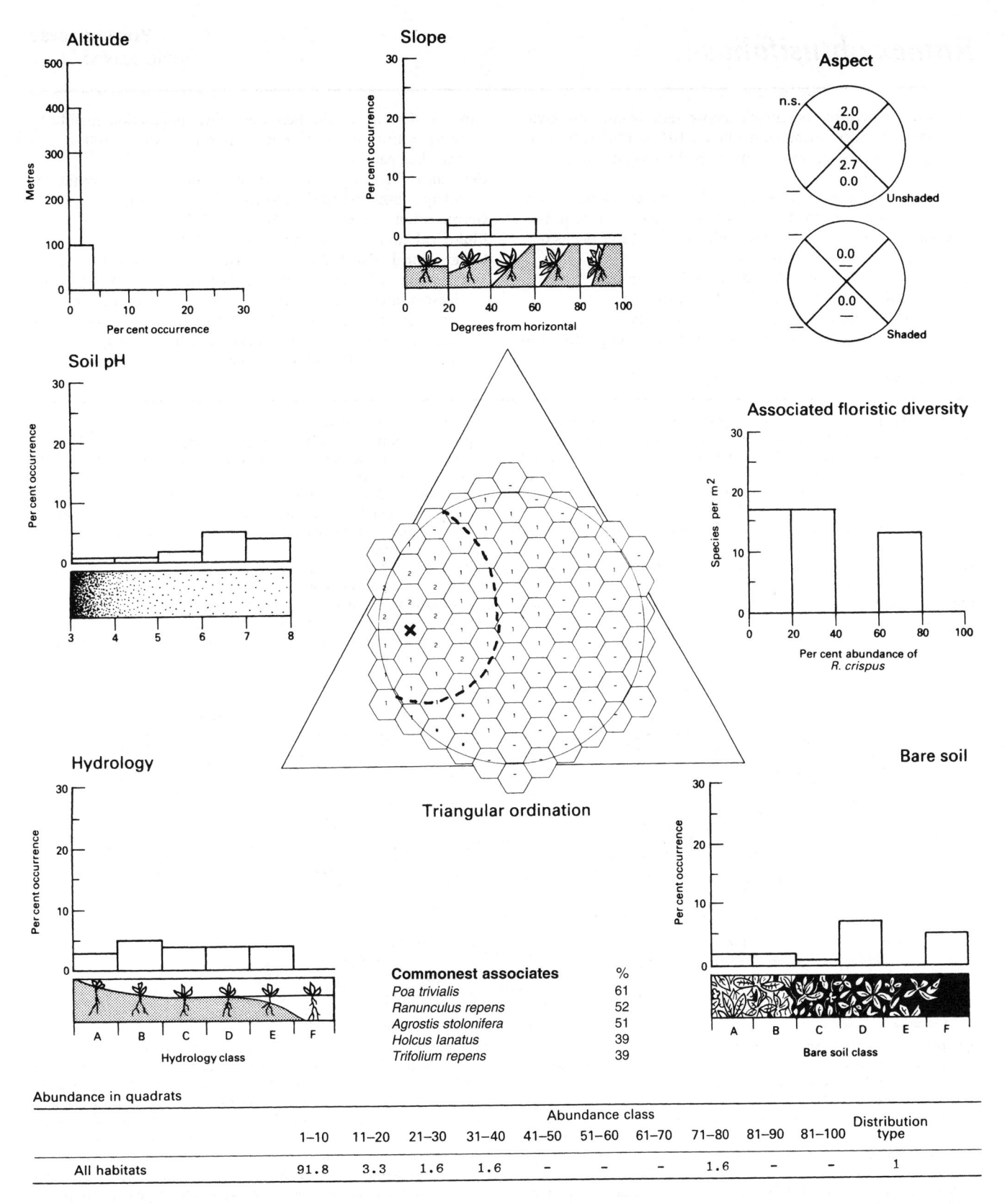

Altitude
Slope
Aspect
Soil pH
Associated floristic diversity
Hydrology
Triangular ordination
Bare soil

Commonest associates	%
Poa trivialis	61
Ranunculus repens	52
Agrostis stolonifera	51
Holcus lanatus	39
Trifolium repens	39

Abundance in quadrats

	1–10	11–20	21–30	31–40	41–50	51–60	61–70	71–80	81–90	81–100	Distribution type
					Abundance class						
All habitats	91.8	3.3	1.6	1.6	–	–	–	1.6	–	–	1

the dispersule to float in sea water for many months (*Biol Fl*). The fruits of other forms of *R.c.* also float but for a lesser period (*Biol Fl*). In a cultivation experiment involving both European and N American populations, Hume and Cavers (1982) found that a majority of variation was associated with phenotypic plasticity, and suggested that populations of *R.c.* were mainly in-breeding. Indeed, in cultivation populations from N America and the British Isles allocate a similar proportion of their resources to reproductive parts, but some differences occur, e.g. in seed size (Hume & Cavers 1983), and populations from ruderal sites, maritime habitats and tidal mud of rivers appear genetically distinct in a number of ecologically important respects (Akeroyd & Briggs 1983a,b). Ten hybrids involving *R.c.* are recorded for Britain (Stace 1975).

Current trends *R.c.* is an effective colonist of many artificial habitats and may even be increasing as a result of the continuing high levels of disturbance. Since it is now uncommon on arable land, *R.c.* may be expected to show an increasing tendency towards perenniality.

Rumex obtusifolius L.

Polygonaceae
Broad-leaved Dock

Life form Polycarpic perennial, semi-rosette hemicryptophyte. Shoots erect; leaves ovate, sometimes >10 000 mm^2; with stout, frequently deep, tap-root, sometimes with slight infection by VA mycorrhizas.

Phenology With a small overwintering rosette. Flowers from June onwards and sheds seed from late summer until winter. Some seed may remain in dead inflorescence. Large plants may produce two crops of seed (*Biol Fl*).

Height Foliage mainly <300 mm; flowering stems to 1 m.

Seedling RGR 1.0–1.4 week^{-1}. SLA 27.5 DMC 16.9.

Nuclear DNA amount 3.0 pg; $2n = 40*$ (*BC*); $4x*$.

Established strategy Intermediate between competitor and C-S-R.

Regenerative strategies B_s. See 'Synopsis'; SB IV.

Flowers Green, usually hermaphrodite, pollination mainly by wind, self-fertile (*Biol Fl*); numerous (often >1000) in a branched panicle.

Germinule 1.1 mg, 2.0 × 1.5 mm, nut enclosed within the enlarged inner perianth segments (5.7 × 3.2 mm).

Germination Epigeal; t_{50} 3 days; 15–28°C; L > D.

Biological flora Cavers and Harper (1964b).

Geographical distribution For aggregate species: the British Isles, although less common in N Scotland (100% of British vice-counties), *ABIF* page 198; most of Europe (90% of European territories and 5% naturalized); now introduced to many other parts of the world including Africa, N and S America, Australia and New Zealand.

Habitats Frequent on spoil heaps, particularly those consisting of mineral soil, brick rubble, cinders, manure and sewage waste. Equally widespread on river banks and waysides. Also found in meadows and pastures, and in drier areas within mire. Not persistent in the majority of arable, skeletal and wetland habitats. Observed as an infrequent and impermanent component of disturbed woodland.

Gregariousness Scattered individuals. All frequency values of more than 20 refer to colonies of seedlings or very young plants.

Altitude Most frequent and abundant in lowland areas, but extends to 400 m. [Up to 565 m.]

Slope Found over a wide range of values, but distribution centred on flat and gently sloping ground.

Aspect No clear-cut trends with respect to aspect. The slight N-facing bias recorded here has not been confirmed by subsequent sampling of the same geographical area.

Hydrology Although found associated with topogenous mire, particularly at pond margins heavily trampled by stock, not typically a wetland species, and many records refer to seedlings which are only of casual occurrence.

Soil pH Virtually restricted to soils of pH >4.5.

Bare soil Widely distributed.

△ **ordination** Occurs most frequently in disturbed, productive vegetation types. Absent from undisturbed and infertile habitats.

Associated floristic diversity Typically found in sites of intermediate species-richness.

Habitats

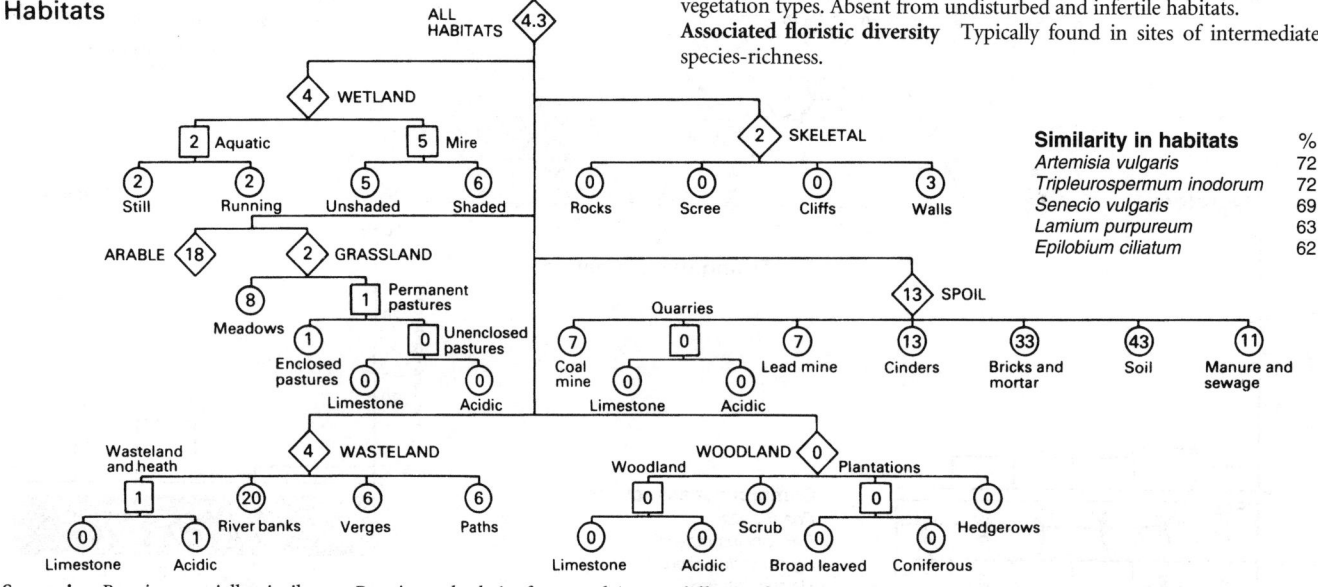

Similarity in habitats	%
Artemisia vulgaris	72
Tripleurospermum inodorum	72
Senecio vulgaris	69
Lamium purpureum	63
Epilobium ciliatum	62

Synopsis *R.o.* is essentially similar to *R. crispus*, both in form and in ecology and the two frequently hybridize. *R.o.* is rather less ruderal, typically more robust, exclusively perennial and usually does not flower until its second year (*Biol Fl*). *R.o.* also exhibits a greater capacity for vegetation dominance (Cideciyan & Malloch 1982). Although classified as one of the commonest weeds by Holm *et al.* (1977) it is clearly much less important than *R.c.* worldwide, and has the narrower geographical range. Also, *R.o.* is probably less wide-ranging ecologically. In Britain the species is designated under the 1959 Weeds Act, and is probably now more troublesome than *R.c.* as a weed of pastures, particularly around field edges and in trampled gateways. It is also more common than *R.c.* in the infrequently cut grass swards of roadsides. *R.o.* is generally refused by cattle, sheep and horses. However, seed number and weight are affected by the grazing of the beetle *Gastrophysa viridula* De Geer. and *R.c.* is even more susceptible (Bentley *et al.* 1980). Established plants of *R.o.* survive mowing (*Biol Fl*) and show a greater tendency than those of *R.c.* to produce two crops of seed. For both *R.c.* and *R.o.*, stems cut down 14 days after anthesis produce seed with a germination percentage not significantly different from that of controls (Weaver & Cavers 1980). Seeds of *R.o.* mature less rapidly than those of the more ruderal *R.c.* (Weaver & Cavers 1980). In addition to the mechanisms for dispersal already described, the spiny teeth on the perianth enclosing the fruit facilitate dispersal in animal fur and in clothing. Seed germinates mainly in spring and forms a persistent bank of buried seeds (*Biol Fl*). *R.o.* often forms compact patches, but not of sufficient size to permit effective vegetative spread. However, pieces of underground stem and, in spring, root fragments readily regenerate after ploughing (*Biol Fl*). Seeds are polymorphic with respect to germination behaviour (Cavers & Harper 1966). *R.o.* frequently grows close to *Urtica dioica* and juice from the leaves is customarily applied to the skin to relieve nettle stings. Variants recorded from Europe include var. *microcarpus* Dierb. and var. *transiens* (Simonk.) Kubat. These are naturalized in S England. Within the British Isles hybrids are recorded with 11 other species.

Current trends Well established in a wide range of artificial habitats, and perhaps increasing.

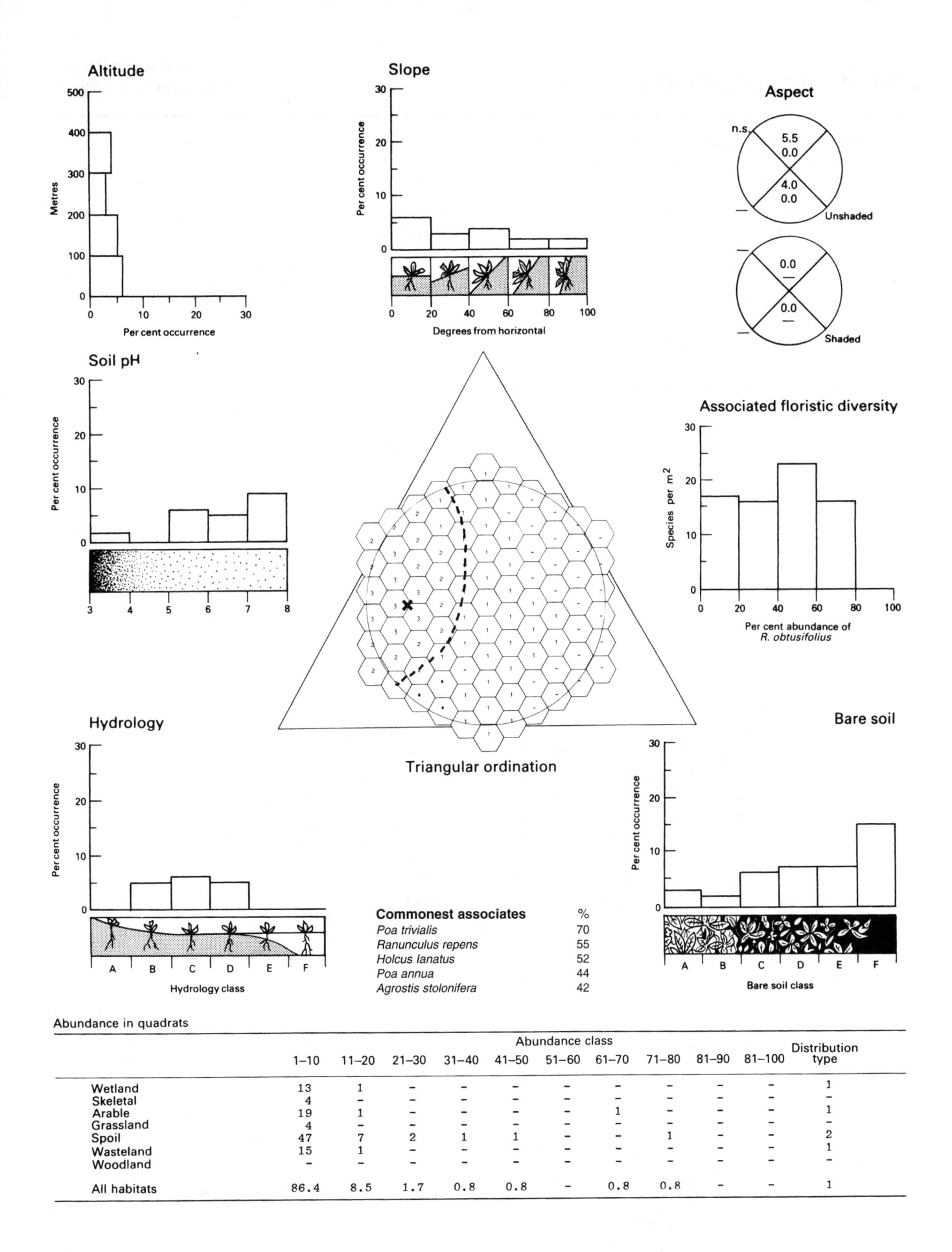

Altitude

Slope

Aspect

Soil pH

Associated floristic diversity

Species per m²

Per cent abundance of *R. obtusifolius*

Hydrology

Triangular ordination

Commonest associates	%
Poa trivialis	70
Ranunculus repens	55
Holcus lanatus	52
Poa annua	44
Agrostis stolonifera	42

Bare soil

Abundance in quadrats

	1–10	11–20	21–30	31–40	41–50	51–60	61–70	71–80	81–90	81–100	Distribution type
Wetland	13	1	–	–	–	–	–	–	–	–	1
Skeletal	4	–	–	–	–	–	–	–	–	–	–
Arable	19	1	–	–	–	–	1	–	–	–	1
Grassland	4	–	–	–	–	–	–	–	–	–	–
Spoil	47	7	2	1	1	–	–	1	–	–	2
Wasteland	15	1	–	–	–	–	–	–	–	–	1
Woodland	–	–	–	–	–	–	–	–	–	–	–
All habitats	86.4	8.5	1.7	0.8	0.8	–	0.8	0.8	–	–	1

Sagina procumbens L.

Caryophyllaceae
Procumbent Pearlwort

Life-form Polycarpic perennial, semi-rosette hemicryptophyte or more rarely summer- or perhaps even winter-annual therophyte. Stems mostly prostrate; leaves linear, <10 mm^2; roots shallow, non-mycorrhizal.
Phenology Winter green. Spring shoot growth followed by flowering from May to September and seed set from June to October. Leaves short-lived.
Height Foliage <10 mm; flowering stems to 20 mm.
Seedling RGR No data. SLA 19.2. DMC 21.7.
Nuclear DNA amount No data; $2n = 22^*$ (*BC*); $2x^*$.
Established strategy Intermediate between ruderal and C-S-R.
Regenerative strategies B$_s$, (V). Regenerative biology is complex (see 'Synopsis'); SB IV.

Flowers Green, hermaphrodite, homogamous, often self-pollinated; often solitary.
Germinule 0.02 mg, 0.5 × 0.25 mm, seed in a 20-135-seeded dehiscent capsule (Salisbury 1964).
Germination Epigeal; t_{50} 3 days; 11-39°C; L > D.
Biological flora None.
Geographical distribution The British Isles (100% of British vice-counties), *ABIF* page 167; Europe (97% of European territories); Asia, Greenland, N America.

Habitats Occurs at high frequencies on spoil heaps of cinders, building rubble, soil, mining waste, etc., and in waste places such as paths, verges, banks of rivers and streams. Found at lower frequency in grassland and in arable fields. Also present in open habitats such as walls, rocks and cliffs, and on marshy ground. Commonly observed on lawns. Absent from woodland and aquatic habitats.
Gregariousness Intermediate.
Altitude Widely distributed up to 430 m. [Up to 1150 m.]
Slope Occurs over the full range in slope.
Aspect No consistent trends.
Hydrology Found occasionally in soligenous mire and on exposed mud beside reservoirs, but absent from submerged habitats.
Soil pH On all except acidic soils of pH <4.5.
Bare soil Found in habitats with widely different amounts of bare soil.
△ **ordination** Wide-ranging, but distribution centred on vegetation associated with moderately fertile habitats subject to relatively high levels of disturbance.
Associated floristic diversity Typically intermediate.
Synopsis *S.p.* is a herb of appressed growth habit and moss-like appearance, found in a wide range of moist, disturbed, moderately fertile habitats. As a consequence of the low growth form, *S.p.* is susceptible to competition from the majority of other pasture species, and is particularly

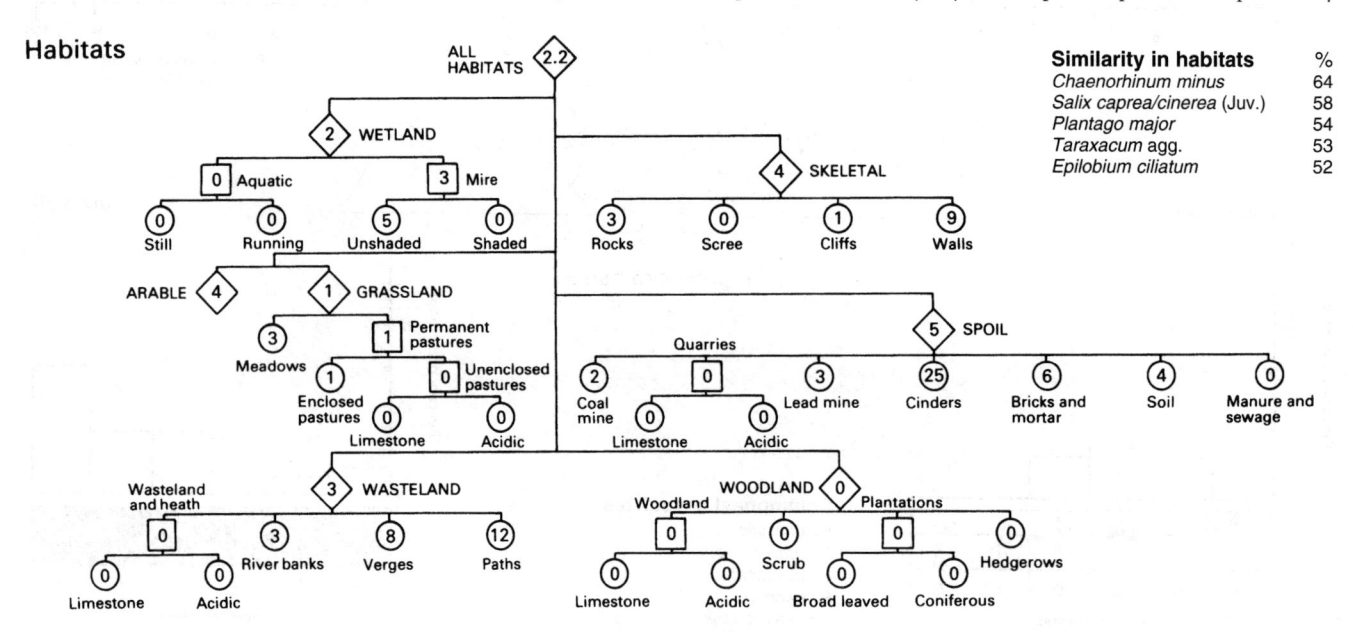

Habitats

Similarity in habitats	%
Chaenorhinum minus	64
Salix caprea/cinerea (Juv.)	58
Plantago major	54
Taraxacum agg.	53
Epilobium ciliatum	52

abundant in lawns, where the growth of other species is debilitated by the effects of close and frequent mowing. *S.p.* is equally frequent in sites with disturbed soil, such as spoil heaps with high exposure of bare ground. In pasture *S.p.* is little grazed by stock (Milton 1933). *S.p.* is tolerant of trampling and colonizes cracks between paving stones (Segal 1969). *S.p.* is also a common garden weed, often occurring in lightly shaded situations. Regeneration by a combination of seedling establishment and vegetative means occasionally results in the formation of local monocultures of up to 0.25 m^2. Vegetative regeneration involves the rooting of prostrate stems, which we suspect may form new plants if detached. As with many other ruderal species, the season over which flowering and seed set may occur is long. Seeds are numerous, minute and highly mobile, and may germinate in autumn or in spring. A large bank of buried seed may be formed (Odum 1978, UCPE unpubl.). The longevity of established plants of *S.p.* varies. In highly disturbed sites *S.p.* may be short-lived and behaves as a summer annual on reservoir margins. However, the species appears to be relatively long-lived in some grassland sites, and Austin (1980) observed plants which appeared to persist for *c.* 8 years in a lawn. He has also noted a change, for no apparent reason, from a long- to a short-life duration within a lawn population of *S.p.* Plants adopt a weakly ascending habit in light shade. We suspect that, like several other lawn weeds (Warwick & Briggs 1979), *S.p.* is both phenotypically plastic and genotypically variable; the species is very variable in Europe (*FE1*). Although mainly self-pollinated, hybrids between *S.p.* and other *Sagina* spp. have been reported (Stace 1975), and *S.p.* tends to be replaced by other species of *Sagina* in dry, maritime, heathland and montane environments.
Current trends The seed of *S.p.* is both persistent in the soil and readily dispersed by human activities. *S.p.* is a common and probably increasing colonist of artificial habitats.

Altitude
Slope
Aspect

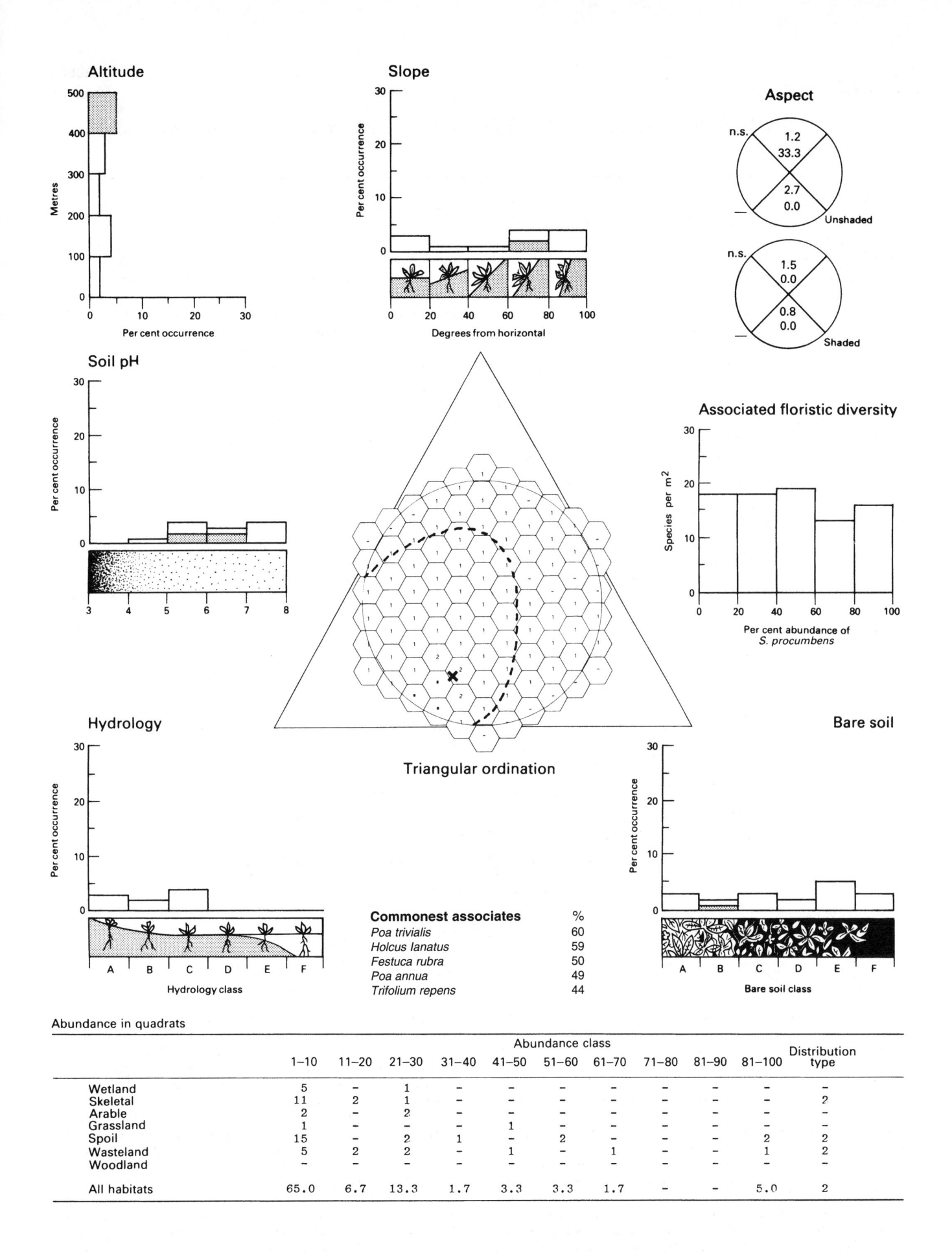

	n.s.	1.2	
		33.3	
	—	2.7	Unshaded
		0.0	

	n.s.	1.5	
		0.0	
	—	0.8	Shaded
		0.0	

Soil pH

Associated floristic diversity

Hydrology

Triangular ordination

Commonest associates	%
Poa trivialis	60
Holcus lanatus	59
Festuca rubra	50
Poa annua	49
Trifolium repens	44

Bare soil

Abundance in quadrats

	Abundance class										Distribution type
	1–10	11–20	21–30	31–40	41–50	51–60	61–70	71–80	81–90	81–100	
Wetland	5	–	1	–	–	–	–	–	–	–	–
Skeletal	11	2	1	–	–	–	–	–	–	–	2
Arable	2	–	2	–	–	–	–	–	–	–	–
Grassland	1	–	–	–	1	–	–	–	–	–	–
Spoil	15	–	2	1	–	2	–	–	–	2	2
Wasteland	5	2	2	–	1	–	1	–	–	1	2
Woodland	–	–	–	–	–	–	–	–	–	–	–
All habitats	65.0	6.7	13.3	1.7	3.3	3.3	1.7	–	–	5.0	2

Salix spp.

Salicaceae
Sallow

Includes *S. caprea* L., *S. cinerea* L. and related hybrids. Field data analysed in this account refer to seedlings and small saplings

Life-form Deciduous, much-branched shrub or more rarely small tree. Leaves ovate, tomentose beneath, typically <1000 mm^2; roots with ecto- or more rarely VA mycorrhizas.
Phenology Flowers March to April, before the leaves. Seed shed from May to June before the leaves are fully expanded. Leaves shed in autumn.
Height Foliage and flowers to 10 m.
Seedling RGR *S.cin.*: 1.0-1.4 week^{-1}. *S.cap.*: SLA 12.4. DMC 31.1.
Nuclear DNA amount *S.cin.*: 0.9 pg; $2n = 76$ (*FE1*); $4x$. *S.cap.*: $2n = 38^*, 76^*$ (*BC*); $2x^*, 4x^*$. Base number of family polyploid in origin.
Established strategy Intermediate between competitor and stress-tolerant competitor.
Regenerative strategies W, S, (V). Seed germinates immediately after shedding; seedling establishment restricted to vegetation gaps. Twigs detached as a result of disturbance readily re-root in moist soils to form new plants; SB I.
Flowers Usually dioecious, mainly insect-pollinated, but some wind-pollination possible (Meikle 1984) and visited by birds (Kay & Stevens 1986); >100 in a dense yellowish male or greenish female catkin.
Germinule 0.09 mg, 1.1 × 0.7 mm, seed short-lived (6 months) and enveloped by hairs *c*. 5 mm long; wind-dispersed from a dehiscent, many-seeded capsule.
Germination Epigeal; t_{50} 1 day.
Biological flora None.
Geographical distribution The British Isles (*S.cap.* 99% of British vice-counties), *ABIF* pages 240-242; most of Europe (*S.cap.* 77% of European territories; *S.cin.* 77% and 3% naturalized).

Habitats Seedlings and saplings are most frequent on various types of spoil tips, particularly cinder heaps, railway ballast, limestone quarry debris and demolition sites. Willows are also widely observed in gravel pits. Also common in mire and on walls, and occurs as an occasional colonist of river banks. Only rarely observed at woodland margins. Absent from pasture, arable and aquatic habitats.

Gregariousness Usually sparse to intermediate, but occasionally forming extensive carpets of seedlings under *Salix* bushes.
Altitude Suitable habitats are most frequent and abundant in the lowland region. [Up to 850 m (Meikle 1984).]
Slope Juveniles found over a wide range of slopes.
Aspect Seedlings and saplings significantly more frequent and abundant on unshaded S-facing slopes.
Hydrology Recorded from topogenous and soligenous mire, but not from aquatic habitats.

Habitats

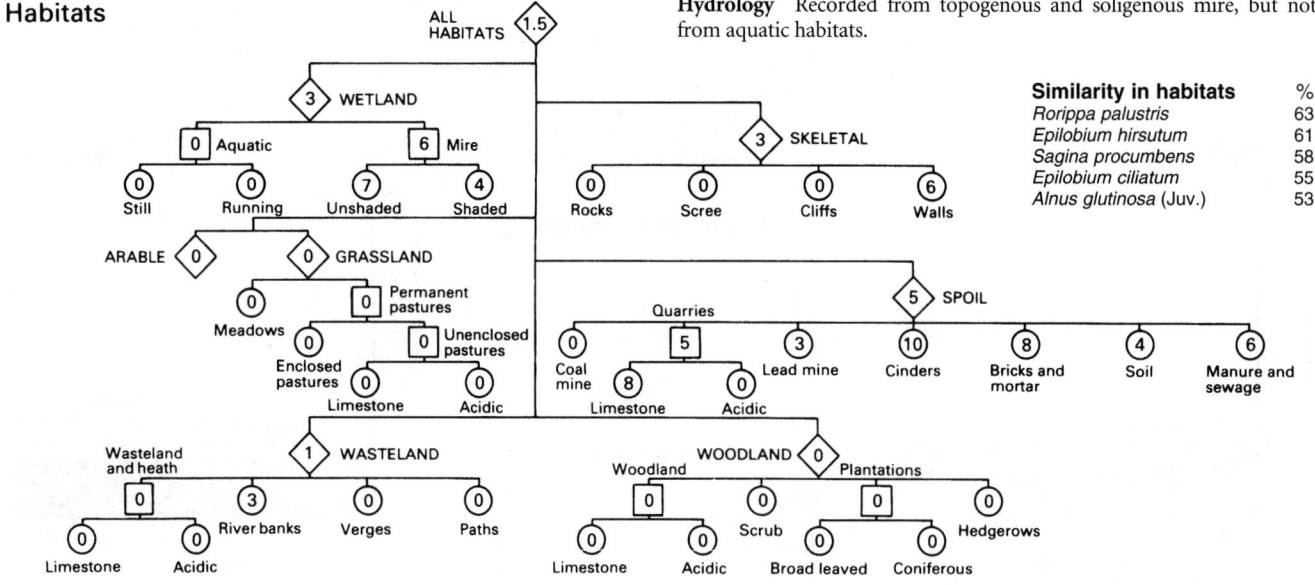

Similarity in habitats	%
Rorippa palustris	63
Epilobium hirsutum	61
Sagina procumbens	58
Epilobium ciliatum	55
Alnus glutinosa (Juv.)	53

Soil pH Juveniles widely distributed, although recorded more frequently on soils of pH >6.0.
Bare soil A widespread distribution, with highest frequencies associated with intermediate levels of exposed soil.
△ **ordination** Juveniles widely distributed, centred on productive vegetation subject to some disturbance. Absent only from vegetation types subject to either intense stress or disturbance.
Associated floristic diversity Typically intermediate, but very variable.
Synopsis *S.* spp. is a deciduous shrub or small tree with diffuse-porous wood and early canopy emergence. As a seedling or young sapling *S.* spp. occurs on bared ground in a wide range of moist but little-grazed habitats. This broad distribution in disturbed habitats stems from the capacity of *S.* spp., despite being dioecious, to produce an abundance of minute, wind-dispersed seeds. These are short-lived (germinating exceedingly rapidly over a wide range of temperatures in sunlight and in darkness) and do not form a persistent seed bank. In *S.cap.* male and female bushes tend to occur in equal numbers; in *S.cin.* there is a bias towards females (Kay & Stevens 1986). *S.* spp. are tolerant of cutting, and the capacity of detached twigs to re-root and sprout, forming new plants, also may be of importance. It is not known whether, like *S. fragilis* L. (Zimmerman *et al.* 1985), *S.* spp. have primordial adventitious roots in their twigs. The ecological limits of the mature plants are much narrower than those of juveniles. Thus, as canopy components, *S.* spp. are typically associated with moist soils and are more frequent on N-facing slopes. *S.* spp. tends to occur on relatively fertile soils, and produces a dense canopy. Litter is generally only shortly persistent. Seedlings have a high growth rate relative to other woody species (Grime & Hunt 1975), and establishment from seed appears restricted to unshaded sites. Plants are also capable of growing from a coppiced stump at the rate of 50 mm day^{-1} and may attain a height of 3-4 m in a single season (Rackham. 1980). The mature plants occur mainly as bushes or as open

Altitude
Slope
Aspect
P < 0.01

0.8 / 0.0 / 4.0 / 11.1 — Unshaded

0.0 / — / 0.0 / — — Shaded

Soil pH

Associated floristic diversity

Hydrology

Triangular ordination

Bare soil

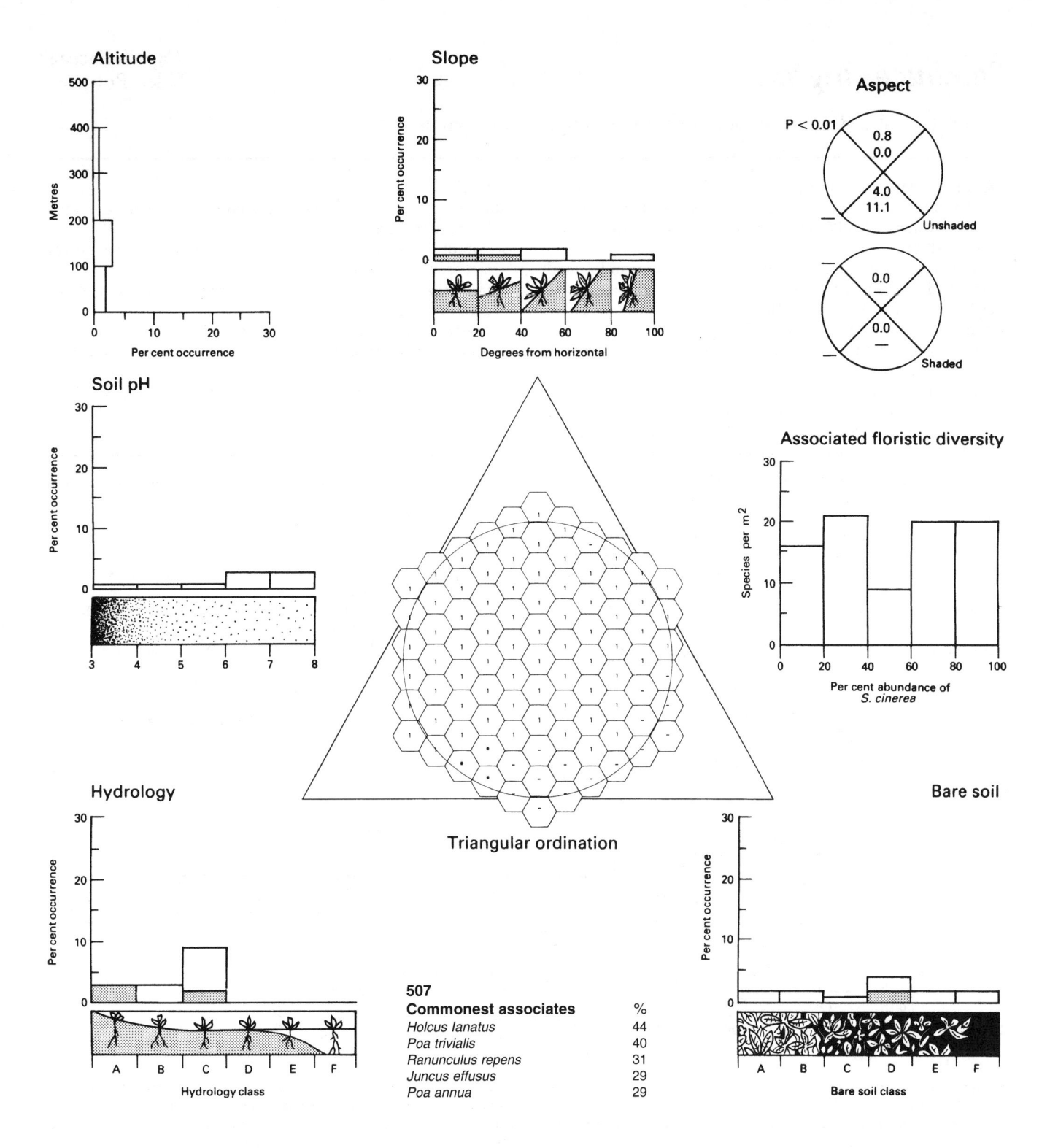

507
Commonest associates	%
Holcus lanatus	44
Poa trivialis	40
Ranunculus repens	31
Juncus effusus	29
Poa annua	29

scrub on wasteland, spoil or in mire, and they are much more frequent as canopy species in scrub than in woodlands. Other studies (Tansley 1953) illustrate that the invasion by *S.* spp. may represent the first stage in the development of secondary woodland or of fen carr, and the distribution of *S.* spp. outlined above is consistent with the premise that *S.* spp. is a seral species of developing woodland or carr. However, as is the case for *Crataegus monogyna*, the plant may also form long-persistent thickets, particularly in upland areas with well-established scrub. Leaves of *S.cin.* are relatively unpalatable to the snail *Helix aspersa* Muller (Wratten et al. 1981) and willows in general harbour a large number of insects (Southwood 1961). *S.cin.* ssp. *oleifolia* Macreight (*S. atrocinerea* Brot.) is probably the most common willow in N England, and is found on stream-sides, marshes, woodland edges and hedgerows (Meikle 1984). Outside Britain the taxon appears restricted to France, Spain and Portugal (Meikle 1984). *S.cap.* is also widespread and shows some tendency to exploit drier, more-calcareous sites, including demolition sites (Meikle 1984). Hybrids between the two taxa are apparently common, and both segregates form ecotypes. Thus, *S.cin.* ssp. *cinerea* is found in base-rich mire, particularly in East Anglia, and *S.cap.* var. *sphacelata* (Sm.) Wahlenb. is a shrub of the Scottish Highlands, extending to 850 m (Meikle 1984).

Current trends The two segregates of *S.* spp. are effective colonists of bared ground, and are probably increasing. The high level of disturbance in the contemporary landscape is presumably also tending further to erode the ecological barriers between *S.cap.* and *S.cin.*, promoting additional hybridization between the two taxa.

Sambucus nigra L.

Caprifoliaceae
Elder, Bourtree

Field data analysed in this account refer to seedlings or small saplings

Life-form Deciduous, much-branched shrub or, more rarely, small tree with deeply furrowed bark. Leaves pinnate, slightly hairy, often >10 000 mm^2; roots with or without VA mycorrhizas.
Phenology Buds break early in spring. Flowers June to July and sets seed from August to October. Leaves shed in autumn.
Height Foliage and flowers to 10 m.
Seedling RGR No data. SLA 18.3. DMC 24.2.
Nuclear DNA amount 3.1 pg; 2n = 36 (*FE4*); 2x, (4x for family).
Established strategy Competitor.
Regenerative strategies S. Reproduces by seed, which germinates in spring. There is no information concerning the presence or absence of a buried seed bank. Cuttings root freely; SB II.
Flowers White, hermaphrodite, usually insect-pollinated; several hundred in a umbel-like cyme.
Germinule 3.4 mg, 3.3 × 2.0 mm, seed enclosed within an up to 5-seeded black drupe.
Germination Epigeal; evidence of a chilling requirement for germination.
Biological flora Atkinson and Atkinson (2002).
Geographical distribution The British Isles (100% of British vice-counties), *ABIF* page 597; Europe except the extreme N and W (77% of European territories and 8% naturalized); W Asia, N Africa.

Habitats A frequent component of hedges. Seedlings and young saplings are frequent in unshaded habitats such as demolition sites, river and stream banks, and manure and sewage spoil. They are also frequent in scrub and a variety of other wooded habitats. Occasionally recorded as seedlings in skeletal habitats, shaded mire and arable land. Absent from aquatic and grassland habitats. [A colonist of shingle beaches and sand dunes (Tansley 1953).]
Gregariousness Sparse

Altitude Juveniles predominantly lowland, extending to 300 m. [Up to 440 m.]
Slope Widely distributed.
Aspect Juveniles slightly more frequent on both shaded and unshaded S facing slopes.
Hydrology Not a wetland species, but seedlings occur occasionally in soligenous and topogenous mire.
Soil pH Wide-ranging, but nearly all records are from above pH 4.0.
Bare soil Juveniles particularly frequent at high exposures.
△ **ordination** Distribution of juveniles very wide-ranging, but centred on fertile habitats subject to slight disturbance.
Associated floristic diversity Typically low to intermediate.

Similarity in habitats	%
Rubus fruticosus agg.	59
Urtica dioica	58
Digitalis purpurea	57
Galium aparine	57
Anisantha sterilis	57

Habitats

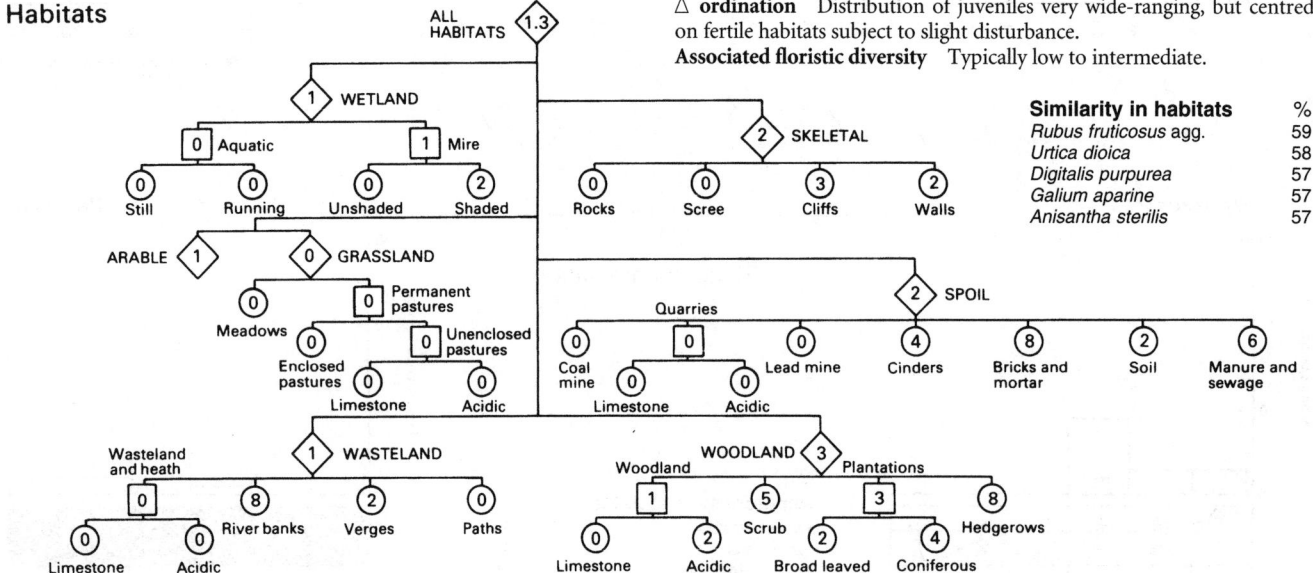

Synopsis *S.n.* is a short-lived shrub, or small tree to 10 m, with a rough bark. The wood of *S.n.* is diffuse-porous, and bud-break is early. Indeed, leaves may be formed during warmer spells during winter, and are often subsequently killed by severe frosts. *S.n.* has a Sub-Atlantic distribution (Birks 1973). In unshaded habitats the species often produces an abundant crop of berries which are avidly removed by birds as soon as they become ripe. Consequently, seeds are widely distributed and seedlings are recorded from a greater diversity of habitats than the adult plant. The occurrence of seedlings and the establishment of saplings appear to be consistently associated with bared ground. Seedlings appear to be sensitive to competition from polycarpic perennial herbs and to grazing by stock. Once established *S.n.* is seldom grazed; the species contains cyanogenic glycosides (Cooper & Johnson 1984), and tissue has an unpleasant odour when crushed. The leaves are toxic to mammals, and stands of *S.n.* are characteristic of rabbit warrens. Edwards and Wratten (1985) also suggest that *S.n.* possesses inducible defences against foliar predation by insects. Litter is only shortly persistent. The species is characteristic of well-drained, base-rich sites of high fertility, and is often found with *Urtica dioica* and *Galium aparine* in farmyards and in other sites close to human habitation. Long-established colonies may occur on rabbit warrens and other fertile derelict land without any evidence of replacement by taller species. Elsewhere succession tends to lead to the formation of woodland (e.g. Watt 1934). *S.n.* is characteristic of more-disturbed and fertile woodlands. Thus, although not one of the most frequent woodland species, *S.n.* is recorded as the third most common volunteer species both in broad-leaved and in coniferous plantations in the Sheffield region. Like *Crataegus monogyna*, the shrub may persist in shaded sites, but few flowers and fruits are formed. Thus, the regeneration of the species is heavily dependent upon plants growing in woodland margins and clearings. No seed bank has been reported. The species is very resilient when coppiced, and cuttings appear to root readily. *S.n.* is a common plant of recent, species-poor hedgerows (Pollard *et al.* 1975).

Current trends A common and increasing colonist of fertile derelict land.

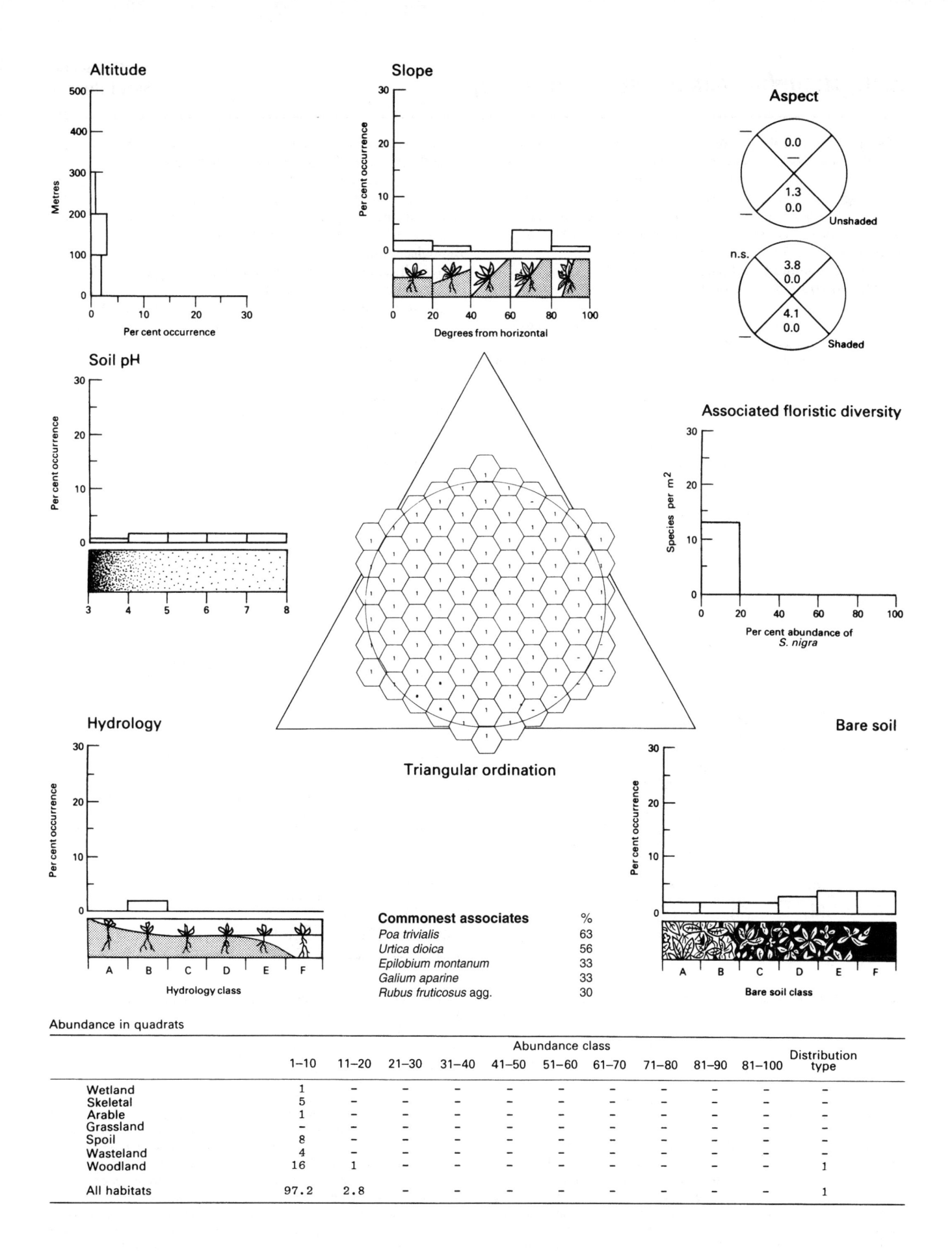

Altitude
Slope
Aspect
Soil pH
Associated floristic diversity
Hydrology
Triangular ordination
Bare soil

Commonest associates	%
Poa trivialis	63
Urtica dioica	56
Epilobium montanum	33
Galium aparine	33
Rubus fruticosus agg.	30

Abundance in quadrats

	\multicolumn{10}{c}{Abundance class}	Distribution type
	1–10	11–20	21–30	31–40	41–50	51–60	61–70	71–80	81–90	81–100	
Wetland	1	–	–	–	–	–	–	–	–	–	–
Skeletal	5	–	–	–	–	–	–	–	–	–	–
Arable	1	–	–	–	–	–	–	–	–	–	–
Grassland	–	–	–	–	–	–	–	–	–	–	–
Spoil	8	–	–	–	–	–	–	–	–	–	–
Wasteland	4	–	–	–	–	–	–	–	–	–	–
Woodland	16	1	–	–	–	–	–	–	–	–	1
All habitats	97.2	2.8	–	–	–	–	–	–	–	–	1

Sanguisorba minor ssp. minor Scop.

Rosaceae
Salad Burnet

Life-form Polycarpic perennial, semi-rosette hemicryptophyte with woody stock. Stem erect; leaves mainly basal, pinnate, <2000 mm^2; deeply tap-rooted and with VA mycorrhizas.
Phenology Winter green. Flowers in June and July and sets seed in July to August. Large summer leaves die back in autumn.
Height Foliage mainly basal, <200 mm; flowers to 400 mm.
Seedling RGR 0.5-0.9 week^{-1}. SLA 21.9. DMC 33.1.
Nuclear DNA amount 1.1 pg; $2n = 28^*$ (*BC*); $2x^*$ ($4x$ for family).
Established strategy Intermediate between stress-tolerator and C-S-R.
Regenerative strategies ?B$_s$. Regenerative biology requires further study (see 'Synopsis').
Flowers Greenish, upper often female, middle and lower hermaphrodite, protandrous (Nordborg 1967), wind-pollinated; usually <30 in a dense capitulum.
Germinule 1.8 mg, 3.1 × 1.6 mm, achene, 1-2 (rarely 3) enclosed in a dry hard fruit (4.11 mg, 3.2 × 2.1 mm).
Germination Epigeal; t_{50} 6 days; 8-30°C; germination rate increased after dry storage at 20°C; L > D.
Biological flora None.
Geographical distribution Mainly in England, particularly the S, becoming rarer to the N and W. Scarce in Scotland and Ireland (64% of British vice-counties), *ABIF* page 337; S, W and C Europe extending northwards to Sweden, and only a casual in NE Europe (77% of European territories and 8% naturalized); N America.

Habitats Almost exclusively restricted to limestone substrata. Occurs predominantly in pastures and wasteland, but also recorded from scree, lead-mine spoil, ungrazed grassland, rock outcrops, cinders and cliffs. Observed more rarely on waysides and in open scrub, where it is perhaps a relict of former grassland. Absent from arable and aquatic habitats.
Gregariousness Intermediate.
Altitude Extending to 380 m. Most-frequently associated with altitudes between 200 and 300 m on carboniferous limestone dalesides. [Up to 500 m.]
Slope Most frequent on carboniferous limestone dalesides on slopes in the range 20-60°.
Aspect Frequent on both N- and S-facing slopes.
Hydrology Absent from wetland habitats.
Soil pH Abundant in the pH range 5.0-8.0, but extending down to pH 4.0 on surface-leached limestone soils.
Bare soil Associated mainly with intermediate amounts of exposed soil.
△ ordination Confined to vegetation associated with relatively undisturbed, unproductive situations.
Associated floristic diversity High, particularly at higher frequencies of *S.m.*

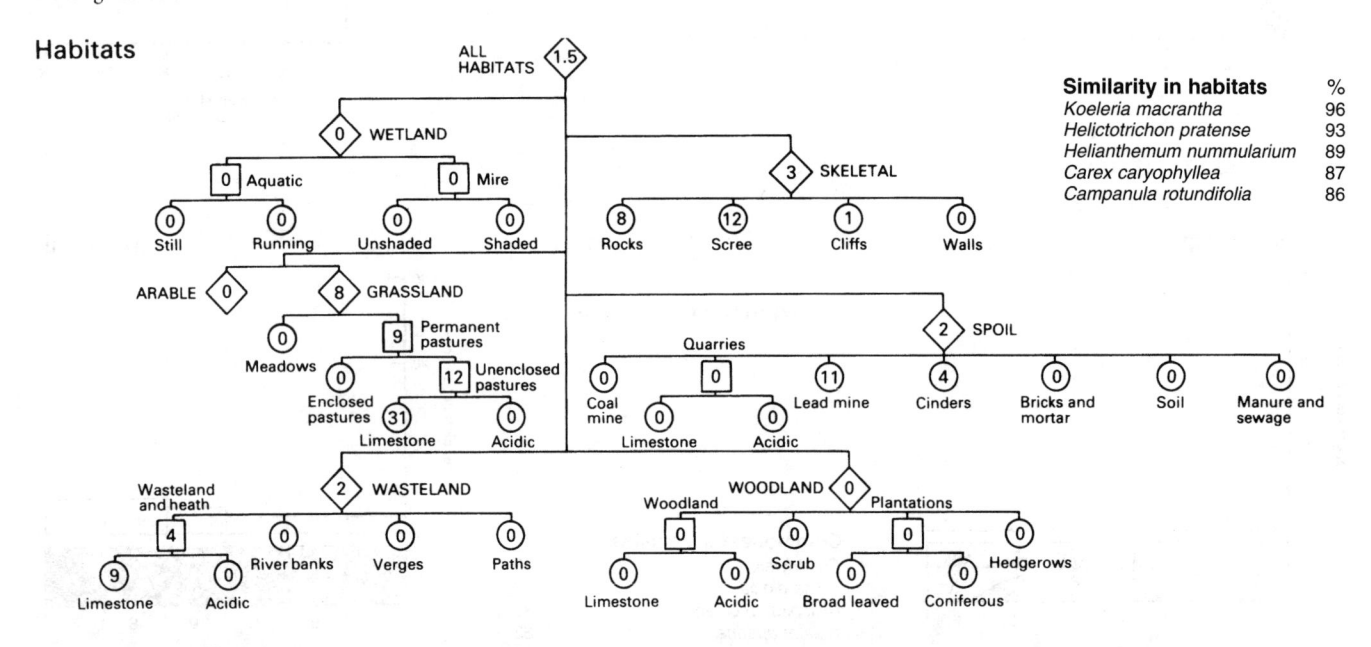

Habitats

Similarity in habitats	%
Koeleria macrantha	96
Helictotrichon pratense	93
Helianthemum nummularium	89
Carex caryophyllea	87
Campanula rotundifolia	86

Synopsis *S.m. is* a tap-rooted, rosette-forming herb particularly characteristic of species-rich pastures on both dry and moist, relatively infertile, calcareous soils. In the PGE *S.m.* only occurs on unfertilized, unproductive plots. Calcium and magnesium contents of the leaves are unusually high (SIEF). *S.m.* maintains a small rosette of overwintering leaves, and expands the shoot biomass slowly in the spring, reaching a peak in summer. This phenological pattern, which contrasts with species such as *Festuca ovina*, with which *S.m.* often grows, may relate to the capacity of the long tap-root (Anderson 1927) to provide access to moisture in deep soil crevices during the period of summer growth. In turf microcosms roots of *S.m.* became heavily colonized by VA mycorrhizas and infection caused up to 5-fold increases in seedling yield (Grime, Mackey, Hillier & Read 1987). Where calcareous soil accumulates a surface layer of acidic humus, larger, older, individuals of *S. minor* with roots mainly exploiting the calcareous subsoil may be observed in situations where seedling establishment is no longer possible (Grime 1963b). *S.m.* survives grazing both as a seedling by gastropods (King 1977c) and as a mature plant by sheep and rabbits. Leaves were formerly used in salads. Because of the low canopy and lack of lateral vegetative spread, *S.m.* is a poor competitor. Thus, although *S.m.* is occasionally represented by large individuals in abandoned pastures, the species is not persistent in tall turf. *S.m.* appears to be susceptible to burial and is infrequent on ant hills (King 1977c). Shoots regenerate from buds situated on the stock, and it is not unusual to find small multiple rosettes arising. In view of the extended longevity of *S.m.* it seems likely that occasional vegetative offspring and recruitments from seed may be sufficient to sustain population levels (cf. *Primula veris*). Seeds germinate in spring (Muller 1978, Nordborg 1967) or autumn (King 1977c). Seed set is relatively low, and often as few as 4 fruits per capitulum contain viable seed (King 1977c). It is not known whether a persistent seed bank is formed. Fruits are large. Thus, seedlings have some capacity to etiolate in response to shading, and at the second true leaf stage the radicle may attain 60 mm in length (King 1977c). Nevertheless, fatalities are high

Altitude

Slope

Aspect

Soil pH

Associated floristic diversity

Triangular ordination

Hydrology

Absent from wetland

Bare soil

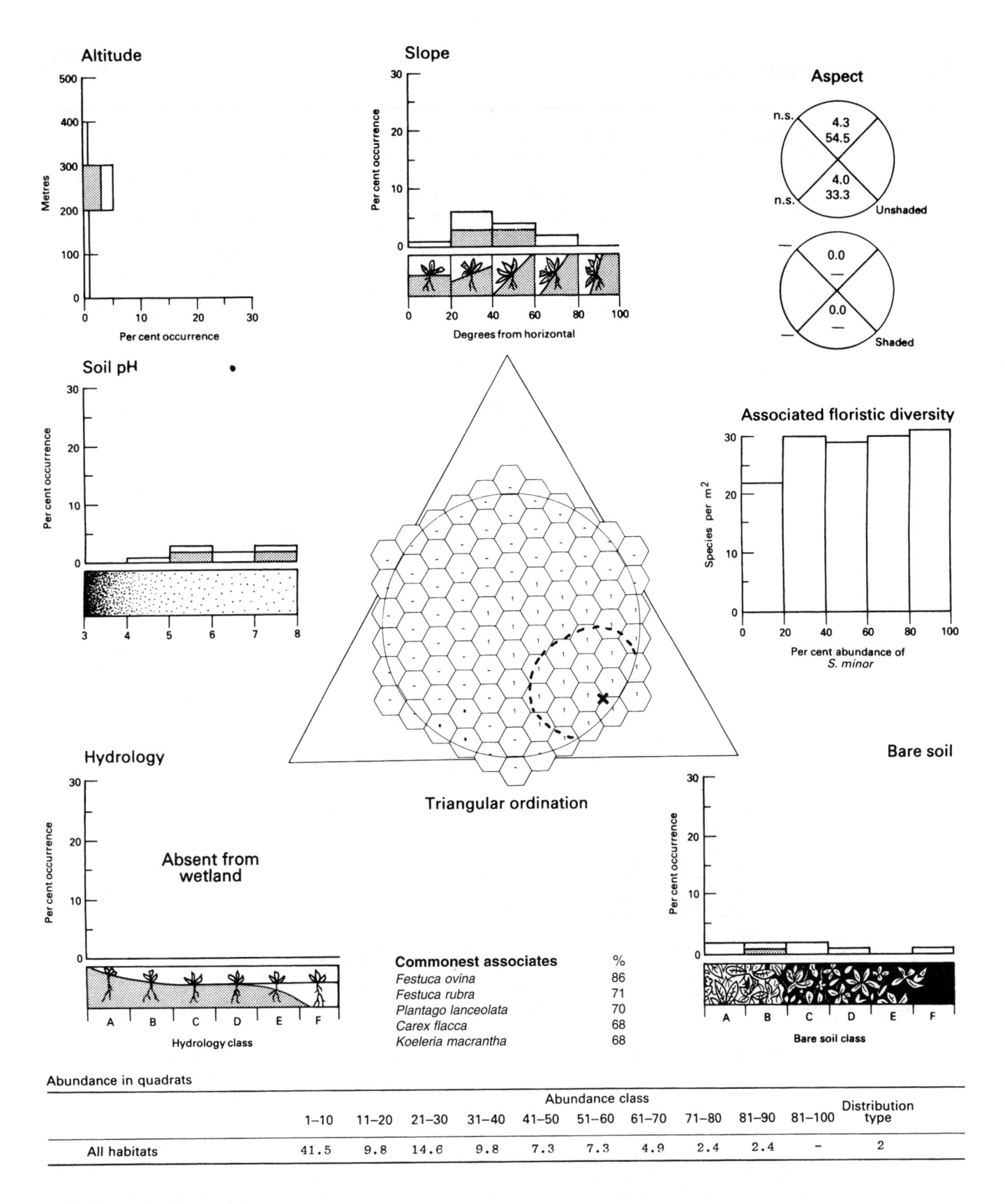

Commonest associates	%
Festuca ovina	86
Festuca rubra	71
Plantago lanceolata	70
Carex flacca	68
Koeleria macrantha	68

Abundance in quadrats

	1–10	11–20	21–30	31–40	41–50	51–60	61–70	71–80	81–90	81–100	Distribution type
All habitats	41.5	9.8	14.6	9.8	7.3	7.3	4.9	2.4	2.4	–	2

during seedling establishment (King 1977c). Although seeds may survive ingestion by rabbits (King 1977c), *S.m.* has an extremely limited capacity to disperse to new sites. However, the species may be found occasionally on railway banks on non-calcareous strata. The extent to which *S.m.* forms ecotypes is uncertain, but the difference between low-growing and taller populations has some genetic basis (Nordborg 1967), and five other ssp. have been recognized in Europe (*FE2*). One variant formerly grown for fodder and a native of S Europe, ssp. *muricata* Briq., is established locally in Britain and, in particular, may be observed in relatively tall calcareous grassland on railway banks. Ssp. *m.* ($2n = 28, 56$) is taller and has larger fruits (dispersule 13.0 mg) than ssp. *minor*.

Current trends Mainly restricted to species-rich, calcareous pastures, and decreasing, particularly in lowland areas.

Sanicula europaea L.

Apiaceae
Sanicle

Life-form Polycarpic perennial, rosette hemicryptophyte with stout stock. Stems erect; leaves rounded, palmately lobed, sometimes >2000 mm^2; roots with VA mycorrhizas.
Phenology Winter green with some leaf replacement in spring. Flowers mainly from May to July and seed set August to September.
Height Foliage usually <300 mm; flowers to 600 mm.
Seedling RGR <0.5 week^{-1}. SLA 34.0. DMC 19.8.
Nuclear DNA amount No data; $2n = 16$ (*FE2*); $2x$.
Established strategy Intermediate between stress-tolerator and C-S-R.
Regenerative strategies S, V. Regenerates by seed which germinates in spring (Roberts 1986a). There is no evidence of a persistent seed bank. Has some capacity for vegetative spread (Inghe & Tamm 1985); SB II.
Flowers White, insect-pollinated (Knuth 1906); in subglobose umbels of shortly stalked male flowers and 3-6 sessile hermaphrodite flowers (Tutin 1980).
Germinule 2.97 mg, 4.0 × 2.8 mm, mericarp with hooked bristles (2 mericarps per ovary).
Germination Epigeal; a chilling requirement is suspected.
Biological flora None, but see Inghe and Tamm (1985).
Geographical distribution The British Isles, but absent from much of Scotland (98% of British vice-counties), *ABIF* page 455; much of Europe except the N and E margins, but only on mountains in the Mediterranean region (79% of European territories); Asia and Africa.

Habitats Restricted to woodland habitats and to shaded waysides.
Gregariousness Sparse to intermediate.
Altitude Extends up to 300 m. [Up to 500 m.]
Slope Not recorded from skeletal habitats or from steep slopes.
Aspect Distribution shows a N-facing bias.
Hydrology Characteristic of moist soils, but absent from wetland habitats.
Soil pH Most frequent on calcareous soils and in the pH range 6.0-7.5. Not recorded below pH 4.5.
Bare soil Wide-ranging.

△ **ordination** Insufficient data, but some evidence that the species is associated with undisturbed and relatively infertile conditions.
Associated floristic diversity Typically low to intermediate.
Synopsis *S.e.* is a slow-growing, winter-green, perennial herb, characteristic of sites where the growth of potential dominants is restricted by heavy shade and probably also by nutrient stress. The species is long-lived, with a half-life of between 59 and 360 years (Inghe & Tamm 1985). Thus, *S.e.* may live for as long as the trees above it (Inghe & Tamm 1985). The species is very shade-tolerant, a fact illustrated by the N-facing bias of its woodland sites and its ability to persist under the canopy of *Taxus*

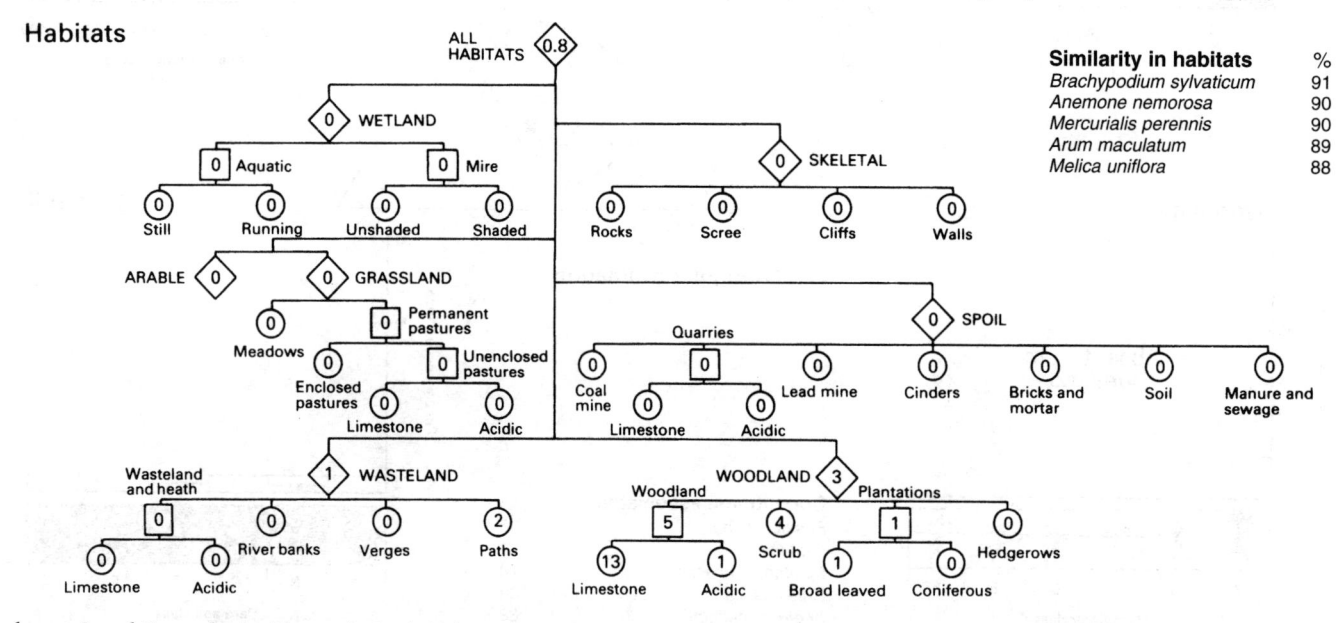

Habitats

Similarity in habitats	%
Brachypodium sylvaticum	91
Anemone nemorosa	90
Mercurialis perennis	90
Arum maculatum	89
Melica uniflora	88

baccata L. and *Fagus sylvatica*. The species is relatively sensitive to burial by beech litter. *S.e.* is susceptible to summer drought (Inghe & Tamm 1985) and, consistent with this, the species is characteristic of clayey calcareous soils, and in N England is particularly common on the heavier soils of the magnesian limestone and Keuper marl. Species with a similar edaphic restriction include *Carex sylvatica* Hudson and *Viola reichenbachiana* Jordan ex Boreau. Leaves are long-lived and of low palatability to unspecialized herbivores (Inghe & Tamm 1985); each year the older leaves are replaced during the spring flush of growth. The species has only a limited capacity to produce daughter rosettes but, because of the low mortality of mature ramets, bifurcation of the rhizome may play an important role in regeneration (Inghe & Tamm 1985). Indeed, even though the rhizome gradually breaks down and no extensive rhizome system is formed, Inghe and Tamm (1985) suggest that *S.e.* is in many ways comparable with a *Trifolium repens* 'growing in slow motion'. Plants do not flower until 8-16 years old (Inghe & Tamm 1985) and often produce less than 100 seeds year^{-1}, which germinate in spring. No persistent seed bank has been recorded in field studies (Donelan & Thompson 1980) or pot experiments (Roberts 1986a). The pioneering studies of plant demography conducted on permanent plots in woodland in Sweden (Tamm 1956, Inghe & Tamm 1985) confirm the existence of a bank of persistent seedlings in this species. The mericarps of *S.e.* are covered in hooked bristles and are dispersed by humans and animals. Within upland Britain, where *S.e.* is scarce, there is evidence of long-distance dispersal and Peterken (1981), perhaps rather surprisingly in view of the population biology of the species, classified *S.e.* as a fast-colonizer of Lincolnshire secondary woodland. Rackham (1980) confirms that *S.e.* has some colonizing ability in secondary woodland in E England. *S.e.* is one of the few British representatives of the subfamily Hydrocotyloideae (Tutin 1980).

Current trends Uncertain, but perhaps decreasing through the destruction of broad-leaved woodland habitats.

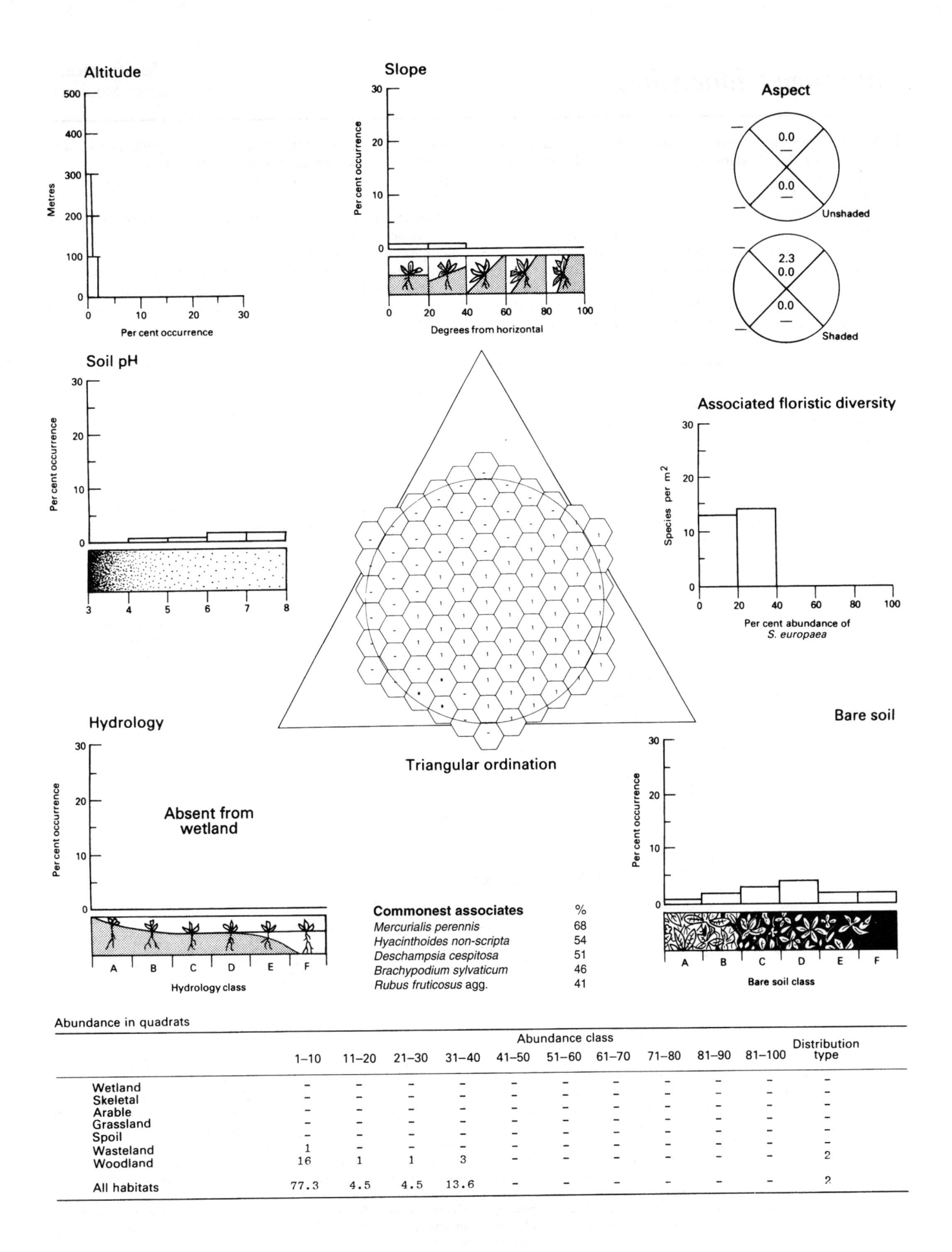

Altitude

Slope

Aspect

Unshaded

Shaded

Soil pH

Associated floristic diversity

Per cent abundance of *S. europaea*

Hydrology

Absent from wetland

Triangular ordination

Bare soil

Commonest associates	%
Mercurialis perennis	68
Hyacinthoides non-scripta	54
Deschampsia cespitosa	51
Brachypodium sylvaticum	46
Rubus fruticosus agg.	41

Abundance in quadrats

	Abundance class										Distribution type
	1–10	11–20	21–30	31–40	41–50	51–60	61–70	71–80	81–90	81–100	
Wetland	–	–	–	–	–	–	–	–	–	–	–
Skeletal	–	–	–	–	–	–	–	–	–	–	–
Arable	–	–	–	–	–	–	–	–	–	–	–
Grassland	–	–	–	–	–	–	–	–	–	–	–
Spoil	–	–	–	–	–	–	–	–	–	–	–
Wasteland	1	–	–	–	–	–	–	–	–	–	2
Woodland	16	1	1	3	–	–	–	–	–	–	
All habitats	77.3	4.5	4.5	13.6	–	–	–	–	–	–	2

Saxifraga tridactylites L.

Saxifragaceae
Rue-leaved Saxifrage

Life-form Semi-rosette, winter-annual therophyte. Stems erect; leaves ovate, lobed, glandular, hairy, often <50 mm^2; roots non-mycorrhizal.

Phenology Winter annual. Flowers from April to June and sets seed from June to July.

Height Foliage mainly <50 mm; flowers to 150 mm.

Seedling RGR No data. SLA. No data. DMC 10.7.

Nuclear DNA amount No data; $2n = 22$ (*FE1*); ?$2x$.

Established strategy Stress-tolerant ruderal.

Regenerative strategies S, B$_s$. Regenerates entirely by seed, which germinates in autumn. Seed shows some persistence in the soil; SB III.

Flowers White, hermaphrodite, usually self-pollinated (Knuth 1906); either solitary or in a cymose inflorescence.

Germinule 0.01 mg, 0.5 × 0.3 mm, seed; often >100 in each dehiscent capsule.

Germination Epigeal; t_{50} 7 days; 7-25°C; germination rate increased after dry storage at 20°C; L > D.

Biological flora Webb (1950).

Geographical distribution Lowland Britain, but rare in Scotland (82% of British vice-counties), A*BIF* page 319; Europe except for the extreme N (85% of European territories); extending to W Iran.

Habitats Recorded from calcareous rock outcrops, limestone quarry spoil and locally on cinders. Also observed on walls, cliffs and lead-mine and cinder spoil. [Found on semi-stabilized sand dunes.]

Gregariousness Sparse.

Altitude Largely confined to limestone strata, and altitudinal range restricted accordingly. [Up to 730 m.]

Slope Widely distributed.

Aspect More frequently recorded from unshaded N-facing slopes. This unexpected result is based on a small number of field records. Greater intensity of sampling (Hodgson unpubl.) reveals that, in common with other winter annuals, *S.t.* has a marked S-facing bias.

Hydrology Absent from wetlands.

Soil pH Largely restricted to calcareous soils, and found only in the pH range 6.0-8.0.

Bare soil Only recorded from rocky or cindery open sites with little exposed soil [but also found on sand dunes with higher exposures].

△ **ordination** Distribution confined to relatively infertile, disturbed habitats.

Associated floristic diversity Often high at sites with large amounts of soil, e.g. some limestone outcrops, but usually low on walls.

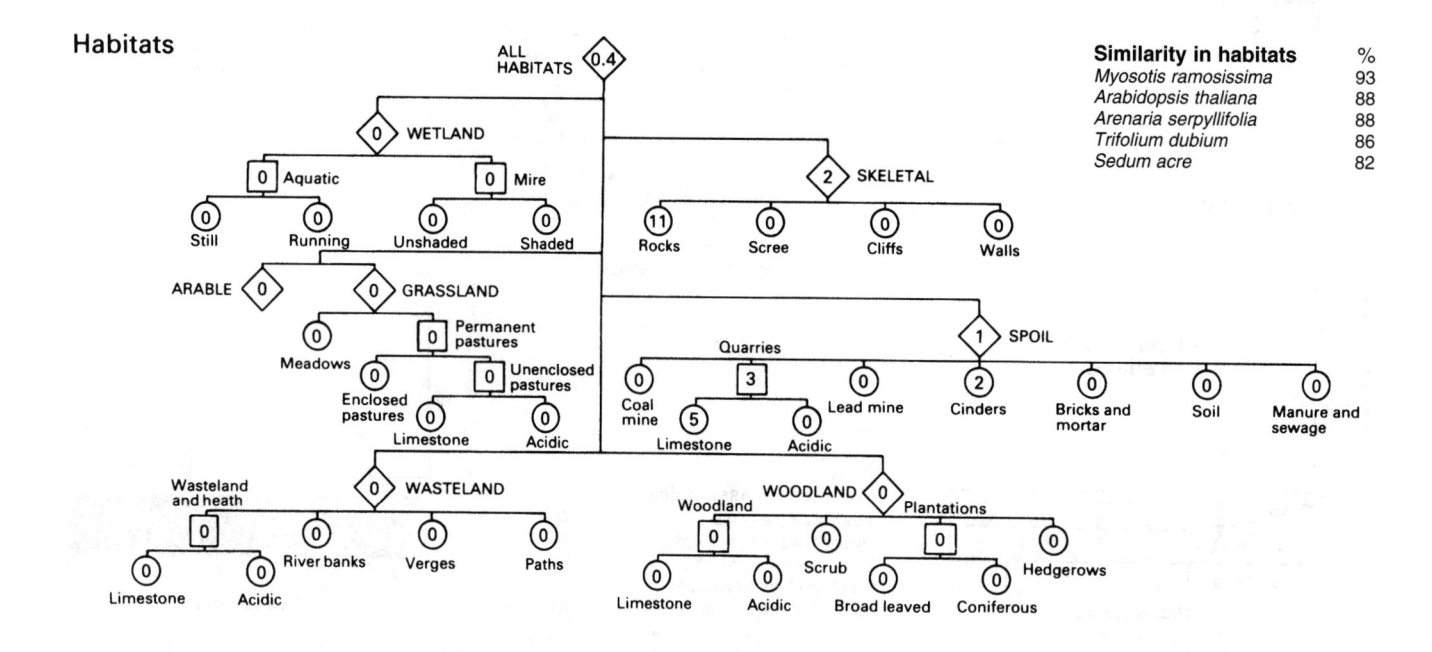

Habitats

Similarity in habitats	%
Myosotis ramosissima	93
Arabidopsis thaliana	88
Arenaria serpyllifolia	88
Trifolium dubium	86
Sedum acre	82

Synopsis *S.t.* is a diminutive, winter-annual herb of dry, nutrient-deficient, calcareous, rocky or sandy habitats in which the cover of perennials is restricted by summer drought. The species is narrowly restricted to 'semi-permanent' winter-annual communities on rock outcrops, crevices in walls and sand dunes, and in habitat and phenology (see Pemadasa & Lovell 1975) closely resembles *Erophila verna* (L.) DC. Both *S.t.* and *E.v.* are late-germinating and early-flowering, and are presumably associated with the most drought-prone microsites. *S.t.* differs primarily in its much smaller seeds, which are released in summer; these are prevented from immediate germination by an after-ripening requirement and germinate in autumn (Ratcliffe 1961, Pemadasa & Lovell 1975). Seedlings perennate as very small rosettes. Flowering is induced by vernalization, but there is no absolute requirement for long days (Clark 1969; Pemadasa & Lovell 1974b) and flowering occurs rapidly in early spring. A persistent seed bank is formed (Thompson & Grime 1979). Seed appears to be mainly dispersed close to the parent plant (Pemadasa & Lovell 1974a). However, the species is capable of wider dispersal, e.g. along railway tracks, and the frequent occurrence of *S.t.* on walls suggest that the minute seeds are capable of movement in air currents. A perennial hybrid between *S.t.* and *S. hypnoides* L. has been recorded once for the British Isles (Stace 1975).

Current trends Not under immediate threat, but infrequent as a colonist of artificial habitats. Confined to local refugia and probably decreasing slowly.

Altitude
Slope
Aspect

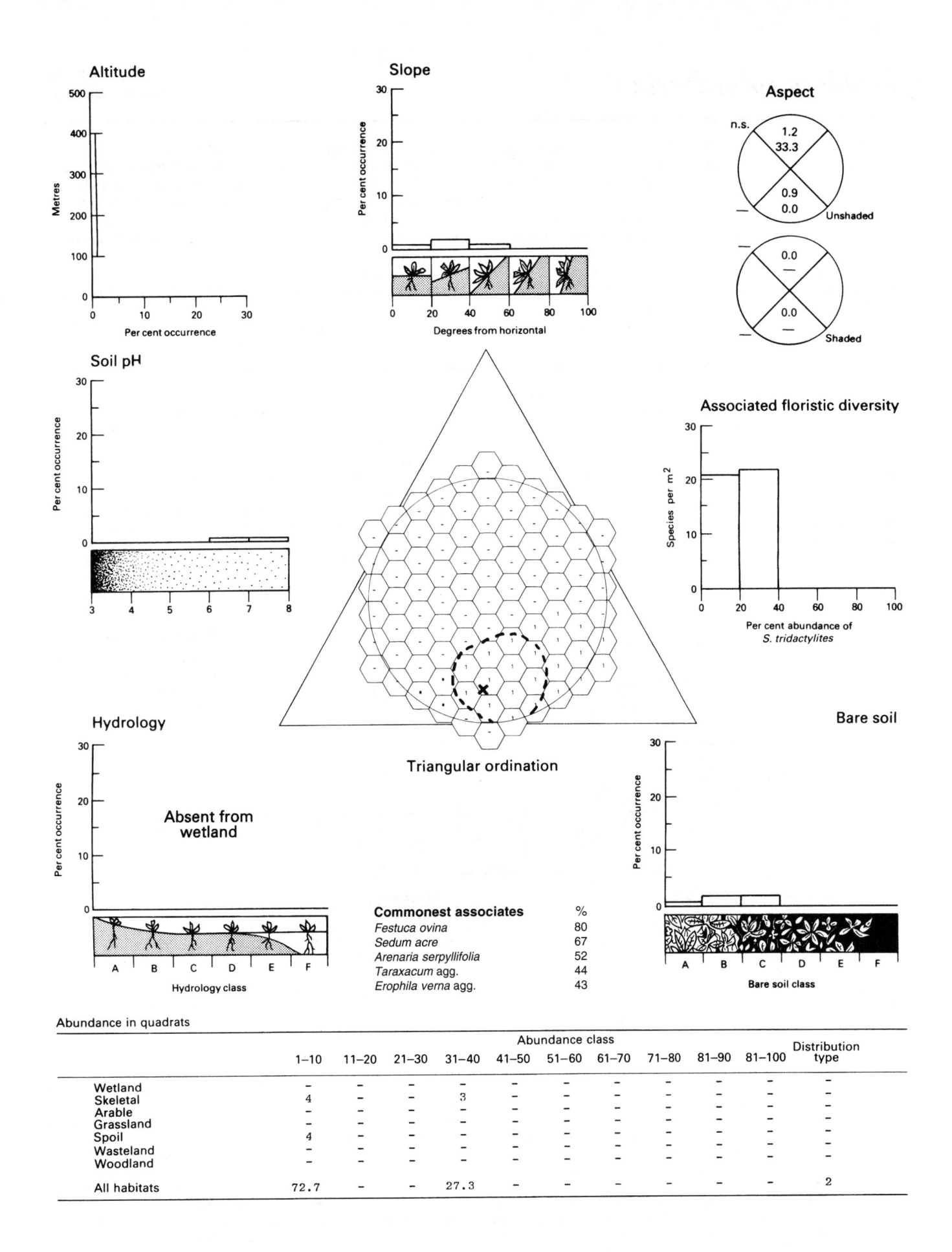

	n.s.	1.2	
		33.3	
	—	0.9	Unshaded
		0.0	

		0.0	
	—	—	
		0.0	Shaded
		—	

Soil pH

Associated floristic diversity

Hydrology

Absent from wetland

Triangular ordination

Commonest associates	%
Festuca ovina	80
Sedum acre	67
Arenaria serpyllifolia	52
Taraxacum agg.	44
Erophila verna agg.	43

Bare soil

Abundance in quadrats

	1–10	11–20	21–30	31–40	41–50	51–60	61–70	71–80	81–90	81–100	Distribution type
Wetland	–	–	–	–	–	–	–	–	–	–	–
Skeletal	4	–	–	3	–	–	–	–	–	–	–
Arable	–	–	–	–	–	–	–	–	–	–	–
Grassland	–	–	–	–	–	–	–	–	–	–	–
Spoil	4	–	–	–	–	–	–	–	–	–	–
Wasteland	–	–	–	–	–	–	–	–	–	–	–
Woodland	–	–	–	–	–	–	–	–	–	–	–
All habitats	72.7	–	–	27.3	–	–	–	–	–	–	2

Scabiosa columbaria L.

Dipsacaceae
Small Scabious

Life-form Monocarpic to polycarpic perennial, semi-rosette hemicryptophyte with deep tap-root. Stock branching; stems erect; leaves hairy, ovate, or pinnatifid, usually <1000 mm^2; roots with VA mycorrhizas.
Phenology Overwinters as a rosette of small leaves produced in autumn. Flowers July to August and sets seed August to October. Leaves long-lived.
Height Foliage mainly <100 mm; flowers to 700 mm.
Seedling RGR 1.0-1.4 week^{-1}. SLA 19.8. DMC 17.2.
Nuclear DNA amount 2.3 pg; $2n = 16*$ (*BC*); $2x*$.
Established strategy Intermediate between stress-tolerator and stress-tolerant ruderal.

Regenerative strategies S. See 'Synopsis'; ?SB II.
Flowers Florets lilac, homogamous to protandrous, insect-pollinated; usually <100 aggregated into terminal capitula.
Germinule 1.32 mg, 2.6 × 1.5 mm, indehiscent dry fruit with a ring of straight spines.
Germination Epigeal; t_{50} 11 days; some seed with a chilling requirement (During *et al.* 1985); L > D.
Biological flora None.
Geographical distribution Much of the British Isles, but absent from Ireland and most of Scotland (48% of British vice-counties), *ABIF* page 607; Europe except the extreme N (64% of European territories); W Asia and N Africa.

Habitats A strictly calcicolous species of semi-natural grassland, both grazed and ungrazed, scree and quarry spoil. Also found on cliffs, rock outcrops and lead-mine spoil. Observed on waysides and railway ballast, but absent from arable, wetland and woodland habitats.
Gregariousness Sparse to intermediate.
Altitude Most frequent in sites on carboniferous limestone between 200 and 350 m. [Up to 610 m.]
Slope Wide-ranging, but most frequent on moderate to steep slopes on the carboniferous limestone.
Aspect Slightly more frequent and abundant on N-facing slopes.
Hydrology Absent from wetlands.
Soil pH Largely absent from soils of pH <5.5.
Bare soil Most frequent and abundant in sites with low exposures of bare soil. Habitats often contain some bare rock.
△ **ordination** Distribution centred on unproductive, relatively undisturbed communities. Absent from productive vegetation and from heavily disturbed environments.
Associated floristic diversity Typically intermediate to species-rich.
Synopsis *S.c.* is a winter-green, rosette-forming herb largely restricted to short turf and rocky habitats on dry or moist infertile calcareous soils. *S.c.* has been described as an 'hapaxanth' by Verkaar *et al.* (1983). Populations under garden cultivation are also short-lived (*c.* 5 years), and floriferous if cultivated on fertile soils, but field observations suggest that plants develop much more slowly and often persist for >10 years in the Derbyshire dales (Rorison pers. comm.). Low stature and lack of lateral vegetative spread restrict the capacity of *S.c.* to exploit tall or productive communities. *S.c.*

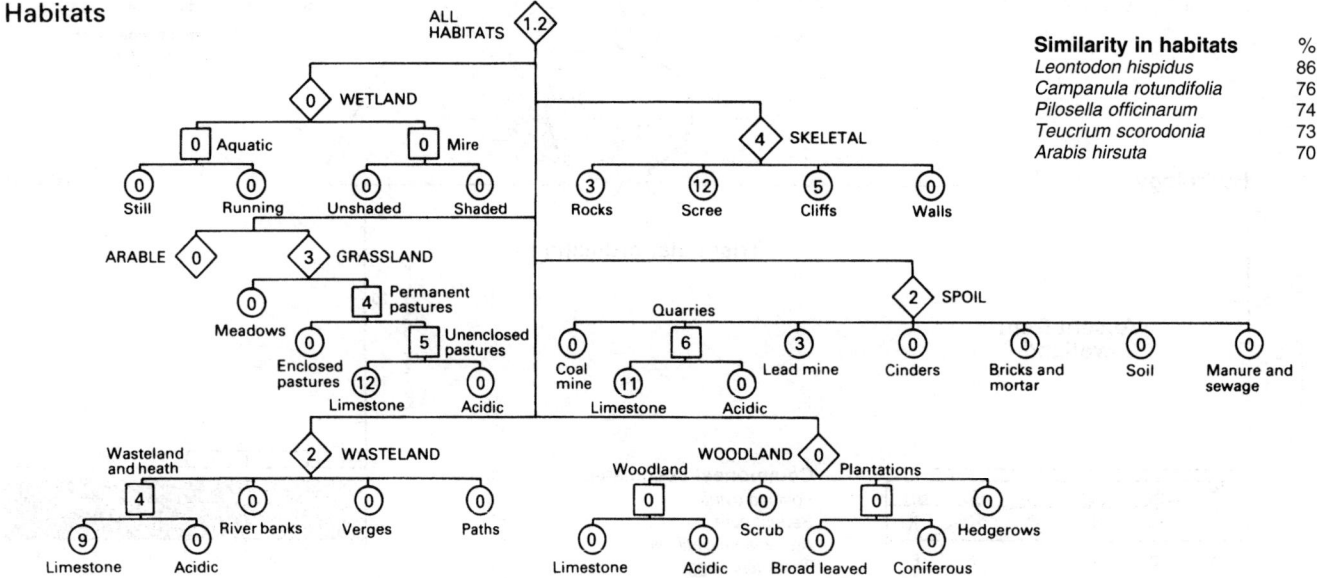

Habitats

Similarity in habitats	%
Leontodon hispidus	86
Campanula rotundifolia	76
Pilosella officinarum	74
Teucrium scorodonia	73
Arabis hirsuta	70

ALL HABITATS 1.2
WETLAND 0
- Aquatic 0: Still 0, Running 0
- Mire 0: Unshaded 0, Shaded 0
SKELETAL 4
- Rocks 3, Scree 12, Cliffs 5, Walls 0
ARABLE 0
GRASSLAND 3
- Meadows 0
- Permanent pastures 4
- Unenclosed pastures 5
- Enclosed pastures: Limestone 12, Acidic 0
SPOIL 2
- Quarries: Coal mine 0, Limestone 11, Lead mine 3, Acidic 0
- Cinders 0, Bricks and mortar 0, Soil 0, Manure and sewage 0
WASTELAND 2
- Wasteland and heath: Limestone 9, Acidic 0
- River banks 0, Verges 0, Paths 0
- (sub) 4
WOODLAND 0
- Woodland 0: Limestone 0, Acidic 0
- Plantations 0
- Scrub 0, Broad leaved 0, Coniferous 0, Hedgerows 0

---

occurs both in grazed and in burned sites. The species is particularly associated with microsites of lower vegetation density in calcareous grassland (Verkaar *et al.* 1983), and this capacity for sustained growth during summer may be related to the long tap-root (Anderson 1927) which provides access to moisture in deep soil crevices. In turf microcosms, *S.c.* became colonized by VA mycorrhizas and seedlings showed up to 5-fold increases in yield as a result of infection (Grime, Mackey, Hillier & Read 1987). In Derbyshire populations several rosettes frequently arise from the basal stock, but *S.c.* forms only very small compact patches. The association of *S.c.* with open vegetation appears to be linked to its dependence upon seed for regeneration. Although visited more rarely than *Centaurea* spp. by insects (Lack 1982a,b), in ungrazed sites *S.c.* produces abundant seed which germinates mainly in autumn. However, flowering and fruiting heads are often heavily grazed by rabbits, in which case the average number of seeds per plant may only be in the order of 20-60 (During *et al.* 1985). Mortality rates of seedlings are low (During *et al.* 1985), but seed predation may be high. A persistent seed bank has been reported (Schenkefeld & Verkaar 1984), but on the basis of laboratory germination characteristics we suspect that seed longevity is short. Fruits vary greatly in size within the same capitulum. Seedlings have a relatively similar growth rate irrespective of seed size (Rorison 1973). Fruits are spiny and have been described as specialized for wind dispersal (*CTW*); this interpretation is not confirmed by our own field observations, and *S.c.* is an ineffective colonist of new sites. The absence of *S.c.* from Ireland may relate to poor dispersal during Postglacial colonizing episodes. The species exhibits much variation in shape and degree of dissection of the leaves and, according to the observations of During *et al.* (1985) and Rorison (pers. comm.), variation also occurs in life-history. Heavy-metal-tolerant races have been recorded in Derbyshire populations (Shaw 1984). *S.c.* is part of a taxonomically complex grouping, and two further ssp. are recognized within Europe (*FE4*).

Current trends Largely restricted to old or semi-natural vegetation. Decreasing and near extinction in some lowland areas.

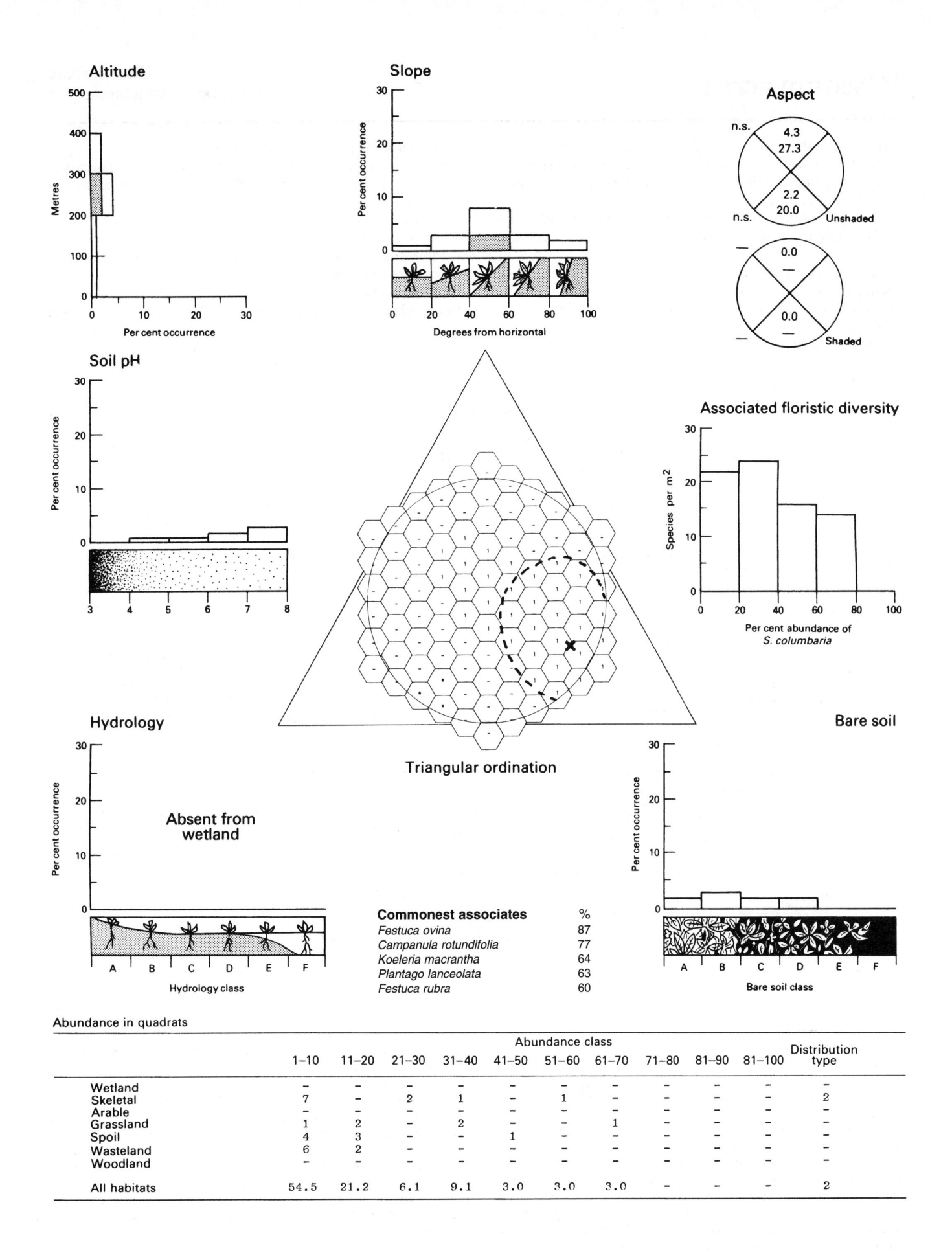

Altitude

Slope

Aspect

	n.s.	4.3	
n.s.	27.3	2.2	Unshaded
		20.0	

		0.0	
—		—	
		0.0	Shaded
—		—	

Soil pH

Associated floristic diversity

Species per m² vs Per cent abundance of *S. columbaria*

Hydrology

Absent from wetland

Hydrology class: A B C D E F

Triangular ordination

Bare soil

Bare soil class: A B C D E F

Commonest associates %
Festuca ovina 87
Campanula rotundifolia 77
Koeleria macrantha 64
Plantago lanceolata 63
Festuca rubra 60

Abundance in quadrats

	1–10	11–20	21–30	31–40	41–50	51–60	61–70	71–80	81–90	81–100	Distribution type
Wetland	–	–	–	–	–	–	–	–	–	–	–
Skeletal	7	–	2	1	–	1	–	–	–	–	2
Arable	–	–	–	–	–	–	–	–	–	–	–
Grassland	1	2	–	2	–	–	1	–	–	–	–
Spoil	4	3	–	–	1	–	–	–	–	–	–
Wasteland	6	2	–	–	–	–	–	–	–	–	–
Woodland	–	–	–	–	–	–	–	–	–	–	–
All habitats	54.5	21.2	6.1	9.1	3.0	3.0	3.0	–	–	–	2

(*)Sedum acre L.
Crassulaceae
Wall-pepper, Biting Stonecrop

Life-form Polycarpic perennial chamaephyte. Stems creeping, forming a mat with ascending or erect non-flowering and flowering branches; leaves ovoid, succulent; roots often shallow, without mycorrhizas.

Phenology Winter green. Plants flower June to July and seed set from July to October. Flowering stems annual. Leaves long-lived.

Height Foliage and flowers to 100 mm.

Seedling RGR 0.5–0.9 week^{-1}. SLA 15.8. DMC 6.4.

Nuclear DNA amount No data; $2n = 40, 48, 80^*$ (*BC, FE1*), $5x$, $6x$, $10x^*$.

Established strategy Intermediate between stress-tolerator and stress-tolerant ruderal.

Regenerative strategies V, ?B$_s$. Regeneration by seed requires further study (see 'Synopsis'); ?SB III.

Flowers Yellow, hermaphrodite, protandrous, insect-pollinated, sometimes selfed (Knuth 1906); up to >10 in a cymose inflorescence.

Germinule 0.03 mg, 0.9×0.5 mm, seed often $c.$ 20 in each of five dehiscent follicles.

Germination Epigeal; t_{50} 3 days; 13–27°C; germination rate increased after dry storage at 20°C; L > D.

Biological flora None.

Geographical distribution The British Isles, but less common in Scotland and Ireland (99% of British vice-counties), *ABIF* page 311; Europe (85% of European territories); N and W Asia, N Africa; naturalized in N America.

Habitats Mainly in open stony habitats such as rock outcrops, stabilized scree, quarry waste, walls and cliffs. Also recorded locally at the sites of old lead-mine workings, and in dry calcareous pastures and on wasteland. Observed on cinders of railway banks and soil heaps, apparently as a garden escape. [Frequent on semi-stabilized sand dunes.]

Gregariousness Typically sparse to intermediate, but potentially patch-forming on sand dunes.

Altitude Observed from 5 m to 400 m, but suitable habitats are most frequent on the carboniferous limestone between 200 m and 300 m. [Up to 460 m.]

Slope Wide-ranging, but particularly frequent on the steeper slopes of skeletal habitats.

Aspect Slightly more frequent on S-facing slopes.

Hydrology Absent from wetlands.

Soil pH Only recorded on near-neutral soils.

Bare soil Often present in sites with much exposed rock, but absent from habitats with a high exposure of bare soil.

△ **ordination** Excluded from fertile and heavily disturbed situations. Distribution centred on undisturbed, unproductive vegetation.

Associated floristic diversity Low in extremely skeletal habitats, but occurrences extend into species-rich pastures, rock outcrops [and sand dunes].

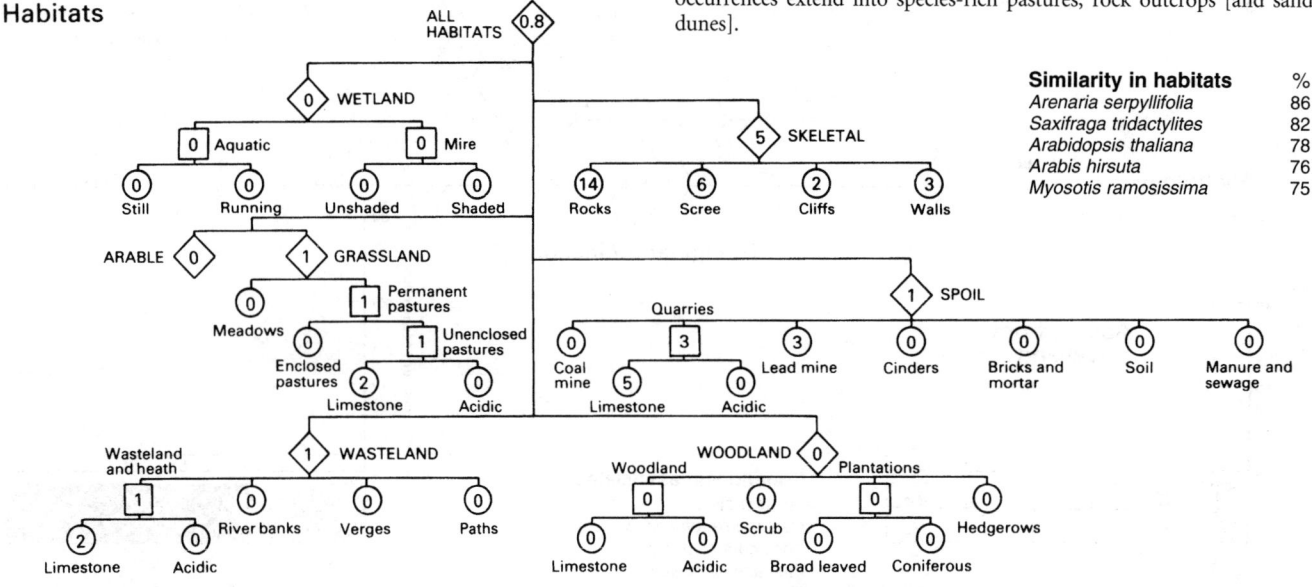

Similarity in habitats	%
Arenaria serpyllifolia	86
Saxifraga tridactylites	82
Arabidopsis thaliana	78
Arabis hirsuta	76
Myosotis ramosissima	75

Habitats

Synopsis *S.a.* is a slow-growing, leaf succulent forming tight cushions which expand vegetatively over shallow infertile soil and bare rock. *S.a.* is the most widespread British succulent. Under droughted conditions carbon fixation shifts from the normal C$_3$ type to crassulaean acid metabolism (CAM) (Schuber & Kluge 1981). However, the extent of the carbon gain resulting from CAM appears slight (Schuber & Kluge 1981) and the importance of CAM to the drought resistance exhibited by *S.a.* is uncertain. Not withstanding, the species is particularly characteristic of dry sand dunes and steeply sloping, S-facing exposures of rock. On sand dunes *S.a.* can survive a limited degree of burial. *S.a.* is strongly protected against herbivory. The foliage is highly acrid and the plant also contains alkaloids which are toxic to stock (Cooper & Johnson 1984). *S.a.* is almost exclusively calcicolous in the Sheffield region. In W Britain a comparable ecological niche on more-acidic soils is occupied by the morphologically similar *S. anglicum* Hudson. In addition to patch formation by creeping stems, vegetative spread may occur from detached shoots, or even leaves, which under suitably moist conditions root to form new colonies. Numerous small seeds are produced during summer. These show increased germination following laboratory storage, and probably germinate in autumn. *S. album* L. is known to form a persistent seed bank and the same is predicted for *S.a*. This is a morphologically variable species (*FE1*), but the extent of ecotypic differentiation between native populations is not clear. The species achieves occasional escapes from cultivation on rockeries, walls and cottage roofs, and may be observed locally in situations such as the cinders adjacent to disused railway stations. It is not known to what extent these 'domestic' populations differ genetically and ecologically from the native stock. *S.a.* is part of a complex in which 11 species have been recognized (*FE1*).

Current trends Despite its frequent cultivation in gardens and the fact that the seeds are produced in large numbers in suitable habitats, the species is not an early colonist of artificial habitats, and may be decreasing.

Altitude
Slope
Aspect

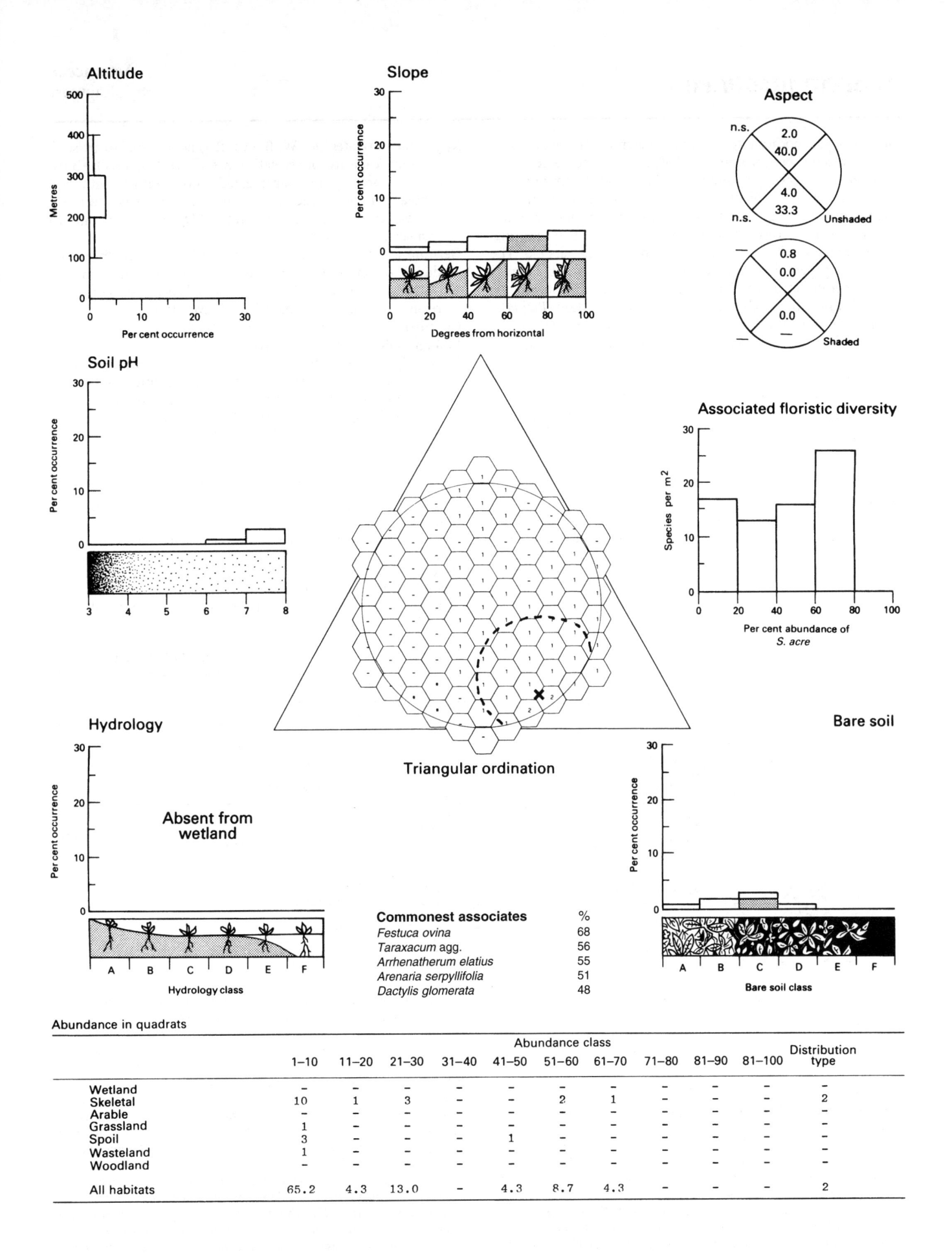

	n.s.	2.0	
		40.0	
		4.0	
n.s.		33.3	Unshaded

	—	0.8	
		0.0	
		0.0	
	—	—	Shaded

Soil pH

Associated floristic diversity

Per cent abundance of *S. acre*

Triangular ordination

Hydrology

Absent from wetland

Commonest associates

	%
Festuca ovina	68
Taraxacum agg.	56
Arrhenatherum elatius	55
Arenaria serpyllifolia	51
Dactylis glomerata	48

Bare soil

Abundance in quadrats

	1–10	11–20	21–30	31–40	41–50	51–60	61–70	71–80	81–90	81–100	Distribution type
Wetland	–	–	–	–	–	–	–	–	–	–	–
Skeletal	10	1	3	–	–	2	1	–	–	–	2
Arable	–	–	–	–	–	–	–	–	–	–	–
Grassland	1	–	–	–	–	–	–	–	–	–	–
Spoil	3	–	–	–	1	–	–	–	–	–	–
Wasteland	1	–	–	–	–	–	–	–	–	–	–
Woodland	–	–	–	–	–	–	–	–	–	–	–
All habitats	65.2	4.3	13.0	–	4.3	8.7	4.3	–	–	–	2

Senecio jacobaea L.

Asteraceae
Ragwort

Life-form Monocarpic to polycarpic perennial (rarely annual), semi-rosette hemicryptophyte (*Biol Fl*) with a short erect stock. Stems erect; leaves pinnatifid, often hairy, <4000 mm^2; roots to 150 mm in depth (*Biol Fl*), with or without VA mycorrhizas.

Phenology Seeds germinate mainly in autumn and seedlings overwinter in a leafy condition. Some seeds germinate in spring. Flowering usually occurs in plants which are 2 years old. The flowering period commences in June and may extend until October in defoliated plants. Ripe seed dispersed from August until winter, by which time the shoot is usually dead (*Biol Fl*).

Height In non-flowering plants foliage is situated close to the ground. Flowering shoots to 1.5 m with cauline leaves up to 1 m.

Seedling RGR 1.0–1.4 week^{-1}. SLA 27.4. DMC 11.4.

Nuclear DNA amount 4.5 pg; 2n = 40*, 80 (*FE4, BC*); 4x*, 8x.

Established strategy Intermediate between C-S-R and competitive-ruderal.

Regenerative strategies W, B$_s$ (V). Regenerates mainly by seed, but some capacity for vegetative spread (see 'Synopsis'); SB III.

Flowers Florets yellow, aggregated into capitula, with >50 tubular hermaphrodite disk florets and >10 female, outer ray florets, ligulate, insect-pollinated (*Biol Fl*); often >100 capitula in a terminal corymb.

Germinule 0.05 mg, 2.0 × 0.5 mm, achene, inner, from disk florets, dispersed with a pappus *c*. 6 mm in length when ripe; outer, from ray florets, heavier, without dispersal structures and retained on plant after maturity (McEvoy 1984).

Germination Epigeal; t_{50} 3 days; 7–28°C; L > D.

Biological flora Harper and Wood (1957).

Geographical distribution The British Isles, but most abundant in the W and SW (100% of British vice-counties), *ABIF* page 653; most of Europe, although rare in the extreme N and S (74% of European territories and 3% naturalized); W Asia and N Africa; introduced into N and S America, New Zealand and Australia.

Habitats Widely dispersed, but restricted to habitats with at least a little bared ground. Particularly abundant in rocky habitats (outcrops, scree, limestone quarry spoil) and in overgrazed pasture. Widespread on road verges, wasteland, spoil tips, paths, cliffs and walls. Almost absent from wetland and woodland, and of only casual occurrence on arable land. [Often a conspicuous plant of sand-dune grassland.]

Gregariousness Mainly represented by scattered individuals.

Altitude Over a wide range of values, reaching 420 m. [Up to 670 m.]

Slope Distributed over the complete range in slope.

Aspect No aspect preference apparent.

Hydrology *S.j.* has a relatively deep root system; habitats include moist rather than waterlogged sites.

Habitats

Similarity in habitats	%
Leontodon hispidus	85
Linum catharticum	82
Pilosella officinarum	81
Festuca rubra	76
Scabiosa columbaria	70

ALL HABITATS 6.3
- WETLAND 1
  - Aquatic 0
    - Still 0
    - Running 0
  - Mire 1
    - Unshaded 1
    - Shaded 0
- ARABLE 4
  - Meadows 0
- GRASSLAND 11
  - Permanent pastures 13
  - Unenclosed pastures 12
  - Enclosed pastures 5
    - Limestone 31
    - Acidic 0
- SKELETAL 14
  - Rocks 25
  - Scree 37
  - Cliffs 9
  - Walls 7
- SPOIL 13
  - Quarries
    - Coal mine 4
    - Limestone 25 (45)
    - Acidic 0
  - Lead mine 17
  - Cinders 8
  - Bricks and mortar 8
  - Soil 16
  - Manure and sewage 6
- WASTELAND 8
  - Wasteland and heath 6
    - Limestone 11
    - Acidic 3
  - River banks 3
  - Verges 18
  - Paths 10
- WOODLAND 1
  - Woodland 1
    - Limestone 2
    - Acidic 0
  - Scrub 2
    - Broad leaved 0
    - Coniferous 0
  - Plantations 0
  - Hedgerows 0

Soil pH Most frequent on base-rich soil of pH >7.0 and rare below pH 5.0.

Bare soil Associated with a wide range of exposures.

△ **ordination** Distribution centred on vegetation associated with moderate levels of disturbance and relatively unproductive conditions, but not excluded from any vegetation type except where disturbance is minimal.

Associated floristic diversity Often associated with relatively species-rich vegetation.

Synopsis *S.j.* is typically a 'biennial', but often takes more than 2 years to flower. Otzen (1977) has suggested that as flowering individuals have significant carbohydrate reserves in the stem bases and roots, this may allow the species to behave also as a facultative perennial. Up to 30 000 achenes may be produced by a single plant (Van der Meijden & Van der Waals-Kooi 1979) and the distribution of *S.j.* is limited by a dependence upon vegetation gaps for colonization by seed. Thus, *S.j.* is characteristic of open habitats and is a noxious weed of overgrazed grassland. Germination occurs mainly during autumn, with subsidiary cohorts of seedlings appearing in spring and summer. *S.j.* forms a buried seed bank in some circumstances but seed is only briefly persistent (Roberts 1986a). Most seed is dispersed only short distances from the parent plant (*Biol Fl*), but field evidence suggests that *S.j.* is highly mobile. The seed survives ingestion by sheep but not birds (*Biol Fl*). Survival in relatively undisturbed habitats may be effected by vigorous recolonization in bared areas formerly occupied by individuals of *S.j.* (McEvoy 1984). Plants also have the capacity to regenerate from root fragments, which may be important in disturbed habitats (*Biol Fl*). New basal branching may occur, particularly following damage, to give several rosettes which may become independent (*Biol Fl*). *S.j.* is highly toxic to cattle and horses, and is avoided if alternative food is available (*Biol Fl*). The poisonous principles are pyrrolizidine alkaloids which are not destroyed by drying or storage (Cooper & Johnson 1984). The species is also unpalatable to rabbits and may be prominent

Altitude
Slope
Aspect
Soil pH
Associated floristic diversity
Hydrology
Triangular ordination
Bare soil

Commonest associates	%
Festuca rubra	70
Plantago lanceolata	54
Festuca ovina	53
Dactylis glomerata	52
Arrhenatherum elatius	45

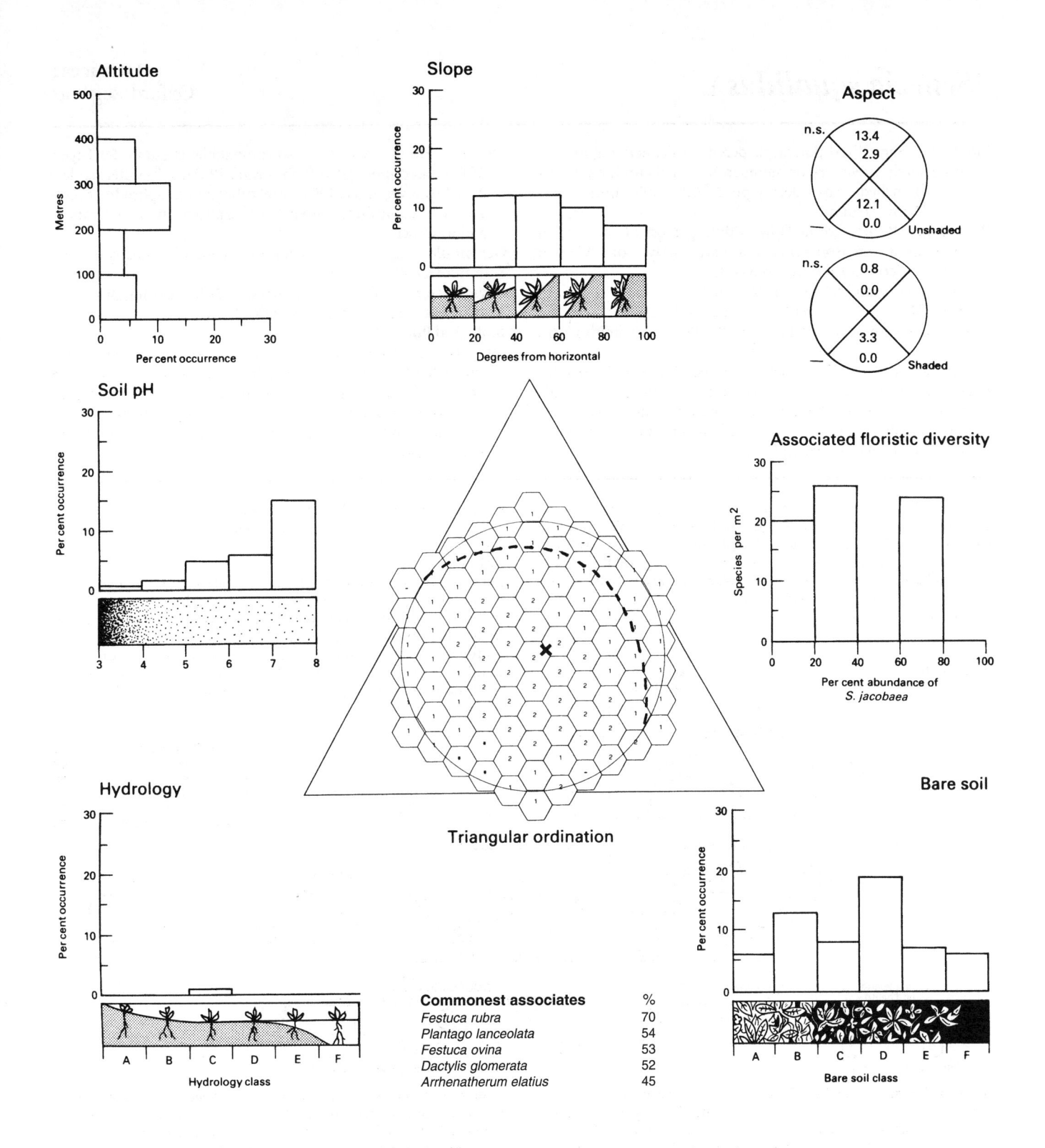

near rabbit warrens. *S.j.* is grazed by sheep, sometimes with harmful effects, and often persists in sheep pastures as non-flowering rosettes for many years. Shoots of *S.j.* are often defoliated and flowering is suppressed by the caterpillars of the cinnabar moth (*Tyria jacobaeae* L.), which has been used in the biological control of *S.j.* in countries where *S.j.* has become a serious weed. *S.j.* shows a number of morphologically distinct variants (*Biol Fl*). There is seed polymorphism (see 'Germinule') and a degree of heterophylly (stem leaves are much more dissected than those of rosettes). The existence of ecotypes has received taxonomic recognition. Thus, Kadereit and Sell (1986) have separated certain rayless, low-growing coastal populations in the Scottish and Irish coasts as ssp. *dunensis* (Dumort.) Kadereit and P. D. Sell, and other populations from dunes and sea cliffs with swollen stem bases as ssp. *jacobaea* var. *condensatus* Druce. In wetter sites *S.j.* is replaced by *S. aquaticus* Hill, with which it hybridizes.

Current trends *S.j.* undoubtedly owes its present abundance to an ability to exploit artificial habitats. Management to eliminate, or at least contain, *S.j.* in pastures is a legal requirement under the Weeds Act of 1959 and, although still common, *S.j.* is now only abundant in poor pasture and wasteland, particularly on sandy or other freely drained soils where the presence of many gaps in the vegetation allows the establishment of seedlings. *S.j.* is thus probably decreasing in many, but not all, areas (see Forbes 1982).

Senecio squalidus L.

Asteraceae
Oxford Ragwort

Life-form Short-lived polycarpic perennial hemicryptophyte or more rarely a winter- or summer-annual therophyte. Stems erect, branched; leaves deeply pinnatifid, <4000 mm²; roots with VA mycorrhizas.
Phenology Winter green. Flowers during most of the year in some urban environments, but particularly from May to October. Seed set from June onwards.
Height Foliage and flowers to 300 mm.
Seedling RGR 2.0–2.4 week^{-1}. SLA 21.2. DMC 11.7.
Nuclear DNA amount 1.8 pg, $2n = 20^*$ (Bennett & Smith 1976); $2x^*$.
Established strategy Intermediate between C-S-R and ruderal.
Regenerative strategies W. Regenerates by means of wind-dispersed seeds, which germinate on bare soil and in vegetation gaps probably mainly in spring. Seeds are only briefly persistent in the soil (Roberts 1986a); ?SB I.

Flowers Florets yellow, self-incompatible (Ingram & Taylor 1982), insect-pollinated (Gibbs *et al.* 1975); *c.* 13 outer ligulate female flowers and *c.* 100 central tubular hermaphrodite florets in each capitulum; often >10 capitula in a corymbose inflorescence.
Germinule 0.21 mg, 2.0 × 0.8 mm, achene with a pappus *c.* 5.5 mm in length.
Germination Epigeal; t_{50} 2 days; 7–36°C; germination rate increased after dry storage at 20°C; L > D.
Biological flora None.
Geographical distribution A European endemic. Naturalized throughout S Britain to Lancashire and Yorkshire, local in S Scotland and Ireland (39% of British vice-counties), *ABIF* page 655; locally abundant in N and C Europe, mainly in the mountains in S Europe (33% of European territories and 13% naturalized).

Habitats Mainly on open ground in urban areas, particularly on demolition sites and cinders, alongside railway tracks, soil heaps and coalmine spoil. Less frequent on rocky habitats such as outcrops, cliffs, walls and quarry spoil, and occasionally on road verges and at the edges of paths and pavements. Absent from grassland, woodland and wetland habitats, and only a casual on arable land.
Gregariousness Sparse.

Slope Distributed over the full range of slope.
Aspect Wide-ranging but abundance tends to be greater on S-facing slopes.
Hydrology Absent from wetlands.
Soil pH Frequency of occurrence and abundance reaching maxima on soils of pH >7.0, and largely absent from soils of pH <4.5.
Bare soil Widely distributed, but evidence of an association with habitats having a high proportion of bare ground.
△ **ordination** Distribution centred on vegetation associated with highly disturbed, fertile habitats (but see 'Synopsis'). Some plants recorded in less-disturbed vegetation, but none present in sites dominated by stress-tolerators.

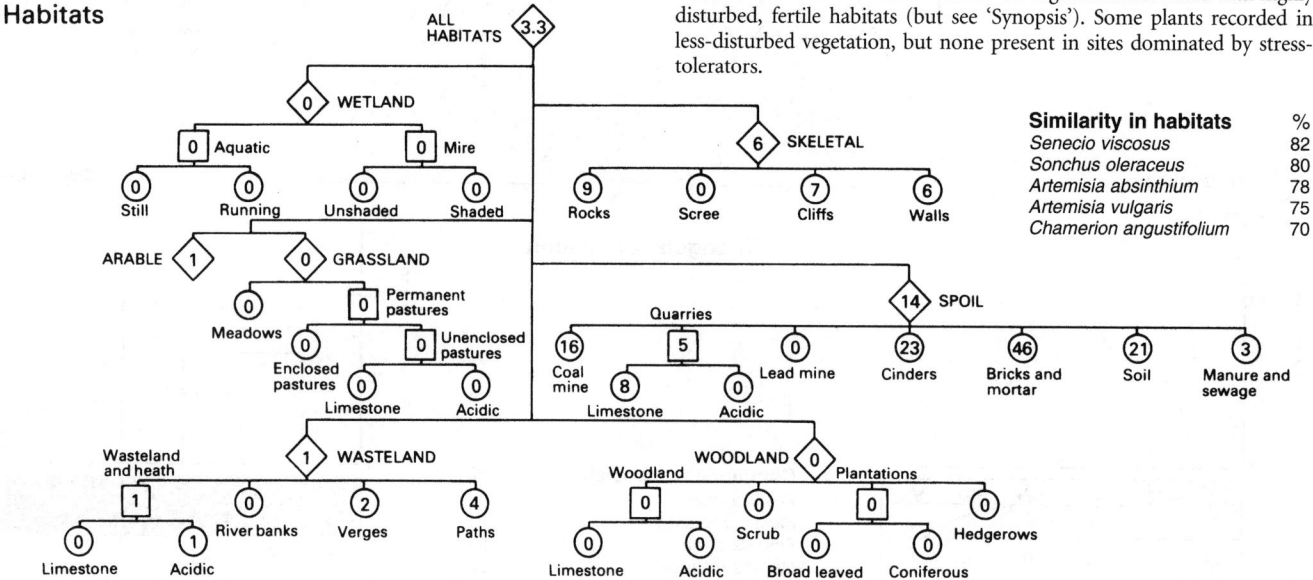

Habitats

Similarity in habitats	%
Senecio viscosus	82
Sonchus oleraceus	80
Artemisia absinthium	78
Artemisia vulgaris	75
Chamerion angustifolium	70

Associated floristic diversity Typically low to intermediate.
Synopsis *S.s.* is typically a short-lived perennial, fast-growing, bushy, winter-green herb which is a native of disturbed areas in C and S Europe (*FE4*, Oberdorfer 1979). The species was first recorded from Britain as an established alien on walls at Oxford in the late 18th century (Kent 1956). Its subsequent spread, particularly since reaching the railway at Oxford in *c.* 1879, has been rapid, and the plumed seeds of this species have been widely dispersed, particularly in the gusts formed by passing trains and through the long-distance transport of railway ballast (Kent 1960). Related in part to its habitat range (which includes demolition sites, cinders and ballast beside the railway and cracks at the edge of pavements), *S.s.* has a lowland distribution. We suspect that the distribution in Britain is also restricted by climatic factors; persistence may be favoured by the warmer winter temperatures associated with certain urban habitats. Thus, *S.s.* has become the most characteristic species of areas of urban dereliction in lowland S and C Britain. *S.s.* also colonizes some naturally-occurring rocky habitats. In all of these sites the lateral spread of perennials is restricted by the rocky nature of the substrata. The species shows an association with soils of high pH, a reflection in part of the large proportion of records from sites contaminated by cement and mortar. *S.s.* is absent from grazed, regularly mown, woodland and wetland sites. The species regenerates entirely by seed, and plants on average produce *c.* 10 000 seeds year^{-1} (Salisbury 1964). The rapid geographical spread of this species is possibly related to its capacity to produce abundant flowers and seeds over much of the year. Seeds exhibit only limited persistence in the soil. Plants vary markedly in leaf shape, even within single populations, and the level of this variation is much greater than that of the native plants of C Europe (*FE4*). *S.s.* hybridizes with *S. viscosus* and *S. vulgaris*, and is one parent of the allohexaploid *S. cambrensis* Rosser (Stace 1975, and see account for *S. vulgaris*).
Current trends Probably still increasing in geographical range and in abundance.

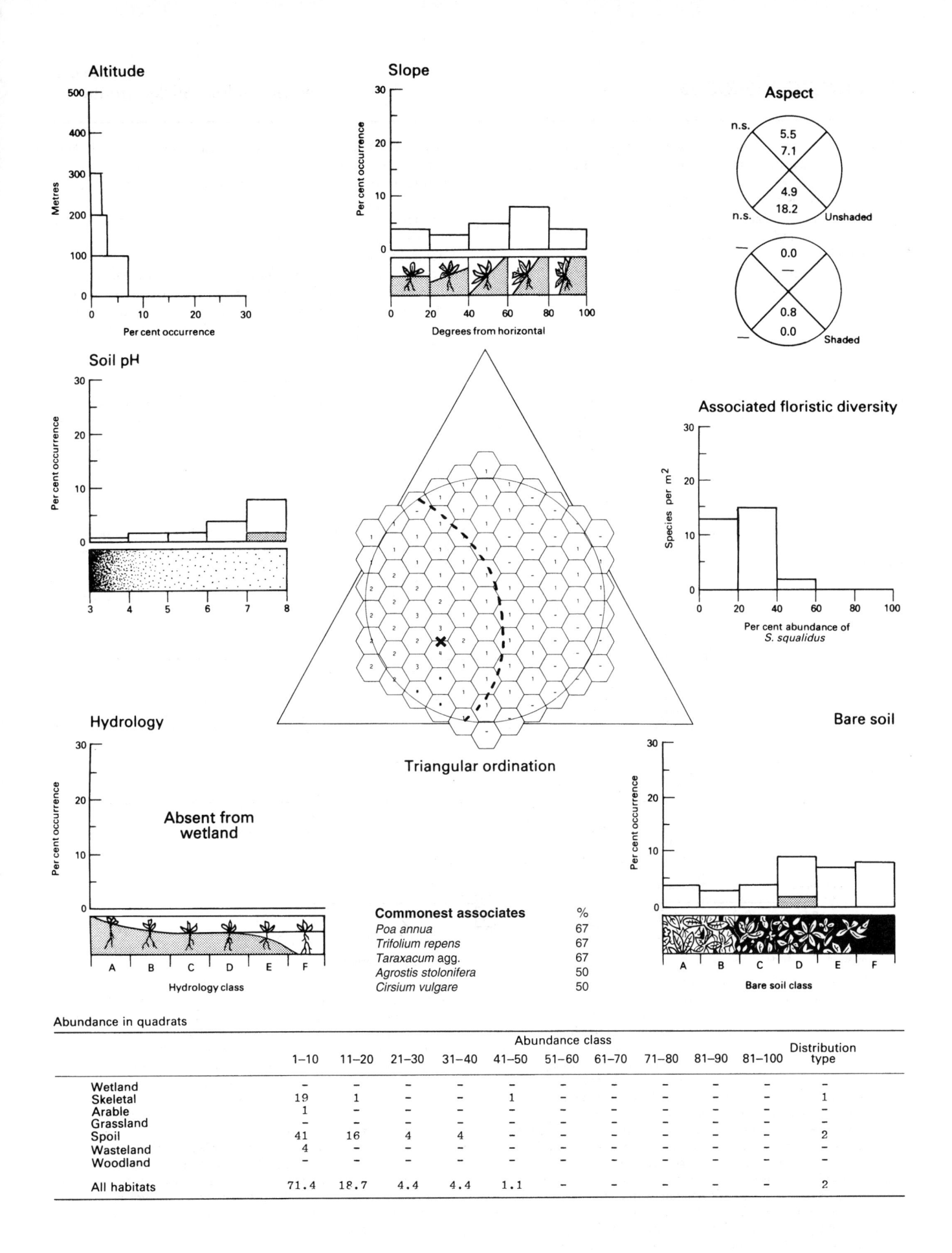

Altitude

Slope

Aspect

	n.s.	5.5	
n.s.		7.1	
		4.9	Unshaded
		18.2	

	—	0.0	
		—	
		0.8	
	—	0.0	Shaded

Soil pH

Associated floristic diversity

Per cent abundance of *S. squalidus*

Hydrology

Absent from wetland

Triangular ordination

Commonest associates %
Poa annua 67
Trifolium repens 67
Taraxacum agg. 67
Agrostis stolonifera 50
Cirsium vulgare 50

Bare soil

Abundance in quadrats

	1–10	11–20	21–30	31–40	41–50	51–60	61–70	71–80	81–90	81–100	Distribution type
Wetland	–	–	–	–	–	–	–	–	–	–	–
Skeletal	19	1	–	–	1	–	–	–	–	–	1
Arable	1	–	–	–	–	–	–	–	–	–	–
Grassland	–	–	–	–	–	–	–	–	–	–	–
Spoil	41	16	4	4	–	–	–	–	–	–	2
Wasteland	4	–	–	–	–	–	–	–	–	–	–
Woodland	–	–	–	–	–	–	–	–	–	–	–
All habitats	71.4	18.7	4.4	4.4	1.1	–	–	–	–	–	2

Senecio viscosus L.

Asteraceae
Stinking Groundsel, 'Sticky Groundsel'

Life-form A viscid, summer-annual therophyte. Stems erect, branched; leaves obovate, pinnatifid, <4000 mm^2; roots with VA mycorrhizas.

Phenology Seeds germinate in spring. Flowers and sets seed from July to October. Established plant killed by autumn frosts.

Height Foliage and flowers to 600 mm.

Seedling RGR No data. SLA 22.5. DMC 12.0.

Nuclear DNA amount No data; $2n = 40^*$ (*BC*); $4x^*$.

Established strategy Intermediate between ruderal and competitive-ruderal.

Regenerative strategies W, B$_s$.

Flowers Florets yellow, with a high selfing ability (Gibbs *et al.* 1975); *c.* 12 outer, ligulate, female and *c.* 90 central, tubular, hermaphrodite, florets in each capitulum; often >20 capitula in a corymbose inflorescence.

Germinule 0.60 mg, 3.5 × 0.5 mm, achene with a pappus *c.* 5.5 mm in length.

Germination Epigeal; t_{50} 4 days; 11-31°C; germination rate increased after dry storage; L > D.

Biological flora None.

Geographical distribution Doubtfully native in lowland Great Britain, introduced in Ireland (58% of British vice-counties), *ABIF* page 657; Europe northwards to Scandinavia, although absent from much of SE Europe (56% of European territories and 15% naturalized); introduced in N America.

Habitats Most common on cinder tips, railway ballast and building rubble, but also recorded on other types of urban spoil and rock outcrops. Observed on screes and roadsides. Absent from grassland, woodland, wetland and arable habitats. [Occurs on shingle ridges by the sea, the only semi-natural habitat with which *S.v.* is commonly associated.

Gregariousness Intermediate.

Altitude Distribution in the Sheffield region extends almost to the British altitudinal limit of 310 m.

Slope Majority of occurrences on flat or moderately sloping sites.

Aspect No evidence of an aspect preference.

Hydrology Absent from wetlands.

Soil pH Over a wide range of values, but absent from more-acidic soils.

Bare soil Always associated with open habitats with high exposures of either bare soil or bare spoil.

△ **ordination** Distribution centred on vegetation associated with highly disturbed habitats (see 'Synopsis').

Associated floristic diversity Rather low and reaching a minimum at the highest frequencies of *S.v.*, probably reflecting the specialized nature of the cinder tips on which some of the largest populations occur.

Synopsis *S.v.* is an erect, summer-annual herb which is introduced in Ireland and is only doubtfully native in Britain. The only semi-natural habitat with which *S.v.* is frequently associated is maritime shingle. The species is most frequently recorded from urban spoil, particularly cinder tips, railway ballast and demolition sites, all providing substrates on which consolidation by large perennials is restricted. Occurrences on more-

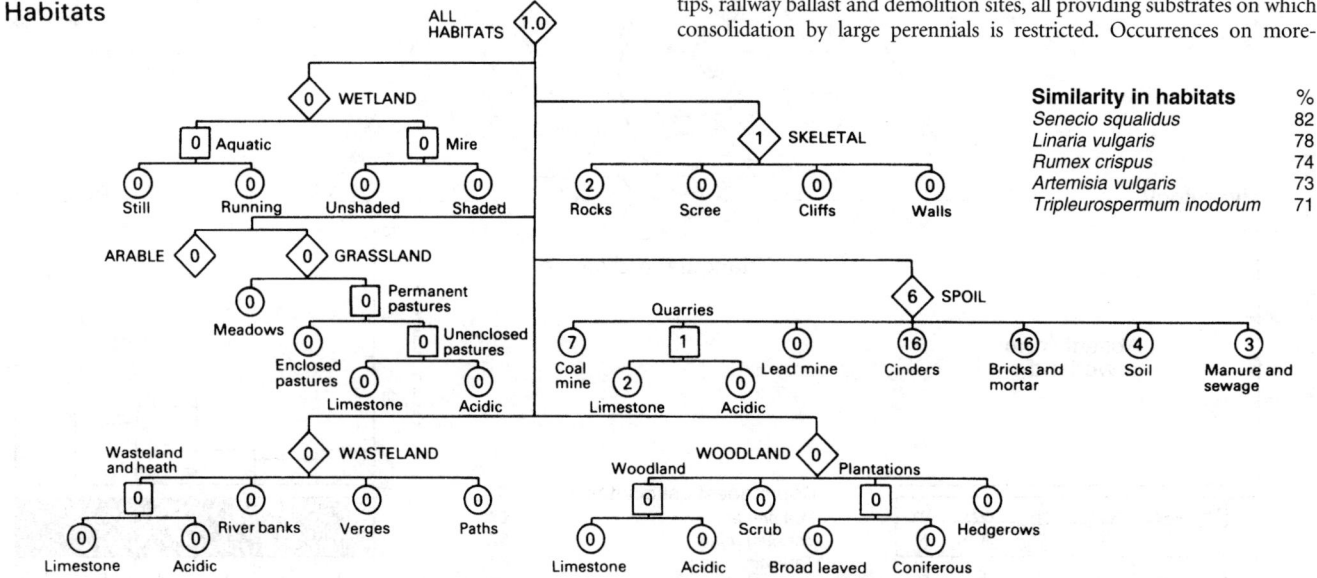

Habitats

Similarity in habitats	%
Senecio squalidus	82
Linaria vulgaris	78
Rumex crispus	74
Artemisia vulgaris	73
Tripleurospermum inodorum	71

natural soils are less frequent and, within the Sheffield region, they are largely restricted to the freely-drained Bunter sandstone. The level of fertility of the sites exploited by *S.v.* is considerably lower than that indicated by the triangular ordination, where a misleading impression has been created by the transient presence of large numbers of seedlings and stunted juveniles of weedy species in many of the urban cinder-tip habitats exploited by *S.v. S.v.* is unrecorded from grazed, shaded or waterlogged habitats. The species is foetid and clothed in viscid hairs. These characteristics tend to deter herbivory by insects (Palmblad 1968) and are presumably effective against some other predators. *S.v.* regenerates entirely by means of wind-dispersed seeds, which germinate in spring, and an average plant may produce 6000 seeds (Salisbury 1964). Seeds produced late in the season are frequently non-viable (Salisbury 1964), and the species has a lowland distribution, with a limit at 310 m in the British Isles (Wilson 1956). The degree of persistence of seed in the soil requires investigation. The dispersules (achene and pappus) are less buoyant in air than those of many members of the Asteraceae (Salisbury 1964). However, the species is an effective colonist of disturbed artificial habitats, and may be dispersed by air currents along roadsides and railways. Transport by humans is implicated by the fact that new sites have been reported up to 360 km from the nearest known localities (Salisbury 1964). Genetic differences between populations have been detected, and a dwarf, early-flowering, form occurs in some shingle sites (Akeroyd *et al.* 1978). The hybrid with *S. squalidus*, which is sterile and annual (see Stace 1975), occurs occasionally on demolition sites, on refuse and cinder tips, and in other disturbed urban habitats. The hybrid with *S. sylvaticus* L. has also been reported (Stace 1975).

Current trends *S.v.*, which was first recorded in Britain in 1666, has spread dramatically in the last 70 years (Salisbury 1964). From a distribution centred on SE England in 1900 the species has extended over much of the UK (Salisbury 1964, *ABIF*). Dispersal was perhaps facilitated by the development of a modern network of roads and railways. *S.v.* is probably still increasing.

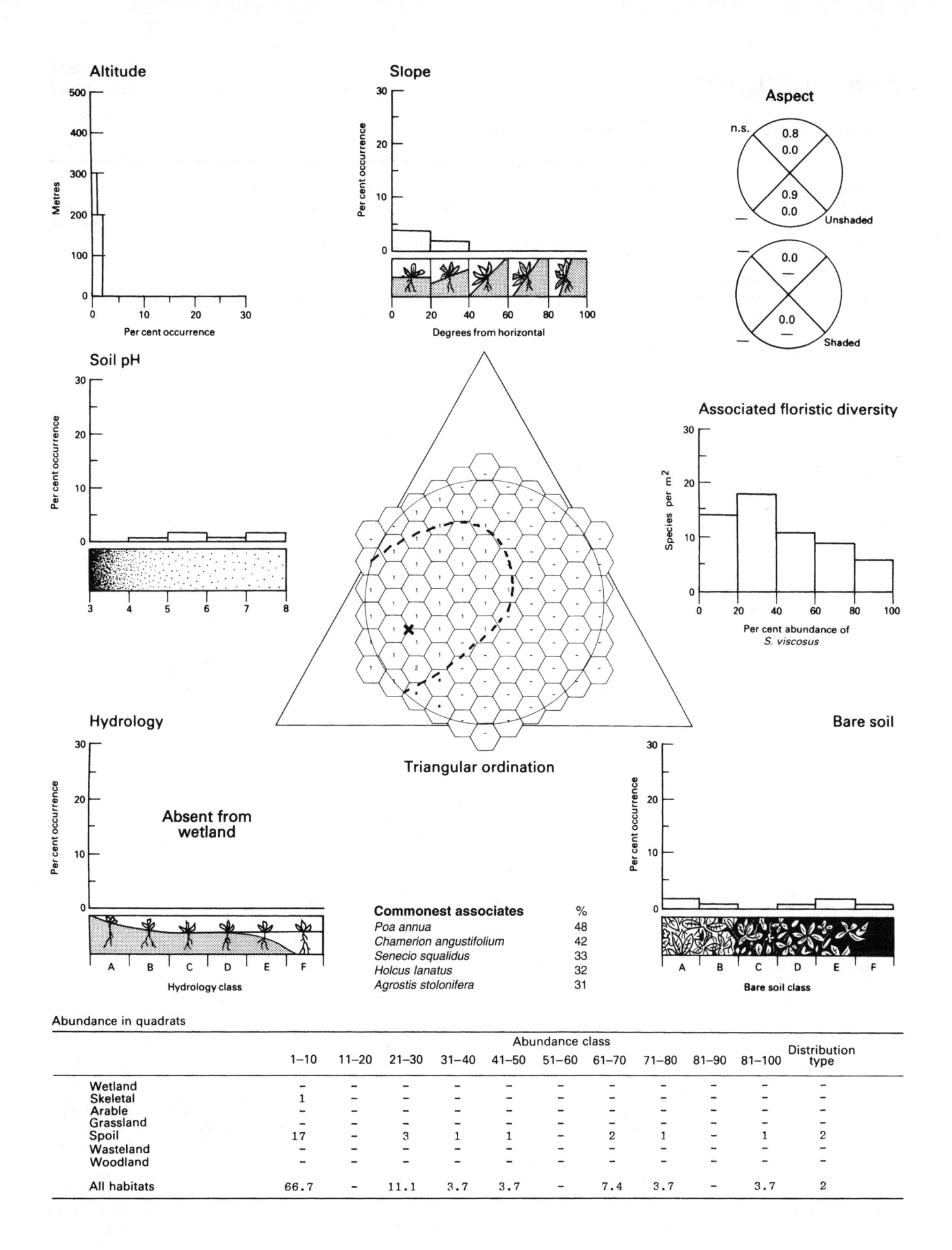

Altitude

Slope

Aspect

n.s. 0.8 / 0.0 / 0.9 / 0.0 — Unshaded

— / 0.0 / 0.0 / — Shaded

Soil pH

Associated floristic diversity

Species per m² vs Per cent abundance of *S. viscosus*

Hydrology

Absent from wetland

Triangular ordination

Bare soil

Commonest associates

	%
Poa annua	48
Chamerion angustifolium	42
Senecio squalidus	33
Holcus lanatus	32
Agrostis stolonifera	31

Abundance in quadrats

	1–10	11–20	21–30	31–40	41–50	51–60	61–70	71–80	81–90	81–100	Distribution type
Wetland	–	–	–	–	–	–	–	–	–	–	
Skeletal	1	–	–	–	–	–	–	–	–	–	
Arable	–	–	–	–	–	–	–	–	–	–	
Grassland	–	–	–	–	–	–	–	–	–	–	
Spoil	17	–	3	1	1	–	2	1	–	1	2
Wasteland	–	–	–	–	–	–	–	–	–	–	
Woodland	–	–	–	–	–	–	–	–	–	–	
All habitats	66.7	–	11.1	3.7	3.7	–	7.4	3.7	–	3.7	2

Senecio vulgaris L.

Asteraceae
Groundsel

Field records include var. *hibernicus* Syme (see 'Synopsis')

Life-form Summer- or winter-annual therophyte. Stems erect, often branched; leaves oblong, pinnatifid, often cottony, <2000 mm^2; roots with or without VA mycorrhizas.

Phenology Variable, since germination may occur throughout most of the year. Plants capable of overwintering and flowers may be seen throughout the year, but particularly from April to October. Seed set mainly from May to October.

Height Foliage and flowers to 450 mm.

Seedling RGR 0.5-0.9 week^{-1}. SLA 25.8. DMC 8.4.

Nuclear DNA amount 3.2 pg, $2n = 40*$ (*BC*, Bennett & Smith 1976); $4x*$.

Established strategy Ruderal.

Regenerative strategies W, B$_s$. Regenerates exclusively by means of wind-dispersed seed. Ungerminated seeds have some capacity to persist in the soil; SB III.

Flowers Florets yellow, tubular, hermaphrodite, usually self-fertilized (Gibbs *et al.* 1975); *c.* 40 in each capitulum; often >40 capitula in a corymbose inflorescence.

Germinule 0.25 mg, 1.8 × 0.5 mm, achene, with a pappus *c.* 2.5 mm in length.

Germination Epigeal; t_{50} 3 days; 11-28°C; germination rate increased after dry storage at 20°C; L > D.

Biological flora None.

Geographical distribution Abundant throughout the British Isles except in the Highlands of Scotland (100% of British vice-counties), *ABIF* page 656; Europe, although only as a casual in the extreme N (97% of European territories); Asia and N Africa.

Habitats Found in a range of disturbed habitats including spoil heaps, building rubble, manure heaps and sewage waste and arable land, particularly among broad-leaved crops. Occasionally present in various other habitats including paths, stream sides, limestone-quarry heaps and rock outcrops. Rarely found in shaded habitats and absent from wetlands. A familiar garden weed. [Grows on unconsolidated sand dunes and maritime shingle ridges.]

Gregariousness Intermediate.

Altitude Extends to 400 m. [Up to 530 m.]

Slope Wide-ranging, but only infrequently associated with steep slopes.

Aspect Widely distributed but slightly more frequent on S-facing slopes.

Hydrology Absent from wetlands.

Soil pH Most commonly associated with soils of pH >6.0. Absent from acidic soils of pH 4.5.

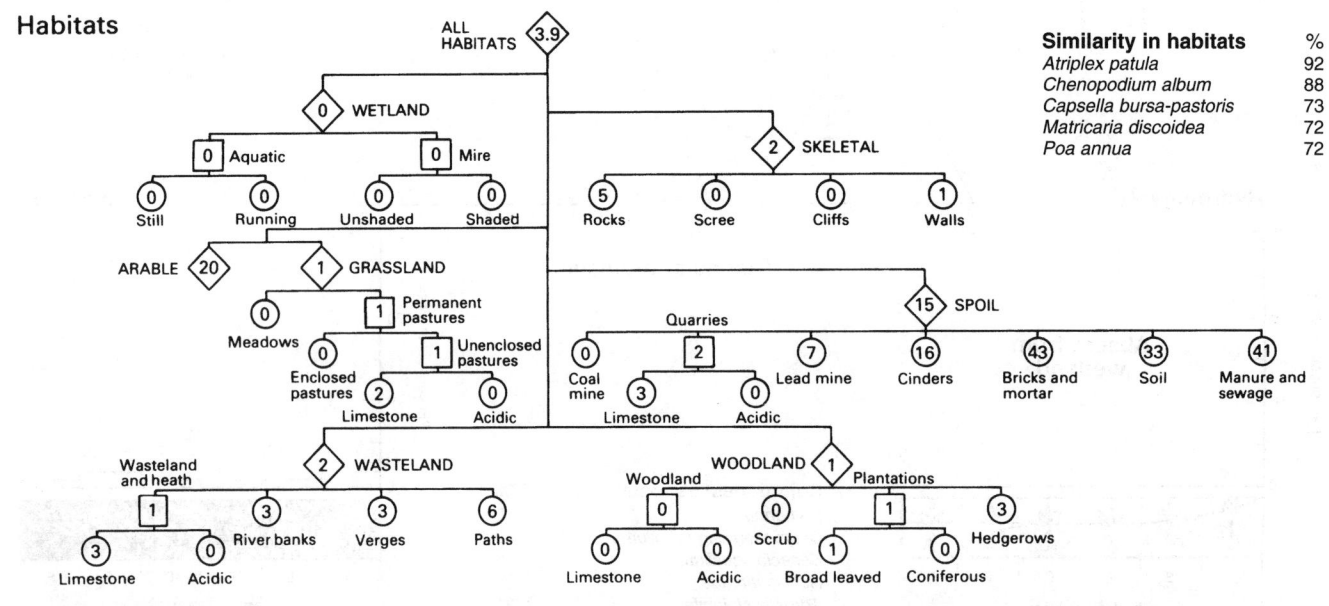

Habitats

Similarity in habitats	%
Atriplex patula	92
Chenopodium album	88
Capsella bursa-pastoris	73
Matricaria discoidea	72
Poa annua	72

Bare soil Mainly restricted to sites with large amounts of either bare soil or exposed spoil.

△ **ordination** Distribution centred on vegetation associated with highly disturbed but productive habitats. Evidence of some capacity to occur locally in less-disturbed fertile situations. Excluded from unproductive habitats.

Associated floristic diversity Typically low to intermediate.

Synopsis *S.v.* is an annual herb which exploits impermanent sites such as soil and manure heaps, as well as disturbed habitats with greater continuity, such as garden plots and arable fields (particularly those with broad-leaved crops). *S.v.* frequently occurs in mixed populations with *Poa annua*. However, unlike *P.a.*, *S.v.* is a relatively tall plant, and is unable to exploit regularly grazed, mown and trampled sites. *S.v.* is also obligately annual. The success of *S.v.* in exploiting fertile, unshaded, disturbed habitats appears to be related to the association between a short life-span and the capacity to germinate over much of the year; seeds may be produced within 6 weeks, and up to three generations may be formed each year *(WCH)*. The species displays a high potential for seed production and an average plant produces over 1000 seeds (Salisbury 1942). Although *S.v.* thrives on soils of high mineral-nutrient status, the species is also capable of regenerating under conditions of nutrient stress and drought; exceptionally stunted plants on skeletal habitats such as rock outcrops or cinder tips may produce a single capitulum and set <40 seeds. Although there is flexibility in seed production, reproductive effort expressed as the percentage of the total weight of the plant allocated to seeds tends to be maintained at a constant level (Harper & Ogden 1970). Seeds of *S.v.* are effectively dispersed by wind, by humans and by other agencies. The small plumed seeds are buoyant in air currents, and the hairs of the pappus are sufficiently viscid to adhere to animals or to clothing (Ridley 1930). Achenes can also survive ingestion by birds (Ridley 1930). The seeds exhibit limited persistence in the soil (Brenchley & Warington 1930, Roberts & Feast 1972) and are strongly responsive to changes in light

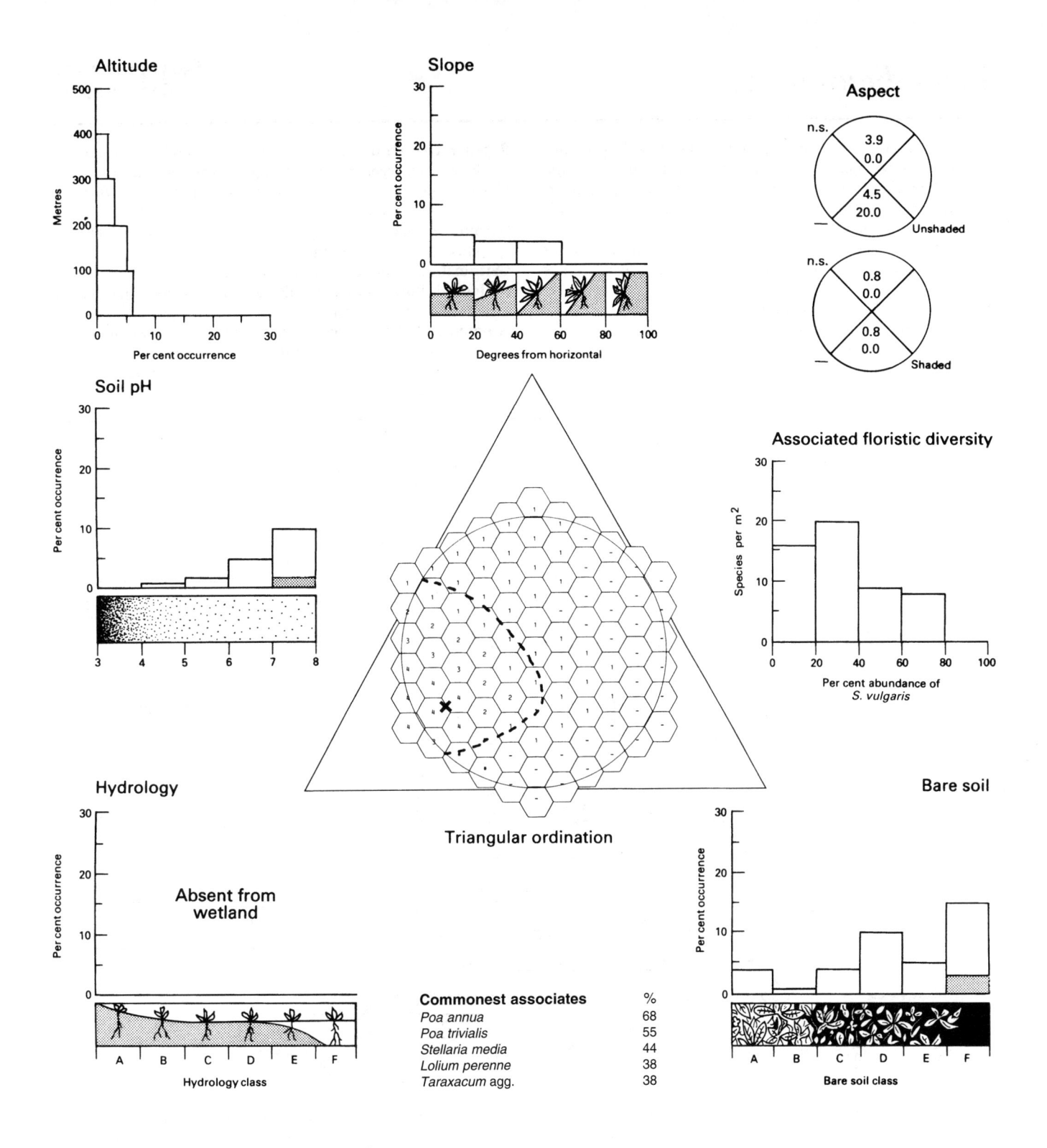

Commonest associates	%
Poa annua	68
Poa trivialis	55
Stellaria media	44
Lolium perenne	38
Taraxacum agg.	38

intensity and quality (Popay & Roberts 1970a,b, Hilton 1983). Although *S.v.* is essentially in-breeding, ecotypic differentiation has occurred. Herbicide resistance, coupled with reduced growth rate, has been reported (Warwick 1980, Holt & Radosevich 1983), and some populations from industrial areas have been reported to be more resistant to 'acid rain' (Hodgkin & Briggs 1981), while those from road verges are more salt-tolerant (Briggs 1978). Var. *hibernicus*, which may have arisen from the introgression of *S. squalidus* into *S.v.* appears to be increasing (Stace 1977; Ingram *et al.* 1980). This taxon shows a higher degree of out-breeding than typical *S.v.* (Marshall & Abbott 1984), produces more seed (Oxford & Andrew 1977) and shows greater restriction to urban spoil and to roadside habitats than typical forms. In addition, it is particularly conspicuous in towns and villages at upland locations above the altitudinal range of *S.s.* Another rayed form (var. *denticulatus* (O. F. Mueller) Hyl.) is a winter annual, has a lower reproductive potential, shows marked seed dormancy (Kadereit 1984) and, except in Mediterranean regions where it occurs on mountains (Kadereit 1984), has a maritime distribution. *S.v.* occasionally forms sterile hybrids with *S.s.*, and the fertile allohexaploid between the two, *S. cambrensis* Rossar, is endemic to Britain.

Current trends An effective colonist of a wide range of fertile artificial habitats; remains exceedingly common.

Silene dioica (L.) Clairv.

Caryophyllaceae
Red Campion

Life-form Polycarpic or perhaps rarely monocarpic perennial protohemicryptophyte or chamaephyte with creeping stock. Non-flowering shoots, numerous decumbent, flowering shoots erect; leaves ovate, hairy, typically <2000 mm², roots mainly shallow (*Biol Fl*), non-mycorrhizal.
Phenology Winter green. Shoots elongate in spring and flowers appear mainly in May and June. Seed shed gradually from June onwards.
Height Foliage largely <600 mm; flowers to 900 mm.
Seedling RGR 1.0-1.4 week^{-1}. SLA 31.6. DMC 11.2.
Nuclear DNA amount 5.4 pg; $2n = 24$ (*FE1*); $2x$.
Established strategy C-S-R.

Regenerative strategies B_s. Further research is needed to characterize regeneration in *S.d.* (see 'Synopsis'); SB III.
Flowers Rose-coloured, dioecious, insect-pollinated (*Biol Fl*); usually <10 in a dichasium.
Germinule 0.69 mg, 1.3 × 1.1 mm, seed, often *c*. 200 in a dehiscent capsule (Thompson 1981).
Germination Epigeal; t_{50} 2 days; 7-31°C; L > D.
Biological flora Baker (1947) as *Melandrium dioicum* (L. emend.) Coss. & Germ.
Geographical distribution Britain but relatively scarce in Ireland (96% of British vice-counties), *ABIF* page 178; Europe, although rare in the S (67% of European territories); N Africa and W Asia.

Habitats Predominantly a species of damp, fertile soils in lightly shaded situations in woodland, scrub, hedgerows, broad-leaved plantations and mire and on river and stream banks. Occasional in open habitats on rocks, cliffs and walls, and among mining waste and other types of spoil heaps. Absent from arable and aquatic habitats. [Found on sea cliffs and mountain ledges (*Biol Fl*).]
Gregariousness Sparse to intermediate.
Altitude More common in lowland areas but extends to 400 m. [Up to 1070 m.]
Slope Wide-ranging.
Aspect In shaded habitats slightly more frequent, and significantly more abundant in S-facing situations. In the open, abundant only on N-facing slopes.
Hydrology In wetland, most frequent on gently sloping ground and excluded from waterlogged habitats.
Soil pH Frequent within the pH range 3.5-8.0.
Bare soil Widely distributed, but most frequently associated with habitats having intermediate amounts of bare ground.
△ **ordination** A diffuse pattern with highest frequency of occurrence in woodland vegetation dominated by vernal geophytes (i.e. contours at base of diagram), scattered occurrences in tall-herb communities (weakly-defined satellites towards top of the diagram) and pronounced inability to exploit conditions of extreme stress or disturbance.
Associated floristic diversity Typically low to intermediate.

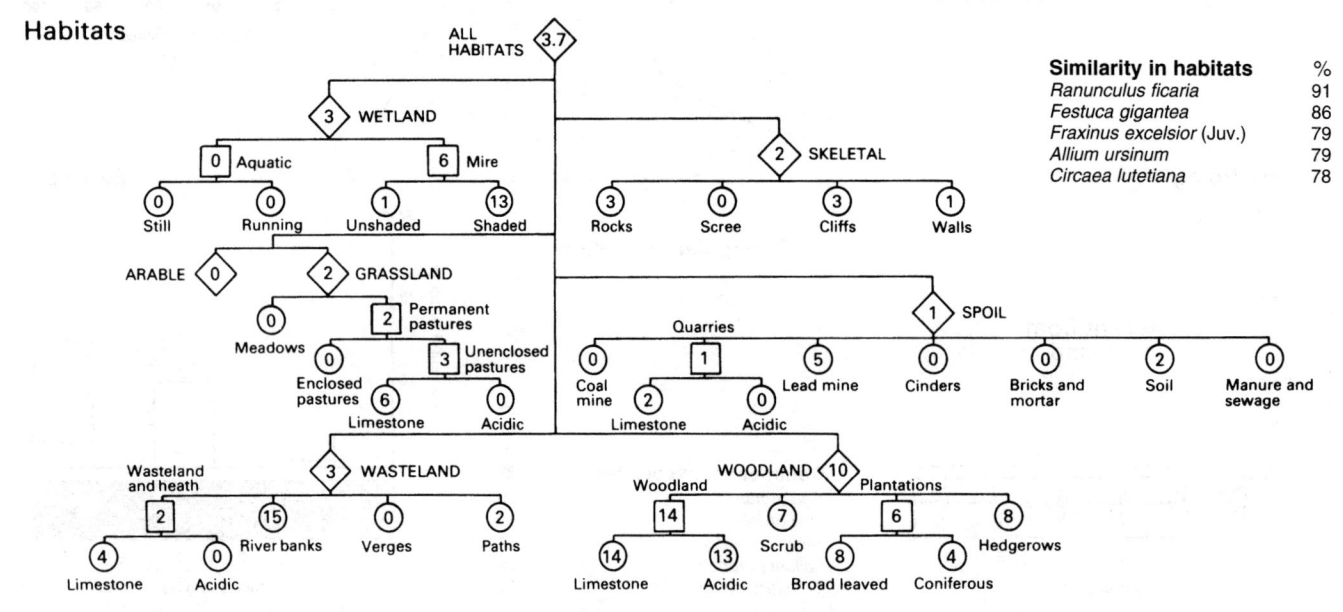

Habitats

Similarity in habitats	%
Ranunculus ficaria	91
Festuca gigantea	86
Fraxinus excelsior (Juv.)	79
Allium ursinum	79
Circaea lutetiana	78

Synopsis *S.d* is a potentially fast-growing, winter-green herb in which the leaf canopy is situated close to the ground. The species is most typically associated with relatively fertile sites where the vigour of potential dominants is reduced by shade and by the occasional disturbance which results from factors such as soil creep on steep slopes or flooding on river banks. *S.d* is perhaps typically a short-lived polycarpic perennial which may flower in its first year (*Biol Fl*). The species is a mesophytic long-day plant (*Biol Fl*), is largely confined to lightly shaded habitats and can survive in a non-flowering condition in deep shade (*Biol Fl*). However, the species is most prominent during the open phase of the coppice cycle (*Biol Fl*) and flowers most profusely at woodland margins. In moist upland and maritime environments *S.d* often grows on rock ledges and at other sites which are inaccessible to stock. In these open sites *S.d* is frequently found on highly calcareous soils. However, in calcareous woodland *Myosotis sylvatica* Hoffm., a species of closely similar life-history, morphology and ecology, may partially replace *S.d*. In lowland Britain the distribution of *S.d.* is biased towards non-calcareous strata. The foliage of *S.d* contains high concentrations of P (*SIEF*). The species is susceptible to frost damage and to both drought and waterlogging (*Biol Fl*). Vegetative spread by means of short stolons is largely ineffectual (*Biol Fl*). In contrast, seed set is frequently good. Flowers are often infected by the smut *Ustilago violacea* (Pers.) Fuckel. Salisbury (1942) suggests that individuals may produce an average of 7000 seeds. Our own studies in Northern England suggest that less than half this number may be more realistic, but seed production is nevertheless prodigious in comparison with that of most other perennial woodland herbs. Pot experiments strongly indicate tendencies for seed persistence and early spring germination (Roberts 1986a). At a relatively fine scale, plants of the two sexes appear to differ in their ecology and distribution. Male plants normally produce more leaves and flowers than females (Cox 1981), and tend to occur in greater numbers within the

Altitude

Slope

Aspect

Soil pH

Associated floristic diversity

Hydrology

Triangular ordination

Bare soil

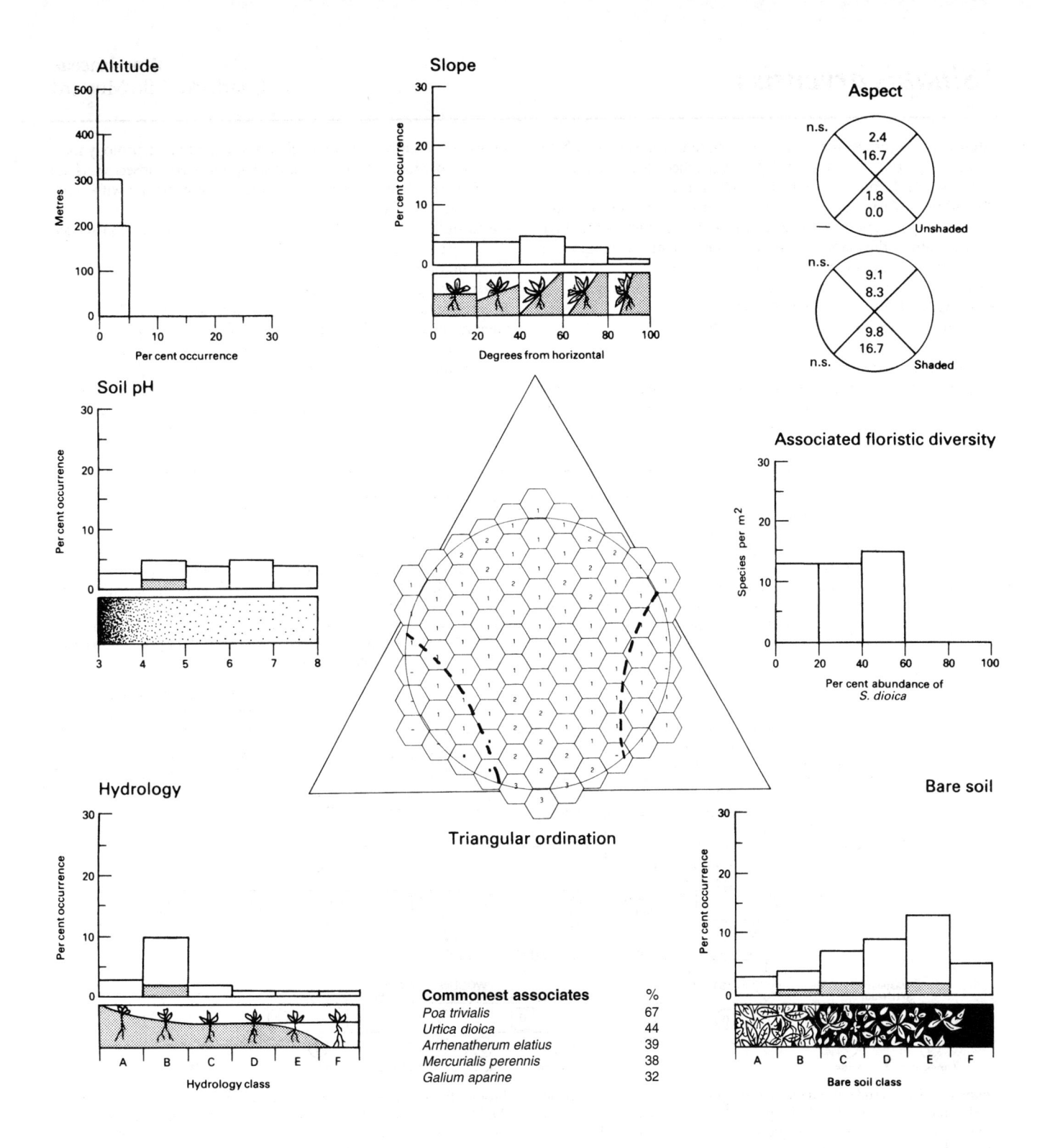

Commonest associates	%
Poa trivialis	67
Urtica dioica	44
Arrhenatherum elatius	39
Mercurialis perennis	38
Galium aparine	32

population (Kay & Stevens 1986). However, female plants grown adjacent to male plants often produce more shoots and seeds, while the size of the male plant is reduced (Cox 1981). Further, they occupy slightly different temporal niches. Thus, females are more prominent in the June canopy while by August, male and female plants contribute equally (Cox 1981). Seed size is very variable within populations (Thompson 1981), and Thompson suggests that the production of many small, relatively disadvantaged seeds may serve to reduce the level of insect predation, which for *S.d.* is high. Ecotypic differentiation in plant form (Turesson 1922, 1925), in heavy-metal tolerance (Prat 1934) and in germination characteristics (Thompson 1975) have been described for European populations, and in Britain a montane variety with seeds almost three times as heavy as normal was formerly described as ssp. *zetlandica* (Compton) Clapham. Hybrid swarms with *S. latifolia* Poiret, a short-lived species of disturbed wasteland, occur in disturbed lowland habitats where the two species overlap (see Stace 1975).

Current trends As a result of deforestation and agriculture, *S.d* is a decreasing species in many lowland areas and is showing increasing hybridization with *S.l*. However, *S.d* is not under immediate threat in upland habitats.

+*Sinapis arvensis* L.

Brassicaceae
Charlock, Wild Mustard

Life-form Summer- (or winter)-annual, semi-rosette therophyte with tap-root. Shoot erect, often branched; leaves lyrate to lanceolate, hispid, occasionally >4000 mm^2.
Phenology Variable, but mainly germinates in spring with flowering in June and July and seed set from August until after the death of the plant (*Biol Fl(a)*). Young plants resulting from summer germination may survive the winter.
Height Foliage to 500 mm overtopped by the inflorescence.
Seedling RGR No data. SLA 29.7. DMC 11.0.
Nuclear DNA amount 0.8 pg; $2n = 18*(BC)$; $2x*$.
Established strategy Intermediate between ruderal and competitive-ruderal.
Regenerative strategies B_s. Regeneration entirely dependent upon seed. Germination in spring and to a lesser extent in autumn or winter. A persistent seed bank is formed (*Biol Fl(a)*); SB IV.

Flowers Yellow, hermaphrodite, homogamous or protogynous, cross-pollinated by insects and self-pollinated if insect visits fail (*Biol Fl(a)*); in a racemose inflorescence generally with >100 flowers per plant.
Germinule 1.15 mg, 1.6 × 1.6 mm, seed, in an 8-13-seeded siliqua (*FE1*).
Germination Epigeal; t_{50} 2 days; germination rate increased after dry storage at 20°C; L > D.
Biological flora (a) Fogg (1950), (b) Mulligan and Bailey (1975).
Geographical distribution The British Isles, but probably not native (100% of British vice-counties), *ABIF* page 281; Europe, but less common in the N and perhaps only native in Mediterranean region (89% of European territories and 5% naturalized); N Africa and SW Asia; introduced in N and S America, S Africa, Australia and New Zealand.

Habitats A common arable weed found also in recently disturbed artificial habitats such as soil heaps and cinder tips, and beside paths and railways. Also frequently observed on disturbed roadsides. Absent or of casual occurrence in shaded, waterlogged and skeletal habitats, and in grazed and ungrazed grasslands.
Gregariousness Usually occurring as isolated individuals.
Altitude More common in lowland areas. Infrequent above 300 m, where the species may be dependent upon regular re-introduction for survival. [Up to 450 m.]

Slope Absent from the steepest slopes.
Aspect Insufficient data.
Hydrology Absent from wetlands.
Soil pH Absent from acidic soils.
Bare soil Distribution correlated with high exposures of bare soil.
Δ ordination Clear evidence of a strong association with highly disturbed fertile habitats.
Associated floristic diversity Usually in vegetation of intermediate species diversity. Although a potentially tall and robust plant, the data are not sufficient to assess its capacity to exclude other annual spp.

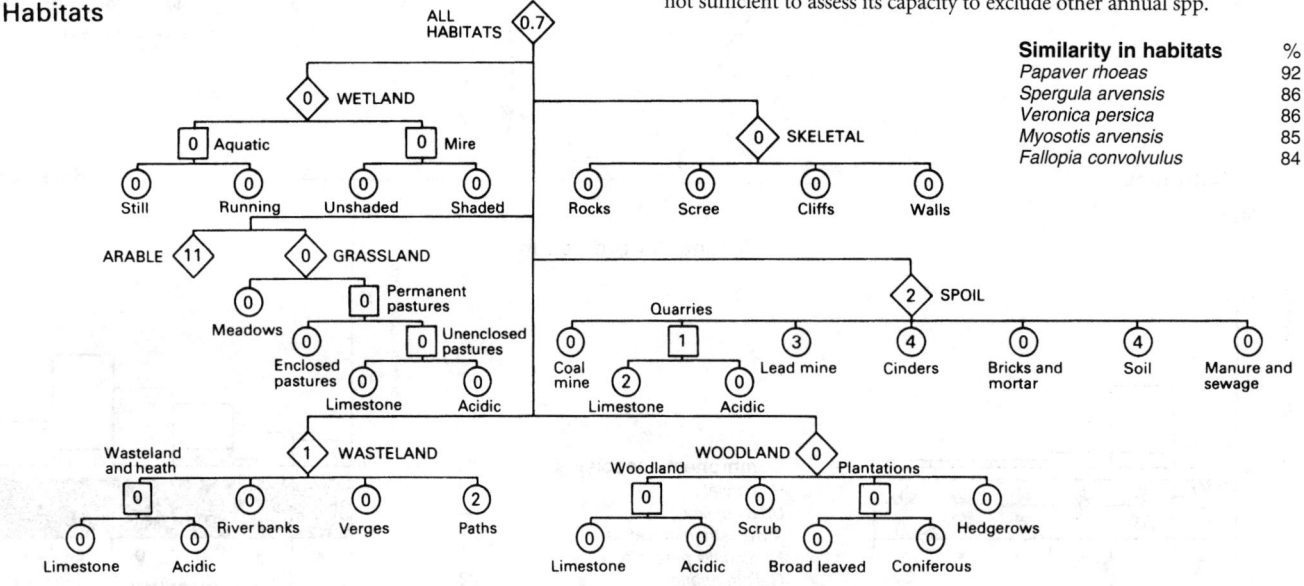

Similarity in habitats	%
Papaver rhoeas	92
Spergula arvensis	86
Veronica persica	86
Myosotis arvensis	85
Fallopia convolvulus	84

Habitats

Synopsis *S.a.* is a ruderal and arable weed, probably originating from the Mediterranean region (*FE1*). In keeping with the Mediterranean origins, seed set in *S.a.* appears to be greatly influenced by climatic factors (Edwards 1980). *S.a.* is regarded as palatable to stock and has been included in human diets, but seeds are very poisonous to livestock (*Biol Fl(b)*). The species is sensitive to mechanical damage (*Biol Fl(a)*) and is seldom recorded from grazed habitats. *S.a.* takes a minimum of 6 weeks to flower and, due to restricted seed output and mobility, does not exploit sites of transient disturbance as effectively as species such as *Poa annua* and *Senecio vulgaris*. Seeds germinate in spring and to a lesser extent in autumn and winter (*Biol Fl(a)*), and germination of seed populations may be partially inhibited by a leaf canopy (Gorski *et al.* 1977). Colonization of much of Europe coincident with the spread of agriculture has probably been assisted by a number of features of its biology. Some seeds are usually still in the siliqua at harvest time, and *S.a.* has been spread as an impurity along with seeds of crop plants (*WCH*). Seeds may pass undamaged through the intestines of birds, which are important as dispersal agents (*Biol Fl(b)*). The long-lived seed bank enables survival in sites subject to infrequent cultivation and disturbance. It has been estimated (Edwards 1980) that at certain sites one reproductive phase every 11 years is sufficient to maintain the seed bank. Like other ruderals, *S.a.* shows marked phenotypic plasticity. In unfavourable habitats an individual may yield as few as 16 seeds, while in robust plants up to 25 000 may be produced (*Biol Fl(b)*). A number of varieties have been recognized (*Biol Fl(a)*) but the extent of ecotypic differentiation in *S.a.* requires investigation.
Current trends Fogg (1950) emphasized that formerly *S.a.* was a serious weed, particularly of cereals. This is no longer the case (Chancellor & Froud-Williams 1984), and *S.a.* is recorded more frequently from broadleaved than from cereal crops in the Sheffield region. This decrease, like that of many arable weeds, stems from the widespread use of weed-killers (Edwards 1980), and to some extent from the improved purity of crop seeds. A further reduction in the abundance of *S.a.* is predicted.

Altitude
Slope
Aspect
Unshaded
Shaded

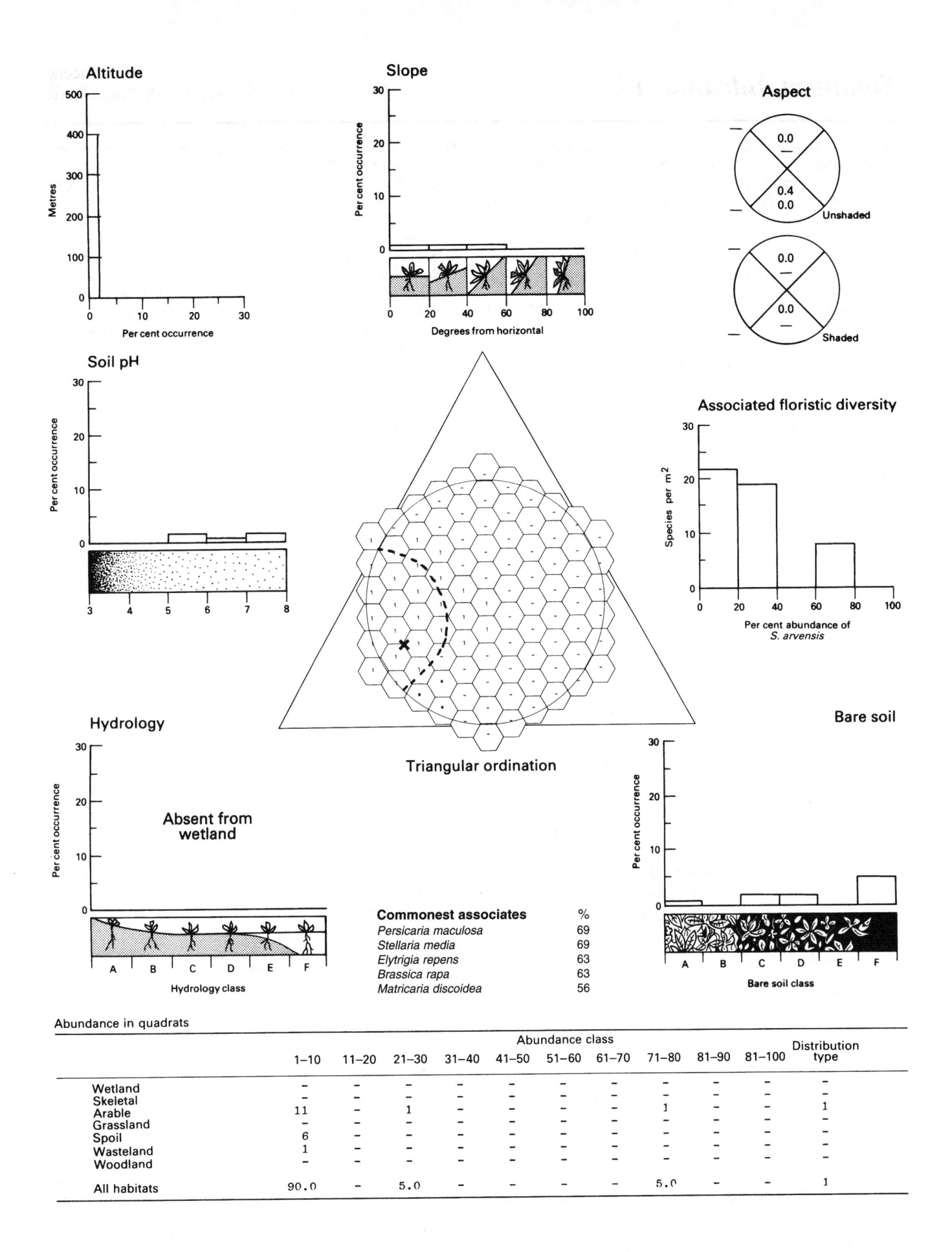

Soil pH
Associated floristic diversity
Species per m^2
Per cent abundance of *S. arvensis*

Hydrology
Absent from wetland

Triangular ordination

Commonest associates	%
Persicaria maculosa	69
Stellaria media	69
Elytrigia repens	63
Brassica rapa	63
Matricaria discoidea	56

Bare soil

Abundance in quadrats

	\multicolumn{10}{c	}{Abundance class}	Distribution type
	1–10	11–20	21–30	31–40	41–50	51–60	61–70	71–80	81–90	81–100	
Wetland	–	–	–	–	–	–	–	–	–	–	–
Skeletal	–	–	–	–	–	–	–	–	–	–	–
Arable	11	–	1	–	–	–	–	1	–	–	1
Grassland	–	–	–	–	–	–	–	–	–	–	–
Spoil	6	–	–	–	–	–	–	–	–	–	–
Wasteland	1	–	–	–	–	–	–	–	–	–	–
Woodland	–	–	–	–	–	–	–	–	–	–	–
All habitats	90.0	–	5.0	–	–	–	–	5.0	–	–	1

Solanum dulcamara L.
Solanaceae
Bittersweet, Woody Nightshade

Life-form Woody perennial phanerophyte or chamaephyte with scrambling stems; leaves ovate, often deeply lobed, sometimes hairy, <2000 mm^2; roots with VA mycorrhizas.

Phenology Shoot extension and leaf production occur in spring. Plants flower from June to September and berries ripen from August to October.

Height Foliage and flowers usually <1 m.

Seedling RGR No data. SLA 28.9. DMC 18.6.

Nuclear DNA amount 2.3 pg; $2n = 24^*$ (*BC*); $2x^*$.

Established strategy Intermediate between C-S-R and competitor.

Regenerative strategies V, ?B$_s$. Attached (or detached) vegetative stems in contact with the ground may root, giving rise to large clonal patches. Also regenerates in spring by seed (see 'Synopsis'); ?SB II.

Flowers Purple, hermaphrodite, cross-pollinated by insects (Knuth 1906); often *c.* 20 in axillary cymes.

Germinule 1.49 mg, 2.0 × 1.9 mm, seed in a *c.* 25-seeded berry (Salisbury 1942).

Germination Epigeal; t_{50} 2 days; 19-35°C; dormancy broken by widely fluctuating temperatures or by chilling; L = D (Roberts & Lockett 1977).

Biological flora None.

Geographical distribution Common in England and Wales, but absent from much of Scotland and Ireland (92% of British vice-counties), *ABIF* page 488; most of Europe except the extreme N (85% of European territories); Asia, N Africa; introduced in N America.

Habitats Primarily a wetland species occurring most consistently in shaded mire. Also frequent on river and stream banks, in unshaded mire and, to a lesser extent, in shaded aquatic habitats. Also found in moist, shaded hedgerows; infrequently on spoil, in skeletal habitats and on waysides. Absent from meadows, pasture and arable habitats. [Occurs on shingle beaches (*CTW*).]

Gregariousness Potentially stand-forming even at low frequencies.

Altitude Confined to altitudes of less than 300 m and only abundant below 200 m. [Up to 310 m.]

Slope Like most wetland species, largely restricted to gentle slopes.

Aspect Only recorded from N-facing slopes, in both shaded and unshaded sites.

Hydrology Frequent in topogenous mire, particularly in sites close to open water, but only rarely present in soligenous mire. Also found in shallow water to 450 m.

Soil pH Restricted to soils of pH 4.0, and most occurrences on soils of neutral pH.

Bare soil Widely distributed; most frequently associated with habitats having intermediate exposures of bare soil.

△ **ordination** Distribution centred on fertile, moderately disturbed vegetation types.

Habitats

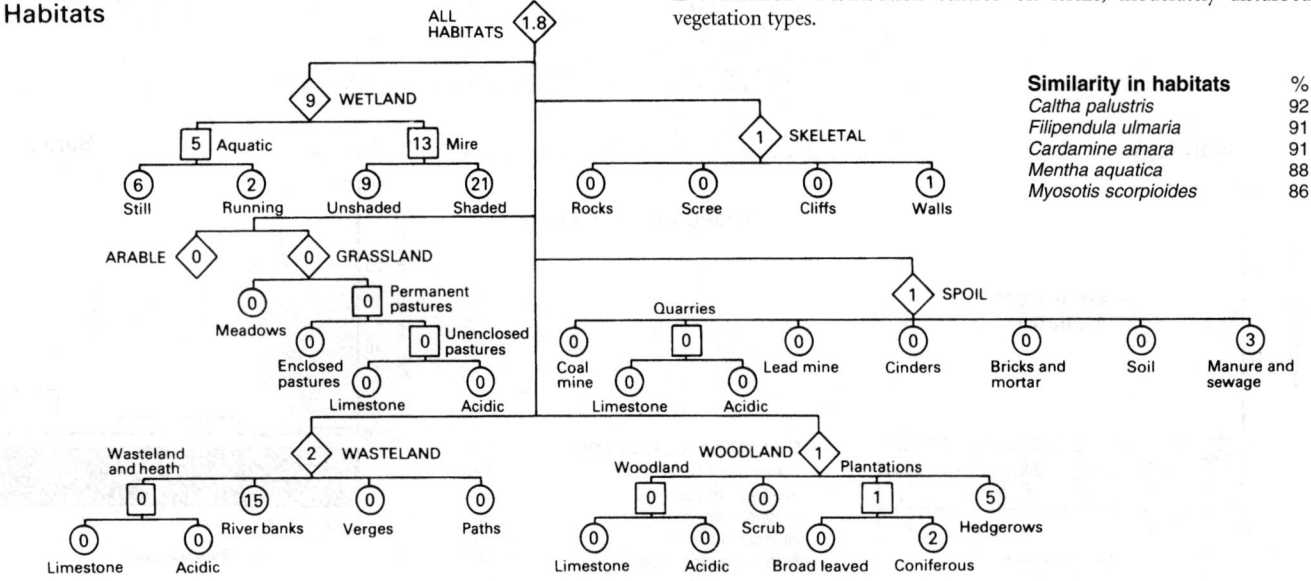

Similarity in habitats	%
Caltha palustris	92
Filipendula ulmaria	91
Cardamine amara	91
Mentha aquatica	88
Myosotis scorpioides	86

Associated floristic diversity Low, particularly at high frequencies of *S.d*.

Synopsis *S.d.* is a scrambling woody plant of fertile habitats, many situated in wetland. The species, which lives at least 20 years (Salisbury 1942), is capable of dominance within the herb layer of vegetation shaded by trees, but in unshaded habitats it may be suppressed by taller-growing herbaceous species. The Solanaceae show their greatest diversity in the tropics (Cronquist 1981) and, consistent with this family trend, *S.d.* is largely confined to southern and lowland Britain and extends only to altitudes of 310 m (Wilson 1956). *S.d.* is strongly defended against herbivory through the presence of glycoalkaloids, and only the fully ripe berry is not highly toxic (Cooper & Johnson 1984). Regeneration in situ is mainly by vegetative means. Stems in contact with the ground root readily, and large clonal patches may result. Pieces of stem are also capable of regeneration, and the frequency of *S.d* on river banks may be a reflection of its ability to form new plants in this way. Plants growing in shade often produce fewer flowers and fruits but Salisbury (1942) gives a mean value of 1400 seeds per plant. Seeds are dispersed by birds within many-seeded berries and are probably only of importance in colonizing new sites. Seedlings emerge in spring from down to 50 mm in the soil (Roberts & Lockett 1977); there is conflicting evidence concerning the capacity to develop a persistent seed bank. Further research is required to resolve this problem. Prostrate, fleshy-leaved forms may constitute a maritime ecotype (*CTW*), and outside Britain sun and shade ecotypes have been identified (Gauhl 1976, Clough *et al.* 1979). Races also occur in Europe, differing in the alkaloid composition of their above-ground parts and in their geographical distribution (Mathe & Mathe 1978).

Current trends Uncertain. Exploits fertile habitats but appears to be a poor colonist of new sites.

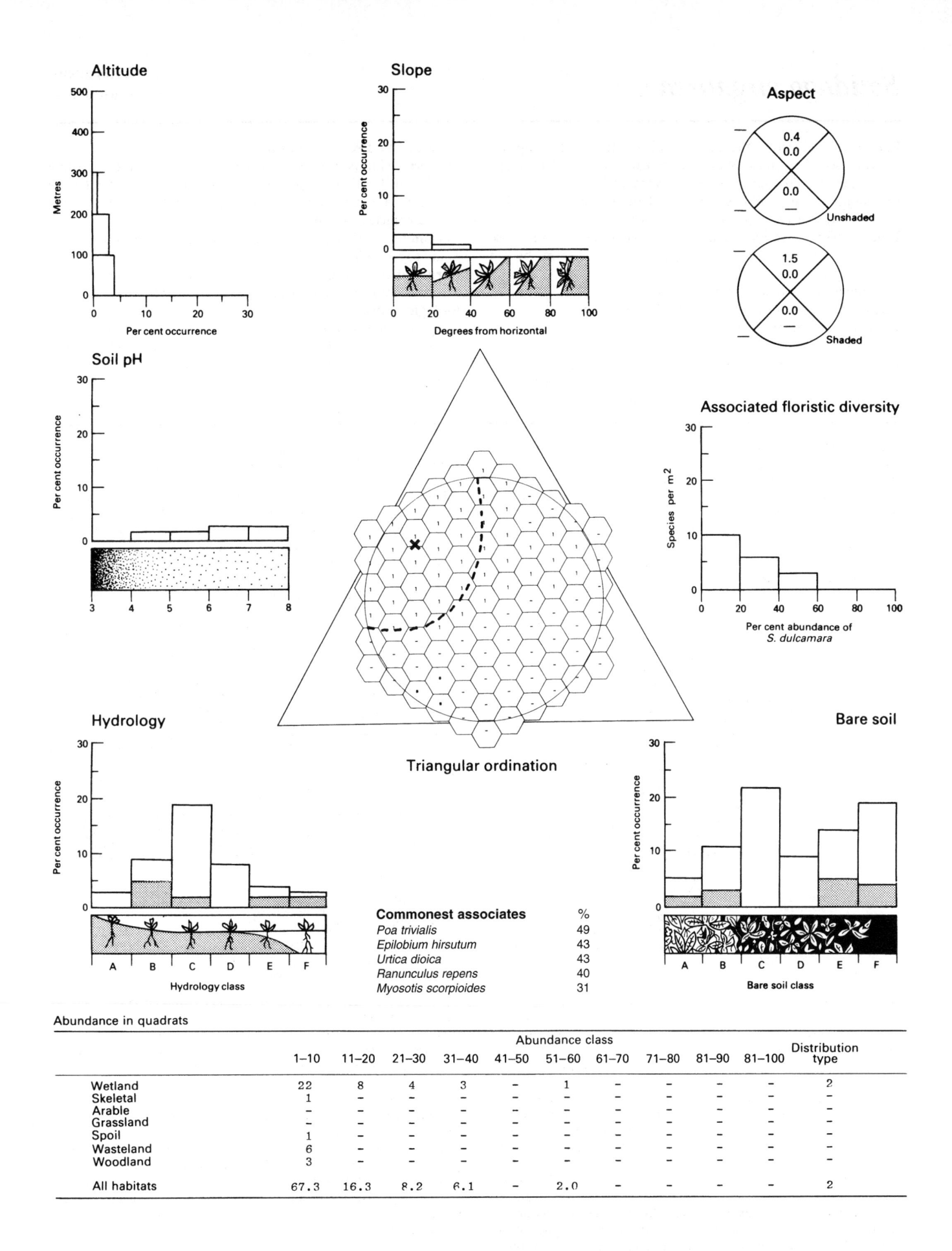

Altitude

Slope

Aspect

Unshaded: 0.4 / 0.0 / 0.0 / —

Shaded: 1.5 / 0.0 / 0.0 / —

Soil pH

Associated floristic diversity

Hydrology

Triangular ordination

Bare soil

Commonest associates

	%
Poa trivialis	49
Epilobium hirsutum	43
Urtica dioica	43
Ranunculus repens	40
Myosotis scorpioides	31

Abundance in quadrats

	1–10	11–20	21–30	31–40	41–50	51–60	61–70	71–80	81–90	81–100	Distribution type
Wetland	22	8	4	3	–	1	–	–	–	–	2
Skeletal	1	–	–	–	–	–	–	–	–	–	–
Arable	–	–	–	–	–	–	–	–	–	–	–
Grassland	–	–	–	–	–	–	–	–	–	–	–
Spoil	1	–	–	–	–	–	–	–	–	–	–
Wasteland	6	–	–	–	–	–	–	–	–	–	–
Woodland	3	–	–	–	–	–	–	–	–	–	–
All habitats	67.3	16.3	8.2	6.1	–	2.0	–	–	–	–	2

Solidago virgaurea L.

Asteraceae
Golden-rod

Life-form Polycarpic perennial, semi-rosette hemicryptophyte with ascending stock. Shoots erect; leaves ovate to lanceolate, typically <1000 mm^2; roots with VA mycorrhizas.

Phenology Leaves not overwintering. Flowers July to September and seed set from September to October.

Height Non-flowering plant, a small basal rosette; otherwise foliage to 550 mm; overtopped to *c.* 750 mm by an inflorescence.

Seedling RGR No data. SLA 22.9. DMC 22.7.

Nuclear DNA amount No data; $2n = 18^*$ (*BC*); $2x^*$.

Established strategy Intermediate between stress-tolerator and C-S-R.

Regenerative strategies W. Further studies of regeneration by seed are required. Effective regeneration is by seed since, although the stock may become multi-stemmed, the plant retains a tufted growth form; SB I.

Flowers Florets yellow, insect-pollinated, self-incompatible (Burton 1980), aggregated into capitula with *c.* 6 outer, female ligulate and *c.* 6 inner, hermaphrodite florets; *c.* >100 capitula in a panicle or raceme.

Germinule 0.52 mg, 2.8 × 0.8 mm, achene with a pappus of hairs up to 5 mm in length.

Germination Epigeal; t_{50} 5 days; percentage germination considerably reduced by dry storage at 20°C; L > D.

Biological flora None.

Geographical distribution The British Isles, but rare in C and E England (99% of British vice-counties), *ABIF* page 637; most of Europe, although absent from the S (85% of European territories); Asia; introduced in N America.

Habitats Most frequent in rocky or open habitats such as quarry spoil, scree, cliffs and outcrops. Found occasionally on wasteland and on quarry spoil, with some relatively rare occurrences on lead-mine heaps and demolition sites. Observed, too, on walls, railway cuttings and at the edge of woodland. Absent from pastures, arable land and wetlands. [Found on mountain ledges and on sheltered sea cliffs (Burnett 1964).]

Gregariousness Sparse.

Altitude Suitable habitats are more frequent in upland areas. Observed up to 400 m. [Up to 1100 m.]

Slope Found over a wide range of slopes.

Aspect Recorded more frequently from N-facing slopes.

Hydrology Absent from wetlands.

Soil pH Predominantly associated with calcareous conditions, but recent studies (Hodgson unpubl.) indicate that the pH distribution is bimodal; in addition to a peak above pH 7.0 there is another small peak at pH 3.5-4.5. [In some regions typically associated with acidic soils.]

Bare soil Usually on open ground, but typically in sites with high exposures of bare rock rather than of bare soil.

△ **ordination** Records indicate that *S.v.* is found in undisturbed vegetation types of low to moderate productivity. The occurrences in moderately productive vegetation appear to be a reflection of the ability of *S.v.* to colonize recently abandoned quarry spoil.

Habitats

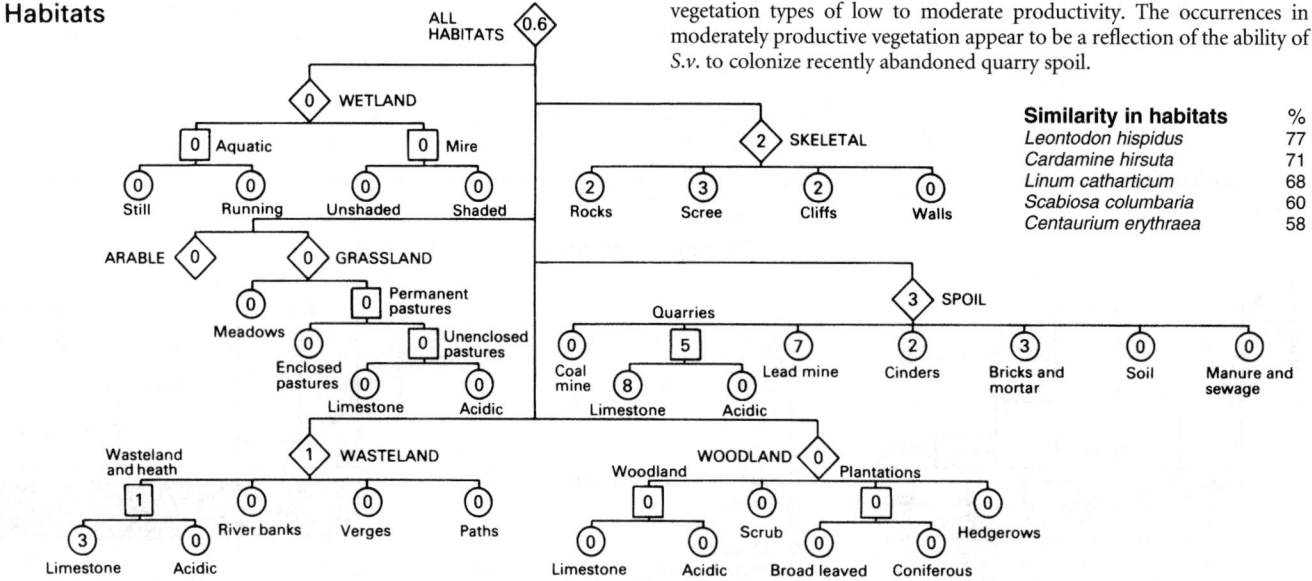

Similarity in habitats	%
Leontodon hispidus	77
Cardamine hirsuta	71
Linum catharticum	68
Scabiosa columbaria	60
Centaurium erythraea	58

Associated floristic diversity Varying widely, but typically intermediate.

Synopsis *S.v.* is a semi-rosette herb with a stout stock. *S.v.* is mainly restricted to infertile habitats, and achieves local abundance on cliffs and outcrops. Plants produce on average *c.* 4000 small, wind-dispersed seeds (Salisbury 1942). This feature has contributed significantly to the ability of *S.v.* to colonize quarries and railway cuttings. At the other extreme of its ecological range *S.v.* is also found at wood margins; here the species flowers less regularly. *S.v.* does not persist, however, in tall herb communities. Regeneration is by seed, and the rarity of *S.v.* in pasture is presumably a reflection of the vulnerability to grazing of the long flowering spike. The species occurs across a wide range in soil pH values. In lowland parts of the region *S.v.* tends to be calcifuge, whereas in the uplands the species occurs on both highly calcareous and highly acidic soils. Consistent with this trend, *S.v.* shows a predominantly calcifuge distribution in SE England (see Jermyn 1974, Hall 1980, Philp 1982). The wide ecological range of this out-breeding species appears to depend to a considerable extent on its capacity to form ecotypes. Both British and European populations are highly polymorphic and, for example, dwarf montane populations with large capitula may merit taxonomic separation (*CTW*, *FE4*). Many ecotypes have also been described for Europe (Turesson 1925, Holmgren 1968). In particular, ecotypes from shaded and unshaded habitats differed markedly in their response to light intensity; in particular, exposure to high irradiance caused a differential amount of chlorophyll-bleaching in shade ecotypes (Bjorkman & Holmgren 1963). *S.v.* hybridizes with the N American garden escapee, *S. canadensis* L. (Burton 1980).

Current trends *S.v.* is essentially restricted to infertile habitats, and its capacity to colonize new artificial habitats is limited, despite the large number of wind-dispersed propagules which each plant may produce. As a result of habitat destruction *S.v.* is now close to extinction in many lowland areas. However, the species is under no immediate threat in its upland habitats.

Altitude
(Per cent occurrence vs Metres)

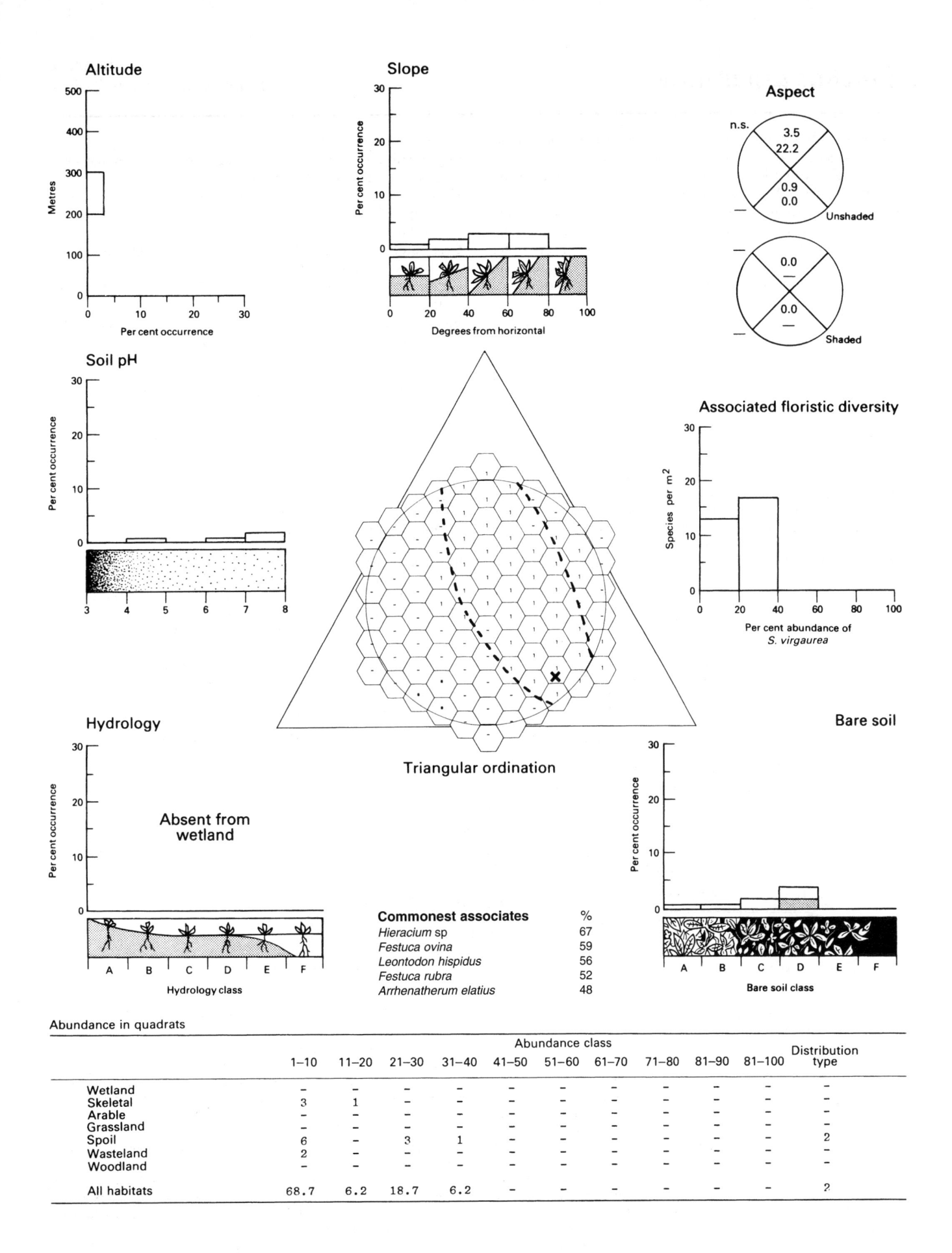

Slope
(Per cent occurrence vs Degrees from horizontal)

Aspect

	n.s.	3.5	
		22.2	
	—	0.9	Unshaded
		0.0	

		0.0	
	—		
		0.0	Shaded
	—		

Soil pH
(Per cent occurrence vs pH 3–8)

Associated floristic diversity
(Species per m² vs Per cent abundance of *S. virgaurea*)

Hydrology
Absent from wetland
(Per cent occurrence vs Hydrology class A–F)

Triangular ordination

Commonest associates

	%
Hieracium sp	67
Festuca ovina	59
Leontodon hispidus	56
Festuca rubra	52
Arrhenatherum elatius	48

Bare soil
(Per cent occurrence vs Bare soil class A–F)

Abundance in quadrats

	\multicolumn{10}{c}{Abundance class}	Distribution type
	1–10	11–20	21–30	31–40	41–50	51–60	61–70	71–80	81–90	81–100	
Wetland	–	–	–	–	–	–	–	–	–	–	–
Skeletal	3	1	–	–	–	–	–	–	–	–	–
Arable	–	–	–	–	–	–	–	–	–	–	–
Grassland	–	–	–	–	–	–	–	–	–	–	–
Spoil	6	–	3	1	–	–	–	–	–	–	2
Wasteland	2	–	–	–	–	–	–	–	–	–	–
Woodland	–	–	–	–	–	–	–	–	–	–	–
All habitats	68.7	6.2	18.7	6.2	–	–	–	–	–	–	2

Sonchus asper (L.) Hill

Asteraceae
Spiny Milk- or Sow-Thistle

Life-form Winter- or summer-annual, semi-rosette therophyte. Stems stout, often branched; leaves ovate, sometimes pinnatifid, with prickly margins, <10 000 mm^2; tap-rooted and sometimes with VA mycorrhizas.

Phenology Autumn-germinating plants overwinter as rosettes; spring-germinating plants as achenes. Growth continues until plants are killed through disturbance, competition or winter frosts. Flowers and fruits produced from May onwards in autumn-germinating and June onwards in spring-germinating plants (*Biol Fl(a)*). Leaves short-lived.

Height Foliage and flowers to 1500 mm. Large leaves all below 1000 mm.

Seedling RGR No data. SLA 24.5. DMC 12.4.

Nuclear DNA amount 3.7 pg; 2*n* = 18* (Bennett & Smith 1976, BC); (2*x**).

Established strategy Intermediate between ruderal and competitive-ruderal.

Regenerative strategies W, B$_s$. Reproduces solely by means of wind-dispersed seed (see 'Synopsis'); SB III.

Flowers Florets yellow, ligulate, visited by insects, self-compatible (*Biol Fl(b)*); >100 in each capitulum; capitula in few-flowered cymose umbels.

Germinule 0.32 mg, 2.5 × 1.2 mm, achene with a deciduous pappus *c*. 8 mm.

Germination Epigeal; t_{50} 2 days; 9-35°C; germination rate increased after dry storage at 20°C L > D.

Biological flora (a) Lewin (1948a), (b) Hutchinson *et al.* (1984).

Geographical distribution Lowland Britain (100% of British vice-counties), *ABIF* page 624; Europe northwards to Scandinavia (92% of European territories); N and W Asia, N Africa; introduced worldwide as a weed of cultivated land.

Habitats Occurs in a range of mainly artificial disturbed habitats. Of frequent occurrence on spoil, particularly limestone quarry waste and heaps of mineral soil. Occasional on limestone outcrops, disturbed wasteland, arable land and waysides. Recorded locally on exposed mud at the margins of ponds. Less frequent than *S. oleraceus* as a garden weed. Absent from pasture and woodland.

Gregariousness Sparse.

Altitude Recorded up to 360 m, but often only of casual occurrence at upland sites. [Up to 400 m.].

Slope Largely absent from slopes >40°.

Aspect More common on S-facing slopes.

Hydrology Persistent on moist rather than waterlogged soils.

Soil pH Not recorded below pH 5.0.

Bare soil Widely distributed, but more common in habitats having a high proportion of bare soil.

△ **ordination** Present in vegetation subject to intermediate levels of disturbance and various levels of productivity.

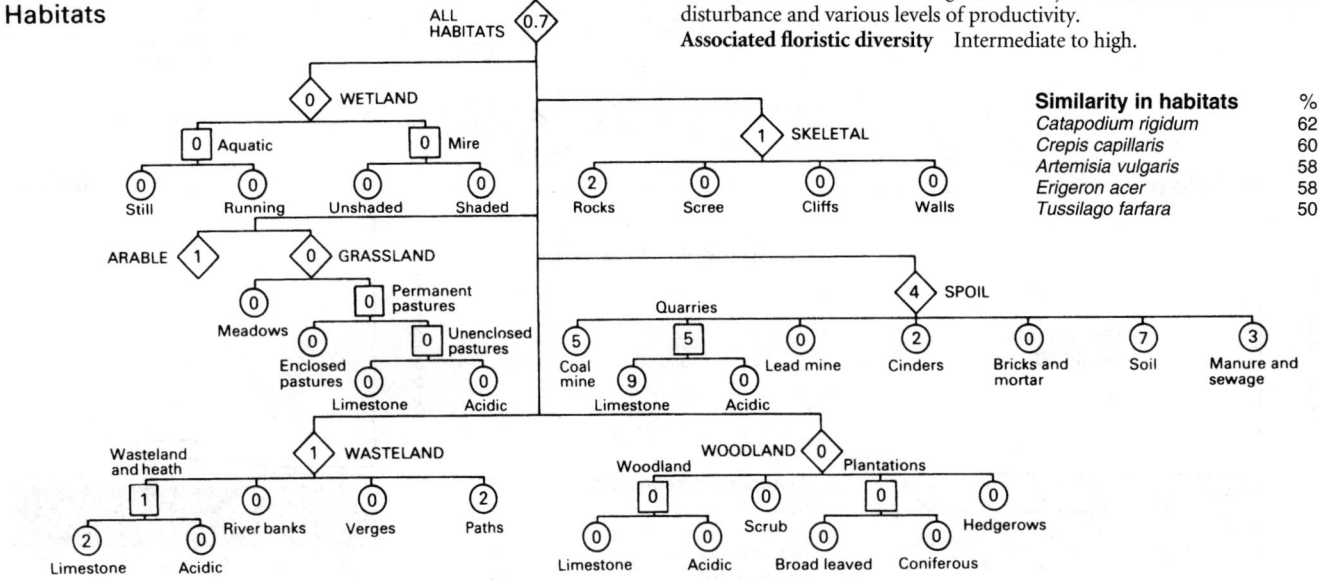

Associated floristic diversity Intermediate to high.

Habitats

Similarity in habitats	%
Catapodium rigidum	62
Crepis capillaris	60
Artemisia vulgaris	58
Erigeron acer	58
Tussilago farfara	50

Synopsis *S.a.* is a ruderal species with numerous wind-dispersed seeds (over 1500 on a robust plant (*Biol Fl(a)*). The species is particularly common in transient artificial habitats, where it exhibits considerable variation in size. *S.a.* often develops a rosette of leaves and a tap-root in autumn, and produces a flowering stem during the following summer, but spring-germinating individuals also occur and may function as summer annuals. *S.a.* reaches a larger size than many strict ruderals and takes longer to set seed (*c*. 10 weeks in spring-germinating plants (*Biol Fl(b)*), and is rarely an important arable weed. Despite their prickles, the leaves of *S.a.* are palatable to grazing stock, and *S.a.* seldom reaches maturity in heavily grazed areas. *S.a.* regenerates solely by means of seeds, which germinate on open ground during autumn or remain dormant until the following year (*Biol Fl(a)*). A persistent seed bank is reported. The plumed achenes are buoyant in air and are widely dispersed. Thus, a high proportion of the occurrences in the Sheffield region refer to isolated seedlings originating from seed dispersed into habitats where there is low probability of survival to flowering. This may explain the heterogeneity of the list of species 'similar in habitats'. The ecological significance, if any, of variation within the species remains to be elucidated (*Biol Fl(b)*). The ecology and geographical distribution of *S.a.* is remarkably similar to that of *S.o.*, with which it hybridizes (Stace 1975). The slight differences between the two are described under *S.o.* A further biennial subspecies of *S.a.* is recorded within Continental Europe (FE4).

Current trends *S.a.* was formerly regarded as a troublesome arable weed, even more so than *S.o.* (*Biol Fl(a)*). Although generally less abundant than *S.o.*, *S.a.* appears to be increasing. The species is now less common in farmland, but remains an effective colonist and is apparently still extending its geographical range (*Biol Fl(b)*) in other recent artificial habitats, particularly those in urban areas.

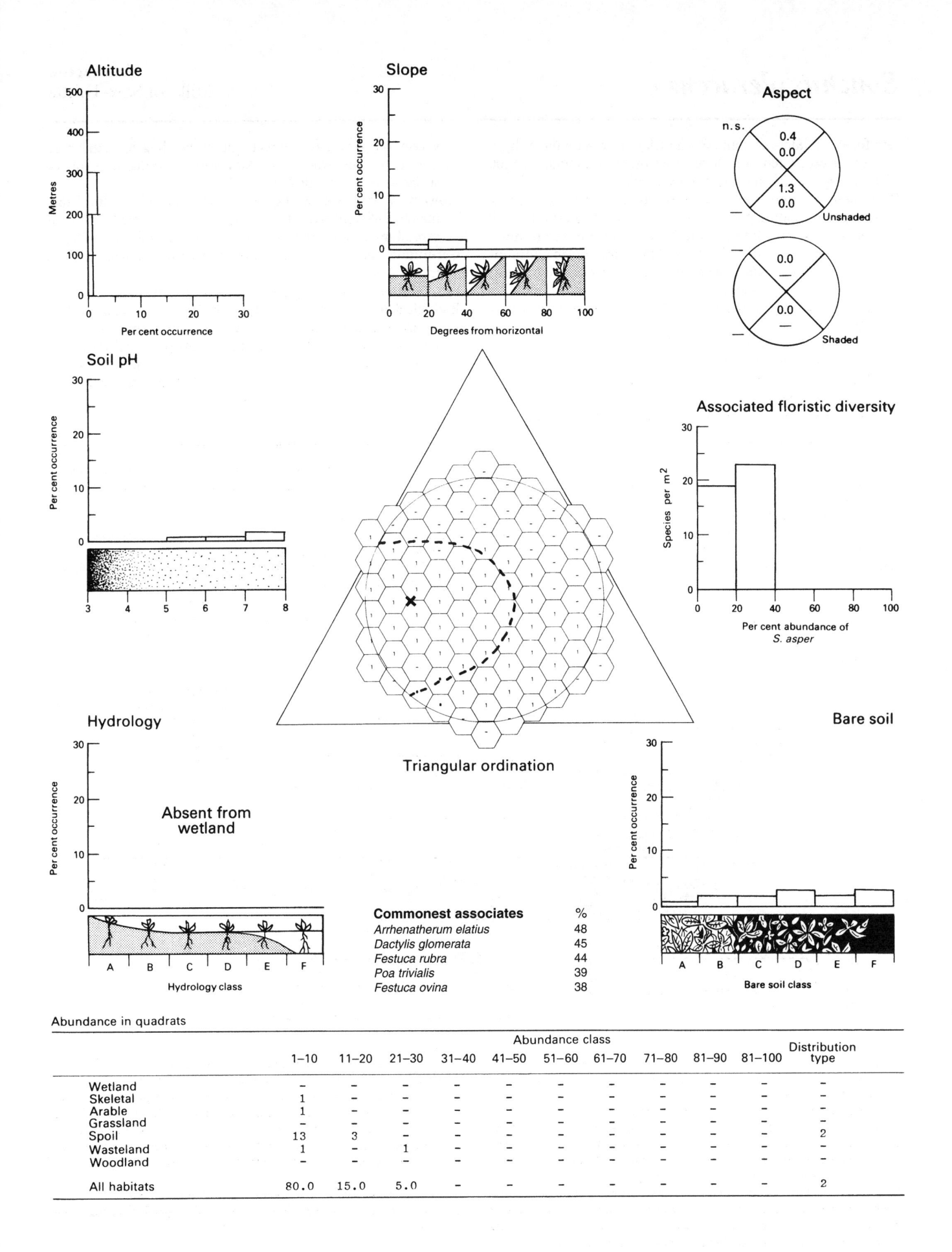

Altitude
Slope
Aspect
Soil pH
Associated floristic diversity
Triangular ordination
Hydrology
Absent from wetland

Bare soil

Commonest associates	%
Arrhenatherum elatius	48
Dactylis glomerata	45
Festuca rubra	44
Poa trivialis	39
Festuca ovina	38

Abundance in quadrats

	\multicolumn{9}{c}{Abundance class}	Distribution type
	1–10	11–20	21–30	31–40	41–50	51–60	61–70	71–80	81–90	81–100	
Wetland	–	–	–	–	–	–	–	–	–	–	–
Skeletal	1	–	–	–	–	–	–	–	–	–	–
Arable	1	–	–	–	–	–	–	–	–	–	–
Grassland	–	–	–	–	–	–	–	–	–	–	–
Spoil	13	3	–	–	–	–	–	–	–	–	2
Wasteland	1	–	1	–	–	–	–	–	–	–	–
Woodland	–	–	–	–	–	–	–	–	–	–	–
All habitats	80.0	15.0	5.0	–	–	–	–	–	–	–	2

Sonchus oleraceus L.

Asteraceae
Milk- or Sow-Thistle

Life-form Winter- or summer-annual, semi-rosette therophyte. Stems stout, often branched; leaves often pinnatifid, <10 000 mm^2; tap-rooted and with VA mycorrhizas.
Phenology Autumn-germinating plants overwinter as rosettes; spring-germinating plants as achenes. Growth continues until plants are killed through disturbance, competition or winter frosts. Flowers and fruits from May onwards in autumn-germinating and June onwards in spring-germinating plants (*Biol Fl(a)*). Leaves short-lived.
Height Foliage and flowers to 1500 m. Large leaves all below 1000 mm.
Seedling RGR No data. SLA 34.9. DMC 11.8.
Nuclear DNA amount 3.3 pg; $2n = 32^*$ (*BC*); $4x^*$.
Established strategy Intermediate between ruderal and competitive-ruderal.
Regenerative strategies W, B$_s$. Regenerates exclusively by seed in vegetation gaps, either directly after shedding in autumn or during the following year. May form a transient bank of surface-lying seeds (*Biol Fl(b)*); SB III.
Flowers Florets yellow, ligulate, visited by insects, probably mainly self-fertilized (*Biol Fl(a)*); >100 aggregated into each capitulum; capitula in few-flowered cymose umbels.
Germinule 0.27 mg, 3.2×0.9 mm, achene with a pappus of *c.* 8 mm.
Germination Epigeal; t_{50} 3 days; 6–34°C; L > D.
Biological flora (a) Lewin (1948b), (b) Hutchinson *et al.* (1984).
Geographical distribution Lowland Britain (100% of British vice-counties), *ABIF* page 623; Europe northwards to Scandinavia (92% of European territories); N and W Asia, N Africa; introduced worldwide as a weed of cultivated land.

Habitats Frequently recorded from demolition sites, soil heaps, limestone quarries and arable land. Also widespread on disturbed road verges, cliffs, rock outcrops, walls, path edges and river banks, particularly in areas with a high level of human activity. Also recorded as non-persistent seedlings in a range of habitats including wetland, enclosed pasture and woodland. [Recorded from maritime cliffs (*Biol Fl(a)*).]
Gregariousness Usually represented by isolated individuals.

Altitude More common at lower altitudes where suitable habitats are more abundant, and often of casual occurrence in the uplands. [Up to 370 m.]
Slope Wide-ranging.
Aspect Significantly more common on unshaded S-facing slopes.
Hydrology Only casual occurrences in wetlands.
Soil pH Most commonly found on soils of neutral to high pH, and largely absent from acidic soils.
Bare soil Associated with a wide range of values, a reflection of the ability of *S.o.* to colonize not only small bare areas within vegetation which normally has a closed canopy, but also highly disturbed sites such as arable land.

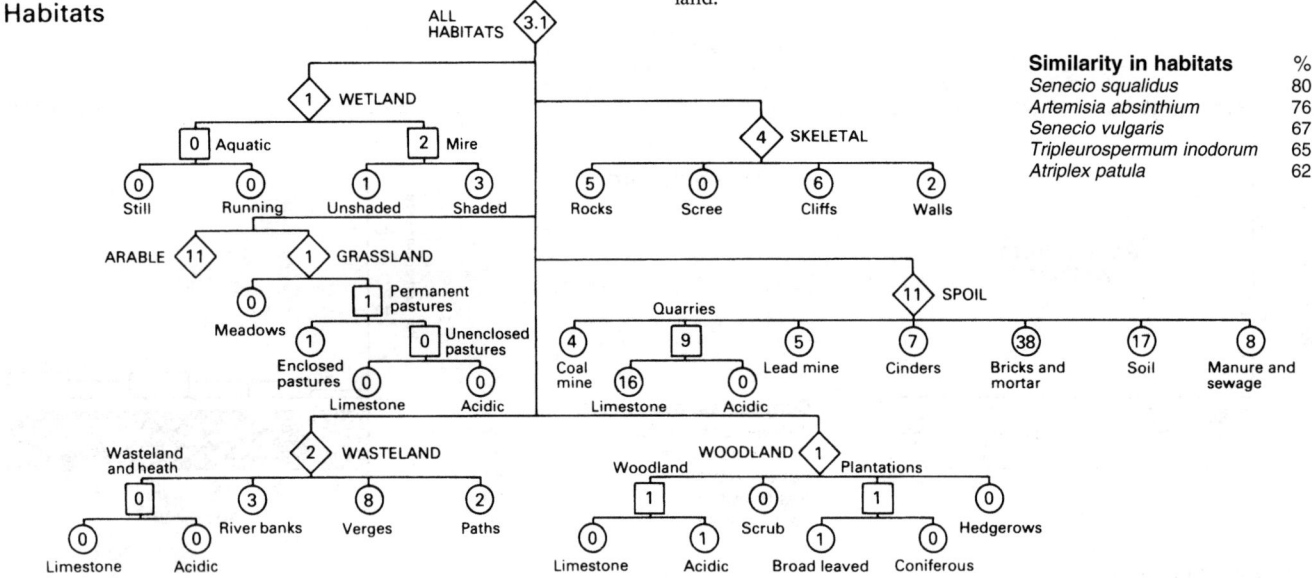

Habitats

Similarity in habitats	%
Senecio squalidus	80
Artemisia absinthium	76
Senecio vulgaris	67
Tripleurospermum inodorum	65
Atriplex patula	62

△ **ordination** Distribution centred on vegetation of fertile but highly disturbed habitats. Absent from undisturbed habitats and also from unproductive situations.
Associated floristic diversity Typically intermediate.
Synopsis The distinctive features of *S.o.* are essentially the same as those of *S. asper*. *S.o.* differs in that it has a slightly more restricted geographical distribution, tending to be replaced by *S.a.* at the N edge of its range (*Biol Fl(b)*). The species is strongly confined to impermanent habitats and, in the Sheffield region, does not share the tendency of *S.a.* to occur in some winter-annual communities on limestone outcrops and dry sandy soils. In *S.o.* the pappus is usually more securely attached to the ripe achene, which is slightly longer and thinner. There appears to be a greater capacity in *S.o.* for ungerminated viable seeds to persist on or near the soil surface (*Biol Fl(a,b)*). *S.o.* is widely distributed, and as a result is classified as one of the world's worst weeds (Holm *et al.* 1977). Like *S.a.*, *S.o.* is palatable but lacks spiny leaves and was formerly eaten as salad (*Biol Fl(b)*). *S.o.*(4x) is believed to have arisen through hybridization between *S.a.* (2x) and the S European *S. tenerrimus* L. (2x) (*FE4*). Some of the variability of the species may relate to its hybrid origin (*Biol Fl(b)*). The existence of ecotypes is suspected (*Biol Fl(a)*).
Current trends Although perhaps less frequent on arable land than formerly, *S.o.* is a common plant of many other artificial habitats, and appears to be increasing.

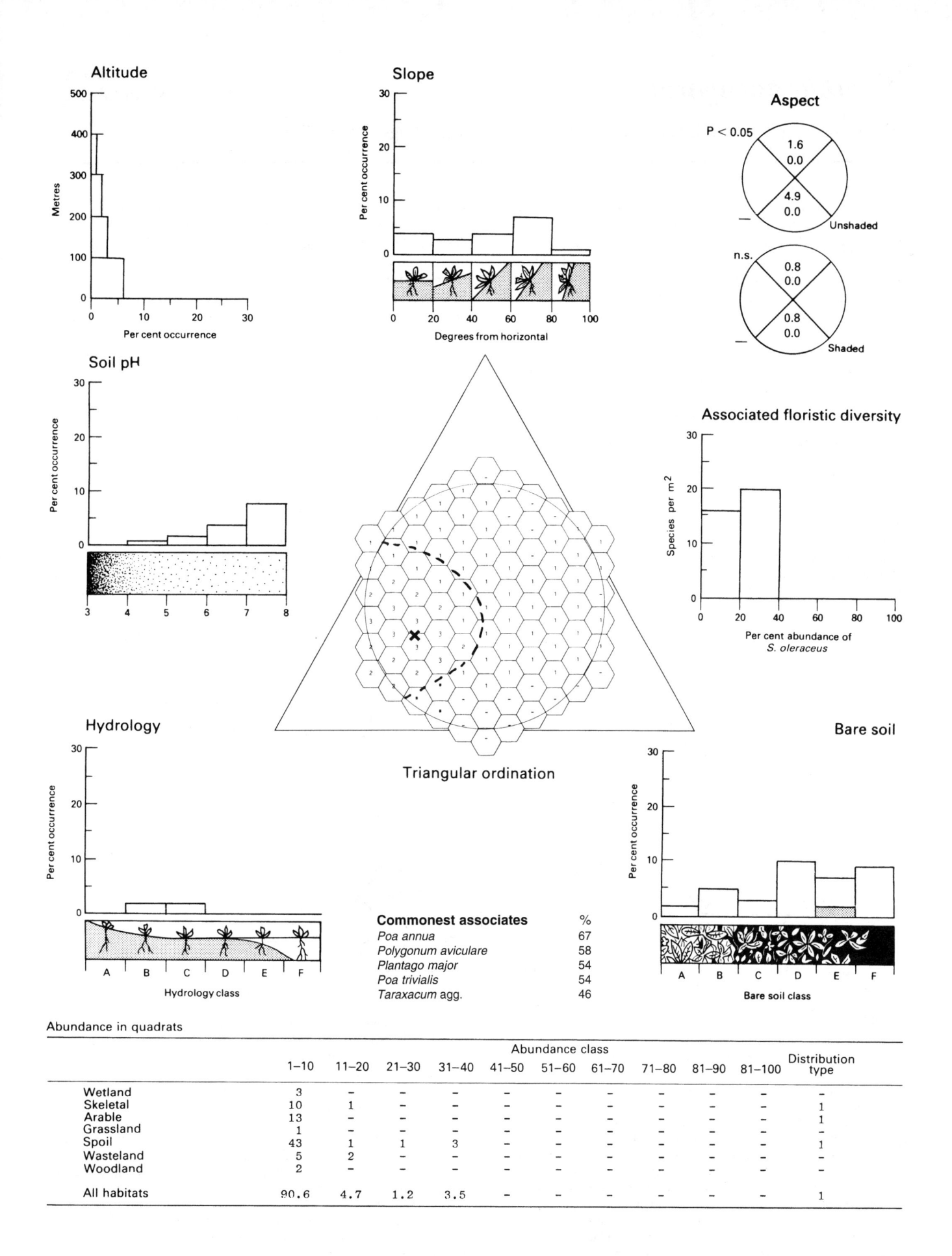

Altitude

Slope

Aspect

$p < 0.05$

	1.6
0.0	
	4.9
0.0	

Unshaded

n.s.

	0.8
0.0	
	0.8
0.0	

Shaded

Soil pH

Associated floristic diversity

Hydrology

Triangular ordination

Commonest associates %
Poa annua 67
Polygonum aviculare 58
Plantago major 54
Poa trivialis 54
Taraxacum agg. 46

Bare soil

Abundance in quadrats

	Abundance class									Distribution type
	1–10	11–20	21–30	31–40	41–50	51–60	61–70	71–80	81–90	81–100	
Wetland	3	–	–	–	–	–	–	–	–	–	–
Skeletal	10	1	–	–	–	–	–	–	–	–	1
Arable	13	–	–	–	–	–	–	–	–	–	1
Grassland	1	–	–	–	–	–	–	–	–	–	–
Spoil	43	1	1	3	–	–	–	–	–	–	1
Wasteland	5	2	–	–	–	–	–	–	–	–	–
Woodland	2	–	–	–	–	–	–	–	–	–	–
All habitats	90.6	4.7	1.2	3.5	–	–	–	–	–	–	1

# (*)*Sorbus aucuparia* L.

Rosaceae
Rowan, Mountain Ash

Field records analysed in this account refer exclusively to seedlings and small saplings

Life-form Deciduous tree with narrow crown and ascending branches. Leaves pinnate, hairy at first, typically <5000 mm^2; roots with VA or rarely ecto-mycorrhizas.
Phenology Buds break in early spring. Flowers in May and June and sets seed in September to October. Leaves shed in autumn.
Height Foliage and flowers to 15 m.
Seedling RGR No data. SLA 13.9. DMC. No data.
Nuclear DNA amount No data; $2n = 34$ (*FE2*); $2x$. Subfamily also of polyploid origin.
Established strategy Stress-tolerant competitor.
Regenerative strategies S. Regenerates entirely by seed, which germinates in spring. No buried seed bank has been reported; SB II.
Flowers White, hermaphrodite, insect-pollinated or selfed (Proctor 1973); <100 flowers in a compound corymb.
Germinule 4.0×2.0 mm, seed in a few-seeded berry (2.58 mg, 11.3×7.8 mm).
Germination Epigeal; dormancy broken by chilling.
Biological flora Raspé, Findlay and Jacquemart (2000).
Geographical distribution The British Isles, although rare and possibly not native in some lowland C and E counties of England (99% of British vice-counties), *ABIF* page 358; most of Europe (82% of European territories); N Asia.

Habitats Seedlings and saplings mainly restricted to wooded sites, particularly on non-calcareous strata, but also recorded from skeletal habitats including lead mines and wasteland.
Gregariousness Sparse.
Altitude Juveniles particularly frequent in gritstone woodlands over 200 m, and observed to 400 m. [Up to 900 m (Barclay & Crawford 1984).]
Slope Widely distributed.
Aspect No bias detected in unshaded sites but in shaded sites juveniles recorded more frequently on N-facing slopes.
Hydrology Absent from wetlands.

Soil pH Mainly restricted to soils of pH <5.5, but some records up to pH 7.0.
Bare soil Juveniles not recorded from sites with 100% bare soil, but otherwise wide-ranging.
△ **ordination** Insuffient data.
Associated floristic diversity Low.
Synopsis *S.a.* is a small deciduous tree or shrub with diffuse porous wood, 15-20 m in height and living a maximum of 150 years (Bernatzky 1978). Bud-break occurs relatively early in spring. Seedlings and mature individuals are concentrated mainly in skeletal habitats, e.g. crevices in rock outcrops, and in woodland. *S.a.* extends on rocky ground to over 900 m, higher than any of the other British trees (Barclay & Crawford 1984). The buds of montane populations show a considerable resistance to

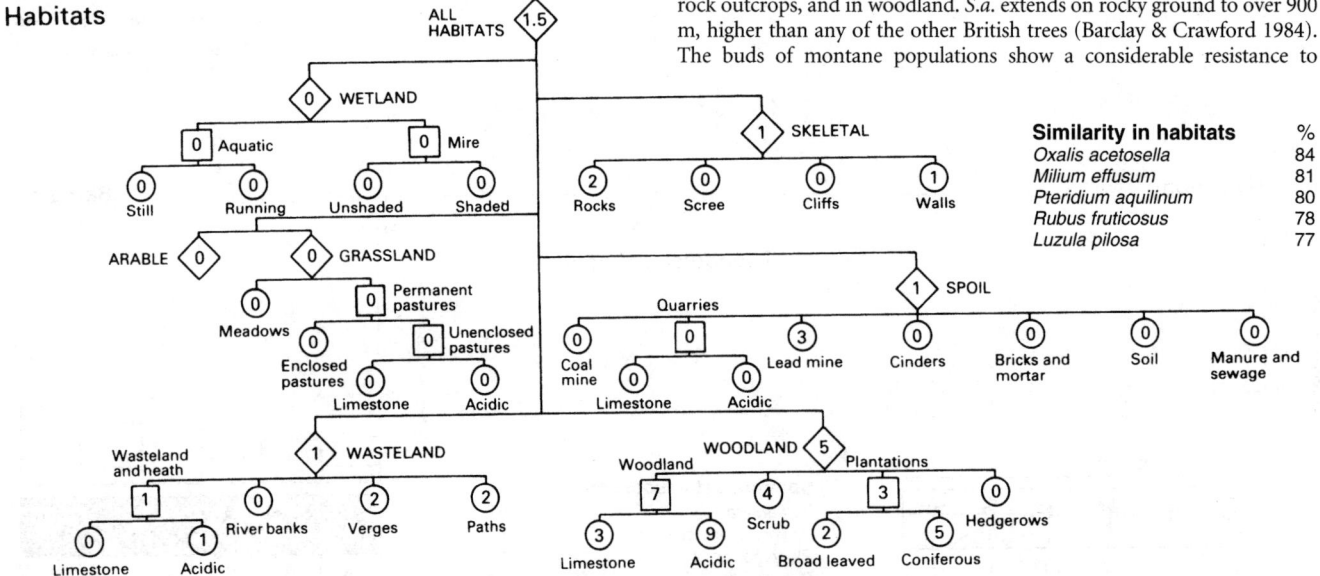

Habitats

Similarity in habitats	%
Oxalis acetosella	84
Milium effusum	81
Pteridium aquilinum	80
Rubus fruticosus	78
Luzula pilosa	77

desiccation (Barclay & Crawford 1982). In wooded habitats the species occurs more frequently in woodland than in scrub, suggesting that establishment by seed and the subsequent development of the tree usually take place under shaded conditions. Pigott (1983) considers the small seedling to be exceptionally shade-tolerant. The species is wide-ranging with respect to soil pH, but is more frequent on acidic soils, often occurring rather sparsely with oak or birch. *S.a.* is characteristic of well-drained soils, and does not accumulate persistent litter. Indeed, leaves show a high palatability to the snail *Helix aspersa* Muller, and *S.a.* supports a relatively species-poor insect fauna (Wratten *et al.* 1981). However, *S.a.* appears to possess a mechanism of inducible defence against foliar predation by insects (Edwards & Wratten 1985). Establishment in woodland, and presumably also in skeletal habitats, appears to be adversely affected by the presence of grazing stock (Pigott 1983). *S.a.* regenerates entirely by seed, which decreases in viability with altitude (Barclay & Crawford 1984). Seeds germinate in spring, have a chilling requirement for germination, and are dispersed within a red fleshy fruit by birds and mammals. Where suitable habitats are adjacent to mountain streams they are also dispersed by water (Ridley 1930). No persistent seed bank has been detected within the Sheffield region, but Hill (1979) suggests that seeds have considerable longevity in the soil. This observation, if confirmed, would make *S.a.* the only tree in the British flora characterized by a long-persistent seed bank. The species is perhaps most similar in ecological distribution to *Betula* spp., since both are relatively short-lived trees most characteristic of acidic soils and extending to high altitudes. In regenerative biology, however, the two clearly differ. *B.* spp. produce vast numbers of small, wind-dispersed seeds and colonize relatively unshaded areas, whereas *S.a.* is capable of regeneration in more-shaded sites and produces fewer, larger, bird-dispersed seeds. *S.a.* is represented by a further four ssp. in Continental Europe, and hybridization between *S.a.* and other taxa has greatly increased the taxonomic complexity of the genus Sorbus in Britain (see Stace 1975).
Current trends Uncertain. Not under any immediate threat, and probably increasing in some lowland areas as a result of a widespread use for amenity or in gardens (and formerly for fruit or 'to frustrate witches' (Rackham 1980)).

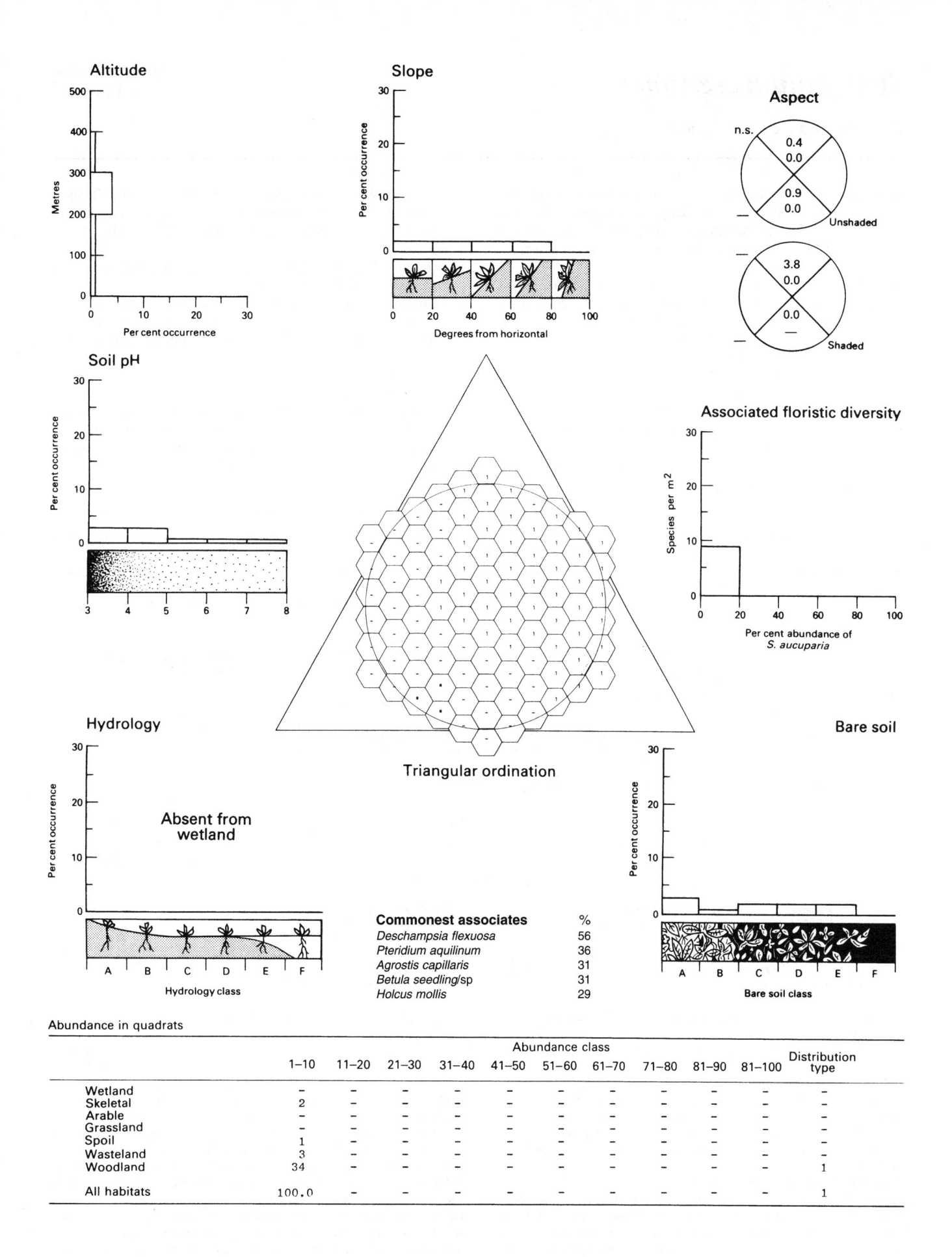

Altitude

Slope

Aspect

n.s. Unshaded

	0.4
—	0.0
0.9	
0.0	—

Shaded

	3.8
—	0.0
0.0	—

Soil pH

Associated floristic diversity

Hydrology

Absent from wetland

Triangular ordination

Commonest associates %
Deschampsia flexuosa 56
Pteridium aquilinum 36
Agrostis capillaris 31
Betula seedling/sp 31
Holcus mollis 29

Bare soil

Abundance in quadrats

	\multicolumn{10}{c	}{Abundance class}	Distribution type
	1–10	11–20	21–30	31–40	41–50	51–60	61–70	71–80	81–90	81–100	
Wetland	–	–	–	–	–	–	–	–	–	–	–
Skeletal	2	–	–	–	–	–	–	–	–	–	–
Arable	–	–	–	–	–	–	–	–	–	–	–
Grassland	–	–	–	–	–	–	–	–	–	–	–
Spoil	1	–	–	–	–	–	–	–	–	–	–
Wasteland	3	–	–	–	–	–	–	–	–	–	–
Woodland	34	–	–	–	–	–	–	–	–	–	1
All habitats	100.0	–	–	–	–	–	–	–	–	–	1

Sparganium erectum L.

Sparganiaceae
Bur-reed

Divisible into ssp. (see 'Synopsis')

Life-form Polycarpic perennial, rhizomatous geophyte, hydrophyte or helophyte, producing corms. Stems erect; leaves mainly basal, linear, sometimes >10 000 mm^2, erect or rarely floating, produced from a corm; roots relatively shallow; VA mycorrhizas recorded.

Phenology Shoot growth commences in spring from corms. Flowers occurs from July to September; fruits ripen from September to November and may be damaged by the frost. Leaves relatively long-lived but die in autumn along with rhizomes and roots which are also renewed annually.

Height Foliage and flowers to 1 m.

Seedling RGR No data. SLA 17.1. DMC 14.7.

Nuclear DNA amount 1.1 pg; 2n = 30 (*FE5*); 2x.

Established strategy Intermediate between competitor and competitive-ruderal.

Regenerative strategies (V), ?B$_s$. Regeneration biology is complex (see 'Synopsis'); ?SB IV.

Flowers Greenish, monoecious, protogynous, wind-pollinated; aggregated into many-flowered capitula, with male capitula borne above female capitula on each branch of the inflorescence.

Germinule 18.71 mg, 9.3 × 5.9 mm, dry drupaceous fruit (minus exocarp 13.1 mg, 3.9 × 3.2 mm).

Germination Cotyledon partially enclosed within seed; hardcoat dormancy; germination occurs only under water or in a saturated atmosphere (*Biol Fl*); germination also promoted by chilling.

Biological flora Cook (1962).

Geographical distribution Most of Britain, although scarce in upland regions (99% of British vice-counties), *ABIF* page 805, *CSABF* pages 144-5; Europe (89% of European territories); N temperate regions; closely allied but possibly distinct species are found in Australia and New Zealand (*Biol Fl*).

Habitats An emergent aquatic of still or slowly-moving water in ponds, ditches, lakes and canals, and the backwaters of rivers and streams. Also occurs in unshaded or shaded mire, usually close to the water's edge or on floating rafts of vegetation rooted in liquid mud.

Gregariousness A stand-forming species, often capable of forming a near-monoculture at frequency values of <70%.

Altitude Suitable habitats are largely confined to river valleys and lowland areas. Extends to 335 m. [Up to 490 m (*Biol Fl*).]

Slope Confined to waterside habitats, usually of gentle slope.

Aspect Not applicable.

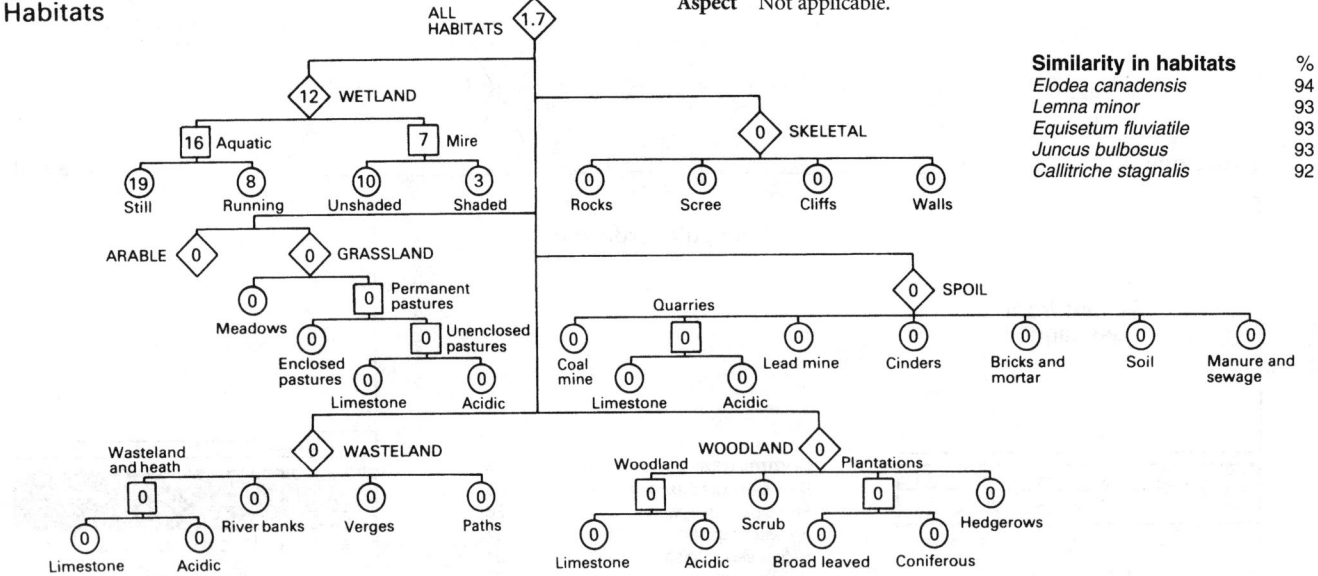

Habitats

Similarity in habitats	%
Elodea canadensis	94
Lemna minor	93
Equisetum fluviatile	93
Juncus bulbosus	93
Callitriche stagnalis	92

Hydrology Grows at the edge of still or gently flowing water. Most common and most abundant in shallow water, but extending to depths of 600 mm [Up to l m (*Biol Fl*)] and also found in mire beside water and in silted-up ponds.

Soil pH Associated most frequently with substrates in the pH range 6.5-7.0, but scattered occurrences on all but the most acidic soils.

Bare soil Mostly recorded from aquatic sites which have no bare soil or litter. In mire, stands of *S.e.* often accumulate persistent litter.

Δ **ordination** Present only in productive vegetation types with low to moderate disturbance.

Associated floristic diversity Low.

Synopsis *S.e.* is a long-lived, robust, stand-forming, emergent aquatic or marsh species resembling a diminutive *Typha latifolia* in form, phenology and habitat range. Each plant is capable of extending 2 m laterally each year (*Biol Fl*). *S.e.* grows most vigorously in shallow water (*c.* 100 mm), but is rarely found in fast-flowing sites and is intolerant of wave action (*Biol Fl*). *S.e.* is typically rooted in loose, often anaerobic, silt and has well-developed air spaces in its roots, stems and leaves (*Biol Fl*). *S.e.* is also found growing less vigorously in mire beside open water, but the species is sensitive to drought and is confined to situations where the water table is above the level of the roots (*c.* 100 mm below the surface) for most of the year (*Biol Fl*). New shoots examined in early spring were found to contain high concentrations of fructose (Hendry & Brocklebank unpubl.). *S.e.* occurs on a wide range of soils and can survive low levels of salinity (*Biol Fl*). *S.e.* grows poorly in shade and produces few flowering stems. The species appears unable to compete effectively with the taller *Phragmites australis* and *T.l.* Despite forming extensive and persistent stands, the clones are rapidly fragmented and each ramet is short-lived (*Biol Fl*). Rhizomes survive only one growing season and corms, which give rise either to vegetative or to flowering shoots and are monocarpic, survive for a

Altitude

Slope

Aspect

Soil pH

Associated floristic diversity

Triangular ordination

Hydrology

Commonest associates %
Lemna minor 58
Agrostis stolonifera 42
Galium palustre 29
Holcus lanatus 26
Juncus effusus 26

Bare soil

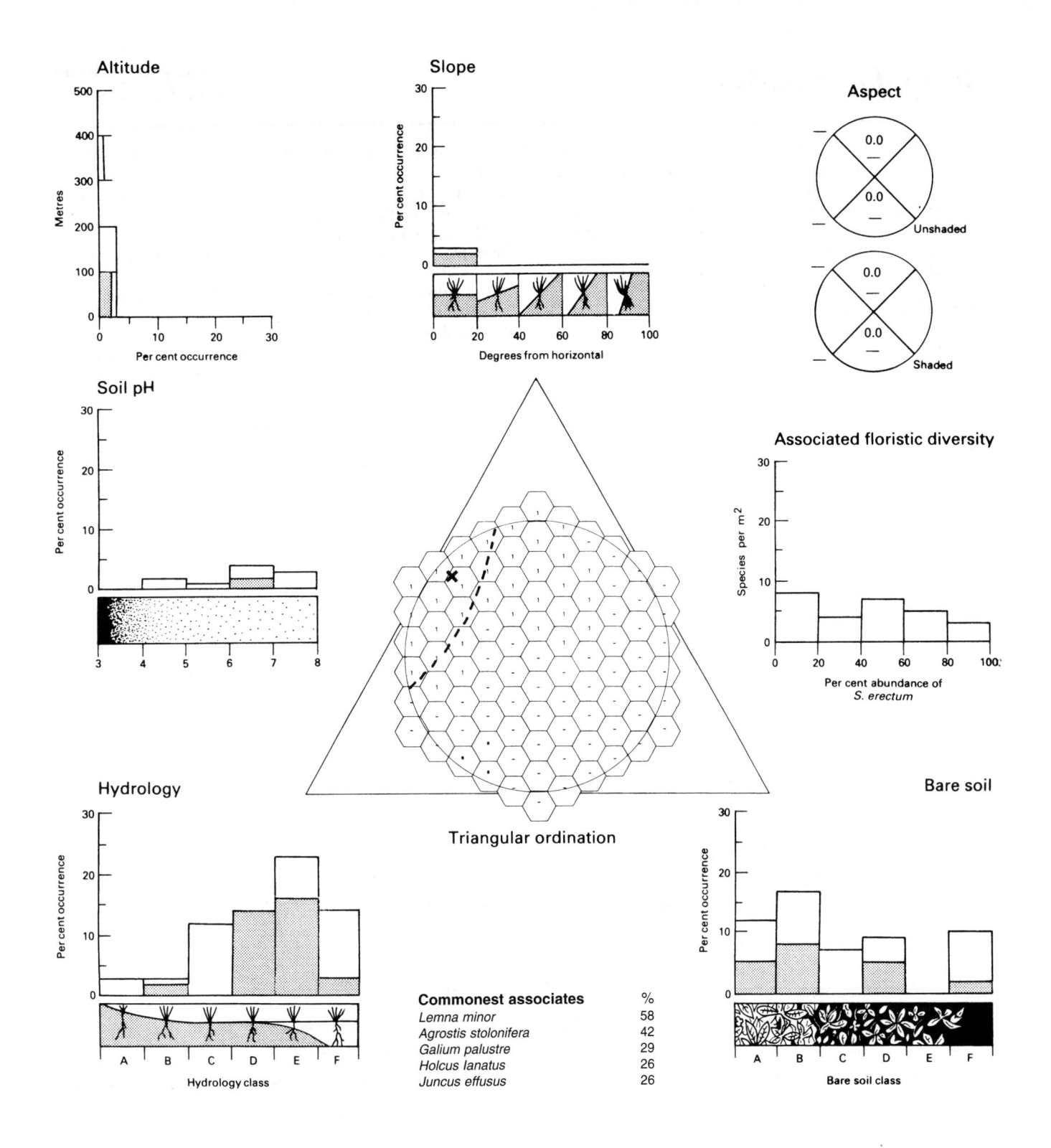

maximum of 3 years (*Biol Fl*). Corms in winter and rhizome pieces in summer may be transported in water, allowing new sites to be colonized (*Biol Fl*). Apart from vegetative fragmentation S.e. also regenerates, albeit rarely, by seeds produced at the rate of up to 720 per 500 flowering head (*Biol Fl*). The fruit has a spongy exocarp and may float in water for 1 year (*Biol Fl*). The endocarp is also hard and the micropile is plugged, and seeds pass undamaged through the gut of waterfowl (*Biol Fl*). Fruits may also be transported externally by birds (*Biol Fl*). Seedlings establish under water or on wet mud, and are sensitive to competition, even from algae. Establishment is not observed in turbid water, and seedlings, unlike their parents, tend not to survive in eutrophic habitats (*Biol Fl*). It seems likely that regeneration by seed is of primary significance in the colonization of new sites. Fruits are relatively long-lived under a variety of conditions (*Biol Fl*) and the possibility of a 'floating seed bank' requires investigation. Three subspecies are recognized, differing in fruit characters and geographical distribution (*FE5*). They do not appear to be ecologically distinct (*Biol Fl*) and have not been separated here. Hybrids between two of the subspecies have also been reported (Stace 1975).

Current trends Uncertain. Although S.e. has the capacity to exploit highly eutrophic aquatic systems, colonization of new land-locked sites is slow, presumably because dispersal by birds on which S.e. appears to rely is less effective than the wind-borne dispersal of species such as *T.l.*

+*Spergula arvensis* L.

Caryophyllaceae
Spurrey, Corn Spurrey

Life-form Summer-annual therophyte. Stems prostrate to erect, often branching from the base; leaves fleshy, linear, with glandular hairs, <10 mm²; root system with or without VA mycorrhizas.

Phenology Seeds germinate in April. Flowers from June to August and sets seed *c.* 2 weeks after flowering. Plants normally die in early autumn (*Biol Fl*). Under favourable conditions two generations may occur within a season. Leaves short-lived.

Height Foliage and flowers to 300 mm (usually less).

Seedling RGR No data. SLA 12.5. DMC 14.5.

Nuclear DNA amount 2.1 pg; $2n = 18^*$ (*BC*); $2x^*$.

Established strategy Intermediate between ruderal and stress-tolerant ruderal.

Regenerative strategies B_s. Exclusively by seed, mainly in spring but seed also incorporated into a persistent seed bank (*Biol Fl*); SB IV.

Flowers White, hermaphrodite, homogamous, usually at least 97% self-pollinated, but also insect-pollinated (*Biol Fl*); often *c.* 20 in a cymose panicle.

Germinule 0.42 mg, 1.5 × 1.3 mm, seed in a dehiscent capsule with up to *c.* 25 seeds.

Germination Epigeal; t_{50} 1 day; (5-35°C; germination rate increased after dry storage at 20°C; L = D.

Biological flora New (1961).

Geographical distribution Most of the British Isles, *ABIF* page 172; Europe (92% of European territories); virtually cosmopolitan.

Habitats Primarily an arable weed of both broad-leaved and cereal crops, particularly on light, sandy soils. Also recorded from other disturbed habitats such as spoil heaps. Frequent in sand and gravel pits and on disturbed sandy wasteland, and during the first year may occur in newly sown grasslands. Occasionally found on sandy silt exposed during summer around the margins of reservoirs. Atypical of trampled sites. [Var. *nana* (Linton) Druce occurs in short cliff-top turf and sand dunes in the Channel Islands.]

Gregariousness Intermediate.

Altitude No records at sites above 400 m. [Up to 460 m.]

Slope Occurs almost entirely on flat ground and gentle slopes.

Aspect Widely distributed.

Hydrology Occasionally found beside reservoirs on moist ground exposed during summer but absent from waterlogged habitats.

Soil pH Most commonly found on soils in the pH range 5.0-6.0. Since the majority of arable soils have pH values above this range it is clear that *S.a.* shows a calcifugous tendency.

Bare soil Most commonly associated with habitats having a high proportion of bare soil.

△ ordination Distribution confined to vegetation associated with highly disturbed but fertile conditions.

Associated floristic diversity Typically high.

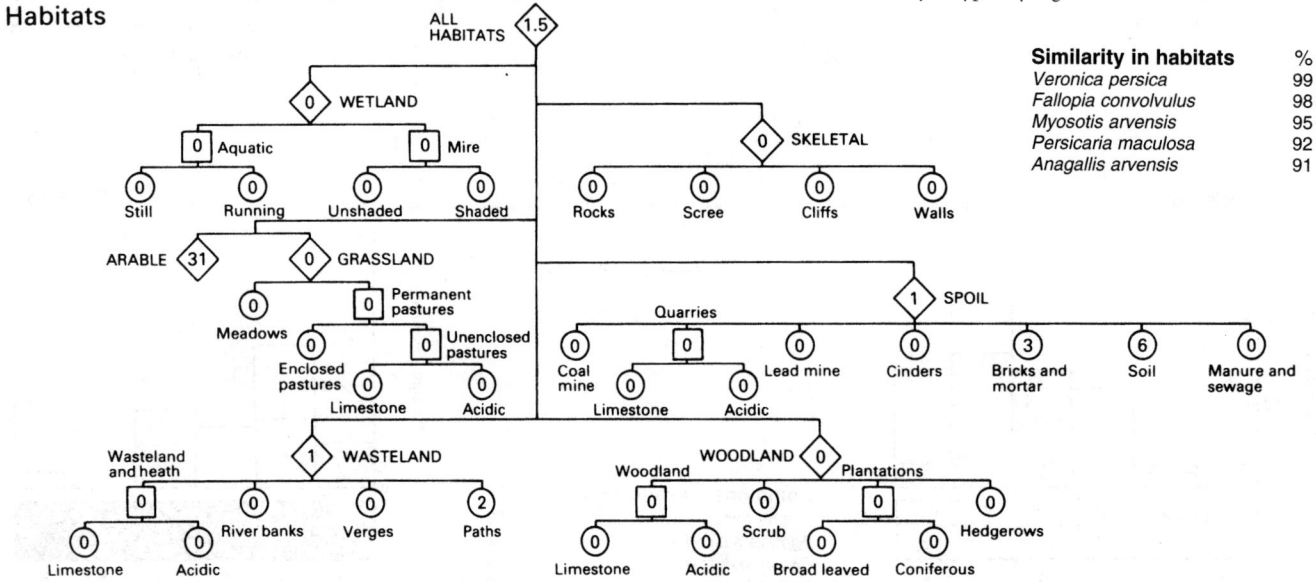

Habitats

Similarity in habitats	%
Veronica persica	99
Fallopia convolvulus	98
Myosotis arvensis	95
Persicaria maculosa	92
Anagallis arvensis	91

Synopsis *S.a.* is a slender summer-annual herb found in Britain on arable land and in other fertile, disturbed habitats. *S.a.* is thought to have been a common weed of flax from the Iron Age onwards (*Biol Fl*). Whether its form and life-history are best suited to this crop is uncertain. Today, *S.a.* is cosmopolitan in distribution, and is classified by Holm *et al.* (1977) as one of the world's worst weeds, particularly of cereals. *S.a.* is regarded as an important weed in 25 crops and in 33 countries. Unlike most other common British arable weeds, the occurrences of *S.a.* are most frequent on surface-leached sandy soils of pH 6.0 or less. *S.a.* responds to defoliation by producing new flowering shoots from the base (*Biol Fl*), but is not tolerant of trampling. *S.a.* is palatable to sheep, cattle and poultry, and has been grown for fodder (Holm *et al.* 1977). Seeds have, in historical times, also been eaten by humans (Holm *et al.* 1977). Regeneration is entirely by seed, which germinates mainly in spring. Seed production is both rapid and prolific. Robust plants may produce as many as 7500 seeds; these may be released within 10 weeks of germination, and two generations may be accommodated within a growing season. Germination may be inhibited by a leaf canopy (Gorski *et al.* 1977). Like many other arable species, *S.a.* forms a large persistent seed bank, which may reach 23 million seeds ha^{-1} (*Biol Fl*). Seeds appear to be capable of extended survival in the soil (Odum 1978) and can withstand ingestion by many mammals and birds (Holm *et al.* 1977). The species is often transported to new sites on agricultural machinery (Holm *et al.* 1977). In common with most ruderals, *S.a.* is phenotypically highly plastic (*Biol Fl*). Genotypic differences are also apparent. Seeds with papillae tend to germinate at higher temperatures and lower moisture tension (New & Herriot 1981), and tend to be replaced by non-papillate forms towards the N and W (*Biol Fl*). Hairiness of the shoot varies in a similar manner (*Biol Fl*).

Current trends Since arable soils with pH values below 6.5 are now infrequent, *S.a.* is probably decreasing, despite remaining a locally common plant of disturbed sandy soils. The proportion of densely hairy plants may be increasing and this may be related to changes in climate (New 1978).

Altitude
Slope
Aspect

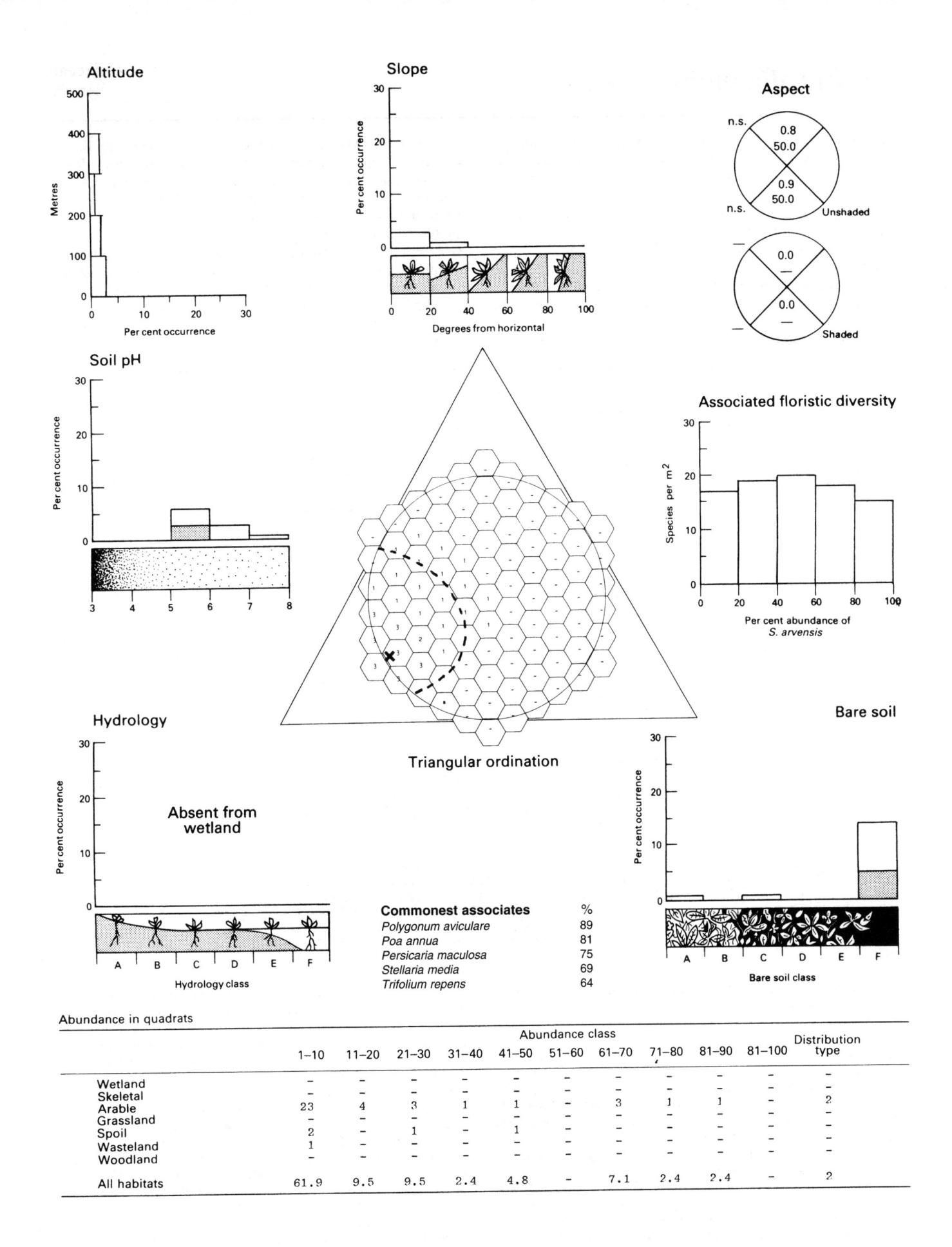

	n.s.	0.8	
		50.0	
n.s.		0.9	Unshaded
		50.0	

	—	0.0	
	—	0.0	Shaded

Soil pH

Associated floristic diversity

Per cent abundance of *S. arvensis*

Hydrology

Absent from wetland

Triangular ordination

Commonest associates   %
Polygonum aviculare   89
Poa annua   81
Persicaria maculosa   75
Stellaria media   69
Trifolium repens   64

Bare soil

Abundance in quadrats

	\multicolumn{10}{c}{Abundance class}	Distribution type
	1–10	11–20	21–30	31–40	41–50	51–60	61–70	71–80	81–90	81–100	
Wetland	–	–	–	–	–	–	–	–	–	–	–
Skeletal	–	–	–	–	–	–	–	–	–	–	–
Arable	23	4	3	1	1	–	3	1	1	–	2
Grassland	–	–	–	–	–	–	–	–	–	–	–
Spoil	2	–	1	–	1	–	–	–	–	–	–
Wasteland	1	–	–	–	–	–	–	–	–	–	–
Woodland	–	–	–	–	–	–	–	–	–	–	–
All habitats	61.9	9.5	9.5	2.4	4.8	–	7.1	2.4	2.4	–	2

Stachys officinalis (L.) Trevisan

Lamiaceae
Betony

Life-form Polycarpic perennial, semi-rosette hemicryptophyte with a short woody rhizome. Stems erect; leaves mostly basal, oblong, hairy or glabrous, <4000 mm^2; roots with VA mycorrhizas.
Phenology Winter green. Flowers June to September and sheds seed from July onwards.
Height Foliage mainly <100 mm; flowers to 600 mm.
Seedling RGR No data. SLA 17.8 DMC 24.2.
Nuclear DNA amount 9.0 pg; $2n = 16*$ (*BC*); $2x*$.
Established strategy Intermediate between stress-tolerator and C-S-R.
Regenerative strategies V, S. Seeds germinate in the spring and do not form a persistent seed bank (Roberts 1986a); SB II.
Flowers Purple, hermaphrodite, homogamous, protandrous, insect-pollinated; usually <50 in a spike-like inflorescence.
Germinule 1.37 mg, 2.6 × 1.8 mm, nutlet (4 per ovary).
Germination Epigeal; t_{50} 7 days; 12-25°C.
Biological flora None.
Geographical distribution Widespread in England and Wales but largely absent from Scotland and Ireland (63% of British vice-counties), *ABIF* page 507; most of Europe from S Scandinavia to C Spain, but absent from the Mediterranean region (72% of European territories and 3% naturalized); N Africa.

Habitats Recorded from grazed and ungrazed grassland on calcareous strata, and also frequently observed on non-calcareous strata. Observed also from waysides, woodland margins and, more rarely, mire. Absent from aquatic, arable, spoil and most skeletal habitats.
Gregariousness Typically occurring as scattered individuals.
Altitude Wide-ranging to 310 m. [Up to 460 m.]
Slope Absent from most skeletal habitats and from steep slopes. Particularly characteristic of the brow of the limestone dale-sides.
Aspect No consistent trends.
Hydrology Not a wetland species, although observed occasionally at the edge of soligenous mire.
Soil pH Restricted to sites above pH 4.0, and particularly characteristic of soils of intermediate pH within the range 5.0-6.5.
Bare soil Restricted to closed vegetation with little exposed soil.
△ **ordination** Confined to vegetation types associated with undisturbed and unproductive conditions.
Associated floristic diversity Typically species-rich.

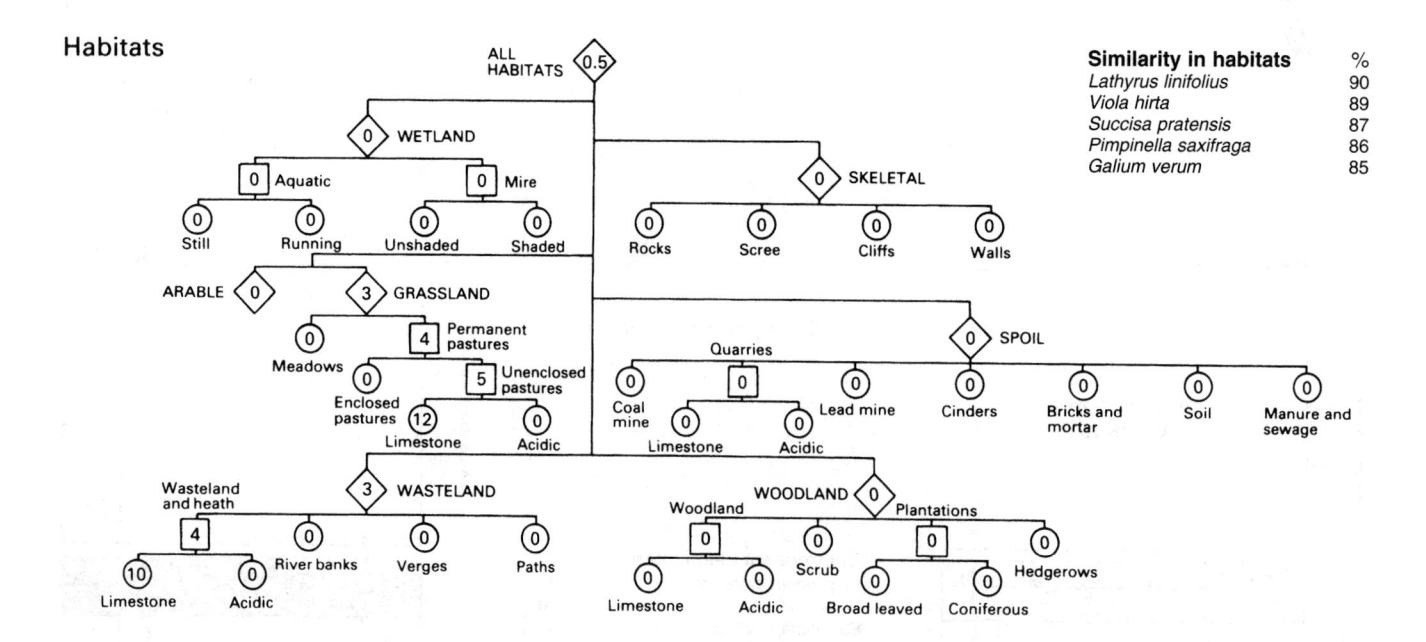

Habitats

Similarity in habitats	%
Lathyrus linifolius	90
Viola hirta	89
Succisa pratensis	87
Pimpinella saxifraga	86
Galium verum	85

Synopsis *S.o.* is a slow-growing, rosette-forming herb which is mainly restricted to short turf on infertile soils where potential dominants are restricted in vigour either by a low intensity of grazing or cutting or by burning. Like *Lathyrus linifolius*, with which it often grows, *S.o.* is most commonly found on soils which are mildly acidic (pH 5.0-6.5) at least in the superficial horizons. In laboratory experiments *S.o.* exhibits high sensitivity to lime-chlorosis on calcareous soils (Grime & Hodgson 1969). *S.o.* tends to be associated with relatively deep clayey soils, and is usually absent from waterlogged and from droughted situations. In a field experiment (Grime 1960) individuals of *S.o.* transplanted to a shallow calcareous soil in Derbyshire suffered irreversible wilting during summer droughts, and had much lower survivorship than transplants on a neighbouring brown earth naturally colonized by the species. At low altitudes *S.o.* persists at woodland margins, and experimental studies (Grime 1966) also indicate a degree of shade-tolerance. *S.o.* shows a pronounced summer peak in both biomass and leaf canopy and thus has a somewhat different phenology from that of many of the grasses with which it is associated. Plants grown continuously at low temperatures (<15°C) produce dark green leaves with small laminae; these are quite unlike the large leaves produced at warm temperatures (Mason 1976). We suspect that *S.o.* is long-lived, and that vegetative spread, although restricted in extent, is more important in the development of populations than regeneration by seed. Seed matures late in the year and is released slowly. In a recent experiment, Roberts (1986a) found that germination was delayed until spring, and was not associated with the formation of a persistent seed bank. The seed appears to be poorly dispersed, and *S.o.* is virtually restricted to old or semi-natural vegetation. Lack of mobility during a Postglacial colonizing episode may be partly responsible for the present rarity of *S.o.* in Ireland. Various extracts of the leaves were formerly used in home remedies for coughs, stomach upsets and kidney, bladder and spleen complaints (Usher 1974).
Current trends Decreasing, particularly in lowland areas.

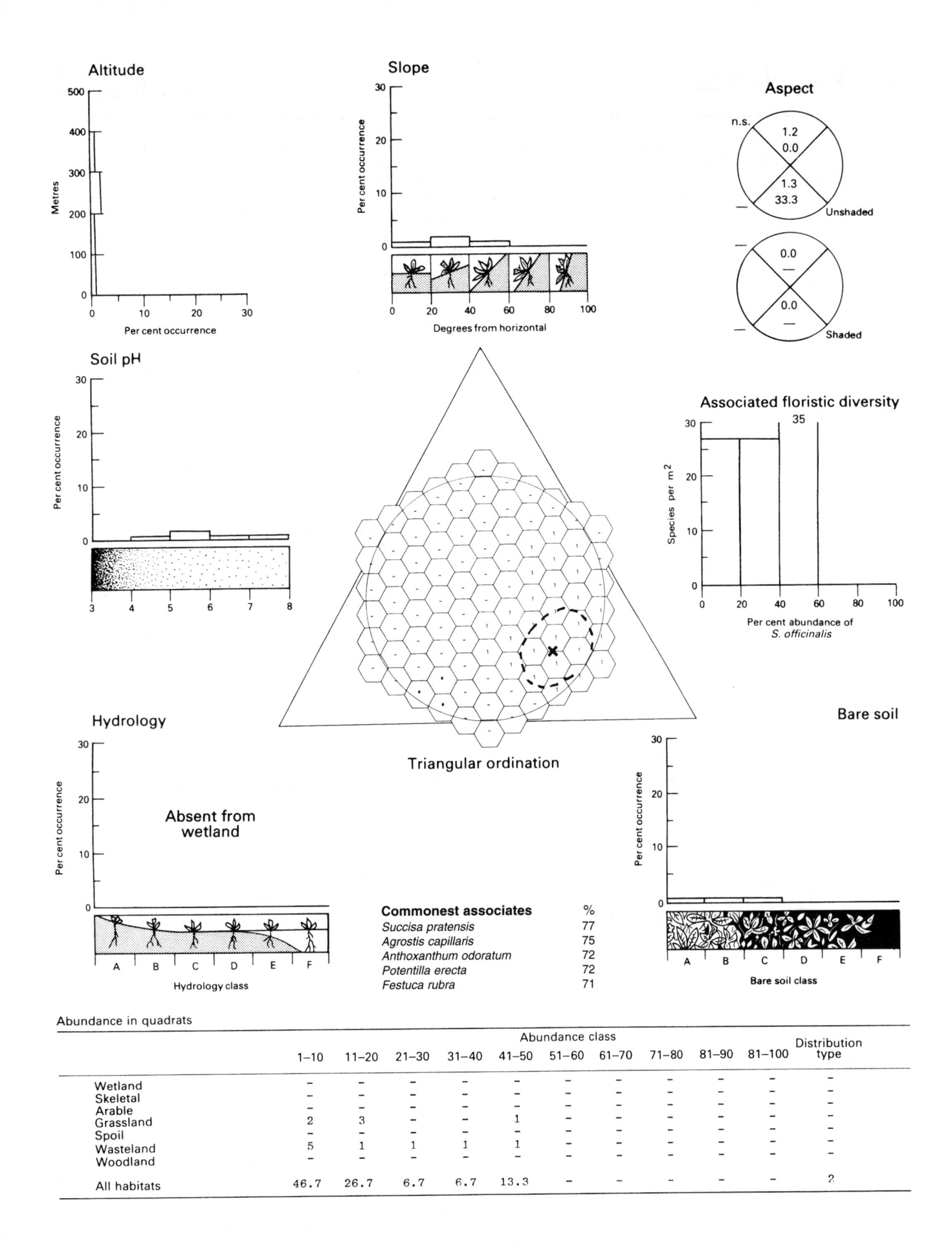

Altitude

Slope

Aspect

	n.s.	1.2
	0.0	1.3
	—	33.3 Unshaded

		0.0
	—	—
		0.0
	—	Shaded

Soil pH

Associated floristic diversity

Hydrology

Absent from wetland

Triangular ordination

Commonest associates	%
Succisa pratensis	77
Agrostis capillaris	75
Anthoxanthum odoratum	72
Potentilla erecta	72
Festuca rubra	71

Bare soil

Abundance in quadrats

	1–10	11–20	21–30	31–40	41–50	51–60	61–70	71–80	81–90	81–100	Distribution type
Wetland	–	–	–	–	–	–	–	–	–	–	–
Skeletal	–	–	–	–	–	–	–	–	–	–	–
Arable	–	–	–	–	–	–	–	–	–	–	–
Grassland	2	3	–	–	1	–	–	–	–	–	–
Spoil	–	–	–	–	–	–	–	–	–	–	–
Wasteland	5	1	1	1	1	–	–	–	–	–	–
Woodland	–	–	–	–	–	–	–	–	–	–	–
All habitats	46.7	26.7	6.7	6.7	13.3	–	–	–	–	–	2

Stachys sylvatica

Lamiaceae
Hedge Woundwort

Life-form Rhizomatous, polycarpic perennial protohemicryptophyte. Stems erect; leaves ovate, hairy, <4000 mm^2; roots with VA mycorrhizas.

Phenology Shoots appear in spring. Flowers from July to August and sets seed from July to October, after which shoots die.

Height Foliage usually <600 mm, overtopped by flowers.

Seedling RGR No data. SLA 33.4. DMC 20.7.

Nuclear DNA amount No data; $2n = 66*$ (*BC*); $8x*$.

Established strategy Intermediate between competitor and competitive-ruderal.

Regenerative strategies (V), B$_s$. Regenerates mainly vegetatively, but also by seed (see 'Synopsis'); ?SB IV.

Flowers Purple, hermaphrodite, insect-pollinated, self-compatible (Wilcock & Jones 1974); usually <50 in a whorled inflorescence.

Germinule 1.4 mg, 2.0 × 1.5 mm, nutlet (4 per ovary).

Germination Epigeal; t_{50} 1 day; dormancy broken by chilling; L > D.

Biological flora None.

Geographical distribution The British Isles, but less common in N Scotland and Ireland (99% of British vice-counties), *ABIF* page 509; Europe, although rare in the Mediterranean region (82% of European territories); central Asia.

Habitats Particularly characteristic of lightly shaded habitats, and recorded most regularly from hedgerows, road verges, banks of rivers and streams, heaps of fertile mineral soil and disturbed fertile soils in open woodland and scrub. Also occurs in shaded mire and occasionally on cliffs, walls and the margins of arable fields. Absent from pasture and aquatic habitats.

Gregariousness Capable of forming patches.

Altitude Extending up to 400 m, but more common at low altitudes. [Up to 490 m.]

Slope Wide-ranging.

Aspect Widely-distributed.

Hydrology Recorded from moist ground at the edge of both soligenous and topogenous mire.

Soil pH Most frequent in the pH range 6.0-8.0, and absent below pH 4.0.

Bare soil Varying widely, but more frequent at sites with large exposures.

△ ordination Most frequent in vegetation associated with fertile, disturbed habitats, but exhibiting a rather diffuse pattern.

Associated floristic diversity Typically intermediate.

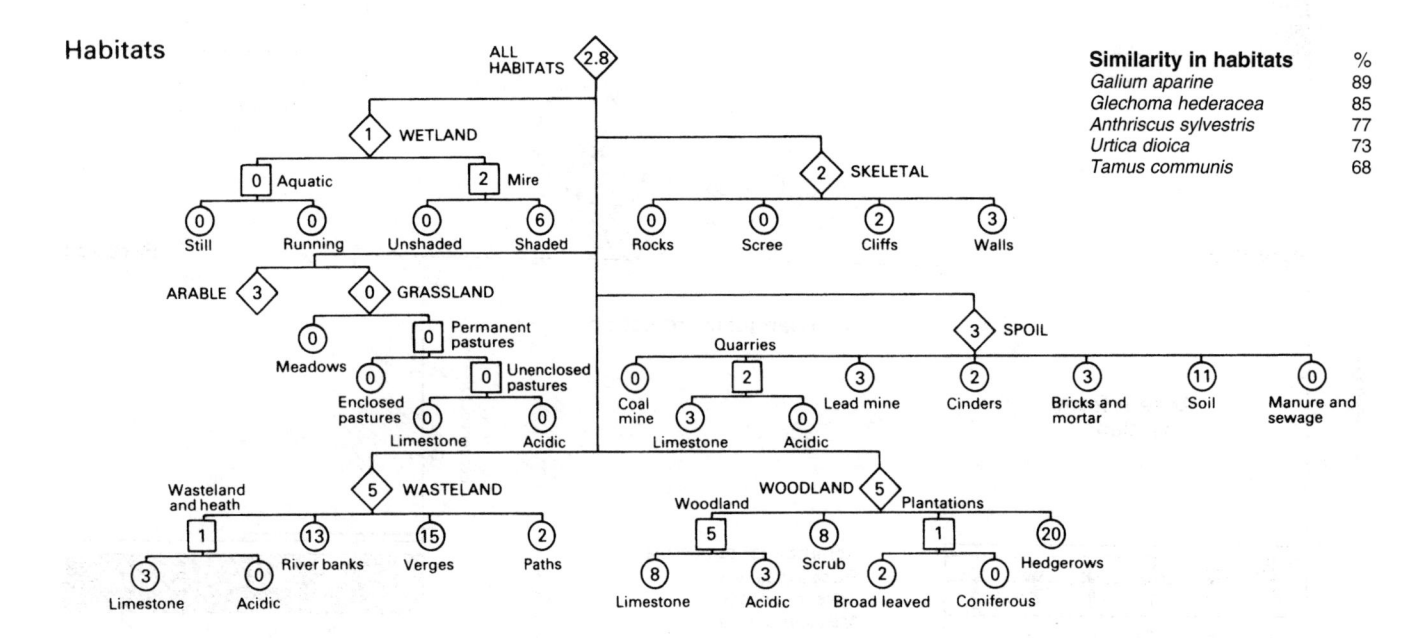

Habitats

Similarity in habitats	%
Galium aparine	89
Glechoma hederacea	85
Anthriscus sylvestris	77
Urtica dioica	73
Tamus communis	68

Synopsis *S.s.* is a rhizomatous, stand-forming herb of moist fertile sites, where the vigour of taller potential dominants is reduced by disturbance and often also by the presence of light-to-moderate shade. Thus, *S.s.* is particularly characteristic of hedgerows, road verges, river banks, the edge of woodland rides and floodplains. The species also occurs on infrequently-cut roadsides, but despite the fact that its leaves are glandular and foetid when crushed, it is absent from pasture. Individual parts of the rhizome system are short-lived and shoots are, in addition, proliferous (Berko 1979). Colonies of *S.s.* are perhaps similar to those of *Ranunculus repens* in terms of the dynamic turnover of ramets. This capacity for intensive vegetative reproduction ensures high competitive ability and facilitates the active foraging for optimal sites within a changing environment (Berko 1979). Effective regeneration is probably mainly by vegetative means, and *S.s.* forms large clonal patches as a result of rhizome growth. The species also regenerates freely from rhizome pieces. Thus, *S.s.* can persist in vegetable gardens provided that the weeding regime is not too severe. Long-distance dispersal of the largely sterile *S.* × *ambigua* Sm., (*S. palustris* L. × *S.s.*) occurs regularly as a result of human activities, and introductions of *S.s.* by similar means are probable. *S.s.* is a rapid colonizer of secondary woodland (Peterken 1981). Although *S.s.* exhibits a degree of shade tolerance, the species has a peak of biomass in summer, and lacks both overwintering shoots and pronounced vernal growth. This leads us to suspect that the capacity for effective dispersal and early establishment (often low in many woodland species) are more critical to the success of *S.s.* in exploiting shaded habitats than phenological specialization (see also Peterken 1981). The conditions required for establishment by seed have been little studied. However, seeds which form a persistent seed bank are produced more consistently in unshaded sites, and appear to lack any well-defined mechanism for long-distance dispersal.

Current trends A frequent wayside plant apparently favoured by modern forestry practices, and probably increasing.

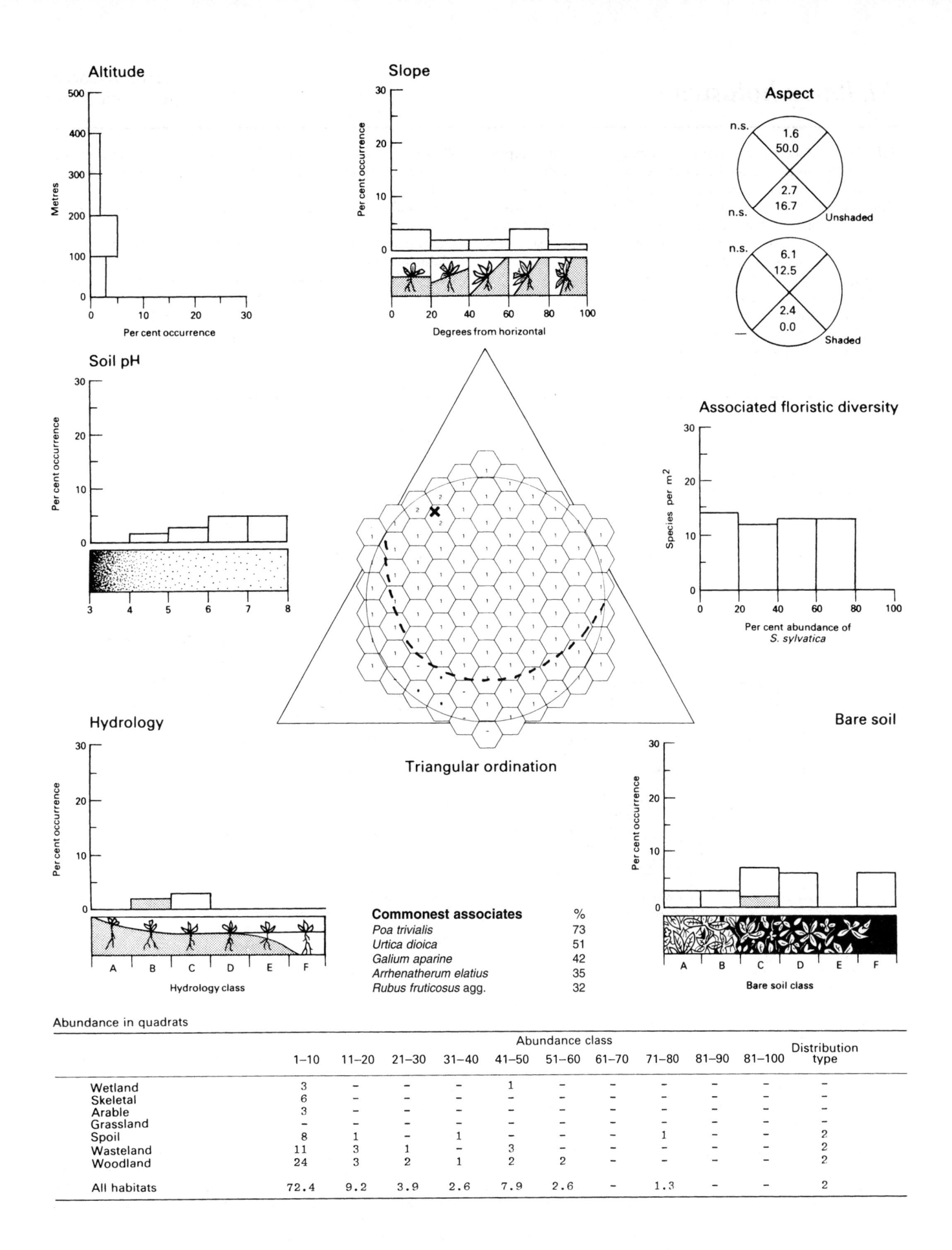

Altitude

Slope

Aspect

	n.s.	1.6	
		50.0	
	n.s.	2.7	Unshaded
		16.7	

	n.s.	6.1	
		12.5	
	—	2.4	Shaded
		0.0	

Soil pH

Associated floristic diversity

Hydrology

Triangular ordination

Bare soil

Commonest associates %
Poa trivialis 73
Urtica dioica 51
Galium aparine 42
Arrhenatherum elatius 35
Rubus fruticosus agg. 32

Abundance in quadrats

	Abundance class										Distribution type
	1–10	11–20	21–30	31–40	41–50	51–60	61–70	71–80	81–90	81–100	
Wetland	3	–	–	–	1	–	–	–	–	–	–
Skeletal	6	–	–	–	–	–	–	–	–	–	–
Arable	3	–	–	–	–	–	–	–	–	–	–
Grassland	–	–	–	–	–	–	–	–	–	–	–
Spoil	8	1	–	1	–	–	–	1	–	–	2
Wasteland	11	3	1	–	3	–	–	–	–	–	2
Woodland	24	3	2	1	2	2	–	–	–	–	2
All habitats	72.4	9.2	3.9	2.6	7.9	2.6	–	1.3	–	–	2

Stellaria holostea L.

Caryophyllaceae
Greater Stitchwort

Life-form Polycarpic perennial chamaephyte with creeping stock. Stems ascending; leaves lanceolate, with rough margins, usually <500 mm^2; roots usually lacking VA mycorrhizas.
Phenology Winter green. The flush of spring growth is accompanied by flowering from April to June. Seed set from June to July. Leaves relatively long-lived, with those surviving the winter replaced in spring.
Height Foliage to *c.* 500 mm; flowers to 600 mm.
Seedling RGR No data. SLA 21.1. DMC 17.8.
Nuclear DNA amount 2.9 pg; $2n = 26^*$ (*BC*); $2x^*$.
Established strategy C-S-R.
Regenerative strategies V, ?S Regeneration from seed requires further study (see 'Synopsis').

Flowers White, hermaphrodite, protandrous or homogamous, insect- or self-pollinated (Knuth 1906); often *c.* 10 in a dichasium.
Germinule 3.7 mg (Salisbury 1942), 3.0 × 2.3 mm, seed in a dehiscent <10-seeded capsule.
Germination Epigeal; freshly-collected seeds dormant.
Biological flora None.
Geographical distribution Most of the British Isles, but less common in Scotland and Ireland (99% of British vice-counties), *ABIF* page 159; Europe northwards to Finland, rare in the Mediterranean (74% of European territories); N Africa.

Habitats Predominantly a plant of hedgerows and lightly-shaded woodland habitats and scrub. Occasionally found on road verges and in unmanaged grassland, and observed on outcrops, soil heaps and in pastures. Absent from wetlands.
Gregariousness Capable of forming patches.
Altitude Observed to 310 m. [Up to 920 m.]
Slope Not recorded from skeletal habitats, and consequently not associated with steep slopes.
Aspect Occurring widely on both N- and S-facing slopes.

Hydrology Absent from wetlands.
Soil pH Occurrences scattered over a wide range of soil pH.
Bare soil Mainly associated with habitats having a low proportion of bare soil.
△ **ordination** A diffuse pattern, suggesting that the species occurs in association with a wide range of vegetation types. This is consistent with the fact that most of its occurrences are in an ecotone habitat – hedgerows. Intolerant of severely disturbed conditions.
Associated floristic diversity Typically low to intermediate.

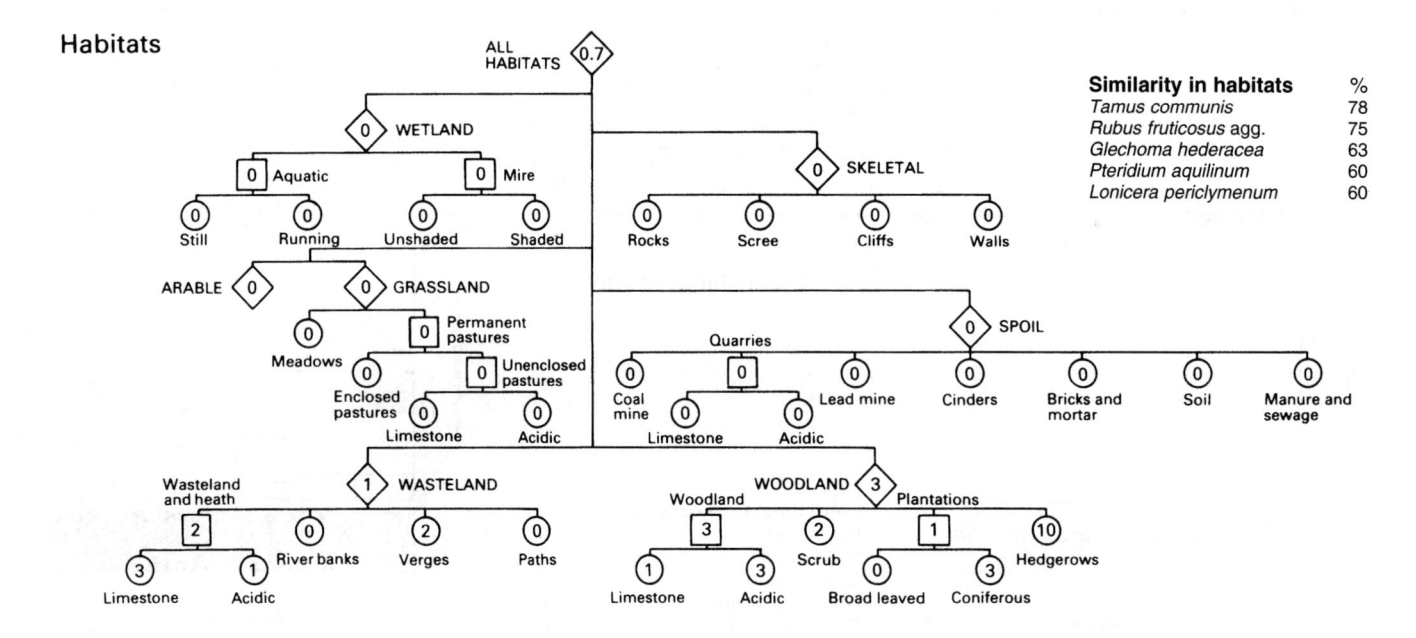

Habitats

Similarity in habitats	%
Tamus communis	78
Rubus fruticosus agg.	75
Glechoma hederacea	63
Pteridium aquilinum	60
Lonicera periclymenum	60

Synopsis A species about which we know little and which deserves further study. *S.h.* is a low-growing, scrambling, winter-green herb, and is typical of sites where the growth of potential dominants is restricted by moderate shade. As in many species of habitats subject to summer shade, *S.h.* shows much early season growth, and flowers in spring. However, *S.h.* maintains foliage throughout the year, and the species is one of the few herbaceous plants to achieve its highest frequency of occurrence in hedgerows. In the uplands *S.h.* is frequent in unshaded sites, and the low tolerance of heavy shade is confirmed experimentally by Rebele *et al.* (1982) and Werner *et al.* (1982). Despite an association with a wide range of soil types, *S.h.* is largely absent from highly calcareous soils and from the most freely-drained sands. It is most abundant on moist, mildly acidic, moderately infertile soils, where it is frequently associated with *Holcus mollis*. The leaves of *S.h.* contain unusually low concentrations of Fe and Mn (*SIEF*). *S.h.* is rarely found in heavily-grazed sites, and appears to be ecologically similar to, but less calcicolous and more shade-tolerant than *Cruciata laevipes*. Shoots are often supported by the surrounding herbage; when this dies back the shoot drops to the soil surface and may form adventitious roots in vegetation gaps. In this way *S.h.* may form large clonal patches. Detached shoots appear capable of regeneration in the field, and thus infrequent disturbance may both serve to check the growth of potential dominants and effect vegetative spread. The 'patchy' distribution of *S.h.* suggests that dispersal by seed does not occur as consistently as regeneration by vegetative means. However, the mechanism and importance of regeneration by the relatively large seed require further investigation. In Lincolnshire *S.h.* is most frequent in ancient woodland (Peterken 1981) and in the Sheffield region, too, *S.h.* does not colonize recent artificial habitats.
Current trends Probably decreasing.

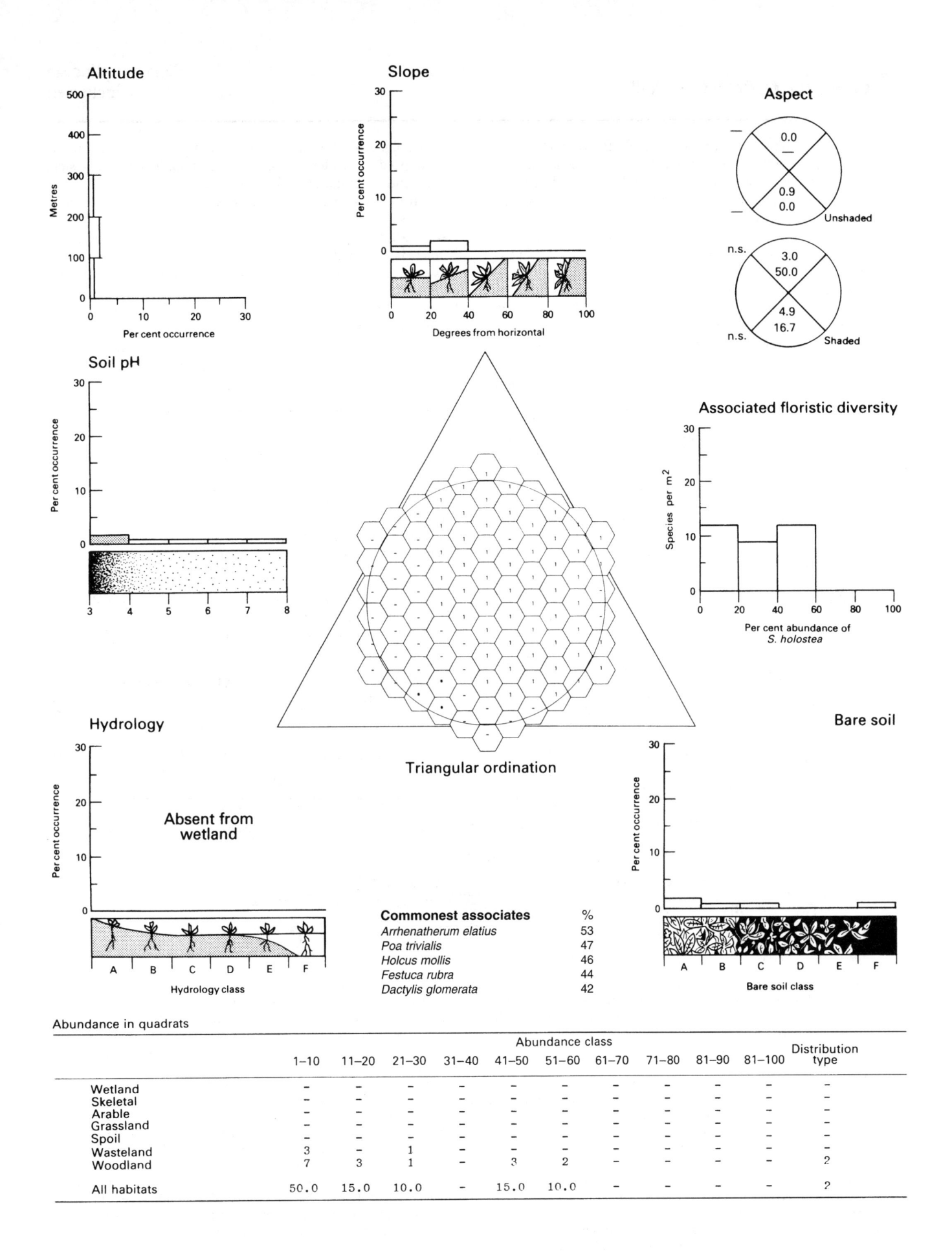

Altitude

Slope

Aspect

Soil pH

Associated floristic diversity

Hydrology

Absent from wetland

Triangular ordination

Bare soil

Commonest associates %
Arrhenatherum elatius 53
Poa trivialis 47
Holcus mollis 46
Festuca rubra 44
Dactylis glomerata 42

Abundance in quadrats

	\multicolumn{10}{c}{Abundance class}	Distribution type
	1–10	11–20	21–30	31–40	41–50	51–60	61–70	71–80	81–90	81–100	
Wetland	–	–	–	–	–	–	–	–	–	–	–
Skeletal	–	–	–	–	–	–	–	–	–	–	–
Arable	–	–	–	–	–	–	–	–	–	–	–
Grassland	–	–	–	–	–	–	–	–	–	–	–
Spoil	–	–	–	–	–	–	–	–	–	–	–
Wasteland	3	–	1	–	3	2	–	–	–	–	2
Woodland	7	3	1	–	3	2	–	–	–	–	2
All habitats	50.0	15.0	10.0	–	15.0	10.0	–	–	–	–	2

Stellaria media (L.) Vill.

Caryophyllaceae
Chickweed

Life-form Winter- or summer-annual therophyte. Stems branched, decumbent or ascending; leaves almost glabrous, ovate, >250 mm²; roots sometimes with VA mycorrhizas.

Phenology Spring-germinating plants may flower and set seed from May onwards, but tend to senesce during late summer. Autumn-germinating plants may survive the winter, and *S.m.* can be seen in flower throughout the year. Leaves short-lived.

Height Usually <300 mm.

Seedling RGR 2.0-2.4 week^{-1}. SLA 54.0. DMC 8.4.

Nuclear DNA amount 2.1 pg; $2n = 40^*$, 42, 44* (*BC*, *FE1*), $4x^*$.

Established strategy Ruderal.

Regenerative strategies B_s. Almost exclusively by seed (see 'Synopsis'); SB IV.

Flowers White, hermaphrodite, homogamous, usually self-pollinated, but insect pollination also possible; flowers produced in winter usually cleistogamous (*Biol Fl(a)*); stem branches terminating in few-flowered cymes.

Germinule 0.35 mg, 1.3×1.0 mm, seed in a capsule containing *c.* 8 seeds.

Germination Epigeal; t_{50} 2 days; 10-28°C; germination rate increased after dry storage at 20°C. Loss of dormancy may be slower in autumn- than in spring-germinating seeds (*Biol Fl(b)*); L = D.

Biological flora (a) Turkington *et al.* (1980), (b) Sobey (1981).

Geographical distribution The British Isles (100% of British vice-counties), *ABIF* page 158; Europe except for the extreme N of Scandinavia (100% of European territories); now cosmopolitan.

Habitats Predominantly an arable weed of both cereal and broad-leaved crops. Also common on manure heaps, sewage waste, soil heaps, demolition sites and river banks. Recorded locally in meadows, on waysides and in recently-established plantations. Found at low frequencies in a variety of other habitats including woodland, hedgerows, open habitats, such as cliffs and walls, and mire. In many shaded habitats *S.m.* is of casual occurrence and does not reach maturity. [May be abundant on cliffs and other coastal sites where sea birds breed. Also occasionally found on shingle beaches (*Biol Fl(b)*).]

Gregariousness Intermediate.

Altitude Most common in areas with a high incidence of arable farming (below 200 m) but extending to 400 m. [Up to 950 m.]

Slope Associated with a wide range of values, but most frequent on comparatively level ground.

Aspect Widely distributed.

Hydrology Occurs on moist, but not submerged or waterlogged, soils.

Soil pH Most frequent and only abundant on soils of pH >5.0.

Bare soil Frequency and abundance positively associated with proportion of bare ground.

△ **ordination** Many records from vegetation associated with disturbed fertile conditions; comparatively scarce elsewhere.

Associated floristic diversity Most typical of intermediate values.

Similarity in habitats	%
Persicaria maculosa	88
Chenopodium album	88
Polygonum aviculare	84
Capsella bursa-pastoris	82
Atriplex patula	82

Habitats

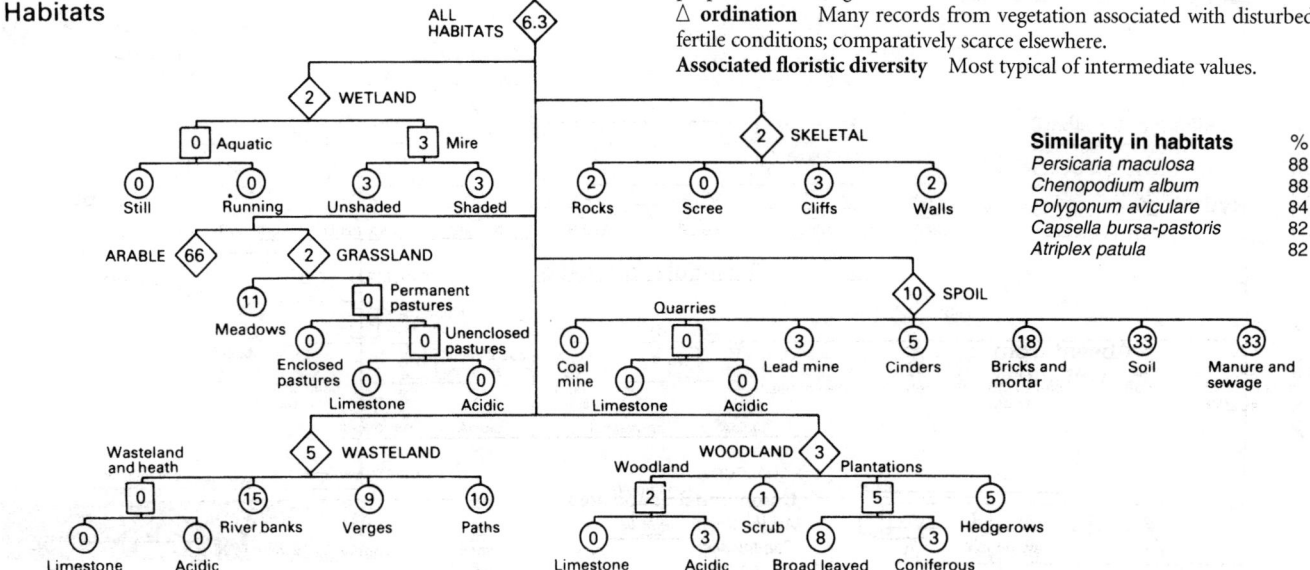

Synopsis *S.m.* is a short-lived annual herb. Although primarily an arable weed capable of severe effects on crop yield, the species also occurs in a wide range of other habitats, but always in association with recently disturbed, fertile soils. *S.m.* is recorded in 50 countries, and is classified as one of the world's worst weeds by Holm *et al.* (1977). *S.m.* is attuned to arable land through several mechanisms. Seed is produced rapidly (within 5 weeks) and abundantly (up to 13 000 per plant) (*Biol Fl(b)*). The species has a high growth rate and shows a swift response to fertilizer additions, sometimes accumulating levels of nitrate which are toxic to stock (*Biol Fl(a)*). Leaves have exceptionally high levels of Na and contain high P, Mg and Fe (*SIEF*). Despite its wide climatic range, *S.m.* thrives best in cool moist environments (*Biol Fl(b)*), and in the Sheffield region is noticeably more abundant and luxuriant during spring and autumn than in summer. *S.m.* is moderately persistent in heavy shade under crops (*Biol Fl(b)*). *S.m.* is palatable to stock and slugs, and has been used as a salad plant (*Biol Fl(b)*, Dirzo 1980). *S.m.* is extremely variable, both phenotypically and genotypically. Summer and winter phenotypes with different growth forms occur. *S.m.* also shows variation in seed characters, particularly with respect to dormancy and persistence in the soil (*Biol Fl(b)*). Seeds are released without a light requirement for germination, but this has been shown to be capable of development after burial (Wesson & Wareing 1969, Taylorson 1970). Germination occurs throughout the year, with peaks in spring and autumn, and two, or even three, generations of plants maybe produced in a year (*Biol Fl(b)*). *S.m.* forms a persistent seed bank. Cuttings root readily in disturbed sites, allowing a limited amount of vegetative regeneration (*Biol Fl(b)*). Seed is widely dispersed in the faeces of birds and animals, and on farm machinery or with crops. *S. neglecta* Weihe ($2n = 22$), a taller, often short-lived perennial of moist woodland, and *S. pallida* (Dumort.) Pire, a smaller, very early flowering winter annual mainly of sandy soils, both occur in Britain and are perhaps best regarded as ssp. of *S.m.* (*Biol Fl(b)*). Two further ssp. are recorded from Europe (*FE1*).

Current trends Another 'follower of humans', likely to remain common as long as disturbed, fertile habitats are created.

Altitude
Slope
Aspect

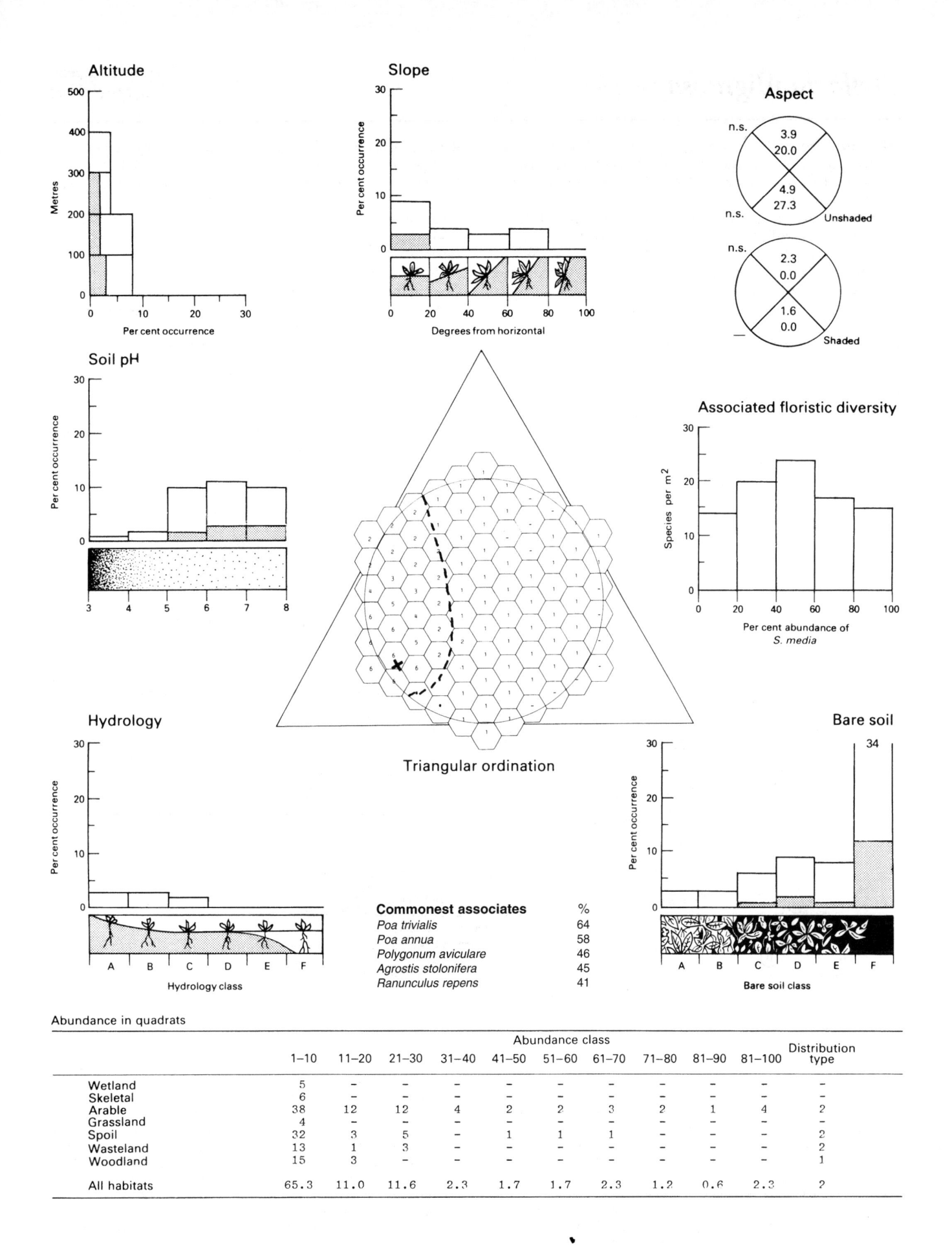

	n.s.	3.9	n.s.
		20.0	
		4.9	
	n.s.	27.3	Unshaded

	n.s.	2.3	n.s.
		0.0	
		1.6	
	—	0.0	Shaded

Soil pH

Associated floristic diversity

Species per m^2 vs Per cent abundance of *S. media*

Hydrology

Hydrology class: A B C D E F

Triangular ordination

Commonest associates %
Poa trivialis 64
Poa annua 58
Polygonum aviculare 46
Agrostis stolonifera 45
Ranunculus repens 41

Bare soil

Bare soil class: A B C D E F

Abundance in quadrats

	1–10	11–20	21–30	31–40	41–50	51–60	61–70	71–80	81–90	81–100	Distribution type
Wetland	5	–	–	–	–	–	–	–	–	–	–
Skeletal	6	–	–	–	–	–	–	–	–	–	–
Arable	38	12	12	4	2	2	3	2	1	4	2
Grassland	4	–	–	–	–	–	–	–	–	–	–
Spoil	32	3	5	–	1	1	1	–	–	–	2
Wasteland	13	1	3	–	–	–	–	–	–	–	2
Woodland	15	3	–	–	–	–	–	–	–	–	1
All habitats	65.3	11.0	11.6	2.3	1.7	1.7	2.3	1.2	0.6	2.3	2

Stellaria uliginosa Murray

Caryophyllaceae
Bog Stitchwort

Life-form Normally a polycarpic perennial helophyte with creeping stock and numerous decumbent and ascending shoots. Leaves lanceolate, often <250 mm^2; roots shallow, non-mycorrhizal.

Phenology Winter green. Flowers in May and June before a majority of the shoot growth has taken place, and sets seed in June and July. Leaves relatively short-lived.

Height Foliage and flowers <100 mm, or greater if plant supported by surrounding vegetation.

Seedling RGR No data. SLA 38.2. DMC 15.5.

Nuclear DNA amount 1.5 pg; $2n = 24*$ (*BC*); $2x*$.

Established strategy Intermediate between ruderal and C-S-R.

Regenerative strategies (V), B$_s$. Effective regeneration by vegetative means and by seed (see 'Synopsis'); ?SB IV.

Flowers White, hermaphrodite, insect- or probably more usually self-pollinated; in <10-flowered cymes.

Germinule 0.06 mg, 0.8 × 0.5 mm, seed in many-seeded dehiscent capsule.

Germination Epigeal; t_{50} 3 days; 6-30°C; L > D.

Biological flora None.

Geographical distribution Most of the British Isles (100% of British vice-counties), *ABIF* page 160; Europe northwards to Norway, but only on mountains in the S (72% of European territories); temperate Asia, N America and New Zealand.

Habitats A wetland species most characteristic of unshaded mire. Also found on river banks, in shaded mire and as an emergent or floating aquatic of stream margins. Occasionally recorded in damp woodland and as a temporary colonist of pond dredgings. Observed on cliffs and walls beside water.

Gregariousness Intermediate or rarely stand-forming.

Altitude Widely distributed, but most common above 400 m. [Up to 1010 m.]

Slope As with most wetland species, concentrated on flat ground and gentle slopes.

Aspect No significant pattern detected.

Hydrology Particularly frequent in soligenous mire, but also widespread in topogenous mire and in very shallow water at the margins of streams.

Soil pH Most frequent in the pH range 5.0-7.0. Absent below pH 4.5, and rare on calcareous strata.

Bare soil Highest frequency of occurrence and abundance associated with habitats with intermediate amounts of bare ground.

△ **ordination** Distribution centred on vegetation associated with fertile but moderately disturbed conditions.

Associated floristic diversity Typically intermediate.

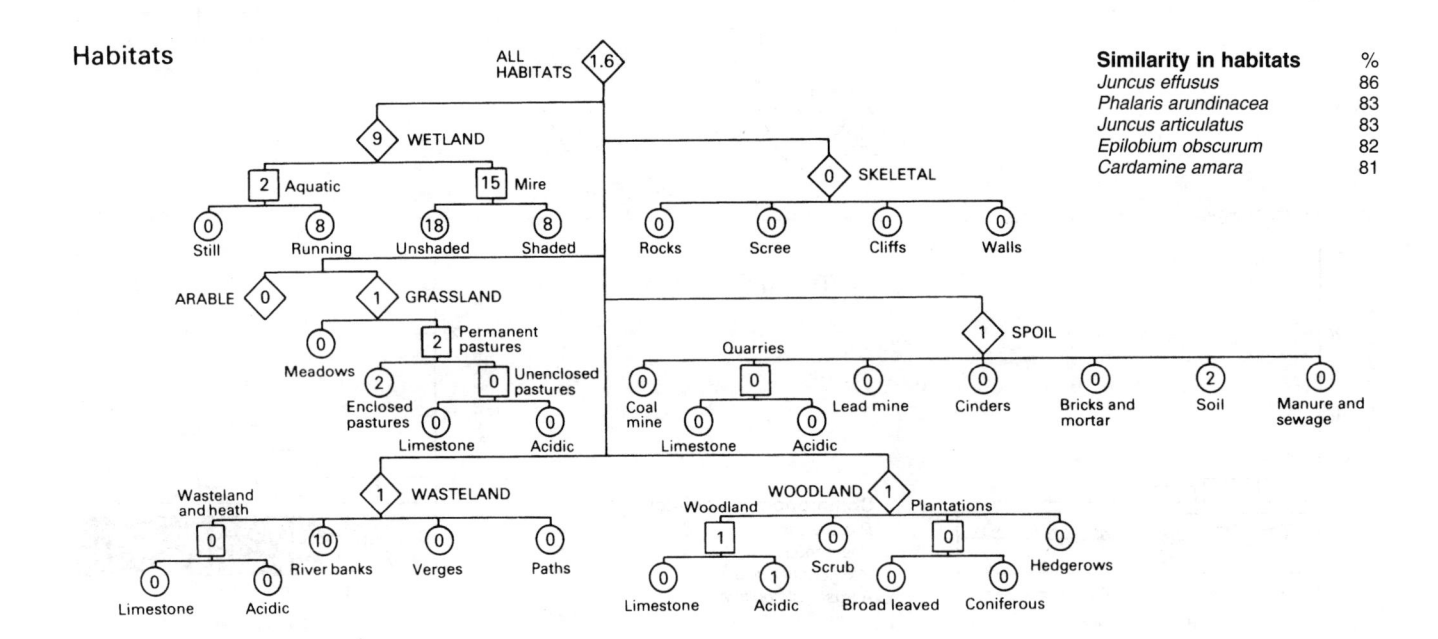

Habitats

Similarity in habitats	%
Juncus effusus	86
Phalaris arundinacea	83
Juncus articulatus	83
Epilobium obscurum	82
Cardamine amara	81

Synopsis *S.u.* is a prostrate mat-forming herb of fertile wetlands and is essentially an opportunist colonizer of disturbed sites. In Japan, at least, the species may flower within 5 months and seed set takes a further 3 weeks (Ankei 1982). Thus, the species often colonizes bare mud in advance of potentially more-dominant species, and may survive in trampled mire and on river banks subject to erosion. *S.u.* also displays a modest degree of shade tolerance although, in shade, flowering is often suppressed and the species may form a sparse understorey beneath taller vegetation. In tall patchy vegetation such as stands of *Juncus effusus*, ascending stems, supported by the surrounding vegetation, may protrude through gaps in the canopy. *S.u.* also forms floating rafts in small upland streams, often intermixed with species such as *Holcus mollis* and *Montia fontana* L. The species is most characteristic of mildly acidic sites, and in the Sheffield region has not been recorded from calcareous mire. Plants observed in autumn have a creeping habit and numerous narrower leaves of lighter hue. Vegetative regeneration by means of prostrate rooted stems enables *S.u.* to form extensive clones. Detached shoots readily reroot and are, we suspect, important in the colonization of river banks and mire beside open water. In unshaded sites seed production is often prolific, and a large persistent seed bank may be formed (Thompson & Grime 1979, UCPE unpubl.). Annual and perennial ecotypes of this inbreeding species have been identified in Japan (Ankei 1982).

Current trends A species of fertile and disturbed wetlands, which may be increasing as a result of human activities.

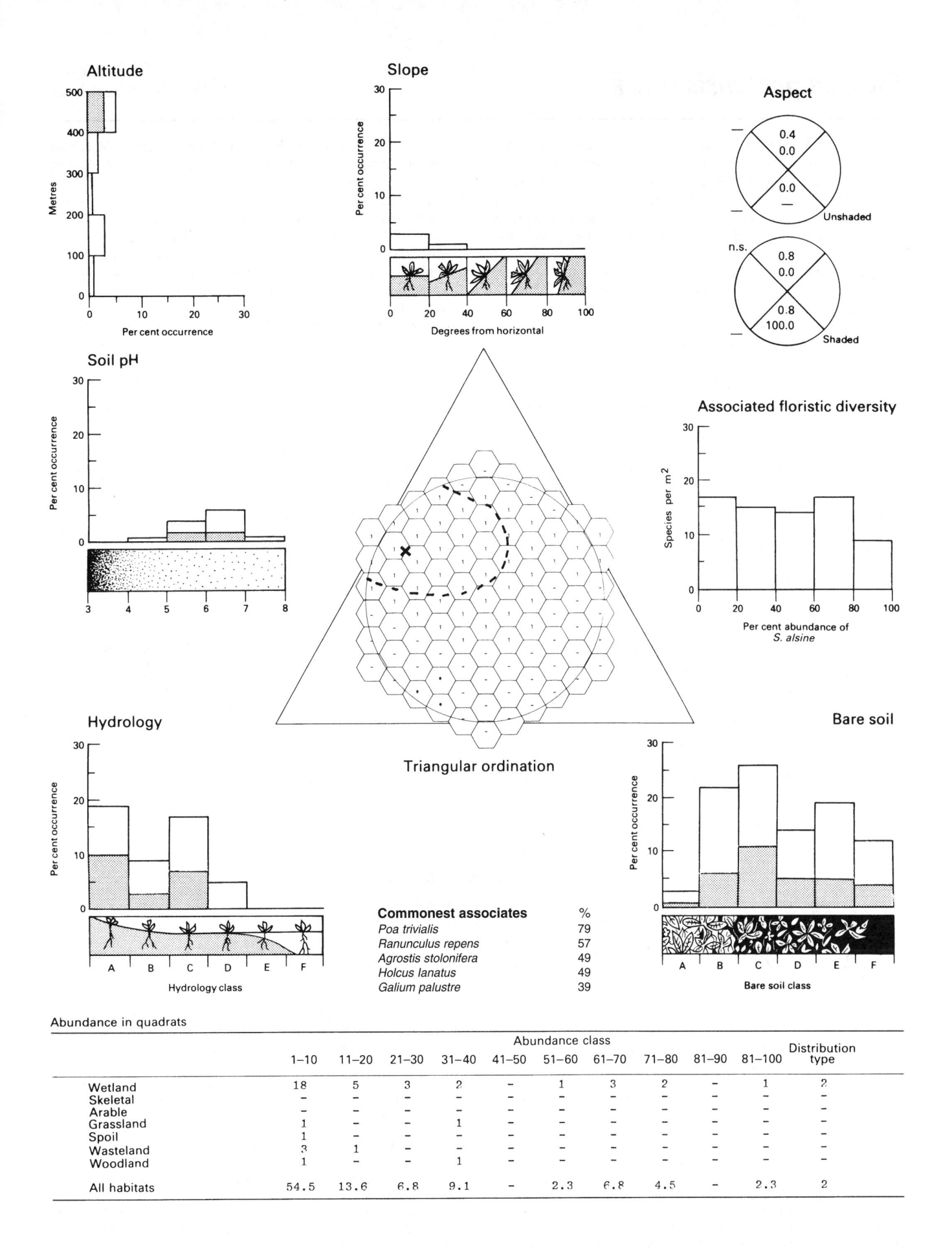

Altitude

Slope

Aspect

Soil pH

Associated floristic diversity

Species per m^2 vs Per cent abundance of *S. alsine*

Hydrology

Hydrology class: A B C D E F

Triangular ordination

Commonest associates	%
Poa trivialis	79
Ranunculus repens	57
Agrostis stolonifera	49
Holcus lanatus	49
Galium palustre	39

Bare soil

Bare soil class: A B C D E F

Abundance in quadrats

	1–10	11–20	21–30	31–40	41–50	51–60	61–70	71–80	81–90	81–100	Distribution type
Wetland	18	5	3	2	–	1	3	2	–	1	2
Skeletal	–	–	–	–	–	–	–	–	–	–	–
Arable	–	–	–	–	–	–	–	–	–	–	–
Grassland	1	–	–	1	–	–	–	–	–	–	–
Spoil	1	–	–	–	–	–	–	–	–	–	–
Wasteland	3	1	–	–	–	–	–	–	–	–	–
Woodland	1	–	–	1	–	–	–	–	–	–	–
All habitats	54.5	13.6	6.8	9.1	–	2.3	6.8	4.5	–	2.3	2

Succisa pratensis Moench

Dipsacaceae
Devil's-bit Scabious

Life-form Polycarpic perennial, semi-rosette hemicryptophyte with short vertical rhizome. Stems erect; leaves ovate-lanceo-late, slightly hairy, usually <4000 mm^2; roots with VA mycorrhizas.
Phenology Winter green. Flowers from July to October and fruit mature within 1 month. Some fruit is shed while green, some is retained until winter (*Biol Fl*). Leaves survive for up to 1 year (*Biol Fl*).
Height Foliage typically appressed to ground, but sometimes extending to 200 mm; flowering shoots to 1 m.
Seedling RGR 0.5–0.9 week^{-1}. SLA 17.9. DMC 17.8.
Nuclear DNA amount 5.5 pg; $2n = 20*$ (*BC*), $2x*$.
Established strategy Intermediate between stress-tolerator and C-S-R.
Regenerative strategies S. Regenerates almost entirely by seed in spring. No buried seed bank has been reported. Vegetative spread through the production of lateral shoots is possible but rare (*Biol Fl*); SB II.
Flowers Usually purple, gynodioecious, normally insect-pollinated although self-pollination possible (*Biol Fl*); c. 100 in terminal hemispherical heads, fewer in lateral heads.
Germinule 1.54 mg, 4.5 × 1.6 mm, indehiscent dry fruit, shed within a persistent epicalyx and overtopped by a spiky calyx.
Germination Epigeal; t_{50} 10 days; 12–31°C; L = D.
Biological flora Adams (1955).
Geographical distribution The British Isles (100% of British vice-counties), *ABIF* page 607; most of Europe except for the extreme N and parts of the Mediterranean (77% of European territories); NW Africa; introduced in N America.

Habitats Mainly restricted to relatively unproductive damp pastures and ungrazed grassland. Less frequent in mire, and on waysides and lead-mine spoil. Occasionally observed at the margin of woodland, particularly along woodland rides. Not recorded from aquatic, arable and skeletal habitats. [Found on sea cliffs and other maritime habitats, and in montane grassland.]
Gregariousness Intermediate.
Altitude Extends to 350 m. Upland bias reflects a strong association with the carboniferous limestone. [Up to 970 m.]
Slope Only present on slopes of <50°.
Aspect Mostly on unshaded slopes of N-facing aspect; this trend very evident in grasslands (*Ecol Atl*).
Hydrology Found primarily in soligenous mire and also observed on tussocks of other species.
Soil pH Attains maximum frequency in the pH range 5.5–6.5. No records from sites of pH outside the range 3.5–7.5.
Bare soil Shows a clear association with habitats having little or no bare soil.
△ ordination Distribution centred on vegetation associated with relatively infertile and undisturbed conditions.
Associated floristic diversity Typically species-rich.

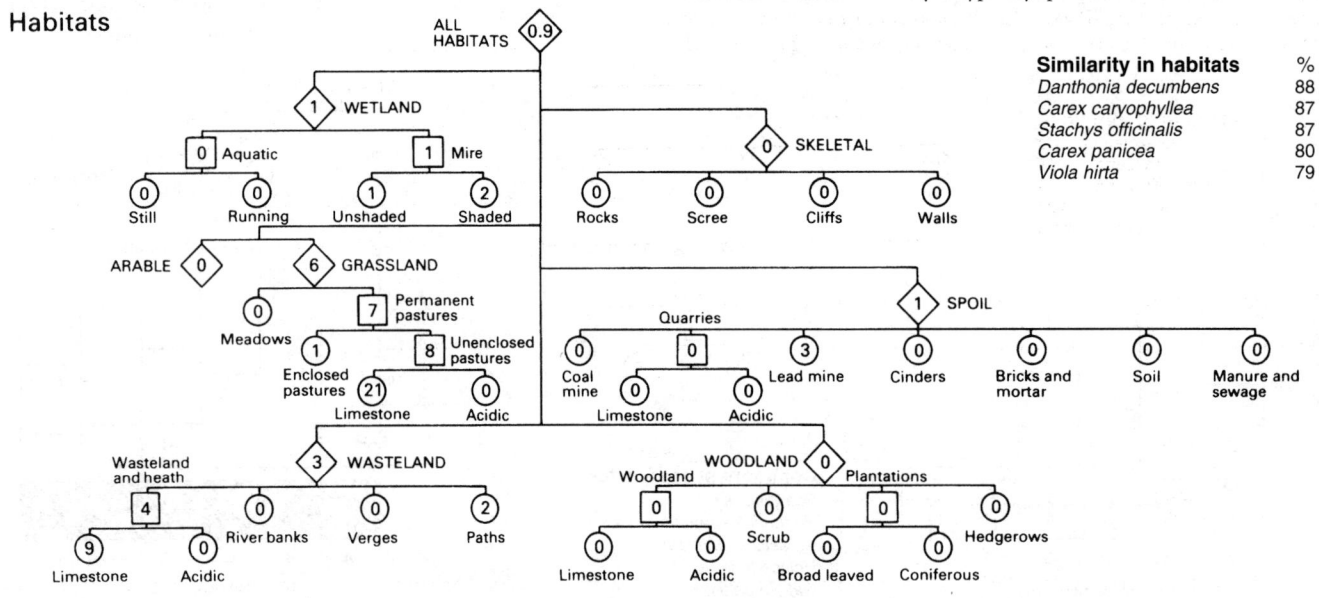

Habitats

Similarity in habitats	%
Danthonia decumbens	88
Carex caryophyllea	87
Stachys officinalis	87
Carex panicea	80
Viola hirta	79

Synopsis *S.p.* is a slow-growing, winter-green, perennial, rosette-forming herb which exploits sites where the growth of potential dominants is restricted by a low level of soil fertility and often also by grazing. The species appears to be long-lived and often does not flower until its fourth year (*Biol Fl*). The root system of *S.p.* is composed of spreading, rather superficial roots (*Biol Fl*) and the species does not occur in droughted habitats. Though strongly associated with continuously moist habitats, *S.p.* is seldom found in waterlogged sites other than in soligenous mire or on tussocks of other species. In unproductive grassland, wood margins and maritime and montane habitats, *S.p.* is characteristic of soils of intermediate pH. However, *S.p.* is also found on a wide range of other soil types, including calcareous mineral soils, but the species is infrequent on freely-drained sandy soils and on soils of low pH. In grazed habitats the leaves of *S.p.* often escape predation, and *S.p.* is tolerant of light trampling. Regeneration is mainly by seed (*Biol Fl*) which germinates in spring and, as far as is known, *S.p.* does not develop a persistent seed bank. The seed has no well-defined dispersal mechanism and is in fact poorly dispersed (*Biol Fl*). *S.p.*, normally an out-breeding species, shows a modest degree of phenotypic and genotypic variation. Dwarf plants with short flowering stems from maritime habitats usually, but not always, increase their size on cultivation (*Biol Fl*). In species-rich short turf, in which *S.p.* reaches maximum abundance, the species tends to be represented by phenotypes with low rosettes (*Biol Fl*). In taller vegetation ascending leaves with longer petioles are observed (*Biol Fl*).
Current trends *S.p.* mainly exploits infertile pasture; as a result of its poor ability to colonize new artificial habitats it is largely restricted to semi-natural vegetation. As a result of drainage and high fertilizer input, *S.p.* is decreasing, dramatically so in lowland habitats where it is becoming increasingly restricted to linear habitats such as railway banks, ditch banks and woodland rides.

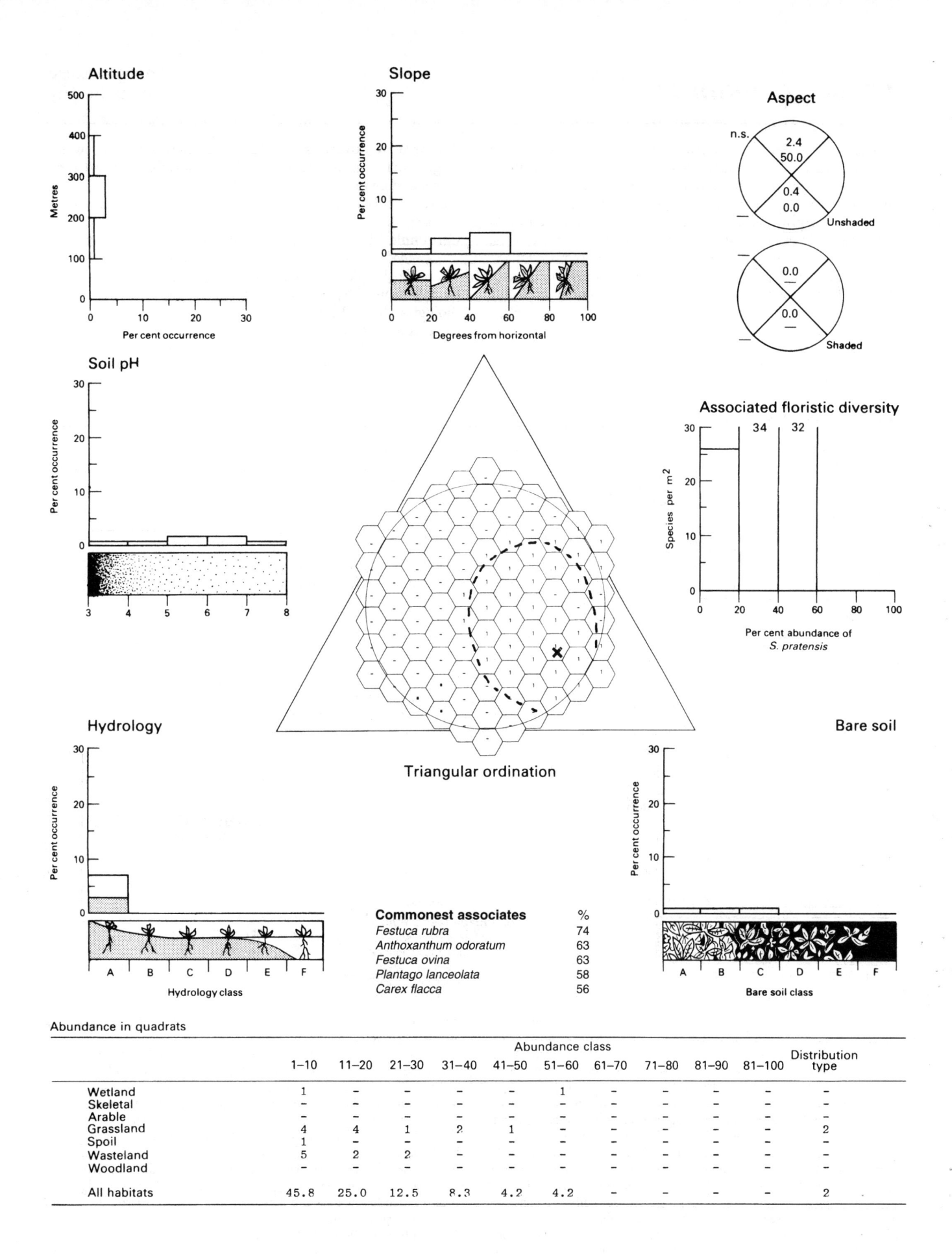

Altitude

Slope

Aspect

	n.s.	2.4	
		50.0	
	—	0.4	Unshaded
		0.0	

	—	0.0	
	—	0.0	Shaded

Soil pH

Associated floristic diversity

Hydrology

Triangular ordination

Bare soil

Commonest associates   %
Festuca rubra   74
Anthoxanthum odoratum   63
Festuca ovina   63
Plantago lanceolata   58
Carex flacca   56

Abundance in quadrats

	1–10	11–20	21–30	31–40	41–50	51–60	61–70	71–80	81–90	81–100	Distribution type
Wetland	1	–	–	–	–	1	–	–	–	–	–
Skeletal	–	–	–	–	–	–	–	–	–	–	–
Arable	–	–	–	–	–	–	–	–	–	–	–
Grassland	4	4	1	2	1	–	–	–	–	–	2
Spoil	1	–	–	–	–	–	–	–	–	–	–
Wasteland	5	2	2	–	–	–	–	–	–	–	–
Woodland	–	–	–	–	–	–	–	–	–	–	–
All habitats	45.8	25.0	12.5	8.3	4.2	4.2	–	–	–	–	2

Tamus communis L.

Dioscoreaceae
Black Bryony

Life-form Polycarpic perennial geophyte with subterranean tuber. Stems twining, supported by surrounding vegetation; leaves ovate, usually <6000 mm^2; roots with VA mycorrhizas.
Phenology Shoots elongate underground during winter and emerge in April or May. Flowers from May to July. Berries do not turn red until *c.* 100 days after fertilization of the ovaries, and usually persist on the plant until at least November. Shoots die back in late autumn (*Biol Fl*). Leaves relatively long-lived.
Height Immature plants persisting under a woodland canopy frequently <100 mm. Foliage and flowers of mature plant sometimes >2 m.
Seedling RGR No data. SLA 27.8. DMC 15.1.
Nuclear DNA amount No data; $2n = 48$ (*FE5*); $4x$ for family.
Established strategy Intermediate between competitor and C-S-R.

Regenerative strategies S. Almost entirely by seeds, which germinate in the spring and are capable of short-term persistence in the soil (*Biol Fl*); SB II.
Flowers Green, dioecious, insect-pollinated; 10-30 in axillary racemes.
Germinule 18.33 mg, 3.2 × 3.2 mm, seed dispersed within a 1-6-seeded red berry during autumn or winter. Berries become more palatable to birds as winter progresses (*Biol Fl*).
Germination Hypogeal; dormancy broken by chilling (*Biol Fl*).
Biological flora Burkhill (1944).
Geographical distribution Extends northwards to N England (46% of British vice-counties), *ABIF* page 834; S, C and W Europe (59% of European territories and 3% naturalized); N Africa.

Habitats A plant of hedgerows and woodland, particularly on calcareous soils. Rarely found, and often impermanent, in unshaded habitats. Occasionally observed in shaded skeletal habitats. Absent from pasture and wetland. [Sometimes occurs on sheltered sea cliffs (*Biol Fl*).]
Gregariousness Occurs only as isolated individuals.
Altitude A lowland species extending to *c.* 300 m, both locally and in Britain as a whole (*Biol Fl*).
Slope Not recorded from slopes >50°.

Aspect Slightly more frequent on S-facing slopes.
Hydrology Absent from wetlands, like many other deeply rooted species.
Soil pH Mainly recorded on soils above pH 6.0, with isolated occurrences on mildly acidic soils.
Bare soil Found over a wide range of values. Sites with little bare soil often have a covering of tree litter.
Δ **ordination** Confined to relatively undisturbed vegetation. Most abundant in habitats of moderately high fertility.
Associated floristic diversity Typically low to intermediate.

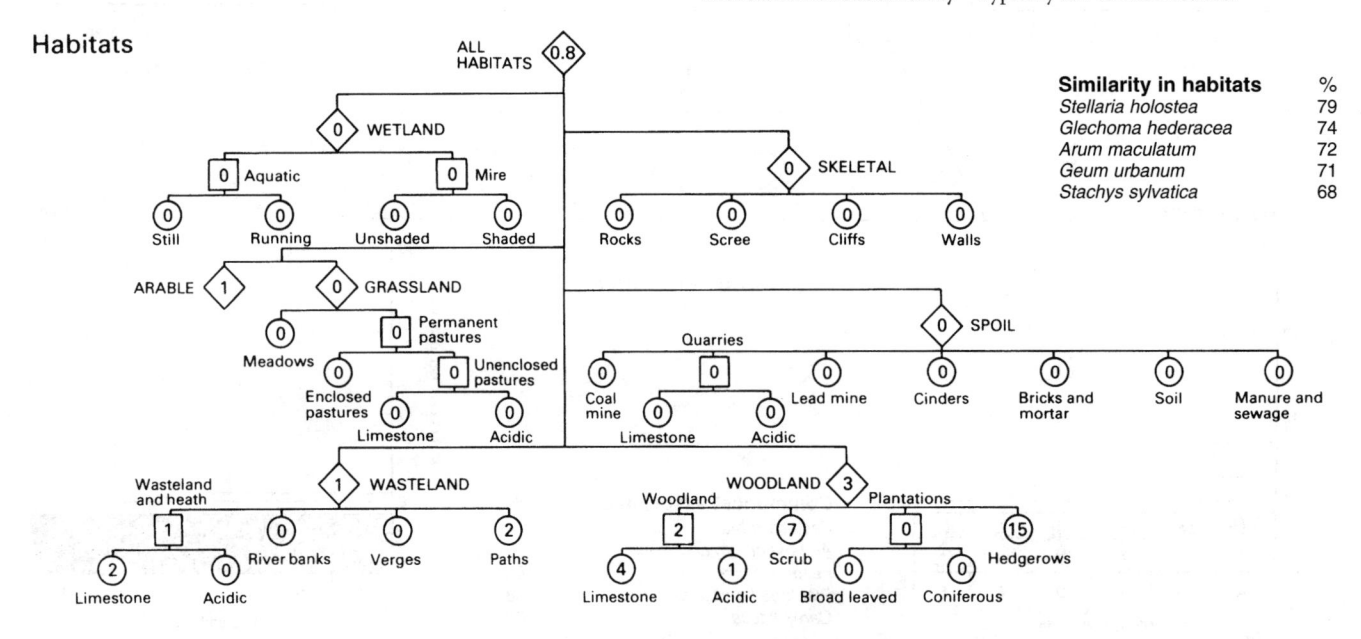

Habitats

Similarity in habitats	%
Stellaria holostea	79
Glechoma hederacea	74
Arum maculatum	72
Geum urbanum	71
Stachys sylvatica	68

Synopsis *T.c.* has long twining stems and a large underground tuber which in extreme cases may have a fresh weight of 10-15 kg (*Biol Fl*). *T.c.* is the only British member of the predominantly tropical Dioscoreaceae, the yam family (Cronquist 1981). The species is virtually restricted to lowland areas, and reaches its N distributional limit in Britain. *T.c.* occupies two distinct habitats. In hedgerows and woodland margins large plants occur and seed-set is generally high, whereas in woodlands plants normally persist in a non-flowering, suppressed state, but are presumably released when gaps form in the tree canopy. The location of the tuber 50 mm or more below the soil surface, and the relatively deep root system, appear to restrict the capacity of *T.c.* to exploit waterlogged soils, and probably explain its virtual absence from shallow soils. Although the leaves and shoots may be used as a vegetable (Dahlgren *et al.* 1985), the plant is little predated and when eaten raw may cause fatalities to humans and stock (Cooper & Johnson 1984). *T.c.* is normally dioecious and populations tend to contain more male than female plants (Kay & Stevens 1986). Effective regeneration is by means of seeds which germinate in spring (*Biol Fl*). Hence, the survival of hedgerow and to a lesser extent woodland margin populations is essential for the success of *T.c.* Large plants of *T.c.* may be 20 years old or more and plants do not flower until at least 5 years old (*Biol Fl*). Seeds retained on the plant until late winter do not germinate in the subsequent spring and remain dormant for a further year (*Biol Fl*). Seeds are enclosed within berries, and may be distributed by birds (*Biol Fl*). However, *T.c.* tends to be a poor colonist of new sites and to be principally associated with those species-rich (older) hedgerows. No persistent seed bank has been detected. Occasionally, when tubers are broken, each half will form a new plant (*Biol Fl*). Otherwise there is no vegetative regeneration.
Current trends Probably decreasing.

Altitude
Slope
Aspect

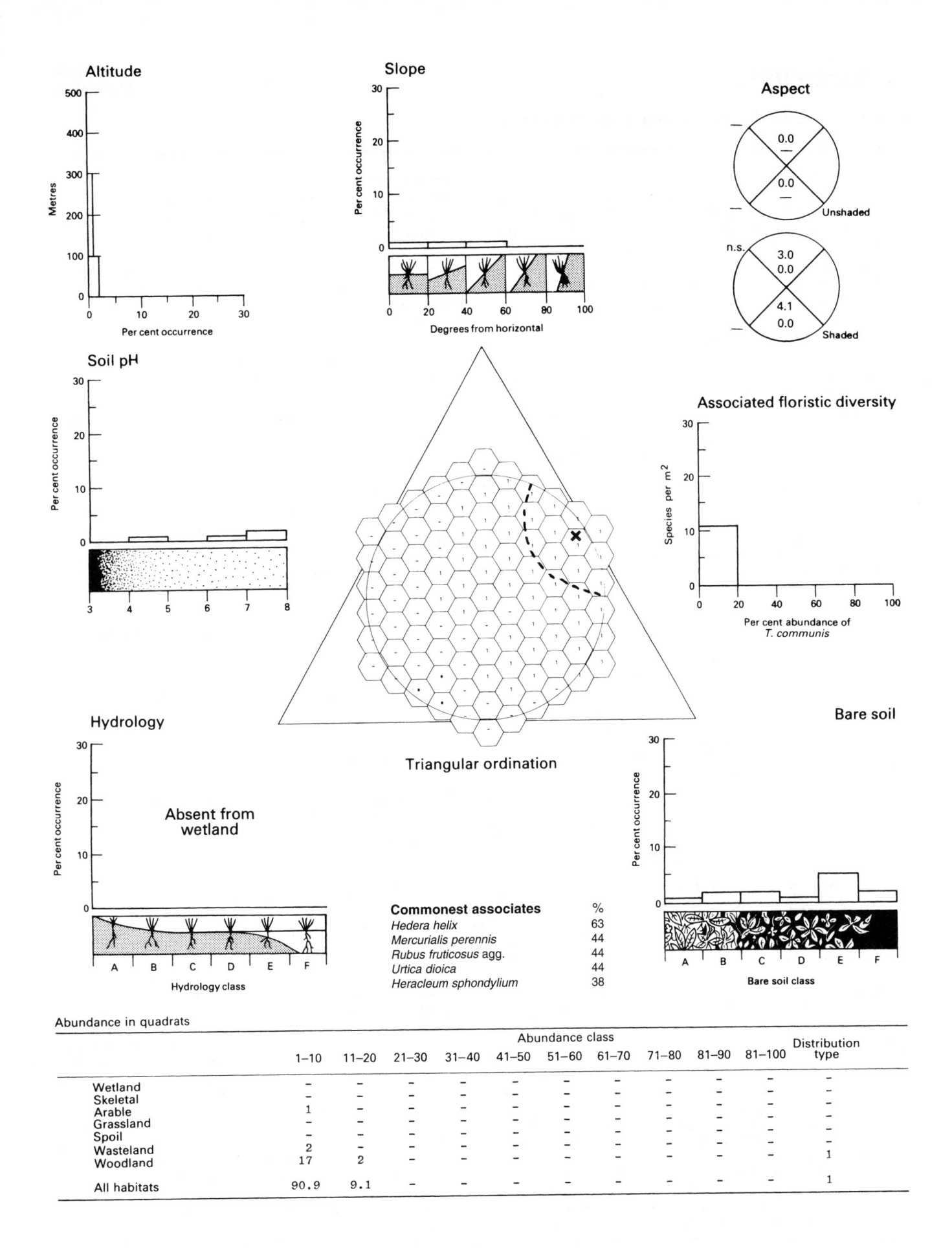

	0.0
0.0	Unshaded

	3.0	
n.s.		
	4.1	Shaded

Soil pH

Associated floristic diversity

Hydrology

Absent from wetland

Commonest associates	%
Hedera helix	63
Mercurialis perennis	44
Rubus fruticosus agg.	44
Urtica dioica	44
Heracleum sphondylium	38

Triangular ordination

Bare soil

Abundance in quadrats

	Abundance class										Distribution type
	1–10	11–20	21–30	31–40	41–50	51–60	61–70	71–80	81–90	81–100	
Wetland	–	–	–	–	–	–	–	–	–	–	–
Skeletal	–	–	–	–	–	–	–	–	–	–	–
Arable	1	–	–	–	–	–	–	–	–	–	–
Grassland	–	–	–	–	–	–	–	–	–	–	–
Spoil	–	–	–	–	–	–	–	–	–	–	–
Wasteland	2	–	–	–	–	–	–	–	–	–	1
Woodland	17	2	–	–	–	–	–	–	–	–	
All habitats	90.9	9.1	–	–	–	–	–	–	–	–	1

Taraxacum agg.

Asteraceae
Dandelion

A taxonomically difficult grouping (see 'Synopsis')

Life-form Polycarpic perennial, rosette hemicryptophyte. Scapes decumbent to erect; leaves entire to pinnately lobed, usually <10 000 mm^2; stoutly tap-rooted and with VA mycorrhizas.
Phenology A small rosette of leaves overwinters. Flowers from May to October but mainly April to June. Most seed set from May to June.
Height Foliage often close to the ground surface and not exceeding 350 mm; flowers to 400 mm.
Seedling RGR 1.0–1.4 week^{-1}. SLA 37.3 DMC 13.7.
Nuclear DNA amount 2.6 pg (Bennett & Smith 1976); $2n = c.$ 16*, *c.* 24*, 32*, 40*, 48* (Richards 1972), $2x$*–$6x$*.
Established strategy Variable; most commonly intermediate between ruderal and C-S-R.

Regenerative strategies W. See 'Synopsis'; SB I.
Flowers Florets yellow, hermaphrodite, ligulate, visited by insects but mostly apomictic; >100 florets in each solitary terminal capitulum. Sexual species are mainly self-incompatible (Richards 1972).
Germinule 0.64 mg, 4.0 × 1.3 mm, achene dispersed with a pappus *c.* 15 mm long. Fruit smaller in some taxa.
Germination Epigeal; t_{50} 3 days; 7–34°C; L > D.
Biological flora None, but see Zarzycki (1976) and Sterk *et al.* 1983.
Geographical distribution The British Isles, *ABIF* page 626; Europe and the rest of the N hemisphere.

Habitats A common constituent of all but aquatic habitats. Particularly frequent in meadows and pastures and on waysides, wasteland and spoil, particularly building rubble. Recorded at lower frequencies from arable land, woods, hedgerows, wetlands, scree slopes, walls and cliffs. A troublesome garden weed. [Some taxa are characteristic of sand dunes and cliff tops, and others are restricted to high mountains over 900 m.]
Gregariousness Intermediate.
Altitude Widely distributed. [Up to 1220 m.]
Slope Wide-ranging.

Aspect Widely distributed (see 'Synopsis').
Hydrology Often present in soligenous mire (see 'Synopsis').
Soil pH Most frequent on soils of pH >7.0; little recorded below pH 4.5.
Bare soil Wide-ranging (see 'Synopsis').
△ **ordination** A very widespread distribution, but centred on vegetation associated with fertile, moderately disturbed habitats, the type of environment exploited by Section *Taraxacum*, the most common group (see 'Synopsis').
Associated floristic diversity Typically intermediate to species-rich.

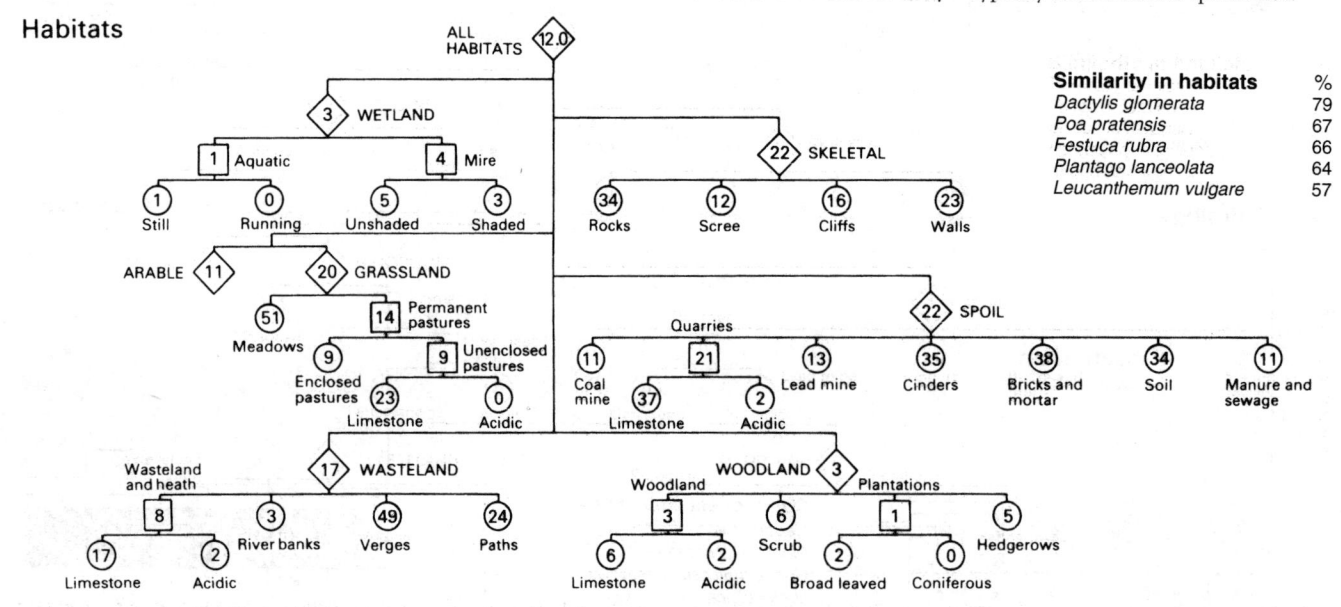

Habitats

Similarity in habitats	%
Dactylis glomerata	79
Poa pratensis	67
Festuca rubra	66
Plantago lanceolata	64
Leucanthemum vulgare	57

Synopsis *T.* agg. is a winter-green, rosette-forming polycarpic perennial with a long stout tap-root, and is found in a wide range of habitats. The aggregate is well known as a troublesome weed of gardens, waysides and pasture, but equally extends onto sand dunes, mountains and into wetland. This diversity is related to the existence of many ecotypes. Thus, *T.* agg., a taxonomically difficult grouping, has been divided into a number of Sections. Over 150 microspecies have been recognized in Britain, and there are possibly 1500 in Europe (*FE4*). Based on Richards (1972), Sterk *et al.* (1983) and extremely limited field observations, the following generalizations may be made. Plants from Section *Taraxacum* are robust and are mainly found in fertile, disturbed, artificial habitats, and include the taxa which are familiar garden and pasture weeds. Those from Section *Erythrosperma* Dahlst. are associated with dry sandy or calcareous soils, account for the occurrences in the portions of the triangular ordination diagram corresponding to less-fertile soils and probably have a bias towards S-facing slopes. They are generally much smaller in all their parts than taxa from Section *T*. Taxa from Section *Spectabilia sensu lato* (see Richards & Haworth 1984) occur in moist, often less-fertile and less disturbed sites. Most wetland records refer to this group, which tends to be more common on N-facing slopes. Other Sections are associated with montane habitats, with sand dunes or with mire. In moist habitats in Holland a reduction in the intensity of land management has been shown to result in the gradual replacement of Section *T.* by Sections *S.* and *Palustria* Dahlst. In drier habitats Section *E.* becomes more prominent (Oosterveld 1978). The tap-root often penetrates deeply and permits the exploitation of sites where mineral subsoil is overlain with debris such as building rubble or paving stones. In rocky or sandy habitats subject to summer drought the tap-root appears to be of critical importance in affording access to subsoil moisture. In pastures *T.* agg. is among the species preferred by stock; it is capable of rapid recovery after defoliation, and is also resilient under trampling damage. In part this resilience is due to the presence of contractile roots which pull the apical meristem 10–20 mm below the soil surface (Sterk *et*

Altitude

Slope

Aspect

Soil pH

Commonest associates	%
Festuca rubra	60
Dactylis glomerata	52
Poa trivialis	41
Holcus lanatus	40
Plantago lanceolata	39

Associated floristic diversity

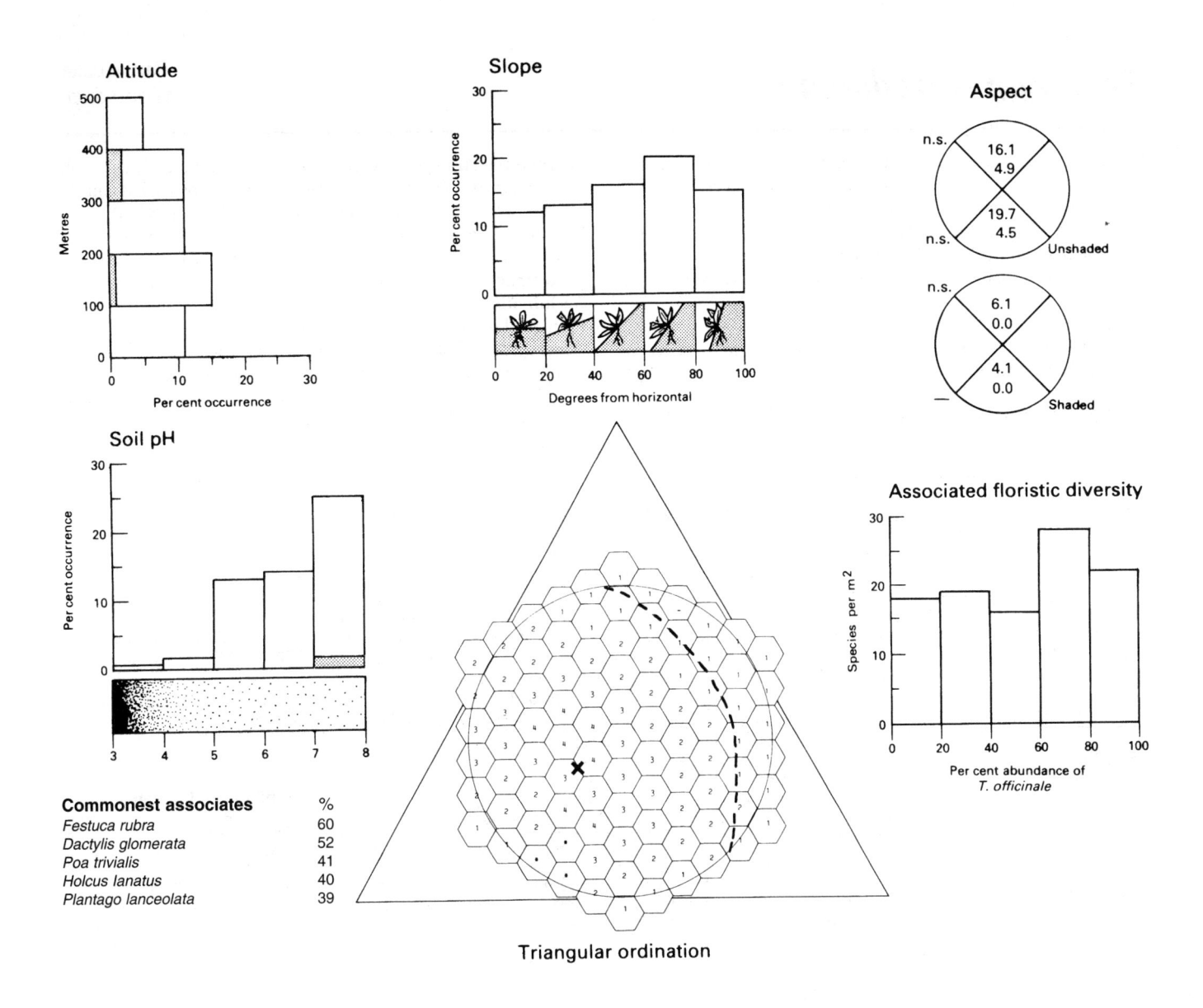

Triangular ordination

Abundance in quadrats

	1–10	11–20	21–30	31–40	41–50	51–60	61–70	71–80	81–90	81–100	Distribution type
					Abundance class						
All habitats	88.2	6.6	1.8	1.2	0.6	0.9	0.3	–	0.3	–	1

al. 1983). Early growth of foliage and flowering, before appreciable growth of many grasses has occurred, may also be a key feature in the success of this species in pasture (Sterk et al. 1983). Fruits of *T.* agg. are buoyant in air, released in large numbers during early summer (up to 2000 or more per plant (Salisbury 1942) or c. 5000 m^{-2} in pastures (Sterk et al. 1983)), and are capable of rapid germination over a wide range of temperatures. This combination enables *T.* agg. to be a highly effective colonizing species of sites with exposed soil. *T.* agg. is also effectively dispersed through human activities, and alien taxa appear to have been introduced from Continental Europe. Only two native taxa are diploid and sexual; the remainder are apomictic (see Richards 1972). Thus, regeneration by seed is essentially an asexual process. No persistent seed bank is formed, but in the event of disturbance *T.* agg. may regenerate by means of fragments of tap-root. Large plants, particularly after close grazing, may also form multiple rosettes. Plants are phenotypically plastic, so much so that only those collected in early summer from typical habitats can be identified with any certainty (Richards 1972). As implied already from the wide ecological range of *T.* agg., populations differ genotypically in the extent of their allocation of resources to vegetative growth and to seed (Solbrig &

Simpson 1974, 1977) and in a variety of characters related to growth and morphology, including RGR (Roetman & Sterk 1986). Microspecies from less-strongly fertilized and less-heavily grazed pastures tend to have a broad ecological range; those of fertile, intensively grazed habitats are more specialized and are probably among the most recently evolved taxa (Sterk et al. 1983). Hybridization occurs between diploid sexual taxa, and pollen-bearing apomictic species (Stace 1975), and this has been important during episodes of speciation (Richards 1972). In Europe an estimated 10% of taxa are diploid, 45% triploid, 28% tetraploid, 5% pentaploid and c. 11% aneuploid (Doll 1977).

Current trends Overall the aggregate species is probably still increasing, particularly in Section *T.* which includes the majority of populations exploiting fertile and disturbed lowland habitats. The status of other Sections is less clear but some which exploit fens and water-meadows, particularly Section *Palustria* Dahlst., are clearly declining (Richards 1972). Sexuality is not uncommon in Sect. *T.* (see Sterk et al. 1983) and the evolution of new taxa capable of exploiting disturbed fertile habitats is doubtless proceeding.

Teucrium scorodonia L.

Lamiaceae
Wood Sage

Life-form Rhizomatous, polycarpic perennial proto-hemicryptophyte. Stems erect; leaves ovate, hairy, typically <2000 mm^2; roots with VA mycorrhizas.
Phenology Flowers July to September; seeds ripen from September onwards, but later flowers seldom set seed (*Biol Fl*). Most shoots die back in autumn but some may overwinter with senescing leaves.
Height Foliage to 400 mm overtopped by flowers.
Seedling RGR 1.0–1.4 week^{-1}. SLA 12.0. DMC 37.8.
Nuclear DNA amount 2.4 pg; $2n = 32*, 34*$ (*BC*); $2x*$. (High generic base numbers common in Lamiaceae.)
Established strategy Intermediate between stress-tolerant competitor and C-S-R.

Regenerative strategies V, B$_s$. Forms extensive but diffuse patches by means of rhizome growth. Also regenerates by seed, mainly in spring, and may produce a persistent seed bank; ?SB II-IV.
Flowers Green, hermaphrodite, protandrous, partially self-incompatible; *c.* 20–400 in terminal racemes (*Biol Fl*).
Germinule 0.87 mg, 1.5 × 1.0 mm, nutlet (4 per ovary).
Germination Epigeal; t_{50} 11 days; L > D.
Biological flora Hutchinson (1968).
Geographical distribution A European endemic. The British Isles, but less common in parts of E England and C Ireland (99% of British vice-counties), *ABIF* page 517; S, W and C Europe, N to Norway and E to Poland and Yugoslavia (49% of European territories and 5% naturalized).

Habitats Most abundant on screes and other rocky habitats such as quarry spoil, cliffs and outcrops, and even in old pastures and on lead-mine heaps and other types of spoil usually associated with a rocky substratum. Also occurs on road verges and in woodland margins and scrub. [Recorded from sand dunes, montane habitats and heaths (*Biol Fl*).]
Gregariousness Intermediate.
Altitude Most abundant on the carboniferous limestone from 200 to 300 m but extends to 320 m. [Up to 600 m (*Biol Fl*).]
Slope Wide-ranging, but showing a particular association with sloping ground.
Aspect More frequent and significantly more abundant on unshaded S-facing slopes, a reflection of its strong association with unshaded limestone screes.
Hydrology Generally absent from wetlands.
Soil pH Wide-ranging, with two peaks between 4.0 and 5.0 and >7.0, a reflection of the known existence of edaphic ecotypes (*Biol Fl*).
Bare soil Found over a wide range of values.
△ **ordination** Distribution centred on vegetation associated with infertile and relatively undisturbed habitats. Excluded from heavily disturbed habitats.
Associated floristic diversity Relatively low, declining in habitats such as scree slopes where *T.s.* reaches maximum abundance.

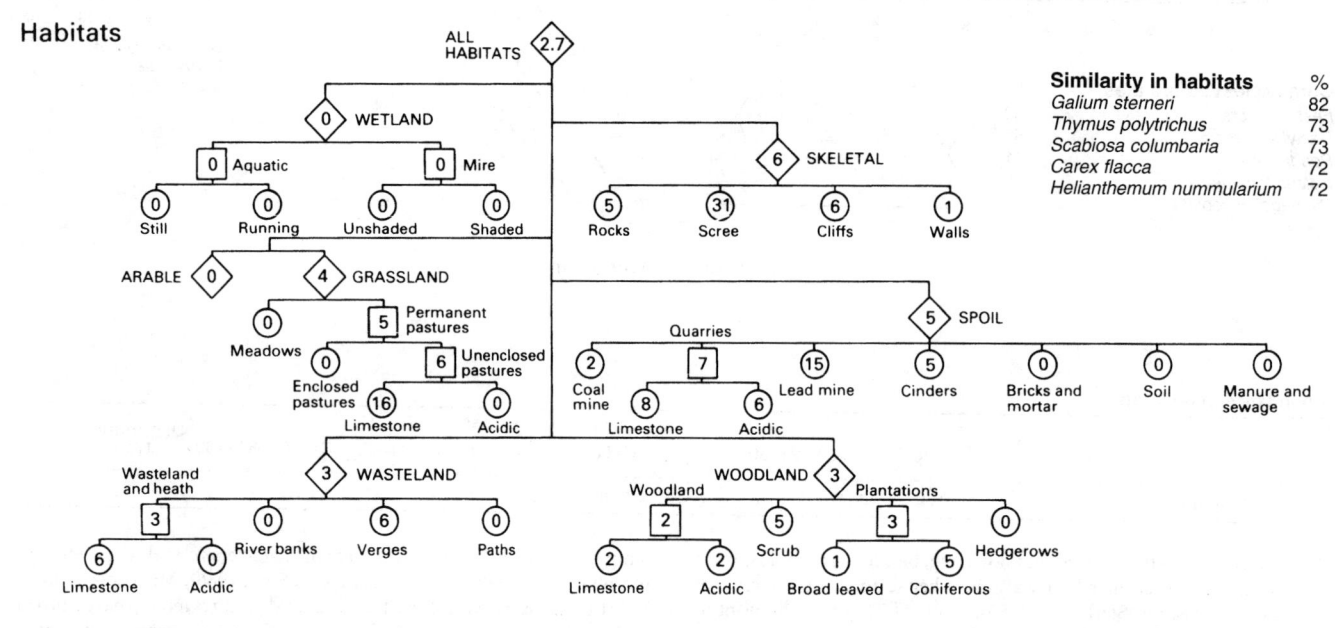

Habitats

Similarity in habitats	%
Galium sterneri	82
Thymus polytrichus	73
Scabiosa columbaria	73
Carex flacca	72
Helianthemum nummularium	72

Synopsis *T.s.* is a long-lived, erect, somewhat woody herb forming extensive patches by means of rhizome growth (*Biol Fl*). The species has a rather slow growth rate and, despite producing robust ascending shoots, tends not to form a dense canopy, and is excluded by species with taller or more-consolidated clonal populations. Consequently, *T.s.* is mainly restricted to open infertile habitats, and is particularly associated with scree and steeper slopes, which may be inherently unstable. *T.s.* is usually found in well-drained mineral soils, but is less frequent in areas with high summer temperatures and is sensitive to severe drought (*Biol Fl*). The widespread occurrence of *T.s.* on scree appears to be in part a result of the continuously moist soil maintained by the 'mulch effect' of talus, even on S-facing slopes. The species is sensitive to grazing and trampling. *T.s.* is eaten by sheep, but not cows, and only survives in lightly grazed pasture. Physiological tolerance of shade has been recognized (Mousseau 1977), but even woodland ecotypes of *T.s.* are only moderately shade-tolerant and the species is much more frequent in open scrub than dense woodland (*Biol Fl*). The species is a European endemic with an Atlantic distribution. *T.s.* regenerates both vegetatively by rhizome proliferation and by seed. The seeds, which may form a persistent seed bank (UCPE unpubl.), have no well-defined dispersal mechanism, although they may be moved by ants. Regeneration by seed is infrequent in shaded sites, and fatalities in calcareous soils result mainly from summer drought and winter frost; on acidic soils fungal attack appears critical to seedling establishment (*Biol Fl*). *T.s.* exploits a wide range of well-drained soils. There is ecotypic differentiation between populations exploiting acidic and those on calcareous soils, and between plants in shaded and unshaded habitats (*Biol Fl*). *T.s.* is rather uniform morphologically (*Biol Fl*) except in the Mediterranean region, where other ssp. are recognized (*FE3*).
Current trends Restricted to infertile habitats. Probably underdispersed to suitable new sites, and decreasing.

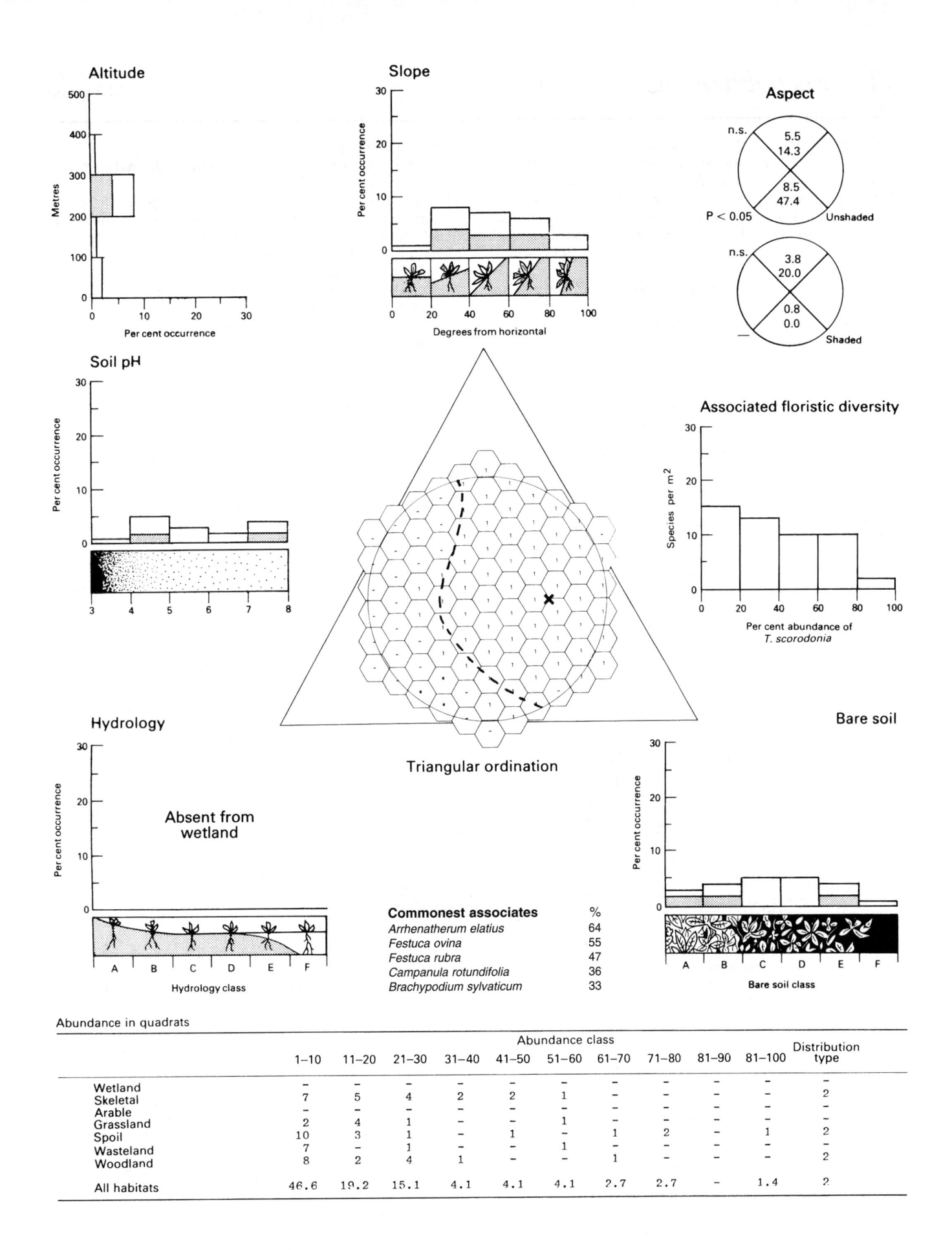

Altitude

Slope

Aspect

n.s. 5.5 / 14.3 / 8.5 / 47.4   Unshaded
P < 0.05

n.s. 3.8 / 20.0 / 0.8 / 0.0   Shaded

Soil pH

Associated floristic diversity

(Species per m² vs Per cent abundance of *T. scorodonia*)

Hydrology

Absent from wetland

Triangular ordination

Bare soil

Commonest associates	%
Arrhenatherum elatius	64
Festuca ovina	55
Festuca rubra	47
Campanula rotundifolia	36
Brachypodium sylvaticum	33

Abundance in quadrats

	1–10	11–20	21–30	31–40	41–50	51–60	61–70	71–80	81–90	81–100	Distribution type
Wetland	–	–	–	–	–	–	–	–	–	–	–
Skeletal	7	5	4	2	2	1	–	–	–	–	2
Arable	–	–	–	–	–	–	–	–	–	–	–
Grassland	2	4	1	–	–	1	–	–	–	–	–
Spoil	10	3	1	–	1	–	1	2	–	1	2
Wasteland	7	–	1	–	1	–	–	–	–	–	2
Woodland	8	2	4	1	–	–	1	–	–	–	2
All habitats	46.6	19.2	15.1	4.1	4.1	4.1	2.7	2.7	–	1.4	2

Thymus polytrichus A. Kern. ex Borbás

Lamiaceae
Wild Thyme

Life-form Woody perennial chamaephyte. Stems prostrate; leaves ovate, leathery, sometimes hairy, <10 mm^2; primary root deep, adventitious roots shallow, with VA mycorrhizas.
Phenology Evergreen. Growth commences in spring, and runners also elongate in autumn. Flowers from May and July and sheds seed from July onwards (*Biol Fl*). Leaves long-lived, usually lasting more than a season. Growth ceases in October (*Biol Fl*).
Height Very low creeping vegetative stems; even flower heads usually <30 mm.
Seedling RGR 0.5–0.9 week^{-1}. SLA 13.6. DMC. No data.
Nuclear DNA amount 2.8 pg; $2n = c.$ 50*, 51, 54*, (*BC*), $4x$*.

Established strategy Stress-tolerator.
Regenerative strategies V, B$_s$. Regenerates both vegetatively and by seed (see 'Synopsis'); SB IV.
Flowers Purple, gynodioecious but a majority of plants hermaphrodite, protandrous, insect pollination often leads to selfing (*Biol Fl*); often *c.* 40 aggregated into a terminal head.
Germinule 0.11 mg, 0.5 × 0.5 mm nutlet, (up to 4 per ovary).
Germination Epigeal; t_{50} 3 days; 9–28°C; L > D.
Biological flora Pigott (1955) (as *T. drucei* Ronniger).
Geographical distribution Much of the British Isles (95% of British vice-counties), *ABIF* page 524; S, W and C Europe (50% of European territories).

Habitats Occurs mainly in dry grassland and on screes, lead-mine spoil and rock outcrops, particularly on shallow, nutrient-deficient calcareous soils. Observed very locally on surface-leached calcareous soils, and on road verges with calcareous sandy soils. [Occurs in montane areas on heaths, river gravels, wet rocks and soligenous mire; in lowland areas of N and W occurs in some acidic grasslands, and in maritime areas on calcareous sand dunes, cliff turf and sea heaths (*Biol Fl*). Also frequent on ant-hills in calcareous grassland (King 1977c).]
Gregariousness Potentially carpet-forming.
Altitude Most common from 200–300 m, reflecting its abundance on carboniferous limestone. Extends to 400 m. [Up to 1200 m (*Biol Fl*).]

Slope Favours steep rocky slopes from 20 to 60°.
Aspect Widely distributed, but in grasslands both frequency and abundance are higher on S-facing slopes (*Ecol Atl*).
Hydrology Absent from wetland in survey area, but see 'Habitats'.
Soil pH Only recorded above pH 5.0 and most common in the pH range 7.5–8.0. [Shows a lesser restriction to higher pH values in N and W of Britain, and occurs down to pH 3.2 (*Biol Fl*).]
Bare soil Habitats of *T.p.* vary considerably in quantity of exposed soil, but the species is less frequent and abundant at the extremes of the range.
△ **ordination** Confined to vegetation associated with infertile relatively undisturbed situations.
Associated floristic diversity Often high.
Synopsis *T.p.* is a long-lived, slow-growing, prostrate, mat-forming, evergreen undershrub. The species is rapidly submerged by taller and faster-growing herbs, and consequently is restricted to open, often rocky,

Habitats

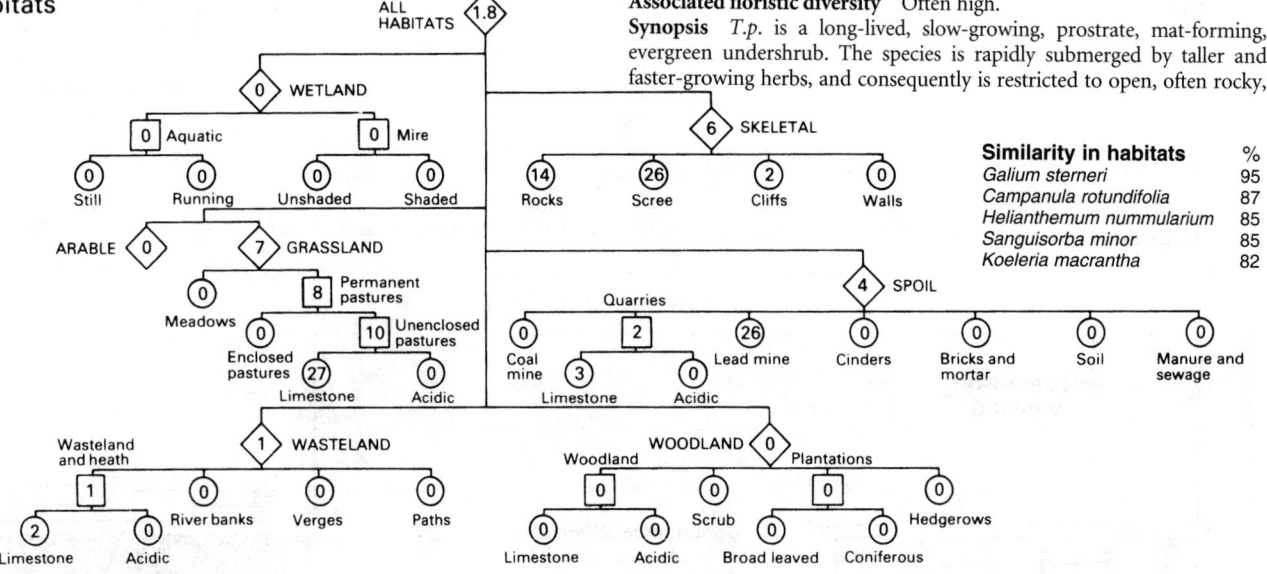

Similarity in habitats	%
Galium sterneri	95
Campanula rotundifolia	87
Helianthemum nummularium	85
Sanguisorba minor	85
Koeleria macrantha	82

habitats and short turf, extending from sea level (cliffs and sand dunes) to mountain sides. *T.p.* is most characteristic of sites which are droughted in summer (*Biol Fl*). Although the shoot is xeromorphic, *T.p.* is essentially a drought-avoiding species and the deep primary root system, which may on occasion descend to 2 m, may reach sources of water unavailable to associated species (*Biol Fl*). The seedling root may reach 40 mm in length before the first true leaves develop, a capacity which must be important in drought avoidance (*Biol Fl*). The leaves of *T.p.* are generally avoided by grazing animals, and the relative abundance of *T.p.*, which is not shade-tolerant, is enhanced by grazing of taller turf components (*Biol Fl*). *T.p.* is a strict calcicole over much of S, E and SE England, but extends into a wider range of habitats including acidic soils in the N and W (*Biol Fl*). Seedlings, establishing upon acidic soil and subject to some root stunting, are vulnerable to summer droughts. This may explain the capacity of *T.p.* to colonize more-acidic sites in areas of heavier rainfall and higher humidity, i.e. in the N and W (*Biol Fl*). Nevertheless, even here *T.p.* is usually associated with elevated levels of exchangeable Ca, Mg or K (*Biol Fl*) and may, for example, grow from calcite veins in mountain rocks or in turf subject to salt spray. *T.p.* is a poor colonist, and is most characteristic

of semi-natural habitats. *Tp.* regenerates both vegetatively and by seed. Seed is shed in autumn or may be retained on the plant over winter, and forms a persistent seed bank. At higher latitudes viable seed may be produced only during warmer, sunnier years (*Biol Fl*), and here regeneration by vegetative means probably assumes greater importance. *T.p.* produces extensive mats by long creeping branches bearing adventitious roots. These mats may break up with age. The production of long running stems, tolerant of shallow burial in the soil, is particularly important in the colonization of ant hills which are a favoured habitat in S England (King 1977a) and in exploiting sand dunes, where *T.p.* can survive at least 50 mm of sand deposition and contributes to the stabilization of dune blow-outs (*Biol Fl*). A heavy-metal-tolerant race of *T.p.* in Derbyshire has been described (Shaw 1984). The species is morphologically variable in Europe where 5 ssp. have been recognized (*FE3*). One European ssp. includes plants with $2n = 28$, $2x$ (*FE3*).
Current trends *T.p.* is rapidly decreasing in many lowland areas. However, in upland regions where pressures of land-use are less intense, *T.p.* is probably declining more slowly, and a restricted capacity to colonize quarries and other artificial habitats is evident.

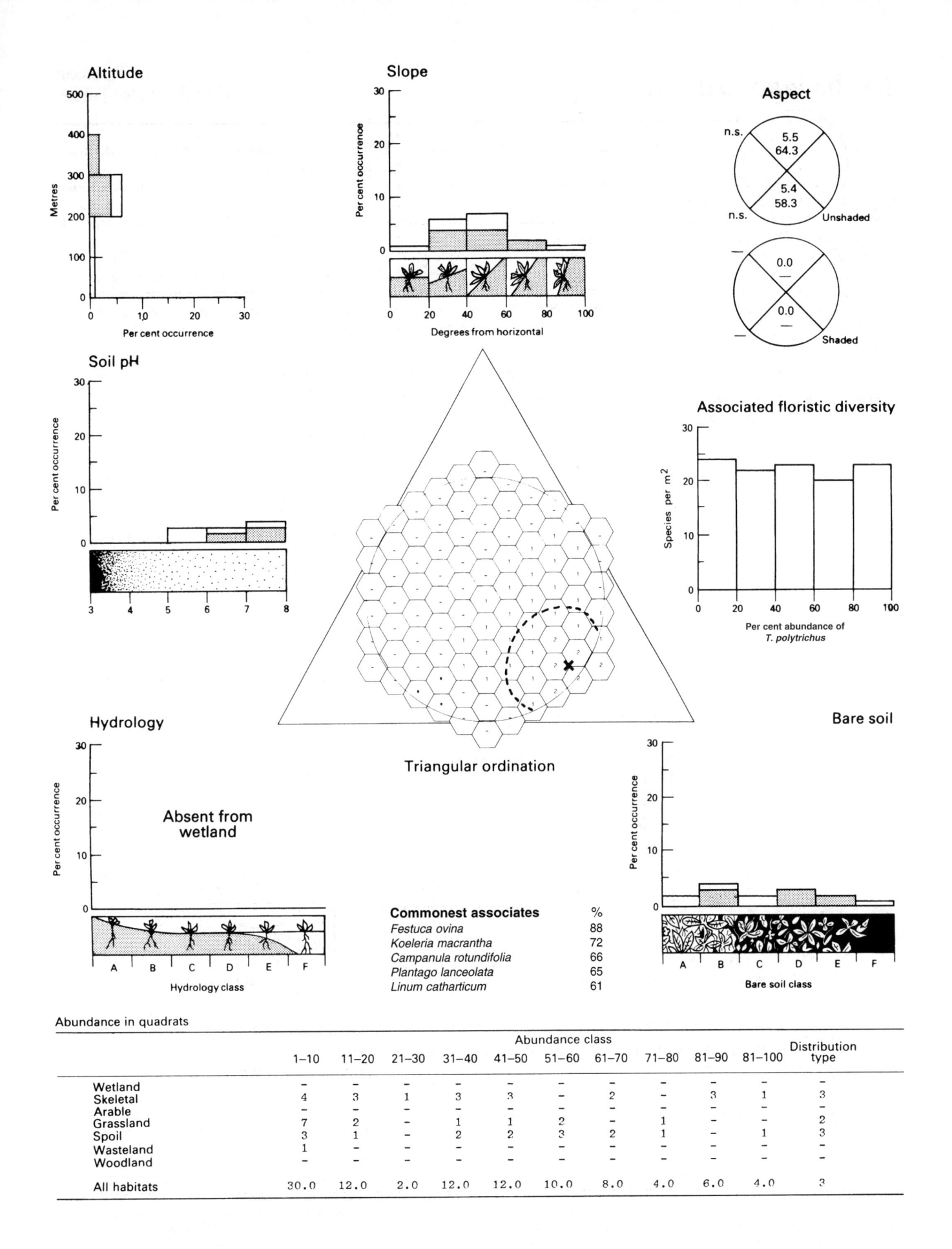

Altitude

Slope

Aspect

	n.s.	5.5	
n.s.		64.3	
		5.4	
		58.3	Unshaded

	—	0.0	
—		—	
		0.0	
		—	Shaded

Soil pH

Associated floristic diversity

Species per m²

Per cent abundance of *T. polytrichus*

Hydrology

Absent from wetland

Triangular ordination

Commonest associates %
Festuca ovina 88
Koeleria macrantha 72
Campanula rotundifolia 66
Plantago lanceolata 65
Linum catharticum 61

Bare soil

Abundance in quadrats

	\multicolumn{10}{c	}{Abundance class}	Distribution type
	1–10	11–20	21–30	31–40	41–50	51–60	61–70	71–80	81–90	81–100	
Wetland	–	–	–	–	–	–	–	–	–	–	–
Skeletal	4	3	1	3	3	–	2	–	3	1	3
Arable	–	–	–	–	–	–	–	–	–	–	–
Grassland	7	2	–	1	1	2	–	1	–	–	2
Spoil	3	1	–	2	2	3	2	1	–	1	3
Wasteland	1	–	–	–	–	–	–	–	–	–	–
Woodland	–	–	–	–	–	–	–	–	–	–	–
All habitats	30.0	12.0	2.0	12.0	12.0	10.0	8.0	4.0	6.0	4.0	3

593

Torilis japonica (Houtt.) DC.

Apiaceae
Upright Hedge-parsley

Life-form Semi-rosette, possibly winter-annual therophyte or perhaps rarely biennial with a slender tap-root. Stems erect; leaves much divided, hairy, typically <2000 mm^2.
Phenology Seeds germinate mainly in autumn and a rosette of leaves overwinters. Flowers July to August and seed set from September onwards.
Height Foliage <1 m; flowers to 1.25 m.
Seedling RGR No data. SLA 28.1. DMC 23.1.
Nuclear DNA amount 2.1 pg; $2n = 16$ (*FE2*); $2x$.
Established strategy Intermediate between C-S-R and competitive ruderal.
Regenerative strategies S. See 'Synopsis'; ?SB II.

Flowers Pale pink or purplish-white, hermaphrodite, insect-pollinated; c. 50 in each umbel.
Germinule 1.98 mg, 4.0 × 2.4 mm, dry mericarp with hooked spines (2 mericarps per ovary).
Germination Epigeal; t_{50} 5 days; <5-25°C; dormancy broken by chilling; L > D (Baskin & Baskin 1975).
Biological flora None.
Geographical distribution The British Isles except for the N of Scotland (98% of British vice-counties), *ABIF* page 477; Europe except the extreme N and much of the Mediterranean region (79% of European territories); temperate Asia, N Africa; introduced in N America.

Habitats Occurs in a wide range of habitats including rock outcrops, disturbed soil, quarry heaps and other spoil. Also frequent at woodland margins, in hedgerows, on waysides and in pastures. Observed locally on wasteland. Absent from wetlands.
Gregariousness Typically sparse.
Altitude Observed up to 300 m. [Up to 415 m.]
Slope Widely distributed, but only infrequently observed on the steepest slopes.
Aspect Exhibits a S-facing bias.
Hydrology Absent from wetland habitats.
Soil pH Most characteristic of sites of pH >5.5.
Bare soil Wide-ranging.

△ ordination A scattered distribution, but most frequently recorded from moderately disturbed habitats of intermediate or low productivity.
Associated floristic diversity Typically species-rich; less so in shaded habitats.
Synopsis *T.j.* is a late-flowering umbellifer found in a range of habitats, including areas on steep roadsides and ditch banks where the growth of perennials has been reduced by soil creep, and at more-stable sites on limestone rock outcrops, where summer drought and low levels of nutrients restrict productivity. Although broadly similar in habitat range to the diminutive winter annuals (see 'Similarity in habitats'), *T.j.* is more robust and persists longer into summer; this may be related to its tendency to exploit pockets of deep soil and its modest potential for competing with perennial plants. The species also occurs along woodland margins and in hedgerows, where populations show a capacity to persist in tall but open vegetation dominated by perennial species. *T.j.* has an erect habit and is

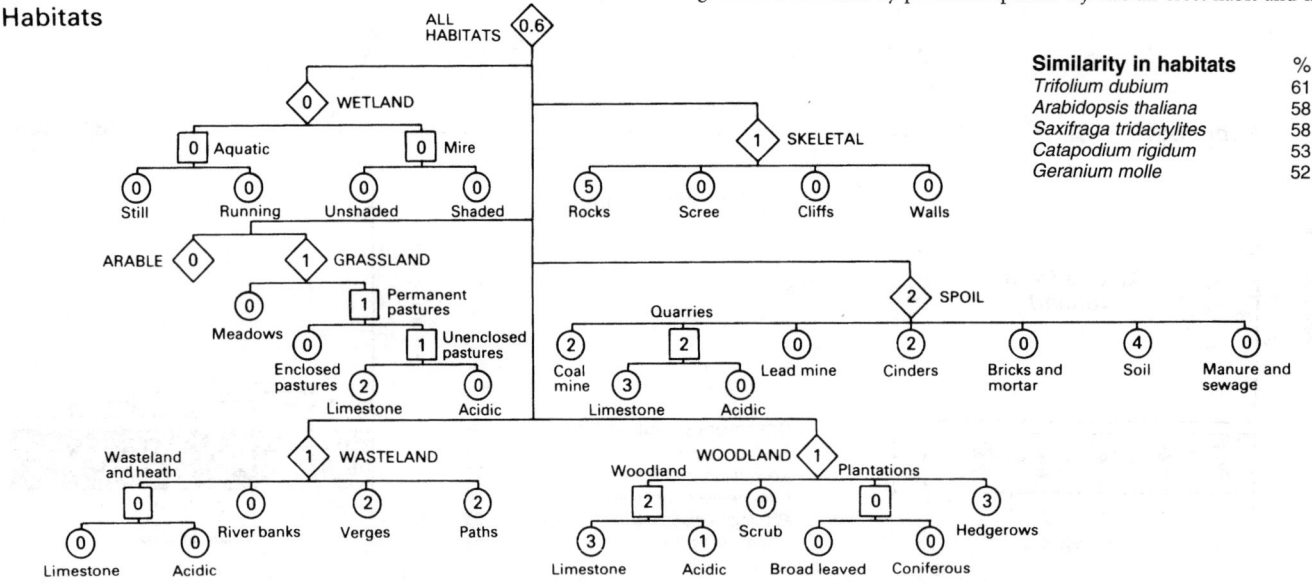

Habitats

Similarity in habitats	%
Trifolium dubium	61
Arabidopsis thaliana	58
Saxifraga tridactylites	58
Catapodium rigidum	53
Geranium molle	52

seldom recorded from cut or grazed habitats. It has a predominantly lowland distribution, reaching only 415 mm in the British Isles (Wilson 1956) and tends to be most abundant on calcareous strata. The species is dependent entirely upon the relatively large seeds for regeneration. Detailed studies of the germination behaviour of *T.j.* have been undertaken in N America by Baskin and Baskin (1975). Seed is retained on the plant after the death of the parent, and falls to the ground in the late summer and autumn. Some seed germinates to produce plants which flower in the following year, but seed which fails to germinate in autumn, mainly because it is retained longer on the plant, is induced into dormancy by low winter temperatures and does not germinate until the following autumn. Thus, *T.j.* has a transient seed bank in which the maximum length of the dormancy period appears to be 12 months. However, in a field experiment in S England Roberts (1979) found that germination was delayed until spring and, consistent with this, Grime *et al.* (1981), working in the Sheffield region, showed that the seeds have a chilling requirement. Clearly, these observations are at variance with the N American studies and with field observations from the Sheffield region that young plants overwinter. Additional work on the germination behaviour of this species, collected from various geographical locations and vegetation types, is required. The mericarps of *T.j.* have hooked bristles and can be widely dispersed by animals and humans. Dwarf plants resembling *T. arvensis* (Hudson) Link, which occur in cereal crops and fruit after harvest-time, may be genetically distinct. Otherwise, nothing is known of the extent of ecotypic variation in this wide-ranging diploid species.
Current trends A mobile and perhaps increasing species.

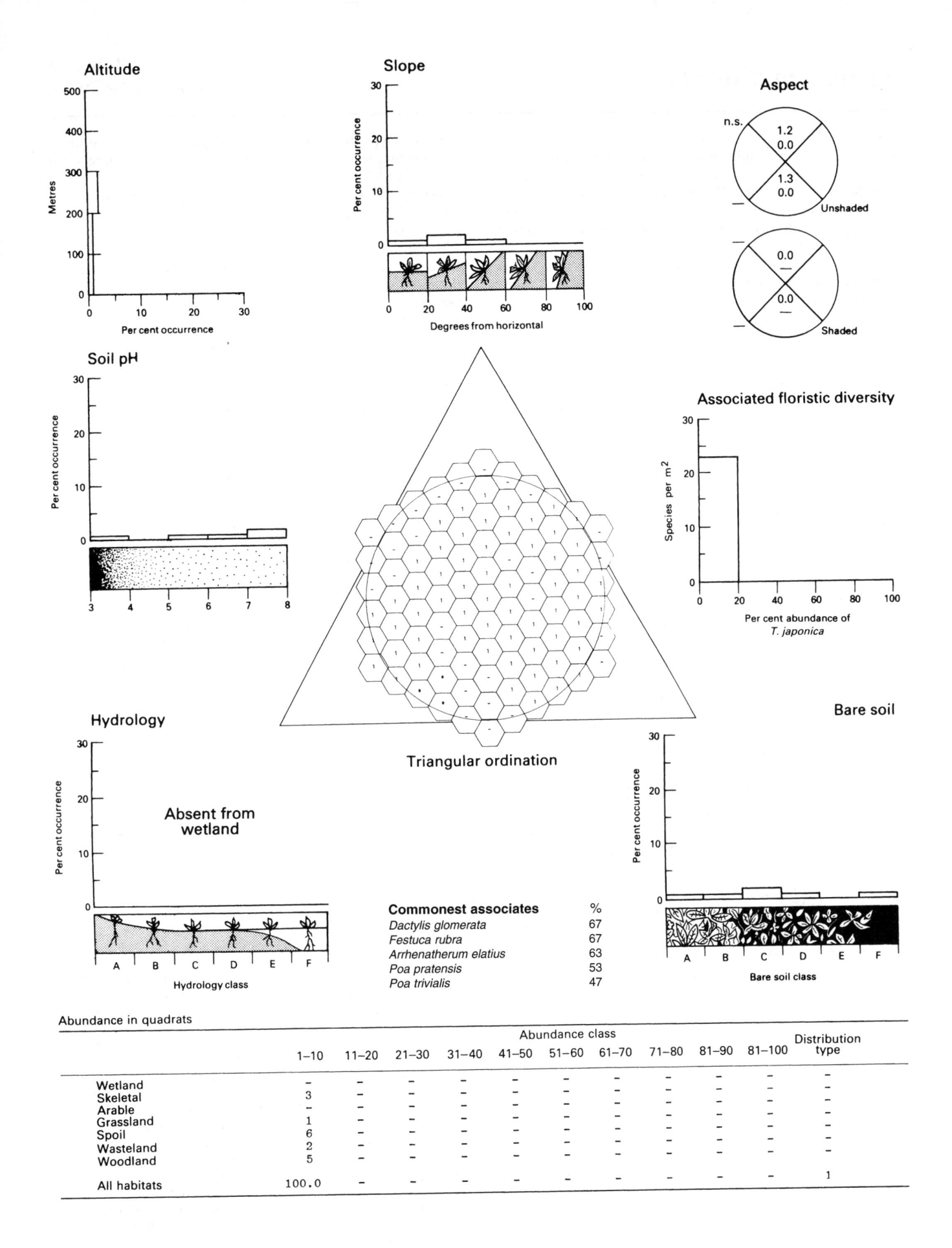

Altitude

Per cent occurrence vs *Metres* (0–500)

Slope

Per cent occurrence vs *Degrees from horizontal* (0–100)

Aspect

Unshaded: n.s. 1.2 / 0.0 / 1.3 / 0.0

Shaded: — 0.0 / 0.0 — / —

Soil pH

Per cent occurrence vs pH (3–8)

Associated floristic diversity

Species per m² vs *Per cent abundance of T. japonica*

Hydrology

Absent from wetland

Hydrology class: A B C D E F

Triangular ordination

Bare soil

Per cent occurrence vs Bare soil class A–F

Commonest associates

	%
Dactylis glomerata	67
Festuca rubra	67
Arrhenatherum elatius	63
Poa pratensis	53
Poa trivialis	47

Abundance in quadrats

	1–10	11–20	21–30	31–40	41–50	51–60	61–70	71–80	81–90	81–100	Distribution type
Wetland	–	–	–	–	–	–	–	–	–	–	–
Skeletal	3	–	–	–	–	–	–	–	–	–	–
Arable	–	–	–	–	–	–	–	–	–	–	–
Grassland	1	–	–	–	–	–	–	–	–	–	–
Spoil	6	–	–	–	–	–	–	–	–	–	–
Wasteland	2	–	–	–	–	–	–	–	–	–	–
Woodland	5	–	–	–	–	–	–	–	–	–	–
All habitats	100.0	–	–	–	–	–	–	–	–	–	1

Trifolium dubium Sibth.

Fabaceae
Lesser Yellow Trefoil

Life-form Winter- or perhaps more rarely summer-annual therophyte. Stems procumbent or ascending; leaves usually hairy, ternate, <500 mm^2; roots nodulated and with VA mycorrhizas.
Phenology Seedlings usually overwinter. Flowers mainly May to June in droughted and May to October in non-droughted sites. Seeds ripen about 1 month after flowering. Shoots die back in autumn or earlier.
Height Foliage and flowers up to 250 mm.
Seedling RGR No data. SLA 32.9. DMC 19.9.
Nuclear DNA amount 2.1 pg; $2n = 28, 32$ (*FE2*); $4x$.
Established strategy Intermediate between ruderal and stress-tolerant ruderal.

Regenerative strategies S, B$_s$. Regenerates by autumn-germinating seed. A persistent buried seed bank has been reported (Charlton 1977); SB III.
Flowers Flowers yellow, hermaphrodite, self-pollinated (Fryxell 1957); 3-16 in each racemose head (*FE2*).
Germinule 0.32 mg, 1.5 × 0.9 mm, seed in a 1-seeded pod, of 2.5-3 mm.
Germination Epigeal; t_{50} 2 days; hard-coat dormancy; L = D.
Biological flora None.
Geographical distribution The British Isles, but rarer in N and NW Scotland (100% of British vice-counties), *ABIF* page 397; most of Europe except the extreme N (72% of European territories and 5% naturalized); N Africa.

Habitats Found in two distinct types: (a) open habitats, particularly rock outcrops, but also quarry spoil, soil heaps and cinders, and (b) grassland sites such as meadows, waysides and wasteland. Also frequently observed as a weed of lawns. Our records refer mainly to the former type.
Gregariousness Intermediate.
Altitude Most frequent and only abundant at <300 m, but extends to 350 m. [Up to 490 m.]
Slope Recorded most frequently from moderate slopes, but absent from the steepest inclines.
Aspect Recorded with greatest frequency on unshaded S-facing slopes.
Hydrology Absent from wetlands.

Soil pH Only recorded from sites of pH >5.0, and observed only sparingly below this value.
Bare soil Widely distributed, but most frequently recorded from sites with open vegetation but low to moderate exposure of bare soil. A considerable amount of bare rock or bare spoil (e.g. cinders) is often also present.
△ **ordination** Distribution centred on vegetation associated with conditions of low to moderate fertility and subject to some disturbance. Absent from undisturbed and highly productive vegetation types.
Associated floristic diversity Typically species-rich.

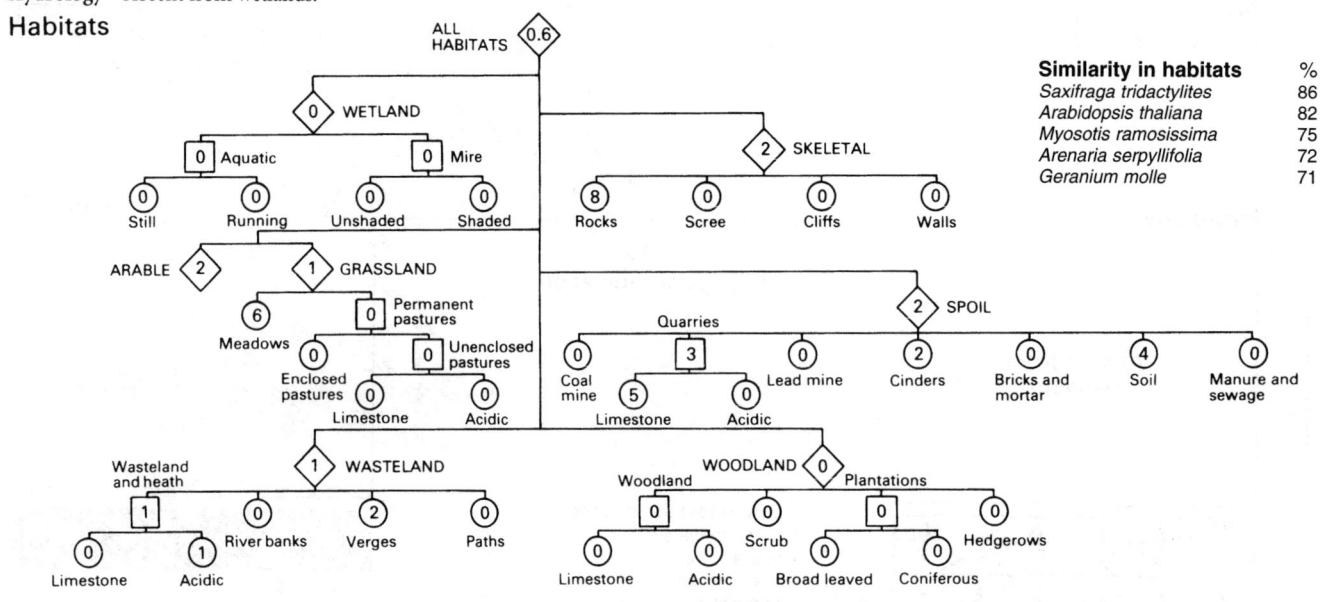

Habitats

Similarity in habitats	%
Saxifraga tridactylites	86
Arabidopsis thaliana	82
Myosotis ramosissima	75
Arenaria serpyllifolia	72
Geranium molle	71

Synopsis *T.d.* is a creeping annual legume found in a wide range of relatively infertile grasslands and open habitats. Observations in N England suggest that *T.d.* is usually a winter annual, rather than a summer annual as stated in *CTW*. Nitrogen-fixing root nodules are formed in conjunction with *Rhizobium trifolii*, but the capacity to fix nitrogen on sand and mine waste is inferior to that of most common legumes (Bradshaw & Chadwick 1980). At the ecological extremes of its distribution *T.d.* appears to exploit two very different types of habitat. On calcareous rock outcrops and dry sandy soils *T.d.* grows in winter-annual communities and has a restricted flowering period due to the onset of summer drought. Here *T.d.* is among the later-flowering winter annuals. It has a relatively deep tap-root which may afford some capacity for drought avoidance. In lawns, a habitat in which the species is not usually droughted, *T.d.* continues to set seed until autumn. Survival here appears to be related to the capacity of *T.d.* to adopt a low growth form. This habit is less effective against grazing, however, and *T.d.* is largely absent from pasture, although it does occur occasionally in unproductive meadows. *T.d.* regenerates entirely by seed, and the presence of a buried seed bank has been reported (Charlton 1977). The quantity of *T.d.* present as a contaminant of agricultural seed stocks of *Trifolium repens* is carefully controlled (*WCH*) but, bearing in mind the vast quantity of *T.r.* sown, this source is probably sufficient to have a significant effect on the distribution of *T.d.* Only one cultivar of this low-yielding species has been grown on a small scale (Spedding & Diekmahns 1972). Plant size and morphology depend to a great extent upon management and site fertility, and in the absence of severe drought populations of *T.d.* on rock outcrops flower throughout summer. The extent to which phenotypic plasticity is complemented by ecotypic differentiation is not known. The wide-ranging distribution of *T.d* overlaps with those of the related but more ecologically restricted *T. micranthrum* Viv. on lawns and *T. campestre* Schreber on outcrops, but there are strong breeding barriers between most species of *Trifolium*, and hybrids have not been reported (Stace 1975).
Current trends *T.d.* is an effective colonist of many habitats of intensively exploited landscapes. Probably increasing, at least within lawns and other sown grasslands.

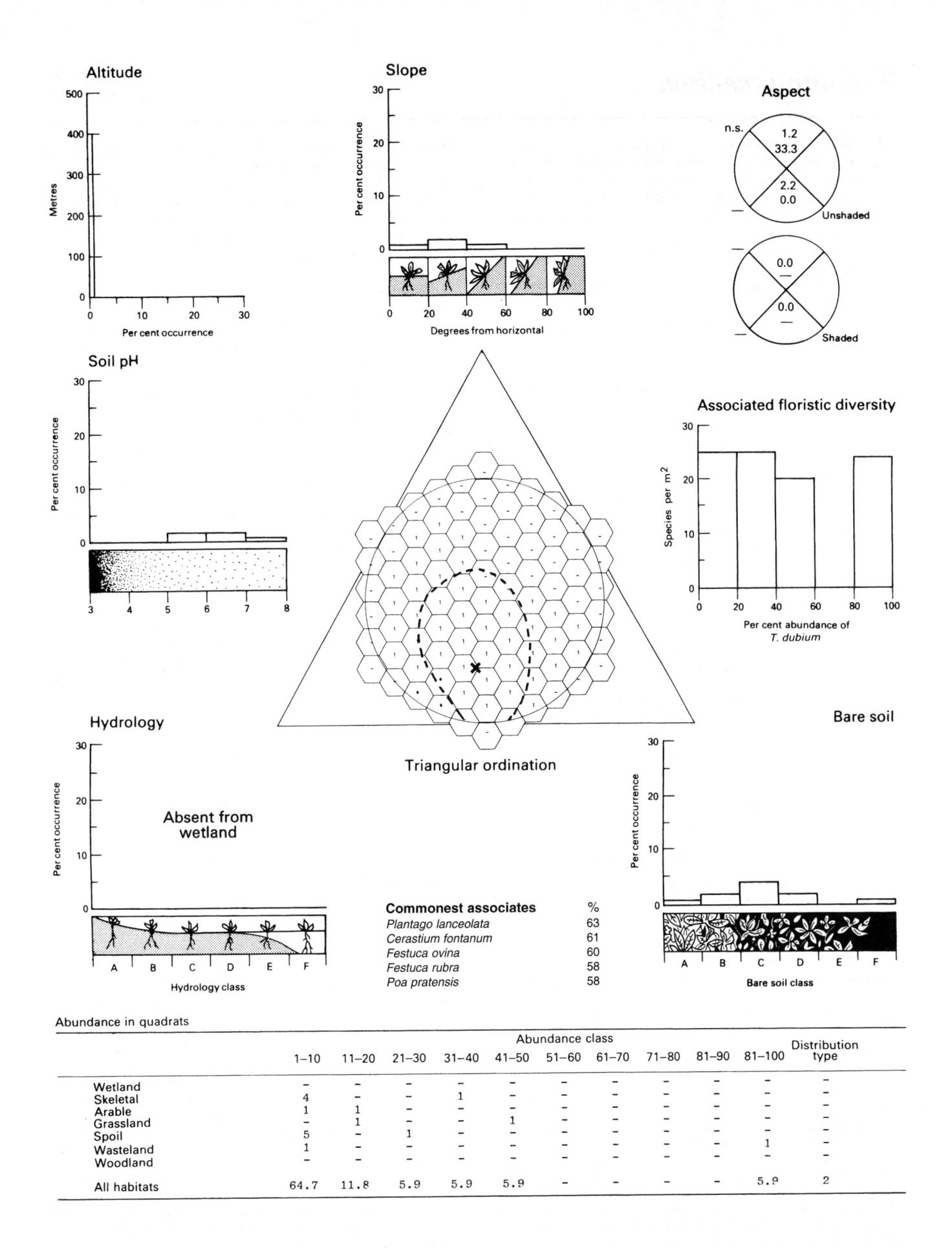

Altitude

Slope

Aspect

	n.s.	1.2	
		33.3	
	2.2		
	—	0.0	Unshaded

	—		
	0.0		
	—		
	0.0		
	—		Shaded

Soil pH

Associated floristic diversity

Hydrology

Absent from wetland

Triangular ordination

Commonest associates %
Plantago lanceolata 63
Cerastium fontanum 61
Festuca ovina 60
Festuca rubra 58
Poa pratensis 58

Bare soil

Abundance in quadrats

	1–10	11–20	21–30	31–40	41–50	51–60	61–70	71–80	81–90	81–100	Distribution type
Wetland	–	–	–	–	–	–	–	–	–	–	–
Skeletal	4	–	–	1	–	–	–	–	–	–	–
Arable	1	1	–	–	–	–	–	–	–	–	–
Grassland	–	1	–	–	1	–	–	–	–	–	–
Spoil	5	–	1	–	–	–	–	–	–	–	–
Wasteland	1	–	–	–	–	–	–	–	–	1	–
Woodland	–	–	–	–	–	–	–	–	–	–	–
All habitats	64.7	11.8	5.9	5.9	5.9	–	–	–	–	5.9	2

Trifolium medium L.

Fabaceae
Zig-zag Clover

Life-form Rhizomatous, polycarpic perennial protohemicryptophyte. Shoots ascending; leaves hairy, ternate, <2000 mm^2; roots with VA mycorrhizas and root nodules.

Phenology Flowers June to September. Fruits ripe in October, and some may remain on the plant until winter. Flowering shoots die back in autumn.

Height Foliage and flowers typically <250 mm.

Seedling RGR 0.5-0.9 week^{-1}. SLA 21.9. DMC 26.3.

Nuclear DNA amount 7.1 pg; $2n = c.$ 70, 73*-84* (*FE2*, Mowforth 1985); 10*x**-12**x*.

Established strategy Intermediate between stress-tolerant competitor and C-S-R.

Regenerative strategies V, ?B$_s$. Regenerative biology requires further study (see 'Synopsis').

Flowers Purple, hermaphrodite, self-incompatible (Duke 1981, Merker 1984), insect-pollinated; often *c.* 50 in a cymose head.

Germinule 2.34 mg, 1.9 × 1.5 mm, seed in a dehiscent 1-seeded pod which is enclosed within a spiny calyx up to *c.* 9 mm.

Germination Epigeal; t_{50} 1 day; <5-36°C; hard-coat dormancy; L = D.

Biological flora None, but see Zarzycki (1976).

Geographical distribution The British Isles, but rather local, rare in East Anglia (96% of British vice-counties), *ABIF* page 398; Europe except the extreme N and S (79% of European territories); naturalized in N America.

Habitats A species of derelict and lightly grazed grasslands and unmanaged habitats such as old grassy quarry spoil and coal-mine heaps. Also observed on waysides, lead-mine spoil and beside railways. Absent from wetland, skeletal and heavily-shaded habitats.

Gregariousness Potentially stand-forming.

Altitude More frequent in lowland areas, but observed up to 310 m. [Up to 460 m.]

Slope Absent from skeletal habitats, and therefore also from steep slopes.

Aspect Recorded as significantly more-frequent and abundant on S-facing slopes. Further studies (Hodgson *et al.* unpubl.) suggest that this pattern is less consistently shown off the magnesian limestone.

Hydrology Absent from wetlands.

Soil pH Not recorded from strongly acidic soils of pH <4.5. Most abundant at pH values within the range 5.0-6.0 but equally frequent on calcareous and non-calcareous strata.

Bare soil Largely restricted to closed communities with little exposed soil.

△ **ordination** Distribution centred on undisturbed vegetation types of moderate productivity. Infrequent in communities subject to disturbance.

Associated floristic diversity Typically intermediate or species-rich.

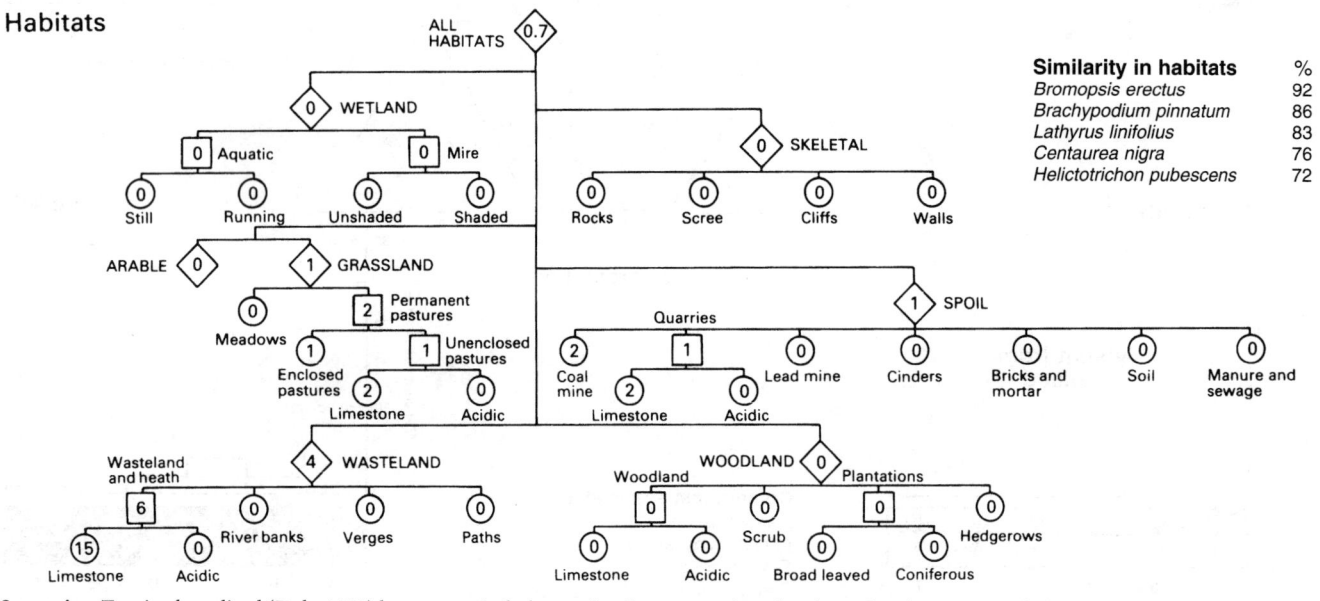

Habitats

Similarity in habitats	%
Bromopsis erectus	92
Brachypodium pinnatum	86
Lathyrus linifolius	83
Centaurea nigra	76
Helictotrichon pubescens	72

Synopsis *T.m.* is a long-lived (Duke 1981) legume particularly associated with grassland on heavy soils of intermediate fertility. Nitrogen-fixing root nodules are formed in conjunction with *Rhizobium trifold*. *T.m.* is a robust species producing long straggling shoots which are capable of emergence through the tall canopies of derelict grasslands. The species is the only legume consistently to achieve a large biomass in communities dominated by *Brachypodium pinnatum*. However, in some situations, particularly in stands of *Arrhenatherum elatius*, *T.m.* may be represented by a few suppressed individuals. The possibility that the balance between *T.m.* and its associated grasses experiences cyclical fluctuations which depend upon the changing nitrogen level of the soil (as is the case in *Lolium perenne* and *Trifolium repens*) deserves investigation. *T.m.* is tolerant of frost and is very winter-hardy (Duke 1981). The species is also moderately tolerant of shade (Duke 1981) and is occasionally found at woodland margins where it is often subject to sublethal attack by mildew in autumn. On the magnesian limestone of South Yorkshire *T.m.* is particularly associated with soils of intermediate pH which are neither droughted nor waterlogged. *T.m.* forms extensive clonal patches by means of rhizomes. Breakdown of these rhizomes results in the isolation of daughter colonies. *T.m.* also regenerates by seeds, which, according to Muller (1978), germinate in autumn. Other species of *Trifolium* which are common in Britain form a persistent seed bank and on this basis one is predicted for *T.m.* *T.m.* is considered self-incompatible (Duke 1981), and some clones produce little seed; this may partly explain the tendency of the species to be associated with older grassland habitats. *T.m.* varies in growth form according to habitat, and Duke (1981) has suggested that it should be possible to select types suitable for grazing and others for hay production. The degree of variation in *T.m.* in the British Isles is clearly less than that in C and S Europe, where *T.m.* is divided into four ssp.

Current trends *T.m.* has in the past colonized artificial habitats such as railway banks, roadsides, coal-mine spoil and cinders but shows little capacity to exploit habitats of more-recent origin. The species is decreasing, particularly in lowland areas. Selection for a higher degree of self-compatability is expected.

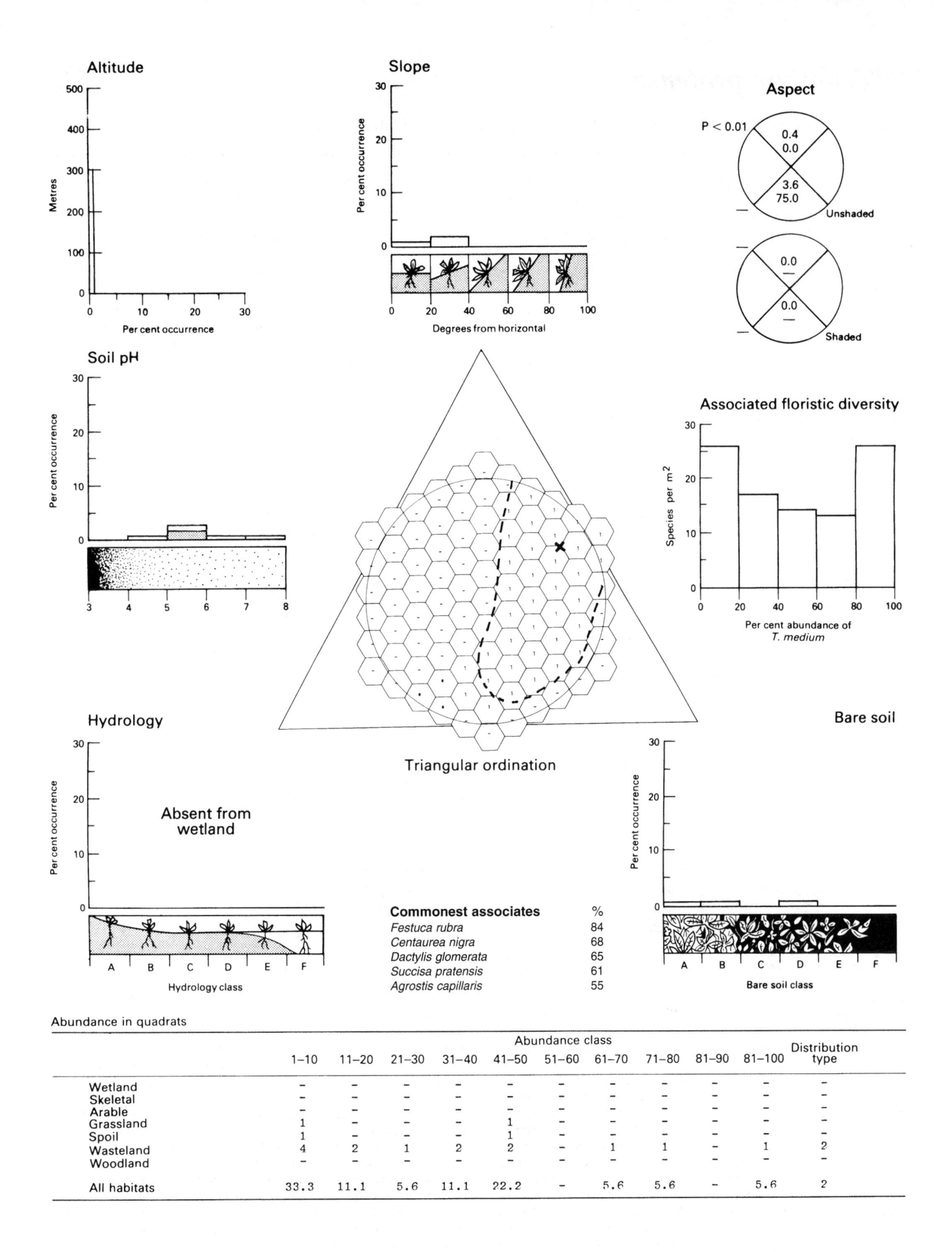

Altitude

Slope

Aspect

$P < 0.01$

	0.4
	0.0
3.6	75.0

Unshaded

	0.0
	0.0

Shaded

Soil pH

Associated floristic diversity

Triangular ordination

Hydrology

Absent from wetland

Bare soil

Commonest associates	%
Festuca rubra	84
Centaurea nigra	68
Dactylis glomerata	65
Succisa pratensis	61
Agrostis capillaris	55

Abundance in quadrats

	\multicolumn{10}{c	}{Abundance class}	Distribution type
	1–10	11–20	21–30	31–40	41–50	51–60	61–70	71–80	81–90	81–100	
Wetland	–	–	–	–	–	–	–	–	–	–	–
Skeletal	–	–	–	–	–	–	–	–	–	–	–
Arable	–	–	–	–	–	–	–	–	–	–	–
Grassland	1	–	–	–	1	–	–	–	–	–	–
Spoil	1	–	–	–	1	–	–	–	–	–	–
Wasteland	4	2	1	2	2	–	1	1	–	1	2
Woodland	–	–	–	–	–	–	–	–	–	–	–
All habitats	33.3	11.1	5.6	11.1	22.2	–	5.6	5.6	–	5.6	2

(*)*Trifolium pratense* L.

Fabaceae
Red Clover

Includes var. *pratense* and var. *sativum* Sturm., the latter of cultivated origin

---

Life-form Tufted, polycarpic perennial protohemicryptophyte. Stems erect or decumbent; leaves hairy, ternate, usually <1000 mm^2; roots with VA mycorrhizas and nodules.

Phenology Winter green. Onset of spring growth relatively late. Flowers from May to September. Seed ripe from July onwards and some may be retained on plant until early winter. Flowering shoots die back in autumn.

Height Foliage typically <300 mm, but leafy flowering stems may reach 600 mm.

Seedling RGR <0.5 week^{-1}. SLA 27.9. DMC 19.9.

Nuclear DNA amount 1.3 pg; $2n = 14*$ (*BC*); $2x*$. $4x$ plants are also cultivated (ADAS 1984).

Established strategy C-S-R.

Regenerative strategies S, B$_s$. By seed which appears to germinate in autumn. A persistent seed bank has been reported (Chippindale & Milton 1934); SB III.

Flowers Purple, hermaphrodite, insect-pollinated, virtually self-sterile in the absence of insects (Knuth 1906); often *c*. 100 in a corymbose head.

Germinule 1.35 mg, 2.1 × 1.5 mm, seed in a 1-seeded dehiscent pod.

Germination Epigeal; t_{50} 1 day; 5-35°C; hard-coat dormancy; germination rate also increased by dry storage at 20°C; L = D.

Biological flora None, but see Zarzycki (1976).

Geographical distribution The British Isles (100% of British vice-counties), *ABIF* page 398; Europe except for parts of the extreme N and S (92% of European territories); W Asia; introduced in N and S America and New Zealand.

Habitats Commonly employed in agriculture. Most frequent in older sown grassland, meadows and pastures. Widespread in other artificial habitats such as spoil heaps and waysides, and in unimproved pasture. Recorded from rock outcrops and unshaded mire but otherwise absent from skeletal and wetland habitats. On arable land either a casual or sown with other species such as *Brassica rapa* L. Absent from woodland. In many of the more fertile and disturbed habitats it is suspected that var. *s*. is the only segregate present.

Gregariousness Sparse to intermediate.

Altitude Widely distributed up to 360 m. [Up to 850 m.]

Slope Absent from skeletal habitats, and consequently not present on the steepest slopes.

Aspect Most records are from unshaded N-facing slopes.

Hydrology Typically a plant of continuously moist habitats, but only a transient constituent of vegetation on waterlogged soils.

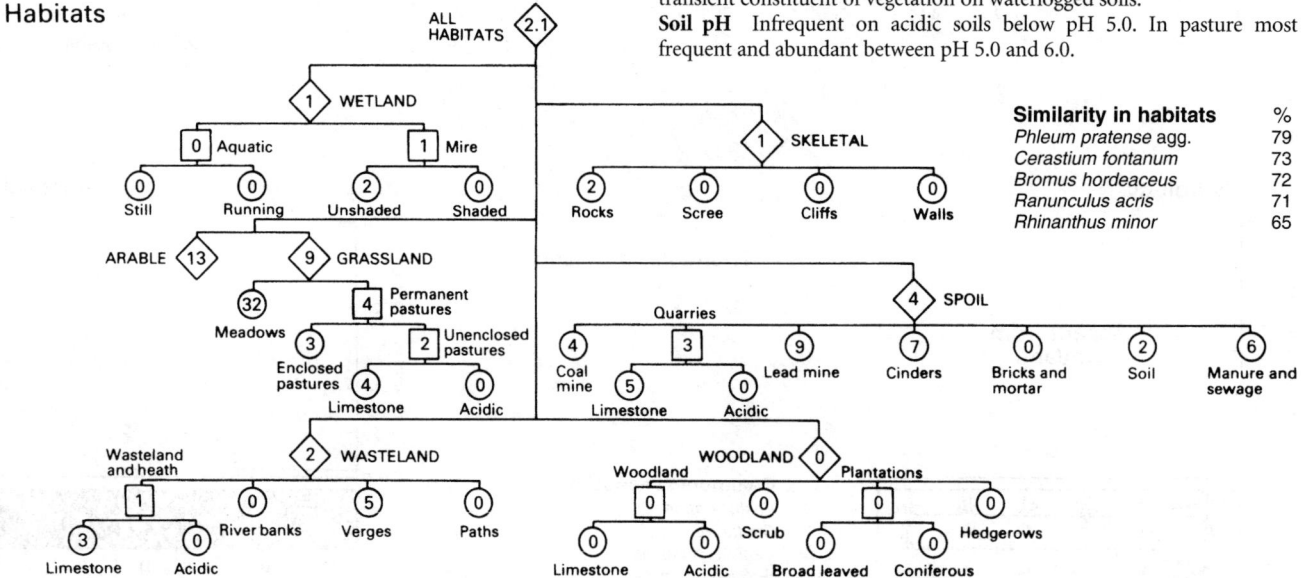

Soil pH Infrequent on acidic soils below pH 5.0. In pasture most frequent and abundant between pH 5.0 and 6.0.

Similarity in habitats	%
Phleum pratense agg.	79
Cerastium fontanum	73
Bromus hordeaceus	72
Ranunculus acris	71
Rhinanthus minor	65

Bare soil Associated with a wide range of values.

Δ **ordination** Recorded in a wide variety of vegetation types with distribution centred on productive, disturbed vegetation.

Associated floristic diversity Varying widely with habitat type, but generally associated with intermediate or species-rich vegetation.

Synopsis *T.p.* is an agriculturally important, tufted, winter-green, grassland legume with a pronounced summer peak in above-ground biomass. Nitrogen-fixing root nodules are formed in conjunction with *Rhizobium trifolii*. *T.p.* is strongly suppressed by nitrogenous fertilizer, whether applied as ammonium salts, sodium nitrate or farmyard manure (*PGE*). Records include var. *p*., a native, long-lived perennial of semi-natural habitats and hay meadows of low fertility and var. *s*, a shorter-lived (often 2-3 years) more erect, robust taxon cultivated mainly for hay and silage in short-term leys (*FE2*, Spedding & Diekmahns 1972, ADAS 1983a). We suspect that var. *s*., which is a frequent escapee from cultivation, is the more common, but the exact ecological distribution of the two varieties has not been determined. *T.p.* is found in relatively moist but freely-drained soils, and var. *p*. appears to be particularly associated with slightly acidic soils (pH 5.0-6.0). Both varieties are intolerant of heavy grazing and trampling, var. *s*. perhaps more so because of its robust erect habit. *T.p.* is a poor competitor in fertile undisturbed habitats. Hence var. *p*. is characteristic of lightly grazed turf and old hay-meadows while var. *s*., is a commonly sown plant of fertile grassland. *T.p.*, which is of tufted habit, is dependent upon seed for regeneration and a persistent seed bank is recorded (Chippindale & Milton 1934). Seeds may survive ingestion by cattle and horses (Ridley 1930). Of the cultivated strains tetraploids have a lower seed-setting ability (Frame 1976). Estimation of the extent to which *T.p.* is an effective colonist of new sites is confounded by the fact that >200 t of seed are sown annually (MAFF 1980-4). The degree of persistence of colonies formed from escaped cultivars is not known, nor have the ecological consequences of interbreeding between native and cultivated varieties been investigated. *Tp.* shows considerable genetic variation both

Altitude
Slope
Aspect

	n.s.	3.1
	12.5	
	0.9	Unshaded
—	0.0	

	—	0.0	
	0.0	—	Shaded

Soil pH

Associated floristic diversity

Hydrology

Triangular ordination

Commonest associates	%
Plantago lanceolata	77
Festuca rubra	75
Trifolium repens	69
Dactylis glomerata	61
Holcus lanatus	59

Bare soil

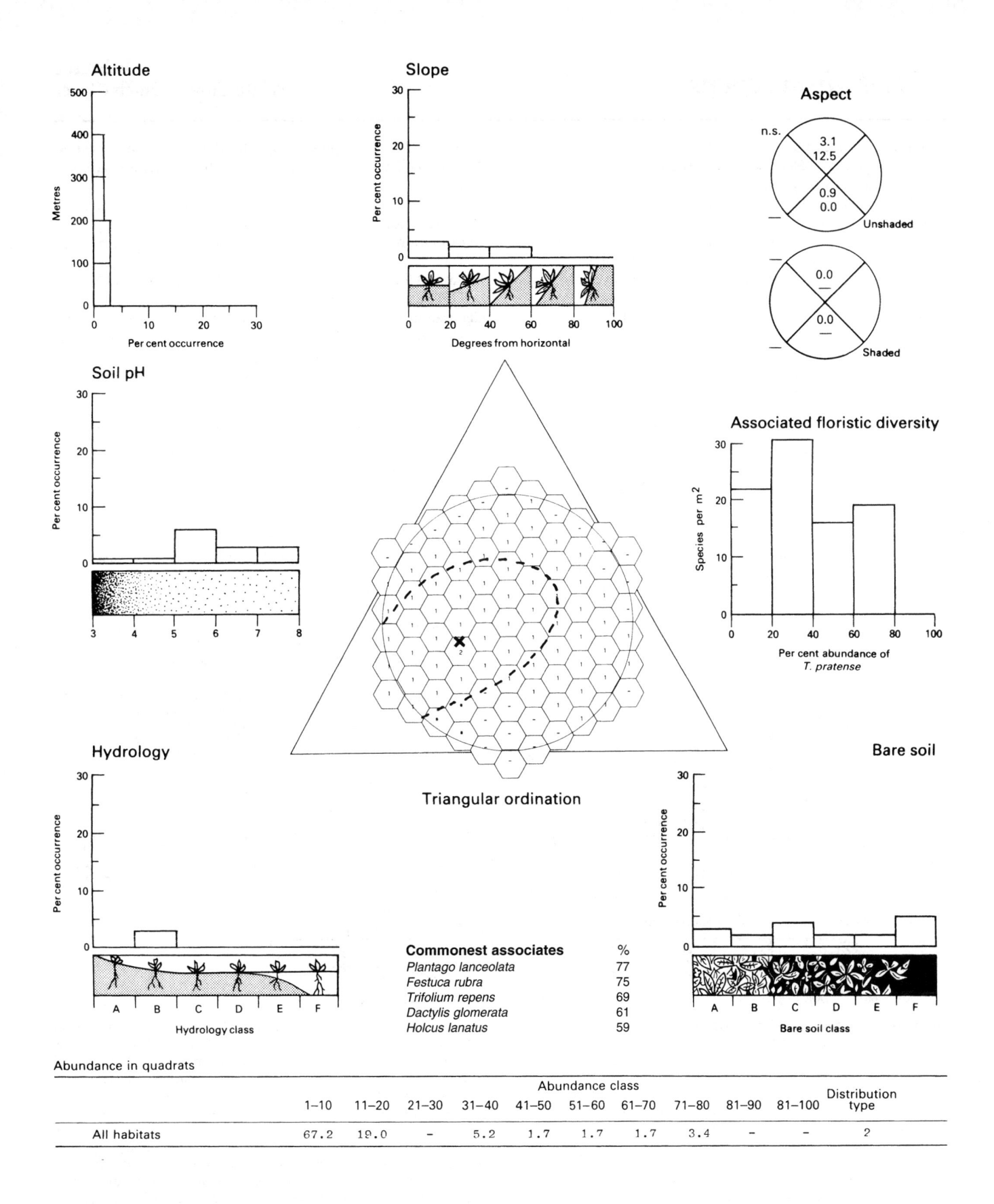

Abundance in quadrats

	1–10	11–20	21–30	31–40	41–50	51–60	61–70	71–80	81–90	81–100	Distribution type
All habitats	67.2	19.0	–	5.2	1.7	1.7	1.7	3.4	–	–	2

within native European populations (*FE2*) and within cultivars, which include early, intermediate- and late-flowering types (Spedding & Diekmahns 1972). Ten cultivars (including 5 tetraploids) are recommended for cultivation (ADAS 1984). *T.p.* also shows potential for the reclamation of derelict land (Dancer *et al.* 1977). Heavy-metal-tolerant populations have been recorded in Derbyshire (Shaw 1984).

Current trends Var. *p.* is presumably decreasing, whereas var. *s.* is held at an artificially high level through its use in agriculture. With the fragmentation of older grassland systems, we expect (a) that hybridization between native and cultivated populations will be increasingly likely and (b) that selection will favour genotypes which combine persistence with the capacity to exploit relatively fertile habitats.

(*)*Trifolium repens* L.

Fabaceae
White Clover, Dutch Clover

Life-form Polycarpic perennial chamaephyte or hemicryptophyte. Stems creeping; leaves ternate, usually <1000 mm², with long petioles; roots predominantly shallow and adventitious with VA mycorrhizas and nodules.

Phenology Winter green. Vegetative growth starts in April (later than in many pasture grasses). Flowers mainly in June and July, and seeds take a further month to ripen (*Biol Fl(a)*).

Height Foliage usually less than 100 mm, slightly exceeded by the inflorescence.

Seedling RGR 1.0-1.4 week^{-1}. SLA 38.8. DMC 17.9.

Nuclear DNA amount 2.0 pg; $2n = 32^*$ (*BC*); $4x^*$.

Established strategy Intermediate between C-S-R and competitive-ruderal.

Regenerative strategies (V), B_s. (See 'Synopsis'); SB IV.

Flowers White or pink, hermaphrodite; mainly outbreeding, insect pollinated (*Biol Fl(a)*); often c. 50 in an axillary racemose head.

Germinule 0.56 mg, 1.0 × 1.0 mm, seed enclosed within a 3-6-seeded pod.

Germination Epigeal; t_{50} 1 day; <5-30°C; hard-coat dormancy; L = D.

Biological flora (a) Burdon (1983), (b) Turkington and Burdon (1983), see also Zarzycki (1976).

Geographical distribution The British Isles (100% of British vice-counties), *ABIF* page 394; Europe except the extreme N (97% of European territories); N and W Asia, N Africa; introduced in S Africa, N and S America and E Asia.

Habitats Wide-ranging. Most common in meadows and pastures, and on road verges and paths. Other common habitats include arable land (mainly as seedlings), soil heaps, lead-mine spoil and demolition sites. Occurs at lower frequencies on wasteland and various other types of spoil, and in skeletal habitats and wetlands. Almost totally excluded from woodland areas. [Also found in montane and dune grassland (*Biol Fl(a)*).]

Gregariousness Intermediate.

Altitude Mainly below 400 m. [Up to over 900 m (*Biol Fl(a)*).]

Slope Most frequent and abundant on flat and gently sloping terraces, and generally absent from steep slopes; indicating both a positive association with fertile agricultural grassland and low frequency in skeletal habitats.

Aspect Slightly more frequent and abundant on N-facing slopes.

Hydrology A dryland species recorded occasionally from tussocks of other species and from soligenous mire.

Soil pH Shows a wide tolerance, but most common on soils of pH 5.0-8.0.

Bare soil Occurs in association with widely different amounts of exposed soil.

△ **ordination** Distribution wide-ranging, though centred on vegetation associated with fertile habitats subject to moderate disturbance.

Associated floristic diversity Typically intermediate.

Similarity in habitats	%
Cynosurus cristatus	85
Bellis perennis	78
Phleum pratense agg.	77
Lolium perenne	76
Cerastium fontanum	74

Habitats

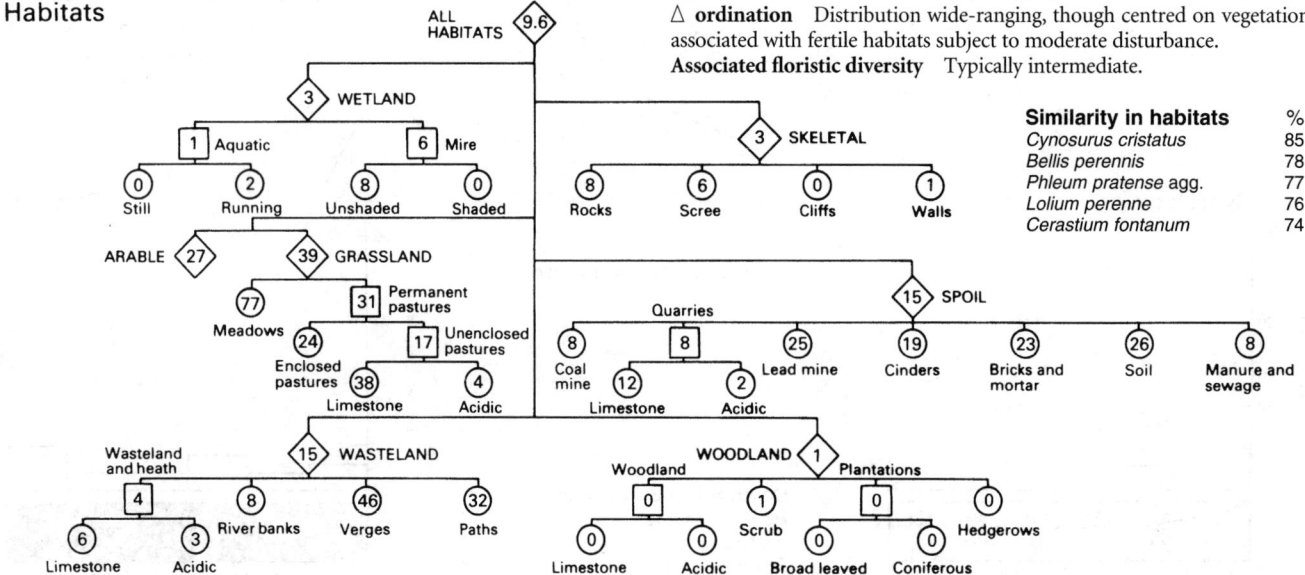

ALL HABITATS 9.6
- WETLAND 3
  - Aquatic 1
    - Still 0
    - Running 2
  - Mire 6
    - Unshaded 8
    - Shaded 0
- SKELETAL 3
  - Rocks 8
  - Scree 6
  - Cliffs 0
  - Walls 1
- ARABLE 27
- GRASSLAND 39
  - Meadows 77
  - Permanent pastures 31
  - Enclosed pastures 24
  - Unenclosed pastures 17
  - Limestone 38
  - Acidic 4
- SPOIL 15
  - Quarries
    - Coal mine 8
    - Limestone 8 / 12
    - Acidic 2
  - Lead mine 25
  - Cinders 19
  - Bricks and mortar 23
  - Soil 26
  - Manure and sewage 8
- WASTELAND 15
  - Wasteland and heath 4
    - Limestone 6
    - Acidic 3
  - River banks 8
  - Verges 46
  - Paths 32
- WOODLAND 1
  - Woodland 0
    - Limestone 0
    - Acidic 0
  - Plantations 0
    - Scrub 1
    - Broad leaved 0
    - Coniferous 0
  - Hedgerows 0

Synopsis *T.r.* is a low-growing but far-creeping, stoloniferous legume of greatest abundance in moist, fertile habitats. In common with the majority of British legumes, *T.r.* is intolerant of shade. *T.r.* is rapidly suppressed in tall vegetation. In many sites nitrogen fixation by *T.r.* often creates soil conditions which encourage invasion and temporary dominance by grasses. This suppresses *T.r.* and eventually leads to reduced N status and conditions favouring renewed vigour of *T.r.* (*Biol Fl(a)*). *T.r.* is intolerant of drought and severe frosts. *T.r.* has been important in fodder since the 17th century, and is by far the most important pasture legume in Britain (c. 900 t are sown annually (MAFF 1980-84)) and 75 cultivars are listed for Europe (*Biol Fl(a)*). The value of *T.r.* to agriculture stems from the capacity to fix nitrogen in habitats where other major nutrients are not limiting (*Biol Fl(a)*). The morphology and phenology also complement that of *Lolium perenne*, with which it often grows. The creeping shoots allow colonization of gaps between tufts of *L.p.* and many other pasture grasses with a consolidated growth form. In keeping with the low nuclear DNA amount, *T.r.* grows predominantly during summer in contrast with the spring and autumn peaks for *L.p.* (*Biol Fl(a)*). In addition, *T.r.* is tolerant of heavy grazing, trampling and cutting. *T.r.* regenerates almost exclusively by rooted stolons in closed communities to form large diffuse clonal patches but colonizes new sites mainly by seed, which germinates mainly in spring and may form a potentially long-lived persistent seed bank (Odum 1978). Seeds can survive ingestion by a range of animals and birds (*Biol Fl(a)*). Although the foliage of *T.r.* is highly palatable to stock, some genotypes contain cyanogenic glucosides, which afford varying degrees of protection against invertebrate herbivores and small mammals and, when consumed in large quantities, can disrupt the rumen of domestic animals (Corkill 1952, Jones 1962). Cyanogenetic phenotypes are less common in cold climates, and at high altitudes show evidence of poorer flowering and susceptibility to frost and drought (Daday 1954a,b, Foulds & Grime 1972a,b). The wide ecological amplitude of *T.r.* in many temperate zones (*Biol Fl(b)*) is correlated with genetic variation in morphology, growth rate, phenotypic plasticity in growth form, and with widespread sowing of agricultural cultivars of *T.r.* and the association with several strains of the

Altitude
Slope
Aspect

	n.s.	10.6	
	37.0		
	9.0	15.0	Unshaded
	n.s.		

	—	0.0	
	—		
	0.8	0.0	Shaded
	—		

Soil pH

Associated floristic diversity

Hydrology

Triangular ordination

Bare soil

Commonest associates	%
Festuca rubra	71
Holcus lanatus	62
Plantago lanceolata	49
Dactylis glomerata	48
Poa trivialis	46

Abundance in quadrats

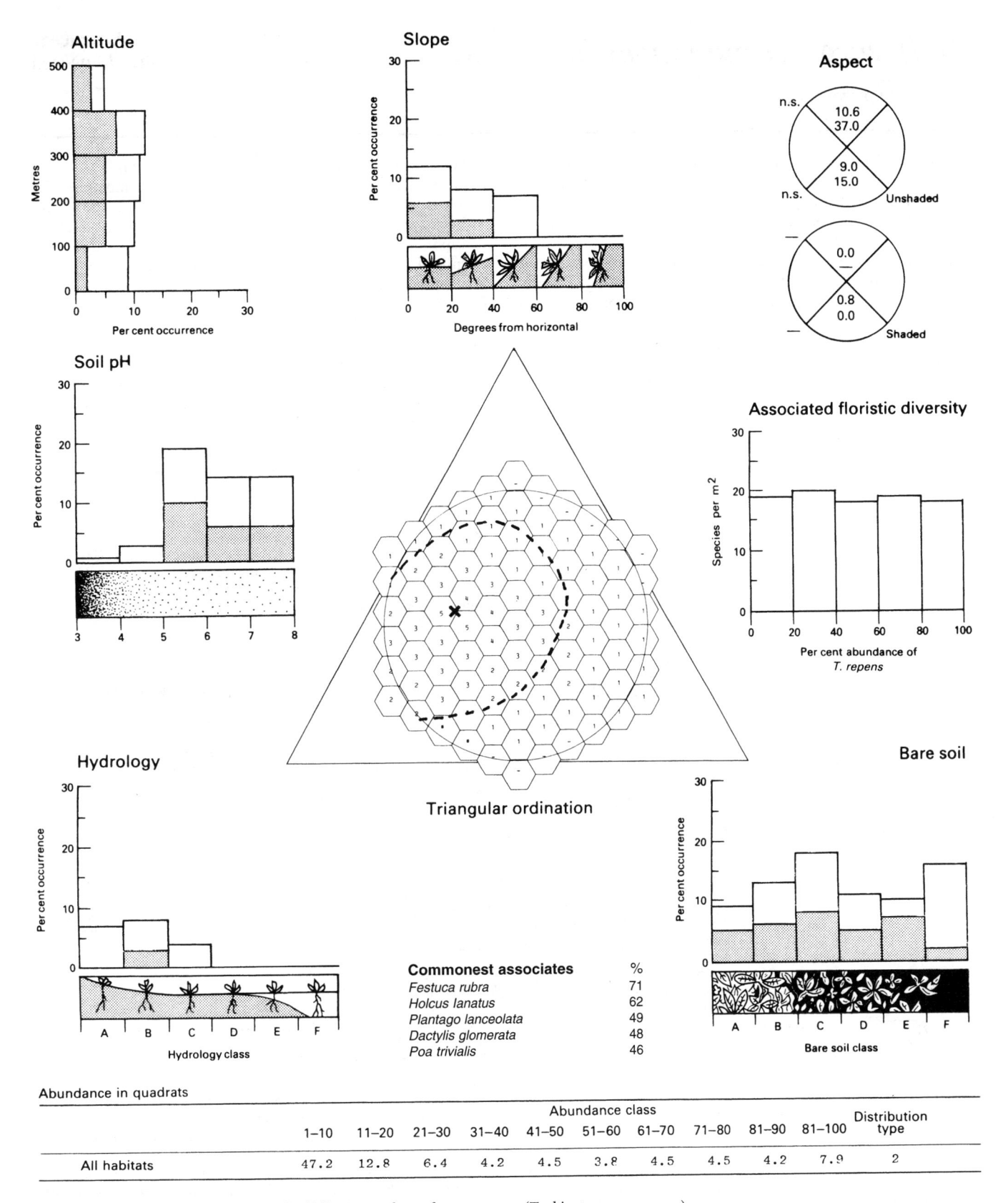

	Abundance class										Distribution type
	1–10	11–20	21–30	31–40	41–50	51–60	61–70	71–80	81–90	81–100	
All habitats	47.2	12.8	6.4	4.2	4.5	3.8	4.5	4.5	4.2	7.9	2

nitrogen-fixing *Rhizobium trifolii* (*Biol Fl(a)*). A number of ssp. are recognized in Europe (*FE2*). Evidence has been obtained of local variation in *T.r.* coinciding with changes in the identity of the grass species with which *T.r.* is associated in the field (Turkington *et al.* 1979). More-recent studies suggest that these differences are not genetic in origin, and it now seems likely that they may arise from changes in the phenotype induced by particular strains of *Rhizobium trifolii*, the distribution of which may be influenced by the rhizosphere associated with particular turf grasses (Turkington pers. comm.).

Current trends Sowings for agriculture and more recently for amenity purposes (*Biol Fl(a)*) maintain the abundance of *T.r.* at an artificially high level. The popularity of *T.r.* is unlikely to wane, since there are few native legumes with as great a capacity to persist in fertile habitats. A greater use of indigenous forms which are shorter, more persistent and drought-tolerant is predicted (*Biol Fl(a)*).

+*Tripleurospermum inodorum* Schultz Bip.

Matricaria perforata Merat

Asteraceae
Scentless Mayweed

Life-form Typically a winter- or summer-annual therophyte, exceptionally behaving as a short-lived, polycarpic perennial, semi-rosette, hemicryptophyte. Stems usually erect, often branched; leaves oblong, finely divided, often <500 mm^2; roots with VA mycorrhizas.

Phenology Variable since germination may occur throughout most of the year. Plant capable of overwintering. Flowers mainly from July to September and sets seed from August to October.

Height Foliage and flowers to 600 mm.

Seedling RGR No data. SLA 23.1. DMC 15.8.

Nuclear DNA amount 3.9 pg; $2n = 18^*, 36^*$ (Kay 1969); $2x^*, 4x^*$.

Established strategy Ruderal.

Regenerative strategies B$_s$. Exclusively by seed on bare ground or in vegetation gaps either directly after shedding in autumn or after incorporation into a persistent seed bank; SB III.

Flowers Florets insect-pollinated, largely self-incompatible (Kay 1969); *c.* 25 white, marginal ray florets and *c.* 400 yellow hermaphrodite disk-florets in each solitary, terminal capitulum.

Germinule 0.29 mg, 2.0 × 1.1 mm, achene with resin glands, pappus reduced to a scaley rim.

Germination Epigeal; t_{50} 6 days; 13-35°C; germination rate increased after dry storage at 20°C; L > D.

Biological flora Kay (1994).

Geographical distribution Throughout lowland Britain (100% of British vice-counties), *ABIF* page 651; N and C Europe (77% of European territories); W Asia. British distribution includes *T. maritimum* (L.) Koch.

Habitats Most common as a weed of arable land. Also frequent on various types of disturbed spoil, particularly demolition sites, soil heaps and cinder tips. Occasionally found on disturbed road verges and in lightly trampled areas on paths. Observed on wasteland and in urban and agricultural areas on cliffs, outcrops and walls. A casual of wetlands and of disturbed ground in pasture or woodland. [Replaced by *T.m.* in maritime habitats (Kay 1972).]

Gregariousness Intermediate.

Altitude More abundant in lowland areas, but occurs up to 400 m. [Up to 530 m.]. Since *T.i.* often occurs in impermanent habitats, the altitudinal limit at which populations are viable is difficult to assess.

Slope Found over the full range of slope.

Aspect Widely distributed.

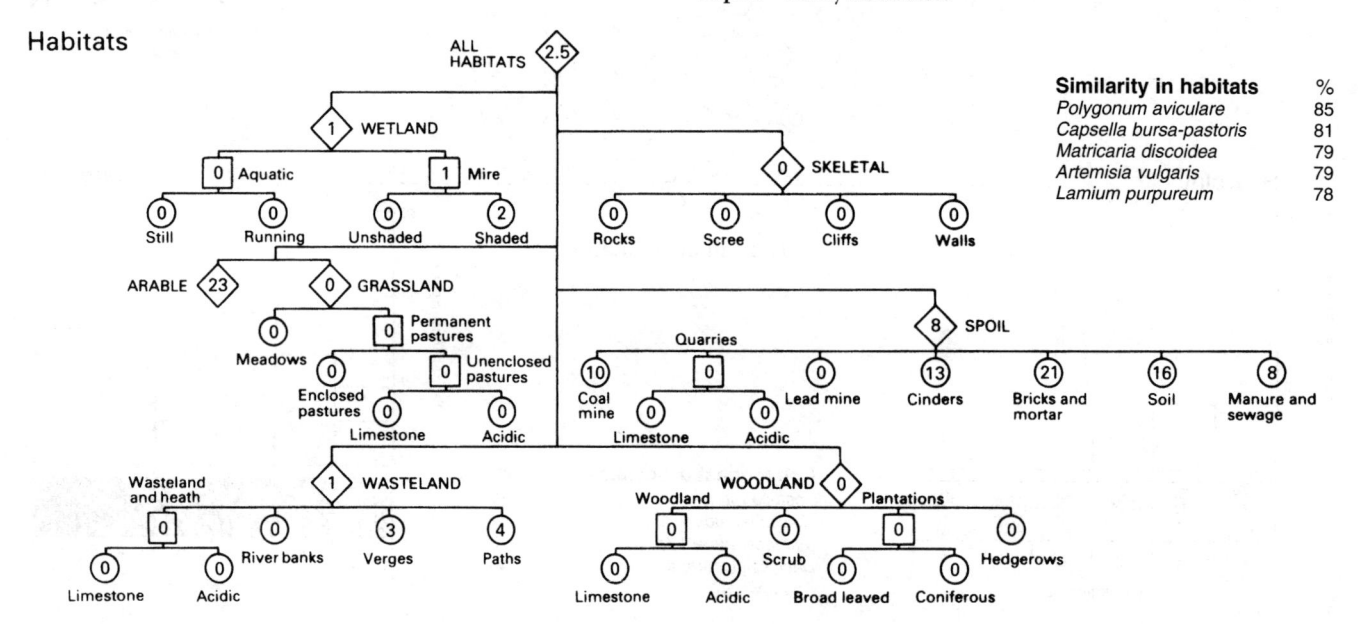

Habitats

Similarity in habitats	%
Polygonum aviculare	85
Capsella bursa-pastoris	81
Matricaria discoidea	79
Artemisia vulgaris	79
Lamium purpureum	78

Hydrology In wetland observed only as a transient colonist of wet disturbed ground near coal-mine heaps and gravel pits.

Soil pH Most abundant on soils of pH >5.5, and not recorded from acidic soils of pH <4.5.

Bare soil Most frequent and abundant in habitats having a large proportion of bare soil.

△ **ordination** Distribution centred on vegetation associated with highly disturbed but fertile habitats. Absent from the most productive vegetation and from stressed environments.

Associated floristic diversity Typically intermediate.

Synopsis *T.i.* is a short-lived herb which exploits a range of disturbed and relatively fertile artificial habitats. *T.i.* is intolerant both of dense shade and of waterlogging, and the erect growth form renders the plant susceptible to grazing and frequent mowing. The persistence of *T.i.* on ballast beside railways may be related to its resistance to herbicides (Salisbury 1964). Like many other common ruderals, the species exploits both 'new' open sites (e.g. soil heaps) and continuously disturbed habitats (e.g. arable). Success appears to be related to a high potential for seed production: in an open, fertile habitat an individual may produce more than 10 000 seeds, while stunted individuals with a single capitulurn may form as few as 100 viable seeds. Seed is dispersed effectively by human and other agencies, including animals; *T.i.* is widely distributed during harvesting and as an impurity in grass seed (*WCH*), and the achenes are eaten by birds and may survive ingestion by stock (Ridley 1930). Seed shows little innate dormancy, but a persistent seed bank is formed and confers the capacity for seed germination over much of the year (but particularly in spring and autumn (*WCH*)). Unlike many weeds, however, *T.i.* is a long-day, relatively slow-maturing plant (Roberts & Feast 1974)

Altitude
Slope
Aspect

n.s.	2.4 / 33.3
n.s.	2.7 / 16.7 Unshaded

—	0.0 / —
—	0.0 / — Shaded

Soil pH

Associated floristic diversity

Hydrology

Triangular ordination

Bare soil

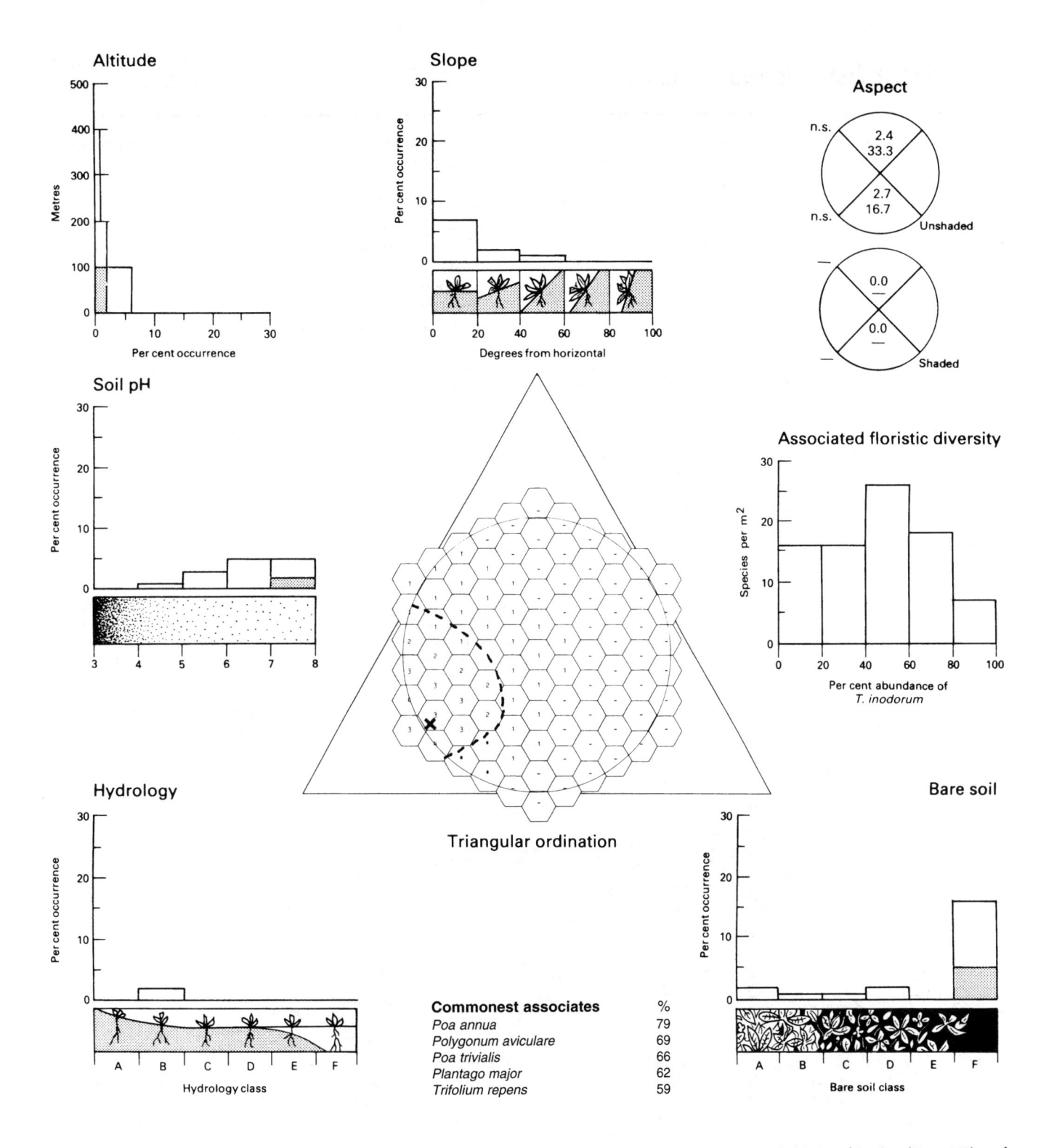

Commonest associates %
Poa annua 79
Polygonum aviculare 69
Poa trivialis 66
Plantago major 62
Trifolium repens 59

and flowering is mainly restricted to the period July to September. This inflexibility may partially explain why *T.i.* reaches maximum frequency in sites with some habitat continuity, e.g. arable fields, and on those areas of spoil heaps where the rate of vegetation recolonization is slow. Although normally an annual, a small percentage of the populations on demolition sites and cinder tips may be short-lived, polycarpic perennials. *T.i.* is usually strongly self-incompatible, and viable populations consist of several individuals (Kay 1969); this may further explain the restriction to sites with some habitat continuity. The degree of ecotypic differentiation in this outbreeding species is not certain. There is no evidence of genetically differentiated winter- and summer-annual forms (Roberts & Feast 1974). However, the mean size of achenes varies from 1.5 × 0.7 mm in S and E England to 2.2 × 1.2 mm in Scotland (Kay 1972); some populations from cereal fields may assume a prostrate habit in cultivation (Kay 1972), and differences in resistance to herbicides have been reported (Ellis & Kay 1975). Native British populations of *T.i.*, like those from W Europe, appear diploid and the tetraploids, which are sometimes more vigorous but otherwise ecologically indistinguishable, occur only as a rare introduction from within their native range in E and N Europe (Kay 1969). Maritime populations have recently been separated from *T.i.* as *T. maritimum* (Kay 1972). These, too, are diploid and self-incompatible and introgression with *T.i.* is widespread near the coast (Stace 1975). Hybrids with *Anthemis cotula* L. have been reported from Britain (Stace 1975).

Current trends A common colonist of artificial habitats and perhaps still increasing.

Trisetum flavescens (L.) Beauv.

Poaceae
Yellow Oat-grass

Life-form Loosely tufted, polycarpic perennial protohemicryptophyte. Shoots erect; leaves linear, hairy, up to 1000 mm^2; roots with VA mycorrhizas.
Phenology Flowers in May and June and seed shed in July and August. Leaves relatively short-lived; shoots dying back in autumn replaced by short, overwintering tillers.
Height Foliage predominantly <250 mm; flowering shoots to 500 mm.
Seedling RGR No information. SLA 20.1. DMC 30.7.
Nuclear DNA amount 4.7 pg; $2n = 28$ (*FE5*); $4x$.
Established strategy C-S-R.
Regenerative strategies S, V. Mainly by freshly shed seed in autumn. No persistent seed bank has been reported. Also has a restricted capacity for lateral vegetative spread, forming small clumps; SB I.
Flowers Yellow-green, hermaphrodite, wind-pollinated, self-incompatible (Fryxell 1957); >100 aggregated into a loose panicle (2-4 per spikelet).
Germinule 0.18 mg, 1.2 × 0.4 mm, caryopsis dispersed with an attached lemma and an hygroscopic bristle (5.7 × 0.7 mm).
Germination Epigeal; t_{50} 2 days; <5-34°C.
Biological flora Dixon (1995).
Geographical distribution Generally distributed in the British Isles, but less common in the N and W (94% of British vice-counties), *ABIF* page 768; throughout Europe to S Scandinavia (82% of European territories and 10% naturalized); temperate Asia; introduced in N America.

Habitats *T.f.* is a frequent constituent of commonly occurring types of base-rich grassland, both managed (meadows, pastures and road verges) and unmanaged (wasteland and quarry spoil). Also frequent in certain skeletal habitats such as rock outcrops, scree slopes and cliffs. Infrequent in shaded habitats, only a casual in arable land, and absent from wetlands.
Gregariousness Sparse to intermediate.
Altitude Up to 400 m, but particularly frequent on the carboniferous limestone between 200 and 400 m. [Up to 500 m.]
Slope Widely distributed.

Habitats

Aspect There is a slight bias towards S-facing slopes; this trend is particularly evident in grasslands (*Ecol Atl*).
Hydrology Absent from wetlands.
Soil pH Most abundant on soils of pH >6.0, and largely absent below pH 5.0.
Bare soil Found both in closed vegetation such as meadows and in habitats with a moderate amount of exposed soil, e.g. rock outcrops.
Δ ordination Widespread in distribution, *T.f.* is absent only from heavily disturbed and from dense, productive vegetation types. Distribution centred on vegetation in which the intensity of competition from other species is limited by moderate levels of nutrient stress and disturbance.
Associated floristic diversity Most commonly associated with vegetation of intermediate to high species-richness.

Similarity in habitats	%
Plantago lanceolata	85
Festuca rubra	75
Lotus corniculatus	71
Linum catharticum	71
Pilosella officinarum	69

Synopsis *T.f.* is a fairly tall, tufted grass of dry grassland and, to a lesser extent, rocky habitats, particularly on base-rich soils. *T.f.* resembles a 'small *Arrhenatherum elatius*' in habit and in producing a midsummer peak in above-ground biomass. In unmanaged grassland the species is usually restricted to sites where the growth of more-productive species is checked by, for example, periodic burning. *T.f.* also tends to be absent from severely droughted habitats. In common with other grasses producing erect nodal stems, *T.f.* appears to be susceptible to heavy grazing by stock. The leaves are palatable to cattle and the stems also are consumed by sheep (Spedding & Diekmahns 1972). Towards the base of the plant long reflexed hairs are often present in abundance on the leaf sheaths; there is some evidence (Grime *et al.* 1968) that such hairs restrict the activities of invertebrate herbivores. *T.f.* is susceptible to trampling. In Britain *T.f.* is associated with a range of soil types, moisture and fertility levels, and a variety of management regimes. *T.f.* appears to show only modest specialization towards any ecological factor or turf structure, and we conclude that the species is a 'congenital subordinate', i.e. never more than a minor component of grassland communities, irrespective of management or turf height, but capable of persisting through fluctuations of environment or management which are sufficient to bring about drastic changes in the abundance of the more-specialized potential dominants. This is consistent with Brenchley and Warington (1958), who described *T.f.* as 'a very insignificant member of all the associations in which it occurs, except occasionally on the limed sections of plots receiving farmyard manure' (*PGE*). Perhaps because *T.f.* is such a minor species, its ecological distribution has been little studied. Early seed set and almost immediate germination has been documented in sites subject to summer drought (Thompson 1977), and this may be expected to confer an advantage in hayfields. No persistent seed bank has been reported. Populations tolerant of heavy metals have been detected in the Derbyshire flora (Shaw 1984). There is considerable taxonomic complexity and genetic diversity in Europe as a whole, and two additional subspecies have been recognized, including one with $2n = 12$ ($2x$) (*FE5*).
Current trends *T.f.* is no longer included in agricultural seed mixtures and is now restricted to less intensively managed grassland. *T.f.* is not among the most frequent colonists of artificial habitats, and is probably decreasing slightly.

Altitude
Slope
Aspect

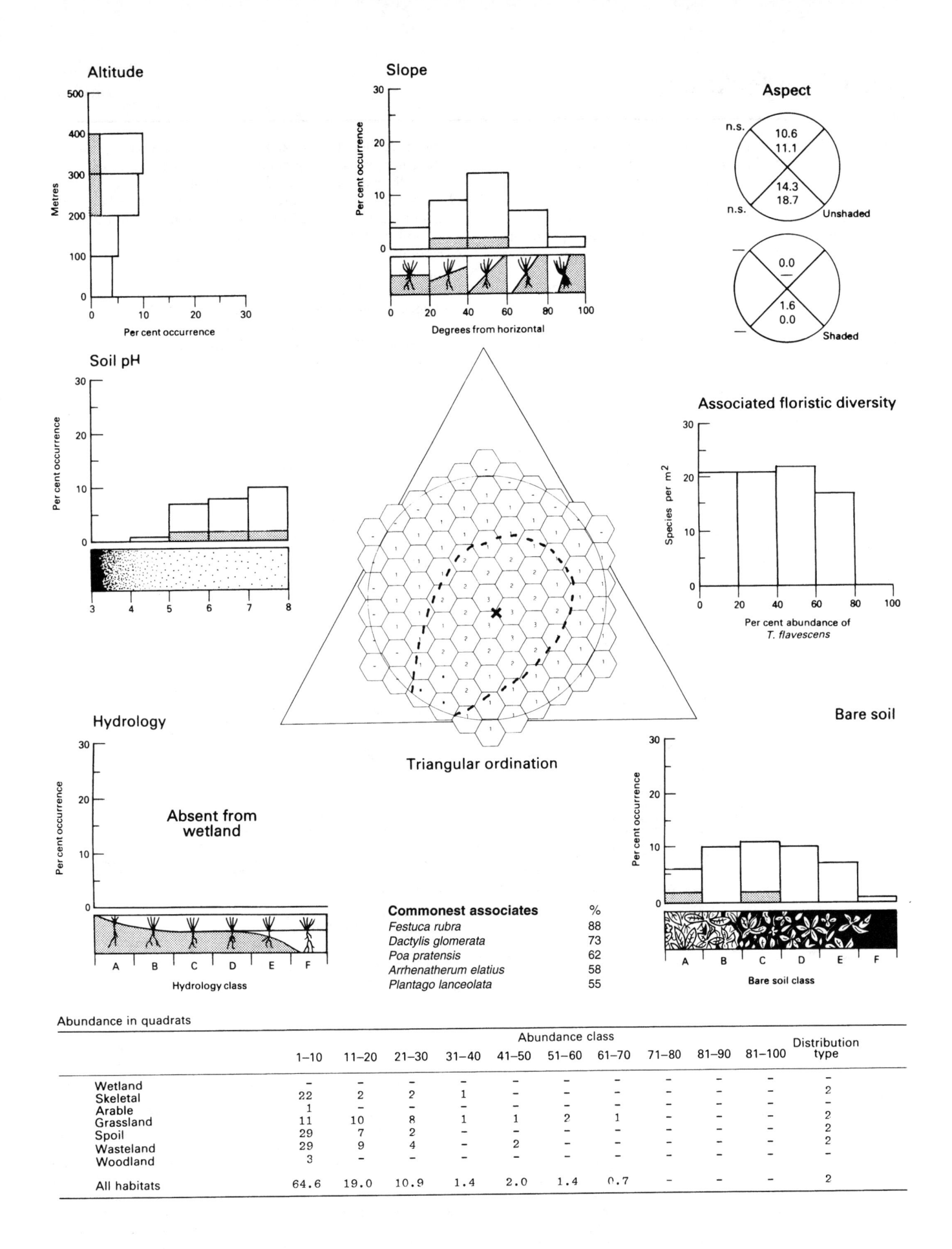

n.s. 10.6 / 11.1 / 14.3 / 18.7 n.s. / Unshaded

— / 0.0 / — / 1.6 / 0.0 / — Shaded

Soil pH

Associated floristic diversity

Hydrology

Absent from wetland

Triangular ordination

Bare soil

Commonest associates	%
Festuca rubra	88
Dactylis glomerata	73
Poa pratensis	62
Arrhenatherum elatius	58
Plantago lanceolata	55

Abundance in quadrats

	1–10	11–20	21–30	31–40	41–50	51–60	61–70	71–80	81–90	81–100	Distribution type
Wetland	–	–	–	–	–	–	–	–	–	–	–
Skeletal	22	2	2	1	–	–	–	–	–	–	2
Arable	1	–	–	–	–	–	–	–	–	–	–
Grassland	11	10	8	1	1	2	1	–	–	–	2
Spoil	29	7	2	–	–	–	–	–	–	–	2
Wasteland	29	9	4	–	2	–	–	–	–	–	–
Woodland	3	–	–	–	–	–	–	–	–	–	–
All habitats	64.6	19.0	10.9	1.4	2.0	1.4	0.7	–	–	–	2

Tussilago farfara L.

Asteraceae
Coltsfoot

Life-form Polycarpic perennial geophyte with long rhizomes bearing short branches, each of which terminates in a rosette of leaves. Scapes erect; leaves rounded, with woolly hairs beneath, sometimes >4000 mm^2; roots with VA mycorrhizas.

Phenology There is some winter growth of the rhizome (Leuchs 1961). Flowers February to April. Fruits formed April to June, during which time the leaves emerge. Leaves relatively long-lived but die back in autumn.

Height Foliage up to 200 mm; flowers to 100 mm (to 300 mm in fruit).

Seedling RGR 1.0-1.4 week^{-1}. SLA 18.6. DMC 13.9.

Nuclear DNA amount 4.6 pg; $2n = 60^*$ (*BC*); $2x^*$ (6x for the tribe).

Established strategy Intermediate between competitor and competitive-ruderal.

Regenerative strategies (V), W, S. Forms extensive patches by means of short-lived rhizomes which die back leaving isolated daughter plants. Seed germinates immediately in spring and is short-lived in the field (Bakker 1960); SB I.

Flowers Florets yellow, insect- or self-pollinated (Bakker 1960); 20-40 male and 200-300 ligulate female florets in a terminal capitulum (Salisbury 1964).

Germinule 0.25 mg, 3.0 × 0.5 mm, achene with a pappus of hairs up to 14 mm.

Germination Epigeal; t_{50} <1 day; <5-34°C; L = D.

Biological flora None, but see Bakker (1960), Myerscough and Whitehead (1966, 1967), Ogden (1974).

Geographical distribution The British Isles (100% of British vice-counties), *ABIF* page 659; Europe northwards to Scandinavia (89% of European territories); N and W Asia, N Africa; introduced in N America.

Habitats Particularly common on various types of spoil. Also found in soligenous mire, on river banks, road verges and path margins, and in lightly shaded hedgerows and woodland margins. May become established in scree and non-persistent seedlings are found in other skeletal habitats and in arable fields. In upland areas *T.f.* is observed on wet and eroding shale cliffs. [Found in maritime shingle and sand dunes.]

Gregariousness Capable of forming stands at low frequencies.

Altitude Observed to 550 m. [Up to 1070 m.]

Slope Found over a wide range of slopes.

Aspect Widely distributed with respect to aspect.

Hydrology Occurs on moist ground adjacent to topogenous mire in sites that are not normally flooded. Also observed in soligenous mire, and sometimes in sites where there is a trickle of water flowing over the mire surface.

Soil pH Found over a wide range of values but infrequent in soils of pH <4.5.

Bare soil Occurs in association with the full range of values.

△ **ordination** Distribution centred on vegetation associated with moderately disturbed and fertile habitats. However, *T.f.* does occur in a wide range of communities, though many of the records refer to seedlings rather than established plants.

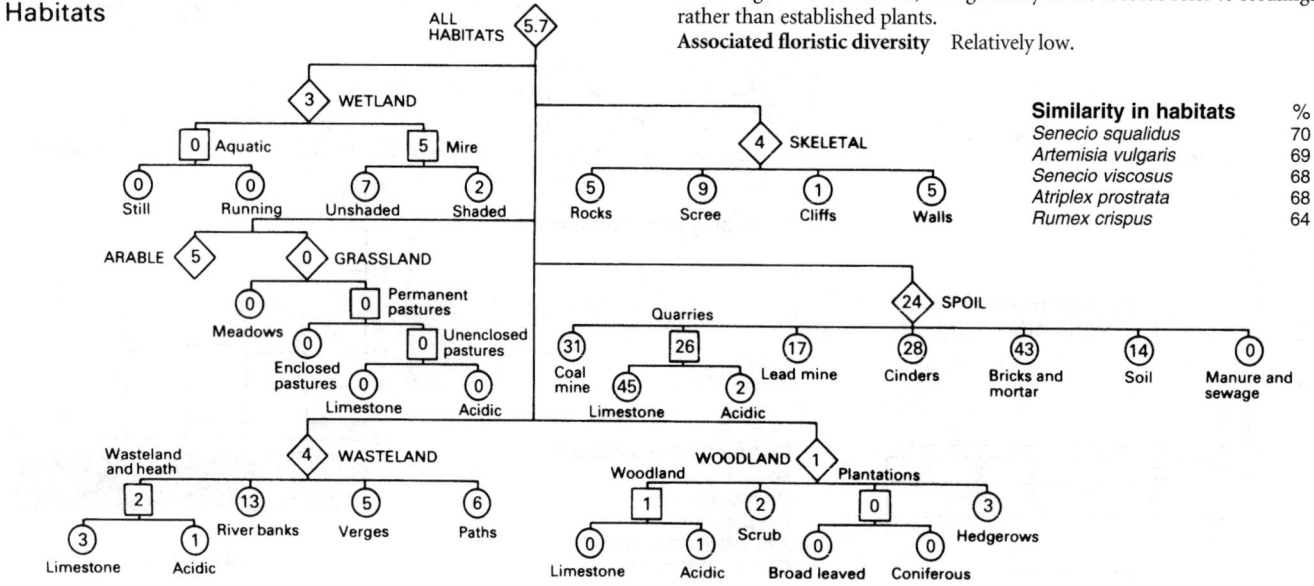

Associated floristic diversity Relatively low.

Habitats

Similarity in habitats	%
Senecio squalidus	70
Artemisia vulgaris	69
Senecio viscosus	68
Atriplex prostrata	68
Rumex crispus	64

Synopsis *T.f.* is a rhizomatous, stand-forming perennial which colonizes bared ground. Extensive lateral spread underground, coupled with the shade created by the large, entire, radical leaves contribute to its subsequent capacity for dominance of relatively short (<200 mm) perennial vegetation. In urban areas *T.f.* is most frequent on spoil tips and demolition sites. As a result of its low canopy height, the persistence of *T.f.* varies according to the productivity of the site. In fertile habitats *T.f.* tends to be replaced eventually by taller dominants, while in less-fertile sites, such as coal-mine spoil, *T.f.* may persist for many years. Rhizomes are produced at various depths in the soil, and this may be important in allowing *T.f.* to survive in such semi-natural habitats as stream banks and boulder-clay cliffs, where soil movement may occur annually (Myerscough & Whitehead 1967). However, because of its robust stature and long-lived leaf canopy, *T.f.* is vulnerable to the effects of regular grazing, cutting and trampling. The species is little affected by waterlogging, but requires exposed mineral soil for seedling establishment. Early spring flowering from buds initiated in autumn, followed by vegetative growth in summer and die-back over winter constitutes an unusual phenological pattern within the British flora and may be related to the ability of *T.f.* to exploit sub-alpine 'snow patch' vegetation in Europe (Ogden 1974). The levels of Na, Ca and Mg in the leaves are unusually high, but the concentrations of N and P are low (*SIEF*). In its ability to colonize disturbed artificial habitats *T.f.* is most similar to *Chamerion angustifolium*. On heavy clay and on calcareous soils, *T.f.* is the more frequent, despite the fact that *C.a.* is a taller and more effective potential dominant; this observation is in agreement with those of Myerscough and Whitehead (1966, 1967). Seed is short-lived and under natural conditions over 50% is non-viable within 2 months (Bakker 1960). However, seeds are produced in abundance during spring

Altitude

Slope

Aspect

n.s. | 9.8
12.0
n.s. | 9.9
13.6 | Unshaded

— | 0.0
0.0
— | — Shaded

Soil pH

Associated floristic diversity

Hydrology

Triangular ordination

Commonest associates	%
Festuca rubra	58
Holcus lanatus	36
Arrhenatherum elatius	35
Poa trivialis	34
Poa pratensis	34

Bare soil

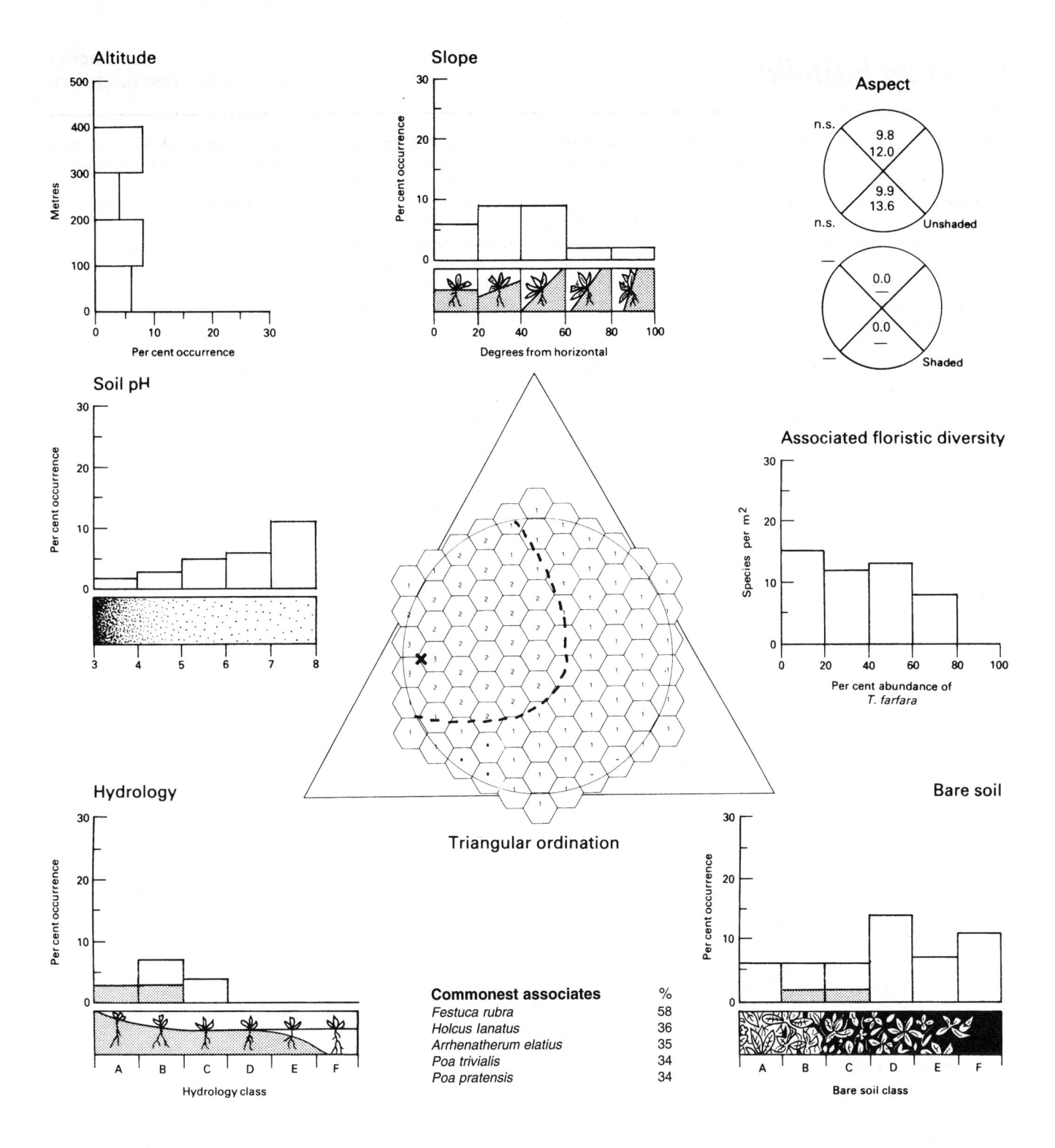

and early summer, with as many as 1000 in each tuft of flower heads (Salisbury 1964, Bostock & Benton 1979). The small-plumed achenes are extremely buoyant in air (Sheldon & Burrows 1973) and may be carried up to 4 km from the parent plant (Bakker 1960). As a result, *T.f.* is a frequent colonist of bare ground in early summer by means of freshly-shed seed (Ogden 1974). Germination is exceptionally rapid, even on substrates with a low water content (Bostock 1978) and at low temperatures (Grime *et al.* 1981). Subsequent establishment of the seedling is also speedy in unshaded sites where plants may flower in their second year (Ogden 1974). Established colonies increase in size mainly by vegetative means, and large clonal patches may be formed through the expansion of short-lived rhizomes by up to 1 m year^{-1} (Bakker 1960). Rhizomes may be dispersed in soil by humans in the same way as those of *Fallopia japonica*. However, the balance between vegetative and sexual reproduction varies according to the stage of colonization. Thus, in diffuse patches a greater proportion of resources are allocated to rhizomes, while in crowded stands of *T.f.* more flowers are produced (Ogden 1974). The existence of ecotypes of this ecologically wide-ranging species is suspected but has not been demonstrated.

Current trends A species which has increased dramatically due to the artificial creation of disturbed habitats. Probably still increasing but, in view of the short-lived nature of its seed and the limited persistence of the adult plant in grazed sites and tall-herb communities, *T.f.* is potentially vulnerable in the event of a reduction in the level of disturbance within the landscape.

(*)*Typha latifolia* L.

Typhaceae
Cat's-tail, Great Reedmace

Life-form Rhizomatous, polycarpic perennial hydrophyte. Shoots erect; leaves typically <30 000 mm^2; roots predominantly shallow.
Phenology Shoots elongate in late spring. Flowers in June and July and sets seed from September onwards. Shoots die in autumn, but some seed may overwinter on dead inflorescences.
Height Foliage and flowers to 2.5 m.
Seedling RGR No data. SLA 14.5. DMC 18.6.
Nuclear DNA amount 0.6 pg; 2n = 30 (*FE5*); 2x.
Established strategy Competitor.
Regenerative strategies V, W, B$_s$. (See 'Synopsis'); ?SB III
Flowers Flowers unisexual, wind-pollinated, protogynous but usually self-pollinated (Krattinger 1975); several thousand on a terminal spadix, the male above, the female below. The female part of the inflorescence contains numerous sterile flowers (see Muller-Doblies 1970).
Germinule 0.03 mg, 1.4 × 0.3 mm, dry fruit with basal hairs of *c*. 8 mm in length.
Germination Epigeal; t_{50} 3 days, 15-37°C; germination rate increased by a number of factors (see 'Synopsis'); L > D.
Biological flora Grace and Harrison (1986).
Geographical distribution The British Isles, but less frequent in the N and W (94% of British vice-counties), *ABIF* page 807; most of Europe (87% of European territories); many parts of the world from the Arctic Circle to 30° S except for C and S Africa, S Asia and Australia.

Habitats A wetland species most frequent at the edge of ponds and ditches, either in shallow water or in mire. Less frequent in shaded mire. Also observed in wet subsidence areas associated with coal-mining, and rather rarely beside slow-moving rivers.
Gregariousness Stand-forming; achieving dominance at low-frequency values.
Altitude Extends to 350 m but suitable sites are more frequent in the lowland. [Up to 500 m.]
Slope Like most other wetland species, *T.l.* is infrequent on steep slopes.
Aspect Insufficient data.

Hydrology Typically a waterside species, occurring in shallow water or on exposed mud, but found in various types of topogenous mire. *T.l.* occasionally colonizes bare soligenous mire on coal heaps. Reaches maximum abundance in 11-25 mm of water, and not recorded at depths of >600 mm.
Soil pH Mainly restricted to sites above pH 5.5, and unrecorded below pH 4.5.
Bare soil Found typically at high exposures of either bare soil or exposed water.
△ **ordination** A potential dominant which is largely confined to productive and relatively undisturbed vegetation.
Associated floristic diversity Low, a reflection of the tendency of *T.l.* to form mono-dominant stands.

Habitats

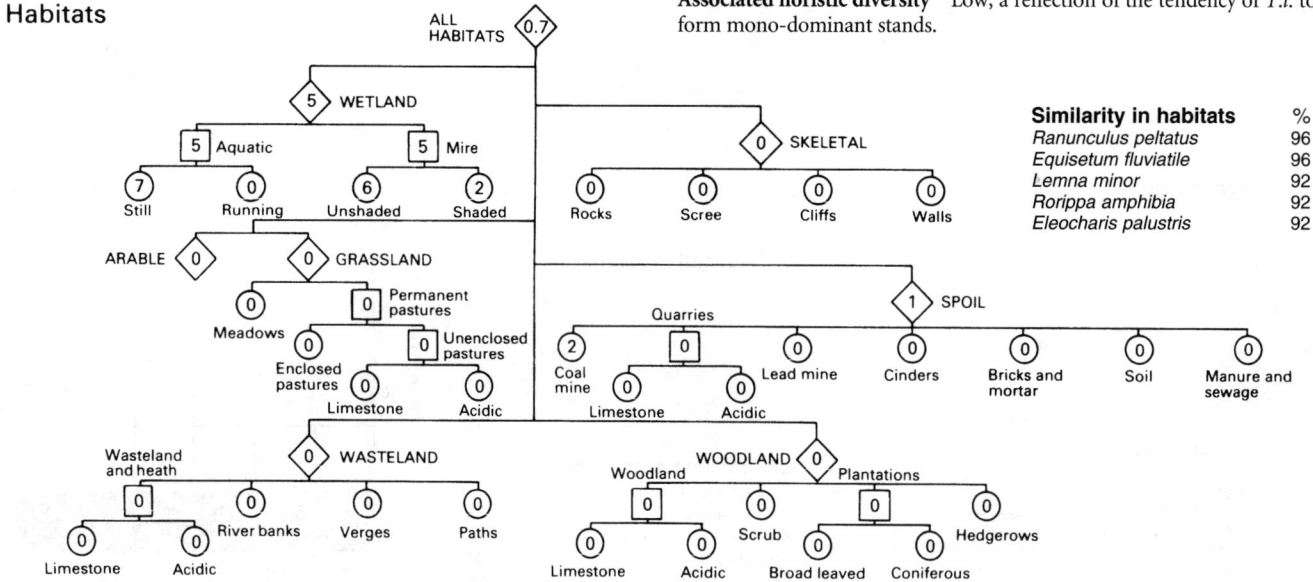

Similarity in habitats	%
Ranunculus peltatus	96
Equisetum fluviatile	96
Lemna minor	92
Rorippa amphibia	92
Eleocharis palustris	92

Synopsis *T.l.* is a robust dominant of ungrazed and uncut eutrophic wetlands. Although common in mire, locally *T.l.* achieves maximum abundance as an emergent aquatic in shallow water and, like other species exploiting this habitat, has well-developed aerenchyma (Smirnoff & Crawford 1983). Unusually among dominants, *T.l.* produces large, ascending basal or sub-basal leaves and, despite its high productivity (Fiala 1971), the species has a rather open leaf canopy. The foliage of plants sampled in the Sheffield region (*SIEF*) contains relatively low levels of N, P and Fe, whereas the concentrations of Na and Mn are often unusually high. *T.l.* produces an extensive rhizome system, and offshoots are produced throughout the growing season (Fiala 1971). *T.l.* forms large clonal patches which may expand at up to 4 m year^{-1} (Fiala 1971). Detached rhizome pieces readily re-root. These may float in water until beached, and have the potential to form new colonies. Each inflorescence produces thousands of minute fruits, and these provide the main mechanism of colonization. They may be transported by wind under dry conditions and may float in water, but seeds are released and sink once the pericarp comes into contact with water (Krattinger 1975). Seeds exhibit hard-coat dormancy, and may be induced to germinate by scarification. Germination is said to be promoted by light in combination with low oxygen concentrations or, at low light intensities, by fluctuating temperatures (Sifton 1959). A persistent seed bank is also recorded (Leck & Gravelime 1979). Regeneration by seed appears not to occur within established colonies of *T.l.* (McNaughton 1968). *T.l.* is grown for ornament and may sometimes be a garden escape (see Allen 1966). The female part of the inflorescence is receptive to pollen for 4 weeks and pollen, which is released in tetrads, is only effectively dispersed in high winds (Krattinger 1975). Thus, despite its protogyny, *T.l.* is mainly inbreeding. Nevertheless, in N America a range of ecotypes has been identified (McNaughton 1966, Grace & Wetzel 1983). *T.l.* is widely distributed on nutrient-rich mineral soils and may occur in peaty water. Under calcareous or saline conditions, however, *T.l.* may be replaced by *Phragmites australis*. *T.l.* hybridizes with *T. angustifolia* L.

Altitude
Slope
Aspect
Soil pH
Associated floristic diversity
Hydrology
Triangular ordination
Bare soil

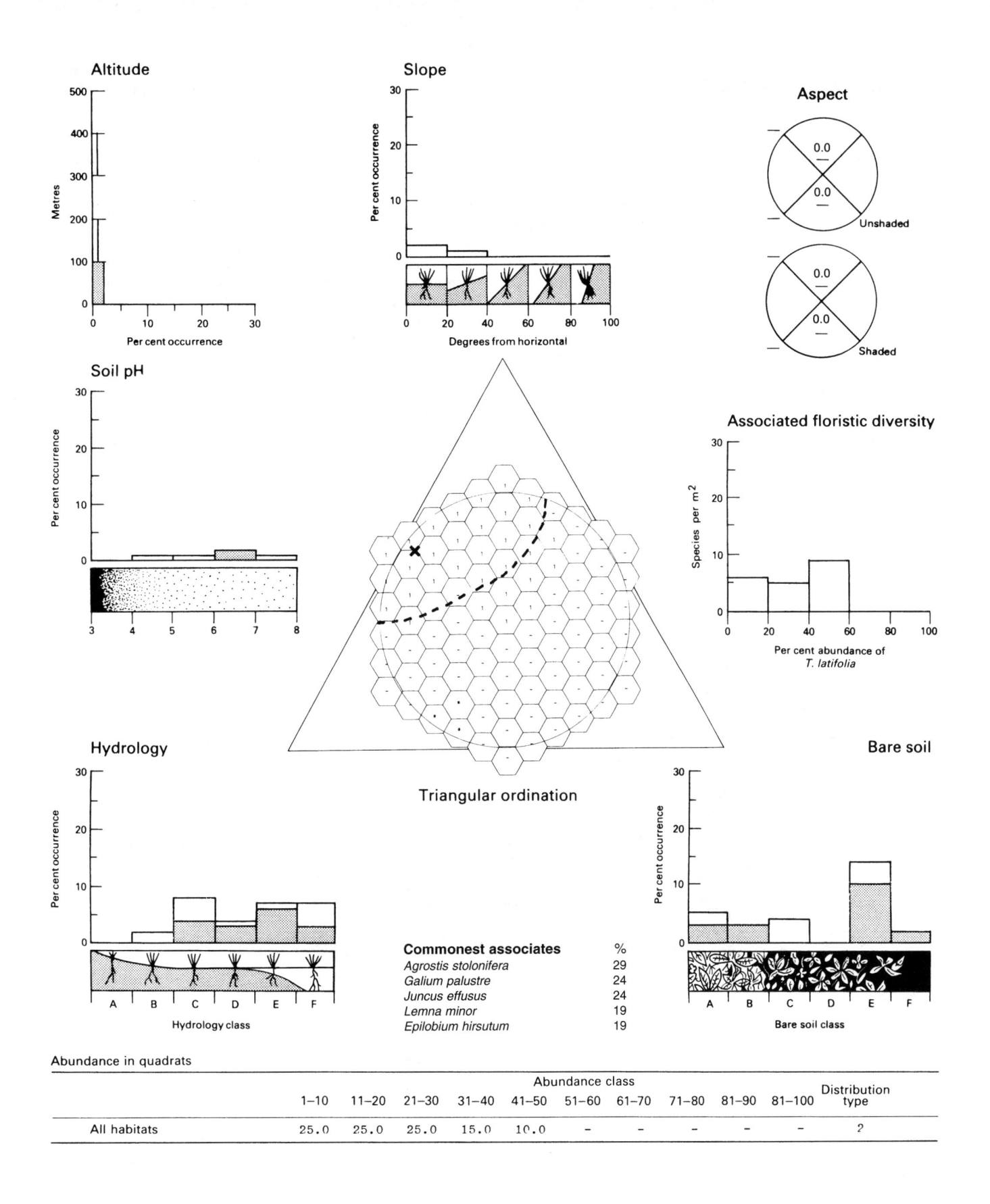

Commonest associates %
Agrostis stolonifera 29
Galium palustre 24
Juncus effusus 24
Lemna minor 19
Epilobium hirsutum 19

Abundance in quadrats

					Abundance class						Distribution type
	1–10	11–20	21–30	31–40	41–50	51–60	61–70	71–80	81–90	81–100	
All habitats	25.0	25.0	25.0	15.0	10.0	–	–	–	–	–	2

(Stace 1975). It has been suggested that in water more than 150 mm deep *T.l.* tends to give way to *T. angustifolia*, which is taller than *T.l.* and exhibits a greater allocation to sexual reproduction (Grace & Wetzel 1981). Other American workers (Fossett & Calhoun 1952) suggest that *T.a.* replaces *T.l.* under saline conditions; both observations are consistent with features of the British distributions of the two species. The species has potential for use as a feed crop (Morton 1975).

Current trends One of the commonest semi-aquatic dominants in Britain, and a very effective colonist even of landlocked sites; *T.l.* is one of the few wetland species which is clearly increasing.

(*)*Ulex europaeus* L.

Fabaceae
Furze, Gorse, Whin

Field data analysed in this account refer to seedlings and small saplings

Life-form Much-branched evergreen shrub. Stems, except for the oldest, green; leaves present during the seedling stage, but thereafter are found only in a reduced form as spines or scales; roots with VA mycorrhizas and N-fixing nodules.
Phenology Evergreen. Flowers mainly January to March in S, April to May in N and sets seed after a further 2 months (T. T. Elkington pers. comm.).
Height Up to 2 m.
Seedling RGR 0.5–0.9 week^{-1}. SLA 9.6. DMC. No data.
Nuclear DNA amount 5.4 pg; $2n = 96$ (*FE2*); $6x$, ($12x$ for family).
Established strategy Stress-tolerant competitor.
Regenerative strategies B_s. Regenerates entirely by seed, which germinates mainly in spring. Forms a persistent seed bank (Moss 1960); SB IV.
Flowers Yellow, hermaphrodite, usually insect-pollinated; 1-3 in axillary clusters.
Germinule 6.2 mg, 3.3 × 2.3 mm, seed explosively discharged from a 2-6-seeded pod.
Germination Epigeal; t_{50} 9 days; hard-coat dormancy; L = D.
Biological flora None, but see bibliography by Gaynor and MacCarter (1981).
Geographical distribution The British Isles (98% of British vice-counties), *ABIF* page 407; W Europe and naturalized in much of C Europe (26% of European territories and 18% naturalized); introduced into many parts of the world.

Habitats Seedlings and saplings are usually found near mature bushes of *U.e.*, and are recorded from wasteland, unenclosed pasture, river banks, cinder tips and quarry spoil. They are also observed on waysides, railway banks and, more rarely, in skeletal habitats. Seldom found in woodland habitats other than at the very edge. However, *U.e.* can form a low scrub, and under these conditions seedlings may be abundant. Absent from wetland and arable habitats. [Particularly characteristic of a number of maritime cliff-top communities in W Britain, and also occurs on old sand dunes (T. T. Elkington pers. comm.).]
Gregariousness Typically sparse, but seedlings capable of reaching high densities after fire.
Altitude Observed to 300 m. [Up to 640 m.]
Slope Seedlings largely absent from skeletal habitats, and therefore also from the steepest slopes.
Aspect Slightly more frequent on S-facing slopes.
Hydrology Absent from wetland habitats.
Soil pH Widely distributed, but most frequent within the pH range 4.0-6.0.
Bare soil Juveniles most frequently associated with areas containing a moderate exposure of bare soil.

Habitats

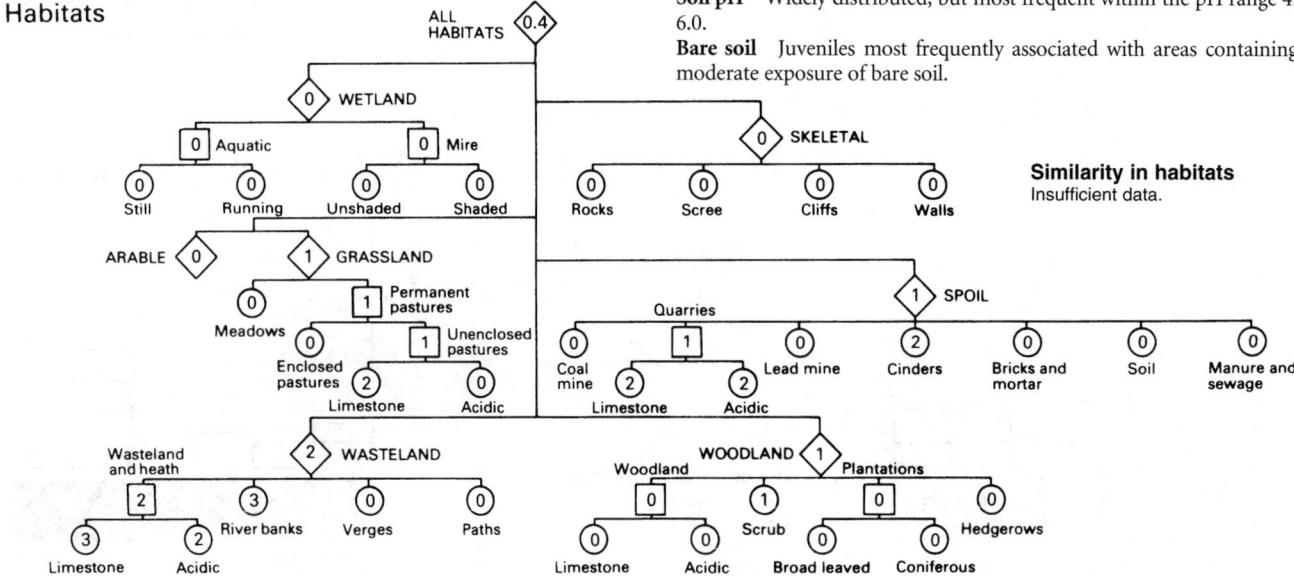

Similarity in habitats Insufficient data.

△ **ordination** Juveniles largely restricted to infrequently disturbed sites of intermediate productivity.
Associated floristic diversity Variable, but most typically intermediate.
Synopsis *U.e.* is a tall colonizing shrub up to 2 m in height, with an Atlantic distribution. Most roots are superficial, but *U.e.* also has a deep tap-root (Zabkiewicz 1976). The species forms nitrogen-fixing root nodules and, unlike those of most herbaceous species, these are markedly perennial (Pate 1961). *U.e.* is most characteristic of roadsides, railway banks, wasteland, derelict pasture and seacliffs on infertile soils, and has been planted in certain areas as hedging. The species is not persistent under the heavy shade of other species. As a seedling the plant forms trifoliate leaves, but these are replaced by woody spines or scales in the established plant. Both short days and lowlight intensities (but not high humidity) reduce the growth of the shoot and prolong the duration of the leafy juvenile phase (Bieniek & Millington 1968). The species frequently forms impenetrable thickets with persistent spiny litter. Studies in New Zealand (Egunjobi 1971) suggest that the amount of litter shed each year corresponds to about half the annual growth increment, and *U.e.* may frequently represent a fire hazard (Beckett & Beckett 1979). The seed is relatively large, and although seeds are explosively discharged and may be dispersed by ants (Weiss 1909), the seedlings are mostly restricted to the vicinity of mature bushes. For this reason the field data presented for juveniles in this account are also relevant to the distribution and ecology of the adult bush. In young bushes the erect stems provide a dense canopy and often have no vegetation beneath them. After 15-20 years bushes become leggy (Tubbs 1974) and the branches more spreading, leaving a central, little-shaded area. *Betula* spp. frequently colonize and eventually replace degenerating gorse bushes. However, gorse may be maintained by controlled burning (Tubbs 1974). The presence of *U.e.* often indicates a site of former disturbance, particularly by fire. The seed has a hard coat, which may afford it some protection from high temperatures and, seeds have been rendered germinable after heating to 88°C (Moss 1959). Zabkiewicz and Gaskin

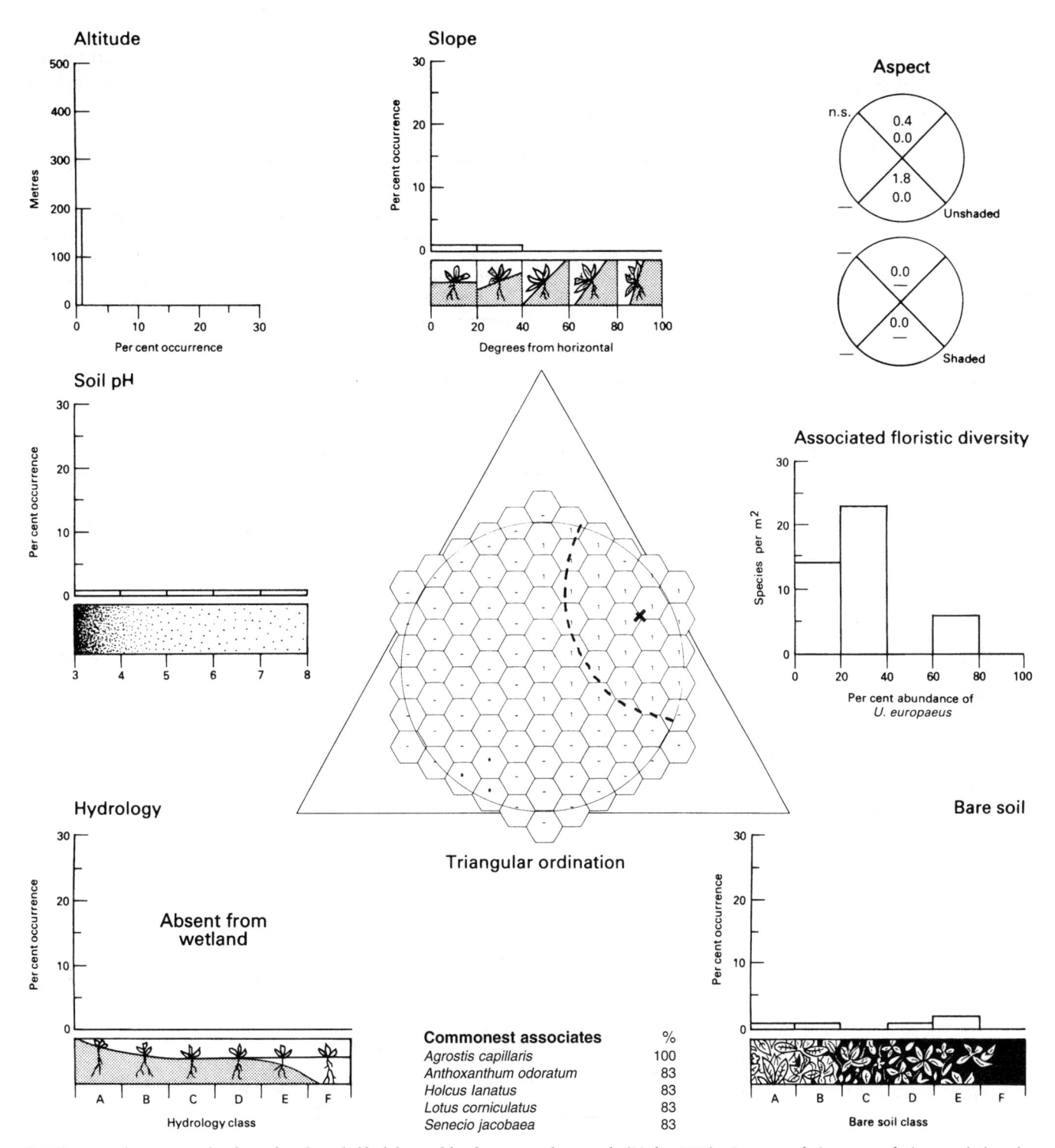

(1978) report that as a result of one fire about half of the seed bank was killed and the remainder subsequently germinated. Seed may persist for up to 28 years in the soil (Moss 1960), and areas may become carpeted with *U.e.* seedlings, and finally a gorse thicket, following a fire. Germination of seeds and survival of seedlings is lower under a gorse canopy than in the open (Ivens 1978). The species does occur on shallow calcareous soil, but is much more characteristic of mildly acidic soils within the pH range 4.0–6.0. The soil around the roots is frequently more acidic than the surrounding soil (Grubb & Suter 1971). *U.e.* is relatively sensitive to lime-chlorosis when grown on calcareous soil (Grime & Hodgson 1969). Formerly, after removal of spines, *U.e.* was widely used for fodder and for other purposes (see Lucas 1960), and is still an important source of food for cattle and ponies on marginal land (Tubbs 1974). Young sprouting shoots may be severely predated by rabbits (Coombe & Frost 1956). The biological control of *U.e.* in New Zealand, using the seed-eating weevil *Apion ulicis* Forster, was unsuccessful even though 90% of seed was destroyed (Little 1960). Intra- and inter-population variation in morphology is largely plastic, but shows a genetic component (Millener 1961, 1962). On the infertile, highly acidic soils of heathland and moorland *U.e.* is replaced in the N and W by *U. gallii* Planch., and in the SE mainly by *U. minor* Roth. Hybrids between *U.e.* and *U.g.* occur in the SW, where the flowering period of the mainly spring-blooming *U.e.* and the autumnal *U.g.* overlap (Stace 1975, T. T. Elkington pers. comm.). A further tetraploid ssp. is recorded from Europe (*FE2*).

Current trends Probably still increasing as a result of the dereliction of poor pasture and by colonizing motorway verges and the margins of woodland rides in plantations on acidic soils. An increase in grassland invasion has also been reported following myxamatosis (Thomas 1960, 1963). Has some potential in land reclamation as a result of its persistent growth under conditions of extreme infertility (Jefferies *et al.* 1981) and its capacity to fix atmospheric nitrogen (Skeffington & Bradshaw 1980).

(*)*Ulmus glabra* Hudson

Ulmaceae
Wych Elm

Field data analysed in this account refer to seedlings and small saplings

Life-form Deciduous tree with widely spreading branches. Leaves ovate, hairy, sometimes >5000 mm^2; roots with VA mycorrhizas.
Phenology Flowers February to April, before the leaves. Seed set May to July. Leaves shed in autumn.
Height Foliage and flowers to 40 m.
Seedling RGR No data. SLA 13.2. DMC. No data.
Nuclear DNA amount 2.1 pg; $2n = 28$ (*FEI*); $2x$. (High generic base numbers common in Ulmaceae.)
Established strategy Intermediate between competitor and stress-tolerant competitor.
Regenerative strategies W, S. By seeds, which germinate soon after shedding and are not incorporated into a persistent seed bank. Shows little capacity for suckering; SB I.
Flowers Reddish, hermaphrodite, protandrous, wind-pollinated, but also visited by insects, normally self-incompatible (Richens 1983); in up to c. 25-flowered clusters on 1-year-old twigs.
Germinule 3.5 mg, 5.0 × 3.5 mm, seed dispersed within a flattened winged fruit (9.93 mg, 20.6 × 12.0 mm).
Germination Epigeal; t_{50} 7 days, seed soon losing viability (Richens 1983); germinates more rapidly at low light intensities; L > D.
Biological flora None, but see Richens (1983).
Geographical distribution The British Isles, but more common in the N and W, *ABIF* page 124; most of Europe except the extreme N and S (72% of European territories); W Asia and N Africa (possibly introduced).

Habitats Seedlings and saplings are found in unshaded rocky and spoil habitats, particularly limestone quarry heaps, lead-mine spoil and cliffs. Equally frequent in shaded habitats, especially limestone woodland and scrub. Widely distributed in other habitats, but conspicuously absent from grazed and ungrazed grassland, and from aquatic habitats.
Gregariousness Seedlings sometimes occurring as locally dense populations.
Altitude Distribution predominantly lowland, and seedlings not recorded above 300 m. [Up to 535 m.]
Slope Widely distributed.
Aspect Juveniles more frequent on S-facing slopes, significantly so in unshaded habitats.
Hydrology Not a wetland species, but seedlings recorded from topogenous and soligenous mire.
Soil pH Distributed over a wide range of values, but infrequent below pH 6.0.
Bare soil Seedlings mainly restricted to sites with a discontinuous cover of herbaceous vegetation. Most sites have moderate exposures of both bare soil and tree litter.
△ ordination Juveniles have a very wide-ranging distribution, in part a reflection of the very large numbers of seeds released.
Associated floristic diversity Typically intermediate.

Similarity in habitats	%
Fraxinus excelsior (Juv.)	68
Betula spp. (Juv.)	66
Melica uniflora	64
Bromopsis ramosa	64
Hedera helix	62

Habitats

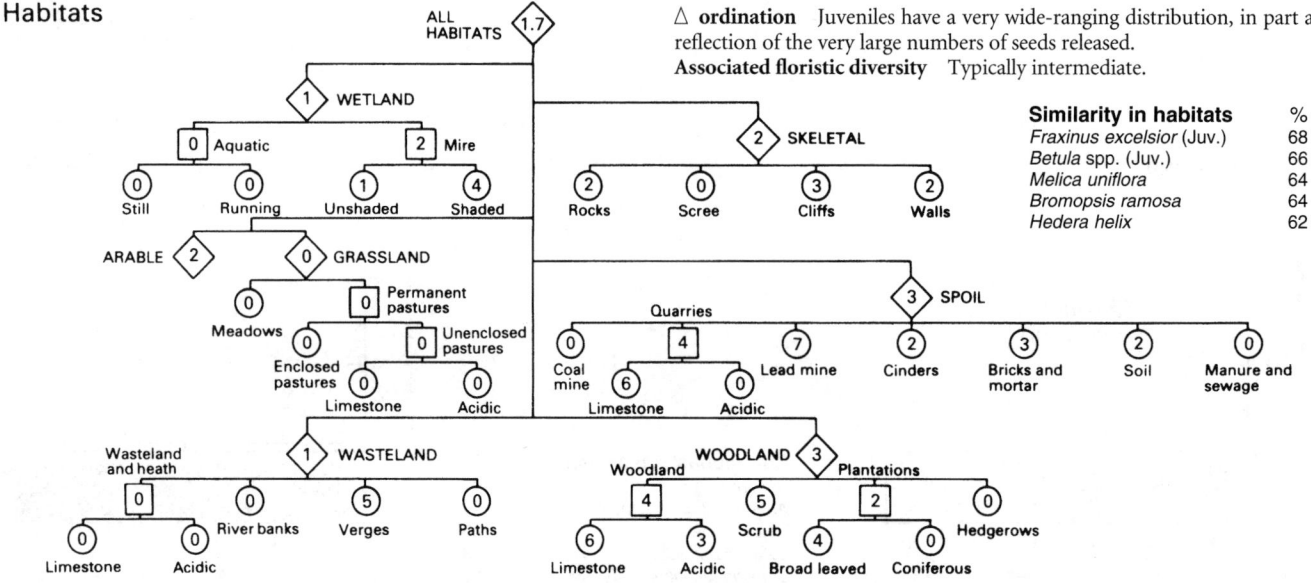

ALL HABITATS 1.7
- WETLAND 1
  - Aquatic 0
    - Still 0
    - Running 0
  - Mire 2
    - Unshaded 1
    - Shaded 4
- ARABLE 2
- GRASSLAND 0
  - Permanent pastures 0
    - Meadows 0
    - Enclosed pastures 0
      - Limestone 0
      - Acidic 0
  - Unenclosed pastures 0
- SKELETAL 2
  - Rocks 2
  - Scree 0
  - Cliffs 3
  - Walls 2
- SPOIL 3
  - Quarries 3
    - Coal mine 0
    - Lead mine 7
      - Limestone 6
      - Acidic 0
  - Cinders 2
  - Bricks and mortar 3
  - Soil 2
  - Manure and sewage 0
- WASTELAND 1
  - Wasteland and heath 0
    - Limestone 0
    - Acidic 0
  - River banks 0
  - Verges 5
  - Paths 0
- WOODLAND 3
  - Woodland 4
    - Limestone 6
    - Acidic 3
  - Scrub 5
  - Plantations 2
    - Broad leaved 4
    - Coniferous 0
  - Hedgerows 0

Synopsis In recent years *U.g.* has been eliminated from large areas of Britain by the devastating effects of Dutch Elm disease. Bud break is rather earlier than in other ring-porous trees. On theoretical grounds this would render vessels liable to cavitation during cold springs (see Lechowicz 1984). The effect of this unusual feature on the distribution of *U.g.* is uncertain, but it is interesting to note that, unlike most other common trees, *U.g.* did not form monocultures within the Sheffield region. In addition the species was vulnerable to drought in East Anglia (Rackham 1980), and was most frequent in the N and W of Britain. All species of elm contributed only c. 11 000 ha to broad-leaved woodland (<2% of total) (Steele & Peterken 1982). Trees over 200 years old were reported in the Derbyshire Dales (Merton 1970). *U.g.* is tolerant of coppicing. The species regularly produced an abundance of winged fruits which are widely wind-dispersed and which are capable of germinating without delay. As a result seedlings occur in a wider range of habitats than the adult tree. In N England *U.g.* occurred as scattered trees on moist, relatively fertile soils, and was particularly common in mixed woodland on limestone. Less frequently the species is recorded from scrub and even occasionally as free-standing trees and in hedges. Saplings are mainly found under the shade of other trees, but sites with extensive regeneration were not observed within the region. The species was a frequent associate of *Fraxinus excelsior* in secondary woodland of the Derbyshire Dales and seedlings regenerated within ashwoods since they are much more tolerant of shade than *F.e.* (Merton 1970). As a result, the proportion of elm in ashwoods tended to increase and ash-elm woods develop (Merton 1970). *U.g.* was also a consistent associate of *Tilia* spp. in ancient woodlands (Pigott 1969). Observations indicate that predation of seeds and mortality of seedlings is extremely high. Furthermore, a high proportion of the seeds shed, particularly those released early, may not be viable (Beckett & Beckett 1979). No persistent seed bank is formed. *U.g.* does not normally sucker, and thus is more vulnerable to Dutch elm disease than *U. procera* Salisb., which may survive by means of root suckers (Rackham 1980). *U.g.* produces a dense canopy,

Altitude

Slope

Aspect

$P < 0.05$

0.8	0.0
3.1	0.0

Unshaded

n.s.

0.8	0.0
1.6	0.0

Shaded

Soil pH

Associated floristic diversity

Hydrology

Triangular ordination

Commonest associates %
Poa trivialis — 46
Fraxinus excelsior — 30
Festuca rubra — 28
Geranium robertianum — 24
Rubus fruticosus agg. — 24

Bare soil

Abundance in quadrats

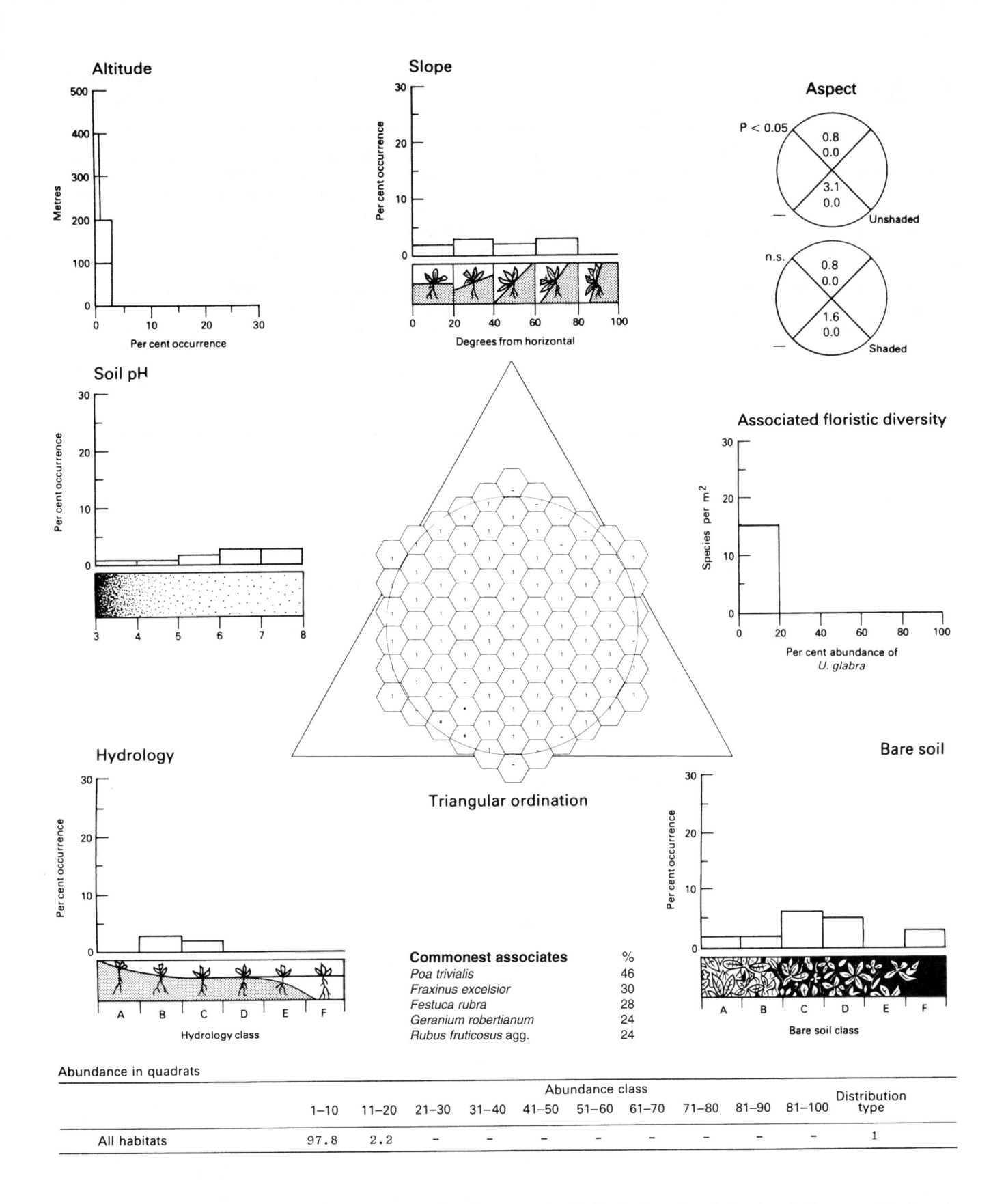

	1–10	11–20	21–30	31–40	41–50	51–60	61–70	71–80	81–90	81–100	Distribution type
All habitats	97.8	2.2	–	–	–	–	–	–	–	–	1

and litter is short-lived. Ssp. *glabra*, with broad leaves, occurs mainly in the S, and ssp. *montana* (Stokes) Lindquist, in the N. *U.g.* hybridizes with the two other native British species (*Excur Fl*, Stace 1975). However, there is little overlap between *U.g.* and the other native or naturalized species, *U.p.* Richens (1983) does not subdivide *U.g.* and only recognizes one other species, *U. minor* Miller, and hybrids. The latter is restricted to lowland coppices, ornamental woodland and to hedgerows. The number of insects associated with elms is moderately low for a deciduous tree (Southwood 1961), implying that the leaves are relatively palatable (see Wratten *et al.* 1981). Elm leaves were formerly fed to animals, but in the Middle Ages its forestry began to be superseded by that of *Acer pseudoplatanus*, which has wood which is easier to work (Richens 1983).

Current trends Decimated by the widespread and severe effects of Dutch elm disease.

Urtica dioica L.

Urticaceae
Stinging Nettle

Life-form Rhizomatous, polycarpic perennial chamaephyte (or, in severe winters, protohemicryptophyte). Stem erect; leaves ovate, <4000 mm², usually with stinging hairs; roots usually lacking mycorrhizas.

Phenology Young shoots, produced in autumn, may overwinter, though some are killed. Flowers June to July and seed shed from August onwards. Leaves relatively short-lived.

Height Foliage and flowers to 1.5 m.

Seedling RGR 2.0-2.4 week^{-1}. SLA 36.7. DMC 21.3.

Nuclear DNA amount 3.1 pg; 2n = 48, 52* (*BC, FE1*); 4x*.

Established strategy Competitor.

Regenerative strategies V, B$_s$. See 'Synopsis'; SB IV.

Flowers Green, dioecious, sometimes with a few flowers of the opposite sex or hermaphrodite, wind-pollinated (*Biol Fl(a)*); in many-flowered axillary spike-like panicles with several thousand flowers on each fertile stem.

Germinule 0.19 mg, 1.3 × 0.9 mm, achene shed enclosed within the persistent perianth.

Germination Epigeal; t_{50} 6 days; 22-35°C; L > D.

Biological flora (a) Greig-Smith (1948a), (b) Bassett *et al.* (1977).

Geographical distribution The British Isles (100% of British vice-counties), *ABIF* page 127; widespread and probably native throughout Europe and Asia from the Mediterranean to the Arctic regions (92% of European territories), and adventive in other temperate regions; India, China, Australia, New Zealand, N and S Africa and N and W America. Absent from the Tropics.

Habitats Recorded from all terrestrial habitats identified in the diagram below, but particularly abundant on the banks of rivers, streams and ditches, unmanaged road verges and hedgerows, shaded mire and recently-disturbed habitats such as manure heaps, sewage waste and heaps of mineral soil at construction sites. On fertile soils the species is persistent under moderate shade. On arable land and in other highly-disturbed habitats *U.d.* usually occurs as isolated seedlings. [Recorded from guano-enriched sea cliffs and sand dunes, limestone pavement, shingle beaches and montane areas (*Biol Fl(a)*).]

Gregariousness Potentially stand-forming.

Altitude Declining with increasing altitude, a reflection of the decrease in the area of fertile habitats in upland areas but reaching 430 m locally. [Up to 840 m.]

Slope Wide-ranging, but most abundant below 40%.

Aspect No consistent trends.

Hydrology Mainly on moist rather than waterlogged soils, but may occur in soligenous mire in shaded habitats.

Soil pH Frequency and abundance high over the pH range 5.0-8.0.

Bare soil A wide range of occurrences consistent with the colonizing ability of the species and its capacity to establish dense clonal patches with shortly persistent stem litter.

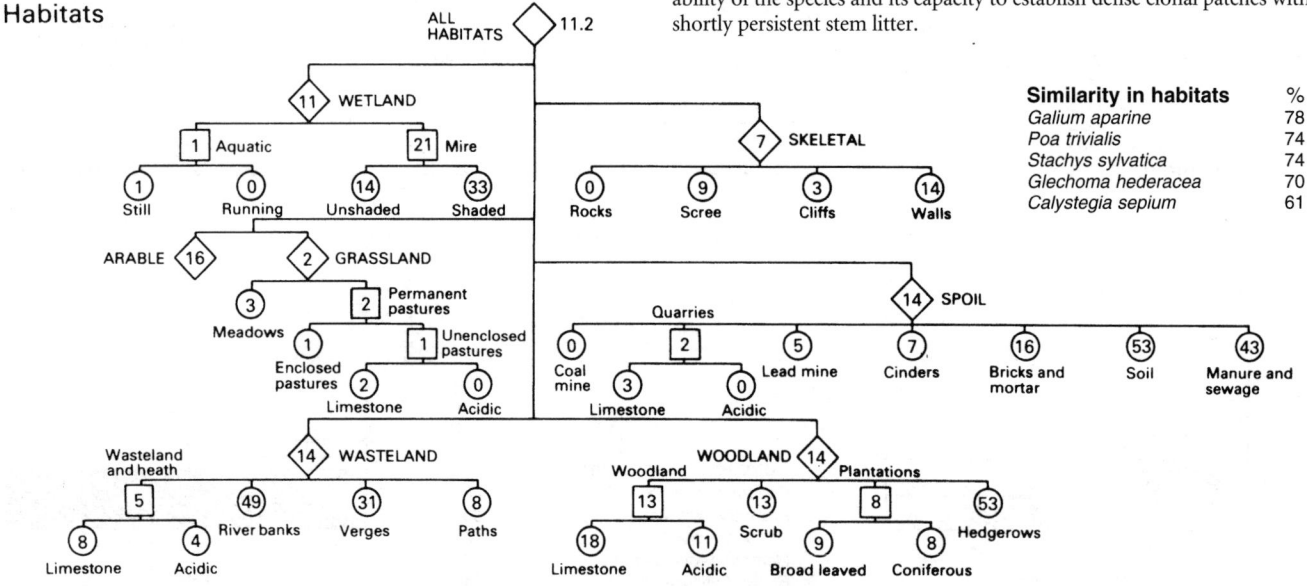

Habitats

Similarity in habitats	%
Galium aparine	78
Poa trivialis	74
Stachys sylvatica	74
Glechoma hederacea	70
Calystegia sepium	61

△ **ordination** Distribution centred on vegetation associated with fertile, relatively undisturbed conditions, but strong evidence of colonizing ability in productive, heavily-disturbed vegetation. Excluded from very unproductive vegetation.

Associated floristic diversity Relatively low, and declining progressively with increasing abundance, and dominance, of *U.d.*

Synopsis *U.d.* is a dioecious, tall rhizomatous herb with stinging hairs. In summer the dense, rapidly-ascending canopy of *U.d.* often precludes the growth of other herbaceous plants, and this, coupled with the impact of relatively persistent stem litter, often causes the species to form monospecific stands. However, because the shoots die back completely in winter, species such as the scrambling annual *Galium aparine*, the low-growing *Poa trivialis* and stem litter-exploiting bryophytes, e.g. *Brachythecium rutabulum* Hedw., are able to benefit from the seasonal availability (between autumn and spring) of relatively high light intensity. *U.d.* is probably only native in fen carr (*Biol Fl(a)*); elsewhere it is a 'follower of humans' (*Biol Fl(a)*), restricted to moist, fertile habitats by features such as its high potential growth rate and its high requirement for mineral nutrients (Pigott & Taylor 1964). Leaves contain unusually high concentrations of N, Ca, Mg and Fe (*SIEF*). *U.d.* is a long-day plant and may need up to 16h daylength for flowering (J. G. Hodgson and R. E. Booth, unpubl.). Flowering is strongly inhibited by drought (Boot *et al.* 1986), but the species shows a modest capacity for drought-hardening and desiccation resistance. A marked increase in specific leaf area in response to shading is observed in plants established under moderately dense woodland canopies (Wheeler 1981), although *U.d.* is intolerant of dense shade and exhibits poor flowering and a truncated shoot phenology in woodland habitats (Al-Mufti *et al.* 1977). *U.d.* is suppressed by repeated cutting, but occurs in pastures, where, perhaps because of the stinging hairs, it is rarely eaten by cattle (*Biol Fl(b)*). However, old leaves are palatable to invertebrates (Grime *et al.* 1970), and the young shoots have been used as a green vegetable. *U.d.* forms large patches by means of

Altitude
Slope
Aspect

n.s. 9.8	16.0
7.6	23.5 Unshaded

n.s. 11.4	33.3
14.6	16.7 Shaded

Soil pH

Associated floristic diversity

Hydrology

Triangular ordination

Bare soil

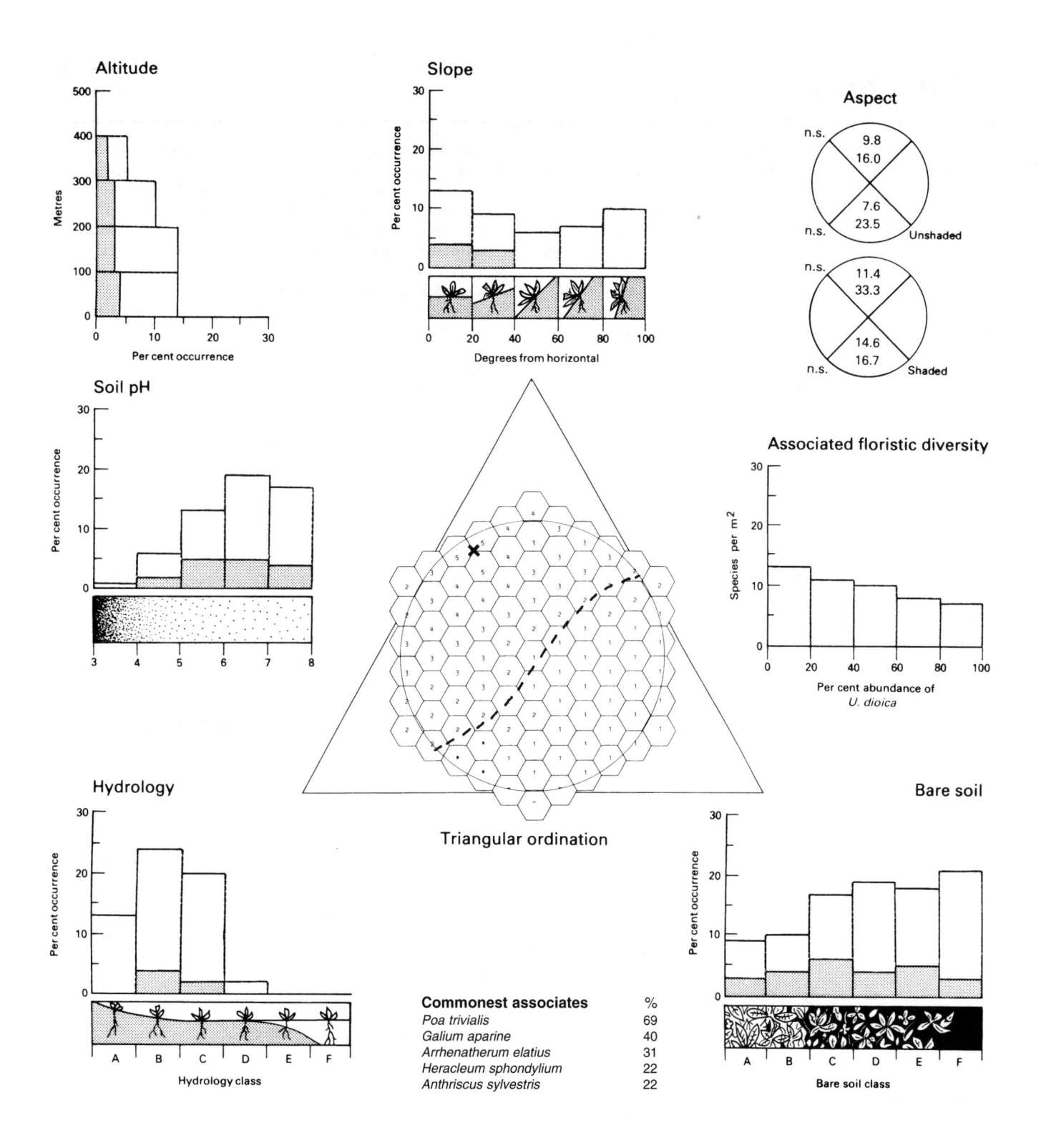

Commonest associates %
Poa trivialis 69
Galium aparine 40
Arrhenatherum elatius 31
Heracleum sphondylium 22
Anthriscus sylvestris 22

rhizome growth, which may increase by 450 mm year^{-1} (*Biol Fl(a)*). Rhizomes broken by digging, or other disturbances, readily re-root to form new colonies. Despite being dioecious and forming large clonal patches, *U.d.* also produces great quantities of seed and accumulates large and persistent banks of buried seeds which have the potential for long-term survival (Odum 1978). Germination is stimulated by light and fluctuating temperatures, and occurs mainly on open and disturbed ground. Thus, many of the sites with a continuous vegetation cover in which *U.d.* persists as established clones are unsuitable for regeneration by seed. Seeds survive ingestion by a variety of animals and may be carried far (*Biol Fl(a)*). Seeds and rhizome pieces are also transported in soil and, for this reason, despite the lack of a well-defined dispersal mechanism, the species is highly mobile. *U.d.* shows considerable genetically-based variation, and includes populations, var. *subinermis* Uechtritz, virtually devoid of stinging hairs (Pollard & Briggs 1982). Ssp. *gracilis* (Ait.) Selander, which is found in N Europe and N America, is monoecious (*Biol Fl(b)*).

Current trends A very effective colonist of newly-disturbed sites, which, once established, is very persistent and invasive. Appears to be increasing in abundance.

+*Urtica urens* L.

Urticaceae
Small Nettle

Life-form Summer-annual therophyte. Stems erect, often branched; leaves ovate, <2000 mm², with stinging hairs; roots without VA mycorrhizas.

Phenology Seeds germinate in spring and plants flower from June onwards until killed by autumn frosts. Seed shed from July onwards. Plants from later-germinating seed occasionally overwinter in sheltered sites.

Height Stems bearing leaves and flowers to 600 mm.

Seedling RGR No data. SLA 21.0. DMC 21.4.

Nuclear DNA amount 0.6 pg; $2n = 24^*, 26, 52$ (*BC, FE1*), $2x^*, 4x$.

Established strategy Intermediate between ruderal and competitive-ruderal.

Regenerative strategies B_s. Regenerates entirely by seeds, which germinate in autumn or more usually in spring. Seeds incorporated into a persistent seed bank; SB IV.

Flowers Green, monoecious, wind-pollinated; in axillary racemes of numerous female and few male flowers (*FE1*).

Germinule 0.50 mg, 2.1×1.5 mm, achene shed enclosed within the persistent perianth.

Germination Epigeal; t_{50} 6 days; maximum germination associated with low irradiance; L < D.

Biological flora Greig-Smith (1948b).

Geographical distribution Much of Britain, particularly in the E (99% of British vice-counties), *ABIF* page 127; Europe except in the extreme N (97% of European territories); temperate Asia, N Africa, Australia, New Zealand and N and S America; probably adventive outside Europe and temperate Asia.

Habitats Occurs mainly on arable land, particularly amongst broad-leaved arable crops such as sugar beet and potatoes. Observed at lower frequency in farm yards, gardens, demolition sites and on waste ground. Not recorded from wetland, woodland, skeletal habitats or pasture. [Found on sand near the coast and on coastal headlands (*Biol Fl*).]

Gregariousness Plants usually widely spaced. Large specimens may occupy 0.25 m² and high-frequency values refer to sites with dense populations of seedlings or stunted plants.

Altitude Most frequent on low-lying arable land, but has been observed locally to 310 m. [Up to 500 m.] Exact altitudinal limit uncertain, since has been observed as a non-persistent introduction.

Slope Most of the suitable habitats are restricted to flat ground and gentle slopes.

Aspect Insufficient data.

Hydrology Absent from wetland.

Soil pH Occurring mainly on soils of around pH 7.0.

Bare soil Maximum occurrence and greatest abundance in habitats with a high proportion of bare soil.

△ **ordination** Distribution restricted to vegetation associated with disturbed productive conditions.

Associated floristic diversity Intermediate.

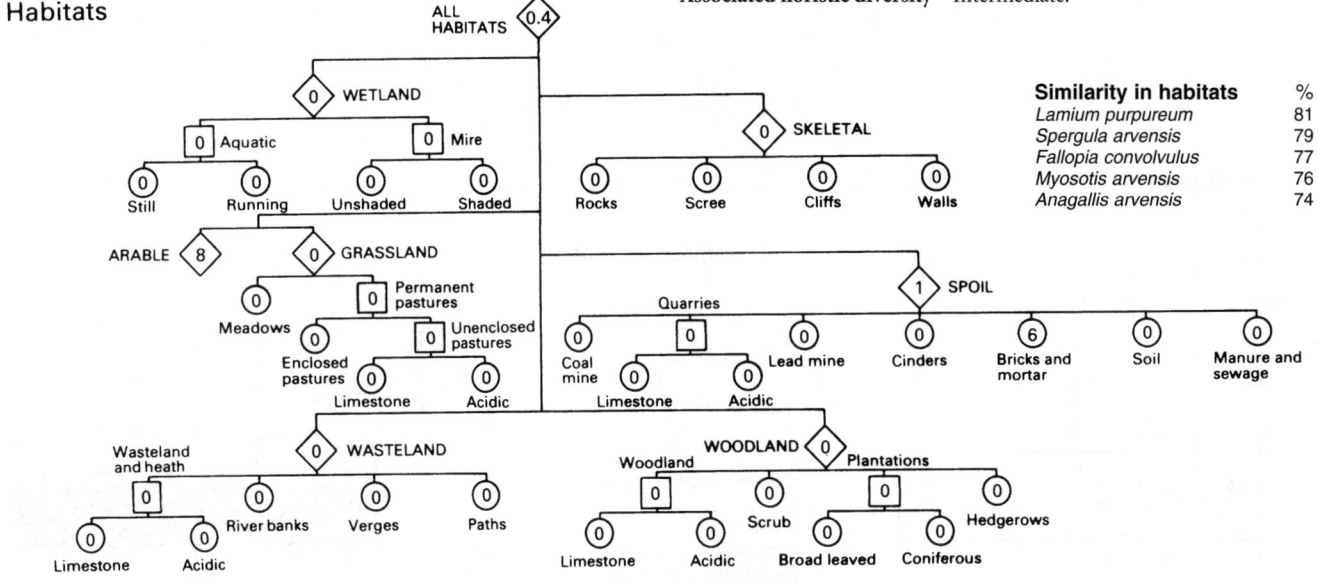

Habitats

Similarity in habitats	%
Lamium purpureum	81
Spergula arvensis	79
Fallopia convolvulus	77
Myosotis arvensis	76
Anagallis arvensis	74

Synopsis *U.u.* is an often tall, potentially robust, summer-annual herb of fertile, disturbed habitats, particularly on sandy soils. The species, regarded by Webb (1985) as doubtfully native, is a 'follower of humans' and is most common in arable areas in which broad-leaved crops are grown. In morphology and habitat range *U.u.* resembles *Chenopodium album*. *U.u.* also occurs in some semi-natural sandy habitats near the coast and is often more frequent in coastal than in inland areas (*Biol Fl*). *U.u.* is usually absent from habitats that are grazed, cut or heavily shaded. The importance of stinging hairs is obscure, since *U.u.* does not normally occur in areas subject to heavy grazing pressure. Under favourable conditions *U.u.* has the potential to become a large plant producing abundant seed (Salisbury 1964). In common with many arable weeds, the plant may flower when very small in less-suitable sites. This is consistent with experimental findings (Boot *et al.* 1986) showing that reproductive effort is maintained in plants severely stunted by drought. The species regenerates entirely by seed, and effective seedling establishment is probably initiated by spring germination. To an unusual extent, freshly-shed seeds are strongly inhibited by high irradiance (Grime & Jarvis 1976). On sandy soils this inhibition could provide a mechanism preventing germination in sites likely to be subject to desiccation during seedling establishment. Seeds persist in the soil, especially when deeply buried in uncultivated soil (Roberts & Feast 1972). Seeds may survive ingestion by animals (*Biol Fl*) and are probably transported in a manner similar to that of *U. dioica*. Despite the absence of well-defined dispersal mechanisms, both *U.u.* and *U.d.* appear to be highly mobile species, frequently occurring as casuals outside their main geographical ranges.

Current trends *U.u.* was not considered an important arable weed by Brenchley (1920), but has probably increased since that time. Now declining on arable land, but its general range may be expanding in response to the creation of an increasing diversity of fertile, disturbed artificial habitats.

Altitude

Slope

Aspect

Unshaded

Shaded

Soil pH

Associated floristic diversity

Species per m^2

Per cent abundance of *U. urens*

Hydrology

Absent from wetland

Hydrology class

Triangular ordination

Commonest associates	%
Poa annua	73
Stellaria media	61
Matricaria discoidea	58
Capsella bursa-pastoris	50
Senecio vulgaris	47

Bare soil

Bare soil class

Abundance in quadrats

	1–10	11–20	21–30	31–40	41–50	51–60	61–70	71–80	81–90	81–100	Distribution type
Wetland	–	–	–	–	–	–	–	–	–	–	–
Skeletal	–	–	–	–	–	–	–	–	–	–	–
Arable	6	1	1	1	–	–	–	–	–	–	–
Grassland	–	–	–	–	–	–	–	–	–	–	–
Spoil	1	1	–	–	–	–	–	–	–	–	–
Wasteland	–	–	–	–	–	–	–	–	–	–	–
Woodland	–	–	–	–	–	–	–	–	–	–	–
All habitats	63.6	18.2	9.1	9.1	–	–	–	–	–	–	2

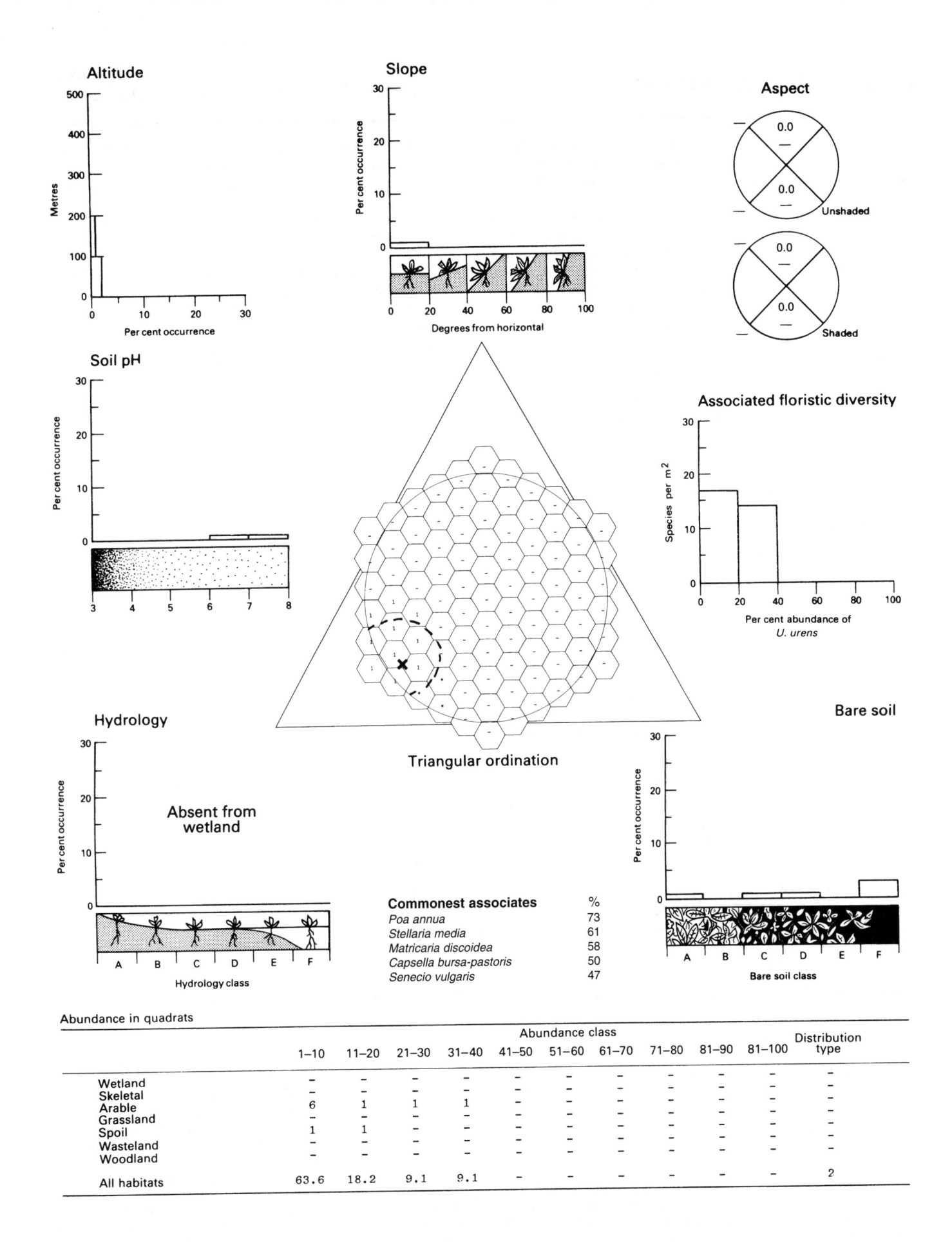

Vaccinium myrtillus L.

Ericaceae
Bilberry, Blueberry, Whortleberry, Huckleberry

Life-form Dwarf, shrubby chamaephyte with rhizomes 150-200 mm below the soil surface. Shoots green, woody, branched; leaves leathery, ovate, often <250 mm^2; roots with ericoid mycorrhizas.

Phenology New shoot growth commences in spring. Flowers, formed the previous winter, are produced at two periods, spring and early summer. Berries formed within 1 month of flowering (*Biol Fl*). Leaves shed in autumn although some leaves on low sheltered shoots may overwinter.

Height Foliage and flowers usually <600 mm.

Seedling RGR 0.5-0.9 week^{-1}. SLA 17.3. DMC. No data.

Nuclear DNA amount No data; 2n = 24* (*BC*); 2x*.

Established strategy Stress-tolerant competitor.

Regenerative strategies V, ?B$_s$. Forms extensive patches as a result of rhizome growth. More information needed to assess the role of seed regeneration. Seedlings are infrequent and establishment from seed is probably restricted to bare areas (*Biol Fl*). Buried seeds persist in some circumstances (Grandstrom 1982); ?SB IV.

Flowers Pink, hermaphrodite, weakly protandrous, insect- or occasionally self-pollinated; borne singly or in pairs on 1-year-old twigs (*Biol Fl*).

Germinule 0.26 mg, 1.3 × 0.8 mm, seed dispersed by birds from a *c.* 18-seeded purple-black berry.

Germination Epigeal; t$_{50}$ 11 days; 12-27°C; L = D.

Biological flora Ritchie (1956).

Geographical distribution Upland Britain, but absent from much of E England and infrequent in C Ireland (93% of British vice-counties), *ABIF* page 294; most of Europe, although only on mountains in the S (77% of European territories); N Asia.

Habitats Mainly on acidic soils in pastures, on heaths and on moorland, and as a local dominant in the field layer of open woodlands. Also occurs on quarry spoil, rock outcrops, cliffs and plantations, and at low frequencies in well-drained peat bogs. Only seedlings have been recorded from arable land. Absent from aquatic habitats. [Also found on scree, boulder slopes and a range of other montane habitats (*Biol Fl*).]

Gregariousness Stand-forming.

Altitude Largely restricted to upland areas extending to 550 m and most abundant above 400 m. [Up to 1300 m (*Biol Fl*).]

Slope Associated with a wide range of slopes, but infrequent on very steeply sloping terrain.

Aspect In unshaded situations recorded significantly more frequently on N-facing slopes, but in grassland significantly more abundant on S-facing slopes (*Biol Atl*). Within shaded habitats *V.m.* is slightly more frequent on S-facing slopes, but slightly more abundant on N-facing slopes. The significance of this distribution pattern is uncertain.

Hydrology Not a wetland plant, but occasionally recorded from sloping, better-drained sites at the edge of ombrogenous mire.

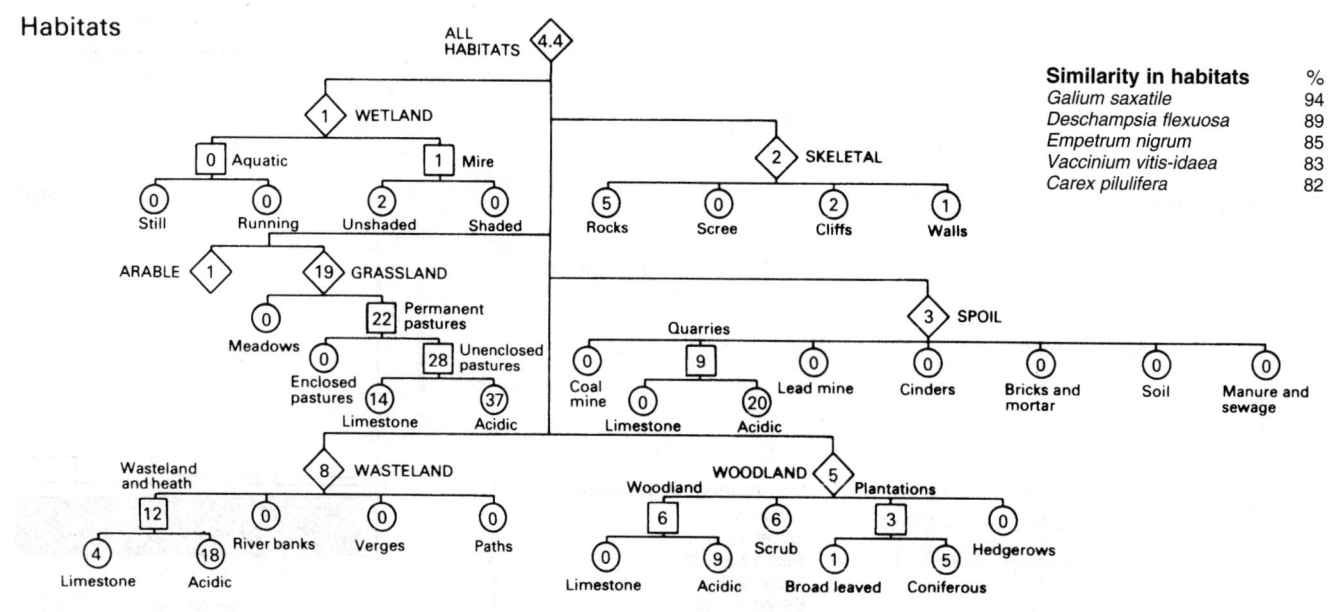

Habitats

Similarity in habitats	%
Galium saxatile	94
Deschampsia flexuosa	89
Empetrum nigrum	85
Vaccinium vitis-idaea	83
Carex pilulifera	82

Soil pH Strongly associated with highly acidic soils of pH <4.5, and not recorded at pH >6.0.

Bare soil Mainly found in habitats with little exposed soil.

Δ ordination Distribution confined to vegetation associated with undisturbed habitats. Centred on relatively unproductive vegetation.

Associated floristic diversity Extremely species-poor.

Synopsis *V.m.* is a strongly rhizomatous, long-lived, slow-growing, dwarf, deciduous shrub which, with the exception of very local sites in N-facing limestone grasslands, behaves as a strict calcifuge and often occurs on peaty soils. The leaves usually contain high concentrations of Mn and Al (*SIEF*). *V.m.* is more tolerant of shade than *Calluna vulgaris*, and in woodlands and scrub often forms a continuous layer consisting of large clonal patches. In these, as in other habitats, the angled shoots, although deciduous, provide a considerable photosynthetic area during winter and early spring. *V.m.* exhibits a N bias in its British distribution (*ABF*) which appears to be largely determined by the availability of siliceous substrata. The species is also especially common in woods of N aspect in the survey area and in S England appears to be largely restricted to humid or shaded sites. *V.m.* is relatively frost-sensitive, and in more-extreme climates may have a distribution which is dependent upon snow protection (*Biol Fl*). *V.m.* is tolerant of sheep grazing, and in pastures a low canopy may be established by the development of phenotypes with a very high density of short erect shoots. Release from grazing usually leads to the formation of tall phenotypes with many erect robust shoots forming a dense, elevated leaf canopy in summer (Pigott 1983). The rhizome system is 150-200 mm below the soil surface, and consequently *V.m.* is much less vulnerable to burning than *C.v.* The depth of the rhizome system may also contribute to the poor tolerance of waterlogged conditions; in ombrogenous mire *V.m.* is restricted to better-drained areas. *V.m.* regenerates mainly by means of rhizome growth to form over many years clonal patches up to 15 m in diameter (*Biol Fl*). Investigations in the Sheffield region and elsewhere in Britain (Hill & Stevens 1981) have failed to detect persistent seeds, but

Altitude

Slope

Aspect
P < 0.01

11.8	63.3
4.0	88.9

n.s. — Unshaded

5.3	57.1
7.3	33.3

n.s. — Shaded

Soil pH

Associated floristic diversity

Hydrology

Triangular ordination

Bare soil

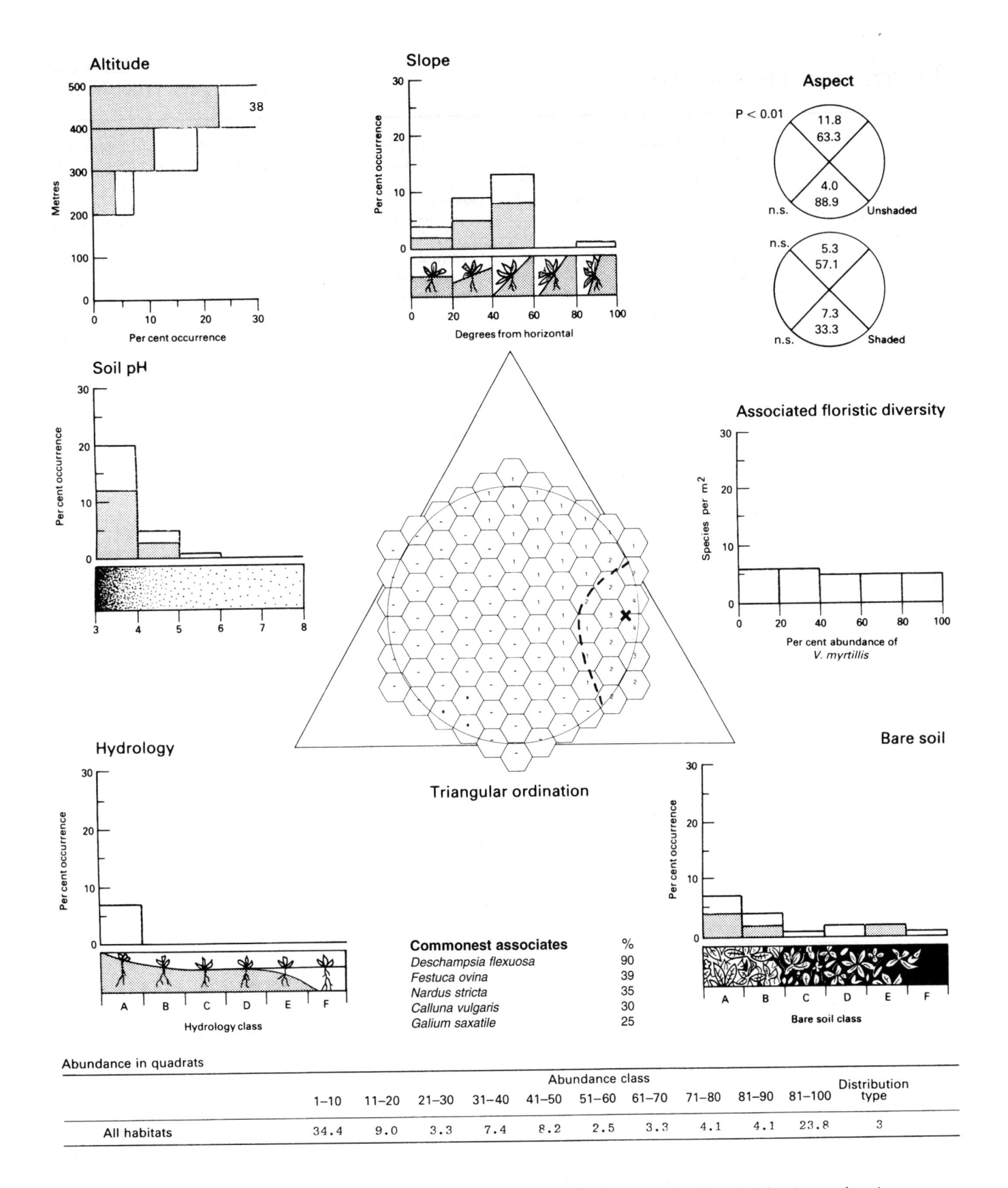

Commonest associates %
- Deschampsia flexuosa — 90
- Festuca ovina — 39
- Nardus stricta — 35
- Calluna vulgaris — 30
- Galium saxatile — 25

Abundance in quadrats

	1–10	11–20	21–30	31–40	41–50	51–60	61–70	71–80	81–90	81–100	Distribution type
All habitats	34.4	9.0	3.3	7.4	8.2	2.5	3.3	4.1	4.1	23.8	3

buried viable seeds have been reported in boreal forest areas in N Sweden (Grandstrom 1982). More research is needed to establish the general significance of seed persistence in the population biology of *V.m.* Seedling establishment is slow and infrequent (*Biol Fl*), and is probably restricted to areas of bare soil. Regeneration from seed is likely to be very uncommon at high altitudes, since few flowers are formed in habitats above 1000 m (*Biol Fl*).

Current trends As a result of factors such as increased grazing pressure, *V.m.* may be increasing in upland areas at the expense of *C.v.* However, in lowland habitats *V.m.* has decreased to a marked extent, a change which is at least partly attributable to habitat destruction.

Vaccinium vitis-idaea L.

Ericaceae
Cowberry, Red Whortleberry

Life-form Dwarf shrub (nanophanerophyte or woody chamaephyte) with numerous rhizomes located 100-200 mm below the soil surface. Stems woody, erect, branched; leaves thick, leathery, ovate, <250 mm^2; roots with ericoid mycorrhizas.

Phenology Evergreen. Flowers in spring and again in early summer or, at higher altitudes, only once and later. Seed set within 1 month of flowering (*Biol Fl*).

Height Foliage and flowers to 300 mm.

Seedling RGR <0.5 week^{-1}. SLA 9.0. DMC. No data.

Nuclear DNA amount No data; $2n = 24$ (*FE3*); $2x$.

Established strategy Intermediate between stress-tolerator and stress-tolerant competitor.

Regenerative strategies V, ?B$_s$. More information needed to assess the role of seedling regeneration. Regenerates mainly by rhizome growth to form large clonal patches. Establishment by seed is slow and infrequent (*Biol Fl*). One report (Grandstrom 1982) suggests that some seeds may persist in the soil.

Flowers Pink, hermaphrodite, homogamous, insect- or self-pollinated; an habitual in-breeder (*Biol Fl*), with flowers in a 4-12-flowered raceme.

Germinule 0.26 mg, 1.9 × 1.1 mm, seed dispersed by birds within a *c.* 7-seeded berry.

Germination Epigeal; t_{50} 11 days; 18-28°C; L > D.

Biological flora Ritchie (1955).

Geographical distribution Restricted to upland Britain and most frequent in the S Pennines, Cumbria and the Scottish Highlands (61% of British vice-counties), *ABIF* page 294; N and C Europe, extending S in mountainous areas (69% of European territories); N America.

Habitats Occurs in upland habitats on acidic soils, including sheep pastures and moorlands, disused sandstone quarries, open woodland and coniferous plantations. Also present in peat bogs, and recorded locally from skeletal habitats. [Occurs also in a range of montane communities on soils of low pH (Gimingham 1972).]

Gregariousness Stand-forming.

Altitude Confined to uplands habitats above 200 m. Formerly found in lowland areas (Howitt & Howitt 1963). [Up to *c.* 1200 m (*Biol Fl*).]

Slope Mainly recorded from sloping ground.

Aspect No consistent trends.

Hydrology Not a wetland species but occurs on better drained peat associated with eroding ombrogenous mire.

Soil pH Strongly calcifuge, observed up to pH 6.8, and rarely found on soils of pH >4.5.

Bare soil Most records are from habitats having little exposed soil but large quantities of persistent litter.

△ **ordination** Strictly confined to vegetation of relatively unproductive sites.

Associated floristic diversity Very low.

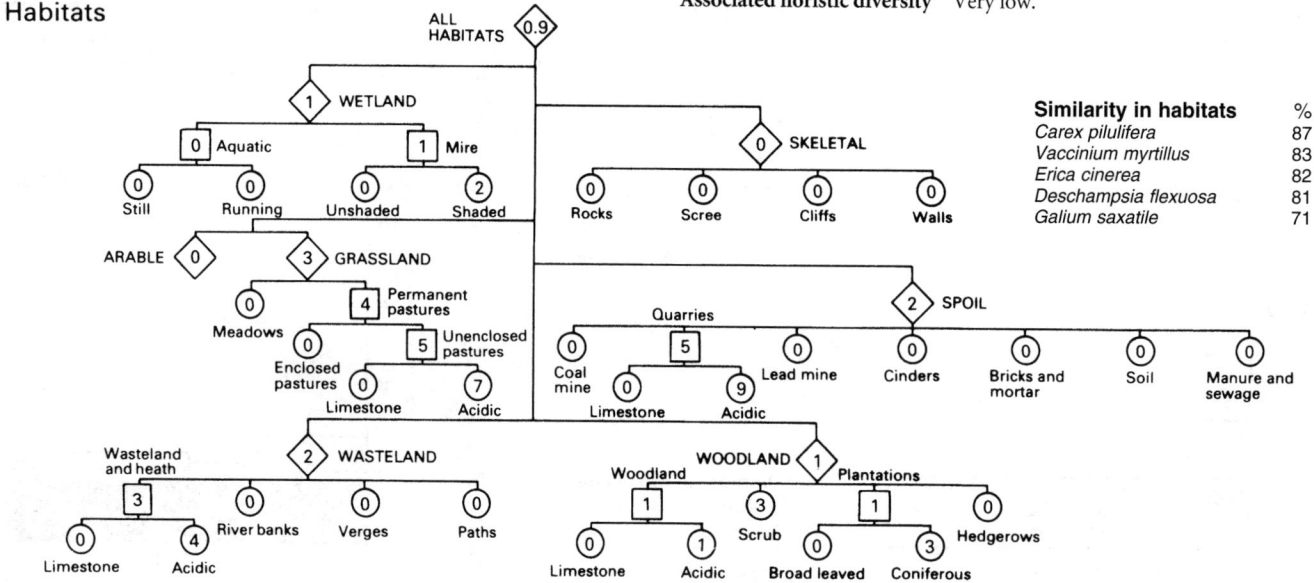

Habitats

Similarity in habitats	%
Carex pilulifera	87
Vaccinium myrtillus	83
Erica cinerea	82
Deschampsia flexuosa	81
Galium saxatile	71

Synopsis *V.v-i.* is, like *V. myrtillus*, with which it commonly occurs, a long-lived, slow-growing, dwarf rhizomatous shrub of acidic peaty soils. However, *V.v-i.* differs from *V.m.* in a number of important respects. The leaves are evergreen; this may enable *V.v-i.* to maximize production per unit of limiting mineral nutrient in nutrient-deficient soils (see Small 1972). *V.v-i.* also has very tough leaves and is little grazed (*Biol Fl*). *V.v-i.* tends to be restricted to upland areas and seldom occurs below 200 m (*Biol Fl*). The species is at the S edge of its British distribution within the region (*ABIF*) and is classified by Matthews (1937) as an Arctic-Alpine species, rather than as a plant characteristic of the Continental Northern Element as is the case for *V.m.* Germination of fresh seed requires high temperatures, a feature characteristic of many northern species (cf. *Eriophorum angustifolium*). *V.v-i.* is more drought-tolerant than *V.m.* (Gimingham 1972), and within the Sheffield region comes closest to forming monocultures on well-drained S-facing slopes. *V.v-i.* is shorter than *V.m.* and may form an understory in stands of *V.m.* In many other aspects *V.v-i.* and *V.m.* are similar. Each forms extensive long-lived clones by means of underground rhizomes which, in each species, confer tolerance of burning and lead to susceptibility to waterlogging. At high altitudes each is particularly associated with sites protected in winter by snow cover (Gimingham 1972). In both, regeneration by seed is infrequent, particularly at high altitudes, and seedling establishment is slow (*Biol Fl*). *V.v-i.* also is tolerant of shade but, unlike *V.m.*, reputedly exhibits its maximum vegetative and reproductive performance in pine woods (*Biol Fl*). The reasons for the greater abundance of *Vm.* in deciduous woodland and in the oceanic montane areas of N Wales require investigation. *V.m.* and *V.v-i.* hybridize within unshaded habitats of the Sheffield region and elsewhere. Another ssp. with an Arctic distribution is recorded (*FE3*) and is the subject of a *Biol Fl* by Hall and Shay (1981).

Current trends *V.v-i.* is now extinct within lowland Britain. In upland areas the species may have increased at the expense of *Calluna vulgaris* because of its greater tolerance of uncontrolled burning and low palatability in grazed sites.

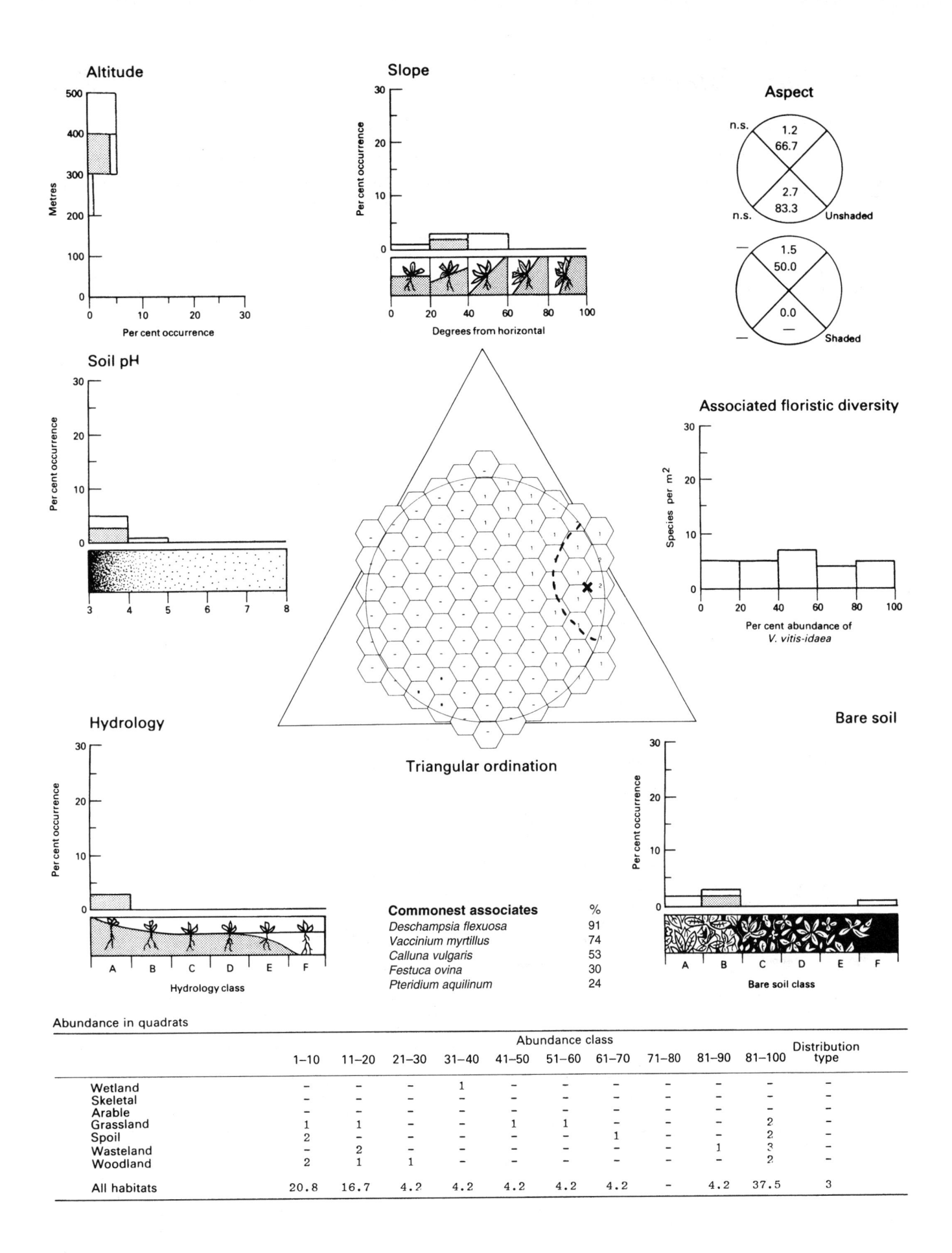

Altitude

Slope

Aspect

	1.2	
n.s.	66.7	
	2.7	
n.s.	83.3	Unshaded

	1.5	
—	50.0	
	0.0	
—	—	Shaded

Soil pH

Associated floristic diversity

Hydrology

Triangular ordination

Bare soil

Commonest associates	%
Deschampsia flexuosa	91
Vaccinium myrtillus	74
Calluna vulgaris	53
Festuca ovina	30
Pteridium aquilinum	24

Abundance in quadrats

	Abundance class										Distribution type
	1–10	11–20	21–30	31–40	41–50	51–60	61–70	71–80	81–90	81–100	
Wetland	–	–	–	1	–	–	–	–	–	–	–
Skeletal	–	–	–	–	–	–	–	–	–	–	–
Arable	–	–	–	–	–	–	–	–	–	–	–
Grassland	1	1	–	–	1	1	–	–	–	2	–
Spoil	2	–	–	–	–	–	1	–	–	2	–
Wasteland	–	2	–	–	–	–	–	–	1	3	–
Woodland	2	1	1	–	–	–	–	–	–	2	–
All habitats	20.8	16.7	4.2	4.2	4.2	4.2	4.2	–	4.2	37.5	3

Valeriana officinalis L.

Valerianaceae
Valerian

Including ssp. *collina* (Wallr.) Nyman and ssp. *sambucifolia* (Mikan fil) Gelak

---

Life-form Polycarpic perennial, usually stoloniferous, semi-rosette hemicryptophyte. Shoots erect; leaves pinnate, the lower often >4000 mm^2; roots relatively shallow, often with VA mycorrhizas.

Phenology Flowers from June to August and sheds seed in July to August. Shoots die back in autumn.

Height Foliage up to 220 mm on non-flowering shoots; flowering stems may reach 1500 mm, with most of leaf area in the lower half.

Seedling RGR No data. SLA 29.1. DMC 14.0.

Nuclear DNA amount No data; ssp. *c.* 2*n* = 28*, 4*x** (*BC, FE4*); ssp. *s.* 2*n* = 56*, 8*x** (*BC, FE4*).

Established strategy C-S-R.

Regenerative strategies V, S. Regenerative biology is complex (see 'Synopsis'); ?SB II.

Flowers Pink, hermaphrodite, insect-pollinated, self-incompatible (Skalinska 1947); >100 in a terminal cymose inflorescence.

Germinule 0.95 mg, 3.6 × 1.5 mm, a 1-seeded nut with stout pappus; ssp. *c.* fruit 2-4 mm, ssp. *s.* 4.5 mm.

Germination Epigeal; t_{50} 10 days; 19-27°C; germination rate increases after dry storage at 20°C; L > D.

Biological flora None, but see Sprague (1949).

Geographical distribution The British Isles (99% of British vice-counties), *ABIF* page 604; Europe except for the extreme S (79% of European territories); temperate Asia to Japan. See also 'Synopsis'.

Habitats Occurs mainly in rough grassland. Also represented locally on limestone quarry heaps, roadsides and river banks. Occasionally in shaded skeletal habitats, mire and woodland. Observed in little-grazed pasture, but absent from arable land, aquatic and most spoil habitats. (See 'Synopsis' for habitat differences between the two ssp.)

Gregariousness Intermediate.

Altitude Extends up to 400 m, but most abundant on carboniferous limestone (between 200 and 300 m). [Up to 800 m.]

Slope Wide-ranging.

Aspect Slightly more frequent on N-facing slopes, but, unexpectedly for a species of moist soils, recorded more abundantly from S-facing slopes.

Hydrology Mainly associated with soligenous mire within the Sheffield region. Occurrences in topogenous mire appear restricted to marginal, relatively well-drained sites.

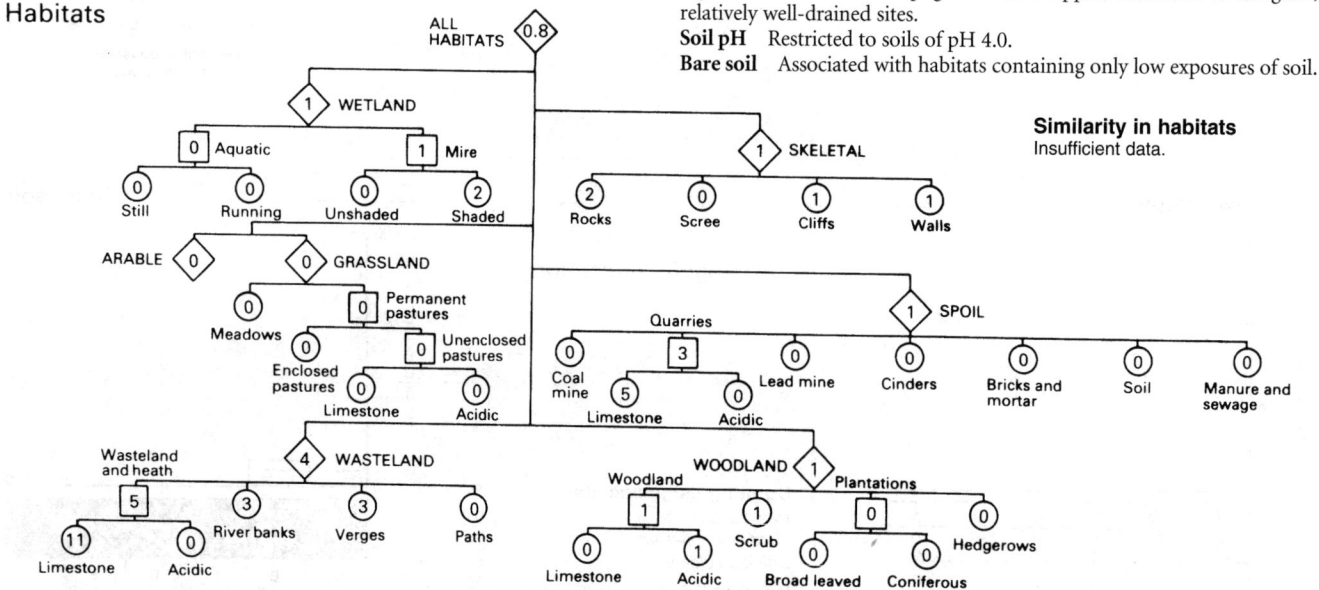

Soil pH Restricted to soils of pH 4.0.

Bare soil Associated with habitats containing only low exposures of soil.

Similarity in habitats Insufficient data.

Habitats

[Habitat hierarchy diagram: ALL HABITATS 0.8 → WETLAND 1 (Aquatic 0: Still 0, Running 0; Mire 1: Unshaded 0, Shaded 2); ARABLE 0; GRASSLAND 0 (Permanent pastures 0: Meadows 0, Enclosed pastures 0; Unenclosed pastures 0: Limestone 0, Acidic 0); WASTELAND 4 (Wasteland and heath 5: Limestone 11, Acidic 0; River banks 3, Verges 3, Paths 0); SKELETAL 1 (Rocks 2, Scree 0, Cliffs 1, Walls 1); Quarries 3 (Coal mine 0, Limestone 5, Lead mine 0, Acidic 0); SPOIL 1 (Cinders 0, Bricks and mortar 0, Soil 0, Manure and sewage 0); WOODLAND 1 (Woodland 1: Limestone 1, Acidic 1; Plantations 0: Scrub, Broad leaved 0, Coniferous 0, Hedgerows 0)]

△ **ordination** A rather diffuse pattern, but centred on vegetation associated with moderately fertile and relatively undisturbed conditions.

Associated floristic diversity Typically intermediate.

Synopsis *V.o.* is a mesophytic herb with a basal tuft consisting of a few, ascending compound leaves, and produces a tall flowering shoot in which the lower leaves are the largest. The low shoot density and modest height of leaf canopy, coupled with the very restricted capacity for lateral vegetative spread, limit the capacity of *V.o.* for vegetation dominance. The shoot consists of a small number of succulent compound leaves; this renders *V.o.* vulnerable to grazing and, consequently, the species is infrequent in pastures. *V.o.* often sets seed in lightly shaded habitats, and we suspect that its survival in tall vegetation involves a modest degree of shade tolerance. It also seems likely that the persistence of *V.o.* depends upon the operation of factors, e.g. occasional grazing, mowing or fire, which limit the vigour of potential dominants. Occurrences over the range from moist to waterlogged base-rich habitats owe much to the existence of two cytotypes which have been afforded subspecific rank. *V.o.* shows considerable genotypic variation at each level of ploidy (Skalinska 1947, Sprague 1949). All or most records from the Sheffield region appear to refer to ssp. *s.*, which is widely distributed in Britain and occurs in N, NC and EC Europe (Skalinska 1947, *FE4*). Ssp. *s.*, an octoploid, occurs typically in wetter habitats within the range of ssp. *c.*, but in the N occupies the whole range of sites exploited by the species (Sprague 1949). In this ssp. plants regenerate vegetatively by both epigeal and hypogeal stolons (*FE4*), which are produced in autumn, albeit very slowly (Sprague 1944, 1949). Regeneration by means of bulbils formed in the axils of the leaves, or by wind-dispersed seed, is infrequent (Sprague 1944, 1949). In common with many other wetland species, the seeds exhibit a high temperature requirement for germination (Grime *et al.* 1981). This suggests that germination will be a spring event, although further research is needed to define the conditions favourable for seedling establishment. Ssp. *c.*, a tetraploid, is in Britain distributed mainly S of a line between the Wash and

Altitude

Slope

Aspect

	n.s.	2.0	
		20.0	
n.s.	0.9	50.0	Unshaded

	—	0.0	
	—	—	
	—	0.0	Shaded

Soil pH

Associated floristic diversity

Triangular ordination

Hydrology

Bare soil

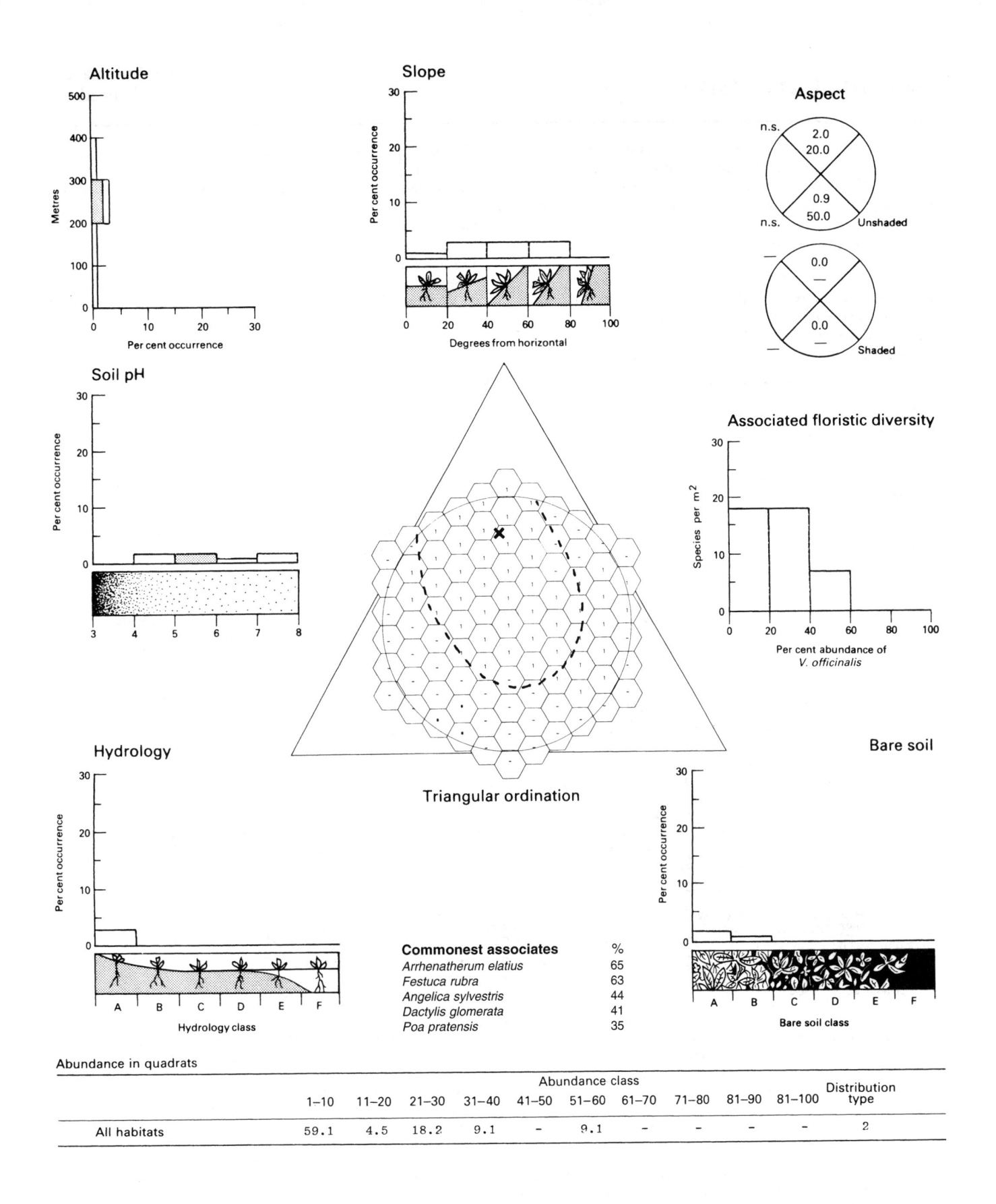

Commonest associates	%
Arrhenatherum elatius	65
Festuca rubra	63
Angelica sylvestris	44
Dactylis glomerata	41
Poa pratensis	35

Abundance in quadrats

	1–10	11–20	21–30	31–40	41–50	51–60	61–70	71–80	81–90	81–100	Distribution type
All habitats	59.1	4.5	18.2	9.1	–	9.1	–	–	–	–	2

the Bristol Channel, and occurs in W and C Europe (Skalinska 1947, FE4). The seeds are smaller and the epigeal stolons (and sometimes also the hypogeal stolons) are absent (FE4). Stolons may be shorter, reducing still further the capacity for vegetative spread. The ssp. is also largely restricted to drier habitats, particularly those on calcareous soils (Sprague 1949). The limits of the two ssp. are confused by the existence of a wide range of intermediates (Sprague 1949), and in Europe a further ssp., *officinalis* (with $2n = 14; 2x$), is recorded (FE4).

Current trends Infrequent as a colonist of recent artificial habitats, and presumably decreasing.

Veronica arvensis L.

Scrophulariaceae
Wall Speedwell

Life-form Semi-rosette, winter- or more rarely summer-annual therophyte. Stems usually erect, often branched; leaves ovate, hairy, <250 mm^2; roots with VA mycorrhizas.
Phenology Typically a winter annual. Flowers mainly in April to June and seed normally set from June to July.
Height Foliage and flowers typically <100 mm or more rarely to 250 mm.
Seedling RGR 1.0-1.4 week^{-1}. SLA 29.1. DMC 13.2.
Nuclear DNA amount No data; $2n = 16$ (*FE3*); $2x$.
Established strategy Stress-tolerant ruderal.
Regenerative strategies B$_s$. Regenerates by seed, which typically germinates in autumn and forms a persistent seed bank; SB III.

Flowers Blue, hermaphrodite, probably often selfed; in many-flowered racemes.
Germinule 0.11 mg, 1.1 × 0.8 mm, seed; *c.* 15 in each dehiscent capsule.
Germination Epigeal; t_{50} 5 days; 7-23°C; germination rate increased after dry storage at 20°C; L > D.
Biological flora None.
Geographical distribution The British Isles (100% of British vice-counties), *ABIF* page 556; most of Europe except NE and SE (92% of European territories and 3% naturalized); C and W Asia, N Africa; naturalized in N America.

Habitats Common in both (a) relatively open rocky habitats, e.g. outcrops, limestone pasture, quarries, cinders and walls and dry sandy soils, and (b) disturbed habitats such as arable fields, grass leys and waysides. Also observed on anthills. Absent from woodland and wetland habitats. [Occurs on fixed dunes.]
Gregariousness Typically sparse.
Altitude Extends to 400 m. [Up to 820 m.]
Slope Recorded or observed from all slopes.
Aspect Distribution shows a S-facing bias.
Hydrology Absent from wetlands.

Soil pH Mainly restricted to neutral soils in the pH range 6.0-8.0, and seldom observed below pH 5.0.
Bare soil Widely distributed, but particularly frequent in sites with higher exposures.
△ **ordination** Restricted to vegetation experiencing some degree of disturbance with contours centred on relatively infertile, moderately disturbed conditions.
Associated floristic diversity Very variable. On walls generally low; in some grassland and winter-annual communities high.

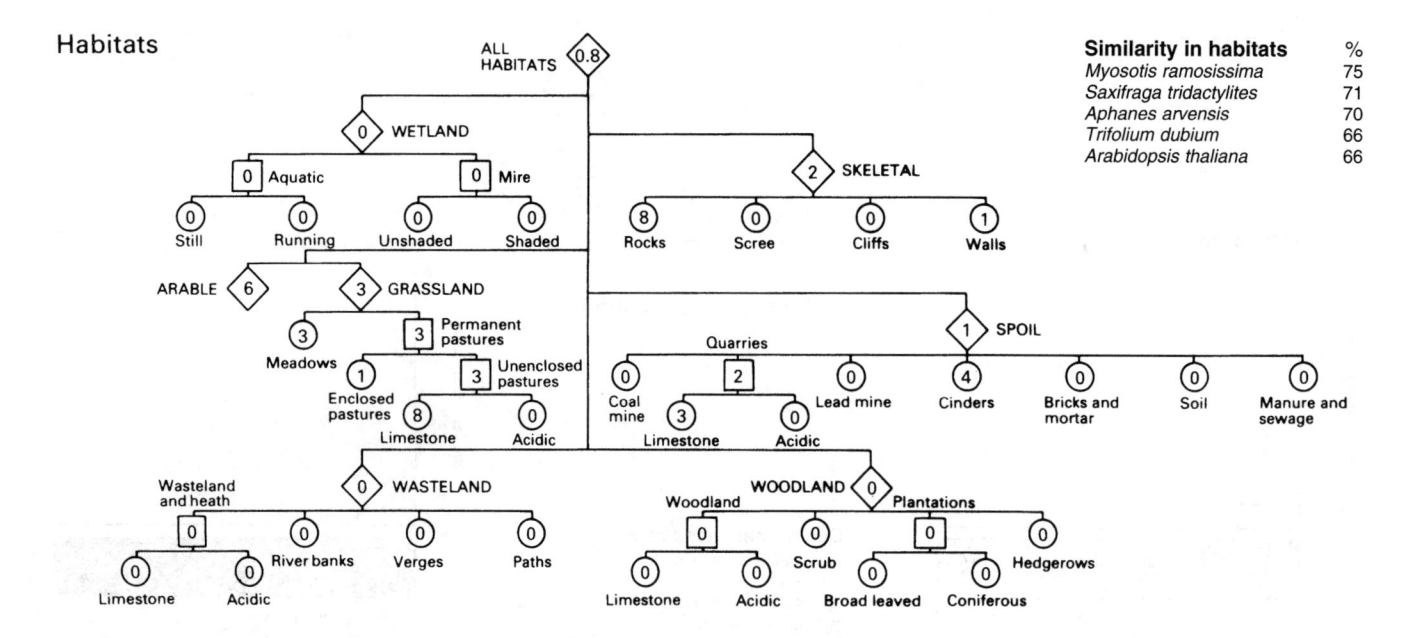

Habitats

Similarity in habitats	%
Myosotis ramosissima	75
Saxifraga tridactylites	71
Aphanes arvensis	70
Trifolium dubium	66
Arabidopsis thaliana	66

Synopsis *V.a.* is a small, facultatively winter- or summer-annual herb found in habitats where the growth of perennials is restricted by drought, nutrient deficiency, disturbance or by a combination of these factors. Like *Aphanes arvensis* and *Arenaria serpyllifolia*, the species extends from droughted, nutrient-deficient calcareous rock outcrops and dry sandy soils to fertile but highly disturbed arable fields, particularly those with a winter-sown cereal crop (Chancellor & Froud-Williams 1984). In addition to exploiting sites which remain open for many years, the species also occupies transient areas of exposed soil in grassland and wasteland. *V.a.* is generally absent from grazed sites, and does not persist in waterlogged or woodland habitats. Germination requires relatively high levels of soil moisture. The species requires both vernalization and long days (Ratcliffe 1961) and is one of the later-flowering winter annuals. Consequently, the species tends to be absent from the most drought-prone microsites. A majority of seeds have an after-ripening requirement, and germination in summer is further inhibited by sensitivity to low moisture and high soil temperatures (Baskin & Baskin 1983). Hence, although seeds are typically shed in summer, germination does not occur until autumn or the following spring (Baskin & Baskin 1983). Germination is inhibited by darkness (Baskin & Baskin 1983) and by the presence of a leaf canopy (King 1975), and the species forms a persistent seed bank. Under optimal conditions seed production by large plants may reach 17 000 (Harris & Lovell 1980a). Seeds can survive ingestion by cattle (Ridley 1930). Despite the tendency of seeds to come to rest close to the parent (Pemadasa & Lovell 1974a), *V.a.* is capable of long-distance dispersal, often apparently through human agency. The minute, saucer shaped seeds are buoyant in air currents and, as in the case of *Saxifraga tridactylites*, local dispersal by wind facilitates colonization of wall-tops. Stem pieces root readily but regeneration by vegetative means does not occur in the field (Harris & Lovell 1980b). The extent of ecotypic differentiation in this ecologically wide-ranging species is unknown.
Current trends Uncertain, but not under any immediate threat.

Altitude
Slope
Aspect

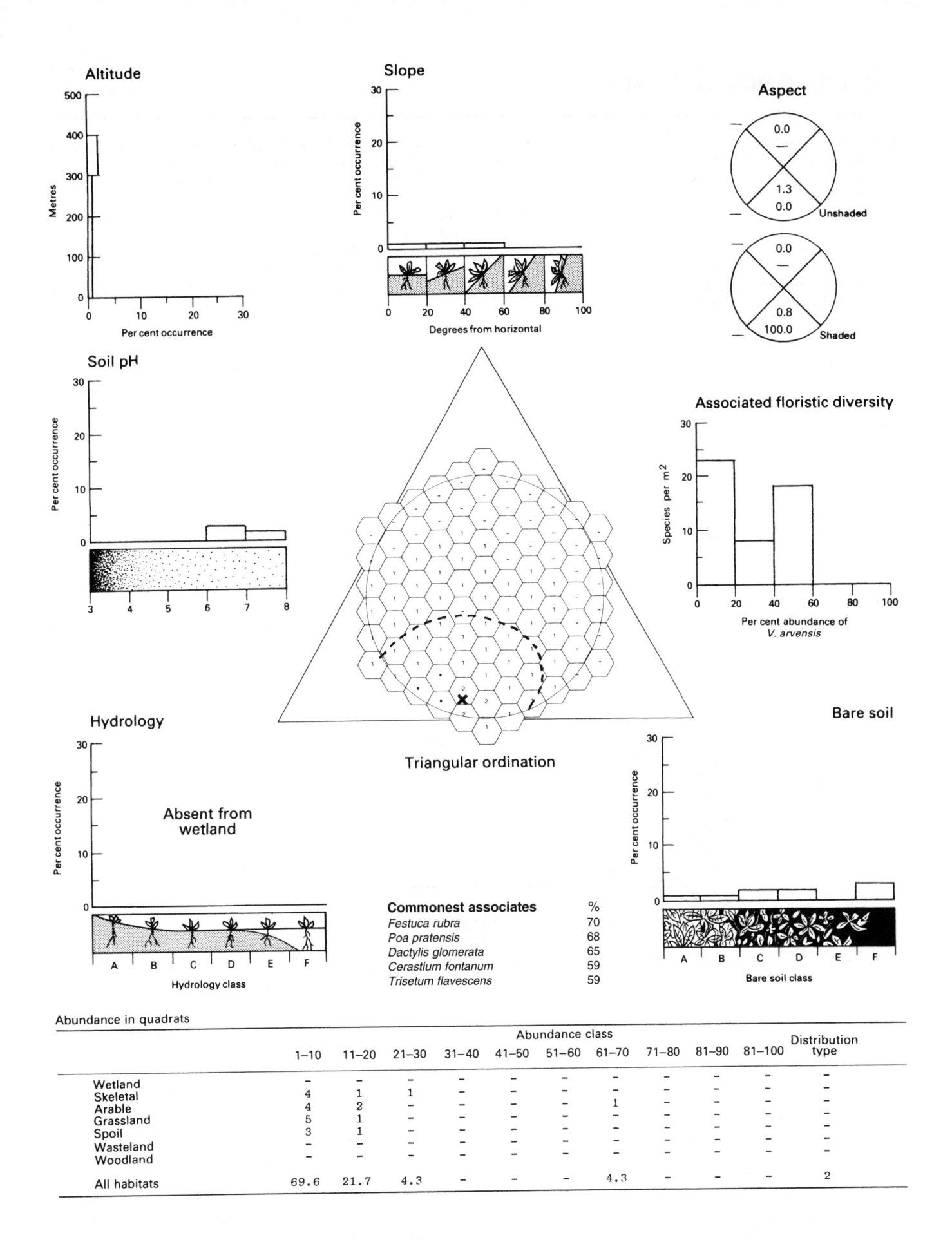

Unshaded

	0.0	
—		—
	1.3	
	0.0	

Shaded

	0.0	
—		—
	0.8	
	100.0	

Soil pH

Associated floristic diversity

Species per m²

Per cent abundance of *V. arvensis*

Hydrology

Absent from wetland

Triangular ordination

Bare soil

Commonest associates %
Festuca rubra 70
Poa pratensis 68
Dactylis glomerata 65
Cerastium fontanum 59
Trisetum flavescens 59

Abundance in quadrats

	\multicolumn{10}{c}{Abundance class}	Distribution type
	1–10	11–20	21–30	31–40	41–50	51–60	61–70	71–80	81–90	81–100	
Wetland	–	–	–	–	–	–	–	–	–	–	–
Skeletal	4	1	1	–	–	–	–	–	–	–	–
Arable	4	2	–	–	–	–	1	–	–	–	–
Grassland	5	1	–	–	–	–	–	–	–	–	–
Spoil	3	1	–	–	–	–	–	–	–	–	–
Wasteland	–	–	–	–	–	–	–	–	–	–	–
Woodland	–	–	–	–	–	–	–	–	–	–	–
All habitats	69.6	21.7	4.3	–	–	–	4.3	–	–	–	2

Veronica beccabunga L.

Scrophulariaceae
Brooklime

Life-form Polycarpic perennial helophyte or protohemicryptophyte. Stems much branched, creeping and rooting at the base then ascending; leaves succulent, ovate, <2000 mm²; roots shallow, without mycorrhizas.
Phenology Winter green. Flowers mainly from May to July and sets seed from July to October.
Height Foliage and flowers normally <200 mm.
Seedling RGR No data. SLA 23.7. DMC 13.7.
Nuclear DNA amount 1.6 pg; $2n = 18^*, 36$ (*BC, FE3*); $2x^*, 4x$.
Established strategy Competitive-ruderal.
Regenerative strategies (V.), B_s. Effective regeneration by vegetative means and by seed (see 'Synopsis'); SB IV.

Flowers Blue, hermaphrodite, protogynous, insect-pollinated or often selfed (*CTW*); <10 in axillary racemes.
Germinule 0.34 mg, 1.8 × 1.2 mm, seed in a dehiscent, many-seeded capsule.
Germination Epigeal; t_{50} 5 days; <5-24°C; L > D.
Biological flora None.
Geographical distribution The British Isles except for the Scottish Highlands (100 % of British vice-counties), *ABIF* page 554; Europe except the extreme N and S (87% of European territories); N Africa and temperate Asia.

Habitats An emergent, semi-aquatic species, particularly common along the margins of streams. Frequent in both shaded and unshaded mire. Elsewhere recorded as a temporary colonist of pond dredgings, and observed on river banks towards the water's edge.
Gregariousness Capable of forming patches.
Altitude Mainly in the lowlands but observed to 340 m. [Up to 850 m.]
Slope As with most wetland species, largely restricted to gentle slopes.
Aspect No data.
Hydrology Most frequent and abundant in shallow water (<100 mm deep) at the margin of streams, but also widespread in stream-side mire.

Less frequent beside ponds and in soligenous or topogenous mire which is not adjoining water.
Soil pH All records fall within the pH range 6.0-8.0, with none below pH 5.0.
Bare soil Insufficient data.
△ **ordination** Distribution centred on vegetation associated with fertile and moderately disturbed conditions.
Associated floristic diversity Typically species-poor and decreasing at higher frequencies of *V.b.*, suggesting a capacity for dominance (but see 'Synopsis').

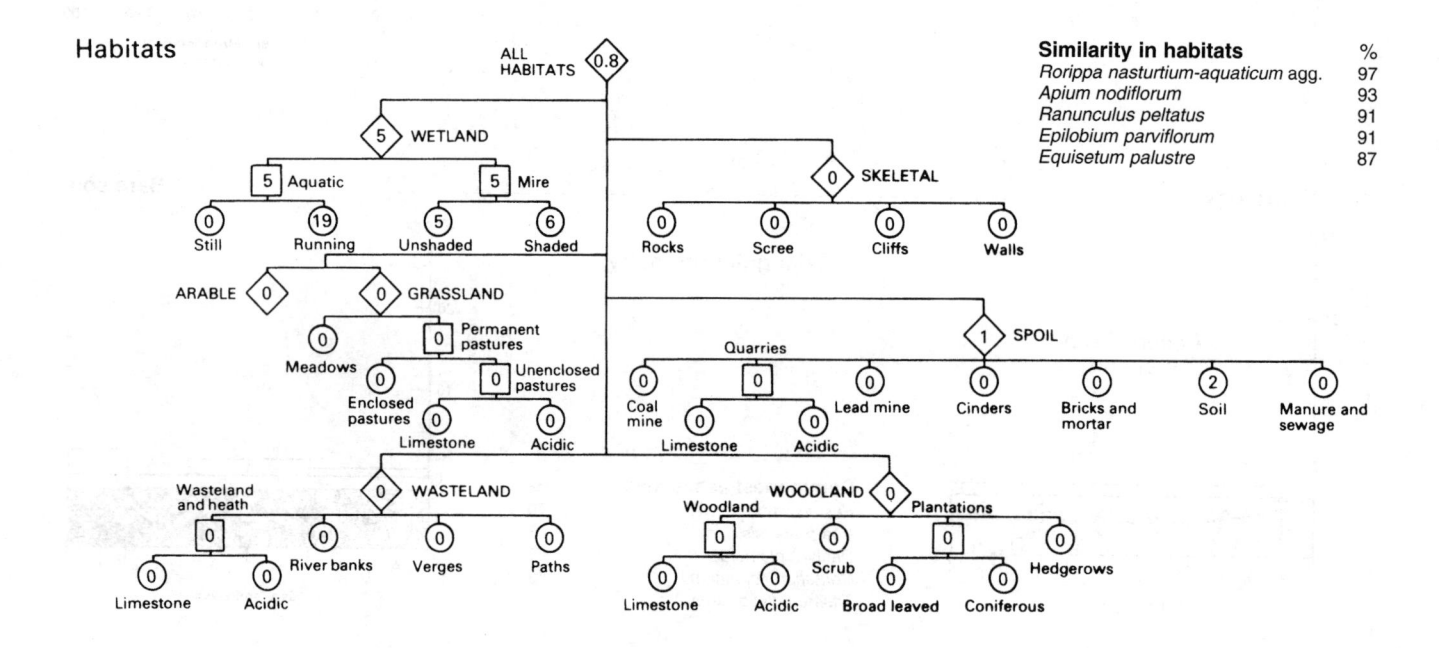

Habitats

Similarity in habitats	%
Rorippa nasturtium-aquaticum agg.	97
Apium nodiflorum	93
Ranunculus peltatus	91
Epilobium parviflorum	91
Equisetum palustre	87

Synopsis *V.b.* is a winter-green, patch-forming herb typically straddling the boundary between mire and aquatic habitats, particularly where the water is flowing. *V.b.* is restricted to fertile sites where disturbances such as winter-flooding, water currents, trampling or grazing restrict the growth of taller perennials. *V.b.* extends into calcareous sites, but is absent from acidic peats. In many respects the ecology of *V.b.* is very similar to that of *Rorippa nasturtium-aquaticum* (agg.). Both form extensive clonal patches by means of prostrate rooted stems. In each, shoot pieces readily root, and regeneration along stream courses by means of detached fragments appears to be a frequent event. Both produce large quantities of small, readily germinable seeds, and in the case of *V.b.* a persistent seed bank has been demonstrated (UCPE unpubl.). A major difference between the two lies in their tolerance of submergence; whereas in *V.b.* a high proportion of the canopy consists of floating or emergent stems rooted to the bank, *N.o.* is strongly amphibious and exhibits considerable plasticity in growth form and leaf shape. A further contrast with *R.a-a.* arises from the fact that *V.b.* occurs more frequently in shaded habitats and persists to a greater extent than *R.a-a.* in taller stream-side vegetation. The seeds of *V.b.* are mucilaginous and adhesive (Ridley 1930) and the species is an effective colonist, even in land-locked sites.
Current trends *V.b.* may have become more abundant beside upland streams in response to increased eutrophication. In lowland areas, where changes to wetland systems have been more marked, the status of *V.b.* is less certain and the species may be decreasing.

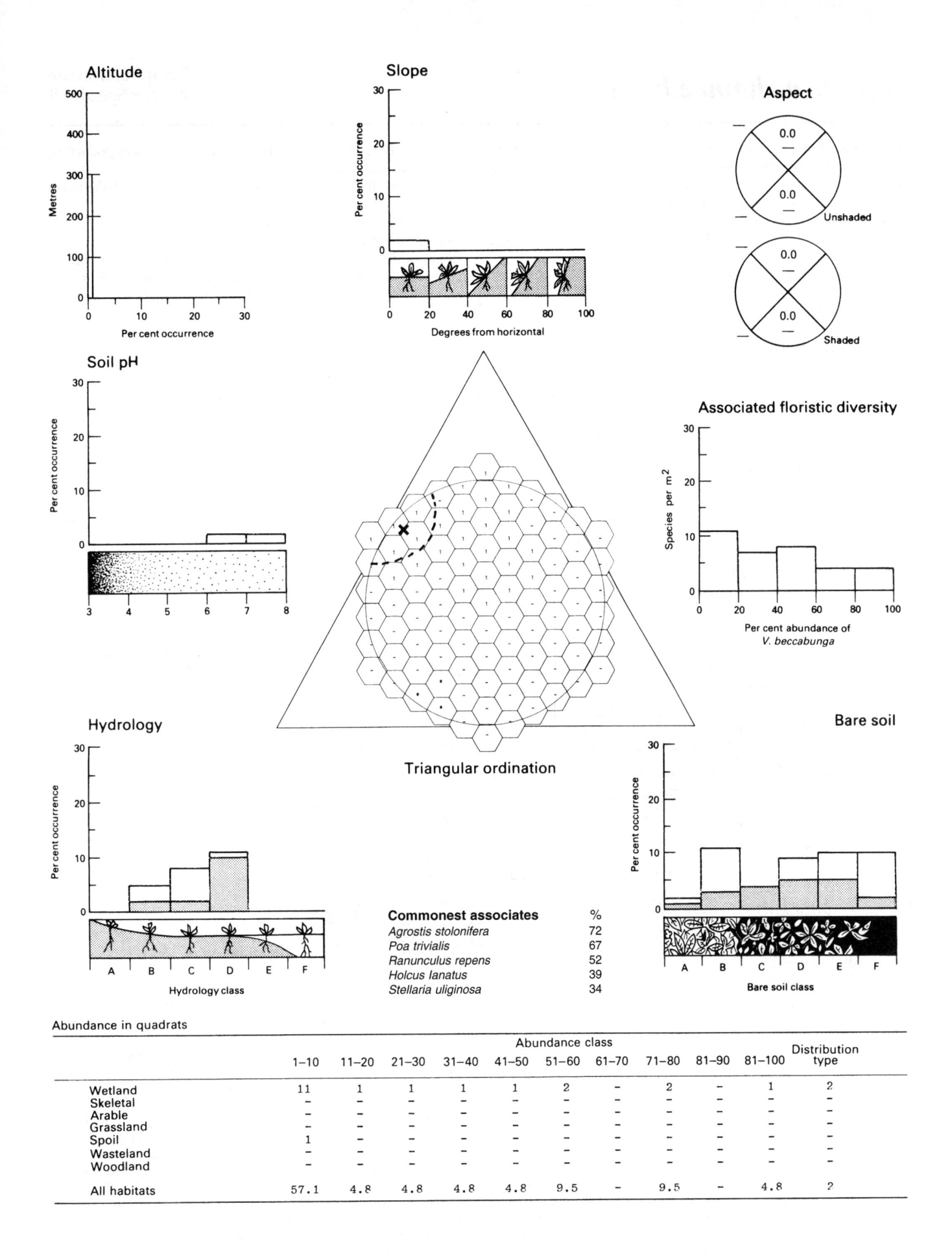

Altitude

Slope

Aspect

Unshaded

Shaded

Soil pH

Associated floristic diversity

Species per m²

Per cent abundance of *V. beccabunga*

Hydrology

Hydrology class

Triangular ordination

Commonest associates	%
Agrostis stolonifera	72
Poa trivialis	67
Ranunculus repens	52
Holcus lanatus	39
Stellaria uliginosa	34

Bare soil

Bare soil class

Abundance in quadrats

	1–10	11–20	21–30	31–40	41–50	51–60	61–70	71–80	81–90	81–100	Distribution type
Wetland	11	1	1	1	1	2	–	2	–	1	2
Skeletal	–	–	–	–	–	–	–	–	–	–	–
Arable	–	–	–	–	–	–	–	–	–	–	–
Grassland	–	–	–	–	–	–	–	–	–	–	–
Spoil	1	–	–	–	–	–	–	–	–	–	–
Wasteland	–	–	–	–	–	–	–	–	–	–	–
Woodland	–	–	–	–	–	–	–	–	–	–	–
All habitats	57.1	4.8	4.8	4.8	4.8	9.5	–	9.5	–	4.8	2

Veronica chamaedrys L.

Scrophulariaceae
Germander Speedwell

Life-form Polycarpic perennial chamaephyte. Stems prostrate towards their base, then ascending; leaves hairy, ovate, often <500 mm²; roots shallow, often with VA mycorrhizas.
Phenology Winter green. New shoots elongate early in spring. Flowers from April to July and sets seed from August onwards.
Height Foliage normally <100 mm, although some flowering shoots may exceed this height.
Seedling RGR No data. SLA 27.1. DMC 29.8.
Nuclear DNA amount No data; $2n = 32$ (*FE3*); $4x$.
Established strategy Intermediate between C-S-R and stress-tolerator.
Regenerative strategies V, B_s. Regeneration from seed is insufficiently studied (see 'Synopsis').

Flowers Blue, hermaphrodite, homogamous, insect-pollinated; 10–20 in axillary racemes.
Germinule 0.18 mg, 1.3×1.0 mm, seed; several in each dehiscent capsule.
Germination Epigeal; t_{50} 7 days; 9–28°C; L > D.
Biological flora None, but see Zarzycki (1976).
Geographical distribution The British Isles (100% of British vice-counties), *ABIF* page 553; Europe except the extremes of N and S (77% of European territories and 3% naturalized); N and W Asia; naturalized in N America.

Habitats Most frequent in limestone pastures, but also commonly recorded from screes, meadows, limestone quarry spoil and waysides. Less frequent on lead-mine spoil, wasteland, soil heaps, outcrops and river banks. Uncommon in open woodland and unshaded mire. Absent from aquatic and arable habitats.
Gregariousness Intermediate.
Altitude Suitable habitats are more abundant in upland areas and *V.c.* therefore shows an upland bias extending to 410 m. [Up to 820 m.]
Slope Not recorded from cliffs or from the steepest slopes.
Aspect Slightly more frequent and abundant on N-facing slopes in unshaded situations. In shaded habitats all records are from S-facing slopes.
Hydrology Not a wetland species but commonly found in damp grasslands.
Soil pH Most frequent and abundant on soils of pH >6.0, and absent from sites of pH <4.5.
Bare soil Occurrences wide-ranging, but some evidence of an association with habitats with low or intermediate exposures.
△ **ordination** Distribution centred on vegetation from which more-robust species are excluded by some disturbance and moderate infertility. Absent from heavily disturbed habitats, but otherwise displaying a fairly broad range.

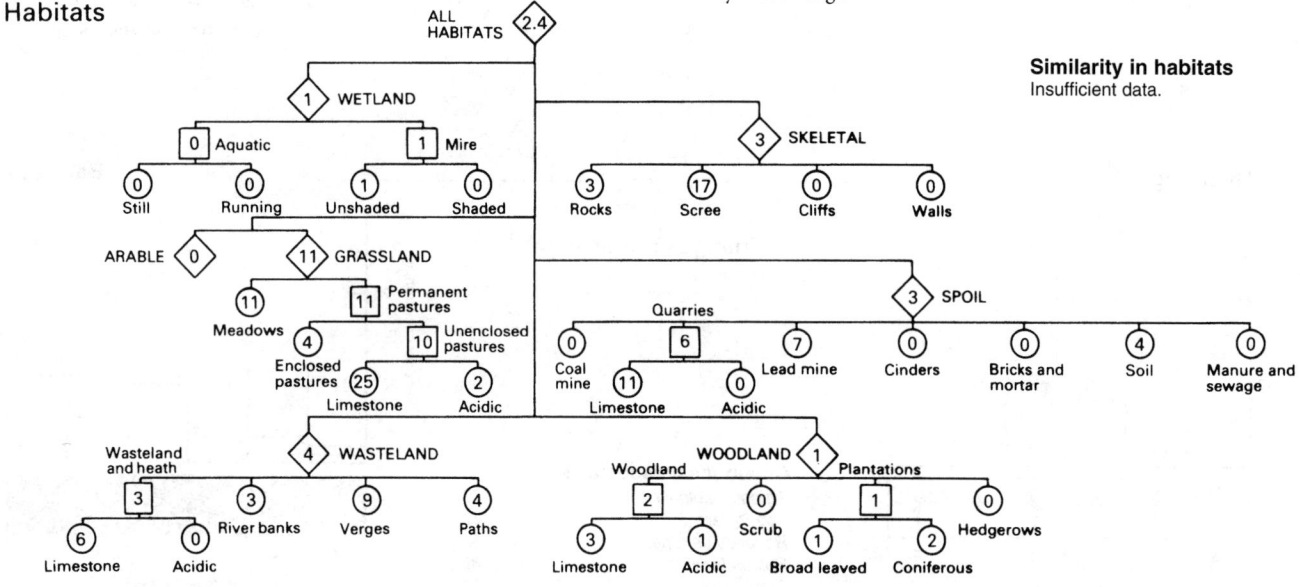

Habitats

Similarity in habitats
Insufficient data.

Synopsis *V.c.* is a creeping, shallow-rooted, perennial herb of moist, base-rich, usually rather infertile soils. *V.c.* is locally frequent in broken short turf in lightly grazed unproductive pastures, where it tends to colonize gaps in the vegetation by means of the stolons, which have the capacity to root along their entire length. *V.c.* is particularly common on stabilized scree where the opportunities for rooting are localized and leaves are often subtended over bare rock at some distance from the roots. A consistent feature of all sites colonized by *V.c.* is that they are liable to some disturbance, and hence present gaps which *V.c.* may exploit by vegetative growth. Suitable sites also include mown hedgebanks and roadsides. In fertile pasture *V.c.* is less common, and gap colonization is effected by larger, faster-growing species such as *Ranunculus repens* and *Trifolium repens*. *V.c.* is occasionally observed in a non-flowering state in open woodland and a degree of shade tolerance is further suggested by the persistence of *V.c.* as an understorey in some types of taller grassland. The plant contains appreciable quantities of mannitol (Hodgson 1973). *V.c.* flowers early, and rapid shoot growth appears restricted to autumn and spring; the significance of this phenology required further study. Apart from the capacity to spread by means of stolons, *V.c.* may also regenerate vegetatively from shoots detached as a result of disturbance (Harris & Lovell 1980b, Hodgson unpubl.). This may be an important mechanism of spread in trampled pastures and on roadsides. As in many other perennial species of less-productive habitats, regeneration by seed appears to be comparatively rare. Persistent seeds have been reported. It is not known when the seed normally germinates. The small flattened seeds survive ingestion by cattle (Ridley 1930). Some morphological plasticity is evident in field populations; plants growing in short turf are appressed closely to the ground, and stolons are rooted along most of their length. In taller vegetation, however, *V.c.* adopts a more ascending growth form. A ssp. with $2n = 16$ ($2x$) also occurs elsewhere in Europe (*FE3*).
Current trends *V.c.* is not a threatened species but it may be expected to fall in abundance as the quantity of unproductive pasture declines further.

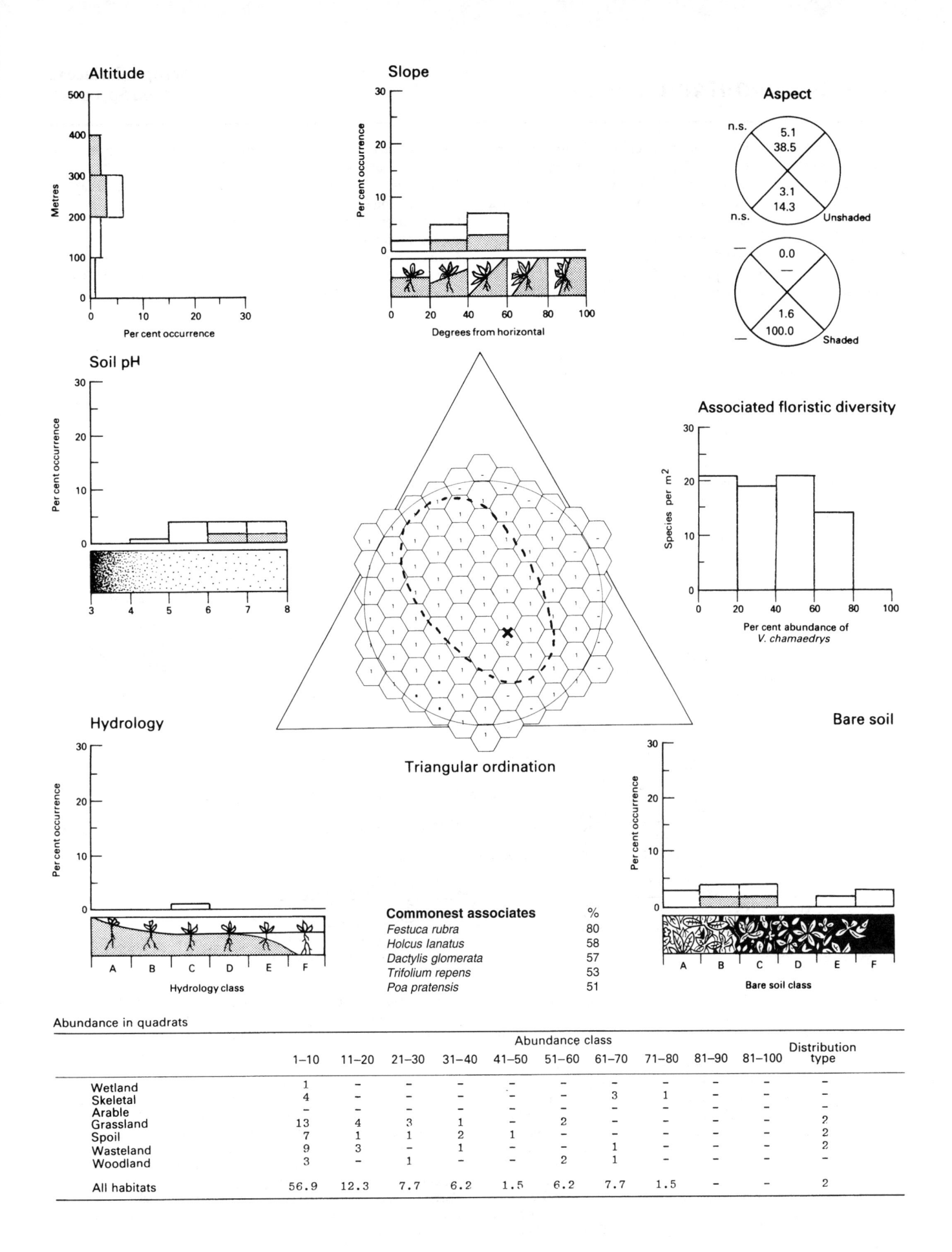

Altitude

Slope

Aspect

n.s. | 5.1 | n.s.
38.5
3.1 | 14.3 | Unshaded

— | 0.0 | —
—
1.6 | 100.0 | Shaded

Soil pH

Associated floristic diversity

Hydrology

Triangular ordination

Bare soil

Commonest associates	%
Festuca rubra	80
Holcus lanatus	58
Dactylis glomerata	57
Trifolium repens	53
Poa pratensis	51

Abundance in quadrats

	1–10	11–20	21–30	31–40	41–50	51–60	61–70	71–80	81–90	81–100	Distribution type
Wetland	1	–	–	–	–	–	–	–	–	–	–
Skeletal	4	–	–	–	–	–	3	1	–	–	–
Arable	–	–	–	–	–	–	–	–	–	–	–
Grassland	13	4	3	1	–	2	–	–	–	–	2
Spoil	7	1	1	2	1	–	–	–	–	–	2
Wasteland	9	3	–	1	–	–	1	–	–	–	2
Woodland	3	–	1	–	–	2	1	–	–	–	–
All habitats	56.9	12.3	7.7	6.2	1.5	6.2	7.7	1.5	–	–	2

Veronica montana L.

Scrophulariaceae
Wood Speedwell

Life-form Polycarpic perennial chamaephyte. Stems creeping with adventitious roots, to 0.5 m in length, the ends of the branches ascending; leaves ovate, hairy, <1000 mm^2.
Phenology Winter green. Flowers from April to July and sets seed from July to August.
Height Foliage and flowers to 100 mm.
Seedling RGR No data. SLA 22.9. DMC 27.2.
Nuclear DNA amount 1.8 pg; $2n = 18$ (*FE3*); $2x$.
Established strategy Intermediate between stress-tolerant ruderal and C-S-R.
Regenerative strategies V, B$_s$. Mainly by vegetative means (see 'Synopsis'); ?SB IV.

Flowers Lilac-blue, hermaphrodite, insect-pollinated; in <10-flowered axillary racemes.
Germinule 0.34 mg, 1.7×1.6 mm, seed dispersed from a capsule containing several seeds.
Germination Epigeal; dormancy broken by chilling.
Biological flora None.
Geographical distribution Most of the British Isles, less common in N Scotland and Ireland (85% of British vice-counties), *ABIF* page 554; W, C and S Europe (69% of European territories).

Habitats In moist woodland, particularly in seepage areas, on stream terraces, in shaded mire and on shaded river banks. Often beside woodland rides, but only rarely observed in unshaded habitats. Absent from aquatic, arable and most spoil habitats.
Gregariousness Intermediate to patch-forming.
Altitude Almost extends to the British altitudinal limit of 320 m.
Slope Not recorded from the steepest slopes.
Aspect Distribution shows a N-facing bias.
Hydrology Typical of moist clayey soils, and extending into soligenous mire.

Soil pH Frequency and abundance greatest on soils within the pH range 4.5-6.5; not observed at pH values >7.5.
Bare soil Wide-ranging.
△ ordination Associated mainly with unproductive and undisturbed vegetation, but found also in more-productive vegetation types subject to slight disturbance. Absent from fertile undisturbed sites.
Associated floristic diversity Typically low to intermediate.

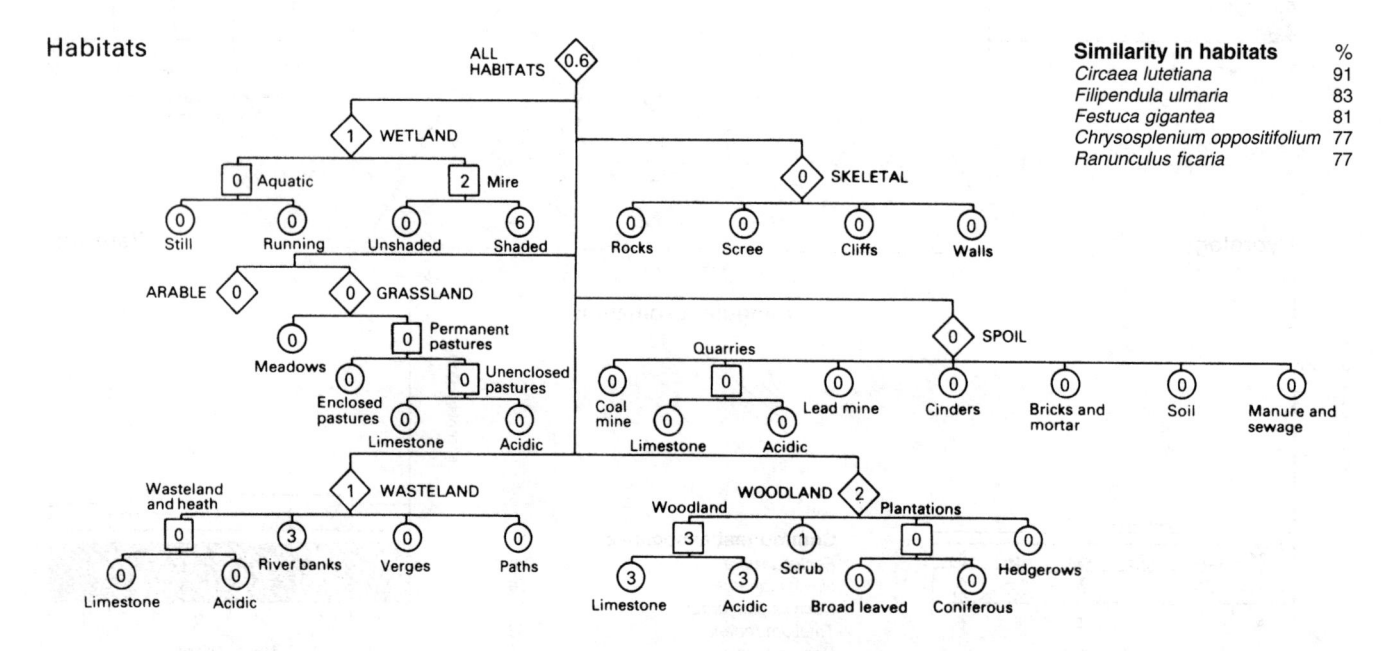

Habitats

Similarity in habitats	%
Circaea lutetiana	91
Filipendula ulmaria	83
Festuca gigantea	81
Chrysosplenium oppositifolium	77
Ranunculus ficaria	77

Synopsis *V.m.* is a sprawling, winter-green herb which exploits moist, often clayey soils in sites where the vigour of dominant species is suppressed by shade and, often, by a degree of disturbance. Mineral nutrient stress may also be an important factor in some habitats. *V.m.* is shade-tolerant and has been observed to flower over a wide range in irradiance. In competition experiments conducted in garden plots with and without summer shade (Al-Mufti 1978), *V.m.* was comparatively resistant to shade and better able to maintain its status as a subordinate component in vegetation where shade had reduced the vigour of potential dominants such as *Holcus mollis*. The species is most floriferous in lightly-shaded sites, and in upland areas is occasionally observed in unshaded habitats. *V.m.* is shallow-rooted and it is uncertain to what extent the N-facing bias in its distribution is related to a high level of shade tolerance or to a requirement for unusually moist conditions. Although extending onto calcareous strata within the region, *V.m.* is most abundant on mildly acidic soils. Further, the restriction of *V.m.* to sites below 320 m in altitude (Wilson 1956) suggests that the distribution of *V.m.* is strongly influenced by climatic factors. *V.m.* regenerates by means of prostrate rooted stems to form extensive clonal patches. It is particularly frequent on stream banks and woodland rides, and appears to spread considerable distances along these linear habitats by means of regenerating shoot fragments. Seed set is often low and seeds, which have a small embryo relative to their internal volume, have a chilling requirement for germination (Wolfenden & Causton 1979). A persistent seed bank has been reported. The seed appears to be poorly dispersed and, with the exception of stream-sides and woodland rides, sites to which *V.m.* appears to be restricted have extended habitat continuity. In Lincolnshire, too, *V.m.* tends to be associated with ancient woodlands (Peterken 1981).
Current trends Perhaps decreasing as a result of woodland destruction but on heavy soils appears to be stable along woodland rides in broadleaved plantations.

Altitude
Slope
Aspect

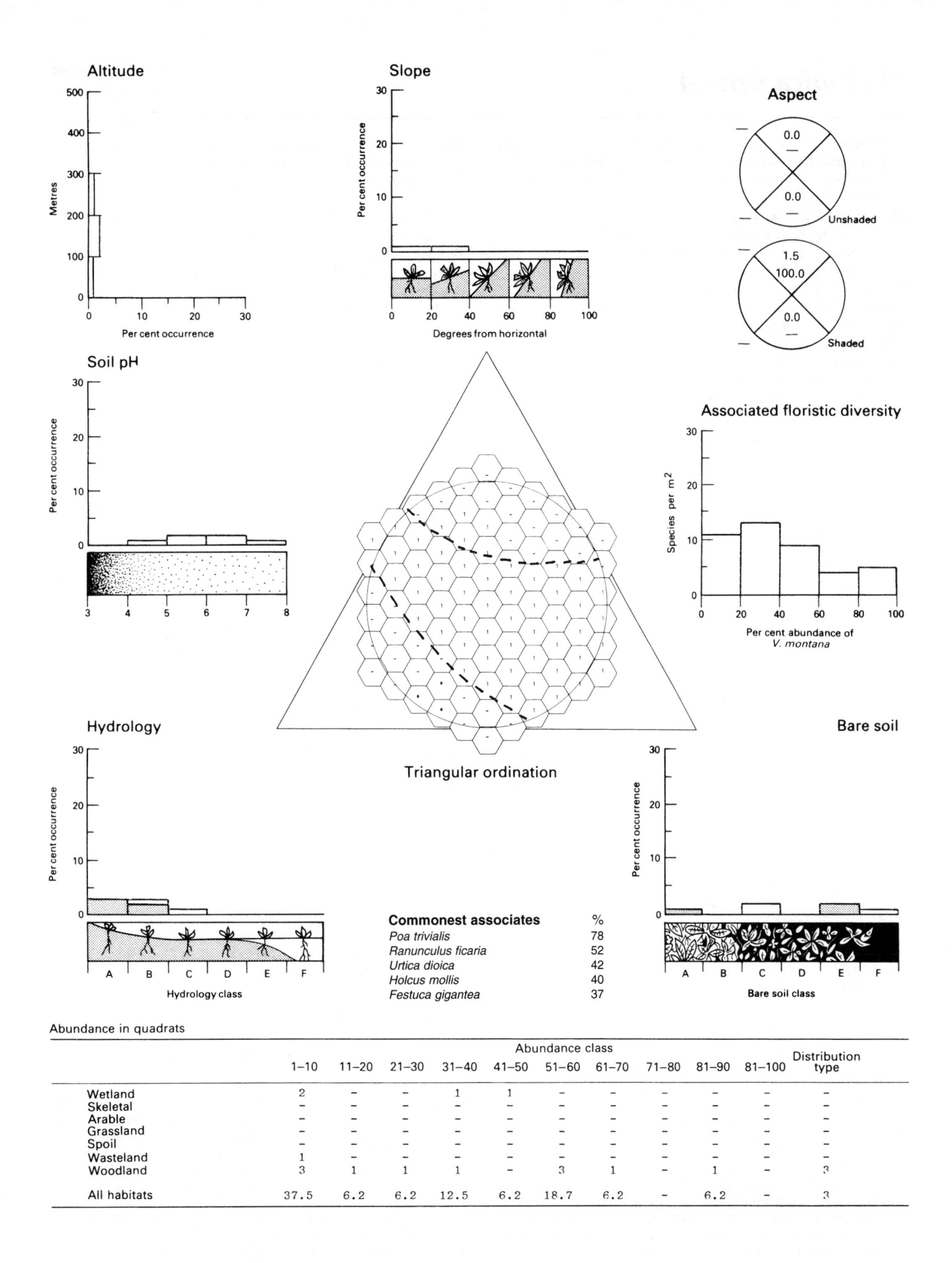

Soil pH

Associated floristic diversity

Species per m^2 vs Per cent abundance of *V. montana*

Aspect values: Unshaded 0.0 / 0.0; Shaded 1.5 / 100.0 / 0.0

Hydrology

Hydrology class: A B C D E F

Triangular ordination

Commonest associates	%
Poa trivialis	78
Ranunculus ficaria	52
Urtica dioica	42
Holcus mollis	40
Festuca gigantea	37

Bare soil

Bare soil class: A B C D E F

Abundance in quadrats

	1–10	11–20	21–30	31–40	41–50	51–60	61–70	71–80	81–90	81–100	Distribution type
Wetland	2	–	–	1	1	–	–	–	–	–	–
Skeletal	–	–	–	–	–	–	–	–	–	–	–
Arable	–	–	–	–	–	–	–	–	–	–	–
Grassland	–	–	–	–	–	–	–	–	–	–	–
Spoil	–	–	–	–	–	–	–	–	–	–	–
Wasteland	1	–	–	–	–	–	–	–	–	–	–
Woodland	3	1	1	1	–	3	1	–	1	–	3
All habitats	37.5	6.2	6.2	12.5	6.2	18.7	6.2	–	6.2	–	3

Veronica persica Poiret

Scrophulariaceae
Large Field Speedwell

Life-form Winter- to summer-annual therophyte. Stems procumbent, sometimes with adventitious roots towards their base; leaves ovate, hairy, <500 mm²; roots with VA mycorrhizas.

Phenology Variable, since germination may occur throughout most of the year. Plants capable of overwintering, and flower throughout the year but particularly from May to October. Seed set mainly from June to October.

Height Foliage and flowers usually <50 mm.

Seedling RGR 1.5-1.9 week^{-1}. SLA 40.1. DMC 12.0.

Nuclear DNA amount 1.5 pg; $2n = 28$ (*FE3*); $4x$.

Established strategy Ruderal.

Regenerative strategies B_s, V. Regenerates by seed which germinates either directly or after incorporation into a persistent seed bank. May exhibit a limited capacity for vegetative spread (see 'Synopsis'); ?SB IV.

Flowers Blue, hermaphrodite, visited by various insects, but often selfed (*CTW*); solitary in leaf axils.

Germinule 0.52 mg, 1.5 × 0.8 mm, seed; in capsule with *c.* 7 seeds.

Germination Epigeal; t_{50} 3 days; 7-35°C; germination rate increased after dry storage at 20°C; L > D.

Biological flora None.

Geographical distribution Native of SW Asia; now naturalized throughout the British Isles (100% of British vice-counties) *ABIF* page 558; S and C Europe northwards to Scandinavia (naturalized in 92% of European territories); also a weed of N and S America, much of Asia, parts of Africa and New Zealand (Holm *et al.* 1977).

Habitats Largely restricted to arable land, particularly cereal fields, but found more sporadically in a range of other disturbed habitats including soil heaps, demolition sites and waysides. Also rarely observed, perhaps only as a casual, in winter-annual communities on rock outcrops.

Gregariousness Sparse to intermediate.

Altitude Largely restricted to the lowland arable region, but observed up to 340 m, the altitudinal limit of its British distribution.

Slope Largely restricted to flat or gently sloping ground.

Aspect Widely distributed with respect to aspect.

Hydrology Absent from wetlands.

Soil pH Most frequent within the pH range 6.0-8.0, and not recorded below pH 5.0.

Bare soil Most frequent and abundant at high exposures.

△ **ordination** Distribution largely restricted to fertile disturbed habitats.

Associated floristic diversity Typically intermediate to species-rich.

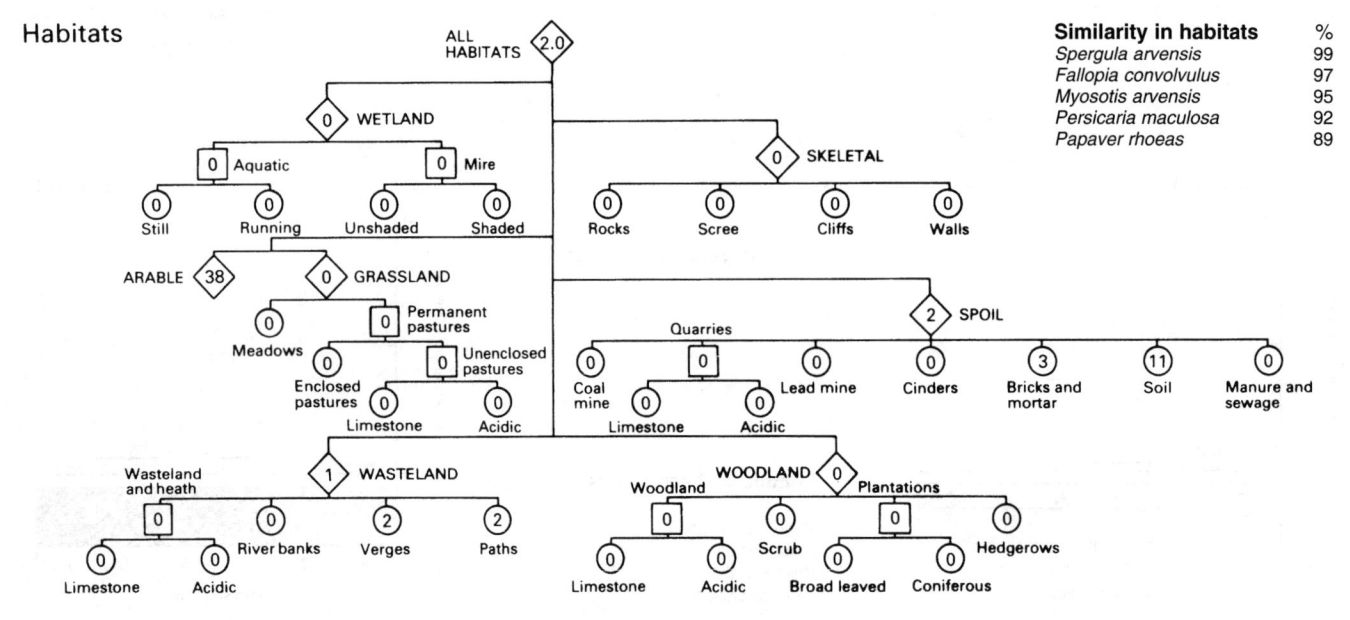

Habitats

Similarity in habitats	%
Spergula arvensis	99
Fallopia convolvulus	97
Myosotis arvensis	95
Persicaria maculosa	92
Papaver rhoeas	89

Synopsis *V.p.* is a fast-growing, annual herb of disturbed but fertile soils, with procumbent stems. It occurs on fallow arable land or beneath the crop canopy (usually of cereals). The species is strongly suppressed in shade (Fitter & Ashmore 1974). Flowers and fruits are formed sequentially in the axils of the leaves. The species is a native of SW Asia, and was first recorded from Europe in *c.* 1800 and from the British Isles in 1825 (*FE3*, Salisbury 1964). Subsequently the species has spread rapidly, perhaps mainly as an impurity in agricultural seeds, in fodder and in manure (Salisbury 1964). *V.p.* is now common throughout the British Isles. It has a high potential seed production, and an average plant may produce >6000 seeds (Harris & Lovell 1980a). The species also forms a persistent seed bank, and the seeds, which show little innate dormancy, are capable of germination during every month of the year (*WCH*). Further, two generations may be produced in a single season. Although capable of exploiting both transient and semi-permanent disturbed habitats, the species is mainly recorded as an arable and garden weed. Its tendency to be restricted to almost continuously disturbed sites may be related to its inferior colonizing potential compared with many other ruderals. Though the seed output per plant is comparable, the seeds are dispersed from capsules held in close contact with the soil, with the result that many appear to be directly incorporated into a buried seed bank. The large seeds are specialized for dispersal by ants, but transport by human activity is more important in the agricultural and horticultural systems exploited by most contemporary populations. Shoot fragments regenerate if placed on moist garden soil or when cultivated in the laboratory (Harris & Lovell 1980b, Hodgson unpubl.), and vegetative regeneration is potentially a significant process in frequently disturbed situations. An unusual degree of herbicide resistance has been recorded for certain populations (Grignac 1978). Two possibly native species (but see Webb 1985), *V. agrestis* L. ($2n = 28$) and *V. polita* Fries ($2n = 14, 18$), also exploit arable land. Both are less robust than *V.p.*,

Altitude
Slope
Aspect
Soil pH
Associated floristic diversity
Triangular ordination
Hydrology
Absent from wetland
Bare soil

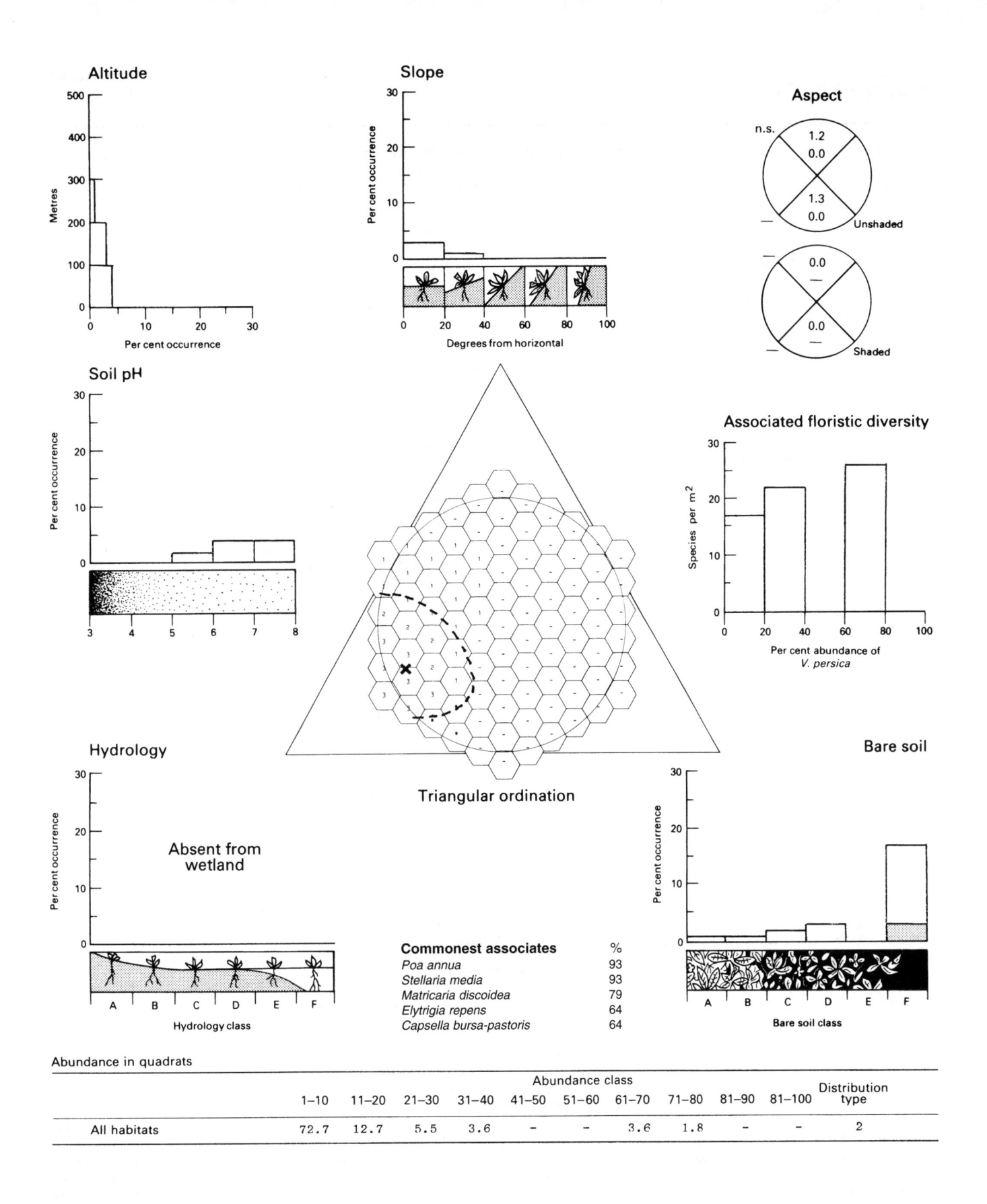

Commonest associates	%
Poa annua	93
Stellaria media	93
Matricaria discoidea	79
Elytrigia repens	64
Capsella bursa-pastoris	64

Abundance in quadrats

	\multicolumn{10}{c}{Abundance class}	
	1–10	11–20	21–30	31–40	41–50	51–60	61–70	71–80	81–90	81–100	Distribution type
All habitats	72.7	12.7	5.5	3.6	–	–	3.6	1.8	–	–	2

and the rate of dry-matter production of *V.a.* is less than that of *V.p.* (Harris & Lovell 1980a). Within the Sheffield area the altitudinal limit of *V.a.* is higher. In the lowland, but not the upland, part of the region this species is also restricted to non-calcareous strata. *V.p.* is more strictly a lowland plant, and is most frequent on calcareous clays. The two species appear to overlap ecologically and often occur in association with *V.p.* It is not clear which features currently limit their effectiveness as arable weeds.

Current trends As a result of the application of herbicides, and other methods of weed control, *V.p.* may be decreasing. However, the species is likely to remain a common arable and garden weed.

Vicia cracca L.

Fabaceae
Tufted Vetch

Life-form Shortly-rhizomatous, polycarpic perennial protohemicryptophyte. Shoots scrambling; leaves hairy, pinnate, usually <2000 mm² with a tendril; roots with VA mycorrhizas and nodules.

Phenology Flowers from June to August and sheds seed from August to September. Shoots die back in autumn.

Height Foliage and flowers up to 1 m or occasionally even 2 m.

Seedling RGR No data. SLA 25.9. DMC 23.9.

Nuclear DNA amount 11.5 pg; $2n = 14, 27, 28*, 30$ (*BC*); $2x, 4x*$.

Established strategy Intermediate between competitor and C-S-R.

Regenerative strategies ?B_s, V. Requires further study (see 'Synopsis').

Flowers Blue-purple, hermaphrodite, insect-pollinated; c. 10-30 in axillary racemes.

Germinule 14.29 mg, 2.8 × 2.6 mm, seed in a 2-6-seeded dehiscent pod.

Germination Hypogeal; t_{50} 10 days; hard-coat dormancy.

Biological flora Aarssen *et al.* 1986, see also Zarzycki (1976).

Geographical distribution The British Isles (100% of British vice-counties), *ABIF* page 380; N Europe (64% of European territories); Asia; introduced in N America.

Habitats Typically a plant of the rough grassland associated with wasteland, road verges, river banks, hedgerows, vegetated scree slopes, cinder tips, unimproved pasture and demolition sites. Only a casual on arable land. Observed in mire and meadows, and at woodland margins. Absent from aquatic and most skeletal habitats.

Gregariousness Sprawling individuals may form small patches.

Altitude Observed up to 380 m, but most frequent and only abundant between 100 and 200 m, despite an apparent abundance of suitable forms of land use above this altitude. [Up to 450 m.]

Slope Virtually absent from skeletal habitats, and for this reason absent from the steepest slopes.

Aspect Widely distributed.

Hydrology Essentially a dryland species, but sometimes observed on moist ground at the margin of topogenous mire.

Soil pH Absent from acidic soils below pH 4.5.

Bare soil Distribution biased towards habitats having relatively little exposed soil.

△ **ordination** Absent from disturbed habitats, with distribution centred on vegetation associated with undisturbed and highly fertile situations.

Associated floristic diversity Typically intermediate, but found in species-poor vegetation; stunted plants have been observed in species-rich meadows with up to 43 spp. m⁻².

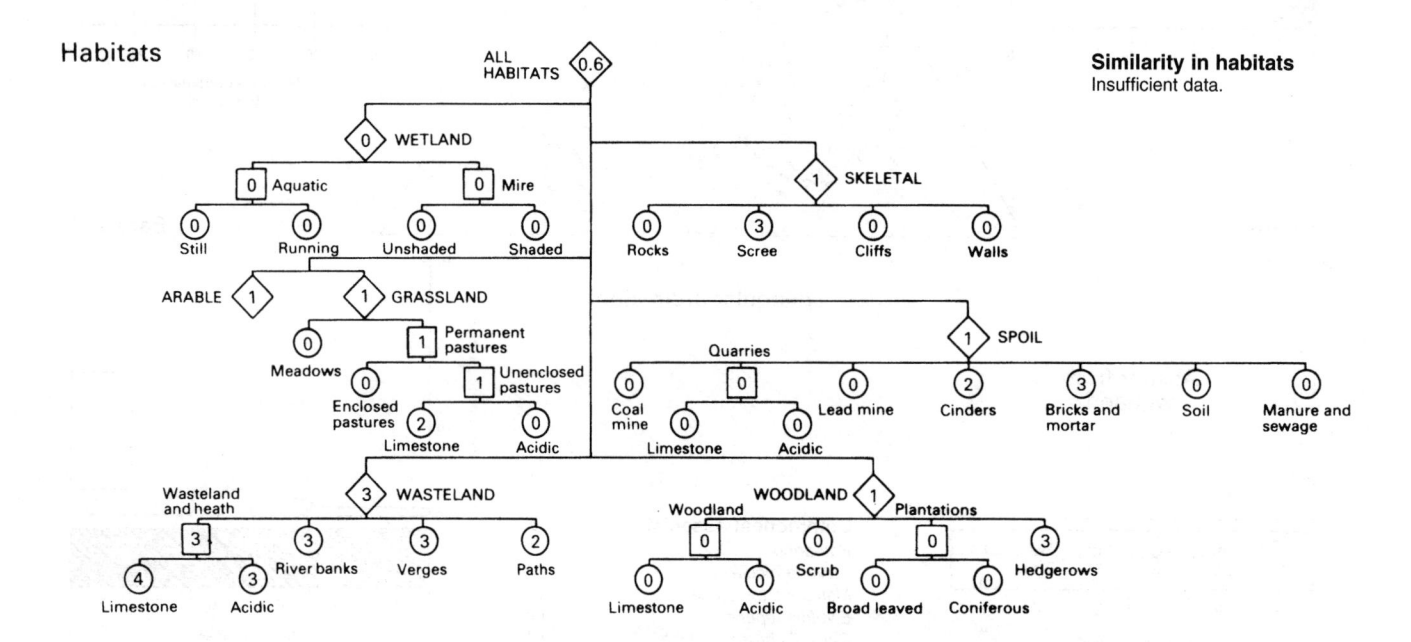

Habitats

Similarity in habitats
Insufficient data.

Synopsis A species much in need of further study (see *V. sepium*). *V.c.* is a large scrambling legume supported on the surrounding vegetation by tendrils. Nitrogen-fixing root nodules are formed in conjunction with *Rhizobium leguminosarum*. *V.c.*, which is found on relatively fertile, moist soils, is capable of forming small clonal patches and produces a moderately dense canopy. However, its capacity for dominance is restricted by its dependence upon other species for support. Studies of the dynamic interactions between *V.c.* and its supporting species are required to understand the factors allowing *V.c.* to attain local dominance. Robust plants (to 2 m) are often found at the margins of hedgerows. *V.c.* has the capacity to persist in a non-flowering condition in shaded sites, and in tall stands such as those of *Phragmites australis V.c.* frequently occurs as stunted individuals. The species is virtually absent from pasture, and its sensitivity to grazing is presumably related to the erect growth form. However, *V.c.* persists in meadows, including those which are grazed later in the year. *V.c.* has a very limited capacity for vegetative spread below ground, and plants may produce 1-25 stems from a short rhizome (Chrtkova-Zertova 1973a). *V.c.* regenerates primarily by means of the large seeds, many of which are said to germinate in autumn, soon after shedding (Muller 1978). Seed persistence has been reported in a minority of the sites occupied by the species. Much of the considerable variability with which *V.c.* is associated represents phenotypic plasticity (Chrtkova-Zertova 1973a). Within Europe *V.c.* is included within a critical group of four species (*FE2*). *V.c.* occurs both as a diploid and a tetraploid in Europe (*FE2*). The diploid has a narrow geographical and ecological range, whereas the tetraploid, which is presumably of more-recent origin, is widespread and is probably the only level of ploidy occurring in N Europe (Chrtkova-Zertova 1973).

Current trends Not an efficient colonist of new artificial habitats, and probably decreasing.

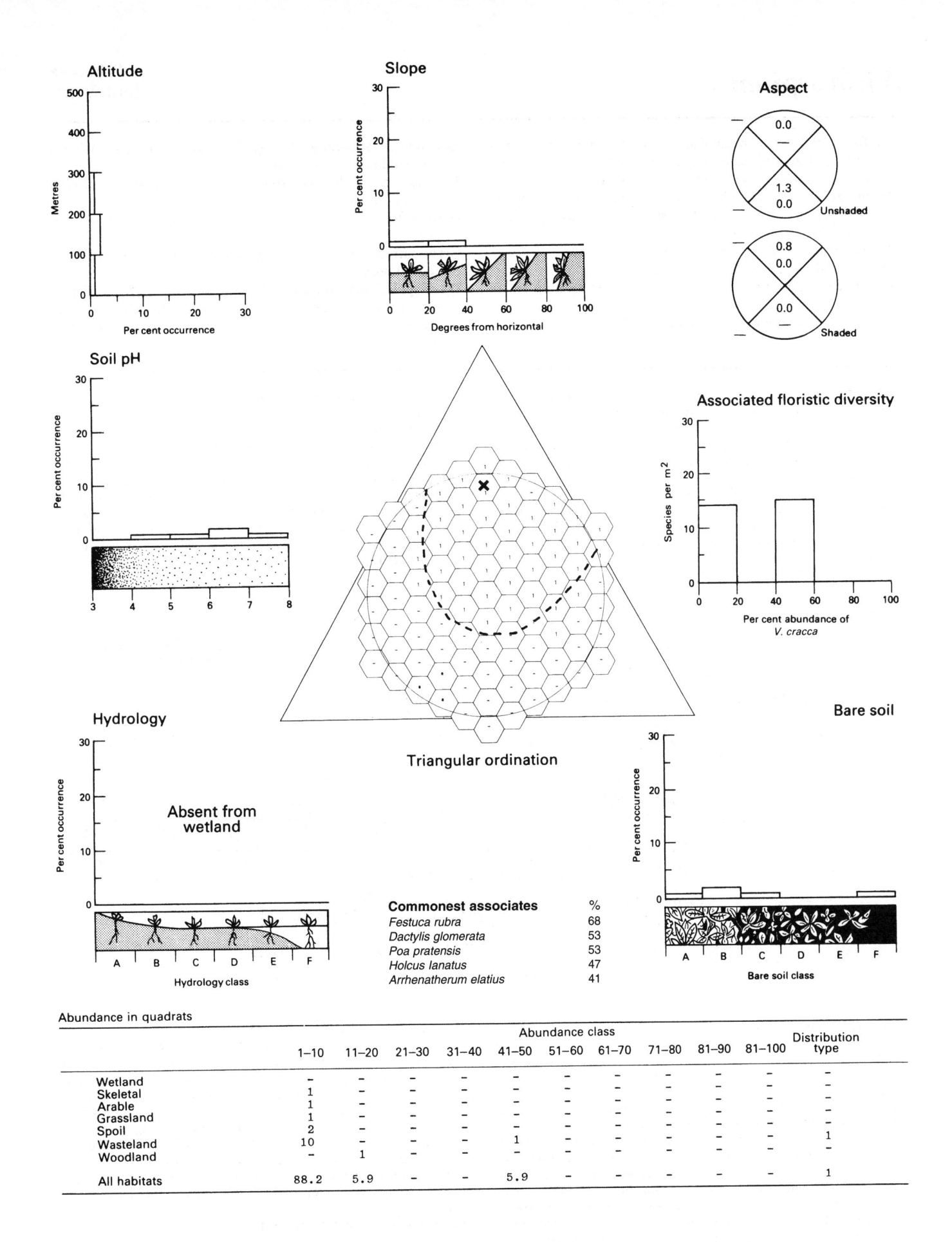

Altitude

Slope

Aspect

Unshaded

0.0	—
1.3	0.0

Shaded

0.8	—
0.0	0.0

Soil pH

Associated floristic diversity

Hydrology

Absent from wetland

Triangular ordination

Commonest associates %
Festuca rubra 68
Dactylis glomerata 53
Poa pratensis 53
Holcus lanatus 47
Arrhenatherum elatius 41

Bare soil

Abundance in quadrats

	Abundance class										Distribution type
	1–10	11–20	21–30	31–40	41–50	51–60	61–70	71–80	81–90	81–100	
Wetland	–	–	–	–	–	–	–	–	–	–	–
Skeletal	1	–	–	–	–	–	–	–	–	–	–
Arable	1	–	–	–	–	–	–	–	–	–	–
Grassland	1	–	–	–	–	–	–	–	–	–	–
Spoil	2	–	–	–	1	–	–	–	–	–	1
Wasteland	10	–	–	–	1	–	–	–	–	–	–
Woodland	–	1	–	–	–	–	–	–	–	–	–
All habitats	88.2	5.9	–	–	5.9	–	–	–	–	–	1

Vicia sepium L.

Fabaceae
Bush Vetch

Life-form Polycarpic perennial, protohemicryptophyte. Stems climbing; leaves pinnate, hairy, with tendrils, usually <4000 mm^2; roots with VA mycorrhizas and nodules.

Phenology Flowers from May to August, particularly during the earlier part of that period, and sets seed from July to September. Shoots die back in autumn.

Height Foliage and flowers to *c*. 750 mm.

Seedling RGR No data. SLA 39.7. DMC 18.6.

Nuclear DNA amount 9.4 pg, $2n = 14$ (Bennett & Smith 1976); $2x$.

Established strategy Intermediate between C-S-R and competitor.

Regenerative strategies V, ?B$_s$. Regeneration biology has been little studied (see 'Synopsis').

Flowers Purple, hermaphrodite, insect-pollinated; 2-6 in each axillary raceme.

Germinule 26.0 mg, 3.5 × 3.0 mm, in a dehiscent 3-7-seeded pod.

Germination Hypogeal; t_{50} 2 days; hard-coat dormancy; L = D.

Biological flora None.

Geographical distribution The British Isles (100% of British vice-counties), *ABIF* page 382; most of Europe, although rare in the Mediterranean region (79% of European territories); W Asia.

Habitats Most commonly recorded in tall moist grassland which is either unmanaged (wasteland) or infrequently cut (road verges). Found less frequently in hedgerows, woodland margins and pastures. Recorded for soil heaps and coal-mine tips, and also observed on limestone quarry spoil, lead mines and on cinder tips. Absent from skeletal habitats, wetland and arable land.

Gregariousness Often forming small patches.

Altitude Found over a wide range of values, but suitable habitats are more frequent at higher altitudes. Extends to 350 m. [Up to 820 m.]

Slope Absent from skeletal habitats and therefore restricted to flat ground or moderate slopes.

Aspect Slightly more frequent and only recorded as abundant on unshaded N-facing slopes.

Hydrology Absent from wetlands.

Soil pH Absent from acidic soils below pH 4.5.

Bare soil Associated most typically with sites having little bare soil, but some woodland, hedgerow and disturbed sites have large exposures.

△ ordination Distribution centred on relatively productive and slightly disturbed vegetation, but also extends into undisturbed, relatively unproductive vegetation types. Absent from vegetation occupied by competitive dominants, and excluded from heavily disturbed habitats.

Associated floristic diversity Typically intermediate.

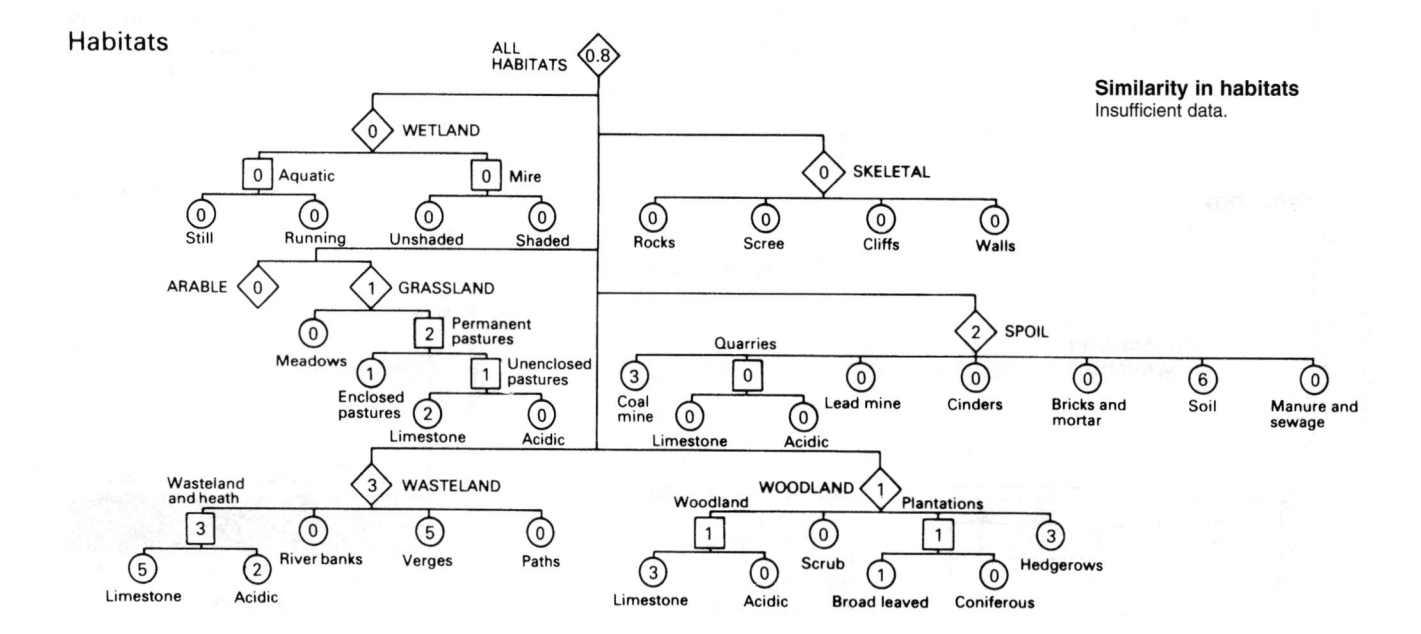

Habitats

Similarity in habitats
Insufficient data.

Synopsis Many aspects of the ecology of *V.s.* require further study. *V.s.* is a scrambling legume with tendrils. The species forms nitrogen-fixing root nodules in association with *Rhizobium leguminosarum*. In lowland Britain *V.s.* is particularly characteristic of hedgerows and woodland margins, while in upland areas the species is frequent on unshaded road verges and ungrazed dale-sides. Thus, *V.s.* appears to be a 'semi-shade' plant, although this may be a reflection of moisture demand rather than shade tolerance. *V.s.* rarely occurs in tall, unmanaged vegetation, tending instead to grow where the vigour of potential dominants is checked either by disturbance, such as that imposed by the infrequent cutting of roadsides, or by lower productivity, as in an unfertilized dale-side grassland. *V.s.* forms relatively long-lived patches but, as with most other native legumes, it is not known whether this local monopoly of the leaf canopy persists over long periods or whether the fixation of nitrogen alters the equilibrium between *V.s.* and its competitors. The possibility that the balance between *V.s.* and associated grasses experiences cyclical fluctuations dependent upon the changing nitrogen level of the soil (as is the case for *Lolium perenne* and *Trifolium repens*) requires investigation. *V.s.* has a limited capacity for vegetative spread, and can produce flowers and seeds over a major part of the summer. However, it is not clear whether regeneration by vegetative means or by seed is more important, and no convincing evidence of a persistent seed bank is available. As with most native legumes, there is a poor understanding of the mechanisms and significance of hard-coat dormancy under British climatic conditions. The large seeds of *V.s.* appear to be ineffectively dispersed, and *V.s.* is seldom a colonist of new artificial habitats.

Current trends Probably decreasing.

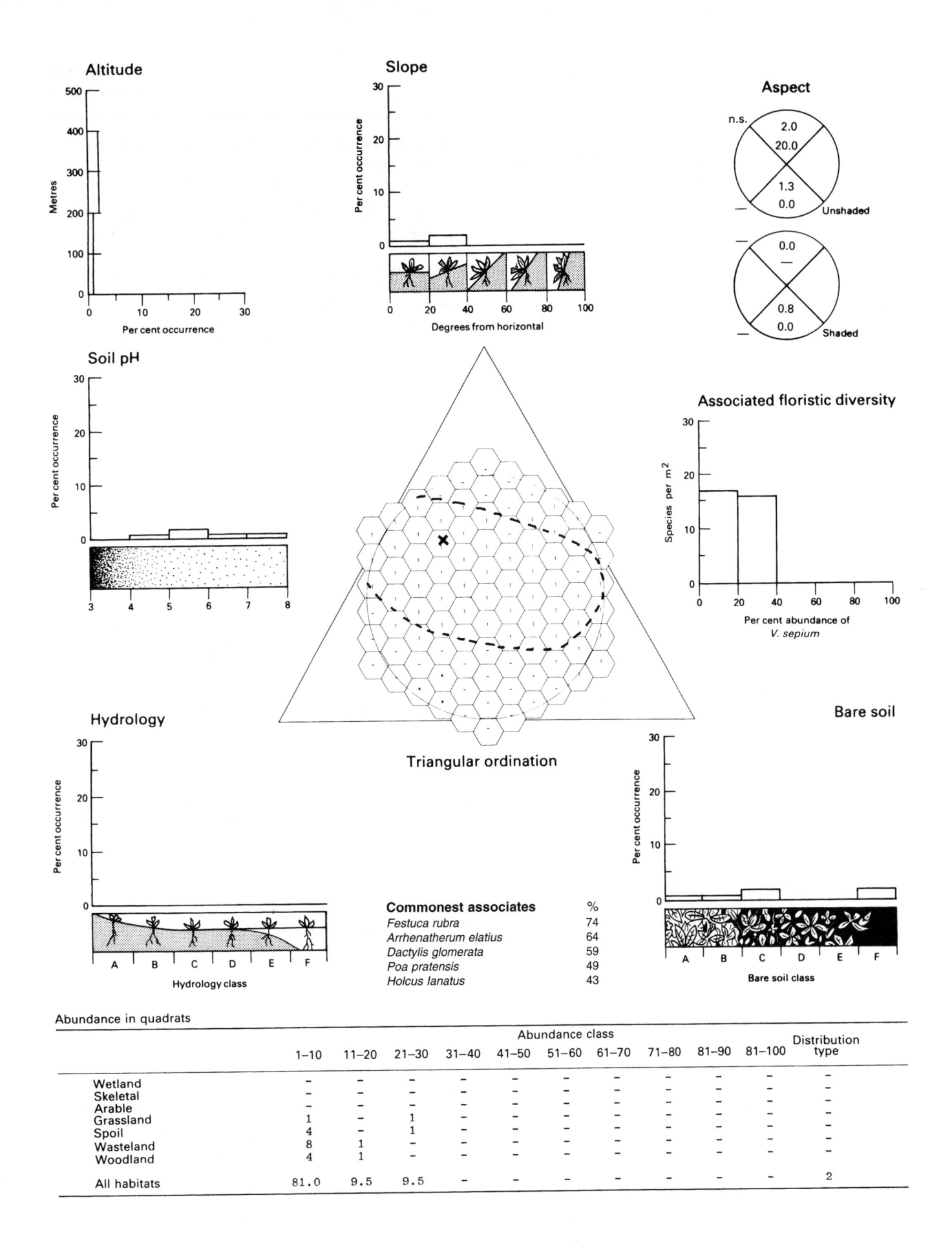

Altitude

Slope

Aspect

n.s. | 2.0
20.0 |
1.3 |
— | 0.0   Unshaded

— | 0.0
— |
0.8 |
— | 0.0   Shaded

Soil pH

Associated floristic diversity

Hydrology

Triangular ordination

Bare soil

Commonest associates	%
Festuca rubra	74
Arrhenatherum elatius	64
Dactylis glomerata	59
Poa pratensis	49
Holcus lanatus	43

Abundance in quadrats

	\multicolumn{10}{c}{Abundance class}	Distribution type
	1–10	11–20	21–30	31–40	41–50	51–60	61–70	71–80	81–90	81–100	
Wetland	–	–	–	–	–	–	–	–	–	–	–
Skeletal	–	–	–	–	–	–	–	–	–	–	–
Arable	–	–	–	–	–	–	–	–	–	–	–
Grassland	1	–	1	–	–	–	–	–	–	–	–
Spoil	4	–	1	–	–	–	–	–	–	–	–
Wasteland	8	1	–	–	–	–	–	–	–	–	–
Woodland	4	1	–	–	–	–	–	–	–	–	–
All habitats	81.0	9.5	9.5	–	–	–	–	–	–	–	2

Viola hirta L.

Violaceae
Hairy Violet

Life-form Rhizomatous, polycarpic perennial, rosette hemicryptophyte. Stock short erect; leaves ovate, hairy, typically <2000 mm²; roots with VA mycorrhizas.
Phenology Winter green. Plant produces chasmogamous flowers in April and cleistogamous ones later during summer (Beattie 1969). Seed set June to August. Leaves relatively long-lived.
Height Foliage and flowers usually <150 mm.
Seedling RGR No data. SLA 20.5. DMC 23.6.
Nuclear DNA amount No data; $2n = 20^*$ (*BC*); $2x^*$.
Established strategy Intermediate between stress-tolerator and C-S-R.

Regenerative strategies V, ?S. Regeneration biology is complex (see 'Synopsis'); SB II.
Flowers Blue-violet, hermaphrodite, insect-pollinated, followed towards the end of the flowering season by cleistogamous flowers without petals; both types solitary and axillary.
Germinule 2.81 mg, 4.0 × 1.4 mm, seed, several in each dehiscent capsule.
Germination Epigeal; t_{50} 4 days; dormancy broken by chilling.
Biological flora None.
Geographical distribution Mainly restricted to S England, rare in the N and W (54% of British vice-counties), *ABIF* page 225; most of Europe (74% of European territories and 3% naturalized); USSR, Central Asia.

Habitats A species of grazed and ungrazed limestone grassland. Also found to a lesser extent on scree, waysides and quarry spoil, and in open woodland and scrub. Absent from wetlands and arable habitats.
Gregariousness Intermediate.
Altitude Observed up to 310 m. [Up to 600 m.]
Slope Mainly restricted to limestone dale-sides and hence distribution biased towards moderate slopes.
Aspect Distribution has a S-facing bias, significantly so in grasslands (*Ecol Atl*).

Hydrology Absent from wetland habitats.
Soil pH Absent on soils of pH <4.0, and most frequent and only abundant above pH 7.0.
Bare soil Wide-ranging, but abundant only in habitats which have little bare soil.
△ ordination Distribution centred on vegetation associated with undisturbed sites of low to intermediate fertility.
Associated floristic diversity Species-rich.

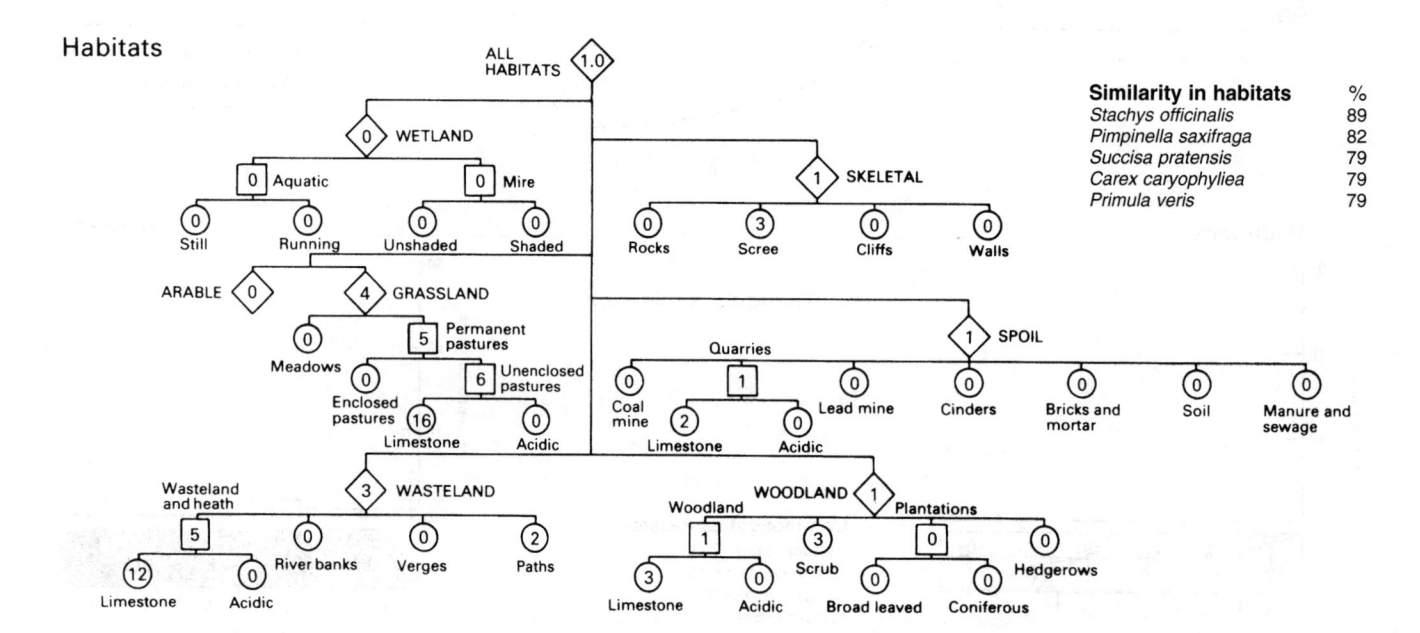

Habitats

Similarity in habitats	%
Stachys officinalis	89
Pimpinella saxifraga	82
Succisa pratensis	79
Carex caryophyllea	79
Primula veris	79

Synopsis *V.h.* is a winter-green herb of short turf on calcareous, relatively infertile soils and is most frequent in S England (*ABF*). Despite a marked bias towards S-facing slopes *V.h.* is restricted to less-droughted sites; this may be related in part to the fact that the root system is relatively shallow (Anderson 1927). These various features lead us to suspect that the distribution of *V.h.* is determined to a considerable extent by climatic factors. *V.h.* has some tolerance of shade, since it occurs infrequently in open woodland and scrub and the leaves also often experience some shading within the grassland canopy. However, *V.h.* does not normally persist in tall grassland. *V.h.* is one of the first species to flower in calcareous turf, but the species presents its maximum leaf canopy during summer, and spring- and summer-produced leaves are recognizably different in that the latter are considerably larger (*FE2*). The floral and regenerative biology of *V.h.* is similar to that of *V. riviniana*. However, since the seeds are passively discharged *V.h.* is dependent upon ants for dispersal (Beattie & Lyons 1975). At least within the Sheffield region, *V.h.* has extremely limited mobility. It has been suggested that ants' nests, to which seeds are taken, represent a favourable environment for seedling establishment (Culver & Beattie 1980). *V.h.* also regenerates vegetatively through the production of a branched rhizome. Regeneration by seed involves high fatalities and vegetative spread is slow. More research is needed to establish the relative importance of these two mechanisms of regeneration in natural populations. A small, late-flowering variant was formerly given subspecific rank (*CTW*), otherwise the extent of ecotypic variation is uncertain. Despite its comparatively narrow habitat range, the distribution of *V.h.* overlaps that of *V.r.* to a limited extent. In shaded habitats local contact also takes place with *V. odorata* L. (with which it hybridizes) and with *V. reichenbachiana* Jord.
Current trends The species is an indicator of long-established, semi-natural turf, and is decreasing.

Altitude
Slope
Aspect

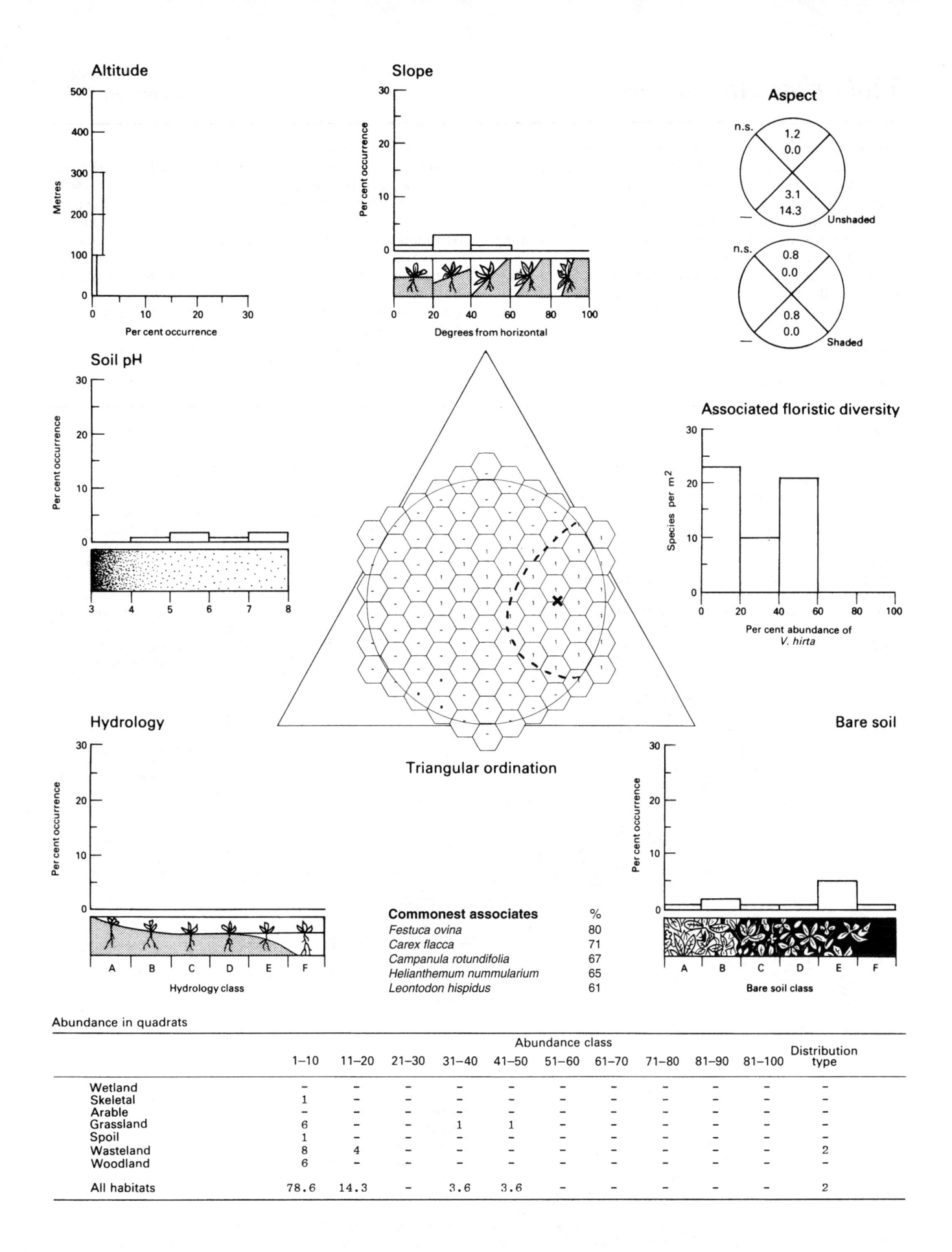

Unshaded
n.s.	1.2
0.0	3.1
	14.3

Shaded
n.s.	0.8
0.0	0.8
	0.0

Soil pH

Associated floristic diversity

Species per m² vs Per cent abundance of *V. hirta*

Hydrology

Triangular ordination

Bare soil

Commonest associates %
Festuca ovina 80
Carex flacca 71
Campanula rotundifolia 67
Helianthemum nummularium 65
Leontodon hispidus 61

Abundance in quadrats

	1–10	11–20	21–30	31–40	41–50	51–60	61–70	71–80	81–90	81–100	Distribution type
Wetland	–	–	–	–	–	–	–	–	–	–	–
Skeletal	1	–	–	–	–	–	–	–	–	–	–
Arable	–	–	–	–	–	–	–	–	–	–	–
Grassland	6	–	–	1	1	–	–	–	–	–	–
Spoil	1	–	–	–	–	–	–	–	–	–	2
Wasteland	8	4	–	–	–	–	–	–	–	–	2
Woodland	6	–	–	–	–	–	–	–	–	–	–
All habitats	78.6	14.3	–	3.6	3.6	–	–	–	–	–	2

Viola riviniana Reichenb.

Violaceae
Common Violet

Life-form Perennial hemicryptophyte with a short stock. Shoot branched, spreading or erect; leaves ovate, sometimes hairy, usually <2000 mm²; roots with VA mycorrhizas.
Phenology Winter green. Chasmogamous flowers open from April to June and cleistogamous flowers from June to September (West 1930). Seed set from June to September. Leaves relatively long-lived.
Height Foliage and flowers generally <200 mm.
Seedling RGR 0.5–0.9 week^{-1}. SLA 18.4. DMC 29.6.
Nuclear DNA amount 2.7 pg; $2n = 35^*, 40^*, 45, 46^*, 47^*$ (*BC, FE2*); c. $4x^*$.
Established strategy Intermediate between stress-tolerator and C-S-R.
Regenerative strategies V, S. Regeneration biology is complex (see 'Synopsis'); SB II.
Flowers Violet, hermaphrodite, insect-pollinated followed towards the end of the flowering season by cleistogamous flowers without petals; both types solitary and axillary. At intermediate times flowers of intermediate structure may be found (West 1930).
Germinule 1.01 mg, 1.9 × 1.5 mm, seed explosively dispersed from a *c.* 20-seeded capsule (Valentine 1950). Seeds from exposed sites 1.0–1.3 mg; from woodland sites 1.4–2.1 mg (Valentine 1941).
Germination Epigeal; dormancy broken by chilling; L > D.
Biological flora None.
Geographical distribution The British Isles (100% of British vice-counties), *ABIF* page 226; Europe except for the SE (79% of European territories); N Africa.

Habitats Recorded most abundantly in short, grazed calcareous turf and on limestone scree, but also common on wasteland and lead-mine spoil. In lowland parts of Britain *V.r.* is more characteristically a woodland plant. Also recorded from waysides, river banks, cinder tips, mire, rock outcrops and walls. Only a casual on arable land, and absent from aquatic habitats.
Gregariousness Intermediate.
Altitude Wide-ranging, extending to 400 m. [Up to 1020 m.]
Slope Occurs over a wide range of slopes, but is largely absent from the steepest.
Aspect In both shaded and unshaded habitats more frequent and abundant on N-facing slopes.
Hydrology Associated with moist rather than waterlogged soils.

Soil pH Wide-ranging, extending from highly calcareous to mildly acidic soils, but infrequent below pH 4.0.
Bare soil Occurring in both grassland turf with low exposures of soil and woodland, where the amounts of bare soil may be high.
△ **ordination** Widely scattered distribution centred on vegetation associated with relatively undisturbed, unproductive conditions.
Associated floristic diversity Varying according to habitat; typically intermediate in woodland; species-rich in grassland.
Synopsis *V.r.* is a long-lived, slow-growing, small, winter-green herb mainly restricted to infertile habitats. Exceptionally among British herbs, *V.r.* exploits open habitats, particularly limestone screes and woodland herb layers. *V.r.* is equally common on highly calcareous and on mildly acidic soils, and the leaves tend to contain high concentrations of Mg and low levels of P (*SIEF*). The species is tolerant of burning, and in pastures tends to be little grazed. The niche of *V.r.* in grassland habitats, where it is generally a minor component, is poorly understood, but foliage tends to

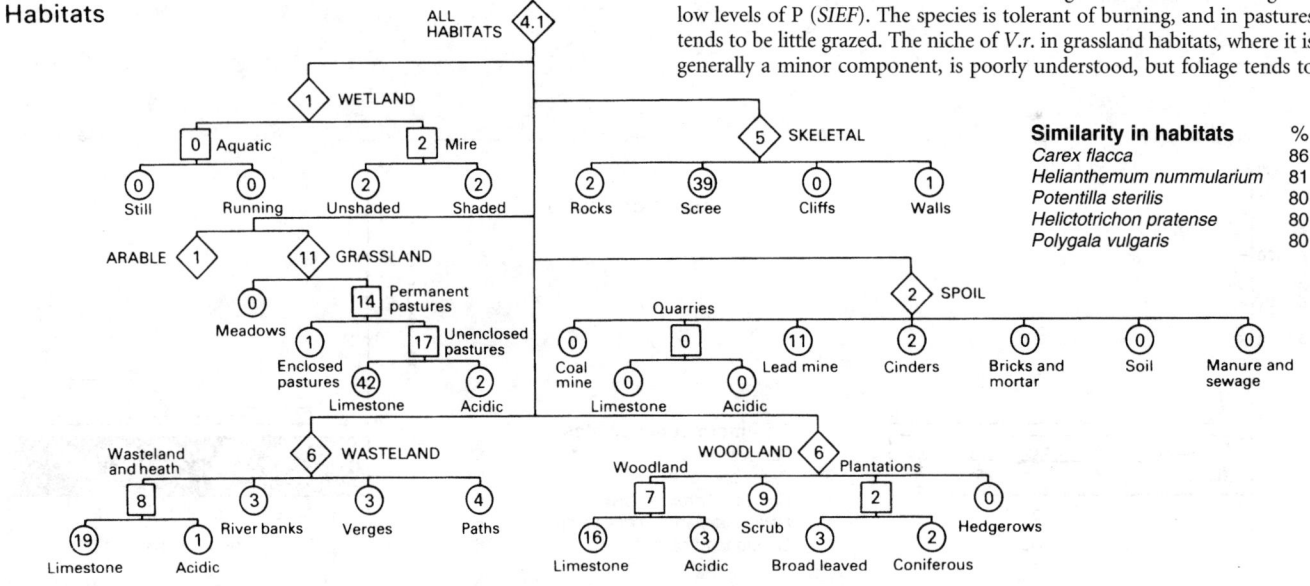

Habitats

Similarity in habitats	%
Carex flacca	86
Helianthemum nummularium	81
Potentilla sterilis	80
Helictotrichon pratense	80
Polygala vulgaris	80

Habitat occupancy diagram: ALL HABITATS 4.1; WETLAND 1 (Aquatic: Still 0, Running 0; Mire 2: Unshaded 0, Shaded 0); SKELETAL 5 (Rocks 2, Scree 39, Cliffs 0, Walls 1); ARABLE 1; GRASSLAND 11 (Meadows 0; Permanent pastures 14: Enclosed pastures 1 — Limestone 42, Acidic 2; Unenclosed pastures 17); SPOIL 2 (Quarries 0: Coal mine 0, Limestone 0; Lead mine 11; Cinders 2; Bricks and mortar 0; Soil 0; Manure and sewage 0); WASTELAND 6 (Wasteland and heath 8: Limestone 19, Acidic 1; River banks 3; Verges 3; Paths 4); WOODLAND 6 (Woodland 7: Limestone 16, Acidic 3; Scrub 9: Broad leaved 3, Coniferous 2; Plantations 2; Hedgerows 0).

occupy lower layers of the leaf canopy and the species appears to be shade-tolerant. In woodland *V.r.* is usually scarce in areas with a high density of vernal species, but may be a relatively conspicuous component on shallow skeletal soils. In transplant experiments (Sydes & Grime 1981b) *V.r.* was found to be capable of surviving moderate depositions of persistent tree leaf litter, mainly by marked extension growth of the petioles. *V.r.* produces both insect-pollinated and, later, cleistogamous flowers (Beattie 1969). A majority of the seeds produced appears to originate from the latter. The seeds, which are heavily predated, are shed by explosive dehiscence and are subsequently dispersed by ants (Beattie & Lyons 1975). Seeds have a chilling requirement which, in calcareous grassland, may result in the appearance of large, even-aged populations of seedlings in the spring (Hillier 1984). Vegetative regeneration through adventitious sprouting from the roots is also possible in some plants (Valentine 1949). Despite the fact that *V.r.* regenerates mainly by vegetative means and from cleistogamous flowers, the species is variable in chromosomal complement. Ecotypic differences have been described, although their relationship to cytology is obscure. Populations occurring in woodlands are genetically and physiologically different from those established in unshaded habitats (Valentine 1941). In woodland habitats on heavy, usually base-rich, soils, *V.r.* often occurs with *V. reichenbachiana* Jord. However, the latter flowers slightly earlier, produces fewer and heavier seeds (Valentine 1950) and is, we suspect, even more shade-tolerant than *V.r.* The two species occasionally hybridize (Stace 1975). In dry, often sandy habitats on acidic soils *V.r.* shows a tendency to be replaced by the later-flowering *V. canina* ssp. *montana* (L.) Hartman, with which it also hybridizes (Stace 1975).
Current trends Seed dispersal by ants is ineffective for long-distance dispersal to new artificial sites. Particularly in lowland areas, the species is probably decreasing, the more so in grassland than in woodland.

Altitude
Slope
Aspect

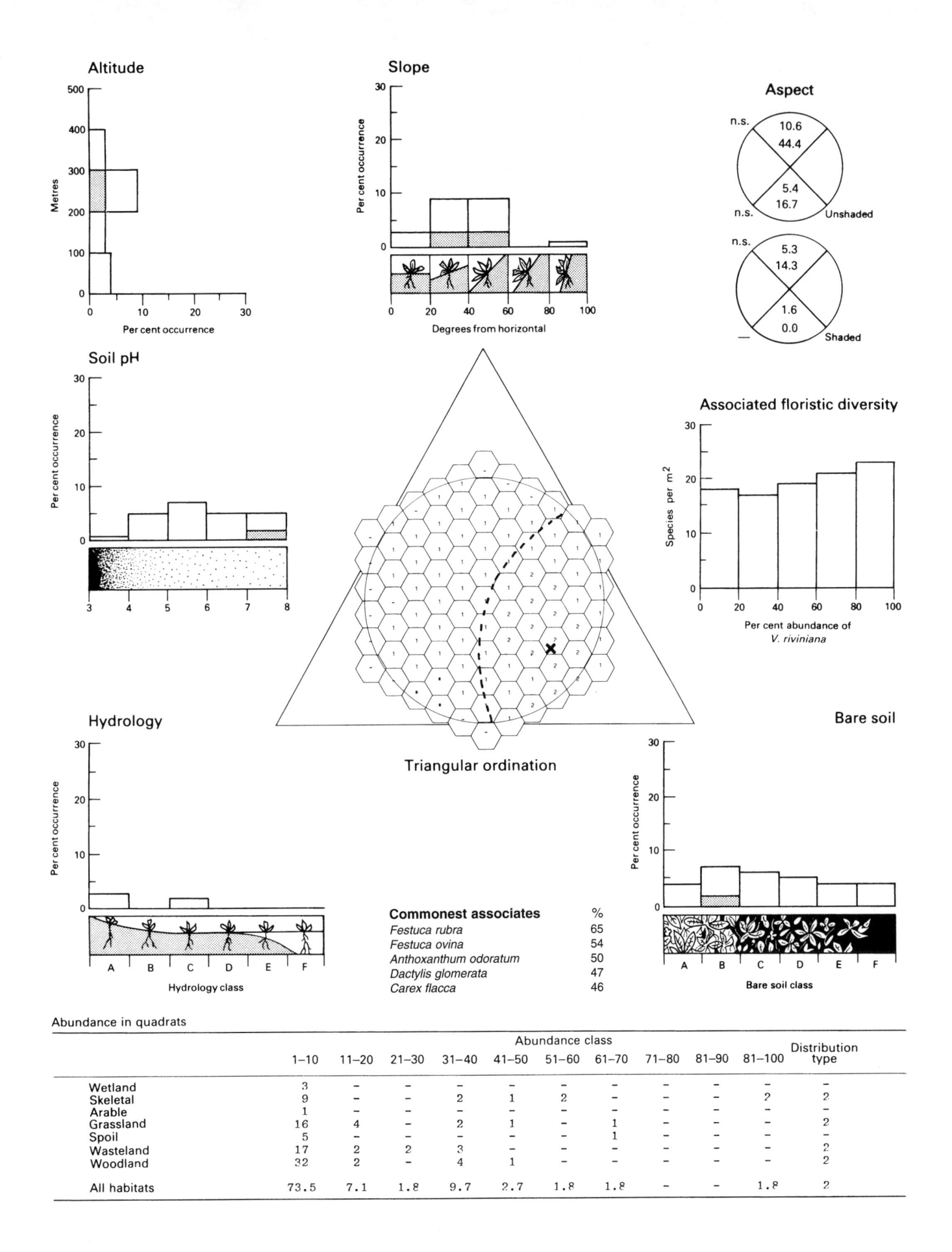

	n.s.	10.6	
	44.4		
n.s.	5.4		Unshaded
	16.7		

	n.s.	5.3	
	14.3		
n.s.	1.6		Shaded
—	0.0		

Soil pH
Associated floristic diversity
Hydrology
Triangular ordination
Bare soil

Commonest associates %
Festuca rubra 65
Festuca ovina 54
Anthoxanthum odoratum 50
Dactylis glomerata 47
Carex flacca 46

Abundance in quadrats

	1–10	11–20	21–30	31–40	41–50	51–60	61–70	71–80	81–90	81–100	Distribution type
Wetland	3	–	–	–	–	–	–	–	–	–	–
Skeletal	9	–	–	2	1	2	–	–	–	2	2
Arable	1	–	–	–	–	–	–	–	–	–	–
Grassland	16	4	–	2	1	–	1	–	–	–	2
Spoil	5	–	–	–	–	–	1	–	–	–	–
Wasteland	17	2	2	3	–	–	–	–	–	–	2
Woodland	32	2	–	4	1	–	–	–	–	–	2
All habitats	73.5	7.1	1.8	9.7	2.7	1.8	1.8	–	–	1.8	2

7 TABLES OF ATTRIBUTES

Introduction

This chapter sets out in tabular form a collection of standardized autecological information for the more common vascular species of the British flora. In addition to providing summaries for the 281 species which are the subjects of accounts in Chapter 6, the tables include data on a further 540 species, which are either (a) recorded in at least 30% of 10 km squares in England, Wales, Scotland, N. Ireland or Eire or (b) widespread and increasing aliens or (c) colonists of saline road verges from maritime habitats. Thus, autecological information is provided for a total of 822 species including the more widespread species in each of the five countries in the British Isles. Only the flora of maritime and montane habitats are poorly represented.

The choice of species was made after reference to Preston *et al.* (2002a) and 320 new species have been added to the ecological summary tables since the first edition. Ecological information for these additional species (and for *Carlina vulgaris*, *Phragmites australis* and *Fallopia japonica*, which were recorded from fewer than ten quadrats in the UCPE Survey II, see p. 18) is derived from the unpublished data of Hodgson *et al.* (Survey III, see p. 18). For completeness we also include one species, *Sedum anglicum*, that was not recorded during fieldwork in the Sheffield area, and which did not reach 30% in any country. This species had been included in the first edition, where different selection criteria had been used. We have attempted to update and revise data from the first edition to take into account ecological advances in the last 19 years and our experiences of using the data in ecological analyses. The quantity and quality of information available still varies considerably according to species. As part of the extended explanations in Chapter 6 we mentioned that certain species are variable in particular biological characteristics, and that field distributions with respect to habitat features such as slope and aspect may be too strongly affected by geographical location to allow reliable generalizations at a national scale. Despite these problems, it has been possible to present data for 25 characters which, apart from caveats included in the explanatory notes, can be applied with some confidence to the species observed in the British Isles; these relate to ecology (Table 7.1), attributes of the established phase (Table 7.2) and attributes of the regenerative phase (Table 7.3). The significance and use of each of the 26 characters is discussed in Chapter 5.

Explanation of Tables

Nomenclature

Nomenclature follows Stace (1997) except for the following species groupings: *Betula* spp. (*B. pendula*, *B. pubescens* and hybrids; the species are also separated in relation to their ecological distribution on the basis of recent field observations), *Festuca ovina* (also includes *F. filiformis*), *Medicago sativa* (relates to spp. *sativa* only), *Poa pratensis* agg. (*P. pratensis* sensu stricto and *P. humilis*; the other segregate, *P. angustifolia*, can generally be identified vegetatively and is treated separately), *Quercus* agg. (*Q. petraea*, *Q. robur* and their hybrid; the individual species are also separated as with *Betula*), *Rosa* spp. (*R.* section caninae DC), *R. canina* (includes *R. caesia* Smith), *R. mollis* (includes *R. sherardii* Davies and *R. tomentosa* Smith), *Salix cinerea* agg. (*S. aurita*, *S. caprea*, *S. cinerea* and hybrids; the individual species are also separated as with *Betula*) and *Taraxacum* agg. (divided into nine sections in Stace (1997)).

One column of symbols precedes the scientific name. Species which are the subject of Accounts in Chapter 6 are indicated by a '+' and those included in the summary tables of the first edition by a '–'. Only species included for the first time in this edition lack a symbol. Only in Table 7.1 are the authorities for the scientific names cited.

Ordering of species

Species lists are divided into three groups: (a) herbs, and woody species up to 1.5 m in height, (b) woody species exceeding 1.5 m in height, and (c) pteridophytes. Within each group species are arranged in alphabetical order.

(Table 7.1) Ecological attributes

For most species only one set of data is given. However, in the case of individual trees and shrubs ecological attributes are, wherever possible, presented separately for seedlings and small saplings, 'juveniles', and for the more mature individuals, that contribute to the canopy.

(1) Native or introduced

In the first edition we subdivided species solely into two groupings, native in the British Isles or introduced. Here, both in the summary tables and the main accounts, we use the newer system adopted by Preston *et al.* (2002a) which uses the archaeological record and ecological criteria to provide a provisional assessment of the status of all species in the British flora. The work represents a significant advance in our knowledge of the British flora and the subdivision of aliens into two groupings, archaeophyte (naturalised before AD 1500) and neophyte (more recent), is especially valuable. The categories of Preston *et al.* are identified as follows: nat (native), arch (archaeophyte), neo (neophyte).

(2) Habitat range

The habitat key (Fig. 3.2, p. 21) has been used as a basis for identifying the habitat range of each species. First, an estimate of the abundance in each of the primary habitat groups [wetland (divided here into aquatic and mire), skeletal, arable, pasture, spoil, wasteland and woodland] is provided. We also identify maritime species that are colonising saline road verges as a separate category. Second, the terminal habitat in which the species is most frequent is indicated.

(a) Abundance in primary habitats The percentage frequency of the species in each major group is compared with the percentage frequency of the species in the overall survey. The primary habitats appear as lozenge-shaped boxes in the 'Habitats' diagrams of the Species Accounts. The symbols used are as follows: 5, very common and characteristic of the particular habitat (percentage frequency >4 times the overall value); 4, common within habitat (>2-4 times the overall value); 3, widespread in the habitat >0.5-2 times the overall value); 2, infrequent and uncharacteristic of the habitat (0.25-0.5 times the overall value); 1, largely absent from habitat or confined to an uncommon variant (<0.25 of the overall value). In 'Woodland' in Survey II and 'Wasteland' in Survey III these primary habitat groups each constitute over 25% of all records, rendering the '5' rating an impossibility. It was decided that all species in which at least 80% of records fall within a single primary habitat should also be classified as '5'. For species for which there are insufficient field records (e.g. *Hypericum androsaemum*, *Salix repens* and *Sedum anglicum*) only the identity of the most characteristic major habitat is indicated.

(b) Commonest terminal habitat The terminal habitats appear as circular boxes in the 'Habitats' diagrams of the Species Accounts. The commonest terminal habitat for the subject species is identified as follows: AQUATp (lakes, canals, ponds and ditches), AQUATr (rivers and streams), ARABLE (arable), BRICK (bricks and mortar rubble), CANALB (canal banks, +), CINDER (cinder tips and cindery railway tracks), CLIFF (cliffs), COAL (coal-mine spoil), GARDEN (flower beds, vegetable plots, horticultural nurseries, +), HEDGE (hedgerows), LAWN (lawns, +), LEAD (lead-mine spoil), MANURE (manure and sewage spoil), MEADOW (meadows), MIREs (shaded mire), MlREu (unshaded mire), OUTCRPu (unshaded rock outcrop), OUTCRPs (shaded rock outcrop), PASTa (pasture on acidic strata), PASTe (enclosed pasture), PASTl (pasture on limestone strata), PATH (paths), PLANTb (broadleaved plantations), PLANTc (coniferous plantations), PONDBK (banks of lakes, canals, ponds and ditches, +), QRYa (quarry spoil on acidic strata), QYRl (limestone quarry spoil), RD/RLY (road verges/railway banks, +), RDVRGEs (shaded road verge, +), RDVRGEu (unshaded road verge), RIVBANKs (shaded river bank, +), RIVBNKu (unshaded river and stream banks), RLYBNK (railway banks, +), SCREEs (shaded limestone scree, +), SCREEu (unshaded limestone scree), SCRUB (scrub), SNDPIT (sand and gravel pits, +), SOIL (soil heaps), WALL (walls), WASTEa (wasteland on acidic strata), WASTEd (dry sandy wasteland droughted during summer, +), WASTEl (wasteland on calcareous strata), WOODa (woodland on acidic strata), WOODl (woodland on limestone strata), WOODle (edge of limestone woodland, +). Habitats suffixed '+' are minor habitats which, because of their infrequent occurrence, were excluded from the habitat classification of Survey II (Fig. 3.2, p. 21). These data are particularly useful for categorizing species of narrow ecological range (e.g. *Chaenorhinum minus*, associated with cinders, and *Lemna minor*, associated with lakes, canals, ponds and ditches). However, three grassland species (*Euphrasia officinalis*, *Leontodon autumnalis* and *Rumex acetosa*), and seedlings of *Ulmus glabra*, were most commonly recorded from lead-mine spoil even though they are characteristic of a range of base-rich habitats. In order to prevent the impression that these species are 'metallophytes' and largely restricted to soils contaminated with heavy metals, the habitat in which these species are second most common is substituted. Rather anomalous data were obtained for two wetland species which are persistent after drainage. *Eriophorum vaginatum* was recorded most commonly in pasture on limestone strata and *Lotus pedunculatus* most commonly in wasteland on limestone. In both species the second most common habitat (unshaded mire) has been promoted to first place. Four species that do not normally reach maturity in their most common habitat (*Athyrium filix-femina*, cliffs and walls, *Dryopteris filix-mas*, walls, *Equisetum palustre*, rivers and streams, and *Ulex europaeus*, river banks) have been similarly treated. For each of these ten species the habitat code is preceded by an asterisk to warn that the information has been modified in this manner. These examples illustrate that data concerning the most common terminal habitat could, if utilized in isolation, lead to misinterpretation. Their use in conjunction with the distribution of the species in major habitats is advised.

N.B. The range and diversity of major habitats differ according to region. For example, in areas where (unlike that sampled here) there is little limestone, species such as *Carex panicea* would be regarded solely as a wetland species. Here it is well represented in 'Wasteland' and 'Grassland' and only scores highly for 'Mire' through the inclusion of data from Survey III.

(3) Soil pH

The data take the form of a number indicating the median surface soil pH value for the species to nearest 0.5 units, followed by a letter indicating the number of pH units that are included within the interquartile range. Thus, '7.0a' would indicate that the median pH for the species is between 6.75 and 7.25 and its interquartile range is ≤ 1 pH unit. In contrast, '3.5c' would indicate a median pH value between 3.25 and 3.75 and an interquartile range of >2 pH units. A few species exploit both strongly acidic and highly calcareous soils and thus possess a bimodal pH distribution. These are additionally identified by an asterisk. Where based upon less than ten values, median pH is given in parentheses and interquartile range is omitted.

In the first edition values were 'corrected' to take into account that some pH classes were more frequently encountered than others. This had the unfortunate result of assigning many species of calcareous grasslands to a pH class of 5 (pH 5.0-6.0). The value is, however, simply a mathematical artefact of an unbalanced distribution of soil pH within the UCPE dataset. Many agricultural and other artificial habitats, which constitute a major part of the survey database, are largely restricted to soils of pH >6.0 but semi-natural communities, and their constituent species, occur over a much wider range of pH. The values presented here are 'uncorrected' for the distribution of pH values and are, therefore, a more accurate reflection of the field distribution of the species within our survey area.

N.B. pH range varies according to region. Some species that are narrowly restricted to soils of high pH in SE England extend onto more acidic soils in the N and W.

(4) Floristic diversity

The mean number of species m^{-2} with which each species is associated has been calculated by reference to survey data from the Sheffield region. The classes recognized are as follows: 1, 10.0 species, or fewer; 2, 10.1-14.0; 3, 14.1-18.0; 4, 18.1-22.0; 5, >22.0.

(5) Distribution in Europe

Since the publication of the first edition Preston and Hill (1997) have provided a detailed classification of the geographical range of native British species in Europe and western Asia and we have used this to update our estimates of the longitudinal and latitudinal range of species. Preston and Hill identify the eastern, longitudinal limit of species as follows: *Oceanic* (restricted to the Atlantic zone at the western margin of Europe, denoted here as longitudinal zone 1); *Suboceanic* (extending further east to western parts of Mediterranean, Central Europe or Sweden, found in zones 1 and 2), *European* (mainly European and not extending east of 60° East; 1-3), *Eurosiberian* (main distribution reaches easterly limit between 60° East and 120° East; 1-4), *Eurasian* (easterly limit of distribution 120°E; 1-5), *Circumpolar* (found in Europe, Asia and N. America; 1-6). Preston and Hill also describe the following latitudinal subdivisions from north to south: *Arctic-montane* (distribution mainly north of, or altitudinally above, the tree line, our zone 1); *Boreal-montane* (northern or montane coniferous zone, 2); *Temperate* (cool-temperate deciduous forest zone, 3); *Southern* (warm-temperate deciduous forest zone, including the Mediterranean, 4). Two of Preston and Hill's phytogeographical groupings have an essentially 'L-shaped' geographical distribution and, therefore, do not lend themselves to subdivision in terms of longitude and latitude. These are 'Mediterranean-Atlantic', restricted to the Mediterranean zone and the Atlantic margin of Europe, and 'Submediterranean-subatlantic', similar but with a broader range extending into south west parts of Central Europe. These are identified as M-A and S-s respectively in both longitudinal and latitudinal columns.

(6) Present status

The extent to which the abundance of species is changing in

response to modern methods of land-use has been estimated (see p. 40).

N.B. The estimation of changing abundance is extremely difficult since documentary evidence is at best fragmentary. Telfer *et al.* (2002) are, therefore, to be commended for developing a mathematical basis for such assessments as a forerunner to the analysis of the distributional data collected for the *New Atlas of the British & Irish Flora* (Preston *et al.* 2002a). Unfortunately, however, this change index, is fundamentally flawed. Its value critically depends on both the abundance of the species (common species always have low values for change) and taxonomy (species that have in the past been under-recorded because of taxonomic difficulties are generally classified as increasing: in many cases such an assessment is counter-intuitive) - see Hodgson (2003). As a result analyses of change to the British flora using the change index (e.g. Preston *et al.* 2002b) are similarly suspect. Important modifications need to be made to the change index (see Hodgson 2003). Until these have been effected, we feel that our data from the first edition of *Comparative Plant Ecology* provide a more acceptable assessment of the present status of species than the values for change index presented in the *New Atlas of the British & Irish Flora* (Preston *et al.* 2002a). Species that are thought to be increasing are represented by '+1' and those considered to be declining by '-1'. Species of uncertain status and those showing no clear increase or decrease are given '0'.

(7) Red Data Book

The publication of the new Red Data Book (Cheffings *et al.* 2005) represents a milestone in British conservation. For the first time the conservation status of all native species and long-established aliens (archaeophytes; see Section 1 above) is assessed. Despite the general excellence of the Red Data Book we disagree strongly on one point. The choice of species for inclusion relies upon a provisional classification of the British flora made using a patently inadequate archaeological and historical record and ecological criteria that are inevitably subjective (see Preston *et al.* 2002a, Hodgson *et al.* 2004). Except where a special case can be made, 'more recent' aliens, neophytes, are treated as unworthy of conservation. We regard the protection of the natural and semi-natural habitats of our forbears and their constituent species as the key issue and would prefer to see all rare and declining species of these habitats protected rather than to make decisions based upon the fragmentary historical record for the British flora. Our view is very much in line with the 'inclusive' approach adopted in *A Nature Conservation Review* (Ratcliffe 1977) and in earlier editions of the *Red Data Book* (Perring & Farrell, 1977, 1983). Nevertheless, because of their importance, we include IUCN categories from the Red Data Book. These are identified as follows: 0, least concern; 1, near threatened; 2, vulnerable; 3, endangered. The most extreme categories, critically endangered and extinct, did not apply to species in our list. Species for which no value is available, are either neophytes or widespread hybrids. *Onobrychis viciifolia*, though classified as near threatened, is frequently sown (as more robust genotypes) in amenity grassland. Our data relate to recent introductions and status is identified as 1*.

(Table 7.2) *Attributes of the established phase*

(8) Life-history

The life-cycles of species have been classified with reference to Clapham *et al.* (1962), Tutin *et al.* (1964-80) and our own field observations. The scheme used is as follows: As, summer annual; Aw, winter annual; B, usually biennial; M, monocarpic perennial (of duration 2 years or, usually, more); P, polycarpic perennial. Some variable species were classified into more than one group; in such cases the most common form of life-cycle is listed first. In the case of *Plantago coronopus*, the symbol M+ indicates that the species is typically monocarpic, but is also capable of being either a winter annual or a polycarpic perennial. Only a minority of plants of *Lemna minor*, *L. gibba* and *L. trisulca* overwinter and these species are denoted as (P). For a few obligate or facultative annuals we are uncertain as to the main germination period and these species are identified by 'A' without a suffix.

(8) Established strategy

An assessment is provided of the established strategy, *sensu* Grime (1974), of the subject species (see p. 33). Where data are available, the methodology of Hodgson *et al.* (1999) has been used to classify species. Abbreviations are as follows: primary strategies C, competitor; R, ruderal; S, stress-tolerator; secondary strategies CR, competitive-ruderal; SC, stress-tolerant competitor; SR, stress-tolerant ruderal; CSR, C-S-R strategist. Strategy types intermediate between these seven are also recognized, e.g. CR/CSR.

(9) Specific leaf area

Specific leaf area (SLA) is expressed as mm^2 leaf area mg leaf mass^{-1} and values relate to material collected from unshaded field conditions. Using unpublished data collected by members of UCPE and supplemented by measurements by the Department of Archaeology at Sheffield University the following classes of SLA are separated: 1, <15; 2, 15-20; 3, 20-25; 4, 25-30; 5, >30; -, no data. There is good correspondence at the species level between values for UK and those from continental W Europe (J. G. Hodgson unpubl.). For a few species, where no UK data are available, we, therefore, present data from this source and identify values by an asterisk.

(10) Leaf dry matter content

Leaf dry matter content (DMC) is expressed as a percentage of the fresh weight of the leaf. The sources of data are the same as those used for specific leaf area (above) and the following classes of DMC are separated: 1, <15; 2, 15-20; 3, 20-25; 4, 25-30; 5, >30; -, no data. As with specific leaf area a small quantity of European data are included, again identified with an asterisk.

(11) Life-form

The life-form *sensu* Raunkiaer (1934), is abstracted from Clapham *et al.* (1962). Using this information, supplemented by field observations, the following classes have been separated: Ph, phanerophyte (woody plant with buds >250 mm above soil surface); Ch, herbaceous chamaephyte (plant with buds not in contact with but <250 mm above the soil surface); Chw, woody chamaephytes; H, hemicryptophyte (herb with buds at soil level); G, geophyte (herb with buds below the soil surface); Hel, helophyte (marsh plant); Hyd, hydrophyte (aquatic plant); Th, therophyte (plant passing the unfavourable season as seeds); Wet, wide-ranging wetland species (facultatively a helophyte or a hydrophyte). Monocarpic and biennial species are classified according to their vegetative overwintering condition, though they could equally well have been classified as therophytes.

(12) Canopy structure

The following classes are recognized from Clapham *et al.* (1962) and from field observations: B, basal (leaves confined to a basal rosette, or to a prostrate stem); S, semi-basal (stems erect or ascending, leafy but with the largest leaves towards their base); L, leafy (stems erect or ascending with no basal rosette, leaves of approximately equal size all the way up the stem); -, leaves small, reduced to scales or spines, with the stem as the main photosynthetic organ). In addition, for the smaller semi-basal and leafy species, those whose canopy does not exceed 100 mm, the suffix 'b' follows the category while the most robust semi-basal or basal species, with large leaves (>10000 mm^2), are identified as 'LA'. Monocarpic species (e.g. *Cirsium vulgare*), which spend their juvenile phase as a rosette and subsequently develop a leafy stem, have been classified as semi-basal species. *Hieracium* spp. includes both semi-basal and leafy taxa and is denoted as V (for various).

A number of additional categories have been used for aquatic species to identify the characteristic positions of leaves or, in the case of *Lemna* spp., the photosynthetic thallus, relative to the water surface: F, floating (some or all of the leaves floating on the water surface); U, underwater (all leaves submerged). For emergent aquatic species leaves are typically all aerial and no symbol is given but emergent aquatic with submerged, floating and emergent leaves are identified as 'E'. For *Callitriche* species, *Persicaria amphibia*, *Potamogeton polygonifolius* and *Sparganium emersum*, information is included on both terrestrial and aquatic leaf forms.

(13) Canopy height
The maximum height of the leaf canopy has been assessed from field observations, and the following broad classes have been constructed: -1, submerged aquatics; 0 floating/emergent aquatics; 1, foliage ≤100 mm in height; 2, 101-299 mm; 3, 300-599 mm; 4, 600-999 mm; 5, 1.0-3.0 m; 6, 3.1-6.0 m; 7, 6.1-15.0 m; 8, ≥15.0 m; V, various (*Hieracium* spp.).

(14) Lateral spread
Based upon field observations, the following classes have been recognized: 1, therophytes (i.e. lateral spread of exceedingly limited extent and duration); 2, perennials with compact unbranched rhizomes or forming small tussocks ≤100 mm in diameter); 3, perennials with rhizomatous systems or tussocks attaining 100-250 mm; 4, perennials attaining diameter of 251-1000 mm; 5, perennials attaining a diameter of >1000 mm; V, various (*Hieracium* spp.).

N.B. Patch size relates to the dimensions of connected branching systems of stems (excluding the leaf canopy). Ancient, fragmented, clonal colonies are often considerably larger (Harberd 1961, 1962, 1967), but estimates of their size are unavailable.

(15) Mycorrhizas
The frequency and nature of mycorrhizal infection, as described by Harley and Harley (1986), have been summarized for each species as follows: Normally mycorrhizal (75% or more records report infection): EC, with ectomycorrhizas; ER, with ericoid mycorrhizas; OR, with orchid mycorrhizas; VA, with vesicular-arbuscular mycorrhizas. Intermediate: +, 26-74% of records report infection with VA mycorrhizas. Non-mycorrhizal: -, 25% or less of records report mycorrhizas; HP, non-mycorrhizal hemiparasite. No information: ?

N.B. This classification of species is only approximate. Although data have been collated from a wide variety of sources, for most species few records are available. The most widespread infection within our region is by vesicular-arbuscular mycorrhizas. These have been shown to be a genetically heterogeneous grouping with different mycorrhizal strains associated with different plant species (van der Heijden *et al.* 1998). To date no functional classification of VA mycorrhizas comparable to that presented here for plant species has been described.

(16) Leaf phenology
The phenology of the leaf canopy, is based upon plants observed in the Sheffield region. The following classes have been recognized. Canopy seasonal, (S): Sa, aestival (duration of canopy spring to autumn); Sh, hibernal (mainly autumn to early summer); Sv, vernal (winter to spring). Canopy evergreen (E): Ea, always evergreen; Ep, partially evergreen (species evergreen in some habitats and not in others, or species evergreen only in mild winters, or leaves slowly but incompletely senescing over winter, or overwintering with small leafy shoots, formed in autumn). In *Hieracium* spp., denoted V (for various), some taxa are wintergreen and others are not.

N.B. The wintergreenness of leaves is very much determined by climate and the leaves of some species may be more persistent in S England and less persistent in Scotland than stated here.

(17) Flowering time and duration
Here the time of first flowering and its duration (from Clapham *et al.* 1981 and field observations) is presented. The month of first flowering is abbreviated to its first three letters and is immediately followed by the span, in months, of the flowering period, e.g. 'Jun3' refers to a species flowering from June to August.

N.B. Like leaf phenology, flowering time shows variation according to region.

(Table 7.3) *Attributes of the regenerative phase*

(18) Regenerative strategies
These estimates are based on laboratory and field studies supplemented by data from the literature. They are presented using the format adopted for the Accounts. The regenerative strategies, which are described more fully on pages 33–36, are abbreviated as follows: S, seasonal regeneration by seed; Sv, seasonal regeneration by vegetative means (offsets soon independent of parent); V, lateral vegetative spread (offsets remaining attached to the parent for a long period, usually for more than one growing season); (V), denotes instances where the period of attachment is intermediate between those of Sv and V, although in most instances we have insufficient demographic data to distinguish fully between Sv and V; W, regeneration involving numerous widely-dispersed seeds or spores; Bs, a persistent bank of buried seeds or spores; ?, strategies of regeneration by seed are uncertain.

N.B. Readers using this information are strongly advised to refer to the rationale concerning the identification of regenerative strategies (pp. 47-50) before attempting to interpret the data presented under this heading.

(19) Soil seed bank
The classes of seed bank, which are illustrated in Figure 2.7 on page 15, were reassessed with the help of Ken Thompson. The main reference source was Thompson *et al.* (1997a) but a supplementary additional dataset (Ken Thompson and Renée Bekker, unpublished data) was also used. The classes of seed bank may be summarized as follows: type 1, transient seed banks present during the summer and germinating synchronously in autumn; 2, transient seed bank present during winter and germinating synchronously in late winter or spring; 3, a small amount of seed persists in the soil, often for >5 years but concentrations of seed in the soil are only high after seed has just been shed; 4, there is a large bank of long persistent seeds in the soil throughout the year. This last category is difficult to separate from class 3 as it requires seasonal sampling, and often high replication. Only for a few very common species can a class 4 seed bank be assigned with confidence and, in practice, 'class 3' probably contains species that more correctly should be ascribed to class 4.

N.B. As in the Accounts, only species with seed banks 1-3 are classified as showing seasonal regeneration (S). Regeneration involving numerous wind-dispersed seeds or spores, (W), is, however, often strongly seasonal. Further, although the proportion of seeds germinating at one time (relative to the proportion which remain dormant) is low, a strong seasonal bias in the timing of germination is often evident in species with a Type 4 seed bank (e.g. *Polygonum aviculare*, p. 482).

(20) Agency of dispersal
The extent of dispersal and the nature of the dispersal mechanism are clearly critical in determining the distribution of species, particularly in a disturbed and changing landscape such as that occupying most of Lowland Britain. Unfortunately, although this topic is too important to ignore, most of the information available on dispersal is anecdotal. It is hoped that the inadequacy of the data presented here will stimulate the production of a more complete appraisal of the mobility of species and of their most important vectors of transport. Until then, readers are advised to utilize these data with caution and in conjunction with the assessment of the extent to which the species is currently increasing or decreasing (attribute 5). What is clear to date is that many of the species, which are becoming increasingly abundant in the British landscape, lack obvious specialization for long-distance dispersal (Salisbury 1953, Grime 1987). The possibility must be considered that this phenomenon is due in part to increased movement of soil-borne seed on motor vehicles and transport of vegetables and potted plants (Hodkinson and Thompson 1997). Types of dispersal to which the seeds of each species appear to be specialized are assessed by reference to Ridley (1930) and from observations of the morphology of seeds and the habitats in which they are shed. The following abbreviations are used: ANIM, dispersal by means of animals; *dispersal a direct consequence of food gathering*: ANIMi, an ingested berry; ANIMn, a nut or related type of hard-coated dispersule; ANIMe, seed with an elaiosome (dispersed short distances by ants); *dispersal adhesive*: ANIMa, dispersule with an awn, or with spiny calyx teeth; ANIMb, an adhesive burr; ANIMm, dispersule adhesive through the secretion of mucilage); AQUAT, dispersal by water, either by means of buoyant seeds or of floating seedlings; WIND, wind-dispersed (WINDm, dispersule minute – orchid seeds and fern spores; WINDp, dispersule plumed or wrapped in woolly hairs; WINDc, seeds small and shed from a capsule, held above the level of the surrounding vegetation; WINDw, seeds winged or strongly flattened); UNSP, unspecialized, with morphological features facilitating dispersal absent or undetected (includes UNSPag, unspecialized but dispersed widely as a result of agricultural practices). In some instances the symbols 'c' for capsule and 'w' for winged or flattened seeds are used in conjunction with UNSP. This occurs where seeds of a morphological character which favours wind dispersal are also associated with a growth habit which renders long-distance dispersal by wind unlikely, e.g. in the case of *Veronica montana*, UNSPcw, where fruits are borne close to the ground in sheltered woodland sites. In such cases, and also where there is explosive discharge of seeds, the main significance of dispersal is likely to lie in its movement of progeny away only from the immediate vicinity of the parent. Question marks are used in some instances to indicate uncertainty, and an asterisk indicates that dispersal of vegetative shoots or root fragments may be as, or more, important for the spread of the species than dispersal of seeds. Species which do not normally produce seed are identified by '–'. In some cases more than one mechanism is cited, and the following abbreviations have been adopted: AN/AQ, by means of animals or water; AQ/WI, by means of water or wind. We suspect that in some instances hard-coat dormancy, a character included in (24) 'Germination requirements', allows seeds to survive ingestion by animals.

(21) Dispersule and germinule form
The following classes are identified: Fr, dispersule and germinule a fruit (or part of a fruit, e.g. nutlet or mericarp); Sd, dispersule and germinule a seed; Sp, dispersule and germinule a spore; F/S, dispersule a fruit, germinule a seed (as in berries and other fleshy fruits); f/S, germinule a seed, dispersed either within a fruit or as a seed (e.g. *Atriplex patula*); Bul, dispersule vegetative with flowers replaced by bulbils; X, seeds, or bulbils, not produced.

(22) Dispersule weight
This describes the weight of seed, achene or other indehiscent germinule collected from the Sheffield region and dried at room temperature (excepting the minute seeds of orchids). The following classes are identified: S, too small to be measured easily (orchid seeds and fern spores); 1, weight ≤0.20 mg; 2, 0.21-0.50 mg; 3, 0.51-1.00 mg; 4, 1.01-2.00 mg; 5, 2.01-10.00 mg; 6, ≥10 mg; X, no seed produced; -, no information available.

(23) Dispersule shape
The following classes have been separated: 1, length/breadth ratio ≤1.5; 2, 1.5-2.5; 3, ≥2.5; -, seed normally absent.

(24) Germination requirements
The treatments required to achieve a high percentage of germination have been determined by UCPE for many common species (see pp. 51-53). Supplementary data from the literature are also included. Treatments are abbreviated as follows: Chill, chilling; Dry, dry storage at room temperature; Fluct, fluctuating temperatures; Freeze, alternate freezing and thawing treatments; Heat, heat treatment; Orchid, fungal symbiont (usually a *Rhizoctonia* spp.) necessary for establishment in the field; Scar, scarification (many species with hard-coat dormancy also respond to chilling); Warm, warm moist incubation; Wash, water-washing to remove inhibitor in seed coat. Species capable of immediate germination are identified as '-' and those which rarely or never produce seed are indicated as 'X'. Where germination requirement differs between collections, e.g. *Arctium minus*, or where seeds are polymorphic in their germination characteristics even on the same plant, e.g. *Chenopodium album*, the various treatments resulting in germination are separated by a '/'. Where several alternative dormancy-breaking mechanisms are effective for a species they are separated by a comma. For a few species two treatments in combination are required, e.g. *Hyacinthoides non-scripta* (warm moist incubation following by chilling, Warm+Chill). 'Unclassified' species lack the capacity for immediate germination, but the dormancy-breaking mechanism(s) have not been identified. This grouping is probably dominated by species with an, as yet, unrecognized chilling requirement. Where parentheses are present the final germination percentage was low (less than 50%) and the time to 50% germination high (more than 20 days). The use of a '?' indicates that confirmation of the dormancy-breaking mechanism is required.

(25) Family
The family name is abbreviated to its first three letters except where this leads to ambiguity (as in the cases of Polygalaceae and Polygonaceae). The abbreviations used are as follows:
Ace, Aceraceae; Ado, Adoxaceae; Ali, Alismataceae; Api, Apiaceae; Apo, Apocynaceae; Aqu, Aquifoliaceae; Ara, Araceae; Arl, Araliaceae; Asp, Aspleniaceae; Ast, Asteraceae; Bal, Balsiminaceae; Ber, Berberidaceae; Bet, Betulaceae; Ble, Blechnaceae; Bor, Boraginaceae; Bra, Brassicaceae; Bud, Budlejaceae; But, Butomaceae; Cal, Callitrichaceae; Cam, Campanulaceae; Can, Cannabaceae; Cap, Caprifoliaceae; Car, Caryophyllaceae; Cel, Celastraceae; Cer, Ceratophyllaceae; Che, Chenopodiaceae; Cis, Cistaceae; Clu, Clusiaceae; Con, Convolvulaceae; Cor, Cornaceae; Cra, Crassulaceae; Cuc, Cucurbitaceae; Cyp, Cyperaceae; Den, Dennstaedtiaceae Dio, Dioscoreaceae; Dip, Dipsacaceae; Dro, Droseraceae; Dry, Dryopteridaceae; Emp, Empetraceae; Equ, Equisetaceae; Eri, Ericaceae; Eup, Euphorbiaceae; Fab, Fabaceae; Fag, Fagaceae; Fum, Fumariaceae; Gen, Gentianaceae; Ger, Geraniaceae; Gro, Grossulariaceae; Hal, Haloragaceae; Hoc,

Hippocastanaceae; Hur, Hippuridaceae; Hyd, Hydrocharitaceae; Iri, Iridaceae; Jce, Juncaceae; Jcg, Juncaginaceae; Lam, Lamiaceae; Lem, Lemnaceae; Len, Lentibulariaceae; Lil, Liliaceae; Lin, Linaceae; Lyc, Lycopodiaceae; Lyt, Lythraceae; Mal, Malvaceae; Men, Menyanthaceae; Nym, Nymphaeaceae; Ole, Oleaceae; Ona, Onagraceae; Oph, Ophioglossaceae; Orc, Orchidaceae; Oro Orobanchaceae; Osm, Osmundaceae; Oxa, Oxalidaceae; Pap, Papaveraceae; Pga, Polygalaceae; Pgo, Polygonaceae; Pla, Plantaginaceae; Poa, Poaceae; Por, Portulacaceae; Pot, Potamogetonaceae; Ppo, Polypodiaceae; Pri, Primulaceae; Ran, Ranunculaceae; Res, Resedaceae; Rha, Rhamnaceae; Ros, Rosaceae; Rub, Rubiaceae; Sal, Salicaceae; Sax, Saxifragaceae; Scr, Scrophulariaceae; Sol, Solanaceae; Spa, Sparganiaceae; Tax, Taxaceae; The, Thelypteridaceae; Thy, Thymelaeaceae; Typ, Typhaceae; Ulm, Ulmaceae; Urt, Urticaceae; Val, Valerianaceae; Ver, Verbenaceae; Vio, Violaceae; Vis, Viscaceae; Woo, Woodsiaceae; Zan, Zannichelliaceae.

Table 7.1 Ecological attributes of species (symbols and notes are explained in the preceding text).

Species and authority	Status	Aquatic	Mire	Skeletal	Arable	Pasture	Spoil	Wasteland	Woodland	Maritime	Commonest terminal habitat	Soil pH	Floristic diversity	Longitudinal range	Latitudinal range	Present status	Red Data book	(Sp.)
(a) Herbs and woody species up to 1.5 m tall																		
+ *Achillea millefolium* L.	nat	1	1	3	3	4	3	4	1	0	PASTe	6.5b	4	1-5	2-3	0	0	Am
- *Achillea ptarmica* L.	nat	1	4	1	1	4	1	3	1	0	MIREu	5.5b	4	1-5	2-3	-1	0	Ap
Acorus calamus L.	neo	4	4	1	1	1	1	1	1	0	AQUATp	6.5a	1	-	-	0		Ac
Adoxa moschatellina L.	nat	1	3	3	1	1	1	2	5	0	WOODa	6.0b	2	1-6	2-3	-1	0	Am
- *Aegopodium podagraria* L.	arch	1	1	1	3	2	2	4	4	0	SOIL	7.0a	2	-	-	1	0	Ap
- *Aethusa cynapium* L.	nat	1	1	1	5	2	3	3	1	0	ARABLE	7.0a	4	1-3	3	0	0	Ac
- *Agrimonia eupatoria* L.	nat	1	1	1	1	3	1	5	2	0	RDVRGEu	6.5b	4	1-4	3-4	-1	0	Ae
Agrimonia procera Wallr.	nat	1	1	1	1	4	1	5	1	0	WASTEl	6.0b	3	1-3	3	-1	0	Ap
+ *Agrostis canina* L.	nat	1	5	3	1	2	1	3	1	0	MIREu	4.5b	2	1-6	2-3	-1	0	Ac
+ *Agrostis capillaris* L.	nat	1	1	3	2	4	3	3	2	0	PASTe	5.5c	2	1-4	2-3	-1	0	Ac
- *Agrostis gigantea* Roth	arch	1	1	1	5	3	4	3	1	0	ARABLE	7.0a	3	1-5	3-4	0	0	Ag
+ *Agrostis stolonifera* L.	nat	1	3	2	4	3	3	3	1	0	ARABLE	7.0a	2	1-6	2-4	1	0	As
+ *Agrostis vinealis* Schreber	nat	1	3	1	1	4	2	5	2	0	WASTEl	4.5b	1	1-3	3	-1	0	Av
- *Aira caryophyllea* L.	nat	1	1	3	2	2	4	4	1	0	CINDER	6.0b	3	1-3	3-4	0	0	Ac
+ *Aira praecox* L.	nat	1	1	3	1	3	4	3	1	0	QRYa	5.5b	2	1-2	3-4	0	0	Ap
- *Ajuga reptans* L.	nat	1	3	2	1	2	4	1	2	4	0	SCRUB	6.5b	4	1-3	3	-1	0	Ar
- *Alchemilla vulgaris* L. agg.	nat	1	1	2	2	4	3	3	3	0	PASTl	6.5b	5	-	-	-1	0	Av
+ *Alisma plantago-aquatica* L.	nat	4	4	1	1	2	1	1	1	0	MIREu	6.5a	1	1-6	2-4	0	0	Ap
+ *Alliaria petiolata* (M. Bieb.) Cavara & Grande	nat	1	1	3	1	2	3	5	3	0	RIVBNKu	7.0a	3	1-3	3	0	0	Ap
+ *Allium ursinum* L.	nat	1	2	1	1	2	1	2	5	0	WOODl	6.5b	1	1-3	3	-1	0	Au
- *Allium vineale* L.	nat	1	1	5	1	2	3	4	1	0	OUTCRPu	7.0a	3	1-3	3	-1	0	Av
+ *Alopecurus geniculatus* L.	nat	2	5	1	3	3	3	1	1	0	MIREu	6.5a	2	1-3	2-3	0	0	Ag
- *Alopecurus myosuroides* Hudson	arch	1	1	1	5	1	3	1	1	0	WASTEa	7.0a	2	1-3	3-4	1	0	Am
+ *Alopecurus pratensis* L.	nat	1	3	2	1	4	3	2	2	0	MEADOW	6.0b	3	1-4	2-3	-1	0	Ap
Amsinckia micrantha Suksd.	neo	1	1	1	5	1	1	2	3	0	WASTEa	6.5a	2	-	-	1		Am
- *Anacamptis pyramidalis* (L.) Rich.	nat	1	1	3	1	4	5	3	1	0	QRYl	7.0a	5	1-3	3-4	-1	0	Ap
+ *Anagallis arvensis* L.	nat	1	1	2	5	2	3	1	1	0	ARABLE	7.0a	4	1-4	3-4	0	0	Aa
Anagallis tenella (L.) L.	nat	1	5	1	1	1	1	1	1	0	MIREu	6.5a	5	1	3-4	-1	0	At
- *Anchusa arvensis* (L.) M. Bieb.	arch	1	1	1	5	2	3	3	1	0	ARABLE	6.5b	4	1-4	3	0	0	Aa
+ *Anemone nemorosa* L.	nat	1	1	1	1	3	1	1	5	0	WOODl	6.5b	2	1-4	3	0	0	An
+ *Angelica sylvestris* L.	nat	1	4	1	1	2	3	4	3	0	RIVBNKu	6.5b	3	1-4	2-3	0	0	As
+ *Anisantha sterilis* (L.) Nevski	arch	1	1	4	4	3	3	3	3	0	HEDGE	7.0a	2	1-3	3-4	0	0	As
Antennaria dioica (L.) Gaertner	nat	1	1	1	1	4	3	1	1	0	PASTl	7.0b	5	1-5	2-3	-1	0	Ad
Anthemis arvensis L.	arch	1	1	1	5	1	3	5	1	0	RLYBNK	6.5b	3	1-3	3-4	-1	3	Aa
Anthemis cotula L.	arch	1	2	1	5	1	3	3	1	0	ARABLE	6.5b	4	1-3	3-4	-1	2	Ac
+ *Anthoxanthum odoratum* L.	nat	1	3	3	1	4	3	3	2	0	PASTl	5.5b	3	1-4	2-4	-1	0	Ao
Anthriscus caucalis M. Bieb.	nat	1	1	4	4	1	3	2	4	0	HEDGE	6.5b	3	1-3	3	-1	0	Ac
+ *Anthriscus sylvestris* (L.) Hoffm.	nat	1	1	3	2	3	3	4	3	0	RDVRGEu	7.0a	3	1-5	2-3	0	0	As
- *Anthyllis vulneraria* L.	nat	1	1	4	1	3	4	3	1	0	OUTCRPu	7.0a	4	1-3	2-3	-1	0	Av
Antirrhinum majus L.	neo	1	1	5	3	1	3	3	2	0	WALL	7.0a	1	-	-	1		Am
Apera interrupta (L.) P. Beauv.	neo	1	1	3	1	5	3	3	1	0	CINDER	7.0b	1	1-4	3-4	1		Ai
+ *Aphanes arvensis* L.	nat	1	1	3	5	3	3	3	1	0	ARABLE	7.0a	4	1-3	3	0	0	Aa
Aphanes australis Rydb.	nat	1	1	1	3	1	2	5	1	0	WASTEd	6.5b	1	1-3	3	-1	0	Aa
Apium inundatum (L.) Reichb. f.	nat	4	4	1	1	1	1	1	1	0	AQUATp	6.0a	1	1-2	3	-1	0	Ai
+ *Apium nodiflorum* (L.) Lagasca	nat	4	4	1	1	2	1	2	1	0	AQUATr	7.0a	1	1-4	3-4	0	0	An
Aquilegia vulgaris L.	nat	1	2	3	1	3	2	3	4	0	WOODl	7.0a	4	1-3	3	-1	0	Av
+ *Arabidopsis thaliana* (L.) Heynh.	nat	1	1	4	1	3	3	3	1	0	OUTCRPu	6.5a	4	1-4	3	1	0	At
Arabis caucasica Willd. ex Schldl.	neo	1	1	5	1	1	2	1	1	0	CLIFF	7.0a	2	-	-			Ac

Table 7.1

Species and authority	Status	Aquatic	Mire	Skeletal	Arable	Pasture	Spoil	Wasteland	Woodland	Maritime	Commonest terminal habitat	Soil pH	Floristic diversity	Longitudinal range	Latitudinal range	Present status	Red Data book	(Sp.)
+ *Arabis hirsuta* (L.) Scop.	nat	1	1	5	1	3	3	3	1	0	SCREEu	7.0a	4	1-6	2-3	-1	0	*Ah*
- *Arctium lappa* L.	arch	1	3	1	1	2	1	5	3	0	RIVBNKu	6.5a	2	1-4	3	1	0	*Al*
+ *Arctium minus* (Hill) Bernh.	nat	1	1	1	3	2	4	4	3	0	SOIL	7.0a	3	1-5	3	0	0	*Am*
Arctostaphylos uva-ursi (L.) Sprengel	nat	1	1	5	1	5	1	1	1	0	WASTEa	[4.0]	1	1-6	2	-1	0	*Au*
+ *Arenaria serpyllifolia* L.	nat	1	1	4	2	2	4	3	1	0	OUTCRPu	7.0a	4	1-4	3-4	1	0	*As*
- *Armoracia rusticana* P. Gaertner, Meyer & Scherb.	arch	1	1	1	3	2	4	4	1	0	RDVRGEu	7.0a	2	-	-	0	0	*Ar*
+ *Arrhenatherum elatius* (L.) P. Beauv. ex J.S. Presl & C. Presl	nat	1	1	4	2	3	3	3	3	0	SCREEu	7.0b	2	1-3	3	1	0	*Ae*
+ *Artemisia absinthium* L.	arch	1	1	3	1	2	4	3	1	0	BRICK	7.0a	3	1-4	3	1	0	*Aa*
Artemisia verlotiorum Lamotte	neo	1	1	1	1	2	4	3	1	0	WASTEa	7.0a	1	-	-	1	0	*Av*
+ *Artemisia vulgaris* L.	nat	1	1	2	4	2	4	3	1	0	SOIL	7.0a	3	1-4	3	1	0	*Av*
+ *Arum maculatum* L.	nat	1	1	1	1	2	1	1	5	0	WOODl	7.0a	1	1-3	3	0	0	*Am*
Aster lanceolatus Willd.	neo	1	1	1	1	3	3	4	1	0	WASTEa	7.0b	1	-	-	1		*Al*
Aster novi-belgii L. agg.	neo	1	3	1	1	2	3	4	1	0	WASTEa	7.0b	2	-	-	1		*An*
Atriplex littoralis L.	nat	1	1	1	1	1	1	5	1	1	RDVRGEu	[7.5]	1	1-6	3	1	0	*Al*
+ *Atriplex patula* L.	nat	1	2	1	5	2	4	2	1	0	MANURE	7.0b	3	1-4	2-4	1	0	*Ap*
+ *Atriplex prostrata* Boucher ex DC.	nat	1	3	1	1	2	5	3	1	0	BRICK	7.0a	2	1-4	2-4	1	0	*Ap*
Avena fatua L.	arch	1	1	1	5	1	2	3	1	0	WASTEa	6.0b	3	-	-	0	0	*Af*
Ballota nigra L.	arch	1	1	3	3	1	3	3	4	0	HEDGE	7.0a	2	1-3	3-4	0	0	*Bn*
- *Barbarea vulgaris* R. Br.	nat	1	3	1	1	2	3	5	1	0	RIVBNKu	7.0a	4	1-4	3	0	0	*Bv*
Bassia scoparia (L.) Voss	neo	1	1	1	1	1	1	5	1	1	WASTEa	[7.5]	1	-	-	1		*Bs*
+ *Bellis perennis* L.	nat	1	2	3	3	4	2	3	3	1	MEADOW	7.0a	4	1-3	3	1	0	*Bp*
- *Berula erecta* (Hudson) Cov.	nat	4	4	1	1	2	1	1	1	1	AQUATp	6.5a	1	1-3	3	-1	0	*Be*
Bidens cernua L.	nat	2	5	2	1	1	1	1	1	1	PONDBK	6.5a	2	1-6	3	-1	0	*Bc*
Bidens tripartita L.	nat	2	5	2	1	1	2	1	1	1	PONDBK	6.5b	3	1-5	3	-1	0	*Bt*
Blackstonia perfoliata (L.) Hudson	nat	1	3	4	1	1	4	3	1	1	QRYl	7.0a	5	S-s	S-s	-1	0	*Bp*
+ *Brachypodium pinnatum* (L.) P. Beauv.	nat	1	1	2	1	2	3	5	2	0	WASTEl	7.0a	4	1-4	3	0	0	*Bp*
+ *Brachypodium sylvaticum* (Hudson) P. Beauv.	nat	1	1	3	1	3	2	1	4	0	WOODl	7.0a	2	1-3	3	-1	0	*Bs*
- *Brassica rapa* L.	arch	1	1	2	4	2	3	4	1	0	RIVBNKu	6.5b	3	-	-	1	0	*Br*
+ *Briza media* L.	nat	1	3	3	1	4	3	4	1	0	PASTl	7.0b	5	1-3	3	-1	0	*Bm*
+ *Bromopsis erecta* (Hudson) Fourr.	nat	1	1	1	1	3	3	5	1	0	WASTEl	7.0a	4	1-3	3	-1	0	*Be*
Bromopsis inermis (Leysser) Holub	neo	1	1	1	3	1	1	5	1	0	WASTEa	7.0a	1	-	-	1		*Bi*
+ *Bromopsis ramosa* (Hudson) Holub	nat	1	1	3	1	2	2	3	4	0	WOODl	7.0b	2	1-3	3	0	0	*Br*
+ *Bromus hordeaceus* L.	nat	1	1	3	3	4	3	2	1	0	MEADOW	6.5a	4	1-3	3-4	0	0	*Bh*
Bromus lepidus O. Holmb.	neo	1	1	1	1	3	3	5	1	0	WASTEa	7.0a	3	-	-	-1		*Bl*
Bryonia dioica Jacq.	nat	1	1	1	3	1	2	3	5	0	HEDGE	6.0b	1	S-s	S-s	-1	0	*Bd*
Butomus umbellatus L.	nat	5	1	1	1	1	1	1	1	0	AQUATp	6.5a	1	1-4	3	1	0	*Bu*
Calamagrostis canescens (Wigg.) Roth	nat	1	4	1	1	1	1	4	3	0	MIREu	4.5b	1	1-4	2-3	-1	0	*Cc*
Calamagrostis epigejos (L.) Roth	nat	1	2	1	1	1	3	3	3	0	CINDER	6.5c	1	1-5	2-3	-1	0	*Ce*
Callitriche hamulata Kutz. ex Koch	nat	4	4	1	1	1	1	1	1	0	AQUATp	6.0b	1	1-2	2-3	-1	0	*Ch*
Callitriche obtusangula Le Gall	nat	5	2	1	1	1	1	1	1	0	AQUATp	7.0a	1	1-2	3-4	-1	0	*Co*
Callitriche platycarpa Kutz.	nat	5	3	1	1	1	1	1	1	0	AQUATp	6.5a	1	1-3	3	-1	0	*Cp*
+ *Callitriche stagnalis* Scop.	nat	4	4	1	2	2	1	1	1	0	AQUATr	6.5a	1	1-3	3	1	0	*Cs*
+ *Calluna vulgaris* (L.) Hull	nat	1	3	3	1	4	3	3	1	0	QRYa	3.5a	1	1-3	2-3	-1	0	*Cv*
+ *Caltha palustris* L.	nat	1	5	2	1	2	1	2	1	0	MIREs	6.5b	2	1-6	1-3	-1	0	*Cp*
Calystegia pulchra Brummitt & Heyw.	neo	1	1	1	1	3	3	3	3	0	HEDGE	6.5b	1	-	-	1		*Cp*
+ *Calystegia sepium* (L.) R. Br.	nat	1	3	1	1	2	4	4	1	0	RIVBNKu	7.0a	2	1-6	3	1	0	*Cs*
Calystegia silvatica (Kit.) Griseb.	neo	1	1	1	1	1	4	3	3	0	WASTEa	7.0a	1	-	-	1		*Cs*
Campanula latifolia L.	nat	1	1	1	1	1	1	2	5	0	WOODl	7.0a	2	1-3	3	-1	0	*Cl*
Campanula rapunculoides L.	neo	1	1	1	3	1	1	5	3	0	RDVRGEu	7.0a	2	-	-	1		*Cr*
+ *Campanula rotundifolia* L.	nat	1	1	4	1	4	3	3	2	0	PASTl	7.0b*	4	1-6	2-3	-1	0	*Cr*
Campanula trachelium L.	nat	1	1	4	1	1	3	2	5	0	WOODl	7.0a	3	1-3	3	-1	0	*Ct*
+ *Capsella bursa-pastoris* (L.) Medikus	nat	1	1	1	5	2	3	3	1	0	ARABLE	7.0a	4	1-4	2-4	1	0	*Cb*
+ *Cardamine amara* L.	nat	1	5	1	1	2	1	3	3	0	MIREs	6.5a	2	1-3	3	-1	0	*Ca*
+ *Cardamine flexuosa* With.	nat	1	4	3	1	2	1	3	3	0	MIREs	6.0b	2	1-3	3	1	0	*Cf*
+ *Cardamine hirsuta* L.	nat	1	1	3	1	3	4	3	1	0	QRYl	7.0b	3	1-4	3-4	1	0	*Ch*
+ *Cardamine pratensis* L.	nat	1	5	1	1	4	1	1	1	0	MIREu	6.0b	3	1-6	1-3	-1	0	*Cp*
- *Carduus crispus* L.	nat	1	1	2	2	2	3	4	4	0	RIVBNKu	7.0a	3	1-4	3	0	0	*Cc*
Carduus nutans L.	nat	1	1	4	3	1	3	4	1	0	WASTEl	6.5a	4	1-3	3	-1	0	*Cn*
Carex acuta L.	nat	2	5	1	1	1	1	1	1	0	MIREu	5.5b	1	1-4	2-3	-1	0	*Ca*
+ *Carex acutiformis* Ehrh.	nat	3	5	1	1	2	1	3	1	0	RIVBNKu	6.5b	1	1-4	3	-1	0	*Ca*
Carex arenaria L.	nat	1	1	1	1	3	5	1	1	1	CINDER	[6.0]	1	1-3	3	1	0	*Ca*

TABLES OF ATTRIBUTES

Table 7.1

Species and authority	Status	Aquatic	Mire	Skeletal	Arable	Pasture	Spoil	Wasteland	Woodland	Maritime	Commonest terminal habitat	Soil pH	Floristic diversity	Longitudinal range	Latitudinal range	Present status	Red Data book	(Sp.)
- Carex binervis Smith	nat	1	4	1	1	4	1	3	2	0	PASTa	4.0a	2	1	3	-1	0	Cb
+ Carex caryophyllea Latour.	nat	1	1	3	1	4	3	4	1	0	PASTl	6.5b	5	1-4	3	-1	0	Cc
Carex curta Gooden.	nat	1	5	1	1	1	1	1	1	0	MIREu	4.5a	1	1-6	2	-1	0	Cc
Carex dioica L.	nat	1	5	1	1	1	1	1	1	0	MIREu	7.0a	4	1-6	1-2	-1	0	Cd
Carex distans L.	nat	1	4	1	1	3	1	4	1	1	MIREu	6.5a	4	1-3	3-4	-1	0	Cd
Carex disticha Hudson	nat	1	5	1	1	3	1	3	1	0	MIREu	6.0b	3	1-4	3	-1	0	Cd
- Carex echinata Murray	nat	1	5	1	1	2	1	1	1	0	MIREu	5.5b	4	1-3	2-3	-1	0	Ce
Carex elata All.	nat	5	1	1	1	1	1	1	1	0	AQUATp	[5.0]	1	1-5	3	-1	0	Ce
+ Carex flacca Schreber	nat	1	3	3	1	4	3	4	2	0	SCREEu	7.0a	5	1-3	3-4	-1	0	Cf
- Carex hirta L.	nat	1	4	1	1	4	2	3	1	0	MIREu	6.5b	4	1-3	3	-1	0	Ch
- Carex hostiana DC.	nat	1	5	1	1	4	1	1	1	0	MIREu	7.0a	5	1-3	3	-1	0	Ch
Carex laevigata Smith	nat	1	5	3	1	1	1	1	1	0	MIREu	5.0b	3	1	3	-1	0	Cl
Carex muricata L.	nat	1	2	2	1	1	1	5	1	0	RD/RLY	5.5b	2	1-4	3-4	-1	0	Cm
+ Carex nigra (L.) Reichard	nat	1	5	1	3	4	1	3	1	0	MIREu	5.5b	2	1-4	2-3	-1	0	Cn
- Carex otrubae Podp.	nat	1	5	1	1	2	1	5	1	0	MIREu	6.5a	3	1-4	3-4	0	0	Co
- Carex ovalis Gooden.	nat	1	5	2	1	4	1	3	1	0	MIREu	6.0b	4	1-4	2-3	-1	0	Co
- Carex pallescens L.	nat	1	3	1	1	4	1	2	1	0	PASTa	5.0a	5	1-4	2-3	-1	0	Cp
+ Carex panicea L.	nat	1	5	1	1	4	1	5	1	0	PASTl	6.0b	5	1-3	2-3	-1	0	Cp
Carex paniculata L.	nat	1	5	1	1	1	1	1	1	0	MIREu	6.0b	2	1-3	3	-1	0	Cp
- Carex pendula Hudson	nat	1	5	1	1	2	1	1	5	0	MIREs	6.5a	2	1-3	3-4	0	0	Cp
+ Carex pilulifera L.	nat	1	1	3	1	4	3	3	3	0	PASTa	4.0a	1	1-3	3	-1	0	Cp
Carex pseudocyperus L.	nat	4	5	1	1	1	1	1	1	0	MIREu	5.5b	1	1-4	3	-1	0	Cp
Carex pulicaris L.	nat	1	5	1	1	5	3	1	1	0	*MIREu	6.0b	5	1-2	3	-1	0	Cp
- Carex remota L.	nat	1	5	2	1	2	1	1	4	0	MIREs	6.0b	2	1-3	3	0	0	Cr
Carex riparia Curtis	nat	4	4	1	1	1	1	1	1	0	AQUATp	6.5a	1	1-4	3	-1	0	Cr
Carex rostrata Stokes	nat	3	5	1	1	1	1	1	1	0	MIREu	5.5b	1	1-6	2-3	-1	0	Cr
Carex spicata Hudson	nat	1	3	1	1	3	2	3	1	0	RDVRGEu	6.5a	4	1-3	3	-1	0	Cs
- Carex sylvatica Hudson	nat	1	2	1	1	3	1	3	5	0	SCRUB	6.5b	3	1-5	3	-1	0	Cs
Carex vesicaria L.	nat	2	5	1	1	1	1	1	1	0	MIREs	6.0b	1	1-6	2-3	-1	0	Cv
Carex viridula subsp. brachyrrhyncha (Celak.) B. Schmid	nat	1	5	3	1	1	1	1	1	0	MIREu	7.0a	4	1-6	2-3	-1	0	Cv
- Carex viridula subsp. oedocarpa (Andersson) B. Schmid	nat	1	5	3	1	1	1	1	1	0	MIREu	6.0b	4	1-6	2-3	-1	0	Cv
+ Carlina vulgaris L.	nat	1	1	5	1	2	4	3	1	0	QRYl	7.0a	4	1-4	3	-1	0	Cv
Catabrosa aquatica (L.) P. Beauv.	nat	5	3	1	1	1	1	1	1	0	AQUATp	6.5a	1	1-3	2-3	-1	0	Ca
+ Catapodium rigidum (L.) C.E. Hubb.	nat	1	1	4	1	2	5	1	1	0	QRYl	7.0a	4	S-s	S-s	1	0	Cr
Centaurea montana L.	neo	1	1	1	1	1	4	5	1	0	WASTEa	7.0a	2	-	-	1		Cm
+ Centaurea nigra L.	nat	1	3	3	1	3	3	4	1	0	WASTEl	7.0b	4	1-2	3	0	0	Cn
+ Centaurea scabiosa L.	nat	1	1	4	1	2	3	4	1	0	WASTEl	7.0a	3	1-4	3	-1	0	Cs
+ Centaurium erythraea Rafn	nat	1	3	3	1	2	3	4	1	0	WASTEl	7.0b	4	1-3	3-4	1	0	Ce
- Centranthus ruber (L.) DC.	neo	1	1	5	1	2	4	3	1	0	CLIFF	7.0a	1	-	-	1		Cr
Cerastium arvense L.	nat	1	1	4	3	1	1	5	1	0	RDVRGEu	6.5b	4	1-6	2-3	-1	0	Ca
Cerastium diffusum Pers.	nat	1	1	1	1	1	5	3	1	1	CINDER	7.0a	2	1-3	3	0	0	Cd
+ Cerastium fontanum Baumg.	nat	1	3	3	3	4	3	3	1	0	MEADOW	6.5b	4	1-4	2-3	1	0	Cf
- Cerastium glomeratum, Thuill.	nat	1	1	3	4	2	4	3	1	0	OUTCRPu	7.0b	4	1-3	3-4	1	0	Cg
- Cerastium semidecandrum L.	nat	1	1	5	1	3	3	4	1	0	OUTCRPu	7.0a	4	1-3	3	0	0	Cs
- Cerastium tomentosum L.	neo	1	2	5	1	2	3	3	1	0	OUTCRPu	7.0b	2	-	-	1		Ct
- Ceratocapnos claviculata (L.) Liden	nat	1	1	1	1	2	1	3	5	0	WOODa	3.5a	1	1	3	0	0	Cc
Ceratochloa carinata (Hook. & Arn.) Tutin	neo	1	1	1	1	1	3	4	1	0	RDVRGEu	[7.0]	1	-	-	1		Cc
Ceratophyllum demersum L.	nat	5	1	1	1	1	1	1	1	0	AQUATp	6.5a	1	1-6	3-4	-1	0	Cd
+ Chaenorhinum, minus (L.) Lange	arch	1	1	1	1	2	5	1	1	0	CINDER	7.0a	3	1-3	3	1	0	Cm
- Chaerophyllum temulum L.	nat	1	1	5	2	1	2	5	1	0	HEDGE	7.0a	3	1-3	3	0	0	Ct
+ Chamerion angustifolium (L.) Holub	nat	1	3	1	2	1	4	3	3	0	CINDER	6.5c	2	1-6	2-3	1	0	Ca
- Chelidonium majus L.	arch	1	1	3	1	2	4	4	1	0	RDVRGEu	7.0a	2	1-5	3	1	0	Cm
+ Chenopodium album L.	nat	1	1	1	5	2	4	1	1	0	ARABLE	7.0a	3	1-5	2-4	-1	0	Ca
- Chenopodium bonus-henricus L.	arch	1	1	1	1	4	3	5	1	0	RDVRGEu	7.0a	2	-	-	-1	2	Cb
Chenopodium polyspermum L.	arch	1	1	1	5	1	3	2	1	0	GARDEN	6.5b	2	1-4	3	1	0	Cp
+ Chenopodium rubrum L.	nat	1	4	3	3	2	5	1	1	0	MANURE	7.0a	2	1-4	3	1	0	Cr
- Chrysanthemum segetum L.	arch	1	1	1	5	2	3	1	1	0	ARABLE	6.0b	3	1-3	3-4	-1	2	Cs
Chrysosplenium alternifolium L.	nat	1	4	3	1	1	1	1	2	4	MIREs	6.5b	2	1-6	1-3	-1	0	Ca
+ Chrysosplenium oppositifolium L.	nat	1	5	3	1	2	1	3	3	0	MIREs	6.5b	2	1-2	3	-1	0	Co
Cicerbita macrophylla (Willd.) Wallr.	neo	1	1	1	3	1	3	5	3	0	RDVRGEu	7.0a	1	-	-	1		Cm
Cichorium intybus L.	arch	1	1	1	1	3	1	5	1	0	RD/RLY	6.5a	2	1-4	3-4	1	0	Ci

652   TABLES OF ATTRIBUTES

		Abundance in the primary habitats														
Species and authority	Status	Aquatic	Mire	Skeletal	Arable	Pasture	Spoil	Wasteland	Woodland	Maritime	Commonest terminal habitat	Soil pH	Floristic diversity	Longitudinal range	Latitudinal range	Present status	Red Data book	(Sp.)
+ Circaea lutetiana L.	nat	1	3	1	1	2	1	1	4	0	MIREs	6.0b	2	1-3	3	1	0	Cl
Cirsium acaule (L.) Scop.	nat	1	1	2	1	5	3	3	1	0	QRYl	7.0a	5	1-3	3	-1	0	Ca
+ Cirsium arvense (L.) Scop.	nat	1	2	2	4	3	4	3	1	0	COAL	7.0b	3	1-5	3	1	0	Ca
Cirsium dissectum (L.) Hill	nat	1	4	1	1	3	1	4	1	0	MIREu	[5.0]	3	1	3	-1	0	Cd
Cirsium heterophyllum (L.) Hill	nat	1	1	1	1	3	1	5	3	0	PASTl	6.5a	3	1-4	2	-1	0	Ch
+ Cirsium palustre (L.) Scop.	nat	1	5	2	1	4	2	3	1	0	PASTl	6.0b	4	1-4	2-3	0	0	Cp
+ Cirsium vulgare (Savi) Ten.	nat	1	1	3	3	3	4	3	1	0	PASTl	7.0b	4	1-4	3	1	0	Cv
Cladium mariscus (L.) Pohl	nat	3	5	1	1	1	1	1	1	0	MIREu	[5.5]	1	1-4	3-4	-1	0	Cm
Claytonia perfoliata Donn ex Willd.	neo	1	1	5	3	1	3	3	3	0	HEDGE	6.0a	2	-	-	1	0	Cp
Claytonia sibirica L.	neo	1	1	3	1	1	1	1	5	0	RIVBNKs	4.5b	2	-	-	1	0	Cs
Clinopodium acinos (L.) Kuntze	nat	1	1	5	3	1	2	1	1	0	OUTCRPu	7.0a	4	1-3	3	-1	2	Ca
- Clinopodium vulgare L.	nat	1	1	4	1	2	4	3	2	0	QRYl	7.0a	4	1-6	3	0	0	Cv
Cochlearia danica L.	nat	1	1	1	1	1	4	3	1	1	WASTEa	[7.0]	2	1	3	1	0	Cd
Coeloglossum viride (L.) Hartman	nat	1	1	1	1	3	3	3	1	0	QRYl	7.0a	5	1-6	2	-1	2	Cv
- Conium maculatum L.	arch	1	1	3	3	2	3	4	3	0	RIVBNKu	7.0a	3	1-4	3-4	1	0	Cm
+ Conopodium majus (Gouan) Loret	nat	1	1	1	1	4	1	3	3	0	PASTl	6.0b	3	1	3	-1	0	Cm
+ Convolvulus arvensis L.	nat	1	1	2	5	2	3	3	2	0	ARABLE	7.0b	3	1-4	3-4	0	0	Ca
Conyza canadensis (L.) Cronq.	neo	1	1	1	1	1	5	3	1	0	CINDER	7.0a	3	-	-	1	0	Cc
Coronopus didymus (L.) Smith	neo	1	1	1	5	1	1	3	1	0	GARDEN	7.0a	2	-	-	1	0	Cd
- Coronopus squamatus (Forsskal) Asch.	arch	1	1	1	4	4	3	1	1	0	PATH	7.0a	2	1-3	3-4	-1	0	Cs
Cotoneaster horizontalis Decne.	neo	1	1	5	1	3	3	2	1	0	OUTCRPu	7.0a	2	-	-	1	0	Ch
Crassula helmsii (Kirk) Cockayne	neo	5	3	1	1	1	1	1	1	0	AQUATp	[6.0]	1	-	-	1	0	Ch
Crepis biennis L.	nat	1	1	3	1	3	1	5	1	0	WASTEl	7.0a	4	1-3	3	0	0	Cb
+ Crepis capillaris (L.) Wallr.	nat	1	1	3	2	3	4	3	1	0	QRYl	7.0b	4	1-3	3	1	0	Cc
Crepis paludosa (L.) Moench	nat	1	4	5	1	2	1	1	1	0	MIREs	6.0b	3	1-3	2-3	-1	0	Cp
Crepis vesicaria L.	neo	1	1	3	1	3	3	5	1	0	OUTCRPu	7.5a	3	-	-	1	0	Cv
+ Cruciata laevipes Opiz	nat	1	2	2	1	3	3	4	2	0	WASTEl	6.5b	3	1-4	3	-1	0	Cl
- Cymbalaria muralis P. Gaertner, Meyer & Scherb.	neo	1	1	5	1	2	3	1	1	0	WALL	7.0a	1	-	-	0	0	Cm
Cynoglossum officinale L.	nat	1	1	1	1	5	4	2	1	0	WASTEa	6.5b	2	1-4	3	-1	1	Co
+ Cynosurus cristatus L.	nat	1	3	2	2	4	2	3	1	0	MEADOW	6.5b	4	1-3	3	-1	0	Cc
+ Dactylis glomerata L.	nat	1	1	3	3	3	3	3	2	0	MEADOW	7.0b	3	1-4	3-4	0	0	Dg
- Dactylorhiza fuchsii (Druce) Soó	nat	1	4	2	1	4	3	4	2	0	PASTl	7.0b	5	1-4	3	0	0	Df
- Dactylorhiza incarnata (L.) Soó	nat	1	5	1	1	4	1	1	1	0	MIREu	[7.0]	3	1-4	2-3	-1	0	Di
- Dactylorhiza maculata (L.) Soó	nat	1	5	1	1	2	1	1	1	0	MIREu	5.0b	3	1-4	2-3	-1	0	Dm
- Dactylorhiza praetermissa (Druce) Soó	nat	1	4	1	1	2	3	3	1	0	MIREu	6.5a	4	1	3	1	0	Dp
Dactylorhiza purpurella (Stephenson & T.A. Stephenson) Soó	nat	1	4	1	1	1	3	1	1	0	MIREu	[7.0]	3	1	2	1	0	Dp
+ Danthonia decumbens (L.) DC.	nat	1	3	2	1	4	1	4	1	0	PASTl	5.5c*	4	1-3	3	-1	0	Dd
Daphne laureola L.	nat	1	1	3	1	1	1	1	5	0	WOODl	7.0a	1	S-s	S-s	-1	0	Dl
+ Daucus carota L.	nat	1	1	3	1	2	3	4	1	5	WASTEl	7.0a	5	1-4	3-4	-1	0	Dc
+ Deschampsia cespitosa (L.) P. Beauv.	nat	1	3	2	3	3	3	3	3	0	WOODl	6.0b	3	1-6	1-3	0	0	Dc
+ Deschampsia flexuosa (L.) Trin.	nat	1	2	3	1	4	3	3	3	0	PASTa	3.5a	1	1-3	2-3	-1	0	Df
+ Digitalis purpurea L.	nat	1	2	3	3	2	3	2	4	0	PLANTb	4.5c	2	1-2	3-4	-1	0	Dp
Diplotaxis muralis (L.) DC.	neo	1	1	3	4	1	4	3	1	0	GARDEN	7.0a	2	-	-	0	0	Dm
Diplotaxis tenuifolia (L.) DC.	arch	1	1	1	1	1	4	3	1	0	WASTEa	7.0a	2	1-3	3	1	0	Dt
Dipsacus fullonum L.	nat	1	1	2	3	1	4	3	1	0	WASTEa	7.0a	3	1-3	3	1	0	Df
Doronicum pardalianches L.	neo	1	1	1	1	1	1	4	5	0	RIVBNKs	6.5b	1	-	-	1	0	Dp
Drosera rotundifolia L.	nat	1	5	1	1	1	1	2	1	0	MIREu	4.5b	2	1-6	2-3	-1	0	Dr
Echium vulgare L.	nat	1	1	2	3	1	4	3	1	0	CINDER	7.0b	3	1-4	3	-1	0	Ev
+ Eleocharis palustris (L.) Roemer & Schultes	nat	4	5	1	1	2	1	1	1	0	MIREu	6.5a	1	1-5	2-4	-1	0	Ep
Eleocharis quinqueflora (F. Hartmann) O. Schwarz	nat	1	5	1	1	1	1	1	1	0	MIREu	7.0a	4	1-3	2-3	-1	0	Eq
Eleogiton fluitans (L.) Link	nat	5	3	1	1	1	1	1	1	0	AQUATp	6.0a	1	1	3-4	-1	0	Ef
+ Elodea canadensis Michaux	neo	5	1	1	1	2	1	1	1	0	AQUATp	6.5a	1	-	-	-1	0	Ec
Elodea nuttallii (Planchon) H. St. John	neo	5	2	1	1	1	1	1	1	0	AQUATp	6.5a	1	-	-	1	0	En
+ Elymus caninus (L.) L.	nat	1	1	4	1	2	1	2	4	0	WOODl	6.5a	2	1-4	2-3	-1	0	Ec
+ Elytrigia repens (L.) Desv. ex Nevski	nat	1	1	1	5	3	3	3	2	0	ARABLE	7.0b	2	1-4	2-4	1	0	Er
+ Empetrum nigrum subsp. nigrum L.	nat	1	3	3	1	4	2	4	1	0	PASTa	3.5a	1	1-6	1-2	0	0	En
Epilobium brunnescens (Cockayne) Raven & Engelhorn	neo	1	1	4	1	1	5	1	1	0	QRYa	5.0b	2	-	-	1	0	Eb
+ Epilobium ciliatum Raf.	neo	1	3	3	1	3	4	3	3	0	CINDER	6.5b	3	-	-	1	0	Ec
+ Epilobium hirsutum L.	nat	1	4	3	1	2	3	3	1	0	RIVBNKu	7.0a	2	1-5	3-4	1	0	Eh

TABLES OF ATTRIBUTES

653

Table 7.1

Species and authority	Status	Aquatic	Mire	Skeletal	Arable	Pasture	Spoil	Wasteland	Woodland	Maritime	Commonest terminal habitat	Soil pH	Floristic diversity	Longitudinal range	Latitudinal range	Present status	Red Data book	(Sp.)
+ Epilobium montanum L.	nat	1	1	4	1	2	3	3	3	0	QRYl	7.0a	3	1-3	3	0	0	Em
+ Epilobium obscurum Schreber	nat	1	5	3	1	2	1	1	1	0	MIREu	6.0b	3	1-3	3	0	0	Eo
+ Epilobium palustre L.	nat	1	5	1	1	2	1	1	1	0	MIREu	6.0b	3	1-6	2-3	-1	0	Ep
+ Epilobium parviflorum Schreber	nat	1	5	2	1	2	3	1	1	0	QRYl	6.5a	3	1-3	3	1	0	Ep
Epilobium roseum Schreber	nat	1	3	4	3	1	3	2	1	0	WASTEa	7.0a	3	1-4	3	0	0	Er
Epilobium tetragonum L.	nat	1	1	3	3	1	3	5	1	0	WASTEa	6.5b	3	1-4	3	1	0	Et
- Epipactis helleborine (L.) Crantz	nat	1	1	1	1	2	3	1	5	0	SCRUB	7.0b	2	1-5	3	0	0	Eh
Epipactis palustris (L.) Crantz	nat	1	5	1	1	1	1	1	1	0	MIREu	[7.0]	5	1-4	3	-1	0	Ep
+ Erica cinerea L.	nat	1	1	3	1	4	4	3	1	0	QRYa	3.5a	1	1	3	-1	0	Ec
- Erica tetralix L.	nat	1	5	1	1	3	1	2	1	0	MIREu	4.0a	1	1-2	3	-1	0	Et
+ Erigeron acer L.	nat	1	1	4	1	2	5	1	1	0	QRYl	7.0a	4	1-6	2-3	0	0	Ea
+ Eriophorum angustifolium Honck.	nat	1	5	3	1	3	1	1	1	0	MIREu	4.5c	3	1-6	1-3	0	0	Ea
+ Eriophorum vaginatum L.	nat	1	5	1	1	4	1	1	1	0	*MIREu	3.5a	1	1-6	1-2	0	0	Ev
- Erodium cicutarium (L.) L'Her.	nat	1	1	4	3	4	3	4	1	0	OUTCRPu	6.5b	4	1-4	3-4	0	0	Ec
- Erophila verna (L.) DC.	nat	1	1	5	2	2	3	3	1	0	OUTCRPu	7.0a	4	1-4	3-4	-1	0	Ev
Erysimum cheiranthoides L.	arch	1	1	1	5	1	4	2	1	0	CINDER	7.0a	2	-	-	0	0	Ec
- Erysimum cheiri (L.) Crantz	arch	1	1	5	1	2	1	1	1	0	CLIFF	[7.0]	1	-	-	1	0	Ec
- Eupatorium cannabinum L.	nat	1	4	3	1	2	3	3	1	0	MIREs	7.0a	3	1-3	3	0	0	Ec
Euphorbia cyparissias L.	neo	1	1	1	1	1	3	5	1	0	WASTEa	[7.0]	3	1-3	3	1		Ec
Euphorbia esula L.	neo	1	1	1	1	1	3	5	1	0	WASTEl	[7.0]	1	-	-	1		Ee
Euphorbia exigua L.	arch	1	1	1	5	1	1	2	1	0	ARABLE	7.0a	4	1-3	3-4	-1	1	Ee
- Euphorbia helioscopia L.	arch	1	1	1	5	2	3	3	1	0	ARABLE	7.0b	5	1-5	3-4	0	0	Eh
- Euphorbia peplus L.	arch	1	1	1	5	2	4	3	2	0	SOIL	7.0a	5	1-3	3-4	0	0	Ep
+ Euphrasia officinalis L. agg.	nat	1	1	3	1	4	4	3	1	0	*PASTl	7.0a	4	-	-	-1	0	Eo
+ Fallopia convolvulus (L.) A. Love	arch	1	1	1	5	2	3	1	1	0	ARABLE	6.5b	3	1-4	2-4	-1	0	Fc
+ Fallopia japonica (Houtt.) Ronse Decraene	neo	1	1	1	1	2	4	4	3	0	RDVRGEu	7.0a	1	-	-	1		Fj
Fallopia sachalinensis (F. Schmidt ex Maxim.) Ronse Decraene	neo	1	1	1	1	1	3	5	3	0	WASTEl	6.5c	1	-	-	1		Fs
- Festuca arundinacea Schreber	nat	1	3	3	1	3	3	3	2	0	WASTEl	7.0a	4	1-4	3-4	0	0	Fa
+ Festuca gigantea (L.) Villars;	nat	1	3	2	1	2	1	3	4	0	RIVBNKu	6.5b	2	1-3	3	0	0	Fg
+ Festuca ovina L.	nat	1	1	3	1	4	3	3	1	0	PASTl	6.0c*	3	1-5	2-3	-1	0	Fo
+ Festuca pratensis Hudson	nat	1	3	1	1	4	1	4	1	0	MEADOW	6.5a	4	1-3	2-3	-1	0	Fp
+ Festuca rubra L.	nat	1	3	3	2	3	3	4	1	0	RDVRGEu	7.0b	3	1-6	1-3	-1	0	Fr
Filago minima (Smith) Pers.	nat	1	1	1	1	1	2	5	1	0	WASTEd	5.5b	3	1-3	3	-1	0	Fm
Filago vulgaris Lam.	nat	1	1	4	1	1	3	5	1	0	WASTEd	6.5b	4	1-3	3-4	-1	1	Fv
+ Filipendula ulmaria (L.) Maxim.	nat	1	4	1	1	2	1	3	3	0	MIREs	6.5b	2	1-5	2-3	0	0	Fu
Filipendula vulgaris Moench	nat	1	1	3	1	5	1	5	1	0	PASTl	7.0a	5	1-4	3	-1	0	Fv
Foeniculum vulgare Miller	arch	1	1	1	1	2	3	5	1	1	WASTEa	7.0a	2	-	-	0	0	Fv
Fragaria × ananassa (Weston) Lois., Vilm., Nois. & J. Deville	neo	1	1	1	1	1	2	5	1	0	RD/RLY	5.5b	2	-	-	1		Fa
+ Fragaria vesca L.	nat	1	1	3	1	3	3	3	3	0	SCREEu	7.0a	4	1-4	3	-1	0	Fv
- Fumaria muralis Sonder ex Koch	nat	1	1	1	5	2	3	3	1	0	ARABLE	[7.0]	5	1	3-4	-1	0	Fm
- Fumaria officinalis L.	arch	1	1	2	5	2	4	3	1	0	ARABLE	7.0a	3	1-3	3-4	-1	0	Fo
Galanthus nivalis L.	neo	1	1	2	1	1	2	2	5	0	WOODa	6.5c	1	1-3	3-4	0		Gn
Galega officinalis L.	neo	1	1	1	3	1	4	4	1	0	WASTEa	7.0a	2	-	-	1		Go
+ Galeopsis tetrahit L.	nat	1	3	2	5	2	2	3	3	0	ARABLE	6.0b	2	1-3	2-3	-1	0	Gt
Galinsoga parviflora Cav.	neo	1	1	1	5	1	3	4	1	0	GARDEN	7.0a	2	-	-	1		Gp
Galinsoga quadriradiata Ruiz Lopez & Pavón	neo	1	1	1	5	1	4	1	1	0	GARDEN	6.5a	3	-	-	1		Gq
+ Galium aparine L.	nat	1	3	1	4	2	2	3	3	0	HEDGE	6.5a	2	1-3	3	1	0	Ga
Galium mollugo L.	nat	1	1	3	1	2	3	3	1	0	WASTEl	7.0a	3	1-3	2-3	0	0	Gm
- Galium odoratum (L.) Scop.	nat	1	2	3	1	2	1	2	5	0	SCRUB	6.5b	2	1-3	3	-1	0	Go
+ Galium palustre L.	nat	1	5	1	1	2	1	2	1	0	MIREu	6.0b	3	1-4	2-3	-1	0	Gp
+ Galium saxatile L.	nat	1	3	1	1	4	2	3	3	0	PASTa	4.0b	1	1-2	3	-1	0	Gs
+ Galium sterneri Ehrend.	nat	1	1	4	1	4	3	2	1	0	SCREEu	7.0a	4	1-2	2	-1	0	Gs
- Galium uliginosum L.	nat	1	5	1	1	2	1	1	1	0	MIREu	6.0a	4	1-5	2-3	-1	0	Gu
+ Galium verum L.	nat	1	1	3	1	4	1	4	1	0	PASTl	6.5b	5	1-5	2-3	0	0	Gv
Genista anglica L.	nat	1	2	1	1	3	1	5	1	0	WASTEa	4.5b	2	1	3	-1	1	Ga
Genista tinctoria L.	nat	1	1	2	1	4	3	3	1	0	WASTEa	6.5c	4	1-3	3	-1	0	Gt
- Gentianella amarella (L.) Boerner	nat	1	2	3	1	4	4	2	1	0	QRYl	7.0a	5	1-6	2-3	0	0	Ga
- Geranium columbinum L.	nat	1	1	5	1	2	1	3	1	0	OUTCRPu	7.0a	5	1-3	3	-1	0	Gc
- Geranium dissectum L.	arch	1	1	4	3	2	3	4	1	0	RDVRGEu	7.0a	4	1-3	3-4	1	0	Gd
Geranium endressii Gay	neo	1	1	1	1	5	1	3	1	0	WASTEl	[6.5]	1	-	-	1		Ge
- Geranium lucidum L.	nat	1	1	5	1	2	3	2	2	0	OUTCRPs	7.0a	3	S-s	S-s	0	0	Gl

Abundance in the primary habitats

Species and authority	Status	Aquatic	Mire	Skeletal	Arable	Pasture	Spoil	Wasteland	Woodland	Maritime	Commonest terminal habitat	Soil pH	Floristic diversity	Longitudinal range	Latitudinal range	Present status	Red Data book	(Sp.)
+ *Geranium molle* L.	nat	1	1	4	4	3	3	3	1	0	OUTCRPu	7.0a	5	1-3	3-4	0	0	*Gm*
Geranium pratense L.	nat	1	1	3	2	3	3	5	2	0	WASTEl	7.0b	3	1-5	2-3	0	0	*Gp*
Geranium pusillum L.	nat	1	1	5	3	1	1	3	1	0	WASTEd	7.0a	4	1-4	3	0	0	*Gp*
- *Geranium pyrenaicum* Burman f.	neo	1	1	1	1	2	4	4	1	0	RDVRGEu	7.0b	2	1-3	3	0		*Gp*
+ *Geranium robertianum* L.	nat	1	2	4	1	2	3	3	3	0	SCREEu	7.0a	2	1-3	3	1	0	*Gr*
Geranium × magnificum N. Hylander	neo	1	1	1	1	1	4	3	1	0	WASTEa	[6.5]	1	-	-	1		*Gm*
- *Geum rivale* L.	nat	1	3	3	1	4	1	3	4	0	SCRUB	6.5a	4	1-4	2-3	-1	0	*Gr*
+ *Geum urbanum* L.	nat	1	1	3	1	2	1	2	5	0	WOODl	6.5a	2	1-4	3	0	0	*Gu*
+ *Glechoma hederacea* L.	nat	1	3	2	1	2	3	3	4	0	HEDGE	6.5a	2	1-5	2-3	0	0	*Gh*
- *Glyceria declinata* Breb.	nat	3	5	1	1	2	1	2	1	0	MIREu	6.0a	2	1-2	3	0	0	*Gd*
+ *Glyceria fluitans* (L.) R. Br.	nat	4	5	1	1	3	1	2	1	0	AQUATp	6.0a	1	1-3	3	0	0	*Gf*
+ *Glyceria maxima* (Hartman) O. Holmb.	nat	3	5	1	1	2	1	3	1	0	MIREu	6.5b	2	1-6	3	-1	0	*Gm*
- *Glyceria notata* Chevall.	nat	3	5	1	1	2	1	1	1	0	MIREu	6.5a	2	1-3	3	0	0	*Gn*
Glyceria × pedicellata F. Towns.	nat	3	5	1	1	1	1	1	1	0	MIREu	7.0a	1	-	-	-1		*Gp*
Gnaphalium sylvaticum L.	nat	1	1	1	1	1	2	5	1	0	WASTEa	5.5b	3	1-4	2-3	-1	3	*Gs*
- *Gnaphalium uliginosum* L.	nat	1	5	1	4	2	3	1	1	0	MIREu	5.5b	3	1-5	2-3	1	0	*Gu*
Groenlandia densa (L.) Fourr.	nat	5	1	1	1	1	1	1	1	0	AQUATp	6.5a	1	1-3	3	-1	2	*Gd*
- *Gymnadenia conopsea* (L.) R. Br.	nat	1	3	3	1	4	4	3	1	0	QRYl	7.0a	5	1-5	2-3	-1	0	*Gc*
+ *Hedera helix* L.	nat	1	1	3	1	2	1	1	4	0	WOODl	7.0b	1	1-3	3-4	0	0	*Hh*
+ *Helianthemum nummularium* (L.) Miller	nat	1	1	4	1	4	3	3	1	0	PASTl	7.0a	5	1-3	3	-1	0	*Hn*
+ *Helictotrichon pratense* (L.) Besser	nat	1	1	4	1	4	3	3	1	0	PASTl	7.0b	5	1-3	3	-1	0	*Hp*
+ *Helictotrichon pubescens* (Hudson) Pilger	nat	1	1	3	1	4	2	4	1	0	WASTEl	6.5a	5	1-3	3	-1	0	*Hp*
Heracleum mantegazzianum Sommier & Levier	neo	1	1	1	1	1	1	5	4	0	WASTEa	6.5b	2	-	-	1		*Hm*
+ *Heracleum, sphondylium* L.	nat	1	1	2	3	3	3	4	3	0	WASTEl	7.0b	3	1-5	2-3	1	0	*Hs*
- *Hesperis matronalis* L.	neo	1	2	2	1	2	1	3	5	0	RIVBNKu	7.0a	2	-	-	0		*Hm*
+ *Hieracium* sp. L.	nat	1	1	3	1	2	4	3	1	0	QRYl	7.0b*	3	-	-	0	0	*Hs*
- *Hippuris vulgaris* L.	nat	5	3	1	1	2	1	1	1	0	AQUATp	6.5a	1	1-6	2-3	-1	0	*Hv*
Hirschfeldia incana (L.) Lagr.-Fossat	neo	1	1	1	1	1	4	3	1	0	BRICK	7.0a	3	-	-	1		*Hi*
+ *Holcus lanatus* L.	nat	1	3	3	3	4	4	3	2	0	PASTe	6.5b	3	1-3	3-4	1	0	*Hl*
+ *Holcus mollis* L.	nat	1	3	2	1	3	2	3	3	0	WOODa	5.0c	1	1-3	3	0	0	*Hm*
Hordeum jubatum L.	neo	1	1	1	1	1	2	5	1	1	RDVRGEu	7.5a	2	-	-	1		*Hj*
+ *Hordeum murinum* L.	nat	1	1	3	3	1	4	3	3	0	PATH	7.0a	2	1-4	3-4	1	0	*Hm*
Hordeum secalinum Schreber	nat	1	1	1	1	5	1	5	1	0	RDVRGEu	6.5a	2	1-3	3	-1	0	*Hs*
Hottonia palustris L.	nat	5	3	1	1	1	1	1	1	0	AQUATp	6.5a	1	1-3	3	-1	0	*Hp*
- *Humulus lupulus* L.	nat	1	3	1	1	2	1	3	5	0	HEDGE	6.0c	1	1-4	3	0	0	*Hl*
Hyacinthoides hispanica (Miller) Rothm.	neo	1	1	1	1	1	1	5	5	0	WOODa	[6.5]	2	-	-	1		*Hh*
+ *Hyacinthoides non-scripta* (L.) Chouard ex Rothm.	nat	1	1	1	1	2	1	1	5	0	PLANTb	4.5c	1	1	3	0	0	*Hn*
+ *Hydrocotyle vulgaris* L.	nat	2	5	3	1	2	1	1	1	0	MIREu	5.5b	1	1-2	3-4	-1	0	*Hv*
- *Hypericum androsaemum* L.	nat	-	-	-	-	-	-	5	0					1-2	3-4	-1	0	*Ha*
Hypericum elodes L.	nat	4	3	1	1	1	1	1	1	0	AQUATp	[5.0]	3	1	3	-1	0	*He*
+ *Hypericum hirsutum* L.	nat	1	1	3	1	3	2	2	3	0	SCREEu	7.0a	3	1-4	3	0	0	*Hh*
- *Hypericum humifusum* L.	nat	1	2	2	1	3	2	4	1	0	PATH	5.5b	2	1-3	3	0	0	*Hh*
- *Hypericum maculatum* Crantz	nat	1	1	1	1	3	4	4	1	0	WASTEa	6.5b	3	1-3	2-3	1	0	*Hm*
+ *Hypericum perforatum* L.	nat	1	1	3	3	3	3	4	1	0	WASTEl	7.0a	4	1-4	3-4	1	0	*Hp*
- *Hypericum pulchrum* L.	nat	1	1	4	1	4	1	3	3	0	PASTa	5.0b	4	1-2	3	-1	0	*Hp*
- *Hypericum tetrapterum* Fries	nat	1	5	3	1	2	3	2	1	0	MIREu	6.5a	4	1-3	3	0	0	*Ht*
+ *Hypochaeris radicata* L.	nat	1	1	3	2	3	4	4	1	0	QRYa	6.0c	3	1-3	3-4	1		*Hr*
+ *Impatiens glandulifera* Royle	neo	1	3	1	1	2	1	3	3	0	RIVBNKu	6.0b	2	-	-	1		*Ig*
Impatiens parviflora DC.	neo	1	3	3	1	1	3	3	3	0	RDVRGEs	7.0b	1	-	-	1		*Ip*
+ *Inula conyzae* (Griess.) Meikle	nat	1	1	3	1	3	4	4	1	0	QRYl	7.0a	4	1-3	3	1	0	*Ic*
Inula helenium L.	arch	1	1	1	1	1	1	5	1	0	WASTEl	7.0b	2	-	-	-1	0	*Ih*
- *Iris foetidissima* L.	nat	1	1	1	1	1	1	1	5	0	WOODl	[5.5]	1	1-2	3-4	-1	0	*If*
Iris germanica L.	neo	1	1	1	1	1	1	5	1	0	RLYBNK	[6.0]	1	-	-	1		*Ig*
- *Iris pseudacorus* L.	nat	2	5	1	1	2	1	1	1	0	MIREs	6.0b	1	1-3	3-4	-1	0	*Ip*
+ *Isolepis setacea* (L.) R. Br.	nat	1	5	3	1	2	1	3	2	0	MIREu	6.5b	4	1-4	3	-1	0	*Is*
Jasione montana L.	nat	1	1	3	1	1	3	5	1	0	WASTEa	5.0b	2	1-3	3	-1	0	*Jm*
- *Juncus acutiflorus* Ehrh. ex Hoffm.	nat	1	5	1	1	2	1	1	1	0	MIREu	5.5b	3	1-3	3	-1	0	*Ja*
+ *Juncus articulatus* L.	nat	2	5	1	1	3	1	3	1	0	MIREu	6.5b	3	1-4	3-4	-1	0	*Ja*
+ *Juncus bufonius* L.	nat	1	4	1	5	2	3	1	1	0	MIREu	6.5b	3	1-6	2-4	1	0	*Jb*
+ *Juncus bulbosus* L.	nat	4	5	1	1	2	1	2	1	0	AQUATp	5.5b	2	1-3	2-3	0	0	*Jb*
- *Juncus conglomeratus* L.	nat	1	5	1	1	3	1	3	3	0	MIREu	5.5a	4	1-3	3	0	0	*Jc*
+ *Juncus effusus* L.	nat	1	5	1	1	3	1	3	3	0	MIREu	5.5b	2	1-3	3-4	1	0	*Je*

TABLES OF ATTRIBUTES

Table 7.1

Species and authority	Status	Aquatic	Mire	Skeletal	Arable	Pasture	Spoil	Wasteland	Woodland	Maritime	Commonest terminal habitat	Soil pH	Floristic diversity	Longitudinal range	Latitudinal range	Present status	Red Data book	(Sp.)
- *Juncus inflexus* L.	nat	1	5	1	1	3	2	2	1	0	MIREu	6.5a	3	1-4	3-4	0	0	Ji
+ *Juncus squarrosus* L.	nat	1	3	3	1	4	1	4	1	0	PATH	4.0b	1	1-2	3	0	0	Js
Juncus subnodulosus Schrank	nat	2	5	1	1	1	1	1	1	0	MIREu	6.5a	3	1-3	3-4	-1	0	Js
Juncus tenuis Willd.	neo	1	3	1	1	1	4	3	1	0	PATH	6.5a	3	-	-	1		Jt
- *Knautia arvensis* (L.) Coulter	nat	1	1	2	2	4	3	4	1	0	WASTEl	7.0b	4	1-4	3	-1	0	Ka
+ *Koeleria macrantha* (Ledeb.) Schultes	nat	1	1	4	1	4	3	3	1	0	PASTl	6.5b	5	1-6	3	-1	0	Km
Lactuca virosa L.	nat	1	1	1	1	1	4	5	1	0	WASTEa	[7.5]	4	1-2	3-4	1	0	Lv
+ *Lamiastrum galeobdolon* (L.) Ehrend. & Polatschek	nat	1	2	1	1	2	1	1	5	0	WOODl	6.5b	1	1-3	3	-1	0	Lg
+ *Lamium album* L.	arch	1	1	3	3	2	4	3	1	0	SOIL	7.0a	2	1-5	2-3	1	0	La
Lamium amplexicaule L.	arch	1	1	1	5	1	3	2	1	0	GARDEN	6.5a	3	1-4	3-4	1	0	La
Lamium hybridum Villars	arch	1	1	3	5	1	2	4	1	0	ARABLE	7.0a	3	1-3	3	-1	0	Lh
Lamium maculatum (L.) L.	neo	1	1	4	1	1	1	3	1	0	RDVRGEs	[6.5]	1	-	-	1	0	Lm
+ *Lamium purpureum* L.	arch	1	1	1	5	2	4	1	1	0	ARABLE	7.0b	4	1-3	3	1	0	Lp
+ *Lapsana communis* L.	nat	1	1	3	5	2	3	1	3	0	ARABLE	7.0a	3	1-3	3	0	0	Lc
Lathraea squamaria L.	nat	1	1	1	1	3	1	1	5	0	WOODl	6.0b	1	1-3	3	-1	0	Ls
Lathyrus latifolius L.	neo	1	1	1	1	1	3	4	4	0	RLYBNK	7.0a	1	-	-	1		Ll
+ *Lathyrus linifolius* (Reichard) Bassler	nat	1	1	1	1	4	2	5	1	0	WASTEl	5.0b	4	1-3	3	-1	0	Ll
+ *Lathyrus pratensis* L.	nat	1	3	1	2	4	3	4	1	0	MEADOW	6.5b	4	1-4	2-3	-1	0	Lp
Lemna gibba L.	nat	5	1	1	1	1	1	1	1	0	AQUATp	6.5a	1	1-3	3-4	-1	0	Lg
+ *Lemna minor* L.	nat	5	3	2	1	2	1	1	1	0	AQUATp	6.5a	1	1-6	3-4	1	0	Lm
- *Lemna trisulca* L.	nat	5	1	1	1	2	1	1	1	0	AQUATp	6.5a	1	1-6	3	-1	0	Lt
+ *Leontodon autumnalis* L.	nat	1	3	3	3	3	4	3	1	0	*BRICK	6.5b	3	1-3	2-3	1	0	La
+ *Leontodon hispidus* L.	nat	1	1	4	1	3	4	3	1	0	QRYl	7.0a	4	1-3	3	0	0	Lh
- *Leontodon saxatilis* Lam.	nat	1	3	3	2	4	2	3	1	0	RDVRGEu	7.0b	2	1-2	3-4	0	0	Ls
- *Lepidium campestre* (L.) R. Br.	arch	1	1	1	1	3	3	4	1	0	RLYBNK	6.5a	2	1-3	3	-1	0	Lc
- *Lepidium draba* L.	neo	1	1	1	3	2	3	5	1	0	RLYBNK	7.0a	2	-	-	1		Ld
- *Lepidium heterophyllum* Benth.	nat	1	1	1	1	2	5	3	1	0	CINDER	6.5b	2	1	3-4	1	0	Lh
Lepidium ruderale L.	arch	1	1	1	1	1	4	3	1	0	PATH	[7.5]	2	1-4	3	1	0	Lr
+ *Leucanthemum vulgare* Lam.	nat	1	1	3	3	4	3	3	1	0	QRYl	7.0a	4	1-4	2-3	-1	0	Lv
Leucanthemum × superbum (Bergmans ex J. Ingram) D.H. Kent	neo	1	2	1	1	2	3	3	1	0	WASTEa	7.0b	2	-	-	1		Ls
Linaria purpurea (L.) Miller	neo	1	1	3	1	1	5	2	1	0	BRICK	7.5a	2	-	-	1		Lp
Linaria repens (L.) Miller	arch	1	1	1	1	1	5	2	1	0	CINDER	7.0b	2	1-2	3	1	0	Lr
+ *Linaria vulgaris* Miller	nat	1	1	2	1	2	5	1	1	0	CINDER	6.5b	2	1-5	2-3	1	0	Lv
+ *Linum catharticum* L.	nat	1	1	4	1	3	4	3	1	0	QRYl	7.0a	4	1-3	3	0	0	Lc
- *Listera ovata* (L.) R. Br.	nat	1	2	1	1	3	3	4	4	0	SCRUB	7.0a	4	1-4	2-3	-1	0	Lo
Lithospermum officinale L.	nat	1	1	3	1	1	1	3	5	0	WOODle	7.0b	3	1-4	3	-1	0	Lo
Littorella uniflora (L.) Asch.	nat	5	5	1	1	1	1	1	1	0	AQUATp	5.5a	1	1-2	3	-1	0	Lu
Lolium multiflorum Lam.	neo	1	1	1	4	5	3	3	2	0	MEADOW	7.0b	3	-	-	0		Lm
+ *Lolium perenne* L.	nat	1	1	3	4	4	3	3	1	0	MEADOW	7.0a	3	1-3	3-4	1	0	Lp
+ *Lonicera periclymenum* L.	nat	1	2	1	1	2	1	1	5	0	WOODl	4.0b	1	1-2	3-4	1	0	Lp
+ *Lotus corniculatus* L.	nat	1	1	3	1	4	4	4	1	0	PASTl	7.0b	4	1-5	3-4	1	0	Lc
+ *Lotus pedunculatus* Cav.	nat	1	5	1	1	3	1	4	1	0	*MIREu	5.5b	4	1-3	3	-1	0	Lp
Lupinus × regalis Bergmans	neo	1	1	2	1	1	4	5	1	0	RLYBNK	6.5b	2	-	-	1		Lr
+ *Luzula campestris* (L.) DC.	nat	1	2	2	1	4	3	4	1	0	PASTl	5.5b	3	1-3	3	0	0	Lc
- *Luzula multiflora* (Ehrh.) Lej.	nat	1	4	2	1	4	3	3	1	0	QRYa	4.5b	3	1-6	1-3	-1	0	Lm
+ *Luzula pilosa* (L.) Willd.	nat	1	1	3	1	2	1	1	5	0	PLANTc	4.0c	1	1-4	2-3	0	0	Lp
- *Luzula sylvatica* (Hudson) Gaudin	nat	1	1	4	1	2	1	1	5	0	WOODa	5.0b	2	1-3	3	-1	0	Ls
- *Lychnis flos-cuculi* L.	nat	1	5	1	1	3	1	2	1	0	MIREu	6.0b	4	1-4	3	-1	0	Lf
- *Lycopus europaeus* L.	nat	1	5	2	1	2	1	2	1	0	PONDBK	6.5b	2	1-4	3	-1	0	Le
- *Lysimachia nemorum* L.	nat	1	4	1	1	3	1	2	4	0	MIREs	5.5b	3	1-2	3	-1	0	Ln
- *Lysimachia nummularia* L.	nat	1	5	1	1	4	1	2	3	0	MIREu	6.5b	3	-	-	-1	0	Ln
Lysimachia punctata L.	neo	1	4	1	1	1	3	3	1	0	WASTEa	6.5b	3	-	-	1		Lp
- *Lysimachia vulgaris* L.	nat	1	5	1	1	3	2	2	1	0	MIREs	5.5b	1	1-5	3	-1	0	Lv
- *Lythrum portula* (L.) D. Webb	nat	2	5	1	1	2	1	1	1	0	MIREu	5.5a	2	1-3	3	-1	0	Lp
- *Lythrum salicaria* L.	nat	2	4	1	1	2	1	2	1	0	MIREu	6.0b	1	1-5	3	-1	0	Ls
- *Malva moschata* L.	nat	1	1	4	1	3	3	4	1	0	OUTCRPu	7.0b	3	1-3	3	-1	0	Mm
- *Malva neglecta* Wallr.	arch	1	1	3	1	3	3	4	1	0	RDVRGEu	7.0a	2	1-3	3	-1	0	Mn
- *Malva sylvestris* L.	arch	1	1	4	3	2	4	4	1	0	OUTCRPu	7.0a	3	1-4	3-4	1	0	Ms
+ *Matricaria discoidea* DC.	neo	1	1	1	5	2	4	3	1	0	ARABLE	7.0a	3	-	-	1		Md
Matricaria recutita L.	arch	1	1	1	5	1	3	3	1	0	ARABLE	7.0a	3	1-3	3-4	0	0	Mr
Meconopsis cambrica (L.) Viguier	nat	1	1	4	1	1	3	1	4	0	WOODa	[7.0]	1	1	2-3	0	0	Mc
+ *Medicago lupulina* L.	nat	1	1	4	4	3	4	4	1	0	OUTCRPu	7.0a	4	1-4	3	0	0	Ml

			Abundance in the primary habitats														
Species and authority	Status	Aquatic	Mire	Skeletal	Arable	Pasture	Spoil	Wasteland	Woodland	Maritime	Commonest terminal habitat	Soil pH	Floristic diversity	Longitudinal range	Latitudinal range	Present status	Red Data book	(Sp.)
- *Medicago sativa* L.	neo	1	1	2	1	2	3	5	1	0	RDVRGEu	7.0a	2	1-4	3-4	1	0	*Ms*
- *Melampyrum pratense* L.	nat	1	1	1	1	4	3	1	5	0	WOODa	3.5a	1	1-4	2-3	-1	0	*Mp*
+ *Melica uniflora* Retz.	nat	1	1	3	1	2	1	1	5	0	WOODl	7.0a	1	1-3	3	-1	0	*Mu*
Melilotus albus Medikus	neo	1	1	1	1	1	4	3	1	0	WASTEa	7.5a	2	-	-	1		*Ma*
Melilotus altissimus Thuill.	arch	1	1	1	1	1	4	5	1	0	WASTEa	7.0a	3	-	-	1	0	*Ma*
Melilotus officinalis (L.) Lam.	neo	1	1	1	1	1	5	3	1	0	WASTEa	7.0a	2	-	-	1		*Mo*
+ *Mentha aquatica* L.	nat	2	5	1	1	2	1	1	1	0	MIREs	6.5a	2	1-3	3	-1	0	*Ma*
- *Mentha arvensis* L.	nat	1	5	3	4	2	1	2	3	0	MIREu	6.0b	3	1-6	2-3	0	0	*Ma*
- *Mentha spicata* L.	arch	1	1	2	3	4	4	4	1	0	SOIL	7.0a	2	-	-	1	0	*Ms*
Mentha × piperita L.	neo	1	5	2	1	1	1	1	1	0	RIVBNKu	6.0a	2	-	-	-1		*Mp*
Mentha × verticillata L.	nat	1	5	1	1	3	1	1	1	0	MIREu	6.0b	3	-	-	0		*Mv*
Mentha × villosa Hudson	neo	1	1	1	1	3	3	5	1	0	WASTEa	7.0a	2	-	-	1		*Mv*
- *Menyanthes trifoliata* L.	nat	3	5	1	1	2	1	1	1	0	MIREu	5.0b	2	1-6	2-3	-1	0	*Mt*
Mercurialis annua L.	arch	1	1	1	5	1	3	3	1	0	WASTEa	7.0a	3	S-s	S-s	1	0	*Ma*
+ *Mercurialis perennis* L.	nat	1	1	2	1	2	2	1	5	0	WOODl	7.0a	2	1-3	3	-1	0	*Mp*
+ *Milium effusum* L.	nat	1	1	1	1	2	1	1	5	0	SCRUB	4.5b	1	1-6	2-3	-1	0	*Me*
- *Mimulus guttatus* DC.	neo	2	4	1	1	2	1	1	1	0	RIVBNKu	6.5a	2	-	-	1		*Mg*
Mimulus × robertsii Silverside	neo	4	4	3	1	1	1	1	1	0	MIREu	6.5a	1	-	-	1		*Mr*
+ *Minuartia verna* (L.) Hiern	nat	1	1	1	1	2	5	1	1	0	LEAD	6.5a	2	1-5	2	-1	1	*Mv*
+ *Moehringia trinervia* (L.) Clairv.	nat	1	1	3	1	2	1	1	5	0	WOODl	6.5c	1	1-3	3	-1	0	*Mt*
+ *Molinia caerulea* (L.) Moench	nat	1	5	2	1	4	1	3	1	0	PASTa	4.0c*	1	1-4	2-3	-1	0	*Mc*
- *Montia fontana* L.	nat	2	4	1	2	2	1	1	1	0	MIREu	6.0a	2	1-3	2-3	0	0	*Mf*
+ *Mycelis muralis* (L.) Dumort.	nat	1	1	5	1	2	2	1	3	0	OUTCRPs	7.0a	3	1-3	3	0	0	*Mm*
+ *Myosotis arvensis* (L.) Hill	arch	1	1	1	5	3	2	1	1	0	ARABLE	7.0b	4	1-4	2-3	0	0	*Ma*
- *Myosotis discolor* Pers.	nat	1	3	2	2	4	2	4	1	0	WASTEd	6.0b	4	1-3	3	0	0	*Md*
- *Myosotis laxa* subsp. *cespitosa* (Schultz) N. Hylander ex Nordh.	nat	2	5	1	1	2	1	1	1	0	MIREu	6.5b	3	1-6	2-3	0	0	*Ml*
+ *Myosotis ramosissima* Rochel	nat	1	1	4	1	3	4	3	1	0	OUTCRPu	6.5a	5	1-3	3-4	-1	0	*Mr*
+ *Myosotis scorpioides* L.	nat	2	5	1	1	2	1	1	1	0	MIREs	6.5a	2	1-4	3	0	0	*Ms*
- *Myosotis secunda* Al. Murray	nat	4	5	1	1	2	1	1	1	0	AQUATr	5.5a	3	1	3	-1	0	*Ms*
Myosotis sylvatica Hoffm.	nat	1	3	1	1	2	3	4	1	0	WOODl	7.0a	3	1-5	3	0	0	*Ms*
Myosoton aquaticum (L.) Moench	nat	1	4	1	1	1	1	3	3	0	PONDBK	6.5b	2	1-4	3	-1	0	*Ma*
Myriophyllum alterniflorum DC.	nat	5	1	1	1	1	1	1	1	0	AQUATp	[6.0]	1	1-2	2-3	-1	0	*Ma*
- *Myriophyllum spicatum* L.	nat	5	1	1	1	2	1	1	1	0	AQUATp	6.5a	1	1-5	3	-1	0	*Ms*
+ *Myrrhis odorata* (L.) Scop.	neo	1	1	1	3	2	1	4	4	0	RIVBNKu	6.5b	2	-	-	0		*Mo*
Narcissus pseudonarcissus L.	nat	1	1	1	1	1	1	3	5	0	WOODa	4.0c	1	1-2	3-4	-1	0	*Np*
+ *Nardus stricta* L.	nat	1	3	1	1	4	2	3	1	0	PASTa	3.5b	1	1-3	2-3	0	0	*Ns*
Narthecium ossifragum (L.) Hudson	nat	1	5	1	1	1	1	1	1	0	MIREu	4.5b	2	1	2-3	-1	0	*No*
Neottia nidus-avis (L.) Rich.	nat	1	1	1	1	1	1	1	5	0	WOODl	6.5c	1	1-4	3	-1	1	*Nn*
- *Nuphar lutea* (L.) Smith	nat	5	2	1	1	2	1	1	1	0	AQUATp	6.5a	1	1-4	2-3	-1	0	*Nl*
Nymphaea alba L.	nat	5	1	1	1	1	1	1	1	0	AQUATp	6.5b	1	1-3	3	0	0	*Na*
- *Odontites vernus* (Bellardi) Dumort.	nat	1	1	1	4	3	2	4	1	0	RDVRGEu	7.0a	4	1-5	3	-1	0	*Ov*
Oenanthe aquatica (L.) Poiret	nat	2	5	1	1	2	1	1	1	0	AQUATp	7.0a	2	1-4	3	-1	0	*Oa*
- *Oenanthe crocata* L.	nat	1	4	1	1	2	1	2	1	0	MIREu	[6.5]	1	1-2	3-4	-1	0	*Oc*
- *Oenanthe fistulosa* L.	nat	4	4	1	1	2	1	1	1	0	AQUATp	6.5a	1	1-3	3	-1	2	*Of*
Oenothera biennis L.	neo	1	1	1	1	1	4	5	1	0	WASTEa	[6.5]	3	-	-	0		*Ob*
Oenothera cambrica Rostanski	neo	1	1	1	1	1	5	2	1	0	COAL	6.5b	3	-	-	1		*Oc*
Oenothera glazioviana Micheli ex C. Martius	neo	1	1	1	1	1	5	3	1	0	WASTEa	7.5a	2	-	-	1		*Og*
- *Onobrychis viciifolia* Scop.	nat	1	1	4	1	1	1	5	1	0	WASTEl	7.0a	3	1-4	3	1	1	*Ov*
- *Ononis repens* L.	nat	1	1	4	1	3	4	4	1	0	OUTCRPu	7.0b	4	1-3	3	-1	0	*Or*
- *Ononis spinosa* L.	nat	1	1	4	1	4	3	1	1	0	WASTEa	[7.0]	3	1-4	3-4	-1	0	*Os*
Ophrys apifera Hudson	nat	1	1	5	1	4	3	1	1	0	QRYl	7.0a	5	S-s	S-s	1	0	*Oa*
- *Orchis mascula* (L.) L.	nat	1	2	1	4	1	3	5	1	0	PASTl	6.5a	4	1-3	3	-1	0	*Om*
- *Orchis morio* L.	nat	1	1	1	1	4	5	1	1	0	PASTl	[7.0]	5	1-3	3	-1	1	*Om*
+ *Origanum vulgare* L.	nat	1	1	3	1	3	4	4	1	0	QRYl	7.0a	4	1-5	3-4	0	0	*Ov*
Ornithogalum angustifolium Boreau	neo	1	1	4	1	1	1	5	4	0	WOODa	6.0b	2	1-3	3-4	0		*Oa*
Ornithopus perpusillus L.	nat	1	1	1	3	3	1	5	1	0	WASTEd	6.0b	3	1-2	3	-1	0	*Op*
+ *Oxalis acetosella* L.	nat	1	2	3	1	2	1	1	5	0	SCREEu	4.5c	1	1-5	2-3	-1	0	*Oa*
Oxalis corniculata L.	neo	1	1	1	5	1	1	1	1	0	GARDEN	[7.0]	2	-	-	0		*Oc*
Oxalis debilis Kunth	neo	1	1	1	5	1	1	3	1	0	GARDEN	6.5a	1	-	-	1		*Od*
Oxalis exilis Cunn.	neo	1	1	4	5	3	3	1	1	0	GARDEN	6.5a	1	-	-	1		*Oe*
- *Papaver argemone* L.	arch	1	1	3	5	2	3	3	3	0	ARABLE	7.0a	4	1-3	3-4	-1	2	*Pa*
- *Papaver dubium* L.	arch	1	1	1	3	2	4	4	1	0	RLYBNK	6.5b	3	1-4	3-4	1	0	*Pd*

Table 7.1

Species and authority	Status	Aquatic	Mire	Skeletal	Arable	Pasture	Spoil	Wasteland	Woodland	Maritime	Commonest terminal habitat	Soil pH	Floristic diversity	Longitudinal range	Latitudinal range	Present status	Red Data book	(Sp.)
+ *Papaver rhoeas* L.	arch	1	1	1	5	2	3	3	1	0	ARABLE	7.0a	4	1-3	3-4	-1	0	*Pr*
Papaver somniferum L.	arch	1	1	1	1	1	4	5	1	0	CINDER	7.0b	3	-	-	1	0	*Ps*
- *Parietaria judaica* L.	nat	1	1	5	1	2	2	2	3	0	WALL	7.0a	1	S-s	S-s	-1	0	*Pj*
Paris quadrifolia L.	nat	1	1	1	1	1	1	3	5	0	WOODl	7.0a	1	1-4	2-3	-1	0	*Pq*
Parnassia palustris L.	nat	1	3	3	1	5	1	1	1	0	*MIREu	7.0a	5	1-6	2-3	-1	0	*Pp*
Pastinaca sativa L.	nat	1	1	3	1	1	5	1	1	0	WASTEl	7.0b	3	1-4	3	1	0	*Ps*
- *Pedicularis sylvatica* L.	nat	1	3	1	1	4	1	1	1	0	PASTa	5.0b	5	1-3	3	-1	0	*Ps*
- *Pentaglottis sempervirens* (L.) Tausch ex L. Bailey	neo	1	1	1	1	2	3	3	5	0	RDVRGEu	6.5a	1	-	-	-1		*Ps*
+ *Persicaria amphibia* (L.) Gray	nat	5	5	1	1	2	1	1	1	0	MIREu	6.0b	1	1-6	2-3	-1	0	*Pa*
Persicaria bistorta (L.) Samp.	nat	1	2	1	1	2	1	3	5	0	WOODa	6.0c	2	1-5	2-3	-1	0	*Pb*
- *Persicaria hydropiper* (L.) Spach	nat	2	5	1	3	2	1	1	1	0	MIREu	6.0a	3	1-6	3	1	0	*Ph*
- *Persicaria lapathifolia* (L.) Gray	nat	1	3	1	5	2	3	2	1	0	ARABLE	6.5a	3	1-6	3-4	0	0	*Pl*
+ *Persicaria maculosa* Gray	nat	1	3	1	5	2	3	1	1	0	ARABLE	6.5b	3	1-5	3	-1	0	*Pm*
- *Petasites fragrans* (Villars) C. Presl	neo	1	1	1	1	2	3	4	4	0	RDVRGEu	6.5a	1	-	-	0	0	*Pf*
+ *Petasites hybridus* (L.) P. Gaertner, Meyer & Scherb.	nat	1	3	3	3	2	2	3	3	0	RIVBNKu	7.0a	2	1-3	3	0	0	*Ph*
+ *Phalaris arundinacea* L.	nat	2	5	3	1	2	1	3	1	0	RIVBNKu	6.5b	1	1-6	2-3	0	0	*Pa*
Phleum bertolonii DC.	nat	1	1	4	2	4	3	3	1	0	PASTl	6.5b	5	1-3	3-4	0	0	*Pb*
+ *Phleum pratense* L.	nat	1	3	1	5	4	3	3	1	0	MEADOW	6.5b	3	1-4	3	0	0	*Pp*
+ *Phragmites australis* (Cav.) Trin. ex Steudel	nat	2	5	1	1	2	1	2	1	0	MIREu	6.0b	1	1-6	2-4	-1	0	*Pa*
Picris echioides L.	arch	1	1	1	1	3	1	5	1	0	RDVRGEu	7.0a	3	1-3	3-4	1	0	*Pe*
Picris hieracioides L.	nat	1	1	4	1	1	4	4	1	0	CINDER	7.0a	3	1-5	3	1	0	*Ph*
Pilosella aurantiaca (L.) F. Schultz & Schultz-Bip.	neo	1	1	1	1	2	4	4	1	0	LAWN	6.5b	3	-	-	1		*Pa*
+ *Pilosella officinarum* F. Schultz & Schultz-Bip.	nat	1	1	3	1	3	4	3	1	0	PASTl	7.0b	4	1-3	3	-1	0	*Po*
Pimpinella major (L.) Hudson	nat	1	1	2	1	1	3	4	4	0	WOODle	7.0a	4	1-3	3	-1	0	*Pm*
+ *Pimpinella saxifraga* L.	nat	1	1	3	1	4	3	4	1	0	PASTl	7.0a	5	1-4	3	-1	0	*Ps*
Pinguicula vulgaris L.	nat	1	4	5	1	1	1	1	1	0	MIREu	6.5a	4	1-6	2	-1	0	*Pv*
- *Plantago coronopus* L.	nat	1	1	3	1	2	1	5	1	1	PATH	6.5a	4	1-4	3-4	-1	0	*Pc*
+ *Plantago lanceolata* L.	nat	1	1	3	3	4	3	3	1	0	MEADOW	7.0b	4	1-4	3-4	1	0	*Pl*
+ *Plantago major* L.	nat	1	3	3	4	3	3	4	1	0	PATH	7.0a	3	1-5	2-4	1	0	*Pm*
Plantago media L.	nat	1	1	3	1	3	4	3	1	0	PASTl	7.0a	5	1-5	3	-1	0	*Pm*
- *Platanthera chlorantha* (Custer) Reichb.	nat	1	1	1	1	4	3	3	3	0	PASTl	5.0a	5	1-3	3	-1	1	*Pc*
Poa angustifolia L.	nat	1	1	2	1	1	3	5	1	0	QRYl	7.0a	4	1-6	3-4	0	0	*Pa*
+ *Poa annua* L.	nat	1	2	1	5	3	4	3	1	0	PATH	7.0b	3	1-4	2-4	1	0	*Pa*
Poa compressa L.	nat	1	1	4	1	1	4	2	1	0	CINDER	7.0a	3	1-3	3	0	0	*Pc*
- *Poa nemoralis* L.	nat	1	1	3	1	2	1	1	5	0	OUTCRPs	7.0b	1	1-6	2-3	-1	0	*Pn*
+ *Poa pratensis* L.	nat	1	2	3	2	3	3	4	1	0	RDVRGEu	7.0b	3	1-6	2-4	-1	0	*Pp*
+ *Poa trivialis* L.	nat	1	3	3	3	3	3	3	3	0	MEADOW	6.5b	2	1-4	2-4	1	0	*Pt*
- *Polygala serpyllifolia* Hose	nat	1	3	2	1	4	3	1	1	0	PASTa	4.5a	3	1-2	3	-1	0	*Ps*
+ *Polygala vulgaris* L.	nat	1	1	3	1	4	3	4	1	0	PASTl	7.0a	5	1-3	3	-1	0	*Pv*
Polygonatum × hybridum Bruegger	neo	1	1	1	1	1	1	5	4	0	RDVRGEu	[7.0]	1	-	-	1		*Ph*
+ *Polygonum aviculare* L.	nat	1	2	1	5	2	3	3	1	0	ARABLE	7.0b	3	1-6	2-4	1	0	*Pa*
Potamogeton berchtoldii Fieber	nat	5	2	1	1	1	1	1	1	0	AQUATp	6.5a	1	1-6	2-3	-1	0	*Pb*
+ *Potamogeton crispus* L.	nat	5	1	1	1	2	1	1	1	0	AQUATp	6,5a	1	1-5	3-4	-1	0	*Pc*
+ *Potamogeton natans* L.	nat	5	2	1	1	2	1	1	1	0	AQUATp	6.5a	1	1-6	2-3	-1	0	*Pn*
Potamogeton pectinatus L.	nat	5	1	1	1	1	1	1	1	0	AQUATp	7.0a	1	1-6	2-4	-1	0	*Pp*
Potamogeton perfoliatus L.	nat	5	1	1	1	1	1	1	1	0	AQUATp	6.5a	1	1-6	2-3	-1	0	*Pp*
Potamogeton polygonifolius Pourret	nat	4	4	1	1	1	1	1	1	0	AQUATr	6.0a	1	1-2	3	-1	0	*Pp*
Potamogeton pusillus L.	nat	5	1	1	1	1	1	1	1	0	AQUATp	7.0a	1	1-6	3-4	-1	0	*Pp*
- *Potentilla anglica* Laich.	nat	1	4	1	1	2	1	4	1	0	RDVRGEu	6.0b	4	1-3	3	0	0	*Pa*
Potentilla anserina L.	nat	1	4	1	3	4	2	3	1	0	MIREu	6.5b	3	1-6	2-3	0	0	*Pa*
+ *Potentilla erecta* (L.) Rausch.	nat	1	3	3	1	4	1	4	1	0	PASTl	5.0b	3	1-4	2-3	-1	0	*Pe*
- *Potentilla palustris* (L.) Scop.	nat	2	5	1	1	2	1	1	1	0	MIREu	5.0a	2	1-6	2-3	-1	0	*Pp*
Potentilla recta L.	neo	1	1	1	1	1	1	5	1	0	RDVRGEu	[6.5]	1	-	-	1		*Pr*
- *Potentilla reptans* L.	nat	1	3	2	1	3	3	4	2	0	RDVRGEu	7.0a	3	1-4	3-4	1	0	*Pr*
+ *Potentilla sterilis* (L.) Garcke	nat	1	1	3	1	4	3	3	3	0	PASTl	6.5b	5	1-2	3	-1	0	*Ps*
+ *Primula veris* L.	nat	1	1	1	1	4	2	4	1	0	PASTl	7.0a	5	1-4	3	-1	0	*Pv*
- *Primula vulgaris* Hudson	nat	1	3	3	1	3	2	3	5	0	SCRUB	6.5b	3	1-3	3	-1	0	*Pv*
+ *Prunella vulgaris* L.	nat	1	3	3	3	4	3	3	3	0	MEADOW	6.5b	5	1-6	2-4	1	0	*Pv*
- *Pseudofumaria lutea* (L.) Borkh.	neo	1	1	5	1	2	1	1	1	0	WALL	[7.0]	1	-	-	0		*Pl*

			Abundance in the primary habitats														
Species and authority		Status	Aquatic	Mire	Skeletal	Arable	Pasture	Spoil	Wasteland	Woodland	Maritime	Commonest terminal habitat	Soil pH	Floristic diversity	Longitudinal range	Latitudinal range	Present status	Red Data book	(Sp.)
Puccinellia distans (Jacq.) Parl.		nat	1	3	1	3	1	3	4	1	1	RDVRGEu	7.5a	1	1-4	2-3	1	0	*Pd*
- *Pulicaria dysenterica* (L.) Bernh.		nat	1	5	1	2	3	2	3	1	0	MIREu	6.5b	3	1-4	3-4	0	0	*Pd*
+ *Ranunculus acris* L.		nat	1	3	1	3	4	3	3	2	0	MEADOW	6.5b	4	1-5	1-3	-1	0	*Ra*
Ranunculus aquatilis L.		nat	5	3	1	1	1	1	1	1	0	AQUATp	7.0a	1	1-3	3	-1	0	*Ra*
- *Ranunculus auricomus* L.		nat	1	1	1	1	4	2	4	5	0	WOODl	7.0a	3	1-3	2-3	-1	0	*Ra*
+ *Ranunculus bulbosus* L.		nat	1	1	2	1	4	3	3	1	0	MEADOW	6.5b	5	1-3	3-4	-1	0	*Rb*
Ranunculus circinatus Sibth.		nat	5	1	1	1	1	1	1	1	0	AQUATp	6.5a	1	1-5	3	-1	0	*Rc*
+ *Ranunculus ficaria* L.		nat	1	3	1	3	1	3	3	4	0	RIVBNKu	6.5b	2	1-3	3-4	0	0	*Rf*
+ *Ranunculus flammula* L.		nat	2	5	1	1	1	2	3	1	0	MIREs	5.5b	1	1-3	3	-1	0	*Rf*
Ranunculus hederaceus L.		nat	4	5	1	1	1	1	1	1	0	MIREu	6.5b	1	1-2	3-4	-1	0	*Rh*
Ranunculus lingua L.		nat	3	5	1	1	1	1	1	1	0	MIREu	[7.0]	1	1-4	3	0	0	*Rl*
Ranunculus omiophyllus Ten.		nat	5	3	1	1	1	1	1	1	0	AQUATr	6.0a	1	1-2	3-4	-1	0	*Ro*
+ *Ranunculus peltatus* Schrank		nat	4	4	1	1	2	1	1	1	0	AQUATp	6.0b	1	1-3	2-4	-1	0	*Rp*
+ *Ranunculus penicillatus* (Dumort.) Bab.		nat	5	2	3	1	2	1	1	1	0	AQUATr	7.0a	1	1-3	3	-1	0	*Rp*
+ *Ranunculus repens* L.		nat	1	4	1	4	4	3	3	1	0	MEADOW	6.5b	3	1-5	2-3	1	0	*Rr*
+ *Ranunculus sceleratus* L.		nat	3	5	1	1	2	3	2	1	0	MIREu	6.5a	2	1-6	2-3	1	0	*Rs*
Ranunculus trichophyllus Chaix		nat	5	3	1	1	1	1	1	1	0	AQUATp	6.5a	1	1-6	1-3	-1	0	*Rt*
- *Raphanus raphanistrum* L.		arch	1	1	1	5	2	3	3	1	0	ARABLE	6.0b	4	1-3	3-4	-1	0	*Rr*
Rapistrum rugosum (L.) Bergeret		neo	1	1	1	1	1	4	1	1	0	CINDER	[7.0]	1	-	-		0	*Rr*
Reseda lutea L.		nat	1	1	1	1	1	5	1	1	0	CINDER	6.5a	2	1-3	3-4	1	0	*Rl*
- *Reseda luteola* L.		arch	1	1	2	2	2	5	3	1	0	COAL	7.0a	1	1-4	3-4	1	0	*Rl*
+ *Rhinanthus minor* L.		nat	1	2	3	1	4	1	1	1	0	MEADOW	6.5b	5	1-3	2-3	-1	0	*Rm*
- *Ribes uva-crispa* L.		neo	1	1	1	1	2	1	1	5	0	WOODl	6.0c	1	1-3	3	0		*Ru*
Rorippa amphibia (L.) Besser		nat	3	5	1	1	1	1	1	1	0	AQUATp	6.5a	2	1-5	3	-1	0	*Ra*
Rorippa microphylla (Boenn.) N. Hylander ex Á. Löve & D. Löve		nat	4	4	1	1	1	1	1	1	0	AQUATr	6.5a	1	-	-	-1	0	*Rm*
+ *Rorippa nasturtium-aquaticum* (L.) Hayek		nat	4	4	1	1	2	1	1	1	0	AQUATr	6.5a	1	1-4	3-4	-1	0	*Rn*
+ *Rorippa palustris* (L.) Besser		nat	1	5	3	3	2	3	1	1	0	MIREu	6.5b	2	1-6	2-3	1	0	*Rp*
Rorippa sylvestris (L.) Besser		nat	2	3	1	5	1	1	2	1	0	GARDEN	6.5b	2	1-3	3	1	0	*Rs*
- *Rubus caesius* L.		nat	1	2	3	1	2	3	3	4	0	SCRUB	7.0a	2	1-4	3	-1	0	*Rc*
Rubus chamaemorus L.		nat	1	3	1	1	5	1	1	1	0	PASTa	3.5a	1	1-6	2	-1	0	*Rc*
+ *Rubus fruticosus* L. agg.		nat	1	2	2	2	1	3	3	4	0	HEDGE	6.5c	1	1-3	3-4	0	0	*Rf*
- *Rubus idaeus* L.		nat	1	2	3	1	2	1	3	4	0	PLANTb	6.0c	2	1-6	2-3	0	0	*Ri*
Rubus saxatilis L.		nat	1	1	4	1	1	1	3	5	0	SCREEs	7.0a	4	1-5	2-3	-1	0	*Rs*
+ *Rumex acetosa* L.		nat	1	3	2	3	4	3	3	1	0	*MEADOW	6.0b	3	1-4	2-3	-1	0	*Ra*
+ *Rumex acetosella* L.		nat	1	1	3	3	3	3	4	2	0	COAL	5.5c	1	1-4	2-4	-1	0	*Ra*
- *Rumex conglomeratus* Murray		nat	1	5	1	1	3	1	2	1	0	MIREu	6.5b	3	1-4	3-4	1	0	*Rc*
+ *Rumex crispus* L.		nat	1	3	1	3	2	4	3	1	0	CINDER	6.5b	3	1-4	3-4	1	0	*Rc*
- *Rumex hydrolapathum* Hudson		nat	3	5	1	1	2	1	1	1	0	CANALB	6.0c	1	1-3	3	-1	0	*Rh*
+ *Rumex obtusifolius* L.		nat	1	3	2	5	2	4	3	1	0	SOIL	7.0a	3	1-3	3	1	0	*Ro*
- *Rumex sanguineus* L.		nat	1	4	1	2	3	1	3	4	0	SCRUB	6.5a	3	1-3	3	-1	0	*Rs*
Ruscus aculeatus L.		nat	1	1	1	1	1	1	1	5	0	WOODl	[6.5]	1	S-s	S-s	-1	0	*Ra*
- *Sagina apetala* Ard.		nat	1	1	3	2	2	4	3	1	0	CINDER	7.0b	4	1-3	3-4	1	0	*Sa*
Sagina maritima G. Don		nat	1	1	1	1	1	1	5	1	1	RDVRGEu	[7.5]	1	1-3	3-4	-1	0	*Sm*
- *Sagina nodosa* (L.) Fenzl		nat	1	4	3	1	2	3	1	1	0	MIREu	7.0a	4	1-4	2-3	-1	0	*Sn*
+ *Sagina procumbens* L.		nat	1	3	3	3	2	4	3	1	0	CINDER	6.5b	3	1-4	2-3	1	0	*Sp*
Sagittaria sagittifolia L.		nat	5	1	1	1	1	1	1	1	0	AQUATp	6.5a	1	1-4	2-3	-1	0	*Ss*
- *Salix repens* L.		nat	1	5	1	1	1	4	1	1	0	MIREu	[5.5]	3	1-4	2-3	-1	0	*Sr*
Sambucus ebulus L.		arch	1	1	1	1	4	3	4	3	0	WASTEl	7.0a	2	1-3	3-4	-1	0	*Se*
Samolus valerandi L.		nat	1	5	1	1	1	1	1	1	0	MIREu	6.5a	1	1-6	3-4	-1	0	*Sv*
+ *Sanguisorba minor* subsp. *minor* Scop.		nat	1	1	3	1	4	3	3	1	0	PASTl	7.0a	5	1-4	3-4	-1	0	*Sm*
Sanguisorba minor subsp. *muricata* (Grernli) Briq.		neo	1	1	4	1	1	1	4	1	0	RD/RLY	7.0a	3	-	-		0	*Sm*
Sanguisorba officinalis L.		nat	1	3	2	1	5	1	3	3	0	WASTEa	6.0b	4	1-6	2-3	-1	0	*So*
+ *Sanicula europaea* L.		nat	1	1	1	1	2	1	1	5	0	WOODl	6.5b	2	1-3	3	-1	0	*Se*
- *Saponaria officinalis* L.		arch	1	1	1	1	2	4	5	1	0	WASTEa	6.5b	1	1-3	3	0	0	*So*
Saxifraga granulata L.		nat	1	1	4	3	5	3	3	1	0	PASTl	6.5a	5	1-3	3	-1	0	*Sg*
+ *Saxifraga tridactylites* L.		nat	1	1	5	1	2	4	1	1	0	OUTCRPu	7.0a	4	1-3	3-4	-1	0	*St*
Saxifraga × *urbium* D. Webb		neo	1	1	4	1	1	1	1	5	0	RIVBNKu	7.0c	1	-	-		1	*Su*
+ *Scabiosa columbaria* L.		nat	1	1	4	1	4	3	3	1	0	PASTl	7.0a	4	1-3	3	-1	0	*Sc*
- *Schoenoplectus lacustris* (L.) Palla		nat	5	3	1	1	1	2	1	1	0	AQUATp	6.5a	1	1-4	2-4	-1	0	*Sl*
Schoenoplectus tabernaemontani (C. Gmelin) Palla		nat	4	5	1	1	1	1	1	1	0	MIREu	6.5a	1	1-5	3-4	0	0	*St*

TABLES OF ATTRIBUTES

Table 7.1

				Abundance in the primary habitats										
Species and authority		Status	Aquatic	Mire	Skeletal	Arable	Pasture	Spoil	Wasteland	Woodland	Maritime	Commonest terminal habitat	Soil pH	Floristic diversity	Longitudinal range	Latitudinal range	Present status	Red Data book	(Sp.)
	Schoenus nigricans L.	nat	1	5	1	1	1	1	1	1	1	MIREu	7.0a	4	1-4	3-4	-1	0	*Sn*
	Scirpus sylvaticus L.	nat	1	5	1	1	1	1	1	1	0	MIREs	6.0b	2	1-4	3	-1	0	*Ss*
-	*Scleranthus annuus* L.	nat	1	1	1	5	3	4	4	1	0	SNDPIT	6.0b	4	1-3	3	-1	3	*Sa*
-	*Scrophularia auriculata* L.	nat	1	4	2	1	2	3	3	3	0	RIVBNKu	7.0a	3	1-2	3-4	0	0	*Sa*
-	*Scrophularia nodosa* L.	nat	1	3	1	1	2	4	3	4	0	SCRUB	7.0b	3	1-4	3	1	0	*Sn*
-	*Scutellaria galericulata* L.	nat	1	5	3	1	2	1	1	1	0	MIREs	6.0b	2	1-4	2-3	0	0	*Sg*
	Scutellaria minor Hudson	nat	1	5	1	1	1	1	1	1	0	MIREu	5.5b	5	1-2	3-4	-1	0	*Sm*
	Securigera varia (L.) Lassen	neo	1	1	1	1	1	5	1	1	0	WASTEa	[6.0]	1	-	-	0		*Sv*
+	*Sedum acre* L.	nat	1	1	5	1	3	3	1	1	0	OUTCRPu	7.0a	3	1-3	3	-1	0	*Sa*
	Sedum album L.	arch	1	1	5	1	1	3	2	1	0	OUTCRPu	7.0a	2	S-s	S-s	1	0	*Sa*
-	*Sedum anglicum* Hudson	nat	-	-	5	-	-	-	-						1	3	-1	0	*Sa*
	Sedum rupestre L.	neo	1	1	5	1	1	3	5	1	0	OUTCRPu	7.0a	3	-	-		1	*Sr*
	Sedum spurium M. Bieb.	neo	1	1	5	1	1	1	5	1	0	OUTCRPu	[5.5]	1	-	-		1	*Ss*
-	*Sedum telephium* L.	nat	1	1	5	1	2	1	3	1	0	OUTCRPu	6.5a	3	1-5	3	-1	0	*St*
-	*Senecio aquaticus* Hill	nat	1	5	1	1	3	1	3	1	0	MIREu	6.5b	4	1-3	3	-1	0	*Sa*
	Senecio erucifolius L.	nat	1	2	1	1	3	3	5	1	0	WASTEl	7.0b	4	1-4	3	0	0	*Se*
+	*Senecio jacobaea* L.	nat	1	1	4	2	3	4	3	1	0	QRYl	7.0b	4	1-4	3	-1	0	*Sj*
+	*Senecio squalidus* L.	neo	1	1	3	1	2	5	3	1	0	BRICK	7.0a	2	-	-		1	*Ss*
-	*Senecio sylvaticus* L.	nat	1	1	1	2	2	3	5	1	0	WASTEd	6.0b	3	1-3	3	-1	0	*Ss*
+	*Senecio viscosus* L.	neo	1	1	2	1	2	5	1	1	0	CINDER	7.0b	2	1-3	3	1		*Sv*
+	*Senecio vulgaris* L.	nat	1	1	2	5	2	4	2	1	0	BRICK	7.0a	3	1-3	3-4	1	0	*Sv*
	Serratula tinctoria L.	nat	1	3	1	1	3	1	4	3	0	WASTEl	6.5b	2	1-3	3	-1	0	*St*
-	*Sherardia arvensis* L.	nat	1	1	5	4	2	3	1	3	0	OUTCRPu	7.0a	5	1-3	3-4	-1	0	*Sa*
	Silaum silaus (L.) Schinz & Thell.	nat	1	3	1	1	3	2	5	1	0	WASTEl	6.0b	4	1-4	3	-1	0	*Ss*
+	*Silene dioica* (L.) Clairv.	nat	1	3	2	1	2	1	3	4	0	RIVBNKu	6.5b	2	1-3	2-3	-1	0	*Sd*
-	*Silene latifolia* Poiret	arch	1	1	3	4	3	3	4	1	0	RDVRGEu	7.0a	3	1-4	3-4	1	0	*Sl*
	Silene noctiflora L.	arch	1	1	1	5	1	3	5	1	0	ARABLE	[7.0]	2	1-3	3	-1	2	*Sn*
-	*Silene vulgaris* Garcke	nat	1	1	3	2	2	4	4	1	0	CINDER	7.0a	2	1-5	3-4	-1	0	*Sv*
+	*Sinapis arvensis* L.	arch	1	1	1	5	2	3	2	1	0	ARABLE	7.0a	4	1-4	3	-1	0	*Sa*
-	*Sisymbrium altissimum* L.	neo	1	1	1	1	2	5	3	1	0	SOIL	7.0a	2	-	-		1	*Sa*
	Sisymbrium loeselii L.	neo	1	1	1	1	1	5	1	1	0	WASTEa	[7.5]	2	-	-		1	*Sl*
-	*Sisymbrium officinale* (L.) Scop.	arch	1	1	1	4	2	4	4	1	0	RDVRGEu	7.0a	3	1-3	3-4	1	0	*So*
	Sisymbrium orientale L.	neo	1	1	1	3	1	5	2	1	0	BRICK	7.5a	3	-	-		1	*So*
+	*Solanum dulcamara* L.	nat	2	5	1	1	2	1	3	3	0	MIREs	6.5b	1	1-5	3-4	0	0	*Sd*
	Solanum nigrum L.	nat	1	1	1	5	3	3	3	1	0	WASTEa	7.0b	2	1-5	3-4	0	0	*Sn*
	Solanum physalifolium Rusby	neo	1	1	1	5	1	1	1	1	0	ARABLE	[6.0]	1	-	-		1	*Sp*
	Solidago canadensis L.	neo	1	1	1	1	3	3	5	1	0	WASTEa	6.5a	2	-	-		1	*Sc*
	Solidago gigantea Aiton	neo	1	1	1	1	1	4	4	3	0	RDVRGEu	7.0a	1	-	-		1	*Sg*
+	*Solidago virgaurea* L.	nat	1	1	3	1	2	4	3	3	1	QRYl	7.0b*	3	1-5	2-3	-1	0	*Sv*
-	*Sonchus arvensis* L.	nat	1	3	1	4	2	3	4	1	0	RDVRGEu	7.0a	3	1-4	3	0	0	*Sa*
+	*Sonchus asper* (L.) Hill	nat	1	1	2	3	2	5	3	1	0	QRYl	7.0a	4	1-3	3-4	1	0	*Sa*
+	*Sonchus oleraceus* L.	nat	1	1	3	4	2	4	3	1	0	BRICK	7.0a	3	1-3	3-4	1	0	*So*
-	*Sparganium emersum* Rehmann	nat	5	3	1	1	2	1	1	1	0	AQUATp	6.5a	1	1-6	2-3	-1	0	*Se*
+	*Sparganium erectum* L.	nat	4	4	1	1	2	1	1	1	0	AQUATp	6.5a	1	1-6	3	0	0	*Se*
+	*Spergula arvensis* L.	arch	1	1	1	5	2	3	1	1	0	ARABLE	6.0b	3	1-4	2-4	-1	2	*Sa*
	Spergularia marina (L.) Griseb.	nat	1	1	1	1	1	1	5	1	1	RDVRGEu	[7.5]	1	1-6	3-4	1	0	*Sm*
-	*Spergularia rubra* (L.) J.S. Presl & C. Presl	nat	1	3	1	3	2	4	3	1	0	CINDER	5.5b	3	1-3	3-4	1	0	*Sr*
-	*Stachys arvensis* (L.) L.	arch	1	1	1	5	2	3	3	1	0	ARABLE	7.0b	4	1-2	3-4	-1	1	*Sa*
+	*Stachys officinalis* (L.) Trev. St. Leon	nat	1	1	1	1	4	1	5	1	0	PASTl	5.5b	5	1-3	3	-1	0	*So*
	Stachys palustris L.	nat	1	4	1	4	3	1	3	2	0	PONDBK	6.5a	2	1-6	2-3	0	0	*Sp*
+	*Stachys sylvatica* L.	nat	1	2	3	3	2	3	3	3	0	HEDGE	7.0a	3	1-4	3	1	0	*Ss*
-	*Stellaria graminea* L.	nat	1	3	1	1	4	1	4	2	0	WASTEa	6.0b	4	1-4	2-3	-1	0	*Sg*
+	*Stellaria holostea* L.	nat	1	1	1	1	2	1	3	5	0	HEDGE	6.0b	2	1-4	3	-1	0	*Sh*
+	*Stellaria media* (L.) Villars	nat	1	1	2	5	2	3	3	2	0	ARABLE	6.5b	3	1-5	2-4	1	0	*Sm*
	Stellaria neglecta Weihe	nat	1	1	1	1	1	1	1	4	0	MIREs	6.5b	1	1-3	3	-1	0	*Sn*
	Stellaria pallida (Dumort.) Crepin	nat	1	1	4	3	3	2	4	3	0	WASTEd	6.5b	2	1-4	3-4	-1	0	*Sp*
+	*Stellaria uliginosa* Murray	nat	1	5	1	1	3	1	3	1	0	MIREu	6.0a	3	1-3	3	1	0	*Su*
+	*Succisa pratensis* Moench	nat	1	3	1	1	4	1	4	1	0	PASTl	6.5b	5	1-4	3	-1	0	*Sp*
-	*Symphytum officinale* L.	nat	1	3	1	1	2	1	4	2	0	RIVBNKu	7.0a	2	1-3	3	0	0	*So*
	Symphytum orientale L.	neo	1	1	1	1	1	1	4	4	0	HEDGE	[7.0]	1	-	-		0	*So*
-	*Symphytum × uplandicum* Nyman	neo	1	2	1	1	3	3	4	3	0	RDVRGEu	7.0a	1	-	-		0	*Su*
+	*Tamus communis* L.	nat	1	1	1	1	1	1	3	5	0	HEDGE	7.0a	2	S-s	S-s	-1	0	*Tc*
-	*Tanacetum parthenium* (L.) Schultz-Bip.	arch	1	1	4	2	2	4	3	1	0	WALL	7.0a	3	-	-		0	*Tp*
-	*Tanacetum vulgare* L.	nat	1	1	1	1	2	5	4	1	0	CINDER	6.5b	2	1-5	2-3	0	0	*Tv*

					Abundance in the primary habitats												
Species and authority	Status	Aquatic	Mire	Skeletal	Arable	Pasture	Spoil	Wasteland	Woodland	Maritime	Commonest terminal habitat	Soil pH	Floristic diversity	Longitudinal range	Latitudinal range	Present status	Red Data book	(Sp.)
+ *Taraxacum* agg. Weber	nat	1	1	3	3	3	3	3	2	0	MEADOW	7.0a	3	1-6	2-4	1	0	Ta
Tellima grandiflora (Pursh) Douglas ex Lindley	neo	1	1	5	1	1	1	1	4	0	RIVBNKs	[6.5]	1	-	-	1		Tg
+ *Teucrium scorodonia* L.	nat	1	1	4	1	3	3	3	3	0	SCREEu	6.5c*	2	1-2	3-4	-1	0	Ts
Thalictrum flavum L.	nat	1	4	1	1	5	1	3	1	0	PONDBK	6.5a	2	1-4	2-3	-1	0	Tf
- *Thlaspi arvense* L.	arch	1	1	1	5	2	3	4	1	0	ARABLE	7.0a	3	1-5	3	-1	0	Ta
+ *Thymus polytrichus* A. Kerner ex Borbas	nat	1	1	4	1	4	3	1	1	0	PASTl	7.0a	5	1-3	2-3	-1	0	Tp
+ *Torilis japonica* (Houtt.) DC.	nat	1	1	3	1	3	4	3	3	0	OUTCRPu	7.0a	5	1-5	3	1	0	Tj
- *Tragopogon pratensis* L.	nat	1	1	3	1	3	3	4	1	0	RDVRGEu	7.0a	3	1-4	3	0	0	Tp
Trichophorum cespitosum (L.) Hartman	nat	1	4	1	1	4	1	3	1	0	MIREu	3.5a	1	1-6	2	-1	0	Tc
Trientalis europaea L.	nat	1	5	1	1	1	1	1	1	0	MIREu	[3.5]	1	1-6	2	-1	0	Te
- *Trifolium arvense* L.	nat	1	1	2	1	2	4	4	1	0	SNDPIT	6.5	3	1-4	3-4	-1	0	Ta
- *Trifolium campestre* Schreber	nat	1	1	4	1	2	3	4	1	0	OUTCRPu	7.0a	4	1-4	3-4	-1	0	Tc
+ *Trifolium dubium* Sibth.	nat	1	1	4	4	3	4	3	1	0	OUTCRPu	6.5a	5	1-3	3	1	0	Td
Trifolium fragiferum L.	nat	1	4	1	1	5	1	1	1	0	PASTl	[7.0]	4	1-4	3-4	-1	0	Tf
- *Trifolium hybridum* L.	neo	1	3	1	1	3	4	4	1	0	RDVRGEu	7.0b	3	-	-	0		Th
+ *Trifolium medium* L.	nat	1	1	1	1	3	3	5	1	0	WASTEl	6.5b	4	1-4	2-3	-1	0	Tm
Trifolium micranthum Viv.	nat	1	1	1	1	5	1	3	1	0	LAWN	7.0b	3	S-s	S-s	-1	0	Tm
+ *Trifolium pratense* L.	nat	1	1	2	5	4	3	3	1	0	MEADOW	7.0a	4	1-4	3	0	0	Tp
+ *Trifolium repens* L.	nat	1	3	1	4	4	3	3	1	0	PASTe	6.5b	3	1-4	2-3	-1	0	Tr
Trifolium striatum L.	nat	1	1	5	1	3	1	3	1	0	OUTCRPu	6.5a	5	1-3	3-4	-1	0	Ts
- *Triglochin palustre* L.	nat	1	5	1	1	2	1	1	1	0	MIREu	6.5a	4	1-6	2-3	-1	0	Tp
+ *Tripleurospermum inodorum* (L.) Schultz-Bip.	arch	1	1	1	5	2	4	2	1	0	ARABLE	7.0a	3	1-4	3	1	0	Ti
+ *Trisetum flavescens* (L.) P. Beauv.	nat	1	1	3	1	4	3	4	1	0	MEADOW	7.0a	4	1-3	3	-1	0	Tf
Trollius europaeus L.	nat	1	1	2	1	1	1	4	5	0	SCRUB	7.0a	5	1-3	2	-1	0	Te
+ *Tussilago farfara* L.	nat	1	3	3	3	2	5	3	1	0	QRYl	7.0b	3	1-4	2-3	1	0	Tf
Typha angustifolia L.	nat	4	3	1	1	1	1	1	1	0	AQUATp	6.5a	1	1-4	3	-1	0	Ta
+ *Typha latifolia* L.	nat	4	4	1	1	2	2	1	1	0	AQUATr	6.5a	1	1-6	3-4	1	0	Tl
Ulex minor Roth	nat	1	1	1	1	1	1	5	1	0	WASTEa	4.0b	1	1	3-4	-1	0	Um
- *Umbilicus rupestris* (Salisb.) Dandy	nat	1	1	5	1	2	1	1	1	0	WALL	[5.5]	1	M-A	M-A	-1	0	Ur
+ *Urtica dioica* L.	nat	1	3	3	3	2	3	3	3	0	SOIL	6.5b	2	1-4	2-3	1	0	Ud
+ *Urtica urens* L.	arch	1	1	1	5	2	3	1	1	0	ARABLE	7.0a	3	1-4	3-4	-1	0	Uu
+ *Vaccinium myrtillus* L.	nat	1	2	1	2	4	2	3	3	0	PASTa	3.5a	1	1-4	3	0	0	Vm
Vaccinium oxycoccos L.	nat	1	5	1	1	1	1	1	1	0	MIREu	4.5b	1	1-6	2	-1	0	Vo
+ *Vaccinium vitis-idaea* L.	nat	1	3	1	1	4	3	3	3	0	QRYa	3.5a	1	1-6	1-2	0	0	Vv
Valeriana dioica L.	nat	1	5	1	1	2	1	2	2	0	MIREu	6.5a	5	1-3	3	-1	0	Vd
+ *Valeriana officinalis* L.	nat	1	3	3	1	2	3	5	2	0	WASTEl	6.5b	3	1-5	2-3	-1	0	Vo
Valerianella locusta (L.) Laterr.	nat	1	1	4	1	3	3	3	1	0	OUTCRPu	7.0a	4	1-3	3	-1	0	Vl
Verbascum nigrum L.	nat	1	1	1	1	1	5	3	1	0	WASTEl	7.0a	3	1-4	3	-1	0	Vn
- *Verbascum thapsus* L.	nat	1	1	3	1	2	5	3	1	0	CINDER	7.0b	3	1-4	3	0	0	Vt
Verbena officinalis L.	arch	1	1	1	1	1	4	1	1	0	WASTEa	[7.0]	3	1-5	3-4	-1	0	Vo
- *Veronica agrestis* L.	arch	1	1	1	5	2	3	3	1	0	ARABLE	7.0b	3	1-3	3	-1	0	Va
- *Veronica anagallis-aquatica* L.	nat	3	5	1	1	2	1	1	1	0	MIREu	7.0a	1	1-5	3-4	-1	0	Va
+ *Veronica arvensis* L.	nat	1	1	4	5	4	3	1	1	0	PASTl	7.0a	4	1-3	3-4	0	0	Va
+ *Veronica beccabunga* L.	nat	3	5	1	1	2	1	1	1	0	AQUATr	6.5a	1	1-4	3	0	0	Vb
Veronica catenata Pennell	nat	4	5	1	1	1	1	1	1	0	MIREu	6.5a	2	1-6	3	-1	0	Vc
+ *Veronica chamaedrys* L.	nat	1	1	3	1	4	3	3	2	0	PASTl	6.5b	4	1-4	2-3	-1	0	Vc
- *Veronica filiformis* Smith	neo	1	1	1	1	4	1	3	1	0	RDVRGEu	7.0a	3	-	-	1		Vf
- *Veronica hederifolia* L.	arch	1	1	3	5	3	1	2	4	0	WOODl	7.0a	2	1-3	3-4	1	0	Vh
+ *Veronica montana* L.	nat	1	3	1	1	2	1	2	4	0	MIREs	6.0b	1	1-3	3	-1	0	Vm
- *Veronica officinalis* L.	nat	1	1	3	1	4	3	3	3	0	QRYa	5.5b	5	1-3	2-3	-1	0	Vo
+ *Veronica persica* Poiret	neo	1	1	1	5	2	3	1	1	0	ARABLE	7.0a	3	-	-	-1		Vp
Veronica polita Fries	neo	1	1	1	5	2	3	1	1	0	ARABLE	7.0a	2	1-4	3-4	-1		Vp
- *Veronica scutellata* L.	nat	1	5	1	1	2	1	1	1	0	MIREu	6.0b	1	1-4	2-3	-1	0	Vs
- *Veronica serpyllifolia* L.	nat	1	3	1	3	4	1	4	3	0	RDVRGEu	6.5b	4	1-6	2-3	-1	0	Vs
+ *Vicia cracca* L.	nat	1	3	2	3	3	3	5	1	0	WASTEl	6.5b	2	1-5	2-3	-1	0	Vc
- *Vicia hirsuta* (L.) Gray	nat	1	1	3	2	3	3	5	1	0	WASTEd	6.5a	3	1-3	3	0	0	Vh
- *Vicia sativa* L.	nat	1	1	2	3	3	3	5	1	0	WASTEd	6.5b	3	1-3	3-4	0	0	Vs
+ *Vicia sepium* L.	nat	1	2	1	1	3	3	4	3	0	SOIL	6.5b	3	1-4	2-3	-1	0	Vs
Vicia tenuifolia Roth	neo	1	1	1	1	3	5	1	1	0	WASTEa	[7.5]	3	-	-	1		Vt
- *Vicia tetrasperma* (L.) Schreber	nat	1	1	1	1	1	3	5	1	0	WASTEa	6.0b	4	1-3	3	0	0	Vt
Vinca major L.	neo	1	1	1	1	1	1	5	4	0	HEDGE	7.0a	1	-	-	0		Vm
Vinca minor L.	arch	1	1	3	1	1	1	1	5	0	WOODl	6.5a	1	-	-	1	0	Vm

TABLES OF ATTRIBUTES

Table 7.1

			Abundance in the primary habitats														
Species and authority		Status	Aquatic	Mire	Skeletal	Arable	Pasture	Spoil	Wasteland	Woodland	Maritime	Commonest terminal habitat	Soil pH	Floristic diversity	Longitudinal range	Latitudinal range	Present status	Red Data book	(Sp.)
-	*Viola arvensis* Murray	arch	1	1	1	5	2	3	3	1	0	ARABLE	6.5b	3	1-4	3	0	0	Va
	Viola canina L.	nat	1	1	1	1	1	1	5	1	0	WASTEa	5.5b	3	1-4	2-3	-1	0	Vc
+	*Viola hirta* L.	nat	1	1	2	1	4	1	4	3	0	PASTl	7.0a	5	1-4	3	-1	0	Vh
-	*Viola odorata* L.	nat	1	1	1	1	3	1	2	5	0	SCRUB	7.0a	2	1-3	3	-1	0	Vo
-	*Viola palustris* L.	nat	1	5	1	1	2	1	1	1	0	MIREu	5.0b	2	1-3	2-3	-1	0	Vp
	Viola reichenbachiana Jordan ex Boreau	nat	1	1	3	1	1	1	1	5	0	WOODl	7.0a	2	1-3	3	-1	0	Vr
+	*Viola riviniana* Reichb.	nat	1	1	3	1	4	2	3	3	0	PASTl	6.5b	3	1-3	3	-1	0	Vr
-	*Viola tricolor* L.	nat	1	1	1	5	2	4	3	1	0	ARABLE	6.0b	3	1-3	3	-1	1	Vt
	Viscum album L.	nat	1	1	1	1	5	1	3	4	0	WOODa	6.5b	1	1-3	3	-1	0	Va
-	*Vulpia bromoides* (L.) Gray	nat	1	1	3	1	2	4	4	1	0	WASTEd	6.5b	3	S-s	S-s	0	0	Vb
	Vulpia myuros (L.) C. Gmelin	neo	1	1	1	1	1	5	3	1	0	CINDER	7.0b	2	1-4	3-4	0	0	Vm
	Wahlenbergia hederacea (L.) Reichb.	nat	1	5	1	1	1	1	1	1	0	MIREu	5.5b	4	1	3-4	-1	1	Wh
	x *Festulolium loliaceum* (Hudson) P. Fourn.	nat	1	3	1	1	5	1	3	1	0	PASTa	6.5b	4	-	-	-1		Fl
	Zannichellia palustris L.	nat	5	1	1	1	1	1	1	1	0	AQUATp	7.0a	1	1-6	3-4	-1	0	Zp

(b) Woody species more than 1.5 m tall

-	*Acer campestre* L.	nat	1	1	1	1	2	1	1	5	0	SCRUB	5.0c	1	1-3	3	-1	0	Ac
	Acer platanoides L.	neo	1	1	1	1	1	3	2	5	0	WOODl	6.0c	2	-	-	1		Ap
+	*Acer pseudoplatanus* L.	neo	1	3	2	1	2	1	1	5	0	WOODl	6.0c	1	-	-	1		Ap
(+)	ditto (juvenile)		1	2	1	1	2	2	3	5	0	PLANTb	6.0c	1	-	-			Ap
	Aesculus hippocastanum L.	neo	1	1	1	1	1	1	1	5	0	WOODa	[6.5]		-	-	0		Ah
+	*Alnus glutinosa* (L.) Gaertner	nat	1	4	1	1	2	1	1	4	0	MIREs	6.0b	1	1-4	3	-1	0	Ag
(+)	ditto (juvenile)		1	4	4	1	2	1	3	2	0	RIVBNKu	6.0b	3	-	-			Ag
+	*Betula* sp. L.	nat	1	3	2	1	2	1	1	5	0	SCRUB	5.0c	1	1-4	2-3	1	0	Bs
(+)	ditto (juvenile)		1	2	3	2	2	3	2	3	0	SCRUB	5.0c	2	-	-			Bs
	Betula pendula Roth	nat	1	2	1	1	1	3	3	5	0	SCRUB	[5.0]		1-4	2-3	1	0	Bp
	Betula pubescens Ehrh.	nat	1	3	1	1	1	2	2	5	0	SCRUB	[5.0]		1-4	2-3	1	0	Bp
	Buddleja davidii Franchet	neo	1	1	1	1	1	5	4	1	0	CINDER	[7.0]	3	-	-	1		Bd
	Carpinus betulus L.	nat	1	3	1	1	1	1	1	5	0	WOODa	[5.0]	1	1-3	3	0	0	Cb
	Castanea sativa Miller	arch	1	1	1	1	1	1	1	5	0	WOODa	[5.0]	1	-	-	1	0	Cs
	Clematis vitalba L.	nat	1	1	3	1	1	3	3	3	0	HEDGE	7.0b	1	1-3	3	1	0	Cv
	Cornus sanguinea L.	nat	1	1	3	1	1	1	2	5	0	SCRUB	7.0a	1	1-3	3	-1	0	Cs
-	*Corylus avellana* L.	nat	1	1	1	1	2	1	1	5	0	SCRUB	7.0a	1	1-3	3	-1	0	Ca
+	*Crataegus monogyna* Jacq.	nat	1	1	2	1	2	1	1	5	0	HEDGE	7.0b	1	1-3	3	1	0	Cm
(+)	ditto (juvenile)		1	1	3	1	3	3	3	3	0	PASTl	7.0b	3	-	-			Cm
-	*Cytisus scoparius* (L.) Link	nat	1	1	1	1	2	4	4	1	0	RD/RLY	6.0b	1	1-3	3	0	0	Cs
-	*Euonymus europaeus* L.	nat	1	1	1	1	2	1	3	5	0	SCRUB	[7.0]	1	1-3	3	-1	0	Ee
+	*Fagus sylvatica* L.	nat	1	1	1	1	2	1	1	5	0	PLANTb	4.0b	1	1-3	3	1	0	Fs
(+)	ditto (juvenile)		1	1	1	1	2	1	1	5	0	PLANTb	4.0b	1	-	-			Fs
+	*Fraxinus excelsior* L.	nat	1	2	2	1	2	1	1	5	0	WOODl	7.0b	1	1-3	3	1	0	Fe
(+)	ditto (juvenile)		1	3	3	1	3	3	3	4	0	WOODl	7.0b	2	-	-			Fe
-	*Ilex aquifolium* L.	nat	1	1	2	1	2	1	1	5	0	WOODa	4.5c	1	1-2	3-4	-1	0	Ia
-	*Ligustrum vulgare* L.	nat	1	2	3	1	2	1	1	5	0	SCRUB	7.0a	1	1-3	3	-1	0	Lv
	Lupinus arboreus Sims	neo	1	5	3	1	1	1	4	1	0	WASTEa	5.5b	3	-	-	1		La
	Mahonia aquifolium (Pursh) Nutt.	neo	1	1	4	3	1	1	1	5	0	WOODa	6.0b	1	-	-	1		Ma
-	*Malus sylvestris* (L.) Miller	nat	1	1	1	1	2	1	1	5	0	HEDGE	[7.0]	2	1-3	3	-1	0	Ms
-	*Populus tremula* L.	nat	1	1	1	1	2	1	1	5	0	SCRUB	[6.5]	2	1-5	2-3	-1	0	Pt
	Populus × *canescens* (Aiton) Smith	neo	1	3	1	3	3	1	3	3	0	WOODa	6.0b	3	-	-	1		Pc
-	*Prunus avium* (L.) L.	nat	1	1	1	1	2	1	1	5	0	WOODa	6.0c	1	1-3	3	-1	0	Pa
	Prunus padus L.	nat	1	2	1	1	1	1	1	5	0	HEDGE	6.5b	1	1-5	2-3	-1	0	Pp
-	*Prunus spinosa* L.	nat	1	1	1	1	2	1	1	5	0	SCRUB	6.5b	2	1-3	3	0	0	Ps
+	*Quercus* agg. L.	nat	1	1	2	1	2	3	3	5	0	WOODa	4.0c	2	1-3	3		0	Qa
(+)	ditto (juvenile)		1	1	3	1	2	2	2	4	0	PLANTb	4.0c	1	-	-			Qa
	Quercus cerris L.	neo	1	1	1	1	1	1	5	1	0	WOODa	[5.0]	2	-	-	1		Qc
	Quercus petraea (Mattuschka) Liebl.	nat	1	1	1	1	2	1	3	5	0	WOODa	[4.0]		1-3	3	0	0	Qp
	Quercus robur L.	nat	1	1	1	1	2	1	1	5	0	WOODa	[5.0]		1-3	3	0	0	Qr
	Rhamnus cathartica L.	nat	1	1	5	1	2	3	3	3	0	SCRUB	7.0a	5	1-4	3	-1	0	Rc
-	*Rhododendron ponticum* L.	neo	1	1	3	3	3	5	3	5	0	SCRUB	[3.5]	1	-	-	1		Rp
	Ribes nigrum L.	neo	1	4	1	1	1	1	3	4	0	MIREs	6.5b	1	1-4	2-3	0	0	Rn
	Ribes rubrum L.	nat	1	3	2	1	1	1	1	5	0	MIREs	5.5a	1	1-2	3	0	0	Rr
-	*Rosa* sp. L.	nat	1	1	2	1	2	1	1	5	0	HEDGE	7.0b	3	-	-	0	0	Rs
(+)	ditto (juvenile)		1	1	3	3	3	3	3	3	0	OUTCRPu	7.0b	4	-	-			Rs
	Rosa arvensis Hudson	nat	1	1	2	1	1	1	3	5	0	SCRUB	6.5a	3	1-3	3	-1	0	Ra

			\multicolumn{8}{c	}{Abundance in the primary habitats}							
Species and authority		Status	Aquatic	Mire	Skeletal	Arable	Pasture	Spoil	Wasteland	Woodland	Maritime	Commonest terminal habitat	Soil pH	Floristic diversity	Longitudinal range	Latitudinal range	Present status	Red Data book	(Sp.)
	Rosa canina L.	nat	1	1	3	1	1	3	3	3	0	HEDGE	7.0a	4	1-3	3	0	0	*Rc*
	Rosa mollis Smith	nat	1	1	1	1	5	1	4	3	0	SCRUB	6.5a	5	1-3	2-3	-1	0	*Rm*
	Salix alba L.	arch	2	4	1	1	1	1	1	1	3	RIVBNKu	[6.5]		1-4	3-4	0	0	*Sa*
	Salix aurita L.	nat	2	5	1	1	1	1	1	2	0	MIREs	[5.0]		1-3	2-3	1	0	*Sa*
+	*Salix caprea/cinerea* agg.	nat	1	4	1	1	2	1	1	4	0	SCRUB	6.5b	2	1-5	2-3	1	0	*Sc*
(+)	ditto (juvenile)		1	4	3	1	2	4	1	1	0	CINDER	6.5b	3	-	-			*Sc*
	Salix caprea L.	nat	1	2	1	1	1	3	3	3	0	SCRUB	[6.5]		1-5	2-3	1	0	*Sc*
	Salix cinerea L.	nat	1	4	2	1	1	3	3	2	0	SCRUB	[6.5]	3	1-4	2-3	1	0	*Sc*
-	*Salix fragilis* L.	arch	3	4	1	1	2	1	1	1	0	MIREs	7.0b	1	1-4	3	0	0	*Sf*
-	*Salix purpurea* L.	nat	1	5	1	1	2	1	1	4	0	MIREs	[6.5]		1-4	3	1	0	*Sp*
-	*Salix viminalis* L.	arch	1	5	1	1	2	1	1	3	0	MIREs	6.5b	3	1-5	3	-1	0	*Sv*
+	*Sambucus nigra* L.	nat	1	2	2	1	2	1	1	5	0	PLANTb	6.5b	1	1-3	3	1	0	*Sn*
(+)	ditto (juvenile)		1	1	3	3	2	3	3	3	0	BRICK	6.5b	2	-	-			*Sn*
+	*Sorbus aucuparia* L.	nat	1	1	2	1	2	1	2	4	0	WOODa	4.0c	1	1-5	2-3	0	0	*Sa*
(+)	ditto (juvenile)		1	1	2	1	2	1	2	5	0	WOODa	4.0c	1	-	-			*Sa*
	Symphoricarpos albus (L.) S.F. Blake	neo	1	1	1	1	1	4	4	1	0	SCRUB	7.0b	5	-	-	1		*Sa*
	Taxus baccata L.	nat	1	1	5	1	1	1	1	1	0	WOODl	7.0a	2	1-3	3	0	0	*Tb*
+	*Ulex europaeus* L.	nat	1	2	1	1	3	3	4	1	0	*WASTEl	6.0c	2	1	3	1	0	*Ue*
	Ulex gallii Planchon	nat	1	3	2	1	3	4	3	1	0	WASTEa	4.0a	1	1	3	-1	0	*Ug*
+	*Ulmus glabra* Hudson	nat	1	2	2	1	2	1	1	5	0	WOODl	7.0b	1	1-3	3	-1	0	*Ug*
(+)	ditto (juvenile)		1	3	3	3	2	3	2	3	0	*QRYl	7.0b	3	-	-			*Ug*
-	*Ulmus procera* Salisb.	nat	1	1	1	1	2	1	1	5	0	WOODa	6.5a	1	1-3	3	-1	0	*Up*
-	*Viburnum opulus* L.	nat	1	3	1	1	1	1	1	5	0	WOODl	7.0b	1	1-6	3	-1	0	*Vo*

(c) Pteridophytes

-	*Asplenium adiantum-nigrum* L.	nat	1	1	5	1	2	1	1	1	0	WALL	7.0a	1	1-3	3	-1	0	*Aa*
+	*Asplenium ruta-muraria* L.	nat	1	1	5	1	2	2	1	1	0	WALL	7.0a	1	1-6	3	-1	0	*Ar*
+	*Asplenium trichomanes* L.	nat	1	1	5	1	2	3	1	1	0	CLIFF	7.0a	2	1-6	3-4	-1	0	*At*
+	*Athyrium filix-femina* (L.) Roth	nat	1	3	5	1	2	1	1	3	0	*MIREs	5.0c	1	1-6	2-3	-1	0	*Af*
-	*Blechnum spicant* (L.) Roth	nat	1	1	5	1	4	1	3	3	0	OUTCRPu	3.5a	1	1-3	3	-1	0	*Bs*
	Ceterach officinarum Willd.	nat	1	1	5	1	1	1	1	1	0	WALL	7.5a	1	S-s	S-s	-1	0	*Co*
+	*Cystopteris fragilis* (L.) Bernh.	nat	1	1	5	1	2	1	1	1	0	WALL	7.0a	1	1-6	1-3	0	0	*Cf*
-	*Dryopteris affinis* (Lowe) Fraser-Jenkins	nat	1	1	5	1	2	3	2	4	0	OUTCRPs	4.0c	1	1-3	3	-1	0	*Da*
	Dryopteris carthusiana (Villars) H.P. Fuchs	nat	1	5	1	1	2	1	1	4	0	MIREs	4.0b	1	1-4	2-3	-1	0	*Dc*
+	*Dryopteris dilatata* (Hoffm.) A. Gray	nat	1	3	3	1	2	1	1	4	0	PLANTc	4.0b	1	1-3	3	1	0	*Dd*
+	*Dryopteris filix-mas* (L.) Schott	nat	1	1	4	1	2	1	1	3	0	*WOODl	6.0c	1	1-6	3	-1	0	*Df*
+	*Equisetum arvense* L.	nat	1	4	1	3	3	4	3	1	0	RIVBNKu	6.5b	2	1-6	1-3	1	0	*Ea*
+	*Equisetum fluviatile* L.	nat	3	5	1	1	2	1	1	1	0	AQUATp	6.0b	1	1-6	2-3	-1	0	*Ef*
+	*Equisetum palustre* L.	nat	1	5	2	1	2	1	1	1	0	*MIREu	6.5a	2	1-6	2-3	0	0	*Ep*
	Equisetum sylvaticum L.	nat	1	5	1	1	1	3	2	3	0	MIREu	5.5b	3	1-6	2-3	-1	0	*Es*
	Equisetum telmateia Ehrh.	nat	1	4	1	1	1	2	3	4	0	MIREs	6.5b	2	1-3	3-4	-1	0	*Et*
	Gymnocarpium dryopteris (L.) Newman	nat	1	1	5	1	1	1	1	1	0	OUTCRPs	4.5a	1	1-6	2-3	-1	0	*Gd*
	Huperzia selago (L.) Bernh. ex Schrank & C. Martius	nat	1	3	2	1	4	2	1	1	0	QRYa	5.0b	3	1-6	1-2	-1	0	*Hs*
	Lycopodium clavatum L.	nat	1	1	1	1	3	3	4	1	0	WASTEa	4.5a	1	1-6	2-3	-1	0	*Lc*
	Oreopteris limbosperma (Bellardi ex All.) Holub	nat	1	3	3	1	3	2	3	3	0	RIVBNKu	4.0b	2	1-3	3	-1	0	*Ol*
	Ophioglossum vulgatum L.	nat	1	3	1	1	4	1	4	3	0	WASTEl	6.0b	4	1-6	3	-1	0	*Ov*
	Osmunda regalis L.	nat	2	4	1	1	1	1	3	4	0	MIREu	[4.5]	1	1-2	3-4	-1	0	*Or*
	Phegopteris connectilis (Michaux) Watt	nat	1	5	1	1	1	1	1	1	0	OUTCRPs	4.5a	1	1-6	2-3	-1	0	*Pc*
-	*Phyllitis scolopendrium* (L.) Newman	nat	1	1	5	1	2	1	1	1	0	CLIFF	7.0a	1	1-3	3	0	0	*Ps*
-	*Polypodium vulgare* L.	nat	1	1	5	1	2	1	1	1	0	CLIFF	6.5b	1	-	-	-1	0	*Pv*
	Polystichum aculeatum (L.) Roth	nat	1	1	5	1	1	1	1	3	0	OUTCRPs	7.0a	1	1-5	3	-1	0	*Pa*
	Polystichum setiferum (Forsskal) T. Moore ex Woynar	nat	1	1	5	1	1	1	1	5	0	RIVBNKs	6.0a	1	S-s	S-s	-1	0	*Ps*
+	*Pteridium aquilinum* (L.) Kuhn	nat	1	2	1	1	3	2	3	4	0	PLANTc	4.0c	1	1-6	3	1	0	*Pa*

TABLES OF ATTRIBUTES

Table 7.2 Attributes of the established phase of species (symbols and notes are explained in the preceding text).

Species	Life history	Established strategy	SLA Class	DMC Class	Life form	Canopy structure	Canopy height	Lateral spread	Mycorrhizas	Leaf phenology	Flowering time and duration	(Sp.)
(a) Herbs and woody species up to 1.5 m tall												
+ Achillea millefolium	P	CSR	4	2	Ch	S	2	5	VA	Ea	Jun3	*Am*
- Achillea ptarmica	P	SC/CSR	3	3	H/Hel	L	3	3	-	Sa	Jul2	*Ap*
Acorus calamus	P	C/CR	2	3	Wet	LA	4	5	-	Sa	May3	*Ac*
Adoxa moschatellina	P	SR	5	1	G	Sb	1	4	-	Sv	Apr2	*Am*
- Aegopodium podagraria	P	CR	3	3	H	LA	2	5	VA	Sa	May3	*Ap*
- Aethusa cynapium	As	R/CR	5	2	Th	L	5	1	VA	Sa	Jul2	*Ac*
- Agrimonia eupatoria	P	CSR	2	5	H	LA	3	2	VA	Sa	Jun3	*Ae*
Agrimonia procera	P	CSR	2	5	H	LA	2	2	?	Sa	Jun3	*Ap*
+ Agrostis canina	P	SR/CSR	5	4	Hel	Lb	1	5	VA	Ea	Jun2	*Ac*
+ Agrostis capillaris	P	CSR	5	4	H	L	2	4	VA	Ea	Jun3	*Ac*
- Agrostis gigantea	P	CR	4	4	H	L	2	5	VA	Ea	Jun3	*Ag*
+ Agrostis stolonifera	P	CR	5	4	H	L	2	5	+	Ea	Jul2	*As*
+ Agrostis vinealis	P	CSR	4	5	H	L	2	2	VA	Ea	Jun3	*Av*
- Aira caryophyllea	Aw	SR	4	3	Th	Sb	1	1	+	Sh	May1	*Ac*
+ Aira praecox	Aw	SR	3	4	Th	Sb	1	1	+	Sh	Apr2	*Ap*
- Ajuga reptans	P	R/CSR	5	2	H	Sb	1	4	VA	Ea	May3	*Ar*
- Alchemilla vulgaris	P	S/CSR	2	5	H	S	3	2	VA	Ea	Jun4	*Av*
+ Alisma plantago-aquatica	P	R/CR	5	1	Wet	LA	3	2	-	Sa	Jun3	*Ap*
+ Alliaria petiolata	A/M	CR	5	1	H	S	5	1	-	Sh	Apr3	*Ap*
+ Allium ursinum	P	SR	5	1	G	B	2	2	VA	Sv	Apr3	*Au*
- Allium vineale	P	SR	2	2	G	L	4	2	VA	Sh	Jun2	*Av*
+ Alopecurus geniculatus	P	CR	5	2	Hel	Lb	1	5	VA	Ea	Jun2	*Ag*
Alopecurus myosuroides	Aws	R	4	4	Th	S	3	1	VA	Ep	Jun2	*Am*
+ Alopecurus pratensis	P	C/CSR	4	4	H	L	2	3	VA	Ea	Apr3	*Ap*
Amsinckia micrantha	Asw	R/CR	4	1	Th	S	3	1	?	Ep	Apr5	*Am*
- Anacamptis pyramidalis	P	SR/CSR	5	1	G	S	2	2	OR	Sa	Jun3	*Ap*
+ Anagallis arvensis	Asw	R/SR	5	1	Th/Ch	L	2	1	VA	Ep	Jun3	*Aa*
Anagallis tenella	P	SR	5	1	Hel/H/Ch	B	1	4	VA	Ep	Jun3	*At*
- Anchusa arvensis	Aws	R/CR	3	1	Th/H	S	3	1	?	Ep	Jun4	*Aa*
- Anemone nemorosa	P	SR	4	3	G	S	3	3	VA	Sv	Mar3	*An*
+ Angelica sylvestris	M/P	C/CR	3	3	H/Hel	LA	3	2	VA	Sa	Jul3	*As*
+ Anisantha sterilis	Aws	R/CR	5	3	Th	S	3	1	-	Sh	May3	*As*
Antennaria dioica	P	SR/CSR	2	5	Ch	Sb	1	3	VA	Ea	Jun2	*Ad*
Anthemis arvensis	Aws	R/SR	2	2	Th	L	2	1	VA	Sh	Jun2	*Aa*
Anthemis cotula	As	R	3	1	Th	L	3	1	?	Ep	Jul3	*Ac*
+ Anthoxanthum odoratum	P	SR/CSR	4	3	H	L	2	2	VA	Ea	Apr3	*Ao*
Anthriscus caucalis	Aw	R/CR	5	2	Th	S	3	1	?	Sh	May2	*Ac*
+ Anthriscus sylvestris	P	C/CR	3	3	H	LA	3	2	-	Sh	Apr3	*As*
- Anthyllis vulneraria	P/M	SR/CSR	2	2	H	S	2	2	VA	Ea	Jun4	*Av*
Antirrhinum majus	P	CSR	3	2	H	L	3	2	?	Ep	Jul3	*Am*
Apera interrupta	Aw	SR	4	5	Th	S	2	1	?	Sh	Jun2	*Ai*
+ Aphanes arvensis	Aws	R/SR	4	4	Th	Lb	1	1	VA	Sh	Apr7	*Aa*
Aphanes australis	Aw	SR	4	3	Th	Lb	1	1	?	Sh	Apr7	*Aa*
Apium inundatum	P	SR	-	-	Hyd	B	1	4	?	Ea	Jun3	*Ai*
+ Apium nodiflorum	P	C/CR	5	1	Wet	L	3	5	-	Ea	Jul2	*An*
Aquilegia vulgaris	P	CSR	3	4	H	S	3	2	VA	Sa	May2	*Av*
+ Arabidopsis thaliana	Aws	SR	5	1	Th	Sb	1	1	+	Sh	Apr2	*At*
Arabis caucasica	P	CSR	2	1	Ch	Lb	1	3	?	Ea	Mar3	*Ac*
+ Arabis hirsuta	P/M	S/CSR	3	2	H	Sb	1	2	-	Ea	Jun3	*Ah*
- Arctium lappa	M	C/CR	3	3	H	LA	5	1	+	Sa	Jul3	*Al*
+ Arctium minus	M	C/CR	3	2	H	LA	5	1	VA	Sa	Jul3	*Am*
Arctostaphylos uva-ursi	P	S/SC	1	5	Chw	L	2	5	ER	Ea	May3	*Au*
+ Arenaria serpyllifolia	Aws	SR	3	3	Th	L	2	1	-	Sh	May4	*As*
- Armoracia rusticana	P	C/CSR	3	2	H	LA	3	3	+	Sa	May2	*Ar*
+ Arrhenatherum elatius	P	C/CSR	3	5	H	L	5	4	+	Ep	Jun2	*Ae*
+ Artemisia absinthium	P	C/CSR	3	3	Ch	L	3	2	-	Ea	Jul2	*Aa*
Artemisia verlotiorum	P	C	2	5	H	L	5	5	?	Ep	Oct2	*Av*
+ Artemisia vulgaris	P	C/CR	3	4	H	L	4	3	VA	Ep	Jul3	*Av*
+ Arum maculatum	P	SR	5	1	G	LA	3	2	VA	Sv	Apr2	*Am*
Aster lanceolatus	P	C/CSR	2	4	H	L	5	5	?	Sa	Aug3	*Al*
Aster novi-belgii	P	C/CSR	4	2	H	L	5	4	VA	Sa	Aug3	*An*
Atriplex littoralis	As	CR	2	1	Th	L	5	1	VA	Sa	Jul2	*Al*
+ Atriplex patula	As	R/CR	2	2	Th	S	4	1	VA	Sa	Jun5	*Ap*

Species	Life history	Established strategy	SLA Class	DMC Class	Life form	Canopy structure	Canopy height	Lateral spread	Mycorrhizas	Leaf phenology	Flowering time and duration	(Sp.)
+ *Atriplex prostrata*	As	R/CR	2	1	Th	S	4	1	-	Sa	Jul3	*Ap*
Avena fatua	As	R/CR	2	3	Th	S	4	1	VA	Sv	Jul3	*Af*
Ballota nigra	P	CR/CSR	2	3	H	L	4	2	-	Ep	Jun5	*Bn*
- *Barbarea vulgaris*	P/M	R/CSR	5	1	H	S	3	2	-	Ea	May4	*Bv*
Bassia scoparia	As	C/CR	2	3	Th	L	5	1	?	Sa	Aug3	*Bs*
+ *Bellis perennis*	P	R/CSR	5	1	H	B	1	3	VA	Ea	Mar12	*Bp*
- *Berula erecta*	P	CR	5	1	Wet	L	4	5	-	Ea	Jul3	*Be*
Bidens cernua	As	CR	5	1	Th	L	3	1	?	Sa	Jul3	*Bc*
Bidens tripartita	As	CR	5	1	Th	L	3	1	VA	Sa	Jul3	*Bt*
Blackstonia perfoliata	A?	SR	2	3	Th	S	2	1	VA	Ea	Jun5	*Bp*
+ *Brachypodium pinnatum*	P	SC	3	5	H	S	2	5	VA	Ea	Jul1	*Bp*
+ *Brachypodium sylvaticum*	P	SC/CSR	5	5	H	S	3	3	VA	Ea	Jul1	*Bs*
- *Brassica rapa*	B/A	CR	2	1	Th/Ch	S	3	1	-	Ea	May4	*Br*
+ *Briza media*	P	S/CSR	2	5	H	Sb	1	3	VA	Ea	Jun2	*Bm*
+ *Bromopsis erecta*	P	SC/CSR	1	5	H	S	2	3	VA	Ea	Jun2	*Be*
Bromopsis inermis	P	C	3	4	H	L	5	5	VA	Sa	Jun1	*Bi*
+ *Bromopsis ramosa*	P	CSR	3	4	H	L	3	2	?	Ea	Jul2	*Br*
+ *Bromus hordeaceus*	Aw	R/CR	3	4	Th	L	3	1	+	Sh	May3	*Bh*
Bromus lepidus	Aw	R/CR	-	-	Th	L	3	1	?	Sh	May3	*Bl*
Bryonia dioica	P	C/CSR	5	1	G	L	6	2	VA	Sa	May5	*Bd*
Butomus umbellatus	P	CR	1	2	Hyd	LA	4	4	-	Sa	Jul3	*Bu*
Calamagrostis canescens	P	C/SC	3	5	H/Hel	L	5	5	VA	Sa	Jun2	*Cc*
Calamagrostis epigejos	P	C/SC	1	5	H	S	4	5	VA	Sa	Jul2	*Ce*
Callitriche hamulata	P	R/SR	5	2	Wet	Lb,F	1	3	VA	Ea	Apr6	*Ch*
Callitriche obtusangula	P	R/CR	-	-	Wet	Lb,F	1	3	?	Ea	May5	*Co*
Callitriche platycarpa	P	CR	5	1	Wet	Lb,F	1	4	?	Ea	Apr7	*Cp*
+ *Callitriche stagnalis*	P	R/CR	5	1	Wet	Lb,F	1	4	?	Ea	May5	*Cs*
+ *Calluna vulgaris*	P	S/SC	1	-	Chw/Ph	L	4	4	ER	Ea	Aug2	*Cv*
+ *Caltha palustris*	P	CSR	2	2	Hel	LA	2	2	VA	Sa	Apr3	*Cp*
Calystegia pulchra	P	C/CR	3	3	G	L	5	5	?	Sa	Jul3	*Cp*
+ *Calystegia sepium*	P	C/CR	5	2	G	L	5	5	VA	Sa	Jun4	*Cs*
Calystegia silvatica	P	C/CR	3	3	G	L	5	5	?	Sa	Jul3	*Cs*
Campanula latifolia	P	C/CSR	5	2	H	L	4	2	?	Sa	Jul2	*Cl*
Campanula rapunculoides	P	CR/CSR	5	2	H	S	2	5	VA	Sa	Jul3	*Cr*
+ *Campanula rotundifolia*	P	S/CSR	3	4	H	S	1	3	VA	Ea	Jul3	*Cr*
Campanula trachelium	P	CSR	5	2	H	L	4	2	-	Sa	Jul3	*Ct*
+ *Capsella bursa-pastoris*	Asw	R	3	1	Th	Sb	1	1	-	Ep	Mar12	*Cb*
+ *Cardamine amara*	P	CR	5	1	Hel	L	3	5	-	Ea	May2	*Ca*
+ *Cardamine flexuosa*	A/P	R/SR	5	1	H/Hel	S	2	1	-	Ea	Apr6	*Cf*
+ *Cardamine hirsuta*	Aws	SR	5	1	Th	S	1	1	-	Sh	Apr5	*Ch*
+ *Cardamine pratensis*	P	R/CSR	2	2	H/Hel	S	2	2	-	Ea	Apr3	*Cp*
- *Carduus crispus*	M	CR	3	1	H	S	5	1	VA	Ea	Jun3	*Cc*
Carduus nutans	M	R/CR	2	1	H	S	3	1	?	Ea	May4	*Cn*
Carex acuta	P	C/SC	-	-	Hel	LA	5	5	-	Ep	May2	*Ca*
+ *Carex acutiformis*	P	C/SC	2	5	Hel	LA	5	5	-	Ep	Jun2	*Ca*
Carex arenaria	P	SC/CSR	1	5	H	L	3	-	5	Ea	Jun2	*Ca*
- *Carex binervis*	P	S	1	5	Hel/H	S	2	3	?	Ea	Jun1	*Cb*
+ *Carex caryophyllea*	P	S	3	5	H	Sb	1	3	-	Ea	Apr2	*Cc*
Carex curta	P	S/CSR	4	5	Hel	S	3	2	-	Sa	May2	*Cc*
Carex dioica	P	S	1	5	Hel	Sb	1	4	-	Ea	May1	*Cd*
Carex distans	P	S/CSR	-	-	Hel	S	2	3	-	Ea	May2	*Cd*
Carex disticha	P	C/CSR	2	5	Hel	S	4	5	-	Ep	Jun2	*Cd*
- *Carex echinata*	P	S/CSR	1	5	Hel/H	Sb	1	2	-	Ea	May2	*Ce*
Carex elata	P	SC	2	5	Wet	S	4	4	?	Sa	May2	*Ce*
+ *Carex flacca*	P	S	2	5	H	S	2	4	+	Ea	May2	*Cf*
- *Carex hirta*	P	C/CSR	2	3	H/Hel	S	3	5	-	Sa	May2	*Ch*
- *Carex hostiana*	P	S/CSR	2	5	Hel/H	S	2	2	-	Ea	Jun1	*Ch*
Carex laevigata	P	SC/CSR	2	4	Hel	S	3	3	?	Ea	Jun1	*Cl*
Carex muricata	P	S/CSR	2	5	H	S	2	3	+	Ea	May3	*Cm*
+ *Carex nigra*	P	S/SC	1	5	Hel/H	S	3	4	+	Ep	May3	*Cn*
- *Carex otrubae*	P	C/CSR	2	5	Hel	S	4	3	-	Ea	Jun2	*Co*
- *Carex ovalis*	P	CSR	3	5	H/Hel	S	3	2	?	Ea	Jun1	*Co*
- *Carex pallescens*	P	S/CSR	3	4	H	S	2	2	-	Ea	May2	*Cp*
+ *Carex panicea*	P	S/CSR	2	4	Hel/H	S	2	4	+	Ea	May2	*Cp*
Carex paniculata	P	C/SC	2	4	Hel	S	4	4	-	Ea	May2	*Cp*

Table 7.2

Species	Life history	Established strategy	SLA Class	DMC Class	Life form	Canopy structure	Canopy height	Lateral spread	Mycorrhizas	Leaf phenology	Flowering time and duration	(Sp.)
− *Carex pendula*	P	C/SC	1	4	Hel/H	LA	4	4	−	Ea	May2	*Cp*
+ *Carex pilulifera*	P	S	3	5	H	Sb	1	2	−	Ea	May2	*Cp*
Carex pseudocyperus	P	C/CSR	2	4	Hel	LA	4	3	?	Ea	May2	*Cp*
Carex pulicaris	P	S	1	5	Hel/H	Sb	1	3	−	Ea	May2	*Cp*
− *Carex remota*	P	CSR	4	3	Hel	S	3	3	+	Ea	Jun1	*Cr*
Carex riparia	P	C/SC	1	5	Wet	LA	5	5	−	Ep	May2	*Cr*
Carex rostrata	P	SC	1	5	Wet	S	4	5	−	Ea	Jun1	*Cr*
Carex spicata	P	CSR	2	4	H	S	3	3	?	Ea	Jun2	*Cs*
− *Carex sylvatica*	P	S/CSR	3	5	H	Sb	1	2	−	Ea	May3	*Cs*
Carex vesicaria	P	SC/CSR	1	−	Hel	S	4	5	?	Ea	Jun2	*Cv*
Carex viridula subsp. *brachyrrhyncha*	P	S/CSR	2	5	Hel	Sb	1	2	−	Ea	May2	*Cv*
− *Carex viridula* subsp. *oedocarpa*	P	S/CSR	2	5	Hel	Sb	1	2	−	Ea	Jun1	*Cv*
+ *Carlina vulgaris*	M	SR	1	3	H	S	3	1	VA	Ep	Jul4	*Cv*
Catabrosa aquatica	P	CR	5	1	Hel	L	2	5	?	Ea	Jun3	*Ca*
+ *Catapodium rigidum*	Aws	SR	4	5	Th	Sb	1	1	VA	Sh	May2	*Cr*
Centaurea montana	P	C/CSR	5	1	H	S	3	4	?	Ep	May3	*Cm*
+ *Centaurea nigra*	P	CSR	4	2	H	S	2	2	VA	Sa	Jun4	*Cn*
+ *Centaurea scabiosa*	P	SC/CSR	3	1	H	S	2	2	VA	Sa	Jul3	*Cs*
+ *Centaurium erythraea*	Aw	SR	3	3	Th	Sb	1	1	VA	Ea	Jun5	*Ce*
− *Centranthus ruber*	P	C/CSR	3	1	H	L	3	2	−	Ea	Jun3	*Cr*
Cerastium arvense	P	SR/CSR	2	2	Ch	La	1	4	−	Ea	Apr5	*Ca*
Cerastium diffusum	Aw	SR	4	1	Th	Sb	1	1	?	Sh	Apr4	*Cd*
+ *Cerastium fontanum*	P/A	R/CSR	4	1	Ch/Th	Lb	1	2	−	Ea	Apr6	*Cf*
− *Cerastium glomeratum*	Aws	R/SR	3	1	Th	L	2	1	?	Sh	Apr6	*Cg*
− *Cerastium semidecandrum*	Aw	SR	3	1	Th	Sb	1	1	?	Sh	Apr2	*Cs*
− *Cerastium tomentosum*	P	CSR	3	−	Ch	L	2	4	?	Ea	May4	*Ct*
− *Ceratocapnos claviculata*	As	R/SR	5	1	Th	L	4	1	?	Sa	Jun4	*Cc*
Ceratochloa carinata	P	CR	3	3	H	S	4	2	?	Ep	Jun3	*Cc*
Ceratophyllum demersum	P	CR	3	1	Hyd	L	−1	3	?	Ea	Jul3	*Cd*
+ *Chaenorhinum minus*	As	R/SR	3	1	Th	L	2	1		−Sa	May6	*Cm*
− *Chaerophyllum temulum*	M	CR	4	3	H	LA	4	1	+	Sh	Jun2	*Ct*
+ *Chamerion angustifolium*	P	C	2	3	G	L	4	5	VA	Sa	Jul3	*Ca*
− *Chelidonium majus*	P	CR/CSR	5	2	H	S	4	2	−	Ea	May4	*Cm*
+ *Chenopodium album*	As	CR	2	2	Th	L	5	1	−	Sa	Jul4	*Ca*
− *Chenopodium bonus-henricus*	P	C/CSR	4	2	H	S	3	2	−	Sa	May3	*Cb*
Chenopodium polyspermum	As	R/CR	4	2	Th	L	3	1	?	Sa	Jul4	*Cp*
+ *Chenopodium rubrum*	As	CR	3	1	Th	L	4	1	−	Sa	Jul3	*Cr*
− *Chrysanthemum segetum*	As	R	4	1	Th	L	2	1	VA	Sa	Jun3	*Cs*
Chrysosplenium alternifolium	P	R/CSR	4	1	Hel/H	S	3	4	−	Sa	Apr4	*Ca*
+ *Chrysosplenium oppositifolium*	P	CR/CSR	3	1	Ch/Hel	L	1	4	−	Ea	Apr4	*Co*
Cicerbita macrophylla	P	C/CSR	4	2	H	LA	3	5	?	Sa	Jul3	*Cm*
Cichorium intybus	P	CSR	5	1	H	S	3	2	VA	Sa	Jul4	*Ci*
+ *Circaea lutetiana*	P	CR	5	2	G	L	2	5	VA	Sa	Jun3	*Cl*
Cirsium acaule	P	SC/CSR	1	2	H	B	2	5	VA	Sa	Jul3	*Ca*
+ *Cirsium arvense*	P	C	2	1	G	L	4	5	VA	Sa	Jul3	*Ca*
Cirsium dissectum	P	SC/CSR	3	1	Hel	S	2	5	?	Sa	Jun3	*Cd*
Cirsium heterophyllum	P	SC/CSR	4	1	H	LA	2	5	VA	Sa	Jul2	*Ch*
+ *Cirsium palustre*	M	CR/CSR	2	1	H/Hel	S	5	1	+	Ea	Jul3	*Cp*
+ *Cirsium vulgare*	M	CR	1	1	H	LA	5	1	+	Ea	Jul4	*Cv*
Cladium mariscus	P	SC	1	5	Wet	S	5	5	+	Ea	Jul2	*Cm*
Claytonia perfoliata	Aw	R/SR	5	1	Th	S	2	1	?	Sv	May3	*Cp*
Claytonia sibirica	P	R/CSR	5	1	Th/H	Sb	1	2	VA	Ep	Apr4	*Cs*
Clinopodium acinos	P/Aw	SR/CSR	4	3	Ch/Th	La	1	1	?	Ep	May5	*Ca*
− *Clinopodium vulgare*	P	SC/CSR	3	4	H	L	4	2	+	Ep	Jul3	*Cv*
Cochlearia danica	Aw	SR	2	1	Th	Sb	1	1	?	Sv	Mar4	*Cd*
Coeloglossum viride	P	SR	3	1	G	Sb	1	2	OR	Sa	Jun3	*Cv*
− *Conium maculatum*	B	C/CR	4	3	H	LA	5	1	VA	Ea	Jun2	*Cm*
+ *Conopodium majus*	P	SR	3	2	G	Sb	1	2	VA	Sv	May2	*Cm*
+ *Convolvulus arvensis*	P	CR	5	1	G	L	4	5	VA	Sa	Jun4	*Ca*
Conyza canadensis	Aws	R/CR	5	2	Th	L	3	1	VA	Ep	Aug2	*Cc*
Coronopus didymus	As	R	5	1	Th	B	2	1	?	Sa	Jul3	*Cd*
− *Coronopus squamatus*	Aw/M	R	2	1	Th	B	1	1	?	Ep	Jun4	*Cs*
Cotoneaster horizontalis	P	SC	1	5	Ph	L	4	5	?	Sa	Jun2	*Ch*
Crassula helmsii	P	CR	5	1	Wet	Lb	1	5	?	Ea	Jul3	*Ch*
Crepis biennis	M	R/CSR	4	1	H	S	3	1	VA	Ea	Jun2	*Cb*

Species	Life history	Established strategy	SLA Class	DMC Class	Life form	Canopy structure	Canopy height	Lateral spread	Mycorrhizas	Leaf phenology	Flowering time and duration	(Sp.)
+ *Crepis capillaris*	Aw	R/SR	5	1	Th	S	1	1	VA	Sh	Jun2	*Cc*
Crepis paludosa	P	CSR	5	1	Hel	L	3	3	VA	Sa	Jul3	*Cp*
Crepis vesicaria	Aw	R/CR	4	2	Th	S	3	1	?	Sh	May3	*Cv*
+ *Cruciata laevipes*	P	CSR	5	3	H	L	3	3	VA	Ea	May2	*Cl*
− *Cymbalaria muralis*	P	R/CSR	5	1	Ch	B	1	4	VA	Ea	May5	*Cm*
Cynoglossum officinale	M	CR	2	2	H	S	3	1	?	Ea	Jun3	*Co*
+ *Cynosurus cristatus*	P	R/CSR	4	3	H	Sb	1	2	VA	Ea	Jun3	*Cc*
+ *Dactylis glomerata*	P	C/CSR	4	3	H	S	3	3	VA	Ea	May3	*Dg*
− *Dactylorhiza fuchsii*	P	SR	3	1	Hel/G	L	3	2	OR	Sa	Jun2	*Df*
− *Dactylorhiza incarnata*	P	SR	-	-	G/Hel	L	3	2	OR	Sa	May3	*Di*
− *Dactylorhiza maculata*	P	SR	3	1	G/Hel	L	3	2	OR	Sa	Jun2	*Dm*
Dactylorhiza praetermissa	P	SR	4	1	Hel/G	L	3	2	OR	Sa	Jun2	*Dp*
Dactylorhiza purpurella	P	SR	-	-	Hel/G	L	2	2	OR	Sa	Jun2	*Dp*
+ *Danthonia decumbens*	P	S/CSR	3	5	H	S	1	2	VA	Ea	Jul1	*Dd*
Daphne laureola	P	SC	-	-	Ph	L	4	2	VA	Ea	Feb3	*Dl*
+ *Daucus carota*	M	SR/CSR	4	4	H	S	2	1	VA	Ea	Jun2	*Dc*
+ *Deschampsia cespitosa*	P	SC/CSR	2	5	H	S	3	4	VA	Ea	Jun3	*Dc*
+ *Deschampsia flexuosa*	P	S/SC	2	5	H	S	2	4	VA	Ea	Jun2	*Df*
+ *Digitalis purpurea*	P/M	SR/CSR	1	3	H	LA	4	2	+	Ea	Jun4	*Dp*
Diplotaxis muralis	A/P	R	4	1	Th	Sb	1	1	?	Ea	Jun4	*Dm*
Diplotaxis tenuifolia	P	R/CSR	4	1	H/Ch	S	2	2	?	Ep	May5	*Dt*
Dipsacus fullonum	M	CR	3	2	H	LA	4	1	+	Ep	Jul2	*Df*
Doronicum pardalianches	P	C/CR	5	1	H	S	2	4	?	Sa	May3	*Dp*
Drosera rotundifolia	P	SR	5	1	Hel	B	1	2	+	Sa	Jun2	*Dr*
Echium vulgare	M	R/CSR	1	1	H	S	3	1	VA	Ea	Jun4	*Ev*
+ *Eleocharis palustris*	P	C/CSR	1	5	Hel	-	3	5	-	Sa	May3	*Ep*
Eleocharis quinqueflora	P	S/SC	1	5	Hel	-	2	4	-	Ep	Jun2	*Eq*
Eleogiton fluitans	P	SR/CSR	-	-	Wet	B	1	5	?	Ep	Jun4	*Ef*
+ *Elodea canadensis*	P	CR	4	3	Hyd	U	-1	5	-	Ep	Jun4	*Ec*
Elodea nuttallii	P	CR	4	4	Hyd	U	-1	5	?	Ep	May6	*En*
+ *Elymus caninus*	P	SC/CSR	4	5	H	L	4	2	VA	Ea	Jul1	*Ec*
+ *Elytrigia repens*	P	C/CR	3	4	H	L	3	5	VA	Ea	Jun4	*Er*
+ *Empetrum nigrum* subsp. *nigrum*	P	S/SC	1	5	Chw/Ph	L	2	5	ER	Ea	Apr2	*En*
Epilobium brunnescens	P	R/CSR	5	2	H	B	1	4	?	Ea	May6	*Eb*
+ *Epilobium ciliatum*	P	R/CSR	5	2	H/Ch	L	4	2	VA	Ea	Jul2	*Ec*
+ *Epilobium hirsutum*	P	C	4	3	Hel/H	L	5	5	VA	Ep	Jul2	*Eh*
+ *Epilobium montanum*	P	R/CSR	4	3	H/Ch	L	3	2	VA	Ea	Jun3	*Em*
+ *Epilobium obscurum*	P	R/CSR	5	1	H/Ch	L	3	2	VA	Ea	Jul2	*Eo*
+ *Epilobium palustre*	P	S/CSR	5	2	Hel	L	3	2	VA	Sa	Jul2	*Ep*
+ *Epilobium parviflorum*	P	CSR	5	2	H/Ch	L	3	2	VA	Ea	Jul2	*Ep*
Epilobium roseum	P	CR/CSR	5	2	H/Ch	L	4	2	VA	Ea	Jul2	*Er*
Epilobium tetragonum	P	R/CSR	2	3	H/Ch	L	3	2	VA	Ea	Jul2	*Et*
− *Epipactis helleborine*	P	S/CSR	4	-	G	L	3	2	OR	Sa	Jul4	*Eh*
Epipactis palustris	P	S/CSR	3	2	G/Hel	L	2	3	OR	Sa	Jun3	*Ep*
+ *Erica cinerea*	P	S/SC	1	5	Chw/Ph	L	3	3	ER	Ea	Jul2	*Ec*
− *Erica tetralix*	P	S/SC	1	5	Chw/Ph	L	3	3	ER	Ea	Jul3	*Et*
+ *Erigeron acer*	M/P	SR/CSR	2	2	Th/H	Sb	1	1	-	Ep	Jul2	*Ea*
+ *Eriophorum angustifolium*	P	S/SC	1	5	Hel	S	3	5	-	Ep	May2	*Ea*
+ *Eriophorum vaginatum*	P	S/SC	1	5	Hel	L	4	3	+	Ep	Apr2	*Ev*
− *Erodium cicutarium*	Aws	R/CSR	4	2	Th	S	2	1	+	Ep	Jun4	*Ec*
− *Erophila verna*	Aw	SR	5	1	Th	B	1	1	-	Ea	Mar4	*Ev*
Erysimum cheiranthoides	Asw	R	3	2	Th	L	3	1	-	Ep	Jun3	*Ec*
− *Erysimum cheiri*	P	CSR	1	2	Ch	L	2	2	?	Ea	Apr3	*Ec*
− *Eupatorium cannabinum*	P	C	5	2	Hel/H	L	5	3	VA	Sa	Jul3	*Ec*
Euphorbia cyparissias	P	CSR	3	4	H	L	2	5	VA	Ep	May4	*Ec*
Euphorbia esula	P	C/CSR	-	-	H	L	3	5	?	Ep	May3	*Ee*
Euphorbia exigua	As	SR	4	3	Th	L	2	1	?	Ep	Jun5	*Ee*
− *Euphorbia helioscopia*	As	R	5	2	Th	L	3	1	+	Sa	May6	*Eh*
− *Euphorbia peplus*	As	R	5	2	Th	L	2	1	+	Sa	Apr8	*Ep*
+ *Euphrasia officinalis*	As	SR	1	3	Th	L	2	1	Hp	Sa	Jun4	*Eo*
+ *Fallopia convolvulus*	As	R/CR	4	2	Th	L	3	1	-	Sa	Jul4	*Fc*
+ *Fallopia japonica*	P	C	2	4	G	L	5	5	?	Sa	Aug3	*Fj*
Fallopia sachalinensis	P	C	3	4	G	L	5	5	-	Sa	Aug2	*Fs*
− *Festuca arundinacea*	P	SC/CSR	2	3	H	S	3	4	VA	Ea	Jun3	*Fa*
+ *Festuca gigantea*	P	CSR	4	3	H	S	3	2	-	Ea	Jun2	*Fg*

TABLES OF ATTRIBUTES

Table 7.2

Species	Life history	Established strategy	SLA Class	DMC Class	Life form	Canopy structure	Canopy height	Lateral spread	Mycorrhizas	Leaf phenology	Flowering time and duration	(Sp.)
+ *Festuca ovina*	P	S	1	5	H	S	2	3	VA	Ea	May3	*Fo*
+ *Festuca pratensis*	P	CSR	3	4	H/Ch	S	2	2	VA	Ea	Jun1	*Fp*
+ *Festuca rubra*	P	CSR	2	4	H	S	2	4	VA	Ea	May3	*Fr*
Filago minima	Aw	SR	5	3	Th	L	2	1	?	Sh	Jun4	*Fm*
Filago vulgaris	Aw	SR	5	1	Th	L	3	1	?	Sh	Jun3	*Fv*
+ *Filipendula ulmaria*	P	C/SC	2	5	Hel/H	LA	5	4	+	Sa	Jun3	*Fu*
Filipendula vulgaris	P	S/CSR	3	4	H	S	2	3	?	Sa	May4	*Fv*
Foeniculum vulgare	P	SC/CSR	1	3	H	LA	4	3	?	Sa	Jul4	*Fv*
Fragaria × ananassa	P	CR/CSR	-	-	H	B	3	5	?	Ea	May3	*Fa*
+ *Fragaria vesca*	P	S/CSR	2	5	H	B	2	4	VA	Ea	Apr3	*Fv*
- *Fumaria muralis*	As	R/CR	4	2	Th	L	3	1	?	Sa	May6	*Fm*
- *Fumaria officinalis*	As	R	5	1	Th	L	3	1	-	Sa	May6	*Fo*
Galanthus nivalis	P	SR	4	1	G	B	2	2	VA	Sv	Jan3	*Gn*
Galega officinalis	P	C/CSR	4	2	H	L	5	3	?	Sa	Jun2	*Go*
+ *Galeopsis tetrahit*	As	CR	4	2	Th	L	4	1	-	Sa	Jul3	*Gt*
Galinsoga parviflora	As	R	4	2	Th	L	4	1	VA	Sa	May6	*Gp*
Galinsoga quadriradiata	As	R	5	1	Th	L	3	1	VA	Sa	May6	*Gq*
+ *Galium aparine*	Aws	CR	4	1	Th	L	5	1	+	Sh	Apr3	*Ga*
Galium mollugo	P	C/CSR	4	1	H	L	4	3	+	Ep	Jun4	*Gm*
- *Galium odoratum*	P	SC/CSR	4	3	G/H	L	3	4	+	Ea	May2	*Go*
+ *Galium palustre*	P	CR/CSR	5	2	Hel	L	4	4	+	Ep	Jun2	*Gp*
+ *Galium saxatile*	P	S/CSR	4	2	H	Lb	1	4	+	Ea	Jun3	*Gs*
+ *Galium sterneri*	P	S/CSR	3	3	H	Lb	1	3	?	Ea	Jun2	*Gs*
- *Galium uliginosum*	P	CSR	3	2	Hel	L	3	4	-	Ep	Jul2	*Gu*
+ *Galium verum*	P	SC/CSR	3	3	H	L	2	4	VA	Ea	Jul2	*Gv*
Genista anglica	P	S/SC	1	4	Ph	L	3	4	?	Sa	May2	*Ga*
Genista tinctoria	P	SC	2	3	Ph	L	4	3	VA	Ea	Jul3	*Gt*
- *Gentianella amarella*	M	SR	2	3	H	S	2	1	VA	Sa	Aug3	*Ga*
- *Geranium columbinum*	Aw	R/SR	4	3	Th	S	3	1	?	Sh	Jun2	*Gc*
- *Geranium dissectum*	Aws	R/CR	4	3	Th	L	3	1	VA	Sh	May4	*Gd*
Geranium endressii	P	C/CSR	5	2	H	S	3	3	?	Sa	Jun2	*Ge*
- *Geranium lucidum*	Aw	R/SR	4	2	Th	S	3	1	+	Sh	May4	*Gl*
+ *Geranium molle*	Aws	R/CSR	4	2	Th	S	2	1	+	Sh	Apr6	*Gm*
Geranium pratense	P	C/CSR	3	4	H	LA	3	2	VA	Sa	Jun4	*Gp*
Geranium pusillum	Aw	R/CSR	4	2	Th	S	3	1	?	Sh	Jun4	*Gp*
- *Geranium pyrenaicum*	P	CSR	4	2	H	S	3	2	?	Ea	Jun3	*Gp*
+ *Geranium robertianum*	A/P	R/CSR	5	2	Th/H	S	3	1	+	Ea	May5	*Gr*
Geranium × magnificum	P	C/CSR	-	-	H	S	3	3	?	Sa	Jun3	*Gm*
- *Geum rivale*	P	S/CSR	3	3	H	S	3	3	VA	Ea	May5	*Gr*
+ *Geum urbanum*	P	CR/CSR	3	5	H	S	3	2	VA	Ea	Jun3	*Gu*
+ *Glechoma hederacea*	P	CR/CSR	5	2	H	S	2	5	+	Ea	Mar3	*Gh*
- *Glyceria declinata*	P	CR	5	2	Hel	L	2	4	?	Ea	Jun4	*Gd*
+ *Glyceria fluitans*	P	CR	5	3	Wet	L	3	4	-	Ea	May4	*Gf*
+ *Glyceria maxima*	P	C	3	3	Wet	L	4	5	-	Ep	Jul2	*Gm*
- *Glyceria notata*	P	CR	5	3	Wet	L	4	4	+	Ea	May2	*Gn*
Glyceria × pedicellata	P	CR	-	-	Hel	L	2	4	?	Ea	Jun2	*Gp*
Gnaphalium sylvaticum	P	SR/CSR	2	3	H	S	3	3	VA	Ea	Jul3	*Gs*
- *Gnaphalium uliginosum*	As	SR	5	1	Th	L	2	1	VA	Sa	Jul2	*Gu*
Groenlandia densa	P	CR	5	1	Hyd	U	-1	4	?	Sa	May5	*Gd*
- *Gymnadenia conopsea*	P	SR	2	1	G	S	2	2	OR	Sa	Jun3	*Gc*
+ *Hedera helix*	P	SC	1	5	Chw/Ph	L	8	5	VA	Ea	Sep3	*Hh*
+ *Helianthemum nummularium*	P	S	2	4	Chw	L	2	3	EC	Ea	Jun2	*Hn*
+ *Helictotrichon pratense*	P	SC/CSR	1	5	H	S	2	3	VA	Ea	Jun1	*Hp*
+ *Helictotrichon pubescens*	P	S/CSR	3	4	H	L	2	2	VA	Ea	Jun2	*Hp*
Heracleum mantegazzianum	M	C/CR	5	2	H	LA	5	1	?	Sa	Jun2	*Hm*
+ *Heracleum sphondylium*	P/M	C/CSR	3	3	H	LA	3	2	+	Sa	Jun4	*Hs*
- *Hesperis matronalis*	P/M	CR/CSR	5	1	H	L	3	2	?	Ea	May3	*Hm*
+ *Hieracium sp.*	P	S/CSR	-	-	H/G	V		V	VA	V	May6	*Hs*
- *Hippuris vulgaris*	P	CR	5	1	Hyd	U,Lb	0	5	?	Ea	Jun2	*Hv*
Hirschfeldia incana	Aw	R/CR	2	2	Th	S	2	1	?	Ep	May6	*Hi*
+ *Holcus lanatus*	P	CSR	5	3	H	L	3	3	VA	Ep	Jun2	*Hl*
+ *Holcus mollis*	P	C/CSR	5	3	H	L	2	5	VA	Ea	Jun2	*Hm*
Hordeum jubatum	P	R/CSR	4	3	H	Sb	1	3	?	Ep	Jun3	*Hj*
Hordeum murinum	Aw	R	5	2	Th	L	3	1	VA	Ep	Jun2	*Hm*
Hordeum secalinum	P	R/CSR	5	3	H	L	2	3	?	Ea	Jun2	*Hs*

Species	Life history	Established strategy	SLA Class	DMC Class	Life form	Canopy structure	Canopy height	Lateral spread	Mycorrhizas	Leaf phenology	Flowering time and duration	(Sp.)
Hottonia palustris	P	CR	-	-	Hyd	U,Lb	1	5	+	Ea	May2	Hp
- Humulus lupulus	P	C	2	4	H	L	6	5	+	Sa	Jul2	Hl
Hyacinthoides hispanica	P	SR	4	1	G	B	2	2	?	Sv	May1	Hh
+ Hyacinthoides non-scripta	P	SR	3	1	G	B	2	2	VA	Sv	Apr3	Hn
+ Hydrocotyle vulgaris	P	R/CSR	5	1	Hel	B	2	5	+	Sa	Jun3	Hv
- Hypericum androsaemum	P	SC/CSR	-	-	Ph	L	4	2	?	Ep	Jun3	Ha
Hypericum elodes	P	SC/CSR	3	3	Hel	L	2	5	?	Ep	Jun4	He
+ Hypericum hirsutum	P	SC/CSR	4	4	H	L	4	2	-	Ep	Jul2	Hh
- Hypericum humifusum	P	SR/CSR	4	3	Ch	Lb	1	3	-	Ep	Jun4	Hh
- Hypericum maculatum	P	CR/CSR	4	4	H	L	3	3	VA	Ep	Jun3	Hm
+ Hypericum perforatum	P	CR/CSR	2	5	H	L	3	4	VA	Ep	Jun4	Hp
- Hypericum pulchrum	P	S/CSR	3	3	H	L	3	2	VA	Ea	Jun3	Hp
- Hypericum tetrapterum	P	R/CSR	5	2	Hel	L	3	2	+	Sa	Jun4	Ht
+ Hypochaeris radicata	P	CSR	3	1	H	B	1	2	VA	Ep	Jun4	Hr
+ Impatiens glandulifera	As	CR	5	1	Th	L	5	1	+	Sa	Jul4	Ig
Impatiens parviflora	As	CR	5	1	Th	L	4	1	VA	Sa	Jul5	Ip
+ Inula conyzae	M/P	SR/CSR	1	3	H	S	3	1	VA	Ea	Jul3	Ic
Inula helenium	P	C/CSR	2	4	H	S	4	2	?	Sa	Jul2	1h
Iris foetidissima	P	C/SC	1	5	G	LA	4	4	VA	Ea	May3	If
Iris germanica	P	C/CSR	-	-	G	LA	3	4	VA	Ep	May2	Ig
- Iris pseudacorus	P	C/CSR	2	2	G/Hel	LA	5	5	-	Ea	May3	Ip
- Isolepis setacea	P/A	R/CSR	4	3	Hel	B	1	2	?	Ea	May3	Is
Jasione montana	M/P	SR/CSR	4	2	H	Sb	1	2	+	Ea	May4	Jm
- Juncus acutiflorus	P	SC	1	3	Hel/H	S	4	5	-	Sa	Jul3	Ja
+ Juncus articulatus	P	CR/CSR	5	2	Hel/H	S	3	4	VA	Ea	Jun4	Ja
+ Juncus bufonius	As	R/SR	2	1	Th	Sb	1	1	+	Sa	May5	Jb
+ Juncus bulbosus	P	S/SR	2	3	Wet	Sb	1	5	-	Ea	Jun4	Jb
- Juncus conglomeratus	P	C/SC	1	4	H/Hel	-	4	3	?	Ea	May3	Jc
+ Juncus effusus	P	C/SC	1	4	Hel	-	5	4	-	Ea	Jun3	Je
- Juncus inflexus	P	SC	1	5	Hel	-	5	4	VA	Ea	Jun3	Ji
+ Juncus squarrosus	P	S/CSR	1	5	H	B	1	3	+	Ea	Jun2	Js
Juncus subnodulosus	P	SC	1	4	Hel/H	S	4	5	?	Sa	Jul3	Js
Juncus tenuis	P	R/CSR	1	5	H	S	2	4	?	Ep	Jun4	Jt
- Knautia arvensis	P	CSR	3	2	H	S	3	2	VA	?S	Jul3	Ka
+ Koeleria macrantha	P	S	1	5	H	Lb	1	2	VA	Ea	Jun2	Km
Lactuca virosa	Aw	CR	4	2	Th	L	5	1	?	Sh	Jul3	Lv
+ Lamiastrum galeobdolon	P	S/SC	3	4	Ch	L	2	4	+	Ea	May2	Lg
+ Lamium album	P	CR	5	2	H	L	3	3	VA	Ea	May8	La
Lamium amplexicaule	Aws	R	3	1	Th	L	2	1	VA	Ep	Apr5	La
Lamium hybridum	Aws	R	4	1	Th	L	3	1	?	Ep	Mar8	Lh
Lamium maculatum	P	CSR	3	3	H	L	3	5	?	Ea	May6	Lm
+ Lamium purpureum	Aws	R	5	1	Th	S	3	1	VA	Sh	Mar8	Lp
+ Lapsana communis	Aws	R/CR	5	1	Th	S	3	1	VA	Ep	Jul3	Lc
Lathraea squamaria	P	-	-	-	G	-		4	-	Sv	Apr2	Ls
Lathyrus latifolius	P	C/CSR	2	3	H	L	5	2	?	Sa	Jun4	Ll
Lathyrus linifolius	P	S/CSR	3	4	H	L	3	3	VA	Sa	Apr4	Ll
+ Lathyrus pratensis	P	CSR	3	4	H	L	4	3	VA	Sa	May4	Lp
Lemna gibba	(P)	R	5	1	Hyd	F	0	2	-	Ep	Jun2	Lg
+ Lemna minor	(P)	CR	5	1	Hyd	F	0	2	-	Ep	Jun2	Lm
- Lemna trisulca	(P)	SR	2	2	Hyd	U	-1	2	-	Ep	May3	Lt
+ Leontodon autumnalis	P	R/CSR	4	2	H	B	1	2	VA	Ea	Jun5	La
+ Leontodon hispidus	P	CSR	4	1	H	B	2	3	VA	Sa	Jun4	Lh
- Leontodon saxatilis	P/M	SR/CSR	3	1	H	B	1	2	?	Ea	Jun4	Ls
- Lepidium campestre	Aw/B	R/SR	3	3	Th	L	3	1	?	Sh	May4	Lc
- Lepidium draba	P	CR	2	2	H/G	L	3	5	?	Sa	May2	Ld
- Lepidium heterophyllum	P	SR/CSR	5	2	H	L	2	3	?	Ea	May4	Lh
Lepidium ruderale	Aw	R/SR	3	1	Th	S	3	1	-	Sh	May3	Lr
+ Leucanthemum vulgare	P	CR/CSR	3	1	H	L	3	2	VA	Ea	Jun3	Lv
Leucanthemum × superbum	P	C/CSR	2	1	H	L	3	2	?	Ep	Jun3	Ls
Linaria purpurea	P	CR/CSR	2	2	H	L	4	2	?	Ep	Jun3	Lp
Linaria repens	P	SR/CSR	2	2	H	L	3	4	+	Ep	Jun4	Lr
+ Linaria vulgaris	P	CR/CSR	3	2	H	L	3	4	VA	Ep	Jul4	Lv
+ Linum catharticum	B/A	SR	5	3	Ch/Th	Lb	1	1	+	Ea	Jun4	Lc
- Listera ovata	P	SR	3	1	G	L	2	3	OR	Sv	Jun2	Lo
Lithospermum officinale	P	SC/CSR	3	2	H	L	4	3	-	Sa	Jun2	Lo

TABLES OF ATTRIBUTES

Table 7.2

Species	Life history	Established strategy	SLA Class	DMC Class	Life form	Canopy structure	Canopy height	Lateral spread	Mycorrhizas	Leaf phenology	Flowering time and duration	(Sp.)
Littorella uniflora	P	SR/CSR	1	1	Wet	B	1	5	VA	Ea	Jun3	*Lu*
Lolium multiflorum	As/P	R/CR	3	4	Th	L	4	1	VA	Ea	May5	*Lm*
+ *Lolium perenne*	P	CR/CSR	4	3	H	S	2	3	VA	Ea	May4	*Lp*
+ *Lonicera periclymenum*	P	SC	2	4	Chw/Ph	L	6	5	VA	Ep	Jun4	*Lp*
+ *Lotus corniculatus*	P	S/CSR	4	2	H	L	2	2	VA	Sa	Jun4	*Lc*
+ *Lotus pedunculatus*	P	C/CSR	3	3	Hel	L	3	4	VA	Sa	Jun3	*Lp*
Lupinus × *regalis*	P	C/CSR	4	2	H	LA	4	3	?	Sa	May3	*Lr*
+ *Luzula campestris*	P	S/CSR	3	3	H	Sb	1	3	+	Ea	Mar4	*Lc*
- *Luzula multiflora*	P	S/CSR	3	3	H/Hel	Sb	1	3	?	Ea	May2	*Lm*
+ *Luzula pilosa*	P	S/CSR	4	3	H	Sb	1	3	-	Ea	Apr3	*Lp*
- *Luzula sylvatica*	P	SC	2	4	H	S	2	4	-	Ea	May2	*Ls*
- *Lychnis flos-cuculi*	P	CSR	4	1	Hel	S	2	3	+	Ea	May2	*Lf*
- *Lycopus europaeus*	P	C/CR	5	3	Hel/H	L	4	3	+	Sa	Jun4	*Le*
- *Lysimachia nemorum*	P	R/CSR	5	2	Hel/Ch	Lb	1	4	VA	Ea	May5	*Ln*
- *Lysimachia nummularia*	P	R/CSR	5	2	Ch/Hel	Lb	1	4	VA	Ea	Jun3	*Ln*
Lysimachia punctata	P	C/CSR	3	3	H	L	4	3	?	Sa	May4	*Lp*
- *Lysimachia vulgaris*	P	C	5	2	Hel/H	L	5	3	VA	Sa	May2	*Lv*
- *Lythrum portula*	As	R/SR	3	4	Th	Lb	1	1	?	Sa	Jun5	*Lp*
- *Lythrum salicaria*	P	C/CR	5	2	Hel/H	L	4	2	+	Sa	Jun3	*Ls*
- *Malva moschata*	P	C/CSR	3	2	H	L	4	2	?	Ea	Jul2	*Mm*
- *Malva neglecta*	P/A	R/CR	3	3	Th/H	L	2	4	?	Ep	Jun4	*Mn*
- *Malva sylvestris*	P	CR	3	3	H	L	4	3	VA	Ea	Jun4	*Ms*
+ *Matricaria discoidea*	Asw	R	4	1	Th	S	2	1	VA	Sa	Jun2	*Md*
Matricaria recutita	Asw	R	5	1	Th	L	3	1	+	Ep	Jun2	*Mr*
Meconopsis cambrica	P	CSR	5	1	H	S	2	2	?	Ep	Jun3	*Mc*
+ *Medicago lupulina*	A/P	R/CSR	4	2	Th/H	L	2	1	VA	Ea	May4	*Ml*
- *Medicago sativa*	P	C/CSR	2	4	H	L	4	2	VA	Ea	Aug2	*Ms*
- *Melampyrum pratense*	As	R/SR	5	3	Th	L	3	1	Hp	Sa	May6	*Mp*
+ *Melica uniflora*	P	S/SC	5	-	H	L	4	5	-	Sa	May2	*Mu*
Melilotus albus	Aw/B	CR	5	2	Th/H	L	4	1	VA	Ep	Jul2	*Ma*
Melilotus altissimus	B/P	CR/CSR	2	3	H	L	4	1	?	Ep	Jun3	*Ma*
Melilotus officinalis	B/P	CR/CSR	2	3	H	L	4	1	VA	Ep	Jul3	*Mo*
+ *Mentha aquatica*	P	C/CR	5	2	Hel/H	L	4	5	VA	Ea	Jul3	*Ma*
- *Mentha arvensis*	P	CR	4	2	H/Hel	L	3	4	VA	Ea	May6	*Ma*
Mentha spicata	P	C/CR	3	4	H	L	3	4	?	Ea	Aug2	*Ms*
Mentha × *piperita*	P	C/CR	3	2	H/Hel	L	4	4	+	Ea	Aug3	*Mp*
- *Mentha* × *verticillata*	P	C/CR	5	1	Hel/H	L	4	4	?	Ea	Jul4	*Mv*
Mentha × *villosa*	P	C/CR	5	1	H	L	5	4	?	Ep	Aug2	*Mv*
- *Menyanthes trifoliata*	P	SC/CSR	3	3	Hel	B	2	4	-	Sa	Jul2	*Mt*
Mercurialis annua	Asw	R/CR	4	2	Th	L	3	1	VA	Ep	Jul4	*Ma*
+ *Mercurialis perennis*	P	SC	4	3	H	L	3	5	+	Ep	Feb4	*Mp*
+ *Milium effusum*	P	S/CSR	5	4	H	L	3	2	+	Ea	Jun1	*Me*
- *Mimulus guttatus*	P	CR	5	1	Hel/H	S	3	3	?	Ea	Jul3	*Mg*
Mimulus × *robertsii*	P	CR	5	1	Wet	L	3	3	?	Ea	Jun3	*Mr*
+ *Minuartia verna*	P	S	2	2	Ch	Lb	1	2	-	Ea	May5	*Mv*
+ *Moehringia trinervia*	A/P	R/CSR	5	2	Th/Ch	L	2	1	-	Sa	May2	*Mt*
+ *Molinia caerulea*	P	SC	3	5	H/Hel	S	4	4	VA	Sa	Jun3	*Mc*
- *Montia fontana*	P/A	SR, CR	3	2	Th/Wet	Lb	1	3	?	Ea	May6	*Mf*
+ *Mycelis muralis*	P	CSR	4	2	H	S	3	2	+	Ea	Jul3	*Mm*
+ *Myosotis arvensis*	Aws	R	4	1	Th	S	2	1	-	Sh	Apr6	*Ma*
- *Myosotis discolor*	Aw	SR	4	1	Th	Sb	1	1	VA	Sh	May5	*Md*
- *Myosotis laxa* subsp. *cespitosa*	As	R	5	1	Hel/H	S	2	1	VA	Sa	May4	*Ml*
+ *Myosotis ramosissima*	Aw	SR	4	1	Th	Sb	1	1	+	Sh	Apr3	*Mr*
+ *Myosotis scorpioides*	P	CR	5	1	Hel	S	2	3	+	Ea	May5	*Ms*
- *Myosotis secunda*	P	CR	4	1	Wet	S	2	3	?	Ea	May4	*Ms*
Myosotis sylvatica	P	R/CSR	5	1	H	S	2	2	-	Ea	May2	*Ms*
Myosoton aquaticum	P	CR	5	1	Hel/H/Ch	L	4	2	-	Ea	Jul2	*Ma*
Myriophyllum alterniflorum	P	SR	-	-	Hyd	U	-1	5	-	Ea	May4	*Ma*
- *Myriophyllum spicatum*	P	CR	5	2	Hyd	U	-1	5	?	Ea	Jun2	*Ms*
+ *Myrrhis odorata*	P	C/CSR	5	3	H	LA	5	3	VA	Sa	May2	*Mo*
Narcissus pseudonarcissus	P	SR	2	1	G	B	2	2	VA	Sv	Mar2	*Np*
+ *Nardus stricta*	P	S/SC	1	5	H	S	3	2	VA	Ea	Jun3	*Ns*
Narthecium ossifragum	P	S/CSR	4	2	Hel	B	1	5	VA	Sa	Jul3	*No*
Neottia nidus-avis	P		-	-	G	-		2	OR	Sa	Jun2	*Nn*
- *Nuphar lutea*	P	C/CR	2	1	Hyd	F	0	5	?	Ea	Jun3	*Nl*

670   TABLES OF ATTRIBUTES

Species	Life history	Established strategy	SLA Class	DMC Class	Life form	Canopy structure	Canopy height	Lateral spread	Mycorrhizas	Leaf phenology	Flowering time and duration	(Sp.)
Nymphaea alba	P	C/CR	1	2	Hyd	F	0	5	?	Ea	Jun3	*Na*
- Odontites vernus	As	R/CR	2	2	Th	L	3	1	Hp	Sa	Jun3	*Ov*
Oenanthe aquatica	As/B	R/CR	5	2	Hel	S	3	1	-	Sa	Jun4	*Oa*
- Oenanthe crocata	P	C/CR	5	1	G	LA	4	2	-	Ea	Jun2	*Oc*
- Oenanthe fistulosa	P	SR/CSR	5	1	Wet	S	3	3	?	Ea	Jul3	*Of*
Oenothera biennis	M	R/CSR	2	2	H	S	3	1	VA	Ep	Jun4	*Ob*
Oenothera cambrica	M	CR/CSR	3	2	H	S	3	1	?	Ep	Jun4	*Oc*
Oenothera glazioviana	M	R/CSR	3	2	H	S	3	1	VA	Ep	Jun4	*Og*
Onobrychis viciifolia	P	CSR	2	3	H	S	3	2	VA	Ea	Jun3	*Ov*
- Ononis repens	P	SC/CSR	4	3	H/Ch	L	3	4	VA	Sa	Jun4	*Or*
Ononis spinosa	P	SC/CSR	3	3	H/Ch	L	3	3	VA	Sa	Jun4	*Os*
Ophrys apifera	M	SR	4	1	G	Sb	1	2	OR	Sh	Jun2	*Oa*
- Orchis mascula	P	SR	2	1	G	S	2	2	OR	Sv	Apr3	*Om*
- Orchis morio	P	SR	-	-	G	S	2	2	OR	Sh	May2	*Om*
+ Origanum vulgare	P	SC/CSR	3	3	Ch/H	L	4	3	VA	Ep	Jul3	*Ov*
Ornithogalum angustifolium	P	SR	1	1	G	B	2	2	VA	Sv	May3	*Oa*
Ornithopus perpusillus	Aw	SR/CSR	5	2	Th	Sb	1	1	?	Sh	May5	*Op*
+ Oxalis acetosella	P	S/SR	5	1	H/Ch	B	1	2	VA	Ea	Apr2	*Oa*
Oxalis corniculata	A/P	R	5	2	Th/Ch	L	2	3	VA	Sa	Jun4	*Oc*
Oxalis debilis	P	R	5	-	G	B	1	2	?	Sa	Jul3	*Od*
Oxalis exilis	A/P	SR	5	2	Th/H	B	1	3	VA	Ep	May6	*Oe*
- Papaver argemone	Asw	R/SR	4	2	Th	S	2	1	?	Sh	Jun2	*Pa*
- Papaver dubium	Asw	R	3	2	Th	S	2	1	?	Sh	Jun2	*Pd*
+ Papaver rhoeas	Asw	R	5	1	Th	S	2	1	-	Sa	Jun3	*Pr*
Papaver somniferum	Asw	R/CR	4	1	Th	S	3	1	-	Sh	Jul2	*Ps*
- Parietaria judaica	P	CR/CSR	5	2	H	B	4	3	+	Ea	Jun5	*Pj*
Paris quadrifolia	P	SR/CSR	5	2	G	L	3	4	VA	Sv/Sa	May4	*Pq*
Parnassia palustris	P	SR/CSR	5	1	H	Sb	1	2	+	Sa	Jul4	*Pp*
Pastinaca sativa	M	CR	5	1	H	LA	3	1	VA	Sa	Jul2	*Ps*
- Pedicularis sylvatica	M	SR	1	2	H	S	2	1	Hp	Sa	Apr4	*Ps*
- Pentaglottis sempervirens	P	C/CSR	4	1	H	LA	4	3	?	Ea	May2	*Ps*
+ Persicaria amphibia	P	C/CR	2	3	Wet	L,F	4	5	-	Sa	Jul3	*Pa*
Persicaria bistorta	P	C/CSR	5	2	G	LA	2	5	VA	Sa	Jun3	*Pb*
- Persicaria hydropiper	As	R/CR	5	1	Th	L	4	1	-	Sa	Jul3	*Ph*
- Persicaria lapathifolia	As	CR	3	3	Th	L	4	1	-	Sa	Jun5	*Pl*
+ Persicaria maculosa	As	R/CR	3	2	Th	L	4	1	+	Sa	Jun5	*Pm*
- Petasites fragrans	P	C/CR	4	2	G	B	2	4	?	Ep	Jan3	*Pf*
+ Petasites hybridus	P	C	2	3	G	LA	5	4	+	Sa	Mar3	*Ph*
+ Phalaris arundinacea	P	C	4	4	Hel	L	5	5	VA	Sa	Jun2	*Pa*
Phleum bertolonii	P	SR/CSR	4	4	H/G	L	2	3	VA	Ea	Jun1	*Pb*
+ Phleum pratense	P	CR/CSR	5	4	H	L	3	3	VA	Ea	Jun2	*Pp*
+ Phragmites australis	P	C	2	5	Wet	L	5	4	VA	Sa	Aug2	*Pa*
Picris echioides	A/P	CR	4	1	Th/H	S	3	1	VA	Ea	Jun5	*Pe*
Picris hieracioides	M/P	R/CSR	5	1	H	S	3	2	VA	Ea	Jul3	*Ph*
Pilosella aurantiaca	P	R/CSR	3	2	H	B	1	4	?	Ea	Jun4	*Pa*
+ Pilosella officinarum	P	S/CSR	3	2	H	B	1	4	VA	Ea	May2	*Po*
Pimpinella major	P	CSR	4	3	H	S	3	2	+	Sa	Jun2	*Pm*
+ Pimpinella saxifraga	P	SR/CSR	2	4	H	Sb	1	2	VA	Ea	Jul2	*Ps*
Pinguicula vulgaris	P	SR	3	1	Hel	B	1	2	-	Sa	May3	*Pv*
Plantago coronopus	M+	SR/CSR	2	1	H	B	1	1	VA	Ea	May3	*Pc*
+ Plantago lanceolata	P	CSR	3	2	H	B	2	2	VA	Ea	Apr5	*Pl*
+ Plantago major	P	R/CSR	5	1	H	LA	2	2	VA	Ep	Jun4	*Pm*
Plantago media	P	S/CSR	3	1	H	B	1	4	VA	Ep	May4	*Pm*
- Platanthera chlorantha	P	SR	3	1	G	L	2	2	OR	Sv	May3	*Pc*
Poa angustifolia	P	SC/CSR	2	5	H	S	4	3	?	Ea	May3	*Pa*
+ Poa annua	A/P	R	5	3	Th/H	S	2	1	VA	Ep	Mar12	*Pa*
Poa compressa	P	SR/CSR	3	3	H	S	2	4	+	Ea	Jun3	*Pc*
- Poa nemoralis	P	SR/CSR	5	3	H	S	4	2	+	Ea	Jun3	*Pn*
+ Poa pratensis	P	CSR	3	5	H	S	2	3	+	Ea	May3	*Pp*
+ Poa trivialis	P	R/CSR	5	2	H/Ch	Sb	1	2	+	Ea	Jun1	*Pt*
- Polygala serpyllifolia	P	S/CSR	3	3	Ch	Lb	1	2	VA	Ea	May4	*Ps*
+ Polygala vulgaris	P	S/CSR	3	4	Ch	Lb	1	2	VA	Ea	May3	*Pv*
Polygonatum × hybridum	P	SC	3	2	G	L	4	4	?	Sa	May2	*Ph*
+ Polygonum aviculare	As	R	4	3	Th	L	4	1	+	Sa	Jul5	*Pa*
Potamogeton berchtoldii	P	R	5	1	Hyd	U	-1	4	?	Sa	Jun4	*Pb*

TABLES OF ATTRIBUTES

Table 7.2

Species	Life history	Established strategy	SLA Class	DMC Class	Life form	Canopy structure	Canopy height	Lateral spread	Mycorrhizas	Leaf phenology	Flowering time and duration	(Sp.)
+ *Potamogeton crispus*	P	CR	5	1	Hyd	U	-1	4	-	Sh	May3	*Pc*
+ *Potamogeton natans*	P	CR	4	3	Hyd	F	0	5	-	Sa	Jun4	*Pn*
Potamogeton pectinatus	P	CR	5	2	Hyd	U	-1	5	?	Sa	May5	*Pp*
Potamogeton perfoliatus	P	CR	5	1	Hyd	U	-1	5	?	Sa	Jun4	*Pp*
Potamogeton polygonifolius	P	S/CSR	3	3	Hyd	F,B	1	5	-	Sa	May6	*Pp*
Potamogeton pusillus	P	R	5	1	Hyd	U	-1	4	-	Sa	Jun4	*Pp*
- *Potentilla anglica*	P	SR/CSR	3	4	Hel/H	Lb	1	5	VA	Ep	Jun4	*Pa*
- *Potentilla anserina*	P	CR/CSR	3	4	Hel/H	L	2	5	VA	Ep	Jun3	*Pa*
+ *Potentilla erecta*	P	S/CSR	3	5	H	S	2	3	+	Sa	Jun4	*Pe*
- *Potentilla palustris*	P	SC/CSR	2	5	Hel	L	3	5	+	Sa	May3	*Pp*
Potentilla recta	P	C/CSR	4	3	H	S	3	2	?	Sa	Jun2	*Pr*
- *Potentilla reptans*	P	CR/CSR	4	3	H	B	2	5	VA	Ep	Jun4	*Pr*
+ *Potentilla sterilis*	P	SR/CSR	3	3	H	Sb	1	3	VA	Ea	Feb4	*Ps*
+ *Primula veris*	P	S/CSR	2	3	H	B	1	2	VA	Ea	Apr2	*Pv*
- *Primula vulgaris*	P	S/CSR	4	2	H	B	2	2	VA	Ea	Dec6	*Pv*
+ *Prunella vulgaris*	P	CSR	5	2	H	Sb	1	3	VA	Ea	Jun4	*Pv*
- *Pseudofumaria lutea*	P	R/CSR	3	2	H	L	2	2	-	Ea	May4	*Pl*
Puccinellia distans	P	R/CSR	5	2	H	Sb	1	3	+	Ea	Jun2	*Pd*
- *Pulicaria dysenterica*	P	C/CSR	4	1	Hel	L	3	4	VA	Sa	Aug2	*Pd*
+ *Ranunculus acris*	P	CSR	3	2	H	S	2	2	VA	Ea	May3	*Ra*
Ranunculus aquatilis	A/P	R	5	2	Th/Hyd	F	0	1	?	Ep	May2	*Ra*
- *Ranunculus auricomus*	P	SR	5	1	H/G	S	2	2	VA	Sv	Apr2	*Ra*
+ *Ranunculus bulbosus*	P	SR	2	2	H	Sb	1	2	VA	Sh	May2	*Rb*
Ranunculus circinatus	P	CR/CSR	-	-	Hyd	U	-1	4	?	Ep	May4	*Rc*
+ *Ranunculus ficaria*	P	SR	5	1	G	S	2	2	+	Sv	Mar3	*Rf*
+ *Ranunculus flammula*	P	CR/CSR	2	2	Hel	S	2	4	+	Ea	May4	*Rf*
Ranunculus hederaceus	As	R	5	1	Hel	B	1	4	?	Ep	Jun4	*Rh*
Ranunculus lingua	P	C/CR	-	-	Hel	S	4	3	VA	Ep	Jun4	*Rl*
Ranunculus omiophyllus	P	R/CSR	5	1	Th/Hyd	B	1	4	?	Ep	Apr7	*Ro*
+ *Ranunculus peltatus*	P/A	R/CR	5	1	Hyd	F	0	1	?	Ep	May4	*Rp*
+ *Ranunculus penicillatus*	P	C/CR	5	1	Hyd	U	-1	5	?	Ea	May4	*Rp*
+ *Ranunculus repens*	P	CR	4	2	Hel/H	S	2	5	VA	Ea	May2	*Rr*
+ *Ranunculus sceleratus*	Asw	R/CR	5	2	Th/Hel	S	3	1	+	Sa	May5	*Rs*
Ranunculus trichophyllus	A/P	R	4	1	Th/Hyd	U	-1	1	?	Ep	May2	*Rt*
- *Raphanus raphanistrum*	Asw	R/CR	3	1	Th	S	3	1	-	Sa	May5	*Rr*
Rapistrum rugosum	Aw	R/CR	3	2	Th	S	2	1	?	Sh	May5	*Rr*
Reseda lutea	P	SR/CSR	2	2	H	S	3	2	+	Ea	Jun3	*Rl*
- *Reseda luteola*	B/P	R/CR	2	2	H	S	4	1	-	Ea	Jun3	*Rl*
+ *Rhinanthus minor*	As	R/SR	2	2	Th	L	3	1	Hp	Sa	May4	*Rm*
- *Ribes uva-crispa*	P	SC	2	-	Ph	L	4	5	VA	Sa	Jun3	*Ru*
Rorippa amphibia	P	C/CR	5	1	Hel	LA	3	4	?	Ep	Jun3	*Ra*
Rorippa microphylla	P	CR	5	1	Wet	L	3	5	?	Ea	May6	*Rm*
+ *Rorippa nasturtium-aquaticum*	P	CR	5	1	Wet	L	3	5	-	Ea	May3	*Rn*
+ *Rorippa palustris*	As	R	5	1	Th	S	3	1	-	Sa	Jun4	*Rp*
Rorippa sylvestris	P	CR	5	1	Hel/H	S	3	5	?	Ep	Jun3	*Rs*
- *Rubus caesius*	P	SC	2	5	Ph	L	4	5	VA	Sa	Jun4	*Rc*
Rubus chamaemorus	P	SC/CSR	1	4	H	B	1	5	-	Sa	Jun3	*Rc*
+ *Rubus fruticosus*	P	SC	1	5	Ph	L	6	5	VA	Ep	Jun4	*Rf*
- *Rubus idaeus*	P	SC	2	5	Ph	L	5	5	VA	Sa	May3	*Ri*
Rubus saxatilis	P	S/CSR	3	5	H	B	2	5	VA	Sa	Jun3	*Rs*
+ *Rumex acetosa*	P	CSR	4	1	H	S	2	2	-	Ea	May2	*Ra*
+ *Rumex acetosella*	P	SR/CSR	3	1	H/G	Sb	1	4	-	Ea	May3	*Ra*
- *Rumex conglomeratus*	P/M	CR/CSR	3	2	H/Hel	S	4	2	-	Ea	May2	*Rc*
+ *Rumex crispus*	P/A	CR/CSR	4	1	H	S	3	2	+	Ea	May6	*Rc*
- *Rumex hydrolapathum*	P	C/CSR	4	2	Wet	LA	5	2	?	Ep	Jul3	*Rh*
+ *Rumex obtusifolius*	P	C/CSR	4	2	H	LA	2	2	-	Ea	Jun5	*Ro*
- *Rumex sanguineus*	P	CR/CSR	4	3	H/Hel	S	4	2	?	Ea	Jun3	*Rs*
Ruscus aculeatus	P	S/SC	1	5	Ph	L	4	5	VA	Ea	Jan4	*Ra*
- *Sagina apetala*	Aw	SR	3	3	Th	Sb	1	1	-	Sh	May4	*Sa*
Sagina maritima	Aw	SR	1	2	Th	Sb	1	1	?	Sh	Apr4	*Sm*
- *Sagina nodosa*	P	S/SR	1	5	Hel/H/Ch	Sb	1	2	?	Ea	Jul3	*Sn*
+ *Sagina procumbens*	P/A	R/CSR	2	3	H/Th	Sb	1	2	-	Ea	May5	*Sp*
Sagittaria sagittifolia	P	CR	4	2	Hyd	E	0	4	?	Sa	Jul2	*Ss*
- *Salix repens*	P	SC	1	5	Ph/Chw/Hel	L	5	5	EC	Sa	Apr2	*Sr*
Sambucus ebulus	P	C	3	2	H	L	4	5	?	Sa	Jul2	*Se*

Species	Life history	Established strategy	SLA Class	DMC Class	Life form	Canopy structure	Canopy height	Lateral spread	Mycorrhizas	Leaf phenology	Flowering time and duration	(Sp.)
Samolus valerandi	P	SR/CSR	3	2	Hel/H	S	2	4	VA	Ea	Jun3	*Sv*
+ *Sanguisorba minor*	P	S/CSR	2	5	H	S	2	2	VA	Ea	Jun2	*Sm*
Sanguisorba minor subsp. *muricata*	P	CSR	3	5	H	S	3	2	?	Ea	Jun3	*Sm*
Sanguisorba officinalis	P	C/CSR	4	3	H	S	3	2	VA	Ea	Jun4	*So*
+ *Sanicula europaea*	P	S/CSR	5	2	H	B	2	2	VA	Ea	May5	*Se*
− *Saponaria officinalis*	P	C/CR	5	2	H	L	4	4	−	Ea	Jul3	*So*
Saxifraga granulata	P	SR/CSR	5	1	H	Sb	1	2	+	Sv	Apr3	*Sg*
+ *Saxifraga tridactylites*	Aw	SR	−	1	Th	Sb	1	1	VA	Sh	Apr3	*St*
Saxifraga × *urbium*	P	S/CSR	1	−	H	B	1	4	?	Ea	Jun2	*Su*
+ *Scabiosa columbaria*	M/P	S/SR	2	2	H	Sb	1	2	VA	Ea	Jul2	*SC*
− *Schoenoplectus lacustris*	P	C/SC	2	2	Hyd	−	5	4	?	Ea	Jun2	*Sl*
Schoenoplectus tabernaemontani	P	C/SC	−	−	Wet	−	5	4	?	Ea	Jun2	*St*
Schoenus nigricans	P	S/SC	1	5	Hel	S	4	3	+	Ea	May2	*Sn*
Scirpus sylvaticus	P	C	4	2	Hel	S	4	5	−	Sa	Jun2	*Ss*
− *Scleranthus annuus*	As	SR	2	2	Th	La	1	1	−	Sa	Jun3	*Sa*
− *Scrophularia auriculata*	P	C/CR	5	1	Hel/H	L	3	2	VA	Sa	Jun4	*Sa*
Scrophularia nodosa	P	C/CR	4	2	H	L	3	2	+	Sa	Jun4	*Sn*
− *Scutellaria galericulata*	P	CR/CSR	5	2	Hel	L	3	3	VA	Sa	Jun4	*Sg*
Scutellaria minor	P	S/CSR	−	−	Hel	L	3	3	?	Sa	Jul4	*Sm*
Securigera varia	P	C/CSR	3	3	H	L	3	4	−	Sa	Jun3	*Sv*
+ *Sedum acre*	P	S/SR	2	1	Ch	Lb	1	3	−	Ea	Jun2	*Sa*
Sedum album	P	S	1	−	Ch	Lb	1	3	−	Ea	Jun3	*Sa*
− *Sedum anglicum*	P	S/SR	−	−	Ch	Lb	1	3	?	Ea	Jun3	*Sa*
Sedum rupestre	P	S/SR	1	1	Ch	Lb	1	3	−	Ea	Jun3	*Sr*
Sedum spurium	P	S/SR	1	−	Ch	Lb	1	4	?	Ea	Jul2	*Ss*
− *Sedum telephium*	P	S/CSR	3	1	H	L	3	3	−	Sa	Jul3	*St*
− *Senecio aquaticus*	M/P	R/CR	5	1	H/Hel	S	3	1	?	Ea	Jul2	*Sa*
Senecio erucifolius	P	SC/CSR	2	2	H	L	4	4	VA	Ep	Jul2	*Se*
+ *Senecio jacobaea*	M/P	CR/CSR	4	1	H	S	4	1	VA	Ea	Jun5	*Sj*
+ *Senecio squalidus*	P/A	R/CSR	3	1	H/Th	L	2	2	VA	Ea	Jun7	*Ss*
− *Senecio sylvaticus*	As	R/CR	4	1	Th	L	4	1	+	Sa	Jul3	*Ss*
+ *Senecio viscosus*	As	R/CR	3	1	Th	L	3	1	VA	Sa	Jul4	*Sv*
+ *Senecio vulgaris*	Asw	R	4	1	Th	L	3	1	+	Ep	Apr12	*Sv*
Serratula tinctoria	P	SC/CSR	3	3	H	S	3	2	VA	Ea	Jul3	*St*
− *Sherardia arvensis*	Aw	R/SR	3	2	Th	L	3	1	?	Sh	May6	*Sa*
Silaum silaus	P	S/CSR	2	4	H	S	3	2	VA	Sa	Jun3	*Ss*
+ *Silene dioica*	P/M	CSR	5	1	H/Ch	L	3	2	−	Ea	May2	*Sd*
− *Silene latifolia*	P/A	R/CR	5	1	H	L	4	2	−	Ea	May5	*Sl*
Silene noctiflora	Asw	R/CR	2	2	Th	L	3	1	?	Ep	Jul3	*Sn*
− *Silene vulgaris*	P	CSR	3	1	H	L	3	2	−	Sa	Jun3	*Sv*
+ *Sinapis arvensis*	Asw	R/CR	4	1	Th	S	3	1	?	Sa	May3	*Sa*
− *Sisymbrium altissimum*	As	R/CR	2	1	Th	S	4	1	?	Sa	Jun3	*Sa*
Sisymbrium loeselii	Aw	CR	3	2	Th	S	3	1	−	Ep	Jun3	*Sl*
− *Sisymbrium officinale*	Aw/B	R/CR	4	1	Th	S	3	1	+	Ep	Jun2	*So*
Sisymbrium orientale	Aw	R/CR	3	1	Th	L	3	1	−	Sh	Jun3	*So*
+ *Solanum dulcamara*	P	C/CSR	4	2	Chw/Ph	L	4	5	VA	Sa	Jun4	*Sd*
Solanum nigrum	As	R/CR	5	1	Th	L	3	1	VA	Sa	Jul3	*Sn*
Solanum physalifolium	As	R/CR	4	1	Th	L	3	1	?	Sa	Jul4	*Sp*
Solidago canadensis	P	C	4	3	H	L	4	5	?	Sa	Aug3	*SC*
Solidago gigantea	P	C	3	5	H	L	4	5	?	Sa	Jul2	*Sg*
+ *Solidago virgaurea*	P	S/CSR	3	3	H	S	3	2	VA	Sa	Jul3	*Sv*
− *Sonchus arvensis*	P	C/CR	3	1	H	L	5	4	+	Sa	Jul4	*Sa*
+ *Sonchus asper*	Aws	R/CR	3	1	Th	S	4	1	+	Ep	May6	*Sa*
+ *Sonchus oleraceus*	Aws	R/CR	5	1	Th	S	4	1	VA	Ep	May6	*So*
− *Sparganium emersum*	P	CR	4	1	Wet	F,B	3	4	?	Sa	Jun2	*Se*
+ *Sparganium erectum*	P	C/CR	2	1	Wet	LA	4	4	VA	Sa	Jul3	*Se*
+ *Spergula arvensis*	As	R/SR	1	1	Th	L	2	1	−	Sa	Jun3	*Sa*
Spergularia marina	As	R/SR	3	1	Th	Lb	1	1	+	Ep	Jun4	*Sm*
− *Spergularia rubra*	Asw	SR	3	1	Th	Lb	1	1	?	Sa	May5	*Sr*
− *Stachys arvensis*	Asw	R/SR	5	1	Th	L	2	1	?	Ep	Apr8	*Sa*
+ *Stachys officinalis*	P	S/CSR	2	3	H	Sb	1	2	VA	Ea	Jun4	*So*
− *Stachys palustris*	P	C/CR	4	3	Hel/G	L	4	4	VA	Sa	Jul3	*Sp*
+ *Stachys sylvatica*	P	C/CR	5	3	H	L	3	4	VA	Sa	Jul2	*Ss*
− *Stellaria graminea*	P	CSR	5	2	H	L	2	4	−	Ea	May4	*Sg*
+ *Stellaria holostea*	P	CSR	3	2	Ch	L	3	5	−	Ea	Apr3	*Sh*

Table 7.2

Species	Life history	Established strategy	SLA Class	DMC Class	Life form	Canopy structure	Canopy height	Lateral spread	Mycorrhizas	Leaf phenology	Flowering time and duration	(Sp.)
+ *Stellaria media*	Aws	R	5	1	Th	L	2	1	-	Ep	Mar12	*Sm*
Stellaria neglecta	Aw	R/CR	5	1	Th	L	4	1	?	Sh	Apr4	*Sn*
Stellaria pallida	Aw	R/SR	5	1	Th	Lb	1	1	-	Sv	Mar3	*Sp*
+ *Stellaria uliginosa*	P	R/CSR	5	2	Hel	Lb	1	3	-	Ea	May2	*Su*
+ *Succisa pratensis*	P	S/CSR	2	2	H	Sb	1	2	VA	Ea	Jul4	*Sp*
- *Symphytum officinale*	P	C/CSR	3	1	H	LA	5	3	VA	Sa	May2	*So*
Symphytum orientale	P	C/CR	4	1	H	S	3	2	?	Sa	Apr2	*So*
- *Symphytum × uplandicum*	P	C/CSR	5	1	H	LA	5	3	?	Sa	Jun3	*Su*
+ *Tamus communis*	P	C/CSR	4	2	G	L	5	2	VA	Sa	May3	*Tc*
- *Tanacetum parthenium*	P	CR/CSR	5	2	H	L	3	2	+	Ea	Jul2	*Tp*
- *Tanacetum vulgare*	P	C/CSR	3	3	H	L	4	3	-	Ea	Jul3	*Tv*
+ *Taraxacum* agg.	P	R/CSR	5	1	H	B	2	2	VA	Ea	Mar8	*Ta*
Tellima grandiflora	P	CR/CSR	5	3	H	S	2	2	?	Ea	May2	*Tg*
+ *Teucrium scorodonia*	P	SC/CSR	1	5	H	L	3	3	VA	Ep	Jul3	*Ts*
Thalictrum flavum	P	C/CSR	3	3	Hel/H	LA	4	4	?	Sa	Jul2	*Tf*
- *Thlaspi arvense*	Asw	R	5	1	Th	L	2	1	-	Sa	May3	*Ta*
+ *Thymus polytrichus*	P	S	1	-	Chw	Lb	1	4	?	Ea	May3	*Tp*
+ *Torilis japonica*	Aw/B	CR/CSR	4	3	Th	S	4	1	?	Ep	Jul2	*Tj*
- *Tragopogon pratensis*	P/M	CR/CSR	2	2	H/G	S	3	2	VA	Sa	Jun2	*Tp*
Trichophorum cespitosum	P	S/SC	-	-	H/Hel	B	3	3	+	Ea	May2	*Tc*
Trientalis europaea	P	S/CSR	-	-	G/Hel	L	2	5	VA	Sa	Jun2	*Te*
- *Trifolium arvense*	Aw	R/SR	2	3	Th	L	2	1	?	Sh	Jun4	*Ta*
- *Trifolium campestre*	Aw	R/SR	5	3	Th	L	2	1	?	Sh	Jun4	*Tc*
+ *Trifolium dubium*	Aws	R/SR	5	2	Th	L	2	1	VA	Sh	May6	*Td*
Trifolium fragiferum	P	CR/CSR	4	3	H/Hel	B	1	5	-	Ea	Jul3	*Tf*
- *Trifolium hybridum*	P	CSR	3	4	H	S	3	2	VA	Ea	Jun4	*Th*
+ *Trifolium medium*	P	SC/CSR	3	4	H	L	2	4	VA	Ep	Jun4	*Tm*
Trifolium micranthum	Aw	R/SR	4	2	Th	Lb	1	1	?	Ep	Jun2	*Tm*
+ *Trifolium pratense*	P	CSR	4	2	H	L	2	3	VA	Ea	May5	*Tp*
+ *Trifolium repens*	P	CR/CSR	5	2	H/Ch	B	1	4	VA	Ea	Jun4	*Tr*
Trifolium striatum	Aw	R/SR	3	4	Th	Lb	1	1	?	Sh	May3	*Ts*
- *Triglochin palustre*	P	SR/CSR	5	2	Hel/H	B	3	4	-	Sa	Jun3	*Tp*
+ *Tripleurospermum inodorum*	Aws	R	3	2	Th	S	3	1	VA	Ep	Jul4	*Ti*
+ *Trisetum flavescens*	P	CSR	3	5	H	L	2	3	VA	Ea	May2	*Tf*
Trollius europaeus	P	SC/CSR	4	2	H	S	3	2	VA	Sa	Jun3	*Te*
+ *Tussilago farfara*	P	C/CR	2	1	G	LA	2	5	VA	Sa	Feb3	*Tf*
Typha angustifolia	P	C/SC	1	4	Hyd	LA	5	5	VA	Sa	Jun2	*Ta*
+ *Typha latifolia*	P	C	1	2	Wet	LA	5	5	?	Sa	Jun2	*Tl*
Ulex minor	P	SC	1	5	Ph	-	4	5	?	Ea	Jul3	*Um*
- *Umbilicus rupestris*	P	S/CSR	3	1	H	Sb	1	2	?	Ea	Jun3	*Ur*
+ *Urtica dioica*	P	C	5	3	Ch/H	L	5	4	-	Ep	Jun2	*Ud*
+ *Urtica urens*	As	R/CR	3	3	Th	L	3	1	-	Sa	Jun4	*Uu*
+ *Vaccinium myrtillus*	P	SC	2	5	Chw	L	3	5	ER	Sa	Apr3	*Vm*
Vaccinium oxycoccos	P	S/SR	1	5	Chw/Hel	B	1	5	ER	Ea	Jun3	*Vo*
+ *Vaccinium vitis-idaea*	P	S/SC	1	5	Chw/Ph	L	2	4	ER	Ea	Jun3	*Vv*
Valeriana dioica	P	SR/CSR	5	1	Hel/H	S	3	4	-	Ep	May2	*Vd*
+ *Valeriana officinalis*	P	CSR	4	1	H	LA	3	2	+	Sa	Jun3	*Vo*
- *Valerianella locusta*	Aw	SR	5	1	Th	L	3	1	?	Sh	Apr3	*Vl*
Verbascum nigrum	M/P	C/CSR	2	2	H	LA	4	1	VA	Ea	Jun5	*Vn*
- *Verbascum thapsus*	M	SR/CSR	2	2	H	LA	5	1	VA	Ea	Jun3	*Vt*
Verbena officinalis	P	CSR	2	3	H	S	2	2	VA	Ep	Jul3	*Vo*
- *Veronica agrestis*	Asw	R/SR	5	1	Th	Lb	1	1	+	Ep	Apr12	*Va*
- *Veronica anagallis-aquatica*	P/M	CR	5	1	Hel/H	L	2	2	VA	Ea	Jun3	*Va*
+ *Veronica arvensis*	Aws	SR	4	1	Th	Sb	1	1	VA	Sh	Apr3	*Va*
+ *Veronica beccabunga*	P	CR	3	1	Hel/H	L	2	4	-	Ea	May3	*Vb*
Veronica catenata	P/As	R/CR	5	1	Wet	L	3	2	?	Ea	Jun3	*Vc*
+ *Veronica chamaedrys*	P	S/CSR	4	4	Ch	Lb	1	3	+	Ea	Apr4	*Vc*
- *Veronica filiformis*	P	R/CSR	5	1	Ch	B	1	4	VA	Ea	Apr3	*Vf*
- *Veronica hederifolia*	As	R/SR	3	1	Th	L	2	1	?	Sv	Mar3	*Vh*
+ *Veronica montana*	P	SR/CSR	3	4	Ch	Lb	1	4	?	Ea	Apr4	*Vm*
+ *Veronica officinalis*	P	SR/CSR	2	4	Ch	Lb	1	3	+	Ea	May4	*Vo*
+ *Veronica persica*	Aws	R	5	1	Th	Lb	1	1	VA	Ep	Apr12	*Vp*
- *Veronica polita*	A	R/SR	4	1	Th	Lb	1	1	?	Ep	Apr12	*Vp*
- *Veronica scutellata*	P	R/CSR	5	3	Hel/H	L	3	2	?	Ea	Jun3	*Vs*
- *Veronica serpyllifolia*	P	R/CSR	3	3	Ch	Lb	1	3	VA	Ea	Mar8	*Vs*

Species	Life history	Established strategy	SLA Class	DMC Class	Life form	Canopy structure	Canopy height	Lateral spread	Mycorrhizas	Leaf phenology	Flowering time and duration	(Sp.)
+ Vicia cracca	P	C/CSR	4	3	H	L	4	4	VA	Sa	Jun3	Vc
- Vicia hirsuta	Aw	R/CR	5	3	Th	L	2	1	VA	Sh	Jun3	Vh
- Vicia sativa	Aw	R/CR	3	3	Th	L	5	1	VA	Sh	May5	Vs
+ Vicia sepium	P	C/CSR	5	2	H	L	4	4	VA	Sa	May4	Vs
Vicia tenuifolia	P	C/CSR	3	3	H	L	4	3	?	Sa	May3	Vt
Vicia tetrasperma	Aw	R/CR	3	4	Th	L	3	1	VA	Sh	May4	Vt
Vinca major	P	C/SC	1	3	Chw	L	4	5	VA	Ea	Apr3	Vm
Vinca minor	P	SC	3	2	Chw	L	2	5	VA	Ea	Mar3	Vm
- Viola arvensis	As	R	3	2	Th	S	3	1	+	Ep	Apr7	Va
Viola canina	P	S/CSR	2	3	H	S	2	2	VA	Ea	Apr3	Vc
+ Viola hirta	P	S/CSR	3	3	H	B	2	2	VA	Ea	Apr2	Vh
- Viola odorata	P	CSR	2	4	H	B	2	4	VA	Ea	Feb3	Vo
- Viola palustris	P	SR/CSR	5	2	Hel	B	2	4	VA	Sa	May4	Vp
Viola reichenbachiana	P	S/CSR	-	-	H	S	2	2	VA	Ea	Mar3	Vr
+ Viola riviniana	P	S/CSR	2	4	H	S	2	2	VA	Ea	Apr3	Vr
- Viola tricolor	A/P	R/SR	3	2	Th/H	S	2	1	VA	Ep	Apr6	Vt
Viscum album	P	S/SC	1	3	Ph	L	3	4	-	Ea	Mar3	Va
- Vulpia bromoides	Aw	SR	3	3	Th	S	2	1	?	Sh	May3	Vb
Vulpia myuros	Aw	R/SR	3	3	Th	S	2	1	?	Sh	May3	Vm
Wahlenbergia hederacea	P	SR/CSR	5	1	Hel	B	1	4	?	Ea	Jul2	Wh
× Festulolium loliaceum	P	CSR	1	4	H	S	2	3	?	Ea	Jun3	Fl
Zannichellia palustris	P	CR	5	1	Hyd	U	-1	4	?	Ep	May4	Zp
(b) Woody species more than 1.5 m tall												
- Acer campestre	P	SC	1	5	Ph	L	7	5	VA	Sa	May2	Ac
Acer platanoides	P	C/SC	2	5	Ph	L	8	5	VA	Sa	Apr2	Ap
+ Acer pseudoplatanus	P	C/SC	1	5	Ph	L	8	5	VA	Sa	Apr3	Ap
Aesculus hippocastanum	P	SC	1	5	Ph	L	8	5	VA	Sa	May2	Ah
+ Alnus glutinosa	P	SC	1	5	Ph/Hel	L	8	5	EC	Sa	Feb3	Ag
+ Betula sp.	P	C/SC	-	-	Ph/Chw	L	8	5	EC	Sa	Apr2	Bs
Betula pendula	P	C/SC	1	5	Ph	L	8	5	EC	Sa	Apr2	Bp
Betula pubescens	P	C/SC	-	-	Ph	L	8	5	EC	Sa	Apr2	Bp
Buddleja davidii	P	C	1	-	Ph	L	6	5	-	Sa	Jun5	Bd
Carpinus betulus	P	SC	-	-	Ph	L	8	5	EC	Sa	Apr2	Cb
Castanea sativa	P	SC	1	-	Ph	L	8	5	EC	Sa	Jul1	Cs
Clematis vitalba	P	SC	2	3	Ph	L	8	2	VA	Sa	Jul2	Cv
Cornus sanguinea	P	SC	1	5	Ph	L	6	5	VA	Sa	Jun2	Cs
- Corylus avellana	P	SC	2	5	Ph	L	6	5	EC	Sa	Jan4	Ca
+ Crataegus monogyna	P	SC	1	5	Ph	L	7	5	EC	Sa	May2	Cm
- Cytisus scoparius	P	SC	2	5	Ph	L	5	5	VA	Ep	May2	Cs
- Euonymus europaeus	P	SC	2	5	Ph	L	6	5	VA	Sa	May2	Ee
+ Fagus sylvatica	P	SC	1	5	Ph	L	8	5	EC	Sa	Apr2	Fs
+ Fraxinus excelsior	P	C/SC	1	5	Ph	L	8	5	VA	Sa	Apr2	Fe
- Ilex aquifolium	P	SC	1	5	Ph	L	7	5	VA	Ea	May4	Ia
- Ligustrum vulgare	P	SC	1	5	Ph	L	6	5	VA	Ep	Jun2	Lv
Lupinus arboreus	P	SC	-	-	Ph	L	5	4	?	Ep	Jun4	La
Mahonia aquifolium	P	SC	1	5	Ph	L	5	3	VA	Ea	Feb4	Ma
Malus sylvestris	P	SC	2	5	Ph	L	6	5	VA	Sa	May1	Ms
- Populus tremula	P	SC	-	-	Ph	L	8	5	EC	Sa	Feb2	Pt
Populus × canescens	P	SC	-	-	Ph	L	8	5	EC	Sa	Mar1	Pc
- Prunus avium	P	SC	1	5	Ph	L	8	5	VA	Sa	Apr2	Pa
Prunus padus	P	SC	1	5	Ph	L	7	5	VA/EC	Sa	May1	Pp
- Prunus spinosa	P	SC	1	5	Ph	L	6	5	+	Sa	Mar3	Ps
+ Quercus agg.	P	SC	1	5	Ph	L	8	5	EC	Sa	Apr2	Qa
Quercus cerris	P	SC	1	5	Ph	L	8	5	EC	Sa	May1	Qc
Quercus petraea	P	SC	1	5	Ph	L	8	5	EC	Sa	Apr2	Qp
Quercus robur	P	SC	1	5	Ph	L	8	5	EC	Sa	Apr2	Qr
Rhamnus cathartica	P	SC	2	5	Ph	L	6	5	VA	Sa	May2	Rc
- Rhododendron ponticum	P	SC	1	5	Ph	L	5	5	ER	Ea	May2	Rp
Ribes nigrum	P	SC	-	-	Ph/Hel	L	5	5	VA	Sa	Apr2	Rn
Ribes rubrum	P	SC	1	5	Ph	L	5	5	VA	Sa	Apr2	Rr
- Rosa sp.	P	SC	-	-	Ph	L	5	5	VA	Sa	May3	Rs
Rosa arvensis	P	SC	3	5	Ph	L	5	5	VA	Sa	Jun2	Ra
Rosa canina	P	SC	2	5	Ph	L	5	5	VA	Sa	Jun2	Rc

TABLES OF ATTRIBUTES

Table 7.2

Species	Life history	Established strategy	SLA Class	DMC Class	Life form	Canopy structure	Canopy height	Lateral spread	Mycorrhizas	Leaf phenology	Flowering time and duration	(Sp.)
Rosa mollis	P	SC	-	-	Ph	L	5	5	VA	Sa	Jun2	Rm
Salix alba	P	C/SC	-	-	Ph/Hel	L	8	5	EC	Sa	Apr2	Sa
Salix aurita	P	SC	1	5	Ph	L	5	5	EC	Sa	Apr1	Sa
+ Salix caprea/cinerea agg.	P	C/SC	-	-	Ph/Hel	L	7	5	EC	Sa	Mar2	SC
Salix caprea	P	C/SC	1	5	Ph	L	7	5	EC	Sa	Mar2	SC
Salix cinerea	P	C/SC	1	5	Ph/Hel	L	7	5	EC	Sa	Mar2	SC
- Salix fragilis	P	C/SC	1	5	Ph/Hel	L	8	5	EC	Sa	Apr2	Sf
- Salix purpurea	P	C/SC	1	5	Ph/Hel	L	5	5	EC	Sa	Mar2	Sp
- Salix viminalis	P	C/SC	1	5	Ph/Hel	L	6	5	EC	Sa	Apr2	Sv
+ Sambucus nigra	P	C	2	3	Ph	L	7	5	+	Sa	Jun2	Sn
+ Sorbus aucuparia	P	SC	1	5	Ph	L	7	5	VA	Sa	May2	Sa
Symphoricarpos albus	P	C/SC	2	-	Ph	L	5	5	+	Sa	Jun4	Sa
Taxus baccata	P	SC	1	5	Ph	L	8	5	VA	Ea	Mar2	Tb
+ Ulex europaeus	P	SC	1	5	Ph	-	5	5	VA	Ea	Mar3	Ue
Ulex gallii	P	SC	1	5	Ph	-	5	5	-	Ea	Jul3	Ug
+ Ulmus glabra	P	C/SC	1	5	Ph	L	8	5	VA	Sa	Feb3	Ug
- Ulmus procera	P	C/SC	1	4	Ph	L	8	5	VA	Sa	Feb2	Up
- Viburnum opulus	P	SC	2	-	Ph	L	6	5	VA	Sa	Jun2	Vo
(c) Pteridophytes												
- Asplenium adiantum-nigrum	P	S	2	4	H	B	3	2	-	Ea	Jun5	Aa
+ Asplenium ruta-muraria	P	S	1	4	H	B	1	2	-	Ea	Jun3	Ar
+ Asplenium trichomanes	P	S	1	5	H	B	2	2	-	Ea	Aug3	At
+ Athyrium filix-femina	P	C/SC	1	4	H	LA	5	3	VA	Sa	Jul2	Af
- Blechnum spicant	P	S/CSR	1	4	H	B	4	2	VA	Ea	Jul4	Bs
Ceterach officinarum	P	S	1	5	H	B	1	2	-	Ea	Apr7	Co
+ Cystopteris fragilis	P	SR/CSR	3	3	H	B	2	2	VA	Sa	Jun3	Cf
- Dryopteris affinis	P	SC/CSR	3	4	H	LA	5	2	?	Ep	Jul3	Da
Dryopteris carthusiana	P	SC/CSR	3	5	Hel/H	LA	4	4	VA	Ep	Jul3	Dc
+ Dryopteris dilatata	P	SC/CSR	3	4	H	LA	5	2	VA	Ep	Jun4	Dd
+ Dryopteris filix-mas	P	SC	3	4	H	LA	5	2	VA	Ep	Aug4	Df
+ Equisetum arvense	P	CR	-	-	G/Hel	-	4	5	-	Sa	Apr1	Ea
+ Equisetum fluviatile	P	SC/CSR	-	-	Wet	-	5	5	-	Sa	Jun2	Ef
+ Equisetum palustre	P	CR/CSR	-	-	G/Hel	-	3	5	-	Sa	Jun2	Ep
Equisetum sylvaticum	P	CR/CSR	-	-	G/Hel	-	4	5	-	Sa	Apr2	Es
Equisetum telmateia	P	C/CSR	-	-	G/Hel	-	5	5	VA	Sa	Apr1	Et
Gymnocarpium dryopteris	P	S/CSR	5	2	G	B	3	3	VA	Sa	Jul2	Gd
Huperzia selago	P	S/SR	1	-	Ch	L	3	2	VA	Ea	Jun3	Hs
Lycopodium clavatum	P	S	5	5	Ch	Sb	1	5	VA	Ea	Jun4	Lc
Oreopteris limbosperma	P	SC/CSR	1	4	H	LA	4	2	VA	Sa	Jul2	Ol
Ophioglossum vulgatum	P	SR	3	1	G	B	1	4	VA	Sv/Sa	May4	Ov
Osmunda regalis	P	C/SC	5	3	Hel/H	L	5	4	VA	Sa	Jun3	Or
Phegopteris connectilis	P	S/CSR	2	4	G	B	3	3	VA	Sa	Jun3	Pc
- Phyllitis scolopendrium	P	S/CSR	1	3	H	LA	3	2	+	Ea	Aug3	Ps
- Polypodium vulgare	P	S/CSR	1	4	G/Ch	B	3	3	-	Ea	Aug9	Pv
Polystichum aculeatum	P	SC/CSR	1	5	H	LA	3	2	VA	Ep	Jul2	Pa
Polystichum setiferum	P	SC/CSR	2	4	H	LA	3	2	?	Ep	Jul2	Ps
+ Pteridium aquilinum	P	C	2	4	G	LA	5	5	VA	Sa	Aug3	Pa

Table 7.3 Attributes of the regenerative phase of species (symbols and notes are explained in the preceding text).

	Species	Regenerative strategies	Seed bank	Agency of dispersal	Dispersule and germinule form	Dispersule weight	Dispersule shape	Germination requirements	Family	(Sp.)
	(a) Herbs and woody species up to 1.5 m tall									
+	Achillea millefolium	V, ?S	1	WINDw	Fr	1	2	-	Ast	Am
-	Achillea ptarmica	V, ?S	1	WINDw	Fr	2	2	-	Ast	Ap
	Acorus calamus	V	-	*-	X	X	-	X	Ara	Ac
	Adoxa moschatellina	(V), S	2	*ANIMi	Fr	?	2	?Chill	Ado	Am
-	Aegopodium podagraria	V, S	2	*UNSP	Fr	4	3	Chill	Api	Ap
-	Aethusa cynapium	S, Bs	3	UNSPag	Fr	4	1	Chill	Api	Ac
-	Agrimonia eupatoria	S, V	2	ANIMb	Fr	5	2	Chill	Ros	Ae
	Agrimonia procera	S, V	?2	ANIMb	Fr	5	2	?Chill	Ros	Ap
+	Agrostis canina	V, Bs	3	*ANIMa	Fr	1	3	Dry	Poa	Ac
+	Agrostis capillaris	V, Bs	3	UNSP	Fr	1	2	-	Poa	Ac
-	Agrostis gigantea	V, Bs	3	*UNSP	Fr	1	2	-	Poa	Ag
+	Agrostis stolonifera	V, Bs	3	*UNSP	Fr	1	2	-	Poa	As
+	Agrostis vinealis	V, Bs	3	UNSP	Fr	1	3	-	Poa	Av
-	Aira caryophyllea	S, Bs	3	ANIMa	Fr	1	2	Dry	Poa	Ac
+	Aira praecox	S, ?Bs	?3	ANIMa	Fr	1	3	Dry	Poa	Ap
	Ajuga reptans	V	?2	*ANIMe	Fr	4	2	Unclassified	Lam	Ar
-	Alchemilla vulgaris	V, ?S	2	ANIMa	Fr	2	1	Chill	Ros	Av
+	Alisma plantago-aquatica	V, Bs	3	AN/AQ	Fr	2	1	Scar	Ali	Ap
+	Alliaria petiolata	S	?2	UNSP	Sd	5	3	Chill	Bra	Ap
+	Allium ursinum	(V), S	2	UNSPc	Sd	5	1	Warm+Chill	Lil	Au
-	Allium vineale	Sv	1	*-	Bul	X	-	X;Dry(bulbil)	Lil	Av
+	Alopecurus geniculatus	V, Bs	?4	*UNSP	Fr	2	2	Dry	Poa	Ag
	Alopecurus myosuroides	Bs	3	ANIMa	Fr	4	3	Dry	Poa	Am
+	Alopecurus pratensis	?V, S	1	UNSPag	Fr	3	2	-	Poa	Ap
	Amsinckia micrantha	?Bs	?3	UNSPag	Fr	4	2	-	Bor	Am
-	Anacamptis pyramidalis	W	?	WINDcm	Sd	S	3	Orchid	Orc	Ap
+	Anagallis arvensis	Bs	4	UNSPag	Sd	2	1	Chill	Pri	Aa
	Anagallis tenella	V, Bs	3	*UNSPc	Sd	3	1	Unclassified	Pri	At
	Anchusa arvensis	Bs	3	ANIMe	Fr	5	2	Unclassified	Bor	Aa
+	Anemone nemorosa	V, S	2	*ANIMa	Fr	3	3	?Chill	Ran	An
+	Angelica sylvestris	S	2	AQUAT	Fr	4	1	Chill	Api	As
+	Anisantha sterilis	S	2	ANIMa	Fr	5	3	-	Poa	As
	Antennaria dioica	V, W	?1	*WINDp	Fr	1	2	Unclassified	Ast	Ad
	Anthemis arvensis	Bs	3	UNSP	Fr	3	2	Scar	Ast	Aa
	Anthemis cotula	Bs	3	UNSP	Fr	3	2	Scar	Ast	Ac
+	Anthoxanthum odoratum	S	?1	ANIMa	Fr	2	2	-	Poa	Ao
	Anthriscus caucalis	S, Bs	?3	ANIMb	Fr	4	2	Dry	Api	Ac
+	Anthriscus sylvestris	(V), S	2	UNSP	Fr	5	2	Chill	Api	As
-	Anthyllis vulneraria	?Bs	?	?WIND	F/S	5	1	Scar	Fab	Av
	Antirrhinum majus	Bs	3	WINDcw	Sd	1	1	-	Scr	Am
	Apera interrupta	S, Bs	3	ANIMa	Fr	1	3	Dry	Poa	Ai
+	Aphanes arvensis	S, Bs	3	UNSPag	Fr	1	2	Dry	Ros	Aa
	Aphanes australis	S, Bs	3	UNSPag	Fr	1	2	Dry	Ros	Aa
	Apium inundatum	(V), ?S	?	*AQUAT	Fr	3	3	Unclassified	Api	Ai
+	Apium nodiflorum	V, ?S	?	*AQUAT	Fr	2	2	-	Api	An
	Aquilegia vulgaris	S	?2	UNSP	Sd	3	1	Chill	Ran	Av
+	Arabidopsis thaliana	S, Bs	3	WINDw	Sd	1	1	Dry	Bra	At
	Arabis caucasica	V, ?Bs	?3	UNSP	Sd	2	1	Not studied	Bra	Ac
+	Arabis hirsuta	S, Bs	3	UNSP	Sd	1	2	-	Bra	Ah
-	Arctium lappa	S, Bs	3	ANIMb	Fr	6	2	-	Ast	Al
+	Arctium minus	S, Bs	3	ANIMb	Fr	6	2	-/Scar	Ast	Am
	Arctostaphylos uva-ursi	V, ?S	?1	*ANIMb	Fr	5	2	Not studied	Eri	Au
+	Arenaria serpyllifolia	S, Bs	3	WINDc	Sd	1	1	-	Car	As
-	Armoracia rusticana	V	-	*-	X	X	-	X	Bra	Ar
+	Arrhenatherum elatius	(V), S	1	ANIMa	Fr	5	2	-	Poa	Ae
	Artemisia absinthium	S, Bs	3	UNSP	Fr	1	2	Dry	Ast	Aa
	Artemisia verlotiorum	V	-	*-	X	X	2	X	Ast	Av
+	Artemisia vulgaris	V, S, Bs	3	UNSP	Fr	1	3	Dry	Ast	Av
+	Arum maculatum	V, ?S	2	ANIMi	F/S	6	1	Chill	Ara	Am
	Aster lanceolatus	V, W	1	WINDp	Fr	2	3	Not studied	Ast	Al
	Aster novi-belgii	V, W	1	WINDp	Fr	2	3	Dry	Ast	An
	Atriplex littoralis	Bs	3	UNSP	US	3	1	Chill	Che	Al
+	Atriplex patula	Bs	?4	UNSP	f/S	4	1	Chill	Che	Ap
+	Atriplex prostrata	Bs	4	UNSP	f/S	3	1	Chill	Che	Ap

Table 7.3

	Species	Regenerative strategies	Seed bank	Agency of dispersal	Dispersule and germinule form	Dispersule weight	Dispersule shape	Germination requirements	Family	(Sp.)
	Avena fatua	S, Bs	3	ANIMa	Fr	6	3	Dry/Scar	Poa	*Af*
	Ballota nigra	V, Bs	3	UNSP	Fr	3	2	Unclassified	Lam	*Bn*
-	*Barbarea vulgaris*	Bs	3	WINDw	Sd	3	1	Scar	Bra	*Bv*
	Bassia scoparia	S, Bs	3	UNSP	f/S	4	2	-	Che	*Bs*
+	*Bellis perennis*	V	2	UNSP	Fr	1	2	-	Ast	*Bp*
-	*Berula erecta*	(V), ?S	?1	*AQUAT	Fr	2	1	-	Api	*Be*
	Bidens cernua	S, Bs	3	ANIMa	Fr	4	3	Scar	Ast	*Bc*
	Bidens tripartita	S, Bs	3	ANIMa	Fr	5	3	Scar	Ast	*Bt*
	Blackstonia perfoliata	S, Bs	3	WINDc	Sd	1	2	-	Gen	*Bp*
+	*Brachypodium pinnatum*	V, ?S	?2	*UNSP	Fr	5	3	-	Poa	*Bp*
+	*Brachypodium sylvaticum*	V, S	?2	ANIMa	Fr	3	3	Dry	Poa	*Bs*
-	*Brassica rapa*	Bs	3	UNSP	Sd	4	1	-	Bra	*Br*
+	*Briza media*	V, S	1	UNSP	Fr	2	2	-	Poa	*Bm*
+	*Bromopsis erecta*	V, S	1	ANIMa	Fr	5	2	-	Poa	*Be*
	Bromopsis inermis	V, ?S	1	*UNSP	Fr	5	3	-	Poa	*Bi*
+	*Bromopsis ramosa*	S	2	ANIMa	Fr	5	3	Chill	Poa	*Br*
+	*Bromus hordeaceus*	S	1	ANIMa	Fr	5	3	-	Poa	*Bh*
	Bromus lepidus	S	1	ANIMa	Fr	5	3	-	Poa	*Bl*
	Bryonia dioica	V, S	1	ANIMi	Sd	6	2	Unclassified	Cuc	*Bd*
	Butomus umbellatus	(V)	1	*AQIWI	Sd	1	3	Unclassified	But	*Bu*
	Calamagrostis canescens	V, W	?2	*WINDp	Fr	1	3	Unclassified	Poa	*Cc*
	Calamagrostis epigejos	V, W	?2	*WINDp	Fr	1	3	Unclassified	Poa	*Ce*
	Callitriche hamulata	V, Bs	3	*UNSP	f/S	1	3	Not studied	Cal	*Ch*
	Callitriche obtusangula	V, Bs	3	*UNSP	f/S	1	3	Not studied	Cal	*Co*
	Callitriche platycarpa	V, Bs	3	*UNSP	f/S	1	3	Not studied	Cal	*Cp*
+	*Callitriche stagnalis*	V, Bs	3	*UNSP	f/S	1	3	-	Cal	*Cs*
+	*Calluna vulgaris*	Bs	4	WINDcm	Sd	1	1	-	Eri	*Cv*
+	*Caltha palustris*	V, ?S	2	*AQUAT	Sd	3	2	Chill	Ran	*Cp*
	Calystegia pulchra	(V), ?Bs	3	*UNSP	Sd	6	1	Scar	Con	*Cp*
+	*Calystegia sepium*	(V), Bs	4	*UNSP	Sd	6	1	Scar	Con	*Cs*
	Calystegia silvatica	(V), ?Bs	3	*UNSP	Sd	6	1	Scar	Con	*Cs*
	Campanula latifolia	?Bs	?3	WINDc	Sd	2	2	Unclassified	Cam	*Cl*
	Campanula rapunculoides	V, ?Bs	?3	*WINDc	Sd	2	2	Not studied	Cam	*Cr*
+	*Campanula rotundifolia*	V, Bs	3	WINDc	Sd	1	2	-	Cam	*Cr*
	Campanula trachelium	Bs	3	WINDc	Sd	1	2	Unclassified	Cam	*Ct*
+	*Capsella bursa-pastoris*	Bs	4	WINDw	Sd	1	2	Chill,Scar	Bra	*Cb*
+	*Cardamine amara*	V, Bs	3	*UNSP	Sd	2	1	Dry	Bra	*Ca*
+	*Cardamine flexuosa*	S, Bs	3	WINDw	Sd	1	2	-	Bra	*Cf*
+	*Cardamine hirsuta*	S, Bs	3	WINDw	Sd	1	1	Dry	Bra	*Ch*
+	*Cardamine pratensis*	(V), Bs	3	*WINDw	Sd	3	2	-	Bra	*Cp*
-	*Carduus crispus*	W, ?Bs	?3	WINDp	Fr	5	2	-	Ast	*Cc*
	Carduus nutans	W, ?Bs	?3	WINDp	Fr	5	2	Unclassified	Ast	*Cn*
	Carex acuta	V, Bs	3	*AQUAT	Fr	2	1	Not studied	Cyp	*Ca*
+	*Carex acutiformis*	V, ?Bs	?3	*AQUAT	Fr	4	2	?Dry	Cyp	*Ca*
	Carex arenaria	V, ?Bs	?3	*UNSP	Fr	4	2	Dry	Cyp	*Ca*
-	*Carex binervis*	V, ?Bs	?3	*UNSP	Fr	4	2	?Chill	Cyp	*Cb*
+	*Carex caryophyllea*	V, ?Bs	?3	*UNSP	Fr	4	2	Chill	Cyp	*Cc*
	Carex curta	V, ?Bs	?3	UNSP	Fr	2	2	Unclassified	Cyp	*Cc*
	Carex dioica	V, ?Bs	?3	UNSP	Fr	2	2	Not studied	Cyp	*Cd*
	Carex distans	V, ?Bs	?3	UNSP	Fr	4	2	Not studied	Cyp	*Cd*
	Carex disticha	V, ?Bs	?3	*UNSP	Fr	4	2	Not studied	Cyp	*Cd*
-	*Carex echinata*	?Bs	?3	AQUAT	Fr	3	1	Dry/Chill	Cyp	*Ce*
	Carex elata	V, ?Bs	?3	AQUAT	Fr	3	1	Dry	Cyp	*Ce*
+	*Carex flacca*	V, Bs	3	*UNSP	Fr	2	1	Chill	Cyp	*Cf*
-	*Carex hirta*	V, ?Bs	?3	*UNSP	Fr	5	2	?Chill	Cyp	*Ch*
-	*Carex hostiana*	?Bs	?3	UNSP	Fr	4	1	?Chill	Cyp	*Ch*
	Carex laevigata	V, ?Bs	?3	AQUAT	Fr	4	1	Unclassified	Cyp	*Cl*
	Carex muricata	V, ?Bs	?3	UNSP	Fr	4	1	Dry	Cyp	*Cm*
+	*Carex nigra*	V, Bs	3	*UNSP	Fr	3	2	Chill	Cyp	*Cn*
-	*Carex otrubae*	V, Bs	3	AQUAT	Fr	4	1	Dry	Cyp	*Co*
	Carex ovalis	Bs	3	UNSP	Fr	2	1	Dry	Cyp	*Co*
	Carex pallescens	Bs	3	UNSP	Fr	4	1	?Chill	Cyp	*Cp*
+	*Carex panicea*	V, Bs	3	*AQUAT	Fr	4	1	Chill	Cyp	*Cp*
	Carex paniculata	V, ?Bs	3	AQUAT	Fr	3	1	?Chill	Cyp	*Cp*
-	*Carex pendula*	V, ?Bs	3	UNSP	Fr	3	2	?Chill	Cyp	*Cp*

Species	Regenerative strategies	Seed bank	Agency of dispersal	Dispersule and germinule form	Dispersule weight	Dispersule shape	Germination requirements	Family	(Sp.)
+ Carex pilulifera	V, Bs	4	ANIMe	Fr	4	1	?Chill	Cyp	Cp
Carex pseudocyperus	V, ?Bs	?3	AQUAT	Fr	4	1	?Chill	Cyp	Cp
Carex pulicaris	V, ?Bs	?3	UNSP	Fr	4	2	?Chill	Cyp	Cp
- Carex remota	Bs	3	AQUAT	Fr	2	1	Chill	Cyp	Cr
Carex riparia	V, ?Bs	?3	*AQUAT	Fr	4	2	?Chill	Cyp	Cr
Carex rostrata	V, ?Bs	?3	*AQUAT	Fr	3	1	?Chill	Cyp	Cr
Carex spicata	V, ?Bs	?3	UNSP	Fr	5	1	?Chill	Cyp	Cs
- Carex sylvatica	Bs	3	UNSP	Fr	4	2	?Chill	Cyp	Cs
Carex vesicaria	V, ?Bs	?3	*AQUAT	Fr	4	1	?Chill	Cyp	Cv
Carex viridula subsp. brachyrrhyncha	V, ?Bs	?3	AQUAT	Fr	3	1	?Chill	Cyp	Cv
- Carex viridula subsp. oedocarpa	V, ?Bs	?3	AQUAT	Fr	3	1	?Chill	Cyp	Cv
+ Carlina vulgaris	S	2	WINDp	Fr	4	3	-	Ast	Cv
Catabrosa aquatica	V, ?Bs	?	*AQUAT	Fr	2	2	-	Poa	Ca
+ Catapodium rigidum	S	?1	UNSP	Fr	1	3	-	Poa	Cr
Centaurea montana	V, ?S	?	UNSP	Fr	5	2	Not studied	Ast	Cm
+ Centaurea nigra	V, S	2	UNSP	Fr	5	2	Dry	Ast	Cn
+ Centaurea scabiosa	S	2	WINDp	Fr	5	2	-	Ast	Cs
+ Centaurium erythraea	S, Bs	3	WINDc	Sd	1	1	-	Gen	Ce
- Centranthus ruber	W	?	WINDp	Fr	4	3	-	Val	Cr
Cerastium arvense	V, Bs	3	*UNSPc	Sd	1	1	-	Car	Ca
Cerastium diffusum	S, ?Bs	?3	UNSPc	Sd	1	1	Dry	Car	Cd
+ Cerastium fontanum	(V), Bs	3	UNSPc	Sd	1	1	-	Car	Cf
- Cerastium glomeratum	S, Bs	3	WINDc	Sd	1	1	Dry	Car	Cg
- Cerastium semidecandrum	S, Bs	3	WINDc	Sd	1	1	Dry	Car	Cs
- Cerastium tomentosum	V, ?Bs	?3	*WINDc	Sd	2	1	-	Car	Ct
- Ceratocapnos claviculata	S	?	ANIMe	Sd	4	1	-	Fum	Cc
Ceratochloa carinata	?	?1	UNSP	Fr	6	3	Not studied	Poa	Cc
Ceratophyllum demersum	V, ?S	?1	*AQ/AN	Fr	?	1	Not studied	Cer	Cd
+ Chaenorhinum minus	S, ?Bs	?3	WINDcm	Sd	1	2	Chill	Scr	Cm
- Chaerophyllum temulum	S	?2	UNSP	Fr	5	3	Chill	Api	Ct
+ Chamerion angustifolium	V, W	1	WINDp	Sd	1	3	Dry	Ona	Ca
- Chelidonium majus	?S, Bs	3	ANIMe	Sd	3	1	Chill	Pap	Cm
+ Chenopodium album	Bs	4	UNSP	f/S	4	1	-/Chill,Dry	Che	Ca
- Chenopodium bonus-henricus	Bs	3	UNSP	f/S	4	1	-	Che	Cb
Chenopodium polyspermum	Bs	3	UNSP	f/S	2	1	Dry	Che	Cp
+ Chenopodium rubrum	Bs	4	UNSP	f/S	1	1	Fluct	Che	Cr
- Chrysanthemum segetum	Bs	3	UNSPag	Fr	4	1	Scar	Ast	Cs
Chrysosplenium alternifolium	(V) ?Bs	?	*UNSP	Sd	1	2	Unclassified	Sax	Ca
+ Chrysosplenium oppositifolium	(V), ?Bs	?	*UNSP	Sd	1	1	-	Sax	Co
Cicerbita macrophylla	V	-	* -	X	X	-	X	Ast	Cm
Cichorium intybus	?	?2	UNSP	Fr	3	2	Dry	Ast	Ci
+ Circaea lutetiana	(V), S	2	*ANIMb	F/S	3	3	?Chill	Ona	Cl
Cirsium acaule	V, W	1	*WINDp	Fr	5	2	Unclassified	Ast	Ca
+ Cirsium arvense	V, W, Bs	3	*WINDp	Fr	4	2	-/Unclassified	Ast	Ca
Cirsium dissectum	V, W	1	*WINDp	Fr	4	2	Scar	Ast	Cd
Cirsium heterophyllum	V, W	1	*WINDp	Fr	5	2	Unclassified	Ast	Ch
+ Cirsium palustre	W, Bs	3	WINDp	Fr	4	2	-	Ast	Cp
+ Cirsium vulgare	W, Bs	?3	WINDp	Fr	5	2	Dry	Ast	Cv
Cladium mariscus	V, ?	?	*UNSP	Fr	5	2	Unclassified	Cyp	Cm
Claytonia perfoliata	S, ?Bs	?3	UNSP	Sd	3	1	Scar	Por	Cp
Claytonia sibirica	S, ?Bs	?3	UNSP	Sd	3	1	Unclassified	Por	Cs
Clinopodium acinos	S, ?Bs	3	UNSP	Fr	2	2	Dry	Lam	Ca
- Clinopodium vulgare	V, ?Bs	3	UNSP	Fr	2	1	Dry	Lam	Cv
Cochlearia danica	?Bs	3	UNSP	Sd	1	2	-	Bra	Cd
Coeloglossum viride	W	?	WINDcm	Sd	S	3	Orchid	Orc	Cv
Conium maculatum	S	2	AQUAT	Fr	5	1	?Dry	Api	Cm
- Conopodium majus	S	2	UNSP	Fr	5	3	Chill	Api	Cm
+ Convolvulus arvensis	(V), Bs	4	*UNSP	Sd	6	2	Scar	Con	Ca
Conyza canadensis	W, Bs	3	WINDp	Fr	1	2	-	Ast	Cc
Coronopus didymus	Bs	3	UNSP	Sd	2	1	(Chill)	Bra	Cd
- Coronopus squamatus	Bs	4	UNSP	F/S	3	2	Unclassified	Bra	Cs
Cotoneaster horizontalis	S	?2	ANIMi	Fr	5	1	Unclassified	Ros	Ch
Crassula helmsii	S, ?Bs	?3	*AQUAT	Sd	1	2	Not studied	Cra	Ch
Crepis biennis	W, S	?1	WINDp	Fr	4	3	-	Ast	Cb
+ Crepis capillaris	W, ?Bs	?1	WINDp	Fr	2	3	-	Ast	Cc

TABLES OF ATTRIBUTES

Table 7.3

	Species	Regenerative strategies	Seed bank	Agency of dispersal	Dispersule and germinule form	Dispersule weight	Dispersule shape	Germination requirements	Family	(Sp.)
	Crepis paludosa	V, W	?2	WINDp	Fr	3	3	Unclassified	Ast	*Cp*
	Crepis vesicaria	W, S	?1	WINDp	Fr	2	3	-	Ast	*Cv*
+	*Cruciata laevipes*	(V), S	1	*UNSP	Fr	5	1	Chill	Rub	*Cl*
-	*Cymbalaria muralis*	?Bs	?3	*UNSPc	Sd	1	2	?Chill	Scr	*Cm*
	Cynoglossum officinale	S	?2	ANIMb	Fr	6	1	Unclassified	Bor	*Co*
+	*Cynosurus cristatus*	S	1	UNSPag	Fr	3	3	-	Poa	*Cc*
+	*Dactylis glomerata*	S	2	UNSPag	Fr	3	2	Dry	Poa	*Dg*
-	*Dactylorhiza fuchsii*	W	?	WINDcm	Sd	S	3	Orchid	Orc	*Df*
-	*Dactylorhiza incarnata*	W	?	WINDcm	Sd	S	3	Orchid	Orc	*Di*
-	*Dactylorhiza maculata*	W	?	WINDcm	Sd	S	3	Orchid	Orc	*Dm*
	Dactylorhiza praetermissa	W	?	WINDcm	Sd	S	3	Orchid	Orc	*Dp*
	Dactylorhiza purpurella	W	?	WINDcm	Sd	S	3	Orchid	Orc	*Dp*
+	*Danthonia decumbens*	Bs	4	ANIMe	Fr	3	2	Dry	Poa	*Dd*
	Daphne laureola	S	2	ANIMi	Fr	6	2	Unclassified	Thy	*Dl*
+	*Daucus carota*	?S, Bs	3	ANIMb	Fr	3	1	Chill	Api	*Dc*
+	*Deschampsia cespitosa*	S, Bs	3	ANIMa	Fr	2	3	-	Poa	*Dc*
+	*Deschampsia flexuosa*	V, S	1	*ANIMa	Fr	2	2	-	Poa	*Df*
+	*Digitalis purpurea*	S, Bs	3	WINDc	Sd	1	2	-	Scr	*Dp*
	Diplotaxis muralis	Bs	3	UNSP	Sd	2	1	Scar	Bra	*Dm*
	Diplotaxis tenuifolia	Bs	3	UNSP	Sd	2	1	Scar	Bra	*Dt*
	Dipsacus fullonum	?Bs	?2	UNSP	Fr	5	3	Dry	Dip	*Df*
	Doronicum pardalianches	V, W	?	*WlNDp	Fr	2	1	-	Ast	*Dp*
	Drosera rotundifolia	(V), ?Bs	3	WINDcmw	Sd	1	3	Unclassified	Dro	*Dr*
	Echium vulgare	?Bs	3	UNSP	Fr	5	2	Unclassified	Bor	*Ev*
+	*Eleocharis palustris*	V, ?Bs	3	*ANIMb	Fr	3	2	Chill	Cyp	*Ep*
	Eleocharis quinqueflora	V	?	*ANIMb	Fr	3	2	Unclassified	Cyp	*Eq*
	Eleogiton fluitans	V	?	*ANIMb	Fr	1	2	Unclassified	Cyp	*Ef*
+	*Elodea canadensis*	V	-	* -	X	X	-	X	Hyd	*Ec*
	Elodea nuttallii	V	-	* -	X	X	-	X	Hyd	*En*
+	*Elymus caninus*	S	?2	ANIMa	Fr	5	3	-	Poa	*Ec*
+	*Elytrigia repens*	(V)	1	*UNSP	Fr	5	3	Fluct	Poa	*Er*
+	*Empetrum nigrum* subsp. *nigrum*	V, S	1	ANIMi	F/S	3	1	Chill	Emp	*En*
	Epilobium brunnescens	W, Bs	?3	*WINDp	Sd	1	3	-	Ona	*Eb*
+	*Epilobium ciliatum*	W, Bs	3	WINDp	Sd	1	3	-	Ona	*Ec*
+	*Epilobium hirsutum*	V, W, Bs	3	*WINDp	Sd	1	2	-	Ona	*Eh*
+	*Epilobium montanum*	W, Bs	3	WINDp	Sd	1	2	-	Ona	*Em*
+	*Epilobium obscurum*	W, Bs	3	WINDp	Sd	1	2	-	Ona	*Eo*
+	*Epilobium palustre*	W, Bs	3	WINDp	Sd	1	3	-	Ona	*Ep*
+	*Epilobium parviflorum*	W, Bs	3	WINDp	Sd	1	2	-	Ona	*Ep*
	Epilobium roseum	W, Bs	3	WINDp	Sd	3	3	-	Ona	*Er*
	Epilobium tetragonum	W, Bs	3	WINDp	Sd	1	3	-	Ona	*Et*
-	*Epipactis helleborine*	W	?	WINDcm	Sd	S	3	Orchid	Orc	*Eh*
	Epipactis palustris	V, W	?	*WINDcm	Sd	S	3	Orchid	Orc	*Ep*
+	*Erica cinerea*	Bs	4	WINDcm	Sd	1	2	Chill,Heat,Scar	Eri	*Ec*
-	*Erica tetralix*	Bs	3	WINDcm	Sd	1	2	?Dry	Eri	*Et*
+	*Erigeron acer*	W	1	WINDp	Fr	1	3	-	Ast	*Ea*
+	*Eriophorum angustifolium*	V, W	?2	*WINDp	Fr	2	3	-	Cyp	*Ea*
+	*Eriophorum vaginatum*	W	?2	WINDp	Fr	4	2	-	Cyp	*Ev*
-	*Erodium cicutarium*	S, ?Bs	?3	ANIMa	Sd	4	3	Scar	Ger	*Ec*
	Erophila verna	Bs	3	WINDw	Sd	1	1	Dry	Bra	*Ev*
	Erysimum cheiranthoides	Bs	3	UNSP	Sd	2	1	Scar	Bra	*Ec*
	Erysimum cheiri	?	?	WINDw	Sd	4	1	-	Bra	*Ec*
-	*Eupatorium cannabinum*	W, Bs	3	WINDp	Fr	2	3	Unclassified	Ast	*Ec*
	Euphorbia cyparissias	V, Bs	3	*ANIMe	Sd	?	1	Not studied	Eup	*Ec*
	Euphorbia esula	V, Bs	3	*ANIMe	Sd	5	1	Not studied	Eup	*Ee*
	Euphorbia exigua	Bs	3	ANIMe	Sd	3	2	(Dry)	Eup	*Ee*
-	*Euphorbia helioscopia*	Bs	3	ANIMe	Sd	5	1	(Dry)	Eup	*Eh*
-	*Euphorbia peplus*	Bs	3	ANIMe	Sd	2	2	Dry	Eup	*Ep*
+	*Euphrasia officinalis*	S	?2	WINDc	Sd	1	2	Chill	Scr	*Eo*
+	*Fallopia convolvulus*	Bs	4	UNSPag	Fr	4	2	Chill	Pgo	*Fc*
+	*Fallopia japonica*	V	-	*UNSP	Fr	3	2	(Dry)	Pgo	*Fj*
	Fallopia sachalinensis	V	-	*UNSP	Fr	3	2	Not studied	Pgo	*Fs*
-	*Festuca arundinacea*	V, S	1	*ANIMa	Fr	4	3	-	Poa	*Fa*
+	*Festuca gigantea*	S	2	ANIMa	Fr	5	3	Chill	Poa	*Fg*
+	*Festuca ovina*	V, S	1	ANIMa	Fr	2	3	-	Poa	*Fo*

Species	Regenerative strategies	Seed bank	Agency of dispersal	Dispersule and germinule form	Dispersule weight	Dispersule shape	Germination requirements	Family	(Sp.)
+ Festuca pratensis	V, S	1	ANIMa	Fr	4	3	-	Poa	Fp
+ Festuca rubra	V, S	1	*ANIMa	Fr	3	3	-	Poa	Fr
Filago minima	Bs	3	WINDp	Fr	1	3	-	Ast	Fm
Filago vulgaris	Bs	3	WINDp	Fr	1	2	Dry	Ast	Fv
+ Filipendula ulmaria	V	?4	*AQUAT	Fr	3	2	-	Ros	Fu
Filipendula vulgaris	V, S	?2	UNSP	Fr	4	2	Dry	Ros	Fv
Foeniculum vulgare	V, S	?1	UNSP	Fr	4	2	Unclassified	Api	Fv
Fragaria × ananassa	V, ?Bs	?3	*ANIMi	Fr	2	1	Unclassified	Ros	Fa
+ Fragaria vesca	V, Bs	3	*ANIMi	Fr	2	1	-	Ros	Fv
- Fumaria muralis	?Bs	3	UNSPag	Sd	5	1	?Chill	Fum	Fm
- Fumaria officinalis	Bs	3	ANIMe	Sd	5	2	?Chill	Fum	Fo
Galanthus nivalis	V, S	2	UNSPc	Sd	5	3	Not studied	Lil	Gn
Galega officinalis	V, ?Bs	3	UNSP	Sd	3	1	Scar	Fab	Go
+ Galeopsis tetrahit	Bs	3	UNSPag	Fr	5	1	Chill	Lam	Gt
Galinsoga parviflora	Bs	3	?ANIMa	Fr	1	2	-	Ast	Gp
Galinsoga quadriradiata	?Bs	3	?ANIMa	Fr	1	2	-	Ast	Gq
+ Galium aparine	S	2	ANIMb	Fr	5	1	Chill	Rub	Ga
Galium mollugo	V, ?Bs	?2	UNSP	Fr	3	1	Dry	Rub	Gm
- Galium odoratum	V, S	2	*ANIMb	Fr	5	1	?Chill	Rub	Go
+ Galium palustre	V, Bs	4	AQUAT	Fr	3	1	-	Rub	Gp
+ Galium saxatile	V, Bs	3	UNSP	Fr	3	1	-	Rub	Gs
+ Galium sterneri	V, Bs	3	UNSP	Fr	2	1	-	Rub	Gs
- Galium uliginosum	V, ?Bs	3	UNSP	Fr	2	1	Unclassified	Rub	Gu
+ Galium verum	V	?1	*UNSP	Fr	2	1	-	Rub	Gv
Genista anglica	V, ?Bs	3	ANIMe	Sd	5	1	Scar	Fab	Ga
Genista tinctoria	V, ?Bs	3	ANIMe	Sd	5	1	Scar	Fab	Gt
- Gentianella amarella	?	?	WINDc	Sd	1	1	Unclassified	Gen	Ga
- Geranium columbinum	S, ?Bs	3	ANIMa	f/S	5	1	Scar	Ger	Gc
- Geranium dissectum	S, ?Bs	3	ANIMa	f/S	5	1	Scar	Ger	Gd
- Geranium endressii	V, ?Bs	3	*ANIMa	f/S	3	2	Scar	Ger	Ge
- Geranium lucidum	S, ?Bs	?3	ANIMa	f/S	1	2	Scar	Ger	Gl
+ Geranium molle	S, ?Bs	3	ANIMa	f/S	4	1	Scar	Ger	Gm
Geranium pratense	V, ?Bs	3	ANIMa	f/S	5	2	Scar	Ger	Gp
Geranium pusillum	?Bs	3	ANIMa	f/S	3	2	Scar	Ger	Gp
- Geranium pyrenaicum	?Bs	3	ANIMa	f/S	4	2	Scar	Ger	Gp
+ Geranium robertianum	?Bs	3	ANIMa	f/S	4	1	Scar	Ger	Gr
Geranium × magnificum	V	-	*-	X	X	-	X	Ger	Gm
- Geum rivale	S	2	ANIMa	Fr	4	3	-	Ros	Gr
+ Geum urbanum	S	2	ANIMa	Fr	3	3	-	Ros	Gu
+ Glechoma hederacea	V, Bs	3	*UNSP	Fr	3	2	Dry	Lam	Gh
- Glyceria declinata	V, Bs	3	*UNSP	Fr	4	2	Dry	Poa	Gd
+ Glyceria fluitans	V, Bs	3	*UNSP	Fr	4	2	Dry	Poa	Gf
+ Glyceria maxima	V, Bs	3	*AQUAT	Fr	3	2	Dry	Poa	Gm
- Glyceria notata	V, Bs	3	*UNSP	Fr	3	2	-	Poa	Gn
Glyceria × pedicellata	V	-	*-	X	X	-	X	Poa	Gp
Gnaphalium sylvaticum	V, Bs	3	WINDp	Fr	1	3	-	Ast	Gs
- Gnaphalium uliginosum	Bs	3	WINDp	Fr	1	2	-	Ast	Gu
- Groenlandia densa	V, ?	?	*AQ/AN	Fr	4	1	Scar	Pot	Gd
- Gymnadenia conopsea	W	?	WINDcm	Sd	S	3	Orchid	Orc	Gc
+ Hedera helix	V, S	1	*ANIMi	F/S	6	2	-	Arl	Hh
+ Helianthemum nummularium	?Bs	3	ANIMm	Sd	4	2	Scar	Cis	Hn
+ Helictotrichon pratense	S	1	ANIMa	Fr	5	3	-	Poa	Hp
+ Helictotrichon pubescens	(V), S	1	ANIMa	Fr	4	3	-	Poa	Hp
Heracleum mantegazzianum	S	2	WINDw	Fr	6	1	Chill	Api	Hm
+ Heracleum sphondylium	S	2	WINDw	Fr	5	1	Chill	Api	Hs
- Hesperis matronalis	?	?	UNSP	Sd	4	3	(-)	Bra	Hm
Hieracium sp.	V, W	?1	WINDp	Fr	2	3	-	Ast	Hs
- Hippuris vulgaris	(V), ?	?	*AQUAT	Fr	3	2	-/Scar	Hur	Hv
Hirschfeldia incana	?Bs	3	UNSP	Sd	2	1	Dry	Bra	Hi
+ Holcus lanatus	V, S, Bs	3	UNSP	Fr	2	2	-	Poa	Hl
+ Holcus mollis	V	?	*ANIMa	Fr	2	3	Unclassified	Poa	Hm
Hordeum jubatum	?S, ?Bs	3	ANIMa	Fr	3	3	Dry	Poa	Hj
Hordeum murinum	S	?1	ANIMa	Fr	5	3	-	Poa	Hm
Hordeum secalinum	V, S	?1	ANIMa	Fr	5	3	Dry	Poa	Hs
Hottonia palustris	V, ?	?	*UNSPc	Sd	1	2	Unclassified	Pri	Hp

Table 7.3

	Species	Regenerative strategies	Seed bank	Agency of dispersal	Dispersule and germinule form	Dispersule weight	Dispersule shape	Germination requirements	Family	(Sp.)
−	Humulus lupulus	V, ?	?	WINDw	Fr	5	1	Unclassified	Can	Hl
	Hyacinthoides hispanica	V, S	2	UNSPc	Sd	5	1	Warm+Chill	Lil	Hh
+	Hyacinthoides non-scripta	V, S	2	UNSPc	Sd	5	1	Warm+Chill	Lil	Hn
+	Hydrocotyle vulgaris	V, ?Bs	?3	*AQUAT	Fr	2	2	Unclassified	Api	Hv
−	Hypericum androsaemum	?Bs	3	ANIMi	F/S	1	2	Unclassified	Clu	Ha
	Hypericum elodes	V, ?Bs	3	*WINDc	Sd	1	2	Not studied	Clu	He
+	Hypericum hirsutum	V, Bs	4	WINDc	Sd	1	3	Dry,Wash	Clu	Hh
−	Hypericum humifusum	Bs	4	UNSPc	Sd	1	2	-	Clu	Hh
	Hypericum maculatum	Bs	4	WINDc	Sd	1	2	Wash	Clu	Hm
+	Hypericum perforatum	V, Bs	4	*WINDc	Sd	1	2	Wash	Clu	Hp
	Hypericum pulchrum	Bs	3	WINDc	Sd	1	2	Dry	Clu	Hp
	Hypericum tetrapterum	Bs	3	WINDc	Sd	1	2	Dry	Clu	Ht
+	Hypochaeris radicata	V, W	1	WINDp	Fr	3	3	-	Ast	Hr
+	Impatiens glandulifera	S	2	AQUAT	Sd	5	1	Chill	Bal	Ig
	Impatiens parviflora	S	2	UNSP	Sd	5	1	Chill	Bal	Ip
+	Inula conyzae	W, ?Bs	2	WINDp	Fr	2	3	-	Ast	Ic
	Inula helenium	W, ?Bs	?	WINDp	Fr	5	3	Unclassified	Ast	Ih
	Iris foetidissima	W, ?Bs	?3	UNSP	Sd	6	1	Unclassified	Iri	If
	Iris germanica	W, ?Bs	?3	UNSP	Sd	6	2	Unclassified	Iri	Ig
−	Iris pseudacorus	V, S	?2	*AQUAT	Sd	6	2	Scar	Iri	Ip
−	Isolepis setacea	Bs	3	UNSP	Fr	1	1	-	Cyp	Is
	Jasione montana	?Bs	3	UNSP	Sd	1	2	Unclassified	Cam	Jm
−	Juncus acutiflorus	V, Bs	4	*ANIMm	Sd	1	2	-	Jce	Ja
+	Juncus articulatus	V, Bs	4	*ANIMm	Sd	1	2	-	Jce	Ja
	Juncus bufonius	Bs	3	ANIMm	Sd	1	1	Dry	Jce	Jb
+	Juncus bulbosus	V, Bs	3	*ANIMm	Sd	1	3	-	Jce	Jb
−	Juncus conglomeratus	V, Bs	3	*ANlMm	Sd	1	2	Dry	Jce	Jc
+	Juncus effusus	V, Bs	4	*ANIMm	Sd	1	2	-	Jce	Je
	Juncus inflexus	V, Bs	4	*ANIMm	Sd	1	2	-	Jce	Ji
+	Juncus squarrosus	V, Bs	3	*ANIMm	Sd	1	2	-	Jce	Js
	Juncus subnodulosus	V, Bs	3	*ANIMm	Sd	1	2	-	Jce	Js
	Juncus tenuis	Bs	3	ANIMm	Sd	1	2	Not studied	Jce	Jt
−	Knautia arvensis	S	?1	ANIMa	Fr	5	2	Unclassified	Dip	Ka
+	Koeleria macrantha	S	1	UNSP	Fr	2	3	-	Poa	Km
	Lactuca virosa	W, Bs	3	WINDp	Fr	3	3	-	Ast	Lv
+	Lamiastrum galeobdolon	V, S	2	*ANIMe	Fr	4	2	Chill	Lam	Lg
+	Lamium album	V, Bs	3	*ANIMe	Fr	5	2	Unclassified	Lam	La
	Lamium amplexicaule	Bs	3	ANIMe	Fr	3	2	Unclassified	Lam	La
	Lamium hybridum	?Bs	3	ANIMe	Fr	3	2	-	Lam	Lh
	Lamium maculatum	V, ?Bs	3	*ANIMe	Fr	4	2	Not studied	Lam	Lm
+	Lamium purpureum	Bs	4	ANIMe	Fr	3	2	Dry	Lam	Lp
+	Lapsana communis	Bs	3	UNSPag	Fr	4	3	Dry	Ast	Lc
	Lathraea squamaria	V, ?S	?	UNSP	Sd	3	1	Unclassified	Oro	Ls
	Lathyrus latifolius	V, ?Bs	3	*UNSP	Sd	6	1	Scar	Fab	Ll
+	Lathyrus linifolius	V, ?Bs	3	*UNSP	Sd	6	1	Scar	Fab	Ll
+	Lathyrus pratensis	V, ?Bs	3	*UNSP	Sd	6	1	Scar	Fab	Lp
	Lemna gibba	(V)	-	* -	X	X	-	X	Lem	Lg
+	Lemna minor	(V)	-	* -	X	X	-	X	Lem	Lm
+	Lemna trisulca	(V)	-	* -	X	X	-	X	Lem	Lt
+	Leontodon autumnalis	W	?1	WINDp	Fr	3	3	Dry	Ast	La
+	Leontodon hispidus	V, W	?1	WINDp	Fr	3	3	-	Ast	Lh
−	Leontodon saxatilis	W	?1	WINDp	Fr	2	3	-	Ast	Ls
−	Lepidium campestre	Bs	3	ANIMm	Sd	5	2	Dry	Bra	Lc
	Lepidium draba	(V), Bs	3	*UNSP	Sd	4	1	Dry	Bra	Ld
	Lepidium heterophyllum	?Bs	3	ANIMm	Sd	4	2	Unclassified	Bra	Lh
	Lepidium ruderale	?Bs	3	ANIMm	Sd	1	2	Dry	Bra	Lr
+	Leucanthemum vulgare	V, S, Bs	3	UNSPag	Fr	2	3	-	Ast	Lv
	Leucanthemum × superbum	V, S, ?Bs	?3	*UNSP	Fr	2	3	Dry	Ast	Ls
	Linaria purpurea	V,Bs	3	WINDcw	Sd	1	1	Unclassified	Scr	Lp
	Linaria repens	V,Bs	3	WINDcw	Sd	1	1	Unclassified	Scr	Lr
+	Linaria vulgaris	V, Bs	3	*WINDcw	Sd	1	1	Dry	Scr	Lv
+	Linum catharticum	S, Bs	3	UNSP	Sd	1	2	Chill	Lin	Lc
−	Listera ovata	V, W	?	WINDcm	Sd	S	3	Orchid	Orc	Lo
	Lithospermum officinale	V, ?S	?	UNSP	Fr	5	1	Unclassified	Bor	Lo
	Littorella uniflora	V, ?	?3	*UNSP	Fr	3	3	Unclassified	Pla	Lu

Species	Regenerative strategies	Seed bank	Agency of dispersal	Dispersule and germinule form	Dispersule weight	Dispersule shape	Germination requirements	Family	(Sp.)
Lolium multiflorum	S	1	ANIMa	Fr	5	3	-	Poa	*Lm*
+ *Lolium perenne*	S	1	UNSPag	Fr	4	3	-	Poa	*Lp*
+ *Lonicera periclymenum*	V, S	1	*ANIMi	F/S	5	1	Chill	Cap	*Lp*
+ *Lotus corniculatus*	Bs	3	*UNSP	Sd	4	1	Scar	Fab	*Lc*
+ *Lotus pedunculatus*	V, Bs	3	*UNSP	Sd	2	1	Scar	Fab	*Lp*
Lupinus × regalis	Bs	3	UNSP	Sd	6	1	Scar	Fab	*Lr*
+ *Luzula campestris*	V, Bs	3	*ANIMe	Sd	3	2	-	Jce	*Lc*
- *Luzula multiflora*	V, Bs	3	ANIMe	Sd	3	2	Dry	Jce	*Lm*
+ *Luzula pilosa*	V, S, Bs	3	ANIMe	Sd	3	1	-	Jce	*Lp*
- *Luzula sylvatica*	V, Bs	3	*ANIMe	Sd	3	2	Dry	Jce	*Ls*
- *Lychnis flos-cuculi*	V, Bs	3	WINDc	Sd	2	1	-	Car	*Lf*
- *Lycopus europaeus*	V	?1	AQ/AN	Fr	2	1	Fluct	Lam	*Le*
- *Lysimachia nemorum*	V, Bs	3	*UNSPc	Sd	2	1	?Chill	Pri	*Ln*
- *Lysimachia nummularia*	V	-	*UNSPc	Sd	X	-	X	Pri	*Ln*
Lysimachia punctata	V	-	* -	X	3	2	X	Pri	*Lp*
- *Lysimachia vulgaris*	V	2	*AQUATc	Sd	2	2	Chill	Pri	*Lv*
- *Lythrum portula*	Bs	3	UNSPc	Sd	1	2	-	Lyt	*Lp*
- *Lythrum salicaria*	Bs	3	AQUAT	Sd	1	2	-	Lyt	*Ls*
- *Malva moschata*	Bs	3	UNSP	Fr	5	1	Scar	Mal	*Mm*
- *Malva neglecta*	Bs	3	UNSP	Fr	4	1	Scar	Mal	*Mn*
- *Malva sylvestris*	Bs	3	UNSP	Fr	5	1	Scar	Mal	*Ms*
+ *Matricaria discoidea*	Bs	3	UNSP	Fr	1	2	Dry	Ast	*Md*
Matricaria recutita	Bs	3	UNSP	Fr	1	2	Dry	Ast	*Mr*
- *Meconopsis cambrica*	?Bs	3	WINDc	Sd	1	1	Not studied	Pap	*Mc*
+ *Medicago lupulina*	Bs	4	UNSP	f/S	5	1	Scar	Fab	*Ml*
- *Medicago sativa*	Bs	3	UNSPag	f/S	5	1	Scar	Fab	*Ms*
- *Melampyrum pratense*	S	2	ANIMe	Sd	5	3	Chill	Scr	*Mp*
+ *Melica uniflora*	V, S	2	*ANIMe	Fr	5	2	Chill	Poa	*Mu*
- *Melilotus albus*	Bs	3	UNSP	Sd	3	1	Scar	Fab	*Ma*
- *Melilotus altissimus*	Bs	3	UNSP	Sd	5	1	Scar	Fab	*Ma*
- *Melilotus officinalis*	Bs	3	UNSP	Sd	4	1	Scar	Fab	*Mo*
+ *Mentha aquatica*	V, Bs	3	*AQUAT	Fr	1	2	Chill,Dry	Lam	*Ma*
- *Mentha arvensis*	V, Bs	4	*AQUAT	Fr	2	1	-	Lam	*Ma*
- *Mentha spicata*	V, ?	?	*UNSP	Fr	X	X	X	Lam	*Ms*
Mentha × piperita	V	-	* -	X	X	-	X	Lam	*Mp*
- *Mentha × verticillata*	V	-	*UNSP	Fr	2	-	X (part sterile)	Lam	*Mv*
Mentha × villosa	V	-	* -	X	X	-	X	Lam	*Mv*
- *Menyanthes trifoliata*	V	2	*AQUAT	Sd	5	1	Scar	Men	*Mt*
Mercurialis annua	Bs	3	ANIMe	Sd	4	1	Unclassified	Eup	*Ma*
+ *Mercurialis perennis*	V, S	2	*ANIMe	Sd	5	1	Chill	Eup	*Mp*
+ *Milium effusum*	V, Bs	3	UNSP	Fr	4	2	-	Poa	*Me*
- *Mimulus guttatus*	V, Bs	3	*AQUAT	Sd	1	1	-	Scr	*Mg*
Mimulus × robertsii	V	-	* -	X	X	-	X	Scr	*Mr*
+ *Minuartia verna*	V, Bs	3	WINDc	Sd	1	1	-	Car	*Mv*
+ *Moehringia trinervia*	?Bs	3	UNSPc	Sd	2	1	Dry	Car	*Mt*
+ *Molinia caerulea*	V, ?Bs	3	UNSP	Fr	3	2	Chill	Poa	*Mc*
- *Montia fontana*	V, ?Bs	3	UNSP	Sd	2	1	Scar	Por	*Mf*
+ *Mycelis muralis*	W	1	ANIMp	Fr	2	3	-	Ast	*Mm*
+ *Myosotis arvensis*	S, Bs	3	ANIMa	Fr	2	2	Dry	Bor	*Ma*
- *Myosotis discolor*	S, ?Bs	3	ANIMa	Fr	1	1	Dry	Bor	*Md*
- *Myosotis laxa* subsp. *cespitosa*	S, ?Bs	3	AQUAT	Fr	2	1	Dry	Bor	*Ml*
+ *Myosotis ramosissima*	S, Bs	3	ANIMa	Fr	1	1	Dry	Bor	*Mr*
+ *Myosotis scorpioides*	V, Bs	3	*UNSP	Fr	2	2	Dry	Bor	*Ms*
- *Myosotis secunda*	V, Bs	3	*UNSP	Fr	2	1	Dry	Bor	*Ms*
Myosotis sylvatica	V, ?Bs	3	UNSP	Fr	3	2	Unclassified	Bor	*Ms*
Myosoton aquaticum	V, ?Bs	?3	UNSPc	Sd	2	1	Unclassified	Car	*Ma*
Myriophyllum alterniflorum	(V)	?	*UNSP	Fr	?	2	Not studied	Hal	*Ma*
- *Myriophyllum spicatum*	(V)	?	*UNSP	Fr	4	2	-/Scar	Hal	*Ms*
+ *Myrrhis odorata*	S	2	UNSP	Fr	6	3	Chill	Api	*Mo*
Narcissus pseudonarcissus	?V, S	2	UNSPc	Sd	5	1	Not studied	Lil	*Np*
+ *Nardus stricta*	V, S	2	ANIMa	Fr	2	3	Chill	Poa	*Ns*
Narthecium ossifragum	V, S	2	*WINDcw	Sd	1	3	Unclassified	Lil	*No*
Neottia nidus-avis	W	?	WINDcm	Sd	S	3	Orchid	Orc	*Nn*
- *Nuphar lutea*	V, ?S	?2	*AQUAT	Sd	6	1	?Chill	Nym	*Nl*
Nymphaea alba	V, ?S	?2	*AQUAT	Sd	?	1	Not studied	Nym	*Na*

TABLES OF ATTRIBUTES

Table 7.3

	Species	Regenerative strategies	Seed bank	Agency of dispersal	Dispersule and germinule form	Dispersule weight	Dispersule shape	Germination requirements	Family	(Sp.)
-	Odontites vernus	S, Bs	3	WINDc	Sd	2	2	Chill	Scr	Ov
	Oenanthe aquatica	S	?2	AQUAT	Fr	5	3	Unclassified	Api	Oa
	Oenanthe crocata	?	?2	AQUAT	Fr	5	3	Dry	Api	Oc
	Oenanthe fistulosa	V	?2	*AQ/AN	Fr	5	1	(-)	Api	Of
	Oenothera biennis	Bs	3	WINDc	Sd	3	2	-	Ona	Ob
	Oenothera cambrica	Bs	3	WINDc	Sd	3	2	Unclassified	Ona	Oc
	Oenothera glazioviana	Bs	3	WINDc	Sd	2	2	Unclassified	Ona	Og
	Onobrychis viciifolia	V, Bs	3	UNSP	Sd	6	1	Scar	Fab	Ov
	Ononis repens	V, ?Bs	3	UNSP	Sd	5	1	Scar	Fab	Or
	Ononis spinosa	V, ?Bs	3	UNSP	Sd	5	1	Scar	Fab	Os
	Ophrys apifera	W	?	WINDcm	Sd	S	3	Orchid	Orc	Oa
-	Orchis mascula	W	?	WINDcm	Sd	S	3	Orchid	Orc	Om
	Orchis morio	W	?	WINDcm	Sd	S	3	Orchid	Orc	Om
+	Origanum vulgare	V, Bs	4	UNSP	Fr	1	1	-	Lam	Ov
	Ornithogalum angustifolium	(V), S	2	UNSPc	Sd	?	1	No studied	Lil	Oa
	Ornithopus perpusillus	Bs	3	UNSP	F/S	4	1	Scar	Fab	Op
+	Oxalis acetosella	V, S, ?Bs	3	*ANIMm	Sd	4	2	Chill	Oxa	Oa
	Oxalis corniculata	V, Bs	3	*ANIMm	Sd	2	2	-	Oxa	Oc
	Oxalis debilis	(V)	-	*	X	X	-	X	Oxa	Od
	Oxalis exilis	V, Bs	3	*ANIMm	Sd	1	2	Not studied	Oxa	Oe
-	Papaver argemone	Bs	3	WINDc	Sd	1	2	Scar	Pap	Pa
-	Papaver dubium	Bs	3	WINDc	Sd	1	1	Chill	Pap	Pd
+	Papaver rhoeas	Bs	4	WINDc	Sd	1	1	Chill	Pap	Pr
	Papaver somniferum	?Bs	3	WINDc	Sd	2	1	Dry	Pap	Ps
	Parietaria judaica	?	?	ANIMa	Fr	1	1	Dry	Urt	Pj
	Paris quadrifolia	V, S	2	*ANIMi	Sd	5	2	?Chill	Lil	Pq
	Parnassia palustris	?V, S	?2	WINDc	Sd	2	3	Unclassified	Sax	Pp
	Pastinaca sativa	S, ?Bs	3	WINDw	Fr	5	1	Unclassified	Api	Ps
-	Pedicularis sylvatica	S, ?Bs	3	ANIMe	Sd	4	2	Chill	Scr	Ps
-	Pentaglottis sempervirens	?	?	UNSP	Fr	5	2	Dry	Bor	Ps
+	Persicaria amphibia	(V), ?Bs	3	*AQUAT	Fr	5	1	Chill	Pgo	Pa
	Persicaria bistorta	V, Bs	3	*UNSP	Fr	5	1	Unclassified	Pgo	Pb
-	Persicaria hydropiper	Bs	3	AQUAT	Fr	4	2	Chill	Pgo	Ph
-	Persicaria lapathifolia	Bs	3	UNSPag	Fr	5	1	Chill,Dry	Pgo	Pl
+	Persicaria maculosa	Bs	4	UNSPag	Fr	5	1	Chill,Dry	Pgo	Pm
	Petasites fragrans	V	-	* -	X	X	-	X	Ast	Pf
+	Petasites hybridus	V, W	1	*WINDp	Fr	2	3	-	Ast	Ph
+	Phalaris arundinacea	V, ?Bs	?2	*AQUAT	Fr	3	2	Dry,Scar	Poa	Pa
	Phleum bertolonii	V, S, ?Bs	?3	ANIMa	Fr	1	2	Unclassified	Poa	Pb
+	Phleum pratense	S, ?Bs	?3	ANIMa	Fr	2	2	-	Poa	Pp
+	Phragmites australis	V, W	?1	*WINDp	Fr	1	2	Unclassified	Poa	Pa
	Picris echioides	W	?	WINDp	Fr	3	3	-	Ast	Pe
	Picris hieracioides	W, ?Bs	3	WINDp	Fr	4	3	Dry	Ast	Ph
	Pilosella aurantiaca	V, W	?1	*WINDp	Fr	1	3	(-)	Ast	Pa
+	Pilosella officinarum	V, W	?1	*WINDp	Fr	1	3	Dry	Ast	Po
	Pimpinella major	S	2	UNSP	Fr	4	2	Unclassified	Api	Pm
+	Pimpinella saxifraga	S	2	UNSP	Fr	4	2	Chill	Api	Ps
	Pinguicula vulgaris	V, ?W	?1	WINDm	Sd	1	3	Unclassified	Len	Pv
-	Plantago coronopus	Bs	3	ANIMm	Sd	1	2	-	Pla	Pc
+	Plantago lanceolata	V, Bs	3	ANIMm	Sd	4	2	Dry	Pla	Pl
+	Plantago major	Bs	4	ANIMm	Sd	2	2	Chill	Pla	Pm
	Plantago media	V, Bs	3	ANIMm	Sd	2	2	-	Pla	Pm
-	Platanthera chlorantha	W	?	WINDm	Sd	S	3	Orchid	Orc	Pc
	Poa angustifolia	V, Bs	3	*UNSP	Fr	1	3	-	Poa	Pa
+	Poa annua	V, S, Bs	3	UNSPag	Fr	2	2	-	Poa	Pa
	Poa compressa	V, ?Bs	3	UNSP	Fr	2	3	Dry	Poa	Pc
-	Poa nemoralis	?Bs	3	UNSP	Fr	2	3	Dry	Poa	Pn
+	Poa pratensis	V, Bs	3	*UNSPag	Fr	2	3	-	Poa	Pp
+	Poa trivialis	V, Bs	3	UNSPag	Fr	1	3	-	Poa	Pt
-	Polygala serpyllifolia	?S	2	ANIMe	Sd	4	2	Chill	Pga	Ps
+	Polygala vulgaris	?S	2	ANIMe	Sd	4	2	Chill	Pga	Pv
	Polygonatum × hybridum	V	-	* -	X	X	-	X	Lil	Ph
+	Polygonum aviculare	Bs	4	UNSPag	Fr	4	2	Chill	Pgo	Pa
	Potamogeton berchtoldii	(V), S	1	*AQ/AN	Fr	4	1	Scar	Pot	Pb
+	Potamogeton crispus	(V), S	1	*AQ/AN	Fr	5	2	Scar	Pot	Pc

	Species	Regenerative strategies	Seed bank	Agency of dispersal	Dispersule and germinule form	Dispersule weight	Dispersule shape	Germination requirements	Family	(Sp.)
+	*Potamogeton natans*	(V), ?	?1	*AQ/AN	Fr	5	1	Scar	Pot	*Pn*
	Potamogeton pectinatus	(V), S	?1	*AQ/AN	Fr	5	1	Scar	Pot	*Pp*
	Potamogeton perfoliatus	(V), S	?1	*AQ/AN	Fr	4	1	Scar	Pot	*Pp*
	Potamogeton polygonifolius	(V), S	?1	*AQ/AN	Fr	4	1	Scar	Pot	*Pp*
	Potamogeton pusillus	(V), S	?1	*AQ/AN	Fr	4	1	Scar	Pot	*Pp*
-	*Potentilla anglica*	V, ?Bs	3	*UNSP	Fr	2	1	Unclassified	Ros	*Pa*
-	*Potentilla anserina*	V	?2	*UNSP	Fr	3	1	Unclassified	Ros	*Pa*
+	*Potentilla erecta*	V, Bs	4	UNSP	Fr	3	1	Warm	Ros	*Pe*
-	*Potentilla palustris*	V	?2	*UNSP	Fr	2	1	Unclassified	Ros	*Pp*
	Potentilla recta	V, ?Bs	?3	UNSP	Fr	1	1	-	Ros	*Pr*
-	*Potentilla reptans*	V, Bs	3	*UNSP	Fr	2	1	Unclassified	Ros	*Pr*
+	*Potentilla sterilis*	V, Bs	3	UNSP	Fr	3	1	-	Ros	*Ps*
+	*Primula veris*	V	?2	WINDc	Sd	3	1	Chill,Wash	Pri	*Pv*
-	*Primula vulgaris*	Bs	3	ANIMe	Sd	2	1	Wash	Pri	*Pv*
+	*Prunella vulgaris*	(V), Bs	3	*ANIMm	Fr	3	2	-	Lam	*Pv*
	Pseudofumaria lutea	?S	?	ANIMe	Sd	4	1	Unclassified	Fum	*Pl*
	Puccinellia distans	Bs	3	UNSP	Fr	2	3	Unclassified	Poa	*Pd*
-	*Pulicaria dysenterica*	V, W	?3	WINDp	Fr	1	3	?Dry	Ast	*Pd*
+	*Ranunculus acris*	V	?2	ANIMa	Fr	4	1	H	Ran	*Ra*
	Ranunculus aquatilis	V, ?Bs	?3	*AQUAT	Fr	3	1	Unclassified	Ran	*Ra*
-	*Ranunculus auricomus*	?	2	UNSP	Fr	5	1	?Chill	Ran	*Ra*
+	*Ranunculus bulbosus*	Bs	3	ANIMa	Fr	5	1	Dry	Ran	*Rb*
	Ranunculus circinatus	V, Bs	?	*AQUAT	Fr	1	1	Unclassified	Ran	*Rc*
+	*Ranunculus ficaria*	Sv, S	2	*ANIMe	Fr	4	2	Chill	Ran	*Rf*
+	*Ranunculus flammula*	V, Bs	3	*AQUAT	Fr	2	1	Dry,Freeze	Ran	*Rf*
	Ranunculus hederaceus	?Bs	?3	*AQUAT	Fr	2	2	Unclassified	Ran	*Rh*
	Ranunculus lingua	V, ?	?	*AQUAT	Fr	5	2	Unclassified	Ran	*Rl*
	Ranunculus omiophyllus	V, ?Bs	?3	*AQUAT	Fr	2	2	Unclassified	Ran	*Ro*
+	*Ranunculus peltatus*	V, ?Bs	?3	*AQUAT	Fr	2	1	Unclassified	Ran	*Rp*
+	*Ranunculus penicillatus*	V, ?	?	*AQUAT	Fr	2	2	Unclassified	Ran	*Rp*
+	*Ranunculus repens*	(V), Bs	4	*AQ/AN	Fr	5	2	(Dry)	Ran	*Rr*
+	*Ranunculus sceleratus*	Bs	3	*AQUAT	Fr	1	2	-	Ran	*Rs*
	Ranunculus trichophyllus	V, ?	?	*AQUAT	Fr	2	1	Unclassified	Ran	*Rt*
-	*Raphanus raphanistrum*	Bs	3	UNSPag	Sd	5	1	Unclassified	Bra	*Rr*
	Rapistrum rugosum	?	?	UNSP	Sd	4	1	Not studied	Bra	*Rr*
	Reseda lutea	Bs	3	WINDc	Sd	3	1	Unclassified	Res	*Rl*
-	*Reseda luteola*	Bs	3	WINDc	Sd	2	1	Unclassified	Res	*Rl*
+	*Rhinanthus minor*	S, ?Bs	?3	WINDcw	Sd	5	1	Chill	Scr	*Rm*
-	*Ribes uva-crispa*	S	2	ANIMi	F/S	5	1	Chill	Gro	*Ru*
	Rorippa amphibia	V, Bs	3	*AQUAT	Sd	1	1	Unclassified	Bra	*Ra*
	Rorippa microphylla	V, Bs	3	*AQUAT	Sd	1	1		Bra	*Rm*
+	*Rorippa nasturtium-aquaticum*	V, Bs	3	*AQUAT	Sd	1	1	-/Dry	Bra	*Rn*
+	*Rorippa palustris*	S, Bs	3	AQUAT	Sd	1	1	Dry,Fluct,Scar	Bra	*Rp*
	Rorippa sylvestris	V, Bs	3	*UNSP	Sd	1	1	Unclassified	Bra	*Rs*
-	*Rubus caesius*	V, ?	2	*ANIMi	Fr	5	2	Scar/Chill	Ros	*Rc*
	Rubus chamaemorus	V, ?	2	*ANIMi	Fr	5	2	Unclassified	Ros	*Rc*
+	*Rubus fruticosus*	V, Bs	4	*ANIMi	Fr	5	1	Scar/Chill	Ros	*Rf*
-	*Rubus idaeus*	V, Bs	4	*ANIMi	Fr	4	2	Scar/Chill	Ros	*Ri*
	Rubus saxatilis	V, ?	?2	*ANIMi	Fr	5	1	Unclassified	Ros	*Rs*
+	*Rumex acetosa*	V, S	?2	WINDw	Fr	3	1	-	Pgo	*Ra*
+	*Rumex acetosella*	V, Bs	4	*UNSP	Fr	2	2	Dry	Pgo	*Ra*
-	*Rumex conglomeratus*	?Bs	3	AQUAT	Fr	4	1	-	Pgo	*Rc*
+	*Rumex crispus*	(V), Bs	4	UNSP	Fr	4	2	Dry	Pgo	*Rc*
	Rumex hydrolapathum	?Bs	3	AQUAT	Fr	5	1	-	Pgo	*Rh*
+	*Rumex obtusifolius*	Bs	4	ANIMa	Fr	4	1	-	Pgo	*Ro*
-	*Rumex sanguineus*	Bs	3	UNSP	Fr	4	1	Dry	Pgo	*Rs*
	Ruscus aculeatus	V, S	2	*ANIMi	Sd	6	1	Not studied	Lil	*Ra*
	Sagina apetala	Bs	3	WINDc	Sd	1	1	Dry	Car	*Sa*
	Sagina maritima	Bs	3	WINDc	Sd	1	1	Not studied	Car	*Sm*
	Sagina nodosa	Sv, Bs	3	*UNSPc	Sd	1	1	Unclassified	Car	*Sn*
+	*Sagina procumbens*	(V), Bs	4	UNSPc	Sd	1	2	-	Car	*Sp*
	Sagittaria sagittifolia	V, ?	?	*AN/AQ	Fr	4	1	Scar	Ali	*Ss*
-	*Salix repens*	V, W	1	*WINDp	Sd	1	2	-	Sal	*Sr*
	Sambucus ebulus	V, ?	?2	*ANIMi	F/S	4	2	Unclassified	Cap	*Se*
	Samolus valerandi	?V, Bs	3	UNSPc	Sd	1	1	Dry	Pri	*Sv*

TABLES OF ATTRIBUTES

Table 7.3

	Species	Regenerative strategies	Seed bank	Agency of dispersal	Dispersule and germinule form	Dispersule weight	Dispersule shape	Germination requirements	Family	(Sp.)
+	Sanguisorba minor	?Bs	?2	UNSP	Fr	4	2	Dry	Ros	Sm
	Sanguisorba minor subsp. muricata	?Bs	?2	UNSP	Fr	6	2	Unclassified	Ros	Sm
	Sanguisorba officinalis	V, ?	?1	UNSP	Fr	5	2	Unclassified	Ros	So
+	Sanicula europaea	V, S	2	ANIMb	Fr	5	1	?Chill	Api	Se
-	Saponaria officinalis	V, ?	1	*WINDc	Sd	4	1	Chill,Scar	Car	So
	Saxifraga granulata	Sv, Bs	3	*WINDc	Sd	1	2	Not studied	Sax	Sg
+	Saxifraga tridactylites	S, Bs	3	WINDc	Sd	1	2	Dry	Sax	St
	Saxifraga × urbium	V	-	*WINDc	Sd	X	-	X	Sax	Su
+	Scabiosa columbaria	S	2	ANIMa	Fr	4	1	-/Chill	Dip	Sc
-	Schoenoplectus lacustris	V, ?S	?1	*ANIMa	Fr	4	2	Unclassified	Cyp	Sl
	Schoenoplectus tabernaemontani	V, ?S	?1	*ANIMa	Fr	4	2	Unclassified	Cyp	St
	Schoenus nigricans	V, ?	?	UNSP	Fr	2	2	Unclassified	Cyp	Sn
	Scirpus sylvaticus	V, Bs	3	*ANIMa	Fr	1	1	Unclassified	Cyp	Ss
	Scleranthus annuus	Bs	3	UNSP	Fr	4	3	Unclassified	Car	Sa
	Scrophularia auriculata	S, Bs	3	AQ/WIc	Sd	1	2	-	Scr	Sa
	Scrophularia nodosa	Bs	3	WINDc	Sd	1	2	-	Scr	Sn
	Scutellaria galericulata	V, ?	?1	AQUAT	Fr	3	1	Unclassified	Lam	Sg
	Scutellaria minor	V, ?Bs	?3	AQUAT	Fr	1	1	Not studied	Lam	Sm
	Securigera varia	V, Bs	3	UNSP	Sd	5	1	Scar	Fab	Sv
+	Sedum acre	V, Bs	3	*WINDcm	Sd	1	2	Dry	Cra	Sa
	Sedum album	V, Bs	3	*WINDcm	Sd	1	3	Unclassified	Cra	Sa
-	Sedum anglicum	V, Bs	3	*WINDcm	Sd	1	2	Unclassified	Cra	Sa
	Sedum rupestre	V, Bs	3	*WINDcm	Sd	1	3	Dry	Cra	Sr
	Sedum spurium	V, Bs	3	*WINDcm	Sd	1	3	Not studied	Cra	Ss
	Sedum telephium	V, Bs	3	WINDcm	Sd	1	3	Dry	Cra	St
-	Senecio aquaticus	W	1	WINDp	Fr	2	3	-	Ast	Sa
	Senecio erucifolius	V, W	1	WINDp	Fr	2	3	-	Ast	Se
+	Senecio jacobaea	(V), W, Bs	3	WINDp	Fr	1	3	-	Ast	Sj
+	Senecio squalidus	W, ?Bs	?2	WINDp	Fr	2	2	Dry	Ast	Ss
-	Senecio sylvaticus	W, Bs	3	WINDp	Fr	2	3	Dry	Ast	Ss
+	Senecio viscosus	W, Bs	3	WINDp	Fr	3	3	Dry	Ast	Sv
+	Senecio vulgaris	W, Bs	3	WINDp	Fr	2	3	Dry	Ast	Sv
-	Serratula tinctoria	V, W	1	WINDp	Fr	5	3	Unclassified	Ast	St
-	Sherardia arvensis	S, Bs	3	ANIMa	Fr	4	2	Dry	Rub	Sa
	Silaum silaus	S	2	UNSP	Fr	4	2	Unclassified	Api	Ss
+	Silene dioica	Bs	3	WINDc	Sd	3	1	-	Car	Sd
-	Silene latifolia	Bs	3	WINDc	Sd	3	1	Dry	Car	Sl
	Silene noctiflora	Bs	3	WINDc	Sd	4	1	Dry	Car	Sn
-	Silene vulgaris	Bs	3	WINDc	Sd	4	1	Dry	Car	Sv
+	Sinapis arvensis	Bs	4	UNSPag	Sd	4	1	Dry	Bra	Sa
-	Sisymbrium altissimum	Bs	3	WINDw	Sd	2	1	Scar	Bra	Sa
	Sisymbrium loeselii	Bs	3	WINDw	Sd	1	1	Dry	Bra	Sl
-	Sisymbrium officinale	Bs	3	WINDw	Sd	2	1	Scar	Bra	So
	Sisymbrium orientale	Bs	3	WINDw	Sd	2	1	Scar	Bra	So
+	Solanum dulcamara	(V), ?Bs	?3	*ANIMi	F/S	4	1	Chill,Fluct	Sol	Sd
	Solanum nigrum	Bs	3	ANIMi	F/S	3	1	Chill	Sol	Sn
	Solanum physalifolium	Bs	3	ANIMi	F/S	4	1	Chill, Dry	Sol	Sp
	Solidago canadensis	V, W	2	WINDp	Fr	1	3	Unclassified	Ast	Sc
	Solidago gigantea	V, W	2	*WINDp	Fr	1	3	Unclassified	Ast	Sg
+	Solidago virgaurea	W	?2	WINDp	Fr	3	3	-	Ast	Sv
-	Sonchus arvensis	V, Bs	3	*WINDp	Fr	3	2	Chill	Ast	Sa
+	Sonchus asper	W, Bs	3	WINDp	Fr	2	2	Dry	Ast	Sa
+	Sonchus oleraceus	W, Bs	3	WINDp	Fr	2	3	-	Ast	So
+	Sparganium emersum	(V)	?	*AQ/AN	Fr	5	2	Chill,Scar	Spa	Se
+	Sparganium erectum	(V)	?1	*AQ/AN	Fr	6	2	Chill,Scar	Spa	Se
+	Spergula arvensis	Bs	4	WINDc	Sd	2	1	Dry	Car	Sa
	Spergularia marina	Bs	3	UNSPc	Sd	1	1	Not studied	Car	Sm
-	Spergularia rubra	?Bs	3	UNSPc	Sd	1	1	-	Car	Sr
-	Stachys arvensis	Bs	4	UNSPag	Fr	3	1	Dry	Lam	Sa
+	Stachys officinalis	V, S, ?Bs	?3	UNSP	Fr	4	1	-	Lam	So
-	Stachys palustris	V, Bs	3	*AQUAT	Fr	4	1	Unclassified	Lam	Sp
+	Stachys sylvatica	(V), Bs	3	*ANIMa	Fr	4	1	Chill	Lam	Ss
	Stellaria graminea	V, ?Bs	3	UNSPc	Sd	2	1	Dry	Car	Sg
+	Stellaria holostea	V, ?S	2	*UNSPc	Sd	5	1	?Chill	Car	Sh
+	Stellaria media	Bs	4	UNSPag	Sd	2	1	Dry	Car	Sm

	Species	Regenerative strategies	Seed bank	Agency of dispersal	Dispersule and germinule form	Dispersule weight	Dispersule shape	Germination requirements	Family	(Sp.)
	Stellaria neglecta	Bs	3	UNSPc	Sd	4	1	Unclassified	Car	*Sn*
	Stellaria pallida	Bs	3	UNSPag	Sd	1	1	Dry	Car	*Sp*
+	*Stellaria uliginosa*	(V), Bs	3	*UNSPc	Sd	1	2	-	Car	*Su*
+	*Succisa pratensis*	S	2	ANIMa	Fr	4	3	-	Dip	*Sp*
-	*Symphytum officinale*	V, ?	?1	AQ/ANe	Fr	6	2	Not studied	Bor	*So*
	Symphytum orientale	V, ?	?1	ANIMe	Fr	5	2	Not studied	Bor	*So*
-	*Symphytum × uplandicum*	V, ?	?1	?AQUAT	Fr	5	2	Unclassified	Bor	*Su*
+	*Tamus communis*	S	2	ANIMi	F/S	6	1	Chill	Dio	*Tc*
-	*Tanacetum parthenium*	?	?	UNSP	Fr	1	3	-	Ast	*Tp*
-	*Tanacetum vulgare*	V	2	*UNSP	Fr	1	3	Dry	Ast	*Tv*
+	*Taraxacum* agg.	V, W	2	*WINDp	Fr	3	3	-	Ast	*Ta*
	Tellima grandiflora	V, ?Bs	?3	WINDc	Sd	1	2	Not studied	Sax	*Tg*
+	*Teucrium scorodonia*	V, Bs	3	*UNSP	Fr	3	1	-	Lam	*Ts*
	Thalictrum flavum	V, S	2	*UNSP	Fr	3	1	Unclassified	Ran	*Tf*
-	*Thlaspi arvense*	Bs	4	WINDw	Sd	3	1	Fluct, Scar	Bra	*Ta*
+	*Thymus polytrichus*	V, Bs	4	*UNSP	Fr	1	1	-	Lam	*Tp*
+	*Torilis japonica*	S, ?Bs	?3	ANIMb	Fr	4	2	Chill	Api	*Tj*
-	*Tragopogon pratensis*	W	1	WINDp	Fr	6	3	Dry	Ast	*Tp*
	Trichophorum cespitosum	V, ?S	2	ANIMa	Fr	2	2	Unclassified	Cyp	*Tc*
	Trientalis europaea	V, ?S	?2	*UNSP	Sd	3	1	Not studied	Pri	*Te*
-	*Trifolium arvense*	S, Bs	3	ANIMa	US	2	1	Scar	Fab	*Ta*
-	*Trifolium campestre*	S, Bs	3	ANIMa	US	2	2	Scar	Fab	*Tc*
+	*Trifolium dubium*	S, Bs	3	ANIMa	f/S	2	2	Scar	Fab	*Td*
	Trifolium fragiferum	(V), Bs	3	*ANIMa	Sd	4	1	Scar	Fab	*Tf*
	Trifolium hybridum	Bs	3	ANIMa	f/S	3	1	Scar	Fab	*Th*
+	*Trifolium medium*	V, Bs	3	*ANIMa	f/S	5	1	Scar	Fab	*Tm*
	Trifolium micranthum	S, Bs	3	ANIMa	Sd	2	1	Scar	Fab	*Tm*
+	*Trifolium pratense*	S, Bs	3	ANIMa	f/S	4	1	Scar	Fab	*Tp*
+	*Trifolium repens*	(V), Bs	4	*ANIMa	f/S	3	1	Scar	Fab	*Tr*
	Trifolium striatum	S, Bs	3	ANIMa	Sd	2	1	Scar	Fab	*Ts*
	Triglochin palustre	V, ?	?1	ANIMa	f/S	2	3	Unclassified	Jcg	*Tp*
+	*Tripleurospermum inodorum*	S, Bs	3	UNSPag	Fr	2	2	Dry	Ast	*Ti*
+	*Trisetum flavescens*	V, S	1	ANIMa	Fr	1	3	-	Poa	*Tf*
	Trollius europaeus	S	2	WINDc	Sd	4	3	Unclassified	Ran	*Te*
+	*Tussilago farfara*	(V), W, S	1	*WINDp	Fr	2	3	-	Ast	*Tf*
	Typha angustifolia	V, W, Bs	3	*WINDp	Fr	1	3	Unclassified	Typ	*Ta*
+	*Typha latifolia*	V, W, Bs	3	*WINDp	Fr	1	3	Fluct,Scar	Typ	*Tl*
	Ulex minor	Bs	3	ANIMe	Sd	5	1	Scar	Fab	*Um*
	Umbilicus rupestris	?	?	WINDcm	Sd	1	3	Dry	Cra	*Ur*
+	*Urtica dioica*	V, Bs	4	*ANIMa	Fr	1	1	-	Urt	*Ud*
+	*Urtica urens*	Bs	4	ANIMa	Fr	2	1	-	Urt	*Uu*
+	*Vaccinium myrtillus*	V, ?S	?2	*ANIMi	F/S	2	2	-	Eri	*Vm*
	Vaccinium oxycoccos	V, ?S	?2	*ANIMi	F/S	2	2	Unclassified	Eri	*Vo*
+	*Vaccinium vitis-idaea*	V, ?S	?2	*ANIMi	F/S	2	2	-	Eri	*Vv*
	Valeriana dioica	V, S	1	*WINDpw	Fr	4	2	Unclassified	Val	*Vd*
+	*Valeriana officinalis*	V, S	1	WINDpw	Fr	3	2	Dry	Val	*Vo*
-	*Valerianella locusta*	S	2	UNSP	Fr	3	2	Dry	Val	*Vl*
-	*Verbascum nigrum*	Bs	3	WINDc	Sd	1	2	-	Scr	*Vn*
-	*Verbascum thapsus*	Bs	4	WINDc	Sd	1	2	Dry	Scr	*Vt*
-	*Verbena officinalis*	Bs	3	UNSP	Fr	2	3	Unclassified	Ver	*Vo*
-	*Veronica agrestis*	Bs	3	ANIMe	Sd	2	2	Dry	Scr	*Va*
-	*Veronica anagallis-aquatica*	(V), Bs	3	WINDc	Sd	1	2	-	Scr	*Va*
+	*Veronica arvensis*	Bs	3	UNSP	Sd	1	1	Dry	Scr	*Va*
+	*Veronica beccabunga*	(V), Bs	4	*ANIMm	Sd	2	1	-	Scr	*Vb*
	Veronica catenata	(V), Bs	3	WINDc	Sd	1	2	-	Scr	*Vc*
+	*Veronica chamaedrys*	V, Bs	3	*UNSPcw	Sd	1	1	-	Scr	*Vc*
-	*Veronica filiformis*	V	-	*-	X	X	-	X	Scr	*Vf*
-	*Veronica hederifolia*	Bs	3	ANIMe	Sd	5	2	Chill	Scr	*Vh*
+	*Veronica montana*	V, Bs	3	*UNSPcw	Sd	2	1	Chill	Scr	*Vm*
-	*Veronica officinalis*	V, Bs	3	*UNSPcw	Sd	1	1	Dry	Scr	*Vo*
+	*Veronica persica*	Bs	3	UNSPag	Sd	3	2	Dry	Scr	*Vp*
-	*Veronica polita*	Bs	3	UNSPag	Sd	2	2	Unclassified	Scr	*Vp*
-	*Veronica scutellata*	V, ?Bs	3	*UNSPcw	Sd	1	1	Dry	Scr	*Vs*
-	*Veronica serpyllifolia*	V, Bs	3	*UNSPcw	Sd	1	2	Dry	Scr	*Vs*
+	*Vicia cracca*	V, ?Bs	?3	*UNSP	Sd	6	1	Scar	Fab	*Vc*

Table 7.3

	Species	Regenerative strategies	Seed bank	Agency of dispersal	Dispersule and germinule form	Dispersule weight	Dispersule shape	Germination requirements	Family	(Sp.)
-	Vicia hirsuta	S, Bs	3	UNSP	Sd	5	1	Scar	Fab	Vh
-	Vicia sativa	S, Bs	3	UNSP	Sd	6	1	Scar	Fab	Vs
+	Vicia sepium	V, ?Bs	?3	*UNSP	Sd	6	1	Scar	Fab	Vs
	Vicia tenuifolia	V, ?Bs	?3	*UNSP	Sd	6	1	Scar	Fab	Vt
	Vicia tetrasperma	S, Bs	3	UNSP	Sd	5	1	Scar	Fab	Vt
	Vinca major	V	-	* -	Fr	X	3	X	Apo	Vm
	Vinca minor	V	-	* -	Fr	X	3	X	Apo	Vm
	Viola arvensis	Bs	3	UNSPag	Sd	2	2	Unclassified	Vio	Va
	Viola canina	V, ?Bs	3	ANIMe	Sd	3	2	Chill	Vio	Vc
+	Viola hirta	V, ?S	2	ANIMe	Sd	5	3	Chill	Vio	Vh
-	Viola odorata	V, ?Bs	3	*ANIMe	Sd	5	1	?Chill	Vio	Vo
-	Viola palustris	V, S	2	*ANIMe	Sd	3	2	Chill	Vio	Vp
	Viola reichenbachiana	V, S	2	ANIMe	Sd	3	1	Chill	Vio	Vr
+	Viola riviniana	V, S	2	ANIMe	Sd	4	1	Chill	Vio	Vr
-	Viola tricolor	Bs	3	ANIMm	Sd	3	2	Dry	Vio	Vt
	Viscum album	?S	?2	ANIMi	F/S	5	1	Not studied	Vis	Va
	Vulpia bromoides	S	1	ANIMa	Fr	2	3	Dry	Poa	Vb
	Vulpia myuros	S, Bs	1	ANIMa	Fr	3	3	Dry	Poa	Vm
	Wahlenbergia hederacea	V, Bs	3	*UNSPc	Sd	1	2	Not studied	Cam	Wh
	× Festulolium loliaceum	V	-	* -	X	X	-	X	Poa	Fl
	Zannichellia palustris	V, Bs	3	*AQ/AN	Fr	2	3	Unclassified	Zan	Zp

(b) Woody species more than 1.5 m tall

	Species	Regenerative strategies	Seed bank	Agency of dispersal	Dispersule and germinule form	Dispersule weight	Dispersule shape	Germination requirements	Family	(Sp.)
-	Acer campestre	W, S	2	WINDw	Fr	6	3	Chill	Ace	Ac
	Acer platanoides	W, S	2	WINDw	Fr	6	3	Chill	Ace	Ap
+	Acer pseudoplatanus	W, S	2	WINDw	Fr	6	3	Chill	Ace	Ap
	Aesculus hippocastanum	S	2	ANIMn		6	1	Chill	Hoc	Ah
+	Alnus glutinosa	W	2	AQ/WIw	Fr	4	1	Dry	Bet	Ag
+	Betula sp.	W, S	2	WINDw	Fr	1	1	-	Bet	Bs
	Betula pendula	W, S	2	WINDw	Fr	1	1	-	Bet	Bp
	Betula pubescens	W	2	WINDw	Fr	1	1	-	Bet	Bp
	Buddleja davidii	W, Bs	3	WINDcm	Sd	1	3	-	Bud	Bd
	Carpinus betulus	W, S	2	WINDw	F/S	6	1	Not studied	Bet	Cb
	Castanea sativa	S	2	ANIMn	Fr	6	1	Not studied	Fag	Cs
	Clematis vitalba	S	2	WINDp	Fr	4	3	Chill	Ran	Cv
	Cornus sanguinea	S	?2	ANIMi	Sd	6	1	Not studied	Cor	Cs
-	Corylus avellana	V, S	2	ANIMn	Fr	6	1	Chill	Bet	Ca
+	Crataegus monogyna	S	?2	ANIMi	Fr	6	2	Warm+Chill	Ros	Cm
-	Cytisus scoparius	Bs	3	ANIMe	Sd	5	1	Scar	Fab	Cs
-	Euonymus europaeus	S	2	ANIMie	F/S	6	2	Chill	Cel	Ee
+	Fagus sylvatica	S	2	ANIMn	Fr	6	1	Chill	Fag	Fs
+	Fraxinus excelsior	W, S	2	WINDw	Fr	6	3	Chill	Ole	Fe
	Ilex aquifolium	S	?2	ANIMi	Fr	6	2	Scar+Chill	Aqu	Ia
	Ligustrum vulgare	V, S	2	ANIMi	F/S	6	2	Chill	Ole	Lv
	Lupinus arboreus	?Bs	?3	UNSP	Sd	6	1	Scar	Fab	La
	Mahonia aquifolium	S	?2	ANIMi	Fr	6	1	Not studied	Ber	Ma
	Malus sylvestris	S	2	ANIMi	F/S	6	1	Chill	Ros	Ms
	Populus tremula	V, W	1	WINDp	Sd	1	2	-	Sal	Pt
	Populus × canescens	V, W	1	WINDp	Sd	X		X	Sal	Pc
	Prunus avium	V, S	2	ANIMi	Fr	6	1	Chill	Ros	Pa
	Prunus padus	V, S	2	ANIMi	Fr	6	1	Not studied	Ros	Pp
-	Prunus spinosa	V, S	2	ANIMi	Fr	6	1	Chill	Ros	Ps
+	Quercus agg.	S	1	ANIMn	Fr	6	2	Chill	Fag	Qa
	Quercus cerris	S	1	ANIMn	Fr	6	2	Not studied	Fag	Qc
	Quercus petraea	S	1	ANIMn	Fr	6	2	Chill	Fag	Qp
	Quercus robur	S	1	ANIMn	Fr	6	2	Chill	Fag	Qr
	Rhamnus cathartica	S	?2	ANIMi	Sd	6	2	Not studied	Rha	Rc
	Rhododendron ponticum	V, W	?1	WINDcm	Sd	1	3	-	Eri	Rp
	Ribes nigrum	V, S	?2	ANIMi	F/S	5	2	Not studied	Gro	Rn
	Ribes rubrum	V, S	?2	ANIMi	F/S	5	2	Not studied	Gro	Rr
-	Rosa sp.	V, S	2	ANIMi	Fr	6	2	?Warm+Chill/Chill	Ros	Rs
	Rosa arvensis	V, S	2	ANIMi	Fr	5	2	?Warm+Chill/Chill	Ros	Ra
	Rosa canina	V, S	2	ANIMi	Fr	5	2	?Warm+Chill/Chill	Ros	Rc
	Rosa mollis	V, S	2	ANIMi	Fr	5	2	?Warm+Chill/Chill	Ros	Rm
	Salix alba	(V), W, S	1	WINDp	Sd	1	2		Sal	Sa

	Species	Regenerative strategies	Seed bank	Agency of dispersal	Dispersule and germinule form	Dispersule weight	Dispersule shape	Germination requirements	Family	(Sp.)
	Salix aurita	(V), W, S	1	WINDp	Sd	1	2	-	Sal	*Sa*
+	*Salix caprea/cinerea* agg.	(V), W, S	1	WINDp	Sd	1	2	-	Sal	*Sc*
	Salix caprea	(V), W, S	1	WINDp	Sd	1	2	-	Sal	*Sc*
	Salix cinerea	(V), W, S	1	WINDp	Sd	1	2	-	Sal	*Sc*
-	*Salix fragilis*	(V), W, S	1	WINDp	Sd	1	2	-	Sal	*Sf*
-	*Salix purpurea*	(V), W, S	1	WINDp	Sd	1	2	-	Sal	*Sp*
-	*Salix viminalis*	(V), W, S	1	WINDp	Sd	1	2	-	Sal	*Sv*
+	*Sambucus nigra*	S, ?Bs	?3	ANIMi	F/S	5	2	?Chill	Cap	*Sn*
+	*Sorbus aucuparia*	S	2	ANIMi	F/S	5	2	Chill	Ros	*Sa*
	Symphoricarpos albus	V, S	?2	ANIMi	F/S	5	2	Not studied	Cap	*Sa*
	Taxus baccata	S	1	ANIMi	F/S	6	1	Not studied	Tax	*Tb*
+	*Ulex europaeus*	Bs	4	ANIMe	Sd	5	1	Scar	Fab	*Ue*
	Ulex gallii	Bs	3	ANIMe	Sd	5	1	Scar	Fab	*Ug*
+	*Ulmus glabra*	W, S	2	WINDw	Fr	5	1	-	Ulm	*Ug*
-	*Ulmus procera*	V, W, S	2	WINDw	Fr	5	1	-	Ulm	*Up*
-	*Viburnum opulus*	S	2	ANIMi	F/S	5	1	Chill	Cap	*Vo*
(c) Pteridophytes										
-	*Asplenium adiantum-nigrum*	W	?	WINDm	Sp	S	1	-	Asp	*Aa*
+	*Asplenium ruta-muraria*	W	?	WINDm	Sp	S	1	-	Asp	*Ar*
+	*Asplenium trichomanes*	W	?	WINDm	Sp	S	1	-	Asp	*At*
+	*Athyrium filix-femina*	V, W	?	WINDm	Sp	S	2	-	Woo	*Af*
-	*Blechnum spicant*	W	?	WINDm	Sp	S	1	?-	Ble	*Bs*
	Ceterach officinarum	V, W	?	WINDm	Sp	S	1	Not studied	Asp	*Co*
+	*Cystopteris fragilis*	W	?	WINDm	Sp	S	2	-	Woo	*Cf*
-	*Dryopteris affinis*	V, W, ?Bs	?	WINDm	Sp	S	1	-	Dry	*Da*
	Dryopteris carthusiana	V, W, ?Bs	?	WINDm	Sp	S	1	-	Dry	*Dc*
+	*Dryopteris dilatata*	V, W, ?Bs	?	WINDm	Sp	S	1	?-	Dry	*Dd*
+	*Dryopteris filix-mas*	V, W, ?Bs	?3	WINDm	Sp	S	1	-	Dry	*Df*
+	*Equisetum arvense*	V, W, S	1	WINDm	Sp	S	1	-	Equ	*Ea*
+	*Equisetum fluviatile*	V, W, S	1	WINDm	Sp	S	1	-	Equ	*Ef*
+	*Equisetum palustre*	V, W, S	1	WINDm	Sp	S	1	-	Equ	*Ep*
	Equisetum sylvaticum	V, W, S	1	WINDm	Sp	S	1	-	Equ	*Es*
	Equisetum telmateia	V, W, S	1	WINDm	Sp	S	1	-	Equ	*Et*
	Gymnocarpium dryopteris	V, W	?	WINDm	Sp	S	1	Not studied	Woo	*Gd*
	Huperzia selago	(V), W	1	*WINDm	Sp	S	1	Not studied	Lyc	*Hs*
	Lycopodium clavatum	V, W, S	1	WINDm	Sp	S	1	Not studied	Lyc	*Lc*
	Oreopteris limbosperma	V, W	?	WINDm	Sp	S	1	Not studied	The	*Ol*
	Ophioglossum vulgatum	V, W, S	1	WINDm	Sp	S	1	Not studied	Oph	*Ov*
	Osmunda regalis	V, W, S	?	WINDm	Sp	S	1	Not studied	Osm	*Or*
	Phegopteris connectilis	V, W	?	WINDm	Sp	S	1	Not studied	The	*Pc*
-	*Phyllitis scolopendrium*	W	?	WINDm	Sp	S	1	-	Asp	*Ps*
-	*Polypodium vulgare*	V, W	?	WINDm	Sp	S	1	-	Ppo	*Pv*
	Polystichum aculeatum	V, W	?	WINDm	Sp	S	1	Not studied	Dry	*Pa*
	Polystichum setiferum	V, W	?	WINDm	Sp	S	1	Not studied	Dry	*Ps*
+	*Pteridium aquilinum*	V, W, ?Bs	?3	WINDm	Sp	S	1	-	Den	*Pa*

REFERENCES

AARSSEN, L. W. 1981. The biology of Canadian weeds. 50: Hypochoeris radicata L. Canadian Journal of Plant Science 61, 365-381.

AARSSSEN, L.W., I.V. HALL and K.I.N. JENSEN 1986. The biology of Canadian weeds. 76: Vicia angustifolia L., V. cracca L., V. sativa L., V. tetrasperma (L.) Shreb. and V. villosa Roth. Canadian Journal of Plant Science 66, 711-737.

ADAMS, A.W. 1955. Biological Flora of the British Isles: Succisa pratensis Moencha. Journal of Ecology 43, 709-718.

ADAS 1983a. Ministry of Agriculture, Fisheries and Food, Booklet 2054. Red clover. Alnwick: MAFF Publications.

ADAS 1983b. Ministry of Agriculture, Fisheries and Food, Leaflet 777. Barren brome. Alnwick: MAFF Publications.

ADAS 1983c. Ministry of Agriculture, Fisheries and Food, Leaflet 190. Bracken and its control. Alnwick: MAFF Publications.

ADAS 1984. National Institute of Agricultural Botany, Grass seed mixtures - list of recommended varieties. Alnwick: MAFF Publications.

AERTS, R. and M.J. VAN DER PEIJIL 1993. A simple model to explain the dominance of low-productive perennials in nutrient-poor habitats. Oikos 66, 144-146.

AKEROYD, J.R. and D. BRIGGS 1983a. Genecological studies of garden Rumex crispus. 1: Experiments using transplanted material. New Phytologist 94, 309-323.

AKEROYD, J.R. and D. BRIGGS 1983b. Genecological studies of garden Rumex crispus. 2: Variation in plants grown from wild collected seed. New Phytologist 94, 325-343.

AKEROYD, J.R., S.I. WARWICK and D. BRIGGS 1978. Variation in four populations of Senecio viscosus as revealed by a cultivation experiment. New Phytologist 81, 391-400.

AL-FARRAJ, M.M. 1983. The influence of environmental conditions on the growth of wetland plants with special reference to Epilobium hirsutum L. and E. palustre L. Ph.D. Thesis, University of Sheffield.

ALLEN, G.P. 1966. The botany, ecology, agronomy and control of Poa trivialis L., rough stalked meadow grass. Agricultural Research Council Weed Research Organisation, Technical Report no. 6.

ALLEN, T.F.H. and T.B. STARR 1982. Hierarchy: perspectives for ecological complexity. Chicago: University of Chicago Press.

AL-MASHHADANI, Y.D. 1980. Experimental investigations of competition and allelopathy in herbaceous plants. Ph.D. Thesis, University of Sheffield.

AL-MUFTI, M.M. 1978. A quantitative and phenological study of the herbaceous vegetation in the deciduous woodlands at Totley (South Yorkshire). Ph.D. Thesis, University of Sheffield.

AL-MUFTI, M.M., C.L. SYDES, S.B. FURNESS, J.P.GRIME and S.R. BAND 1977. A quantitative analysis of shoot phenology and dominance in herbaceous vegetation. Journal of Ecology 65, 759-792.

ALVEY, N.G., C.F. BANFIELD, R.I. BAXTER, J.C. GOWER, W.J. KRZANOWSKI, P.W. LANE, P.K. LEECH, J.A. NELDER, R.W. PAYNE, K.M. PHELPS, C.E. ROGERS, G.J.S. ROSS, H.R. SIMPSON, A.D. TODD, R.W.M. WEDDERBURN and G.M. WILKINSON 1980. GENSTAT: A general statistical program. Oxford: Numerical Algorithms Group Ltd.

AMOR, R.L. and R.G. RICHARDSON 1980. The biology of Australian weeds. 2: Rubus fruticosus agg. Journal of the Australian Institute of Agricultural Science 46, 87-97.

ANDERSON, P. and D. SHIMWELL 1981. Wild flowers and other plants of the Peak District; an ecological study. Ashbourne: Moorland Publishing.

ANDERSON, M. 1979. The development of plant habitats under exotic forest crops. In: Ecology and design in amenity land management, S.E. Wright and G.P. Buckley (eds), 87-108. Wye, Kent: Wye College and Recreation Ecology Research Group.

ANDERSON, V.L. 1927. The water economy of the chalk flora. (Studies of the vegetation of the English chalk 5.) Journal of Ecology 15, 72-129.

ANGEVINE, M.W. 1983. Variations in the demography of natural populations of the wild strawberries Fragaria vesca and F. virigiana. Journal of Ecology 71, 959-974.

ANKEI, T. 1982. Habitat gradients and reproductive habits of the seven Stellaria species in Japan. Botanical Magazine (Tokyo) 95, 35-48.

ANTONOVICS, J. 1972. Population dynamics of the grass Anthoxanthum odoratum on a zinc mine. Journal of Ecology 60, 351-365.

ANTONOVICS, J. 1976. The population genetics of mixtures. In: Plant Relations in Pastures, J.R. Wilson (ed.), 233-252. Melbourne: CSIRO.

ANTONOVICS, J. AND R.B. PRIMACK 1982. Experimental ecological genetics in Plantago. 6: The demography of seedling transplants of P. lanceolata. Journal of Ecology 70, 55-75.

ANTONOVICS, J., A.D. BRADSHAW and R.G. TURNER 1971. Heavy metal tolerance in plants. Advances in Ecological Research 7, 1-85.

ARMSTRONG, W. 1964. Oxygen diffusion from the roots of some British bog plants. Nature 204, 801-802.

ARMSTRONG, W. and D.J. BOATMAN 1967. Some field observations relating the growth of bog plants to conditions of soil aeration. Journal of Ecology 55, 101-110.

ARNOLD, H.R., L. FARRELL and F.H. PERRING 1978. Atlas of ferns of the British Isles, A.C. Jermy (ed.). Botanical Society of the British Isles and the British Pteridological Society.

ARNOLD, R.M. 1981. Population dynamics and seed dispersal of Chaenorrhinum minus on railroad cinder ballast. American Midland Naturalist 106, 80-91.

ATKINS, D.P., I.C. TRUEMAN, C.B. CLARKE and A.D. BRADSHAW 1982. The evolution of lead tolerance of Festuca rubra on a motorway verge. Environmental Pollution (Series A) 27, 234-241.

ATKINSON, D. and A.W. DAVISON 1971. The effects of phosphorus deficiency on the growth of Epilobium montanum, L. New Phytologist 70, 789-797.

AUSTIN, M.P. 1968. Pattern in a Bromus erectus dominated community. Journal of Ecology 56, 197-218.

AUSTIN, M.P. 1980. An exploratory analysis of grassland dynamics; an example of a lawn succession. Vegetatio 43, 87-94.

AVDULOV, N.P. 1931. Karyo-systematische Untersuchtungen der Familie Gramineen. Bulletin of Applied Botany 44, suppl. 43, 1-428.

AYRES, P. 1977. The growth of Arrhenatherum elatius var. bulbosum (Willd.) Spenn. in spring barley, as influenced by cultivation. Weed Research 17, 422-428.

BAILEY, J.P. and A.P. CONOLLY 1985. Chromosome numbers of some alien Reynoutria species in the British Isles. Watsonia 15, 270-271.

BAKER, A.J.M. 1981. Accumulators and excluders - strategies in the response of plants to heavy metals. Journal of Plant Nutrition 3, 643-654.

BAKER, H.G. 1947. Biological Flora of the British Isles: Melandrium dioicum (L. emend.) Coss. & Germ. Journal of Ecology 35, 283-292.

BAKER, H.G. 1972. Seed weight in relation to environmental conditions in California. Ecology 53, 997-1010.

BAKKER, D. 1960. A comparative life-history study of Cirsium arvense and Tussilago farfara, the most troublesome arable weeds in the newly reclaimed polders of the former Zuider Zee. In: The biology of weeds, J.L. Harper (ed.), 205-222. Oxford: Blackwell Scientific.

BAKKER, D. 1966. On life forms of hapaxanths in the Dutch flora. Wentia 15, 13-24.

BAKKER, J.P. 1989. Nature management by grazing and cutting. Dordrecht: Kluwer.

BAKSHI, T.S. and R.T. COUPLAND 1960. Vegetative propagation in Linaria vulgaris. Canadian Journal of Botany 38, 243-249.

BALLARD, L.A.T. 1973. Physical barriers to germination. Seed Science and Technology 1, 285-303.

BALME, O.E. 1953. Edaphic and vegetational zoning on the carboniferous limestone of the Derbyshire Dales. Journal of Ecology 41, 331-344.

BANNISTER, P. 1965. Biological Flora of the British Isles: Erica cinerea L. Journal of Ecology 53, 527-542.
BANNISTER, P. 1978. Flowering and shoot extension in heath plants of different geographical origin. Journal of Ecology 66, 117-131.
BARCLAY, A.M. and R.M.M. CRAWFORD 1982. Winter desiccation stress and resting bud viability in relation to high altitude survival in Sorbus aucuparia. Flora 172, 21-34.
BARCLAY, A.W. and R.M.M. CRAWFORD 1984. Seedling emergence in the Rowan (Sorbus aucuparia) from an altitudinal gradient. Journal of Ecology 72, 627-636.
BARLING, D.M. 1959. Biological studies in Poa angustifolia. Watsonia 4, 147-168.
BARLING, D.M. 1962. Studies in the biology of Poa subcaerulea Sm. Watsonia 5, 163-173.
BARR, C.J., BUNCE R.G.H., CLARKE, R.T., FULLER, R.M., FURSE, M.T., GILLESPIE, M.K., GROOM, G.B., HALLAM, C.J., HORNUNG, M., HOWARD, D.C. and M.J. NESS 1993. Countryside Survey 1990. London: H.M.S.O.
BASKIN, C.C. and J.M. BASKIN 1998. Seeds: Ecology, Biogeography and Evolution of Dormancy and Germination. San Diego: Academic Press.
BASKIN, J.M. and C.C. BASKIN 1975. Ecophysiology of seed dormancy and germination in Torilis japonica in relation to its life cycle strategy. Bulletin of the Torrey Botanical Club 102, 67-72.
BASKIN, J.M. and C.C. BASKIN 1978. A contribution to the germination ecology of Rumex crispus. Bulletin of the Torrey Botanical Club 105, 278-281.
BASKIN, J.M. and C.C. BASKIN 1983. Germination ecology of Veronica arvensis. Journal of Ecology 71, 57-68.
BASKIN, J.M. and C.C. BASKIN 1984. Role of temperature in regulating timing of germination in soil seed reserves of Lamium purpureum. Weed Research 24, 341-349.
BASKIN, J.M. and C.C. BASKIN 1985. Life cycle ecology of annual plant species of cedar glades of Southeastern United States. In: The population structure of vegetation, J. White (ed.), 371-398. Dordrecht: Junk.
BASSETT, I.J. and C.W. CROMPTON 1978. The biology of Canadian weeds. 32: Chenopodium album L. Canadian Journal of Plant Science 58, 1061-1072.
BASSETT, I.J., C.W. CROMPTON and D.W. WOODLAND 1977. The biology of Canadian weeds. 21: Urtica dioica L. Canadian Journal of Plant Science 57, 491-498.
BEADLE, N.C.W. 1966. Soil phosphate and its role in molding segments of the Australian flora and vegetation, with special reference to xeromorphy and sclerophylly. Ecology 46, 992-1007.
BEAN, W.J. 1970. Trees and shrubs hardy to the British Isles, 8th edn. Vol. 1: Species A-C. London: John Murray.
BEATTIE, A.J. 1969. The floral biology of three species of Viola. New Phytologist 68, 1187-1201.
BEATTIE, A.J. and N. LYONS 1975. Seed dispersal in Viola, adaptations and strategies. American Journal of Botany 62, 714-722.
BECKETT, K. and G. BECKETT 1979. Planting native trees and shrubs. Norwich: Jarrold & Sons.
BEDDOWS, A.R. 1931. Self and cross fertility and flowering habits of certain herbage grasses and legumes. Bulletin H12, Welsh Plant Breeding Station.
BEDDOWS, A.R. 1959. Biological Flora of the British Isles: Dactylis glomerata L. Journal of Ecology 47, 223-239.
BEDDOWS, A.R. 1961. Biological Flora of the British Isles: Holcus lanatus L. Journal of Ecology 49, 421-430.
BEDDOWS, A.R. 1967. Biological Flora of the British Isles: Lolium perenne L. Journal of Ecology 55, 567-587.
BEDDOWS, A.R. 1971. The inter- and intra-specific relationships of Holcus lanatus L. and H. mollis. Botanical Journal of the Linnean Society 64, 183-198.
BEDDOWS, A.R. 1973. Biological Flora of the British Isles: Lolium multiflorum Lam. Journal of Ecology 61, 587-600.
BELL, J.N.B. and J.H. TALLIS 1973. Biological Flora of the British Isles: Empetrum nigrum L. Journal of Ecology 61, 289-305.
BELL, P.R. 1985. Introduction: The essential role of the Pteridophyta in the study of land plants. PRS Edinburgh 86B, 1-4.
BENNETT, M.D. 1971. The duration of meiosis. Proceedings of the Royal Society of London 178B, 277-299.
BENNETT, M.D. 1972. Nuclear DNA content and minimum generation time in herbaceous plants. Proceedings of the Royal Society of London 181B, 109-135.
BENNETT, M.D. 1976. DNA amount, latitude and crop plant distribution. Environmental and Experimental Botany 16, 93-108.
BENNETT, M.D. 1987. Ecological implications of interspecific variation in genomic form. In: Frontiers of comparative plant ecology, I.H. Rorison, J.P. Grime, R. Hunt, G.A.F. Hendry and D.H. Lewis (eds). London: Academic Press.
BENNETT, M.D. and I.J. LEITCH 2004. Plant DNA C-values database (release 3.0, Dec. 2004) http://www.rbgkew.org.uk/cval/homepage.html.
BENNETT, M.D. and J.P. SMITH 1976. Nuclear DNA amounts in angiosperms. Philosophical Transactions of the Royal Society B274, 227-274.
BENNETT, M.D., J.P. SMITH and J.S. HESLOP-HARRISON 1982. Nuclear DNA amounts in angiosperms. Proceedings of the Royal Society of London B216, 179-199.
BENTLEY, S., J.B. WHITTAKER and A.J.C. MALLOCH 1980. Field experiments on the effects of grazing by a Chrysomelid beetle (Gastrophysa viridula) on seed production and quality in Rumex obtusifolius and R. crispus. Journal of Ecology 68, 671-674.
BERKO, I.M. 1979. Vital cycle of long rhizome vegetative juvenile Stachys sylvatica. Ukrayinski Botanichnyi Zhurnal 36, 147-152.
BERNATZKY, A. 1978. Tree ecology and preservation. Amsterdam: Elsevier.
BHAT, K.K. and P.H. NYE 1973. Diffusion of phosphate to plant roots in soil. 1: Quantitative autoradiography of the depletion zone. Plant and Soil 38, 161-175.
BIENIEK, M.E. and W.F. MILLINGTON 1968. Thorn formation of Ulex europaeus in relation to environmental and endogenous factors. Botanical Gazette 129, 145-150.
BIGGIN, P. 1982. Forestry and bracken. Proceedings of the Royal Society of Edinburgh 81B, 19-28.
BILLINGS, W.D. and H.A. MOONEY 1968. The ecology of arctic and alpine plants. Biological Reviews 43, 481-529.
BIRKS, H.J.B. 1973. Past and present vegetation of the Isle of Skye, a paleological study. Cambridge: Cambridge University Press.
BISHOP, G.F., A.J. DAVY and R.L. JEFFERIES 1978. Demography of Hieracium pilosella in a Breck grassland. Journal of Ecology 66, 615-629.
BJORK, S. 1967. Ecologic investigations of P. communis; studies in theoretic and applied limnology. Folia Limnologica Scandinavica 14, 1-248.
BJORKMAN, O. and P. HOLMGREN 1963. Adaptability of the photosynthetic apparatus to light intensity in ecotypes from exposed and shaded habitats. Physiologia Plantarum 16, 889-914.
BJORKMAN, O. and P. HOLMGREN 1966. Adaptation to light intensity in plants native to shaded and exposed habitats. Physiologia Plantarum 19, 854-859.
BJORKVIST, 1. 1967. Studies in Alisma L. I. Distribution, variation and germination. Opera Botanica 17, 5-128.
BLACK, M. 1969. Light controlled germination of seeds. Symposium of the Society of Experimental Biology 23, 193.
BLACK, M. and P.F. WAREING 1960. Photoperiodism in the light-inhibited seed of Nemophila insignis. Journal of Experimental Botany 11, 28-39.
BLACK, R.F. 1956. Effect of NaCl in water culture on the ion uptake and growth of Atriplex hastata L. Australian Journal of Biological Science 9, 67-80.
BLACK, R.F. 1958. Effect of sodium chloride on leaf succulence and area of Atriplex hastata L. Australian Journal of Botany 6, 306-321.
BLACKMAN, G.E. and A.J. RUTTER 1954. Biological Flora of the British Isles: Endymion non-scriptus (L.) Garcke. Journal of Ecology 42, 629-638.
BLOM, C.W.P.M. 1979. Effects of trampling and soil compaction on the occurrence of some Plantago species in coastal sand dunes. Ph.D. Thesis, University of Nijmegen.
BOCHER, T.W. 1949. Racial divergences in Prunella. New Phytologist 48, 285-314.
BOCHER, T.W. 1963. The study of ecotypical variation in relation to

experimental morphology. Regnum Vegetabile 27, 10-16.
BOCHER, T.W. 1975. Experimental and cytological studies on plant species. 13: Clinopodium vulgare L. Botanisk Tidsskrift 70, 152-179.
BOCHER, T.W. and K. LARSEN 1957. Cytotaxonomical studies in the Chrysanthemum leucanthemum complex. Watsonia 4, 11-16.
BOCHER, T.W., K. LARSEN and K. RAHN 1955. Experimental and cytological studies on plant species. 2: Trifolium Biologiske Skrifter 8, 1-31.
BOLKER, B.M. and S.W. PACALA 1999. Spatial moment equations for plant competition: understanding spatial strategies and the advantages of short dispersal. American Naturalist 153, 575-602.
BONNEMAISON, F. and D.A. JONES 1986. Variation in alien Lotus corniculatus L. 1. Morphological differences between alien and native British plants. Heredity 56, 129-138.
BOOT, R., D.J. RAYNAL and J.P. GRIME 1986. A comparative study of the influence of drought stress on flowering in Urtica dioica and U. urens. Journal of Ecology 74, 485-495.
BORG, P.J. 1971. Ecology of Equisetum palustre in Finland with special reference to its role as a noxious weed. Annals Botanica Fennici 8, 93-141.
BOOTH, R.E. and J.P. GRIME 2003. The effects of genetic impoverishment on plant community diversity. Journal of Ecology 91, 721-730.
BORNKAMM, R. 1961. Zur Konkurrenzkraft von Bromus erectus. Ein sechsjahriger Dauerversuch. Botanische Jahrbucher 80, 466-479.
BORRILL, M. 1958. A biosystematic study of some Glyceria species in Britain. 4: Breeding systems, fertility relationships and general discussion. Watsonia 4, 89-100.
BOSBACH, K., H. HURKA and R. HAASE 1982. The soil seed bank of Capsella bursa-pastoris: its influence on population variability. Flora 172, 47-56.
BOSTOCK, S.J. 1978. Seed germination strategies of five perennial weeds. Oecologia 36, 113-126.
BOSTOCK, S.J. and R.A. BENTON 1979. The reproductive strategies of five perennial weeds. Journal of Ecology 67, 91-107.
BOURDOT, G.W., D.J. SAVILLE and R.J. FIELD 1984. The response of Achillea millefolium to shading. New Phytologist 97, 653-663.
BOX, E.O. 1981. Macroclimate and plant forms; an introduction to predictive modelling in phytogeography. The Hague: Junk.
BRADLEY, M.V. 1954. Cell and nuclear size in relation to polysomaty and the nuclear cycle. American Journal of Botany 41, 398-402.
BRADLEY, R., A.J. BURT and D.J. READ 1982. The biology of mycorrhiza in the Ericaceae. 8: The role of mycorrhizal infection in heavy metal resistance. New Phytologist 91, 197-209.
BRADSHAW, A.D. 1958. Natural hybridisation of Agrostis tenuis Sibth. and A. stolonifera L. New Phytologist 57, 66.
BRADSHAW, A.D. 1959a. Population differentiation in Agrostis tenuis. 1: Morphological differentiation. New Phytologist 58, 208-227.
BRADSHAW, A.D. 1959b. Population differentiation in Agrostis tenuis. 3: Populations in varied environments. New Phytologist 59, 92-103.
BRADSHAW, A.D. 1987. Comparison - its scope and limits. In: Frontiers of comparative plant ecology, I.H. Rorison, J.P. Grime, R. Hunt, G.A.F. Hendry and D.H. Lewis (eds), 3-22. London: Academic Press.
BRADSHAW, A.D. and M.J. CHADWICK 1980. The restoration of land. Oxford: Blackwell Scientific.
BRADSHAW, A.D., M.J. CHADWICK, D. JOWETT and R.W. SNAYDON 1964. Experimental investigations into the mineral nutrition of several grass species 4. Nitrogen level. Journal of Ecology 52, 665-676.
BRENCHLEY, W.E. 1918. Buried weed seeds. Journal of Agricultural Science 9, 1-31.
BRENCHLEY, W.E. 1920. Weeds of farmland. London: Longman.
BRENCHLEY, W.E. and K. WARINGTON 1930. The weed seed population of arable soil. 1: Numerical estimation of viable seeds and observations on their natural dormancy. Journal of Ecology 18, 235-272.
BRENCHLEY, W.E. and K. WARINGTON 1958. The Park Grass plots at Rothamsted 1856-1949. Harpenden: Rothamsted Experimental Station.

BRIGGS, D. 1978. Genecological studies of salt-tolerance in groundsel (Senecio vulgaris L.) with particular reference to roadside habitats. New Phytologist 81, 381-389.
BRITTEN, R.J. and E.H. DAVIDSON 1971. Repetitive and non-repetitive DNA sequences and a speculation on the origins of evolutionary novelty. Quarterly Review of Biology 46, 111-133.
BRITTON, C.E. 1923. Centaurea scabiosa L. Varieties and a hybrid. Reprints of the Botanical Society Exchange Club of the British Isles 6, 767-773.
BROOKER, R.W. and T.V. CALLAGHAN 1998. The balance between positive and negative plant interactions and its relationship to environmental gradients: a model. Oikos 81, 196-207.
BROWN, A.H.F. and L. OOSTERHUIS 1981. The role of buried seeds in coppice woods. Biological Conservation 21, 329-330.
BROWN, J.M.B. 1953. Studies on British beechwoods. Forestry Commission Bulletin 20. London: HMSO.
BROWNLEE, C., J.A. DUDDRIDGE, A. MALIBARI and D.J. READ 1983. The structure and function of mycelial systems of ectomycorrhizal roots with special reference to their role in forming inter-plant connections and providing pathways for assimilate and water transport. Plant and Soil 71, 433-443.
BRUNSBERG, K. 1977. Biosystematics of the Lathyrus pratensis complex. Opera Botanica 41, 1-78.
BRYANT, J.D. and P.S. KUROPAT 1980. Selection of winter forage by subarctic browsing vertebrates; the role of plant chemistry. Annual Review of Ecological Systems 11, 261-285.
BUCKLAND, S.M. and J.P. GRIME 2000. The effects of trophic structure and soil fertility on the assembly of plant communities: a microcosm approach. Oikos 91, 336-352.
BUCKLAND, S.M., J.P. GRIME, K. THOMPSON and J.G. HODGSON 1997. A comparison of responses to the extreme drought of 1995 in Northern England. Journal of Ecology 85, 875-882.
BUDD, E.G. 1970. Seasonal germination patterns of Poa trivialis and subsequent plant behaviour. Weed Research 10, 243-249.
BUNCE R.G.H., BARR, C.J., GILLESPIE, M.K., HOWARD, D.C., SCOTT, W.A., SMART, S.M., VAN DE POLL, H.M. and J.W. WATKINS 1999. Vegetation of the British countryside - the Countryside Vegetation System. London: Department of the Environment, Transport and the Regions.
BURDON, J.J. 1983. Biological Flora of the British Isles: Trifolium repens L. Journal of Ecology 71, 307-330.
BURDON, J.J. and J.L. HARPER 1980. Relative growth rates of individual members of a plant population. Journal of Ecology 68, 953-957.
BURKHILL, I.H. 1944. Biological Flora of the British Isles: Tamus communis L. Journal of Ecology 32, 121-129.
BURNETT, J.H. 1964. The vegetation of Scotland. Edinburgh: Oliver & Boyd.
BURTON, R.M. 1980. Solidago x niederedi Khek in Britain. Watsonia 13, 123-124.
BURT-SMITH, G.S., J.P. GRIME and D. TILMAN 1998. Seedling resistance to herbivory as a predictor of relative abundance in a synthesised prairie community. Oikos 81, 99-108.
BUTTERY, B.R. and J.M. LAMBERT 1965. Competition between Glyceria maxima and Phragmites communis in the region of Surlingham Broad. 1: The competition mechanism. Journal of Ecology 53, 163-181.
BUTTERY, B.R., W.T. WILLIAMS and J.M. LAMBERT 1965. Competition between Glyceria maxima and Phragmites communis in the region of Surlingham Broad. 2: The fen gradient. Journal of Ecology 53, 183-195.
BYATT, J.I. 1975. Hybridisation between Crataegus monogyna and Crataegus laevigata in southeastern England. Watsonia 10, 253-264.
CACCIANIGA M., LUZZARO A., PIERCE S., CERIANI R.M., CERABOLINI B. 2006. The functional basis of a primary succession resolved by CSR classification. Oikos 112, 10-20.
CALLAGHAN, T.V. and E. SHEFFIELD 1985. Pteridium aquilinum: weed or resource? PRS Edinburgh 86B, 461.
CALLAWAY, R.M., BROOKER, R.W., CHOLER, P., KIKVIDZE, Z., LORTIE, C.J., MICHALET, R., PAOLINI L., PUGNAIRE, F.I., NEWINGHAM, B. ASCHEHOUG, E.T., ARMAS, C., KIKODZE,

D. and B.J. COOK 2002. Positive interactions among alpine plants increase with stress. Nature 417, 844-848.
CAMPBELL, B.D. and J.P. GRIME 1989. A comparative study of plant responsiveness to the duration of episodes of mineral nutrient enrichment. New Phytologist 112, 261-267.
CAMPBELL, B.D. and J.P. GRIME 1992. An experimental test of plant strategy theory. Ecology 73, 15-29.
CAMPBELL, B.D., J.P. GRIME and J.M.L. MACKEY 1991. A trade-off between scale and precision in resource foraging. Oecologia 87, 532-538.
CAMPBELL, M.H. 1985. Germination, emergence and seedling growth of Hypericum perforatum L. Weed Research 25, 259-266.
CAMPBELL, M.H. and E.S. DELFOSSE 1984. The biology of Australian weeds. 13: Hypericum perforatum L. Journal of the Australian Institute of Agricultural Science 50, 63-73.
CARROLL, C.P. and K. JONES 1962. Cytotaxonomic studies in Holcus. 3: A morphological study of the triploid FI hybrid between H. lanatus and H. mollis. New Phytologist 61, 73-84.
CARTER, R.N. and S.D. PRINCE 1981. Epidemic models used to explain biogeographical distribution limits. Nature 293, 644-645.
CASWELL, H. 1989. Matrix population models. Sunderland: Sinauer.
CASWELL, S.A., I.F. HOLMES and D.A. SPEARS 1984b. Total chlorine in coal seam profiles from the South Staffordshire (Cannock) coalfield. Fuel 63, 782-787.
CASWELL, H. 1989. Matrix population models. Sunderland: Sinauer.
CATLING, P.M. and I. DOBSON 1985. The biology of Canadian weeds. 69: Potamogeton crispus L. Canadian Journal of Plant Science 65, 655-668.
CAVALIER-SMITH, T. 1978. The evolutionary origin and phylogeny of microtubules, mitotic spindles and eukaryote flagella. BioSystems 10, 93-114.
CAVALIER-SMITH, T. 1980. How selfish is DNA? Nature 285, 617-618.
CAVALIER-SMITH, T. 1982. Skeletal DNA and the evolution of genome size. Annual Review of Biophysics and Bioengineering 11, 273-302.
CAVERS, P.B. and J.L. HARPER 1964a. Biological Flora of the British Isles: Rumex crispus L. Journal of Ecology 52, 754-766.
CAVERS, P.B. and J.L. HARPER 1964b. Biological Flora of the British Isles: Rumex obtusifolius L. Journal of Ecology 52, 737-754.
CAVERS, P.B. and J.L. HARPER 1966. Germination polymorphism in Rumex crispus and R. obtusifolius. Journal of Ecology 54, 367-382.
CAVERS, P.B. and J.L. HARPER 1967a. Studies in the dynamics of plant populations. 1: The fate of seeds and transplants introduced into various habitats. Journal of Ecology 55, 59-71.
CAVERS, P.B. and J.L. HARPER 1967b. The comparative biology of closely related species living in the same area. 9: Rumex. Journal of Ecology 55, 73-82.
CAVERS, P.B., J. BASSETT and C.W. CROMPTON 1980. The biology of Canadian weeds. 47: Plantago lanceolata L. Canadian Journal of Plant Science 60, 1269-1282.
CAVERS, P.B., M.I. HEAGY and R.F. KOKRON 1979. The biology of Canadian weeds. 35: Alliaria petiolata (M.Bieb.) Cavara and Grande. Canadian Journal of Plant Science 59, 217-229.
CHABOT, B.F. and H.A. MOONEY 1985. Physiological ecology in North American plant communities. New York: Chapman and Hall.
CHADWICK, M.J. 1960. Biological Flora of the British Isles: Nardus stricta L. Journal of Ecology 48, 255-267.
CHAMPNESS, S.S. and K. MORRIS 1948. The population of buried viable seeds in relation to contrasting pasture and soil types. Journal of Ecology 36, 149-173.
CHANCELLOR, R.J. 1979. Grass seeds beneath pastures. In: Changes in sward composition and productivity. Occasional symposium no. 10., British Grassland Society, A.H. Charles and R.J. Haggar (eds), 147-150. Maidenhead, Berkshire: British Grassland Society.
CHANCELLOR, R.J. 1985. Changes in the weed flora of an arable field cultivated for twenty years. Journal of Applied Ecology 22, 491-501.
CHANCELLOR, R.J. and R.J. FROUD-WILLIAMS 1984. A second survey of cereal weeds in central southern England. Weed Research 24, 29-36.
CHAPIN, F.S. 1980. The mineral nutrition of wild plants. Annual Review of Ecology and Systematics 11, 233-260.
CHAPIN, F.S., G.R. SHAVER and R.A. KEDROWSKI 1986. Environmental controls over carbon, nitrogen and phosphorus fractions in Eriophorum vaginatum in Alaskan tussock tundra. Journal of Ecology 74, 167-195.
CHARLTON, J.F.L. 1977. Establishment of pasture legumes in North Island hill country. 1: Buried seed populations. New Zealand Journal of Experimental Agriculture 5, 211-214.
CHARLTON, W.A. 1973. Studies in the Alismataceae. 2: Inflorescences of Alismataceae. Canadian Journal of Botany 51, 775-789.
CHARNOV, E.L. 1979. Simultaneous hermaphroditism and sexual selection. Proceedings of the National Academy of Sciences of the United States of America 79, 2480-2484.
CHEFFINGS, C.M., L.FARRELL (Eds.), T.D. DINES, R.A. JONES, S.J. LEACH, D.R. McKEAN, D.A. PEARMAN, C.D. PRESTON, F.J. RUMSEY and I. TAYLOR 2005. The Vascular Plant List for Great Britain. Species Status 7, 1-116. Peterborough: Joint Nature Conservation Committee.
CHEN, S.S.C. and K.V. THIMAN 1964. Studies on the germination of light-inhibited seeds of Phacelia tanacetifolia. Israel Journal of Botany 13, 57-73.
CHEPIL, W.S. 1946. Germination of weed seeds. 1: Longevity, periodicity of germination and vitality of seeds in cultivated soil. Science in Agriculture 26, 307-346.
CHERFAS, J. 1985. When is a tree more than a tree? New Scientist 111 No.1461, 42-45.
CHIARIELLO, N., J.C. HICKMAN and H.A. MOONEY 1982. Endomycorrhizal role for interspecific transfer of phosphorus in a community of annual plants. Science 217, 941-943.
CHIPPINDALE, H.G. and W.E.J. MILTON 1934. On the viable seeds present in the soil beneath pastures. Journal of Ecology 22, 508-531.
CHRTKOVA-ZERTOVA, A. 1973a. A cytotaxonomic study of the Vicia cracca complex. 2: The taxa of the Southern part of Northern Europe. Folia Geobotanica et Phytotaxonomica 8, 249-254.
CHRTKOVA-ZERTOVA, A. 1973b. A monographic study of Lotus corniculatus L. 1: Central and Northern Europe. Rozpravy Ceskoslovenske Akademie Ved 83, 194.
CHURCH, A.H. 1908. Types of floral mechanism. Oxford: Clarendon Press.
CIDECIYAN, M.A. and J.C. MALLOCH 1982. Effects of seed size on the germination, growth and competitive ability of Rumex crispus and R. obtusifolius. Journal of Ecology 70, 227-232.
CLAPHAM, A.R. 1953. Human factors contributing to a change in our flora. In: The changing face of Britain, J.E. Lousley (ed.), 26-39. Oxford: Botanical Society of the British Isles.
CLAPHAM, A.R. 1956. Autecological studies and the 'Biological Flora of the British Isles'. Journal of Ecology 44, 1-11.
CLAPHAM, A.R., T.G. TUTIN and E.F. WARBURG 1952. Flora of the British Isles. Cambridge: Cambridge University Press.
CLAPHAM, A.R., T.G. TUTIN and E.F. WARBURG 1962. Flora of the British Isles, 2nd edn. Cambridge: Cambridge University Press.
CLAPHAM, A.R., T.G. TUTIN and E.F. WARBURG 1981. Excursion flora of the British Isles, 3rd edn. Cambridge: Cambridge University Press.
CLAPHAM, A.R, T.G. TUTIN and D.M. MOORE 1987. Flora of the British Isles, 3rd edn. Cambridge: Cambridge University Press.
CLARK, S.C. 1969. Some effects of temperature and photoperiod on growth and floral development in three winter-annuals. New Phytologist 68, 1137-1144.
CLARK, S.C. 1974. Biological Flora of the British Isles: Catapodium rigidum (L.) C.E. Hubbard. Journal of Ecology 62, 937-958.
CLARK, S.C. 1980a. Reproductive and vegetative performance in two winter-annual grasses Catapodium rigidum amd C. marinum. 1: The effects of soil and genotype on reproductive performance in the field and in a growth room. New Phytologist 84, 59-78.
CLARK, S.C. 1980b. Reproductive and vegetative performance in two winter-annual grasses Catapodium rigidum and C. marinum. 2: Leaf demography and its relationship to the production of caryopses. New Phytologist 84, 79-93.
CLATWORTHY, J.N. and J.L. HARPER 1962. The comparative biology of closely related tree species living in the same area. 5 : Inter- and

intra-specific interference within cultures of Lemna spp. and Salvinia natans. Journal of Experimental Botany 13, 307-324.

CLAUSEN, J., KECK, D.D. and HIESEY, W.M. 1940. Experimental studies on the nature of species: I. Effect of varied environments on western North American plants. Carnegie Institute Washington Pub. 520.

CLOUGH, J.M., J.A. TEERI and R.I. ALBERTE 1979. Photosynthetic adaptation of Solanum dulcamara to sun and shade environments. Oecologia 38, 13-21.

CLUTTON-BROCK, T.H. and P.H. HARVEY 1979. Comparison and adaptation. Proceedings of the Royal Society of London 205B, 547-565.

COCKBURN, W. 1985. Variation in photosynthetic acid metabolism in vascular plants; CAM and related phenomena. New Phytologist 101, 3-24.

CODY, W.J. and C.W. CROMPTON 1975. The biology of Canadian weeds. 15: Pteridium. aquilinum (L.) Kuhn. Canadian Journal of Plant Science 55, 1059-1072.

CODY, W.J. and V. WAGNER 1980. The biology of Canadian weeds. 49: Equisetum arvense L. Canadian Journal of Plant Science 61, 123-133.

COLASANTI, R.L. and R. HUNT, 1997. Resource dynamics and plant growth; a self-assembling model for individuals, populations and communities. Functional Ecology 11, 133-145.

COLBRY, V.L. 1953. Factors affecting the germination of reed canary grass. Proceedings of the Association of Official Seed Analysts 43, 50-56.

COLEY, P.D. 1983. Herbivory and defensive characteristics of tree species in a lowland tropical forest. Ecological Monographs 53, 209-233.

COLEY, P.D. 1987. Inter- and intra-specific variation in plant antiherbivore properties. In: Frontiers of comparative plant ecology, I.H. Rorison, J.P. Grime, R. Hunt, G.A.F. Hendry and D.H. Lewis (eds), 251-264. London: Academic Press.

COLEY, P.D., J.P. BRYANT and F.S. CHAPIN 1985. Resource availability and plant antiherbivore defence. Science 230, 895-899.

COMMONER, B. 1964. Roles of deoxyribonucleic acid in inheritance. Nature 202, 960-968.

CONNELL, J.H. 1980. Diversity and the coevolution of competitors, or the ghost of competition past. Oikos 35,131-138.

CONOLLY, A.P. 1977. The distribution and history in the British Isles of some alien species of Polygonum and Reynoutria. Watsonia 11, 291-311.

CONWAY, E. 1949. The autecology of bracken (Pteridium aquilinum Kuhn). The germination of the spore, the development of the prothallus and the young sporophyte. Proceedings of the Royal Society of Edinburgh 163B, 325-343.

CONWAY, E. 1953. Spore and sporeling survival in bracken. Journal of Ecology 41, 289-294.

CONWAY, E. 1957. Spore production in bracken. Journal of Ecology 45, 273-284.

COOK, C.D.K. 1962. Biological Flora of the British Isles: Sparganium erectum L. Journal of Ecology 50, 247-255.

COOK, C.D.K. 1966. A monographic study of Ranunculus subgenus Batrachium (D.C.) A. Gray. Mitteilungen der Botanischen Staatssammlung Munchen 6, 47-237.

COOK, C.D.K. 1969. On the determination of leaf form in Ranunculus aquatilis. New Phytologist 68, 469-480.

COOK, S.A. and M.P. JOHNSON 1968. Adaptation to heterogeneous environments. 1: Variation in heterophylly in Ranunculus flammula L. Evolution 22, 495-516.

COOMBE, D.E. and L.C. FROST 1956. The heaths of the Cornish Serpentine. Journal of Ecology 44, 226-256.

COOPER, A. 1976. The vegetation of carboniferous limestone soils in South Wales. 2: Ecotypic adaptations in response to calcium and magnesium. Journal of Ecology 64, 147-155.

COOPER, M.R. and A.W. JOHNSON 1984. Poisonous plants in Britain and their effects on animals and man. Ministry of Agriculture, Fisheries and Food, Reference Book 161. London: HMSO.

COOPER-DRIVER, G. 1976. Chemotaxonomy and phytochemical ecology of bracken. Botanical Journal of the Linnean Society 73, 35-46.

COOPER-DRIVER, G. 1985. Anti-predation strategies in pteridophytes - a biochemical approach. Proceedings of the Royal Society of Edinburgh 86B, 397-402.

COPE, T.A. and C.A. STACE 1978. The Juncus bufonius L. aggregate in Western Europe. Watsonia 12, 113-128.

COPE, T.A. and C.A. STACE 1985. Cytology and hybridisation in the Juncus bufonius aggregate in Western Europe. Watsonia 15, 309-320.

CORKILL, L. 1952. Cyanogenesis in white clover (Trifolium repens L.) 6. Experiments with high glucoside and glucoside-free strains. New Zealand Journal of Science and Technology 34, 1A-16A.

CORNELISSEN, J.H.C. 1996. An experimental comparison of leaf decomposition rates in a wide range of temperate plant species and types. Journal of Ecology 84, 573-582.

CORNELISSEN, J.H.C. 1999. Leaf structure and defence control litter decomposition rate across species, life forms and continents. New Phytologist 143, 191-200.

COURTNEY, A.D. 1968. Seed dormancy and field emergence in Polygonum aviculare. Journal of Applied Ecology 5, 675-684.

COX, P. 1981. Niche partitioning between sexes of dioecious plants. American Naturalist 117, 295-307.

CRAINE, J.M., J. FARGIONE and S. SUGITA 2005. Supply pre-emption, not concentration reduction, is the mechanism of competition for nutrients. New Phytologist 166, 933-946.

CRAWFORD, R.M.M. 1982. Physiological responses to flooding. In: Encyclopedia of plant physiology 12B, A. Pirson and M.H. Zimmermann (eds), 453-478. Berlin: Springer.

CRAWLEY, M.J. and R.M. MAY 1987. Population dynamics and plant community structure: competition between annuals and perennials. Journal of Theoretical Biology 125, 475-489.

CRESSWELL, E.G. 1982. The developmental origin and ecological consequences of seed germination responses to light. Ph.D. Thesis, University of Sheffield.

CRESSWELL, E.G. and J.P. GRIME 1981. Induction of a light requirement during seed development and its ecological consequences. Nature 291, 583-585.

CRICK, J.C. 1985. The role of plasticity in resource acquisition by higher plants. Ph.D. Thesis, University of Sheffield.

CRONQUIST, A. 1968. The evolution and classification of flowering plants. London: Nelson.

CRONQUIST, A. 1981. An integrated system of classification of flowering plants. New York: Columbia University Press.

CULVER, D.C. and A.J. BEATTIE 1980. The fate of Viola seeds dispersed by ants. American Journal of Botany 67, 710-714.

CUMMING, B.G. 1969. Chenopodium rubrum L. and related species. In: The induction of flowering, L.T. Evans (ed.), 156-185. New York: Cornell University Press.

CUNNINGHAM, S.A., B. SUMMERHAYES and M. WESTOBY 1999. Evolutionary divergences in leaf structure and chemistry, comparing rainfall and soil nutrient gradients. Ecology 69, 569-588.

DADAY, H. 1954a. Gene frequencies in wild populations of Trifolium repens L. 1: Distribution by latitude. Heredity 8, 61-78.

DADAY, H. 1954b. Gene frequencies in wild populations of Trifolium repens L. 2: Distribution by altitude. Heredity 8, 377-384.

DAHLGREN, R.M.T., H.T. CLIFFORD and P.F. YEO 1985. The families of the Monocotyledons; structure, evolution and taxonomy. New York: Springer.

DALE, A. and T.T. ELKINGTON 1974. Variation within Cardamine pratensis L. in England. Watsonia 10, 117.

DALE, H.M. 1974. The biology of Canadian weeds. 5: Daucus carota L. Canadian Journal of Plant Science 54, 673-685.

DANCER, W.S., J.F. HANDLEY and A.D. BRADSHAW 1977. Nitrogen accumulation in kaolin mining wastes in Cornwall, Britain. 2: Forage legumes. Plant and Soil 48, 303-314.

DARLINGTON, C.D. 1965. Cytology. London: Churchill.

DARLINGTON, H.T. and G.P STEINBAUER 1961. The eighty year period for Dr Beal's seed viability experiment. American Journal of Botany 48, 321-325.

DARWIN C. 1859. The origin of species by means of natural selection or the preservation of favoured races in the struggle for life. London: Murray.

DAVIDSON, E.H. and R.J. BRITTEN 1973. Organisation, transcription and regulation in the animal genome. Quarterly Review of Biology 48, 565-613.

DAVIES, M.S. 1984. The response of contrasting populations of Erica cinerea and Erica tetralix to soil type and waterlogging. Journal of Ecology 72, 197-208.

DAVIES, M.S. and R.W. SNAYDON 1974. Physiological differences among populations of Anthoxanthum odoratum. collected from the Park Grass experiment, Rothamsted. 3: Response to phosphates. Journal of Applied Ecology 11, 699-707.

DAVISON, A.W. 1964. Some factors affecting seedling establishment in calcareous soils. Ph.D. Thesis, University of Sheffield.

DAVISON, A.W. 1977. The ecology of Hordeum murinum L. 3: Some effects of adverse climate. Journal of Ecology 65, 523-530.

DAVY, A.J. 1980. Biological Flora of the British Isles: Deschampsia caespitosa (L.) Beauv. Journal of Ecology 68, 1075-1096.

DAWSON, F.H. 1976. The annual production of the aquatic macrophyte Ranunculus penicillatus var. calcareus R.W. Butcher) C.D.K. Cook. Aquatic Botany 2, 51-73.

DAY, R.T., P.A. KEDDY and J. MC NEIL 1988. Fertility and disturbance gradients: a summary model for riverine marsh vegetation. Ecology 69, 1044-1054.

DE ANGELIS, D.L. 1980. Energy flow, nutrient cycling and ecosystem resilience. Ecology 61, 764-771.

DE JONG, T. and P. KLINKHAMER 1986. Population ecology of the biennials Cirsium vulgare and Cynoglossum officinale: an experimental and theoretical approach. Utrecht: Drukkerij Elinkwijk.

DENNIS, R.L.H., J.G. HODGSON, R. GREYNER, T.G. SHREEVE and D.B. ROY 2004. Host plants and butterfly biology: do host plant strategies drive butterfly status? Ecological Entomology 29, 12-26.

DENNY, P. 1980. Solute movement in submerged angiosperms. Biological Review 55, 65-92.

DIAMOND, J.M. 1975. Assembly of species communities. In Ecology and Evolution of communities. Eds. Cody ML and Diamond JM, 342-444. Belknap Press.

DÍAZ, S. and M. CABIDO 1997. Plant functional types and ecosystem function in relation to global change. Journal of Vegetation Science 8, 463-474.

DÍAZ, S., J.G. HODGSON, K. THOMPSON, M. CABIDO, J.H.C. CORNELISSEN, A. JALILI, G. MONTSERRAT-MARTÍ, J.P. GRIME, F. ZARRINKAMAR, Y. ASRI, S.R. BAND, S. BASCONCELO, P. CASTRO-DÍEZ, G. FUNES, B. HAMZEHEE, M. KHOSHNEVIS, N. PÉREZ-HARGUINDEGUY, M.C. PÉREZ-RONTOMÉ, F.A. SHIRVANY, F. VENDRAMINI, S. YAZDANI, R. ABBAS-AZIMI, A. BOGARD, S. BOUSTANI, M. CHARLES, M. DEHGHAN, L. DE TORRES-ESPUNY, J. GUERRERO-CAMPO, A. HYND, G. JONES, E. KOWSARY, F. KAZEMI-SAEED, M. MAESTRO-MARTÍNEZ, A. ROMO-DÍEZ, S. SHAW, B. SIAVASH, P. VILLAR-SALVADOR and M.R. ZAK, 2004. The plant traits that drive ecosystems: evidence from three continents. Journal of Vegetation Science 15, 295-304.

DICKINSON, T.A. 1985. The biology of Canadian weeds. 68 : Crataegus crus-galli L. sensu lato. Canadian Journal of Plant Science 65, 641-654.

DIMBLEBY, G.W. 1953. Natural regeneration of pine and birch on the heather moors of North-East Yorkshire. Journal of Forestry 26, 41-52.

DIRZO, R. 1980. Experimental studies on slug-plant interactions. 1: The acceptability of thirty plant species to the slug, Agriolimax carvanae. Journal of Ecology 68, 981-998.

DIXON, A.F.G. 1971. The role of aphids in wood formation. 1: The role of the sycamore aphid, Drepanosiphum platanoides (Schr.) (Aphididae), on the growth of sycamore, Acer pseudoplatanus (L.). Journal of Applied Ecology 8, 165-179.

DOLL, R. 1977. Grundriss der Evolution der Gattung Taraxacum Zinn. Ph.D. Thesis, University of Berlin.

DOLPH, G.E. and D.L. DILCHER 1980. Variation in leaf size with respect to climate in the tropics of the Western Hemisphere. Bulletin of the Torrey Botanical Club 107, 154-162.

DONALD, C.M. 1958. The interaction of competition for light and for nutrients. Australian Journal of Agricultural Research 9, 421-432.

DONELAN, M. and K. THOMPSON 1980. Distribution of buried viable seeds along a successional series. Biological Conservation 17, 297-311.

DOOLITTLE, W.F. and C. SAPIENZA 1980. Selfish genes, the phenotype paradigm and genome evolution. Nature 284, 601-603.

DORPH-PETERSEN, K. 1924. Examination of the occurrence and vitality of various weed species under different conditions, made at the Danish seed testing station during the years 1896-1923. Report of the Fourth International Seed Testing Congress, 124-138.

DOSIER, L.W. and J.L. RIOPEL 1978. Origin, development and growth of differential trichoblasts in Elodea canadensis. American Journal of Botany 65, 813-822.

DOUST, J.L. 1980. Floral sex ratios in andromonoecious Umbelliferae. New Phytologist 85, 265-273.

DOUST, J.L. and L.L. DOUST 1982. Life-history patterns in British Umbelliferae: A review. Botanical Journal of the Linnean Society 85, 179-194.

DOUST, L.L. 1981a. Population dynamics and local specialisation in a clonal perennial Ranunculus repens. 1: The dynamics of ramets in contrasting habitats. Journal of Ecology 69, 743-755.

DOUST, L.L. 1981b. Population dynamics and local specialisation in a clonal perennial Ranunculus repens. 2: The dynamics of leaves and a reciprocal transplant-replant experiment. Journal of Ecology 69, 757-768.

DOVER, G.A. and R.B. FLAVELL (eds) 1982. Genome evolution. London: Academic Press.

DRING, M.J. 1965. The influence of shaded conditions on the fertility of bracken. British Fern Gazette 9, 222-227.

DUCKETT, J.G. and A.R. DUCKETT 1980. Reproductive biology and population dynamics of wild gametophytes of Equisetum. Botanical Journal of the Linnean Society 80, 1-40.

DUKE, J.A. 1981. Handbook of legumes of world economic importance. New York: Plenum Press.

DURING, H.J., A.J. SCHENKEVELD, H.J. VERKAAR and J.H. WILLEMS 1985. The population structure of vegetation. In: The population structure of vegetation, J. White (ed.), 341-370. Dordrecht: Junk.

DYER, A.F. and P.R.H. HADFIELD 1985. Polymorphism for cyanogenesis in British bracken (Pteridium aquilinum, ssp. aquilinum var. aquilinum). Proceedings of the Royal Society of Edinburgh 86B, 462-464.

EDWARDS, K.E. 1962. The Peak District. London: Collins.

EDWARDS, K.E. (ed.) 1966. Nottingham and its region. British Association for the Advancement of Science. Nottingham: Derry.

EDWARDS, M. 1980. Aspects of the population ecology of charlock. Journal of Applied Ecology 17, 151-171.

EDWARDS, P.J. and S.D. WRATTEN 1982. Wound induced changes in palatability in birch (Betula pubescens Ehrh. ssp. pubescens). American Naturalist 120, 816-818.

EDWARDS, P.J. and S.D. WRATTEN 1985. Induced plant defences against insect grazing; fact or artefact? Oikos 44, 70-74.

EGUNJOBI, J.K. 1971. Ecosystem processes in a stand of Ulex europaeus L. 1: Dry matter production, litter fall and efficiency of solar energy utilisation. Journal of Ecology 59, 31-38.

ELLENBERG, H. 1963. Vegetation Mitteleuropas mit den Alpen. Stuttgart: Eugen Ulmer.

EL-SHEIKH, A.M. 1973. Comparative physiological ecology of Daucus carota ssp. carota and Daucus carota ssp. gummifer. Ph.D. Thesis, University of Reading.

ELIAS, C.O. and M.J. CHADWICK 1979. Growth characteristics of grass and legume cultivars and their potential for land reclamation. Journal of Applied Ecology 16, 537-544.

ELKINGTON, T.T. and L.C. MIDDLEFELL 1972. Population variation within Centaurea nigra L. in the Sheffield region. Watsonia 9, 109-116.

ELLENBERG, H. 1953. Physiologisches und okologisches Verhalten derselben Pflanzenarten. Ebenda 65, 351-362.

ELLENBERG, H. 1978. Vegetation Mitteleuropas mit den Alpen in okologischer Sicht. Stuttgart: Ulmer.

ELLIS, M. and Q.O.N. KAY 1975. Genetic variation in herbicide resistance in scentless mayweed (Tripleurospermum inodorum (L.) Schultz Bip.) 3. Selection for increased resistance to ioxynil,

MCPA and simazine. Weed Research 15, 327-333.
ELLIS, R.P. and B.M.G. JONES 1969. The origin of Cardamine flexuosa with evidence from morphology and geographical distribution. Watsonia 7, 92-103.
ELLIS, W.M., D.M. CALDER and B.T.O. LEE 1970. A diploid population of Poa annua L. from Australia. Experientia 26, 1156.
ELZEBROEK, A.T.G. 1981. The occurrence of Leontodon autumnalis and Leontodon hispidus in grasslands. Meded Vakgroep Landbouwplantenteelt Graslandkunde 59, 1-23.
ENGLEDOW, F. and L. AMEY 1980. Britain's future in farming: Principles of policy for British agriculture. Berkhamsted: Geographical Publications.
ENNOS, R.A. 1985. The mating system and genetic structure in a perennial grass, Cynosurus cristatus L. Heredity 55, 121-126.
ERNST, W.H.O. 1979. Population biology of Allium ursinum in Northern Germany. Journal of Ecology 67, 347-362.
ETHERINGTON, J.R. 1983. Control of germination and seedling morphology by ethene; differential responses related to habitat of Epilobium hirsutum L. and Chamerion angustifolium L. Annals of Botany 52, 653-658.
EVANS, R. 1977. Overgrazing and soil erosion on hill pastures with particular reference, to the Peak District. Journal of the British Grassland Society 32, 65-76.
FALINSKA, K. and E. PIROZNIKOW 1983. Ecological structure of Geranium robertianum L. populations under natural conditions and in the garden. Ekologia Polska 31, 93-121.
FARMER, A.M. and D.H.N. SPENCE 1985. Studies of diurnal acid fluctuations in British isoetid-type submerged aquatic macrophytes. Annals of Botany 56, 347-350.
FAULKNER, J.S. 1973. Experimental hybridisation of North-West European species of Carex section Acutae (Cyperaceae). Botanical Journal of the Linnean Society 67, 233-253.
FEDEROV, A.A. (ed.) 1974. Chromosome numbers of flowering plants. Academy of Sciences USSR Botanical Institute. Koenigstein: Otto Koeltz Science Publishers.
FENNER, M. 1978. Susceptibility to shade in seedlings of colonising and closed turf species. New Phytologist 81, 739-744.
FENNER, M. 1980a. The inhibition of germination of Bidens pilosa seeds by leaf canopy shade in some natural vegetation types. New Phytologist 84, 95-101.
FENNER, M. 1980b. The induction of a light requirement in Bidens pilosa seeds by leaf canopy shade. New Phytologist 84, 103-106.
FENNER, M. 1985. Seed ecology. London: Chapman & Hall.
FIALA, K. 1971. Seasonal changes in the growth of clones of Typha latifolia. Folia Geobotanica et Phytotaxonomica 6, 255-270.
FIALA, K. 1976. Underground organs of Phragmites communis, their growth, biomass and net production. Folia Geobotanica et Phytotaxonomica 11, 225-259.
FIELD, C. and H.A. MOONEY 1986. The photosynthesis-nitrogen relationship in world plants. In: On the economy of plant form and function. T.V. Givnish (ed.), 25-55. Cambridge: Cambridge University Press.
FINCH, R.A. 1967. Natural chromosomal variation in Leontodon. Heredity 22, 359-386.
FINEGAN, B.G. and H.J. HARVEY 1982. The dynamics of chalk quarry vegetation. In: The ecology of quarries; the importance of natural vegetation. Natural Environment Research Council Institute of Terrestrial Ecology symposium No. 11, proceedings of a workshop held at Monks Wood Experimental Station, 23-24 February 1981, 41-46. Abbots Ripton: ITE, Monks Wood Experimental Station.
FIRBANK, L.G., S.M. SMART, H.M. VAN DE POLL, R.G.H. BUNCE, M.O. HILL, D.C. HOWARD, J.W. WATKINS and G.J. STARK 2000. Causes of change in British vegetation. Grange-over-Sands: Institute of Terrestrial Ecology.
FITTER, A.H. 1978. An atlas of the wild flowers of Britain and Northern Europe. London: Collins.
FITTER, A.H. and C.J. ASHMORE 1974. Response of two Veronica species to a simulated woodland light climate. New Phytologist 73, 997-1001.
FITTER, A.H. and R.K.M. HAY 1981. Environmental physiology of plants. London: Academic Press.
FOGG, G.E. 1950. Biological Flora of the British Isles: Sinapis arvensis L. Journal of Ecology 38, 415-429.
FORBES, J.C. 1982. Ragwort survey in North-East Scotland. The Scottish Agricultural Colleges Research and Development Note 5.
FOSSETT, N. and B.M. CALHOUN 1952. Introgression between Typha latifolia and T. angustifolia. Evolution 6, 367-379.
FOSTER, J. 1964. Studies on the population dynamics of the daisy Bellis perennis. Ph.D. Thesis, University of Wales.
FOULDS, W. 1978. Response to soil moisture supply in three leguminous species. 1: Growth, reproduction and mortality. New Phytologist 80, 535-545.
FOULDS, W. and J.P. GRIME 1972a. The influence of soil moisture on the frequency of cyanogenic plants in populations of Trifolium repens and Lotus corniculatus. Heredity 28, 143-146.
FOULDS, W. and J.P. GRIME 1972b. The response of cyanogenic and acyanogenic phenotypes of Trifolium repens to soil moisture supply. Heredity 28, 181-187.
FOWLER, M.C. and T.O. ROBSON 1978. The effects of the food preferences and stocking rates of grass carp Ctenopharyngodon idella on mixed plant communities. Aquatic Botany 5, 261-276.
FRAME, J. 1976. The potential of tetraploid red clover and its role in the United Kingdom. Journal of the British Grassland Society 31, 139-152.
FRANKEL, R. and E. GALUN 1977. Pollination mechanisms, reproduction and plant breeding. Berlin: Springer.
FRASER, H.F. and J.P. GRIME 1999. Interacting effects of herbivory and fertility on a synthesised plant community. Journal of Ecology 87, 514-525.
FRASER, L.H. and J.P. GRIME 1997. Primary productivity and trophic dynamics investigated in a North Derbyshire, UK dale. Oikos 80, 499-508.
FROUD-WILLIAMS, R.J. 1983. The influence of straw disposal and cultivation regime on the population dynamics of Bromus sterilis. Annals of Applied Biology 103, 139-148.
FROUD-WILLIAMS, R.J., J.R. HILTON and J. DIXON 1986. Evidence for an endogenous cycle of dormancy in dry stored seeds of Poa trivialis L. New Phytologist 102, 123-132.
FRYER, J.D. and R.J. MAKEPEACE 1977. Weed control handbook. 1: Principles, 6th edn. Oxford: Blackwell Scientific.
FRYXELL, P.A. 1957. Mode of reproduction of higher plants. Botanical Review 23, 135-233.
FUCHS, C. 1957. Sur le developpement des structures de l'appereil souterrain du Polygonum cuspidatum Sieb. et Zucc. Bulletin de la Societe Botanique de France 104, 141-147.
FUKAMI, T., BEZEMER, T.M., MORTIMER, S.R. and W.H. VAN DER PUTTEN 2005. Species divergence and trait convergence in experimental plant community assembly. Ecology Letters 8, 1283-1492.
FURNESS, S.B. 1980. Ecological investigations of growth and temperature responses in bryophytes. Ph.D. Thesis, University of Sheffield.
FURNESS, S.B. and R.H. HALL 1981. An explanation of the intermittant occurrence of Physcomitrium, sphaericum (Hedw.) Brid. Journal of Bryology 11, 733-742.
GADELLA, T.W.J. 1972. Biosystematic studies in Hieracium pilosella L. and some related species of the subgenus pilosella. Botaniska Notiser 125, 361-369.
GAJEWSKI, W. 1959. Evolution in the genus Geum. Evolution 13, 378-388.
GARDNER, F.P. and W.E. LOOMIS 1953. Floral induction and development in orchard grass. Plant Physiology 28, 201-217.
GARDNER, G. 1976. Light and the growth of ash. In: Light as an ecological factor, G.C. Evans, O. Rackham and R. Bainbridge (eds), 557-563. Oxford: Blackwell Scientific.
GARDNER, G. 1977. The reproductive capacity of Fraxinus excelsior on the Derbyshire limestone. Journal of Ecology 65, 107-118.
GARNIER, E. and G. LAURENT 1994. Leaf anatomy, specific mass and water content in congeneric annual and perennial grass species. New Phytologist 128, 725-736.
GARNIER, E., J. CORTEZ, G. BILLES, M.-L. NAVAS, C. ROUMET, M. DEBUSSCHE, G. LAURENT, A. BLANCHARD, D. AUBRY, A. BELLMANN, C. NEILL, J.-P. and TOUSSAINT 2004. Plant

functional markers capture ecosystem properties during secondary succession. Ecology 85, 2630-2637.
GARNOCK-JONES, B.J. 1986. Floret specialisation, seed production and gender in Artemisia vulgaris L. (Asteraceae, Anthemideae). Botanical Journal of the Linnean Society 92, 285-302.
GARTNER, B.L., F.S. CHAPIN and G.R. SHAVER 1983. Demographic patterns of seedlings establishment and growth of native graminoids in an Alaskan tundra disturbance. Journal of Applied Ecology 20, 965-980.
GATES, D.M. 1962. Energy exchange in the biosphere. New York: Harper & Row.
GATES, D.M. 1968. Transpiration and leaf temperature. Annual Review of Plant Physiology 19, 211-238.
GAUDET, C.L. and P.A. KEDDY 1988. A comparative approach to predicting competitive ability from plant traits. Nature 334, 242-243.
GAUHL, E. 1976. Photosynthetic response to varying light intensity and ecotypes of Solanum dulcamara from shaded and exposed habitats. Oecologia 22, 275-286.
GAYNOR, D.L. and L.E. MacCARTER 1981. Biology, ecology and control of gorse (Ulex europaeus L.); a bibliography. New Zealand Journal of Agricultural Research 24, 123-137.
GEIGER, R. 1957. The climate near the ground. Cambridge, Mass.: Harvard University Press.
GIBBS, P.E. and R.J. GORNALL 1976. A biosystematic study of the creeping spearworts at Loch Leven, Kinross, Scotland. New Phytologist 77, 777-785.
GIBBS, P.E., C. MILNE and M.V. CARRILLO 1975. Correlation between the breeding system of recombinati on index in five species of Senecio. New Phytologist 75, 619-626.
GIBBY, M. and S. WALKER 1977. Further cytogenetic studies and a reappraisal of the diploid ancestry in the Dryopteris carthusiana complex. Fern Gazette 11, 315-324.
GIGON, A. and I.H. RORISON 1972. The response of some ecologically distinct plant species to nitrate and to ammonium nitrogen. Journal of Ecology 60, 92-102.
GIGON, A. and A. LEUTERT 1996. The dynamic keyhole-key model of coexistence to explain diversity of plants in limestone and other grasslands. Journal of Vegetation Science 7, 29-40.
GILL, D.E. 1978. On selection at high population density. Ecology 59, 1289-1291.
GIMINGHAM, C.H. 1960. Biological Flora of the British Isles: Calluna vulgaris (L.) Hull. Journal of Ecology 48, 455-483.
GIMINGHAM, C.H. 1972. Ecology of heathlands. London: Chapman & Hall.
GIVNISH, T.J. 1982. On the adaptive significance of leaf height in forest herbs. American Naturalist 120, 353-381.
GIVNISH, T.J. 1986. On the economy of plant form and function. Cambridge: Cambridge University Press.
GIVNISH, T.J. and G. VERMEIJ 1976. Sizes and shapes in liane leaves. American Naturalist 110, 743-778.
GLIESSMAN, S.R. 1976. Allelopathy in a broad spectrum of environments as illustrated by bracken. Botanical Journal of the Linnean Society 73, 95-104.
GODWIN, H. 1929. The 'sedge' and 'litter' of Wicken Fen. Journal of Ecology 17, 148-160.
GODWIN, H. 1956. The history of the British flora. Cambridge: Cambridge University Press.
GODWIN, H. 1975. History of the British flora. London: Cambridge University Press.
GODWIN, H. 1985. Cambridge and Clare. Cambridge: Cambridge University Press.
GOEBEL, K. 1896. Ueber den Einfluss des Lichtes auf die Gestaltung der Kakteen und anderer, Pflanzen. 2: Die Abhangigkeit der Blattform von Campanula rotundifolia von der Lichtintensitat. Flora 82, 1-37.
GOEBEL, K. 1908. Einleitung in die experimentelle Morphologie der Pflanzen. Leipzig und Berlin.
GOLDSTEIN, D.J. 1981. Errors in microdensitometry. Histochemical Journal 13, 251-267.
GOOD, R. 1974. The geography of flowering plants, 4th edn. London: Longman.
GOODWAY, K.M. 1957. The species problem in Galium pumilum. In: Progress in the study of the British flora, J.E. Lousley (ed.), 116-118. London: Botanical Society of the British Isles.
GORDON, A.G. and D.K. FRASER 1982. The supply and utilisation of seed of native broadleaves. In: Broadleaves in Britain; future management and research, D.C. Malcolm, J. Evans and P.N. Edwards (eds), 161-168. Farnham, Surrey: Institute of Chartered Foresters.
GORSKI, T. 1975. Germination of seeds in the shadow of plants. Physiologia Plantarum 34, 342-346.
GORSKI, T., K. GORSKA and J. NOWICKI 1977. Germination of seeds of various herbaceous species under leaf canopy. Flora 166, 249-259.
GOULD, S.J. and R.C. LEWONTIN 1979. The spandrels of San Marco and the Panglossian paradigm: a critique of the adaptationist programme. Proceedings of the Royal Society of London 205B, 581-598.
GRABHAM, P.W. and J.R. PACKHAM 1983. A comparative study of the bluebell Hyacinthoides non-scripta in two different woodland situations in the West Midlands. Biological Conservation 26, 105-126.
GRACE, J.B. 1977. Plant response to wind. London: Academic Press.
GRACE, J.B. 1987. Climatic tolerance and the distribution of plants. In: Frontiers of comparative plant ecology, I.H. Rorison, J.P. Grime, R. Hunt, G.A.F. Hendry and D.H. Lewis (eds), 113-130. London: Academic Press.
GRACE, J.B. and J.S. HARRISON 1986. The biology of Canadian weeds. 73: Typha latifolia L., T. angustifolia L. and Typha x glauca Godr. Canadian Journal of Plant Science 66, 361-380.
GRACE, J.B. and R.G. WETZEL 1981. Phenotypic and genotypic components of growth and reproduction in Typha latifolia; experimental studies in marshes of differing successional maturity. Ecology 62, 789-801.
GRACE, J.B. and R.G. WETZEL 1983. Variations in growth and reproduction within populations of two rhizomatous plant species: Typha latifolia and T. angustifolia. Oecologia 53, 258-263.
GRACE, J.B. (1990) A clarification of the debate between Grime and Tilman. In Perspectives on plant competition. J.B. Grace and D. Tilman (eds). New York: Academic Press.
GRACE, J.B. 1999. The factors controlling species density in herbaceous plant communities. Perspectives in Plant Ecology, Evolution and Systematics 3, 1-28.
GRADWELL, G. 1974. The effect of defoliators on tree growth. In: The British oak; its history and natural history, M.G. Morris and F.H. Perring (eds), 182-193. Faringdon, Berkshire: E.W. Classey.
GRANDSTROM, A. 1982. Seed banks in five boreal forest stands originating between 1810 and 1963. Canadian Journal of Botany 60, 1815-1821.
GRANT, C.J. and J. WHITEBROOK 1983. Tetraploidy in Chrysanthemum leucanthemum L. in relation to heavy metal pollution. In: Proceedings of the 2nd Kew chromosome conference, P.E. Brandham and M.D. Bennett (eds), 342-343. London: George Allen & Unwin.
GRANT, M.C. and J. ANTONOVICS 1978. Biology of ecologically marginal populations of Anthoxanthum odoratum. 1: Phenetics and dynamics. Evolution 32, 822-838.
GRANT, V. 1963. The origin of adaptation. New York: Columbia University Press.
GRANT, W.F. and G.C. MARTEN 1985. Birdsfoot trefoil. In: Forages: The science of grassland agriculture, 4th edn, M.E. Heath, R.F. Barnes and D.S. Metcalfe (eds), 98-108. Ames: Iowa State University Press.
GRAY, H. 1982. Plant dispersal and colonisation. In: The ecology of quarries, the importance of natural vegetation. Natural Environment Research Council Institute of Terrestrial Ecology Symposium no. 11. Abbots Ripton: ITE, Monks Wood Experimental Station.
GREEN, B.H. 1981. A policy on introductions to Britain. In: The biological aspects of rare plant conservation, H. Synge (ed.), 403-412. Chichester: Wiley.
GREENSLADE, P.J.M. 1972a. Distribution patterns of Prochirus species (Coeleoptera; Staphylindae) in the Solomon Islands. Evolution, Lancaster, Pennsylvania 26, 130-142.
GREENSLADE, P.J.M. 1972b. Evolution in the staphylinid genus

Priochirus (Coleoptera). Evolution, Lancaster, Pennsylvania 26, 203-220.

GREENSLADE, P.J.M. 1983. Adversity selection and the habitat templet. American Naturalist 122, 352-365.

GREIG-SMITH, J. and G.R. SAGAR 1981. Biological causes of local rarity in Carlina vulgaris. In: The biological aspects of rare plant conservation, H. Synge (ed.), 389-400. Chichester: Wiley.

GREIG-SMITH, P. 1948a. Biological Flora of the British Isles: Urtica dioica L. Journal of Ecology 36, 343-351.

GREIG-SMITH, P. 1948b. Biological Flora of the British Isles: Urtica urens L. Journal of Ecology 36, 351-355.

GRIGNAC, P. 1978. The evolution of resistance to herbicides in weedy species. Agro-Ecosystems 4, 377-385.

GRIME, J.P. 1955. A preliminary study of magnesian limestone grasslands east of Sheffield. B.Sc. Thesis, University of Sheffield.

GRIME, J.P. 1960. A study of the ecology of a group of Derbyshire plants with particular reference to their nutrient requirements. Ph.D. Thesis, University of Sheffield.

GRIME, J.P. 1963a. An ecological investigation at a junction between two plant communities in Coombsdale on the Derbyshire limestone. Journal of Ecology 51, 391-402.

GRIME, J.P. 1963b. Factors determining the occurrence of calcifuge species on shallow soils over calcareous substrata. Journal of Ecology 51, 375-391.

GRIME, J.P. 1965. The ecological significance of lime chlorosis. An experiment with two species of Lathyrus. New Phytologist 64, 477-487.

GRIME, J.P. 1965,. Comparative experiments as a key to the ecology of flowering plants. Ecology 45, 513-515.

GRIME, J.P. 1966. Shade avoidance and shade tolerance in flowering plants. In: Light as an ecological factor, R. Bainbridge, G.C. Evans and O. Rackham (eds), 187-207. Oxford: Blackwell Scientific.

GRIME, J.P. 1973. Competitive exclusion in herbaceous vegetation. Nature 242, 344-347.

GRIME, J.P. 1973b. Control of species density in herbaceous vegetation. Journal of Environmental Management 1, 151-167.

GRIME, J.P. 1974. Vegetation classification by reference to strategies. Nature 250, 26-31.

GRIME, J.P. 1977. Evidence for the existence of three primary strategies in plants and its relevance to ecological and evolutionary theory. American Naturalist 111, 1169-1194.

GRIME, J.P. 1978. Interpretation of small scale patterns in the distribution of plant species in space and time. In: Structure and functioning of plant populations, A.H.J. Freysen and J.W. Woldendorp (eds), 101-121. Amsterdam: North-Holland.

GRIME, J.P. 1979. Plant strategies and vegetation processes. Chichester: Wiley.

GRIME, J.P. 1980. An ecological approach to management. In: Amenity grassland: an ecological perspective, I.H. Rorison and R. Hunt (eds), 13-55. Chichester: Wiley.

GRIME, J.P. 1981. Plant strategies in shade. In: Plants and the daylight spectrum, H. Smith (ed.), 159-186. London: Academic Press.

GRIME, J.P., 1983. Prediction of weed and crop response to climate based upon measurements of nuclear DNA content. In: Aspects of Applied Biology. 4: Influence of environmental factors on herbicide performance and crop and weed biology. Wellesbourne: National Vegetable Research Station.

GRIME, J.P. 1984. The ecology of species, families and communities of the contemporary British Flora. New Phytologist 98, 15-33.

GRIME, J.P. 1985a. Towards a functional classification of vegetation. In: The population structure of vegetation. J White (ed.) 503-514. Dordrect: Dr. W. Junk.

GRIME, J.P. 1985b. Dominant and subordinate components of plant communities: implications for succession, stability and diversity. In: Colonisation, succession and stability, A. Gray, P. Edwards and A. Crawley (eds), 413-428. Oxford: Blackwell.

GRIME, J.P. 1986a. The circumstances and characteristics of spoil colonization within a local flora. In: Proceedings of the Royal Society discussion meeting on quantitative aspects of the ecology of biological invasions. M.W. Holdgate (ed.). Philosophical Transactions of the Royal Society of London B314, 637-654.

GRIME, J.P. 1986b. Manipulation of plant species and communities. In: Ecology and design in landscape, 24th symposium of the British Ecological Society, A.D. Bradshaw, E. Thorn and D.A. Goode (eds), 175-194. Oxford: Blackwell.

GRIME, J.P. 1987. The C-S-R model of primary plant strategies: origins, implications and tests. In: Evolutionary plant biology, L.D. Gottlieb and S.K. Jain (eds). London: Chapman & Hall.

GRIME, J.P. 1988. Appendix 39 Memorandum submitted to the Agriculture Committee Second Report. Chernobyl: The Governments Reaction. Volume II. Minutes of Evidence and Appendices, pp.399-403. London, HMSO.

GRIME, J.P. 1993. Ecology sans frontiers. Oikos 68, 385-392.

GRIME, J.P. 1998. Benefits of plant diversity to ecosystems: immediate, filter and founder effects. Journal of Ecology 86, 902-910.

GRIME, J.P. 2001. Plant strategies, vegetation processes and ecosystem properties. 2nd edn. Chichester: Wiley.

GRIME, J.P. 2002. Declining plant diversity: empty niches or functional shifts? Journal of Vegetation Science 13, 457-460.

GRIME, J.P. 2006. Trait convergence and trait divergence in the plant community: mechanisms and consequences. Journal of Vegetation Science 17, 255-260.

GRIME, J.P. and J.M. ANDERSON 1986. Environmental controls over organism activity. In: Forest ecosystems in the Alaskan Taiga. A synthesis of structure and function, K. van Cleve, F.S. Chapin, P.W. Flanagan, L.A. Viereck and C.T. Dyrness (eds), 89-95. Berlin: Springer.

GRIME, J.P. and A.V. CURTIS 1976. The interaction of drought and mineral nutrient stress in calcareous grassland. Journal of Ecology 64, 976-998.

GRIME, J.P. and J.G. HODGSON 1969. An investigation of the significance of lime-chlorosis by means of large-scale comparative experiments. In: Ecological aspects of the mineral nutrition of plants, I.H. Rorison (ed.), 67-99. Oxford: Blackwell Scientific.

GRIME, J.P. and J.G. HODGSON 1987. Botanical contributions to contemporary ecological theory. In: Frontiers of comparative plant ecology, I.H. Rorison, J.P. Grime, R. Hunt, G.A.F. Hendry and D.H. Lewis (eds), 283-296. London: Academic Press.

GRIME, J.P. and R. HUNT 1975. Relative growth-rate; its range and adaptive significance in a local flora. Journal of Ecology 63, 393-422.

GRIME, J.P. and B.C. JARVIS 1976. Shade avoidance and shade tolerance in flowering plants. 2: Effects of light on the germination of species of contrasted ecology. In: Light as an ecological factor. 2: Symposium of the British Ecological Society No. 16, G.C. Evans, R. Bainbridge and O. Rackham (eds), 525-532. Oxford: Blackwell Scientific.

GRIME, J.P. and D.W. JEFFREY 1965. Seedling establishment in vertical gradients of sunlight. Journal of Ecology 53, 621-642.

GRIME, J.P. and P.S. LLOYD 1973. An ecological atlas of grassland plants. Natural Environment Research Council. London: Edward Arnold.

GRIME, J.P. and J.M.L. MACKEY 2002. The role of plasticity in resource capture by plants. Evolutionary Ecology 16, 299-307.

GRIME, J.P. and M.A. MOWFORTH 1982. Variation in genome size - an ecological interpretation. Nature 299, 151-153.

GRIME, J.P., G.M. BLYTHE and J.D. THORNTON 1970. Food selection by the snail Cepaea nemoralis L. In: Animal populations in relation to their food resources, A. Watson (ed.), 73-99. Oxford: Blackwell Scientific.

GRIME, J.P., MASON, G. CURTIS A.V., RODMAN, J., BAND S R., NEAL A. M. and SHAW S. (1981) A comparative study of germination characteristics in a local flora. Journal of Ecology 69, 1017-1059.

GRIME, J.P., J.C. CRICK and J.E. RINCON 1986. The ecological significance of plasticity. In: Plasticity in plants. Proceedings of the Society of Experimental Biology 40th symposium, D.H. Jennings and A.J. Trewevas (eds), 5-29. Cambridge: The Society for Experimental Biology.

GRIME, J.P., J.G. HODGSON and R. HUNT 1988. Comparative plant ecology: a functional approach to common British species. London: Unwin Hyman.

GRIME, J.P., R. HUNT and W.J. KRZANOWSKI 1987. Evolutionary physiological ecology of plants. In: Evolutionary physiological ecology, P. Calow (ed.), 105-126. Cambridge: Cambridge University Press.

GRIME, J.P., S.F. MacPHERSON-STEWART and R.S. DEARMAN 1968. An investigation of leaf palatability using the snail Cepaea nemoralis L. Journal of Ecology 56, 405-420.

GRIME, J.P., J.M.L. SHACKLOCK and S.R. BAND 1985. Nuclear DNA contents, shoot phenology and species coexistence in a limestone grassland community. New Phytologist 100, 435-445.

GRIME, J.P., J.M.L. MACKEY, S.H. HILLIER and D.J. READ 1987. Floristic diversity in a model system using experimental microcosms. Nature 328, 420-422.

GRIME, J.P., J.H.C. CORNELISSEN, K. THOMPSON and J.G. HODGSON 1996. Evidence of a causal connection between anti-herbivore defence and the decomposition rate of leaves. Oikos 77, 489-494.

GRIME, J.P., K. THOMPSON, J.D. FRIDLEY, A.P. ASKEW and M. SHANTA 2007. Responses to fourteen years of simulated climate change. (in preparation).

GRIME, J.P., G. MASON, A-V. CURTIS, J. RODMAN, S.R. BAND, M.A.G. MOWFORTH, A.M. NEAL and S.C. SHAW 1981. A comparative study of germination characteristics in a local flora. Journal of Ecology 69, 1017-1059.

GRIME, J.P., K. THOMPSON, R. HUNT, J.G. HODGSON, J.H.C. CORNELISSEN, I.H. RORISON, G.A.F. HENDRY, T.W. ASHENDEN, A.P. ASKEW, S.R. BAND, R.E. BOOTH, C.C. BOSSARD, B.D. CAMPBELL, J.E.L. COOPER, A. DAVISON, P.L. GUPTA, W. HALL, D.W. HAND, M.A. HANNAH, S.H. HILLIER, D.J. HODKINSON, A. JALILI, Z. LIU, J.M.L. MACKEY, N. MATTHEWS, M.A. MOWFORTH, A.M. NEAL, R.F. SUTTON, D.E. TASKER, P.C. THORPE and J. WJ. READER, K. REILING, W. ROSS-FRASER, R.E. SPENCER, F. SUTTON, D.E. TASKER, P.C. THORPE and J. WHITEHOUSE, 1997. Integrated screening validates primary axes of specialisation in plants. Oikos 79, 259-281.

GRIME, J.P., V.K. BROWN, K. THOMPSON, G.J. MASTERS, S.H. HILLIER, I.P. CLARKE, A.P. ASKEW, D. CORKER and J.P. KIELTY 2000. The response of two contrasted grasslands to simulated climate change. Science 289, 762-765.

GROSS, K.L. 1984. Effect of seed size and growth form on seedling establishment of six monocarpic perennial plants. Journal of Ecology 72, 369-387.

GROSS, R.S. and P.A. WERNER 1983. Probabilites of survival and reproduction relative to rosette size in the common burdock Arctium minus. American Midland Naturalist 109, 184-193.

GROSS, R.S., P.A. WERNER and W.R. HAWTHORNE 1980. The biology of Canadian weeds. 38: Arctium minus (Hill) Bernh. and A. lappa L. Canadian Journal of Plant Science 60, 621-634.

GRUBB, P.J. 1976. A theoretical background to the conservation of ecologically distinct groups of annuals and biennials in the chalk grassland ecosystem. Biological Conservation 10: 53-76.

GRUBB, P.J. 1977. The maintenance of species-richness in plant communities: the importance of the regeneration niche. Biological Reviews 52, 107-145.

GRUBB, P.J. 1982. Control of relative abundance in roadside Arrhenatherum, results of a long-term experiment. Journal of Ecology 70, 845-861.

GRUBB, P.J. 1985. Plant populations and vegetation in relation to habitat, disturbance and competition: problems of generalization. In: The population structure of vegetation, J. White (ed.), 595-621. Dordrecht: Junk.

GRUBB, P.J. 1992. A positive distrust in simplicity - lessons from plant defences and from competition among plants and among animals. Journal of Ecology 80, 585-610.

GRUBB, P.J. and M.B. SUTER 1971. The mechanisms of acidification of soil by Calluna and Ulex and the significance for conservation. In: The scientific management of animal and plant communities for conservation, E. Duffey and A.S. Watt (eds), 115-133. Oxford: Blackwell Scientific.

GRUBB, P.J. and M.B. SUTER 1971. The mechanisms of acidification of soil by Calluna and Ulex and the significance for conservation. In: The scientific management of animal and plant communities for conservation, E. Duffey and A.S. Watt (eds), 115-133. Oxford: Blackwell Scientific.

GRUBB, P.J., H.E. GREEN and R.C.J. MERRIFIELD 1969. The ecology of chalk heath: its relevance to the calcicole-calcifuge and soil acidification problems. Journal of Ecology 57, 175-212.

GRUBB, PJ, D. KELLY and G.F. MITCHLEY 1982. The control of relative abundance in communities of herbaceous plants. In: The Plant Community as a Working Mechanism. E. Newman (ed.). 77-97. Oxford: Blackwell.

GUPPY, H.B. 1912. Studies in seeds and fruits. London: Williams & Norgate.

GUPTA, P.L. and I.H. RORISON 1975. Seasonal differences in the availability of nutrients down a podzolic profile. Journal of Ecology 63, 521-534.

GUSTAFSSON, M. 1976. Evolutionary trends in the Atriplex prostrata group of Scandinavia. Taxonomy and morphological variation. Opera Botanica 39, 1-63.

HAAG, R.W. 1979. The ecological significance of dormancy in some rooted aquatic plants. Journal of Ecology 67, 727-738.

HACKETT, C. 1965. Ecological aspects of the nutrition of Deschampsia flexuosa (L.) Trin. 2: The effects of Al, Ca, Fe, K, Mn, N, P and pH on the growth of seedlings and established plants. Journal of Ecology 53, 315-334.

HAGGAR, R.J. 1971. The significance and control of Poa trivialis in ryegrass pastures. Journal of the British Grassland Society 26, 117-122.

HAGON, M.W. 1971. The action of temperature fluctuations on hard seeds of subterranean clover. Australian Journal of Experimental Agronomy and Animal Husbandry 11, 440-443.

HAGON, M.W. and L.A.T. BALLARD 1969. Reversibility of strophiolar permeability to water in seeds of subterranean clover (Trifolium subterraneum L.). Australian Journal of Biological Science 23, 519-528.

HAINES-YOUNG, R., C.J. BARR, H.I.J. BLACK, D.J. BRIGGS, R.G.H. BUNCE, R.T. CLARKE, A. COOPER, F.H. DAWSON, L.G. FIRBANK, R.M. FULLER, M.T. FURSE, M.K. GILLESPIE, M.O. HILL, M. HORNUNG, D.C. HOWARD, T. McCANN, M.D. MORECROFT, S. PETIT, A.R.G. SIER, S.M. SMART, G.M. SMITH, A.P. STOTT, R.G. STUART and J.W. WATKINS 2000. Accounting for nature: assessing habitats in the UK countryside. London: Department of the Environment, Transport and the Regions.

HASTINGS, A 1980. Disturbance, coexistence, history and competition for space. Theoretical Population Biology 18, 363-373.

HAJAR, A.S.M. 1987. The comparative ecology of Minuartia verna (L.) Hiern and Thlaspi alpestre L. in the Southern Pennines, with special reference to heavy metal tolerance. Ph.D. Thesis, University of Sheffield.

HALL, I.V. and J.M. SHAY 1981. The Biological Flora of Canada. 3: Vaccinium vitis-idaea L. var. minus Lodd., supplementary account. Canadian Field Naturalist 95, 434-464.

HALL, J.B. 1971. Pattern in a chalk grassland community. Journal of Ecology 59, 749-762.

HALL, P.C. 1980. Sussex plant atlas; an atlas of the distribution of wild plants in Sussex. Brighton: Borough of Brighton, Booth Museum of Natural History.

HALLIDAY, G. 1960. Taxonomic and ecological studies in the Arenaria ciliata and Minuartia verna complexes. Ph.D. Thesis, University of Cambridge.

HAM, S.F., J.F. WRIGHT and A.D. BERRY 1982. The effect of cutting on the growth and recession of the freshwater macrophyte Ranunculus penicillatus (Dumort). Journal of Environmental Management 15, 263-271.

HAMBLER, D.J. 1958. Some taxonomic investigations in the genus Rhinanthus. Watsonia 4, 101-116.

HAMLY, D.H. 1932. Softening of the seeds of Melilotus alba. Botanical Gazette 93, 345-375.

HAMMERTON, J.L. 1965. Studies on weed species of the genus Polygonum L. 1: Physiological variation within Polygonum persicaria L. Weed Research 5, 13-26.

HAMMERTON, J.L. 1967. Studies on weed species of the genus Polygonum L. 5: Variation in seed weight, germination behaviour and seed polymorphism in Polygonum persicaria L. Weed Research 7, 331-348.

HANCOCK, J.F. and R.S. BRINGHURST 1978. Interpopulational differentiation and adaptation in the perennial diploid species Fragaria vesca. American Journal of Botany 65, 795-803.

HANDEL, S.N. 1978. New ant dispersed species in the genera Carex, Luzula and Claytonia. Canadian Journal of Botany 56, 2925-2927.

HANSEN, K. and J. JENSEN 1974. Edaphic conditions and plant-soil relationships on roadsides in Denmark. Dansk Botanisk Arkiv 28, No. 3, 1-143.

HARBERD, D.J. 1961. Observations on population structure and longevity of Festuca rubra. New Phytologist 60, 184-210.

HARBERD, D.J. 1962. Some observations on natural clones in Festuca ovina. New Phytologist 61, 85-100.

HARBERD, D.J. 1967. Observations on natural clones in Holcus mollis. New Phytologist 66, 401-408.

HARDING, J.S. 1981. Regeneration of birch (Betula pendula Ehrh. and Betula pubescens Roth. In: The regeneration of oak and beech; a literature review, A.N. Newbold and F.B. Goldsmith, with an addendum on birch by J.S. Harding, 83-112. Discussion Papers in Conservation no. 33. London: University College London.

HARDMAN, N. 1986. Structure and function of repetitive DNA in eukaryotes. Biochemical Journal 234, 1-11.

HARLEY, J.L. 1969. The biology of mycorrhiza, 2nd edn. London: Hill.

HARLEY, J.L. and E.L. HARLEY 1986. A checklist of mycorrhizas in the British Flora. New Phytologist, suppl. 105, 1-102.

HARLEY, R.M. and C.A. BRIGHTON 1977. Chromosome numbers in the genus Mentha. Botanical Journal of the Linnean Society 74, 71-96.

HARMER, R. and J.A. LEE 1978. The germination and viability of Festuca vivipara (L.) SM. plantlets. New Phytologist 81, 745-751.

HARPER, C.W. and K.G. STOTT 1966. The control of Japanese Knotweed, Polygonum cuspidatum Sieb. and Zucc. Long Ashton Agricultural and Horticultural Research Station Annual Report, 268-276.

HARPER, J.L. 1957a. Biological Flora of the British Isles: Ranunculus acris L. Journal of Ecology 45, 289-313.

HARPER, J.L. 1957b. Biological Flora of the British Isles: Ranunculus repens L. Journal of Ecology 45, 314-325.

HARPER, J.L. 1957c. Biological Flora of the British Isles: Ranunculus bulbosus L. Journal of Ecology 45, 325-342.

HARPER, J.L. 1977. Population biology of plants. London: Academic Press.

HARPER, J.L. 1982. After description. In: The plant community as a working mechanism, E.I. Newman (ed.), 11-25. Special Publication No. 1, British Ecological Society. London: Blackwell.

HARPER, J.L. and T.L. OGDEN 1970. The reproductive strategy of higher plants. 1: The concept of strategy with special reference to Senecio vulgaris L. Journal of Ecology 58, 681-698.

HARPER, J.L. and W.A. WOOD 1957. Biological Flora of the British Isles: Senecio jacobaea L. Journal of Ecology 45, 617-637.

HARPER, J.L., P.H. LOVELL and K.G. MOORE 1970. The shapes and sizes of seeds. Ann Review of Ecology and Systematics 1, 327-356.

HARRIS, G.R. and P.H. LOVELL 1980a. Growth and reproductive strategy in Veronica. Annals of Botany 45, 447-458.

HARRIS, G.R. and P.H. LOVELL 1980b. Adventitious root formation in Veronica. Annals of Botany 45, 459-468.

HARRIS, W. 1968. Environmental effects on the sex ratio of Rumex acetosella L. Proceedings of the New Zealand Ecological Society 15, 51-54.

HARTSEMA, A.M. 1961. Influence of temperature on flower formation and flowering of bulbous and tuberous plants. In: Handbuch der Pflanzenphysiologie. 16: Ausenfaktoren in Wachstum und Entwicklung, W. Ruhland (ed.), 123-167. Berlin: Springer.

HARVEY, H.J. and R.C. MEREDITH 1981. Ecological studies on Peucedanum palustre and their implications for conservation management at Wicken Fen, Cambridgeshire. In: The biological aspects of rare plant conservation, H. Synge (ed.) 365-378. Chichester: Wiley.

HASELWANDTER, K. and D.J. READ 1982. The significance of a root-fungus association in two Carex species of high Alpine plant communities. Oecologia 53, 352-354.

HASLAM, S.M. 1972. Biological Flora of the British Isles: Phragmites communis Trin. Journal of Ecology 60, 585-610.

HASLAM, S.M., C. SINKER and P. WOLSELEY 1975. British water plants. Field Studies 4, 243-351.

HAUKIOJA, E. and P. NIEMELA 1976. Does birch defend itself actively against herbivores? Report of Kevu Subarctic Research Station 13, 44-47.

HAUKIOJA, E. and P. NIEMELA 1977. Retarded growth of a geometrid larva after mechanical damage to leaves of the host tree. Annals Zoologica Fennici 14, 48-52.

HAUKIOJA, E. and P. NIEMELA 1979. Birch leaves as a resource for herbivores; seasonal occurrence of increased resistance in foliage after mechanical damage of adjacent leaves. Oecologia 39, 151-159.

HAWKES, J.G. (ed.) 1966. Reproductive biology and taxonomy of vascular plants. Oxford: Pergamon Press.

HAWTHORN, W.R. 1974. The biology of Canadian weeds. 4: Plantago major and P. rugelii. Canadian Journal of Plant Science 54, 383-396.

HAYASHI, 1. 1977. Secondary succession of herbaceous communities in Japan. Japanese Journal of Ecology 27, 191-200.

HEADS, P.A. and J.H. LAWTON 1984. Bracken, ants and extrafloral nectaries 2. The effect of ants on the insect herbivores of bracken. Journal of Animal Ecology 53, 995-1014.

HECTOR, A., B. SCHMID, C. BEIERKUHNLEIN, M.C. CALDEIRA, M. DIEMER, P.G. DIMITRAKOPOULOS, J.A. FINN, H. FREITAS, P.S. GILLER, J. GOOD, R. HARRIS, P. HOGBERG, K. HUSS-DANELL, J. JOSHI, A. JUMPPONEN, C. KORNER, P.W. LEADLEY, M. LOREAU, A. MINNS, C.P.H. MULDER, G. O'DONOVAN, S.J. OTWAY, J.S. PEREIRA, A. PRINZ, D.J. READ, M. SCHERER-LORENZEN, E.D. SCHULZE, A.S.D. SIAMANTZIOURAS, E.M. SPEHN, A.C. TERRY, A.Y. TROUMBIS, F.I. WOODWARD, S. YACHI, J.H. LAWTON, 1999. Plant diversity and productivity experiments in European grasslands. Science 286, 1123-1127.

HEGI, G. 1907-1981. Illustrierte Flora von Mittel-Europa. Munchen: Lehmann and Berlin: Parey.

HEGI, G. 1964. Illustrierte Flora von Mittel-Europa. Bild 4, Teil 3, 1712-1716. Munchen: Carl Hanser.

HELPSPER, H.P.G. and G.A.M. KLERKEN 1984. Germination of Calluna vulgaris (L.) Hull in vitro under different pH conditions. Acta Botanica Neerlandica 33, 347-353.

HENDERSON, D.M. and D. MANN (eds) 1984. Birches. Botanical Society of Edinburgh Symposium. Proceedings of the Royal Society of Edinburgh 85B, 1213.

HENDRY, G.A.F. and K.J. BROCKLEBANK 1985. Iron-induced oxygen radical metabolism in waterlogged plants. New Phytologist 101, 199-206.

HENRY, M., M.H.H. STEVENS, D.E. BUNKER, S.A. SCHNITZER and W.P. CARSON 2004. Establishment limitation reduces species recruitment and species richness as soil resources rise. Journal of Ecology 92, 339-347.

HERMY, M. 1985. Capitalists and proletarians (MacLeod 1894): an early theory of plant strategies. Oikos 44, 364-366.

HESLOP-HARRISON, J.W. 1953. The present position of the rosebay willow-herb. Vasculum 38, 25.

HEUBL, G.R. 1984. Systematische Untersuchungen an Mittel-Europaischen Polygala-Arten. Mitteilungen der Botanischen Staatssamlung Munchen 20, 205-428.

HEUKELS, H. and S.J. VAN OOSTROOM 1968. Flora van Nederland. Groningen: Noordhof.

HEYWOOD, V.H. 1983. Relationships and evolution in the Daucus carota complex. Israeli Journal of Botany 32, 51-65.

HIGGS, D.E.B. and D.B. JAMES 1969. Comparative studies on the biology of upland grasses. 1: Rate of dry-matter production and its control in four grass species. Journal of Ecology 57, 553-563.

HILL, M.O. 1979. The development of a flora in even-aged plantations. In: The ecology of even-aged forest plantations, E.D. Ford, D.C. Malcolm and J.C. Atterson (eds), 175-192. Institute of Terrestrial Ecology.

HILL, M.O. and P.A. STEVENS 1981. The density of viable seed in soils of forest plantations in upland Britain. Journal of Ecology 69, 693-709.

HILLIER, S.H. 1984. A quantitative study of gap recolonisation in two contrasted limestone grasslands. Ph.D. Thesis, University of Sheffield.

HILTON, J.R. 1983. The influence of light on the germination of Senecio vulgaris. New Phytologist 94, 29-37.

HILTON, J.R. 1984. The influence of dry storage temperatures on the response of Bromus sterilis seeds to light. New Phytologist 98, 129-134.

HILTON, J.R., R.H. FROUD-WILLIAMS and J. DIXON 1984. A relationship between phytochrome photoequilibrium and germination of seeds of Poa trivialis from contrasting habitats. New Phytologist 97, 375-379.

HIROSE, T. and T. TATENO 1984. Soil nitrogen patterns induced by colonisation of Polygonum cuspidatum on Mount Fuji, Japan. Oecologia 61, 218-223.

HODGKIN, S.E. and D. BRIGGS 1981. The effect of simulated acid rain on two populations of Senecio vulgaris L. New Phytologist 89, 687-691.

HODGSON, J.F. 1973. Aspects of the carbon nutrition of angiospermous parasites. Ph.D. Thesis, University of Sheffield.

HODGSON, J.G. 1972. A comparative study of seedling root growth with respect to aluminium and iron supply. Ph.D. Thesis, University of Sheffield.

HODGSON, J.G. 1986a. Commonness and rarity in plants with special reference to the Sheffield flora. 1: The identity, distribution and habitat characteristics of the common and rare species. Biological Conservation 36, 199-252.

HODGSON, J.G. 1986b. Commonness and rarity in plants with special reference to the Sheffield flora. 2: The relative importance of climate, soils and land use. Biological Conservation 36, 253-274.

HODGSON, J.G. 1986c. Commonness and rarity in plants with special reference to the Sheffield flora. 3: Taxonomic and evolutionary aspects. Biological Conservation 36, 275-296.

HODGSON, J.G. 1986d. Commonness and rarity in plants with special reference to the Sheffield flora. 4: European context with particular reference to endemism. Biological Conservation 36, 297-314.

HODGSON, J.G. 1987. Growing rare in Britain. New Scientist 113 No. 1547, 38-39.

HODGSON, J.G. 1987b. Why do so few plant species exploit productive habitats? An investigation into cytology, plant strategies and abundance within a local flora. Functional Ecology 1, 243-250.

HODGSON, J.G. 1993. Commonness and rarity in British butterflies. Journal of Applied Ecology 30, 407-427.

HODGSON, J. G. 2003. Change the Change Index? BSBI News 93, 44-48.

HODGSON, J.G. and J.M.L. MACKEY 1986. The ecological specialization of dicotyledonous families within a local flora: some factors constraining optimization of seed size and their possible evolutionary significance. New Phytologist 104, 479-515.

HODGSON, J., G. JONES and C. PALMER 2004 Are some species missing from the new Plant Status Lists for Great Britain? BSBI News 97, 39-43.

HODGSON, J.G., G. MONTSERRAT-MARTÍ, B. CERABOLINI, R.M. CERIANI, M. MAESTRO-MARTÍNEZ, B. PECO, P.J. WILSON, K. THOMPSON, J.P. GRIME, S.R. BAND, A. BOGARD, P. CASTRO-DÍEZ, M. CHARLES, G. JONES, M.C. PÉREZ-RONTOMÉ, M. CACCIANIGA, D. ALARD, J.P. BAKKER, J.H.C. CORNELISSEN, T. DUTOIT, A.P. GROOTJANS, J. GUERRERO-CAMPO, P.L. GUPTA, A. HYND, S. KAHMEN, P. POSCHLOD, A. ROMO-DÍEZ, I.H. RORISON, E. ROSÉN, K.-F. SCHREIBER, J. TALLOWIN, L. DE TORRES ESPUNY and P. VILLAR-SALVADOR 2005b. A functional method for classifying European grasslands for use in joint ecological and economic studies. Basic and Applied Ecology 6, 107-118.

HODGSON, J.G., G. MONTSERRAT-MARTÍ, J. TALLOWIN, K. THOMPSON, S. DÍAZ, M. CABIDO, J.P. GRIME, P.J. WILSON, S.R. BAND, A. BOGARD, R. CABIDO, D. CÁCERES, P. CASTRO-DÍEZ, C. FERRER, M. MAESTRO-MARTÍNEZ, M.C. PÉREZ-RONTOMÉ, M. CHARLES, J.H.C. CORNELISSEN, S. DABBERT, N. PÉREZ-HARGUINDEGUY, T. KRIMLY, F.J. SIJTSMA, D. STRIJKER, F. VENDRAMINI, J. GUERRERO-CAMPO, A. HYND, G. JONES, A. ROMO-DÍEZ, L. DE TORRES ESPUNY, P. VILLAR-SALVADOR and M.R. ZAK 2005c. How much will it cost to save grassland diversity? Biological Conservation 122, 263-273.

HODGSON, J.G., J.P. GRIME, P.J. WILSON, K. THOMPSON and S.R. BAND 2005a. The impacts of agricultural change (1965-1997) on the grassland flora of Central England: Processes and prospects. Basic and Applied Ecology 6, 107-118.

HODGSON, J.G., P.J. WILSON, R. HUNT, J.P. GRIME and K. THOMPSON 1999. Allocating C-S-R plant functional types: a soft approach to a hard problem. Oikos 85, 282-296.

HOFFMAN, G.R., M.B. HOGAN and L.D. STANLEY 1980. Germination of plant species common to reservoir shores in the Northern Great Plains. Bulletin of the Torrey Botanical Club 107, 506-513.

HOLDRIDGE, L.R. 1947. Determination of world formations from simple climatic data. Science 105, 367-368.

HOLM, L.G., J.V. PANCHO, J.P. HERBERGER and D.L. PLUCKNETT 1979. A geographical atlas of world weeds. New York: Wiley.

HOLM, L.G., D.L. PLUCKNETT, J.V. PANCHO and J.P. HERBERGER 1977. The world's worst weeds. Honolulu: University Press of Hawaii.

HOLMGREN, P. 1968. Leaf factors affecting light saturated photosynthesis in ecotypes of Solidago virgaurea from exposed and shaded habitats. Physiologia Plantarum 21, 676-698.

HOLT, J.S. and S.R. RADOSEVICH 1983. Differential growth of two common groundsel (Senecio vulgaris) biotypes. Weed Science 31, 112-120.

HOOPER, D.U. and P.M. VITOUSEK 1997. The effects of plant composition and diversity on ecosystem processes. Science 277, 1302-1305.

HOPE-SIMPSON, J.F. 1940. Studies of the vegetation of the English chalk. 6: Late stages in succession leading to chalk grassland. Journal of Ecology 28, 386-402.

HOPE-SIMPSON, J.F. 1941. Studies of the vegetation of the English chalk. 7: Bryophytes and lichens in chalk grassland, with a comparison of their occurrence in other calcareous grassland. Journal of Ecology 29, 107-116.

HOPKINS, A. 1979. The botanical composition of grasslands in England and Wales: An appraisal of the role of species and varieties. Journal of the Royal Agricultural Society of England 140, 140-150.

HORGAN 1996 The end of science: facing the limits of knowledge in the twilight of the scientific age. New York: Broadway Books.

HOWARD, H.W. and A.G. LYON 1952. Biological Flora of the British Isles: Nasturtium microphyllum Boennigh. ex Rchl. Journal of Ecology 40, 239-245.

HOWARTH, S.E. and J.J. WILLIAMS 1968. Biological Flora of the British Isles: Chrysanthemum leucanthemum, L. Journal of Ecology 56, 585-595.

HOWITT, R.C.L. and B.M. HOWITT 1963. A flora of Nottinghamshire. Nottingham: Derry & Son.

HUBBARD, C.E. 1984. Grasses: A guide to their structure, identification, uses and distribution in the British Isles, 3rd edn (revised by J.C.E. Hubbard). London: Penguin.

HULL, R. 1961. Variation in populations of Bellis perennis. Proceedings of the Botanical Society of the British Isles 4, 269-272.

HULTEN, E. 1958. The Amphi-atlantic plants and their phytogeographical connections. Stockholm: Almqvist & Wiksell.

HULTEN, E. 1964. The circumpolar plants. 1: Vascular cryptogams, conifers and monocotyledons. Stockholm: Almqvist & Wiksell.

HULTEN, E. 1971. The circumpolar plants. 2: Dicotyledons. Stockholm: Almqvist & Wiksell.

HUME, L. and P.B. CAVERS 1982. Geographic variation in a widespread perennial weed. The relative amounts of genetic and environmentally induced variation among Rumex crispus populations. Canadian Journal of Botany 60, 1928-1937.

HUME, L. and P.B. CAVERS 1983. Resource allocation and reproductive and life-history strategies in widespread populations of Rumex crispus. Canadian Journal of Botany 61, 1276-1282.

HUME, L., J. MARTINEZ and K. BEST 1983. The biology of Canadian weeds. 60: Polygonum convolvulus L. Canadian Journal of Plant

Science 63, 959-971.

HUNDT, R. 1966. Okologisch-geobotanische Untersuchungen an Pflanzen der Mitteleuropaischen Wiesenvegetation. Jena: Fischer.

HUNT, R. 1970. Relative growth-rate: Its range and adaptive significance in a local flora. Ph.D. Thesis, University of Sheffield.

HUNT, R. 1979. Plant growth analysis: the rationale behind the use of the fitted mathematical function. Annals of Botany 43, 245-249.

HUNT, R. 1981. The fitted curve in plant growth studies. In: Mathematics and plant physiology, D.A. Rose and D.A. Charles-Edwards (eds), 283-298. London: Academic Press.

HUNT, R. 1982. Plant growth curves: the functional approach to plant growth analysis. London: Edward Arnold.

HUNT, R. 1984. Relative growth rates of cohorts of ramets cloned from a single genet. Journal of Ecology 72, 299-305.

HUNT, R. and C.J. DOYLE 1984. Modelling the partitioning of research effort in ecology. Journal of Theoretical Biology 111, 451-461.

HUNT, R. and P.S. LLOYD 1987. Growth and partitioning In: Frontiers of comparative plant ecology, I.H. Rorison, J.P. Grime, R. Hunt, G.A.F. Hendry and D.H. Lewis (eds), 235-250. London: Academic Press.

HUNT, R, J.F. HOPE-SIMPSON and J. SNAPE 1985. Growth of the dune wintergreen (Pyrola rotundifolia ssp. maritima) at Braunton Burrows in relation to weather factors. International Journal of Biometeorology 29, 327-334.

HUNT, R., HODGSON, J.G., THOMPSON, K., BUNGENER, P., DUNNETT, N.P. and A.P. ASKEW 2004. A tool for deriving a functional signature for herbaceous vegetation. Applied Vegetation Science 7, 163-170.

HURKA, H. and R. HAASE 1982. Seed ecology of Capsella bursa-pastoris; dispersal mechanism and the soil seed bank. Flora 172, 35-46.

HURKA, H., R. KRAUS, T. REINER and K. WOHRMANN 1976. Studies on the flowering of Capsella bursa-pastoris; a partially self-pollinating autogamous species. Plant Systems and Evolution 125, 89-95.

HUSSEIN, F. 1955. Chromosome races in Cardamine pratensis in the British Isles. Watsonia 3, 170-174.

HUSTON, M. and D.L. DE ANGELIS 1994. Competition and co-existence: the effects of resource transport and supply rates. American Naturalist 144; 954-977.

HUSTON, M.A. 1997. Hidden treatments in ecological experiments: re-evaluating the ecosystem function of biodiversity. Oecologia 110, 449-460.

HUSTON, M.A. and T. M. SMITH 1987. Plant succession: life history and competition. American Naturalist 130, 168-198.

HUTCHINGS, M.J. 1978. Standing crop and pattern in pure stands of Mercurialis perennis and Rubus fruticosus in mixed deciduous woodland. Oikos 31, 351-357.

HUTCHINGS, M.J. and J.P. BARKHAM 1976. An investigation of shoot interactions in Mercurialis perennis; a rhizomatous perennial herb. Journal of Ecology 64, 723-744.

HUTCHINGS M.J. and H. DE KROON 1994. Foraging in plants: the role of morphological plasticity in resource acquisition. Advances in Ecological Research 25, 159-238.

HUTCHINSON, C.S. and G.B. SEYMOUR 1982. Biological Flora of the British Isles: Poa annua L. Journal of Ecology 70, 887-901.

HUTCHINSON, G.E. 1951. Copepodology for the ornithologist. Ecology 32, 571-577.

HUTCHINSON, G.E. 1959. Homage to Santa Rosalia or why are there so many kinds of animals? American Naturalist 93, 145-159.

HUTCHINSON, I., J. COLOSI and R.A. LEWIN 1984. The biology of Canadian weeds. 63: Sonchus asper (L.) Hill and S. oleraceus L. Canadian Journal of Plant Science 64, 731-749.

HUTCHINSON, T.C. 1967. Comparative studies of the ability of species to withstand prolonged periods of darkness. Journal of-Ecology 55, 291-299.

HUTCHINSON, T.C. 1968. Biological Flora of the British Isles: Teucrium. scorodonia L. Journal of Ecology 56, 901-911.

IESTEWAART, J.H., R.A. BAREL and M.E. IKELAAR 1984. Male sterility and ecology of Dutch Origanum vulgare populations. Acta Botanica Neerlandica 33, 335-345.

INGHE, O. and C.O. TAMM 1985. Survival and flowering of perennial herbs. 4: The behaviour of Hepatica nobilis and Sanicula europaea on permanent plots during 1943-1981. Oikos 45, 400-420.

INGRAM, R. and L. TAYLOR 1982. The genetic control of a non-radiate condition in Senecio squalidus L. and some observations on the role of ray florets in the Compositae. New Phytologist 91, 749-756.

INGRAM, R., J. WEIR and R.J. ABBOTT 1980. New evidence concerning the origin of inland radiate groundsel, S. vulgaris L. var. hibernicus Syme. New Phytologist 84, 543-546.

INTERGOVERNMENTAL PANEL ON CLIMATE CHANGE. 2001. Climate Change 2001: Synthesis Report. A contribution of Working Groups I, II, and III to the Third Assessment Report of the Intergovernmental Panel on Climate Change. Ed. Watson R.T and Core Working Team. New York: Cambridge University Press.

IVENS, G.W. 1978. Some aspects of seed ecology of gorse. In: Proceedings of the 31st New Zealand weed and pest control conference, M.J. Hartley (ed. 53-57. Palmerston North: New Zealand Weed and Pest Control Society.

JALAS, J. and J. SUOMINEN (eds) 1972-1983. Atlas Florae Europaeae, volumes 1-6. Helsinki: The Committee for Mapping the Flora of Europe together with Societas Biologica Fennica Vanamo.

JALLOQ, M.C. 1974. The invasion of molehills by weeds as a possible factor in the degeneration of reseeded pasture. 1: The buried viable seed population of molehills from four reseeded pastures in W Wales. Journal of Applied Ecology 12, 643-657.

JAMES, D.B. 1962. Growth of Nardus stricta on a calcareous soil. Nature 196, 390-391.

JANSSEN, J.M.G. 1973. Effects of light, temperature and seed age on the germination of the winter annuals Veronica arvensis and Myosotis ramosissima. Oecologia 12, 141-146.

JANSSENS, F., A. PEETERS, J.R.B. TALLOWIN, J.P. BAKKER, R.M. BEKKER, F. FILLAT and M.J.M. OOMES, 1998. Relationship between soil chemical factors and grassland diversity. Plant and Soil 202, 69-78.

JANZEN, D.H. 1986. Lost plants. Oikos 46, 129-131.

JARVIS, P.G. 1964. Interference by Deschampsia flexuosa. Oikos 15, 56 78.

JARVIS, P.G., M.S. JARVIS 1964. Growth rates of woody plants. Physiologia Plantarum 17, 654-666.

JEFFERIES, R.A., A.D. BRADSHAW and P.D. PUTWAIN 1981. Growth, nitrogen accumulation and nitrogen transfer by legume species established on mine spoils. Journal of Applied Ecology 18, 945-956.

JEFFERIES, R.L. and A.J. WILLIS 1964. Studies on the calcicole-calcifuge habit. 2: The influence of calcium on the growth and establishment of four species in soil and sand cultures. Journal of Ecology 52, 691-707.

JEFFERIES, T.A. 1915. The ecology of the purple heath grass (Molinia caerulea). Journal of Ecology 3, 93-109.

JEFFERIES, T.A. 1916. The vegetative anatomy of Molinia caerulea, the purple moor grass. New Phytologist 15, 47-71.

JEFFREYS, H. 1917. On the vegetation of four Durham coal-measure fells. 2: On water supply as an ecological factor. Journal of Ecology 5, 129-154.

JERMY, A.C., A.O. CHATER and R.W. DAVID 1982. Sedges of the British Isles: A new edition of British sedges by A.C. Jermy and T.G. Tutin (illustrated by S. Bownas, G. Dalby, M. Tebbs and J. Webb). BSBI Handbook no. 1. London: Botanical Society of the British Isles.

JERMY, A.C., H.R. ARNOLD, L. FARRELL and F.H. PERRING 1978. Atlas of ferns of the British Isles. London: Botanical Society of the British Isles and The British Pteridological Society.

JERMYN, S.T. 1974. Flora of Essex. Colchester: Essex Naturalists Trust.

JOENJE, W. 1978. Plant colonisation and succession on embanked sandflats. Ph.D. Thesis, University of Groningen.

JOHNSON, H. and M.E. BIONDINI 2001. Root morphological plasticity and nitrogen uptake of 59 species from the Great Plains grasslands, USA. Basic and Applied Ecology 2, 127-143.

JONASSON, S. and F.S. CHAPIN 1985. Significance of sequential leaf development for nutrient balance of the cotton sedge, Eriophorum vaginatum L. Oecologia 67, 511-518.

JONES, D.A. 1962. Selective eating of the acyanogenic form of the plant

Lotus corniculatus L. by various animals. Nature 193, 1109.

JONES, D.A. and R. TURKINGTON 1986. Biological Flora of the British Isles: Lotus corniculatus L. Journal of Ecology 74, 1185-1212.

JONES, E.W. 1945. Biological Flora of the British Isles: Acer pseudoplatanus L. Journal of Ecology 32, 220-237.

JONES, E.W. 1959a. Biological Flora of the British Isles: Quercus petraea (Matt.) Liebl. Journal of Ecology 47, 169-216.

JONES, E.W. 1959b. Biological Flora of the British Isles: Quercus robur L. Journal of Ecology 47, 169-216.

JONES, E.W. 1974. Introduction. In: The British oak; its history and natural history, M.G. Morris and F.H. Perring (eds), 11-12. Faringdon, Berkshire: E.W. Classey.

JONES, K. 1956. Some aspects of plant variation: The grasses. In: Progress in the study of the British Flora, J.E. Lousley (ed.), 45-55. London: Botanical Society of the British Isles.

JONES, K. 1958. Cytotaxonomic studies in Holcus 1. The chromosome complex in Holcus mollis L. New Phytologist 57, 191-210.

JONES, R. 1975. Comparative studies of plant growth and distribution in relation to waterlogging. 8: The uptake of phosphorus by dune and dune-slack plants. Journal of Ecology 63, 109-116.

JONES, R.N. and L.M. BROWN 1976. Chromosome evolution and DNA variation in Crepis. Heredity 36, 91-104.

JUSTICE, O.L. 1941. A study of dormancy in seeds of Polygonum. Cornell University Agricultural Experimental Station Memoir 235, 1-43.

KADEREIT, J.W. 1984. Studies on the biology of Senecio vulgaris ssp. denticulatus. New Phytologist 97, 681689.

KADEREIT, J.W. and P.D. SELL 1986. Variation in Senecio jacobaea L. (Asteraceae) in the British Isles. Watsonia 16, 21-23.

KADMON, R. and A. SHMIDA 1990. Competition in a variable environment: an experimental study in a desert annual population. Israel Journal of Botany 39, 403-412.

KAHLERT, B.R., P. RYSER, and P.J. EDWARDS 2005. Leaf phenology of three dominant limestone grassland plants matching the disturbance regime. Journal of Vegetation Science 16, 433-442.

KANNANGARA, H.W. and R.J. FIELD 1985. Environmental and physiological factors affecting the fate of seeds of yarrow (Achillea millefolium L.) in arable land in New Zealand. Weed Research 25, 87-92.

KAY, Q.O.N. 1969. The origin and distribution of diploid and tetraploid Tripleurospermum inodorum (L.) Schultz Bip. Watsonia 7, 130-141.

KAY, Q.O.N. 1972. Variation in sea mayweed (Tripleurospermum maritimum (L.) Koch.) in the British Isles. Watsonia 9, 81-107.

KAY, Q.O.N. 1985. Hermaphrodite and subhermaphrodites in a reputedly dioecious plant Cirsium arvense (L.) Scop. New Phytologist 100, 457-472.

KAY, Q.O.N. 1987. Comparative ecology of flowering. In: Frontiers of comparative plant ecology, I.H. Rorison, J.P. Grime, R. Hunt, G.A.F. Hendry and D.H. Lewis (eds), 265-282. London: Academic Press.

KAY, Q.O.N. and D.P. STEVENS 1986. The frequency, distribution and reproductive biology of dioecious species in the native flora of Britain and Ireland. Botanical Journal of the Linnean Society 92, 39-64.

KEDDY, P.A. 1990. Competitive hierarchies and centrifugal organization in plant communities. In: Perspectives on Plant Competition (J. Grace and D. Tilman, eds.), 265-289. New York: Academic Press.

KEDDY P.A. 1992. Assembly and response rules: two goals for predictive community ecology. Journal of Vegetation Science 3, 157-164.

KEELEY, J.E. and B.A. MORTON 1982. Distribution of diurnal acid metabolism in submerged aquatic plants outside the genus Isoetes. Photosynthetica 16, 546-553.

KELLY, D. 1989. Short-lived plants of chalk grassland 2: Control of phenology, mortality and fecundity. Journal of Ecology 77, 770-784.

KENDRICK, R.E. and B. FRANKLAND 1969. Photocontrol of germination in Amaranthus caudatus. Planta 85, 326-339.

KENT, D.H. 1956. Senecio squalidus L. in the British Isles 1. Early records (to 1877). Proceedings of the Botanical Society of the British Isles 2, 115-118.

KENT, D.H. 1960. Senecio squalidus L. in the British Isles 2. The spread from Oxford 1879-1939. Proceedings of the Botanical Society of the British Isles 3, 375-379.

KHEYR-POUR, A. 1981. Wide nucleo-cytoplasmic polymorphism for male sterility in Origanum vulgare. Journal of Heredity 72, 5-51.

KIMATA, M. 1983. Comparative studies on the reproductive systems of Cardamine flexuosa, C. impatiens, C. scutata and C. lyrata, Cruciferae. Botanical Magazine Tokyo 96, 299-312.

KING, J. 1960. Observations on the seedling establishment and growth of Nardus stricta in burned callunetum. Journal of Ecology 48, 667-677.

KING, T.J. 1975. Inhibition of seed germination under leaf canopies in Arenaria serpyllifolia, Veronica arvensis and Cerastium holosteoides. New Phytologist 75, 87-90.

KING, T.J. 1977a. The plant ecology of ant-hills in calcareous grasslands. 1: Patterns of species in relation to ant-hills in Southern England. Journal of Ecology 65, 235-256.

KING, T.J. 1977b. The plant ecology of ant-hills in calcareous grasslands. 2: Factors affecting the population sizes of selected species. Journal of Ecology 65, 257-278.

KING, T.J. 1977c. The plant ecology of ant-hills in calcareous grasslands. 3: Factors affecting the population sizes of selected species. Journal of Ecology 65, 279-316.

KINNAIRD, J.W. 1968. Ecology of birch woods. Proceedings of the Botanical Society of the British Isles 7, 181-182.

KINNAIRD, J.W. 1974. Effect of site conditions on the regeneration of birch (Betula pendula and B. pubescens). Journal of Ecology 62, 467-472.

KINNAIRD, J.W. and E. KEMP 1971. Effect of shade on the growth of birch. Nature Conservancy Research in Scotland Report 1968-1970, 32-33.

KINZEL, W. 1920. Frost und Licht als beeinflussende Krafte der Samenkeimung. Stuttgart: E. Ulmer.

KIRCHNER, O., E. LOEW and C. SCHROETER 1908. Lebensgeschichte der Blutenpflanzen Mitteleuropas. Stuttgart.

KJELLSSON, G. 1985a. Seed fate in a population of Carex pilulifera L. 1: Seed dispersal and ant-seed mutualism. Oecologia 67, 416-423.

KJELLSSON, G. 1985b. Seed fate in a population of Carex pilulifera L. 2: Seed predation and its consequences for dispersal and seed bank. Oecologia 67, 424-429.

KLEYER, M. 1999. Distribution of plant functional types along gradients of disturbance intensity and resource supply in an agricultural landscape. Journal of Vegetation Science 10, 697-708.

KLIPHUIS, E. 1967. Cytotaxonomic notes on some Galium species. Acta Botanica Neerlandica 15, 535-538.

KNEVEL, I.C., R.M. BEKKER, J.P. BAKKER and M. KLEYER 2003. Life-history traits of the Northwest European flora: the LEDA database. Journal of Vegetation Science 14, 611-614.

KNIGHT, G.H. 1964. Some factors affecting the distribution of Endymion non-scriptus in Warwickshire woods. Journal of Ecology 52, 405-421.

KNOX, J.P. and A.D. DODGE 1985. Isolation and activity of the photodynamic pigment hypericin. Plant Cell and Environment 8, 19-26.

KNUTH, P. 1906. Handbook of flower pollination. Oxford: Clarendon Press.

KOOTIN-SANWU, M. and S.R.J. WOODELL 1970. The cytology of Caltha palustris; distribution and behaviour of the chromosome races. Caryologia 32, 225-239.

KOOTIN-SANWU, M. and S.R.J. WOODELL 1971. The cytology of Caltha palustris; cytogenetic relationships. Heredity 26, 121-135.

KRATTINGER, K. 1975. Genetic mobility in Typha. Aquatic Botany 1, 57-70.

KREKULE, J. and L. HAJKOVA 1972. The developmental pattern in a group of therophytes. 2: Vernalisation and photoperiodic induction. Flora 161, 121-128.

KWAK, M.M. 1979. Effects of bumblebee visits on the seed set of Pedicularis, Rhinanthus and Melampyrum in the Netherlands. Acta Botanica Neerlandica 28, 177-195.

KYTOVUORI, I. 1969. Epilobium davuricum Fisch. (Onagraceae) in Eastern Fennoscandia compared with Epilobium palustre L. A morphological, ecological and distributional study. Annals Botanica Fennici 6, 35-58.

LAANE, M. 1983. Campanula rotundifolia complex in Norway and cyto-morphological characteristics of diploid and tetraploid groups. Hereditas 99, 21-48.

LACEY, E.P. 1982. Timing of seed dispersal in Daucus carota. Oikos 39, 83-91.

LACEY, E.P. 1986. The genetic and environmental control of reproductive timing in a short-lived monocarpic species Daucus carota (Umbelliferae). Journal of Ecology 74, 73-86.

LACK, A.J. 1976. Competition for pollinators and evolution in Centaurea. New Phytologist 77, 787-792.

LACK, A.J. 1982a. Competition for pollinators in the ecology of Centaurea scabiosa L. and Centaurea nigra L. 1: Variation in flowering time. New Phytologist 91, 297-308. 684.

LACK, A.J. 1982b. Competition for pollinators in the ecology of Centaurea scabiosa L. and Centaurea nigra L. 2: Observations on nectar production. New Phytologist 91, 309-320.

LACK, A.J. 1982c. Competition for pollinators in the ecology of Centaurea scabiosa L. and Centaurea nigra L. 3: Insect visits and the number of successful pollinations. New Phytologist 91, 321-339.

LACK, A.J. and Q.O.N. KAY 1986. Phosphoglucose isomerase (EC 5.3.1.9.) isozymes in diploid and tetraploid Polygala species; evidence for gene duplication and diversification. Heredity 56, 111-118.

LADD, J.M. and J.M. FACELLI 2005. Effects of competition, resource availability and invertebrates on tree seedling establishment. Journal of Ecology 93, 968-977.

LAIBACH, F. 1951. Uber Sommer- und Winterannuelle Rassen von Arabidopsis thaliana (L.) Heynh. Ein Beitrag zur Atiologie der Blutenbildung. Beitrage zur Biologie Pflanzen 28, 173.

LAMARCK, J.B. de. 1778. Flora Francaise 3, 538.

LAMBERS, H. and H. POORTER 1992. Inherent variation in growth rate between higher plants: a search for physiological causes and ecological consequences. Advances in Ecological Research 23, 187-261.

LAMBERT, J.M. 1947. Biological Flora of the British Isles: Glyceria maxima (Hartm.) Holmb. Journal of Ecology 34, 310-344.

LANDSBERG, J., LAVOREL, S. and J. STOL 1999. Grazing response groups among understorey plants in arid rangelands. Journal of Vegetation Science 10, 683-690.

LANDOLT, E. 1982. Distribution pattern and ecophysiological characteristics of the European species of the Lemnaceae. Berichte des Geobotanischen Institutes der ETH, Stiftung Rubel 49, 127-185.

LANGER, R.H.M. 1956. Growth and nutrition of timothy 1. The life history of individual tillers. Annals of Applied Biology 44, 166-187.

LARCHER, W. 1980. Physiological plant ecology. Berlin: Springer.

LAUER, E. 1953. Ueber die Keimtemperatur von Ackerunkrautern und deren Einfluss auf die Zusammensetzung von Unkrautgesellsehaften. Flora 40, 551-595.

LAVOREL, S., LEPART, J., DEBUSSCHE, M., LEBRETON, J.D. and J.L.BEFFY 1994. Small scale disturbance and the maintenance of species diversity in Mediterranean old fields. Oikos 70, 455-473.

LAVOREL, S., S. MCINTYRE, and K. GRIGULIS 1999. Plant response to disturbance in a Mediterranean grassland: how many functional groups? Journal of Vegetation Science 10, 661-67.

LAW, R. 1974. Features of the biology and ecology of Bromus erectus and Brachypodium pinnatum in the Sheffield region. Ph.D. Thesis, University of Sheffield.

LAW, R., A. D. BRADSHAW and P.D. PUTWAIN 1977. Life history variation in Poa annua. Evolution 31, 233-246.

LAWRENCE, M.J. 1976. Variation in natural populations of Arabidopsis thaliana (L.) Heynh. In: The biology and chemistry of the Cruciferae, J.G. Vaughan, A.J. Macleod and B.M.G. Jones (eds), 167-190. London: Academic Press.

LAWTON, J.H. and P.A. HEADS 1984. Bracken, ants and extrafloral nectaries. 1: The components of the system. Journal of Animal Ecology 53, 1015-1031.

LE DEUNFF, Y. 1974. Heterogeneite de la germination des semences de Rumex crispus L. Physiologie Vegetale 9, 201-208.

LECHOWICZ, M.J. 1984. Why do temperate deciduous trees leaf out at different times? Adaptation and ecology of forest communities. American Naturalist 124, 821-842.

LECK, M.A. and K.J. GRAVELINE 1979. The seed bank of a freshwater tidal marsh. American Journal of Botany 66, 1006-1015.

LEES, D.R. 1971. Frequency of pin and thrum plants in a wild population of the cowslip Primula veris. Watsonia 8, 289-291.

LEES, F.A. 1888. The flora of West Yorkshire. London: Lovell Reeve.

LEPS, J., J. OSBORNA-KOSINOVA and M. REJMANEK 1982. Community stability, complexity and species life-history strategies. Vegetatio 50, 53-63.

LESLIE, A.C. and S.M. WALTERS 1983. The occurrence of Lemna minuscula Herter in the British Isles. Watsonia 14, 243-248.

LEVANG-BRILZ, N. and M.E. BIONDINI 2002. Growth rate, root development and nutrient uptake of 55 plant species from the Great Plains grasslands, USA. Plant Ecology 165, 117-144.

LEVIN, D.A. and S.W. FUNDERBURG 1979. Genome size in angiosperms; temperate versus tropical species. American Naturalist 114, 784-795.

LEWIN, R.A. 1948a. Biological Flora of the British Isles: Sonchus asper (L.) Hill. Journal of Ecology 36, 213-216.

LEWIN, R.A. 1948b. Biological Flora of the British Isles: Sonchus oleraceus L. Journal of Ecology 36, 204-216.

LEWIS, J. 1973. Longevity of crop and weed seeds, survival after twenty-years in the soil. Weed Research 13, 179-191.

LEWIS, M.C. 1972. The physiological significance of variation in leaf structure. Science Progress, 60, 25-51.

LEWONTIN, R.C. 1974. The genetic basis of evolutionary change. New York: Columbia University Press.

LHOTSKA, H. 1977. Notes on the ecology of germination in Myrrhis odorata. Folia Geobotanica et Phytotaxonomica 12, 209-213.

LINHART, Y.B. and R.J.-WHELAN 1980. Woodland regeneration in relation to grazing and fencing in Coed Gorswen, North Wales. Journal of Applied Ecology 17, 827-840.

LINTON, D.L. (ed.) 1956. Sheffield and its region. British Association for the Advancement of Science.

LINTON, W.R. 1903. Flora of Derbyshire. London: Bemrose.

LITTLE, E.C.S. 1960. The ecology of some New Zealand woody weeds. In: The biology of weeds, J.L. Harper (ed.), 176-183. Oxford: Blackwell Scientific.

LLOYD, P.S. 1972a. The grassland vegetation of the Sheffield region. 2: Classification of grassland types. Journal of Ecology 60, 739-776.

LLOYD, P.S. 1972b. Effects of fire on a Derbyshire grassland community. Ecology 53, 915-920.

LLOYD, P.S. and C.D. PIGOTT 1967. The influence of soil conditions on the course of succession on the chalk of Southern England. Journal of Ecology 55, 137-146.

LLOYD, P.S., J.P. GRIME and I.H. RORISON 1971. The grassland vegetation of the Sheffield region. 1: General features. Journal of Ecology 59, 863-886.

LLOYD, P.S., V. MOFFETT and D.W. WINDLE 1972. Computer storage and retrieval of botanical survey data. Journal of Applied Ecology 9, 1-10.

LOACH, K. 1968. Seasonal growth and nutrient uptake in a Molinietum. Journal of Ecology 56, 433-444.

LOACH, K. 1970. Shade tolerance in tree seedlings 2. Growth analysis of plants raised under artificial shade. New Phytologist 69. 273-286.

LODGE, R.W. 1959. Biological Flora of the British Isles: Cynosurus cristatus L. Journal of Ecology 47, 511-518.

LODGE, R.W. 1962a. Autecology of Cynosurus cristatus. 1: Habitat studies. Journal of Ecology 50, 63-73.

LODGE, R.W. 1962b. Autecology of Cynosurus cristatus. 2: Ecotypic variation. Journal of Ecology 50, 75-86.

LOENHOED, P.J. van and A.A. STERK 1976. A study of the Juncus bufonius complex in the Netherlands. Acta Botanica Neerlandica 25, 193-204.

LONGMAN, K.A. and M.P. COUTTS 1974. Physiology of the oak tree. In: The British oak; its history and natural history, M.G. Morris and F.H. Perring (eds), 194-221. Faringdon, Berkshire: E.W. Classey.

LOUSLEY, J.E. (ed.) 1953. The changing flora of Britain. Arbroath:

Buncle.
LOUSLEY, J.E. and D.H. KENT 1981. Docks and knotweeds of the British Isles. BSBI Handbook No. 3. London: Botanical Society of the British Isles.
LOVELESS, A.R. 1961. A nutritional interpretation of sclerophylly based on differences in the chemical composition of sclerophyllous and mesophytic leaves. Annals of Botany 25, 168-184.
LOVKVIST, B. 1956. The Cardamine pratensis complex. Outlines of its cyanogenetics and taxonomy Symbolae Botanicae Upsalienses 14, 5-131.
LOVKVIST, B. 1975. Experimental studies in Cardamine amara L. Botaniska Notiser 110, 423-441.
LUCAS, A.T. 1960. Furze, a survey of its uses in Ireland. Dublin: National Museum of Ireland.
McALLISTER, H. 1979. The species of ivy. Ivy Exchange Newsletter 1, 4-6. MACAN, T.T. 1977. Changes in the vegetation of a moorland fish pond in twenty-one years. Journal of Ecology 65, 95-106.
MacARTHUR, R.H. and E.D. WILSON 1967. The theory of island biogeography. Princeton: Princeton University Press.
McCALLISTER, H.A. 1970. Biosystematic studies in Campanula rotundifolia. Transactions and Proceedings of the Botanical Society of Edinburgh 40, 642.
McCREATH, J.B. 1982. Some reflections and projections on bracken control. Proceedings of the Royal Society of Edinburgh 81B, 135-143.
McEVOY, P.B. 1984. Seedling dispersion and the persistence of ragwort (Senecio jacobaea) in a grassland dominated by perennials. Oikos 42, 138-143.
MACGILLIVRAY, C.W., J.P. GRIME and THE ISP TEAM.1995. Testing predictions of resistance and resilience of vegetation subjected to extreme events. Functional Ecology 9, 640-649.
McGRATH, S.P. 1979. Growth and distribution of Holcus lanatus L. populations with reference to nitrogen source and aluminium. Ph.D. Thesis, University of Sheffield.
McGRAW, J.B. 1980. Seed bank size and distribution of seeds in cottongrass tussock tundra, Eagle Creek, Alaska. Canadian Journal of Botany 58, 1607-1611.
MacLEOD, J. 1894. Over de bevruchting der bloemen in het Kempisch gedeelte van Vlaanderen. Deel II. Bot. Jaarboek Dodonaea 6, 119-511.
McNAUGHTON, I.H. and J.L. HARPER 1964. Biological Flora of the British Isles: Papaver rhoeas L. Journal of Ecology 52, 767-779.
McNAUGHTON, S.J. 1966. Ecotype function in the Typha community type. Ecological Monographs 36, 297-325.
McNAUGHTON, S.J. 1968. Autotoxic feedback in relation to germination and seedling growth in Typha latifolia. Ecology 49, 367-369.
McVEAN, D.N. 1952. Ecology of Molinia pastures with particular reference to their invasion by Polytrichum commune. Journal of the British Grassland Society 7. 21-27.
McVEAN, D.N. 1953. Biological Flora of the British Isles: Alnus glutinosa (L.) Gaertn. Journal of Ecology 41, 447-466.
McVEAN, D.N. 1955a. Ecology of Alnus glutinosa (L.) Gaertn. 1: Fruit formation. Journal of Ecology 43, 46-60.
McVEAN, D.N. 1955b. Ecology of Alnus glutinosa (L.) Gaertn. 2: Seed distribution and germination. Journal of Ecology 43, 61-71.
McVEAN, D.N. 1956a. Ecology of Alnus glutinosa (L.) Gaertn. 3: Seedling establishment. Journal of Ecology 44, 195-218.
McVEAN, D.N. 1956b. Ecology of Alnus glutinosa (L.) Gaertn. 4: Root system. Journal of Ecology 44, 219-255.
McVEAN, D.N. 1956c. Ecology of Alnus glutinosa (L.) Gaertn. 5: Notes on some British alder populations. Journal of Ecology 44, 321-330.
McVEAN, D.N. and D.A. RATCLIFFE 1962. Plant communities of the Scottish Highlands. Monograph of the Nature Conservancy No. 1. London: HMSO.
MAFF 1980. Culinary and medicinal herbs. ADAS reference book 325. London: HMSO.
MAFF 1980-84. Stocks of seeds in the hands of seedsmen. Ministry of Agriculture, Fisheries and Food, published annually.
MAGISTAD, O.C. 1925. The aluminium content of the soil solution and its relation to soil reaction and plant growth. Soil Science 20, 181-213.
MAHMOUD, A. 1973. A laboratory approach to ecological studies of the grasses: Arrhenatherum elatius (L.) Beauv. ex J. and C. Presl., Agrostis tenuis Sibth. and Festuca ovina L. Ph.D. Thesis, University of Sheffield.
MAHMOUD, A. and J.P. GRIME 1974. A comparison of negative relative growth rates in shaded seedlings. New Phytologist 73, 1215-1219.
MAHMOUD, A. and J.P. GRIME 1976. An analysis of competitive ability in three perennial grasses. New Phytologist 77, 431-435.
MAHMOUD, A., J.P. GRIME and S.B. FURNESS 1975. Polymorphism in Arrhenatherum elatius (L.) eauv. ex J. and C. Presl. New Phytologist 75, 269-276.
MALLIK, A.U. and C.H. GIMINGHAM 1985. Ecological effects of burning heather. 2: Effects on seed germination and vegetative regeneration. Journal of Ecology 73, 633-644.
MALLIK, A.U., C.H. GIMINGHAM and A.A. RAHMAN 1984. Ecological effects of heather burning. 1: Water infiltration, moisture retention and porosity of surface soil. Journal of Ecology 72, 767-776.
MALLOCH, A.J.C. and O.T. OKUSANYA 1979. An experimental investigation into the ecology of some maritime cliff species. 1: Field observations. Journal of Ecology 67, 283-292.
MANZUR, M.I. and S.P. COURTNEY 1984. Influence of insect damage in fruits of hawthorn Crataegus monogyna on bird foraging and seed dispersal. Oikos 43, 265-270.
MARKS, P.L. 1974. The role of pin cherry (Prunus pensylvanica L.) in the maintenance of stability in northern hardwood ecosystems. Ecological Monographs 44, 73-88.
MARRS, R.H., M.J. HICKS and R.M. FULLER 1986. Losses of lowland heath through succession at four sites in Breckland, East Anglia. Biological Conservation 36, 19-38.
MARSDEN-JONES, E.M. and W.B. TURRILL 1954. British knapweeds, a study in synthetic taxonomy. London: Printed for the Ray Society.
MARSDEN-JONES, E.M. and F.E. WEISS 1959. The genetics and pollination of Anagallis arvensis ssp. arvensis and A. arvensis ssp. foemina. Proceedings of the Linnean Society 171, 27-29.
MARSHALL, D.F. and R.J. ABBOTT 1984. Polymorphism for outcrossing frequency at the ray floret locus in Senecio vulgaris L. 3: Causes. Heredity 53, 145-149.
MARTIN, M.H. 1968. Conditions affecting the distribution of Mercurialis perennis in certain Cambridgeshire woodlands. Journal of Ecology 56, 777-793.
MARUTA, E. 1976. Seedling establishment of Polygonum cuspidatum. on Mount Fuji. Japanese Journal of Ecology 26, 101-105.
MASON, G. 1976. Effects of temperature on the germination and growth of native species, using temperature gradient techniques. Ph.D. Thesis, University of Sheffield.
MASUDA, M. and I. WASHITANI 1990. A comparative ecology of the seasonal schedules for reproduction by seeds in a moist tall grassland community. Functional Ecology 4,169-182.
MATFIELD, B. and J. R. ELLIS 1972. The allopolyploid origin and genomic constitution of Potentilla anglica. Heredity 29, 315-327.
MATHE, I. and I. MATHE Jr 1978. Data to European area of chemical taxa of Solanum dulcamara L. Acta Botanica Academiae Scientiarum Hungaricae 19, 441-451.
MATTHEWS, J.R. 1937. Geographical relationships of the British flora. Journal of Ecology 25, 1-90.
MAW, M.G., A.G. THOMAS and A. STAHEVITCH 1985. The biology of Canadian weeds. 66: Artemisia absinthium L. Canadian Journal of Plant Science 65, 389-400.
MAY, R.M. and J. SEGER 1986. Ideas in ecology. American Scientist 74, 256-267.
MAYNARD SMITH, J, 1982. Evolution and the theory of games. Cambridge: Cambridge University Press.
MEARNS, L.O., R.W. KATZ and S.H. SCHNEIDER 1984. Changes in the probability of extreme high temperature events with changes in global mean temperature. Journal of Climate and Applied Meteorology 23, 1601-1613.
MEEUSE, B.J.D. 1975. Thermogenic respiration in aroids. Annual

Review of Plant Physiology 26, 117-126.
MEIKLE, R.D. 1984. Willows and poplars of Great Britain and Ireland. BSBI Handbook No. 4. London: Botanical Society of the British Isles.
MERKER, A. 1984. Hybrids between Trifolium medium and Trifolium pratense. Hereditas 101, 267-268.
MERTON, L.F.H. 1970. The history and status of woodlands of the Derbyshire limestone. Journal of Ecology 58, 723-744.
MEUSEL, H., E. JAGER and E. WEINERT 1965. Vergleichende Chorologie der zentraleuropaischen Flora. Jena: Fischer.
MILES, J. 1974. A note on the relation between P supply and the abundance of Agrostis canina ssp. montana on a Southern English heath. Journal of Ecology 62, 355-358.
MILLENER, L.H. 1961. Day-length as related to vegetative development in Ulex europaeus L. 1: The experimental approach. New Phytologist 60, 339-354.
MILLENER, L.H. 1962. Day-length as related to vegetative development in Ulex europaeus L 2: Ecotypic variation with latitude. New Phytologist 61, 119-127.
MILTON, W.E.J. 1933. The palatability of the self-establishing species. Empire Journal of Experimental Agriculture 1, 347-360.
MOGFORD, D.J. 1974. Flower colour polymorphism in Cirsium palustre. Heredity 33, 241-256.
MOLES, A.T., D.W. HODSON and C.J. WEBB 2000. Seed size and shape and persistence in the soil in the New Zealand flora. Oikos 89, 541-45.
MOORE, D.M. 1982. Flora Europaea check-list and chromosome index. Cambridge: Cambridge University Press.
MOORE, R.J. 1973. Index of plant chromosome numbers 1967-1971. Regnum Vegetabile 90, 1-539.
MOORE, R.J. 1974. Index of plant chromosome numbers 1972. Regnum Vegetabile 91, 1-108.
MOORE, R.J. 1975. The biology of Canadian weeds 13. Cirsium arvense (L.) Scop. Canadian Journal of Plant Science 55, 1033-1048.
MOORE, R.J. 1977. Index of plant chromosome numbers 1973-197. Regnum Vegetabile 96, 1-257.
MOORE, R.J. and C. FRANKTON 1962. Cytotaxonomic studies in the Tribe Cynaraea (Compositae). Canadian Journal of Botany 40, 281-293.
MORENO-CASASOLA, P. 1985. Patterns of plant species distribution on Mexican coastal dunes along the Gulf of Mexico. Ph.D. Thesis, University of Uppsala.
MORRIS, M.G. 1974. Oak as a habitat for insect life. In: The British oak; its history and natural history, M.G. Morris and F.H. Perring (eds), 274-297. Faringdon, Berkshire: E.W. Classey.
MORRIS, M.G. and F.H. PERRING, (eds) 1974. The British oak; its history and natural. history. Faringdon, Berkshire: Published for the Botanical Society of the British Isles by E.W. Classey.
MORRISON, J. 1979. Botanical change in agricultural grassland in Britain. In: Changes in sward composition and productivity. Occasional symposium No. 10, A.H. Charles and R.J. Hagger (eds), 5-10. Maidenhead, Berkshire: British Grassland Society.
MORTON, J.F. 1975. Cattails (Typha spp.) - weed problem or potential crop? Economic Botany 29, 297-329.
MORTON-BOYD, J., N.F. ROBERTSON, A.F. DYER and J.D. FRYER (eds) 1982. Bracken in Scotland. Proceedings of the Royal Society of Edinburgh 81-82B, 1-143.
MOSS, G.R. 1959. The gorse seed problem. Proceedings of the 12th New Zealand weed control conference, 59-64. Wellington: New Zealand Weed Control Conference.
MOSS, G.R. 1960. Gorse, a weed problem of thousands of acres of farm land. New Zealand Journal of Agriculture 100, 561-567.
MOUSSEAU, M. 1977. Adaptation to photosynthesis of a shade tolerant plant Teucrium scorodonia. In: Geobiology, ecology and planning. Processes of primary plant production. French works on photosynthesis from the International Biological Program, A. Moyse (ed.), 157-181. Paris: GauthierVillars.
MOWFORTH, M.A.G. 1985. Variation in nuclear DNA amounts in flowering plants: An ecological analysis. Ph.D. Thesis, University of Sheffield.
MUKERJI, S.K. 1936. Contributions to the autecology of Mercurialis perennis L. Journal of Ecology 24, 53-81.

MULLER, F.M. 1978. Seedlings of the North-Western European lowland. The Hague: Junk.
MULLER, H. 1883. The fertilization of flowers. London: Macmillan.
MULLER-DOBLIES, D. 1970. Uber die Verwandtschaft von Typha und Sparganium. im Infloreszenz und Blutenbau. Botanische Jahrbucher 89, 451-562.
MULLIGAN, G.A. and L.G. BAILEY 1975. The biology of Canadian weeds 8. Sinapis arvensis L. Canadian Journal of Plant Science 55, 171-183.
MURRAY, B.G. 1974. Breeding system and floral biology in the genus Briza L. (Gramineae). Heredity 33, 285-292.
MURRAY, B.G. 1976. The cytology of the genus Briza 2. Chiasma frequency, polyploidy and interchange heterozygosity. Chromosoma 57, 81-93.
MUSTARD, M.J., STANDING, D.B., AITKENHEAD, M.J., ROBINSON D. and A.J.S. McDONALD 2003. The emergence of primary strategies in evolving plant populations. Evolutionary Ecology Research 5, 1067-1081.
MYERSCOUGH, P.J. 1980. Biological Flora of the British Isles: Epilobium angustifolium. L. Journal of Ecology 68, 1047-1074.
MYERSCOUGH, P.J. and J.K. MARSHALL 1973. Population dynamics of Arabidopsis thaliana strain Westland at different densities and nutrient levels. New Phytologist 72, 595-617.
MYERSCOUGH, P.J. and F.H. WHITEHEAD 1966. Comparative biology of Tussilago farfara L., Chamaenerion angustifolium (L.) Scop., Epilobium montanum L. and Epilobium adenocaulon Hausskn. 1: General biology and germination. New Phytologist 65, 192-210.
MYERSCOUGH, P.J. and F.H. WHITEHEAD 1967. Comparative biology, of Tussilago farfara L., Chamaenerion angustifolium (L.) Scop., Epilobium montanum L. and Epilobium adenocaulon Hausskn. 2: Growth and ecology. New Phytologist 66, 785-823.
NEW, J.K. 1961. Biological Flora of the British Isles: Spergula arvensis L. Journal of Ecology 49, 205-215.
NEW, J.K,. 1978. Change and stability of clines in Spergula arvensis L. (corn spurrey) after twenty years. Watsonia 12, 137-143.
NEW, J.K. and J.C. HERRIOTT 1981. Moisture for germination as a factor affecting the distribution of the seed coat morphs of Spergula arvensis. Watsonia 13, 323-324.
NEWBOLD, A.J. and F.B. GOLDSMITH 1981, with an addendum on birch by J.S. HARDING 1981. The regeneration of oak and beech; a literature review. Discussion papers on conservation No. 33. London: University College London.
NEWMAN, E.I. 1963. Factors controlling the germination date of winter annuals. Journal of Ecology 51, 625-638.
NICHOLSON, G.C. 1983. Studies on the distribution and the relationship between the chromosome races of Ranunculus ficaria in South-East Yorkshire. Watsonia 14, 321-328.
NIEMELA, P., F.M. ARO and F. HAUKIOJA 1979. Birch leaves as a resource for herbivores. Damage-induced increase in leaf phenols with trypsin-inhibiting effects. Report of Kevo Subarctic Research Station 15, 37-40.
NIKOLAJEVSKII, V.G. 1971. Research into the biology of the common reed (Phragmites communis) in the USSR. Folia Geobotanica et Phytotaxonomica 6, 221-290.
NOBLE, I.R. and R.O. SLATYER 1979. The use of vital attributes to predict successional changes in plant communities subject to recurrent disturbances. Vegetatio 43, 5-21.
NORDBORG, G. 1967. The genus Sanguisorba Section Poterium; experimental studies and taxonomy. Opera Botanica 16, 5-166.
NYAHOZA, F., C. MARSHALL and G.R. SAGAR 1973. The inter-relationship between tillers and rhizomes of Poa pratensis L.: an autoradiographic study. Weed Research 13, 304-309.
NYBOM, H. 1980. Germination in Swedish blackberries (Rubus L. subgenus Rubus). Botaniska Notiser 133, 619-631.
NYE, P.H. 1966. The effect of nutrient intensity and buffering power of a soil, and the absorbing power, size and root hairs of a root, on nutrient absorption by diffusion. Plant and Soil 25, 81-105.
NYE, P.H. 1969. The soil model and its application to plant nutrition. In: Ecological aspects of the mineral nutrition of plants, I.H. Rorison (ed.), 105-114. Oxford: Blackwell.

NYE, P.H. and P.B. TINKER 1977. Solute movement in the soil systems. Oxford: Blackwell Scientific Publications.

OBERDORFER, E. 1979. Pflanzenphysiologische Exkursions Flora. Stuttgart: Ulmer.

ODUM, S. 1978. Dormant seeds in Danish ruderal soils: An experimental study of relations between seed bank and pioneer flora. Horstolm Arboretum, Denmark: The Royal Vetinary & Agricultural University.

OGDEN, J. 1974. The reproductive strategy of higher plants. 2: The reproductive strategy of Tussilago farfara. Journal of Ecology 62, 291-324.

OKUSANYA, O.T. 1979a. An experimental investigation into the ecology of some maritime cliff species 2: Germination studies. Journal of Ecology 67, 293-304.

OKUSANYA, O.T. 1979b. An experimental investigation into the ecology of some maritime cliff species 3: Effect of sea water on growth. Journal of Ecology 67, 579-590.

OKUSANYA, O.T. 1979c. An experimental investigation into the ecology of some maritime cliff species 4: Cold sensitivity and competition studies. Journal of Ecology 67, 591-600.

OLMO, E. 1983. Nucleotype and cell size in vertebrates -a review. Basic and Applied Histochemistry 27, 227-256.

OLSEN C. 1958. Iron uptake in different plant species as function of the pH value of the nutrient solution. Physiologia Plantarum 11, 889-905.

OLSZEWSKA, M.J. and R. OLSIECKA 1982. The relationship between 2C DNA content, life cycle type, systematic position and the level of DNA endoreplication in nuclei of parenchyma cells during growth and differentiation of roots in some monocotyledonous species. Biochemica Physiologia Pflanzen 177, 319-336.

OOMES, M.J.M. and W.T. ELBERSE 1976. Germination of six grassland herbs in microsites with different water contents. Journal of Ecology 64, 745-755.

OOSTERVELD, P. 1978. De indicatiewaarde van het genus Taraxacum voor het beheer van grasslanden. Gorteria 9, 188-193.

OREDSSON, A. 1975. Factors possibly influencing the range of shrubby Rubus species in Sweden. 1: Severity of winter. Botaniska Notiser 128, 47-54.

ORGEL, L.E. and F.H.C. CRICK 1980. Selfish DNA: The ultimate parasite. Nature 284, 604-607.

OTTOSSON, J.G. and J.M. ANDERSON 1983. Seasonal and inter-specific variation in the biochemical composition of some British UK fern species and their effects on Spodoptera littoralis larvae. Biological Journal of the Linnean Society 19, 305-320.

OTZEN, D. 1977. Life forms of three Senecio species in relation to accumulation and utilisation of nonstructural carbohydrates. Acta Botanica Neerlandica 26, 401-409.

OVINGTON, J.D. and G. SCURFIELD 1956. Biological Flora of the British Isles: Holcus mollis L. Journal of Ecology 44, 272-280.

OXFORD, G.S. and T. ANDREW 1977. Variations in characters affecting fitness between radiate and non- radiate morphs in natural populations of groundsel (Senecio vulgaris L.). Heredity 38, 367-371.

PACALA, S.W. and D. TILMAN (1994) Limiting similarity in mechanistic and spatial models of plant competition in heterogeneous environments. American Naturalist 143, 222-257.

PACKHAM, J.R. 1978. Biological Flora of the British Isles: Oxalis acetosella L. Journal of Ecology 66, 669-693.

PACKHAM, J.R. 1983. Biological Flora of the British Isles: Lamiastrum galeobdolon (L.). Journal of Ecology 71, 975-997.

PAGE, C.N. 1967. Sporelings of Equisetum arvense in the wild. British Fern Gazette 9, 335-338.

PAGE, C.N. 1976. The taxonomy and phytogeography of bracken - a review. Botanical Journal of the Linnean Society 73, 1-34.

PAGE, C.N. 1982. The ferns of Britain and Ireland. Cambridge: Cambridge University Press.

PAGE, C.N. and M.A. BARKER 1985. Ecology and geography of hybridisation in British and Irish horsetails. Proceedings of the Royal Society of Edinburgh 86B, 265-272.

PALMBLAD, I.G. 1968. Comparative response of closely related species to reduced competition. 1: Senecio sylvaticus and S. viscosus. Canadian Journal of Botany 46, 225-228.

PALMER, J.H. and G.R. SAGAR 1963. Biological Flora of the British Isles: Agropyron repens (L.) Beauv. Journal of Ecology 51, 783-794.

PANETTA, F.D. 1979. Germination and seed survival in the woody weed, groundsel bush Baccharis halimifolia. Australian Journal of Agricultural Research 30, 1067 -1077.

PANIGRAHI, G. 1955. Caltha in the British flora. In: Species studies in the British flora, J.E. Lousley (ed.), 107-110. London: Botanical Society of the British Isles.

PARKHURST, D.F. and O.L. LOUCKS 1972. Optimal leaf size in relation to environment. Journal of Ecology 60, 505-537.

PARRY, J. and B. BUTTERWORTH 1981. Grassland Management. London: Northwood Publications.

PARSONS, R.F. 1968. The significance of growth-rate comparisons for plant ecology. American Naturalist 102, 595-597.

PASTOR, J. and W.M.POST 1986 Influence of climate, soil-moisture, and succession on forest carbon and nitrogen cycles. Biogeochemistry 2, 3-27.

PATE, J.S. 1961. Perennial nodules on native legumes in the British Isles. Nature 192, 367-377.

PEARCE, S., A. VIANELLI, and B. CERABOLINI, 2006. From ancient genes to modern communities: the cellular stress response and the evolution of plant strategies. Functional Ecology 19, 763-776.

PEARSALL, W.H. 1950. Mountains and Moorlands. London: Bloomsbury Books.

PEARSON, P.L. 1967. The experimental taxonomy of Chrysanthemum leucanthemum L. Ph.D. Thesis, University of Durham.

PEART, M.H. 1984. The effects of morphology, orientation and position of grass diaspores on seedling survival. Journal of Ecology 72, 437-453.

PEMADASA, M.A. and P.H. LOVELL 1974a. Interference in populations of some dune annuals. Journal of Ecology 62, 855-868.

PEMADASA, M.A. and P.H. LOVELL 1974b. Factors controlling the flowering time of some dune annuals. Journal of Ecology 62, 869-880.

PEMADASA, M.A. and P.H. LOVELL 1975. Factors controlling germination time of some dune annuals. Journal of Ecology 63, 41-59.

PERRING, F.H. 1959. Topographical gradients of chalk grassland. Journal of Ecology 47, 447-481.

PERRING, F.H. 1960. Climatic gradients of chalk grassland. Journal of Ecology 48, 415-442.

PERRING, F.H. (ed.) 1968. Critical supplement to the Atlas of the British flora. London: Nelson published for the Botanical Society of the British Isles.

PERRING, F.H. and B.G. GARDINER (eds) 1976. The biology of bracken. Botanical Journal of the Linnean Society 73, 1-302.

PERRING, F.H. and S.M. WALTERS (eds) 1976. Atlas of the British flora, 2nd edn. Wakefield: EP Publishing, published for the Botanical Society of the British Isles.

PERRING, F.H. and L. FARRELL 1977. British Red Data Books: 1 Vascular Plants. Lincoln: Society for the Promotion of Nature Conservation.

PERRING, F.H. and L. FARRELL 1983. British Red Data Books: 1 Vascular Plants. 2nd ed. Lincoln: Society for the Promotion of Nature Conservation.

PETERKEN, G.F. 1981. Woodland conservation and management. London: Chapman & Hall.

PETERSEN, R.L. 1985. Towards an appreciation of fern edaphic niche requirements . Proceedings of the Royal Society of Edinburgh 86B, 93-104.

PETERSON, S-R. 1969. Biology of the mouse-ear chickweed Cerastium vulgatum. Michigan Botanist 8, 151-157.

PFITZENMEYER, C.D.C. 1962. Biological Flora of the British Isles: Arrhenatherum elatius (L.) Beauv. ex J. and C. Presl. Journal of Ecology 50, 235-245.

PHILBRICK, C.T. 1984. Pollen tube growth within vegetative tissues of Callitriche (Callitrichaceae). American Journal of Botany 71, 882-886.

PHILIPSON, W.R. 1937. A revision of the British species of the genus Agrostis Linn. Botanical Journal of the Linnean Society 51, 73-151.

PHILLIPS, M.E. 1954. Biological Flora of the British Isles: Eriophorum angustifolium Roth. Journal of Ecology 42, 612-622.

PHILP, E.G. 1982. Atlas of the Kent flora. Maidstone: Kent Field Club.

PIANKA, E.R. 1970. On r- and K-selection. American Naturalist 104, 592-597.

PIGOTT, C.D. 1955. Biological Flora of the British Isles: Thymus drucei Ronniger em. Jalas. Journal of Ecology 43, 369-379.

PIGOTT, C.D. 1958. Biological Flora of the British Isles: Polemonium caeruleum L. Journal of Ecology 46, 507-525.

PIGOTT, C.D. 1968. Biological Flora of the British Isles: Cirsium acaulon. Journal of Ecology 56, 597-612.

PIGOTT, C.D. 1969. The status of Tilia cordata and T. platyphyllos on the Derbyshire limestone. Journal of Ecology 57, 491-504.

PIGOTT, C.D. 1971. Analysis of the response of Urtica dioica to phosphate. New Phytologist 70, 953-966.

PIGOTT, C.D. 1974. The response of plants to climate and climatic change. In: The flora of a changing Britain, F. Perring (ed.), 32-44. London: Classey.

PIGOTT, C.D. 1983. Regeneration of oak-birch woodland following exclusion of sheep. Journal of Ecology 71, 629-646.

PIGOTT, C.D. and J.P. HUNTLEY 1978. Factors controlling the distribution of Tilia cordata at the northern limit of its geographical range. 1: Distribution in north-west England. New Phytologist 81, 429-441.

PIGOTT, C.D. and J.P. HUNTLEY 1980. Factors controlling the distribution of Tilia cordata at the northern limit of its geographical range. 2: History in north-west England. New Phytologist 84, 145-164.

PIGOTT, C.D. and J.P. HUNTLEY 1981. Factors controlling the distribution of Tilia cordata at the northern limit of its geographical range. 3: Nature and cause of seed sterility. New Phytologist 87, 817-839.

PIGOTT, C.D. and K. TAYLOR 1964. The distribution of some woodland herbs in relation to the supply of nitrogen and phosphorus in the soil. Journal of Ecology 52, 175-185.

PITCAIRN, C.E.R. and J. GRACE 1984. The effect of wind on provenances of Molinia caerulea. Annals of Botany 54, 135-143.

POGORELOVA, T.D. and T.A. RABOTNOV 1978. Experience in determining the age of Dryopteris filix-mas specimens. Rastitelnye Resursy (USSR) 14, 601-604.

POLLARD, A.J. 1980. Diversity of metal tolerances in Plantago lanceolata L. from the South-Eastern United States. New Phytologist 86, 109-117.

POLLARD, A.J. and D. BRIGGS 1982. Genecological studies of Urtica dioica. 1: The nature of variation in Urtica dioica. New Phytologist 92, 453-470.

POLLARD, F. and G.W. CUSSANS 1981. The influence of tillage on the weed flora in a succession of winter cereal crops on a sandy loam soil. Weed Research 21, 185-190.

POLLOCK, C.J. 1979. Pathway of fructan synthesis in leaf bases of Dactylis glomerata. Phytochemistry 18, 777-779.

POLLOCK, C.J. and T. JONES 1979. Seasonal patterns of fructan metabolism in forage grasses. New Phytologist 83, 9-15.

PONS, T.L. 1977. An ecophysiological study in the field layer of ash coppice. 2: Experiments with Geum urbanum and Cirsium palustre in different light intensities. Acta Botanica Neerlandica 26, 29-42.

PONS, T.L. 1983. Significance of inhibition of seed germination under the leaf canopy in ash coppice. Plant Cell and Environment 6, 385-392.

PONS, T.L. 1984. Possible significance of changes in the light requirement of Cirsium palustre seeds after dispersal in ash coppice. Plant Cell and Environment 7, 263-268.

POPAY, A.I. and E.H. ROBERTS 1970b. Ecology of Capsella bursa-pastoris and Senecio vulgaris in relation to germination behaviour. Journal of Ecology 58, 123-139.

PORT, G.R. and J.R. THOMPSON 1980. Outbreaks of insect herbivores on plants along motorways in the UK. Journal of Applied Ecology 17, 649-656.

PRAT, S. 1934. Die Erblichkeit der Resistenz gegen Kupfer. Berichte der Deutschen Botanischen Gesellschaft 52, 65-67.

PRESTON, C.D. and M.O. HILL 1997. The geographical relationships of British and Irish vascular plants. Botanical Journal of the Linnean Society 124, 1-120.

PRESTON, C.D., D.A. PEARMAN and T.D. DINES 2002a. New Atlas of the British and Irish Flora. Oxford: Oxford University Press.

PRESTON, C.D., M.G. TELFER, H.R. ARNOLD, P.D. CAREY, J.M. COOPER, T.D. DINES, M.O. HILL, D.A. PEARMAN, D.B. ROY and S.M. SMART 2002b. The Changing Flora of The UK. London: DEFRA.

PRIME, C.T. 1960. Lords and ladies. London: Collins.

PRITCHARD, T. 1960. Race formation in weedy species with special reference to Euphorbia cyparissias L. and Hypericum. perforatum L. In: The biology of weeds, J.L. Harper (ed.), 61-68. Oxford: Blackwell Scientific.

PROCTOR, M.C.F. 1956. Biological Flora of the British Isles: Helianthemum. chamaecistus Mill. Journal of Ecology 44, 683-688. .

PROCTOR, M.C.F. 1958. Ecological and historical factors in the distributions of the British Helianthemum species. Journal of Ecology 46, 349-371.

PROCTOR, M.C.F. and P. YEO 1973. The pollination of flowers. London: Collins.

PROFFITT, G.W.H. 1985. The biology and ecology of purple-moor grass Molinia caerulea (L.) Moench. with special reference to the root system. Ph.D. Thesis, University of Aberystwyth.

PUTWAIN, P.D. and J.L. HARPER 1970. Studies in the dynamics of plant populations. 3: The influence of associated species on populations of Rumex acetosa and R. acetosella in grassland. Journal of Ecology 58, 251-264.

PUTWAIN, P.D. and J.L. HARPER 1972. Studies in the dynamics of plant populations. 5: Mechanisms governing the sex ratios in Rumex acetosa and R. acetosella. Journal of Ecology 60, 113-129.

QUINLIVAN, B.J. 1961. The effect of constant and fluctuating temperature in the permeability of the hard seeds of some legume species. Australian Journal of Agricultural Research 12, 1009-1022.

QUINLIVAN, B.J. 1968. Seed coat impermeability in the common annual legume pasture species of Western Australia. Australian Journal of Experimental Agriculture and Animal Husbandry 8, 695-701.

QUINLIVAN, B.J. and A.J. MILLINGTON 1962. The effect of a Mediterranean summer environment on the permeability of hard seeds of subterranean clover. Australian Journal of Agricultural Research 13, 377-387.

RABOTNOV, T.A. 1978. On coenopopulations of plants reproducing by seeds. In: Structure and functioning of plant populations, A.H.J. Freysen and J.T. Voldendorp (eds), 1-26. Amsterdam: NorthHolland.

RABOTNOV, T.A. 1983. Types of plant strategies. Ekologiya 3, 3-12.

RABOTNOV, T.A. 1985. Dynamics of plant coenotic populations. In: Structure of vegetation, J. White (ed.), 121-142. Dordrecht: Junk.

RACKHAM, O. 1980. Ancient woodland: Its history, vegetation and uses in England. London: Edward Arnold.

RAMENSKII, L.G. 1938. Introduction to the geobotanical study of complex vegetation. Moscow: Selkozgiz.

RANDALL, R.E. 1974. Rorippa islandica (Oeder) Borbas sensu stricta in the British Isles. Watsonia 10, 80-82.

RATCLIFFE, D. 1961. Adaptation to habitat in a group of annual plants. Journal of Ecology 49, 187-203.

RATCLIFFE, D.A. (ed.) 1977. A nature conservation review. Cambridge: Cambridge University Press.

RATCLIFFE, D.A. 1984. Post-mediaeval and recent changes in British vegetation: the culmination of human influence. New Phytologist 98, 73-100.

RAUNKIAER, C. 1934. The life forms of plants and statistical plant geography; being the collected papers of C. Raunkiaer, translated into English by H.G. Carter, A.G. Tansley and Miss Fansboll. Oxford: Clarendon Press.

RAVEN, P.H. 1963. Circaea in the British Isles. Watsonia 5, 262-272.

RAWES, M. and R. HOBBS 1979. Management of semi-natural blanket bog in the North Pennines. Journal of Ecology 67, 789-807.

READ, D.J. and R. BAJWA 1985. Some nutritional aspects of the biology of the ericaceous mycorrhizas. Proceedings of the Royal Society of Edinburgh 85B, 317-332.

READ, D.J., R.FRANCIS and R.D. FINLAY 1985. Mycorrhizal mycelia and nutrient cycling in plant communities. In: Ecological

interactions in soil, A.H. Fitter (ed.), 193-217. Special Publication No. 4 of the British Ecological Society. Oxford: Blackwell.
READER, P.M. and T.R.E. SOUTHWOOD 1981. The relationship between palatability to invertebrates and the successional status of a plant. Oecologia 51, 271-275.
REBELE, F., P. WERNER and R. BORNKAMM 1982. Wirkung von Lichtquantitat und Lichtqualitat auf die Konkurrenz zwischen der Schattenpflanze Lamium galeobdolon (L.) Crantz und der Halbschattenpflanze Stellaria holostea L. Flora Jena 172, 251-266.
REES, W.J. and G.H. SIDRAK 1955. Plant growth on fly ash. Nature 176, 352.
REICH, P.B., M.B. WALTERS and D.S. ELLESWORTH 1992. Leaf life-span in relation to leaf, plant and stand characteristics among diverse ecosystems. Ecological Monographs 62, 365-392.
REJMANEK, M. 1976. Centres of species diversity and centres of species diversification. Evolutionary Biology 9, 393-408.
REJMANKOVA, E. 1975. Comparison of Lemna gibba and Lemna.minor from the production ecological viewpoint. Aquatic Botany 1, 423-428.
REUSCH, T.B.H., EHLERS, A., HAMMERLI, A. and B. WORM 2005. Ecosystem recovery after climatic extremes enhanced by genotypic diversity. Proceedings of the National Academy of Sciences of the United States of America 102, 2826-2831.
RICHARDS, A.J. 1972. The Taraxacum flora of the British Isles. Watsonia, suppl. to vol. 9.
RICHARDS, A.J. 1986. Plant breeding systems. London: George Allen & Unwin.
RICHARDS, A.J. and C.C. HAWARTH 1984. Further new species of Taraxacum from the British Isles. Watsonia 15, 85-94.
RICHARDS, P.W. and A.R. CLAPHAM 1941. Biological Flora of the British Isles: Juncus effusus L. Journal of Ecology 29, 375-380.
RICHENS, R.H. 1983. Elm. Cambridge: Cambridge University Press.
RIDLEY, H.N. 1930. The dispersal of plants throughout the world. Ashford: Reeve.
RITCHIE, J.C. 1955. Biological Flora of the British Isles: Vaccinium. vitis-idaea L. Journal of Ecology 43, 701-708.
RITCHIE, J.C. 1956. Biological Flora of the British Isles: Vaccinium, myrtillus L. Journal of Ecology 44, 291-299.
ROBERTS, H.A. 1970. Viable weed seeds in cultivated soils. Report of the National Vegetable Research Station 25-38.
ROBERTS, H.A. 1979. Periodicity of seedling emergence and seed survival in some Umbelliferae. Journal of Applied Ecology 16, 195-201.
ROBERTS, H.A. (ed.) 1982. Weed control handbook, 7th edn. Vol. 1: Principles. Oxford: Blackwell Scientific.
ROBERTS, H.A. 1986a. Seed persistence in soil and seasonal emergence in plant species from different habitats. Journal of Applied Ecology 23, 639-656.
ROBERTS, H.A. 1986b. Persistence of seeds of some grass species in cultivated soil. Grass and Forage Science 41, 273-276.
ROBERTS, H.A. and J.E. BODDRELL 1983. Seed survival and periodicity of seedling emergence in ten species of annual weeds. Annals of Applied Biology 102, 523-532.
ROBERTS, H.A. and P.M. FEAST 1972. Fate of seeds of some annual weeds in different depths of cultivated and undisturbed soil. Weed Research 12, 316-324.
ROBERTS, H.A. and P.M. FEAST 1974. Observations on the time of flowering in mayweeds. Journal of Applied Ecology 11, 223-229.
ROBERTS, H.A. and P.M. LOCKETT 1977. Temperature requirements for germination of dry-stored, cold-stored and buried seeds of Solanum dulcamara L. New Phytologist 79, 505-510.
ROBERTS, H.A. and J.E. NEILSON 1981. Changes in the soil seed bank of four long term crop/herbicide experiments. Journal of Applied Ecology 18, 661-668.
ROBINSON, D. and I.H. RORISON 1983. A comparison of the responses of Lolium perenne L., Holcus lanatus L. and Deschampsia flexuosa (L.) Trin. to a localised supply of nitrogen. New Phytologist 94, 263-273.
ROBSON, N.K.B. 1981. Studies in the genus Hypericum L. (Guttiferae). 2: Characters of the genus. Bulletin of the British Museum of Natural History (Botany) 8, 55-226.
RODEWALD-RUDESCU, L. 1974. Das Schilfrohr (die Binnengewasser 27). Stuttgart: Schweizerbartsche.
RODWELL, J.S. (Ed.) 1992. British plant communities. Vol. 3: Grasslands and montane communities. Cambridge: Cambridge University Press.
ROELOFS, J.G.M. 1983. Impact of acidification and eutrophication on macrophyte communities in soft waters in the Netherlands. 1: Field observations. Aquatic Botany 17, 139-155.
ROETMAN, E. and A.A. STERK 1986. Growth of microspecies of different sections of Taraxacum in climatic chambers. Acta Botanica Neerlandica 35, 5-22.
ROGERS, R.W. and H.T. CLIFFORD 1993. The taxonomic and evolutionary significance of leaf longevity. New Phytologist 123, 811-821.
ROGERS, S. 1969. Studies on British poppies 1. Some observations on the reproductive biology of the British species of Papaver. Watsonia 7, 55-63.
ROLLIN, P. and G. MAIGNAN 1967. Phytochrome and the photoinhibition of germination. Nature 214, 741-742.
ROLLO, C.D., J.D. MacFARLANE and B.S. SMITH 1985. Electrophoretic and allometric variation in burdock (Arctium spp.); hybridisation and its ecological implications. Canadian Journal of Botany 63, 1255-1302.
RORISON, I.H. 1960. The calcicole-calcifuge problem. 2: The effects of mineral nutrition on seedling growth in solution culture. Journal of Ecology 48, 679-688.
RORISON, I.H. 1973. Seed ecology - present and future. In: Seed ecology, W. Heydecker (ed.), 497-519. London: Butterworths.
RORISON, I.H. 1985. Nitrogen source and tolerance of Deschampsia flexuosa, Holcus lanatus and Bromus erectus to aluminium during seedling growth. Journal of Ecology 73, 83-90.
RORISON, I.H., P.L. GUPTA and R. HUNT 1986. Local climate topography and plant growth in Lathkill Dale NNR. 2: Growth and nutrient uptake within a single season. Plant, Cell and Environment 9, 57-64.
ROSE, P.Q. 1980. Ivies. Poole: Blandford Press.
ROSS, M.D. 1973. Inheritance of self-incompatibility in Plantago lanceolata. Heredity 30, 169-176.
ROTHERA, S.L. and A.J. DAVY 1986. Polyploidy in habitat differentiation in Deschampsia cespitosa. New Phytologist 102, 449-468.
RUSKIN, J. 1872. Arartra Pentelici. Six lectures on the elements of sculpture, given before the university of Oxford in Michaelmas term, 1870. London: Smith, Elder and Co.
RUTTER, A.J. 1955. The composition of wet-heath vegetation in relation to the water-table. Journal of Ecology 43, 507-543.
RYMER, L. 1976. The history and ethnobotany of bracken. Botanical Journal of the Linnean Society 73, 157-176.
RYSER, P. and H. LAMBERS 1995. Root and leaf attributes accounting for the performance of fast- and slow-growing grasses at different nutrient supply. Plant and Soil 170, 251-265.
SAGAR, G.R. and J.L. HARPER 1964a. Biological Flora of the British Isles: Plantago lanceolata L. Journal of Ecology 52, 211-221.
SAGAR, G.R. and J.L. HARPER 1964b. Biological Flora of the British Isles: Plantago major L. Journal of Ecology 52, 189-205.
SALA O.E., LAUENROTH, W.K. and PARTON W.J. 1992. Long term soil water dynamics in the shortgrass steppe. Ecology 73, 1175 1181.
SALISBURY, E.J. 1916. The emergence of the aerial organs in woodland plants. Journal of Ecology 4, 121-218.
SALISBURY, E.J. 1942. The reproductive capacity of plants. London: George Bell.
SALISBURY, E.J. 1952. Downs and dunes: their plant life and its environment. London: George Bell.
SALISBURY, E.J. 1953. A changing flora as shown in the study of weeds of arable land and waste places. In: The changing flora of Britain, J.E. Lousley (ed.), 30-139. Arbroath: Buncle.
SALISBURY, E.J. 1964. Weeds and aliens, 2nd edn. London: Collins.
SALISBURY, E.J. 1965. The reproduction of Cardamine pratensis and C. palustris particularly in relation to their specialised foliar vivipary and its deflexion of the constraints of natural selection: Proceedings of the Royal Society of Biology 163, 321-342.
SALISBURY, E.J. 1967. The reproduction and germination of Limosella aquatica. Annals of Botany 31, 147-162.

SALTER, P.J. and J.E. GOODE 1967. Crop responses to water at different stages of growth. Research Review 2. East Malling, Kent: Commonwealth Bureau of Horticulture and Plantation Crops.

SARGENT, C., O. MOUNTFORD and D. GREENE 1986. The distribution of Poa angustifolia L. in Britain. Watsonia 16, 31-36.

SARUKHAN, J. 1974. Studies on plant demography; Ranunculus repens, R. bulbosus and R. acris 2: Reproductive strategies and seed population dynamics. Journal of Ecology 62, 151-177.

SARUKHAN, J. and M. GADGIL 1974. Studies on plant demography; Ranunculus repens, R. bulbosus and R. acris. 3: A mathematical model encorporating multiple modes of reproduction. Journal of Ecology 62, 921-936.

SARUKHAN, J, and J.L. HARPER 1973. Studies on plant demography; Ranunculus repens, R. bulbosus and R. acris. 1: Population flux and survivorship. Journal of Ecology 61, 675-716.

SCHENKEVELD, A.J. and H.J. VERKAAR 1984. The ecology of short-lived forbs in chalk grasslands; distribution of germinative seeds and its significance for seedling emergence. Journal of Biogeography 11, 251-260.

SCHMID, B. 1985a. Clonal growth in grassland perennials 2: Growth form and fine-scale colonising ability. Journal of Ecology 73, 809-818.

SCHMID, B. 1985b. Clonal growth in grassland perennials 3: Genetic variation and plasticity between and within populations of Bellis perennis and Prunella vulgaris. Journal of Ecology 73, 819-830.

SCHMID, B. and J.L. HARPER 1985. Clonal growth in grassland perennials. 1: Density and pattern dependent competition between plants of different growth forms. Journal of Ecology 73, 793-808.

SCHNELLER, J.J. 1975. Studies on Swiss ferns in particular on the group of Dryopteris filix-mas 3: Ecological studies. Berichte der Schweizerischen Botanischen Gesellsehaft 85, 110-159.

SCHOTSMAN, H.D. 1958. Notes on Callitriche hermaphroditica Jusl. Acta Botanica Neerlandica 7, 519-523.

SCHUBER, M. and M. KLUGE 1981. In-situ studies on crassulacean acid metabolism in Sedum acre and S. mite. Oecologia 50, 82-87.

SCHULZE, E.D. and H.A. MOONEY (eds) 1993. Biodiversity and ecosystem function. Berlin, Springer-Verlag.

SCHULTZ, M.R. and R.M. KLEIN 1963. Effects of visible and ultra-violet. radiation on the germination of Phacelia tanacetifolia. American Journal of Botany 50, 430-434.

SCOTT, A.C., E.G. CHALONER and S. PATTERSON 1985. Evidence on pteridophyte-arthropod interactions in the fossil record. Proceedings of the Royal Society of Edinburgh 86B, 133-140.

SCULTHORPE, C.D. 1967. The biology of aquatic vascular plants. London: Edward Arnold.

SCURFIELD, G. 1954. Biological Flora of the British Isles: Deschampsia flexuosa (L.) Trin. Journal of Ecology 42, 225-233.

SCURFIELD, G. 1959. The ashwoods of the Derbyshire carboniferous limestone in Monks Dale. Journal of Ecology 47, 357-369.

SEGAL, S. 1969. Ecological notes on wall vegetation. The Hague: Junk.

SEILACHER, A. 1970. Arbeitskonzept zur Konstruktionsmorphologie. Lethaia 3, 393-396.

SHAMSI, S.R.A. 1970. Comparative biology of Epilobium hirsutum and Lythrum salicaria. Ph.D. Thesis, University of London.

SHAMSI, S.R.A. and F.H. WHITEHEAD 1974. Comparative ecophysiology of Epilobium hirsutum and Lythrum salicaria. 1: General biology, distribution and germination. Journal of Ecology 62, 279-290.

SHAMSI, S.R.A. and F.H. WHITEHEAD 1977. Comparative ecophysiology of Epilobium hirsutum and Lythrum salicaria. 4: Effects of temperature and interspecific competition and concluding discussion. Journal of Ecology 65, 71-84.

SHANTA, M., GRIME, J.P. and K. THOMPSON 2007. Contrasted effects of vascular plants and bryophytes on diversity: the humped-back model revisited. Journal of Vegetation Science (in prep.).

SHAVER, G.R., F.S. CHAPIN and B.L. GARTNER 1986. Factors limiting seasonal growth and peak biomass accumulation in Eriophorum vaginatum in Alaskan tussock tundra. Journal of Ecology 74, 257-278.

SHAW, M.W. 1974. The reproductive characteristics of oak. In: The British oak; its history and natural history, M.G. Morris and F.H. Perring (eds), 162-181. Faringdon, Berkshire: E.W. Classey.

SHAW, S.C. 1984. Ecophysiological studies on heavy metal tolerance in plants colonising Tideslow Rake, Derbyshire. Ph.D. Thesis, University of Sheffield.

SHELDON, J.C. 1974. The behaviour of seeds in soil 3: The influence of seed morphology and the behaviour of seedlings on the establishment of plants from surface-lying seeds. Journal of Ecology 62, 47-66.

SHELDON, J.C. and F.M. BURROWS 1973. The dispersal effectiveness of the achene pappus units of selected compositae in steady winds with convection. New Phytologist 72, 665-675.

SHILDRICK, J. 1980. Turfgrass manual. Bingley: Sports Turf Research Institute.

SHILDRICK, J. 1984. Turfgrass manual. Bingley: Sports Turf Research Institute.

SHIRREFFS, D.A. 1985. Biological Flora of the British Isles: Anemone nemorosa L. Journal of Ecology 73, 1005-1020.

SHUGART, H.H. 1984. A theory of forest dynamics. New York: Springer Verlag.

SHUGART, H.H. 1997. Plant and ecosystem functional types. In: plant functional types. T. M. Smith, H. H. Shugart and F. I. Woodward (Eds.) 20-43 Cambridge: Cambridge University Press.

SIDHU, S.S. and P.B. CAVERS 1977. Maturity-dormancy relationships in attached and detached seeds of Medicago lupulina L. (Black medick). Botanical Gazette 138, 174-182.

SIFTON, H.B. 1959. The germination of light-sensitive seeds of Typha latifolia. Canadian Journal of Botany 37, 719-739.

SILVERTOWN, J.W. 1980. Leaf-canopy induced seed dormancy in a grassland flora. New Phytologist 85, 109-118.

SILVERTOWN, J.W. 1983. The distribution of plants in limestone pavement. Tests of species interaction and niche separation against null hypotheses. Journal of Ecology 71, 819-828.

SILVESTRE, S. 1972. Estudio taxonomico de los generos Conopodium. Koch y Bunium L. en la Peninsula Iberica 1: Parte experimental. Lagascalia 2, 143-173.

SILVESTRE, S. 1973. Estudio taxonomico de los generos Conopodium Koch y Bunium L. en la Peninsula Iberica 2: Parte sistematica. Lagascalia 3, 3-48.

SIMMONDS, N.W. 1945. Biological Flora of the British Isles: Polygonum persicaria. Journal of Ecology 33, 121-131.

SIMPSON, D.A. 1984. A short history of the introduction and spread of Elodea Michx in the British Isles. Watsonia 15, 1-9.

SIMPSON, D.A. 1986. Taxonomy of Elodea Michx. in the British Isles. Watsonia 16, 1-14.

SKALINSKA, M. 1947. Polyploidy in Valeriana officinalis Linn. in relation to its ecology and distribution. Botanical Journal of the Linnean Society 53, 159-186.

SKEFFINGTON, R.A. and A.D. BRADSHAW 1980. Nitrogen fixation by plants grown on reclaimed china clay waste. Journal of Applied Ecology 17, 469-477.

SLATYER, R.O. 1970. Comparative photosynthesis, growth and transpiration of two species of Atriplex. Planta 93, 175-189.

SLEDGE, W.A. 1959. Yorkshire Naturalists' Union Excursion in 1959. Askern V.C. 63, July 4th, flowering plants. The Naturalist, 135-136.

SMALL, E. 1972. Ecological significance of four critical elements in plants of raised sphagnum peat bogs. Ecology 53, 498-503.

SMIRNOFF, N. and R.M.M. CRAWFORD 1983. Variation in the structure and response to flooding of root aerenchyma in some wetland plants. Annals of Botany 51, 237-249.

SMITH, D. 1967. Carbohydrates in grasses. 2: Sugar and fructosan composition of the stem bases of bromegrass and timothy at several growth stages and in different plant parts at anthesis. Crop Science 7, 62-67.

SMITH, P.M. 1968. The Bromus mollis aggregate in Britain. Watsonia 6, 327-344.

SMITH, P.M. 1981. Ecotypes and subspecies in annual brome-grasses. Botanische Jahrbucher 102, 497-509.

SNAYDON, R.W. and A.D. BRADSHAW 1961. Differential response to calcium within the species Festuca ovina. New Phytologist 60, 219-234.

SNAYDON, R.W. and M.S. DAVIES 1972. Rapid population

differentiation in a mosaic environment 2: Morphological variation in Anthoxanthum odoratum. Evolution 26, 390-405.

SOBEY, D.G. 1981. Biological Flora of the British Isles: Stellaria media (L.) Vill. Journal of Ecology 69, 311-335.

SOLBRIG, O.T. and B.B. SIMPSON 1974. Components of regulation of a population of dandelions in Michigan. Journal of Ecology 62, 473-486.

SOLBRIG, O.T. and B.B. SIMPSON 1977. A garden experiment on competition between biotypes of the common dandelion (Taraxacum officinale). Journal of Ecology 65, 427-430.

SOUTHWOOD, T.R.E. 1961. The number of species of insect associated with various trees. Journal of Animal Ecology 30, 1-8.

SOUTHWOOD, T.R.E 1976. Bionomic strategies and population parameters. In: Theoretical ecology, principles and applications, R.M. May (ed.), 26-48. Oxford: Blackwell.

SOUTHWOOD, T.R.E. 1977. Habitat, the templet for ecological strategies. Journal of Animal Ecology 46, 337-365.

SOWTER, F.A. 1949. Biological Flora of the British Isles: Arum maculatum L. Journal of Ecology 37, 207-219.

SPEDDING, C.R.W. and E.W. DIEKMAHNS (eds) 1972. Grasses and legumes in British agriculture. Bulletin of the Commonwealth Bureau of Pastures and Field Crops no. 49. Farnham: Royal Commonwealth Agricultural Bureau.

SPENCE, D.H.N. 1964. The macrophytic vegetation of freshwater lochs, swamps and associated fens. In: The vegetation of Scotland, J.H. Burnett (ed.), 306-425. Edinburgh: Oliver & Boyd.

SPENCE, D.H.N. and J. CHRYSTAL 1970. Photosynthesis and zonation of freshwater macrophytes. 1: Depth distribution and shade tolerance. New Phytologist 69, 205-215.

SPRAGUE, T.A. 1944. Field studies on Valeriana officinalis Linn. in the Cotswold hills. Proceedings of the Linnean Society 155, 93-104.

SPRAGUE, T.A. 1949. Forms of Valeriana officinalis. In: British flowering plants and modern systematic methods, A.J. Wilmott (ed.), 67-71. London: Botanical Society of the British Isles.

SPRENT, J.I. 1984. Nitrogen fixation. In: Advanced plant physiology, M.B. Wilkins (ed.), 249-276. London: Pitman.

STACE, C.A. 1961. Some studies in Calystegia; compatibility and hybridisation in C. sepium and C. sylvatica. Watsonia 5, 88-105.

STACE, C.A. 1965. Some studies in Calystegia (Convolvulaceae). 2: Observations on the floral biology of the British island taxa. Proceedings of the Botanical Society of the British Isles 6, 21-31.

STACE C.A. 1975. Introductory. In: hybridisation and the flora of the British Isles, C.A. Stace (ed.), 1-90. London: Academic Press.

STACE, C.A. 1977. The origin of radiate Senecio vulgaris L. Heredity 39, 383-388.

STACE, C.A. 1997. New Flora of the British Isles, Second Edition. Cambridge: Cambridge University Press.

STANIFORTH, R.J. and P.B. CAVERS 1979. Field and laboratory germination responses of achenes of Polygonum lapathifolium, P. pensylvanicum and P. persicaria. Canadian Journal of Botany 57, 877-885.

STAUDT, G. 1962. Taxonomic studies in the genus Fragaria. Typification of Fragaria spp. known at the time of Linnaeus. Canadian Journal of Botany 40, 869-886.

STEARNS, S.C. 1976. Life-history tactics; a review of the ideas. Quarterly Review of Biology 51, 3-47.

STEBBINS, G.L. 1951. Natural selection and the differentiation of angiosperm families. Evolution 5, 299-324.

STEBBINS, G.L. 1956. Cytogenetics and evolution of the grass family. American Journal of Botany 43, 890-905.

STEBBINS, G.L. 1971. Chromosomal evolution in higher plants. London: Edward Arnold.

STEBBINS, G.L. 1974. Flowering plants; evolution above the species level. London: Edward Arnold.

STEBBINS, G.L. 1980. Rarity of plant species: a synthetic viewpoint. Rhodora 82, 77-86.

STEELE, R.C. and G.F. PETERKEN 1982. Management objectives for broadleaved woodland conservation. In: Broadleaves in Britain; future management and research, D.C. Malcolm, J. Evans and P.N. Edwards (eds), 91-103. Farnham, Surrey: Institute of Chartered Foresters.

STEINBAUER, G.P. and B. GRIGSBY 1958. Dormancy and germination characteristics of the seeds of sheep sorrel, Rumex acetosella. Proceedings of the Association of Official Seed Analysts 48, 118.

STERK, A.A., M.C. GROENHART and J.F.A. MOOREN 1983. Ecology of some microspecies of Taraxacum in the Netherlands. Acta Botanica Neerlandica 32, 385-415.

STERN, R.C. 1982. The use of sycamore in British forestry. In: Broadleaves in Britain; future management and research, D.C. Malcolm, J. Evans and P.N. Edwards (eds), 85-87. Farnham, Surrey: Institute of Chartered Foresters.

STEWART, F. and J. GRACE 1984. An experimental study of hybridisation between Heracleum mantegazzianum Somm. & Levier and H. sphondylium L. ssp. sphondylium (Umbelliferae). Watsonia 15, 73-83.

STIEPERAERE, H. and C. TIMMERMAN 1983. Viable seeds in the soils of some parcels of reclaimed and unreclaimed heath in the Flemish District of Northern Belgium. Bulletin de la Societe Royale de Botanique de Belgique 116, 62-73.

STOKES, P. 1952. A physiological study of embryo development in Heracleum sphondylium L. 1: The effect of temperature on embryo development. Annals of Botany 16, 441-447.

STRANDHEDE, S.O. 1965. Chromosome studies in Eleocharis, subser. palustre. 3: Observations on western European taxa. Opera Botanica 9, 2.

STUBBS, W.J. and J.B. WILSON 2004. Evidence for limiting similarity in a sand dune community. Journal of Ecology 92, 557-567.

STYLES, B.T. 1962. The taxonomy of Polygonum aviculare and its allies in Britain. Watsonia 4, 177-214.

SYDES, C. 1981. Investigations into the effects of tree leaf litter on herbaceous vegetation. Ph.D. Thesis, University of Sheffield.

SYDES, C. and J.P. GRIME 1981a. Effects of tree leaf litter on herbaceous vegetation in deciduous woodland 1: Field investigations. Journal of Ecology 69, 237-248.

SYDES, C. and J.P. GRIME 1981b. Effects of tree leaf litter on herbaceous vegetation in deciduous woodland 2: An experimental investigation. Journal of Ecology 69, 249-262.

SYDES, C.L. 1984. A comparative study of leaf demography in limestone grassland. Journal of Ecology 72, 331-345.

SYDES, C.L. and J.P. GRIME 1984. A comparative study of root development using a simulated rock crevice. Journal of Ecology 72, 937-946.

SZUJKO-LACZA, J. and G. FEKETE 1974. Examination of development and growth of Brachypodium sylvaticum and Euphorbia cyparissias in parkwood. Acta Botanica Academiae Scientiarum Hungaricae 20, 147-158.

TAKHTAJAN, A. 1969. Flowering plants, origin and dispersal. Edinburgh: Oliver & Boyd.

TAMM, C.O. 1956. Further observations on the survival and flowering of some perennial herbs. Oikos 7, 273292.

TAMM, C.O. 1972. Survival and flowering of some perennial herbs. 3: The behaviour of Primula veris in permanent plots. Oikos 23, 159-166.

TANSLEY, A.G. 1917. On competition between Galium saxatile L. and G. sylvestre Poll. on.different types of soil. Journal of Ecology 5, 173-179.

TANSLEY, A.G. and R.S. ADAMSON 1925. Studies of the vegetation of the English Chalk. III . The chalk grasslands of the Hampshire-Sussex Border. Journal of Ecology 13, 177-223.

TANSLEY, A.G. 1939. The British Islands and their vegetation. London: Cambridge University Press.

TANSLEY, A.G. 1953. The British Isles and their vegetation. Cambridge: Cambridge University Press.

TASCHEREAU, P.M. 1985. Taxonomy of Atriplex species indigenous to the British Isles. Watsonia 15, 183-209.

TAYLOR, F.J. 1956. Biological Flora of the British Isles: Carex flacca Schreb. Journal of Ecology 44, 281-290.

TAYLOR, K. and B. MARKHAM 1978. Biological Flora of the British Isles: Ranunculus ficaria L. Journal of Ecology 66, 1011-1031.

TAYLORSON, R.B. 1970. Changes in dormancy and viability of weed seeds in soils. Weed Science 18, 265-269.

TAYLORSON, R.B. and H.A. BORTHWICK 1969. Light filtration by foliar canopies: significance for light controlled weed seed

germination. Weed Science 17, 48-51.
TELFER, M.G., C.D. PRESTON and P. ROTHERY 2002. A general method for calculating relative change in range size from biological atlas data. Biological Conservation 107, 99-109.
TELTSCHEROVA, L. and S. HEJNY 1973. The germination of some Potamogeton species from South Bohemian fish ponds. Folia Geobotanica et Phytotaxonomica 8, 231-239.
THOMAS, A.S. 1960. Changes in vegetation since the advent of myxomatosis. Journal of Ecology 48, 287-306.
THOMAS, A.S. 1963. Further changes in vegetation since the advent of myxomatosis. Journal of Ecology 51, 151-186.
THOMMEN, G.H. and D.F. WESTLAKE 1981. Factors affecting the distribution of populations of Apium nodiflorum and Nasturtium officinale in small chalk streams. Aquatic Botany 11, 21-36.
THOMPSON, K. 1977. An ecological investigation of germination responses to diurnal fluctuations in temperature. Ph.D. Thesis, University of Sheffield.
THOMPSON, K. 1986. Small-scale heterogeneity in an acid grassland seed bank. Journal of Ecology 74, 733-738.
THOMPSON, K. 1987a. The resource ratio concept and the meaning of competition. Functional Ecology 1, 297-303.
THOMPSON, K. 1987b. Seeds and seed banks. In: Frontiers of comparative plant ecology, I.H. Rorison, J.P. Grime, R. Hunt, G.A.F. Hendry and D.H. Lewis (eds), 23-34. London: Academic Press.
THOMPSON, K. 1994. Predicting the fate of temperate species in response to human disturbance and global change. In: Biodiversity, temperate ecosystems and global change, T.J.B. Boyle and C.E.B. Boyle (eds), 61-76. Berlin: Springer-Verlag.
THOMPSON, K. and A. JONES 1999. Human population density and prediction of local plant extinction in Britain. Conservation Biology 15, 1-6.
THOMPSON, K. and J.P. GRIME 1979. Seasonal variation in the seed banks of herbaceous species in ten contrasting habitats. Journal of Ecology 67, 893-921.
THOMPSON, K. and J.P. GRIME 1983. A comparative study of germination responses to diurnally-fluctuating temperatures. Journal of Applied Ecology 20, 141-156.
THOMPSON, K. and J.P. GRIME 1988. Competition reconsidered- a reply to Tilman. Functional Ecology 2, 114-116.
THOMPSON, K. and A.J.A. STEWART 1981. The measurement and meaning of reproductive effort in plants. American Naturalist 117, 205-211.
THOMPSON, K. and J.C. WHATLEY 1983. Germination responses of naturally buried weed seeds to diurnal temperature fluctuations. In: Aspects of applied biology. 4: Influence of environmental factors on herbicide performance and crop and weed biology. Wellesbourne: National Vegetable Research Station.
THOMPSON, K., J.P. GRIME and G. MASON 1977. Seed germination in response to diurnal fluctuations of temperature. Nature 67, 147-149.
THOMPSON, K., A.P. ASKEW, J.P. GRIME, N.P. DUNNETT and A.J. WILLIS 2005. Biodiversity, ecosystem function and plant traits in mature and immature plant communities. Functional Ecology 19, 355-358.
THOMPSON, K., J.P. BAKKER and R.M. BEKKER 1997a. The Soil Seed Banks of North West Europe: Methodology, Density and Longevity. Cambridge: Cambridge University Press.
THOMPSON, K., S.R. BAND and J.G. HODGSON 1993. Seed size and shape predict persistence in soil. Functional Ecology 7, 236-241.
THOMPSON, K., J.A. PARKINSON, S.R. BAND and R.E. SPENCER 1997b. A comparative study of leaf nutrient concentrations in a regional herbaceous flora. New Phytologist 136, 679-689.
THOMPSON, K., S.H. HILLIER, J.P. GRIME, C.C. BOSSARD and S.R. BAND, 1996. A functional analysis of a limestone grassland community. Journal of Vegetation Science 7, 371-380.
THOMPSON, P.A. 1968. Germination of Caryophyllaceae at low temperatures in relation to geographical distribution. Nature 217, 1156-1157.
THOMPSON, P.A. 1970. Germination of Caryophyllaceae in relation to their geographical distribution in Europe. Annals of Botany 34, 427-449.
THOMPSON, P.A. 1975. Characterisation of the germination responses of Silene dioica (L.) Clairv. populations from Europe. Annals of Botany 39, 1-19.
THOMPSON, P.A. 1980. Germination strategy of a woodland grass Milium effusum L. Annals of Botany 46, 593-602.
THOMPSON, P.A. 1981. Variation in seed size within populations of Silene dioica (L.) Clairv. in relation to habitat. Annals of Botany 47, 623-634.
THOMPSON, P.A. and S.A. COX 1978. Germination of the bluebell Hyacinthoides non-scripta in relation to its distribution and habitat. Annals of Botany 42, 51-62.
THURSTON, J.M. 1960. Dormancy in weed seeds. In: The biology of weeds, J. Harper (ed.), 69-82. Oxford: Blackwell Scientific.
THURSTON, J.M. 1969. The effects of liming and fertilizers on the botanical composition of permanent grassland, and on the yield of hay. In: Ecological aspects of the mineral nutrition of plants, I.H. Rorison (ed.), 3-10. Oxford: Blackwell.
THURSTON, J.M. and E.D. WILLIAMS 1968. Growth of perennial grass weeds in relation to the cereal crop. Proceedings of the 9th British weed control conference, Vol. 3, 1115-1123. London: British Crop Protection Council.
TILMAN, D. 1976. Ecological competition between algae: experimental confirmation of resource-based competition theory. Science 192, 463-465.
TILMAN, D. 1982. Resource competition and community structure. Princeton: Princeton University Press.
TILMAN, D. 1987. On the meaning of competition and the mechanisms of competitive superiority. Functional Ecology 1, 304-315.
TILMAN, D. 1988. Plant strategies and the dynamics and structure of plant communities. Princeton: Princeton University Press.
TILMAN, D. 1994. Competition and biodiversity in spatially structured habitats. Ecology 78, 81-92.
TILMAN, D. and J.A. DOWNING 1994. Biodiversity and stability in grasslands. Nature 367, 363-365.
TIMMS, E.W. and CLAPHAM, A.R. 1940. Jointed rushes of the Oxford district. New Phytologist 39, 1-16.
TINKER, P.B. and P.H. NYE 2000. Solute movement in the rhizosphere. Oxford, Oxford University Press.
TOOLE, V.K. 1973. Effects of light, temperature and their interactions on the germination of seeds. Seed Science and Technology 1, 339-396.
TRIPATHI, R.S. and J.H. HARPER 1973. The comparative biology of Agropyron repens and A. caninum. Journal of Ecology 61, 353-358.
TUBBS, C.R. 1974. Heathland management in the New Forest, Hampshire. Biological Conservation 6, 303-306.
TUCKER, J.J. and A.H. FITTER 1981. Ecological studies at Askham Bog Nature Reserve. 2: The tree population of Far Wood. Naturalist 106, 3-14.
TUKEY, J.W. 1977. Exploratory data analysis. Reading, Mass.: Addison-Wesley.
TURESSON, G. 1922. The genotypical response of the plant species to the habitat. Hereditas 3, 211.
TURESSON, G. 1925. The plant species in relation to habitat and climate. Contribution to the knowledge of genecological units. Hereditas 6, 147-236.
TURKINGTON, R. and L.W. AARSSEN 1983. Biological Flora of the British Isles: Hypochoeris radicata L. Journal of Ecology 71, 999-1022.
TURKINGTON, R. and J.J. BURDON 1983. The biology of Canadian weeds. 54: Trifolium repens L. Canadian Journal of Plant Science 63, 243-266.
TURKINGTON, R. and P.B. CAVERS 1979. The biology of Canadian weeds. 33: Medicago lupulina L. Canadian Journal of Plant Science 59, 99-110.
TURKINGTON, R. and G.D. FRANKO 1980. The biology of Canadian weeds. 41: Lotus corniculatus L. Canadian Journal of Plant Science 60, 965-979.
TURKINGTON, R., N.C. KENKEL and G.D. FRANKO 1980. The

biology of Canadian weeds. 42: Stellaria media (L.) Vill. Canadian Journal of Plant Science 60, 981-902.

TURKINGTON, R., M.A. CAHN, A. VARDY and J.L. HARPER 1979. The growth, distribution and neighbour relationship of Trifolium repens in a permanent pasture. 3: The establishment and growth of T. repens in natural and perturbed sites. Journal of Ecology 67, 231-243.

TURNER, J.H. 1933. The viability of seeds. Kew Bulletin 6, 257-269.

TUTIN, T.G. 1957. Biological Flora of the British Isles: Allium ursinum, L. Journal of Ecology 45, 1003-1010.

TUTIN, T.G. 1980. Umbellifers of the British Isles. BSBI Handbook No. 2. London: Botanical Society of the British Isles.

TUTIN, T.G., V.H. HEYWOOD, N.A. BURGES, D.H. VALENTINE, S.M. WALTERS and D.A. WEBB (eds) with the assistance of P.W. BALL and A.O. CHATER 1964. Flora Europaea. Vol. 1: Lycopodiaceae to Platanaceae. Cambridge: Cambridge University Press.

TUTIN, T. G. , V. H. HEYWOOD, N. A. BURGES, D. M. MOORE, D.H. VALENTINE' S.M. WALTERS and D.A. WEBB (eds) with the assisiance of P.W. BALL, A.O. CHATER and I.K FERGUSON 1968. Flora Europaea. Vol. 2: Rosaceae to Umbelliferae. Cambridge: Cambridge University Press.

TUTIN, T.G. , V.H. HEYWOOD, N.A. BURGES, D.M. MOORE, D.H. VALENTINE, S.M. WALTERS and D.A. WEBB (eds) with the assistance of P.W. BALL, A.O. CHATER, R A DEFILIPPS, I.K. FERGUSON and I.B.K. RICHARDSON 1978. Flora Europaea. Vol. 3: Diapensiaceae to Myoporaceae. Cambridge: Cambridge University Press.

TUTIN, T.G., V.H. HEYWOOD, N.A. BURGES, D.M. MOORE, D.H. VALENTINE, S.M. WALTERS and D.A. WEBB (eds) with the assistance of A.O. CHATER, R.A. DEFILIPPS and I.B.K. RICHARDSON 1976. Flora Europaea. Vol . 4: Plantaginaceae to Compositae (and Rubiaceae). Cambridge: Cambridge University Press.

TUTIN, T.G., V.H. HEYWOOD, N.A. BURGES, D.M. MOORE, D.H. VALENTINE, S.M. WALTERS and D.A. WEBB (eds) with the assistance of A.O. CHATER and I.B.K. RICHARDSON 1980. Flora Europaea. Vol. 5: Alismataceae to Orchidaceae (Monocotyledones). London: Cambridge University Press.

TYLER, B. M. BORRILL and K. CHORLTON 1978. Studies in Festuca. Observations on germination and seedling cold tolerance in diploid Festuca pratensis apennina and tetraploid Festuca pratensis apennina in relation to their altitudinal distribution. Journal of Applied Ecology 15, 219-226.

UBSDELL, R.A.E. 1976a. Studies on variation and evolution in Centaurium erythraea Rafn. and C. littorale (D. Turner) Gilmour in the British Isles. 1: Taxonomy and biometrical-studies. Watsonia 11, 7-31.

UBSDELL, R.A.E. 1976b. Studies on variation and evolution in Centaurium erythraea Rafn. and C. littorale (D. Turner) Gilmour in the British Isles. 2: Cytology. Watsonia 11, 33-43.

UBSDELL, R.A.E. 1979. Studies on variation and evolution in Centaurium erythraea Rafn. and C. littorale (D. Turner) Gilmour in the British Isles. 3: Breeding systems, floral biology and general discussion. Watsonia 12, 225-232.

URBANSKA-WORYTKIEWICZ, K. 1975. The applicability of polyphenolic data to systematic problems in the Lemnaceae. Aquatic Botany 1, 377-394.

URBANSKA-WORYTKIEWICZ, K. 1980. Reproductive strategies in a hybridogenous population of Cardamine. Oecologia Plantarum. 1. 137-150.

USHER, G. 1974. A dictionary of plants used by man. London: Constable.

VAARTAJA, O. 1952. Forest humus quality and light conditions as factors influencing damping-off. Phytopathology 42, 501-506.

VALENTINE, D.H. 1941. Variation in Viola riviniana Rchb. New Phytologist 40, 189-209.

VALENTINE, D.H. 1949. Vegetative and cytological variation in Viola riviniana. In: British flowering plants and modern systematic methods, A.J. Wilmott (ed.), BSBI Conference Report No. 1, 48. London: Botanical Society of the British Isles.

VALENTINE, D.H. 1950. The experimental taxonomy of two species of Viola. New Phytologist 49, 193-212.

VALENTINE, D.H. 1978. Ecological criteria in plant taxonomy. 1: Essays in plant taxonomy, H.E. Street (ed.), 1-18. London: Academic Press.

VALENTINE, D.H. 1980. Ecotypic and polymorphic variation in Centaurea scabiosa. Watsonia 13, 103-109.

VAN ANDEL, J. and F. VERA 1977. Reproductive allocation in Senecio sylvaticus and Chamaenerion angustifolium in relation to mineral nutrition. Journal of Ecology 65, 747-758.

VAN BAALEN, J. 1982. Population biology of plants in woodland clearings. Ph.D. Thesis, Free University of Amsterdam.

VAN BREEMAN, A.A.M. and B.H. VAN LEEUWEN 1983. The seed-bank of three short-lived monocarpic species; Cirsium vulgare (Compositae), Echium vulgare and Cynoglossum officinale (Boraginaceae). Acta Botanica Neerlandica 32, 245.

VAN DEN BERGH, J.P. 1979. Changes in the composition of mixed populations of grassland species. Commissioned for the study of vegetation of the Royal Botanical Society of the Netherlands.

VAN DER DIJK, S.J. 1981. Nitrogen metabolism in higher plants in different environments. Ph.D. Thesis, University of Groningen.

VAN DER HEIJDEN, M.G.A., J.N. KLIRONOMOS, M. URSIC, P. MOUTOGLIS, R. STREITWOLF-ENGEL, T. BOLLER, A. WIEMKEN and I.R. SANDERS 1998. Mycorrhizal fungal diversity determines plant biodiversity, ecosystem variability and productivity. Nature 396, 69-72.

VAN DER MEIJDEN, E. and R.E. VAN DER WAALS-KOOI 1979. The population ecology of Senecio jacobaea in a sand-dune system. 1: Reproductive strategy and the biennial habit. Journal of Ecology 67, 131-153.

VAN DER PIJL, L. 1972. Principles of dispersal in higher plants. Berlin: Springer.

VAN DER TOORN, J. 1969. Ecological differentiation of Phragmites communis. Acta Botanica Neerlandica 18, 484-485.

VAN DER TOORN, J. 1972. Variability of Phragmites australis (Cav.) Trin. ex. Stendel in relation to the environment. Ph.D. Thesis, University of Groningen.

VAN DER TOORN, J. 1980. On the ecology of Cotula coronopifolia and Ranunculus sceleratus L. Geographical distribution, habitat and field observations. Acta Botanica Neerlandica 29, 385-396.

VAN DER TOORN, J. and H.J. TEN HOVE 1982. On the ecology of Cotula coronopifolia and Ranunculus sceleratus L. 2: Experiments on germination, seed longevity and seedling survival. Oecologia Plantarum 3, 409-418.

VAN DER VALK, A.G. and C.B. DAVIS 1976. The seed banks of prairie glacial marshes. Canadian Journal of Botany 54, 1832-1838.

VAN DER WERF, A., VAN NUENEN, M., VISSER, A.J. and H. LAMBERS 1993. Contribution of physiological and morphological plant traits to a species' competitive ability at high and low nitrogen supply. Oecologia 94, 434-440.

VAN GILS, H. and P. HUITS 1978. Standplaats, stengelhoogte en levensduur van Inula conyza DC. in Nederlande. Gorteria 9, 93-103.

VAN GROENENDAEL, J. 1985. Selection for different life histories in Plantago lanceolata. Ph.D. Thesis, University of Nijmegen.

VAN LEEUWEN, B.H. 1981. The role of pollination in the ecology of the monocarpic species Cirsium palustre and Cirsium vulgare. Oecologia 51, 28-32.

VAN'T HOF, J. and A.H. SPARROW 1963. A relationship between DNA content, nuclear volume and minimum mitotic cycle time. Proceedings of the National Academy of Sciences USA 49, 897-902.

VARAPOLOUS, A. 1979. Breeding systems in Myosotis scorpioides. 1: Self-incompatibility. Heredity 42, 149-157.

VENDRAMINI, F., S. DÍAZ, D.E. GURVICH, P.J. WILSON, K. THOMPSON and J.G. HODGSON 2002. Leaf traits as indicators of resource-use in floras with succulent species. New Phytologist 154, 147-157.

VERKAAR, H.J. and A.J. SCHENKEVELD 1984a. On the ecology of short-lived forbs in chalk grassland; life history characteristics. New Phytologist 98, 659-672.

VERKAAR, H.J. and A.J. SCHENKEVELD 1984b. On the ecology of short-lived forbs in chalk grassland; semelparity and seed output

of some species in relation to various levels of nutrient supply. New Phytologist 98, 673-682.
VERKAAR, H.J., A.J. SCHENKEVELD and J.M. BRAND 1983. On the ecology of short-lived forbs in chalk grasslands; microsite tolerances in relation to vegetation structure. Vegetatio 52, 91-102.
VILE, D., E. GARNIER, B. SHIPLEY, G. LAURENT, M.-L. NAVAS, C. ROUMET, S. LAVOREL, S. DIAZ, J.G. HODGSON, F. LLORET, G.F. MIDGLEY, H. POORTER, M.C. RUTHERFORD, P.J. WILSON and I.J. WRIGHT 2005. Specific leaf area and dry matter content estimate thickness in laminar leaves. Annals of Botany 96, 1129-1136.
VINCENT, E.M. and P.B. CAVERS 1978. The effects of wetting and drying on the subsequent germination of Rumex crispus. Canadian Journal of Botany 56, 2207-2217.
VITOUSEK, P.M., H.A. MOONEY, J. LUBCHENKO and J.M. MELILLO 1997. Human domination of Earth's ecosystems. Science 277, 494-499.
VOSE, P.D. 1962. Delayed germination in reed-canary grass Phalaris arundinaceae. Annals of Botany 26, 197-206.
WADE, K.M. 1981a. Experimental studies on the distribution of the sexes of Mercurialis perennis L. 2: Transplanted populations under different canopies in the field. New Phytologist 87, 439-446.
WADE, K.M. 1981b. Experimental studies on the distribution of the sexes of Mercurialis perennis L. 3: Transplanted populations under light screens. New Phytologist 87, 447-455.
WADE, K.M., R.A. ARMSTRONG and S.R.J. WOODELL 1981. Experimental studies on the distribution of the sexes of Mercurialis perennis L. 1: Field observations and canopy removal experiments. New Phytologist 87, 431-438.
WALTERS, S.M. 1949. Biological Flora of the British Isles: Eleocharis palustris (L.) R. Br. em. R. and S. Journal of Ecology 37, 194-202.
WALTERS, S.M. 1968. Betula L. in Britain. Proceedings of the Botanical Society of the British Isles 7, 179-180.
WARDLE, P. 1961. Biological Flora of the British Isles: Fraxinus excelsior L. Journal of Ecology 49, 739-751.
WARDLE, D.A., K.I BONNER. and K.S NICHOLSON. 1997. Biodiversity and plant litter: experimental evidence that does not show support for the view that enhanced species-richness improves ecosystem function. Oikos 79, 247-258.
WAREING, P.F. 1966. Ecological aspects of seed dormancy and germination. In: Reproductive biology and taxonomy of vascular plants, J.G. Hawkes (ed.), 39-51. Oxford: Pergamon Press.
WARWICK, S.I. 1979. The biology of Canadian weeds 37: Poa annua L. Canadian Journal of Plant Science 59, 1053-1066.
WARWICK, S.I. 1980. Differential growth between and within triazine-resistant and triazine- succeptible biotypes of Senecio vulgaris L. Weed Research 20, 299-304.
WARWICK, S.I. and L. BLACK 1982. The biology of Canadian weeds. 52: Achillea millefolium L. Canadian Journal of Plant Science 62, 163-182.
WARWICK, S.I. and D. BRIGGS 1979. The genecology of lawn weeds. 3: Cultivation experiments with Achillea millefolium L., Bellis perennis L., Plantago lanceolata L., Plantago, major L. and Prunella vulgaris L. collected from lawns and contrasting grassland habitats. New Phytologist 83, 509-536.
WARWICK, S.I. and D. BRIGGS 1980a. The genecology of lawn weeds. 4: Adaptive significance of variation in Bellis perennis L. as revealed in a transplant experiment. New Phytologist 85, 275-288.
WARWICK, S.I. and D. BRIGGS 1980b. The genecology of lawn weeds. 5: The adaptive significance of different growth habit in lawn and roadside populations of Plantago major L. New Phytologist 85, 289-300.
WATSON, P.J. 1958. The distribution in Britain of diploid and tetraploid races within the Festuca ovina group. New Phytologist 57, 11-18.
WATSON, P.J. 1969. Evolution in closely adjacent plant populations. 6: An entomophilous species Potentilla erecta in two contrasting habitats. Heredity 24, 407-422.
WATSON, W.C.R. 1958. Handbook of the Rubi of Great Britain and Ireland. Cambridge: Cambridge University Press.
WATT, A.S. 1934. The vegetation of the Chiltern Hills, with special reference to the beechwoods and their seral relationships. Journal of Ecology 22, 445-507.
WATT, A.S. 1940. Contributions to the ecology of bracken. 1: The rhizome. New Phytologist 39, 401-422.
WATT, A.S. 1947. Pattern and process in the plant community. Journal of Ecology 35, 1-22.
WATT, A.S. 1960. Population changes in acidophilous grass-heath in Breckland 1936-1957. Journal of Ecology 48, 605-629.
WATT, A.S. 1962. The effect of excluding rabbits from grassland 'A' (xerobrometum) in Breckland. Journal of Ecology 50, 181-198.
WATT, A.S. 1976. The ecological status of bracken. Botanical Journal of the Linnean Society 73, 217-240.
WATT, A.S. 1981. A comparison of grazed and ungrazed grassland in East Anglian Breckland. Journal of Ecology 69, 499-508.
WATT, T.A. 1978. The biology of Holcus lanatus L. (Yorkshire fog) and its significance in grassland. Herbage Abstracts 48, 195-204.
WATT, T.A. and R.S. HAGGAR 1980. The effect of height of water table on the growth of Holcus lanatus with reference to Lolium perenne. Journal of Applied Ecology 17, 423-430.
WCISLO, H. 1977. Chromosome numbers in the genus Polygonum sensu-lato in Poland. Acta Biologica Cracovienska Series Botanica (Poland) 20, 153-166.
WEAVER, S.E. and P.B. CAVERS 1980. Reproductive effort in two perennial weed species in different habitats. Journal of Applied Ecology 17, 505-513.
WEAVER, S.E. and W.R. RILEY 1982. The biology of Canadian weeds. 53: Convolvulus arvensis L. Canadian Journal of Plant Science 62, 461-472.
WEBB, D.A. 1950. Biological Flora of the British Isles: Saxifraga L. genus. Journal of Ecology 38, 185-194.
WEBB, D.A. 1985. What are the criteria for presuming native status? Watsonia 15, 231-236.
WEBSTER, J.R. 1962. The composition of wet-heath vegetation in relation to aeration of the ground- water and soil. 1: Field studies of ground-water and soil aeration in several communities. Journal of Ecology 50, 619-637.
WEBSTER, M.M. 1978. Flora of Moray, Nairn and East Inverness: botanical vice-counties 95, Elgin and 96, Easterness. Aberdeen: Aberdeen University Press.
WEBSTER, S.D. 1986. Two natural hybrids in Ranunculus L. subgenus Batrachium (DC.) Gray. Watsonia 16, 25-30.
WEIHER, E. and P. KEDDY 1999. Ecological assembly rules: perspectives, advances, retreats. Cambridge: Cambridge University Press.
WEIMARCK, A. 1968. Self-incompatibility in the Gramineae. Hereditas 60, 157-166.
WEIN, R.W. 1973. Biological Flora of the British Isles: Eriophorum vaginatum L. Journal of Ecology 61, 601-615.
WEIN, R.W. and D.A. MacLEAN 1973. Cotton grass (Eriophorum vaginatum) germination requirements and colonising potential in the Arctic. Canadian Journal of Botany 51, 2509-2513.
WEISS, F.E. 1909. The dispersal of the seeds of gorse and broom by ants. New Phytologist 8, 81-89.
WELCH, D. 1966. Biological Flora of the British Isles: Juncus squarrosus L. Journal of Ecology 54, 535-548.
WELLS, T.C.E. 1971. A comparison of the effects of sheep grazing and mechanical cutting on the structure and botanical composition of chalk grassland. In: The scientific management of animal and plant communities for conservation, E. Duffey and A.S. Watt (eds), 497-515. Oxford: Blackwell Scientific.
WELLS, T.C.E., J. SHEAIL, D.F. BALL and L.K. WARD 1976. Ecological studies on the Porton Ranges: Relationships between vegetation, soils and land-use history. Journal of Ecology 64, 589-626.
WENT, F.W. 1949. Ecology of desert plants. 2: The effect of rain and temperature on germination and growth. Ecology 30, 1-13.
WERNER, P.A. 1977. Colonisation success of a biennial plant species; experimental field studies of species cohabitation and replacement. Ecology 58, 840-849.
WERNER, P.A. and R. RIOUXE 1977. The biology of Canadian weeds. 24: Agropyron repens (L.) Beauv. Canadian Journal of Plant Science 57, 905-919.

WERNER, P.A., F. REBELE and R. BORNKAMM 1982. Effects of light intensity and light quality on the growth and competition of the shadow plant Lamium galeobdolon (L.) Crantz and the half-shadow plant Stellaria holostea L. Flora Jena 172, 235-249.

WESSON, G. and P.F. WAREING 1969. The role of light in the germination of naturally occurring populations of buried weed seeds. Journal of Experimental Botany 20, 402-413.

WEST, G. 1930. Cleistogamy in Viola riviniana with especial reference to its cytological aspects. Annals of Botany 44, 87-109.

WEST, R.G. 1969. Pleistocene geology and biology. London: Longmans.

WESTOBY, M. 1998. A leaf-height-seed (LHS) plant ecology strategy scheme. Plant and Soil 199, 213-227.

WESTON, R.L. 1964. Nitrogen nutrition in Atriplex hastata L. Plant and Soil 20, 251-259.

WHEELER, B.D. 1975. Phytosociological studies on rich fen systems in England and Wales. Ph.D. Thesis, University of Durham.

WHEELER, B.D. 1980. Plant communities of rich-fen systems in England and Wales. 3: Fen meadow, fen grassland and fen woodland communities and contact communities. Journal of Ecology 68, 761-788.

WHEELER, B.D. 1983. Vegetation, nutrients and agricultural land use in a North Buckinghamshire valley fen. Journal of Ecology 71, 529-544.

WHEELER, B.D. and K.E. GILLER 1982. Status of aquatic macrophytes in an undrained area of fen in the Norfolk Broads, England. Aquatic Botany 12, 277-296.

WHEELER, K.G.R. 1981. A study of morphological and physiological differences between woodland and pastureland clones of the great stinging nettle, Urtica dioica L. Ph.D. Thesis, University of Exeter.

WHITE, R.A. 1963a. Tracheary elements of the ferns. 1: Factors which influence tracheid length; correlation of length with evolutionary divergence. American Journal of Botany 50, 447-455.

WHITE, R.A. 1963b. Tracheary elements of the ferns. 2: Morphology of tracheary elements; conclusions. American Journal of Botany 50, 514-522.

WHITE, T.R.C. 1993. The inadequate environment. Berlin: Springer-Verlag.

WHITTAKER, R.H. 1965. Dominance and diversity in land plant communities. Science 147, 250-260.

WHITTAKER, R.H. 1975. Communities and ecosystems. New York: Macmillan.

WHITTAKER, R.H. and D. GOODMAN 1979. Classifying species according to their demographic strategy. 1: Population fluctuations and environmental heterogeneity. American Naturalist 113, 185-200.

WHITTINGHAM, J. and D.J. READ 1982. Vesicular arbuscular mycorrhiza in natural vegetation systems. 3: Nutrient transfer between plants with mycorrhizal inter-connections. New Phytologist 90, 277-284.

WHITWORTH, J.W. and J.J. MUZIK 1967. Differential response of selected clones of bindweed to 2,4-D. Weeds 15, 275-280.

WIDLECHNER, M.P. 1983. Historical and phenological observations on the spread of Chaenorrhinum minus across North America. Canadian Journal of Botany 61, 179-187.

WIEGERS, J. 1985. Succession in fen woodland ecosystems in the Dutch Laf district. Dissertationes Botanicae 86, 152.

WIGLEY, T.M.L. 1985. Impact of extreme events. Nature 316, 106-107.

WILBUR, H.M., D.W. TINKLE and J.P. COLLINS 1974. Environmental certainty, trophic level and resource availability in life history evolution. American Naturalist 108, 805-817.

WILCOCK, C.C. and B.M.G. JONES 1974. The identification and origin of Stachys x ambigua. Watsonia 10, 139-148.

WILKINS, D.A. 1963. Plasticity and establishment in Euphrasia. Annals of Botany 27, 533-552.

WILLEMS, J.H. 1983. Species composition and aboveground phytomass in chalk grassland with different management. Vegetatio 52, 171-180.

WILLEMS, J.H. 1985. Growth form spectra and species diversity in permanent grassland plots with different management. Munstersche Geographische Arbeiten 20, 35-43.

WILLIAMS, E.D. 1973. A comparison of the growth and competition behaviour of seedlings and plants from rhizomes of Agropyron repens and Agrostis gigantea. Weed Research 13, 422-429.

WILLIAMS, J.T. 1963. Biological Flora of the British Isles: Chenopodium album (L.). Journal of Ecology 51, 711-725.

WILLIAMS, J.T. 1969. Biological Flora of the British Isles: Chenopodium rubrum (L.). Journal of Ecology 57, 831-841.

WILLIAMS, J.T. and J.L. HARPER 1965. Seed polymorphism and germination. 1: The influence of nitrates and low temperatures on the germination of Chenopodium album. Weed Research 5, 141-150.

WILLIAMSON, P. 1976. Above-ground primary production of chalk grassland allowing for leaf death. Journal of Ecology 64, 1059-1075.

WILLIS, A.J., B.F. FOLKES, J.F. HOPE-SIMPSON and E.W. YEMM 1959. Braunton Burrows; the dune system and its vegetation. Journal of Ecology 47, 249-288.

WILLMOT, A.J. 1985. Population dynamics of woodland Dryopteris in Britain. Proceedings of the Royal Society of Edinburgh 86B, 307-314.

WILLSON, M.F. and N. BURLEY 1983. Mate choice in plants, tactics, mechanisms and consequences. Monographs in Population Biology No. 19. Princeton, New Jersey: Princeton University Press.

WILSON, A. 1956. Altitudinal range of British plants, 2nd edn. Suppl. to North West Naturalist.

WILSON, J.Y. 1959a. Vegetative reproduction in the bluebell, Endymion non-scriptus. New Phytologist 58, 155-163.

WILSON, J.Y. 1959b. Verification of the breeding system in the bluebell, Endymion non-scriptus (L.) Garcke. Annals of Botany 23, 201-203.

WILSON, P.J., K. THOMPSON, and J.G. HODGSON 1999. SLA and leaf dry matter content as alternative predictors of plant strategies. New Phytologist 143, 155-162.

WINN, A.A. 1985. Effect of seed size and microsite on seedling emergence of Prunella vulgaris in four habitats. Journal of Ecology 73, 831-840.

WISHART, D. 1978. Clustan user manual, 3rd edn. Inter-University/Research Councils Series Report No. 47. Edinburgh University: Program Library Unit.

WISHEU, I.C. and Keddy, P.A. 1992. Competition and centrifugal organization of ecological communities: theory and tests. Journal of Vegetation Science 3: 147-156.

WOLFENDEN, E.A. and D.R. CAUSTON 1979. The germination of some woodland herbaceous species under laboratory conditions. A multi-factorial study. New Phytologist 83, 549-558.

WOODELL, S.R.J. and M. KOOTIN-SANWU 1971. Intraspecific variation in Caltha palustris. New Phytologist 70, 173-186.

WOODWARD, F.I. 1987. Climate and plant distribution. Cambridge: Cambridge University Press.

WRATTEN, S.D., P. GODDARD and P.J. EDWARDS 1981. British trees and insects; the role of palatability. American Naturalist 118, 916-919.

WRIGHT, I.J., P.B. REICH, M. WESTOBY, D.D. ACKERLY, Z. BARUCH, F. BONGERS, J. CAVENDER-BARES, T. CHAPIN, J.H.C. CORNELISSEN, M. DIEMER, J. FLEXAS, E. GARNIER, P.K. GROOM, J. GULIAS, K. HIKOSAKA, B.B. LAMONT, T. LEE, W. LEE, C. LUSK, J.J. MIDGLEY, M.L. NAVAS, U. NIINEMETS, J. OLEKSYN, N. OSADA, H. POORTER, P. POOT, L. PRIOR, V.I. PYANKOV, C. ROUMET, S.C. THOMAS, M.G. TJOELKER, E.J. VENEKLAAS and R. VILLAR 2004. The worldwide leaf economics spectrum. Nature 428, 821-827.

WU, L. and S. JAIN 1980. Self-fertility and seed set in natural populations of Anthoxanthum odoratum L. Botanical Gazette 141, 300-304.

YAPP, R.H. 1912. Spiraea ulmaria L. and its bearing on the problem of xeromorphy in marsh plants. Annals of Botany 26, 815-869.

YEO, P.F. 1961. Germination, seedlings and the formation of haustoria in Euphrasia. Watsonia 5, 11-22.

YEO, P.F. 1964. The growth of Euphrasia in cultivation. Watsonia 6, 1-24.

YEO, P.F. 1966. The breeding relationships of some European Euphrasiae. Watsonia 6, 216-245.

YOUNG, J.E. 1985. Some effects of temperature on germination and protonemal growth in Asplenium ruta-muraria and A. trichomanes. Proceedings of the Royal Society of Edinburgh 86B,

454-455.

ZABKIEWICZ, J.A. 1976. The ecology of gorse and its relevance to New Zealand forestry. In: The use of herbicides in forestry in New Zealand. New Zealand Forest Service, Forest Research Institute symposium 18, 63-68. Rotorua: Forest Research Institute.

ZABKIEWICZ, J.A. and R.E. GASKIN 1978. Effect of fire on gorse seeds. In: Proceedings of the 31st New Zealand weed and pest control conference, M.J. Hartley (ed.), 47-52. Palmerston North: New Zealand Weed and Pest Control Society Inc.

ZANDEE, M. 1981. Studies in the Juncus articulatus L. J. acutiflorus Ehrh.-J. anceps Laharpe-J. alpinus vill. aggregate. Proceedings Koniklijke Nederlandse Akadamie von Wettenschappen C 84, 243-254.

ZARZYCKI, K. 1976. Ecodiagrams of common vascular plants in the Pieniny Mountains (the Polish West Carpathians). Part 2: Ecodiagrams of selected grassland species. Fragmenta Floristica et Geobotanica 22, 500-528.

ZIMMERMANN, J.K. and M.J. LECHOWITZ 1982. Responses to moisture stress in male and female plants of Rumex acetosella L. (Polygonaceae). Oecologia 53, 305-309.

ZIMMERMANN, R.H., W.P. HACKETT and K.P. PHARIS 1985. Hormonal aspects of phase change and precocious flowering. In: Encyclopaedia of plant physiology, K.P. Pharis and D.M. Reid (eds), 79-115. Berlin: Springer.

Subject Index

In the case of a subject occurring within the text of a Species Account only the left-hand page of the Account is indexed. However, the subject may also (or only) occur within material which over-runs onto a right-hand page.

Abundance 14, 20, 23, 24, 37, 55, 59, 62, 65, 73, 152, 268, 308, 376
    beneath suppressed competitors 454
    fluctuations in relative 248
    high 290
    local 188, 458, 572, 630
    low 272
    maintained by agriculture 600, 602
    manipulation of relative 62, 65, 72, 73
    maximum 216
    trends in *see* Population trends
    widespread 212, 376, 394, 434
Active foraging *see* Foraging
Adventive species 246
Aestivation
    of bulb 356
    of corm 502
    of dormant apices 484
Aggregate species 320, 372, 482
Agriculture 18, 68
    cultivated species 246, 252, 312, 314, 352, 406, 462, 600, 602
    developments in 68, 438, 524
    intensive methods of 230, 244, 246, 386, 388
    minimum tillage 348
    ploughing and regeneration 464, 472, 512, 530
    species now little used in 238, 306, 412, 434, 476, 606
    unsuitability for 346
    use in 106, 238, 242, 246, 306, 308, 352, 406, 410, 474, 600, 602
        over-reliance on 406
        tonnage sown in 406, 410, 462, 476, 600, 602
    'weeds' of 202, 242, 248, 364, 392, 528, 588
        arable land 84, 90, 108, 114, 120, 130, 132, 146, 176, 208, 210, 214, 222, 230, 264, 280, 298, 316, 318, 348, 354, 386, 388, 400, 418, 438, 452, 474, 478, 482, 456, 520, 528, 554, 558, 564, 566, 572, 580, 604, 618, 626, 634
        broad-leaved crops 316, 454, 456, 482, 554, 558, 572, 618
        cereal crops 316, 388, 452, 478, 482, 626, 634
        grassland 364
        horticulture 520
        pasture 248, 528, 588
    deleterious effects on yield 580
    see also Weed species
    see also Grassland, Meadow, Pasture species

Alien species 260, 268, 300, 366, 550
Allelopathic species 264, 496
Altitude 37, 55
    and edaphic tolerance 556, 562
    and phenology 448
    and shade 390, 426, 490, 496, 632, 638
    effects of 88
    effects on competition 218
    habitats at varying 56, 236
    high
        decreasing seed viability 568
        inability to set seed at 102, 620
        poor performance at 602
        restriction to 622
    low
        bias towards 368, 418
        restriction to 108, 186, 338, 632
        problems of establishment at 160
Altitudinal limits 128, 256, 266, 270, 278, 286, 368, 520, 552, 554, 560, 568, 634
Aluminium
    foliar concentrations of 168, 322, 342, 488, 620
    tolerance of high substrate 250
Amenity
    grassland species 88, 92, 238, 250, 304, 406, 474
    lawn weeds 150, 208, 226, 234, 328, 414, 420, 494, 532, 596
    purposes, species unsuitable 152
    use 82, 308, 382, 568, 602
    tonnage sown in 304, 308, 406, 410, 602
Ammonium ion
    intolerance of 346
    tolerance of 148
Ammonium salts
    increased yield where added 164, 212
    suppression by 84, 392, 500, 600
Anaerobic soil conditions *see* Soils
Ancestral
    specializations 9
    constraints 9
Animal dispersal
    seeds *see* Seed dispersal
    see also Fruit dispersal
Annual *see* Life-cycle
Aquatic
    emergent species 122, 282, 570, 610
    habitats
        destruction of 394, 518, 486
        extension into 358
        species of 90, 96, 122, 166, 178, 186, 260, 282, 374, 394, 460, 454, 508
    populations, derived from land
        form following flooding 454

semi-, species 336, 370, 374, 628
submerged species 122, 260, 484, 510
Aromatic species 134, 136, 348, 418, 424, 448
Arctic habitats, species of 170, 292
Associated floristic diversity *see* Diversity
Associated species
    absence of 144
Aspect preference 37, 57
Attributes *see* Traits

Bare soil
    species exploiting 150, 152, 234, 252, 254, 324, 328, 342, 358, 360, 362, 372, 376, 378, 388, 420, 434, 442, 472, 512, 520, 532, 534, 536, 554, 582, 588, 626
Biennial *see* Life-cycle
Biological
    control, weed species 548, 612
    flora 37, 53
    specialisation, intermediate degree of 380
Biomass, above-ground
    annual increase in 302, 422, 428
    little seasonal change in 142
    peak 126, 220, 310, 426, 492
        in summer 154, 212, 326, 362, 384, 396, 398, 494, 510, 538, 576, 600, 606
    reduced after severe winters 352
    total contribution to 468
        large 570
        small 468
    shoot 156, 262, 380
Bird
    dispersal *see* Seeds
    nesting, species suitable for 340
Bud-break, timing of 82, 152, 232, 314, 568, 614
Bulb-forming species 122, 356
Burned sites
    distribution biased towards 362
    increase in frequency in 202
    minor component of 228, 346, 494
    species of 156, 296, 448, 544
Burning
    controlled
        maintenance by 612
        tolerance of 168, 266
    increase in abundance after 126, 286
    limited tolerance of 242, 250, 292, 360, 378
    protection against 196
    restriction of competitors by 526, 574, 606, 624
    susceptibility to 160, 212, 228, 236, 304, 446

Burning *continued*
    tolerance of 88, 110, 152, 154, 204, 290, 344, 402, 434, 476, 496, 620, 622, 642

CSR model 9-15
Calcicole habit 142, 154, 324, 336, 344, 346, 380, 436, 546, 592
Calcifuge habit 94, 250, 252, 322, 342, 354, 446, 562, 620
Calcium, foliar concentrations of 132, 168, 228, 242, 248, 264, 290, 292, 304, 308, 318, 330, 342, 344, 348, 352, 354, 376, 378, 380, 406, 426, 430, 450, 470, 474, 496, 538, 608, 616
Calcium levels in substrate, high
    inhibition by 378
    requirement for 448, 592
    indifference to 276
Calcium nutrition, variation in 304
Callunetum, establishment in 152
Canals, species of 122, 394, 484, 486
Canopy
    dense, capacity for development of 154, 264, 270, 300, 338, 402, 460, 534, 614, 616, 620, 636
        produced by young bushes 612
    emergence
        early 534
        late 498
    expansion, causing shoot death 228
    exploitation of phase before development of 356, 386
    -forming species 168, 212, 232, 270, 300, 314, 338, 426, 460, 464, 534
    gaps and flowering 408, 586
    germination under 164
    height 236, 450, 608, 624
    increase in abundance under 282
    long-lived 608
    non-flowering shoots and 310
    persistence beneath 334, 404, 414, 586
        of young plants 348
    protrusion through gaps in 314, 582, 598, 602
    sparse 284
        in old bushes 612

    species occupying top part of 394

    suppression by 112
    tall, species successful in 242
    timing of maximum 334, 574, 640
    understorey species 174, 184, 218, 232, 260, 286, 320, 322, 330, 342, 414, 426, 478, 480, 490, 582, 592, 622, 634, 642
'Capitalist species'
Carbohydrates, storage 26
Carbon metabolism
    C3 148, 546
    C4 146
    CAM 260, 484, 506, 546
Carpet-forming species 250, 334, 340, 434
Chilling and germination *see* Seed dormancy breaking
Chloride concentration, high 148
Chloroplasts, absence of 166
Chromosome
    numbers 98, 110, 116, 132, 170, 174, 184, 234, 242, 300, 318, 324, 364, 476
    races 170, 240

Cinder tips, species of 94, 268, 300, 476, 550, 552, 554, 604
Cliff-tops, species of 164, 380
Cliffs, species of 140, 142, 162, 180, 230, 240, 256, 308, 436, 548, 562, 592
Climate
    and cyanogenic phenotypes 602
    change 5
        simulation experiments 73, 74, 75
    in relation to life-cycle 108, 318, 432
    in relation to life-form
Climbing *see* Scrambling
Climatic
    adaptations
        morphological 580
        physiological 168
    agents of disturbance 13
    factors limiting distribution 128, 134, 136, 204, 206, 230, 266, 268, 288, 296, 300, 368, 550, 558, 590, 620, 632, 640
        dependence on snow protection 620, 622
    factors and reproduction 382
    requirements 378, 430, 464
    tolerance 100, 110, 152, 250, 266, 580
Climax vegetation 296, 498
Clonal species *see* Vegetative expansion
Closed vegetation 234, 348
Clump-forming species 242, 250, 378, 386
CLUSTAN statistical package 39
Coal waste *see* Mining spoil
Co-existence
    between ephemerals and perennials 64
    between grassland perennials 66
    mechanisms 64-72
Cold
    damage
        susceptibility to 138, 200, 352, 498, 556, 614
            in spring 434, 464, 498, 614
            to leaves formed in mild weather 536
            to xylem vessels 614
    resistance to 154, 342
    sensitivity 200, 260, 286, 300, 314, 340, 394, 472, 454, 496, 508, 590, 602, 620
        seedlings 98
    tolerance 82, 264, 266, 304, 382, 450, 462, 470, 476, 478, 498, 500, 5
        seedlings 514
Colonisation
    high ability 84, 86, 88, 90, 94, 136, 140, 142, 146, 178, 180, 182, 208, 226, 234, 268, 270, 272, 274, 276, 288, 308, 330, 334, 364, 370, 394, 400, 408, 464, 466, 482, 486, 552, 554, 564, 588, 596, 610, 612, 616, 628
    low ability 110, 188, 194, 228, 284, 390, 422, 444, 468, 490, 544, 592, 624, 636, 638
    over long distances 378, 464, 472
    patterns of 260
    worldwide 464, 472, 558
Colonists
    of base-rich sites 314
    of burned sites 152, 196, 212, 290, 292, 446

    of derelict land 212, 398, 612
    of disturbed sites 146, 196, 220, 226, 268, 284, 292, 316, 334, 342, 358, 368, 376, 384, 412, 420, 528, 552, 582, 616

    of fertile habitats 536
    of forest clearance 152
    of gaps *see* Gaps
    of grassland 232, 474
        closed 498
    of infertile habitats 410, 562, 612
    of landlocked sites 374, 486, 518, 610, 628

    of little grazed sites 314
    of moist habitats 272, 278, 280, 284, 412
    of quarries 152, 154, 196
    of spoil heaps 464
    of waterway systems 218, 260, 310, 338, 366, 394, 424, 444, 458, 460, 464, 454, 484, 508, 518
Common names *see* Nomenclature
Commonest associates 39, 60
Communities 62, 63
Community
    assembly 64
        from the species pool 64
    diversity *see also* Diversity
    dominants 62
        ruderal dominants 62
        competitive dominants 62
        stress-tolerant dominants 62
    dynamics 66-73
    ecology 62
        'the agony of community ecology' 63
    resilience 67, 73
    resistance 67, 73
    structure 62-66, 72
    subordinates 62
        the fitness of subordinates 71
    transients 65
Comparative approach 1
Competition 9, 11, 12, 13
    by roots 152
    from more robust species 362, 518
    in productive and unproductive habitats 12
    interdependence of root and shoot 11, 13
    lack of in disturbed habitats 138
    low levels favouring flowering 372
    response to 84, 106
    seedling sensitivity to 122, 536, 570
    sporeling susceptibility to 254, 284
    tolerance of, by seedlings 82
    tradeoffs between root and shoot 11
    vulnerability to 430, 532
Competitors 9, 13, 46
Competitive
    ability
        effective 300, 338, 458, 498, 576
        moderate 594
        poor 158, 304, 324, 410, 538, 570, 600
Competitive-ruderal strategy
Competitive species
    characteristics of 13, 34, 46
    co-existence with 356, 458, 504
Competitive strategy 13, 34, 46
Conservation
    Nature 2, 3, 14, 68, 77
Cool-season
    exploitation of dry soils 160
    species 150, 264, 478, 580
Coppice cycles
    closed phase, seed banks in 224, 248, 360
    open phase, species of 556
Coppicing 82
    cessation of, effects 498
    intolerance of 296
    response to 110, 314

tolerance of 152, 534, 536, 614
Cost-effectiveness, species 406
Coumarin 116
Crassulacean acid metabolism *see* CAM
Crevices, species of 140, 142, 144, 254
Crop seed impurity *see* Seed
CSR theory 9, 10
    objections 12
Cultivation, 'escapes' from 600
Cushion-forming species 430, 546
Cutting *see* Mowing
Cyclical, vegetation processes 37, 52
Cytological variation 158, 240, 318, 354, 392, 476, 624, 642

Day-length
    for initiation of flowering 216
    for initiation of seed germination 404, 626
    long-day species and 108, 418, 556, 604
    vegetative reproduction and 484
    seed plasticity in response to 214
Deciduous species 82, 102, 152, 232, 296, 434, 498, 614, 620
Decomposition
    rate 6
Defoliation
    infrequent, success where 242
    rapid recovery following 588
    sensitivity to 116
    severe, by Lepidoptera 498, 548
    survival of 482, 572
    tolerance of 110, 438
        severe 474
    see also Grazing, Mowing, Predation
Demographic
    approach 2
    studies 23, 470, 500, 512, 540
Demolition sites, species of 134, 146, 148, 212, 268, 534, 550, 552, 604, 608
Depletion zones, in soil 13
Depth-sensing mechanism in seeds 52, 216
Derelict
    environments
        reclamation of, species for 61, 410, 600, 612
        species of 132, 134, 148, 212, 214, 222, 278, 398, 522, 536, 550, 552, 554
Desiccation
    adaptation to 506
    sensitivity to 382
    tolerance of 342, 568, 616
Diseases *see* Pathogens
Dispersal *see* Fruit
Distribution
    arctic-alpine 622
    atlantic 114, 252, 356, 590, 612
    coastal 618
    comparison of juvenile and adult 102, 612
    continental
    cosmopolitan 108, 572
    disjunct 430
    montane 430, 444, 422
    north-western 240, 294, 498, 614
    northern bias 292, 620
    northern limits of 200, 586
    oceanic 286, 430
    related to land use 136, 470
    related to shade and altitude 390, 426, 490, 496, 632, 638
    related to soils and climate
    restricted 520

sex differential in 426, 556
south-eastern 296
southern 368, 456, 514, 520, 550, 560, 624, 640
southern limits of 266, 292, 476, 622
Sub-Atlantic 168, 218, 228, 408
warm temperate 498
widespread 94, 174, 216, 250, 332, 400, 534, 566, 624
see also Species distribution
Disturbance 9, 13
    abundance following extensive 196, 252
    filter 64
    restriction to sites of 634
    high, in woodlands 152, 416
    inability to establish under 214
    inability to exploit due to germination biology 482
    intermittent, frequency under 138, 210, 252, 274, 372
    light, species of 462
        mechanised, vegetative spread 166, 392
        sensitivity to 558
    microsites due to 292, 494
    moderate, tolerance of 372
    promoting species richness 63
    rapid re-establishment after 456
    recovery following 372
    regeneration favoured by 314, 358
    restriction of competitors by 252, 258, 262, 268, 278, 320, 330, 332, 334, 360, 362, 370, 374, 384, 392, 400, 442, 506, 526, 556, 576, 578, 626, 628, 632, 638
    restriction of perennials by 180, 186, 234, 388, 438, 556, 594
    seasonally unpredictable 330
    severe, absence from 214, 216
    soil movement and sporelings 254
    spatially unpredictable
    species indicative of human 484
    susceptibility to 458
    survival of repeated 418
    unpredictable, survival of 452
    vulnerability to reduced 608
    vulnerability to sustained 138
    vulnerability to unpredictable 118
Ditch banks, species of 144, 224, 276, 284, 358, 370, 454, 584, 594
Ditches, species of 96, 122, 166, 280, 282, 336, 374, 394, 424, 442, 486, 508, 514, 520
Diversity, floristic *see* Species-richness
    associated floristic diversity 39, 58
    decreases in 3, 154
    low, 536
        under forestry monocultures 296
    reduced 3
DNA amount *see* Nuclear DNA
Dominance
    capacity for 86, 88, 104, 112, 132, 144, 154, 168, 212, 234, 254, 256, 260, 270, 292, 296, 300, 344, 370, 388, 394, 458, 460, 510, 530, 560, 610, 616
    in relative short vegetation 88, 90, 100, 250, 354, 356, 608
    in shade 426, 522, 560
    local 192, 220, 222, 242, 248, 356, 392, 522, 636
    seasonal 100, 356, 484

susceptibility to 126, 238, 244, 246, 258, 326, 436, 484, 494
    in seedlings 212, 422
    transitory 232, 286, 392
Dominant species
    coexistence with 380, 442, 502, 616
    giving mechanical support 318
    inability to coexist with 448
    restricted by disturbance 252, 374, 392
    restricted by grazing 358
    restricted by fertility 358, 362, 374
    restricted by low pH 374
    restricted by mowing 392
    see also Subordinate spp.
Drainage, effects of 96, 122, 170, 224, 248, 274, 276, 290, 320, 336, 370, 412, 464, 506, 508, 584
Drained habitats, persistence in 386, 454
Drought
    and subsoil moisture 198, 204, 246, 342, 538, 592, 596
    avoidance
        mechanisms 94, 120, 342, 436, 502, 592, 596
    causing growth check 200
    causing shift in C metabolism 546
    colonisation of spaces due to 420
    destruction by 140
    dioecious responses to 526
    distribution affected by 322, 640
    flowering inhibition under 616
    resistance 310, 546
    restriction of perennials by 234, 288, 328, 438, 440, 502, 542, 594, 626
    causing seedling mortality 130
    sensitivity 100, 106, 118, 124, 138, 238, 300, 382, 404, 412, 426, 430, 466, 478, 540, 556, 570, 590, 602, 614
        in seedlings 94, 102, 152, 342, 488
        in seeds 498
    tolerance 6, 84, 92, 120, 140, 242, 302, 364, 400, 554
Dry storage *see* Seed
Dutch Elm Disease *see* Pathogens

Ecosystems 62, 64, 75
    carbon storage 64, 77
    functioning 64
    plant drivers of 75
    plant <-> ecosystem feedbacks 76
    tempo 76, 77
    the mass ratio hypothesis 73
Ecotone species 360
Ecotypic:
    Differentiation
        basis of 204, 234
        lack of 150
        significance of 136, 500, 564
Edaphic
    Ecotypes 304, 352
    tolerance
        wide-ranging 156, 166, 174, 182, 186, 188, 212, 218, 242, 290, 304, 312, 320, 336, 352, 362, 382, 396, 414, 422, 424, 438, 442, 456, 570, 578, 584, 588, 590, 606, 626
    see also Soil type
Edible species *see* Food
Endemic species 166, 218, 228, 294, 590
Ephemeral species 13, 146, 318, 328
Established phase
    attributes of 34

SUBJECT INDEX 721

Established phase *continued*
    strategies of *see* Strategies and Plant
        functional types
Eutrophication, consequences of 86, 96, 122,
    258, 260, 270, 274, 276, 282, 336,
    370, 394, 424, 442, 486, 506, 510,
    518, 570, 628
Evergreen habit 168, 250, 266, 286, 340, 342,
    422
Evolution
    and 'channelling' 9
    of ecotypes 516, 588
        rapid 472
    of inbreeding populations 158, 304
Extinction
    approaching in lowland areas 344, 266, 342,
        380, 544, 562, 622
    approaching in native habitats 462
Extreme events 73
    resistance and resilience 73

Fallow ground, occurrence on 452, 634
Farming *see* Agriculture
Fen carr, species of 534, 616
Fen-meadows, species of 186, 588
Fern species 140, 142, 144, 240, 254, 256, 496
Ferrous iron toxicity
    susceptibility to 276, 310, 318, 412, 426
    tolerance of 90, 186, 270, 370, 376, 434, 506
Fertility *see* Productivity
    soil *see* Soil fertility
Fertilizers
    effects of addition of 84, 88, 106, 584
    effects on species richness 67, 68

Field surveys 3, 17
Fire hazard, litter 612
Floating rafts, species forming 582
Flood hollows, species of 276
Flooding
    ability to re-anchor following 520
    facilitating vegetative spread 258, 274, 458
    restriction of competitors by 178, 252, 268,
        274, 320, 442, 444, 556, 628
    root response to 270, 460
    seedling vulnerability to 464
    tolerance of 98, 104, 180, 444, 460
    unaffecting overwintered seed 366
Floodplains
    dispersal along 236
    species of 144, 334, 576
    terraces 310
Floral
    Biology 640
    structure 108 138
Floristic diversity *see* Diversity
Flower characteristics 36, 50, 94, 200, 260, 418
    colour
        influence on pollination 224
        variation within populations 366
    formed during previous years 292
    heterostyly, proportions in 492
    persistence 230 *see also* Pollination,
        Reproduction
Flowering
    before leaf expansion 426, 458
    conditions for 128, 266
    delayed 198, 224, 226, 246, 626
    early 184, 188, 218, 266, 292, 328, 334, 372,
        380, 414, 426, 490, 492, 502, 504,
        542, 578, 588, 630, 640
    free flowering species 450

in first season 238, 462, 470, 528
in shade, capacity for 366, 632
inhibition
    at high altitudes 620
    by caterpillar infestation 548
    by drought 616
    by shade *see* Shade
initiation of 216
late 200
late summer 206
period 440
    extended 328, 348, 372, 430, 488, 532,
        550, 638
    inflexible 604
    restricted by drought 596
    short 380
prevented by grazing 392
probability of in next season 528
rapid 208, 364, 542, 582
rates of 124, 394
    of male and female inflorescences 526
season 150
secondary 200
shoot, height of 320, 350
shy 454
stimulated by nitrogen 466
Food plant 214, 222, 228, 266, 400, 558, 586,
    616
    culinary herbs 134, 444
    flavouring substance 84
    historic use 340
    in salads 98, 400, 518, 538, 566, 580
Forage species 406, 410
Foraging
    by roots and shoots 13, 90, 220, 576
    scale and precision 64
    tempo 64
Forestry *see* Woodland
Frond characteristics 256
Frost *see* Cold
Fructan concentration 84, 100, 150, 174, 242,
    264, 356, 404, 406
Fructose concentration in shoots 206, 348, 356,
    570
Fruiting characteristics 82, 266, 314, 468, 484,
    538, 570
Fruits
    dispersal
        by mammals 128
        by birds 358, 408, 484, 528, 570, 586
        by water 358, 444, 484, 528, 570, 610
        by wind 152, 290, 314, 588, 610
    life span 570
    palatability 522
    production
        abundant 126, 534, 536, 582, 588, 610
        low, in shade 408, 522
Functional classification
    of plants 3
    types 5
Functional shifts 5, 14, 64
    and human population density 14

Gametophyte generation 284
Gaps
    colonization 86, 116, 134, 160, 232, 242,
        278, 314, 358, 602
    in woodland canopy 314

    left by frost action 292
    left by overgrazing 290, 292, 328, 378, 548,
        630

    left by trampling 278, 328, 378
    limited ability to exploit 176, 462
    detection of 130, 216
    exploitation of 184, 248, 252, 320, 404, 544,
        630
    species ill equipped for 482
    seasonal regeneration in
    seedling establishment in 128, 152, 168,
        206, 314, 342, 360, 362, 368, 398,
        398, 466, 472, 548
    temporary expansion into 494
    vegetative spread encouraged by presence of
        512
Garden
    escapes 212, 300, 444, 546, 562, 610
    habitats
        absence from 316
        ornamental species 212, 256, 296, 382,
            544, 546, 568, 610
        'weeds' of 124, 172, 176, 180, 182, 220,
            230, 268, 272, 274, 280, 386, 388,
            504, 532, 554, 576, 588, 634
Garlic odour 98, 100
Genetic
    diversity 116, 246, 296, 430, 480, 606
    heterogeneity 366
    impoverishment 72
    mosaic, in individual trees 498
    uniformity in populations 260
    variation 8
Genome size *see* Nuclear DNA amount
Genotypes
    ecological differences between 464, 466

    fitness of foreign 296
    morphologically distinct 506
    recognition of 376
    resilient following defoliation 132
Genotypic
    control, erect/prostrate form 472

    variation
        due to formerly more extensive
            populations 256
        in growth rate 190, 554, 588, 602
        in morphology 240, 342, 470, 472, 506,
            538, 556, 602, 612
        in physiology 642
        in response to shade 642
    polymorphism *see* Polymorphism
Geographical
    distribution 37
    isolation, reduced 476
    origins 82, 118
    races 326, 362
    range
        increasing 550, 564
        occurrence as casuals outside 616, 618
        restricted 378, 636
        variation in germination
            biology over 428, 572
Geology, of underlying strata
    Bunter sandstone, species of 552
    Keuper marls, species of 540
    Limestone strata
        carboniferous, species of 160, 204, 324,
            336, 368
        confinement to 344
        frequent on 312, 436, 440
        magnesian 204, 278, 288, 464, 540, 598
Millstone grit 94, 218, 290
Siliceous strata, species of 620

Germination *see* Seed
Germinule *see also* seed, spore
'Ghost of competition past' 63
Glacial relict species 430
Glucose concentration 206
Gold, accumulation of 284
Grassland
  acidic, species of 250
  ancient 404, 480
  burned sites, distribution biased towards 362, 448
  damp, species of 184, 192, 194, 224, 228, 434
  derelict
    species of 132, 154, 160, 242, 248, 326, 352, 612
    species suppressed in 158, 304, 322, 342, 410, 466, 468
  destruction of older systems 304, 306, 308
  disturbed, species of 328
  dry, species of 204, 246, 606
  floristic diversity in 67, 68
  invaded after myxomatosis 612
  limestone
    floristic diversity in 24
  lowland 160
  managed 158, 248, 406
  montane 476
  mown
    absence from 130, 230, 594
    increased frequency in 202
    species of 500
  old established 110, 158, 188, 202, 244, 342, 344, 346, 476, 598
    indicator species of 468, 640
    seedling establishment in 466
  open 204, 246, 324, 328, 368, 404, 626
  productive 406, 600
    persistence in 410
  ridge and furrow, niches in 500
  rough, species of 348, 612
  semi-natural 244, 342, 410, 492
  short 364
  tall
    inability to persist in 306, 342, 410, 466, 640
    persistence in 346
    species of 360, 368
  ungrazed, species of 246
  unmanaged 368, 606
    low persistence in 406
  unproductive 326, 390, 398, 410, 466, 476, 594, 596, 612
    major constituent of 446
  upland 236
  wet 412
  xeric 342
Gravel pits, species of 86, 206, 268, 276, 284, 412
Grazed sites
  ability to establish in 232, 378
  fruit production in 396
  infrequent in 204, 214, 216, 252, 452
  persistence in 202, 258, 364, 524
  restriction to 238
  seed production in 410
Grazing
  close
    favoured by 494
    susceptibility to 380
  detrimental to regeneration 498
  effects of cessation 132, 410

effects of damage by, severe 244
elimination by 310
favoured by increased 292
heavy
  colonisation of bare areas left by 226, 266, 278, 290, 292, 472, 500, 548
  effects of 160, 242, 250, 236
  exploitation of sites of 500, 502
  inability to persist under 306, 354
  poor growth under 462
  susceptibility to 168, 600, 606
  vegetative regeneration under 364
light
  species abundant where 246, 346, 630
  species restricted to situations of 392
  survival of 590
  tolerance of 292, 390, 412, 466
pressure
  resulting in appressed forms 150, 584
  resulting in displacement 270
  resulting in reduced stature 344
  resulting in semi-prostration 132, 470
prevention of flowering by 392
protection from by other species 276, 314
  by low growth-habit 158, 596
  by odour 330
  by siliceous stem characteristics 284
  by spines 222, 224, 226
  by toxic principles 356
relaxation of
  leading to tall phenotypes 620
  prominence in sites subject to 236, 326, 354
  vulnerability to 188, 410, 492
  woodlands attributable to 314
restriction of competitors by 184, 186, 188, 192, 194, 208, 236, 250, 266, 276, 284, 320, 358, 400, 480, 502, 574, 584, 592, 624, 628
seedling sensitivity to 82, 314, 536
species exploited for 460
susceptibility to 116, 122, 126, 132, 164, 172, 196, 246, 200, 270, 274, 278, 314, 236, 348, 350, 352, 356, 408, 424, 444, 458, 516, 562, 590, 604, 608, 624, 636
tolerance of 84, 88, 104, 106, 110, 112, 132, 176, 242, 290, 406, 420, 434, 474, 476, 518, 602, 620
  by seedlings 498
value to stock 434
see also Predation
Gregariousness 37, 55
Growth
  dependent on high rainfall 314
  early in season 578, 588
  effects of temperature on 84, 228, 394, 490
  -form *see* Morphology inhibition of, by humus 250
  opportunistic 124
  period, short 476
  rate
    diminution of 156
    varying with habitat and area 494
  spring and autumn peak in 264
  'stored' 45
  summer-peak 154

winter 344, 462, 478
see also Relative growth rate

Habitat
  continuity, restriction to sites of 604
  destruction, effects of 86, 88, 102, 104, 122, 144, 154, 170, 178, 190, 192, 194, 196, 204, 250, 256, 260, 266, 276, 278, 282, 290, 304, 306, 308, 320, 336, 344, 370, 380, 382, 394, 404, 412, 424, 448, 464, 480, 488, 492, 506, 508, 516, 562, 584, 620
  distribution
    in relation to altitude 21, 55
    in relation to slope
    range 90, 112, 116, 154, 212, 272, 312, 474, 500, 592
    exploitation by seedlings relative to adult plant 536, 612, 614
    narrow 294, 360, 458, 640
    rapidly increasing 550
    species similar in 436, 542, 570
    wide 192, 232, 254, 268, 308, 348, 350, 388, 410, 414, 470, 478, 496, 512, 522, 524, 580, 596, 624
Habitats 37, 54
  specialisation of taxa with respect to 294
  species similar in 618
  variation in growth form with 598
Hapaxanth species 544
Hard-coat dormancy *see* Seed
Hay
  crops, contaminants of 470
  meadows, species of 164, 208, 306, 348, 352, 392, 400, 412, 470, 516, 624
  species exploited for 460, 462, 478, 600
Heather canopy, understorey species 322
Heathland, species of 152, 168, 196, 244, 250, 286, 294, 378, 446, 488, 612
Heavy-metal
  accumulator species 430
Hedgerows
  removal of 522
  species of 98, 120, 138, 232, 252, 312, 314, 236, 332, 334, 340, 348, 384, 386, 408, 422, 490, 498, 522, 534, 536, 576, 578, 586, 594, 614, 630, 636, 638
  species planted as 232, 612
Hemiparasitic species 294, 516
Herbicides
  control by 90, 222, 502, 512, 558
  ineffective 230, 318
  occurrence in areas treated by 124
  resistance to 214, 230, 264, 300, 402, 474, 526, 634
    population variation in 604
  restriction of perennials by 210, 268
  selective 452
  specific to 318
Herbivory *see* Predation
Herbs
  culinary *see* Food
  medicinal *see* Medicinal
Heterophyllous species *see* Morphological variation
Host species
  of hemi-parasites 294, 516
Humans
  plants associated with 118, 150, 164, 176, 264, 288, 400, 406, 470, 456, 512, 536, 580, 616, 618
  population increases associated with activities of 582
  species of economic use 152, 496

Humans
  species of economic use *continued*
    for compost 496
    for fuel 464, 496
    for industry 464, 496, 498
    for thatching 464, 496
    see also Agriculture, Fodder, Food plants, Timber
Hump-backed model *see* Species-richness
Hydrology 38, 57

Infections *see* Pathogens
Infertile habitats *see* Soils
Ingestion, survival by seeds of
  by livestock 114, 176, 216, 316, 318, 372, 418, 482, 506, 526, 548, 600, 604
    cattle 148, 320, 328, 372, 438, 468, 524, 526, 626, 630
  by wildlife 392, 428, 470, 456, 538, 602, 618
    birds 176, 216, 298, 316, 318, 482, 456, 484, 486, 526, 528, 554, 558, 570, 572, 602
    earthworms 176
Insect fauna, association with 82, 102, 152, 232, 296, 314, 496, 534, 548, 61
Introduced species 82, 136, 114, 212, 242, 260, 300, 366, 418, 452, 552, 558, 634
Invasive species 54, 152, 154, 300, 366, 376, 496, 548, 588, 612
Iron
  foliar concentrations of 170, 218, 242, 252, 310, 330, 348, 352, 376, 382, 430, 450, 470, 488, 578, 580, 610, 616
  tolerance of low substrate 270

Juveniles, persistent *see also* seedlings

Lake-shore, species of 186, 464, 506
  Lakes, species of 122, 216, 282, 374, 486
Land management
  current deleterious trends in 14, 440
  distribution of ecotypes in relation to 238, 588
  fluctuating, persistence under 352, 606
  intensification, extinctions under 14, 68, 342, 524
  plant size dependent on 596
  regimes, effects of 14, 160, 242, 266, 294, 324, 352
  species favoured by modern 378, 388

Landscape reclamation, species for 102, 118, 412, 420

Lateral spread *see* Vegetative spread

Lawn species *see* Amenity
Lead-mine spoil, species of 84, 88, 198, 324, 430, 524
Lead tolerance 302
Leaf
  arrangement *see* Morphology
  canopy *see* Canopy
  decomposition rate 7, 10
  defence against herbivores 7
  dry matter content 7, 32, 43

  expansion 82, 96, 348, 458, 498
    timing of emergence in trees and saplings 314 *see also* Shoot expansion
  form *see* Morphology
  litter *see* Litter
  longevity 10

long-lived 190, 244, 310, 350
movement, nastic 'sleep' 450
nutrient concentrations 10
palatability 10
persistence
  in shade 310
  in winter 236
replacement
  lack of 356
  rate, slow 244
shed during drought 342
specific leaf area 7, 32, 43
tensile strength 10
Life-cycle
  annual
    ability to function as 176, 208, 268, 316, 318, 386, 388, 420, 432, 438, 582
    obligate 372, 554
    summer- 108, 146, 148, 114, 210, 214, 298, 366, 432, 438, 452, 482, 456, 520, 552, 618
      ability to function as 564, 626
    winter- 94, 108, 114, 124, 130, 164, 182, 200, 206, 234, 328, 440, 484
      ability to function as 420, 438, 596, 626
  biennial 404
  established phase 8
  regenerative (juvenile) phase 8
Life-form 31, 32, 41
Life-history 11, 180, 272, 286, 318, 420, 544
  variation in 544
  variable 198
Light
  bright, toxicity induced under 362
  compensation point 260
  inhibition of seed germination by intensity
    flowering over wide range in 632
    plasticity in response to 522, 562
    response to increased 426
  low, intolerance of 190
    prolonged duration of growth in 612
    tolerance of 240, 426, 450, 508
  microsites, restriction to 198
  patchy, stolon exploitation of 312
  phase
    exploitation of 110, 356, 426
    removal, and oak regeneration 498
  photoperiod, leaf production dependent on 508
Lime chlorosis
  species exhibiting 322, 378, 390, 496, 526, 574, 612
Limestone
  dust, resistance to 340
  pavement 64
Liming
  abundance increased by 158
  inhibition of establishment by 212
  suppression by 84
'Limiting similarity' 63
Linnaean taxonomy 9
Literature searches
Litter (leaf, grass, frond, stem etc.)
  accumulation, species producing 24, 67, 144, 154, 212, 232, 296, 300, 496
  breakdown, rate of 152
  capacity for emergence through 354, 356, 426, 428, 478
  grass, submergence beneath 322
  harvested as animal bedding 434

non-persistent 314, 568, 614
persistent 168, 248, 296, 300, 308, 376, 464, 612
restriction to sites with no 302, 432
seed germination in sites with 262, 356
seedling establishment
  inhibited by 162, 356, 432
  promoted by 498
shortly persistent 534, 536
suppression by 154, 262, 300, 432, 460, 496
tolerance of 100, 110, 382, 642
see also Leaf litter
Livestock carrying capacity 68
  and farm income 68
  and conservation of species diversity 68
'Lost plants' 65

Magnesium
  foliar concentrations of 250, 310, 356, 382, 450, 538, 580, 608, 616, 642
  requirement for high levels of 592
Man *see* Human
Manganese
  foliar concentrations of 168, 252, 322, 356, 488, 578, 610, 620
  tolerance
    of high levels of 250
    of low levels of 270
Manure heaps, species of 146, 148, 216, 554
Maritime habitats, species of 90, 134, 148, 200, 308, 330, 380, 410, 482, 506, 528, 552, 554, 556, 560, 584, 604
Mast seeding 102, 296, 498
  and absence of spring frosts 296
Mat-forming species 192, 304, 322, 324, 326, 582, 592
Maturation, age of 128, 246, 604
Meadow habitats
  fertiliser application to 516
  re-sowing of 416
  species of 104, 106, 118, 132, 150, 242, 306, 392, 462, 494, 500, 512, 524, 636
Medicinal plants 84, 96, 134, 252, 518, 530, 574, 624
Microspecies, species divided into 350, 522, 588
Migration, barriers to 82
Milk
  reduction of yield 284
  spp. tainting 84, 134, 246, 518
Mineral *see* Nutrient
Mining spoil
  reclamation *see* Landscape
  species of 148, 206, 250, 268, 276, 300, 350, 596, 598, 608
Mining subsidence, species associated with 454
Minor species *see* Subordinate
Mire habitats
  adjacent to open water 424, 442, 454, 628
  absence of species from 192
  grazed, species of 370
  ombrogenous 290
  shaded
    absence of species from 358
    species of 172, 180, 218, 310
  soligenous 158, 166, 178, 184, 186, 190, 192, 194, 218, 224, 244, 266, 274, 276, 278, 282, 284, 290, 320, 336, 358, 374, 412, 424, 434, 488, 506, 584
  topogenous 170, 192, 224, 276, 278, 358, 412, 460

transient occurrence in 486
undisturbed, species of 370
Mobility, species
  High 152, 268, 288, 302, 364, 372, 374, 396, 398, 438, 442, 484, 534, 548, 582, 588, 594, 616, 618
  Low 110, 188, 302, 426, 480, 574, 640
Moist environments
  restriction to 366, 370, 450, 616
  species more robust in 478, 488

Moisture
  stress, susceptibility to 404
  subsoil, access to 202
Monocarpic species *see* Seed production
Monocotyledonous species
Monocultures, species forming 172, 212, 270, 296, 300, 338, 354, 358, 366, 460, 464, 510, 532, 616, 622
Monopolistic species 62
Moorlands, species of 266, 286, 374, 434, 480, 612
Morphological
  variation
    association with management 596
    between dioecious plants 314
    between populations 456
    flower colour, in 230, 468
    genetic in origin 240, 342, 470, 472, 506, 538, 556, 612
    induced by grazing and mowing 494
    in response to canopy height 476
    in response to lack of grazing 344, 556
    in response to submergence 358, 486, 506, 508, 514
    in seed size, induced by shade 494
    intra- and inter-population 612
    leaf
      heterophyllous species 166, 174, 478, 486, 506, 508, 548
      in response to illumination 174, 314, 340, 382, 426, 450, 560, 616
      seasonal patterns in 140
      shape 230, 424, 468, 544, 548, 550
    root, in response to flooding 270
    see also seed polymorphism
Morphologically
  dissimilar forms of species 374
  distinct ecotypes 482
  distinct taxa 308, 350
  indistinguishable populations 324
  intermediate species 346
  similar species 280, 518
  (of) bark
Mowing
  effects on relative abundance 338
  aquatic macrophytes 510
  inability to persist under 308, 478
  persistence under 364
  prevention of flowering by 392
  restriction of competitors by 194, 208, 284, 348, 392, 400, 532, 574, 624, 638
  survival of intermittent 452, 460, 462
  susceptibility to 164, 172, 212, 238, 250, 270, 310, 316, 356, 446, 458, 604, 608, 616, 618
  tolerance of 98, 112, 120, 134, 146, 176, 338, 376, 396, 406, 420, 444, 482, 494, 530, 534, 602
Mulch effect of talus 132, 156, 590

Mycorrhizas
  association with 138, 174, 308, 472
  capacity to bind heavy metals 168
  colonisation by 242, 326, 398, 466, 538, 544
  enhancement of seedling yield 152, 158, 174, 202, 326, 398, 466, 470, 538, 544
  lack of 96, 188
  no benefit gained from 524
  requirement for in seedlings 206

Native
  populations, loss of 462
Naturalised species 260, 444, 474, 614, 618
Nectaries, extrafloral 496
Niche 65
  differentiation and species coexistence 67
  exposure by disturbance 63, 64, 65
  foraging 71
  genetically-differentiated 72
  regeneration 72
  spatial 69
Nitrogen
  enrichment of soil by legumes 410
  -fixation, capacity for 102, 390, 392, 410, 412, 420, 596, 598, 600, 602, 612, 636, 638
  -fixing bacteria, changes induced by strains of 574
  foliar concentration of 170, 212, 218, 228, 248, 250, 308, 310, 344, 352, 376, 382, 426, 430, 444, 470, 474, 608, 610, 616
  requirement 394
  -rich habitats, species of 214, 216
  soil levels, cyclical 406, 598, 602, 638
  utilisation of low levels 148, 250, 300, 266
Nitrogenous fertiliser
  and internal nitrate accumulation 580
  spp. increased by 164
  suppression by 116, 410, 470, 600
Nomenclature 31, 41, 108, 240
Nuclear DNA amount 26, 33, 44, 65, 66, 70
  intraspecific variation in 474
Nutrient
  -deficiency, tolerance of 152, 266, 276, 402, 554, 622, 626
  levels, sensitivity to soil 92, 116
  releases by burning 286
    seasonal pulses in 292, 496
  recycling, efficiency at
  requirement
    high 264, 354, 616
    low 250, 622
  stress, restriction of competitors by 188, 310, 320, 328, 402, 480, 632
  turnover rates 406
Nutritionally-extreme habitats tolerance of 118

Odour 84, 138, 330, 426, 444, 536
  Acridity 546
  foetid Species 552, 576
Orchids, ecology of 138
Osmosis, role of mannitol in 516
Overwintering, ability for 104, 260
  as dormant apices 484
  as dormant buds 486
Oxygen requirements
  for germination 610
  for nitrogen fixation 412

Palatability
  leaf
    relatively low 190, 202, 246, 248, 304, 314 326, 420, 446, 490, 494, 526 540, 590
    very low 160, 222, 224, 226, 232, 284, 294, 452, 486, 488, 498, 500, 502, 530, 548, 522
Paleobotanical research, species used in 340
Pasture, species sown for 238, 306, 406, 474, 476, 602
Pasture habitats
  derelict, species of 232, 612
  species of 84, 88, 184, 208, 238, 304, 308, 322, 324, 406, 410, 414, 474, 476, 488, 524, 616
  infertile
    destruction of 190, 630
  old, species of 416, 502
  permanent, species of 494, 500, 502
  ploughed, species decreasing in 492
  tolerance of 524
Patch-forming species 88, 90, 110, 150, 166, 178, 208, 358, 374, 412, 414, 424, 434, 446, 450, 480, 488, 494, 526, 546, 590, 628, 638
Pathogens
  attack by 54, 152, 260, 460, 498, 556, 590, 598
  Dutch Elm Disease 614
  fungal, of flowers 460, 556
  interactions, genetic mosaics 498
Paths, species exploiting 372, 472, 474, 504, 532
Perennation, means of 116
Perennial vegetation
  dominance *see* Dominant
  establishment in 348
  persistence in 280, 318
Perennial *see* Life-cycle
Perenniality, evolution towards 528
Persistence *see* Species persistence
pH
  bimodal distribution in 216, 304, 434
  most common
    at high values 148, 152, 278, 550
    at intermediate values 376, 390, 490, 584, 598
    at low values 250, 572
    at wide-ranging values 568
  restriction at low 374
Phenolic substances, presence of 152
Phenology 23, 31, 42
  Complementary 458, 478, 484, 602, 616
  cool season 118, 264, 406, 462
  facilitating co-existence
    with arable species 264, 386
    with taller species 218, 338, 386, 478, 602
  flowering 426, 488
  leaf
  shoot 23, 144, 162, 202, 204, 304, 306, 396, 408, 426, 478
    relationship with genome size
  species similar in 542 570
  tolerance of grazing due to 462
  variable 182
  variations with altitude in 448
  vernal 100, 132, 138, 158, 170, 180, 228, 318, 356, 390, 426, 504
  winter green

Phenotypes
    cyanogenic 602
    descriptions of 166, 408
Phenotypic plasticity see
    Plasticity
Phosphate
    foliar concentration of 132, 168, 170, 212,
        218, 228, 248, 250, 252, 310, 318,
        348, 356, 382, 406, 426, 444, 450,
        470, 474, 496, 556, 580, 608, 610,
        642
    requirement for 394
    utilisation of low levels of 266
Photobleaching 156
Photosynthate, allocation of 348
Photosynthetic
    activity
        spring and autumn peaks in 264
        winter 510
    apparatus 280
        stem 258, 284, 376, 420
    rate limited by water speeds 510
    utilisation of bicarbonate ion 260, 484
Phylogeny 15, 41
Physiological variation, in shade 642
Phytochrome 224
Plantations see Woodland
Plant functional types
    'acquisitive' 12
    CSR model of 9, 33, 46
        and ecosystem functioning 76, 77
        and the retention of radionucleides 76
        and butterfly biology 77
    'Capitalists' 11
    competitors 46
    global 8
    hierarchy of 8
    intermediate strategies 47
    intra-populational 8
    intraspecific variation in 15
    local 8
    of established phase 33, 46
    of regenerative phase 33, 35, 48
        vegetative expansion 48
        seasonal 48
        persistent seeds 25, 48, 66, 73
        persistent juveniles 512
        widely-dispersed seeds or spores 49
    'Proletarians' 11
    primary 9, 33, 46
    regional 8
    'retentive' 12
    ruderals 47
    stress-tolerators 47
Plasticity
    morphogenetic 470
    phenotypic 8, 214, 236, 242, 260, 322, 410,
        424, 430, 464, 470, 474, 476, 478,
        456, 492, 508, 524, 528, 532, 558,
        572, 580, 584, 588, 636
        ecological importance of 494, 602
        enabling exploitation of different habitats
            374, 454, 506
        extreme 454
        floret number 150, 164
        flower number 206
        growth form 84, 88, 96, 122, 140, 150,
            168, 184, 204, 212, 242, 374, 452,
            522, 602
        high 148, 164, 370, 484, 604
        N-fixing bacteria, responses to 602
        in different habitats 374, 454, 506

    in response to
        canopy height 630
        cutting 396, 494, 524
        disturbance 396
        litter 642
        grazing 236, 396, 470, 456, 494, 584,
            620, 630
        management 348, 524, 596
        shade 212, 248, 282, 322, 426, 490,
            522, 532, 616
        site fertility 596
        submergence 486
        temperature 494, 574
        trampling 396, 446, 456
    interaction with genetic diversity 530,
        596
    leaf characteristics 248, 310, 450, 454,
        490, 506
    life cycle 474
    relating to climate 572, 602
    reproductive effort 216, 294
    resource allocation, rhizome and root
        450
    seed characteristics 214, 558, 572
    seed production 124, 452
    shoot 470
    size 124, 206, 294, 358, 450, 596
    stem dimensions 258
    terrestrial and aquatic forms 506, 508
Ploidy levels 140, 142, 248, 380, 394, 400, 410,
    430, 466, 594, 604, 624
    aneuploid series 476
    ecological differences associated with 174,
        504, 636
Poisonous plants 100, 340, 360, 362, 426, 450,
    502
Pollination
    influence of flower colour on 224
    heterostyly 492
    insect 204, 366, 642
    mechanism 138, 294
    self 126, 164, 166, 224, 226, 532
    timing of 102
    wind 166, 610
    see also Flowering, Reproduction
Pollution
    acid rain
        effects on wetland species 374
        resistance to 554
    atmospheric, effects on 140
    indicator species 216
    sulphur dioxide, distribution and 142
    water, effects of 510, 518
Polycarpic species see Seed production
Polymorphism see Morphological variation,
    seed polymorphism
Pond-margins, species of 166, 216, 282, 372,
    374, 442, 484, 486, 508
Population
    biology
        seed persistence in 620
    changes, evolutionary 116
    densities, regulation of 198
    dynamics 108, 124, 330, 614
    expansion 352
    fluctuations in 130, 466
        cyclical, dependent on soil N 406, 598,
            602, 638
    maintenance
        role of disturbance in 416
        role of seed regeneration in 468
    recruitment from gardens 382

    resurgence on motorway verges 492, 612
    size, annual variations in 374
    structure 112, 330, 348
    trends
        decreasing 86, 92, 100, 102, 106, 108,
            110, 116, 126, 140, 142, 144, 154,
            156, 158, 160 170, 174, 178, 186,
            204, 214, 228, 236, 238, 242, 258,
            262, 294, 304, 302, 310, 312, 316,
            324, 326, 342, 344, 346, 360, 390,
            410, 422, 450, 452, 454, 456, 464,
            466, 468, 484, 490, 500, 502, 542,
            558, 572, 578, 586, 590, 614, 630,
            634, 636, 638, 640
        due to modern farming practice
            392, 400, 606, 634
        increasing 90, 96, 98, 118, 124, 128, 130,
            132, 136, 110, 142, 148, 150, 152,
            166, 176, 180, 182, 200, 206, 212,
            220, 226, 232, 234, 254, 264, 270,
            292, 296, 314, 318, 348, 352, 362,
            364, 366, 368, 372, 376, 384, 394,
            396, 402, 406, 418, 432, 472, 474,
            482, 494, 504, 512, 514, 520, 534,
            536, 550, 552, 564, 566, 576, 588,
            594, 596, 604, 610, 612, 616
        as a result of human activities 582
        in lowland areas 216, 386
        in upland areas 184, 216, 266, 356,
            374, 378, 416, 446, 496, 498, 514,
            520, 622
        with changed use of land 210, 216,
            270, 274, 280, 330, 350, 362, 478,
            530, 532, 608, 618
        with relaxation of grazing 154
        under forest clearance 314

        uncertain 94, 104, 112, 120, 122, 134,
            138, 164, 162, 242, 272, 274, 284,
            286, 288, 302, 328, 330, 336, 340,
            350, 354, 356, 374, 398, 404, 408,
            414, 420, 428, 436, 438, 460, 456,
            522, 540, 560, 568, 570, 626
Populations
    adapted to climate of site of origin 286

    distribution within dioecious 556, 586

    former genetic variation of 256

    homeostasis in 346, 380
    isolated
        cytologically 324
        geographically 324
        topographically 324
    showing male sterility 448
    variation between 124, 314, 364, 552
        according to habitat and area 488, 494
        in regeneration 494, 588
Potassium, requirement for high 592
Predation 118, 158
    avoidance of 376, 378, 400, 492, 502, 504,
        584, 586, 592, 620, 642
    by low growth habit 470, 490, 492
    by cattle 208, 248, 356, 378, 512, 518
    by grouse 168, 266
    by insects 232, 498, 530, 548, 556, 568
    by rabbits 154, 156, 250, 538, 612
    by rodents 498
    by sheep 156, 250, 356, 512, 538
    by slugs 404, 538
    by squirrels 498

deterrents to 96, 102, 186, 198, 222, 224,
        226, 232, 248, 254, 284, 356, 360,
        394, 410, 426, 450, 512, 536, 546,
        552, 560, 602
    by strategy of masting 296
    chemical 496, 552
    inducible 536, 568
    physical barriers 552, 606, 616
        stinging hairs 616
        thorns 232
        tough leaves 622
    detrimental effects of 498
    flowers 470
    foliar 152, 232, 498, 548
    fruits 232, 544
    seedlings 130
    seeds 128, 196, 198, 202, 204, 226, 296, 314,
        348, 422, 498, 502, 512, 544, 556,
        614, 616, 642
  species subject to little 340
Predictive value, of plant attributes
Productivity
  filter 64

  persistence varying with 414, 608

  response to increased 188, 200

  sites of high
    species of 106, 270, 372, 372, 452, 454,
        456, 472, 482, 496, 512, 528, 532,
        536, 554, 560, 572, 588, 602, 604,
        616, 618, 634
    exclusion from 244, 400
  sites of intermediate
    restriction of competitors in 358
    species of 236, 370, 488, 500
  sites of low
    persistence in 498, 612
    restriction of competitors in 186, 192,
        194, 258, 262, 276, 278, 282, 288,
        312, 330, 360, 362, 370, 374, 400,
        402, 412, 416, 422, 424, 428, 502,
        506, 540, 584, 626, 530, 638
    restriction to 126, 174, 188, 206, 244,
        276, 288, 292, 304, 344, 360, 362,
        390, 398, 404, 422, 446, 468, 480,
        562, 642
'Proletarian strategy' 11
Prostrate
  ecotypes 560, 604
  species 322, 342, 364, 370, 420, 482, 582,
        592, 634
Protein concentration 348
Prothallus, ferns, ecology of 142, 240, 254, 256
  by-passing of vulnerable stage of 496
Pteridophyta see Fern species

Quadrat record 24
Quarries, species of 130, 152, 154, 158, 196,
        198, 200, 288, 334, 340, 362, 398,
        404, 416, 430, 440, 448, 466, 480,
        562, 592

Radionucleides 76
r and K selection 11
Railway
  ballast, species of 124, 210, 230, 268, 312,
        362, 402, 546, 550, 552, 604
Rare species 194, 204, 256, 324, 444, 492

Rarity and abundance in vascular plants 15
  and ancestral specialization 15
  and historical factors 15
Reductionist approach 1
Refugia, species restricted to 210, 342, 430,
        440, 542
Regeneration
  by seed
    associated with disturbance events 314,
        490
    flexibility in 252, 270, 364, 470, 474
    high mortality rate during 640
    at southern limits of range 382
    during colonisation of new sites 312,
        376, 378, 446, 460, 468, 518, 522,
        560, 570, 602
    in population maintenance 468

    in unshaded habitats 408
    infrequent 218, 228, 266, 322, 358, 414,
        424, 426, 454, 486, 422, 624
    in established vegetation 378,
        442, 522, 616
    low importance relative to vegetative
        reproduction 150, 172, 178, 186,
        188, 192, 194, 258, 264, 336, 354,
        446, 506
    rare 192, 326, 338, 458, 484, 510, 570,
        630
    seasonal 35, 48
  by spores 140, 142, 144, 240, 496
    importance in colonization 496
  vegetative see Vegetative reproduction
Regenerative
  biology 118, 138, 208, 280, 408, 640
  failure see also Reproductive capacity, poor
    flexibility 49, 200
  phase
    attributes of 88, 90, 116, 162, 266, 270,
        376, 416, 508
  strategies 47
    involving a bank of persistent juveniles
        35, 540
    involving a persistent seed bank 25, 35,
        48, 73
    involving numerous widely-dispersed
        seeds or spores 35, 49
    seasonal regeneration in vegetation gaps
        35, 48
    vegetative expansion 35, 48
  versatility 49, 268
Relative growth-rate 32, 42, 82, 84, 88, 90, 102,
        106, 112, 116, 118, 138, 152, 158,
        174, 220, 244, 250, 252, 268, 270,
        272, 3 44, 382, 394
  high 298, 318, 352, 478, 526, 550, 556, 570,
        580, 616, 634
  relationship with nuclear DNA amount
        45
  seedling 348, 534, 544
  low 304, 308, 310, 322, 342, 348, 380, 382,
        390, 398, 404, 450, 488, 494, 498,
        540, 546, 574, 590, 592, 620, 622,
        642
  variation in 190, 554, 588, 602
Reproduction
  asexual
    apomixis 120, 256, 350, 446, 466, 476,
        522, 588
  sexual

    cleistogamy 244, 450, 642

    dioecious, species 222, 260, 300, 426,
        524, 526, 534, 586, 616
    inbreeding species 176, 244, 268, 272,
        336, 456, 528, 554, 610
      reproductive depression in 442
    outbreeding species 174, 184, 202, 222,
        264, 304, 336, 396, 398, 412, 448,
        470, 554, 562, 584
    selection for see Selection
    self compatability 204, 222
    self-incompatibility 172, 178, 232, 402,
        460, 488, 598, 604
  vegetative see Vegetative reproduction
  see also Pollination, Regeneration by seed
Reproductive
  capacity 340, 354
    high 152
    poor 624, 640
  effort
    associated with pulses of mineral
        nutrient release 286
    constancy in 554
    maintained under conditions of mineral
        nutrient stress 452
    maintained under drought 618
    plastic 216
  maturity 138
    failure to reach 144
  output, variable 146
  phase, frequency of, required to maintain
        seed bank 558
  potential 330, 554
Reservoir margins, exploitation of 104, 358,
        372, 374, 424, 454, 508, 520
Resource
  allocation between rhizome and root 450
    between seed and vegetative
        reproduction 382, 494, 532, 588,
        608, 610, 638, 640
    between structural/ reproductive tissue
        298
    differentials in dioecious plants 526, 556
    to seed production/ trunk size 314
  capture, rates of 264, 270
    and spatial separation of organs 340,
        630
  depletion 13
Respiration
  rate limited by water flow 510
  thermogenic 138
Rhizomatous species 88, 92, 110, 132, 144, 154,
        172, 174, 192, 194, 230, 236, 240,
        250, 254, 264, 270, 280, 282, 284,
        290, 300, 308, 338, 376, 384, 390,
        392, 424, 426, 442, 450, 458, 460,
        464, 476, 454, 486, 496, 570, 576,
        590, 598, 608, 610, 616, 620, 622
River bank habitats, species of 98, 102, 112,
        118, 144, 170, 172, 186, 218, 224,
        252, 262, 268, 270, 274, 300, 310,
        330, 336, 366, 442, 444, 458, 460,
        454, 556, 560, 576, 582
River beds, seasonally exposed, species 278
Rivers
  pollution, indicator species 216
  species of 122, 180, 460, 484, 510
Road verges
  population resurgence on 492
  species of 84, 98, 108, 118, 128, 132, 136,
        154, 158, 160, 164, 168, 172, 230,
        234, 242, 280, 286, 288, 308, 236,
        300, 328, 334, 348, 350, 384, 392,

Road verges
    Species of *continued*
        398, 482, 490, 346, 530, 554, 576, 588, 594, 598, 612, 630, 638
    winter salted, species of 148, 264, 554
Rock outcrops, species of 94, 108, 130, 142, 164, 182, 200, 204, 206, 208, 218, 230, 234, 256, 328, 342, 360, 380, 386, 430, 436, 448, 476, 542, 554, 556, 562, 594, 596, 626
Roots
    abscission of 458
    aerenchyma production in 460, 464, 570, 610
    capacity to conduct oxygen 190, 282, 570
    capacity to exploit subsoil 198, 204, 284, 328, 342, 348, 350, 356, 362, 364, 368, 410, 420, 448, 468, 476, 526, 538, 544, 592, 596
    depth of 102, 104, 136, 138, 154, 160, 172, 222, 230, 264, 270, 280, 284, 296, 304, 314, 324, 326, 342, 346, 398, 404, 426, 434, 460, 466, 470, 586
        deep, causing vulnerability to waterlogging 356, 586, 620, 622
        shallow 450, 478, 488, 490, 512, 632, 640
    expansion of 356
    formation of 340, 518
    parasites of 294
    physiology of 102
    plasticity under low nitrogen 250
    shoot ratio 152
    system
        adventitious, buttressing by 366
        aquatic species 510, 570
        at surface in waterlogged soils 352
        contractile 100, 138, 356, 588
        extensive 402
        of spreading superficial roots 584
        stunted
            on acidic soils 592
        tap 226, 246, 328, 342, 348, 350, 362, 398, 420, 444, 468, 482, 528, 538, 544, 564, 588, 596, 612
        produced in seedlings 498
        providing anchorage in wetland 444
Rosette-forming species 84, 126, 150, 166, 176, 180, 184, 198, 206, 212, 224, 226, 246, 252, 310, 350, 364, 368, 378, 388, 396, 398, 440, 466, 468, 470, 472, 492, 502, 524, 538, 544, 548, 562, 564, 574, 584, 588
Ruderal
    characteristics 13, 34, 35, 47, 168, 256, 482, 482, 518, 564
    habitats, species of 12, 13, 148, 452
    species 35, 214, 200, 234, 294, 330, 348, 366, 474, 532, 558, 604, 634
    strategy 13, 33, 47

Salt-marshes, species of 148, 216
Salt-tolerance 90, 148, 166, 216, 308, 372, 412, 418, 464, 482, 570
Sampling
    bias 418
    techniques 18-20
Sand-dunes, species of 108, 164, 194, 198, 200, 206, 288, 308, 326, 362, 476, 542, 546, 548, 588, 592
Sand-pits, species of 94, 206, 440
Saplings *see* Seedlings

Scarification *see* Seed dormancy
Scrambling species 318, 236, 320, 340, 392, 408, 560, 578, 636, 638
Scree habitats, species of 132, 156, 256, 308, 312, 314, 324, 330, 480, 590, 530, 642
Screening experiments 6, 7, 10, 11, 26-29
Scrub vegetation, species of 232, 312, 332, 414, 534, 568, 590, 614, 620, 640
Sea shore, species of 148
Sea spray, species of soils subjected to 380
Secondary thickening, species with 284
Sections, species divided into 350, 588
Seed
    country of origin 296
    crop, impurities of 108, 114, 146, 164, 204, 208, 214, 246, 316, 318, 328, 352, 364, 388, 478, 482, 456, 524, 558, 596, 604, 634
    dispersal 84, 88, 94, 98, 102, 110, 126, 130, 132, 176, 506
    see also seed mobility
        along transport corridors 136, 210, 268, 236, 396, 542, 550, 552, 632
        animal 84, 96, 114, 146, 148, 150, 162, 220, 246, 258, 266, 296, 302, 318, 330, 332, 376, 392, 418, 438, 456, 512, 514, 530, 540, 554, 580, 594, 604, 616, 628
        ants 174, 196, 244, 382, 386, 414, 416, 422, 426, 432, 480, 590, 612, 634, 640, 642
        bird 108, 138, 146, 148, 186, 214, 230, 232, 266, 296, 312, 340, 456, 486, 498, 528, 536, 558, 560, 568, 570, 572, 580, 586, 604
    buoyancy
        in air 226, 528, 552, 554, 564, 588, 608, 626
        in water 570
    capacity for 360
        limited 538
    censer mechanism 452
    distance 384
        long 200, 246, 288, 350, 368, 378, 418, 452, 498, 532, 540, 550, 552, 564, 608, 626
        short 198, 382, 392, 446, 480, 542, 548, 612, 642
    effectiveness 152, 232, 376
        poor 178, 222, 226, 236, 342, 382, 426, 500, 544, 574, 584, 632, 638, 642
    explosive 180, 182, 184, 328, 366, 612, 642
    from capsules at soil level 634
    human 94, 136, 146, 148, 150, 162, 114, 152, 196, 220, 246, 298, 302, 316, 318, 330, 306, 350, 352, 400, 432, 438, 470, 472, 482, 512, 524, 526, 530, 532, 540, 552, 554, 588, 594, 604, 626, 634
    ineffective 178, 638, 642
    in crop seeds 114, 298, 316, 318, 352, 388, 470, 472, 478, 482, 524, 558, 580, 596, 604, 634
    in fodder 392, 524, 534
    in manure 634
    by mammals and birds 214, 232, 498, 568, 572
    mechanisms of 412, 634
        lacking well-defined 470, 500, 502, 576, 584, 590, 616, 590
    in mud 370, 482
    on machinery and vehicles 150, 176, 196, 418, 482, 572, 580
    poor 222, 226, 236, 342, 382, 426, 500, 544, 574, 584, 632
    soil contamination 386, 526, 550, 616, 618
    temporal *see* time of shedding
    by trains 210, 542, 550
    under-dispersed species 422, 590
    by water 96, 102, 112, 170, 172, 194, 302, 310, 316, 318, 320, 330, 366, 424, 460, 484, 486, 528, 568
    widespread 212, 314, 328, 350, 352, 370, 536, 564, 604
    by wind 84, 102, 136, 168, 176, 246, 252, 268, 272, 274, 276, 278, 288, 292, 348, 350, 364, 396, 398, 436, 458, 464, 534, 542, 544, 550, 552, 554, 562, 564, 588, 608, 614, 624, 626, 466
    see also Ingestion
    dormancy 386, 446, 554, 564, 586, 594
        annually re-imposed 298, 386, 404, 482
        breaking mechanisms 51, 94, 98, 124, 196, 342
        after- ripening requirement 51, 130, 182, 440, 542, 626
        short 200
        chilling requirement 51, 98, 102, 112, 118, 152, 170, 176, 228, 232, 298, 318, 348, 356, 366, 404, 468, 482, 498, 516, 522, 568, 594, 632, 642
        cold treatment 51, 306
        dry storage 51, 462
        hard-coat dormancy 51, 96, 232, 328, 330, 342, 484, 486, 522, 610, 638
        heat treatment 286
        lack of 164, 226, 304
        light requirement 52, 53, 108, 136, 208, 252, 352, 478, 498, 514, 520, 554, 580, 616, 626
        scarification 51, 146, 222, 390, 460, 610
        stratification 460
        temperature fluctuations 51, 146, 178, 216, 352, 442, 460, 506, 514, 520, 610, 616
    double 522
    epicotyl 498
    induced by winter temperatures 594
    lacking 336, 478, 518, 634
    population variations in 478
    secondary 482, 500
    variation in 580
dry weight 226, 470, 456
ecology 326
embryo
    characteristics 118, 314, 348
    dormancy 232, 314
    poorly differentiated 444, 516
    size 50, 112, 228, 314, 444, 468, 632
germination 36, 51, 82, 82, 90, 102
    behaviour 146, 176, 386, 442, 530, 594
        complex 440, 484
        conservative 440
        flexibility in 180, 352
        opportunist 440
        variable 214, 386

      over geographical range 428
  characteristics 51-53, 168, 246, 356, 388, 528
      ecotypic differentiation in 556
  cues stimulating 51, 52, 506, 610, 616
      day length 404, 626
      nitrate 526
      oxygen 610
      sunlight 52, 344
  in darkness 53, 152
  delayed 51, 222, 232, 244, 424, 432, 438, 444, 466, 488, 518, 522, 542, 574, 586, 594, 624
  direct 234, 262, 266, 270, 278, 356, 380, 424, 458, 474, 514, 518, 534, 588, 606, 614, 628
  early 294, 298, 376, 556
  environmental control of 108, 124
  epigeal 53
  hypogeal 53, 390, 498
  inhibition 108
      by canopy shade 53, 130, 136, 152, 224, 252, 396, 478, 460, 572, 612, 626
      by darkness 52, 626
      by high temperatures 432, 626
      by light 52, 114, 618
      by litter 262
      by low moisture 68, 626
  inhibitors, present in coat 314, 360, 362
  in pods still attached to parent plant 390
  moisture requirements for 110, 114, 130, 172, 356, 626
  period
      protracted 104, 146, 150, 208, 252, 318, 328, 330, 364, 386, 418, 474, 554, 580, 604, 634
      restricted 210, 404
  phenology of 324
  rapid 53, 234, 352, 400, 478, 534, 588, 608
  rate of 52, 270, 428
  season of
      autumn/winter 90, 94, 102, 108, 120, 126, 132, 158, 164, 182, 214, 250, 262, 270, 286, 308, 344, 346, 356, 368, 388, 390, 400, 438, 440, 466, 474, 478, 532, 542, 544, 546, 548, 564, 598, 636
      in southern areas 428
      spring/summer 82, 98, 112, 114, 118, 124, 136, 146, 148, 154, 166, 168, 170, 178, 184, 186, 198, 204, 214, 222, 228, 230, 236, 262, 266, 270, 292, 296, 298, 302, 304, 310, 314, 316, 320, 340, 342, 360, 362, 366, 370, 378, 382, 388, 390, 392, 400, 402, 404, 408, 412, 422, 424, 434, 442, 444, 448, 452, 454, 456, 458, 460, 480, 482, 494, 506, 516, 518, 526, 530, 532, 540, 552, 556, 560, 568, 572, 584, 586, 602, 618, 624
      at northern limits 428
      uncertain 188, 196, 414, 490, 492, 630
      variable 84, 96, 172, 174, 176, 202, 226, 388, 514, 538, 558
  synchronous 118, 304, 348, 366, 380, 468, 642
  temperature requirement for 150, 180, 204, 230, 244, 250, 270, 272, 298, 318, 344, 352, 356, 380, 396, 414, 434, 478, 488, 588, 608, 622, 624, 626
longevity 212, 340, 458, 544, 570, 608, 616
maturation
  early 478
  late 244, 574
mixtures, commercial 116, 238, 242
mobility 152, 156, 162, 208, 210, 236, 242
morphology *see* Morphology
mortality 640
  by drought 296, 498
  by frost 498
plasticity 572, 580
  in response to day length 214
polymorphism 146, 148, 214, 482, 456, 528, 530, 548, 556, 580
predation *see* Predation
production 166, 248, 388, 400, 414, 416, 466, 476, 498
  abundant 136, 168, 202, 204, 206, 208, 210, 224, 226, 234, 238, 270, 278, 286, 290, 292, 310, 318, 362, 364, 366, 370, 376, 378, 386, 418, 428, 430, 434, 448, 452, 458, 486, 492, 500, 502, 514, 520, 522, 532, 534, 544, 548, 550, 552, 554, 562, 580, 588, 608, 614, 616, 618, 626, 628
  age of first 296, 352, 498
  allocation of resources to 314
  at higher altitudes 592
  early in season 180, 298, 472, 588
  extended period of 638
  flexibility in 554
  in grazed sites 410, 472
  low 138, 156, 114, 178, 200, 296, 298, 354, 436, 540, 558
  mast years 102, 296, 498
  moderate 410, 560, 570
  monocarpic spp. 198, 288, 330, 368
  plastic 216, 452, 474, 558
  potential for 402
      high 482, 604, 634
  prolific 268, 288, 316, 352, 360, 384, 420, 432, 460, 472, 556, 572, 582
  quantity of 84, 102, 108, 116, 126, 130, 152, 198, 246, 252, 272, 276, 350, 388, 440
  rapid 176, 234, 420, 554, 572, 580
  sustained at expense of vegetative development 420
  temperature requirement for 116
  timing of 436, 546, 608, 638
  variable 152, 176, 182, 236, 328, 404, 498, 528
  years of maximum 296, 498
retention on plant 400, 472, 456, 586, 592
ripening 164, 488
-set 88, 322, 336
  at high altitudes 102, 154, 446
  early 386, 418, 490, 502, 524, 582, 606
  factors reducing 232, 310
  high 242, 524, 556
  in cold conditions 102
  in shade 366, 450
      reduced 332, 334, 340, 416
  low 172, 186, 222, 300, 334, 338, 402, 446, 464, 504, 538, 632
      in grazed sites 184, 392, 398, 400
  modest 392
  rare 394
  timing of 104, 110, 118, 124, 210, 214, 234, 272, 304, 420, 428, 532, 564, 596
  variable 296
shape 36
size 114, 120, 132, 158, 162, 168, 204, 218, 244, 336, 380, 498
  in cross-pollinated individuals 224, 226
  large 236, 316, 330, 348, 366, 388, 390, 410, 412, 422, 426, 444, 476, 480, 516, 538, 568, 594, 612, 636
  minute 206, 210, 234, 286, 360, 362, 370, 532, 534, 542
  small 6, 430, 448, 628
  variable 146, 184, 224, 246, 472, 494, 528, 556, 604
species overwintering as 366
time of shedding 182, 198, 202, 306, 308, 314, 352, 404, 460, 488, 542, 572, 574, 588, 592
  before hay harvest 516
  flexibility in 470
  synchronisation with leaf fall 498
viability 152, 154, 170, 172, 296, 426, 498, 514, 538, 552, 608, 614
  relationship with altitude 568
weight 50
Seed banks
  absence of 48, 102, 106, 110, 112, 114, 118, 132, 154, 156, 158, 164, 212, 220, 228, 232, 236, 238, 250, 262, 294, 296, 302, 306, 336, 340, 344, 354, 358, 364, 380, 426, 458, 462, 466, 480, 494, 510, 516, 520, 524, 534, 536, 540, 560, 574, 584, 588, 606, 614
  buried 214, 218, 298, 324, 376, 422, 448, 452, 470, 474, 488, 532, 596, 620, 634
  'floating' 570
  insubstantial 202, 242, 266
  persistent 88, 90, 96, 98, 194, 108, 120, 124, Seed banks 126, 128, 130, 136, 140, 168, 170, 176, 178, 180, 182, 184, 192, 196, 194, 206, 208, 216, 222, 224, 322, 322, 326, 328, 338, 324, 360, 362, 370, 372, 374, 276, 378, 386, 400, 402, 410, 412, 414, 416, 418, 420, 424, 428, 430, 432, 434, 438, 472, 478, 456, 500, 506, 512, 514, 526, 528, 530, 540, 544, 546, 548, 558, 566, 572, 576, 582, 580, 590, 592, 600, 602, 604, 610, 612, 616, 618, 626, 628, 634
      variation in 580
  predicted 454, 508, 598
  requirement for maintaining 558
  shortly persistent 314, 318, 334, 388, 398, 500, 548, 550, 554, 564
  suspected 148, 440, 442, 460, 518
  transient 84, 94, 210, 264, 308, 346, 396, 406, 594
  uncertain 150, 152, 170, 186, 188, 226, 258, 288, 290, 316, 342, 384, 392, 436, 446, 462, 464, 476, 486, 490, 492, 438, 552, 568, 630, 632, 636, 638
Seedling
  colonisation, infrequent 422
  damage by frosts, reduction of 434
  dispersal by water 372, 376

Seedling *continued*
  distribution
    relative to adults 612
    relative to canopy species 296
  establishment 82, 84, 102, 122, 132, 138, 154, 290, 348, 352, 368, 396
    adverse effects of livestock 568
    at higher altitudes 160
    conditions required for 576, 624
      high rainfall 296, 466
    conditions unsuitable for 570
    constraints on 314
    dependence on seed reserves in 390
    factors conducive to 324
    in ants nests 640
    on bare ground 376, 502, 536, 608, 620
    in closed perennial communities 348
    in compacted soil 472
    in disturbed sites 292, 376
    in dry, exposed microsites 494
    in established grassland, rare 466
    in heavily grazed sites 152, 232, 266, 502
    in shade 152, 296, 568, 614
    in ungrazed sites 314
    in unshaded sites 160, 224, 266, 340, 416, 534
    under leaf litter, favoured by 498
      inhibition by 162, 356, 432
    on wet mud 570
    poor 500
    prevention by understorey spp. 314
    rapid 230, 608
    sites unsuitable for 314
    slow 398, 528, 620, 622
    soil depth favouring 114
    strategy 204
    timing of 268
    uncommon 426
    under water
      capacity for 570
      inability for 376
  morphology 612
  mortality 130, 152, 158, 246, 294, 296, 330, 468, 538, 544, 614
    due to climatic factors 590
    due to flooding 464
    due to fungal attack 590
    due to shade 152, 612
    in close proximity to parent 356, 612
    in sites with deep litter 356
    risk of 224
  persistence 112, 250, 348, 540
    in darkness 250
    prolonged by low light levels 612
  shade sensitivity 314
  size 298
  susceptibility
    to damping off 314
    to desiccation 314, 342
    survival 246, 330
    of winter conditions 474
  yield 242
  see also Sporeling
Selection
  pressures 600
    promoting inbreeding 452, 598
    promoting outbreeding 448
  r and K 11
Seral species 102, 314
Sex ratio 314, 524, 526
  existence of females only 260
  proportion of male to female individuals 506

534
Shade-adapted species 138
Shade-avoidance strategy 100, 120, 228
Shade
  association with altitude 390, 426, 490, 496, 632, 638
  association with aspect preference 408, 632
  capacity for dominance in 560
  capacity to etiolate in response to, as seedlings 538
  cast by 128, 296, 608
  deep, absence from 156, 230, 490, 518
    disappearance under 188
    less abundant in 162
    low tolerance of 578, 604, 616
    survival of 220, 240, 450
    tolerance of 382, 540, 568
  dry matter production in 450
  ecotypes, response to light 562
  extension into 126, 132
  failure to prosper in 226, 246, 570
  flowering, inhibited by 232, 250, 340, 354, 376, 382, 408, 434, 522, 526, 536, 556, 560, 562, 570, 582, 586, 636
  fruiting inhibited by 232, 522, 536, 560
  fungal attack under 190, 354
  light
    flowering maximised in 632
    species of 252, 256, 272, 334, 354, 384, 444, 556
    tolerance of 532
  margin species 360
  moderate 302, 316, 332, 366, 424
  patchy, species of 334
  persistence in 84, 108, 118, 196, 208, 224, 232, 314, 434, 526, 536, 580, 586, 636
    greater in male plants 526
    inability for 612, 626
    of leaves 310
    of seedlings 296
  plants 44
  population variation and 642
  reduced ability to regenerate 498, 590
  response to 112, 616, 642
  response to release from 334, 586
  restricted tolerance of 352, 376, 458, 576, 582, 640
  restriction of competitors by 178, 180, 186, 220, 262, 274, 282, 312, 320, 330, 334, 388, 422, 428, 444, 556, 576, 578, 632
  restriction to 220, 620
    seed production in 340, 450, 494
  sensitivity to 116, 130, 238, 314, 324, 482, 602
  species achieving prominence in 426
  suppression by 128, 586, 634
  tolerance of 82, 86, 98, 102, 106, 142, 174, 180, 184, 188, 224, 250, 254, 272, 286, 330, 398, 408, 414, 426, 436, 574, 620, 622, 628, 632, 614
    intermediate 322, 332, 366
    moderate 478, 480, 494, 590, 598, 624, 630, 638
    seedlings 296, 568, 614
    variation in 524
  topographic 236
  vigour, reduced by 250, 312, 354, 394, 504, 582
  vulnerability to 198, 206, 466, 378
Shaded sites

absence from 136, 194, 358, 516, 552, 554
  species of 180, 218, 272, 274, 356, 382
Shingle beaches, species of 330
Shoot
  biomass see Biomass
  capacity to diffuse oxygen into roots 190, 570
  density 86, 426, 624
  emergence, characteristics of 356, 404, 426
  expansion 110, 118, 132, 156, 230, 266, 348, 400, 406
    delayed 188, 398, 414
    rapid 160, 178, 310, 390, 630
    reduced by low light intensity 612
    timing of 96, 158, 308, 324, 344, 356, 380, 458, 504, 630
  penetration, of asphalt 300
  phenology see Phenology
Shrubby species 232, 266, 314, 342, 534, 536, 568, 612, 620, 622
Similarity in habitats *see also* Associated species
Skeletal habitats
  species of 140, 240, 272, 330, 404, 554, 568, 642
Slope 37, 56
Sodium
  foliar concentrations of 170, 218, 242, 252, 330, 352, 356, 376, 378, 470, 504, 580, 608, 610
  tolerance of high substrate 148
'Soft tests' 7
Soil type
  acidic
    associated species of 57, 86, 88, 92, 102, 110, 116, 122, 168, 196, 212, 224, 250, 252, 266, 276, 282, 286, 292, 322, 352, 354, 358, 376, 378, 408, 416, 428, 434, 446, 450, 488, 496, 524, 526, 568, 574, 582, 620, 622
    in upland areas 188, 192
    non-association with 186, 314
  alluvial 338, 366, 458
  anaerobic
    failure to persist in 224
    survival of 352
    tolerance of 104, 282, 338, 570
  base-poor 194, 374, 380
  base-rich 162, 206, 242, 262, 302, 308, 330, 336, 354, 370, 376, 422, 426, 434, 468, 480, 490, 536, 606, 630, 642
  brown earths 354, 488, 496, 524
  calcareous
    absence from 114, 166, 274, 276, 320, 336, 634
    associated species of 57, 110, 120, 122, 126, 140, 142, 154, 156, 158, 160, 182, 188, 190, 194, 198, 200, 204, 206, 210, 236, 246, 288, 294, 312, 344, 346, 324, 326, 360, 368, 370, 380, 398, 404, 420, 422, 424, 430, 448, 466, 468, 480, 492, 538, 540, 542, 544, 584, 592, 624, 614
    vigour reduced on 488
  chalk 362, 522
  clay, associated species of 120, 158, 212, 310, 360, 416, 428, 498, 522, 512, 574, 608, 632
  compacted 418
  deep, associated species of 228, 284, 438, 444, 496, 574
  depth, uneven exploitation of 312
  dry, associated species of 160, 162, 196,

198, 200, 256, 286, 288, 294, 328, 362, 364, 368, 402, 438, 440, 448, 466, 476, 538, 542, 588, 596, 526
eroded
   species associated with 430
   species unaffected by 366
exhausted, indicator species 158, 470
fertility
   high, associated species of 98, 102, 104, 106, 112, 114, 128, 138, 146, 164, 166, 172, 174, 180, 214, 216, 220, 230, 238, 264, 270, 274, 314, 316, 318, 338, 348, 352, 366, 372, 384, 386, 388, 406, 458, 472, 454, 482, 456, 512, 528, 534, 536, 554, 560, 576, 580, 588, 602, 604, 616, 618, 634, 636
   low, associated species of 110, 154, 156, 158, 160, 174, 188, 196, 198, 200, 206, 244, 246, 250, 276, 288, 290, 292, 294, 304, 328, 342, 344, 466, 350, 360, 362, 368, 378, 380, 390, 398, 404, 410, 414, 420, 422, 446, 468, 526, 538, 542, 544, 546, 574, 630, 640
intermediate base status 326
leached 196, 326, 572
microsites, exploitation of 120, 124, 126, 130, 542
mull humus 57, 152
non-calcareous strata, less abundant on 190, 278
peaty, species of 192, 266, 292, 320, 358, 370, 414, 500, 526, 620, 622
   pH *see* pH
podzolic, species of 250, 354, 488
poorly drained, species of 248, 512
rocky
   associated species of 94, 124, 126, 312, 362, 542
   restriction of competitors by 210, 526, 550
sandy, species. of 90, 94, 120, 124, 130, 136, 182, 234, 328, 350, 362, 364, 414, 438, 440, 526, 542, 548, 572, 596, 618, 626
shallow, species. of 160, 182, 288, 312, 438, 546
waterlogged, species. of 198, 224, 292, 338, 376
well-drained, species of 138, 168, 182, 256, 262, 286, 314, 342, 356, 362, 384, 422, 444, 446, 502, 536, 548, 568, 590, 600, 346,
see also Edaphic, pH
Soil-less habitats, extension over 340, 630
Speciation 522, 588
Species
   co-existence 65-72
   composition 72
     patchy 578
     scattered 116, 254, 286, 306, 302, 314, 438, 614
     see also Stand-forming
   -richness *see also* Diversity
     conservation 68
     control of 65-72
     exceptionally high 24, oss 2, 14
     low 12, 46
Specific leaf area 7, 32, 43
Spoil heaps

species of 90, 134, 136, 148, 168, 172, 226, 368, 400, 402, 410, 420, 482, 532, 554, 604, 608, 466,
Spore-bank 140, 142, 144, 240, 254, 496
Spore
   dispersal 144, 284
     by ingestion by insects 496
     by wind 140, 240, 496
   germination
     light requirement for 254
     rate of 140, 142, 240
     temperature requirement for 140
   production 140, 142, 144, 240, 254, 256, 496
   viability 280, 496
Sporeling
   establishment 254, 256, 282, 284, 496
   mortality 254
   recruitment 254
Stand-forming species 100, 122, 154, 186, 220, 258, 264, 270, 282, 292, 300, 536, 570, 576, 608, 612, 616, 308, 310, 336, 354, 356, 362, 366, 370, 354, 394, 424, 426, 460
Standing crop 24, 25, 510
Stems, short-lived 518, 522
Stoloniferous species 58, 88, 90, 104, 178, 218, 220, 236, 268, 270, 274, 278, 312, 326, 382, 414, 466, 474, 478, 490, 512, 602, 630
'Stored growth' 45
Strategic range 58, 522
Strategies *see* Plant functional types
Stratification *see* Seed dormancy
Stream-banks, species of 180, 302, 330, 358, 370, 534, 608, 628, 632
Streams, species of 122, 192, 510
Stress 13
Submergence
   adaptation to 506
   tolerance of 628
     in winter 372, 442, 484
Subordinate component species 174, 192, 208, 284, 302, 306, 324, 346, 384, 450, 476, 478, 606, 632, 642
Succession 102, 152, 232, 252, 268, 296, 368, 534, 536, 614
   birch, replacement of degenerating gorse bushes 612
   hawthorn, invasion associated with 340
Succulence, species exhibiting 44, 148, 400, 514, 546, 624
Summer annual species *see* Life-cycle
Sun and shade leaves 43
Support tissue 44
Surveys *see* Vegetation surveys

Tall-herb communities
   absence from 414, 538, 562, 640
   capacity to persist under 334, 476, 594, 628, 636
   species of 170, 270, 310, 318, 292, 426, 504, 576, 616, 624
   suppression by 398, 560, 590, 602
   understorey, species. of 302, 358, 426, 478
Taxonomic study, species in need of 500
Timber management *see* Woodland management
Toxic principles
   alkaloids 284, 452, 460, 546, 548
   cyanogenic glycosides 186, 410, 496, 536, 602

digitalin 252
glycoalkaloids 560
glycosides 356
hypericin 360, 362
oxalate 394, 456, 524, 526
oxalic acid 450
protoanemonin 500, 504, 506, 514
Toxicity 108, 136, 138, 170, 214, 232, 252, 254, 280, 284, 296, 300, 516, 524, 528, 536, 546, 548, 316, 334, 340, 404, 424, 426, 450, 460, 456, 496, 498, 500, 504, 506, 514
   berries 560
   fatal 368, 586
   reduction of animal yield 500
   seeds 558
Trackways, species of 372, 406, 418, 472, 482
Tradeoffs
   a definition 6
   between 'aquisitive' and 'retentive' 10, 12, 63
   between 'fast and leaky' and 'slow and tight' 63, 76
   between resistance and resilience 73
   between root and shoot 9, 11
   in life-history 9
   in resource dynamics 9, 63, 76
   in plant design 5
Traits of plants 6, 7, 10, 11, 34, 63
   effects on communities 63
     conferring resilience 73
     conferring resistence 73
     convergent and divergent 63, 64
   effects on ecosystems 63
     carbon storage 77
     decomposition 64
     nutrient cycling 64, 76
     productivity 63, 64, 76
     trophic interactions 64, 76, 77
'soft' 7
Trampled habitats
   adoption of prostrate form in 472, 456
   inability to exploit 554
   plasticity of growth form in 446
   species of 150, 278, 378, 396, 472, 474
Trampling
   persistence under moderate 342, 418
   resistance to 196, 244, 370, 494
   restriction of competitors by 418, 628
   sensitivity to 106, 118, 146, 148, 164, 208, 424, 452, 496, 590
   strategy for avoidance of 504
   tolerance of 86, 154, 176, 304, 376, 400, 406, 446, 474, 476, 482, 532, 584, 588, 602
Transient habitats, species of 438, 472, 514, 520, 554, 564, 566, 634
Transient constituent species 462, 552, 582
Transpirational water loss 156
Tree species 82, 102, 152, 296, 314, 498, 534, 536, 568, 614
Triangular model 9
Triangular ordination 38, 58
Tuber-forming species 504, 586
Turf, species of 90, 92, 190, 304, 308, 312, 364, 396, 404, 414, 420, 434, 462, 466, 468, 470, 472, 480, 492, 544, 574, 584, 592, 630
   species sown for 304, 308, 406, 476, 478
Tussock-forming species 132, 192, 196, 242, 248, 292, 344, 380, 434, 446
Twining habit 172, 230, 298, 408, 586

SUBJECT INDEX

Uncoupling of resource capture and growth 13
Underground
    parts, majority of biomass 282
    reserves 348
Understorey *see* Canopy
Upland habitats
    carpeting species 218
    extension into 272, 276
    grazing value in 434
    populations 160
    species of 142, 168, 240, 266, 376, 378, 426, 434, 446, 622
Urban dereliction *see* Derelict environment
Urban environments
    species of 136, 216, 268, 300, 550, 552, 564 474

Variation 226, 348, 498, 510
    coinciding with associated species 602
    due to hybrid origin 566
    due to introgression 192
    in size 146, 182, 564
        with geographical distribution 196
    interspecific
    intraspecific variation
Varieties
    description of 206, 510
    native x cultivated 600, 602
Vascular tissue
    development in 484
    lacking vessels 394, 486
    vessels in 258
Vegetation
    density 330
    dynamics 66-73
    see also Population dynamics
    gaps *see* Gaps
    management *see* Land
    structure 62-72
    surveys 17-23
    type, wide-ranging 478
Vegetative
    dispersal
        by human activity 576
        by soil contaminated with rhizomes 300, 608, 616
        by water 86, 122, 144, 282, 284, 300, 310, 320, 336, 338, 374, 382, 424, 442, 458, 460, 464, 454, 486, 570, 610, 610, 628
        effective 576
        of whole plants 394
    expansion 35, 48
    parentage, common 458
    reproduction 158, 166, 188, 192, 364, 500
        by adventitious buds 98, 402, 526
        by adventitious roots 578, 592, 642
        by axillary shoots 362
        by branched stock 398
        by buds in leaf axils 118
        by buds on stock 538
            by bulbils 504, 624
        by corms 570
        by creeping stems 172, 358
        by daughter rosettes 252, 466, 472, 540, 548, 588
        by daughter tubers 138
        by detached fronds 240
        by detached inflorescences 518
        detached leaves 182, 184, 546
        detached roots 172, 222, 230, 266, 410, 470, 522, 526, 528, 548, 588

        effected by disturbance 258, 530, 588, 616
        by detached shoots 122, 166, 170, 180, 184, 208, 218, 236, 260, 274, 282, 334, 336, 354, 360, 362, 372, 374, 382, 386, 410, 424, 442, 460, 464, 472, 454, 486, 494, 508, 518, 546, 576, 578, 580, 582, 628, 634
        effected by disturbance 258, 464, 472, 478, 456, 494, 506, 520, 578, 580, 630, 634
        by detached stems 132, 276, 514
        by detached twigs 534, 536
        by dormant apices 484
        by lateral roots 362
        by layering of old branches 342
        by leaflets 184
        by plant fragments 284, 464, 510
        by plants detached by disturbance 312, 430, 512
        by prostrate rooting shoots 352, 386, 408, 506, 518, 532, 628, 632
        by ramets 512, 524, 576
        by rhizome proliferation 84, 88, 110, 132, 144, 154, 172, 184, 186, 192, 248, 258, 264, 270, 282, 290, 300, 308, 310, 338, 354, 376, 378, 384, 390, 392, 422, 424, 442, 446, 450, 454, 458, 460, 464, 486, 496, 540, 570, 576, 590, 598, 608, 610, 616, 620, 636, 640
        by root tubers 504, 586
        by rooted stems 340, 522, 560, 582, 578, 580, 632
        by runners 592
        by stem buds 470
        by stolons 86, 88, 90, 104, 150, 178, 268, 276, 278, 312, 382, 412, 478, 512, 556, 602, 624, 630, 466
        by suckers 522, 614
        by vegetative branching 96
    effective 186, 276, 484
    exclusive 260
    importance at N edge of range 382, 592
    importance in disturbed habitats 548, 634
    importance in grazed habitats 392, 524
    importance in S. Britain 410
    importance in trampled habitats 524
    lack of effective method 106, 116, 126, 292, 332, 348, 480, 556, 586, 626
    limited capacity for 322, 528
    low capacity for 322, 528
    main method of 266, 300, 334, 338, 354, 412, 426, 460, 454, 510, 512, 560, 574, 576, 578
        in shaded habitats 408
        in stable communities 492, 496, 608
    of minor importance 356, 472
    slow 624, 640
    strategies of 512
    vigorous 284, 358
spread
    aggressive 172, 222, 230, 338
    capacity for 170, 220, 274, 304, 494
        lacking 372, 436, 444, 538, 544
        slight 288, 462
    confined 254
    extensive 340
    'fairy ring' 202
    ineffective 406
    lateral 132, 136, 140, 142, 156, 196, 208,

        210, 236, 242, 248, 290, 320, 336, 350, 434, 608
    limited ability for 360, 362, 396, 400, 402, 448, 470, 502, 636, 638
        in dry habitats 398
        little capacity for 368, 624.
    lack of ability for 240, 268, 346, 352, 462
    by movement of rhizome fragments 392
    rapid 334, 454

    resulting from human action 264
    by rooting at nodes 370
    tillering capacity 88, 116, 132
    unimportant 202
    state, persistence in 198, 208
Vernal species *see* Phenology
Vernalisation
    requirement for flowering 124, 130, 114, 200, 440, 542, 626
    requirement for germination 286
Viviparous species 374

Walls, species of 140, 142, 144, 180, 240, 256, 272, 340, 350, 436, 542, 550, 626
Wasteland habitats
    species of 90, 94, 128, 132, 154, 164, 182, 208, 222, 234, 268, 274, 328, 348, 360, 368, 438, 452, 476, 482, 482, 516, 534, 548, 612, 626
Water
    characteristics
        base poor 394, 442, 508
        base rich 486
        brackish 484, 508
            intolerance of 610
        calcareous 260, 394, 484, 628
        eutrophic 216, 260, 484
            capacity to exploit 570, 610
            sensitivity to, seedlings 570
        mesotrophic 260
        peaty 260, 284, 508, 610
        species poorly represented in 484
        stagnant 486
        turbid, failure to establish in 570
    currents
        morphological adaptation to 510
        physiological adaptation to 510
        restriction of competitors by 186, 258, 628
    depth
        light penetration at varying 486
    sites subject to seasonal levels of 374
    exploitation of subsoil 126
    flowing, species of 460, 464, 510, 628
        absence from 570
    level, fluctuations in
        restriction of competitors by 508
        sensitivity to 186, 282
        tolerance of 454, 508
            facilitated by heterophylly 508
            limited 486
    -margin communities 186, 258, 370
    pollution *see* Eutrophication, Pollution
    running, close proximity to 374
    shallow, species of 290, 570, 610
    still 394, 508
    subsoil, access to 202
    substrate, characteristics of 510, 570
Water-conduction *see* Xylem
Waterlogged sites
    absence from 136, 310, 332, 490, 552, 574

inability to exploit 358
inability to persist in 626
Waterlogging
    absence due to prolonged 472
    intolerance of 226, 266, 422, 432, 620, 622
    loss of seed viability due to 296, 498
    sensitivity to 100, 106, 118, 212, 352
    specialisations associated with 270
    survival of intermittent 462
    susceptibility to 238, 356, 426, 556, 602
    tolerance of 84, 102, 418
Water-table
    colonisation of sites above 376
    deep, exploitation of 132
    fluctuating
        characteristic species of 96, 102, 192, 374, 424, 434
        intolerance of 258, 274, 424
        tolerance of 166, 336, 376, 442
        tussock growth form in 434
    lack of root penetration into 314
    rooting below level of 570
    surface 110
Wave action, vulnerability to 282, 486, 570
Weed
    killers see Herbicides
    species 150, 208, 222, 226, 234, 264, 300, 328, 362, 364, 414, 438, 474, 478, 454, 494, 532, 548, 588, 596, 634
        biological control of 548
        control by farming practices 222, 226, 634
        highly invasive 264, 300, 496
        of forestry 152, 168, 496
        of tree nurseries 124, 182, 354, 526
        see also Agricultural weeds
    status 82, 56, 90, 108, 124, 176, 214, 222, 226, 248, 258, 264, 280, 298, 318, 464, 470, 472, 528, 530, 548, 558, 566, 572, 580
Weeding
    restriction of competitors by 268
    vulnerability to the effects of 316, 388

'Weedy' behaviour 254
Wetland habitats 86, 104
    absence from 482, 550
    colonists of 68, 104
    drainage see Drainage
    persistence in 314
    sites transitional with dryland 310, 318, 336, 412, 434, 514, 520
    species of 166, 170, 186, 258, 274, 276, 278, 284, 310, 320, 354, 358, 370, 372, 374, 376, 412, 464, 478, 454, 506, 514, 560, 582, 588, 610, 624
    survival in 190, 474
Wind
    dispersal
        seeds see Seed dispersal
        spores see Spore dispersal
    inability to tolerate strong 296
Winter-annual see Life-cycle
Winter damage, sensitivity to 352
Winter-dormancy 306
Winter- green species see Phenology
Wood, characteristics of 152, 534, 536, 568, 614
Woodland
    ancient 152, 314, 356, 390, 416, 422, 426, 428, 490, 498, 578, 614, 632
    bare soil, sporelings of 254
    broad-leaved, species of 82, 100, 102, 152, 162, 248, 296, 314, 328, 354, 382, 408, 416, 536, 540, 614, 520, 632
    canopy see Canopy
    clearance, species abundant after 212, 314, 416
    clearings, species of 220, 252, 302, 316, 334, 388, 416, 536
    composition, tree species 614
    destruction of habitats 256, 540, 632
    disturbed, species of 416, 432, 536
    floor, dominant species of 254, 256, 356, 426
    forestry, 'weeds' of 152, 168, 496
    grass species of 156, 354, 422, 428, 478
    habitats, species of 98, 100, 110, 118, 138, 144, 156, 220, 232, 272, 332, 340, 354, 356, 382, 450, 496, 568, 620,
    herb-layer
        coppiced 156, 332, 354
        effects of tree species on 82, 296
    limestone
        absence from 356
        species of 232, 256, 314, 340, 614
    lowland 162, 254, 302, 356
    management 82, 138, 162
        changes in 296
        favoured by modern methods 302, 316, 332, 432, 522, 576
        species less vulnerable to 356
    margins, species of 236, 262, 302, 312, 360, 390, 534, 556, 562, 574, 584, 586, 594, 598, 638
    mixed 498, 614
    old established 110, 450
        indicator species of 382
    open 262, 310, 322, 354, 366, 426, 590, 630, 640
    plantations
        coniferous, exploitation of 450, 622
        species comprising 82, 102, 296
        species of 138, 220, 252, 254, 256, 356, 416, 536, 632
    relict species 156, 228
    rides, species of 156, 268, 332, 334, 424, 490, 494, 576, 584, 612, 632
    secondary 98, 162, 220, 232, 262, 302, 314, 330, 332, 334, 340, 384, 408, 432, 534, 540, 576
    species exclusive to 432
    thicket-forming species of 612
    timber species of 296, 314, 498, 614
    undershrub 408, 522
    upland, species of 256, 356, 428

Xeromorphic species 284, 380, 592
Xerophytic species 304
Xylem system 96, 100, 356, 366, 496, 498

Species Index of Generic Binomial and Vernacular names

A bold page number refers to the Species Account of which the species is a subject. Entries in the Tables of Attributes are also highlighted in bold italics.
In the case of a name occurring within the text of a Species Account of some other subject species, only the left-hand page of the Account is indexed. However, the name may also (or only) occur within material which over-runs onto a right-hand page.

Acer
   campestre *662, 675, 688*
   platinoides *662, 675, 688*
   pseudoplatanus 49, 50, 55, 58, **82**, 252, 254, 286, 354, 356, *662, 675, 688*
Achillea
   millefolium 24, 42, **84**, 476, *650, 664, 677*
   ptarmica *650, 664, 677*
Acinos arvensis, see *Clinopodium acinos*
Acorus calamus *650, 664, 677*
Adder's-tongue, see *Ophioglossum vulgatum*
Adoxa moschatellina *650, 664, 677*
Aegopodium podagraria *650, 664, 677*
Aesculus hippocastanum *662, 675, 688*
Aethusa cynapium *650, 664, 677*
Agrimony, see *Agrimonia eupatoria*
   Fragrant, see *A. procera*
Agrimonia
   eupatoria *650, 664, 677*
   procera *650, 664, 677*
Agropyron
   caninum, see *Elymus caninus*
   donianum, see *E. caninus*
   repens, see *Elytrigia repens*
Agrostis
   canina s.l.
      ssp. *canina* see *canina s.s.*
      ssp. *montana* see *vinealis*
   canina s.s. **86**, 92, 224, *650, 664, 677*
   capillaris 10, 24, 29, 46, 50, 57, 58, 61, 66, **88**, 90, 92, 196, 208, 244, 308, 342, 352, 364, 380, 390, 414, 434, 446, 526, *650, 664, 677*
   castellana 88
   gigantea 70, 90, *650, 664, 677*
   spp. 51, 86, 88, 92, 414
   stolonifera 46, 54, 55, 61, 88, **90**, 92, 104, *650, 664, 677*
      var. *palustris* 90
      var. *stolonifera* 90
   tenuis, see *capillaris*
   vinealis 86, **92**, *650, 664, 677*
Aira
   caryophyllea *650, 664, 677*
   praecox 41, 49, 51, 55, 88, **94**, 168, 286, 364, 526, *650, 664, 677*
Ajuga reptans 312, 390, *650, 664, 677*
Alchemilla vulgaris agg. *650, 664, 677*
Alder, see *Alnus glutinosa*
Alisma
   plantago-aquatica **96**, 338, 454, *650, 664, 677*
Alkanet, Green, see *Pentaglottis sempervirens*
Alliaria petiolata 46, **98**, 118, 172, 316, 366, 444, *650, 664, 677*

Allium
   cepa 33
   ursinum 45, 50, **100**, 110, 248, 302, 314, 356, 504, 556, *650, 664, 677*
   vineale *650, 664, 677*
Alnus
   glutinosa 55, **102**, 152, 458, 478, 534, *662, 675, 688*
   incana 102
Alopecurus
   aequalis 104
   myosuroides *650, 664, 677*
   geniculatus **104**, 106, 500, 520, *650, 664, 677*
   pratensis **106**, *650, 664, 677*
Amsinckia micrantha *650, 664, 677*
Anacamptis
   morio, see *Orchis morio*
   pyramidalis *650, 664, 677*
Anagallis
   arvensis **108**, 120, 298, 438, 572, 618, *650, 664, 677*
      ssp. *arvensis* 108
      ssp. *foemina* 108
   tenella *650, 664, 677*
Anchusa arvensis *650, 664, 677*
Anemone nemorosa 24, 48, 57, 62, 100, **110**, 156, 228, 232, 248, 314, 382, 426, 540, *650, 664, 677*
   Wood, see *Anemone nemorosa*
Angelica sylvestris 57, **112**, 366, *650, 664, 677*
   Wild, see *Angelica sylvestris*
Anisantha sterilis 10, 49, 62, **114**, 536, *650, 664, 677*
Antennaria dioica *650, 664, 677*
Anthemis
   arvensis *650, 664, 677*
   cotula *650, 664, 677*
Anthoxanthum odoratum 10, 24, 49, 51, 53, 57, 61, 66, 106, **116**, 194, 414, 502, 524, *650, 664, 677*
Anthriscus
   caucalis *650, 664, 677*
   sylvestris 10, 43, 48, 50, 58, 59, 98, 114, **118**, 318, 332, 348, 576, *650, 664, 677*
      var. *angustisecta* 118
      var. *latisecta* 118
Anthyllis vulneraria *650, 664, 677*
Antirrhinum majus *650, 664, 677*
Apera interrupta *650, 664, 677*
Aphanes
   arvensis **120**, 626, *650, 664, 677*
   australis 94, 120, *650, 664, 677*
   inexspectata, see *australis*

   microcarpa, see *australis*
Apion ulicis 612
Apium
   inundatum 122, *650, 664, 677*
   nodiflorum 57, 60, **122**, 166, 278, 284, 336, 370, 518, 628, *650, 664, 677*
   repens 122
Apple, Crab, see *Malus sylvestris*
Apple-mint, see *Mentha x villosa*
Aquilegia vulgaris *650, 664, 677*
Arabidopsis thaliana 10, 46, 52, 57, 58, **124**, 130, 182, 328, 440, 542, 546, 594, 596, 626, *650, 664, 677*
Arabis
   brownii 126,
   caucasica *650, 664, 677*
   Garden, see *Arabis caucasica*
   hirsuta 53, **126**, 544, 546, *651, 664, 677* *Arctium*
   lappa **128**, *651, 664, 677*
   minus agg. 50, 60, **128**, 384, 649, *651, 664, 677*
      ssp. *minus* 128
      ssp. *nemorosum* 128
      ssp. *pubens* 128
   minus s.s., see *minus* ssp. *minus*
   nemorosum, see *minus* ssp. *nemorosum*
   pubens, see *minus* ssp. *pubens*
Arctostaphylos uva-ursi *651, 664, 677*
Arenaria
   leptoclados, see ssp. *leptoclados* 130
   serpyllifolia 6, 46, 49, 57, 58, **130**, 182, 420, 440, 542, 546, 596, 626, *651, 664, 677*
      ssp. *leptoclados* 130
      var. *macrocarpa* 130
      ssp. *serpyllifolia* 130
Armoracia rusticana *651, 664, 677*
Arrhenatherum elatius 10, 24, 47, 53, 58, 61, 71, 118, **132**, 230, 330, 392, 462, 476, 598, 606, *651, 664, 677*
   ssp. *bulbosum* 132
   ssp. *elatius* 132
Arrowgrass, Marsh, see *Triglochin palustre*
Arrowhead, see *Sagittaria sagittifolia*
Artemisia
   absinthium 56, 57, **134**, 550, 566, *651, 664, 677*
   verlotiorum *651, 664, 677*
   vulgaris **136**, 148, 212, 222, 268, 402, 528, 530, 550, 552, 564, 604, 608, *651, 664, 677*
Arum
   italicum 138

SPECIES INDEX   735

Arum continued
　　maculatum **138**, 156, 232, 332, 382, 408, 426, 432, 540, 586, **651, 664, 677**
Ash, see *Fraxinus excelsior*
　　Mountain, see *Sorbus aucuparia*
Aspen, see *Populus tremula*
Asphodel, Bog, see *Narthecium ossifragum*
Asplenium
　　adiantum-nigrum **663, 676, 689**
　　ruta-muraria 41, 49, 57, 58, 62, **140**, 142, 144, 242, 256, 272, 436, **663, 676, 689**
　　trichomanes 140, **142**, 144, 204, 240, 272, 436, **663, 676, 689**
　　　ssp. *pachyrachis* 142
　　　ssp. *quadrivalens* 142
　　　ssp. *trichomanes* 142
Aster
　　lanceolatus **651, 664, 677**
　　novi-belgii (agg.) **651, 664, 677**
Athyrium filix-femina 140, 142, **144**, 240, 436, 646, **663, 676, 689**
Atriplex
　　hastata, see *prostrata*
　　laciniata 148
　　littoralis 146, **651, 664, 677**
　　patula 51, **146**, 148, 214, 216, 474, 554, 566, 580, 649, **651, 664, 677**
　　prostrata 134, 136, 146, **148**, 216, 222, 352, 402, 528, 608, **651, 664, 677**
Avena fatua **651, 665, 678**
Avens,
　　Water, see *Geum rivale*
　　Wood, see *G. urbanum*
Avenula
　　pratensis, see *Helictotrichon pratense*
　　pubescens, see *H. pubescens*

Ballota nigra **651, 665, 678**
Balsam, Small, see *Impatiens parviflora*
　　Himalayan, see *I. glandulifera*
Barbarea vulgaris **651, 665, 678**
Bassia scoparia **651, 665, 678**
Barley,
　　Foxtail, see *Hordeum jubatum*
　　Meadow, see *H. secalinum*
　　Wall, see *H. murinum*
Bartsia, Red, see *Odontites vernus*
Basil, Wild, see *Clinopodium vulgare*
Bearberry, see *Arctostaphylos uva-ursi*
Bedstraw,
　　Fen, see *Galium uliginosum*
　　Heath, see *G. saxatile*
　　Hedge, see *G. mollugo*
　　Lady's, see *G. verum*
　　Marsh, see *G. palustre* s.l.
　　Sterner's, see *G. sterneri*
　　Yellow, see *G. verum*
Beech, see *Fagus sylvatica*
Bellbine, see *Calystegia sepium*
Bellflower,
　　Creeping, see *Campanula rapunculoides*
　　Giant, see *C. latifolia*
　　Ivy-leaved, see *Wahlenbergia hederacea*
　　Nettle-leaved, see *Campanula trachelium*
Bell-heather, see *Erica cinerea*
Bellis perennis 52, **150**, 208, 238, 406, 462, 494, 502, 602, **651, 665, 678**
Bent,
Black, see *Agrostis gigantea*
　　Creeping, see *A. stolonifera*

Bent-grass,
　　Brown, see *Agrostis canina* & *A. vinealis*
　　Common, see *A. capillaris*
Berula erecta **651, 665, 678**
Betonica officinalis, see *Stachys officinalis*
Betony, see *Stachys officinalis*
Betula
　　agg. 49, 54, 58, 82, 102, **152**, 232, 314, 498, 568, 612, 614, 645, **662, 675, 688**
　　nana 152
　　pendula 31, 152, 645, **662, 675, 688**
　　pubescens 31, 102, 152, 645, **662, 675, 688**
Bidens
　　cernua **651, 665, 678**
　　tripartita **651, 665, 678**
Bilberry, see *Vaccinium myrtillus*
Bilderdykia convolvulus see *Fallopia covolvulus*
Bindweed, see *Convolvulus arvensis*
　　Black, see *Fallopia convolvulus*
　　Hairy, see *Calystegia pulchra*
　　Large, see *C. silvatica*
　　Larger, see *C. sepium*
Birch, see *Betula*. agg
　　Downy, see *B. pubescens*
　　Silver, see *B. pendula*
Bird's-foot, see *Ornithopus perpusillus*
Birdsfoot-trefoil, see *Lotus corniculatus*
　　Large, see *L. pedunculatus*
Bistort,
　　Amphibious, see *Persicaria amphibia*
　　Common, see *P. bistorta*
Bitter-cress,
　　Large, see *Cardamine amara*
　　Hairy, see *Cardamine hirsuta*
　　Wood, see *Cardamine flexuosa*
Bittersweet, see *Solanum dulcamara*
Blackberry, see *Rubus fruticosus* s.l.
Black-grass, see *Alopecurus myosuroides*
Black Hay, see *Medicago lupulina*
Blackstonia perfoliata 206, **651, 665, 678**
Blackthorn, see *Prunus spinosa*
Bladder-fern, Brittle, see *Cystopteris fragilis*
Bladder-sedge, see *Carex vesicaria*
Blechnum spicant **663, 676, 689**
Blinks, see *Montia fontana*
Bluebell, see *Hyacinthoides non-scripta*
　　(Scotland), see *Campanula rotundifolia*
　　Spanish, see *Hyacinthoides hispanica*
Blueberry, see *Vaccinium myrtillus*
Blue-sow-thistle, Common, see *Cicerbita macrophylla*
Bog Cotton, see *Eriophorum angustifolium*
Bogbean, see *Menyanthes trifoliata*
Bog-rush, Black, see *Schoenus nigricans*
Bourtree, see *Sambucus nigra*
Brachypodium
　　pinnatum 10, 24, 50, 55, 58, 59, 60, 66, 110, **154**, 156, 160, 202, 204, 246, 362, 476, 598, **651, 665, 678**
　　　ssp. *rupestre* 154
　　rupestre, see *B. pinnatum*
　　sylvaticum 52, 53, 61, 110, 138, **156**, 228, 314, 422, 426, 432, 540, **651, 665, 678**
Brachythecium rutabulum 616
Bracken, see *Pteridium aquilinum*
Bramble, see *Rubus fruticosus* agg.
　　Stone, see *Rubus saxatilis*
Brassica
　　oleracea 600
　　rapa 600, **651, 665, 678**
　　spp. 388

Briza
　　media 10, 24, 55, 57, 116, **158**, 188, 194, 244, 326, 346, 410, 414, 468, 492, **651, 665, 678**
Brome,
　　Barren, see *Anisantha sterilis*
　　California, see *Ceratochloa carinata*
　　Hairy, see *Bromus ramosa*
　　Hungarian, see *Bromopsis inermis*
　　Soft, see *Bromus hordeaceus*
　　Tall, see *Festuca gigantea*
　　Upright, see *Bromopsis erecta*
　　Wood, see *B. ramosus*
Bromopsis
　　Erecta 6, 10, 24, 29, 42, 45, 51, 60, 66, 71, 154, **160**, 202, 362, 390, 598, **651, 665, 678**
　　inermis **651, 665, 678**
　　ramosus **162**, 262, 302, 332, 340, 614, **651, 665, 678**
Bromus
　　carinatus, see *Ceratochloa carinata*
　　erectus, see *Bromopsis erecta*
　　inermis, see *Bromopsis inermis*
　　hordeaceus 42, 48, **164**, 306, 318, 600, **651, 665, 678**
　　　ssp. *divaricartus* 164
　　　ssp. *ferronii* 164
　　　ssp. *hordaceus* var. *leiostachys* 164
　　　ssp. *thominei* 164
　　hordeaceus x *lepidus*, see x *pseudothominei*
　　lepidus 164, **651, 665, 678**
　　mollis, see *Bromus hordeaceus*
　　ramosus, see *Bromopsis ramosa*
　　sterilis, see *Anisantha sterilis*
　　x *pseudothominei* 164
Brooklime, see *Veronica beccabunga*
Brookweed, see *Samolus valerandi*
Broom, see *Cytisus scoparius*
Bryonia dioica **651, 665, 678**
Bryony,
　　Black, see *Tamus communis*
　　White, see *Bryonia dioica*
Buckthorn, see *Rhamnus cathartica*
Buckler-fern,
　　Broad, see *Dryopteris dilatata*
　　Narrow, see *D. carthusiana*
Buddleja davidii **662, 675, 688**
Bugle, see *Ajuga reptans*
Bugloss, see *Anchusa arvensis*
Bulrush, Lesser, see *Typha angustifolia*
Burdock, see *Arctium minus* agg.
　　Greater, see *A. lappa*
Bur-marigold,
　　Nodding, see *Bidens cernua*
　　Trifid, see *B. tripartita*
Burnet,
　　Fodder, see *Sanguisorba minor* ssp. *muricata*
　　Great, see *S. officinalis*
　　Salad, see *S. minor* ssp. *minor*
Burnet-saxifrage, Greater, see *Pimpinella major*
Bur-reed, see *Sparganium erectum*
　　Unbranched, see *S. emersum*
Butomus umbellatus **651, 665, 678**
Butcher's-broom, see *Ruscus aculeatus*
Butterbur, see *Petasites hybridus*
Buttercup,
　　Bulbous, see *Ranunculus bulbosus*
　　Creeping, see *R. repens*
　　Goldilocks, see *R. auricomus*
　　Meadow, see *R. acris*
Butterfly-bush, see *Buddleja davidii*

Butterfly-orchid, Greater, see *Platanthera chlorantha*
Butterwort, Common, see *Pinguicula vulgaris*

Cabbage,
    Bargeman's, see *Brassica rapa*
    Bastard, see *Rapistrum rugosum*
Calamagrostis
    canescens **651, 665, 678**
    epigejos **651, 665, 678**
Callitriche
    hamulata 166, **651, 665, 678**
    hermaphroditica
    intermedia, see *hamulata*
    obtusangula **651, 665, 678**
    platycarpa 166, **651, 665, 678**
    spp. 648
    stagnalis 57, 60, 122, **166**, 284, 336, 570, **651, 665, 678**
Calluna vulgaris 49, 50, 56, 57, 58, 69, 94, 110, **168**, 196, 250, 266, 286, 292, 364, 620, 622, **651, 665, 678**
Caltha palustris 170, 180, 218, 310, 322, 560, **651, 665, 678**
    var. *radicans* 170
Calystegia
    pulchra 172, **651, 665, 678**
    sepium s.l. 98, 128, 172, 230, 616,
    sepium s.s. **172, 651, 665, 678**
        ssp. *pulchra*, see *pulchra*
        ssp. *roseata* 172
        ssp. *sepium*, see *sepium s.s*
        ssp. *silvatica*, see *silvatica*
    silvatica 172, **651, 665, 678**
    sylvatica see *silvatica*
Campanula
    latifolia **651, 665, 678**
    rapunculoides **651, 665, 678**
    rotundifolia 10, 24, 55, 58, 61, 62, 71, **174**, 304, 324, 380, 404, 410, 466, 538, 544, 592, **651, 665, 678**
    trachelium **651, 665, 678**
Campion,
    Bladder, see *Silene vulgaris*
    Red, see *S. dioica*
    White, see *S. latifolia*
Canary-grass, Reed, see *Phalaris arundinacea*
Capsella
    bursa-pastoris 35, 50, 146, **176**, 214, 386, 418, 474, 482, 554, 580, 604, **651, 665, 678**
    rubella 176
Cardamine
    amara 170, **178**, 180, 320, 442, 462, 582, **651, 665, 678**
    flexuosa 41, 52, 57, 58, 102, 112, **180**, 182, 184, 218, 310, 478, **651, 665, 678**
    hirsuta 52, 124, **182**, 200, 206, 234, 288, 562, **651, 665, 678**
    impatiens 180
    palustris 184
    pratensis 61, 86, 180, **184**, 224, 320, 376, 442, 512, **651, 665, 678**
Cardaria draba, see *Lepidium draba*
Carduus
    acanthoides, see *crispus*
    crispus **651, 665, 678**
    nutans 200, **651, 665, 678**
Carex
    acuta **651, 665, 678**
    acutiformis 58, 62, 178, **186**, 270, **651, 665, 678**
    arenaria 264, **651, 665, 678**
    binervis **652, 665, 678**
    caryophyllea 24, 48, 55, 116, 158, **188**, 242, 326, 344, 380, 414, 468, 480, 492, 538, 584, 640, **652, 665, 678**
    curta **652, 665, 678**
    demissa, see *viridula* ssp. *oedocarpa*
    dioica **652, 665, 678**
    distans **652, 665, 678**
    disticha **652, 665, 678**
    echinata 194, **652, 665, 678**
    elata **652, 665, 678**
    flacca 10, 24, 29, 52, 57, 61, 132, 158, **190**, 194, 324, 342, 590, 642, **652, 665, 678**
    hirta **652, 665, 678**
    hostiana 194, **652, 665, 678**
    laevigata **652, 665, 678**
    lepidocarpa, see *viridula* ssp. *brachyrrhyncha*
    muricata **652, 665, 678**
    nigra 2, 86, **192, 652, 665, 678**
    otrubae **652, 665, 678**
    ovalis **652, 665, 678**
    pallescens **652, 665, 678**
    panicea 47, 57, 184, 190, 192, **194**, 584, 646, **652, 665, 678**
    paniculata **652, 665, 678**
    pendula **652, 666, 678**
    pilulifera 52, 57, 69, 94, 168, **196**, 266, 286, 378, 416, 446, 620, 622, **652, 666, 679**
    pseudocyperus **652, 666, 679**
    pulicaris 69, 170, **652, 666, 679**
    remota **652, 666, 679**
    riparia 186, **652, 666, 679**
    rostrata **652, 666, 679**
    spicata **652, 666, 679**
    sylvatica 540, **652, 666, 679**
    vesicaria **652, 666, 679**
    viridula
        ssp. *brachyrrhyncha* **652, 666, 679**
        ssp. *oedocarpa* 194, **652, 666, 679**
Carlina vulgaris 23, 35, 57, **198**, 294, 312, 430, 645, **652, 666, 679**
Carnation-grass, see *Carex flacca*
Carpinus betulus **662, 675, 688**
Carrot, Wild, see *Daucus carota* ssp. *carota*
Castanea sativa **662, 675, 688**
Catabrosa aquatica 474, **652, 666, 679**
Catapodium
    marina 200
    rigidum 10, 35, 51, 59, 182, **200**, 234, 288, 440, 564, 594, **652, 666, 679**
        ssp. *majus* 200
Catchfly, Night-flowering, see *Silene noctiflora*
Cat's Ear, see *Hypochaeris radicata*
Cat's-tail (Great Reedmace), see *Typha latfolia*
Cat's-tail (Timothy), see *Phleum pratense*
    Smaller, see *P. bertolonii*
Caucasian-stonecrop, see *Sedum spurium*
Celandine,
    Greater, see *Chelidonium majus*
    Lesser, see *Ranunculus ficaria*
Centaurea
    debeauxii
        ssp. *nemoralis*, see *nigra s.l.* ssp. *nemoralis*
        ssp. *thullieri* 202
    montana **652, 666, 679**
    nigra s.l. 24, 42, 154, 160, **202**, 204, 206, 362, 598, **652, 666, 679**
        ssp. *nigra* 176
        ssp. *nemoralis* 176
    scabiosa 10, 24, 41, 50, 57, 202, **204**, 246, 368, **652, 666, 679**
Centaurium
    erythraea 42, **206**, 362, 368, 562, **652, 666, 679**
    spp. 206
Centaury, Common, see *Centaurium erythraea*
Centranthus ruber **652, 666, 679**
Cerastium
    arvense 208,
    fontanum 10, 24, 57, 62, 71, 84, 150, **208**, 470, 494, 500, 524, 600, 602, **652, 666, 679**
    glomeratum **652, 666, 679**
    holosteoides, see *fontanum*
    semidecandrum **652, 666, 679**
    spp. 208
    tomentosum **652, 666, 679**
    vulgatum see *fontanum*
Ceratocapnos claviculata 432, **652, 666, 679**
Ceratochloa carinata **652, 666, 679**
Ceratophyllum demersum **652, 666, 679**
Ceterach officinarum **663, 676, 689**
Chaenorhinum minus 54, 58, **210**, 246, 268, 402, 532, 646, **652, 666, 679**
Chaerophyllum
    temulentum, see following entry
    temulum **652, 666, 679**
Chamaenerion angustifolium, see following entry
Chamerion angustifolium 10, 24, 35, 45, 46, 49, 50, 54, **212**, 220, 270, 402, 550, 608, **652, 666, 679**
Chamomilla
    recutita, see *Matricaria recutita*
    suaveolens, see *M. discoidea*
Chamomile,
    Corn, see *Anthemis arvensis*
    Stinking, see *A. cotula*
Charlock, see *Sinapis arvensis*
Cheiranthus cheiri, see *Erysimum cheiri*
Chelidonium majus **652, 666, 679**
Chenopodium
    album 10, 42, 46, 50, 51, 146, 176, **214**, 216, 264, 316, 554, 580, 618, 649, **652, 666, 679**
    bonus-henricus **652, 666, 679**
    polyspermum **652, 666, 679**
    rubrum 52, 214, **216**, **652, 666, 679**
Cherry, Bird, see *Prunus padus*
    Wild, see *P. avium*
Chervil, Rough, see *Chaerophyllum temulum*
Chestnut, Sweet, see *Castanea sativa*
Chickweed, see *Stellaria media*
    Greater, see *S. neglecta*
    Lesser, see *S. pallida*
    Water, see *Myosoton aquaticum*
Chicory, see *Cichorium intybus*
Chrysanthemum
    leucanthemum, see *Leucanthemum vulgare*
    segetum **652, 666, 679**
Chrysosplenium
    alternifolium 218, **652, 666, 679**
    oppositifolium 170, 180, **218**, 310, 632, **652, 666, 679**
Cicerbita macrophylla **652, 666, 679**
Cichorium intybus **652, 666, 679**
Cinquefoil,
    Creeping, see *Potentilla reptans*
    Marsh, see *P. palustris*
    Sulphur, see *P. recta*

Circaea
 alpina 220
 alpina x *lutetiana*, see x *intermedia*
 lutetiana 220, 302, 504, 556, 632, **653, 666, 679**
 x *intermedia* 220
Cirsium
 acaule 50, 53, **653, 666, 679**
 arvense 51, 136, 148, **222**, 226, 352, **653, 666, 679**
 dissectum **653, 666, 679**
 helenioides, see *heterophyllum*
 heterophyllum **653, 666, 679**
 palustre 57, 86, 222, **224**, 226, 276, 332, **653, 666, 679**
  spp. 224, 226
 vulgare 55, 222, **226**, 648, **653, 666, 679**
Cladium mariscus **653, 666, 679**
Claviceps purpurea 460
Cleavers, see *Galium aparine*
Clematis vitalba 408
Claytonia
 perfoliata **653, 666, 679**
 sibirica **653, 666, 679**
Clematis vitalba 662, 675, 688
Clinopodium
 acinos **653, 666, 679**
 vulgare **653, 666, 679**
Cloudberry, see *Rubus chamaemorus*
Clover
 Alsike, see *Trifolium hybridum*
 Dutch, see *T. repens*
 Hare's-foot, see *T. arvense*
 Knotted, see *T. striatum*
 Red, see *T. pratense*
 Strawberry, see *T. fragiferum*
 White, see *T. repens*
 Zig-zag, see *T. medium*
Clubmoss,
 Fir, see *Huperzia selago*
 Stag's-horn, see *Lycopodium clavatum*
Club-rush,
 Bristle, see *Isolepis setacea*
 Common, see *Schoenoplectus lacustris*
  Floating, see *Eleogiton fluitans*
  Grey, see *Schoenoplectus tabernaemontani*
  Wood, see *Scirpus sylvaticus*
Cochlearia danica **653, 666, 679**
Cock's-foot, see *Dactylis glomerata*
Coeloglossum viride **653, 666, 679**
Coltsfoot, see *Tussilago farfara*
Columbine, see *Aquilegia vulgaris*
Comfrey,
 Common, see *Symphytum officinale*
 Russian, see *S.* x *uplandicum*
 White, see *S. orientale*
Conium maculatum **653, 666, 679**
Conopodium majus **228**, 504, **653, 666, 679**
Convolvulus arvensis 51, 114, **230**, 264, **653, 666, 679**
  var. *linearifolius* 230
Conyza canadensis 288, **653, 666, 679**
Cornbine, see *Convolvulus arvensis*
Cornflower, Perennial, see *Centaurea montana*
Cornsalad, Common, see *Valerianella locusta*
Cornus sanguinea 662, 675, 688
Coronilla varia, see *Securigera varia*
Coronopus
 didymus **653, 666, 679**
 squamatus **653, 666, 679**
Corydalis
 claviculata, see *Ceratocapnos claviculata*

Climbing, see *Ceratocapnos claviculata*
 lutea, see *Pseudofumaria lutea*
 Yellow, see *Pseudofumaria lutea*
Corylus avellana 662, 675, 688
Cotoneaster
 horizontalis **653, 666, 679**
 Wall, see *Cotoneaster horizontalis*
Cotton-grass, see *Eriophorum vaginatum*
 Common, see *E. angustifolium*
Cotula coronopifolia
Couch,
 Bearded, see *Elymus caninus*
 Onion, see *Arrhenatherum elatius* ssp. *bulbosum*
Couch-grass, see *Elytrigia repens*
Cowberry, see *Vaccinium vitis-idaea*
Cowslip, see *Primula veris*
Cow-wheat, Common, see *Melampyrum pratense*
Cranberry, see *Vaccinium oxycoccos*
Crane's-bill,
 Cut-leaved, see *Geranium dissectum*
 Dove's-foot, see *G. molle*
 French, see *G. endressii*
 Hedgerow, see *G. pyrenaicum*
 Long-stalked, see *G. columbinum*
 Meadow, see *G. pratense*
 Purple, see *G.* x *magnificum*
 Shining, see *G. lucidum*
 Small-flowered, see *G. pusillum*
Crassula helmsii **653, 666, 679**
Crataegus
 laevigata 232
 monogyna 152, **232**, 314, 534, 536, **662, 675, 688**
 oxyacanthoides, see *laevigata*
Creeping-Jenny, see *Lysimachia nummularia*
Crepis
 biennis **653, 666, 679**
 capillaris 182, 200, **234**, 288, 350, 400, 564, **653, 667, 679**
  var. *agrestis* 234
  var. *capillaris* 234
 paludosa **653, 667, 680**
 vesicaria **653, 667, 680**
Cress, Hoary, see *Lepidium draba*
Crosswort, see *Cruciata laevipes*
Crowberry, see *Empetrum nigrum* ssp. *nigrum*
Crowfoot, see *Ranunculus repens*
 Celery-leaved, see *R. sceleratus*
 Ivy-leaved, see *R. hederaceus*
 Round-leaved, see *R. omiophyllus*
Cruciata laevipes 58, 60, **236**, **653, 667, 680**
Ctenopharyngodon idella 486
Cuckoo Flower, see *Cardamine pratensis*
Cuckoo-pint, see *Arum maculatum*
Cudweed,
 Common, see *Filago vulgaris*
 Heath, see *Gnaphalium sylvaticum*
 Marsh, see *G. uliginosum*
 Small, see *Filago minima*
Currant,
 Black, see *Ribes nigrum*
 Red, see *R. rubrum*
Cymbalaria muralis **653, 667, 680**
Cynoglossum officinale **653, 667, 680**
Cynosurus cristatus 66, 106, 150, **238**, 306, 406, 462, 494, 602, **653, 667, 680**
Cystopteris fragilis 49, 55, 57, 140, 142, 144, **240**, 256, 272, 432, **663, 676, 689**
Cytisus scoparius 662, 675, 688

Dactylis glomerata 10, 24, 29, 60, 61, 66, 238, **242**, 308, 348, 352, 400, 470, 476, 588, **653, 667, 680**
Dactylorhiza
 fuchsii **653, 667, 680**
 incarnata **653, 667, 680**
 maculata **653, 667, 680**
 majalis
  ssp. *praetermissa*, see *praetermissa*
  ssp. *purpurella*, see *purpurella*
 praetermissa **653, 667, 680**
 purpurella **653, 667, 680**
 viridis, see *Coeloglossum viride*
Daffodil, see *Narcissus pseudonarcissus*
Daisy, see *Bellis perennis*
 Moon, see *Leucanthemum vulgare*
 Ox-eye, see *L. vulgare*
 Shasta, see *L.* x *superbum*
Dandelion, see *Taraxacum* agg.
Danthonia decumbens 35, 42, 48, 49, 50, 51, 53, **244**, 414, 492, 584, **653, 667, 680**
 ssp. *decipiens* 244
 ssp. *decumbens* 244
Daphne laureola **653, 667, 680**
Daucus carota
 ssp. *carota* 24, 204, 210, **246**, **653, 667, 680**
 ssp. *gummifer* 246
 ssp. *sativus* 246
Dead-nettle,
 Cut-leaved, see *Lamium hybridum*
 Hen-bit, see *L. amplexicaule*
 Red, see *L. purpureum*
 Spotted, see *L. maculatum*
 White, see *L. album*
Deergrass, see *Trichophorum cespitosum*
Deschampsia
 cespitosa 44, 50, 51, 57, **248**, **653, 667, 680**
 flexuosa 6, 57, 58, 61, 66, 88, 168, 196, **250**, 286, 292, 322, 434, 450, 526, 622, **653, 667, 680**
Dewberry, see *Rubus caesius*
Desmazeria rigida, see *Catapodium rigidum*
Digitalis purpurea 50, 53, 55, **252**, 296, 432, 536, **653, 667, 680**
Diplotaxis
 muralis **653, 667, 680**
 tenuifolia **653, 667, 680**
Dipsacus fullonum 128, **653, 667, 680**
Dock
 Broad-leaved, see *Rumex obtusifolius*
 Clustered, see *R. conglomeratus*
 Curled, see *R. crispus*
 Water, see *R. hydrolapathum*
 Wood, see *R. sanguineus*
Doddering Dillies, see *Briza media*
Dog's tail, Crested, see *Cynosurus cristatus*
Dog-violet,
 Early, see *Viola reichenbachiana*
 Heath, see *V. canina*
Dogwood, see *Cornus sanguinea*
Doronicum pardalianches **653, 667, 680**
Draba muralis 53
Drepanosiphum platanoides 82
Dropwort, see *Filipendula vulgaris*
Drosera rotundifolia **653, 667, 680**
Dryas octopetala 10
Dryopteris
 affinis 256, **663, 676, 689**
 assimilis, see *expansa*
 borreri, see *affinis*
 carthusiana **663, 676, 689**

dilatata 56, 144, 240, **254**, 256, 416, ***663, 676, 689***
expansa 254
filix-mas 35, 54, 102, 140, 144, 240, 254, **256**, 272, 646, ***663, 676, 689***
intermedia 254
Duck's meat, see *Lemna minor*
Duckweed, see *Lemna minor*
  Fat, see *L. gibba*
  Ivy-leaved, see *L. trisulca*

Earthnut, see *Conopodium majus*
Echium vulgare ***653, 667, 680***
Eggs-and-Bacon, see *Lotus corniculatus*
Elder, see *Sambucus nigra*
  Dwarf, see *S. ebulus*
Elecampane, see *Inula helenium*
Eleocharis
  palustris
    ssp. *palustris* **258**, 282, 338, 454, 610, ***653, 667, 680***
    ssp. *vulgaris* 258
  quinqueflora 194, ***653, 667, 680***
  uniglumis 258
Eleogiton fluitans ***653, 667, 680***
Elm,
  English, see *Ulmus procera*
  Wych, see *U. glabra*
Elodea
  canadensis **260**, 484, 486, 570, ***653, 667, 680***
  nuttallii 260, ***653, 667, 680***
  spp. 44
Elymus
  caninus 142, 144, 162, 240, 256, **262**, 340, 436, ***653, 667, 680***
  repens, see following entry
Elytrigia repens 42, 55, 61, 70, 90, 118, 154, 230, **264**, 384, 476, ***653, 667, 680***
Empetrum
  nigrum
    ssp. *hermaphroditum* 266
    ssp. *nigrum* 56, 194, **266**, 322, 378, 446, 620, ***653, 667, 680***
Endymion non-scriptus, see *Hyacinthoides non-scripta*
Epilobium
  adenocaulon, see *ciliatum*
  angustifolium, see *Chamerion*
  brunnescens ***653, 667, 680***
  ciliatum 210, **268**, 270, 272, 530, 532, 534, ***653, 667, 680***
  hirsutum 10, 46, 49, 51, 54, 102, 112, 218, **270**, 274, 276, 278, 310, 458, 478, ***653, 667, 680***
  montanum 57, 256, 268, 270, **272**, 274, ***654, 667, 680***
  nerterioides, see *brunnescens*
  obscurum 218, **274**, 276, 278, 520, ***654, 667, 680***
  palustre 86, 90, 274, **276**, 278, 376, 520, ***654, 667, 680***
  parviflorum 276, **278**, 518, 628, ***654, 667, 680***
  roseum ***654, 667, 680***
  spp. 49, 268, 270, 274
  tetragonum ***654, 667, 680***
Epipactis
  helleborine ***654, 667, 680***
  palustris ***654, 667, 680***
Equisetum
  arvense 172, **280**, ***663, 676, 689***

fluviatile 60, 260, **282**, 284, 484, 486, 508, 570, 610, ***663, 676, 689***
palustre 122, 166, 274, 278, 280, 282, **284**, 336, 370, 518, 646, ***663, 676, 689***
  spp. 280, 284
sylvaticum ***663, 676, 689***
telmateia ***663, 676, 689***
variegatum 284
Erica
  cinerea 50, 57, 69, 94, 168, 196, 250, 266, **286**, 622, ***654, 667, 680***
  tetralix ***654, 667, 680***
Erigeron
  acer 10, 182, 200, 234, **288**, 368, 448, 564, ***654, 667, 680***
  canadensis see *Conyza canadensis*
Eriophorum
  angustifolium 52, **290**, 292, 622,
  vaginatum 10, 250, 266, **292**, 434, 646, ***654, 667, 680***
    ssp. *spissum* 292
Erodium cicutarium ***654, 667, 680***
Erophila verna 124, 182, 440, 542, ***654, 667, 680***
Erysimum
  cheiranthoides ***654, 667, 680***
  cheiri ***654, 667, 680***
Euonymus europaeus ***662, 675, 688***
Eupatorium cannabinum ***654, 667, 680***
Euphorbia
  cyparissias ***654, 667, 680***
  esula ***654, 667, 680***
  exigua 108, ***654, 667, 680***
  helioscopia ***654, 667, 680***
  peplus ***654, 667, 680***
Euphrasia
  confusa 294
  nemorosa 294
  officinalis agg. 49, **294**, 430, 524, 646, ***654, 667, 680***
Evening-primrose,
  Common, see *Oenothera biennis*
  Large-flowered, see *O. glazioviana*
  Small-flowered, see *O. cambrica*
Everlasting-pea, Broad-leaved, see *Lathyrus latifolius*
Eyebright, see *Euphrasia officinalis*

Fagus sylvatica 47, 49, 50, 62, 82, 252, **296**, 356, 498, 540, ***662, 675, 688***
Fallopia
  convolvulus 60, 108, 120, 146, 230, **298**, 388, 438, 452, 456, 558, 572, 618, 634, ***654, 667, 680***
  japonica 23, 42, 46, 48, 55, 60, 62, 128, 172, **300**, 608, 645, ***654, 667, 680***
    var. *compacta* 300
  sachalinensis 300, 366, ***654, 667, 680***
False-brome,
  Heath, see *Brachypodium pinnatum*
  Slender, see *B. sylvaticum*
Fat Hen, see *Chenopodium album*
Fen-sedge, Great, see *Cladium mariscus*
Fennel, see *Foeniculum vulgare*
Fern,
  Beech, see *Phegopteris connectilis*
  Hard, see *Blechnum spicant*
  Lemon-scented, see *Oreopteris limbosperma*
  Male, see *Dryopteris filix-mas*
  Oak, see *Gymnocarpium dryopteris*
  Royal, see *Osmunda regalis*
  Scaly Male, see *Dryopteris affinis*

Fescue,
  Creeping, see *Festuca rubra* ssp. *rubra*
  Giant, see *F. gigantea*
  Meadow, see *F. pratensis*
  Rat's-tail, see *Vulpia myuros*
  Red, see *Festuca rubra* ssp. *rubra*
  Sheep's, see *F. ovina*
  Squirrel-tail, see *Vulpia bromoides*
  Tall, see *Festuca arundinacea*
Festuca
  arundinacea 274, 306, ***654, 667, 680***
  filiformis 304, 645
  gigantea 100, 112, 162, 220, 248, **302**, 306, 366, 458, 504, 556, 632, ***654, 667, 680***
  nigrescens, see *rubra* ssp. *commutata*
  ovina 10, 29, 41, 42, 47, 51, 57, 59, 61, 66, 69, 70, 116, 132, 160, 244, 250, **304**, 308, 342, 344, 380, 414, 428, 434, 446, 538, 645, ***654, 668, 680***
  pratensis 24, 106, 164, 302, **306**, 392, 462, 516, ***654, 668, 681***
    ssp. *apennina* 306
    ssp. *pratensis* 306
  rubra 10, 24, 25, 46, 50, 52, 57, 60, 61, 66, 88, 132, 194, 242, 304, **308**, 352, 410, 430, 470, 476, 548, 588, 606, ***654, 668, 681***
    ssp. *arctica* 308
    ssp. *commutata* 308
    ssp. *litoralis* 308
    ssp. *pruinosa* 308
    ssp. *rubra* 304, 308
    spp. 302
  tenuifolia, see *filiformis*
  vivipara 304
Festuca x *Lolium*
  F. pratensis x *L. perenne*, see x *Festulolium loliaceum*
Feverfew, see *Tanacetum parthenium*
Fiddleneck, Common, see *Amsinckia micrantha*
Field Speedwell,
  Large, see *Veronica persica*
  Green, see *V. agrestis*
  Grey, see *V. polita*
Figwort,
  Common, see *Scrophularia nodosa*
  Water, see *S. auriculata*
Filago
  minima ***654, 668, 681***
  vulgaris ***654, 668, 681***
Filipendula
  ulmaria 112, 170, 180, 218, **310**, 424, 426, 560, 632, ***654, 668, 681***
  vulgaris ***654, 668, 681***
Fiorin, see *Agrostis stolonifera*
Fireweed, see *Chamerion*
Flax, Purging, see *Linum catharticum*
Fleabane,
  Blue, see *Erigeron acer*
  Canadian, see *Conyza canadensis*
  Common, see *Pulicaria dysenterica*
Flote-grass, see *Glyceria fluitans*
  Hybrid, see *G.* x *pedicellata*
  Plicate, see *G. notata*
  Small, see *G. declinata*
Foeniculum vulgare ***654, 668, 681***
Forget-me-not,
  Changing, see *Myosotis discolor*
  Common, see *M. arvensis*
  Creeping, see *M. secunda*
  Early, see *M. ramosissima*

SPECIES INDEX   739

Forget-me-not, *continued*
   Tufted, see *M. laxa* ssp. *cespitosa*
   Water, see *M. scorpioides*
   Wood, see *M. sylvatica*
Fox-and-cubs, see *Pilosella aurantiaca*
Foxglove, see *Digitalis purpurea*
Fox-sedge, False, see *Carex otrubae*
Foxtail,
   Marsh, see *Alopecurus geniculatus*
   Meadow, see *A. pratensis*
Fragaria
   vesca **312**, 360, 490, **654, 668, 681**
   x *ananassa* 312, **654, 668, 681**
Frankia spp. 102
Fraxinus excelsior 49, 58, 82, 100, 232, **314**, 382, 556, 614, **662, 675, 688**
Fringe-cups, see *Tellima grandiflora*
Fritillaria meleagris 44
Fumaria
   muralis **654, 668, 681**
   officinalis **654, 668, 681**
Fumitory, Common, see *Fumaria officinalis*
Furze, see *Ulex europaeus*

Galanthus nivalis **654, 668, 681**
Galega officinalis **654, 668, 681**
Galeopsis
   bifida 316
   tetrahit sl 50, **316**, 348, 388, 444, **654, 668, 681**
   tetrahit ss 316
Gallant Soldier, see *Galinsoga parviflora*
Galinsoga
   ciliata, see *G. quadriradiata*
   parviflora **654, 668, 681**
   quadriradiata **654, 668, 681**
Galium
   aparine 10, 42, 46, 50, 58, 59, 61, 62, 98, 114, 118, 316, **318**, 334, 348, 444, 536, 576, 616, **654, 668, 681**
   cruciata, see *Cruciata laevipes*
   elongatum, see *palustre* ssp. *elongatum*
   fleurotti 324
   mollugo **326, 654, 668, 681**
   odoratum **654, 668, 681**
   palustre 61, 178, 184, 224, 270, **320**, 376, 424, 442, 512, **654, 668, 681**
      ssp. *elongatum* 320
      ssp. *palustre* 320
   pumilum 324
   pusillum 324
   saxatile 56, 57, 58, 66, 250, 266, 320, **322**, 324, 446, 620, 622, **654, 668, 681**
   sterneri 55, 174, 294, 312, 320, **324**, 342, 590, 592, **654, 668, 681**
   sylvestre, see *sterneri*
   uliginosum 320, **654, 668, 681**
   verum 24, 42, **326**, 346,468 574, **654, 668, 681**
      ssp. *verum* 326
Garlic, Hedge, see *Alliaria petiolata*
Gastrophysa viridula 530
Genista
   anglica **654, 668, 681**
   tinctoria **654, 668, 681**
Gentian, Autumn, see *Gentianella amarella*
Gentianella amarella **654, 668, 681**
Geranium
   columbinum **654, 668, 681**
   dissectum 328, **654, 668, 681**
   endressii **654, 668, 681**
   ibericum x *platypetalum*, see x *magnificum*

   lucidum **654, 668, 681**
   molle 49, 51, 124, 234, **328**, 594, 596, **655, 668, 681**
   pratense **655, 668, 681**
   purpureum 330
   pusillum 328, **655, 668, 681**
   pyrenaicum **655, 668, 681**
   robertianum 58, 61, 132, 328, **330, 655, 668, 681**
   x *magnificum* **655, 668, 681**
Geum
   rivale 332, **655, 668, 681**
   urbanum 52, 58, 138, 162, **332**, 340, 408, 422, 432, 586, **655, 668, 681**
Gipsywort, see *Lycopus europaeus*
Glaucobactus spp. 522
Glechoma hederacea 318, **334**, 576, 578, 586, 616, **655, 668, 681**
Globe-flower, see *Trollius europaeus*
Glyceria
   declinata 336, **655, 668, 681**
   fluitans 60, 122, 166, 284, **336**, 370, **655, 668, 681**
   fluitans x *notata*, see x *pedicellata*
   maxima 1, **338**, 460, 464, 454, **655, 668, 681**
   notata 336, **655, 668, 681**
   x *pedicellata* 336, **655, 668, 681**
   plicata, see *notata*
Gnaphalium
   sylvaticum **655, 668, 681**
   uliginosum **655, 668, 681**
Goat's-beard, see *Tragopogon pratensis*
Goat's-rue, see *Galega officinalis*
Golden-rod, see *Solidago virgaurea*
   Canadian, see *S. canadensis*
   Early, see *S. gigantea*
Golden Saxifrage, see *Chrysosplenium oppositifolium*
   Alternate-leaved, see *C. alternifolium*
Good King Henry, see *Chenopodium bonus-henricus*
Gooseberry, see *Ribes uva-crispa*
Goosefoot,
   Many-seeded, see *Chenopodium polyspermum*
   Red, see *C. rubrum*
Goosegrass, see *Galium aparine*
Gorse, see *Ulex europaeus*
   Dwarf, see *U. minor*
   Western, see *U. gallii*
Grass,
   Fern, see *Catapodium rigidum*
   Heath, see *Danthonia decumbens*
   Lop, see *Bromus hordeaceus*
   Quaking, see *Briza media*
   Tor, see *Brachypodium pinnatum*
Grass of Parnassus, see *Parnassia palustris*
Greenweed, Dyer's, see *Genista tinctoria*
Groenlandia densa **655, 668, 681**
Gromwell, Common, see *Lithospermum officinale*
Ground-elder, see *Aegopodium podagraria*
Groundsel, see *Senecio vulgaris*
   Heath, see *S. sylvaticus*
   Sticky, see *S. viscosus*
   Stinking, see *S. viscosus*
Guelder-rose, see *Viburnum opulus*
Gymnadenia conopsea **655, 668, 681**
Gymnocarpium dryopteris **663, 676, 689**

Hardheads, see *Centaurea nigra*

Hair-grass,
   Crested, see *Koeleria macrantha*
   Early, see *Aira praecox*
   Silver, see *A. caryophyllea*
   Wavy, see *Deschampsia flexuosa*
Harebell, see *Campanula rotundifolia*
Hart's-tongue, see *Phyllitis scolopendrium*
Hare's Taff, see *Eriophorum vaginatum*
Hawkbit,
   Autumn, see *Leontodon autumnalis*
   Lesser, see *L. saxatilis*
   Rough, see *L. hispidus*
Hawk's-beard,
   Beaked, see *Crepis vesicaria*
   Marsh, see *C. paludosa*
   Rough, see *C. biennis*
   Smooth, see *C. capillaris*
Hawkweed, see *Hieracium*
   Mouse-ear, see *Pilosella officinarum*
Hawthorn, see *Crataegus monogyna*
Hazel, see *Corylus avellana*
Heath, Cross-leaved, see *Erica tetralix*
Heather, see *Calluna vulgaris*
Hedera
   helix 162, 262, 332, **340**, 408, 450, 614, **655, 668, 681**
   hibernica 340
Hedge-parsley, Upright, see *Torilis japonica*
Helianthemum
   apenninum 342
   canum 342
   chamaecistus see *following entry*
   nummularium 10, 24, 51, 52, 53, 188, 190, 324, **342**, 344, 380, 480, 538, 590, 592, 642, **655, 668, 681**
Helianthus annuus 10
Helictotrichon
   pratense 10, 24, 42, 56, 57, **344**, 188, 190, 342, 380, 480, 538, **655, 668, 681**
   pubescens 24, 116, **346**, 158, 326, 390, 468, 492, 598, **655, 668, 681**
Heliotrope, Winter, see *Petasites fragrans*
Helix aspersa 152, 232, 314
Helleborine,
   Broad-leaved, see *Epipactis helleborine*
   Marsh, see *E. palustris*
Hemlock, see *Conium maculatum*
Hemp-agrimony, see *Eupatorium cannabinum*
Hemp-nettle, Common, see *Galeopsis tetrahit*
Heracleum
   mantegazzianum 348, 366, **655, 668, 681**
   sphondylium 43, 48, 50, 51, 61, 118, **348**, **655, 668, 681**
      ssp. *sibiricum* 348
      ssp. *sphondylium* 348
Herb Bennet, see *Geum urbanum*
Herb Paris, see *Paris quadrifolia*
Herb Robert, see *Geranium robertianum*
Hesperis matronalis **655, 668, 681**
Hieracium
   aurantiacum, see *Pilosella aurantiaca*
   pilosella, see *P. officinarum*
   spp 24, 49, 234, **350**, 364, 364, 400, 420, 466, 648, **655, 668, 681**
   umbellatum 350
Hippuris vulgaris **655, 668, 681**
Hirschfeldia incana **655, 668, 681**
Hogweed, see *Heracleum sphondylium*
   Giant, see *H. mantegazzianum*
Holcus
   lanatus 10, 25, 29, 35, 49, 51, 54, 57, 61, 66, 222, **352**, 354, **655, 668, 681**

mollis 54, 82, 252, 352, **354**, 496, 526, 578, 582, 632, **655, 668, 681**
Holly, see *Ilex aquifolium*
Honeysuckle, see *Lonicera periclymenum*
Hop, see *Humulus lupulus*
Hordeum
    jubatum **655, 668, 681**
    murinum 49, 53, **655, 668, 681**
    secalinum **655, 668, 681**
    spp. 264,
Horehound, Black, see *Ballota nigra*
Hornbeam, see *Carpinus betulus*
Hornwort, Rigid, see *Ceratophyllum demersum*
Horse-chestnut, see *Aesculus hippocastanum*
Horse-radish, see *Armoracia rusticana*
Horsetail,
    Common, see *Equisetum arvense*
    Great, see *E. telmateia*
    Marsh, see *E. palustre*
    Water, see *E. fluviatile*
    Wood, see *E. sylvaticum*
Hottonia palustris **655, 669, 681**
Hound's-tongue, see *Cynoglossum officinale*
Huckleberry, see *Vaccinium myrtillus*
Humulus lupulus **655, 669, 682**
Huperzia selago **663, 676, 689**
Hyacinthoides
    hispanica (including *hispanica* x *non-scripta*) 356, **655, 669, 682**
    non-scripta 24, 42, 45, 51, 57, 61, 82, 112, 152, 228, 252, 254, 296, 354, **356**, 408, 416, 450, 498, 522, 649, **655, 669, 682**
Hydrocotyle vulgaris 338, **358**, 454, **655, 669, 682**
Hymenophyllum spp. 44
Hypericum
    androsaemum 646, **655, 669, 682**
    elodes **655, 669, 682**
    hirsutum 312, **360, 655, 669, 682**
    humifusum **655, 669, 682**
    maculatum **362, 655, 669, 682**
        ssp. *obtusiusculum* 362
    perforatum 50, 53, 71, 154, 160, 202, 204, 206, 360, **362**, 368, 402, **655, 669, 682**
    pulchrum 69, **390, 655, 669, 682**
    tetrapterum **655, 669, 682**
Hypochaeris
    glabra 364
    radicata 35, 41, 55, 88, 94, **364**, 526, **655, 669, 682**
Hypochoeris see *Hypochaeris*

Ilex aquifolium **662, 675, 688**
Impatiens
    glandulifera 35, 48, 49, 50, 62, 98, 112, 172, 316, 444, 458, **366, 655, 669, 682**
    parviflora **655, 669, 682**
Inula
    conyza, see following entry
    conyzae 200, 206, 288, 362, **368**, 448, **655, 669, 682**
    helenium **655, 669, 682**
Iris
    Bearded, see *Iris germanica*
    foetidissima **655, 669, 682**
    germanica **655, 669, 682**
    pseudacorus **655, 669, 682**
    Stinking, see *I. foetidissima*
    Yellow, see *I. pseudacorus*

Iron-root, see *Atriplex patula*
Isolepis setacea **655, 669, 682**
Ivy, see *Hedera helix*
    Ground, see *Glechoma hederacea*

Jack-by-the-hedge, see *Alliaria petiolata*
Jasione montana **655, 669, 682**
Juncus
    acutiflorus 370, **655, 669, 682**
    ambiguus 372
    articulatus 49, 52, 90, 274, 278, 284, **370**, 582, **655, 669, 682**
    bufonius 166, **372, 655, 669, 682**
    bulbosus 166, 370, **374**, 570, **655, 669, 682**
    conglomeratus 376, **655, 669, 682**
    effusus 49, 52, 61, 84, 184, 224, 276, 320, 358, **376**, 434, 512, 582, **655, 669, 682**
        var. *compactus* 376
    inflexus 376, **656, 669, 682**
    kochii, see *bulbosus*
    minutulus, see *bufonius*
    ranarius, see *ambiguus*
    spp. 50
    squarrosus 47, 266, **378, 656, 669, 682**
    subnodulosus 370, **656, 669, 682**
    tenuis **656, 669, 682**

Keck, see *Anthriscus sylvestris*
Kingcup, see *Caltha palustris*
Knapweed,
    Greater, see *Centaurea scabiosa*
    Lesser, see *C. nigra* s.l.
Knautia arvensis 24, **656, 669, 682**
Knawel Annual, see *Scleranthus annuus*
Knotgrass, see *Polygonum aviculare*
Knotweed,
    Giant, see *Fallopia sachalinensis*
    Japanese, see *F. japonica*
Kochia scoparia, see *Bassia scoparia*
Koeleria
    cristata, see *macrantha*
    glauca, see *macrantha*
    macrantha 10, 24, 47, 51, 58, 61, 71, 174, 188, 304, 344, **380**, 538, 592, **656, 669, 682**
    vallesiana 380

Lactuca virosa **656, 669, 682**
Lady-fern, see *Athyrium filix-femina*
Lady's Mantle, see *Alchemilla vulgaris*
Lady's Smock, see *Cardamine pratensis*
Lamiastrum galeobdolon
    ssp. *flavidum* 382
    ssp. *galeobdolon* 382
    ssp. *montanum* 24, 41, 57, 100, 138, 220, 248, 314, **382**, 428, 504, **656, 669, 682**
Lamium
    album 128, **384, 656, 669, 682**
    amplexicaule **656, 669, 682**
    hybridum **656, 669, 682**
    maculatum **656, 669, 682**
    purpureum 128, 134, **386**, 530, 604, **656, 669, 682**
Lapsana communis 51, 230, 264, 316, **388**, **656, 669, 682**
    ssp. *intermedia* 388
Lathraea squamaria **656, 669, 682**
Lathyrus
    latifolius **656, 669, 682**
    linifolius 60, 160, 228, 346, **390**, 392, 574,

    598, **656, 669, 682**
    montanus, see *linifolius*
    pratensis 164, 352, **392, 656, 669, 682**
Lemna
    gibba 394, 486, 647, **656, 669, 682**
    minor 61, 260, **394**, 484, 486, 508, 570, 646, 647, **656, 669, 682**
    minuscula, 394
    trisulca 647, **656, 669, 682**
    spp. 648
Leontodon
    autumnalis 40, **396**, 430, 646, **656, 669, 682**
        ssp. *pratensis* 396
    hispidus 10, 24, 53, 61, 126, 350, 368, **398**, 404, 448, 466, 492, 544, 548, 562, **656, 669, 682**
    saxatilis 398, **656, 669, 682**
    taraxacoides, see *L. saxatilis*
Leopard's-bane, see *Doronicum pardalianches*
Lepidium
    campestre **656, 669, 682**
    draba **656, 669, 682**
    heterophyllum **656, 669, 682**
    ruderale **656, 669, 682**
Lettuce,
    Greater, see *Lactuca virosa*
    Wall, see *Mycelis muralis*
Leucanthemum
    lacustre x *maximum*, see x *superbum*
    maximum, see x *superbum*
    vulgare 60, 234, 242, 350, **400**, 420, 448, 588, **656, 669, 682**
    x *superbum* **656, 669, 682**
Ligustrum vulgare **662, 675, 688**
Linaria
    purpurea **656, 669, 682**
    repens 402, **656, 669, 682**
    vulgaris 210, 212, 362, **402**, 552, **656, 669, 682**
Ling, see *Calluna vulgaris*
Linum catharticum 55, 174, 398, **404**, 410, 466, 470, 548, 562, 606, **656, 669, 682**
Listera ovata **656, 669, 682**
Lithospermum officinale **656, 669, 682**
Littorella uniflora 506, **656, 670, 682**
Logfia minima, see *Filago minima*
Lolium
    multiflorum 406, **656, 670, 683**
    perenne 10, 24, 29, 40, 44, 45, 61, 66, 106, 238, 242, 302, 306, 308, 396, **406**, 472, 598, 602, 638, **656, 670, 683**
    ssp. *multiflorum*, see *multiflorum*
    ssp. *perenne*, see *perenne*
Lonicera periclymenum 340, 356, **408**, 428, 450, 498, **656, 670, 683**
Londonpride, see *Saxifraga* x *urbium*
Loosestrife,
    Dotted, see *Lysimachia punctata*
    Yellow, see *L. vulgaris*
Lords-and-Ladies, see *Arum maculatum*
Lotus
    corniculatus 10, 24, 50, 53, 57, 61, 84, **410**, 412, 466, 606, **656, 670, 683**
    pedunculatus 410, **412**, 646, **656, 670, 683**
    uliginosus, see *pedunculatus*
Lousewort, see *Pedicularis sylvatica*
Lucerne, see *Medicago sativa*
Lupin,
    Garden, see *Lupinus* x *regalis*
    Tree, see *L. arboreus*

Lupinus
    arboreus   *662, 675, 688*
    arboreus x *polyphyllus*, see x *regalis*
    x *regalis*   *656, 670, 683*
Luzula
    campestris   66, 88, 116, **414**, *656, 670, 683*
    forsteri   416
    multiflora   414, *656, 670, 683*
        ssp. *congesta*   414
        ssp. *hibernica*   414
        ssp. *multiflora*   414
    pilosa   52, 53, 254, 296, **416**, 496, 498, 568, *656, 670, 683*
    sylvatica   *656, 670, 683*
Lychnis flos-cuculi   *656, 670, 683*
Lycopodium clavatum   *663, 676, 689*
Lycopersicum esculentum   216
Lycopus europaeus   *656, 670, 683*
Lysimachia
    nemorum   *656, 670, 683*
    nummularia   *656, 670, 683*
    punctata   *656, 670, 683*
    vulgaris   *656, 670, 683*
Lythrum
    portula   *656, 670, 683*
    salicaria   *656, 670, 683*

Madder, Field, see *Sherardia arvensis*
Mahonia aquifolium   *662, 675, 688*
Mallow,
    Common, see *Malva sylvestris*
    Dwarf, see *M. neglecta*
Malus sylvestris   *662, 675, 688*
Malva
    moschata   *656, 670, 683*
    neglecta   *656, 670, 683*
    sylvestris   *656, 670, 683*
Maple,
    Field, see *Acer campestre*
    Great, see *A. pseudoplatanus*
    Norway, see *A. platanoides*
Mare's-tail, see *Hippuris vulgaris*
Marguerite, see *Leucanthemum vulgare*
Marigold,
    Corn, see *Chrysanthemum segetum*
    Marsh, see *Caltha palustris*
Marjoram, Wild, see *Origanum vulgare*
Marsh-orchid,
    Early, see *Dactylorhiza incarnata*
    Northern, see *D. purpurella*
    Southern, see *D. praetermissa*
Marshwort, Lesser, see *Apium inundatum*
Mat-grass, see *Nardus stricta*
Matricaria
    discoidea   42, 47, 51, 55, 176, **418**, 472, 474, 482, 554, 604, *656, 670, 683*
    matricarioides, see *discoidea*
    recutita   *656, 670, 683*
    perforata, see *Tripleurospermum inodorum*
Mayweed,
    Rayless, see *Matricaria discoidea*
    Scented, see *M. recutita*
    Scentless, see *Tripleurospermum inodorum*
Meadow-grass
    Annual, see *Poa annua*
    Flattened, see *P. compressa*
    Narrow-leaved, see *P. angustifolia*
    Rough, see *P. trivialis*
    Rough-stalked, see *P. trivialis*
    Smooth-stalked, see *P. pratensis*
    Spreading, see *P. humilis*
    Wood, see *P. nemoralis*

Meadow-rue, Common, see *Thalictrum flavum*
Meadow-sweet, see *Filipendula ulmaria*
Meconopsis cambrica   *656, 670, 683*
Medicago
    lupulina   50, 52, 57, 130, 234, 328, 350, 400, **420**, 440, *656, 670, 683*
    sativa ssp. *sativa*   645, *657, 670, 683*
Medick, Black, see *Medicago lupulina*
Melampyrum pratense   *657, 670, 683*
Melandrium
    album, see *Silene latifolia*
    dioicum, see *S. dioica*
Melica uniflora   40, 110, 156, 162, 332, 340, **422**, 426, 432, 540, 614, *657, 670, 683*
Melick, Wood, see *Melica uniflora*
Melilotus
    alba, see following entry
    albus   *657, 670, 683*
    altissima, see following entry
    altissimus   *657, 670, 683*
    officinalis   *657, 670, 683*
Melilot,
    Ribbed, see *Melilotus officinalis*
    Tall, see *M. altissimus*
    White, see *M. albus*
Mentha
    aquatica   170, 218, 310, 320, **424**, 442, *657, 670, 683*
    aquatica x *arvensis*, see x *verticillata*
    aquatica x *spicata*, see x *piperita*
    arvensis   424, *657, 670, 683*
    x *piperita*   *657, 670, 683*
    spicata x *suaveolens*, see x *villosa*
    spicata   *657, 670, 683*
    x *verticillata*   424, *657, 670, 683*
    x *villosa*   *657, 670, 683*
Menyanthes trifoliata   *657, 670, 683*
Mercurialis
    annua   *657, 670, 683*
    perennis   35, 55, 57, 61, 62, 110, 138, 156, 228, 232, 262, 314, 332, 340, 382, **426**, *657, 670, 683*
Mercury
    Annual, see *Mercurialis annua*
    Dog's, see *M. perennis*
Mespilus germanica
Michaelmas-daisy, see *Aster novi-belgii*
    Narrow-leaved, see *A. lanceolatus*
Mignonette, Wild, see *Reseda lutea*
Milfoil, see *Achillea millefolium*
Milk-Thistle, see *Sonchus oleraceus*
    Spiny, see *S. asper*
Milkwort,
    Common, see *Polygala vulgaris*
    Heath, see *P. serpyllifolia*
Milium effusum   51, 53, 152, 408, **428**, 450, 496, 522, 568, *657, 670, 683*
Millet, Wood, see *Milium effusum*
Mimulus
    guttatus   *657, 670, 683*
    guttatus x *luteus*, see x *robertsii*
    x *robertsii*   *657, 670, 683*
Mint,
    Corn, see *Mentha arvensis*
    Spear, see *M. spicata*
    Water, see *M. aquatica*
    Whorled, see *M.* x *verticillata*
Minuartia verna   47, 58, 59, 294, **430**, 524, *657, 670, 683*
Mistletoe, see *Viscum album*

Moehringia trinervia   50, 422, **432**, *657, 670, 683*
Molinea
    caerulea   52, 57, 192, 322, 378, **434**, 446, *657, 670, 683*
        ssp. *arundinacea*   434
    littoralis, see *caerulea* ssp. *arundinacea*
Monkeyflower, see *Mimulus guttatus*
    Hybrid, see *M.* x *robertsii*
Montia
    fontana   166, 218, 504, 582, *657, 670, 683*
    perfoliata, see *Claytonia perfoliata*
    sibirica, see *Claytonia sibirica*
Moor-grass Purple, see *Molinia caerulea*
Moschatel, see *Adoxa moschatellina*
Mountain Everlasting, see *Antennaria dioica*
Mouse-ear,
    Dark-green, see *Cerastium diffusum*
    Field, see *C. arvense*
    Little, see *C. semidecandrum*
    Sticky, see *C. glomeratum*
Mouse-ear Chickweed, Common, see *Cerastium fontanum*
Mugwort, see *Artemisia vulgaris*
    Chinese, see *A. verlotiorum*
Mullein,
    Dark, see *Verbascum nigrum*
    Great, see *V. thapsus*
Mustard,
    Garlic, see *Alliaria petiolata*
    Hedge, see *Sisymbrium officinale*
    Hoary, see *Hirschfeldia incana*
    Treacle, see *Erysimum cheiranthoides*
    Wild, see *Sinapis arvensis*
Musk-mallow, see *Malva moschata*
Mycelis muralis   49, 58, 140, 142, 144, 240, 256, 272, **436**, *657, 670, 683*
Myosotis
    arvensis   53, 60, 108, 298, 388, **438**, 452, 456, 558, 572, 618, 634, *657, 670, 683*
        var. *arvensis*   438
        var. *sylvestris*   438
    discolor   *657, 670, 683*
    laxa ssp. *cespitosa*   442, *657, 670, 683*
    ramosissima   41, 51, 58, 120, 124, 130, 328, 420, **440**, 542, 546, 596, 626, *657, 670, 683*
        ssp. *globularis*   440
        ssp. *ramosissima*   440
    scorpioides   61, 178, 184, 270, 320, 424, **442**, 560, *657, 670, 683*
    secunda   442, *657, 670, 683*
    sylvatica   556, *657, 670, 683*
Myosoton aquaticum   *657, 670, 683*
Myriophyllum
    alterniflorum   *657, 670, 683*
    spicatum   *657, 670, 683*
Myrrhis odorata   50, 98, 118, 172, 316, 318, 366, **444**, 458, *657, 670, 683*

Narcissus pseudonarcissus   *657, 670, 683*
Nardus stricta   42, 47, 56, 66, 196, 266, 304, 322, 378, 434, **446**, *657, 670, 683*
Narthecium ossifragum   *657, 670, 683*
Nasturtium
microphyllum, see *Rorippa microphylla*
microphyllum x *officinale*, see *R. microphylla* x *nasturtium-aquaticum*
officinale, see *R. nasturtium-aquaticum*
Navelwort, see *Umbilicus rupestris*
Neottia nidus-avis   *657, 670, 683*

Nettle,
  Small, see *Urtica urens*
  Stinging, see *U. dioica*
Nightshade,
  Black, see *Solanum nigrum*
  Enchanter's, see *Circaea lutetiana*
  Green, see *Solanum physalifolium*
  Woody, see *S. dulcamara*
Nipplewort, see *Lapsana communis*
Nuphar lutea *657, 670, 683*
Nymphaea alba *657, 671, 683*

Oak, see *Quercus* agg.
  Pedunculate, see *Q. robur*
  Sessile, see *Q. petraea*
  Turkey, see *Q. cerris*
Oat,
  Meadow, see *Helictotrichon pratense*
  Wild, see *Avena fatua*
Oat-grass, see *Arrhenatherum elatius*
  Hairy, see *Helictotrichon pubescens*
  Yellow, see *Trisetum flavescens*
Odontites
  verna, see following entry
  vernus *657, 671, 684*
Oenanthe
  aquatica *657, 671, 684*
  crocata *657, 671, 684*
  fistulosa *657, 671, 684*
Oenothera
  biennis *657, 671, 684*
  cambrica *657, 671, 684*
  erythrosepala, see *glazioviana*
  glazioviana *657, 671, 684*
  rubricaulis, see *cambrica*
Onion, Wild, see *Allium vineale*
Onobrychis viciifolia 647, *657, 671, 684*
Ononis
  repens *657, 671, 684*
  spinosa *657, 671, 684*
Ophioglossum vulgatum *663, 676, 689*
Ophrys
  apifera *657, 671, 684*
  insectifera 390
Orchardgrass, see *Dactylis glomerata*
Orache,
  Common, see *Atriplex patula*
  Grass-leaved, see *A. littoralis*
  Hastate, see *A. prostrata*
Orchid,
  Bee, see *Ophrys apifera*
  Bird's-nest, see *Neottia nidus-avis*
  Early-purple, see *Orchis mascula*
  Fragrant, see *Gymnadenia conopsea*
  Frog, see *Coeloglossum viride*
  Green-winged, see *Orchis morio*
  Pyramidal, see *Anacamptis pyramidalis*
Orchis
  mascula *657, 671, 684*
  morio *657, 671, 684*
Oregon Grape, see *Mahonia aquifolium*
Oreopteris limbosperma *663, 676, 689*
Origanum vulgare 10, 50, 206, 288, 368, 400, **448**, *657, 671, 684*
Ornithogalum
  angustifolium *657, 671, 684*
  umbellatum, see *angustifolium*
Ornithopus perpusillus 94, *657, 671, 684*
Orpine, see *Sedum telephium*
Osier, see *Salix viminalis*
Osmunda regalis *663, 676, 689*
Oxalis

  acetosella 41, 57, 228, 416, 428, **450**, *657, 671, 684*
  corniculata *657, 671, 684*
  corymbosa, see *debilis*
  debilis *657, 671, 684*
  exilis *657, 671, 684*
Oxtongue,
  Bristly, see *Picris echioides*
  Hawkweed, see *P. hieracioides*

Paigle, see *Primula veris*
Pansy,
  Field, see *Viola arvensis*
  Wild, see *V. tricolor*
Papaver
  argemone *657, 671, 684*
  dubium 452, *657, 671, 684*
  dubium ssp. *lecoqii* 452
  rhoeas 49, 51, 60, 70, 438, **452**, 558, 634, *658, 671, 684*
  somniferum *658, 671, 684*
  spp. 690
Parietaria judaica *658, 671, 684*
Paris quadrifolia *658, 671, 684*
Parnassia palustris 194, *658, 671, 684*
Parsley,
  Bur, see *Anthriscus caucalis*
  Cow, see *A. sylvestris*
  Fool's, see *Aethusa cynapium*
Parsley Piert, see *Aphanes arvensis*
  Slender, see *A. australis*
Parsnip,
  Cow, see *Heracleum sphondylium*
  Wild, see *Pastinaca sativa*
Pastinaca sativa *658, 671, 684*
Pearlwort,
  Annual, see *Sagina apetala*
  Knotted, see *S. nodosa*
  Procumbent, see *S. procumbens*
  Sea, see *S. maritima*
Pedicularis sylvatica *658, 671, 684*
Pellitory-of-the-Wall, see *Parietaria judaica*
Penny-cress, Field, see *Thlaspi arvense*
Pentaglottis sempervirens *658, 671, 684*
Pennywort, see *Hydrocotyle vulgaris*
Pentaglottis sempervirens
Peppermint, see *Mentha x piperita*
Pepper-saxifrage, see *Silaum silaus*
Pepperwort,
  Field, see *Lepidium campestre*
  Narrow-leaved, see *L. ruderale*
  Smith's, see *L. heterophyllum*
Periwinkle,
  Greater, see *Vinca major*
  Lesser, see *V. minor*
Persicaria
  amphibia 57, 338, 376, **454**, 610, 648, *658, 671, 684*
  bistorta *658, 671, 684*
  hydropiper, *658, 671, 684*
  lapathifolia 456, *658, 671, 684*
  maculosa 60, 70, 214, 298, 388, 482, **456**, 572, 580, 634, *658, 671, 684*
  Pale, see *Persicaria lapathifolia*
Petasites
  fragrans *658, 671, 684*
  hybridus 24, 46, 55, 58, 98, 102, 218, 316, 366, 444, **458**, *658, 671, 684*
Peucedanum palustre 112
Phalaris arundinacea 35, 46, 59, 102, 178, 270, **460**, 582, *658, 671, 684*
  ssp. *rotgesii* 460

Phegopteris connectilis *663, 676, 689*
Phleum
  bertolonii 462, *658, 671, 684*
  pratense 106, 150, 164, 208, 238, 306, 406, **462**, 500, 502, 516, 600, 602, *658, 671, 684*
  ssp. *bertolonii*, see *bertolonii*
Phragmites
  australis 23, 48, 316, 338, 460, **464**, 570, 610, 645, *658, 671, 684*
  communis, see *australis*
Phyllitis scolopendrium 142, *663, 676, 689*
Picris
  echioides *658, 671, 684*
  hieracioides *658, 671, 684*
Pigmyweed, New Zealand, see *Crassula helmsii*
Pignut, see *Conopodium majus*
Pilosella
  aurantiaca *658, 671, 684*
  officinarum 6, 10, 42, 59, 126, 304, 350, 398, 404, 410, **466**, 544, 548, 606, *658, 671, 684*
Pimpernel,
  Bog, see *Anagallis tenella*
  Scarlet, see *A. arvensis* ssp. *arvensis*
  Yellow, see *Lysimachia nemorum*
Pimpinella
  major *658, 671, 684*
  saxifraga 24, 50, 158, 202, 326, 346, 390, **468**, 574, 640, *658, 671, 684*
Pineapple Weed, see *Matricaria discoidea*
Pinguicula vulgaris 194, *658, 671, 684*
Pink-sorrel, Large-flowered, see *Oxalis debilis*
Plantago
  coronopus 647, *658, 671, 684*
  lanceolata 10, 24, 51, 53, 60, 61, 62, 84, 208, 242, 308, **470**, 472, 476, 588, 606, *658, 671, 684*
  major ssp. *major* 61, 406, 418, **472**, 474, 532, *658, 671, 684*
  media 24, *658, 671, 684*
Plantain,
  Buck's-horn, see *Plantago coronopus*
  Hoary, see *P. media*
  Rat-tail, see *P. major* ssp. *major*
  Water, see *Alisma plantago-aquatica*
Platanthera chlorantha *658, 671, 684*
Ploughman's Spikenard, see *Inula conyzae*
Poa
  angustifolia 476, 645, *658, 671, 684*
  annua 10, 40, 42, 46, 58, 61, 176, 242, 418, 472, **474**, 554, 558, *658, 671, 684*
  Annual, see *Poa annua*
  Hard, see *Catapodium rigidum*
  humilis 476, 645
  compressa *658, 671, 684*
  nemoralis *658, 671, 684*
  pratensis 24, 58, 60, 61, 242, 308, 346, 348, **476**, 588, 645, *658, 671, 684*
  subcaerulea, see *humilis*
  trivialis 10, 25, 42, 46, 48, 54, 57, 61, 62, 66, 71, 174, 262, 384, **478**, 484, 616, *658, 671, 684*
Policeman's Helmet, see *Impatiens glandulifera*
Polygala
  serpyllifolia 480, *658, 671, 684*
  vulgaris 56, 58, 158, 188, 190, 342, 344, **480**, 642, *658, 671, 684*
  ssp. *collina* 480
Polygonatum x hybridum *658, 671, 684*
Polygonum
  amphibium, see *Persicaria amphibia*

SPECIES INDEX

Polygonum continued
    arenastrum 482
    aviculare s.l. 47, 51, 61, 146, 176, 404, 418, 472, **482**, 580, 604, 648, **658, 671, 684**
    aviculare s.s. 482
    bistorta, see *Persicaria bistorta*
    convolvulus, see *Fallopia convolvulus*
    cuspidatum, see *F. japonica*
    hydropiper, see *Persicaria hydropiper*
    lapathifolium, see *P. lapathifolia*
    persicaria see *P. maculosa*
    spp. 146
Polypodium vulgare **663, 676, 689**
Polypody, see *Polypodium vulgare*
Polypogon monspeliensis 90
Polystichum
    aculeatum **663, 676, 689**
    setiferum **663, 676, 689**
Polytrichum
    commune 86
    juniperinum 94
Pond-sedge,
    Great, see *Carex riparia*
    Lesser, see *C. acutiformis*
Pondweed,
    Bog, see *Potamogeton polygonifolius*
    Broad-leaved, see *P. natans*
    Canadian, see *Elodea canadensis*
    Curled, see *Potamogeton crispus*
    Fennel, see *P. pectinatus*
    Horned, see *Zannichellia palustris*
    Lesser, see *Potamogeton pusillus*
    Opposite-leaved, see *Groenlandia densa*
    Perfoliate, see *Potamogeton perfoliatus*
    Small, see *P. berchtoldii*
Poplar, Grey, see *Populus* x *canescens*
Poppy,
    Field, see *Papaver rhoeas*
    Long-headed, see *P. dubium*
    Opium, see *P. somniferum*
    Prickly, see *P. argemone*
    Welsh, see *Meconopsis cambrica*
Populus
    tremula **662, 675, 688**
    x *canescens* **662, 675, 688**
Potamogeton
    berchtoldii **658, 671, 684**
    coloratus 486
    crispus 260, **484**, 486, **658, 672, 684**
    natans 260, 484, **486**, 508, **658, 672, 685**
    pectinatus **658, 672, 685**
    perfoliatus **658, 672, 685**
    polygonifolius 648, **658, 672, 685**
    pusillus **658, 672, 685**
    spp. 57, 486
Potato, see *Solanum tuberosum*
Potentilla
    anglica 488, **658, 672, 685**
    anserina **658, 672, 685**
    erecta 58, 61, 69, 70, 88, 224, 390, **488, 658, 672, 685**
    palustris **658, 672, 685**
    recta **658, 672, 685**
    reptans 488, **658, 672, 685**
    sterilis 190, 312, 360, 480, **490**, 642, **658, 672, 685**
Poterium sanguisorba, see *Sanguisorba minor* ssp. *minor*
Primrose, see *Primula vulgaris*
Primula
    veris 24, 35, 47, 55, 58, 244, 312, 480, **492**, 538, 640, **658, 672, 685**
    ssp. *veris* 492
    vulgaris 390, **658, 672, 685**
Privet, Wild, see *Ligustrum vulgare*
Prunella vulgaris 150, 312, **494, 658, 672, 685**
Prunus
    avium **662, 675, 688**
    padus **662, 675, 688**
    spinosa **662, 675, 688**
Pseudofumaria lutea **658, 672, 685**
Pteridium aquilinum 42, 46, 48, 55, 57, 58, 59, 62, 254, 284, 322, 354, 356, 416, **496**, 522, 568, 578, **663, 676, 689**
Puccinellia distans **659, 672, 685**
Pulicaria dysenterica **659, 672, 685**
Purple-loosestrife, see *Lythrum salicaria*
Purslane,
    Pink, see *Claytonia sibirica*
    Water, see *Lythrum portula*
Pyrola rotundifolia 54

Quercus
    agg. 49, 50, 53, 55, 62, 82, 152, 252, 290, 356, 416, **498**, 568, 645, **662, 675, 688**
    cerris **662, 675, 688**
    petraea 31, 47, 498, 645, **662, 675, 688**
    robur 31, 498, 645, **662, 675, 688**
    x *rosacea* 498

Radish, Wild, see *Raphanus raphanistrum*
Ragged Robin, see *Lychnis flos-cuculi*
Ragwort, see *Senecio jacobaea*
    Hoary, see *S. erucifolius*
    Marsh, see *S. aquaticus*
    Oxford, see *S. squalidus*
Ramping-fumitory, Common, see *Fumaria muralis*
Ramsons, see *Allium ursinum*
Ranunculus
    acris 106, 208, 392, **500**, 502, 512, 524, 600, **659, 672, 685**
    ssp. *acris* 500
    ssp. *borealis* 500
    aquatilis 508, **659, 672, 685**
    aquatilis x *fluitans* 510
    auricomus **659, 672, 685**
    bulbosus 45, 96, 116, 228, 392, 494, 500, **502**, 512, 516, **659, 672, 685**
    circinatus 508, **659, 672, 685**
    ficaria 24, 33, 41, 48, 58, 62, 110, 220, 228, 248, 298, 458, **504**, 556, 632, **659, 672, 685**
    ssp. *bulbifer* 504
    ssp. *ficaria* 504
    flammula 260, 274, 276, 336, 370, 484, **506, 659, 672, 685**
    ssp. *flammula* 506
    ssp. *minimus* 506
    ssp. *scoticus* 506
    fluitans 508, 510
    hederaceus **659, 672, 685**
    lenormandii, see *omiophyllus*
    lingua **659, 672, 685**
    omiophyllus **659, 672, 685**
    peltatus **508**, 510, 610, **659, 672, 685**
    penicillatus 57, **510**, 518, **659, 672, 685**
    ssp. *penicillatus* 510
    var. *vertumnus* 510
    ssp. *pseudofluitans* 510
    pseudofluitans, see *penicillatus*
    repens 35, 48, 57, 61, 184, 478, 500, 502, **512**, 576, 630, **659, 672, 685**
    reptans 506
    sceleratus 57, **514**, 520, **659, 672, 685**
    trichophyllus 508, 510, **659, 672, 685**
Raphanus raphanistrum **659, 672, 685**
Rapistrum rugosum **659, 672, 685**
Raspberry, see *Rubus idaeus*
Ray-grass, see *Lolium perenne*
Redshank, see *Persicaria maculosa*
Reed, see *Phragmites australis*
    Common, see *P. australis*
Reed-grass, see *Glyceria maxima* & *Phalaris arundinacea*
Reedmace, Great, see *Typha latifolia*
Reseda
    lutea **659, 672, 685**
    luteola **659, 672, 685**
Restharrow,
    Common, see *Ononis repens*
    Spiny, see *O. spinosa*
Reynoutria
    japonica, see *Fallopia japonica*
    sachalinense, see *F. sachalinensis*
Rhamnus cathartica **662, 675, 688**
    catharicus, see preceeding entry
Rheum rhaponticum 128
Rhinanthus
    minor 42, 51, 106, 164, 306, 462, 502, **516**, 600, **659, 672, 685**
    ssp. *minor* 516
    ssp. *stenophyllus* 516
Rhizobium
    leguminosarum 390, 392, 636
    lupine 410, 412
    meliloti 420
    trifolii 596, 598, 600, 602
Rhizoctonia spp. 649
Rhododendron ponticum **662, 675, 688**
Rhododendron, see *Rhododendron ponticum*
Ribes
    nigrum **662, 675, 688**
    rubrum **662, 675, 688**
    uva-crispa **659, 672, 685**
Ribwort, see *Plantago lanceolata*
Rock Cress, Hairy, see *Arabis hirsuta*
Rocket,
    Eastern, see *Sisymbrium orientale*
    False London, see *S. loeselii*
    Tall, see *S. altissimum*
Rockrose, Common, see *Helianthemum nummularium*
Rorippa
    amphibia 520, **659, 672, 685**
    islandica s.l., see *palustris*
    microphylla 518, **659, 672, 685**
    microphylla x *nasturtium-aquaticum* 518
    nasturtium-aquaticum 60, **518, 659, 672, 685**
    palustris 49, 50, 57, 216, 514, **520**, 534, **659, 672, 685**
    sylvestris **659, 672, 685**
Rosa
    arvensis **662, 675, 688**
    caesia 645
    canina **663, 675, 688**
    mollis **663, 676, 688**
    rubiginosa 645
    sherardii 645
    spp. 645, **663, 676, 688**
    tomentosa 645
Rose,
    Dog, see *Rosa canina*

Field, see *R. arvensis*
Rowan, see *Sorbus aucuparia*
Rubus
    caesius 522, **659, 672, 685**
    chamaemorus **659, 672, 685**
    fruticosus agg. 31, 48, 54, 61, 82, 254, 354, 356, 408, 428, 450, 496, **522**, 536, 568, 578, **659, 672, 685**
    idaeus **659, 672, 685**
    inermis 522
    saxatilis **659, 672, 685**
    ulmifolius, see *inermis*
Rumex
    acetosa 10, 35, 42, 57, 84, 208, 276, 294, 392, 430, 500, **524**, 646, **659, 672, 685**
    acetosella 88, 94, 364, **526, 659, 672, 685**
        ssp. *acetosella* var. *tenuifolius* 526
        ssp. *pyrenaicus* 526
    alpinus (= *pseudoalpinus*) 444
    conglomeratus **659, 672, 685**
    crispus 32, 49, 51, 52, 148, 268, 524, **528**, 530, 552, 608, **659, 672, 685**
        ssp. *littoreus* 528
    hydrolapathum **659, 672, 685**
    obtusifolius 32, 42, 136, 268, 384, 524, 528, **530, 659, 672, 685**
        var. *microcarpus* 530
        var. *transiens* 530
    sanguineus **659, 672, 685**
Ruscus aculeatus **659, 672, 685**
Rush, Blunt-flowered, see *Juncus subnodulosus*
    Bulbous, see *Juncus bulbosus*
    Compact, see *J. conglomeratus*
    Flowering, see *Butomus umbellatus*
    Hard, see *Juncus inflexus*
    Heath, see *J. squarrosus*
    Jointed, see *J. articulatus*
    Sharp-flowered, see *J. acutiflorus*
    Slender, see *J. tenuis*
    Soft, see *J. effusus*
    Toad, see *J. bufonius*
Rustyback, see *Ceterach officinarum*
Rye-grass, see *Lolium perenne*
    Italian, see *L. multiflorum*
    Perennial, see *L. perenne*

Sage, Wood, see *Teucrium scorodonia*
Sagina
    apetala **659, 672, 685**
    maritima **659, 672, 685**
    nodosa **659, 672, 685**
    procumbens 52, 208, 210, **532**, 534, **659, 672, 685**
Sagittaria sagittifolia **659, 672, 685**
Sainfoin, see *Onobrychis viciifolia*
Salix
    alba **663, 676, 688**
    atrocinerea, see *cinerea* ssp. *oleifolia*
    aurita 645, **663, 676, 689**
    caprea 49, 58, 532, 534, 645, **663, 676, 689**
        var. *sphacelata* 534
    caprea/cinerea agg. **534**, 645, **663, 676, 689**
    cinerea s.l. 43, 102, 532, 534, 645, **663, 676, 689**
    cinerea s.s
        ssp. *cinerea* 534
        ssp. *oleifolia* 534
    fragilis 102, **663, 676, 689**
    purpurea **663, 676, 689**
    repens 646, **659, 672, 685**
    viminalis **663, 676, 689**

Saltmarsh-grass, Reflexed, see *Puccinellia distans*
Sambucus
    ebulus **659, 672, 685**
    nigra 49, 58, 114, **536**, 663, 676, 689
Samolus valerandi **659, 673, 685**
Sandwort,
    Spring, see *Minuartia verna*
    Three-nerved, see *Moehringia trinerva*
    Thyme-leaved, see *Arenaria serpyllifolia*
Sanguisorba
    minor
        ssp. *minor* 24, 42, 50, 56, 174, 188, 304, 344, 380, 468, **538**, 592, **659, 673, 686**
        ssp. *muricata* 538, **659, 673, 686**
    officinalis **659, 673, 686**
Sanicle, see following entry
Sanicula europaea 35, 47, 48, 49, 57, 62, 100, 110, 156, 382, 422, 426, **540, 659, 673, 686**
Saponaria officinalis **659, 673, 686**
Saxifraga
    granulata **659, 673, 686**
    hypnoides 542
    tridactylites 51, 62, 124, 130, 200, 234, 328, 420, 440, **542**, 546, 594, 596, 626, **659, 673, 686**
    x *urbium* **659, 673, 686**
Saxifrage,
    Burnet, see *Pimpinella saxifraga*
    Meadow, see *Saxifraga granulata*
    Rue-leaved, see *S. tridatylites*
Saw-wort, see *Serratula tinctoria*
Scabiosa columbaria 24, 42, 56, 57, 126, 344, 398, 448, 468, 524, **544**, 548, 562, 590, **659, 673, 686**
Scabious,
    Devil's-bit, see *Succisa pratensis*
    Field, see *Knautia arvensis*
    Small, see *Scabiosa columbaria*
Schoenoplectus
    lacustris **659, 673, 686**
    tabernaemontani **659, 673, 686**
Schoenus nigricans **660, 673, 686**
Scirpus
    cespitosus, see *Trichophorum cespitosum*
    fluitans, see *Eleogiton fluitans*
    lacustris, see *Schoenoplectus lacustris*
    tabernaemontani, see *S. tabernaemontani*
    setaceus, see *Isolepis setacea*
    sylvaticus **660, 673, 686**
Scleranthus annuus **660, 673, 686**
Scrophularia
    aquatica, see *auriculata*
    auriculata **660, 673, 686**
    nodosa **660, 673, 686**
Scurvygrass, Danish, see *Cochlearia danica*
Scutch, see *Elytrigia repens*
Scutellaria
    galericulata **660, 673, 686**
    minor **660, 673, 686**
Securigera varia **660, 673, 686**
Sea-spurrey, Lesser, see *Spergularia marina*
Sedge,
    Bottle, see *Carex rostrata*
    Brown, see *C. disticha*
    Carnation, see *C. panicea*
    Common Yellow, see *C. viridula* ssp. *oedocarpa*
    Common, see *C. nigra*
    Cyperus, see *C. pseudocyperus*

Dioecious, see *C. dioica*
Distant, see *C. distans*
Flea, see *C. pulicaris*
Glaucous, see *C. flacca*
Green-ribbed, see *C. binervis*
Hairy, see *C. hirta*
Long-stalked Yellow, see *C. viridula* ssp. *brachyrrhyncha*
Oval, see *C. ovalis*
Pale, see *C. pallescens*
Pendulous, see *C. pendula*
Pill-headed, see *C. pilulifera*
Prickly, see *C. muricata*
Remote, see *C. remota*
Sand, see *C. arenaria*
Smooth-stalked, see *C. laevigata*
Spiked, see *C. spicata*
Spring, see *C. caryophyllea*
Star, see *C. echinata*
Tawny, see *C. hostiana*
Tufted, see *C. elata*
White, see *C. curta*
Sedum
    acre 41, 44, 47, 57, 59, 62, 124, 126, 130, 328, 542, **546, 660, 673, 686**
    album 546, **660, 673, 686**
    anglicum 546, 645, 646, **660, 673, 686**
    reflexum, see *rupestre*
    rupestre **660, 673, 686**
    spurium **660, 673, 686**
    telephium **660, 673, 686**
Self-heal, see *Prunella vulgaris*
Senecio
    aquaticus 548, **660, 673, 686**
    cambrensis 550, 554
    erucifolius **660, 673, 686**
    jacobaea 42, 126, 128, 132, 308, 398, 404, 466, **548, 660, 673, 686**
        ssp. *dunensis* 548
        ssp. *jacobaea* var. *condensatus* 548
    squalidus 53, 55, 56, 57, 58, 134, 136, 148, 212, 226, 268, 528, **550**, 552, 554, 566, 608, **660, 673, 686**
    sylvaticus 552, **660, 673, 686**
    viscosus 58, 59, 134, 136, 148, 210, 212, 268, 402, 528, 550, **552**, 608, **660, 673, 686**
    vulgaris 33, 35, 42, 47, 58, 146, 176, 214, 216, 226, 530, 550, **554**, 558, 566, **660, 673, 686**
        var. *denticulatus* 554
        var. *hibernicus* 554
Serratula tinctoria 53, **660, 673, 686**
Shaggy Soldier, see *Galinsoga quadriradiata*
Sheep's-bit, see *Jasione montana*
Shepherd's Purse, see *Capsella bursa-pastoris*
Shepherd's Weatherglass, see *Anagallis arvensis* ssp. *arvensis*
Sherardia arvensis **660, 673, 686**
Shield-fern,
    Hard, see *Polystichum aculeatum*
    Soft, see *P. setiferum*
Shoreweed, see *Littorella uniflora*
Sieglingia decumbens, see *Danthonia decumbens*
Silaum silaus **660, 673, 686**
Silene
    alba, see *latifolia*
    dioica 53, 58, 162, 220, 262, 302, 354, **556, 660, 673, 686**
        ssp. *zetlandica* 556
    latifolia 556, **660, 673, 686**
    noctiflora **660, 673, 686**

SPECIES INDEX

Silene continued
   nutans 53
   vulgaris **660, 673, 686**
Silky-bent, Dense, see *Apera interrupta*
Silverweed, see *Potentilla anserina*
Sinapis arvensis 108, 214, 386, 452, 456, **558, 660, 673, 686**
Sisymbrium
   altissimum **660, 673, 686**
   loeselii **660, 673, 686**
   officinale **660, 673, 686**
   orientale **660, 673, 686**
Skullcap, see *Scutellaria galericulata*
   Lesser, see *S. minor*
   Small-reed,
      Purple, see *Calamagrostis canescens*
      Wood, see *C. epigejos*
Snapdragon, see *Antirrhinum majus*
Sneezewort, see *Achillea ptarmica*
Snowberry, see *Symphoricarpos albus*
Snowdrop, see *Galanthus nivalis*
Snow-in-summer, see *Cerastium tomentosum*
Soapwort, see *Saponaria officinalis*
Soft-brome, Slender, see *Bromus lepidus*
Soft-grass, Creeping, see *Holcus mollis*
Solanum
   dulcamara 112, 170, 178, 180, 218, 270, 310, 442, **560, 660, 673, 686**
   nigrum **660, 673, 686**
   physalifolium **660, 673, 686**
   sarrachoides, see *physalifolium*
   tuberosum 388
Solidago
   canadensis 562, **660, 673, 686**
   gigantea **660, 673, 686**
   virgaurea 182, 312, 398, **562, 660, 673, 686**
Solomon's Seal, Garden, see *Polygonatum* x *hybridum*
Sonchus
   arvensis **660, 673, 686**
   asper **564, 660, 673, 686**
   oleraceus 134, 550, **566, 660, 673, 686**
   tenerrimus 566
Sorbus
   aucuparia 354, 416, 428, 450, 496, 498, 522, **568, 663, 676, 689**
Sorrel,
   Common, see *Rumex acetosa*
   Sheep's, see *R. acetosella*
   Wood, see *Oxalis acetosella*
Sow-Thistle, see *Sonchus oleraceus*
   Perennial, see *S. arvensis*
Sparganium
   emersum 648, **660, 673, 686**
   erectum 166, 260, 484, 486, 508, **570, 660, 673, 686**
Spearwort,
   Greater, see *Ranunculus lingua*
   Lesser, see *R. flammula*
Speedwell,
   Germander, see *Veronica chamaedrys*
   Heath, see *V. officinalis*
   Ivy-leaved, see *V. hederifolia*
   Marsh, see *V. scutellata*
   Slender, see *V. filiformis*
   Thyme-leaved, see *V. serpyllifolia*
   Wall, see *V. arvensis*
   Wood, see *V. montana*
Spergula arvensis 60, 108, 120, 230, 298, 388, 438, 452, 500, 558, **572**, 618, 634, **660, 673, 686**
   var. *nana* 572

Spergularia
   marina **660, 673, 686**
   rubra **660, 673, 686**
Sphagnum spp. 86, 192, 290, 304, 358, 488
Spike-rush,
   Common, see *Eleocharis palustris* ssp. *vulgaris*
   Few-flowered, see *E. quinqueflora*
Spindle, see *Euonymus europaeus*
Spirodela polyrhiza 394
Spleenwort,
   Black, see *Asplenium adiantum-nigrum*
   Maidenhair, see *A. trichomaes*
Spodoptera littoralis
Spotted-orchid,
   Common, see *Dactylorhiza fuchsii*
   Heath, see *D. maculata*
Spring Beauty, see *Claytonia perfoliata*
Spurge,
   Cypress, see *Euphorbia cyparissias*
   Dwarf, see *E. exigua*
   Leafy, see *E. esula*
   Petty, see *E. peplus*
   Sun, see *E. helioscopia*
Spurge-laurel, see *Daphne laureola*
Spurrey, see *Spergula arvensis*
   Corn, see *S. arvensis*
   Sand, see *Spergularia rubra*
Squirrel
St. John's Wort,
   Common, see *H. perforatum*
   Hairy, see *H. hirutum*
   Imperforate, see *H. maculatum*
   Marsh, see *H. elodes*
   Slender, see *H. pulchrum*
   Square-stalked, see *H. tetrapterum*
   Trailing, see *H. humifusum*
Stachys
   arvensis **660, 673, 686**
   officinalis 60, 244, 326, 346, 390, 468, **574, 584**, 640, **660, 673, 686**
   palustris 576, **660, 673, 686**
   palustris x *sylvatica*, see x *ambigua*
   sylvatica 114, 118, 220, 318, 334, 348, **576, 586**, 616, **660, 673, 686**
   x *ambigua* 576
Star-of-Bethlehem, see *Ornithogalum angustifolium*
Stellaria
   alsine, see *uliginosa*
   graminea **660, 673, 686**
   holostea 40, 236, 330, **578**, 586, **660, 673, 686**
   media 40, 42, 47, 58, 61, 146, 176, 214, 432, 482, 456, **580, 660, 674, 686**
   neglecta 580, **660, 674, 687**
   pallida 580, **660, 674, 687**
   uliginosa 50, 90, 184, 274, 370, 376, 512, **582, 660, 674, 687**
Stitchwort,
   Bog, see *Stellaria alsine*
   Greater, see *S. holostea*
   Lesser, see *S. graminea*
Stonecrop,
   Biting, see *Sedum acre*
   English, see *S. anglicum*
   Reflexed, see *S. rupestre*
   White, see *S. album*
Stork's-bill, Common, see *Erodium cicutarium*
Strawberry
   Barren, see *Potentilla sterilis*
   Garden, see *Fragaria* x *ananassa*

   Wild, see *F. vesca*
Succisa pratensis 47, 53, 57, 69, 244, 468, 574, **584**, 640, **660, 674, 687**
Summer-cypress, see *Bassia scoparia*
Sundew, Round-leaved, see *Drosera rotundifolia*
Sweet Cicely, see *Myrrhis odorata*
Sweet-flag, see *Acorus calamus*
Swine-cress, see *Coronopus squamatus*
   Lesser, see *C. didymus*
Sycamore, see *Acer pseudoplatanus*
Symphoricarpos albus **663, 676, 689**
Symphytum
   officinale **660, 674, 687**
   orientale **660, 674, 687**
   x *uplandicum* **660, 674, 687**

Tamus communis 56, 57, 114, 334, 340, 408, 576, **586, 660, 674, 687**
Tanacetum
   parthenium **660, 674, 687**
   vulgare **660, 674, 687**
Tansy, see *Tanacetum vulgare*
Taraxacum
   agg. 24, 31, 46, 60, 61, 242, 262, 476, 532, **588**, 645, **661, 674, 687**
   sect. *Erythrosperma* 588
   sect. *Palustria* 588
   sect. *Spectabilia* 588
   sect. *Taraxacum* 588
Tare,
   Hairy, see *Vicia hirsuta*
   Smooth, see *V. tetrasperma*
Taxus baccata 540, **663, 676, 689**
Teasel, Wild, see *Dipsacus fullonum*
Teesdalia nudicaulis 94
Tellima grandiflora **661, 674, 687**
Teucrium scorodonia 132, 312, 324, 360, 544, **590, 661, 674, 687**
Thale Cress, see *Arabidopsis thaliana*
Thalictrum flavum **661, 674, 687**
Thelypteris
   limbosperma, see *Oreopteris limbosperma*
   phegopteris, see *Phegopteris connectilis*
Thistle,
   Carline, see *Carlina vulgaris*
   Creeping, see *Cirsium arvense*
   Dwarf, see *C. acaule*
   Field, see *C. arvense*
   Marsh, see *C. palustre*
   Meadow, see *C. dissectum*
   Melancholy, see *C. heterophyllum*
   Musk, see *Carduus nutans*
   Spear, see *Cirsium vulgare*
   Welted, see *Carduus crispus*
Thlaspi caerulescens 44, 430
   arvense **661, 674, 687**
Thyme,
   Basil, see *Clinopodium acinos*
   Wild, see *Thymus polytrichus*
Thymus
   drucei, see *polytrichus*
   polytrichus 10, 47, 49, 57, 174, 188, 294, 304, 324, 342, 404, 430, 590, **592, 661, 674, 687**
   praecox, see *polytrichus*
Tilia
   cordata 53
   spp. 82, 614
Timothy, see *Phleum pratense*
Toadflax,
   Ivy-leaved, see *Cymbalaria muralis*

Pale, see *Linaria repens*
Purple, see *L. purpurea*
Small, see *Chaenorhinum minus*
Yellow, see *Linaria vulgaris*
Tomato, see *Lycopersicum esculentum*
Toothwort, see *Lathraea squamaria*
Torilis
    arvensis 594
    japonica 50, 60, 234, **594, 661, 674, 687**
Tormentil,
    Common, see *Potentilla erecta*
Trailing, see *P. anglica*
Tragopogon pratensis 53, **661, 674, 687**
Traveller's Joy, see *Clematis vitalba*
Trefoil,
    Hop, see *Trifolium campestre*
    Lesser Yellow, see *T. dubium*
    Slender, see *T. micranthum*
Trichophorum cespitosum **661, 674, 687**
Trientalis europaea **661, 674, 687**
Trifolium
    arvense **661, 674, 687**
    campestre 596, **661, 674, 687**
    dubium 57, 120, 124, 234, 328, 440, 542, 594, **596, 626, 661, 674, 687**
    fragiferum **661, 674, 687**
    hybridum **661, 674, 687**
    medium 24, 60, 154, 160, 202, 246, 288, **598, 661, 674, 687**
    micranthum 596, **661, 674, 687**
    pratense 24, 164, 392, **600, 661, 674, 687**
        var. *pratense* 600
        var. *sativum* 600
    repens 29, 40, 48, 61, 66, 150, 208, 238, 406, 450, 494, 500, 516, 540, 596, 598, **602, 630, 661, 674, 687**
    spp. 596, 598
    striatum **661, 674, 687**
Triglochin palustris **661, 674, 687**
Tripleurospermum
    inodorum 51, 136, 226, 386, 418, 482, 530, 566, **604, 661, 674, 687**
    maritimum 604
Trisetum flavescens 24, 51, 308, 346, 380, 470, **606, 661, 674, 687**
Trollius europaeus **661, 674, 687**
Tufted-sedge, Slender, see *Carex acuta*
Tussilago farfara 35, 49, 57, 212, 222, 280, 352, 402, 528, 564, **608, 661, 674, 687**
Tussock-sedge Greater, see *Carex paniculata*
Tutsan, see *Hypericum androsaemum*
Twayblade Common, see *Listera ovata*
Twitch, see *Elytrigia repens*
Typha
    angustifolia 610, **661, 674, 687**
    x *glauca*
    latifolia 258, 338, 454, 464, 508, 510, 570, **610, 661, 674, 687**
Tyria jacobaea 548

Ulex
    europaeus 62, **612, 646, 663, 676, 689**
    gallii 612, **663, 676, 689**
    minor 612, **661, 674, 687**
Ulmus
    glabra 152, **614, 646, 663, 676, 689**
        ssp. *glabra* 614
        ssp. *montana* 614
    minor 614
    procera 614, **663, 676, 689**
Umbilicus rupestris **661, 674, 687**

Urtica
    dioica 10, 24, 35, 42, 46, 49, 61, 62, 71, 136, 264, 310, 318, 334, 384, 478, 530, 536, 576, **616, 618, 661, 674, 687**
        ssp. *gracilis* 616
        var. *subinermis* 616
    urens 35, 42, 52, 386, **618, 661, 674, 687**
Ustilago violacea 528, 556

Vaccinium
    myrtillus 35, 48, 61, 62, 66, 168, **620, 661, 674, 687**
    oxycoccos **661, 674, 687**
    vitis-idaea 52, 53, 56, 196, 250, 266, 286, 620, **622, 661, 674, 687**
Valerian, see *Valeriana officinalis*
    Marsh, see *V. dioica*
    Red, see *Centranthus ruber*
Valeriana
    dioica 194, **661, 674, 687**
    officinalis 154, **624, 661, 674, 687**
        ssp. *collina* 624
        ssp. *officinalis* 624
        ssp. *sambucifolia* 624
Valerianella locusta **661, 674, 687**
Verbascum
    nigrum **661, 674, 687**
    thapsus **661, 674, 687**
Vernal-grass, Sweet, see *Anthoxanthum odoratum*
Verbena officinalis **661, 674, 687**
Veronica
    agrestis 634, **661, 674, 687**
    anagallis-aquatica **661, 674, 687**
    arvensis 120, 124, 440, **626, 661, 674, 687**
    beccabunga 60, 90, 122, 278, 442, 518, **628, 661, 674, 687**
    chamaedrys 158, **630, 661, 674, 687**
    filiformis **661, 674, 687**
    hederifolia **661, 674, 687**
    montana 57, 220, 302, **632, 649, 661, 674, 687**
    officinalis **661, 674, 687**
    persica 49, 58, 60, 108, 120, 230, 298, 388, 438, 452, 456, 558, 572, **634, 661, 674, 687**
    polita 634, **661, 674, 687**
    scutellata **661, 674, 687**
    serpyllifolia **661, 674, 687**
    spp. 630
Vervain, see *Verbena officinalis*
Vetch,
    Bitter, see *Lathyrus linifolius*
    Bush, see *Vicia sepium*
    Common, see *V. sativa*
    Crown, see *Securigera varia*
    Fine-leaved, see *Vicia tenuifolia*
    Kidney, see *Anthyllis vulneraria*
    Tufted, see *Vicia cracca*
Vetchling, Meadow, see *Lathyrus pratensis*
Viburnum opulus **663, 676, 689**
Vicia
    angustifolium, see *sativa* ssp. *nigra*
    cracca 50, 60, **636, 661, 675, 687**
    hirsuta **661, 675, 688**
    sativa ssp. *nigra* 234, **661, 675, 688**
    sepium 60, 128, 636, **638, 661, 675, 688**
    tenuifolia **661, 675, 688**
    tetrasperma **661, 675, 688**
Vinca
    major **661, 675, 688**
    minor **661, 675, 688**
Viola
    arvensis **662, 675, 688**
    canina ssp. *montana* 642, **662, 675, 688**
    hirta 24, 188, 228, 344, 574, 584, **640, 662, 675, 688**
    lutea 390
    odorata 640, **662, 675, 688**
    palustris **662, 675, 688**
    reichenbachiana 540, 640, **662, 675, 688**
    riviniana 47, 62, 190, 490, **642, 662, 675, 688**
    tricolor **662, 675, 688**
Violet,
    Common, see *Viola riviniana*
    Dame's, see *Hesperis matronalis*
    Hairy, see *Viola hirta*
    Marsh, see *V. palustris*
    Sweet, see *V. odorata*
Viper's Bugloss, see *Echium vulgare*
Viscum album **662, 675, 688**
Vulpia
    bromoides **662, 675, 688**
    myuros **662, 675, 688**
    spp 308

Wahlenbergia hederacea **662, 675, 688**
Wallflower, see *Erysimum cheiri*
Wall-pepper, see *Sedum acre*
Wall-rocket
    Annual, see *Diplotaxis muralis*
    Perennial, see *D. tenuifolia*
Wall-rue, see *Asplenium ruta-muraria*
Water-cress, see *Rorippa nasturtium-aquaticum* agg
    Narrow-fruited, see *Rorippa microphylla*
    Fool's, see *Apium nodiflorum*
Water-crowfoot,
    Common, see *Ranunculus aquatilis* & *R. peltatus*
    Fan-leaved, see *R. circinatus*
    Thread-leaved, see *R. trichophyllus*
Water-dropwort,
    Fine-leaved, see *Oenanthe aquatica*
    Hemlock, see *O. crocata*
    Tubular, see *O. fistulosa*
Water-lily,
    White, see *Nymphaea alba*
    Yellow, see *Nuphar lutea*
Water-milfoil,
    Alternate, see *Myriophyllum alterniflorum*
    Spiked, see *M. spicatum*
Water-parsnip, Lesser, see *Berula erecta*
Water-pepper, see *Persicaria hydropiper*
Water-speedwell,
    Blue, see *Veronica anagallis-aquatica*
    Pink, see *V. catenata*
Water-starwort,
    Blue-fruited, see *Callitriche obtusangula*
    Common, see *C. stagnalis*
    Intermediate, see *C. hamulata*
    Various-leaved, see *C. platycarpa*
Water-violet, see *Hottonia palustris*
Water-weed, Nuttall's, see *Elodea nuttallii*
Weld, see *Reseda luteola*
Whin, see *Ulex europaeus*
    Petty, see *Genista anglica*
White-rot, see *Hydrocotyle vulgaris*
Whitlowgrass, Common, see *Erophila verna*
Whorl-grass, see *Catabrosa aquatica*

Whortleberry, see *Vaccinium myrtillus*
    Red, see *V. vitis-idaea*
Willow,
    Crack, see *Salix fragilis*
    Creeping, see *S. repens*
    Eared, see *S. aurita*
    Goat, see *S. caprea*
    Grey, see *S. cinerea*
    Purple, see *S. purpurea*
White, see *S. alba*
Willow-herb
    American, see *Epilobium ciliatum*
    Broad-leaved, see *E. montanum*
    Dull-leaved, see *E. obscurum*
    Great Hairy, see *E.. hirsutum*
    Lesser Hairy, see *E. parviflorum*
    Marsh, see *E. palustre*
    New Zealand, see *E. brunnescens*
    Pale, see *E. roseum*
    Rose-bay, see *Chamerion*
    Square-stalked, see *Epilobium tetragonum*
Winter-cress, see *Barbarea vulgaris*
Wintergreen, Chickweed, see *Trientalis europaea*
Woodruff, see *Galium odoratum*
Woodrush,
    Field, see *Luzula campestris*
    Great, see *L. sylvatica*
    Hairy, see *L. pilosa*
    Heath, see *L. multiflora*
Wood-sedge, see *Carex sylvatica*
Wormwood, see *Artemisia absinthium*
Woundwort,
    Field, see *Stachys arvensis*
    Hedge, see *S. sylvatica*
    Marsh, see *S. palustris*

x *Festulolium loliaceum*   306, **662, 675, 688**

Yarrow, see *Achillea millefolium*
Yellow Archangel, see *Lamiastrum galeobdolon*
Yellow-cress,
    Creeping, see *Rorippa sylvestris*
    Greater, see *R. amphibia*
    Marsh, see *R. palustris*
Yellow-rattle, see *Rhinanthus minor*
Yellow-wort, see *Blackstonia perfoliata*
Yellow-sorrel,
    Least, see *Oxalis exilis*
    Procumbent, see *O. corniculata*
Yew, see *Taxus baccata*
Yorkshire Fog, see *Holcus lanatus*

Zannichellia palustris   **662, 675, 688**
Zea mays   10

The Botanical Society of the British Isles (BSBI)

- Has a membership of about 3,000 amateur and professional botanists
- Is the leading charitable society promoting the study and enjoyment of British and Irish wild plants
- Maintains a network of 152 Vice-county recorders, a central database of threatened plants and a panel of referees for difficult plants
- Maintains a comprehensive scientific database that enables the list of British and Irish plants to be kept up to date
- Carries out national surveys and publishes the results, notably survey for the *New Atlas of the British and Irish Flora*, 2002 and *Change in the British Flora 1987 - 2004*
- Publishes authoritative identification handbooks on difficult plants such as sedges and roses
- Promotes the publication of local floras and county rare plant registers
- Holds field meetings and conferences to bring botanists together, whether amateur or professional
- Publishes journals to enable members to share their observations and the results of their studies
- Encourages the training of botanists of all ages

More information will be found on the BSBI web site at www.bsbi.org.uk

Membership and other enquiries may be sent to
BSBI Honorary General Secretary,
c/o Department of Botany,
The Natural History Museum, Cromwell Road,
London, SW7 5BD.